CALCULUS
Graphical, Numerical, Algebraic
Single Variable Version

Ross L. Finney

George B. Thomas
Massachusetts Institute of Technology

Franklin Demana
The Ohio State University

Bert K. Waits
The Ohio State University

 ADDISON-WESLEY PUBLISHING COMPANY
Reading, Massachusetts • Menlo Park, California • New York
Don Mills, Ontario • Wokingham, England • Amsterdam • Bonn • Sydney
Singapore • Tokyo • Madrid • San Juan • Milan • Paris

About The Cover

Why a trillium on the cover of a calculus text?

The name *trillium* generally describes 25 to 30 species of wildflowers that are native to North America and eastern Asia. Trilliums, which derive their name from the Latin word for three, develop with all their parts in multiples of three. They bear three leaves and, just above these, a flower with three green sepals, three petals, six stamens, and a three-chambered pistil with three stigmas. The plant bears a berry fruit, often with three to six angles.

The trillium provides a natural representation of the threefold focus of this text. The text consistently uses graphical, numerical, and algebraic (GNA) representations to explore problem situations. This theme is woven throughout the text material and is most obvious in the numerous GNA Explorations.

 The GNA icon is also represented graphically on the cover with the image of a graphing utility-generated graph of the polar equation $r = -\sin 3\theta$.

Sponsoring Editor	Jerome Grant	*Art Consultant*	Joe Vetere
Associate Editor	Kathleen A. Manley	*Manufacturing Manager*	Roy Logan
Editorial Assistant	Cristina Malinn	*Manufacturing Coordinator*	Evelyn Beaton
Development Editors	Elka Block, Joe Will	*Text Designer*	Nancy Blodget
Editorial Assistant	Amy Branowicki	*Copy Editor*	Barbara Willette
Managing Editor	Karen Guardino	*Proofreader*	Laura K. Michaels,
Production Supervisor	Jennifer Brownlow Bagdigian		Michaels Communications
Production Coordinator	Patricia A. Oduor	*Art Coordinator*	Connie Hulse
Prepress Manager	Sarah McCracken	*Production Services*	Sandra Rigney
Prepress Buyer	Caroline Fell Skurat	*Technical Illustration*	Tech Graphics
Cover Design	Peter Blaiwas	*Cover Photograph*	Dick Morton
Marketing Manager	Andrew Fisher	*Composition*	ETP Services, Inc.

Library of Congress Cataloging-in-Publication Data

Calculus : graphical, numerical, algebraic of a single variable / Ross
  L. Finney . . . [et al.].
    p. cm.
  Includes index.
  ISBN 0-201-56902-7
  1. Calculus.   I. Finney, Ross L.
QA303.C1755 1994
515—dc20   93-37037
                         CIP

1 2 3 4 5 6 7 8 9 10 VH 9796959493

About The Authors

Ross L. Finney

Ross Finney received his undergraduate degree and Ph.D. from the University of Michigan at Ann Arbor. He taught at the University of Illinois at Urbana–Champaign from 1966 to 1980 and the Massachusetts Institute of Technology (MIT) from 1980 to 1990. Currently, he is involved in a number of mathematical organizations. Dr. Finney worked as a consultant for Education Development Center in Newton, Massachusetts. He directed the Undergraduate Mathematics and Its Applications Project (UMAP) from 1977 to 1984 and was the founding editor of the *UMAP Journal*. In 1984, he traveled with a Mathematical Association of America (MAA) delegation to China on a teacher education project through People to People International.

Dr. Finney has coauthored a number of Addison-Wesley textbooks, including *Calculus, Calculus and Analytic Geometry,* and *Elementary Differential Equations with Linear Algebra.*

George B. Thomas, Jr.

George Thomas is the author of one of the first mathematics textbooks published by Addison-Wesley, *Calculus and Analytic Geometry*—now a best-selling calculus text. Dr. Thomas received his M.S. and B.S. from the State College of Washington and his Ph.D. from Cornell University. He taught at Cornell University for three years and at MIT for 38 years, was a visiting Associate Professor at Stanford University for a year while on sabbatical from MIT, and was a lecturer at Birla Institute of Technology and Science in India.

Dr. Thomas has written and coauthored a number of texts for Addison-Wesley, including *Calculus, Calculus and Analytic Geometry, Probability: A First Course, Probability with Statistical Applications,* and *Elements of Calculus and Analytic Geometry.*

Franklin D. Demana

Frank Demana received his Master's degree in Mathematics and his Ph.D. from Michigan State University. Currently, he is Professor of Mathematics at The Ohio State University. As an active supporter of the use of technology to teach and learn mathematics, he is codirector of the very successful C^2PC (Calculator and Computer Precalculus) technology-enhanced curriculum revision project now operating in over 1000 high schools and colleges. He has been the director and co-director of over $10 million of National Science Foundation (NSF) and foundational grant activities over the past 10 years. Along with frequent presentations at professional meetings, he has published a variety of articles in the areas of computer and calculator-enhanced mathematics instruction. Dr. Demana is also the cofounder (with Bert Waits) of the annual International Conference on Technology in Collegiate Mathematics (ICTCM).

Dr. Demana coauthored *Essential Algebra: A Calculator Approach, Transition to College Mathematics, Precalculus: A Graphing Approach, College Algebra and Trigonometry: A Graphing Approach, College Algebra: A Graphing Approach, Precalculus: Functions and Graphs,* and *Intermediate Algebra: A Graphing Approach.*

Bert K. Waits

Bert Waits received his Ph.D. from The Ohio State University in 1969 and is currently Professor of Mathematics there. Dr. Waits is codirector of the C^{2}PC project along with Frank Demana and has been codirector or principal investigator on several large NSF projects. Dr. Waits has published articles in over 50 nationally recognized professional journals. He frequently gives invited lectures, workshops, and minicourses at national meetings of the MAA and the National Council of Teachers of Mathematics (NCTM) on how to use computer technology to enhance the teaching and learning of mathematics. Recently, he has given invited presentations at the International Congress on Mathematical Education (ICME 6 and 7) in Budapest (1988) and Quebec (1992), the International Conference on the Teaching of Mathematics by Applications (ICTMA 5) in Noordwijkerhout, The Netherlands (1991), and colleges and universities in Moscow, Zürich, Nice, Combria, Göteborg, Tübingen, Edinburgh, Birmingham, Lancaster, London, Lisbon, and Melbourne.

Dr. Waits coauthored *Precalculus: A Graphing Approach, College Algebra and Trigonometry: A Graphing Approach, College Algebra: A Graphing Approach, Precalculus: Functions and Graphs,* and *Intermediate Algebra: A Graphing Approach.*

Preface

Audience and Prerequisites

This book explores all concepts necessary for the standard calculus sequence. Its purpose, in addition to making it possible to learn calculus, is to teach students how to use calculus effectively and to show how knowing calculus can pay off in any profession they decide to enter. The applications described within the text are real, and their presentations are self-contained; students will not need any experience in the fields from which the applications are drawn. The prerequisites are exposure to algebra, geometry, and trigonometry; Chapter 1 reviews this material.

Philosophy

Calculus: Graphical, Numerical, Algebraic grew out of a strong conviction that incorporating technology into instruction makes the successful study of calculus realistic for *all* students, better prepares them for further study in mathematics and science, and turns calculus into a *pump* and not a *filter*. It builds on the 10-year experience of Demana and Waits in the Calculator and Computer Precalculus (C²PC) project. This project pioneered the use of a computer/visualization-based approach to mathematics while integrating the use of numerical, graphical, and algebraic techniques. Finney and Thomas bring many years of experience from an applications-based method of teaching and learning. The combination of these four authors allows for a richer approach than would have been possible for any of the four alone.

The approach of this textbook is consistent with the call by national mathematics organizations for reform in calculus teaching and learning that takes advantage of today's technological tools. While doing so, the text balances the use of these tools with the well-established approaches.

Computer-generated numerical, visual, and symbolic mathematics is revolutionizing the teaching and learning of calculus. The content of calculus is changing by reducing time spent on paper-and-pencil methods and increasing time spent on concepts, problem solving, and applications. Teaching methods are also dramatically changing by moving toward an investigative, exploratory approach. The computer used to facilitate this exploration can be an inexpensive pocket computer with built-in software (graphing calculator) or a desktop computer with computer algebra graphing software. Today, *any* classroom can become a computer laboratory with student use of graphing calculators.

The text was built on this philosophy, which can be summarized by the following points. The text requires students to:

- Do a problem algebraically (using only paper and pencil), then *support* the results numerically and/or graphically (with a graphing calculator or computer);

- Do a problem numerically and/or graphically (with a graphing calculator or computer), then *confirm* the results algebraically (using paper and pencil);

- Do a problem numerically and/or graphically because paper-and-pencil and paper methods are impractical (too time consuming) or impossible!

This Text and the Evolution of the Calculus Course

Calculus Reform

Our approach to calculus reform is realistic. It does not require drastic change on the part of the instructor or the student. Fundamentally, we believe in the principle of *incremental change*. This principal forms the basis of our approach to calculus curriculum reform and the related integration of graphing utility technology used in this textbook. We believe in a balanced approach to the teaching and learning of calculus when it comes to paper-and-pencil methods versus calculator or computer methods. In this textbook, students are exposed to the utility and value of numerical and graphical methods while they are learning traditional techniques and developing facility with, and appreciation for, the power of algebraic methods.

We have taken a familiar body of calculus material and made the assumption that *every* student has at least an inexpensive graphing calculator for both in-class activities and homework. Of course, students with access to powerful computers and software like Derive®, Mathematica®, and Maple® will easily be able to use this textbook as well.

This textbook is still evolving. As pocket computers become more powerful and less costly and as the calculus community comes to more agreement on the content of the modern calculus course, the traditional calculus table of contents will change more dramatically. For example, less time will be spent on paper-and-pencil differentiation and integration techniques. In fact, instructors using *this edition* can easily choose to spend less time on paper-and-pencil differentiation and integration techniques. We encourage any and all feedback. It is through your suggestions that this text will continue to evolve. Please also feel free to send us suggestions for future changes (such as content coverage and organization) and additions (such as new Explorations, Exploration Bits, and exercises).

Graphical, Numerical, and Algebraic Themes

We use multiple representations to explore problem situations. First, we find an algebraic representation of the problem. Then we find a complete graph of the algebraic representation and determine which values of the domain of the algebraic representation make sense in the problem situation. Technology allows us to move easily among these representations and exploit connections.

The concept of function is central to the study of calculus. A function can be thought of as a process that produces outputs for associated inputs. The notion can be represented *numerically* as a table of input-output pairs, *graphically* as a plot of output versus input, and *symbolically* as an algebraic representation. With technology, we are no longer restricted to a purely algebraic approach to calculus; now we can also approach concepts of calculus numerically and graphically.

Furthermore, modern research on learning suggests that students have different learning styles. Some mathematics students respond better to visual representations, some to numerical representations, and others to algebraic representations.

With all of this in mind, we strive to use these three themes—Graphical, Numerical, and Algebraic (GNA)—to explore problem situations. Moreover, the text emphasizes the connections between the graphical, numerical, and algebraic approaches, building a richer understanding of calculus.

Graphing Utilities

Graphing calculators and computer graphing utility software permit the focus of calculus to be on problem solving and exploration. The use of a graphing utility is required with this textbook. Depending on the particular technology used, it may be necessary to enter certain "toolbox" programs on the graphing utility. These useful programs are provided in the *Technology Resource Manuals for Calculus*, available from your instructor or for purhase separately.

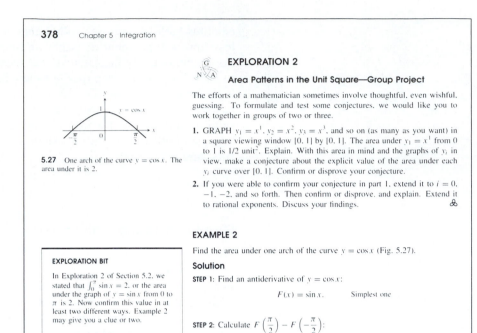

378   Chapter 5   Integration

G
N A   **EXPLORATION 2**

Area Patterns in the Unit Square—Group Project

The efforts of a mathematician sometimes involve thoughtful, even wishful, guessing. To formulate and test some conjectures, we would like you to work together in groups of two or three.

1. GRAPH $y_1 = x^1$, $y_2 = x^2$, $y_3 = x^3$, and so on (as many as you want) in a square viewing window [0, 1] by [0, 1]. The area under $y_1 = x^1$ from 0 to 1 is 1/2 unit^2. Explain. With this area in mind and the graphs of y_i in view, make a conjecture about the explicit value of the area under each y_i curve over [0, 1]. Confirm or disprove your conjecture.

2. If you were able to confirm your conjecture in part 1, extend it to $i = 0$, -1, -2, and so forth. Then confirm or disprove, and explain. Extend it to rational exponents. Discuss your findings.

5.27   One arch of the curve $y = \cos x$. The area under it is 2.

EXPLORATION BIT

In Exploration 2 of Section 5.2, we stated that $\int_0^{\pi} \sin x = 2$, or the area under the graph of $y = \sin x$ from 0 to π is 2. Now confirm this value in at least two different ways. Example 2 may give you a clue or two.

EXAMPLE 2

Find the area under one arch of the curve $y = \cos x$ (Fig. 5.27).

Solution

STEP 1: Find an antiderivative of $y = \cos x$:

$$F(x) = \sin x. \qquad \text{Simplest one}$$

STEP 2: Calculate $F\left(\dfrac{\pi}{2}\right) - F\left(-\dfrac{\pi}{2}\right)$:

$$F\left(\frac{\pi}{2}\right) - F\left(-\frac{\pi}{2}\right) = \sin\left(\frac{\pi}{2}\right) - \sin\left(-\frac{\pi}{2}\right) = 1 - (-1) = 2.$$

The area is 2 square units.   ≡

Features

Balanced Approach.   The text strives to use multiple representations—Graphical, Numerical, and Algebraic (GNA)—to explore problem situations. Establishing connections between these approaches builds a richer understanding of calculus.

Current Trends in Teaching and Learning.   Recognizing the current goals in teaching and learning mathematics, the text encourages discovery, critical thinking, working in groups, writing, and verbalization in its exercise sets and Explorations.

Graphical, numerical, algebraic explorations.   The GNA Explorations make the student an active partner in the learning process, helping the student to explore and "do" mathematics in an exciting and creative way that was simply unavailable until recently. They help to develop an appreciation of the power of a graphing utility and actually teach the student how to apply the technology. They also help the instructor in the role of facilitator and guide to a high-interest, stimulating, and balanced classroom approach. Explorations are marked throughout the text with the special icon shown here. Shaded "G" (graphical), "N" (numerical), and/or "A" (algebraic) suggest the mode(s) for approaching the Exploration. A list of the Explorations can be found in an appendix at the back of the book.

Exploration bits.   These are related to the GNA Explorations in that they also provide an opportunity for active experimentation with challenges that generally are adjunct to the text development.

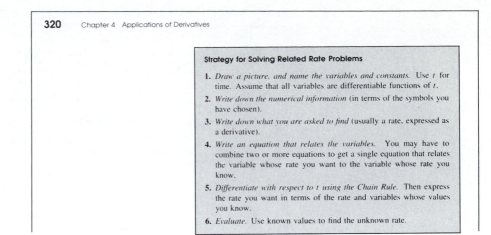

320   Chapter 4 Applications of Derivatives

Strategy for Solving Related Rate Problems

1. *Draw a picture, and name the variables and constants.* Use t for time. Assume that all variables are differentiable functions of t.
2. *Write down the numerical information* (in terms of the symbols you have chosen).
3. *Write down what you are asked to find* (usually a rate, expressed as a derivative).
4. *Write an equation that relates the variables.* You may have to combine two or more equations to get a single equation that relates the variable whose rate you want to the variable whose rate you know.
5. *Differentiate with respect to t using the Chain Rule.* Then express the rate you want in terms of the rate and variables whose values you know.
6. *Evaluate.* Use known values to find the unknown rate.

Problem Solving. Problem solving is developed and encouraged throughout the text. Practical applications and situations (many with sources cited) are included. Students learn how to evaluate a problem critically, formulate a strategy, solve the problem, and analyze and interpret the results.

Strategy boxes   help students to develop general strategies for solving problems.

Sidelight boxes   throughout the margins provide commentary on technology, mathematical development, mathematical guidance, interesting facts, and historical perspectives. The core text does not depend on this material; it is provided to give students a broader understanding of calculus.

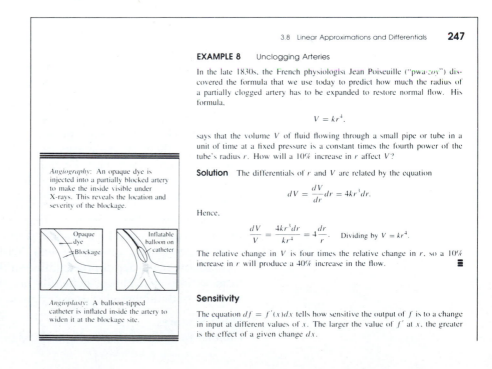

3.8 Linear Approximations and Differentials   **247**

EXAMPLE 8   Unclogging Arteries

In the late 1830s, the French physiologist Jean Poiseuille ("pwazoy") discovered the formula that we use today to predict how much the radius of a partially clogged artery has to be expanded to restore normal flow. His formula,

$$V = kr^4,$$

says that the volume V of fluid flowing through a small pipe or tube in a unit of time at a fixed pressure is a constant times the fourth power of the tube's radius r. How will a 10% increase in r affect V?

Solution   The differentials of r and V are related by the equation

$$dV = \frac{dV}{dr} dr = 4kr^3 dr.$$

Hence,

$$\frac{dV}{V} = \frac{4kr^3 dr}{kr^4} = 4\frac{dr}{r}. \quad \text{Dividing by } V = kr^4.$$

The relative change in V is four times the relative change in r, so a 10% increase in r will produce a 40% increase in the flow. ▄

Sensitivity

The equation $df = f'(x)dx$ tells how sensitive the output of f is to a change in input at different values of x. The larger the value of f' at x, the greater is the effect of a given change dx.

Angiography: An opaque dye is injected into a partially blocked artery to make the inside visible under X-rays. This reveals the location and severity of the blockage.

Opaque dye — Blockage

Inflatable balloon on catheter

Angioplasty: A balloon-tipped catheter is inflated inside the artery to widen it at the blockage site.

89. *Inflation and the CPI.* You have just seen a newspaper headline saying that the Consumer Price Index rose 4% last year. Assuming that this was caused by a constant inflation rate, what was that rate?

90. *Runaway inflation.* At the end of 1988, the Consumer Price Index in Brazil was increasing at the continuous annual rate of 800%. (Yes, it was.) How many days does it take prices to double at this rate? To find out, solve the equation

$$p_0 e^{8t} = 2p_0$$

for t, and convert your answer to days, rounding your answer to the nearest day.

Brazil's hyperinflation

Year-to-year change in Consumer Price Index; monthly data in percent

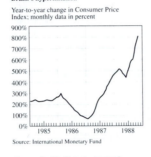

Source: International Monetary Fund

(*Source: Wall Street Journal*, Thursday December 8, 1988, p. 1.)

91. *The purchasing power of the dollar.* How long will it take the purchasing power of the dollar to fall to half its present value if the inflation rate holds at a steady $r\%$?

92. *Oil depletion.* Suppose that the amount of oil pumped from one of the canyon wells in Whitter, California, decreases at the continuous rate of 10% a year. When will the well's output fall to a fifth of its present value?

93. *Continuous price discounting.* To encourage buyers to place 100-unit orders, your firm's sales department applies a continuous discount that makes the unit price a function $p(x)$ of the number of units x ordered. The discount decreases the price at the rate of $0.01 per unit ordered. The price per unit for a 100-unit order is $p(100) = \$20.09$.
  a) Find $p(x)$ by solving the following initial value problem:

  Differential equation: $\dfrac{dp}{dx} = -\dfrac{1}{100}\,p$,

  Initial condition: $p(100) = 20.09$.

  b) Find the unit price $p(10)$ for a 10-unit order and the unit price $p(90)$ for a 90-unit order.

c) The sales department has asked you to find out whether it is discounting so much that the firm's revenue, $r(x) = x \cdot p(x)$, will actually be less for a 100-unit order than, say, for a 90-unit order. Reassure them by showing that r has its maximum value at $x = 100$.
  d) Graph the revenue function $r(x) = xp(x)$ for $0 \leq x \leq 200$.

94. *Subindices of the Consumer Price Index.* There are many "subindices" of the Consumer Price Index—separate indices for food, rent, medical care, and so on. Each index has its own inflation rate. From March 1987 to March 1988, the costs of food, rent, and medical care rose at 3%, 5%, and 6.4%, respectively. At these rates,
  a) how long does it take for costs to increase 50%?
  b) how long does it take for costs to double? (Use the Rule of 70.)

Variable Inflation Rates

When the inflation rate k varies with time instead of being constant, the formula $p = p_0 e^{kt}$ in Eq. (11) no longer gives the solution of the equation $dp/dt = kp$. The corrected formula is

$$p(t) = p_0 e^{\int_0^t k(\tau)\,d\tau} \qquad (12)$$

Use this formula to answer the questions in Exercises 95 and 96.

95. Suppose that $p_0 = 100$ and that $k(t) = 0.04/(1 + t)$, a rate that starts at 4% when $t = 0$ but decreases steadily as the years pass.
  a) Find $\int_0^9 k(\tau)\,d\tau$.
  b) Use Eq. (12) to find the value of $p(t)$ when $t = 9$ yr.
  c) Determine $p(9)$, assuming that the inflation rate was the (constant) average of $k(\tau)$ for the 9 years. Compare your solutions in parts (b) and (c), and discuss.

96. Suppose that $p_0 = 100$ and that the inflation rate is $k(t) = 1 + 1.3t$ (as it was in Brazil during the first few months of 1987).
  a) What does the formula in Eq. (12) look like in this case?
  b) Find $p(1)$, $p(2)$, and the 1-yr and 2-yr percentage increases in the associated Consumer Price Index.

97. The population of rabbits in a certain area is given by

$$P(t) = \frac{1000}{1 + e^{4.8 - 0.7t}}.$$

where t is the number of months after a few rabbits are released.
  a) Estimate the initial number of rabbits released.
  b) What is the maximum possible number of rabbits that this area can sustain?
  c) When, if ever, will the number of rabbits be 700? 1200? Give reasons for your answers.
  d) When will the rate at which the rabbits reproduce be a maximum? What is this rate?

Applications. Applications in the text are real (many with sources cited), and their presentations are self-contained. Students will not need any experience with the fields from which the applications are drawn. Applications both motivate the mathematics and provide an intuitive foundation for further study in these fields.

Exercises sets. There are more than 6000 exercises in the text. Each set of exercises is carefully graded to run from routine at the beginning to more challenging toward the end. Within this framework, the exercises follow the order of presentation in the text. Applications, coming from various disciplines, generally are titled for easy reference. Answers to odd-numbered exercises appear in the back of the text.

Other Features

Chapter overview. Each chapter begins with a preview of the material in the chapter, its significance in calculus, and its relationship to previous and forthcoming material.

Examples. The text contains more than 900 examples to help students grasp the material developed.

Theorems and definitions are boxed and shaded for easy reference.

Key equations are boxed for easy reference.

Supplements

Technology Resource Manuals
Each manual contains keystroke-level instructions for using the technological tool for calculus, as well as concrete examples and machine-specific calculus "toolbox" programs. Once entered, these programs allow you to perform specific calculus tasks at the touch of a few buttons! While not required for use with the text (much of the material can be found in the manufacturer's Owner's Guide), they are recommended as a time-saver. Separate manuals available for TI, Casio, Sharp, and HP graphing calculators and also for computer algebra systems.

Analyzer★ for Macintosh®. This interactive software allows for manipulation of variables and equations. It can graph a function of a single variable and overlay graphs of other functions. It can also differentiate, integrate, and find roots, maxima, minima, and inflection points, as well as vertical asymptotes. It can compose functions, graph polar and parametric equations, display families of curves, and make animated sequences with changing parameters.

Calculus Explorer for IBM®-PC® and compatibles. Contains an assortment of calculus utility programs. The Explorer is highly interactive and allows for the manipulation of variables and equations. The Explorer provides user-friendly operation through an easy-to-use menu-driven system. An accompanying manual includes sections covering each program.

For the Instructor

Answer Book. Contains brief answers for all exercises in the text.

Instructor's Solutions Manual. This supplement contains the worked-out solutions for all exercises, as well as answers to selected Explorations, from the text.

Printed Test Item Bank. Provides a convenient collection of test questions. Includes 30 items per text section, covering the section's learning objectives.

For the Student
Student's solutions manual. Contains carefully worked-out solutions to all of the odd-numbered exercises in the text. In two volumes.

Acknowledgments

This text has been thoroughly class-tested by more than 100 instructors and 5000 students in its preliminary edition. Class-testers, both students and instructors, provided feedback, which was incorporated into this edition.

In-depth reviewers.   The following instructors provided detailed, page-by-page feedback throughout the development process. We would like to thank them for their diligence and assistance.

Carol Benson, University High School

Steven Bianco, University of North Carolina at Wilmington

John Brunsting, Hinsdale Central High School

Barbara Collins, McCullough High School

William Coroscio, Elmira South Side High School

Arthur Finco, Indiana University, Purdue University
   at Fort Wayne

Marsha Finkel-Babadi, University of North Florida

Gregory Foley, Sam Houston State University

Frank Greene, Essex Community College

John Hanna, Teaneck High School

Zenas Hartvigson, University of Colorado at Denver

Jane B. Heinmiller, Capital University

Kiyoshi Igusa, Brandeis University

Cathy Jahr, Martin Westview High School

Luella Johnson, Buffalo State College

Helmer Junghans, Montgomery College

Douglas A. Klumpe, West Lafayette Junior/Senior High School

James Lang, Valencia Community College–East

Alan Lipp, Williston Northampton School

Judith Massey, Livingston University

Mary McCammon, Penn State University

Connie Meek, Craig High School

Darrell Minor, Columbus State Community College

Robert Mizer, Upper Arlington High School

Gary Phillips, Oakton Community College

James L. Smith, Muskingum College

Thomas F. Sweeney, Russell Sage College

Carol Swindell, Cookeville High School

Dixie Trollinger, Mainland High School

Gerald White, Western Illinois University

Debra Wilson, Spokane Educational Service District 101

Ben F. Zirkle, Virginia Western Community College

Survey Respondents.   The following instructors and their students provided feedback via chapter-by-chapter, midterm, and end-of-term surveys. We would like to thank them for their reports.

Bill Barkland, University of Colorado at Denver

Brenda P. Batten, Patrick Henry Academy

Harriet Briscoe, Brookwood High School
Joseph G. Brown, Ferrum College
James Carpenter, Iona College
Pam Caudill, Spencer County High School
Terry Chen, University of Colorado at Denver
Frances Coleman, Ackerman & Weir High School
Loyce Collenback, Robert E. Lee High School
Lynette Corley, Granby High School
James Corsica, Prairie School
Carolyn Cox, Milford Mill High School
Marlin DeWeerdt, Hempstead High School
Robert Fitzsimmons, Iona College
Carl Gatje, Western Branch High School
Albert Goetz, Ramaz Upper School
Samuel Gough, East Mecklenburg High School
Robert P. Hammond, Deerfield Academy
Wanda S. Henry, Deerfield Academy
Bruce H. Hoelter, Raritan Valley Community College
Larry Kaber, Flathead High School
Ellen Kamischke, Interlochen Arts Academy
Mike Koehler, Blue Valley North High School
James Kozman, Franklin Heights High School
Iris Lynnette Lopez, Universidad de Puerto Rico–Arecibo
David Love, Odessa High School
G. Ray Meester, Horizon Senior High School
Carroll Rabon, Ferrum College
Ivan Schukei, University of South Carolina–Beaufort
Dave Slomer, Winton Woods High School
Donald J. Sparks, Kennesaw State College
Jeff Spring, Scotch Plains–Fanwood High School
James Stones, Spring Woods High School
Eddie Warren, University of Texas at Arlington
Harlan Weber, Sheboygan South High School
Peggy Wielenberg, Prairie School
Sharon Wiest, University of Colorado at Denver
Roland Young, Deerfield Academy
Edward Zeidman, Essex Community College

We would particularly like to express our gratitude to Ray Barton of Olympus High School in Salt Lake City, Utah, for his advice and ongoing help in developing the manuscript.

We want to express our special appreciation to our supplements coordinator, Penny Dunham of Muhlenberg College in Allentown, Pennsylvania.

Special thanks also go to David and Mary Winter of Michigan State University in East Lansing, Michigan, for their work providing the answers for the text and the solutions manual.

We also want to express our gratitude to Guillermo Barberena III of Eastern Hills High School in Fort Worth, Texas, Tommy Eads of North Lamar High School in Paris, Texas, and Charlie Reno of Euclid High School in Euclid, Ohio, for their generous advice and help.

Thanks go out to Sherrie Lowery for typing the manuscript and to Renee Hartshorn, Amy Edwards, and Greg Ferrar, who prepared the artwork for the manuscript.

We appreciate all the assistance from the staff at Addison-Wesley. We particularly appreciate the expert assistance and advice provided by Joe Will and Elka Block, Development Editors for this edition. We would also like to thank Jenny Bagdigian, Andrew Fisher, Jerome Grant, Kathy Manley, and the rest of the Finney/Thomas/Demana/ Waits book team: Peter Blaiwas, Amy Branowicki, Barbara Brooks, Susan Howard, Connie Hulse, Aaron Klebanoff, Trisha Mack, Cristina Malinn, Pat Oduor, Laurie Petrycki, and Joe Vetere for their hard work in bringing this text to fruition.

Accuracy

Addison-Wesley is committed to publishing high quality educational materials. Rigorous editing and proofreading procedures ensure the highest level of accuracy possible. This commitment to quality motivates every step of the publishing process. We hope you are satisfied with this text, and we welcome any response you may have.

As is traditional at this point in a Preface, we acknowledge our full responsibility for any errors that remain in the text. With the enhanced accuracy/quality control process we and Addison-Wesley have put into place with this text, we feel confident that we have indeed made this text as error free as is humanly possible. Your Addison-Wesley representative will be happy to describe the process to you.

Some of our colleagues have suggested that since we are so confident about the book's accuracy, we should go one step further and offer payment for any remaining errors that are found. With this in mind, but primarily because we want to detect any remaining errors quickly and correct them in subsequent printings, we hereby are offering a reward of $5 per *mathematical* error to the first instructor or student who reports the error. Any mathematical error that has follow-through effects will be counted as two errors only. Please report any errors to us in care of:

Bert Waits
P.O. Box 3135
Columbus, OH 43210 USA

RLF	Monterey, California
GBT, Jr.	State College, Pennsylvania
FDD	Columbus, Ohio
BKW	Columbus, Ohio

Prologue: Using Graphing Utilities

Devices that automatically produce graphs are called **graphing utilities** or **graphers**. The most practical graphing utilities today are graphing calculators. Larger computers with graphing software are also graphing utilities.

In this book, we assume that you have a grapher to use at all times. Because there are a variety of quality graphers available, we also assume that the ones used—even within one calculus class—are likely to differ.

Different graphers use different terminology and different keying or menu selection steps. We try to cover the differences among graphing calculators in the *Technology Resource Manuals for Calculus* (we'll call it the *Resource Manual*) written as a supplement to accompany this book. Nonetheless, we still had to choose a single terminology to use herein. In each instance, we considered the possibilities and made what we feel is the best "generic" choice possible even though it sometimes matches what is used by a particular grapher. Other than implying our preferences for terminology in those individual instances, we do not recommend one grapher over another. Indeed, most of the ones we have used are fine products, and the owner of any of them should be able to develop expertise needed for this book fairly efficiently.

The following lists are arranged approximately in the order in which the terms appear in the book. This means that as you proceed through the book, you should become familiar with each list from the top down.

Basic Terminology:
(You may wish to note equivalent terminology for your calculator in the margin.)

Home screen   The calculator screen for performing calculations, entering data, editing, and so forth.

Viewing window (or view, or window)   The portion of the coordinate plane shown in the display screen of the grapher. The *standard viewing window* has view dimensions [−10, 10] by [−10, 10].

View dimensions (or window dimensions)   The minimum to maximum *x*-values and the minimum to maximum *y*-values in the viewing window. We often speak of a viewing window with view dimensions [*x*Min, *x*Max] by [*y*Min, *y*Max].

Pixel   A small rectangle on the display screen that can be illuminated to represent a point in the coordinate plane. In a rectangular viewing window, the pixels are arranged in a rectangular array of rows and columns.

x*Min*, x*Max*   The minimum and maximum *x* values, respectively, in a viewing window. The first screen coordinate of each pixel in the leftmost pixel column is *x*Min. The first screen coordinate of each pixel in the rightmost pixel column is *x*Max. When a function $y = f(x)$ plots across the *N* pixel columns of the viewing window, *x* takes on values from *x*Min to *x*Max in δx increments, where

$$\delta x = \frac{x\text{Max} - x\text{Min}}{N - 1}.$$

y*Min*, y*Max*   The minimum and maximum *y*-values, respectively, in a viewing window. The second screen coordinate of each pixel in the bottom pixel row is *y*Min. The second screen coordinate of each pixel in the top pixel row is *y*Max.

Screen coordinates   The coordinate pair that names a screen pixel. If a viewing window has *N* pixel columns and *M* pixel rows, then, from left to right, the pixel columns have

$$x\text{Min} + (i - 1)\,\delta x \qquad (i = 1, 2, \ldots, N)$$

as their first screen coordinates. From bottom to top, the pixel rows have

$$y\text{Min} + (j - 1)\delta y \qquad (j = 1, 2, \ldots, M)$$

as their second screen coordinates, where

$$\delta x = \frac{x\text{Max} - x\text{Min}}{N - 1} \qquad \text{and} \qquad \delta y = \frac{y\text{Max} - y\text{Min}}{M - 1}.$$

If a pixel is part of the graph of $y = f(x)$, (for example, as in TRACE), its first screen coordinate is x_i as computed above. Its second coordinate is $f(x_i)$.

Scale unit   The unit length on a coordinate axis.

x-*scale*, y-*scale*   The distances between scale marks on the *x*-axis and the *y*-axis, respectively. If *x*-scale = 1, the marks on the *x*-axis will show the scale unit.

Cursor control(s)   The device(s) that we use to move the screen cursor from pixel to pixel.

t*Min*, t*Max*   In parametric mode, the minimum and maximum values, respectively, of the parameter *t*.

t-*step*   The increment used as a parameter t changes from tMin to tMax.

Grapher Procedures:

(You can consider each term that we use here simply as a name for *any* "toolbox" procedure that you can program into your calculator to do the indicated task. You can create your own program or use one from the *Resource Manual*. You may wish to note the program name that you use beside the name that we use.)

GRAPH (or PLOT).   Draws a graph in the viewing window.

SQUARE.   Makes the scale unit appear to be the same length on each axis. This results in a graph with no scale distortion.

TRACE   Allows you to move the cursor from pixel to pixel along a graph. The x-value that shows on the screen is the first screen coordinate of the selected pixel. The y-value is the associated function value.

ZOOM-IN,ZOOM-OUT   ZOOM-IN magnifies a portion of the coordinate plane so that you can see finer detail. ZOOM-OUT reduces a view to be part of a larger view. The effects are much like those of zooming in and zooming out with a camcorder.

SOLVE (or ROOT)   Finds a real-number solution of an equation of the form $f(x) = 0$ or, equivalently, finds a real-number zero of f, or an x-intercept of its graph. Applied to $(f - g)(x) = 0$, this procedure finds the x-coordinate of a solution of $f(x) = g(x)$ or, equivalently, finds the x-coordinate of a point of intersection of the graphs of f and g.

STORE   Stores a constant, a variable, an expression, and so on, in an assigned internal location.

NDER   Computes the numerical derivative of a function.

NDER 2   Computes the numerical second derivative of a function by applying NDER twice, that is, NDER 2 (f) = NDER (NDER (f)).

NINT   Computes the numerical integral of a function.

Screen Formats:

(If a particular format is unavailable, you may skip activities using that format.)

Dot and Connected   In dot format, only computed pixels are illuminated. In connected format, computed pixels are illuminated, and "in-between" pixels are also illuminated to give the graph more of a visual "connected" effect, an effect that may be desirable with a continuous function.

Sequential and Simultaneous   For two functions y_1 and y_2, the graph of y_1 can be plotted first, followed by the graph of y_2 (sequential), or the graphs of y_1 and y_2 can be plotted at the same time (simultaneous).

Axes On and Axes Off   We usually want to plot with the axes shown (axes on). Sometimes, we want to see better what happens along an axis, so we "hide" the axes pixels (axes off).

Contents

CHAPTER 12   **913**

Appendixes   **APP-1**

Prerequisites for Calculus

OVERVIEW   This chapter reviews the most important things you need to know to start learning calculus. It also introduces the use of a graphing utility as a tool to investigate mathematical concepts and ideas, to provide support for analytic work, and to solve problems with numerical approximations when analytic methods fail or are impractical. The emphasis is on functions and graphs, the main building blocks of calculus.

In calculus, functions are the major tools for describing the real world in mathematical terms, from temperature variations to planetary motions, from business cycles to brain waves, and from population growth to heartbeat patterns. Many functions have particular importance because of the kind of behavior they describe. For example, trigonometric functions describe cyclic, repetitive activity; exponential and logarithmic functions describe growth and decay; and polynomial functions can approximate these and most other functions. We will examine some of these functions in this chapter and meet others in later chapters.

1.1   Coordinates and Graphs in the Plane

To assign coordinates to a point in a plane, we start with two number lines that cross at their zero points at right angles. Each line represents the real numbers, which are the numbers that can be represented by decimals. Figure 1.1 on the following page shows the usual way of drawing the lines, with one line horizontal and the other vertical. The horizontal line is called the **x-axis** and the vertical line the **y-axis**. The point at which the lines cross is the **origin**.

On the x-axis, the positive number a lies a units to the right of the origin, and the negative number $-a$ lies a units to the left of the origin. On the y-axis, the positive number b lies b units above the origin, while the negative number $-b$ lies b units below the origin. The marks on each axis show the *scale* used for that axis. Each scale is uniform along its entire axis, but quite often it is helpful—even necessary—to make the **x-scale** and **y-scale** different sizes.

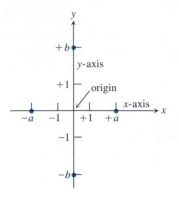

1.1 The intersecting x- and y-axes determine a plane called the *Cartesian plane* or *xy-plane*. The two axes can show different scales, but on each axis the scaling is symmetric about the origin.

The coordinates defined here are often called *Cartesian* coordinates, after their chief inventor, René Descartes (1596–1650).

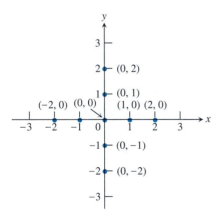

1.3 Axis points can be labeled in two ways.

With the axes in place, we assign a pair (a, b) of real numbers to each point P in the plane. The number a is the number at the foot of the perpendicular from P to the x-axis. The number b is the number at the foot of the perpendicular from P to the y-axis. Figure 1.2 shows the construction. The notation (a, b) is read "a b."

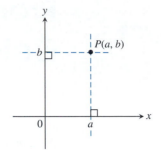

1.2 The point P corresponds to the pair (a, b) with a and b marking where the perpendiculars from P meet the x- and y-axes, respectively.

The construction can be reversed to assign a point in the plane to each ordered pair of real numbers. The point assigned to the pair (a, b) is the point where the perpendicular to the x-axis at a crosses the perpendicular to the y-axis at b. Thus the assignment of number pairs is a one-to-one correspondence between the points of the plane and the set of all ordered pairs of real numbers. Every point has a pair and every pair has a point, so to speak.

The number a from the x-axis is the **x-coordinate** of P. The number b from the y-axis is the **y-coordinate** of P. The pair (a, b) is the **coordinate pair** of the point P. It is an **ordered pair**, with the x-coordinate first and y-coordinate second. To show that P has the coordinate pair (a, b), we sometimes write the P and (a, b) together: $P(a, b)$.

The points on the coordinate axes now have two kinds of numerical labels: single numbers from the axes and paired numbers from the plane. They match up as suggested in Fig. 1.3. In particular, as you can see, every point on the x-axis has y-coordinate zero, and every point on the y-axis has x-coordinate zero. The origin is the point $(0, 0)$.

Directions and Quadrants

Motion from left to right along or parallel to the x-axis is said to be in the **positive x-direction**. Motion from right to left is in the **negative x-direction**. Along or parallel to the y-axis, the positive direction is up, and the negative direction is down.

The origin separates the axes into the **positive x-axis** to its right, the **negative x-axis** to its left, the **positive y-axis** above it, and the **negative y-axis** below it. The axes divide the rest of the plane into four regions called quadrants, numbered I, II, III, and IV (Fig. 1.4).

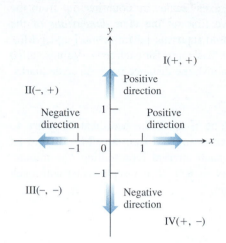

1.4 Directions along or parallel to the axes. Roman numerals label the quadrants.

Intervals

On an axis, or on any number line, the set of all real numbers that lie *strictly between* two fixed numbers a and b is an **open interval**; a and b are called **endpoints** of the interval. The interval is "open" at each end because it contains neither of its endpoints. Intervals that contain both endpoints are **closed**. Intervals that contain one endpoint but not both are **half-open** (Fig. 1.5). Half-open intervals could just as well be called half-closed, but no one seems to call them that. Any point of an interval other than the endpoints is called an **interior point**.

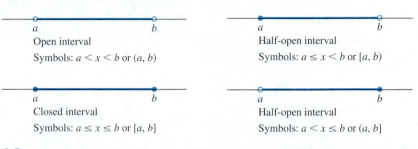

1.5 Depicting intervals. The open dot in the picture and the parenthesis in the symbol mean that the endpoint *is not* part of the interval. The solid dot in the picture and the bracket in the symbol mean that the endpoint *is* part of the interval.

Intervals that have two endpoints are **finite intervals.** The number line itself and rays on the number line are called **infinite intervals** (Fig. 1.6).

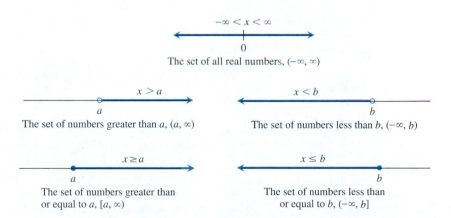

1.6 Infinite intervals. The symbol ∞ (infinity) in the notation is used merely for convenience; it is not to be taken as a suggestion that there is a number ∞.

A Viewing Window on the Coordinate Plane

The points in the coordinate plane that correspond to a set of ordered pairs is a **graph** of the set. Whether we show a graph on paper or on the display screen of a graphing utility, the picture represents a small portion of the coordinate plane. We will call that portion the *viewing window* or *view* (Fig. 1.7 on the following page).

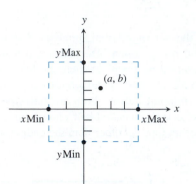

1.7 The dashed rectangle shows a viewing window with view dimensions [xMin, xMax] by [yMin, yMax]. The scale marks are spaced according to values of x-scale and y-scale that we enter into the grapher. When x-scale and y-scale are both 1, the scale marks on each axis show the scale unit. See the *Resource Manual* or your *Owner's Guide* to find how to adjust viewing dimensions in your grapher.

SCALING

When we set *x-scale* on a grapher, we are setting the distance between the scale marks that appear along the x-axis or along the top or bottom of the viewing window. We use *scale unit* to stand for the length 1. Thus, if x-scale is 1, the scale marks will be 1 unit apart; they will actually show the scale unit. Similar statements can be made for *y-scale*. When scales are set so that the *lengths* of the scale units on both axes are the same (whether marked or not), the viewing window is *square* in the sense that scale distortions have been removed.

On paper, we usually mark the axes' scales by counting out from the origin. On an electronic grapher, we first set the *view dimensions* of the viewing window in terms of two closed intervals [xMin, xMax] and [yMin, yMax]. These establish the size of the *scale unit* for each axis. Values called *x-scale* and *y-scale* in the grapher establish the distance between scale marks.

EXAMPLE 1

The graph in Fig. 1.8 shows how long it takes the heartbeat to return to normal after a person runs. The x-axis is marked from an xMin of 0 to an xMax of 10. The x-scale is 1 with each interval representing one minute. The y-axis has yMin = 80 and yMax = 200. The y-scale is 10 with each interval representing 10 heartbeats/min.

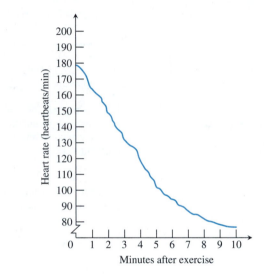

1.8 How the heartbeat returns to a normal rate after running. (Adapted from James F. Fixx's *The Complete Book of Running*, New York: Random House, 1977.)

When the scale units on the axes are the same length in a viewing window, one unit of distance up and down in the plane will then *look* the same as one unit of distance right and left. As on a surveyor's map or a scale drawing, line segments that are supposed to have the same length will look as if they do. The measure of angles will also appear to be correct. Circles will look like circles and squares like squares. When we make the scale units the same length on the axes, we say that we *square* the viewing window.

EXPLORATION 1

The Viewing Window on Your Grapher

To do the following, be sure you first understand the meaning of (a) a square viewing window and (b) a scale unit.

1. Find how to set the view dimensions on your grapher. Then set [xMin, xMax] = [−10, 10] and [yMin, yMax] = [−10, 10], the view

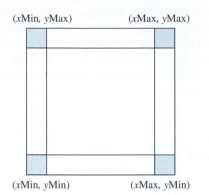

(xMin, yMax) (xMax, yMax)

(xMin, yMin) (xMax, yMin)

1.9 The screen coordinates of the four corner pixels.

AGREEMENT

In this book, the phrase *sketch a graph* will mean to use pencil and paper. *GRAPH* (all capitals) will suggest (but not necessarily require) the use of a grapher. All other phrases, such as *graph* (not all capitals) or *draw a graph*, will mean to use a method of your choice.

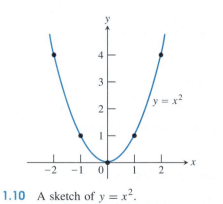

1.10 A sketch of $y = x^2$.

dimensions of the *standard viewing window*. Use the cursor controls to move the cursor left, right, up, and down, and observe the *screen coordinates* of the cursor location. What are the screen coordinates of the four corners of the viewing window?

2. Is your viewing window square? Why?

3. What are your values for x-scale and y-scale? Do the scale marks in your viewing window show either scale unit? Explain.

4. How does the viewing window change if you change x-scale and y-scale?

5. Square your viewing window. Be prepared to explain your method and give a convincing argument that it works.   ✿

The viewing window of a graphing utility consists of a rectangular array of lights called *pixels*. Each pixel can be named with a pair of screen coordinates. A pixel on the border of the viewing window has at least one of xMin (left side), xMax (right side), yMin (bottom), or yMax (top) as a coordinate (Fig. 1.9). Thus, if you know how many rows and columns of pixels there are in your window, you can find how the screen coordinates are spaced.

Graphs of Equations

The points in the xy-plane whose coordinates *satisfy* an equation like $y = x^2$ make up the **graph of the equation**. Graphs provide a way to visualize equations.

EXAMPLE 2 Sketch a graph of the equation $y = x^2$ using pencil and paper.

Solution

STEP 1: Make a table of xy-pairs that satisfy the equation $y = x^2$.

x	$y = x^2$
-2	4
-1	1
0	0
1	1
2	4

STEP 2: Sketch coordinate axes with x-scale $= 1$ and y-scale $= 1$. Let [xMin, xMax] be about $[-3, 3]$ and [yMin, yMax] be about $[-1, 5]$. Then plot the points (x, y) whose coordinates appear in the table.

STEP 3: Sketch a smooth curve through the plotted points. Label the curve with its equation (Fig. 1.10).   ≡

EXAMPLE 3 Graph the equation $y = x^2$ using a graphing utility.

STEP 1: Enter the equation $y = x^2$.

STEP 2: Set the view dimensions of the viewing window.

STEP 3: Then GRAPH (Fig. 1.11).   ≡

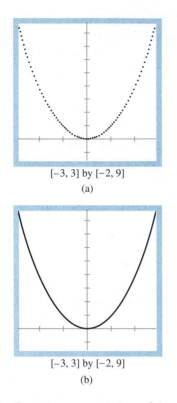

[−3, 3] by [−2, 9]

(a)

[−3, 3] by [−2, 9]

(b)

1.11 *Computer representations* of the graph of $y = x^2$ in (a) dot format and (b) connected format.

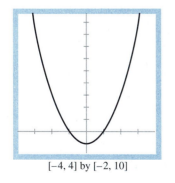

[−4, 4] by [−2, 10]

1.12 A computer representation of the graph of $y = x^2 - 1$ suggests x-intercepts of −1 and 1 and a y-intercept of −1. For strong support, TRACE along the graph and watch the screen coordinates as you cross the axes.

The graphs in Examples 2 and 3 resemble each other but are not identical. The one shown from the grapher in Fig. 1.11(a) is formed (*dot format*) by pixels using pixel x-coordinates and calculated y-coordinates. Some graphers have an option (*connected format*) that fills in from dot to dot to suggest more of a smooth curve (Fig. 1.11b). Grapher viewing windows shown in this book will usually be in connected format.

Intercepts and Graphic Support

Our use of technology will follow two major themes.

1. Do algebraically (using algebra) or analytically (using calculus), *support graphically.*

2. Do graphically, *confirm* algebraically or analytically.

As an example, consider the points on the graph of an equation where the graph touches, or crosses, the axes. These points are the **intercepts.** We can find the (algebraic) **x-intercepts** by setting y equal to 0 in the equation and solving for x. We find the (algebraic) **y-intercepts** by setting x equal to 0 and solving for y. We can *support graphically* our algebraic findings by graphing the equation and seeing where the graph meets the axes.

EXAMPLE 4

Using algebra, find the intercepts of the graph of the equation $y = x^2 - 1$. Support your answer graphically.

Solution

The x-intercepts:

$$y = x^2 - 1 \qquad \text{Write the given equation.}$$

$$0 = x^2 - 1 \qquad \text{Let } y = 0.$$

$$x^2 = 1$$

$$x = 1, -1 \qquad \text{Solve for } x.$$

The y-intercepts:

$$y = x^2 - 1 \qquad \text{Write the given equation again.}$$

$$y = (0)^2 - 1 \qquad \text{Let } x = 0.$$

$$y = -1 \qquad \text{Solve for } y\text{—not much to do in this case.}$$

The graph of $y = x^2 - 1$ (Fig. 1.12) visually supports x-intercepts of 1 and −1 and a y-intercept of −1. Using TRACE on the graph gives even stronger support. ≡

We could have turned the above steps around by first finding the intercepts graphically (a conjecture) and then confirming algebraically. The algebra could be as shown in Example 4 or simply as substitution of the suspected values into the equation.

Complete Graphs

Many graphs will extend beyond a viewing window no matter how great the magnitude of xMin, xMax, yMin, and yMax. For other graphs, the scales will be so great that some behavior will actually be "hidden" within the curve that appears in the viewing window. Therefore, when we are asked to "Draw *complete* graphs," it is understood that the portion in the viewing window must suggest *all the important features* of the graph. A complete graph will contain no "hidden behavior."

The graph $y = x^2 - 1$ in Fig. 1.12 is complete. It suggests all the intercepts, a lowest point at $(0, -1)$, symmetry about the y-axis, and that the graph increases without bound as we move away from $x = 0$.

How will we know for sure that a graph is complete? The answer lies in calculus, as we will see in Chapter 4. There we will learn to use a marvelous mathematical tool called a *derivative* to find the exact shape for a great many curves. We will see that a modern role of differential calculus is to analyze hidden behavior.

In the meantime, we will use what we have learned previously to draw complete graphs. For a *line*, a graph that shows the intercepts is complete. For a *quadratic equation*, a graph (a parabola) is complete if it suggests the vertex, the two branches, and the intercepts. For a *cubic equation*, a complete graph suggests, in general, a local high point, a local low point, and the intercepts. Possible complete graphs of quadratics and cubics are shown in Figs. 1.13 and 1.14.

$a < 0$ $a > 0$

1.13 Possible complete graphs of $y = ax^2 + bx + c, a \neq 0$.

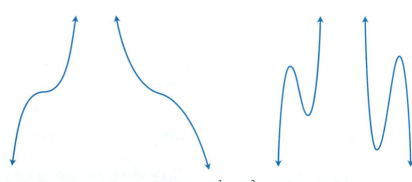

1.14 Possible complete graphs of $y = ax^3 + bx^2 + cx + d, a \neq 0$.

EXAMPLE 5

Draw a complete graph of $y = x^3 - 4x$. Use the graph to estimate the intercepts and the coordinates of any local high or low points. Confirm the intercepts algebraically.

Solution   The graph of $y = x^3 - 4x$ in Fig. 1.15 is complete according to Fig. 1.14. It appears from Fig. 1.15 that $y = 0$, $x = -2$, $x = 0$, and $x = 2$ are the intercepts. We confirm this algebraically as follows.

The y-intercept:

$$y = x^3 - 4x$$
$$y = 0^3 - 4(0) \qquad \text{Let } x = 0.$$
$$y = 0$$

The x-intercepts:

$$y = x^3 - 4x$$
$$0 = x^3 - 4x \qquad \text{Let } y = 0.$$
$$0 = x(x^2 - 4)$$
$$0 = x(x - 2)(x + 2)$$
$$x = 0, 2, -2$$

There appears to be a local high point near $x = -1$. We can estimate its coordinates by reading directly from the graph or by placing the cursor near the high point and reading the coordinates of the cursor directly from the screen to be about $(-1.2, 3.1)$. Similarly, the coordinates of the local low point near $x = 1$ are about $(1.2, -3.1)$.   ☰

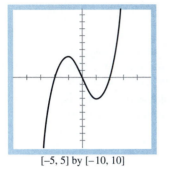

[−5, 5] by [−10, 10]

1.15   A complete graph of $y = x^3 - 4x$.

EXAMPLE 6

Draw a complete graph of $y = x^3 - 2x^2 + x - 30$.

Solution   A complete graph must look like one of the four possibilities in Fig. 1.14. Fig. 1.16 gives three views of the graph of the equation. Only part (c) looks like one of the four in Fig. 1.14.

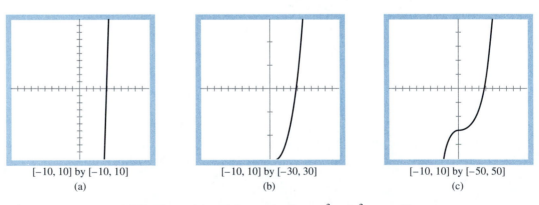

[−10, 10] by [−10, 10]   [−10, 10] by [−30, 30]   [−10, 10] by [−50, 50]
(a)   (b)   (c)

1.16   Three views of the graph of $y = x^3 - 2x^2 + x - 30$.

Knowing what a complete graph should look like and seeing the relatively flat portion of the graph in Fig. 1.16(c) should alert us to the possibility that some important features are hidden from view. In Fig. 1.17, we ZOOM-IN on the flat portion and see a local high point and a local low point. By using the scale marks, we can estimate from Fig. 1.17 that the high point is about $(0.3, -29.8)$ and the low point is about $(1, -30)$.

Notice that the *y-intercept* is $y = -30$. We have to look back to Fig. 1.16(c) to estimate the *x*-intercept to be about 4. Thus, it takes *both* Figs. 1.16(c) and 1.17 to constitute a complete graph of the equation.

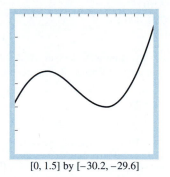

[0, 1.5] by [−30.2, −29.6]

1.17 Using ZOOM-IN on the view in Fig. 1.16(c) gives this closer look at the flat portion of the graph of $y = x^3 - 2x^2 + x - 30$. To help you understand ZOOM-IN, compare the view dimensions and the scales on the axes here with those in Fig. 1.16(c).

Exercises 1.1 _____

In Exercises 1 and 2, determine which points lie in the indicated viewing window.

1. Viewing window: $[-5, 5]$ by $[-10, 10]$
Points: $(-6, 2)$, $(0, 6)$, $(5, -5)$, $(-4, 12)$

2. Viewing window: $[0, 10]$ by $[0, 100]$
Points: $(1.5, 17.5)$, $(0.5, -1)$, $(90, 8)$, $(10, 100)$

In Exercises 3 and 4, determine a viewing window that contains the three indicated points.

3. Points: $(-17, 3)$, $(21, -12)$, $(18, 76)$

4. Points: $(-42, 31)$, $(8, -6)$, $(53, -89)$

In Exercises 5–8, view dimensions for a grapher are given. Choose appropriate values for *x*-scale and *y*-scale. Explain why your choices are reasonable.

5. $[-10, 50]$ by $[-50, 50]$ **6.** $[-1, 1.5]$ by $[1, 2]$

7. $[-1, 0]$ by $[-100, 100]$ **8.** $[-50, 150]$ by $[-2, 2]$

Find the intercepts and sketch a graph of each equation in Exercises 9–14. Remember, "sketch" means to graph without using a graphing utility.

9. $y = x + 1$ **10.** $y = -x + 1$
11. $y = -x^2$ **12.** $y = 4 - x^2$
13. $x = -y^2$ **14.** $x = 1 - y^2$

In Exercises 15–20, which view gives the best complete graph of the indicated equation?

15. $y = 2x^2 - 40x + 150$
  a) $[-5, 5]$ by $[-5, 5]$ **b)** $[-10, 10]$ by $[-10, 10]$
  c) $[10, 30]$ by $[-50, 100]$
  d) $[-100, 100]$ by $[-10, 10]$
  e) $[-10, 30]$ by $[-100, 100]$

16. $y = -3x^2 + 9x - 20$
  a) $[-10, 10]$ by $[-10, 10]$
  b) $[-10, 10]$ by $[-100, 100]$
  c) $[2, 10]$ by $[-100, 10]$
  d) $[-10, 10]$ by $[-3000, 3000]$
  e) $[-300, 300]$ by $[-100, 100]$

17. $y = 20 + 9x - x^3$
  a) $[-10, 10]$ by $[-10, 10]$
  b) $[-1, 10]$ by $[-30, 40]$
  c) $[-5, 0]$ by $[0, 100]$ **d)** $[-5, 5]$ by $[-500, 1000]$
  e) $[-10, 10]$ by $[-50, 50]$

18. $y = x^3 - x + 15$
  a) $[-5, 5]$ by $[-10, 10]$
  b) $[-10, 10]$ by $[-100, 100]$
  c) $[-10, 10]$ by $[-30, 30]$
  d) $[-5, 5]$ by $[-5, 15]$
  e) $[-100, 100]$ by $[-100, 100]$

19. $y = 3x - 800$
a) $[0, 500]$ by $[0, 500]$ b) $[260, 270]$ by $[-10, 10]$
c) $[-10, 10]$ by $[-810, -790]$
d) $[-10, 10]$ by $[-10, 10]$
e) $[-100, 500]$ by $[-1000, 500]$

20. $y = -2x^2 + 500$
a) $[-10, 10]$ by $[-10, 10]$
b) $[240, 260]$ by $[-10, 10]$
c) $[-10, 10]$ by $[490, 510]$
d) $[-100, 400]$ by $[-200, 800]$
e) $[0, 400]$ by $[0, 400]$

Draw a complete graph of the equation in Exercises 21–26. Use the graph to estimate the intercepts. Confirm the intercepts algebraically.

21. $y = 3x - 5$ **22.** $y = 4 - 5x$
23. $y = 10 + x - 2x^2$ **24.** $y = 2x^2 - 2x - 12$
25. $y = 2x^2 - 8x + 3$ **26.** $y = -3x^2 - 6x - 1$

Draw a complete graph of the equation in Exercises 27–34. Use the graph to estimate the intercepts and the coordinates of any local high or low points.

27. $y = x^2 + 4x + 5$ **28.** $y = -3x^2 + 12x - 8$
29. $y = 12x - 3x^3$ **30.** $y = 2x^3 - 2x$
31. $y = -x^3 + 9x - 1$ **32.** $y = x^3 - 4x + 3$
33. $y = x^3 + 2x^2 + x + 5$ **34.** $y = 2x^3 - 5.5x^2 + 5x - 5$

Exercises 35–40 refer to a grapher that uses N horizontal pixels and M vertical pixels on its display screen. (You can replace N and M with the values for your grapher, or you can use $N = 127$ and $M = 63$ if you wish.) Using (x_i, y_j) for the screen coordinates associated with each pixel, the four corner points have the following coordinates:

$$(x_1, y_M) = (x\text{Min}, y\text{Max}) \quad (x_N, y_M) = (x\text{Max}, y\text{Max})$$

$$(x_1, y_1) = (x\text{Min}, y\text{Min}) \quad (x_N, y_1) = (x\text{Max}, y\text{Min})$$

35. To find the pixel with screen coordinates (x_i, y_j), we count i pixels to the right and j pixels up beginning in the lower left corner where $(x_1, y_1) = (x\text{Min}, y\text{Min})$. Show that

$$(x_i, y_j) = (x\text{Min} + (i - 1)\delta x, y\text{Min} + (j - 1)\delta y)$$

where

$$\delta x = \frac{x\text{Max} - x\text{Min}}{N - 1}, \qquad \delta y = \frac{y\text{Max} - y\text{Min}}{M - 1}.$$

(The symbol δ is the lowercase Greek letter "delta," and a symbol like δx is read "delta x." It is a single symbol and not "δ times x." Note that δx is the horizontal distance between the midpoints of any two adjacent pixels.)

36. Let (a, b) be a point in the coordinate plane and (x_i, y_j) the corresponding screen coordinate.
a) Explain why the error in using x_i as an approximation to a is at most $\delta x/2$.
b) Explain why the error in using y_j as an approximation to b is at most $\delta y/2$.

In Exercises 37–40, determine a and b so that the screen coordinates of the $[-10, a]$ by $[-10, b]$ viewing window have the indicated property.

37. $\delta x = \delta y = 1$ **38.** $\delta x = \delta y = 0.1$
39. $\delta x = 0.5, \delta y = 2$ **40.** $\delta x = 2, \delta y = 10$

Decompression Stops

Scuba divers often have to make decompression stops on their way to the surface after deep or long dives. A stop, which may vary from a few minutes to more than an hour, provides time for the safe release of nitrogen and other gases absorbed by the tissues and blood while the body was under pressure. Dives to 33 ft or less do not require decompression stops. Dives to greater depths can be made without return stops if the diver does not stay down too long. The graph in Fig. 1.18 shows the longest times that a diver breathing compressed air may spend at various depths and still surface directly (at 60 ft/min). The times shown are total lengths of the dives, not just the times spent at maximum depth.

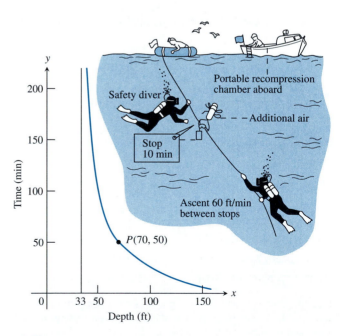

1.18 The coordinates of the points on the curve show how long scuba divers breathing compressed air can safely stay below the surface without a decompression stop. (Data from *U.S. Navy Diving Manual*, NAVSKIPS 250-538.)

The coordinates of the point P in Fig. 1.18 are $x = 70$ and $y = 50$, so a diver going to 70 ft (but no deeper) can return safely without stopping if the total dive time is 50 min or less. Decompression stops are needed for dives plotted above or to the right of the curve because their times exceed the corresponding limits set for their depths. Decompression stops are not needed for dives plotted on the curve or for dives plotted below or to the left of it because the lengths

of these dives do not exceed the limits for their depths. Exercises 41–42 are about these time limits.

41. You have been working at 100 ft below the surface for 1 hr. Do you need a decompression stop on the way up?

42. Which of the following dives need decompression stops and which do not?

   a) (40, 100)   **b)** (100, 40)
   c) (70, 100)   **d)** (50, 50)

1.2   Slope, and Equations for Lines

One of the many reasons calculus has proved so useful over the years is that it is the right mathematics for relating a quantity's rate of change to its graph. Explaining this relationship is one of the goals of this book. Our basic plan is first to define what we mean by the slope of a line. Then we define the slope of a curve at each point on the curve. Later we will relate the slope of a curve to a rate of change. Just how this is done will become clear as the book goes on. Our first step is to find a practical way to calculate the slopes of lines.

INCREMENTS

The symbols Δx and Δy are read "delta x" and "delta y." They denote net changes or *increments* in the variables x and y. The symbol Δ is the capital Greek letter "dee" for "difference." Neither Δx nor Δy denotes multiplication; Δx is not "delta times x," nor is Δy "delta times y." (See Exercise 35 of Section 1.1.)

Slopes of Nonvertical Lines

We calculate the slope of a line from changes in coordinates. The result is a number that helps us visualize how the line is oriented in the coordinate plane. If the (vertical) change in y-coordinates is large in comparison to the (horizontal) change in x-coordinates, the slope is great and the line is steep. Every *non*vertical line has a slope. The calculations will show why vertical lines are an exception.

To begin, let L be a nonvertical line in the plane. Let $P_1(x_1, y_1)$ and $P_2(x_2, y_2)$ be two points on L (Fig. 1.19). We call $\Delta y = y_2 - y_1$ the **rise** from P_1 to P_2 and $\Delta x = x_2 - x_1$ the **run** from P_1 to P_2. Since L is not vertical, $\Delta x \neq 0$, and we may define the **slope** of L to be $\Delta y/\Delta x$, the amount of rise per unit of run. It is conventional to denote the slope by the letter m.

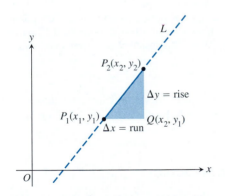

1.19 The slope of the line is

$$m = \frac{\Delta y}{\Delta x} = \frac{\text{rise}}{\text{run}}.$$

DEFINITION

The **slope** of a nonvertical line is $m = \dfrac{\text{rise}}{\text{run}} = \dfrac{\Delta y}{\Delta x} = \dfrac{y_2 - y_1}{x_2 - x_1}$.

The slope of a line depends only on how steeply the line rises or falls and not on the points we use to calculate it (Fig. 1.20 on the following page). A line that rises as x increases (Fig. 1.21a) has positive slope. A line that falls as x increases (Fig. 1.21b) has negative slope. A horizontal line has slope zero. The points on it all have the same y-coordinate, so $\Delta y = 0$.

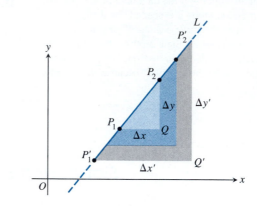

1.20   The slope of a line is the same no matter what points we use to calculate it. Because triangles $P_1 Q P_2$ and $P_1' Q' P_2'$ are similar, $\Delta y / \Delta x = \Delta y' / \Delta x'$.

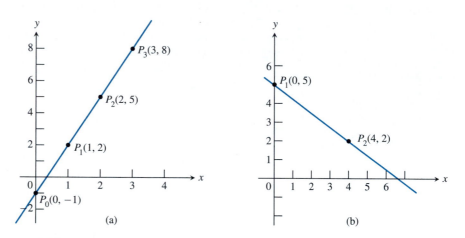

(a)   (b)

1.21   **(a)** The slope of the line is $m = \Delta y / \Delta x = (5 - 2)/(2 - 1) = 3$. This means that $\Delta y = 3 \Delta x$ for every change of position on the line. (Compare Δy and Δx from P_1 to P_2 and from P_2 to P_3.) **(b)** The slope of the line is $m = \Delta y / \Delta x = (2 - 5)/(4 - 0) = -3/4$. This means that y decreases 3 units every time x increases 4 units.

The formula $m = \Delta y / \Delta x$ does not apply to vertical lines because Δx is zero along a vertical line. We express this by saying *vertical lines have no slope* or *the slope of a vertical line is undefined*.

Lines That Are Parallel or Perpendicular

Parallel lines form equal angles with the x-axis. Hence, if they are not vertical, parallel lines have the same slope. Conversely, two lines with equal slopes form equal angles with the x-axis and are therefore parallel (Fig. 1.22).

If neither of two perpendicular lines L_1 and L_2 is vertical, their slopes m_1 and m_2 are related by the equation $m_1 m_2 = -1$. Figure 1.23 shows why.

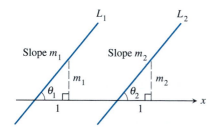

1.22   If $L_1 \| L_2$, then $\theta_1 = \theta_2$ and $m_1 = m_2$. Conversely, if $m_1 = m_2$, then $\theta_1 = \theta_2$ and $L_1 \| L_2$.

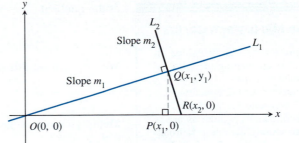

1.23 If $L_1 \perp L_2$, then triangles QPO and RPQ are similar. Therefore,

$$m_1 m_2 = \frac{y_1 - 0}{x_1 - 0} \cdot \frac{0 - y_1}{x_2 - x_1} = \frac{x_2 - x_1}{y_1 - 0} \cdot \frac{0 - y_1}{x_2 - x_1} = -1.$$

Be sure you know a reason for each equation above.

EXAMPLE 1

The slope of a line perpendicular to a line of slope 3/4 is $-4/3$, the *negative reciprocal* of 3/4.   ≡

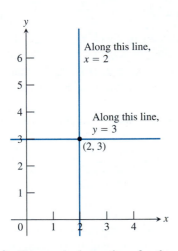

1.24 The standard equations for the horizontal and vertical lines through the point (2, 3) are $x = 2$ and $y = 3$.

Equations for Horizontal and Vertical Lines

The standard equations for the horizontal and vertical lines through a point (a, b) are simply $y = b$ and $x = a$, respectively (Fig. 1.24). A point (x, y) lies on the horizontal line through (a, b) if and only if $y = b$. It lies on the vertical line through (a, b) if and only if $x = a$.

Point-Slope and Slope-Intercept Equations

To write an equation for a line L that is not vertical, it is enough to know its slope m and the coordinates of a point $P_1(x_1, y_1)$ on it. If $P(x, y)$ is any other point on L (Fig. 1.25), then $x \neq x_1$, and we can write the slope of L

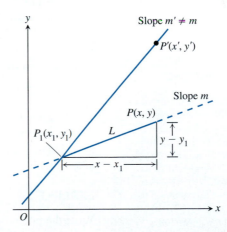

1.25 If L is the line through $P_1(x_1, y_1)$ whose slope is m, then other points $P(x, y)$ lie on this line *if and only if* slope $PP_1 = m$. This fact gives us the point-slope equation for L.

as the quotient

$$\frac{y - y_1}{x - x_1}.$$

We can then set this expression equal to m to get

$$\frac{y - y_1}{x - x_1} = m. \tag{1}$$

Multiplying both sides by $x - x_1$ gives us the more useful equation

$$y - y_1 = m(x - x_1). \tag{2}$$

Equation (2) is an equation for L, as we can check right away. Every point (x, y) on L satisfies the equation—even the point (x_1, y_1). What about the points not on L? If $P'(x', y')$ is a point not on L (Fig. 1.25), then the slope m' of $P'P_1$ is different from m, and the coordinates x' and y' of P' do not satisfy Eqs. (1) and (2).

If we take $(x_1, y_1) = (0, b)$, so that b is the y-intercept of the line (Fig. 1.26), we find that

$$y - b = m(x - 0).$$

When rearranged this takes on a second useful form,

$$y = mx + b.$$

This form matches the desired form for entering the equation into a grapher. It also displays the slope m and the y-intercept b, which can provide us with immediate algebraic confirmation that the graph shown in the viewing window is correct.

RAILROAD AND HIGHWAY GRADES

Civil engineers calculate the slope of a roadbed by calculating the ratio of the distance it rises or falls to the distance it runs horizontally. They call this ratio the **grade** of the roadbed, usually written as a percentage. Along the coast, railroad grades are usually less than 2%. In the mountains, they may go as high as 4%. Highway grades are usually less than 5%.

In analytic geometry, we calculate slopes the same way, but we usually do not express them as percentages.

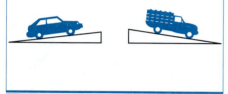

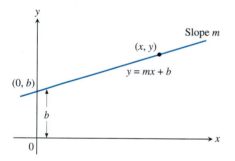

1.26 The line with slope m and y-intercept b has equation $y = mx + b$.

DEFINITION

The equation

$$y - y_1 = m(x - x_1) \tag{3}$$

is the **point-slope equation** of the line that passes through the point (x_1, y_1) with slope m. The equation

$$y = mx + b \tag{4}$$

is the **slope-intercept equation** of the line with slope m and y-intercept b.

In a viewing window, a line with slope m will look like it has slope m only if the viewing window is square, that is, only if the scale unit is the same length on each axis. Most graphers can SQUARE the viewing window automatically.

EXAMPLE 2

Write the point-slope equation for the line that passes through the point $(2, 3)$ with slope $-3/2$. Then, show the equation in slope-intercept form. Sketch a complete graph of the line. Support with a grapher using a square viewing window.

SQUARE VIEWING WINDOWS

A "square viewing window" refers to the scale unit on the x- and y-axes being identical, *not* to the shape of the grapher display. In fact, in the rectangular display of a graphing calculator, $[x\text{Min}, x\text{Max}]$ will always be larger than $[y\text{Min}, y\text{Max}]$ when the viewing window is square. Try showing the graph of $y = (-3/2)x + 6$ in a square viewing window on your grapher. Compare your view with Fig. 1.27(b).

Solution

$$y - y_1 = m(x - x_1)$$ Start with the general point-slope equation, Eq. (3).

$$y - 3 = -\frac{3}{2}(x - 2)$$ Take $(x_1, y_1) = (2, 3)$ and $m = -3/2$.

$$y = -\frac{3}{2}x + 6$$ Rewrite in slope-intercept form.

Sketch the graph (Fig. 1.27a). Notice that $y = 6$ is the y-intercept and $x = 4$ is the x-intercept. Then support your sketch with a grapher in which you SQUARE the viewing window (Fig. 1.27b).

EXPLORATION BITS

1. In general, viewing windows with

$$x\text{-scale} = y\text{-scale}$$

need *not* be square. Why?

2. In any square viewing window on your grapher, find

$$\frac{x\text{Max} - x\text{Min}}{y\text{Max} - y\text{Min}}.$$

Explain why this ratio is constant. How does it relate to the pixels in your display?

3. If you could magnify Fig. 1.27(b) to see the pixels on some graphers, they would be arranged in a 2-1-2-1 pattern. Why?

1.27 Complete graphs of the line $y = (-3/2)x + 6$ of Example 2: (a) sketched and (b) grapher generated with a square viewing window. Here x-scale $= y$-scale $= 1$, and the scale marks show the scale unit. The scale unit on the x- and y-axes are identical, and the view is true to scale. Note here how the orientations of the two lines agree with each other and with the slope of $-3/2$. How would the grapher view change if we changed x-scale or y-scale?

Figure (a): graph of $y = -\frac{3}{2}x + 6$ passing through point $(2, 3)$, with y-intercept 6 and x-intercept 4.

Figure (b): [−2, 10] by [−2, 10]

EXPLORATION 1

Do Algebraically, Support Graphically

Write an equation for the line through $(-2, -1)$ and $(3, 4)$. Support graphically.

1. Write the point-slope equation for the line.

2. Change the equation to slope-intercept form.

3. Enter $y = mx + b$ on your grapher using the values for m and b found in part 2. GRAPH. Tell what you should look for in the graph that will support your algebra. (Did you find it necessary to SQUARE?)

The General Linear Equation

The equation

$$Ax + By = C \quad (A \text{ and } B \text{ not both zero}) \qquad (5)$$

is called the **general linear equation** because its graph is a line for all A and B and because every line has an equation in this form.

EXAMPLE 3

Find the slope and y-intercept of the line $8x + 5y = 20$. Then draw a complete graph in a square viewing window.

Solution   Solve the equation for y to put the equation in slope-intercept form.

$$8x + 5y = 20$$

$$5y = -8x + 20$$

$$y = -\frac{8}{5}x + 4.$$

The slope is $m = -8/5$. The y-intercept is $b = 4$. On a graphing utility, enter either $y = -(8/5)x + 4$ or $y = -1.6x + 4$. Set the view dimensions to include the intercepts. Then GRAPH and SQUARE your viewing window. Compare your view with Fig. 1.28.

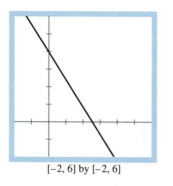

$[-2, 6]$ by $[-2, 6]$

1.28   A complete graph of $8x + 5y = 20$ in a square viewing window.

Generally speaking, it is easier to *sketch* graphs of lines than to use a grapher. We can plot two points (particularly the intercepts) and sketch the line through them, or we can plot one point and sketch the line through it to have the desired slope. We can use a grapher to support our result. Conversely, if we draw a line on a grapher, we can use what we know about its algebraic form to confirm that the picture is accurate.

EXPLORATION BIT

To *quick sketch* the graph of a line, locate the intercepts:
Let $x = 0$, find y;
Let $y = 0$, find x;
and draw the line. Try it for

a) $8x + 5y = 20$

b) $-38x + 60y = 285.$

Support each sketch with your grapher.

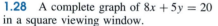

EXPLORATION 2

Do Graphically, Confirm Algebraically

1. Draw complete graphs of the lines $y = mx + 3$ in the same square viewing window for $m = 0.5, 1, 2,$ and 4. What appears to be true about the lines? Give an algebraic reason to confirm your answer.

2. Draw complete graphs of the lines $y = -3x + b$ in the same square viewing window for $b = -4, -1, 2,$ and 5.5. What appears to be true about the lines? Give an algebraic reason to confirm your answer.

3. For a more dramatic effect, change your grapher format from *sequential* to *simultaneous*, then GRAPH.

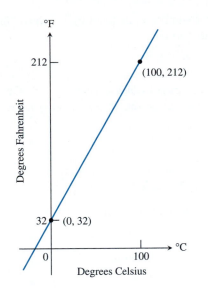

°F

212

Degrees Fahrenheit

(100, 212)

32 — (0, 32)

0   100   °C

Degrees Celsius

1.29 Fahrenheit versus Celsius temperature.

EXAMPLE 4   Celsius versus Fahrenheit

The standard equation for converting Celsius temperature to Fahrenheit temperature is a slope-intercept equation. If we plot Fahrenheit temperature against Celsius temperature in a coordinate plane, the points we plot always lie in a line (Fig. 1.29). The line passes through the point (0, 32) because $F = 32$ when $C = 0$. It also passes through the point (100, 212) because $F = 212$ when $C = 100$.

This is enough information to make a formula for F in terms of C. The line's slope is

$$m = \frac{212 - 32}{100 - 0} = \frac{180}{100} = \frac{9}{5}.$$

The F-intercept of the line is

$$b = 32.$$

The resulting slope-intercept equation for the line,

$$F = \frac{9}{5}C + 32,$$

is the formula we seek.

The following summarizes our algebraic and graphical understandings about lines.

Slope

1. The slope of the line through $P_1(x_1, y_1)$ and $P_2(x_2, y_2)$, $x_1 \neq x_2$, is

$$m = \frac{\text{rise}}{\text{run}} = \frac{\Delta y}{\Delta x} = \frac{y_2 - y_1}{x_2 - x_1}.$$

2. Vertical lines have no slope.

3. Horizontal lines have slope 0.

4. For lines that are neither horizontal nor vertical, it is handy to remember:

   a) lines are parallel $\Leftrightarrow m_2 = m_1$;

   b) lines are perpendicular $\Leftrightarrow m_1 m_2 = -1$, or $m_2 = -1/m_1$. (The symbol $\Leftrightarrow$ is read "if and only if.")

Equations for Lines

$x = a$	Vertical line through (a, b)
$y = b$	Horizontal line through (a, b)
$y - y_1 = m(x - x_1)$	Point-slope equation
$y = mx + b$	Slope-intercept equation
$Ax + By = C$	General linear equation
	(A and B not both zero)

Absolute Value

The **absolute value** or **magnitude** of a number x, denoted by $|x|$ (read "the absolute value of x"), is defined by the formula

$$|x| = \begin{cases} x & \text{if } x \geq 0, \\ -x & \text{if } x < 0. \end{cases}$$

The vertical lines in the symbol $|x|$ are called *absolute value bars*. On a number line, $|x|$ is the distance of x from the origin. It tells us how far it is from 0 to x but not the direction.

EXPLORATION 3

Understanding Algebraic Definitions

To evaluate $|x|$, the definition says that we must compare x with 0.

$$|3| = 3 \text{ because } 3 > 0. \qquad |-5| = -(-5) = 5 \text{ because } -5 < 0.$$

1. What can you say about $|-x|$? (How is it related to x?)
2. What can you say about $-|x|$?
3. Why is $|x|$ always ≥ 0?
4. What can you say about $|x|^2$? about $|-x|^2$?
5. $\sqrt{x^2}$ stands for the positive square root of x^2. How is $\sqrt{x^2}$ related to $|x|$? Explain.

When we do arithmetic with absolute values, we can always use the following rules:

Arithmetic with Absolute Values

1. $|-a| = |a|$ A number and its negative have the same absolute value.

2. $|ab| = |a|\,|b|$ The absolute value of a product is the product of the absolute values.

3. $\left|\dfrac{a}{b}\right| = \dfrac{|a|}{|b|}$ The absolute value of a quotient is the quotient of the absolute values.

EXAMPLE 5

a) $|-\sin x| = |\sin x|$

b) $|-2(x+5)| = |-2|\,|x+5| = 2|x+5|$

c) $\left|\dfrac{3}{x}\right| = \dfrac{|3|}{|x|} = \dfrac{3}{|x|}$

The absolute value of a sum of two numbers is never larger than the sum of their absolute values. When we put this in symbols, we get the important Triangle Inequality.

The Triangle Inequality

$$|a + b| \leq |a| + |b| \text{ for all numbers } a \text{ and } b.$$

The number $|a + b|$ is less than $|a| + |b|$ if a and b have different signs. In all other cases, $|a + b|$ equals $|a| + |b|$:

$$|-3 + 5| = |2| = 2 < |-3| + |5| = 8$$
$$|3 + 5| = |8| = 8 = |3| + |5|$$
$$|-3 + 0| = |-3| = 3 = |-3| + |0|$$
$$|-3 - 5| = |-8| = 8 = |-3| + |-5|.$$

Notice that the absolute value bars in expressions such as $|-3 + 5|$ also work like parentheses: We do the arithmetic inside *before* we take the absolute value.

Distance on the Number Line

The numbers $|a - b|$ and $|b - a|$ are always equal because

$$|a - b| = |(-1)(b - a)| = |-1|\ |b - a| = |b - a|.$$

Both $|a - b|$ and $|b - a|$ give the distance (but not the direction) from a to b on the number line (Fig. 1.30).

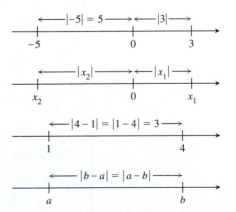

1.30 Absolute values give distances between points on the number line.

Number-Line Distance

$|a - b| = |b - a|$ for all numbers a and b. This number is the distance between a and b on the number line.

Distance in the Coordinate Plane

The coordinates of two points in the xy-plane can tell us how far apart the two points are. (See Fig. 1.31.)

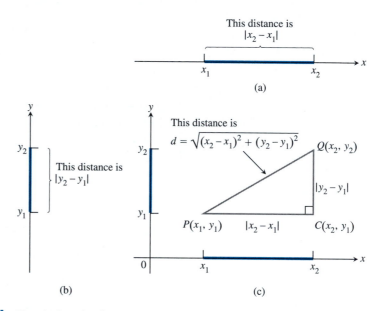

1.31 To calculate the distance between $P(x_1, y_1)$ and $Q(x_2, y_2)$, apply the Pythagorean Theorem to triangle PCQ.

If the points lie on a horizontal line, the distance between them is the usual number-line distance $|x_1 - x_2|$ or $|x_2 - x_1|$ between their x-coordinates (Fig. 1.31a). If the points lie on a vertical line, the distance between them is the usual number-line distance $|y_1 - y_2|$ or $|y_2 - y_1|$ between their y-coordinates (Fig. 1.31b).

If the line joining the points is not parallel to either coordinate axis, we calculate the distance between the points with the Pythagorean theorem (Fig. 1.31c). The resulting formula works for the other cases as well, so there is only one formula to remember:

Coordinate-Plane Distance; The Distance Formula

The distance between $P(x_1, y_1)$ and $Q(x_2, y_2)$ is

$$d = \sqrt{|x_2 - x_1|^2 + |y_2 - y_1|^2} = \sqrt{(x_2 - x_1)^2 + (y_2 - y_1)^2}.$$

EXAMPLE 6

The distance between $P(-1, 2)$ and $Q(3, 4)$ is

$$\sqrt{(3 - (-1))^2 + (4 - 2)^2} = \sqrt{(4)^2 + (2)^2} = \sqrt{20} = \sqrt{4 \cdot 5} = 2\sqrt{5}. \quad \blacksquare$$

Distance from a Point to a Line

To calculate the distance from a point $P(x_1, y_1)$ to a line L, we find the point $Q(x_2, y_2)$ at the foot of the perpendicular from P to L and calculate the distance from P to Q. The next example shows how this is done.

EXAMPLE 7

Find the distance from the point $P(2, 1)$ to the line $L: y = x + 2$.

Solution   Because the distance from a point to a line is the length of the perpendicular segment from the point to the line, we solve the problem in three steps (Fig. 1.32): (1) Find an equation for the line L' through P perpendicular to L; (2) find the point Q where L' meets L; and (3) calculate the distance between P and Q.

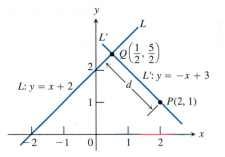

1.32   The distance d from P to L is measured along the line L' perpendicular to L. We can calculate it from the coordinates of P and Q.

STEP 1: We find an equation for the line L' through $P(2, 1)$ perpendicular to L. The slope of $L: y = x + 2$ is $m = 1$. The slope of L' is therefore $m' = -1/1 = -1$. We set $(x_1, y_1) = (2, 1)$ and $m = -1$ in Eq. (3) to find L':

$$y - 1 = -1(x - 2)$$

$$y = -x + 2 + 1$$

$$y = -x + 3.$$

STEP 2: Find the point Q by solving the equations for $L: y = x + 2$ and $L': y = -x + 3$ simultaneously. To find the x-coordinate of Q, we equate the two expressions for y:

$$x + 2 = -x + 3$$

$$2x = 1$$

$$x = \frac{1}{2}.$$

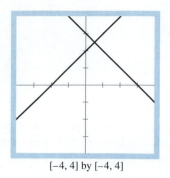

[−4, 4] by [−4, 4]

1.33   Support: Checking screen coordinates suggests that the point of intersection of the graphs of $y_1 = x + 2$ and $y_2 = -x + 3$ is $(0.5, 2.5)$.

We can now obtain the y-coordinate by substituting $x = 1/2$ in the equation for either line. We choose $y = x + 2$ arbitrarily and find

$$y = \frac{1}{2} + 2 = \frac{5}{2}.$$

The coordinates of Q are $(1/2, 5/2)$, which is supported by the graphs of the two lines in a square viewing window (Fig. 1.33).

STEP 3: We calculate the distance between $P(2, 1)$ and $Q(1/2, 5/2)$:

$$d = \sqrt{\left(2 - \frac{1}{2}\right)^2 + \left(1 - \frac{5}{2}\right)^2} = \sqrt{\left(\frac{3}{2}\right)^2 + \left(-\frac{3}{2}\right)^2} = \sqrt{\frac{18}{4}} = \frac{3}{2}\sqrt{2}.$$

The distance from P to L is $(3/2)\sqrt{2}$, or 2.121 to three decimal places.

Exercises 1.2

In Exercises 1–4, compute the rise (Δy) and the run (Δx) for the line segment AB.

1. $A(1, 2), B(-1, -1)$ **2.** $A(-3, 2), B(-1, -2)$

3. $A(-3, 1), B(-8, 1)$ **4.** $A(0, 4), B(0, -2)$

Plot the points A and B in Exercises 5–8. Then find the slope (if any) of the line they determine. Also find the slope (if any) of the lines perpendicular to line AB.

5. $A(1, -2), B(2, 1)$ **6.** $A(-2, -1), B(1, -2)$

7. $A(2, 3), B(-1, 3)$ **8.** $A(1, 2), B(1, -3)$

In Exercises 9–14, find the distance between the given points.

9. $(1, 0)$ and $(0, 1)$ **10.** $(2, 4)$ and $(-1, 0)$

11. $(2\sqrt{3}, 4)$ and $(-\sqrt{3}, 1)$ **12.** $(2, 1)$ and $(1, -1/3)$

13. (a, b) and $(0, 0)$ **14.** $(0, y)$ and $(x, 0)$

Find the absolute values in Exercises 15–20.

15. $|-3|$ **16.** $|2 - 7|$

17. $|-2 + 7|$ **18.** $|1.1 - 5.2|$

19. $|(-2)3|$ **20.** $\left|\frac{2}{-7}\right|$

In Exercises 21–24, find an equation for (a) the vertical line and (b) the horizontal line through the given point.

21. $(2, 3)$ **22.** $(-1, 4/3)$

23. $(0, -\sqrt{2})$ **24.** $(-\pi, 0)$

In Exercises 25–30, write an equation for the line that passes through the point P and has slope m.

25. $P(1, 1), m = 1$ **26.** $P(1, -1), m = -1$

27. $P(-1, 1), m = 1$ **28.** $P(-1, 1), m = -1$

29. $P(0, b), m = 2$ **30.** $P(a, 0), m = -2$

In Exercises 31–36, find an equation for the line through the two points.

31. $(0, 0), (2, 3)$ **32.** $(1, 1), (2, 1)$

33. $(1, 1), (1, 2)$ **34.** $(-2, 0), (-2, -2)$

35. $(-2, 1), (2, -2)$ **36.** $(1, 3), (3, 1)$

In Exercises 37–42, write an equation for the line with the given slope m and y-intercept b. Draw a complete graph.

37. $m = 3, b = -2$ **38.** $m = -1, b = 2$

39. $m = 1, b = \sqrt{2}$ **40.** $m = -1/2, b = -3$

41. $m = -5, b = 2.5$ **42.** $m = 1/3, b = -1$

In Exercises 43–48, find the x- and y-intercepts of the line. Then use the intercepts to sketch a complete graph of the line. Support your sketch with a graphing utility.

43. $3x + 4y = 12$ **44.** $x + y = 2$

45. $4x - 3y = 12$ **46.** $2x - y = 4$

47. $y = 2x + 4$ **48.** $x + 2y = -4$

In Exercises 49 and 50, find the x- and y-intercepts of the line. Draw a complete graph of the line.

49. $\dfrac{x}{3} + \dfrac{y}{4} = 1$ **50.** $\dfrac{x}{-2} + \dfrac{y}{3} = 1$

In Exercises 51 and 52, find the x- and y-intercepts of the line.

51. $\dfrac{x}{a} + \dfrac{y}{b} = 1$ **52.** $\dfrac{x}{a} + \dfrac{y}{b} = 2$

In Exercises 53–58, find an equation for the line through P perpendicular to L. Graph each pair of lines in a square viewing window. Then find the distance from P to L.

53. $P(0, 0), L : y = -x + 2$

54. $P(0, 0), L : x + \sqrt{3}y = 3$

55. $P(1, 2)$, $L : x + 2y = 3$

56. $P(-2, 2)$, $L : 2x + y = 4$

57. $P(3, 6)$, $L : x + y = 3$

58. $P(-2, 4)$, $L : x = 5$

In Exercises 59–62, find an equation for the line through P parallel to L. Draw a complete graph of each pair of lines.

59. $P(2, 1)$, $L : y = x + 2$

60. $P(0, 0)$, $L : y = 3x - 5$

61. $P(1, 0)$, $L : 2x + y = -2$

62. $P(1, 1)$, $L : x + y = 1$

Coordinates of points on a number line are specified in Exercises 63 and 64. Use absolute value notation, and write an expression for the distance between the points.

63. a) x and 3   **b)** x and -2

64. a) y and -1.3 **b)** y and 5.5

In Exercises 65 and 66, write a sentence involving distance which is equivalent to the given algebraic sentence.

65. $|x - 5| = 1$ **66.** $|x + 3| = 5$

67. Do not fall into the trap $|-a| = a$. This equation does not hold for all values of a.
   a) Find a value of a for which $|-a| \neq a$.
   b) For what values of a does the equation $|-a| = a$ hold?

68. For what values of x does $|1 - x|$ equal $1 - x$? For what values of x does it equal $x - 1$?

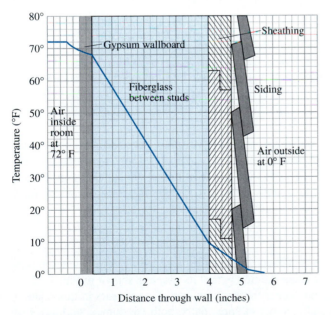

1.34   The temperature changes in the wall in Exercise 69. (Source: *Differentiation*, by W. U. Walton et al., Project CALC, Education Development Center, Inc., Newton, Mass. (1975), p. 25.)

Applications

69. *Insulation.* By measuring slopes in Fig. 1.34, find the temperature change in degrees per inch for the following:
   a) gypsum wall board
   b) fiberglass insulation
   c) wood sheathing

70. *Insulation.* Which of the materials listed in Exercise 69 is the best insulator? The poorest? Explain.

71. *Pressure under water.* The pressure p experienced by a diver under water is related to the diver's depth d by an equation of the form $p = kd + 1$ (k a constant). When $d = 0$ meters, the pressure is 1 atmosphere. The pressure at 100 meters is about 10.94 atmospheres. Find the pressure at 50 meters.

72. *Reflected light.* A ray of light comes in along the line $x + y = 1$ above the x-axis and reflects off the x-axis. The angle of departure is equal to the angle of arrival. Write an equation of the line along which the departing light travels.

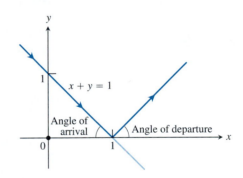

73. *Fahrenheit versus Celsius.* We found a relationship between Fahrenheit temperature and Celsius temperature in Example 4.
   a) Is there a temperature at which a Fahrenheit thermometer and a Celsius thermometer give the same reading? If so, what is it?
   b) GRAPH $y_1 = (9/5)x + 32$, $y_2 = x$, and $y_3 = (5/9)(x - 32)$ in the same viewing window. Explain what you see in the window and how this is related to part (a).

74. *The Mt. Washington Cog Railway.* The steepest part of the Mt. Washington Cog Railway in New Hampshire has a phenomenal 37.1% grade. At this point, the passengers in the front of the car are 14 ft above those in the rear. About how far apart are the front and rear rows of seats?

75. A car starts from point P at time $t = 0$ and travels at 45 mph.
   a) Write an algebraic expression $d(t)$ for the distance the car travels from P.
   b) Graph $y = d(t)$.
   c) What is the slope of the graph in part (b)? What does it have to do with the car?

d) Create a scenario in which t could have negative values in $y = d(t)$.

e) Create a scenario in which the y-intercept of $y = d(t)$ could be 30.

76. A car starts from point P at time $t = 0$ and travels at 55 mph.

a) Write an algebraic expression $d(t)$ for the distance the car travels from P.

b) Graph $y = d(t)$.

c) What is the slope of the graph in part (b)? What does it have to do with the car?

d) Create a scenario in which t could have negative values in $y = d(t)$.

e) Create a scenario in which the y-intercept of $y = d(t)$ could be 40.

Geometry

77. Three different parallelograms have vertices at $(-1, 1)$, $(2, 0)$, and $(2, 3)$. Sketch them and give the coordinates of the missing vertices.

78. For what value of k is the line $2x + ky = 3$ perpendicular to the line $x + y = 1$? For what values of k are the lines parallel?

79. Find the line that passes through the point $(1, 2)$ and the point of intersection of the lines $x + 2y = 3$ and $2x - 3y = -1$.

80. Show that the distance from (x_1, y_1) to the line $Ax + By + C = 0$ is

$$\frac{|Ax_1 + By_1 + C|}{\sqrt{A^2 + B^2}}.$$

1.3 Relations, Functions, and Their Graphs

Points in the coordinate plane correspond to ordered pairs of numbers. Sets of ordered pairs are *relations*. Certain relations are *functions*. A picture of the points in the xy-plane that correspond to the ordered pairs of a relation is the *graph* of the relation. This is the picture that we show in a viewing window.

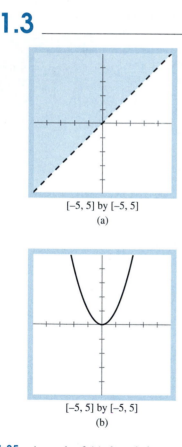

1.35 A graph of (a) the relation $y > x$ and (b) the relation $y = x^2$. The domain of either is made up of the points of the x-axis that are in vertical line with the points of the graph. The range is made up of the points of the y-axis in horizontal line with the points of the graph.

> **DEFINITIONS**
>
> A **relation** is a set of ordered pairs. The set of first entries of the ordered pairs is the **domain** (D) of the relation. The set of second entries is the **range** (R) of the relation. The points in the coordinate plane that correspond to the ordered pairs of a relation form the **graph of the relation**.

In most of our work, relations are usually defined by formulas, equations, or other mathematical sentences and have domains and ranges that are sets of real numbers. Unless otherwise stated in this textbook, the domain of a relation defined by using x and y is taken to be the *largest* set of real x-values for which there are corresponding real y-values.

EXAMPLE 1

a) The inequality $y > x$ defines the set of ordered pairs (x, y) where the second entry is greater than the first entry. Its graph is shown in Fig. 1.35(a). Because *any* real number is the first member of some pairs in this relation and also the second member of some pairs, both the domain and range of this relation are the set of real numbers.

b) The equation $y = x^2$ defines the set of (x, y) pairs where the second entry is the square of the first entry. Its graph is shown in Fig. 1.35(b). The domain is the set of all real numbers, and the range is the nonnegative real numbers.   ≡

For a relation to be a function, each domain element is paired with a single range element. Note that this does *not* mean that there is only one range element. It means that no two range elements are paired with the same domain element.

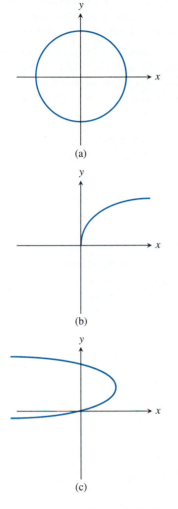

(a)

(b)

(c)

1.36 The graphs for Example 2. Only one of these is the graph of a function. However, all three can be graphed by using a pair of parametric equations for each. Consult the *Resource Manual* for details about parametric graphing.

> **DEFINITION**
>
> A **function** is a relation that assigns a single element of R to each element of D.

Ways to Recognize a Function

A working definition of a function is that it is a device that assigns an *output* to every allowable *input*. The inputs make up the domain of the function. The outputs make up the range. Then a function is distinguished as a relation for which each input from the domain is assigned exactly one output from the range; no more, no less.

A way to distinguish a function from a relation that is not a function is to think of how the graphs of each must differ. A function can have no x-value paired with more than one y-value. This means that its graph cannot have two points in the same vertical line.

> **Vertical Line Test for a Function**
>
> If every vertical line in the xy-plane intersects the graph of a relation in at most one point, then the relation is a function of x.

EXAMPLE 2

Which of the relations graphed in Fig. 1.36 are functions?

Solution   The relation graphed in Fig. 1.36(b) is a function. The relations graphed in Figs. 1.36(a) and (c) are not functions because they fail the Vertical Line Test.   ≡

EXPLORATION 1

Graphing Functions and Relations—Parametric Representation

A grapher can be put into *function mode* to graph functions or into *parametric mode* to graph functions and relations.

In function mode, we enter the equation "$y = \ldots$," set the view dimensions, and then GRAPH. The grapher pairs each (pixel) domain value with a single range value. The graph automatically passes the Vertical Line Test for a function.

1. Enter the functions $y_1 = 2x$ and $y_2 = x^2$. Set the dimensions to include the origin. Then GRAPH. The graphs will appear in either sequential or simultaneous format. Change to the other format and GRAPH again.

2. We can graph in either dot format or connected format. Explain why a graph of a function always passes the Vertical Line Test in dot format but can appear to fail the Vertical Line Test in connected format. As an example, check both formats for $y = \sqrt[3]{x - 4} + 3$ in a $[-10, 10]$ by $[0, 5]$ viewing window.

In parametric mode, graphing relations can be both useful and fun. We enter the equations $x(t) = \ldots$, $y(t) = \ldots$, set a range and increment size for the t values, set the view dimensions, and then GRAPH.

3. Put your grapher in parametric mode. Complete the equations

$$x(t) = \ldots, \qquad y(t) = \ldots,$$

using t as the variable. For example, $x(t) = t^2 - 4t + 1$ and $y(t) = t$. This defines the ordered pairs (x, y) in terms of t. Assign values to tMin, tMax, and t-step, then GRAPH. This will cause t to take on values from tMin to tMax in increments of t-step. The corresponding (x, y) pairs will be computed, and those within the view dimensions will have their pixels illuminated in the viewing window.

By creative choices of the *parametric equations*, "$x(t) = \ldots$," and "$y(t) = \ldots$," we can show a variety of pictures in the viewing window. But it takes a lot of practice, and it helps to know about functions. As we proceed, you will learn much more about functions, and about parametric representation and its use. In the meantime, try graphing the following in the standard viewing window.

4. $x_1(t) = 8 \cos t$, $y_1(t) = 8 \sin t$

5. $x_2(t) = 8 \sin 2t \cos t$, $y_2(t) = 8 \sin 2t \sin t$

6. $x_3(t) = t \cos t$, $y_3(t) = \sqrt{t} \sin t$

Real-Valued Functions of a Real Variable

Functions can model real-life situations in which the values of one variable are related to the values of another. For example:

> Pressure in a boiler depends on steam temperature.
>
> Area of a circle depends on the radius.
>
> Population depends on food supply.

In each of these examples, the value of one variable quantity, which we could call y, *depends on* the value of another variable quantity, which we could call x. Then we can call y the **dependent variable** of the function and x the **independent variable** (or **argument**). In most of our work, the range for y and the domain for x are sets of real numbers; the function is called a

real-valued function of a real variable, we say "y is a function of x," and we symbolize these words using notation invented by Euler:

$$y = f(x) \qquad (\text{read "} y \text{ equals } f \text{ of } x \text{"}).$$

With this notation, we say that f is the name of the function, y or $f(x)$ is the value of the function at x, the function is defined by an equation in x and y, and the graph of f is the graph of the equation. Also, with y isolated on the left-hand side of the equation, the equation is in a form that is ready for entering into most graphing utilities.

You should also be aware, however, that the choice of f, x, and y is somewhat arbitrary. To say that boiler pressure is a function of steam temperature, it may be more useful to write $p = f(T)$. To say that the area of a circle is a function of its radius, we can write Area $= A(r)$. All the terminology mentioned above can be adapted to this new notation, and the new notation itself can be easily adapted to the notation used by a grapher.

Identifying the Domain and Range

Identifying the domain and range of a function is often essential for understanding the function and its graph, particularly when we are concerned with whether a graph is complete. For a function defined by an equation, we must keep two restrictions in mind. First, we *never divide by 0.* When we see $y = 1/x$, we must think $x \neq 0$. Zero is not in the domain of the function. When we see $y = 1/(x - 2)$, we must think "$x - 2 \neq 0$, so $x \neq 2$."

The second restriction is that we will deal exclusively with real-valued functions (except for a very short while later in the book). We may therefore have to restrict our domains when we have square roots (or fourth roots or other *even* roots). If $y = \sqrt{1 - x^2}$, we should think "$1 - x^2 \geq 0$, so x^2 must not be greater than 1. The domain must not extend beyond the interval $-1 \leq x \leq 1$."

With the above in mind, how do we identify the domain and range of a function? One very powerful way is to gain clues from a graph and then use algebraic techniques to confirm what the graph suggests.

EXAMPLE 3

In each of the following functions, the domain is taken to be the largest set of real x-values for which the equation gives real y-values.

Function	Domain	Range
$y = f(x) = \dfrac{1}{x}$	$x \neq 0$	$y \neq 0$
$y = f(x) = \sqrt{x}$	$x \geq 0$	$y \geq 0$
$y = f(x) = \sqrt{4 - x}$	$x \leq 4$	$y \geq 0$
$y = f(x) = \sqrt{1 - x^2}$	$-1 \leq x \leq 1$	$0 \leq y \leq 1$

The equation, or formula, $y = 1/x$ gives a real y-value for every x except $x = 0$. Thus, the domain of the function $y = 1/x$ is the set of all real numbers different from 0 as supported by the complete graph in Fig. 1.38(a). This graph also supports that the range of $y = 1/x$ is the set of all real

---

AGREEMENT

As we mentioned at the start of this section, if the domain of a function is not stated explicitly, then you should assume it to be the largest set of x-values for which the equation gives real y-values. If we wish to exclude values from the domain, we must say so. The function defined by $y = x^2$ has domain $(-\infty, \infty)$ (Fig. 1.35b). The function defined by $y = x^2$, $x \geq 0$, has domain $[0, \infty)$. Its domain has been *restricted* (Fig. 1.37).

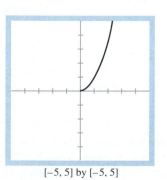

$[-5, 5]$ by $[-5, 5]$

1.37 A complete graph of $y = x^2$ with domain restricted to $x \geq 0$.

numbers different from 0. To confirm this algebraically requires that we observe that every real number $y \neq 0$ occurs (output) as a reciprocal of a real number (input) x. For example, 3 is the output associated with input 1/3.

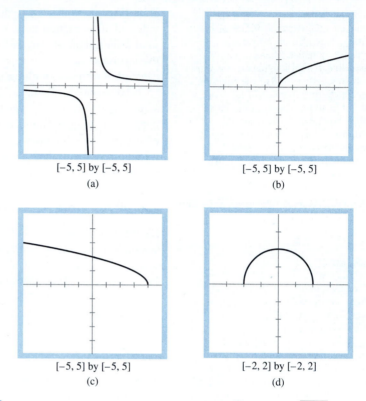

[−5, 5] by [−5, 5]	[−5, 5] by [−5, 5]
(a)	(b)

[−5, 5] by [−5, 5]	[−2, 2] by [−2, 2]
(c)	(d)

1.38 Complete graphs of (a) $y = 1/x$, (b) $y = \sqrt{x}$, (c) $y = \sqrt{4-x}$, and (d) $y = \sqrt{1-x^2}$. Do you get graphs as complete as these on your grapher?

The graph of $y = \sqrt{x}$ in Fig. 1.38(b) suggests that both the domain and the range of this function are the set of nonnegative real numbers. The domain can be confirmed by observing that the formula $y = \sqrt{x}$ gives a real y-value only when x is positive or zero. The range can be confirmed by observing that every nonnegative real number occurs as a y-value.

The graph of $y = \sqrt{4-x}$ in Fig. 1.38(c) is the graph of $y = \sqrt{x}$ reflected through the y-axis and then shifted right 4 units. The domain of $y = \sqrt{4-x}$ appears to be $x \leq 4$, and the range $y \geq 0$. The domain can be confirmed by observing that $4-x$ cannot be negative, that is, $0 \leq 4-x$ or $x \leq 4$.

The graph of $y = \sqrt{1-x^2}$ in Fig. 1.38(d) suggests that the domain is $-1 \leq x \leq 1$ and the range is $0 \leq y \leq 1$. The domain is confirmed by observing that the formula $y = \sqrt{1-x^2}$ gives a real y-value for $1-x^2 \geq 0$, that is, for $-1 \leq x \leq 1$. Confirming the range is more difficult. We need to observe that $-1 \leq x \leq 1$ means that $0 \leq x^2 \leq 1$, which gives us $-1 \leq -x^2 \leq 0$, or $0 \leq 1-x^2 \leq 1$. Thus, $0 \leq \sqrt{1-x^2} \leq 1$, or $0 \leq y \leq 1$.

In Chapters 3 and 4 we will see how to use calculus to confirm that the graphs shown with Example 3 are complete.

Graphing

We already know what complete graphs of linear, quadratic, and cubic equations look like. On the next page is a reference table, Table 1.1, of complete graphs for some of these and for other basic functions.

EXAMPLE 4

Make a conjecture about the domain and range of $y = f(x) = \sqrt[3]{4 - x^2}$. Confirm algebraically.

Solution   The graph in Fig. 1.39(a) suggests that the domain is all real numbers and the range is all real numbers less than or equal to about 1.6. Knowing that $\sqrt[3]{x}$ is defined for all real numbers confirms the domain. To confirm the range, we can start with the graph of $y = 4 - x^2$ (Fig. 1.39b) and see that $4 - x^2 \le 4$ for all x. Then we observe that $\sqrt[3]{4 - x^2} \le \sqrt[3]{4}$, which is 1.5874 to four decimal places, and so we feel reasonably confident that $y \le \sqrt[3]{4}$ is the range. To confirm that *any* value y_0 with $y_0 \le \sqrt[3]{4}$ is in the range of the function, we first observe that $y_0^3 \le 4$. Then, working backward from $y_0 = \sqrt[3]{4 - x_0^2}$, we find that the values $x_0 = \pm\sqrt{4 - y_0^3}$ (which are real because $y_0^3 \le 4$) will give $f(x_0) = y_0$.

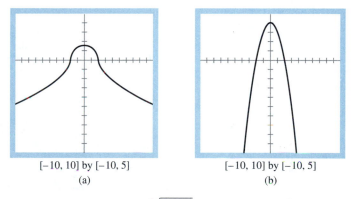

$[-10, 10]$ by $[-10, 5]$ $[-10, 10]$ by $[-10, 5]$
(a)   (b)

1.39   Complete graphs of (a) $y = \sqrt[3]{4 - x^2}$ and (b) $y = 4 - x^2$.

Symmetry

We can use coordinate formulas to describe important symmetries in the coordinate plane. Figure 1.40 shows how this is done and suggests the symmetry test summarized on the next page.

EXAMPLE 5

Symmetric points:

$$P(5, 2) \text{ and } Q(-5, 2) \qquad \text{symmetric about the } y\text{-axis}$$
$$P(5, 2) \text{ and } R(5, -2) \qquad \text{symmetric about the } x\text{-axis}$$
$$P(5, 2) \text{ and } S(-5, -2) \qquad \text{symmetric about the origin}$$

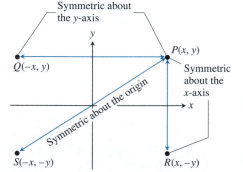

1.40   The coordinate formulas for symmetry with respect to the axes and origin in the coordinate plane.

TABLE 1.1 Some Useful Complete Graphs

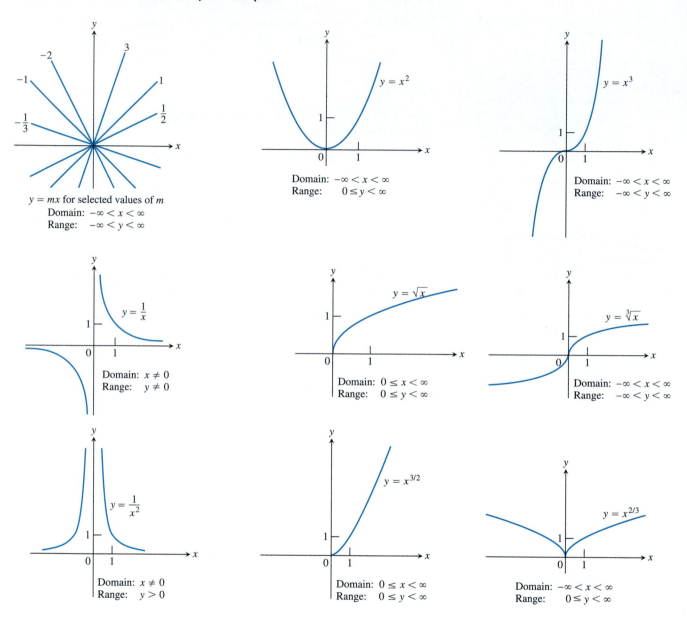

$y = mx$ for selected values of m
Domain: $-\infty < x < \infty$
Range: $-\infty < y < \infty$

$y = x^2$
Domain: $-\infty < x < \infty$
Range: $0 \le y < \infty$

$y = x^3$
Domain: $-\infty < x < \infty$
Range: $-\infty < y < \infty$

$y = \dfrac{1}{x}$
Domain: $x \ne 0$
Range: $y \ne 0$

$y = \sqrt{x}$
Domain: $0 \le x < \infty$
Range: $0 \le y < \infty$

$y = \sqrt[3]{x}$
Domain: $-\infty < x < \infty$
Range: $-\infty < y < \infty$

$y = \dfrac{1}{x^2}$
Domain: $x \ne 0$
Range: $y > 0$

$y = x^{3/2}$
Domain: $0 \le x < \infty$
Range: $0 \le y < \infty$

$y = x^{2/3}$
Domain: $-\infty < x < \infty$
Range: $0 \le y < \infty$

Tests for Symmetry in Graphs

1. *Symmetry about the y-axis:* If (x, y) is on the graph, then $(-x, y)$ is on the graph (Fig. 1.41a on the following page).

2. *Symmetry about the x-axis:* If (x, y) is on the graph, then $(x, -y)$ is on the graph (Fig. 1.41b).

3. *Symmetry about the origin:* If (x, y) is on the graph, then $(-x, -y)$ is on the graph (Fig. 1.41c)

EXPLORATION 2

Supporting Symmetry

The Tests for Symmetry can be supported nicely on a grapher using parametric equations.

1. Test 1 in parametric representation says:

If (x_1, y_1) is on the graph, then $(x_2 = -x_1, y_2 = y_1)$ is on the graph.

We know that $y = x^2$ is symmetric about the y-axis (Fig. 1.41a). The points of $y = x^2$ can be represented parametrically as $x = t$, $y = t^2$. (Think of the pairs you get for $t = 1, 2, 3,$ and so on.)

In parametric mode on your grapher, let:

$$x_1 = t, \qquad\qquad y_1 = t^2.$$

$$x_2 = -x_1(= -t), \qquad y_2 = y_1(= t^2).$$

GRAPH (x_1, y_1) and (x_2, y_2) with $t\text{Min} = -10$, $t\text{Max} = 10$, and t-step $= 0.1$. TRACE will put the cursor on one of the graphs. Then use your cursor control and hop back and forth between graphs (for the same value of t). You should get a strong visual and numerical impression of what is meant by symmetry about the y-axis.

2. Use parametric equations to check the symmetry of each. If needed, you will have to develop parametric representations for symmetry tests 2 and 3 on your own.

a) $x = y^2$ **b)** $y = x^3$ **c)** $y = 1/x^2$

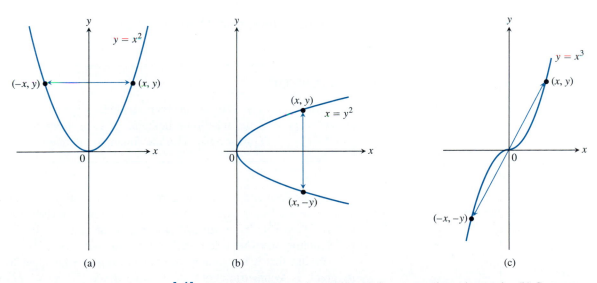

(a)   (b)   (c)

1.41 Tests for symmetry in graphs. (a) Symmetry about the y-axis. (b) Symmetry about the x-axis. (c) Symmetry about the origin.

The following example suggests one way to confirm each type of symmetry algebraically.

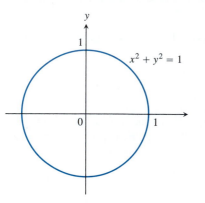

1.42 The graph of the equation $x^2 + y^2 = 1$ is the circle of radius 1, centered at the origin. It is symmetric about both axes and about the origin.

EXAMPLE 6

The graph of $x^2 + y^2 = 1$ has all three of the symmetries listed above (Fig. 1.42).

1. Symmetry about the y-axis:

(x, y) on the graph $\Rightarrow x^2 + y^2 = 1$ $\Rightarrow$ means "implies."

$\Rightarrow (-x)^2 + y^2 = 1$ $x^2 = (-x)^2$

$\Rightarrow (-x, y)$ on the graph.

2. Symmetry about the x-axis:

(x, y) on the graph $\Rightarrow x^2 + y^2 = 1$

$\Rightarrow x^2 + (-y)^2 = 1$ $y^2 = (-y)^2$

$\Rightarrow (x, -y)$ on the graph.

3. Symmetry about the origin:

(x, y) on the graph $\Rightarrow x^2 + y^2 = 1$

$\Rightarrow (-x)^2 + (-y)^2 = 1$

$\Rightarrow (-x, -y)$ on the graph.

Even Functions and Odd Functions

> **DEFINITION**
>
> A function $y = f(x)$ is an **even** function of x if $f(-x) = f(x)$ for every x in the function's domain. It is an **odd** function of x if $f(-x) = -f(x)$ for every x in the function's domain.

The names *even* and *odd* come from powers of x. If y is an even power of x, as in $y = x^2$ or $y = x^4$, it is an even function of x (because $(-x)^2 = x^2$ and $(-x)^4 = x^4$). If y is an odd power of x, as in $y = x$ or $y = x^3$, it is an odd function of x (because $(-x)^1 = -x$ and $(-x)^3 = -x^3$).

Saying that a function $y = f(x)$ is even is equivalent to saying that its graph is symmetric about the y-axis. Since $f(-x) = f(x)$, the point (x, y) lies on the curve if and only if the point $(-x, y)$ lies on the curve (Fig. 1.43a).

Saying that a function $y = f(x)$ is odd is equivalent to saying that its graph is symmetric with respect to the origin. Since $f(-x) = -f(x)$, the point (x, y) lies on the curve if and only if the point $(-x, -y)$ lies on the curve (Fig. 1.43b).

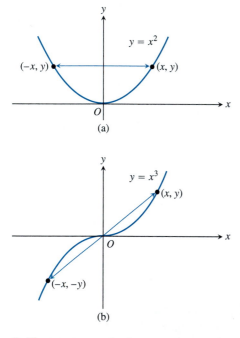

1.43 (a) The graph of an even function is symmetric about the y-axis. (b) The graph of an odd function is symmetric about the origin. A function cannot be symmetric about the x-axis. Why?

EXPLORATION 3

Seeing Even, Odd Functions

The graph of an even function is symmetric about the y-axis. The graph of an odd function is symmetric about the origin. Explain how you could use a grapher to support the following:

EXPLORATION BIT

If your grapher has both keys $\boxed{\wedge}$ (for exponents) and $\boxed{\sqrt[3]{}}$, then GRAPH

$$y_1 = x^{1/3} \qquad \text{and} \qquad y_2 = \sqrt[3]{x},$$

the function in Exploration 3, part 2. Are your results identical?

1. A conjecture that a function is even. Illustrate with $y = 4 - x^2$.
2. A conjecture that a function is odd. Illustrate with $y = \sqrt[3]{x}$.

(*Hint:* For parts 1 and 2, use parametric equations. See Exploration 2.) ✌

EXAMPLE 7

Even, Odd, and Neither

$f(x) = x^2$	Even function: $(-x)^2 = x^2$ for all x Symmetry about y-axis (Fig. 1.44a)
$f(x) = x^2 + 1$	Even function: $(-x)^2 + 1 = x^2 + 1$ for all x Symmetry about the y-axis (Fig. 1.44a)
$f(x) = x$	Odd function: $(-x) = -(x)$ for all x Symmetry about the origin (Fig. 1.44b)
$f(x) = x + 1$	Not odd: $f(-x) = -x + 1$, but $-f(x) = -x - 1$; the two are not equal Not even, either: $(-x) + 1 \neq x + 1$ (Fig. 1.44b)

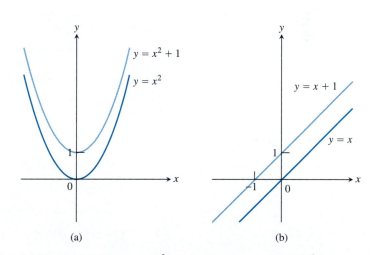

(a)   (b)

1.44 (a) When we add 1 to $y = x^2$, the resulting function $y = x^2 + 1$ is still even, and its graph is still symmetric about the y-axis. (b) When we add 1 to $y = x$, the resulting function $y = x + 1$ is no longer odd. The symmetry about the origin is lost. ≡

Library Functions—Absolute Value

Most graphing utilities have a library of "built-in" functions. These are special functions that either are used so often that it is a matter of efficiency to have them built in or are simply too difficult for the user to enter by formula each time one is needed. The *absolute value function*, $y = |x|$ or $y = \text{abs}(x)$, is a library function on most graphers. Its graph (Fig. 1.45a) lies along the line $y = x$ when $x \geq 0$ and along the line $y = -x$ when $x < 0$. The viewing window (Fig. 1.45b) supports the sketch of $y = |x|$. Note the sharp turn in the graph at the point ($x = 0$) where x changes sign.

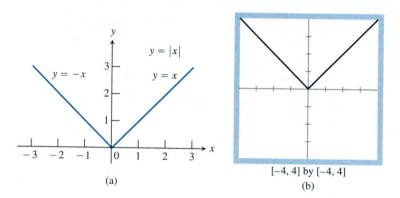

[−4, 4] by [−4, 4]
(b)

1.45 A complete graph of the absolute value function $|x|$ obtained (a) by sketch and (b) by grapher.

EXAMPLE 8

Draw a complete graph of $f(x) = |x + 1| + |x - 3|$ with a graphing utility and confirm algebraically.

Solution If *abs* is the absolute value function on the graphing utility, enter $y = \text{abs}(x + 1) + \text{abs}(x - 3)$. Then GRAPH. (See Fig. 1.46.)

To confirm algebraically, we first note from the graph that something happens when $x = -1$ and $x = 3$. These, in fact, are the points where the expressions inside the absolute value bars change sign. These points divide the x-axis into intervals on which we can write absolute-value-free functions for $f(x)$:

For $x < -1$:   Here, $x + 1 < 0$ and $x - 3 < 0$, so

$$f(x) = |x + 1| + |x - 3|$$
$$= -(x + 1) - (x - 3)$$
$$= -x - 1 - x + 3 = -2x + 2.$$

For $-1 \leq x \leq 3$:   Here, $x + 1 \geq 0$ and $x - 3 \leq 0$, so

$$f(x) = |x + 1| + |x - 3|$$
$$= x + 1 - (x - 3)$$
$$= x + 1 - x + 3 = 4.$$

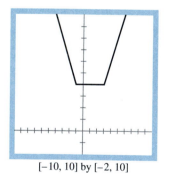

[−10, 10] by [−2, 10]

1.46 A complete graph of $f(x) = |x + 1| + |x - 3|$.

For $x > 3$:   Here, $x + 1 > 0$ and $x - 3 > 0$, so

$$f(x) = |x + 1| + |x - 3|$$

$$= x + 1 + x - 3 = 2x - 2.$$

Thus,

$$f(x) = \begin{cases} -2x + 2, & (x < -1) \\ 4, & (-1 \leq x \leq 3) \\ 2x - 2, & (x > 3) \end{cases}$$

which agrees "piece by piece" with the picture in the viewing window. ≣

Notice that the complete graph in Fig. 1.46 suggests that the equation $|x + 1| + |x - 3| = 0$ has no solution and that the function $f(x) = |x + 1| + |x - 3|$ achieves an *absolute minimum* value, namely 4, for infinitely many values of x in the interval $-1 \leq x \leq 3$.

Integer-Valued Functions

The **greatest integer function**, $y = [\, x \,]$, appears on most graphers as the library function $y = \text{int}\,(x)$. The notation $[x]$ or $\text{int}\,(x)$ represents the greatest integer less than or equal to x, sometimes called *the greatest integer in x*. Because each real number corresponds to only one greatest integer, the greatest integer in x is a function of x. Fig. 1.47 shows a graph of $y = [x]$.

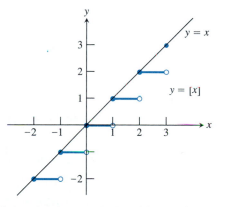

1.47   The graph of $y = [x]$ (or $y = \lfloor x \rfloor$) and its relation to the line $y = x$. As the figure shows, $[x]$ is less than or equal to x, so it provides an *integer floor* for x. Functions like $y = [x]$, with different constant values over adjacent intervals, are called *step functions*.

EXAMPLE 9   Values of $y = [x]$

Positive	$[1.9] = 1, [2.0] = 2, [2.4] = 2$
Zero	$[0.5] = 0, [0] = 0$
Negative	$[-1.2] = -2, [-0.5] = -1$

Notice that if x is negative, $[x]$ may have a larger absolute value than x has.

≣

EXPLORATION 4

Library Functions

On a graphing utility, some library functions appear on the keyboard. Less popular ones may be found in menus.

1. What library functions can you find on the keyboard of your grapher? What ones can you find in menus? Is one of them the greatest integer function? If *yes*, how is it symbolized?

2. We assume that *int* names the greatest integer function on a grapher. Enter $y = \text{int } x$ and GRAPH. Explain what you see in the viewing window. Try graphing in dot and connected mode. Which is better?

   The graph of $y = [x]$ suggests *integer floors* (Fig. 1.47) below the graph of $y = x$. Sometimes $y = [x]$ is called the **integer floor function** and denoted $y = \lfloor x \rfloor$. The companion notation $y = \lceil x \rceil$ is used to denote the result of rounding x up to the nearest integer and is called the **integer ceiling function**. $\lceil x \rceil$ is the least integer greater than or equal to x (Fig. 1.48).

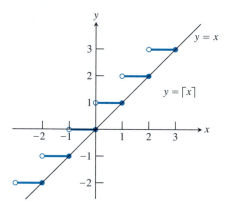

1.48 The graph of $y = \lceil x \rceil$ and its relation to the line $y = x$. As the figure shows, $\lceil x \rceil$ is greater than or equal to x, so it provides an integer ceiling for x. A graph of postal rates for first-class letters resembles the graph of an integer ceiling function. Why?

3. Is there a library function for $y = \lceil x \rceil$ on your grapher? If not, can you write a formula for $y = \lceil x \rceil$ using *int*? In either case, GRAPH the integer ceiling function. Explain what you see.

   Notice, for later reference, that the integer floor (greatest integer) and integer ceiling functions exhibit points called *discontinuities*, where the functions jump from one value to the next without taking on the intermediate values. They jump like this at every integer value of x.

Functions Defined in Pieces

While some functions are defined by single formulas, others are defined by applying different formulas to different parts of their domains.

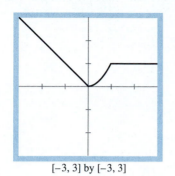

[−3, 3] by [−3, 3]

1.49 A complete graph of the function defined in pieces in Example 10. Functions like the one graphed here are called *piecewise defined functions*.

EXAMPLE 10

The values of the function

$$y = f(x) = \begin{cases} -x & (x < 0) \\ x^2 & (0 \le x \le 1) \\ 1 & (x > 1) \end{cases}$$

are given by the formula $y = -x$ when $x < 0$, by the formula $y = x^2$ when $0 \le x \le 1$, and by the formula $y = 1$ when $x > 1$. The function is *just one function,* however, whose domain is the entire real line (Fig. 1.49). ≡

EXPLORATION 5

Piecewise Defined Functions

Some graphers provide ways to graph piecewise defined functions, others have features that allow you to devise your own method.

1. Investigate whether your grapher can show piecewise defined functions. If your *Owner's Guide* or *Resource Manual* gives no explicit way, can you improvise a method using the grapher's features?

2. On paper, write a piecewise defined function for each graph shown. If you were successful in part 1, apply the method you found, and support your result on the grapher.

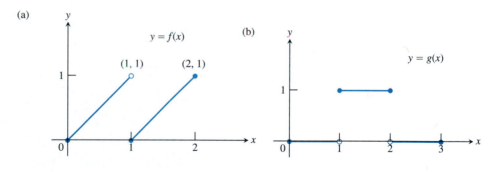

3. A sentence that expresses a relation, such as $x < 0$, can be true for some values of x and false for others. Some graphers will allow you to include a relation within an equation. The graphers then will replace it with value 0 or value 1 according to whether the relation is false or true, respectively.

$(0 < 1) + (0 > -2) = 1 + 1 = 2$ Both relations are true.

$y = |x| = -x(x < 0) + x(x \ge 0)$ The absolute value function

$y = -x(x < 0) + x^2(0 \le x)(x \le 1)$

$\quad + 1(x > 1).$ The function of Example 10

EXPLORATION BIT

Most piecewise defined functions should be GRAPHED in dot format. Why?

With this grapher feature how, if possible, can each of the following functions be written with a single formula?

a) The function for part 2(a) above.

b) The function for part 2(b) above.

c) The function $y = -|x|$.

d) The integer ceiling function.

Sums, Differences, Products, and Quotients

The sum $f + g$ of two functions of x is itself a function of x, defined at any point x that lies in both domains. The same holds for the differences $f - g$ and $g - f$, the product $f \cdot g$, and the quotients f/g and g/f, as long as we exclude any points that require division by zero.

EXAMPLE 11

Find the sum, differences, product, and quotients of the functions f and g defined by the formulas

$$f(x) = \sqrt{x} \qquad \text{and} \qquad g(x) = \sqrt{1 - x}.$$

Function	Formula	Domain
f	$f(x) = \sqrt{x}$	$0 \le x$
g	$g(x) = \sqrt{1 - x}$	$x \le 1$
$f + g$	$(f + g)(x) = f(x) + g(x) = \sqrt{x} + \sqrt{1 - x}$	$0 \le x \le 1$ (The intersection of the domains of f and g)
$f - g$	$(f - g)(x) = f(x) - g(x) = \sqrt{x} - \sqrt{1 - x}$	$0 \le x \le 1$
$g - f$	$(g - f)(x) = g(x) - f(x) = \sqrt{1 - x} - \sqrt{x}$	$0 \le x \le 1$
$f \cdot g$	$(f \cdot g)(x) = f(x)g(x) = \sqrt{x(1 - x)}$	$0 \le x \le 1$
f/g	$\dfrac{f}{g}(x) = \dfrac{f(x)}{g(x)} = \sqrt{\dfrac{x}{1 - x}}$	$0 \le x < 1$ ($x = 1$ excluded)
g/f	$\dfrac{g}{f}(x) = \dfrac{g(x)}{f(x)} = \sqrt{\dfrac{1 - x}{x}}$	$0 < x \le 1$ ($x = 0$ excluded)

Fig. 1.50(a) shows complete graphs of the functions $f(x) = \sqrt{x}$ and $g(x) = \sqrt{1 - x}$. Notice that the domain of f is $0 \le x$ and the domain of g is $x \le 1$. Fig. 1.50(b) shows a complete graph of $f + g$ and gives visual support that the domain of $f + g$ is $0 \le x \le 1$, the intersection of the domains of f and g.

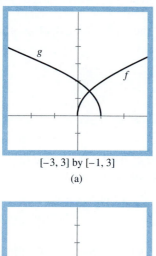

[−3, 3] by [−1, 3]

(a)

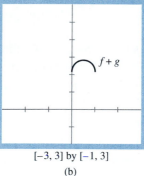

[−3, 3] by [−1, 3]

(b)

1.50 Complete graphs of (a) f and g and (b) $f + g$, where $f(x) = \sqrt{x}$ and $g(x) = \sqrt{1-x}$.

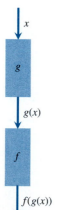

1.51 Two functions can be composed when the range of the first lies in the domain of the second.

EXPLORATION 6

Combining Functions

Some graphers allow us to enter functions y_1 and y_2 and then combine them to form new functions.

1. Enter $y_1 = \sqrt{x}$ and $y_2 = \sqrt{1-x}$. Then enter $y_3 = y_1 + y_2$. GRAPH y_1, y_2, and y_3 in a $[-3, 3]$ by $[-1, 3]$ viewing window. Compare your view with Fig. 1.50.

2. Set your viewing dimensions to be $[-0.5, 1.5]$ by $[-1.5, 1.5]$. Define y_4, y_5, y_6, y_7, and y_8 to be $y_1 - y_2$, $y_2 - y_1$, $y_1 \cdot y_2$, y_1/y_2, and y_2/y_1, respectively, to match the list in the table above. Graph each of y_4 through y_8.

3. Of the eight functions, y_1 through y_8, which combinations might be interesting to view at the same time? Explain why. &

Composition of Functions

Suppose that the outputs of a function g can be used as inputs of a function f. We can then link g and f to form a new function whose inputs are the inputs of g and whose outputs are the numbers $f(g(x))$, as in Fig. 1.51. We say that the function $f(g(x))$ (pronounced "f of g of x") is the *composite* of g and f. It is made by composing g and f in the order of first g, then f. The usual "stand-alone" notation for this composite is $f \circ g$, which is read as "f of g." Thus, the value of $f \circ g$ at x is $(f \circ g)(x) = f(g(x))$.

DEFINITION

The **composition $f \circ g$** of the functions f and g is defined by

$$(f \circ g)(x) = f(g(x)).$$

The domain of $f \circ g$ consists of those x's for which $g(x)$ is in the domain of f.

EXAMPLE 12

Find a formula for $f(g(x))$ if $g(x) = x^2$ and $f(x) = x - 7$. Then find the value of $f(g(2))$.

Solution To find $f(g(x))$, we replace x in the formula $f(x) = x - 7$ by the expression given for $g(x)$:

$$f(x) = x - 7$$

$$f(g(x)) = g(x) - 7 = x^2 - 7.$$

We then find the value of $f(g(2))$ by substituting 2 for x:

$$f(g(2)) = (2)^2 - 7 = 4 - 7 = -3.$$

■

EXPLORATION BIT

In Example 12, we compute
$(f \circ g)(2) = f(g(2))$. Now you
compute $(g \circ f)(2)$ and compare with
$(f \circ g)(2)$.

In the notation for composite functions, the parentheses tell which function comes first:

The notation $f(g(x))$ says "first g, then f." To calculate $f(g(2))$, calculate $g(2)$ and then apply f.

The notation $g(f(x))$ says "first f, then g." To calculate $g(f(2))$, calculate $f(2)$ and then apply g.

EXPLORATION 7

Composing Functions

Some graphers allow a function such as y_1 to be used as a variable (argument) in another function. In other words, we can compose functions.

1. Enter the functions $g(x) = x^2$ and $f(x) = x - 7$ of Example 12 as $y_1 = x^2$ and $y_2 = x - 7$. Enter $y_3 = y_1 - 7$ and $y_4 = y_2^2$. Which of y_3 and y_4 corresponds to $f \circ g$? To $g \circ f$?

2. GRAPH y_3 and y_4. Make a conjecture about $f \circ g$ and $g \circ f$. Confirm your conjecture algebraically by finding formulas for $f(g(x))$ and $g(f(x))$.

3. Explore function composition on your own. Investigate such questions as the following. Feel free to create your own questions.

   a) Are there two functions f and g so that $f \circ g = g \circ f$?

   b) For a given function f, is there a function g so that $f \circ g = g \circ f$?

   c) Are there two functions f and g with nonstraight graphs for which the graph of $f \circ g$ is straight?

   d) What is a reasonable definition of composition $f \circ g \circ h$? For three functions? For four functions?

Exercises 1.3

In Exercises 1–4, identify which of the relations graphed are functions. Explain your answer.

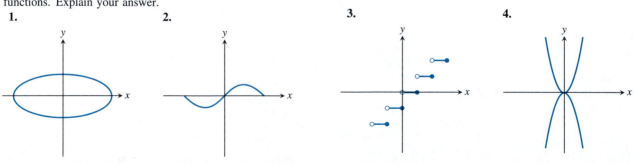

1.   **2.**   **3.**   **4.**

For each given point $P(x, y)$ in Exercises 5–12, use symmetry tests to find the point Q that is (a) symmetric to P across the x-axis, (b) symmetric to P across the y-axis, and (c) symmetric to P through the origin.

5. $(3, 1)$

6. $(-2, 2)$

7. $(-2, 1)$

8. $(-1, -1)$

9. $(1, -\sqrt{2})$

10. $(-\sqrt{3}, -\sqrt{3})$

11. $(0, \pi)$

12. $(2, 0)$

In Exercises 13–20, find the domain and range of each function. Support your answer with a graphing utility.

13. $y = 2 + \sqrt{x - 1}$

14. $y = -3 + \sqrt{x + 4}$

15. $y = -\sqrt{-x}$

16. $y = \sqrt{-x}$

17. $y = 2\sqrt{3 - x}$

18. $y = -3\sqrt{2 - x}$

19. $y = \dfrac{1}{x - 2}$

20. $y = \dfrac{1}{x + 2}$

Use a graph to find the domain and range of each function in Exercises 21–30. What symmetries described in this section, if any, does each graph have?

21. $y = x^2 - 9$

22. $y = 4 - x^2$

23. $y = \sqrt[3]{x - 3}$

24. $y = -2\sqrt[3]{x + 2}$

25. $y = 1 + \sqrt[3]{2 - x}$

26. $y = -2 + 5\sqrt[3]{4 - x}$

27. $y = -\dfrac{1}{x}$

28. $y = -\dfrac{1}{x^2}$

29. $y = 1 + \dfrac{1}{x}$

30. $y = 1 + \dfrac{1}{x^2}$

31. Consider the function $y = 1/\sqrt{x}$.
   a) Can x be negative?
   b) Can $x = 0$?
   c) What is the domain of the function?

32. Consider the function $y = \sqrt{(1/x) - 1}$.
   a) Can x be negative?
   b) Can $x = 0$?
   c) Can x be greater than 1?
   d) What is the domain of the function?

Determine whether each function in Exercises 33-42 is even, odd, or neither. Try to answer without writing anything (except the answer).

33. $y = x^3$

34. $y = x^4$

35. $y = x + 2$

36. $y = x + x^2$

37. $y = x^2 - 3$

38. $y = x + x^3$

39. $y = \dfrac{1}{x^2 - 1}$

40. $y = \dfrac{1}{x - 1}$

41. $y = \dfrac{x}{x^2 - 1}$

42. $y = \dfrac{x^2}{x^2 - 1}$

Test each equation in Exercises 43–50 to find what symmetries its graph has. Then draw a complete graph of the equation.

43. $y = -x^2$

44. $x = 4 - y^2$

45. $y = 1/x^2$

46. $y = 1/(x^2 + 1)$

47. $xy = 1$

48. $xy^2 = 1$

49. $x^2 y^2 = 1$

50. $x^2 + 4y^2 = 1$

Draw a complete graph of each function in Exercises 51–62.

51. $y = |x + 3|$

52. $y = |2 - x|$

53. $y = \dfrac{|x|}{x}$

54. $y = \dfrac{|x - 1|}{x - 1}$

55. $y = \dfrac{x - |x|}{2}$

56. $y = \dfrac{x + |x|}{2}$

57. a) $f(x) = \begin{cases} 3 - x, & x \le 1 \\ 2x, & 1 < x \end{cases}$
   b) Compute $f(0), f(1), f(2.5)$.

58. a) $f(x) = \begin{cases} 1/x, & x < 0 \\ x, & 0 \le x \end{cases}$
   b) Compute $f(-1), f(0), f(\pi)$.

59. a) $f(x) = \begin{cases} 1, & x < 5 \\ 0, & 5 \le x \end{cases}$
   b) Compute $f(0), f(5), f(6)$.

60. a) $f(x) = \begin{cases} 1, & x < 0 \\ \sqrt{x}, & x \ge 0 \end{cases}$
   b) Compute $f(-1), f(0), f(5)$.

61. a) $f(x) = \begin{cases} 4 - x^2, & x < 1 \\ \dfrac{3}{2}x + \dfrac{3}{2}, & 1 \le x \le 3 \\ x + 3, & x > 3 \end{cases}$
   b) Compute $f(0.5), f(1), f(3), f(4)$.

62. a) $f(x) = \begin{cases} x^2, & x < 0 \\ x^3, & 0 \le x \le 1 \\ 2x - 1, & x > 1 \end{cases}$
   b) Compute $f(-1), f(0), f(1), f(2.5)$.

63. Find formulas for the functions graphed in the following figures.

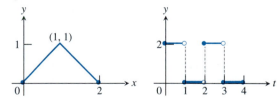

64. Find formulas for the functions graphed in the following figures.

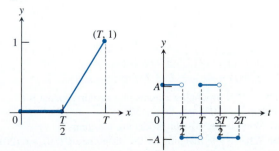

65. For what values of x does (a) $[x] = 0$? (b) $\lceil x \rceil = 0$?

66. Does $[x]$ ever equal $\lceil x \rceil$? Explain.

67. Graph each function over the given interval.
  a) $y = x - [x], -3 \le x \le 3$
  b) $y = [x] - \lceil x \rceil, -3 \le x \le 3$

68. *Integer parts of decimals.* When x is positive or zero, $[x]$ is the integer part of the decimal representation of x. What is the corresponding description of $\lceil x \rceil$ when x is negative or zero?

Draw a complete graph of each function in Exercises 69–72. Confirm algebraically by expressing the function without absolute value symbols.

69. $f(x) = |x + 1| + 2|x - 3|$ *Hint*: Use three intervals.

70. $f(x) = |x + 2| + |x - 1|$

71. $f(x) = |x| + |x - 1| + |x - 3|$ *Hint*: Use four intervals.

72. $f(x) = |x + 2| + |x| + |x + 1|$

In Exercises 73 and 74 find the domains of f and g; then find the corresponding domains and complete graphs of

$$f + g, f - g, f \cdot g, f/g, \text{ and } g/f.$$

73. $f(x) = x \quad g(x) = \sqrt{x - 1}$

74. $f(x) = \sqrt{x + 1}, \quad g(x) = \sqrt{x - 1}$

75. If $f(x) = x + 5$ and $g(x) = x^2 - 3$, find the following:
  a) $f(g(0))$
  b) $g(f(0))$
  c) $f(g(x))$
  d) $g(f(x))$
  e) $f(f(-5))$
  f) $g(g(2))$
  g) $f(f(x))$
  h) $g(g(x))$

76. If $f(x) = x + 1$ and $g(x) = x - 1$, find the following:
  a) $f(g(0))$
  b) $g(f(0))$
  c) $f(g(1))$
  d) $g(f(1))$
  e) $f(g(x))$
  f) $g(f(x))$

77. Copy and complete the following table.

	$g(x)$	$f(x)$	$f \circ g(x)$
a)	$x - 7$	$\sqrt{x}$	?
b)	$x + 2$	$3x$	?
c)	?	$\sqrt{x - 5}$	$\sqrt{x^2 - 5}$
d)	$\dfrac{x}{x - 1}$	$\dfrac{x}{x - 1}$	?
e)	?	$1 + \dfrac{1}{x}$	x
f)	$\dfrac{1}{x}$	?	x

78. a) If $f(x) = 1/x$, find $f(x + 2) - f(2)$.
  b) If $F(t) = 4t - 3$, find $F(t + 1) - F(1)$.

79. The cost of producing x items of a certain product is $C(x) = 0.001x^3 - 0.05x^2 + 2.6x + 50$.
  a) What is the cost of producing ten items?
  b) What is the meaning of the difference $C(30) - C(20)$?

80. Compare the domains and ranges of the functions $y = \sqrt{x^2}$ and $y = (\sqrt{x})^2$.

81. Compare the graphs of $y = \sqrt{x^2}$ and $y = |x|$. Explain what you observe.

82. Find $f(x)$ if $g(x) = \sqrt{x}$ and $(g \circ f)(x) = |x|$.

83. Find $g(x)$ if $f(x) = x^2 + 2x + 1$ and $(g \circ f)(x) = |x + 1|$.

84. Find functions $f(x)$ and $g(x)$ whose composites satisfy the two equations

$$(g \circ f)(x) = |\sin x| \text{ and } (f \circ g)(x) = (\sin \sqrt{x})^2.$$

85. *The best location for a factory assembly table.* (adapted from *Fantastiks of Mathematiks*, Cliff Sloyer, Janson Publications, Inc., Providence, R.I., 1986.) Because of a design change, the parts produced by three machines along a factory aisle (shown here as the x-axis) are to go to a nearby table for assembly before they undergo further processing. Each assembly takes one part from each machine, and there is a fixed cost per foot for moving each part. As the plant's production engineer, you have been asked to find a location for the assembly table that will keep the total cost of moving the parts at a minimum.

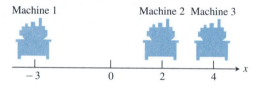

To solve the problem, you let x represent the table's location and look for the value of x that minimizes the sum

$$d(x) = |x + 3| + |x - 2| + |x - 4|$$

of the distances from the table to the three machines. Since the cost of moving the parts to the assembly table is proportional to the total distance the parts travel, any value of x that minimizes d will minimize the cost.

Complete the job now by graphing $d(x)$ to find its smallest value. Then say where you would put the table.

86. *Best location (continuation of Exercise 85).* You solved the table location problem in Exercise 85 so well that your manager has asked you to solve a similar problem at a neighboring plant. This time there are four machines instead of

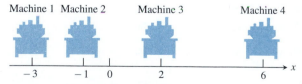

three and the cost is proportional to

$$d(x) = |x + 3| + |x + 1| + |x - 2| + |x - 6|.$$

Where should the assembly table go now?

87. *Best location (continuation of Exercise 86).* As the result of another design change, the assembly in the plant in Exercise 86 is to use twice as many parts from Machine 1 as before and three times as many parts from Machine 3 as before. The total cost of moving parts from the four machines to the assembly table is now proportional to the "weighted" distance

$$d(x) = 2|x + 3| + |x + 1| + 3|x - 2| + |x - 6|.$$

What is the minimum value of this new function? Where should the table go?

1.4 ——————— Geometric Transformations: Shifts, Reflections, Stretches, and Shrinks

In this section, we show how to change an equation to move its graph up or down or right or left, stretch it or shrink it, or reflect it through the x- or y-axis in the coordinate plane. Each of these changes is called a **geometric transformation** of the previous graph. We develop transformations using parametric form because this form allows for "natural" representations of the important ideas. Then we convert our results to the standard representations of the formulas for shifts, stretches, shrinks, and reflections. We practice with parabolas and other basic graphs, but our methods apply to other curves as well.

How to Shift a Graph

The function $y = x^2$ can be represented in parametric form by

$$x_1(t) = t, \qquad y_1(t) = t^2. \qquad (x_1, y_1) = (t, t^2) \qquad (1)$$

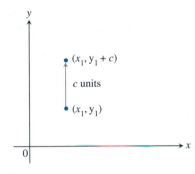

1.52 To shift a graph straight up, we add a positive constant to its second coordinates.

The graph has ordered pairs of the form (t, t^2). Shifting the graph c units straight up (Fig. 1.52) results in ordered pairs of the form $(t, t^2 + c)$. Parametric equations for the new graph are

$$x_2(t) = x_1(t) = t, \qquad y_2(t) = y_1(t) + c = t^2 + c. \qquad (2)$$

The parametric Eqs. (2) represent the function $y_2 = t^2 + c = x_2^2 + c$, or $y = x^2 + c$, the standard representation for a vertical shift.

EXPLORATION 1

Seeing Vertical Shifts

1. Graph the equations $y_1 = x^2$ and $y_2 = x^2 + 2$ in function mode. Graph the parametric Eqs. (1) and (2) with $c = 2$. Both pairs of graphs show a vertical shift of 2 units.

2. If one graph can be found by shifting another, the two graphs are the same shape. Explain why the graphs of $y_1 = x^2$ and $y_2 = x^2 + 2$ appear to have different shapes in the same viewing window. Using TRACE and the cursor control to hop from one graph to the other can give a clue.

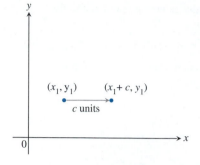

1.53 To shift a graph to the right, we add a positive constant to its first coordinates.

To shift the parabola $y = x^2$ to the right by c units, we must add c to each x-coordinate (Fig. 1.53). We get

$$x_3(t) = x_1(t) + c = t + c, \qquad y_3(t) = y_1(t) = t^2, \tag{3}$$

or ordered pairs of the form $(t + c, t^2)$. Because $t = x_3 - c$, the parametric Eqs. (3) represent the function $y_3 = t^2 = (x_3 - c)^2$, or $y = (x - c)^2$, the standard representation for a horizontal shift.

EXPLORATION 2

Seeing Horizontal Shifts

Support graphically each of the following. You may wish to use TRACE.

1. Show that parametric Eqs. (3) shift the graph of Eqs. (1) to the right by c units. Let $c = 2, 4,$ and 6.

2. Show that $y_2 = (x - c)^2$ in function mode shifts the graph of $y_1 = x^2$ to the right by c units. Let $c = 2, 4,$ and 6.

By similar steps, we can show that $y = x^2 - c$ is a shift of $y = x^2$ *down* c units and $y = (x + c)^2$ is a shift of $y = x^2$ *to the left* c units.

IN CASE YOU FORGET

If you know what the graph of f looks like and you have to graph $y = f(x - c)$, you can ask yourself, "What makes $x - c = 0$?" It is $x = c$. And you can think of taking hold of the graph of f where $x = 0$ and moving the graph, without distortion, to where $x = c$.

Shift Formulas ($c > 0$)

Vertical shifts of the graph of $y = f(x)$:

$$y = f(x) + c \qquad \text{Shifts the graph of } f \text{ up } c \text{ units.}$$

$$y = f(x) - c \qquad \text{Shifts the graph of } f \text{ down } c \text{ units.}$$

Horizontal shifts of the graph of $y = f(x)$:

$$y = f(x - c) \qquad \text{Shifts the graph of } f \text{ right } c \text{ units.}$$

$$y = f(x + c) \qquad \text{Shifts the graph of } f \text{ left } c \text{ units.}$$

Note in particular that the parametric representations for horizontal shifts are "natural" (add to shift to the right, subtract to shift to the left), but the standard representations above are "backwards" (add to shift to the left, subtract to shift to the right). If you don't understand why this is so, study again the steps used above to find the function $y = (x - c)^2$.

EXAMPLE 1

Complete graphs of $y = x^2 - 4$ and $y = x^2 + 4$ are obtained by shifting the graph of $y = x^2$ down 4 units and up 4 units, respectively (Fig. 1.54a).

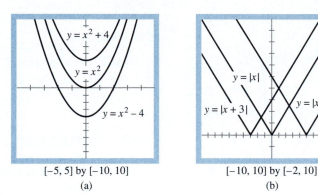

[-5, 5] by [-10, 10]   [-10, 10] by [-2, 10]
(a)   (b)

1.54 Complete graphs of (a) $y = x^2$, $y = x^2 - 4$, and $y = x^2 + 4$ and (b) $y = |x|$, $y = |x + 3|$, and $y = |x - 5|$.

EXAMPLE 2

Complete graphs of $y = |x + 3|$ and $y = |x - 5|$ are obtained by shifting the graph of $y = |x|$ to the left 3 units and to the right 5 units, respectively (Fig. 1.54b).

EXAMPLE 3

Describe how the graph of $y = [1/(x + 2)] - 3$ can be sketched from the graph of $y = 1/x$. Find the domain and range of $y = [1/(x + 2)] - 3$.

Solution

Sketch $y = 1/x$.

Shift left 2 units to get $y = 1/(x + 2)$.

Shift down 3 units to get $y = [1/(x + 2)] - 3$.

or

Sketch $y = 1/x$.

Shift down 3 units to get $y = [1/x] - 3$.

Shift left 2 units to get $y = [1/(x + 2)] - 3$.

The domain is all $x \neq 2$ and the range is all $y \neq -3$. The results are supported by the viewing windows in Fig. 1.55.

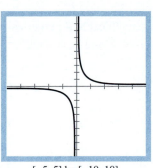

[-5, 5] by [-10, 10]
(a)

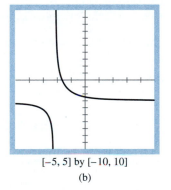

[-5, 5] by [-10, 10]
(b)

1.55 Complete graphs of (a) $y = 1/x$ and (b) $y = [1/(x + 2)] - 3$.

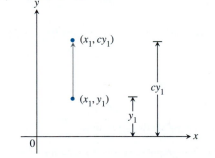

How to Stretch or Shrink a Graph

To stretch the parabola $y = x^2$ or $(x_1, y_1) = (t, t^2)$ vertically by a factor of c ($c > 1$), we must multiply each y-coordinate by c (Fig. 1.56).

$$x_4(t) = x_1(t) = t, \qquad y_4(t) = cy_1(t) = ct^2. \tag{4}$$

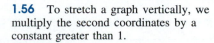

1.56 To stretch a graph vertically, we multiply the second coordinates by a constant greater than 1.

Thus, stretching the graph vertically by a factor of c gives ordered pairs of the form (t, ct^2). The parametric Eqs. (4) represent the function $y_4 = ct^2 = cx_4{}^2$, or $y = cx^2$, the standard representation for a vertical stretch.

By similar steps, we can show that when $0 < c < 1$, $y = cx^2$ shrinks the graph by a factor of c.

EXAMPLE 4

A TRACE on the graphs in Fig. 1.57 gives graphical and numerical support that $y = 2x^2$ stretches $y = x^2$ by a factor of 2 and $y = (1/2)x^2$ shrinks $y = x^2$ by a factor of 1/2. ≡

EXPLORATION 3

Following a Model

We have seen how to shift, stretch, and shrink the graph of $y = x^2$ vertically. We have seen how to shift the graph of $y = x^2$ horizontally. The steps have been essentially the same. Now, we ask you to follow our model to show how to stretch or shrink the graph of $y = x^2$ horizontally.

To stretch $y = x^2$ horizontally by a factor of 2, follow these steps.

1. Begin by asking how the coordinates $(x_1, y_1) = (t, t^2)$ are changed in a horizontal stretch by a factor of 2 (Fig. 1.58).
2. Write the parametric Eqs. (5) to give this change.

$$x_5(t) = \ldots, \qquad y_5(t) = \ldots \qquad (5)$$

3. Combine the parametric Eqs. (5) into a single equation of the form $y = f(x)$. You should get $y = (x/2)^2$.
4. Follow the very same model and show that $y = (2x)^2$ shrinks $y = x^2$ horizontally by a factor of 1/2.
5. Follow the model and show how $y = (cx)^2$ changes the graph of $y = x^2$ horizontally when $0 < c < 1$ and when $c > 1$. ⚘

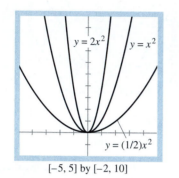

1.57 Complete graphs of $y = x^2$, $y = 2x^2$, and $y = 0.5x^2$.

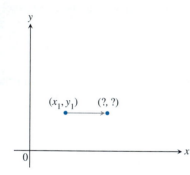

1.58 How are the coordinates (x, y) changed in a horizontal stretch by a factor of 2?

For $c > 0$ in general, $y = cf(x)$ is a *vertical* stretch or shrink; $y = f(cx)$ is a *horizontal* shrink or stretch.

Stretch or Shrink Formulas ($c > 0$)

Vertical stretches or shrinks of the graph of $y = f(x)$:

$y = cf(x), c > 1$ — Stretches the graph vertically by a factor of c.

$y = cf(x), 0 < c < 1$ — Shrinks the graph vertically by a factor of c.

Horizontal shrinks or stretches of the graph of $y = f(x)$:

$y = f(cx), c > 1$ — Shrinks the graph horizontally by a factor of $1/c$.

$y = f(cx), 0 < c < 1$ — Stretches the graph horizontally by a factor of $1/c$.

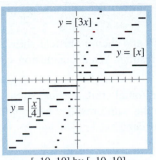

[−10, 10] by [−10, 10]

1.59 Complete graphs of $y = [x]$, $y = [x/4]$, and $y = [3x]$.

Again, we note that although the parametric representations for horizontal stretches and shrinks seem "natural" (multiply by $c > 1$ to stretch, by $0 < c < 1$ to shrink), the standard representations shown above seem "backwards" (multiply by $c > 1$ to shrink, by $0 < c < 1$ to stretch).

EXAMPLE 5

The graph of $y = [x/4] = [(1/4)x]$ is a horizontal stretch of the graph of $y = [x]$ by a factor of 4. The graph of $y = [3x]$ is a horizontal shrink of the graph of $y = [x]$ by a factor of 1/3 (Fig. 1.59). ≣

EXAMPLE 6

Describe how the graph of $y = 2x^3 - 5$ can be sketched from the graph of $y = x^3$.

Solution

Sketch $y = x^3$.

Stretch vertically by a factor of 2 to get $y = 2x^3$.

Shift down 5 units to get $y = 2x^3 - 5$.

The results are supported by Fig. 1.60.

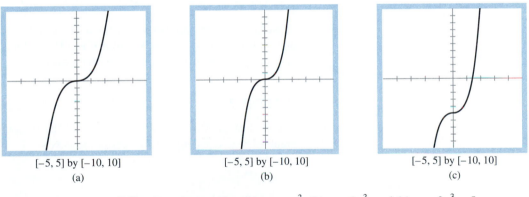

[−5, 5] by [−10, 10]
(a)

[−5, 5] by [−10, 10]
(b)

[−5, 5] by [−10, 10]
(c)

1.60 Complete graphs of (a) $y = x^3$, (b) $y = 2x^3$, and (c) $y = 2x^3 - 5$. ≣

EXPLORATION 4

Proving and Disproving—Group Project

A graphing utility is a laboratory tool. We can perform experiments with it. If the experiments show a pattern, we can make a conjecture about the pattern that we try to confirm with algebra. If an experiment shows an idea to be not true, it serves as a counterexample to the conjecture based on that idea. We would like you to work in groups of two or three on the following:

1. In Example 3, we found that we could apply a vertical shift and a horizontal shift in either order and get the same result. Can we conclude that

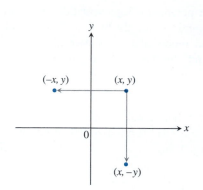

$(-x, y)$ (x, y)

0   x

$(x, -y)$

1.61 To reflect a graph across the y-axis, we multiply the first coordinates by -1. To reflect a graph across the x-axis, we multiply the second coordinates by -1.

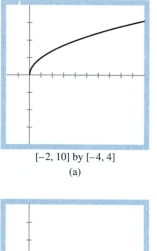

$[-2, 10]$ by $[-4, 4]$

(a)

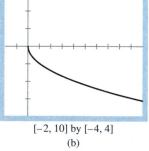

$[-2, 10]$ by $[-4, 4]$

(b)

1.62 Complete graphs of (a) $y_1 = \sqrt{x}$ and (b) $y_2 = -\sqrt{x}$, the reflection of y_1 across the x-axis.

a vertical shift and a horizontal shift in either order will always give the same result? Explain.

2. In Example 6, we apply a vertical stretch and then a vertical shift. If we reverse the order, do we get the same result? What conclusion can we draw?

3. For the transformations vertical shift, horizontal shift, vertical stretch, and horizontal stretch, which pairs will give the same result when applied in either order? (There are six pairs to consider.) ♣

How to Reflect a Graph

To reflect the graph of $y = f(x)$ across the y-axis, we multiply each x-coordinate by -1. To reflect across the x-axis, we multiply each y-coordinate by -1 (Fig. 1.61).

$$x_1(t) = t, \qquad y_1(t) = f(t).$$

$$x_2(t) = -x_1(t) = -t, \qquad y_2(t) = y_1(t) = f(t). \qquad \text{Reflect across } y\text{-axis} \qquad (6)$$

$$x_3(t) = x_1(t) = t, \qquad y_3(t) = -y_1(t) = -f(t). \qquad \text{Reflect across } x\text{-axis} \qquad (7)$$

From Eqs. (6), we find that $y_2 = f(t) = f(-x_2)$, or $y = f(-x)$, is a reflection of $y = f(x)$ across the y-axis. From Eqs. (7), we find $y_3 = -f(t) = -f(x_3)$, or $y = -f(x)$, is a reflection across the x-axis.

EXAMPLE 7

A complete graph of $y = -\sqrt{x}$ (Fig. 1.62b) is obtained by reflecting the graph of $y = \sqrt{x}$ (Fig. 1.62a) across the x-axis. ≡

EXPLORATION 5

Seeing Reflections

1. Draw the graphs of $y_1 = \sqrt{x}$ and $y_2 = -\sqrt{x}$ in the same viewing window. Use TRACE and the cursor control to hop from one graph to the other.

2. Draw the graphs of $y_1 = \sqrt[4]{x}$ and $y_2 = \sqrt[4]{-x}$ in the same viewing window. Use TRACE and the cursor control to hop from one graph to the other. Explain what happens. Can you demonstrate the desired reflection using parametric equations?

3. Which of the above shows reflection across the x-axis? The y-axis? Recall what you know from science and explain why "reflection" is a good word to use when describing how the graphs are related. ♣

Reflection Formulas

With respect to the y-axis

$$y = f(-x)$$ Reflects the graph of f across the y-axis.

With respect to the x-axis

$$y = -f(x)$$ Reflects the graph of f across the x-axis.

The following list summarizes the standard representations for shifting, stretching, shrinking, and reflecting a graph. This information helps us use familiar graphs to help sketch graphs of more complicated equations. It also helps us know the basic shape to expect when graphing a more complicated equation on a graphing utility. Note that the basic shape is the graph of $y = f(x)$.

Summary: Graphing $y = a\,f(bx + c) + d = a\,f(b(x + c/b)) + d$

The graph of $y = af(x) + d$ can be sketched from the graph of $y = f(x)$ as follows:

1. Stretch or shrink the graph vertically by the factor $|a|$. Then reflect across the x-axis if $a < 0$.

2. Shift the graph vertically $|d|$ units: upward if $d > 0$, downward if $d < 0$.

The graph of $y = f(bx + c) = f(b(x + c/b))$ can be sketched from the graph of $y = f(x)$ as follows:

3. Stretch or shrink the graph horizontally by the factor $1/|b|$. Then reflect across the y-axis if $b < 0$.

4. Shift the graph horizontally $|c/b|$ units: to the left if $c/b > 0$, to the right if $c/b < 0$.

The graph of $y = af(bx + c) + d = af(b(x + c/b)) + d$ can be sketched from the graph of $y = f(x)$ by using a combination of steps 1–4.

The Parabola $y = ax^2 + bx + c$

The graph of every equation of the form $y = ax^2 + bx + c$ can be obtained by applying a sequence of geometric transformations (shifts, stretches, shrinks, reflections) to the graph of $y = x^2$. This follows because $y = ax^2 + bx + c$ can be written in the form $y = a(x + h)^2 + k$.

EXPLORATION 6

Analyzing a Parabola

We write $y = ax^2 + bx + c$ in the form $y = a(x + h)^2 + k$ by *completing the square*. We do it here for a specific case (left column) and in general (right column). The key step is in the third line, where we add and subtract the square of 1/2 of the coefficient of x.

$$y = -2x^2 + 12x - 5 \qquad\qquad y = ax^2 + bx + c$$

$$= -2(x^2 - 6x) - 5 \qquad\qquad = a\left(x^2 + \frac{b}{a}x\right) + c$$

$$= -2(x^2 - 6x + 9 - 9) - 5 \qquad = a\left(x^2 + \frac{b}{a}x + \frac{b^2}{4a^2} - \frac{b^2}{4a^2}\right) + c$$

$$= -2(x^2 - 6x + 9) + 18 - 5 \qquad = a\left(x^2 + \frac{b}{a}x + \frac{b^2}{4a^2}\right) - \frac{b^2}{4a} + c$$

$$= -2(x - 3)^2 + 13 \qquad\qquad = a\left(x + \frac{b}{2a}\right)^2 + c - \frac{b^2}{4a}$$

1. Look at the last equation in the right column. Give a convincing argument for each of the following.

    a) The vertex of the parabola occurs at $x = -b/2a$.

    b) The parabola is symmetrical about the line $x = -b/2a$.

    c) The parabola opens upward or downward depending on the sign of a.

2. Support your answers to part 1 graphically for $y = -2x^2 + 12x - 5$.

3. Use completing the square to write $y = -2x^2 + 12x - 5$ in the form $y = a(x + h)^2 + k$. Then tell how the graph of $y = -2x^2 + 12x - 5$ can be obtained from the graph of $y = x^2$ by using shifts, stretches, and shrinks.   ☙

EXPLORATION BIT: A SPECIAL CASE

The equation

$$y = ax^2 + bx + c$$

can be written as

$$y = a(x + \frac{b}{2a})^2 + c - \frac{b^2}{4a}.$$

This has the form

$$y = A(x + B)^2 + C$$

and thus is a special case of the form

$$y = af(bx + c) + d.$$

The graph of

$$y = A(x + B)^2 + C$$

can be obtained from the graph of $y = x^2$ by applying

1. a vertical stretch or shrink by the factor $|A|$ followed by a reflection across the x-axis if $A < 0$,
2. a horizontal shift of $|B|$ units, and
3. a vertical shift of $|C|$ units.

Explain how these three facts are connected to the information summarized for the graph of $y = ax^2 + bx + c$ given at the right.

The Graph of $y = ax^2 + bx + c$

- A parabola that opens in the positive y direction if $a > 0$ and in the negative y direction if $a < 0$.

- The axis of symmetry is $x = -\dfrac{b}{2a}$.

- The vertex is at $(\dfrac{-b}{2a},\ c - \dfrac{b^2}{4a})$.

The Parabola $x = ay^2 + by + c$

If we interchange x and y in the equation for a parabola to obtain $x = ay^2 + by + c$, then the three facts about the graph of $y = ax^2 + bx + c$ shown above remain true when you interchange the x's and y's in each statement. Since the parabola opens to the left or right, however, the Vertical Line Test tells us that the graph is not the graph of a function. Nonetheless, we can graph the relation using parametric equations.

EXPLORATION 7

Graphing a Relation

The relations $(x_1(t), y_1(t))$ and $(x_2(t), y_2(t))$ described below with parametric equations can also be represented as $y = x^2$ and $x = y^2$, respectively. Note how the values of x and y are interchanged from (x_1, y_1) to (x_2, y_2).

$$x_1(t) = t, \qquad\qquad y_1(t) = t^2.$$
$$x_2(t) = y_1(t) = t^2, \qquad y_2(t) = x_1(t) = t.$$

1. GRAPH $(x_2(t), y_2(t))$. Your result should match Fig. 1.63. Does the graph support the facts about the graph of $x = ay^2 + by + c$ suggested in the paragraph preceding this Exploration for this particular case, $x = y^2$? Explain.

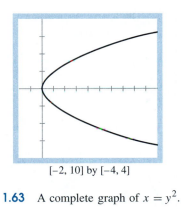

[−2, 10] by [−4, 4]

1.63 A complete graph of $x = y^2$.

2. GRAPH $x = y^2$ in function mode by expressing $x = y^2$ as two functions of y in terms of x.

3. The paragraph preceding this Exploration suggests three facts about the graph of $x = ay^2 + by + c$. Illustrate these facts using $x = 3y^2 - 12y + 11$. First, GRAPH $x = 3y^2 - 12y + 11$ in parametric mode. Also GRAPH this relation in function mode, using two functions of y in terms of x. (To get the two functions, complete the square in y and then solve for y.) Explain your conclusions and confirm algebraically.

Exercises 1.4

1. The graph of $y = x^2$ is shifted to two new positions. Write equations for the new graphs.

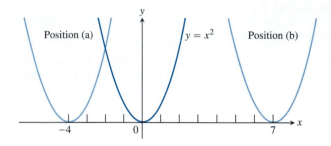

2. The graph of $y = x^2$ is shifted to two new positions. Write equations for the new graphs.

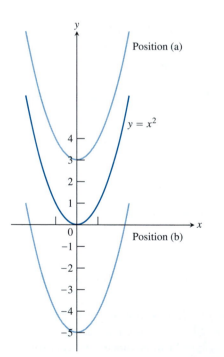

3. Match the equations listed to the graphs shown.
  a) $y + 4 = (x - 1)^2$ **b)** $y - 2 = (x - 2)^2$
  c) $y - 2 = (x + 2)^2$ **d)** $y + 2 = (x + 3)^2$

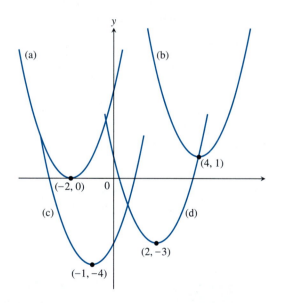

4. The graph of $y = x^2$ is shifted to four new positions. Write an equation for each new graph.

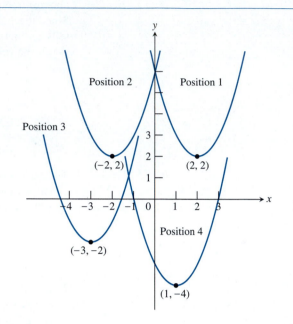

In Exercises 5–14, describe how each graph can be obtained from the graph of $f(x) = |x|, g(x) = 1/x, h(x) = \sqrt{x}$, or $k(x) = x^3$.

5. $y = |x + 4| - 3$ **6.** $y = |x - 3| + 2$

7. $y = 3\sqrt{-x}$ **8.** $y = -0.2\sqrt{-x}$

9. $y = \dfrac{2}{x} - 3$ **10.** $y = -\dfrac{0.5}{x} + 1$

11. $y = -0.5(x - 3)^3 + 1$ **12.** $y = 3\sqrt{2 - x} - 5$

13. $y = \dfrac{1}{x-2} + 3$ **14.** $y = \dfrac{2}{x-3} - 5$

In Exercises 15–24, sketch a complete graph of each function. Support your answer with a grapher. Determine the domain and range of the function.

15. $y = 4\sqrt[3]{2-x} - 5$ **16.** $y = 0.5|x+3| - 4$

17. $y = 5\sqrt[3]{-x} - 1$ **18.** $y = -2\sqrt[4]{1-x} + 3$

19. $y = -\dfrac{1}{(x+3)^2} + 2$ **20.** $y = \dfrac{1}{(x-2)^2}$

21. $y = -2(x-1)^{2/3} + 1$ **22.** $y = (x+2)^{3/2} + 2$

23. $y = 2[1-x]$ **24.** $y = [x-2] + 0.5$

Exercises 25–36 specify the order in which transformations are to be applied to the graph of the given equation. Give an equation for the transformed graph in each case. Support your answer with a grapher.

25. $y = x^2$, vertical stretch by 3, shift up 4

26. $y = x^2$, shift up 4, vertical stretch by 3

27. $y = \dfrac{1}{x}$, shift down 2, vertical shrink by 0.2

28. $y = \dfrac{1}{x}$, vertical shrink by 0.2, shift down 2

29. $y = |x|$, shift left 2, vertical stretch by 3, shift up 5

30. $y = |x|$, shift right 3, vertical shrink by 0.3, shift down 1

31. $y = x^3$, reflect through x-axis, vertical shrink by 0.8, shift right 1, shift down 2

32. $y = x^3$, vertical stretch by 2, reflect through x-axis, shift left 5, shift down 6

33. $y = \sqrt{x}$, reflect through y-axis, vertical stretch by 5, shift left 6, shift up 5

34. $y = \sqrt[4]{x}$, vertical shrink by 0.7, reflect through y-axis, shift right 8, shift down 7

35. $y = \sqrt{3x}$, horizontal stretch by 2, shift up 1

36. $y = 4|x|$, horizontal shrink by 0.5, shift down 3

37. Use a grapher to compare a vertical stretch of $y = x^2$ by a factor of 4 with a horizontal shrink by a factor of 0.5. Algebraically confirm your observation. Generalize.

38. Use a grapher to compare a vertical stretch of $y = |x|$ by a factor of 2 with a horizontal shrink by a factor of 0.5. Algebraically confirm your observation. Generalize.

39. Are the graphs of the equations found in Exercises 25 and 26 the same? Explain any differences.

40. Are the graphs of the equations found in Exercises 27 and 28 the same? Explain any differences.

41. Let $f(x) = \sqrt[3]{x}$.
 a) Describe how the graph of $y = \sqrt[3]{-x}$ can be obtained from the graph of f.
 b) Describe how the graph of $y = -\sqrt[3]{x}$ can be obtained from the graph of f.

c) Compare the graphs of the functions in parts (a) and (b). Explain any similarities or differences.

42. Let $f(x) = \sqrt{x}$.
 a) Describe how the graph of $y = \sqrt{-x}$ can be obtained from the graph of f.
 b) Describe how the graph of $y = -\sqrt{x}$ can be obtained from the graph of f.
 c) Compare the graphs of the functions in parts (a) and (b). Explain any similarities or differences.

43. The line $y = mx$ is shifted vertically to make it pass through the point $(0, b)$. What is the line's new slope-intercept equation?

44. The line $y = mx$ is shifted horizontally and vertically to make it pass through the point (x_0, y_0). What is the line's new point-slope equation?

In Exercises 45–48, the graph of g is obtained by applying, in order, the given transformations to the graph of $f(x) = |x|$. Find the points on the graph of g that correspond under the sequence of transformations to the points $(-1, 1)$, $(0, 0)$, and $(1, 1)$ on the graph of f.

45. Shift right 3, shift up 2

46. Shift down 4, shift left 1

47. Vertical stretch by 2, reflect through x-axis

48. Vertical shrink by 0.3, shift up 4

A complete graph of f is pictured. Sketch a complete graph of each function in Exercises 49–56.

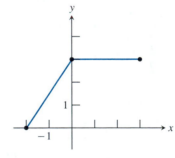

49. $y = f(-x)$ **50.** $y = -f(x)$

51. $y = f(x-2)$ **52.** $y = f(x+3)$

53. $y = 0.5f(x) - 3$ **54.** $y = -3f(x) + 2$

55. $y = -2f(x+1) + 3$ **56.** $y = 0.2f(x-1) - 1$

57. Sketch a complete graph of $y = \dfrac{2x+1}{x-2}$. (*Hint:* Use long division to write $y = 2 + \dfrac{5}{x-2}$.)

58. Sketch a complete graph of $y = \dfrac{x+1}{x+3}$. (*Hint:* See Exercise 57).

In Exercises 59–62, determine the vertex and axis of symmetry, and sketch a complete graph of each parabola. Support your work with a grapher.

59. $y = -2x^2 + 12x - 11$ **60.** $y = 3x^2 + 12x + 7$

61. $y = 4x^2 + 20x + 19$ **62.** $y = -4x^2 + 12x - 3$

In Exercises 63–68, use completing the square to rewrite the equation in the form $x = a(y+h)^2 + k$. Then describe how the graph of the equation can be obtained from the graph of $x = y^2$ and sketch a complete graph. Support your work by graphing the equation (a) in parametric mode and (b) in function mode using two functions of y.

63. $x = y^2 - 6y + 11$ **64.** $x = y^2 + 4y + 1$

65. $x = 2y^2 + 4y + 1$ **66.** $x = -3y^2 + 12y - 7$

67. $x = -2y^2 + 12y - 13$ **68.** $x = 4y^2 + 16y + 9$

69. Let $f(x) = x^2$ and suppose g is obtained by applying the following transformations in some order to f: shift up 3, shift right 2, reflect through the y-axis. How many different graphs can be obtained for g?

1.5

Solving Equations and Inequalities Graphically

The standard algebraic techniques for solving equations and inequalities, such as the quadratic formula and factoring, are really not very useful for solving most equations or inequalities (ones like $x^3 - 2x^2 - 7 = 0$, $\sin x > x$, etc.). In this section, we show how to use a graphing utility to solve equations and inequalities graphically with prescribed accuracy.

Unless we state otherwise, **solving an equation** will mean finding all its real number solutions. Every equation involving a single variable x can be put in the equivalent form $f(x) = 0$. To use the graph of $y = f(x)$ to help solve the equation $f(x) = 0$, we must find where the graph crosses the x-axis, that is, where $y = 0$ (Fig. 1.64).

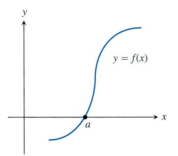

1.64   The real number $x = a$ is a solution to the equation $f(x) = 0$ if and only if the point $(a, 0)$ is on the graph of $y = f(x)$.

Solving Equations Graphically Using ZOOM

A grapher procedure called ZOOM can be used to find solutions of equations to a high degree of accuracy regardless of the complexity of the equation. The idea is to "trap" the x-intercept of $y = f(x)$ in a decreasing sequence of viewing rectangles, each new one contained within the previous one, until the last viewing rectangle has enlarged a small enough portion of the graph so that the value of x can be read to the accuracy desired and within the limits of machine precision (Fig. 1.65).

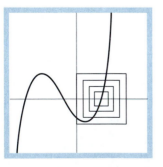

1.65   ZOOM-IN or ZOOM-BOX creates a nested sequence of rectangles that fill our viewing windows with magnifications of the tiny region we are interested in exploring.

ERROR OF A SOLUTION

If we estimate $x = a$ as a solution of an equation when x-scale $= r$, we say that a is a solution with *error of at most r*. Thus, our last estimate of 2.206 in Example 1 is a solution with error of at most 0.01.

EXAMPLE 1

Solve the equation $x^3 - 2x^2 = 1$.

Solution Rewrite the equation in the form $x^3 - 2x^2 - 1 = 0$, and find a complete graph of $f(x) = x^3 - 2x^2 - 1$ (Fig. 1.66). Because the graph in Fig. 1.66(a) is complete, we see that the equation $x^3 - 2x^2 - 1 = 0$ has only one real solution. Because x-scale in Fig. 1.66(a) is 1, we can estimate the solution to be 2.2.

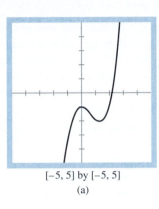

[−5, 5] by [−5, 5]

(a)

[2, 3] by [−1, 1]

(b)

[2.2, 2.3] by [−0.1, 0.1]

(c)

1.66 (a) A complete graph of $f(x) = x^3 - 2x^2 - 1$ that shows all real solutions to $x^3 - 2x^2 - 1 = 0$. The x-scale marks are 1 unit apart (they show the scale unit). (b) A magnified portion of the viewing window from a ZOOM-IN. The x-scale marks are 0.1 unit apart. (c) Another magnification. Here the x-scale marks are only 0.01 unit apart.

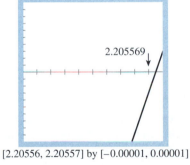

2.205569

[2.20556, 2.20557] by [−0.00001, 0.00001]

1.67 A graph of $f(x) = x^3 - 2x^2 - 1$ that displays its real zero with an error of at most 0.000001.

In Fig. 1.66(b), we have ZOOMED-IN in to magnify a portion of the viewing window. Here x-scale is 0.1, so we can estimate the solution to be 2.20. Another ZOOM-IN (Fig. 1.66c) shows an x-scale of 0.01, and we can estimate the solution to be 2.206. We can continue to ZOOM-IN until we reach the limit of our grapher's precision. ≣

EXAMPLE 2

Solve the equation $x^3 - 2x^2 = 1$ with an error of at most 0.000001.

We use **SOLVE** to stand for a graphing utility procedure that solves the equation $f(x) = 0$.

Solution We continue the process of Example 1 until we reach a viewing window in which we can distinguish the scale marks when x-scale $= 0.000001$ (Fig. 1.67). The graph crosses the x-axis between 2.205569 and 2.205570. We may say the solution is 2.2055694 with an error of at most 0.000001. We could also use SOLVE to find that the solution is 2.2055694304 with 11-digit accuracy. ≣

Most graphing utilities can provide answers, found in various ways, that are accurate to nine or ten significant digits. For this textbook, we make the following agreements. The accuracy agreement will work well most of the time. Exceptions occur when numbers are very large or very small.

Most equations (and inequalities) cannot be solved by using traditional algebraic techniques. The graphical solution method, however, works with a large number of such equations. We will make use of this advantage of graphical and numerical methods over and over again in this book.

EXPLORATION 1

Solving an Equation

Finding the one real solution for the equation

$$x - 3 + \ln x = 0$$

illustrates the need for the accuracy agreements above. (The expression "ln x" stands for the built-in natural logarithm function. Its meaning will be discussed in more detail in the next section and later in the book.)

1. Set x-scale $= 0.01$ and ZOOM-IN until you see the graph of $f(x) = x - 3 + \ln x$ cross the x-axis between consecutive x-scale marks.

2. Read a three-decimal-place solution with error of at most 0.01 for $f(x) = 0$.

3. Find a seven-decimal-place solution with error of at most 0.000001 for $f(x) = 0$.

4. Use SOLVE (if available) to solve $f(x) = 0$.

Polynomials

An *n***th-degree polynomial**, a **polynomial function**, and a **polynomial equation** have the respective forms

$$a_n x^n + a_{n-1} x^{n-1} + \ldots + a_1 x + a_0, \qquad \text{Polynomial (expression)}$$

$$f(x) = a_n x^n + a_{n-1} x^{n-1} + \ldots + a_1 x + a_0, \qquad \text{Polynomial function}$$

$$a_n x^n + a_{n-1} x^{n-1} + \ldots + a_1 x + a_0 = 0, \qquad \text{Polynomial equation}$$

where n is a positive integer and the a_i's are real numbers with $a_n \neq 0$. The equation of Example 1 is a polynomial equation of degree 3, and the equation of Exploration 1 is not a polynomial equation because of the term $\ln x$.

Throughout the history of mathematics, men and women have dedicated countless hours—sometimes even their lives—to the study of polynomials. One major objective was to find all solutions of a polynomial equation, that is, all the zeros of the corresponding polynomial function. That problem has been completely solved in theory, but in practice the results were rather limited and difficult to work with until computers were applied.

Solving linear equations is easy. Solving quadratic equations is a little more difficult with the quadratic formula providing the complete solution. Beyond that, arguably the best result is the Rational Zeros Theorem, which allows us to find algebraically all rational number zeros of a polynomial equation with rational coefficients.

RATIONAL ZEROS THEOREM

Suppose all the coefficients in the polynomial equation

$$f(x) = a_n x^n + a_{n-1} x^{n-1} + \cdots + a_1 x + a_0 = 0 \qquad (a_n \neq 0, a_0 \neq 0)$$

are integers.

If $x = c/d$ is a rational zero of f, where c and d have no common factors, then

- c is a factor of the constant term a_0, and
- d is a factor of the leading coefficient a_n.

The Rational Zeros Theorem is one of the best results available for solving polynomial equations, and students for years have applied it to equations that were especially constructed for the theorem. Unfortunately, equations that result from real applications rarely have rational number solutions. Now, however, polynomial equations can be completely solved by using (1) what we know about a complete graph of a polynomial function and (2) numerical methods that will identify each real solution as accurately as we wish.

We will show more on this in the coming chapters, but for the present we will review the use of the Rational Zeros Theorem and show how computer graphing enhances the application of algebraic procedures for finding exact solutions of polynomial equations.

We recall that an algebraic way to solve a polynomial equation is to factor the polynomial into a product of polynomials, each of degree 1 or 2. We recall also that $x = a$ is a zero of a polynomial if and only if $(x - a)$ is a factor. The Rational Zeros Theorem supplies a finite list of candidates for the zeros that are rational numbers.

Some graphing utilities can even find the complex number zeros of polynomials.

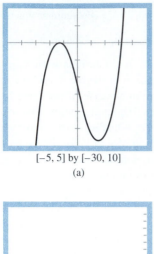

[−5, 5] by [−30, 10]

(a)

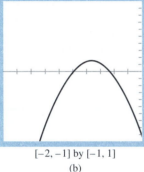

[−2, −1] by [−1, 1]

(b)

1.68 A complete graph of $f(x) = 2x^3 - x^2 - 14x - 12$. Two views are needed because we are not sure whether view (a) hides any behavior, namely how many x-intercepts there are.

WHICH METHOD TO USE?

In real applications, the first method shown here for solving $f(x) = 0$ is rarely useful because most equations do not have exact solutions, or they have exact solutions that are tedious to find. For our work, you will need to use methods 1 and 2. You may use method 3 if available to you. *Note:* In method 2(b), effective programmed SOLVE procedures first draw the graph to prompt the user to determine a good initial approximation of a root.

EXAMPLE 3

Solve $2x^3 - x^2 - 14x - 12 = 0$.

Solution We use the Rational Zeros Theorem to find the 16 possible solution candidates c/d.

c: $\pm 1, \pm 2, \pm 3, \pm 4, \pm 6, \pm 12$ *c* must be a factor of the constant term -12.

d: $\pm 1, \pm 2$ *d* must be a factor of the leading coefficient 2.

c/d: $\pm 1, \pm 2, \pm 3, \pm 4, \pm 6, \pm 12, \pm \dfrac{1}{2}, \pm \dfrac{3}{2}$.

Then, instead of computing $f(c/d)$ for all 16 candidates, we find a complete graph of f (Fig. 1.68) and see that any real zeros must be between -2 and -1 or 3 and 4. This knocks out 15 of the 16 candidates, leaving only $-3/2$ to be checked. Direct computation shows that $f(-3/2) = 0$, so we know that $(x + 3/2)$, or $(1/2)(2x + 3)$, is a factor of f. Thus, by division,

$$2x^3 - x^2 - 14x - 12 = (2x + 3)(x^2 - 2x - 4) = 0.$$

We use the quadratic formula to find the other two real solutions:

$$x = \frac{2 \pm \sqrt{4 + 16}}{2} = \frac{2 \pm 2\sqrt{5}}{2} = 1 \pm \sqrt{5}.$$

≡

How to Solve $f(x) = 0$

There are several general methods for solving the equation $f(x) = 0$.

1. Find the exact solutions of $f(x) = 0$ algebraically (often by factoring).

2. Draw a complete graph of $y = f(x)$.

  a) Use ZOOM-IN to solve $f(x) = 0$ to the desired accuracy.

  b) Use SOLVE or your own programmed procedure to numerically solve $f(x) = 0$. Record the solution to the desired accuracy.

3. Use a Computer Algebra System like *Derive*® or *Mathematica*® to apply a combination of methods 1 and 2 to solve the equation $f(x) = 0$.

Solving Inequalities Graphically Using ZOOM

Every inequality involving a single variable x can be put into one of the four forms $f(x) < 0$, $f(x) \leq 0$, $f(x) > 0$, $f(x) \geq 0$. For most functions f that we will study, the solutions to these inequalities involve intervals of real numbers. To use the graph of f to help solve one of the four inequalities, say $f(x) < 0$, we must find the intervals where the graph lies below the x-axis. Since, in general, we are not able to determine the endpoints of a solution interval *exactly* with a grapher, *solve an inequality* will mean to find solution intervals with *endpoint error* of at most 0.01. When we need to be more careful, we will be so.

EXAMPLE 4

Solve the inequality $x^3 - 2x^2 < 1$.

Solution   Rewrite the inequality in the form $x^3 - 2x^2 - 1 < 0$, and let $f(x) = x^3 - 2x^2 - 1$. This is the function of Example 1, whose complete graph is shown in Fig. 1.66. We know that the graph crosses the x-axis at 2.206 with an error of at most 0.01, so the solution of the inequality is $-\infty < x < 2.206$ (with endpoint error of at most 0.01). Notice the Rational Zeros Theorem provides no help in finding zeros of $f(x) = x^3 - 2x^2 - 1$ because the *exact* real solution is not rational.   ≡

EXAMPLE 5

From Example 3, we know that $f(x) = 2x^3 - x^2 - 14x - 12 \geq 0$ when $-1.5 \leq x \leq 1 - \sqrt{5}$ or $x \geq 1 + \sqrt{5}$. If we solved the inequality with a grapher, we could say that the solution is $[-1.5, -1.236] \cup [3.236, \infty)$ with endpoint error of at most 0.01.   ≡

How to Combine Graphical, Algebraic, and Numerical Methods

Solving a problem efficiently usually requires a combination of graphical, algebraic, and numerical methods. We find the following steps helpful.

Steps for Solving a Problem

1. Find an algebraic representation, $y = f(x)$, involving the variables.
2. Draw a complete graph of $y = f(x)$.
3. Find the domain and range of $y = f(x)$.
4. Determine the values of x (and y) that make sense in the problem situation.
5. Draw a graph of the problem situation.
6. Solve the problem using appropriate combinations of numerical, graphical, and algebraic methods.

EXAMPLE 6

A fence for a rectangular garden with one side against an existing wall is constructed by using 50 feet of fencing. What is the maximum area that can be enclosed?

Solution

1. *Algebraic representation*. We let x be the length of one side of the fence perpendicular to the existing wall (Fig. 1.69). The length of the side parallel to the wall is $50 - 2x$. Then the relationship between the area, A,

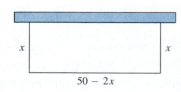

1.69   A garden enclosed by using an existing wall and 50 feet of fence.

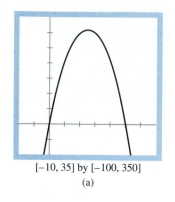

[−10, 35] by [−100, 350]

(a)

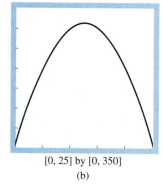

[0, 25] by [0, 350]

(b)

1.70   (a) A complete graph of

$$A(x) = x(50 - 2x).$$

(b) A graph of the garden fence problem situation $A(x) = x(50 - 2x), 0 \le x \le 25$.

of the garden and x is

$$y = A(x) = x(50 - 2x).$$

2. *Complete graph.* A complete graph of the function $y = A(x)$ is shown in Fig. 1.70(a).

3. *Domain and range.* The domain of $y = A(x)$ is all real numbers. The range is $y \le 312.5$, because 312.5 is the y-coordinate of the vertex of the parabola.

4. *Values of x that make sense.* The value of x can never be less than 0 nor more than 25 (one fenced side).

5. *Graph of the problem situation.* A graph of the problem situation

$$A(x) = x(50 - 2x) \qquad (0 \le x \le 25)$$

is shown in Fig. 1.70(b).

6. *Solve the problem.* The coordinates of the vertex of the parabola are (12.5, 312.5). The x-coordinate 12.5 is the side length in feet that produces maximum fenced-in area of 312.5 ft^2.   ≡

EXAMPLE 7

Pure acid is added to 40 ounces of a 30% acid mixture. How much pure acid should be added to produce a mixture that is 75% acid?

Solution

1. *Algebraic representation.* Let x be the number of ounces of pure acid added.

$$(0.3)40 = 12 \text{ ounces of pure acid in original mixture}$$

$$x + 12 = \text{ounces of pure acid in final mixture}$$

$$x + 40 = \text{ounces of final mixture}$$

$$C(x) = \frac{x + 12}{x + 40} = \text{concentration of acid in final mixture}$$

2. *Complete graph.* A graph of $C(x)$ is shown in Fig. 1.71(a) on the following page. We can use long division to rewrite $C(x)$:

$$C(x) = \frac{x + 12}{x + 40} = 1 - \frac{28}{x + 40} = 1 - 28 \cdot \left(\frac{1}{x + 40}\right) \qquad (1)$$

This form shows how the graph of $C(x)$ can be obtained from the graph of $y = 1/x$, so we know the graph shown is complete.

3. *Domain and range.* The domain is all real numbers different from −40. Equation (1) provides algebraic confirmation that the range is all real numbers different from 1, since $28/(x + 40)$ can never be 0.

4. *Values of x that make sense.* The problem situation gives no limitation on the amount of acid that is added. Therefore, $0 \le x < \infty$.

5. *Graph of the problem situation.* A graph of the problem situation, with $0 \le x < \infty$, is shown in Fig. 1.71(b).

6. *Solve the problem.* C represents the concentration (in decimal form). The amount of pure acid that must be added to produce a 75% mixture is given

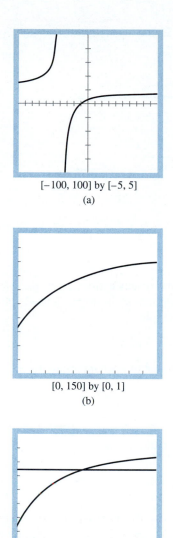

[−100, 100] by [−5, 5]
(a)

[0, 150] by [0, 1]
(b)

[0, 150] by [0, 1]
(c)

1.71 (a) A complete graph of

$$y = (x + 12)/(x + 40).$$

(b) A graph of the problem situation of Example 7. (c) Graphs of $y = (x + 12)/(x + 40)$ and $y = 0.75$ showing their point of intersection. The x-coordinate of this point is the solution to the problem in Example 7.

by the value of x for which $C(x) = 0.75$. Graphically (Fig. 1.71c), this is the x-coordinate of the point of intersection of

$$y = C(x) = \frac{x + 12}{x + 40} \qquad \text{and} \qquad y = 0.75.$$

Using ZOOM-IN shows $x = 72$ with an error of at most 0.01. We can confirm the solution algebraically:

$$C(x) = 0.75$$

$$\frac{x + 12}{x + 40} = 0.75$$

$$x + 12 = 0.75x + 30$$

$$0.25x = 18$$

$$x = 72$$

Thus, the solution that we found graphically using ZOOM-IN is exact. So 72 ounces of pure acid must be added to 40 ounces of a 30% acid solution to produce a mixture that is 75% acid. ≡

Equations and Inequalities with Absolute Values

To solve graphically an equation or inequality that contains absolute values requires *no* new techniques. To solve such equations and inequalities algebraically, we write equivalent equations or inequalities without absolute values and then solve as usual.

EXAMPLE 8

Solve the equation $|2x - 3| = 7$ algebraically, and support your work graphically.

Solution The equation says that $2x - 3 = \pm 7$, so there are two possibilities:

$$2x - 3 = 7 \qquad \text{or} \qquad 2x - 3 = -7 \qquad \text{Equivalent equations without absolute values}$$

$$2x = 10 \qquad\qquad\qquad 2x = -4 \qquad \text{Solve as usual.}$$

$$x = 5 \qquad\qquad\qquad\quad x = -2$$

The equation $|2x - 3| = 7$ has two solutions: $x = 5$ or $x = -2$.

The x-intercepts of the graph of the function $y = |2x - 3| - 7$ in Fig. 1.72 appear to be $x = 5$ and $x = -2$, supporting the reasonableness of the solutions found algebraically. ≡

Absolute Values and Intervals

Recall (Section 1.2) that $|a - b|$ is the distance between a and b on the number line. This connection between absolute value and distance gives us a nice way to represent intervals.

The inequality $|a| < 5$ says that the distance from a to the origin, $|a - 0|$, is less than 5. This is the same as saying that a lies between -5 and 5 on the number line. In symbols,

$$|a| < 5 \Leftrightarrow -5 < a < 5.$$

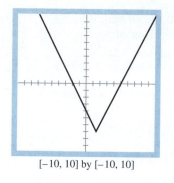

[−10, 10] by [−10, 10]

1.72   A complete graph of $y = |2x - 3| - 7$.

The set of numbers a with $|a| < 5$ is the open interval from -5 to 5 (Fig. 1.73a).

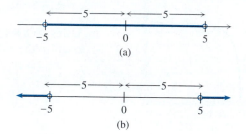

(a)

(b)

1.73   (a) The inequality $|a| < 5$ means $-5 < a < 5$. (b) The inequality $|a| > 5$ means $(-\infty, -5) \cup (5, \infty)$.

The inequality $|a| > 5$ says that the distance from a to the origin is greater than 5. This is the same as saying that a lies to the left of -5 or to the right of 5 on the number line. In symbols,

$$|a| > 5 \qquad \Leftrightarrow a < -5 \text{ or } a > 5.$$

The set of numbers a with $|a| > 5$ is the union of open intervals: $(-\infty, -5) \cup (5, \infty)$ (Fig. 1.73b).

Relation between Intervals and Absolute Values

If D is any positive number, then

$$|a| < D \Leftrightarrow -D < a < D, \qquad \text{The interval } (-D, D) \qquad (2)$$

$$|a| \leq D \Leftrightarrow -D \leq a \leq D, \qquad \text{The interval } [-D, D] \qquad (3)$$

$$|a| > D \Leftrightarrow a < -D \text{ or } a > D, \qquad \text{The intervals } (-\infty, -D) \cup (D, \infty) \qquad (4)$$

$$|a| \geq D \Leftrightarrow a \leq -D \text{ or } a \geq D. \qquad \text{The intervals } (-\infty, -D] \cup [D, \infty) \qquad (5)$$

EXAMPLE 9

Solve the inequality $|x - 5| < 9$ algebraically, and support your work graphically.

Solution

$$|x - 5| < 9$$

$$-9 < x - 5 < 9 \qquad \text{Equation (2) with } a = x - 5 \text{ and } D = 9$$

$$-9 + 5 < x < 9 + 5 \qquad \text{Adding a positive number to both sides of an inequality gives an equivalent inequality. Adding 5 here isolates the } x.$$

$$-4 < x < 14.$$

1.74 $|x - 5| < 9$ means $-4 < x < 14$.

The steps we just took are reversible, so the values of x that satisfy the inequality $|x-5| < 9$ are the numbers in the interval $-4 < x < 14$ (Fig. 1.74). The graph of $y = |x - 5| - 9$ in Fig. 1.75 supports the solution found algebraically because the graph appears to be below the x-axis for $-4 < x < 14$.

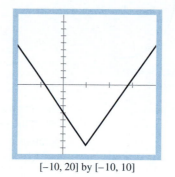

[−10, 20] by [−10, 10]

1.75 A complete graph of $y = |x - 5| - 9$.   ≡

EXAMPLE 10

Solve the inequality $\left|5 - \dfrac{2}{x}\right| < 1$.

Solution

$$\left|5 - \frac{2}{x}\right| < 1$$

$$-1 < 5 - \frac{2}{x} < 1 \qquad \text{Equation (2) with } a = (5 - 2/x) \text{ and } D = 1$$

$$-6 < -\frac{2}{x} < -4 \qquad \begin{array}{l}\text{Subtracting a positive number, in this}\\\text{case 5, from both sides of an inequality gives an}\\\text{equivalent inequality.}\end{array}$$

$$4 < \frac{2}{x} < 6 \qquad \begin{array}{l}\text{Multiplying both sides of an inequality}\\\text{by } -1 \text{ reverses the inequality.}\end{array}$$

$$2 < \frac{1}{x} < 3 \qquad \text{Divide by 2.}$$

$$\frac{1}{3} < x < \frac{1}{2}. \qquad \begin{array}{l}\text{Take reciprocals. When the numbers involved}\\\text{have the same sign, taking reciprocals reverses}\\\text{an inequality.}\end{array}$$

The original inequality holds if and only if x lies between 1/3 and 1/2.   ≡

EXAMPLE 11

Solve the inequality $|3/(x - 2)| < 1$ algebraically, and support your work graphically.

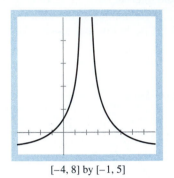

[−4, 8] by [−1, 5]

1.76 A complete graph of $y = |3/(x − 2)| − 1$.

Solution The following inequalities are equivalent provided that $x \neq 2$:

$$\left| \frac{3}{x − 2} \right| < 1$$

$$\frac{3}{|x − 2|} < 1 \qquad \left| \frac{a}{b} \right| = \frac{|a|}{|b|}$$

$$\frac{|x − 2|}{3} > 1 \qquad \text{Taking reciprocals reverses the inequality.}$$

$$|x − 2| > 3 \qquad \text{Multiply by 3.}$$

$$x − 2 < −3 \text{ or } x − 2 > 3 \qquad \text{Equation (4) with } a = x − 2 \text{ and } D = 3.$$

$$x < −1 \text{ or } x > 5 \qquad \text{Adding 2.}$$

The original inequality holds if and only if $x < −1$ or $x > 5$.

The graph of $y = |3/(x − 2)| − 1$ in Fig. 1.76 supports the solution because it appears to be below the x-axis for x in the intervals $(−\infty, −1)$ and $(5, \infty)$. ≡

EXAMPLE 12

Describe the interval $−3 < x < 5$ with an absolute value inequality of the form $|x − x_0| < D$.

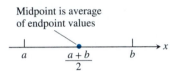

1.77 The midpoint value in the interval (a, b) is found by averaging the endpoint values.

Solution We average the endpoint values to find the interval's midpoint (Fig. 1.77):

$$\text{midpoint } x_0 = \frac{−3 + 5}{2} = \frac{2}{2} = 1.$$

The midpoint lies 4 units from each endpoint. The interval therefore consists of the points that lie within 4 units of the midpoint, or the points x with

$$|x − 1| < 4. \qquad ≡$$

Exercises 1.5

Solve the equations in Exercises 1–4 algebraically. Support your answers graphically.

1. $6x^2 + 5x − 6 = 0$ **2.** $x^2 − 4x + 1 = 0$

3. $4x^3 − 16x^2 + 15x + 2 = 0$

4. $x^3 − x^2 + x = 0$

Solve the equations in Exercises 5–8 graphically. Confirm your answers algebraically.

5. $2x^2 + 7x − 4 = 0$ **6.** $9x^2 − 6x = 4$

7. $18x^3 − 3x^2 − 14x = 4$

8. $x^3 − 2x^2 + 2x = 0$

In Exercises 9–10, find a sequence of four viewing windows containing each solution. Choose each sequence to permit the solutions to be read with errors of at most 0.1, 0.01, 0.001, and 0.0001.

9. $x^3 − 2x^2 − 5x + 5 = 0$

10. $x^3 − 4x + 1 = 0$

Solve the equations in Exercises 11–16.

11. $9x^2 − 6x + 5 = 0$ **12.** $x^2 + x − 12 = 0$

13. $2x^3 − 8x^2 + 3x + 9 = 0$

14. $10 + x + 2x^2 = x^3$

15. $x^3 − 21x^2 + 111x = 71$

16. $x^3 + 19x^2 + 90x + 52 = 0$

17. If $2 < x < 6$, which of the following statements about x are true and which are false?

a) $0 < x < 4$ b) $0 < x - 2 < 4$

c) $1 < \dfrac{x}{2} < 3$ d) $\dfrac{1}{6} < \dfrac{1}{x} < \dfrac{1}{2}$

e) $1 < \dfrac{6}{x} < 3$ f) $|x - 4| < 2$

g) $-6 < -x < 2$ h) $-6 < -x < -2$

18. If $-1 < y - 5 < 1$, which of the following statements about y are true and which are false?

a) $4 < y < 6$ b) $|y - 5| < 1$

c) $y > 4$ d) $y < 6$

e) $0 < y - 4 < 2$ f) $2 < \dfrac{y}{2} < 3$

g) $\dfrac{1}{6} < \dfrac{1}{y} < \dfrac{1}{4}$ h) $-6 < y < -4$

Solve the equations in Exercises 19–24.

19. $|x| = 2$ **20.** $|x - 3| = 7$

21. $|2x + 5| = 4$ **22.** $|1 - x| = 1$

23. $|8 - 3x| = 9$ **24.** $\left| \dfrac{x}{2} - 1 \right| = 1$

The inequalities in Exercises 25–30 define intervals. Describe each interval with inequalities that do not involve absolute values.

25. $|y - 1| \le 2$ **26.** $|y + 2| < 1$

27. $|3y - 7| < 2$ **28.** $\left| \dfrac{y}{3} \right| \le 10$

29. $|1 - y| < \dfrac{1}{10}$ **30.** $\left| \dfrac{7 - 3y}{2} \right| < 1$

Describe the intervals in Exercises 31–34 with absolute value inequalities of the form $|x - x_0| < D$. It may help to draw a picture of the interval first.

31. $3 < x < 9$ **32.** $-3 < x < 9$

33. $-5 < x < 3$ **34.** $-7 < x < -1$

Solve the inequalities in Exercises 35–38 algebraically. Support your answers graphically.

35. $|x - 5| < 2$ **36.** $\left| \dfrac{3x}{2} \right| < 5$

37. $\left| \dfrac{4}{x - 1} \right| \le 2$ **38.** $|3x + 2| > 3$

Solve the inequalities in Exercises 39–44 graphically. Confirm your answers algebraically.

39. $|x + 3| \le 5$ **40.** $\left| \dfrac{2x}{5} \right| \le 1$

41. $\left| \dfrac{2}{x + 3} \right| < 1$ **42.** $|3x + 2| \ge 1$

43. $|2 - 3x| < 4$ **44.** $\left| 5 - \dfrac{x}{2} \right| \le 1$

Solve the inequalities in Exercises 45–50.

45. $x^2 + 3x - 10 \le 0$ **46.** $4x^2 - 8x + 5 > 0$

47. $x^3 - 6x^2 + 5x + 6 \le 0$

48. $x^3 - 2x^2 - 5x + 20 > 0$

49. $x^3 - 4x^2 + 3.99x > 0$

50. $-x^3 + 0.2x^2 + 18.14x > -28.8$

Use ZOOM-IN to find the coordinates of the vertex of each parabola in Exercises 51–52, with an error of at most 0.01. Confirm your answers algebraically.

51. $y = 30x - x^2$ **52.** $y = -x^2 + 22x - 21$

53. Consider the function $f(x) = x^3 - 2x^2 - 1$ of **Example 1**.

a) Draw the graph of f in the following viewing windows: $[-5, 5]$ by $[-5, 5]$, $[2, 3]$ by $[-5, 5]$, $[2.2, 2.3]$ by $[-5, 5]$, and $[2.2, 2.3]$ by $[-0.1, 0.1]$.

b) Explain why it is necessary to change yMin and yMax when we ZOOM-IN on an x-intercept.

54. Solve the equation $x^3 - 2x = 2 \cos x$.

55. Explain how the Rational Zeros Theorem can be applied to a polynomial equation with noninteger rational coefficients such as $2x^3 + \dfrac{1}{2}x^2 - \dfrac{2}{3}x + 1 = 0$.

56. Explain why $ax^3 + bx^2 + cx + d = 0$ (a, b, c, d real, $a \ne 0$) will always have at least one real zero. Explain how to find all real zeros of any cubic polynomial equation.

57. One hundred feet of fencing is used to enclose a rectangular garden. Let x be the length of one side of the garden.

a) Find an algebraic representation $A(x)$ for the area of the garden.

b) Draw a complete graph of $y = A(x)$.

c) What are the domain and range of $y = A(x)$?

d) What values of x make sense in the problem situation? Draw a graph of the problem situation.

e) Use a graph to determine the dimensions of the garden if the area is 500 ft². Confirm your answer algebraically.

f) Determine the possible values of x if the area of the garden is to be less than 500 ft².

58. One hundred feet of fencing is used to enclose three sides of a rectangular pasture. The side of a barn closes off the fourth side. Let x be the length of one side of the fence perpendicular to the barn.

a) Find an algebraic representation $A(x)$ for the area of the pasture.

b) Draw a complete graph of $y = A(x)$.

c) What are the domain and range of $y = A(x)$?

d) What values of x make sense in the problem situation? Draw a graph of the problem situation.

e) Use a graph to determine the dimensions of the pasture if the area is 500 ft². Confirm your answer algebraically.

f) Determine the possible values of x if the area of the pasture is to be less than 500 ft².

59. An 8.5- by 11-in. piece of paper contains a picture with uniform border. The distance from the edge of the paper to the picture is x inches on all sides.

a) Write the area A of the picture as a function of x.

b) Draw a complete graph of $y = A(x)$.

c) What are the domain and range of $y = A(x)$?

d) What values of x make sense in the problem situation? Which portion of the graph in part (b) is a graph of the problem situation?

e) Determine the width of the border if the area of the picture is 60 in².

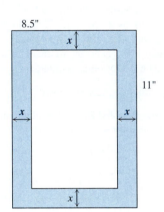

8.5"

11"

60. A 20- by 70-ft swimming pool is surrounded by a walk of uniform width. The distance from the outer edge of the walk to the pool is x feet on all sides.

a) Write the area A of the sidewalk as a function of x.

b) Draw a complete graph of $y = A(x)$.

c) What are the domain and range of $y = A(x)$?

d) What values of x make sense in the problem situation? Which portion of the graph in part (b) is a graph of the problem situation?

e) Determine the width of the sidewalk if its area is 500 ft².

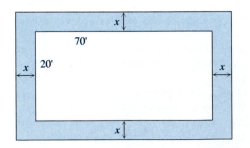

70'

20'

61. A money box contains 50 coins in dimes and quarters. Let x be the number of dimes in the box.

a) Find an algebraic representation $V(x)$ for the value of the coins in the box.

b) Draw a complete graph of $y = V(x)$.

c) What are the domain and range of $y = V(x)$?

d) What values of x make sense in the problem situation? Which portion of the graph in part (b) is a graph of the problem situation?

e) Determine the number of each coin in the box if the value of the coins is $9.20.

f) Repeat part (e) if the value of the coins is $6.25.

62. Sherrie invests $10,000, part at 6.5% simple interest and the remainder at 8% simple interest. Let x be the amount she invests at 6.5% interest.

a) Find an algebraic representation $I(x)$ for the total interest received in one year.

b) Draw a complete graph of $y = I(x)$.

c) What are the domain and range of $y = I(x)$?

d) What values of x make sense in the problem situation? Which portion of the graph in part (b) is a graph of the problem situation?

e) Determine the amount invested at each rate if Sherrie receives $766.25 interest in one year.

63. *The open box problem.*   Equal squares of side length x are removed from each corner of a 20- by 25-in. piece of cardboard. The sides are turned up to form a box with no top.

a) Write the volume V of the box as a function of x.

b) Draw a complete graph of $y = V(x)$.

c) What are the domain and range of $y = V(x)$?

d) What values of x make sense in the problem situation? Draw a graph of the problem situation.

e) Use ZOOM-IN to determine x so that the resulting box has maximum possible volume. What is the maximum possible volume?

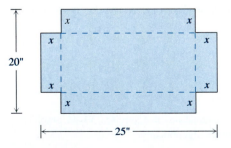

20"

25"

64. Repeat Exercise 63 for a 25- by 30-in. piece of cardboard.

65. *The box with lid problem..*   Two congruent squares are removed from one end of a rectangular 8.5- by 11-in. piece of cardboard. Two congruent rectangles are removed from the other end as shown.

a) Determine the length of the two rectangles to be removed so that when the cardboard is folded along the dashed lines, the box formed has a lid that exactly covers the top.

b) Find an algebraic representation $V(x)$ for the volume of the box with lid.

c) Draw a complete graph of $y = V(x)$. What are the domain and range of the function?

d) What values of x make sense in the problem situation? Draw a graph of the problem situation.

e) Determine the side length x of the squares to remove to construct a box with volume 25 in.^{3}

f) Determine x so that the resulting box has maximum volume. What is the maximum volume?

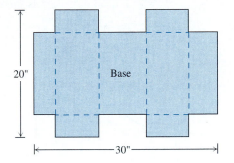

The sheet is then unfolded.

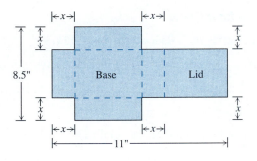

66. *The suitcase box with lid problem.* A rectangular 20- by 30-in. sheet of cardboard is folded in half to form a 20- by 15-in. rectangle from which four congruent corner squares of side length x are removed.

a) Show that folding along the dashed lines forms a box with lid with volume $V(x) = 2x(20 - 2x)(15 - 2x)$.

b) Draw a complete graph of $y = V(x)$. What are the domain and range?

c) What values of x make sense in the problem situation? Draw a graph of the problem situation.

d) Determine x so that the resulting box has volume 300 in.^{3}

e) Determine x so that the resulting box has maximum volume. What is the maximum volume?

1.6 _____ Relations, Functions, and Their Inverses

We recall from Section 1.3 that a relation is simply a set of ordered pairs. A function is a relation with a special condition, namely that no two of its ordered pairs can have the same first element. A nice example of a relation that is *not* a function is one whose graph is a circle, which clearly has a pair of points (many pairs, in fact) with the same first coordinate.

Equations for Circles in the Plane

> **DEFINITION**
>
> A **circle** is the set of points in a plane whose distance from a fixed point in the plane is a constant. The fixed point is the **center** of the circle. The constant distance is the **radius** of the circle.

To write an equation for the circle of radius a centered at the point $C(h, k)$, we let $P(x, y)$ denote a typical point on the circle (Fig. 1.78). The statement that CP equals a then becomes

$$\sqrt{(x - h)^2 + (y - k)^2} = a,$$ The length $CP = a$ using the Distance Formula

or

$$(x - h)^2 + (y - k)^2 = a^2.$$ Both sides squared   (1)

If $CP = a$, then Eq. (1) holds. If Eq. (1) holds, then $CP = a$. Equation (1) is therefore an equation for the circle.

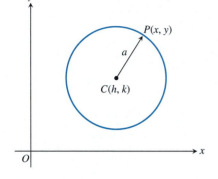

1.78 The standard equation for the circle shown here is $(x - h)^2 + (y - k)^2 = a^2$.

The Standard Equations for a Circle

$$(x - h)^2 + (y - k)^2 = a^2$$ Radius a, centered at (h, k) (2)

$$x^2 + y^2 = a^2$$ Radius a, centered at $(0, 0)$ (3)

The circle of radius 1 unit centered at the origin ($x^2 + y^2 = 1$) has a special name, *the unit circle*.

EXAMPLE 1

The standard equation for the circle of radius 2 centered at the point $(3, 4)$ is

$$(x - 3)^2 + (y - 4)^2 = (2)^2 \qquad \text{or} \qquad (x - 3)^2 + (y - 4)^2 = 4.$$

There is no need to square out the x and y terms in this equation. In fact, it is better not to do so. The present form reveals the circle's center and radius.

EXAMPLE 2

Find the center and radius of the circle. Determine its domain and range. Sketch a graph.

$$x^2 + y^2 - 2x + 10y = -16$$

Solution Complete the square for the terms involving x and the terms involving y.

$$(x^2 - 2x) + (y^2 + 10y) = -16$$

$$(x^2 - 2x + 1) + (y^2 + 10y + 25) = -16 + 1 + 25$$

$$(x - 1)^2 + (y + 5)^2 = 10$$

Comparing

$$(x - h)^2 + (y - k)^2 = a^2 \qquad \text{with} \qquad (x - 1)^2 + (y + 5)^2 = 10$$

shows that

$$-h = -1 \qquad \text{or} \qquad h = 1$$
$$-k = 5 \qquad \text{or} \qquad k = -5$$
$$a^2 = 10 \qquad \text{or} \qquad a = \sqrt{10}.$$

The center is the point $(h, k) = (1, -5)$. The radius is $a = \sqrt{10}$. The domain is $[1 - \sqrt{10}, 1 + \sqrt{10}]$. The range is $[-5 - \sqrt{10}, -5 + \sqrt{10}]$. A sketch is shown in Fig. 1.79.

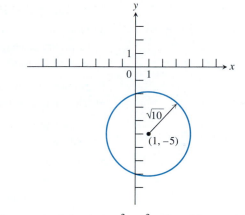

1.79   A complete graph of the circle $x^2 + y^2 - 2x + 10y = -16$, or $(x - 1)^2 + (y + 5)^2 = 10$.

Notice that the circle $(x - h)^2 + (y - k)^2 = a^2$ is the same as the circle $x^2 + y^2 = a^2$ with its center shifted from the origin to the point (h, k). The shift formulas that we have been using for graphs and functions (Section 1.4) apply to equations of any kind. Shifts to the right and up are accomplished by subtracting positive values of h and k from x and y, respectively. Shifts to the left and down are accomplished by subtracting negative values of h and k from x and y, respectively.

EXAMPLE 3

If the circle $x^2 + y^2 = 25$ is shifted two units to the left and three units up, its new equation is

$$(x - (-2))^2 + (y - 3)^2 = 25,$$

or

$$(x + 2)^2 + (y - 3)^2 = 25.$$

As Eq. (2) says it should be, this is the equation of the circle of radius 5 centered at $(h, k) = (-2, 3)$.

EXPLORATION 1

Drawing a Circle, Part 1

Since a circle is a relation and not a function, it is easiest to graph by using parametric equations. The clue to the form of the parametric equations is in the basic trigonometric relationship $\cos^2 x + \sin^2 x = 1$, which we will review in the next section. Therefore, put your grapher in parametric (and radian) mode and let

$$x(t) = \cos t, \qquad y(t) = \sin t.$$

1. Figure 1.80 suggests how to set tMin and tMax. The fact that it is to be a unit circle suggests how to set xMin, xMax, yMin, and yMax. Make these Fig. 1.80 settings, and then GRAPH. Does your graph make sense?

2. Change your t-step. If it was large, make it small. If it was small, make it large. GRAPH and explain what you see.

3. Does your viewing window show a circle or an oval? Can you change it to the other shape by changing the view dimensions? Can you find a fast way to do this?

In Section 1.4, we learned how to transform graphs "naturally" using parametric equations. Those methods will give you clues to the following.

4. Draw the circle with center $(0, 0)$ and radius 4. (You have to stretch the unit circle.)

5. Draw the circle with center $(3, -4)$ and radius 1. (You have to shift the unit circle.)

6. Draw the circle $(x - 3)^2 + (y + 4)^2 = 16$.

The points that lie inside the circle $(x - h)^2 + (y - k)^2 = a^2$ are the points less than a units from (h, k). They satisfy the inequality

$$(x - h)^2 + (y - k)^2 < a^2.$$

They make up the region we call the **interior** of the circle (Fig. 1.81).

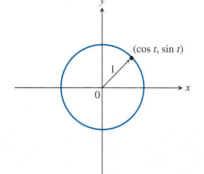

1.80 The points on the unit circle have coordinates $(\cos t, \sin t)$ for $0 \le t \le 2\pi$.

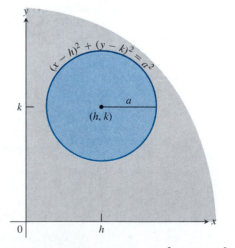

1.81 The interior and exterior of the circle $(x - h)^2 + (y - k)^2 = a^2$.

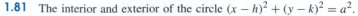

The circle's **exterior** consists of the points that lie more than a units from (h, k). These points satisfy the inequality

$$(x - h)^2 + (y - k)^2 > a^2.$$

EXAMPLE 4

Inequality	Region
$x^2 + y^2 < 1$	Interior of the unit circle
$x^2 + y^2 \leq 1$	Unit circle plus its interior
$x^2 + y^2 > 1$	Exterior of the unit circle
$x^2 + y^2 \geq 1$	Unit circle plus its exterior

EXPLORATION 2

Drawing a Circle, Part 2

Since a circle is a relation and not a function, it cannot be graphed directly in function mode. We can, however, solve $x^2 + y^2 = 1$ for y to get

$$y^2 = 1 - x^2 \qquad \text{or} \qquad y = \pm\sqrt{1 - x^2}.$$

Each of the formulas

$$y = \sqrt{1 - x^2} \qquad \text{or} \qquad y = -\sqrt{1 - x^2}$$

defines a function of x. Drawing a graph of the two functions in the same viewing window produces a complete graph of the unit circle.

1. Try it. Be sure to choose a square viewing window containing the unit circle. Figure 1.82 shows what your viewing window might look like.

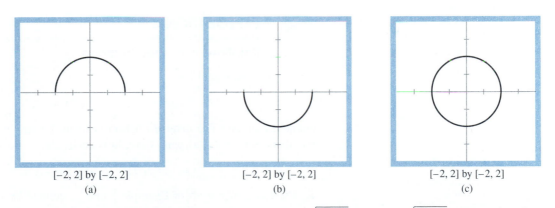

$[-2, 2]$ by $[-2, 2]$	$[-2, 2]$ by $[-2, 2]$	$[-2, 2]$ by $[-2, 2]$
(a)	(b)	(c)

1.82 Complete graphs of (a) $y = \sqrt{1 - x^2}$, (b) $y = -\sqrt{1 - x^2}$, and (c) $x^2 + y^2 = 1$.

2. Stay in function mode. Draw a complete graph of the circle $(x - 3)^2 + (y + 4)^2 = 16$.

3. Draw a complete graph of the *ellipse*, $\dfrac{x^2}{4} + \dfrac{y^2}{9} = 1$.

4. Draw a complete graph of the *hyperbola*, $\dfrac{x^2}{16} - \dfrac{y^2}{9} = 1$.

Inverse Relations and Inverse Functions

Associated with each relation is its *inverse relation*, namely, the relation obtained by switching the members of each ordered pair. A function is a relation, so each function has an inverse. The questions that we are concerned with include whether the inverse of a relation is a function and, in particular, whether the inverse of a function is a function.

> **CAUTION**
>
> The inverse relation, R^{-1}, is a set of ordered pairs. Do not make the mistake of thinking $R^{-1} = 1/R$, one reason being that $1/R$ does not even have a meaning. A little later in this section you will have the opportunity to learn why the word *inverse* for R^{-1} is *not* ill-chosen.

> **DEFINITION**   Inverse Relation
>
> Let R be a relation. The **inverse relation** R^{-1} of R consists of all those ordered pairs (b, a) for which (a, b) belongs to R. So the domain of R^{-1} = the range of R and the range of R^{-1} = the domain of R.

The circle $x^2 + y^2 = 1$ defines a relation that is not a function. If we interchange x and y, we get the same equation, so this relation is its own inverse.

EXAMPLE 5

Find the inverse relation for $y = x^2$.

Solution   If we interchange x and y in $y = x^2$, we get $x = y^2$. So $y = x^2$ and $x = y^2$ are inverse relations. Notice that their graphs are symmetric with respect to the line $y = x$ (Fig. 1.83).   ≡

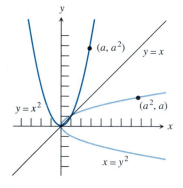

1.83 Complete graphs of the inverse relations $y = x^2$ and $x = y^2$ and the line $y = x$. Notice that for each point (a, a^2) on the graph of $y = x^2$, (a^2, a) is a point on the graph of $x = y^2$.

EXPLORATION 3

Relations and Their Inverses

The parametric equations

$$x_1(t) = \ldots, \qquad y_1(t) = \ldots$$

define a relation. The inverse relation is found by simply interchanging the members of each (x, y) pair. Look how easily this is done:

$$x_2(t) = y_1(t), \qquad y_2(t) = x_1(t).$$

1. The function $y = x^2$ of Example 5 can be defined by the equations

$$x_1(t) = t, \qquad y_1(t) = t^2.$$

Enter these equations in your grapher. Then enter parametric equations for the inverse relationship. GRAPH and compare your graphs with Fig. 1.83.

2. In the same square viewing window and using *simultaneous* format, GRAPH the two relations of Example 5 *and* the line $y = x$. Watch how they appear in the viewing window. Describe what you see.

When Are Inverse Relations Also Functions?

We are especially interested in functions whose inverse relations are also functions. Notice that the inverse relation $x = y^2$ of the function $y = x^2$ is not a function because its graph in Fig. 1.83 fails the Vertical Line Test. Suppose the graph of the inverse relation R^{-1} of a relation R satisfies the Vertical Line Test for functions. This means that no two ordered pairs in R^{-1} have the same first element, or no two ordered pairs in R have the same second element.

Horizontal Line Test That R^{-1} Is a Function

The inverse relation R^{-1} of the relation R is a function if and only if every horizontal line intersects the graph of R in at most one point.

The two Line Tests combined tell us when a relation and its inverse are both functions (Fig. 1.84). The domain elements and range elements are uniquely paired; they are paired *one-to-one* (sometimes written "1-1"). A function f is said to be *one-to-one* if and only if for every pair x_1 and x_2 of distinct values in the domain of f, it is also true that $f(x_1)$ and $f(x_2)$ are distinct.

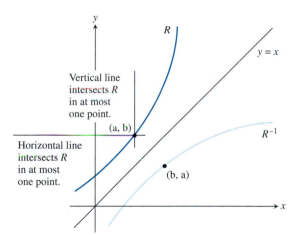

1.84 For a relation R to be a function, any vertical line intersects the graph of R in at most one point. For R^{-1} to be a function, any horizontal line intersects the graph of R in at most one point. If both Line Tests are passed, then both R and R^{-1} are functions, and they are said to be one-to-one.

One-to-One Test That f^{-1} Is a Function

The inverse f^{-1} of a function f is a function if and only if f is a one-to-one function.

Combining the Horizontal Line and One-to-One Tests, we obtain a graphical test of one-to-oneness of a function.

Graphical Test That f Is One-to-One

The function f is one-to-one if and only if every horizontal line intersects the graph of f in at most one point.

To summarize, the graph of a one-to-one function will have no two points in line either horizontally or vertically. Therefore, we can conclude that the inverse of a one-to-one function is also a function.

EXPLORATION 4

Exploring Inverses

1. Draw graphs and use the Horizontal Line Test to determine whether these functions have inverses that are also functions.

$$\textbf{a)} \quad y = x^3 \qquad \textbf{b)} \quad y = \frac{1}{x^2} \qquad \textbf{c)} \quad y = -2x + 4$$

2. Support your conclusions by graphing the inverse of each and using the Vertical Line Test.

The **identity function** $f(x) = x$, or $y = x$, is the function that assigns each number to itself.

Since each output of a one-to-one function comes from just one input, a one-to-one function can be reversed to send the outputs back to the inputs from which they came (Fig. 1.85). Thus, the result of composing f and f^{-1} in either order is the identity function (Fig. 1.86).

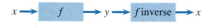

1.85 The inverse of a one-to-one function f sends every output of f back to the input from which it came.

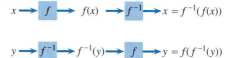

1.86 If $y = f(x)$ is a one-to-one function, then $f^{-1}(f(x)) = x$ and $f(f^{-1}(y)) = y$. Each of the composites $f^{-1} \circ f$ and $f \circ f^{-1}$ is the identity function on its domain. Can you think of another instance (or two) in which you combine something and its inverse and get an identity?

This gives us a way to test whether two functions f and g are inverses of one another. Compute $f \circ g$ and $g \circ f$. If both composites are identity functions, then f and g are inverses of one another; otherwise, they are not. For example, if f squares every number in its domain, g had better take square roots or it isn't the inverse of f.

If the inverse of a function $y = f(x)$ is also a function, we can sometimes find an explicit rule for f^{-1} in terms of x, as Example 6 illustrates.

EXAMPLE 6

Find the inverse of $y = -2x + 4$, expressed as a function of x. Check that the two functions are inverses by verifying that their composites are the identity function.

Solution

STEP 1: Switch x and y: $x = -2y + 4$.

STEP 2: Solve for y in terms of x:

$$x = -2y + 4$$
$$2y = -x + 4$$
$$y = -\frac{1}{2}x + 2.$$

The inverse of the function $f(x) = -2x + 4$ is the function $f^{-1}(x) = -\frac{1}{2}x + 2$. To check, we verify that both composites give the identity function:

$$f^{-1}(f(x)) = -\frac{1}{2}(-2x + 4) + 2 = x - 2 + 2 = x,$$

$$f(f^{-1}(x)) = -2\left(-\frac{1}{2}x + 2\right) + 4 = x - 4 + 4 = x.$$

≡

EXPLORATION 5

Inverses of Linear Functions

Answer parts 1 and 2 based on what you know about the graphs.

1. Do all linear functions ($y = ax + b$, $a \neq 0$) have inverses that are functions? Explain.

2. For a linear function that has an inverse function, what kind of function is the inverse?

Confirm your answers to parts 1 and 2 algebraically.

3. What must be true for a linear function $f(x) = mx + b$ to have an inverse?

4. Find an explicit equation for the inverse. How is the information from part 3 used? Explain how your result confirms part 2.

5. Confirm that the function $f(x) = mx + b$ and the inverse that you found are indeed inverses by composing.

6. If you enter the inverse functions y_1 and y_2 in your grapher in function mode, how can you enter y_3 to be the composition of y_1 and y_2? What would you anticipate the graph of y_3 to look like? Support your answers using a specific example of inverse linear functions y_1 and y_2.

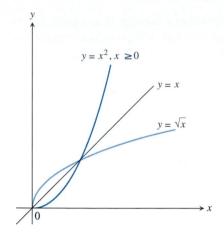

1.87 The functions $y = \sqrt{x}$ and $y = x^2, x \geq 0$, are inverses of one another. Their graphs are symmetric with respect to the line $y = x$.

We can sometimes restrict the domain of a function whose inverse is not a function and create a new function whose inverse is a function.

EXAMPLE 7

Show that the inverse of the function $y = x^2, x \geq 0$, is a function, and express the inverse as a function of x.

Solution   The unrestricted function, $y = x^2$, is not one-to-one, so its inverse is not a function. The restricted function $y = x^2, x \geq 0$, is one-to-one (Fig. 1.87). Thus, its inverse is a function.

STEP 1: Switch x and y: $y = x^2, x \geq 0$, becomes $x = y^2, y \geq 0$.

STEP 2: Solve for y in terms of x:

$$\sqrt{x} = \sqrt{y^2} = |y| = y \qquad \text{because } y \geq 0.$$

The inverse of the function $y = x^2, x \geq 0$, is the function $y = \sqrt{x}$ (Fig. 1.87).

≡

Exponential and Logarithmic Functions

A graphing utility can bring out the scientist in a mathematician. We can take functions about which we have very little prior knowledge and learn much about them from looking at their graphs. The ideas that we get from the viewing window must, of course, remain pure conjecture until we are able to confirm or deny them mathematically.

The equation $y = a^x, a > 0$, identifies a family of functions called *exponential functions*. At present, we know what 2^x means for some special values of x. We will find in Chapter 7 that 2^x (or $a^x, a > 0$) can be given precise definition for all real numbers x (Fig. 1.88). Until then, we will be content with using calculators to find the values of such functions and graphers to learn about them visually.

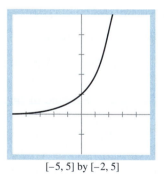

[−5, 5] by [−2, 5]

1.88 A graph of $y = 2^x$. Try showing this graph on your grapher. Use ZOOM-IN to find a value for $2^{\sqrt{3}}$. Can you come close to 3.32199708548, a value we found on one calculator?

> ### DEFINITION   Exponential Function
>
> Let a be a positive real number other than 1. The function $f(x) = a^x$ whose domain is $(-\infty, \infty)$ and whose range is $(0, \infty)$ is the **exponential function with base** a.

EXPLORATION 6

The Function $y = a^x$

Set viewing dimensions of $[-5, 5]$ by $[-2, 5]$.

1. Graph the functions $y = a^x$, for $a = 2, 3, 5$, in the same viewing window.

2. For what values of x is it true that $2^x > 3^x > 5^x$?

3. For what values of x is it true that $2^x < 3^x < 5^x$?

4. For what values of x is it true that $2^x = 3^x = 5^x$?

5. Graph the functions $y = (1/a)^x$, for $a = 2, 3, 5$. Repeat parts 2–4 above.

6. Graph the functions $y = a^{-x}$, for $a = 2, 3, 5$. Repeat parts 2–4 above.

7. Compare the graphs of $y = (1/a)^x$ and $y = a^{-x}$; of $y = a^x$ and $y = (1/a)^x$ for $a = 2, 3, 5$. What would you conjecture for arbitrary $a > 0$?

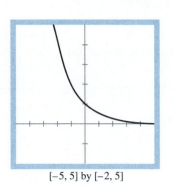

[−5, 5] by [−2, 5]

1.89 A complete graph of the exponential function $y = 0.5^x = 2^{-x}$.

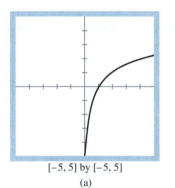

[−5, 5] by [−5, 5]

(a)

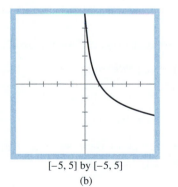

[−5, 5] by [−5, 5]

(b)

1.90 Complete graphs of (a) $y = \log_2 x$, the inverse of $y = 2^x$ (Fig. 1.88), and (b) $y = \log_{0.5} x$, the inverse of $y = 0.5^x$ (Fig. 1.89).

From Exploration 6, we made the following conjectures (which we will eventually confirm as true).

EXPONENTIAL FUNCTIONS

- If $a > 1$, the graph of $y = a^x$ has a shape like the graph of $y = 2^x$.

- If $0 < a < 1$, the graph of $y = a^x$ has a shape like the graph of $y = (1/2)^x = 0.5^x = 2^{-x}$ (Fig. 1.89).

- If $a \neq b$, then $a^x = b^x$ if and only if $x = 0$.

- $y = (1/a)^x$ and $y = a^{-x}$ define the same function. The graph of either is the reflection of the graph of $y = a^x$ across the y-axis.

Notice that the two principal models $y = 2^x$ and $y = (1/2)^x = 2^{-x}$ of exponential functions are one-to-one functions because their graphs satisfy the Horizontal Line Test (see Figs. 1.88 and 1.89). So the inverse relation of an exponential function is also a function. If we interchange x and y in $y = a^x$, we get $x = a^y$. Because we have no technique to solve for y in terms of x, we make the following definition.

DEFINITION **Logarithmic Function**

Let a be a positive real number other than 1. The function $f(x) = \log_a x$ with domain $(0, \infty)$ and range $(-\infty, \infty)$ is the inverse of the exponential function $y = a^x$ and is called the **logarithmic function with base a**.

$$y = \log_a x \Leftrightarrow x = a^y$$

The number $\log_a x$ is the **logarithm of x to the base a**.

Figure 1.90 shows complete graphs of $y = \log_2 x$ and $y = \log_{0.5} x$, which are the inverse functions of $y = 2^x$ and $y = 0.5^x$, respectively. In Example 8, we will show how to graph $y = \log_a x$ for any a ($a > 0, a \neq 1$).

There are several immediate consequences of the definition of the logarithmic function as the inverse of the exponential function: $f(x) = a^x$, $f^{-1}(x) = \log_a x$.

$$a^0 = 1 \qquad \Leftrightarrow \qquad \log_a 1 = 0 \qquad\qquad (4)$$

$$a^1 = a \qquad \Leftrightarrow \qquad \log_a a = 1 \qquad\qquad (5)$$

$$x = (f \circ f^{-1})(x) = a^{\log_a x} \qquad\qquad (6)$$

$$x = (f^{-1} \circ f)(x) = \log_a a^x \qquad\qquad (7)$$

Equations (4)–(7) have established the following properties.

PROPERTIES OF LOGARITHMS

Let $a > 0$ and $a \neq 1$. Then

1. $\log_a 1 = 0$
2. $\log_a a = 1$
3. $a^{\log_a x} = x$ for every positive real number x.
4. $\log_a a^x = x$ for every real number x.

Properties of Logarithms 1–4 will help us establish the following additional properties.

PROPERTIES OF LOGARITHMS (CONTINUED)

Let a, r, and s be positive real numbers with $a \neq 1$. Then

5. $\log_a rs = \log_a r + \log_a s$
6. $\log_a \dfrac{r}{s} = \log_a r - \log_a s$
7. $\log_a r^c = c \log_a r$ for every real number c.

Proof   We prove Property 5 and leave Properties 6 and 7 for the exercises.

$$\log_a rs = \log_a(a^{\log_a r} \cdot a^{\log_a s}) \qquad \text{Log Property 3}$$

$$\log_a rs = \log_a(a^{\log_a r + \log_a s}) \qquad \text{Property of Exponents}$$

$$\log_a rs = \log_a r + \log_a s \qquad \text{Log Property 4}$$

Properties 1–7 allow us to prove the Change of Base formula. With it, we can graph logarithmic functions having any base.

Change of Base Formula

Let a and b be positive real numbers with $a \neq 1$ and $b \neq 1$. Then

$$\log_b x = \frac{\log_a x}{\log_a b}.$$

Proof

$$y = \log_b x$$

$$x = b^y \qquad \text{Equivalent exponential form}$$

$$\log_a x = \log_a b^y$$

$$\log_a x = y \log_a b \qquad \text{Log Property 7}$$

$$y = \frac{\log_a x}{\log_a b}.$$

≡

Special Bases for Logarithms

Traditionally, calculators, including graphing calculators, have the logarithm functions with base e and base 10 built in. The number e is irrational and is important enough that it likely has its own key on your calculator. Check it out, and you will see that to ten significant digits,

$$e = 2.718281828.$$

We write $\log_e x$ as $\ln x$ and omit the base when using base 10:

$$\log_e x = \ln x, \log_{10} x = \log x.$$

The function $y = \ln x$ is usually called the **natural logarithm function**. You may find the function $y = \log x$ called the **common logarithm function**. We will learn much more about the number e, the natural logarithm function, and logarithmic functions in general in Chapter 7.

EXAMPLE 8

Use a graphing utility to draw a complete graph of $f(x) = \log_5 x$. Find the domain and range.

Solution   We use the Change of Base Formula to rewrite $f(x)$ using either log or ln to enter f.

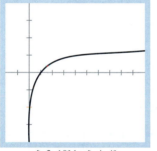

[−2, 10] by [−4, 4]

1.91   A complete graph of $f(x) = \log_5 x$ using either of the representations $f(x) = \log x / \log 5$ or $f(x) = \ln x / \ln 5$.

$$f(x) = \log_5 x = \frac{\log x}{\log 5} = \frac{\ln x}{\ln 5}$$

Figure 1.91 gives a complete graph of f. Notice that the domain of f is $(0, \infty)$ and the range of f is $(-\infty, \infty)$. All functions $y = \log_a x$ have the same domain and range.

≡

EXPLORATION 7

Mentally "Estimate" the Graph

Use what you know about shifts, stretches, shrinks, and reflections to give you an idea of what each of the following graphs below should look like. Then use a grapher to support your conjecture and determine its domain and range.

1. The graph of $f(x) = -1.5(2^{x+1}) - 3$ based on the graph of $y = 2^x$.

2. The graph of $f(x) = \log_3(2 - x)$ based on the graph of $y = \log_3 x$. ⚬

Exercises 1.6

Which of the functions graphed in Exercises 1–4 are one-to-one?

1.

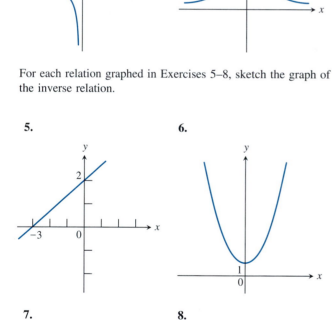

2.

3.

4.

For each relation graphed in Exercises 5–8, sketch the graph of the inverse relation.

5.

6.

7.

8.

In Exercises 9–12, find an equation for the circle with the given center $C(h, k)$ and radius a. Then sketch a complete graph of the circle.

9. $C(0, 2), a = 2$

10. $C(-2, 0), a = 3$

11. $C(3, -4), a = 5$

12. $C(1, 1), a = \sqrt{2}$

Write equations for the circles in Exercises 13–16.

13.

14.

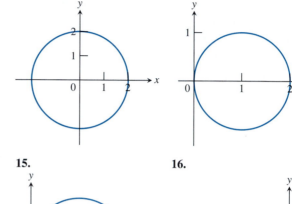

15.

16.

Find the center and radius of the circles in Exercises 17–20. Determine each domain and range. Sketch a complete graph.

17. $x^2 + y^2 - 6x + 8y = -16$

18. $x^2 + y^2 + 2x - 4y = 11$

19. $x^2 + y^2 + 4x + 6y + 8 = 0$

20. $x^2 + y^2 - 2x - 8y + 10 = 0$

Use a grapher to obtain complete graphs of the equations in Exercises 21 and 22.

21. $16x^2 - 9y^2 = 144$

22. $4x^2 + 9y^2 = 36$

Describe the regions defined by the inequalities and pairs of inequalities in Exercises 23 and 24.

23. a) $x^2 + y^2 > 1$
   b) $x^2 + y^2 < 4$
   c) the inequalities in parts (a) and (b) together

24. a) $x^2 + y^2 \geq 1$
   b) $x^2 + y^2 \leq 4$
   c) the inequalities in parts (a) and (b) together

25. Write an inequality that describes the points that lie inside the circle with center $C(-2, -1)$ and radius $a = \sqrt{6}$.

26. Write an inequality that describes the points that lie outside the circle with center $C(-4, 2)$ and radius $a = 4$.

Which of each function in Exercises 27–32 has an inverse that is also a function? Explain and demonstrate how you can support your answers with a grapher in parametric mode.

27. $y = \dfrac{3}{x-2} - 1$

28. $y = x^2 + 5x$

29. $y = x^3 - 4x + 6$

30. $y = x^3 + x$

31. $y = \ln x^2$

32. $y = 2^{3-x}$

In Exercises 33–44, find $f^{-1}(x)$ and show that $f \circ f^{-1}(x) = f^{-1} \circ f(x) = x$. Draw a complete graph of f and f^{-1} in the same viewing window.

33. $f(x) = 2x + 3$

34. $f(x) = 5 - 4x$

35. $f(x) = x^3 - 1$

36. $f(x) = 2 - x^3$

37. $f(x) = x^2 + 1, x \geq 0$

38. $f(x) = x^2, x \leq 0$

39. $f(x) = -(x-2)^2, x \leq 2$

40. $f(x) = (x+1)^2, x \geq -1$

41. $f(x) = \dfrac{1}{x^2}, x > 0$

42. $f(x) = \dfrac{1}{x^3}, x \neq 0$

43. $f(x) = \dfrac{2x+1}{x+3}, x \neq -3$

44. $f(x) = \dfrac{x+3}{x-2}, x \neq 2$

Draw a complete graph of each function in Exercises 45–52.

45. $y = 2 \log_3 (x - 4) - 1$

46. $y = -3 \log_5 (2 - x) + 1$

47. $y = -3 \log_{0.5} (x + 2) + 2$

48. $y = 2 \log_{0.2} (3 - x) + 1$

49. $y = 5(e^{3x}) + 2$

50. $y = 3(e^{2-x}) - 1$

51. $y = -2(3^x) + 1$

52. $y = -5(2^{-x+1}) + 3$

Determine the domain and range and describe how the graphs of each equation in Exercises 53–58 can be obtained from the graph of an appropriate exponential function $y = a^x$, logarithmic function $y = \log_a x$, or circle $x^2 + y^2 = a^2$.

53. $y = -3 \log (x + 2) + 1$

54. $y = 2 \ln (3 - x) - 4$

55. $y = 2(3^{1-x}) + 1.5$

56. $y = -3(5^{x-2}) + 3$

57. $(x + 3)^2 + (y - 5)^2 = 9$

58. $(x - 6)^2 + (y + 1)^2 = 25$

In Exercises 59–62, graph f, f^{-1}, and the line $y = x$ in the same square viewing window.

59. $f(x) = 2^x$

60. $f(x) = 0.5^x$

61. $f(x) = \log_3 x$

62. $f(x) = \log_{0.3} x$

Solve the equations in Exercises 63–66.

63. $e^x + e^{-x} = 3$

64. $2^x + 2^{-x} = 5$

65. $\log_2 x + \log_2 (4 - x) = 0$

66. $\log x + \log (3 - x) = 0$

67. Show that the graph of the inverse R^{-1} of a relation R can be obtained by the following two-step process: Rotate the graph of R $90°$ counterclockwise about the origin, and then reflect the resulting graph through the y-axis. (*Hint:* The path of a point under this process is $(a, b) \to (-b, a) \to (b, a)$.)

68. Prove Properties of Logarithms 6 and 7.

69. Use the Change of Base Formula to show that $\log_b a = \dfrac{1}{\log_a b}$

70. Suppose $a \neq 0, b \neq 1$, and $b > 0$. Determine the domain and range of each function.
a) $y = a(b^{c-x}) + d$
b) $y = a \log_b (x - c) + d$

1.7 —————— A Review of Trigonometric Functions

In surveying, navigation, and astronomy, we measure angles in degrees, but in calculus, it is usually best to use radians. We will see why in Section 3.5. In the present section, we use radians and degrees together so that you can practice relating the two. We also review the trigonometry that you will need for calculus and its applications.

Radian Measure

The **radian measure** of the angle ACB at the center of the unit circle (Fig. 1.92) equals the length of the arc that the angle cuts from the unit circle.

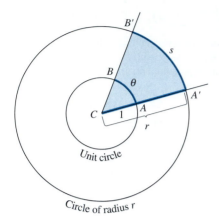

1.92 The radian measure of angle ACB is the length θ of arc AB on the unit circle centered at C. The value of θ can be found from any other circle, however, as the ratio of s to r.

If angle ACB cuts an arc $A'B'$ from a second circle centered at C, then circular sector $A'CB'$ will be similar to circular sector ACB. In particular,

$$\frac{\text{Length of arc } A'B'}{\text{Radius of second circle}} = \frac{\text{Length of arc } AB}{\text{Radius of first circle}}. \qquad (1)$$

In the notation of Fig. 1.92, Eq. (1) says that

$$\frac{s}{r} = \frac{\theta}{1} = \theta \qquad \text{or} \qquad \theta = \frac{s}{r}.$$

Notice that the units of length for s and r cancel out and that radian measure is a dimensionless number.

The equation $\theta = s/r$ tells us that when we know s and r, we can calculate the angle's radian measure as the ratio of arc length to radius. Dividing both sides of $\theta = s/r$ by 2π gives $\theta/2\pi = s/2\pi r$. This equation is part of the following proportions, which are easily remembered:

$$\frac{\text{degree measure}}{360°} = \frac{\text{radian measure}}{2\pi} = \frac{\text{arc length}}{\text{circumference}}$$

EXAMPLE 1   Conversions Between Degree and Radian Measure

For 45°, $\qquad \dfrac{45°}{360°} = \dfrac{x}{2\pi} \Rightarrow x = \dfrac{45 \cdot 2\pi}{360} \text{ rad} \Rightarrow x = \dfrac{\pi}{4} \text{ rad}.$

For 90°, $\qquad \dfrac{90°}{360°} = \dfrac{x}{2\pi} \Rightarrow x = \dfrac{90 \cdot 2\pi}{360} \text{ rad} \Rightarrow x = \dfrac{\pi}{2} \text{ rad}.$

For $\dfrac{\pi}{6}$ rad, $\qquad \dfrac{\pi/6}{2\pi} = \dfrac{x}{360°} \Rightarrow x = \dfrac{(\pi/6) \cdot 360}{2\pi} \text{ deg} \Rightarrow x = 30°.$

For $\dfrac{\pi}{3}$ rad, $\qquad \dfrac{\pi/3}{2\pi} = \dfrac{x}{360°} \Rightarrow x = \dfrac{(\pi/3) \cdot 360}{2\pi} \text{ deg} \Rightarrow x = 60°.$

Remembering that $2\pi = 360°$, $\pi = 180°$, $\pi/2 = 90°$, $\pi/3 = 60°$, $\pi/4 = 45°$, and $\pi/6 = 30°$ can be useful (Fig. 1.93). For all other conversions, we use a calculator.

EXPLORATION BIT

Show that the numerical value of θ (Fig. 1.92) in radians is twice the numerical value of the area of the "pie slice" cut in the unit circle.

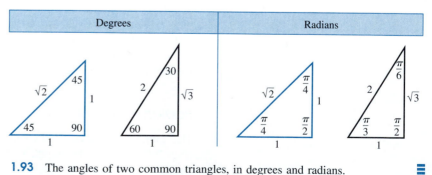

1.93 The angles of two common triangles, in degrees and radians.

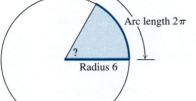

1.94 What is the radian measure of this angle? See Example 3.

EXAMPLE 2 Conversions Using a Calculator

For 26°, $\quad \dfrac{26°}{360°} = \dfrac{x}{2\pi} \Rightarrow x = \dfrac{26 \cdot 2\pi}{360} \text{ rad} \Rightarrow x = 0.454 \text{ rad}.$

For 2 rad, $\quad \dfrac{2}{2\pi} = \dfrac{x}{360°} \Rightarrow x = \dfrac{2 \cdot 360}{2\pi} \text{ deg} \Rightarrow x = 114.592°.$

EXAMPLE 3

An acute angle whose vertex lies at the center of a circle of radius 6 subtends an arc of length 2π (Fig. 1.94). We find the angle's radian measure as follows:

$$\frac{\text{radian measure}}{2\pi} = \frac{\text{arc length}}{\text{circumference}} = \frac{2\pi}{2\pi r} = \frac{2\pi}{2\pi(6)} \Rightarrow \text{radian measure} = \frac{\pi}{3}.$$

EXAMPLE 4

An angle of $3\pi/4$ radians lies at the center of a circle of radius 8. We find the length of the arc it subtends as follows:

$$\frac{\text{arc length}}{2\pi(8)} = \frac{3\pi/4}{2\pi} \Rightarrow \text{arc length} = 6\pi.$$

EXAMPLE 5

The length of the arc subtended by a central angle of 100° in a circle of radius 4 can be found as follows:

$$\frac{\text{arc length}}{2\pi(4)} = \frac{100°}{360°} \Rightarrow \text{arc length} = \frac{100(2\pi)(4)}{360} = 6.981.$$

When angles are used to describe counterclockwise rotations, our measurements can go arbitrarily far beyond 2π radians or 360°. Similarly, angles that describe clockwise rotations can have negative measures of all sizes (Fig. 1.95).

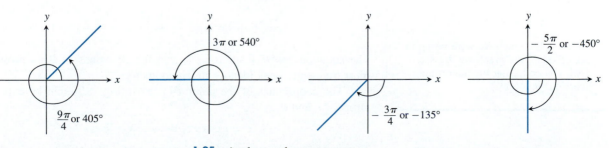

1.95 Angles can have any measure.

The Six Basic Trigonometric Functions

When an angle of measure θ is placed in *standard position* at the center of a circle of radius r (Fig. 1.96), the six basic trigonometric functions of θ are defined in the following way:

Sine: $\quad \sin \theta = \dfrac{y}{r}$, $\qquad$ Cosecant: $\quad \csc \theta = \dfrac{r}{y}$,

Cosine: $\quad \cos \theta = \dfrac{x}{r}$, $\qquad$ Secant: $\quad \sec \theta = \dfrac{r}{x}$, $\qquad$ (2)

Tangent: $\quad \tan \theta = \dfrac{y}{x}$, $\qquad$ Cotangent: $\quad \cot \theta = \dfrac{x}{y}$.

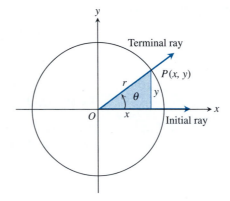

1.96 An angle θ in standard position.

As you can see, $\tan \theta$ and $\sec \theta$ are not defined if $x = 0$. In terms of radian measure, this means that they are not defined when θ is $\pm\pi/2, \pm 3\pi/2$, ... Similarly, $\cot \theta$ and $\csc \theta$ are not defined for values of θ for which $y = 0$, namely, $\theta = 0, \pm\pi, \pm 2\pi, \ldots$.

Notice also that

$$\tan \theta = \frac{\sin \theta}{\cos \theta}, \qquad \csc \theta = \frac{1}{\sin \theta},$$

$$\sec \theta = \frac{1}{\cos \theta}, \qquad \cot \theta = \frac{1}{\tan \theta}$$

whenever the quotients on the right-hand sides are defined.

Because $x^2 + y^2 = r^2$ (Pythagorean Theorem),

$$\cos^2 \theta + \sin^2 \theta = \frac{x^2}{r^2} + \frac{y^2}{r^2} = \frac{x^2 + y^2}{r^2} = 1.$$

The equation $\cos^2 \theta + \sin^2 \theta = 1$, true for all values of θ, is probably the most frequently used identity in trigonometry.

The coordinates of the point $P(x, y)$ in Fig. 1.96 can be expressed in terms of r and θ as

$x = r \cos \theta \quad$ Because $x/r = \cos \theta$

$y = r \sin \theta \quad$ Because $y/r = \sin \theta$. $\qquad$ (3)

EXPLORATION BIT

One of the fun things about using a grapher is that mistakes can lead to further explorations and insights. For example, in an attempt to graph

$$y = (\sin x)^2 + (\cos x)^2,$$

we entered

$$y = \sin x^2 + \cos x^2.$$

Describe the graph we were trying to see. Explain the graph we did see. (Are you collecting the interesting functions you find?)

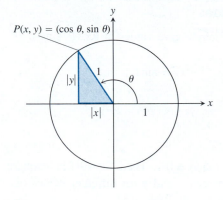

1.97 The acute reference triangle for an angle θ.

Finding Cosines and Sines

If the circle in Fig. 1.96 has radius $r = 1$ unit, Eqs. (3) simplify to

$$x = \cos\theta, \qquad y = \sin\theta.$$

We can therefore find the values of cosine and sine geometrically from the acute reference triangle made by dropping a perpendicular from the point $P(x, y)$ to the x-axis (Fig. 1.97). The numerical values of x and y are read from the triangle's sides. The signs of x and y are determined by the quadrant in which the triangle lies.

EXAMPLE 6

Find the cosine and sine of $-\pi/4$ radians.

Solution

STEP 1: Draw the angle in standard position and write in the lengths of the sides of the reference triangle (Fig. 1.98).

STEP 2: Find the coordinates of the point P where the angle's terminal ray cuts the circle:

$$\cos\left(-\frac{\pi}{4}\right) = x\text{-coordinate of } P$$

$$= \frac{\sqrt{2}}{2}.$$

$$\sin\left(-\frac{\pi}{4}\right) = y\text{-coordinate of } P$$

$$= -\frac{\sqrt{2}}{2}.$$

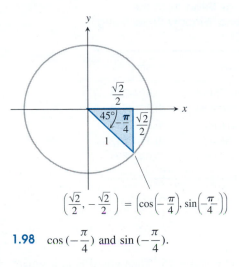

1.98 $\cos\left(-\dfrac{\pi}{4}\right)$ and $\sin\left(-\dfrac{\pi}{4}\right)$.

Table 1.2 gives sine, cosine, and tangent values for selected values of θ. These are found by using methods similar to the ones in Example 6. It is useful to remember these values—or at least how they were found. For other values of trigonometric functions, we will use a calculator.

TABLE 1.2   Exact Values of $\sin\theta$, $\cos\theta$, $\tan\theta$ for Selected Values of θ

θ (degrees)	-180	-135	-90	-45	0	45	90	135	180
θ (radians)	$-\pi$	$-3\pi/4$	$-\pi/2$	$-\pi/4$	0	$\pi/4$	$\pi/2$	$3\pi/4$	π
$\sin\theta$	0	$-\sqrt{2}/2$	-1	$-\sqrt{2}/2$	0	$\sqrt{2}/2$	1	$\sqrt{2}/2$	0
$\cos\theta$	-1	$-\sqrt{2}/2$	0	$\sqrt{2}/2$	1	$\sqrt{2}/2$	0	$-\sqrt{2}/2$	-1
$\tan\theta$	0	1	*	-1	0	1	*	-1	0

* Not defined.

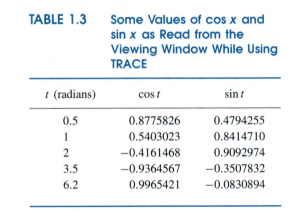

EXPLORATION 1

Finding Sines and Cosines

The parametric equations for the unit circle give values of $(\cos\theta, \sin\theta)$.

$$x(t) = \cos t, \qquad y(t) = \sin t$$

Set your grapher in radian mode.

> **CAUTION**
>
> When using calculators to compute trigonometric values, be sure your calculator is set in the appropriate mode (degree or radian).

1. Enter $t\text{Min} = 0$, $t\text{Max} = 2\pi$, and t-step = 0.1. Then GRAPH in a square viewing window. Use TRACE to read x and y for different values of t (Fig. 1.99). Can you match the entries in Table 1.3?

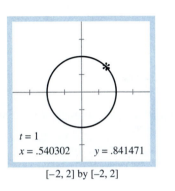

$t = 1$
$x = .540302 \qquad y = .841471$

[−2, 2] by [−2, 2]

1.99 A complete graph of the unit circle $x^2 + y^2 = 1$ obtained from a grapher in parametric mode. The values of $x = \cos 1$ and $y = \sin 1$ are shown using TRACE.

TABLE 1.3 Some Values of cos x and sin x as Read from the Viewing Window While Using TRACE

t (radians)	$\cos t$	$\sin t$
0.5	0.8775826	0.4794255
1	0.5403023	0.8414710
2	−0.4161468	0.9092974
3.5	−0.9364567	−0.3507832
6.2	0.9965421	−0.0830894

Set your grapher in degree mode.

2. Enter $t\text{Min} = 0$, $t\text{Max} = 360$, and t-step = 15. Then GRAPH in a square viewing window. For what values of t are $\cos t$ and $\sin t$ being computed?

3. Change $t\text{Min}$ to -180 and $t\text{Max}$ to 180. Then GRAPH. Explain what you see in the viewing window. Use TRACE, and compare with the exact values in Table 1.2.

4. Are there other $t\text{Min}$ and $t\text{Max}$ values that give the graph in part 3? Explain.

5. Return to radian mode. Repeat parts 3 and 4 using radian equivalents of -180 and 180.

6. How can you obtain values of $\cos\theta$ and $\sin\theta$ for values of θ in $0 \le \theta \le 4\pi$? Explain.

Periodicity

When an angle of measure θ and an angle of measure $\theta + 2\pi$ are in standard position, their terminal rays coincide. The two angles therefore have the same trigonometric function values:

$$\cos(\theta + 2\pi) = \cos\theta \quad \sin(\theta + 2\pi) = \sin\theta \quad \tan(\theta + 2\pi) = \tan\theta$$
$$\cot(\theta + 2\pi) = \cot\theta \quad \sec(\theta + 2\pi) = \sec\theta \quad \csc(\theta + 2\pi) = \csc\theta \tag{4}$$

Similarly, $\cos(\theta - 2\pi) = \cos\theta$, $\sin(\theta - 2\pi) = \sin\theta$, and so on.

From another point of view, Eqs. (4) tell us that if we start at any particular value $\theta = \theta_0$ and let θ increase or decrease steadily, we see the values of the trigonometric functions repeat over regular intervals. We describe this behavior by saying that the six basic trigonometric functions are *periodic*.

DEFINITION

A function $f(x)$ is **periodic** if there is a positive number p such that $f(x + p) = f(x)$ for every value of x. The smallest such value of p is the **period** of f.

EXAMPLE 7

Equations (4) tell us that the six basic trigonometric functions are periodic. A view of each function in a window with $[x\text{Min}, x\text{Max}] = [-2\pi, 2\pi]$ suggests that the cosine, sine, secant, and cosecant functions have period 2π and the tangent and cotangent functions have period π.   ≡

EXPLORATION BITS

1. Give graphical support to the suggestion in Example 7 about the periods of the six basic trigonometric functions.

2. Give a convincing argument about the periods of the six basic trigonometric functions using Eqs. (2) and the unit circle.

3. Show that the word "positive" is not needed in the definition of *periodic* function. If "positive" is removed from the definition, how should the definition of period be changed?

The importance of periodic functions stems from the fact that much of the behavior we study in science is periodic. Brain waves and heartbeats are periodic, as are household voltage and electric current. The electromagnetic field that heats food in a microwave oven is periodic, as are cash flows in seasonal businesses and the behavior of rotational machinery. The seasons are periodic—so is the weather. The phases of the moon are periodic, as are the motions of the planets. There is strong evidence that the ice ages are periodic, with a period of 90,000–100,000 years.

Why are trigonometric functions so important in the study of things periodic? The answer lies in a surprising and beautiful theorem from advanced calculus that says that every periodic function that we want to use in mathematical modeling can be written as an algebraic combination of sines and cosines. Thus, once we learn the calculus of sines and cosines, we will know everything we need to know to model the mathematical behavior of periodic phenomena.

EXPLORATION 2

Modeling Circular Motion

A wheel of radius 4 is rotating counterclockwise once every six seconds.

1. Show that the position of a certain point on the wheel is modeled by

$$x(t) = 4\cos(\pi t/3), \qquad y(t) = 4\sin(\pi t/3)$$

where t is time in seconds.

2. Use TRACE to model the motion of the point.

3. Find the position of the point 3 seconds, 6 seconds, and 9 seconds after the wheel starts to revolve. Where is the point at $t = 0$ seconds?

4. At what values of t is the point at $(4, 0)$?

5. At what values of t is the point at $(2\sqrt{2}, 2\sqrt{2})$?

Graphs of Trigonometric Functions

When we graph trigonometric functions in the coordinate plane, we usually denote the independent variable (radians) by x instead of θ. Figure 1.100

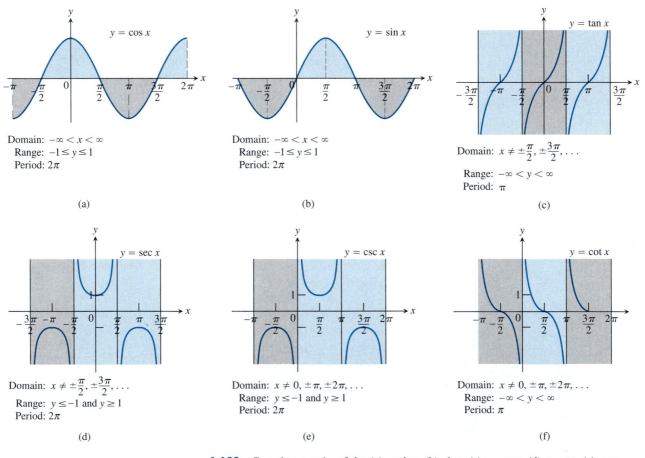

1.100 Complete graphs of the (a) cosine, (b) sine, (c) tangent, (d) secant, (e) cosecant, and (f) cotangent functions using radian measure.

shows sketches of the six trigonometric functions. It is a good exercise for you to compare these with what you see in a grapher viewing window.

Notice that the graph of

$$\tan x = \frac{\sin x}{\cos x}$$

"blows up" whenever x nears an odd-integer multiple of $\pi/2$. These are the points for which $\sin x = 1$ and $\cos x = 0$. Notice, too, how the periodicity of the sine, cosine, and tangent appears in the graphs. Choose any starting point, and each graph repeats after an interval of length equal to the period.

For a graph of a periodic function to be complete, it must show at least one period. If the graph is to be viewed by someone else, you need to specify the length of the period.

EXPLORATION BITS

1. There are infinitely many periodic functions with period smaller than 2π. Can you find two?

2. For any number a, can you find a periodic function with period a? (*Hint:* Recall how to stretch and shrink a graph.)

EXPLORATION 3

Unwrapping Trigonometric Functions

Set your grapher in radian mode and enter the parametric equations

$$x_1(t) = \cos t, \qquad y_1(t) = \sin t,$$

$$x_2(t) = t, \qquad y_2(t) = \sin t$$

with $t\text{Min}=0$, $t\text{Max}=2\pi$, $[x\text{Min}, x\text{Max}]=[-1.5, 2\pi]$, and $[y\text{Min}, y\text{Max}] = [-2, 2]$,

1. Before you GRAPH (in *simultaneous* format), try to predict what will happen in the viewing window. Then GRAPH. Describe what you see in the viewing window. You may wish to make the viewing window square.

2. Use TRACE and compare the y-values on the two graphs. Explain.

3. Set $x\text{Max} = 4\pi$, and repeat part 2.

4. Repeat parts 1–3 for $y_2 = \cos t, \tan t, \csc t, \sec t$, and $\tan t$.

Odd versus Even

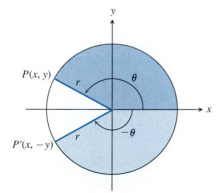

1.101 Angles of opposite sign.

Figure 1.101 shows two angles that have the same magnitude but opposite signs. By symmetry, the points where the terminal rays cross the circle have the same x-coordinate, and their y-coordinates differ only in sign. Hence,

$$\cos(-\theta) = \frac{x}{r} = \cos\theta, \qquad \text{The cosine is an even function.} \qquad (5a)$$

$$\sin(-\theta) = \frac{-y}{r} = -\sin\theta. \qquad \text{The sine is an odd function.} \qquad (5b)$$

Equation (5a) means that reflecting the graph of $f(x) = \cos x$ across the y-axis leaves it unchanged ($f(-x) = f(x)$). Equation (5b) means that reflecting the graph of $g(x) = \sin x$ across the y-axis is the same as reflecting it across the x-axis ($g(-x) = -g(x)$).

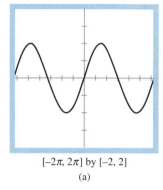

$[-2\pi, 2\pi]$ by $[-2, 2]$
(a)

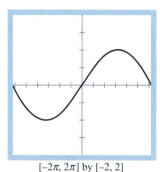

$[-2\pi, 2\pi]$ by $[-2, 2]$
(b)

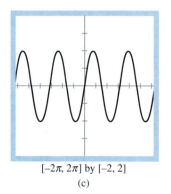

$[-2\pi, 2\pi]$ by $[-2, 2]$
(c)

1.102 Complete graphs of (a) $y = \sin x$ showing two periods of length 2π, (b) $y = \sin (1/2)x$ showing one period of length 4π, and (c) $y = \sin 2x$ showing four periods of length π.

EXAMPLE 8

$$\cos \left(-\frac{\pi}{3}\right) = \cos \frac{\pi}{3} = \frac{1}{2}$$

$$\sin \left(-\frac{\pi}{3}\right) = -\sin \frac{\pi}{3} = -\frac{\sqrt{3}}{2}$$

As for the other basic trigonometric functions, the secant is even and the cosecant, tangent, and cotangent are odd. For the secant and tangent,

$$\sec (-\theta) = \frac{1}{\cos (-\theta)} = \frac{1}{\cos \theta} = \sec \theta,$$

$$\tan (-\theta) = \frac{\sin (-\theta)}{\cos (-\theta)} = \frac{-\sin \theta}{\cos \theta} = -\tan \theta.$$

Similar calculations show that the cotangent and cosecant are odd, as we ask you to verify in Exercises 87 and 88.

EXPLORATION 4

"Seeing" Odd and Even

1. Choose an even trigonometric function f, and prepare to graph $f(x)$ and $f(-x)$ simultaneously. Predict what you will see in the viewing window. Then GRAPH and explain what you see.

2. Repeat part 1 for an odd function.

Geometric Transformations of Trigonometric Graphs

Our rules for shifting, stretching, shrinking, and reflecting the graph of a function apply to the trigonometric functions. The rules are summarized in Section 1.4. The following diagram is to remind you of the controlling parameters.

Vertical stretch or shrink;
reflection across x-axis   Vertical shift

$$y = a\, f(b(x+c))+d$$

Horizontal stretch or shrink;   Horizontal shift
reflection across y-axis

For example, to horizontally stretch the graph of $\sin x$, we multiply x by a number b between 0 and 1. The graph is stretched by the factor $1/b$.

EXAMPLE 9

A complete graph of $y = \sin (1/2)x$ is obtained (Fig. 1.102b) by horizontally stretching the graph of $y = \sin x$ by a factor of $1/(1/2) = 2$.

To horizontally shrink the graph of $\sin x$, we multiply x by a number b greater than 1. The graph is shrunk by the factor $1/b$.

EXAMPLE 10

A complete graph of $y = \sin 2x$ is obtained (Fig. 1.102c) by horizontally shrinking the graph of $y = \sin x$ by a factor of 1/2.

Shift Formulas

EXPLORATION BIT

Examples 9 and 10 suggest how you can find the period of the function $y = \sin bx$ (or even of $y = a \sin(b(x + c)) + d$). The clue is to remember that the number b causes a horizontal stretch or shrink by a factor $1/|b|$.

a) If the graph of a periodic function is stretched horizontally, what happens to the period?

b) If the graph of a periodic function is shrunk horizontally, what happens to the period?

c) Make a conjecture about the period of $y = \sin bx$. Be as complete as possible.

d) Is your conjecture consistent with the period $|B|$ in Example 11?

The shift formulas provide another connection between $\cos x$ and $\sin x$. If you look again at Fig. 1.100, you will see that the cosine curve is the same as the sine curve shifted $\pi/2$ units to the left. Also, the sine curve is the same as the cosine curve shifted $\pi/2$ units to the right. In symbols,

$$\sin\left(x + \frac{\pi}{2}\right) = \cos x, \qquad \cos\left(x - \frac{\pi}{2}\right) = \sin x.$$

Figure 1.103(a) shows the cosine shifted to the left $\pi/2$ units to become the reflection of the sine curve through the x-axis. Figure 1.103(b) shows the sine curve shifted $\pi/2$ units to the right to become the reflection of the cosine curve through the x-axis. In symbols,

$$\cos\left(x + \frac{\pi}{2}\right) = \sin(-x) = -\sin x, \qquad \sin\left(x - \frac{\pi}{2}\right) = -\cos x.$$

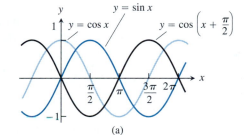

(a)

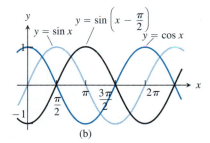

(b)

1.103   (a) The reflection of the sine as a shifted cosine. (b) The reflection of the cosine as a shifted sine.

EXPLORATION BIT

In the function $y = a \sin x$ (or even in $y = a \sin(b(x + c)) + d$), the size, $|a|$ units, of the vertical stretch is the **amplitude** of the function. GRAPH $y = a \sin x$ using different values for a. Does changing the sign of a change the amplitude? Does changing the amplitude change the period? Explain.

EXAMPLE 11

The builders of the Trans-Alaska Pipeline used insulated pads to keep the pipeline heat from melting the permanently frozen soil beneath. To design the pads, it was necessary to take into account the variation in air temperature throughout the year. The variation was represented in the calculations by a *general sine function* of the form

$$f(x) = A \sin\left[\frac{2\pi}{B}(x - C)\right] + D,$$

where $|A|$ is the *amplitude*, $|B|$ is the *period*, C is the *horizontal shift,* and D is the *vertical shift* (Fig. 1.104).

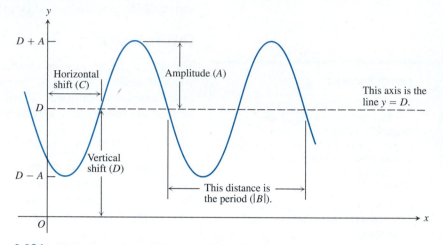

1.104   The general sine curve $y = A \sin[(2\pi/B)(x - C)] + D$, shown for $A, B, C,$ and D positive.

Figure 1.105 shows how we can use such a function to represent temperature data. The data points in the figure are plots of the mean air temperature for Fairbanks, Alaska, based on records of the National Weather Service from 1941 to 1970. The sine function used to fit the data is

$$f(x) = 37 \sin \left[\frac{2\pi}{365} (x - 101) \right] + 25,$$

where f is temperature in degrees Fahrenheit and x is the number of the day counting from the beginning of the year. The fit is remarkably good.

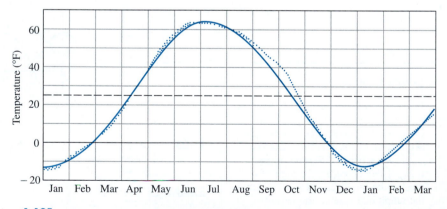

1.105   Normal mean air temperature at Fairbanks, Alaska, plotted as data points. The approximating sine function is

$$f(x) = 37 \sin \left[\frac{2\pi}{365} (x - 101) \right] + 25.$$

(*Source:* "Is the Curve of Temperature Variation a Sine Curve?" by B. M. Lando and C. A. Lando, *The Mathematics Teacher*, 7:6, Fig. 2, p. 535 (September 1977).)

Angle Sum and Difference Formulas

As you may recall from an earlier course,

$$\cos (A + B) = \cos A \cos B - \sin A \sin B,$$
$$\sin (A + B) = \sin A \cos B + \cos A \sin B.$$

The Angle Sum formulas   (6)

These formulas hold for all angles A and B.

If we replace B by $-B$ in Eqs. (6) we get

$$\cos (A - B) = \cos A \cos (-B) - \sin A \sin (-B)$$
$$= \cos A \cos B - \sin A(- \sin B)$$
$$= \cos A \cos B + \sin A \sin B,$$
$$\sin (A - B) = \sin A \cos (-B) + \cos A \sin (-B)$$
$$= \sin A \cos B + \cos A(- \sin B)$$
$$= \sin A \cos B - \cos A \sin B$$

The Angle Difference formulas   (7)

Sinusoids

The sine function used in Example 11 is an example of a *sinusoid*.

DEFINITION

A **sinusoid** is a function that can be written in the form

$$f(x) = a \sin (bx + c) + d,$$

where a, b, c, and d are real numbers.

As was suggested earlier, we can model periodic behavior in nature and science by using an algebraic combination of sines and cosines. In turn, certain combinations of sines and cosines can be expressed as a single sine function, a sinusoid, as follows.

Sinusoids

For all real numbers a and b, there are real numbers A and α such that

$$a \sin x + b \cos x = A \sin (x + \alpha).$$

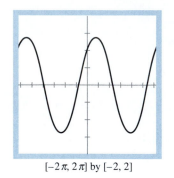

[$-2\pi, 2\pi$] by [$-2, 2$]

1.106 The graph of $f(x) = \sin x + \cos x$ appears to be a sinusoid.

EXPLORATION 5

Refining a Conjecture

A graph of $f(x) = \sin x + \cos x$ is shown in Fig. 1.106. It seems reasonable to conjecture that the curve is a sinusoid. But can we make the conjecture more specific? Recall what a, b, c, and d tell us about shifting, stretching, shrinking, and reflecting the graph of $\sin x$ to obtain the graph of

$$f(x) = a \sin (bx + c) + d.$$

1. Graph $f(x) = \sin x + \cos x$.

2. Make a conjecture about the amplitude of f. Use ZOOM-IN to help you. Make a conjecture about a.

3. Determine three consecutive x-intercepts. Make a conjecture about the period, the horizontal shift, and then about b and c.

4. Make a conjecture about the vertical shift. Make a conjecture about d.

5. Our original conjecture was that $f(x)$ seemed to be a sine function. Now we can be more specific. Let $y_1 = f(x)$ and $y_2 =$ the sine function with the values a, b, c, and d conjectured above. Compare the graphs of y_1 and y_2. Can you refine our original conjecture that the curve is a sinusoid?

6. If you graph $y_1 - y_2 + 1$, what would you like to see in the viewing window?

EXAMPLE 12   Working Backwards to Confirm Algebraically

Confirm that $f(x) = \sin x + \cos x$ is a sinusoid (a statement that we hope was your refined conjecture above).

Solution

We want
$$\sin x + \cos x = a \sin (bx + c) + d$$
$$= a(\sin bx \cos c + \cos bx \sin c) + d$$
$$= (a \cos c) \sin bx + (a \sin c) \cos bx + d.$$

We are done if we let $d = 0$, $b = 1$, and we can choose a and c so that

$$a \cos c = 1 \qquad \text{and} \qquad a \sin c = 1.$$

This would mean

$$(a \cos c)^2 + (a \sin c)^2 = 2$$
$$a^2 = 2.$$

So we choose $a = \sqrt{2}$ and c so that $\cos c = 1/\sqrt{2} = \sin c$; that is, we let $c = \pi/4$. Thus, with $a = \sqrt{2}$, $b = 1$, $c = \pi/4$, and $d = 0$,

$$a \sin (bx + c) + d = \sqrt{2} \sin (x + \pi/4).$$

We can use the Angle Sum formula for $\sin (A + B)$ to show that

$$\sqrt{2} \sin (x + \pi/4) = \sin x + \cos x.$$

≡

EXPLORATION BIT

If you have not done so already as a result of Exploration 5, now is a good time to show graphically that

$$\sin x + \cos x = \sqrt{2} \sin (x + \pi/4).$$

Double-Angle (Half-Angle) Formulas

As you will see later, it is sometimes possible to simplify a calculation by changing trigonometric functions of θ into trigonometric functions of 2θ. There are four basic formulas for doing this, called **Double-Angle formulas**. The first two come from setting A and B equal to θ in the Angle Sum formulas.

$$\cos 2\theta = \cos^2 \theta - \sin^2 \theta$$
$$\sin 2\theta = 2 \sin \theta \cos \theta$$

Two Double-Angle formulas (8)

The other two double-angle formulas come from the equations

$$\cos^2 \theta + \sin^2 \theta = 1, \qquad \cos^2 \theta - \sin^2 \theta = \cos 2\theta.$$

We add to get

$$2 \cos^2 \theta = 1 + \cos 2\theta,$$

subtract to get

$$2 \sin^2 \theta = 1 - \cos 2\theta,$$

and divide by 2 to get

THE HALF-ANGLE FORMULAS

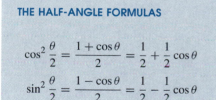

$$\cos^2 \frac{\theta}{2} = \frac{1 + \cos \theta}{2} = \frac{1}{2} + \frac{1}{2} \cos \theta$$
$$\sin^2 \frac{\theta}{2} = \frac{1 - \cos \theta}{2} = \frac{1}{2} - \frac{1}{2} \cos \theta$$

$$\cos^2 \theta = \frac{1 + \cos 2\theta}{2}$$
$$\sin^2 \theta = \frac{1 - \cos 2\theta}{2}.$$

The other two Double-Angle formulas. (9)

When θ is replaced by $\theta/2$ in Eqs. (9) the resulting formulas are called **Half-Angle formulas**. Some books refer to Eqs. (9) by this name as well.

Additional information and trigonometric formulas are available in Appendix 1 of this book and in standard mathematics reference books such as the *CRC Standard Mathematical Tables* from CRC Press, Inc. Support for all the trigonometric identities, or even a check on whether an equation is an identity, can be given graphically by comparing graphs of the left- and right-hand sides of the equations.

Inverse Trigonometric Functions

Notice that none of the six basic trigonometric functions graphed in Fig. 1.100 is one-to-one. So the inverses of these functions are not functions. However, in each case the domain can be restricted to produce a new function whose inverse is a function.

EXAMPLE 13

Each of the following restricted functions is one-to-one:

a) $y = \sin x, \left(-\dfrac{\pi}{2} \le x \le \dfrac{\pi}{2} \right)$ Fig. 1.107(a)

b) $y = \cos x, (0 \le x \le \pi)$ Fig. 1.107(b)

c) $y = \tan x, \left(-\dfrac{\pi}{2} < x < \dfrac{\pi}{2} \right)$ Fig. 1.107(c)

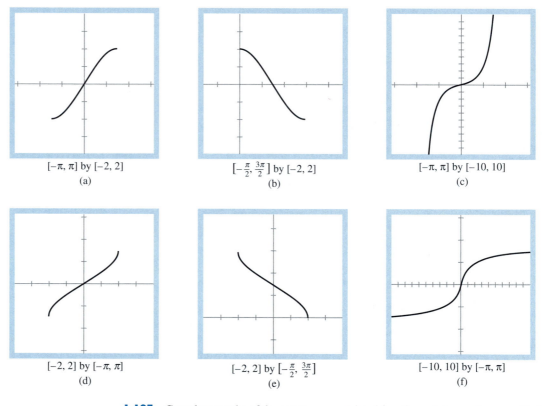

$[-\pi, \pi]$ by $[-2, 2]$
(a)

$[-\frac{\pi}{2}, \frac{3\pi}{2}]$ by $[-2, 2]$
(b)

$[-\pi, \pi]$ by $[-10, 10]$
(c)

$[-2, 2]$ by $[-\pi, \pi]$
(d)

$[-2, 2]$ by $[-\frac{\pi}{2}, \frac{3\pi}{2}]$
(e)

$[-10, 10]$ by $[-\pi, \pi]$
(f)

1.107 Complete graphs of the one-to-one restricted functions: (a) $y = \sin x$, $-\pi/2 \le x \le \pi/2$, (b) $y = \cos x$, $0 \le x \le \pi$, and (c) $y = \tan x$, $-\pi/2 < x < \pi/2$, and of their inverses (d) $y = \sin^{-1} x$, $-1 \le x \le 1$, (e) $y = \cos^{-1} x$, $-1 \le x \le 1$, and (f) $y = \tan^{-1} x$, $-\infty < x < \infty$.

The inverses of the restricted sine, cosine, and tangent functions are called the inverse sine, inverse cosine, and inverse tangent functions, respectively. We define these three inverse functions now and define the inverse secant, inverse cosecant, and inverse cotangent in Chapter 7.

DEFINITIONS

The **inverse sine function**, denoted by $y = \sin^{-1} x$, or $y = \arcsin x$, is the function with domain $[-1, 1]$ and range $[-\pi/2, \pi/2]$ such that $x = \sin y$ (Fig. 1.107d).

The **inverse cosine function**, denoted by $y = \cos^{-1} x$, or $y = \arccos x$, is the function with domain $[-1, 1]$ and range $[0, \pi]$ such that $x = \cos y$ (Fig. 1.107e).

The **inverse tangent function**, denoted by $y = \tan^{-1} x$, or $y = \arctan x$, is the function with domain $(-\infty, \infty)$ and range $(-\pi/2, \pi/2)$ such that $x = \tan y$ (Fig. 1.107f).

EXPLORATION 6

Graphing Inverse Trigonometric Functions

1. Use parametric mode to graph the restricted sine function. (*Hint:* Use tMin and tMax to restrict the domain of $\sin t$.) Then graph the inverse sine function. Compare your graphs with Figs. 1.07(a) and 1.07(d).

2. Scientific and graphing calculators have the inverse trigonometric functions built in as part of their function library. Overlay the graph of the library function $y = \sin^{-1} x$ in the viewing window of part 1.

3. Repeat parts 1 and 2 for the restricted cosine and tangent functions.

The inverse trigonometric functions are useful for finding the possibilities for angles when a trigonometric value is known.

EXAMPLE 14

Solve each equation for x. Support graphically.

a) $\sin x = 0.7$,

b) $\tan x = -2$.

Solution

a) Notice that $\theta = \sin^{-1}(0.7)$ gives one angle that is a solution to this equation. We can find that

$$\theta = \sin^{-1}(0.7) = 0.7754$$

with a calculator. The keystrokes needed to access this value are usually $\boxed{\text{2nd}}$ $\boxed{\text{SIN}}$, $\boxed{\text{INV}}$ $\boxed{\text{SIN}}$, or $\boxed{\text{SHIFT}}$ $\boxed{\text{SIN}}$.

Since θ is in the first quadrant, $\pi - \theta$ is in Quadrant II (Fig. 1.108) and $\sin(\pi - \theta) = \sin\theta$. Thus, $\pi - \theta$ is another solution to this equation. Any other solution to this equation is coterminal with either θ or $\pi - \theta$.

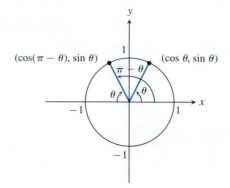

1.108 If θ is in the first quadrant, then $\pi - \theta$ is in the second quadrant, and $\sin(\pi - \theta) = \sin\theta$.

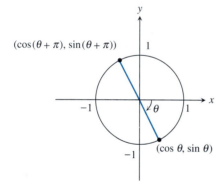

1.110 If θ is in the fourth quadrant, then $\theta + \pi$ is in the second quadrant and $\tan\theta = \tan(\theta + \pi)$.

Thus, every solution has one of the two forms:

$$\theta + 2k\pi = 0.7754 + 2k\pi, \qquad \text{or}$$

$$(\pi - \theta) + 2k\pi = 2.3662 + 2k\pi, \qquad \text{where } k \text{ is any integer.}$$

Figure 1.109(a) supports these answers.

b) First, we find that

$$\theta = \tan^{-1}(-2) = -1.1071$$

is one solution that is in Quadrant IV because $-\pi/2 < \theta < 0$. The angle $\theta + \pi$ in Quadrant II (Fig. 1.110) will have the same tangent value, so $\tan(\theta + \pi) = \tan\theta$, and every solution is of the form

$$\theta + k\pi = -1.1071 + k\pi, \qquad \text{where } k \text{ is any integer.}$$

Figure 1.109(b) supports these answers.

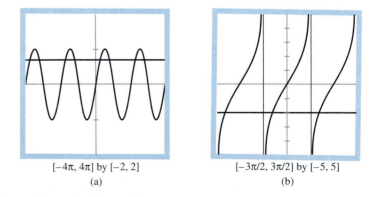

[−4π, 4π] by [−2, 2]   [−3π/2, 3π/2] by [−5, 5]

(a)   (b)

1.109 (a) The solutions to $\sin x = 0.7$ in four periods of $\sin x$. (b) The solutions to $\tan x = -2$ in three periods of $\tan x$.

Exercises 1.7

Exercises 1–4 give angles in degrees. Change them to radians.
1. $510°$ 2. $120°$ 3. $-42°$ 4. $-150°$

Exercises 5–8 give angles in radians. Change them to degrees.

5. 6.2 6. $-\dfrac{\pi}{6}$ 7. -2 8. $\dfrac{3\pi}{4}$

In Exercises 9–14, the angle lies at the center of a circle and subtends an arc of the circle. Find the missing angle measure, circle radius, or arc length.

	Angle	Radius	Arc length
9.	$\dfrac{5\pi}{8}$	2	?
10.	$75°$	10	?
11.	$\dfrac{\pi}{6}$	?	3π
12.	$175°$	?	10
13.	?	14	7
14.	?	6	$\dfrac{3\pi}{2}$

Exercises 15–20 give angles in radians. Find the sine, cosine, tangent, cotangent, secant, and cosecant of each angle (when defined).

15. a) $\dfrac{\pi}{3}$ b) $-\dfrac{\pi}{3}$ 16. a) 2.5 b) -2.5

17. a) 6.5 b) -6.5 18. a) 3.7 b) -3.7

19. a) $\dfrac{\pi}{2}$ b) $\dfrac{3\pi}{2}$ 20. a) 0 b) π

Give the measure of the angles in Exercises 21–24 in radians and degrees. Give exact values whenever possible.

21. $\sin^{-1}(0.5)$ 22. $\sin^{-1}\left(-\dfrac{\sqrt{2}}{2}\right)$

23. $\tan^{-1}(-5)$ 24. $\cos^{-1}(0.7)$

Explain how to use a grapher in parametric mode to compute $\sin t$ and $\cos t$ for the values of t specified in Exercises 25 and 26.

25. $0, 0.5, 1, 1.5, \ldots, 6$

26. $0°, 5°, 10°, \ldots, 360°$

27. Choose appropriate viewing windows to support the graphs of $y = \sin x$, $y = \cos x$, and $y = \tan x$ given in Fig. 1.100.

28. Choose appropriate viewing windows to support the graphs of $y = \sec x$, $y = \csc x$, and $y = \cot x$ given in Fig. 1.100.

29. Choose appropriate viewing windows to display two complete periods of $y = \sec x$, $y = \csc x$, and $y = \cot x$ in degree mode.

30. Choose appropriate viewing windows to display two complete periods of $y = \sin x$, $y = \cos x$, and $y = \tan x$ in degree mode.

In Exercises 31–34, draw the graphs of both functions over the given intervals of x-values.

31. $y = \sin x$ and $y = \csc x$, $-\pi \le x \le \pi$

32. $y = \cos x$ and $y = \sec x$, $-\dfrac{3\pi}{2} \le x \le \dfrac{3\pi}{2}$

33. $y = \cos 4x$ and $y = \cos x$, $0 \le x \le 2\pi$

34. $y = \sin \dfrac{x}{4}$ and $y = \sin x$, $0 \le x \le 8\pi$

Determine the amplitude, period, horizontal shift, and vertical shift and draw a complete graph of the function in Exercises 35–38.

35. $y = 2\cos \dfrac{x}{3}$ 36. $y = 2\cos 3x$

37. $y = \cot\left(2x + \dfrac{\pi}{2}\right)$ 38. $y = 3\cos\left(x + \dfrac{\pi}{4}\right) - 2$

In Exercises 39–42, determine the period, domain, and range, and draw a complete graph of the function. Describe how the function's graph can be obtained from the graph of one of the six basic trigonometric functions.

39. $y = 3\csc(3x + \pi) - 2$ 40. $y = 2\sin(4x + \pi) + 3$

41. $y = -3\tan(3x + \pi) + 2$

42. $y = 2\sin\left(2x + \dfrac{\pi}{3}\right)$

Solve the equations in Exercises 43–50.

43. $\cos x = -0.7$ 44. $\sin x = 0.2$

45. $\tan x = 4$ 46. $\sin x = -0.2$

47. $\sin x = 0.2x$ 48. $\cos x = 0.2x$

49. $\sin x = \ln x$, $x > 0$ 50. $\cos x = -0.23x^2 + 0.5$

Solve the inequalities in Exercises 51 and 52.

51. $\sin x > 0.2x$ 52. $\cos x > 0.2x$

Use parametric mode to draw a complete graph of each circle in Exercises 53–56. Specify the viewing rectangle and the range of values for the parameter used.

53. $x^2 + y^2 = 5$ 54. $x^2 + y^2 = 4$

55. $(x - 2)^2 + (y + 3)^2 = 9$

56. $(x + 1)^2 + (y + 3)^2 = 16$

Show that the functions in Exercises 57–60 are sinusoids, $a\sin(bx + c)$. Use a graph to conjecture the values of a, b, and c. Confirm algebraically.

57. $y = 2\sin x + 3\cos x$ 58. $y = \sin x + \sqrt{3}\cos x$

59. $y = \sin 2x + \cos 2x$ 60. $y = 2\sin 3x + 2\cos 3x$

61. Which equations have the same graph? Confirm algebraically.
   a) $y = \sin x$ b) $y = \cos x$
   c) $y = \sin(-x)$ d) $y = \cos(-x)$
   e) $y = -\sin x$ f) $y = -\cos x$

62. Which equations have the same graph? Confirm algebraically.

a) $y = \sin\left(x + \dfrac{\pi}{2}\right)$ **b)** $y = \sin\left(x - \dfrac{\pi}{2}\right)$

c) $y = \cos\left(x + \dfrac{\pi}{2}\right)$ **d)** $y = \cos\left(x - \dfrac{\pi}{2}\right)$

e) $y = \cos(x + \pi)$ **f)** $y = \cos(x - \pi)$

g) $y = \sin(x + \pi)$ **h)** $y = \sin(x - \pi)$

Two More Useful Identities

Verify the following identities. Support graphically.

63. $\sec^2\theta = 1 + \tan^2\theta$ **64.** $\csc^2\theta = 1 + \cot^2\theta$

65. What symmetries do the graphs of cosine, sine, and tangent have?

66. What symmetries do the graphs of secant, cosecant, and cotangent have?

67. Consider the function $y = \sqrt{(1 + \cos 2x)/2}$.
a) Can x take on any real value?
b) How large can $\cos 2x$ become? How small?
c) How large can $(1 + \cos 2x)/2$ become? How small?
d) What are the domain and range of $y = \sqrt{(1 + \cos 2x)/2}$?

68. Consider the function $y = \tan(x/2)$.
a) What values of $x/2$ must be excluded from the domain of $\tan(x/2)$?
b) What values of x must be excluded from the domain of $\tan(x/2)$?
c) What values does $y = \tan(x/2)$ assume on the interval $-\pi < x < \pi$?
d) What are the domain and range of $y = \tan(x/2)$?

69. *Temperature in Fairbanks, Alaska.* Find the (a) amplitude, (b) period, (c) horizontal shift, and (d) vertical shift of the general sine function

$$f(x) = 37\sin\left[\frac{2\pi}{365}(x - 101)\right] + 25.$$

70. *Temperature in Fairbanks, Alaska.* Use the equation in Exercise 69 to approximate the answers to the following questions about the temperature in Fairbanks, Alaska, shown in Fig. 1.105. Assume that the year has 365 days.
a) What are the highest and lowest mean daily temperatures shown?
b) What is the average of the highest and lowest mean daily temperature shown? Why is this average the vertical shift of the function?

71. What happens if you take $A = B$ in Eqs. (7)? Do these results agree with some things you already know?

72. What happens if you take $B = \pi/2$ in Eqs. (7)? Do these results agree with some things you already know?

73. What happens if you take $B = \pi/2$ in Eqs. (6)? Do these results agree with some things you already know?

74. What happens if you take $B = \pi$ in Eqs. (6)? In Eqs. (7)?

75. Evaluate $\cos 15°$ as $\cos(45° - 30°)$.

76. Evaluate $\sin 75°$ as $\sin(45° + 30°)$.

77. Evaluate $\sin\dfrac{7\pi}{12}$ (radians) as $\sin\left(\dfrac{\pi}{4} + \dfrac{\pi}{3}\right)$.

78. Evaluate $\cos\dfrac{10\pi}{24}$ (radians) as $\cos\left(\dfrac{\pi}{4} + \dfrac{\pi}{6}\right)$.

Use the Double-Angle formulas to find the exact function values in Exercises 79–82 (angles in radians).

79. $\cos^2\dfrac{\pi}{8}$ **80.** $\cos^2\dfrac{\pi}{12}$

81. $\sin^2\dfrac{\pi}{12}$ **82.** $\sin^2\dfrac{\pi}{8}$

83. Use graphs to support the Double-Angle formulas, Eqs. (8) and (9).

84. Prove that for all real numbers a and b, there are real numbers A and α so that $a\sin x + b\cos x = A\sin(x + \alpha)$.

85. *The tangent sum formula.* The standard formula for the tangent of the sum of two angles is

$$\tan(A + B) = \frac{\tan A + \tan B}{1 - \tan A \tan B}.$$

Derive the formula by writing $\tan(A + B)$ as

$$\frac{\sin(A + B)}{\cos(A + B)}$$

and applying Eqs. (6) and (7).

86. Derive a formula for $\tan(A - B)$ by replacing B by $-B$ in the formula for $\tan(A + B)$ in Exercise 85.

Even vs. Odd

87. a) Show that $\cot x$ is an odd function of x.
b) Show that the quotient of an even function and an odd function is always odd (on their common domain).
c) Describe how the graph of $y = \cot(-x)$ can be obtained from the graph of $y = \cot x$.

88. a) Show that $\csc x$ is an odd function of x.
b) Show that the reciprocal of an odd function (when defined) is odd.
c) Describe how the graph of $y = \csc(-x)$ can be obtained from the graph of $y = \csc x$.

89. a) Show that the product $y = \sin x \cos x$ is an odd function of x.
b) Show that the product of an even function and an odd function is always odd (on their common domain).

90. a) Show that the function $y = \sin^2 x$ is an even function of x (even though the sine itself is odd).
b) Show that the square of an odd function is always even.
c) Show that the product of any two odd functions is even (on their common domain).

In Exercises 91 and 92, draw a complete graph that shows exactly one period.

91. $f(x) = \sin(60x)$ **92.** $f(x) = \cos(60\pi x)$

Angles of Inclination

The **angle of inclination** of a line that crosses the x-axis is the smallest angle that we find when we measure counterclockwise from the x-axis around the point of intersection (see figure at right). The angle of inclination of a horizontal line is taken to be $0°$. Thus, angles of inclination may have any measure from $0°$ up to but not including $180°$.

93. Show that the tangent of the angle of inclination of a line in the coordinate plane is the slope of the line.

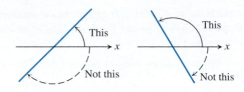

In Exercises 94–97, find the angle of inclination of the line

94. that contains the points $(1, 5)$ and $(3, 1)$.

95. that contains the points $(-1, 2)$ and $(3, 5.5)$.

96. given by $y = 2.5x$.

97. given by $2x - 6y = 7$.

Chapter 1 Review Questions

1. How do you find the distance between two points in the xy-plane? Between a point and a line in the plane? Give examples.

2. Explain the difference between "sketch a graph" and "draw a graph." What are the basic steps in sketching a graph of an equation in x and y? Illustrate them. What does it mean for a graph to be complete? Can a grapher be used to graph any equation in x and y? Give examples.

3. Distinguish between "screen coordinates" and "Cartesian coordinates." What are viewing dimensions? What is a scale unit? What is a square viewing window?

4. What coordinate tests determine whether a graph in the xy-plane is symmetric with respect to the coordinate axes or the origin? Give examples.

5. How can you write the equation for a line if you know the coordinates of two points on the line? The slope of the line and the coordinates of one point on the line? The slope of the line and the y-intercept? Give examples.

6. What are the standard equations for lines perpendicular to the coordinate axes?

7. How are the slopes of mutually perpendicular lines related? Give examples.

8. What is a function? Give examples. How do you use a graphing utility to graph a real-valued function of a real variable? What do you have to be careful about?

9. Name some typical functions and draw their graphs.

10. What is an even function? An odd function? What symmetries do the graphs of such functions have? Give examples. Give an example of a function that is neither odd nor even.

11. When is it possible to compose one function with another? Give examples of composites and their values at various points. Does the order in which functions are composed ever matter?

12. How can you write an equation for a circle in the xy-plane if you know its radius a and the coordinates (h, k) of its center? Give examples.

13. What inequality is satisfied by the coordinates of the points that lie inside the circle of radius a centered at (h, k)? What inequality is satisfied by the coordinates of the points that lie outside the circle?

14. The graph of a function $y = f(x)$ in the xy-plane is shifted 5 units to the left, vertically stretched by a factor of 2, reflected through the x-axis, and then shifted 3 units straight up. Write an equation for the new graph.

15. What are parabolas? What are their typical equations?

16. What does it mean to solve an equation or inequality with an error of at most 0.01? Give examples.

17. Explain the meaning of algebraic representation of a problem situation, graphical representation of a problem situation, graph of a problem situation. Give examples.

18. How do you convert between degree measure and radian measure? Give examples.

19. Graph the six basic trigonometric functions as functions of radian measure. What symmetries do the graphs have?

20. What does it mean for a function $y = f(x)$ to be periodic? Give examples of functions with various periods. Name some real-world phenomena that we model with periodic functions.

21. List the Angle Sum and Difference formulas for the sine and cosine functions.

22. List the four basic Double-Angle formulas for sines and cosines.

23. Define the function $y = |x|$. Give examples of numbers and their absolute values. How are $|-a|$, $|ab|$, $|a/b|$, and $|a+b|$ related to $|a|$ and $|b|$?

24. How are absolute values used to describe intervals of real numbers?

25. What is the inverse of a relation? How are the graphs, domains, and ranges of relations and their inverses related? Give an example.

26. When is the inverse of a relation a function? How do you tell when functions are inverses of one another? Give examples.

Chapter 1 Practice Exercises

In Exercises 1–4, find the points that are symmetric to the given point (a) across the x-axis, (b) across the y-axis, and (c) across the origin.

1. $(1, 4)$ **2.** $(2, -3)$ **3.** $(-4, 2)$ **4.** $(-2, -2)$

Test the equations in Exercises 5–8 to find out whether their graphs are symmetric with respect to the axes or the origin.

5. **a)** $y = x$ **b)** $y = x^2$

6. **a)** $y = x^3$ **b)** $y = x^4$

7. **a)** $x^2 - y^2 = 4$ **b)** $x - y = 4$

8. **a)** $y = x^{1/3}$ **b)** $y = x^{2/3}$

Find equations for the vertical and horizontal lines through the points in Exercises 9–12.

9. $(1, 3)$ **10.** $(2, 0)$ **11.** $(0, -3)$ **12.** (x_0, y_0)

In Exercises 13–20, write an equation for the line that passes through point P with slope m. Then use the equation to find the line's intercepts and graph the line.

13. $P(2, 3), m = 2$ **14.** $P(2, 3), m = 0$

15. $P(1, 0), m = -1$ **16.** $P(0, 1), m = -1$

17. $P(1, -6), m = 3$ **18.** $P(-2, 0), m = 1$

19. $P(-1, 2), m = -\dfrac{1}{2}$ **20.** $P(3, 1), m = \dfrac{1}{3}$

In Exercises 21–24, find an equation for the line through the two points.

21. $(-2, -2), (1, 3)$ **22.** $(-3, 6), (1, -2)$

23. $(2, -1), (4, 4)$ **24.** $(3, 3), (-2, 5)$

In Exercises 25–28, find an equation for the line with the given slope m and y-intercept b.

25. $m = \dfrac{1}{2}, b = 2$ **26.** $m = -3, b = 3$

27. $m = -2, b = -1$ **28.** $m = 2, b = 0$

In Exercises 29–32, (a) find an equation for the line through P parallel to L; (b) then find an equation for the line through P perpendicular to L and (c) the distance from P to L.

29. $P(6, 0), L: 2x - y = -2$

30. $P(3, 1), L: y = x + 2$

31. $P(4, -12), L: 4x + 3y = 12$

32. $P(0, 1), L: y = -\sqrt{3}x - 3$

Sketch a complete graph of each equation in Exercises 33–40. Give the domain and range. Support your answer with a grapher.

33. $y = 2x - 3$ **34.** $y = |x| - 2$

35. $y = 2|x - 1| - 1$ **36.** $y = \sec x$

37. $y = \cos x$ **38.** $y = [x]$

39. $x = -y^2$ **40.** $y = -2 + \sqrt{1 - x}$

Find the domain and range and draw a complete graph of each function in Exercises 41–46.

41. $f(x) = x^3 + 8x^2 + x - 37$

42. $f(x) = -1 + \sqrt[3]{1 - x}$

43. $f(x) = \log_7(x - 1) + 1$

44. $f(x) = 3^{2-x} + 1$

45. $f(x) = |x - 2| + |x + 3|$

46. $f(x) = \dfrac{|x - 2|}{x - 2}$

In Exercises 47–50, describe how the graph of f can be obtained from the graph of g.

47. $f(x) = -2(x - 1)^3 + 5, g(x) = x^3$

48. $f(x) = 2\ln(-x + 1) + 3, g(x) = \ln x$

49. $f(x) = 3\sin(3x + \pi), g(x) = \sin x$

50. $f(x) = -2\sqrt[4]{3 - x} + 5, g(x) = \sqrt[4]{x}$

Exercises 51–54 specify the order in which transformations are to be applied to the graph of the given equation. Give an equation for the transformed graph in each case.

51. $y = x^2$, vertical stretch by 2, reflect through x-axis, shift right 2, shift up 3

52. $x = y^2$, horizontal shrink by 0.5, reflect through y-axis, shift left 3, shift down 2

53. $y = \dfrac{1}{x}$, vertical stretch by 3, shift left 2, shift up 5

54. $x^2 + y^2 = 1$, shift left 3, shift up 5

A complete graph of f is shown. Sketch a complete graph of each function in Exercises 55–58.

55. $y = f(-x)$

56. $y = -f(x)$

57. $y = -2f(x + 1) + 1$

58. $y = 3f(x - 2) - 2$

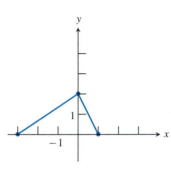

Determine the vertex and axis of symmetry and sketch a complete graph of each parabola in Exercises 59–60. Support your work with a grapher.

59. $y = -x^2 + 4x - 1$ **60.** $x = 2y^2 + 8y + 3$

In Exercises 61–66, say whether each function is even, odd, or neither.

61. a) $y = \cos x$ **b)** $y = -\cos x$ **c)** $y = 1 - \cos x$

62. a) $y = \sin x$ **b)** $y = -\sin x$ **c)** $y = 1 - \sin x$

63. a) $y = x^2 + 1$ **b)** $y = x$ **c)** $y = x(x^2 + 1)$

64. a) $y = x^3$ **b)** $y = -x$ **c)** $y = -x^4$

65. a) $y = \sec x$ **b)** $y = \tan x$ **c)** $y = \sec x \tan x$

66. a) $y = \csc x$ **b)** $y = \cot x$ **c)** $y = \csc x \cot x$

67. Graph the function $y = x - [x]$. Is the function periodic? If so, what is its period?

68. Graph the function $y = \lceil x \rceil - [x]$. Is the function periodic? If so, what is its period?

Graph the functions in Exercises 69–72.

69. $y = \begin{cases} \sqrt{-x}, & -4 \le x \le 0 \\ \sqrt{x}, & 0 < x \le 4 \end{cases}$

70. $y = \begin{cases} -x - 2, & -2 \le x \le -1 \\ x, & -1 < x \le 1 \\ -x + 2, & 1 < x \le 2 \end{cases}$

71. $y = \begin{cases} \sin x, & 0 \le x \le 2\pi \\ 0, & 2\pi < x \end{cases}$

72. $y = \begin{cases} \cos x, & 0 \le x \le 2\pi \\ 0, & 2\pi < x \end{cases}$

Write formulas for the piecewise functions graphed in Exercises 73 and 74.

73.   **74.**

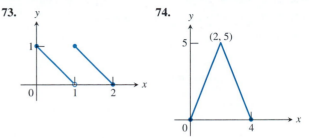

In Exercises 75 and 76, find the domains and ranges of f, g, $f + g$, $f \cdot g$, f/g, g/f Also, find the domains and ranges of the composites $f \circ g$ and $g \circ f$.

75. $f(x) = \dfrac{1}{x}, g(x) = \dfrac{1}{\sqrt{x}}$

76. $f(x) = \sqrt{x}, g(x) = \sqrt{1 - x}$

In Exercises 77–80, write an equation for the circle with the given center (h, k) and radius a.

77. $(h, k) = (1, 1), a = 1$ **78.** $(h, k) = (2, 0), a = 5$

79. $(h, k) = (2, -3), a = \dfrac{1}{2}$ **80.** $(h, k) = (-3, 0), a = 3$

In Exercises 81–84 identify the centers and radii of the circles.

81. $(x - 3)^2 + (y + 5)^2 = 16$

82. $x^2 + (y - 5)^2 = 2$

83. $x^2 + y^2 + 2x - 14y = 71$

84. $x^2 + y^2 + 8x + 2y = 64$

In Exercises 85 and 86 use inequalities to describe the regions.

85. a) The interior of the circle of radius 1 centered at the origin.

    b) The region consisting of the circle and its interior.

86. a) The exterior of the circle of radius 2 centered at the point (1, 1).

    b) The region consisting of the circle and its exterior.

Solve the equations in Exercises 87–90 algebraically. Support your answer with a grapher.

87. $|x - 1| = \dfrac{1}{2}$ **88.** $|2 - 3x| = 1$

89. $\left| \dfrac{2x}{5} + 1 \right| = 7$ **90.** $\left| \dfrac{5 - x}{2} \right| = 7$

Describe the intervals in Exercises 91–94 with inequalities that do not involve absolute values.

91. $|x + 2| \le \dfrac{1}{2}$ **92.** $|2x - 7| \le 3$

93. $\left| y - \dfrac{2}{5} \right| < \dfrac{3}{5}$ **94.** $\left| 8 - \dfrac{y}{2} \right| < 1$

Solve the equations in Exercises 95–98.

95. $x^3 - 7x^2 + 12x - 2 = 0$

96. $4x^3 - 10x^2 + 9 = 0$

97. $2 + \log_3(x - 2) + \log_3(3 - x) = 0$

98. $\sin x = -0.7$

Solve the inequalities in Exercises 99–102 algebraically. Support your answer graphically.

99. $|1 - 2x| < 3$

100. $\left| \dfrac{2x - 1}{5} \right| \le 1$

101. $\left| \dfrac{3}{x - 2} \right| < 1$

102. $|2 - 3x| > 1$

Solve the inequalities in Exercises 103 and 104.

103. $x^3 - 7x^2 + 12x < 2$
(See Exercise 95.)

104. $4x^3 - 10x^2 + 9 \ge 0$
(See Exercise 96.)

105. Change from degrees to radians:
  a) $30°$ **b)** $22°$ **c)** $-130°$ **d)** $-150°$

106. Change from radians to degrees:
  a) $\dfrac{3\pi}{2}$ **b)** -0.9 **c)** 2.75 **d)** $-\dfrac{5\pi}{4}$

Find the sine, cosine, tangent, cotangent, secant, and cosecant of each angle in Exercises 107 and 108. The angles are given in radian measure.

107. a) 1.1 **b)** -1.1 **c)** $\dfrac{2\pi}{3}$ **d)** $-\dfrac{2\pi}{3}$

108. a) $\dfrac{\pi}{4}$ **b)** $-\dfrac{\pi}{4}$ **c)** 2.7 **d)** -2.7

109. Graph the following functions on the same set of axes over the interval $0 \le x \le 2\pi$.
  a) $y = \cos 2x$ **b)** $y = 1 + \cos 2x$ **c)** $y = \cos^2 x$

110. Graph the following functions on the same set of axes over the interval $0 \le x \le 2\pi$.
  a) $y = \cos 2x$ **b)** $y = -\cos 2x$
  c) $y = 1 - \cos 2x$ **d)** $y = \sin^2 x$

111. Find $\cos^2 \dfrac{\pi}{6}$
  a) by finding $\cos \dfrac{\pi}{6}$ and squaring.
  b) by using a Double-Angle formula.

112. Find $\sin^2 \dfrac{\pi}{4}$
  a) by finding $\sin \dfrac{\pi}{4}$ and squaring.
  b) by using a Double-Angle formula.

In Exercises 113 and 114, find $f^{-1}(x)$, and show that $f \circ f^{-1}(x) = f^{-1} \circ f(x) = x$. Draw a complete graph of f and f^{-1} in the same viewing window.

113. $f(x) = 2 - 3x$

114. $f(x) = (x + 2)^2, x \ge -2$

Draw a graph of the inverse relation of each function in Exercises 115 and 116. Is the inverse relation a function?

115. $y = x^3 - x$

116. $y = \dfrac{x + 2}{x - 1}$

Give the measure of the angles in Exercises 117 and 118 in radians and degrees.

117. $\sin^{-1}(0.7)$

118. $\tan^{-1}(-2.3)$

Draw complete graphs of the functions in Exercises 119–122. Explain why your graphs are complete.

119. $y = |\cos x|$

120. $y = \dfrac{\cos x + |\cos x|}{2}$

121. $y = \dfrac{|\cos x| - \cos x}{2}$

122. $y = \dfrac{\cos x - |\cos x|}{2}$

123. A 100-in. piece of wire is cut into two pieces. Each piece of wire is used to make a square wire frame. Let x be the length of one piece of the wire.
  a) Determine an algebraic representation $A(x)$ for the total area of the two squares.
  b) Draw a complete graph of $y = A(x)$.
  c) What are the domain and range of $A(x)$?
  d) What values of x make sense in the problem situation? Which portion of the graph in part (b) is a graph of the problem situation?
  e) Use a graph to determine the length of the two pieces of wire if the total area is 400 in.^{2} Confirm your answer algebraically.
  f) Use a graph to determine the length of the two pieces of wire if the total area is a maximum.

124. Equal squares of side length x are removed from each corner of a 20- by 30-in. piece of cardboard, and the sides are turned up to form a box with no top.
  a) Write the volume V of the box as a function of x.
  b) Draw a complete graph of $y = V(x)$.
  c) What are the domain and range of $y = V(x)$?
  d) What values of x make sense in the problem situation? Draw a graph of the problem situation.
  e) Use ZOOM-IN to determine x so that the box has maximum volume. What is the maximum volume?
  f) Use a graph to determine the dimensions of a box with volume 750 in.^{3}

125. Draw a complete graph of the equation $|x| + |y| = 1$.

2

Limits and Continuity

OVERVIEW This chapter shows how limits of function values are defined and calculated.

Calculus is built on the concept of limit. The rules for calculating limits are straightforward, and most of the limits we need can be found by using one or more of direct substitution, graphing, calculator approximation, or algebra. Proving that the calculation rules always work or that approximations are accurate, however, is a more subtle affair that requires a formal definition of limit. We present this definition in Section 2.6 and show there how it is used to justify the rules.

One of the most important uses of limits in calculus is to test functions for continuity. Continuous functions are widely used in science because they serve to model an enormous range of natural behavior. In Section 2.2 we will see what makes continuous functions special. We shall work mainly with continuous functions in this book.

2.1 _____ Limits

HOW THE BALANCE GROWS

$S_0 = 100$

$S_1 = 100(1 + 0.06/k)$

$S_2 = 100(1 + 0.06/k)^2$

$S_3 = S_2 + (0.06/k)S_2$

$\quad = S_2(1 + 0.06/k)$

$\quad = 100(1 + 0.06/k)^2(1 + 0.06/k)$

$\quad = 100(1 + 0.06/k)^3$

$\quad \vdots$

$S_k = 100(1 + 0.06/k)^k$

Suppose you have \$100 to invest at a fixed annual rate of 6% compounded k times a year. If $k = 4$, the interest is added to your account four times per year. If $k = 12$, the interest is compounded monthly, and so forth. It would seem that the greater the value of k, the better your investment. But watch:

The interest earned at the end of the first of k compounding periods ($1/k$th of a year) is $(0.06/k)(100)$, and the balance in your account is $S_1 = 100 + (0.06/k)(100) = 100\,(1 + 0.06/k)$. The interest added at the end of the second period is $(0.06/k)S_1$, and the balance in your account after two periods is $S_2 = S_1 + (0.06/k)S_1 = 100\,(1 + 0.06/k)^2$. It can be shown by using mathematical induction that the amount in your account at the end of one year is

$$S_k = 100\left(1 + \frac{0.06}{k}\right)^k. \tag{1}$$

105

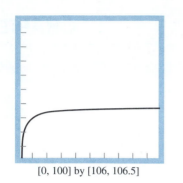

[0, 100] by [106, 106.5]

2.1   The value $S = 100(1 + 0.06/k)^k$ of a $100 investment after one year with interest compounded k times a year. Notice that S increases relatively rapidly as k increases from 1 to 5 but shows very little increase as k increases further.

EXPLORATION BIT

The graph in Fig. 2.1 does not show how an investment grows over time, but rather the value of an investment after one year for different numbers of compounding periods. To see how an investment grows over time, graph

$$y(x) = P(1 + i)^x,$$

entering specific values for the principal (P) and interest rate (i). To compare investment growth for two interest rates, for example, enter the two equations, $y_1(x)$ and $y_2(x)$, with the different values for i. Then GRAPH in the same viewing window (simultaneously, if available). If you also want to see the effects of different numbers of compounding periods, use the equation

$$y(x) = P\left(1 + \frac{i}{k}\right)^{xk}.$$

The graph in Fig. 2.1 shows the amount in your account (y-axis) at the end of a year for k compounding periods (x-axis) during the year. For one compounding period ($k = 1$) the investment will grow to $106. For $k = 2$, 3, and 4 the investment will grow a few more cents. For any value of k, no matter how large, the graph suggests that the investment will never get above $106.20. There is a limit to how much the investment can grow. Later we will see that the limit is $100e^{0.06}$. (As in Section 1.6, $e \approx 2.718281828$.) Here is a situation in which an understanding of limit can help us to be wiser investors.

In fact, we live in a world of limits. There is a limit to

- how far a bat can propel a softball
- how fast we can brake for a red light
- how much weight we can lift
- how long a battery will last
- the position of an oscillating body that is slowing down
- the population that a particular environment will support.

And this is where a graphing utility and calculus come in. Most of the limits that interest us can be viewed as numerical limits to values of functions. As you will see, a grapher can suggest the limits, and calculus can give the right mathematics for confirming the limiting values of functions analytically.

Examples of Limits

To start off on a slightly different track, one of the important things to know about a function f is how its outputs will change when the inputs change. If the inputs get closer and closer to some specific value c, for example, will the outputs get closer to some specific value L? If they do, we want to know that, because it means we can control the outputs by controlling the inputs.

To talk sensibly about this, we need the language of limits. We develop this in two stages. First, we define limit informally, look at examples, and learn the calculation rules that have been discovered over the years to be the most useful ones to know. Then, in Section 2.6, when we have worked with enough examples for the formal definition to make sense, we define limit more precisely and examine the mathematics behind our calculations.

INFORMAL DEFINITION OF LIMIT

Given real numbers c and L, if the values $f(x)$ of a function f approach or equal L as the values of x approach (but do not equal) c, we say that f has limit L as x approaches c. We write

$$\lim_{x \to c} f(x) = L,$$

(read "The limit of f of x as x approaches c equals L").

EXAMPLE 1

As the graph (Fig. 2.2) or the following calculator approximations suggest,

$$\lim_{x \to 2}(3x - 1) = 5$$

Values of $f(x) = 3x - 1$ as $x \to 2$

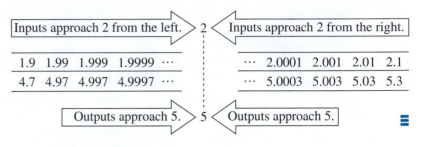

y

$y = 3x - 1$

5

0

x

−1

2

2.2 As the values of x approach 2, the values of $y = 3x - 1$ approach 5. The word *approach* means that we can make *all* y-values be within a certain distance of 5 (and this distance can be made smaller and smaller) by keeping all the x-values within a certain distance of 2.

EXPLORATION 1

Support Graphically

It turns out that many limits are easy to determine. Some can be found by substitution as is the case for any linear function. For instance, in Example 1, $\lim_{x \to 2}(3x - 1) = 3(2) - 1 = 5$.

Try substitution to find the following limits. Support the limits you find graphically.

1. $\lim_{x \to -3}(x + 4)$ **2.** $\lim_{x \to 0}(x + 4)$ **3.** $\lim_{x \to 1}2x^2$

4. $\lim_{x \to -2}2x^2$ **5.** $\lim_{x \to 2}4x^2$ **6.** $\lim_{x \to -3}4x^2$

Use your results from parts 1–6. Make a conjecture about each limit.

7. $\lim_{x \to c}(x + 4)$ **8.** $\lim_{x \to c}2x^2$ **9.** $\lim_{x \to c}4x^2$

Unfortunately, some limits cannot be found by immediate substitution, as Example 2 illustrates.

EXAMPLE 2

Determine $\lim_{x \to 0}\left[(x + 3)^2 - 9\right]/x$. Confirm algebraically.

Solution If we try to substitute 0 for x, we might conclude that the limit does not exist because 0/0 is undefined. To analyze this limit, however, we need concern ourselves with the values of the fraction only for values of x *different from but close* to 0. The graph (Fig. 2.3) suggests that the limit *does exist*, indeed that

$$\lim_{x \to 0}\frac{(x + 3)^2 - 9}{x} = 6.$$

This limit can be confirmed algebraically as follows. When $x \neq 0$,

$$\frac{(x + 3)^2 - 9}{x} = \frac{x^2 + 6x + 9 - 9}{x} = \frac{x^2 + 6x}{x} = \frac{x(x + 6)}{x} = x + 6.$$

Thus, $\lim_{x \to 0}[(x + 3)^2 - 9]/x = \lim_{x \to 0}(x + 6) = 6$.

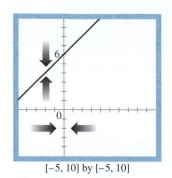

[−5, 10] by [−5, 10]

2.3   This view suggests that

$$f(x) = \frac{(x+3)^2 - 9}{x} \to 6$$

as $x \to 0$ even though $f(0)$ is undefined. TRACE and watch what the screen shows for y when $x = 0$. (To check $f(0)$, you may have to adjust your view dimensions to get $x = 0$ as a pixel coordinate. Do you know how to do this?)

Example 2 illustrates an important fact about limits: The limit of a function $f(x)$ as x approaches c *never* depends on what happens when $x = c$. The limit, if it exists at all, is entirely determined by the values of f when $x \neq c$. In Example 2 the quotient $[(x+3)^2 - 9]/x$ is not even defined at $x = 0$, nor does f ever have the value 6. Yet it does have a limit as x approaches 0, and this limit is 6.

Properties of Limits

We could make a huge list of functions and their limits, but we would be missing the boat if we did, because there is a more constructive (and much easier) way to proceed. For instance, once we know the limits for two particular functions as x approaches some value c, we automatically know the limit of their sum—it is the sum of their limits. Similarly, the limit of the difference of two functions is the difference of their limits, and so on, as described in the following theorem.

THEOREM 1

a) If f is the *constant function* $f(x) = k$ (the function whose outputs have the constant value k), then for any value of c,

$$\lim_{x \to c} f(x) = \lim_{x \to c}(k) = k.$$

b) If f is the *identity function* $f(x) = x$, then for any value of c,

$$\lim_{x \to c} f(x) = \lim_{x \to c}(x) = c.$$

c) Properties of Limits   If L_1 and L_2 are real numbers and $\lim_{x \to c} f(x) = L_1$ and $\lim_{x \to c} g(x) = L_2$, then

  1. *Sum Rule:* $\lim[f(x) + g(x)] = L_1 + L_2$.

  2. *Difference Rule:* $\lim[f(x) - g(x)] = L_1 - L_2$.

  3. *Product Rule:* $\lim f(x) \cdot g(x) = L_1 \cdot L_2$.

  4. *Constant Multiple Rule:* $\lim k \cdot f(x) = k \cdot L_1$ (any number k).

  5. *Quotient Rule:* $\lim \dfrac{f(x)}{g(x)} = \dfrac{L_1}{L_2}$ provided that $L_2 \neq 0$.

The limits are all taken as $x \to c$.

In words, Theorem 1(c) loosely says:

1. The limit of the sum of two functions is the sum of their limits.

2. The limit of the difference of two functions is the difference of their limits.

3. The limit of a product of two functions is the product of their limits.

4. The limit of a constant times a function is the constant times the limit of the function.

5. The limit of a quotient of two functions is the quotient of their limits, provided that the limit of the denominator is not zero.

FORMAL LIMITS

In Section 2.6 we will be more formal about limits. There you will have the opportunity to explore a formal proof of Theorems 1(a) and 1(b). A formal proof of Theorem 1(c), presented in Appendix 2, should then make sense to you also.

Informally, we can paraphrase Theorem 1(c) in terms that make it highly reasonable: When x is close to c, $f(x)$ is close to L_1 and $g(x)$ is close to L_2. Then we naturally think that $f(x) + g(x)$ is close to $L_1 + L_2$; $f(x) - g(x)$ is close to $L_1 - L_2$; $f(x)g(x)$ is close to $L_1 L_2$; $kf(x)$ is close to kL_1, and $f(x)/g(x)$ is close to L_1/L_2 if L_2 is not zero.

What keeps this discussion from being a proof is that the word *close* is vague. Phrases like *arbitrarily close to* and *sufficiently close to* might seem at first to improve the argument, but what are really needed are the formal definitions and arguments that were developed for the purpose by the great mathematicians of the nineteenth century. You will see what we mean in Section 2.6.

In the meantime, here are some examples of what Theorem 1 can do for us.

EXAMPLE 3

We know from Theorems 1(a) and 1(b) that $\lim_{x \to c} k = k$ and $\lim_{x \to c} x = c$. The various parts of Theorem 1(c) now let us combine these results to calculate other limits:

a) $\displaystyle\lim_{x \to c} x^2 = \lim_{x \to c} x \cdot x = c \cdot c = c^2$ Product Rule

b) $\displaystyle\lim_{x \to c}(x^2 + 5) = \lim_{x \to c} x^2 + \lim_{x \to c} 5$ Sum Rule from (a)

$\qquad\qquad = c^2 + 5$

c) $\displaystyle\lim_{x \to c} 4x^2 = 4 \lim_{x \to c} x^2$ Constant Multiple Rule from (a)

$\qquad\quad = 4c^2$

d) $\displaystyle\lim_{x \to c}(4x^2 - 3) = \lim_{x \to c} 4x^2 - \lim_{x \to c} 3$ Difference Rule from (c)

$\qquad\qquad = 4c^2 - 3$

e) $\displaystyle\lim_{x \to c} x^3 = \lim_{x \to c} x^2 \cdot x = c^2 \cdot c = c^3$ Product Rule and (a)

f) $\displaystyle\lim_{x \to c}(x^3 + 4x^2 - 3) = \lim_{x \to c} x^3 + \lim_{x \to c}(4x^2 - 3)$ Sum Rule (e) and (d)

$\qquad\qquad\qquad = c^3 + 4c^2 - 3$

g) $\displaystyle\lim_{x \to c} \frac{x^3 + 4x^2 - 3}{x^2 + 5} = \frac{\lim_{x \to c}(x^3 + 4x^2 - 3)}{\lim_{x \to c}(x^2 + 5)}$ Quotient Rule (f) and (b)

$\qquad\qquad\qquad = \frac{c^3 + 4c^2 - 3}{c^2 + 5}.$ ≡

Example 3 shows the remarkable strength of Theorem 1. From the two simple observations that $\lim_{x \to c} k = k$ and $\lim_{x \to c} x = c$ we can immediately work our way to limits of all polynomial functions and most **rational functions** (ratios of polynomials). As in part (f) of Example 3, the limit of any polynomial $f(x)$ as x approaches c is $f(c)$, the number we get when we substitute c for x. As in part (g) of Example 3, the limit of the ratio $f(x)/g(x)$ of two polynomials is $f(c)/g(c)$, provided that $g(c)$ is different from 0.

THEOREM 2   Limits of Polynomial Functions Can Be Found by Substitution

If $f(x) = a_n x^n + a_{n-1} x^{n-1} + \cdots + a_0$ is any polynomial function, then
$$\lim_{x \to c} f(x) = f(c) = a_n c^n + a_{n-1} c^{n-1} + \cdots + a_0.$$

EXAMPLE 4

a) $\lim\limits_{x \to 3}[x^2(2-x)] = \lim\limits_{x \to 3}(2x^2 - x^3) = 2(3)^2 - (3)^3 = 18 - 27 = -9.$

b) Same limit, found another way:
$$\lim_{x \to 3}[x^2(2-x)] = \lim_{x \to 3} x^2 \cdot \lim_{x \to 3}(2-x) = (3)^2 \cdot (2-3) = 9 \cdot (-1) = -9.$$
≡

THEOREM 3   (Many But Not All) Limits of Rational Functions Can Be Found by Substitution

If $f(x)$ and $g(x)$ are polynomials, then
$$\lim_{x \to c} \frac{f(x)}{g(x)} = \frac{f(c)}{g(c)} \qquad \text{(provided that } g(c) \neq 0\text{)}.$$

EXAMPLE 5

$$\lim_{x \to 2} \frac{x^2 + 2x + 4}{x + 2} = \frac{(2)^2 + 2(2) + 4}{2 + 2} = \frac{12}{4} = 3.$$
≡

In Example 5 we use Theorem 3 to find the limit of $f(x)/g(x)$ because the value of the denominator, $g(x) = x + 2$, is different from zero when $x = 2$. However, many interesting limits cannot be found by substitution. In Exploration 2, the denominator is zero when $x = 5$, so we cannot use Theorem 3 directly. We can rewrite the fraction $f(x)/g(x)$ to find the limit in this case.

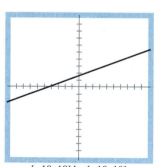

[−10, 10] by [−10, 10]

2.4 The graph suggests that
$$\frac{x^2 - 25}{3(x - 5)}$$
has a limit at $x = 5$. What is it? The graph also suggests that the limit is the same as the limit of a particular linear function. Which one? Some graphers may show a "hole" at $x = 5$. This happens when $x = 5$ is a screen coordinate. Can you make this happen with your grapher?

EXPLORATION 2

Confirm Algebraically

When the limit of the denominator of a rational function $f(x)/g(x)$ is zero, we cannot apply Theorem 3. We can use a grapher, however, to see whether there may be a limit, what it might be, and possibly a clue on how to confirm the limit algebraically (Fig. 2.4).

Estimate $\lim_{x \to 5} \left[\left(x^2 - 25 \right) / (3(x - 5)) \right]$ graphically. Then rewrite $(x^2 - 25)/(3(x - 5))$ into a more familiar, equivalent form (when $x \neq 5$)—one for which you know how to find the limit as $x \to 5$ (x approaches 5).

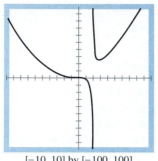

[−10, 10] by [−100, 100]

2.5 Notice that as $x \to 2$, the values of $f(x) = (x^3 - 1)/(x - 2)$ do not approach any real number. How does the graph in your viewing window compare to the one shown here around $x = 2$?

EXPLORATION BIT

In Exploration 2 the denominator and numerator $\to 0$, and the limit of the fraction exists. In Example 6 the denominator $\to 0$, but the numerator $\not\to 0$, and the limit of the fraction does not exist. On your book cover or in some other prominent place, write two (very) *loose* conjectures that you can make on the basis of these two examples. Keep the conjectures in front of you, and firm them up as you explore more examples.

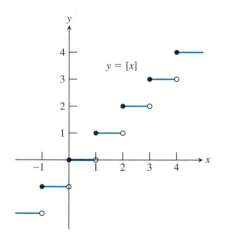

2.6 At each integer the greatest integer function $y = [x]$ (also known as the integer floor function) has different right-hand and left-hand limits.

In both Example 2 and Exploration 2 we found the limit of a function of x as x approaches a value at which the function is undefined. Functions, however, need not have a limit at such points, as the next example illustrates.

EXAMPLE 6

Use a graph to show that $\lim_{x \to 2} \left(x^3 - 1 \right) /(x - 2)$ does not exist.

Solution   Notice that the denominator is 0 when x is replaced by 2. The graph in Fig. 2.5 of $f(x) = (x^3 - 1)/(x - 2)$ strongly suggests that as $x \to 2$ (x approaches 2) from either side, the function values do not get close to any one number. This, in turn, suggests that $\lim_{x \to 2}(x^3 - 1)/(x - 2)$ does not exist. ≡

 EXPLORATION 3

Seeing What Doesn't Exist

A grapher can suggest what a limit is. It can also suggest when a limit doesn't exist. Use a grapher to examine what happens to the values $f(x)$ as $x \to c$ for the following functions. If it appears that a limit exists, tell what it seems to be. Otherwise, explain why a limit likely does not exist on the basis of your present understanding of what a limit is.

1. $f(x) = \dfrac{x^3 - 1}{x - 1}$ as $x \to 1$ 2. $f(x) = \dfrac{x^2 + 1}{x + 1}$ as $x \to -1$

3. $f(x) = \sin \dfrac{1}{x}$ as $x \to 0$ 4. $f(x) = [x]$ as $x \to 3$ ⚹

Right-hand Limits and Left-hand Limits

Sometimes the values of a function f tend to different limits as x approaches a number c from opposite sides. When this happens, we call the limit of f as x approaches c from the right the **right-hand limit** of f at c and the limit as x approaches c from the left the **left-hand limit** of f at c.

The notation for the right-hand limit is

$$\lim_{x \to c^+} f(x). \quad \text{The limit of } f(x) \text{ as } x \text{ approaches } c \text{ from the right.}$$

The $(+)$ is there to say that x approaches c through values to the right of c on the number line.

The notation for the left-hand limit is

$$\lim_{x \to c^-} f(x). \quad \text{The limit of } f(x) \text{ as } x \text{ approaches } c \text{ from the left.}$$

The $(-)$ is there to say that x approaches c through values to the left of c on the number line.

EXAMPLE 7

The greatest integer function $f(x) = [x]$ has different right-hand and left-hand limits at each integer, as we can see in Fig. 2.6,

$$\lim_{x \to 3^+} [x] = 3 \qquad \text{and} \qquad \lim_{x \to 3^-} [x] = 2.$$

The limit of $[x]$ as x approaches an integer n from the right is n, while the limit as x approaches n from the left is $n - 1$.   ≡

One-sided Limits and Two-sided Limits

We sometimes call $\lim_{x \to c} f(x)$ the **two-sided limit** of f at c to distinguish it from the **one-sided** right-hand and left-hand limits of f at c. If the two one-sided limits of f exist at c and are equal, their common value is the two-sided limit of f at c. Conversely, if the two-sided limit of f at c exists, the two one-sided limits exist and have the same value as the two-sided limit.

ON THE FAR SIDE

If f is not defined to the left of c, then f does not have a left-hand limit at c. A similar statement can be made if f is not defined to the right of c.

THEOREM 4

A function $f(x)$ has a limit as x approaches c if and only if the right-hand and left-hand limits at c exist and are equal. In symbols,

$$\lim_{x \to c} f(x) = L \quad \Leftrightarrow \quad \lim_{x \to c^+} f(x) = L \quad \text{and} \quad \lim_{x \to c^-} f(x) = L.$$

Thus, $[x]$ of Example 7 does not have a limit as $x \to 3$. The implications in Theorem 4 will be proved formally in Section 2.6, but you can see what is happening if you look at the next examples.

EXAMPLE 8

All the following statements about the function $y = f(x)$ graphed in Fig. 2.7 are true.

At $x = 0$: $\displaystyle\lim_{x \to 0^+} f(x) = 1.$

At $x = 1$: $\displaystyle\lim_{x \to 1^-} f(x) = 0$ even though $f(1) = 1,$

$\displaystyle\lim_{x \to 1^+} f(x) = 1,$

f has no limit as $x \to 1$. (The right- and left-hand limits at 1 are not equal. We say that the $\lim_{x \to 1} f(x)$ does not exist.)

At $x = 2$: $\displaystyle\lim_{x \to 2^-} f(x) = 1,$

$\displaystyle\lim_{x \to 2^+} f(x) = 1,$

$\displaystyle\lim_{x \to 2} f(x) = 1$ even though $f(2) = 2.$

At $x = 3$: $\displaystyle\lim_{x \to 3^-} f(x) = \lim_{x \to 3^+} f(x) = 2 = f(3) = \lim_{x \to 3} f(x).$

At $x = 4$: $\displaystyle\lim_{x \to 4^-} f(x) = 1.$

At noninteger values of c between 0 and 4, f has a limit as $x \to c$.   ≡

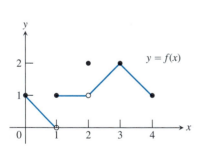

2.7 The graph of the function

$$f(x) = \begin{cases} -x + 1, & 0 \le x < 1, \\ 1, & 1 \le x < 2, \\ 2, & x = 2, \\ x - 1, & 2 < x \le 3, \\ -x + 5, & 3 < x \le 4, \end{cases}$$

of Example 8.

EXAMPLE 9

Show algebraically that $\lim_{x\to 0}|x| = 0$. Support graphically.

Solution We prove that $\lim_{x\to 0}|x| = 0$ by showing that the right-hand and left-hand limits are both 0:

$$\lim_{x\to 0^+} |x| = \lim_{x\to 0^+} x = 0. \qquad |x| = x \quad \text{if} \quad x > 0$$

$$\lim_{x\to 0^-} |x| = \lim_{x\to 0^-} (-x) \qquad |x| = -x \quad \text{if} \quad x < 0$$

$$= -\lim_{x\to 0^-} x \qquad \text{A special case of Theorem 1(c) 4}$$

$$= -0$$

$$= 0.$$

The graph of $f(x) = |x|$ in Fig. 2.8 supports $\lim_{x\to 0}|x| = 0$. ▤

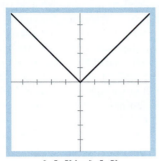

[−5, 5] by [−5, 5]

2.8 The values of $f(x) = |x|$ appear to approach 0 as $x \to 0$.

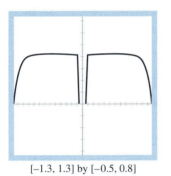

[−1.3, 1.3] by [−0.5, 0.8]

2.9 The grapher used for this view of $f(x) = (1 - \cos x^6)/x^{12}$ does not have enough precision to produce a correct graph near $x = 0$. You may find it interesting to ZOOM-IN around the origin and the "ends" of the graph. Another interesting view occurs in a $[-0.2, 0.2]$ by $[0.3, 0.7]$ viewing window. In Chapter 7 we will confirm analytically that $\lim_{x\to 0} f(x) = 0.5$.

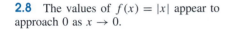

EXPLORATION 4

Complete Graphs

Graphers are very useful in mathematics but cannot be used blindly. To use a grapher for information about a function, including clues as to limiting values, it is essential that viewing-window graphs be complete. GRAPH

$$f(x) = \frac{1 - \cos x^6}{x^{12}}$$

in the standard viewing window and ZOOM-IN to investigate its behavior as $x \to 0$. The view (Fig. 2.9) and TRACE may incorrectly suggest that $f(x) = 0$ for x near 0. Compute $f(x)$ for $x = 0.0001$ and a few other values of x near 0. If your grapher suggests misleading information around 0, you should conclude that the grapher is inadequate for f near 0 (as consideration of the denominator of f might suggest) and that analytic mathematics is *essential* for analyzing f in this trouble spot. ⚘

Exercises 2.1

Find the limits in Exercises 1–14 by substitution and support with a grapher.

1. $\lim_{x\to 2} 2x$

2. $\lim_{x\to 0} 2x$

3. $\lim_{x\to 1} (3x - 1)$

4. $\lim_{x\to 1/3} (3x - 1)$

5. $\lim_{x\to -1} 3x(2x - 1)$

6. $\lim_{x\to -1} 3x^2(2x - 1)$

7. $\lim_{x\to -2} (x + 3)^{171}$

8. $\lim_{x\to -4} (x + 3)^{1994}$

9. $\lim_{x\to 1} (x^3 + 3x^2 - 2x - 17)$

10. $\lim_{x\to -2} (x^3 - 2x^2 + 4x + 8)$

11. $\lim_{x\to -1} \dfrac{x + 3}{x^2 + 3x + 1}$

12. $\lim_{y\to 2} \dfrac{y^2 + 5y + 6}{y + 2}$

13. $\lim_{y\to -3} \dfrac{y^2 + 4y + 3}{y^2 - 3}$

14. $\lim_{x\to -1} \dfrac{x^3 - 5x + 7}{-x^3 + x^2 - x + 1}$

Explain why substitution does not work to find the limits in Exercises 15–18. Find the limits if they exist.

15. $\lim_{x\to -2} \sqrt{x - 2}$

16. $\lim_{x\to 0} \dfrac{1}{x^2}$

17. $\lim_{x\to 0} \dfrac{|x|}{x}$

18. $\lim_{x\to 0} \dfrac{(4 + x)^2 - 16}{x}$

In Exercises 19–26, find the limits graphically. Then confirm algebraically.

19. $\lim\limits_{x\to 1} \dfrac{x-1}{x^2-1}$

20. $\lim\limits_{x\to -5} \dfrac{x^2+3x-10}{x+5}$

21. $\lim\limits_{t\to 1} \dfrac{t^2-3t+2}{t^2-1}$

22. $\lim\limits_{t\to 2} \dfrac{t^2-3t+2}{t^2-4}$

23. $\lim\limits_{x\to 2} \dfrac{2x-4}{x^3-2x^2}$

24. $\lim\limits_{x\to 0} \dfrac{5x^3+8x^2}{3x^4-16x^2}$

25. $\lim\limits_{x\to 0} \dfrac{\frac{1}{2+x}-\frac{1}{2}}{x}$

26. $\lim\limits_{x\to 0} \dfrac{(2+x)^3-8}{x}$

Investigate $\lim_{x\to 0} f(x)$ in Exercises 27–32 by making tables of values.

a)

x	-0.1	-0.01	-0.001	-0.0001	$\ldots$
$f(x)$	?	?	?	?	$\ldots$

b)

x	0.1	0.01	0.001	0.0001	$\ldots$
$f(x)$	?	?	?	?	$\ldots$

On the basis of the tables, state what you believe the limit to be. Support graphically.

27. $f(x) = x\sin\dfrac{1}{x}$

28. $f(x) = \dfrac{1}{x}\sin x$

29. $f(x) = \sin\dfrac{1}{x}$

30. $f(x) = \dfrac{10^x-1}{x}$

31. $f(x) = \dfrac{2^x-1}{x}$

32. $f(x) = x\sin(\ln|x|)$

33. We have seen that there is little difference in the value of an investment when interest is compounded a few times or many times during a year. Use your grapher and Eq. (1). Find the limit on the value of a \$100 investment as the number of compounding periods within a year increases. Does frequent compounding offer any real advantage to an investor?

34. Set $[0, 1000]$ by $[106, 106.5]$ view dimensions. For each value of i, GRAPH $f(x) = 100(1+i/x)^x$ and $g(x) = 100e^i$ in the same window. Where x is about 365, ZOOM-IN enough on both graphs so that you can see that the graphs are different. Find the difference between $f(365)$ and $g(365)$. Instead of \$100, how much money would you have to invest for each value of i to have a "real" difference that you could pocket?

a) $i = 0.06$

b) $i = 0.08$

35. GRAPH $f(x) = [(x+2)^2-4]/x$. Magnify the graph around the point $(0, 4)$ as much as possible. What do you observe? Confirm your observation algebraically.

36. GRAPH $f(x) = [(x+1)^3-1]/x$. Magnify the graph around point $(0, 3)$ as much as possible. What do you observe? Confirm your observation algebraically.

37. GRAPH $f(x) = (x^3-1)/(x-2)$. Use narrow, tall viewing windows to magnify the graph near $x = 2$. What do you notice about the values of f as x approaches 2 from the right? From the left? How could you have reached similar conclusions without using a graphing utility? (A *narrow, tall viewing window* is one with a small horizontal view dimension compared to a large vertical view dimension such as $[2, 2.1]$ by $[-2000, 2000]$ to the right of $x = 2$, or $[1.99, 2]$ by $[-2000, 2000]$ to the left of $x = 2$.)

38. GRAPH $f(x) = \dfrac{1-x^3}{x-2}$. Use tall, narrow viewing windows (see Exercise 37) to magnify the graph near $x = 2$. What do you notice about the values of f as x approaches 2 from the right? From the left? How could you have reached similar conclusions without using a graphing utility?

39. Which of the following statements are true of the function $y = f(x)$ graphed here?

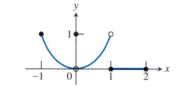

a) $\lim\limits_{x\to -1^+} f(x) = 1$

b) $\lim\limits_{x\to 0^-} f(x) = 0$

c) $\lim\limits_{x\to 0^-} f(x) = 1$

d) $\lim\limits_{x\to 0^-} f(x) = \lim\limits_{x\to 0^+} f(x)$

e) $\lim\limits_{x\to 0} f(x)$ exists

f) $\lim\limits_{x\to 0} f(x) = 0$

g) $\lim\limits_{x\to 0} f(x) = 1$

h) $\lim\limits_{x\to 1} f(x) = 1$

i) $\lim\limits_{x\to 1} f(x) = 0$

j) $\lim\limits_{x\to 2^-} f(x) = 2$

40. Which of the following statements are true of the function graphed here?

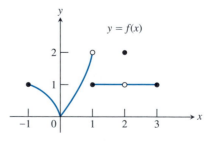

a) $\lim\limits_{x\to -1^+} f(x) = 1$

b) $\lim\limits_{x\to 2} f(x)$ does not exist.

c) $\lim\limits_{x\to 2} f(x) = 2$

d) $\lim\limits_{x\to 1^-} f(x) = 2$

e) $\lim\limits_{x \to 1^+} f(x) = 1$ **f)** $\lim\limits_{x \to 1} f(x)$ does not exist.

g) $\lim\limits_{x \to 0^+} f(x) = \lim\limits_{x \to 0^-} f(x)$

h) $\lim\limits_{x \to c} f(x)$ exists at every c in $(-1, 1)$.

i) $\lim\limits_{x \to c} f(x)$ exists at every c in $(1, 3)$.

Refer to Exploration 5 in Section 1.3 for how to view the piecewise-defined functions in Exercises 41–48 with a grapher.

41. Let $f(x) = \begin{cases} 3 - x, & x < 2; \\ \dfrac{x}{2} + 1, & x > 2. \end{cases}$

a) Determine a complete graph of f.
b) Find $\lim\limits_{x \to 2^+} f(x)$ and $\lim\limits_{x \to 2^-} f(x)$.
c) Does $\lim\limits_{x \to 2} f(x)$ exist? If so, what is it? If not, why not?

42. Let $f(x) = \begin{cases} 3 - x, & x < 2, \\ 2, & x = 2, \\ \dfrac{x}{2}, & x > 2. \end{cases}$

a) Determine a complete graph of f.
b) Find $\lim\limits_{x \to 2^+} f(x)$ and $\lim\limits_{x \to 2^-} f(x)$.
c) Does $\lim\limits_{x \to 2} f(x)$ exist? If so, what is it? If not, why not?

43. Let $f(x) = \begin{cases} \dfrac{1}{x - 1}, & x < 1, \\ x^3 - 2x + 5, & x \geq 1. \end{cases}$

a) Determine a complete graph of f.
b) Find $\lim\limits_{x \to 1^+} f(x)$ and $\lim\limits_{x \to 1^-} f(x)$.
c) Does $\lim\limits_{x \to 1} f(x)$ exist? If so, what is it? If not, why not?

44. Let $f(x) = \begin{cases} \dfrac{1}{2 - x}, & x < 2, \\ 5 - x^2, & x \geq 2. \end{cases}$

a) Determine a complete graph of f.
b) Find $\lim\limits_{x \to 2^+} f(x)$ and $\lim\limits_{x \to 2^-} f(x)$.
c) Does $\lim\limits_{x \to 2} f(x)$ exist? If so, what is it? If not, why not?

45. Let $f(x) = \begin{cases} a - x^2, & x < 2, \\ x^2 + 5x - 3, & x \geq 2. \end{cases}$
For what values of a does $\lim\limits_{x \to 2} f(x)$ exist?

46. Let $f(x) = \begin{cases} x^3 - 4x, & x < -1, \\ 2x + a, & x \geq -1. \end{cases}$
For what values of a does $\lim\limits_{x \to -1} f(x)$ exist?

47. a) Determine a complete graph of $f(x) = \begin{cases} x^3, & x \neq 1, \\ 0, & x = 1. \end{cases}$

b) Find $\lim\limits_{x \to 1^-} f(x)$ and $\lim\limits_{x \to 1^+} f(x)$.
c) Does $\lim\limits_{x \to 1} f(x)$ exist? If so, what is it? If not, why not?

48. a) Determine a complete graph of $f(x) = \begin{cases} 1 - x^2, & x \neq 1, \\ 2, & x = 1. \end{cases}$

b) Find $\lim\limits_{x \to 1^+} f(x)$ and $\lim\limits_{x \to 1^-} f(x)$.

c) Does $\lim\limits_{x \to 1} f(x)$ exist? If so, what is it? If not, why not?

Determine a complete graph of the two functions in Exercises 49 and 50. Then answer these questions.

a) At what points c in the domain of f does $\lim\limits_{x \to c} f(x)$ exist?
b) At what points does only the left-hand limit exist?
c) At what points does only the right-hand limit exist?

49. $f(x) = \begin{cases} \sqrt{1 - x^2} & \text{if} \quad 0 \leq x < 1, \\ 1 & \text{if} \quad 1 \leq x < 2, \\ 2 & \text{if} \quad x = 2. \end{cases}$

50. $f(x) = \begin{cases} x & \text{if} \quad -1 \leq x < 0, \text{ or } 0 < x \leq 1, \\ 1 & \text{if} \quad x = 0, \\ 0 & \text{if} \quad x < -1, \text{ or } x > 1. \end{cases}$

Find the limits of the greatest integer function in Exercises 51–54.

51. $\lim\limits_{x \to 0^+} [x]$ **52.** $\lim\limits_{x \to 0^-} [x]$

53. $\lim\limits_{x \to 0.5} [x]$ **54.** $\lim\limits_{x \to 2^-} [x]$

Find the limits in Exercises 55 and 56.

55. $\lim\limits_{x \to 0^+} \dfrac{x}{|x|}$ **56.** $\lim\limits_{x \to 0^-} \dfrac{x}{|x|}$

Let a be any real number. Find the limits in Exercises 57 and 58.

57. $\lim\limits_{x \to a^+} \dfrac{|x - a|}{x - a}$ **58.** $\lim\limits_{x \to a^-} \dfrac{|x - a|}{x - a}$

59. Suppose $\lim\limits_{x \to c} f(x) = 5$ and $\lim\limits_{x \to c} g(x) = 2$. Find

a) $\lim\limits_{x \to c} f(x)g(x)$ **b)** $\lim\limits_{x \to c} 2f(x)g(x)$

60. Suppose $\lim\limits_{x \to 4} f(x) = 0$ and $\lim\limits_{x \to 4} g(x) = 3$. Find

a) $\lim\limits_{x \to 4} (g(x) + 3)$ **b)** $\lim\limits_{x \to 4} xf(x)$

c) $\lim\limits_{x \to 4} g^2(x)$ **d)** $\lim\limits_{x \to 4} \dfrac{g(x)}{f(x) - 1}$

61. Suppose $\lim\limits_{x \to b} f(x) = 7$ and $\lim\limits_{x \to b} g(x) = -3$. Find

a) $\lim\limits_{x \to b} (f(x) + g(x))$ **b)** $\lim\limits_{x \to b} f(x) \cdot g(x)$

c) $\lim\limits_{x \to b} 4g(x)$ **d)** $\lim\limits_{x \to b} f(x)/g(x)$

62. Suppose $\lim\limits_{x \to -2} p(x) = 4$, $\lim\limits_{x \to -2} r(x) = 0$, and $\lim\limits_{x \to -2} s(x) = -3$. Find
a) $\lim\limits_{x \to -2}(p(x) + r(x) + s(x))$
b) $\lim\limits_{x \to -2} p(x) \cdot r(x) \cdot s(x)$

Draw a graph and determine the limits in Exercises 63–72.

63. $\lim\limits_{x \to 0} x \sin x$ **64.** $\lim\limits_{x \to 0} \dfrac{\sin x}{x}$

65. $\lim\limits_{x \to 0} x^2 \sin x$ **66.** $\lim\limits_{x \to 0} \dfrac{1}{x} \sin \dfrac{1}{x}$

67. $\lim\limits_{x \to 0} x^2 \sin \dfrac{1}{x}$

68. $\lim\limits_{x \to 0} (1 + x)^{3/x}$

69. $\lim\limits_{x \to 0} (1 + x)^{4/x}$

70. $\lim\limits_{x \to 1} \dfrac{\ln (x^2)}{\ln x}$

71. $\lim\limits_{x \to 0} \dfrac{2^x - 1}{x}$

72. $\lim\limits_{x \to 0} \dfrac{3^x - 1}{x}$

73. Consider the function $f(x) = (1 - \cos x^6)/x^{12}$ of Exploration 4.

   a) Reproduce the graph in Fig. 2.9 in both dot and connected mode. Use TRACE to investigate $f(x)$ for x near 0.

   b) ZOOM-IN around $x = 0$, and GRAPH in both dot and connected mode.

   c) Discuss your findings.

Find the limits in Exercises 74–76 graphically. Does your grapher suggest incorrect information?

74. $\lim\limits_{x \to 0} \dfrac{1 - \cos x^{15}}{x^{30}}$

75. $\lim\limits_{x \to 3\pi/2} \dfrac{(1 + \sin x)^{20}}{\left(x - \frac{3\pi}{2}\right)^{40}}$

76. $\lim\limits_{x \to \pi} \dfrac{(1 + \cos x)^{20}}{(x - \pi)^{40}}$

77. *Challenge: limits and geometry.* Let $P = (a, a^2)$ be a point on the parabola $y = x^2$, $a > 0$. Let O denote the origin and $(0, b)$ denote the y-intercept of the perpendicular bisector of line segment OP. Evaluate $\lim\limits_{P \to O} b$. Confirm your answer analytically. Support with a grapher.

2.2 _____ Continuous Functions

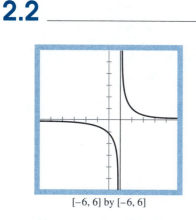

[−6, 6] by [−6, 6]

2.10   Graphers will sometimes connect the two branches of the graph of $y = 1/(x - 1)$, suggesting that the function is defined and continuous at $x = 1$. To avoid "spikes" like this, some graphers allow us to turn off the *connected format* and view graphs in *dot format* to get a better idea of whether the function is really continuous.

Most graphers can plot points (*dot format*). Some can illuminate pixels between plotted points to suggest an unbroken curve (*connected format*). For functions, the connected format basically assumes that outputs vary *continuously* with inputs and do not jump from one value to another without taking on all values in between. (See Fig. 2.10.)

Continuous functions are the functions that we normally use in the equations that describe numerical relations in the world around us. They are the functions we use to find a planet's closest approach to the sun or the peak concentration of antibodies in blood plasma. They are also the functions that we use to describe how a body moves through space or how the speed of a chemical reaction changes with time. In fact, so many observable processes proceed continuously that throughout the eighteenth and nineteenth centuries it rarely occurred to anyone to look for any other kind of behavior. It came as quite a surprise when the physicists of the 1920s discovered that the vibrating atoms in a hydrogen molecule can oscillate only at discrete energy levels, that light comes in particles, and that, when heated, atoms emit light in discrete frequencies and not in continuous spectra.

As a result of these and other discoveries, and because of the heavy use of discrete functions in computer science and statistics, the issue of continuity has become one of practical as well as theoretical importance. As scientists, we need to know when continuity is called for, what it is, and how to test for it.

The Definition of Continuity

A function $y = f(x)$ whose graph can be sketched over any interval of its domain with one continuous motion of the pencil is an example of a **continuous function**. The height of the graph over the interval varies continuously with x. At each interior point of the function's domain, like the point c in

Fig. 2.11, the function value $f(c)$ is the limit of the function values on either side; that is,

$$f(c) = \lim_{x \to c} f(x).$$

The function value at an endpoint of the domain is also the limit of the nearby function values. At the left endpoint a and at the right endpoint b in Fig. 2.11,

$$f(a) = \lim_{x \to a^+} f(x) \qquad \text{and} \qquad f(b) = \lim_{x \to b^-} f(x).$$

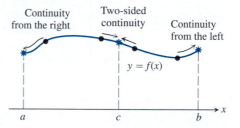

Continuity
from the right

Two-sided
continuity

Continuity
from the left

$y = f(x)$

a $\qquad$ c $\qquad$ b

2.11 Continuity at points a, b, and c for a function $y = f(x)$ that is continuous on the interval $[a, b]$.

To be specific, let us look at the function in Fig. 2.12, whose limits we investigated in Example 8 in Section 2.1.

EXAMPLE 1

The function in Fig. 2.12 is continuous at every point in its domain $[0, 4]$ except $x = 1$ and $x = 2$. At these points there are breaks in the graph. Note the relation between the limit of f and the value of f at each point of the function's domain.

Points of discontinuity:

At $x = 1$: $\quad \lim_{x \to 1} f(x)$ does not exist.

At $x = 2$: $\quad \lim_{x \to 2} f(x) = 1$, but $1 \neq f(2)$.

Points at which f is continuous:

At $x = 0$: $\quad \lim_{x \to 0^+} f(x) = f(0)$.

At $x = 4$: $\quad \lim_{x \to 4^-} f(x) = f(4)$.

At every point $0 < c < 4$ except $x = 1, 2$: $\quad \lim_{x \to c} f(x) = f(c)$. $\quad$ ≡

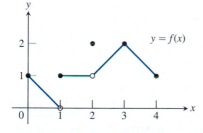

$y = f(x)$

2.12 The function is continuous on $[0, 4]$ except for discontinuities at $x = 1$ and $x = 2$.

We now come to the formal definitions of continuity at a point in a function's domain. In the definitions we distinguish between continuity at an endpoint (which involves a one-sided limit) and continuity at an interior point (which involves a two-sided limit).

DEFINITIONS

Continuity at an Interior Point   A function $y = f(x)$ is continuous at an interior point c of its domain if

$$\lim_{x \to c} f(x) = f(c).$$

Continuity at an Endpoint   A function $y = f(x)$ is continuous at a left endpoint a or a right endpoint b of its domain if

$$\lim_{x \to a^+} f(x) = f(a) \qquad \text{or} \qquad \lim_{x \to b^-} f(x) = f(b), \text{ respectively.}$$

Continuous Function   A function is continuous if it is continuous at each point of its domain.

Discontinuity at a Point   If a function f is not continuous at a point c, we say that f is discontinuous at c and call c a point of discontinuity of f. Note that c may or may not be in the domain of f.

To check for continuity at a point, we apply the Continuity Test.

AN EQUIVALENT TEST

The following set of conditions is equivalent to the conditions in the Continuity Test for continuity at $x = c$:

1. $f(c)$ exists.

2. $\lim_{x \to c^+} f(x)$ and $\lim_{x \to c^-} f(x)$ exist.

3. $\lim_{x \to c^+} f(x) = \lim_{x \to c^-} f(x) = f(c)$.

When $\lim_{x \to c^-} f(x) = f(c)$ or $\lim_{x \to c^+} f(x) = f(c)$, we say that f is *continuous from the left at c or continuous from the right at c,* respectively.

THE CONTINUITY TEST

The function $y = f(x)$ is continuous at $x = c$ if and only if *all three* of the following statements are true:

1. $f(c)$ exists   (c is in the domain of f).

2. $\lim_{x \to c} f(x)$ exists   (f has a limit as $x \to c$).

3. $\lim_{x \to c} f(x) = f(c)$ (the limit value equals the function value).

The limit is to be two-sided if c is an interior point of the domain of f; it is to be the appropriate one-sided limit if c is an endpoint of the domain.

EXAMPLE 2

When applied to the function $y = f(x)$ of Example 1 at the points $x = 0, 1, 2, 3,$ and 4, the Continuity Test gives the following results. (The graph of f is reproduced here as Fig. 2.13.)

a) f is continuous at $x = 0$ because

   i) $f(0)$ exists   (it equals 1),

   ii) $\lim_{x \to 0^+} f(x) = 1$ (f has a limit as $x \to 0^+$),

   iii) $\lim_{x \to 0^+} f(x) = f(0)$ (the limit value equals the function value).

b) f is discontinuous at $x = 1$ because $\lim_{x \to 1} f(x)$ does not exist. The function fails part 2 of the test. (The right-hand and left-hand limits exist at $x = 1$, but they are not equal.)

c) f is discontinuous at $x = 2$ because $\lim_{x \to 2} f(x) \neq f(2)$. The function fails part 3 of the test. This type of discontinuity is called a *removable*

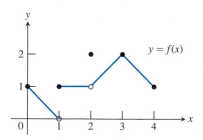

2.13   The function is continuous on $[0, 1)$, discontinuous at 1, continuous on $(1, 2)$, discontinuous at 2, and continuous on $(2, 4]$.

discontinuity because it is possible to redefine f at $x = 2$, that is, $f(2) = 1$ to make f continuous at $x = 2$. Notice that this is not possible at $x = 1$.

d) f is continuous at $x = 3$ because

   i) $f(3)$ exists   (it equals 2),

   ii) $\lim_{x \to 3} f(x) = 2$ (f has a limit as $x \to 3$),

   iii) $\lim_{x \to 3} f(x) = f(3)$ (the limit value equals the function value).

e) f is continuous at $x = 4$ because

   i) $f(4)$ exists   (it equals 1),

   ii) $\lim_{x \to 4^-} f(x) = 1$ (f has a limit as $x \to 4^-$),

   iii) $\lim_{x \to 4^-} f(x) = f(4)$ (the limit value equals the function value).

TYPES OF DISCONTINUITIES

Discontinuities may be classified in various ways. When the classification is based on what the graph looks like at the discontinuity, the names used include

- removable discontinuity,
- jump discontinuity,
- oscillating discontinuity,
- infinite discontinuity.

As you encounter discontinuities, try to apply these names or whatever names are selected for use by your class.

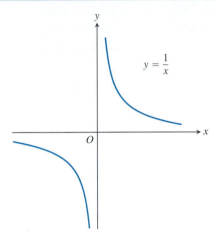

2.14 The function $y = 1/x$ is continuous at every value of x except $x = 0$.

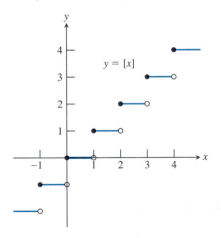

2.15 The greatest integer function $y = [x]$ is discontinuous at every integer (but continuous at every other point).

EXAMPLE 3

Polynomial functions are continuous at every point. In Section 2.1 we saw that $\lim_{x \to c} f(x) = f(c)$ for any polynomial $f(x)$ at any point $x = c$.

EXAMPLE 4

The reciprocal function $y = 1/x$ (Fig. 2.14) is continuous because it is continuous at every point of its domain. It is not continuous on the set of real numbers, however, because it is not defined at $x = 0$ and thus fails part 1 of the Continuity Test at $x = 0$. This is a special case of Example 5.

EXAMPLE 5

Rational functions are continuous at every point of their domains. In Section 2.1 we saw that $\lim_{x \to c} f(x)/g(x) = f(c)/g(c)$ at every point where $g(c) \neq 0$. Discontinuities occur where $g(c) = 0$.

EXAMPLE 6

The *absolute value function* $y = |x|$ is continuous at every value of x. It agrees with the continuous (polynomial) function $y = x$ if $x \geq 0$. It agrees with the continuous function $y = -x$ if $x < 0$. And, finally, $\lim_{x \to 0} |x| = |0|$, as we saw in Example 9 in Section 2.1.

EXAMPLE 7

The *greatest integer function*, or *floor function*, $y = [x]$ (Fig. 2.15) is discontinuous at every integer. At every integer the function fails to have a limit and so fails part 2 of the Continuity Test.

EXAMPLE 8

The *exponential functions, logarithmic functions, trigonometric functions,* and *radical functions* like $y = a^x$, $y = \log_a x$, $y = \cos x$, and $y = \sqrt[n]{x}$ (n is a positive integer greater than 1), respectively, are continuous at every point of their domains. These facts are suggested by exploring their graphs and will be proved in later chapters.

Algebraic Combinations of Continuous Functions

As you may have guessed, algebraic combinations of continuous functions are continuous wherever they are defined. The relevant theorem is this:

THEOREM 5   **Algebraic Properties of Continuous Functions**

If the functions f and g are continuous at $x = c$, then the following combinations are continuous at $x = c$:

1. Sums: $f + g$
2. Differences: $f - g$
3. Products: $f \cdot g$
4. Constant multiples: $k \cdot g$ (any number k)
5. Quotients: f/g (provided that $g(c) \neq 0$)

Proof Theorem 5 is a special case of the Property of Limits Theorem, Theorem 1(c) in Section 2.1. If the latter were restated for the continuous functions f and g, it would say that if $\lim_{x \to c} f(x) = f(c)$ and $\lim_{x \to c} g(x) = g(c)$, then

1. $\lim_{x \to c} [f(x) + g(x)] = f(c) + g(c)$,
2. $\lim_{x \to c} [f(x) - g(x)] = f(c) - g(c)$,
3. $\lim_{x \to c} f(x)g(x) = f(c)g(c)$,
4. $\lim_{x \to c} kf(x) = kf(c)$ (any number k),
5. $\lim_{x \to c} \dfrac{f(x)}{g(x)} = \dfrac{f(c)}{g(c)}$ (provided that $g(c) \neq 0$).

In other words, the limits as $x \to c$ of the functions in (1)–(5) exist and equal the function values at $x = c$. Therefore, each function fulfills the three requirements of the continuity test at any interior point $x = c$ of its domain. Similar arguments with right-hand and left-hand limits establish the theorem for continuity at endpoints.   ≣

EXAMPLE 9

The functions

$$f(x) = x^{14} + 20x^4 \qquad \text{and} \qquad g(x) = 5x(2 - x) + 1/(x^2 + 1)$$

are continuous at every value of x. The function

$$h(x) = \frac{x + 3}{x^2 - 3x - 10} = \frac{x + 3}{(x - 5)(x + 2)}$$

is continuous at every value of x except $x = 5$ and $x = -2$.   ≣

EXPLORATION 1

Removing Discontinuities

Visually, a removable discontinuity appears as a one-point gap in the graph of a function. In reality, it can *never* be "seen." This may be *represented* in a sketch by an open dot. It can be shown in a viewing window by an unlit pixel if the horizontal view dimensions are set in a certain way.

We know that

$$f(x) = \frac{x^3 - 7x - 6}{x^2 - 9} = \frac{x^3 - 7x - 6}{(x+3)(x-3)}$$

has neither 3 nor -3 in its domain. The graph of f (Fig. 2.16) suggests that the discontinuity at $x = 3$ is removable by defining $f(3)$ to be 10/3. (Use ZOOM and TRACE to check this.) It also suggests that we look for $x - 3$ to be a factor of the numerator. Why?

$$f(x) = \frac{x^3 - 7x - 6}{x^2 - 9} = \frac{x^3 - 7x - 6}{(x+3)(x-3)} = \frac{(x-3)(x+1)(x+2)}{(x+3)(x-3)}$$

Thus,

$$\lim_{x \to 3} \frac{x^3 - 7x - 6}{x^2 - 9} = \lim_{x \to 3} \frac{(x-3)(x+1)(x+2)}{(x+3)(x-3)}$$

$$= \lim_{x \to 3} \frac{(x+1)(x+2)}{(x+3)}$$

$$= \frac{(3+1)(3+2)}{3+3} = \frac{20}{6} = 10/3$$

The *extended function*

$$g(x) = \begin{cases} \dfrac{x^3 - 7x - 6}{x^2 - 9} & \text{if } x \neq 3, \\ 10/3 & \text{if } x = 3, \end{cases}$$

is continuous at $x = 3$ because $\lim_{x \to 3} g(x)$ exists and equals $g(3)$. The function g is the **continuous extension** of the original function f to the point $x = 3$.

For each of the following, tell how to remove a discontinuity and extend the function.

1. $f(x) = \dfrac{x^2 + x - 6}{x^2 - 4}$ **2.** $f(x) = \left(1 + \dfrac{1}{x}\right)^x$

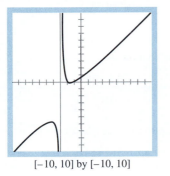

[−10, 10] by [−10, 10]

2.16 A graph of
$$f(x) = \frac{x^3 - 7x - 6}{x^2 - 9}.$$

We know that 3 and -3 are not in the domain of f, so there are discontinuities at $x = 3$ and $x = -3$. The graph of f, however, *appears* continuous at $x = 3$, suggesting that the discontinuity at $x = 3$ is removable.

EXPLORATION BIT

Figure 2.13 shows an open dot representing a removable discontinuity at $x = 2$. A viewing window will not show a removable discontinuity unless the point of discontinuity has an x-coordinate that matches a pixel or screen x-coordinate. For the function f in Exploration 1, can you set [xMin, xMax] so that a one-pixel hole shows in the graph of f?

Composites of Continuous Functions

All composites of continuous functions are continuous. This means that composites like

$$y = \sin(x^2) \qquad \text{and} \qquad y = |\cos x|$$

are continuous at every point at which they are defined. The idea is that if $f(x)$ is continuous at $x = c$ and $g(x)$ is continuous at $x = f(c)$, then $g \circ f$ is continuous at $x = c$ (Fig. 2.17).

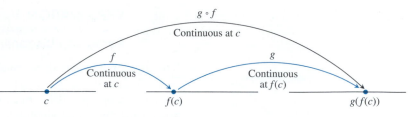

2.17   Composites of continuous functions are continuous.

> **THEOREM 6**
>
> If f is continuous at c and g is continuous at $f(c)$, then the composite $g \circ f$ is continuous at c.

For an outline of a proof

EXAMPLE 10

Make a conjecture about the continuity of $y = \left| \dfrac{x \sin x}{x^2 + 2} \right|$. Give a convincing argument to support your conjecture.

Solution   The graph (Fig. 2.18) of $y = \left| (x \sin x)/(x^2 + 2) \right|$ suggests that the function is continuous at every value of x. By letting

$$g(x) = |x| \qquad \text{and} \qquad f(x) = \frac{x \sin x}{x^2 + 2},$$

we see that y is the composite function $g \circ f$. The function f is continuous by Theorem 5, the function g by Example 6, and their composite $g \circ f$ by Theorem 6.   ≣

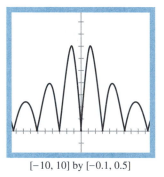

[−10, 10] by [−0.1, 0.5]

2.18   The graph suggests that $y = \left| (x \sin x)/(x^2 + 2) \right|$ is continuous.

If a composite function $g \circ f$ is continuous at a point $x = c$, its limit as $x \to c$ is $g(f(c))$. (See Fig. 2.17.)

EXAMPLE 11

a) $\lim_{x \to 0} |1 + \cos x| = |1 + \cos 0| = |1 + 1| = 2$

$$\lim_{x \to c} (g \circ f)(x) = g(f(c))$$

b) $\lim_{x \to 1^+} \sin \sqrt{x - 1} = \sin \sqrt{1 - 1} = \sin 0 = 0$ ≣

Important Properties of Continuous Functions

We study continuous functions because they are useful in mathematics and its applications. A function that is continuous on a closed interval $[a, b]$ assumes every value between $f(a)$ and $f(b)$. This concept justifies the use of ZOOM-IN to solve many equations and inequalities as presented in Chapter 1.

In addition, a function that is continuous at every point of a closed interval $[a, b]$ has an *absolute maximum* value and an *absolute minimum*

ZOOM-IN TO SOLVE

If a view of the graph of a continuous function f shows part of the graph below and part above the x-axis, then the continuity of f assures us that there is some point in our viewing window where the graph meets the axis, that is, where $f(x) = 0$. If we ZOOM-IN to always include graph points above and below the x-axis, then we can close in on a solution of $f(x) = 0$ to a desired degree of accuracy.

value on this interval. We always look for these values when we graph a function, and we shall see the role they play in problem solving (Chapter 4) and in the development of the integral calculus (Chapters 5 and 6).

Finally, it turns out that every continuous function is some other function's *derivative*, as we shall see in Chapter 5. The ability to recover a function from information about its derivative is one of the great powers given to us by calculus. Thus, given a formula $v(t)$ for the velocity of a moving body as a continuous function of time, we shall be able, with the calculus of Chapters 3, 4, and 5, to produce a formula $s(t)$ that tells how far the body has traveled from its starting point at any instant and whose derivative is $v(t)$.

MAXIMA AND MINIMA

If $f(x_0) \geq f(x)$ for all x in the domain D of f, then $f(x_0)$ is the (**absolute**) **maximum** value of f on D. If $f(x_0) \geq f(x)$ for all x in a subset S of D, and x_0 is in D, then $f(x_0)$ is a **local maximum** value of f on S.

If $f(x_0) \leq f(x)$ for all x in the domain D of f, then $f(x_0)$ is the (**absolute**) **minimum** value of f on D. If $f(x_0) \leq f(x)$ for all x in a subset S of D, and x_0 is in D, then $f(x_0)$ is a **local minimum** value of f on S.

In practice, S is usually taken to be a small interval about x_0.

THEOREM 7   The Max-Min Theorem for Continuous Functions

If f is continuous at every point of a closed interval $[a, b]$, then f takes on both a maximum value M and a minimum value m somewhere in that interval. That is, for some numbers x_1 and x_2 in $[a, b]$ we have $f(x_1) = m$, $f(x_2) = M$, and $m \leq f(x) \leq M$ at every other point x of the interval.

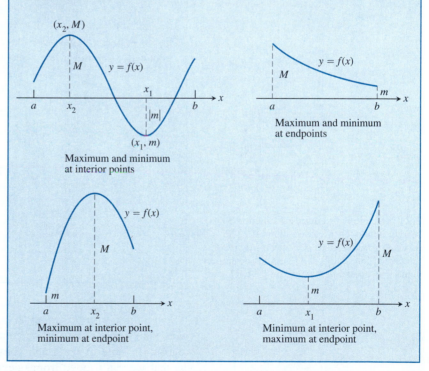

Maximum and minimum at interior points

Maximum and minimum at endpoints

Maximum at interior point, minimum at endpoint

Minimum at interior point, maximum at endpoint

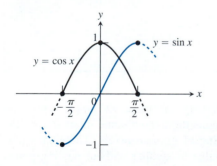

2.19 $\sin x$ and $\cos x$ for $-\pi/2 \leq x \leq \pi/2$. Theorem 7 says that a continuous function on a closed interval takes on a maximum and a minimum value somewhere in the closed interval. Either extreme value could occur at more than one point.

EXAMPLE 12

On the closed interval $\left[-\dfrac{\pi}{2}, \dfrac{\pi}{2}\right]$ the cosine takes on a maximum value of 1 (once) and a minimum value of 0 (twice). The sine takes on a maximum value of 1 and a minimum value of -1, one time each (Fig. 2.19).   ≣

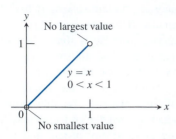

2.20 Theorem 7 requires a function to be continuous on a *closed* interval. The function $y = x$ is continuous on the *open* interval (0, 1) but has no largest value and no smallest value on (0, 1).

EXAMPLE 13

On an open interval a continuous function need not have either a maximum or a minimum value. The function $f(x) = x$ (Fig. 2.20) has neither a largest nor a smallest value on the interval $0 < x < 1$. ≡

EXAMPLE 14

Even a single point of discontinuity can keep a function from having either a maximum or a minimum value on a closed interval. The function

$$y = \begin{cases} x + 1, & -1 \le x < 0, \\ 0 & x = 0, \\ x - 1, & 0 < x \le 1, \end{cases}$$

is continuous at every point of the interval $-1 \le x \le 1$ except $x = 0$, yet its graph over the interval (Fig. 2.21) has neither a highest nor a lowest point. ≡

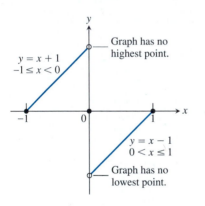

2.21 Theorem 7 requires a function to be *continuous* on a closed interval. The function graphed here is *not continuous* on the closed interval $[-1, 1]$ and has no largest value and no smallest value on $[-1, 1]$.

THEOREM 8   The Intermediate Value Theorem for Continuous Functions

A function $y = f(x)$ that is continuous on a closed interval $[a, b]$ takes on every value between $f(a)$ and $f(b)$. In other words, if y_0 is between $f(a)$ and $f(b)$, then $y_0 = f(c)$ for some c in $[a, b]$.

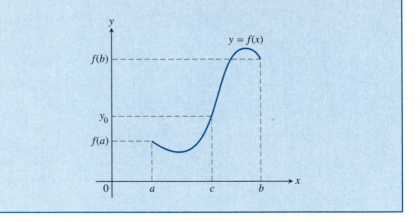

The proofs of Theorems 7 and 8 require a detailed knowledge of the real number system, and we shall not give them here.

A consequence for graphing: connectivity Suppose we graph a function $y = f(x)$ that we know to be continuous throughout some interval I on the x-axis. In spite of the graph consisting of discrete pixels, Theorem 8 tells us that we can't move from one y-value to another without passing points on the graph that take on all the y-values in between. The graph of f over I consists of a single, unbroken curve and is said to be **connected**. We know that it cannot have separate branches like the graph of $y = 1/x$ and that we can ZOOM-IN and *never* find "holes" like the one in the graph of $y = (x^2 - 1)/(x - 1)$ or "jumps" as in the graph of the greatest integer function.

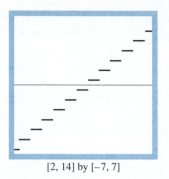

[2, 14] by [−7, 7]

2.22 The graph of a continuous function never "steps" across the x-axis the way this step function does. If we know that a function is continuous and its graph appears to cross the x-axis, then we know that there indeed is an x-intercept.

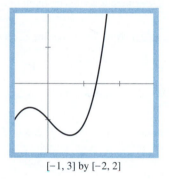

[−1, 3] by [−2, 2]

2.23 The graph of $f = x^3 - x - 1$ crosses the x-axis between $x = 1$ and $x = 2$.

A consequence for solving equations Knowing that a function is continuous contributes to knowing that a graph in a viewing window is complete. Suppose that $f(x)$ is continuous at every point of a closed interval $[a, b]$ and that $f(a)$ and $f(b)$ differ in sign. Because zero lies between $f(a)$ and $f(b)$, there is at least one number c between a and b where $f(c) = 0$. In other words, if f is continuous on $[a, b]$ and $f(a)$ and $f(b)$ differ in sign, then the equation $f(x) = 0$ has at least one solution in the open interval (a, b); the graph of f really has an x-intercept where it clearly crosses the x-axis. This is the principle that underlies our method of solving equations graphically using ZOOM-IN in Section 1.5. (See Fig. 2.22.)

EXAMPLE 15

Is any real number exactly 1 less than its cube?

Solution Any such number must satisfy the equation $x = x^3 - 1$ or, equivalently, $x^3 - x - 1 = 0$. Hence, we are looking for a zero value of the function $f(x) = x^3 - x - 1$. Figure 2.23 suggests that the solution is about 1.3. In Exercise 47, we ask you to find the solution with greater accuracy using ZOOM-IN. ≡

Concluding Remarks

For any function $y = f(x)$, it is important to distinguish between continuity at $x = c$ and having a limit as $x \to c$. The limit, $\lim_{x \to c} f(x)$, is where the function values are heading as $x \to c$. Continuity is the property of arriving at the point where the $f(x)$ has been heading when x actually gets to c. (Someone is home when you get there, so to speak.) If the limit is what you expect as $x \to c$ and the number $f(c)$ is what you get when $x = c$, then the function is continuous at c if you get what you expect.

Finally, remember the test for continuity at a point:

1. Does $f(c)$ exist?
2. Does $\lim_{x \to c} f(c)$ exist?
3. Does $\lim_{x \to c} f(x) = f(c)$?

For f to be continuous at $x = c$, all three answers must be *yes*.

Exercises 2.2 _____

Exercises 1–6 are about the function f defined as follows and whose graph is shown.

$$f(x) = \begin{cases} x^2 - 1, & -1 \le x < 0; \\ 2x, & 0 \le x < 1; \\ 1, & x = 1; \\ -2x + 4, & 1 < x < 2; \\ 0, & 2 < x \le 3. \end{cases}$$

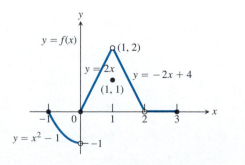

1. a) Does $f(-1)$ exist?

b) Does $\lim_{x \to -1^+} f(x)$ exist?

c) Does $\lim_{x \to -1^+} f(x) = f(-1)$?

d) Is f continuous at $x = -1$?

2. a) Does $f(1)$ exist?

b) Does $\lim_{x \to 1} f(x)$ exist?

c) Does $\lim_{x \to 1} f(x) = f(1)$?

d) Is f continuous at $x = 1$?

3. a) Is f defined at $x = 2$? (Look at the definition of f.)

b) Is f continuous at $x = 2$?

4. At what values of x is f continuous?

5. a) What is the value of $\lim_{x \to 2} f(x)$?

b) Can a function g be defined to make g a continuous extension of f to the point $x = 2$? If so, give g. If not, explain.

6. How should h be defined to make h a continuous extension of f to the point $x = 1$?

At which points are the functions in the following exercises in Section 2.1 continuous?

7. Exercise 39   **8.** Exercise 40

9. Exercise 41   **10.** Exercise 42

11. Exercise 43   **12.** Exercise 44

13. Exercise 47   **14.** Exercise 48

15. Let $f(x) = \begin{cases} 0, & x < 0, \\ 1, & 0 \le x \le 1, \\ 0, & 1 < x. \end{cases}$

a) Determine a complete graph of f.

b) At what points is the function continuous?

16. Let $f(x) = \begin{cases} 1, & x < 0, \\ \sqrt{1 - x^2}, & 0 \le x \le 1, \\ x - 1, & x > 1. \end{cases}$

a) Determine a complete graph of f.

b) Is f continuous? Explain.

Find the points, if any, at which the functions in Exercises 17–30 are *not* continuous.

17. $y = \dfrac{1}{x - 2}$ **18.** $y = \dfrac{1}{(x + 2)^2}$

19. $y = \dfrac{x + 1}{x^2 - 4x + 3}$ **20.** $y = \dfrac{x + 3}{x^2 - 3x - 10}$

21. $y = \dfrac{x^3 - 1}{x^2 - 1}$ **22.** $y = \dfrac{1}{x^2 + 1}$

23. $y = |x - 1|$ **24.** $y = |2x + 3|$

25. $y = \dfrac{\cos x}{x}$ **26.** $y = \dfrac{|x|}{x}$

27. $y = \sqrt{2x + 3}$ **28.** $y = \sqrt[4]{3x - 1}$

29. $y = \sqrt[3]{2x - 1}$ **30.** $y = \sqrt[5]{2 - x}$

31. The function $f(x)$ is defined by $f(x) = (x^2 - 1)/(x - 1)$ when $x \ne 1$ and by $f(1) = 2$. Is f continuous at $x = 1$? Explain.

32. Define $g(3)$ so that $g(x) = (x^2 - 9)/(x - 3)$ is continuous at $x = 3$.

33. Define $h(2)$ so that $h(x) = (x^2 + 3x - 10)/(x - 2)$ is continuous at $x = 2$.

34. Define $f(1)$ so that $f(x) = (x^3 - 1)/(x^2 - 1)$ is continuous at $x = 1$.

35. Define $g(4)$ so that $g(x) = (x^2 - 16)/(x^2 - 3x - 4)$ is continuous at $x = 4$.

36. How should g be defined to make g a continuous extension of f to the point $x = 2$ in Fig. 2.12?

37. What value should be assigned to a to make the function

$$f(x) = \begin{cases} x^2 - 1, & x < 3, \\ 2ax, & x \ge 3, \end{cases}$$

continuous at $x = 3$? Determine a complete graph of f for this value of a.

38. What value should be assigned to b to make the function

$$g(x) = \begin{cases} x^3, & x < 1/2, \\ bx^2, & x \ge 1/2, \end{cases}$$

continuous at $x = 1/2$? Determine a complete graph of g for this value of b.

Use the properties of continuous functions in Theorem 5 and the definition of continuous function. Explain why the limits in Exercises 39–42 exist, and give their values. Support your results with a graphing utility.

39. $\lim_{x \to 0} \sec x$ **40.** $\lim_{x \to 0} \tan x$

41. $\lim_{x \to 0}[(1 + \cos x)/2]$ **42.** $\lim_{x \to 0} \sin\left(\dfrac{\pi}{2} \cos(\tan x)\right)$

In Exercises 43 and 44, investigate the limits using tables of values for $(x, f(x))$ near $x = 0$. On the basis of the tables, state what you believe each limit to be. Support your answer with a graphing utility.

43. $\lim_{x \to 0} \cos\left(1 - \dfrac{\sin x}{x}\right)$ **44.** $\lim_{x \to 0} \sin\left(\dfrac{1 - \cos x}{x}\right)$

45. Let $f(x) = x^3 + 4$. Find c in $[-2, 4]$ for which $f(c) = 2$.

46. Let $f(x) = 2 - x^3$. Find c in $[-2, 2]$ for which $f(c) = 5$.

47. Let $f(x) = x^3 - x - 1$ (see Example 15).

a) Determine the solution to $f(x) = 0$ with an error of at most 10^{-8}.

b) It can be shown that the exact solution in part (a) is

$$\left(\dfrac{\sqrt{69}}{18} + \dfrac{1}{2}\right)^{1/3} + \left(\dfrac{1}{2} - \dfrac{\sqrt{69}}{18}\right)^{1/3}.$$

Evaluate this answer with a calculator, and compare with the value determined in part (a).

48. Let $f(x) = x^3 - 2x + 2$.
    a) Determine the solution to $f(x) = 0$ with an error of at most 10^{-4}.
    b) It can be shown that the exact solution in part (a) is

$$\left(\frac{\sqrt{57}}{9} - 1\right)^{1/3} - \left(\frac{\sqrt{57}}{9} + 1\right)^{1/3}.$$

    Evaluate this answer with a calculator, and compare with the value determined in part (a).

49. At what values of x (if any) does the function in Fig. 2.12 take on its maximum value? Does the function take on a minimum value? Explain.

50. At what values of x (if any) does the function in Fig. 2.23 take on a maximum value? A minimum value?

51. Does the function $y = x^2$ have a maximum value on the open interval $-1 < x < 1$? A minimum value? Explain.

52. On the closed interval $0 \le x \le 1$, the greatest integer function $y = [x]$ takes on a minimum value $m = 0$ and a maximum value $M = 1$. It does so even though it is discontinuous at $x = 1$. Does this violate the Max-Min Theorem? Why?

53. A continuous function $y = f(x)$ is known to be negative at $x = 0$ and positive at $x = 1$. Why does the equation $f(x) = 0$ have at least one solution between $x = 0$ and $x = 1$? Illustrate with a sketch.

54. Assuming $y = \cos 3x$ to be continuous, show that the equation $\cos 3x = x$ has at least one solution. (*Hint:* Show that the equation $\cos 3x - x = 0$ has at least one solution.)

55. Show that $e^{-x} = x$ has at least one solution.

56. *Group discussion.* You hear a calculus teacher say, "A one-point gap on a graph is *not* visible." Write a paragraph explaining what you think the teacher means. Explain why you sometimes see a one-point gap on a grapher.

2.3

The Sandwich Theorem and (sin θ)/θ

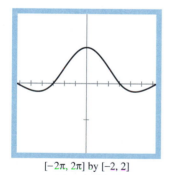

$[-2\pi, 2\pi]$ by $[-2, 2]$

2.24 The graph suggests that

$$\lim_{\theta \to 0} \frac{\sin\theta}{\theta} = 1,$$

even though $(\sin\theta)/\theta$ is undefined at $\theta = 0$.

A useful fact in calculus is that $\lim_{\theta \to 0}(\sin\theta)/\theta = 1$ when θ is measured in radians. (Unless explicitly stated otherwise, trigonometric arguments will be in radians.) This beautiful and simple result turns out to be the key to measuring the rates at which all trigonometric functions of θ change their values as θ changes, as we shall see in Chapter 3.

    Figure 2.24 provides support that $\lim_{\theta \to 0}(\sin\theta)/\theta = 1$. To confirm it, we introduce and apply a powerful theorem called the Sandwich Theorem.

THEOREM 9   **The Sandwich Theorem**

Suppose that

$$g(x) \le f(x) \le h(x)$$

for all $x \ne c$ in some interval about c and that

$$\lim_{x \to c} g(x) = \lim_{x \to c} h(x) = L.$$

Then

$$\lim_{x \to c} f(x) = L.$$

    We have included a proof of the Sandwich Theorem in Appendix 2. The idea of this theorem is that if the values of f are sandwiched between the values of two functions that approach L, then the values of f also approach L (Fig. 2.25).

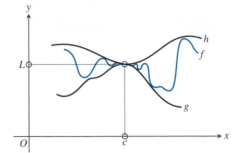

2.25 The Sandwich Theorem. Sandwiching f between g and h forces the limiting value of f to be between the limiting values of g and h.

THE SANDWICH THEOREM IN ACTION, PART 1

GRAPH

$$y_1 = (\sin \theta)/\theta,$$

$$y_2 = \cos \theta, \text{ and}$$

$$y_3 = 1$$

simultaneously.

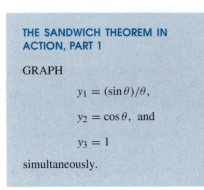

$$\text{Sector area} = \frac{\theta}{2\pi} \cdot \text{Circle area}$$

$$= \frac{\theta}{2\pi} \cdot \pi r^2$$

$$= \frac{1}{2} r^2 \theta \quad \text{(Usual form)}$$

$$= \frac{\theta}{2} \quad \text{(If } r = 1)$$

2.26 When θ is measured in radians (and not degrees), the formula for the area of a sector of a unit circle is $A = \theta/2$, as shown above.

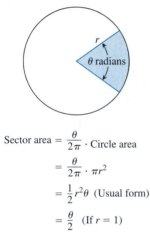

2.27 Area $\triangle OAP <$ area sector $OAP <$ area $\triangle OAT$.

In the proof that $\lim_{\theta \to 0}(\sin \theta)/\theta = 1$, we draw on a formula from geometry that says that the area cut from a unit circle by a central angle of θ radians is $\theta/2$. Figure 2.26 shows where this formula comes from.

We also use the fact that $\cos \theta$ approaches 1 as θ approaches 0, which follows from $\cos \theta$ being a continuous function, hence $\lim_{\theta \to 0} \cos \theta = \cos 0 = 1$. Then we sandwich $(\sin \theta)/\theta$ between the functions 1 and $\cos \theta$—two functions that approach 1 as θ approaches 0. The Sandwich Theorem tells us that $(\sin \theta)/\theta$ must approach 1 as well.

Proof That $\lim_{\theta \to 0} \dfrac{\sin \theta}{\theta} = 1$ Our plan is to show that as $\theta \to 0$, the right-hand and left-hand limits are both 1. We will then know that the two-sided limit is 1 as well.

To show that the right-hand limit is 1, we begin with the values of θ that are positive and less than $\pi/2$ (Fig. 2.27). We compare the areas of $\triangle OAP$, sector OAP, and $\triangle OAT$ and note that

$$\text{Area } \triangle OAP < \text{Area sector } OAP < \text{Area } \triangle OAT.$$

We can express these areas in terms of θ as follows:

$$\text{Area } \triangle OAP = \frac{1}{2}\text{base} \times \text{height} = \frac{1}{2}(1)(\sin \theta) = \frac{1}{2}\sin \theta,$$

$$\text{Area sector } OAP = \frac{1}{2}r^2\theta = \frac{1}{2}(1)^2\theta = \frac{\theta}{2} \quad \text{(Fig. 2.26)},$$

$$\text{Area } \triangle OAT = \frac{1}{2}\text{base} \times \text{height} = \frac{1}{2}(1)(\tan \theta) = \frac{1}{2}\tan \theta,$$

so that

$$\frac{1}{2}\sin \theta < \frac{1}{2}\theta < \frac{1}{2}\tan \theta.$$

We multiply the three terms by the positive quantity $2/\sin \theta$:

$$1 < \frac{\theta}{\sin \theta} < \frac{1}{\cos \theta}.$$

Next we take reciprocals, which reverses the inequalities:

$$\cos \theta < \frac{\sin \theta}{\theta} < 1.$$

Because $\cos\theta$ approaches 1 as θ approaches 0, the Sandwich Theorem tells us that

$$\lim_{\theta\to 0^+}\frac{\sin\theta}{\theta}=1. \qquad (1)$$

This limit is a right-hand limit because we have been dealing with values of θ between 0 and $\pi/2$, but we obtain the same limit for $(\sin\theta)/\theta$ as θ approaches 0 from the left. For if $\theta=-\alpha$ and α is positive, then

$$\frac{\sin\theta}{\theta}=\frac{\sin(-\alpha)}{-\alpha}=\frac{-\sin(\alpha)}{-\alpha}=\frac{\sin\alpha}{\alpha}.$$

Therefore,

$$\lim_{\theta\to 0^-}\frac{\sin\theta}{\theta}=\lim_{\alpha\to 0^+}\frac{\sin\alpha}{\alpha}=1. \qquad (2)$$

Together, Eqs. (1) and (2) imply that $\lim_{\theta\to 0}(\sin\theta)/\theta=1$ to conclude the proof. ▤

Because $f(\theta)=(\sin\theta)/\theta$ is undefined at $\theta=0$, f is discontinuous there. But because $\lim_{\theta\to 0}(\sin\theta)/\theta=1$, the discontinuity is removable. Thus, the extended function

$$F(\theta)=\begin{cases}\dfrac{\sin\theta}{\theta}, & \theta\neq 0,\\[2mm] 1, & \theta=0,\end{cases}$$

is continuous at every real number θ.

Knowing the limit of $(\sin\theta)/\theta$ and that $F(\theta)$ is continuous will be very useful in calculating a number of related limits.

EXAMPLE 1

If $a\neq 0$,

$$\lim_{x\to 0}\frac{\sin ax}{ax}=\lim_{x\to 0}F(ax) \qquad \text{A composite function}$$

$$=F(a\cdot 0) \qquad \begin{array}{l}\text{A composite of continuous}\\\text{functions is continuous.}\end{array}$$

$$=F(0)$$

$$=1 \qquad\qquad\qquad ▤$$

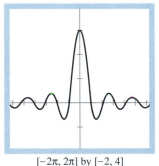

$[-2\pi, 2\pi]$ by $[-2, 4]$

2.28 The graph of $f(x)=(\sin 3x)/x$ suggests that $f(x)\to 3$ as $x\to 0$.

EXAMPLE 2

Find $\lim_{x\to 0}(\sin 3x)/x$ graphically. Confirm algebraically.

Solution The graph of $f(x)=(\sin 3x)/x$ in Fig. 2.28 suggests that $\lim_{x\to 0}(\sin 3x)/x=3$. Algebraically,

$$\lim_{x\to 0}\frac{\sin 3x}{x}=\lim_{x\to 0}3\frac{\sin 3x}{3x} \qquad \text{Multiply by 3/3}$$

$$=3\cdot\lim_{x\to 0}\frac{\sin 3x}{3x} \qquad \text{Constant Multiple Rule}$$

$$=3\cdot 1 \qquad\qquad\qquad \text{Example 1}$$

$$=3 \qquad\qquad\qquad\qquad ▤$$

EXAMPLE 3

Find $\lim_{x \to 0} \dfrac{\tan x}{x}$ graphically. Confirm algebraically.

Solution The graph of $f(x) = (\tan x)/x$ (Fig. 2.29) suggests that $\lim_{x \to 0}(\tan x)/x = 1$.

$$\lim_{x \to 0} \frac{\tan x}{x} = \lim_{x \to 0} \frac{\sin x}{x} \frac{1}{\cos x} \qquad \tan x = \frac{\sin x}{\cos x}$$

$$= \lim_{x \to 0} \frac{\sin x}{x} \lim_{x \to 0} \frac{1}{\cos x} \qquad \text{Property of Limits}$$

$$= 1 \cdot 1$$

$$= 1 \qquad \blacksquare$$

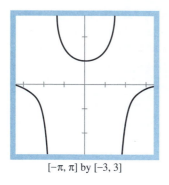

$[-\pi, \pi]$ by $[-3, 3]$

2.29 The graph of $f(x) = (\tan x)/x$ suggests that $f(x) \to 1$ as $x \to 0$.

EXPLORATION 1

Confirm Algebraically

Find each limit graphically. Then confirm algebraically using the limit of $(\sin \theta)/\theta$.

1. $\lim_{x \to 0} \dfrac{\sin 2x}{\sin 3x}$

2. $\lim_{x \to 0} \dfrac{\sin 5x}{\sin 8x}$

3. Make a conjecture about $\lim_{x \to 0}(\sin ax / \sin bx)$. Then prove your conjecture true or find a counterexample to prove it false.

Exercises 2.3

Find the limits in Exercises 1–16 graphically and confirm algebraically.

1. $\lim\limits_{x \to 0} \dfrac{1}{\cos x}$

2. $\lim\limits_{x \to 0} (2 \sin x + 3 \cos x)$

3. $\lim\limits_{x \to 0} \dfrac{1 + \sin x}{1 + \cos x}$

4. $\lim\limits_{x \to 0} \dfrac{x^2 + 1}{1 - \sin x}$

5. $\lim\limits_{x \to 0} \dfrac{x}{\sin x}$

6. $\lim\limits_{x \to 0} \dfrac{x}{\tan x}$

7. $\lim\limits_{x \to 0} \dfrac{\sin 2x}{x}$

8. $\lim\limits_{x \to 0} \dfrac{x}{\sin 3x}$

9. $\lim\limits_{x \to 0} \dfrac{\tan 2x}{2x}$

10. $\lim\limits_{x \to 0} \dfrac{\tan 2x}{x}$

11. $\lim\limits_{x \to 0} \dfrac{\sin x}{2x^2 - x}$

12. $\lim\limits_{x \to 0} \dfrac{x + \sin x}{x}$

13. $\lim\limits_{x \to 0} \dfrac{\sin^2 x}{x}$

14. $\lim\limits_{x \to 0} \dfrac{\tan^2 x}{2x}$

15. $\lim\limits_{x \to 0} \dfrac{3 \sin 4x}{\sin 3x}$

16. $\lim\limits_{x \to 0} \dfrac{\tan 5x}{\tan 2x}$

17. Let $f(x) = \dfrac{\tan 3x}{\sin 5x}$.

a) Estimate $\lim_{x \to 0} f(x)$ graphically. (*Hint:* ZOOM-IN around $x = 0$).

b) Compare the function values of f near 0 to the limit found in part (a).

c) Find the exact value of $\lim_{x \to 0} f(x)$ algebraically.

18. Let $f(x) = \dfrac{\cot 3x}{\csc 2x}$.

a) Estimate $\lim_{x \to 0} f(x)$ graphically. (*Hint:* ZOOM-IN around $x = 0$).

b) Compare the function values of f near 0 to the limit found in part (a).

c) Find the exact value of $\lim_{x \to 0} f(x)$ algebraically.

19. *Sandwich Theorem.* The inequality

$$1 - \frac{x^2}{6} < \frac{\sin x}{x} < 1$$

holds when x is measured in radians and $-1 < x < 1$. Use this inequality to calculate $\lim_{x \to 0}(\sin x)/x$. In Chapter 9 we shall see where this inequality comes from.

Sandwich Theorem. As we saw in the proof that $\lim_{\theta \to 0}(\sin \theta)/\theta = 1$ and again in Exercise 19, we can sometimes use the Sandwich Theorem to find the limit of a fraction whose numerator and denominator both approach zero. Exercises 20–22 contain three other examples in which the inequalities come from infinite series (Chapter 9) and are true for radian values of x close to zero. In each exercise:

a) Apply the Sandwich Theorem graphically by graphing the three functions in the inequality simultaneously.

b) Give the value of $\lim_{x \to 0} f(x)$ suggested by the graphs.

c) Confirm the limit analytically by applying the Sandwich Theorem.

20. $1 - \dfrac{x^2}{6} < f(x) = \dfrac{x \sin x}{2 - 2\cos x} < 1$

21. $\dfrac{1}{2} - \dfrac{x^2}{24} < f(x) = \dfrac{1 - \cos x}{x^2} < \dfrac{1}{2}$

22. $1 < f(x) = \dfrac{\tan x}{x} < 1 + x^2$

23. Show that $\lim_{x \to 0}(\cos x)/x$ does not exist.

24. The area formula $A = (1/2)r^2\theta$ derived in Fig. 2.26 for radian measure has to be changed if the angle is measured in degrees. What should the new formula be?

25. Find $\lim_{\theta \to 0}(\sin \theta)/\theta$ if θ is measured in degrees.

Investigate $\lim_{x \to 0} f(x)$ in Exercises 26–29 by making tables of values. On the basis of the tables, state what you believe the limit to be. Support graphically.

26. $f(x) = \dfrac{\sin 2x}{x}$

27. $f(x) = \dfrac{\tan 3x}{x}$

28. $f(x) = \dfrac{\cos x - 1}{x}$

29. $f(x) = \dfrac{x - \sin x}{x^2}$

30. Give a geometric proof (similar to the one given for $\lim_{\theta \to 0}\dfrac{\sin \theta}{\theta}$) to show that $\lim_{\theta \to 0}\dfrac{1 - \cos \theta}{\theta} = 0$.

2.4

Limits Involving Infinity

Although there is no real number *infinity*, the word *infinity* is useful for describing how some functions behave when their domains or ranges exceed all bounds. In this section we describe what it means for the values of a function to approach infinity and what it means for a function $f(x)$ to have a limit as x approaches infinity. Our presentation continues to be informal. In Section 2.6 we define the limits involving infinity more precisely.

THE SYMBOL ∞

The symbol ∞, read "infinity," does not represent any real number. We cannot use ∞ in arithmetic in the usual way, but it is convenient to be able to say things like "the limit of $1/x$ as x approaches infinity is 0." Here, "x approaches infinity" means that x *increases without bound*.

Functions with Finite Limits as $x \to \pm\infty$

Our strategy is again the one that we used in Section 2.1. We find the limits of two "basic" functions as $x \to \infty$ and $x \to -\infty$, and then, to find everything else, we use a theorem about limits of algebraic combinations. In Section 2.1 the basic functions were the constant function $y = k$ and the identity function $y = x$. Here, the basic functions are $y = k$ and the reciprocal function $y = 1/x$.

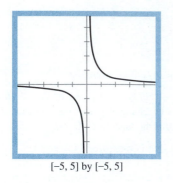

[−5, 5] by [−5, 5]

2.30 The graph of $f(x) = 1/x$.

EXAMPLE 1

$$\lim_{x \to \infty} \frac{1}{x} = 0 \qquad \text{and} \qquad \lim_{x \to -\infty} \frac{1}{x} = 0.$$

The function

$$f(x) = \frac{1}{x}$$

is defined for all real numbers except $x = 0$. As its graph suggests (Fig. 2.30),

a) $1/x \to 0$ as $x \to \infty$ in the following sense: No matter how small a positive number E you name, we can find a value x_0 large enough so that for all x to the right of x_0, the values of $1/x$ will be smaller than E.

b) $1/x \to 0$ as $x \to -\infty$ in the following sense: No matter how small a positive number E you name, we can find a value x_0 large enough so that for all x to the left of $-x_0$, the values of $|1/x|$ will be smaller than E.

We summarize these facts by writing:

a) as $x \to \infty$, $1/x \to 0$ and $\lim_{x \to \infty} \dfrac{1}{x} = 0$.

b) as $x \to -\infty$, $1/x \to 0$ and $\lim_{x \to -\infty} \dfrac{1}{x} = 0$.

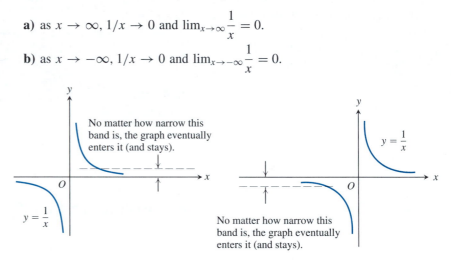

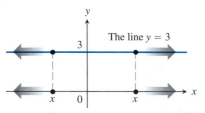

2.31 The values of $f(x) = 3$ remain constant no matter how large $|x|$ becomes.

EXAMPLE 2

If $f(x) = 3$ is the constant function whose outputs have the constant value 3 (Fig. 2.31), then

$$\lim_{x \to \infty} f(x) = \lim_{x \to \infty} (3) = 3$$

$$\lim_{x \to -\infty} f(x) = \lim_{x \to -\infty} (3) = 3.$$

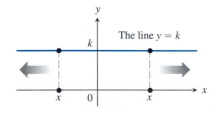

2.32 The values of $f(x) = k$ remain constant no matter how large $|x|$ becomes.

EXAMPLE 3

What we saw in Example 2 holds for any number k: If f is the constant function with $f(x) = k$ (Fig. 2.32), then

$$\lim_{x \to \infty} f(x) = \lim_{x \to \infty} (k) = k$$

$$\lim_{x \to -\infty} f(x) = \lim_{x \to -\infty} (k) = k.$$

Believe it or not, we now have enough specific information to calculate the limits of a wide variety of rational functions as $x \to \pm\infty$. To do so, we simply use the limit properties listed in the following theorem.

THEOREM 10

a) If f is the constant function $f(x) = k$, then

$$\lim_{x \to \infty} f(x) = \lim_{x \to \infty} (k) = k.$$

b) If f is the reciprocal function $f(x) = 1/x$, then

$$\lim_{x \to \infty} f(x) = \lim_{x \to \infty} \frac{1}{x} = 0.$$

c) Properties of Finite Limits as $x \to \infty$ If L_1 and L_2 are real numbers and $\lim_{x \to \infty} f(x) = L_1$ and $\lim_{x \to \infty} g(x) = L_2$, then

1. *Sum Rule:* $\qquad\qquad\qquad \lim_{x \to \infty} [f(x) + g(x)] = L_1 + L_2.$

2. *Difference Rule:* $\qquad\quad\; \lim_{x \to \infty} [f(x) - g(x)] = L_1 - L_2.$

3. *Product Rule:* $\qquad\qquad\; \lim_{x \to \infty} f(x) \cdot g(x) = L_1 \cdot L_2.$

4. *Constant Multiple Rule:* $\lim_{x \to \infty} k \cdot f(x) = k \cdot L_1$ (any number k).

5. *Quotient Rule:* $\lim_{x \to \infty} \dfrac{f(x)}{g(x)} = \dfrac{L_1}{L_2}$ (provided $L_2 \neq 0$).

These properties hold for $x \to -\infty$ as well.

These properties are just like the properties we stated in Section 2.1 for limits as $x \to c$, and we use them the same way.

EXAMPLE 4

$$\lim_{x \to \infty} \left(5 + \frac{1}{x} \right) = \lim_{x \to \infty} 5 + \lim_{x \to \infty} \frac{1}{x} \qquad \text{Sum Rule}$$

$$= 5 + 0 = 5. \qquad \text{Theorems 10(a), 10(b)}$$

EXAMPLE 5

$$\lim_{x \to -\infty} \frac{4}{x^2} = \lim_{x \to -\infty} 4 \cdot \lim_{x \to -\infty} \frac{1}{x} \cdot \lim_{x \to -\infty} \frac{1}{x} \qquad \text{Product Rule}$$

$$= 4 \cdot 0 \cdot 0 = 0. \qquad \text{Theorems 10(a), 10(b)}$$

THE SYMBOL $\pm\infty$

Using "$\pm\infty$" in a sentence gives us two statements "for the price of one." For example,

$$\lim_{x \to \pm\infty} f(x) = L$$

says that

$$\lim_{x \to \infty} f(x) = L \; and \; \lim_{x \to -\infty} f(x) = L.$$

EXAMPLE 6

$$\lim_{x \to \pm\infty} \frac{\sin x}{x} = 0$$

Notice first that $\sin x$ lies between -1 and 1. Therefore, for all positive values of x,

$$-\frac{1}{x} \leq \frac{\sin x}{x} \leq \frac{1}{x},$$

and for all negative values of x,

$$-\frac{1}{x} \geq \frac{\sin x}{x} \geq \frac{1}{x}.$$

THE SANDWICH THEOREM IN ACTION, PART 2

The Sandwich Theorem, stated in Section 2.3 for limits as $x \to c$, also holds for limits as $x \to \pm\infty$. You can see this version of the Sandwich Theorem in action if you GRAPH

$$y_1 = (\sin x)/x,$$

$$y_2 = -(1/x), \text{ and}$$

$$y_3 = 1/x$$

simultaneously.

Because $-1/x$ and $1/x$ both approach 0 as $x \to \infty$, a modified version of the Sandwich Theorem tells us that $(\sin x)/x$ approaches 0 as well (Fig. 2.33).

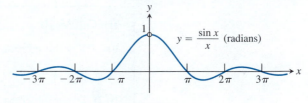

NOT TO SCALE

2.33 The graph of $y = (\sin x)/x$ oscillates about the x-axis. The amplitude of the oscillation decreases toward zero as $x \to \pm\infty$.

Lim $f(x) = \infty$ or lim $f(x) = -\infty$

As is suggested by the behavior of $1/x$ as $x \to 0$ (Fig. 2.34) or x^2 as $x \to \infty$, we sometimes want to say such things as

$$\lim_{x \to c} f(x) = \infty, \qquad \lim_{x \to c^+} f(x) = \infty, \qquad \lim_{x \to c^-} f(x) = \infty,$$

$$\lim_{x \to \infty} f(x) = \infty, \qquad \lim_{x \to -\infty} f(x) = \infty.$$

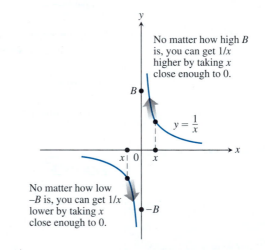

No matter how high B is, you can get $1/x$ higher by taking x close enough to 0.

No matter how low $-B$ is, you can get $1/x$ lower by taking x close enough to 0.

2.34 As $x \to 0^+$ and $x \to 0^-$, the limits of $f(x) = (1/x)$ are ∞ and $-\infty$, respectively.

None of the limits exist in the sense of Section 2.1. However, in every instance, we mean that the value of $f(x)$ eventually exceeds any positive real number B. That is, for any real number B, no matter how large, the values of f eventually satisfy the condition

$$f(x) > B.$$

Similarly, we write

$$\lim_{x \to c} f(x) = -\infty, \qquad \lim_{x \to c^+} f(x) = -\infty, \qquad \lim_{x \to c^-} f(x) = -\infty$$

to say that no matter how large the absolute value of the negative number $-B$ may be, the values of f eventually satisfy the condition

$$f(x) < -B.$$

EXAMPLE 7

$\lim_{x \to 0^+} 1/x = \infty$ because for any positive number B, no matter how great, we can find positive values of x so close to 0 that $1/x$ will be greater than B. Similarly, $\lim_{x \to 0^-} 1/x = -\infty$ because for any positive number B, no matter how large, we can find negative values of x so close to 0 that $1/x$ will be less than $-B$ (Fig. 2.34). ≡

End Behavior Models

We call the function $y = 0$ an end behavior model for $f(x) = 1/x$ in the sense that $y = 0$ is a simpler function that behaves virtually the same way as f for $|x|$ large.

LIMITS OF AND AT INFINITY

You should be able to describe informally the meanings of

a) these two statements:

$$\lim_{x \to \pm\infty} f(x) = L;$$

b) these two statements:

$$\lim_{x \to c} f(x) = \pm\infty;$$

c) and these four statements:

$$\lim_{x \to \pm\infty} f(x) = \pm\infty.$$

> **DEFINITION**
>
> The function g is an **end behavior model** of the function f if
>
> **1.** $\lim_{x \to \pm\infty} f/g = 1$ when $g(x) \neq 0$ for $|x|$ large, or
>
> **2.** $\lim_{x \to \pm\infty} f(x) = 0$ when $g(x) = 0$.

EXAMPLE 8

From Example 6 we see that $g(x) = 0$ is an end behavior model for the function $f(x) = (\sin x)/x$. (See Fig. 2.33.) ≡

LEFT- AND RIGHT-END BEHAVIOR

If $\lim_{x \to \infty} f(x)$ and $\lim_{x \to -\infty} f(x)$ are different, we can extend the idea of end behavior model to those of left-end behavior model and right-end behavior model.

Polynomial Function End Behavior

Polynomial functions $f(x) = a_n x^n + a_{n-1} x^{n-1} + \cdots + a_1 x + a_0$ of degree 1 or higher have the property that

$$\lim_{x \to \infty} f(x) = \infty \text{ or } -\infty,$$

and

$$\lim_{x \to -\infty} f(x) = \infty \text{ or } -\infty.$$

It turns out that for $|x|$ large, the values of $f(x)$ and its leading term $a_n x^n$ are approximately the same in the sense that their ratio is close to 1. More precisely, $y = a_n x^n$ is an end behavior model of f, that is,

$$\lim_{x \to \pm\infty} \frac{f(x)}{a_n x^n} = 1 \qquad (a_n \neq 0).$$

REMINDER

The *degree* of $a_n x^n + a_{n-1} x^{n-1} + \cdots + a_1 x + a_0$, $a_n \neq 0$, is n, the largest exponent. The *leading coefficient* is a_n. The *leading term* is $a_n x^n$.

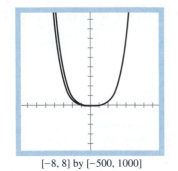

[−8, 8] by [−500, 1000]

2.35 $y = 3x^4$ is an end behavior model of $f(x) = 3x^4 - 2x^3 + 3x^2 - 5x + 6$.

EXAMPLE 9

Show that $y = 3x^4$ is an end behavior model of
$f(x) = 3x^4 - 2x^3 + 3x^2 - 5x + 6$.

Solution   The graphs of $y = 3x^4$ and f in Fig. 2.35 are nearly identical. This is visual evidence that the end behavior of f and $y = 3x^4$ are the same. Algebraically,

$$\lim_{x \to \pm\infty} \frac{f(x)}{3x^4} = \lim_{x \to \pm\infty} \frac{3x^4 - 2x^3 + 3x^2 - 5x + 6}{3x^4}$$

$$= \lim_{x \to \pm\infty} \left(1 - \frac{2}{3x} + \frac{1}{x^2} - \frac{5}{3x^3} + \frac{2}{x^4} \right)$$

$$= 1. \qquad \blacksquare$$

The algebraic techniques of Example 9 and mathematical induction can be used to establish the following theorem.

THEOREM 11   Polynomial End Behavior Model

If $f(x) = a_n x^n + a_{n-1} x^{n-1} + \cdots + a_1 x + a_0 \ (a_n \neq 0)$, then $y = a_n x^n$ is an end behavior model of f.

For $n \geq 1$, there are really only four types of polynomial end behavior models:

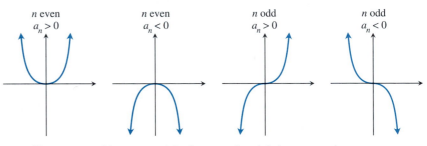

The steepness of the curves and the flatness at the origin increase as n increases.

EXPLORATION 1

Understanding End Behavior

GRAPH $y_1 = 2x^2 + 3x$ and its end behavior model $y_2 = 2x^2$ in the same viewing window. Adjust the view dimensions, or use TRACE, to see large x and y coordinates and decide whether the two functions are "alike" as $x \to \infty$. Use TRACE and the cursor control to compare values of the two functions as $x \to \infty$.

1. Do the two functions get "farther apart"? Consider $y_1 - y_2$. Confirm algebraically and support graphically that the two functions are getting farther apart at each x when $x \to \infty$.

2. Do the two functions get "closer together"? Consider y_1/y_2. Confirm algebraically and support graphically that the two functions are getting closer together as $x \to \infty$.

3. Express your understanding of "end behavior model" in your own words.

&

EXPLORATION BIT

For horizontal asymptotes, we can use TRACE to explore the values of a function as $x \to \pm\infty$. Can we use TRACE to explore for vertical asymptotes? Explain.

Horizontal and Vertical Asymptotes

If the graph of a function appears indistinguishable from a line as we move away from the origin, we say that the line is an *asymptote* of the graph and the graph *approaches the line asymptotically*. For the function $f(x) = 1/x$, both axes are asymptotes of the graph (Fig. 2.36).

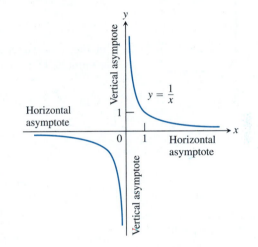

2.36 The coordinate axes are asymptotes of both branches of the graph of $y = 1/x$.

DEFINITION

A line $y = b$ is a **horizontal asymptote** of the graph of a function $y = f(x)$ if either

$$\lim_{x \to \infty} f(x) = b \qquad \text{or} \qquad \lim_{x \to -\infty} f(x) = b.$$

A line $x = a$ is a **vertical asymptote** of the graph of a function $y = f(x)$ if either

$$\lim_{x \to a^-} f(x) = \pm\infty \qquad \text{or} \qquad \lim_{x \to a^+} f(x) = \pm\infty.$$

EXAMPLE 10

The graph of $y = \tan^{-1} x$ in Fig. 2.37 suggests that a graph can have two horizontal asymptotes. The asymptotes $g(x) = -\pi/2$ and $h(x) = \pi/2$ also happen to be left- and right-end behavior models, respectively, for $f(x) = \tan^{-1} x$.

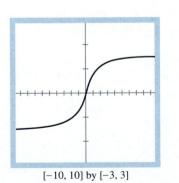

[−10, 10] by [−3, 3]

2.37 The function $y = \tan^{-1} x$ has two horizontal asymptotes.

Asymptotic Behavior of Rational Functions

Rational functions (ratios of polynomials) provide many examples of asymptotic behavior. To find horizontal asymptotes, we check how the graphs leave the viewing window to the right and left ($x \to \pm\infty$). To find vertical asymptotes, we look for the graphs to go off the screen at the top or bottom ($f(x) \to \pm\infty$).

We can study rational functions systematically by comparing the degrees of the numerator and denominator.

EXAMPLE 11   A Rational Function in Which the Numerator and Denominator Have the Same Degree

The graph of

$$f(x) = \frac{-x}{7x + 4}$$

in Fig. 2.38 suggests a horizontal asymptote of $y = -0.14$ and a vertical asymptote of $x = -0.57$. Algebraically,

$$\lim_{x \to \pm\infty} \frac{-x}{7x + 4} = \lim_{x \to \pm\infty} \frac{-1}{7 + (4/x)}$$

> Divide the numerator and denominator by the highest power of x in the denominator, in this case x.

$$= \frac{-1}{7 + 0}$$

> Theorem 10

$$= -\frac{1}{7}$$

$$= -0.142\ldots$$

Because $\lim_{x \to +\infty} f(x) = -1/7$, the line $y = -1/7$ is a horizontal asymptote. Notice also that $7x + 4 = 0$ when $x = -4/7 (\approx -0.57)$, the function values are unbounded as $x \to -4/7$, and the line $x = -4/7$ is a vertical asymptote. ▣

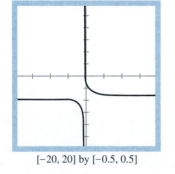

[−20, 20] by [−0.5, 0.5]

2.38 The values of

$$f(x) = \frac{-x}{7x + 4}$$

appear to approach a limit as $x \to \infty$ and as $x \to -\infty$. Using TRACE, we estimate the limit to be −0.142.

EXAMPLE 12   Another Rational Function in Which the Numerator and Denominator Have the Same Degree

The graph of $f(x) = (2x^2 - x + 3)/(3x^2 + 5)$ in Fig. 2.39 suggests that $\lim_{x \to \infty} f(x)$ and $\lim_{x \to -\infty} f(x)$ are both about 0.65. Algebraically,

$$\lim_{x \to \pm\infty} \frac{2x^2 - x + 3}{3x^2 + 5} = \lim_{x \to \pm\infty} \frac{2 - (1/x) + (3/x^2)}{3 + (5/x^2)}$$

> Divide the numerator and denominator by x^2.

$$= \frac{2 - 0 + 0}{3 + 0}$$

> Theorem 10

$$= \frac{2}{3}$$

$$= 0.66\ldots$$

Because $\lim_{x \to \pm\infty} f(x) = 2/3$, the line $y = 2/3$ is a horizontal asymptote. Notice that there are no vertical asymptotes because the denominator, $3x^2 + 5$, can never equal 0 for any real value of x. ▣

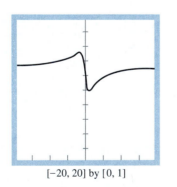

[−20, 20] by [0, 1]

2.39 The values of $f(x) = \dfrac{2x^2 - x + 3}{3x^2 + 5}$
appear to approach a limit as $x \to \infty$ and as $x \to -\infty$. TRACE can help us to estimate the limit.

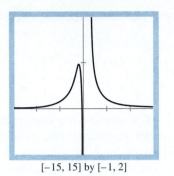

2.40 The values of $f(x) = \dfrac{5x + 2}{2x^3 - 1}$ appear to approach 0 as $x \to \infty$ and as $x \to -\infty$.

EXAMPLE 13   A Rational Function in Which the Degree of the Numerator Is Less Than the Degree of the Denominator

The graph of $f(x) = (5x+2)/(2x^3-1)$ in Fig. 2.40 suggests that $\lim_{x \to \infty} f(x) = \lim_{x \to -\infty} f(x) = 0$. Algebraically,

$$\lim_{x \to \pm\infty} \frac{5x + 2}{2x^3 - 1} = \lim_{x \to \pm\infty} \frac{(5/x^2) + (2/x^3)}{2 - (1/x^3)} \qquad \text{Divide the numerator and denominator by } x^3.$$

$$= \frac{0 + 0}{2 - 0} \qquad \text{Theorem 10}$$

$$= 0$$

The line $y = 0$ (x-axis) is a horizontal asymptote.   ≡

In Example 13 the graph suggests there may be a vertical asymptote somewhere between $x = 0$ and $x = 2$. We can search for it algebraically as follows.

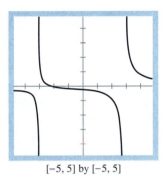

2.41 The graph of

$$f(x) = \frac{x^2 - 4}{x^3 - 2x^2 - 9x + 18}$$

appears to have $x = -3$ and $x = 3$ as vertical asymptotes.

EXPLORATION 2

Finding Vertical Asymptotes

For a rational function, vertical asymptotes occur at points where the denominator is 0. The graph of $f(x) = (x^2 - 4)/(x^3 - 2x^2 - 9x + 18)$ (see Fig. 2.41) suggests vertical asymptotes $x = -3$ and $x = 3$. This in turn suggests that we look for $x + 3$ and $x - 3$ to be factors of the denominator:

$$f(x) = \frac{x^2 - 4}{x^3 - 2x^2 - 9x + 18} = \frac{x^2 - 4}{(x + 3)(x - 3)(x - 2)}.$$

The factorization of the denominator shows that $x = 2$ is not in the domain of f. However, the graph suggests that there is a removable discontinuity at $x = 2$, which we can see algebraically by recognizing that $x - 2$ is a factor of $x^2 - 4$:

$$\frac{x^2 - 4}{(x + 3)(x - 3)(x - 2)} = \frac{(x + 2)(x - 2)}{(x + 3)(x - 3)(x - 2)} = \frac{x + 2}{(x + 3)(x - 3)}$$

(when $x \neq 2$).

1. For f defined above, tell how to remove the discontinuity at $x = 2$.

For f defined above, give a convincing argument for each of the following. (It may help to think of

$$\frac{x + 2}{(x + 3)(x - 3)} \qquad \text{as} \qquad \frac{x + 2}{x + 3} \cdot \frac{1}{x - 3} \qquad \text{and as} \qquad \frac{1}{x + 3} \cdot \frac{x + 2}{x - 3}$$

or to make tables of values of f.)

2. $\lim_{x \to 3^+} f(x) = \infty$ **3.** $\lim_{x \to 3^-} f(x) = -\infty$

4. $\lim_{x \to -3^+} f(x) = \infty$ **5.** $\lim_{x \to -3^-} f(x) = -\infty$

Thus, if the numerator and denominator of a rational function have no common factors, then factors of the form $x - a$ of the denominator correspond

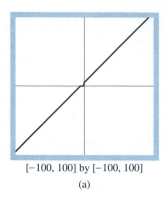

[−100, 100] by [−100, 100]

(a)

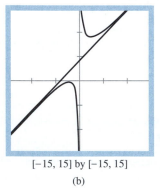

[−15, 15] by [−15, 15]

(b)

2.42 (a) The large viewing window shows that $y = x$ is an *end behavior model* of

$$f(x) = \frac{x^2 + 4x + 5}{x}.$$

(b) The close-up view shows the *end behavior asymptote* $y = x + 4$. The graph of f approaches the line $x + 4$ as $x \to \pm \infty$.

to vertical asymptotes of the form $x = a$. Find the vertical asymptotes for these rational functions.

6. $f(x) = \dfrac{x^3 - 2x + 1}{x^2 - 1}$ **7.** $f(x) = \dfrac{5x + 2}{2x^3 - 1}$ (See Example 13.)

Rational Function End Behavior

Now we consider the limit as $x \to \infty$ of a rational function $f(x) = g(x)/h(x)$ where *the degree of the numerator is greater than the degree of the denominator*. Since the leading terms of the polynomials g and h provide end behavior models for g and h, it is reasonable to conjecture that the quotient of the leading terms provides an end behavior model for f.

EXAMPLE 14

The function $x^2/x = x \, (x \neq 0)$ is an end behavior model for $f(x) = \dfrac{x^2 + 4x + 5}{x}$ because

$$\lim_{x \to \pm\infty} \frac{\dfrac{x^2 + 4x + 5}{x}}{\dfrac{x^2}{x}} = \lim_{x \to \pm\infty} \frac{x^2 + 4x + 5}{x^2} = 1.$$

Away from 0, x^2/x is the same as the function x. The graphs of f and x in a large viewing window (Fig. 2.42a) illustrate that x is an end behavior model of f. ≡

For the function f in Example 14 it can be shown that any function $y = x + a$ is an end behavior model for f. So an end behavior model is not unique. However, there is a special end behavior model that can be found by using division:

$$\frac{x^2 + 4x + 5}{x} = x + 4 + \frac{5}{x}.$$

Among the end behavior models $x + a$ of f it can be shown that $y = x + 4$ best approximates f (Fig. 2.42b). The line $y = x + 4$ is sometimes called a *slant asymptote* of f. We will call it the *end behavior asymptote* to distinguish it among all the possible end behavior models of f.

EXPLORATION BIT

Writing f in the form

$$x + 4 + \frac{5}{x}$$

can help you to *sketch* the graph of f. Think of the graph of $x + 4$, the graph of $5/x$, and how to combine them into the graph of their sum. Figure 2.42(b) can give you a clue.

> **DEFINITION**   End Behavior Asymptote
>
> Let $f(x) = p(x)/h(x)$ be a rational function and $q(x)$ and $r(x)$ be the quotient and remainder when $p(x)$ is divided by $h(x)$. That is,
>
> $$f(x) = q(x) + \frac{r(x)}{h(x)} \qquad \text{with} \qquad \deg r < \deg h.$$
>
> The graph of q is called the **end behavior asymptote** of f.

In general, q is an end behavior model of f. It can be shown that among the end behavior models of f, q gives the best approximation of f, thus earning the right for its graph to be called an asymptote. Note that an end behavior asymptote need not be a line, but if q is a constant, including 0, then its graph is indeed a horizontal asymptote.

EXPLORATION 3

End Behavior Models and Asymptotes

By comparing leading terms of the numerator and denominator, we see that $y = x^3/x = x^2 (x \neq 0)$ is an *end behavior model* for

$$f(x) = \frac{p(x)}{h(x)} = \frac{x^3 - 4x^2 + 2x + 8}{x - 2}.$$

1. Support this by showing y and f in the same "large" viewing window, $[-100, 100]$ by $[-10{,}000, 10{,}000]$.

2. Rewrite $f(x)$ as $q(x) + \dfrac{r(x)}{h(x)}$ with $\deg r < \deg h$ to find q, the *end behavior asymptote* of f. Show q and f simultaneously in the $[-8, 8]$ by $[-10, 20]$ viewing window. You will find the effect rather dramatic! Predict what will happen if you ZOOM-OUT.

Give an end behavior model and the end behavior asymptote of the rational function.

3. $f(x) = \dfrac{x^3 - 10x^2 + x + 50}{x - 2}$ 4. $f(x) = \dfrac{2x^3 + x^2 - 27x - 31}{x^2 - x - 12}$

5. $f(x) = \dfrac{x^4 + 3x^3 - 9x^2 - 27x + 2}{x + 3}$

A simple procedure for identifying the behavior of a rational function as $x \to \pm\infty$ is this: Compare the degree of the numerator and the degree of the denominator. If the degree "on top" is less, the limit as $x \to \infty$ is 0. If the degrees are equal, the limit is the ratio of the leading coefficients. If the degree "on top" is greater, divide to find the polynomial that gives the end behavior asymptote.

SUMMARY FOR RATIONAL FUNCTIONS

For the rational function $f(x) = p(x)/h(x)$,

If:	*then:*	*and:*
$\deg p < \deg h$	$\displaystyle\lim_{x \to \pm\infty} f(x) = 0$	$y = 0$ is (horizontal) end behavior asymptote.
$\deg p = \deg h$	$\displaystyle\lim_{x \to \pm\infty} f(x) = k$	$y = k$ is (horizontal) end behavior asymptote.
$\deg p > \deg h$	$\displaystyle\lim_{x \to \pm\infty} f(x) = \pm\infty$	$y =$ polynomial is end behavior asymptote.

Changing Variables with Substitutions

Sometimes a change of variable can turn an unfamiliar expression into one whose limit we know how to find. Here are two examples.

EXAMPLE 15

$$\lim_{x \to \infty} \sin \frac{1}{x} = \lim_{\theta \to 0^+} \sin \theta \quad \text{Substitute } \theta = 1/x. \text{ Then } \theta \to 0^+ \text{ as } x \to \infty.$$

$$= 0. \qquad \text{Sin } \theta \text{ is continuous.} \qquad \blacksquare$$

EXAMPLE 16

Determine the limit graphically. Confirm algebraically using substitution.

$$\lim_{x \to \pm\infty} \left(1 + \frac{2}{x}\right)\left(\cos \frac{1}{x}\right)$$

Solution   The graph of $f(x) = [1 + (2/x)][\cos(1/x)]$ in Fig. 2.43 suggests that the limit is 1. Algebraically,

$$\lim_{x \to \pm\infty} \left(1 + \frac{2}{x}\right)\left(\cos \frac{1}{x}\right) = \lim_{\theta \to 0}(1 + 2\theta)(\cos \theta) \qquad \begin{array}{l}\text{Substitute } \theta = \dfrac{1}{x}. \\ \text{Then } \theta \to 0 \\ \text{as } x \to \pm\infty.\end{array}$$

$$= \lim_{\theta \to 0}(1 + 2\theta) \lim_{\theta \to 0} \cos \theta \qquad \text{Product Rule}$$

$$= 1 \cdot 1 = 1. \qquad \begin{array}{l}\text{The functions} \\ \text{are continuous.}\end{array} \qquad \blacksquare$$

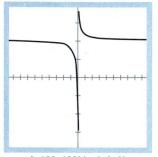

[−100, 100] by [−2, 2]

2.43   The values of

$$f(x) = \left(1 + \frac{2}{x}\right)\left(\cos \frac{1}{x}\right)$$

approach 1 as $x \to \pm\infty$. Use TRACE to estimate the limit.

Exercises 2.4

Use algebra to find the limits of the functions defined by the expressions in Exercises 1–12 (a) as $x \to \infty$ and (b) as $x \to -\infty$. Support graphically.

1. $\dfrac{2x + 3}{5x + 7}$

2. $\dfrac{2x^3 + 7}{x^3 - x^2 + x + 7}$

3. $\dfrac{x + 1}{x^2 + 3}$

4. $\dfrac{3x + 7}{x^2 - 2}$

5. $\dfrac{3x^2 - 6x}{4x - 8}$

6. $\dfrac{x^4}{x^3 + 1}$

7. $\dfrac{1}{x^3 - 4x + 1}$

8. $\dfrac{10x^5 + x^4 + 31}{x^6}$

9. $\dfrac{-2x^3 - 2x + 3}{3x^3 + 3x^2 - 5x}$

10. $\dfrac{-x^4}{x^4 - 7x^3 + 7x^2 + 9}$

11. $\left(\dfrac{-x}{x + 1}\right)\left(\dfrac{x^2}{5 + x^2}\right)$

12. $\left(\dfrac{2}{x} + 1\right)\left(\dfrac{5x^2 - 1}{x^2}\right)$

Find the limits in Exercises 13–20 by a convincing method of your choice.

13. $\lim_{x \to 2^+} \dfrac{1}{x - 2}$

14. $\lim_{x \to 2^-} \dfrac{1}{x - 2}$

15. $\lim_{x \to 2^+} \dfrac{x}{x - 2}$

16. $\lim_{x \to 2^-} \dfrac{x}{x - 2}$

17. $\lim_{x \to -3^+} \dfrac{1}{x + 3}$

18. $\lim_{x \to -3^-} \dfrac{1}{x + 3}$

19. $\lim_{x \to -3^+} \dfrac{x}{x + 3}$

20. $\lim_{x \to -3^-} \dfrac{x}{x + 3}$

Find the end behavior asymptote algebraically in Exercises 21–30 and support graphically. Find all vertical asymptotes.

21. $f(x) = \dfrac{x - 2}{2x^2 + 3x - 5}$

22. $T(y) = \dfrac{2y + 3}{4 - y^2}$

23. $g(x) = \dfrac{3x^2 - x + 5}{x^2 - 4}$

24. $f(x) = \dfrac{2 - 3x^2}{5 + 2x - 6x^2}$

25. $f(x) = \dfrac{x^2 - 2x + 3}{x + 2}$

26. $f(x) = \dfrac{x^2 - 3x - 7}{x + 3}$

27. $g(x) = \dfrac{x^3 - 2x + 1}{x - 2}$

28. $g(x) = \dfrac{x^4 - 2x^2 - x + 3}{x^2 - 4}$

29. $f(x) = \dfrac{x}{x^2+3} - \dfrac{x^2+2}{1+x-x^2}$

30. $g(x) = \dfrac{x^2-4}{x+1} + \dfrac{x}{x^2-3x+2}$

In Exercises 31–34, use graphs to find the limits.

31. a) $\lim_{x\to 2^+}\dfrac{1}{x^2-4}$ **b)** $\lim_{x\to 2^-}\dfrac{1}{x^2-4}$

c) $\lim_{x\to -2^+}\dfrac{1}{x^2-4}$ **d)** $\lim_{x\to -2^-}\dfrac{1}{x^2-4}$

32. a) $\lim_{x\to 1^+}\dfrac{x}{x^2-1}$ **b)** $\lim_{x\to 1^-}\dfrac{x}{x^2-1}$

c) $\lim_{x\to -1^+}\dfrac{x}{x^2-1}$ **d)** $\lim_{x\to -1^-}\dfrac{x}{x^2-1}$

33. a) $\lim_{x\to -2^+}\dfrac{x^2-1}{2x+4}$ **b)** $\lim_{x\to -2^-}\dfrac{x^2-1}{2x+4}$

34. a) $\lim_{x\to 0^+}\left(x^2+\dfrac{4}{x}\right)$ **b)** $\lim_{x\to 0^-}\left(x^2+\dfrac{4}{x}\right)$

Find the limits in Exercises 35–42 by a convincing method of your choice.

35. $\lim_{x\to 0^+}\dfrac{[x]}{x}$ **36.** $\lim_{x\to 0^-}\dfrac{[x]}{x}$

37. $\lim_{x\to\infty}\dfrac{|x|}{|x|+1}$ **38.** $\lim_{x\to -\infty}\dfrac{x}{|x|}$

39. $\lim_{x\to 0^+}\dfrac{1}{\sin x}$ **40.** $\lim_{x\to 0^-}\dfrac{1}{\sin x}$

41. $\lim_{x\to(\pi/2)^+}\dfrac{1}{\cos x}$ **42.** $\lim_{x\to(\pi/2)^-}\dfrac{1}{\cos x}$

43. Let $f(x)=\begin{cases}\dfrac{1}{x}, & x<0,\\ -1, & x\geq 0.\end{cases}$
Find $\lim f(x)$ as $x\to -\infty, 0^-, 0^+$, and ∞.

44. Let $f(x)=\begin{cases}\dfrac{x-2}{x-1}, & x\leq 0,\\ \dfrac{1}{x^2}, & x>0.\end{cases}$
Find $\lim f(x)$ as $x\to -\infty, 0^-, 0^+$, and ∞.

Find the limits in Exercises 45–50.

45. $\lim_{x\to\infty}\left(2+\dfrac{\sin x}{x}\right)$ **46.** $\lim_{x\to -\infty}\dfrac{\sin x}{x}$

47. $\lim_{x\to\infty}\left(1+\cos\dfrac{1}{x}\right)$ **48.** $\lim_{x\to\infty}x\sin\dfrac{1}{x}$

49. $\lim_{x\to\infty}\dfrac{\sin 2x}{x}$ **50.** $\lim_{x\to\infty}\dfrac{\cos(1/x)}{1+(1/x)}$

Use the Sandwich Theorem for limits at ∞ to find the limits in Exercises 51 and 52.

51. Find $\lim_{x\to\infty}f(x)$ and $\lim_{x\to -\infty}f(x)$ if
$$\dfrac{2x^2}{x^2+1}<f(x)<\dfrac{2x^2+5}{x^2}.$$

52. *The greatest integer function.* Find $\lim_{x\to\infty}[x]/x$ and $\lim_{x\to -\infty}[x]/x$ given that
$$\dfrac{x-1}{x}<\dfrac{[x]}{x}\leq 1 \quad (x\neq 0).$$

53. Draw complete graphs of the following functions in the same viewing window: $2x, 2x^3, 2x^5, 2x^7$. Compare their limits as $x\to\pm\infty$ and the steepness of the graphs. How can you distinguish their behavior? Explain.

54. Draw complete graphs of the following functions in the same viewing window: $-3x^2, -3x^4, -3x^6, -3x^8$. Compare their limits as $x\to\pm\infty$ and the steepness of the graphs. How can you distinguish their behavior? Explain.

55. Show that $y=-1/7$ is an end behavior model for the function
$$f(x)=-\dfrac{x}{7x+4}$$
of Example 11. (*Hint:* Show that $\lim_{x\to\pm\infty}f(x)/(-1/7)=1$.)

56. Show that $y=2/3$ is an end behavior model for the function
$$f(x)=\dfrac{2x^2-x+3}{3x^2+5}$$
of Example 12.

In Exercises 57 and 58, sketch a graph of a function $y=f(x)$ with domain the largest subset of real numbers that satisfies the stated conditions.

57. $\lim_{x\to 1}f(x)=2$, $\lim_{x\to 5^-}f(x)=\infty$,
$\lim_{x\to\infty}f(x)=-1$, $\lim_{x\to -\infty}f(x)=-\infty$

58. $\lim_{x\to 2}f(x)=-1$, $\lim_{x\to 4^+}f(x)=-\infty$,
$\lim_{x\to\infty}f(x)=\infty$, $\lim_{x\to -\infty}f(x)=2$

In Exercises 59–62, find the limits (or state that they do not exist) of f, g, and fg as $x\to c$.

59. $f(x)=\dfrac{1}{x}, g(x)=x, c=0$

60. $f(x)=-\dfrac{2}{x^3}, g(x)=4x^3, c=0$

61. $f(x)=\dfrac{3}{x-2}, g(x)=(x-2)^3, c=2$

62. $f(x)=\dfrac{5}{(3-x)^4}, g(x)=(x-3)^2, c=3$

63. Let $\lim_{x\to c}f(x)=0$ and $\lim_{x\to c}g(x)=\infty$. Give examples to show that $\lim_{x\to c}(fg)$ can be 0, finite and nonzero, or infinite.

64. Let L be a real number, $\lim_{x\to c}f(x)=L$, and $\lim_{x\to c}g(x)=\pm\infty$. Can $\lim_{x\to c}(f\pm g)$ be determined? Explain.

65. Prove Theorem 11.

66. Use graphs to find the limits.

a) $\lim\limits_{x\to\infty} \dfrac{\ln(x+1)}{\ln x}$.

b) $\lim\limits_{x\to\infty} \dfrac{\ln(x+999)}{\ln x}$.   Compare with part (a).

c) $\lim\limits_{x\to\infty} \dfrac{\ln x^2}{\ln x}$.

d) $\lim\limits_{x\to\infty} \dfrac{\ln x}{\log x}$.

In Exercises 67–74, find $\lim_{x\to\infty} f(x)$ and $\lim_{x\to-\infty} f(x)$. The behavior of the functions will be explained in Chapter 7.

67. $f(x) = \left(1 + \dfrac{1}{x}\right)^x$

68. $f(x) = \left(1 + \dfrac{5}{x}\right)^x$

69. $f(x) = \left(1 + \dfrac{0.07}{x}\right)^x$

70. $f(x) = \left(1 + \dfrac{1}{x}\right)^{1/x}$

71. $f(x) = xe^{-x}$

72. $f(x) = x^2 e^{-x}$

73. $f(x) = xe^x$

74. $f(x) = x^2 e^x$

75. Explain why there is no value L (not even $\pm\infty$) for which $\lim_{\theta\to\infty} \sin\theta = L$.

76. *Group discussion.* Discuss the number of horizontal and vertical asymptotes possible for a rational function. Illustrate with examples. Use the function of Example 13 as the example for its type of rational function.

2.5 _____   Controlling Function Outputs: Target Values

We sometimes want the outputs of a function $y = f(x)$ to lie near a particular target value y_0. This need can come about in different ways. A gas-station attendant, asked for $5.00 worth of gas, will try to pump the gas to the nearest cent. A mechanic grinding a 3.385-in. cylinder bore knows to not let the bore vary from this value by more than 0.002 in. A pharmacist making ointments will measure the ingredients to the nearest milligram.

So the question becomes: How accurate do our machines and instruments have to be to keep the outputs within useful bounds? When we express this question with mathematical symbols, we ask: How closely must we control x to keep $y = f(x)$ within an acceptable interval about some particular target value y_0? In this section we use a graphing utility to give us clues to the answers for specific examples. Then we confirm the answers algebraically. In Section 2.6 we use ideas from this section to give a general answer with the formal definition of limit.

Controlling Function Outputs as $x \to c$

EXPLORATION 1

Aiming at the Target

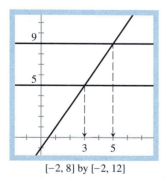

[-2, 8] by [-2, 12]

2.44   Keeping x between 3 and 5 will keep $y_1 = 2x - 1$ between $y_2 = 5$ and $y_3 = 9$.

Controlling a linear function   For the *linear function* $y = 2x - 1$, $y = 7$ when $x = 4$. How close must we hold x to $x_0 = 4$ for y to be within 2 units of $y_0 = 7$? In other words, we want to find x values around 4 that give y values between 5 and 9.

The viewing window (Fig. 2.44) shows graphs of $y_1 = 2x - 1$, $y_2 = 5$, and $y_3 = 9$. A TRACE on $y_1 = 2x - 1$ shows that $5 < y < 9$ when $3 < x < 5$. Confirming algebraically, we get

$$3 < x < 5 \Rightarrow 6 < 2x < 10$$
$$\Rightarrow 5 < 2x - 1 < 9.$$

For the given function $y = f(x)$ and the given point x_0, use your grapher to find x values near x_0 that will give you y-values within 1 unit of $y_0 = 5$. Confirm algebraically.

1. $y = 4x - 3, x_0 = 2$

2. $y = x + 6, x_0 = -1$

3. $y = -2x + 7, x_0 = 1$

4. Explain how you could use ZOOM-BOX to find the values around x_0.

Controlling a square root function For the *square root function* $y = \sqrt{3x - 2}$, $y = 2$ when $x = 2$. Find an interval about $x_0 = 2$ that will give y values within 0.2 unit of $y_0 = 2$ (Fig. 2.45).

TRACE or ZOOM-BOX shows that $1.8 < y < 2.2$ when $1.75 < x < 2.28$. Algebraically,

$$1.75 < x < 2.28 \Rightarrow 3(1.75) < 3x < 3(2.28)$$

$$\Rightarrow 3(1.75) - 2 < 3x - 2 < 3(2.28) - 2$$

$$\Rightarrow a = \sqrt{3(1.75) - 2} < \sqrt{3x - 2} < \sqrt{3(2.28) - 2} = b.$$

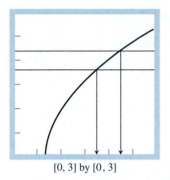

[0, 3] by [0, 3]

2.45 Keeping x between 1.75 and 2.28 will keep $y_1 = \sqrt{3x - 2}$ between $y_2 = 1.8$ and $y_3 = 2.2$.

A calculator shows $a = 1.803$ and $b = 2.2$, which means that the x-values in $(1.75, 2.28)$ hit the target interval $(1.8, 2.2)$ around $y_0 = 2$.

For the given function $y = f(x)$ and the given points x_0 and y_0, use your grapher to find x values near x_0 that will give you y-values within 0.1 unit of y_0. Confirm algebraically.

5. $y = \sqrt{2x + 1}, x_0 = 4, y_0 = 3$ **6.** $y = \sqrt{x + 3}, x_0 = -2, y_0 = 1$

7. $y = \sqrt{7 - 2x}, x_0 = 3, y_0 = 1$

For the function $y = 2x - 1$ we found that $3 < x < 5 \Rightarrow 5 < 2x - 1 < 9$. We can convert an interval like $3 < x < 5$ to one that is symmetric about its midpoint (Section 1.5) as follows.

First the midpoint $x_0 = (3 + 5)/2 = 4$. Then,

$$3 < x < 5 \Rightarrow 3 - 4 < x - 4 < 5 - 4$$

$$\Rightarrow -1 < x - 4 < 1$$

$$\Rightarrow |x - 4| < 1.$$

When we aim for a target around y_0, we try to find the interval around x_0 to be symmetric and as large as reasonably possible. The answers you provide for the exercises of this section should use this convention. If the interval we find is skewed, then we take the largest one inside it that is symmetric. Its x-values, of course, will still hit the y-value target.

EXAMPLE 1   Controlling a Quadratic Function

Observe that if $y = x^2$ and $x = 1$, then $y = 1$. How close to $x_0 = 1$ must we hold x to be sure that $y = x^2$ lies within 0.4 unit of $y_0 = 1$? Estimate graphically and confirm algebraically.

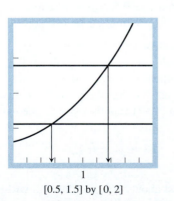

1

[0.5, 1.5] by [0, 2]

2.46 Keeping x between $\sqrt{0.6}$ (or 0.78) and $\sqrt{1.4}$ (or 1.18) will keep $y_1 = x^2$ between $y_2 = 0.6$ and $y_3 = 1.4$.

Solution $|y - 1| < 0.4$ is equivalent to $0.6 < y < 1.4$. The viewing window (Fig. 2.46) shows graphs of $y_1 = x^2$, $y_2 = 0.6$, and $y_3 = 1.4$. We know that $y_1 = 0.6$ when $x = \sqrt{0.6}$ and $y_1 = 1.4$ when $x = \sqrt{1.4}$. Thus,

$$\sqrt{0.6} < x < \sqrt{1.4} \Rightarrow 0.6 < y_1 < 1.4.$$

Algebraically (for x near 1),

$$|y - 1| < 0.4 \Leftrightarrow |x^2 - 1| < 0.4$$

$$\Leftrightarrow -0.4 < x^2 - 1 < 0.4$$

$$\Leftrightarrow 0.6 < x^2 < 1.4$$

$$\Leftrightarrow \sqrt{0.6} < x < \sqrt{1.4} \qquad \begin{array}{l} \text{For } a, b, \text{ and } c \text{ nonnegative,} \\ a < b < c \Rightarrow \sqrt{a} < \sqrt{b} < \sqrt{c}. \end{array}$$

▰

> **TERMINOLOGY**
>
> For any point x_0 on a number line, the interval from $x_0 - r$ to $x_0 + r$ may be called the *interval centered at* x_0 *with radius* r. This terminology applies nicely to symmetric intervals. Do you understand why?

By using decimals, the inequality $\sqrt{0.6} < x < \sqrt{1.4}$ in Example 1 becomes $0.7745\ldots < x < 1.1832\ldots$. If we want to give the x-interval using hundredths, we play safe and round to *shorten* the x-interval at each endpoint. Thus, to hundredths,

$$0.78 < x < 1.18 \Rightarrow \sqrt{0.6} < x < \sqrt{1.4}$$

$$\Rightarrow 0.6 < y_1 < 1.4.$$

But notice that the interval $0.78 < x < 1.18$ is not symmetric about $x_0 = 1$.

EXAMPLE 2

Find a symmetric interval about $x_0 = 1$ for which $y = x^2$ lies within 0.4 unit of $y_0 = 1$.

Solution For the interval $0.78 < x < 1.18$ from Example 1, the distance 0.18 from $x_0 = 1$ to the right endpoint 1.18 is smaller than the distance 0.22 from $x_0 = 1$ to the left endpoint 0.78. We use the smaller distance, 0.18, as the radius for the symmetric interval about 1,

$$|x - 1| < 0.18.$$

▰

EXAMPLE 3   Controlling a Trigonometric Function

Observe that if $y = \sin x$ and $x = \pi/6$, then $y = 0.5$. How close to $x_0 = \pi/6$ must we hold x to be sure that $y = \sin x$ lies within 0.2 unit of $y_0 = 0.5$?

Solution $|\sin x - 0.5| < 0.2$ is equivalent to $0.3 < \sin x < 0.7$. The viewing window (Fig. 2.47) shows graphs of $y_1 = \sin x$, $y_2 = 0.3$, and $y_3 = 0.7$. We know that $y_1 = 0.3$ when $x = \sin^{-1} 0.3$. (*Note:* x is near $x_0 = \pi/6$; thus it is between 0 and 1.) Similarly, $y_1 = 0.7$ when $x = \sin^{-1} 0.7$. The interval $(\sin^{-1} 0.3, \sin^{-1} 0.7)$ rounds *safely* to the smaller interval $(0.31, 0.77)$, thus

$$0.31 < x < 0.77 \quad \Rightarrow \quad 0.3 < y_1 < 0.7.$$

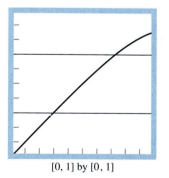

[0, 1] by [0, 1]

2.47 Keeping x between 0.31 and 0.77 will keep $y_1 = \sin x$ between $y_2 = 0.3$ and $y_3 = 0.7$.

Because $\pi/6 \approx 0.52$, a symmetric interval about $\pi/6$ of radius 0.2 will have endpoints of about 0.32 and 0.72, safely within the interval $0.31 < x < 0.77$. Hence, for the symmetric interval $|x - \pi/6| < 0.2$ centered on $x_0 = \pi/6$, we can be sure that the values of $y = \sin x$ lie within 0.2 unit of $y_0 = 0.5$.

▰

EXAMPLE 4   Controlling a Rational Function

If $f(x) = (x + 1)/(x - 2)$, then $f(3) = 4$. Find an interval about $x_0 = 3$ for which the values of f lie within 0.1 unit of 4. Estimate graphically, and confirm algebraically.

Solution   We need to determine the values of x for which $3.9 < f(x) < 4.1$. Figure 2.48 shows a complete graph of f, and Fig. 2.49 shows the graphs of f, $y_2 = 3.9$, and $y_3 = 4.1$ near the point $(3, 4)$. To algebraically determine the values of x for which $f(x) = 3.9$ and $f(x) = 4.1$, we need to solve the equations

$$\frac{x + 1}{x - 2} = 3.9 \qquad \text{and} \qquad \frac{x + 1}{x - 2} = 4.1.$$

The solutions, rounded safely, are 3.03 and 2.97, respectively. Thus, $f(x) = (x + 1)/(x - 2)$ is within 0.1 unit of 4 if $2.97 < x < 3.03$, or $|x - 3| < 0.03$.

Study these two sequences of steps algebraically. The one on the left shows another way to find the interval around $x_0 = 3$. The one on the right confirms that the interval found is the desired interval.

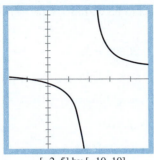

[−2, 5] by [−10, 10]

2.48   A complete graph of $f(x) = (x + 1)/(x - 2)$.

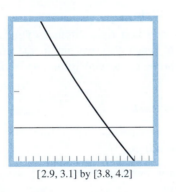

[2.9, 3.1] by [3.8, 4.2]

2.49   Keeping x between 2.97 and 3.03 will keep

$$f(x) = (x + 1)/(x - 2)$$

between 3.9 and 4.1.

Finding interval about x_0

$$\left| \frac{x + 1}{x - 2} - 4 \right| < 0.1$$

$$3.9 < \frac{x + 1}{x - 2} < 4.1$$

$$3.9 < 1 + \frac{3}{x - 2} < 4.1$$

$$2.9 < \frac{3}{x - 2} < 3.1$$

$$\frac{1}{2.9} > \frac{x - 2}{3} > \frac{1}{3.1}$$

$$\frac{1}{3.1} < \frac{x - 2}{3} < \frac{1}{2.9}$$

$$\frac{3}{3.1} < x - 2 < \frac{3}{2.9}$$

$$\frac{3}{3.1} + 2 < x < \frac{3}{2.9} + 2$$

$$2.967 \ldots < x < 3.034 \ldots$$

$$2.97 < x < 3.03$$

(rounding safely)

$$\therefore |x - 3| < 0.03$$

(symmetric about $x_0 = 3$)

Confirming interval about x_0

$$|x - 3| < 0.03$$

$$2.97 < x < 3.03$$

$$0.97 < x - 2 < 1.03$$

$$\frac{3}{0.97} > \frac{3}{x - 2} > \frac{3}{1.03}$$

$$\frac{3}{1.03} < \frac{3}{x - 2} < \frac{3}{0.97}$$

$$1 + \frac{3}{1.03} < 1 + \frac{3}{x - 2} < 1 + \frac{3}{0.97}$$

$$1 + \frac{3}{1.03} < \frac{x + 1}{x - 2} < 1 + \frac{3}{0.97}$$

$$\frac{3}{1.03} - 3 < \frac{x + 1}{x - 2} - 4 < \frac{3}{0.97} - 3$$

$$-0.087 \ldots < f(x) - 4 < 0.092 \ldots$$

(which is inside the interval)

$$-0.1 < f(x) - 4 < 0.1$$

$$\therefore |f(x) - 4| < 0.1$$

The examples that follow should give you some appreciation of the value of the calculator.

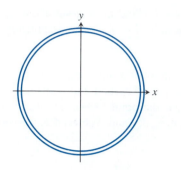

2.50 A circle with radius between 9.73 cm and 9.82 cm will have area within 1% of 300 cm². The circles shown here are not drawn to scale.

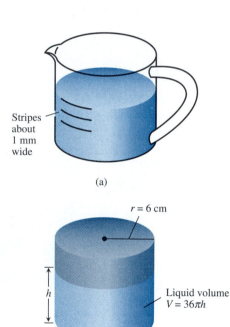

Stripes about 1 mm wide

(a)

$r = 6$ cm

h

Liquid volume $V = 36\pi h$

(b)

2.51 (a) A typical 1-liter measuring cup. (b) The measuring cup modeled as a right circular cylinder of radius $r = 6$ cm. To get a liter of water to the nearest 1%, how accurately must we measure h?

EXAMPLE 5   Controlling the Area of a Circle

What must the radius of a circle be for its area to be within 1% of 300 square units?

Solution We want $297 < \pi r^2 < 303$,   1% of $300 = 3$.

or
$$297/\pi < r^2 < 303/\pi$$

or
$$\sqrt{297/\pi} < r < \sqrt{303/\pi}$$

or
$$9.723\ldots < r < 9.8207\ldots$$

or
$$9.73 < r < 9.82 \qquad \text{(rounding safely)}$$

Thus, a circle with radius between 9.73 and 9.82 will have area within the allowable deviation of 300 (Fig. 2.50).

EXAMPLE 6   Why the Stripes on a 1-Liter Kitchen Measuring Cup Are About a Millimeter Wide

The interior of a typical 1-L measuring cup is a right circular cylinder of radius 6 cm (Fig. 2.51). The volume of water that we put in the cup is therefore a function of the level h in centimeters to which the cup is filled, the formula being

$$V = \pi (6)^2 h = 36\pi h.$$

How closely do we have to measure h to measure out 1 liter of water (1000 cm³) with an error of less than 1% (10 cm³)?

Solution In terms of V and h, we want to know in what interval of values to hold h to make V satisfy the inequality

$$|V - 1000| = |36\pi h - 1000| < 10.$$

To find out, we change

$$|36\pi h - 1000| < 10$$

to
$$-10 < 36\pi h - 1000 < 10$$

to
$$990 < 36\pi h < 1010$$

to
$$\frac{990}{36\pi} < h < \frac{1010}{36\pi}$$

to
$$8.76 < h < 8.93. \qquad \text{(rounding safely)}$$

The interval in which we should hold h, then, is about 0.17 cm (1.7 mm) wide. Thus, with a cup having stripes that are 1 millimeter (even 1.5 mm) wide, we can expect to measure a liter of water with an accuracy of better than 1%, which is more than enough accuracy for our cooking.

Controlling Function Outputs as $x \to \infty$

Suppose $\lim_{x \to \infty} f(x)$ is finite, say y_0. We can control the function values $f(x)$ to be within an acceptable interval around y_0 by controlling the values of x. This time we need to find a positive number N beyond which the x values will have y values within the acceptable interval.

A similar statement can be made about a limit as $x \to -\infty$.

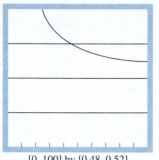

[0, 100] by [0.48, 0.52]

2.52 "Watch" $f \to 1/2$ by drawing y_1, y_2, y_3, and y_4 simultaneously. We see the graph of f cross over into the [0.49, 0.51] interval just beyond $x = 50$. We can confirm algebraically that keeping $x > 50$ will keep $f(x) = (x+1)/(2x)$ between 0.49 and 0.51.

EXAMPLE 7

$\lim_{x\to\infty} \dfrac{x+1}{2x} = \dfrac{1}{2}$. How large must we choose x so that the values of $f(x) = (x+1)/(2x)$ are within 0.01 unit of 1/2? Estimate graphically and confirm algebraically.

Solution   Figure 2.52 shows the graphs $y_1 = f$, $y_2 = 0.5$, $y_3 = 0.49$, and $y_4 = 0.51$. As $x \to \infty$, the graph of f appears to cross into the target interval [0.49, 0.51] and stay there, approaching 0.5 through values larger than 0.5. ZOOM-IN or SOLVE shows that the crossover into the target interval happens just past $x = 50$.

Algebraically, if $x > 50$, then $0 < \dfrac{1}{x} < \dfrac{1}{50}$. So

$$\left| f(x) - \frac{1}{2} \right| = \left| \frac{x+1}{2x} - \frac{1}{2} \right| = \left| \frac{x+1-x}{2x} \right| = \frac{1}{2x} < \frac{1}{2(50)} = 0.01.$$

Exercises 2.5

1. Which of the following statements are equivalent to the statement $2 < x < 8$?

a) $0 < x < 6$ **b)** $1 < x - 1 < 7$

c) $1 < \dfrac{x}{2} < 4$ **d)** $\dfrac{1}{8} < \dfrac{1}{x} < \dfrac{1}{2}$

e) $x > 8$ **f)** $|x - 5| < 3$

g) $4 < x < 10$ **h)** $-8 < -x < -2$

2. Which of the following statements are equivalent to the statement $-1 < y - 3 < 1$?

a) $2 < y < 4$ **b)** $|y - 3| < 1$

c) $y > 2$ **d)** $y < 4$

e) $0 < y - 2 < 2$ **f)** $1 < \dfrac{y}{2} < 2$

g) $\dfrac{1}{4} < \dfrac{1}{y} < \dfrac{1}{2}$ **h)** $-4 < y < -2$

In Exercises 3–10, match each absolute value inequality with the interval that it determines.

3. $|x + 3| < 1$ **a)** $-2 < x < 1$

4. $|x - 5| < 2$ **b)** $-1 < x < 3$

5. $\left| \dfrac{x}{2} \right| < 1$ **c)** $3 < x < 7$

6. $|1 - x| < 2$ **d)** $-\dfrac{5}{2} < x < -\dfrac{3}{2}$

7. $|2x - 5| \le 1$ **e)** $-2 < x < 2$

8. $|2x + 4| < 1$ **f)** $-4 \le x \le 4$

9. $\left| \dfrac{x-1}{5} \right| \le 1$ **g)** $-4 < x < -2$

     h) $2 \le x \le 3$

10. $\left| \dfrac{2x+1}{3} \right| < 1$ **i)** $-4 \le x \le 6$

The inequalities in Exercises 11–18 define intervals. Describe each interval with an inequality that does not involve absolute values.

11. $|y - 2| \le 5$ **12.** $|y + 3| < 1$

13. $|2y - 5| < 1$ **14.** $|2y + 5| < 1$

15. $\left| \dfrac{y}{2} - 1 \right| \le 1$ **16.** $\left| 2 - \dfrac{y}{2} \right| < \dfrac{1}{2}$

17. $|2 - y| < \dfrac{1}{5}$ **18.** $\left| \dfrac{5 - 2y}{3} \right| < 1$

Describe each interval in Exercises 19–22 with an absolute value inequality of the form $|x - x_0| < D$. It may help to draw a picture of the interval first.

19. $1 < x < 8$ **20.** $-2 < x < 7$

21. $-4 < x < 1$ **22.** $-8 < x < -1$

23. How close to $x_0 = -1$ must we hold x to be sure that $y = x^2$ lies within 0.5 unit of $y_0 = 1$? Estimate graphically, and confirm algebraically. (See Example 1.)

24. Find an interval in $[\pi/2, \pi]$ for which $y = \sin x$ lies within 0.2 unit of $y_0 = 0.5$. (See Example 3.)

25. Find an interval in $[0, \pi/2]$ for which $y = \cos x$ lies within 0.2 unit of $y_0 = 0.4$.

26. Find an interval in $[0, \pi/2]$ for which $y = \tan x$ lies within 0.2 unit of $y_0 = 2$.

27. Find an interval in $[-\pi/2, 0]$ for which $y = \cos x$ lies within 0.2 unit of $y_0 = 0.4$.

28. Find an interval in $[-\pi/2, 0]$ for which $y = \tan x$ lies within 0.2 unit of $y_0 = -2$.

Each of Exercises 29–36 gives a function $y = f(x)$, a number E, and a target value y_0. In what interval must we hold x in each case to be sure that $y = f(x)$ lies within E units of y_0? Estimate graphically, and confirm algebraically.

29. $y = x^2$, $E = 0.1$, $y_0 = 100$, $x_0 > 0$

30. $y = x^2$, $E = 0.1$, $y_0 = 100$, $x_0 < 0$

31. $y = \sqrt{x - 7}$, $E = 0.1$, $y_0 = 4$

32. $y = \sqrt{19 - x}$, $E = 1$, $y_0 = 3$

33. $y = 120/x$, $E = 1$, $y_0 = 5$

34. $y = 1/(4x)$, $E = 1/2$, $y_0 = 1$

35. $y = \dfrac{3 - 2x}{x - 1}$, $E = 0.1$, $y_0 = -3$

36. $y = \dfrac{3x + 8}{x + 2}$, $E = 0.1$, $y_0 = 1$

Each of Exercises 37–42 gives a function $y = f(x)$, a number E, and a target value y_0. In what interval must we hold x in each case to be sure that $y = f(x)$ lies within E units of y_0?

37. $y = x^2 - 5$, $E = 0.5$, $y_0 = 11$, $x_0 > 0$

38. $y = x^2 - 5$, $E = 0.5$, $y_0 = 11$, $x_0 < 0$

39. $y = x^3 - 9x$, $E = 0.2$, $y_0 = 5$, x_0 near -3

40. $y = x^4 - 10x^2$, $E = 0.2$, $y_0 = -5$, x_0 near 1

41. $y = e^x$, $E = 0.1$, $y_0 = 0.5$

42. $y = \ln x$, $E = 0.1$, $y_0 = 2$

Each of Exercises 43–46 gives a function $y = f(x)$, a number E, a point x_0, and a target value y_0. In what interval about x_0 must we hold x in each case to be sure that $y = f(x)$ lies within E units of y_0? Describe the interval with an absolute value inequality of the form $|x - x_0| < D$.

43. $y = x + 1$, $E = 0.5$, $x_0 = 3$, $y_0 = 4$

44. $y = 2x - 1$, $E = 1$, $x_0 = -2$, $y_0 = -5$

45. $y = 2x^2 + 1$, $E = 0.2$, $x_0 = 1$, $y_0 = 3$

46. $y = \sqrt{2x - 3}$, $E = 0.2$, $x_0 = 3.5$, $y_0 = 2$

In Example 1 we asked how close to $x_0 = 1$ we have to be so that $y = x^2$ is within 0.4 of $y_0 = 1$. We know there is an answer because

$$\lim_{x \to x_0} x^2 = y_0, \quad \text{that is,} \quad \lim_{x \to 1} x^2 = 1.$$

In Exercises 47–50, write the limit statements that are suggested by each group of examples or exercises.

47. Examples 1–4.

48. Exercises 23–28.

49. Exercises 29–36.

50. Exercises 37–46.

51. *Grinding engine cylinders.* Before contracting to grind engine cylinders to a cross-sectional area of 9 in^2, you want to know how much deviation from the ideal cylinder diameter of $x_0 = 3.385$ in. you can allow and still have the area come within 0.01 in^2 of the required 9 in^2. To find out, you let $A = \pi(x/2)^2$ and look for the interval in which you must hold x to make $|A - 9| \le 0.01$. What interval do you find?

52. *Manufacturing electrical resistors.* Ohm's law for electrical circuits, like the one shown here, states that $V = RI$. In this equation, V is a constant voltage, I is the current in amperes, and R is the resistance in ohms. Your firm has been asked to supply the resistors for a circuit in which V will be 120 volts and I is to be 5 ± 0.1 amperes. In what interval does R have to lie for I to be within 0.1 ampere of the target value $I_0 = 5$?

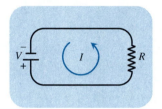

Each of Exercises 53–56 gives a function $y = f(x)$. How large must we choose x to be so that the values of f are within 0.01 of $\lim_{x \to \infty} f(x)$? Estimate graphically, and confirm algebraically.

53. $f(x) = \dfrac{3x + 1}{x - 2}$

54. $f(x) = \dfrac{2x + 5}{5x - 7}$

55. $f(x) = \dfrac{2x^2 - x + 2}{x^2 - 4}$

56. $f(x) = \dfrac{3x^3 - x + 1}{2x^3 + 5}$

57. What must be true about the values of x for $f(x) = \sin x$ to be within 0.1 unit of $\sqrt{2}/2$? Give as complete an answer as you can.

58. The interior of a 1-L measuring cup is a right circular cylinder of radius 5 cm. How wide should the stripes on the side be to measure one liter of water with an error of less than 1%?

Controlling function outputs as $x \to -\infty$. We need to find a positive number N so that for $x < -N$, we keep $y = f(x)$ within an acceptable interval about y_0. Each of Exercises 59–62 gives a function $y = f(x)$. How small must we choose x to be so that the values of f lie within 0.01 of $\lim_{x \to -\infty} f(x)$? Estimate graphically, and confirm algebraically.

59. $f(x) = \dfrac{x - 3}{2x + 1}$

60. $f(x) = \dfrac{3x + 4}{2x - 1}$

61. $f(x) = \dfrac{x - 4}{1 - 3x}$

62. $f(x) = \dfrac{5 - 2x}{x + 1}$

Defining Limits Formally with Epsilons and Deltas

We have just spent five sections using a grapher to understand and estimate limits and using algebra to calculate them. For calculating limits, our basic tools included the substitution theorems, the Sandwich Theorem, and the Properties of Limits Theorem that gave the calculation rules for sums, differences, products, and quotients. With these tools in hand, we started with sensible assumptions about the limits of constant functions and the identity function and worked our way to limits of rational functions. In every case the calculations were straightforward, and the results made sense.

The only problem is that we do not yet know why the theorems that we used to get our results are true. The entire calculus depends on these theorems, and we haven't a clue about why they hold. If we were to try to prove them now, however, we would quickly realize that we never said what a limit really is (except by example) and that we do not have a definition good enough to establish even the simple facts that

$$\lim_{x \to x_0} (k) = k \qquad \text{and} \qquad \lim_{x \to x_0} (x) = x_0.$$

To establish these facts and to understand why the limit theorems hold, we must take one final step and define *limit* formally. In this section, we take that step.

As it turns out, the mathematics that we need to make the notion of limit precise enough to be useful is the same mathematics that we used in Section 2.5 to study target values of functions. There, we had a function $y = f(x)$, a target value y_0, and a bound E on the amount of error we could allow the output values y to have. We wanted to know how close we had to keep x to a particular value x_0 so that y would lie within E units of y_0. In symbols, we were asking for a value of D that would make the inequality $|x - x_0| < D$ imply the inequality $|y - y_0| < E$. The number D described the amount by which x could differ from x_0 and still give y-values that approximated y_0 with an error less than E.

In the limit discussions that follow, we shall use the traditional Greek letters δ (delta) and ϵ (epsilon) in place of the English letters D and E. These are the letters that Cauchy and Weierstrass used in their pioneering work on continuity in the nineteenth century. In their arguments, δ meant *différence* (French for "difference") and ϵ meant *erreur* (French for "error").

As you read along, please keep in mind that the purpose of this section is not to calculate limits of particular functions. We already know how to do that. The purpose here is to develop a technical definition of limit that is good enough to establish the limit theorems on which our calculations depend.

The Definition of Limit

Suppose we are watching the values $f(x)$ of a function as x approaches x_0 (without x taking on the value x_0 itself). What do we have to know about the values of f to say that they have a particular number L as their limit? What observable pattern in their behavior would guarantee their eventual approach to L?

Certainly we want to be able to say that $f(x)$ stays within one tenth of a unit of L as soon as x stays within a certain radius r_1 of x_0, as shown here:

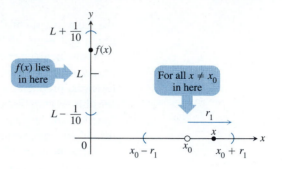

EXPLORATION BITS

To see examples of functions that are "jittery" near $x = 0$, GRAPH

a) $y = \sin(1/x)$,

b) $y = e^{\sin(1/x)}$.

c) $y = (1/x)\sin(1/x)$.

Group Project
Explain why the behavior of each is jittery near $x = 0$. Then create your own example of a jittery function—without using $1/x$.

But that in itself is not enough because as x continues on its course toward x_0, what is to prevent $f(x)$ from jittering about within the interval from $L - 1/10$ to $L + 1/10$ without tending toward L?

We need to say also that as x continues toward x_0, the number $f(x)$ has to get still closer to L. We might say this by requiring $f(x)$ to lie within $1/100$ of a unit of L for all values of x within some smaller radius r_2 of x_0:

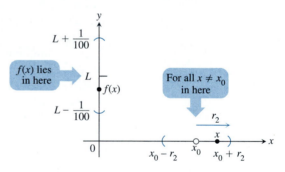

But this is not enough either. What if $f(x)$ skips about within the interval from $L - 1/100$ to $L + 1/100$ from then on, without heading toward L?

We had better require that $f(x)$ lie within $1/1000$ of a unit of L after a while. That is, for all values of x within some still smaller radius r_3 of x_0, all the values of $y = f(x)$ should lie in the interval

$$L - \frac{1}{1000} < y < L + \frac{1}{1000},$$

as shown here:

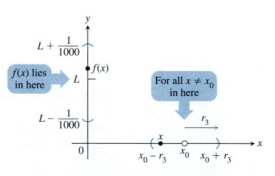

This still does not guarantee that $f(x)$ will now move toward L as x approaches x_0. Even if $f(x)$ has not skipped about before, it might start now. We need more.

We need to require that for *every* interval about L, no matter how small, we can find an interval of numbers about x_0 whose f-values all lie within that interval about L. In other words, given any positive radius ϵ about L, there should exist some positive radius δ about x_0 such that for all x within δ units of x_0 (except $x = x_0$ itself) the values $y = f(x)$ lie within ϵ units of L:

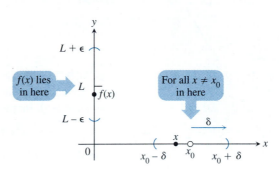

If f satisfies these requirements, we will say that

$$\lim_{x \to x_0} f(x) = L.$$

Here, at last, is a mathematical way to say "the closer x gets to x_0, the closer $y = f(x)$ must get to L."

DEFINITION

The **limit** of $f(x)$ as x approaches x_0 is the number L if the following criterion holds:

Given any radius $\epsilon > 0$ about L there exists a radius $\delta > 0$ about x_0 such that for all x,

$$0 < |x - x_0| < \delta \qquad \text{implies that} \qquad |f(x) - L| < \epsilon.$$

EXPLORATION BIT

Now you should be able to give an ϵ-δ definition of a function being continuous at a point. Try it.

If it turns out that $f(x_0) = L$, then the inequality $0 < |x - x_0| < \delta$ can be replaced by $|x - x_0| < \delta$ because the inequality $|f(x) - L| < \epsilon$ is automatically satisfied for $x = x_0$.

To return to the notions of error and difference, we might think of machining something like a generator shaft with great accuracy. We try for diameter L, but since nothing is perfect, we must be satisfied to get the diameter $f(x)$ somewhere between $L - \epsilon$ and $L + \epsilon$. The δ is the measure of how accurate our control setting for x must be to guarantee this ϵ-accuracy in the diameter of the shaft.

Testing the Definition

Whenever someone proposes a new definition, it is a good idea to determine whether it gives results that are consistent with past experience. For instance, our experience tells us that as x approaches 1, the number $5x - 3$ approaches $5 - 3 = 2$. If our new definition were to lead to some other result, we would want to throw out the definition and look for a new one. The following example and Exploration are included in part to show that the definition of limit gives the kinds of results we want.

EXAMPLE 1   Testing the Definition

Show that
$$\lim_{x \to 1}(5x - 3) = 2.$$

Solution   In the definition of limit we set $f(x) = 5x - 3$, $x_0 = 1$, and $L = 2$. To show that $\lim_{x \to 1}(5x - 3) = 2$, we need to show that for any number $\epsilon > 0$, there exists a number $\delta > 0$ such that for all x,

$$0 < |x - 1| < \delta \quad \Rightarrow \quad |(5x - 3) - 2| < \epsilon. \tag{1}$$

To find a suitable value for δ, we note that

$$|(5x - 3) - 2| < \epsilon$$

$$\Leftrightarrow |5x - 5| < \epsilon$$

$$\Leftrightarrow 5|x - 1| < \epsilon$$

$$\Leftrightarrow |x - 1| < \epsilon/5.$$

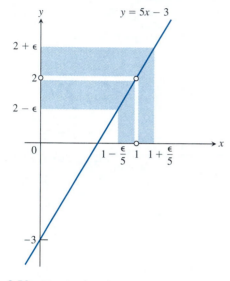

2.53 For the function $f(x) = 5x - 3$ we find that $|x - 1| < \epsilon/5$ will guarantee $|f(x) - 2| < \epsilon$.

Because the above inequalities are equivalent, the last line here tells us that the original ϵ-inequality, and hence the implication in Eq. (1), will hold if we choose $\delta = \epsilon/5$. (See Fig. 2.53.)

The value $\delta = \epsilon/5$ is not the only value that will make the implication in Eq. (1) hold. Any smaller positive δ will do as well. The definition does not ask for a "best" δ, just one that will work.   ≡

EXPLORATION 1

Understanding Analytic Definitions

Theorems 1(a) and 1(b) in Section 2.1 state that

$$\lim_{x \to c}(k) = k \qquad \text{and} \qquad \lim_{x \to c}(x) = c,$$

where (k) represents the constant function $f(x) = k$ and (x) represents the identity function $g(x) = x$. To prove these two theorems formally, you can test the definition directly.

1. For $f(x) = k$ you must find some positive number δ such that $|x - c| < \delta \Rightarrow |k - k| < \epsilon$. This is easier than it looks!

2. For $g(x) = x$ you must find $\delta > 0$ such that $|x - c| < \delta \Rightarrow |x - c| < \epsilon$. This is as easy as it looks!

3. Write an explanation of why $f(x) = k$ and $g(x) = x$ are continuous functions using the definition of continuous function.   ⚇

Finding Deltas for Given Epsilons

In Section 2.5 we were given numerical ϵ's and asked to find numerical δ's so that x values within δ units of a particular value, x_0, yielded y values within ϵ units of a particular value, y_0.

To prove a limit mathematically, we have to find a δ interval about x_0 for *every* ϵ interval about y_0. Thus δ frequently must be expressed in terms of ϵ.

Study carefully the next three examples, all about $f(x) = 1/x$. In the first, we find a numerical δ for a given numerical ϵ one more time. In the second, we find a numerical δ algebraically. In Example 4, we find a δ that will work for any given ϵ. As a check, the numerical results (Example 2) should fit the general solution (Example 4).

EXAMPLE 2

If $f(x) = 1/x$, then $f(0.5) = 2$. Let $\epsilon = 0.01$. Use a grapher to find a value $\delta > 0$ for which $|x - 0.5| < \delta$ implies that $|f(x) - 2| < e$.

Solution We are asked to find a value $\delta > 0$ for which $|x-0.5| < \delta$ implies that $f(x)$ is within 0.01 unit of 2, that is, $f(x)$ is between 1.99 and 2.01. Figure 2.54 shows the graphs of $y_1 = f(x) = \dfrac{1}{x}$, $y_2 = 1.99$, and $y_3 = 2.01$. We can ZOOM, TRACE, or SOLVE to see that if $0.498 < x < 0.502$, or $|x - 0.5| < 0.002$, then $|f(x) - 2| < 0.01$. ≡

EXAMPLE 3

Now find a value $\delta > 0$ algebraically for which $|x - 0.5| < \delta$ implies that $|f(x) - 2| < 0.01$.

Solution We want to find an interval of x values around 0.5 for which $|f(x) - 2| < 0.01$. To do this, we change

$$|f(x) - 2| < 0.01$$

to
$$\left|\frac{1}{x} - 2\right| < 0.01$$

to
$$-0.01 < \frac{1}{x} - 2 < 0.01$$

to
$$1.99 < \frac{1}{x} < 2.01$$

to
$$\frac{1}{2.01} < x < \frac{1}{1.99} \qquad 0 < a < b < c \Rightarrow \frac{1}{c} < \frac{1}{b} < \frac{1}{a}$$

to
$$\frac{1}{2.01} - 0.5 < x - 0.5 < \frac{1}{1.99} - 0.5$$

to
$$-0.0024875\cdots < x - 0.5 < 0.0025125\ldots$$

to
$$-0.00248 < x - 0.5 < 0.00251 \qquad \text{Rounding safely}$$

to
$$-0.00248 < x - 0.5 < 0.00248 \qquad \text{Choosing the smaller radius}$$

to
$$|x - 0.5| < 0.00248.$$

We could have chosen an even smaller $\delta = 0.002$ to have been in agreement with Example 2. To complete a mathematical proof that $\delta = 0.00248$ works, we must show that $|x - 0.5| < \delta$ implies that $|f(x) - 2| < \epsilon$. (See Exercise 41.) ≡

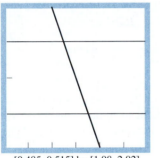

[0.485, 0.515] by [1.98, 2.02]

2.54 Keeping x between 0.498 and 0.502 will keep $y_1 = 1/x$ between $y_2 = 1.99$ and $y_3 = 2.01$.

EXAMPLE 4

Finally, find a value $\delta > 0$ for which $|x - 0.5| < \delta$ implies that $|f(x) - 2| < \epsilon$ for any $\epsilon > 0$, no matter how small ϵ is.

Solution   We want to find an interval of x values around 0.5 for which $|f(x) - 2| < \epsilon$. To do this, we write the equivalent inequalities

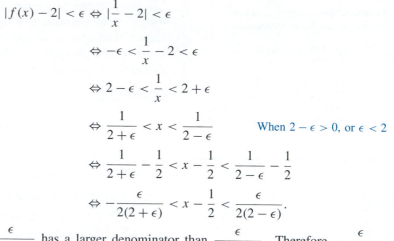

$$|f(x) - 2| < \epsilon \Leftrightarrow |\frac{1}{x} - 2| < \epsilon$$

$$\Leftrightarrow -\epsilon < \frac{1}{x} - 2 < \epsilon$$

$$\Leftrightarrow 2 - \epsilon < \frac{1}{x} < 2 + \epsilon$$

$$\Leftrightarrow \frac{1}{2 + \epsilon} < x < \frac{1}{2 - \epsilon} \qquad \text{When } 2 - \epsilon > 0, \text{ or } \epsilon < 2$$

$$\Leftrightarrow \frac{1}{2 + \epsilon} - \frac{1}{2} < x - \frac{1}{2} < \frac{1}{2 - \epsilon} - \frac{1}{2}$$

$$\Leftrightarrow -\frac{\epsilon}{2(2 + \epsilon)} < x - \frac{1}{2} < \frac{\epsilon}{2(2 - \epsilon)}.$$

$\frac{\epsilon}{2(2 + \epsilon)}$ has a larger denominator than $\frac{\epsilon}{2(2 - \epsilon)}$. Therefore, $\frac{\epsilon}{2(2 + \epsilon)} < \frac{\epsilon}{2(2 - \epsilon)}$. We choose the smaller radius $\delta = \frac{\epsilon}{2(2 + \epsilon)}$, and we can show that $|x - 0.5| < \delta \Rightarrow |f(x) - 2| < \epsilon$ by essentially working backwards through the above inequalities. Note that at one point in finding this δ, we required that ϵ had to be smaller than 2. Thus, this δ "works" when $\epsilon < 2$. Whenever $\epsilon \geq 2$, we can use $\delta = 1/4$. (See Exercise 43.) ▤

The solution in Example 4 may seem rather neat and concise, but it took heaps of scratch paper and several blind alleys to finally find the δ. Finding δ in Example 3 for a *specific* case before finding a *general* value for δ in Example 4 illustrates a problem-solving skill that helped us immensely.

Locally Straight Functions

In Examples 1 and 4 we found δ's that corresponded to arbitrary ϵ's. Generally, finding a δ corresponding to an arbitrary ϵ can be very difficult because δ usually depends on ϵ in a very complicated way. In Chapter 3, we introduce the idea of differentiable functions. Functions that are differentiable are *locally straight*, that is, as we ZOOM-IN, the graph of the function will appear to be a straight line. We can estimate the relationship between δ and ϵ at $x = a$ where $\lim_{x \to a} f(x) = L$ and f is locally straight.

Figure 2.55 shows the graph of such a function near $x = a$ between the horizontal lines $y = L + \epsilon$ and $y = L - \epsilon$. It turns out that if f is differentiable at $x = a$, then f is continuous at $x = a$ so that $L = f(a)$. We choose δ as indicated in Fig. 2.55. This is the largest that δ can be so that $|x - a| < \delta$ implies that $|f(x) - L| < \epsilon$. Let m be the slope of the straight line graph of f in Fig. 2.55. Because m may be negative, we have $|m| = \epsilon/\delta$ and $\delta = \epsilon/|m|$. An appropriate estimate for m gives an appropriate estimate for δ as illustrated in the next example.

EXPLORATION BIT

In Example 2, for $\epsilon = 0.01$ we found $\delta = 0.002$. Using $\epsilon = 0.01$ in Example 4, we find that $\delta = \frac{\epsilon}{2(2 + \epsilon)} = 0.00248\dots$. Note that the δ-interval from Example 2 fits inside the δ-interval of Example 4. What is the implication of this?

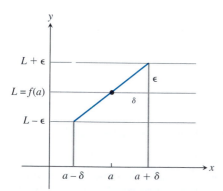

2.55 A ZOOM-IN view of a function f that is locally straight and $\lim_{x \to a} f(x) = L$.

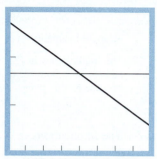

[1.98, 2.02] by [0.65, 0.68]

2.56 A ZOOM-IN view of $f(x) = x/(x^2 - 1)$ and $y = 2/3$. We use two points on the graph of f to estimate the slope of the line suggested by the graph. Then we use that slope to tell us what δ must be for any small ϵ. Because we have zoomed-in so that the graph appears straight between the two chosen points, we know that ϵ is sufficiently small for the value $\delta = (5/3)\epsilon$ found in Example 5 to be correct. Certainly the δ-interval about $x = 2$ must be small enough to avoid the vertical asymptote at $x = 1$.

EXAMPLE 5

If $f(x) = x/(x^2 - 1)$, then $\lim_{x\to 2} f(x) = 2/3$. For $\epsilon > 0$, estimate $\delta > 0$ so that $|x - 2| < \delta$ implies that $|f(x) - 2/3| < \epsilon$.

Solution   Figure 2.56 gives a ZOOM-IN view near the point (2, 2/3) of the graph of $y = 2/3$ and the graph of f that appears to be a straight line. We can find the coordinates of two points on the graph of f in Fig. 2.56 to estimate the slope of the line to be $m = -0.5559$. Thus, $\epsilon/\delta = |m|$ or $\delta = \epsilon/|m|$ (see Fig. 2.55). To be safe, we choose a smaller value for δ by replacing $|m|$ by 0.6, and we see that $\delta = \epsilon/0.6$, or $\delta = (5/3)\epsilon$.   ≡

How Limit Theorems Are Proved

Although we shall not ask you to prove limit theorems yourself, we want to show how a typical proof goes, if only to support our claim that having a precise definition of limit now makes it possible for us to prove the limit theorems on which calculus depends. Our example will be the proof of the Sum Rule for limits (the first part of Theorem 1(c) from Section 2.1). You can find a proof of the rest of Theorem 1 in Appendix 2.

> **THEOREM 1**   **(Revisited)**
>
> **Part c(1):**
>
> If $\lim_{x\to x_0} f(x) = L_1$ and $\lim_{x\to x_0} g(x) = L_2$, then $\lim_{x\to x_0}(f(x) + g(x)) = L_1 + L_2$.

Proof   To show that $\lim_{x\to x_0}(f(x) + g(x)) = L_1 + L_2$, we must show that for any $\epsilon > 0$, there exists a $\delta > 0$ such that for all x,

$$0 < |x - x_0| < \delta \quad \Rightarrow \quad |f(x) + g(x) - (L_1 + L_2)| < \epsilon.$$

Suppose, then, that ϵ is a positive number. The number $\epsilon/2$ is positive, too, and because $\lim_{x\to x_0} f(x) = L_1$, we know that there is a $\delta_1 > 0$ such that for all x,

$$0 < |x - x_0| < \delta_1 \quad \Rightarrow \quad |f(x) - L_1| < \frac{\epsilon}{2}. \qquad (2)$$

Because $\lim_{x\to x_0} g(x) = L_2$, there is also a $\delta_2 > 0$ such that for all x,

$$0 < |x - x_0| < \delta_2 \quad \Rightarrow \quad |g(x) - L_2| < \frac{\epsilon}{2}. \qquad (3)$$

Now, either δ_1 equals δ_2 or it doesn't. If δ_1 equals δ_2, the implications in Eqs. (2) and (3) both hold true for their common value δ. Taken together, Eqs. (2) and (3) then say that for all x, $0 < |x - x_0| < \delta$ implies that

$$|f(x) + g(x) - (L_1 + L_2)|$$

$$= |(f(x) - L_1) + (g(x) - L_2)|$$

$$\leq |(f(x) - L_1)| + |(g(x) - L_2)| \qquad \text{Triangle inequality}$$

$$< \frac{\epsilon}{2} + \frac{\epsilon}{2} \qquad \begin{array}{l} \text{The implications in (2)} \\ \text{and (3) both hold for } \delta \\ \text{because } \delta = \delta_1 = \delta_2. \end{array}$$

$$< \epsilon.$$

If $\delta_1 \neq \delta_2$, let δ be the smaller of δ_1 and δ_2. The implications in (2) and (3) then both hold for all x such that $0 < |x - x_0| < \delta$. As before,

$$|f(x) + g(x) - (L_1 + L_2)| < \epsilon.$$

Either way, we know that given any $\epsilon > 0$, there exists a $\delta > 0$ such that for all x,

$$0 < |x - x_0| < \delta \quad \Rightarrow \quad |f(x) + g(x) - (L_1 + L_2)| < \epsilon.$$

According to the ϵ-δ definition of limit, then,

$$\lim_{x \to x_0} (f(x) + g(x)) = L_1 + L_2. \qquad \blacksquare$$

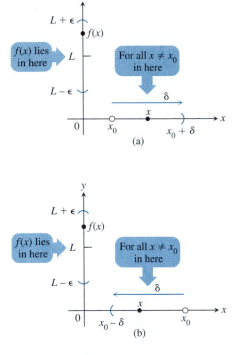

2.57 The (a) right-hand limit and (b) left-hand limit as $x \to x_0$.

The Relation between One-sided and Two-sided Limits

The formal definitions of right-hand and left-hand limits go like this:

DEFINITION

Right-hand Limit: $\lim_{x \to x_0^+} f(x) = L$

The limit of $f(x)$ as x approaches x_0 from the right (Fig. 2.57a) is the number L if the following criterion holds:

Given any radius $\epsilon > 0$ about L, there exists a radius $\delta > 0$ to the right of x_0 such that for all x,

$$x_0 < x < x_0 + \delta \qquad \text{implies that} \qquad |f(x) - L| < \epsilon. \qquad (4)$$

Left-hand Limit: $\lim_{x \to x_0^-} f(x) = L$

The limit of $f(x)$ as x approaches x_0 from the left (Fig. 2.57b) is the number L if the following criterion holds:

Given any radius $\epsilon > 0$ about L, there exists a radius $\delta > 0$ to the left of x_0 such that for all x,

$$x_0 - \delta < x < x_0 \qquad \text{implies that} \qquad |f(x) - L| < \epsilon. \qquad (5)$$

By comparing Eqs. (4) and (5) with the corresponding line in the limit definition, we can see the relation between the one-sided limits just defined and the two-sided limit defined earlier. If we subtract x_0 from the δ-inequalities in Eqs. (4) and (5), they become

$$0 < x - x_0 < \delta \quad \Rightarrow \quad |f(x) - L| < \epsilon \tag{6}$$

and

$$-\delta < x - x_0 < 0 \quad \Rightarrow \quad |f(x) - L| < \epsilon. \tag{7}$$

Together, Eqs. (6) and (7) say the same thing as

$$0 < |x - x_0| < \delta \quad \Rightarrow \quad |f(x) - L| < \epsilon$$

in the definition of limit. In other words, $f(x)$ has limit L at x_0 if and only if the right-hand and left-hand limits of f at x_0 exist and equal L.

Limits Involving Infinity

There are several formal definitions of limits involving infinity. They are suggested by the following two:

DEFINITIONS

Limit at Infinity: The limit of $f(x)$ as x increases without bound $(x \to \infty)$ is the number L,

$$\lim_{x \to \infty} f(x) = L,$$

if the following holds:

Given any radius $\epsilon > 0$ about L, there exists a number $N > 0$ such that for all x,

$$x > N \quad \text{implies that} \quad |f(x) - L| < \epsilon.$$

Limit of Infinity: The values of $f(x)$ as x approaches x_0 are unboundedly large,

$$\lim_{x \to x_0} f(x) = \infty,$$

if the following holds:

Given any number $N > 0$, there exists a radius $\delta > 0$ about x_0 such that for all x,

$$0 < |x - x_0| < \delta \quad \text{implies that} \quad f(x) > N.$$

LIMITS OF AND AT INFINITY

You should be able to state or extend the given definitions to cover all limits involving infinity:

a) $\lim_{x \to \pm\infty} f(x) = L$;

b) $\lim_{x \to x_0} f(x) = \pm\infty$;

c) $\lim_{x \to \pm\infty} f(x) = \pm\infty$;

d) the various related one-sided limits.

Examples 6 and 7 show how to apply the definitions.

EXAMPLE 6

Confirm that $\lim_{x\to\infty} \dfrac{2x+3}{x+1} = 2$.

Solution   For any $\epsilon > 0$, we want to find a number $N > 0$ so that

$$\left| \frac{2x+3}{x+1} - 2 \right| < \epsilon$$

when $x > N$. We work backwards, assuming that ϵ is small to begin with, say $\epsilon < 1$:

$$\left| \frac{2x+3}{x+1} - 2 \right| < \epsilon \Leftrightarrow \left| \frac{2x+3-2x-2}{x+1} \right| < \epsilon$$

$$\Leftrightarrow \left| \frac{1}{x+1} \right| < \epsilon$$

$$\Leftrightarrow \frac{1}{x+1} < \epsilon \qquad \text{\color{blue}Assuming } x \text{ \color{blue}is to be}$$
$$\text{\color{blue}large and positive.}$$

$$\Leftrightarrow \frac{1}{\epsilon} < x + 1$$

$$\Leftrightarrow \frac{1}{\epsilon} - 1 = \frac{1-\epsilon}{\epsilon} < x.$$

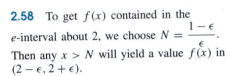

2.58 To get $f(x)$ contained in the e-interval about 2, we choose $N = \dfrac{1-\epsilon}{\epsilon}$. Then any $x > N$ will yield a value $f(x)$ in $(2-\epsilon, 2+\epsilon)$.

So we choose $N = \dfrac{1-\epsilon}{\epsilon}$. Then $x > N$ implies that $x > \dfrac{1-\epsilon}{\epsilon}$, and we can reverse the steps above to conclude that $\left| \dfrac{2x+3}{x+1} - 2 \right| < \epsilon$. (See Fig. 2.58.)   ▇

EXAMPLE 7

Confirm that $\lim_{x\to -1^{+}} \dfrac{2x+3}{x+1} = \infty$.

Solution   For any number $N > 0$, we want to find an interval $-1 < x < -1 + \delta$ on which $\dfrac{2x+3}{x+1} > N$. Again, we work backwards, assuming that N is large to begin with, say $N > 2$:

$$\frac{2x+3}{x+1} > N \Leftrightarrow 2x+3 > N(x+1) \qquad \text{\color{blue}}x > -1 \Rightarrow x+1 > 0$$

$$\Leftrightarrow x(2-N) > N - 3$$

$$\Leftrightarrow x < \frac{N-3}{2-N} \qquad \text{\color{blue}}N > 2 \Rightarrow 2 - N < 0$$

$$\Leftrightarrow x + 1 < \frac{N-3}{2-N} + 1 = \frac{-1}{2-N} = \frac{1}{N-2}.$$

So we choose $\delta = \dfrac{1}{N-2}$. Then $-1 < x < -1 + \delta$ implies that $0 < x + 1 < \delta = \dfrac{1}{N-2}$, and we can reverse the steps above to conclude that $\dfrac{2x+3}{x+1} > N$.   ▇

Exercises 2.6

In Exercises 1–4, sketch the interval (a, b) on the x-axis with the point x_0 inside. Then find the largest value of $\delta > 0$ such that $|x - x_0| < \delta$ implies that $a < x < b$.
1. $a = 1$, $b = 7$, $x_0 = 5$ **2.** $a = 1$, $b = 7$, $x_0 = 2$
3. $a = -7/2$, $b = -1/2$, $x_0 = -3$
4. $a = -7/2$, $b = -1/2$, $x_0 = -3/2$

Use the graphs in Exercises 5–10 to find a $\delta > 0$ such that for all x,

$$0 < |x - x_0| < \delta \quad \Rightarrow \quad |f(x) - L| < \epsilon.$$

5.

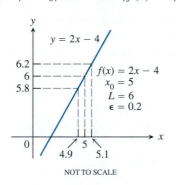

$$y = 2x - 4$$
$$f(x) = 2x - 4$$
$$x_0 = 5$$
$$L = 6$$
$$\epsilon = 0.2$$

NOT TO SCALE

6.

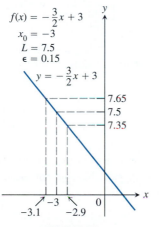

$$f(x) = -\frac{3}{2}x + 3$$
$$x_0 = -3$$
$$L = 7.5$$
$$\epsilon = 0.15$$
$$y = -\frac{3}{2}x + 3$$

NOT TO SCALE

7.

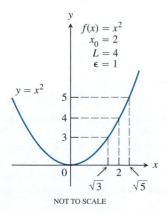

$$f(x) = x^2$$
$$x_0 = 2$$
$$L = 4$$
$$\epsilon = 1$$
$$y = x^2$$

NOT TO SCALE

8.

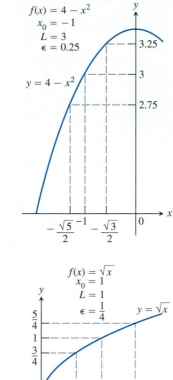

$$f(x) = 4 - x^2$$
$$x_0 = -1$$
$$L = 3$$
$$\epsilon = 0.25$$
$$y = 4 - x^2$$

9.

$$f(x) = \sqrt{x}$$
$$x_0 = 1$$
$$L = 1$$
$$\epsilon = \frac{1}{4}$$
$$y = \sqrt{x}$$

10.

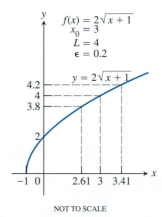

$$f(x) = 2\sqrt{x + 1}$$
$$x_0 = 3$$
$$L = 4$$
$$\epsilon = 0.2$$
$$y = 2\sqrt{x + 1}$$

NOT TO SCALE

Each of Exercises 11–18 gives a function $f(x)$, a point x_0, and a positive number ϵ. Find $L = \lim_{x \to x_0} f(x)$. Then find a number $\delta > 0$ such that for all x,

$$0 < |x - x_0| < \delta \quad \Rightarrow \quad |f(x) - L| < \epsilon.$$

11. $f(x) = 2x + 3, x_0 = 1, \epsilon = 0.01$

12. $f(x) = 3 - 2x, x_0 = 3, \epsilon = 0.02$

13. $f(x) = \dfrac{x^2 - 4}{x - 2}, x_0 = 2, \epsilon = 0.05$

14. $f(x) = \dfrac{x^2 + 6x + 5}{x + 5}, x_0 = -5, \epsilon = 0.05$

15. $f(x) = \sqrt{x - 7}, x_0 = 11, \epsilon = 0.01$

16. $f(x) = \sqrt{1 - 5x}, x_0 = -3, \epsilon = 0.5$

17. $f(x) = 4/x, x_0 = 2, \epsilon = 0.4$

18. $f(x) = 4/x, x_0 = 1/2, \epsilon = 0.04$

In Exercises 19 and 20, find the largest $\delta > 0$ such that for all x,

$$0 < |x - 4| < \delta \quad \Rightarrow \quad |f(x) - 5| < \epsilon.$$

19. $f(x) = 9 - x;\ \epsilon = 0.01, 0.001, 0.0001$, arbitrary $\epsilon > 0$

20. $f(x) = 3x - 7;\ \epsilon = 0.003, 0.0003$, arbitrary $\epsilon > 0$

In Exercises 21–24, evaluate and confirm the limits using the definitions of limits involving infinity.

21. $\lim_{x \to \infty} \dfrac{x + 2}{x + 1}$

22. $\lim_{x \to \infty} \dfrac{x^2}{2x^2 - 1}$

23. $\lim_{x \to -1^+} \dfrac{x + 2}{x + 1}$

24. $\lim_{x \to (\sqrt{2}/2)^+} \dfrac{x^2}{2x^2 - 1}$

The functions in Exercises 25–32 are locally straight. Let $L = \lim_{x \to x_0} f(x), \epsilon > 0$, and estimate $\delta > 0$ so that $|x - x_0| < \delta$ implies that $|f(x) - L| < \epsilon$. You may assume that ϵ is small, say $\epsilon < 0.01$.

25. $f(x) = \sin x, x_0 = 1$

26. $f(x) = \tan x, x_0 = 1$

27. $f(x) = \cos x, x_0 = 1$

28. $f(x) = \sec x, x_0 = 4$

29. $f(x) = x^3 - 4x, x_0 = 0.5$

30. $f(x) = 9x - x^3, x_0 = 2.5$

31. $f(x) = \dfrac{x}{x^2 - 4}, x_0 = -1$

32. $f(x) = \dfrac{2x}{5 - x^2}, x_0 = -1$

33. Given $\epsilon > 0$, find an interval $I = (5, 5 + \delta), \delta > 0$, such that if x lies in I, then $\sqrt{x - 5} < \epsilon$. What limit is being verified?

34. Given $\epsilon > 0$, find an interval $I = (4 - \delta, 4), \delta > 0$, such that if x lies in I, then $\sqrt{4 - x} < \epsilon$. What limit is being verified?

35. Graph the function

$$f(x) = \begin{cases} 4 - 2x, & x < 1, \\ 6x - 4, & x \geq 1. \end{cases}$$

Then, given $\epsilon > 0$, find the largest δ for which $f(x)$ lies between $y = 2 - \epsilon$ and $y = 2 + \epsilon$ for x in the interval $I = (1 - \delta, 1 + \delta)$.

36. Let $f(x) = |x - 5|/(x - 5)$. Find the set of x-values for which

$$1 - \epsilon < f(x) < 1 + \epsilon, \quad \text{for } \epsilon = 4, 2, 1, \text{ and } 1/2.$$

37. Define what it means to say that $\lim_{x \to 2} f(x) = 5$.

38. Define what it means to say that $\lim_{x \to 0} g(x) = k$.

39. Suppose $0 < \epsilon < 4$. Find the largest $\delta > 0$ with the property that $|x - 2| < \delta$ implies that $|x^2 - 4| < \epsilon$. What limit is being verified? What happens to δ as ϵ decreases toward 0? Graph δ as a function of ϵ.

40. Suppose $0 < \epsilon < 1$. Find the largest $\delta > 0$ with the property that $|x - 3| < \delta$ implies that $\left| \dfrac{2}{x - 1} - 1 \right| < \epsilon$. What limit is being verified? What happens to δ as ϵ decreases toward 0? Graph δ as a function of ϵ.

41. Let $f(x) = 1/x$. Show that $|x - 0.5| < 0.00248$ implies that $|f(x) - 2| < 0.01$. (See Example 3.)

42. Let $0 < \epsilon < 2$.

a) Use graphs to support $\dfrac{1}{2 + \epsilon} < \dfrac{1}{2} < \dfrac{1}{2 - \epsilon}$.

b) Show that for $\epsilon > 2$, the inequality statement in part (a) does not hold. Which part fails?

43. Let $f(x) = 1/x$ and

$$\delta = \begin{cases} \dfrac{\epsilon}{2(2 + \epsilon)}, & \text{if } \epsilon < 2, \\ 1/4, & \text{if } \epsilon \geq 2. \end{cases}$$

Show that $|x - 0.5| < \delta$ implies that $|f(x) - 2| < \epsilon$. (See Example 4 and Exercise 42.)

44. Let

$$f(x) = \begin{cases} x^3 + 1.001, & x \geq 0, \\ x^3 + 0.009, & x < 0. \end{cases}$$

a) Draw a complete graph of f.

b) Find $\lim_{x \to 0} f(x)$ if it exists.

c) If the limit in part (b) does not exist, show how the implication

$$0 < |x - x_0| < \delta \Rightarrow |f(x) - L| < \epsilon$$

fails.

45. Suppose

$$f(x) = \begin{cases} x, & \text{if } x = 1/n \text{ for every natural number } n, \\ 0, & \text{otherwise.} \end{cases}$$

a) Graph $f(x)$.

b) Does $\lim_{x \to 0} f(x)$ exist? If so, find the limit. Explain.

46. Note that $\lim_{x \to -1} |x + 1|/(x + 1)$ does not exist. Prove the limit is not 1 with an ϵ-δ argument. That is, show that there is an $\epsilon > 0$ such that for each $\delta > 0$ there is always an x_0 with $0 < |x_0 + 1| < \delta$ but $||x + 1|/(x + 1) - 1| \geq \epsilon$. (*Hint:* Choose a value for δ, say $\delta = 2$. Then find an ϵ that makes the last inequality hold. Finally, complete the argument for any arbitrary δ.)

Review Questions

1. You have been asked to calculate the limit of a function $f(x)$ as x approaches a finite number c. What theorems are available for calculating the limit? Give examples to show how the theorems are used.

2. What is the relation between one-sided and two-sided limits? How is this relation sometimes used to calculate a limit or to prove that a limit does not exist? Give examples.

3. You used a graphing calculator to estimate $\lim_{\theta\to 0}(\sin\theta)/\theta$ and found the graph of $f(\theta) = (\sin\theta)/\theta$ to be the horizontal line $y = 0.0174532925 \approx \pi/180$ in this figure. Explain what you did wrong.

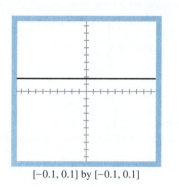

$[-0.1, 0.1]$ by $[-0.1, 0.1]$

4. What is the procedure for finding the limit of a rational function of x as x approaches $\pm\infty$? When is the limit zero? Finite and different from zero? Infinite?

5. What test can you apply to find out whether a function $y = f(x)$ is continuous at point $x = c$? Give examples of functions that are continuous at $x = 0$. Give examples of functions that fail to be continuous at $x = 0$ for various reasons. (They don't have to be examples from the book. You may make up your own.)

6. What can be said about the continuity of polynomial functions and rational functions?

7. What can be said about the continuity of composites of continuous functions?

8. What are the important theorems about continuous functions? Can functions that are not continuous be expected to have the properties guaranteed by these theorems? Give examples.

9. What are the formal definitions of (two-sided) limit, right-hand limit, and left-hand limit?

10. Discuss how a grapher can be used to *help* determine the limit of a function. Explain why this is not conclusive.

11. Define horizontal and vertical asymptote.

12. Give an example of a function with a removable discontinuity.

13. Define end behavior and end behavior model for a function.

14. Define end behavior asymptote for a rational function.

15. How are absolute values used to describe intervals of real numbers?

16. Show by example how absolute values are used to control function values.

17. State the formal definitions of limits of infinity and limits at infinity.

Practice Exercises

In Exercises 1–6, decide whether the limits exist on the basis of the graph of $y = f(x)$ shown. The domain of f is the set of real numbers.

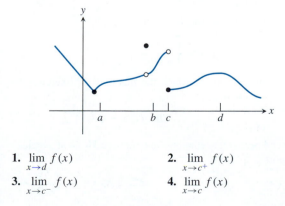

1. $\lim\limits_{x\to d} f(x)$

2. $\lim\limits_{x\to c^+} f(x)$

3. $\lim\limits_{x\to c^-} f(x)$

4. $\lim\limits_{x\to c} f(x)$

5. $\lim\limits_{x\to b} f(x)$

6. $\lim\limits_{x\to a} f(x)$

In Exercises 7–10, decide whether $f(x)$ whose graph is shown above is continuous at the following points.

7. $x = a$

8. $x = b$

9. $x = c$

10. $x = d$

Find the limits in Exercises 11–30. Some limits may not exist.

11. $\lim\limits_{x\to -2} x^2(x + 1)$

12. $\lim\limits_{x\to 3} (x + 2)(x - 5)$

13. $\lim\limits_{x\to 3} \dfrac{x - 3}{x^2}$

14. $\lim\limits_{x\to -1} \dfrac{x^2 + 1}{3x^2 - 2x + 5}$

15. $\lim\limits_{x\to -2} \left(\dfrac{x}{x + 1}\right)\left(\dfrac{3x + 5}{x^2 + x}\right)$

16. $\lim\limits_{x \to 1} \left(\dfrac{1}{x+1} \right) \left(\dfrac{x+6}{x} \right) \left(\dfrac{3-x}{7} \right)$

17. $\lim\limits_{x \to 4} \sqrt{1 - 2x}$

18. $\lim\limits_{x \to 5} \sqrt[4]{9 - x^2}$

19. $\lim\limits_{x \to 1} \dfrac{x^2 - 1}{x - 1}$

20. $\lim\limits_{x \to -5} \dfrac{x^2 + 3x - 10}{x + 5}$

21. $\lim\limits_{x \to 2} \dfrac{x - 2}{x^2 + x - 6}$

22. $\lim\limits_{x \to 1} \dfrac{x^2 - 2x + 1}{x^3 - 2x^2 + x}$

23. $\lim\limits_{x \to 0} \dfrac{(1+x)(2+x) - 2}{x}$

24. $\lim\limits_{x \to 0} \dfrac{\frac{1}{2+x} - \frac{1}{2}}{x}$

25. $\lim\limits_{x \to \infty} \dfrac{2x + 3}{5x + 7}$

26. $\lim\limits_{x \to -\infty} \dfrac{2x^2 + 3}{5x^2 + 7}$

27. $\lim\limits_{x \to -\infty} \dfrac{x^2 - 4x + 8}{3x^3}$

28. $\lim\limits_{x \to \infty} \dfrac{1}{x^2 - 7x + 1}$

29. $\lim\limits_{x \to -\infty} \dfrac{x^2 - 7x}{x + 1}$

30. $\lim\limits_{x \to \infty} \dfrac{x^4 + x^3}{12x^3 + 128}$

Find the limits in Exercises 31–34.

31. $\lim\limits_{x \to 3^+} \dfrac{1}{x - 3}$

32. $\lim\limits_{x \to 3^-} \dfrac{1}{x - 3}$

33. $\lim\limits_{x \to 0^+} \dfrac{1}{x^2}$

34. $\lim\limits_{x \to 0^-} \dfrac{1}{|x|}$

Find the limits in Exercises 35–38.

35. $\lim\limits_{x \to 0} \dfrac{\sin 2x}{4x}$

36. $\lim\limits_{x \to 0} \dfrac{x + \sin x}{x}$

37. $\lim\limits_{x \to 0} \dfrac{\sin^3 2x}{x^3}$

38. $\lim\limits_{x \to 0} \dfrac{2 \csc 5x}{\csc 3x}$

39. Let $f(x) = \dfrac{\sec 2x \csc 9x}{\cot 7x}$.

   a) Estimate $\lim_{x \to 0} f(x)$ graphically.

   b) Compare the function values of f near 0 to the limit found in part (a).

   c) Describe (informally, no proof required) the behavior of f near 0. (*Hint:* Restrict x to $[-0.2, 0.2]$.)

   d) Find the exact value of $\lim_{x \to 0} f(x)$ algebraically.

40. Let $f(x) = \dfrac{\csc 5x \sec 8x}{\csc 8x \sec 3x}$.

   a) Estimate $\lim_{x \to 0} f(x)$ graphically.

   b) Compare the function values of f near 0 to the limit found in part (a).

   c) Describe (informally, no proof required) the behavior of f near 0. (*Hint:* Restrict x to $[-0.2, 0.2]$.)

   d) Find the exact value of $\lim_{x \to 0} f(x)$ algebraically.

In Exercises 41 and 42 find the limits.

41. a) $\lim_{x \to -2^+} \dfrac{x + 3}{x + 2}$
   b) $\lim_{x \to -2^-} \dfrac{x + 3}{x + 2}$

42. a) $\lim_{x \to 2^+} \dfrac{x - 1}{x^2(x - 2)}$
   b) $\lim_{x \to 2^-} \dfrac{x - 1}{x^2(x - 2)}$

   c) $\lim_{x \to 0^+} \dfrac{x - 1}{x^2(x - 2)}$
   d) $\lim_{x \to 0^-} \dfrac{x - 1}{x^2(x - 2)}$

43. Let

$$f(x) = \begin{cases} 1, & x \le -1, \\ -x, & -1 < x < 0, \\ 1, & x = 0, \\ -x, & 0 < x < 1, \\ 1, & x \ge 1. \end{cases}$$

   a) Determine a complete graph of f.

   b) Find the right-hand and left-hand limits of f at -1, 0, and 1.

   c) Does f have a limit as x approaches -1? 0? 1? If so, what is it? If not, why not?

   d) At which of the points $x = -1, 0, 1$, if any, is f continuous?

44. Repeat Exercise 43 for the function

$$f(x) = \begin{cases} 0, & x \le -1, \\ |2x|, & -1 < x < 1, \\ 0, & x = 1, \\ 1, & x > 1. \end{cases}$$

45. Let $f(x) = \begin{cases} |x^3 - 4x|, & x < 1, \\ x^2 - 2x - 2, & x \ge 1. \end{cases}$

   a) Determine a complete graph of f.

   b) Find the right-hand and left-hand limits of f at 1.

   c) Does f have a limit as x approaches 1? If so, what is it? If not, why not?

   d) At what points is f continuous? Why?

   e) At what points is f not continuous? Why?

46. Let $f(x) = \begin{cases} 1 - \sqrt{3 - 2x}, & x < 3/2, \\ 1 + \sqrt{2x - 3}, & x \ge 3/2. \end{cases}$

   a) Determine a complete graph of f.

   b) Find the right-hand and left-hand limits of f at 3/2.

   c) Does f have a limit as x approaches 3/2? If so, what is it? If not, why not?

   d) At what points is f continuous? Why?

   e) At what points is f not continuous? Why?

47. Let $f(x) = \begin{cases} -x, & x < 1, \\ x - 1, & x > 1. \end{cases}$

   a) Graph f.

   b) Find the right-hand and left-hand limits of f at $x = 1$.

   c) What value, if any, should be assigned to $f(1)$ to make f continuous at $x = 1$?

48. Repeat Exercise 47 for the function

$$f(x) = \begin{cases} 3x^2, & x < 1, \\ 4 - x^2, & x > 1. \end{cases}$$

Find the points, if any, at which the functions in Exercises 49 and 50 are *not* continuous.

49. $f(x) = \dfrac{x + 1}{4 - x^2}$

50. $f(x) = \sqrt[3]{3x + 2}$

Find the end behavior asymptotes in Exercises 51–54 algebraically and support graphically.

51. $f(x) = \dfrac{2x + 1}{x^2 - 2x + 1}$

52. $g(x) = \dfrac{2x^2 + 5x - 1}{x^2 + 2x}$

53. $h(x) = \dfrac{x^3 - 4x^2 + 3x + 3}{x - 3}$

54. $T(x) = \dfrac{x^4 - 3x^2 + x - 1}{x^3 - x + 1}$

55. Suppose that $f(x)$ and $g(x)$ are defined for all x and that $\lim_{x \to c} f(x) = -7$ and $\lim_{x \to c} g(x) = 0$. Find the limit as $x \to c$ of the following functions.

  a) $3f(x)$　　**b)** $(f(x))^2$　　**c)** $f(x) \cdot g(x)$

  d) $\dfrac{f(x)}{g(x) - 7}$　　**e)** $\cos(g(x))$　　**f)** $|f(x)|$

56. Suppose that $f(x)$ and $g(x)$ are defined for all x and that $\lim_{x \to 0} f(x) = \frac{1}{2}$ and $\lim_{x \to 0} g(x) = \sqrt{2}$. Find the limits as $x \to 0$ of the following functions.

  a) $-g(x)$　　**b)** $g(x) \cdot f(x)$　　**c)** $f(x) + g(x)$

  d) $1/f(x)$　　**e)** $x + f(x)$　　**f)** $\dfrac{f(x) \cdot \sin x}{x}$

57. Use the inequality

$$0 \le \left| \sqrt{x} \sin \frac{1}{x} \right| \le \sqrt{x}$$

to find $\lim_{x \to 0} \sqrt{x} \sin (1/x)$.

58. Use the inequality

$$0 \le \left| x^2 \sin \frac{1}{x} \right| \le x^2$$

to find $\lim_{x \to 0} x^2 \sin (1/x)$.

Given that $\lim_{x \to \infty} (\sin x)/x = \lim_{x \to \infty} (\cos x)/x = 0$, find the limits in Exercises 59 and 60.

59. $\lim_{x \to \infty} \dfrac{x + \sin x}{x}$　　　**60.** $\lim_{x \to \infty} \dfrac{x + \sin x}{x + \cos x}$

Use the Sandwich Theorem to find the limits in Exercises 61 and 62.

61. $\lim_{x \to \infty} \left(\dfrac{\sin x}{\sqrt{x}} \right)$　　　**62.** $\lim_{x \to \infty} \dfrac{\cos x}{\sqrt{x}}$

63. Let $f(x) = \begin{cases} \dfrac{x^2 + 2x - 15}{x - 3}, & x \ne 3, \\ k, & x = 3. \end{cases}$

  What value, if any, should be assigned to k to make f continuous at $x = 3$?

64. With the help of a graphing utility, determine $\lim_{x \to 0^+} x^x$. Why must this limit be approached from the right-hand side?

65. Study $f(x) = (2^x + 3^x)^{1/x}$ near $x = 0$ to determine each limit.

  a) $\lim_{x \to 0^-} f(x)$
  b) $\lim_{x \to 0^+} f(x)$
  c) Does $\lim_{x \to 0} f(x)$ exist? Explain.

66. Let $f(x) = \begin{cases} \dfrac{\sin x}{2x}, & x \ne 0, \\ k, & x = 0. \end{cases}$

  What value, if any, should be assigned to k to make f continuous at $x = 0$?

67. The function $y = 1/x$ does not take on either a maximum or a minimum on the interval $0 < x < 1$, even though the function is continuous on this interval. Does this contradict the Max-Min Theorem for continuous functions? Why?

68. What are the maximum and minimum values of the function $y = |x|$ on the interval $-1 \le x < 1$? Note that the interval is not closed. Is this consistent with the Max-Min Theorem for continuous functions? Why?

69. True or false? If $y = f(x)$ is continuous, with $f(1) = 0$ and $f(2) = 3$, then f takes on the value 2.5 at some point between $x = 1$ and $x = 2$. Explain.

70. Show that there is at least one value of x for which $x + \cos x = 0$.

71. How do you know (for sure) that $x + \log x = 0$ has at least one solution? Explain.

The Definition of Limit

72. Define what it means to say that

$$\lim_{x \to 1} f(x) = 3.$$

73. Define what it means to say that

$$\lim_{x \to 0} \frac{\sin x}{x} = 1.$$

Wrong descriptions of limit. Show by example that the statements in Exercises 74 and 75 are wrong.

74. The number L is the limit of $f(x)$ as x approaches x_0 if $f(x)$ gets closer to L as x approaches x_0.

75. The number L is the limit of $f(x)$ as x approaches x_0 if, given any $\epsilon > 0$, there is a value of x for which $|f(x) - L| < \epsilon$.

Each of Exercises 76–81 gives a function $y = f(x)$, a number E, and a target value y_0. In each case, find an interval of x-values for which $y = f(x)$ lies within E units of y_0. (Recall our agreement to try to choose the interval as large as possible.) Then find an interval which can be described with an absolute value inequality of the form $|x - x_0| < D$.

	$f(x)$	E	y_0
76.	$\sqrt{x + 2}$	1	4
77.	$\sqrt{\dfrac{x + 1}{2}}$	1/2	1
78.	$\dfrac{x - 1}{x - 3}$	0.1	2
79.	$\dfrac{x - 1}{x - 3}$	0.1	-2
80.	$x^3 - 4x$	0.1	4
81.	$x^3 - 4x$	0.1	$1 (-1 < x_0 < 0)$

82. The function $f(x) = 2x - 3$ is continuous at $x = 2$. Given a positive number ϵ, how small must δ be for $|x - 2| < \delta$ to imply that $|f(x) - 1| < \epsilon$?

83. The function $f(x) = |x|$ is continuous at $x = 0$. Given a positive number ϵ, how small must δ be for $|x - 0| < \delta$ to imply that $|f(x) - 0| < \epsilon$?

In Exercises 84 and 85, evaluate and confirm the limits using the definitions of limits involving infinity.

84. $\lim_{x \to \infty} \dfrac{1 - 2x}{3x - 1}$

85. $\lim_{x \to (1/3)^+} \dfrac{1 - 2x}{3x - 1}$

Each of Exercises 86–89 gives a function $f(x)$, a point x_0, and a positive number ϵ. Find $L = \lim_{x \to x_0} f(x)$. Then find a number $\delta > 0$ such that for all x,

$$0 < |x - x_0| < \delta \Rightarrow |f(x) - L| < \epsilon.$$

86. $f(x) = 5x - 10$, $x_0 = 3$, $\epsilon = 0.05$

87. $f(x) = 5x - 10$, $x_0 = 2$, $\epsilon = 0.05$

88. $f(x) = \sqrt{x - 5}$, $x_0 = 9$, $\epsilon = 1$

89. $f(x) = \sqrt{2x - 3}$, $x_0 = 2$, $\epsilon = 1/2$

The functions in Exercises in 90 and 91 are locally straight at x_0. Let $L = \lim_{x \to 0} f(x)$, $\epsilon > 0$, and estimate $\delta > 0$ so that $|x - x_0| < \delta$ implies that $|f(x) - L| < \epsilon$. You may assume that ϵ is small, say $\epsilon < 0.01$.

90. $f(x) = \dfrac{x^2 - x}{x + 2}$, $x_0 = 5$ **91.** $f(x) = \dfrac{x - 1}{x^2 + 3x}$, $x_0 = -5$

92. *Controlling the flow from a draining tank.* Torricelli's Law says that if you drain a tank like the one shown, the rate y at which the water runs out is a constant times the square root of the water's depth. As the tank drains, x decreases, and so does y, but y decreases less rapidly than x. The value of the constant depends on the size of the exit valve.

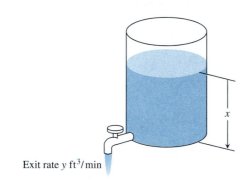

Exit rate y ft^3/min

Suppose that for the tank in question, $y = \sqrt{x}/2$. You are trying to maintain a constant exit rate of $y_0 = 1$ ft^3/min by refilling the tank with a hose from time to time. How deep must you keep the water to hold the rate to within 0.2 ft^3/min of $y_0 = 1$? Within 0.1 ft^3/min of $y_0 = 1$? In other words, in what interval must you keep x to hold y within 0.2 (or 0.1) units of $y_0 = 1$?

Remark: What if we want to know how long it will take the tank to drain if we do not refill it? We cannot answer such a question with the usual equation Time = Amount/Rate, because the rate changes as the tank drains. We could always open the valve, sit down with a watch, and wait; but with a large tank or a reservoir, that might take hours or even days. With calculus, we will be able to find the answer in just a minute or two, as you will see if you do Exercise 47 in Section 4.7.

93. *Dimension changes in equipment.* As you probably know, most metals expand when heated and contract when cooled, and people sometimes have to take this into account in their work. Boston and Maine Railroad crews try to lay track at temperatures as close to 65°F as they can so that the track won't expand too much in the summer or shrink too much in the winter. Surveyors have to correct their measurements for temperature when they use steel measuring tapes.

The dimensions of a piece of laboratory equipment are often so critical that the machine shop in which it is made has to be held at the same temperature as the laboratory where a part is to be installed. And once the piece is installed, the laboratory must continue to be held at that temperature.

A typical aluminum bar that is 10 cm wide at 70°F will be

$$y = 10 + (t - 70) \times 10^{-4}$$

centimeters wide at a nearby temperature t. As t rises above 70, the bar's width increases; as t falls below 70, the bar's width decreases.

Suppose you had a bar like this made for a gravity-wave detector you were building. You need the width of the bar to stay within 0.0005 cm of the ideal 10 cm. How close to 70°F must you maintain the temperature of your laboratory to achieve this? In other words, how close to $t_0 = 70$ must you keep t to be sure that y lies within 0.0005 of $y_0 = 10$?

94. The equation $1000x^2 + x - 10^{-15} = 0$ has meaning in chemistry. (See *The Mathematics Teacher*, Vol. 85(6), p. 462.) Confirm that $x = 0$ is not a solution. Find the solution, both exactly and approximately. Explain your results, and discuss the error in your approximation.

3

Derivatives

OVERVIEW   Derivatives are the functions we use to measure the rates at which things change. We define derivatives as limiting values of average changes, just as we define slopes of curves as limiting values of slopes of secants. Now that we can calculate limits, we can calculate derivatives.

The notion of derivative is one of the most important ideas in calculus. Any subject area that uses calculus, and most subjects do, has applications of derivatives. This chapter shows how to accurately approximate the derivative of a given function using the numerical derivative and how to determine exact derivatives quickly, without having to do tedious limit calculations. Graphs of numerical derivatives will be used to support the determination of exact derivatives. We will see that the ability to use a computer to graph numerical derivatives is as useful as determining exact derivatives.

3.1 Slopes, Tangent Lines, and Derivatives

In this section, we get our first view of the role calculus plays in describing how rapidly things change. Our point of departure is the coordinate plane of Descartes and Fermat. The plane is the natural place to draw curves and calculate the slopes of lines, and it is from the slopes of lines that we find the slopes of curves. Once we can do that, we can do two really important things: we can find tangent lines for curves, and we can find formulas for rates of change.

These formulas for the rates at which functions change define new functions called *derivatives*. In this section, we develop the ideas of slope and derivative and learn how to calculate derivatives with limits.

Average Rates of Change

We encounter average rates of change in such forms as average speeds (distance traveled divided by elapsed time, say, in miles per hour), growth rates of populations (in percent per year), and average monthly rainfall (in inches per month). The **average rate of change** in a quantity over a period of time is the amount of change divided by the time it takes.

Experimental biologists often want to know the rates at which populations grow under controlled laboratory conditions. Figure 3.1 shows data from a fruit fly–growing experiment, the setting for our first example.

EXAMPLE 1   The Average Growth Rate of a Laboratory Population

The graph in Fig. 3.1 shows how the number of fruit flies *(Drosophila)* grew in a controlled 50-day experiment. The graph was made by counting flies at regular intervals, plotting a point for each count, and drawing a smooth curve through the plotted points.

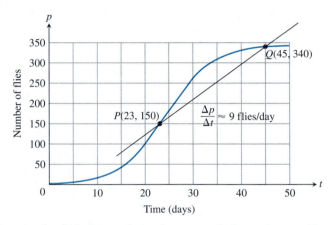

3.1 Growth of a fruit fly population in a controlled experiment. (*Source:* A. J. Lotka, *Elements of Mathematical Biology.* Dover, New York (1956), p. 69.)

There were 150 flies on day 23 and 340 flies on day 45. This gave an increase of $340 - 150 = 190$ flies in $45 - 23 = 22$ days. The average rate of change in the population from day 23 to day 45 was therefore

Average rate of change: $\dfrac{\Delta p}{\Delta t} = \dfrac{340 - 150}{45 - 23} = \dfrac{190}{22} = 8.64,$

or about 9 flies per day.

This average rate of change is also the slope of the secant line through the two points $P(23, 150)$ and $Q(45, 340)$ on the population curve. (A line through two points on a curve is called a **secant** to the curve.) We can calculate the slope of the secant PQ from the coordinates of P and Q:

Secant slope: $\dfrac{\Delta p}{\Delta t} = \dfrac{340 - 150}{45 - 23} = \dfrac{190}{22} = 8.64.$

As the above suggests, *we can always think of an average rate of change as the slope of a secant line.*   ≡

In addition to knowing the average rate at which the population grew from day 23 to day 45, we may also want to know how fast the population was growing on day 23 itself. To find out, we can watch the slope of the secant PQ change as we back Q along the curve toward P. The results for four positions of Q are shown in Fig. 3.2.

In terms of geometry, what we see as Q approaches P along the curve is this: The secant PQ approaches the tangent line AB that we drew by eye

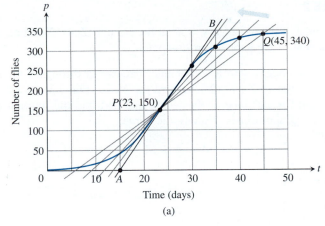

Q	Slope of $PQ = \Delta p/\Delta t$ (flies/day)
$(45, 340)$	$(340 - 150)/(45 - 23) \approx 8.64$
$(40, 330)$	$(330 - 150)/(40 - 23) \approx 10.59$
$(35, 310)$	$(310 - 150)/(35 - 23) \approx 13.33$
$(30, 265)$	$(265 - 150)/(30 - 23) \approx 16.43$

(b)

3.2 (a) Four secants to the fruit fly graph of Fig. 3.1, through the point $P(23, 150)$.
(b) The slopes of the four secants.

at P. This means that within the limitations of our drawing, the slopes of the secants approach the slope of the tangent, which we calculate from the coordinates of A and B to be

$$\frac{350 - 0}{35 - 15} = 17.5 \text{ flies/day.}$$

In terms of population change, what we see as Q approaches P is this: The average growth rates for increasingly smaller time intervals approach the slope of the tangent to the curve at P (17.5 flies per day). The slope of the tangent line is therefore the number we take as the rate at which the fly population was changing on day $t = 23$.

Defining Slopes and Tangent Lines

The moral of the fruit fly story would seem to be that we should define the rate at which the value of the function $y = f(x)$ is changing with respect to x at any particular value $x = x_1$ to be the slope of the tangent to the curve $y = f(x)$ at $x = x_1$. But how are we to define the tangent line at an arbitrary point P on the curve and find its slope from the formula $y = f(x)$? The problem here is that we only know one point. Our usual definition of slope requires two points.

The solution that Fermat finally found in 1629 proved to be one of that century's major contributions to calculus. We still use his method of defining tangents to produce formulas for slopes of curves and rates of change:

1. We start with what we *can* calculate, namely, the slope of a secant through P and a point Q nearby on the curve.

2. We find the limiting value of the secant slope (if it exists) as Q approaches P along the curve.

3. We define the *slope of the curve* at P to be this number and define the *tangent to the curve* at P to be the line through P with this slope.

WHAT IS A TANGENT TO A CURVE?

It is hard to overestimate how important the answer to this question was to the scientists of the early seventeenth century. In optics, the angle at which a ray of light strikes the surface of a lens may be defined in terms of the tangent to the surface. In physics, the direction of a body's motion at any point of its path is along the tangent to the path. In geometry, the angle between two intersecting curves is the angle between their tangents at the point of intersection. Follow along as we build a mathematical definition of tangent that fits all these situations.

WORKING DEFINITIONS

We will turn these working definitions of slope and tangent into mathematical definitions in just a little while.

EXAMPLE 2

Find the slope of the parabola $y = x^2$ at the point $P(2, 4)$. Write an equation for the tangent to the parabola at this point.

Solution   We begin with a secant line that passes through $P(2, 4)$ and a neighboring point $Q(2 + h, (2 + h)^2)$ on the curve (Fig. 3.3). We then write an expression for the slope of the secant line and find the limiting value of this slope as Q approaches P along the curve.

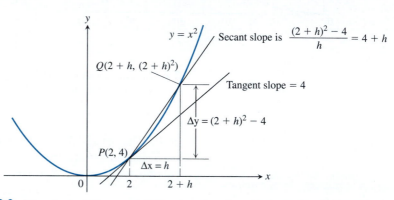

3.3   The slope of the secant PQ approaches 4 as Q approaches P along the curve.

The slope of the secant PQ is

$$\text{Secant slope} = \frac{\Delta y}{\Delta x} = \frac{(2 + h)^2 - 2^2}{h}$$

$$= \frac{h^2 + 4h + 4 - 4}{h}$$

$$= \frac{h^2 + 4h}{h} = h + 4.$$

The limit of the secant slope as Q approaches P along the curve is

$$\lim_{Q \to P} (\text{secant slope}) = \lim_{h \to 0} (h + 4) = 4.$$

EXPLORATION BIT

Graph $y_1 = x^2$ and $y_2 = 4(x - 2) + 4$. ZOOM-IN at (2, 4) several times. Explain what you see.

Thus, the slope of the parabola at $P(2, 4)$ is 4. The tangent to the parabola at P is the line through $P(2, 4)$ with slope $m = 4$:

$$y - 4 = 4(x - 2) \quad \text{Point-slope equation}$$

$$y = 4x - 8 + 4$$

$$y = 4x - 4.$$ ≣

The mathematics that we used to find the slope of the parabola $y = x^2$ at the point $P(2, 4)$ will find the slope of the parabola at *any* point on the curve. Here's how it works.

EXAMPLE 3

Find the slope of the parabola $y = x^2$ at any point on the curve.

Solution Let $P(a, a^2)$ be the point. In the notation of Fig. 3.4, the slope of the secant line through P and any nearby point $Q(a + h, (a + h)^2)$ is

$$\text{Secant slope} = \frac{\Delta y}{\Delta x} = \frac{(a + h)^2 - a^2}{h}$$

$$= \frac{a^2 + 2ah + h^2 - a^2}{h}$$

$$= \frac{2ah + h^2}{h} = 2a + h.$$

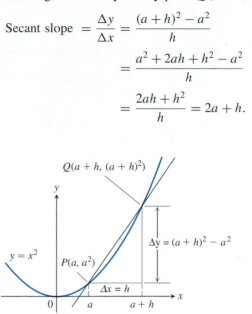

3.4 The slope of the secant PQ is

$$\frac{(a + h)^2 - a^2}{h} = 2a + h.$$

The limit of the secant slope as Q approaches P along the curve and $h \to 0$ is

$$\lim_{Q \to P} (\text{secant slope}) = \lim_{h \to 0} (2a + h) = 2a.$$

Thus, the slope of the parabola at $P(a, a^2)$ is $2a$. Because a can be any value of x, we have that the slope at any point (x, x^2) of the parabola is $m = 2x$. When $x = 2$, for example, the slope is $m = 2(2) = 4$, as in Example 2. ≡

EXPLORATION 1

Equations of Tangent Lines

To write an equation of a line, we need one point on the line, the slope of the line, and the point-slope formula $y - y_1 = m(x - x_1)$ (Section 1.2). The function $m = 2x$ gives the slope of the parabola $y = x^2$ at each point (x, x^2) on the parabola and hence the slope of the line tangent to the parabola at each point (x, x^2).

1. Find equations for the tangents to the curve $y = x^2$ at the points $(1, 1)$, $(-1/2, 1/4)$, and a point of your choice.

2. Support your results graphically by graphing each tangent line in the same viewing window with $y = x^2$. Before you GRAPH, explain what you would expect to see. ✤

The Derivative of a Function

The equation $m = 2x$ defines a function that gives the slope of the parabola $y = x^2$ at x. The function $m = 2x$ is the derivative of the function $y = x^2$.

To find the derivative of an arbitrary function $y = f(x)$ (when the function has one—we'll come back to that), we simply repeat for f the steps we took in Examples 2 and 3 for x^2. We start with an arbitrary point $P(x, f(x))$ on the graph of f, as in Fig. 3.5. The slope of the secant line through P and a nearby point $Q(x + h, f(x + h))$ is then

$$\text{Secant slope} = \frac{\Delta y}{\Delta x} = \frac{f(x + h) - f(x)}{h}.$$

The slope of the graph at P is the limit of the secant slope as Q approaches P along the graph. We find this limit from the formula for f by calculating the limit

$$\lim_{h \to 0} \frac{f(x + h) - f(x)}{h}.$$

This limit is itself a function of x. We denote it by f' ("f prime") and call it the derivative of f. Its domain is a subset of the domain of f. For most of the functions in this book, f' will be defined at all or all but a few of the points where f is defined.

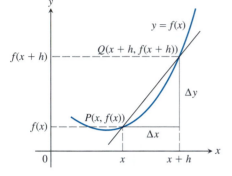

3.5 The slope of the secant line PQ is

$$\frac{\Delta y}{\Delta x} = \frac{f(x + h) - f(x)}{(x + h) - x} =$$

$$\frac{f(x + h) - f(x)}{h}.$$

DEFINITIONS

Derivative of a Function The **derivative of a function** f is the function f' whose value at x is defined by the equation

$$f'(x) = \lim_{h \to 0} \frac{f(x + h) - f(x)}{h}. \tag{1}$$

The fraction $(f(x+h) - f(x))/h$ is the **difference quotient** for f at x.

Differentiable at a Point A function f is **differentiable at a point** a if a is in the domain of f'. We say, "f has a derivative at a."

Differentiable Function A function that is differentiable at every point of its domain is a **differentiable function**.

Slope of a Curve and **Tangent to a Curve** When the number $f'(a)$ exists, it is called the **slope of the curve** $y = f(x)$ at $x = a$. The line through the point $(a, f(a))$ with slope $f'(a)$ is the **tangent to the curve** at a.

The most common notations for the derivative of a function $y = f(x)$, besides $f'(x)$, are

y' ("y prime")   (Nice and brief.)

$\dfrac{dy}{dx}$ ("dy dx")   (Names the variables and has a "d" for derivative.)

$\dfrac{df}{dx}$ ("df dx")   (Emphasizes the function's name.)

$D_x(f)$ ("Dx of f")   (Emphasizes the idea that taking the derivative is an operation performed on f.)

$\dfrac{d}{dx}(f)$ ("ddx of f")   (Ditto.)

We also read dy/dx as "the derivative of y with respect to x" and df/dx as "the derivative of f with respect to x." See Fig. 3.6.

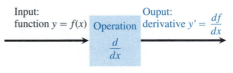

Input: Ouput:
function $y = f(x)$ Operation derivative $y' = \dfrac{df}{dx}$
                     $\dfrac{d}{dx}$

3.6 The operation *taking a derivative* defines a function d/dx whose domain is a set of functions and whose range is a set of functions.

EXPLORATION 2

Confirming a Definition

We now have two definitions for the *slope of a line* $y = mx + b$: the number m and, at each point, the derivative of the function $f(x) = mx + b$. Whenever we have two ways of defining an idea, it is good practice to confirm that the two definitions agree.

1. For the line $y = mx + b$, we want to show that $f'(x) = m$ at every x. Study the following equations, and explain why each is true.

$$\lim_{h \to 0} \frac{f(x+h) - f(x)}{h} = \lim_{h \to 0} \frac{m(x+h) + b - (mx + b)}{h}$$

$$= \lim_{h \to 0} \frac{mh}{h}$$

$$= \lim_{h \to 0} m$$

$$= m.$$

2. What can you conclude from part 1? Explain.

3. Interpret the definition of *tangent to a curve* when the curve is a line. That is, tell what is meant by a *tangent to a line*.

Typical Exact Derivative Calculations

EXAMPLE 4

Use Eq. (1) to find $f'(x)$ if $f(x) = 1/x$. Then find the slope of the curve $y = f(x)$ at $x = 0.5$, and write an equation for the tangent there. GRAPH the curve and tangent together.

Solution We take $f(x) = 1/x$, $f(x + h) = 1/(x + h)$, and form the difference quotient

$$\frac{f(x+h) - f(x)}{h} = \frac{\dfrac{1}{x+h} - \dfrac{1}{x}}{h}$$

$$= \frac{1}{h} \cdot \frac{x - (x+h)}{x(x+h)}$$

$$= \frac{1}{h} \cdot \frac{-h}{x(x+h)} = \frac{-1}{x(x+h)}.$$

We then take the limit as $h \to 0$:

$$f'(x) = \lim_{h \to 0} \frac{f(x+h) - f(x)}{h} = \lim_{h \to 0} \frac{-1}{x(x+h)} = \frac{-1}{x(x+0)} = -\frac{1}{x^2}.$$

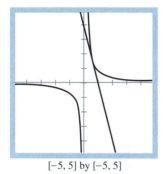

[−5, 5] by [−5, 5]

3.7 The graphs of $y_1 = 1/x$ and its tangent line $y_2 = -4(x - 0.5) + 2$ at $x = 0.5$.

The slope of the curve at $x = 0.5$ is $f'(0.5) = -1/(0.5)^2 = -4$. Since $f(0.5) = 2$, the equation of the tangent line at $x = 0.5$ is $y - 2 = -4(x - 0.5)$, or $y = -4x + 4$. See Fig. 3.7. ≡

EXAMPLE 5

Use Eq. (1) to find $f'(x)$ if $f(x) = x^3$.

Solution

$$f'(x) = \lim_{h \to 0} \frac{(x+h)^3 - x^3}{h}$$

$$= \lim_{h \to 0} \frac{(x^3 + 3hx^2 + 3h^2x + h^3) - x^3}{h}$$

$$= \lim_{h \to 0} (3x^2 + 3hx + h^2)$$

$$= 3x^2.$$ ≡

Exploration 3 lists steps for finding the derivative using the definition. Then it applies the steps to show that

$$\text{if} \quad f(x) = \sqrt{x}, \quad \text{then} \quad f'(x) = \frac{1}{2\sqrt{x}}.$$

EXPLORATION 3

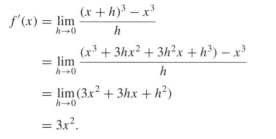

Finding the Derivative Using the Definition; The Derivative of $\sqrt{x}$

We use the following steps to calculate $f'(x)$ directly from the limit definition.

STEP 1: Write out $f(x)$ and $f(x + h)$.

STEP 2: Write the difference quotient $[f(x+h) - f(x)]/h$, replacing $f(x+h)$ and $f(x)$ with their values from Step 1.

STEP 3: Manipulate the expression algebraically into a form for which you can find the limit as $h \to 0$. This is usually the challenging step.

STEP 4: Take the limit as $h \to 0$.

Now let's try putting a lot of ideas together for the function $y_1 = \sqrt{x}$.

1. Show that the derivative of $y_1 = \sqrt{x}$ is $y_2 = 1/(2\sqrt{x})$. For Step 3 above, multiply the numerator and denominator by $\sqrt{x+h} + \sqrt{x}$.

2. Find an equation for the tangent to the curve $y_1 = \sqrt{x}$ at $x = 1.5$. Use $y_1'(1.5)$ for slope. Support your result graphically. (See Fig. 3.8.)

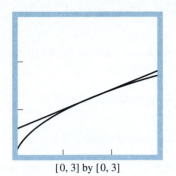

[0, 3] by [0, 3]

3.8 The curve $y = \sqrt{x}$ and its tangent line, $y = 0.41x + 0.61$, at $x = 1.5$.

Connections Between Differentiable Functions and Continuous Functions

We now show that a differentiable function is continuous. The converse is not true. There are continuous functions that are not differentiable. In Chapter 5, we will see that every continuous function is a derivative but there are derivatives that are not continuous.

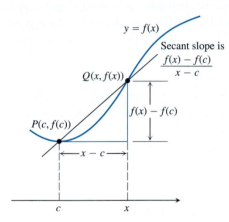

3.9 Figure for the proof that a function is continuous at every point at which it has a derivative.

THEOREM 1

If f has a derivative at $x = c$, then f is continuous at $x = c$.

Proof Our task is to show that $\lim_{x \to c} f(x) = f(c)$ or, equivalently, that

$$\lim_{x \to c}[f(x) - f(c)] = 0.$$

To this end, we have point $P(c, f(c))$ on the graph of f, and we let $Q(x, f(x))$ be a point nearby (Fig. 3.9). The slope of the secant PQ is

$$\text{Secant slope} = \frac{f(x) - f(c)}{x - c}.$$

By definition, the derivative of f at c is the limiting value of this slope as Q approaches P along the curve, which means in this case the limit as $x \to c$:

$$f'(c) = \lim_{x \to c} \frac{f(x) - f(c)}{x - c}. \tag{2}$$

Why should the mere existence of this limit imply that $[f(x) - f(c)] \to 0$ as $x \to c$? Because, with the denominator $x - c$ going to zero, the quotient can have a finite limit only if the numerator goes to zero at the same time. This is exactly what we find if we apply the Limit Product Rule from Section 2.1:

$$\lim_{x \to c}\left[f(x) - f(c)\right] = \lim_{x \to c}\left[(x - c)\frac{f(x) - f(c)}{x - c}\right] \quad \text{Divide and multiply by } x - c.$$

$$= \lim_{x \to c}(x - c) \cdot \lim_{x \to c}\frac{f(x) - f(c)}{x - c} \quad \text{Limit Product Rule}$$

$$= 0 \cdot f'(c) = 0. \qquad \blacksquare$$

EXPLORATION BIT

Show that

$$f'(c) = \lim_{h \to 0}\frac{f(c+h) - f(c)}{h}$$

and

$$f'(c) = \lim_{x \to c}\frac{f(x) - f(c)}{x - c}$$

are equivalent by letting $h = x - c$.

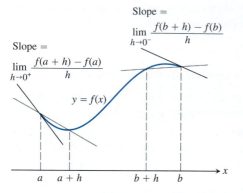

3.10 Derivatives at endpoints are one-sided limits.

Differentiable on a Closed Interval—One-sided Derivatives

A function $y = f(x)$ is *differentiable on a closed interval* $[a, b]$ if it has a derivative at every interior point and if the limits

$$\lim_{h \to 0^+} \frac{f(a + h) - f(a)}{h} \qquad \text{Right-hand derivative at } a$$

$$\lim_{h \to 0^-} \frac{f(b + h) - f(b)}{h} \qquad \text{Left-hand derivative at } b$$

exist at the endpoints. In the right-hand derivative, h is positive and $a + h$ approaches a from the right. In the left-hand derivative, h is negative and $b + h$ approaches b from the left (Fig. 3.10).

Right-hand and left-hand derivatives may be defined at any point of a function's domain.

The usual relation between one-sided and two-sided limits holds for derivatives. A function has a (two-sided) derivative at a point if and only if the function's right-hand and left-hand derivatives are defined and are equal at that point.

EXPLORATION BIT

GRAPH $y_1 = |x|$ and explain the two limits in Example 6 using your graph.

EXAMPLE 6

The function $f(x) = |x|$ has no derivative with respect to x at $x = 0$ even though it has a derivative with respect to x at every other point. The reason is that the right-hand and left-hand derivatives of $f(x) = |x|$ are not equal at $x = 0$:

$$\lim_{h \to 0^+} \frac{|0 + h| - |0|}{h} = \lim_{h \to 0^+} \frac{|h|}{h}$$

$$= \lim_{h \to 0^+} \frac{h}{h} \qquad \text{Because } h > 0$$

$$= \lim_{h \to 0^+} 1 = 1,$$

$$\lim_{h \to 0^-} \frac{|0 + h| - |0|}{h} = \lim_{h \to 0^-} \frac{|h|}{h}$$

$$= \lim_{h \to 0^-} \frac{-h}{h} \qquad |h| = -h \text{ because } h < 0.$$

$$= \lim_{h \to 0^-} (-1) = -1.$$

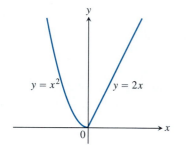

3.11 The "left-hand tangent" at $x = 0$ would have slope 0. The "right-hand tangent" at $x = 0$ would have slope 2. Hence there is no tangent at 0. The derivative does not exist at 0.

Using a slope argument, we can let $P = (0, 0)$ and $Q = (x, f(x))$. If $x > 0$, then the slope of secant PQ is 1. If $x < 0$, the slope of secant $PQ = -1$. There is not a single limiting value to the slopes of the secant lines as $x \to 0$. $f'(0)$ does not exist, and there is no tangent line to the graph of f at $x = 0$.

EXAMPLE 7

Show that the following function has no derivative at $x = 0$:

$$y = \begin{cases} x^2, & x \le 0, \\ 2x, & x > 0. \end{cases}$$

Solution There is no derivative at the origin because the right-hand and left-hand derivatives are different there. The slope at $x = 0$ of the parabola on the left (Fig. 3.11) is $m = 2(0) = 0$ (from Example 3). The slope at $x = 0$ of the line on the right is 2. ≡

Examples 6 and 7 are examples of continuous functions that are not differentiable, counterexamples to the converse of Theorem 1.

Exercises 3.1

In Exercises 1–4, estimate the slope of the curve (in y-units per x-unit) at the point with the indicated x-coordinate. Be careful; x-scale and y-scale may not equal 1 in the viewing window shown.

1. a) $x = 2$ **b)** $x = 4$

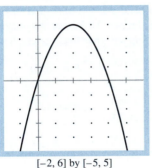

[−2, 6] by [−5, 5]

2. a) $x = -1$ **b)** $x = 0$

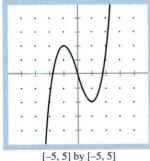

[−5, 5] by [−5, 5]

3. a) $x = 0.5$ **b)** $x = 4$

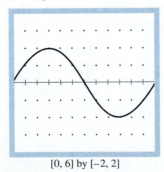

[0, 6] by [−2, 2]

4. a) $x = -1$ **b)** $x = 1$

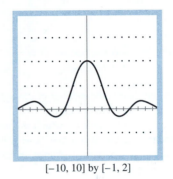

[−10, 10] by [−1, 2]

5. The viewing window below shows the Fahrenheit temperature in Fairbanks, Alaska, for a typical 365-day period from January 1 to December 31. Answer the following questions by estimating slopes on the graph in degrees per day. For the purpose of estimation, assume that each month has 30 days.

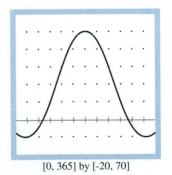

[0, 365] by [-20, 70]

a) On about what date is the temperature increasing at the fastest rate? What is the rate?

b) Do there appear to be days on which the temperature's rate of change is zero? If so, which ones?

c) During what period is the temperature's rate of change positive? Negative?

6. The viewing window below shows the number of hours of daylight in Fairbanks, Alaska, on each day for a typical 365-day period from January 1 to December 31. Answer the following questions by estimating slopes on the graph in hours per day. For the purpose of estimation, assume that each month has 30 days.

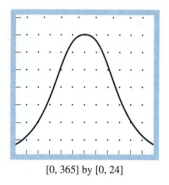

[0, 365] by [0, 24]

a) On about what date is the amount of daylight increasing at the fastest rate? What is that rate?

b) Do there appear to be days on which the rate of change in the amount of daylight is zero? If so, which ones?

c) On what dates is the rate of change in the number of daylight hours positive? Negative?

In Exercises 7–20, use Eq. (1) to find the derivative $dy/dx = f'(x)$ of the function $y = f(x)$. Then find the slope of the curve $y = f(x)$ at $x = 3$, and write an equation for the tangent line there.

7. $y = 2x^2 - 5$
8. $y = x^2 - 6x$

9. $y = 2x^2 - 13x + 5$
10. $y = -3x^2 + 4x$

11. $y = \dfrac{2}{x}$
12. $y = \dfrac{1}{x+1}$

13. $y = \dfrac{x}{x+1}$
14. $y = \dfrac{1}{2x+1}$

15. $y = x + \dfrac{9}{x}$
16. $y = x - \dfrac{1}{x}$

17. $y = 1 + \sqrt{x}$
18. $y = \sqrt{x+1}$

19. $y = \sqrt{2x}$
20. $y = \sqrt{2x+3}$

In Exercises 21–24, find an equation for the tangent line to the curve at the given point. Then GRAPH the curve and tangent in the same viewing window.

21. $y = 4 - x^2$, $(-1, 3)$
22. $y = (x-1)^2 + 1$, $(1, 1)$

23. $y = \sqrt{x}$, $(1, 1)$
24. $y = \dfrac{1}{x^2}$, $(-1, 1)$

In Exercises 25–30, use the alternate derivative formula

$$f'(c) = \lim_{x \to c} \frac{f(x) - f(c)}{x - c}$$

from the proof of Theorem 1 to find the derivative of f at the given value of c.

25. $f(x) = x^2 - x + 1$, $c = 1/2$

26. $f(x) = -3x^2 + 7x + 5$, $c = 2$

27. $f(x) = \dfrac{1}{x+2}$, $c = -1$

28. $f(x) = \dfrac{1}{(x-1)^2}$, $c = 2$

29. $f(x) = \dfrac{1}{\sqrt{x}}$, $c = 4$

30. $f(x) = \dfrac{1}{\sqrt{2x+13}}$, $c = -2$

Compare the right-hand and left-hand derivatives to show that the functions in Exercises 31–34 are not differentiable at the indicated point P. Support your findings graphically.

31. $f(x) = \begin{cases} x^2, & x < 0 \\ x, & x \geq 0 \end{cases}$ $\quad P = (0, 0)$

32. $f(x) = \begin{cases} 2, & x < 1 \\ 2x, & x \geq 1 \end{cases}$ $\quad P = (1, 2)$

33. $f(x) = \begin{cases} \sqrt{x}, & x \leq 1 \\ 2x - 1, & x > 1 \end{cases}$ $\quad P = (1, 1)$

34. $f(x) = \begin{cases} x, & x \leq 1 \\ \dfrac{1}{x}, & x > 1 \end{cases}$ $\quad P = (1, 1)$

In Exercises 35–38, do the following:

a) Find the derivative $f'(x)$ of the given function $f(x)$.

b) GRAPH $y_1 = f(x)$, and $y_2 = f'(x)$ in the same viewing window.

Then answer these questions:

c) For what values of x, if any, is $f'(x)$ positive? Zero? Negative?

d) Over what intervals of x-values, if any, does the function $f(x)$ increase as x increases? Decrease as x increases? How is this connected with what you found in part (c)? (We say more about this connection in Chapter 4.)

35. $y = -x^2$
36. $y = -\dfrac{1}{x}$

37. $y = \dfrac{x^3}{3}$
38. $y = \dfrac{x^4}{4}$

3.2 _____

Numerical Derivatives

In this section we learn how to approximate derivatives numerically. We also explore graphs of derivatives.

The NDER Procedure

Some graphing utilities can compute derivatives by using rules of differentiation. Some can approximate derivatives by applying a numerical method. We assume that your grapher can do the latter and accurately approximate the derivative of any differentiable function at most values in its domain. For a function f we will denote the *numerical derivative at a point a* and the *numerical derivative* (as a function), respectively, as follows:

$$\text{NDER} f(a) \text{ and NDER} f(x).$$

Sometimes we will use NDER $(f(x), a)$ for NDER $f(a)$ when we want to emphasize both the function *and* the point.

Our graphing calculator uses the *symmetric* difference quotient

$$\frac{f(a+h) - f(a-h)}{2h}$$

to determine NDER $f(a)$. (See Fig. 3.12.) It can be proven that

$$\lim_{h \to 0} \frac{f(a+h) - f(a-h)}{2h}$$

is equal to $f'(a)$, whenever $f'(a)$ exists.

How close to 0 do we want h for the approximation of $f'(a)$? Most graphers allow the user to specify h (consult your *Owner's Guide*), and for the symmetric difference method, the specified value becomes the grapher value of h. That is,

$$f'(a) \approx \frac{f(a+h) - f(a-h)}{2h}.$$

For our purposes, we use $h = 0.01$. Thus,

$$\text{NDER} f(a) = \frac{f(a + 0.01) - f(a - 0.01)}{0.02}.$$

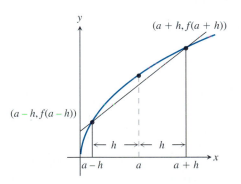

3.12 The symmetric difference quotient uses two points whose x-coordinates, $a - h$ and $a + h$, are symmetrically positioned on either side of a.

EXAMPLE 1

If $f(x) = x^2$ and $h = 0.01$, the grapher computes NDER $(x^2, 2)$ as follows:

$$\text{NDER} (x^2, 2) = \frac{(2.01)^2 - (1.99)^2}{0.02}$$

$$= 4,$$

exactly what we found in Example 2 of Section 3.1 for the slope of $y = x^2$ at $x = 2$.   ≡

It was shown in Section 3.1 that $D_x(x^3) = 3x^2$. So if $f(x) = x^3$, $f'(2) = 12$. Compare this result with NDER $(x^3, 2)$.

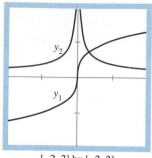

[−2, 2] by [−2, 2]

3.13 For the function $y_1 = \sqrt[3]{x}$ whose graph is shown, NDER $(\sqrt[3]{x}, 0) = 21.54$ on one graphing calculator. The graph of $y_2 = $ NDER $\sqrt[3]{x}$ is also shown above. It suggests correctly that $D_x \sqrt[3]{x}$ does not exist at $x = 0$. Can you give a convincing argument that this, in fact, is true? *Hint:* What is the slope of the tangent line at $x = 0$?

In the exercises you will see that NDER $f(d)$ gives the exact value of $f'(d)$ for $f(x) = ax^2 + bx + c$.

EXPLORATION BITS

1. In Example 3, compute NDER $f(-1)$ using $h = 0.1, 0.01$, and 0.001. Watch how the values of NDER $f(-1)$ get closer and closer to the more precise value shown for $f'(-1)$.

2. GRAPH the function f in Example 3 and see whether you can spot any places where there seems to be hidden behavior. If you suspect hidden behavior, can you support your suspicion by a quick analysis of the expression for f?

Agreement: In our work, we will always use $h = 0.01$ for NDER, and all displays of computations using NDER will be accurate to a reasonable number of decimal places, usually two or three.

EXAMPLE 2

If $f(x) = x^3$, compute NDER$(x^3, 2)$, the numerical derivative of x^3 at $x = 2$.

Solution   Using $h = 0.01$,

$$\text{NDER}(x^3, 2) = \frac{(2.01)^3 - (1.99)^3}{0.02}$$
$$= 12.0001. \qquad \blacksquare$$

Example 2 suggests that NDER is very accurate when $h = 0.01$. This is usually the case, although it is possible for NDER to produce some very inaccurate, even wrong, results (Fig. 3.13). Recognizing these instances is one of the skills that you will develop as you learn *differential calculus*, the calculus of derivatives.

In this book, we will use NDER to find derivatives with our graphing utility. We assume that you will use NDER when specifically requested to compute NDER $f(a)$. Otherwise, you may use your grapher's version of $f'(x)$. With NDER, our hope is that you will come to appreciate its power when you have to study the derivative of a very complicated function, as the next example illustrates.

EXAMPLE 3

Let

$$f(x) = \frac{2^{x+1} - \sin\left(\dfrac{x+1}{x-2}\right)}{\sqrt[3]{x^3 - 2x + 7}}.$$

Find an equation for the tangent to the curve $y = f(x)$ at $x = -1$.

Solution   We find $f(-1) = 0.5$ and NDER $f(-1) = 0.492413961911$. Thus, at $(-1, 0.5)$, the slope is approximately 0.49. An equation for the tangent line is

$$y - 0.5 = 0.49(x - (-1))$$
$$y - 0.5 = 0.49x + 0.49$$
$$y = 0.49x + 0.99.$$

To support this conclusion, we can graph f and the line $y = 0.49x + 0.99$ in the same viewing window and ZOOM-IN at $(-1, 0.5)$ to see whether the line and the curve seem to be tangent. $\blacksquare$

Using the rules that will be presented later in this book, the exact derivative f' of the function in Example 3 could be determined, and we could find that, to 12 decimal places, $f'(-1) = 0.492406923613$. But finding f' exactly is a very tedious process. You can compare NDER $f(-1)$ found in Example 3 with $f'(-1)$ shown here and see that the NDER approximation is quite accurate. In most applications, an accurate approximation of a derivative is all that is needed. We are safe in saying that the line $y = 0.49x + 0.99$ is "very close" to the exact tangent line.

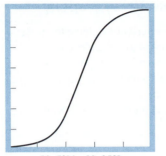

[0, 50] by [0, 350]

3.14 The graph of

$$P(t) = \frac{350}{1 + 8^{(25-t)/10}}.$$

A curve like this is called a *logistic growth curve*. Can you give an intuitive algebraic argument why it has this shape? Can you give an argument based on the real-life situation that it models?

AN "ALTERNATIVE TO NDER"

Graphing

$$y = \frac{f(x + 0.01) - f(x - 0.01)}{0.02}$$

is equivalent to graphing $y = \text{NDER } f(x)$ if NDER is not readily available on your grapher.

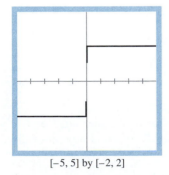

[−5, 5] by [−2, 2]

3.15 The graph of $y = \text{NDER } |x|$.

The fruit fly problem revisited The graph of $P(t) = 350/(1 + 8^{(25-t)/10})$ in Fig. 3.14 is a very close approximation to the graph of the fruit fly population in Fig. 3.1. It is easy to compute that NDER $(P(t), 23) = 17.43$. This compares very favorably with the slope estimate of 17.5 given in Section 3.1, the rate at which the fly population was changing on day 23.

Graphs of Derivatives

In this text we assume that your electronic grapher can produce a graph of the numerical derivative, that is, a graph of $y = \text{NDER } f(x)$. We will see that the ability to obtain an accurate graph of $y = \text{NDER } f(x)$ is often more useful than determining the exact derivative.

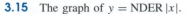

EXPLORATION 1

Making Conjectures

1. GRAPH $y = \text{NDER } f(x)$ for each of the following cubic functions:

$$f_1(x) = 3x^3 - 2x^2 + 5x - 6, \qquad f_2(x) = -x^3 + x^2 + 2x + 5,$$

$$f_3(x) = x^3 - 4x - 5, \qquad f_4(x) = -2x^3 - 3x^2 + 5x - 10.$$

2. If $f(x)$ is a cubic polynomial, NDER $f(x)$ is a polynomial function. Make a conjecture about the degree of NDER $f(x)$ based on the graphs in part 1.

3. GRAPH $y = \text{NDER } f(x)$ for each of the following quartic functions:

$$f_1(x) = 4x^4 - 3x^3 + 2x^2 - x + 10, \qquad f_2(x) = -x^4 + 2x^3 - 3x + 5,$$

$$f_3(x) = 2x^4 + x^2 - 2x + 5, \qquad f_4(x) = -2x^4 + x^3 - 2x^2.$$

4. If $f(x)$ is a quartic polynomial, NDER $f(x)$ is a polynomial function. Make a conjecture about the degree of NDER $f(x)$ based on the graphs in part 2.

5. Make a conjecture about the derivative of a polynomial function of degree n.

In Example 6 of Section 3.1, we showed that $f(x) = |x|$ has no derivative at $x = 0$. NDER gives *incorrect* information about the derivative of $f(x) = |x|$ at $x = 0$ because NDER $(|x|, 0) = 0$. However, the graph of $y = \text{NDER } |x|$ in Fig. 3.15 does correctly suggest that

$$f'(x) = \begin{cases} -1, & x < 0, \\ 1, & x > 0. \end{cases}$$

We will see in Exercise 46 that there is hidden behavior near $x = 0$ in Fig. 3.15. The graph should alert us that $f'(0)$ does not exist because of the following important property about derivatives.

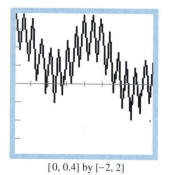

[0, 0.4] by [−2, 2]

3.16 The graph of the first four terms of Weierstrass's function given above. In the Exercises, you will graph the first eight terms. Where on this graph do you think hidden behavior occurs?

Derivatives Have the Intermediate Value Property

It comes in handy now and then to know that derivatives have the intermediate value property: If f has a derivative at every point of a closed interval $[a, b]$, then f' assumes every value between $f'(a)$ and $f'(b)$. We will refer to the property briefly in Chapter 4 but will make no attempt to prove it. There are proofs in more advanced texts.

For $f(x) = |x|$, suppose we conclude from Fig. 3.15 and NDER $(|x|, 0) = 0$ that

$$f'(x) = \begin{cases} -1, & x < 0, \\ 0, & x = 0, \\ 1, & x > 0. \end{cases}$$

Then we can apply the intermediate value property to f' on $[-1, 1]$ and conclude that f' takes on all values between $f'(-1) = -1$ and $f'(1) = 1$. This is clearly false.

This property of derivatives allows a partial answer to the question: When is a function that is defined on an interval the derivative of some other function throughout that interval? The partial answer is: Only when it has the intermediate value property. No step function, for example, is the derivative of some other function.

When Does a Function *Not* Have a Derivative at a Point?

As we know, a function has a derivative at a point x_0 if the slopes of the secant lines through $P(x_0, f(x_0))$ and a nearby point Q on the graph approach a limit as Q approaches P. If the secants fail to approach a limiting position as Q approaches P, the derivative does not exist. Typically, a function whose graph is otherwise smooth will fail to have a derivative at a point where the graph has

1. A *corner* where the one-sided derivatives differ, as in

$$f(x) = \begin{cases} x^3 + 6x^2 + 12x, & x < 0, \\ -x^2, & x \geq 0, \end{cases}$$

at $x = 0$. (See Exercise 39.)

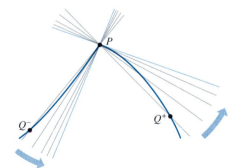

2. A *discontinuity* (below left) that makes one of the one-sided derivatives infinite (here the left derivative). This could happen at a point of discontinuity such as at $x = 0$ for

$$f(x) = \begin{cases} -3\sqrt{|x|}, & x < 0, \\ 3 - 0.2x^2, & x \geq 0. \end{cases} \qquad \text{(See Exercise 40.)}$$

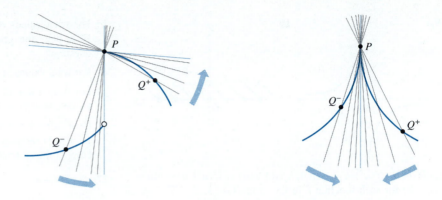

3. A *vertical tangent* (above right) where a graph is vertical "for an instant" as is the graph of $f(x) = -5\sqrt[3]{|x|}$ at $x = 0$. (See Exercise 41.)

Part 2 above illustrates the contrapositive of Theorem 1: If f is not continuous at $x = c$, then $f'(c)$ does not exist. (The contrapositive of an "if P, then Q" theorem is the true statement "if not Q, then not P.") Recall the Continuity Test from Section 2.2. Now, if any part of the test fails at $x = c$, then f is not continuous at that point *and* it is not differentiable there either. Here are the three ways in which the test can fail:

1. $f(c)$ does not exist. *Example:* $f = 1/x$ is not differentiable at $x = 0$.

2. $\lim\limits_{x \to c} f(x)$ does not exist. *Example:* Illustrated in part 2 above.

3. $\lim\limits_{x \to c} f(x) \neq f(c)$. *Example:* Think of $f(c)$ defining a removable discontinuity.

What Functions Are Differentiable?

Most of the functions that we have worked with so far are differentiable. Polynomials are differentiable, as are rational functions and trigonometric functions. Composites of differentiable functions are differentiable, and so are sums, differences, products, powers, and the quotients of differentiable functions, where defined. We will explain all this as the chapter continues.

Exercises 3.2

In Exercises 1–6, use NDER to find an equation for the tangent line to the curve at the given point. Then graph the curve and the tangent line in the same viewing window.

1. $y = x^2 + 1$, $x = 2$.

2. $y = 2x^3 - 5x - 2$, $x = 1.5$.

3. $y = \sqrt{4 - x^2}$, $x = -1$.

4. $y = (x - 1)^3 + 1$, $x = 2.5$.

5. $y = \dfrac{x^2 - 4}{x^2 + 1}$, $x = 2$.

6. $y = x \sin x$, $x = 2$.

Which of the graphs in Exercises 7–10 suggest a function $y = f(x)$ that is

a) Continuous at every point of its domain?
b) Differentiable at every point of its domain?
c) Both (a) and (b)?
d) Neither (a) nor (b)?

Explain in each case.

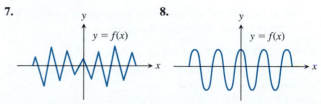

7.

8.

9.

10.

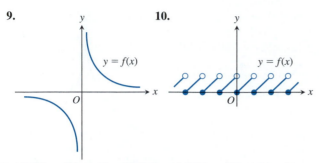

Let $D(h) = (f(a+h)-f(a))/h$ and $S(h) = (f(a+h)-f(a-h))/(2h)$ for each function f in Exercises 11–18.

a) For $a = 2$ and $a = 0$, complete a table of values comparing $D(h)$ (difference quotient) and $S(h)$ (symmetric difference quotient) for $h = -0.1, 0.1, -0.01, 0.01, -0.001, 0.001, -0.0001, 0.0001, -10^{-15}, 10^{-15}$.

b) Make a conjecture about the exact value of $f'(a)$, or state that it does not exist. Which of $D(h)$ or $S(h)$ is closer to your conjecture for the same value of h?

c) Explain your results in part (a) when $h = \pm 10^{-15}$.

11. $f(x) = 3x^2 - 2x$

12. $f(x) = 2x^3 + 4x$

13. $f(x) = 1/x^2$

14. $f(x) = 1/x$

15. $f(x) = |x|$

16. $f(x) = |x - 2|$

17. $f(x) = \sqrt{4x - x^2}$

18. $f(x) = \sqrt{4 - x^2}$

In Exercises 19–22, explain why you cannot use (a) $S(h)$ or (b) $D(h)$ to approximate the indicated derivative.

19. $f'(0)$ in Exercise 13

20. $f'(0)$ in Exercise 14

21. $f'(0)$ in Exercise 15

22. $f'(2)$ in Exercise 16

In Exercises 23 and 24, explain why you cannot use $S(h)$ to approximate the indicated derivative.

23. $f'(0)$ in Exercise 17

24. $f'(2)$ in Exercise 18

25. Let $f(x) = \sqrt[3]{\dfrac{x^2 \sin{(\tan x)}}{x^2 \cdot 2^x + x^5 - 4x^2}}$.

a) Use NDER to approximate $f'(-1)$, $f'(0)$, $f'(1.5)$, and $f'(3.5)$.

b) Explain why you must interpret the results of part (a) carefully. (*Hint:* GRAPH $y = f(x)$ in a $[-5, 5]$ by $[-0.5, 0.5]$ window, and consider the domain of f.)

26. Let $g(x) = \dfrac{x^8 - 2x + 5}{\left(x^2 + 3\right)^4 (2 + \sin x)}$.

a) Use NDER to approximate $g'(-2)$, $g'(0)$, $g'(1.5)$, and $g'(5)$.

b) How confident are you about the results of part (a)? Explain. (*Hint:* Investigate the graph of g.)

Consider the functions $y = f(x)$ in Exercises 27–32.

a) GRAPH $y_1 = f(x)$ and $y_2 = $ NDER y_1 in the same viewing window.

b) For what values of x, if any, does y_1' fail to exist? Why? How does the graph of $y_2 = $ NDER y_1 help answer this question?

c) For what values of x, if any, is y_1' positive? Zero? Negative?

d) For what value of x, if any, is the slope of the line tangent to the curve $y_1 = f(x)$ positive? Zero? Negative?

e) Over what intervals of x-values, if any, does the function $y_1 = f(x)$ increase as x increases? Decrease as x increases? How is this connected with what you found in part (c)?

27. $f(x) = -x^2$

28. $f(x) = -\dfrac{1}{x}$

29. $f(x) = \sqrt[3]{x - 2}$

30. $f(x) = \sqrt[3]{2 - x}$

31. $f(x) = \sqrt{1 - x}$

32. $f(x) = \sqrt[4]{x - 1}$

33. GRAPH the first eight terms of the Weierstrass function in the standard viewing window. ZOOM-IN several times. How wiggly and bumpy is this graph? Specify a viewing window in which the displayed portion of the graph is smooth.

34. GRAPH the first six terms of the Weierstrass function in the standard viewing window. ZOOM-IN several times. How wiggly and bumpy is this graph? Specify a viewing window in which the displayed portion of the graph is smooth.

For each function f in Exercises 35–37, do the following.

a) Draw the graph of $y = $ NDER $f(x)$ in the $[-3, 3]$ by $[-5, 5]$ viewing window.

b) Describe the graph in part (a). Write an algebraic representation for $y = $ NDER $f(x)$.

c) Compute NDER $(f(x), 0)$. Find $f'(0)$, or explain why it does not exist.

35. $f(x) = \begin{cases} -x^2, & x < 0, \\ 4 - x^2, & x \geq 0. \end{cases}$

36. Let $f(x) = \begin{cases} x^2, & x < 0, \\ x, & x \geq 0. \end{cases}$

37. Let $f(x) = \begin{cases} -x^2/2, & x < 0, \\ x^2/2, & x \geq 0. \end{cases}$

38. Let $f(x) = ax^2 + bx + c, a \neq 0$.

a) Show that $f'(x) = 2ax + b$.

b) Show that the symmetric difference quotient $[f(x+h) - f(x-h)]/(2h) = 2ax + b$ when $h \neq 0$.

c) Explain why NDER $f(x)$ is identical to $f'(x)$.

In Exercises 39–41, GRAPH $y_1 = f(x)$ and $y_2 = $ NDER $f(x)$ in the same viewing window. Find $f'(0)$, or explain why it does not exist.

39. $f(x) = \begin{cases} x^3 + 6x^2 + 12x, & x < 0 \\ -x^2, & x \geq 0 \end{cases}$

40. $f(x) = \begin{cases} -3\sqrt{|x|}, & x < 0 \\ 3 - 0.2x^2, & x \geq 0 \end{cases}$

41. $f(x) = -5\sqrt[3]{|x|}$

42. a) Draw the graphs of $y_1 = a^x$ and $y_2 = \text{NDER}(a^x)$ in the $[-1, 2]$ by $[-2, 8]$ viewing window for $a = 0.5$, $a = 0.75$, $a = 1$, $a = 1.5$, $a = 2$ and $a = 3$.

b) Determine a value for a so that the graphs of y_1 and y_2 in part (a) are identical.

c) Find a nontrivial function $y = f(x)$ with the property that $f(x) = f'(x)$ for each value of x in the domain of f. ($f(x) = 0$ is an example of a trivial function for which $f = f'$.)

43. a) Draw the graph of $y_1 = \sin x$ and $y_2 = \text{NDER}(\sin x)$ in the $[-10, 10]$ by $[-2, 2]$ viewing window.

b) Make a conjecture about $D_x(\sin x)$.

c) Test your conjecture by graphing your conjecture and $y_2 = \text{NDER}(\sin x)$ in the same viewing window.

44. Let $f(x) = \sqrt[5]{x}$. What are $f'(0)$ and NDER $(f, 0)$? Explain.

45. Let $f(x) = |x|$. What are $f'(0)$ and NDER $(f, 0)$? Explain.

46. Let $f(x) = |x|$ and $g(x) = (f(x+0.01) - f(x-0.01))/0.02$.
a) Show that f is a continuous function.
b) Show that g is a continuous function.

c) Draw the graph of g in the $[-10, 10]$ by $[-2, 2]$ viewing window.

d) Does this graph contradict the continuity of g at $x = 0$? (*Hint:* ZOOM-IN around $x = 0$.)

e) Write g as a piecewise function.

47. Show that the function

$$f(x) = \begin{cases} 0, & -1 \le x < 0, \\ 1, & 0 \le x \le 1, \end{cases}$$

is not the derivative of any function on the interval $-1 \le x \le 1$. (*Hint:* Does f have the intermediate value property on the interval?)

48. Let $f(x) = \sin x$. Use the definition of the derivative to compute $f'(0)$.

49. Show that the greatest integer function $y = [x]$ is not the derivative of any function throughout the interval $-\infty < x < \infty$.

50. Let $S_f(h) = [f(x+h) - f(x-h)]/2h$ be the symmetric difference quotient that is used to compute NDER f. Show that if f is a continuous function, then $S_f(h)$ is a continuous function.

3.3 Differentiation Rules

The process of calculating a derivative is called differentiation. The goal of this section is to show how to differentiate functions rapidly—without having to apply the definition each time. It will then be an easy matter to calculate the velocities, accelerations, and other important rates of change that we will encounter in Section 3.4.

Integer Powers, Multiples, Sums, and Differences

The first rule of differentiation is that the derivative of every constant function is zero. In short,

---

RULE 1   Derivative of a Constant

If c is a constant, then

$$\frac{d}{dx}(c) = 0.$$

Proof of Rule 1   If $f(x) = c$ is a function with a constant value c, then

$$\lim_{h \to 0} \frac{f(x+h) - f(x)}{h} = \lim_{h \to 0} \frac{c - c}{h} = \lim_{h \to 0} 0 = 0. \qquad \blacksquare$$

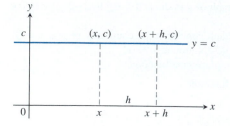

3.17 The slope of the graph of $y = c$, a constant, is zero at every point.

Rule 1 is another way to say that the values of constant functions never change and that the slope of a horizontal line is zero at every point (Fig. 3.17).

EXPLORATION 1

Making Conjectures

1. Let $f(x) = x^3$. GRAPH $y = \text{NDER}\,(x^3)$ in a $[-5, 5]$ by $[-5, 5]$ viewing window. Overlay graphs of the form $y = nx^2$ for $n = 1, 2, 3, 4, 5$. Which graph appears to match the graph of the derivative of f?

2. Try a similar idea for the derivative of $g(x) = x^4$. Overlay graphs of $y = nx^3$ for various values of n on the graph of $y = \text{NDER}\,(x^4)$. Which graph appears to match the graph of the derivative of g?

3. Make a conjecture about $f'(x)$. Make a conjecture about $g'(x)$. Make a conjecture about $h(x) = x^n$.

Your last conjecture above could easily have been the following.

EXPLORATION BIT

Give a proof of Rule 2 by mathematical induction.

> **RULE 2   Power Rule for Positive Integer Powers of x**
>
> If n is a positive integer, then
>
> $$\frac{d}{dx}(x^n) = nx^{n-1}.$$
>
> As a special case, we have
>
> $$\frac{d}{dx}(x) = \frac{d}{dx}(x^1) = 1 \cdot x^0 = 1.$$

Proof of Rule 2   We set $f(x) = x^n$ and find the limit as h approaches zero of

$$\frac{f(x+h) - f(x)}{h} = \frac{(x+h)^n - x^n}{h}.$$

Since n is a positive integer, we can apply the algebra formula

$$a^n - b^n = (a - b)(a^{n-1} + a^{n-2}b + \cdots + ab^{n-2} + b^{n-1})$$

to the numerator ($a = x + h$ and $b = x$) and obtain

$$\frac{f(x+h) - f(x)}{h} = \frac{(x+h)^n - x^n}{h}$$

$$= \frac{(h)[(x+h)^{n-1} + (x+h)^{n-2}x + \cdots + (x+h)x^{n-2} + x^{n-1}]}{h}$$

$$= \underbrace{[(x+h)^{n-1} + (x+h)^{n-2}x + \cdots + (x+h)x^{n-2} + x^{n-1}]}_{n \text{ terms, each with limit } x^{n-1} \text{ as } h \to 0}.$$

Hence

$$\frac{d}{dx}(x^n) = \lim_{h \to 0} \frac{f(x+h) - f(x)}{h} = nx^{n-1}. \qquad \blacksquare$$

To apply the Power Rule, we subtract 1 from the original exponent (n) and multiply the result by n. Thus, the derivatives of $x^2, x^3, x^4, \ldots,$ are $2x^1, 3x^2, 4x^3, \ldots,$ respectively.

RULE 3   The Constant Multiple Rule

If u is a differentiable function of x and c is a constant, then

$$\frac{d}{dx}(cu) = c\frac{du}{dx}.$$

As special cases, we have

$$\frac{d}{dx}(-u) = \frac{d}{dx}(-1 \cdot u) = -1 \cdot \frac{d}{dx}(u) = -\frac{du}{dx}$$

$$\frac{d}{dx}(cx^n) = cnx^{n-1}, \qquad \text{when } n \text{ is a positive integer.}$$

Proof of Rule 3

$$\frac{d}{dx}cu = \lim_{h \to 0}\frac{cu(x+h) - cu(x)}{h} \qquad \text{Derivative definition applied to } f(x) = cu(x)$$

$$= c\lim_{h \to 0}\frac{u(x+h) - u(x)}{h} \qquad \text{Limit property}$$

$$= c\frac{du}{dx}. \qquad u \text{ is differentiable.} \qquad \blacksquare$$

Rule 3 says that if a differentiable function is multiplied by a constant, then its derivative is multiplied by the same constant. Geometrically, if a graph is stretched vertically and possibly reflected about the x-axis, then the tangent at each point is changed in the same manner.

EXAMPLE 1

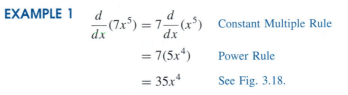

$$\frac{d}{dx}(7x^5) = 7\frac{d}{dx}(x^5) \qquad \text{Constant Multiple Rule}$$

$$= 7(5x^4) \qquad \text{Power Rule}$$

$$= 35x^4 \qquad \text{See Fig. 3.18.}$$

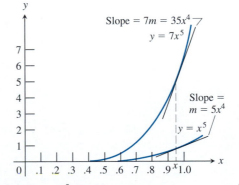

3.18 The graphs of $y = x^5$ and the stretched curve $y = 7x^5$. Multiplying the y-coordinates by 7 multiplies the slopes by 7. $\blacksquare$

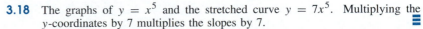

The next rule says that the derivative of the sum or difference of two differentiable functions is the sum or difference of their derivatives.

RULE 4   The Sum and Difference Rule

If u and v are differentiable functions of x, then their sum and difference are differentiable at every point where u and v are both differentiable. At such points,

$$\textbf{1.} \quad \frac{d}{dx}(u+v) = \frac{du}{dx} + \frac{dv}{dx}, \qquad \textbf{2.} \quad \frac{d}{dx}(u-v) = \frac{du}{dx} - \frac{dv}{dx}.$$

Similar equations hold for more than two functions, as long as the number of functions involved is finite.

Proof of Rule 4   To prove part 1, we apply the derivative definition with $f(x) = u(x) + v(x)$:

$$\frac{d}{dx}[u(x) + v(x)] = \lim_{h \to 0} \frac{[u(x+h) + v(x+h)] - [u(x) + v(x)]}{h}$$

$$= \lim_{h \to 0} \left[\frac{u(x+h) - u(x)}{h} + \frac{v(x+h) - v(x)}{h} \right]$$

$$= \lim_{h \to 0} \frac{u(x+h) - u(x)}{h} + \lim_{h \to 0} \frac{v(x+h) - v(x)}{h}$$

$$= \frac{du}{dx} + \frac{dv}{dx}.$$

The proof of part 2 is similar.   ≡

Rule 4 lets us differentiate any polynomial term by term.

EXAMPLE 2

a)
$$y = x^4 + 12x$$

$$\frac{dy}{dx} = \frac{d}{dx}(x^4) + \frac{d}{dx}(12x)$$

$$= 4x^3 + 12$$

b)
$$y = \frac{7x^2}{3} - 5$$

$$\frac{dy}{dx} = \frac{d}{dx}\left(\frac{7x^2}{3}\right) - \frac{d}{dx}(5)$$

$$= \frac{7}{3} \cdot 2x - 0 = \frac{14}{3}x$$

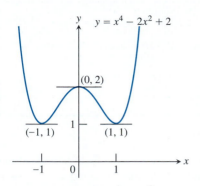

3.19 The curve $y = x^4 - 2x^2 + 2$ and its horizontal tangents. The tangents were located by setting dy/dx equal to zero and solving for x, as in Example 3.

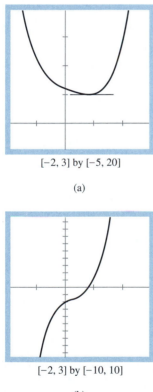

$[-2, 3]$ by $[-5, 20]$

(a)

$[-2, 3]$ by $[-10, 10]$

(b)

3.20 Complete graphs of
(a) $f(x) = x^4 - x^3 + x^2 - 2x + 6$
and (b) $f'(x) = 4x^3 - 3x^2 + 2x - 2$.
$f'(x) = 0$ at $x = 0.852$, so the curve for f has a horizontal tangent (shown) at $(0.852, f(0.852)) = (0.852, 4.930)$.

c)
$$y = x^3 + 3x^2 - 5x + 1$$
$$\frac{dy}{dx} = \frac{d}{dx}(x^3) + \frac{d}{dx}(3x^2) - \frac{d}{dx}(5x) + \frac{d}{dx}(1)$$
$$= 3x^2 + 3 \cdot 2x - 5 + 0$$
$$= 3x^2 + 6x - 5$$

EXAMPLE 3 Finding Horizontal Tangents

Does the curve $y = x^4 - 2x^2 + 2$ have any horizontal tangents? If so, where?

Solution The horizontal tangents, if any, occur where the slope dy/dx is zero. To find these points, we

1. Calculate dy/dx:
$$\frac{dy}{dx} = \frac{d}{dx}(x^4 - 2x^2 + 2) = 4x^3 - 4x.$$

2. Solve the equation $\frac{dy}{dx} = 0$ for x:
$$4x^3 - 4x = 0$$
$$4x(x^2 - 1) = 0$$
$$x = 0, 1, -1.$$

The curve in question has horizontal tangents at $x = 0, 1$, and -1. The corresponding points on the curve (found from the equation $y = x^4 - 2x^2 + 2$) are $(0, 2)$, $(1, 1)$, and $(-1, 1)$. See Fig. 3.19.

The successful solution of Example 3 depends on correctly applying the differentiation rules to $y = f(x)$ or being able to work graphically with NDER $f(x)$. It also depends on our ability to solve either equation $dy/dx = 0$ or NDER $f(x) = 0$. In many problems, an algebraic solution of $dy/dx = 0$ can be difficult, while a graphical solution of $dy/dx = 0$ or NDER $f(x) = 0$ is relatively easy using one of the methods discussed in Section 1.5.

EXAMPLE 4

Where does the curve $f(x) = x^4 - x^3 + x^2 - 2x + 6$ have horizontal tangent lines, if any?

It is easy to determine that $f'(x) = 4x^3 - 3x^2 + 2x - 2$. However, solving the equation $f'(x) = 0$ is another matter. It can be shown that there are no rational zeros using the Rational Zeros Theorem. It can also be shown that there is only one real (but irrational) solution. Using SOLVE, we can find that $f'(x) = 0$ when $x = 0.852$ (Fig. 3.20(b)). Thus there is only one horizontal tangent to f, and it occurs at the point $(0.852, f(0.852))$ as shown in Fig. 3.20(a).

Products

While the derivative of the sum of two functions is the sum of their derivatives and the derivative of the difference of two functions is the difference of their derivatives, the derivative of the product of two functions is *not* the product of their derivatives. The derivative of a product is a sum of *two* products, as we now explain.

> **RULE 5   The Product Rule**
>
> The product of two differentiable functions u and v is differentiable, and
>
> $$\frac{d}{dx}(uv) = u\frac{dv}{dx} + v\frac{du}{dx}.$$

EXPLORATION BIT

Find a counterexample to show why

$$(uv)' \neq u' \cdot v'.$$

In fact, prove that if

$$(uv)' = u' \cdot v'$$

in general, then the derivative of *every* differentiable function is zero.

As with the Sum and Difference Rules, the Product Rule is understood to hold only at values of x where u and v both have derivatives. At such a value of x, the derivative of the product uv is u times the derivative of v plus v times the derivative of u.

Proof of Rule 5

$$\frac{d}{dx}(uv) = \lim_{h \to 0} \frac{u(x+h)v(x+h) - u(x)v(x)}{h}.$$

To change this fraction into an equivalent one that contains difference quotients for the derivatives of u and v, we subtract and add $u(x+h)v(x)$ in the numerator. Then

$$\frac{d}{dx}(uv) = \lim_{h \to 0} \frac{u(x+h)v(x+h) - u(x+h)v(x) + u(x+h)v(x) - u(x)v(x)}{h}$$

$$= \lim_{h \to 0}\left[u(x+h)\frac{v(x+h) - v(x)}{h} + v(x)\frac{u(x+h) - u(x)}{h}\right]$$

$$= \lim_{h \to 0} u(x+h) \cdot \lim_{h \to 0}\frac{v(x+h) - v(x)}{h} + v(x) \cdot \lim_{h \to 0}\frac{u(x+h) - u(x)}{h}.$$

As h approaches zero, $u(x+h)$ approaches $u(x)$ because u, being differentiable at x, is continuous at x. The two fractions approach the values of du/dx at x and dv/dx at x.

$$\frac{d}{dx}(uv) = u\frac{dv}{dx} + v\frac{du}{dx}.$$

EXPLORATION BIT

Support Example 5 graphically by entering

$$y_1 = (x^2 + 1)(x^3 + 3),$$

$$y_2 = 5x^4 + 3x^2 + 6x,$$

$$y_3 = \text{NDER } y_1.$$

Then GRAPH y_2 and y_3 and compare. What should you expect to see?

EXAMPLE 5

Find the derivative of $y = (x^2 + 1)(x^3 + 3)$.

Solution   From the Product Rule with

$$u = x^2 + 1, \qquad v = x^3 + 3,$$

we find

$$\frac{d}{dx}[(x^2 + 1)(x^3 + 3)] = (x^2 + 1)(3x^2) + (x^3 + 3)(2x)$$

$$= 3x^4 + 3x^2 + 2x^4 + 6x$$

$$= 5x^4 + 3x^2 + 6x.$$

This particular example can be done as well (perhaps better) by multiplying out the original expression for y and differentiating the resulting polynomial. We do that now as a check. From

$$y = (x^2 + 1)(x^3 + 3) = x^5 + x^3 + 3x^2 + 3,$$

we obtain

$$\frac{dy}{dx} = 5x^4 + 3x^2 + 6x,$$

in agreement with our first calculation.

There are times, however, when the Product Rule *must* be used, as is shown in Exercise 68.

Quotients

Just as the derivative of the product of two differentiable functions is not the product of their derivatives, the derivative of the quotient of two functions is not the quotient of their derivatives. What happens instead is this:

RULE 6   The Quotient Rule

At a point where $v \neq 0$, the quotient $y = u/v$ of two differentiable functions is differentiable, and

$$\frac{d}{dx}\left(\frac{u}{v}\right) = \frac{v\dfrac{du}{dx} - u\dfrac{dv}{dx}}{v^2}.$$

As with the earlier combination rules, the Quotient Rule holds only at values of x at which u and v both have derivatives.

Proof of Rule 6

$$\frac{d}{dx}\left(\frac{u}{v}\right) = \lim_{h \to 0} \frac{\dfrac{u(x+h)}{v(x+h)} - \dfrac{u(x)}{v(x)}}{h}$$

$$= \lim_{h \to 0} \frac{v(x)u(x+h) - u(x)v(x+h)}{hv(x+h)v(x)}$$

To change the last fraction into an equivalent one that contains the difference quotients for the derivatives of u and v, we subtract and add $v(x)u(x)$ in the numerator. This allows us to continue with

$$\frac{d}{dx}\left(\frac{u}{v}\right) = \lim_{h \to 0} \frac{v(x)u(x+h) - v(x)u(x) + v(x)u(x) - u(x)v(x+h)}{hv(x+h)v(x)}$$

$$= \lim_{h \to 0} \frac{v(x)\dfrac{u(x+h) - u(x)}{h} - u(x)\dfrac{v(x+h) - v(x)}{h}}{v(x+h)v(x)}.$$

Taking the limit in the numerator and denominator now gives the Quotient Rule.

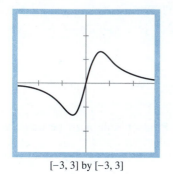

[−3, 3] by [−3, 3]

3.21 The graphs of

$$y_1 = \frac{4x}{(x^2 + 1)^2}$$

and

$$y_2 = \text{NDER} \frac{x^2 - 1}{x^2 + 1}$$

appear to be identical. When you use your grapher to check that graphs are identical, how can you be sure that one of the graphs is not entirely outside the viewing window or hidden by the x-axis?

EXAMPLE 6

Find the derivative of $y = f(x) = \dfrac{x^2 - 1}{x^2 + 1}$. Support graphically.

Solution   We apply the Quotient Rule with $u = x^2 - 1$ and $v = x^2 + 1$:

$$\frac{dy}{dx} = \frac{(x^2 + 1) \cdot 2x - (x^2 - 1) \cdot 2x}{(x^2 + 1)^2}$$

$$= \frac{2x^3 + 2x - 2x^3 + 2x}{(x^2 + 1)^2}$$

$$= \frac{4x}{(x^2 + 1)^2}.$$

The graphs of $y_1 = f'(x)$ calculated above and of $y_2 = \text{NDER}\ f(x)$ are shown in Fig. 3.21. The fact that they appear to be identical provides strong graphical support that our calculations are correct.   ▤

Negative Integer Powers of x

The rule for differentiating negative powers of x is the same as the rule for differentiating positive powers of x.

> **RULE 7   Power Rule for Negative Integer Powers of x**
>
> If n is a negative integer and $x \neq 0$, then
>
> $$\frac{d}{dx}(x^n) = nx^{n-1}.$$

Proof of Rule 7   The proof uses the Quotient Rule in a clever way. If n is a negative integer, then $n = -m$, where m is a positive integer. Hence, $x^n = x^{-m} = 1/x^m$, and

$$\frac{d}{dx}(x^n) = \frac{d}{dx}\left(\frac{1}{x^m}\right)$$

$$= \frac{x^m \cdot \dfrac{d}{dx}(1) - 1 \cdot \dfrac{d}{dx}(x^m)}{(x^m)^2} \qquad \begin{array}{l}\text{Quotient Rule} \\ \text{with } u = 1 \text{ and } v = x^m\end{array}$$

$$= \frac{0 - mx^{m-1}}{x^{2m}} \qquad \begin{array}{l}\text{Since } m > 0, \\ \dfrac{d}{dx}(x^m) = mx^{m-1}.\end{array}$$

$$= -mx^{-m-1}$$

$$= nx^{n-1}. \qquad \text{Changing back to } n \qquad ▤$$

EXAMPLE 7

$$\frac{d}{dx}\left(\frac{1}{x}\right) = \frac{d}{dx}(x^{-1}) = (-1)x^{-2} = -\frac{1}{x^2}$$

$$\frac{d}{dx}\left(\frac{4}{x^3}\right) = 4\frac{d}{dx}(x^{-3}) = 4(-3)x^{-4} = -\frac{12}{x^4}$$

$\equiv$

EXAMPLE 8

Find an equation for the tangent to the curve

$$y = x + \frac{2}{x}$$

at the point (1, 3). Support graphically.

Solution NDER $(x+2/x, 1) = -1.00020002$. This suggests that the slope at $x = 1$ is exactly -1, which we confirm analytically.

The slope of the curve is

$$\frac{dy}{dx} = \frac{d}{dx}(x) + 2\frac{d}{dx}\left(\frac{1}{x}\right)$$

$$= 1 + 2\left(-\frac{1}{x^2}\right) \qquad \text{From Example 7}$$

$$= 1 - \frac{2}{x^2}.$$

The slope at $x = 1$ is

$$\left.\frac{dy}{dx}\right|_{x=1} = \left[1 - \frac{2}{x^2}\right]_{x=1} = 1 - 2 = -1.$$

The line through (1, 3) with slope $m = -1$ is

$$y - 3 = (-1)(x - 1) \qquad \text{Point-slope equation}$$

$$y = -x + 1 + 3$$

$$y = -x + 4.$$

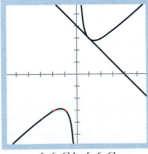

[-6, 6] by [-6, 6]

3.22 The graph of $y = -x + 4$ appears to be tangent to the graph of $y = x + 2/x$ at the point (1, 3).

The equation for the tangent to the curve $y = x + 2/x$ at the point (1, 3) is $y = -x + 4$. Figure 3.22 provides graphical support. $\equiv$

 EXPLORATION 2

Be Careful about Conclusions Based on a Picture

Graphs provide strong *support* for algebraic manipulations and analytic methods. However, we must be careful about any *conclusions* that we try to make from what we see in a viewing window.

1. GRAPH $y = x$ and $y = \sin x$ in the $[-0.1, 0.1]$ by $[-0.1, 0.1]$ viewing window. Are the graphs identical? What conclusions can you make? What conclusion can you *not* make? Explain.

2. GRAPH $y = \sin kx$ in the $[-6, 6]$ by $[-2, 2]$ viewing window with $k = N$, the number of columns of pixels on your grapher. Change k

by increments of 10 in both directions and GRAPH. Are the graphs correct? Explain.

Second and Higher Order Derivatives

The derivative

$$y' = \frac{dy}{dx}$$

is the *first derivative* of y with respect to x. The first derivative may also be a differentiable function of x. If so, its derivative,

$$y'' = \frac{dy'}{dx} = \frac{d}{dx}\left(\frac{dy}{dx}\right) = \frac{d^2y}{dx^2},$$

is called the *second derivative* of y with respect to x. If y'' ("y double prime") is differentiable, its derivative,

$$y''' = \frac{dy''}{dx} \qquad \text{("y triple prime"),}$$

is the *third derivative* of y with respect to x. The names continue as you imagine they would, with

$$y^{(n)} = \frac{d}{dx}y^{(n-1)} \qquad \text{("y super n")}$$

denoting the *nth derivative* of y with respect to x.

EXPLORATION 3

Finding Higher Order Derivatives

Find *all* higher order derivatives of the third-degree polynomial $y = x^3 - 3x^2 + 2$. Begin with y'. Then find y'', y''', $y^{(4)}$, and so on. (Might this take forever?!) What can you say, in general, about the higher order derivatives of a polynomial function of degree n?

For the *numerical* second derivative, we use NDER2. So NDER2 $f(x) = $ NDER (NDER $f(x)$). We simplify this notation further as follows:

Let $y_1 = $ NDER $f(x)$ First derivative

Then $y_2 = $ NDER y_1.   Second derivative

EXAMPLE 9

Let $f(x) = x2^{-x}$. GRAPH f, NDER f, and NDER2 f. Use a $[-1, 10]$ by $[-0.2, 0.6]$ viewing window.

Solution   Study the notation that follows. We enter

$y_1 = x2^{-x}$ The function

$y_2 = $ NDER y_1 First derivative

$y_3 = $ NDER y_2 Second derivative

The graphs are shown in Fig. 3.23.

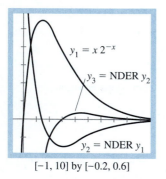

$y_1 = x\, 2^{-x}$

$y_3 = $ NDER y_2

$y_2 = $ NDER y_1

$[-1, 10]$ by $[-0.2, 0.6]$

3.23 The graphs of

$$y_1 = x\, 2^{-x},$$

$$y_2 = \text{NDER}x2^{-x} = \text{NDER}\, y_1,$$

and

$$y_3 = \text{NDER2}\, (x\, 2^{-x})$$

$$= \text{NDER}\, (\text{NDER}\, y_1) = \text{NDER}\, y_2.$$

Notice how the *x*-intercept of the graph of NDER y_1 suggests where the graph of y_1 could have a horizontal tangent *and* how the *x*-intercept of the graph of NDER2 y_1 suggests where the graph of NDER y_1 could have a horizontal tangent.

EXPLORATION 4

Differentiation Routes, Part 1

The choice of which rules to use in finding a derivative can make a difference in how much work you do.

1. Find the derivative of $y = \dfrac{(x-1)(x^2-2x)}{x^4}$ in two ways:

   a) Use the Quotient Rule.

   b) Expand the numerator, divide by x^4, then use the Sum and Power Rules.

2. Compare your results.

Two points have to be made:

- When paper and pencil manipulations are lengthy, it is best to work neatly in an organized step-by-step fashion so that others and perhaps you yourself can check easily through your work.

- A derivative found by two different methods may have two different forms. You should be able to show that they are equivalent. ⚛

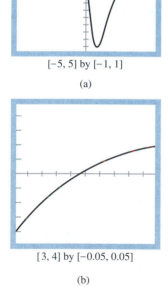

[−5, 5] by [−1, 1]

(a)

[3, 4] by [−0.05, 0.05]

(b)

3.24 Two views of

$$\text{NDER2}\left(\frac{5x}{x^2+4}\right).$$

Neither view shows a complete graph.

EXPLORATION 5

Differentiation Routes, Part 2

The choice of whether to use rules of differentiation and of algebra or a graphing utility can make a difference in how much work you do.

1. For $f(x) = \dfrac{5x}{x^2+4}$, find where $f''(x) > 0$.

   a) Use the Quotient Rule twice, and solve the inequality $f''(x) > 0$ using algebra.

   b) Graph NDER2 $f(x)$, and solve NDER2 $f(x) > 0$ by finding where the graph of NDER2 f is *above* the x-axis.

2. Compare your results.

3. Which method do you prefer, 1(a) or 1(b)? Explain why.

4. Now try part 1 for $g(x) = (2x-5)/(3x^2+4)$.

Two points have to be made:

- When you used algebra in part 1(a), you were lucky that the numerator of $f''(x)$ was factorable, thereby allowing you to find the zeros of f'' easily. How did you fare with g in part 4?

- When you use a graphing utility, you must be sure that you have a complete graph. Figure 3.24 shows two views of the graph of NDER2 f. Neither is a complete graph. In Fig. 3.24(a), it is not clear whether there are one, two, three, or more zeros. Seeing another view and using ZOOM-OUT will completely resolve this issue and complete the graph. ⚛

Exercises 3.3

In Exercises 1–40, use the rules of differentiation to find the requested derivatives.

In Exercises 1–12, find dy/dx and d^2y/dx^2.

1. $y = x$

2. $y = -x$

3. $y = -x^2 + 3$

4. $y = \dfrac{x^3}{3} - x$

5. $y = 2x + 1$

6. $y = x^2 + x + 1$

7. $y = \dfrac{x^3}{3} + \dfrac{x^2}{2} + x$

8. $y = 1 - x + x^2 - x^3$

9. $y = x^4 - 7x^3 + 2x^2 + 15$

10. $y = 5x^3 - 3x^5$

11. $y = 4x^2 - 8x + 1$

12. $y = \dfrac{x^4}{4} - \dfrac{x^3}{3} + \dfrac{x^2}{2} - x + 3$

Find all derivatives of the functions in Exercises 13–16.

13. $y = x^2 - x$

14. $y = \dfrac{x^3}{3} + \dfrac{x^2}{2} - 5$

15. $y = \dfrac{x^4}{2} - \dfrac{3}{2}x^2 - x$

16. $y = \dfrac{x^5}{120}$

In Exercises 17–24, find dy/dx. Find each derivative in two ways: (a) by applying the Product Rule and (b) by multiplying the factors to produce a sum of simpler terms to differentiate.

17. $y = (x + 1)(x^2 + 1)$

18. $y = (x + 1)(3 - x^2)$

19. $y = (x - 1)(x^2 + x + 1)$

20. $y = \left(x + \dfrac{1}{x}\right)\left(x - \dfrac{1}{x}\right)$

21. $y = (3x - 1)(2x + 5)$

22. $y = (5 - 3x)(4 - 2x)$

23. $y = x^2(x^3 - 1)$

24. $y = x^2\left(x + 5 + \dfrac{1}{x}\right)$

In Exercises 25–40, find dy/dx. Support your answer by graphing $y_1 = dy/dx$ and $y_2 = \text{NDER}\,(y)$ in the same viewing window.

25. $y = \dfrac{x - 1}{x + 7}$

26. $y = \dfrac{2x + 5}{3x - 2}$

27. $y = \dfrac{x^3 + 7}{x}$

28. $y = \dfrac{x^2 + 5x - 1}{x^2}$

29. $y = \dfrac{(x - 1)(x^2 + x + 1)}{x^3}$

30. $y = \dfrac{(x^2 + x)(x^2 - x + 1)}{x^4}$

31. $y = (1 - x)(1 + x^2)^{-1}$

32. $y = (2x - 7)^{-1}(x + 5)$

33. $y = \dfrac{x^2}{1 - x^3}$

34. $y = \dfrac{x^2 - 1}{x^2 + x - 2}$

35. $y = \dfrac{10}{\sqrt{x} - 4}$

36. $y = \dfrac{x}{2\sqrt{x} - 7}$

37. $y = \dfrac{\sqrt{x} - 1}{\sqrt{x} + 1}$

38. $y = \dfrac{1 + x - 4\sqrt{x}}{x}$

39. $y = \dfrac{1}{\left(x^2 - 1\right)\left(x^2 + x + 1\right)}$

40. $y = \dfrac{(x + 1)(x + 2)}{(x - 1)(x - 2)}$

In Exercises 41–46, find $y' = dy/dx$ and $y'' = d^2y/dx^2$. Support your answer by graphing $y_1 = dy/dx$ and $y_2 = \text{NDER}\,(y)$ in the same viewing window and by graphing $y_3 = d^2y/dx^2$ and $y_4 = \text{NDER2}\,(y)$ in the same viewing window.

41. $y = \dfrac{3}{x^2}$

42. $y = -\dfrac{1}{x}$

43. $y = \dfrac{5}{x^4}$

44. $y = -\dfrac{3}{x^7}$

45. $y = x + 1 + \dfrac{1}{x}$

46. $y = \dfrac{12}{x} - \dfrac{4}{x^3} + \dfrac{1}{x^4}$

For each function $y = f(x)$ in Exercises 47–50 use NDER to determine the equation of the tangent line at $(a, f(a))$. GRAPH both the function and the tangent line in the same viewing window.

47. $f(x) = x3^{-0.2x}$, $\quad a = 1$

48. $f(x) = \dfrac{\sin x}{x}$, $\quad a = \pi$

49. $f(x) = \dfrac{x + 3}{x^3 - 2x + 5}$, $\quad a = 0$

50. $f(x) = \sqrt[3]{\dfrac{x - 1}{x^2 + 5}}$, $\quad a = 2$

In Exercises 51–56, determine complete graphs for f' and f'' for each function $y = f(x)$.

51. $y = x \sin x$

52. $y = x^2 \sin x$

53. $y = \dfrac{2^x}{x^2 - 1}$

54. $y = \dfrac{3^x}{4 - x^2}$

55. $y = \sqrt[3]{\dfrac{x + 3}{x - 5}}$

56. $y = \sqrt[3]{\dfrac{x + 1}{x^2 + 2}}$

57–62. For each function in Exercises 51–56, solve $f'(x) = 0$ and $f''(x) > 0$.

63. Let f be the function $f(x) = (2x - 5)/(3x^2 + 4)$. Determine $f''(x)$. Try to solve $f''(x) > 0$ *exactly*. Solve $f''(x) > 0$ using any appropriate method.

64. Use the definition of derivative (given in Section 3.1, Eq. (1)) to show that

a) $\dfrac{d}{dx}(x) = 1$.

b) $\dfrac{d}{dx}(-u) = -\dfrac{du}{dx}$.

65. Use the product rule to show that $\dfrac{d}{dx}c \cdot f(x) = c \cdot \dfrac{d}{dx}f(x)$ for any constant c.

66. Devise a rule for $\dfrac{d}{dx}\left(\dfrac{1}{f(x)}\right)$.

67. Suppose u and v are functions of x that are differentiable at $x = 0$ and that

$$u(0) = 5, \quad u'(0) = -3, \quad v(0) = -1, \quad v'(0) = 2.$$

Find the values of the following derivatives at $x = 0$.

a) $\dfrac{d}{dx}(uv)$ **b)** $\dfrac{d}{dx}\left(\dfrac{u}{v}\right)$

c) $\dfrac{d}{dx}\left(\dfrac{v}{u}\right)$ **d)** $\dfrac{d}{dx}(7v - 2u)$

68. Suppose u and v are functions of x that are differentiable at $x = 2$ and that $u(2) = 3$, $u'(2) = -4$, $v(2) = 1$, and $v'(2) = 2$. Find the values of the following derivatives at $x = 2$.

a) $\dfrac{d}{dx}(uv)$ **b)** $\dfrac{d}{dx}\left(\dfrac{u}{v}\right)$

c) $\dfrac{d}{dx}\left(\dfrac{v}{u}\right)$ **d)** $\dfrac{d}{dx}(3u - 2v + 2uv)$

69. Which of the following numbers is the slope of the line tangent to the curve $y = x^2 + 5x$ at $x = 3$?

  a) 24   **b)** $-5/2$ **c)** 11   **d)** 8

70. Which of the following numbers is the slope of the line $3x - 2y + 12 = 0$?

  a) 6   **b)** 3   **c)** 3/2   **d)** 2/3

In Exercises 71–76, support your answers graphically.

71. Find the equation of the line perpendicular to the tangent to the curve $y = x^3 - 3x + 1$ at the point $(2, 3)$.

72. Find the tangents to the curve $y = x^3 + x$ at the points where the slope is 4. What is the smallest slope on the curve? At what value of x does the curve have this slope?

73. Find the points on the curve $y = 2x^3 - 3x^2 - 12x + 20$ where the tangent is parallel to the x-axis.

74. Find the x- and y-intercepts of the line that is tangent to the curve $y = x^3$ at the point $(-2, -8)$.

75. Find the tangents to *Newton's Serpentine*,

$$y = \frac{4x}{x^2 + 1}$$

at the origin and the point $(1, 2)$.

76. Find the tangent to the *Witch of Agnesi*

$$y = \frac{8}{4 + x^2}$$

at the point $(2, 1)$. (There is a nice story about the name of this curve in the historical note on Maria Agnesi in Chapter 10.)

When we work with functions of a single variable in mathematics, we often call the independent variable x and the dependent variable y. Applied fields use many different letters, however. Here are some examples. In these cases, the exact derivative is very useful.

77. *Cylinder pressure.*   If the gas in a cylinder is maintained at a constant temperature T, the pressure P is related to the volume V by a formula of the form

$$P = \frac{nRT}{V - nb} - \frac{an^2}{V^2},$$

in which a, b, n, and R are constants. Find dP/dV.

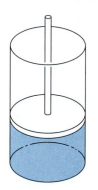

78. *Free fall.*   When a rock falls from rest near the surface of the earth, the distance it covers during the first few seconds is given by the equation

$$s = 4.9t^2.$$

In this equation, s is the distance in meters, and t is the elapsed time in seconds. Find ds/dt and d^2s/dt^2.

79. *The body's reaction to medicine.*   The reaction of the body to a dose of medicine can often be represented by an equation of the form

$$R = M^2\left(\frac{C}{2} - \frac{M}{3}\right),$$

where C is a positive constant and M is the amount of medicine absorbed in the blood. If the reaction is a change in blood pressure, R is measured in millimeters of mercury. If the reaction is a change in temperature, R is measured in degrees, and so on.

Find dR/dM. This derivative, as a function of M, is called the sensitivity of the body to the medicine. In Chapter 4, we shall see how to find the amount of medicine to which the body is most sensitive. (*Source: Some Mathematical Models in Biology*, Revised Edition, December 1967, PB-202 364, p. 221; distributed by N.T.I.S., U.S. Department of Commerce.)

80. Let K, m, n, a, and b be positive constants. Explain why horizontal tangent lines to the graphs of $y = Ka^x/(m + nb^x)$ and $y = a^x/(m + nb^x)$ have the same x-coordinate. Determine the x-coordinates of all points where the tangent lines are horizontal for $a = 2, m = 2, n = 1$, and $b = 3$. Draw complete graphs for $K = 1, K = 2$, and $K = 3$ in the same viewing window. Describe a technique, suggested by this exercise, that you could employ using a graphing calculator to help you investigate a function.

81. Show that if f is an even function, then f' is an odd function.

82. Show that if f is an odd function, then f' is an even function.

Exercises 83 and 84 refer to the function $f(x) = y_1 = x2^{-x}$ of Example 9 and Fig. 3.23.

83. a) Find the x-intercept of $y_2 = $ NDER y_1 shown in Fig. 3.23.
   b) Find the coordinates of the local maximum of f shown in Fig. 3.23.
   c) Compare the x-coordinate of the point in part (b) with the x-intercept in part (a).

84. a) Find the x-intercept of $y_3 = $ NDER $2f$ shown in Fig. 3.23.
   b) Find the coordinates of the local minimum of $y_2 = $ NDER y_1 shown in Fig. 3.23.
   c) Compare the x-coordinate of the point in part (b) with the x-intercept in part (a).

85. *Third derivatives.* Most graphing calculators that have a numerical derivative procedure (NDER) allow only one "nested" computation. This permits quick numerical computation of second derivatives as NDER2 $f = $ NDER (NDER f). However, NDER (NDER (NDER f)) is *not* allowed. Let $y_1 = $ NDER (NDER f) for $f(x) = x^5 - 3x^4 + x^3 - 6x^2 + 7x - 5$.
   a) Compute the maximum error in using y_1 as an estimate for $y = f''(x)$ for $-10 \leq x \leq 10$. Explain how you arrived at your solution.
   b) Use the symmetric difference quotient to determine a function y_2 that closely approximates $y = f^{(3)}(x)$.
   c) Compute the maximum error in using y_2 as an estimate for $y = f^{(3)}(x)$ for $-10 \leq x \leq 10$. Explain how you arrived at your solution.

3.4 — Velocity, Speed, and Other Rates of Change

In this section, we see how derivatives provide the mathematics we need to understand the way in which things change in the world around us. With derivatives, we can describe the rates at which water reservoirs empty, populations change, rocks fall, the economy evolves, and an athlete's blood sugar varies with exercise. We begin with free fall, the kind of fall that takes place in a vacuum near the surface of Earth.

Free Fall

Near the surface of the earth, all bodies fall with the same constant acceleration. The distance a body falls after it is released from rest is a constant multiple of the square of the time elapsed. At least, that is what happens when the body falls in a vacuum, where there is no air to slow it down. The square-of-time rule also holds for dense, heavy objects like rocks, ball bearings, and steel tools during the first few seconds of their fall through air, before their velocities build up to where air resistance begins to matter. When air resistance is absent or insignificant and the only force acting on a falling body is the force of gravity, we call the way in which the body falls *free fall*.

The equation we write to say that the distance an object falls from rest is proportional to the square of the time elapsed is

$$s = \frac{1}{2}gt^2.$$

In this equation, s is distance, t is time, and g, as we will see in a moment, is the constant acceleration given to an object by the force of gravity.

t (seconds) s (feet)

$t = 0$ ─── 0
        ─── 10
$t = 1$ ─── 20
        ─── 30
        ─── 40
        ─── 50
$t = 2$ ─── 60
        ─── 70
        ─── 80
        ─── 90
        ─── 100
        ─── 110
        ─── 120
        ─── 130
        ─── 140
$t = 3$ ─── 150

3.25 Distance fallen by a ball bearing released from rest at $t = 0$ sec.

The Gravitational Constant of Acceleration, g

The value of g in the equation $s = (1/2)gt^2$ depends on the units used to measure t and s. With t in seconds (the usual unit),

$$g = 32 \text{ ft/sec}^2 \qquad s = \frac{1}{2}(32)t^2 = 16t^2 \quad (s \text{ in feet}),$$

$$g = 9.80 \text{ m/sec}^2 \qquad s = \frac{1}{2}(9.80)t^2 = 4.9t^2 \ (s \text{ in meters}),$$

$$g = 980 \text{ cm/sec}^2 \qquad s = \frac{1}{2}(980)t^2 = 490t^2 \ (s \text{ in centimeters}).$$

The abbreviation ft/sec^2 is read "feet per second squared" or "feet per second per second." The other units for g are "meters per second squared" and "centimeters per second squared."

EXAMPLE 1

Figure 3.25 shows the free fall of a heavy ball bearing released from rest at time $t = 0$. During the first 2 sec, the ball falls

$$s(2) = 16(2)^2 = 16 \cdot 4 = 64 \text{ ft.}$$

EXAMPLE 2

How long did it take the ball bearing in Fig. 3.25 to fall the first 100 feet?

Solution   Because distance is in feet, we use $g = 32$ in the free-fall equation:

$$s = 16t^2.$$

To find the time it took the ball bearing to cover the first 100 ft, we substitute $s = 100$ and solve for t.

$$100 = 16t^2$$

$$t^2 = \frac{100}{16}$$

$$t = \frac{5}{2} \qquad \text{Time increases from } t = 0 \text{, so we ignore the negative root.}$$

It took the ball 2.5 seconds to fall the first 100 feet.

EXPLORATION 1

Linear Animation

Let c be a constant. The parametric equations

$$x(t) = c, \qquad y(t) = f(t)$$

will illuminate pixels along the vertical line $x = c$. As t takes on values from tMin to tMax in increments of t-step, different $(c, y(t))$ pixels illuminate and can give an illusion of up or down movement along the line $x = c$. Since the t-values occur in equal increments, t can be thought of as representing time. The motion on the screen then can simulate that of a moving object.

1. Try dropping the ball bearing of Example 1 as follows. In parametric mode, enter $x(t) = 2$ and $y(t) = 16t^2$. Then GRAPH, and explain what happens. Can you make the pixels simulate the *falling* ball bearing? What adjustment can you make to slow down the fall? To speed up the fall? How can you represent a ball bearing falling (a) on Earth, (b) on the moon, and (c) on Jupiter, all on the same screen? (The free-fall equations for the moon and Jupiter are $s = 2.6t^2$ and $s = 37.2t^2$, respectively.)

The parametric equations

$$x(t) = g(t), \qquad y(t) = c,$$

will illuminate pixels along the horizontal line $y = c$. The following problem is a typical algebra problem. Try modeling it on your grapher.

2. Train 1 leaves city A at 10:00 A.M. heading towards city B, 300 miles to the east, at 45 miles per hour. At 11:00 A.M., train 2 leaves city A for city B at 55 miles per hour. Which train will complete the trip first?

3. Model the problem of part 2 if train 2 leaves city B heading for city A.

EXPLORATION BIT

Try to simulate *nonlinear* motion in the xy-plane.

When we TRACE a parametric graph in the xy-plane, the cursor is controlled by the values of t from tMin to tMax in increments of t-step. The viewing window will display the three values $t, x(t)$, and $y(t)$.

EXAMPLE 3

Graph the parametric equations $x(t) = 1$, $y(t) = -16t^2$ in the $[0, 3]$ by $[-200, 0]$ viewing window for $0 \le t \le 4$ with t-step $= 0.01$. Use TRACE to approximate the time it takes the ball bearing in Fig. 3.25 to fall 100 feet, 125 feet, 150 feet, and 180 feet (Fig. 3.26).

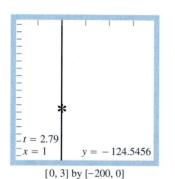

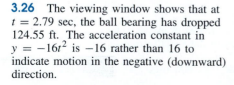

[0, 3] by [−200, 0]

3.26 The viewing window shows that at $t = 2.79$ sec, the ball bearing has dropped 124.55 ft. The acceleration constant in $y = -16t^2$ is -16 rather than 16 to indicate motion in the negative (downward) direction.

Solution   The ball bearing falls 100 feet in about 2.5 seconds, 125 feet in about 2.79 seconds, 150 feet in about 3.06 seconds, and 180 feet in about 3.35 seconds.   ■

Velocity

Suppose we have a body moving along a coordinate line and we know that its position at time t is $s = f(t)$.

Position at time t...   and at time $t +$ t

$s = f(t)$ $s + \Delta s = f(t + \Delta t)$

As the body moves along, it has a velocity at each particular instant, and we want to find out what that velocity is. The information that we seek must somehow be contained in the formula $s = f(t)$, but how do we find it?

We reason like this: In the interval from any time t to the slightly later time $t + \Delta t$, the body moves from position $s = f(t)$ to position

$$s + \Delta s = f(t + \Delta t).$$

The body's net change in position, or *displacement*, for this short time interval is

$$\Delta s = f(t + \Delta t) - f(t).$$

The body's average velocity for the time interval is Δs divided by Δt.

DEFINITION

The **average velocity** of a body moving along a line from position $s = f(t)$ to position $s + \Delta s = f(t + \Delta t)$ is

$$v_{av} = \frac{\text{displacement}}{\text{travel time}} = \frac{\Delta s}{\Delta t} = \frac{f(t + \Delta t) - f(t)}{\Delta t}.$$

To find the body's velocity at the exact instant t, we take the limit of the average velocity over the interval from t to $t + \Delta t$ as the interval gets shorter and shorter and Δt shrinks to zero. Here is where the derivative comes in. As we now know, this limit is the derivative of f with respect to t.

DEFINITION

Instantaneous velocity is the derivative of position, or distance. If the position function of a body moving along a line is $s = f(t)$, the body's (instantaneous) velocity at time t is

$$v(t) = \frac{ds}{dt} = \lim_{\Delta t \to 0} \frac{f(t + \Delta t) - f(t)}{\Delta t}.$$

CONVENTION

When people use the word *velocity* by itself, they usually mean instantaneous velocity.

EXAMPLE 4

Figure 3.27 on the following page is a time-to-distance graph of a 1989 Ford Thunderbird SC. The slope of the secant PQ is the average speed for the 10-second interval from $t = 5$ to $t = 15$ sec, in this case 32 m/sec or 115 km/h. The slope of the tangent at P is the speedometer reading at $t = 5$ sec, about 20 m/sec or 72 km/h. The car's top speed is 235 km/h (146 mph). (*Source: Car and Driver,* March 1989.)

The free-fall equation for distance is $s = (1/2)gt^2$. The free-fall equation for velocity is

$$v = gt,$$

which is found as follows:

$$v = \frac{ds}{dt} = \frac{d}{dt}\left(\frac{1}{2}gt^2\right) = \frac{1}{2}g\,\frac{d}{dt}(t^2) = \frac{1}{2}g(2t) = gt.$$

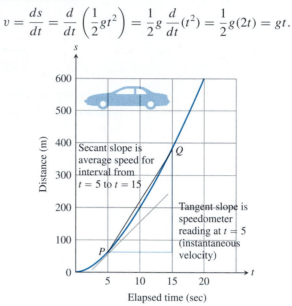

3.27 The distance a 1989 Ford Thunderbird SC can travel in time t. The speed at time t can be read from the car's speedometer and matches the slope of the curve for that value of t.

For the different values of g, we have $v = 32t$ (ft/sec), $v = 9.8t$ (m/sec), and $v = 980t$ (cm/sec).

EXAMPLE 5

For the falling ball bearing of Example 1 (Fig. 3.25), the velocity at t seconds after release is $32t$ ft/sec.

$$\text{At } t = 2: \qquad v = 32(2) = 64 \text{ ft/sec,}$$

$$\text{At } t = 3: \qquad v = 32(3) = 96 \text{ ft/sec.}$$

The parametric equations

$$x(t) = c, \qquad y(t) = f(t) \tag{1}$$

will illuminate pixels in a viewing window of a graphing utility along the vertical line $x = c$. This allows us to model vertical motion dynamically.

The parametric equations

$$x(t) = t, \qquad y(t) = f(t) \tag{2}$$

will illuminate the points $(t, f(t))$, which are precisely the points $(x, f(x))$ of the function $y = f(x)$. Thus, the parametric Eqs. (2) provide another way to graph the function $y = f(x)$.

So if $f(t)$ describes the height of a body at time t, graphing the parametric Eqs. (1) and (2) in the same viewing window will show the vertical motion of the body as well as a graph of its height plotted against time.

EXAMPLE 6

A dynamite blast blows a heavy rock straight up with a launch velocity of 160 ft/sec (about 109 mph) (Fig. 3.28). It reaches a height of $s(t) = 160t - 16t^2$ ft after t sec. Solve parts (a)–(d) graphically.

a) Simulate the moving rock.

b) Graph the rock's height as a function of time.

c) How high does the rock go? How long is it in the air?

d) How fast is the rock going when it is 256 ft above the ground?

e) Confirm the results of parts (c) and (d) analytically.

Solution

a) The parametric equations

$$x_1(t) = 3, \qquad y_1(t) = 160t - 16t^2 \tag{3}$$

will show the path of the rock along the vertical line $x = 3$ (Fig. 3.29).

b) The parametric equations

$$x_2(t) = t, \qquad y_2(t) = 160t - 16t^2 \tag{4}$$

will plot the rock's height as a function of time (Fig. 3.29). What should we set for [tMin, tMax]? We can use tMin = 0 to represent the time of the blast. For tMax, we try different values and watch what happens when we graph Eqs. (3) and (4) simultaneously. (Try different values of tMax yourself.)

c) Using TRACE, we find the rock's maximum height to be 400 ft, and it hits the ground after 10 seconds.

d) We can also TRACE to find that $y = 256$ when $t = 2$ *or* 8. The velocity of the rock is given by

$$v(t) = \frac{ds}{dt} = 160 - 32t \quad \text{ft/sec.}$$

So $v(2) = 96$ ft/sec, and $v(8) = -96$ ft/sec. This means that the velocity of the rock when it is 256 ft above the ground is 96 ft/sec (on the way up) and -96 ft/sec (on the way down).

e) To confirm part (c) analytically, we observe that the maximum height occurs when $v(t) = 160 - 32t = 0$, that is, when $t = 5$. At $t = 5$, $s(5) = 160(5) - 16(5)^2 = 400$.

To confirm part (d) analytically, we observe that $s(t) = 160t - 16t^2 = 256$ exactly when $t = 2$ and $t = 8$ (by factoring). ≡

In Example 6, the velocity function can be graphed parametrically by

$$x_3(t) = t, \qquad y_3(t) = 160 - 32t.$$

Graphs of the path of the rock, the height function, and the velocity function are shown in Fig. 3.30.

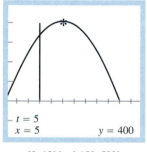

3.28 A sketch of the situation in Example 6.

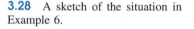

[0, 12] by [-150, 500]

3.29 The simulation, and the graph of height against time. The rock's movement is simulated along the line $x = 3$, but we see only its upward flight. How could we also see its downward flight? On the graph, TRACE shows y to have a maximum value 400 at $x = t = 5$ sec.

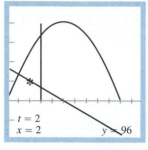

[0, 12] by [−150, 500]

3.30 The path of the rock, the graph of the height function, and the graph of the velocity function. Notice that $v > 0$ but decreasing as the rock is rising, $v = 0$ when the height is maximum, and $v < 0$ with $|v|$ increasing as the rock falls back to Earth. Downward velocity is negative because of how we set up the coordinate system. Since s measures height above the ground, changes in s are positive as the rock rises and negative as the rock falls.

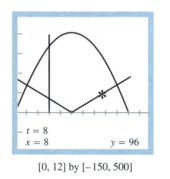

[0, 12] by [−150, 500]

3.31 The path of the rock, the graph of the height function, and the graph of the speed function, a nonnegative quantity.

$a(t) = v'(t) = s''(t)$

Acceleration is the derivative of velocity. Velocity is the derivative of distance or position. Acceleration is the derivative of the derivative—the second derivative—of distance or position.

Speed

If we drive over to a friend's house and back at, say, 30 mph, the speedometer will show 30 on the way over, but it will not show −30 on the way back. The speedometer always shows speed, and speed is the absolute value of velocity. Speed measures the rate at which our position changes, regardless of direction.

> **DEFINITION**
>
> **Speed** is the absolute value of velocity.

EXAMPLE 7

When the rock in Example 6 passed the 256-ft mark, its forward speed was 96 ft/sec on the way up and $|-96| = 96$ ft/sec again on the way down. The graph of the speed function $y = |v(t)| = |160 - 32t|$ is V-shaped (Fig. 3.31).

Acceleration

In studies of motion along a coordinate line, we usually assume that the body's position function $s = f(t)$ has a second derivative as well as a first. The first derivative gives the body's velocity as a function of time; the second derivative gives the body's *acceleration*. Thus, the velocity is how fast the position is changing, and the acceleration is how fast the velocity is changing. The acceleration tells how quickly the body picks up or loses speed.

> **DEFINITION**
>
> **Acceleration** is the derivative of velocity. If a body's position at time t is $s = f(t)$, then the body's acceleration at time t is
>
> $$a = \frac{dv}{dt} = \frac{d^2s}{dt^2}.$$

EXAMPLE 8

The acceleration of the rock in Example 6 is

$$a = \frac{dv}{dt} = \frac{d}{dt}(160 - 32t) = 0 - 32 = -32 \text{ ft/sec}^2.$$

The negative sign confirms that the acceleration is downward, in the negative s direction. Whether the rock is going up or down, it is subject to the same constant downward pull of gravity.

EXPLORATION 2

Horizontal Motion

In Examples 6–8, we analyzed vertical motion. Now try applying the ideas to horizontal motion. On your graphing utility, enter the following parametric equations. Use a $[-4, 6]$ by $[-3, 6]$ viewing window with $[t\text{Min}, t\text{Max}] = [0, 5]$ and t-step $= 0.05$.

$$x_1(t) = 4t^3 - 16t^2 + 15t, \qquad y_1(t) = 2; \qquad \text{Graph 1}$$

$$x_2(t) = x_1(t), \qquad y_2(t) = t; \qquad \text{Graph 2}$$

$$x_3(t) = t, \qquad y_3(t) = x_1(t). \qquad \text{Graph 3}$$

Draw Graph 1 first, then draw Graphs 1 and 2 *simultaneously*. In Graph 1, the (x_1, y_1) pairs begin at $(0, 2)$ and travel along $y = 2$, reversing direction a couple of times en route. In Graph 2, the (x_2, y_2) pairs show the back-and-forth nature of the (x_1, y_1) path (Fig. 3.32). TRACE shows that the (x_1, y_1) "particle" reverses direction at $t = 0.60$ when its distance from $(0, 2)$ is 4.104.

1. When does the particle reverse direction again? Where is it?

2. Describe the velocity of the particle during its journey.

3. Draw Graphs 1 and 3 simultaneously. What does Graph 3 show?

4. Confirm the motion of the particle by analyzing the function for Graph 3 as completely as you can.

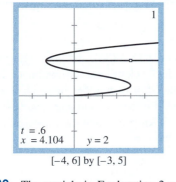

$t = .6$
$x = 4.104$ $y = 2$

$[-4, 6]$ by $[-3, 5]$

3.32 The particle in Exploration 2 moves along the line $y_1 = 2$. At $t = 0.6$, as shown here, it has gone as far to the right as it can for the time being and will begin moving in the other direction. Its back-and-forth motion is suggested by the second graph shown.

Other Rates of Change

Average and instantaneous rates give us the right language for many other applications.

EXAMPLE 9

Suppose that the number of gallons $g(t)$ of water in a reservoir at time t(min) can be regarded as a differentiable function of t. If the volume of water changes by the amount Δg in the time interval Δt, then

$$\frac{\Delta g}{\Delta t} \text{ (gal/min)} = \begin{array}{l} \text{average rate of change} \\ \text{for the time interval,} \end{array}$$

$$\frac{dg}{dt} = \lim_{\Delta t \to 0} \frac{\Delta g}{\Delta t} = \begin{array}{l} \text{instantaneous rate} \\ \text{of change at time } t. \end{array}$$

Although it is natural to think of rates of change in terms of motion and time, there is no need to be so restrictive. We can define the *average* rate of change of any function over any interval of its domain as the change in the function divided by the length of the interval. We can then go on to define the *instantaneous* rate of change as the limit of average change as the length of the interval goes to zero.

INSTANTANEOUS CHANGE

It is conventional to use the word *instantaneous* in the definition of *rate of change* even when x does not represent time.

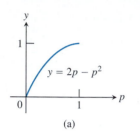

(a)

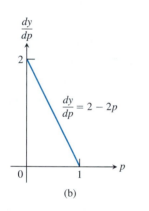

(b)

3.33 (a) The frequency $y = 2p - p^2$ of smooth skin in garden peas as a function of the prevalence p of the dominant smooth-skin gene. (b) The graph of dy/dp shows, among other things, how much y will change in response to a small change in p—a lot if p is near 0, but not much if p is near 1.

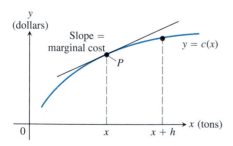

3.34 Weekly steel production: $c(x)$ is the cost of producing x tons in a week. The cost of producing an additional h tons is $c(x + h) - c(x)$.

DEFINITIONS   Rates of Change

The **average rate of change** of a function $f(x)$ over the interval from x to $x + h$ is

$$\text{Average rate of change} = \frac{f(x + h) - f(x)}{h}.$$

The **(instantaneous) rate of change** of f at x is the derivative

$$f'(x) = \lim_{h \to 0} \frac{f(x + h) - f(x)}{h},$$

provided that the limit exists.

EXAMPLE 10

The Austrian monk Gregor Johann Mendel (1822–1884), working with garden peas and other plants, provided the first scientific explanation of hybridization. His careful records showed that if p (a number between 0 and 1) is the frequency of the gene for smooth skin in peas (dominant) and $(1 - p)$ is the frequency of the gene for wrinkled skin in peas, then the proportion of smooth-skinned peas, in the population at large is

$$y = 2p(1 - p) + p^2 = 2p - p^2.$$

The graph of y versus p in Fig. 3.33(a) suggests that the value of y is more sensitive to a change in p when p is small than it is to a change in p when p is large. Indeed, this is borne out by the derivative graph in Fig. 3.33(b), which shows that dy/dp is close to 2 when p is near 0 and close to 0 when p is near 1.   ≡

Derivatives in Economics

Economists often call the derivative of a function the **marginal value** of the function.

EXAMPLE 11   Marginal Cost

Suppose it costs a company $c(x)$ dollars to produce x tons of steel in a week. It costs more to produce $x + h$ tons in a week, and the cost difference, divided by h, is the average increase in cost per ton:

$$\frac{c(x + h) - c(x)}{h} = \text{average increase in cost.}$$

The limit of the ratio as $h \to 0$ is the marginal cost when x tons of steel are produced:

$$c'(x) = \lim_{h \to 0} \frac{c(x + h) - c(x)}{h} = \text{ marginal cost.}$$

How are we to interpret this derivative? First of all, it is the slope of the graph of c at the point marked P in Fig. 3.34. But there is more.

Figure 3.35 on the following page shows an enlarged view of the curve and its tangent at P. We can see that if the company, currently producing x tons in a week, increases production by one ton, then the additional cost

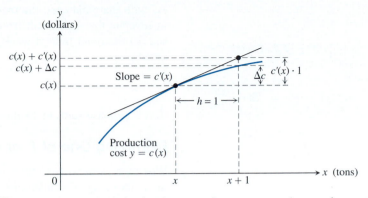

3.35 As weekly steel production increases from x to $x + 1$ tons, the cost curve rises by the amount Δc. The tangent line rises by the amount slope · run $= c'(x) \cdot 1 = c'(x)$. Since $\Delta c / h \approx c'(x)$, we have $\Delta c \approx c'(x)$ when $h = 1$.

$\Delta c = c(x + 1) - c(x)$ of producing that one ton is approximately $c'(x)$. That is,

$$\Delta c \approx c'(x) \qquad \text{when} \qquad h = 1.$$

Herein lies the economic importance of marginal cost. It estimates the cost of producing one unit beyond the present production level. ≡

EXAMPLE 12   Marginal Cost (Continued)

Suppose it costs

$$c(x) = x^3 - 6x^2 + 15x + 100$$

dollars to produce x stoves and your shop is currently producing 10 stoves a day. About how much extra will it cost to produce one more stove a day?

Solution   The cost of producing one more stove a day when 10 are produced is about $c'(10)$. Since

$$c'(x) = \frac{d}{dx}(x^3 - 6x^2 + 15x + 100)$$

$$= 3x^2 - 12x + 15,$$

$$c'(10) = 3(100) - 12(10) + 15$$

$$= 195.$$

The additional cost will be about $195 if you produce 11 stoves a day. ≡

EXAMPLE 13   Marginal Revenue

If

$$r(x) = x^3 - 3x^2 + 12x$$

gives the dollar revenue from selling x thousand heavy-duty bolts, the marginal revenue when x thousand are sold is

$$r'(x) = \frac{d}{dx}(x^3 - 3x^2 + 12x)$$

$$= 3x^2 - 6x + 12.$$

The actual additional cost to increase production from 10 to 11 stoves in Example 12 is

$$C(11) - C(10) = 870 - 650$$

$$= \$220.$$

As with marginal cost, the marginal-revenue function estimates the increase in revenue that will result from selling one additional unit. If you currently sell 10 thousand bolts a week, you can expect your revenue to increase by about

$$r'(10) = 3(100) - 6(10) + 12$$

$$= \$252$$

if you increase sales to 11 thousand bolts a week.   ≡

> The actual revenue increase in Example 13 is
>
> $$r(11) - r(10) = \$280.$$

Estimating One of f' or f from the Other

When we gather data in the laboratory or in the field, we are often recording the pairs (x, y) of a function f. We might be recording volume and pressure of a gas, the time and size of a population, or the sales and profits of a business. To see what the function looks like, we can plot the data points and fit them with a curve and possibly a formula. Even if we can find no formula for f, however, we can still learn something about the derivative f' *and* about a function F whose derivative is f.

EXAMPLE 14

It has been observed that the daily marginal profit (derivative of the profit) of a company selling x widgets per day is given by the graph of $y = M(x)$ shown in Fig. 3.36.

This graph shows that the profit function, $P(x)$, is increasing very slowly for each additional unit sold when x is between 0 and 40 and between 160 and 200. (Why?) It also shows the maximum marginal profit occurs when sales are about 106 units, a point where P would have its steepest slope.

≡

We will find much more information about *antiderivative* functions like P in the next several chapters.

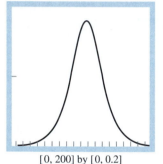

[0, 200] by [0, 0.2]

3.36 A marginal profit graph. Marginal profit = marginal revenue − marginal cost. Marginal profit estimates the increase in profit from selling one additional unit.

EXAMPLE 15

On April 23, 1988, the human-powered airplane *Daedalus* flew a record-breaking 119 km from Crete to the island of Santorini in the Aegean Sea, southeast of mainland Greece. During the 6-hour endurance tests before the flight, researchers monitored the prospective pilots' blood-sugar concentrations. The concentration graph for one of the athlete-pilots is shown in Fig. 3.37(a), where the concentration in milligrams/deciliter is plotted against time in hours.

The graph is made of line segments connecting data points. The constant slope of each segment gives an estimate of the derivative of the concentration between measurements. We calculated the slopes from the coordinate grid and plotted the derivative as a step function in Fig. 3.37(b). To make the plot for the first hour, for instance, we observed that the concentration increased from about 79 mg/dL to 93 mg/dL. The net increase was $\Delta y = 93 - 79 = 14$ mg/dL. Dividing this by $\Delta t = 1$ hr gave the rate of change:

$$\frac{\Delta y}{\Delta t} = \frac{14}{1} = 14 \text{ mg/dL per hour.}$$

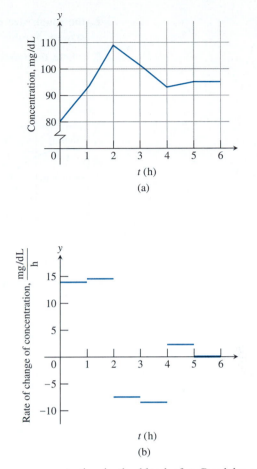

3.37 (a) The sugar concentration in the blood of a *Daedalus* pilot during a 6-hour preflight endurance test. (b) The derivative of the pilot's blood-sugar concentration shows how rapidly the concentration rose and fell during various portions of the test. (*Source:* Ethan R. Nadel and Steven R. Bussolari, "The Daedalus Project: Physiological Problems and Solutions," *American Scientist*, Vol. 76, No. 4, July–August 1988, p. 358.)

When we have so many data that the graph we get by connecting the data points looks like a smooth curve, we may also wish to plot the derivative as a smooth curve. The next example shows how this is done.

EXAMPLE 16

Graph the derivative of the function f in Fig. 3.38(a) on the following page.

Solution First, we draw a pair of coordinate axes, marking the horizontal axis in x-units and the vertical axis in slope units (Fig. 3.38(b)). Next, we estimate the slope of the graph of f in y-units per x-unit at frequent intervals, plotting the corresponding points against the new axes. We then connect the plotted points with a smooth curve.

From the graph of $y' = f'(x)$, we can see at a glance:

1. where f's rate of change is positive, negative, or zero;

2. the rough size of the growth rate at any x and its size in relation to the size of $f(x)$;

3. where the rate of change itself is increasing or decreasing.   ≡

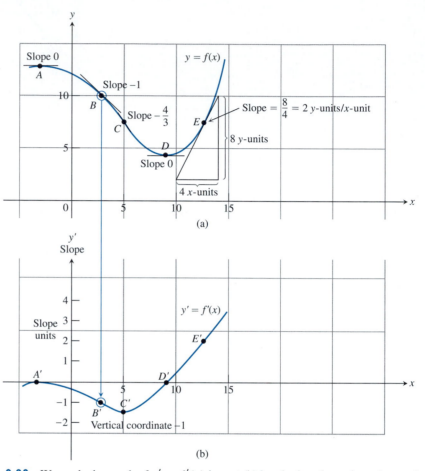

(a)

(b)

3.38   We made the graph of $y' = f'(x)$ in part (b) by plotting slopes from the graph of $y = f(x)$ in part (a). The vertical coordinate, -1, of B' is the slope at B, and so on. The graph of $y' = f'(x)$ is a visual record of how the slope of f changes with x.

Exercises 3.4

The equations in Exercises 1–6 give the position $s = f(t)$ of a particle moving along the line $y = 3$; s is in meters and t is in seconds.

a) Use parametric mode to simulate the motion of the particle for the indicated values of t. Describe the path of the particle.

b) Determine the position of the particle at $t = 0$, $t = 1$, $t = 2$, and $t = 3$ seconds.

c) Determine where and at what time the direction of the par-

ticle changes (if it does). Also determine the velocity and acceleration at these times.

d) Determine the total distance the particle travels for the specified t interval.

e) Simultaneously GRAPH the path of the particle and its position as a function of time.

f) Simultaneously GRAPH the path of the particle, the velocity function $y = v(t) = s'(t)$, and the acceleration function

$y = a(t) = s''(t)$. When is the particle at rest?

1. $s = t^2 - 3t + 2, \quad 0 \le t \le 5$

2. $s = 5 + 3t - t^2, \quad 0 \le t \le 5$

3. $s = t^3 - 6t^2 + 7t - 3, \quad 0 \le t \le 5$

4. $s = 4 - 7t + 6t^2 - t^3, \quad 0 \le t \le 5$

5. $s = t \sin t, \quad 0 \le t \le 15$

6. $s = 5 \sin \left(\dfrac{2t}{\pi} \right), \quad 0 \le t \le 2\pi$

7. The equations for free fall at the surfaces of Mars and Jupiter (s in meters, t in seconds) are: Mars, $s = 1.86t^2$; Jupiter, $s = 11.44t^2$. How long would it take a rock falling from rest to reach a velocity of 16.6 m/sec (about 60 km/h) on each planet?

8. A rock thrown vertically upward from the surface of the moon at a velocity of 24 m/sec (about 86 km/h) reaches a height of $s = 24t - 0.8t^2$ meters in t seconds.
   a) Find the rock's velocity and acceleration as a function of time. (The acceleration in this case is the acceleration of gravity on the moon.)
   b) How long did it take the rock to reach its highest point?
   c) How high did the rock go?
   d) How long did it take the rock to reach half its maximum height?
   e) How long was the rock aloft?

9. Devise a grapher simulation of the problem situation in Exercise 8. Use it to support the answers obtained analytically.

10. On Earth, in the absence of air, the rock in Exercise 8 would reach a height of $s = 24t - 4.9t^2$ meters in t seconds. How high would the rock go?

11. A 45-caliber bullet fired straight up from the surface of the moon would reach a height of $s = 832t - 2.6t^2$ feet after t seconds. On Earth, in the absence of air, its height would be $s = 832t - 16t^2$ feet after t seconds. How long would it take the bullet to get back down in each case?

12. Devise a grapher simulation of the problem situation in Exercise 11. Use it to support the answers obtained analytically.

13. When a bactericide was added to a nutrient broth in which bacteria were growing, the bacterium population continued to grow for a while but then stopped growing and began to decline. The size of the population at time t (hours) was $b(t) = 10^6 + 10^4 t - 10^3 t^2$. Find the growth rates at $t = 0$, $t = 5$, and $t = 10$ hours.

14. The number of gallons of water in a tank t minutes after the tank has started to drain is $Q(t) = 200(30 - t)^2$. How fast is the water running out at the end of 10 min? What is the average rate at which the water flows out during the first 10 min?

15. *Marginal cost.* Suppose that the dollar cost of producing x washing machines is $c(x) = 2000 + 100x - 0.1x^2$.
   a) Find the average cost of producing 100 washing machines.
   b) Find the marginal cost when 100 washing machines are produced.
   c) Show that the marginal cost when 100 washing machines are produced is approximately the cost of producing one more washing machine after the first 100 have been made, by calculating the latter cost directly.

16. *Marginal revenue.* Suppose the weekly revenue in dollars from selling x custom-made office desks is
$$r(x) = 2000 \left(1 - \frac{1}{x + 1} \right).$$
   a) Draw a complete graph of r. What values of x make sense in this problem situation?
   b) Find the marginal revenue when x desks are sold.
   c) Use the function $r'(x)$ to estimate the increase in revenue that will result from increasing sales from 5 desks a week to 6 desks a week.
   d) Find the limit of $r'(x)$ as $x \to \infty$. How would you interpret this number?

17. The position of a body at time t sec is $s = t^3 - 6t^2 + 9t$ m. Find the body's acceleration each time the velocity is zero.

18. A body's velocity at time t sec is $v = 2t^3 - 9t^2 + 12t - 5$ m/sec. Find the body's speed each time the acceleration is zero.

19. Determine complete graphs of the following parametric equations.
   a) $x(t) = 3t - 5 \sin t, \quad y(t) = 3t - 5 \cos t$
   b) $x(t) = \dfrac{6 \cos t}{4 - 3 \cos t}, \quad y(t) = \dfrac{6 \sin t}{4 - 3 \cos t}$
   c) $x(t) = 2 - 7 \cos t, \quad y(t) = -2 + 3 \cos t$

20. Which of the graphs in Exercise 19 are graphs of functions?

21. Let $f'(x) = 3x^2$.
   a) Compute the derivatives of $g(x) = x^3$, $h(x) = x^3 - 2$, and $t(x) = x^3 + 3$.
   b) Graph the numerical derivatives of g, h, and t.
   c) Describe the *family* of functions, $f(x)$, that have the property that $f'(x) = 3x^2$.
   d) Is there a unique f such that $f'(x) = 3x^2$ and $f(0) = 0$? What is it?
   e) Is there a unique g such that $g'(x) = 3x^2$ and $g(0) = 3$? What is it?

22. The monthly profit (in thousands of dollars) of a software company is given by
$$P(x) = \frac{10}{1 + 50 \cdot 2^{5 - 0.1x}},$$
where x is the number of software packages sold.
   a) Draw a complete graph of $y = P(x)$.
   b) What values of x make sense in the problem situation?

c) Draw a graph of $y = P'(x)$ (Use $y =$ NDER $P(x)$.) Compare with the graph of $y = M(x)$ in Example 14.

d) What is the profit when the marginal profit is maximum? What is the marginal profit when 50 units are produced? 100 units, 125 units, 150 units, 175 units, and 300 units?

e) What is $\lim\limits_{x \to \infty} P(x)$? What is the maximum profit possible?

f) Is there a practical explanation to the maximum profit question? Explain your reasoning.

23. When a model rocket is launched, the propellant burns for a few seconds, accelerating the rocket upward. After burnout, the rocket coasts upward for a while and then begins to fall. A small explosive charge pops out a parachute shortly after the rocket starts downward. The parachute slows the rocket to keep it from breaking when it lands. This graph shows velocity data from the flight.

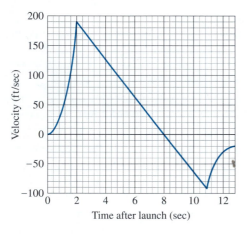

Use the data to answer the following.

a) How fast was the rocket climbing when the engine stopped?

b) For how many seconds did the engine burn?

c) When did the rocket reach its highest point? What was its velocity then?

d) When did the parachute pop out? How fast was the rocket falling then?

e) How long did the rocket fall before the parachute opened?

f) When was the rocket's acceleration greatest? When was the acceleration constant?

24. *Pisa by parachute (continuation of Exercise 23).* A few years ago, Mike McCarthy parachuted 179 ft from the top of the Tower of Pisa. Make a rough sketch to show the shape of the graph of his downward velocity during the jump.

Exercises 25 and 26 are about the graphs in Fig. 3.39. The graphs in part (a) show the numbers of rabbits and foxes in a small arctic population. They are plotted as functions of time for 200 days. The number of rabbits increases at first, as the rabbits reproduce. But the foxes prey on the rabbits, and as the number of foxes increases, the rabbit population levels off and

then drops. Figure 3.39(b) shows the graph of the derivative of the rabbit population. We made it by plotting slopes, as in Example 16.

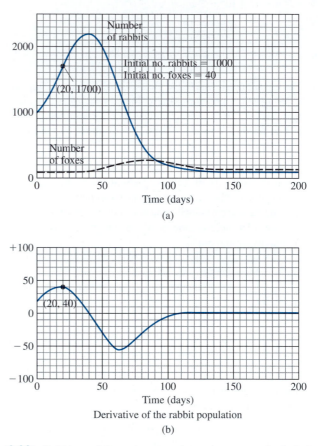

3.39 Rabbits and foxes in an arctic predator–prey food chain. (*Source: Differentiation,* by W. U. Walton et al., Project CALC, Education Development Center, Inc., Newton, Mass. (1975), p. 86.)

25. a) What is the value of the derivative of the rabbit population when the number of rabbits is largest? Smallest?

b) What is the size of the rabbit population when its derivative is largest? Smallest?

c) In what units should the slopes of the rabbit and fox population curves be measured?

26. Clearly there cannot be a fractional number of rabbits. Explain then, why the graphs appear to be continuous, and also how they should be interpreted.

Match the graphs of the functions in Exercises 27–30 with the graphs of the derivatives shown here:

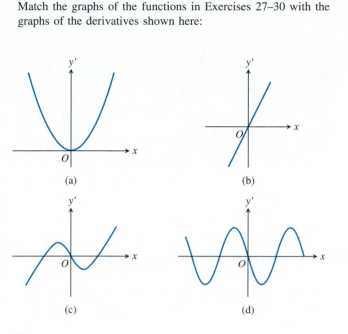

(a)

(b)

(c)

(d)

27.

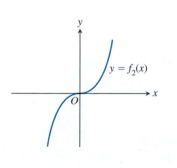

28.

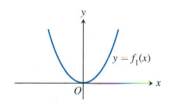

29.

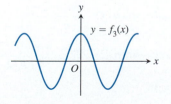

30.

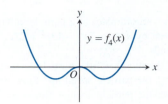

31. The graph of the function $y = f(x)$ shown here is made of line segments joined end to end.

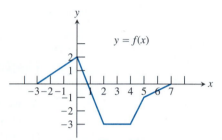

a) Graph the function's derivative.

b) At what values of x between $x = -3$ and $x = 7$ is the derivative not defined?

32. *Fruit flies (Example 1, Section 3.1 continued).* Populations starting out in closed environments grow slowly at first, when there are relatively few members, then more rapidly as the number of reproducing individuals increases and resources are still abundant, then slowly again as the population reaches the carrying capacity of the environment.

a) Use the graphical technique of Example 16 to graph the derivative of the fruit fly population introduced in Section 3.1. The graph of the population is reproduced below. What units should be used on the horizontal and vertical axes for the derivative's graph?

b) During what days does the population seem to be increasing fastest? Slowest?

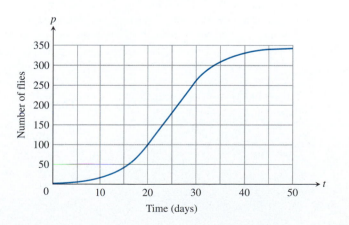

33. In Exploration 2, at what time is the particle at the point (5, 2)?

34. The position (*x*-coordinate) of a particle moving on the line $y = 2$ is given by $s(t) = 2t^3 - 13t^2 + 22t - 5$ where *t* is time in seconds.

a) Describe the motion of the particle for $t \geq 0$.
b) When does the particle speed up? Slow down?
c) When does the particle change direction?
d) When is the particle at rest?
e) Describe the velocity and speed of the particle.
f) When is the particle at the point (5, 2)?

The data in Exercises 35 and 36 give the coordinates *s* of a moving body for various values of *t*. Plot *s* versus *t* on coordinate paper, and sketch a smooth curve through the given points. Assuming that this smooth curve represents the motion of the body, estimate the velocity at (a) $t = 1.0$; (b) $t = 2.5$, (c) $t = 3.5$.

35.

t (sec)	0	0.5	1.0	1.5	2.0	2.5	3.0	3.5	4.0
s (ft)	12.5	26	36.5	44	48.5	50	48.5	44	36.5

36.

t (sec)	0	0.5	1.0	1.5	2.0	2.5	3.0	3.5	4.0
s (ft)	3.5	−4	−8.5	−10	−8.5	−4	3.5	14	27.5

3.5   Derivatives of Trigonometric Functions

As we mentioned in Section 1.7, trigonometric functions are important because so many of the phenomena about which we want information are periodic (heart rhythms, earthquakes, tides, weather). Continuous periodic functions can always be expressed in terms of sines and cosines, so the derivatives of sines and cosines play a key role in describing and predicting important changes. This section shows how to differentiate the six basic trigonometric functions.

For $y_1 = \sin x$, graphs of $y_2 = \text{NDER}\, y_1$ and $y_3 = \cos x$ in the same viewing window of a grapher strongly suggest that the derivative of the sine function is the cosine function. We now confirm this analytically.

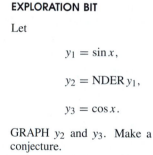

EXPLORATION BIT

Let

$$y_1 = \sin x,$$

$$y_2 = \text{NDER}\, y_1,$$

$$y_3 = \cos x.$$

GRAPH y_2 and y_3. Make a conjecture.

A New Limit

First, we have

$$\frac{\cos h - 1}{h} = -\frac{1 - \cos h}{h}$$

$$= -\frac{1 - \cos 2(h/2)}{2(h/2)}$$

$$= -\frac{\sin^2(h/2)}{h/2} \qquad \text{Section 1.7, half-angle formula}$$

$$= -\sin(h/2) \cdot \frac{\sin(h/2)}{h/2}.$$

Then,

$$\lim_{h \to 0} \frac{\cos h - 1}{h} = \lim_{h \to 0} \left(-\sin(h/2) \cdot \frac{\sin(h/2)}{h/2}\right)$$

$$= \lim_{h \to 0}(-\sin(h/2)) \cdot \lim_{h \to 0} \frac{\sin(h/2)}{h/2}$$

$$= 0 \cdot 1 = 0.$$

The Derivative of the Sine

The derivative of $y = \sin x$ is the limit

$$\frac{dy}{dx} = \lim_{h \to 0} \frac{\sin(x+h) - \sin x}{h}$$

$$= \lim_{h \to 0} \frac{\sin x \cos h + \cos x \sin h - \sin x}{h} \qquad \text{Section 1.7, angle sum formula}$$

$$= \lim_{h \to 0} \frac{\sin x(\cos h - 1) + \cos x \sin h}{h}$$

$$= \lim_{h \to 0} \sin x \cdot \lim_{h \to 0} \frac{\cos h - 1}{h} + \lim_{h \to 0} \cos x \cdot \lim_{h \to 0} \frac{\sin h}{h}$$

$$= \sin x \cdot 0 + \cos x \cdot 1$$

$$= \cos x.$$

In short,

$$\frac{d}{dx} \sin x = \cos x.$$

Now, with sine differentiable, we know that sine and its derivative obey all the differentiation rules. We also know that $\sin x$ is continuous. The same holds for other trigonometric functions in this section. The differentiation rules apply for each one. Each one is differentiable at every point in its domain and is therefore continuous at every point in its domain.

The Derivative of the Cosine

The graphs of $y_2 = \text{NDER} \cos x$ and $y_3 = \sin x$ strongly suggest that

$$\frac{d}{dx} \cos x = -\sin x.$$

You will confirm this analytically in the exercises.

$$\frac{d}{dx} \cos x = -\sin x.$$

Notice the negative sign. The derivative of the sine is the cosine, but the derivative of the cosine is the opposite of the sine. We'll say, "The derivative of the cosine is *minus* the sine" because the nice rhythm of the wording makes it easier to remember.

EXPLORATION BIT

Exercises 24 and 25 of Section 2.3 suggest one reason why we use radian measure in calculus. Here's another: The derivative of $\sin x$ is $\cos x$ only when x is in radians. Can you find the derivative of $\sin x$ when x is in degrees?

EXPLORATION BIT

Let

$$y_1 = \cos x,$$

$$y_2 = \text{NDER } y_1,$$

$$y_3 = \sin x.$$

GRAPH y_2 and y_3. Make a conjecture.

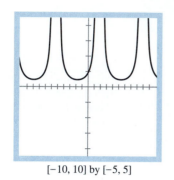

[−10, 10] by [−5, 5]

3.40 To support our analytic work in Example 1(b), we view the graphs of

$$y_1 = \text{NDER} \frac{\cos x}{1 - \sin x}$$

and

$$y_2 = \frac{1}{1 - \sin x}.$$

They appear to be identical. That's good.

EXAMPLE 1

a) $y = x^2 \sin x$ $\dfrac{dy}{dx} = x^2 \dfrac{d}{dx}(\sin x) + 2x \sin x$ Product Rule

$$= x^2 \cos x + 2x \sin x$$

b) $y = \dfrac{\cos x}{1 - \sin x}$

$$\frac{dy}{dx} = \frac{(1 - \sin x)\dfrac{d}{dx}(\cos x) - \cos x \dfrac{d}{dx}(1 - \sin x)}{(1 - \sin x)^2} \qquad \text{Quotient Rule}$$

$$= \frac{(1 - \sin x)(-\sin x) - \cos x(0 - \cos x)}{(1 - \sin x)^2}$$

$$= \frac{1 - \sin x}{(1 - \sin x)^2} \qquad \sin^2 x + \cos^2 x = 1$$

$$= \frac{1}{1 - \sin x} \qquad \text{See Fig. 3.40.}$$

EXPLORATION 1

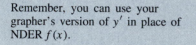

Support Graphically

Use rules of differentiation to find y' for each function $y = f(x)$. Support graphically by comparing the graph of your result and the graph of NDER $f(x)$.

1. $y = x^2 - \sin x$ **2.** $y = \dfrac{\sin x}{x}$

3. $y = 5x + \cos x$ **4.** $y = \sin x \cos x$

> Remember, you can use your grapher's version of y' in place of NDER $f(x)$.

Simple Harmonic Motion

The motion of a weight bobbing up and down on the end of a spring is a *simple harmonic motion*. The next example describes a case in which there are no opposing forces such as air friction to slow the motion down.

EXAMPLE 2

A weight hanging from a spring (Fig. 3.41) is compressed 5 units above its rest position ($s = 0$) and released at time $t = 0$ to bob down and up. Its position at any later time t is

$$s = 5 \cos t.$$

What are its velocity and acceleration at time t?

Solution We have

Position: $s = 5 \cos t,$

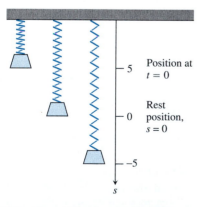

3.41 The weighted spring in Example 2.

Velocity: $\dfrac{ds}{dt} = \dfrac{d}{dt}(5 \cos t) = 5 \dfrac{d}{dt}(\cos t) = -5 \sin t,$

Acceleration: $\dfrac{dv}{dt} = \dfrac{d}{dt}(-5\sin t) = -5\dfrac{d}{dt}(\sin t) = -5\cos t.$

Here is what we can learn from these equations:

1. As time passes, the weight moves down and up between $s = -5$ and $s = 5$ on the s-axis. The amplitude of the motion is 5. The period of the motion is 2π.

2. The function $\sin t$ attains its greatest magnitude, 1, when $\cos t = 0$, as the graphs of the sine and cosine show. Hence, the weight's speed, $|v| = 5\,|\sin t|$, is greatest when $\cos t = 0$, that is, when $s = 0$.

   The weight's speed is zero when $\sin t = 0$. This occurs when $\cos t = \pm 1$, at the endpoints of the interval of motion.

3. The acceleration, $a = -5\cos t$, is zero only at the rest position, where $\cos t = 0$, and the force of gravity and the force from the spring offset each other. When the weight is anywhere else, the two forces are unequal, and acceleration is nonzero. The acceleration is greatest in magnitude at the points farthest from the origin, where $\cos t = \pm 1$.

EXPLORATION 2

Simulating Simple Harmonic Motion

GRAPH simultaneously the parametric equations

$$x_1(t) = -1, \qquad y_1(t) = 5\cos t,$$
$$x_2(t) = t, \qquad y_2(t) = -5\sin t,$$

for $0 \le t \le 3\pi$. TRACE to explore the position and velocity functions from Example 2. Change $x_1(t) = -1$ to $x_1(t) = t$. This allows you to see the up-and-down motion of the spring, just as we saw the horizontal motion of the particle in Section 3.4. GRAPH simultaneously again, and explain what you see.

The Derivatives of the Other Basic Functions

Because $\sin x$ and $\cos x$ are differentiable functions of x, the related functions

$$\tan x = \frac{\sin x}{\cos x}, \qquad\qquad \sec x = \frac{1}{\cos x},$$
$$\cot x = \frac{\cos x}{\sin x}, \qquad\qquad \csc x = \frac{1}{\sin x}$$

are differentiable at every value of x at which they are defined. Their derivatives, calculated from the quotient rule, are given by the following formulas.

$$\frac{d}{dx}\tan x = \sec^2 x \qquad (1) \qquad\qquad \frac{d}{dx}\sec x = \sec x \tan x \qquad (2)$$

$$\frac{d}{dx}\cot x = -\csc^2 x \qquad (3) \qquad\qquad \frac{d}{dx}\csc x = -\csc x \cot x \qquad (4)$$

Notice the negative signs in the equations for the cotangent and cosecant. To show how a typical calculation goes, we derive Eq. (1). The other derivations are left as exercises.

EXAMPLE 3

Find dy/dx if $y = \tan x$.

Solution

$$\frac{d}{dx}\tan x = \frac{d}{dx}\left(\frac{\sin x}{\cos x}\right) = \frac{\cos x \dfrac{d}{dx}(\sin x) - \sin x \dfrac{d}{dx}(\cos x)}{\cos^2 x}$$

$$= \frac{\cos x \cos x - \sin x(-\sin x)}{\cos^2 x} = \frac{\cos^2 x + \sin^2 x}{\cos^2 x}$$

$$= \frac{1}{\cos^2 x} = \sec^2 x$$

≡

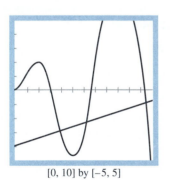

[0, 10] by [−5, 5]

3.42 The graphs of

$$y_1 = x \sin x$$

and its normal line

$$y_2 = 0.3x - 4.23$$

at $x = 4$. If you show this on your grapher, try to (a) graph in a square window that (b) includes the view dimensions [0, 10] by [−5, 5].

EXAMPLE 4

Find the equation of the line that is normal (perpendicular to the tangent) to the curve $y = x \sin x$ at $x = 4$. Support graphically.

Solution   The slope of the tangent to the curve at $(4, 4\sin 4) = (4, -3.03)$ is

$$\text{NDER}\,(x \sin x, 4) = -3.37.$$

The normal (perpendicular) has slope $-1/-3.37 = 0.30$, and the equation of the normal line to the curve at $(4, -3.03)$ is

$$y = 0.30(x - 4) - 3.03 \quad \text{or}$$

$$y = 0.30x - 4.23.$$

The graphs in Fig. 3.42 support this result.

≡

EXAMPLE 5

Find y'' if $y = \sec x$.

Solution

$$y = \sec x$$

$$y' = \sec x \tan x \qquad\qquad \text{(Eq. 2)}$$

$$y'' = \frac{d}{dx}(\sec x \tan x)$$

$$= \sec x \frac{d}{dx}(\tan x) + \tan x \frac{d}{dx}(\sec x)$$

$$= \sec x (\sec^2 x) + \tan x (\sec x \tan x)$$

$$= \sec^3 x + \sec x \tan^2 x$$

≡

EXPLORATION 3

Support Graphically

Use rules of differentiation to find y' for each function $y = f(x)$. Support graphically by comparing the graph of your result with the graph of NDER $f(x)$.

1. $y = 3x + \cot x$

2. $y = 2/\sin x$

Find y'' for each function $y = f(x)$. Support graphically by comparing the graph of your result with the graph of NDER2 $f(x)$.

3. $y = \sec x$

4. $y = 2 \sin x + 3 \cos x$

Exercises 3.5

In Exercises 1–24, find dy/dx. Support your answer by graphing the numerical derivative and comparing with the graph of the exact derivative.

1. $y = 1 + x - \cos x$

2. $y = 2 \sin x - \tan x$

3. $y = \dfrac{1}{x} + 5 \sin x$

4. $y = x^2 - \sec x$

5. $y = \csc x - 5x + 7$

6. $y = 2x + \cot x$

7. $y = x \sec x$

8. $y = x \csc x$

9. $y = x^2 \cot x$

10. $y = 4 - x^2 \sin x$

11. $y = 3x + x \tan x$

12. $y = x \sin x + \cos x$

13. $y = \sin x \sec x$

14. $y = \sec x \csc x$

15. $y = \tan x \cot x$

16. $y = \cos x (1 + \sec x)$

17. $y = \dfrac{4}{\cos x}$

18. $y = 5 + \dfrac{1}{\tan x}$

19. $y = \dfrac{\cos x}{x}$

20. $y = \dfrac{2}{\csc x} - \dfrac{1}{\sec x}$

21. $y = \dfrac{x}{1 + \cos x}$

22. $y = \dfrac{\sin x + \cos x}{\cos x}$

23. $y = \dfrac{\cot x}{1 + \cot x}$

24. $y = \dfrac{\cos x}{1 + \sin x}$

25. Find y'' if $y = \csc x$.

26. Find $y^{(4)} = d^4 y/dx^4$ if
 a) $y = \sin x$,
 b) $y = \cos x$.

In Exercises 27–30, find equations for the lines that are tangent and normal to the curve $y = f(x)$ at the given point $(x, f(x))$. Support your answers graphically using square viewing windows.

27. $y = \sin x, \quad x = 0$

28. $y = \tan x, \quad x = 0$

29. $y = 2 \sin^2 x, \quad x = 2$

30. $y = \dfrac{2 + \cot x}{x}, \quad x = 1$

31. Prove that $D_x \cos x = -\sin x$.

32. Show that the graphs of $y = \sec x$ and $y = \cos x$ have horizontal tangents at $x = 0$.

33. Show that the graphs of $y = \tan x$ and $y = \cot x$ never have horizontal tangents.

Do the graphs of the functions in Exercises 34–37 have any horizontal tangents in the interval $0 \le x \le 2\pi$? If so, where? If not, why not?

34. $y = x + \sin x$

35. $y = 2x + \sin x$

36. $y = x + \cos x$

37. $y = x + 2 \cos x$

38. *Simple harmonic motion.* The equations in parts (a) and (b) give the position $s = f(t)$ of a body moving along a coordinate line. Find each body's velocity, speed, and acceleration at time $t = \pi/4$.

 a) $s = 2 - 2 \sin t$
 b) $s = \sin t + \cos t$

39. Find equations for the lines that are tangent and normal to the curve $y = \sqrt{2} \cos x$ at the point $(\pi/4, 1)$.

40. Find the points on the curve $y = \tan x$, $-\pi/2 < x < \pi/2$, where the tangent is parallel to the line $y = 2x$.

Find equations for the horizontal tangents to the graphs in Exercises 41 and 42.

41. $y = \cot x - \sqrt{2}\csc x$, $0 < x < \pi$

42. $y = \tan x + 3\cot x - 3$, $0 < x < \pi/2$

43. GRAPH $y = \tan x$ and its derivative together over the interval $-\pi/2 < x < \pi/2$.

44. GRAPH $y = \cot x$ and its derivative together for $0 < x < \pi$.

45. Although $\lim\limits_{h \to 0} (1 - \cos h)/h = 0$, it turns out that

$$\lim_{h \to 0} \frac{1 - \cos h}{h^2} \neq 0.$$

Determine the limit. Confirm analytically.

46. Derive Eq. (2) by writing $\sec x = 1/\cos x$ and differentiating with respect to x.

47. Derive Eq. (3) by writing $\cot x = (\cos x)/(\sin x)$ and differentiating with respect to x.

48. Derive Eq. (4) by writing $\csc x = 1/\sin x$ and differentiating with respect to x.

3.6 The Chain Rule

We now know how to differentiate $\sin x$ and $x^2 - 4$, but how do we differentiate a composite like $\sin(x^2 - 4)$? The answer is: with the Chain Rule, which says that the derivative of the composite of two differentiable functions is the product of their derivatives evaluated at appropriately related points. The Chain Rule is probably the most widely used differentiation rule in mathematics. This section describes the rule and how to use it.

Introductory Examples

Some examples will help to show what is going on.

EXAMPLE 1

The function $y = 9x^2 + 6x + 1 = (3x + 1)^2$ is the composite of $y = u^2$ and $u = 3x + 1$. How are the derivatives of these three functions related?

$$\frac{dy}{dx} = \frac{d}{dx}(9x^2 + 6x + 1) = 18x + 6 = 6(3x + 1) = 6u$$

$$\frac{dy}{du} = \frac{d}{du}(u^2) = 2u$$

$$\frac{du}{dx} = \frac{d}{dx}(3x + 1) = 3.$$

Because $6u = 2u \cdot 3$,

$$\frac{dy}{dx} = \frac{dy}{du} \cdot \frac{du}{dx}.$$

EXAMPLE 2

In the gear train in Fig. 3.43, the ratios of the radii of gears A, B, and C are 3:1:2. If gear A turns x times, then gear B turns $u = 3x$ times, and gear C turns $y = u/2 = (3/2)x$ times. In terms of derivatives,

$$\frac{dy}{du} = \frac{1}{2} \qquad\qquad C \text{ turns at half } B\text{'s rate.}$$

$$\frac{du}{dx} = 3 \qquad\qquad B \text{ turns at three times } A\text{'s rate.}$$

In this example, too, we can calculate dy/dx by multiplying dy/du by du/dx:

$$\frac{dy}{dx} = \frac{3}{2} = \frac{1}{2} \cdot 3 = \frac{dy}{du} \cdot \frac{du}{dx}. \qquad \begin{array}{l} C \text{ turns at three-halves } A\text{'s rate,} \\ \text{three-halves of a turn for } A\text{'s one.} \end{array}$$

C: y turns   B: u turns   A: x turns

3.43 When wheel A takes x turns, wheel B takes u turns, and wheel C takes y turns. By comparing circumferences, we see that $dy/du = 1/2$ and $du/dx = 3$. What is dy/dx?

EXPLORATION 1

The Chain Rule

Now you try the following.

1. A particle moves along the line $y = 5u - 2$ in the uy-plane in such a way that its u-coordinate at time t is $u = 3t$ (Fig. 3.44).

Write y as a function of t.

Compute $\dfrac{dy}{dt}$, then $\dfrac{dy}{du}$ and $\dfrac{du}{dt}$. Compare $\dfrac{dy}{dt}$ with $\dfrac{dy}{du} \cdot \dfrac{du}{dt}$.

2. The function $y = 2(2x + 1)^3 + 1$ is the composite of $y = 2u^3 + 1$ and $u = 2x + 1$.

Compute $\dfrac{dy}{dx}$, then $\dfrac{dy}{du}$ and $\dfrac{du}{dx}$. Compare $\dfrac{dy}{dx}$ with $\dfrac{dy}{du} \cdot \dfrac{du}{dx}$.

$y = 5u - 2$

$P(3t, 15t - 2)$

3.44 A particle moving along the line $y = 5u - 2$ in such a way that $u = 3t$.

The Chain Rule

The preceding examples all work because the derivative of a composite $f \circ g$ of two differentiable functions is the product of their derivatives evaluated at appropriate points. This is the observation that we state formally as the Chain Rule. As in Section 1.3, the notation $f \circ g$ ("f of g") denotes the composite of the functions f and g, with f following g. The value of $f \circ g$ at a point x is $(f \circ g)(x) = f(g(x))$.

THE CHAIN RULE (FIRST FORM)

Suppose that $f \circ g$ is the composite of the differentiable functions $y = f(u)$ and $u = g(x)$. Then $f \circ g$ is a differentiable function of x whose derivative at each value of x is

$$(f \circ g)'_{\text{at } x} = f'_{\text{at } u=g(x)} \cdot g'_{\text{at } x}. \tag{1}$$

In short,

$$(f \circ g)'(x) = f'(g(x)) \cdot g'(x). \tag{2}$$

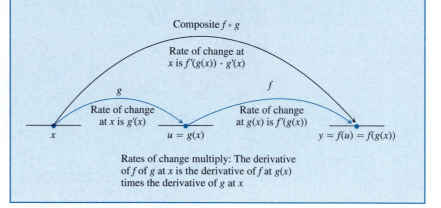

Composite $f \circ g$

Rate of change at x is $f'(g(x)) \cdot g'(x)$

g f

Rate of change at x is $g'(x)$ Rate of change at $g(x)$ is $f'(g(x))$

x $u = g(x)$ $y = f(u) = f(g(x))$

Rates of change multiply: The derivative of f of g at x is the derivative of f at $g(x)$ times the derivative of g at x

---

Equations (1) and (2) name the function involved as well as the dependent and independent variables. Once we know what the functions are, as we usually do in any particular example, we can get by with writing the Chain Rule in a shorter way.

THE CHAIN RULE (SHORTER FORM)

If y is a differentiable function of u and u is a differentiable function of x, then y is a differentiable function of x, and

$$\frac{dy}{dx}\Big|_{\text{at } x} = \frac{dy}{du}\Big|_{\text{at } u(x)} \cdot \frac{du}{dx}\Big|_{\text{at } x}. \tag{3}$$

Equation (3) still tells how each derivative is to be evaluated. When we don't need to be told that, we can get along with an even shorter form.

THE CHAIN RULE (SHORTEST FORM)

If y is a differentiable function of u and u is a differentiable function of x, then

$$\frac{dy}{dx} = \frac{dy}{du} \cdot \frac{du}{dx}. \tag{4}$$

You might think it would be a relatively easy matter to prove the Chain Rule by starting with the derivative definition the way we started the proofs of the Product and Quotient Rules. Unfortunately, this is the way the *hard* proof starts. The (relatively) easy proof starts with an equation that we won't get to until Section 3.8. We have therefore placed the proof of the Chain Rule in Appendix 3, to be looked at later. We'll direct your attention to it when the time comes.

Like different instruments in a doctor's bag, each form of the Chain Rule makes some task a little easier. We shall use them all in the examples that follow. But remember—they all express the same one rule: The derivative of a composite of differentiable functions is the product of their derivatives evaluated at appropriately related points.

EXAMPLE 3

If $f(u) = \sin u$ and $u = g(x) = x^2 - 4$, find $(f \circ g)'$ at $x = 2$.

Solution   Equation (1) for the Chain Rule gives

$$(f \circ g)'_{\text{at } x=2} = f'_{\text{at } u=g(2)} \cdot g'_{\text{at } x=2}$$

$$= \frac{d}{dx}(\sin u)_{\text{at } u=0} \cdot \frac{d}{dx}(x^2 - 4)_{\text{at } x=2}$$

$$= \cos u \, |_{u=0} \cdot 2x|_{x=2}$$

$$= 1 \cdot 4 = 4.$$

EXPLORATION 2

Support Graphically

Let $y_1 = x^2 - 4$ and $y_2 = \sin y_1$. Note that $y_2 = f \circ g$ from Example 3.

1. Graph y_2 and $y_3 = \text{NDER} \, y_2$ in the $[-3, 3]$ by $[-4, 4]$ viewing window. TRACE to explore the values of $(f \circ g)'$ for $-3 \le x \le 3$.

2. How many horizontal tangents are there for $f \circ g$ in $[-3, 3]$?

3. Support the result of Example 3, namely that $(f \circ g)'(2) = 4$. Interpret this value as it relates to $f \circ g$. Does the viewing window support your interpretation? Explain.

EXAMPLE 4

Find dy/dx at $x = 0$ if $y = \cos u$ and $u = \pi/2 - 3x$.

Solution   With Eq. (3) this time,

$$\frac{dy}{dx}\bigg|_{x=0} = \frac{dy}{du}\bigg|_{u=\frac{\pi}{2}} \cdot \frac{du}{dx}\bigg|_{x=0}$$

$$= -\sin u \, |_{u=\frac{\pi}{2}} \cdot (-3) = 3 \sin \frac{\pi}{2} = 3 \cdot 1 = 3.$$

It is more direct in Example 4 to write $y = \cos(\pi/2 - 3x)$. Then

$$\frac{dy}{dx} = -\sin\left(\frac{\pi}{2} - 3x\right) \cdot \frac{d}{dx}\left(\frac{\pi}{2} - 3x\right)$$

$$= 3\sin\left(\frac{\pi}{2} - 3x\right),$$

and at $x = 0$,

$$\frac{dy}{dx} = 3\sin\left(\frac{\pi}{2} - 3 \cdot 0\right) = 3.$$

EXAMPLE 5

Express dy/dx in terms of x if $y = u^3$ and $u = x^2 - 1$.

Solution

$$\frac{dy}{dx} = \frac{dy}{du} \cdot \frac{du}{dx} \qquad \text{Eq. (4)}$$

$$= 3u^2 \cdot 2x$$

$$= 3(x^2 - 1)^2 \cdot 2x \qquad u = x^2 - 1$$

$$= 6x(x^2 - 1)^2$$

Integer Powers of Differentiable Functions

As illustrated in Example 5, if $y = u^n$ and u is a differentiable function of x, then the Chain Rule in the form

$$\frac{dy}{dx} = \frac{dy}{du} \cdot \frac{du}{dx}$$

gives

$$\frac{dy}{dx} = \frac{d}{du}(u^n) \cdot \frac{du}{dx}$$

$$= nu^{n-1}\frac{du}{dx}. \qquad \begin{array}{l}\text{Differentiating } u^n \text{ with respect to } u \text{ itself} \\ \text{gives } nu^{n-1} \text{ by Rules 2 and 7 in Section 3.3.}\end{array}$$

Integer Powers of a Differentiable Function

If u^n is an integer power of a differentiable function $u(x)$, then u^n is differentiable, and

$$\frac{d}{dx}u^n = nu^{n-1}\frac{du}{dx}. \qquad (5)$$

EXAMPLE 6

Let (a) $y = \sin^5 x$ and (b) $y = (2x + 1)^{-3}$. Find dy/dx for each. Support graphically.

Solution

a) $\dfrac{d}{dx}\sin^5 x = 5\sin^4 x\,\dfrac{d}{dx}(\sin x)$ Eq. (5) with
$u = \sin x, n = 5;$

$= 5\sin^4 x\cos x$ See Fig. 3.45(a) for
graphical support.

b) $\dfrac{d}{dx}(2x+1)^{-3} = -3(2x+1)^{-4}\dfrac{d}{dx}(2x+1)$ Eq. (5) with
$u = 2x+1, n = -3$

$= -3(2x+1)^{-4}(2)$

$= -6(2x+1)^{-4}$ See Fig. 3.45(b) for
graphical support.

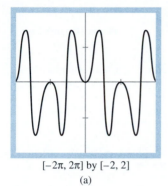

$[-2\pi, 2\pi]$ by $[-2, 2]$
(a)

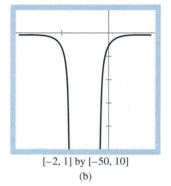

$[-2, 1]$ by $[-50, 10]$
(b)

3.45 The graphs appear the same (a) for $y = 5\sin^4 x\cos x$ and $y = \text{NDER}\sin^5 x$ and (b) for $y = -6(2x+1)^{-4}$ and $y = \text{NDER}\,(2x+1)^{-3}$.

The "Outside-Inside" Rule

It sometimes helps to think about the Chain Rule in the following way. If $y = f(g(x))$, Eq. (2) tells us that

$$\frac{dy}{dx} = f'(g(x))\cdot g'(x).$$

In words, this says: To find dy/dx, differentiate the "outside" function f and leave the "inside" $g(x)$ alone; then multiply by the derivative of the inside. Study how this works in Example 7 with $y = \sin(x^2 + x)$, then $y = \tan\sqrt{x}$.

EXAMPLE 7

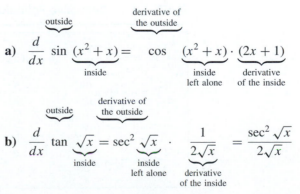

a) $\underset{}{\dfrac{d}{dx}}\ \underbrace{\sin}_{\text{outside}}\ \underbrace{(x^2+x)}_{\substack{\text{inside}}} = \underbrace{\cos}_{\substack{\text{derivative of}\\ \text{the outside}}}\underbrace{(x^2+x)}_{\substack{\text{inside}\\ \text{left alone}}}\cdot\underbrace{(2x+1)}_{\substack{\text{derivative}\\ \text{of the inside}}}$

b) $\dfrac{d}{dx}\ \underbrace{\tan}_{\text{outside}}\ \underbrace{\sqrt{x}}_{\text{inside}} = \underbrace{\sec^2}_{\substack{\text{derivative of}\\ \text{the outside}}}\underbrace{\sqrt{x}}_{\substack{\text{inside}\\ \text{left alone}}}\cdot\underbrace{\dfrac{1}{2\sqrt{x}}}_{\substack{\text{derivative}\\ \text{of the inside}}} = \dfrac{\sec^2\sqrt{x}}{2\sqrt{x}}$

Repeated Use

We sometimes have to use the Chain Rule two or more times to get the job done. Here is an example.

EXAMPLE 8

a)
$$\frac{d}{dx}\cos^2 3x = 2\cos 3x \cdot \frac{d}{dx}(\cos 3x) \qquad \text{Power (Chain) Rule}$$

$$= 2\cos 3x(-\sin 3x)\frac{d}{dx}(3x) \qquad \text{Chain Rule again}$$

$$= 2\cos 3x(-\sin 3x)(3)$$

$$= -6\cos 3x \sin 3x$$

b)
$$\frac{d}{dx}\sin(1+\tan 2x) = \cos(1+\tan 2x) \cdot \frac{d}{dx}(1+\tan 2x) \qquad \text{Chain Rule}$$

$$= \cos(1+\tan 2x) \cdot \sec^2 2x \cdot \frac{d}{dx}(2x) \qquad \text{Chain Rule}$$

$$= 2\cos(1+\tan 2x)\sec^2 2x \qquad \blacksquare$$

> **EXPLORATION BIT**
>
> Let $y_1 = \sin 2x$ and $y_2 = \cos 2x$. GRAPH y_1, y_2, and $y_3 = \text{NDER } y_1$. On the basis of the graphs, make a conjecture about the derivative of $\sin 2x$. Confirm it using the Chain Rule. What other question could be asked?

> **EXPLORATION BITS**
>
> Apply the Chain Rule to find $f'(x)$. Support graphically by comparing $f'(x)$ and NDER $f(x)$ in the same viewing window.
>
> 1. $f(x) = \sin(-x)$
> 2. $f(x) = \tan(1/x)$
> 3. $f(x) = \cos(\cos x)$

Derivative Formulas that Include the Chain Rule

Many of the derivative formulas you will encounter in your scientific work already include the Chain Rule.

If f is a differentiable function of u and u is a differentiable function of x, then substituting $y = f(u)$ in the Chain Rule formula

$$\frac{dy}{dx} = \frac{dy}{du} \cdot \frac{du}{dx}$$

leads to

$$\frac{d}{dx}f(u) = f'(u)\frac{du}{dx}.$$

When we spell this out for the functions whose derivatives we have studied so far, we get the formulas in Table 3.1

TABLE 3.1 Derivative Formulas That Include the Chain Rule

$$\frac{d}{dx}u^n = nu^{n-1}\frac{du}{dx} \qquad (n \text{ an integer})$$

$$\frac{d}{dx}\sin u = \cos u\frac{du}{dx} \qquad\qquad \frac{d}{dx}\sec u = \sec u \tan u\frac{du}{dx}$$

$$\frac{d}{dx}\cos u = -\sin u\frac{du}{dx} \qquad\qquad \frac{d}{dx}\cot u = -\csc^2 u\frac{du}{dx}$$

$$\frac{d}{dx}\tan u = \sec^2 u\frac{du}{dx} \qquad\qquad \frac{d}{dx}\csc u = -\csc u \cot u\frac{du}{dx}$$

Exercises 3.6

In Exercises 1–14, find dy/dx.

1. $y = \sin(3x + 1)$

2. $y = \sin(7 - 5x)$

3. $y = \cos(-x/3)$

4. $y = \cos(\sqrt{3}x)$

5. $y = \tan(2x - x^3)$

6. $y = \tan 5(2x - 1)$

7. $y = x + \sec(x^2 + \sqrt{2})$

8. $y = x \sec(3 - 8x)$

9. $y = -\csc(x^2 + 7x)$

10. $y = \sqrt{x} + \csc(1 - 2x)$

11. $y = 5\cot\left(\dfrac{2}{x}\right)$

12. $y = \cot\left(\pi - \dfrac{1}{x}\right)$

13. $y = \cos(\sin x)$

14. $y = \sec(\tan x)$

In Exercises 15–26, find dy/dx. Support your answer with the graph of $y_1 = $ NDER y.

15. $y = (2x + 1)^5$

16. $y = (4 - 3x)^9$

17. $y = (x^2 + 1)^{-3}$

18. $y = (x + \sqrt{x})^{-2}$

19. $y = \left(1 - \dfrac{x}{7}\right)^{-7}$

20. $y = \left(\dfrac{x}{2} - 1\right)^{-10}$

21. $y = \left(\dfrac{x^2}{8} + x - \dfrac{1}{x}\right)^4$

22. $y = \left(\dfrac{x}{5} + \dfrac{1}{5x}\right)^5$

23. $y = (\csc x + \cot x)^{-1}$

24. $y = -(\sec x + \tan x)^{-1}$

25. $y = \sin^4 x + \cos^{-2} x$

26. $y = \sin^{-5} x - \cos^3 x$

Use the Chain Rule in combination with the Product and Quotient Rules to find dy/dx in Exercises 27–36.

27. $y = x^3(2x - 5)^4$

28. $y = (1 - x)(3x^2 - 5)^5$

29. $y = (4x + 3)^4(x + 1)^{-3}$

30. $y = (2x - 5)^{-1}(x^2 - 5x)^6$

31. $y = \left(\dfrac{\sin x}{1 + \cos x}\right)^2$

32. $y = \left(\dfrac{1 + \cos x}{\sin x}\right)^{-1}$

33. $y = \left(\dfrac{x}{x - 1}\right)^{-3}$

34. $y = \left(\dfrac{x}{x - 1}\right)^2 - \dfrac{4}{x - 1}$

35. $y = \sin^3 x \tan 4x$

36. $y = \cos^4 x \cot 7x$

Find dy/dx in Exercises 37–44. Support your answer with the graph of $y_1 = $ NDER y.

37. $y = \sqrt{\sin x}$

38. $y = \sqrt{\cos x}$

39. $y = 4\sqrt{\sec x + \tan x}$

40. $y = 2\sqrt{\csc x + \cot x}$

41. $y = \dfrac{3}{\sqrt{2x + 1}}$

42. $y = \dfrac{x}{\sqrt{1 + x^2}}$

43. $y = (2x - 6)\sqrt{x + 5}$

44. $y = x\sqrt{x^2 - 2x}$

In Exercises 45–48, find ds/dt.

45. $s = \cos\left(\dfrac{\pi}{2} - 3t\right)$

46. $s = t\cos(\pi - 4t)$

47. $s = \dfrac{4}{3\pi}\sin 3t + \dfrac{4}{5\pi}\cos 5t$

48. $s = \sin\left(\dfrac{3\pi}{2}t\right) + \cos\left(\dfrac{7\pi}{4}t\right)$

In Exercises 49–52, find $dr/d\theta$.

49. $r = \tan(2 - \theta)$

50. $r = \sec 2\theta \tan 2\theta$

51. $r = \sqrt{\theta \sin \theta}$

52. $r = 2\theta\sqrt{\sec \theta}$

In Exercises 53–60, find dy/dx. You will need to use the Chain Rule two or three times in each case.

53. $y = \sin^2(3x - 2)$

54. $y = \sec^2 5x$

55. $y = (1 + \cos 2x)^2$

56. $y = (1 - \tan(x/2))^{-2}$

57. $y = \sin(\cos(2x - 5))$

58. $y = (1 + \cos^2 7x)^3$

59. $y = \cot\sqrt{2x}$

60. $y = \sqrt{\tan 5x}$

Find y'' in Exercises 61–64.

61. $y = \tan x$

62. $y = \cot x$

63. $y = \cot(3x - 1)$

64. $y = 9\tan(x/3)$

In Exercises 65–70, find the value of $(f \circ g)'$ at the given value of x.

65. $f(u) = u^5 + 1, u = g(x) = \sqrt{x}, x = 1$

66. $f(u) = 1 - \dfrac{1}{u}, u = g(x) = \dfrac{1}{1 - x}, x = -1$

67. $f(u) = \cot\dfrac{\pi u}{10}, u = g(x) = 5\sqrt{x}, x = 1$

68. $f(u) = u + \dfrac{1}{\cos^2 u}, u = g(x) = \pi x, x = 1/4$

69. $f(u) = \dfrac{2u}{u^2 + 1}, u = g(x) = 10x^2 + x + 1, x = 0$

70. $f(u) = \left(\dfrac{u - 1}{u + 1}\right)^2, u = g(x) = \dfrac{1}{x^2} - 1, x = -1$

What happens if you can write a function as a composite in different ways? Do you get the same derivative each time? The Chain Rule says you should. Try it with the functions in Exercises 71–74.

71. Find dy/dx if $y = \cos(6x + 2)$ by writing y as a composite with

  a) $y = \cos u$ and $u = 6x + 2$

  b) $y = \cos 2u$ and $u = 3x + 1$.

72. Find dy/dx if $y = \sin(x^2 + 1)$ by writing y as a composite with

  a) $y = \sin(u + 1)$ and $u = x^2$

  b) $y = \sin u$ and $u = x^2 + 1$.

73. Find dy/dx if $y = x$ by writing y as the composite of

  a) $y = (u/5) + 7$ and $u = 5x - 35$

  b) $y = 1 + (1/u)$ and $u = 1/(x - 1)$.

74. Find dy/dx if $y = \sin(\sin(2x))$ by writing y as the composite of

  a) $y = \sin u$ and $u = \sin 2x$

  b) $y = \sin(\sin u)$ and $u = 2x$.

75. Evaluate ds/dt when $s = \cos\theta$ and $d\theta/dt = 5$ when $\theta = 3\pi/2$.

76. Evaluate dy/dt when $y = x^2 + 7x - 5$ and $dx/dt = 1/3$ when $x = 1$.

77. What is the largest value possible for the slope of the curve $y = \sin(x/2)$?

78. Write an equation for the tangent to the curve $y = \sin mx$ at the origin.

79. Find the lines that are tangent and normal to the curve $y = 2\tan(\pi x/4)$ at $x = 1$. Support your answer graphically.

80. *Orthogonal curves.* Two curves are said to cross at right angles if their tangents are perpendicular at the crossing point. The technical word for "crossing at right angles" is *orthogonal*. Show that the curves $y = \sin 2x$ and $y = -\sin(x/2)$ are orthogonal at the origin. Draw both graphs and both tangents in a square viewing window.

81. Suppose that functions f and g and their derivatives have the following values at $x = 2$ and $x = 3$:

x	$f(x)$	$g(x)$	$f'(x)$	$g'(x)$
2	8	2	$\frac{1}{3}$	-3
3	3	-4	2π	5

Evaluate the derivatives with respect to x of the following combinations at the given value of x.

a) $2f(x)$ at $x = 2$ **b)** $f(x) + g(x)$ at $x = 3$

c) $f(x) \cdot g(x)$ at $x = 3$ **d)** $f(x)/g(x)$ at $x = 2$

e) $f(g(x))$ at $x = 2$ **f)** $\sqrt{f(x)}$ at $x = 2$

g) $1/g^2(x)$ at $x = 3$

h) $\sqrt{f^2(x) + g^2(x)}$ at $x = 2$

82. Suppose that the functions f and g and their derivatives with respect to x have the following values at $x = 0$ and $x = 1$:

x	$f(x)$	$g(x)$	$f'(x)$	$g'(x)$
0	1	1	5	1/3
1	3	-4	$-1/3$	$-8/3$

Evaluate the derivatives with respect to x of the following combinations at the given value of x:

a) $5f(x) - g(x)$, $x = 1$ **b)** $f(x)g^3(x)$, $x = 0$

c) $\dfrac{f(x)}{g(x) + 1}$, $x = 1$ **d)** $f(g(x))$, $x = 0$

e) $g(f(x))$, $x = 0$ **f)** $(g(x) + f(x))^{-2}$, $x = 1$

g) $f(x + g(x))$, $x = 0$

83. *Running machinery too fast.* Suppose that a piston is moving straight up and down and that its position at time t sec is

$$s = A\cos(2\pi bt),$$

with A and b positive. The value of A is the amplitude of the motion, and b is the frequency (number of times the piston moves up and down each second). What effect does doubling the frequency have on the piston's velocity and acceleration? (Once you find out, you will know why machinery breaks when you run it too fast.)

84. Use parametric mode to simulate the motion of the piston in Exercise 83. Simultaneously graph the path of the piston and the velocity, both as a function of time t.

85. *Temperatures in Fairbanks, Alaska.* The equation that approximates the average temperature (°F) on day t in Fairbanks, Alaska, during a typical 365-day year is

$$y = 37\sin\left[\frac{2\pi}{365}(t - 101)\right] + 25.$$

a) Draw a complete graph of $y = f(t)$. Assume that $t = 0$ is January 1.

b) On what day is the temperature increasing the fastest?

c) About how many degrees per day is the temperature increasing when it is increasing at its fastest?

86. *Daylight hours in Fairbanks, Alaska.* The equation that approximates the number of hours of daylight on day t in Fairbanks, Alaska, during a typical 365-day year is

$$y = 8.5\sin\left(\frac{2\pi(t - 83)}{365}\right) + 12.5.$$

a) Draw a complete graph of $y = f(t)$. Assume that $t = 0$ is January 1.

b) On what day is the number of hours of daylight increasing the fastest?

c) About how many hours per day is the number of daylight hours increasing when it is increasing at its fastest?

3.7   Implicit Differentiation and Fractional Powers

Graphing Curves of the Form $F(x, y) = 0$

The equations $x^2 + y^2 = 64$, $y^2 = x$, and $y^5 + \sin xy = 0$ define relations that are *not* functions. The equation $y = x^3 + 2x - 3$ defines a function.

Each can be written in the form

$$F(x, y) = 0.$$

For example, $x^2 + y^2 = 64$ is

$$F(x, y) = 0 \qquad \text{where} \qquad F(x, y) = x^2 + y^2 - 64,$$

and $y = x^3 + 2x - 3$ is $F(x, y) = 0$, where $F(x, y) = y - x^3 - 2x + 3$.

Some equations of the form $F(x, y) = 0$ can be graphed in parametric mode as we first used it in Chapter 1.

EXPLORATION 1

Graphing Relations in Parametric Mode

We have already seen how any function $y = f(x)$ can be represented by the parametric equations

$$x(t) = t, \qquad y(t) = f(t).$$

1. To review, try graphing $y = x^3 - 2x$ using parametric equations.

The trigonometric identities

$$\cos^2 t + \sin^2 t = 1, \qquad \sec^2 t - \tan^2 t = 1$$

suggest implicitly-defined relationships that can be represented parametrically.

GRAPH each pair of parametric equations in a square viewing window. Describe the result. Confirm your description by relating x and y algebraically.

2. $x(t) = \cos t, \quad y(t) = \sin t$ 3. $x(t) = 3\cos t, \quad y(t) = 3\sin t$

4. $x(t) = 4\cos t, \quad y(t) = 2\sin t$ 5. $x(t) = \sec t, \quad y(t) = \tan t$

6. $x(t) = 3\sec t, \quad y(t) = 3\tan t$ 7. $x(t) = 4\sec t, \quad y(t) = 2\tan t$

Implicit Differentiation

Many equations $F(x, y) = 0$, such as $y^5 + \sin xy = 0$ (see Exercise 57), define relationships between x and y but do not let us solve for y *explicitly* in terms of x. The graph of $F(x, y) = 0$ shows that the relationship is not a function of x if any vertical line intersects the graph more than once (Fig. 3.46 on the following page). However, various parts of the graph, such as arcs AB, BC, and CD, may well be the graphs of functions of x, and we can consider the relationship to be a function "a piece at a time." We say that the function has been defined *implicitly* by $F(x, y) = 0$. In this way, it will make sense to talk about y being a differentiable function of x, and we will be able to find dy/dx using a technique called *implicit differentiation*.

When may we expect the functions defined by an equation $F(x, y) = 0$ to be differentiable? The answer is: when their graphs are smooth enough to have a nonvertical tangent at almost every point, as they will, for instance, if the formula for F is an algebraic combination of powers of x and y (a theorem from advanced mathematics).

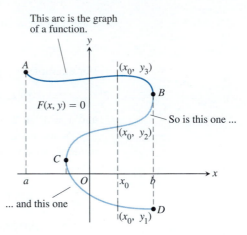

3.46 As a whole, the curve $F(x, y) = 0$ is not the graph of a function of x. Some of the vertical lines that cross it intersect it more than once. However, the curve can be divided into separate arcs that *are* the graphs of functions of x.

EXAMPLE 1

The graph of $F(x, y) = x^2 + y^2 - 1 = 0$ is the circle $x^2 + y^2 = 1$. Taken as a whole, the circle is not the graph of any single function of x (Fig. 3.47). Each x in the interval $-1 < x < 1$ gives two values of y, namely, $y = \sqrt{1 - x^2}$ and $y = -\sqrt{1 - x^2}$.

EXPLORATION BIT

Give a convincing argument why neither of the functions whose graphs are the arcs AB and BC in Fig. 3.46 have a one-sided derivative at the domain endpoint b.

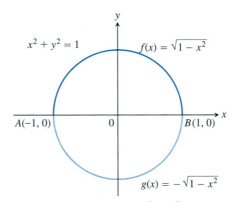

3.47 The graph of the equation $F(x, y) = x^2 + y^2 - 1 = 0$ is the complete circle $x^2 + y^2 = 1$. The upper semicircle AB is the graph of the function $f(x) = \sqrt{1 - x^2}$. The lower semicircle AB is the graph of $g(x) = -\sqrt{1 - x^2}$. Can you show that f' and g' match dy/dx found in Example 1?

The upper and lower semicircles are the graphs of the functions $f(x) = \sqrt{1 - x^2}$ and $g(x) = -\sqrt{1 - x^2}$. These functions are differentiable (except at 1 and -1) because they are composites of differentiable functions. The quickest way to find their derivatives, however, is not to differentiate the square-root formulas but to differentiate both sides of the original equation, treating y as a differentiable but otherwise unknown function of x:

$$x^2 + y^2 = 1$$

$$\frac{d}{dx}(x^2) + \frac{d}{dx}(y^2) = \frac{d}{dx}(1)$$

$$2x + 2y\frac{dy}{dx} = 0$$

$$\frac{dy}{dx} = -\frac{x}{y}.$$

This formula for dy/dx is simpler than either of the formulas that we would get by differentiating f and g and holds for all points on the curve above or below the x-axis. It is also easy to evaluate at any such point. At $(\sqrt{2}/2, \sqrt{2}/2)$, for instance,

$$\frac{dy}{dx} = -\frac{\sqrt{2}/2}{\sqrt{2}/2} = -1.$$

To calculate the derivatives of other implicitly defined functions, we simply proceed as in Example 1: We treat y as a differentiable (but otherwise unknown) function of x and apply the already familiar rules of differentiation to differentiate both sides of the defining equation, then solve for dy/dx. This procedure is called *implicit differentiation*.

EXAMPLE 2

Find dy/dx if $2y = x^2 + \sin y$.

Solution

$$2y = x^2 + \sin y$$

$$\frac{d}{dx}(2y) = \frac{d}{dx}(x^2) + \frac{d}{dx}(\sin y) \qquad \text{Differentiate both sides with respect to } x.$$

$$2\frac{dy}{dx} = 2x + \cos y\frac{dy}{dx}$$

$$2\frac{dy}{dx} - \cos y\frac{dy}{dx} = 2x \qquad \text{Collect terms with } dy/dx.$$

$$(2 - \cos y)\frac{dy}{dx} = 2x \qquad \text{Factor out } dy/dx \ldots$$

$$\frac{dy}{dx} = \frac{2x}{2 - \cos y} \qquad \ldots \text{ and divide.}$$

In the exercises, we will ask you to graph the equation $2y = x^2 + \sin y$ and dy/dx.

IMPLICIT DIFFERENTIATION TAKES FOUR STEPS

STEP 1: Differentiate both sides of the equation with respect to x.

STEP 2: Collect the terms with dy/dx on one side of the equation.

STEP 3: Factor out dy/dx.

STEP 4: Solve for dy/dx by dividing.

EXPLORATION 2

Another Approach

Looking at a relationship $F(x, y) = 0$ parametrically can provide an alternative approach to finding dy/dx. It will be established later in the text that

$$\frac{dy}{dx} = \frac{dy}{dt} \cdot \frac{dt}{dx} = \frac{dy}{dt} \cdot \frac{1}{\dfrac{dx}{dt}}.$$

1. Using this rule and the parametric equations for a circle,

$$x(t) = \sin t, \qquad y(t) = \cos t,$$

show that $dy/dx = -(x/y)$ as found in Example 1.

2. Figure 3.48 shows the circle and the graphs of the derivatives for the upper half and the lower half. Tell which is which. Explain the end behavior of the derivatives. Duplicate, if you can, this viewing window on your grapher. If you can't, explain why.

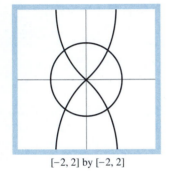

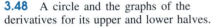
[−2, 2] by [−2, 2]

3.48 A circle and the graphs of the derivatives for its upper and lower halves.

Lenses, Tangents, and Normal Lines

In the law that describes how light changes direction as it enters a lens, the important angles are the angles the light makes with the line perpendicular to the surface of the lens at the point of entry (angles A and B in Fig. 3.49). This line is called the *normal to the surface* at the point of entry. In a profile view of a lens like the one in Fig. 3.49, the normal is the line perpendicular to the tangent to the profile curve at the point of entry.

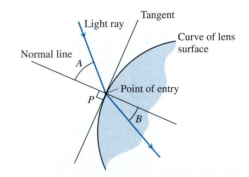

3.49 The profile or cutaway view of a lens, showing the bending (refraction) of a light ray as it passes through the lens surface.

The profiles of lenses are often described by quadratic curves like the one in Fig. 3.50. When they are, we can use implicit differentiation to find the tangents and normals.

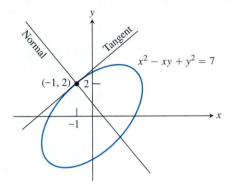

3.50 The graph of $x^2 - xy + y^2 = 7$ is an ellipse. The slope of the curve at the point $(-1, 2)$ is $(dy/dx)|_{(-1,2)} = 4/5$.

EXAMPLE 3

Find the tangent and normal to the curve $x^2 - xy + y^2 = 7$ at the point $(-1, 2)$. (See Fig. 3.50.)

Solution We first use implicit differentiation to find dy/dx:

$$x^2 - xy + y^2 = 7$$

$$\frac{d}{dx}(x^2) - \frac{d}{dx}(xy) + \frac{d}{dx}(y^2) = \frac{d}{dx}(7) \qquad \text{Differentiate both sides with respect to } x \ldots$$

$$2x - \left(x\frac{dy}{dx} + y\frac{dx}{dx}\right) + 2y\frac{dy}{dx} = 0 \qquad \ldots \text{ treating } xy \text{ as a product and } y^2 \text{ as a power.}$$

$$(2y - x)\frac{dy}{dx} = y - 2x \qquad \text{Collect terms.}$$

$$\frac{dy}{dx} = \frac{y - 2x}{2y - x} \qquad \text{Solve as usual.}$$

We then evaluate the derivative at $x = -1$, $y = 2$ to obtain

$$\left.\frac{dy}{dx}\right|_{(-1,2)} = \left.\frac{y - 2x}{2y - x}\right|_{(-1,2)}$$

$$= \frac{2 - 2(-1)}{2(2) - (-1)} = \frac{4}{5}.$$

The tangent to the curve at $(-1, 2)$ is

$$y - 2 = \frac{4}{5}(x - (-1))$$

$$y = \frac{4}{5}x + \frac{14}{5}.$$

The normal to the curve at $(-1, 2)$ is

$$y - 2 = -\frac{5}{4}(x + 1)$$

$$y = -\frac{5}{4}x + \frac{3}{4}.$$

Using Implicit Differentiation to Find Derivatives of Higher Order

Implicit differentiation can also produce derivatives of higher order. Here is an example.

EXAMPLE 4

Find d^2y/dx^2 if $2x^3 - 3y^2 = 7$.

Solution   To start, we differentiate both sides of the equation with respect to x to find $y' = dy/dx$:

$$2x^3 - 3y^2 = 7$$

$$\frac{d}{dx}(2x^3) - \frac{d}{dx}(3y^2) = \frac{d}{dx}(7)$$

$$6x^2 - 6yy' = 0$$

$$x^2 - yy' = 0$$

$$y' = \frac{x^2}{y}, \qquad \text{when } y \neq 0.$$

We now apply the Quotient Rule to find y'':

$$y'' = \frac{d}{dx}\left(\frac{x^2}{y}\right) = \frac{2xy - x^2 y'}{y^2} = \frac{2x}{y} - \frac{x^2}{y^2}y'.$$

Finally, we substitute $y' = x^2/y$ to express y'' in terms of x and y:

$$y'' = \frac{2x}{y} - \frac{x^2}{y^2}\left(\frac{x^2}{y}\right) = \frac{2x}{y} - \frac{x^4}{y^3}, \qquad \text{when } y \neq 0.$$

> **EXPLORATION BIT**
>
> Find d^2y/dx^2 for each function y defined implicitly.
>
> 1. $2y^3 - 3x^2 = 9$
> 2. $x^3 + xy + y^3 = 3$

Fractional Powers of Differentiable Functions

We know that the Power Rule

$$\frac{d}{dx}u^n = nu^{n-1}\frac{du}{dx}$$

holds when n is an integer. Our goal now is to show that it holds when n is a fraction. We will then be able to differentiate functions like

$$y = x^{4/3} \qquad \text{and} \qquad y = (\cos x)^{-1/5}$$

that were beyond our reach before.

POWER RULE FOR FRACTIONAL EXPONENTS

If n is any rational number, then

$$\frac{d}{dx}x^n = nx^{n-1}, \tag{1}$$

provided that $x \neq 0$ if $n - 1 < 0$ (i.e., $n < 1$).

If n is a rational number and u is a differentiable function of x, then u^n is a differentiable function of x and

$$\frac{d}{dx}u^n = nu^{n-1}\frac{du}{dx}, \tag{2}$$

provided that $u \neq 0$ if $n < 1$.

HELGA VON KOCH'S SNOWFLAKE CURVE (1904)

Start with an equilateral triangle, calling it Curve 1. On the middle third of each side, build an equilateral triangle pointing outward. Then erase the old middle thirds. Call the expanded curve Curve 2. Now put equilateral triangles, again pointing them outward, on the middle thirds of the sides of Curve 2. Erase the old middle thirds to make Curve 3. Repeat the process, as shown, to define an infinite sequence of plane curves. The limit curve of the sequence is Koch's snowflake curve.

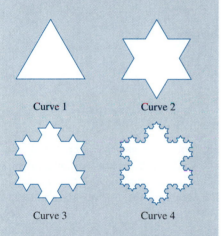

Curve 1   Curve 2

Curve 3   Curve 4

The snowflake curve is too rough to have a tangent at any point. In other words, the equation $F(x, y) = 0$ defining the curve does not define y as a differentiable function of x or x as a differentiable function of y at any point. Continuous curves that fail to have a tangent anywhere, such as the snowflake curve or the graph of Weierstrass's function of Section 3.2, play an important role in chaos theory, in part because there is no way to measure the length of such a curve. We'll see what length has to do with derivatives when we get to Section 6.4, where we'll encounter the snowflake curve again.

Proof of the Power Rule for Fractional Exponents We prove Eq. (1) first and then apply the Chain Rule to get Eq. (2).

To prove Eq. (1), let p and q be integers with $q > 0$, and suppose that $y = x^{p/q}$. Then

$$y^q = x^p.$$

This equation is an algebraic combination of powers of x and y, so the advanced theorem that we mentioned earlier assures us that y is a differentiable function of x. Since p and q are integers (for which we already have the Power Rule), we can differentiate both sides of the equation with respect to x and obtain.

$$qy^{q-1}\frac{dy}{dx} = px^{p-1}.$$

Hence, if $y \neq 0$,

$$\frac{dy}{dx} = \frac{p}{q}\frac{x^{p-1}}{y^{q-1}} = \frac{p}{q}\frac{x^{p-1}}{(x^{p/q})^{q-1}} = \frac{p}{q}\frac{x^{p-1}}{x^{p-p/q}} = \frac{p}{q}x^{(p/q)-1}.$$

This proves Eq. (1).

To prove Eq. (2), we let $y = u^{p/q}$ and apply the Chain Rule in the form

$$\frac{dy}{dx} = \frac{dy}{du} \cdot \frac{du}{dx}.$$

From Eq. (1), $(d/du)\left(u^{p/q}\right) = (p/q)u^{(p/q)-1}$. Hence,

$$\frac{dy}{dx} = \frac{p}{q}u^{(p/q)-1}\frac{du}{dx}$$

and we're done.   ≡

The restrictions $x \neq 0$ if $n < 1$ and $u \neq 0$ if $n < 1$ are part of the Power Rule to protect against inadvertent attempts to divide by zero. There is nothing mysterious about these restrictions. They come up quite naturally in practice, as the next example shows.

EXAMPLE 5

a) $\dfrac{d}{dx}(x^{1/2}) = \dfrac{1}{2}x^{-1/2} = \dfrac{1}{2\sqrt{x}}$ Eq. (1) with $n = \dfrac{1}{2}$

  function   derivative defined
  defined for $x \geq 0$ for $x > 0$

b) $\dfrac{d}{dx}(x^{1/5}) = \dfrac{1}{5}x^{-4/5}$ Eq. (1) with $n = \dfrac{1}{5}$

  function   derivative not
  defined for all x defined at $x = 0$

c) $\dfrac{d}{dx}(1 - x^2)^{1/2} = \dfrac{1}{2}(1 - x^2)^{-1/2}(-2x)$ Eq. (2) with $u = 1 - x^2$
                        and $n = \dfrac{1}{2}$

  function defined
  for $-1 \leq x \leq 1$

  $= \dfrac{-x}{(1 - x^2)^{1/2}}$ derivative defined
                  only for $-1 < x < 1$

EXPLORATION 3

Points of Nondifferentiability

For each of the three functions in Example 5, graph $y = f(x)$ and $y = \text{NDER}\,f(x)$ in the same viewing window. Explain in terms of tangent lines the behavior of NDER $f(x)$ at the points in the domain of f that are not in the domain of NDER $f(x)$, that is, the points where f is not differentiable.

The derivatives of the functions $x^{4/3}$ and $(\cos x)^{-1/5}$ are defined wherever the functions themselves are defined, as we see in the next example.

EXAMPLE 6

a) $\dfrac{d}{dx}x^{4/3} = \dfrac{4}{3}x^{1/3}$

b) $\dfrac{d}{dx}(\cos x)^{-1/5} = -\dfrac{1}{5}\cos x^{-6/5}\dfrac{d}{dx}(\cos x)$

    $= -\dfrac{1}{5}(\cos x)^{-6/5}(-\sin x)$

    $= \dfrac{1}{5}\sin x(\cos x)^{-6/5}$

EXAMPLE 7

Find the derivative of $y = x^{3/2}$ at $x = 0$.

Solution   When the graph of a function stops abruptly at a point, as the graph of $y = x^{3/2}$ does at $x = 0$ (Fig. 3.51), we calculate its derivative as a one-sided limit. The Power Rule still applies, giving in this case

$$\left.\frac{dy}{dx}\right|_{x=0} = \left.\frac{3}{2}x^{1/2}\right|_{x=0} = 0.$$

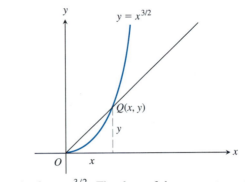

3.51   The graph of $y = x^{3/2}$. The slope of the curve at $x = 0$ is
$$\lim_{Q \to O} m_{OQ} = \lim_{x \to 0^+}\left(\frac{y}{x}\right) = 0.$$

We can see why this equation holds by looking at the geometry of the curve. The slope of a typical secant line through the origin and a point $Q(x, y)$ on the curve is

$$m_{OQ} = \frac{y - 0}{x - 0} = \frac{x^{3/2}}{x} = x^{1/2}.$$

As Q approaches the origin from the right, m_{OQ} approaches zero, in agreement with the result from the Power Rule.   ≡

Exercises 3.7

Find dy/dx in Exercises 1–26.

1. $y = x^{9/4}$

2. $y = x^{-3/5}$

3. $y = \sqrt[3]{x}$

4. $y = \sqrt[4]{x}$

5. $y = (2x + 5)^{-1/2}$

6. $y = (1 - 6x)^{2/3}$

7. $y = x\sqrt{x^2 + 1}$

8. $y = \dfrac{x}{\sqrt{x^2 + 1}}$

9. $x^2y + xy^2 = 6$

10. $x^3 + y^3 = 18xy$

11. $2xy + y^2 = x + y$

12. $x^3 - xy + y^3 = 1$

13. $x^2y^2 = x^2 + y^2$

14. $(3x + 7)^2 = 2y^3$

15. $y^2 = \dfrac{x - 1}{x + 1}$

16. $x^2 = \dfrac{x - y}{x + y}$

17. $y = \sqrt{1 - \sqrt{x}}$

18. $y = 3(2x^{-1/2} + 1)^{-1/3}$

19. $y = 3(\csc x)^{3/2}$

20. $y = [\sin (x + 5)]^{5/4}$

21. $x = \tan y$

22. $x = \sin y$

23. $x + \tan (xy) = 0$

24. $x + \sin y = xy$

25. $y \sin \left(\dfrac{1}{y}\right) = 1 - xy$

26. $y^2 \cos \left(\dfrac{1}{y}\right) = 2x + 2y$

27. Which of the following could be true if $f''(x) = x^{-1/3}$?

a) $f(x) = \dfrac{3}{2}x^{2/3} - 3$

b) $f(x) = \dfrac{9}{10}x^{5/3} - 7$

c) $f'''(x) = -\dfrac{1}{3}x^{-4/3}$

d) $f'(x) = \dfrac{3}{2}x^{2/3} + 6$

28. Which of the following could be true if $g''(t) = 1/t^{3/4}$?

a) $g'(t) = 4\sqrt[4]{t} - 4$ **b)** $g'''(t) = -4/\sqrt[4]{t}$

c) $g(t) = t - 7 + (16/5)t^{5/4}$

d) $g'(t) = (1/4)t^{1/4}$

In Exercises 29–34, use implicit differentiation to find dy/dx and then d^2y/dx^2.

29. $x^2 + y^2 = 1$ **30.** $x^{2/3} + y^{2/3} = 1$

31. $y^2 = x^2 + 2x$ **32.** $y^2 + 2y = 2x + 1$

33. $y + 2\sqrt{y} = x$ **34.** $xy + y^2 = 1$

In Exercises 35–44, find the lines that are (a) tangent and (b) normal to the curve at the given point.

35. $x^2 + xy - y^2 = 1$, $(2, 3)$

36. $x^2 + y^2 = 25$, $(3, -4)$

37. $x^2y^2 = 9$, $(-1, 3)$

38. $y^2 - 2x - 4y - 1 = 0$, $(-2, 1)$

39. $6x^2 + 3xy + 2y^2 + 17y - 6 = 0$, $(-1, 0)$

40. $x^2 - \sqrt{3}xy + 2y^2 = 5$, $(\sqrt{3}, 2)$

41. $2xy + \pi \sin y = 2\pi$, $(1, \pi/2)$

42. $x \sin 2y = y \cos 2x$, $(\pi/4, \pi/2)$

43. $y = 2 \sin(\pi x - y)$, $(1, 0)$

44. $x^2 \cos^2 y - \sin y = 0$, $(0, \pi)$

45. Assume that the equation $2xy + \pi \sin y = 2\pi$ defines y as a differentiable function of x. Evaluate dy/dx when $x = 1$ and $y = \pi/2$.

46. Find an equation for the tangent to the curve $x \sin 2y = y \cos 2x$ at the point $(\pi/4, \pi/2)$.

47. *The eight curve.*

  a) Find the slopes of the figure-eight–shaped curve

$$y^4 = y^2 - x^2$$

at the two points shown on the graph that follows.

  b) Use parametric mode and the two pairs of parametric equations

$$x_1(t) = \sqrt{y^2 - y^4}, \quad y_1(t) = t,$$

$$x_2(t) = -\sqrt{y^2 - y^4}, \quad y_2(t) = t,$$

to show the curve in a viewing window.

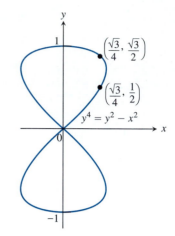

48. *The Cissoid of Diocles (dates from about 200 B.C.).*

  a) Find equations for the tangent and normal to the Cissoid of Diocles,

$$y^2(2 - x) = x^3,$$

at the point $(1, 1)$ as pictured below.

  b) Use parametric mode to reproduce the curve and the tangent and normal lines at $(1, 1)$.

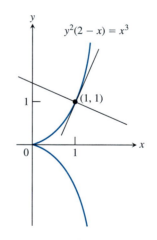

49. a) Confirm that $(-1, 1)$ is on the curve defined by $x^3y^2 = \cos(\pi y)$.

  b) Use part (a) to find the slope of the line tangent to the curve at $(-1, 1)$.

50. a) Show that the relation

$$y^3 - xy = -1$$

cannot be a function of x by showing that there is more than one possible y-value when $x = 2$.

  b) On a small enough square with center $(2, 1)$, the part of the graph of the relation within the square will define a function $y = f(x)$. For this function, find $f'(2)$ and $f''(2)$.

51. Find the two points where the curve $x^2 + xy + y^2 = 7$ crosses the x-axis, and show that the tangents to the curve at these points are parallel. What is the common slope of these tangents?

52. Find points on the curve $x^2 + xy + y^2 = 7$ (a) where the tangent is parallel to the x-axis and (b) where the tangent is parallel to the y-axis. (In the latter case, dy/dx is not defined, but dx/dy is. What value does dx/dy have at these points?)

53. *Orthogonal curves.* Two curves are *orthogonal* at a point of intersection if their tangents there cross at right angles. Show that the curves $2x^2 + 3y^2 = 5$ and $y^2 = x^3$ are orthogonal at $(1, 1)$ and $(1, -1)$. Use parametric mode to draw the curves and to show the tangent lines.

54. The position of a body moving along a coordinate line at time t is $s = \sqrt{1 + 4t}$, with s in meters and t in seconds. Find the body's velocity and acceleration when $t = 6$ sec.

55. The velocity of a falling body is $v = k\sqrt{s}$ meters per second (k a constant) at the instant the body has fallen s meters from its starting point. Show that the body's acceleration is constant.

56. Use parametric mode to GRAPH the curve given by the equation in Example 2. Show that it is a function. Find its domain. GRAPH its derivative.

57. Consider the equation $y^5 + \sin xy = 0$.

a) Show that $-1 \leq y \leq 1$.

b) Show that $xy = \sin^{-1}(-y^5) + 2k\pi$ or $xy = \pi - \sin^{-1}(-y^5) + 2k\pi$, k any integer.

c) GRAPH the first relation in part (b) parametrically for $k = 0$ by setting

$$x(t) = (1/t)\sin^{-1}(-t^5), \quad y(t) = t.$$

What are the domain and range of this relation?

d) GRAPH the relation in part (b) parametrically for $k = 1$. What are the domain and range of this relation?

58. Consider the relation $x = (\sin^{-1}(-y^5))/y$ graphed in Exercise 57.

a) Find x when $y = -1/2$.

b) Find dy/dx using analytic methods and compute its value at $y = -1/2$.

c) Draw the tangent line to the graph in Exercise 57(c) at $y = -1/2$.

59. Refer to Example 3. There are two points at which dy/dx does not exist. Find them.

60. Draw complete graphs of $y = x^{1/3}$ and its derivative. Find the domain and range of each function.

3.8 _____ Linear Approximations and Differentials

Sometimes we can approximate complicated functions with simpler ones that give the accuracy we want for specific applications. It is important to know how to do this, and in this section we study the simplest of the useful approximations. For reasons that will be clear in a moment, the approximation is called a *linearization*.

We also introduce a new symbol, dx, for an increment in a variable x. This symbol is called the *differential* of x. In the physical sciences, it is used more frequently than Δx. In mathematics, differentials are used to estimate changes in function values, as we shall see toward the end of this section.

> Contrary to what you may think, we have not been using dx as a symbol already. We have been using "dx" only as *part* of the derivative symbol d/dx.

EXPLORATION 1

Seeing a Local Approximation

Set your ZOOM factor at 4. In the standard viewing window, GRAPH $f(x) = x^2 - x - 3$ and the line tangent at $(2, f(2))$, namely, $y - f(2) = f'(2)(x - 2)$. ZOOM-IN at the point $(2, f(2))$. ZOOM-IN a second time, then a third time. Explain what you see.   ☙

Linearizations Are Linear Functions

As we can see in Fig. 3.52, the tangent to a curve $y = f(x)$ lies close to the curve near the point of tangency. For a small interval to either side, the y-values along the tangent line give good approximations to the y-values on the curve. If we ZOOM-IN at the point $(a, f(a))$, the curve and its tangent line will appear to coincide. We say that the graph of f is *locally straight* at $x = a$.

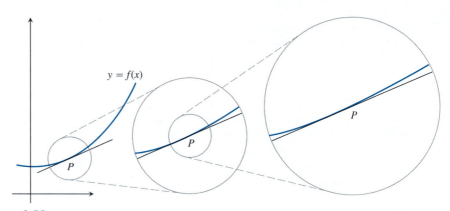

3.52 Successive magnifications show a close fit between a curve and its tangent line.

To simplify work that sometimes is necessary with a function f whose graph is locally straight near a point, we propose to replace the formula for f over the small interval around a by the formula for its tangent line. In the notation of Fig. 3.53, the tangent passes through the point $P(a, f(a))$ with slope $f'(a)$ so its point-slope equation is

$$y - f(a) = f'(a)(x - a),$$

or

$$y = f(a) + f'(a)(x - a).$$

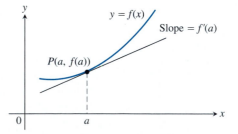

3.53 The equation of the tangent line is $y = f(a) + f'(a)(x - a)$.

Thus, the tangent line is the graph of the function

$$L(x) = f(a) + f'(a)(x - a).$$

For as long as the line remains close to the graph of f, $L(x)$ will give a good approximation to $f(x)$.

> **DEFINITIONS** Linearization and Standard Linear Approximation
>
> If $y = f(x)$ is differentiable at $x = a$, then
>
> $$L(x) = f(a) + f'(a)(x - a) \qquad (1)$$
>
> is the **linearization** of f at a. The approximation
>
> $$f(x) \approx L(x)$$
>
> is the **standard linear approximation** of f at a, and we say the graph of f is **locally straight** at a.

EXAMPLE 1

Find the linearization of $f(x) = \sqrt{1 + x}$ at $x = 0$.

Solution We evaluate Eq. (1) for $f(x) = \sqrt{1 + x}$ and $a = 0$.
  The derivative of f is

$$f'(x) = \frac{1}{2}(1 + x)^{-1/2} = \frac{1}{2\sqrt{1 + x}}.$$

Its value at $x = 0$ is $1/2$. We substitute this along with $a = 0$ and $f(0) = 1$ into Eq. (1):

$$L(x) = f(a) + f'(a)(x - a) = 1 + \frac{1}{2}(x - 0) = 1 + \frac{x}{2}.$$

The linearization of $\sqrt{1 + x}$ at $x = 0$ is $L(x) = 1 + \dfrac{x}{2}$. See Fig. 3.54.

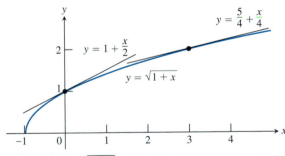

3.54 The graphs of $y = \sqrt{1 + x}$ and its linearizations at $x = 0$ and $x = 3$.

EXPLORATION 2

Approximations

Example 1 shows the linearization of $f(x) = \sqrt{1 + x}$ at $x = 0$ is $1 + x/2$.

1. Show that the linearization of f at $x = 3$ is $\dfrac{5}{4} + \dfrac{x}{4}$ (Fig. 3.54).

  How close are the approximations $1 + x/2$ and $5/4 + x/4$ to the values of $\sqrt{1 + x}$ at $x = 0$ and $x = 3$, respectively? Use a calculator to complete these tables. Then comment on what you observe.

2.

x	0.2	0.05	0.005
$\sqrt{1+x}$	?	?	?
$1 + \dfrac{x}{2}$	?	?	?

x	3.2	3.05	3.005
$\sqrt{1+x}$	?	?	?
$\dfrac{5}{4} + \dfrac{x}{4}$	?	?	?

3. Use a calculator to show that the linearization of f at $x = 0$ is not a good approximation of f at $x = 3$. Support your argument with visual evidence by extending the sketch in Fig. 3.54.

Do not be misled by our calculations here into thinking that whatever we do with a linearization is better done with a calculator. In practice, we would never use a linearization to find the value of a particular square root. That is not what linearizations are for. The utility of the linearization in Example 1 lies in its ability to replace the complicated formula $\sqrt{1+x}$ by a simpler formula. If we have to work with $\sqrt{1+x}$ for values of x close to 0 and can tolerate the small amount of error involved, we can safely work with $1 + x/2$ instead. Of course, we then need to know just how much error there really is. We shall look at that in a moment, but the full answer won't come until Chapter 9.

EXAMPLE 2

The most important linearization for replacing roots and powers is

$$(1 + x)^k \approx 1 + kx \qquad \text{(any number } k) \tag{2}$$

for x near zero and any number k. (See Exercise 22.) For instance, when x is numerically small,

$$\sqrt{1+x} \approx 1 + \frac{x}{2}. \qquad \textcolor{blue}{\text{Example 1}}$$

Also,

$$\frac{1}{1-x} = (1-x)^{-1} \approx 1 + (-1)(-x) = 1 + x$$

$$\sqrt[3]{1 + 5x^4} = (1 + 5x^4)^{1/3} \approx 1 + \frac{1}{3}(5x^4) = 1 + \frac{5}{3}x^4$$

$$\frac{1}{\sqrt{1-x^2}} = (1 - x^2)^{-1/2} \approx 1 + \left(-\frac{1}{2}\right)(-x^2) = 1 + \frac{1}{2}x^2. \qquad \blacksquare$$

> **GENERALIZATION**
>
> Equation (2) can be extended to
>
> $$(1 + f(x))^k \approx 1 + kf(x)$$
>
> for any function f whose values are near 0 when x is near 0.

EXPLORATION 3

Support Graphically

In the same viewing window, GRAPH the third or fourth function of Example 2 and its approximation. Explain what you see and why the approximation is not a linearization. ZOOM-IN at $(0, f(0))$ a few times until the graphs appear indistinguishable. TRACE away from $(0, f(0))$ until you can find a value for x where the cursor control shows different values for the function and its approximation. How far from 0 can you get?

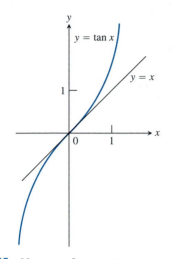

3.55 Near $x = 0$, $\tan x \approx x$.

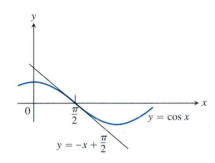

3.56 The graphs of $y = \cos x$ and its linearization at $x = \pi/2$. Do you have a hunch what the linearization equation is? Can you confirm it algebraically?

Trigonometric functions have delightfully simple linearizations at the origin. $L(x) = 0 + 1(x - 0) = x$. Near $x = 0$, $\tan x \approx x$ (Fig. 3.55).

EXAMPLE 3

Find the linearization of $f(x) = \tan x$ at $x = 0$.

Solution We use the equation

$$L(x) = f(a) + f'(a)(x - a)$$

with $f(x) = \tan x$ and $a = 0$. Since

$$f(0) = \tan(0) = 0, \qquad f'(0) = \sec^2(0) = 1,$$

we have

EXPLORATION 4

Make a Conjecture

In Exercises 13 and 14, you will be asked to find the linearizations of $\sin x$ and $\cos x$ at $x = 0$. The local straightness of their graphs should give you a clue.

1. What do you expect the linearizations of $\sin x$ and $\cos x$ to be at $x = 0$? (You can ZOOM-IN at $(0, f(0))$ on your grapher to get an idea, but by now you may be able to do this in the viewing window of your mind!)

2. Make a conjecture about the linearization of $f(x) = \cos x$ at $x = \pi/2$. (See Fig. 3.56.) Use algebra to confirm your conjecture.

Estimating Change with Differentials

Suppose we know the value of a differentiable function $f(x)$ at a particular point x_0 and want to predict how much this value will change if we move nearby to the point $x_0 + h$. If h is small, f and its linearization L at x_0 will change by nearly the same amount. Since the values of L are always simple to calculate, calculating the change in L gives a practical way to estimate the change in f.

In the notation of Fig. 3.57 on the following page, the change in f is

$$\Delta f = f(x_0 + h) - f(x_0).$$

The corresponding change in L is

$$\begin{aligned}\Delta L &= L(x_0 + h) - L(x_0)\\ &= f(x_0) + f'(x_0)[(x_0 + h) - x_0] - f(x_0)\\ &= f'(x_0)h.\end{aligned}$$

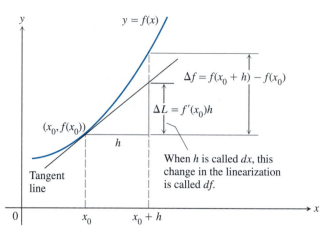

3.57 If h is small, the change in the linearization of f is nearly the same as the change in f.

The formula for Δf is usually as hard to work with as the formula for f. The formula for ΔL, however, is always simple to work with. As you can see, the change in L is just a constant times h.

The change $\Delta L = f'(x_0)h$ is usually described by using the suggestive notation of *differentials* as in the form of Eq. (3) in the following definition.

<table>
<tr>
<td>

ON NOTATION

Δf: a change in f;
df: a change in the linearization
  of f.
Remember, f and the *linearization of f*
denote two functions. The differential
of f, df, refers to change in the
linearization of f, not to change in f.

</td>
<td>

DEFINITIONS

If $f(x)$ is a differentiable function at x_0, dx is a change in x and df is the corresponding change in the linearization of f so that

$$df = f'(x_0)dx, \qquad (3)$$

then dx and df are **differentials of x and f**, respectively. Note that in Eq. (3), the value of df depends on both x_0 and dx.

</td>
</tr>
</table>

If $y = f(x)$ and we divide both sides of the equation $dy = f'(x)dx$ by dx, we obtain the familiar equation

$$\frac{df}{dx} = f'(x).$$

This equation now says that we may regard the derivative function df/dx as a quotient of differentials. In many calculations, it is convenient to be able to think this way. For example, in writing the Chain Rule as

$$\frac{dy}{dx} = \frac{dy}{du} \cdot \frac{du}{dx},$$

we can think of the derivatives on the right as quotients in which the du's cancel to produce the fraction on the left. This gives a quick check on whether we remembered the rule correctly.

EXAMPLE 4

The radius of a circle increases from an initial value of $r_0 = 10$ by an amount $dr = 0.1$ (Fig. 3.58). Estimate the corresponding increase in the circle's area $A = \pi r^2$ by calculating dA. Compare dA with the true change ΔA.

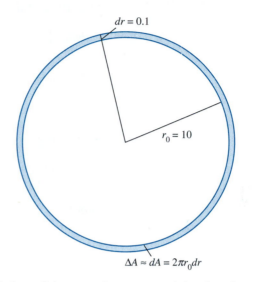

$dr = 0.1$

$r_0 = 10$

$\Delta A \approx dA = 2\pi r_0 dr$

3.58 When dr is small in comparison to r_0, as it is when $dr = 0.1$ and $r_0 = 10$, the differential $dA = 2\pi r_0 dr$ gives a good estimate of ΔA.

Solution To calculate dA, we apply Eq. (3) to the function $A = \pi r^2$:

$$dA = A'(r_0)dr = 2\pi r_0 dr.$$

We then substitute the values $r_0 = 10$ and $dr = 0.1$:

$$dA = 2\pi(10)(0.1) = 2\pi.$$

The estimated change is 2π square units.

A direct calculation of ΔA gives

$$\Delta A = \pi(10.1)^2 - \pi(10)^2 = (102.01 - 100)\pi = \underbrace{2\pi}_{dA} + \underbrace{0.01\pi}_{\text{error}}.$$

The error in the estimate dA is 0.01π square unit. As a percentage of the circle's original area, the error is quite small, as we can see from the following calculation:

$$\frac{\text{Error}}{\text{Original area}} = \frac{0.01\pi}{100\pi} = 0.01\%.$$

Absolute, Relative, and Percentage Change

What is the difference again between Δf and df? The increment Δf is the change in f; the differential df is the change in the linearization of f. Unlike Δf, the differential df is always simple to calculate, and it gives a good estimate of Δf when the change in x is small.

As we move from x_0 to a nearby point, we can describe the corresponding change in the value of f in three ways:

	True (or Exact)	Estimate (or Approximation)
Absolute change	Δf	df
Relative change	$\dfrac{\Delta f}{f(x_0)}$	$\dfrac{df}{f(x_0)}$
Percentage change	$\dfrac{\Delta f}{f(x_0)} \times 100$	$\dfrac{df}{f(x_0)} \times 100$

EXAMPLE 5

Estimate the percentage change that will occur in the area of a circle if its radius increases from $r_0 = 10$ units to 10.1 units.

Solution   From the preceding table, we have

$$\text{Estimated percentage change} = \frac{dA}{A(r_0)} \times 100.$$

With $dA = 2\pi$ (from Example 4) and $A(r_0) = 100\pi$, the formula gives

$$\frac{dA}{A(r_0)} \times 100 = \frac{2\pi}{100\pi} \times 100 = 2\%.$$

EXAMPLE 6

Suppose Earth were a perfect sphere and we determined its radius to be 3959 ± 0.1 miles. Estimate the effect the tolerance of ± 0.1 would have on our calculation of Earth's surface area.

Solution   The surface area of a sphere of radius r is $S = 4\pi r^2$. The uncertainty in the calculation of S that arises from measuring r with a tolerance of dr miles is about

$$dS = \left(\frac{dS}{dr}\right) dr = 8\pi r \ dr.$$

With $r = 3959$ and $dr = 0.1$,

$$dS = 8\pi (3959)(0.1) = 9950 \text{ square miles}$$

to the nearest square mile, which is about the area of the state of Maryland. In absolute terms, this might seem like a large error. However, 9950 mi^2 is a relatively small error when compared to the calculated surface area of Earth:

$$\frac{dS}{\text{Calculated } S} = \frac{9950}{4\pi(3959)^2} \approx \frac{9950}{196,961,284} \approx 0.005\%.$$

EXPLORATION BIT

Find more precise bounds for Earth's surface area by calculating ΔS for $r = 3959 \pm 0.1$. That is, calculate $4\pi(3958.9)^2$ and $4\pi(3959.1)^2$. Then compare your results with dS.

EXAMPLE 7

About how accurately should we measure the radius r of a sphere to calculate the surface area $S = 4\pi r^2$ within 1% of its true value?

Solution   We want any inaccuracy in our measurement to be small enough to make the corresponding increment ΔS in the surface area satisfy the inequality

$$|\Delta S| \le \frac{1}{100} S = \frac{4\pi r^2}{100}.$$

We replace ΔS in this inequality with

$$dS = \left(\frac{dS}{dr}\right) dr = 8\pi r \, dr.$$

This gives

$$|8\pi r \, dr| \le \frac{4\pi r^2}{100} \quad \text{or} \quad |dr| \le \frac{1}{8\pi r} \cdot \frac{4\pi r^2}{100} = \frac{1}{2}\frac{r}{100}.$$

We should measure the radius with an error dr that is no more than 0.5% of the true value.

EXAMPLE 8   Unclogging Arteries

In the late 1830s, the French physiologist Jean Poiseuille ("pwa·*zoy*") discovered the formula that we use today to predict how much the radius of a partially clogged artery has to be expanded to restore normal flow. His formula,

$$V = kr^4,$$

says that the volume V of fluid flowing through a small pipe or tube in a unit of time at a fixed pressure is a constant times the fourth power of the tube's radius r. How will a 10% increase in r affect V?

Solution   The differentials of r and V are related by the equation

$$dV = \frac{dV}{dr}dr = 4kr^3 dr.$$

Hence,

$$\frac{dV}{V} = \frac{4kr^3 dr}{kr^4} = 4\frac{dr}{r}. \qquad \text{Dividing by } V = kr^4.$$

The relative change in V is four times the relative change in r, so a 10% increase in r will produce a 40% increase in the flow.

Angiography: An opaque dye is injected into a partially blocked artery to make the inside visible under X-rays. This reveals the location and severity of the blockage.

Angioplasty: A balloon-tipped catheter is inflated inside the artery to widen it at the blockage site.

Sensitivity

The equation $df = f'(x)dx$ tells how sensitive the output of f is to a change in input at different values of x. The larger the value of f' at x, the greater is the effect of a given change dx.

EXAMPLE 9

You want to calculate the height of a bridge from the equation $s = 16t^2$ by timing how long it takes a heavy stone that you drop to splash into the water below. How sensitive will your calculation be to a 0.1-sec error in measuring the time?

Solution The size of ds in the equation

$$ds = 32t\, dt$$

depends on how big t is. If $t = 2$ sec, the error caused by $dt = 0.1$ is only

$$ds = 32(2)(0.1) = 6.4 \text{ ft.}$$

Three seconds later, at $t = 5$ sec, the error caused by the same dt is

$$ds = 32(5)(0.1) = 16 \text{ ft.}$$

The Error in the Approximation $\Delta f \approx f'(a)\Delta x$

How well does the quantity $f'(a)\Delta x$ estimate the true increment $\Delta f = f(a + \Delta x) - f(a)$? We measure the error by subtracting one from the other:

$$
\begin{aligned}
\text{Approximation error} &= \Delta f - f'(a)\Delta x \\
&= f(a + \Delta x) - f(a) - f'(a)\Delta x \\
&= \underbrace{\left(\frac{f(a + \Delta x) - f(a)}{\Delta x} - f'(a) \right)}_{\epsilon} \Delta x, \\
&= \epsilon \cdot \Delta x.
\end{aligned}
$$

As $\Delta x \to 0$, the difference quotient

$$\frac{f(a + \Delta x) - f(a)}{\Delta x}$$

approaches $f'(a)$ (remember the definition of $f'(a)$), so the quantity in parentheses becomes a very small number (which is why we called it ϵ). In fact,

$$\epsilon \to 0 \quad \text{as} \quad \Delta x \to 0.$$

Thus, when Δx is small, the approximation error $\epsilon \Delta x$ is smaller still.

$$\underbrace{\Delta f}_{\substack{\text{true} \\ \text{change}}} = \underbrace{f'(a)\Delta x}_{\substack{\text{estimated} \\ \text{change}}} + \underbrace{\epsilon \Delta x}_{\text{error}}$$

While we do not know exactly how small the error is and will not be able to make much progress on this front until later in Chapter 9, there is something worth noting here, namely, the *form* taken by the equation.

> If $y = f(x)$ is differentiable at $x = a$ and x changes from a to $a + \Delta x$, the change Δy in f is given by an equation of the form
>
> $$\Delta y = f'(a)\Delta x + \epsilon \Delta x, \tag{4}$$
>
> in which $\epsilon \to 0$ as $\Delta x \to 0$.

Surprising as it may seem, just knowing the form of Eq. (4) enables us to bring the proof of the Chain Rule to a successful conclusion. You can find out what we mean by turning to Appendix 3.

Formulas for Differentials

Every formula like

$$\frac{d(u + v)}{dx} = \frac{du}{dx} + \frac{dv}{dx}$$

has a corresponding differential formula like

$$d(u + v) = du + dv$$

that comes from multiplying both sides by dx.

To find dy when y is a differentiable function of x, we may either find dy/dx and multiply by dx or use one of the formulas in Table 3.2.

TABLE 3.2   Formulas for Differentials

$$d(c) = 0$$

$$d(cu) = c\,du$$

$$d(u + v) = du + dv$$

$$d(uv) = u\,dv + v\,du$$

$$d\left(\frac{u}{v}\right) = \frac{v\,du - u\,dv}{v^2}$$

$$d(u^n) = nu^{n-1}du$$

$$d(\sin u) = \cos u\,du$$

$$d(\cos u) = -\sin u\,du$$

$$d(\tan u) = \sec^2 u\,du$$

$$d(\cot u) = -\csc^2 u\,du$$

$$d(\sec u) = \sec u \tan u\,du$$

$$d(\csc u) = -\csc u \cot u\,du$$

EXAMPLE 10

a) $d(3x^2 - 6) = 6x\,dx$

b) $d(\cos 3x) = -(\sin 3x)d(3x) = -3\sin 3x\,dx$

c)
$$d\frac{x}{(x + 1)} = \frac{(x + 1)dx - x\,d(x + 1)}{(x + 1)^2}$$

$$= \frac{x\,dx + dx - x\,dx}{(x + 1)^2}$$

$$= \frac{dx}{(x + 1)^2}$$

Notice that a differential on one side of an equation always calls for a differential on the other side of the equation. Thus, we never have $dy = 3x^2$ but, instead, $dy = 3x^2\,dx$.

Exercises 3.8

In Exercises 1–6, find the linearization $L(x)$ of $f(x)$ at $x = a$.

1. $f(x) = x^4$ at $x = 1$

2. $f(x) = x^{-1}$ at $x = 2$

3. $f(x) = x^3 - x$ at $x = 1$

4. $f(x) = x^3 - 2x + 3$ at $x = 2$

5. $f(x) = \sqrt{x}$ at $x = 4$

6. $f(x) = \sqrt{x^2 + 9}$ at $x = -4$

You want linearizations that will replace the functions in Exercises 7–12 over intervals that include the given points x_0. To make your subsequent work as simple as possible, you want to center each linearization not at x_0 but at a nearby integer $x = a$ at which the given function and its derivative are easy to evaluate. What linearization do you use in each case?

7. $f(x) = x^2 + 2x$, $x_0 = 0.1$

8. $f(x) = x^{-1}$, $x_0 = 0.6$

9. $f(x) = 2x^2 + 4x - 3$, $x_0 = -0.9$

10. $f(x) = 1 + x$, $x_0 = 8.1$

11. $f(x) = \sqrt[3]{x}$, $x_0 = 8.5$

12. $f(x) = \dfrac{x}{x+1}$, $x_0 = 1.3$

In Exercises 13–18, find the linearization $L(x)$ of the given function at $x = a$. Then graph f and L together near $x = a$.

13. $f(x) = \sin x$ at $x = 0$

14. $f(x) = \cos x$ at $x = 0$

15. $f(x) = \sin x$ at $x = \pi$

16. $f(x) = \cos x$ at $x = -\pi/2$

17. $f(x) = \tan x$ at $x = \pi/4$

18. $f(x) = \sec x$ at $x = \pi/4$

19. Use the linearization $(1 + x)^k \approx 1 + kx$ to find linear approximations of the following functions for values of x near zero. Graph each function and its linearization in the $[-2, 2]$ by $[-2, 2]$ viewing window.

a) $(1 + x)^2$

b) $\dfrac{1}{(1 + x)^5}$

c) $\dfrac{2}{1 - x}$

d) $(1 - x)^6$

e) $3(1 + x)^{1/3}$

f) $\dfrac{1}{\sqrt{1 + x}}$

20. Use the linearization $(1 + x)^k \approx 1 + kx$ to estimate and compare with a calculator value.

a) $(1.002)^{100}$

b) $\sqrt[3]{1.009}$

21. Find the linearization of $f(x) = \sqrt{x + 1} + \sin x$ at $x = 0$. How is it related to the individual linearizations for $\sqrt{x + 1}$ and $\sin x$?

22. We know from the Power Rule that the equation

$$\frac{d}{dx}(1 + x)^k = k(1 + x)^{k-1}$$

holds for every rational number k. In Section 7.3, we shall show that it holds for every irrational number as well. Assuming this result for now, verify Eq. (2) by showing that the linearization of $f(x) = (1 + x)^k$ at $x = 0$ is $L(x) = 1 + kx$ for any number k.

23. a) Compute $\sqrt{2}$, $\sqrt{\sqrt{2}}$, $\sqrt{\sqrt{\sqrt{2}}}$ and so forth. At each step, show that the decimal part of the square root is

about half the decimal part of its radicand. Informally, we might write: $\sqrt{1.x} \approx 1.(x/2)$.

b) Write a paragraph explaining what is happening in part (a) using the linearization of $\sqrt{1 + x}$ at $x = 0$. (See Example 1.)

c) Make a conjecture about what the numbers in part (a) are approaching if the process of taking square roots is continued. Use the linearization of $\sqrt{1 + x}$ at $x = 0$ to give a convincing argument for your conjecture.

d) Repeat parts (a)–(c), replacing 2 by other numbers greater than 1.

24. *Continuation of Exercise 23.* One way to describe the halving of the decimal parts of the square roots in Exercise 23 is to say that the number-line distance between 1 and the square root is approximately halved each time.

a) What happens if you start with a positive number that is less than 1 instead of greater than 1? Try it with 0.5.

b) Do the successive square roots approach a number? Can you use the linearization of $\sqrt{1 + x}$ at x to explain?

c) Extend the ideas in this and the previous exercise to include $\sqrt[10]{2}$ and $\sqrt[10]{0.5}$.

In Exercises 25–30, each function $f(x)$ changes value when x changes from x_0 to $x_0 + dx$. Find

a) the change $\Delta f = f(x_0 + dx) - f(x_0)$;

b) the value of the estimate $df = f'(x_0)dx$; and

c) the error $|\Delta f - df|$.

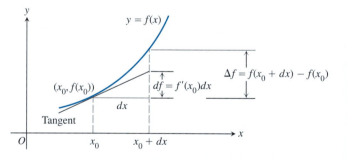

25. $f(x) = x^2 + 2x$, $x_0 = 0$, $dx = 0.1$

26. $f(x) = 2x^2 + 4x - 3$, $x_0 = -1$, $dx = 0.1$

27. $f(x) = x^3 - x$, $x_0 = 1$, $dx = 0.1$

28. $f(x) = x^4$, $x_0 = 1$, $dx = 0.1$

29. $f(x) = x^{-1}$, $x_0 = 0.5$, $dx = 0.1$

30. $f(x) = x^3 - 2x + 3$, $x_0 = 2$, $dx = 0.1$

In Exercises 31–36, write a differential formula that estimates the given change in volume or surface area.

31. The change in the volume $V = (4/3)\pi r^3$ of a sphere when the radius changes from r_0 to $r_0 + dr$.

32. The change in the surface area $S = 4\pi r^2$ of a sphere when the radius changes from r_0 to $r_0 + dr$.

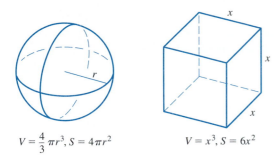

$V = \frac{4}{3}\pi r^3, S = 4\pi r^2$ $\qquad$ $V = x^3, S = 6x^2$

33. The change in the volume $V = x^3$ of a cube when the edge lengths change from x_0 to $x_0 + dx$.

34. The change in the surface area $S = 6x^2$ of a cube when the edge lengths change from x_0 to $x_0 + dx$.

35. The change in the volume $V = \pi r^2 h$ of a right circular cylinder when the radius changes from r_0 to $r_0 + dr$ and the height does not change.

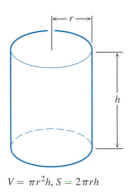

$V = \pi r^2 h, S = 2\pi rh$

36. The change in the lateral surface area $S = 2\pi rh$ of a right circular cylinder when the height changes from h_0 to $h_0 + dh$ and the radius does not change.

37. The radius of a circle is increased from 2.00 to 2.02 m.
   a) Estimate the resulting change in area.
   b) Express the estimate as a percentage of the circle's original area.

38. The diameter of a tree was 10 in. During the following year, the circumference grew 2 in. About how much did the tree's diameter grow? The tree's cross-sectional area?

39. The edge of a cube is measured as 10 cm with an error of 1%. The cube's volume is to be calculated from this measurement. Estimate the percentage error in the volume calculation.

40. About how accurately should you measure the side of a square to be sure of calculating the area to within 2% of its true value?

41. The diameter of a sphere is measured as 100 ± 1 cm, and the volume is calculated from this measurement. Estimate the percentage error in the volume calculation.

42. Estimate the allowable percentage error in measuring the diameter d of a sphere if the volume is to be calculated correctly to within 3%.

43. The height and radius of a right circular cylinder are equal, so the cylinder's volume is $V = \pi h^3$. The volume is to be calculated from a measurement of h and must be calculated with an error of no more than 1% of the true value. Find approximately the greatest error that can be tolerated in the measurement of h, expressed as a percentage of h.

44. a) About how accurately must the interior diameter of a 10-m-high cylindrical storage tank be measured to calculate the tank's volume to within 1% of its true value?
   b) About how accurately must the tank's exterior diameter be measured to calculate the amount of paint it will take to paint the side of the tank to within 5% of the true amount?

45. A manufacturer contracts to mint coins for the federal government. How much variation dr in the radius of the coins can be tolerated if the coins are to weigh within 1/1000 of their ideal weight? Assume that the thickness does not vary.

46. *Continuation of Example 8.* By what percentage should r be increased to increase V by 50%?

47. *Continuation of Example 9.* Show that a 5% error in measuring t will cause about a 10% error in calculating $s(t)$ from the equation $s = 16t^2$.

48. *The effect of flight maneuvers on the heart.* The amount of work done by the heart's main pumping chamber, the left ventricle, is given by the equation

$$W = PV + \frac{Vpv^2}{2g},$$

where W is the work per unit time, P is the average blood pressure, V is the volume of blood pumped out during the unit of time, p is the density of the blood, v is the average velocity of the exiting blood, and g is the acceleration of gravity.

When P, V, p, and v remain constant, W becomes a function of g, and the equation takes the simplified form

$$W = a + \frac{b}{g} \qquad (a, b \text{ constant}) \qquad (5)$$

As a member of NASA's medical team, you want to know how sensitive W is to apparent changes in g caused by flight maneuvers, and this depends on the initial value of g. As part of your investigation, you decide to compare the effect on W of a given change dg on the moon, where $g = 5.2$ ft/sec^2, with the effect the same change dg would have on Earth, where $g = 32$ ft/sec^2. You use Eq. (5) to find the ratio of dW_{moon} to dW_{Earth}. What do you find?

Using Linearizations to Solve Equations

49. Let $g(x) = \sqrt{x} + \sqrt{1+x} - 4$.

   a) Show that $g(3) < 0$ and $g(4) > 0$ to confirm (by the Intermediate Value Theorem, Section 2.2) that the equation $g(x) = 0$ has a solution between $x = 3$ and $x = 4$.

   b) To estimate the solution of $g(x) = 0$, replace the square roots by their linearizations at $x = 3$, and solve the resulting linear equation.

   c) Check your estimate in the original equation.

   d) GRAPH $g(x)$ and solve $g(x) = 0$ using your graphing utility. Compare with part (b).

   e) Find the exact solutions to $g(x) = 0$.

50. Carry out the following steps to estimate the solution of $2\cos x = \sqrt{1+x}$.

   a) Let $f(x) = 2\cos x - \sqrt{1+x}$. Show that $f(0) > 0$ and $f(\pi/2) < 0$ to confirm that $f(x)$ has a zero between 0 and $\pi/2$.

   b) Find the linearizations of $2\cos x$ at $x = \pi/4$ and $\sqrt{1+x}$ at $x = 0.69$.

   c) To estimate the solution of the original equation, replace $2\cos x$ and $\sqrt{1+x}$ by their linearizations from part (b), and solve the resulting linear equation for x. Check your estimate in the original equation.

   d) Solve the original equation using your graphing utility.

   e) Can you find exact solutions to the original equation?

Differentials

In Exercises 51–62, find dy.

51. $y = x^3 - 3x$ **52.** $y = x\sqrt{1-x^2}$

53. $y = 2x/(1+x^2)$ **54.** $y = (3x^2 - 1)^{3/2}$

55. $y + xy - x = 0$ **56.** $xy^2 + x^2y - 4 = 0$

57. $y = \sin(5x)$ **58.** $y = \cos(x^2)$

59. $y = 4\tan(x/2)$ **60.** $y = \sec(x^2 - 1)$

61. $y = 3\csc(1 - (x/3))$ **62.** $y = 2\cot\sqrt{x}$

63. Let $f(x) = \sqrt{1+x}$.

   a) ZOOM-IN on the graph of f at $x = 0$ until it appears to be a straight line.

   b) Compute the slope of the straight line graph of f in part

(a). Compare with $f'(0)$.

   c) Graph both f and its linearization in the final window of part (a).

64. Repeat Exercise 63 with $f(x) = x/(x^2 - 4)$ at $x = 3$.

65. Show that the approximation of $\sqrt{1+x}$ by its linearization at the origin must improve as $x \to 0$ by showing that

$$\lim_{x \to 0} \frac{\sqrt{1+x}}{1 + (x/2)} = 1.$$

66. Show that the approximation of $\tan x$ by its linearization at the origin must improve as $x \to 0$ by showing that

$$\lim_{x \to 0} \frac{\tan x}{x} = 1.$$

67. *The linearization is the best linear approximation.* Suppose that $y = f(x)$ is differentiable at $x = a$ and that $g(x) = m(x - a) + c$ (m and c constants). If the error $E(x) = f(x) - g(x)$ were small enough near $x = a$, we might think of using g as a linear approximation of f instead of the linearization $L(x) = f(a) + f'(a)(x - a)$. Show that if we impose on g the conditions:

   a) $E(a) = 0$,   The error is zero at $x = a$.

   b) $\displaystyle\lim_{x \to a} \frac{E(x)}{x - a} = 0$,   The error is negligible when compared with $(x - a)$.

then $g(x) = f(a) + f'(a)(x - a)$. Thus, the linearization gives the only linear approximation whose error is both zero at $x = a$ and negligible in comparison with $(x - a)$.

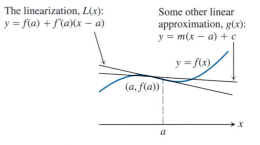

The linearization, $L(x)$:
$y = f(a) + f'(a)(x - a)$

Some other linear approximation, $g(x)$:
$y = m(x - a) + c$

$y = f(x)$

$(a, f(a))$

a

Review Questions

1. What is a derivative? A right-hand derivative? A left-hand derivative? How are they related? Give examples.

2. What geometric significance do derivatives have?

3. How is the differentiability of a function at a point related to its continuity (if any) at a point?

4. Give some examples of differentiable functions.

5. Give an example of a continuous function that is not differentiable.

6. Explain why all differentiable functions are continuous.

7. Explain the meaning of NDER $f(x)$. Is NDER $f(x) = f'(x)$ always true? Illustrate with examples.

8. When can a function typically *not* have a derivative at a point? Illustrate with graphs.

9. What rules do you know for calculating derivatives? Give examples.

10. Explain how the three formulas

    a) $\dfrac{d(x^n)}{dx} = nx^{n-1}$,

    b) $\dfrac{d(cu)}{dx} = c\dfrac{du}{dx}$,

    c) $\dfrac{d(u+v)}{dx} = \dfrac{du}{dx} + \dfrac{dv}{dx}$

    let us differentiate any polynomial.

11. Explain how the graph of $y = $ NDER $f(x)$ can be used to support the analytic computation of $D_x f(x)$.

12. What formula do we need, in addition to the three listed in Exercise 10, to differentiate rational functions?

13. What is a second derivative? A third derivative? How many derivatives do the functions you know have? Give examples.

14. Give an example of a function that is continuous and differentiable but not twice differentiable.

15. When a body moves along a coordinate line and its position $s(t)$ is a differentiable function of t, how do you define the body's velocity, speed, and acceleration? Give an example.

16. Besides velocity, speed, and acceleration, what other rates of change are found with derivatives?

17. What are the derivatives of the six basic trigonometric functions? How does their calculation depend on radian measure?

18. When is the composite of two functions differentiable at a point? What do you need to know to calculate its derivative there? Give examples.

19. What is implicit differentiation, and what is it good for? Give examples.

20. What is the linearization $L(x)$ of a function $f(x)$ at a point $x = a$? What is required of f at a for the linearization to exist? How are linearizations used? Give examples.

21. If x moves from x_0 to a nearby value $x_0 + dx$, how do we estimate the corresponding change in the value of a differentiable function $f(x)$? How do we estimate the relative change? The percentage change? Give an example.

22. How are derivatives expressed in differential notation? Give examples.

Practice Exercises

In Exercises 1–34, find dy/dx analytically. Support with a graph of NDER y.

1. $y = x^5 - \dfrac{1}{8}x^2 + \dfrac{1}{4}x$

2. $y = 3 - 7x^3 + 3x^7$

3. $y = (x+1)^2(x^2 + 2x)$

4. $y = (2x - 5)(4 - x)^{-1}$

5. $y = 2\sin x \cos x$

6. $y = \sin x - x \cos x$

7. $y = \dfrac{x}{x+1}$

8. $y = \dfrac{2x+1}{2x-1}$

9. $y = (x^3 + 1)^{-4/3}$

10. $y = (x^2 - 8x)^{-1/2}$

11. $y = \cos(1 - 2x)$

12. $y = \cot \dfrac{2}{x}$

13. $y = (x^2 + x + 1)^3$

14. $y = \left(-1 - \dfrac{x}{2} - \dfrac{x^2}{4}\right)^2$

15. $y = \sqrt{2u + u^2}, u = 2x + 3$

16. $y = \dfrac{-u}{1+u}, u = \dfrac{1}{x}$

17. $xy + y^2 = 1$

18. $xy + 2x + 3y = 1$

19. $x^2 + xy + y^2 - 5x = 2$

20. $x^3 + 4xy - 3y^2 = 2$

21. $5x^{4/5} + 10y^{6/5} = 15$

22. $\sqrt{xy} = 1$

23. $y^2 = \dfrac{x}{x+1}$

24. $y^2 = \sqrt{\dfrac{1+x}{1-x}}$

25. $y^2 = \dfrac{(5x^2 + 2x)^{3/2}}{3}$

26. $y = \dfrac{3}{(5x^2 + 2x)^{3/2}}$

27. $y = \sqrt{x} + 1 + \dfrac{1}{\sqrt{x}}$

28. $y = x\sqrt{2x + 1}$

29. $y = \sec(1 + 3x)$

30. $y = \sec^2(1 + 3x)$

31. $y = \cot x^2$

32. $y = x^2 \cos 5x$

33. $y = \sqrt{\dfrac{1-x}{1+x^2}}$

34. $y^2 = \dfrac{x^2 - 1}{x^2 + 1}$

35. a) GRAPH the function

$$f(x) = \begin{cases} x, & 0 \le x \le 1, \\ 2 - x, & 1 < x \le 2. \end{cases}$$

b) Is f continuous at $x = 1$?

c) Is f differentiable at $x = 1$? Explain.

36. a) Find the values of the left-hand and right-hand derivatives of

$$f(x) = \begin{cases} \sin 2x, & x \le 0, \\ mx, & x > 0 \quad (m \text{ constant}) \end{cases}$$

at $x = 0$.

b) For what value of m, if any, is f differentiable at $x = 0$.

37. Find the points on the curve $y = 2x^3 - 3x^2 - 12x + 20$ where the tangent is parallel to the x-axis.

38. The line normal to the curve $y = x^2 + 2x - 3$ at $(1, 0)$ intersects the curve at what other point?

39. The position at time $t \ge 0$ of a particle moving along the x-axis is

$$s(t) = 10 \cos (t + \pi/4).$$

a) Use parametric mode on a graphing utility to simulate the motion of the particle.

b) What is the particle's starting position $(t = 0)$?

c) What are the points farthest to the left and right of the origin reached by the particle?

d) Find the particle's velocity and acceleration at the points in part (b).

e) When does the particle first reach the origin? What are its velocity, speed, and acceleration then?

40. On Earth, you can shoot a paper clip 64 ft straight up into the air with a rubber band. In t seconds after firing, the paper clip is $s = 64t - 16t^2$ ft above your hand.

a) Use parametric mode on a graphing utility to simulate the position of the paper clip.

b) How long does it take the paper clip to reach its maximum height? With what velocity does it leave your hand?

c) On the moon, the same force will send the paper clip to a height of $s(t) = 64t - 2.6t^2$ ft in t seconds. About how long will it take the paper clip to reach its maximum height, and how high will it go?

41. Suppose two balls are falling from rest at a certain height in centimeters above the ground. Use the equation $s = 490t^2$ to answer the following questions:

a) How long did it take the balls to fall the first 160-cm? What was their average velocity for the period?

b) How fast were the balls falling when they reached the 160-cm mark? What was their acceleration then?

42. The following data give the coordinates s of a moving body for various values of t. Plot s versus t on coordinate paper, and sketch a smooth curve through the given points. Assuming that this smooth curve represents the motion of the body, estimate the velocity at (a) $t = 1.0$; (b) $t = 2.5$; (c) $t = 2.0$.

t (in sec)	0	0.5	1.0	1.5	2.0	2.5	3.0	3.5	4.0
s (in ft)	10	38	58	70	74	70	58	38	10

43. The following graphs show the distance traveled (miles), velocity (mph), and acceleration (mph/sec) for each second of a 2-minute automobile trip. Which graph shows
a) distance?   **b)** velocity?   **c)** acceleration?

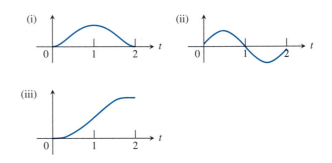

44. The following graph shows the position $s(t)$ at time t of a truck traveling on a highway. The truck starts at $t = 0$ and returns 15 hours later at $t = 15$.

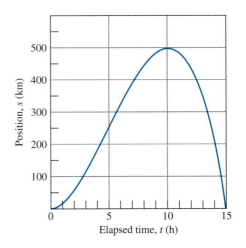

a) Use the technique described at the end of Section 3.4 to graph the truck's velocity $v = ds/dt$ for $0 \le t \le 15$. Then repeat the process with the velocity curve to graph the truck's acceleration dv/dt.

b) When is the truck going the slowest? The fastest?

c) Suppose $s(t) = 15t^2 - t^3$. Graph ds/dt and d^2s/dt^2, and compare your graphs with those in part (a).

45. Use the following information to graph the function $y = f(x)$ for $-1 \le x \le 6$.

i) The graph of f is made of line segments joined end to end.

ii) The graph starts at the point $(-1, 2)$.

iii) The derivative of f, where defined, is the step function shown in the following graph.

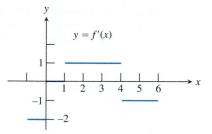

46. Repeat Exercise 45, supposing that the graph starts at $(-1, 0)$ instead of $(-1, 2)$.

For Exercises 47 and 48, graph $f'(x)$ given the graph of $f(x)$.

47.

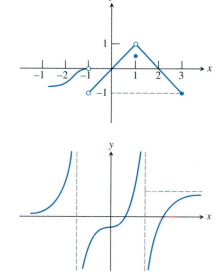

48.

49. If a hemispherical bowl of radius 10 in. is filled with water to a depth of x in., the volume of water is given by $V = \pi[10 - (x/3)]x^2$. Find the rate of increase of the volume per inch increase of depth.

50. A bus will hold 60 people. The fare charged (p dollars) is related to the number x of people who use the bus by the law $p = [3 - (x/40)]^2$. Write an expression for the total revenue $r(x)$ per trip received by the bus company. What number of people per trip will make the marginal revenue dr/dx equal to zero? What is the corresponding fare? (This is the fare that maximizes the revenue, so the bus company should probably rethink its fare policy.)

In Exercises 51 and 52, find

a) all points with horizontal tangents and

b) the tangent to the curve at the indicated point P.

51. $y = 4 + \cot x - 2 \csc x, P = (\pi/2, 2)$

52. $y = 1 + \sqrt{2} \csc x + \cot x, P = (\pi/4, 4)$

53. The graph of $y = \sin(x - \sin x)$ appears to have horizontal tangents at the x-axis. Does it?

54. The following figure shows a boat 1 km offshore, sweeping the shore with a search light. The light turns at the constant rate $d\theta/dt = -3/5$ radian per second. (This rate is called the light's *angular velocity*.)

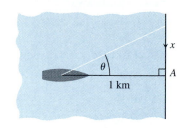

a) Express x (see the figure) in terms of θ.

b) Differentiate both sides of the equation that you obtained in part (a) with respect to t. Then substitute $d\theta/dt = -3/5$. This will express dx/dt (the rate at which the light moves along the shore) as a function of θ.

c) How fast (m/sec) is the light moving along the shore when it reaches point A?

d) How many revolutions per minute is 0.6 radian per second?

55. Suppose that functions f and g and their derivatives have the following values at $x = 0$ and $x = 1$:

x	$f(x)$	$g(x)$	$f'(x)$	$g'(x)$
0	1	1	5	1/3
1	3	−4	−1/3	−8/3

Find the derivatives of the following combinations at the given value of x.

a) $5f(x) - g(x)$, $x = 1$

b) $f(x)g^3(x)$, $x = 0$

c) $\dfrac{f(x)}{g(x) + 1}$, $x = 1$

d) $f(g(x))$, $x = 0$

e) $g(f(x))$, $x = 0$

f) $(x + f(x))^{3/2}$, $x = 1$

g) $f(x + g(x))$, $x = 0$

56. Suppose that $f(x) = x^2$ and $g(x) = |x|$. Then the composites

$$(f \circ g)(x) = |x|^2 = x^2 \qquad \text{and} \qquad (g \circ f)(x) = |x^2| = x^2$$

are both differentiable at $x = 0$ even though g is not differentiable at $x = 0$. Does this contradict the Chain Rule? Explain.

57. If the identity $\sin(x + a) = \sin x \cos a + \cos x \sin a$ is differentiated with respect to x, is the resulting equation also an identity? Does this "principle" apply to the equation $x^2 - 2x - 8 = 0$? Explain.

58. Find dy/dt at $t = 0$ if $y = 3 \sin 2x$ and $x = t^2 + \pi$.

59. Find ds/du at $u = 2$ if $s = t^2 + 5t$ and $t = (u^2 + 2u)^{1/3}$.

60. Find dw/ds at $s = 0$ if $w = \sin(\sqrt{r} - 2)$ and $r = 8 \sin(s + \pi/6)$.

61. Find the points where the tangent to the curve $y = \sqrt{x}$ at $x = 4$ crosses the coordinate axes.

62. What horizontal line crosses the curve $y = \sqrt{x}$ at a $45°$ angle?

63. Find the lines that are (i) tangent and (ii) normal to the curve at the given point.
a) $x^2 + 2y^2 = 9$ at $(1, 2)$
b) $x^3 + y^2 = 2$ at $(1, 1)$
c) $xy + 2x - 5y = 2$ at $(3, 2)$

64. Which of the following statements could be true if $f''(x) = x^{1/3}$?

i. $f(x) = \dfrac{9}{28} x^{7/3} + 9$ **ii.** $f'(x) = \dfrac{9}{28} x^{7/3} - 2$

iii. $f'(x) = \dfrac{3}{4} x^{4/3} + 6$ **iv.** $f(x) = \dfrac{3}{4} x^{4/3} - 4$

a) i only   **b)** iii only

c) ii and iv only   **d)** i and iii only

65. The designer of a 30-ft-diameter spherical hot-air balloon wishes to suspend the gondola 8 ft below the bottom of the balloon. Two of the cables are shown running from the top edges of the gondola to their points of tangency, $(-12, -9)$ and $(12, -9)$. How wide must the gondola be?

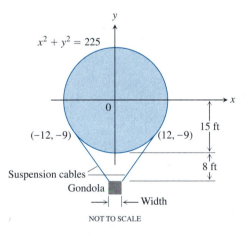

$x^2 + y^2 = 225$

$(-12, -9)$ $(12, -9)$

15 ft

8 ft

Suspension cables

Gondola

Width

NOT TO SCALE

66. *What determines the fundamental frequency of a vibrating piano string?* We measure the frequencies at which wires vibrate in cycles (trips back and forth) per second. The unit of measure is a *hertz*: 1 cycle per second. Middle A on a piano has a frequency of 440 hertz. For any given wire, the fundamental frequency y is a function of four variables:
   r: the radius of the wire
   l: the length
   d: the density of the wire
   T: the tension (force) holding the wire taut.

With r and l in centimeters, d in grams per cubic centimeter, and T in dynes (it takes about 100,000 dynes to lift an apple), the fundamental frequency of the wire is

$$y = \frac{1}{2rl} \sqrt{\frac{T}{\pi d}}.$$

If we keep all the variables fixed except one, then y can be alternatively thought of as four different functions of one variable, $y(r)$, $y(l)$, $y(d)$, and $y(T)$. How would changing each variable then affect the string's fundamental frequency? To find out, calculate $y'(r)$, $y'(l)$, $y'(d)$, and $y'(T)$.

67. Find d^2y/dx^2 by implicit differentiation:

a) $x^3 + y^3 = 1$ **b)** $y^2 = 1 - \dfrac{2}{x}$

68. a) By differentiating $x^2 - y^2 = 1$ implicitly, show that $dy/dx = x/y$.
b) Then show that $d^2y/dx^2 = -1/y^3$.

69. Find d^2y/dx^2 if
a) $y = \sqrt{2x + 7}$ **b)** $x^2 + y^2 = 1$

70. If $y^3 + y = 8x - 6$, find d^2y/dx^2 at the point $(1, 1)$.

71. Find the linearizations of
a) $\tan x$ at $x = -\pi/4$ **b)** $\sec x$ at $x = -\pi/4$.
Graph the curves and linearizations together.

72. A useful linear approximation to

$$\frac{1}{1 + \tan x}$$

at $x = 0$ can be obtained by combining the approximations

$$\frac{1}{1 + x} \approx 1 - x \quad \text{and} \quad \tan x \approx x$$

to get

$$\frac{1}{1 + \tan x} \approx 1 - x.$$

Show that this is the standard linear approximation of $1/(1 + \tan x)$.

73. Let $f(x) = \sqrt{1 + x} + \sin x - 0.5$.
a) Show that $f(-\pi/4) < 0$ and $f(0) > 0$, to confirm that the equation $f(x) = 0$ has a solution between $-\pi/4$ and 0.
b) To estimate the solution of $f(x) = 0$, replace $\sqrt{1 + x}$ and $\sin x$ by their linearizations at $x = 0$, and solve the resulting linear equation.
c) Check your estimate in the original equation.
d) Solve $f(x) = 0$ using your graphing utility. What are the exact solutions?

74. Let

$$f(x) = \frac{2}{1-x} + \sqrt{1+x} - 3.1.$$

a) Show $f(0) < 0$ and $f(0.5) > 0$ to confirm that the equation $f(x) = 0$ has a solution between $x = 0$ and $x = 0.5$.
b) To estimate the solution of the equation $f(x) = 0$, replace $2/(1-x)$ and $\sqrt{1+x}$ by their linearizations at $x = 0$, and solve the resulting linear equation.
c) Check your estimate in the original equation.
d) Solve $f(x) = 0$ using your graphing utility. What are the exact solutions?

75. Write a formula that estimates the change that occurs in the volume of a right circular cone when the radius changes from r_0 to $r_0 + dr$ and the height does not change.

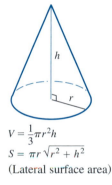

$$V = \frac{1}{3}\pi r^2 h$$
$$S = \pi r \sqrt{r^2 + h^2}$$
(Lateral surface area)

76. Write a formula that estimates the change that occurs in the lateral surface area of a cone when the height changes from h_0 to $h_0 + dh$ and the radius does not change.

77. a) How accurately should you measure the edge of a cube to be reasonably sure of calculating the cube's surface area with an error of no more than 2%?
b) Suppose the edge is measured with the accuracy required in part (a). About how accurately can the cube's volume be calculated from the edge measurement? To find out, estimate the percentage error in the volume calculation that would result from using the edge measurement.

78. The circumference of the equator of a sphere is measured as 10 cm with a possible error of 0.4 cm. The measurement is then used to calculate the radius. The radius is then used to calculate the surface area and volume of the sphere. Es-

timate the percentage errors in the calculated values of (a) the radius, (b) the surface area, and (c) the volume.

79. To find the height of a tree, you measure the angle from the ground to the treetop from a point 100 ft away from the base. The best figure you can get with the equipment at hand is $30° \pm 1°$. About how much error could the tolerance of $\pm 1°$ create in the calculated height? Remember to work in radians.

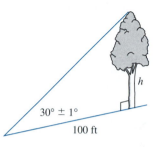

30° ± 1°

100 ft

80. To find the height of a lamppost, you stand a 6-ft pole 20 ft from the lamp and measure the length a of its shadow, finding it to be 15 ft, give or take an inch. Calculate the height of the lamppost from the value $a = 15$, and estimate the possible error in the result.

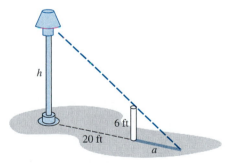

h

6 ft

20 ft

a

81. The volume $y = x^3$ of a cube with edges of length x increases an amount Δy when x increases by an amount Δx. Draw the cube with side lengths x, and carefully overlay the cube with side lengths $x + \Delta x$. Then, show with your sketch how to represent Δy geometrically as the sum of the volumes of three slabs of dimensions x by x by Δx, three bars of dimensions x by Δx by Δx, and one cube of dimensions Δx by Δx by Δx. The differential formula $dy = 3x^2 dx$ estimates the change in y with the three slabs.

4

Applications of Derivatives

OVERVIEW   Graphs reveal much about the behavior of functions. However, confirmation of what we see and believe to be true in a viewing window must still come from calculus. In the past, when virtually all graphing was done by hand—sometimes very laboriously—derivatives were the key tool used for sketching the graphs of polynomial, rational, radical, and transcendental functions (such as the trigonometric, exponential, and logarithmic functions). Now the derivative is the key instrument used for confirming the information portrayed in a graph and predicting any information hidden within the graph. Indeed, the derivative is the key to establishing that a graph is complete.

Recall that we also associate derivatives with the rates at which functions change. If we know the rate at which a function is changing, this chapter will show that we can often calculate the rates at which functions closely related to it are changing at the same time. If we know the derivative of a function and the value of the function at a particular point, this chapter will show how to find the function. The key to recovering functions from their derivatives is the Mean Value Theorem, a theorem whose corollaries provide the gateway to *integral calculus*, which we will begin to study in Chapter 5.

4.1 _____ Maxima, Minima, and the Mean Value Theorem

Differential calculus is the mathematics of working with derivatives. One of the things we do with derivatives is confirm where functions take on their maximum and minimum values. In this section, we lay the theoretical groundwork for finding these *extreme values* and establish the First Derivative Test for determining when functions are increasing or decreasing on an interval.

We also introduce the Mean Value Theorem, one of the most influential theorems in calculus.

Maxima and Minima—Local and Absolute

Figure 4.1 shows a point c where a function $y = f(x)$ with domain $[a, b]$ has a maximum value. If we move to either side of c, the function values get smaller, and the curve falls away. When we take in more of the curve, however, we find that f assumes an even larger value at d. Thus, $f(c)$ is not the *absolute* maximum value of f on the interval $[a, b]$ but only a *local* maximum value.

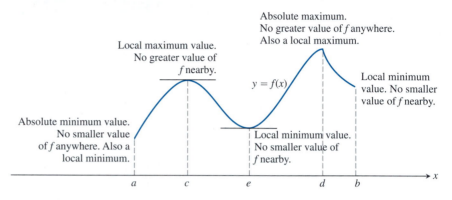

4.1 How to classify maxima and minima.

DEFINITIONS

A function f has a **local maximum** value at an interior point c of its domain if $f(x) \leq f(c)$ for all x in some open interval I containing c. The function has an **absolute maximum** value at c if $f(x) \leq f(c)$ for all x in the domain.

Similarly, f has a **local minimum** value at an interior point c of its domain if $f(x) \geq f(c)$ for all x in an open interval I containing c. The function has an **absolute minimum** value at c if $f(x) \geq f(c)$ for *all* x in the domain.

The definitions of a function's maximum values and minimum values, also called **extreme values**, are extended to endpoints of the function's domain by requiring the intervals I to be appropriate half-open intervals containing the endpoints.

Notice that an absolute maximum is also a local maximum because it is the largest value in its immediate neighborhood as well as overall. Hence, a list of all local maxima will include the absolute maximum if there is one. Similarly, an absolute minimum, when it exists, is also a local minimum. A list of all local minima will include the absolute minimum if there is one.

The First Derivative Theorem

In Figure 4.1, two extreme values of f occur at endpoints of the function's domain, one occurs at a point where f' fails to exist, and two occur at interior points where $f' = 0$. This is typical for a function defined on a closed interval. As we will prove shortly, a function's first derivative, if it exists, is always zero at an interior point where the function has a local extreme value. Hence, the only places where a function f can ever have an extreme value are

1. interior points where f' is zero,
2. interior points where f' does not exist, and
3. endpoints of the function's domain.

When the function is defined on a closed interval

DEFINITION

A point in the domain of a function f at which $f' = 0$ or f' does not exist is a **critical point** of f.

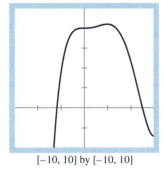

[−10, 10] by [−10, 10]

4.2 The critical points of

$$f(x) = x^5 - 5x^4 + 5x^3 + 20$$

on $[-3, 3]$ are $x = 0, x = 1$, and $x = 3$. The function f can have extreme values only at these points and the point $x = -3$. On the basis of the graph of f shown here and your knowledge about the graph of a fifth-degree polynomial function, where do the extreme values of f appear to be?

Thus, the only points where a function can take on extreme values are critical points and endpoints. We will see the importance of this observation as the chapter continues.

EXAMPLE 1

Determine the critical points of $f(x) = x^5 - 5x^4 + 5x^3 + 20$ on the interval $[-3, 3]$.

Solution $f'(x) = 5x^4 - 20x^3 + 15x^2$, which exists everywhere on $[-3, 3]$, so the critical points occur only where $f'(x) = 0$. We see that

$$f'(x) = 5x^4 - 20x^3 + 15x^2$$
$$= 5x^2(x^2 - 4x + 3)$$
$$= 5x^2(x - 1)(x - 3) = 0$$

at $x = 0, 1$, and 3, all of which are within $[-3, 3]$.

Therefore, the critical points of f are 0, 1, and 3. See Fig. 4.2. ≡

EXPLORATION 1

Finding Extreme Values

REMINDER

You may replace NDER f by your grapher's version of f' or, when appropriate, compute f' analytically.

1. GRAPH $y_1 = f(x) = x^3 - 12x$. For what values of x does it appear that f has extreme values? Support your response using ZOOM and TRACE. Support your response by graphing $y_2 = $ NDER y_1 in the same viewing window and locating the critical points of y_1 graphically and analytically.

2. GRAPH $y_3 = $ abs(y_1). Discuss the extreme values. Include comments on the behavior of $y_4 = $ NDER y_3 near the values of x where extreme values occur.

3. GRAPH $y_5 = y_3$ with domain restricted to $-1 < x < 4$. Discuss the extreme values.

> **THEOREM 1**   **The First Derivative Theorem for Local Extreme Values**
>
> If a function f has a local maximum or a local minimum value at an interior point c of its domain and if f' exists at c, then
>
> $$f'(c) = 0.$$

Proof   You may not have seen an argument like the one we are about to use, so we will explain its form first. We want to show that $f'(c) = 0$, and our plan is to do that indirectly by showing first that $f'(c)$ cannot be positive and second that $f'(c)$ cannot be negative either. Why does that show that $f'(c) = 0$? Because in the entire real-number system, only one number is neither positive nor negative, and that number is zero.

To be specific, suppose f has a local maximum value at $x = c$, so that $f(x) \le f(c)$ for all values of x near c (Fig. 4.3). Since c is an interior point of f's domain, the limit

$$\lim_{x \to c} \frac{f(x) - f(c)}{x - c}$$

defining $f'(c)$ is two-sided. This means that the right-hand and left-hand limits both exist at $x = c$, and both equal $f'(c)$.

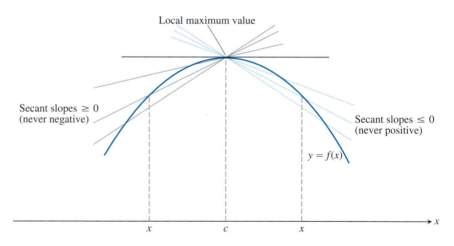

4.3   A curve with a local maximum value. The slope at c, simultaneously the limit of nonpositive numbers and nonnegative numbers, is zero.

Let's examine these limits separately. To the right of c, $f(x) \le f(c)$, so $f(x) - f(c) \le 0$. Also $x - c > 0$, so $(f(x) - f(c))/(x - c) \le 0$. Thus,

$$\lim_{x \to c^+} \frac{f(x) - f(c)}{x - c} \le 0. \tag{1}$$

Similarly, to the left of c, $f(x) - f(c) \le 0$, and $x - c < 0$. As a result, $(f(x) - f(c))/(x - c) \ge 0$, and

$$\lim_{x \to c^-} \frac{f(x) - f(c)}{x - c} \ge 0. \tag{2}$$

The inequality in (1) says that $f'(c)$ cannot be greater than zero, whereas (2) says that $f'(c)$ cannot be less than zero. So $f'(c) = 0$.

This proves the theorem for local maximum values. To prove it for local minimum values, simply replace f by $-f$ and run through the argument again. ≡

Rolle's Theorem

There is strong geometric evidence that between any two points at which a smooth curve crosses the x-axis there is a point on the curve where the tangent is horizontal. A 300-year-old theorem of Michel Rolle (1652–1719) assures us that this is indeed the case.

THEOREM 2   Rolle's Theorem

Suppose that $y = f(x)$ is continuous at every point of the closed interval $[a, b]$ and differentiable at every point of its interior (a, b). If

$$f(a) = f(b) = 0,$$

then there is at least one number c between a and b at which

$$f'(c) = 0.$$

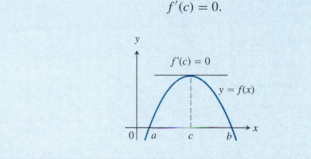

Proof of Rolle's Theorem   We know from Section 2.2 that a continuous function defined on a closed interval assumes absolute maximum and minimum values on the interval. The question is, where? By Theorem 1, we deduce that there are only three kinds of places to look:

1. At interior points where f' is zero,

2. At interior points where f' does not exist, and

3. At the endpoints of the function's domain, in this case a and b.

By hypothesis, f has a derivative at every interior point. That rules out the second possibility above, leaving us with interior points where $f' = 0$ and with the two endpoints a and b.

If either the maximum or the minimum occurs at a point c inside the interval, then $f'(c) = 0$ by Theorem 1, and we have found a point for Rolle's Theorem.

If both the maximum and the minimum occur at the endpoints a and b where f is zero, then zero is the maximum value of f as well as the minimum value of f. In other words, for every value of x,

$$0 = \min(f) \le f(x) \le \max(f) = 0.$$

MORE ON ROLLE'S THEOREM

For suitably defined f, the contrapositive of Rolle's Theorem tells us:

If $f'(x) \neq 0$ for all x, then f has no more than one zero.
In other words, the graph of f crosses the x-axis no more than once.

EXPLORATION BIT

Give an example of a function that is defined on a closed interval $[a, b]$, continuous and differentiable on the open interval (a, b), but for which Rolle's Theorem does not apply. In other words, show that continuity on $[a, b]$ is a necessary condition for Rolle's Theorem.

So f is zero throughout the interval, and because f has a constant value, its derivative is zero throughout the interval. In other words, $f'(c) = 0$ at every interior point. Either way, we find a point c in (a, b) where $f'(c)$ is zero. This concludes the proof. ≡

EXAMPLE 2

The polynomial function

$$f(x) = \frac{x^3}{3} - 3x$$

graphed in Fig. 4.4 is continuous at every point of the interval $-3 \leq x \leq 3$ and differentiable at every point of the interval $-3 < x < 3$. Since $f(-3) = f(3) = 0$, Rolle's Theorem says that f' must be zero at least once in the open interval between $a = -3$ and $b = 3$. In fact, $f'(x) = x^2 - 3$ is zero twice in this interval, once at $x = -\sqrt{3}$ and again at $x = \sqrt{3}$.

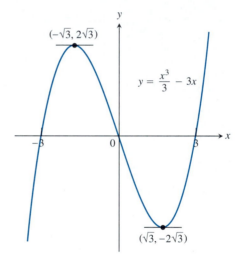

4.4 As predicted by Rolle's Theorem, this smooth curve has horizontal tangents between the points at which it crosses the x-axis. ≡

EXAMPLE 3

As the function $f(x) = 1 - |x|$ shows (Fig. 4.5), the differentiability of f is essential to Rolle's Theorem. If we allow even one interior point in (a, b) where f is not differentiable, there may be no horizontal tangent to the curve. ≡

The Mean Value Theorem

The Mean Value Theorem is a slanted version of Rolle's Theorem. You will see what we mean if you look at Fig. 4.6. The figure shows the graph of a differentiable function f defined on an interval $a \leq x \leq b$. There is a point on the curve where the tangent is parallel to the secant AB.

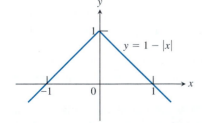

4.5 This curve has no horizontal tangent between the points where it crosses the x-axis.

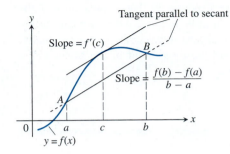

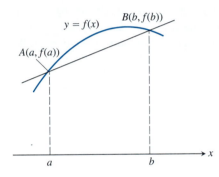

4.6 Geometrically, the Mean Value Theorem says that somewhere between A and B, the curve has at least one tangent parallel to secant AB.

4.7 The graph of f, and the secant AB over the interval $a \le x \le b$.

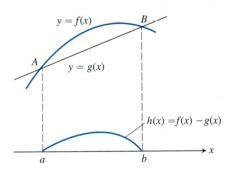

4.8 The secant AB in Fig. 4.6 is the graph of the function $g(x)$. The function $h(x) = f(x) - g(x)$ gives the vertical distance between the graphs of f and g at x.

In Rolle's Theorem, the secant AB is the x-axis, and $f'(c) = 0$. Here the secant AB meets the curve above a and b, and $f'(c)$ is the slope of the secant.

THEOREM 3   The Mean Value Theorem

If $y = f(x)$ is continuous at every point of the closed interval $[a, b]$ and differentiable at every point of its interior (a, b), then there is at least one number c between a and b at which

$$f'(c) = \frac{f(b) - f(a)}{b - a}. \tag{3}$$

Proof If we graph f over $[a, b]$ and draw the line through the endpoints $A(a, f(a))$ and $B(b, f(b))$, the figure we get, Fig. 4.7, resembles the one we drew for Rolle's Theorem. The difference is that the line AB need not be the x-axis because $f(a)$ and $f(b)$ may not be zero. We cannot apply Rolle's Theorem directly to f, but we can apply it to the function that measures the vertical distance between the graph of f and the line AB. This, it turns out, will tell us what we want to know about the derivative of f.

The line AB is the graph of the function

$$g(x) = f(a) + \frac{f(b) - f(a)}{b - a}(x - a)$$

(point–slope equation), and the formula for the vertical distance between the graphs of f and g at x is

$$h(x) = f(x) - g(x) = f(x) - f(a) - \frac{f(b) - f(a)}{b - a}(x - a). \tag{4}$$

Figure 4.8 shows the graphs of f, g, and h together.

The function h satisfies the hypotheses of Rolle's Theorem on the interval $[a, b]$. It is continuous on $[a, b]$ because f and g are. Both $h(a)$ and $h(b)$ are zero because the graphs of f and g pass through A and B.

Therefore, $h' = 0$ at some point c between a and b. To see what this says about f', we differentiate both sides of Eq. (4) with respect to x and set $x = c$. This gives

$$h'(x) = f'(x) - \frac{f(b) - f(a)}{b - a} \qquad \text{Derivative of Eq. (4)} \ldots$$

$$h'(c) = f'(c) - \frac{f(b) - f(a)}{b - a} \qquad \ldots \text{ with } x = c$$

$$0 = f'(c) - \frac{f(b) - f(a)}{b - a} \qquad h'(c) = 0$$

$$f'(c) = \frac{f(b) - f(a)}{b - a}, \qquad \text{Rearranged}$$

which is what we set out to prove. ≡

The importance of the Mean Value Theorem lies in the mathematical conclusions that come from Eq. (3), one of which we will see in a moment.

A peculiarity of the Mean Value Theorem is that it tells us that the number c exists but doesn't tell how to find it. In some cases, we can satisfy our curiosity algebraically, and we can almost always satisfy it graphically. Being able to identify c, however, is not the main point of the Mean Value Theorem; its importance lies elsewhere.

EXPLORATION 2

The Mean Value Theorem

1. The function $f(x) = x^2$ is continuous on [0, 2] and differentiable on (0, 2). Since $f(0) = 0$ and $f(2) = 4$, the Mean Value Theorem says that there is some point c in (0, 2) for which $f'(c) = (4-0)/(2-0) = 2$. (See Fig. 4.9.) Explain how you could find c graphically and algebraically.

2. Now create a function of your own that is continuous on [a, b] and differentiable on (a, b). You specify the a and b. Identify the value of $f'(c)$. Then find c algebraically or graphically, or both ways if possible. GRAPH simultaneously

$$y_1 = f(x),$$

$$y_2 = \frac{f(b) - f(a)}{b - a}(x - a) + f(a),$$

$$y_3 = f'(c)(x - c) + f(c).$$

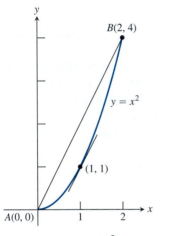

4.9 The function $f(x) = x^2$ over [0, 2] has a tangent at $c = 1$ that is parallel to the secant AB through $A(0, 0)$ and $B(2, 4)$.

Explain what you see in the viewing window.

EXAMPLE 4

The function $y = \sqrt{1 - x^2}$ (Fig. 4.10) satisfies the hypotheses (and conclusions) of Rolle's Theorem and the Mean Value Theorem on the interval $-1 \le x \le 1$. It is continuous on the closed interval, and its derivative

$$y' = \frac{-x}{\sqrt{1 - x^2}}$$

is defined at every interior point. Notice that $y' = 0$ at $x = 0$ where the graph has a horizontal tangent. Notice also that the function is not differentiable at $x = -1$ and $x = 1$. It does not need to be for the theorems to apply.

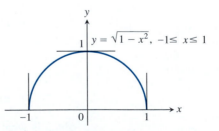

4.10 The function $y = \sqrt{1 - x^2}$ satisfies the hypotheses (and conclusions) of Rolle's Theorem and the Mean Value Theorem on the interval [−1, 1] despite the presence of vertical tangents at $x = 1$ and $x = -1$.

Physical Interpretation

If we think of $(f(b) - f(a))/(b - a)$ as the average change in f over $[a, b]$ and $f'(c)$ as an instantaneous change, then the Mean Value Theorem says that the instantaneous change at some interior point must equal the average change over the entire interval.

EXAMPLE 5

If a car takes 8 sec to drive 352 ft, its average velocity for the 8-sec interval is $352/8 = 44$ ft/sec, or 30 mph. At some point during the acceleration, the theorem says, the speedometer must have read exactly 30.

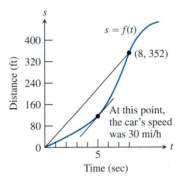

The First Corollary—A Test for Increasing and Decreasing

The Mean Value Theorem is famous for, among other things, three important corollaries. The first, which we will get to in a moment, says exactly when graphs rise and fall. The second, which we will come to in Section 4.7, says that only constant functions can have zero derivatives. The third, also in Section 4.7, says that functions with identical derivatives must differ at most by a constant value.

 ## EXPLORATION 3

Making a Conjecture

GRAPH $y_1 = x^3 - 2x$ and $y_2 = $ NDER y_1 in the same viewing window. Find the portions of the domain over which the graph of y_1 rises. What do you notice about y_2? Do the same for the portion of the domain over which the graph of y_1 falls. Make a conjecture that connects the idea of the increasing and decreasing behavior of a function f with the behavior of the function NDER f? Test your conjecture with another function. ✿

   The conjecture that you formed above, we hope, is the corollary to which we are coming. To understand the corollary, we first need precise definitions of *increasing* and *decreasing*.

DEFINITIONS

A function $f(x)$ defined throughout an interval I is said to **increase** on I if, for any two points x_1 and x_2 in I,

$$x_1 < x_2 \Rightarrow f(x_1) < f(x_2).$$

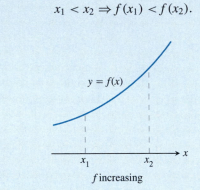

$y = f(x)$

x_1　　x_2

x

f increasing

Similarly, a function $f(x)$ defined throughout an interval I is said to **decrease** on I if, for any two points x_1 and x_2 in I,

$$x_1 < x_2 \Rightarrow f(x_1) > f(x_2).$$

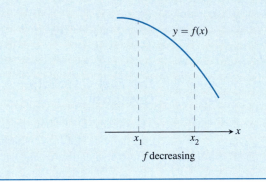

$y = f(x)$

x_1　　x_2

x

f decreasing

COROLLARY 1 **The First Derivative Test for Increasing and Decreasing: f increases when $f' > 0$ and decreases when $f' < 0$.**

Suppose that f is continuous at each point of the closed interval $[a, b]$ and differentiable at each point of its interior (a, b). If $f' > 0$ at each point of (a, b), then f increases throughout $[a, b]$. If $f' < 0$ at each point of (a, b), then f decreases throughout $[a, b]$.

Proof Let x_1 and x_2 be any two numbers in $[a, b]$ with $x_1 < x_2$. Apply the Mean Value Theorem to f on $[x_1, x_2]$:

$$f(x_2) - f(x_1) = f'(c)(x_2 - x_1)$$

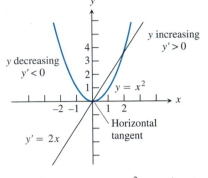

4.11 The graphs of $y = x^2$ and $y' = 2x$.

for some c between x_1 and x_2. The sign of the right-hand side of this equation is the same as the sign of $f'(c)$ because $x_2 - x_1$ is positive. Therefore,

$$f(x_2) > f(x_1) \qquad \text{if } f'(x) \text{ is positive on } (a, b)$$

(f is increasing), and

$$f(x_2) < f(x_1) \qquad \text{if } f'(x) \text{ is negative on } (a, b)$$

(f is decreasing).

EXAMPLE 6

For the function $y = x^2$, the derivative $y' = 2x$ is negative on $(-\infty, 0)$, so y is decreasing on $(-\infty, 0]$. Also, y' is positive on $(0, \infty)$, so y is increasing on $[0, \infty)$. At $x = 0, y' = 0$ and the tangent to the curve is horizontal (Fig. 4.11).

EXAMPLE 7

Let $f(x) = x^3 - 4x$ (Fig. 4.12a). Determine (a) the intervals on which f is increasing and the intervals on which f is decreasing, (b) the local extreme values of f, and (c) the absolute maximum and absolute minimum of f on $[0, 3]$.

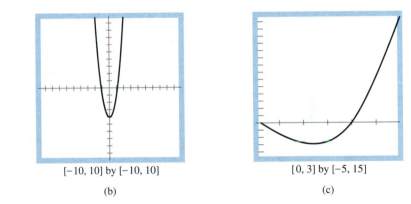

[−10, 10] by [−10, 10]	[−10, 10] by [−10, 10]	[0, 3] by [−5, 15]
(a)	(b)	(c)

4.12 (a) A complete graph of $f(x) = x^3 - 4x$. (b) A complete graph of $f'(x) = 3x^2 - 4$. (c) The graph of $f(x) = x^3 - 4x$ on $[0, 3]$.

Solution

a) The domain of both f and $f'(x) = 3x^2 - 4$ is $(-\infty, \infty)$. Thus, the only possible local extreme values of f occur where $f'(x) = 0$, namely, at $x = \sqrt{4/3}$ and $x = -\sqrt{4/3}$. We can see from the graph of f' in Fig. 4.12(b) that $f'(x) > 0$ for both $x < -\sqrt{4/3} = -1.15$ and $x > \sqrt{4/3} = 1.15$, and $f'(x) < 0$ for $-1.15 < x < 1.15$. Thus, f increases in $(-\infty, -1.15]$, decreases in $[-1.15, 1.15]$, and increases in $[1.15, \infty)$.

b) The increasing–decreasing behavior of f determined in part (a) means that f has a local maximum of 3.08 at $x = -1.15$ and a local minimum of -3.08 at $x = 1.15$. In the next section, we will formalize a procedure for confirming local extrema.

c) The value of f at $x = 0$ is 0, and the value of f at $x = 3$ is 15. So on $[0, 3]$, f has an absolute minimum of -3.08 at $x = 1.15$, a local maximum of 0 at $x = 0$, and an absolute maximum of 15 at $x = 3$ (Fig. 4.12c).

Exercises 4.1

In Exercises 1–8, determine the critical points of the function.

1. $f(x) = x^3 + x^2 - 8x + 5$

2. $f(x) = x^3 - 2x^2 - 15x + 2$

3. $F(x) = \sqrt[3]{x}, -1 \le x \le 8$

4. $F(x) = \sin x, -\frac{\pi}{2} \le x \le \frac{5\pi}{6}$

5. $g(x) = \sqrt{3 + 2x - x^2}$

6. $g(x) = \sqrt{5 - 4x - x^2}$

7. $h(x) = \begin{cases} x^2, & x \le 0 \\ 2x, & x > 0 \end{cases}$

8. $h(x) = \begin{cases} -\dfrac{x^2}{2}, & x < 0 \\ \dfrac{x^2}{2}, & 0 \le x < 2 \\ 2x - 2, & x \ge 2 \end{cases}$

9. a) Draw a complete graph of each polynomial function. On a number line, mark the zeros of the polynomial together with the zeros of its derivative.

i) $y = x^2 + 8x + 15$

ii) $y = x^3 - 3x^2 + 4$

iii) $y = x^3 - 33x^2 + 216x$

iv) $y = -x^3 + 4x - 2$

On the basis of these four trials, make a conjecture about how the zeros of a polynomial and the zeros of its derivative are related. Do you believe the conjecture to be true? Give a convincing argument.

b) Use Rolle's Theorem to prove that between every two zeros of the polynomial

$$x^n + a_{n-1}x^{n-1} + \cdots + a_1 x + a_0,$$

there lies a zero of the polynomial

$$nx^{n-1} + (n-1)a_{n-1}x^{n-2} + \cdots + a_1.$$

10. The function

$$y = f(x) = \begin{cases} x & \text{if } 0 \le x < 1, \\ 0 & \text{if } x = 1, \end{cases}$$

is zero at $x = 0$ and at $x = 1$. Its derivative at every point between 0 and 1 is $y' = 1$, and so y' is never zero between 0 and 1. Why doesn't that contradict Rolle's Theorem?

11. a) Draw a complete graph of $y = \sin x$. On a number line, mark its zeros together with the zeros of its derivative.

b) Use Rolle's Theorem to prove that between every two zeros of $y = \sin x$, there is a zero of $y = \cos x$.

12. a) Draw a complete graph of $y = (x^2 - 4x + 2)/(x + 1)$. On a number line, mark its zeros together with the zeros of its derivative.

b) Can you use Rolle's Theorem to prove the existence of each zero of the derivative of y?

13. Let $f(x) = |x^3 - 9x|$.

a) Determine a complete graph of f.

b) Does $f'(0)$ exist? Explain.

c) Does $f'(-3)$ exist? Explain.

d) Determine all local extrema of f.

e) Are parts (b)–(d) in conflict with Theorem 1? Explain.

f) Determine the intervals on which f is increasing and the intervals on which f is decreasing.

14. Let $g(x) = (x - 2)^{2/3}$.

a) Determine a complete graph of g.

b) Does $g'(2)$ exist? Explain.

c) Determine all local extrema of g.

d) Are parts (b) and (c) in conflict with Theorem 1? Explain.

e) Determine the intervals on which g is increasing and the intervals on which g is decreasing.

In Exercises 15–22, determine the local extrema of the function, the intervals on which the function is increasing and the intervals on which the function is decreasing, and the absolute maximum and absolute minimum of the function on the specified interval.

15. $f(x) = 5x - x^2$, $[0, 6]$

16. $f(x) = x^2 - x - 12$, $[-4, 4]$

17. $f(x) = \sqrt{x - 4}$, $[4, 8]$

18. $f(x) = 4 - \sqrt{x + 2}$, $[-2, 5]$

19. $y(x) = x^4 - 10x^2 + 9$, $[-3, 3]$

20. $g(x) = -x^4 + 5x^2 - 4$, $[-3, 3]$

21. $h(x) = x^3 - 2x - 2\cos x$, $[-5, 5]$

22. $h(x) = \sin x + 2\cos 2x$, $[0, 2\pi]$

The Mean Value Theorem

In Exercises 23–26, find the value(s) of c that satisfies the equation

$$\frac{f(b) - f(a)}{b - a} = f'(c)$$

in the conclusion of the Mean Value Theorem for the functions and intervals shown.

23. $f(x) = x^2 + 2x - 1, \quad 0 \le x \le 1$

24. $f(x) = x^{2/3}, \quad 0 \le x \le 1$

25. $f(x) = x + \dfrac{1}{x}, \quad \dfrac{1}{2} \le x \le 2$

26. $f(x) = \sqrt{x - 1}, \quad 1 \le x \le 3$

27. *Speeding.* A trucker handed in a ticket at a toll booth showing that in 2 hours the truck had covered 159 mi on a toll road on which the speed limit was 65 mph. The trucker was cited for speeding. Why?

28. *Temperature change.* It took 20 sec for a thermometer to rise from $10°F$ to $212°F$ when it was taken from a freezer and placed in boiling water. Explain why somewhere along the way the mercury was rising at exactly $10.1°F$/sec.

29. *Triremes.* Classical accounts tell us that a 170-oar trireme (ancient Greek or Roman warship) once covered 184 sea miles in 24 hours. Explain why at some point during this feat the trireme's speed exceeded 7.5 knots.

30. *Running the marathon.* A marathoner ran the 26.2-mi New York City Marathon in 2.2 hours. Show that at least twice, the marathoner was running at exactly 11 mph.

31. Sketch the graph of a differentiable function with a local minimum value that is greater than one of its local maximum values.

32. Suppose that $y = f(x)$ is continuous on $[a, b]$, differentiable on (a, b) and $f(a) = f(b) = k$. Use Rolle's Theorem to prove there is at least one number c between a and b at which $f'(c) = 0$.

For the functions and intervals specified in Exercises 33–36, find the value(s) of c that satisfy the extension of Rolle's Theorem given in Exercise 32.

33. $f(x) = x \sin x, [-4, 4]$ **34.** $f(x) = x^2 \cos x, [-5, 5]$

35. $f(x) = x^2 - 2x, [-2, 4]$ **36.** $f(x) = 3x - x^2, [0.5, 2.5]$

37. Show that $y = 1/x$ decreases on any interval on which it is defined.

38. Show that $y = 1/x^2$ increases on any interval to the left of the origin and decreases on any interval to the right of the origin.

39. Suppose the derivative of a differentiable function $f(x)$ is never zero on the interval $0 \le x \le 1$. Show that $f(0) \ne f(1)$.

40. Show that for any numbers a and b

$$|\sin b - \sin a| \le |b - a|.$$

41. Suppose that f is differentiable for $a \le x \le b$ and that $f(b) < f(a)$. Show that f' is negative at some point between a and b.

42. Suppose that a function f is continuous on $[a, b]$ and differentiable on (a, b). Suppose also that $f(a)$ and $f(b)$ have opposite signs and $f' \ne 0$ between a and b. Show that f has exactly one zero between a and b.

Use Exercise 42 to show analytically that the equations in Exercises 43-46 have exactly one solution in the given interval.

43. $x^4 + 3x + 1 = 0, \quad -2 \le x \le -1$

44. $-x^3 - 3x + 1 = 0, \quad 0 \le x \le 1$

45. $x - \dfrac{2}{x} = 0, \quad 1 \le x \le 3$

46. $2x - \cos x = 0, \quad -\pi \le x \le \pi$

47. Suppose that $f(0) = 3$ and that $f'(x) = 0$ for all x. Use the Mean Value Theorem to show that $f(x)$ must be 3 for all x.

48. Suppose that $f'(x) = 2$ and that $f(0) = 5$. Use the Mean Value Theorem to show that $f(x) = 2x + 5$ at every value of x.

For the functions $y = f(x)$ graphed in Exercises 49 and 50, estimate where f' is (a) positive, (b) negative, and (c) zero.

49.

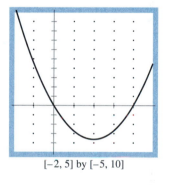

$[-2, 5]$ by $[-5, 10]$

50.

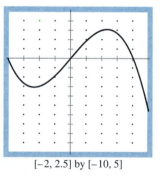

$[-2, 2.5]$ by $[-10, 5]$

51. Find a function and real numbers a and b $(a < b)$ so that f is continuous on $[a, b]$, differentiable on (a, b), *not* differentiable on $[a, b]$, and has a tangent line at every point of $[a, b]$.

52. Suppose the graph of a differentiable function $y = f(x)$ passes through the point $(1, 1)$. Sketch the graph if
a) $f'(x) > 0$ for $x < 1$ and $f'(x) < 0$ for $x > 1$.
b) $f'(x) < 0$ for $x < 1$ and $f'(x) > 0$ for $x > 1$.
c) $f'(x) > 0$ for $x \ne 1$.
d) $f'(x) < 0$ for $x \ne 1$.

53. Sketch one possible graph of $y = g(x)$ if $g(-2) = 1$ and $g'(x)$ is the function graphed in Exercise 49.

54. Sketch one possible graph of $y = g(x)$ if $g(-2) = 1$ and $g'(x)$ is the function graphed in Exercise 50.

55. *Challenge.*   Consider the graph of $y = f(x) = xe^{-x}$ shown here. For each $a > 0$, think of the rectangle drawn in the first quadrant under the graph of f as suggested.

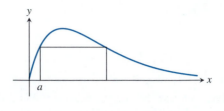

a) Determine a where the rectangle has area zero.

b) Compute the area of the rectangle for $a = 0.5, 0.8, 1.0, 1.2, 1.5$.

c) Determine a so that the rectangle has maximum area. What is the area? This may require nonstandard numerical techniques.

4.2   Predicting Hidden Behavior

Computer-drawn graphs of most functions that appear in this textbook are usually very reliable. In this section we will see how to use calculus to confirm completeness of graphs determined technologically and to predict behavior that is hidden from view on a computer graph. The verification that the graph really looks like what's on the screen and an analysis of any hidden behavior must come from calculus. The computer can only suggest what *might* be true.

We also introduce the concepts of concavity of graphs and of points where the sense of concavity changes, commonly called points of inflection of graphs. We will see that points of inflection are easy to locate by using a graphing utility even though they may be difficult to *see* in a viewing window because of the pixel nature of the graphs and the local straightness which shows if we ZOOM-IN.

The First Derivative

When we know that a function has a derivative at every point of an interval, we also know that it is continuous throughout the interval (Section 3.1) and that its graph over the interval is connected (Section 2.2). Thus, the graphs of $y = \sin x$ and $y = \cos x$ remain unbroken however far extended, as do the graphs of polynomials. The graphs of $y = \tan x$ and $y = 1/x^2$ are not connected only at points where the functions are undefined. On every interval that avoids these points, the functions are differentiable, so they are continuous and have connected graphs.

We gain additional information about the shape of a function's graph when we know where the function's first derivative is positive, negative, or zero. For, as we saw in Section 4.1, this tells us where the graph is rising, falling, or possibly has a horizontal tangent (Fig. 4.13.).

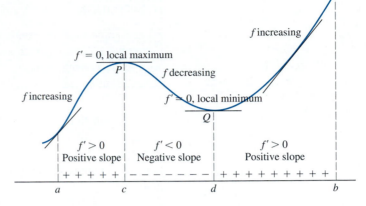

4.13 The function $y = f(x)$ increases on (a, c) where $f' > 0$, decreases on (c, d) where $f' < 0$, and increases again on (d, b). The transitions are marked on the graph by horizontal tangents.

EXPLORATION BIT

Give the view dimensions of a window that hides some behavior of the function. Identify a small interval in which the hidden behavior occurs. Describe the behavior that you find when you ZOOM-IN.

1. $y = 3x^3 - 5x^2 + 2x - 20$

2. $y = x^2 + 1/(100 - 150x)$

3. $y = 2^{\sin(1/x)}$

There are two things to watch out for here. A function may have a local maximum or minimum without its graph having a horizontal tangent (Fig. 4.14), and a graph may have a horizontal tangent without its function having a local maximum or minimum (Fig. 4.15a).

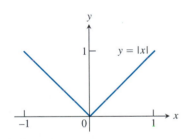

4.14 The graph of $y = |x|$ shows a minimum value for the function at $x = 0$ without having a horizontal tangent there.

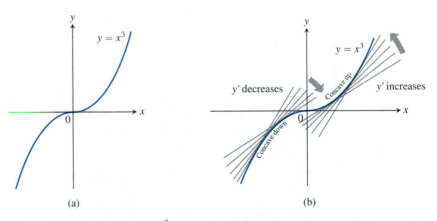

(a)   (b)

4.15 (a) The graph of $y = x^3$ has a horizontal tangent at the origin, but the function has no maximum or minimum there. (b) As x increases, the slopes of the tangents to the curve $y = x^3$ on $(-\infty, 0)$ are positive and decreasing; the slopes of the tangents to the curve on $(0, \infty)$ are positive and increasing.

The Second Derivative Test for Concavity

As you can see in Fig. 4.15(b), the function $f(x) = x^3$ increases as x increases, but portions of the curve $y = x^3$ increase in different ways. Looking at tangents as we scan from left to right, we see that the slope of the curve decreases on the interval $(-\infty, 0)$ and then increases on the interval $(0, \infty)$. We say that the curve $y = x^3$ is *concave down* on $(-\infty, 0)$ and *concave up* on $(0, \infty)$.

DEFINITION

The graph of a differentiable function $y = f(x)$ is **concave up** on an interval where y' is increasing and **concave down** on an interval where y' is decreasing.

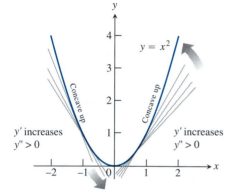

4.16 The graph of $y = x^2$. As x increases, the slopes of the tangents to the curve on $(-\infty, 0)$ are negative and increasing; the slopes of the tangents to the curve on $(0, \infty)$ are positive and increasing.

If a function $y = f(x)$ has a second derivative as well as a first, we can apply Corollary 1 of the Mean Value Theorem (Section 4.1) to conclude that y' decreases if $y'' < 0$ and increases if $y'' > 0$. We therefore have a test that we can apply to the formula $y = f(x)$ to determine the concavity of its graph.

The Second Derivative Test for Concavity

The graph of a twice-differentiable function $y = f(x)$ is

concave down on any interval where $y'' < 0$,

concave up on any interval where $y'' > 0$.

EXAMPLE 1

The curve $y = x^2$ (Fig. 4.16) is concave up on the entire x-axis because its second derivative $y'' = 2$ is always positive. ≡

EXAMPLE 2

The curve $y = x^3$ in Fig. 4.15 is concave down on $(-\infty, 0)$, where its second derivative $y'' = 6x$ is negative. It is concave up on $(0, \infty)$, where its second derivative $y'' = 6x$ is positive. ≡

THE GRAPHER ADVANTAGE

As you do Exploration 1, you will find that you have plenty of time to "follow the action" in the viewing window. But it is still much faster than if you had to sketch the graphs with pencil and paper and much more interesting. And think of the incredible amount of calculation going on that you don't have to do! If you don't watch while the grapher computes, keep in mind that you can always distinguish y_1 from y_3 by using TRACE.

EXPLORATION 1

Concavity

1. Enter $y_1 = x^2$, $y_2 = \text{NDER}\, y_1$, and $y_3 = \text{NDER}\, y_2$. GRAPH y_1 and y_3 simultaneously. Watch that y_1 remains concave up wherever y_3 is above the x-axis. (See Example 1.)

2. Repeat part 1 with $y_1 = x^3$. Watch how the concavity of y_1 changes as y_3 crosses the x-axis. (See Example 2.)

3. Repeat part 1 with a complete graph of $y_1 = 2x^5 - 9x^4 - 52x^3 + 141x^2 + 230x - 312$. (For your choice of xMin and xMax, what clue does the first term $2x^5$ give you about a first choice for yMin and yMax?)

   a) How many times does the concavity change?

   b) Make a conjecture and give a convincing argument about the number of times concavity changes in the graph of an nth-degree polynomial.

c) ZOOM-IN to check whether concavity is changing at integer values of x. Can you make a conjecture?

4. Explore and discuss the concavity of $y = \sin x$.

Points of Inflection

When we graph the position of a moving body as a function of time, we keep in mind that the first derivative is velocity and the second derivative is acceleration. The concavity of the position function changes when the second derivative changes sign, that is, when acceleration changes to be in the opposite direction. The moving body begins speeding up or slowing down at such *points of inflection*.

Points of inflection are also important in business applications. The growth of an individual company, population or a new product often follows a *logistic* or *life cycle* curve like the one shown in Fig. 4.17. For example, sales of a new product will generally grow slowly at first, then the sales will experience a rapid growth. Eventually, the growth of sales slows down again. The function f in Fig. 4.17 is increasing. Its rate of increase, f', is at first increasing ($f'' > 0$) up to the point of inflection, and then its rate of increase, f', is decreasing ($f'' < 0$). This is, in a sense, the opposite of what happens in Fig. 4.15(b).

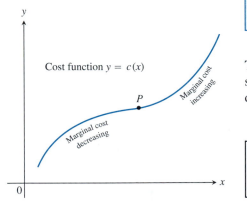

4.17 A logistic curve.

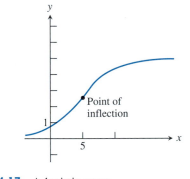

4.18 A point of inflection shows where the derivative reaches a maximum or minimum. On a typical cost curve, the point of inflection separates the interval of decreasing marginal cost from the interval of increasing marginal cost. This is the point where the marginal cost is least. For a position function, a point of inflection shows where velocity has peaked or has reached a minimum.

> **DEFINITION**
>
> A point where the graph of a function has a tangent line and where the concavity changes is called a **point of inflection.**

Thus a point of inflection on a curve is a point where y'' is positive on one side and negative on the other. At such a point, y'' is either zero (because derivatives have the intermediate value property) or undefined.

> At a point of inflection on the graph of a twice-differentiable function, $y'' = 0$.

Inflection points have important applications in some areas of economics.

EXAMPLE 3

Suppose that the function $y = c(x)$ in Fig. 4.18 is the total cost of producing x units of something. The point of inflection at P is then the point at which the marginal cost (the approximate cost of producing one more unit) changes from decreasing to increasing. ≡

EXAMPLE 4

The graph of the simple harmonic motion $y = \sin t$ shown here changes concavity at $t = 0$ and $t = \pi$, where the acceleration $y'' = -\sin t$ is zero. (To review simple harmonic motion see Section 3.5.)

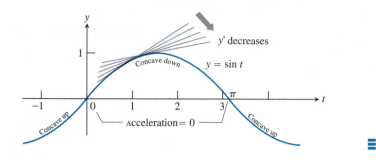

Identifying points of inflection for $y = f(x)$ visually or ZOOMING-IN on one in a viewing window can be difficult (see Exercise 56). Generally, however, there is a way to use a graphing utility to find such points quite easily. We simply find solutions for $y'' = f''(x) = 0$ or NDER2 $f(x) = 0$ using ZOOM-IN or SOLVE. We must keep in mind that y'' could be zero at a point that is not a point of inflection and that a point of inflection may occur where y'' fails to exist.

EXAMPLE 5   No inflection where $y'' = 0$

The curve $y = x^4$ (Fig. 4.19) has no inflection point at $x = 0$. Even though $y'' = 12x^2$ is zero there, y'' does not change sign.

EXAMPLE 6   An inflection point where y'' does not exist.

The curve $y = x^{1/3}$ (Fig. 4.20) has a point of inflection at $x = 0$, but y'' does not exist there. The formulas for y' and y'' are

$$y' = \frac{1}{3}x^{-2/3}, \qquad y'' = -\frac{2}{9}x^{-5/3}.$$

The curve is concave up for $x < 0$, where $y'' > 0$ and y' is increasing. It is concave down for $x > 0$, where $y'' < 0$ and y' is decreasing.

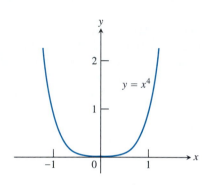

4.19 The graph of $y = x^4$ has no inflection point at the origin, even though $y''(0) = 0$.

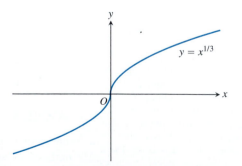

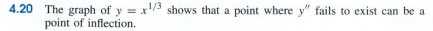

4.20 The graph of $y = x^{1/3}$ shows that a point where y'' fails to exist can be a point of inflection.

Confirming Computer-generated Graphs

In Chapter 1, we found hidden behavior in our graph of $f(x) = x^3 - 2x^2 + x - 30$ (Fig. 4.21a). This is supported by the graph of $f'(x) = 3x^2 - 4x + 1$ (Fig. 4.21b). Notice that $f'(x) = (3x - 1)(x - 1)$ is positive over $(-\infty, 1/3) \cup (1, \infty)$, negative on $(1/3, 1)$, and zero at $x = 1/3$ and $x = 1$. This *sign pattern* for f' (Fig. 4.22) tells us that the curve for f rises as x approaches 1/3 from the left to a local maximum at $x = 1/3$, then falls to a local minimum at $x = 1$.

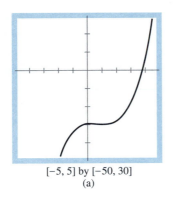
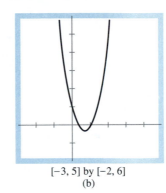
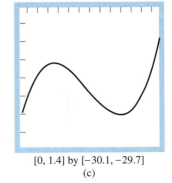

$[-5, 5]$ by $[-50, 30]$ $[-3, 5]$ by $[-2, 6]$ $[0, 1.4]$ by $[-30.1, -29.7]$
(a)   (b)   (c)

4.21 (a) A view of $f(x) = x^3 - 2x^2 + x - 30$ that suggests hidden behavior where the graph appears flat. (b) A graph of $f'(x) = 3x^2 - 4x + 1$ shows f' is negative on $(1/3, 1)$ to support our hunch of hidden behavior. (c) A magnified view of f showing the hidden behavior.

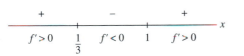

4.22 The sign pattern for

$$f'(x) = 3x^2 - 4x + 1$$
$$= (3x - 1)(x - 1)$$

from Fig. 4.21(b).

A close-up view (Fig. 4.21c) reveals the hidden behavior.

The First Derivative Test for Local Extreme Values

Determining the signs and the zeros of f' helps to identify local extreme values of f.

The First Derivative Test for Local Extreme Values

If $f'(c) = 0$ and the sign of f' changes at c, then f has a local extreme value at $x = c$:

f has a local maximum if f' changes from $+$ to $-$.

f has a local minimum if f' changes from $-$ to $+$.

Now we summarize how the derivative can help us to confirm that a graph generated by computer or sketched by hand is complete.

A Complete Graph of a Function f; Applying the Derivative

To determine a complete graph of a function f, we establish its domain and intercepts, its intervals of continuity and differentiability, its points of discontinuity and nondifferentiability, and its end behavior. Then, to use derivatives, we do the following:

1. Find f' and f''.
2. Find where f' is positive, negative, and zero to confirm where f is increasing and decreasing and has extreme values.
3. Find where f'' is positive, negative, and zero to confirm where f is concave up and concave down and has points of inflection.
4. Compare the information found above with our graph (viewing window or sketch), and resolve conflicting information.

The order, difficulty, and importance of these tasks will vary from function to function and depend on what you really want to know about the function.

EXPLORATION 2

Supporting and Confirming

Themes throughout this book include the following:

Do algebraically and support graphically or numerically.

Do graphically or numerically and confirm algebraically.

Do using a method of your choice (often involving a combination of algebraic, numerical, and graphical steps).

The power of these themes, particularly the third one, is wonderfully illustrated in the way they can be used to create and confirm a complete graph. As you do the following, pay attention to how you use—hand-in-hand—your grapher skills to support and your mathematical skills to confirm.

1. Draw and confirm a complete graph of

$$f(x) = x^3 - 2x^2 - 7x + 2.$$

Carefully list all the information you find about the function, including f', f'', and the information they provide. Make a *large* sketch of the graph (you can copy from your grapher), and mark the sketch to illustrate the known information. Be prepared to defend the information you show.

2. Draw and confirm a complete graph of
   a) $f(x) = -20x^3 + 273x^2 - 1242x + 1883.$
   b) $f(x) = x^4 - 6x^2 + 8x + 1.$

The mathematics for confirming a complete graph takes on new meaning for the position function of a particle moving along a line in the coordinate plane. (Look again at Exploration 2, Section 3.4.)

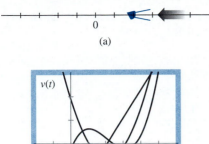

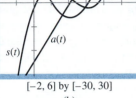

[−2, 6] by [−30, 30]

(b)

4.23 (a) A particle moving on a number line and (b) graphs of the functions that describe its motion.

Distance from origin:

$$s(t) = 2t^3 - 13t^2 + 22t - 5.$$

Velocity: $v(t) = 6t^2 - 26t + 22$.
Acceleration: $a(t) = 12t - 26$.
The particle reaches an extreme distance from the origin at $t = 1.15$, where it stops ($v = 0$) and begins traveling to the left. Figure (a) shows the particle at maximum speed heading to the left, at a distance 1.98 from the origin. Where is this information in Figure (b)?

Particle motion

Moving right	Moving left	Moving right
+	−	+

$v > 0$ 1.15 $v < 0$ 3.18 $v > 0$ → x

v decreasing v increasing
| − | + |

$a < 0$ $\dfrac{13}{6}$ $a > 0$ → x

4.24 The particle of Example 7 moves to the right, slowing down until it stops when $t = 1.15$. Then it moves to the left, first speeding up until $t = 13/6$, then slowing down until it stops when $t = 3.18$. For $t > 3.18$, it moves to the right, speeding up.

EXAMPLE 7

A particle is moving on the x-axis (or any number line). Its position $x(t)$, or distance from the origin $s(t)$, at time t is given by

$$x(t) = s(t) = 2t^3 - 13t^2 + 22t - 5.$$

Analyze the motion of the particle.

Solution To simulate the motion on a horizontal line see Exploration 2, Section 3.4. The velocity and acceleration of the particle are given by

$$v(t) = s'(t) = 6t^2 - 26t + 22$$

$$a(t) = v'(t) = s''(t) = 12t - 26.$$

The moving particle is suggested in Fig. 4.23(a), and complete graphs of $s(t)$, $v(t)$, and $a(t)$ are shown in Fig. 4.23(b). Refer to these graphs as you continue.

$a(t) = 0$ for $t = 13/6$. For $t < 13/6$, $a(t) = v'(t)$ is negative, and $v(t)$ decreases. For $t > 13/6$, $a(t) = v'(t)$ is positive, and $v(t)$ increases. Because $v(t) = 0$ for $t = 1.15$ and 3.18, $v(t) = s'(t)$ is positive for t outside of (1.15, 3.18) and negative for t within (1.15, 3.18).

It follows that $s(t)$ is increasing for t outside of (1.15, 3.18) and decreasing for t within (1.15, 3.18). This means that the particle is moving to the right for t in $(-\infty, 1.15)$, to the left for t in (1.15, 3.18), and then to the right again for t in $(3.18, \infty)$.

We can summarize the motion of the particle with a diagram (Fig. 4.24) or with a table as follows:

Time	a(t)	v(t)	Speed	Direction
$t < 1.15$	Negative	Positive, decreasing	Slowing	To the right
$t = 1.15$	Negative	0	0	Stopped
$1.15 < t < 13/6$	Negative	Negative, decreasing	Gaining	To the left
$t = 13/6$	0	Minimum	Maximum	To the left
$13/6 < t < 3.18$	Positive	Negative, increasing	Slowing	To the left
$t = 3.18$	Positive	0	0	Stopped
$t > 3.18$	Positive	Positive, increasing	Gaining	To the right

What causes the change in v? Answer: The acceleration. When acceleration is negative, it acts "against" positive v to slow down the particle, bring it to a stop, and then start it moving in the opposite direction. When acceleration becomes positive, it acts against the negative movement to bring the particle to a halt again and then start it moving again to the right.

If we encounter a function whose derivative is unknown or is very complicated, we can use NDER f and NDER2 f.

EXAMPLE 8

Analyze graphically the logistic curve

$$f(x) = \frac{1}{0.2 + 2^{-0.5x}}.$$

Solution   The graphs of f, f' (NDER f), and f'' (NDER2 f) are shown in Fig. 4.25. We see that f' is always positive, which means that f is always increasing. We see that the rate of increase of f, namely, f', increases to a maximum and then decreases, a characteristic of logistic curves. We see that f'' changes sign at $x = 4.64$, so there is a point of inflection at 4.64, and we can find it to be (4.64, 2.50).

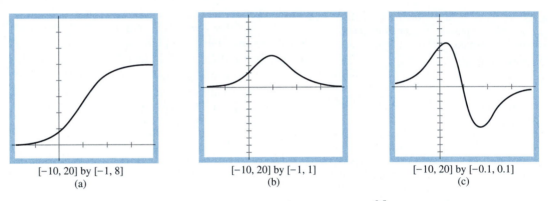

| $[-10, 20]$ by $[-1, 8]$ | $[-10, 20]$ by $[-1, 1]$ | $[-10, 20]$ by $[-0.1, 0.1]$ |
| (a) | (b) | (c) |

4.25   Graphs of (a) $y_1 = f(x) = 1/(0.2 + 2^{-0.5x})$, (b) $y_2 = $ NDER y_1, and (c) $y_3 = $ NDER2 y_1. Ways to analytically *calculate* derivatives of exponential functions will be established later in the text.   ≡

The Second Derivative Test for Local Extreme Values

Instead of looking at how the sign of f' changes at a point where $f' = 0$, we can often use the following test to determine whether there is a local maximum or minimum at the point.

The Second Derivative Test for Local Extreme Values

If $f'(c) = 0$ and $f''(c) < 0$, then f has a local maximum at $x = c$.

If $f'(c) = 0$ and $f''(c) > 0$, then f has a local minimum at $x = c$.

Notice that the test requires us to know f'' only at c itself and not in an interval about c. This makes the test easy to apply. That's the good news. The bad news is that the test fails if $f''(c) = 0$, if $f''(c)$ fails to exist, or if $f''(c)$ is hard to find both analytically *and* numerically.

EXPLORATION 3

When the Second Derivative Test Does Not Apply

Show that the Second Derivative Test may not be applied when $f''(c) = 0$:
Find examples of functions f and points c for which $f'(c) = 0 = f''(c)$, but
for which

a) f has a maximum value at c,

b) f has a minimum value at c, or

c) f has neither a maximum nor a minimum value at c.

For each of your examples, explain how the First Derivative Test for Extreme
Values can be applied instead.   &

EXAMPLE 9

Find all maxima and minima of the following function analytically:

$$y = x^3 - 3x + 2.$$

Support graphically.

Solution The function is differentiable at every point of its domain
$(-\infty, \infty)$, so there are no endpoints to consider. Therefore, extreme values
occur only where the first derivative,

$$y' = 3x^2 - 3 = 3(x - 1)(x + 1),$$

equals zero. The critical points are $x = 1$ and $x = -1$. The second derivative,

$$y'' = 6x,$$

is positive at $x = 1$ and negative at $x = -1$. Hence, $y(1) = 0$ is a local
minimum value, and $y(-1) = 4$ is a local maximum value (Fig. 4.26.)

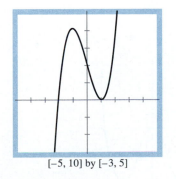

[−5, 10] by [−3, 5]

4.26 A complete graph of $y = x^3 - 3x + 2$.

≡

Summary: What Derivatives Confirm About Functions and Graphs

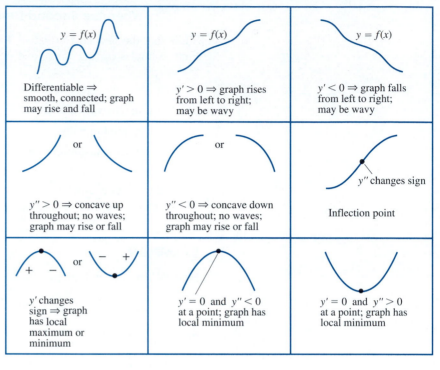

$y = f(x)$	$y = f(x)$	$y = f(x)$
Differentiable ⟹ smooth, connected; graph may rise and fall	$y' > 0$ ⟹ graph rises from left to right; may be wavy	$y' < 0$ ⟹ graph falls from left to right; may be wavy
or	or	y'' changes sign
$y'' > 0$ ⟹ concave up throughout; no waves; graph may rise or fall	$y'' < 0$ ⟹ concave down throughout; no waves; graph may rise or fall	Inflection point
or		
y' changes sign ⟹ graph has local maximum or minimum	$y' = 0$ and $y'' < 0$ at a point; graph has local minimum	$y' = 0$ and $y'' > 0$ at a point; graph has local minimum

Exercises 4.2

For each function f graphed in Exercises 1 and 2, identify where the derivative f' is 0, positive, and negative.

1.

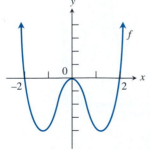

For each function f whose derivative is graphed in Exercises 3 and 4, identify the intervals on which the graph of f is rising or falling. Identify the local extreme values of f.

3.

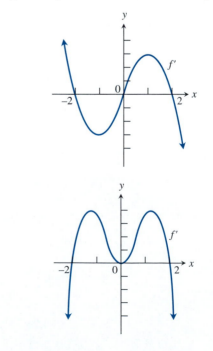

2.

4.

In Exercises 5–32, show a complete graph and identify the inflection points, local maximum and minimum values, and the intervals on which the graph is rising, falling, concave up, and concave down. We suggest you do exercises from the three groups in the order given.

Do Exercises 5–10 analytically, then support graphically.

5. $y = x^2 - x - 1$ **6.** $y = 4x^2 + 8x + 1$
7. $y = x^3 - 6x^2 + 9x + 1$ **8.** $y = -2x^3 + 6x^2 - 3$
9. $y = 2x^4 - 4x^2 + 1$ **10.** $y = x^4 - 2x^2$

Do Exercises 11–16 graphically, then confirm analytically.

11. $y = 2x^3 - 5x^2 + 4x + 10$

12. $y = 4x^3 + 21x^2 + 36x - 20$

13. $y = 3x^4 - x^2 - 10$

14. $y = 20 + 2x^2 - 9x^4$

15. $y = x + \sin x, \quad 0 \le x \le 2\pi$

16. $y = x - \sin x, \quad 0 \le x \le 2\pi$

Do Exercises 17–32 using a method of your choice.

17. $y = x^4 - 8x^2 + 4x + 2$

18. $y = -x^4 + 4x^3 - 4x + 1$

19. $y = -x^4 + 2x^2 - 3x - 2$

20. $y = 2x^4 - x^2 - 3x + 5$

21. $y = 2x^{1/5} + 3$ **22.** $y = 5 - x^{1/3}$

23. $y = 3x^{1/2} - 1$ **24.** $y = 2x^{1/4}$

25. $y = \dfrac{5}{1 + 2^{1-0.5x}}$ **26.** $y = \dfrac{-7}{1 + 3^{1-0.5x}}$

27. $f(x) = \begin{cases} x^2 + 1, & x \ge 0 \\ 3 - x^2, & x < 0 \end{cases}$

28. $f(x) = \begin{cases} 2 - x^2, & x \ge 1 \\ 2x, & x < 1 \end{cases}$

29. $y = x^{2/3}(3 - x)$ **30.** $y = x^{3/4}(5 - x)$

31. $y = x^{1/3}(x - 4)$ **32.** $y = x^{1/4}(x + 3)$

Analyze the motion of the particle moving on the x-axis with distance from the origin given by each function in Exercises 33–36. Support the analysis by graphing the parametric equations

$$x_1(t) = s(t), \quad y_1(t) = 2, \text{ and}$$

$$x_2(t) = s(t), \quad y_2(t) = t,$$

and using TRACE.

33. $x(t) = s(t) = t^2 - 4t + 3$ **34.** $x(t) = s(t) = 6 - 2t - t^2$
35. $x(t) = s(t) = t^3 - 3t + 3$ **36.** $x(t) = s(t) = 3t^2 - 2t^3$

In Exercises 37 and 38, the derivative of the function $y = f(x)$ is given. At what points, if any, does the graph of f have a local minimum, local maximum, or point of inflection?

37. $y' = (x - 1)^2(x - 2)$

38. $y' = (x - 1)^2(x - 2)(x - 4)$

Find the local maximum and minimum values of the functions in Exercises 39 and 40.

39. $y = x + \dfrac{1}{x}$ **40.** $y = \dfrac{x}{2} + \dfrac{1}{2x - 1}$

41. If $f(x)$ is a differentiable function and $f'(c) = 0$ at an interior point c of f's domain, must f have a local maximum or minimum at $x = c$? Explain.

42. If $f(x)$ is a twice-differentiable function and $f''(c) = 0$ at an interior point c of f's domain, must the graph of f have an inflection point at $x = c$? Explain.

43. *Quadratic curves.* True or false? A quadratic curve $y = ax^2 + bx + c$ never has an inflection point. Explain.

44. *Cubic curves.* True or false? A cubic curve $y = ax^3 + bx^2 + cx + d$, $a \ne 0$, always has one inflection point. Explain.

Velocity and Acceleration

Each graph in Exercises 45 and 46 is the graph of the position function $y = s(t)$ of a body moving back and forth on a coordinate line. At approximately what times is each body's (a) velocity equal to zero? (b) acceleration equal to zero?

45.

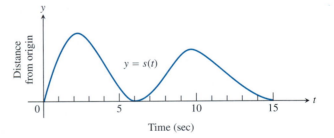

46.

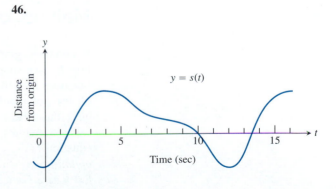

47. Sketch a smooth curve $y = f(x)$ through the origin with the properties that $f'(x) < 0$ for $x < 0$ and $f'(x) > 0$ for $x > 0$.

48. Sketch a smooth curve $y = f(x)$ through the origin with the properties that $f''(x) < 0$ for $x < 0$ and $f''(x) > 0$ for $x > 0$.

49. Sketch a continuous curve $y = f(x)$ having the following characteristics:

$f(-2) = 8,$ $\qquad f'(2) = f'(-2) = 0,$
$f(0) = 4,$ $\qquad f'(x) < 0$ for $|x| < 2,$
$f(2) = 0,$ $\qquad f''(x) < 0$ for $x < 0,$
$f'(x) > 0$ for $|x| > 2,$ $\quad f''(x) > 0$ for $x > 0.$

50. Sketch a continuous curve $y = f(x)$ with the following properties. Label coordinates where possible.

x	y	Curve
$x < 2$		Falling, concave up
2	1	Horizontal tangent
$2 < x < 4$		Rising, concave up
4	4	Inflection point
$4 < x < 6$		Rising, concave down
6	7	Horizontal tangent
$x > 6$		Falling, concave down

Linearizations at Inflection Points

Linearizations fit particularly well at points of inflection. You will see what we mean if you graph the pairs of functions in Exercises 51–54.

51. $f(x) = \sin x$ and its linearization $L(x) = x$ at $x = 0$.

52. $f(x) = \sin x$ and its linearization $L(x) = -x + \pi$ at $x = \pi$.

53. *Newton's serpentine.* $f(x) = 4x/x^2 + 1$ and its linearization $L(x) = 4x$ at $x = 0$.

54. *Newton's serpentine.* $f(x) = 4x/(x^2 + 1)$ and its linearization $L(x) = -x/2 + 3\sqrt{3}/2$ at the point $(\sqrt{3}, \sqrt{3})$.

55. Let $y_1 = f(x) = x^3 - 9x$.

a) Show that f has a point of inflection at $(0, 0)$.

b) Determine m and b so that $y_2 = mx + b$ is the equation of the tangent line to the graph of f at $(a, f(a))$.

c) Let $a = 1$. GRAPH y_1 and y_2 simultaneously and ZOOM-IN on the point $(1, f(1))$ in the graph of f. Is the graph of y_1 above or below the graph of y_2? Explain.

d) Repeat part (c) with $a = 2, 3, -1, -2, -3$.

56. The function $f(x) = x^3 - 2x - 2\cos x$ has a point of inflection in $-1 < x < 0$.

a) Try to find the coordinates of this point of inflection by using ZOOM-IN. Explain why this procedure does not work.

b) Find the coordinates of the point of inflection by using $y = $ NDER2 f, and ZOOM-IN or SOLVE.

c) *Group Discussion.* Can you *confirm* part (b) analytically? Explain.

A ball is hit at an angle of elevation α and with initial velocity v_0. Neglecting air resistance, the position of the ball in Cartesian coordinates at time t is

$$(x(t), y(t)) = (v_0 t \cos \alpha, v_0 t \sin \alpha - 16t^2).$$

57. Use your grapher in parametric mode to simulate the motion of the ball hit with $v_0 = 90$ ft/sec and $\alpha = 40°, 50°, 60°, 75°, 80°, 85°$.

58. Determine formulas in terms of v_0 and α for the length of time the ball is in the air, the maximum height of the ball, and the range of the ball (the distance from the origin to the point of impact on level ground).

59. Determine the angle α so that the maximum height of the ball is equal to the range. Is this angle independent of v_0?

4.3 Polynomial Functions, Newton's Method, and Optimization

A complete graph has no important behavior hidden from view. In Section 1.1, we suggested what complete graphs of linear, quadratic, and cubic functions look like. Now we are equipped to establish what a complete graph of any polynomial function looks like. Because complete graphs also require that we know about the intercepts, we will look at a numerical method called Newton's method, or the Newton-Raphson method, for solving an equation $f(x) = 0$. We will close the section with applications involving *optimization*—finding values of x that give maximum or minimum values of $f(x)$ where f is a function that models the real situation.

Polynomial Functions

To help understand polynomial functions, we state two theorems about the zeros or roots of polynomials. You can find proofs in books on complex-number analysis or advanced algebra. The first theorem is simply to remind you that zeros of polynomials can be complex numbers and that complex zeros appear in pairs. For review, you can check that both $1 + 2i$ and $1 - 2i$ are zeros of the polynomial $x^2 - 2x + 5$.

A *zero* of a function f is a *root* of the equation $f(x) = 0$ and vice versa.

THEOREM 4

If the complex number $a + bi$ is a zero of a polynomial with real coefficients, then its complex conjugate $a - bi$ is also a zero.

Some polynomials such as $x^2 - 2x + 5$ or, even simpler, $x^2 + 1$, have no real-number zeros—their graphs never cross the x-axis. The next theorem says that every nonconstant polynomial has the same number of zeros as its degree. Two equivalent forms of the theorem are given.

THEOREM 5

Fundamental Theorem of Algebra (First Form) Any polynomial function with degree greater than 0 has at least one zero.

Fundamental Theorem of Algebra (Second Form) Any polynomial function with degree $n (n > 0)$ can be written as

$$f(x) = a(x - c_1)(x - c_2) \cdots (x - c_n),$$

where the c_i are its zeros.

For either form, a zero may be a real number or a nonreal complex number.

Notice that the example

$$f(x) = (x + 2)(x + 2)(x - 1)(x - 1)(x - 1)$$

$$= (x + 2)^2(x - 1)^3$$

shows that the c_i in Theorem 5 need not be distinct. In this case, we say that -2 is a zero of *multiplicity 2* and 1 is zero of *multiplicity 3*. Counting multiplicities, we can conclude from Theorem 5 that a polynomial of degree n has n zeros. The payoff for us is that a polynomial of degree n has *at most n real zeros*, and if there are fewer than n real zeros, then by Theorem 4, the number is one of $n - 2, n - 4$, and so on.

Theorems 4 and 5 also let us deduce the possible number of extreme values a polynomial can have. We know that a polynomial function f of degree n is differentiable on $(-\infty, \infty)$. Therefore, its extreme values can occur only at points where $f' = 0$, and f' is of degree $n - 1$.

THEOREM 6

The number of extrema of a polynomial function f of degree n can be $n - 1$, $n - 3$, and so forth.

Partial Proof   If f has an extreme value at $x = a$, then $f'(a) = 0$, and so a is a real zero of f'. Thus, a count of the number of real zeros possible for f' will tell us how many extrema are possible. Because f is a polynomial of degree n, the derivative f' is a polynomial of degree $n - 1$. By Theorem 5, f' has $n - 1$ zeros, and some may not be real. For every complex zero that we discard from our count, we must also discard its conjugate (Theorem 4). Thus, the possible number of real zeros of f' is $n - 1$, $(n - 1) - 2 = n - 3$, and so on.

The proof is "partial" because we have not argued, for example, why a real zero of multiplicity 2 also reduces the number of possible extreme points by two, and not just one. (See Exercise 70.)

Now we ask you to confirm some information about the graphs that are possible for the general polynomial functions of degrees 3 and 4.

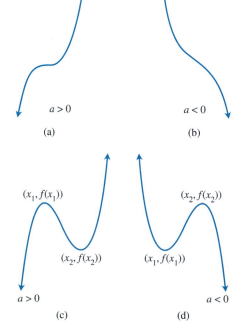

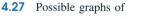

4.27   Possible graphs of

$$f(x) = ax^3 + bx^2 + cx + d.$$

EXPLORATION 1

Confirming Graphs of Cubic Polynomials

By Theorem 6, a graph of the cubic polynomial

$$f(x) = ax^3 + bx^2 + cx + d, \, a \neq 0,$$

has two or no extreme values. The possible graphs of a cubic are shown in Fig. 4.27. We note that f is continuous and differentiable on $(-\infty, \infty)$ and that

$$f'(x) = 3ax^2 + 2bx + c,$$
$$f''(x) = 6ax + 2b.$$

1. Using f'', explain why the graph of any cubic has one point of inflection and thus has opposite concavity on either side of that point.

2. Next, consider f'. Explain why f' might not change sign on $(-\infty, \infty)$ and what this confirms about the graphs of f in Figs. 4.27(a) and 4.27(b). Explain why f' could change sign on $(-\infty, \infty)$ and, if it does, why it would change sign twice. Tell what this confirms about the graphs of f in Figs. 4.27(c) and 4.27(d).

3. Give examples of cubic functions whose graphs have the shapes in Fig. 4.27. Use the information from parts 1 and 2 to help you build the examples.

EXPLORATION 2

Confirming Graphs of Quartic Polynomials— Group Project

By Theorem 6, a graph of the quartic polynomial

$$f(x) = ax^4 + bx^3 + cx^2 + dx + e, \qquad a \neq 0,$$

has three or one extreme values. We note that f is continuous and differentiable on $(-\infty, \infty)$, and possible graphs are shown in Figs. 4.28 and 4.29. To confirm that these are the possible graphs, we would like you to work in groups of two or three.

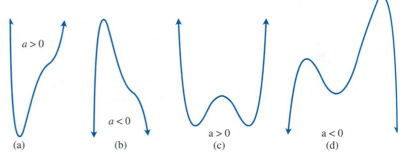

4.28 Possible graphs of $f(x) = ax^4 + bx^3 + cx^2 + dx + e$ when f'' changes sign and (a) or (b) f' has one or two real zeros and (c) or (d) f' has three distinct real zeros.

1. First, show that $f'(x)$ is a cubic function and $f''(x)$ is a quadratic function.

2. Next, consider the case in which f'' never changes sign. Decide what this tells you about f'. (Use the results of Exploration 1.) Use your information about f'' and f' to make conclusions about f. Match your conclusions to our graphs of the quartic polynomial.

3. Then consider what happens if f'' changes sign. Here is where you have to organize your work very carefully as you relate information about f'', f', and f. Again, use the results of Exploration 1. Prepare a presentation of your results to clearly show that you have covered all possibilities.

4. Illustrate your results with the following quartic polynomials or with quartic polynomials that you construct.

   a) $x^4 - 4x^3 + 6x^2 - 4x + 6$

   b) $2x^4 - 8x + 6$

   c) $3x^4 - 16x^3 + 24x^2 - 12$

   d) $3x^4 - 16x^3 + 24x^2 - 6x - 4$

EXPLORATION BIT

Some graphers and CAS can find all the zeros of a polynomial. If your grapher has this option, try it with part 4 of Exploration 2.

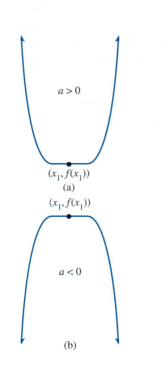

$(x_1, f(x_1))$
(a)

$(x_1, f(x_1))$

(b)

4.29 Possible graphs of

$$f(x) = ax^4 + bx^3 + cx^2 + dx + e$$

when f'' does not change sign.

If we are not concerned with details about concavity, we can describe the possible graphs of polynomials with less effort. For example, suppose that g is a polynomial of degree 4 with leading coefficient $a < 0$ (so $\lim_{|x| \to \infty} g(x) = -\infty$). Then there must be an interval $(-\infty, c)$ on which

AN ITERATIVE ALGORITHM

It is customary to call a specified sequence of computational steps like the one in Newton's method an *algorithm*. When an algorithm proceeds by repeating a given set of steps over and over, using the answer from the previous step as the input for the next, the algorithm is called *iterative*, and each repetition is called an *iteration*. Newton's method is one of the fastest algorithms known for finding a root of an equation.

TERMINOLOGY

The numbers x_n of Newton's method are said to be defined *recursively*.

Remember, linearizations are linear functions and therefore are easy to work with. The value of a linearization in Newton's method is that it is easy to find its zero.

g is increasing. Because g cannot be increasing throughout $(-\infty, \infty)$, the graph must rise to a local maximum and then begin to decrease. It may decrease for all $x > c$. This means that g has one extreme value and its graph looks like Fig. 4.28(b) or 4.29(b). On the other hand, the graph may fall until it hits a local minimum and then rise again. Because of the right-end behavior, the graph must eventually fall again. This means that g has at least three local extrema, and because of Theorem 6, we know that it has exactly three local extreme values. The graph of g would look like Fig. 4.29(d).

Newton's Method

The numerical technique called Newton's method for solving an equation $f(x) = 0$ uses a sequence of linearizations of f near a point where f is zero. Each linearization has a zero that usually gets closer and closer to the actual zero of f. Newton's method is the method used by many calculators to calculate zeros because it *converges* so fast. This means that for this *iterative algorithm*, the number of *iterations* needed is small.

The Procedure for Newton's Method

1. Guess a first approximation of a root for the equation $f(x) = 0$. A graph of $y = f(x)$ will help.

2. Use the first approximation to get a second, the second to get a third, and so on. To go from the nth approximation x_n to the next approximation x_{n+1}, use the formula

$$x_{n+1} = x_n - \frac{f(x_n)}{f'(x_n)}, \tag{1}$$

where $f'(x_n)$ is the derivative of f at x_n.

EXAMPLE 1

Use Newton's method to solve $x^3 + 3x + 1 = 0$.

Solution   With $f'(x) = 3x^2 + 3$, Eq. (1) becomes

$$x_{n+1} = x_n - \frac{x^3{}_n + 3x_n + 1}{3x^2{}_n + 3}.$$

The graph of $f(x) = x^3 + 3x + 1$ (Fig. 4.30) suggests -0.3 to be a reasonable first approximation for x_0. Thus, we get

$$x_0 = -0.3,$$

$$x_1 = -0.322324159021,$$

$$x_2 = -0.322185360251,$$

$$x_3 = -0.322185354626.$$

The x_n for $n \geq 4$ all appear to be equal to x_3 on the graphing calculator that we used to calculate the above values. We conclude that x_3 is the solution.

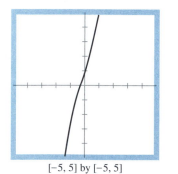

[−5, 5] by [−5, 5]

4.30 A complete graph of $f(x) = x^3 + 3x + 1$. Using ZOOM-IN or SOLVE shows that a zero occurs at $x = -0.322185354626$. Compare this to the result found by Newton's method in Example 1.

EXPLORATION 3

Newton's Method

The process described here is based on Eq. (1) and provides a clue to how Newton's method can be programmed into your grapher.
Let

$$y_1 = f(x) \qquad \text{and} \qquad y_2 = \text{NDER} f(x).$$

Then on your home screen, let x_0 be your first approximation of a root for $f(x) = 0$, and

STORE $x_0 \to x$ ENTER .

Then

STORE $x - (y_1/y_2) \to x$ ENTER repeatedly.

The first STORE- ENTER *initializes* x with your first guess, x_0. The second STORE- ENTER uses the value stored in x as the value of x in $x - (y_1/y_2)$ to compute a new value to STORE in x. Pressing ENTER repeatedly then generates x values iteratively, each being a better approximation of a root of f than the last. You can sit back, relax, and (usually) watch the values converge to the root until they get so close that they stop changing.

1. Find a root for $f(x) = x^3 + 3x + 1$. Let your first approximation be -0.3. Compare your results to those in Example 1.

2. Repeat part 1 but with a different first approximation.

3. Repeat part 1 to find zeros of other functions f. For each f, compare your result from Newton's method above and your result when you GRAPH f and use ZOOM-IN or SOLVE.

4. Write a SOLVE program for your grapher. Base it on Newton's method.

&

Solving $f(x) = 0$ using graphing features such as ZOOM-IN or SOLVE is faster than using Newton's method, and we don't have to remember the algorithm. The important understanding related to Newton's method is that it is typical of the numerical techniques that are built into calculators.

To truly understand the processes that govern the performance of such electronic devices, we have to spend time understanding their background algorithms. In advanced courses in numerical methods, we would research and study ways to approximate advanced mathematical ideas with relatively simple arithmetic, algebraic, and analytic techniques, being always on the lookout for a way to build the better algorithm.

What Is the Theory behind Newton's Method?

It is this: We use the tangent to approximate the graph of $y = f(x)$ near the point $P(x_n, y_n)$, where $y_n = f(x_n)$ is small, and we let x_{n+1} be the value

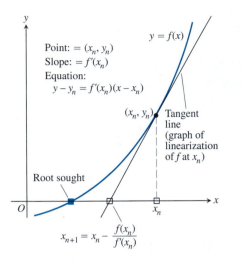

4.31   The geometry of the successive steps of Newton's method. From x_n, we go up to the curve and follow the tangent line down to find x_{n+1}.

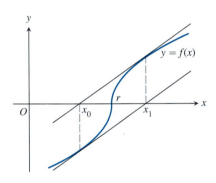

4.32   The graph of the function

$$f(x) = \begin{cases} \sqrt{x - r} & \text{for } x \geq r, \\ -\sqrt{r - x} & \text{for } x < r. \end{cases} \quad (4)$$

If we begin Newton's method with $x_0 = r - h$, we get $x_1 = r + h$. Successive approximations go back and forth between these two values, and Newton's method fails to converge.

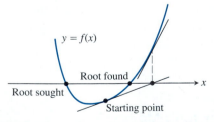

4.33   Newton's method may miss the root you want if you start too far away.

of x where that tangent line crosses the x-axis. (We assume that the slope $f'(x_n)$ of the tangent is not zero.) The equation of the tangent is

$$y - y_n = f'(x_n)(x - x_n).$$

We put $y_n = f(x_n)$ and $(x, y) = (x_{n+1}, 0)$ into this equation, then solve for x_{n+1} to get

$$x_{n+1} = x_n - \frac{f(x_n)}{f'(x_n)}.$$

See Fig. 4.31.

From another point of view, every time we replace the graph of f by one of its tangent lines, we are replacing the function by one of its linearizations L. We then solve $L(x) = 0$ to estimate the solution of $f(x) = 0$.

Strengths and Limitations of Newton's Method

Newton's method does not work if $f'(x_n) = 0$. In that case, choose a new starting point.

Newton's method does not always converge. For instance, successive approximations a and b can go back and forth between these two values, and no amount of iteration will bring us any closer to the root r. (See Fig. 4.32.)

If Newton's method does converge, it converges to a root of $f(x)$. However, the method may converge to a root that is different from the expected one if the starting value is not close enough to the root sought. Figure 4.33 shows how this might happen.

Optimization

In the mathematical models in which we use differentiable functions to describe the things that interest us, optimization means finding where some function has its greatest or smallest value. What is the size of the most profitable production run? What is the best shape for an oil can? What is the stiffest beam we can cut from a 12-in. log?

Most applications call for finding the absolute maximum value or absolute minimum value of a function that is continuous on a closed interval $[a, b]$ and differentiable on the open interval (a, b). The Max-Min Theorem in Section 2.2 assures us that these extreme values exist. Our observation in Section 4.1 that extreme values can occur only at critical points or the endpoints a and b usually limits our candidates to a small number that we can check one by one.

Optimization Example from Mathematics

EXAMPLE 2   Products of numbers

Find two numbers whose sum is 20 and whose product is as large as possible.

Solution   If one number is x, the other is $(20 - x)$. Their product is

$$f(x) = x(20 - x) = 20x - x^2.$$

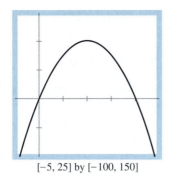

[−5, 25] by [−100, 150]

4.34   A complete graph of
$$f(x) = x(20 - x).$$

Figure 4.34 shows a complete graph of f. We can see from the graph that there is a maximum, and from what we know about parabolas, it occurs at $x = 10$.

To confirm, we need to evaluate f at the critical points and the endpoints of its domain. There are no endpoints. The first derivative,

$$f'(x) = 20 - 2x,$$

is defined for all values of x and is zero only at $x = 10$ to agree with our graphical result.

The value of f at this one critical point is

$$f(10) = 100,$$

which we conclude is the maximum value of f.

The two numbers we seek—the solution to the problem—are $x = 10$ and $20 - x = 20 - 10 = 10$.

Optimization Example from Industry

EXAMPLE 3   Metal fabrication

An open-top box is to be made by cutting congruent squares of side length x from the corners of a 20- by 25-in. sheet of tin and bending up the sides (Fig. 4.35). Find how large the cut-out squares should be so that the box can hold as much as possible. Find the resulting maximum volume.

Of the three approaches:

Do algebraically, support graphically;

Do graphically, confirm algebraically;

Do using a method of your choice (often involving a combination of algebraic and graphical steps).

Examples 2 and 3 illustrate the third approach.

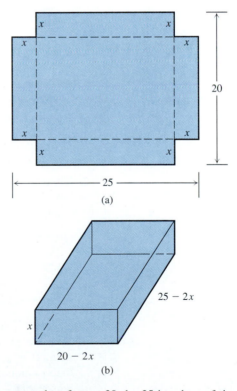

4.35   To make an open box from a 20- by 25-in. sheet of tin, (a) squares are cut from the corners, and (b) the sides are bent upward. What value of x gives the largest volume? Give a convincing argument why the largest volume does not occur for $x = 0$ or 10.

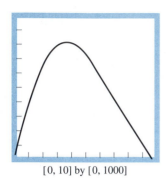

[0, 10] by [0, 1000]

4.36 The graph of

$$V(x) = x(20 - 2x)(25 - 2x)$$

over its domain, [xMin, xMax] = [0, 10]. You can estimate or ZOOM-IN to see that the maximum occurs at about (3.7, 820.5).

Solution The height of the box is x, and the other two dimensions are $(20 - 2x)$ and $(25 - 2x)$. The volume of the box is

$$V(x) = x(20 - 2x)(25 - 2x) = 4x^3 - 90x^2 + 500x.$$

Because the tin is 20 in. wide and the cut-out corners are square, the only values of x that make sense in the problem are $0 \le x \le 10$. Figure 4.36 shows the graph of V with [xMin, xMax] = [0, 10] and suggests that the maximum value of V occurs at about $x = 3.7$ and is about 820.5.

To confirm, we need to evaluate V at the critical points and endpoints. The first derivative,

$$V'(x) = 12x^2 - 180x + 500,$$

is defined for all x and is zero at

$$\frac{180 - \sqrt{180^2 - 48(500)}}{24} = 3.68 \quad \text{and} \quad \frac{180 + \sqrt{180^2 - 48(500)}}{24} = 11.32.$$

Only $x = 3.68$ lies in the domain [0, 10] of V. The values of V at this one critical point and two endpoints are

$$\text{Critical point value:} \quad V(3.68) = 820.53,$$

$$\text{Endpoint values:} \quad V(0) = 0, V(10) = 0.$$

Thus, cut-out squares that are 3.68 in. on a side give a maximum volume of 820.53 in^3. ≡

LIMITING OUR SEARCH FOR OPTIMUM VALUES

To find the absolute maximum and minimum values of a continuous function on a closed interval, evaluate the function at the critical points and endpoints, and take the largest and smallest of these values.

Strategy for Solving Max-Min Problems

1. *Draw a picture of the problem situation.* Label the parts that are important in the problem.

2. *Write an algebraic representation.* Represent, preferably as a function f of a single variable, the quantity whose extreme value you want.

3. *Draw a complete graph of $y = f(x)$.* Find the domain and range of f. Determine the values of x and y that make sense in the problem situation.

4. *Draw a graph of the problem situation.* Use the values of x and y from Step 3 as a guide for [xMin, xMax] and [yMin, yMax].

5. *Confirm the critical points.* Find where f' is zero or fails to exist.

6. *Find the extreme value and state the solution.* Select the critical point(s) or endpoint(s) where f reaches the desired extreme. State the solution of the original problem. Check that the solution is reasonable.

Exercises 4.3

In Exercises 1–8, use Newton's method to find the solutions.

1. $x^2 + x - 1 = 0$. Confirm your answers using the quadratic formula.

2. $x^3 + x - 1 = 0$. Support your answers using ZOOM-IN.

3. $x^4 + x - 3 = 0$ **4.** $2x - x^2 + 1 = 0$

5. $x^4 - 2x^3 - x^2 - 2x + 2 = 0$

6. $2x^4 - 4x^2 + 1 = 0$ **7.** $9 - \dfrac{3}{2}x^2 + x^3 - \dfrac{1}{4}x^4 = 0$

8. $x^5 - 5x^3 + 4x + 5 = 0$

9. Use Newton's method to find the positive fourth root of 2 by solving the equation $x^4 - 2 = 0$. Start with $x_0 = 1$.

10. Use Newton's method to find the negative fourth root of 2 by solving the equation $x^4 - 2 = 0$. Start with $x_0 = -1$.

11. Suppose your first guess in using Newton's method is lucky in the sense that x_0 is a root of $f(x) = 0$. What happens to x_1 and later approximations?

12. You plan to estimate $\pi/2$ to five decimal places by solving the equation $\cos x = 0$ by Newton's method. Does it matter what your starting value is? Explain.

13. *Oscillation.* Show that if $h > 0$, applying Newton's method to

$$f(x) = \begin{cases} \sqrt{x}, & x \ge 0 \\ \sqrt{-x}, & x < 0 \end{cases}$$

leads to $x_1 = -h$ if $x_0 = h$ and to $x_1 = h$ if $x_0 = -h$. Draw a picture that shows what is going on.

14. *Approximations that get worse and worse.* Apply Newton's method to $f(x) = x^{1/3}$ with $x_0 = 1$, and calculate x_1, x_2, x_3, and x_4. Find a formula for $|x_n|$. What happens to $|x_n|$ as $n \to \infty$? Draw a picture that shows what is going on.

In Exercises 15–36, show a complete graph and identify the inflection points, local maximum and minimum values, and the intervals on which the graph is rising, falling, concave up, and concave down. Indicate the number of real roots. We suggest you do exercises from the three groups in the order given.

Do Exercises 15–20 analytically, then support graphically.

15. $y = x^3 - 3x^2 + 5x - 4$ **16.** $y = \dfrac{1}{3}x^3 - 2x^2 + 4x + 8$

17. $y = x^3 - 2x^2 - 3x + 8$ **18.** $y = x^3 + 10x^2 - 23x + 12$

19. $y = 12 + x - 4x^3$ **20.** $y = 2x^3 - x^2 - 14x - 12$

Do Exercises 21–24 graphically, then confirm analytically.

21. $y = 20 - 3x - \dfrac{1}{3}x^3$ **22.** $y = -9x^3 + 4x - 15$

23. $y = x^4 + x^2 + x + 8$

24. $y = x^4 - 8x^3 + 17x^2 - 10x - 1$

Do Exercises 25–36 using a method of your choice.

25. $y = 4x^3 - 17x^2 + 8x - 1$

26. $y = -x^3 - 3x^2 + 10x + 3$

27. $y = 10 + \dfrac{8}{3}x^3 - x^4$

28. $y = \dfrac{1}{3}x^4 + \dfrac{4}{3}x^3 + 2x^2 + 2x - 10$

29. $y = x^4 - 2x^2 + x + 20$

30. $y = 20 - 4x + 2x^2 + 3x^3 - \dfrac{9}{4}x^4$

31. $y = -x^4 - 7x^3 - 9x^2 + 7x + 12$

32. $y = x^4 - 8x^3 + 14x^2 + 8x - 5$

33. $y = -x^5 - x^4 + 11x^3 + 9x^2 - 18x + 5$

34. $y = x^5 - \dfrac{5}{2}x^4 - 5x^2 + 1$

35. $y = 3x^5 + \dfrac{15}{4}x^4 + 10x^3 + \dfrac{15}{2}x^2 + 15x + 1$

36. $y = \dfrac{1}{5}x^5 - \dfrac{9}{4}x^4 + \dfrac{19}{3}x^3 + \dfrac{9}{2}x^2 - 16x + 1$

37. The sum of two nonnegative numbers is 20. Find the numbers if the sum of their squares is to be as large as possible.

38. Show that among all rectangles with an 8-ft perimeter, the one with the largest area is a square.

39. You are planning to make an open rectangular box from an 8-in. × 15-in. piece of cardboard by cutting squares from the corners and folding up the sides. What are the dimensions of the box of largest volume you can make this way?

40. A rectangular plot of farmland will be bounded on one side by a river and on the other three sides by a single-strand electric fence. With 800 m of wire at your disposal, what is the largest area you can enclose?

41. The figure below shows a rectangle inscribed in an isosceles right triangle whose hypotenuse is 2 units long.
 a) Express the y-coordinate of P in terms of x. (You might start by writing an equation for the line AB.)

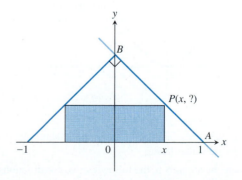

 b) Express the area of the rectangle in terms of x.
 c) What is the largest area the rectangle can have?

42. A rectangle has its base on the x-axis and its upper two vertices on the parabola $y = 12 - x^2$. What is the largest area the rectangle can have?

In Exercises 43 and 44, the height of an object moving vertically is given by $s(t)$ with s in feet and t in seconds. Find (a) the object's velocity when $t = 0$, (b) its maximum height, and (c) its velocity when $s = 0$.

43. $s = -16t^2 + 100t + 200$ **44.** $s = -16t^2 + 96t + 112$

45. The U.S. Postal Service will accept a box for domestic shipment only if the sum of the length and girth (distance around) does not exceed 108 in. Find the dimensions of the largest acceptable box with a square end.

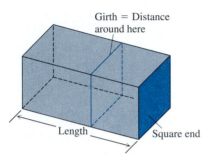

Girth = Distance around here

Length   Square end

46. *The sonobuoy problem.* From C. O. Wilde's *The Contraction Mapping Principle*, UMAP Unit 326 (Arlington, Mass.: COMAP, Inc.). In submarine location problems, it is often necessary to find the submarine's closest point of approach (CPA) to a sonobuoy (sound detector) in the water. Suppose that the submarine travels on a parabolic path $y = x^2$ and that the buoy is located at the point $(2, -1/2)$.

a) Show that the value of x that minimizes the square of the distance, and hence the distance, between the points (x, x^2) and $(2, -1/2)$ in the figure below is a solution of the equation $x = 1/(x^2 + 1)$.

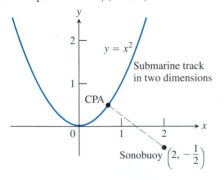

$y = x^2$

Submarine track in two dimensions

CPA

Sonobuoy $\left(2, -\dfrac{1}{2}\right)$

b) Use Newton's method to solve the equation in part (a) to five decimal places.

47. Compare the answers to the following two construction problems.

a) A rectangular sheet of perimeter 36 cm and dimensions x cm by y cm is to be rolled into the cylinder shown in part (a) of the figure below. What values of x and y give the largest volume?

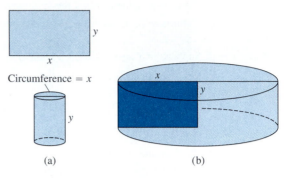

Circumference = x

(a)   (b)

b) The rectangular sheet of perimeter 36 cm and dimensions x by y is to be revolved about one of the sides of length y to sweep out the cylinder in part (b) of the figure above. What values of x and y give the largest volume?

48. A right triangle whose hypotenuse is $\sqrt{3}$ meters long is revolved about one of its legs to generate a right circular cone. Find the radius, height, and volume of the cone of greatest volume that can be made in this way.

49. A spherical floating buoy has a radius of 1 meter and a density 1/4 that of sea water. By Archimedes' Law, the weight of the buoy is equal to the weight of the displaced water. The volume of the displaced water is equal to the volume of the submerged portion of the buoy. Find the depth x (see figure below) that the buoy sinks in sea water.

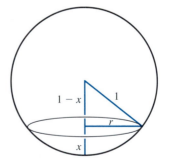

$1 - x$ 1

r

x

50. Repeat Exercise 49 assuming that the buoy has a density 1/3 that of sea water.

51. What values of a and b cause $f(x) = x^3 + ax^2 + bx$ to have

a) a local maximum at $x = -1$ and a local minimum at $x = 3$?

b) a local minimum at $x = 4$ and a point of inflection at $x = 1$?

52. Suppose that at time $t \geq 0$ the position of a particle moving on the x-axis is $x = (t - 1)(t - 4)^4$.

a) When is the particle at rest?

b) During what time interval does the particle move left?

c) What is the fastest the particle goes while moving left?

53. Find the volume of the largest right circular cone that can be inscribed in a sphere of radius 3.

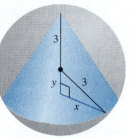

54. *The strength of a beam.* The strength of a rectangular beam is proportional to its width times the square of its depth. Find the dimensions of the strongest beam that can be cut from a 12-in.-diameter log.

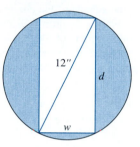

55. *The stiffness of a beam.* The stiffness of a rectangular beam is proportional to its width times the cube of its depth. Find the dimensions of the stiffest beam that can be cut from a 12-in.-diameter log. Compare your answers with the dimensions of the beam in Exercise 54.

56. *Chemical reactions.* Metallic tin, when kept below 13° C for a while, becomes brittle and crumbles to a gray powder. Tin objects eventually crumble to this gray powder spontaneously if kept in a cold climate for years. The Europeans who saw the tin organ pipes in their churches crumble away years ago called the change *tin pest* because it seemed to be contagious. And indeed it was, for the gray powder is a catalyst for its own formation.

A *catalyst* for a chemical reaction is a substance that controls the rate of the reaction without undergoing any permanent change in itself. An *autocatalytic reaction* is one whose product is a catalyst for its own formation. Such a reaction may proceed slowly at first if the amount of catalyst present is small, and slowly again at the end when most of the original substance is used up. But in between, when both the substance and its product are abundant, the reaction proceeds at a faster pace.

In some cases, it is reasonable to assume that the rate $v = dx/dt$ of the reaction is proportional both to the amount of the original substance and to the amount of product. That is, v may be seen as a function of x alone, and

$$v = kx(a - x) = kax - kx^2,$$

where

$x = $ the amount of product,

$a = $ the amount of substance at the beginning, and

$k = $ a positive constant.

At what value of x does the rate v have a maximum? What is the maximum value of v?

57. *Logistic growth.* Many natural populations grow at a rate directly proportional to their current size when small. Because of competition for resources, however, the rate of growth ultimately decreases until the population attains its maximum size, known as its *carrying capacity*. Population biologists call this type of growth *logistic* and often model it by

$$\frac{dR}{dt} = rp \left(\frac{K - p}{K} \right),$$

where dR/dt is the population growth rate, p is the population size, and r and K are positive constants representing the instantaneous growth rate of the population (at low population sizes) and the carrying capacity, respectively. Determine the population size, in terms of the carrying capacity, at which the population is growing most rapidly.

Medicine

58. *How we cough.* When we cough, the trachea (TRAY-kee-uh, windpipe) contracts to increase the velocity of the air going out. This raises the questions of how much it should contract to maximize the velocity and whether it really contracts that much when we cough.

Under reasonable assumptions about the elasticity of the tracheal wall and about how the air near the wall is slowed by friction, the average flow velocity v can be modeled by

$$v = c(r_0 - r)r^2 \text{ cm/sec}, \quad \frac{r_0}{2} \leq r \leq r_0,$$

where r_0 is the rest radius of the trachea in centimeters and c is a positive constant whose value depends in part on the length of the trachea.

Show that v has its maximum value when $r = (2/3)r_0$, that is, when the trachea is about 33% contracted. The remarkable fact is that X ray photographs confirm that the trachea contracts about this much during a cough.

59. *Sensitivity to medicine (continuation of Exercise 79, Section 3.3).* Find the amount of medicine to which the body is most sensitive by finding the value of M that maximizes the derivative dR/dM.

60. The derivative of a polynomial is

$$p'(x) = (x + 1)(x - 1)^2(x - 5)^3.$$

What is the degree of p? Discuss the graph of p. Use technology to determine where the graph of p is concave up.

61. *Curves that are almost flat near the root.* Some curves are so flat that, owing to the finiteness of computers, Newton's method stops some distance from the root. Try Newton's method on $f(x) = (x-1)^{40}$ with a starting value of $x_0 = 2$. How close does the computer come to the root $x = 1$?

62. *Finding a root different from the one sought.* All three roots of $f(x) = x^4 - x^2$ can be found by starting Newton's method near $x = -\sqrt{2}/2$.

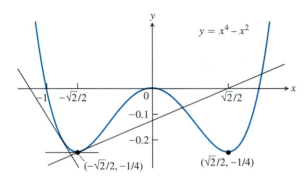

$y = x^4 - x^2$

$(-\sqrt{2}/2, -1/4)$ $(\sqrt{2}/2, -1/4)$

The method will find the root $x = -1$ if x_0 is far enough to the left of $-\sqrt{2}/2$. If x_0 is too close to $-\sqrt{2}/2$, the computer will encounter a zero slope or a value of x_1 that is too large to handle. There is a zone just to the right of $-\sqrt{2}/2$ where values of x_0 lead to $x = 1$ instead of $x = 0$.

Likewise, selected values of x_0 near $\sqrt{2}/2$ will lead to $x = 1$, $x = 0$, $x = -1$, or no root at all.

Try it.

63. a) Explain why the following four statements ask for the same information:

i) Find the roots of $f(x) = x^3 - 3x - 1$.

ii) Find the x-coordinates of the intersections of the curve $y = x^3$ with the line $y = 3x + 1$.

iii) Find the x-coordinates of the points where the curve $y = x^3 - 3x$ crosses the horizontal line $y = 1$.

iv) Find the values of x where the derivative of $g(x) = (1/4)x^4 - (3/2)x^2 - x + 5$ equals zero.

b) Sketch the graph of $f(x) = x^3 - 3x - 1$ over the interval $-2 \le x \le 2$.

c) Find the roots of $f(x) = x^3 - 3x - 1$ to five decimal places.

64. The curve $y = \tan x$ crosses the line $y = 2x$ somewhere between $x = 0$ and $x = \pi/2$. Where?

65. Estimate π to five decimal places by applying Newton's method to the equation $\tan x = 0$ with $x_0 = 3$. Remember to use radians.

66. *Locating a planet.* To calculate a planet's space coordinates, we have to solve equations like $x = 1 + 0.5 \sin x$. Graphing the function $f(x) = x - 1 - 0.5 \sin x$ suggests

that the function has a root near $x = 1.5$. Use Newton's method to find the solution accurate to three decimal places.

67. *The suitcase problem (continuation of Exercise 66, Section 1.5).* A 24- by 36-in. sheet of cardboard is folded in half to form a 24- by 18-in. rectangle. Then four squares of equal side length x are removed from each corner of the folded rectangle. The sheet is unfolded, and the six tabs are folded up to form a box with sides and a lid.

a) Draw a picture of the situation. Refer back to Section 1.5 if needed.

b) Write an algebraic representation $V(x)$ for the volume of the box.

c) Draw a complete graph of $y = V(x)$. Find the domain and range of V. Determine the values of x (and y) that make sense in the problem situation.

d) Draw a graph of the problem situation.

e) Find x so that the volume of the box is maximum. Find the maximum volume. Confirm your result.

f) Find x so that the volume of the box is 1120 in³. Confirm, if possible. Find the exact value of x, if possible.

g) Write a paragraph discussing the issues that arise in part (e).

68. *The box with lid problem (continuation of Exercise 65, Section 1.5).* A piece of cardboard measures 10 in. by 15 in. Two equal squares are removed from the corners of a 10-in. side. Two equal rectangles are removed from the other corners so that the tabs can be folded to form a rectangular box with lid.

a) Draw a picture of the situation. Refer back to Section 1.5 if needed.

b) Write an algebraic representation $V(x)$ for the volume of the box.

c) Draw a complete graph of $y = V(x)$. Find the domain and range of V. Determine the values of x (and y) that make sense in the problem situation.

d) Draw a graph of the problem situation.

e) Find x so that the volume of the box is maximum. Find the maximum volume. Confirm your result.

69. Repeat Exercise 68. Begin by removing the two equal squares from the corners of a 15-in. side.

70. Refer to Theorem 6, and let $f'(x) = (x - c)^n g(x)$ with $g(c) \ne 0$.

a) If $n = 2k$, show that f does not have an extreme value at $x = c$. Explain why this reduces the number of extrema of f by $2k$.

b) If $n = 2k + 1$, show that f does have an extreme value at $x = c$. Explain why this reduces the number of possible extrema of f by $2k$.

71. Suppose f and g are polynomials and f' and g' have the same real roots with the same multiplicities. What do the graphs of f and g have in common? Try some examples first.

72. Prove that if a polynomial of degree 4 has three local extrema, then there are real numbers a and c and four distinct real numbers $x_1, x_2, x_3,$ and x_4 such that

$$f(x) = a(x - x_1)(x - x_2)(x - x_3)(x - x_4) + c.$$

An airplane is flying at altitude H when it begins its descent to an airport runway that is at horizontal ground distance L from the plane as shown. Assume that the landing path of the airplane is the cubic polynomial function $y = ax^3 + bx^2 + cx + d$ where $y(-L) = H$ and $y(0) = 0$.

73. What is dy/dx at $x = 0$?

74. What is dy/dx at $x = -L$?

75. Use the values for dy/dx at $x = 0$ and $x = -L$ together with $y(0) = 0$ and $y(-L) = H$ to show that

$$y(x) = H \left[2 \left(\frac{x}{L} \right)^3 + 3 \left(\frac{x}{L} \right)^2 \right].$$

76. GRAPH the landing path for $H = 25{,}000$ feet and $L = 90$ miles. Specify a viewing window that shows only the landing path.

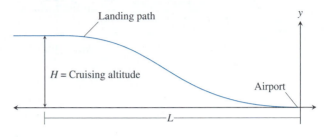

4.4 Rational Functions and Economics Applications

In this section, we use calculus and graphing utilities to provide more detail about rational functions. Applications whose models involve rational functions are introduced. We also illustrate how calculus can make a contribution to economics.

Complete Graphs of Rational Functions

Let $f(x) = p(x)/h(x)$ be a rational function with $p(x)$ and $h(x)$ polynomials with no common factors. In Section 2.4, we rewrote f in the form

$$f(x) = q(x) + \frac{r(x)}{h(x)}, \qquad \deg r < \deg h, \qquad (1)$$

where q and r are the quotient and remainder, respectively, when p is divided by h. The zeros of h give vertical asymptotes of f, and q is the end behavior asymptote of f. This means that, *except* near the vertical asymptotes of f, q is a very good approximation of f.

Graphs produced by graphing utilities are usually very accurate. This means that we can save time normally spent in algebraic computation of derivatives by letting our grapher calculate and GRAPH f' and f'' (or NDER f and NDER2 f) and then using these graphs to support the information suggested by the graph of the function. We illustrate what we mean in Example 1.

EXAMPLE 1

Draw a complete graph of

$$f(x) = \frac{x^4 + x^3 - 6x^2 + 6}{x^2 + x - 6} = x^2 + \frac{6}{(x - 2)(x + 3)}.$$

Find the points of inflection and the local extreme values, and identify the intervals on which the graph is rising, falling, concave up, and concave down.

Solution   Two views (Fig. 4.37) help to clarify the behavior of f. We can use SOLVE or ZOOM-IN to find zeros of f at $x = -2.85$ and $x = -1$. The points of discontinuity of f are the zeros, $x = 2$ and $x = -3$, of its denominator—points at which the graph has vertical asymptotes. Otherwise, f is everywhere differentiable. The right-hand side of the equation above is in the form

$$q(x) + r(x)/h(x),$$

which tells us that x^2 is the end behavior asymptote.

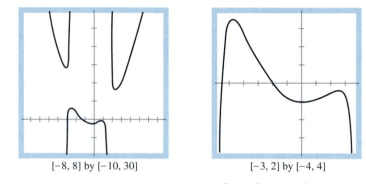

$[-8, 8]$ by $[-10, 30]$ $[-3, 2]$ by $[-4, 4]$

4.37   Two views of the graph of $y_1 = (x^4 + x^3 - 6x^2 + 6)/(x^2 + x - 6)$.

We note that f seems to be rising ($\nearrow$) on four intervals and falling ($\searrow$) on four intervals. We also note that f seems to have three local minimum values and two local maximum values. The graph of f appears to be concave up over three intervals and concave down over two. There appear to be two points of inflection, both in the interval $(-3, 2)$.

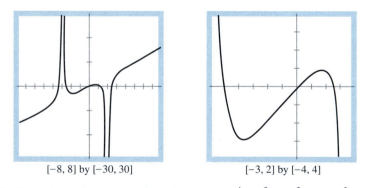

$[-8, 8]$ by $[-30, 30]$ $[-3, 2]$ by $[-4, 4]$

4.38   Two views of $y_2 = \text{NDER } y_1$ where $y_1 = (x^4 + x^3 - 6x^2 + 6)/(x^2 + x - 6)$.

Two views of the graph of f' (Fig. 4.38) and one view of f'' (Fig. 4.39) support our conclusions and give more information about the graph of f. We note that f' has five zeros and f'' has two zeros at which each changes sign. These zeros (found by using grapher techniques) and the vertical asymptotes identify nine points that divide the domain of f into intervals for which we can find useful information about the behavior of f as summarized in the following chart.

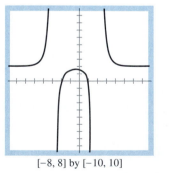

$[-8, 8]$ by $[-10, 10]$

4.39   A complete graph of $y_3 = \text{NDER2 } y_1$ where $y_1 = (x^4 + x^3 - 6x^2 + 6)/(x^2 + x - 6)$.

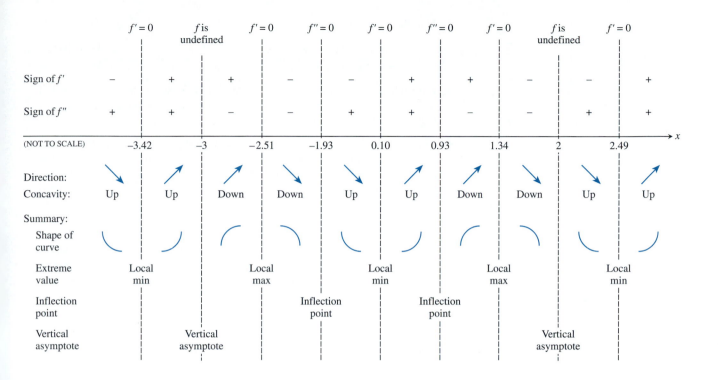

	$f'=0$	f is undefined	$f'=0$	$f''=0$	$f'=0$	$f''=0$	$f'=0$	f is undefined	$f'=0$
Sign of f'	$-$	$+$	$+$	$-$	$-$	$+$	$+$	$-$	$-$	$+$
Sign of f''	$+$	$+$	$-$	$-$	$+$	$+$	$-$	$-$	$+$	$+$

(NOT TO SCALE) -3.42 -3 -2.51 -1.93 0.10 0.93 1.34 2 2.49

Direction:									
Concavity:	Up	Up	Down	Down	Up	Up	Down	Down	Up	Up
Summary:									
Shape of curve									
Extreme value	Local min		Local max		Local min		Local max		Local min
Inflection point				Inflection point		Inflection point			
Vertical asymptote		Vertical asymptote						Vertical asymptote	

EXPLORATION BIT

For f in Example 1, give a convincing argument for each of the following:

a) $f(x) \to \infty$ as $x \to -3^-$

b) $f(x) \to -\infty$ as $x \to -3^+$

c) $f(x) \to -\infty$ as $x \to 2^-$

d) $f(x) \to \infty$ as $x \to 2^+$

We can confirm the information in the chart by calculating

$$f'(x) = 2x - \frac{6(2x+1)}{(x^2+x-6)^2}$$

and

$$f''(x) = 2 + \frac{12(3x^2+3x+7)}{(x^2+x-6)^3}$$

and analyzing these two functions. We can use the information in the chart to confirm the complete graph of f.   ≡

EXPLORATION 1

Complete Graph of a Rational Function

Draw and confirm a complete graph of

$$f(x) = \frac{3x+2}{2x^2+5x-3}.\qquad \text{See Fig. 4.40(a).}$$

1. Establish the intercepts and domain of f. Describe the behavior of f near the points that are not in the domain. Put $f(x)$ in the form of Eq. (1). Use what you have learned about rational functions from Section 2.4 to confirm the end behavior of f.

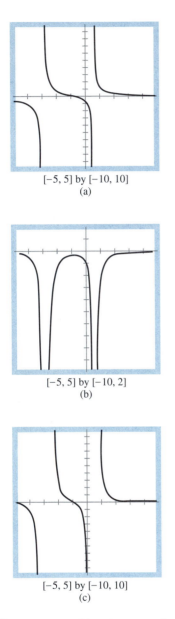

[−5, 5] by [−10, 10]
(a)

[−5, 5] by [−10, 2]
(b)

[−5, 5] by [−10, 10]
(c)

4.40 (a) Our candidate for a complete graph of

$$f(x) = (3x + 2)/(2x^2 + 5x - 3).$$

For support, we also graphed (b) NDER f and (c) NDER2 f. When you have confidence in recognizing complete graphs, the graphs of NDER f and NDER2 f are usually enough to draw the necessary conclusions about the graph of f, and you can avoid the tedious computation of exact derivatives.

2. Use the quotient rule to show

$$f'(x) = \frac{-(6x^2 + 8x + 19)}{(2x^2 + 5x - 3)^2} \quad \text{and} \quad f''(x) = \frac{2(12x^3 + 24x^2 + 114x + 107)}{(2x^2 + 5x - 3)^3}.$$

3. Find where f' is positive, negative, and zero. Use this information to tell where f is increasing and decreasing and has extreme values. A graph of NDER f (Fig. 4.40b) gives a clue of what to look for.

4. Find where f'' is positive, negative, and zero. Use this information to tell about the concavity of f and its points of inflection. In this case, the analysis of f'' becomes difficult (but it can be done), and you should feel comfortable using a graph of NDER2 f (Fig. 4.40c) to make the desired conclusions. ✣

The following guidelines help to summarize how we can analyze a rational function and know that its graph is complete. As before, the order, difficulty, and importance of these tasks will vary depending on the function and what we want to know about it.

A Complete Graph of a Rational Function f

To determine a complete graph of a rational function f, we do the following:

1. Establish the intercepts and domain of f, noting the vertical asymptotes. Find the limits from the left and right at each vertical asymptote. Find and visualize (or sketch) the end behavior asymptote.

2. Find f' and f'', or GRAPH NDER f and NDER2 f.

3. Find where f' is positive, negative, and zero to confirm where f is increasing and decreasing and has extreme values.

4. Find where f'' is positive, negative, and zero to confirm where f is concave up and concave down and has points of inflection.

5. Compare the information found above with our graph (viewing window or sketch) and resolve conflicting information.

Optimization Examples from Industry

EXAMPLE 2   Product Design

You have been asked to design a one-liter oil can shaped like a right circular cylinder. What dimensions will use the least material? Support your answer with a graph.

Solution   We picture the can as a right circular cylinder with height h and radius r (Fig. 4.41). If r and h are measured in centimeters and the volume is expressed as 1000 cm^3, then r and h are related by the equation

$$\pi r^2 h = 1000. \quad \text{1000 cm}^3 = 1\text{L}. \tag{2}$$

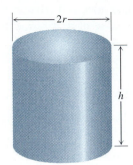

4.41 This one-liter can uses the least material when $h = 2r$ (Example 2).

How can we interpret the phrase "least material"? One possibility is to ignore the thickness of the material and the waste in manufacturing. Then we ask for dimensions r and h that make the total surface area

$$A = \underbrace{2\pi r^2}_{\substack{\text{cylinder} \\ \text{ends}}} + \underbrace{2\pi rh}_{\substack{\text{cylinder} \\ \text{wall}}} \tag{3}$$

as small as possible while satisfying the constraint $\pi r^2 h = 1000$. (Exercise 34 describes one way in which we might take waste into account.)

What kind of oil can do we expect? Not a tall, thin one like a 6-ft pipe, nor a short, wide one like a covered pizza pan. We expect something in between.

We are not quite ready to find critical points because Eq. (3) gives A as a function of two variables and our procedure calls for A to be a function of a single variable. However, Eq. (2) can be solved to express either r or h in terms of the other.

Solving for h is easier, so we take

$$h = \frac{1000}{\pi r^2}.$$

This changes the formula for A to

$$A = 2\pi r^2 + 2\pi rh = 2\pi r^2 + 2\pi r \frac{1000}{\pi r^2} = 2\pi r^2 + \frac{2000}{r}.$$

Thus, we have a mathematical model that shows A as a rational function of r (instead of x). Our mathematical goal is to find the value of $r > 0$ that minimizes the value of A. As support that we are heading in the right direction, we GRAPH $y = A(r)$ (Fig. 4.42) and see that there indeed appears to be a minimum value of A for r somewhere between 5 and 6.

Since A is differentiable on $r > 0$, an interval with no endpoints, it can have a minimum value only where its first derivative is zero.

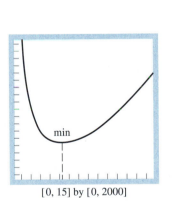

[0, 15] by [0, 2000]

4.42 A complete graph of

$$A(r) = 2\pi r^2 + 2000/r$$

for the domain $r > 0$ is concave up, has $r = 0$ as a vertical asymptote on the left, and has $2\pi r^2$ as its right-end behavior asymptote. What would be the shapes of the oil cans for $r \to 0$ and $r \to \infty$?

$$A = 2\pi r^2 + \frac{2000}{r}$$

$$\frac{dA}{dr} = 4\pi r - \frac{2000}{r^2} \qquad \text{Find } dA/dr.$$

$$4\pi r - \frac{2000}{r^2} = 0 \qquad \text{Set it equal to zero.}$$

$$4\pi r^3 = 2000 \qquad \text{Solve for } r.$$

$$r = \sqrt[3]{\frac{500}{\pi}} = 5.42. \qquad \text{Critical point}$$

Now, with a fairly high degree of confidence, we confirm that a minimum for A occurs at this value of r. We investigate d^2A/dr^2:

$$\frac{dA}{dr} = 4\pi r - \frac{2000}{r^2},$$

$$\frac{d^2A}{dr^2} = 4\pi + \frac{4000}{r^3}.$$

This derivative is positive throughout the interval $r > 0$. Hence, by the Second Derivative Test for Extreme Values, the value of A at $r = \sqrt[3]{500/\pi}$ is indeed an absolute minimum.

When $r = \sqrt[3]{500/\pi}$,

$$h = \frac{1000}{\pi r^2} = 2\sqrt[3]{500/\pi} = 2r. \qquad \text{After some arithmetic}$$

Thus, the most efficient one-liter can has its height equal to the diameter, so

$$r = 5.42 \ \text{cm} \qquad \text{and} \qquad h = 10.84 \ \text{cm}. \qquad \blacksquare$$

EXAMPLE 3   Maximizing Profit

Suppose a manufacturer can sell x items a week for a revenue of $r(x) = 200x - 0.01x^2$ cents, and it costs $c(x) = 50x + 20{,}000$ cents to make x items. Is there a most profitable number of items to make each week? If so, find what it is.

Solution   Profit is revenue minus cost, so the weekly profit on x items is

$$p(x) = r(x) - c(x) = 150x - 0.01x^2 - 20{,}000.$$

We want to find the maximum value (if any) of p on the open interval $x > 0$. The graph of p (Fig. 4.43) is a parabola that opens down, so p has an absolute maximum value. It occurs where

$$\frac{dp}{dx} = 150 - 0.02x = 0,$$

which means at

$$x = \frac{150}{0.02} = 7500.$$

To answer the questions, then, there *is* a production level for maximum profit, and that level is $x = 7500$ units per week. $\blacksquare$

Cost and Revenue in Economics

Here we want to point out two of the many places where calculus makes an important contribution to economic theory. The first has to do with the relationship between profit, revenue (money received), and cost. Suppose that

$$r(x) = \text{the revenue from selling } x \text{ items,}$$

$$c(x) = \text{the cost of producing the } x \text{ items,}$$

$$p(x) = r(x) - c(x) = \ \text{the profit from selling } x \text{ items.}$$

The marginal revenue and marginal cost at this production level (x items) are

$$\frac{dr}{dx} = \text{marginal revenue,}$$

$$\frac{dc}{dx} = \text{marginal cost.}$$

The first theorem is about the relationship of p to these derivatives.

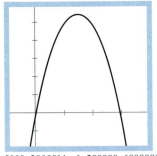

[−5000, 20000] by [−200000, 600000]

4.43   A complete graph of

$$p(x) = 150x - 0.01x^2 - 20{,}000.$$

Note that $p(0) = -c(0) = -20{,}000$. This represents *fixed costs* that the manufacturer incurs even if no items are made.

> ### THEOREM 7
>
> Maximum profit (if any) occurs at a production level at which marginal revenue equals marginal cost.

Proof We assume that $r(x)$ and $c(x)$ are differentiable for all $x > 0$, so if $p(x) = r(x) - c(x)$ has a maximum value, it occurs at a production level at which $p'(x) = 0$. Since $p'(x) = r'(x) - c'(x)$, $p'(x) = 0$ implies that

$$r'(x) - c'(x) = 0 \qquad \text{or} \qquad r'(x) = c'(x).$$

This concludes the proof (Fig. 4.44). ∎

THE LEAST AND THE GREATEST

Many problems of the seventeenth century that motivated the development of the calculus were maximum and minimum problems. Often these problems came from research in physics, such as finding the maximum range of a cannon. Galileo showed that the maximum range of a cannon is obtained with a firing angle of 45 degrees above the horizontal. He also found formulas for predicting maximum heights reached by projectiles fired at various angles to the ground. Pierre de Fermat worked on other problems of maxima and minima, culminating in his principle of least time (see Section 4.5 Exercise 68). This was generalized by Sir William Hamilton in his principle of least action—one of the most powerful underlying ideas in physics.

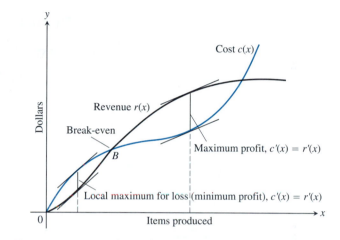

4.44 The graph of a typical cost function starts concave down and later turns concave up. It crosses the revenue curve at the break-even point B. To the left of B, the company operates at a loss. To the right, the company operates at a profit, the maximum profit occurring where $c'(x) = r'(x)$. Farther to the right, cost exceeds revenue (perhaps because of inefficiencies in handling a high rate of production combined with a "saturated" marketplace), and production levels become unprofitable again.

What guidance do we get from this theorem? We know that a production level at which $p'(x) = 0$ need not be a level of maximum profit. It might be a level of minimum profit, for example. But if we are making financial projections for our company, we should look for production levels at which marginal cost seems to equal marginal revenue. If there is a most profitable production level, it will be one of these.

EXAMPLE 4

Suppose that

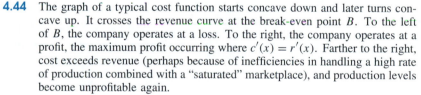

where x represents thousands of units. Is there a production level that maximizes profit? If so, what is it?

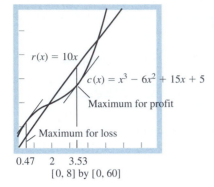

0.47 2 3.53

[0, 8] by [0, 60]

4.45 The cost and revenue curves for Example 4.

Solution

$$r(x) = 10x, \quad c(x) = x^3 - 6x^2 + 15x + 5 \qquad \text{Find } r'(x) \text{ and } c'(x).$$

$$r'(x) = 10, \quad c'(x) = 3x^2 - 12x + 15$$

$$3x^2 - 12x + 15 = 10 \qquad \text{Set them equal.}$$

$$3x^2 - 12x + 5 = 0 \qquad \text{Rearrange.}$$

Using the quadratic formula, we find that $x = 2 \pm \sqrt{21}/3$, that is, $x = 0.472$ or $x = 3.528$. The possible production levels for maximum profit are $x = 0.472$ thousand units or $x = 3.527$ thousand units. The graphs in Fig. 4.45 show $x = 3.528$ to be a point of maximum profit and $x = 0.472$ to be a local maximum for loss. ≡

EXPLORATION 2

Minimizing Average Cost

Another way to look for optimal production levels is to look for levels that minimize the *average cost* of the units produced. If $c(x)$ is the cost of producing x items ($x > 0$), then $A(x) = c(x)/x$ is the average cost per item of producing the x items.

1. Use differentiation on $A(x) = c(x)/x$ to find candidates for values of x that minimize average cost. (Find where $A' = 0$.) Summarize your result in the form of a theorem.

2. Suppose $c(x) = x^3 - 6x^2 + 15x + 5$ is a cost function where x represents thousands of units produced. Use two different methods to find any candidate(s) for values of x that minimize average cost. For one method, use your theorem from part 1. For the second method, give a straightforward analysis of $A(x)$. Use your grapher to support your work.

3. Confirm that one of the values of x that you found in part 2 does indeed minimize average cost. &

Modeling Discrete Phenomena with Differentiable Functions

In case you are wondering how we can use differentiable functions $c(x)$ and $r(x)$ to describe the cost and revenue that come from producing a number of items x that can only be an integer, here is the rationale.

When x is large, we can reasonably fit the cost and revenue data with smooth curves $c(x)$ and $r(x)$ that are defined not only at integer values of x but at the values in between. Once we have these differentiable functions, which are supposed to behave like the real cost and revenue when x is an integer, we can apply calculus to come to conclusions about their values. We then translate these mathematical conclusions into inferences about the real world that we hope will have predictive value. When they do, as is the case with the economic theory here, we say that the functions give a good model of reality.

What do we do when our calculus tells us that the best production level is a value of x that isn't an integer, as it did in Example 4 when it said that $x = 2 + \sqrt{21}/3$ thousand units would be the production level for maximum profit? The answer is to use the nearest "convenient" integer. For $x = 2 + \sqrt{21}/3$ thousand, we might use 3528, or 3530, or 3500, depending perhaps on a "convenience factor" such as whether the items will be shipped in boxes of 12, 10, or 100.

Exercises 4.4

In Exercises 1–26, show a complete graph, and identify the inflection points, local maximum and minimum values, and the intervals on which the graph is rising, falling, concave up, and concave down. We suggest that you do exercises from the three groups in the order given.

Do Exercises 1–6 analytically, then support graphically.

1. $y = \dfrac{x^2 - 1}{x}$

2. $y = \dfrac{x^2 + 4}{2x}$

3. $y = \dfrac{x^4 + 1}{x^2}$

4. $y = \dfrac{x^3 + 1}{x^2}$

5. $y = \dfrac{x}{x^2 - 4}$

6. $y = \dfrac{x - 1}{x^3 - 2x^2}$

Do Exercises 7–12 graphically, then confirm analytically.

7. $y = \dfrac{1}{x^2 - 1}$

8. $y = \dfrac{x^2}{x^2 - 1}$

9. $y = -\dfrac{x^2 - 2}{x^2 - 1}$

10. $y = \dfrac{x^2 - 4}{x^2 - 2}$

11. $y = \dfrac{x^2 - 4}{x - 1}$

12. $y = -\dfrac{x^2 - 4}{x + 1}$

Do Exercises 13–26 using a method of your choice.

13. $y = x^2 - x + \dfrac{1}{x + 2}$

14. $y = x + \dfrac{4}{x^2}$

15. $y = \dfrac{x + 1}{x^2 + 1}$

16. $y = \dfrac{x^2 + 1}{x^3 - 4x}$

17. $y = \dfrac{2x^3 + 3x^2 - 2x + 1}{x^2 + 2x}$

18. $y = \dfrac{3x^3 - 7x^2 - 7x - 1}{x^2 - 3x}$

19. $y = \dfrac{2x^3 - 4x^2 + 3}{x - 2}$

20. $y = \dfrac{-x^3 - 3x^2 + x + 5}{x + 3}$

21. $y = \dfrac{x^4 - 3x^2 + 4x + 1}{x^2 + x - 2}$

22. $y = \dfrac{2x^4 + 3x^3 + x^2 + x - 3}{x^2 + x}$

23. $y = \dfrac{2x^5 - 18x^3 + 2}{x^2 - 9}$

24. $y = \dfrac{x^5 + 3x^4 - 11x^3 - 3x^2 + 9x - 4}{x^2 + 3x - 10}$

25. $y = \dfrac{8}{x^2 + 4}$ (Agnesi's witch)

26. $y = \dfrac{4x}{x^2 + 4}$ (Newton's serpentine)

27. What is the smallest perimeter possible for a rectangle whose area is 16 in.²?

28. You are planning to make an open rectangular box with a square base that will hold a volume of 50 ft³. What are the dimensions of the box with minimum surface area?

29. A 216-m² rectangular pea patch is to be enclosed by a fence and divided into two equal parts by another fence parallel to one of the sides. What dimensions for the outer rectangle will require the smallest total length of fence? How much fence will be needed?

30. *The lightest steel holding tank.* Your iron works has contracted to design and build a 500-ft³, square-based, open-top, rectangular steel holding tank for a paper company. The tank is to be made by welding half-inch-thick stainless steel plates together along their edges. As the production engineer, your job is to find dimensions for the base and height that will make the tank weigh as little as possible. What dimensions do you tell the shop to use? Explain.

31. *Catching rainwater.* An 1125-ft³ open-top rectangular tank with a square base x ft on a side and y ft deep is to be built with its top flush with the ground to catch runoff water. The costs associated with the tank involve not only the material from which the tank is made but also an excavation charge proportional to the product xy. If the total cost is

$$c = 5(x^2 + 4xy) + 10xy,$$

what values of x and y will minimize it?

32. You are designing a poster to contain 50 in.² of printing with margins of 4 in. each at top and bottom and 2 in. at each side. What overall dimensions will minimize the amount of paper used?

33. What are the dimensions of the lightest open-top right circular cylindrical can that will hold a volume of 1000 cm^3? Compare the result here with the result in Example 2.

34. You are designing 1000-cm^3 right circular cylindrical cans whose manufacture will take waste into account. There is no waste in cutting the aluminum for the sides, but the tops and bottoms of radius r cm will be cut from squares that measure $2r$ cm on a side. The total amount of aluminum used up by each can will therefore be

$$A = 8r^2 + 2\pi rh$$

rather than the $A = 2\pi r^2 + 2\pi rh$ in Example 2. In Example 2 the ratio of h to r for the most economical cans was 2 to 1. What is the ratio now?

35. A rectangular sheet of $8\frac{1}{2} \times 11$-in. paper is placed on a flat surface, and one of the corners is lifted up and placed on the opposite longer edge as shown. (The other corners are held in their original positions.) With all four corners now held fixed, the paper is smoothed flat. Make the length of the crease as small as possible (call the length L).

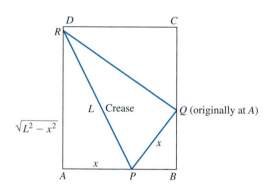

a) Try it with paper.
b) Show that $L^2 = 2x^3/(2x - 8.5)$.
c) Minimize L^2.
d) Find the minimum value of L.

36. Let $f(x)$ and $g(x)$ be the differentiable functions graphed as shown. Point c is the point where the vertical distance between the curves is the greatest. Show that the tangents to the curves at $x = c$ have to be parallel.

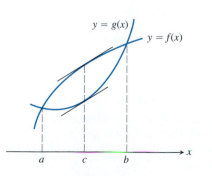

37. What value of a makes

$$f(x) = x^2 + \frac{a}{x}$$

have
a) a local minimum at $x = 2$?
b) a point of inflection at $x = 1$?

38. Show that

$$f(x) = x^2 + \frac{a}{x}$$

cannot have a local maximum for any value of a.

39. Show that the function $y = -(6x^2 + 8x + 19)$, the numerator of f' of Exploration 1, is less than 0 for all x.

40. Show that the function $f(x) = x^2 - x + 1$ is never negative.

Business

41. It costs a manufacturer c dollars each to manufacture and distribute a certain item. If the items sell at x dollars each, the number sold is given by $N = a/(x - c) + b(100 - x)$, where a and b are certain positive constants. What selling price will bring a maximum profit?

42. You operate a tour service that offers the following rates:
a) $200 per person if 50 people (the minimum number to book the tour) go on the tour.
b) For each additional person, up to a maximum of 80 people total, everyone's charge is reduced by $2.

It costs you $6000 (a fixed cost) plus $32 per person to conduct the tour. How many people does it take to maximize your profit?

43. *The best quantity to order.* One of the formulas for inventory management says that the average weekly cost of ordering, paying for, and holding merchandise is

$$A(q) = \frac{km}{q} + cm + \frac{hq}{2},$$

where q is the quantity that you order when things run low (shoes, radios, brooms, or whatever the item might be), k is the cost of placing an order (the same, no matter how often you order), c is the cost of one item (a constant), m is the number of items sold each week (a constant), and h is the weekly holding cost per item (a constant that takes into account things such as space, utilities, insurance, and security). Your job, as the inventory manager for your store, is to find the quantity that will minimize $A(q)$. What is it? (The formula that you get for the answer is called the *Wilson lot size formula.*)

44. *Continuation of Exercise 43.* Shipping costs sometimes depend on order size. When they do, it is more realistic to replace k by $k + bq$, the sum of k and a constant multiple of q. What is the most economical quantity to order now?

Economics

45. Use Theorem 7 to show that if $r(x) = 6x$ and $c(x) = x^3 - 6x^2 + 15x$ are your revenue and cost functions, then your operation will never be profitable and the best you can do is break even (have revenue equal cost).

46. Suppose $c(x) = x^3 - 20x^2 + 20,000x + 1000$ is the cost of manufacturing x items. Use Exploration 2 to find a production level that will minimize the average cost of making x items.

47. Let $f(x)$ be a rational function.
 a) Show that the domains of f, f', f'', and so forth are the same.
 b) Show that the vertical asymptotes of f, f', f'', and so forth are the same.

48. Let $f(x)$ be a rational function with $q(x)$ its end behavior

asymptote. Show that $q'(x)$ and $q''(x)$ are the end behavior asymptotes of f' and f'', respectively.

49. You have been asked to design an oil can shaped like a right cylinder (including both top and bottom) with a specified volume. Determine analytically what dimensions will use the least material.

50. A can shaped like a right cylinder (including both top and bottom) with volume 1 ft^3 is to be made from material costing \$0.80 per ft^2. A vertical seam and the top and bottom are soldered at a cost of \$0.20 per linear ft. Find the dimensions of the can of minimum cost and the minimum cost graphically, and confirm analytically.

4.5 _____ Radical and Transcendental Functions

In this section, we use calculus and graphing utilities to expand our understanding about radical and transcendental functions and some of their applications. We use the numerical derivative feature of graphing utilities, even for functions whose derivatives we are not yet ready to calculate exactly.

Radical Functions

The domain and range of the **radical function** $y = \sqrt[n]{x}$ is $[0, \infty)$ if n is even and $(-\infty, \infty)$ if n is odd. The graphs of $y = \sqrt[n]{x}$ with n even all have behavior similar to that of the graph of $y = \sqrt{x}$, which is illustrated in Example 1.

EXAMPLE 1

Draw a complete graph of $f(x) = \sqrt{x}$. Confirm its behavior analytically.

Solution Figure 4.46(a) gives a complete graph of f. It suggests that f is increasing in its domain $[0, \infty)$ with no local extreme values. The graph of f appears to be concave down on the domain of f with no points of inflection. The graphs of f' and f'' in Figs. 4.46(b) and 4.46(c) support these observations.

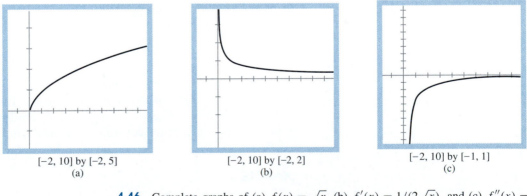

[−2, 10] by [−2, 5]   [−2, 10] by [−2, 2]   [−2, 10] by [−1, 1]
(a)   (b)   (c)

4.46 Complete graphs of (a) $f(x) = \sqrt{x}$, (b) $f'(x) = 1/(2\sqrt{x})$, and (c) $f''(x) = -1/(4\sqrt{x^3})$.

Differentiating, we have

$$f'(x) = \frac{1}{2}x^{-1/2} = \frac{1}{2\sqrt{x}} \quad \text{and}$$

$$f''(x) = -\frac{1}{4}x^{-3/2} = -\frac{1}{4\sqrt{x^3}}.$$

Neither derivative exists at $x = 0$. Also $f' > 0$ in its domain $(0, \infty)$, and $f'' < 0$ in its domain $(0, \infty)$. So f has no extreme values nor points of inflection.

The graphs of $y = \sqrt[n]{x}$ with n odd all have behavior similar to that of the graph of $y = \sqrt[3]{x}$, which is illustrated in Example 2.

EXAMPLE 2

Draw a complete graph of $f(x) = \sqrt[3]{x}$. Confirm its behavior analytically.

Solution   Figure 4.47(a) gives a complete graph of f. It suggests that f is increasing in its domain $(-\infty, \infty)$ with no local extreme values. The graphs of f' (Fig. 4.47b) and f'' (Fig. 4.47c) support this behavior of f.

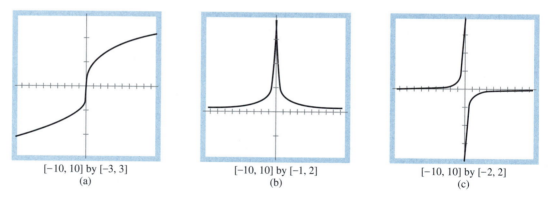

[−10, 10] by [−3, 3]   [−10, 10] by [−1, 2]   [−10, 10] by [−2, 2]
(a)   (b)   (c)

4.47   Complete graphs of (a) $f(x) = \sqrt[3]{x}$, (b) $f'(x)$, and (c) $f''(x)$.

Differentiating, we have

$$f'(x) = \frac{1}{3}x^{-2/3} = \frac{1}{3\sqrt[3]{x^2}} \quad \text{and}$$

$$f''(x) = -\frac{2}{9}x^{-5/3} = \frac{-2}{9\sqrt[3]{x^5}}.$$

The equation for f' tells us that $f' > 0$ for $x \neq 0$, confirming that f is always increasing and thus has no extreme values. The equation for f'' tells us that $f'' > 0$ for $x < 0$ and $f'' < 0$ for $x > 0$, confirming that f is concave up in $(-\infty, 0)$ and concave down in $(0, \infty)$ and has a point of inflection at $x = 0$.

EXAMPLE 3

Find the absolute maximum and minimum values of $y = x^{2/3}$ on the interval $-2 \leq x \leq 3$ analytically. Support your answers graphically.

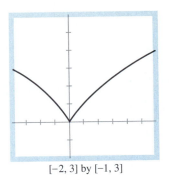

[-2, 3] by [-1, 3]

4.48 The graph of $f(x) = x^{2/3}$ showing extreme values on $-2 \le x \le 3$.

EXPLORATION BIT

We must be careful when graphing functions of the form $y = x^{m/n}$ with graphing utilities. For $y = x^{2/3}$, for example, entering the exponent in the form 2/3 may produce an incorrect graph. Try this with your graphing utility, and compare with the graph of Example 3. If you get the wrong graph, try entering the function in the forms $y = (x^2)^{1/3}$ and $y = (x^{1/3})^2$.

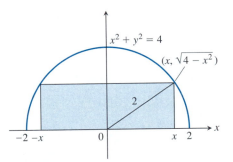

4.49 For Example 4, visualize a rectangle inscribed in a semicircle as shown with x ranging from 0 to 2. What is the area of the rectangle when $x = 0$? when $x = 2$? Can you give a convincing argument that a largest rectangle must exist for some x with $0 \le x \le 2$?

Solution We evaluate the function $f(x) = x^{2/3}$ at the critical points and endpoints and take the largest and smallest of these values.

The first derivative,

$$f'(x) = \frac{2}{3}x^{-1/3} = \frac{2}{3\sqrt[3]{x}},$$

has no zeros and is undefined at $x = 0$. The values of the function at this one critical point and at the endpoints are

Critical point value: $f(0) = 0$

Endpoint values: $f(-2) = (-2)^{2/3} = 4^{1/3} = 1.59,$

$$f(3) = (3)^{2/3} = 9^{1/3} = 2.08.$$

We conclude that the function's maximum value is $9^{1/3}$, taken on at $x = 3$. The minimum value is 0, taken on at $x = 0$. The complete graph of f in Fig. 4.48 supports these conclusions. ≡

EXAMPLE 4   Geometry

Rectangles of different sizes can be inscribed in a semicircle of radius 2. Find the dimensions of the inscribed rectangle with largest area. Support your work graphically.

Solution To describe the dimensions of the rectangle, we place the circle and rectangle in the coordinate plane (Fig. 4.49). The length, height, and area of the rectangle can then be expressed in terms of the position x of the lower right-hand corner:

Length: $2x,$

Height: $\sqrt{4 - x^2},$

Area: $2x \cdot \sqrt{4 - x^2}.$

Our mathematical goal is to find the maximum value of the continuous function

$$A(x) = 2x\sqrt{4 - x^2}$$

on the interval $0 \le x \le 2$. To support our work so far, we GRAPH $y = A(x)$ (Fig. 4.50) and see that there appears to be a maximum value of A near $x = 1.4$. The graph suggests and the situation confirms that the maximum value does not occur at the endpoints 0 and 2 of the domain, so we have only to find and check values of A at any critical points in (0, 2).

The derivative

$$\frac{dA}{dx} = \frac{-2x^2}{\sqrt{4 - x^2}} + 2\sqrt{4 - x^2}$$

is equal to zero when

$$\frac{-2x^2}{\sqrt{4-x^2}} + 2\sqrt{4-x^2} = 0$$

$$-2x^2 + 2(4-x^2) = 0$$

$$8 - 4x^2 = 0$$

$$x^2 = 2$$

$$x = \pm\sqrt{2}.$$

For $0 < x < 2$, the only critical point is $\sqrt{2}$. The value of A at this critical point is

$$A(\sqrt{2}) = 2\sqrt{2}\sqrt{4-2} = 4.$$

The area has a maximum value of 4 when the rectangle is $x = \sqrt{2}$ units high and $2x = 2\sqrt{2}$ units long. This is supported quite nicely by our view (Fig. 4.50b) of the graph of A where an extreme value appears to occur at about $x = 1.4$. ≡

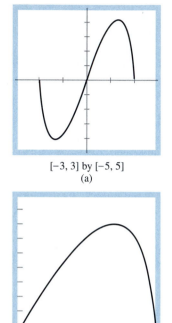

[−3, 3] by [−5, 5]
(a)

[0, 2] by [0, 5]
(b)

4.50 (a) A complete graph of $A(x) = 2x\sqrt{4-x^2}$. (b) The graph of A on [0, 2].

EXPLORATION 1

The Largest Cone

A cone of height h and radius r is constructed from a flat, circular disk of radius 4 in. by removing a sector AOC of arc length x in. and then connecting the edges AO and CO. Find the arc length x that will produce the cone of maximum volume.

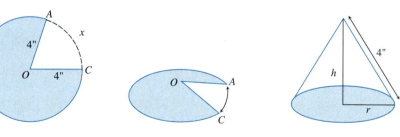

1. Show that

$$r = \frac{8\pi - x}{2\pi}, \qquad h = \sqrt{16 - r^2},$$

and

$$V(x) = \frac{\pi}{3}\left(\frac{8\pi - x}{2\pi}\right)^2 \sqrt{16 - \left(\frac{8\pi - x}{2\pi}\right)^2}.$$

2. Show that the domain of $y = V(x)$ is $0 \le x \le 16\pi$. Draw a complete graph of $y = V(x)$. (*Hint*: To simplify your keying, let $y_1 = (8\pi - x)/2\pi$, $y_2 = \sqrt{16 - y_1^2}$, and so on.)

3. Explain why only $0 \le x \le 8\pi$ makes sense in the problem situation, and draw a graph of the problem situation.

EXPLORATION BIT

For the value of x found in Exploration 1, compute r, h, and the ratio h/r. Do you recognize the ratio? Make a conjecture about the exact value of the ratio. (See Exercise 72 for an extension of Exploration 1.)

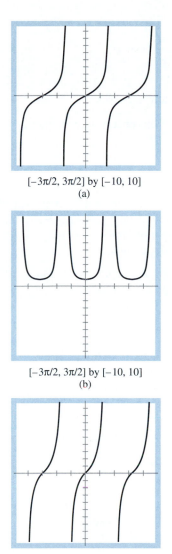

$[-3\pi/2, 3\pi/2]$ by $[-10, 10]$
(a)

$[-3\pi/2, 3\pi/2]$ by $[-10, 10]$
(b)

$[-3\pi/2, 3\pi/2]$ by $[-10, 10]$
(c)

4.51 Complete graphs of (a) $f(x) = \tan x$, (b) $f'(x)$, and (c) $f''(x)$. A complete graph must show at least one period. Each of these complete graphs shows three.

4. Use a grapher (ZOOM-IN or SOLVE) to show that $x = 4.612$ in. produces the cone with maximum volume of 25.796258 in.^{3} Explain why you can determine V with error at most 10^{-6} by determining x with error at most 10^{-3}.

5. Confirm your results in part 4 analytically. (*Hint*: Use $V(x) = (1/3)\pi r^2 h$, $h^2 + r^2 = 4^2$, and the Chain Rule.)

Trigonometric Functions

In Chapter 1 and in prior courses, you studied the graphs of $y = af(bx + c) + d$, where f is any one of the six basic trigonometric functions: $\sin x$, $\cos x$, $\tan x$, $\csc x$, $\sec x$, $\cot x$. In the exercises, you will use calculus to confirm and deepen your understanding of these functions.

EXAMPLE 5

Draw a complete graph of $f(x) = \tan x$. Confirm its behavior analytically.

Solution We know that f is a periodic function with period π, and it is undefined at $\pi/2 + k\pi$ for all integers k. The complete graph of f in Fig. 4.51(a) shows three periods and suggests asymptotes at $x = \pi/2 + k\pi$. The derivatives of f (Figs. 4.51b and 4.51c) must also be periodic with period π and exhibit asymptotic behavior at $x = \pi/2 + k\pi$. We can calculate the derivatives

$$f'(x) = \sec^2 x, \qquad \text{and} \qquad f''(x) = 2\sec x \sec x \tan x = 2\sec^2 x \tan x.$$

Now $f'(x) > 0$ for all x, so f is increasing on $(-\pi/2, \pi/2)$ and on any interval $(-\pi/2 + k\pi, \pi/2 + k\pi)$. Also, f has no local extrema. The graph of f'' shows that f is concave down on $(-\pi/2, 0)$ and all the intervals $(-\pi/2 + k\pi, 0 + k\pi)$; f is concave up on $(0, \pi/2)$ and all the intervals $(0 + k\pi, \pi/2 + k\pi)$. ≡

We can conclude from Example 5 that for a graph of a periodic function to be complete, it is enough to describe its behavior in one period.

EXPLORATION 2

Confirming Analytically

Draw a complete graph of $f(x) = \tan x - 3\cos x$.

1. Find the period. Since the period of $\tan x$ is π and the period of $\cos x$ is 2π, the period of f must divide 2π. Why? A graph will suggest this period (telling you which divisor of 2π) and also will suggest values for a better choice of [xMin, xMax].

2. GRAPH f on [xMin, xMax]. Tell where the graph appears to be rising and falling, is concave up and concave down, and exhibits asymptotic behavior. Tell what appears to be true about local extreme values of f and points of inflection. GRAPH f' and f''. How do these graphs support what you know about f from its graph?

3. Show that the derivatives of f are

$$f'(x) = \sec^2 x + 3 \sin x$$

and

$$f''(x) = 2 \sec^2 x \tan x + 3 \cos x.$$

Give convincing analytic arguments for your observations in part 2. ☘

EXAMPLE 6

A sphere of radius 20 cm is cut so that a cylindrical plug with spherical caps at each end can be removed as illustrated in Fig. 4.52. Determine the dimensions of the cylinder (without the caps) of largest possible volume that can be removed from the sphere. What is its volume?

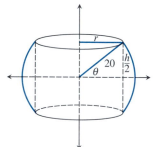

4.52 The sphere of Example 6.

Solution Let r, h, and θ be as shown in Fig. 4.52. The volume of the cylinder is $V = \pi r^2 h$. Notice that $\cos \theta = r/20$ and $\sin \theta = (h/2)/20$. Thus, $r = 20 \cos \theta, h = 40 \sin \theta$, and $V = 16,000\pi \sin \theta \cos^2 \theta$, where $0 \leq \theta \leq \pi/2$.

Our intuition suggests that there is a cylinder with greatest volume as θ ranges from 0 to $\pi/2$. Our mathematical goal is to find the value of θ where this maximum volume occurs and then the maximum value itself. That we are on the right track is supported by the graph of V in the viewing window (Fig. 4.53), which shows a maximum volume of about 19,500 cm^3 when θ is 0.6.

We use calculus to confirm these results by finding where $V'(\theta) = 0$.

$$V'(\theta) = 16,000\pi \,(\cos^3 \theta - 2 \sin^2 \theta \cos \theta) = 0$$

$$2 \sin^2 \theta \cos \theta = \cos^3 \theta$$

$$\tan^2 \theta = 1/2 \qquad \text{Divide by } 2\cos^3 \theta \neq 0 \text{ in } (0, \pi/2)...$$

$$\tan \theta = 1/\sqrt{2}. \qquad \text{because } \tan \theta > 0 \text{ in } (0, \pi/2).$$

$$\theta = \tan^{-1}(1/\sqrt{2}) = 0.6154\ldots$$

Finally, for this value of θ, $r = 16.33$, $h = 23.09$, and $V = 19,347.19$ cm^3, the dimensions and volume of the largest cylinder. ▤

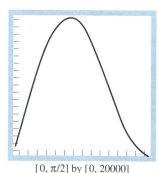

[0, π/2] by [0, 20000]

4.53 The graph of

$$V(\theta) = 16,000\pi \sin \theta \cos^2 \theta$$

in $0 \leq \theta \leq \pi/2$.

We have to build our experience with and our confidence in NDER. Later in this book, we will encounter functions for which we *have to* place our faith in NDER that it gives a good approximation of the derivative.

Exponential and Logarithmic Functions

In Section 1.6, we introduced the exponential function $y = a^x (a > 0, a \neq 1)$ and the logarithmic function $y = \log_a x \, (a > 0, a \neq 1)$ as inverses of each other. The domain of $y = a^x$ is $(-\infty, \infty)$, and the range is $(0, \infty)$. This means that the domain of $y = \log_a x$ is $(0, \infty)$ and its range is $(-\infty, \infty)$.

Now we will look at graphical behavior involving these functions. Although we don't know the formulas yet for finding their derivatives, we can look at numerical derivatives and give arguments that the graphs are complete.

EXAMPLE 7

The function $y = a^x (a > 1)$ has behavior similar to $y = 2^x$. The function $y = a^x (0 < a < 1)$ has behavior similar to $y = 0.5^x$. Graph (a) $y = 2^x$ and (b) $y = 0.5^x$, and give convincing arguments that the graphs are complete.

Solution   For $y = 2^x$, graphs of y, y', and y'' are shown in Fig. 4.54. The graphs of y' and y'' allow us to conclude that 2^x is increasing and concave up on $(-\infty, \infty)$ and that there are no local extrema or points of inflection.

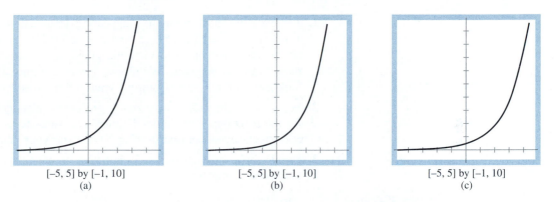

| $[-5, 5]$ by $[-1, 10]$ | $[-5, 5]$ by $[-1, 10]$ | $[-5, 5]$ by $[-1, 10]$ |
| (a) | (b) | (c) |

4.54   Complete graphs of (a) $y_1 = 2^x$, (b) $y_2 = \text{NDER } y_1$, and (c) $y_3 = \text{NDER } y_2$. It turns out that y_2 and y_3 are vertical shrinks of the graph of y_1, so you may find it instructive to graph all three simultaneously.

For $y = 0.5^x$, graphs of y, y', and y'' are shown in Fig. 4.55. The graphs of y' and y'' allow us to conclude that 0.5^x is decreasing and concave up on $(-\infty, \infty)$ and that there are no local extrema or points of inflection.   ≡

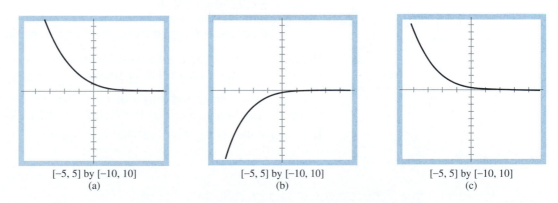

| $[-5, 5]$ by $[-10, 10]$ | $[-5, 5]$ by $[-10, 10]$ | $[-5, 5]$ by $[-10, 10]$ |
| (a) | (b) | (c) |

4.55   Complete graphs of (a) $y_1 = 0.5^x$, (b) $y_2 = \text{NDER } y_1$, and (c) $y_3 = \text{NDER } y_2$. How do the graphs in Fig. 4.54(a) and Fig. 4.55(a) compare? How can you test this?

The graph of $g(x) = 0.5^x$ is the reflection in the y-axis of the graph of $f(x) = 2^x$. This is because

$$g(x) = 0.5^x = \left(\frac{1}{2}\right)^x = 2^{-x} = f(-x).$$

Later we will see that $f'(x) = (\ln 2)2^x$ and $f''(x) = (\ln 2)^2 2^x$. This explains why the graphs of f' and f'' are vertical shrinks of the graph of f. We will also see that $g'(x) = (\ln 0.5)0.5^x$ and $g''(x) = (\ln 0.5)^2 0.5^x$. This together with the fact that $\ln 0.5 < 0$ explains the graphs in Figs. 4.55(b) and 4.55(c).

EXPLORATION 3

Exploring Exponential Functions

1. Three of the most commonly appearing irrational numbers are $\sqrt{2} = 1.41421\ldots,\ \pi = 3.14159\ldots,$ and $e = 2.71828\ldots.$ You perhaps have never before attempted to view the graph of $y = a^x$ where a is one of these irrational numbers. Try it now, but first predict what the graph will look like with a sketch. Compare your grapher result with your sketch.

2. For $y = 2^x$, the graphs of y and y' are nearly identical. It is a fact that for the functions $y = a^x\,(a > 1)$, there is a value for a for which $y = y'$. See whether you can find it using NDER and your graphing utility.

3. Draw graphs of each function, and give convincing arguments that the graphs are complete.

   a) $f(x) = xe^{-x}$ **b)** $f(x) = \dfrac{e^x + e^{-x}}{2}$

REMINDER

To graph $y = \log_a x$ with a graphing utility, we need to use the change of base formula

$$\log_a x = \frac{\log_b x}{\log_b a}$$

established in Chapter 1. Enter $y = \log_a x$ as either

$$y = \log_a x = \frac{\log_{10} x}{\log_{10} a} \quad \text{or}$$

$$y = \log_a x = \frac{\ln x}{\ln a}.$$

Caution: Some graphing utilities use log, which means logarithm base 10, for ln, which means logarithm base e.

Later we will establish connections between derivatives of functions that are inverses of each other. For now, we can get a good idea of a complete graph of a logarithmic function using numerical derivatives.

EXPLORATION 4

Complete Graphs of Logarithmic Functions

The function $y = \log_a x\,(a > 1)$ has behavior similar to $\log_2 x$. The function $y = \log_a x\,(0 < a < 1)$ has behavior similar to $y = \log_{0.5} x$.

1. Graph (a) $y = \log_2 x$ and (b) $y = \log_{0.5} x$, and give convincing arguments that the graphs are complete. Use NDER and NDER2 as needed.

2. Compare the graphs of $y_1 = 2^x$ and $y_2 = \log_2 x$.

Exercises 4.5

In Exercises 1–48 show a complete graph and identify the inflection points, local maximum and minimum values, and the intervals on which the graph is rising, falling, concave up, and concave down—in only one period if the function is periodic. We suggest that you do exercises from the three groups in the order given.

Do Exercises 1–10 analytically, then support graphically.

1. $y = \sqrt[3]{x}$ **2.** $y = x^{4/3}$

3. $y = x^{3/2}$ **4.** $y = \sqrt[5]{1-x}$

5. $y = \sqrt{2x+3}$ **6.** $y = \sqrt[4]{x-2}+5$

7. $y = 2\sin(3x+5)+3$ **8.** $y = -3\cos(2x+\pi)+1$

9. $y = \sin 3x + \cos 3x$ **10.** $y = 5 \sin 2x + 3 \cos 2x$

Do Exercises 11–18 graphically, then confirm analytically.

11. $y = x^{5/3}$ **12.** $y = x^{5/4}$

13. $y = \sqrt{2x - 3}$ **14.** $y = \sqrt[3]{3 - x} + 5$

15. $y = \sqrt[4]{x - 4} - 2$ **16.** $y = \sqrt[3]{x^2 - 3x}$

17. $y = 3 \csc (2x + \pi) - 5$ **18.** $y = -2 \sec (3x + \pi) + 7$

Do Exercises 19–48 using a method of your choice.

19. $y = x^2 + \sqrt[3]{x}$ **20.** $y = x^{2/3} - x$

21. $y = (x + 2)^{3/5}$ **22.** $y = (3 - x)^{2/3}$

23. $y = \sec 2x$ **24.** $y = \tan (3x + \pi)$

25. $y = 2e^{3x+1} + 5$ **26.** $y = 2^{x-3} + 5$

27. $y = 3 \log_2 (x + 1)$ **28.** $y = 2 \log_3 (2 - x)$

29. $y = \sqrt{x^2 - 2x}$ **30.** $y = \sqrt[3]{x^3 - 4x}$

31. $y = \sec x \tan x$ **32.** $y = \csc x \cot x$

33. $y = \sin 2x + \cos 3x$ **34.** $y = \sin \dfrac{x}{2} + \sin \dfrac{x}{3}$

35. $y = \cot x + 4 \sin x$ **36.** $y = 3 \sin x - \cot x$

37. $y = x \ln x$ **38.** $y = x^2 \ln x$

39. $y = x^2 \ln |x|$ **40.** $y = \dfrac{\ln x}{x}$

41. $y = 2^{x^2 - 1}$ **42.** $y = 3^{-x^2}$

43. $y = x2^{-x^2}$ **44.** $y = x3^x$

45. $y = \dfrac{x}{\sin x}$ on $-4\pi \le x \le 4\pi$

46. $y = x^2 \sin x$ on $-4\pi \le x \le 4\pi$

47. $y = \dfrac{x - \sin x}{x^2 + x}$ **48.** $y = \dfrac{1 - \cos x}{x^2 - 2x}$

Find the absolute maximum and minimum of each function in Exercises 49–52 on the specified interval.

49. $y = x^{3/5}$ on $-5 \le x \le 5$

50. $y = x^{4/3}$ on $-3 \le x \le 4$

51. $y = e^x \sin x$ on $-6 \le x \le 6$

52. $y = e^{-x} \cos x$ on $-5 \le x \le 5$

The *hyperbolic functions*—built-in on some calculators—are important functions with important applications. They describe the motions of waves in elastic solids, the shapes of hanging electric power lines, and the temperature distributions in metal cooling fins. You were given the opportunity to investigate the hyperbolic cosine,

$$\cosh x = \frac{e^x + e^{-x}}{2},$$

in Exploration 3. In Exercises 53–57, investigate the other five by drawing a complete graph of each. (These functions will be studied in more detail later.)

53. Hyperbolic sine of x : $\sinh x = \dfrac{e^x - e^{-x}}{2}$

54. Hyperbolic tangent of x : $\tanh x = \dfrac{\sinh x}{\cosh x} = \dfrac{e^x - e^{-x}}{e^x + e^{-x}}$

55. Hyperbolic cotangent of x : $\coth x = \dfrac{\cosh x}{\sinh x} = \dfrac{e^x + e^{-x}}{e^x - e^{-x}}$

56. Hyperbolic secant of x : $\operatorname{sech} x = \dfrac{1}{\cosh x} = \dfrac{2}{e^x + e^{-x}}$

57. Hyperbolic cosecant of x : $\operatorname{csch} x = \dfrac{1}{\sinh x} = \dfrac{2}{e^x - e^{-x}}$

58. The sum of two nonnegative numbers is 20. Find the numbers if one number plus the square root of the other is to be as large as possible.

59. What is the largest possible area for a right triangle whose hypotenuse is 5 cm long?

60. Two sides of a triangle have lengths a and b, and the angle between them is θ. What value of θ will maximize the triangle's area? (*Hint:* $A = (1/2)ab \sin \theta$.)

61. Find the largest possible value of $s = 2x + y$ if x and y are side lengths in a right triangle whose hypotenuse is $\sqrt{5}$ units long.

62. You are planning to close off a corner of the first quadrant with a line segment 20 units long running from $(a, 0)$ to $(0, b)$. Show that the area of the triangle enclosed by the segment is largest when $a = b$.

63. The trough is to be made to the dimensions shown. Only the angle θ can be varied. What value of θ will give the trough its maximum volume?

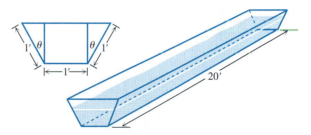

64. How close does the curve $y = \sqrt{x}$ come to the point $(3/2, 0)$?

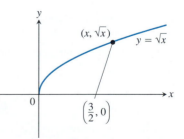

65. How close does the semicircle $y = \sqrt{16 - x^2}$ come to the point $(1, \sqrt{3})$?

66. Is the function $f(x) = 3 + 4\cos x + \cos 2x$ ever negative? How do you know?

67. Is the function $f(x) = (3^x + 3^{-x} - 2)/2$ ever negative? How do you know?

68. Fermat's principle in optics states that light always travels from one point to another along a path that minimizes the travel time. Figure 4.56 shows light from a source A reflected by a plane mirror to a receiver at point B. Show that for the light to obey Fermat's principle, the angle of incidence must equal the angle of reflection.

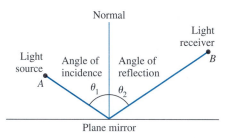

4.56 In studies of light reflection, the angles of incidence and reflection are measured from the line normal to the reflecting surface. Exercise 68 asks you to show that if light obeys Fermat's "least-time" principle, then $\theta_1 = \theta_2$.

69. Jane is 2 miles offshore in a boat and wishes to reach a coastal village 6 miles down a straight shoreline from the point nearest the boat. She can row 2 mph and can walk 5 mph. Where should she land her boat to reach the village in the least amount of time?

70. Find the dimensions of a right circular cylinder of maximum possible volume that can be inscribed in a sphere of radius 10 cm.

71. Rework Example 4 by writing the area of the inscribed rectangle as a function of the angle θ that the line segment from the origin to the upper right-hand corner of the rectangle makes with the positive x-axis.

72. *Generalized cone problem.* A cone of height h and radius r is constructed from a flat, circular disk of radius a inches as described in Exploration 1.

a) Determine an algebraic representation of the volume V of the cone in terms of x and a.

b) Find r and h in the cone of maximum volume for $a = 4$, 5, 6, 8.

c) Find a simple relationship between r and h that is independent of a for the cone of maximum volume. Show your steps and explain.

d) Confirm your generalization in part (c).

73. A rectangle has its base on the x-axis and its upper two vertices on the curve $y = 8\cos(0.3x)$. What is the largest area the rectangle can have?

74. Let $P(x, a)$ and $Q(-x, a)$ be two points on the upper half of the ellipse

$$\frac{x^2}{100} + \frac{(y-5)^2}{25} = 1$$

centered at $(0, 5)$. A triangle RST is formed by using the tangent lines to the ellipse at P and Q as shown.

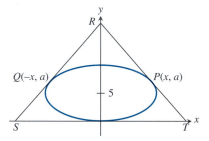

a) Show that the area of the triangle is

$$A(x) = -f'(x)\left[x - \frac{f(x)}{f'(x)}\right]^2,$$

where $y = f(x)$ is the function representing the upper half of the ellipse.

b) Draw a complete graph of $y = A(x)$. What values of x make sense in the problem situation? How are the asymptotes of the graph related to the problem situation?

c) Determine the height of the triangle with minimum area. How is it related to the y-coordinate of the center of the ellipse?

d) *Generalization.* Repeat parts (a)–(c) for the ellipse

$$\frac{x^2}{C^2} + \frac{(y-B)^2}{B^2} = 1$$

centered at $(0, B)$. Confirm that the triangle has minimum area when its height is $3B$. You may find it convenient to use CAS.

4.6 _____ Related Rates of Change

How fast does the radius change when you blow air into a spherical soap bubble at the rate of 10 cm^3/sec? How fast does the water level drop when a cylindrical tank is drained at the rate of 3 L/sec?

Questions like these ask us to calculate a rate that we cannot measure directly from a rate that we can. To do so, we write an equation that relates the variables involved and differentiate it to get an equation that relates the rate we seek to the rate we know.

EXAMPLE 1   The Soap Bubble

How fast does the radius of a spherical soap bubble change when you blow air into it at the rate of 10 cm^3/sec?

Solution   We are given the rate at which the volume is changing and are asked for the rate at which the radius is changing.

We think abstractly at first, picturing the bubble as a sphere whose volume V and radius r are differentiable functions of time t. The equation that relates V and r is

$$V = \frac{4}{3}\pi r^3. \tag{1}$$

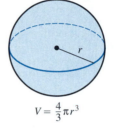

$$V = \frac{4}{3}\pi r^3$$

To find an equation that relates the rate we seek (dr/dt) to the rate we know (dV/dt), we differentiate both sides of Eq. (1) with respect to t using the Chain Rule:

$$\frac{dV}{dt} = \frac{d}{dt}\left(\frac{4}{3}\pi r^3\right) = \frac{4}{3}\pi \frac{d}{dt}(r^3) = \frac{4}{3}\pi \cdot 3r^2 \frac{dr}{dt} = 4\pi r^2 \frac{dr}{dt}. \tag{2}$$

We are told that

$$\frac{dV}{dt} = 10. \qquad \text{Air is blown in at the rate of 10 cm}^3\text{/sec.}$$

We substitute this value in Eq. (2) and solve for dr/dt:

$$10 = 4\pi r^2 \frac{dr}{dt}$$

$$\frac{dr}{dt} = \frac{10}{4\pi r^2}. \tag{3}$$

We see from Eq. (3) that the rate at which r changes at any particular time depends on how big r is at the time. When r is small, dr/dt will be large; when r is large, dr/dt will be small:

$$\text{At } r = 1 \text{ cm}: \quad \frac{dr}{dt} = \frac{10}{4\pi} \approx 0.8 \text{ cm/sec,}$$

$$\text{At } r = 10 \text{ cm}: \quad \frac{dr}{dt} = \frac{10}{400\pi} \approx 0.008 \text{ cm/sec.}$$

SETTING THE VIEW DIMENSIONS

To set the view dimensions for problems like those in Exploration 1, it is helpful to think of the constraints of the physical situation. For example, t (time) should begin at 0. Also, because V is proportional to r^3, it is sensible in part 2 to make $V\text{Max} = (r\text{Max})^3$.

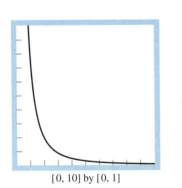

[0, 10] by [0, 1]

4.57 A graph of r' of Example 1 plotted against time t. How do you interpret this graph for t near 0? for $t \to \infty$?

EXPLORATION 1

Visualizing Relationships

Parametric mode on a grapher can help us to visualize relationships in related rate problems *when we are able to represent the related variables as functions of time, t.* In Example 1, V is increasing at the rate of 10 cm³/sec. We can think of V (in cubic centimeters) as a function of t (in seconds), namely, $V(t) = 10t$, assuming that $V = 0$ when $t = 0$.

1. Write r in terms of t by replacing V with $10t$ in $V = (4/3)\pi r^3$ and solving for r.

2. GRAPH the parametric equations

$$x_1(t) = r(t), \qquad y_1(t) = V(t).$$

Explain the graph that results. (*Hint:* What variable is represented by each axis?)

3. Now graph each of the following, and explain the graph that results in terms of the radius or volume of the soap bubble. (*Hint:* One of your graphs should match Fig. 4.57. Why?)

$$x_2(t) = t, \qquad y_2(t) = y_1(t).$$
$$x_3(t) = t, \qquad y_3(t) = x_1(t).$$
$$x_4(t) = t, \qquad y_4(t) = \text{NDER}\, x_1(t).$$

4. Define parametric equations $(x_5(t), y_5(t))$ whose graph shows the relationship between r' and r given in Eq. (3).

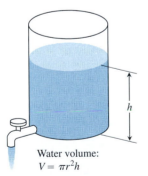

Water volume:
$V = \pi r^2 h$

4.58 When water flows out of the cylindrical tank, the volume of water decreases. The "cylinder" of water keeps the same radius but loses height. Example 2 shows a way to find how fast the water level drops when the water flows out at a constant rate.

EXAMPLE 2   The Cylindrical Tank

How fast does the water level drop when a cylindrical tank is drained at the rate of 3 liters/sec?

Solution   We draw a picture of a partially filled cylindrical tank, calling its radius r and the height of the water h (Fig. 4.58) with both r and h measured in centimeters. We call the volume of water in the tank V.

The radius r is a constant, but V and h change with time. We think of V and h as differentiable functions of time and use t to represent time. The derivatives dV/dt and dh/dt give the rates at which V and h change. We are told that

$$\frac{dV}{dt} = -3000 \text{ cm}^3/\text{sec}, \qquad \text{The tank is drained at the rate of } 3 \text{ liters/sec} = 3000 \text{ cm}^3 \text{sec,}$$

and we are asked to find

$$\frac{dh}{dt}. \qquad \text{How fast does the water level drop?}$$

To answer this question, we first write an equation that relates V and h:

$$V = \pi r^2 h. \qquad \text{The tank is cylindrical.}$$

We then differentiate both sides with respect to t to get an equation that relates dV/dt to dh/dt:

$$\frac{dV}{dt} = \pi r^2 \frac{dh}{dt}. \qquad \text{Radius } r \text{ is a constant.}$$

We substitute the known value $dV/dt = -3000$ and solve for dh/dt:

$$\frac{dh}{dt} = -\frac{3000}{\pi r^2}. \qquad \begin{array}{l}\text{Height } h \text{ is decreasing} \\ \text{because } dh/dt \text{ is negative.}\end{array}$$

The water level is dropping at the constant rate of $3000/\pi r^2$ cm/sec.   ≡

EXAMPLE 3   A Rising Balloon

A hot-air balloon, rising straight up from a level field, is tracked by a range finder 500 ft from the lift-off point. At the moment the range finder's elevation angle is $\pi/4$, the angle is increasing at the rate of 0.14 radians/min. How fast is the balloon rising?

Solution   We answer the question in six steps.

STEP 1: *Draw a picture, and name the variables and constants* (Fig. 4.59). The variables in the picture are

$\theta = $ the angle the range finder makes with the ground (radians),

$y = $ the height of the balloon (feet).

We let t represent time (in minutes) and assume θ and y to be differentiable functions of t. The one constant in the picture is the distance from the range finder to the lift-off point (500 ft).

STEP 2: *Write down the additional numerical information:*

$$\frac{d\theta}{dt} = 0.14 \text{ rad/min} \qquad \text{when} \quad \theta = \frac{\pi}{4}.$$

STEP 3: *Write down what we are asked to find.* We are asked to find dy/dt when $\theta = \pi/4$.

STEP 4: *Write an equation that relates the variables.* An equation that relates y to θ is

$$\frac{y}{500} = \tan\theta \qquad \text{or} \qquad y = 500\tan\theta.$$

STEP 5: *Differentiate with respect to t using the Chain Rule.* The result tells how dy/dt (which we want) is related to $d\theta/dt$ (which we know):

$$\frac{dy}{dt} = 500 \sec^2\theta \frac{d\theta}{dt}.$$

STEP 6: *Evaluate with $\theta = \pi/4$ and $d\theta/dt = 0.14$ to find dy/dt:*

$$\frac{dy}{dt} = 500(\sqrt{2})^2(0.14) = (1000)(0.14) = 140. \qquad \sec\frac{\pi}{4} = \sqrt{2}.$$

At the moment in question, the balloon is rising at the rate of 140 ft/min.   ≡

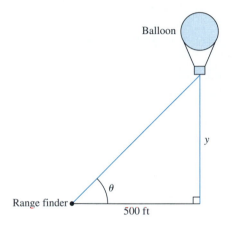

Balloon

y

θ

Range finder

500 ft

4.59 If $d\theta/dt = 0.14$ when $\theta = \pi/4$, what is the value of dy/dt when $\theta = \pi/4$? See Example 3.

Strategy for Solving Related Rate Problems

1. *Draw a picture, and name the variables and constants.* Use t for time. Assume that all variables are differentiable functions of t.

2. *Write down the numerical information* (in terms of the symbols you have chosen).

3. *Write down what you are asked to find* (usually a rate, expressed as a derivative).

4. *Write an equation that relates the variables.* You may have to combine two or more equations to get a single equation that relates the variable whose rate you want to the variable whose rate you know.

5. *Differentiate with respect to t using the Chain Rule.* Then express the rate you want in terms of the rate and variables whose values you know.

6. *Evaluate.* Use known values to find the unknown rate.

An unwritten step in solving related-rate problems—or any problem—is that you should be alert for ways to support your work with a grapher if support is needed. We showed some ways in which it could be done for Example 1. In Example 2, since dh/dt was a constant, grapher support would have been trivial.

EXAMPLE 4   The Conical Tank

Water runs into a conical tank at the rate of 9 ft³/min. The tank stands vertex down and has a height of 10 ft and a base radius of 5 ft. How fast is the water level rising when the water is 6 ft deep?

Solution   We carry out the steps of the basic strategy.

STEP 1: *Picture and variables.* We draw a picture of a partially filled conical tank (Fig. 4.60). The variables in the problem are

$V = $ volume (ft³) of water in the tank at time t (min),

$x = $ radius (ft) of the surface of the water at time t,

$y = $ depth (ft) of water in the tank at time t.

We assume V, x, and y to be differentiable functions of t. The constants are the dimensions of the tank.

STEP 2: *Numerical information.* At the time in question,

$$y = 6 \text{ ft}, \qquad \frac{dV}{dt} = 9 \text{ ft}^3/\text{min}.$$

STEP 3: *Find:* $\dfrac{dy}{dt}$.

STEP 4: *Relate the variables.*

$$V = \frac{1}{3}\pi x^2 y \qquad \text{Cone volume formula}$$

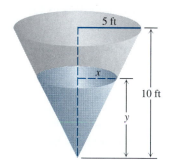

4.60 The conical tank in Example 4.

This equation involves x as well as V and y. Because no information is given about x and dx/dt at the time in question, we need to eliminate x. Using similar triangles (Fig. 4.60) gives us a way to express x in terms of y:

$$\frac{x}{y} = \frac{5}{10} \quad \text{or} \quad x = \frac{y}{2}.$$

Therefore,

$$V = \frac{1}{3}\pi\left(\frac{y}{2}\right)^2 y = \frac{\pi}{12}y^3. \tag{4}$$

STEP 5: *Differentiate with respect to t.* We differentiate Eq. (4) to get

$$\frac{dV}{dt} = \frac{\pi}{12}\cdot 3y^2\frac{dy}{dt} = \frac{\pi}{4}y^2\frac{dy}{dt}.$$

We then solve for dy/dt to express the rate we want (dy/dt) in terms of the rate we know (dV/dt):

$$\frac{dy}{dt} = \frac{4}{\pi y^2}\frac{dV}{dt}.$$

STEP 6: *Evaluate.* We evaluate with $y = 6$ and $dV/dt = 9$:

$$\frac{dy}{dt} = \frac{4}{\pi(6)^2}\cdot 9 = \frac{1}{\pi} = 0.32 \text{ ft/min.}$$

At the moment in question, the water level is rising by about 0.32 ft/min.

≡

EXAMPLE 5   Truck Convoys

Two truck convoys leave a depot. Convoy A travels east at 40 mph, and convoy B travels north at 30 mph. (a) How fast is the distance between the convoys changing 6 min later, when convoy A is 4 mi from base and convoy B is 3 mi from base? (b) How fast is the distance changing at any time?

Solution

a) We carry out the steps of the basic strategy.

STEP 1: *Picture and variables.* We picture the convoys in the coordinate plane, using the positive x-axis as the eastbound highway and the positive y-axis as the northbound highway (Fig. 4.61, on the following page). We let t represent time and set

$$x(t) = \text{ position of convoy A,}$$
$$y(t) = \text{ position of convoy B,}$$
$$s(t) = \text{ distance between convoys.}$$

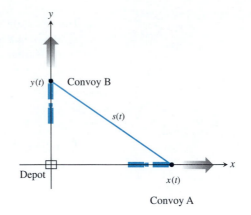

4.61   If you know where the convoys are and how fast they are moving, you can calculate how fast the distance between them is growing (Example 5).

We assume x, y, and s to be differentiable functions of t.

STEP 2: *Numerical information.* At the time in question,

$$x = 4 \text{ mi}, \qquad y = 3 \text{ mi}, \qquad \frac{dx}{dt} = 40 \text{ mph}, \qquad \frac{dy}{dt} = 30 \text{ mph}.$$

STEP 3: *Find:* $\dfrac{ds}{dt}$.

STEP 4: *Relate the variables.*

$$s^2 = x^2 + y^2 \qquad \text{Pythagorean Theorem; the equation } s = \sqrt{x^2 + y^2}$$
$$\text{would also work.}$$

STEP 5: *Differentiate with respect to t.*

$$2s\frac{ds}{dt} = 2x\frac{dx}{dt} + 2y\frac{dy}{dt}$$

$$\frac{ds}{dt} = \frac{1}{s}\left(x\frac{dx}{dt} + y\frac{dy}{dt}\right).$$

STEP 6: *Evaluate.* Let $x = 4$, $y = 3$, $(dx/dt) = 40$, $(dy/dt) = 30$:

$$\frac{ds}{dt} = \frac{1}{\sqrt{4^2 + 3^2}}(4(40) + 3(30)) = \frac{1}{5}(160 + 90) = \frac{250}{5} = 50.$$

At the moment in question, the distance between the convoys is growing at the rate of 50 mph.

b) Using the formula $d = rt$, we can write $x = 40t$ and $y = 30t$. Thus,

$$s = \sqrt{x^2 + y^2} = \sqrt{(40t)^2 + (30t)^2} = 50t.$$

So the distance between convoys is changing at the constant rate of 50 mph.

If we substitute $s = 50t$, $x = 40t$, $y = 30t$, $dx/dt = 40$, and $dy/dt = 30$ into

$$\frac{ds}{dt} = \frac{1}{s}\left(x\frac{dx}{dt} + y\frac{dy}{dt}\right),$$

we find that $ds/dt = 50$, so the techniques of this section produce the same answer as the formula $d = rt$. ▣

EXPLORATION BIT

A dynamic model of Example 5 is possible by using parametric mode on your grapher. Try to create such a model that will "show" the truck convoys traveling away from the depot. Exploration 2 can give you a clue on how to do this.

EXPLORATION 2

Modeling Related Motion—The Sliding Ladder

Parametric mode on a grapher can be used to model the motion of moving objects when the motion of each can be expressed as a function of time t. Consider, for example, a 13-ft ladder leaning against a wall. Suppose that the base of the ladder slides away from the wall at the constant rate of 3 ft/sec. How fast does the top of the ladder move down the wall?

1. The motion of the two ends of the ladder can be represented as

$$x_1(t) = 3t, \qquad y_1(t) = 0;$$

$$x_2(t) = 0, \qquad y_2(t) = \sqrt{13^2 - (3t)^2}.$$

   Explain.

2. What values of t make sense in this situation? What view dimensions are appropriate for a grapher viewing window?

3. Use simultaneous format. (You may wish to also use dot format.) Describe the motions you see when you GRAPH. Include an observation about the apparent rates.

4. Find analytically the rates at which the top of the ladder is moving down the wall at $t = 0.5$, 1, 1.5, and 2 sec.

SEEING THE SLIDING LADDER

To see the action in part 3 of Exploration 2, it may be helpful to "hide" the coordinate axes or adjust your parametric equations so that the wall and ground are moved away from the axes. Also, to control the speed of the action, you may want to adjust t-step.

EXAMPLE 6   Relief from a Heart Attack

A heart attack victim has been given a blood vessel dilator drug to lower the pressure against which the heart has to pump. For a short while after the drug is administered, the radii of the affected blood vessels will increase at about 1% per minute. According to Poiseuille's Law, $V = kr^4$ (Section 3.8, Example 8), what percentage rate of increase can we expect in the blood flow over the next few minutes (all other things being equal)?

Solution

STEP 1: *Picture and variables.* We really don't need a picture, and the variables r and V are already named. It remains only to assume that r and V are differentiable functions of time t.

STEP 2: *Numerical information.*

$$\frac{dr}{dt} = 0.01r \qquad r \text{ increases at 1\% of } r \text{ per min.}$$

$$= \frac{1}{100}r$$

STEP 3: *Find.* The rate of increase in V expressed as a percentage of V, that is, the number P such that

$$\frac{dV}{dt} = \frac{P}{100}V.$$

STEP 4: *Relate the variables.* $V = kr^4$.

STEP 5: *Differentiate* to find how dV/dt is related to dr/dt:

$$\frac{dV}{dt} = 4kr^3 \frac{dr}{dt}.$$

STEP 6: *Evaluate* by substituting $(dr/dt) = (1/100)r$

$$\frac{dV}{dt} = 4kr^3 \left(\frac{1}{100} r \right)$$

$$= \frac{4}{100} kr^4$$

$$= \frac{4}{100} V \qquad V = kr^4.$$

During the next few minutes, the blood flow will increase at the rate of $P = 4\%$ per minute. ≡

Exercises 4.6

1. The radius r and area $A = \pi r^2$ of a circle are differentiable functions of t. Write an equation that relates dA/dt to dr/dt.

2. The radius r and surface area $S = 4\pi r^2$ of a sphere are differentiable functions of t. Write an equation that relates dS/dt to dr/dt.

3. The radius r and volume $V = (1/3)\pi r^2 h$ of a right circular cone are differentiable functions of t. How is dV/dt related to dr/dt if h is constant?

4. The height h and volume $V = (1/3)\pi r^2 h$ of a right circular cone are differentiable functions of t. How is dV/dt related to dh/dt if r is constant?

5. The power P (watts) of an electric circuit is related to the circuit's resistance R (ohms) and current I (amperes) by the equation $P = RI^2$.
   a) How is dP/dt related to dR/dt and dI/dt?
   b) How is dP/dt related to dI/dt if R is constant?

6. Let $x(t)$ and $y(t)$ be differentiable functions of t, and let $s = \sqrt{x^2 + y^2}$ be the distance between the points $(x, 0)$ and $(0, y)$ in the xy-plane.
   a) How is ds/dt related to dx/dt and dy/dt?
   b) How is dx/dt related to dy/dt if s is constant?

7. If x, y, and z are the lengths of the edges of a rectangular box, the common length of the box's diagonals is $s = \sqrt{x^2 + y^2 + z^2}$. If x, y, and z are differentiable functions of t, how is ds/dt related to dx/dt, dy/dt, and dz/dt?

8. If a and b are the sides of a triangle and include an angle of measure θ, then the area of the triangle is $A = (1/2)ab \sin \theta$. If a, b, and θ are differentiable functions of t, how is dA/dt related to da/dt, db/dt, and $d\theta/dt$?

9. *Heating a plate.* As a circular plate of metal is heated in an oven, its radius increases at a rate of 0.01 cm/min. At what rate is the plate's area increasing when the radius is 50 cm?

10. *Changing voltage.* Ohm's Law for electrical circuits like the one shown here states that $V = IR$, where V is the voltage, I is the current in amperes, and R is the resistance in ohms. Suppose that V is increasing at the rate of 1 volt/sec while I is decreasing at the rate of 1/3 amp/sec. Let t denote time in seconds.
    a) What is the value of dV/dt?
    b) What is the value of dI/dt?
    c) What equation relates dR/dt to dV/dt and dI/dt?
    d) Find the rate at which R is changing when $V = 12$ volts and $I = 2$ amp. Is R increasing or decreasing?

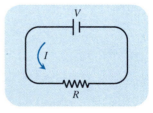

11. *Changing dimensions in a rectangle.* The length l of a rectangle is decreasing at the rate of 2 cm/sec while the width w is increasing at the rate of 2 cm/sec. When $l = 12$ cm and $w = 5$ cm, find the rates of change of (a) the area, (b) the perimeter, and (c) the lengths of the diagonals of the rectangle. Which of these quantities are decreasing and which are increasing?

12. *Changing dimensions in a rectangular box.* Suppose that the edge lengths x, y, and z of a closed rectangular box are changing at the following rates:

$$\frac{dx}{dt} = 1 \text{ m/sec}, \quad \frac{dy}{dt} = -2 \text{ m/sec}, \quad \frac{dz}{dt} = 1 \text{ m/sec}.$$

Find the rates at which the box's (a) volume, (b) surface area, and (c) diagonal length $s = \sqrt{x^2 + y^2 + z^2}$ are changing at the instant when $x = 4$, $y = 3$, and $z = 2$.

13. *Commercial air traffic.* Two commercial jets at 40,000 ft are flying at 520 mi/h along straight-line courses that cross at right angles. How fast is the distance between the airplanes closing when airplane A is 5 mi from the intersection point and airplane B is 12 mi from the intersection point? How fast is the distance closing at any time?

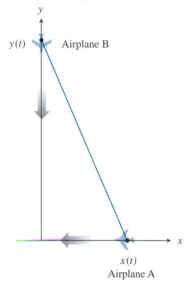

$y(t)$ Airplane B

$x(t)$
Airplane A

14. *A sliding ladder.* A 13-ft ladder is leaning against a house when its base starts to slide away. By the time the base is 12 ft from the house, the base is moving at the rate of 5 ft/sec. How fast is the top of the ladder sliding down the wall then? How fast is the area of the triangle formed by the ladder, wall, and ground changing?

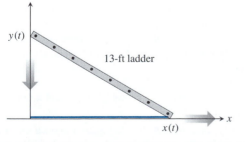

$y(t)$

13-ft ladder

$x(t)$

15. *A shrinking lollipop.* A spherical Tootsie Roll Pop that you are enjoying is giving up volume at a steady rate of 0.08 mL/min. How fast will the radius be decreasing when the Tootsie Roll Pop is 20 mm across?

16. *Boring a cylinder.* The mechanics at Lincoln Automotive are reboring a 6-in.-deep cylinder to fit a new piston. The machine that they are using increases the cylinder's radius one-thousandth of an inch every 3 min. How rapidly is the cylinder volume increasing when the bore (diameter) is 3.80 in.?

17. *A growing sand pile.* Sand falls from a conveyor belt at the rate of 10 m³/min onto the top of a conical pile. The height of the pile is always three-eighths of the base diameter. How fast are the (a) height and (b) radius changing when the pile is 4 m high? Answer in centimeters per minute.

18. *A draining conical reservoir.* Water is flowing at the rate of 50 m³/min from a shallow concrete conical reservoir (vertex down) of base radius 45 m and height 6 m.
   a) How fast is the water level falling when the water is 5 m deep?
   b) How fast is the radius of the water's surface changing then? Answer in centimeters per minute.

19. *A draining hemispherical reservoir.* Water is flowing at the rate of 6 m³/min from a reservoir shaped like a hemispherical bowl of radius 13 m. Answer the following questions, given that the volume of water in a hemispherical bowl of radius R is $V = (\pi/3)y^2(3R - y)$ when the water is y units deep.
   a) How fast is the water level falling when the water is 8 m deep?
   b) What is the radius of the water's surface when the water is y units deep?
   c) How fast is the radius of the water's surface changing when the water is 8 m deep? Answer in centimeters per minute.

20. *A growing raindrop.* Suppose that a drop of mist is a perfect sphere and that, through condensation, the drop picks up moisture at a rate proportional to its surface area. Show that under these circumstances the drop's radius increases at a constant rate.

21. *The radius of an inflating balloon.* A spherical balloon is inflated with helium at the rate of 100π ft³/min. How fast is the balloon's radius increasing at the instant the radius is 5 ft? How fast is the surface area increasing?

22. *Hauling in a dinghy.* A dinghy is pulled toward a dock by a rope from the bow through a ring on the dock 6 ft above the bow. If the rope is hauled in at the rate of 2 ft/sec, how fast is the boat approaching the dock when 10 ft of rope are out?

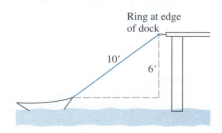

Ring at edge
of dock

10′

6′

23. *A balloon and a bicycle.* A balloon is rising vertically above a level, straight road at a constant rate of 1 ft/sec. Just when the balloon is 65 ft above the ground, a bicycle passes under it, going 17 ft/sec. How fast is the distance between the bicycle and balloon increasing 3 sec later?

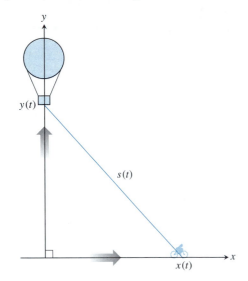

24. *Making coffee.* Coffee is draining from a conical filter into a cylindrical coffeepot at the rate of 10 in.³/min.

a) How fast is the level in the pot rising when the coffee in the cone is 5 in. deep?

b) How fast is the level in the cone falling then?

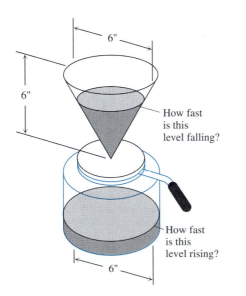

25. *Blood flow.* Cold water has the effect of contracting the blood vessels in the hands, and the radius of a typical vein might decrease at the rate of 20%/min. According to Poiseuille's Law, $V = kr^4$ (see Example 6), at what percentage rate can we expect the volume of blood flowing through that vein to decrease?

26. *Cardiac output.* In the late 1860s, Adolf Fick, a professor of physiology in the Faculty of Medicine in Würtzberg, Germany, developed the method that we use today for measuring how much blood your heart pumps in a minute. Your cardiac output as you read this sentence is probably about 7 L/min. At rest, it is likely to be a bit under 6 L/min. If you are a trained marathon runner running a marathon, your cardiac output can be as high as 30 L/min.

Your cardiac output can be calculated with the formula

$$y = \frac{Q}{D},$$

where Q is the number of milliliters of CO_2 you exhale in a minute and D is the difference between the CO_2 concentration (mL/L) in the blood pumped to the lungs and the CO_2 concentration in the blood returning from the lungs. With $Q = 233$ mL/min and $D = 97 - 56 = 41$ mL/L,

$$y = \frac{233 \text{ mL/min}}{41 \text{ mL/L}} \approx 5.68 \text{ L/min},$$

close to the 6 L/min that most people have at rest. (Data courtesy of J. Kenneth Herd, M.D., Quillan College of Medicine, East Tennessee State University.)

Suppose that when $Q = 233$ and $D = 41$, we also know that D is decreasing at the rate of 2 units a minute but that Q remains unchanged. What is happening to the cardiac output?

27. *Moving along a parabola.* A particle moves along the parabola $y = x^2$ in the first quadrant in such a way that its x-coordinate increases at a steady 10 m/sec. How fast is the angle of inclination θ of the line joining the particle to the origin changing when $x = 3$ m?

28. *Moving along another parabola.* A particle moves from right to left along the parabola $y = \sqrt{-x}$ in such a way that its x-coordinate (measured in meters) decreases at the rate of 8 m/sec. How fast is the angle of inclination θ of the line joining the particle to the origin changing when $x = -4$?

29. *Cost, revenue, and profit.* A company can manufacture x items at a cost of $c(x)$ dollars, a sales revenue of $r(x)$ dollars, and a profit of $p(x) = r(x) - c(x)$ dollars (everything in thousands). Find the rates of change of cost, revenue, and profit for the following values of x and dx/dt.

a) $r(x) = 9x$, $c(x) = x^3 - 6x^2 + 15x$, and $dx/dt = 0.1$ when $x = 2$

b) $r(x) = 70x$, $c(x) = x^3 - 6x^2 + 45/x$, and $dx/dt = 0.05$ when $x = 1.5$

30. *A moving shadow.* A 6-ft-tall man walks at the rate of 5 ft/sec toward a street light that is 16 ft above the ground. At what rate is the tip of his shadow moving? At what rate is the length of his shadow changing when he is 10 ft from the base of the light?

31. *Another moving shadow.* A light shines from the top of a pole 50 ft high. A ball is dropped from the same height from a point 30 ft away from the light. How fast is the shadow of the ball moving along the ground 1/2 sec later? (Assume that the ball falls a distance $s = 16t^2$ ft in t sec.)

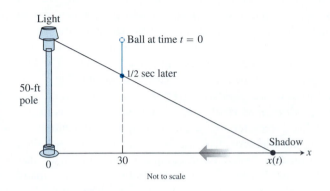

Light

Ball at time $t = 0$

1/2 sec later

50-ft pole

Shadow

x

0   30 $x(t)$

Not to scale

32. *Flying a kite.* A girl flies a kite at a height of 300 ft, the wind carrying the kite horizontally away from her at a rate of 25 ft/sec. How fast must she let out the string when the kite is 500 ft away from her?

33. *A melting ice layer.* A spherical iron ball 8 in. in diameter is coated with a layer of ice of uniform thickness. If the ice melts at the rate of 10 in^3/min, how fast is the thickness of the ice decreasing when it is 2 in. thick? How fast is the outer surface area of ice decreasing?

34. *Highway patrol.* A highway patrol airplane flies 3 mi above a level, straight road at a steady speed of 120 mi/h. The pilot sees an oncoming car and with radar determines that the line-of-sight distance from the airplane to the car is 5 mi and decreasing at the rate of 160 mi/h. Find the car's speed along the highway.

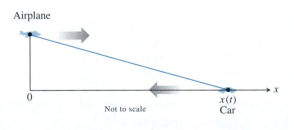

Airplane

0

x

$x(t)$
Car

Not to scale

35. *A building's shadow.* On a morning when the sun will pass directly overhead, the shadow of an 80-ft-tall building on level ground is 60 ft long. At the moment in question, the angle θ that the sun makes with the ground is increasing at the rate of 0.27°/min. At what rate is the shadow decreasing? (Remember to use radians. Express your answer in inches per minute, to the nearest tenth.)

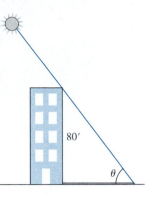

80′

θ

36. *Walkers.* A and B are walking on straight streets that meet at right angles. A approaches the intersection at 2 m/sec; B moves away from the intersection at 1 m/sec. At what rate is the angle θ changing when A is 10 m from the intersection and B is 20 m from the intersection? Express your answer in degrees per minute to the nearest degree.

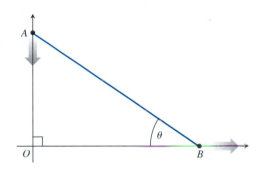

A

θ

O B

37. *Ships.* Two ships are heading straight away from a point O along routes that make a 120° angle. Ship A moves at 14 knots (nautical miles per hour—a nautical mile is 2000 yd). Ship B moves at 21 knots. How fast are the ships moving apart when $OA = 5$ and $OB = 3$? How fast are the ships moving apart at any time?

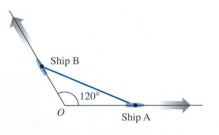

Ship B

120°

O Ship A

38. *The second hand.* At what rate is the distance between the tip of the second hand and the 12 o'clock mark changing when the second hand points to 4 o'clock?

Antiderivatives, Initial Value Problems, and Mathematical Modeling

One of the early accomplishments of calculus was predicting the future position of a moving body from one of its known locations and a formula for its velocity function. Today we view this as one of a number of occasions on which we recover a function from one of its known values and a formula for its rate of change. It is a routine process today, thanks to calculus, to calculate a factory's future output from its present output and production rate or to predict a population's future size from its present size and growth rate.

The process that we use in this section to find a formula for a function from one of its known values and its derivative $f(x)$ has two steps. The first is to find a formula that gives all the functions that could possibly have f as a derivative. These functions are *antiderivatives of f*, and the formula that gives them all is the *general antiderivative of f*. The second step is to use the known function value to pick the one antiderivative that we want.

In general, finding one antiderivative can be a very difficult and often impossible task. Even modern sophisticated computer algebra systems that are designed to produce antiderivatives can fail simply because for many functions, there are no exact formulas for an antiderivative. The good news, however, is that such antiderivatives are no longer mysterious and inaccessible because we have numerical techniques that can approximate and produce their graphs. One may be in your graphing utility. If so, then you can use it to support your work in this section.

The other good news is that finding all the antiderivatives (once we have one) is incredibly easy, thanks to the remaining two corollaries of the Mean Value Theorem of Section 4.1.

This section begins with these two corollaries. It then shows how to "reverse" known differentiation formulas to find general antiderivatives and uses this technique to solve a number of differential equations. It closes with a brief discussion of mathematical modeling, the process by which we use mathematics to learn about reality.

The Second and Third Corollaries of the Mean Value Theorem

The first corollary of the Mean Value Theorem in Section 4.1 gave the First Derivative Test for a function being increasing or decreasing. The second corollary says that only constant functions can have zero derivatives. The third says that functions with identical derivatives differ only by a constant.

> **COROLLARY 2**   If $F' = 0$, then F is a constant.
>
> If $F'(x) = 0$ for all x in (a, b), then there is a constant C such that $F(x) = C$ for all x in (a, b).

Corollary 2 is the converse of the rule that says the derivative of a constant is zero. While the derivatives of nonconstant functions may be zero

The hypotheses of the Mean Value Theorem require that a function be continuous on a closed interval and differentiable on the corresponding open interval. Here, x_1 and x_2 are chosen *within* an open interval (a, b) on which F is differentiable.

at single points, the only functions whose derivatives are zero throughout an entire interval are the functions that are constant on the interval.

Proof of Corollary 2   We want to show that F has a constant value throughout the interval (a, b). The way we do so is to show that if x_1 and x_2 are any two points in (a, b), then $F(x_1) = F(x_2)$.

Suppose that x_1 and x_2 are two points in (a, b), numbered from left to right so that $x_1 < x_2$. Then F satisfies the hypotheses of the Mean Value Theorem on the closed interval $[x_1, x_2]$. It is differentiable at every point of the interval and hence continuous at every point of the interval. Therefore,

$$\frac{F(x_2) - F(x_1)}{x_2 - x_1} = F'(c)$$

at some point c between x_1 and x_2. Since $F' = 0$ throughout (a, b), this translates into

$$\frac{F(x_2) - F(x_1)}{x_2 - x_1} = 0, \qquad F(x_2) - F(x_1) = 0, \qquad \text{and} \qquad F(x_1) = F(x_2).$$

EXPLORATION 1

Making a Conjecture

1. Let $y_1 = \sqrt[3]{x + 1}$, $y_2 = \sqrt[3]{x + 1} + 5$, $y_3 = \sqrt[3]{x + 1} - 6$, $y_4 = \sqrt[3]{5x + 1}$, and $y_5 = 4\sqrt[3]{x + 1}$. Compare the graphs of NDER y_i for $i = 1, 2, 3, 4, 5$ in a $[-10, 10]$ by $[-2, 2]$ viewing window.

2. Let $y_1 = e^x$, $y_2 = e^x + 2$, $y_3 = e^x - 3$, $y_4 = e^{3x}$, and $y_5 = 2e^x$. Compare the graphs of NDER y_i, for $i = 1, 2, 3, 4, 5$ in a $[-5, 5]$ by $[-5, 5]$ viewing window.

3. Make a conjecture about functions that have the same derivative based on your observations in parts 1 and 2.

Your conjecture may have resembled Corollary 3 of the Mean Value Theorem.

A GRAPHICAL INTERPRETATION

The graphical implication of Corollary 3 is that a graph of an antiderivative in a viewing window is a vertical shift of each graph of the other antiderivatives. This is important to remember when you are expecting (hoping) to be supported with a particular graph and a different graph appears. It may be that the particular graph that you want is simply a shift up or down.

COROLLARY 3   **Functions with the same derivative differ only by a constant.**

If $F'(x) = G'(x)$ at each point x in (a, b), then there is a constant C such that

$$F(x) = G(x) + C$$

for all x in (a, b).

Proof of Corollary 3   Since $F'(x) = G'(x)$ at each point of (a, b), the derivative of the function $H = F - G$ at each point is

$$H'(x) = F'(x) - G'(x) = 0.$$

Therefore H has a constant value C throughout (a, b) (from Corollary 2), so $F(x) - G(x) = C$. That is, $F(x) = G(x) + C$ for all x in (a, b).

Corollary 3 says that the only way two functions can have identical rates of change throughout an interval is for their values on the interval to differ by a constant. For example, we know that the derivative of the function x^2 is $2x$. Therefore, every other function whose derivative is $2x$ is given by the formula $x^2 + C$ for some value of C. No other functions have $2x$ as their derivative.

Finding Antiderivatives

Once we have found one antiderivative F of a function f, Corollary 3 tells us how to represent all the other antiderivatives of f as a collection of functions that we call the *general antiderivative* of f.

DEFINITIONS

A function F is an **antiderivative** of a function f over an interval I if $F'(x) = f(x)$ at every point of I.

If F is an antiderivative of f, then the family of functions $F(x) + C$ (C any real number), is the **general antiderivative** of f over the interval I. The constant C is called the **arbitrary constant**.

The use of the capital F for an antiderivative of f is conventional, even though F and f are pronounced the same way in normal speech. To distinguish between the two, we suggest saying *cap eff* for F and *little eff* for f.

In scientific work, we find many antiderivatives that we need by reversing derivative formulas that we already know. Once we have a particular antiderivative F, we know the general antiderivative is F plus an arbitrary constant. The next example shows what we mean. In this example and the others of this section, the interval I will be the natural domain of f unless we say otherwise.

EXAMPLE 1

Function $f(x)$	General Antiderivative $F(x) + C$	Reversed Derivative Formula
$\cos x$	$\sin x + C$	$\dfrac{d}{dx}(\sin x) = \cos x$
$\cos 2x$	$\dfrac{\sin 2x}{2} + C$	$\dfrac{d}{dx}\left(\dfrac{\sin 2x}{2}\right) = \cos 2x$
$3x^2$	$x^3 + C$	$\dfrac{d}{dx}(x^3) = 3x^2$
$\dfrac{1}{2\sqrt{x}}$	$\sqrt{x} + C$	$\dfrac{d}{dx}(\sqrt{x}) = \dfrac{1}{2\sqrt{x}}$
$\dfrac{1}{x^2}$	$-\dfrac{1}{x} + C$	$\dfrac{d}{dx}\left(-\dfrac{1}{x}\right) = \dfrac{1}{x^2}$

Some useful general rules are listed in Example 2.

EXAMPLE 2   Some Useful General Rules

	Function	General Antiderivative	Source
1.	$k\dfrac{du}{dx}$ (k constant)	$ku + C$	Constant Multiple Rule
2.	$\dfrac{du}{dx} + \dfrac{dv}{dx}$	$u + v + C$	Sum Rule
3.	$\dfrac{du}{dx} - \dfrac{dv}{dx}$	$u - v + C$	Difference Rule
4.	x^n $(n \neq -1)$	$\dfrac{x^{n+1}}{n+1} + C$	Power Rule
5.	$\sin kx$	$-\dfrac{\cos kx}{k} + C$	Chain Rule
6.	$\cos kx$	$\dfrac{\sin kx}{k} + C$	Chain Rule

Examples 3–5 show how to apply the rules in Example 2. Check the results in one of two ways: Take the derivative of the general antiderivative to see whether it agrees with the function, or graph the numerical derivative of the general antiderivative (use any value for C) to see whether it is the same as the graph of the function.

EXAMPLE 3   We Can Find Antiderivatives Term by Term.

Function	General Antiderivative	Source
$10x$	$5x^2 + C$	(Constant Multiple and Power Rules)
$10x - x^2$	$5x^2 - \dfrac{x^3}{3} + C$	(... along with the Difference Rule ...)
$10x - x^2 + 2$	$5x^2 - \dfrac{x^3}{3} + 2x + C$	(... and Sum Rule ...)

EXAMPLE 4   Fractional Powers Are Handled in the Same Way as Integer Powers.

Function	General Antiderivative	Source
$\sqrt{x} = x^{1/2}$	$\dfrac{x^{3/2}}{3/2} + C = \dfrac{2}{3}x^{3/2} + C$	(Power Rule with $n = 1/2$)
$\dfrac{1}{\sqrt{x}} = x^{-1/2}$	$\dfrac{x^{1/2}}{1/2} + C = 2x^{1/2} + C$	(Power Rule with $n = -1/2$)

EXAMPLE 5   The k in Rules 5 and 6 Can Be Any Real Number $\neq 0$.

Function	General Antiderivative	Source
$6 \sin 3x$	$6 \cdot \dfrac{-\cos 3x}{3} + C = -2 \cos 3x + C$	(Rule 5 with $k = 3$)
$5 \cos \dfrac{x}{2}$	$5 \cdot \dfrac{\sin(x/2)}{1/2} + C = 10 \sin \dfrac{x}{2} + C$	(Rule 6 with $k = \dfrac{1}{2}$)
$\cos 2\pi x$	$\dfrac{\sin 2\pi x}{2\pi} + C$	(Rule 6 with $k = 2\pi$)

EXAMPLE 6

Find the general antiderivative of $f(x) = x^{2/3}$. Support graphically.

Solution   Rule 4 says that the general antiderivative is

$$y = \frac{x^{5/3}}{5/3} + C = \frac{3}{5} x^{5/3} + C.$$

Let $C = 0$, and the antiderivative is $y = F(x) = (3/5) x^{5/3}$. For this F,

$$F'(x) = \frac{5}{3} \cdot \frac{3}{5} x^{2/3}$$

$$= x^{2/3} = f(x),$$

and the graph of NDER F agrees with the graph of f (Fig. 4.62).

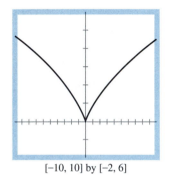

[−10, 10] by [−2, 6]

4.62 Complete graphs of f and NDER F of Example 6 appear to be identical.

Differential Equations and Initial-Value Problems

The problem of finding a function y of x when we know its derivative

$$\frac{dy}{dx} = f(x)$$

and its value y_0 at a particular point x_0 is called an **initial-value problem.** We solve it in two steps. First we find the general antiderivative of f,

$$y = F(x) + C.$$

Then we use the *initial condition* that $y = y_0$ when $x = x_0$ to find the correct value of C. In this general case, $y_0 = F(x_0) + C$, so $C = y_0 - F(x_0)$, and the solution of the initial-value problem is the function

$$y = F(x) + (y_0 - F(x_0)).$$

This function has the correct derivative because

$$\frac{dy}{dx} = \frac{d}{dx} F(x) + \frac{d}{dx}(y_0 - F(x_0)) = f(x) + 0 = f(x).$$

It also has the correct value when $x = x_0$ because

$$y \Big|_{x=x_0} = F(x_0) + (y_0 - F(x_0)) = y_0.$$

An equation like

$$\frac{dy}{dx} = f(x) \tag{1}$$

that has a derivative in it is called a **differential equation.** Equation (1) gives dy/dx as a function of x. A more complicated differential equation might involve y as well as x:

$$\frac{dy}{dx} = 2xy^2.$$

It might also involve higher-order derivatives:

$$\frac{d^2y}{dx^2} - \frac{dy}{dx} + 5y = 3.$$

When we solve an algebraic equation, $f(x) = 0$, we find the values for x that make the equation true. When we solve a differential equation, $dy/dx = f(x)$, we find the functions y that make the equation true. The general antiderivative $y = F(x) + C$ of the function $f(x)$ is called the **general solution** of the differential equation $dy/dx = f(x)$. In an initial-value problem, the **initial condition** that $y = y_0$ when $x = x_0$ yields the **particular solution** $y = F(x) + (y_0 - F(x_0))$.

GRAPHING THE SOLUTION

The toolbox program EULERG (pronounced *oiler-gee*) can graph the solution to an initial-value problem. You can use it with Examples 7–10 and compare the graphs in your viewing windows to the solutions in the examples. Details about this approach are in Section 8.7 and the *Resource Manual.*

EXAMPLE 7   Finding Velocity from Acceleration

The acceleration of gravity near the surface of Earth is 9.8 m/sec². This means that the velocity v of a body falling freely in a vacuum changes at the rate of

$$\frac{dv}{dt} = 9.8 \text{ m/sec}^2.$$

If the body is dropped from rest, what will its velocity be t seconds after it is released?

Solution   In mathematical terms, we want to solve the initial-value problem that consists of

The differential equation: $\dfrac{dv}{dt} = 9.8$,

The initial condition: $v = 0$ when $t = 0$.

To solve it, we first use what we know about antiderivatives to find the general solution of the differential equation $dv/dt = 9.8$:

$$v = 9.8t + C.$$

Reversed derivative formula:
$$\frac{d}{dt}(9.8t) = 9.8$$

Then we use the initial condition to find the correct value of C for this particular problem:

$$v = 9.8t + C,$$
$$0 = 9.8(0) + C, \qquad v = 0 \text{ when } t = 0.$$
$$C = 0.$$

The velocity of the falling body t seconds after release is

$$v(t) = 9.8t \text{ m/sec.}$$

EXAMPLE 8 Finding a Curve from Its Slope Function and a Point

Find the curve whose slope at the point (x, y) is $3x^2$ if the curve is required to pass through the point $(1, -1)$.

Solution In mathematical language, we are asked to solve the initial-value problem that consists of

The differential equation: $\dfrac{dy}{dx} = 3x^2,$

The initial condition: $y = -1$ when $x = 1.$

To solve it, we first use what we know about antiderivatives to find the general solution of the differential equation:

$$y = x^3 + C.$$

Reverse derivative formula:
$$\frac{d}{dx}(x^3) = 3x^2$$

Then we substitute $x = 1$ and $y = -1$ to find C:

$$-1 = (1)^3 + C,$$
$$C = -2.$$

The curve that we want is $y = x^3 - 2$ (Fig. 4.63). ≣

Some problems require us to solve two or more differential equations in a row. Here is an example.

EXAMPLE 9

A heavy projectile is fired straight up from a platform 10 ft above the ground, with an initial velocity of 160 ft/sec. Assume that the only force affecting the projectile during its flight is gravity, which produces a downward acceleration of 32 ft/sec². Find an equation for the projectile's height above the ground as a function of time t if $t = 0$ when the projectile is fired.

Solution To model the problem, we draw a figure (Fig. 4.64) and let $s(t)$ denote the projectile's height above the ground at time t. We assume s to be a twice-differentiable function of t, so that

$$v = \frac{ds}{dt} \qquad \text{and} \qquad a = \frac{dv}{dt} = \frac{d^2s}{dt^2}.$$

Since gravity acts in the negative s direction, the direction of decreasing s in our model, the initial-value problem to solve is

The differential equation: $\dfrac{dv}{dt} = -32$ ft/sec²,

The initial conditions: $v(0) = 160$ ft/sec, $s(0) = 10$ ft.

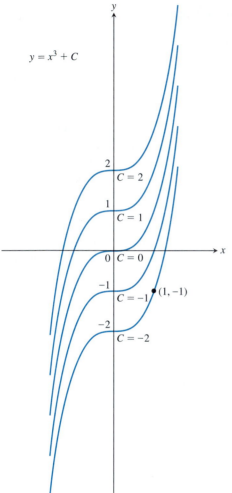

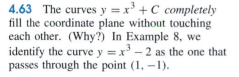

4.63 The curves $y = x^3 + C$ *completely fill the coordinate plane without touching each other. (Why?) In Example 8, we identify the curve $y = x^3 - 2$ as the one that passes through the point $(1, -1)$.

ESTABLISHING MARGIN DIRECTION

By choosing *down* to be the negative direction, we must choose dv/dt to be negative.

We find v from the equation $dv/dt = -32$:

$$v = \frac{ds}{dt} = -32t + C_1 \qquad \text{General antiderivative of } -32$$

and s from the equation $ds/dt = -32t + C_1$:

$$s = -16t^2 + C_1 t + C_2. \qquad \text{General antiderivative of } -32t + C_1$$

The appropriate values of C_1 and C_2 are determined by the initial conditions:

$$C_1 = v(0) = 160, \qquad C_2 = s(0) = 10.$$

The projectile's height above the ground at time t is

$$s(t) = -16t^2 + 160t + 10.$$

Notice that the formula for s agrees with the one used in prior courses and in Section 3.4.

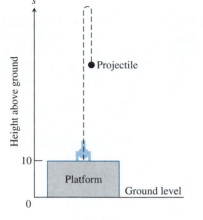

4.64 The sketch for modeling the projectile motion in Example 9.

Solution Curves

The graph of a solution of a differential equation is called a **solution curve.** The curves $y = x^3 + C$ in Fig. 4.63 are solution curves of the differential equation $dy/dx = 3x^2$. Often we cannot find explicit formulas $y = f(x) + C$ for the solution curves of a differential equation $dy/dx = f(x, y)$. We can use the tools of calculus, however, to obtain a sketch of the solution curves as the next example illustrates. In Chapter 5, we will see that we can obtain a particular solution curve using a toolbox program or a built-in grapher feature.

EXAMPLE 10

Sketch each of the following.

a) Solution curves of the differential equation

$$y' = \frac{1}{x^2 + 1}.$$

b) The solution curve of the initial-value problem

$$\text{Differential equation:} \quad y' = \frac{1}{x^2 + 1},$$

$$\text{Initial condition:} \quad y = 0 \quad \text{when} \quad x = 0.$$

c) Apply SLOPEFLD to

$$y' = \frac{1}{x^2 + 1}.$$

SLOPEFIELDS

The toolbox program SLOPEFLD gives a picture that approximates a family of solution curves to $dy/dx = f(x, y)$. The program computes slopes and draws small tangent lines at a lattice of points (x, y) in the viewing window. These small tangent lines approximate the solution curves.

Solution

a) The domain of y' (and therefore y) is $(-\infty, \infty)$. We note that $y' > 0$ for all x, so the solution curves are increasing on $(-\infty, \infty)$ and have no extreme values.

Because

$$y'' = \frac{d}{dx}\left(\frac{1}{x^2 + 1}\right) = \frac{-2x}{\left(x^2 + 1\right)^2},$$

all solution curves are concave down ($y'' < 0$) in $(0, \infty)$ and concave up ($y'' > 0$) in $(-\infty, 0)$ and have a point of inflection at $x = 0$.

Because

$$\lim_{|x| \to \infty} y' = \lim_{|x| \to \infty} \frac{1}{x^2 + 1} = 0,$$

the solution curves have end behavior models whose tangent lines are nearly horizontal. Thus, we have the general shape of the curve (Fig. 4.65a).

To sketch several of the solution curves, we note that y' is 1 at $x = 0$, so we sketch the general shape in the coordinate plane to cross the y-axis with slope 1. (We include a portion of the tangent line to help guide the sketch. To continue the sketch, we estimate other points on this first solution curve near $x = 0$, compute the tangent lines at those points, and use the tangent lines to guide us.)

Then, because the solution curves completely fill the coordinate plane, we sketch others to be "parallel" to the first one drawn (Fig. 4.65b).

b) The solution curve that satisfies $y = 0$ when $x = 0$ is the one in Fig. 4.65(b) that passes through the origin.

c) The SLOPEFLD output closely approximates the family of solution curves (Fig. 4.65c).

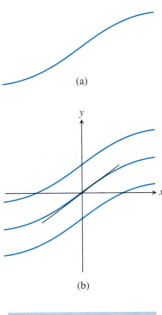

(a)

(b)

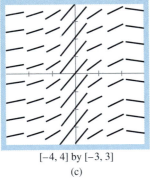

[−4, 4] by [−3, 3]

(c)

4.65 (a) The general shape of a solution curve of $y' = 1/(x^2 + 1)$ in Example 10. There is a point of inflection at $x = 0$ where the slope of the curve is 1. The curve is concave up to the left of 0, concave down to the right of 0, and flattens out as $|x| \to \infty$. (b) To sketch several of the solution curves, sketch the general shape and then sketch others to be "parallel" to the general shape. (c) Output from SLOPEFLD.

Mathematical Modeling

The development of a mathematical model usually takes four steps: First we observe something in the real world (a ball bearing falling from rest or the trachea contracting during a cough, for example) and construct a system of mathematical variables and relationships that imitate some of its important features. We build a mathematical metaphor for what we see. Next we apply (usually) existing mathematics to the variables and relationships in the model to draw conclusions about them. After that, we translate the mathematical conclusions into information about the system under study. Finally, we check the information against observation to see whether the model has predictive value. We also investigate the possibility that the model applies to other systems. The really good models are the ones that lead to conclusions that are consistent with observation, that have predictive value and broad application, and that are not too hard to use.

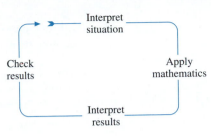

The natural cycle of mathematical imitation, deduction, interpretation, and comparison is shown in the following diagrams.

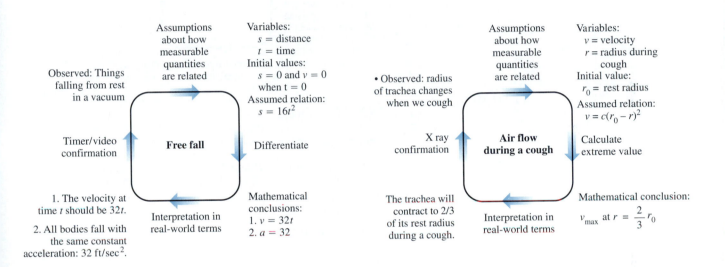

Computer Simulation

When a system that we want to study is complicated, we can sometimes experiment first to see how the system behaves under different circumstances. But if this is not possible (the experiments might be expensive, time consuming, or dangerous), we might run a series of simulated experiments on a computer—experiments that behave like the real thing, without the disadvantages. For example, we have simulated the motion of objects using the parametric graphing feature of our graphing utility (really a computer). Thus, we might model the effects of atomic war, the effect of waiting a year longer to harvest trees, the effect of crossing particular breeds of cattle, or the effect of reducing atmospheric ozone by 1%, all without having to pay the consequences or wait to see how things work out.

We also bring computers in when the model that we want to use has too many calculations to be practical any other way. NASA's space-flight models are run on computers—they have to be to generate course corrections on time. If you want to model the behavior of galaxies that contain billions and billions of stars, a computer offers the only possible way. One of the most spectacular computer simulations in recent years, carried out by Alar Toomre at MIT, explained a peculiar galactic shape that was not consistent with our previous ideas about how galaxies are formed. The galaxies had acquired their odd shapes, Toomre deduced, by passing through one another.

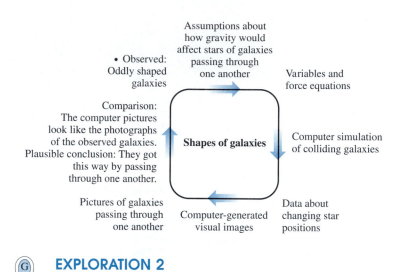

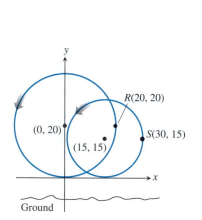

4.66 The two Ferris wheels of Exploration 2.

EXPLORATION BIT

Some large fairs feature a "multiple Ferris wheel" that has two smaller wheels attached to the outside of a larger wheel. Another popular (horizontal) amusement ride is *The Scrambler*, which is essentially three wheels attached to the outside of a larger wheel. Can you model the motion of either of these rides with parametric equations?

EXPLORATION 2

Computer Simulation—Two Ferris Wheels

Two Ferris wheels with radii 20 ft and 15 ft are next to each other and centered at (0, 20) and (15, 15), respectively (Fig. 4.66). The larger wheel is rotating at 1 revolution every 12 sec, and the smaller wheel at 1 revolution every 8 sec. Renee, on the larger Ferris wheel, and Sherrie, on the smaller Ferris wheel, are at the points $R(20, 20)$ and $S(30, 15)$, respectively, when $t = 0$.

1. Assume that the two Ferris wheels are in the same plane. Show that Renee's and Sherrie's positions on the Ferris wheels at any time are given parametrically by

$$x_1(t) = 20\cos\left(\frac{\pi t}{6}\right), \qquad y_1(t) = 20 + 20\sin\left(\frac{\pi t}{6}\right) \qquad (2)$$

$$x_2(t) = 15 + 15\cos\left(\frac{\pi t}{4}\right), \quad y_2(t) = 15 + 15\sin\left(\frac{\pi t}{4}\right) \qquad (3)$$

Which equations are Renee's? Which are Sherrie's? GRAPH and compare with Fig. 4.66. (What might make your graph more realistic?)

2. TRACE and use the cursor to explore the positions of Renee and Sherrie on the Ferris wheels at different times. When is the distance between them a maximum? A minimum? What are the maximum and minimum distances? Can you confirm them analytically? Explain.

3. Let $y = D(t)$ be the distance between Renee and Sherrie at time t and

$$x_3(t) = t, \qquad y_3(t) = D(t).$$

Explain what you see when you GRAPH. Do the minimum and maximum distances apart agree with your observations in part 2?

4. When is the distance between them increasing the fastest? When is it decreasing the fastest? Explain how you found your answers.

Models in Biology

You may have noticed that we haven't mentioned models in biology yet. The reason is that most mathematical models of life processes use either exponential functions or logarithms, functions whose exact derivatives are not found until Chapter 7. Typical of the models that we will study there is the model for unchecked bacterial growth. The basic assumption is that at any time t, the rate dy/dt at which the population is changing is proportional to the number $y(t)$ of bacteria present. If the population's original size is y_0, this leads to the initial-value problem

$$\text{Differential equation:} \quad \frac{dy}{dt} = ky$$

$$\text{Initial condition:} \quad y = y_0 \quad \text{when} \quad t = 0.$$

As you will see, the solution turns out to be $y = y_0 e^{kt}$, so the modeling cycle looks like this:

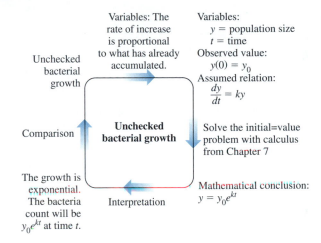

This model is one of the really good models that we talked about earlier, because it applies to so many of the phenomena that we want to forecast and understand: cell growth, heat transfer, radioactive decay, the flow of electrical current, and the accumulation of capital by compound interest, to mention only a few. We will see how all of this works by the time we are through with Chapter 7.

Exercises 4.7

Find the general antiderivatives of the functions in Exercises 1–18. Do as many as you can without writing anything down (except the answer). Then support your answers with a graphing utility.

1. a) $2x$ **b)** x^2 **c)** $x^2 - 2x + 1$

2. a) $6x$ **b)** x^5 **c)** $x^5 - 6x + 3$

3. a) $-3x^{-4}$ **b)** x^{-4} **c)** $x^{-4} + 2x + 3$

4. a) $2x^{-3}$ **b)** $\dfrac{x^{-3}}{2} + x^2$ **c)** $-x^{-3} + x - 1$

5. a) $\dfrac{1}{x^2}$ **b)** $\dfrac{5}{x^2}$ **c)** $2 - \dfrac{5}{x^2}$

6. a) $-\dfrac{2}{x^3}$ **b)** $\dfrac{1}{2x^3}$ **c)** $x^3 - \dfrac{1}{x^3}$

7. a) $\dfrac{3}{2}\sqrt{x}$ **b)** $4\sqrt{x}$ **c)** $x^2 - 4\sqrt{x}$

8. a) $\dfrac{4}{3}\sqrt[3]{x}$ **b)** $\dfrac{1}{3\sqrt[3]{x}}$ **c)** $\sqrt[3]{x} + \dfrac{1}{\sqrt[3]{x}}$

9. a) $\dfrac{2}{3}x^{-1/3}$ **b)** $\dfrac{1}{3}x^{-2/3}$ **c)** $-\dfrac{1}{3}x^{-4/3}$

10. a) $\frac{1}{2}x^{-1/2}$ **b)** $-\frac{1}{2}x^{-3/2}$ **c)** $-\frac{3}{2}x^{-5/2}$

11. a) $-\sin 3x$ **b)** $3 \sin x$ **c)** $3 \sin x - \sin 3x$

12. a) $\pi \cos \pi x$ **b)** $\frac{\pi}{2} \cos \frac{\pi x}{2}$ **c)** $\cos \frac{\pi x}{2}$

13. a) $\sec^2 x$ **b)** $5 \sec^2 5x$ **c)** $\sec^2 5x$

14. a) $\csc^2 x$ **b)** $7 \csc^2 7x$ **c)** $\csc^2 7x$

15. a) $\sec x \tan x$ **b)** $2 \sec 2x \tan 2x$ **c)** $4 \sec 2x \tan 2x$

16. a) $\csc x \cot x$ **b)** $8 \csc 4x \cot 4x$ **c)** $\csc 4x \cot 4x$

17. $(\sin x - \cos x)^2$ (*Hint:* $2 \sin x \cos x = \sin 2x$)

18. $(1 + 2 \cos x)^2$ (*Hint:* $2 \cos^2 x = 1 + \cos 2x$)

19. Suppose that $1 - \sqrt{x}$ is an antiderivative of $f(x)$ and that $x + 2$ is an antiderivative of $g(x)$. Find the *general* antiderivatives of the following functions.

   a) $f(x)$ **b)** $g(x)$ **c)** $-f(x)$

   d) $-g(x)$ **e)** $f(x) + g(x)$ **f)** $3f(x) - 2g(x)$

   g) $x + f(x)$ **h)** $g(x) - 4$

20. Repeat Exercise 19, assuming that e^x is an antiderivative of $f(x)$ and that $x \sin x$ is an antiderivative of $g(x)$.

21. Which of the following graphs shows the solution of the initial-value problem

$$\frac{dy}{dx} = 2x, \qquad x = 4 \qquad \text{when} \quad x = 1?$$

Give reasons for your answer.

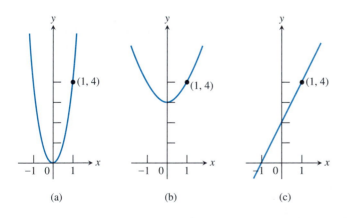

(a)   (b)   (c)

22. Which of the following graphs shows the solution of the initial-value problem

$$\frac{dy}{dx} = -x, \qquad y = 1 \qquad \text{when} \quad x = -1?$$

Give reasons for your answer.

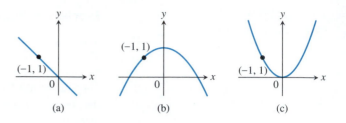

(a)   (b)   (c)

Solve the initial-value problems in Exercises 23–34 for y as a function of x. If possible, support your answers with a graphing utility.

23. $\frac{dy}{dx} = 2x - 7$, $y = 0$ when $x = 2$

24. $\frac{dy}{dx} = 10 - x$, $y = -1$ when $x = 0$

25. $\frac{dy}{dx} = x^2 + 1$, $y = 1$ when $x = 0$

26. $\frac{dy}{dx} = x^2 + \sqrt{x}$, $y = 1$ when $x = 1$

27. $\frac{dy}{dx} = -5/x^2$, $x > 0$; $y = 3$ when $x = 5$

28. $\frac{dy}{dx} = \frac{1}{x^2} + x$, $x > 0$; $y = 1$ when $x = 2$

29. $\frac{dy}{dx} = 3x^2 + 2x + 1$, $y = 0$ when $x = 1$

30. $\frac{dy}{dx} = 9x^2 - 4x + 5$, $y = 0$ when $x = -1$

31. $\frac{dy}{dx} = 1 + \cos x$, $y = 4$ when $x = 0$

32. $\frac{dy}{dx} = \cos x + \sin x$, $y = 1$ when $x = \pi$

33. $\frac{d^2y}{dx^2} = 2 - 6x$, $y = 1$ and $\frac{dy}{dx} = 4$ when $x = 0$

34. $\frac{d^3y}{dx^3} = 6$; $y = 5$, $\frac{dy}{dx} = 0$, and $\frac{d^2y}{dx^2} = -8$ when $x = 0$

Exercises 35 and 36 give the velocity and initial position of a body moving along a coordinate line. Find the body's position at time t. Simulate the motion with a grapher in parametric mode.

35. $v = 9.8t$, $s = 10$, when $t = 0$

36. $v = \sin t$, $s = 0$, when $t = 0$

Exercises 37 and 38 give the acceleration, initial velocity, and initial position of a body moving along a coordinate line. Find the body's position at time t. Simulate the motion with a grapher in parametric mode.

37. $a = 32$, $v = 20$, and $s = 0$ when $t = 0$

38. $a = \sin t$, $v = -1$, and $s = 1$ when $t = 0$

39. Find the curve in the xy-plane that passes through the point $(9, 4)$ and whose slope at each point is $3\sqrt{x}$.

40. a) Find a function $y = f(x)$ with the following properties:

    i) $\dfrac{d^2y}{dx^2} = 6x$.

    ii) Its graph in the xy-plane passes through the point $(0, 1)$ and has a horizontal tangent there.

  b) How many functions like this are there? How do you know?

41. *Revenue from marginal revenue.* Suppose that the marginal revenue when x thousand units are sold is

$$\frac{dr}{dx} = 3x^2 - 6x + 12$$

dollars per unit. Find the revenue function $r(x)$ given that there is no revenue if no units are sold.

42. *Cost from marginal cost.* Suppose that the marginal cost of manufacturing an item when x thousand items are produced is

$$\frac{dc}{dx} = 3x^2 - 12x + 15$$

dollars per item. Find the cost function $c(x)$ if $c(0) = 400$.

43. On the moon, the acceleration of gravity is 1.6 m/sec². If a rock is dropped into a crevasse, how fast will it be going just before it hits bottom 30 sec later?

44. A rocket lifts off the surface of Earth with a constant acceleration of 20 m/sec². How fast will the rocket be going 1 min later?

45. With approximately what velocity do you enter the water if you dive from a 10-m platform? (Use $g = 9.8$ m/sec².)

46. The acceleration of gravity near the surface of Mars is 3.72 m/sec². If a rock is blasted straight up from the surface with an initial velocity of 93 m/sec (about 208 mi/h), how high does it go? (*Hint:* When is the velocity zero?)

47. *How long will it take a tank to drain?* If we open a valve to drain the water from a cylindrical tank, the water will flow fast when the tank is full but slow down as the tank drains. It turns out that the rate at which the water level drops is proportional to the square root of the water's depth. In the notation of the diagram this means that $\dfrac{dy}{dt} = -k\sqrt{y}.$ (4)

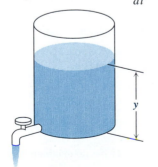

The value of k depends on the acceleration of gravity and the cross-sectional areas of the tank and drain hole. Equation (4)

has a negative sign because y decreases with time. To solve Eq. (4), rewrite it as

$$\frac{1}{\sqrt{y}}\frac{dy}{dt} = -k,\qquad (5)$$

and carry out the following steps.
a) Find the general antiderivative of each side of Eq. (5).
b) Set the antiderivatives in part (a) equal, and combine their arbitrary constants into a single arbitrary constant. (Nothing is achieved by having two when one will do.) This will give an equation that relates y directly to t.

48. *Continuation of Exercise 47.* (a) Suppose t is measured in minutes and $k = 1/10$. Find y as a function of t if $y = 9$ ft when $t = 0$. (b) How long does it take the tank to drain if the water is 9 ft deep to start with?

49. *Two ferris wheels (continued from Exploration 2).* Let $y = D(t)$ be the distance between Renee and Sherrie.

a) Show that $y = D(t)$ is a periodic function, and determine its period.

b) Draw a complete graph of $y = D(t)$.

c) Find the maximum and minimum distances between Renee and Sherrie and the first time each value occurs.

d) Explain how you would try to confirm the results in part (c) analytically. Can you do it?

50. Repeat Exercise 49 assuming that Renee and Sherrie start at $(0, 0)$ and $(15, 0)$, respectively, when $t = 0$.

51. The graph below is that of a function $y = f(x)$ that solves one of the following initial-value problems. Which one? How do you know?
a) $dy/dx = 2x$, $y(1) = 0$,   **b)** $dy/dx = x^2$, $y(1) = 1$,
c) $dy/dx = 2x + 2$, $y(1) = 1$,   **d)** $dy/dx = 2x$, $y(1) = 1$.

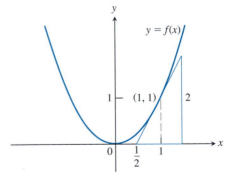

52. Give a convincing argument that

$$y = \frac{2}{3}(x^4 + x^2 + 3)^{3/2} + 2$$

is not the solution for the initial-value problem

$$dy/dx = (x^4 + x^2 + 3)^{1/2}(4x^3 + 2x), \; y(0) = 2.$$

Use the technique described in Example 10 to sketch solution curves of the differential equations in Exercises 53–56. Then solve the differential equations, and graph the solutions to support your work. Also support graphically with SLOPEFLD.

53. $\dfrac{dy}{dx} = 2x$

54. $\dfrac{dy}{dx} = -2x + 2$

55. $\dfrac{dy}{dx} = 1 - 3x^2$

56. $\dfrac{dy}{dx} = x^2$

Use the technique described in Example 10 to sketch the solution curve of each initial-value problem in Exercises 57–60. Support graphically using SLOPEFLD.

57. $\dfrac{dy}{dx} = \sqrt{1 + x^4}, \quad y = 1 \quad$ when $x = 0$

58. $\dfrac{dy}{dx} = \dfrac{1}{\sqrt{1 - x^2}}, \quad -1 < x < 1; y = 0 \quad$ when $x = 0$

59. $\dfrac{dy}{dx} = \dfrac{x}{x^2 + 1}, y = 0 \quad$ when $x = 0$

60. $\dfrac{dy}{dx} = \dfrac{1}{x^2 + 1} - 1, \quad y = 1 \quad$ when $x = 0$

61. Use SLOPEFLD to approximate the solution curves to $dy/dt = 0.001y(100 - y)$ in a $[0, 100]$ by $[0, 100]$ viewing window. Then sketch the particular solution $y = f(t)$ that satisfies $f(0) = 10$.

Chapter 4 Review Questions

1. What does it mean for a function $y = f(x)$ to have an absolute or local maximum or minimum value?

2. How do you find the local and absolute maximum and minimum values of a function $y = f(x)$?

3. What are the hypotheses and conclusion of Rolle's Theorem? How does the theorem sometimes help you to tell how many solutions an equation has in a given interval?

4. What are the hypotheses and conclusion of the Mean Value Theorem? What physical interpretation does the theorem sometimes have? Give an example.

5. This chapter gives three important corollaries of the Mean Value Theorem. State each one and describe how it is used.

6. How do you test a function to find out where its graph is concave up or concave down? What is an inflection point? What physical significance do inflection points sometimes have?

7. State the First Derivative Test for Local Extreme Values.

8. State the Second Derivative Test for Local Extreme Values.

9. List the steps that you would take to confirm a computer-generated graph of a function. How does calculus tell you the shape of the graph between plotted points? Give an example.

10. Give a general description of the class of polynomial functions. Indicate the possible number of real zeros and the possible number of local extrema.

11. Describe Newton's method for solving equations. Give an example. What is the theory behind the method? What are some of the things to watch out for when you use the method?

12. Describe how you would solve max-min problems. Illustrate with an example.

13. What guidance do you get from calculus about finding production levels that maximize profit? That minimize average manufacturing cost?

14. Describe how you would sketch the graph of a rational function.

15. Indicate a reasonable way to describe the behavior of a periodic function.

16. Describe how you would use a graphing utility to draw the graph of $f(x) = \log_a x$.

17. Describe how you would solve related rate problems. Illustrate with an example.

18. What is an antiderivative of a function $y = f(x)$? When a function has an antiderivative, how do we find its general antiderivative? Illustrate with an example.

19. What general rules can you call on to help find antiderivatives? Show, by example, how they are used.

20. What is an initial-value problem? How do you solve one? Illustrate with an example.

Chapter 4 Practice Exercises

1. Show that $y = x/(x + 1)$ increases on every interval in its domain.

2. Show that $y = \sin^2 t - 3t$ decreases on every interval in its domain.

3. Show that $y = x^3 + 2x$ has no maximum or minimum values.

4. Does $f(x) = x^3 + 2x + \tan x$ have any local maximum or minimum values?

5. If $f'(x) \leq 2$ for all x, what is the most f can increase on the interval $0 \leq x \leq 6$?

6. Show that the equation $x^4 + 2x^2 - 2 = 0$ has exactly one solution on the interval $0 \leq x \leq 1$.

In Exercises 7 and 8, suppose that the first derivative of $y = f(x)$ is as given. At what points, if any, does the graph of f have a local maximum, local minimum, or point of inflection?

7. $y' = 6(x + 1)(x - 2)^2$

8. $y' = 6x(x + 1)(x - 2)$

9. At which of the five points on the graph of $y = f(x)$ shown here (a) are y' and y'' both negative? (b) is y' negative and y'' positive?

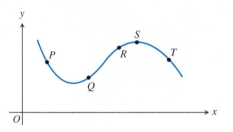

10. Here is the graph of the fruit fly population again. On approximately what day did the population's growth rate change from increasing to decreasing?

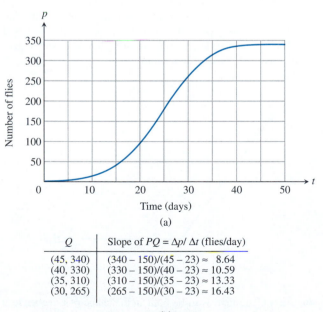

(a)

Q	Slope of $PQ = \Delta p / \Delta t$ (flies/day)
(45, 340)	$(340 - 150)/(45 - 23) \approx 8.64$
(40, 330)	$(330 - 150)/(40 - 23) \approx 10.59$
(35, 310)	$(310 - 150)/(35 - 23) \approx 13.33$
(30, 265)	$(265 - 150)/(30 - 23) \approx 16.43$

(b)

In Exercises 11–30, show a complete graph and identify the inflection points, local maximum and minimum values, and the intervals on which the graph is rising, falling, concave up, and concave down. We suggest that you do exercises from the three groups in the order given.

Do Exercises 11–14 analytically, then support graphically.

11. $y = -x^3 - 3x^2 - 4x - 2$

12. $y = x^3 - 9x^2 - 21x - 11$

13. $y = \sqrt[3]{x - 2}$

14. $y = \sqrt[4]{1 - x}$

Do Exercises 15 and 16 graphically, then confirm analytically.

15. $y = 1 + x - x^2 - x^4$

16. $y = \dfrac{2}{3}x^3 + 5x + 20$

Do Exercises 17–30 using a method of your choice.

17. $y = -\dfrac{8}{3}x^3 + 4x^2 - 2x - 12$

18. $y = -x^4 + 4x^3 - 4x^2 + x + 20$

19. $y = x^4 - \dfrac{8}{3}x^3 - \dfrac{x^2}{2} + 1$

20. $y = 4x^5 + 5x^4 + \dfrac{20}{3}x^3 + 4$

21. $y = -x^5 + \dfrac{7}{3}x^3 + 5x^2 + 4x + 2$

22. $y = \dfrac{1}{5}x^5 + \dfrac{3}{2}x^4 + \dfrac{5}{3}x^3 - 6x^2 + 3x + 1$

23. $y = \dfrac{5 - 4x + 4x^2 - x^3}{x - 2}$

24. $y = \dfrac{3x^3 - 5x^2 - 11x - 11}{x^2 - 2x - 3}$

25. $y = \log_3 |x|$

26. $y = e^{x-1} - x$

27. $y = x \log (x - 2)$

28. $y = \sin 3x + \cos 4x$

29. $y = \sqrt[4]{x - x^2}$

30. $y = \sinh (x + 2)$

31. Use Newton's method to find where the curve $y = -x^3 + 3x + 4$ crosses the x-axis. Support graphically.

32. Use Newton's method to solve the equation $\sec x = 4$ on the interval $0 \leq x \leq \pi/2$. Support graphically.

33. Use Newton's method to solve the equation $2 \cos x - \sqrt{1 + x} = 0$.

34. Find the approximate values of r_1 through r_4 in the factorization

$$8x^4 - 14x^3 - 9x^2 + 11x - 1 = 8(x - r_1)(x - r_2)(x - r_3)(x - r_4).$$

35. *Estimating reciprocals without division.* Newton's method in Section 4.3 can be used to estimate the reciprocal of a positive number a without ever dividing by a, by taking $f(x) = 1/x - a$. For example, if $a = 3$, the function involved is $f(x) = 1/x - 3$.

   a) Graph $y = 1/x - 3$. Where does the graph cross the x-axis?

   b) Show that the recursion formula in Newton's method in this case is

   $$x_{n+1} = x_n(2 - 3x_n),$$

   so indeed there is no division.

36. Show that the equation $x^3 + x - 1 = 0$ has exactly one solution, and use Newton's method to find it to three decimal places.

37. Find the maximum and minimum values of $f(x) = 10 + 20x - 11x^2 - 8x^3 - x^4$ on $-6 \le x \le 1$, and say where they are assumed.

38. Find the maximum and minimum of $f(x) = \sqrt{x} + \cos x$ on $0 \le x \le 11$, and say where they are assumed.

39. If the perimeter of the circular sector shown here is 100 ft, what values of r and s will give the sector the greatest area?

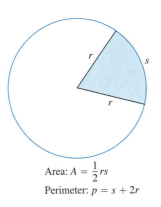

Area: $A = \dfrac{1}{2}rs$

Perimeter: $p = s + 2r$

40. An isosceles triangle has its vertex at the origin and its base parallel to the x-axis with the vertices above the axis on the curve $y = 27 - x^2$. Find the largest area the triangle can have.

41. Find the dimensions of the largest open storage bin with a square base and vertical sides that can be made from 108 ft^2 of sheet steel. (Neglect the thickness of the steel, and assume that there is no waste.)

42. You are to design an open-top rectangular stainless-steel vat. It is to have a square base and a volume of 32 ft^3, to be welded from quarter-inch plate, and weigh no more than necessary. What dimensions do you recommend?

43. Find the height and radius of the largest right circular cylinder that can be put in a sphere of radius $\sqrt{3}$. (See following diagram.)

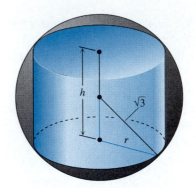

44. The figure shows two right circular cones, one upside down inside the other. The two bases are parallel, and the vertex of the smaller cone lies at the center of the larger cone's base. What values of r and h will give the smaller cone the largest possible volume?

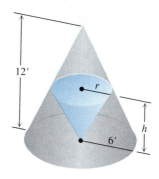

45. A drilling rig 12 mi off shore is to be connected by a pipe to a refinery on shore, 20 mi down the coast from the rig. If underwater pipe costs \$40,000 per mile and land-based pipe costs \$30,000 per mile, what values of x and y give the least expensive connection?

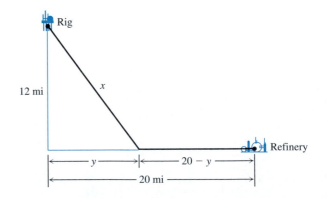

46. An athletic field is to be built in the shape of a rectangle x units long capped by semicircular regions of radius r at the two ends. The field is to be bounded by a 400-m running track. What values of x and r will give the rectangle the largest possible area?

47. Your company can manufacture x hundred grade A tires and y hundred grade B tires a day, where $0 \le x \le 4$ and

$$y = \frac{40 - 10x}{5 - x}.$$

Your profit on Grade A tires is twice your profit on grade B tires. What is the most profitable number of each kind of tire to make?

48. The positions of two particles on the s-axis are $s_1 = \sin t$ and $s_2 = \sin(t + \pi/3)$.

   a) What is the farthest apart the particles ever get?

   b) When, if ever, do the particles collide?

49. An open-top rectangular box is constructed from a 10- by 16-in. piece of cardboard by cutting squares of equal side length from the corners and folding up the sides. Find analytically the dimensions of the box of largest volume and the maximum volume. Support your results graphically.

50. a) Repeat Exercise 67 in Section 4.3 for a 22- by 34-in. piece of cardboard. In part (f), replace 1120 in.^{3} by 400 in.^{3}

   b) Repeat Exercise 68 in Section 4.3 for a 12- by 20-in. piece of cardboard. There are two possibilities.

Describe the motion of the particle moving on the x-axis with distance from the origin given by the functions in Exercises 51 and 52. Simulate the motion with a grapher in parametric mode.

51. $s(t) = t^3 + t^2 - 6t + 5$

52. $s(t) = 3 + 4t - 3t^2 - t^3$

53. The radius of a circle is changing at the rate of $-2/\pi$ m/sec. At what rate is the circle's area changing when $r = 10$ m?

54. The coordinates of a particle moving in the metric xy-plane are differentiable functions of time t with $dx/dt = -1$ m/sec and $dy/dt = -5$ m/sec. How fast is the particle approaching the origin as it passes through the point $(5, 12)$?

55. The volume of a cube is increasing at the rate of 1200 cm^3/min at the instant its edges are 20 cm long. At what rate are the edges changing at that instant?

56. A point moves smoothly along the curve $y = x^{3/2}$ in the first quadrant in such a way that its distance from the origin increases at the constant rate of 11 units per second. Find dx/dt when $x = 3$.

57. Water drains from the conical tank shown in the following diagram at the rate of 5 ft^3/min.

   a) What is the relation between the variables h and r?

   b) How fast is the water level dropping when $h = 6$ ft?

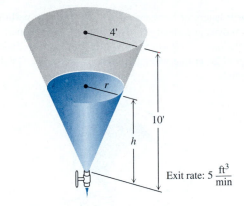

Exit rate: $5 \dfrac{ft^3}{min}$

58. Two cars are approaching an intersection along straight highways that cross at right angles, car A moving at 36 mi/h and car B at 50 mi/h. At what rate is the straight-line distance between the cars changing when car A is 5 mi and car B is 12 mi from the intersection? At what rate is the distance changing at any time?

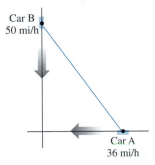

59. You are videotaping a race from a stand 132 ft from the track, following a car that is traveling at 180 mi/h (264 ft/sec). How fast will your camera angle θ be changing when the car is right in front of you? A half-second later?

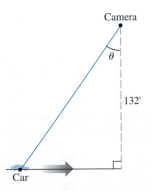

60. As telephone cable is pulled from a large spool to be strung from the telephone poles along a street, it unwinds from the spool in layers of constant radius. If the truck pulling the cable moves at a steady 6 ft/sec (a touch over 4 mi/h), use the equation $s = r\theta$ to find how fast (in radians per second) the spool is turning when the layer of radius 1.2 ft is being unwound?

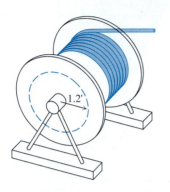

61. The formula $F(x) = 3x + C$ gives a different function for each value of C. All of these functions, however, have the same derivative with respect to x, namely, $F'(x) = 3$. Are these the only differentiable functions whose derivative is 3? Could there be any others? Explain.

62. Show that

$$\frac{d}{dx}\left(\frac{x}{x+1}\right) = \frac{d}{dx}\left(-\frac{1}{x+1}\right)$$

even though

$$\frac{x}{x+1} \neq -\frac{1}{x+1}.$$

Doesn't this contradict Corollary 3 of the Mean Value Theorem? Explain.

63. Find the general antiderivatives of the following functions.

a) 0 b) 1 c) x

d) x^2 e) x^{10} f) x^{-2}

g) x^{-5} h) $x^{5/2}$ i) $x^{4/3}$

j) $x^{3/4}$ k) $x^{1/2}$ l) $x^{-1/2}$

m) $x^{-3/7}$ n) $x^{-7/3}$

64. Find the general antiderivatives of the following functions.

a) $\sin x$ b) $\cos x$ c) $\sec x \tan x$

d) $-\csc^2 x$ e) $\sec^2 x$ f) $-\csc x \cot x$

Find the general antiderivatives of the functions in Exercises 65–80. Then support your answer with a graphing utility.

65. $3x^2 + 5x - 7$

66. $\dfrac{1}{x^2} + x + 1$

67. $\sqrt{x} + \dfrac{1}{\sqrt{x}}$

68. $\sqrt[3]{x} + \sqrt[4]{x}$

69. $3\cos 5x$

70. $8\sin(x/2)$

71. $3\sec^2 3x$

72. $4\csc^2 2x$

73. $\dfrac{1}{2} - \cos x$

74. $3x^5 + 16\cos 8x$

75. $\sec\dfrac{x}{3}\tan\dfrac{x}{3} + 5$

76. $1 - \csc x \cot x$

77. $\tan^2 x$ (*Hint:* $\tan^2 x = \sec^2 x - 1$.)

78. $\cot^2 x$ (*Hint:* $\cot^2 x = \csc^2 x - 1$.)

79. $2\sin^2 x$ (*Hint:* $2\sin^2 x = 1 - \cos 2x$.)

80. $\sin^2 x - \cos^2 x$ (*Hint:* $\cos^2 x - \sin^2 x = \cos 2x$.)

Solve the initial-value problems in Exercises 81–86. If possible, support your answer with a graphing utility.

81. $\dfrac{dy}{dx} = 1 + x + \dfrac{x^2}{2}$, $y = 1$ when $x = 0$

82. $\dfrac{dy}{dx} = 4x^3 - 21x^2 + 14x - 7$, $y = 1$ when $x = 1$

83. $\dfrac{dy}{dx} = \dfrac{x^2 + 1}{x^2}$, $y = -1$ when $x = 1$

84. $\dfrac{dy}{dx} = \left(x + \dfrac{1}{x}\right)^2$, $y = 1$ when $x = 1$

85. $\dfrac{d^2 y}{dx^2} = -\sin x$, $y = 0$ and $\dfrac{dy}{dx} = 1$ when $x = 0$

86. $\dfrac{d^2 y}{dx^2} = \cos x$, $y = -1$ and $\dfrac{dy}{dx} = 0$ when $x = 0$

87. Does any function $y = f(x)$ satisfy all of the following conditions? If so, what is it? If not, why not?
   a) $d^2 y/dx^2 = 0$ for all x
   b) $dy/dx = 1$ when $x = 0$
   c) $y = 0$ when $x = 0$

88. Find an equation for the curve in the xy-plane that passes through the point $(1, -1)$ if its slope at x is always $3x^2 + 2$.

89. You sling a shovelful of dirt up from the bottom of a 17-ft hole with an initial velocity of 32 ft/sec. Is that enough speed to get the dirt out of the hole, or had you better duck?

90. The acceleration of a particle moving along a coordinate line is $d^2 s/dt^2 = 2 + 6t$ m/sec^2. At $t = 0$, the velocity is 4 m/sec. Find the velocity as a function of t. Then find how far the particle moves during the first second of its trip, from $t = 0$ to $t = 1$.

Sketch the solution curves of the initial-value problems in Exercises 91 and 92. Support graphically with the toolbox program SLOPEFLD.

91. $\dfrac{dy}{dx} = 4 - \sqrt{x^2 + 3}$, $y = 0$ when $x = 0$

92. $\dfrac{dy}{dx} = \sqrt{x^2 + 1} - 1$, $y = 1$ when $x = 0$

5

Integration

OVERVIEW   This chapter introduces the second main branch of calculus, the branch called integral calculus. Integral calculus is the mathematics that we use to find lengths, areas, and volumes, to calculate the average values of functions; and to predict future population sizes and future costs of living. In this chapter, we set the stage for these and other applications.

The development of integral calculus starts from the calculation of areas by a technique that leads to a natural definition of area as a limit of finite sums. The limits used to define areas are special cases of a kind of limit called a *definite integral*. Presenting the properties of definite integrals, developing numerical methods of computing definite integrals, and applying the numerical methods with a graphing calculator are central goals of this chapter.

The single most important concept in this chapter is the connection between definite integrals and derivatives. The discovery of this connection (called the Fundamental Theorem of Calculus) by Leibniz and Newton turned calculus into the most important application of mathematics in the world.

5.1 _____ Calculus and Area

Integral calculus is the mathematics that we use to define and calculate the areas of regions like the cross sections of machine parts and airplane wings for which no standard area formulas are known. This section explains what calculus and area have to do with one another and shows how to calculate the areas of regions like the one in Fig. 5.1.

Regions Bounded by Curves

To find the area of a triangle, we use the formula $A = (1/2)bh$, area equals one-half base times height. To find the areas of more general polygonal regions, we can divide them into triangles, then add the areas of the triangles (Fig. 5.2). But we get stuck if we try to calculate the area of a circle this way. No matter how many triangles we draw inside the circle, their straight

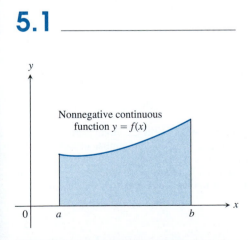

Nonnegative continuous function $y = f(x)$

5.1   We can use integral calculus to find the areas of regions like the one shaded here.

347

edges never quite match the curve of the circle, and some of the circle's interior remains uncovered.

The Greeks of the fifth century B.C. overcame this problem by filling the circle with an infinite sequence of increasingly fine regular polygons, exhausting the circle's area, so to speak, step by step (Fig. 5.3a). They then took the circle's area to be the limit of the areas of these polygons. A decreasing sequence of circumscribed polygons would have worked as well (Fig. 5.3b).

The difficulty with applying this approach to more general curves is not the involvement of limits (at least, that's not a difficulty for us). Rather, it is the complication associated with finding workable formulas for the areas of the inscribed polygons, which in an arbitrary curve can assume irregular shapes. We can avoid this difficulty if, instead of working in the abstract plane of Euclidean geometry, we work in the coordinate plane of Descartes and Fermat. Then we can approximate the region under a curve with rectangles whose numerical dimensions, and hence areas, are given by the curve itself.

You will see what we mean if you look at the curve and rectangles in Fig. 5.4(a). The rectangle areas, added, approximate the area between the curve $y = f(x)$ and the x-axis over the interval from $x = a$ to $x = b$. The area of each rectangle, base times height, is the base length times some particular function value, a value we can find from the formula $y = f(x)$.

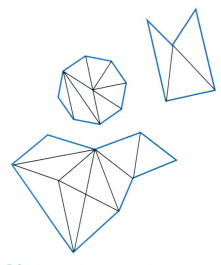

5.2 We find the areas of plane regions with polygonal boundaries by dividing the regions into triangles. The answer is the same for every triangulation.

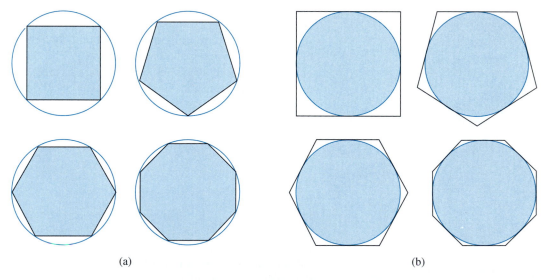

(a)   (b)

5.3 The areas of the (a) inscribed regular polygons or (b) the circumscribed regular polygons have the area of the circle as a limit as the number of sides of the polygons increase without bound. One sequence approaches the area of the circle from below, the other from above. Approximations like the ones shown here were the basis of the method used in classical Greek times to find the area of a circle.

Notice in Fig. 5.4(b) how the approximations improve as the rectangles become thinner and more numerous. With each refinement, we get closer to

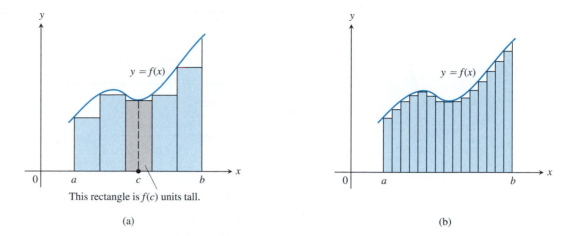

(a)   (b)

5.4 (a) If we approximate the region under the curve $y = f(x)$ from $x = a$ to $x = b$ with inscribed rectangles that reach from the x-axis up to the curve, then the height of each rectangle is the value of f at some point along the rectangle's base. (b) The more rectangles we use, the better the approximation becomes (provided that the rectangles in general become narrower as we go along).

filling up the region whose area we want to find. To finish the job, all we need is

1. A way to write formulas for sums of large numbers of terms, and

2. A way to find the numerical limits of such sums as the number of terms tends to infinity (when the limits exist, that is).

As soon as we know how to take these two steps, we will be able to define and calculate all the areas we want.

What we find, when we take these steps, will also be surprising. We will be able to calculate much more than just area. And thanks to a great breakthrough in connecting integration and differentiation discovered by Leibniz and Newton, the calculations will be easier. Furthermore, using a graphing utility, we will be able to illustrate what is going on in an instructive and unusual manner, sometimes creating pictures that have rarely, if ever, been seen before.

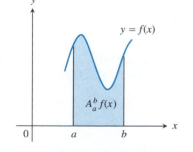

5.5 For a nonnegative continuous function $y = f(x)$, the area under its graph and above the x-axis from a to b is $A_a^b f(x)$.

The Area under the Graph of a Nonnegative Continuous Function

Here is a preview of how we will be able to calculate area once we have made the necessary mathematical arrangements. We all know what we want area to be like, so let us suppose for the moment that the forthcoming mathematical definition in Section 5.2 gives us everything we want. (It will.) Also, we will consider the function $y = x^2$ as an example of a nonnegative continuous function like the one in Fig. 5.5. We will let $A_a^b f(x)$ denote the area of the region bounded by the curve $y = f(x)$, the x-axis, and the lines $x = a$ and $x = b$. We will call it *the area under the curve from a to b*.

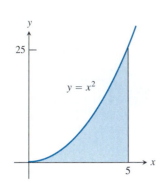

5.6 The area of the shaded region = $A_0^5 x^2$.

A PIXEL APPROXIMATION

You could approximate $A_0^5 \left(x^2 \right)$ in a viewing window by counting the pixels under the graph of $y = x^2$ from 0 to 5 and multiply by the area of one pixel. (However, we aren't recommending that you try this!)

$$A_a^b f(x) = \text{the area under the curve from } a \text{ to } b.$$

EXAMPLE 1

For $f(x) = x^2$, the area under the curve from 0 to 5 is $A_0^5(x^2)$ (Fig. 5.6).

■

The Rectangle Approximation Method (RAM)

How can we find an explicit value for the area in Example 1? We can do what was suggested in Fig. 5.4, namely, use a *rectangle approximation method* (RAM) to approximate the area with rectangles of known area and then consider what happens to the total area of the rectangles as we view more and more of them with smaller and smaller widths.

First, we establish some notation. We partition the interval $[a, b]$ into n subintervals by selecting $n - 1$ points, say, $x_1, x_2, \ldots, x_{n-1}$, between a and b. To make the notation consistent, we usually denote a by x_0 and b by x_n. The set

$$P = \{x_0, x_1, \ldots, x_{n-1}, x_n\}$$

is then called a **partition** of $[a, b]$.

The partition P defines n closed **subintervals**

$$[x_0, x_1], [x_1, x_2], \ldots, [x_{n-1}, x_n].$$

The typical closed subinterval $[x_{k-1}, x_k]$ is called the **kth subinterval** of P.

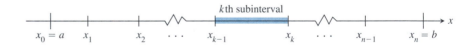

The length of the kth subinterval is $\Delta x_k = x_k - x_{k-1}$.

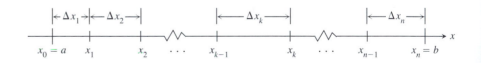

If P is chosen so that its points are equally spaced, then $\Delta x_i = (b-a)/n$ for $i = 1$ to n; that is, all subintervals are the same length (Fig. 5.7).

5.7 With the points x_i equally spaced, $\Delta x = (b - a)/n$ is the width of each subinterval. Notice that $x_0 = a, x_1 = a + \Delta x, x_2 = a + 2\Delta x, \ldots, x_k = a + k\Delta x, \ldots, x_n = a + n\Delta x = b$.

Now we let f be a nonnegative function on $[a, b]$ (remember, we are illustrating with $f(x) = x^2$), and we stand rectangles on each subinterval so that they reach up to the graph of f. But how high should we make these rectangles? For each $[x_{k-1}, x_k]$, we have three suitable choices for what we want to do. We can choose the value of f at the left endpoint, $f(x_{k-1})$, the value of f at the right endpoint, $f(x_k)$, or the value of f at the midpoint, $f((x_{k-1} + x_k)/2)$, to be the height of each rectangle.

For a partition of $[a, b]$ into n subintervals of equal length, we call the RAM results using the left endpoints $\text{LRAM}_n f$ (Fig. 5.8a), the right endpoints $\text{RRAM}_n f$ (Fig. 5.8b), and the midpoints $\text{MRAM}_n f$ (Fig. 5.8c).

PRONUNCIATION AND READING

For "$\text{LRAM}_n f$," we say *el ram sub n of f*. To help you get used to the idea of what it represents, you may wish initially to read "$\text{LRAM}_n f$" as the "left-RAM_n *value of f*." Similar comments apply to $\text{RRAM}_n f$ and $\text{MRAM}_n f$ (the *mid-RAM* value).

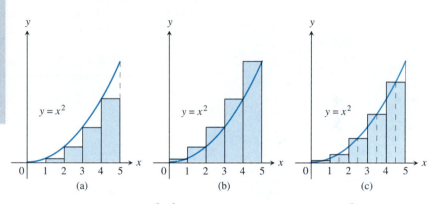

5.8 Approximations to $A_0^5(x^2)$, the area under the curve $y = x^2$ from $x = 0$ to $x = 5$, using (a) $\text{LRAM}_5(x^2)$, (b) $\text{RRAM}_5(x^2)$, and (c) $\text{MRAM}_5(x^2)$.

EXAMPLE 2

For $f(x) = x^2$, the partition shown in Fig. 5.8 gives $\Delta x_i = (5 - 0)/5 = 1$. We numerically approximate the area under the graph of f with LRAM, RRAM, and MRAM by adding the areas of the five rectangles.

For each RAM,

$$\text{Area of each rectangle} = \text{height} \cdot \text{width}$$

$$= f(x_i) \cdot \Delta x_i.$$

Thus,

$$\text{LRAM}_5(x^2) = 0^2 \cdot 1 + 1^2 \cdot 1 + 2^2 \cdot 1 + 3^2 \cdot 1 + 4^2 \cdot 1$$

$$= 0 + 1 + 4 + 9 + 16 = 30$$

$$\text{RRAM}_5(x^2) = 1^2 \cdot 1 + 2^2 \cdot 1 + 3^2 \cdot 1 + 4^2 \cdot 1 + 5^2 \cdot 1 = 55$$

$$\text{MRAM}_5(x^2) = \left(\frac{1}{2}\right)^2 \cdot 1 + \left(\frac{3}{2}\right)^2 \cdot 1 + \left(\frac{5}{2}\right)^2 \cdot 1 + \left(\frac{7}{2}\right)^2 \cdot 1 + \left(\frac{9}{2}\right)^2 \cdot 1$$

$$= 41.25$$

ESTIMATES AND ACTUAL AREA

Later we will show that the actual area under the curve $y = x^2$ from 0 to 5 is 41 2/3. Notice that $\text{LRAM}_5(x^2)$ is an *under*estimate, $\text{RRAM}_5(x^2)$ is an *over*estimate and $\text{MRAM}_5(x^2)$ is the best estimate. In the exercises, you will be asked why LRAM_n and $\text{RRAM}n$ are always underestimates and overestimates, respectively, of $A_0^5(x^2)$.

A key to understanding a mathematical definition of area is to understand the answer to this question: How can $\text{LRAM}_n f$, $\text{RRAM}_n f$ and $\text{MRAM}_n f$

be made to more closely approximate $A_a^b f$, the actual area under the nonnegative function $f(x)$ from $x = a$ to $x = b$? The answer is to make n larger, visualize what is happening to the rectangles, and watch how the RAM values converge to a single number using a computer to do the arithmetic.

EXPLORATION 1

Computing RAM Values

On your graphing utility, enter a RAM program for computing $\text{LRAM}_n f$, $\text{RRAM}_n f$, and $\text{MRAM}_n f$. Run the program for $f(x) = x^2$, and verify the values in Table 5.1. (A simple RAM program for most graphing calculators is shown in the *Resource Manual* that accompanies this textbook.)

RAM CONVERGENCE

In Exploration 1, the LRAM values approach $A_0^5 (x^2)$ *from below* and the RRAM values approach *from above* because x^2 is an increasing function on [0, 5]. For other functions $f \geq 0$ on [0,5], LRAM and RRAM will still approach $A_0^5 f$, but not necessarily from one particular side. Which one of LRAM, RRAM, or MRAM converges on $A_0^5 f$ the fastest will also vary from function to function.

TABLE 5.1   RAM Computations for $A_0^5 x^2$

n	$\text{LRAM}_n f$	$\text{MRAM}_n f$	$\text{RRAM}_n f$
5	30	41.25	55
10	35.625	41.5625	48.125
25	39.2	41.65	44.2
50	40.425	41.6625	42.925
100	41.04375	41.665625	42.29375
1000	41.6041875	41.66665625	41.7291875

Notice that for $f(x) = x^2$ in Table 5.1, the LRAM and RRAM values approach 41 2/3 with the LRAM values increasing from below and the RRAM values decreasing from above. Notice that MRAM approaches 41 2/3 also, but much more rapidly. In fact, on [0, 5],

$$\lim_{n \to \infty} \text{LRAM}_n(x^2) = \lim_{n \to \infty} \text{RRAM}_n(x^2) = \lim_{n \to \infty} \text{MRAM}_n(x^2) = 41\frac{2}{3}.$$

Now it is sensible to make the following definition.

DEFINITION   Area under a Curve

If $y = f(x)$ is nonnegative and continuous over a closed interval $[a, b]$, then the **area** of the region between the graph of f and the x-axis from $x = a$ to $x = b$ is

$$A_a^b f(x) = \lim_{n \to \infty} \text{LRAM}_n f(x) = \lim_{n \to \infty} \text{RRAM}_n f(x) = \lim_{n \to \infty} \text{MRAM}_n f(x).$$

We will also call this the **area under the curve $y = f(x)$ from a to b**.

The above definition makes sense only if the limits mentioned make sense and are equal. We included in the definition the requirement that f be continuous because, as we will see later, if f is continuous, these limits indeed do exist and are equal.

EXAMPLE 3

Estimate the area under $f(x) = x^2 \sin x$ from 0 to 3 using the RAM with $n = 5, 10, 25, 50$, and 100.

Solution   First, we must verify that $x^2 \sin x \geq 0$ on $[0, 3]$. Our viewing window (Fig. 5.9) supports that it is. A simple algebraic analysis, which we leave for you to do, confirms that it is.

Next we apply our RAM computer program to obtain Table 5.2.

TABLE 5.2   RAM Approximations for $A_0^3 \, (x^2 \sin x)$.

n	$\text{LRAM}_n \, f$	$\text{MRAM}_n \, f$	$\text{RRAM}_n \, f$
5	5.15480...	5.89668...	5.91685...
10	5.52574...	5.80684...	5.90676...
25	5.69078...	5.78150...	5.84319...
50	5.73614...	5.77787...	5.81235...
100	5.75701...	5.77696...	5.79511...
1000	5.77475...	5.77667...	5.77856...

We will see in Chapter 8 (with a lot of effort) that the *exact* area is $-7 \cos 3 + 6 \sin 3 - 2$ or 5.77666752456 to 12 digits. Notice that $\text{MRAM}_{100} f(x)$ is "correct" to three decimal places while $\text{MRAM}_{1000} f(x)$ is correct to four decimal places!

Sigma Notation and Algebra Rules for Finite Sums

The three RAMs that we have seen so far and a more general RAM that we will see in the next section suggest that we need notation that will simplify working with sums of large numbers of terms. The symbol that we use to indicate sums is Σ (pronounced "sigma"), the Greek letter for S.

EXAMPLE 4

The sum	In sigma notation	One way to read the notation
$a_1 + a_2$	$\displaystyle\sum_{k=1}^{2} a_k$	The sum of a sub k from k equals 1 to k equals 2.
$a_1 + a_2 + a_3$	$\displaystyle\sum_{k=1}^{3} a_k$	The sum of a sub k from k equals 1 to k equals 3.
$a_1 + a_2 + a_3 + a_4$	$\displaystyle\sum_{k=1}^{4} a_k$	The sum of a sub k from k equals 1 to k equals 4.
$a_1 + a_2 + \cdots + a_n$	$\displaystyle\sum_{k=1}^{n} a_k$	The sum of a sub k from k equals 1 to k equals n.

[0, 3] by [0, 5]

5.9   The graph of $f(x) = x^2 \sin x$ on $0 \leq x \leq 3$.

As you can see, the notation remains compact no matter how many terms are being added. It is just what we need for writing sums that involve millions and millions of terms.

There are many different ways to read the notation $\sum_{k=1}^{n} a_k$, all equally good. Some people say, "Summation from k equals 1 to n of a sub k." Others say, "Summation a k from k equals 1 to n." Still others, "The sum of the a sub k's as k goes from 1 to n," and so on. Take your pick.

DEFINITIONS　**Sigma Notation for Finite Sums**

The symbol

$$\sum_{k=1}^{n} a_k$$

denotes the sum of the n terms

$$a_1 + a_2 + \cdots + a_{n-1} + a_n.$$

The variable k is the **index of summation**. The values of k run through the integers from 1 to n. The a's are the **terms** of the sum; a_1 is the first term, a_2 is the second term, a_k is the **kth term**, and a_n is the nth and last term. The number 1 is the **lower limit of summation**; the number n is the **upper limit of summation**.

Here are some numerical examples.

EXAMPLE 5

The sum in sigma notation	The sum written out— one term for each value of k	The value of the sum
$\displaystyle\sum_{k=1}^{5} k$	$1+2+3+4+5$	15
$\displaystyle\sum_{k=1}^{3} (-1)^k k$	$(-1)^1(1)+(-1)^2(2)+(-1)^3(3)$	$-1+2-3=-2$
$\displaystyle\sum_{k=1}^{2} \frac{k}{k+1}$	$\dfrac{1}{1+1}+\dfrac{2}{2+1}$	$\dfrac{1}{2}+\dfrac{2}{3}=\dfrac{7}{6}$

≡

EXPLORATION 2

Understanding Notation

1. Show that

$$\sum_{k=1}^{3} \sin\left(\frac{k\pi}{2}\right) = 0.$$

2. Here are two more sums to evaluate. Note that the lower limit of summation tells you to use an integer other than 1 for k in the first term.

a) Show that

$$\sum_{k=0}^{2} \frac{1}{2^k} = \frac{7}{4}.$$

b) Show that

$$\sum_{k=-3}^{-1} (k+1) = -3.$$

When you work with finite sums, you can always use the following rules.

INFINITE SUMS

In Chapter 9, we will study the *infinite sum*

$$\sum_{k=1}^{\infty} a_k = \lim_{n \to \infty} \sum_{k=1}^{n} a_k.$$

We will see that this limit exists for the first sum in the Exploration Bit above but not for the second sum.

Algebra Rules for Finite Sums

1. Constant Multiple Rule: $\displaystyle\sum_{k=1}^{n} ca_k = c \cdot \sum_{k=1}^{n} a_k$ (any number c).

2. Constant Value Rule: $\displaystyle\sum_{k=1}^{n} a_k = n \cdot c$ if a_k has the constant value c.

3. Sum Rule: $\displaystyle\sum_{k=1}^{n}(a_k + b_k) = \sum_{k=1}^{n} a_k + \sum_{k=1}^{n} b_k.$

4. Difference Rule: $\displaystyle\sum_{k=1}^{n}(a_k - b_k) = \sum_{k=1}^{n} a_k - \sum_{k=1}^{n} b_k.$

There are no surprises in this list of rules, but the formal proofs require a technique called mathematical induction (Appendix 4).

EXAMPLE 6

a) $\displaystyle\sum_{k=1}^{n} -a_k = \sum_{k=1}^{n} -1 \cdot a_k = -1 \cdot \sum_{k=1}^{n} a_k = -\sum_{k=1}^{n} a_k$ Constant Multiple Rule

b) $\displaystyle\sum_{k=1}^{3}(k+4) = \sum_{k=1}^{3} k + \sum_{k=1}^{3} 4$ Sum Rule

$$= (1+2+3) + (3 \cdot 4) = 6 + 12 = 18$$ Constant Value Rule

c) $\displaystyle\sum_{k=1}^{n}(k - k^2) = \sum_{k=1}^{n} k - \sum_{k=1}^{n} k^2$ Difference Rule

Standard Formulas for Sums

Over the years, people have discovered a variety of formulas for the values of finite sums. The most famous of these are the formula for the sum of the first n integers (which Gauss discovered at age 5) and the formulas for the sums of the squares and cubes of the first n integers.

The first n integers: $\displaystyle\sum_{k=1}^{n} k = \frac{n(n+1)}{2}.$ (1)

The first n squares: $\displaystyle\sum_{k=1}^{n} k^2 = \frac{n(n+1)(2n+1)}{6}.$ (2)

The first n cubes: $\displaystyle\sum_{k=1}^{n} k^3 = \left(\frac{n(n+1)}{2}\right)^2.$ (3)

Notice the relationship between the first sum and the third.

EXAMPLE 7

a) $\displaystyle\sum_{k=1}^{5} k = 1 + 2 + \cdots + 5 = \frac{5(5+1)}{2} = \frac{5 \cdot 6}{2} = 15$

b) $\displaystyle\sum_{k=1}^{5} k^2 = 1 + 4 + \cdots + 25 = \frac{5(6)(2 \cdot 5 + 1)}{6} = 5 \cdot 11 = 55$

c) $\displaystyle\sum_{k=1}^{5} k^3 = 1 + 8 + \cdots + 125 = \left(\frac{5(5+1)}{2}\right)^2 = 15^2 = 225$ ▤

> **EXPLORATION BIT**
>
> Note that Example 7(b) is $\text{RRAM}_5(x^2)$ on $[0, 5]$. Can you express any of the other sums in Examples 7 and 8 as RAM values?

EXAMPLE 8

Using Algebra Rules and the values from Example 7, we have

a) $\displaystyle\sum_{k=1}^{5} \frac{k^3}{3} = \frac{1}{3}\sum_{k=1}^{5} k^3 = \frac{225}{3} = 75,$

b) $\displaystyle\sum_{k=1}^{5} (k - k^2) = \sum_{k=1}^{5} k - \sum_{k=1}^{5} k^2 = 15 - 55 = -40.$ ▤

Computing Area

We return now to the RAM for a preview of things to come. Using summation notation, we have

$$\text{LRAM}_n \, f(x) = f(x_0) \cdot \Delta x + f(x_1) \cdot \Delta x + \cdots + f(x_{n-1}) \cdot \Delta x$$

$$= \sum_{k=0}^{n-1} f(x_k) \cdot \Delta x$$

$$\text{RRAM}_n \, f(x) = f(x_1) \cdot \Delta x + f(x_2) \cdot \Delta x + \cdots + f(x_n) \cdot \Delta x$$

$$= \sum_{k=1}^{n} f(x_k) \cdot \Delta x$$

$$\text{MRAM}_n \, f(x) = f\left(\frac{x_0 + x_1}{2}\right) \cdot \Delta x + f\left(\frac{x_1 + x_2}{2}\right) \cdot \Delta x + \cdots$$

$$+ f\left(\frac{x_{n-1} + x_n}{2}\right) \cdot \Delta x$$

$$= \sum_{k=1}^{n} f\left(\frac{x_{k-1} + x_k}{2}\right) \cdot \Delta x,$$

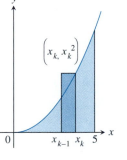

5.10 In computing $\text{RRAM}_n(x^2)$, the kth subinterval has length $5/n$ and $x_k = k(5/n)$. The rectangle over the partition interval $[x_{k-1}, x_k]$ has width $5/n$ and height $[k(5/n)]^2$.

where $\{x_0, x_1, \ldots, x_{n-1}, x_n\}$ is a partition of $[a, b]$ into intervals of length $\Delta x = (b - a)/n$.

Using RRAM with $f(x) = x^2$ on $[0, 5]$, we will compute $A_0^5(x^2) = \lim_{n \to \infty} \text{RRAM}_n(x^2)$. The key here is to recognize that $\Delta x = (5 - 0)/n = 5/n$ is the width of each of the n rectangles and that the right-hand endpoint of the kth interval $[x_{k-1}, x_k]$ is $x_k = 0 + k\Delta x = k(5/n)$. (See Fig. 5.10.) Thus,

$$\text{RRAM}_n\left(x^2\right) = \sum_{k=1}^{n} (x_k)^2 \cdot \Delta x$$

$$= \sum_{k=1}^{n} \left(k \cdot \frac{5}{n}\right)^2 \cdot \frac{5}{n}$$

$$= \sum_{k=1}^{n} k^2 \cdot \left(\frac{5}{n}\right)^2 \cdot \frac{5}{n}$$

$$= \left(\frac{5}{n}\right)^3 \sum_{k=1}^{n} k^2 \qquad \text{Constant Multiple Rule}$$

$$= \left(\frac{5}{n}\right)^3 \frac{n(n + 1)(2n + 1)}{6} \qquad \text{Sum of the first } n \text{ squares} \quad (4)$$

$$= \frac{125}{6} \cdot \frac{n(n + 1)(2n + 1)}{n^3}$$

$$= \frac{125}{6} \cdot \frac{2n^3 + 3n^2 + n}{n^3}$$

$$= \frac{125}{6}\left(2 + \frac{3}{n} + \frac{1}{n^2}\right)$$

EXPLORATION BIT

Equation (5) confirms our earlier statement that
$$\lim_{n \to \infty} \text{RRAM}_n(x^2) = 41 \ 2/3.$$
Now confirm that
$$\lim_{n \to \infty} \text{LRAM}_n(x^2) = 41 \ 2/3 \text{ and}$$
$$\lim_{n \to \infty} \text{MRAM}_n(x^2) = 41 \ 2/3.$$

Now, taking the limit of this sum as $n \to \infty$ gives

$$
\begin{aligned}
\lim_{n \to \infty} \text{RRAM}_n(x^2) &= \lim_{n \to \infty} \left[\frac{125}{6} \left(2 + \frac{3}{n} + \frac{1}{n^2} \right) \right] \\
&= \frac{125}{6} \lim_{n \to \infty} \left(2 + \frac{3}{n} + \frac{1}{n^2} \right) \\
&= \frac{125}{6} \cdot (2 + 0 + 0) \\
&= \frac{125}{3} = 41\frac{2}{3}.
\end{aligned}
\tag{5}
$$

Now we change notation from $f(x) = x^2$ to $f(t) = t^2$ and ask the question: Can we determine a formula for $A_0^x(t^2)$ for any value of $x > 0$? The answer is that we surely can. All that is required is that we replace 5 by x in the derivation of $A_0^5(t^2)$. Thus, Eq. (4),

$$
\text{RRAM}_n(t^2) = \left(\frac{5}{n} \right)^3 \frac{n(n+1)(2n+1)}{6},
$$

becomes

$$
\begin{aligned}
\text{RRAM}_n(t^2) &= \left(\frac{x}{n} \right)^3 \frac{n(n+1)(2n+1)}{6} \\
&= \frac{x^3}{n^3} \frac{2n^3 + 3n^2 + n}{6} \\
&= \frac{x^3}{6} \left(2 + \frac{3}{n} + \frac{1}{n^2} \right).
\end{aligned}
$$

As $n \to \infty$, $\text{RRAM}_n(t^2) \to x^3/3$ for any positive value of x. That is, the area under the curve $f(t) = t^2$ from 0 to x is *exactly* $x^3/3$. So

$$
A_0^x(t^2) = \frac{x^3}{3}, \qquad x \geq 0.
$$

As a tantalizing glimpse of what will come in Section 5.3, note that

$$
A_0^x(t^2) = \frac{x^3}{3} \qquad \text{and} \qquad D_x \left(\frac{x^3}{3} \right) = x^2,
$$

an illustration of one of the most remarkable connections in mathematics!

What Lies Ahead

If all we wanted to do was find areas, we would be nearly done now, but the idea of approximating things with small manageable pieces the way we approximated regions with rectangles extends to thousands of other situations. We calculate the volumes of large objects by slicing them like loaves of bread and adding up the volumes of the slices. We find the lengths of curves by approximating small pieces of the curve with line segments and adding the lengths of the line segments. The idea extends to finding the areas of surfaces and the forces against dams and to calculating how much work it takes to serve a tennis ball or lift a satellite into orbit. We will see these examples and many more in subsequent chapters.

Exercises 5.1

For each function $y = f(x)$ in Exercises 1–6, consider the area of the region between the graph of $y = f(x)$ and the x-axis from $x = a$ to $x = b$.

a) Make sketches illustrating the RAM for LRAM$_5$, RRAM$_5$, and MRAM$_5$ showing the five approximating rectangles.

b) Write out by hand LRAM$_5$, RRAM$_5$, and MRAM$_5$, and compute each sum.

1. $y = 6 - x^2$ from $x = 0$ to $x = 2$

2. $y = x^2 + 2$ from $x = -3$ to $x = 2$

3. $f(x) = x + 1$ from $x = 0$ to $x = 5$

4. $f(x) = 5 - x$ from $x = 0$ to $x = 5$

5. $f(x) = 2x^2$ from $x = 0$ to $x = 5$

6. $f(x) = x^2 + 2$ from $x = 1$ to $x = 6$

For each function f in Exercises 7–14, estimate the area of the region between the graph of f and the x-axis from $x = a$ to $x = b$ using the RAM for $n = 10$, 100, and 1000. Do all three RAMs: LRAM$_n f$, RRAM$_n f$, and MRAM$_n f$. First verify that each function is nonnegative on the specified interval $[a, b]$.

7. $f(x) = x^2 - x + 3$ from $x = 0$ to $x = 3$

8. $f(x) = 2x^2 - 5x + 6$ from $x = -1$ to $x = 4$

9. $f(x) = 2x^3 + 3$ from $x = 0$ to $x = 5$

10. $f(x) = x^3 + x^2 + 2x + 3$ from $x = -1$ to $x = 3$

11. $f(x) = \sin x$ from $x = 0$ to $x = \pi$

12. $f(x) = \cos x$ from $x = 0$ to $x = \dfrac{\pi}{2}$

13. $f(x) = e^{-x^2}$ from $x = -5$ to $x = 5$

14. $f(x) = 2 + \dfrac{\sin x}{x}$ from $x = -3$ to $x = 4$

15. Make a conjecture about the *exact* area of the region described in Exercises 3, 5, 7, 9, and 11.

16. Make a conjecture about the *exact* area of the region described in Exercises 4, 6, 8, 10, and 12.

17. Confirm the values in Table 5.1.

18. Confirm the values in Table 5.2.

Write the sums in Exercises 19–28 without sigma notation. Then evaluate them.

19. $\displaystyle\sum_{k=1}^{4} \frac{1}{k}$

20. $\displaystyle\sum_{k=1}^{4} \frac{12}{k}$

21. $\displaystyle\sum_{k=1}^{3} (k+2)$

22. $\displaystyle\sum_{k=1}^{5} (2k-1)$

23. $\displaystyle\sum_{k=0}^{4} \frac{k}{4}$

24. $\displaystyle\sum_{k=-2}^{2} 3k$

25. $\displaystyle\sum_{k=1}^{4} \cos k\pi$

26. $\displaystyle\sum_{k=1}^{3} \sin \frac{\pi}{k}$

27. $\displaystyle\sum_{k=1}^{4} (-1)^k$

28. $\displaystyle\sum_{k=1}^{4} (-1)^{k+1}$

29. Which of the following express $1 + 2 + 4 + 8 + 16 + 32$ in sigma notation?

a) $\displaystyle\sum_{k=1}^{6} 2^{k-1}$ b) $\displaystyle\sum_{k=0}^{5} 2^{k}$ c) $\displaystyle\sum_{k=-1}^{4} 2^{k+1}$

30. Which formula is not equivalent to the others?

a) $\displaystyle\sum_{k=-1}^{1} \frac{(-1)^k}{k+2}$ b) $\displaystyle\sum_{k=0}^{2} \frac{(-1)^k}{k+1}$

c) $\displaystyle\sum_{k=1}^{3} \frac{(-1)^k}{k}$ d) $\displaystyle\sum_{k=2}^{4} \frac{(-1)^{k-1}}{k-1}$

Express the sums in Exercises 31–36 in sigma notation.

31. $1 + 2 + 3 + 4 + 5 + 6$ 32. $1 + 4 + 9 + 16$

33. $\dfrac{1}{2} + \dfrac{1}{4} + \dfrac{1}{8} + \dfrac{1}{16}$ 34. $1 + \dfrac{1}{2} + \dfrac{1}{3} + \dfrac{1}{4} + \dfrac{1}{5}$

35. $\dfrac{1}{5} - \dfrac{2}{5} + \dfrac{3}{5} - \dfrac{4}{5} + \dfrac{5}{5}$ 36. $-\dfrac{1}{5} + \dfrac{2}{5} - \dfrac{3}{5} + \dfrac{4}{5} - \dfrac{5}{5}$

Use algebra and the formulas in Eqs. (1)–(3) to evaluate the sums in Exercises 37–44.

37. $\displaystyle\sum_{k=1}^{10} k$

38. $\displaystyle\sum_{k=1}^{7} 2k$

39. $\displaystyle\sum_{k=1}^{6} -k^2$

40. $\displaystyle\sum_{k=1}^{6} (k^2 + 5)$

41. $\displaystyle\sum_{k=1}^{5} k(k-5)$

42. $\displaystyle\sum_{k=1}^{7} (2k-8)$

43. $\displaystyle\sum_{k=1}^{100} k^3 - \sum_{k=1}^{99} k^3$

44. $\left(\displaystyle\sum_{k=1}^{7} k\right)^2 - \sum_{k=1}^{7} k^3$

45. Suppose that $\displaystyle\sum_{k=1}^{n} a_k = -5$ and $\displaystyle\sum_{k=1}^{n} b_k = 6.$ Evaluate:

a) $\displaystyle\sum_{k=1}^{n} 3a_k$

b) $\displaystyle\sum_{k=1}^{n} \frac{b_k}{6}$

c) $\displaystyle\sum_{k=1}^{n} (a_k + b_k)$

d) $\displaystyle\sum_{k=1}^{n} (a_k - b_k)$

e) $\displaystyle\sum_{k=1}^{n} (b_k - 2a_k)$

46. Suppose that $\displaystyle\sum_{k=1}^{n} a_k = 0$ and $\displaystyle\sum_{k=1}^{n} b_k = 1.$ Evaluate:

a) $\displaystyle\sum_{k=1}^{n} 8a_k$

b) $\displaystyle\sum_{k=1}^{n} 250b_k$

c) $\displaystyle\sum_{k=1}^{n} (a_k + 1)$

d) $\displaystyle\sum_{k=1}^{n} (b_k - 1)$

Write the first five terms of each sum in Exercises 47–50. Then use the built-in capability of your graphing calculator or the PARTSUMT program to evaluate each sum.

47. $\displaystyle\sum_{k=1}^{100} \frac{6k}{k+1}$

48. $\displaystyle\sum_{k=1}^{100} \frac{k-1}{k}$

49. $\displaystyle\sum_{k=1}^{500} k(k-1)(k-2)$

50. $\displaystyle\sum_{k=0}^{500} (1-k)(2-k)$

51. Use a summation formula to find the number of boxes in this supermarket display.

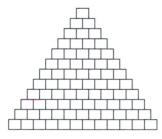

Use arithmetic rules and the formula for the value of $\displaystyle\sum_{k=1}^{n} k$ to establish the formulas in Exercises 52 and 53.

52. $\displaystyle\sum_{k=1}^{n} (2k-1) = n^2$

53. $\displaystyle\sum_{k=1}^{n} k + \sum_{k=1}^{n-1} k = n^2$

54. In Example 2 we found $\mathrm{LRAM}_5(x^2)$ and $\mathrm{RRAM}_5(x^2)$ to be an underestimate and an overestimate, respectively, of $A_0^5(x^2)$. Show that, on [0,5], $\mathrm{LRAM}_n(x^2)$ and $\mathrm{RRAM}_n(x^2)$ are always underestimates and overestimates, respectively of $A_0^5(x^2)$.

55. Show that on every interval $[a, b]$ with $0 < a < b$, $\mathrm{LRAM}_n(1/x) > \mathrm{RRAM}_n(1/x)$ for all positive integers n.

56. Assume that f is an increasing, nonnegative function on $[a, b]$. What can you conclude about the relationships among $\mathrm{LRAM}_n f$, $\mathrm{RRAM}_n f$, and $A_a^b f$?

57. Assume that f is a decreasing, nonnegative function on $[a, b]$. What can you always conclude about the relationships among $\mathrm{LRAM}_n f$, $\mathrm{RRAM}_n f$ and $A_a^b f$?

58. a) Refer to Example 2. Show that $\mathrm{RRAM}_5(x^2) = \mathrm{LRAM}_5(x^2) + 1 \cdot 5^2 - 1 \cdot 0^2$ on [0, 5].

b) Show that

$$\mathrm{RRAM}_n f = \mathrm{LRAM}_n f + f(x_n) \cdot \Delta x - f(x_0) \cdot \Delta x$$

for any nonnegative function f and any partition of $[a, b]$ into n subintervals of equal length.

59. Prove or disprove the following statement:

$\mathrm{MRAM}_n f$ is the average of $\mathrm{LRAM}_n f$ and $\mathrm{RRAM}_n f$.

Try it for $f(x) = x^2$.

60. Explain why in Exercise 11, $\mathrm{LRAM}_n \sin x = \mathrm{RRAM}_n \sin x$ for $n = 10, 100,$ and 1000.

61. Show that if f is nonnegative on $[a, b]$ and the line $x = (a + b)/2$ is a line of symmetry of the graph of $y = f(x)$, then $\mathrm{LRAM}_n f = \mathrm{RRAM}_n f$ for any positive integer n.

Find a formula in terms of n for $\mathrm{RRAM}_n f$ for each of the functions $y = f(x)$ in Exercises 62–67. Then determine $\lim_{n \to \infty} \mathrm{RRAM}_n f$ to find the *exact* area under the curve $y = f(x)$ from $x = a$ to $x = b$.

62. $f(x) = x + 1$ from $x = 0$ to $x = 5$ (Exercise 3)

63. $f(x) = 2x^2$ from $x = 0$ to $x = 5$ (Exercise 5)

64. $f(x) = x^2 + 2$ from $x = 1$ to $x = 6$ (Exercise 6)

65. $f(x) = x^2 - x + 3$ from $x = 0$ to $x = 3$ (Exercise 7)

66. $f(x) = 2x^3 + 3$ from $x = 0$ to $x = 5$ (Exercise 9)

67. $f(x) = x^3 + x^2 + 2x + 3$ from $x = -1$ to $x = 3$ (Exercise 10)

68. Compute $\mathrm{LRAM}_n(x^2)$ for $x = 0$ to $x = 5$. Show that $\lim_{n \to \infty} \mathrm{LRAM}_n(x^2)$ is exactly 125/3.

69. Show that

a) $\displaystyle\sum_{k=1}^{n} (2k-1)^2 = \frac{n(2n-1)(2n+1)}{3}$

b) $\displaystyle\sum_{k=1}^{n} (2k-1)^3 = n^2(2n^2 - 1)$

Hint: First show that $\displaystyle\sum_{k=1}^{n} (2k-1)^2 = \sum_{k=1}^{2n} k^2 - \sum_{k=1}^{n} (2k)^2$.

70. Use the result of Exercise 69(a) to show that $\lim_{n \to \infty} \text{MRAM}_n (x^2) = 125/3$ in the area computation of $A_0^5(x^2)$.

71. Show that $\lim_{n \to \infty} \text{LRAM}_n (x^3)$ is exactly 156.25 in the area computation of $A_0^5(x^3)$.

72. Use the result of Exercise 69(b) to show that $\lim_{x \to \infty} \text{MRAM}_n (x^3)$ is exactly 156.25 in the area computation of $A_0^5(x^3)$.

73. Find a formula for $A_0^x(t^3)$ for $x \geq 0$.

Proofs without Words

Exercises 74–76 present informal pictorial proofs of summation formulas. Explain what is going on in each proof.

74.

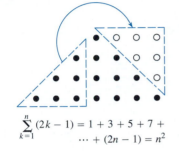

$$\sum_{k=1}^{n} (2k - 1) = 1 + 3 + 5 + 7 + \cdots + (2n - 1) = n^2$$

75.

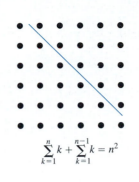

$$\sum_{k=1}^{n} k + \sum_{k=1}^{n-1} k = n^2$$

76.

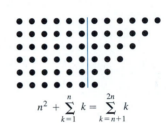

$$n^2 + \sum_{k=1}^{n} k = \sum_{k=n+1}^{2n} k$$

5.2 Definite Integrals

In this section, we develop the mathematics that supports the area calculations in Section 5.1. We do this by defining a limit of sums called the definite integral of a function $y = f(x)$ over an interval $[a, b]$, a limit that exists whenever f is continuous, regardless of the numerical signs of the values of f. In the special case in which f is nonnegative, the definite integral of f from a to b is also the number that we call the area under the curve $y = f(x)$ from a to b.

RIEMANN SUM

A "Riemann sum" is named in honor of the German mathematician Georg Friedrich Bernhard Riemann (1826–1866), who studied the limits of such sums.

Riemann Sums

The RAM that was used for finding area in the previous section involved summing the areas of rectangles. These were special cases of a more general sum called a *Riemann sum*. The limit of a Riemann sum is central to the understanding of integral calculus.

We begin with an arbitrary continuous function $y = f(x)$ defined over a closed interval $a \le x \le b$. Like the function graphed in Fig. 5.11, it may have negative values as well as positive values.

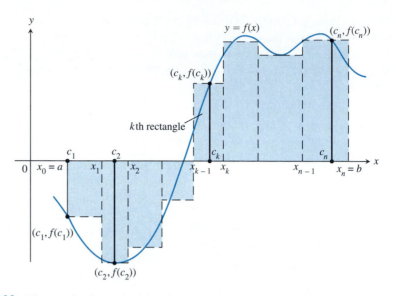

5.11 The graph of a typical function $y = f(x)$ over a closed interval $[a, b]$. The rectangles for the partition $\{x_0, x_1, \ldots, x_n\}$ of $[a, b]$ approximate the region between the graph of the function and the x-axis. Note that the subintervals formed by the partition can be of varying lengths.

We then use the partition, $P = \{x_0, x_1, \ldots, x_n\}$ of $[a, b]$ to define n subintervals $[x_0, x_1], [x_1, x_2], \ldots, [x_{n-1}, x_n]$ of $[a, b]$. In each subinterval $[x_{k-1}, x_k]$, we select some point c_k. On each subinterval, we stand a vertical rectangle that reaches from the x-axis to touch the curve at $(c_k, f(c_k))$.

If $f(c_k)$ is positive, the number $f(c_k) \Delta x_k = $ height $\times$ base is the area of the rectangle. If $f(c_k)$ is negative, then $f(c_k) \Delta x_k$ is the negative of the area. In any case, we add the n products $f(c_k) \Delta x_k$ to form the sum

$$S_P = \sum_{k=1}^{n} f(c_k) \Delta x_k.$$

This sum, which depends on P and the choice of the numbers c_k, is called a **Riemann sum for f on the interval $[a, b]$**.

EXAMPLE 1

Find three different Riemann sums for $f(x) = \sin \pi x$ on the interval $[0, 3/2]$ using the partition $P = \{0, 1/2, 1, 3/2\}$.

Solution The partition P gives us subintervals of equal lengths:

$$\Delta x_k = \frac{1}{2}, \qquad k = 1, 2, 3.$$

We choose c_k to be the subinterval midpoints (Fig. 5.12):

$$c_1 = \frac{1}{4}, \qquad c_2 = \frac{3}{4}, \qquad c_3 = \frac{5}{4}.$$

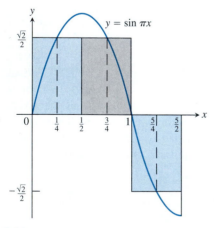

5.12 To calculate a Riemann sum for

$$f(x) = \sin \pi x \quad \text{on} \quad [0, 3/2]$$

in Example 1, we choose the points $c_k = 1/4, 3/4, 5/4$ in the subintervals of the partition $\{0, 1/2, 1, 3/2\}$. This Riemann sum is the same as $\text{MRAM}_3 \, f(x)$.

The corresponding Riemann sum is

$$\sum_{k=1}^{3} f(c_k)\Delta x_k = \sum_{k=1}^{3} \sin(\pi c_k)\cdot\frac{1}{2}$$

$$= \frac{1}{2}\sum_{k=1}^{3}\sin(\pi c_k) = \frac{1}{2}\left(\sin\frac{\pi}{4} + \sin\frac{3\pi}{4} + \sin\frac{5\pi}{4}\right)$$

$$= \frac{1}{2}\left(\frac{\sqrt{2}}{2} + \frac{\sqrt{2}}{2} - \frac{\sqrt{2}}{2}\right) = \frac{\sqrt{2}}{4}.$$

If instead of choosing the c_k's to be midpoints we choose each c_k to be the left endpoint of its subinterval (Fig. 5.13), then

$$c_1 = 0, \qquad c_2 = \frac{1}{2}, \qquad c_3 = 1,$$

and the corresponding Riemann sum is

$$\sum_{k=1}^{3}\sin(\pi c_k)\Delta x_k = \frac{1}{2}\sum_{k=1}^{3}\sin(\pi c_k)$$

$$= \frac{1}{2}\left(\sin 0 + \sin\frac{\pi}{2} + \sin\pi\right)$$

$$= \frac{1}{2}(0 + 1 + 0) = \frac{1}{2}.$$

If we choose each c_k to be the right endpoint of its subinterval (Fig. 5.14), then

$$c_1 = \frac{1}{2}, \qquad c_2 = 1, \qquad c_3 = \frac{3}{2},$$

and the corresponding Riemann sum is

$$\sum_{k=1}^{3}\sin(\pi c_k)\Delta x_k = \frac{1}{2}\left(\sin\frac{\pi}{2} + \sin\pi + \sin\frac{3\pi}{2}\right) = \frac{1}{2}(1 + 0 - 1) = 0.\ \blacksquare$$

EXPLORATION BIT

In Fig. 5.12, why are the rectangles standing on [0, 1/2] and [1/2, 1] the same height?

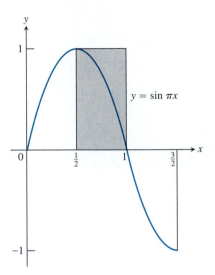

5.13 For another Riemann sum using the same partition, we could choose $c_k = 0, 1/2, 1$, the left endpoints of the subintervals. This Riemann sum is the same as LRAM$_3$ $f(x)$.

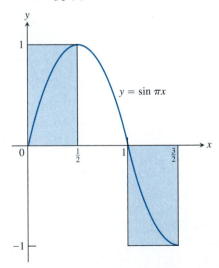

5.14 For a third Riemann sum using the same partition, we could choose $c_k = 1/2, 1, 3/2$, the right endpoints of the subintervals. This Riemann sum is the same as RRAM$_3$ $f(x)$.

EXPLORATION 1

The RAM Riemann Sums

As suggested by Example 1, the sums MRAM$_n$, LRAM$_n$, and RRAM$_n$ are Riemann sums although the choices of the c_k's for each value of n are quite specific.

1. Compute the Riemann sum RRAM$_n$ f for $f(x) = x^3$ on each interval:
   a) [0, 1]   **b)** [0, 5]   **c)** [0, x]
Hints: Use the formula for $\sum_{k=1}^{n} k^3$ in Section 5.1. For part (c), you will have to distinguish between two uses for x. One way to do this is to replace one of the x's with t.

2. What is $A_0^x(t^3)$ exactly?

What happens to the Riemann sums as the number of points in the partition increases and the partition becomes finer? As Fig. 5.15 suggests, the rectangles involved overlap the region between the curve and the x-axis with increasing accuracy, and we should find the sums approaching a limiting value of some kind. To make this idea precise, we need to define what it means for partitions to become finer and for Riemann sums to have a limit. We accomplish this with the definitions that follow.

> **ANIMATING RAM**
>
> Use the toolbox program AREA to animate the RAM for $f(x) = x^3 - 3x^2 - x + 3$ on $[0, 4]$.

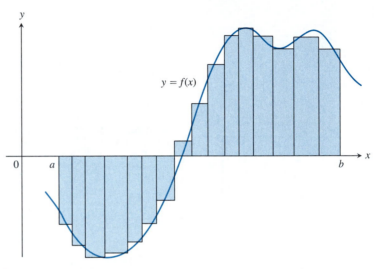

5.15   The curve of Fig. 5.11 with rectangles from a finer partition.

> **DEFINITION**
>
> The **norm** of a partition P is the length of the partition's longest subinterval. It is denoted by $\|P\|$.

EXAMPLE 2

Find the norm of the partition $P = \left\{ 0, \dfrac{1}{4}, \dfrac{2}{3}, 1, \dfrac{3}{2}, 2 \right\}$ of the interval $[0, 2]$.

Solution   The subintervals in the partition are

$$\left[0, \frac{1}{4} \right], \quad \left[\frac{1}{4}, \frac{2}{3} \right], \quad \left[\frac{2}{3}, 1 \right], \quad \left[1, \frac{3}{2} \right], \quad \left[\frac{3}{2}, 2 \right].$$

The lengths of the subintervals are

$$\Delta x_1 = \frac{1}{4}, \qquad \Delta x_2 = \frac{5}{12}, \qquad \Delta x_3 = \frac{1}{3}, \qquad \Delta x_4 = \frac{1}{2}, \qquad \Delta x_5 = \frac{1}{2}.$$

The norm of the partition is 1/2, the longest of these lengths. As you can see, there are two subintervals of length 1/2. There can be more than one longest subinterval. ≡

The way we say that successive partitions of an interval become finer is to say that their norms approach zero.

DEFINITIONS   **The Limit of Riemann Sums**

Let $f(x)$ be a function defined on a closed interval $[a, b]$. We say that the **limit** of the Riemann sums $\sum_{k=1}^{n} f(c_k)\Delta x_k$ on $[a, b]$ as $\|P\| \to 0$ is the number I if the following condition is satisfied:

Given any positive number ϵ, there exists a positive number δ such that for every partition P of $[a, b]$,

$$\|P\| < \delta \quad \text{implies that} \quad \left| \sum_{k=1}^{n} f(c_k)\Delta x_k - I \right| < \epsilon$$

for any choice of the numbers c_k in the subintervals $[x_{k-1}, x_k]$.

The Definite Integral as a Limit of Riemann Sums

If the above limit exists, we write

$$\lim_{\|P\| \to 0} \sum_{k=1}^{n} f(c_k)\Delta x_k = I.$$

We call I the **definite integral** of f over $[a, b]$, we say that f is **integrable** over $[a, b]$, and we say that Riemann sums of f on $[a, b]$ approach the number I. We usually denote the limit I by the more suggestive notation

$$\int_{a}^{b} f(x)\, dx,$$

which is read "the integral of f from a to b." Thus,

$$\lim_{\|P\| \to 0} \sum_{k=1}^{n} f(c_k)\Delta x_k = \int_{a}^{b} f(x)\, dx.$$

Despite the potential for variety in the Riemann sums $\Sigma f(c_k)\Delta x_k$ as the partitions change and the c_k's are chosen at random in the intervals of each new partition, the sums always have a limit as $\|P\| \to 0$ when f is continuous on $[a, b]$. The existence of this limit, of the definite integral of a continuous function on a closed interval, blithely assumed by the mathematicians of the seventeenth and eighteenth centuries, was finally established, once and for all, by Georg Riemann in 1854. You can find a current version of Riemann's proof in most advanced calculus books.

> **THEOREM 1** **The Existence of Definite Integrals**
>
> All continuous functions are integrable. That is, if a function $y = f(x)$ is continuous on an interval $[a, b]$, then its definite integral over $[a, b]$ exists.

Theorem 1 says nothing about *how* to calculate definite integrals. Except for a few special cases, doing that takes another theorem, and we will get to it in Section 5.4. We do know how to find the limit of some Riemann sums for the particular functions $f(x) = ax$, $f(x) = ax^2$, and $f(x) = ax^3$, and thus we can compute definite integrals for these special cases. However, this process is tedious at best.

Finally, Theorem 1 speaks only about continuous functions. Many discontinuous functions are integrable, as we will see later in this section.

Terminology of Integration

There is a fair amount of terminology to learn in connection with definite integrals: The symbol $\int$ is an **integral sign.** When we find the value of $\int_a^b f(x)\,dx$, we say that we have **evaluated the integral** and that we have **integrated** f from a to b. We call $[a, b]$ the **interval of integration.** The numbers a and b are the **limits of integration**, a being the **lower limit of integration** and b the **upper limit of integration.** The function f is the **integrand** of the integral, and the variable x is the **variable of integration**.

While the integral of f from a to b is usually denoted by $\int_a^b f(x)\,dx$, its value over any particular interval depends on the function and not on the letter that we choose to represent its independent variable. If we decide to use t or u instead of x, we simply write the integral as

$$\int_a^b f(t)\,dt \qquad \text{or} \qquad \int_a^b f(u)\,du \qquad \text{instead of} \qquad \int_a^b f(x)\,dx.$$

No matter how we write the integral, it is still the same number, defined as a limit of Riemann sums. Since it does not matter what letter we use, the variable of integration is called a **dummy variable.**

THE SYMBOL $\int$

Leibniz established the use of the symbol $\int$. He chose it because it resembled the S in the German word for *summation*. As you proceed, you will find that $\int$ will be used in different ways. You should remember that the form $\int_a^b f(x)\,dx$ represents a number.

EXAMPLE 3

Express the limit of Riemann sums

$$\lim_{\|P\| \to 0} \sum_{k=1}^n (3c_k^2 - 2c_k + 5)\Delta x_k$$

as an integral if P denotes a partition of the interval $[-1, 3]$.

Solution The function being evaluated at c_k in each term of the sum is $f(x) = 3x^2 - 2x + 5$. The interval being partitioned is $[-1, 3]$. The limit is therefore the integral of f from -1 to 3:

$$\lim_{\|P\| \to 0} \sum_{k=1}^n (3c_k^2 - 2c_k + 5)\Delta x_k = \int_{-1}^3 (3x^2 - 2x + 5)\,dx. \qquad \blacksquare$$

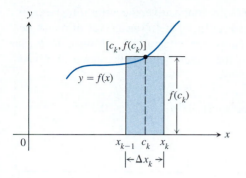

5.16 A term of a Riemann sum

$$\sum f(c_k)\Delta x_k$$

for a nonnegative function f is the area of a rectangle such as the one shown.

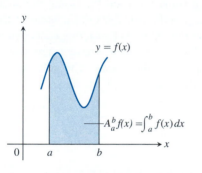

5.17 The area under the graph of f from a to b is defined and calculated as an integral.

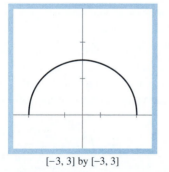

$[-3, 3]$ by $[-3, 3]$

5.18 A square viewing window on $y = \sqrt{4 - x^2}$. To confirm that the graph is a semicircle, we see that $y = \sqrt{4 - x^2}$ is the same as $y^2 = 4 - x^2$, or $x^2 + y^2 = 4$, with $y \geq 0$.

The Definite Integral and Area

If an integrable function $y = f(x)$ is nonnegative throughout an interval $[a, b]$, each term $f(c_k)\Delta x_k$ is the area of a rectangle reaching from the x-axis up to the curve $y = f(x)$. (See Fig. 5.16.) The Riemann sum

$$\sum_{k=1}^{n} f(c_k)\Delta x_k,$$

which is the sum of the areas of these rectangles, gives an estimate of the area of the region between the curve and the x-axis from a to b. Since the rectangles give an increasingly good approximation of the region as we use subdivisions with smaller and smaller norms, we call the limiting value

$$\lim_{\|P\| \to 0} \sum f(c_k)\Delta x_k = \int_a^b f(x)\,dx$$

the area under the curve.

> **DEFINITION**   Area Under a Curve (Revisited)
>
> If $y = f(x)$ is nonnegative and integrable over a closed interval $[a, b]$, then the **area under the curve $y = f(x)$ from a to b** is the integral of f from a to b:
>
> $$A = \int_a^b f(x)\,dx.$$

In Section 5.1, we used the symbol $A_a^b f$ to denote the area under the graph of a nonnegative continuous function f from a to b (Fig. 5.17). We can now see precisely how A_a^b is defined. According to the definition above,

$$A_a^b f(x) = \int_a^b f(x)\,dx.$$

EXAMPLE 4

Find the value of the integral

$$\int_{-2}^{2} \sqrt{4 - x^2}\,dx$$

by regarding it as the area under the graph of an appropriately chosen function.

Solution   We recognize $f(x) = \sqrt{4 - x^2}$ as a function whose graph is a semicircle of radius 2, which is supported in our grapher viewing window (Fig. 5.18). The area between the semicircle and the x-axis from -2 to 2 is

$$\text{Area} = \frac{1}{2} \cdot \pi r^2 = \frac{1}{2}\pi(2)^2 = 2\pi.$$

Because the area is also the value of the integral of f from -2 to 2,

$$\int_{-2}^{2} \sqrt{4 - x^2}\,dx = 2\pi.$$

■

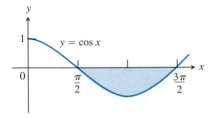

5.19 Because $f(x) = \cos x$ is nonpositive on $[\pi/2, 3\pi/2]$, the integral of f is a negative number. The area of the shaded region is the opposite of this integral,

$$\text{Area} = -\int_{\pi/2}^{3\pi/2} \cos x \, dx.$$

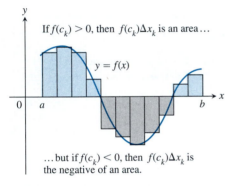

5.20 An integrable function f with negative as well as positive values.

NET AREA

Sometimes $\int_a^b f(x) \, dx$ is called the *net area* of the region determined by the curve $y = f(x)$ and the x-axis between $x = a$ and $x = b$.

If an integrable function $y = f(x)$ is nonpositive, the terms $f(c_k)\Delta x_k$ in the Riemann sums for f over an interval $[a, b]$ are all negatives of rectangle areas. The limit of the Riemann sums, the integral of f from a to b, is therefore the negative of the area of the region between the graph of f and the x-axis (Fig. 5.19).

$$\int_a^b f(x) \, dx = - \text{ (the area)} \qquad \text{if} \quad f(x) \le 0.$$

Or, turning this around,

$$\boxed{\text{Area} = -\int_a^b f(x) \, dx \qquad \text{when} \quad f(x) \le 0.}$$

If an integrable function $y = f(x)$ has both positive and negative values on an interval $[a, b]$, then the Riemann sums for f on $[a, b]$ add the areas of the rectangles that lie above the x-axis to the negatives of the areas of the rectangles that lie below the x-axis, as in Fig. 5.20. The resulting cancellation reduces the sums, so their limiting value is a number whose magnitude is less than the total area between the curve and the x-axis. The value of the integral is the area above the axis minus the area below the axis.

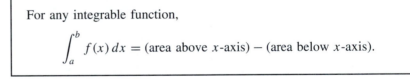

For any integrable function,

$$\int_a^b f(x) \, dx = (\text{area above } x\text{-axis}) - (\text{area below } x\text{-axis}).$$

EXPLORATION 2

Making Conjectures

It is a fact (which we will revisit) that $\int_0^\pi \sin x = 2$. With that information, what you know from above about integrals and areas, what you know about geometric transformations, and sometimes a bit of intuition, make a good guess and give a convincing argument for the value for each of the following. The appropriate graphs can give you clues.

1. $\displaystyle\int_0^{2\pi} \sin x \, dx$ **2.** $\displaystyle\int_0^{\pi/2} \sin x \, dx$ **3.** $\displaystyle\int_0^{\pi} (\sin x + 2) \, dx$

4. $\displaystyle\int_0^{\pi} 2 \sin x \, dx$ **5.** $\displaystyle\int_0^{\pi+2} \sin(x - 2) \, dx$ **6.** $\displaystyle\int_0^{2\pi} \sin(x/2) \, dx$

7. Consider each of the above, and decide whether it suggests a way to extend your thinking to making a conjecture about $\int_0^b f(x) \, dx$ for any b. For each one that does, make a conjecture, and give a convincing argument.

Constant Functions

Integrals of constant functions are always easy to evaluate. Over a closed interval, they are simply the constant times the length of the interval (Fig. 5.21).

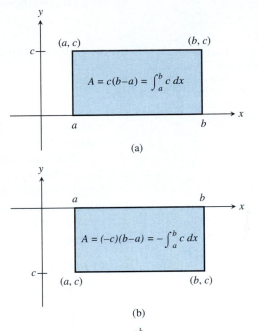

5.21 (a) If c is a positive constant, then $\int_a^b c\,dx$ is the area of the rectangle shown. (b) If c is negative, then $\int_a^b c\,dx$ is the opposite of the area of the rectangle.

THEOREM 2

If $f(x) = c$ has the constant value c on the interval $[a, b]$, then

$$\int_a^b f(x)\,dx = \int_a^b c\,dx = c(b - a).$$

Proof   The Riemann sums of f on $[a, b]$ all have the constant value

$$\sum f(c_k)\Delta x_k = \sum c \cdot \Delta x_k$$

$$= c \cdot \sum \Delta x_k \qquad \text{Constant multiple rule for sums}$$

$$= c(b - a). \qquad \text{The sum of } \Delta x_k\text{'s} = \text{the length of } [a, b] = b - a.$$

The limit of these sums, the integral to which they converge, therefore has this value also.   ≡

EXAMPLE 5

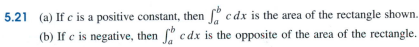

$$\int_1^4 5\,dx = 5(4 - 1) = 5(3) = 15 \qquad\qquad ≡$$

Evaluating Definite Integrals Numerically with Technology—The NINT Procedure

We assume that your graphing utility has a simple method (a few keystrokes or a program) for approximating definite integrals. We use the notation

$$\text{NINT}\,(f(x), a, b)$$

to denote a calculator or computer approximation of $\int_a^b f(x)\,dx$. And we write

$$\int_a^b f(x)\,dx = \text{NINT}\,(f(x), a, b)$$

with the understanding that the right-hand side of the equation is an approximation of the left-hand side.

There are many methods for numerically approximating $\int_a^b f(x)\,dx$. Thus, NINT $(f(x), a, b)$ values will vary depending on the utility you are using. However, most graphing utilities give results that are very accurate, usually to five or six significant digits. Consult your *Owner's Guide*. We usually will report answers that are meaningful for the problem under consideration.

EXAMPLE 6

Evaluate numerically (a) $\int_{-1}^2 x \sin x\,dx$, and (b) $\int_0^5 e^{-x^2}\,dx$.

Solution

a) NINT $(x \sin x, -1, 2) = 2.04275977886$, or 2.043.

b) NINT $(e^{-x^2}, 0, 5) = 0.88622692545$, or 0.886. ≡

We eventually will be able to confirm that the exact value for Example 6(a) is $-2\cos 2 + \sin 2 - \cos 1 + \sin 1$, which is 2.04275977886 to 11 decimal places. This is comforting, to say the least. It is *not* comforting, however, to know that no explicit *exact* value has ever been found for Example 6(b). The best anyone can do is to numerically approximate this definite integral. Here, technology is *essential*!

Discontinuous Integrable Functions

A function f is said to be **bounded** on the interval $[a, b]$ if there are real numbers m and M such that $m \le f(x) \le M$ for all x in $[a, b]$. Continuous functions are bounded on closed intervals, as the Max-Min Theorem for Continuous Functions of Section 2.2 shows.

If a function is bounded and has only a finite number of discontinuities on the interval $[a, b]$, then it is integrable on the interval. (Actually, the number of discontinuities need not be finite, but we will not deal with this here.) So bounded functions with a finite number of "jump" or "removable" discontinuities are integrable.

EXAMPLE 7

The region determined by the curve $y = |x|/x$ and the x-axis between $x = -1$ and $x = 2$ has the shape of two rectangles, one below the x-axis and one

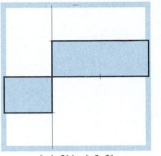

[−1, 2] by [−2, 2]

5.22 A discontinuous integrable function:

$$\int_{-1}^{2} (|x|/x) \, dx =$$

−(area below x-axis) + (area above x-axis).

above (Fig. 5.22). Using the idea of net area, we have

$$\int_{-1}^{2} \frac{|x|}{x} \, dx = -1 + 2 = 1.$$

Check, using your grapher, that NINT $(|x|/x, -1, 2) = 1$.

EXPLORATION 3

Discontinuous Integrands

1. Show that $f(x) = (x^2 - 4)/(x - 2)$ has a removable discontinuity on $[0, 3]$ and that $g(x) = x + 2$ is the continuous extension of f on $[0, 3]$.

2. Use area to show that $\int_{0}^{3} \frac{x^2 - 4}{x - 2} \, dx = 10.5$.

3. Use area to show that $\int_{0}^{5} [x] \, dx = 10$.

Exercises 5.2

In Exercises 1–4, make three sketches of the graph of each function $f(x)$ over the given interval. Partition the interval into four subintervals of equal length. Then add to your sketches the rectangles associated with the Riemann sum $\sum_{k=1}^{4} f(c_k)\Delta x_k$, given that c_k is (a) the left endpoint, (b) the right endpoint, and (c) the midpoint of the kth subinterval. Support your sketch with the toolbox program AREA.

1. $f(x) = x^2 - 1, \quad [0, 2]$

2. $f(x) = -x^2, \quad [0, 1]$

3. $f(x) = \sin x, \quad [-\pi, \pi]$

4. $f(x) = \sin x + 1, \quad [-\pi, \pi]$

5. Find the norm of partition $P = \{0, 1.2, 1.5, 2.3, 2.6, 3\}$.

6. Find the norm of partition $P = \{-2, -1.6, -0.5, 0, 0.8, 1\}$.

Express the limits in Exercises 7–14 as definite integrals.

7. $\lim_{\|P\|\to 0} \sum_{k=1}^{n} c_k^2 \Delta x_k$, where P is a partition of $[0, 2]$

8. $\lim_{\|P\|\to 0} \sum_{k=1}^{n} 2c_k^3 \Delta x_k$, where P is a partition of $[-1, 0]$

9. $\lim_{\|P\|\to 0} \sum_{k=1}^{n} (c_k^2 - 3c_k) \Delta x_k$, where P is a partition of $[-7, 5]$

10. $\lim_{\|P\|\to 0} \sum_{k=1}^{n} \frac{1}{c_k} \Delta x_k$, where P is a partition of $[1, 4]$

11. $\lim_{\|P\|\to 0} \sum_{k=1}^{n} \frac{1}{1 - c_k} \Delta x_k$, where P is a partition of $[2, 3]$

12. $\lim_{\|P\|\to 0} \sum_{k=1}^{n} \sqrt{4 - c_k^2} \Delta x_k$, where P is a partition of $[0, 1]$

13. $\lim_{\|P\|\to 0} \sum_{k=1}^{n} \cos c_k \Delta x_k$, where P is a partition of $[0, 4]$

14. $\lim_{\|P\|\to 0} \sum_{k=1}^{n} \sin^3 c_k \Delta x_k$, where P is a partition of $[-\pi, \pi]$

In Exercises 15–20, use your grapher or your prior knowledge of the graph to check whether $f(x) \geq 0$ or $f(x) \leq 0$ from $x = a$ to $x = b$. Then, use an integral to express the area between the x-axis and the graph of f from a to b. Use NINT $(f(x), a, b)$ to evaluate the integral.

15. $f(x) = x^2 - 4, a = 0, b = 2$

16. $f(x) = 9 - x^2, a = 0, b = 3$

17. $f(x) = \sqrt{25 - x^2}, a = 0, b = 5$

18. $f(x) = \sqrt{36 - 4x^2}, a = -3, b = 3$

19. $f(x) = \tan x, a = 0, b = \pi/4$

20. $f(x) = \cos x, a = \pi/2, b = 3\pi/2$

For Exercises 21–26, compute $\text{LRAM}_n f$, $\text{RRAM}_n f$, and $\text{MRAM}_n f$ for $n = 100$ and $n = 1000$ on the given interval $[a, b]$. Compare with $\text{NINT}(f(x), a, b)$.

21. $f(x) = x^2 e^{-x^2}$, $[0, 3]$ **22.** $f(x) = \sin(x^2)$, $[0, 2\pi]$

23. $f(x) = \dfrac{\sin x}{x}$, $[1, 10]$ **24.** $f(x) = \dfrac{1 - \cos x}{x^2}$, $[1, 2\pi]$

25. $f(x) = x \sin x$, $[-1, 2]$ **26.** $f(x) = e^{-x^2}$, $[0, 5]$

For Exercises 27–30, use $\text{NINT}(f(x), a, b)$ to compute the definite integral $\int_a^b f(x)\,dx$. Graph each function $y = f(x)$ for $a \le x \le b$. Use area to explain the value of the integral.

27. $\displaystyle\int_{-1}^{3} (2 - x - 5x^2)\,dx$ **28.** $\displaystyle\int_0^3 x^2 e^{-x^2}\,dx$

29. $\displaystyle\int_0^{2\pi} \sin(x^2)\,dx$ **30.** $\displaystyle\int_1^{10} \dfrac{\sin x}{x}\,dx$

In Exercises 31–34, find the exact value of each integral by regarding it as the area under the graph of an appropriately chosen function and using an area formula from plane geometry. Compare the exact value with a NINT computation.

31. $\displaystyle\int_{-1}^{1} \sqrt{1 - x^2}\,dx$ **32.** $\displaystyle\int_0^2 \sqrt{4 - x^2}\,dx$

33. $\displaystyle\int_{-1}^{1} (1 - |x|)\,dx$ **34.** $\displaystyle\int_{-1}^{1} (1 + \sqrt{1 - x^2})\,dx$

35. Use the formula for $\displaystyle\sum_{k=1}^{n} (2k - 1)^3$ given in Exercise 69(b) from Section 5.1 to compute the Riemann sums MRAM_n for $f(x) = x^3$ on $[0, 5]$ and $[0, a]$. Find the limit of each Riemann sum.

36. Find the exact value of the area between $f(x) = x^3 + 1$ and the x-axis from $x = 0$ to $x = 2$ by finding the limit of a Riemann sum.

Exercises 37–40 involve integrals of discontinuous functions. Sketch a graph of the integrand, and identify the discontinuities. Use area to evaluate the integrals. Support with NINT.

37. $\displaystyle\int_{-2}^{3} \dfrac{|x|}{x}\,dx$ **38.** $\displaystyle\int_{-6}^{5} 2\,[|x - 3|]\,dx$

39. $\displaystyle\int_{-3}^{4} \dfrac{x^2 - 1}{x + 1}\,dx$ **40.** $\displaystyle\int_{-5}^{6} \dfrac{9 - x^2}{x - 3}\,dx$

41. a) What is the continuous extension of $y = \dfrac{\sin x}{x}$? Use NINT to compute.

b) $\displaystyle\int_0^3 \dfrac{\sin x}{x}\,dx$ **c)** $\displaystyle\int_{-3}^{3} \dfrac{\sin x}{x}\,dx$

Note: You may obtain a "division by zero error" on your grapher in part (b) or (c). If so, changing an endpoint slightly

(say, 3 to 2.99999999) will allow your grapher to complete computations.

42. a) What is the continuous extension of $y = \dfrac{1 - \cos x}{x^2}$? Use NINT to compute.

b) $\displaystyle\int_0^3 \dfrac{|x|}{x} \left(\dfrac{1 - \cos x}{x^2} \right) dx$

c) $\displaystyle\int_{-3}^{3} \dfrac{|x|}{x} \left(\dfrac{1 - \cos x}{x^2} \right) dx$

43. In Exercise 41, explain why

$$\int_0^3 \dfrac{\sin x}{x}\,dx = \dfrac{1}{2} \int_{-3}^{3} \dfrac{\sin x}{x}\,dx.$$

44. In Exercise 42, explain why

$$\int_{-3}^{3} \dfrac{|x|}{x} \left(\dfrac{1 - \cos x}{x^2} \right) dx = 0.$$

45. Compare $\text{NINT}(y_i(x), 0, 2)$ with $\text{NINT}(y_i(x), 0, 1) + \text{NINT}(y_i(x), 1, 2)$ for each of $y_1(x) = x^2$, $y_2(x) = x \sin x$, and $y_3(x) = e^x$. Make a conjecture based on what you find.

46. Compare $\int_0^3 x^2 e^{-x}\,dx$ and $\int_0^{10} x^2 e^{-x}\,dx$. Explain analytically how you know that $\int_0^3 x^2 e^{-x}\,dx < \int_0^{10} x^2 e^{-x}\,dx$.

47. Let $f(x) = 2x - x^2$, and assume a, b, and c are either 1 or -1. Determine a, b, and c so that

$$a \int_{-1}^{0} f(x)\,dx + b \int_0^2 f(x)\,dx + c \int_2^3 f(x)\,dx$$

is as large as possible.

48. Let f be a continuous function on the interval $[a, b]$. Write a paragraph that explains why $\int_a^b f(x)\,dx$ may be called the *net area* of the region determined by the curve $y = f(x)$ and the x-axis between $x = a$ and $x = b$.

49. Let $S_n = \dfrac{1^2}{n^3} + \dfrac{2^2}{n^3} + \cdots + \dfrac{(n-1)^2}{n^3} + \dfrac{n^2}{n^3}$. Interpret S_n as a Riemann sum and find $\lim_{n\to\infty} S_n$ as a definite integral. (*Hint:* Consider that

$$S_n = \left[\left(\dfrac{1}{n}\right)^2 + \left(\dfrac{2}{n}\right)^2 + \cdots + \left(\dfrac{n-1}{n}\right)^2 + \left(\dfrac{n}{n}\right)^2 \right] \left(\dfrac{1}{n}\right)$$

and compare with $\text{RRAM}_n(x^2)$ on $[0, 1]$.)

In Exercises 50–53, interpret S_n as a Riemann sum and find $\lim_{n\to\infty} S_n$ as a definite integral. (See Exercise 49.)

50. $S_n = \displaystyle\sum_{k=1}^{n} \dfrac{8k^2}{n^3}$ **51.** $S_n = \displaystyle\sum_{k=1}^{n} \dfrac{n + k}{n^2}$

52. $S_n = \displaystyle\sum_{k=1}^{n} \dfrac{1}{n + k}$ **53.** $S_n = \displaystyle\sum_{k=1}^{n} \dfrac{(2n + k)^2}{n^3}$

5.3 _____ Definite Integrals and Antiderivatives

In this section, we continue our discussion of definite integrals, introduce the concept of average value of a function, and conclude with a preview of the most remarkable result of calculus discovered independently by Newton and Liebniz.

Useful Rules for Working with Integrals

We often want to add and subtract definite integrals, multiply them by constants, and compare them with other definite integrals. We do this with the Rules for Definite Integrals in Table 5.3. All the rules except the first two follow from the way in which integrals are defined as limits of Riemann sums. The sums have these properties, so their limits do too. For example,

TABLE 5.3   Rules for Definite Integrals

If f and g are integrable functions on $[a, b]$ and $[b, c]$, then

1. *Zero Rule:* $\displaystyle\int_a^a f(x)\,dx = 0.$ A definition

2. *Reversing Limits of Integration Rule:* $\displaystyle\int_b^a f(x)\,dx = -\int_a^b f(x)\,dx.$ The sign changes. Also a definition.

3. *Constant Multiple Rule:* $\displaystyle\int_a^b kf(x)\,dx = k\int_a^b f(x)\,dx.$ (any number k)

3a. *Special Case:* $\displaystyle\int_a^b -f(x)\,dx = -\int_a^b f(x)\,dx.$ Take $k = -1$ in Rule 3.

4. *Sum Rule:* $\displaystyle\int_a^b [f(x) + g(x)]\,dx = \int_a^b f(x)\,dx + \int_a^b g(x)\,dx.$

5. *Difference Rule:* $\displaystyle\int_a^b [f(x) - g(x)]\,dx = \int_a^b f(x)\,dx - \int_a^b g(x)\,dx.$

6. *Domination Rule:* $g(x) \geq f(x)$ on $[a, b]$ $\Rightarrow$ $\displaystyle\int_a^b g(x)\,dx \geq \int_a^b f(x)\,dx.$

6a. *Special Case:* $f(x) \geq 0$ on $[a, b]$ $\Rightarrow$ $\displaystyle\int_a^b f(x)\,dx \geq 0.$

7. *Max-Min Rule:* $\underbrace{\min f \cdot (b-a)}_{\text{Lower bound}} \leq \displaystyle\int_a^b f(x)\,dx \leq \underbrace{\max f \cdot (b-a)}_{\text{Upper bound}},$

8. *Interval Addition Rule:* $\displaystyle\int_a^b f(x)\,dx + \int_b^c f(x)\,dx = \int_a^c f(x)\,dx.$

9. *Interval Subtraction Rule:* $\displaystyle\int_b^c f(x)\,dx = \int_a^c f(x)\,dx - \int_a^b f(x)\,dx.$ Rule 8 in another form.

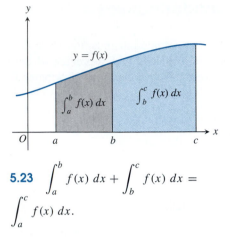

5.23 $\displaystyle\int_a^b f(x)\,dx + \int_b^c f(x)\,dx =$

$\displaystyle\int_a^c f(x)\,dx.$

Similarly,

$$\int_b^c f(x)\,dx =$$

$$\int_a^c f(x)\,dx - \int_a^b f(x)\,dx.$$

Rule 3 says that the integral of k times a function is k times the integral of the function. This is true because

$$\int_a^b kf(x)\,dx = \lim_{\|P\|\to 0}\sum_{i=1}^n kf(c_i)\Delta x_i$$

$$= \lim_{\|P\|\to 0} k\sum_{i=1}^n f(c_i)\Delta x_i$$

$$= k\lim_{\|P\|\to 0}\sum_{i=1}^n f(c_i)\Delta x_i = k\int_a^b f(x)\,dx.$$

Rule 1 is a definition. We want the integral of a function over an interval of zero length to be zero. Rule 2 is also a definition. It lets us show the limits of integration in either order.

Rules 4 and 5 infer that $f + g$ and $f - g$ must also be integrable on $[a, b]$, which is easily shown. Rule 6a uses the fact that the definite integral of the zero function is zero. In Rule 7, min f and max f refer to values of f on $[a, b]$. Rules 8 and 9 (Fig. 5.23) infer that integrability on $[a, b]$ and $[b, c]$ ensures integrability on $[a, c]$, all of which can also be shown.

You should be able to state these rules in English sentences. For example, Rule 6a says that integrals of nonnegative functions are nonnegative.

EXAMPLE 1

Suppose that f, g, and h are integrable, that

$$\int_{-1}^1 f(x)\,dx = 5, \qquad \int_1^4 f(x)\,dx = -2, \qquad \int_{-1}^1 h(x)\,dx = 7,$$

and that $g(x) \geq f(x)$ on $[-1, 1]$. Then

1. $\displaystyle\int_4^1 f(x)\,dx = -\int_1^4 f(x)\,dx = -(-2) = 2,$ Rule 2

2. $\displaystyle\int_{-1}^1 [2f(x) + 3h(x)]\,dx = 2\int_{-1}^1 f(x)\,dx + 3\int_{-1}^1 h(x)\,dx$
$= 2(5) + 3(7) = 31,$ Rules 4 and 3

3. $\displaystyle\int_{-1}^1 [f(x) - h(x)]\,dx = 5 - 7 = -2,$ Rule 5

4. $\displaystyle\int_{-1}^1 g(x)\,dx \geq 5,$ Because $g(x) \geq f(x)$ on $[-1, 1]$ — Rule 6

5. $\displaystyle\int_{-1}^4 f(x)\,dx = \int_{-1}^1 f(x)\,dx + \int_1^4 f(x)\,dx = 5 + (-2) = 3.$ Rule 8

The Average Value of a Function

If we divide the inequality in the Max-Min Rule for definite integrals by $(b - a)$, we get the inequality

$$\min f \leq \frac{1}{b-a}\int_a^b f(x)\,dx \leq \max f.$$

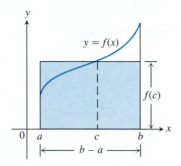

5.24 The value $f(c)$ in the Mean Value Theorem is, in a sense, the average (or *mean*) height of f on $[a, b]$. When $f \geq 0$, the area of the shaded rectangle

$$f(c)(b - a) = \int_a^b f(x)\, dx,$$

is the area under the graph of f from a to b.

MAKING THE CONNECTION

Here's how the average value of a function connects to the familiar idea of the average value of a set of n numbers. The average of a set of n functional values $f(c_1), f(c_2), \ldots, f(c_{n-1}), f(c_n)$, is

$$\frac{\displaystyle\sum_{i=1}^{n} f(c_i)}{n}.$$

Considering such an average on the partitions P of $[a, b]$ with subintervals of equal length Δx and taking the limit as $\|P\| \to 0$, we have

$$\lim_{\|P\| \to 0} \frac{\displaystyle\sum_{i=1}^{n} f(c_i)}{n}$$

$$= \lim_{\|P\| \to 0} \frac{\displaystyle\sum_{i=1}^{n} f(c_i)\Delta x}{n\,\Delta x}$$

$$= \frac{1}{b - a} \lim_{\|P\| \to 0} \sum_{i=1}^{n} f(c_i)\Delta x$$

$$= \frac{1}{b - a} \int_a^b f(x)\, dx.$$

If f is continuous, the Intermediate Value Theorem in Section 2.2 says that f must assume every value between min f and max f. In particular, f must assume the value

$$\frac{1}{b - a} \int_a^b f(x)\, dx.$$

THEOREM 3   The Mean Value Theorem for Definite Integrals

If f is continuous on the closed interval $[a, b]$, then, at some point c in the interval $[a, b]$,

$$f(c) = \frac{1}{b - a} \int_a^b f(x)\, dx. \tag{1}$$

The number on the right-hand side of Eq. (1) is called the mean value or average value of f on the interval $[a, b]$. We will see some of the applications of average values in Section 5.4.

DEFINITION

The **average value** of f on $[a, b]$ is $\dfrac{1}{b - a} \displaystyle\int_a^b f(x)\, dx.$

Notice that the average value of f on $[a, b]$ is the integral of f divided by the length of the interval. If f is continuous and nonnegative on $[a, b]$, its average value suggests an "average height" of the graph of f. Note that the rectangle on $[a, b]$ whose height is $f(c)$ has area

$$f(c)(b - a) = \int_a^b f(x)\, dx,$$

which is the area under the graph of f on $[a, b]$ (Fig. 5.24). ≡

EXPLORATION 1

Visualizing the Average Value

1. Find the average value $f(c)$ of $f(x) = \sqrt{4 - x^2}$ on the interval $[-2, 2]$. Use the value of $\int_{-2}^2 \sqrt{4 - x^2}$ from Example 4 of Section 5.2.

2. GRAPH $y_1 = f(x)$ and $y_2 = f(c)$. Can you "see" the rectangle with height $f(c)$ whose area is equal to the area under the semicircle? (If you want to get "fancy" and show the complete rectangle including the sides $x = -2$ and $x = 2$, you can GRAPH parametrically.) Can you match up portions of the rectangle with portions of the semicircle that have the same area?

3. Find c.

NOTATION ALERT

We could use either representation

$$A_a^x f(x) \text{ or } \int_a^x f(x)\,dx$$

to define this new function. Regardless of which we use, however, we must keep in mind that both representations display a dual use of x. Thus, it is a good idea to use the "dummy" nature of the second x here and view these as $A_a^x f(t)$ and $\int_a^x f(t)\,dt$, respectively. Be sure that you understand this dual use of x.

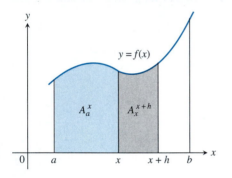

5.25 The area of the two shaded regions combined is

$$A_a^{x+h} = A_a^x + A_x^{x+h}.$$

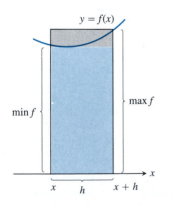

5.26 The area under the curve $y = f(x)$ and over the interval $[x, x+h]$ can be trapped between $h \cdot \min f$ and $h \cdot \max f$ on the interval.

Connecting Differential and Integral Calculus

We have two equal representations for area under the graph of a nonnegative continuous function from a to b, namely,

$$A_a^b f \quad \text{and} \quad \int_a^b f.$$

Now we will use one of these to define a function that will ultimately suggest the link between the calculus of derivatives and the calculus of integrals.

The new function is $A_a^x f$, where f remains a given nonnegative continuous function on an interval $[a, b]$. This new function $A_a^x f$ (or A_a^x) denotes the area of the region that lies between the curve for f and the x-axis, from the point a to a point x somewhere between a and b (Fig. 5.25). The area A_a^x, whose value changes with each new value of x, is a function of x.

If we move from x to a nearby point $x + h$, the amount of area that we add is A_x^{x+h}, the area under the curve from x to $x + h$. These two areas combine to give the area from a to $x + h$, so

$$A_a^x + A_x^{x+h} = A_a^{x+h},$$

and

$$A_x^{x+h} = A_a^{x+h} - A_a^x.$$

Now, the area under the curve from x to $x + h$ can be trapped between the areas of two rectangles of base length h, as shown in Fig. 5.26: The height of the shorter rectangle is $\min f$, the minimum value of f on the interval from x to $x + h$. The area of this rectangle is therefore $h \cdot \min f$ (base times height). The height of the taller rectangle is $\max f$, the maximum value of f on the interval from x to $x + h$. The area of this rectangle is therefore $h \cdot \max f$.

We record the observation that the area under the curve from x to $x + h$ lies between the areas of these rectangles by writing the following inequality:

$$h \cdot \min f \leq A_a^{x+h} - A_a^x \leq h \cdot \max f.$$

Dividing by h gives

$$\min f \leq \frac{A_a^{x+h} - A_a^x}{h} \leq \max f.$$

The fraction in the middle of this inequality is Fermat's difference quotient for the derivative of the area function A_a^x. It is the value of the function at $x + h$ minus the value of the function at x, all divided by h. We can therefore calculate the derivative of the area function at x by finding the limit of this quotient as h goes to zero.

As h goes to zero, the interval from x to $x + h$ gets shorter and shorter. As it does so, the values of $\max f$ and $\min f$ both approach the value of f at x (remember that f is continuous). Hence, by the Sandwich Theorem of Section 2.3, the difference quotient approaches $f(x)$ as well. In symbols,

$$\lim_{h \to 0} \frac{A_a^{x+h} - A_a^x}{h} = f(x).$$

We are thus led to the astonishing conclusion that when f is a nonnegative continuous function of x, the area under its graph from a to x is a differentiable function of x whose derivative at x is $f(x)$:

$$\frac{d}{dx} A_a^x = f(x). \tag{2}$$

THE CONNECTION, PART 1

In the representation using the integral sign, Eq. (2) can be written

$$\frac{d}{dx} \int_a^x f(t)\,dt = f(x).$$

This is one form of the remarkable connection between derivatives and integrals.

Among other things, this means that we can find an explicit formula for A_a^x whenever we have an explicit formula for $f(x)$ and are able to solve the following *initial value problem:*

Differential equation: $\quad \dfrac{d}{dx} A_a^x = f(x)$

Initial condition: $\quad A_a^x = 0 \quad$ when $\quad x = a.$

The initial condition comes from the observation that the area under the graph from a point a to the same point a is zero. To solve the initial value problem, we find an antiderivative $F(x)$ of $f(x)$ to get

$$A_a^x = F(x) + C. \tag{3}$$

We then find the correct value of C from the initial condition by setting x equal to a:

$$A_a^a = F(a) + C \qquad \text{\textcolor{blue}{$x = a$ in Eq. (3).}}$$

$$0 = F(a) + C \qquad \text{\textcolor{blue}{$A_a^x = 0$ when $x = a$.}}$$

$$C = -F(a). \qquad \text{\textcolor{blue}{Solved for C}}$$

The area under the curve $y = f(x)$ from a to x is therefore

$$A_a^x = F(x) - F(a). \tag{4}$$

Here, then, is the amazing relation between area and differential calculus. We calculate area with antiderivatives.

THE CONNECTION, PART 2

In the representation using the integral sign, Eq. (4) can be written

$$\int_a^x f(t)\,dt = F(x) - F(a).$$

In particular,

$$\int_a^b f(t)\,dt = F(b) - F(a).$$

This is another form of the connection between derivatives and integrals. In the next section, we will confirm that both forms hold for *any* continuous function f.

How to Find the Area under the Graph of a Nonnegative Continuous Function $y = f(x)$ from $x = a$ to $x = b$

STEP 1: Find an antiderivative $F(x)$ of $f(x)$. (Any antiderivative will do.)

STEP 2: Calculate the number $F(b) - F(a)$. This number will be $A_a^b f$, the area under the curve from a to b.

There are three practical questions to face here: How do we know f *has* an antiderivative, how do we *find* one when f does, and why will *any* antiderivative do? We will take care of the existence in Section 5.4. As for finding antiderivatives, there is no need to worry. We will get better at that as we learn more calculus, and we will be able to use our graphing utility for some of the more difficult functions.

And why will any antiderivative do? Consider the general antiderivative of f that we call $F(x) + C$. Evaluating at a and b and then subtracting shows that

$$(F(b) + C) - (F(a) + C) = F(b) + C - F(a) - C = F(b) - F(a)$$

with the constant canceling out.

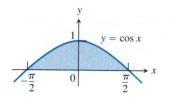

5.27 One arch of the curve $y = \cos x$. The area under it is 2.

EXPLORATION 2

Area Patterns in the Unit Square—Group Project

The efforts of a mathematician sometimes involve thoughtful, even wishful, guessing. To formulate and test some conjectures, we would like you to work together in groups of two or three.

1. GRAPH $y_1 = x^1$, $y_2 = x^2$, $y_3 = x^3$, and so on (as many as you want) in a square viewing window $[0, 1]$ by $[0, 1]$. The area under $y_1 = x^1$ from 0 to 1 is 1/2 unit^2. Explain. With this area in mind and the graphs of y_i in view, make a conjecture about the explicit value of the area under each y_i curve over $[0, 1]$. Confirm or disprove your conjecture.

2. If you were able to confirm your conjecture in part 1, extend it to $i = 0$, -1, -2, and so forth. Then confirm or disprove, and explain. Extend it to rational exponents. Discuss your findings.

EXAMPLE 2

Find the area under one arch of the curve $y = \cos x$ (Fig. 5.27).

Solution

STEP 1: Find an antiderivative of $y = \cos x$:

$$F(x) = \sin x. \qquad \text{Simplest one}$$

STEP 2: Calculate $F\left(\dfrac{\pi}{2}\right) - F\left(-\dfrac{\pi}{2}\right)$:

$$F\left(\frac{\pi}{2}\right) - F\left(-\frac{\pi}{2}\right) = \sin\left(\frac{\pi}{2}\right) - \sin\left(-\frac{\pi}{2}\right) = 1 - (-1) = 2.$$

The area is 2 square units.   ▣

EXPLORATION BIT

In Exploration 2 of Section 5.2, we stated that $\int_0^\pi \sin x = 2$, or the area under the graph of $y = \sin x$ from 0 to π is 2. Now confirm this value in at least two different ways. Example 2 may give you a clue or two.

Exercises 5.3

In Exercises 1–16, use your grapher or your prior knowledge of the graph to check that the function is nonnegative on the given interval. If it is, then use antiderivatives to find the area under the graph on the given interval. Support your result using NINT.

1. $y = x^2$, $[0, 2]$

2. $y = x^2$, $[-2, 1]$

3. $y = \sqrt{x}$, $[0, 4]$

4. $y = 1/x^2$, $[1/2, 2]$

5. $y = 2 - \sqrt{x}$, $[0, 4]$

6. $y = 1 - x^2$, $[-1, 1]$

7. $y = x^3 - 3x^2 + 4$, $[-1, 2]$

8. $y = 2x^4 - 4x^2 + 2$, $[-1, 1]$

9. $y = \cos x$, $[0, \pi/2]$

10. $y = 2 \cos 2x$, $[0, \pi/4]$

11. $y = \sin 2x$, the interval under any one arch

12. $y = (1/2) \sin x$, the interval under any one arch

13. $y = \cos \pi x$, the interval under any one arch

14. $y = 1 + \sin x$, $[0, \pi]$

15. $y = \sec^2 x$, $[-\pi/4, \pi/3]$

16. $y = \sec x \tan x$, $[0, \pi/3]$

17. Let b be any positive number and n any rational number other than -1. Find a formula for the area under the curve $y = x^n$ from $x = 0$ to $x = b$.

18. Whenever we find a new way to calculate something, it is a good idea to be sure that the new and old ways agree on the objects to which they both apply. If you use an antiderivative to find the area of the following triangle, will you still get $A = (1/2)bh$? Try it and find out.

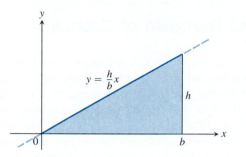

$$y = \frac{h}{b}x$$

19. Suppose that f and g are continuous and that

$$\int_1^2 f(x)\,dx = -4, \quad \int_1^5 f(x)\,dx = 6, \quad \int_1^5 g(x)\,dx = 8.$$

Use the rules in Table 5.3 to find

a) $\displaystyle\int_2^2 g(x)\,dx$

b) $\displaystyle\int_5^1 g(x)\,dx$

c) $\displaystyle\int_1^2 3f(x)\,dx$

d) $\displaystyle\int_2^5 f(x)\,dx$

e) $\displaystyle\int_1^5 [f(x) - g(x)]\,dx$

f) $\displaystyle\int_1^5 [4f(x) - g(x)]\,dx.$

20. Suppose f and h are continuous and that

$$\int_1^9 f(x)\,dx = -1, \quad \int_7^9 f(x)\,dx = 5, \quad \int_7^9 h(x)\,dx = 4.$$

Use the rules in Table 5.3 to find

a) $\displaystyle\int_1^9 -2f(x)\,dx$

b) $\displaystyle\int_7^9 [f(x) + h(x)]\,dx$

c) $\displaystyle\int_7^9 [2f(x) - 3h(x)]\,dx$

d) $\displaystyle\int_9^1 f(x)\,dx$

e) $\displaystyle\int_1^7 f(x)\,dx$

f) $\displaystyle\int_9^7 [h(x) - f(x)]\,dx.$

21. Suppose that f is continuous and that

$$\int_1^2 f(x)\,dx = 5, \quad \int_1^3 f(z)\,dz = 2.$$

Find

a) $\displaystyle\int_1^2 f(u)\,du,$

b) $\displaystyle\int_2^1 f(t)\,dt,$

c) $\displaystyle\int_2^3 f(y)\,dy.$

22. Suppose that f is continuous and that

$$\int_0^3 f(x)\,dx = 3, \quad \int_0^4 f(z)\,dz = 7.$$

Find

a) $\displaystyle\int_0^3 f(t)\,dt,$

b) $\displaystyle\int_4^0 f(w)\,dw,$

c) $\displaystyle\int_3^4 f(y)\,dy.$

In Exercises 23–28, find the average value of f on $[a, b]$.

23. $f(x) = 2 - x^2$ on $[-3, 5]$

24. $f(x) = x^3 - 1$ on $[-1, 3]$

25. $f(x) = \sqrt{x}$ on $[0, 3]$

26. $f(x) = \cos^2 x$ on $[-\pi, \pi]$

27. $f(x) = x \sin x$ on $[0, 5]$

28. $f(x) = 2e^{-x^2}$ on $[1, 4]$

29. Use Rule 7 in Table 5.3 to find upper and lower bounds for the value of

$$\int_0^1 \frac{1}{1 + x^2}\,dx.$$

30. *Continuation of Exercise 29.* Use Rule 7 in Table 5.3 to find upper and lower bounds for the values of

$$\int_0^{1/2} \frac{1}{1 + x^2}\,dx \quad \text{and} \quad \int_{1/2}^1 \frac{1}{1 + x^2}\,dx.$$

Then add these to arrive at an improved estimate of

$$\int_0^1 \frac{1}{1 + x^2}\,dx.$$

Support with a NINT computation.

In Exercises 31–34, the value of $\int_a^b f(x)\,dx$ is given. Is there a point c in (a, b) such that $f(c)(b - a) = \int_a^b f(x)\,dx$? If so, find it. Support your result with a grapher by drawing the curve and the corresponding rectangle.

31. $\displaystyle\int_0^1 (x^3 + 1)\,dx = \frac{5}{4}$

32. $\displaystyle\int_0^3 (9 - x^2)\,dx = 18$

33. $\displaystyle\int_0^1 \frac{1}{x^2 + 1}\,dx = \frac{\pi}{4}$

34. $\displaystyle\int_{-1/2}^{1/2} \frac{1}{\sqrt{1 - x^2}}\,dx = \frac{\pi}{3}$

35. The average value of $f(t)$ on $[-1, 2]$ is 5, and the average value of $f(t)$ on $[2, 7]$ is 3. Can you determine the average value of $f(t)$ on $[-1, 7]$? If so, what is it? Give reasons for your answers.

36. The average value of $g(s)$ on $[-5, -1]$ is 10, and the average value of $g(s)$ on $[-5, 10]$ is 50. Can you determine the average value of $g(s)$ on $[-1, 10]$? If so, what is it? Give reasons for your answers.

37. Suppose that f is continuous and that

$$\int_1^2 f(x)\,dx = 4.$$

Show that $f(x) = 4$ at least once on the interval $[1, 2]$.

38. Show that the value of

$$\int_0^1 \sin^2 x\,dx$$

cannot possibly be 2.

5.4 _____ The Fundamental Theorem of Calculus

We know that definite integrals can be used to define area. We will see that integrals are also used to define arc length, volume, force and work. The list of applications of the definite integral is almost endless. This section presents the discovery by Newton and Leibniz of the astonishing connections between integration and differentiation that started the mathematical development that fueled the scientific revolution for the next 200 years and constitutes what is still regarded as the most important computational discovery in the history of mathematics. We introduce the Fundamental Theorem of Calculus. The first part of the theorem says that the definite integral of a continuous function is a differentiable function of its upper limit of integration and tells us what the value of that derivative is. The second part tells us that the definite integral of a continuous function from a to b can be found from any one of the function's antiderivatives F as the number $F(b) - F(a)$.

The Fundamental Theorem, Part 1

If $f(t)$ is an integrable function, its integral from any fixed number a to another number x defines a function F whose value at x is

$$F(x) = \int_a^x f(t)\, dt.$$

The variable x in the function F is the upper limit of integration of an integral, but the function is just like any other function. For each value of the input x, there is an output $\int_a^x f$, namely, the value of the integral of f from a to x.

The formula

$$F(x) = \int_a^x f(t)\, dt$$

gives an important way to define new functions in science and provides an especially useful way to describe solutions of differential equations (more about this later). The reason for our mentioning the formula now is that this formula makes the connection between integrals and derivatives. For if f is any continuous function whatever, we will show that F is a differentiable function of x and, even more important, that its derivative, dF/dx, is f itself! That is, at every value of x,

$$\frac{d}{dx} \int_a^x f(t)\, dt = f(x). \tag{1}$$

If you were being sent to a desert island and could take your grapher and only one equation with you, Eq. (1) might well be your choice. It says that the differential equation $dF/dx = f$ has a solution for any continuous function f. It says that every continuous function f is the derivative of some other function, namely, $\int_a^x f(t)\, dt$. It says that every continuous function has an antiderivative. And it says that the processes of integration and differentiation are inverses of one another. Equation (1) is so important that we call it the first part of the Fundamental Theorem of Calculus.

SEKI KOWA

Seki Kowa (1642–1708) was born into a samurai warrior family in Tokyo, Japan, but he was adopted by the family of an accountant. He showed his great mathematical aptitude as a child prodigy. In 1674, Seki Kowa published solutions to 15 problems thought to be unsolvable up to that time. This gave him a reputation throughout Japan as a brilliant mathematician. He was also an inspirational teacher.

Among his contributions were an improved method of solving higher-degree equations, the use of determinants in solving simultaneous equations, and a seventeenth-century form of calculus native to Japan, known as *yenri*. It is difficult to know the entire extent of his work, because the samurai code demanded great modesty. Seki Kowa is credited with awakening the scientific spirit in Japan that continues to thrive today.

> ### THEOREM 4   The Fundamental Theorem of Calculus, Part 1
>
> If f is continuous on $[a, b]$, then the function
>
> $$F(x) = \int_a^x f(t)\, dt$$
>
> has a derivative at every point in $[a, b]$, and
>
> $$\frac{dF}{dx} = \frac{d}{dx} \int_a^x f(t)\, dt = f(x).$$

Proof   We prove Theorem 4 by applying the definition of derivative directly to the function $F(x)$. This means writing out Fermat's difference quotient,

$$\frac{F(x + h) - F(x)}{h},$$

and showing that its limit as $h \to 0$ is the number $f(x)$.

When we replace $F(x + h)$ and $F(x)$ by their defining integrals, the numerator above becomes

$$F(x + h) - F(x) = \int_a^{x+h} f(t)\, dt - \int_a^x f(t)\, dt.$$

The Interval Subtraction Rule for integrals (Table 5.3, Section 5.3) simplifies this difference to

$$\int_x^{x+h} f(t)\, dt,$$

so that the difference quotient above becomes

$$\frac{F(x + h) - F(x)}{h} = \frac{1}{h}[F(x + h) - F(x)] = \frac{1}{h} \int_x^{x+h} f(t)\, dt. \qquad (2)$$

According to the Mean Value Theorem for definite integrals, Theorem 3 in Section 5.3, the value of the entire expression on the right-hand side of Eq. (2) is one of the values taken on by f in the interval joining x and $x + h$. That is, for some number c in this interval,

$$\frac{1}{h} \int_x^{x+h} f(t)\, dt = f(c). \qquad (3)$$

We can therefore find out what happens to $(1/h)$ times the integral as $h \to 0$ by watching what happens to $f(c)$ as $h \to 0$.

What does happen to $f(c)$ as $h \to 0$? As $h \to 0$, the endpoint $x + h$ approaches x, taking c along with it like a bead on a wire:

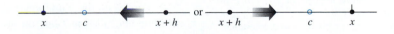

So c approaches x, and since f is continuous at x, $f(c)$ approaches $f(x)$:

$$\lim_{h \to 0} f(c) = f(x). \qquad (4)$$

Going back to the beginning, then, we have

$$\frac{dF}{dx} = \lim_{h \to 0} \frac{F(x+h) - F(x)}{h} \qquad \text{The definition of derivative}$$

$$= \lim_{h \to 0} \frac{1}{h} \int_x^{x+h} f(t)\, dt \qquad \text{Eq. (2)}$$

$$= \lim_{h \to 0} f(c) \qquad \text{Eq. (3)}$$

$$= f(x). \qquad \text{Eq. (4)}$$

This concludes the proof. ▤

Graphing the function $\int_a^x f(t)\, dt$

GRAPHING NINT f

Some graphers can graph the numerical integral, $y = \text{NINT } f$, directly. Others will require a toolbox program such as the one called NINTGRAF that is provided in the *Resource Manual*.

It turns out, as we will see, that $F(x) = \int_a^x f$ can sometimes be determined explicitly as a function of x. But this happens in relatively few cases, and these few special cases make up much of what has been traditionally studied in integral calculus. The good news this time is that computer technology can show us what $F(x)$ looks like for many functions f for which an explicit form of F is unknown or is known not to exist!

Just like the function $\int_a^x f$ is defined in terms of the definite integral of f at varying values of x in $[a, b]$, so the graphing utility feature NINT f—the numerical integral introduced in Section 5.2—can light up pixels for varying values of x in $[a, b]$ and thus show a graph that approximates the graph of $\int_a^x f$.

We use NINT $(f(x), a, b)$ for the numerical integral of f from a to b, so we use NINT $f = \text{NINT } (f(t), a, x)$ to approximate $\int_a^x f$. We will graph $y = \text{NINT } f$ to learn about $\int_a^x f$. Since explicit forms of $\int_a^x f$ are not known for many functions f, we will be seeing graphs that have rarely, and sometimes never, been seen before.

EXPLORATION BIT

The "speed" of graphing $y = \text{NINT } f$ is sometimes affected by the NINT tolerance setting. GRAPH

$$y = \text{NINT} \cos x$$

in the $[-10, 10]$ by $[-2, 2]$ viewing window for tolerances of 1, 0.001, and 0.00001. What do you observe to be the most significant difference in the three trials?
Note: Although we use a tolerance of 10^{-5} to *evaluate* with NINT, we find that a tolerance of 1 is usually accurate enough to *graph* with NINT.

EXPLORATION 1

Using NINT to Make a Conjecture

If $f(x) = \cos x$, then the function $F(x) = \int_a^x f(t)\, dt = \int_a^x \cos t\, dt$ has an explicit form. GRAPH the function NINT $(\cos t, 0, x)$, and make a conjecture about an explicit form for $\int_a^x \cos t\, dt$. ✛

Part 1 of the Fundamental Theorem of Calculus assures us that *any* continuous function f has an antiderivative, and we even have a definition, $F(x) = \int_a^x f(t)$, for the antiderivative. However, the only functions F that have an *explicit form* are those for which f has an explicit antiderivative. The next example shows an explicit form for one function $\int_a^x f$ and how this is supported by the graph of NINT f.

EXAMPLE 1

Compare the graph of $y_1 = F(x) = \int_0^x t^2\, dt$ with the graph of $y_2 = x^3/3$.

Solution The graphs of $y_1 = \text{NINT}\ (t^2, 0, x)$ and $y_2 = x^3/3$ are shown in the same viewing window in Fig. 5.28. We TRACE on the graphs and see that the graphs are virtually identical. This gives strong support that an explicit form for $\int_0^x t^2 dt$ is $x^3/3$; that is, $\int_0^x t^2 dt = x^3/3$. ≡

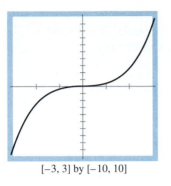

[−3, 3] by [−10, 10]

5.28 The graphs of

$$y_1 = \text{NINT}\ (t^2, 0, x)$$

and

$$y_2 = \frac{x^3}{3}$$

appear to be identical.

EXPLORATION 2

The Power of Numerical Techniques

Example 1 strongly suggests that $x^3/3$ is an explicit form for $\int_0^x t^2 dt$. We can confirm analytically that $D_x(x^3/3) = x^2$. But do you *really* believe that $D_x(\int_0^x t^2 dt) = x^2$ as the Fundamental Theorem says it does? Some graphing utilities can give strong support that this is indeed true. The process demonstrates the power of the numerical techniques operating in the background on the utility.

1. Enter $y_1 = \text{NINT}\ (t^2, 0, x)$, $y_2 = \text{NDER}\ (y_1, x)$, and $y_3 = x^2$. GRAPH y_2 and y_3. Before looking at the viewing window, predict what you should see according to the Fundamental Theorem.

2. TRACE on the graphs and explain what you do see. ⚘

We find the results to Exploration 2 to be rather remarkable. To appreciate what happens in part 1 of the Exploration, think of what the computer is doing to create the graph of $y_2 = \text{NDER}\ (y_1, x) = \text{NDER}\ (\text{NINT}\ (t^2, 0, x), x)$. For each value of x (127 of them on the grapher we use), two approximations of the definite integral are determined, since, by definition,

$$\text{NDER}\ (\text{NINT}\ (t^2, 0, x), x) = \frac{\text{NINT}\ (t^2, 0, x+h) - \text{NINT}\ (t^2, 0, x-h)}{2h}.$$

It shouldn't be too surprising that it took a little longer for the graphs to be completed in the viewing window!

You can run similar supportive tests on both parts of Example 2.

HOW TO MAKE SLOW PLOTS FAST!

If your grapher can graph NINT $(f(t), a, x)$ directly, try graphing it in parametric mode to significantly speed up the plot. Let

$$x_1(t) = t,$$

$$y_1(t) = \text{NINT}(f(t), a, t),$$

and adjust t-step to control the plotting speed. For example, in function plotting mode we obtained the graph in Example 1 in 30 seconds, but in parametric mode, with t-step = 0.5, we obtained the same graph in 3 seconds!

EXAMPLE 2

$$\frac{d}{dx} \int_{-\pi}^{x} \cos t\, dt = \cos x \qquad \text{Theorem 4 with } f(t) = \cos t$$

$$\frac{d}{dx} \int_{0}^{x} \frac{1}{1+t^2}\, dt = \frac{1}{1+x^2} \qquad \text{Theorem 4 with } f(t) = \frac{1}{1+t^2} \quad ≡$$

Examples 3 and 4 show other applications of the Fundamental Theorem, Part 1.

EXAMPLE 3

Find dy/dx if $y = \displaystyle\int_{1}^{x^2} \cos t \, dt$.

Solution   Notice that the upper limit of integration is not x but x^2. To find dy/dx, we must therefore treat y as the composite of

$$y = \int_{1}^{u} \cos t \, dt \qquad \text{and} \qquad u = x^2$$

and apply the Chain Rule:

$$\frac{dy}{dx} = \frac{dy}{du} \cdot \frac{du}{dx} \qquad\qquad \text{Chain Rule}$$

$$= \frac{d}{du} \int_{1}^{u} \cos t \, dt \cdot \frac{du}{dx} \qquad \text{Substitute the formula for } y.$$

$$= \cos u \cdot \frac{du}{dx} \qquad\qquad \text{Theorem 4 with } f(t) = \cos t$$

$$= \cos x^2 \cdot 2x \qquad\qquad u = x^2.$$

$$= 2x \cos x^2. \qquad\qquad \text{Usual form} \qquad \blacksquare$$

EXPLORATION 3

Extending the Fundamental Theorem

As illustrated in Example 3, The Fundamental Theorem, Part 1, can be extended as follows:

> If f and v are continuous on $[a, b]$, then the function
> $$F(x) = \int_{a}^{v(x)} f(t) \, dt$$
> has a derivative at every point in $[a, b]$, and $F'(x) = f(v(x))v'(x)$.

Use your graphing utility to support this extended form of the Fundamental Theorem by checking one particular example. Let

$$F(x) = \int_{1}^{x^2} \sqrt{t} \sin t \, dt.$$

1. Use NINT, and GRAPH $y_1 = F(x)$ for $-1.7 \le x \le 1.7$.

2. Compute F' analytically. (*Hint:* See Example 3.)

3. Compare the graphs of $y_2 = $ NDER F and $y_3 = F'$. Do your graphs support the extended form of the Fundamental Theorem shown above? Explain.   ✿

EXPLORATION BIT

Extend the Fundamental Theorem one step further than in Exploration 3. Give meaning to the function

$$F(t) = \int_{u(x)}^{v(x)} f(t) \, dt.$$

(*Hint:* Use Table 5.3, Rule 8.) Test your extension. Let

$$F(t) = \int_{x^2}^{x^3} \sin t \, dt,$$

and compare the graphs of NDER F and F' found analytically, as in Exploration 3.

EVALUATING $\int_a^b f(x)dx$

To evaluate $\int_a^b f(x)\,dx$ exactly:

1. Find an antiderivative F of f.
2. Calculate the number $F(b)-F(a)$.

This number will be the exact value of $\int_a^b f(x)\,dx$. Any antiderivative will do.

EXAMPLE 4

Show that the function

$$y = \int_0^x \tan t \, dt + 5$$

solves the initial value problem

Differential equation: $\dfrac{dy}{dx} = \tan x,$

Initial condition: $y = 5$ when $x = 0.$

Solution   The function satisfies the differential equation because

$$\frac{d}{dx}\left(\int_0^x \tan t \, dt + 5\right) = \tan x + 0 = \tan x. \qquad \text{Theorem 4 with } f(t) = \tan t$$

It fulfills the initial condition because

$$y(0) = \int_0^0 \tan t \, dt + 5 = 0 + 5. \qquad \text{Table 5.3, Rule 1} \qquad \blacksquare$$

The Fundamental Theorem, Part 2 (Integral Evaluation Theorem)

We now come to the second part of the Fundamental Theorem of Calculus, the part that describes how to evaluate definite integrals.

NOTATION

The usual notation for the number $F(b) - F(a)$ is $F(x)\big]_a^b$ or $[F(x)]_a^b$, depending on whether F has one or more terms. This notation provides a compact "recipe" for the evaluation.

Write $F(x)\big]_a^b$ for $F(b) - F(a)$ when $F(x)$ has a single term.

Write $[F(x)]_a^b$ for $F(b) - F(a)$ when $F(x)$ has more than one term.

> **THEOREM 4 (Continued)   The Fundamental Theorem of Calculus, Part 2**
>
> If f is continuous at every point of $[a, b]$ and F is any antiderivative of f on $[a, b]$, then
>
> $$\int_a^b f(x) \, dx = F(b) - F(a). \qquad (5)$$
>
> This part of the Fundamental Theorem is also called the **Integral Evaluation Theorem.**

Proof   Part 1 of the Fundamental Theorem tells us that an antiderivative of f exists, namely,

$$G(x) = \int_a^x f(t) \, dt.$$

Thus, if F is *any* antiderivative of f, then $F(x) = G(x) + C$ for some constant C (by Corollary 3 of the Mean Value Theorem for derivatives, Section 4.7).

Evaluating $F(b) - F(a)$, we have

$$F(b) - F(a) = [G(b) + C] - [G(a) + C]$$

$$= G(b) - G(a)$$

$$= \int_a^b f(t) \, dt - \int_a^a f(t) \, dt = \int_a^b f(t) \, dt - 0 = \int_a^b f(t) \, dt.$$

This establishes Eq. (5) and concludes the proof. $\blacksquare$

Theorem 4 says that to evaluate the definite integral of a continuous function f from a to b, all we need do is find an antiderivative F of f and calculate the number $F(b) - F(a)$.

EXAMPLE 5

Evaluate $\int_{-1}^{3}(x^3 + 1)\,dx$. Support by finding the numerical integral, NINT $(x^3 + 1, -1, 3)$.

Solution

$$\int_{-1}^{3}(x^3 + 1)\,dx = \left[\frac{x^4}{4} + x\right]_{-1}^{3} \qquad D_x\left(\frac{x^4}{4} + x\right) = x^3 + 1$$

$$= \left(\frac{81}{4} + 3\right) - \left(\frac{1}{4} - 1\right)$$

$$= \frac{80}{4} + 4 = 24$$

Using a calculator, NINT $(x^3 + 1, -1, 3) = 24$. ≡

EXAMPLE 6

Evaluate each integral, if possible.

a) $\displaystyle\int_{0}^{3} \frac{x^2 - 4}{x - 2}\,dx$ **b)** $\displaystyle\int_{-1}^{2} \frac{dx}{x}$

Solution

a) The function $f(x) = (x^2 - 4)/(x - 2)$ has a removable discontinuity on [0, 3], so f is integrable, but Theorem 4 does not apply. We can, however, apply Theorem 4 to $g(x) = x + 2$, the continuous extension of f.

$$\int_{0}^{3} \frac{x^2 - 4}{x - 2}\,dx = \int_{0}^{3}(x + 2)\,dx = \left[\frac{x^2}{2} + 2x\right]_{0}^{3} = 10.5$$

b) The function $f(x) = 1/x$ is discontinuous at 0 on $[-1, 2]$, so Theorem 4 does not apply. The discontinuity is unbounded; it is not removable, nor is it a jump, so f is not integrable by our current understanding. ≡

Evaluating Definite Integrals

To evaluate a definite integral, we sometimes have the choice to

> Do algebraically and support with a graphing utility,
>
> Do with a graphing utility and confirm algebraically, or
>
> Do using a method of our choice.

In many cases, however, we will not be able to find explicit antiderivatives for reasons that will become clear later. In these cases, our only "choice" will be to apply the numerical approximation, NINT.

EXAMPLE 7

Evaluate $\int_{1}^{10\pi} x \cos x\,dx$.

Solution No explicit antiderivative is readily apparent. However,

$$\text{NINT}(x \cos x, 1, 10\pi) = -0.381773290\underline{682} = -0.382.$$

The exact solution is

$$[x \sin x + \cos x]_1^{10\pi} = (10\pi \sin 10\pi + \cos 10\pi) - (\sin 1 + \cos 1)$$

$$= 1 - \sin 1 - \cos 1$$

$$= -0.381773290\underline{676} \quad \text{(to 12 decimal places)}. \quad \blacksquare$$

Example 7 shows how incredibly accurate NINT can be. For all practical purposes, the NINT approximation is just as useful as the exact solution (what would you do with $1 - \sin 1 - \cos 1$ other than approximate it?) and usually more quickly found. So at this point, can we say that we have completed our study of definite integrals?

Not quite.

Although use of a calculator definite integral key is valuable and often appropriate, it is essential to come out of this course equally well versed in the mathematics and the supportive use of technology. It is important to know mathematical techniques for finding exact solutions. It is also important to become sensitive to the limitations of technology.

We must learn to be comfortable with and appreciate the high accuracy of most NINT f approximations as we become comfortable with the various functions that we study. We must also learn when not to trust NINT f. For example, we will know to be suspicious if a machine's numerical methods return a value for NINT $(f(x), a, b)$ when we know that $\int_a^b f(x)\,dx$ does not exist!

Viewing Graphs Never Seen Before

One of the most exciting advantages gained with technology, and one that continues to amaze us, is that we can now visualize relationships that have never been seen before. For almost 200 years, the efforts in integral calculus that have focused on finding antiderivatives have produced a relatively small set of functions whose antiderivatives are explicit. Most integrable functions, such as the important

$$f(x) = e^{-x^2} \quad \text{and} \quad g(x) = \frac{\sin x}{x},$$

do not have antiderivatives that *anyone* can write down explicitly. Now, using computer graphing, we can see (almost touch and feel) most antiderivatives that we know exist for *any* continuous function.

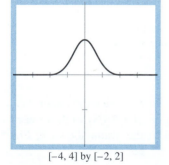

[−4, 4] by [−2, 2]

5.29 The graph of

$$y = f(x) = e^{-x^2}.$$

We can use $f(x)$ and its graph to deduce information about the function $F(x)$ where $F'(x) = f(x)$.

EXPLORATION 4

Confirming an Unseen Graph

The function $f(x) = e^{-x^2}$ (Fig. 5.29) is central to probability theory. Try as hard as you can, however, and you will not be able to find an explicit form for $F(x) = \int_0^x e^{-t^2}\,dt$. None exists! Nevertheless, you can still use your analytic tools to confirm information about F and its "unseen" graph. To do this, it is helpful to recall what derivatives reveal about a function.

1. Where is F increasing and decreasing? Where are its extreme points, if any? (Recall what information F' gives.)

2. What can you say about concavity and the graph of F? What, if anything, does the peak in the graph of f at $x = 0$ tell you? (Recall what information F'' gives and how you learn about F'' in this case.)

3. What, if anything, can you conclude about the end behavior of F?

4. What, if anything, can you conclude about the intercepts of F? ⚛

USING A LIST

If your grapher has the capability to use a *list* as an argument in a function, you can create the viewing window in Fig. 5.30(b) with one NINT line. Simply replace the a in $y = \text{NINT}(f(t), a, x)$ with the list $\{-1, 0, 1\}$. Try it.

EXAMPLE 8

We cannot leave the function $F(x) = \int_0^x e^{-t^2} dt$ of Exploration 4 without showing you what its graph looks like and giving you graphical support for results that you may have found. In fact, we do more than that. We show you in Fig. 5.30 what the graph of $F(x) = \int_a^x e^{-t^2} dt$ looks like for *three* different values of a:

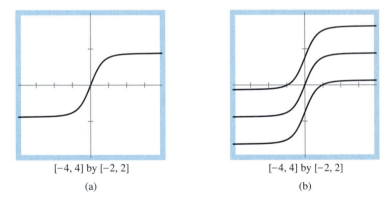

[−4, 4] by [−2, 2]

(a)

[−4, 4] by [−2, 2]

(b)

5.30 Graphs of (a) NINT $(e^{-t^2}, 0, x)$ and (b) NINT (e^{-t^2}, a, x) for $a = -1, 0, 1$. The graph of NDER of each of these functions is shown in Fig. 5.29.

To support that the graphs do indeed show antiderivatives of $f(x) = e^{-x^2}$ for $a = -1, 0, 1$, you can check the graph of NDER (NINT f) for each value of a. Each graph should appear identical to the graph shown in Fig. 5.29. Try it! ≣

EXPLORATION BIT

Run the toolbox program SLOPEFLD for $dy/dx = e^{-x^2}$ in the window $[-4, 4]$ by $[-2, 2]$. Compare with Fig. 5.30(b). Explain the relationship between slopefields of $dy/dx = f(x)$ and graphs of $y = \text{NINT}(f(t), a, x)$.

Graphically, we may say that antiderivatives are vertical shifts of one another and that every vertical shift is an antiderivative. Infinitely many vertical shifts are possible. Numerically, Example 8 reveals another source of infinitely many antiderivatives of a continuous function f, namely, $\int_a^x f(t)\, dt$ for each number a. These antiderivatives are "connected" slopefields.

EXPLORATION 5

Infinitely Many Antiderivatives—Two Sources

1. Give a convincing argument that of all antiderivatives $\int_a^x f(t)\,dt$, the one with y-intercept 0 has $a = 0$.

2. How far apart are the graphs of the two antiderivatives

$$\int_a^x f(t)\,dt \quad \text{and} \quad \int_b^x f(t)\,dt\,?$$

That is, what is the value of the constant by which these antiderivatives differ? Give a convincing argument.

3. $\int_0^x f(t)\,dt + C$ is an antiderivative of f for each value of C. $\int_a^x f(t)\,dt$ is an antiderivative of f for each value of a.

Is there a one-to-one correspondence between antiderivatives of the form $\int_0^x f(t)\,dt + C$ and antiderivatives of the form $\int_a^x f(t)\,dt$? If so, can you describe it. If not, give a convincing argument. ✤

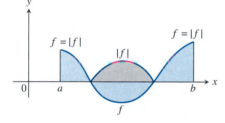

5.31 Wherever $f < 0$, the graph of $|f|$ is the reflection of the graph of f across the x-axis. The region enclosed by the x-axis and the graph of f is congruent to the region enclosed by the x-axis and the graph of $|f|$, so the regions have equal area.

The Area Connection

Because $A_a^b f$ and $\int_a^b f$ are identical when f is continuous and nonnegative, Part 2 of the Fundamental Theorem formalizes the way in which we calculated area with antiderivatives in Section 5.3. The area under the graph of f from $x = a$ to $x = b$ is $\int_a^b f(x)\,dx = F(b) - F(a)$.

When f is continuous and its graph dips below the x-axis in places, NINT gives us an efficient way to find the area between the graph and the x-axis. We simply calculate NINT $|f|$. This works because the graph of $|f|$ either is identical to the graph of f (where $f \geq 0$) or is the reflection of the graph of f across the x-axis (where $f < 0$), so the size of the region bounded by $|f|$ is identical to the size of the region bounded by f (Fig. 5.31).

EXAMPLE 9

Find the area of the region beween the x-axis and the curve

$$y = x^3 - 4x, \qquad -2 \leq x \leq 2.$$

Solution   NINT (abs $(x^3 - 4x), -2, 2) = 8$.

EXPLORATION 6

Confirming Area between a Curve and the x-Axis

To confirm analytically the area between the graph of f and the x-axis, we have to consider f "piecewise." We could follow these steps. (See Fig. 5.32.)

STEP 1: Draw a complete graph of f on $[a, b]$, and find the zeros where f changes sign.

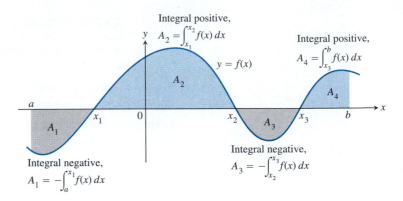

5.32   The area between the graph of f and the x-axis can be found by using NINT $|f|$ and confirmed by finding and summing the indicated A_i values.

STEP 2: Use those zeros to partition $[a, b]$ into subintervals, and integrate f over each subinterval.

STEP 3: Add the absolute values of the results.

1. Confirm analytically the result found in Example 9.
2. For $y_1 = x^3 - 4x$ (Example 9), which of the following will give the area between the graph of f and the x-axis from -2 to 2? Explain.

    a) NINT $(y_1, -2, 2)$

    b) 2 NINT $(y_1, -2, 0)$

    c) NINT $(|y_1|, -2, 2)$

    d) NINT $(y_1, -2, 0) +$ NINT $(y_1, 0, 2)$

    e) NINT $(y_1, -2, 0) -$ NINT $(y_1, 0, 2)$

    f) $|$NINT $(y_1, -2, 0) +$ NINT $(y_1, 0, 2)|$

    g) $|$NINT $(y_1, -2, 0)| + |$NINT $(y_1, 0, 2)|$

3. Let $y_1 = 4 - x^2$ and $y_2 = x^2 - 4$. Describe at least four ways to find the area between the graphs of y_1 and y_2. Give a convincing argument for each.   ⚛

Business Applications

EXAMPLE 10   Cost from Marginal Cost

The fixed cost of starting a manufacturing run and producing the first 10 units is \$200. After that, the marginal cost at x units output is

$$\frac{dc}{dx} = \frac{1000}{x^2}.$$

Find the total cost of producing the first 100 units.

Solution   If $c(x)$ is the cost of x units, then

$$\underbrace{c(100)}_{\substack{\text{Cost of} \\ \text{100 units}}} = \underbrace{200}_{\substack{\text{Startup} \\ \text{first 10}}} + \underbrace{c(100) - c(10)}_{\substack{\text{Cost of units} \\ \text{11--100}}}$$

$$= 200 + \int_{10}^{100} \frac{dc}{dx}\, dx \qquad\qquad \int_{10}^{100} \frac{dc}{dx}\, dx = c(100) - c(10)$$

$$= 200 + \int_{10}^{100} \frac{1000}{x^2}\, dx \qquad\qquad \frac{dc}{dx} = \frac{1000}{x^2}$$

$$= 200 + 1000 \int_{10}^{100} \frac{1}{x^2}\, dx \qquad\qquad \text{Rule 3, Table 5.3}$$

$$= 200 + 1000 \left[-\frac{1}{x} \right]_{10}^{100} = 200 + 1000 \left[-\frac{1}{100} + \frac{1}{10} \right]$$

$$= 200 - 10 + 100 = 290.$$

The total cost of producing the first 100 units is $290.

Average Daily Inventory

The notion of a function's average value is used in economics to study things like average daily inventory. If $I(x)$ is the number of radios, tires, shoes, or whatever product a firm has on hand on day x (we call $I(x)$ an **inventory function**), the average value of I over a time period $a \le x \le b$ is the firm's average daily inventory for the period.

DEFINITION

If $I(x)$ is the number of items on hand on day x, the **average daily inventory** of these items for the period $a \le x \le b$ is

$$I_{av} = \frac{1}{b - a} \int_a^b I(x)\, dx.$$

If h is the dollar cost of holding one item per day, the **average daily holding cost** for the period $a \le x \le b$ is $I_{av} \cdot h$.

EXAMPLE 11

Suppose a wholesaler receives a shipment of 1200 cases of chocolate bars every 30 days. The chocolate is sold to retailers at a steady rate, and x days after the shipment arrives, the inventory of cases still on hand is $I(x) = 1200 - 40x$. Find the average daily inventory. Also find the average daily holding cost for the chocolate if the cost of holding one case is 3¢ a day.

Solution   The average daily inventory is

$$I_{av} = \frac{1}{30 - 0} \int_0^{30} (1200 - 40x)\, dx = \frac{1}{30} \left[1200x - 20x^2 \right]_0^{30} = 600.$$

The average daily holding cost for the chocolate is the dollar cost of holding one case times the average daily inventory. This works out to $18 a day:

$$\text{Average daily holding cost} = (0.03)(600) = 18.$$

Exercises 5.4

In Exercises 1–14, evaluate each integral using Part 2 of the Fundamental Theorem *and* using NINT. Vary the order in which you do the two evaluations.

1. $\int_0^3 (4 - x^2)\, dx$

2. $\int_0^1 (x^2 - 2x + 3)\, dx$

3. $\int_0^1 (x^2 + \sqrt{x})\, dx$

4. $\int_0^5 x^{3/2}\, dx$

5. $\int_1^{32} x^{-6/5}\, dx$

6. $\int_{-2}^{-1} \frac{2}{x^2}\, dx$

7. $\int_0^\pi \sin x\, dx$

8. $\int_0^\pi (1 + \cos x)\, dx$

9. $\int_0^{\pi/3} 2 \sec^2 x\, dx$

10. $\int_{\pi/6}^{5\pi/6} \csc^2 x\, dx$

11. $\int_{\pi/4}^{3\pi/4} \csc x \cot x\, dx$

12. $\int_0^{\pi/3} 4 \sec x \tan x\, dx$

13. $\int_{-1}^1 (r + 1)^2\, dr$

14. $\int_9^4 \frac{1 - \sqrt{u}}{\sqrt{u}}\, du$

(*Hint:* Square first.)

(*Hint:* Divide first.)

In Exercises 15–18, find the total area of the region between the curve and the x-axis.

15. $y = 2 - x, \quad 0 \le x \le 3$

16. $y = 3x^2 - 3, \quad -2 \le x \le 2$

17. $y = x^3 - 3x^2 + 2x, \quad 0 \le x \le 2$

18. $y = x^3 - 4x, \quad -2 \le x \le 2$

In Exercises 19–24, can Theorem 4 be used to evaluate the integral? If it can, then find the value. If it cannot, then explain why not.

19. $\int_{-2}^3 \frac{x^2 - 1}{x + 1}\, dx$

20. $\int_0^5 \frac{9 - x^2}{3x - 9}\, dx$

21. $\int_0^{2\pi} \tan x\, dx$

22. $\int_0^2 \frac{x + 1}{x^2 - 1}\, dx$

23. $\int_{-1}^2 \frac{\sin x}{x}\, dx$

24. $\int_{-2}^3 \frac{1 - \cos x}{x^2}\, dx$

In Exercises 25–28, find the area of the shaded region.

25.

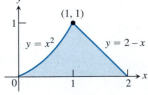

26.

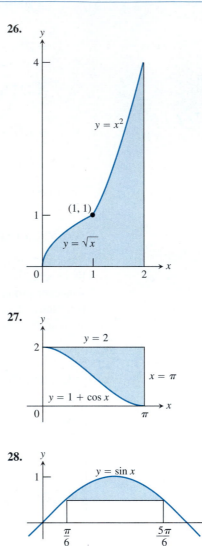

27.

28.

In Exercises 29–32, determine an explicit function for $F(x) = \int_a^x f(t)\, dt$ in terms of x. GRAPH $y = F(x)$ and $y = $ NINT $(f(t), 0, x)$ in the same viewing window. Compare values at $x = 0.5, 1, 1.5, 2,$ and 5.

29. $\int_0^x (t - 2)\, dt$

30. $\int_0^x (t^3 + 1)\, dt$

31. $\int_0^x (t^2 - 3t + 6)\, dt$

32. $\int_0^x 3 \sin t\, dt$

GRAPH in the interval specified in Exercises 33–36. Support with SLOPEFLD.

33. $\displaystyle\int_0^x t^2 \sin t \, dt \qquad$ for $-3 \le x \le 3$

34. $\displaystyle\int_0^x \sqrt{1+t^2}\, dt \qquad$ for $0 \le x \le 5$

35. $\displaystyle\int_0^x 5e^{-0.3t^2}\, dt \qquad$ for $0 \le x \le 5$

36. $\displaystyle\int_0^x t \sin(t^3)\, dt \qquad$ for $0 \le x \le \pi$

In Exercises 37 and 38, let $F(x) = \int_0^x f(t)\, dt$. GRAPH $y =$ NDER $(F(x), x)$, and compare with the graph of $y = f(x)$.

37. $f(x) = 4 - x^2 \qquad$ for $-5 \le x \le 5$

38. $f(x) = x \sin x \qquad$ for $0 \le x \le 2\pi$

In Exercises 39 and 40, find K so that $\int_a^x f(t)\, dt + K = \int_b^x f(t)\, dt$.

39. $f(x) = x^2 - 3x + 1; a = -1; b = 2$

40. $f(x) = \sin^2 x; a = 0; b = 2$

Exercises 41 and 42 involve the power of visualization. In these exercises, solve for x.

41. $\displaystyle\int_0^x e^{-t^2}\, dt = 0.6 \qquad$ **42.** $\displaystyle\int_0^x \frac{\sin t}{t}\, dt = 1.8$

(*Note:* We assume that the integrand is the continuous extension of $y = (\sin t)/t$.)

Find dy/dx in Exercises 43–46.

43. $\displaystyle y = \int_0^x \sqrt{1+t^2}\, dt \qquad$ **44.** $\displaystyle y = \int_1^x \frac{1}{t}\, dt, x > 0$

45. $\displaystyle y = \int_0^{\sqrt{x}} \sin(t^2)\, dt \qquad$ **46.** $\displaystyle y = \int_0^{2x} \cos t \, dt$

Each of the following functions solves one of the initial value problems in Exercises 47–50. Which function solves which problem?

a) $\displaystyle y = \int_1^x \frac{1}{t}\, dt - 3 \qquad$ **b)** $\displaystyle y = \int_0^x \sec t \, dt + 4$

c) $\displaystyle y = \int_{-1}^x \sec t \, dt + 4 \qquad$ **d)** $\displaystyle y = \int_\pi^x \frac{1}{t}\, dt - 3$

47. $\dfrac{dy}{dx} = \dfrac{1}{x}, y(\pi) = -3 \qquad$ **48.** $y' = \sec x, y(-1) = 4$

49. $y' = \sec x, y(0) = 4 \qquad$ **50.** $y' = \dfrac{1}{x}, y(1) = -3$

51. For what value of x is

$$\int_a^x f(t)\, dt$$

sure to be zero?

52. Suppose $\int_1^x f(t)\, dt = x^2 - 2x + 1$. Find $f(x)$. (*Hint:* Differentiate both sides of the equation with respect to x.)

53. Find $f(4)$ if $\int_0^x f(t)\, dt = x \cos \pi x$.

54. Find the linearization of

$$f(x) = 2 + \int_0^x \frac{10}{1+t}\, dt$$

at $x = 0$.

55. Show that if k is a positive constant, then the area between the x-axis and one arch of the curve $y = \sin kx$ is always $2/k$.

56. *Archimedes' area formula for parabolas.* Archimedes (287–212 B.C.), inventor, military engineer, physicist, and the greatest mathematician of classical times, discovered that the area under a parabolic arch like the one shown here is always two-thirds the base times the height.

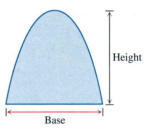

a) Find the area under the parabolic arch

$$y = 6 - x - x^2, \quad -3 \le x \le 2.$$

b) Find the height of the arch. (Where does y have its maximum value?)

c) Show that the area is two-thirds the base times the height.

57. *Cost from marginal cost.* The marginal cost of printing a poster when x posters have been printed is

$$\frac{dc}{dx} = \frac{1}{2\sqrt{x}}$$

dollars. Find (a) $c(100) - c(1)$, the cost of printing posters 2–100; (b) $c(400) - c(100)$, the cost of printing posters 101–400.

58. *Revenue from marginal revenue.* Suppose that a company's marginal revenue from the manufacture and sale of eggbeaters is

$$\frac{dr}{dx} = 2 - 2/(x+1)^2,$$

where r is measured in thousands of dollars and x in thousands of units. How much money should the company expect from a production run of $x = 3$ thousand eggbeaters? To find out, integrate the marginal revenue from $x = 0$ to $x = 3$.

59. Solon Container receives 450 drums of plastic pellets every 30 days. The inventory function (drums on hand as a function of days) is $I(x) = 450 - x^2/2$. Find the average daily inventory. If the holding cost for one drum is 2¢ per day, find the average daily holding cost.

60. Tina's Team Outfitters receives a shipment of 600 cases of athletic socks every 60 days. The number of cases on hand x days after the shipment arrives is $I(x) = 600 - 20\sqrt{15x}$. Find the average daily inventory. If the holding cost for one case is 1/2¢ per day, find the average daily holding cost.

61. Let $f(x) = \dfrac{\cos x}{x}$.

   a) Draw a complete graph of f.

   b) Draw a graph of $g(x) = \int_1^x f(t)\, dt$, $x > 0$.

   c) Explain why $\int_0^x f(t)\, dt$ is undefined.

   d) How are $h(x) = \int_{0.5}^x f(t)\, dt$ and $g(x)$ related? Graph both, and support your answer.

62. Explain how you know that $\int (\sin 2x/x)\, dx$ has no explicit analytic antiderivative? Draw a graph of $f(x) = \int_0^x (\sin 2t/t)\, dt$. How is it related to the graph of $y = \int_0^x (\sin t/t)\, dt$?

In Exercises 63–66, let $F(x) = \int_c^{u(x)} f(t)\, dt$ for the specified c, u, and f.

   a) Find the domain of F.

   b) Determine $F'(x)$ analytically.

   c) Determine the zeros of $F'(x)$, the intervals on which $F'(x)$ is increasing and decreasing, and the local extrema of F'.

   d) Explain how part (c) can be used to sketch a complete graph of $y = F(x)$. Support your sketch by graphing $y = F(x)$.

63. $c = 1$, $u(x) = x^2$, $f(x) = \sqrt{1 - x^2}$

64. $c = 0$, $u(x) = x^2$, $f(x) = \sqrt{1 - x^2}$

65. $c = 1$, $u(x) = 2x$, $f(x) = 1/\sqrt{1 - x^2}$

66. $c = 1$, $u(x) = 3x$, $f(x) = 1/\sqrt{1 - x^2}$

In Exercises 67 and 68, find $F'(x)$ analytically. Support your result by comparing the graphs of NDER F and F'.

67. $F(x) = \displaystyle\int_{x^2}^{x3} \cos (2t)\, dt$

68. $F(x) = \displaystyle\int_{\sin x}^{\cos x} t^2\, dt$

5.5 _____ Indefinite Integrals

Because antiderivatives make it possible to evaluate definite integrals with arithmetic, we will be working with antiderivatives a lot. We therefore need a notation that will make antiderivatives easier to describe and work with. This section introduces the notation and shows how to use it.

The Indefinite Integral of a Function

We call the set of all antiderivatives of a function the indefinite integral of the function, according to the following definition.

> **DEFINITION**
>
> If the function $f(x)$ is a derivative, then the set of all antiderivatives of f is called the **indefinite integral** of f, denoted by the symbols
>
> $$\int f(x)\, dx.$$
>
> As in definite integrals, the symbol $\int$ is called an **integral sign**. The function f is the **integrand** of the integral, and x is the **variable of integration.**

Since every continuous function has an antiderivative (by the Fundamental Theorem), every continuous function has an indefinite integral.

EVALUATING $\int f(x)\,dx$

To evaluate $\int f(x)\,dx$:

1. Find an antiderivative $F(x)$ of $f(x)$.

2. Add C (the constant of integration). Then

$$\int f(x)\,dx = F(x) + C.$$

Once we have found an antiderivative $F(x)$ of a function $f(x)$, the other antiderivatives of f differ from F only by a constant (Corollary 3 of the Mean Value Theorem for Derivatives). We indicate this in the new notation by writing

$$\int f(x)\,dx = F(x) + C.$$

The constant C is called the **constant of integration** or the **arbitrary constant,** and the equation is read, "The indefinite integral of f with respect to x is $F(x) + C$." When we find $F(x) + C$, we say that we have **evaluated** the indefinite integral.

EXAMPLE 1

Evaluate $\int (x^2 - 2x + 5)\,dx$.

Solution

$$\int (x^2 - 2x + 5)\,dx = \underbrace{\frac{x^3}{3} - x^2 + 5x}_{\text{An antiderivative of } f(x) = x^2 - 2x + 5} + \overset{\text{The constant of integration}}{C}$$

The indefinite integral of a function f is a family of functions whose graphs fill the entire plane (over the domain of f). If we calculate the indefinite integral $\int f$ to be $F(x) + C$, then $F(x)$ is a single representative of the family. If we use NINT $(f(t), a, x)$, then NINT f approximates a single representative of the family for each value of a. The $F(x)$ calculated as a particular antiderivative and NINT $(f(t), 0, x)$, where $a = 0$ and whose graph has y-intercept 0, may or may not match.

Hence, to use a grapher to compare F and NINT f, we can GRAPH both to see (a) whether they rest one of top of the other or (b) whether they are vertical shifts of each other. In the latter case, we have to be careful because congruent graphs that are vertical shifts of each other tend to appear not congruent in the small frame of a viewing window.

EXPLORATION 1

Viewing the Indefinite Integral

Suppose the graphs of F and NINT f appear to be different. How can you use the following to support that they both represent $\int f$?

1. TRACE

2. $y = F - \text{NINT } f$
Support graphically that F and NINT f both represent $\int f$ (from Example 1). Explain the support in each case.

3. $y_1 = F(x) = x^3/3 - x^2 + 5x;\ y_2 = \text{NINT } (t^2 - 2t + 5, 0, x)$.

4. $y_1 = F(x) = x^3/3 - x^2 + 5x;\ y_2 = \text{NINT } (t^2 - 2t + 5, 2, x)$. (Do the shapes in the viewing window *appear* to be identical?)

5. Apply SLOPEFLD to $dy/dx = x^2 - 2x + 5$. Explain how the results can be used to support the solution to Example 1.

EXPLORATION BIT

The equations

$$F = \int f$$

and

$$F' = f$$

suggest two methods of support for a differentiation problem and two methods of support for an integration problem using NDER and NINT. State what they are.
(Here is a more unconventional method of support: In part 4 of Exploration 1, ZOOM-OUT (far out) on the two graphs. Explain what you see.)

Formulas and Rules for Indefinite Integrals

The trigonometric derivative and antiderivative formulas from Sections 3.5, 3.6, and 4.7 translate into the integration formulas listed in Table 5.4.

> The integrals of $\tan x$, $\cot x$, $\sec x$, and $\csc x$ are not listed here because the usual formulas for them require logarithms. We know that these functions have indefinite integrals on intervals where they are continuous, but we will have to wait until Chapters 7 and 8 to see what the integrals are.

TABLE 5.4   **Integration Formulas**

1. $\displaystyle\int x^n \, dx = \frac{x^{n+1}}{n+1} + C \quad (n \neq -1)$

2. $\displaystyle\int \sin x \, dx = -\cos x + C$ **3.** $\displaystyle\int \cos x \, dx = \sin x + C$

4. $\displaystyle\int \sin kx \, dx = -\frac{\cos kx}{k} + C$ **5.** $\displaystyle\int \cos kx \, dx = \frac{\sin kx}{k} + C$

6. $\displaystyle\int \sec^2 x \, dx = \tan x + C$ **7.** $\displaystyle\int \csc^2 x \, dx = -\cot x + C$

8. $\displaystyle\int \sec x \tan x \, dx = \sec x + C$ **9.** $\displaystyle\int \csc x \cot x \, dx = -\csc x + C$

EXAMPLE 2   Selected Integrals from Table 5.4

a) $\displaystyle\int x^5 \, dx = \frac{x^6}{6} + C$ Formula 1

b) $\displaystyle\int \sin 2x \, dx = -\frac{\cos 2x}{2} + C$ Formula 4 with $k = 2$

c) $\displaystyle\int \cos \frac{x}{2} \, dx = \int \cos \frac{1}{2}x \, dx = \frac{\sin (1/2)x}{1/2} + C$ Formula 5 with $k = 1/2$

$\displaystyle\qquad\qquad = 2 \sin \frac{x}{2} + C$ ▤

The integration formulas (Table 5.4) hold because in each case the derivative of the function $F(x) + C$ on the right is the integrand $f(x)$ on the left. Finding an explicit form for an integral is often an impossible task, but *confirming* an explicit form, once found, is relatively easy: Differentiate the explicit form. If the derivative is equivalent to the integrand, the explicit form is correct; otherwise, it is wrong.

EXPLORATION 2

Do Graphically, Confirm Analytically

One of

$$y_1 = x \sin x, \qquad y_2 = x \sin x + \cos x, \qquad y_3 = x \sin x - \cos x$$

belongs to the indefinite integral $\int x \cos x \, dx$. You can get a clue as to which one by using NDER or NINT.

1. Demonstrate how both NDER and NINT can help.

2. Confirm the conclusion from part 1 analytically.   ꙮ

Supporting Indefinite Integral Evaluation Graphically

Two of our major themes on the use of technology in calculus:

1. Do analytically, *support* graphically, and

2. Do graphically, *confirm* analytically,

are illustrated repeatedly in our work with integrals and derivatives. If we calculate F, an antiderivative of f, we can support our result by comparing the graphs of F and NINT f, or the graphs of NDER F and f. We can also provide graphical support by applying SLOPEFLD to $dy/dx = f(x)$.

If we GRAPH NINT g or NDER g, the viewing window can give us guidance on what we seek to confirm about $\int g$ or g'. Let's take one more look at $\int x \cos x \, dx$.

EXAMPLE 3

Support graphically: $\displaystyle\int x \cos x \, dx = x \sin x + \cos x + C.$

Solution   There are two ways to do this: GRAPH

$$y_1 = \text{NINT}\ (t \cos t, 0, x) \qquad \text{and} \qquad y_2 = x \sin x + \cos x$$

or

$$y_3 = x \cos x \qquad \text{and} \qquad y_4 = \text{NDER}\ (x \sin x + \cos x).$$

Figure 5.33 shows graphs of y_1 and y_2, graphs that you may have seen in Exploration 2. The graphs appear to resemble each other. You should know how to use TRACE to check that they are related by a vertical shift. (We leave the graphs of y_3 and y_4 for you to check.) ≡

The definitions of indefinite integral and derivative and the Rules for Antiderivatives from Section 4.7 translate directly into Rules for Indefinite Integrals (Table 5.5).

TABLE 5.5   Rules for Indefinite Integrals

1. $\displaystyle\int \frac{dF}{dx}\,dx = F(x) + C$ (C an arbitrary constant)

2. $\displaystyle\frac{d}{dx}\int f(x)\,dx = f(x)$

3. Constant Multiple Rule: $\displaystyle\int k\, f(x)\,dx = k\int f(x)\,dx$ (any number k)

3a. Special Case: $\displaystyle\int -f(x)\,dx = -\int f(x)\,dx$

4. Sum Rule: $\displaystyle\int [f(x) + g(x)]\,dx = \int f(x)\,dx + \int g(x)\,dx$

5. Difference Rule: $\displaystyle\int [f(x) - g(x)]\,dx = \int f(x)\,dx - \int g(x)\,dx$

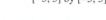

[−5, 5] by [−5, 5]

5.33   Graphs of

$$y_1 = \text{NINT}\ (t \cos t, 0, x)$$

and

$$y_2 = x \sin x + \cos x.$$

It appears that $y_2 = y_1 + C$. If so, what is the value of C? Explain.

EXPLORATION BITS

1. For an integrable function f, what should you expect to see when you GRAPH NDER (NINT f)? For a differentiable function F, what should you expect to see when you GRAPH NINT (NDER F)? Check your answers using y_3 and y_2, respectively, from Example 3.

2. In Section 5.4, we made the statement, "The processes of integration and differentiation are inverses of one another." Write a paragraph explaining the meaning of this statement. Use part 1 to help you.

EXAMPLE 4

$$\frac{d}{dx} \int \tan x \, dx = \tan x \qquad \text{Rule 2 with } f(x) = \tan x$$

We know this even without knowing how to evaluate the integral.   ≡

In $\int f = F(x) + C$, the C is "arbitrary," meaning that it can represent several constants rewritten in the form of one constant.

EXAMPLE 5   Rewriting the Constant of Integration

$$\int 5 \sec x \tan x \, dx = 5 \int \sec x \tan x \, dx \qquad \text{Constant Multiple Rule}$$

$$= 5(\sec x + C) \qquad \text{Table 5.4, Formula 8}$$

$$= 5 \sec x + 5C \qquad \text{First form}$$

$$= 5 \sec x + C_1 \qquad \text{Shorter form}$$

$$= 5 \sec x + C. \qquad \text{Usual form—no subscript} \ ≡$$

What about all the different forms in Example 5? Each one of them gives all the antiderivatives of $f(x) = 5 \sec x \tan x$, so each answer is correct. But the least complicated of the three, and the usual choice, is

$$\int 5 \sec x \tan x \, dx = 5 \sec x + C.$$

The general rule, in practice, is this:

If $F'(x) = f(x)$ and k is a constant, then

$$k \int f(x) \, dx = \int k \, f(x) \, dx = k \, F(x) + C. \qquad (1)$$

EXAMPLE 6

$$\int 8 \cos x \, dx = 8 \sin x + C \qquad \begin{array}{l} \text{Eq. (1) with } k = 8, \, f(x) = \cos x, \\ F(x) = \sin x \end{array} \quad ≡$$

The Sum Rule and Difference Rule allow us to integrate term by term. The arbitrary constants from each integration are rewritten as one constant.

EXAMPLE 7

Evaluate $\int (x^2 - 2x + 5)\,dx$.

Solution

$$\int (x^2 - 2x + 5)\,dx = \int x^2\,dx - \int 2x\,dx + \int 5\,dx$$

$$= \frac{x^3}{3} + C_1 - x^2 + C_2 + 5x + C_3$$

$$= \frac{x^3}{3} - x^2 + 5x + C.$$

In effect, we find the simplest antiderivative for each part, then add on the arbitrary constant of integration at the end. Compare with how we evaluated this integral in Example 1.   ≡

The Integrals of $\sin^2 x$, $\cos^2 x$, and $(\sin x)/x$

We can sometimes use trigonometric identities to transform indefinite integrals that we do not know how to evaluate into indefinite integrals that we do know how to evaluate. Among the examples that you should know about, important because of how frequently they arise in applications, are the integral formulas for $\sin^2 x$ and $\cos^2 x$. Instead of remembering the integration formulas themselves, you should try to remember how they are derived. We show steps that work for $\sin^2 x$ and ask that you follow a similar process for $\cos^2 x$ as an exercise.

EXPLORATION BIT

Apply SLOPEFLD to $dy/dx = \sin^2 x$ and then overlay the graph of $y_1 = x/2 - (\sin 2x)/4$. Explain how this supports the result of Example 8.

EXAMPLE 8

$$\int \sin^2 x\,dx = \int \frac{1 - \cos 2x}{2}\,dx \qquad \text{Because } \sin^2 x = \frac{1 - \cos 2x}{2}$$

$$= \frac{1}{2}\int (1 - \cos 2x)\,dx = \frac{1}{2}\int dx - \frac{1}{2}\int \cos 2x\,dx$$

$$= \frac{1}{2}x - \frac{1}{2}\frac{\sin 2x}{2} + C = \frac{x}{2} - \frac{\sin 2x}{4} + C \qquad\qquad ≡$$

Another important trigonometric integral that occurs in applications is

$$\int \frac{\sin x}{x}\,dx,$$

whose integrand we studied in Section 2.3.

For $F(x) = \int (\sin t)/t\,dt$, we do not have to remember a formula or a process because it has been proved that there is no explicit formula for F in terms of elementary functions. However, we can "see" what F looks like using NINT.

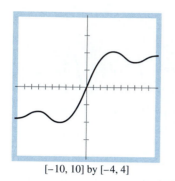

[−10, 10] by [−4, 4]

5.34 The graph of

$$y = \text{NINT}\left(\frac{\sin t}{t}, 0, x\right).$$

Remember that we are really graphing the integral of the continuous extension of $y = (\sin x)/x$.

By choosing dv / dt for the falling body to be positive, we are choosing *down* to be the positive direction in the coordinate plane.

EXAMPLE 9

To see what $F(x) = \int (\sin t)/t \, dt$ looks like, we GRAPH NINT $((\sin t)/t, 0, x)$. (See Fig. 5.34.)

Example 9 shows again the remarkable power of the graphing utility. We don't have an explicit formula for an antiderivative, but we can see how it behaves. In many applications, this suffices quite nicely, as we will see in many examples later. In Chapter 9, we will see that there is a formula for F as an infinite series.

Solving Initial Value Problems with Indefinite Integrals

As you know, we solve initial value problems with antiderivatives. We now look at the solutions in the language of indefinite integration. The first example is based on Example 7 in Section 4.7. You need not look back, however, unless you are interested in the modeling that gave rise to the problem. The solution that we give here is mathematically self-contained.

EXAMPLE 10   Finding Velocity from Acceleration

As a function of elapsed time t, the velocity $v(t)$ of a body falling from rest in a vacuum near the surface of Earth satisfies

Differential equation: $\dfrac{dv}{dt} = 9.8,$ The acceleration is 9.8 m/sec^2.

Initial condition: $v = 0$ when $t = 0$.   The velocity is 0 at the start.

Find v as a function of t.

Solution   We find the general solution of the differential equation by integrating both sides of it with respect to t:

$$\frac{dv}{dt} = 9.8 \qquad \text{Differential equation}$$

$$\int \frac{dv}{dt} \, dt = \int 9.8 \, dt \qquad \text{Integral equation}$$

$$v + C_1 = 9.8t + C_2 \qquad \text{Integrals evaluated}$$

$$v = 9.8t + C_2 - C_1 \qquad \text{Solved for } v$$

$$v = 9.8t + C. \qquad \text{Constants combined as one}$$

This last equation tells us that the body's velocity t seconds into the fall is $9.8t + C$ m/sec for some value of C. What value? We find out from the initial condition:

$$v = 9.8t + C$$

$$0 = 9.8(0) + C \qquad v = 0 \text{ when } t = 0.$$

$$C = 0.$$

Conclusion: The body's velocity t seconds into the fall is

$$v = 9.8t + 0 = 9.8t \text{ m/sec.}$$

In the next example, we have to integrate a second derivative twice to find the function we are after. The first integration,

$$\int \frac{d^2y}{dx^2}\,dx = \frac{dy}{dx} + C,$$

gives the function's first derivative. The second integration gives the function.

EXAMPLE 11

Solve the following initial value problem for y as a function of x:

Differential equation: $\dfrac{d^2y}{dx^2} = 6x - 2,$

Initial conditions: $\dfrac{dy}{dx} = 0 \quad\text{and}\quad y = 10 \quad\text{when}\quad x = 1.$

Solution   We integrate the differential equation with respect to x to find dy/dx:

$$\int \frac{d^2y}{dx^2}\,dx = \int (6x - 2)\,dx$$

$$\frac{dy}{dx} = 3x^2 - 2x + C_1. \qquad \text{Constants of integration combined as } C_1$$

We apply the first initial condition to find C_1:

$$0 = 3(1)^2 - 2(1) + C_1 \qquad \frac{dy}{dx} = 0 \text{ when } x = 1$$

$$C_1 = -3 + 2 = -1.$$

This completes the formula for dy/dx:

$$\frac{dy}{dx} = 3x^2 - 2x - 1.$$

We integrate dy/dx with respect to x to find y:

$$\int \frac{dy}{dx}\,dx = \int (3x^2 - 2x - 1)\,dx$$

$$y = x^3 - x^2 - x + C_2. \qquad \text{Constants of integration combined as } C_2$$

We apply the second initial condition to find C_2:

$$10 = (1)^3 - (1)^2 - 1 + C_2 \qquad y = 10 \text{ when } x = 1.$$

$$10 = -1 + C_2$$

$$C_2 = 11.$$

This completes the formula for y as a function of x:

$$y = x^3 - x^2 - x + 11. \qquad\qquad \blacksquare$$

SUPPORT

One way to support your result in Example 11 is to begin with $y = x^3 - x^2 - x + 11$ and differentiate. Since you used rules of differentiation to find antiderivatives, however, faulty use of a rule could repeat itself, and checking by differentiating might not reveal an error.

A second way to support your result is to let

$$y_1 = x^3 - x^2 - x + 11,$$

$y_2 = \text{NDER } y_1,$ and $y_3 = \text{NDER } y_2.$

You can use views of y_1 and y_2 to check the initial conditions and views of y_3 to check the initial differential equation.

When we find a function by integrating its first derivative, we have one constant of integration, as in Example 10. When we find a function from its second derivative, we have to deal with two constants of integration, one from each integration, as in Example 11. If we were to find a function from its third derivative, we would have to find the values of three constants of integration, and so on. In each case, the values of the constants are determined by the problem's initial conditions. Each time we integrate, we need an initial condition to tell us the value of C.

EXPLORATION 3

Solving an Initial Value Problem

We have seen that a family of antiderivatives of f that are part of the indefinite integral $\int f$ can be represented by $F(x) = \int_a^x f(t) \, dt$ where a represents an arbitrary real number. Suppose we have the initial value problem:

Differential equation: $\dfrac{dy}{dx} = (0.6)2^x,$

Initial condition: $y = -1$ when $x = 0$

and we want to find the value of a, if any, so that $F(x) = \int_a^x f(t) \, dt$ is the solution of the problem (that is, $F(x)$ includes the constant of integration so that F satisfies the initial condition and F' satisfies the given differential equation). Then

$$-1 = F(0) = \int_a^0 f(t) \, dt = -\int_0^a f(t) \, dt, \qquad \text{or} \qquad 1 = \int_0^a f(t) \, dt.$$

1. For $f(x) = (0.6)2^x$, GRAPH $y_1 = \displaystyle\int_0^x f(t) \, dt = \text{NINT } (f(t), 0, x)$. Use ZOOM-IN or SOLVE to find a where $\int_0^a f(t) \, dt = \text{NINT } (f(t), 0, a) = 1$.

2. In part 1, you should find $a = 1.108$, accurate to three decimal places. The solution to the initial value problem, then, is

$$y = \int_{1.108}^x (0.6)2^t \, dt.$$

In a later section, we will find the representation $(0.6)2^x / \ln 2 - 1.866$ to be the solution. For now, GRAPH $y_1 = \text{NINT } ((0.6)2^t, 1.108, x)$ and $y_2 = (0.6)2^x / \ln 2 - 1.866$ for support that they are the same solution of the differential equation. (What would you expect the y-intercept of each to be? Why?)   ∞

Exercises 5.5

Evaluate each integral in Exercises 1–10. Confirm by differentiating your answer and comparing the result with the integrand.

1. $\displaystyle\int x^3 \, dx$

2. $\displaystyle\int 7 \, dx$

3. $\displaystyle\int (x + 1) \, dx$

4. $\displaystyle\int (6 - 6x) \, dx$

5. $\displaystyle\int 3\sqrt{x} \, dx$

6. $\displaystyle\int \frac{4}{x^2} \, dx$

7. $\displaystyle\int x^{-1/3} \, dx$

8. $\displaystyle\int (1 - 4x^{-3}) \, dx$

9. $\displaystyle\int (5x^2 + 2x) \, dx$

10. $\displaystyle\int \left(\frac{x^2}{2} + \frac{x^3}{3} \right) dx$

Evaluate each integral in Exercises 11–20. Support by comparing the graph of NDER of your answer with the graph of the integrand.

11. $\displaystyle\int (2x^3 - 5x + 7) \, dx$

12. $\displaystyle\int (1 - x^2 - 3x^5) \, dx$

13. $\displaystyle\int 2 \cos x \, dx$

14. $\displaystyle\int 5 \sin \theta \, d\theta$

15. $\displaystyle\int \sin\frac{x}{3}\,dx$ **16.** $\displaystyle\int 3\cos 5x\,dx$

17. $\displaystyle\int 3\csc^2 x\,dx$ **18.** $\displaystyle\int \frac{\sec^2 x}{3}\,dx$

19. $\displaystyle\int \frac{\csc x\cot x}{2}\,dx$ **20.** $\displaystyle\int \frac{2}{5}\sec x\tan x\,dx$

Evaluate each integral in Exercises 21–30. Support by comparing the graph of the antiderivative you found and $y = \text{NINT}\,(f(t),\,0,\,x)$ or by using SLOPEFLD.

21. $\displaystyle\int (4\sec x\tan x - 2\sec^2 x)\,dx$

22. $\displaystyle\int \frac{1}{2}(\csc^2 x - \csc x\cot x)\,dx$

23. $\displaystyle\int (\sin 2x - \csc^2 x)\,dx$

24. $\displaystyle\int (2\cos 2x - 3\sin 3x)\,dx$

25. $\displaystyle\int 4\sin^2 y\,dy$ **26.** $\displaystyle\int \frac{\cos^2 x}{7}\,dx$

27. $\displaystyle\int \sin x\cos x\,dx$ (*Hint:* $2\sin x\cos x = \sin 2x$)

28. $\displaystyle\int (1 - \cos^2 t)\,dt$

29. $\displaystyle\int (1 + \tan^2\theta)\,d\theta$ (*Hint:* $1 + \tan^2\theta = \sec^2\theta$)

30. $\displaystyle\int \frac{1 + \cot^2 x}{2}\,dx$

In Exercises 31 and 32, write the equation of the continuous extension of the function in the specified closed interval. Graph $y = \int_0^x f(t)\,dt$ for the continuous extension of f.

31. $f(x) = \dfrac{1 - \cos x}{x^2},\quad -10 \le x \le 10$

32. $f(x) = \left(1 + \dfrac{1}{x}\right)^x,\quad 0 \le x \le 5$

Show that the integral formulas in Exercises 33–36 are correct by showing that the derivative of the right-hand side is the integrand in the integral on the left-hand side. (In Section 5.6 we will see where formulas like these come from.)

33. $\displaystyle\int (7x - 2)^3\,dx = \frac{(7x - 2)^4}{28} + C$

34. $\displaystyle\int \sec^2 5x\,dx = \frac{\tan 5x}{5} + C$

35. $\displaystyle\int \frac{1}{(x + 1)^2}\,dx = -\frac{1}{x + 1} + C$

36. $\displaystyle\int \frac{1}{(x + 1)^2}\,dx = \frac{x}{x + 1} + C$

37. Right or wrong? Say which for each formula.

a) $\displaystyle\int x\sin x\,dx = \frac{x^2}{2}\sin x + C$

b) $\displaystyle\int x\sin x\,dx = -x\cos x + C$

c) $\displaystyle\int x\sin x\,dx = -x\cos x + \sin x + C$

38. Right or wrong? Say which for each formula.

a) $\displaystyle\int (2x + 1)^2\,dx = \frac{(2x + 1)^3}{3} + C$

b) $\displaystyle\int 3(2x + 1)^2\,dx = (2x + 1)^3 + C$

c) $\displaystyle\int 6(2x + 1)^2\,dx = (2x + 1)^3 + C$

Solve each initial value problem in Exercises 39–46 for y as a function of x. GRAPH the solution $y = \text{NINT}\,f$ if analytic solutions are unavailable.

39. Differential equation: $\dfrac{dy}{dx} = 3\sqrt{x}$

Initial condition: $y = 4$ when $x = 9$

40. Differential equation: $\dfrac{dy}{dx} = \dfrac{1}{2\sqrt{x}}$

Initial condition: $y = 0$ when $x = 4$

41. Differential equation: $\dfrac{dy}{dx} = 2^x$

Initial condition: $y = 2$ when $x = 0$

42. Differential equation: $\dfrac{dy}{dx} = \dfrac{1 - \cos x}{x}$

Initial condition: $y = 0.5$ when $x = 0$

43. Differential equation: $\dfrac{d^2y}{dx^2} = 0$

Initial conditions: $\dfrac{dy}{dx} = 2$ and $y = 0$ when $x = 0$

44. Differential equation: $\dfrac{d^2y}{dx^2} = \dfrac{2}{x^3}$

Initial conditions: $\dfrac{dy}{dx} = 1$ and $y = 1$ when $x = 1$

45. Differential equation: $\dfrac{d^2y}{dx^2} = \dfrac{3x}{8}$

Initial conditions: $\dfrac{dy}{dx} = 3$ and $y = 4$ when $x = 4$

46. Differential equation: $\dfrac{d^3y}{dx^3} = 6$

Initial conditions: $\dfrac{d^2y}{dx^2} = -8,\ \dfrac{dy}{dx} = 0$, and

$y = 5$ when $x = 0$

47. *Stopping a car in time.* You are driving along a highway at a steady 60 mph (88 ft/sec) when you see an accident ahead and slam on the brakes. What constant deceleration is required to stop your car in 242 ft? To find out, carry out the following steps:

STEP 1: Solve the initial value problem

Differential equation: $\dfrac{d^2s}{dt^2} = -k$ (k constant)

Initial conditions: $\dfrac{ds}{dt} = 88$ and $s = 0$ when $t = 0$.

Measuring time and distance from when the brakes are applied

STEP 2: Find the value of t that makes $ds/dt = 0$. (The answer will involve k.)

STEP 3: Find the value of k that makes $s = 242$ for the value of t you found in Step 2.

48. *Motion along a coordinate line.* A particle moves along a coordinate line with acceleration $a = d^2s/dt^2 = 15\sqrt{t} - 3/\sqrt{t}$, subject to the conditions that $ds/dt = 4$ and $s = 0$ when $t = 1$. Find

a) the velocity $v = ds/dt$ in terms of t, and

b) the position s in terms of t.

49. *The hammer and the feather.* When Apollo 15 astronaut David Scott dropped a hammer and a feather on the moon to demonstrate that in a vacuum all bodies fall with the same (constant) acceleration, he dropped them from about 4 ft above the ground. The television footage of the event shows the hammer and feather falling more slowly than on Earth, where, in a vacuum, they would have taken only half a second to fall the four feet. How long did it take the hammer and feather to fall the four feet on the moon? To find out, solve the following initial value problem for s as a function of t. Then find the value of t that makes s equal 4.

Differential equation: $\dfrac{d^2s}{dt^2} = 5.2$ ft/sec^2

Initial conditions: $\dfrac{ds}{dt} = 0$ and $s = 0$ when $t = 0$

50. *The standard equation for free fall.* The standard equation for free fall near the surface of every planet is

$$s(t) = \frac{1}{2}gt^2 + v_0 t + s_0,$$

where $s(t)$ is the body's position on the line of fall, g is the planet's (constant) acceleration of gravity, v_0 is the body's initial velocity, and s_0 is the body's initial position. Derive this equation by solving the following initial value problem:

Differential equation: $\dfrac{d^2s}{dt^2} = g$

Initial conditions: $\dfrac{ds}{dt} = v_0$ and $s = s_0$ when $t = 0$.

51. Derive the integration formula

$$\int \cos^2 x \, dx = \frac{x}{2} + \frac{\sin 2x}{4} + C.$$

In Exercises 52–57, consider the curved paths or "tracks" given parametrically by $x(t)$, $y(t)$. Each track begins at $(0, 0)$ and ends at $(\pi, -2)$ for $0 \le t \le 1$. (Contributed by Jerry Johnston.)

A: $x_1(t) = \pi t$, $y_1(t) = -2t$

B: $x_2(t) = \pi t$, $y_2(t) = 2(t - 1)^2 - 2$

C: $x_3(t) = \pi t - \sin(\pi t)$, $y_3(t) = \cos(\pi t) - 1$

D: $x_4(t) = \pi t$, $y_4(t) = -2\sqrt{t}$

A small steel ball is rolled down each track with initial velocity 0. It can be shown that the time T required for the ball to roll down the track is given by

$$T = \frac{1}{\sqrt{g}} \int_0^1 \sqrt{\frac{(x'(t))^2 + (y'(t))^2}{-2y(t)}} \, dt$$

where g is the acceleration due to gravity.

52. Explain why the integrand $y = f(t)$ of T is *not* a continuous function for $0 \le t \le 1$. Can you apply Theorem 4 to evaluate this integral?

53. Graph each track A, B, C, D in the $[0, \pi]$ by $[0, -2]$ window.

54. Graph each integrand $y = f(t)$ of T in the $[0, 1]$ by $[0, 100]$ window. In each case, find $\lim_{t \to 0^+} f(t)$.

55. It can be shown that

$$T = \lim_{x \to 0^+} \frac{1}{\sqrt{g}} \int_x^1 \sqrt{\frac{(x'(t))^2 + (y'(t))^2}{-2y(t)}} \, dt$$

if the limit exists. Find this limit analytically for curves A and C. Support your answers by computing NINT $(f(t), h, 1)$ for $h > 0$ and very small.

56. Investigate T for curves B and D by computing NINT $(f(t), h, 1)$ for $h = 0.1, 0.01, 0.001$, and so forth. Interpret your results in the problem situation.

57. It can be shown (with difficulty) that track C (a path called a *brachistochrone*) is the path of "least time" for paths that always decrease from $(0, 0)$ to $(\pi, -2)$. Explain how your results in Exercise 56 might contradict this fact. It turns out that the "improper" integrals (more in Chapter 8) given by the paths B and D are very difficult to approximate (even with powerful CAS).

5.6

Integration by Substitution—Running the Chain Rule Backward

A change of variable can often turn an unfamiliar integral into one that we can evaluate. The method for doing this is called the substitution method of integration. It is the principal method by which integrals are evaluated. This section shows how and why the method works.

The Generalized Power Rule in Integral Form

When u is a differentiable function of x and n is a rational number different from -1, the Chain Rule tells us that

$$\frac{d}{dx}\left(\frac{u^{n+1}}{n+1}\right) = u^n\frac{du}{dx}.$$

This same equation, from another point of view, says that $u^{n+1}/(n+1)$ is one of the antiderivatives of $u^n(du/dx)$. The set of all antiderivatives of $u^n(du/dx)$ is therefore

$$\int\left(u^n\frac{du}{dx}\right)dx = \frac{u^{n+1}}{n+1} + C. \tag{1}$$

The integral on the left-hand side of this equation is usually written in the simpler "differential" form,

$$\int u^n\,du,$$

obtained by treating the dx's as differentials that cancel. Substituting this into Eq. (1) then gives the following rule.

If u is any differentiable function of x,

$$\int u^n\,du = \frac{u^{n+1}}{n+1} + C \qquad (n \neq -1). \tag{2}$$

Whenever we can put an integral in the form

$$\int u^n\,du$$

with u a differentiable function of x and du the differential of u, we can integrate with respect to u in the usual way to get $[u^{n+1}/(n+1)] + C$ as the value of the integral.

EXAMPLE 1

Evaluate $\int (x+2)^5 \, dx$.

Solution We can put the integral in the form $\int u^5 \, du$ by substituting $u = x + 2$, $du = d(x+2) = dx$. Then

$$\int (x+2)^5 \, dx = \int u^5 \, du \qquad \text{Substitute } u = x + 2, du = dx.$$

$$= \frac{u^6}{6} + C \qquad \text{Integrate, using Eq. (2) with } n = 5.$$

$$= \frac{(x+2)^6}{6} + C. \qquad \text{Replace } u \text{ by } x + 2.$$

EXPLORATION 1

Sometimes You Must "Do Graphically"

1. Use the substitution $u = 1 + x^2$ (hence, $du = 2x \, dx$) to show that $\int \sqrt{1 + x^2} \cdot 2x \, dx = \frac{2}{3}(1 + x^2)^{3/2} + C$. Support this result graphically by comparing the graphs of $y_1 = \text{NINT}\left(\sqrt{1 + t^2} \cdot 2t, 0, x\right)$ and $y_2 = (2/3)(1 + x^2)^{3/2}$. Explain what you see in the viewing window.

2. Use the substitution $u = x^3$ to show that $\int x^2 \sin x^3 \, dx = -1/3 \cos x^3 + C$. Support this result graphically, using $y_1 = x^2 \sin x^3$ and $y_2 = \text{NDER}\left(-(1/3) \cos x^3\right)$. Explain what you see in the viewing window.

3. The integrands in parts 1 and 2 conveniently include factors that make finding an explicit form relatively easy by using substitution. Now, try finding an explicit form for $\int \sqrt{1 + x^2} \, dx$. Failing that (and don't spend more than five minutes looking for one), GRAPH $\text{NINT}\left(\sqrt{1 + t^2}, a, x\right)$ for $a = -4, -2, 0, 2, 4$ to see what five antiderivatives look like.

EXAMPLE 2

$$\int \sqrt{4x - 1} \, dx = \int u^{1/2} \cdot \frac{1}{4} \, du \qquad \begin{array}{l} \text{Substitute } u = 4x - 1, \\ du = 4 \, dx, \dfrac{1}{4} du = dx. \end{array}$$

$$= \frac{1}{4} \int u^{1/2} \, du \qquad \begin{array}{l} \text{The Constant Multiple Rule; the} \\ \text{integral is now in standard form.} \end{array}$$

$$= \frac{1}{4} \cdot \frac{u^{3/2}}{3/2} + C \qquad \text{Integrate, using Eq. (2) with } n = \frac{1}{2}.$$

$$= \frac{1}{6} u^{3/2} + C \qquad \text{Simpler form}$$

$$= \frac{1}{6}(4x - 1)^{3/2} + C. \qquad \text{Replace } u \text{ by } 4x - 1.$$

Trigonometric Integrands

We know that

$$\int \cos x \, dx = \sin x + C.$$

If u is a differentiable function of x, call it $u(x)$, then $\sin u$ is a differentiable function of x, and the Chain Rule tells us that

$$\frac{d}{dx} \sin u(x) = \cos u(x) \frac{du}{dx}.$$

Thus, we can write

$$\int \left(\cos u(x) \frac{du}{dx} \right) dx = \sin u(x) + C,$$

or simply,

$$\int \cos u \, du = \sin u + C. \qquad (3)$$

This equation says that whenever we can put an integral in the form

$$\int \cos u \, du,$$

where u is a function of x, then we can integrate with respect to u in the usual way before replacing u in the result with $u(x)$.

Similarly, we can write

$$\int \sin u \, du = -\cos u + C. \qquad (4)$$

EXAMPLE 3

$$\int \cos (7x + 5) \, dx = \int \cos u \cdot \frac{1}{7} \, du \qquad \text{Substitute } u = 7x + 5,$$
$$du = 7 \, dx, \frac{1}{7} \, du = dx.$$

$$= \frac{1}{7} \int \cos u \, du \qquad \text{With the } \frac{1}{7} \text{ out front, the integral is now in standard form.}$$

$$= \frac{1}{7} \sin u + C \qquad \text{Integrate with respect to } u.$$

$$= \frac{1}{7} \sin (7x + 5) + C \qquad \text{Replace } u \text{ by } 7x + 5. \qquad \blacksquare$$

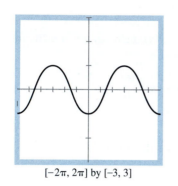

$[-2\pi, 2\pi]$ by $[-3, 3]$

5.35 This is the graph of

$$\int_a^x \sin t\,dt = \text{NINT} \ (\sin t, a, x) = -\cos x$$

for a particular value of a. Can you find a?

 EXPLORATION 2

Solving an Initial Value Problem

1. Predict what a viewing window should show for the particular antiderivative $\int_0^x \sin t\,dt = \text{NINT} \ (\sin t, 0, x)$. (*Hint:* Recall the antiderivative of $\sin x$ and the difference between $\int_0^x f$ and $\int_a^x f$.)

2. Try to find the value of a so that $\int_a^x \sin t\,dt = \text{NINT} \ (\sin t, a, x)$ gives the *familiar* graph of $-\cos x$ (Fig. 5.35). 🙰

The Chain Rule formulas for the derivatives of the tangent, cotangent, secant, and cosecant of a differentiable function u of x lead to the following integrals. In each formula, we assume u to be a differentiable function of x. Each formula can be checked by differentiating the right-hand side with respect to x. In each case, the Chain Rule applies to produce the integrand on the left.

$$\int \sec^2 u \, du = \tan u + C \qquad\qquad (5)$$

$$\int \csc^2 u \, du = -\cot u + C \qquad\qquad (6)$$

$$\int \sec u \tan u \, du = \sec u + C \qquad\qquad (7)$$

$$\int \csc u \cot u \, du = -\csc u + C \qquad\qquad (8)$$

EXAMPLE 4

Find an explicit form for $\int 1/(\cos^2 2x) \, dx$. Support graphically.

Solution

$$\int \frac{1}{\cos^2 2x} \, dx = \int \sec^2 2x \, dx \qquad\qquad \sec 2x = \frac{1}{\cos 2x}$$

$$= \int \sec^2 u \cdot \frac{1}{2} \, du \qquad\qquad \text{Substitute } u = 2x, du = 2\,dx, dx = \frac{1}{2}\,du.$$

$$= \frac{1}{2} \int \sec^2 u \, du$$

$$= \frac{1}{2} \tan u + C \qquad\qquad \text{Integrate, using Eq. (5).}$$

$$= \frac{1}{2} \tan 2x + C \qquad\qquad \text{Replace } u \text{ by } 2x.$$

Figure 5.36 shows a viewing window that supports this result graphically.

∎

$[-0.7, 0.7]$ by $[-4, 4]$

5.36 The graphs of

$$y_1 = \int_0^x \frac{1}{\cos^2 2t} dt$$

$$= \text{NINT} \left(\frac{1}{\cos^2 2t}, 0, x \right)$$

and

$$y_2 = \frac{1}{2} \tan 2x$$

appear to be the same.

The Substitution Method of Integration

The substitutions that we have been using in the explorations and examples are all instances of a general rule:

$$\int f(g(x)) \cdot g'(x)\, dx = \int f(u)\, du \qquad \text{Substitute } u = g(x),$$
$$ du = g'(x)\, dx.$$

$$= F(u) + C \qquad \text{Evaluate by finding an}$$
$$ \text{antiderivative of } f(u).$$
$$ \text{(Any one will do.)}$$

$$= F(g(x)) + C. \qquad \text{Substitute back.}$$

These three steps are the steps of the substitution method of integration.

The Substitution Method of Integration

To evaluate the integral

$$\int f(g(x))g'(x)\, dx$$

when f and g' are continuous functions, carry out the following steps:

STEP 1: Substitute $u = g(x)$ and $du = g'(x)\, dx$ to obtain the integral

$$\int f(u)\, du.$$

STEP 2: Integrate with respect to u.

STEP 3: Replace u by $g(x)$ in the result.

EXAMPLE 5

$$\int (x^2 + 2x - 3)^2 (x + 1)\, dx = \int u^2 \cdot \frac{1}{2}\, du \qquad \begin{array}{l} \text{Substitute } u = x^2 + 2x - 3, \\ du = 2x\, dx + 2\, dx \\ = 2(x + 1)\, dx, \\ \frac{1}{2} du = (x + 1)\, dx. \end{array}$$

$$= \frac{1}{2} \int u^2\, du$$

$$= \frac{1}{2} \cdot \frac{u^3}{3} + C = \frac{1}{6} u^3 + C \qquad \begin{array}{l} \text{Integrate with} \\ \text{respect to } u. \end{array}$$

$$= \frac{1}{6}(x^2 + 2x - 3)^3 + C \qquad \begin{array}{l} \text{Replace } u \text{ by} \\ x^2 + 2x - 3. \end{array}$$

EXAMPLE 6

$$\int \sin^4 x \cos x \, dx = \int u^4 \, du \qquad \text{Substitute } u = \sin x, du = \cos x \, dx.$$

$$= \frac{u^5}{5} + C \qquad \text{Integrate with respect to } u.$$

$$= \frac{\sin^5 x}{5} + C \qquad \text{Replace } u \text{ by } \sin x. \qquad \blacksquare$$

EXPLORATION 3

Substitution Guess and Test

Sometimes, different substitutions are possible when evaluating an integral. Sometimes, no substitution will work. The process is one of intelligent "guess and test." One guideline is to substitute for the most troublesome part of the integrand and see how things work out. Evaluate

$$\int \frac{2z \, dz}{\sqrt[3]{z^2 + 1}}$$

in two ways:

1. Use the substitution $u = z^2 + 1$.
2. Use the substitution $u = \sqrt[3]{z^2 + 1}$.

Substitution in Definite Integrals

To evaluate a *definite* integral by substitution, we can avoid Step 3 of the substitution method. Instead of writing an antiderivative $F(g)$ in terms of x as $F(g(x))$ (Step 3) and then evaluating $F(g(b)) - F(g(a))$, we can leave F in terms of $u = g(x)$ with $d = g(b)$ and $c = g(a)$ and evaluate $F(d) - F(c)$.

THE DEFINITE INTEGRAL FORMULA

The formula for evaluating definite integrals by substitution first appeared in a book by Isaac Barrow (1630–1677), Newton's mathematics teacher at Cambridge University.

Substitution in Definite Integrals

The Formula

$$\int_a^b f(g(x)) \cdot g'(x) \, dx = \int_{g(a)}^{g(b)} f(u) \, du \qquad (9)$$

How to Use It Substitute $u = g(x)$, $du = g'(x) \, dx$, and integrate with respect to u from $u = g(a)$ to $u = g(b)$.

To use the formula, make the same u-substitution you would use to evaluate the corresponding indefinite integral. Then integrate with respect to u from the value u has at $x = a$ to the value u has at $x = b$.

EXAMPLE 7

RECOGNIZING $g'(x)\, dx$

In Example 7, we did not substitute for what appears to be the most troublesome part of the integrand, namely, $\sec^2 x$, because we recognized that the derivative of $\tan x$ is $\sec^2 x$.

$$\int_0^{\pi/4} \tan x \sec^2 x \, dx = \int_0^1 u \, du$$

Substitute $u = \tan x$, $du = \sec^2 x \, dx$. Integrate from $u(0) = \tan 0 = 0$ to $u\left(\dfrac{\pi}{4}\right) = \tan\dfrac{\pi}{4} = 1$.

$$= \left.\frac{u^2}{2}\right]_0^1$$

Evaluate the definite integral.

$$= \frac{(1)^2}{2} - \frac{(0)^2}{2}$$

$$= \frac{1}{2}$$

EXPLORATION 4

Two Routes to the Integral

We do not have to use Eq. (9) if we do not want to. We can always transform the integral as an indefinite integral, integrate, change back to x, and use the original x limits. To show that the the two methods give the same result, evaluate $\int_{-1}^{1} 3x^2\sqrt{x^3 + 1}\, dx$ both ways:

1. Substitute $u = x^3 + 1$, $du = 3x^2\, dx$. Then integrate from $u(-1)$ to $u(1)$.
2. Substitute $u = x^3 + 1$, $du = 3x^2\, dx$. Then find an antiderivative with respect to u. Replace u with $x^3 + 1$, and evaluate using the limits of integration -1 and 1.

Explain how you could use NINT to support your work.

Which method is better—evaluating the transformed integral with transformed limits or transforming back to use the original limits of integration? In Exploration 4, the first method may have seemed easier, but that is not always the case. As a rule, it is best to know both methods and use whichever one seems better at the time.

Does either method always work? Hardly. You should have noticed that in integrals that can be evaluated by using substitution,

OTHER METHODS OF INTEGRATION

There are other methods of integration that we will see later. But each of them is applicable to only a limited set of integrands.

$$\int f(g(x))g'(x) \, dx,$$

the integrand must have the part $g'(x)$. Most integrands that occur in applications lack this part; the ones that are usually given in textbooks are constructed so that you can have some practice with the method. As examples of what we mean, remove the $g'(x)$ part from any of the examples of this section, and try to evaluate the integral.

Finally, we point out that the limits of integration in textbook problems are also selected—contrived, some say—so that our pencil-and-paper arithmetic is easy and does not get in the way of the major ideas. In real-life, the limits of integration are not as considerate. The pencil-and-paper arithmetic is usually brutal. For both of the above reasons, it becomes vital, we feel, that you learn to use the calculator and become much better equipped from this course to make it work for you.

EXPLORATION 5

Integrals in the Real World

Just for fun, here for you to evaluate are three definite integrals based on Exploration 4. First identify how each integral differs from the integral in Exploration 4. Then evaluate each, using pencil and paper or a graphing utility. It's your choice.

1. $\displaystyle\int_{-1}^{1} \sqrt{x^3 + 1}\, dx$

2. $\displaystyle\int_{2.1}^{5.6} 3x^2 \sqrt{x^3 + 1}\, dx$

3. $\displaystyle\int_{2.1}^{5.6} \sqrt{x^3 + 1}\, dx$

(With a NINT tolerance of 0.00001, we found the values to be 1.95, 1542.87, and 27.39 on our graphing calculator. All three took us a total of 80 seconds including keystroking and some shortcuts. By now, we hope that you know some calculator shortcuts and have been able to share them with others.)

Exercises 5.6

Evaluate each indefinite integral in Exercises 1–10 by using the given substitution to reduce the integral to a standard form. Support by graphing the antiderivative and $y = \text{NINT}\,(f(t), 0, x)$ in the same viewing window.

1. $\displaystyle\int \sin 3x\, dx,\, u = 3x$

2. $\displaystyle\int x \sin (2x^2)\, dx,\, u = 2x^2$

3. $\displaystyle\int \sec 2x \tan 2x\, dx,\, u = 2x$

4. $\displaystyle\int \left(1 - \cos \frac{t}{2}\right)^2 \sin \frac{t}{2}\, dt,\, u = 1 - \cos \frac{t}{2}$

5. $\displaystyle\int 28(7x - 2)^3\, dx,\, u = 7x - 2$

6. $\displaystyle\int 4x^3(x^4 - 1)^2\, dx,\, u = x^4 - 1$

7. $\displaystyle\int \frac{9r^2\, dr}{\sqrt{1 - r^3}},\, u = 1 - r^3$

8. $\displaystyle\int 12(y^4 + 4y^2 + 1)^2(y^3 + 2y)\, dy,\, u = y^4 + 4y^2 + 1$

9. $\displaystyle\int \csc^2 2\theta \cot 2\theta\, d\theta$

   a) Using $u = \cot 2\theta$ b) Using $u = \csc 2\theta$

10. $\displaystyle\int \frac{dx}{\sqrt{5x}}$

   a) Using $u = 5x$ b) Using $u = \sqrt{5x}$

Evaluate each definite integral in Exercises 11–18. Support with a NINT computation.

11. $\displaystyle\int_{0}^{1/2} \frac{dx}{(2x + 1)^3}$

12. $\displaystyle\int_{0}^{1} \sqrt{5x + 4}\, dx$

13. $\displaystyle\int_{0}^{\pi/6} \frac{\sin 2x}{\cos^2 2x}\, dx$

14. $\displaystyle\int_{\pi/6}^{\pi/2} \sin^2 \theta \cos \theta\, d\theta$

15. $\displaystyle\int_{-1}^{1} x \sqrt{1 - x^2}\, dx$

16. $\displaystyle\int_{1}^{4} \frac{dy}{2\sqrt{y}(1 + \sqrt{y})^2}$

17. $\displaystyle\int_{-\pi/2}^{\pi/2} \frac{\cos x}{(2 + \sin x)^2}\, dx$

18. $\displaystyle\int_{\pi^2/4}^{\pi^2} \frac{\sin \sqrt{x}}{\sqrt{x}}\, dx$

Evaluate each indefinite integral in Exercises 19–28. Support by comparing the graph of NDER of your answer with the graph of the integrand.

19. $\displaystyle\int \frac{dx}{(1-x)^2}$

20. $\displaystyle\int \frac{4y}{\sqrt{2y^2+1}}\,dy$

21. $\displaystyle\int \sec^2(x+2)\,dx$

22. $\displaystyle\int \sec^2\left(\frac{x}{4}\right)dx$

23. $\displaystyle\int 8r(r^2-1)^{1/3}\,dr$

24. $\displaystyle\int x^4(7-x^5)^3\,dx$

25. $\displaystyle\int \sec\left(\theta+\frac{\pi}{2}\right)\tan\left(\theta+\frac{\pi}{2}\right)d\theta$

26. $\displaystyle\int \sqrt{\tan x}\,\sec^2 x\,dx$

27. $\displaystyle\int \frac{6x^3}{\sqrt[4]{1+x^4}}\,dx$

28. $\displaystyle\int (s^3+2s^2-5s+6)^2(3s^2+4s-5)\,ds$

Evaluate each definite integral in Exercises 29–47 by an analytic method of your choice.

29. a) $\displaystyle\int_0^3 \sqrt{y+1}\,dy$ **b)** $\displaystyle\int_{-1}^0 \sqrt{y+1}\,dy$

30. a) $\displaystyle\int_0^1 r\sqrt{1-r^2}\,dr$ **b)** $\displaystyle\int_{-1}^1 r\sqrt{1-r^2}\,dr$

31. a) $\displaystyle\int_0^{\pi/4} \tan x\,\sec^2 x\,dx$ **b)** $\displaystyle\int_{-\pi/4}^0 \tan x\,\sec^2 x\,dx$

32. a) $\displaystyle\int_0^1 x^3(1+x^4)^3\,dx$ **b)** $\displaystyle\int_{-1}^1 x^3(1+x^4)^3\,dx$

33. a) $\displaystyle\int_0^1 \frac{x^3}{\sqrt{x^4+9}}\,dx$ **b)** $\displaystyle\int_{-1}^0 \frac{x^3}{\sqrt{x^4+9}}\,dx$

34. a) $\displaystyle\int_{-1}^1 \frac{x}{(1+x^2)^2}\,dx$ **b)** $\displaystyle\int_0^1 \frac{x}{(1+x^2)^2}\,dx$

35. a) $\displaystyle\int_0^{\sqrt{7}} x(x^2+1)^{1/3}\,dx$ **b)** $\displaystyle\int_{-\sqrt{7}}^0 x(x^2+1)^{1/3}\,dx$

36. a) $\displaystyle\int_0^{\pi} 3\cos^2 x\,\sin x\,dx$ **b)** $\displaystyle\int_{2\pi}^{3\pi} 3\cos^2 x\,\sin x\,dx$

37. a) $\displaystyle\int_0^{\pi/6} (1-\cos 3x)\,\sin 3x\,dx$

b) $\displaystyle\int_{\pi/6}^{\pi/3} (1-\cos 3x)\,\sin 3x\,dx$

38. a) $\displaystyle\int_0^{\sqrt{3}} \frac{4x}{\sqrt{x^2+1}}\,dx$ **b)** $\displaystyle\int_{-\sqrt{3}}^{\sqrt{3}} \frac{4x}{\sqrt{x^2+1}}\,dx$

39. a) $\displaystyle\int_0^{2\pi} \frac{\cos x}{\sqrt{2+\sin x}}\,dx$ **b)** $\displaystyle\int_{-\pi}^{\pi} \frac{\cos x}{\sqrt{2+\sin x}}\,dx$

40. a) $\displaystyle\int_{-\pi/2}^0 \frac{\sin x}{(3+\cos x)^2}\,dx$ **b)** $\displaystyle\int_0^{\pi/2} \frac{\sin x}{(3+\cos x)^2}\,dx$

41. $\displaystyle\int_0^1 \sqrt{t^5+2t}\,(5t^4+2)\,dt$

42. $\displaystyle\int_0^3 t\sqrt{1+t}\,dt$

43. $\displaystyle\int_0^{\pi/2} \cos^3 2x\,\sin 2x\,dx$

44. $\displaystyle\int_{-\pi/4}^{\pi/4} \tan^2 x\,\sec^2 x\,dx$

45. $\displaystyle\int_0^{\pi} \frac{8\sin t}{\sqrt{5-4\cos t}}\,dt$

46. $\displaystyle\int_0^{\pi/4} (1-\sin 2t)^{3/2}\cos 2t\,dt$

47. $\displaystyle\int_0^1 15x^2\sqrt{5x^3+4}\,dx$

48. Do analytically and support with a NINT computation:

$$\int_0^1 (y^3+6y^2-12y+5)(y^2+4y-4)\,dy.$$

Find the total areas of each shaded region in Exercises 49 and 50.

49.

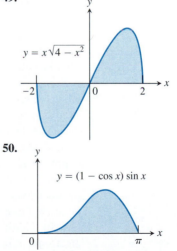

$y = x\sqrt{4-x^2}$

50.

$y = (1-\cos x)\sin x$

Solve each initial value problem in Exercises 51–54.

51. $\dfrac{ds}{dt} = 24t(3t^2-1)^3,\ s=0$ when $t=0$

52. $\dfrac{dy}{dx} = 4x(x^2+8)^{-1/3},\ y=0$ when $x=0$

53. $\dfrac{ds}{dt} = 6\sin(t+\pi),\ s=0$ when $t=0$

54. $\dfrac{d^2s}{dt^2} = -4\sin\left(2t-\dfrac{\pi}{2}\right),\ \dfrac{ds}{dt}=100$ and $s=0$ when $t=0$

Sequences of Substitutions

If you do not know what substitution to make, try reducing the integral step by step, using a trial substitution to simplify the integral a bit, then another to simplify it some more. You will see what we mean if you try the sequences of substitutions in Exercises 55 and 56.

55. $\displaystyle\int_0^{\pi/4} \frac{18 \tan^2 x \sec^2 x}{(2 + \tan^3 x)^2} \, dx$

  a) $u = \tan x$, followed by $v = u^3$, then by $w = 2 + v$

  b) $u = \tan^3 x$ followed by $v = 2 + u$

  c) $u = 2 + \tan^3 x$

56. $\displaystyle\int \sqrt{1 + \sin^2 (x - 1)} \, \sin (x - 1) \cos (x - 1) \, dx$

  a) $u = x - 1$, followed by $v = \sin u$, then by $w = 1 + v^2$

  b) $u = \sin (x - 1)$ followed by $v = 1 + u^2$

  c) $u = 1 + \sin^2 (x - 1)$

57. It looks as if we can integrate $2 \sin x \cos x$ with respect to x in at least three different ways:

a) $\displaystyle\int 2 \sin x \cos x \, dx = \int 2u \, du \quad \left(\begin{matrix} u = \sin x, \\ du = \cos x \, dx \end{matrix} \right)$

$$= u^2 + C_1$$

$$= \sin^2 x + C_1;$$

b) $\displaystyle\int 2 \sin x \cos x \, dx = \int - 2u \, du \quad \left(\begin{matrix} u = \cos x, \\ du = -\sin x \, dx, \\ -du = \sin x \, dx \end{matrix} \right)$

$$= -u^2 + C_2$$

$$= -\cos^2 x + C_2;$$

c) $\displaystyle\int 2 \sin x \cos x \, dx = \int \sin 2x \, dx \quad \left(\begin{matrix} 2 \sin x \cos x \\ = \sin 2x \end{matrix} \right)$

$$= -\frac{\cos 2x}{2} + C_3.$$

Can all three integrations be correct? Graph each antiderivative with $C_i = 0, i = 1, 2, 3$. Explain.

5.7 _____

Numerical Integration: The Trapezoidal Rule and Simpson's Method

Most modern graphing calculators have powerful and accurate numerical differentiation and integration algorithms (like NDER and NINT) built in. Should you come to rely seriously on an electronic aid, then you may find it necessary to learn more thoroughly how it performs its algorithms. You would likely need to understand the limitations of the algorithms. You might even find it necessary to do some formal work in *numerical analysis*.

The Trapezoidal Rule and Simpson's Rule are two relatively simple numerical techniques that enable us to estimate an integral's value with a good degree of accuracy. As we study them, you should note the routine procedures that are required for the approximations, whether they are limited in number in order to be done by pencil on paper or they are to be programmed to be done electronically.

In addition to giving us some background on the internal workings of electronic utilities, these rules also serve as an introduction to *error analysis*—procedures by which we know how much a numerically generated answer could differ from the exact answer.

The Trapezoidal Rule

The Trapezoidal Rule for estimating the value of

$$\int_a^b f(x) \, dx$$

is based on approximating the region between the graph of f and the x-axis with n trapezoids of equal width (Fig. 5.37). The trapezoids have the common base length $h = (b-a)/n$, and the side of each trapezoid runs from the x-axis up (or down) to the curve. Adding the areas of the n trapezoids gives

$$T = \frac{1}{2}(y_0 + y_1)h + \frac{1}{2}(y_1 + y_2)h + \cdots + \frac{1}{2}(y_{n-2} + y_{n-1})h + \frac{1}{2}(y_{n-1} + y_n)h$$

$$= h\left(\frac{1}{2}y_0 + y_1 + y_2 + \cdots + y_{n-1} + \frac{1}{2}y_n\right)$$

$$= \frac{h}{2}(y_0 + 2y_1 + 2y_2 + \cdots + 2y_{n-1} + y_n),$$

where

$$y_0 = f(a), \quad y_1 = f(x_1), \quad \ldots, \quad y_{n-1} = f(x_{n-1}), \quad y_n = f(b).$$

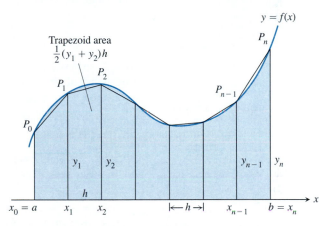

5.37 The Trapezoidal Rule approximates short stretches of curve with line segments. To estimate the area under the curve, we add the areas of the trapezoids formed by joining the ends of these segments to the x-axis.

The Trapezoidal Rule says: Use T to estimate the integral of f from a to b.

THE TRAPEZOIDAL RULE

To approximate

$$\int_a^b f(x)\,dx,$$

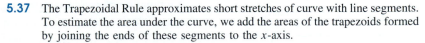

use

$$T = \frac{h}{2}(y_0 + 2y_1 + 2y_2 + \cdots + 2y_{n-1} + y_n) \tag{1}$$

with $[a, b]$ partitioned into n subinervals of length $h = (b-a)/n$.

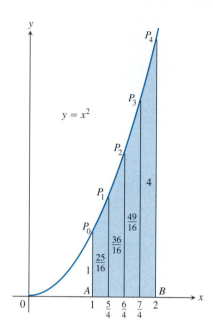

5.38 For a trapezoidal approximation of the area under the graph of $y = x^2$ from $x = 1$ to $x = 2$, we partition the interval $[1, 2]$ into four equal subintervals of length 1/4, the height of each trapezoid. The base lengths of the trapezoids are shown beside each (vertical) base. The sum of the trapezoid areas is 75/32 and is a slight overestimate of the area under the curve.

TABLE 5.6

x	$y = x^2$
1	1
$\dfrac{5}{4}$	$\dfrac{25}{16}$
$\dfrac{6}{4}$	$\dfrac{36}{16}$
$\dfrac{7}{4}$	$\dfrac{49}{16}$
2	4

EXAMPLE 1

Use the Trapezoidal Rule with $n = 4$ to estimate $\displaystyle\int_1^2 x^2\, dx$. Compare the estimate with the value of NINT $(x^2, 1, 2)$ and the exact value.

Solution To find the trapezoidal approximation, we draw a sketch (Fig. 5.38) to help us keep track of the information. We partition $[1, 2]$ into four subintervals of equal length and list the values of $y = x^2$ at each partition point (Table 5.6). The sketch suggests that the trapezoids should give us an overestimate of the area because each trapezoid contains slightly more than the corresponding strip under the curve.

We evaluate Eq. (1) with $n = 4$ and $h = 1/4$ to find

$$T = \frac{h}{2}(y_0 + 2y_1 + 2y_2 + 2y_3 + y_4)$$

$$= \frac{1}{8}\left(1 + 2\left(\frac{25}{16}\right) + 2\left(\frac{36}{16}\right) + 2\left(\frac{49}{16}\right) + 4\right) = \frac{75}{32} = 2.34375.$$

The value of NINT $(x^2, 1, 2) = 2.333$. The value of the integral is

$$\int_1^2 x^2\, dx = \frac{x^3}{3}\Bigg]_1^2 = \frac{8}{3} - \frac{1}{3} = \frac{7}{3}.$$

The T approximation overestimates the area by about half a percent of its true value of 7/3. The percent error is $(2.34375 - 7/3)/(7/3) = 0.00446$.

◼

Simpson's Rule

Simpson's Rule is based on approximating curves with parabolic arcs instead of line segments. The shaded area under the parabola in Fig. 5.39 is

$$A_p = \frac{h}{3}(y_0 + 4y_1 + y_2). \tag{2}$$

Applying this formula successively along a continuous curve $y = f(x)$ from $x = a$ to $x = b$ leads to an estimate of $\int_a^b f(x)\, dx$ that is generally more accurate than T for a given value of h.

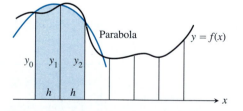

5.39 Simpson's Rule approximates short stretches of curve with parabolas.

We can derive the formula for A_p in the following way. To simplify the algebra, we use the coordinate system shown in Fig. 5.40. The area under the parabola is the same no matter where the y-axis is, as long as we preserve the vertical scale. The parabola has an equation of the form

$$y = Ax^2 + Bx + C,$$

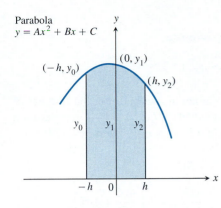

5.40 By integrating from $-h$ to h, the shaded area is found to be

$$A_p = \frac{h}{3}(y_0 + 4y_1 + y_2).$$

SIMPSON'S ONE-THIRD RULE

The rule

$$\text{Area} = \frac{h}{3}(y_0 + 4y_1 + y_2)$$

for calculating area by replacing curves by parabolas (Eq. (2) in the text) was discovered long before Thomas Simpson (1720–1761) was born. It is another of history's beautiful quirks that one of the ablest mathematicians of eighteenth-century England is remembered not for his successful texts and his contributions to mathematical analysis but for a rule that was never his, that he never laid claim to, and that bears his name only because he happened to mention it in one of his books.

---

so the area under it from $x = -h$ to $x = h$ is

$$A_p = \int_{-h}^{h} (Ax^2 + Bx + C)\, dx = \frac{Ax^3}{3} + \frac{Bx^2}{2} + Cx \Big]_{-h}^{h} \tag{3}$$

$$= \frac{2Ah^3}{3} + 2Ch = \frac{h}{3}(2Ah^2 + 6C).$$

Since the curve passes through the three points $(-h, y_0)$, $(0, y_1)$, and (h, y_2), we also have

$$y_0 = Ah^2 - Bh + C, \qquad y_1 = C, \qquad y_2 = Ah^2 + Bh + C,$$

from which we obtain

$$y_0 + 4y_1 + y_2 = Ah^2 - Bh + C + 4C + Ah^2 + Bh + C \tag{4}$$

$$= 2Ah^2 + 6C.$$

Substituting the result from Eq. (4) into Eq. (3) gives

$$A_p = \frac{h}{3}(y_0 + 4y_1 + y_2).$$

Simpson's Rule follows from partitioning $[a, b]$ into an even number of subintervals of equal length h, applying the formula for A_p to successive interval pairs, and adding the results.

SIMPSON'S RULE

To approximate

$$\int_a^b f(x)\, dx,$$

use

$$S = \frac{h}{3}(y_0 + 4y_1 + 2y_2 + 4y_3 + \cdots + 2y_{n-2} + 4y_{n-1} + y_n) \tag{5}$$

with $[a, b]$ partitioned into an *even number*, n, of subintervals of length $h = (b - a)/n$.

The y's in Eq. (5) are the values of $y = f(x)$ at the partition points

$$x_0 = a, \ x_1 = a + h, \ x_2 = a + 2h, \ \ldots, \ x_{n-1} = a + (n-1)h, \ b = x_n$$

(See Fig. 5.41).

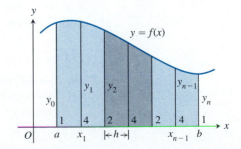

5.41 The y's in Eq. (5) are the values of f at the points of subdivision.

TABLE 5.7

x	$y = 5x^4$
0	0
$\dfrac{1}{4}$	$\dfrac{5}{256}$
$\dfrac{2}{4}$	$\dfrac{80}{256}$
$\dfrac{3}{4}$	$\dfrac{405}{256}$
1	5

EXAMPLE 2

Use Simpson's Rule with $n = 4$ to approximate

$$\int_0^1 5x^4\, dx.$$

Solution   To find the Simpson approximation, we partition the interval of integration into four subintervals and evaluate $f(x) = 5x^4$ at the partition points (Table 5.7). We then evaluate Eq. (5) with $n = 4$ and $h = 1/4$:

$$S = \frac{h}{3}(y_0 + 4y_1 + 2y_2 + 4y_3 + y_4)$$

$$= \frac{1}{12}\left(0 + 4\left(\frac{5}{256}\right) + 2\left(\frac{80}{256}\right) + 4\left(\frac{405}{256}\right) + 5\right)$$

$$= 1.00260416667.$$

We can calculate the exact value of the integral directly:

$$\int_0^1 5x^4\, dx = x^5\Big]_0^1 = 1. \qquad \text{Four subintervals gives an overestimate with error less than 0.3 percent.}$$

The error in the estimate is $1.002604 - 1 = 0.0026041$.   ≡

Error Analysis

The magnitudes of error in the Trapezoidal Rule and Simpson's Rule approximations,

$$E_T = T - \int_a^b f(x)\, dx \qquad \text{and} \qquad E_S = S - \int_a^b f(x)\, dx,$$

respectively, decrease as **step size** (the subinterval length) $h = (b - a)/n$ decreases. For small steps, the graph of f, the upper segments of the trapezoids, and the parabolic approximations closely resemble each other as the local straightness of each might suggest. The slight curvature present in the parabola, however, gives it an edge over the side of the trapezoid as an approximation device. This is reflected in the following two theorems from advanced calculus that put a bound on the error possible from each method.

EXPLORATION BIT

Predict the error in using Simpson's Rule to compute

$$\int_l^u \left(ax^2 + bx + c\right) dx.$$

Explain.

> **The Error Estimate for the Trapezoidal Rule**
>
> If f'' is continuous and M is any upper bound for the values of $|f''|$ on $[a, b]$, then
>
> $$|E_T| \le \frac{b - a}{12} h^2 M. \qquad (6)$$
>
> **The Error Estimate for Simpson's Rule**
> If $f^{(4)}$ is continuous and M is any upper bound for the values of $|f^{(4)}|$ on $[a, b]$, then
>
> $$|E_S| \le \frac{b - a}{180} h^4 M. \qquad (7)$$

Note that to apply these error estimates, we need f'' continuous for the Trapezoidal Rule and $f^{(4)}$, the fourth derivative, continuous for Simpson's Rule. Also, we need to find an upper bound, M, for $|f''|$ or $|f^{(4)}|$. Although theory tells us that there will always be a smallest value M, in practice we can hardly ever find it. Instead, we can find a "safe" value analytically or graphically and go on to *estimate* $|E_T|$ or $|E_S|$ from there. This may seem a bit sloppy, but it works because to make $|E_T|$ or $|E_S|$ small for a given M, we simply make h as small as necessary.

EXAMPLE 3

Find an upper bound for the error in the trapezoid approximation found in Example 1 for

$$\int_1^2 x^2 \, dx.$$

Solution   We first find an upper bound M for the magnitude of the second derivative of $f(x) = x^2$ on the interval $1 \le x \le 2$. Since $f''(x) = 2$ for all x, we may safely take $M = 2$. With $b - a = 1$ and $h = 1/4$, Eq. (6) gives

$$|E_T| \le \frac{b-a}{12} h^2 M = \frac{1}{12} \left(\frac{1}{4} \right)^2 (2) = \frac{1}{96}.$$

This is precisely what we find when we subtract $T = 75/32$ from $\int_1^2 x^2 \, dx = 7/3$, since $|7/3 - 75/32| = |-1/96| = 1/96$. Here we are able to give the error *exactly*, but this is exceptional.   ≡

EXPLORATION 1

Bounding the Error, Part 1

Suppose the Trapezoidal Rule has been used to estimate the value of $\int_0^1 f(x) \, dx$ where $f(x) = x \sin x$, $n = 10$, and $h = (b - a)/n = 1/10$.

1. Show $|E_T| \le \dfrac{1}{1200} M$ for *any* upper bound M of $|f''|$.

2. Find a value for M in two ways (the values need not be the same):

   a) Calculate f'' analytically. Then use the triangle inequality to find a value for M.

   b) Let $y_1 = x \sin x$, $y_2 = \text{NDER}(y_1, x)$, and $y_3 = \text{abs}(\text{NDER}(y_2, x))$. GRAPH y_3 on $[0, 1]$, and select a value for M different from the value for M in part 2(a). Give a visual argument why $|f''| \le M$.

3. Compare the two upper bounds that you get for $|E_T|$ for your two M values from parts 2(a) and 2(b).

4. Let $n = 100$ and $h = 1/100$. Find upper bounds for $|E_T|$ using the values found for M in parts 2(a) and 2(b). Compare the results of parts 3 and 4 and discuss how the integral approximation is affected by the choice of h compared to the choice of M.   ✿

THE TRIANGLE INEQUALITY

The triangle inequality says that for all real numbers, a and b,

$$|a + b| \le |a| + |b|.$$

In Exploration 1, part 2(a), you could use the triangle inequality as follows:

$$f''(x) = 2 \cos x - x \sin x.$$

Therefore,

$$|f''(x)| = |2 \cos x - x \sin x|$$

$$\le |2 \cos x| + |-x \sin x|,$$

and proceed to find an upper bound M of $|f''(x)|$.

EXAMPLE 4

Find an upper bound for the error in the Simpson approximation found in Example 2 for

$$\int_0^1 5x^4 \, dx.$$

Solution   For an upper bound for the error, we first find an upper bound M for the magnitude of the fourth derivative of $f(x) = 5x^4$ on the interval $0 \le x \le 1$. Since the fourth derivative has the constant value $f^{(4)}(x) = 120$, we may safely take $M = 120$. With $(b - a) = 1$ and $h = 1/4$, Eq. (7) then gives

> Compare this upper bound for the error with the actual error 0.002604 in Example 2.

$$|E_S| \le \frac{b-a}{180} h^4 M = \frac{1}{180} \left(\frac{1}{4}\right)^4 (120) = \frac{1}{384} < 0.0026042.$$

Controlling the Error in an Integral Approximation

Exploration 1 suggests that the choice of n, the number of subintervals, gives us better control over the accuracy of an integral approximation than does the bound M on a derivative. The following example shows how we can actually exercise this control.

THE NATURAL LOG OF 2

In Chapter 7, we will see that $\int_1^2 (1/x)\, dx$ is ln 2 (the natural log of 2). Example 5 shows that we can use the Trapezoidal Rule to approximate ln 2 with error of magnitude less than 10^{-4} when we use a partition that gives 41 subintervals.

EXAMPLE 5

How many subintervals (steps) should be used in the Trapezoidal Rule to approximate the integral

$$\int_1^2 \frac{1}{x} \, dx$$

to guarantee an error of absolute value less than 10^{-4}?

Solution   To determine n, the number of subintervals, we use Eq. (6) with

$$b - a = 2 - 1 = 1, \qquad h = \frac{b-a}{n} = \frac{1}{n},$$

$$f''(x) = \frac{d^2}{dx^2}(x^{-1}) = 2x^{-3} = \frac{2}{x^3}.$$

Then

$$|E_T| \le \frac{b-a}{12} h^2 \left(\max |f''(x)|\right) = \frac{1}{12} \left(\frac{1}{n}\right)^2 \left(\max \left|\frac{2}{x^3}\right|\right),$$

where $\max |2/x^3|$ refers to the maximum value of $|2/x^3|$ on the interval $[1, 2]$.

This is one of the rare cases in which we can find the exact value of $\max |f''|$. On $[1, 2]$, $y = 2/x^3$ decreases steadily from a maximum of $y = 2$ to a minimum of $y = 1/4$ (Fig. 5.42). Therefore,

$$|E_T| \le \frac{1}{12} \left(\frac{1}{n}\right)^2 \cdot 2 = \frac{1}{6n^2}.$$

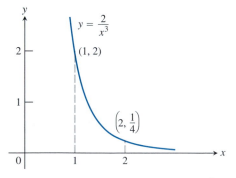

5.42 The continuous function $y = 2/x^3$ has its maximum value on $[1, 2]$ at $x = 1$.

EXPLORATION BIT

Use the TRAP procedure with 40 subintervals to show that

$$\int_1^2 \frac{1}{x}\,dx = 0.69318624$$

so that the actual error is 3.91×10^{-5}. (Consult the *Resource Manual*.)

The error's absolute value will therefore be less than 10^{-4} if

$$\frac{1}{6n^2} < 10^{-4}, \quad \frac{10^4}{6} < n^2, \quad \frac{100}{\sqrt{6}} < n, \quad \text{or} \quad 40.83 < n.$$

The first integer beyond 40.83 is $n = 41$. With $n = 41$ subintervals, we can guarantee calculating $\int_1^2 (1/x)\,dx$ with an error of magnitude less than 10^{-4}. Any larger n will work, too.   ▬

EXPLORATION 2

Bounding the Error, Part 2

In Exploration 1, we used a graphing utility to find a bound M on a derivative in order to bound the error resulting from the Trapezoidal Rule. We could also use a grapher to bound an error estimate directly. Consider

$$\int_{-1}^{1} \sin(x^2)\,dx.$$

There is no explicit formula for an antiderivative of $\sin(x^2)$. So to get trapezoidal and Simpson's approximations for this definite integral, we used our graphing calculator and loaded the procedures TRAP and SIMP from the *Resource Manual*. We let $n = 20$ so that $h = 1/10$ and got

$$\int_{-1}^{1} \sin(x^2)\,dx = 0.6223416 \qquad \text{Trapezoid, } n = 20$$

and

$$\int_{-1}^{1} \sin(x^2)\,dx = 0.6205205. \qquad \text{Simpson, } n = 20$$

How accurate are these results? To answer, try the following:

1. For $f(x) = \sin(x^2)$, calculate $f'(x)$ and $f''(x)$.
2. Let $y_1 = f''(x)$. Let y_2 be the *trapezoidal error function*,

$$y_2 = E_T(x) = \frac{b-a}{12} h^2 |y_1|.$$

   GRAPH y_2, and find its maximum value on $[-1, 1]$. This gives the *smallest* maximum possible error using the Trapezoidal Rule.
3. Let $y_3 = \text{NDER}(y_1, x)$ and $y_4 = \text{NDER}(y_3, x)$. Then let y_5 be the *Simpson's error function*,

$$y_5 = E_S(x) = \frac{b-a}{180} h^4 |y_4|.$$

   GRAPH y_5, and find its maximum value on $[-1, 1]$. This gives the *smallest* maximum possible error using Simpson's Rule.
4. Compare the results of parts 2 and 3.   ⌘

Because of how its algorithm is constructed, it is not possible to use NDER beyond NDER2. Hence to compute the error bound for Simpson's Rule we evaluated f'' directly, then used NDER2 to approximate $f^{(4)}$. In the exercises you will be asked to find $f^{(4)}$ analytically and compare graphs of $E_S(x)$, one using NDER2 (f'', x) and the other using the exact form of $f^{(4)}$.

Which Rule Gives Better Results?

The answer lies in the error-control formulas for the two rules:

$$|E_T| \le \frac{b-a}{12}h^2M, \qquad |E_S| \le \frac{b-a}{180}h^4M.$$

The M's of course mean different things, the first being an upper bound on $|f''|$ and the second an upper bound on $|f^{(4)}|$. But there is more than that going on here. The factor $(b-a)/180$ in the Simpson formula is one-fifteenth of the factor $(b-a)/12$ in the trapezoidal formula. More important still, the Simpson formula has an h^4, while the trapezoidal formula has only an h^2. If h is one-tenth, then h^2 is a hundredth, but h^4 is a ten-thousandth! If both M's are 1, for example, and $b-a=1$, then, with $h=1/10$,

while

$$|E_T| \le \frac{1}{12}\left(\frac{1}{10}\right)^2 \cdot 1 \le \frac{1}{1200},$$

$$|E_S| \le \frac{1}{180}\left(\frac{1}{10}\right)^4 \cdot 1 \le \frac{1}{1,800,000} = \frac{1}{1500} \cdot \frac{1}{1200}.$$

For roughly the same amount of computational effort, we get better accuracy with Simpson's Rule—at least in this case.

The h^2 versus h^4 is the key. If h is less than 1, then h^4 can be significantly smaller than h^2. On the other hand, if h equals 1, there is no difference between h^2 and h^4. If h is greater than 1, the value of h^4 may be significantly larger than the value of h^2. In the latter two cases, the error-control formulas offer little help. We have to go back to the geometry of the curve $y=f(x)$ to see whether line segments or parabolas, if either, are going to give the results we want.

> **CAUTION**
>
> Although decreasing the step size h reduces the error in these integral approximation rules in theory, it may fail to do so in practice. When h is very small, say, $h=10^{-5}$, the rounding errors in the arithmetic may accumulate to such an extent that the error formulas no longer describe what is going on. Shrinking h below a certain size can actually make things worse rather than better. While this will not be an issue in the present book, you should consult a text on numerical analysis for alternative methods should you ever have problems with rounding errors.

Working with Numerical Data

Simpson's and the Trapezoidal Rules enable us to calculate integrals with reasonable accuracy from tables of function values. This is particularly handy when the only information we have about a function is a set of specific values measured in the laboratory or the field.

EXAMPLE 6

A town wants to drain and fill the small swamp shown in Fig. 5.43. The swamp averages 5 ft deep. About how many cubic yards of dirt will it take to fill the area after the swamp is drained?

Solution   To calculate the volume of the swamp, we estimate the surface area and multiply by 5. To estimate the area, we use Simpson's Rule with $h=20$ ft and the y's equal to the distances measured across the swamp, as shown in Fig. 5.43:

$$S = \frac{h}{3}(y_0 + 4y_1 + 2y_2 + 4y_3 + 2y_4 + 4y_5 + y_6)$$

$$= \frac{20}{3}(146 + 488 + 152 + 216 + 80 + 120 + 13) = 8100.$$

The volume is about $(8100)(5) = 40,500 \text{ ft}^3$ or 1500 yd^3.

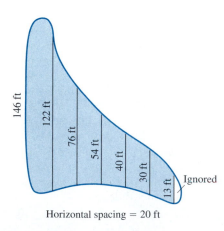

146 ft 122 ft 76 ft 54 ft 40 ft 30 ft 13 ft Ignored

Horizontal spacing = 20 ft

5.43   The swamp in Example 6.

Computer Programs

Table 5.8 shows TRAP and SIMP outputs for $\int_1^2 (1/x)\, dx$ (which is equal to ln 2) from our graphing calculator. For comparison, $\ln 2 = 0.69314718056$.

Notice how Simpson's Rule improves on the Trapezoidal Rule. The Simpson approximation for $n = 50$ rounds accurately to seven places instead of three. The approximation for $n = 100$ is accurate to nine decimal places (billionths)!

TABLE 5.8 **Trapezoidal Rule Approximations (T_n) and Simpson's Rule Approximations (S_n) of ln 2 $= \int_1^2 (1/x)\, dx$.**

n	T_n	|Error| less than ...	S_n	|Error| less than ...
10	.6937714032	.0006242226	.6931502307	.0000030501
20	.6933033818	.0001562012	.6931473747	.0000001941
30	.6932166154	.0000694348	.6931472190	.0000000385
40	.6931862400	.0000390594	.6931471927	.0000000122
50	.6931721793	.0000249988	.6931471856	.0000000050
100	.6931534305	.0000062499	.6931471809	.0000000003

EXPLORATION 3

Summing Sequences

Some graphing calculators or computer software can generate sequences and sum them. Try this if your graphing calculator or software allows. Type the following on the home screen:

$$10 \to N:\ \text{seq}(2/(1 + K\,(1/10)),\, K, 0, N, 1).$$

Press ENTER , and confirm that the terms of the sequence are almost the terms in the $N = 10$ trapezoidal approximation of $\int_1^2 1/x\, dx$, namely,

$$\frac{H}{2}(y_0 + 2y_1 + 2y_2 + \cdots + 2y_9 + y_{10}).$$

Now show that if the *sum* of

$$\text{seq}\,(2/(1 + K * H),\, K, 0, N, 1)$$

is denoted by S, then

$$\frac{H}{2}(y_0 + 2y_1 + 2y_2 + \cdots + 2y_{n-1} + y_n) = \frac{H}{2}(S - 1 - 0.5).$$

Next, enter this on the home screen:

$$10 \to N:\ 1/N \to H: (H/2)(-1-0.5+\text{sum seq}(2/(1+K*H),\, K, 0, N, 1)).$$

Press ENTER, and confirm that you obtain the trapezoid estimate for $\int_1^2 1/x \, dx$ when $N = 10$. Repeat by editing and storing 20, 30, 40, and 50 to N, respectively. Compare with Table 5.8. Also compare with $\ln 2$. Explain. ⚘

We close by showing you the values (Table 5.9) that we found for $\int_1^5 (\sin x)/x \, dx$ by six different calculator methods. It is known that $\int_1^5 (\sin x)/x \, dx = 0.603848$ is accurate to six decimal places, so both Simpson's method with 50 steps and NINT (with a tolerance of 10^{-5}) give results accurate to 10^{-6}.

TABLE 5.9   Approximations of $\int_1^5 \sin x / x \, dx$.

Method	Subintervals	Value
LRAM	50	0.6453898
RRAM	50	0.5627293
MRAM	50	0.6037425
TRAP	50	0.6040595
SIMP	50	0.6038481
NINT	Tol = 0.00001	0.6038482

Exercises 5.7

Use (a) the Trapezoidal Rule and (b) Simpson's Rule to approximate each integral in Exercises 1–6 with $n = 4$. Then (c) find the integral's exact value for comparison.

1. $\int_0^2 x \, dx$ **2.** $\int_0^2 x^2 \, dx$

3. $\int_0^2 x^3 \, dx$ **4.** $\int_1^2 \frac{1}{x^2} \, dx$

5. $\int_0^4 \sqrt{x} \, dx$ **6.** $\int_0^\pi \sin x \, dx$

Use (a) TRAP, (b) SIMP, and (c) the three RAMs with $n = 10$, 100, and 1000 to estimate each integral in Exercises 7–12. Support your results with a NINT computation.

7. $\int_{-1}^3 e^{-x^2} \, dx$ **8.** $\int_2^5 \sqrt{x^2 - 2} \, dx$

9. $\int_{-5}^5 x \sin x \, dx$ **10.** $\int_{-2}^2 \sin x^2 \, dx$

11. $\int_{3\pi/4}^{4.5} \frac{\tan x}{x} \, dx$ **12.** $\int_1^{2\pi} \frac{2 \sin x}{x} \, dx$

13. Use Eq. (6) to estimate the error when you use the Trapezoidal Rule with $n = 10$ to estimate the value of

$$\ln 2 = \int_1^2 \frac{1}{x} \, dx.$$

Compare your answer with the result in Table 5.8.

14. Use Eq. (7) to estimate the error when you use Simpson's Rule with $n = 10$ to estimate the value of

$$\ln 2 = \int_1^2 \frac{1}{x} \, dx.$$

Compare your answer with the result in Table 5.8.

In Exercises 15–20, estimate the minimum number of subintervals needed to approximate each integral with an error of absolute value less than 10^{-4} by (a) the Trapezoidal Rule and (b) Simpson's Rule. Check your results using TRAP and SIMP, and compare with the exact values.

15. $\int_0^2 x \, dx$ **16.** $\int_0^2 x^2 \, dx$ **17.** $\int_0^2 x^3 \, dx$

18. $\int_{1}^{2} \frac{1}{x^2} \, dx$ **19.** $\int_{1}^{4} \sqrt{x} \, dx$ **20.** $\int_{0}^{\pi} \sin x \, dx$

Use Simpson's Rule with $n = 50$ and $n = 100$ to compute values of each integral in Exercises 21–24.

21. $\int_{-1}^{1} 2\sqrt{1 - x^2} \, dx$ The exact value is π.

22. $\int_{0}^{1} \sqrt{1 + x^4} \, dx$ A nonelementary integral that came up in Newton's research

23. $\int_{0}^{\pi/2} \frac{\sin x}{x} \, dx$ To avoid division by zero, you may have to start the integration at a small positive number like 10^{-6} instead of 0.

24. $\int_{0}^{\pi/2} \sin(x^2) \, dx$ An integral associated with the diffraction of light

25. a) Compute both the Trapezoidal Rule and Simpson's Rule estimate of $\int_{\pi/2}^{2\pi} (\sin x)/x \, dx$ for $n = 10$.

   b) Graph the error functions $y = E_T(x)$ and $y = E_S(x)$, and establish the smallest maximum error of each estimate. (See Exploration 2.)

   c) Repeat part (b) for $n = 20$.

   d) Repeat part (b) for $n = 50$.

   e) What is the *exact* value of the definite integral?

26. Repeat Exercise 25 with $\int_{1}^{\pi} x \sin x^2 \, dx$.

27. Repeat Exercise 25 with $\int_{0}^{2} \sqrt{1 + x^4} \, dx$.

28. Let $f(x) = \sin(x^2)$ and $y_1 = f^{(4)}(x)$, the exact fourth derivative of f. Let $y_2 = \text{NDER } 2 \, (f''(x), x)$. GRAPH y_1 and y_2. Compare function values at $x = -0.2$, 0.2, 0.4, and 0.6. What is the smallest maximum error in using y_2 to approximate y_1 in $[-1, 1]$? What does this imply about $E_S(x)$ computations when y_2 is used rather than y_1?

Let $T_n f$ and $S_n f$ denote the approximations of the integral of f using TRAP and SIMP, respectively, with n subintervals.

29. Let $f(x) = x^2 + \sin x$ on $[1, 2]$. Compute $\text{LRAM}_{50} f$, $\text{RRAM}_{50} f$, and $T_{50} f$. Is $2T_{50} f = \text{LRAM}_{50} f + \text{RRAM}_{50} f$? Explain.

30. Let $f(x) = x^3 - \cos x$ on $[1, 2]$. Compute $T_{50} f$, $\text{MRAM}_{25} f$, and $S_{50} f$. Is $S_{50} f = (\text{MRAM}_{25} f + 2T_{50} f)/3$? Explain.

31. Prove that, in general,

$$T_{10} f = \frac{\text{LRAM}_{10} f + \text{RRAM}_{10} f}{2}.$$

32. Prove that, in general,

$$S_{10} f = \frac{\text{MRAM}_5 f + 2T_{10} f}{3}.$$

33. As the fish-and-game warden of your town, you are responsible for stocking the town pond mapped out in the diagram below with fish before fishing season. The average depth of the pond is 20 ft. You plan to start the season with one

fish per 1000 ft³. You intend to have at least 25% of the opening day's fish population left at the end of the season. About what is the maximum number of licenses the town can sell if the average seasonal catch is 20 fish per license?

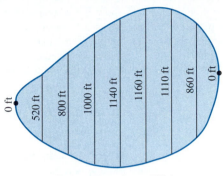

Horizontal spacing = 200 ft

34. The design of a new airplane requires a gasoline tank of constant cross-sectional area in each wing. A scale drawing of a cross-section is shown here. The tank must hold 5000 lb of gasoline that has a density of 42 lb/ft³. Estimate the length of the tank.

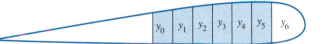

$y_0 = 1.5$ ft, $y_1 = 1.6$ ft, $y_2 = 1.8$ ft, $y_3 = 1.9$ ft,

$y_4 = 2.0$ ft, $y_5 = y_6 = 2.1$ ft,   Horizontal spacing = 1 ft

35. A vehicle's aerodynamic drag is determined in part by its cross-sectional area, and, all other things being equal, engineers try to make this area as small as possible. Use Simpson's Rule to estimate the cross-sectional area of James Worden's solar-powered Solectria car at M.I.T. from the diagram below.

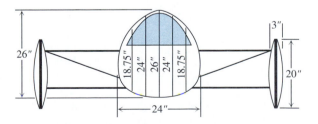

36. *The dye-dilution technique for measuring cardiac output.* Instead of measuring a patient's cardiac output with exhaled carbon dioxide, as in Exercise 26, Section 4.6, a doctor may

prefer to use the dye-dilution technique described here. You start by injecting 5–10 mg of dye into a main vein near the heart. The dye is drawn into the right side of the heart and then pumped through the lungs and out the left side of the heart into the aorta, where its concentration is measured each second as the blood flows past. The data in Table 5.10 and the plot in Fig. 5.44 show the response of a healthy, resting patient to an injection of 5.6 mg of dye.

The patient's cardiac output is calculated by dividing the area under the concentration curve into the number of milligrams of dye and multiplying the result by 60:

$$\text{Cardiac output} = \frac{\text{milligrams of dye}}{\text{area under curve}} \times 60. \qquad (8)$$

You can see why if you check the units in which these quantities are measured. The dye is in milligrams, the area is in (milligrams/liter) × seconds, and

$$\frac{\text{mg}}{\text{mg/L} \cdot \text{sec}} \cdot 60 = \text{mg} \cdot \frac{\text{L}}{\text{mg} \cdot \text{sec}} \cdot 60 = \frac{\text{L}}{\text{sec}} \cdot 60 = \frac{\text{L}}{\text{min}}.$$

a) Use the Trapezoidal Rule and the data in Table 5.10 to estimate the area under the concentration curve in Fig. 5.44.
b) Then use Eq. (8) to calculate the patient's cardiac output.

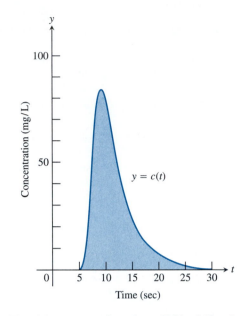

5.44 The dye concentrations from Table 5.10, plotted as a function of time t. The data points have been fitted with a smooth curve. Time is measured with $t = 0$ at the time of injection. The concentration readings are zero at the beginning, while the dye passes through the lungs. They then rise to a maximum at about $t = 9$ seconds and taper exponentially to zero at $t = 30$.

TABLE 5.10   Dye-Dilution Data for Exercise 36

Seconds after injection, t	Dye concentration (adjusted for recirculation), $c(t)$	Seconds after injection, t	Dye concentration (adjusted for recirculation), $c(t)$
1	0	16	18.5
2	0	17	14.5
3	0	18	11.5
4	0	19	9.1
5	0	20	7.3
6	1.5	21	5.7
7	38.0	22	4.5
8	67.0	23	3.6
9	80.0	24	2.8
10	73.0	25	2.3
11	61.0	26	1.8
12	48.0	27	1.4
13	36.0	28	1.1
14	29.0	29	0.9
15	23.0	30	0

Electric Energy Consumption

The Louisiana Power and Light Company tries to forecast the demand for electricity throughout the day so that it can have enough generators on line at any given time to carry the load. Boilers take a while to fire up, so the company has to know in advance what the load is going to be if service is not to be interrupted. Like all power companies, Louisiana Power measures electrical demand in kilowatt-hours (KWH on your electric bill). A 1500-watt space heater, running for 10 hours, for instance, uses 15,000 watt-hours or 15 kilowatt-hours of electricity. A kilowatt, like a horsepower, is a unit of power. A kilowatt-hour is a unit of energy, and energy is what Louisiana Power, despite its name, sells.

Table 5.11 and the graph in Fig. 5.45 show the results of a 1984 residential-load study of Louisiana Power residential customers with electric heating, for typical weekdays and weekend days in January. A typical customer used 56.60 kWh on a weekday that January and 53.46 kWh on a weekend day. At the time of the study, 53.60 kWh cost a customer $3.40.

TABLE 5.11 **Residential Electric Loads (Exercises 37 and 38)**

Hour of day	Weekday kW	Weekend kW
1	1.88	1.69
2	1.88	1.64
3	2.02	1.63
4	2.02	1.73
5	2.25	1.80
6	2.76	1.97
7	3.60	2.25
8	3.66	2.68
9	3.05	3.05
10	2.70	3.05
11	2.38	2.88
12	2.17	2.55
13	2.02	2.25
14	1.82	1.95
15	1.72	1.87
16	1.77	1.83
17	1.97	1.90
18	2.43	2.17
19	2.68	2.46
20	2.75	2.52
21	2.65	2.50
22	2.40	2.57
23	2.21	2.40
24	1.90	2.22

37. Find the weekday energy consumption of a typical residential customer by using the Trapezoidal Rule to estimate the area under the weekday power curve. To save time, you might try using only the data for the even-numbered hours or the data for the odd-numbered hours. How close do you come to Louisiana Power's own estimate if you do that? (If you use the even-numbered hours, be sure to count hour 2 twice so that you cover a complete 24-hr period. If you use odd-numbered hours, use hour 1 twice.)

38. *Continuation of Exercise 37.* Use the Trapezoidal Rule to find the daily weekend energy consumption.

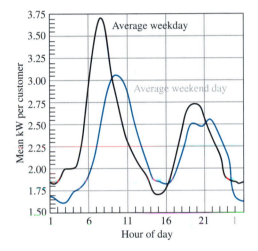

5.45 Louisiana Power and Light Company 1984 residential-load profile, residential customers with electric heating, averaged days for January. The curves plot kilowatts against hours. The areas under the curves give kilowatt-hours. See Table 5.11 for data.

Challenge For Exercises 39–44, consider the curve defined parametrically by $x(t) = \pi t - \sin(\pi t)$, $y(t) = \cos(\pi t) + 1$ for $0 \le t \le 1$.

39. Graph the curve. Show that it defines a function $y = f(x)$. Find the domain and range of f.

40. Determine an approximation for the area under the curve $y = f(x)$ in the first quadrant using the trapezoidal method with the four trapezoids determined by $t = 0.25, 0.5, 0.75, 1$. *Note:* The bases of the trapezoids will not be equal.

41. Repeat Exercise 40 using 10 trapezoids determined by $t = 0.1, 0.2, \ldots, 0.9, 1$.

42. Determine an explicit formula for x in terms of y (the inverse of $y = f(x)$). Interchange x and y, and graph $y = g(x)$. Show that f and g are inverses of each other.

43. Use NINT g to obtain an accurate approximation to the area under the curve $y = f(x)$. Explain why g can be used rather than an explicit formula for f (which can't be determined).

44. Write a grapher program that automates the procedure that you used in Exercises 40 and 41 to approximate the area

under the curve $y = f(x)$. Apply the program using 100 trapezoids determined by $t = 0.01, 0.02, \ldots, 0.99, 1$. Compare your answer with the NINT computation in Exercise 43.

45. *Challenge problem.* Prove the following generalizations of Exercises 31 and 32:

$$T_n f = \frac{\text{LRAM}_n f + \text{RRAM}_n f}{2}$$

$$S_{2n} f = \frac{\text{MRAM}_n f + 2T_{2n} f}{3}.$$

Review Questions

1. How can you use antiderivatives to find areas? Give an example.

2. How are finite sums written in sigma notation? Give examples.

3. What is a partition of an interval? What is the norm of a partition? Give examples.

4. What is a Riemann sum? Give an example.

5. What does $\int_a^b f(x)\,dx$ mean? Is it a number? How is it defined? What is it called? When does it exist?

6. What is the relation between definite integrals and area? Does a definite integral have to represent an area?

7. State 11 rules for working with definite integrals (Table 5.3). Give a specific example of each rule.

8. What is the average value of a function $f(x)$ over an interval $[a, b]$? Give an example.

9. State the Mean Value Theorem for definite integrals.

10. State the two parts of the Fundamental Theorem of Calculus. What are they good for? Illustrate each part with an example.

11. What is an indefinite integral? How are indefinite integrals evaluated? What corollary of the Mean Value Theorem for derivatives makes the evaluation possible?

12. How does integration by substitution work? Does it apply to definite integrals as well as indefinite integrals? Give specific examples.

13. What numerical methods are available for estimating the values of definite integrals that cannot be evaluated directly with antiderivatives? What formulas sometimes help to determine the accuracy of these methods? How do you know whether you are using a step size small enough to get the accuracy you want? Give an example.

14. Explain how to visualize part 1 of the Fundamental Theorem of Calculus.

15. Explain how to use NINT to support a definite integral computation. Why is NINT often more useful than applying part 2 of the Fundamental Theorem of Calculus to evaluate definite integrals?

16. Given that there is no explicit formula in terms of x for $F(x) = \int (\sin 2x)/x\,dx$, we can use a grapher to draw a graph of one antiderivative of $f(x) = (\sin 2x)/x$. How are the graphs of all the antiderivatives of f related to NINT f?

17. Explain how the SLOPEFLD toolbox program can be used to graphically support indefinite integration. Give an example.

Practice Exercises

For each of the following functions $y = f(x)$, consider the area above the x-axis and under the graph of $y = f(x)$ from $x = a$ to $x = b$.

a) Make a sketch illustrating LRAM$_5$, RRAM$_5$, and MRAM$_5$ showing the five approximating rectangles.

b) Write out by hand LRAM$_5$, RRAM$_5$, and MRAM$_5$, and

compute each sum.

1. $f(x) = 6 - x$ for $x = 0$ to $x = 5$

2. $f(x) = x^2 + 2x - 1$ for $x = 1$ to $x = 6$

For Exercises 3–6, estimate the area above the x-axis and under the graph of $y = f(x)$ from $x = a$ to $x = b$ using the RAM for

$n = 10, 100,$ and 1000. Do all three: $\text{LRAM}_n\, f$, $\text{RRAM}_n\, f$, and $\text{MRAM}_n\, f$. First verify that each function is nonnegative on the specified interval $[a, b]$.

3. $f(x) = x^2 + 5$ $\qquad$ for $x = 0$ to $x = 3$

4. $f(x) = x^3 - 2x + 2$ $\qquad$ for $x = 0$ to $x = 2$

5. $f(x) = 2 \sin x$ $\qquad$ from $x = 0$ to $x = \pi$

6. $f(x) = \dfrac{1}{2}e^{-x^2}$ $\qquad$ for $x = -5$ to $x = 5$

Use standard formulas to evaluate each sum in Exercises 7–9.

7. a) $\displaystyle\sum_{k=1}^{10}(k + 2)$ $\qquad$ **b)** $\displaystyle\sum_{k=1}^{10}(2k - 12)$

8. a) $\displaystyle\sum_{k=1}^{6}\left(k^2 - \dfrac{1}{6}\right)$ $\qquad$ **b)** $\displaystyle\sum_{k=1}^{6}k(k + 1)$

9. a) $\displaystyle\sum_{k=1}^{5}(k^3 - 45)$ $\qquad$ **b)** $\displaystyle\sum_{k=1}^{6}\left(\dfrac{k^3}{7} - \dfrac{k}{7}\right)$

10. Evaluate the following sums.

a) $\displaystyle\sum_{k=1}^{3} 2^{k-1}$ $\qquad$ **b)** $\displaystyle\sum_{k=0}^{4}(-1)^k \cos k\pi$

c) $\displaystyle\sum_{k=-1}^{2} k(k + 1)$ $\qquad$ **d)** $\displaystyle\sum_{k=1}^{4}\dfrac{(-1)^{k+1}}{k(k + 1)}$

11. Express the following sums in sigma notation.

a) $1 + 2 + 4 + 8$ $\qquad$ **b)** $1 + \dfrac{1}{3} + \dfrac{1}{9} + \dfrac{1}{27} + \dfrac{1}{81}$

c) $1 - 2 + 3 - 4 + 5$ $\qquad$ **d)** $\dfrac{5}{2} + \dfrac{5}{4} + \dfrac{5}{6}$

Express the limits in Exercises 12 and 13 as definite integrals.

12. $\displaystyle\lim_{\|P\| \to 0}\sum_{k=1}^{n}\dfrac{1}{c_k}\Delta x_k$, where P is a partition of $[1, 2]$.

13. $\displaystyle\lim_{\|P\| \to 0}\sum_{k=1}^{n} e^{c_k}\Delta x_k$, where P is a partition of $[0, 1]$.

Find a formula for $\text{RRAM}_n\, f$ in terms of n for each of the functions $y = f(x)$. Then determine $\lim_{n \to \infty}\text{RRAM}_n\, f$ to find the *exact* area under the curve $y = f(x)$ from $x = a$ to $x = b$.

14. $f(x) = x^2 + 2x + 3$ $\qquad$ for $x = 0$ to $x = 4$

15. $f(x) = 2x^3 + 3x$ $\qquad$ for $x = 0$ to $x = 5$

16. Suppose $\int_{-2}^{2} f(x)\, dx = 4$, $\int_{2}^{5} f(x)\, dx = 3$, $\int_{-2}^{5} g(x)\, dx = 2$. Which, if any, of the following statements are true, and which, if any, are false?

a) $\displaystyle\int_{5}^{2} f(x)\, dx = -3$

b) $\displaystyle\int_{-2}^{5}(f(x) + g(x))\, dx = 9$

c) $f(x) \le g(x)$ on the interval $-2 \le x \le 5$

17. Suppose $\int_{0}^{1} f(x)\, dx = \pi$. Find

a) $\displaystyle\int_{0}^{1} f(t)\, dt$ $\quad$ **b)** $\displaystyle\int_{1}^{0} f(y)\, dy$ $\quad$ **c)** $\displaystyle\int_{0}^{1} -3f(z)\, dz.$

Find the areas of each of the shaded regions in Exercises 18–20 analytically. Support with a NINT computation.

18.

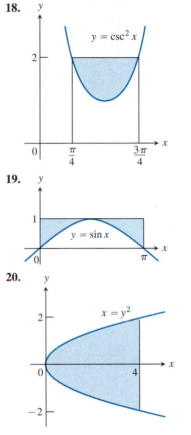

19.

20.

In Exercises 21 and 22, find the total area between the curve and the x-axis.

21. $y = 4 - x,\quad 0 \le x \le 6$

22. $y = \cos x,\quad -\pi \le x \le \pi$

Evaluate each integral in Exercises 23–34 analytically. Support with a NINT computation.

23. $\displaystyle\int_{-1}^{1}(3x^2 - 4x + 7)\, dx$ $\qquad$ **24.** $\displaystyle\int_{0}^{1}(8s^3 - 12s^2 + 5)\, ds$

25. $\displaystyle\int_{1}^{2}\dfrac{4}{x^2}\, dx$ $\qquad$ **26.** $\displaystyle\int_{1}^{27} x^{-4/3}\, dx$

27. $\displaystyle\int_{1}^{4}\dfrac{dt}{t\sqrt{t}}$ $\qquad$ **28.** $\displaystyle\int_{0}^{2} 3\sqrt{4x + 1}\, dx$

29. $\displaystyle\int_{0}^{\pi} \sin 5\theta\, d\theta$ $\qquad$ **30.** $\displaystyle\int_{0}^{\pi} \cos 5t\, dt$

31. $\displaystyle\int_0^{\pi/3} \sec^2\theta \, d\theta$

32. $\displaystyle\int_{\pi/4}^{3\pi/4} \csc^2 x \, dx$

33. $\displaystyle\int_{\pi}^{3\pi} \cot^2 \frac{x}{6} \, dx$

34. $\displaystyle\int_0^{\pi} \tan^2 \frac{\theta}{3} \, d\theta$

Evaluate each integral in Exercises 35–44 by a method of your choice.

35. $\displaystyle\int_0^1 \frac{36 \, dx}{(2x+1)^3}$

36. $\displaystyle\int_1^2 \left(x + \frac{1}{x^2}\right) dx$

37. $\displaystyle\int_{-\pi/3}^0 \sec x \tan x \, dx$

38. $\displaystyle\int_{\pi/4}^{3\pi/4} \csc x \cot x \, dx$

39. $\displaystyle\int_0^{\pi/2} 5(\sin x)^{3/2} \cos x \, dx$

40. $\displaystyle\int_{-1}^1 2x \sin(1 - x^2) \, dx$

41. $\displaystyle\int_4^8 \frac{1}{t} \, dt$

42. $\displaystyle\int_0^2 \frac{2}{x+1} \, dx$

43. $\displaystyle\int_0^2 \frac{x \, dx}{x^2+5}$

44. $\displaystyle\int_0^{\pi} \frac{\cos x}{3 - \sin x} \, dx$

Evaluate each integral in Exercises 45–48 analytically. Support with a NINT computation.

45. $\displaystyle\int_2^3 \left(t - \frac{2}{t}\right)\left(t + \frac{2}{t}\right) dt$ *Hint*: Multiply first.

46. $\displaystyle\int_{-1}^0 (1 - 3w)^2 \, dw$ *Hint*: Square first.

47. $\displaystyle\int_{-4}^0 |x| \, dx$ *Hint*: Write without absolute value first.

48. $\displaystyle\int_{1/2}^4 \frac{x^2 + 3x}{x} \, dx$ *Hint*: Divide first.

Evaluate each integral in Exercises 49 and 50.

49. $\displaystyle\int_{-\pi/2}^{\pi/2} 15 \sin^4 3x \cos 3x \, dx$

50. $\displaystyle\int_0^{\pi/2} \frac{3 \sin x \cos x}{\sqrt{1 + 3\sin^2 x}} \, dx$

Support each integral formula in Exercises 51–54 graphically.

51. $\displaystyle\int \frac{\ln 5x}{x} \, dx = \frac{1}{2}(\ln 5x)^2 + C$

52. $\displaystyle\int x^2 \ln x \, dx = \frac{x^3}{3} \ln x - \frac{x^3}{9} + C$

53. $\displaystyle\int e^x \sin x \, dx = \frac{e^x}{2}(\sin x - \cos x) + C$

54. $\displaystyle\int x e^x \, dx = x e^x - e^x + C$

Determine a graphical representation of each integral in Exercises 55 and 56. Can you find an explicit antiderivative in terms of x?

55. $\displaystyle\int \frac{\sin 3x}{x} \, dx$

56. $\displaystyle\int \frac{1}{4} e^{-x^2/2} \, dx$

57. Solve for x: $\displaystyle\int_0^x (t^3 - 2t + 3) \, dt = 4$.

58. Solve for x : $\displaystyle\int_1^x \frac{1}{2} e^{-t^2} \, dt = 0.05$.

59. Solve for x: $\displaystyle\int_1^x \frac{\cos t}{t} \, dt = 1$.

60. What is the domain of $f(x) = \int_1^x \cos t / t \, dt$? Draw a graph of the function.

61. Which of the following methods could be used successfully to prepare the integral

$$\int 3x^2(x^3 - 1)^5 \, dx$$

for evaluation?

a) Expand $(x^3 - 1)^5$, and multiply the result by $3x^2$ to get a polynomial to integrate term by term.

b) Factor $3x^2$ out front to get an integral of the form

$$3x^2 \int u^5 \, du.$$

c) Substitute $u = x^3 - 1$ to get an integral of the form

$$\int u^5 \, du.$$

62. The substitution $u = \tan x$ gives

$$\int \sec^2 x \tan x \, dx = \int \tan x \cdot \sec^2 x \, dx$$

$$= \int u \, du = \frac{u^2}{2} + C = \frac{\tan^2 x}{2} + C.$$

The substitution $u = \sec x$ gives

$$\int \sec^2 x \tan x \, dx = \int \sec x \cdot \sec x \tan x \, dx$$

$$= \int u \, du = \frac{u^2}{2} + C = \frac{\sec^2 x}{2} + C.$$

Can both integrations be correct? Explain.

63. Suppose that $y = f(x)$ is continuous and positive throughout the interval $[0, 1]$. Suppose also that for every value of x in this interval the area between the graph of f and the subinterval $[0, x]$ is $\sin x$. Find $f(x)$.

64. Suppose that f has a positive derivative for all values of x and that $f(1) = 0$. Which of the following statements must be true of the function

$$g(x) = \int_0^x f(t)\,dt?$$

a) g is a differentiable function of x.

b) g is a continuous function of x.

c) The graph of g has a horizontal tangent at $x = 1$.

d) g has a local maximum at $x = 1$.

e) g has a local minimum at $x = 1$.

f) The graph of g has an inflection point at $x = 1$.

g) The graph of dg/dx crosses the x-axis at $x = 1$.

65. Suppose $F(x)$ is an antiderivative of $f(x) = \sqrt{1 + x^4}$.
Express $\displaystyle\int_0^1 \sqrt{1 + x^4}\,dx$ in terms of F.

66. Show that $y = x^2 + \int_1^x 1/t\,dt + 1$ solves the initial value problem.

Differential equations: $y'' = 2 - \dfrac{1}{x^2}$

Initial conditions: $y = 2$ and $y' = 3$ when $x = 1$.

67. The acceleration of a particle moving back and forth along a line is $d^2s/dt^2 = \pi^2 \cos \pi t$ m/sec². If $s = 0$ m and $v = 8$ m/sec when $t = 0$, find the value of s when $t = 1$.

68. Solve the following initial value problems.

a) Differential equation: $y^{(4)} = \cos x$
Initial conditions: $y = 3$, $y' = 2$, $y'' = 1$,
and $y''' = 0$ when $x = 0$

b) Differential equation: $\dfrac{dy}{dx} = \dfrac{4x}{(1 + x^2)^2}$
Initial condition: $y = 0$ when $x = 0$

69. *Stopping a motorcycle.* The State of Illinois Cycle Rider Safety Program requires riders to be able to brake from 30 mi/h (44 ft/sec) to 0 mi/h in 45 ft. What constant deceleration does it take to do that? To find out, carry out these steps:

STEP 1: Solve the following initial value problem. The answer will involve k.

Differential equation: $\dfrac{d^2s}{dt^2} = -k$

Initial conditions: $ds/dt = 44$ and $s = 0$ when $t = 0$

STEP 2: Find the time t^* when $ds/dt = 0$. The answer will still involve k.

STEP 3: Solve the equation $s(t^*) = 45$ for k.

70. Which of the following graphs shows the solution of the initial value problem

$$\frac{dy}{dx} = 2x, \quad y = 4 \quad \text{when } x = 1?$$

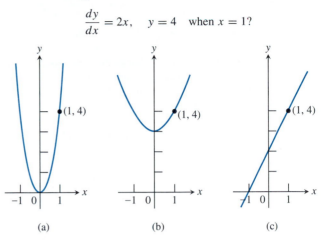

(a)　　　　　(b)　　　　　(c)

71. *Household electricity.* We model the voltage V in our homes with the sine function

$$V = V_{\max} \sin 120\pi t,$$

which expresses V in volts as a function of time t in seconds. The function runs through 60 cycles each second. The number $V_{\max}$ is the **peak voltage**.

To measure the voltage effectively, we use an instrument that measures the square root of the average value of the square of the voltage over a 1-sec interval:

$$V_{\mathrm{rms}} = \sqrt{(V^2)_{\mathrm{av}}}.$$

The subscript "rms" (read the letters separately) stands for "root mean square." It turns out that

$$V_{\mathrm{rms}} = V_{\max}/\sqrt{2}. \tag{1}$$

The familiar phrase "115 volts ac" means that the rms voltage is 115. The peak value, obtained from Eq. (1) as $V_{\max} = 115\sqrt{2}$, is about 163 volts.

a) Find the average value of V^2 over a 1-sec interval. Then find V_{rms}, and verify Eq. (1).

b) The circuit that runs your electric stove is rated 240 volts rms. What is the peak value of the allowable voltage?

72. Compute the average value of the temperature function

$$f(x) = 37 \sin\left[\frac{2\pi}{365}(x - 101)\right] + 25$$

for a 365-day year. This is one way to estimate the annual mean air temperature in Fairbanks, Alaska. The National Weather Service's official figure, a numerical average of the daily normal mean air temperatures for the year, is 25.7°F, which is slightly higher than the average value of $f(x)$. Figure 1.105 shows why.

73. Let f be a function that is differentiable on $[a, b]$. In Chapter 3, we defined the average rate of change of f on $[a, b]$ to be

$$\frac{f(b) - f(a)}{b - a}$$

and the instantaneous rate of change of f at x to be $f'(x)$. In this chapter, we defined the average value of a function. For the new definition of average to be consistent with the old one, we should have

$$\frac{f(b) - f(a)}{b - a} = \text{average value of } f' \text{ on } [a, b].$$

Show that this is the case.

74. Find the average value of

  a) $y = \sqrt{3x}$ over the interval $0 \le x \le 3$,

  b) $y = \sqrt{ax}$ over the interval $0 \le x \le a$.

75. What step size h would you use to be sure of estimating the value of

$$\int_1^3 \frac{1}{x} \, dx$$

by Simpson's Rule with an error of no more than 10^{-4} in absolute value?

76. A brief calculation shows that if $0 \le x \le 1$, then the second derivative of $f(x) = \sqrt{1 + x^4}$ lies between 0 and 8. On the basis of this, about how many subdivisions would you need to estimate the integral of f from 0 to 1 with an error no greater than 10^{-3} in absolute value?

77. A direct calculation shows that

$$\int_0^\pi 2 \sin^2 x \, dx = \pi.$$

How close do you come to this value by using the Trapezoidal Rule with $n = 6$? Simpson's Rule with $n = 6$?

78. You are planning to use Simpson's Rule to estimate the value of the integral

$$\int_1^2 f(x) \, dx$$

with an error magnitude less than 10^{-5}. You have determined that $|f^{(4)}(x)| \le 3$ throughout the interval of integration. How many subintervals should you use to ensure the required accuracy? (Remember that for Simpson's Rule, the number has to be even.)

79. *A new parking lot.* To meet the demand for parking, your town has allocated the area shown below. As the town engineer, you have been asked by the town council to find out whether the lot can be built for $11,000. The cost to clear the land will be $0.10 a square foot, and the lot will cost $2.00 a square foot to pave. Can the job be done for $11,000?

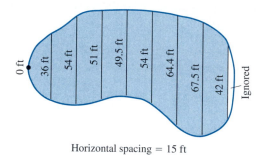

Horizontal spacing = 15 ft

CHAPTER

6

Applications of Definite Integrals

Developing an Integral to Calculate Something

1. Model something that we want to measure in terms of continuous functions on a closed interval $[a, b]$.

2. Form small approximating objects using the continuous functions on the subintervals of $[a, b]$.

3. Add the approximations to form a Riemann sum.

4. Take the limit of the Riemann sum to get an integral, which gives precisely the measurement we want in step 1.

OVERVIEW   The importance of integral calculus stems from the fact that thousands of things that we want to know can be calculated with integrals. This includes areas between curves, volumes and surface areas of solids, lengths of curves, the amount of work it takes to pump liquids up from below ground, the forces against flood gates, and the coordinates of the points where solid objects will balance. We can define all these things in natural ways as limits of Riemann sums of continuous functions on closed intervals. Then we can evaluate these limits by applying techniques of integral calculus including the use of technology.

There is a pattern to how we go about defining and calculating the integrals in our applications, a pattern that, once learned, enables us to define new integrals whenever we need them. An outline of the pattern is given at the left. In this chapter, we look at specific applications, each time following this pattern. In the concluding section, we examine the pattern itself and show how it leads to integrals in new situations.

6.1 _____ Areas between Curves

This section shows how to find the area of a region in the coordinate plane by integrating the functions that define the region's boundaries.

The Basic Formula, Derived from Riemann Sums

Suppose that functions $f_1(x)$ and $f_2(x)$ are continuous and that $f_1(x) \geq f_2(x)$ throughout an interval $a \leq x \leq b$ (Fig. 6.1). The region between the curves $y = f_1(x)$ and $y = f_2(x)$ from a to b may sometimes have a shape that would let us find its area with a formula from geometry (a trapezoid, for example), but we usually have to find the area with an integral instead.

To see what that integral should be, we start by approximating the region with vertical rectangles. Figure 6.2(a) shows a typical approximation, based on a partition $P = \{x_0, x_1, \ldots, x_n\}$ of the interval $[a, b]$. The kth rectangle,

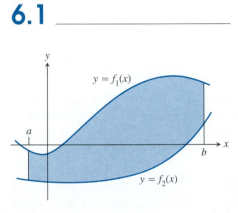

6.1   A typical region between two curves.

433

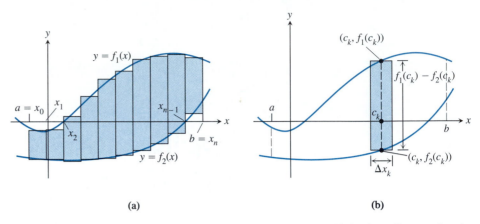

6.2 To develop an integral to calculate area, (a) we think of small approximating rectangles whose long sides are perpendicular to the x-axis. (b) A typical rectangle is $f_1(c_k) - f_2(c_k)$ units high by Δx_k units wide.

shown in detail in Fig. 6.2(b), is Δx_k units wide and runs from the point $(c_k, f_2(c_k))$ on the lower curve to the point $(c_k, f_1(c_k))$ on the upper curve. Its area, height × width, is $(f_1(c_k) - f_2(c_k))\Delta x_k$.

We approximate the area of the region by adding the areas of the n rectangles from a to b:

<table>
<tr><td>

Developing an Integral to Calculate Area

1. Model the region that we want to measure in terms of continuous functions on a closed interval $[a, b]$.

2. Form small approximating rectangles using the continuous functions on the subintervals of $[a, b]$.

3. Add the areas of the small rectangles to form a Riemann sum (Eq. 1).

4. Take the limit of the Riemann sum to get an integral (Eq. 2), which gives precisely the area of the region in step 1.

</td></tr>
</table>

$$\text{Rectangle area sum} = \sum_{k=1}^{n} (f_1(c_k) - f_2(c_k))\Delta x_k. \qquad (1)$$

This sum is a Riemann sum for the difference function $(f_1 - f_2)$ on the closed interval $[a, b]$.

Since f_1 and f_2 are continuous, two things will happen as we subdivide $[a, b]$ more finely and let the norm of the partition go to zero. The rectangles will approximate the region with increasing geometric accuracy, and the Riemann sums in (1) will approach a limit. We therefore define this limit, the definite integral of $(f_1 - f_2)$ from a to b, to be the area of the region.

DEFINITION

If functions f_1 and f_2 are continuous and if $f_1(x) \geq f_2(x)$ throughout the interval $a \leq x \leq b$, then the **area of the region between the curves** $y = f_1(x)$ and $y = f_2(x)$ from a to b is the integral of $(f_1 - f_2)$ from a to b:

$$\text{Area} = \int_a^b (f_1(x) - f_2(x))\, dx. \qquad (2)$$

To apply Eq. (2), we often have a choice of procedures.

> **How to Find the Area between Two Curves (Three Ways)**
>
> **The Analytic Way (A):** *Do analytically. Support with a grapher.*
> **The Graphical Way (G):** *Do with a grapher. Confirm analytically.*
> Key steps for both ways:
>
> 1. Determine which curve is the upper curve, $f_1(x)$, and which curve is the lower curve, $f_2(x)$.
>
>   **A:** Show that $f_1(x) \geq f_2(x)$.
>
>   **G:** Show that the graph of $f_1(x)$ lies above the graph of $f_2(x)$.
>
> 2. Find the limits of integration a and b. This may require finding where the curves intersect as follows:
>
>   **A:** Solve $y = f_1(x)$ and $y = f_2(x)$ simultaneously for x.
>
>   **G:** Use ZOOM-IN or SOLVE.
>
> 3. Integrate $f_1(x) - f_2(x)$ from a to b. The result is the area.
>
>   **A:** Simplify $f_1(x) - f_2(x)$, then integrate.
>
>   **G:** Compute NINT $(f_1(x) - f_2(x), a, b)$.
>
> 4. Support or confirm.
>
>   **A:** Support with a grapher using steps 1G–3G.
>
>   **G:** Confirm analytically using steps 1A–3A.
>
> **The Third Way:** *Do any way you can.*
> This often involves a combination of analytic and graphical steps. We sometimes phrase this third way as *Do the easiest way* or *Just do it.*

EXPLORATION BIT

When your work is analytic, NINT provides one means of support. Other types of support are often possible, and some of them can be more efficient than NINT. For example, if you can "see" a familiar geometric shape, such as a rectangle or parallelogram, that is close in size to the given region, then finding its area will give you a quick check as to whether your analytic result is reasonable. Try this method of support for Example 1 by identifying a familiar shape that resembles the region in Fig. 6.3 and then finding its area. For the geometric shape that you use, should you expect an analytic or NINT result that is a little more or a little less?

The following example is a combination of steps, but the result is essentially analytic.

EXAMPLE 1

Find the area between the curves $y = \cos x$ and $y = -\sin x$ from 0 to $\pi/2$.

Solution

STEP 1: *Viewing graphs.* We graph the curves together (Fig. 6.3). We also sketch a representative rectangle (as shown in Fig. 6.3) to help determine the area function to be integrated. The upper curve is $y_1 = \cos x$, so we take $f_1(x) = \cos x$ in the area formula (Eq. 2). The lower curve is $y_2 = -\sin x$, so $f_2(x) = -\sin x$.

STEP 2: *The limits of integration.* They are already given: $a = 0$ and $b = \pi/2$.

STEP 3: *Integrate.* From step 1,

$$f_1(x) - f_2(x) = \cos x - (-\sin x)$$
$$= \cos x + \sin x,$$

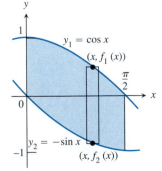

6.3   The graphs of

$$y_1 = \cos x \quad \text{and} \quad y_2 = -\sin x$$

show that $\cos x \geq -\sin x$ on $[0, \pi/2]$.

and

$$\int_0^{\pi/2} (\cos x + \sin x)\, dx = \left[\sin x - \cos x \right]_0^{\pi/2}$$

$$= [1 - 0] - [0 - 1] = 2.$$

The area between the curves is 2.

STEP 4: *Support.* NINT $(y_1 - y_2, 0, \pi/2) = 2.$ ≡

Curves That Intersect

When a region is determined by curves that intersect, the intersection points give the limits of integration.

EXAMPLE 2

Find the area of the region enclosed by the parabola $y = 2 - x^2$ and the line $y = -x$.

Solution

STEP 1: *Viewing graphs.* We graph the curves together (Fig. 6.4). Identifying the upper and lower curves, we take $f_1(x) = 2 - x^2$ and $f_2(x) = -x$.

STEP 2: *The limits of integration.* The x-coordinates of the points where the parabola and line intersect are the limits of integration. We find them by solving the equations $y = 2 - x^2$ and $y = -x$ simultaneously for x:

$$2 - x^2 = -x \qquad \text{Equate } f_1(x) \text{ and } f_2(x).$$

$$x^2 - x - 2 = 0 \qquad \text{Rearrange.}$$

$$(x + 1)(x - 2) = 0 \qquad \text{Factor.}$$

$$x = -1, \quad x = 2 \qquad \text{Solve.}$$

The region runs from $x = -1$ on the left to $x = 2$ on the right. The limits of integration are $a = -1$ and $b = 2$.

STEP 3: *Integrate.* For $y_1 = 2 - x^2$ and $y_2 = -x$, NINT $(y_1 - y_2, -1, 2) = 4.5$. The area of the region is 4.5.

STEP 4: *Confirm.*

$$f_1(x) - f_2(x) = (2 - x^2) - (-x) = 2 - x^2 + x$$

$$= 2 + x - x^2, \qquad \text{\color{blue}Rearranged—a matter of taste}$$

and

$$\int_{-1}^{2} (2 + x - x^2)\, dx = \left[2x + \frac{x^2}{2} - \frac{x^3}{3} \right]_{-1}^{2}$$

$$= \left(4 + \frac{4}{2} - \frac{8}{3} \right) - \left(-2 + \frac{1}{2} + \frac{1}{3} \right)$$

$$= 6 + \frac{3}{2} - \frac{9}{3} = \frac{9}{2} \qquad ≡$$

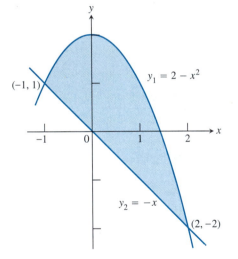

6.4 The two curves

$$y_1 = f_1(x) = 2 - x^2$$

and

$$y_2 = f_2(x) = -x$$

intersect. The limits of integration are the x-coordinates of their points of intersection.

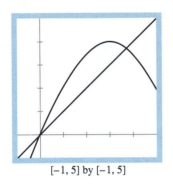

[−1, 5] by [−1, 5]

6.5 The graphs of

$$y_1 = 4\sin\frac{x}{2} \quad \text{and} \quad y_2 = x.$$

To approximate the area between the curves, we can use NINT to evaluate

$$\int_0^{3.79} \left(4\sin\frac{x}{2} - x\right) dx,$$

or a calculator to evaluate

$$\left[-8\cos\frac{x}{2} - \frac{x^2}{2}\right]_0^{3.79},$$

or both and compare results.

EXPLORATION 1

Practical Problems, Part 1

Examples 1 and 2 are written so that the analytic steps are straightforward: The limits of integration and the antiderivatives were easy to find, and the integration arithmetic is simple. In practical situations, most problems are not so conveniently solved, and our choice method of solution becomes essentially steps 1G–3G.

1G. *Viewing graphs.* GRAPH the functions in the same viewing window to find the upper curve y_1 and the lower curve y_2.

2G. *The limits of integration.* Use ZOOM-IN or SOLVE to find the limits, a and b, when they are determined by points of intersection of the curves.

3G. *Integrate.* Compute NINT $(y_1 - y_2, a, b)$ to find the area.

Try these steps to find the area of the region in the first quadrant between the curves $y_1 = 4\sin(x/2)$ and $y_2 = x$ (Fig. 6.5). The caption for Fig. 6.5 provides a check for Step 3. ☙

Boundaries with Changing Formulas

If the formula for one of the bounding curves changes at some point across the region, you may have to add two or more integrals to find the area.

EXAMPLE 3

Find the area of the region in the first quadrant bounded above by the curve $y = \sqrt{x}$ and below by the x-axis and the line $y = x - 2$.

Solution

STEP 1: *Viewing graphs.* We graph the curves together. The entire upper boundary of the region consists of the curve $y = \sqrt{x}$, so $f_1(x) = \sqrt{x}$. The lower boundary consists of two curves, first $y = 0$ for $0 \le x \le 2$ and then $y = x - 2$ for $2 \le x \le 4$. Hence, the formula for $f_2(x)$ changes from $f_2(x) = 0$ for $0 \le x \le 2$ to $f_2(x) = x - 2$ for $2 \le x \le 4$.

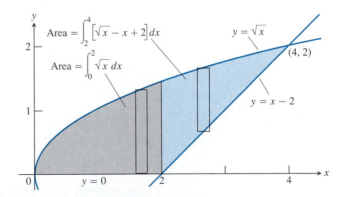

STEP 2: *The limits of integration.* The limits of integration for the pair $f_1(x) = \sqrt{x}$ and $f_2(x) = 0$ are $a = 0$ and $b = 2$.

For the pair $f_1(x) = \sqrt{x}$ and $f_2(x) = x - 2$, the left-hand limit is $a = 2$, and the right-hand limit is the x-coordinate of the upper point where the line crosses the parabola. To find it, we solve the equations $y = \sqrt{x}$ and $y = x - 2$ simultaneously for x:

$$\sqrt{x} = x - 2 \qquad \text{Equate } f_1(x) = \sqrt{x} \text{ and } f_2(x) = x - 2.$$

$$x = (x - 2)^2 = x^2 - 4x + 4 \qquad \text{Square.}$$

$$x^2 - 5x + 4 = 0 \qquad \text{Rearrange.}$$

$$(x - 1)(x - 4) = 0 \qquad \text{Factor.}$$

$$x = 1, \quad x = 4. \qquad \text{Solve.}$$

The value $x = 1$ does not satisfy the equation $\sqrt{x} = x - 2$. It is an extraneous root introduced by squaring. The value $x = 4$ gives our upper limit of integration.

STEP 3: *Integrate.* We have two integrals to evaluate. Their sum is the area.

For $0 \le x \le 2$: $f_1(x) - f_2(x) = \sqrt{x} - 0 = \sqrt{x}$,

For $2 \le x \le 4$: $f_1(x) - f_2(x) = \sqrt{x} - (x - 2) = \sqrt{x} - x + 2$.

Therefore,

$$\text{Area} = \int_0^4 \left[f_1(x) - f_2(x) \right] \, dx = \int_0^2 \sqrt{x} \, dx + \int_2^4 \left(\sqrt{x} - x + 2 \right) \, dx$$

$$= \left[\frac{2}{3} x^{3/2} \right]_0^2 + \left[\frac{2}{3} x^{3/2} - \frac{x^2}{2} + 2x \right]_2^4$$

$$= \frac{2}{3}(2)^{3/2} + \left(\frac{2}{3}(4)^{3/2} - \frac{16}{2} + 8 \right) - \left(\frac{2}{3}(2)^{3/2} - \frac{4}{2} + 4 \right)$$

$$= \frac{2}{3}(8) - 2 = \frac{10}{3}.$$

The area of the region is 10/3.

STEP 4: *Support.* With $y_1 = \sqrt{x}, y_2 = 0$ and $y_3 = x - 2$,

$$\text{Area} = \text{NINT}(y_1 - y_2, 0, 2) + NINT(y_1 - y_3, 2, 4) = 3.333.$$

Integrating with Respect to y

When a region's bounding curves are described by giving x as a function of y, the basic formula changes.

For regions like these,

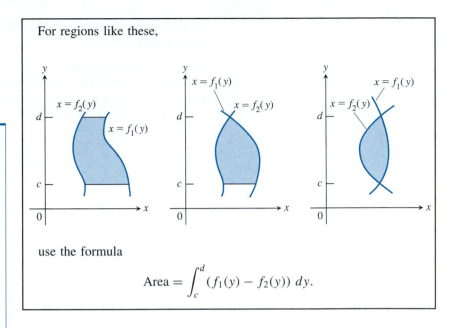

use the formula

$$\text{Area} = \int_c^d (f_1(y) - f_2(y))\ dy.$$

EXPLORATION BIT

Because we have found the area of the region in Fig. 6.6 three times now, we leave step 4 (*Confirm analytically*) in Example 4 to you. You have to evaluate

$$\int_0^2 (y + 2 - y^2)\, dy.$$

After you have obtained a result (area = 10/3, we hope), tell which analytic confirmation was easier, this one or the one in Example 3.

The only difference between this formula and the one in Eq. (2) is that we are now integrating with respect to y instead of x. We can sometimes save time by doing so. The basic steps are the same as before.

EXAMPLE 4

The area in Example 3, found by a single integration with respect to y. Find the area of the region between the curves $x = y^2$ and $x = y + 2$ in the first quadrant.

Solution

STEP 1: *Viewing graphs.* We graph the curves together (Fig. 6.6). The right-hand curve is $x = y + 2$, so $f_1(y) = y + 2$. The left-hand curve is $x = y^2$, so $f_2(y) = y^2$.

STEP 2: *The limits of integration.* The lower limit of integration is $y = 0$. The upper limit is the y-coordinate of the upper point where the line crosses the parabola. We find it by solving the equations $x = y + 2$ and $x = y^2$ simultaneously for y:

$$y + 2 = y^2 \qquad \text{Equate } f_1(y) = y + 2 \text{ and } f_2(y) = y^2.$$

$$y^2 - y - 2 = 0 \qquad \text{Rearrange.}$$

$$(y + 1)(y - 2) = 0 \qquad \text{Factor.}$$

$$y = -1, \quad y = 2. \qquad \text{Solve.}$$

The upper limit of integration is 2. (The value $y = -1$ is for the point of intersection of f_1 and f_2 *below* the x-axis.)

STEP 3: *Integrate.* Using y as the variable of integration,

$$\text{Area} = \text{NINT}\,(y + 2 - y^2, 0, 2) = 3.333. \qquad \blacksquare$$

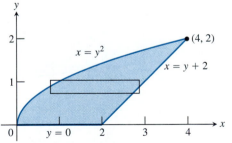

6.6 The region between the graphs of $x = y^2$ and $x = y + 2$ in the first quadrant. Compare with the diagram in Example 3.

Combining Integrals with Formulas from Geometry

Regardless of the method that we choose for finding the area of a region, we have found that viewing a diagram, when possible, is often a desirable early step. Other steps, however, need not be limited to the use of a graphing utility or analytic methods. Indeed, we can bring to bear on a problem *any* mathematics we know.

EXAMPLE 5

If we view the region in Examples 3 and 4 as the area between the curve $y = \sqrt{x}, 0 \leq x \leq 4$, and the *x*-axis, *minus* the area of a triangle with base 2 and height 2, then

$$\text{Area} = \int_0^4 \sqrt{x} \, dx - \frac{1}{2}(2)(2) = \frac{2}{3} x^{3/2} \Big]_0^4 - 2$$

$$= \frac{2}{3}(8) - 0 - 2 = \frac{10}{3}. \qquad \blacksquare$$

Moral of Examples 3–5 We view the curves first. It is sometimes easier to find the area between them by integrating with respect to *y* instead of *x*. The picture may also suggest how familiar geometry formulas could be used to simplify our work.

Exercises 6.1

Find the areas of the shaded regions in Exercises 1–6 analytically. Support with a grapher.

1.

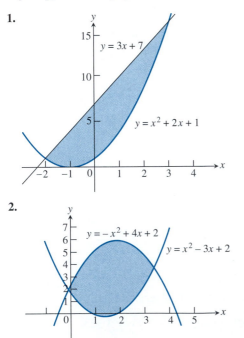

2.

3.

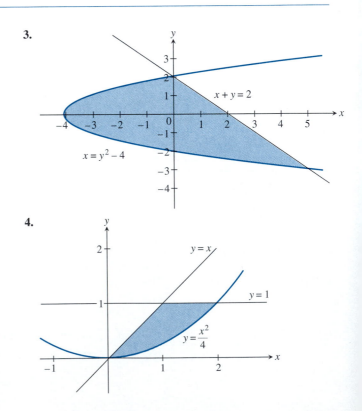

4.

5.

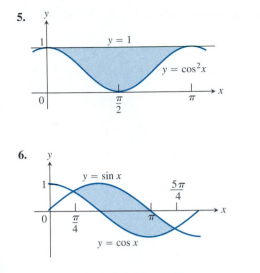

6.

Find the area of the regions enclosed by the lines and curves in Exercises 7–22 any way you can. If you find the area analytically, support with a grapher. If you find the area graphically, confirm analytically if possible.

7. The curve $y = x^2 - 2$ and the line $y = 2$

8. The x-axis and the curve $y = 2x - x^2$

9. The curve $y^2 = x$ and the line $x = 4$

10. The curve $y = 2x - x^2$ and the line $y = -3$

11. The curve $y = x^2$ and the line $y = x$

12. The curve $x = 3y - y^2$ and the line $x + y = 3$

13. The line $y = 2x$ and the curve $y = x^3 + 2x^2 - 3x + 1$

14. Below the line $y = 3 - 2x$ and above the curve $y = 2\cos 2x$ in the first quandrant

15. The curve $y = x^2 - 2x$ and the line $y = x$

16. The curve $x = 10 - y^2$ and the line $x = 1$

17. Above the line $y = 2x$ and below the curve $y = e^{-x^2}$ in the first quadrant

18. The curve $y = e^x$ and the lines $y = -x$ and $x = 2$

19. The line $y = x$ and the curve $y = 2 - (x - 2)^2$

20. The curves $y = 7 - 2x^2$ and $y = x^2 + 4$

21. The line $y = x$ and the curve $y = x^3 - 2x^2 - 3x + 1$ (*Hint*: There are two regions.)

22. The curve $y = 2 - x^2$ and $y = 2\cos 2x$ (*Hint*: There are two regions.)

Find the areas of the shaded regions in Exercises 23–26 graphically. Confirm analytically.

23.

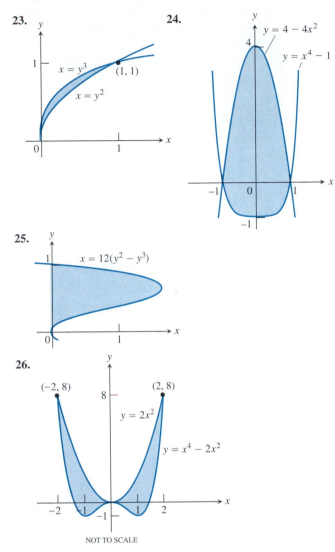

24.

25.

26.

NOT TO SCALE

27. Find the area of the "triangular" region bounded by the y-axis and the curves $y = \sin x$ and $y = \cos x$ in the first quadrant.

28. Find the area of the region between the curve $y = 3 - x^2$ and the line $y = -1$ by integrating with respect to (**a**) x; (**b**) y.

29. The area of the region between the curve $y = x^2$ and the line $y = 4$ is divided into two equal portions by the line $y = c$.
a) Find c by integrating with respect to y. (This puts c into the limits of integration.)
b) Find c by integrating with respect to x. (This puts c into the integrand as well.)

30. The diagram shows triangle AOC inscribed in the region cut from the parabola $y = x^2$ by the line $y = a^2$. Consider the ratio of the area of the triangle to the area of the parabolic region. Visualize a approaching 0, then find the limiting value of this ratio.

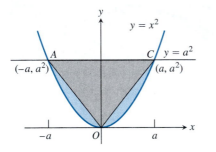

In Exercises 31 and 32, let $R(t)$ be the area of the rectangle and $A(t)$ be the area of the shaded region inside the rectangle and between the curves $y = x^2$ and $y = 2x^2$.

a) Determine $A(t)$ by integrating with respect to y.

b) Determine $A(t)$ by integrating with respect to x.

c) Show that the ratio $A(t)/R(t)$ is constant for all $t > 0$, and determine the constant.

31. The rectangle is bounded by $x = t > 0, y = t^2$, and the coordinate axes.

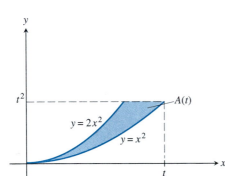

32. The rectangle is bounded by $x = t > 0, y = 2t^2$, and the coordinate axes.

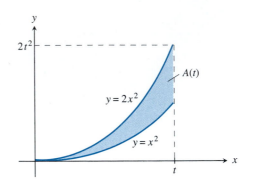

33. Suppose that the area between the continuous curve $y = f(x)$ shown here and the x-axis from $x = a$ to $x = b$ is 4 square units. Find the area between the curves $y = f(x)$ and $y = 2f(x)$ from $x = a$ to $x = b$.

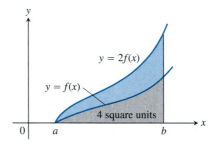

34. Write a paragraph that explains why the first quadrant region determined by $y = 2x, y = e^{-x^2}$, and the x-axis is not bounded.

6.2   Volumes of Solids of Revolution—Disks and Washers

Solids of revolution are solids whose shapes can be generated by revolving plane regions about axes. Thread spools are solids of revolution; so are hand weights and billiard balls. Solids of revolution sometimes have volumes that we can find with formulas from geometry, as we can the volume of the billiard ball. But, likely as not, we want to find the volume of a blimp instead or to predict the weight of a part that we are going to have turned on a lathe.

In cases like these, formulas from geometry are of little help, and we must turn to calculus for the answers. In this section and the next, we show how to find these answers.

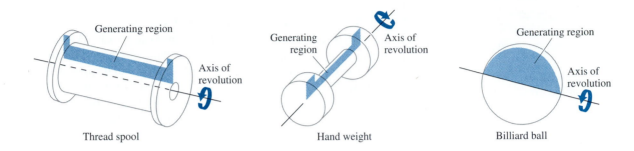

Thread spool

Hand weight

Billiard ball

The Disk Method

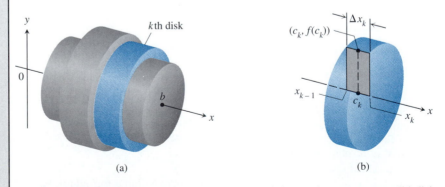

6.7 To visualize solids of revolution better, we show oblique views of the xy-plane and the curve and region that will generate the solid. In developing an integral to calculate the volume of the solid, we then approximate the generating region with rectangles.

As we said, a solid of revolution is generated by revolving a plane region about an axis. We usually assume that the axis lies in the same plane as the region. If we can set things up so that the region is contained between the graph of a continuous function $y = f(x)$, $a \leq x \leq b$, and the x-axis, and the axis of revolution is the x-axis, then we can define and calculate the resulting solid's volume in the following way.

We begin by approximating the region itself with vertical rectangles based on a partition of the closed interval $[a, b]$ (Fig. 6.7). We then imagine the rectangles to be revolved about the x-axis along with the region (Fig. 6.8a). Each rectangle generates a solid disk (Fig. 6.8b). The disks, taken together, approximate the solid of revolution. The approximation is similar to what we would get if we sliced up the original solid like a loaf of bread and reshaped each slice into a disk.

If we focus on a typical rectangle and the disk it generates (Fig. 6.8b), we see that the disk is a right circular cylinder with height Δx_k and radius $f(c_k)$. We can therefore calculate the volume of the disk with the geometry formula $V = \pi r^2 h$:

$$\text{Disk volume} = \pi \times (\text{radius})^2 \times \text{height} = \pi (f(c_k))^2 \Delta x_k.$$

Developing an Integral to Calculate Volume

1. Model the solid that we want to measure in terms of a "generating" continuous function on a closed interval $[a, b]$.

2. Form thin approximating disks using the continuous functions on the subintervals of $[a, b]$.

3. Add the volumes of the thin disks to form a Riemann sum (Eq. 1).

4. Take the limit of the Riemann sum to get an integral (Eq. 2), which gives precisely the volume of the solid in step 1.

(a)

(b)

6.8 (a) The rectangles of Fig. 6.7 revolve around the x-axis to generate solid disks whose volumes, when added, approximate the volume of the solid. (b) The kth disk is shown in an enlarged view. Its volume is base area × height $= \pi [f(c_k)]^2 \Delta x_k$.

How To Apply Eq. (2)

1. Square the expression for the radius function $f(x)$.
2. Multiply by π.
3. Integrate from a to b analytically, or numerically using NINT $(\pi f^2, a, b)$.
4. Support with a grapher, or confirm analytically.

The sum of the volumes of the n disks is

$$\text{Disk volume sum} = \sum_{k=1}^{n} \pi (f(c_k))^2 \Delta x_k. \qquad (1)$$

The expression on the right-hand side of Eq. (1) is a Riemann sum for the function πf^2 on the closed interval $[a, b]$. As the norm of the partition of $[a, b]$ approaches zero, two things happen simultaneously. First, the rectangles approximating the revolved region fit the region with increasing accuracy, and the disks they generate fit the solid of revolution with increasing accuracy. Second, the Riemann sums in Eq. (1) approach the integral of πf^2 from a to b. We therefore define the volume of the solid of revolution to be the value of this integral.

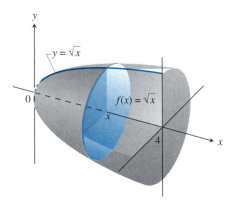

6.9 A slice perpendicular to the axis of the solid in Example 1. The radius at x is $f(x) = \sqrt{x}$.

DEFINITION   **Volume of a Solid of Revolution (about the x-Axis)**

Consider the region between the graph of a continuous function $y = f(x)$ and the x-axis from $x = a$ to $x = b$. The volume of the solid generated by revolving this region about the x-axis is

$$\text{Volume} = \int_a^b \pi (\text{radius})^2 \, dx = \int_a^b \pi (f(x))^2 \, dx. \qquad (2)$$

EXAMPLE 1

The region between the curve $y = \sqrt{x}, 0 \le x \le 4$, and the x-axis is revolved about the x-axis to generate the solid in Fig. 6.9. Find its volume.

Solution

$$\text{Volume} = \int_a^b \pi (\text{radius})^2 \, dx \qquad \text{Eq. 2}$$

$$= \int_0^4 \pi (\sqrt{x})^2 \, dx \qquad \begin{array}{l} \text{The radius function is} \\ f(x) = \sqrt{x}, 0 \le x \le 4. \end{array}$$

$$= \pi \int_0^4 x \, dx = \pi \frac{x^2}{2} \Big]_0^4 = \pi \frac{(4)^2}{2} = 8\pi$$

Support: $\pi \cdot$ NINT $(x, 0, 4) = \pi \cdot 8$. (We could also compute NINT $(\pi x, 0, 4)$ and get $25.133 = 8\pi$.)

EXPLORATION BITS

1. If the continuous function defining the generating region of a solid of revolution has a graph that dips below the x-axis, what modifications, if any, are needed in Eq. (2) for the volume of its solid of revolution? Explain.

2. How would you interpret

$$\int_0^x \pi (f(t))^2 \, dt ?$$

The volume formula in Eq. (2) is consistent with all the standard formulas from geometry. If we use Eq. (2) to calculate the volume of a sphere of radius a, for instance, we get $(4/3)\pi a^3$, just as we should. The next example shows how the calculation goes.

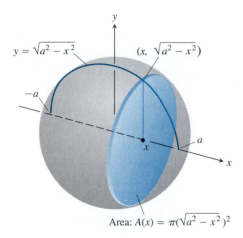

$y = \sqrt{a^2 - x^2}$ $(x, \sqrt{a^2 - x^2})$

Area: $A(x) = \pi(\sqrt{a^2 - x^2})^2$

6.10 The sphere generated by revolving the semicircle $y = \sqrt{a^2 - x^2}$ about the x-axis. The radius at x is $f(x) = \sqrt{a^2 - x^2}$.

EXPLORATION BIT

The volume of the sphere, found in Example 2, is a general result and thus requires an analytic approach rather than graphical or numerical methods. You can, however, use your graphing utility to support the result. Demonstrate how.

EXAMPLE 2

The region enclosed by the semicircle $y = \sqrt{a^2 - x^2}$ and the x-axis is revolved about the x-axis to generate a sphere (Fig. 6.10). Find the volume of the sphere.

Solution

$$\text{Volume} = \int_{-a}^{a} \pi(\text{radius})^2 \, dx \qquad \text{Eq. (2) with limits of integration } -a \text{ and } a$$

$$= \int_{-a}^{a} \pi(\sqrt{a^2 - x^2})^2 \, dx \qquad \text{The radius at } x \text{ is } f(x) = \sqrt{a^2 - x^2}.$$

$$= \pi \int_{-a}^{a} (a^2 - x^2) \, dx = \pi \left[a^2 x - \frac{x^3}{3} \right]_{-a}^{a}$$

$$= \pi \left(a^3 - \frac{a^3}{3} \right) - \pi \left(-a^3 + \frac{a^3}{3} \right)$$

$$= \pi \left(\frac{2a^3}{3} \right) - \pi \left(-\frac{2a^3}{3} \right) = \frac{4}{3} \pi a^3 \qquad \equiv$$

The axis of revolution in the next example is not the x-axis, but the rule for calculating the volume is the same: Integrate $\pi(\text{radius})^2$ between appropriate limits.

EXAMPLE 3

Find the volume generated by revolving the region bounded by $y = \sqrt{x}$ and the lines $y = 1$ and $x = 4$ about the line $y = 1$.

Solution We draw a figure that shows the region and the radius at a typical point on the axis of revolution (Fig. 6.11). The region runs from $x = 1$ to $x = 4$. At each x in the interval $1 \le x \le 4$, the cross-sectional radius is

$$\text{Radius at } x = \sqrt{x} - 1,$$

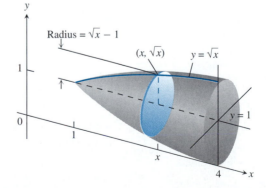

Radius $= \sqrt{x} - 1$

$(x, \sqrt{x})$ $y = \sqrt{x}$

$y = 1$

6.11 The solid swept out by revolving the region bounded by $y = \sqrt{x}$, $x = 4$, and $y = 1$ about the line $y = 1$.

and the integrand in Eq. (2) becomes

$$\pi(\text{radius})^2 = \pi(\sqrt{x} - 1)^2.$$

Therefore,

$$\text{Volume} = \text{NINT}\,(\pi(\sqrt{x} - 1)^2, 1, 4) = 3.665.$$

Confirm:

$$\int_1^4 \pi(\sqrt{x} - 1)^2 \, dx = \pi \int_1^4 (x - 2\sqrt{x} + 1) \, dx$$

$$= \pi \left[\frac{x^2}{2} - 2 \cdot \frac{2}{3}x^{3/2} + x \right]_1^4$$

$$= \frac{7\pi}{6} = 3.665. \qquad \text{Arithmetic omitted} \quad \blacksquare$$

EXPLORATION BIT

Some computer graphing software can produce pictures of solids of revolution that we find are nothing less than stunning. In the next few years, we will also see hand-held graphing utilities that will be able to show quality 3-D visuals. Until then, and in the absence of a computer, we have to rely on our visualization and sketching skills to "see" the solids. A basis for visualizing a solid of revolution is the generating region. We can extend the generating region on a graphing utility to show the complete cross-section of the solid in the xy-plane. Show how to do this for each of Examples 1–4.

Revolving about the y-Axis

To find the volume of the solid generated by revolving the region between a curve $x = f(y)$ and the y-axis from $y = c$ to $y = d$ about the y-axis, use Eq. (2) with x replaced by y.

Volume of a Solid of Revolution (about the y-Axis)

$$\text{Volume} = \int_c^d \pi(\text{radius})^2 \, dy = \int_c^d \pi(f(y))^2 \, dy$$

EXAMPLE 4

The region between the curve $x = 1/\sqrt{y},\, 1 \le y \le 4$, and the y-axis is revolved about the y-axis to generate a solid (Fig. 6.12). Find the volume of the solid.

Solution

$$\text{Volume} = \int_1^4 \pi(\text{radius})^2 \, dy = \int_1^4 \pi \left(\frac{1}{\sqrt{y}} \right)^2 \, dy \qquad \begin{array}{l}\text{The radius function}\\ \text{is } f(y) = 1/\sqrt{y}.\end{array}$$

$$= \int_1^4 \pi \cdot \frac{1}{y} \, dy \qquad = \pi \int_1^4 \frac{1}{y} \, dy,$$

which has an integrand for which we do not yet know an explicit antiderivative. Therefore, we use

$$\text{NINT}\,(\pi/y, 1, 4) = 4.355$$

to get the volume of the solid. (We could also compute $\pi \cdot \text{NINT}\,(1/y, 1, 4)$).

$\blacksquare$

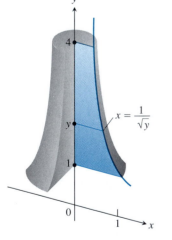

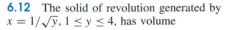

6.12 The solid of revolution generated by $x = 1/\sqrt{y},\, 1 \le y \le 4$, has volume

$$\text{NINT}\,(\pi/y, 1, 4) = 4.355.$$

The Washer Method

If the region we revolve to generate a solid does not border on the axis of revolution,

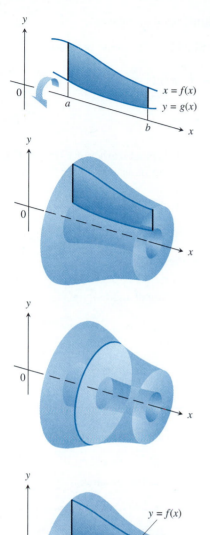

. . . the solid has a hole in it.

The cross-sections perpendicular to the axis of revolution are washers instead of disks.

If we fill in the hole, the volume is

$$\int_a^b \pi (f(x))^2 \ dx.$$

Hole filled in

The volume of the hole itself is

$$\int_a^b \pi (g(x))^2 \ dx,$$

Volume of the hole

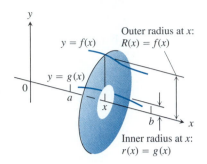

6.13 The typical washer, at x, has outer radius $R(x) = f(x)$ and inner radius $r(x) = g(x)$.

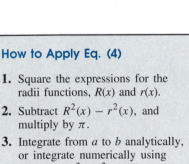

How to Apply Eq. (4)

1. Square the expressions for the radii functions, $R(x)$ and $r(x)$.

2. Subtract $R^2(x) - r^2(x)$, and multiply by π.

3. Integrate from a to b analytically, or integrate numerically using NINT $(\pi(R^2 - r^2), a, b)$.

4. Support with a grapher, or confirm analytically.

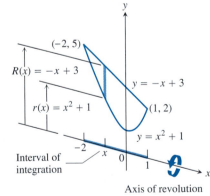

6.14 The region bounded by the curve

$$y = r(x) = x^2 + 1$$

and the line

$$y = R(x) = -x + 3$$

with a thin rectangle whose long sides are perpendicular to the axis of revolution.

and we can find the volume of the original solid as follows:

Volume = (volume with hole filled in) − (volume of hole)

$$= \int_a^b \pi(f(x))^2 \, dx - \int_a^b \pi(g(x))^2 \, dx$$

$$= \int_a^b \pi(f^2(x) - g^2(x)) \, dx. \tag{3}$$

If we return to the cross-sections to look at the typical washer at x (Fig. 6.13), we see that it has two radii, an outer one and an inner one:

Outer washer radius: $R(x) = f(x)$,

Inner washer radius: $r(x) = g(x)$.

Equation (3) can be described in terms of these radii as

$$\text{Volume} = \int_a^b \pi((\text{outer radius})^2 - (\text{inner radius})^2) \, dx$$

$$= \int_a^b \pi(R^2(x) - r^2(x)) \, dx.$$

The Washer Method (about the x-Axis)

$$\text{Volume} = \int_a^b \pi(R^2(x) - r^2(x)) \, dx \tag{4}$$

$$R(x) = \text{outer radius}, \qquad r(x) = \text{inner radius}$$

Notice that the function being integrated in Eq. (4) is $\pi(R^2 - r^2)$, not $\pi(R - r)^2$. Also notice that Eq. (4) turns into the disk-method formula if the inner radius $r(x)$ is zero throughout the interval $a \le x \le b$. The disk-method formula is a special case of what we have here.

Steps for using the washer method are shown in the following example. We apply the steps carefully to remind us of the Riemann sum origin of the integral. We list the steps again after the example.

EXAMPLE 5

The region bounded by the curve $y = x^2 + 1$ and the line $y = -x + 3$ is revolved about the x-axis to generate a solid. Find the volume of the solid.

Solution

STEP 1: Draw the curves and identify the generating region. Draw a thin rectangle across the region (Fig. 6.14) with long sides perpendicular to the axis of revolution.

EXPLORATION BIT

Reproduce the generating region of Example 5 in a viewing window using parametric equations. What role can [*t*Min, *t*Max] play in your picture? Then extend the generating region to show a complete cross-section of the solid in the *xy*-plane.

STEP 2: Find the limits of integration. The limits of integration are the *x*-coordinates of the points where the parabola and line cross. We find them by solving the equations $y = x^2 + 1$ and $y = -x + 3$ simultaneously for *x*:

$$x^2 + 1 = -x + 3$$

$$x^2 + x - 2 = 0$$

$$(x + 2)(x - 1) = 0$$

$$x = -2, \quad x = 1.$$

STEP 3: Find the outer and inner radii of the thin approximating washer that would be swept out by the rectangle if it were revolved about the *x*-axis along with the region (Fig. 6.15). These radii are the distances of the two ends of the rectangle from the axis of revolution, and they give us the radius functions to be used in the volume integral.

Outer radius: $\quad R(x) = -x + 3$

Inner radius: $\quad r(x) \;= x^2 + 1.$

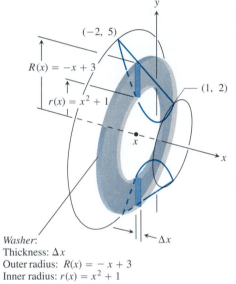

Washer:
Thickness: Δx
Outer radius: $R(x) = -x + 3$
Inner radius: $r(x) = x^2 + 1$

6.15 The volume of the thin washer is

$$(\pi R^2(x) - \pi r^2(x))\Delta x.$$

The sum of the volumes of all the approximating washers is the Riemann sum

$$\Sigma(\pi R^2(x) - \pi r^2(x))\Delta x$$

(shown without subscripts). The limiting value of the Riemann sums as $\Delta x \to 0$ is the integral in Step 4.

STEP 4: Evaluate the volume integral (Eq. 4). Here we evaluate analytically, then support with a grapher.

$$\text{Volume} = \int_a^b \pi(R^2(x) - r^2(x))\,dx \qquad \text{Eq. 4}$$

$$= \int_{-2}^1 \pi((-x+3)^2 - (x^2+1)^2)\,dx \qquad \text{Values from step 3}$$

$$= \int_{-2}^1 \pi(8 - 6x - x^2 - x^4)\,dx \qquad \text{Expressions squared and combined}$$

$$= \pi\left[8x - 3x^2 - \frac{x^3}{3} - \frac{x^5}{5}\right]_{-2}^1 = \frac{117\pi}{5} \qquad \text{Arithmetic omitted}$$

The volume of the solid is $117\pi/5$, or 23.4π.
Support: $\pi \cdot$ NINT $((R(x))^2 - (r(x))^2, -2, 1) = 23.4\pi.$

The steps we took in Example 5 to implement the washer method are these:

> **How to Find Volume by the Washer Method**
>
> **STEP 1:** Draw the curves and identify the generating region. Draw a thin rectangle across the region perpendicular to the axis of revolution.
>
> **STEP 2:** Find the limits of integration.
>
> **STEP 3:** Find the outer and inner radii of the washer (the distances to the ends of the rectangle from the axis of revolution) that would be swept out by the rectangle if it were revolved around the axis of revolution. These give the radius functions to be used in the volume integral.
>
> **STEP 4:** Evaluate the volume integral. Do analytically and support with a grapher, or do with a grapher and confirm analytically.

When we speak of a rectangle being perpendicular or parallel to an axis, we mean that its long sides are perpendicular or parallel, respectively, to the axis.

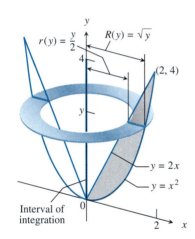

6.16 The region bounded by the parabola $y = x^2$ and the line $y = 2x$. The thin rectangle perpendicular to the axis of revolution sweeps out a washer whose radii can be found from the formulas for the curves.

EXPLORATION BIT

We could revolve the generating region in Fig. 6.16 around either the x-axis or the y-axis. Would the two solids of revolution be congruent? Make a conjecture, and give a convincing argument. Would the two solids have the same volume? Make a conjecture, and give a convincing argument. (This is easy if the solids *are* congruent.)

EXPLORATION 1

Washers about the *y*-Axis

To find the volume of a solid generated by revolving a region about the y-axis, we use the steps listed above and integrate with respect to y instead of x. Try it for the region in the first quadrant bounded by the parabola $y = x^2$ and the line $y = 2x$.

STEP 1: A sketch of the curves, the generating region, and a thin rectangle across the region is shown in Fig. 6.16. Show the region in a viewing window using $y_1 = 2x$ and $y_2 = x^2$. Just for fun, use ZOOM BOX to show a thin rectangle (two or three pixels wide) in your viewing window perpendicular to the axis of revolution.

STEP 2: The sketch suggests that the limits of integration are $y = 0$ and $y = 4$. Support the limits of integration graphically using ZOOM or SOLVE, and confirm algebraically.

STEP 3: Explain why the distances of the ends of the rectangle from the axis of revolution are

$$R(y) = \sqrt{y} \qquad \text{and} \qquad r(y) = \frac{y}{2}.$$

STEP 4: Set up and evaluate the volume integral in two ways, analytically and with a grapher. Compare the results.

If the axis of revolution is not one of the coordinate axes, the rule for finding the volume is still the same: Find formulas for the solid's outer and inner radii, and integrate $\pi(R^2 - r^2)$ between appropriate limits. In the next example, we revolve a region about a line parallel to the y-axis. We would handle a region revolved about a line parallel to the x-axis in a similar way.

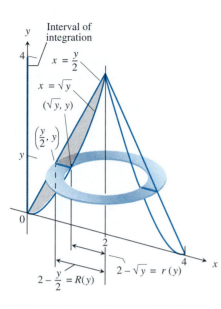

6.17 The inner and outer radii of the washer are measured, as always, as distances from the axis of revolution.

EXAMPLE 6

The region between the parabola $y = x^2$ and the line $y = 2x$ is revolved about the line $x = 2$ parallel to the y-axis. Find the volume swept out.

Solution

STEP 1: Draw the region and a thin rectangle across the region perpendicular to the axis of revolution (Fig. 6.17).

STEP 2: Find the limits of integration. The line crosses the parabola at $(0, 0)$ and $(2, 4)$, so we integrate from $y = 0$ to $y = 4$.

STEP 3: Find the distances of the ends of the rectangle from the axis of revolution. From Fig. 6.17,

$$R(y) = 2 - \frac{y}{2}, \qquad r(y) = 2 - \sqrt{y}.$$

STEP 4: With $y_2 = r(y) = 2 - \sqrt{y}$ and $y_1 = R(y) = 2 - y/2$, integrate with respect to y and find

$$\pi \, \text{NINT} \, (y_1^2 - y_2^2, 0, 4) = 2.667\pi.$$

The volume is 2.667π.

We confirm this numerical integration result analytically:

$$\text{Volume} = \int_0^4 \pi \left(\left(2 - \frac{y}{2}\right)^2 - (2 - \sqrt{y})^2 \right) \, dy$$

$$= \pi \int_0^4 \left(\frac{y^2}{4} - 3y + 4\sqrt{y} \right) \, dy \qquad \text{Expressions squared and combined.}$$

$$= \pi \left[\frac{y^3}{12} - \frac{3y^2}{2} + \frac{8}{3} y^{3/2} \right]_0^4 = \frac{8}{3}\pi \qquad \text{Arithmetic omitted} \quad \equiv$$

EXPLORATION 2

Practical Problems, Part 2

To illustrate analytical techniques, the examples of this section are designed so that calculations do not "get in the way." The antiderivatives are easy to find, and the arithmetic using the limits of integration is straightforward. In practical situations, most problems are not this accommodating.

1. Try to find the volume of the solid formed by revolving the region bounded by the curves

$$y_1 = x + \frac{6}{\sqrt{x + 6}} \qquad \text{and} \qquad y_2 = \frac{2x^2 + 2x + 1}{5}$$

about the x-axis. As Fig. 6.18 seems to suggest, the limits of integration may be "nice." You should find, however, that you still will need to use technology. Why?

2. ZOOM OUT. Have you given a *complete* solution to the problem?

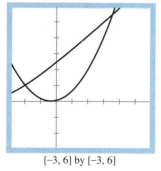

[−3, 6] by [−3, 6]

6.18 The graphs of

$$y_1 = x + \frac{6}{\sqrt{x + 6}}$$

and

$$y_2 = \frac{2x^2 + 2x + 1}{5}.$$

Will the solid of revolution about the x-axis have a hole through it?

Solids of Revolution

We can generate a solid of revolution by revolving a plane region about an axis. To calculate the volume of the solid, we integrate the square of the radius of the solid along an interval on the axis and multiply by π.

If the axis of revolution is the x-axis, we integrate along the x-axis (Example 1).

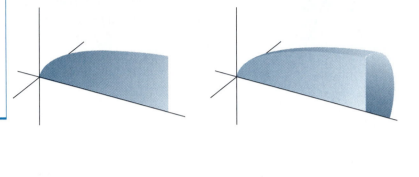

EXPLORATION BIT

Notice that the generating region in Fig. 6.17 is the same as the one in Fig. 6.16. Are the two solids of revolution congruent? Give a convincing argument. Do the two solids have the same volume?

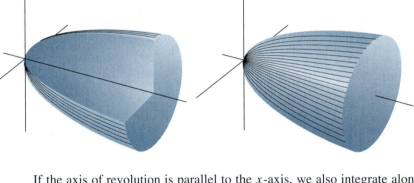

If the axis of revolution is parallel to the x-axis, we also integrate along an interval of the x-axis (Example 3).

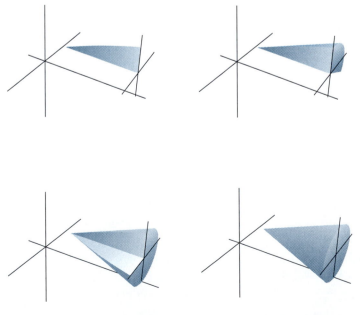

If the axis of revolution is the y-axis, we integrate along an interval of the y-axis (Example 4).

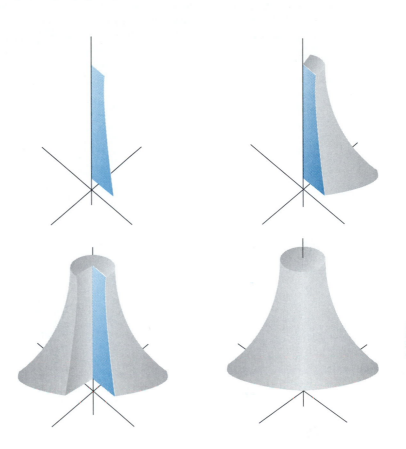

If the axis of revolution is not part of the region being revolved, the solid has a hole in it. We integrate along the coordinate axis parallel to the axis of revolution (Example 5).

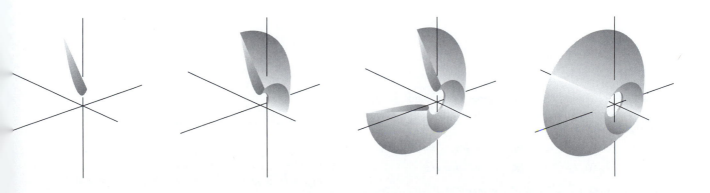

(Generated by *Mathematica*.)

Exercises 6.2

Use the disk method to find the volumes of the solids generated by revolving the regions bounded by the lines and curves in Exercises 1–8 about the x-axis.

1. $x + y = 2$, $y = 0$, $x = 0$

2. $y = x^2$, $y = 0$, $x = 2$

3. $y = \sqrt{9 - x^2}$, $y = 0$

4. $y = x - x^2$, $y = 0$

5. $y = x^3$, $y = 0$, $x = 2$

6. $y = e^x$, $y = 0$, $x = 0$, $x = \ln 2$

7. $y = \sqrt{\cos x}$, $0 \le x \le \pi/2$, $y = 0$, $x = 0$

8. $y = \sec x$, $y = 0$, $x = -\pi/4$, $x = \pi/4$

Use the disk method to find the volumes of the solids generated by revolving about the y-axis the regions bounded by the lines and curves in Exercises 9–16.

9. $y = x/2$, $y = 2$, $x = 0$

10. $x = \sqrt{4 - y}$, $x = 0$, $y = 0$

11. $x = \sqrt{5}y^2$, $x = 0$, $y = -1$, $y = 1$

12. $x = 1 - y^2$, $x = 0$

13. $x = y^{3/2}$, $x = 0$, $y = 2$

14. $x = \sqrt{2 \sin 2y}$, $0 \le y \le \pi/2$, $x = 0$

15. $x = 2/\sqrt{y + 1}$, $x = 0$, $y = 0$, $y = 3$

16. $x = 2/(y + 1)$, $x = 0$, $y = 0$, $y = 1$

Use the washer method to find the volumes of the solids generated by revolving about the x-axis the regions bounded by the lines and curves in Exercises 17–24.

17. $y = x$, $y = 1$, $x = 0$ 18. $y = 2x$, $y = x$, $x = 1$

19. $y = x^2$, $y = 4$, $x = 0$ 20. $y = x^2 + 3$, $y = 4$

21. $y = x^2 + 1$, $y = x + 3$ 22. $y = 4 - x^2$, $y = 2 - x$

23. $y = \sec x$, $y = \sqrt{2}$, $-\pi/4 \le x \le \pi/4$

24. $y = 2/\sqrt{x}$, $y = 2$, $x = 4$

Use the washer method to find the volumes of the solids generated by revolving the regions bounded by the lines and curves in Exercises 25–30 about the y-axis.

25. $y = x - 1$, $y = 1$, $x = 1$ 26. $y = x - 1$, $y = 0$, $x = 4$

27. $y = x^2$, $y = 0$, $x = 2$ 28. $y = x$, $y = \sqrt{x}$

29. The semicircle $x = \sqrt{25 - y^2}$ and the y-axis

30. The semicircle $x = \sqrt{25 - y^2}$ and the line $x = 4$

Find the volume of the solid generated by revolving the shaded region about the indicated axis in Exercises 31–34.

31. The x-axis

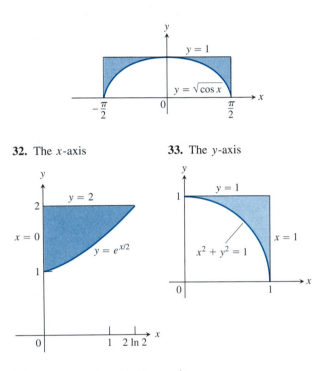

32. The x-axis   33. The y-axis

34. a) The x-axis   b) The y-axis

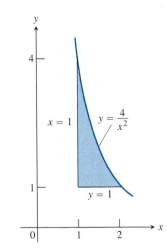

35. Find the volume of the solid generated by revolving the region bounded by $y = \sqrt{x}$ and the lines $y = 2$ and $x = 0$

a) about the x-axis,   b) about the y-axis,
c) about the line $y = 2$,   d) about the line $x = 4$.

36. Find the volume of the solid generated by revolving the triangular region bounded by the lines $y = 2x$, $y = 0$, and $x = 1$
   a) about the line $x = 1$,
   b) about the line $x = 2$.

37. Find the volume of the solid generated by revolving the region bounded by the parabola $y = x^2$ and the line $y = 1$
   a) about the line $y = 1$,
   b) about the line $y = 2$,
   c) about the line $y = -1$.

38. Determine the volume of a right circular cylinder of radius r and height h by revolving the rectangle about the x-axis.

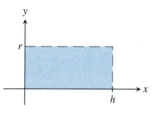

39. Determine the volume of a right circular cone of radius r and height h by revolving the triangle about the y-axis.

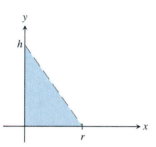

40. *Minimizing a volume.*   The arch $y = \sin x$, $0 \leq x \leq \pi$, is revolved about the line $y = c$ to generate the solid shown here. Find the value of c that minimizes this volume. What is the minimum volume?

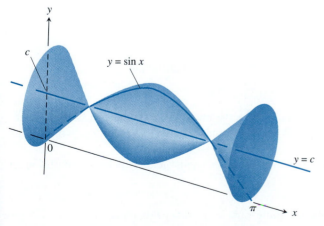

41. a) Solve $5 \cos (0.5x + 1) = x^2 + 1$ for x.
   b) Find the volume of the solid of revolution determined by revolving the region bounded by the curves $y = 5 \cos (0.5x + 1)$ and $y = x^2 + 1$ about the x-axis.

42. Solve Exercise 40 with the arch given by $y = (\sin x)\sqrt{x^2 + 3}$, $0 \leq x \leq \pi$.

43. *Designing a wok.*   You are designing a wok frying pan that will be shaped like a spherical bowl with handles. A bit of experimentation at home persuades you that you can get one that holds about 3 L if you make it 9 cm deep and give the sphere a radius of 16 cm. To be sure, you picture the wok as a solid of revolution and calculate the volume with an integral. Your picture looks like this:

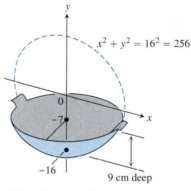

Dimensions in centimeters

What volume do you really get?

44. *Designing a plumb bob.*   Having been asked to design a brass plumb bob that will weigh in the neighborhood of 190 gm, you decide to shape it like the solid of revolution shown here. What is the volume of the solid? If you specify a brass that weighs 8.5 gm/cm³, about how much will the bob weigh?

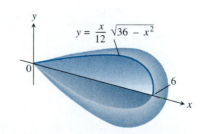

6.3

Cylindrical Shells—An Alternative to Washers

If the rectangular strips that approximate a region being revolved about an axis run parallel to the axis instead of perpendicular to it, they sweep out cylindrical shells. Cylindrical shells are sometimes easier to work with than washers because the formula that they lead to does not require squaring. They also allow us to revolve around one axis and integrate along an interval of the other axis.

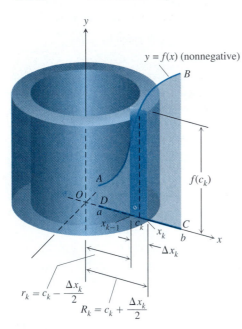

6.19 The solid swept out by revolving region $ABCD$ about the y-axis can be approximated with cylindrical shells like the one shown here.

The Basic Shell Formula

We arrive at the cylindrical shell volume formula in the following way.

Suppose we revolve region $ABCD$ under the graph of f in Fig. 6.19 about the y-axis to generate a solid. To find the volume of the solid, we first approximate the region with vertical rectangles based on a partition of the closed interval $[a, b]$ on which the region stands. The rectangles run parallel to the y-axis, the axis of revolution.

Figure 6.19 shows a typical approximating rectangle. Its dimensions are $f(c_k)$ by Δx_k. The point c_k is chosen to be the midpoint of the interval from x_{k-1} to x_k. Since it does not matter where the c_k's are chosen in their intervals when we find limits of Riemann sums, we are free to choose each c_k as we please. The resulting formula in this case will be less cumbersome if we stay with midpoints. You will see why in just a moment.

Again with reference to Fig. 6.19, the approximating cylindrical shell swept out by revolving the rectangle about the y-axis is a cylinder with a hole through it with these dimensions:

Shell height:	$f(c_k)$
Inner radius:	$r_k = c_k - \dfrac{\Delta x_k}{2}$
Outer radius:	$R_k = c_k + \dfrac{\Delta x_k}{2}$
Base ring area:	$A_k = \pi R_k^2 - \pi r_k^2 = \pi (R_k + r_k)(R_k - r_k)$
	$= \pi (2c_k) \Delta x_k$
Shell volume:	$V_k = \text{base ring area} \times \text{shell height}$
	$= 2\pi c_k f(c_k) \Delta x_k.$

The advantage of choosing c_k to be the midpoint of its interval becomes clear in the formula for A_k. With this choice, $R_k + r_k$ equals $2c_k$; without this choice, it doesn't.

The sum of the volumes of the n shells generated by the partition of $[a, b]$ is

$$\text{Shell volume sum} = \sum_{k=1}^{n} 2\pi c_k f(c_k) \Delta x_k.$$

Developing an Integral to Calculate Volume

1. Model the solid that we want to measure in terms of a "generating" continuous function on a closed interval $[a, b]$.

2. Form thin-walled approximating shells using the continuous functions on the subintervals of $[a, b]$.

3. Add the volumes of the thin shells to form a Riemann sum.

4. Take the limit of the Riemann sum to get an integral, which gives precisely the volume of the solid in step 1.

The volume of the solid swept out by revolving region $ABCD$ about the y-axis is taken to be the limit of the shell volume sums as the norm of the partition of $[a, b]$ goes to zero. If f is continuous, the limit exists and can be found by integrating the product $2\pi x f(x)$ from $x = a$ to $x = b$.

The Shell Method (about the y-Axis)

Suppose $y = f(x)$ is continuous throughout an interval $a \leq x \leq b$ that does not cross the y-axis. Then, the volume of the solid generated by revolving the region between the graph of f and the interval $a \leq x \leq b$ about the y-axis is found by integrating $2\pi x f(x)$ with respect to x from a to b.

$$\text{Volume} = \int_a^b 2\pi \left(\begin{array}{c} \text{shell} \\ \text{radius} \end{array} \right) \left(\begin{array}{c} \text{shell} \\ \text{height} \end{array} \right) dx = \int_a^b 2\pi x f(x) \, dx. \quad (1)$$

One way to remember Eq. (1) is to imagine that a cylindrical shell of average circumference $2\pi x$, height $f(x)$, and thickness dx has been cut along a generating rectangle and rolled flat like a sheet of tin (Fig. 6.20). The sheet is almost a rectangular solid of dimensions $2\pi x$ by $f(x)$ by dx. Hence, the shell's volume is about $2\pi x f(x) \, dx$. Equation (1) says that the volume of the complete solid is the integral of $2\pi x f(x) \, dx$ from a to b.

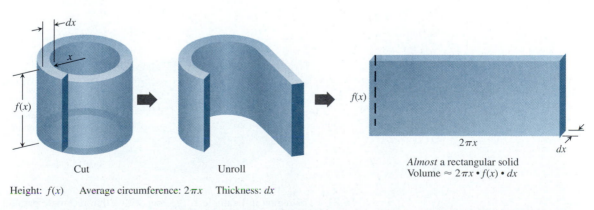

Height: $f(x)$ Average circumference: $2\pi x$ Thickness: dx

Cut   Unroll   *Almost* a rectangular solid Volume $\approx 2\pi x \cdot f(x) \cdot dx$

6.20   A model for remembering the integrand in the formula for cylindrical shells.

How to Find Volume by the Shell Method

STEP 1: Draw the curves and identify the generating region. Draw a thin rectangle across the region parallel to the axis of revolution.

STEP 2: Find the limits of integration.

STEP 3: Identify the rectangle's height (shell height, $f(x)$), width (shell thickness, dx), and distance from the axis of revolution (average shell radius, x). Write the volume of the shell as the product

$$2\pi(\text{shell radius})(\text{shell height})(\text{shell thickness}) = 2\pi x f(x)\,dx.$$

STEP 4: Evaluate the volume integral. Do analytically and support with a grapher, or do with a grapher and confirm analytically.

EXAMPLE 1

The region bounded by the curve $y = \sqrt{x}$, the x-axis, and the line $x = 4$ is revolved about the y-axis to generate a solid. Find the volume of the solid.

Solution

STEP 1: Sketch the region and a thin rectangle across the region parallel to the axis of revolution (Fig. 6.21).

STEP 2: The limits of integration are $x = 0$ and $x = 4$.

STEP 3: The height of the rectangle is $\sqrt{x}$. Its width is dx, and its distance from the axis of revolution is x. The shell that it sweeps out has volume

$$2\pi(\text{shell radius})(\text{shell height})(\text{shell thickness}) = 2\pi x \sqrt{x}\,dx.$$

STEP 4: Evaluate the volume integral in Eq. (1). For this example, we evaluate numerically using NINT and then confirm analytically.

$$\text{Volume} = \int_0^4 2\pi x \sqrt{x}\,dx$$

$$= \text{NINT}\,(2\pi x \sqrt{x}, 0, 4) = 80.425$$

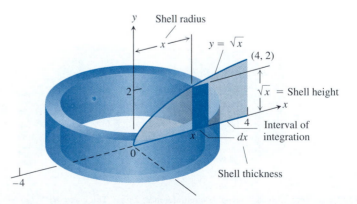

6.21 The region in Example 1, together with a thin rectangle parallel to the axis of revolution (the y-axis). The rectangle sweeps out a cylindrical shell.

The volume is 80.425. We confirm this numerical integration result analytically:

$$\int_0^4 2\pi x \sqrt{x}\; dx = 2\pi \int_0^4 x^{3/2}\; dx$$

$$= 2\pi \left[\frac{2}{5} x^{5/2} \right]_0^4$$

$$= \frac{128\pi}{5} = 80.425. \qquad \text{Arithmetic omitted} \qquad \blacksquare$$

Shells about the *x*-Axis

To use shells to find the volume of a solid generated by revolving a region about the *x*-axis instead of the *y*-axis, use Eq. (1) with *y* in place of *x*. Except for changes in notation, the steps that we take to implement the new formula are the same as before.

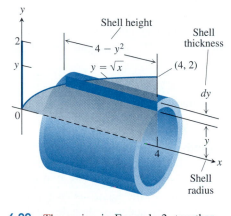

6.22 The region in Example 2, together with a thin rectangle parallel to the axis of revolution.

EXAMPLE 2

The region bounded by the curve $y = \sqrt{x}$, the *x*-axis, and the line $x = 4$ is revolved about the *x*-axis to generate a solid. Find the volume of the solid.

Solution

STEP 1: Sketch the region and a thin rectangle across the region parallel to the *x*-axis, the axis of revolution (Fig. 6.22).

STEP 2: The limits of integration are $y = 0$ and $y = 2$.

STEP 3: The height of the shell swept out by the thin rectangle is $4 - y^2$. The thickness of the shell is dy, and its radius is y. Write the volume of the shell as

$$2\pi(\text{shell radius})\,(\text{shell height})\, dy = 2\pi y(4 - y^2)\, dy.$$

STEP 4: Evaluate the volume integral in Eq. (1) with *y* in place of *x*.

$$\text{Volume} = \int_c^d 2\pi \begin{pmatrix} \text{shell} \\ \text{radius} \end{pmatrix} \begin{pmatrix} \text{shell} \\ \text{height} \end{pmatrix} dy = \int_0^2 2\pi y \left(4 - y^2\right)\; dy$$

$$= 2\pi \int_0^2 (4y - y^3)\; dy = 2\pi \left[2y^2 - \frac{y^4}{4} \right]_0^2 = 8\pi. \qquad \text{Arithmetic omitted}$$

Support: $\pi \cdot \text{NINT}\,(2y(4 - y^2), 0, 2) = 8\pi.$ $\qquad \blacksquare$

Another form of support, of course, is to do the problem a second way and get the same result. Compare the volume found here with the volume found by using the disk method in Section 6.2, Example 1.

EXPLORATION 1

Shifts

If the axis of revolution is a line parallel to one of the coordinate axes, we use the same steps as before. The only added complication is that the expression for the typical shell radius is no longer simply x or y. Figure 6.23 shows the region in the first quadrant bounded by the parabola $y = x^2$, the y-axis, and the line $y = 1$. It is to be revolved about the line $x = 2$ to generate a solid. We would like you to find the volume in two ways.

1. **The shell method:** Sketch the region and a thin rectangle across the region parallel to the line $x = 2$ (the axis of revolution). Determine the limits of integration, $x = a$ to $x = b$. Describe the radius of the shell swept out by the thin rectangle. (If the axis is at 2 and the rectangle is at x, how far is it from the axis to the rectangle?) Describe the rectangle's height from $y = x^2$ up to $y = 1$. Then integrate to find the volume:

$$\text{Volume} = \int_a^b 2\pi \left(\begin{array}{c}\text{shell}\\\text{radius}\end{array}\right) \left(\begin{array}{c}\text{shell}\\\text{height}\end{array}\right) dx.$$

Support your analytic result. Confirm your NINT result.

2. **Using a shift with the shell method:** The volume of the solid described in part 1 will be the same if we first shift the given region 2 units to the left and then revolve about the line $x = 0$ (the y-axis). Confirm that the resulting volume is the same as that found in part 1. (Recall that a shift to the left is accomplished by replacing x by $x + 2$. Both the integrand and the limits of integration are affected.)

EXPLORATION BIT

Suppose, in Exploration 1, that the graph of $y = x^2$ is replaced by the curve

$$y = 1 - \frac{\sin 4x}{4x}.$$

Show that the volume of the solid of revolution is

$$\int_0^b 2\pi (2 - x)\frac{\sin 4x}{4x}\, dx,$$

where b is the x-coordinate of the first point of intersection of $y = 1 - (\sin 4x)/4x$ and $y = 1$ to the right of the y-axis. What must you do to evaluate the volume? Explain how you can do it.

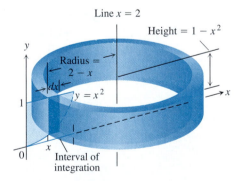

6.23 The volume of a solid of revolution is the same whether the generating region is revolved around $x = 2$ or the region is shifted 2 units to the left and then revolved around $x = 0$ (the y-axis).

Table 6.1 summarizes the methods of finding volumes with washers and shells. All three methods—disk, washer, and shell—for calculating the volume of a solid of revolution always agree. We illustrate this agreement in Example 3.

TABLE 6.1 Washers versus Shells

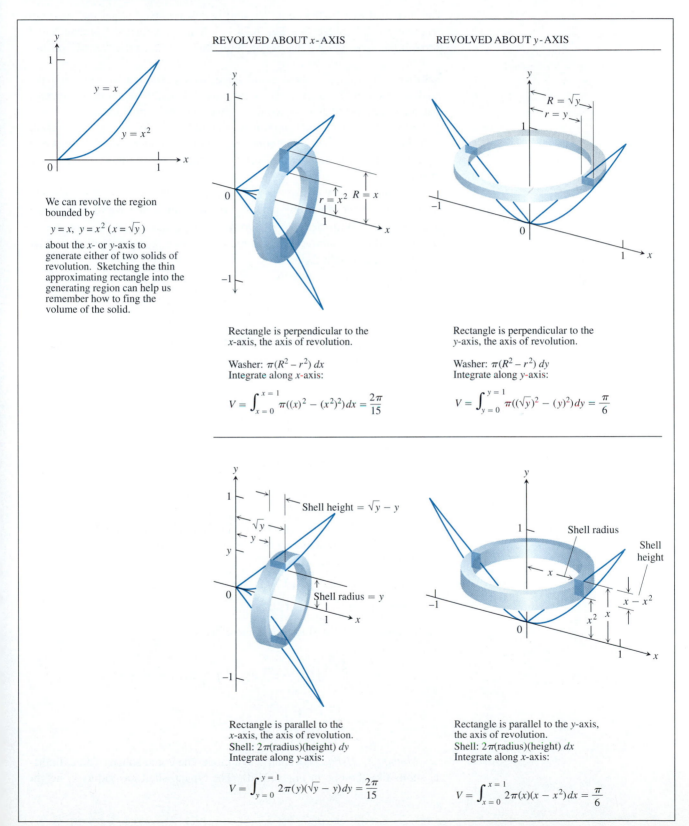

We can revolve the region bounded by

$$y = x, \quad y = x^2 \; (x = \sqrt{y})$$

about the x- or y-axis to generate either of two solids of revolution. Sketching the thin approximating rectangle into the generating region can help us remember how to fing the volume of the solid.

REVOLVED ABOUT x-AXIS

Rectangle is perpendicular to the x-axis, the axis of revolution.

Washer: $\pi(R^2 - r^2)\,dx$
Integrate along x-axis:

$$V = \int_{x=0}^{x=1} \pi((x)^2 - (x^2)^2)\,dx = \frac{2\pi}{15}$$

Rectangle is parallel to the x-axis, the axis of revolution.
Shell: $2\pi(\text{radius})(\text{height})\,dy$
Integrate along y-axis:

$$V = \int_{y=0}^{y=1} 2\pi(y)(\sqrt{y} - y)\,dy = \frac{2\pi}{15}$$

REVOLVED ABOUT y-AXIS

Rectangle is perpendicular to the y-axis, the axis of revolution.

Washer: $\pi(R^2 - r^2)\,dy$
Integrate along y-axis:

$$V = \int_{y=0}^{y=1} \pi((\sqrt{y})^2 - (y)^2)\,dy = \frac{\pi}{6}$$

Rectangle is parallel to the y-axis, the axis of revolution.
Shell: $2\pi(\text{radius})(\text{height})\,dx$
Integrate along x-axis:

$$V = \int_{x=0}^{x=1} 2\pi(x)(x - x^2)\,dx = \frac{\pi}{6}$$

EXAMPLE 3

The disk enclosed by the circle $x^2 + y^2 = 4$ is revolved about the y-axis to generate a solid sphere (see Fig. 6.24.) A hole of diameter 2 is then bored through the sphere along the y-axis. Find the volume of the "cored" sphere.

Solution   The cored sphere could have been generated by revolving the shaded region in Fig. 6.24(a) about the y-axis. Thus, there are three methods that we might use to find the volume: disks, washers, and shells.

Method 1: Disks and subtraction. Figure 6.24(b) shows the solid sphere with the core pulled out. The core is a circular cylinder with spherical end caps. Our plan is to subtract the volume of the core from the volume of the sphere.

We can simplify matters by imagining that the two caps have already been sliced off the sphere by planes perpendicular to the y-axis at $y = \sqrt{3}$ and $y = -\sqrt{3}$. With the caps removed, the truncated sphere has volume T, say. From this we subtract the volume of the hole, a right circular cylinder of radius 1 and height $2\sqrt{3}$.

The volume of the hole is

$$H = \pi (1)^2 \left(2\sqrt{3}\right) = 2\pi \sqrt{3}.$$

The truncated sphere (before drilling) is a solid of revolution whose cross-sections perpendicular to the y-axis are disks. The radius of a typical disk is $\sqrt{4 - y^2}$. Therefore,

$$T = \int_{-\sqrt{3}}^{\sqrt{3}} \pi \,(\text{radius})^2 \; dy = \int_{-\sqrt{3}}^{\sqrt{3}} \pi \left(4 - y^2\right) \; dy = \pi \left[4y - \frac{y^3}{3}\right]_{-\sqrt{3}}^{\sqrt{3}} = 6\pi \sqrt{3}.$$

The volume of the cored sphere is

$$T - H = 6\pi \sqrt{3} - 2\pi \sqrt{3} = 4\pi \sqrt{3}.$$

Method 2: Washers. The cored sphere is a solid of revolution whose cross-sections perpendicular to the y-axis are washers (Fig. 6.24c). The radii of a typical washer are

Outer radius: $R = \sqrt{4 - y^2}$,

Inner radius: $r = 1$.

The volume of the cored sphere is therefore

$$V = \int_{-\sqrt{3}}^{\sqrt{3}} \pi \,(R^2 - r^2) \; dy = \int_{-\sqrt{3}}^{\sqrt{3}} \pi \left(4 - y^2 - 1\right) \; dy$$

$$= \pi \left[3y - \frac{y^3}{3}\right]_{-\sqrt{3}}^{\sqrt{3}} = 4\pi \sqrt{3}.$$

Method 3: Cylindrical shells. We model the cored sphere with cylindrical shells like the one in Fig. 6.24(d). The typical shell has radius x, height

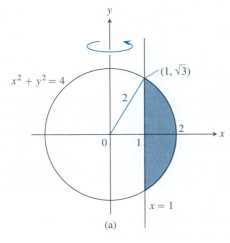

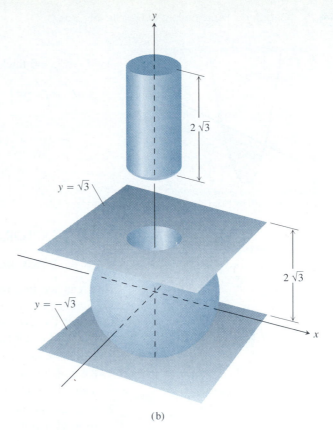

(a)

(b)

The volume of the cored sphere is the volume swept out by the shaded region as it revolves about the y-axis.

With the method of disks, we can subtract the volume of the core from the volume of the sphere, or we can subtract the volume of the hole left by the core from the volume of the truncated sphere.

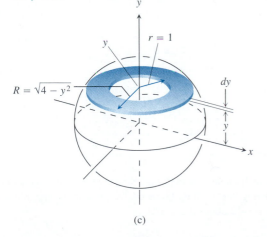

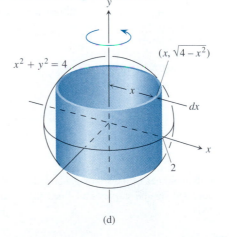

(c)

(d)

With the method of washers, we can calculate the volume of the cored sphere directly by modeling it as a stack of washers perpendicular to the y-axis.

With the method of cylindrical shells, we can calculate the volume of the cored sphere directly by modeling it as a union of cylindrical shells parallel to the y-axis.

6.24 (a) The tinted region generates the solid of revolution. (b) An exploded view showing the sphere with the core removed. (c) A phantom view showing a cross-sectional slice of the sphere with the core removed. (d) Filling the volume with cylindrical shells.

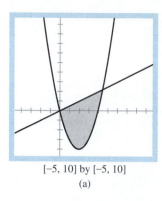

[−5, 10] by [−5, 10]
(a)

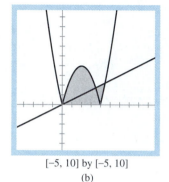

[−5, 10] by [−5, 10]
(b)

6.25 (a) The region between

$$y_1 = \frac{1}{2}x \quad \text{and} \quad y_2 = x^2 - 4x.$$

When this region is revolved about the x-axis, it will generate part of the solid of revolution twice. (b) Reflecting the lower portion of the generating region about the x-axis shows the effective generating region.

$2\sqrt{4 - x^2}$, and thickness dx. The volume of the cored sphere is

$$V = \int_1^2 2\pi \left(\begin{array}{c} \text{shell} \\ \text{radius} \end{array} \right) \left(\begin{array}{c} \text{shell} \\ \text{height} \end{array} \right) dx = \int_1^2 4\pi x \sqrt{4 - x^2} \, dx$$

$$= 4\pi \left[-\frac{1}{3}(4 - x^2)^{3/2} \right]_1^2 \qquad \begin{array}{l} \text{After substituting } u = 4 - x^2, \\ \text{integrating, and substituting back} \end{array}$$

$$= 0 - 4\pi \left[-\frac{1}{3}(4 - 1)^{3/2} \right] = 4\pi\sqrt{3}.$$

EXPLORATION 2

Solids Generated by Overlapping Regions

How can we find the volume of the solid of revolution when the axis of revolution passes through the generating region (Fig. 6.25a)? One way is to reflect the generating region on one side of the axis to the other side (Fig. 6.25b) and then use familiar techniques.

1. Explain how you would find the volume of the solid of revolution (about the x-axis) for the generating region shown in Fig. 6.25(a). Show the actual integration you would use.

2. In your viewing window, show the cross-section in the xy-plane of the solid of revolution for the given generating region.

   a) The region bounded by the graphs of $y_1 = 3 \sin x$ and $y_2 = 2 \cos x$ for $0 \le x \le \pi$ about the x-axis.

   b) The region bounded by the graphs of $y_1 = 2^x$ and $y_2 = 4 - (x - 1)^2$ about the y-axis.

Exercises 6.3

Use the shell method to find the volumes of the solids generated by revolving the regions bounded by the curves and lines in Exercises 1–6 about the y-axis.

1. $y = x$, $y = -x/2$, and $x = 2$

2. $y = \sqrt{x}$, $y = 0$, and $x = 4$

3. $y = x^2 + 1$, $y = 0$, $x = 0$, and $x = 1$

4. $y = 2x - 1$, $y = \sqrt{x}$, and $x = 0$

5. $y = 1/x$, $y = 0$, $x = 1/2$, $x = 2$

6. $y = 1/x^2$, $y = 0$, $x = 1/2$, $x = 2$

Use the shell method to find the volumes of the solids generated by revolving the regions bounded by the curves and lines in Exercises 7–12 about the x-axis.

7. $y = |x|$ and $y = 1$

8. $y = x$, $y = 1$, and $x = 2$

9. $y = \sqrt{x}$, $y = 0$, and $y = x - 2$

10. $y = 2$, $x = -y$, and $x = \sqrt{y}$

11. The parabola $x = 2y - y^2$ and the y-axis

12. The parabola $x = 2y - y^2$ and the line $y = x$

Find the volumes of the solids generated by revolving the shaded regions in Exercises 13–16 about the indicated axes.

13. The y-axis

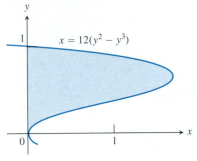

14. The y-axis

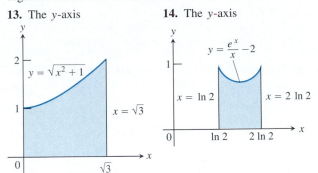

15. The x-axis

16. The x-axis

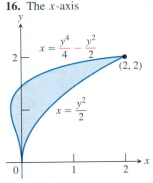

In Exercises 17–26, find the volumes of the solids generated by revolving the regions about the given axes.

17. The triangle with vertices $(1, 1)$, $(1, 2)$, and $(2, 2)$ about

   a) the x-axis.   **b)** the y-axis.

18. The region in the first quadrant bounded by the curve $x = y - y^3$ and the y-axis about

   a) the x-axis.   **b)** the y-axis.

19. The region in the first quadrant bounded by $x = y - y^3$, $x = 1$, and $y = 1$ about

   a) the x-axis.   **b)** the y-axis.

   c) the line $x = 1$.   **d)** the line $y = 1$.

20. The triangular region bounded by the lines $2y = x + 4$, $y = x$, and $x = 0$ about

   a) the x-axis.

b) the y-axis.

c) the line $x = 4$.

d) the line $y = 8$.

21. The region in the first quadrant bounded by $y = x^3$ and $y = 4x$ about

   a) the x-axis.

   b) the line $y = 8$.

22. The region bounded by $y = \sqrt{x}$ and $y = x^2/8$ about

   a) the x-axis.

23. The region bounded by $y = 2x - x^2$ and $y = x$ about

   a) the y-axis.

   b) the line $x = 1$.

24. The region bounded by $y = \sqrt{x}$, $y = 2$, $x = 0$ about

   a) the x-axis.

   b) the y-axis.

   c) the line $x = 4$.

   d) the line $y = 2$.

25. The region bounded by $y = -x + 3$ and $y = x^2 - 3x$ about

   a) the x-axis.

   b) the y-axis.

26. The region bounded by $y = -x^2 + x + 2$ and $y = -x - 1$ about

   a) the x-axis.   **b)** the y-axis.

27. The region enclosed by the circle $x^2 + y^2 = 9$ is revolved about the y-axis to generate a solid sphere. A hole of diameter 2 is then bored through the sphere along the y-axis. Find the volume of the "cored" sphere.

28. The region enclosed by the ellipse $\dfrac{x^2}{4} + \dfrac{y^2}{9} = 1$ is revolved about the y-axis to generate a solid ellipsoid. A hole of diameter 1 is then bored through the ellipsoid along the y-axis. Find the volume of the "cored" ellipsoid.

29. Consider the region in the first quadrant bounded by the curve $y = 1 - (\sin 4x)/4x$ and the line $y = 1$ from the y-axis to their first point of intersection.

   a) Show that the volume of the solid determined by revolving this region about the line $x = 2$ is

$$\int_0^b 2\pi (2 - x)(\sin 4x)/4x \ dx,$$

   where b is the x-coordinate of the first point of intersection.

   b) Determine b given in part (a).

   c) Find the volume of the solid determined in part (a).

30. Suppose $f(x)$ and $g(x)$ are continuous on $[a, c]$ and satisfy the following conditions:

  i) $f(a) = g(a)$, $f(b) = g(b)$, and $f(c) = g(c)$, for $a < b < c$;

  ii) $f(x) > g(x)$ for all x in (a, b);

  iii) $f(x) < g(x)$ for all x in (b, c).

Answer *always true* or *not always true* with a brief explanation to each of the following claims.

  a) The volume of the solid generated by revolving the region between $f(x)$ and $g(x)$ on $[a, b]$ about the x-axis is given by

$$V = \int_a^b \pi (g(x) - f(x))^2 \, dx.$$

  b) The volume of the solid generated by revolving the region between $f(x)$ and $g(x)$ on $[b, c]$ about the line $x = a$ is given by

$$V = \int_b^c 2\pi (x - a)(g(x) - f(x)) \, dx.$$

  c) The volume of the solid generated by revolving the region between $f(x)$ and $g(x)$ on $[a, c]$ about $x = b$ is given by

$$V = \int_a^b 2\pi (b - x)(f(x) - g(x)) \, dx +$$
$$\int_b^c 2\pi (x - b)(g(x) - f(x)) \, dx.$$

6.4 —————— Lengths of Curves in the Plane

We approximate the length of a curved path in the plane as we use a ruler to estimate the length of a curved road on a map, by measuring from point to point with straight-line segments and adding the results. There is a limit to the accuracy of such an estimate, however, imposed in part by how accurately we measure and in part by how many line segments we are willing to use.

With calculus, we can usually do a better job because we can imagine using straight-line segments that are as short as we please, each set of segments making a polygonal path that fits the curve more tightly than before. When we proceed this way, with a smooth enough curve (one that is locally straight), the sums of the lengths of the polygonal paths are Riemann sums with limiting value that we can calculate as an integral. In this section, we will see what that integral is. We shall also see what happens if, instead of being smooth, the curve is Helga von Koch's snowflake curve (Section 3.7).

Developing an Integral to Calculate Length

1. Model the curve that we want to measure in terms of a continuous function on a closed interval $[a, b]$.

2. Form small approximating segments of the curve on the subintervals of $[a, b]$.

3. Add the lengths of the small segments to form a Riemann sum.

4. Take the limit of the Riemann sum to get an integral, which gives a reasonable definition for the length of the curve in step 1.

The Basic Formula

Suppose the curve whose length we want to find is the graph of the function $y = f(x)$ from $x = a$ to $x = b$. We partition the closed interval $[a, b]$ in the usual way and connect the corresponding points on the curve with line segments (Fig. 6.26). The line segments, taken together, form a polygonal path that approximates the curve.

The length of a typical line segment PQ (shown in the figure) is

$$\sqrt{(\Delta x_k)^2 + (\Delta y_k)^2}.$$

The length of the curve is therefore approximated by the sum

$$\sum_{k=1}^n \sqrt{(\Delta x_k)^2 + (\Delta y_k)^2}. \tag{1}$$

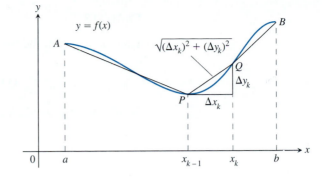

6.26 A typical segment PQ of a polygonal path approximating the curve AB.

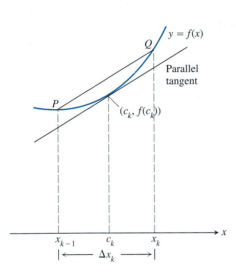

6.27 Enlargement of the arc PQ in Fig. 6.26.

We expect the approximation to improve as the partition $[a, b]$ becomes finer, and we would like to show that the sums in (1) approach a calculable limit as the norm of the partition goes to zero. To show this, we rewrite the sum in (1) in a form to which we can apply the Integral Existence Theorem from Chapter 5. Our starting point, oddly enough, is the Mean Value Theorem for derivatives.

Suppose that f has a derivative that is continuous at every point of $[a, b]$. Then, by the Mean Value Theorem, there is a point $(c_k, f(c_k))$ on the curve between P and Q where the tangent is parallel to the segment PQ (Fig. 6.27). At this point,

$$f'(c_k) = \frac{\Delta y_k}{\Delta x_k},$$

so that

$$\Delta y_k = f'(c_k)\Delta x_k.$$

With this substitution for Δy_k, the sums in (1) take the form

$$\sum_{k=1}^{n} \sqrt{(\Delta x_k)^2 + (f'(c_k)\Delta x_k)^2} = \sum_{k=1}^{n} \sqrt{1 + (f'(c_k))^2}\,\Delta x_k.$$

The sums on the right are Riemann sums for the continuous function $\sqrt{1 + (f'(x))^2}$ on the interval $[a, b]$. They therefore converge to the integral of this function as the norm of the partition of the interval goes to zero. We define this integral to be the length of the curve from a to b.

How to Apply Eq. (2)

1. Find dy/dx, and substitute it into the integrand, or let $y_1 = \text{NDER}\,(y, x)$.

2. Integrate from a to b analytically, or integrate numerically using

$$\text{NINT}\left(\sqrt{1 + y_1{}^2}, a, b\right).$$

3. Support with a grapher or confirm analytically, respectively.

DEFINITION

If the function $y = f(x)$ has a continuous first derivative throughout the interval $a \leq x \leq b$, the **length of the curve $y = f(x)$ from a to b** is the number

$$L = \int_a^b \sqrt{1 + \left(\frac{dy}{dx}\right)^2}\, dx. \qquad (2)$$

EXAMPLE 1

Find the length of the curve

$$y = f(x) = \frac{4\sqrt{2}}{3}x^{3/2} - 1, \qquad 0 \le x \le 1.$$

Solution   We use Eq. (2) with $a = 0$, $b = 1$, and

$$y = \frac{4\sqrt{2}}{3}x^{3/2} - 1,$$

$$\frac{dy}{dx} = \frac{4\sqrt{2}}{3} \cdot \frac{3}{2}x^{1/2} = 2\sqrt{2}x^{1/2},$$

$$1 + \left(\frac{dy}{dx}\right)^2 = 1 + (2\sqrt{2}x^{1/2})^2 = 1 + 8x.$$

The length of the curve from $x = 0$ to $x = 1$ is

$$L = \int_0^1 \sqrt{1 + \left(\frac{dy}{dx}\right)^2}\, dx = \int_0^1 \sqrt{1 + 8x}\, dx \qquad \text{Eq. (2) with } a = 0, \ b = 1$$

$$= \frac{2}{3} \cdot \frac{1}{8}(1 + 8x)^{3/2}\Big]_0^1 = \frac{13}{6}. \qquad \text{Substitute } u = 1 + 8x, \text{ integrate, and replace } u \text{ by } 1 + 8x.$$

Support: Let $y_1 = \text{NDER}\,(f(x), x)$. Then $\text{NINT}\left(\sqrt{1 + y_1{}^2}, 0, 1\right) = 2.167$.

≡

The function in Example 1 was especially chosen so that the calculations would work out nicely. Most curve-length problems that arise in practical applications produce an integrand in Eq. (2) whose antiderivative is impossible to find analytically. We can still solve the problem numerically, however, using technology.

EXAMPLE 2

A portion of a roller coaster track is modeled by the graph of

$$y = 100 + 30\sin 0.03x + 40\cos(0.02x - 1)$$

from $x = 0$ to $x = 628$ ft. (See Fig. 6.28.) Find the length of this portion of track.

Solution   The length from $x = 0$ to $x = 628$ is given by

$$\int_0^{628} \sqrt{1 + (0.9\cos(0.03x) - 0.8\sin(0.02x - 1))^2}\, dx$$

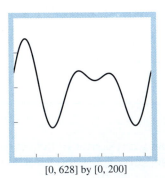

[0, 628] by [0, 200]

6.28 The roller coaster of Example 2 covers a "ground distance" of 628 feet while traveling over 804.7 feet of track.

$$\text{NINT}\left(\sqrt{1 + (0.9\cos(0.03x) - 0.8\sin(0.02x - 1))^2}, 0, 628\right) = 804.702.$$

The length of this portion of track is about 804.7 ft.

≡

Dealing with Discontinuities in *dy/dx*

At a point on a curve where dy/dx fails to exist, dx/dy may exist, and we may be able to find the curve's length by interchanging x and y in Eq. (2). The revised formula looks like this:

$$L = \int_c^d \sqrt{1 + \left(\frac{dx}{dy}\right)^2} \, dy \qquad (3)$$

We could justify Eq. (3) similarly to the way we did Eq. (2).

To use Eq. (3), we express x as a function of y, calculate dx/dy, and proceed as before to square, add 1, take the square root, and integrate.

EXAMPLE 3

Find the length of the curve $y = (x/2)^{2/3}$ from $x = 0$ to $x = 2$.

Solution   The derivative

$$\frac{dy}{dx} = \frac{2}{3}\left(\frac{x}{2}\right)^{-1/3}\left(\frac{1}{2}\right) = \frac{1}{3}\left(\frac{2}{x}\right)^{1/3}$$

is not defined at $x = 0$, so we cannot find the curve's length with Eq. (2).

We therefore rewrite the equation to express x in terms of y:

$$y = \left(\frac{x}{2}\right)^{2/3}$$

$$y^{3/2} = \frac{x}{2} \qquad \text{Raise both sides to the power 3/2.}$$

$$x = 2y^{3/2}. \qquad \text{Solve for } x.$$

From this, we see that the curve whose length we want is also the graph of $x = 2y^{3/2}$ from $y = 0$ to $y = 1$ (Fig. 6.29).

The derivative

$$\frac{dx}{dy} = 2 \cdot \frac{3}{2}y^{1/2} = 3y^{1/2}$$

is continuous throughout the interval $0 \le y \le 1$. We may therefore find the curve's length by setting

$$\left(\frac{dx}{dy}\right)^2 = (3y^{1/2})^2 = 9y$$

in Eq. (3) and integrating from $y = 0$ to $y = 1$:

$$L = \int_c^d \sqrt{1 + \left(\frac{dx}{dy}\right)^2} \, dy = \int_0^1 \sqrt{1 + 9y} \, dy \qquad \text{Eq. 3}$$

$$= \frac{1}{9} \cdot \frac{2}{3}(1 + 9y)^{3/2}\Big]_0^1 \qquad \begin{array}{l}\text{Substitute } u = 1 + 9y,\ du/9 = dy,\\ \text{integrate, and substitute back.}\end{array}$$

$$= \frac{2}{27}(10\sqrt{10} - 1) = 2.268.$$

Support: NINT $\left(\sqrt{1 + (\text{NDER }(2y^{3/2}, y))^2}, 0, 1\right) = 2.268.$ ≡

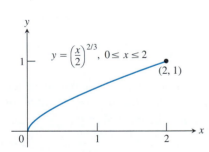

6.29 The graph of $y = (x/2)^{2/3}$ from $x = 0$ to $x = 2$ is also the graph of $x = 2y^{3/2}$ from $y = 0$ to $y = 1$, a function that has a continuous first derivative. We may therefore use Eq. (3) with $x = 2y^{3/2}$ to find the curve's length.

The Short Differential Formula

The equations

$$L = \int_a^b \sqrt{1 + \left(\frac{dy}{dx}\right)^2}\, dx \qquad \text{and} \qquad L = \int_c^d \sqrt{1 + \left(\frac{dx}{dy}\right)^2}\, dy \quad (4)$$

are often written with differentials instead of derivatives. This is done formally by thinking of the derivatives as quotients of differentials and bringing the dx and dy inside the radicals to cancel the denominators. In the first integral, we have

$$\sqrt{1 + \left(\frac{dy}{dx}\right)^2}\, dx = \sqrt{1 + \frac{dy^2}{dx^2}}\, dx = \sqrt{dx^2 + \frac{dy^2}{dx^2}\, dx^2} = \sqrt{dx^2 + dy^2}.$$

In the second integral, we have

$$\sqrt{1 + \left(\frac{dx}{dy}\right)^2}\, dy = \sqrt{1 + \frac{dx^2}{dy^2}}\, dy = \sqrt{dy^2 + \frac{dx^2}{dy^2}\, dy^2} = \sqrt{dx^2 + dy^2}.$$

Thus, the integrals in (4) reduce to a single differential formula:

$$L = \int_a^b \sqrt{dx^2 + dy^2}. \qquad (5)$$

Of course, dx and dy must be expressed in terms of a common variable, and appropriate limits of integration must be found before the integration in Eq. (5) is performed.

We can shorten Eq. (5) still further. Think of dx and dy as two sides of a small triangle whose hypotenuse is

$$ds = \sqrt{dx^2 + dy^2}$$

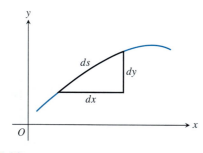

6.30 Diagram for remembering the equation $ds = \sqrt{dx^2 + dy^2}$.

(Fig. 6.30). The differential ds is then regarded as a differential of arc length that can be integrated between appropriate limits to give the length of the curve. With $\sqrt{dx^2 + dy^2}$ set equal to ds, the integral in Eq. (5) simply becomes the integral of ds.

DEFINITION   **The Arc Length Differential and the Differential Formula for Arc Length**

$$ds = \sqrt{dx^2 + dy^2} \qquad L = \int ds$$

Arc length   Differential formula
differential   for arc length

Curves with Infinite Length

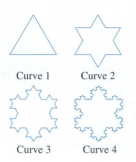

Curve 1   Curve 2

Curve 3   Curve 4

6.31 The first four curves of Helga von Koch's snowflake.

As you may recall from Section 3.7, Helga von Koch's snowflake curve K is the limit curve of an infinite sequence $C_1, C_2, \ldots, C_n, \ldots$ of "triangular" polygonal curves. The first four curves in the sequence are shown in Fig. 6.31.

Each time we introduce a new vertex in the construction process, it remains as a vertex in all subsequent curves and becomes a point on the limit curve K. This means that each C_n is itself a polygonal approximation of K—the endpoints of its sides all belonging to K. The length of K should therefore be the limit of the lengths of the curves C_n. At least, that is what it should be if we stick to the definition of the length that we developed for smooth curves.

What, then, is the limit of the lengths of the curves C_n? If the original equilateral triangle C_1 has sides of length 1, the total length of C_1 is 3. To make C_2 from C_1, we replace each side of C_1 by four segments, each of which is one-third as long as the original side.

To make C_2 from C_1, we do this to each side.

The total length of C_2 is therefore $3(4/3)$. To get the length of C_3, we multiply by $4/3$ again and get $3(4/3)^2$. Similarly, we find the length of C_4 to be $3(4/3)^3$. When we get to C_n, we have a curve of length $3(4/3)^{n-1}$.

Curve Number	1	2	3	$\ldots$	n	$\ldots$
Length	3	$3\left(\dfrac{4}{3}\right)$	$3\left(\dfrac{4}{3}\right)^2$	$\ldots$	$3\left(\dfrac{4}{3}\right)^{n-1}$	$\ldots$

The length of C_{10} is nearly 40, and the length of C_{100} is greater than 7,000,000,000,000. The lengths grow too rapidly to have any finite limit. Therefore, the snowflake curve has no length or, if you prefer, infinite length.

What went wrong? Nothing. The formulas that we derived for length are for the graphs of functions with continuous first derivatives, curves that are smooth enough to have a continuously turning tangent at every point. Helga von Koch's snowflake curve is too rough for that, and our derivative-based formulas do not apply.

EXPLORATION BIT

In Chapter 9, we find that $3(4/3)^{n-1}$ for $n = 1, 2, 3, \ldots$ is a *geometric sequence* of numbers whose terms *increase without bound*. You can confirm our values for C_{10} and C_{100} easily on your calculator. To support our claim that the length of C_n increases without bound, find the smallest value n_O for which the length causes an *overflow error* (number too great) on your calculator. What is the length of C_{n_O-1}?

Exercises 6.4

Find the lengths of the curves in Exercises 1–14 using a method of your choice. Confirm your NINT result analytically, if possible, or support your analytical result using NINT.

1. $y = (1/3)(x^2 + 2)^{3/2}$ from $x = 0$ to $x = 3$

2. $y = x^{3/2}$ from $x = 0$ to $x = 4$

3. $9x^2 = 4y^3$ from $(0, 0)$ to $(2\sqrt{3}, 3)$

4. $y = x^{2/3}$ from $x = 0$ to $x = 4$

5. $y = x^3/3 + 1/(4x)$ from $x = 1$ to $x = 3$
   (*Hint:* $1 + (dy/dx)^2$ is a perfect square.)

6. $y = x^{3/2}/3 - x^{1/2}$ from $x = 1$ to $x = 9$
   (*Hint:* $1 + (dy/dx)^2$ is a perfect square.)

7. $x = y^4/4 + 1/(8y^2)$ from $y = 1$ to $y = 2$
   (*Hint:* $1 + (dx/dy)^2$ is a perfect square.)

8. $x = y^3/6 + 1/(2y)$ from $y = 2$ to $y = 3$
   (*Hint:* $1 + (dx/dy)^2$ is a perfect square.)

9. $y = (e^x + e^{-x})/2$ from $x = -\ln 2$ to $x = \ln 2$
   (*Hint:* $1 + (dy/dx)^2$ is a perfect square.)

10. $x = y^2 - (1/8)\ln y$ from $y = 1$ to $y = 3$
    (*Hint:* $1 + (dx/dy)^2$ is a perfect square.)

11. $y = \sec x$ from $x = -\pi/3$ to $x = \pi/3$

12. $y = \sin x$ from $x = 0$ to $x = 2\pi$

13. $y = (e^x + e^{-x})/2$ from $x = -3$ to $x = 3$

14. $y = (2^x + 2^{-x})/2$ from $x = -3$ to $x = 3$

15. *The length of an astroid.* The graph of the equation $x^{2/3} + y^{2/3} = 1$ is one of a family of curves called *astroids* (not "asteroids") because of their starlike shapes. Find the length of this particular astroid by finding the length of half the first quadrant portion, $y = (1 - x^{2/3})^{3/2}$, $\sqrt{2}/4 \le x \le 1$, and multiplying by 8.

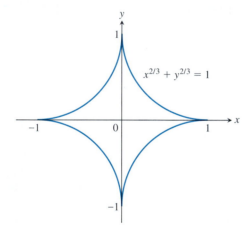

16. Find the length of the curve

$$y = \int_0^x \sqrt{\cos 2t} \, dt$$

from $x = 0$ to $x = \pi/4$ analytically and support with a grapher. (*Hint:* For $0 \le x \le \pi/4$, $1 + \cos 2x = 2\cos^2 x$.)

17. Find a curve through the origin whose length is

$$L = \int_0^4 \sqrt{1 + \frac{1}{4x}} \, dx.$$

(*Hint:* Solve the initial value problem $dy/dx = \sqrt{1/(4x)}$, $y(0) = 0$.)

18. Find a curve through the point $(0, 1)$ whose length from $x = 0$ to $x = 1$ is

$$L = \int_0^1 \sqrt{1 + e^{2x}} \, dx.$$

19. Find a curve through the point $(1, 3)$ whose length from $x = 1$ to $x = 2$ is

$$L = \int_1^2 \sqrt{1 + \frac{1}{x^2}} \, dx.$$

20. Without evaluating either integral, show why

$$2 \int_{-1}^1 \sqrt{1 - x^2} \, dx = \int_{-1}^1 \frac{1}{\sqrt{1 - x^2}} \, dx,$$

and then support with a grapher. (*Hint:* Interpret one integral as an area and the other as a length.) (*Source:* Peter A. Lindstrom, *Mathematics Magazine,* Volume 45, Number 1, January 1972, page 47.)

Many of the curves that we have been working with have unusual formulas because they are especially chosen so that the calculations work out nicely. Most curve-length problems that arise in practical applications are not so nice. They yield a square root $\sqrt{1 + (dy/dx)^2}$ in the arc-length integral whose antiderivative is impossible to find analytically. In fact, this square root is a famous source of nonelementary integrals. Most arc-length integrals have to be evaluated numerically, as in the exercises that follow.

Determine the lengths of the curves in Exercises 21 and 22. Do not try to evaluate the integrals analytically right now. We shall see how to do that when we are in Chapter 8.

21. $y = \ln(1 - x^2)$, $0 \le x \le 1/2$

22. $y = \ln(\cos x)$, $0 \le x \le \pi/3$

23. Your metal-fabrication company is bidding for a contract to make sheets of corrugated iron roofing like the one shown. The cross-sections of the corrugated sheets are to conform to the curve

$$y = \sin \frac{3\pi}{20} x, \quad 0 \le x \le 20 \text{ in.}$$

Original sheet   Corrugated sheet

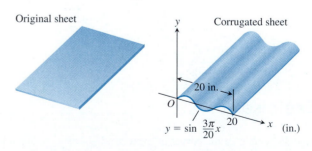

$$y = \sin \frac{3\pi}{20} x$$

If the roofing is to be stamped from flat sheets by a process that does not stretch the material, how wide should the original material be?

24. Your engineering firm is bidding for the contract to construct the tunnel suggested by the figure below. The tunnel is 300 ft long and 50 ft wide at the base. The cross-section is shaped like one arch of the curve $y = 25 \cos (\pi x/50)$. Upon completion, the tunnel's inside surface (excluding the roadway) will be treated with a waterproof sealer that costs \$1.75 per ft^2 to apply. How much will it cost to apply the sealer?

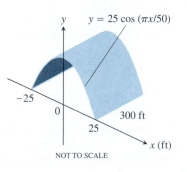

25. An asteroid (not "astroid") follows the path modeled by $y^2 = 0.2x^2 - 1$ for $x > \sqrt{5}$ (x is in units of 10^6 miles). Assume that Earth is at the origin. How many miles will the asteroid travel starting from the point whose x-coordinate is $x = 10$ (million miles), until it reaches its closest point to Earth?

26. Two lanes of a running track are modeled by semiellipses, as shown:

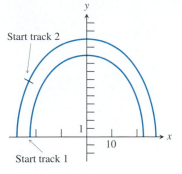

The equation for track 1 is $y = \sqrt{100 - 0.2x^2}$, and the equation for track 2 is $y = \sqrt{150 - 0.2x^2}$. The starting point for track 1 is at the negative x-intercept $(-\sqrt{500}, 0)$. The finish points for both tracks are the positive x-intercepts. Where should the starting point be placed on track 2 so that the two track lengths will be equal?

Note: There are values of x and y where dy/dx or dx/dy are zero. Give a convincing argument that the results obtained using Eq. (3) are correct.

27. The boundary of a garden is modeled by the equation $y^2 = 400(0.1x)^2(2 - 0.1x)$ for $x \geq 0$. Find the length of fence needed to completely enclose the garden.

28. Suppose the boundary of the garden in Exercise 27 is modeled by the equation $y^2 = K(0.1x)^2(2 - 0.1x)x$ and only 80 feet of fencing is needed to enclose it. Determine K.

6.5 Areas of Surfaces of Revolution

When you jump rope, the rope sweeps out a surface in the space around you, a surface called a **surface of revolution**. As you can imagine, the area of this surface depends on the rope's length and on how far away each segment of the rope swings. This section explores the relation between the area of a surface of revolution and the length and reach of the curve that generates it.

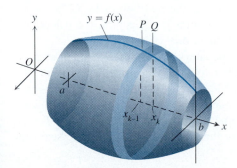

6.32 The surface swept out by revolving the curve $y = f(x), a \leq x \leq b$, about the x-axis is a union of bands like the one swept out by the arc PQ.

The Basic Formula

Suppose we want to find the area of the surface swept out by revolving the graph of the function $y = f(x), a \leq x \leq b$, about the x-axis. We partition the closed interval $[a, b]$ in the usual way and use the points in the partition to divide the graph into short arcs. Figure 6.32 shows a typical arc PQ and the band it sweeps out as part of the surface of revolution.

As the arc PQ revolves about the x-axis, the line segment joining P and Q sweeps out part of a cone whose axis lies along the x-axis (magnified

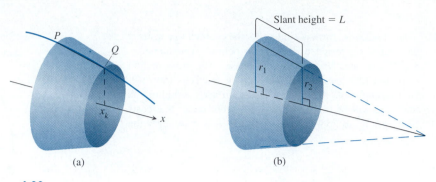

Slant height = L

(a) (b)

6.33 (a) The line segment joining P and Q sweeps out a frustum of a cone with (b) important dimensions shown.

view in Fig. 6.33a). A piece of a cone like this is called a **frustum** of the cone, *frustum* being Latin for "piece." The surface area of the frustum approximates the surface area of the band swept out by the arc PQ.

The surface area of the frustum of a cone (see Fig. 6.33b) is 2π times the average of the base radii times the slant height:

$$\text{Frustum surface area } = 2\pi \cdot \frac{r_1 + r_2}{2} \cdot L = \pi(r_1 + r_2)L. \qquad (1)$$

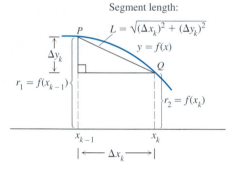

Segment length:
$$L = \sqrt{(\Delta x_k)^2 + (\Delta y_k)^2}$$
$$y = f(x)$$
$$r_1 = f(x_{k-1})$$
$$r_2 = f(x_k)$$
$$x_{k-1} \qquad x_k$$
$$\Delta x_k$$

6.34 The important dimensions associated with the arc and segment PQ.

For the frustum swept out by the segment PQ (Fig. 6.34), this works out to be

$$\text{Frustum surface area } = \pi(f(x_{k-1}) + f(x_k))\sqrt{(\Delta x_k)^2 + (\Delta y_k)^2}.$$

The area of our original surface, being the sum of the areas of the bands swept out by arcs like arc PQ, is approximated by the frustum area sum

$$\sum_{k=1}^{n} \pi(f(x_{k-1}) + f(x_k))\sqrt{(\Delta x_k)^2 + (\Delta y_k)^2}. \qquad (2)$$

We expect the approximation to improve as the partition of $[a, b]$ becomes finer, and we would like to show that the sums in (2) approach a calculable limit as the norm of the partition goes to zero.

To show this, we try to rewrite the sum in (2) as the Riemann sum of some function over the interval from a to b. As in the calculation of arc length, we begin by appealing to the Mean Value Theorem for derivatives.

Suppose as before that f has a derivative that is continuous at every point of $[a, b]$. Then, by the Mean Value Theorem, there is a point $(c_k, f(c_k))$ on the curve between P and Q where the tangent is parallel to the segment PQ (Fig. 6.35). At this point,

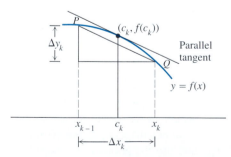

$$(c_k, f(c_k))$$
Parallel tangent
$$y = f(x)$$
$$x_{k-1} \quad c_k \quad x_k$$
$$\Delta x_k$$

6.35 If f' is continuous, the Mean Value Theorem guarantees the existence of a point on arc PQ where the tangent is parallel to segment PQ.

$$f'(c_k) = \frac{\Delta y_k}{\Delta x_k},$$

so

$$\Delta y_k = f'(c_k)\Delta x_k.$$

Developing an Integral to Calculate Surface Area

1. Model the solid whose surface area we want to measure in terms of a "generating" continuous function on a closed interval $[a, b]$.

2. Form small approximating frustums using the continuous function on the subintervals of $[a, b]$.

3. Add the surface areas of the small frustums to form a sum that "behaves" like a Riemann sum as the norm of the partitions of $[a, b]$ goes to zero.

4. Take the limit of the Riemann sum to get an integral, which gives precisely the surface area of the solid in step 1.

With this substitution for Δy_k, the sums in (2) take the form

$$\sum_{k=1}^{n} \pi (f(x_{k-1}) + f(x_k)) \sqrt{(\Delta x_k)^2 + (f'(c_k) \Delta x_k)^2}$$

$$= \sum_{k=1}^{n} \pi (f(x_{k-1}) + f(x_k)) \sqrt{1 + (f'(c_k))^2} \, \Delta x_k. \qquad (3)$$

At this point, there is both good news and bad.

The bad news is that the sums in (3) are not the Riemann sums of any function because the points x_{k-1}, x_k, and c_k are not the same and there is no way to make them the same. The good news is that this does not matter. A theorem from advanced calculus assures us that as the norm of the partition of $[a, b]$ goes to zero, the sums in (3) converge to

$$\int_a^b 2\pi f(x) \sqrt{1 + (f'(x))^2} \, dx$$

just the way we want them to. We therefore define this integral to be the area of the surface swept out by the graph of f from a to b.

DEFINITION

If the function f has a continuous first derivative throughout the interval $a \le x \le b$, the **area of the surface** generated by revolving the curve $y = f(x)$ about the x-axis is the number

$$S = \int_a^b 2\pi y \sqrt{1 + \left(\frac{dy}{dx}\right)^2} \, dx. \qquad (4)$$

Notice that the square root in Eq. (4) is the same one that appears in the formula for arc length. More about that later.

EXAMPLE 1

Find the area of the surface generated by revolving the curve $y = 2\sqrt{x}$, $1 \le x \le 2$, about the x-axis (Fig. 6.36).

Solution   We let $a = 1$, $b = 2$, $y_1 = 2\sqrt{x}$, and $y_2 = $ NDER (y_1, x). Then

$$\text{NINT} \left(2\pi y_1 \sqrt{1 + y_2{}^2}, 1, 2\right) = 19.836.$$

The surface area is about 19.8.

To confirm analytically, we have

$$a = 1, \quad b = 2, \quad y = 2\sqrt{x}, \quad \frac{dy}{dx} = \frac{1}{\sqrt{x}},$$

$$\sqrt{1 + \left(\frac{dy}{dx}\right)^2} = \sqrt{1 + \left(\frac{1}{\sqrt{x}}\right)^2} = \sqrt{1 + \frac{1}{x}} = \sqrt{\frac{x+1}{x}} = \frac{\sqrt{x+1}}{\sqrt{x}}.$$

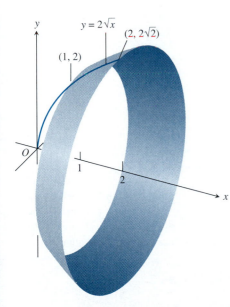

6.36 Example 1 calculates the area of this surface.

How to Apply Eq. (4)

1. Find dy/dx and substitute it and y into the integrand; or let $y_1 = y$ and $y_2 = \text{NDER}\,(y_1, x)$.

2. Integrate

$$2\pi y \sqrt{1 + \left(\frac{dy}{dx}\right)^2},$$

from a to b; or integrate numerically using

$$\text{NINT}\,(2\pi y_1 \sqrt{1 + y_2{}^2}, a, b).$$

3. Support with a grapher or confirm analytically, respectively.

Then

$$\text{Surface area} = \int_a^b 2\pi y \sqrt{1 + \left(\frac{dy}{dx}\right)^2}\, dx \qquad \text{Eq. 4}$$

$$= \int_1^2 2\pi \cdot 2\sqrt{x} \cdot \frac{\sqrt{x+1}}{\sqrt{x}}\, dx$$

$$= 4\pi \int_1^2 \sqrt{x+1}\, dx = 4\pi \cdot \frac{2}{3}(x+1)^{3/2}\Big]_1^2$$

$$= \frac{8\pi}{3}\left(3\sqrt{3} - 2\sqrt{2}\right) = 19.836.$$

Revolution about the y-Axis

If the axis of revolution is the y-axis, we use the formula that we get from interchanging x and y in Eq. (4):

SUPPORT

For grapher support, we can let $y_1 = f(y)$, and then $S =$

$$\text{NINT}(2\pi y_1 \sqrt{1 + (\text{NDER}(y_1, y))^2}, c, d)$$

Note that the variable of differentiation and integration is y. In Example 2, for example, we can let $y_1 = 1 - y$. Try it.

Revolution about the y-Axis

If $x = f(y)$ has a continuous first derivative throughout the interval $c \le y \le d$, the area of the surface S generated by revolving the curve $x = f(y), c \le y \le d$, about the y-axis is

$$S = \int_c^d 2\pi x \sqrt{1 + \left(\frac{dx}{dy}\right)^2}\, dy. \qquad (5)$$

EXAMPLE 2

The line segment $x = 1 - y, 0 \le y \le 1$, is revolved about the y-axis to generate the cone in Fig. 6.37. Find its lateral surface area.

Solution Here we have a calculation that we can also confirm with a formula from geometry:

$$\text{Lateral surface area} = \frac{1}{2}(\text{base circumference})(\text{slant height})$$

$$= \frac{1}{2}(2\pi)(\sqrt{2}) = \pi\sqrt{2}.$$

To see how Eq. (5) gives the same result, we take

$$c = 0, \quad d = 1, \quad x = 1 - y, \quad \frac{dx}{dy} = -1,$$

$$\sqrt{1 + \left(\frac{dx}{dy}\right)^2} = \sqrt{1 + (-1)^2} = \sqrt{2}$$

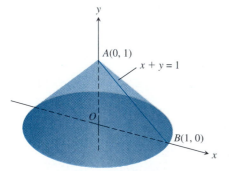

6.37 Revolving line segment AB about the y-axis generates a cone whose lateral surface area we can now calculate in two different ways. See Example 2.

and calculate

$$\text{Surface area} = \int_c^d 2\pi x \sqrt{1 + \left(\frac{dx}{dy}\right)^2} \, dy = \int_0^1 2\pi (1 - y)\sqrt{2} \, dy \quad \text{Eq. 5}$$

$$= 2\pi \sqrt{2} \left[y - \frac{y^2}{2} \right]_0^1 = 2\pi \sqrt{2} \left(1 - \frac{1}{2} \right) = \pi \sqrt{2}.$$

The results agree, as they should.

The Short Differential Formula

The equations

$$S = \int_a^b 2\pi y \sqrt{1 + \left(\frac{dy}{dx}\right)^2} \, dx \qquad \text{and} \qquad S = \int_c^d 2\pi x \sqrt{1 + \left(\frac{dx}{dy}\right)^2} \, dy$$

are often written in terms of the arc length differential $ds = \sqrt{dx^2 + dy^2}$ as

$$S = \int_a^b 2\pi y \, ds \qquad \text{and} \qquad S = \int_c^d 2\pi x \, ds.$$

In the first of these, y is the distance from the x-axis to an element of arc length ds. In the second, x is the distance from the y-axis to an element of arc length ds. In both cases, the integrals have the form

$$S = \int 2\pi (\text{radius})(\text{band width}) = \int 2\pi \rho \, ds,$$

where the Greek letter ρ (rho) represents the radius from the axis of revolution to an element of arc length ds (Fig. 6.38).

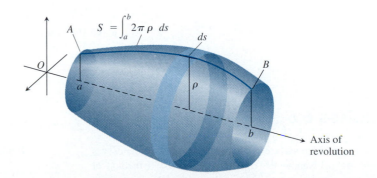

6.38 The area of the surface swept out by revolving arc AB about the axis shown here is $\int_a^b 2\pi \rho \, ds$. The exact expression depends on the formulas for ρ and ds.

If you wish to remember only one formula for surface area, you might make it the short differential formula.

Short Differential Formula

$$S = \int 2\pi \, \rho \, ds$$

In any particular problem, you would then express the radius function ρ and the arc length differential ds in terms of a common variable and supply limits of integration for that variable.

EXAMPLE 3

Find the area of the surface generated by revolving the curve $y = x^3$, $0 \leq x \leq 1/2$, about the x-axis (Fig. 6.39).

Solution   We start with the short differential formula:

$$S = \int 2\pi \, \rho \, ds$$

$$= \int 2\pi \, y \, ds \qquad \text{For revolution about the } x\text{-axis,}$$
$$\text{the radius function is } \rho = y.$$

$$= \int 2\pi \, y \, \sqrt{dx^2 + dy^2}. \qquad ds = \sqrt{dx^2 + dy^2}.$$

We then decide whether to express dy in terms of dx or dx in terms of dy. The original form of the equation, $y = x^3$, makes it easier to express dy in terms of dx, so we continue the calculation with

$$y = x^3, \quad dy = 3x^2 \, dx, \quad \text{and} \quad \sqrt{dx^2 + dy^2} = \sqrt{dx^2 + (3x^2 \, dx)^2}$$

$$= \sqrt{1 + 9x^4} \, dx.$$

With these substitutions, x becomes the variable of integration, and we obtain

$$S = \int_{x=0}^{x=1/2} 2\pi \, y \, \sqrt{dx^2 + dy^2} = \int_{0}^{1/2} 2\pi \, x^3 \sqrt{1 + 9x^4} \, dx.$$

We leave the evaluation of this integral as an exercise.   ▤

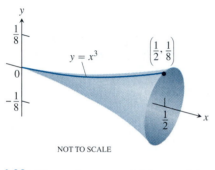

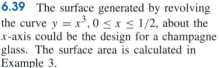

6.39   The surface generated by revolving the curve $y = x^3$, $0 \leq x \leq 1/2$, about the x-axis could be the design for a champagne glass. The surface area is calculated in Example 3.

NOT TO SCALE

Exercises 6.5

Find the areas of the surfaces generated by revolving the curves in Exercises 1–14 about the axes indicated. Use a method of your choice. Confirm your NINT result analytically, if possible, or support your analytic result using NINT.

1. $y = x/2, 0 \leq x \leq 4$, about the x-axis. Also, confirm your result with a formula from geometry, as in Example 2.

2. $y = x/2, 0 \leq x \leq 4$, about the y-axis. Also, confirm your result with a formula from geometry, as in Example 2.

3. $y = x/2 + 1/2, 1 \leq x \leq 3$, about the x-axis. Also, confirm your result with the geometry formula in Eq. (1).

4. $y = x/2 + 1/2, 1 \leq x \leq 3$, about the y-axis. Also, confirm your result with the geometry formula in Eq. (1).

5. $y = x^3/9, 0 \leq x \leq 2$, about the x-axis

6. $x = y^3/3, 0 \leq y \leq 1$, about the y-axis

7. $y = \sqrt{x}, 3/4 \leq x \leq 15/4$, about the x-axis

8. $y = \sqrt{2x - x^2}, 0 \leq x \leq 2$, about the x-axis

9. $y = \sqrt{x + 1}, 1 \leq x \leq 5$, about the x-axis

10. $y = 2\sqrt{4 - x}, 1 \leq x \leq 4$, about the y-axis.

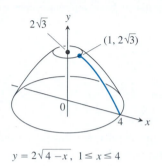

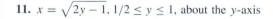

$$y = 2\sqrt{4 - x}, \; 1 \leq x \leq 4$$

11. $x = \sqrt{2y - 1}, 1/2 \leq y \leq 1$, about the y-axis

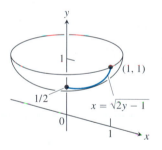

12. $x = (e^y + e^{-y})/2, 0 \leq y \leq \ln 2$, about the y-axis

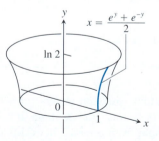

13. $x = y^4/4 + 1/(8y^2), 1 \leq y \leq 2$, about the x-axis
(*Hint:* Express $ds = \sqrt{dx^2 + dy^2}$ in terms of dy, and evaluate the integral $S = \int 2\pi y \, ds$ with appropriate limits.)

14. $y = (1/3)(x^2 + 2)^{3/2}, 0 \leq x \leq 3$, about the y-axis
(*Hint:* Express $ds = \sqrt{dx^2 + dy^2}$ in terms of dx, and evaluate the integral $S = \int 2\pi x \, ds$ with appropriate limits.)

15. Complete Example 3 by evaluating the surface area integral analytically and supporting numerically using NINT.

16. Use an integral to find the surface area of the sphere generated by revolving the semicircle $y = \sqrt{1 - x^2}, -1 \leq x \leq 1$, about the x-axis. Confirm your result with a formula from geometry.

17. Find the area of the surface generated by revolving the curve $y = \cos x, -\pi/2 \leq x \leq \pi/2$, about the x-axis.

18. *The surface of an astroid.* Find the area of the surface generated by revolving the portion of the astroid $x^{2/3} + y^{2/3} = 1$ shown below about the x-axis. (*Hint:* Revolve the first quadrant portion $y = (1 - x^{2/3})^{3/2}, 0 \leq x \leq 1$, about the x-axis, and double your result.)

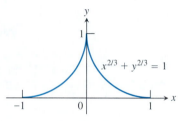

19. *Enameling woks.* Your company has decided to put out a deluxe version of the successful wok that you designed in Exercise 43 of Section 6.2. The plan is to coat it inside with white enamel and outside with blue enamel. Each enamel will be sprayed on 1 millimeter thick before baking. Manufacturing wants to know how much enamel it will take for a production run of 5000 woks. What do you tell them? (Neglect waste and unused material. Answer in liters.)

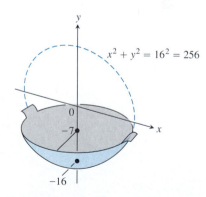

Dimensions in centimeters

20. *Slicing bread.* Did you know that if you cut a spherical loaf of bread into slices of equal width, each slice will have the same amount of crust? To see why, suppose the semicircle $y = \sqrt{r^2 - x^2}$ shown in the following diagram is revolved about the x-axis to generate a sphere. Let AB be an arc of the semicircle that lies above an interval of length h on the x-axis. Show that the area swept out by AB does not depend on the location of the interval. (It does depend on the length of the interval.)

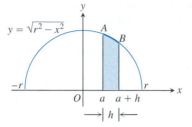

$y = \sqrt{r^2 - x^2}$

21. The nose cone of a space shuttle is modeled by the solid of revolution determined by revolving the region in the first quadrant given by the curve $y = 10x^{1.3/3}$ (where x and y are measured in feet) and the line $x = 40$ about the x-axis. Assume that each of the special heat-resistant tiles covering the nose cone is 1 ft square. About how many tiles are necessary to cover the space shuttle's nose cone?

22. A parabolic mirror modeled by revolving the curve $y = 2\sqrt{x}$ (where x and y are measured in feet), $0 \le x \le c$ about the x-axis needs to have 50 square feet of surface area. Determine c.

6.6 — Work

Developing an Integral to Calculate Work

1. Model the work that we want to measure in terms of continuous force directed along the x-axis on a closed interval $[a, b]$.

2. Form approximations of the small amounts of work done on the subintervals of $[a, b]$.

3. Add the small amounts of work to form a Riemann sum.

4. Take the limit of the Riemann sum to get an integral, which gives precisely the amount of work we want in step 1.

In everyday life, *work* describes any activity that takes muscular or mental effort. In science, however, the term is used in a narrower sense that involves the application of a force to a body and the body's subsequent displacement. This section shows how to calculate work. The applications run from stretching springs and pumping liquids from subterranean tanks to lifting satellites into orbit.

The Constant-Force Formula for Work

We begin with a definition.

> **DEFINITION**
>
> When a body moves a distance d along a straight line as the result of being acted on by a force that has a constant magnitude F in the direction of the motion, the **work** W done by the force in moving the body is F times d:
>
> $$W = Fd. \tag{1}$$

Work is measured in foot-pounds, newton-meters, or whatever force-distance unit is appropriate to the occasion. For example, it takes a force of about 1 newton (1 N) to lift an apple from a table. If we lift it 1 meter, we have done about 1 newton-meter (N·m) of work on the apple.

We can see right away from the definition of work that there is a considerable difference between what we are used to calling work and what this formula says work is. If you push a car down a street, you are doing work, both by our own reckoning and according to Eq. (1). But if you push against the car and the car does not move, Eq. (1) says that you are doing no work, no matter how hard or how long you push.

The Variable-Force Integral Formula for Work

If the force that you apply varies along the way, as it will if you are lifting a leaking bucket or compressing a spring, the formula $W = Fd$ has to be replaced by an integral formula that takes the variation in F into account. It takes calculus to measure the work done by a variable force.

Suppose that the force performing the work varies continuously along a line that we can take to be the x-axis and that the force is represented by the function $F(x)$. We are interested in the work done along an interval from $x = a$ to $x = b$. We partition the closed interval $[a, b]$ in the usual way and choose an arbitrary point c_k in each subinterval $[x_{k-1}, x_k]$.

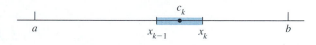

If the subinterval is short enough, F, being continuous, will not vary much from x_{k-1} to x_k. The amount of work done by the force from x_{k-1} to x_k will be nearly equal to $F(c_k)$ times distance Δx_k, as it would be if we could apply Eq. (1). The total work done from a to b is thus approximated by the Riemann sum

$$\sum_{k=1}^{n} F(c_k)\Delta x_k.$$

We expect the approximations to improve as the norm of the partition goes to zero, so we define the work done by the force from a to b to be the integral of F from a to b.

DEFINITION

The **work** done by a continuous force $F(x)$ directed along the x-axis from $x = a$ to $x = b$ is

$$W = \int_a^b F(x)\,dx. \tag{2}$$

We call Eq. (2) the **integral formula for work**.

EXAMPLE 1

A leaky 5-lb bucket is lifted from the ground into the air by a worker pulling in 20 ft of rope at a constant speed (Fig. 6.40). The rope weighs 0.08 lb/ft. The bucket starts with 2 gal of water (16 lb) and leaks at a constant rate. It finishes draining just as it reaches the top. How much work was spent

a) lifting the water alone?

b) lifting the water and bucket together?

c) lifting the water, bucket, and rope?

Solution

a) *The water alone.* The force required to lift the water is the water's weight, which varies steadily from 16 to 0 lb over the 20-ft lift. When the bucket

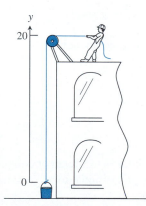

6.40 Because the bucket is leaking steadily and the length of rope being lifted is decreasing steadily, the force applied to raise the bucket with the rope is decreasing steadily.

is x ft off the ground, the water weighs

$$F(x) = 16 \left(\frac{20 - x}{20} \right) = 16 \left(1 - \frac{x}{20} \right) = 16 - \frac{4x}{5} \text{ lb.}$$

Original weight of water

Proportion left at elevation xi

The work done is

$$W = \int_a^b F(x) \, dx \qquad \text{Use Eq. (2) for variable forces.}$$

$$= \int_0^{20} \left(16 - \frac{4x}{5} \right) dx = \left[16x - \frac{2x^2}{5} \right]_0^{20} = 320 - 160 = 160 \text{ ft} \cdot \text{lb.}$$

b) *The water and bucket together.* According to Eq. (1), it takes $5 \times 20 = 100 \text{ ft} \cdot \text{lb}$ to lift a 5-lb weight 20 ft. Therefore,

$$160 + 100 = 260 \text{ ft} \cdot \text{lb}$$

of work was spent lifting the water and bucket together.

c) *The water, bucket, and rope.* Now the total weight at level x is

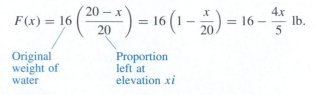

lb/ft$i$ ft

$$F(x) = \left(16 - \frac{4x}{5} \right) + \quad 5 \quad + (0.08)(20 - x).$$

Variable weight of water

Constant weight of bucket

Weight of rope paid out at elevation x

NUMERICAL SOLUTION

To do part (c) numerically, we compute

NINT $\left(16 - \dfrac{4x}{5} + 5 + 0.08 \, (20 - x), 0, 20 \right)$ = 276.

The work lifting the rope is

$$\text{Work on rope} = \int_0^{20} (0.08)(20 - x) \, dx = \int_0^{20} (1.6 - 0.08x) \, dx$$

$$= \left[1.6x - 0.04x^2 \right]_0^{20} = 32 - 16 = 16 \text{ ft} \cdot \text{lb.}$$

The total work for the water, bucket, and rope combined is

$$160 + 100 + 16 = 276 \text{ ft} \cdot \text{lb.} \qquad \equiv$$

Hooke's Law for Springs, *F = kx*

Hooke's Law says that the amount of force F it takes to stretch or compress a spring x length units from its natural length is proportional to x. In symbols,

$$F = kx. \tag{3}$$

EXPLORATION BIT

In Fig. 6.41, what is a geometric representation of W, the work done in compressing the spring?

The number k, measured in force units per unit length, is a constant characteristic of the spring, called the **spring constant.** Hooke's Law (Eq. 3) holds as long as the force doesn't distort the metal in the spring. We will assume that the forces in this section are too small to do that.

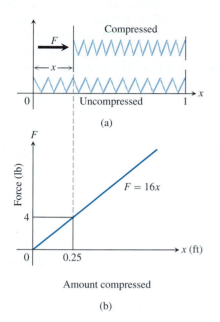

6.41 (a) The more the spring is compressed (or stretched), the greater the force that is being applied. The algebraic relationship is $F = kx$, and (b) the graphical relationship is linear. For Example 2, the relationship is $F = 16x$ (graphed here). The work done is $W = \int_0^{0.25} 16x\,dx$.

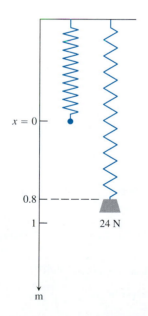

6.42 A 24-newton weight stretches this spring 0.8 m beyond its unstressed length.

EXAMPLE 2

Find the work required to compress a spring from its natural length of 1 ft to a length of 0.75 ft if the spring constant is $k = 16$ lb/ft.

Solution We picture the uncompressed spring laid out along the x-axis with its movable end at the origin and its fixed end at $x = 1$ ft (Fig. 6.41). This enables us to describe the force required to compress the spring from 0 to x with the formula $F = 16x$. As the spring is compressed from 0 to 0.25 ft, the force varies from

$$F(0) = 16 \cdot 0 = 0 \text{ lb} \qquad \text{to} \qquad F(0.25) = 16 \cdot 0.25 = 4 \text{ lb}.$$

The work done by F over this interval is

$$W = \int_a^b F(x)\,dx = \int_0^{0.25} 16x\,dx = 8x^2 \Big]_0^{0.25} = 0.5 \text{ ft} \cdot \text{lb}. \qquad \text{Eq. (2)}$$

EXPLORATION 1

The Work-Force-Length Relationships

Hooke's Law tells us that the amount of force that it takes to stretch or compress a spring is linearly related to the length of the stretch or compression. Knowing one pair of length-force values $(x, F(x))$ other than $(0, 0)$ will reveal the relationship.

1. A spring has a natural length of 1 m. A 24-N weight extends the spring 0.8 m (Fig. 6.42). What's the value of k in $F = kx$?

2. Use the equation from part 1 to tell by how much a 45-N force will stretch the spring.

3. Use the force function from part 1 to tell how much work will be spent in extending the spring by 2 m.

4. Describe, in general, how the work required to stretch or compress a spring is related to the length of the stretch or compression of the spring. What does this mean in nonmathematical terms when you try to stretch or compress a spring?

Pumping Liquids from Containers—Do-it-yourself Integrals

As we stated at the beginning of this chapter, thousands of things can be measured with integrals. We have seen several examples in which Riemann integration is applied. It is desirable that we learn to recognize such situations and be able to build our own Riemann sums—and hence Riemann integrals—for them.

For example, suppose we want to find how much work it takes to pump all or part of the liquid from a container. We can imagine lifting out one thin

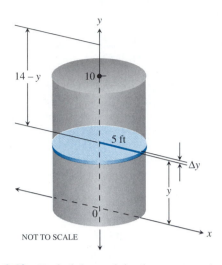

6.43 To find the work it takes to pump the water to a point 4 feet above the top of the tank, find the work needed to lift the water one thin slab at a time, and then sum the work for all the slabs.

EXPLORATION BIT

After you understand Example 3, try it with the positive y-axis pointed down and the origin at the top surface. Then the slab of water at distance y below the origin has to be lifted y ft to the top of the tank and $(y + 4)$ ft to 4 ft above the top.

Developing an Integral to Find Work Done by Pumping

1. Draw a diagram of the container with a coordinate system. Then find the *weight* of a thin horizontal slab of liquid.

2. Find how *far* to lift the slab. Then find the *work* needed to lift the slab.

3. Add the small amounts of work needed for each slab to form a Riemann sum on an interval that corresponds to the bottom and top of the liquid.

4. Take the limit of the Riemann sum to get an integral, which gives precisely the work needed to pump the contents out of the container.

---

horizontal slab of liquid at a time, applying the equation $W = Fd$ to that slab, and then summing the work needed to lift all the slabs. The sum is a Riemann sum, and the integral that we get each time depends on the weight of the liquid and the cross-sectional dimensions of the container, but the way we find the integral is the same for all containers. The next example shows what to do.

EXAMPLE 3

How much work does it take to pump the water from an upright and full right circular cylindrical tank of radius 5 ft and height 10 ft to a level 4 ft above the top of the tank?

Solution   We draw coordinate axes (Fig. 6.43) and imagine the water divided into thin slabs by planes perpendicular to the y-axis at the points of a partition of the interval [0, 10].

The typical slab between the planes at y and $y + \Delta y$ has a volume of approximately

$$\Delta V = \pi(\text{radius})^2(\text{thickness}) = \pi(5)^2 \Delta y = 25\pi \Delta y \text{ ft}^3.$$

The force $F(y)$ required to lift this slab is its weight,

$$F(y) = w\Delta V = 25\pi w \Delta y \text{ lb},$$

where w is the weight of a cubic foot of water. (We can substitute for w later.)

The distance through which $F(y)$ must act is about $(14 - y)$ ft, so the work done in lifting this slab 4 ft above the top of the tank is about

$$\Delta W = 25\pi w (14 - y)\Delta y \text{ ft} \cdot \text{lb}.$$

The work of lifting all the slabs of water is about

$$\sum \Delta W = \sum 25\pi w (14 - y)\Delta y \text{ ft} \cdot \text{lb}.$$

This is a Riemann sum for the function $25\pi w (14 - y)$ on the interval from $y = 0$ to $y = 10$. The work of pumping the tank dry is the limit of these sums as the norm of the partition goes to zero:

$$\text{Work} = \int_0^{10} 25\pi w (14 - y)\, dy = 25\pi\, w \int_0^{10} (14 - y)\, dy$$

$$= 25\pi\, w \left[14y - \frac{y^2}{2} \right]_0^{10} = 25\pi\, w(140 - 50) = 2250\pi\, w$$

$$= 2250\pi\,(62.5) \qquad \text{Water weighs 62.5 lb/ft}^3.$$

$$= 441{,}786 \text{ ft} \cdot \text{lb}. \qquad \text{Nearest ft} \cdot \text{lb}$$

A 1-horsepower pump, rated at 550 ft · lb per second, could empty the tank in a little less than 14 min. ≡

EXPLORATION 2

Build the Integral Yourself

We recognized Example 3 as a variable-force problem and built our own integral to solve it. Now try building an integral on your own. The inverted conical tank in Fig. 6.44 is filled to within 2 ft of the top with salad oil weighing 57 lb/ft^3. How much work does it take to pump the oil to the rim of the tank?

1. A thin horizontal slab of the oil is approximately cylindrical, as suggested in Fig. 6.44. At height y in the tank, the radius of the slab is $(1/2)y$; the thickness of the slab is Δy. What is the cylindrical volume, ΔV, of the slab?

2. What is the force needed to lift the slab? That is, what is the weight of the slab?

3. How far must the slab be lifted to reach the rim of the tank?

4. How much work is done in lifting the slab to the rim of the tank?

5. Write a Riemann sum for the work done in lifting all the slabs.

6. Tell how to go from the Riemann sums to an integral. Write the integral. Evaluate. Give a solution to the problem. Support your solution with a NINT computation.

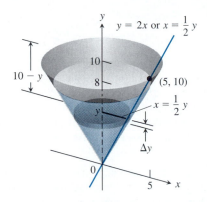

6.44 The force needed to lift each slab of salad oil is the weight of the slab.

Exercises 6.6

1. The worker in Example 1 changed to a larger bucket that held 5 gal (40 lb) of water, but the new bucket had an even larger leak, so it, too, was empty by the time it reached the top. Assuming that the water leaked out at a steady rate, how much work was done lifting the water? (Do not include the rope and bucket.)

2. The bucket in Example 1 is hauled up twice as fast so that there is still 1 gal (8 lb) of water left when the bucket reaches the top. How much work is done lifting the water this time? (Do not include the rope and bucket.)

3. A mountain climber is about to haul up a 50-m length of hanging rope. How much work will it take if each meter of rope weighs 0.74 newton?

4. A model rocket engine burned up its 2-oz fuel cartridge lifting the rocket to 170 ft. Assuming that the fuel burned at a steady rate, how much work was spent just lifting fuel?

5. An electric elevator with a motor at the top has a multistrand cable weighing 4 lb/ft. One hundred eighty feet of cable are paid out when the car is at the first floor and effectively zero feet are out when the car is at the top floor. How much work does the motor do just lifting the cable when it takes the car from the first floor to the top?

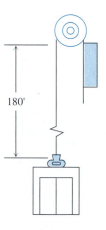

6. A bag of sand that originally weighed 144 lb was lifted at a constant rate. The sand leaked out at a steady rate, and the sand was half gone by the time the bag had been lifted 18 ft. How much work was done in lifting the sand this far? (Neglect the weight of the bag and lifting equipment.)

7. If a force of 6 N stretches a spring 0.4 m beyond its natural length, how much work is done?

8. If a force of 90 N stretches a spring 1 m beyond its natural length, how much work does it take to stretch the spring 5 m beyond its natural length?

9. A 10,000-lb force compressed a spring from its natural length of 12 in. to a length of 11 in. How much work did it do in
a) compressing the spring the first half-inch?
b) compressing the spring the second half-inch?

10. A bathroom scale is compressed 1/16 in. when a 150-lb person stands on it. Assuming that the scale behaves like a spring, how much does someone who compresses the scale 1/8 in. weigh? How much work does it take to compress the scale 1/8 in.?

11. In Example 3, how much work would it take to pump the water to the top rim of the tank (instead of 4 ft higher)?

12. Suppose that instead of being completely full, the tank in Example 3 is only half full. How much work does it take to pump the water that's left to a level 4 ft above the top of the tank?

13. A vertical right circular cylindrical tank measures 30 ft high and 20 ft in diameter. It is full of kerosene weighing 51.2 lb/ft^3. How much work does it take to pump the kerosene to the level of the top of the tank?

14. a) Suppose the cone in Fig. 6.44 contained milk (weight density 64.5 lb/ft^3) instead of salad oil. How much work would it have taken to pump the contents to the rim?
b) How much work would it have taken to pump the oil in Fig. 6.44 to a level 3 ft above the cone's rim?

15. The rectangular tank shown here with its top at ground level is used to catch runoff water. (*Note:* The diagram suggests that you point the y-axis down.)

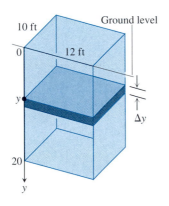

a) How much work does it take to empty the tank by pumping the water back to ground level once the tank is full?
b) If the water is pumped to ground level with a (5/11)-hp motor (work output 250 ft · lb/sec), how long will it take to empty the full tank?
c) Show that the pump in part (b) will lower the water level 10 ft (halfway) during the first 25 min of pumping.

16. The full rectangular cistern (rainwater storage tank) shown here with its top 10 ft below ground level is to be emptied for inspection by pumping its contents to ground level.

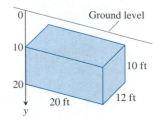

a) How much work will it take to empty the cistern?
b) How long will it take a (1/2)-hp pump, rated at 225 ft · lb/sec, to pump the tank dry?
c) How long will it take the pump in part (b) to empty the tank halfway? (It will be less than half the time required to empty the tank completely.)

17. To design the interior surface of a huge stainless steel tank, you revolve the curve $y = x^2$, $0 \le x \le 4$, about the y-axis. The container (where x and y are measured in meters) is to be filled with sea water, which weighs 10,000 newtons per cubic meter. How much work will it take to empty the tank by pumping the water to the tank's top?

18. We model pumping from spherical containers the way we do from others, with the axis of integration along the vertical axis of the sphere. We drew the following diagram to help find out how much work it takes to pump the water from a full hemispherical bowl of radius 5 ft to a height 4 ft above the top of the bowl. How much work does it take?

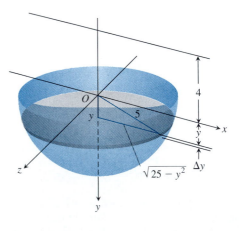

19. The cylindrical tank shown in the following diagram can be filled by pumping water from a lake 15 ft below the bottom of the tank. There are two ways to go about it. One is to pump the water through a hose to a valve in the bottom of the tank. The other is to attach the hose to the rim of the tank and let the water pour in. Which way will be faster? Explain.

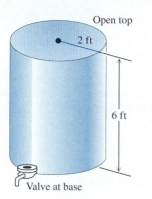

Open top
2 ft
6 ft
Valve at base

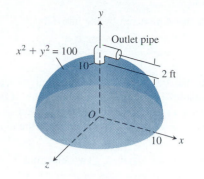

y
Outlet pipe
$x^2 + y^2 = 100$
10
2 ft
O
10   x
z

20. The truncated conical container shown below is full of strawberry milkshake that weighs (4/9) oz/in.^{3}.

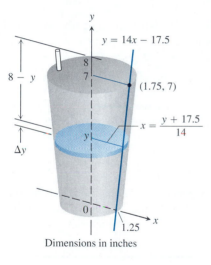

y
$y = 14x - 17.5$
8
8 − y
7
(1.75, 7)
$x = \dfrac{y + 17.5}{14}$
y
Δy
0
1.25
x
Dimensions in inches

As you can see, the container is 7 in. deep, 2.5 in. across at the base, and 3.5 in. across at the top (a standard size at Brigham's in Boston). The straw sticks up an inch above the top. About how much work does it take to suck up the milkshake through the straw (neglecting friction)? Answer in inch-ounces.

21. You are in charge of the evacuation and repair of the storage tank shown in the following diagram. The tank is a hemisphere of radius 10 ft and is full of benzene weighing 56 lb/ft^3. A firm that you contacted says that it can empty the tank for 1/2¢ per foot-pound of work. Find the work required to empty the tank by pumping the benzene to an outlet 2 ft above the top of the tank. If you have $5000 budgeted for the job, can you afford to hire the firm?

22. Your town has decided to drill a well to increase its water supply. As the town engineer, you have determined that a water tower will be necessary to provide the pressure needed for distribution, and you have designed the system shown below. The water is to be pumped from a 300-ft well through a vertical 4-in. pipe into the base of a cylindrical tank 20 ft in diameter and 25 ft high. The base of the tank will be 60 ft above ground. The pump is a 3-hp pump, rated at 1650 ft · lb/sec. How long will it take to fill the tank the first time? (Include the time it takes to fill the pipe.)

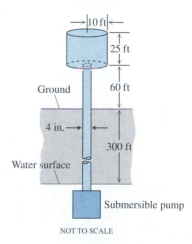

10 ft
25 ft
Ground
60 ft
4 in.
300 ft
Water surface
Submersible pump

NOT TO SCALE

23. *Putting a satellite in orbit.* The strength of Earth's gravitational field varies with the distance r from the planet's center, and the magnitude of the gravitational force experienced by a satellite of mass m during and after launch is

$$F(r) = \frac{m\,MG}{r^2}.$$

Here, $M = 5.975 \times 10^{24}$ kg is Earth's mass, $G = 6.6720 \times 10^{-11}$ Nm^{2}kg^{-2} is the universal gravitational constant, and r is measured in meters. The number of newton-meters of work that it takes to lift a 1000-kg satellite from Earth's surface to a circular orbit 35,780 km above the planet's center is therefore given by the integral

$$\text{Work} = \int_{6,370,000}^{35,780,000} \frac{m\,MG}{r^2}\,dr.$$

Evaluate the integral. The lower limit of integration is Earth's radius in meters at the launch site. (This calculation does not take into account energy spent on the launch vehicle or energy spent bringing the satellite to orbit velocity.)

24. *Forcing electrons together.* Two electrons r meters apart repel each other with a force of

$$F = \frac{23 \times 10^{-29}}{r^2}$$

newtons.

a) Suppose one electron is held fixed at the point $(1, 0)$ on the x-axis (units in meters). How much work does it take to move a second electron along the x-axis from the point $(-1, 0)$ to the origin?

b) Suppose an electron is held fixed at each of the points $(-1, 0)$ and $(1, 0)$. How much work does it take to move a third electron along the x-axis from $(5, 0)$ to $(3, 0)$?

6.7 _____ Fluid Pressures and Fluid Forces

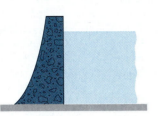

6.45 To withstand the increasing pressure, dams are built thicker approaching the bottom.

We make dams thicker at the bottom than at the top (Fig. 6.45) because the pressure against them increases with depth. The deeper the water, the thicker the dam has to be.

It is a remarkable fact that the pressure at any point on the dam depends only on how far below the water surface the point is and not on how much the dam wall happens to be tilted at that point. The pressure, in pounds per square foot at a point h feet below the surface, is always $62.5h$. The number 62.5 is the weight-density of water in pounds per cubic foot.

The formula, pressure $= 62.5h$, makes sense when you think of the units involved: Pounds per square foot equals pounds per cubic foot times feet:

$$\frac{\text{lb}}{\text{ft}^2} = \frac{\text{lb}}{\text{ft}^3} \times \text{ft}.$$

As you can see, this equation depends only on units and not on what fluid is involved. The pressure h feet below the surface of *any* fluid is the fluid's weight-density times the depth.

WEIGHT-DENSITY

Because the distance to Earth's center of gravity can vary, so can the weight of objects on the planet's surface. At sea level, a cubic foot of water weighs from 62.26 lb at the equator to about 62.59 lb at the poles (which are closer to Earth's center), a variation of about 0.5%. A cubic foot that weighs about 62.4 lb in Melbourne and New York City will weigh about 62.5 lb in Juneau and Stockholm. Typical weight-density values (weight per unit volume) are as follows in lb/ft^3:

Gasoline	42
Mercury	849
Milk	64.5
Molasses	100
Olive oil	57
Sea water	64
Water	62.5

These are the values that we use in this textbook.

The Pressure-Depth Equation

In a fluid that is standing still, the pressure at depth h is the fluid's weight-density times h:

$$p = wh. \qquad (1)$$

In this section, we use the equation $p = wh$ to derive a formula for the total force exerted by a fluid against all or part of a vertical or horizontal containing wall.

The Constant-Depth Formula for Fluid Force

In a container of fluid with a flat horizontal base, the total force exerted by the fluid against the base can be calculated by multiplying the area of the

EXPLORATION BIT

These two containers have the same base area and are filled with water to the same depth.

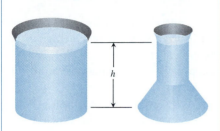

How does the total force of the water on the two bases compare? Give a convincing argument.

base by the pressure at that level. We can do this because total force equals force per unit area (pressure) times area. If F, p, and A are the total force, pressure, and area, then

$$F = \text{total force} = \text{force per unit area} \times \text{ area}$$

$$= \text{pressure} \times \text{area} = pA$$

$$= whA. \qquad p = wh \text{ from Eq. (1)}$$

Total Force on a Constant-Depth Surface

$$F = pA = whA \qquad (2)$$

EXAMPLE 1   The Great Molasses Flood

At 1:00 P.M. on January 15, 1919, an unusually warm day, a 90-ft-high, 90-ft-diameter, cylindrical metal tank, in which the Puritan Distilling Company was storing molasses at the corner of Foster and Commercial streets in Boston's North End, exploded. The molasses flooded into the streets, 30 ft deep, trapping pedestrians and horses, knocking down buildings, and oozing into homes. It was eventually tracked all over town and even into the suburbs on people's shoes. The cleanup went on for weeks.

Given that the molasses weighed 100 lb/ft^3 and assuming that the tank was full, we can use Eq. (2) to compute the total force exerted by the molasses against the base of the tank at the time it burst:

$$\text{Total force } = whA = (100)(90)(\pi(45)^2) = 57{,}255{,}526 \text{ lb.} \qquad \equiv$$

How about the force against the walls of the tank? For example, what was the total force against the bottom foot-wide band of tank wall (Fig. 6.46)?

The area of the band was

$$A = \pi d h = \pi(90)(1) = 90\pi \text{ ft}^2.$$

The tank was 90 ft deep, so the pressure near the bottom was about

$$p = wh = (100)(90) = 9000 \text{ lb/ft}^2.$$

Therefore, the total force against the band was about

$$F = wh\ A = (9000)(90\pi) = 2{,}544{,}690 \text{ lb.}$$

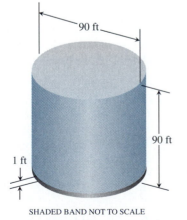

SHADED BAND NOT TO SCALE

6.46 Schematic drawing of the molasses tank in Example 1. How much force did the bottom foot of wall have to withstand when the tank was full? It takes an integral to find out.

But this is not exactly right. The top of the band was 89 ft below the surface, not 90, and the pressure there was less. To find out exactly what the force on the band was, we need to take the variation of the pressure across the band into account, and this means using calculus.

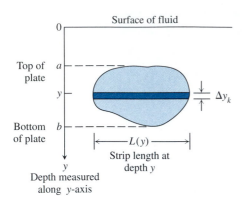

6.47 The force exerted by the fluid against one side of an approximating horizontal strip y units beneath the surface is about

$$\Delta F = (\text{pressure})(\text{area}) = (wy)(L(y)\Delta y).$$

Developing an Integral to Calculate Fluid Force

1. Model the force that we want to measure in terms of a continuous force-per-unit-length "pressure" on a closed interval $[a, b]$.

2. Form approximations of the small amounts of total force on the subintervals of $[a, b]$.

3. Add the small amounts of total force to form a Riemann sum.

4. Take the limit of the Riemann sum to get an integral, which gives precisely the force that we want in step 1.

The Variable-Depth Integral Formula for Force

Suppose we want to find the force against one side of a submerged vertical plate whose surface looks like the shaded region in Fig. 6.47. The region runs from a units below the surface to b units below the surface. We have chosen to measure depth with the y-axis, and the region's width at depth y is $L(y)$.

We partition the closed interval $[a, b]$ in the usual way and imagine the region to be cut into thin horizontal strips by planes perpendicular to the y-axis at the points of the partition. The typical strip from y to $y + \Delta y$ is Δy units wide by $L(y)$ units long. We assume $L(y)$ to be continuous throughout the closed interval $[a, b]$.

The pressure varies across the strip from top to bottom, just as it did in the molasses tank. But, if the strip is narrow enough, the pressure will remain close to its top-edge value wy (w being the weight-density of the liquid that we are working with, and y the depth of the top edge). The total force against one side of the strip will therefore be about

$$\Delta F = (\text{pressure along top edge})(\text{area})$$
$$= (wy)(L(y)\Delta y) = wy\,L(y)\Delta y.$$

The force against the entire wall will be about

$$\sum \Delta F = \sum wy\,L(y)\Delta y. \qquad (3)$$

The sum on the right-hand side of Eq. (3) is a Riemann sum for the continuous function $wy\,L(y)$ on the closed interval $[a, b]$. We expect the approximations to improve as the norm of the partition of $[a, b]$ goes to zero, so we define the total force against the wall to be the limit of these sums as the norm goes to zero.

DEFINITION

Suppose a submerged vertical plate running from depth $y = a$ to depth $y = b$ in a fluid of weight-density w is $L(y)$ units across at depth y, as measured along the plate. Then the **total force** of the fluid against one side of the plate is

$$F = \int_a^b wy\,L(y)\,dy. \qquad (4)$$

We call Eq. (4) the **integral formula for fluid force**.

EXAMPLE 2

A flat triangular plate is submerged vertically, base up, 2 ft below the surface of a swimming pool (Fig. 6.48). Find the fluid force against one side of the plate.

Solution We see from Fig. 6.48 that the plate runs from depth $y = 2$ to depth $y = 5$. At depth y, its width is $L(y) = 2(5 - y)$. Therefore, Eq. (4)

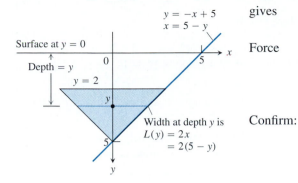

6.48   The important dimensions of the plate in Example 2. Note that the origin of the coordinate system is at the surface of the water with the y-axis pointing down.

gives

$$\text{Force} = \int_a^b wy\,L(y)\,dy$$

$$= \text{NINT}\,((62.5)y(2)(5-y),\,2,\,5) = 1687.5\ \text{lb}. \qquad \text{For water,}\ w = 62.5.$$

Confirm: $$\int_2^5 (62.5)(y)(2)(5-y)\,dy$$

$$= 125\int_2^5 (5y - y^2)\,dy = 125\left[\frac{5}{2}y^2 - \frac{1}{3}y^3\right]_2^5 = 1687.5\ \text{lb} \quad \blacksquare$$

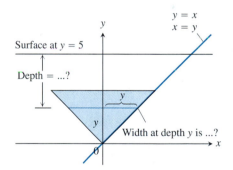

EXPLORATION 1

Choosing a Convenient Coordinate System

A coordinate system with the positive y-axis pointing down is not always the most convenient one to choose for fluid force. It sometimes is better to put the origin at the bottom of the plate instead of at the fluid's surface and have the y-axis point up. This changes the depth factor y in the integrand of Eq. (4) to some other value $D(y)$ while $L(y)$ remains the width of the plate at depth y. The limits of integration are from the bottom of the plate to the top.

Use the coordinate system of Fig. 6.49 for the problem in Example 2, and evaluate the integral for fluid force, $\int_a^b wD(y)L(y)\,dy$.

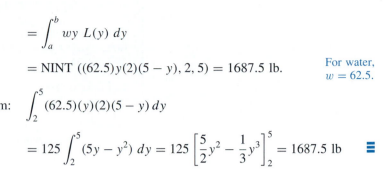

6.49   The origin is at the bottom of the submerged plate with the y-axis pointing up. Compare with Fig. 6.48.

1. What are the limits of integration (the depth values from the bottom of the plate to the top)?

2. What is w?

3. For y as shown in Fig. 6.49, what is the depth $D(y)$?

4. What is $L(y)$?

5. Now evaluate the integral $\int_a^b wD(y)L(y)\,dy$, and compare with Example 2.

> ### Strategy for Finding Fluid Force
>
> Whatever coordinate system you use, you can always find the fluid force against one side of a submerged vertical plate or wall by taking these steps:
>
> 1. Find expressions for the depth and length of a typical thin horizontal strip.
> 2. Multiply their product by the fluid's weight-density w, and integrate over the interval of depths occupied by the plate or wall.

EXAMPLE 3

We can now calculate exactly the force exerted by the molasses against the bottom 1-ft band of the Puritan Distilling Company's storage tank when the tank was full.

The tank was a right circular cylindrical tank, 90 ft high and 90 ft in diameter. Using a coordinate system with the origin at the bottom of the tank and the y-axis pointing up (Fig. 6.50), we find that the typical horizontal strip at level y has

Strip depth: $(90 - y)$,

Strip length: $\pi \times$ tank diameter $= 90\pi$.

The force against the band is therefore

$$\text{Force} = \int_0^1 w \ (\text{depth})(\text{length}) \ dy$$

$$= \int_0^1 100(90 - y)(90\pi) \ dy \qquad \text{For molasses, } w = 100.$$

$$= 9000\pi \int_0^1 (90 - y) \ dy = 9000\pi \left[90y - \frac{1}{2}y^2 \right]_0^1$$

$$= 9000\pi \, (89.5) = 2{,}530{,}553 \text{ lb.}$$

Support:

$$\text{NINT} \, (100(90 - y)(90\pi), 0, 1) = 2{,}530{,}553.$$

As expected, the actual total force of the molasses on the bottom 1-ft band was slightly less than the constant-depth estimate at the end of Example 1.

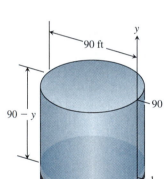

6.50 For y values from 0 to 1, the depth of the strip at the bottom of the molasses tank is $90 - y$. The length of the strip is the constant circumference, 90π.

Exercises 6.7

1. Suppose the triangular plate in Figs. 6.48 and 6.49 is 4 ft beneath the surface instead of 2 ft. What is the force on one side of the plate now?

2. What was the total force against the side wall of the Puritan Distilling Company's molasses tank when the tank was full? Half full? (See Example 1.)

3. The vertical ends of a hog-watering trough are inverted isosceles triangles like the one shown here. What is the

force on each end of the trough when the trough is full? Does it matter how long the trough is?

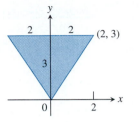

4. What is the force on each end of the trough in Exercise 3 if the water level is lowered 1 ft?

5. The triangular plate shown here is submerged vertically, 1 ft below the surface of the water. Find the force on one side of the plate.

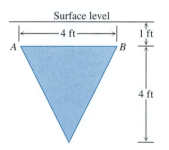

6. The triangular plate in Exercise 5 is revolved 180° about the line AB so that part of it sticks up above the surface. What force does the water exert on one face of the plate now?

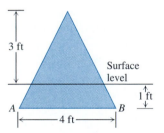

7. A semicircular plate is submerged straight down in the water with its diameter at the surface. Find the force exerted by the water on one side of the plate.

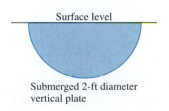

Submerged 2-ft diameter
vertical plate

8. A rectangular fish tank of interior dimensions $2 \times 2 \times 4$ ft is filled to within 2 in. of the top with water. Find the force against the sides and ends of the tank.

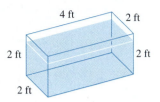

9. The viewing portion of the rectangular glass window in a typical fish tank at the New England Aquarium in Boston, Massachusetts, is 63 in. wide and runs from 0.5 in. below the water's surface to 33.5 in. below the surface. Find the force of sea water against this portion of the window. The weight-density of sea water is 64 lb/ft^3. (In case you were wondering, the glass is 3/4 in. thick.)

10. A rectangular milk carton measures $3.75 \times 3.75 \times 7.75$ in. Find the force of the milk on one side of a full carton.

11. A tank truck hauls milk in a 6-ft-diameter horizontal right circular cylindrical tank. What is the force on the end of the tank when the tank is half full?

12. The cubical metal tank shown below is used to store liquids. It has a parabolic gate, held in place by bolts and designed to withstand a force of 160 lb without rupturing. The liquid that you plan to store has a weight-density of 50 lb/ft^3.

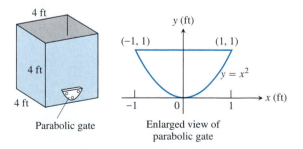

a) What will be the force on the gate when the liquid is 2 ft deep?

b) What is the maximum height to which the container can be filled without exceeding the gate's design limitation?

13. The end plates in the trough shown here were designed to withstand a force of 6667 lb. How many cubic feet of water can the tank hold without exceeding design limitations?

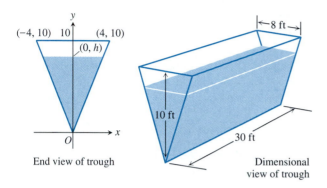

End view of trough

Dimensional view of trough

14. A vertical triangular drain plate is at the bottom of one end of a rectangular swimming pool as shown in the following diagram. Water is running into the pool at the rate of 1000 ft³/hr.

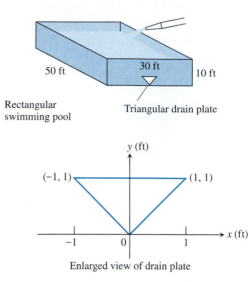

Rectangular swimming pool

Triangular drain plate

Enlarged view of drain plate

a) Find the force against the drain plate after 9 hr of filling.
b) The plate is designed to withstand a force of 520 lb. How high can the pool be filled without exceeding this design limitation?

15. A vertical rectangular plate is submerged in a fluid with its top edge parallel to the fluid's surface. Show that the force on one side of the plate is the average value of the pressure up and down the plate times the area of the plate.

6.8 _____ Centers of Mass

Many structures and mechanical systems behave as if their masses were concentrated at a single point called the center of mass. (See Fig. 6.51.) It is important to know how to locate this point, and it turns out that doing so is basically a mathematical undertaking. We do it with calculus, and this section shows how. For the moment, we will deal only with one- and two-dimensional shapes. Three-dimensional shapes are best done with the so-called multiple integrals of Chapter 14.

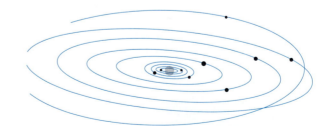

6.51 From the structure of atoms to the structure of the universe, the center of mass is a key element controlling how structural components relate to each other. In our solar system, for example, the planets, asteroids, and comets all revolve about their collective center of mass. (It lies inside the sun.)

Masses along a Line

If we imagine masses m_1, m_2, and m_3 placed on a rigid x-axis, and the axis supported by a fulcrum at the origin, the resulting system might or might not balance.

Each mass m_k exerts a downward force $m_k g$ equal to the magnitude of the mass times the acceleration of gravity. Each of these forces has a tendency to turn the axis about the origin, the way somebody's weight might turn a seesaw. This turning effect, called a **torque**, is measured by multiplying the force $m_k g$ by the signed distance x_k from the mass to the origin. Masses to the left of the origin exert a negative (counterclockwise) torque. Masses to the right of the origin exert a positive (clockwise) torque.

We use the sum of the torques to measure the tendency of a system to rotate about the origin. This sum is called the **system torque:**

$$\text{System torque} = m_1 g x_1 + m_2 g x_2 + m_3 g x_3. \tag{1}$$

The system is in balance if and only if its net torque is zero.

If we factor out the g in Eq. (1), we see that the system torque is

$$g(m_1 x_1 + m_2 x_2 + m_3 x_3).$$

A feature of the environment

A feature of the system

Thus, the torque is the product of the gravitational acceleration g, which is a feature of the environment in which the system happens to reside, and the number $(m_1 x_1 + m_2 x_2 + m_3 x_3)$, which is a feature of the system itself, a constant that stays the same no matter where the system is placed.

The number $(m_1 x_1 + m_2 x_2 + m_3 x_3)$ is called the **moment of the system about the origin:**

$$M_O = \text{Moment of system about origin} = \sum m_k x_k.$$

(We shift to sigma notation here to allow for sums with more terms. If you want a quick way to read $\Sigma m_k x_k$, try "summation $m\,k\,x\,k$".)

We usually want to know where to place the fulcrum to make the system balance, that is, at what point $\overline{x}$ to place it to make the torque zero.

The torque of each mass about the fulcrum in this special location is

$$\text{Torque of } m_k \text{ about } \overline{x} = \left(\begin{array}{c} \text{signed distance} \\ \text{of } m_k \text{ from } \overline{x} \end{array} \right) \cdot \left(\begin{array}{c} \text{downward} \\ \text{force} \end{array} \right)$$

$$= (x_k - \overline{x}) \cdot m_k g.$$

When we write down the equation that says that the sum of these torques is zero, we get an equation that we can solve for $\overline{x}$:

$$\sum (x_k - \overline{x}) m_k g = 0 \qquad \text{Sum of the torques equals zero.}$$

$$g \sum (x_k - \overline{x}) m_k = 0 \qquad \text{Constant Multiple Rule for Sums}$$

$$\sum (m_k x_k - \overline{x} m_k) = 0 \qquad g \text{ divided out, } m_k \text{ distributed}$$

$$\sum m_k x_k - \sum \overline{x} m_k = 0 \qquad \text{Difference Rule for Sums}$$

$$\sum m_k x_k = \overline{x} \sum m_k \qquad \text{Rearranged, Constant Multiple Rule again}$$

$$\overline{x} = \frac{\sum m_k x_k}{\sum m_k}. \qquad \text{Solved for } \overline{x}$$

This last equation tells us to find $\overline{x}$ by dividing the system's moment about the origin by the system's total mass:

$$\overline{x} = \frac{\sum x_k m_k}{\sum m_k} = \frac{\text{System moment about origin}}{\text{System mass}}.$$

The point $\overline{x}$ is called the system's **center of mass.**

Wires and Thin Rods

In many applications, we want to know the center of mass of a rod or a thin strip of metal. In cases like these, in which we can assume the distribution of mass is continuous, the summation signs in our formulas become integrals in a manner we shall now describe.

Imagine a long, thin strip lying along the x-axis from $x = a$ to $x = b$ and cut into small pieces of mass Δm_k by a partition of the interval $[a, b]$.

Each piece is Δx units long and lies approximately x_k units from the origin. Now observe three things.

First, the strip's center of mass $\overline{x}$ is nearly the same as the center of mass of the finite system of point masses that we would get by attaching each mass Δm_k to the point x_k:

$$\overline{x} \approx \frac{\text{System moment}}{\text{System mass}}.$$

Second, the moment of each piece of the strip about the origin is approximately $x_k \Delta m_k$, so the system moment is approximately the sum of the $x_k \Delta m_k$:

$$\text{System moment} \approx \sum x_k \Delta m_k.$$

Third, if the density of the strip at x_k is $\delta(x_k)$, expressed in terms of mass per unit length, and δ is continuous, then Δm_k is approximately equal to $\delta(x_k)\Delta x$ (mass per unit length times length):

$$\Delta m_k \approx \delta(x_k)\Delta x.$$

Combining these three observations gives

$$\bar{x} \approx \frac{\text{System moment}}{\text{System mass}} \approx \frac{\sum x_k \Delta m_k}{\sum \Delta m_k} \approx \frac{\sum x_k \delta(x_k)\Delta x}{\sum \delta(x_k)\Delta x}. \qquad (2)$$

The sum in the numerator of the last quotient in (2) is a Riemann sum for the continuous function $x\delta(x)$ over the closed interval $[a, b]$. The sum in the denominator is a Riemann sum for the function $\delta(x)$ over this interval. We expect the approximations in (2) to improve as the strip is partitioned ever more finely and are led to the equation

$$\bar{x} = \frac{\displaystyle\int_a^b x\delta(x)\,dx}{\displaystyle\int_a^b \delta(x)\,dx}.$$

This is the formula that we use to calculate $\bar{x}$.

Developing an Integral to Calculate Mass of a Thin Rod

1. Model the mass of the thin rod (wire or strip) that we want to measure as a continuous density function $\delta(x)$ on a closed interval $[a, b]$ (in terms of mass per unit length).

2. Form approximations of the masses of small pieces of rod on the subintervals of $[a, b]$.

3. Add the masses of the small pieces of rod to form a Riemann sum.

4. Take the limit of the Riemann sum to get an integral, which gives precisely the mass of the rod in step 1.

In a similar manner, the integral for the moment about the origin follows from Riemann sums. We approximate moments about the origin of small pieces of rod on the subintervals of $[a, b]$. Their sum is the Riemann sum that we seek.

Moment, Mass, and Center of Mass of a Wire, Thin Rod, or Strip along the x-Axis

Moment about the origin:	$M_O = \displaystyle\int_a^b x\delta(x)\,dx$	(3a)
Mass:	$M = \displaystyle\int_a^b \delta(x)\,dx$	(3b)
Center of mass:	$\bar{x} = \dfrac{M_O}{M}$	(3c)

Equation (3c) says that to find the center of mass of a rod or a thin strip, we divide its moment about the origin by its mass.

EXAMPLE 1   A Useful Result

Show that the center of mass of a straight, thin strip or rod of constant density is always located halfway between its two ends.

Solution We model the strip as a portion of the x-axis from $x = a$ to $x = b$ (Fig. 6.52). Our goal is to show that $\bar{x} = (a+b)/2$, the point halfway between a and b.

The key is the density's having a constant value. This enables us to regard the function $\delta(x)$ in the integrals in Eqs. (3) as a constant (call it δ),

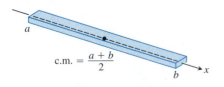

6.52 The center of mass (c.m.) of a straight, thin rod or strip of constant density lies halfway between its ends.

with the result that

$$M_O = \int_a^b \delta x \, dx = \delta \int_a^b x \, dx = \delta \left[\frac{1}{2} x^2 \right]_a^b = \frac{\delta}{2} (b^2 - a^2)$$

$$M = \int_a^b \delta \, dx = \delta \int_a^b dx = \delta \left[x \right]_a^b = \delta(b - a)$$

$$\bar{x} = \frac{M_O}{M} = \frac{\frac{\delta}{2}(b^2 - a^2)}{\delta(b - a)} = \frac{a + b}{2}.$$ The δ's cancel.   ▤

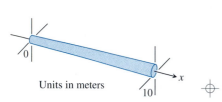

Units in meters

6.53 We can treat a rod of variable thickness as a rod of variable density. See Example 2.

EXAMPLE 2   A Variable Density

The 10-meter-long rod in Fig. 6.53 thickens from left to right so that its density, instead of being constant, is $\delta(x) = 1 + x/10$ kg/m. Find the rod's center of mass.

Solution   The rod's moment about the origin (Eq. 3a) is

$$M_O = \int_0^{10} x \delta(x) \, dx = \int_0^{10} x \left(1 + \frac{x}{10} \right) dx = \int_0^{10} \left(x + \frac{x^2}{10} \right) dx$$

$$= \left[\frac{x^2}{2} + \frac{x^3}{30} \right]_0^{10} = 50 + \frac{100}{3} = \frac{250}{3} \text{ kg} \cdot \text{m}.$$ The units of a moment are mass × length.

The rod's mass (Eq. 3b) is

$$M = \int_0^{10} \delta(x) \, dx = \int_0^{10} \left(1 + \frac{x}{10} \right) dx = \left[x + \frac{x^2}{20} \right]_0^{10} = 10 + 5 = 15 \text{ kg}.$$

The center of mass (Eq. 3c) is located at the point

$$\bar{x} = \frac{M_O}{M} = \frac{250/3}{15} = 5.556 \text{ m}.$$ ▤

SUPPORT

In Example 2, let

$$y_1 = 1 + \frac{x}{10};$$

then

$$M_O = \text{NINT} \, (x y_1, 0, 10) = 83.333,$$

$$M = \text{NINT} \, (y_1, 0, 10) = 15,$$

and their quotient $\bar{x} = 5.556$.

Masses Distributed over a Plane Region

Suppose we have a finite collection of masses located in the coordinate plane, the mass m_k being located at the point (x_k, y_k). (See Fig. 6.54.) The total mass of the system is

System mass: $M = \sum m_k.$

Each mass m_k has a moment about each axis. Its moment about the x-axis is $m_k y_k$, and its moment about the y-axis is $m_k x_k$. The moments of the entire system about the two axes are

Moment about x-axis: $M_x = \sum m_k y_k,$

Moment about y-axis: $M_y = \sum m_k x_k.$

The x-coordinate of the system's center of mass is defined to be

$$\bar{x} = \frac{M_y}{M} = \sum m_k x_k \Big/ \sum m_k.$$ (4)

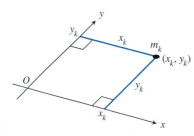

6.54 The mass m_k has a moment about each axis. Note that x_k and y_k are the (perpendicular) distances to the axes.

With this choice of $\overline{x}$, as in the one-dimensional case, the system balances about the line $x = \overline{x}$ (Fig. 6.55).

The y-coordinate of the system's center of mass is defined to be

$$\overline{y} = \frac{M_x}{M} = \sum m_k y_k \bigg/ \sum m_k. \tag{5}$$

With this choice of $\overline{y}$, the system balances about the line $y = \overline{y}$ as well. The torques exerted by the masses about the lines $x = \overline{x}$ and $y = \overline{y}$ cancel out. Thus, as far as balance is concerned, the system behaves as if all its mass were at the single point $(\overline{x}, \overline{y})$. We call this point the system's **center of mass**.

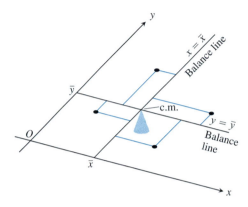

6.55 A two-dimensional array of masses balances on its center of mass (c.m.).

Thin, Flat Plates

In many applications, we need to find the center of mass of a thin, flat plate: a disk of aluminum, say, or a triangular sheet of steel. In such cases, we assume the distribution of mass to be continuous, and the formulas that we use to calculate $\overline{x}$ and $\overline{y}$ contain integrals instead of finite sums. The integrals arise in the following way.

Imagine the plate occupying a region in the xy-plane, cut into thin strips parallel to one of the axes (in Fig. 6.56, the y-axis). The center of mass of a typical strip is $(\tilde{x}, \tilde{y})$. (The symbol ˜ over the x and y is a *tilde*, pronounced to rhyme with "Hilda." Thus, $\tilde{x}$ is read "x tilde.") We treat the strip's mass Δm as if it were concentrated at $(\tilde{x}, \tilde{y})$. The moment of the strip about the y-axis is then $\tilde{x} \Delta m$, and the moment of the strip about the x-axis is $\tilde{y} \Delta m$. Equations (4) and (5) then become

$$\overline{x} = \frac{M_y}{M} = \frac{\sum \tilde{x} \Delta m}{\sum \Delta m}, \qquad \overline{y} = \frac{M_x}{M} = \frac{\sum \tilde{y} \Delta m}{\sum \Delta m}.$$

As in the one-dimensional case, the sums in the numerator and denominator are Riemann sums for integrals and approach these integrals as limiting values as the strips into which the plate is cut become narrower and narrower. We write these integrals symbolically as

$$\overline{x} = \frac{\displaystyle\int \tilde{x}\, dm}{\displaystyle\int dm} \qquad \text{and} \qquad \overline{y} = \frac{\displaystyle\int \tilde{y}\, dm}{\displaystyle\int dm}.$$

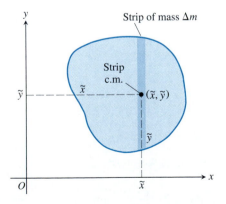

6.56 A plate cut into thin strips parallel to the y-axis. The moment about each axis of a typical strip with mass Δm is equal to the moment of a mass Δm concentrated at the strip's center of mass $(\tilde{x}, \tilde{y})$.

Moments, Mass, and Center of Mass of a Thin Plate Covering a Region in the *xy*-Plane

Moment about the *x*-axis: $M_x = \int \tilde{y}\, dm$

Moment about the *y*-axis: $M_y = \int \tilde{x}\, dm$

Mass: $M = \int dm$

Center of mass: $\bar{x} = \dfrac{M_y}{M}, \qquad \bar{y} = \dfrac{M_x}{M}$

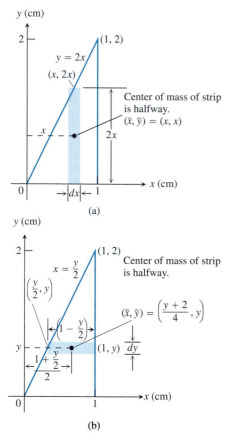

(a)

(b)

6.57   Two ways to model the calculation of the moment M_y of the triangular plate in Example 3.

To evaluate these integrals, we picture the plate in the coordinating plane and sketch a strip of mass parallel to one of the coordinate axes. We then express the strip's mass dm and the coordinates $(\tilde{x}, \tilde{y})$ of the strip's center of mass in terms of x or y. Finally, we integrate $\tilde{y}\, dm$, $\tilde{x}\, dm$, and dm between limits of integration determined by the plate's location in the plane.

EXAMPLE 3

The triangular plate shown in Fig. 6.57, bounded by the lines $y = 0$, $y = 2x$, and $x = 1$, has a constant density of $\delta = 3$ gm/cm^2. Find (a) the plate's moment M_y, about the *y*-axis, (b) the plate's mass M, and (c) the *x*-coordinate of the plate's center of mass.

Solution
Method 1: Vertical strips (Fig. 6.57a).

a) The moment M_y: The typical vertical strip has

center of mass (c.m.): $(\tilde{x}, \tilde{y}) = (x, x)$,

length: $2x$,

width: dx,

area: $dA = 2x\, dx$,

mass: $dm = \delta\, dA = 3 \cdot 2x\, dx = 6x\, dx$,

distance of c.m. from *y*-axis: $\tilde{x} = x$.

The moment of the strip about the *y*-axis is

$$\tilde{x}\, dm = x \cdot 6x\, dx = 6x^2\, dx.$$

The moment of the plate about the *y*-axis is therefore

$$M_y = \int \tilde{x}\, dm = \int_0^1 6x^2\, dx = 2x^3\Big]_0^1 = 2 \text{ gm} \cdot \text{cm}.$$

b) The plate's mass:

$$M = \int dm = \int_0^1 6x\, dx = 3x^2\Big]_0^1 = 3 \text{ gm}.$$

c) The x-coordinate of the plate's center of mass:

$$\bar{x} = \frac{M_y}{M} = \frac{2 \text{ gm} \cdot \text{ cm}}{3 \text{ gm}} = \frac{2}{3} \text{ cm.}$$

By a similar computation, we could find M_x and $\bar{y} = M_x/M$.

Method 2: Horizontal strips (Fig. 6.57b).

a) The moment M_y: The typical horizontal strip has

center of mass (c.m.): $(\tilde{x}, \tilde{y}) = \left(\frac{1}{2} \left(1 + \frac{y}{2} \right), y \right) = \left(\frac{y+2}{4}, y \right),$

length: $1 - \frac{y}{2} = \frac{2 - y}{2},$

width: $dy,$

area: $dA = \frac{2 - y}{2} \, dy,$

mass: $dm = \delta dA = 3 \cdot \frac{2 - y}{2} \, dy,$

distance of c.m. to y-axis: $\tilde{x} = \frac{y + 2}{4}.$

The moment of the strip about the y-axis is

$$\tilde{x} \, dm = \frac{y + 2}{4} \cdot 3 \cdot \frac{2 - y}{2} \, dy = \frac{3}{8}(4 - y^2) \, dy.$$

The moment of the plate about the y-axis is

$$M_y = \int \tilde{x} \, dm = \int_0^2 \frac{3}{8}(4 - y^2) \, dy = \frac{3}{8} \left[4y - \frac{y^3}{3} \right]_0^2$$

$$= \frac{3}{8} \left(\frac{16}{3} \right) = 2 \text{ gm} \cdot \text{ cm.}$$

b) The plate's mass:

$$M = \int dm = \int_0^2 \frac{3}{2}(2 - y) \, dy = \frac{3}{2} \left[2y - \frac{y^2}{2} \right]_0^2 = \frac{3}{2}(4 - 2) = 3 \text{ gm.}$$

c) The x-coordinate of the plate's center of mass:

$$\bar{x} = \frac{M_y}{M} = \frac{2 \text{ gm} \cdot \text{ cm}}{3 \text{ gm}} = \frac{2}{3} \text{ cm.}$$

By a similar computation, we could find M_x and $\bar{y}$.

How to Find a Plate's Center of Mass

1. Picture the plate in the xy-plane.
2. Sketch a strip of mass parallel to one of the coordinate axes, and find its dimensions.
3. Find the strip's mass dm and center of mass $(\tilde{x}, \tilde{y})$.
4. Integrate $\tilde{y}\, dm, \tilde{x}\, dm$, and dm analytically or numerically to find M_x, M_y, and M, respectively.
5. Divide the moments by the mass to calculate $\bar{x}$ and $\bar{y}$.

EXPLORATION 1

Support and Confirm

Example 3 illustrates two methods for finding the center of mass of a thin, flat plate. Thus, this type of problem has a "built-in" method of support and confirmation, namely, solving it by two alternative approaches. Here is one more to try.

A thin plate of constant density δ covers the region bounded above by the parabola $y = 4 - x^2$ and below by the x-axis (Fig. 6.58). Since the plate is symmetric about the y-axis and has constant density, the center of mass lies on the y-axis. This means that $\bar{x} = 0$. It remains to find $\bar{y} = M_x/M$.

1. A trial calculation with horizontal strips (Fig. 6.58a) leads to

$$M_x = \delta \int_0^4 2y \sqrt{4 - y}\; dy.$$

Confirm that this integral is correct, and evaluate it by a method of your choice.

2. Confirm your value for M_x by modeling the distribution of mass with vertical strips (Fig. 6.58b). Evaluate the integral both analytically *and* numerically.

3. Complete the solution by finding $\bar{y}$ and identifying $(\bar{x}, \bar{y})$.

4. Suppose, instead of constant density, the density of the plate at any point (x, y) is $\delta = 2x^2$, twice the distance from the point to the y-axis. Now find its center of mass $(\bar{x}, \bar{y})$.

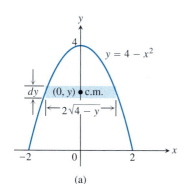

(a)

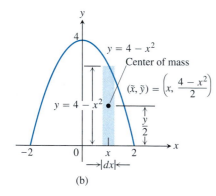

(b)

6.58 We have two ways to model a center-of-mass problem for a plate in the xy-plane: (a) using horizontal strips and (b) using vertical strips. Solving the problem both ways and getting the same result provide strong support that the result is correct.

Centers of Gravity, Homogeneity, Uniformity, and Centroids

As you read elsewhere, you will find some variety in the vocabulary used in connection with centers of mass.

When physicists discuss the effects of a constant gravitational force on a system of masses, they may call the center of mass the **center of gravity**.

Material that has a constant density δ is also said to be **homogeneous**, or to be **uniform**, or to have **uniform density.**

When the density function is constant, it cancels out of the numerator and denominator of the formulas for $\bar{x}$ and $\bar{y}$. This happened in nearly every example in this section. As far as $\bar{x}$ and $\bar{y}$ were concerned, δ might as well have been 1. Thus, when the density is constant, the location of the center of mass is a feature of the geometry of the object and not of the material from which it is made. In such cases, engineers may call the center of mass the **centroid** of the shape, as in "Find the centroid of a triangle or a solid cone." To do so, we just set δ equal to 1 and proceed to find $\bar{x}$ and $\bar{y}$ as before, by dividing moments by masses.

Exercises 6.8

1. Two children are balancing on a seesaw. The 80-lb child is 5 ft from the fulcrum. How far from the fulcrum is the 100-lb child?

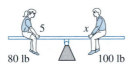

2. The ends of two thin steel rods of equal length are welded together to make a right-angled frame as pictured. Locate the frame's center of mass. (*Hint:* Where is the center of mass of each rod?)

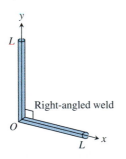

Right-angled weld

Exercises 3–6 give the density functions of thin rods lying along various intervals of the x-axis. Find each rod's moment about the origin and its center of mass.

3. $\delta(x) = 4, 0 \le x \le 2$

4. $\delta(x) = 1 + x/3, 0 \le x \le 3$

5. $\delta(x) = \left(1 + \dfrac{x}{4}\right)^2, 0 \le x \le 4$

6. $\delta(x) = \begin{cases} 2, & 0 \le x \le 3 \\ 1, & 3 < x \le 6 \end{cases}$

In Exercises 7–18, find the center of mass of a thin plate of constant density δ covering the given region.

7. The triangular region bounded below by the x-axis and on the sides by the lines $y = 2x + 2$ and $y = -2x + 2$

8. The region bounded by the parabola $y = x^2$ and the line $y = 4$

9. The region bounded by the y-axis and the curve $x = y - y^3, 0 \le y \le 1$

10. The region bounded by the parabola $y = x - x^2$ and the line $y = -x$

11. The region bounded by the parabola $x = y^2 - y$ and the line $y = x$

12. The region bounded by the parabola $y = 25 - x^2$ and the x-axis

13. The region bounded by the x-axis and the curve $y = \cos x$, $-\pi/2 \le x \le \pi/2$

14. The region between the x-axis and the curve $y = \sec x$, $-\pi/4 \le x \le \pi/4$

15. The region bounded by the parabolas $y = 2x^2 - 4x$ and $y = 2x - x^2$

16. **a)** The region cut from the first quadrant by the circle $x^2 + y^2 = 9$

    b) The region bounded by the x-axis and the semicircle $y = \sqrt{9 - x^2}$. Compare your answer with the answer in part (a).

17. The "triangular" region in the first quadrant between the circle $x^2 + y^2 = 9$ and the lines $x = 3$ and $y = 3$. (*Hint:* Use geometry to find the area.)

18. The region bounded above by the curve $y = 1/x^2$, below by the curve $y = -1/x^2$, and on the left and right by the lines $x = 1$ and $x = 2$.

It can be shown that the centroid of a triangle always lies at the intersection of the medians, one-third of the way from the midpoint of each side toward the opposite vertex. Use this to find the centroids of the triangles whose vertices are given in Exercises 19–22. (*Hint:* Draw each triangle first.)

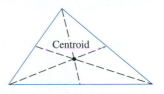

Centroid

19. $(-1, 0), (1, 0), (0, 3)$ **20.** $(0, 0), (1, 0), (0, 1)$

21. $(0, 0), (a, 0), (0, a)$ **22.** $(0, 0), (a, 0), (0, b)$

23. Find the center of mass of a thin plate covering the region between the x-axis and the curve $y = 2/x^2$, $1 \le x \le 2$, if the density is $\delta(x) = x^3$.

24. Find the center of mass of a thin plate covering the region bounded below by the parabola $y = x^2$ and above by the line $y = x$ if the density is $\delta(x) = 12x$.

The Theorems of Pappus

In the third century, an Alexandrian Greek named Pappus discovered two formulas that relate centers of mass to surfaces and volumes of revolution. These formulas, easy to remember, provide useful shortcuts to a number of otherwise lengthy calculations.

Theorem 1

If a plane region is revolved once about an axis in the plane that does not pass through the region's interior, then the volume of the solid swept out by the region is equal to the region's area times the distance traveled by the region's center of mass. In symbols, $V = 2\pi \bar{y} A$.

Theorem 2

If an arc of a plane curve is revolved once about a line in the plane that does not cut through the interior of the arc, then the area of the surface swept out by the arc is equal to the length of the arc times the distance traveled by the arc's center of mass. In symbols, $S = 2\pi \bar{y} L$.

Example 1. The volume of the torus (doughnut) generated by revolving a circle of radius a about an axis in its plane at a distance $b \ge a$ from its center (see Fig. 6.59) is $V = (2\pi b)(\pi a^2) = 2\pi^2 b a^2$. ≡

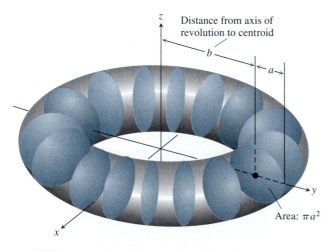

6.59 The volume swept out by the revolving disk is $(2\pi b)\,(\pi a^2)$.

Example 2. The surface area of the torus in Example 1 is

$$S = (2\pi b)(2\pi a) = 4\pi^2 ba. \qquad ≡$$

25. The square region with vertices $(0, 2)$, $(2, 0)$, $(4, 2)$ and $(2, 4)$ is revolved about the x-axis to generate a solid. Find the volume and surface area of the solid.

26. Use a theorem of Pappus to find the volume generated by revolving about the line $x = 5$ the triangular region bounded by the coordinate axes and the line $2x + y = 6$. (The centroid of a triangle lies at the intersection of the medians, one third of the way from the midpoint of each side toward the opposite vertex.)

27. Find the volume of the torus generated by revolving the circle $(x - 2)^2 + y^2 = 1$ about the y-axis.

28. Use the theorems of Pappus to find the lateral surface area and the volume of a right circular cone.

29. Use the second theorem of Pappus and the fact that the surface area of a sphere of radius a is $4\pi a^2$ to find the center of mass of the semicircle $y = \sqrt{a^2 - x^2}$.

30. As found in Exercise 29, the center of mass of the semicircle $y = \sqrt{a^2 - x^2}$ lies at the point $(0, 2a/\pi)$. Find the area of the surface swept out by revolving the semicircle about the line $y = a$.

31. Use the first theorem of Pappus and the fact that the volume of a sphere of radius a is $V = (4/3)\pi a^3$ to find the center of mass of the region enclosed by the x-axis and the semicircle $y = \sqrt{a^2 - x^2}$.

32. As found in Exercise 31, the center of mass of the region enclosed by the x-axis and the semicircle $y = \sqrt{a^2 - x^2}$ lies at the point $(0, 4a/3\pi)$. Find the volume of the solid generated by revolving this region about the line $y = -a$.

33. The region of Exercise 32 is revolved about the line $y = x - a$ to generate a solid. Find the volume of the solid.

34. As found in Exercise 29, the center of mass of the semicircle $y = \sqrt{a^2 - x^2}$ lies at the point $(0, 2a/\pi)$. Find the area of the surface generated by revolving the semicircle about the line $y = x - a$.

6.9

The Basic Idea; Other Modeling Applications

To reiterate the pattern that we followed in the preceding sections: In each section, we wanted to measure something that was modeled or described by one or more continuous functions. In Section 6.1, it was the area between the graphs of two continuous functions. In Section 6.2, it was the volume of the solid defined by revolving the graph of a continuous function about an axis. In Section 6.6, it was the work done by a force directed along the x-axis whose magnitude was given by a continuous function, and so on. In each case, we responded by partitioning the interval on which the function or functions were defined and approximated what we wanted to measure with Riemann sums over the interval. We then used the integral defined by the limit of the Riemann sums to define and calculate what we wanted to measure. You will see what we mean if you look at Table 6.2. Literally thousands of things in biology, chemistry, economics, engineering, finance, geology, medicine, and other fields (the list would fill pages) are modeled and calculated by exactly this process.

In this section we review the process and look at more of the important integrals to which it leads.

TABLE 6.2   The phases of developing an integral to calculate something

Phase 1	Phase 2	Phase 3
We describe or model something that we want to measure in terms of one or more continuous functions defined on a closed interval $[a, b]$.	We partition $[a, b]$ into subintervals of length Δx_k and choose a point c_k in each subinterval.	The approximations improve as the norm of the partition goes to zero.
	We approximate what we want to measure with a finite sum.	The Riemann sums approach a limiting integral.
	We identify the sum as a Riemann sum of a continuous function over $[a, b]$.	We use the integral to define and calculate what we originally wanted to measure.
The area between the curves $y = f_1(x)$, $y = f_2(x)$, on $[a, b]$ when $f_2(x) \le f_1(x)$	$\sum (f_1(c_k) - f_2(c_k)) \Delta x_k$	$\text{Area} = \int_a^b (f_1(x) - f_2(x))\, dx$
The volume of the solid defined by revolving the curve $y = f(x)$, $a \le x \le b$, about the x-axis	$\sum \pi f^2(c_k) \Delta x_k$	$\text{Volume} = \int_a^b \pi f^2(x)\, dx$
The length of a differentiable curve $y = f(x)$ whose derivative is continuous, $a \le x \le b$	$\sum \sqrt{1 + (f'(c_k))^2} \,\Delta x_k$	$\text{Length} = \int_a^b \sqrt{1 + (f'(x))^2}\, dx$
The work done by a variable force $F(x)$ directed along the x-axis from a to b	$\sum F(c_k) \Delta x_k$	$\text{Work} = \int_a^b F(x)\, dx$

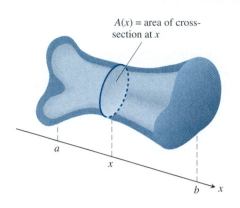

$A(x)$ = area of cross-section at x

6.60 If the area of the cross-section is a continuous function of x, we can find the volume of the solid in the way explained in the text.

Volumes of Arbitrary Solids—Slicing

Now that we can find the areas of regions bounded by smooth curves, we can find the volumes of a wide variety of cylinders. As we will now see, this enables us to define the volumes of many new solids.

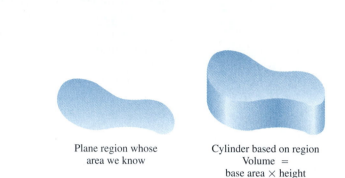

Plane region whose area we know

Cylinder based on region
Volume =
base area × height

Suppose we want to find the volume of a solid like the one shown in Fig. 6.60. The solid lies between planes perpendicular to the x-axis at $x = a$ and $x = b$. Each cross-section of the solid by a plane perpendicular to the x-axis is a region whose area we know how to find. Specifically, at each point x in the closed interval $[a, b]$, the cross-section of the solid is a region $R(x)$ whose area is $A(x)$. This makes A a real-valued function of x. If it is also a continuous function of x, we can use it to define and calculate the volume of the solid as an integral in the following way.

We partition the interval $[a, b]$ in the usual manner and slice the solid, as we would a loaf of bread, by planes perpendicular to the x-axis at the partition points. The kth slice, the one between the planes at x_{k-1} and x_k, has approximately the same volume as the cylinder between these two planes based on the region $R(x_k)$ (Fig. 6.61). The volume of this cylinder is

$$V_k = \text{base area} \times \text{height}$$

$$= A(x_k) \times (\text{distance between the planes at } x_{k-1} \text{ and } x_k)$$

$$= A(x_k)\Delta x_k.$$

The volume of the solid is therefore approximated by the cylinder volume sum

$$\sum_{k=1}^{n} A(x_k)\Delta x_k.$$

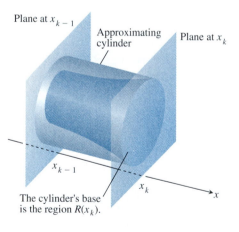

Plane at x_{k-1}

Approximating cylinder

Plane at x_k

x_{k-1}

x_k

The cylinder's base is the region $R(x_k)$.

NOT TO SCALE

6.61 Enlarged view of the slice of the solid between the planes at x_{k-1} and x_k and its approximating cylinder.

This is a Riemann sum for the function $A(x)$ on the closed interval $[a, b]$. We expect the approximations that we get from sums like these to improve as the norm of the partition of $[a, b]$ goes to zero, so we define their limiting integral to be the volume of the solid.

DEFINITION

The **volume** of a solid of known cross-sectional area $A(x)$ from $x = a$ to $x = b$ is the integral of A from a to b,

$$\text{Volume} = \int_a^b A(x)\, dx. \tag{1}$$

Notice that the new formula is consistent with the disk and washer formulas for solids of revolution (a good sign). In the disk formula,

$$V = \int_a^b \pi (f(x))^2\, dx,$$

the cross-section at x is a disk of radius $f(x)$ whose area is

$$A(x) = \pi (f(x))^2.$$

In the washer formula,

$$V = \int_a^b \pi (R^2(x) - r^2(x))\, dx,$$

the cross-section at x is a washer of inner radius $r(x)$ and outer radius $R(x)$ whose area is

$$A(x) = \pi R^2(x) - \pi r^2(x) = \pi (R^2(x) - r^2(x)).$$

To apply Eq. (1), we take the following steps.

The Method of Slicing

STEP 1: Sketch the solid and a typical cross-section.

STEP 2: Find a formula for $A(x)$.

STEP 3: Find the limits of integration.

STEP 4: Integrate $A(x)$ analytically or numerically to find the volume. Support the analytic result with a grapher, or confirm the numeric result analytically, if possible.

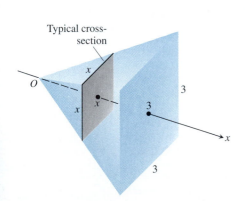

6.62 The cross-sections of the pyramid in Example 1 are squares.

EXAMPLE 1

A 3-m-high pyramid has a square base that is 3 m on a side. The cross-section of the pyramid perpendicular to the altitude x units down from the vertex is a square x units on a side. Find the volume of the pyramid.

Solution

STEP 1: *A sketch.* We draw the pyramid with its altitude along the x-axis and its vertex at the origin and include a typical cross-section (Fig. 6.62).

STEP 2: *A formula for A(x).* The cross-section at x is a square x meters on a side, so its area is

$$A(x) = x^2.$$

STEP 3: *The limits of integration.* The squares go from $x = 0$ to $x = 3$.

STEP 4: *The volume (found analytically):*

$$\text{Volume} = \int_a^b A(x)\, dx \qquad \text{Eq. 1}$$

$$= \int_0^3 x^2\, dx$$

$$= \frac{x^3}{3}\Bigg]_0^3$$

$$= 9.$$

The volume is 9 m³.
Support:

$$\text{Volume} = \text{NINT}\left(x^2, 0, 3\right) = 9.$$

EXAMPLE 2

A curved wedge is cut from a cylinder of radius 3 by two planes. One plane is perpendicular to the axis of the cylinder. The second plane crosses the first plane at a 45° angle at the center of the cylinder. Find the volume of the wedge.

Solution

STEP 1: *A sketch.* We draw the wedge and sketch a typical cross-section perpendicular to the x-axis (Fig. 6.63).

STEP 2: *The formula for $A(x)$.* The cross-section at x is a rectangle of area

$$A(x) = (\text{height})(\text{width}) = (x)(2\sqrt{9 - x^2}) = 2x\sqrt{9 - x^2}.$$

STEP 3: *The limits of integration.* The rectangles run from $x = 0$ to $x = 3$.

STEP 4: *The volume (found numerically):*

$$\text{Volume} = \int_a^b A(x)\, dx \qquad \text{Eq. 1}$$

$$= \text{NINT}\left(2x\sqrt{9 - x^2}, 0, 3\right) = 18.$$

Confirm:

$$\text{Volume} = \int_0^3 2x\sqrt{9 - x^2}\, dx$$

$$= -\frac{2}{3}\left(9 - x^2\right)^{3/2}\Bigg]_0^3 = 18.$$

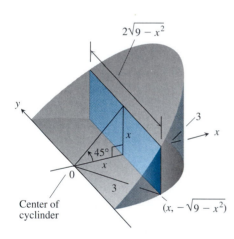

$2\sqrt{9 - x^2}$

$45°$

0

3

Center of cyclinder

$(x, -\sqrt{9 - x^2})$

6.63 The wedge of Example 2, sliced perpendicular to the x-axis. The cross-sections are rectangles.

EXAMPLE 3   Cavalieri's Theorem

Cavalieri's Theorem says that two solids with equal altitudes and identical parallel cross-sections have the same volume (Fig. 6.64). We can see this immediately from Eq. (1) because the cross-sectional area function $A(x)$ is the same in each case.

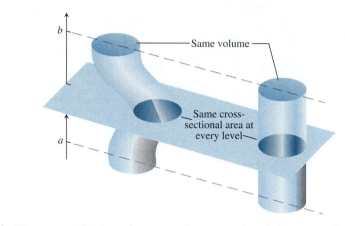

6.64 These two solids have the same volume even though the one on the left looks bigger. You can illustrate this yourself with two stacks of coins.

Position Shift and Distance Traveled

The total distance traveled by a body moving up and down a coordinate line is found by integrating the absolute value of the body's velocity (that is, the body's speed) over the time interval of the motion.

To see why, partition the time interval $a \leq t \leq b$ into subintervals in the usual way, and let Δt_k denote the length of the kth interval. If Δt_k is small enough, the body's velocity $v(t)$ will not change much from t_{k-1} to t_k, and the right-hand endpoint value $v(t_k)$ will give a good approximation of the velocity throughout the interval. Accordingly, the change in the body's position coordinate during the kth time interval will be about

$$v(t_k)\Delta t_k.$$

The change will be positive if $v(t_k)$ is positive and negative if $v(t_k)$ is negative.

In either case, the amount of distance traveled by the body during the kth interval will be about

$$|v(t_k)|\Delta t_k.$$

The total trip distance will be about

$$\sum_{k=1}^{n} |v(t_k)|\Delta t_k. \tag{2}$$

The sum in (2) is a Riemann sum for the speed $|v(t)|$ on the interval $[a, b]$. We expect the approximations that we get from sums like these to improve as the norm of the partition of $[a, b]$ goes to zero. It therefore looks as if we should be able to calculate the total distance traveled by the body by integrating the body's speed from a to b. In practice, this turns out to be just the right thing to do. The mathematical model predicts the distance every time.

BONAVENTURA CAVALIERI (1598–1647)

Cavalieri, a student of Galileo's, discovered that if two plane regions can be arranged to lie over the same interval of the x-axis in such a way that they have identical cross-sections at every point, then the regions have the same area. The theorem (and a letter of recommendation from Galileo) were enough to win Cavalieri a chair at the University of Bologna in 1629. The solid-geometry version in Example 3, which Cavalieri never proved, was given his name by later geometers.

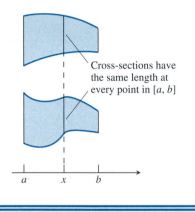

Cross-sections have the same length at every point in $[a, b]$

If $s(t)$ is a body's position on a coordinate line at time t, then

$$\frac{ds}{dt} = v = \text{velocity},$$

and

$$\left|\frac{ds}{dt}\right| = |v| = \text{speed}.$$

$$\text{Distance traveled} = \int_a^b |v(t)| \, dt$$

If we wish instead to predict how far up or down the line from its initial position a body will be when a trip is over, we integrate v instead of its absolute value. To understand why, let $s(t)$ be the body's position at time t, and let F be an antiderivative of v. Then

$$s(t) = F(t) + C$$

for some constant C. The shift in the body's position caused by the trip from $t = a$ to $t = b$ is

$$s(b) - s(a) = (F(b) + C) - (F(a) + C) = F(b) - F(a) = \int_a^b v(t) \, dt.$$

$$\text{Position shift} = \int_a^b v(t) \, dt$$

EXAMPLE 4

A body moving along a line from $t = 0$ to $t = 3\pi/2$ seconds has velocity

$$v(t) = 5 \cos t \text{ m/sec.}$$

Find the total distance traveled and the shift in the body's position.

Solution

$$\text{Distance traveled} = \int_0^{3\pi/2} |5 \cos t| \, dt \qquad \text{Distance is the integral of speed.}$$

$$= \int_0^{\pi/2} 5 \cos t \, dt + \int_{\pi/2}^{3\pi/2} (-5 \cos t) \, dt$$

$$= 5 \sin t \Big]_0^{\pi/2} - 5 \sin t \Big]_{\pi/2}^{3\pi/2} = 5(1 - 0) - 5(-1 - 1)$$

$$= 5 + 10 = 15 \text{ m.}$$

$$\text{Position shift} = \int_0^{3\pi/2} 5 \cos t \, dt \qquad \text{Shift is the integral of velocity.}$$

$$= 5 \sin t \Big]_0^{3\pi/2} = 5(-1) - 5(0) = -5 \text{ m.}$$

During the trip, the body traveled 5 m forward and 10 m backward for a total distance of 15 m. This shifted the body 5 m to the left.

EXPLORATION 1

Parametric Graphing

We can give a visual model of the moving body in Example 4 by using parametric mode.

1. Let

$$x_1(t) = v(t) = 5 \cos t, \qquad\qquad y_1(t) = t; \qquad (3)$$

$$x_2(t) = \text{NINT } (x_1(t), 0, t), \qquad y_2(t) = t; \qquad (4)$$

$$x_3(t) = \text{NINT } (|x_1(t)|, 0, t), \qquad y_3(t) = 2. \qquad (5)$$

Explain how Eqs. (3)–(5) relate to Example 4. (Note that t has two uses in the integrals, one as the dummy variable of integration and the other as the upper limit of the integral.)

2. GRAPH Eqs. (4) and (5) simultaneously. Explain what you see.

3. In Eq. (4), letting $y_2(t) =$ a constant gives a better representation of a position function. Why? What is the advantage of letting $y_2(t) = t$?

4. GRAPH Eqs. (3) and (5) simultaneously in DOT mode with t-step $= 0.1$. Explain how the distance between dots relates to the acceleration of the moving body.

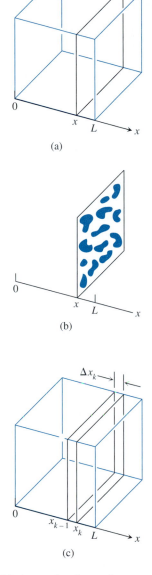

(a)

(b)

Δx_k

(c)

6.65 (a) A sample of granular material in the shape of a cube. (b) A cross-sectional slice at x. (c) The x_kth "slice" of the cube.

Delesse's Rule

As you may know, the sugar in an apple starts turning into starch as soon as the apple is picked, and the longer the apple sits around, the starchier it becomes. You can tell fresh apples from stale by both flavor and consistency.

To find out how much starch is in a given apple, we can look at a thin slice under a microscope. The cross-sections of the starch granules will show up clearly, and it is easy to estimate the proportion of the viewing area they occupy. This two-dimensional proportion will be the same as the three-dimensional proportion of uncut starch granules in the apple itself. The apparently magical quality of these proportions was first discovered by a French geologist, Achille Ernest Delesse, in the 1840s. Its explanation lies in the notion of average value.

Suppose that we want to find the proportion of some granular material in a solid and that the sample we have chosen to analyze is a cube whose edges have length L. We picture the cube (Fig. 6.65) with an x-axis along one edge and imagine slicing the cube with planes perpendicular to points of the interval $[0, L]$. Call the proportion of the area of the slice at x occupied by the granular material of interest (starch, in our apple example) $r(x)$, and assume that r is a continuous function of x.

Now partition the interval $[0, L]$ into subintervals in the usual way. Imagine the cube sliced into thin slices by planes at the subdivision points. The length Δx_k of the kth subinterval is the distance between the planes at

x_{k-1} and x_k. If the planes are close enough together, the sections cut from the grains by the planes will resemble cylinders with bases in the plane at x_k. The proportion of granular material between the planes will therefore be about the same as the proportion of cylinder base area in the plane at x_k, which in turn will be about $r(x_k)$. Thus, the amount of granular material in the slab between the two planes will be about

$$\text{(Proportion)} \times \text{(slab volume)} = r(x_k)L^2\Delta x_k.$$

The amount of granular material in the entire sample cube will be about

$$\sum_{k=1}^{n} r(x_k)L^2\Delta x_k.$$

This sum is a Riemann sum for the function $r(x)L^2$ over the interval $[0, L]$. We expect the approximations by sums like these to improve as the norm of the subdivision of $[0, L]$ goes to zero and therefore expect the integral

$$\int_a^b r(x)L^2 \, dx$$

to give the amount of granular material in the sample cube.

We can then obtain the proportion of granular material in the sample by dividing this amount by the cube's volume L^3. If we have chosen our sample well, this will also be the proportion of granular material in the solid from which the sample was taken. Putting it all together, we get

$$
\begin{aligned}
\begin{array}{l}\text{Proportion of granular}\\ \text{material in solid}\end{array} &= \begin{array}{l}\text{Proportion of granular}\\ \text{material in the sample cube}\end{array} \\[2mm]
&= \frac{\displaystyle\int_0^L r(x)L^2 \, dx}{L^3} \\[4mm]
&= \frac{\displaystyle L^2\int_0^L r(x) \, dx}{L^3} = \frac{1}{L}\int_0^L r(x) \, dx \\[4mm]
&= \text{average value of } r(x) \text{ over } [0, L] \\[2mm]
&= \begin{array}{l}\text{proportion of area occupied by granular}\\ \text{material in a typical cross-section.}\end{array}
\end{aligned}
$$

This is Delesse's Rule. Once we have found $\bar{r}$, the average of $r(x)$ over $[0, L]$, we have found the proportion of granular material in the solid.

In practice, $\bar{r}$ is found by averaging over a number of cross-sections. There are several things to watch out for in the process. In addition to the possibility that the granules cluster in ways that make representative samples difficult to find, there is the possibility that we might not recognize a granule's trace for what it is. Some cross-sections of normal red blood cells look like disks and ovals, but others look surprisingly like outlines of dumbbells. We do not want to dismiss the dumbbells as experimental error as one research group that we know of did a few years ago.

Achille Ernest Delesse was a mining engineer who was interested in determining the composition of rocks. To find out how much of a particular mineral a rock contained, he cut the rock through, polished an exposed face, and covered the face with transparent waxed paper, trimmed to size. He then traced on the paper the exposed portions of the mineral that interested him. After weighing the paper, he cut out the mineral traces and weighed them. The ratio of the weights gave not only the proportion of the surface occupied by the mineral but, more important, the proportion of the rock occupied by the mineral. Delesse described his method in an article entitled "A mechanical procedure for determining the composition of rocks," in the *Annales des Mines*, 13, 1848, pp. 379–388. His method is still used by petroleum geologists today. A two-dimensional analogue of it is used to determine the porosities of the ceramic filters that extract organic molecules in chemistry laboratories and screen out microbes in water purifiers.

Useless Integrals

Some of the integrals that we get from forming Riemann sums do what we want, but others do not. It all depends on how we choose to model the problems that we want to solve. Some choices are good—others are not. Here is an example.

We use the surface area formula

$$\text{Surface area} = \int_a^b 2\pi f(x) \sqrt{1 + \left(\frac{df}{dx}\right)^2} \, dx \qquad (6)$$

because it has predictive value and always gives results that are consistent with information from other sources. In other words, the model that we used to derive the formula was a good one.

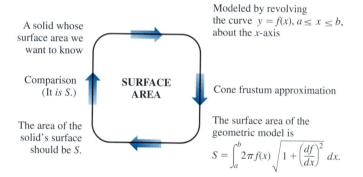

A solid whose surface area we want to know

Modeled by revolving the curve $y = f(x)$, $a \le x \le b$, about the x-axis

Comparison (It *is* S.)

SURFACE AREA

Cone frustum approximation

The area of the solid's surface should be S.

The surface area of the geometric model is

$$S = \int_a^b 2\pi f(x) \sqrt{1 + \left(\frac{df}{dx}\right)^2} \, dx.$$

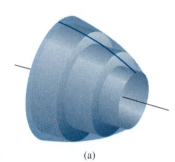

(a)

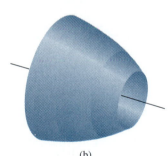

(b)

6.66 Why not use (a) cylindrical bands instead of (b) conical bands to approximate surface area?

Why not find the surface area by approximating with cylindrical bands instead of conical bands, as suggested in Fig. 6.66? The Riemann sums that we get this way converge just as nicely as the ones based on conical bands, and the resulting integral is simpler. Instead of Eq. (6), we get

$$\text{Surface area candidate} = \int_a^b 2\pi f(x) \, dx. \qquad (7)$$

After all, we might argue, we used cylinders to derive good volume formulas, so why not use them again to derive surface-area formulas?

The answer is that the formula in Eq. (7) has no predictive value and almost never gives results that are consistent with experience. The comparison step in the model fails for this formula.

There is a moral here: Just because we end up with a nice-looking integral does not mean that it will do what we want. Constructing an integral is not enough—we have to test it too.

Exercises 6.9

Find the volumes of the solids in Exercises 1–10.

1. The solid lies between planes perpendicular to the x-axis at $x = 0$ and $x = 4$. The cross-sections perpendicular to the axis on the interval $0 \le x \le 4$ are squares whose diagonals run from the parabola $y = -\sqrt{x}$ to the parabola $y = \sqrt{x}$.

2. The solid lies between planes perpendicular to the x-axis at $x = -1$ and $x = 1$. The cross-sections perpendicular to the x-axis between these planes are squares whose diagonals run from the semicircle $y = -\sqrt{1 - x^2}$ to the semicircle $y = \sqrt{1 - x^2}$.

3. The solid lies between planes perpendicular to the x-axis at $x = -1$ and $x = 1$. The cross-sections perpendicular to the axis between these planes are squares with edges running from the semicircle $y = -\sqrt{1 - x^2}$ to the semicircle $y = \sqrt{1 - x^2}$.

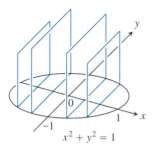

$x^2 + y^2 = 1$

4. The solid lies between the planes perpendicular to the x-axis at $x = -1$ and $x = 1$. The cross-sections perpendicular to the x-axis are circular disks whose diameters run from the parabola $y = x^2$ to the parabola $y = 2 - x^2$.

5. The solid lies between planes perpendicular to the x-axis at $x = 1$ and $x = 2$. The cross-sections perpendicular to the x-axis are circular disks with diameters running from the x-axis up to the curve $y = 2/\sqrt{x}$.

6. The solid lies between planes perpendicular to the x-axis at $x = 0$ and $x = 2$. The cross-sections perpendicular to the x-axis are circular disks with diameters running from the x-axis up to the parabola $y = \sqrt{5}x^2$.

7. The base of the solid is the disk $x^2 + y^2 \le 1$. The cross-sections by planes perpendicular to the y-axis between $y = -1$ and $y = 1$ are isosceles right triangles with one leg in the disk.

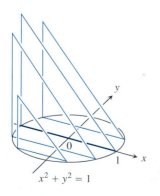

$x^2 + y^2 = 1$

8. The base of the solid is the region between the curve $y = 2\sqrt{\sin x}$ and the interval $0 \le x \le \pi$ of the x-axis. The cross-sections perpendicular to the x-axis are equilateral triangles.

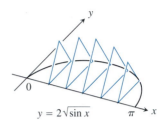

$y = 2\sqrt{\sin x}$ π

9. The solid is a sphere of radius r.

10. The solid is a right circular cone of radius r and height h.

11. A twisted solid is generated as follows: We are given a line L in space and a square of side s in a plane perpendicular to L. One vertex of the square is on L. As this vertex moves a distance h along L, the square turns through a full revolution about L. Find the volume of the solid generated by this motion. What would the volume be if the square had turned two revolutions in moving the same distance along L?

12. Prove Cavalieri's original theorem. (See the historical note in this section.) Assume that each region is the region between the graphs of two continuous functions over the interval $a \le x \le b$.

In Exercises 13–20, the function $v(t)$ is the velocity in meters per second of a body moving along a coordinate line. (a) GRAPH v as a function of t to see where it is positive and negative. Then find (b) the total distance traveled by the body during the given time interval and (c) the shift in the body's position.

13. $v(t) = 5 \cos t, 0 \le t \le 2\pi$

14. $v(t) = \sin \pi t, 0 \le t \le 2$

15. $v(t) = 6 \sin 3t, 0 \le t \le \pi/2$

16. $v(t) = 4 \cos 2t, 0 \le t \le \pi$

17. $v(t) = 49 - 9.8t, 0 \le t \le 10$

18. $v(t) = 8 - 1.6t, 0 \le t \le 10$

19. $v(t) = 6t^2 - 18t + 12 = 6(t - 1)(t - 2), 0 \le t \le 2$

20. $v(t) = 6t^2 - 18t + 12 = 6(t - 1)(t - 2), 0 \le t \le 3$

21. The following graphs are velocity graphs of four bodies moving on coordinate lines. Find the distance traveled by each body and the position shift for the given time interval.

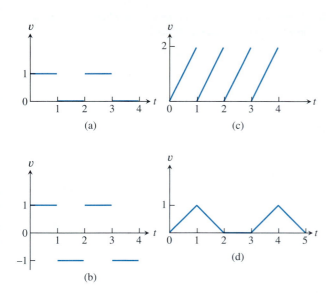

(a)

(c)

(b)

(d)

22. Table 6.3 shows the velocity of a model train engine moving back and forth on a track for 10 sec. Use Simpson's Rule to estimate the resulting position shift and total distance traveled.

23. *Modeling surface area.* The lateral surface area of the cone swept out by revolving the line segment $y = x/\sqrt{3}$, $0 \le x \le \sqrt{3}$, about the x-axis should be (1/2) (base circumference)(slant height) $= (1/2)(2\pi)(2) = 2\pi$. What do you get if you use Eq. (7) with $f(x) = x/\sqrt{3}$?

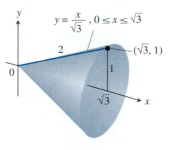

$y = \dfrac{x}{\sqrt{3}}, 0 \le x \le \sqrt{3}$

$(\sqrt{3}, 1)$

24. *Modeling surface area.* The only surface for which Eq. (7) gives the area we want is a cylinder. Show that Eq. (7) gives $S = 2\pi rh$ for the cylinder swept out by revolving the line segment $y = r, 0 \le x \le h$, about the x-axis.

TABLE 6.3 Selected velocities of the model train engine in Exercise 22

Time (sec)	Velocity (in. per sec)	Time (sec)	Velocity (in. per sec)
0	0	6	−11
1	12	7	−6
2	22	8	2
3	10	9	6
4	−5	10	0
5	−13		

Review Questions

1. How do you define and calculate the area between the graphs of two continuous functions? Give an example.

2. How do you define and calculate the volume of a solid of revolution by
   a) the disk method?
   b) the washer method?
   c) the method of cylindrical shells? Give examples.

3. How do you define and calculate the length of the graph of a differentiable function whose derivative is continuous over a closed interval? Give an example. What about functions that aren't continuously differentiable?

4. How do you define and calculate the area of the surface swept out by revolving the graph of a differentiable function $y = f(x)$ whose derivative is continuous, $a \le x \le b$, about

the x-axis? Give an example.

5. How do you define and calculate the work done by a force directed along a portion of the x-axis? How do you calculate the work that it takes to pump liquid from a tank? Give examples.

6. How do you calculate the force exerted by a liquid against a portion of vertical wall? Give an example.

7. How do you locate the center of mass of a straight, narrow rod or strip of material? Give an example. If the density of the material is constant, you can tell right away where the center of mass is. Where is it?

8. How do you locate the center of mass of a thin, flat plate of material? Give an example.

9. Suppose that you know the velocity of a body that will move back and forth along a coordinate line tomorrow from time $t = a$ to time $t = b$. How can you calculate in advance how much the motion will shift the body's position? How do you predict the total distance the body will travel?

10. How do you define and calculate volumes of solids by the method of slicing? Give an example. How is the method of slicing related to the disk and washer methods?

11. What does Delesse's Rule say? Give an example.

12. There is a basic pattern to the way in which we constructed integrals in this chapter. What is it?

Practice Exercises

Find the areas of the regions enclosed by the curves and lines in Exercises 1–14 analytically and support with a grapher.

1. $y = x$, $y = 1/x^2$, $x = 2$

2. $y = x$, $y = 1/\sqrt{x}$, $x = 2$

3. $y = x + 1$, $y = 3 - x^2$

4. $\sqrt{x} + \sqrt{y} = 1$, $x = 0$, $y = 0$

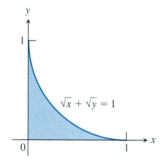

5. $x = 2y^2$, $x = 0$, $y = 3$

6. $4x = y^2 - 4$, $4x = y + 16$

7. $y^2 = 4x$, $y = 4x - 2$

8. $x = 4 - y^2$, $x = 0$, $y = 0$

9. $y = \sin x$, $y = x$, $x = \pi/4$

10. $y = |\sin x|$, $y = 1$, $-\pi/2 \le x \le \pi/2$

11. $y = 2\sin x$, $y = \sin 2x$, $0 \le x \le \pi$

12. $y = 8\cos x$, $y = \sec^2 x$, $-\pi/3 \le x \le \pi/3$

13. The "triangular" region bounded on the left by $x + y = 2$, on the right by $y = x^2$, and above by $y = 2$

14. The "triangular" region bounded on the left by $y = \sqrt{x}$, on the right by $y = 6 - x$, and below by $y = 1$

Find the areas of the shaded regions in Exercises 15 and 16 with a grapher, and confirm analytically, if possible.

15.   **16.**

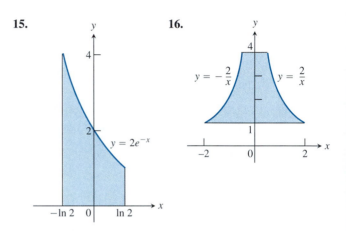

17. Find the volume of the solid generated by revolving the region bounded by the x-axis, the curve $y = 3x^4$, and the lines $x = 1$ and $x = -1$ about
a) the x-axis.
b) the y-axis.

18. Find the volume of the solid generated by revolving the region in the first quadrant bounded by the x-axis, the parabola $y = x^2$, and the line $x = 3$ about **(a)** the x-axis; **(b)** the y-axis.

19. Find the volume of the solid generated by revolving the region bounded on the left by the parabola $x = y^2 + 1$ and on the right by the line $x = 5$ about **(a)** the x-axis; **(b)** the y-axis; **(c)** the line $x = 5$.

20. Find the volume of the solid generated by revolving the region bounded by the parabola $y^2 = 4x$ and the line $y = x$ about **(a)** the x-axis; **(b)** the y-axis; **(c)** the line $x = 4$; **(d)** the line $y = 4$.

21. Find the volume of the solid generated by revolving the "triangular" region bounded by the x-axis, the line $x = \pi/3$, and the curve $y = \tan x$ in the first quadrant about the x-axis.

22. Find the volume of the solid generated by revolving the region bounded by the curve $y = \sin x$ and the lines $x = 0$, $x = \pi$, and $y = 2$ about the line $y = 2$.

23. Find the volume of the solid spindle generated by revolving the region bounded by the x-axis, the curve $y = 1/\sqrt{x}$, and the lines $x = 1$ and $x = 16$ about the x-axis.

24. Find the volume of the solid generated by revolving the region bounded by the curve $y = e^{x/2}$ and the lines $x = \ln 3$ and $y = 1$ about the x-axis.

25. A round hole of radius $\sqrt{3}$ ft is bored through the center of a sphere of radius 2 ft. Find the volume cut out.

26. The profile of a football resembles the ellipse shown here. Find the volume of the football to the nearest cubic inch.

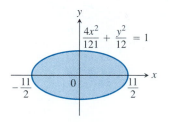

Find the lengths of the curves in Exercises 27–30 analytically and support with a grapher.

27. $y = x^{1/2} - (1/3)x^{3/2}$, $0 \le x \le 3$
   (*Hint*: $1 + (dy/dx)^2$ is a perfect square.)

28. $x = y^{2/3}$, $1 \le y \le 8$

29. $y = x^4/8 + 1/(4x^2)$, $1 \le x \le 3$

30. $y = x^3/2 + 1/(6x)$, $1 \le x \le 2$

In Exercises 31–36, find the areas of the surfaces generated by revolving the curves about the given axes with a grapher, and confirm analytically, if possible.

31. $y = \sqrt{2x + 1}$, $0 \le x \le 12$, x-axis

32. $y = x^3/9$, $-1 \le x \le 1$, x-axis

33. $y = (1/3)x^{3/2} - x^{1/2}$, $0 \le x \le 3$, x-axis

34. $y = e^x + (1/4)e^{-x}$, $-\ln 4 \le x \le \ln 2$, x-axis

35. $y = (1/3)(x^2 + 2)^{3/2}$, $0 \le x \le 1$, y-axis
   (*Hint*: To confirm, express $ds = \sqrt{dx^2 + dy^2}$ in terms of dx, and evaluate $\int 2\pi x \, ds$.)

36. $y = x^2$, $0 \le y \le 2$, y-axis

37. A rock climber is about to haul up 10 kg of equipment that has been hanging on 40 m of rope that weighs 0.8 newton per meter. How much work will it take? (*Hint*: Solve for the rope and equipment separately; then add.)

38. You drove an 800-gal tank truck from the base to the summit of Mt. Washington and discovered on arrival that the tank of water was only half full. You started out with a full tank, climbed at a steady rate, and took 50 min to accomplish the 4750-ft elevation change. Assuming that the water leaked out at a steady rate, how much work was spent in carrying water to the top? Do not count the work done in getting yourself and the truck there. Water weighs 8 lb/gal.

39. If a force of 20 lb is required to hold a spring 1 ft beyond its unstressed length, how much work does it take to stretch the spring this far? How much work does it take to stretch the spring an additional foot?

40. A force of 2 N will stretch a rubber band 2 cm. Assuming that Hooke's Law applies, how far will a 4-N force stretch the rubber band? How much work does it take to stretch the rubber band this far?

41. A reservoir, shaped like an inverted right circular cone 20 ft across the top and 8 ft deep, is full of water. How much work does it take to pump the water to a level 6 ft above the top?

42. *Continuation of Exercise 41.* The reservoir is filled to a depth of 5 ft, and the water is to be pumped to the same level as the top. How much work does it take?

43. The vertical triangular plate shown below on the left is the end plate of a triangular watering trough full of water. What is the force against the end of the plate?

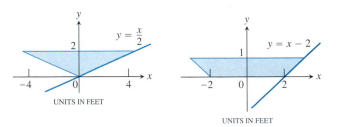

44. The vertical trapezoidal plate shown above on the right is the end plate of a trough of maple syrup weighing 75 lb/ft^3. What is the force against the end of the trough when there are 10 in. of syrup in the trough?

45. A flat vertical gate in the face of a dam is shaped like the parabolic region between the curve $y = 4x^2$ and the line $y = 4$, with measurements in feet. The top of the gate lies 5 ft below the surface of the water. Find the force against the gate.

46. A standard olive oil can measures 5.75 by 3.5 by 10 in. Find the fluid force against the base and each side of the can when the can is full. (Refer to the weight-density table given in Section 6.7.)

47. Find the center of mass of a thin, flat plate of constant density covering the region enclosed by the parabolas $y = 2x^2$ and $y = 3 - x^2$.

48. Find the center of mass of a thin, flat plate of constant density covering the region enclosed by the x-axis, the lines $x = 2$ and $x = -2$, and the parabola $y = x^2$.

49. Find the center of mass of a thin, flat plate of constant density covering the "triangular" region in the first quadrant bounded by the y-axis, the parabola $y = x^2/4$, and the line $y = 4$.

50. Find the center of mass of a thin, flat plate of density $\delta = 3$ covering the region enclosed by the parabola $y^2 = x$ and the line $x = 2y$.

51. Find the centroid of the trapezoid in Exercise 44.

52. Find the centroid of the triangle in Exercise 43.

53. Find the center of mass of a thin plate of constant density covering the region between the curve $y = 1/\sqrt{x}$ and the x-axis from $x = 1$ to $x = 16$.

54. *Continuation of Exercise 53.* Now find the center of mass, assuming that the density is $\delta(x) = 4/\sqrt{x}$.

Find the volumes of the solids in Exercises 55–58.

55. The solid lies between planes perpendicular to the x-axis at $x = 0$ and $x = 1$. The cross-sections perpendicular to the x-axis between these planes are circular disks whose diameters run from the parabola $y = x^2$ to the parabola $y = \sqrt{x}$.

56. The base of the solid is the region in the first quadrant between the line $y = x$ and the parabola $y = 2\sqrt{x}$. The cross-sections of the solid perpendicular to the x-axis are equilateral triangles whose bases stretch from the line to the curve.

57. The solid lies between planes perpendicular to the x-axis at $x = \pi/4$ and $x = 5\pi/4$. The cross-sections between these planes are circular disks whose diameters run from the curve $y = 2 \cos x$ to the curve $y = 2 \sin x$.

58. The solid lies between planes perpendicular to the x-axis at $x = 0$ and $x = 6$. The cross-sections between these planes are squares whose bases run from the x-axis up to the curve $x^{1/2} + y^{1/2} = \sqrt{6}$.

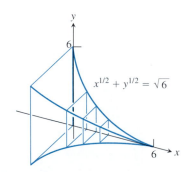

59. The volume of the wedge in Section 6.9, Example 2, could have been found by drawing a picture like the one below, taking cross-sections perpendicular to the y-axis, and integrating with respect to y. Find the volume this way. (*Hint:* What is the shape of each cross-section?)

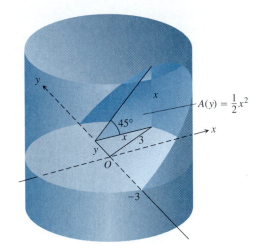

60. A solid lies between planes perpendicular to the x-axis at $x = 0$ and $x = 12$. The cross-sections by planes perpendicular to the x-axis for $0 \le x \le 12$ are circular disks whose diameters run from the line $y = x/2$ to the line $y = x$. Use Cavalieri's Theorem (Section 6.9, Example 3) to explain why the solid has the same volume as a right circular cone with base radius 3 and altitude 12.

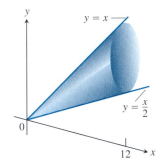

In Exercises 61–64, the function $v(t)$ is the velocity in feet per second of a body moving on a coordinate line. (a) Graph v as a function of t to see where it is positive and negative. Then find (b) the total distance traveled by the body during the given time interval and (c) how much the trip shifted the body's position.

61. $v(t) = t/2 - 1,\ 0 \le t \le 6$

62. $v(t) = t^2 - 8t + 12 = (t - 2)(t - 6),\ 0 \le t \le 6$

63. $v(t) = 5 \cos t,\ 0 \le t \le 3\pi/2$

64. $v(t) = \pi \sin \pi t,\ 0 \le t \le 3/2$

The Calculus of Transcendental Functions

OVERVIEW   Many of the functions that we work with in mathematics and science are inverses of one another. The functions ln x and e^x are probably the most famous function-inverse pair, but there are others that are nearly as important. The trigonometric functions, when suitably restricted, have important inverses that are functions, and there are other useful pairs of logarithmic and exponential functions. Less widely known are the hyperbolic functions and their inverses, algebraic representations that arise when we study hanging cables, heat flow, and the friction encountered by objects falling through the air. In this chapter, we will complete the description of these algebraic representations and look at the kinds of problems they were designed to solve.

We introduced exponential functions in Chapter 1 and worked with them in Chapters 4, 5, and 6. However, we left them undefined and promised a more detailed look in this Chapter. In Section 7.2, we show a definition for an exponential function, $y = e^x$, a function so important that it is often called *the* exponential function. The key to understanding this function is its inverse, the natural logarithm function, so we study that first.

7.1 ——————————— The Natural Logarithm Function

The function $f(x) = 1/x$ is continuous except at $x = 0$. It follows from the Fundamental Theorem of Calculus, Part 1, that $f(x) = 1/x$ has antiderivatives of the form $F(x) = \int_a^x 1/t\, dt$ if *both* a and x are positive numbers. Figure 7.1 displays a family of antiderivatives of $1/x$, namely, NINT $(1/t, a, x)$ for $a = 0.5, 1, 1.5, 2, 2.5,$ and 3.

Although the small viewing window distorts an overall view, we know that the antiderivatives differ by constant values, with each value of a corresponding to a particular antiderivative. In previous sections, we preferred the antiderivative $\int_0^x f$ in which $a = 0$ because its graph passes through the origin. Now, since we are working to the right of 0, we opt for the next best choice and use $\int_1^x 1/t\ dt$. The graph of this function passes through $(1, 0)$

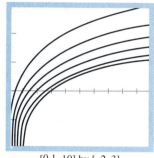

[0.1, 10] by [−2, 3]

7.1 The graphs of

$$y = \text{NINT} \left(\frac{1}{t}, a, x \right)$$

for $a = 0.5, 1, 1.5, 2, 2.5$, and 3. The graphs are vertical shifts of each other, which you can check by using TRACE and the cursor control. Try reproducing these on your graphing utility. Be careful to set $x\text{Min} > 0$ to avoid the discontinuity at $x = 0$. We used $x\text{Min} = 0.1$.

(Why?), and its domain is $(0, \infty)$. We call the function

$$\ln x = \int_1^x 1/t \, dt$$

the *natural logarithm function*, and we are about to find that it has very special properties.

DEFINITION   **The Natural Logarithm Function**

$$\ln x = \int_1^x \frac{1}{t} \, dt \qquad (x > 0)$$

If $x > 1$, then $\ln x$ is the area under the curve $y = 1/t$ from $t = 1$ to $t = x$ (Fig. 7.2). If x is less than 1 (but still positive), $\ln x$ gives the negative of the area under the curve from $t = x$ to $t = 1$. The function is not defined for $x \le 0$. The natural logarithm of 1 itself is zero because

$$\ln 1 = \int_1^1 \frac{1}{t} \, dt = 0. \qquad \text{Upper and lower limits equal}$$

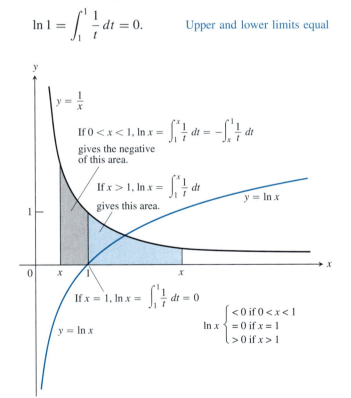

EXPLORATION BITS

1. It is natural, even compelling, for a mathematician to wonder about $\int (1/x) \, dx$ because $1/x = x^{-1}$ is the one "power" function for which the Power Rule of Integration,

   $$\int x^n \, dx = \frac{x^{n+1}}{n+1},$$

   does not apply. Why?

2. Your graphing utility quite likely has a natural logarithm key, $\boxed{\text{LN}}$, or the natural logarithm function built in. Let $y_1 = \ln x$. Then GRAPH $y_2 = \text{NDER}\,(y_1, x)$. What would you expect to see?

7.2 The graph of $y = \ln x$ and its relation to the function $y = 1/x$, $x > 0$. The graph of $\ln x$ rises above the x-axis as x moves from 1 to the right and falls below the axis as x moves from 1 to the left.

Historically, the importance of logarithms came from the improvement they brought to arithmetic. The revolutionary properties of logarithms made possible the calculations behind the great seventeenth-century advances in off-shore navigation and celestial mechanics. Today, calculators do arithmetic far better than the old logarithmic techniques, but logarithms are still useful for reasons other than computation.

Rules of Arithmetic for Logarithms

The following rules hold for any positive numbers a and x and for any exponent n.

1. $\ln ax = \ln a + \ln x$

2. $\ln \dfrac{a}{x} = \ln a - \ln x$

2(a) $\ln \dfrac{1}{x} = -\ln x$ \qquad Rule 2 with $a = 1$

3. $\ln x^n = n \ln x$ \qquad The Exponent Rule

EXPLORATION 1

Support Graphically

We can "see" that the rules of arithmetic for logarithms are true in a grapher viewing window.

1. For your choice of a positive number a, let $y_1 = \ln a$, $y_2 = \ln x$, and $y_3 = \ln ax$.

   a) Support Rule 1 graphically.

   b) Support Rule 2 graphically.

   c) Let $a = 1$, and support Rule 2(a) graphically.

2. Support Rule 3 graphically for $n > 1$; for $0 < n < 1$; for $n < 0$.

3. For your choice of a *positive function* $f(x)$, let $y_1 = \ln f(x)$, $y_2 = \ln x$, and $y_3 = \ln (f(x) \cdot x)$. Make a conjecture about how Rule 1 might generalize, and check out your conjecture. ✦

The Derivative of $y = \ln x$

By the Fundamental Theorem of Calculus, Part 1, in Section 5.4,

$$\frac{d}{dx} \ln x = \frac{d}{dx} \int_1^x \frac{1}{t}\, dt = \frac{1}{x}.$$

For every positive value of x, therefore,

$$\frac{d}{dx} \ln x = \frac{1}{x}.$$

If u is a differentiable function of x whose values are positive, so that $\ln u$ is defined, then applying the Chain Rule

$$\frac{dy}{dx} = \frac{dy}{du} \cdot \frac{du}{dx}$$

to the function $y = \ln u$ gives

$$\frac{d}{dx} \ln u = \frac{d}{du} \ln u \cdot \frac{du}{dx} = \frac{1}{u} \frac{du}{dx}.$$

$$\frac{d}{dx} \ln u = \frac{1}{u} \frac{du}{dx} \qquad (u > 0) \tag{1}$$

EXAMPLE 1

Equation (1) with $u = x^2 + 3$ gives

$$\frac{d}{dx} \ln (x^2 + 3) = \frac{1}{x^2 + 3} \cdot \frac{d}{dx}(x^2 + 3) = \frac{1}{x^2 + 3} \cdot 2x = \frac{2x}{x^2 + 3}.$$

EXPLORATION 2

Some Curious Facts for ln x

Arithmetic Rule 1 for logarithms suggests a curious fact about the natural logarithm function.

We know that $f(ax)$ is a horizontal stretch $(0 < a < 1)$ or shrink $(a > 1)$ of the graph of $f(x)$ by a factor of $1/a$. Also, $f(x) + C$ is a vertical shift. Thus, Rule 1 tells us that the horizontal stretch or shrink $\ln ax$ of $\ln x$ is equivalent to a vertical shift of $\ln x$ by the amount $\ln a$. This means that $\ln ax$ does not change the shape of the graph of $\ln x$ at all, and hence the derivatives of $\ln ax$ and $\ln x$ should be the same.

Use Eq. (1) to show

1. $\dfrac{d}{dx}(\ln 2x) = \dfrac{1}{x}$ 2. $\dfrac{d}{dx}(\ln kx) = \dfrac{1}{x}$

3. If $\ln kx$ is an antiderivative of $1/x$ (part 2) and $\ln x$ is an antiderivative of $1/x$, what can you conclude about $\ln kx$ and $\ln x$. What "natural" question arises from your conclusion? Answer it, and give a convincing argument.

The derivative of $\ln x$ makes possible an unusual—and elegant—proof of the first rule of arithmetic for logarithms, namely, that $\ln ax = \ln a + \ln x$. It starts by showing that $\ln ax$ and $\ln x$ have the same derivative. According to Corollary 3 of the Mean Value Theorem, then, the functions must differ by a constant, which means that

$$\ln ax = \ln x + C$$

for some C. With this much accomplished, it remains only to show that C equals $\ln a$.

Proof That ln ax = ln a + ln x   The equality of the derivatives of $\ln ax$ and $\ln x$ comes from the fact that

$$\frac{d}{dx} \ln ax = \frac{1}{ax} \cdot \frac{d}{dx}(ax) = \frac{1}{ax} \cdot a = \frac{1}{x}, \qquad \text{Eq. (1) with } u = ax$$

In the late 1500s, a Scottish baron, John Napier, invented a device called the *logarithm* that simplified arithmetic by replacing multiplication by addition. The equation that accomplished this was

$$\ln ax = \ln a + \ln x.$$

To multiply two positive numbers a and x, you looked up their logarithms in a table, added the logarithms, found the sum in the body of the table, and read the table backward to find the product ax.

Having the table was the key, of course, and Napier spent the last 20 years of his life working on a table that he never finished (while the astronomer Tycho Brahe waited in vain for the information that he needed to speed his calculations). The table was completed after Napier's death (and Brahe's) by Napier's friend Henry Briggs in London. Base 10 logarithms subsequently became known as Briggs's logarithms (what else?), and some books on navigation still refer to them this way.

Napier also invented an artillery piece that could hit a cow a mile away. Horrified by the weapon's accuracy, he stopped production and suppressed the cannon's design.

which is the derivative of $\ln x$. Therefore,

$$\ln ax = \ln x + C \tag{2}$$

for some constant C. Equation (2) holds for all positive values of x, so it must hold for $x = 1$. Hence,

$$\ln (a \cdot 1) = \ln 1 + C$$

$$\ln a = 0 + C \qquad \text{\color{blue}{ln 1 = 0.}}$$

$$C = \ln a. \qquad \text{\color{blue}{Rearranged}}$$

Substituting $C = \ln a$ in Eq. (2) gives the equation that we wanted to prove:

$$\ln ax = \ln a + \ln x. \tag{3}$$

≡

Proof That $\ln \dfrac{a}{x} = \ln a - \ln x$ We get this from Eq. (3) in two stages. Equation (3) with a replaced by $1/x$ gives

$$\ln \frac{1}{x} + \ln x = \ln \left(\frac{1}{x} \cdot x\right) = \ln 1 = 0,$$

so that

$$\ln \frac{1}{x} = - \ln x.$$

Equation (3) with x replaced by $1/x$ then gives

$$\ln \frac{a}{x} = \ln \left(a \cdot \frac{1}{x}\right) = \ln a + \ln \frac{1}{x} = \ln a - \ln x. \qquad ≡$$

We note that there are proofs of Rule 2(a) embedded in the proof of Rule 2 and the proof of Rule 3 that follows. For Rule 3, there is an added restriction for the moment that the exponent n be a rational number.

Proof That $\ln x^n = n \ln x$ (Assuming n Rational) We use the same-derivative argument again. For all positive values of x,

$$\frac{d}{dx} \left(\ln x^n\right) = \frac{1}{x^n} \cdot \frac{d}{dx}(x^n) \qquad \text{\color{blue}{Eq. (1) with $u = x^n$}}$$

$$= \frac{1}{x^n} nx^{n-1} \qquad \text{\color{blue}{Here is why we need n to be rational, at least for now. We have proved the power rule only for rational exponents.}}$$

$$= n \cdot \frac{1}{x} = \frac{d}{dx}(n \ln x).$$

Since $\ln x^n$ and $n \ln x$ have the same derivative,

$$\ln x^n = n \ln x + C$$

for some constant C. Taking x to be 1 identifies C as zero, and we're done.

≡

One useful consequence of Rule 3 is that $\ln \sqrt[m]{x} = \ln x^{1/m} = (1/m) \ln x$.

EXPLORATION BITS

Even our calculators cannot evaluate a number like $2^{\sqrt{3}}$ exactly. They can only give an approximate value. Calculators work only with rational numbers—for that matter, only with rational numbers having terminating decimal representations. Fortunately, calculators work with finite decimals to a *large* number of places, making their approximate results *very* accurate. So go right ahead and use your calculator to approximate positive numbers raised to irrational powers.

a) For example, you should feel quite comfortable with the calculator result

$$2^{\sqrt{3}} = 3.32199708548.$$

Is this what your calculator shows? (Recall our agreement to use = in general, in place of ≈.)

b) What does your calculator show for these numbers:

$$e^e, e^\pi, \pi^e, \pi^\pi, (\sqrt{\pi})^{\cos e}$$

EXAMPLE 2

$$\ln \sqrt{\cos x} = \frac{1}{2} \ln \cos x \qquad \ln \sqrt[3]{x + 1} = \frac{1}{3} \ln (x + 1)$$

≣

As for using the rule $\ln x^n = n \ln x$ for irrational values of n, go right ahead and do so. It does hold for all n, and there is no need to pretend otherwise. From the point of view of mathematical development, however, we want you to be aware that the rule has not yet been proved. Indeed, we have not yet even given meaning to x^n when, for example, we try to raise a base like 2 to an irrational power like $\sqrt{3}$.

The Graph of $y = \ln x$

The derivative $d(\ln x)/dx = 1/x$ is positive and continuous at every point in the domain of $\ln x$, so $\ln x$ is an increasing function of x. The graph is connected and rises steadily from left to right with a continuously turning tangent. The second derivative

$$\frac{d^2(\ln x)}{dx^2} = \frac{d}{dx}\left(\frac{1}{x}\right) = -\frac{1}{x^2}$$

is always negative, so the graph is concave down throughout. The graph of $y = \ln x$ shown in Figure 7.3 is complete.

The natural logarithm of x tends to ∞ as x tends to ∞ and tends to $-\infty$ as x approaches zero from the right. In short,

$$\lim_{x \to \infty} \ln x = \infty \qquad \text{and} \qquad \lim_{x \to 0^+} \ln x = -\infty.$$

The domain of $\ln x$ is the set of positive real numbers. The range of $\ln x$ is the entire set of real numbers.

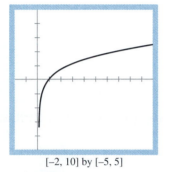

[–2, 10] by [–5, 5]

7.3 A complete graph of $y = \ln x$. Although not readily apparent in the viewing window, $\ln x$ tends to ∞ as x tends to ∞, a fact that we will confirm in Section 8.6.

Logarithmic Differentiation

The derivatives of functions given by formulas that involve products, quotients, and powers can often be found quickly if we take the natural logarithm of both sides before differentiating. This enables us to use the rules of arithmetic for logarithms to simplify the formulas before we go to work. The process, called **logarithmic differentiation**, is illustrated in the next example.

Logarithmic differentiation has five steps:

1. Take logs of both sides and simplify.
2. Differentiate implicitly.
3. Solve for dy/dx.
4. Substitute for y.
5. (Optional) Support by comparing the graphs of dy/dx and NDER (y, x).

EXAMPLE 3

Find dy/dx if

$$y = \frac{(x^2 + 1)(x + 3)^{1/2}}{x - 1} \qquad (x > 1).$$

Solution Notice that $y > 0$ for $x > 1$. We take the natural logarithm of both sides and apply the rules for logarithms:

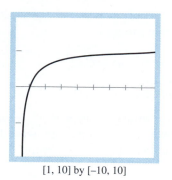

[1, 10] by [−10, 10]

7.4 The graphs of

$$y_2 = \text{NDER}\,(y_1, x) \quad \text{and} \quad y_3 = \frac{dy}{dx}$$

of Example 3 where

$$y_1 = y = \frac{(x^2 + 1)(x + 3)^{1/2}}{x - 1} \quad (x > 1).$$

The graphs appear to be identical.

$$\ln y = \ln \left(\frac{(x^2 + 1)(x + 3)^{1/2}}{x - 1} \right)$$

$$= \ln (x^2 + 1)(x + 3)^{1/2} - \ln (x - 1) \qquad \text{Rule 2}$$

$$= \ln (x^2 + 1) + \ln (x + 3)^{1/2} - \ln (x - 1) \qquad \text{Rule 1}$$

$$= \ln (x^2 + 1) + \frac{1}{2} \ln (x + 3) - \ln (x - 1). \qquad \text{Rule 3}$$

We then take derivatives of both sides with respect to x, using implicit differentiation on the left:

$$\frac{1}{y}\frac{dy}{dx} = \frac{1}{x^2 + 1} \cdot 2x + \frac{1}{2} \cdot \frac{1}{x + 3} - \frac{1}{x - 1}.$$

Next, we solve for dy/dx:

$$\frac{dy}{dx} = y \left(\frac{2x}{x^2 + 1} + \frac{1}{2x + 6} - \frac{1}{x - 1} \right).$$

Then, we substitute for y:

$$\frac{dy}{dx} = \frac{(x^2 + 1)(x + 3)^{1/2}}{x - 1} \left(\frac{2x}{x^2 + 1} + \frac{1}{2x + 6} - \frac{1}{x - 1} \right).$$

Finally, we support our analytic work by letting $y_1 = y$ and comparing the graphs of $y_2 = \text{NDER}\,(y_1, x)$ and $y_3 = dy/dx$. (See Fig. 7.4.) ≡

The Integral $\int \frac{1}{u}\, du$

If u is a positive differentiable function of x, Eq. (1) yields the integral formula

$$\int \frac{1}{u}\, du = \ln u + C \qquad (u > 0) \tag{4}$$

SUPPORT

1. For Example 4, GRAPH and compare

$$y_1 = \text{NINT}\left(\frac{2t}{t^2 + 3}, a, x \right)$$

(your choice of a) and

$$y_2 = \ln \left(x^2 + 3 \right).$$

2. For Example 5, compute

$$\text{NINT}\left(\frac{2x}{x^2 + 3}, 0, 3 \right).$$

Compare the result with the analytic result, ln 4. Use the [LN] key on your grapher to find a value for ln 4.

EXAMPLE 4

$$\int \frac{2x}{x^2 + 3}\, dx = \int \frac{du}{u} \qquad \text{Substitute } u = x^2 + 3,\, du = 2x\, dx.$$

$$= \ln u + C \qquad \text{Eq. (4)}$$

$$= \ln (x^2 + 3) + C \qquad \text{Replace } u \text{ by } x^2 + 3.$$ ≡

EXAMPLE 5

$$\int_0^3 \frac{2x}{x^2 + 3}\, dx = \ln (x^2 + 3) \Big]_0^3 \qquad \begin{array}{l}\text{Antiderivative}\\ \text{from Example 4}\end{array}$$

$$= \ln (3^2 + 3) - \ln (0 + 3)$$

$$= \ln 12 - \ln 3$$

$$= \ln \frac{12}{3} \qquad \text{Logarithm Rule 2}$$

$$= \ln 4$$ ≡

The expression on the right in Eq. (4) could also be written as $\ln |u| + C$ because $u = |u|$ if u is positive.

If u is a negative differentiable function of x, then $-u$ is a positive differentiable function of x, and Eq. (4) applies to give

$$\int \frac{1}{u}\, du = \underbrace{\int \frac{1}{-u}\, d(-u) = \ln(-u) + C}_{\text{Applying Eq. (4)}}.$$

The expression on the right could also be written as $\ln |u| + C$ because $-u = |u|$ if u is negative. Whether u is positive or negative, the integral of $(1/u)\, du$ is $\ln |u| + C$. This removes the restriction that u be positive.

If u is a positive or negative differentiable function, then

$$\int \frac{1}{u}\, du = \ln |u| + C. \tag{5}$$

EXAMPLE 6

SUPPORT

For Example 6, compute

$$\text{NINT}\left(\frac{2x}{x^2 - 5}, 0, 2 \right)$$

and compare with the analytic result $-\ln 5$.

$$\int_0^2 \frac{2x}{x^2 - 5}\, dx = \int_{-5}^{-1} \frac{du}{u} = \ln |u| \Big]_{-5}^{-1} \qquad \text{Substitute } u = x^2 - 5, du = 2x\, dx,$$
$$u(0) = -5, u(2) = -1 \text{ in Eq. (5)}.$$

$$= \ln |-1| - \ln |-5| = \ln 1 - \ln 5 = -\ln 5 \qquad \blacksquare$$

The Integrals of tan x and cot x

Equation (5) enables us at last to integrate the tangent and cotangent functions.

For the tangent

$$\int \tan x\, dx = \int \frac{\sin x}{\cos x}\, dx = \int \frac{-du}{u} \qquad \text{Substitute } u = \cos x,$$
$$du = -\sin x\, dx.$$

$$= -\int \frac{du}{u} = -\ln |u| + C \qquad \text{Eq. (5)}$$

$$= -\ln |\cos x| + C = \ln \frac{1}{|\cos x|} + C \qquad \text{Logarithm Rule 2(a)}$$

$$= \ln |\sec x| + C.$$

For the cotangent,

$$\int \cot x\, dx = \int \frac{\cos x\, dx}{\sin x} = \int \frac{du}{u} \qquad \text{Substitute } u = \sin x,$$
$$du = \cos x\, dx.$$

$$= \ln |u| + C = \ln |\sin x| + C. \qquad \text{Eq. (5)}$$

The general formulas are

$$\int \tan u \, du = -\ln |\cos u| + C = \ln |\sec u| + C \qquad (6)$$

$$\int \cot u \, du = \ln |\sin u| + C = -\ln |\csc x| + C \qquad (7)$$

EXAMPLE 7

$$\int_0^{\pi/6} \tan 2x \, dx = \int_0^{\pi/3} \tan u \cdot \frac{du}{2} \qquad \begin{array}{l} \text{Substitute } u = 2x, \\ dx = du/2, u(0) = 0, \\ u(\pi/6) = \pi/3. \end{array}$$

$$= \frac{1}{2} \int_0^{\pi/3} \tan u \, du$$

$$= \frac{1}{2} \ln |\sec u| \Big]_0^{\pi/3} \qquad \text{Eq. (6)}$$

$$= \frac{1}{2}(\ln 2 - \ln 1) = \frac{1}{2} \ln 2 \qquad \blacksquare$$

SUPPORT

For Example 7, compare

$$\text{NINT}\left(\tan 2x, 0, \frac{\pi}{6}\right)$$

with $(\ln 2)/2$.

Exercises 7.1

Express the logarithms in Exercises 1–8 in terms of $\ln 2$ and $\ln 3$ using the rules of arithmetic for logarithms. For example, $\ln 1.5 = \ln (3/2) = \ln 3 - \ln 2$.

1. $\ln 4/9$

2. $\ln 12$

3. $\ln (1/2)$

4. $\ln (1/3)$

5. $\ln 4.5$

6. $\ln \sqrt[3]{9}$

7. $\ln 3\sqrt{2}$

8. $\ln \sqrt{13.5}$

Find dy/dx in Exercises 9–18. Support graphically.

9. $y = \ln (x^2)$

10. $y = (\ln x)^2$

11. $y = \ln (1/x)$

12. $y = \ln (10/x)$

13. $y = \ln (x + 2)$

14. $y = \ln (2x + 2)$

15. $y = \ln (2 - \cos x)$

16. $y = \ln (x^2 + 1)$

17. $y = \ln (\ln x)$

18. $y = x \ln x - x$

Use logarithmic differentiation to find dy/dx in Exercises 19–30. In each case, assume y to be positive.

19. $y = \sqrt{x(x + 1)}$

20. $y = \sqrt{\dfrac{x}{x + 1}}$

21. $y = \sqrt{x + 3} \sin x$

22. $y = \dfrac{\tan x}{\sqrt{2x + 1}}$

23. $y = x(x + 1)(x + 2)$

24. $y = \dfrac{1}{x(x + 1)(x + 2)}$

25. $y = \dfrac{x + 5}{x \cos x}$

26. $y = \dfrac{x \sin x}{\sqrt{\sec x}}$

27. $y = \dfrac{x\sqrt{x^2 + 1}}{(x + 1)^{2/3}}$

28. $y = \sqrt{\dfrac{(x + 1)^{10}}{(2x + 1)^5}}$

29. $y = \sqrt[3]{\dfrac{x(x - 2)}{x^2 + 1}}$

30. $y = \sqrt[3]{\dfrac{x(x + 1)(x - 2)}{(x^2 + 1)(2x + 3)}}$

Evaluate the integrals in Exercises 31–48. Support your answer with a NINT computation.

31. $\displaystyle\int_{-3}^{-2} \frac{dx}{x}$

32. $\displaystyle\int_{-9}^{-4} \frac{dx}{2x}$

33. $\displaystyle\int_{-1}^{0} \frac{3 \, dx}{3x - 2}$

34. $\displaystyle\int_{-1}^{0} \frac{dx}{2x + 3}$

35. $\displaystyle\int_{3}^{4} \frac{dx}{x - 5}$

36. $\displaystyle\int_{2}^{5} \frac{dx}{1 - x}$

37. $\displaystyle\int_{0}^{3} \frac{2x \, dx}{x^2 - 25}$

38. $\displaystyle\int_{0}^{1} \frac{8x \, dx}{4x^2 - 5}$

39. $\displaystyle\int_{0}^{3} \frac{1}{x + 1} \, dx$

40. $\displaystyle\int_{0}^{4} \frac{2x \, dx}{x^2 + 9}$

41. $\displaystyle\int_{0}^{\pi} \frac{\sin x}{2 - \cos x} \, dx$

42. $\displaystyle\int_{0}^{\pi/3} \frac{4 \sin x}{1 - 4 \cos x} \, dx$

43. $\displaystyle\int_{1}^{2} \frac{2 \ln x}{x} \, dx$

44. $\displaystyle\int_{3}^{4} \frac{5 \ln x}{x} \, dx$

45. $\displaystyle\int_{0}^{\pi/2} \tan \frac{x}{2} \, dx$

46. $\displaystyle\int_{-\pi/2}^{-\pi/4} \cot x \, dx$

47. $\displaystyle\int_{\pi/2}^{\pi} 2\cot\frac{x}{3}\,dx$ **48.** $\displaystyle\int_{0}^{\pi/12} 6\tan 3x\,dx$

Evaluate the integrals in Exercises 49–54.

49. $\displaystyle\int \frac{2\,dx}{x}$ **50.** $\displaystyle\int \frac{dx}{3x+1}$

51. $\displaystyle\int \frac{x\,dx}{x^2+4}$ **52.** $\displaystyle\int \frac{\sin x\,dx}{1+2\cos x}$

53. $\displaystyle\int \tan\frac{x}{3}\,dx$ **54.** $\displaystyle\int \cot 2x\,dx$

Find the limits in Exercises 55–58.

55. $\displaystyle\lim_{x\to\infty} \ln\frac{1}{x}$ **56.** $\displaystyle\lim_{x\to 0^+} \ln\frac{1}{x}$

57. $\displaystyle\lim_{x\to 0} \ln|x|$ **58.** $\displaystyle\lim_{x\to 0} \ln\left|\frac{1}{x}\right|$

59. Find the area between the curve $y = \tan x$ and the x-axis from $x = 0$ to $x = \pi/3$.

60. The region between the curve $y = \sqrt{\cot x}$ and the x-axis from $x = \pi/6$ to $x = \pi/2$ is revolved about the x-axis to generate a solid. Find the volume of the solid.

61. *Increasing functions and decreasing functions.* As we saw in Section 4.1, a function $f(x)$ increases on its domain if for any two points x_1 and x_2 in the domain,

$$x_1 < x_2 \implies f(x_1) < f(x_2).$$

Similarly, a function decreases on its domain if for any two points x_1 and x_2 in the domain,

$$x_1 < x_2 \implies f(x_1) > f(x_2).$$

Show that increasing functions and decreasing functions are one-to-one. That is, show that $x_2 \neq x_1$ always implies that $f(x_2) \neq f(x_1)$. Hence, argue that $\ln x$ is one-to-one.

62. Refer to Example 3. Let

$$y = y_1 = \frac{(x^2+1)(x+3)^{1/2}}{x-1}.$$

a) What is the domain of y?

b) Draw a complete graph of y. Use the graph of $y_2 = $ NDER (y_1, x) to support the completeness of your graph of y.

c) What is the domain of dy/dx?

d) Use the formula obtained for dy/dx in Example 3 for $y_3 = dy/dx$. (Recall that this formula was obtained with the restriction $x > 1$.) Compare the graphs of y_2 and y_3 on the domain found in part (c).

e) Give a reason why the restriction $x > 1$ was placed on y in Example 3.

In Exercises 63–67, (a) draw a complete graph of the function, and (b) use graphs to support that the formula obtained by logarithmic differentiation is also valid where $y < 0$.

63. The function of Exercise 21.

64. The function of Exercise 23.

65. The function of Exercise 25.

66. The function of Exercise 27.

67. The function of Exercise 29.

In Exercises 68 and 69, let f and g be differentiable functions. Use logarithmic differentiation to find a formula for dy/dx. *Hint:*

$$d|u|/dx = (|u|/u)(du/dx).$$

68. $y = fg$. *Hint:* $|y| = |f||g|$.

69. $y = f/g$.

Grapher Support for Indefinite Integrals

We can use the graph of $y_1 = $ NINT $(f(t), a, x)$ for a specific value of a to support the computation of $\int f(x)\,dx$, the indefinite integral of f. We must be careful about points of discontinuity of f on the interval from a to x. In Exercises 70 and 71, graph the function in the specified intervals.

70. Let $\displaystyle f(x) = \int_{0}^{x} \tan t\,dt$.

a) GRAPH f in $[-\pi/2, \pi/2]$. On the basis of this graph, make a conjecture about

$$\lim_{x\to-(\pi/2)^+} f(x) \quad\text{and}\quad \lim_{x\to(\pi/2)^-} f(x).$$

b) GRAPH f in $[-\pi/2, 2]$ by $[-1, 10]$. What happens near $x = 2$? Why?

c) Compare the graph in part (b) with a graph of the antiderivative of $\tan x$.

71. Let $\displaystyle f(x) = \int_{\pi/2}^{x} \cot t\,dt$.

a) GRAPH f in $[0, \pi]$ by $[-5, 5]$. On the basis of this graph, make a conjecture about

$$\lim_{x\to 0^+} f(x) \quad\text{and}\quad \lim_{x\to\pi^-} f(x).$$

b) GRAPH f in $[0, 4]$ by $[-5, 5]$. What happens near $x = 4$? Why?

c) Compare the graph in part (b) with a graph of the antiderivative of $\cot x$.

72. Refer to Example 4.

a) Compare the graphs of

$$y_1 = \text{NINT}\left(\frac{2t}{t^2+3}, a, x\right) \quad\text{and}$$

$$y_2 = \text{NINT}\left(\frac{1}{t}, 1, x^2+3\right).$$

b) Can you find a value of a for which $y_1 = y_2$?

c) Explain why the graph of y_2 also supports the result of Example 4.

7.2 _____ The Exponential Function

You may recall our saying that the exponential model for bacterial growth, reproduced here from Section 4.7, is one of the really good models in science because it applies to so many phenomena that people want to forecast and understand. Whenever you have a quantity y whose rate of change over time is proportional to the amount of y present, you have a function that satisfies the differential equation

$$\frac{dy}{dt} = ky.$$

If, in addition, $y = y_0$ when $t = 0$, the function is none other than the exponential function

$$y = y_0 e^{kt}.$$

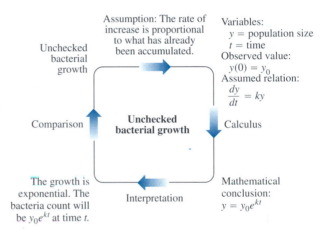

Assumption: The rate of increase is proportional to what has already been accumulated.

Unchecked bacterial growth

Variables:
y = population size
t = time
Observed value:
$y(0) = y_0$
Assumed relation:
$\frac{dy}{dt} = ky$

Comparison

Unchecked bacterial growth

Calculus

The growth is exponential. The bacteria count will be $y_0 e^{kt}$ at time t.

Interpretation

Mathematical conclusion:
$y = y_0 e^{kt}$

The Exponential Function $y = \exp(x)$

As you may recall, a function $y = f(x)$ is one-to-one if different input values for x always give different output values for y. The function $y = x$ is one-to-one because different x-values always give different y-values. The function $y = \sin x$ is not one-to-one because some different x-values give the same y-value. For example, the sines of 0 and π both have the value 0.

The natural logarithm function $y = \ln x$ has derivative $y' = 1/x$, which is positive at every point in the domain of the natural log function. This means that $\ln x$ is increasing on its domain and thus is one-to-one (Section 7.1, Exercise 61). The natural logarithm function therefore has an inverse. We call the inverse the exponential function of x, abbreviated $\exp(x)$.

We are coming at the exponential function from two directions. There is the unfamiliar function

$$y = exp(x)$$

that we define for all real numbers x. Then there is the more familiar

$$y = e^x$$

which, at this time, has meaning only for rational numbers x. We will argue that $\exp(x)$ and e^x are the same for all rational numbers x. Then we will say that e^x has the value $\exp(x)$ for all irrational numbers x. Thus, e^x and $\exp(x)$ will be identical, and we will have a complete definition of $y = e^x$.

DEFINITION

The function $y = \exp(x)$, **the exponential function of x**, is the inverse of the natural logarithm function, $y = \ln x$.

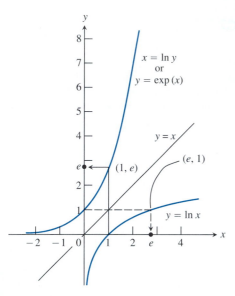

7.5 The graphs of $y = \ln x$ and its inverse $y = \exp(x)$. The number e is the one number whose natural logarithm is 1.

EXPLORATION BIT

Let

$$y_1 = \int_1^x 1/t\,dt, \qquad y_2 = 1,$$

and find the point of intersection of their graphs using a grapher.

EXPLORATION BIT

Show graphically that

$$\ln^{-1} x \neq (\ln x)^{-1}.$$

The Graph of $y = \exp(x)$

When functions are inverses of each other their graphs are reflections of each other across the line $y = x$. (The point (x, y) belongs to one graph, and (y, x) belongs to the other.) Thus, the graph of the exponential function $y = \exp(x)$ can be obtained by reflecting the graph of $y = \ln x$ across the line $y = x$ (Fig. 7.5).

The Number e

The number e is the one number whose natural logarithm is 1. That is, $\ln e = \int_1^e 1/t\,dt = 1$. Because exp and ln are inverse functions, this also means that $\exp(1) = e$ (Fig. 7.5). Although e is not a rational number, it is possible to find its value with a grapher to a large number of decimal places by solving the equation

$$\int_1^x 1/t\,dt = 1.$$

To 15 places,

$$e = 2.7\,1828\,1828\,45\,90\,45.$$

(This layout makes the number easier to remember.)

The Exponential Function $y = e^x$

We raise the number e to a rational power x in the usual way:

$$e^2 = e \cdot e, \qquad e^{-2} = \frac{1}{e^2}, \qquad e^{1/2} = \sqrt{e},$$

and so on. Since e is positive, the number e^x ("e to the x") is always positive. This means that e^x has a logarithm. When we take its logarithm, here is what we find:

$$\ln e^x = x \ln e \qquad \text{Rule 3}$$
$$= x \cdot 1 = x. \qquad \ln e = 1$$

But we also have

$$\ln(\exp(x)) = x$$

because ln and exp are inverses. Since ln is one-to-one and $\ln e^x = \ln(\exp(x))$ when x is rational, we have

$$e^x = \exp(x) \qquad \text{when } x \text{ is rational}$$

This equation provides a way to complete the definition of e^x to include irrational values of x. The function $\exp(x)$ is defined for all x, so we can use it to give a value to e^x at points where e^x had no previous value.

DEFINITION

For every real number x,

$$e^x = \exp(x) = \ln^{-1}(x).$$

TRANSCENDENTAL NUMBERS AND TRANSCENDENTAL FUNCTIONS

The numbers e and π are different from numbers like -2 and $\sqrt{3}$ in that they are not solutions of algebraic equations. Euler called e and π transcendental because, as he said, "they transcend the power of algebraic methods." Today, we call a function $y = f(x)$ *transcendental* if it satisfies no equation of the form

$$P_n(x)y^n + \cdots + P_1(x)y + P_0(x) = 0$$

in which the coefficients $P_0(x)$, $P_1(x)$, ..., $P_n(x)$ are polynomials in x with rational coefficients.

Functions that do satisfy such an equation are called *algebraic*. For instance, $y = 1/\sqrt{(x+1)}$ is algebraic because it satisfies the equation $(x+1)y^2 - 1 = 0$. Polynomials and rational functions are algebraic, and all sums, products, quotients, powers, and roots of algebraic functions are algebraic.

The six basic trigonometric functions are transcendental, as are the inverse trigonometric functions and the exponential and logarithmic functions.

EXPLORATION BIT

On your calculator, check $\ln(e^x)$ for $x = -1, 2, 10$, and 100. Also check $e^{\ln(x)}$ for your choice of values of $x > 0$. What should you expect the result of each computation to be?

TWO USEFUL OPERATING RULES

Equations (1) give us two approaches to dealing with equations that contain logarithms or exponents.

1. To remove logarithms from an equation, exponentiate both sides.
2. To remove exponents from an equation, take the logarithm of both sides.

EXPLORATION 1

Exploring the Exponential Function

The exponential function $y = e^x$ is so important that it usually has a key assigned to it on a hand calculator.

1. GRAPH $y_1 = e^x$. Compare your graph with the sketch in Fig. 7.5.
2. GRAPH $y_2 = \text{NDER }(y_1, x)$. What do you observe?
3. GRAPH $y_3 = \text{NINT }(e^t, 0, x)$. What do you observe? What can you conjecture? (Recall what you know about the family of antiderivatives of a function.)   ⚘

Equations Involving ln x and e^x

Because e^x and $\ln x$ are inverses, their composite is the identity function.

$$y = e^x \Leftrightarrow \ln y = x \qquad y = \ln x \Leftrightarrow e^y = x$$

$$\ln e^x = x \quad \text{(for all } x\text{)} \qquad e^{\ln x} = x \quad \text{(for } x > 0\text{)} \qquad (1)$$

Equations (1) are the two most important rules for combining $\ln x$ and e^x.

EXAMPLE 1

a) $\ln e^2 = 2$

b) $\ln e^{-1} = -1$

c) $\ln \sqrt{e} = \dfrac{1}{2}$

d) $\ln e^{\sin x} = \sin x$

e) $e^{\ln 2} = 2$

f) $e^{\ln(x^2+1)} = x^2 + 1$

g) $e^{3\ln 2} = e^{\ln 2^3} = e^{\ln 8} = 8$ One way

h) $e^{3\ln 2} = (e^{\ln 2})^3 = 2^3 = 8$ Another way   ☰

EXAMPLE 2

Find y if $\ln y = 3t + 5$.

Solution   Exponentiate:

$$e^{\ln y} = e^{3t+5} \qquad e^{\ln y} = y$$

$$y = e^{3t+5}$$

☰

EXAMPLE 3

Find k if $e^{2k} = 10$.

Solution Take the logarithm of both sides:

$$e^{2k} = 10$$

$$\ln e^{2k} = \ln 10 \qquad\qquad \ln e^{2k} = 2k$$

$$2k = \ln 10$$

$$k = \frac{1}{2} \ln 10$$

Rules of Exponents

Even though e^x is defined in a seemingly roundabout way as $\ln^{-1} x$, it obeys the familiar rules of exponents from algebra.

RULES OF EXPONENTS

For all real numbers $x, x_1,$ and x_2,

1. $e^{x_1} \cdot e^{x_2} = e^{x_1 + x_2}$ **2.** $\dfrac{e^{x_1}}{e^{x_2}} = e^{x_1 - x_2}$ **2(a).** $e^{-x} = \dfrac{1}{e^x}$

Proof of Rule 1 Let

$$y_1 = e^{x_1} \qquad \text{and} \qquad y_2 = e^{x_2}. \qquad\qquad (2)$$

Then,

$$\ln y_1 = x_1 \qquad \text{and} \qquad \ln y_2 = x_2 \qquad \text{Take logs of both sides of Eqs. (2).}$$

$$x_1 + x_2 = \ln y_1 + \ln y_2 = \ln y_1 y_2 \qquad \text{Logarithm Rule 1}$$

$$e^{x_1 + x_2} = e^{\ln y_1 y_2} \qquad \text{Exponentiate.}$$

$$= y_1 y_2 = e^{x_1} e^{x_2}. \qquad e^{\ln u} = u$$

Rules 2 and 2(a) follow from Rule 1, as you can show in the exercises.

EXAMPLE 4

a) $e^{x + \ln 2} = e^x \cdot e^{\ln 2} = 2e^x$ Rule 1

b) $\dfrac{e^{2x}}{e} = e^{2x-1}$ Rule 2

c) $e^{-\ln x} = \dfrac{1}{e^{\ln x}} = \dfrac{1}{x}$ Rule 3

AGREEMENT

From now on when we say that f has an inverse, it should be understood that f^{-1} is a function (with domain appropriately restricted, if necessary).

Derivatives of Inverses of Differentiable Functions

In Section 1.6, we saw that the inverse, f^{-1}, of a function f is itself a function if and only if the function f is one-to-one. Increasing functions and decreasing

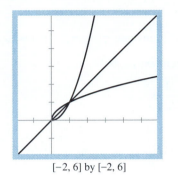

[−2, 6] by [−2, 6]

7.6 The graphs of $y = x^2$, $x \geq 0$ and $y = \sqrt{x}$. Notice the symmetry about the line $y = x$.

functions are one-to-one (Section 7.1, Exercise 61) and hence have inverses. In fact these are the only continuous functions whose inverses are functions.

If two functions are inverses of each other, call them f and f^{-1}, their graphs are symmetric about the line $y = x$. (See Fig. 7.5 again and Fig. 7.6.) This symmetry and the fact that derivatives give the slopes of tangent lines suggest that there should be some relationship between the derivatives f' and $(f^{-1})'$, as indeed there is.

EXPLORATION 2

A Relationship between f' and $(f^{-1})'$

Let $f(x) = x^2$, $x \geq 0$, and $g(x) = f^{-1}(x) = \sqrt{x}$.

1. Find $f'(x)$ and $g'(x)$.

2. Copy and complete this table:

x	$f(x)$	$f'(x)$	$g'(f(x))$
0.5	?	?	?
1	?	?	?
1.5	?	?	?
2	?	?	?
3	?	?	?
4	?	?	?

How are the numbers in the last two columns related?

3. Use $f'(5)$ to predict $g'(25)$. Then verify directly.

4. Make a conjecture about a relationship between $f'(x)$ and $g'(f(x))$ if f and g are inverses.

5. Test the conjecture with $f(x) = \dfrac{2x + 3}{x + 3}$ and its inverse.

A relationship that you may have found in Exploration 2 is that the derivative of a function f evaluated at a and the derivative of the inverse function evaluated at $f(a)$ are reciprocals. We state this with necessary conditions in the following theorem from advanced calculus.

EXPLORATION BIT

It is important that you understand what Theorem 1 *does not* say. It does not say that $g' = 1/f'$. Nor should you confuse $g'(f(x))$ with $(g(f(x)))'$.

a) Verify that the functions g' and $1/f'$ are not equal.

b) Verify that $g'(f(x)) \neq (g(f(x)))'$.

THEOREM 1   The Derivative Rule for Inverses

If f is differentiable at every point of an interval I and df/dx is never zero on I, then $g = f^{-1}$ is differentiable at every interior point of the interval $f(I)$. The value of g' at any particular image point $f(a)$ is the reciprocal of the value of f' at a:

$$g'(f(x)) = \frac{1}{f'(x)}. \tag{3}$$

EXAMPLE 5

Verify Eq. (3) for $f(x) = x^2$, $x > 0$, and its inverse $g(x) = f^{-1}(x) = \sqrt{x}$.

Solution $f'(x) = 2x$, which is never zero on $(0, \infty)$.

$$g'(x) = \frac{1}{2\sqrt{x}}.$$

Thus, $g'(f(x)) = \dfrac{1}{2\sqrt{f(x)}} = \dfrac{1}{2\sqrt{x^2}} = \dfrac{1}{2|x|} = \dfrac{1}{2x} = \dfrac{1}{f'(x)}.$ (See Fig. 7.7.)

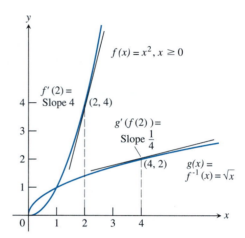

7.7   The graphs of inverse functions have reciprocal slopes at corresponding points. For the functions f and $g = f^{-1}$ of Example 5,

$$g'(f(2)) = \frac{1}{2 \cdot 2} = \frac{1}{4} = \frac{1}{f'(2)}.$$

The slope of f at 2 and the slope of g at $f(2)$ are reciprocals, namely, 4 and 1/4.

EXAMPLE 6

Let $f(x) = \dfrac{2x + 3}{x + 3}$. Find a rule for f^{-1}, and verify Eq. (3) at the points $(-5, f(-5)) = (-5, 3.5)$ and $(3.5, f^{-1}(3.5)) = (3.5, -5)$.

Solution   Fig. 7.8(a) gives a complete graph of

$$y = f(x) = \frac{2x + 3}{x + 3} = 2 - \frac{3}{x + 3}. \tag{4}$$

Notice that

$$\text{Domain } f = (-\infty, -3) \cup (-3, \infty) \quad \text{Why?}$$

and

$$\text{Range } f = (-\infty, 2) \cup (2, \infty). \quad \text{Why?}$$

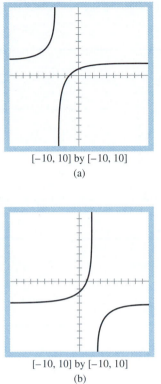

[−10, 10] by [−10, 10]

(a)

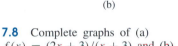

[−10, 10] by [−10, 10]

(b)

7.8 Complete graphs of (a) $f(x) = (2x + 3)/(x + 3)$ and (b) $f^{-1}(x) = (-3x + 3)/(x - 2)$. You may recognize the function f as the one in part 5 of Exploration 2.

Also, f is one-to-one. (Why?) To obtain a rule for f^{-1}, we switch x and y in Eq. (4) and solve for y:

$$y = \frac{2x + 3}{x + 3}$$

$$x = \frac{2y + 3}{y + 3} \qquad \text{Switch } x \text{ and } y.$$

$$xy + 3x = 2y + 3 \qquad \text{Solve for } y.$$

$$(x - 2)y = -3x + 3$$

$$y = \frac{-3x + 3}{x - 2}$$

$$f^{-1}(x) = \frac{-3x + 3}{x - 2} = -3 - \frac{3}{x - 2}$$

A complete graph of f^{-1} is shown in Fig. 7.8(b). Notice that the Domain f^{-1} = Range f and Range f^{-1} = Domain f.

Finally, at the points $(-5, 3.5)$ and $(3.5, -5)$,

$$\text{NDER }(f, -5) = 0.75$$

and

$$\text{NDER }(f^{-1}, 3.5) = 1.33,$$

respectively. These two values are approximately reciprocals of each other. In the exercises, you are asked to confirm that the exact values are 3/4 and 4/3.

≡

The Derivative of $y = e^x$

Because $f(x) = \ln x$ is differentiable at every point of its domain, a straightforward application of Theorem 1 tells us that $f^{-1}(x) = e^x$ is also differentiable at every point of its domain. We can calculate the derivative of e^x as follows:

$$y = e^x$$

$$\ln y = \ln e^x \qquad \text{Take logarithms of both sides.}$$

$$\ln y = x \qquad \text{Because } \ln e^x = x$$

$$\frac{d}{dx} \ln y = \frac{d}{dx} x \qquad \begin{array}{l}\text{Implicit differentiation—assumes} \\ y = e^x \text{ to be differentiable}\end{array}$$

$$\frac{1}{y} \frac{dy}{dx} = 1 \qquad \text{Eq. (1), Section 7.1, with } u = y$$

$$\frac{dy}{dx} = y$$

$$\frac{dy}{dx} = e^x \qquad \text{Replace } y \text{ by } e^x.$$

The conclusion that we draw from this sequence of equations is that the function $y = e^x$ is its own derivative. Never before have we encountered such a nonzero function. No matter how many times we differentiate it, we always get the function back. Constants times e^x are the only other functions to behave this way.

$$\frac{d}{dx}e^x = e^x \tag{5}$$

EXAMPLE 7

$$\frac{d}{dx}(5e^x) = 5\frac{d}{dx}e^x = 5e^x$$

The Chain Rule extends Eq. (5) in the usual way to a more general form.

If u is any differentiable function of x, then
$$\frac{d}{dx}e^u = e^u \frac{du}{dx}. \tag{6}$$

EXAMPLE 8

Equation (6) with $u = -x$:
$$\frac{d}{dx}e^{-x} = e^{-x}\frac{d}{dx}(-x) = e^{-x}(-1) = -e^{-x}.$$

EXAMPLE 9

Equation (6) with $u = \sin x$:
$$\frac{d}{dx}e^{\sin x} = e^{\sin x}\frac{d}{dx}(\sin x) = e^{\sin x} \cdot \cos x.$$

The integral-formula equivalent of Eq. (6) is

$$\int e^u \, du = e^u + C.$$

EXAMPLE 10

$$\int_0^{\ln 2} e^{3x} \, dx = \int_0^{3\ln 2} e^u \cdot \frac{1}{3} \, du \qquad \begin{aligned} &u = 3x, du = 3dx, \frac{1}{3}du = dx, \\ &u(0) = 0, u(\ln 2) = 3\ln 2. \end{aligned}$$

$$= \frac{1}{3}\int_0^{3\ln 2} e^u \, du$$

$$= \frac{1}{3}e^u \Big]_0^{3\ln 2} = \frac{1}{3}[e^{3\ln 2} - e^0] = \frac{1}{3}[8 - 1] = \frac{7}{3}$$

EXAMPLE 11

$$\int_0^{\pi/2} e^{\sin x} \cos x \, dx = e^{\sin x} \Big]_0^{\pi/2} \qquad \text{Antiderivative from Example 9}$$

$$= e^{\sin(\pi/2)} - e^{\sin 0} = e^1 - e^0 = e - 1 \qquad \blacksquare$$

The Law of Exponential Change, $y = y_0 e^{kt}$

In many instances in biology and economics, some positive quantity y grows or decreases at a rate that at any given time t is proportional to the amount that is present. If we also know the initial amount y_0 at time $t = 0$, we can find y by solving the initial value problem

Differential equation: $\quad \dfrac{dy}{dt} = ky$

Initial condition: $\qquad y = y_0 \qquad$ when $t = 0$. $\qquad (7)$

The constant k is positive if y is increasing and negative if y is decreasing.

To solve Eq. (7), we divide through by y to get

$$\frac{1}{y}\frac{dy}{dt} = k$$

and integrate both sides with respect to t to get

$$\ln y = kt + C.$$

We then solve this for y by exponentiating:

$$e^{\ln y} = e^{kt+C}$$

$$y = e^{kt} \cdot e^C$$

$$y = Ae^{kt} \qquad \text{Write } A \text{ for } e^C.$$

We find the value of A from the initial condition:

$$y_0 = Ae^{k(0)} = A \cdot 1 = A.$$

The solution of the initial value problem is $y = y_0 e^{kt}$. We call this equation the Law of Exponential Change.

The Law of Exponential Change

$$y = y_0 e^{kt}$$

LIMITATIONS ON CELL GROWTH

Some limitations are those that make the environment less than ideal, such as inadequate nutrients or oxygen. Other limitations would also include the strength of the cell wall or the space available for the cell to grow.

EXAMPLE 12 The Growth of a Cell

In an ideal environment, the mass m of a cell will grow exponentially, at least early on. Nutrients pass quickly through the cell wall, and growth is limited only by the metabolism within the cell, which in turn depends on the mass of participating molecules. If we make the reasonable assumption that, at each instant of time, the cell's growth rate dm/dt is proportional to the mass that has already been accumulated, then

$$\frac{dm}{dt} = km \qquad \text{and} \qquad m = m_0 e^{kt}.$$

There are limitations, of course, and in any particular case, we would expect this equation to provide reliable information only for values of m below a certain size. ≡

EXAMPLE 13   Birth Rates and Population Growth

Strictly speaking, the number of individuals in a population (of people, plants, foxes, or whatever) is a discontinuous function of time because it takes on discrete values (a step function). However, as soon as the number of individuals becomes large enough, it may safely be described with a continuous and even differentiable function. If we assume that the proportion of reproducing individuals remains constant and assume a constant fertility, then at any instant t, the birth rate is proportional to the number $y(t)$ of individuals present. If, further, we neglect departures, arrivals, and deaths, the growth rate dy/dt will be the same as the birth rate ky. In other words,

$$\frac{dy}{dt} = ky.$$

Once again, we find that $y = y_0 e^{kt}$. ≡

> **LIMITATIONS ON POPULATION**
>
> Factors such as disease or birth control practices affect the value of k. There are also limitations on the size of y. What are some of them?

EXPLORATION 3

Vertical and Horizontal Stretches and Shrinks

The constant a in the function $y = af(bx)$ stretches or shrinks the graph of $y = f(x)$ vertically. The constant b stretches or shrinks the graph of f horizontally. Interpret the *real* meanings of y_0 and a positive k in the population growth model $y = y_0 e^{kt}$, and tell what effects changes in their values have. Support your conclusions with grapher examples. ⚛

The Incidence of Disease

One model for the way in which a disease spreads assumes that the rate dy/dt at which the number of infected people changes is proportional to the number y itself. The more infected people there are, the faster the disease will spread. The fewer there are, the more slowly it will spread. Once again,

$$y = y_0 e^{kt}.$$

EXAMPLE 14

In the course of any given year, the number y of cases of a disease is reduced by 20%. If there are 10,000 cases today, how many years will it take to reduce the number of cases to 1000?

Solution   The equation that we use is $y = y_0 e^{kt}$. There are three things to find:

1. The value of y_0,

2. The value of k,

3. The value of t that makes $y = 1000$.

STEP 1: *The value of y_0.* We are free to count time from anywhere we want. If we start counting from today, then $y = 10,000$ when $t = 0$, so $y_0 = 10,000$. Our equation becomes

$$y = 10,000e^{kt}.$$

STEP 2: *The value of k.* When $t = 1$, the number of cases will be 80% of its present value, or 8000. Hence,

$$10,000e^{k(1)} = 8000$$

$$e^k = 0.8$$

$$\ln e^k = \ln 0.8$$

$$k = \ln 0.8.$$

At any given time t, therefore,

$$y = 10,000e^{(\ln 0.8)t}.$$

Two down and one to go.

STEP 3: *The value of t that makes $y = 1000$.* Set y equal to 1000, and solve for t:

$$10,000e^{(\ln 0.8)t} = 1000$$

$$e^{(\ln 0.8)t} = 0.1 \qquad\qquad \text{Divide by 10,000.}$$

$$(\ln 0.8)t = \ln 0.1 \qquad\qquad \text{Take logs of both sides.}$$

$$t = \frac{\ln 0.1}{\ln 0.8} = 10.32.$$

It will take a little more than 10 years to reduce the number of cases to 1000.

≡

For Example 14, the graph of

$$y = 10,000e^{(\ln 0.8)t}$$

in a [0, 20] by [−2,000, 10,000] viewing window gives a dramatic representation of the decline of the disease. Showing the graph of $y = 1000$ in the same viewing window adds a nice visual touch in support of the analytic solution.

Logistic Growth

An exponential model for growth of population or disease is reasonable only under special assumptions and then only for short periods of time. For example, exponential population growth would eventually exceed Earth's resources to sustain life. The logistic curve

$$P(t) = \frac{k}{1 + e^{a-rt}},$$

where a, k, and r are constants and t represents time, usually provides a better model.

EXPLORATION BIT

For Example 15, is it better to show the graphs of P and P' (Figs. 7.9 and 7.10) in the same viewing window or in different viewing windows? Explain.

EXAMPLE 15

The spread of flu in a certain school is given by

$$P(t) = \frac{100}{1 + e^{3-t}},$$

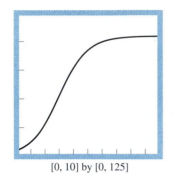

[0, 10] by [0, 125]

7.9 The graph of

$$P(t) = \frac{100}{1 + e^{3-t}},$$

the number of students infected with the flu.

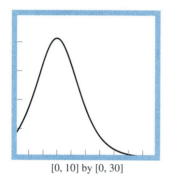

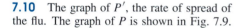

[0, 10] by [0, 30]

7.10 The graph of P', the rate of spread of the flu. The graph of P is shown in Fig. 7.9.

where t is the number of days after students are exposed to infected students (Fig. 7.9).

a) Estimate the initial number of students infected with the flu.

b) What is the maximum number of students who will get the flu?

c) Estimate the number of days it will take for 99 students to become infected.

d) When will the rate at which the flu spreads be a maximum? What is this rate?

Solution

a) $P(t)$ represents the number of students infected with the flu, and $P(0) = 4.74$. Thus, the initial number of students infected can be estimated at 5.

b) The graph of $y = P(t)$ shows that P is always increasing and

$$\lim_{t \to \infty} P(t) = 100,$$

that is, $y = 100$ is a right-end behavior model for P. Thus, the maximum number of students who will be infected is 100.

c) We can use ZOOM-IN or SOLVE to show that $P(t) = 99$ when $t = 7.60$. So 99 students will be infected after eight days.

d) Figure 7.10 shows the graph of P', the rate at which the flu spreads. The maximum value 25 of P' occurs at $t = 3$ days. This means that 25 students per day is the maximum rate of spread of the flu. Notice that the graph of P has a point of inflection at $t = 3$. ≡

Continuously Compounded Interest

If you invest an amount A_0 of money at a fixed annual interest rate r and interest is added to your account k times a year, it turns out that the amount of money that you will have at the end of t years is

$$A(t) = A_0 \left(1 + \frac{r}{k}\right)^{kt}. \tag{8}$$

The money might be added ("compounded," bankers say) monthly ($k = 12$), weekly ($k = 52$), daily ($k = 365$), or even more frequently, say, by the hour or by the minute. In Section 2.1, we viewed the graph of

$$A = 100 \left(1 + \frac{0.06}{k}\right)^{k}$$

and saw how the growth of A in one year was limited regardless of the size of k. In Section 7.5, we will see that $\lim_{k \to \infty}(1 + (r/k))^{k} = e^{r}$, and hence, the growth of an investment can be given by the following formula:

$$A(t) = A_0 e^{rt}.$$

The Continuous Compound Interest Formula

$$A(t) = A_0 e^{rt} \tag{9}$$

Interest paid according to this formula is said to be **compounded continuously**. The number r is called the **continuous interest rate**.

EXAMPLE 16

Suppose you deposit $621 in a bank account that pays 6% compounded continuously. How much money will you have in the account 8 years later?

Solution   We use Eq. (9) with $A_0 = 621, r = 0.06$, and $t = 8$:

$$A(8) = 621e^{(0.06)(8)} = 621e^{0.48} = 1003.58. \qquad \blacksquare$$

Had the bank paid interest quarterly ($k = 4$ in Eq. (8)), the amount in your account would have been $1000.01. Thus, the effect of continuous compounding, as compared with quarterly compounding, is an additional $3.57. A bank might decide it would be worth this additional amount to be able to advertise, "We compound your money every second, night and day—better than that, we compound the interest continuously."

EXPLORATION 4

The Rules of 70 and 72

How long does it take to double money at a continuous interest rate of $r\%$? To find out, set $A_0e^{rt} = 2A_0$, and solve for t:

$$A_0e^{rt} = 2A_0$$

$$e^{rt} = 2$$

$$\ln e^{rt} = \ln 2$$

$$rt = \ln 2, \text{ so } t = \frac{\ln 2}{r} = \frac{100 \ln 2}{100r} \approx \frac{69.3}{100r} \approx \frac{70}{100r} \approx \frac{72}{100r}.$$

Now GRAPH $y_1 = (\ln 2)/r, y_2 = 70/100r$, and $y_3 = 72/100r$ in a $[0, 0.1]$ by $[0, 100]$ viewing window. You will see that the graphs of y_1, y_2, and y_3 are close approximations of each other. Thus, we have two rules of thumb for computing in our heads the number of years it will take money to double at various interest rates compounded continuously.

> **The Rules of 70 and 72**
>
> It takes about $y = 70/i$ or $72/i$ years for money to double at i percent, using whichever of 70 or 72 is easier to divide by i.

For example, at 7%, it will take about 70/7 or 10 years for money to double. Using TRACE in your viewing window will support this result.

The rules can also be used "backward" to estimate the interest rate needed to double money in a given number of years. For example, to double any given amount in 6 years, you need to invest at the rate $i = 72/6 = 12\%$ (as TRACE again will support in your viewing window).

Although developed for continuous compounding, the Rules of 70 and 72 have good predictive value even for quarterly or annual compounding because, as Example 16 suggests, the results of different compounding frequencies are much the same.   ⚘

The higher the CPI, or p, the faster it is changing for a fixed rate of inflation, k.

The Consumer Price Index

As you know, prices often change (they usually go up), and it is important to know what things are likely to cost in the years ahead. The economists in the U.S. Department of Labor measure the cost of living with a number called the **Consumer Price Index (CPI).** This index is a weighted average of the costs of food, clothing, housing, transportation, medical care, personal care, and entertainment.

The current index, set at 100 (an arbitrary choice for convenient arithmetic) in 1984, was 112.1 in March of 1987 and 116.5 in March of 1988. What you could buy for $1.00 in 1984 cost roughly $1.12 in 1987 and $1.17 in 1988.

One of the predictors of price change assumes that the rate dp/dt at which the CPI, or p, changes is proportional to p, so

$$\frac{dp}{dt} = kp. \tag{10}$$

The constant k is called the **continuous rate of inflation.** In newspapers, k is usually expressed as a percent. If $k = 0.04$, for example, the U.S. Bureau of Labor Statistics reports an inflation rate of 4%. The solution of Eq. (10) is

$$p(t) = p_0 e^{kt}. \tag{11}$$

If the CPI is p_0 at time $t = 0$, then t years later, it will be $p(t) = p_0 e^{kt}$.

EXAMPLE 17

The CPI in March 1988 was 116.5. What will its value be 10 years later, in March 1998, if the inflation rate is a constant 4%?

Solution   We take $p_0 = 116.5$ and $k = 0.04$ in Eq. (11) and find $p(10)$ (see Fig. 7.11):

$$p(10) = 116.5 e^{(0.04)(10)} = 173.8$$

What conclusions can we draw from this? One conclusion is that from 1988 to 1998, prices will rise

$$\frac{173.8 - 116.5}{116.5} \times 100 = 49\%.$$

Another is that by 1998, relative to March 1984, when the index was 100, prices will have risen

$$\frac{173.8 - 100}{100} \times 100 = 73.8\%.$$

(Now you see why the index is set at 100 in the base year. Subtracting 100 from any later value automatically gives the percentage change.)

Still another conclusion is that by 1998, it will cost nearly $1.74 to buy what $1.00 bought in 1984.   ≡

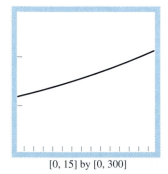

[0, 15] by [0, 300]

7.11   A graph of the Consumer Price Index based on the index for March 1988 and an inflation rate of 4%. Using TRACE on the graph of $p(x) = 116.5 e^{0.04x}$ shows that $p(10) = 173.80$.

EXPLORATION 5

The Consumer Price Index

1. How long does it take the cost of living to increase 50%? To answer the question, assume a steady inflation rate of 4%, and find how long it takes the CPI to increase from p_0 to $1.5p_0$. This is equivalent to asking: When does $p_0e^{0.04t} = 1.5p_0$, or $e^{0.04t} = 1.5$? Find an answer graphically, then confirm algebraically.

2. GRAPH $y_1 = e^{rt}$ for $r = 0.03, 0.04, 0.06$, and 0.10 and $y_2 = 1.5$. Explain what you see in the viewing window in terms of the CPI model.

The Purchasing Power of the Dollar

The higher the Consumer Price Index, the less a dollar will buy. Economists call the number $100/p$ the **purchasing power of the dollar.**

In 1984, when the CPI was 100, the purchasing power of the dollar was $100/100 = 1.00$. Three years later, in 1987, it was $100/112.1 = 0.89$, and a year after that, it was $100/116.5 = 0.86$.

EXAMPLE 18

Assuming the 4% inflation rate of Example 17, what will the purchasing power of the dollar be in 1998?

Solution   We divide the CPI for 1998 into 100:

$$\frac{100}{1998 \text{ CPI}} = \frac{100}{173.8} = 0.58. \qquad \text{CPI value from Example 17}$$

This represents a change of -0.42, or a loss in purchasing power of 42% since 1984.

Exercises 7.2

Use the fact that the functions $y = \ln x$ and $y = e^x$ are inverses of each other to simplify the expressions in Exercises 1–6.

1. $e^{\ln 7}$

2. $e^{-\ln 7}$

3. $\ln e^2$

4. $e^{3\ln 2}$

5. $e^{2+\ln 3}$

6. $e^{-2\ln 3}$

In Exercises 7–12, solve for k.

7. $e^{2k} = 4$

8. $e^{5k} = \dfrac{1}{4}$

9. $100e^{10k} = 200$

10. $100e^k = 1$

11. $2^{k+1} = 3^k$

12. $\sqrt{4^{k-1}} = 3^k$

In Exercises 13–18, solve for t.

13. $e^t = 1$

14. $e^{kt} = \dfrac{1}{2}$

15. $e^{-0.3t} = 27$

16. $e^{-0.01t} = 1000$

17. $2^{e^t} = 2 - t$

18. $e^{-2t} = t + 2$

In Exercises 19–24, solve for y.

19. $\ln y = 2t + 4$

20. $\ln y = -t + 5$

21. $\ln (y - 40) = 5t$

22. $\ln (1 - 2y) = t$

23. $5 + \ln y = 2^{x^2+1}$

24. $\ln (2^y - 1) = x^2 - 3$

Find dy/dx in Exercises 25–34. Support graphically.

25. $y = 2 e^x$

26. $y = e^{2x}$

27. $y = e^{-x}$

28. $y = e^{-5x}$

29. $y = e^{2x/3}$

30. $y = e^{-x/4}$

31. $y = xe^2 - e^x$

32. $y = x^2e^x - xe^x$

33. $y = e^{\sqrt{x}}$

34. $y = e^{(x^2)}$

Evaluate the integrals in Exercises 35–48. Support using a NINT computation.

35. $\int_1^{e^2} \frac{1}{x}\, dx$

36. $\int_1^e \frac{2}{x}\, dx$

37. $\int_{\ln 2}^{\ln 3} e^x\, dx$

38. $\int_{-1}^1 e^{(x+1)}\, dx$

39. $\int_{\ln 3}^{\ln 5} e^{2x}\, dx$

40. $\int_0^{\ln 2} e^{-x}\, dx$

41. $\int_0^1 (1+e^x)e^x\, dx$

42. $\int_{-1}^1 \frac{e^x}{1+e^x}\, dx$

43. $\int_2^4 \frac{dx}{x+2}$

44. $\int_{-1}^0 \frac{8\, dx}{2x+3}$

45. $\int_{-1}^1 2xe^{-x^2}\, dx$

46. $\int_0^1 \frac{x\, dx}{4x^2+1}$

47. $\int_1^4 \frac{e^{\sqrt{x}}\, dx}{2\sqrt{x}}$

48. $\int_e^{e^2} \frac{dx}{x\ln x}$

Evaluate the integrals in Exercises 49–54.

49. $\int 2e^x \cos(e^x)\, dx$

50. $\int 3e^x \sin(e^x)\, dx$

51. $\int \frac{e^x\, dx}{1+e^x}$

52. $\int \frac{dx}{1-e^{-x}}$

53. $\int \frac{\tan(\sqrt{x})\, dx}{\sqrt{x}}$

54. $\int \frac{\cot(\sqrt{2x})\, dx}{\sqrt{x}}$

In Exercises 55–58:

a) Find the inverse f^{-1} of the function f, expressed as a function of x.

b) Graph f and f^{-1} together.

c) Verify Eq. (3) by evaluating df/dx at $x=a$ and df^{-1}/dx at $x=f(a)$.

55. $f(x) = 2x+3, \quad a=-1$

56. $f(x) = 5-4x, \quad a=1/2$

57. $f(x) = \dfrac{1-2x}{x+2}, \quad a=-3$

58. $f(x) = \dfrac{x-5}{x-3}, \quad a=2$

One of the virtues of Eq. (3) is that it enables us to find values of df^{-1}/dx even when we do not have an explicit formula for the derivative. As a case in point, let $f(x) = x^2-4x-3, x>2$, and find the value of df^{-1}/dx at the point specified in Exercises 59 and 60.

59. $x=-3=f(4)$

60. $x=2=f(5)$

61. Let $f(x) = \dfrac{2x+3}{x+3}$. Verify the claims in Example 6 that $f'(-5) = 3/4$ and $(f^{-1})'(3.5) = 4/3$.

Find the limits in Exercises 62–65.

62. $\lim\limits_{x\to\infty} e^{-x}$

63. $\lim\limits_{x\to-\infty} e^{-x}$

64. $\lim\limits_{x\to-\infty} \ln(2+e^x)$

65. $\lim\limits_{x\to\infty} \int_x^{2x} \frac{1}{t}\, dt$

66. Solve for x: $\int_0^x \frac{1}{\sqrt{2\pi}} e^{(-t^2/2)}\, dt = 0.3$

67. Solve for x: $\int_1^x 2^t \ln t\, dt = 0.1$

68. Find the linearization of $f(x) = e^x$ at $x=0$.

69. Find the linearization of $f(x) = x+e^{4x}$ at $x=0$.

70. Find the maximum value of $f(x) = x^2 \ln(1/x)$.

71. Find the maximum and minimum values of the periodic function $f(x) = e^{\sin x}$.

Solve the initial value problems in Exercises 72 and 73.

72. Differential equation: $\dfrac{dy}{dx} = (\cos x)e^{\sin x}$

   Initial condition: $y=0$ when $x=0$

73. Differential equation: $\dfrac{dy}{dx} = 1 + \dfrac{1}{x}$

   Initial condition: $y=3$ when $x=1$

74. A body moves along a coordinate line with acceleration $d^2s/dt^2 = 4/(4-t)^2$. When $t=0$, the body's velocity is 2 m/sec. Find the total distance traveled by the body from time $t=1$ sec to time $t=2$ sec.

75. Show that, for any number $a>1$,

$$\int_1^a \ln x\, dx + \int_0^{\ln a} e^y\, dy = a\ln a.$$

(*Hint:* Study this diagram.)

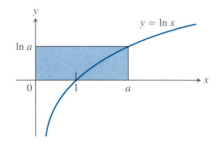

76. *Rules of Exponents.* Let $e^{x_1} = e^{x_1-x_2+x_2}$ and use the first rule of exponents to show that $e^{x_1}/e^{x_2} = e^{x_1-x_2}$.

77. *Rules of Exponents.* Use the second rule of exponents to prove that $e^{-x} = 1/e^x$.

78. *Cholera bacteria.* Suppose that the bacteria in a colony can grow unchecked, by the Law of Exponential Change. The colony starts with 1 bacterium and doubles every half hour. How many bacteria will the colony contain at the end of 24 hr? (Under favorable laboratory conditions, the number of cholera bacteria can double every 30 min. In an infected person, many bacteria are destroyed, but this example helps to explain why a person who feels well in the morning may be dangerously ill by evening.)

79. *Growth of bacteria.* A colony of bacteria is grown under ideal conditions in a laboratory so that the population increases exponentially with time. At the end of 3 hr, there are 10,000 bacteria. At the end of 5 hr, there are 40,000. How many bacteria were present initially?

80. *The incidence of a disease (continuation of Example 14).* Suppose that in any given year, the number of cases can be reduced by 25% instead of 20%.
  a) How long will it take to reduce the number of cases to 1000?
  b) How long will it take to eradicate the disease, that is, to reduce the number of cases to less than 1?

81. *John Napier's question.* John Napier, who invented natural logarithms, was the first person to answer the question: What happens if you invest an amount of money at 100% interest, compounded continuously?
  a) What does happen?
  b) How long does it take to triple your money?
  c) How much can you earn in a year?

82. *Benjamin Franklin's will.* The Franklin Technical Institute of Boston owes its existence to a provision in a codicil to the will of Benjamin Franklin. In part it reads:

> I was born in Boston, New England and owe my first instruction in Literature to the free Grammar Schools established there: I have therefore already considered those schools in my Will. . . . I have considered that among Artisans good Apprentices are most likely to make good citizens. . . . I wish to be useful even after my Death, if possible, in forming and advancing other young men that may be serviceable to their Country in both Boston and Philadelphia. To this end I devote Two thousand Pounds Sterling, which I give, one thousand thereof to the Inhabitants of the Town of Boston in Massachusetts, and the other thousand to the inhabitants of the City of Philadelphia, in Trust and for the Uses, Interests and Purposes hereinafter mentioned and declared.

Franklin's plan was to lend money to young apprentices at 5% interest with the provision that each borrower should pay each year

> . . . with the yearly Interest, one tenth part of the Principal, which sums of Principal and Interest shall be again let to fresh Borrowers. . . . If this plan is executed and succeeds as projected without interruption for one hundred Years, the Sum will then be one hun-

dred and thirty-one thousand Pounds of which I would have the Managers of the Donation to the Inhabitants of the Town of Boston, then lay out at their discretion one hundred thousand Pounds in Public Works. . . . The remaining thirty-one thousand Pounds, I would have continued to be let out on Interest in the manner above directed for another hundred Years. . . . At the end of this second term if no unfortunate accident has prevented the operation the sum will be Four Millions and Sixty-one Thousand Pounds.

It was not always possible to find as many borrowers as Franklin had planned, but the managers of the trust did the best they could; they lent money to medical students as well as to others. At the end of 100 years from the reception of the Franklin gift, in January 1894, the fund had grown from 1000 pounds to almost exactly 90,000 pounds. In 100 years, the original capital had multiplied about 90 times instead of the 131 times Franklin had imagined.

What rate of interest, compounded continuously for 100 years, would have multiplied Benjamin Franklin's original capital by 90?

83. In Benjamin Franklin's estimate that the original 1000 pounds would grow to 131,000 in 100 years, he was using an annual rate of 5% and compounding once each year. What rate of interest per year when compounded continuously for 100 years would multiply the original amount by 131?

84. *The U.S. population.* The Museum of Science in Boston, Massachusetts, displays the running total of the U.S. population. On December 27, 1988, it was increasing the total by 1 every 21 seconds, which works out to an instantaneous rate of 1,502,743 people a year. Working from 246,605,103, the display's population figure for 2:40 P.M. that day, this gives $k = 0.00609$, or 0.609%. Assuming this value of k for the next 10 years, what will the U.S. population be at 2:40 P.M. Boston time on December 27, 1998?

85. *The Rule of 70.* Use the Rule of 70 to answer these questions:
  a) How long does it take to double money at 5% interest? At 7% interest?
  b) What interest rate do you need to double money in 5 years? In 20 years?

86. The Consumer Price Index in May 1993 was 144.4. Assuming a constant inflation rate of 3.1% from then on, about when would you expect the index to double? (*Hint*: Use the Rule of 70.)

87. Repeat Exercise 86, but now use Eq. (10) directly instead of the Rule of 70.

88. *The purchasing power of the dollar.* Assuming a constant inflation rate of 3.1% starting in May 1993, when the Consumer Price Index was 144.4, what will the purchasing power of the dollar be in May 1995? May 1997? May 1999?

89. *Inflation and the CPI.* You have just seen a newspaper headline saying that the Consumer Price Index rose 4% last year. Assuming that this was caused by a constant inflation rate, what was that rate?

90. *Runaway inflation.* At the end of 1988, the Consumer Price Index in Brazil was increasing at the continuous annual rate of 800%. (Yes, it was.) How many days does it take prices to double at this rate? To find out, solve the equation

$$p_0 e^{8t} = 2p_0$$

for t, and convert your answer to days, rounding your answer to the nearest day.

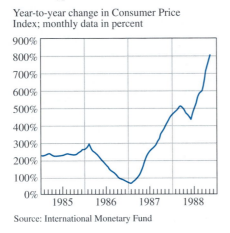

Brazil's hyperinflation

Year-to-year change in Consumer Price Index; monthly data in percent

Source: International Monetary Fund

(*Source* Wall Street Journal,
Thursday December 8, 1988, p. 1.)

91. *The purchasing power of the dollar.* How long will it take the purchasing power of the dollar to fall to half its present value if the inflation rate holds at a steady $r\%$?

92. *Oil depletion.* Suppose that the amount of oil pumped from one of the canyon wells in Whitter, California, decreases at the continuous rate of 10% a year. When will the well's output fall to a fifth of its present value?

93. *Continuous price discounting.* To encourage buyers to place 100-unit orders, your firm's sales department applies a continuous discount that makes the unit price a function $p(x)$ of the number of units x ordered. The discount decreases the price at the rate of $0.01 per unit ordered. The price per unit for a 100-unit order is $p(100) = \$20.09$.

a) Find $p(x)$ by solving the following initial value problem:

Differential equation: $\dfrac{dp}{dx} = -\dfrac{1}{100}p,$

Initial condition: $p(100) = 20.09.$

b) Find the unit price $p(10)$ for a 10-unit order and the unit price $p(90)$ for a 90-unit order.

c) The sales department has asked you to find out whether it is discounting so much that the firm's revenue, $r(x) = x \cdot p(x)$, will actually be less for a 100-unit order than, say, for a 90-unit order. Reassure them by showing that r has its maximum value at $x = 100$.

d) Graph the revenue function $r(x) = xp(x)$ for $0 \le x \le 200$.

94. *Subindices of the Consumer Price Index.* There are many "subindices" of the Consumer Price Index—separate indices for food, rent, medical care, and so on. Each index has its own inflation rate. From March 1987 to March 1988, the costs of food, rent, and medical care rose at 3%, 5%, and 6.4%, respectively. At these rates,

a) how long does it take for costs to increase 50%?

b) how long does it take for costs to double? (Use the Rule of 70.)

Variable Inflation Rates

When the inflation rate k varies with time instead of being constant, the formula $p = p_0 e^{kt}$ in Eq. (11) no longer gives the solution of the equation $dp/dt = kp$. The corrected formula is

$$p(t) = p_0 e^{\int_0^t k(\tau)\, d\tau} \qquad (12)$$

Use this formula to answer the questions in Exercises 95 and 96.

95. Suppose that $p_0 = 100$ and that $k(t) = 0.04/(1+t)$, a rate that starts at 4% when $t = 0$ but decreases steadily as the years pass.

a) Find $\int_0^9 k(\tau)\, d\tau$.

b) Use Eq. (12) to find the value of $p(t)$ when $t = 9$ yr.

c) Determine $p(9)$, assuming that the inflation rate was the (constant) average of $k(\tau)$ for the 9 years. Compare your solutions in parts (b) and (c), and discuss.

96. Suppose that $p_0 = 100$ and that the inflation rate is $k(t) = 1 + 1.3t$ (as it was in Brazil during the first few months of 1987).

a) What does the formula in Eq. (12) look like in this case?

b) Find $p(1)$, $p(2)$, and the 1-yr and 2-yr percentage increases in the associated Consumer Price Index.

97. The population of rabbits in a certain area is given by

$$P(t) = \frac{1000}{1 + e^{4.8 - 0.7t}},$$

where t is the number of months after a few rabbits are released.

a) Estimate the initial number of rabbits released.

b) What is the maximum possible number of rabbits that this area can sustain?

c) When, if ever, will the number of rabbits be 700? 1200? Give reasons for your answers.

d) When will the rate at which the rabbits reproduce be a maximum? What is this rate?

98. The spread of measles in a certain school is given by

$$P(t) = \frac{200}{1 + e^{5.3-t}},$$

where t is the number of days after students are exposed to infected students.

a) Estimate the initial number of students infected with measles.

b) What is the maximum number of students who will get the measles?

c) Estimate the number of days it will take for 150 students

to become infected.

d) When will the rate of spread of measles be at a maximum? What is this rate?

99. Let $f(t) = e^{-0.7t^2}$ and $F(x) = \int_0^x f(t)\,dt$.

a) Estimate $D_x F^{-1}(1)$ using only a value of f and the graph of $y = F(x)$.

b) Support your answer by graphing (in parametric mode) $y = F(x)$ and its inverse, and then estimating the slope of the tangent line to $y = F^{-1}(x)$ at $x = 1$.

7.3

Other Exponential and Logarithmic Functions

While we have not yet devised a way to raise positive numbers to any but rational powers, we have an exception in the number e. The definition $e^x = \ln^{-1} x$ defines e^x for every real value of x, irrational as well as rational. In this section, we show how this good fortune enables us to raise any other positive number to an arbitrary power and thus to define an exponential function $y = a^x$ for any positive number a. We also prove the power rule for differentiation in its final form (good for all exponents) and define and graph functions like x^x and $(\sin x)^{\tan x}$ that involve raising the values of one function to powers given by another.

Just as e^x is but one of many exponential functions, $\ln x$ is one of many logarithmic functions, the others being the inverses of the functions a^x. These logarithmic functions have important applications in science and engineering.

The Function a^x

If a is a positive number and x is any number whatever, we define the function a^x ("a to the x") by the equation

$$a^x = e^{x \ln a}.$$

It is natural to think of $e^{x \ln a}$ as being the same as $(e^{\ln a})^x$, and this, in turn, is $(a)^x$ because $e^{\ln a} = a$.

EXPLORATION BITS

1. Compute and compare 2^3 and $e^{3 \ln 2}$.

2. GRAPH and compare $y_1 = a^x$ and $y_2 = e^{x \ln a}$ for your choice of $a > 0$.

3. Consult your *Owner's Guide* to find whether your grapher has "list" capabilities. If it does, do part 2 for several values of a at one time by replacing a with a list of different values for a.

> **DEFINITION**   **The Function $y = a^x$**
>
> If a is a positive number, then $a^x = e^{x \ln a}$.   (1)

This definition allows us to fill in some of the gaps in our development. The number x^n can now be defined for any positive number x and any real number n:

$$x^n = e^{n \ln x}. \qquad \text{Eq. (1) with } a = x \text{ and } x = n$$

Thus, $\ln x^n = \ln (e^{n \ln x})$

$$= n \ln x, \qquad \begin{array}{l} \ln e^u = u \text{ for any } u, \text{ in} \\ \text{particular for } u = n \ln x. \end{array}$$

**RULES OF EXPONENTS
($a > 0$, ANY x AND y)**

1. $a^x \cdot a^y = a^{x+y}$

2. $\dfrac{a^x}{a^y} = a^{x-y}$

2a. $a^{-x} = \dfrac{1}{a^x}$

3. $(a^x)^y = a^{(xy)} = (a^y)^x$

and the n in the equation $\ln x^n = n \ln x$ no longer has to be a rational number. It can be any real number, as long as $x > 0$.

The exponential function a^x obeys all the standard rules of exponents as listed in the margin. (We omit the proofs.)

The Power Rule for Differentiation—Final Form

Rule 2, together with the definition of x^n as $e^{n \ln x}$, enables us to prove the Power Rule for differentiation in its final form. Differentiating x^n with respect to x gives

$$\frac{d}{dx}x^n = \frac{d}{dx}e^{n \ln x} \qquad \text{Definition of } x^n$$

$$= e^{n \ln x} \cdot \frac{d}{dx}(n \ln x) \qquad \text{Chain Rule for } e^u$$

$$= x^n \cdot \frac{n}{x} \qquad \begin{array}{l}\text{The definition again, and} \\ \text{differentiating } n \ln x\end{array}$$

$$= n x^{n-1}. \qquad \text{Rule 2 of Exponents}$$

In short, as long as $x > 0$,

$$\frac{d}{dx}x^n = n x^{n-1}. \tag{2}$$

The Chain Rule extends Eq. (2) to the Power Rule's final form.

POWER RULE (FINAL FORM)

If u is a positive differentiable function of x and n is any real number, then u^n is a differentiable function of x, and

$$\frac{d}{dx}u^n = nu^{n-1}\frac{du}{dx}.$$

EXAMPLE 1

SUPPORT

For each function f in Example 1, compare NDER f and df/dx.

a) $\qquad \dfrac{d}{dx}x^{\sqrt{3}} = \sqrt{3}x^{\sqrt{3}-1} \qquad\qquad (x > 0)$

b) $\qquad \dfrac{d}{dx}(\sin x)^{\pi} = \pi(\sin x)^{\pi-1}\cos x \qquad (\sin x > 0)$ ≡

EXPLORATION 1

Support and Confirm

1. Draw a complete graph of $f(x) = x^{\sqrt{3}}$, $x > 0$.

2. Support that your graph of f is complete by graphing NDER f and NDER2 f and explaining what they suggest about f.

3. Confirm analytically that your graph of f is complete. ⅋

The Derivatives of a^x and a^u

We know that for a *power function* $f(x) = x^a (x > 0)$, $f'(x) = ax^{a-1}$. Now, what about the *exponential function* $f(x) = a^x (a > 0, a \neq 1)$?

EXPLORATION 2

Exploring the Derivative of $f(x) = a^x$

Let $y_1 = a^x$ and $y_2 = \text{NDER } y_1$.

1. Store 2 into a. GRAPH y_1 and y_2 in $[-5, 5]$ by $[-2, 10]$. Use TRACE to compare $y_1(x)$ and $y_2(x)$ for several values of x.

2. Let $y_3 = y_2/y_1$, $y_4 = e^{y_3}$, and GRAPH y_4.

3. Repeat part 2 replacing a by several other positive numbers. Make a conjecture about y_4.

4. Use your conjecture in part 3 and properties of logarithms to express y_2 in terms of y_1.

We can find the derivative of a^x the way we found the derivative of x^n, starting with the definition $a^x = e^{x \ln a}$:

$$\frac{d}{dx}a^x = \frac{d}{dx}e^{x \ln a}$$

$$= e^{x \ln a} \cdot \frac{d}{dx}(x \ln a) \qquad \text{Chain Rule}$$

$$= a^x \ln a.$$

SUPPORT

Do your conclusions in Exploration 2 support Equation (3)? Now that you know this result, how can you use your grapher to support it?

EXPONENTIAL RULE

If $a > 0$, then

$$\frac{d}{dx}a^x = a^x \ln a. \qquad (3)$$

EXAMPLE 2

$$\frac{d}{dx}3^x = 3^x \ln 3$$

Equation (3) shows why the function e^x is the exponential function preferred in calculus. If $a = e$, then $\ln a = 1$, and Eq. (3) simplifies to

$$\frac{d}{dx}e^x = e^x.$$

Equation (3) also comes in a more general form, based on the Chain Rule. This is shown on the following page.

EXPLORATION BITS

The derivative of a^u can always be found by logarithmic differentiation (Section 7.1), so it is not necessary that we memorize Eq. (4). However, logarithmic differentiation *and* Eq. (4) give us two ways to find a derivative, so one can serve as analytic support of the other, and we can give our graphers a rest! For each function in parts 1 and 2 below, find the derivative in two ways. First, use logarithmic differentiation; second, use Eq. (4). The results should be identical.

1. $y = 3^{-x}$

2. $y = 3^{\sin x}$

3. A third method is to let, in part 1, for example, $3^{-x} = e^{-x \ln 3}$ and then find $D_x(e^{-x \ln 3})$. Try it. (Some people like to follow the rule, "If it's not in base e, get it there.")

EXPONENTIAL RULE (FINAL FORM)

If $a > 0$ and u is a differentiable function of x, then a^u is a differentiable function of x, and

$$\frac{d}{dx}a^u = a^u \ln a \, \frac{du}{dx}. \tag{4}$$

The Complete Graph of a^x

From the formula

$$\frac{d}{dx}a^x = a^x \ln a,$$

we see that the derivative of a^x is positive if $a > 1$ and negative if $0 < a < 1$. Thus, a^x is an increasing function of x if $a > 1$ and a decreasing function of x if $0 < a < 1$. In either case, a^x is one-to-one. The second derivative,

$$\frac{d^2}{dx^2}a^x = \frac{d}{dx}(a^x \ln a) = \ln a \frac{d}{dx}(a^x)$$

$$= \ln a(a^x \ln a) = (\ln a)^2 a^x,$$

is always positive, so the graph is concave up. This confirms the information given in Section 4.5 about complete graphs of $y = a^x$ resembling the graph of $y = 2^x$ if $a > 1$, and the graph of $y = (1/2)^x$ if $0 < a < 1$. Figure 7.12 shows the graphs of $y = 2^x$ (increasing and one-to-one) and $y = (1/2)^x$ (decreasing and one-to-one).

The Integral of a^u

If $a \neq 1$ so that $\ln a \neq 0$, we can divide both sides of Eq. (4) by $\ln a$ to obtain

$$a^u \frac{du}{dx} = \frac{1}{\ln a} \cdot \frac{d}{dx}(a^u).$$

Integrating with respect to x then gives

$$\int a^u \frac{du}{dx} \, dx = \int \frac{1}{\ln a} \cdot \frac{d}{dx}(a^u) \, dx = \frac{1}{\ln a} \int \frac{d}{dx}(a^u) \, dx = \frac{1}{\ln a}a^u + C.$$

Writing the first integral in differential form then gives

$$\int a^u du = \frac{a^u}{\ln a} + C. \tag{5}$$

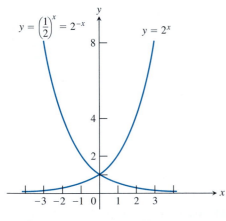

7.12 The graphs of $y = 2^x$ and $y = (1/2)^x$. Both functions are one-to-one. As $x \to \infty$, $2^x \to \infty$ and $(1/2)^x \to 0$. As $x \to -\infty$, $2^x \to 0$ and $(1/2)^x \to \infty$.

EXAMPLE 3

$$\int 2^x dx = \frac{2^x}{\ln 2} + C \qquad \text{Eq. (5) with } a = 2, u = x$$

RULES OF ARITHMETIC FOR BASE a LOGARITHMS

1. $\log_a uv = \log_a u + \log_a v$

2. $\log_a \dfrac{u}{v} = \log_a u - \log_a v$

3. $\log_a u^n = n \log_a u$

CHANGE OF BASE FORMULAS (SECTION 1.6)

$$\log_b x = \frac{\log_a x}{\log_a b}$$

$$\log_b x = \frac{\ln x}{\ln b}$$

EXAMPLE 4

$$\int 2^{\sin x} \cos x \, dx = \int 2^u du = \frac{2^u}{\ln 2} + C = \frac{2^{\sin x}}{\ln 2} + C \qquad u = \sin x \text{ in Eq. (5).}$$

≡

Base a Logarithms and Their Graphs

As we saw earlier, if a is any positive number other than 1, the function a^x is one-to-one and has a nonzero derivative at every point. It therefore has a differentiable inverse. We call the inverse the **base a logarithm of x** and denote it by $\log_a x$.

> **DEFINITION**
>
> The function $y = \log_a x$ is the inverse of the function $y = a^x (a > 0, a \neq 1)$.

This definition for $\log_a x$ is the same as the one given in Section 1.6. There we had to assume understanding about the function $y = a^x$. Now we have given formal meaning to a^x and have analyzed it fairly completely. Consequently, we know much about $\log_a x$. First, because the logarithm and exponential are inverses of one another, their composites in either order give the identity function.

$$\log_a a^x = x \quad \text{(all } x) \qquad a^{\log_a x} = x \quad (x > 0).$$

The rules of arithmetic and the change of base formulas hold for base a logarithms (see margin note above). A complete graph of $\log_a x$ is obtained by reflecting the graph of a^x across the line $y = x$. Thus, the graph of $\log_a x$ resembles the graph of $\log_2 x$ if $a > 1$ and the graph of $\log_{0.5} x$ if $0 < a < 1$ (Figs. 7.13 and 7.14).

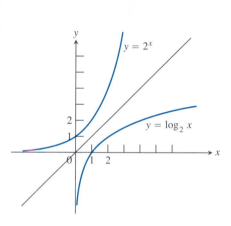

7.13 The graph of 2^x and its inverse, $\log_2 x$.
As $x \to \infty, \log_2 x \to \infty$.
As $x \to 0^+, \log_2 x \to -\infty$.

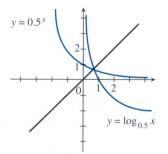

7.14 The graph of $y = 0.5^x$ and its inverse $\log_{0.5} x$.
As $x \to \infty, \log_{0.5} x \to -\infty$.
As $x \to 0^+, \log_{0.5} x \to \infty$.

 ## EXPLORATION 3

Viewing the Function $y = \log_a x$

Your graphing utility may allow you to exponentiate directly using any base, but it probably allows you to find logarithms directly for just two bases, e and 10. Base 10 logarithms are a throwback to the days when logs (short for logarithms) were used to ease pencil and paper computation as calculators are used today. They were easier to work with than base e logarithms because of a natural relationship to our base 10 number system.

1. One way to support the result in Example 3 is to let

$$y_1 = 2^x,$$

$$y_2 = \text{NINT } y_1,$$

$$y_3 = 2^x / \ln 2,$$

and then compare the graphs of y_2 and y_3. Try it.

2. For Example 4, let

$$y_1 = 2^{\sin x} \cos x,$$

$$y_2 = \text{NINT } y_1,$$

$$y_3 = 2^{\sin x} / \ln 2,$$

and compare the graphs of y_2 and y_3.

3. In each of the above, both y_2 and y_3 correspond to specific antiderivatives of the integrand, so they differ by a constant C. Do the graphs appear to differ by a constant? Can you find C in each case? Also, can you find a in each case so that NINT $(y_1, a, x) = y_3$?

If a graphing utility provides only for base e (and base 10), there are two convenient ways to graph $\log_a x$ for any positive a (other than 1).

1. Show how to graph $\log_a x$ using the change of base formula and base e.
2. Show how to graph $\log_a x$ by defining parametric equations $(x_1(t), y_1(t))$ and then letting $(x_2(t), y_2(t)) = (y_1(t), x_1(t))$. Explain.
3. Recall how to stretch, shrink, and reflect a graph. Confirm that the graph of $\log_a x$ is a stretch or shrink of the graph of $\ln x$, sometimes followed by a reflection. Analyze the role of a as it relates to the type of transformation.
4. Confirm a similar relationship (see part 3) between the graphs of $y = a^x$ and $y = e^x$.

The Derivative of $\log_a x$

To find the derivative of $\log_a x$, we first convert it to a natural logarithm:

$$\frac{d}{dx}(\log_a x) = \frac{d}{dx}\left(\frac{\ln x}{\ln a}\right) = \frac{1}{\ln a}\frac{d}{dx}\ln x = \frac{1}{\ln a} \cdot \frac{1}{x}.$$

If $a > 0, a \neq 1$, then

$$\frac{d}{dx}(\log_a x) = \frac{1}{\ln a} \cdot \frac{1}{x}. \tag{6}$$

Equation (6) extends to a more general form:

If $a > 0, a \neq 1$, and u is a positive differentiable function of x, then $\log_a u$ is a differentiable function of x, and

$$\frac{d}{dx}\log_a u = \frac{1}{\ln a} \cdot \frac{1}{u} \cdot \frac{du}{dx}.$$

For the function $y = \log_{10}(3x + 1)$ in Example 5, compare NDER y and dy/dx.

EXAMPLE 5

$$\frac{d}{dx}\log_{10}(3x + 1) = \frac{1}{\ln 10} \cdot \frac{1}{3x + 1} \cdot \frac{d}{dx}(3x + 1) = \frac{3}{(\ln 10)(3x + 1)} \quad \blacksquare$$

Integrals Involving $\log_a x$

To evaluate integrals involving base a logarithms, we convert them to natural logarithms.

EXPLORATION BIT

For Example 6, let

$$y_1 = \frac{\ln x}{x \ln 2},$$

$$y_2 = \text{NINT} y_1,$$

$$y_3 = \frac{\ln 2}{2} (\log_2 x)^2,$$

$$y_4 = \text{NDER} y_3.$$

Which *two* pairs of graphs could you view to support the result of Example 6? Explain. Which comparison do you prefer? Why?

EXPLORATION BIT

Graph $y = \log_{10} x$ in the viewing windows

$[0, 10]$ by $[0, 1]$,

$[0, 100]$ by $[0, 2]$,

$[0, 1000]$ by $[0, 3]$,

$[0, 10{,}000]$ by $[0, 4]$,

$\vdots$

$[0, 10^{10}]$ by $[0, 10]$.

What do you observe? Describe the behavior of logarithmic growth, that is, how changes in x compare to changes in y.

EXPLORATION BIT

Small jumps on the Richter scale sometimes mean the difference between a gentle tremor and a "killer quake." To help you understand why, do the following.

a) Find the value of a that will increase R from 7.8 to 8.8 in Example 7, and interpret your finding.

b) Show how R behaves graphically using your grapher, and interpret your graph(s).

EXAMPLE 6

$$\int \frac{\log_2 x}{x} \, dx = \frac{1}{\ln 2} \int \frac{\ln x}{x} \, dx \qquad \log_2 x = \frac{\ln x}{\ln 2}.$$

$$= \frac{1}{\ln 2} \int u \, du \qquad \text{Let } u = \ln x, \, du = \frac{1}{x} \, dx.$$

$$= \frac{1}{\ln 2} \frac{u^2}{2} + C = \frac{1}{\ln 2} \frac{(\ln x)^2}{2} + C$$

$$= \frac{\ln 2}{2} \frac{(\ln x)^2}{(\ln 2)^2} + C = \frac{\ln 2}{2} (\log_2 x)^2 + C \qquad \equiv$$

Base 10 Logarithms

If $\log_{10} x = y$, then $x = 10^y$. Thus a base 10 logarithm y tells us something about the size of x, namely that $10^{[y]} \le x < 10^{[y]+1}$ where $[y]$ is the greatest integer less than or equal to y. In other words, it tells us whether x is between 10^0 and 10^1 (1 and 10), 10^1 and 10^2 (10 and 100), 10^2 and 10^3 (100 and 1000), and so on, or between 10^{-1} and 10^0 (0.1 and 1), 10^{-2} and 10^{-1} (0.01 and 0.1), and so on.

Base 10 logarithms, often called common logarithms, appear in many scientific formulas. For example, earthquake intensity is often reported on the *Richter scale*. Here the formula is

$$\text{Magnitude } R = \log_{10}\left(\frac{a}{T}\right) + B,$$

where a is the amplitude of the ground motion in microns at the receiving station, T is the period of the seismic wave in seconds, and B is an empirical quantity that allows for the weakening of the seismic wave with increasing distance from the epicenter of the earthquake.

EXAMPLE 7

For an earthquake 10,000 km from the receiving station, $B = 6.8$. If the recorded vertical ground motion is $a = 10$ microns and the period is $T = 1$ second, the earthquake's magnitude is

$$R = \log_{10}\left(\frac{10}{1}\right) + 6.8 = 1 + 6.8 = 7.8.$$

An earthquake of this magnitude does great damage near its epicenter. $\equiv$

The pH scale for measuring the acidity of a solution is a base 10 logarithmic scale. The pH value (hydrogen potential) of the solution is the common logarithm of the reciprocal of the solution's hydronium ion concentration, $[H_3O^+]$:

$$\text{pH} = \log_{10} \frac{1}{[H_3O^+]} = -\log_{10}[H_3O^+].$$

The hydronium ion concentration is measured in moles per liter. Vinegar has a pH of 3, distilled water a pH of 7, sea water a pH of 8.15 and household ammonia a pH of 12. The total scale ranges from about 0.1 for normal hydrochloric acid to 14 for a normal solution of sodium hydroxide.

Another example of the use of common logarithms is the dB ("dee bee") scale for measuring loudness in decibels. If I is the intensity of sound in watts per square meter, the decibel level of the sound is

$$\text{Sound level} = 10 \log_{10}(I \times 10^{12}) \quad \text{dB.} \tag{7}$$

If you ever wondered why doubling the power of your audio amplifier increased the sound level by only a few decibels, Eq. (7) provides the answer. As the following calculation shows, doubling I adds only about 3 dB.

EXAMPLE 8

Doubling I in Eq. (7) adds about 3 dB: Writing log for $\log_{10}$ (a common practice), we have

$$
\begin{aligned}
\text{Sound level with } I \text{ doubled} &= 10 \log(2I \times 10^{12}) &&\quad \text{Eq. (7) with} \\
&&&\quad 2I \text{ for } I \\
&= 10 \log(2 \cdot I \times 10^{12}) \\
&= 10 \log 2 + 10 \log(I \times 10^{12}) \\
&= \text{original sound level} + 10 \log 2 \\
&\approx \text{original sound level} + 3 &&\quad \log_{10} 2 \approx 0.30.
\end{aligned}
$$

TYPICAL ACIDITIES OF FOODS (MOST FOODS ARE ACIDIC WITH pH < 7).	
Food	pH value
Bananas	4.5–4.7
Grapefruit	3.0–3.3
Oranges	3.0–4.0
Limes	1.8–2.0
Milk	6.3–6.6
Soft drinks	2.0–4.0
Spinach	5.1–5.7

TYPICAL SOUND LEVELS	
Threshold of hearing	0 dB
Rustle of leaves	10 dB
Average whisper	20 dB
Quiet automobile	50 dB
Ordinary conversation	65 dB
Pneumatic drill 10 feet away	90 dB
Threshold of pain	120 dB

EXPLORATION BIT

What sound intensity I gives 0 dB, the threshold of hearing? (You can compute it from Eq. (7).) What sound intensity I gives 120 dB, the threshold of pain? How many times greater intensity is the threshold of pain than the threshold of hearing? This wide range of intensities to which the ear is sensitive suggests why a view of the decibel function is not too informative. However, you may wish to have a look at it in the $[10^{-12}, 1]$ by $[0, 150]$ viewing window.

Extending the Power and Exponential Functions

The ability to raise positive numbers to arbitrary real powers makes it possible to define functions like x^x and $x^{\ln x}$ for $x > 0$. A graphing utility makes their graphs readily accessible. And being able to find derivatives by using logarithmic differentiation gives us the ability to confirm that the graphs are complete and thus gain a reasonably good understanding of each function's behavior.

EXAMPLE 9

Draw a complete graph of $y = x^x$ for $x > 0$.

Solution Figure 7.15(a) shows a complete graph of $y_1 = x^x$. Figures 7.15(b) and 7.15(c) show graphs of $y_2 = \text{NDER } y_1$ and $y_3 = \text{NDER2 } y_1$, respectively. The graph of NDER y_1 crosses the x-axis from negative to positive at $x = 0.368$, suggesting that y_1 is decreasing on $(0, 0.368)$ and increasing on $(0.368, \infty)$. The graph of y_1 appears to reach a minimum at $x = 0.368$, the point where $y_2 = 0$. The graph of y_3 appears to be positive everywhere, supporting a minimum value of y_1 at 0.368 and upward concavity of y_1 over its entire domain.

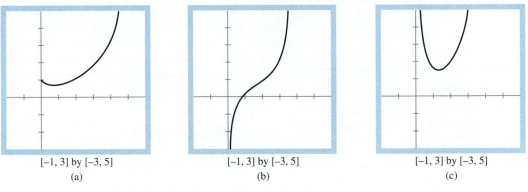

[–1, 3] by [–3, 5]
(a)

[–1, 3] by [–3, 5]
(b)

[–1, 3] by [–3, 5]
(c)

7.15 Complete graphs of (a) $y_1 = x^x, x > 0$, (b) $y_2 = $ NDER y_1, and (c) $y_3 = $ NDER2 y_1.

To confirm this information, we first note that with $x > 0$, $y = x^x$ is positive. Then we find y' and y'' using logarithmic differentiation:

$$y = x^x$$

$$\ln y = \ln x^x = x \ln x$$

$$\frac{1}{y}y' = x \cdot \frac{1}{x} + 1 \cdot \ln x = 1 + \ln x$$

$$y' = y(1 + \ln x) = (1 + \ln x)x^x$$

Thus,

$$y'' = \frac{1}{x}x^x + (1 + \ln x)^2 x^x,$$

using the Product Rule and the formulas for y and y'.

Because $x^x > 0$, it can easily be argued that $y'' > 0$ when $x > 0$. This confirms that the graph of y is concave up in its domain $(0, \infty)$ and that y has a local minimum where $y' = 0$; namely, where

$$(1 + \ln x)x^x = 0$$

$$1 + \ln x = 0$$

$$\ln x = -1$$

$$x = e^{-1} = \frac{1}{e} = 0.368.$$

So f has a local minimum value of 0.692 at $x = e^{-1} = 0.368$. ≡

In the Section 7.5 exercises, you will show that $f(x) = x^x$ has a removable discontinuity at $x = 0$ and $\lim_{x \to 0^+} f'(x) = -\infty$ as suggested by Fig. 7.15(b). Also we note that the function x^x takes on real values for some negative values of x. For example,

$$(-3)^{-3} = \frac{1}{(-3)^3} = -\frac{1}{27}.$$

The value of x^x is real for any negative integer. It is also negative for any negative rational number p/q, where p/q is in reduced form and q is odd. However, x^x has no real value for many negative values of x. For this reason, we will assume that $f(x) > 0$ when we study functions of the form $f(x)^{q(x)}$. That is, the understood domain will be the values of x for which $f(x) > 0$.

EXPLORATION BIT

Demonstrate that x^x is real when x is any negative integer. Demonstrate that x^x is real when x is any negative rational number p/q, where p/q is in reduced form and q is odd. Demonstrate the difficulty that arises for some rational numbers that do not meet this restriction.

EXPLORATION 4

Confirm Analytically

Let $y_1 = x^x$.

1. GRAPH y_1 so that some points are graphed for $x < 0$. (This requires that you set $[x\text{Min}, x\text{Max}]$ in a certain way, as suggested by Exercises 35–40 of Section 1.1.)

   Draw a complete graph of the functions g and h. Use what you know about the complete graph of y_1 to support the completeness of the graphs of g and h. Then confirm analytically that your graphs are complete.

2. $g(x) = |x|^x$

3. $h(x) = |x|^{|x|}$

Exercises 7.3

In Exercises 1–24, find dy/dx.

1. $y = x^\pi$

2. $y = x^{1+\sqrt{2}}$

3. $y = x^{-\sqrt{2}}$

4. $y = x^{1-e}$

5. $y = 8^x$

6. $y = 9^{-x}$

7. $y = 3^{\csc x}$

8. $y = 3^{\cot x}$

9. $y = x^{\ln x}, x > 0$

10. $y = x^{(1/\ln x)}$

11. $y = (x + 1)^x$

12. $y = (x + 2)^{x+2}$

13. $y = x^{\sin x}$

14. $y = (\sin x)^{\tan x}$

15. $y = \log_4 x^2$

16. $y = \log_5 \sqrt{x}$

17. $y = \log_2(3x + 1)$

18. $y = \log_{10} \sqrt{x + 1}$

19. $y = \log_2(1/x)$

20. $y = 1/\log_2 x$

21. $y = \ln 2 \cdot \log_2 x$

22. $y = \log_3(1 + x \ln 3)$

23. $y = \log_{10} e^x$

24. $y = \ln 10^x$

Evaluate the integrals in Exercises 25–40 analytically and with an NINT computation.

25. $\displaystyle\int_0^1 3x^{\sqrt{3}}\, dx$

26. $\displaystyle\int_0^1 x^{\sqrt{2}}\, dx$

27. $\displaystyle\int_0^1 5^x\, dx$

28. $\displaystyle\int_1^e x^{\ln 2 - 1}\, dx$

29. $\displaystyle\int_0^1 \frac{1}{2^x}\, dx$

30. $\displaystyle\int_{-1}^1 2^{(x+1)}\, dx$

31. $\displaystyle\int_{-1}^0 4^{-x} \ln 2\, dx$

32. $\displaystyle\int_{-2}^0 5^{-x}\, dx$

33. $\displaystyle\int_1^{\sqrt{2}} x 2^{x^2}\, dx$

34. $\displaystyle\int_0^{\pi/2} 2^{\cos x} \sin x\, dx$

35. $\displaystyle\int_1^{10} \frac{\log_{10} x}{x}\, dx$

36. $\displaystyle\int_1^4 \frac{\log_2 x}{x}\, dx$

37. $\displaystyle\int_0^2 \frac{\log_2(x + 2)}{x + 2}\, dx$

38. $\displaystyle\int_{1/10}^{10} \frac{\log_{10}(10x)}{x}\, dx$

39. $\displaystyle\int_0^9 \frac{2\log_{10}(x + 1)}{x + 1}\, dx$

40. $\displaystyle\int_2^3 \frac{2\log_2(x - 1)}{x - 1}\, dx$

Evaluate the integrals in Exercises 41–44 analytically.

41. $\displaystyle\int 2^{\sin x}\cos x\,dx$

42. $\displaystyle\int \frac{3^{\sqrt{x}}\,dx}{\sqrt{x}}$

43. $\displaystyle\int \frac{\log_3(x-2)}{x-2}\,dx$

44. $\displaystyle\int \frac{\log_5(2x-1)}{1-2x}\,dx$

Draw a complete graph of the functions in Exercises 45–52. Find the local extrema and inflection points and identify the intervals on which the graphs are rising, falling, concave up, and concave down. Show just one period if the function is periodic.

45. $y = x^{-\sqrt{3}}$

46. $y = x^{\sqrt{7}}$

47. $y = x^{\sqrt{x}}$

48. $y = x^{\ln x}$

49. $y = 2^{\sec x}$

50. $y = 2^{\tan x}$

51. $y = \log_7 \sin x$

52. $y = \log_5 (x+1)^2$

Evaluate the integrals in Exercises 53–56.

53. $\displaystyle\int_1^2 2^{x^2}dx$

54. $\displaystyle\int_0^{\pi/2} 2^{\cos x}dx$

55. $\displaystyle\int_1^3 x^{\ln x}dx$

56. $\displaystyle\int_0^3 x^x dx$

Solve the equations in Exercises 57 and 58 for x.

57. $3^{\log_3 7} + 2^{\log_2 5} = 5^{\log_5 x}$

58. $8^{\log_8 3} - e^{\ln 5} = x^2 - 7^{\log_7 3x}$

Solve the systems of equations in Exercises 59 and 60.

59. $y = x^2$
$y = 2^x$

60. $y = x^{10}$
$y = 10^x$

Find the limits in Exercises 61–64.

61. a) $\displaystyle\lim_{x\to\infty} \log_2 x$

b) $\displaystyle\lim_{x\to\infty} \log_2 (1/x)$

62. a) $\displaystyle\lim_{x\to 0^+} \log_{10} x$

b) $\displaystyle\lim_{x\to 0^+} \log_{10} (1/x)$

63. a) $\displaystyle\lim_{x\to\infty} 3^x$

b) $\displaystyle\lim_{x\to\infty} 3^{-x}$

64. a) $\displaystyle\lim_{x\to -\infty} 3^x$

b) $\displaystyle\lim_{x\to -\infty} 3^{-x}$

Express the ratios in Exercises 65 and 66 as ratios of natural logarithms. Then find the limit of each ratio as $x \to \infty$.

65. a) $\dfrac{\ln x}{\log x}$ See Section 2.4, Exercise 66(d).

b) $\dfrac{\log_2 x}{\log_3 x}$

66. a) $\dfrac{\log_9 x}{\log_3 x}$

b) $\dfrac{\log_{\sqrt{10}} x}{\log_{\sqrt{2}} x}$

Draw a complete graph of the functions in Exercises 67 and 68 in the interval $(0, 6\pi]$.

67. $y = x^{\sin x}$

68. $y = x^{\cos x}$

69. Find the points of intersection and investigate the relationship among the graphs of the following functions for $x > 0$.

a) $y = x$, $y = x^2$, $y = x^{\sqrt{3}}$

b) $y = x^{1.7}$, $y = x^{1.8}$, $y = x^{\sqrt{3}}$

c) $y = x^{1.73}$, $y = x^{1.74}$, $y = x^{\sqrt{3}}$

70. Find dy/dx if

a) $y = 2^{\ln x}$

b) $y = \ln 2^x$

c) $y = \ln x^2$

d) $y = (\ln x)^2$

71. a) Assume that f and g are increasing functions. Show that the composition, $f \circ g$, is an increasing function. (See Section 7.1, Exercise 61.)

b) Use part (a) to show that $f(x) = x^{\sqrt{3}} = e^{\sqrt{3}\ln x}$ is an increasing function on $x > 0$.

72. Show that $\dfrac{d}{dx}(a^x) = a^x$ if and only if $a = e$.

73. *Blood pH.* The pH of human blood normally falls between 7.37 and 7.44. Find the corresponding bounds for the solution's hydronium ion concentration, $[H_3O^+]$.

74. *Brain fluid pH.* The cerebrospinal fluid in the brain has a hydronium ion concentration of about $[H_3O^+] = 4.8 \times 10^{-8}$ moles per liter. What is the pH?

75. *Audio amplifiers.* By what factor k do you have to multiply the intensity I of the sound from your audio amplifier to add 10 dB to the sound level?

76. *Conversion factors.* Show that an equation for converting base 10 logarithms to base 2 logarithms is

$$\log_2 x = \frac{\ln 10}{\ln 2}\log_{10} x.$$

77. *Continuation of Exercise 76.* Show that an equation for converting base a logarithms to base b logarithms is

$$\log_b x = \frac{\ln a}{\ln b}\log_a x.$$

78. (a) Pour warm liquid into a container. Place a thermometer designed to measure hot liquids in the container. Record the lapsed time (in minutes) and temperature at approximately 10-minute intervals until the liquid temperature stabilizes. This may take a few hours. The liquid will cool faster if you can leave it in a refrigerator. (b) Plot the ordered pairs (time, temperature) on a grapher, or use graph paper. (c) If the points in the graph from part (b) were connected with a smooth curve, would this curve look familiar? In Section 7.4 you will learn how to find an algebraic representation for the curve.

7.4 — The Law of Exponential Change Revisited

Suppose a quantity y that we are interested in—velocity, temperature, electrical current, whatever—grows or decreases at a rate that at any given time t is proportional to the amount present. If we also know the amount present at time $t = 0$, call it y_0, we can find y as a function of t by solving the following initial value problem:

$$\text{Differential equation:} \qquad \frac{dy}{dt} = ky \qquad\qquad (1)$$

$$\text{Initial condition:} \qquad y = y_0 \qquad \text{when } t = 0.$$

If y is positive and *increasing*, then k is positive, and Eq. (1) says that the *rate of growth* is proportional to the amount that has accumulated. If y is positive and *decreasing*, then k is negative, and Eq. (1) says that the *rate of decay* is proportional to the amount that remains.

The solution to the initial value problem, as shown in Section 7.2, is

$$y = y_0 e^{kt},$$

the Law of Exponential Change. (See Fig. 7.16.) In Section 7.2, we looked at numerous growth examples. Now we look at some important decay applications.

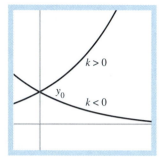

7.16   The Law of Exponential Change,

$$y = y_0 e^{kt},$$

models growth behavior when k is positive and decay behavior when k is negative. Note that when $t = 0$, y has the initial value y_0.

Radioactive Decay

When a radioactive atom loses some of its mass as radiation energy, the remainder of the atom reforms to make an atom of some new substance. This process of radiation and change is called **radioactive decay**, and an element whose atoms go through this process spontaneously is called **radioactive**. Thus, radioactive carbon-14 decays into nitrogen; radium, through a number of intervening radioactive steps, decays into lead.

Experiments have shown that at any given time, the rate at which a radioactive element decays (as measured by the number of nuclei that change per unit time) is approximately proportional to the number of radioactive nuclei present. Thus, the decay of a radioactive element is described by the equation $dy/dt = -ky$. If y_0 is the number of radioactive nuclei present at time zero, the number still present at any later time t will be $y = y_0 e^{-kt}$.

k AS A DECAY CONSTANT

In the radioactive decay equation, we use $-k$ (with $k > 0$) instead of simply k (with $k < 0$) to help remind us that y is decreasing. For radon gas, k is 0.18, and t is measured in days. For radium-226, which used to be painted on watch dials to make them glow at night (a dangerous practice), $k = 4.3 \times 10^{-2}$ and t is measured in years. (Incidentally, the decay of radium in Earth's crust is the source of the radon that we find in our basements.)

Radioactive Decay Equation

$$y = y_0 e^{-kt}$$

The **half-life** of a radioactive element is the time required for half of the radioactive nuclei present in a sample to decay. Because the rate of decay is proportional to the amount of the element present, the time it takes for *any* amount to decay by one-half is constant.

EXPLORATION 1

Radioactive Half-Life

Polonium-210 The effective radioactive lifetime of polonium-210 is so short that we measure it in days rather than years. The number of radioactive atoms remaining after t days in a sample that starts with y_0 radioactive atoms is

$$y = y_0 e^{-4.95 \times 10^{-3} t}.$$

GRAPH

$$y_1 = y_0 e^{-4.95 \times 10^{-3} t} \qquad \text{and} \qquad y_2 = \frac{1}{2} y_0$$

in the same viewing window for your choice of y_0 (Fig. 7.17). Find the value of t at which the graphs intersect. This is the half-life of polonium-210. What happens when you use other values of y_0? ☙

To confirm that the initial amount, y_0, present in a sample has no effect on a half-life, we seek the value of t at which

$$y_0 e^{-kt} = \frac{1}{2} y_0,$$

for this will be the time when the number of radioactive nuclei present equals half the original number. The y_0's cancel in this equation to give

$$e^{-kt} = \frac{1}{2}, \qquad -kt = \ln \frac{1}{2} = -\ln 2, \qquad \text{and} \qquad t = \frac{\ln 2}{k}.$$

This value of t is the half-life of the element. It depends only on the value of k. The fact that the y_0's cancel shows that the half-life does not depend on the initial amount present, y_0.

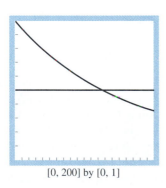

[0, 200] by [0, 1]

7.17 The graphs of the functions

$$y_1 = y_0 e^{-4.95 \times 10^{-3} t}$$

and

$$y_2 = \frac{1}{2} y_0$$

with $y_0 = 1$. The value of t where the graphs intersect is the half-life of polonium-210. We used SOLVE to find the half-life to be about 140 days. Does this agree with your result from Exploration 1?

$$\boxed{\text{Half-life } = \frac{\ln 2}{k} \qquad\qquad (2)}$$

EXAMPLE 1

Confirm that the half-life of polonium-210 is about 140 days. (See Fig. 7.17.)

Solution

$$\text{Half-life} = \frac{\ln 2}{k} \qquad\qquad \text{Eq. (2)}$$

$$= \frac{\ln 2}{4.95 \times 10^{-3}} \qquad\qquad \text{The } k \text{ from polonium's decay equation}$$

$$= 140 \text{ days} \qquad\qquad ▤$$

EXAMPLE 2   Carbon-14

People who do carbon-14 dating use a figure of 5700 years for its half-life (more about carbon-14 dating in the exercises). Find the age of a sample in which 10% of the radioactive nuclei originally present have decayed.

Solution   We use the decay equation $y = y_0 e^{-kt}$. There are two things to find:

1. The value of k

2. The value of t that makes $y_0 e^{-kt} = (0.9) y_0$, or $e^{-kt} = 0.9$

STEP 1: *The value of k:* We find it from the half-life equation:

$$k = \frac{\ln 2}{\text{Half-life}} \qquad \text{Eq. (2) turned around}$$

$$= \frac{\ln 2}{5700} = 1.2 \times 10^{-4}$$

STEP 2: *The value of t that makes $e^{-kt} = 0.9$:*

$$e^{-(\ln 2/5700)t} = 0.9$$

$$-\frac{\ln 2}{5700} t = \ln (0.9) \qquad \text{Logs of both sides}$$

$$t = \frac{-5700 \ln (0.9)}{\ln 2} = 866 \text{ years}$$

The sample is about 866 years old. (See Fig. 7.18.)   ▤

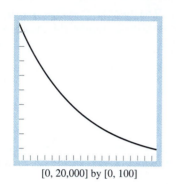

[0, 20,000] by [0, 100]

7.18   A plant contains the same proportion of carbon-14 as the atmosphere and stops taking it in when it dies. Thus, the proportion of carbon-14 in plant remains are a good indicator of how long ago the plant died. This viewing window shows a graph of the decay equation

$$y = y_0 e^{-1.2 \times 10^{-4} t}$$

for carbon-14 with $y_0 = 100$. The y-axis shows the percentage of the radioactive substance remaining. The x-axis (with x-scale set to 1000) shows the time carbon-14 took to decay to that percentage. Similar techniques are used with other elements for dating geologic specimens.

Resistance to a Moving Object

In some cases, it makes sense to assume that the resistance encountered by a moving object, such as a car coasting to a stop, is proportional to the velocity of the object's motion. The more slowly the object moves, the less its forward progress is resisted by the medium through which it passes. We can describe this in mathematical terms if we picture the object as a mass m moving along a coordinate line with position $s(t)$ and velocity $v(t)$ at time t. The magnitude of the resisting force (mass *times* acceleration) opposing the motion is then $m(dv/dt)$, and we can write

$$m\frac{dv}{dt} = -kv \qquad (k > 0) \tag{3}$$

to say that the force decreases in proportion to velocity. If we rewrite Eq. (3) as

$$\frac{dv}{dt} = -\frac{k}{m}v,$$

we can see that the solution is $v = v_0 e^{-(k/m)t}$, where v_0 is the initial velocity.

Resistance to a Moving Object

$$v = v_0 e^{-(k/m)t} \tag{4}$$

EXPLORATION 2

Slowing Down More Slowly

In Eqs. (3) and (4), the constant k depends on the shape of the object and the medium through which it moves. It is independent of other conditions. If there are two different masses of the same shape moving through the same medium, Eq. (3) tells us that the greater mass m will have a lesser speed $|dv/dt|$. That is, it will slow down more slowly.

1. GRAPH $v = 100e^{-(0.5/m)t}$ for $m = 1, 2, 4, 10$ in a [0, 10] by [0, 100] viewing window. Watch the viewing window to see how the increasing values of m cause an object to slow down more slowly.

2. For the v from part 1, GRAPH NINT v to "see" the distances traveled as the object slows down. Notice that as the mass increases, the total distance traveled becomes greater but, in each case, it never exceeds a certain value. (In other words, the object will travel only so far before stopping.)

From Eq. (4) and Exploration 2, we can see that if m is something large, like the mass of a 20,000-ton ore boat in Lake Erie, it will take a long time for the velocity to get near zero. Also we can integrate Eq. (4) to find s.

Suppose a body is coasting to a stop and the only force acting on it is a resistance proportional to its speed. How far will it coast? To find out, we start with Eq. (4) and solve the initial value problem

Differential equation: $\dfrac{ds}{dt} = v_0 e^{-(k/m)t}$,

Initial condition: $\quad s = 0 \quad$ when $t = 0$.

Integrating with respect to t gives

$$s = -\frac{v_0 m}{k} e^{-(k/m)t} + C.$$

Substituting $s = 0$ when $t = 0$ gives

$$0 = -\frac{v_0 m}{k} + C \qquad \text{and} \qquad C = \frac{v_0 m}{k}.$$

The body's position at time t is therefore

$$s(t) = -\frac{v_0 m}{k} e^{-(k/m)t} + \frac{v_0 m}{k} = \frac{v_0 m}{k}(1 - e^{-(k/m)t}). \tag{5}$$

To find how far the body will coast, we find the limit of $s(t)$ as $t \to \infty$. Since $\lim\limits_{t \to \infty} e^{-(k/m)t} = 0$, the limit of $s(t)$ is $v_0 m/k$:

$$\text{Distance coasted} = \frac{v_0 m}{k}. \tag{6}$$

This is an ideal figure, of course. Only in mathematics can time stretch to infinity. The number $v_0 m/k$ is only an upper bound (albeit a useful one). It is true to life in one respect, at least—if m is large, the body will coast a long way before it stops. That is why ocean liners have to be docked by tugboats. Any liner of conventional design entering a slip with enough speed to steer would smash into the pier before it could stop.

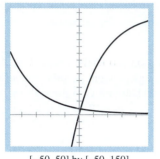

[−50, 50] by [−50, 150]

7.19 Complete graphs for Example 3 of $v(t)$ and $s(t)$. From what you know of v and s, which graph is which?

EXAMPLE 3

Jenny, with a mass of 50 kg, relaxes as she crosses the finish line of an ice skating race. She is traveling 7 m/sec and continues to coast in a straight line.

a) By experiment, it is known that $k = 2.5$ kg/sec. Write an equation for Jenny's velocity v in terms of time t since crossing the finish line.

b) Write an equation for Jenny's distance s from the finish line in terms of time t.

c) Draw complete graphs of $y_1 = v(t)$ and $y_2 = s(t)$.

d) What values of t make sense in the problem situation?

e) Draw complete graphs for the problem situation.

f) When will Jenny be coasting at 1 m/sec?

g) How far will Jenny have coasted in 3 sec?

h) When will Jenny have coasted 100 m from the finish line?

i) Explain when Jenny will come to a stop. Relate the explanation to the graphs of v and s.

Solution

a) From Eq. (4), $v(t) = 7e^{-(2.5/50)t}$.

b) $s(t) = $ NINT v or, from Eq. (5), $s(t) = \dfrac{7 \cdot 50}{2.5}(1 - e^{-(2.5/50)t})$.

c) See Fig. 7.19.

d) $t \geq 0$

e) See Fig. 7.20.

f) Using ZOOM-IN or SOLVE with the graph of $v(t)$ shows that $v = 1$ when $t = 38.9$. Jenny will be coasting at 1 m/sec in about 38.9 sec.

g) $s(3) = 19.5$. In 3 sec, Jenny will have coasted about 19.5 m.

h) Using ZOOM-IN or SOLVE with the graph of $s(t)$ shows that $s = 100$ when $t = 25.1$. Jenny will have coasted 100 m in about 25.1 sec.

i) The graph of v approaches 0 asymptotically while s approaches $v_0 m / k = 140$ asymptotically (from Eq. (6)). In a little over a minute, however, Jenny's speed will be about 0.3 m/sec. It is safe to assume that she is about to stop, if not having stopped already. Thus, Jenny will stop well short of 140 m in about a minute. ≡

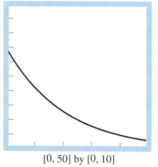

[0, 50] by [0, 10]
(a)

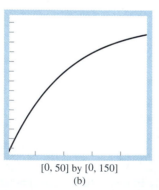

[0, 50] by [0, 150]
(b)

7.20 Complete graphs for Example 3 of (a) $v(t)$ and (b) $s(t)$ for $t \geq 0$.

EXPLORATION 3

Slowing Down to a Stop

With your grapher in parametric mode, let

$$x_1(t) = \frac{7 \cdot 50}{2.5}(1 - e^{-(2.5/50)t}), \ y_1(t) = 100.$$

1. GRAPH the parametric equations in a [0, 150] by [0, 150] viewing window with tMin = 0, tMax = 150, and t-step = 1. Watch the graph being drawn.

Tell how this simulates Jenny's motion (Example 3), including her velocity and acceleration.

2. Let

$$x_2(t) = t, \qquad y_2(t) = x_1(t).$$

Change to dot and simultaneous formats, and change t-step to 2. GRAPH $(x_1(t), y_1(t))$ and $(x_2(t), y_2(t))$. Explain what you see. Explain why the distances between the dots decrease in the graph of (x_1, y_1). Explain why the graph of (x_1, y_1) seems to end near $x = 140$ but the graph of (x_2, y_2) goes all the way across the viewing window. &

Heat Transfer—Newton's Law of Cooling

Hot chocolate left in a tin cup cools to the temperature of the surrounding air. A hot silver ingot immersed in water cools to the temperature of the surrounding water. In situations like these, the rate at which an object's temperature is changing at any given time is roughly proportional to the difference between its temperature and the temperature of the surrounding medium. This observation is called *Newton's Law of Cooling*, although it applies to warming as well. It can be written as an equation in the following way.

If $T(t)$ is the temperature of the object at time t and T_s is the surrounding temperature, then

$$\frac{dT}{dt} = -k(T - T_s). \tag{7}$$

Since T_s is constant, this is the same as $dy/dt = -ky$ with $y = (T - T_s)$. Hence, the solution of Eq. (7) is

$$y = y_0 e^{-kt}$$

or

$$T - T_s = (T_0 - T_s)e^{-kt},$$

where T_0 is the value of T at time zero.

FACTORS AFFECTING k

Here, k depends on the size, shape, surface area, and thermal properties of the object and the thermal properties of the media in contact with the object.

Newton's Law of Cooling

$$T - T_s = (T_0 - T_s)e^{-kt} \tag{8}$$

EXAMPLE 4

A hard-boiled egg at 98°C is put in a sink of 18°C water to cool. After 5 minutes, the egg's temperature is found to be 38°C. Assuming that the water has not warmed appreciably, how much longer will it take the egg to reach 20°C?

Solution We find how long it would take the egg to cool from 98°C to 20°C and subtract the 5 minutes that have already elapsed.

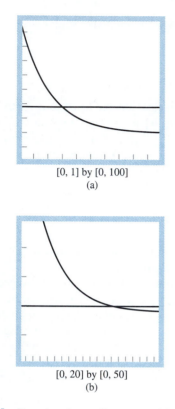

[0, 1] by [0, 100]
(a)

[0, 20] by [0, 50]
(b)

7.21 To solve the cooling egg problem of Example 4 graphically, we find that (a) the graphs of

$$T = 18 + 80e^{-5k} \quad \text{and} \quad T = 38$$

meet at $k = 0.277$. Using this value of k, we find that (b) the graphs of

$$T = 18 + 80e^{-0.277t} \quad \text{and} \quad T = 20$$

meet at $t = 13.3$. These values are confirmed in the analytic solution shown at right.

According to Eq. (8), the egg's temperature t minutes after it is put in the sink is

$$T - 18 = (98 - 18)e^{-kt}, \quad \text{or} \quad T = 18 + 80e^{-kt}.$$

To find k, we use the information that $T = 38$ when $t = 5$. This gives

$$38 = 18 + 80e^{-5k}$$

$$e^{-5k} = \frac{1}{4}$$

$$-5k = \ln\frac{1}{4} = -\ln 4$$

$$k = \frac{1}{5}\ln 4 = 0.277 \quad \text{See Fig. 7.21(a).}$$

The egg's temperature at time t is

$$T = 18 + 80e^{-0.277t}.$$

When will $T = 20$? When

$$20 = 18 + 80e^{-0.277t}$$

$$80e^{-0.277t} = 2$$

$$e^{-0.277t} = \frac{1}{40}$$

$$-0.277t = \ln\frac{1}{40} = -\ln 40$$

$$t = \frac{\ln 40}{0.277} = 13.3 \text{ min.}$$

The egg's temperature will reach 20°C in a little more than 13 minutes after it is put in to cool (Fig. 7.21b). Because it took 5 minutes to reach 38°C, it will take slightly more than 8 additional minutes to reach 20°C. ≡

Exercises 7.4

1. *Human evolution continues.* The analysis of tooth shrinkage by C. Loring Brace and colleagues at the University of Michigan's Museum of Anthropology indicates that human tooth size is continuing to decrease and that the evolutionary process did not come to a halt some 30,000 years ago as many scientists contend. In northern Europeans, for example, tooth size reduction now has a rate of 1% per 1000 years. (Source: *LSA Magazine*, Spring 1989, Volume 12, Number 2, p. 19, Ann Arbor, MI.)

   a) If t represents time in years and y represents tooth size, use the condition that $y = 0.99y_0$ when $t = 1000$ to find the value of k in the equation $y = y_0e^{kt}$. Then use this

   value of k to answer the following questions.
   b) In about how many years will human teeth be 90% of their present size? Support your answer graphically.
   c) What will be our descendants' tooth size 20,000 years from now (as a percentage of our present tooth size)?

2. *Atmospheric pressure.* Earth's atmospheric pressure p is often modeled by assuming that the rate dp/dh at which p changes with the altitude h above sea level is proportional to p. Suppose that the pressure at sea level is 1013 millibars (about 14.7 pounds per square inch) and that the pressure at an altitude of 20 km is 90 millibars.

a) Solve the equation $dp/dh = kp$ (k a constant) to express p in terms of h. Determine the values of k and the constant of integration from the given initial conditions.

b) What is the atmospheric pressure at $h = 50$ km?

c) At what altitude is the pressure equal to 900 millibars? Support your answer graphically.

3. *First-order chemical reactions.* In some chemical reactions, the rate at which the amount of a substance changes with time is proportional to the amount present. For example, the rate of change of δ-glucono lactone into gluconic acid is

$$\frac{dy}{dt} = -0.6y,$$

where t is measured in hours. If 100 grams of δ-glucono lactone are present when $t = 0$, how many grams will be left after the first hour?

4. *The inversion of sugar.* The processing of raw sugar has a step called "inversion" that changes the sugar's molecular structure. Once the process has begun, the rate of change of the amount of raw sugar is proportional to the amount of raw sugar remaining. If 1000 kg of raw sugar reduces to 800 kg of raw sugar during the first 10 hours, how much raw sugar will remain after another 14 hours?

5. *Radon gas.* The decay equation for (radioactive) radon gas is $y = y_0 e^{-0.18t}$, with t in days. About how long will it take the radon in a sealed sample of air to fall to 90% of its original value?

6. *Polonium-210.* The half-life of polonium is 140 days, but your sample will not be useful to you after 95% of the radioactive nuclei present on the day the sample arrives has disintegrated. About how many days after the sample arrives will you be able to use the polonium?

7. For a 66-kg cyclist on a 7-kg bicycle on level ground, the k in Eq. (4) is about 3.9. The cyclist starts coasting at 9 m/sec.

a) About how far will the cyclist coast before reaching a complete stop?

b) How long will it take the cyclist's speed to drop to 1 m/sec?

8. An Iowa class battleship might have mass around 51,000 metric tons (51,000,000 kg) and a k value in Eq. (4) of about 59,000. Suppose such a ship loses power when it is moving at a speed of 9 m/sec.

a) About how far will the ship coast before it is dead in the water?

b) About how long will it take the ship's speed to drop to 1 m/sec?

9. *Cooling cocoa.* Suppose that a cup of cocoa cooled from 90°C to 60°C after 10 minutes in a room whose temperature was 20°C. Use Newton's law of cooling to answer the following questions.

a) How much longer would it take the cocoa to cool to 35°C?

b) Instead of being left to stand in the room, the cup of 90°C cocoa is put in a freezer whose temperature is −15°C. How long will it take the cocoa to cool from 90°C to 35°C?

10. *Body of unknown temperature.* A body of unknown temperature was placed in a room that was held at 30°F. After 10 minutes, the body's temperature was 0°F, and 20 minutes after the body was placed in the room the body's temperature was 15°F. Use Newton's law of cooling to estimate the body's initial temperature.

11. *Surrounding medium of unknown temperature.* A pan of warm water (46°C) was put in a refrigerator. Ten minutes later, the water's temperature was 39°C; 10 minutes after that, it was 33°C. Use Newton's law of cooling to estimate how cold the refrigerator was.

12. *Silver cooling in air.* The temperature of an ingot of silver is 60°C above room temperature right now. Twenty minutes ago, it was 70°C above room temperature. How far above room temperature will the silver be 15 minutes from now? Two hours from now? When will the silver be 10°C above room temperature?

13. *Voltage in a discharging capacitor.* Suppose that electricity is draining from a capacitor at a rate that is proportional to the voltage $V(t)$ across its terminals and that, if t is measured in seconds,

$$\frac{dV}{dt} = -\frac{1}{40}V.$$

Solve this equation for V, using V_0 to denote the value of V when $t = 0$. How long will it take the voltage to drop to 10% of its original value?

14. *The mean life of a radioactive nucleus.* Physicists using the radioactivity equation $y = y_0 e^{-kt}$ call the number $1/k$ the *mean life* of a radioactive nucleus. The mean life of a radon nucleus is about $1/0.18 = 5.6$ days. The mean life of a carbon-14 nucleus is more than 8000 years. Show that 95% of the radioactive nuclei originally present in a sample will disintegrate within three mean lifetimes, that is, by time $t = 3/k$. Thus, the mean life of a nucleus gives a quick way to estimate how long the radioactivity of a sample will last.

Carbon-14 Dating

The half-lives of radioactive elements can sometimes be used to date events from Earth's past. The ages of rocks that are more than 2 billion years old have been measured by the extent of the radioactive decay of uranium (half-life 4.5 billion years!). In a living organism, the ratio of radioactive carbon, carbon-14, to ordinary carbon stays fairly constant during the lifetime of the organism, being approximately equal to the ratio in the organism's surroundings at the time. After the organism's death, however, no new carbon is ingested, and the proportion of carbon-14 in the organism's remains decreases as the carbon-14 decays. Since the half-life of carbon-14 is known to be about 5700 years, it is possible to estimate the age of organic remains by comparing the proportion of carbon-14 that they contain with the proportion assumed to have been in the organism's environment at the time it lived. Archeologists have dated shells (which contain $CaCO_3$), seeds, and wooden artifacts this way. The estimate of 15,500 years for the age of the cave paintings at Lascaux, France, is based on carbon-14 dating. After generations of controversy, the cloth in the Shroud of Turin, long believed to have been the burial cloth of Christ, was shown by carbon-14 dating in 1988 to have been made later than 1200 A.D., probably between the years 1260 and 1390.

15. The charcoal from a tree killed in the volcanic eruption that formed Crater Lake in Oregon contained 44.5% of the carbon-14 found in living matter. About how old is Crater Lake?

16. To see the effect of a relatively small error in the estimate of the amount of carbon-14 in a sample being dated, consider this hypothetical situation:
a) A fossilized bone found in central Illinois in the year 2000 A.D. is found to contain 17% of its original carbon-14 content. Estimate the year the animal died.
b) Repeat part (a) assuming 18% instead of 17%.
c) Repeat part (a) assuming 16% instead of 17%.

d) Use a graphing utility to show the range of time corresponding to the 16%–18% range of carbon-14 content.

17. *Art forgery.* A painting attributed to Vermeer (1632–1675) should have contained no more than 96.2% of its original carbon-14, but it contained 99.5% instead. How many years ago was the painting made?

18. Find an algebraic representation for the smooth curve connecting the points you plotted in Exercise 78 of Section 7.3.

19. Refer to Example 3.
a) Use analytic methods to confirm the results in parts (f) through (h).
b) When will Jenny have coasted 141 m from the finish line? Confirm your answer analytically.

20. Solve the following initial value problem without assuming that y is a positive function.

Differential equation: $\dfrac{dy}{dt} = ky$

Initial condition: $y = y_0$ when $t = 0$

21. A population of honeybees grows at an annual rate equal to 1/4 of the number present when there are no more than 10,000 bees. If there are more than 10,000 but fewer than 50,000 bees, the growth rate is equal to 1/12 of the number present. If there are 5000 bees now, when will there be 25,000 bees?

7.5   Indeterminate Forms and l'Hôpital's Rule

EXPLORATION BIT

The result 0/0 is called *indeterminate* because it is possible to have functions f, g, h, and k so that the limiting value of each is 0 (hence, we get 0/0 by substituting into their quotients) but with different actual limits of f/g and h/k. A result like 0/∞ is not indeterminate because we know that the limit would be 0. Other indeterminate forms are

∞/∞ from f/g,
$\infty \cdot 0$ from $f \cdot g$,
$\infty - \infty$ from $f - g$,

and

$1^\infty, 0^0$, and ∞^0, all from f^g.

Find functions f, g, h, and k to show why these forms are indeterminate. Also, be alert for other indeterminate forms. Keep a list of your findings.

Graphs played an important role in Chapter 2 when we used them to predict limits. Some of the limits could be confirmed algebraically, but often the algebraic process was very tedious. John Bernoulli, in the late seventeenth century, provided a confirming method for the limits of fractions whose numerators and denominators both approach zero. The rule is known today as l'Hôpital's rule, named after Guillaume François Antoine de l'Hôpital (1661–1704), a French nobleman who wrote the first introductory differential calculus text, in which the rule first appeared.

L'Hôpital's Rule gives fast confirmation even when other analytic methods are slow or unavailable. This section introduces the rule, uses it to confirm that $\lim_{x \to \infty} (1 + (1/n))^n = e$, and sets the stage for the growth-rate comparisons in Section 7.6.

The Indeterminate Form 0/0

If functions f and g are continuous at $x = a$ but $f(a) = g(a) = 0$, the limit

$$\lim_{x \to a} \frac{f(x)}{g(x)}$$

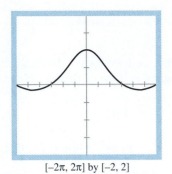

[-2π, 2π] by [-2, 2]

7.22 The graph suggests that the $\lim_{x \to 0} (\sin x)/x = 1$.

cannot be evaluated by substituting $x = a$, since this produces 0/0, a meaningless expression known as an **indeterminate form**.

An important example of this type is the ratio $(\sin x)/x$ studied in Section 2.3. Notice that the graph (Fig. 7.22) suggests that

$$\lim_{x \to 0} \frac{\sin x}{x} = 1.$$

The technique used in Section 2.3 to confirm this limit is very tedious. Derivatives provide an easier way.

The limit

$$f'(a) = \lim_{x \to a} \frac{f(x) - f(a)}{x - a}$$

from which we calculate derivatives always produces the indeterminate form 0/0. Our success in calculating derivatives suggests that we might turn things around and use derivatives to calculate limits that lead to indeterminate forms. For example, knowing the derivative of $\sin x$ would let us find

$$\lim_{x \to 0} \frac{\sin x}{x} = \lim_{x \to 0} \frac{\sin x - \sin 0}{x - 0} = \frac{d}{dx}(\sin x)\Big|_{x=0} = \cos 0 = 1.$$

L'Hôpital's Rule gives an explicit connection between derivatives and limits that lead to the indeterminate form 0/0.

Instead of substituting a into $f(x)/g(x)$ when the given conditions are satisfied, we first find f' and g' and then substitute a into $f'(x)/g'(x)$.

THEOREM 2   **L'Hôpital's Rule (First Form)**

Suppose that $f(a) = g(a) = 0$, that $f'(a)$ and $g'(a)$ exist, and that $g'(a) \neq 0$. Then

$$\lim_{x \to a} \frac{f(x)}{g(x)} = \frac{f'(a)}{g'(a)}.$$

Proof Working backward from $f'(a)$ and $g'(a)$, which are themselves limits, we have

$$\frac{f'(a)}{g'(a)} = \frac{\lim_{x \to a} \dfrac{f(x) - f(a)}{x - a}}{\lim_{x \to a} \dfrac{g(x) - g(a)}{x - a}} = \lim_{x \to a} \frac{\dfrac{f(x) - f(a)}{x - a}}{\dfrac{g(x) - g(a)}{x - a}}$$

$$= \lim_{x \to a} \frac{f(x) - f(a)}{g(x) - g(a)} = \lim_{x \to a} \frac{f(x) - 0}{g(x) - 0} = \lim_{x \to a} \frac{f(x)}{g(x)}. \quad \blacksquare$$

The proof of Theorem 2 has a nice geometric interpretation. If we ZOOM-IN on the graphs of f and g at $(a, f(a)) = (a, g(a)) = (a, 0)$, their graphs appear to be straight lines (Fig. 7.23). We let m_1 and m_2 be the slopes of the lines for f and g, respectively, and for x near a,

$$\frac{f(x)}{g(x)} = \frac{\dfrac{f(x)}{x-a}}{\dfrac{g(x)}{x-a}} = \frac{m_1}{m_2}.$$

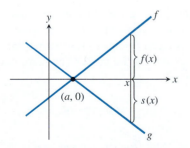

7.23 A ZOOM-IN view in which the graphs of f and g appear to be straight lines.

As $x \to a$, m_1 and m_2 approach $f'(a)$ and $g'(a)$, respectively. Therefore,

$$\lim_{x \to a} \frac{f(x)}{g(x)} = \lim_{x \to a} \frac{m_1}{m_2} = \frac{f'(a)}{g'(a)}.$$

EXPLORATION 1

Avoiding L'Hôpital's Traps

There are a couple "traps" we can fall victim to if we apply l'Hôpital's Rule carelessly or we do not completely understand it. One of them is suggested in part 2 below.

1. Let

$$y_1 = \sin x, \qquad y_2 = x, \qquad y_3 = y_1/y_2, \qquad y_4 = \text{NDER } y_1/\text{NDER } y_2.$$

Support l'Hôpital's Rule for $(\sin x)/x$ by graphing y_3 and y_4. Interpret l'Hôpital's Rule in terms of what you see in the viewing window.

2. Let $y_5 = \text{NDER } y_3$. GRAPH y_3, y_4, and y_5. On the basis of what you see in the viewing window, make a statement about what l'Hôpital's Rule does *not* say. ❧

> Recall that the graph of y_3 has a removable discontinuity at $x = 0$. The graph of y_4 is continuous at $x = 0$, and we can find its limit by substituting 0 for x. Also, be sure that you understand the "trap" suggested by part 2.

EXAMPLE 1

Use graphs to predict the limit of each function as $x \to 0$. Confirm analytically using l'Hôpital's Rule.

a) $y = \dfrac{3x - \sin x}{x}$ **b)** $y = \dfrac{\sqrt{1+x} - 1}{x}$ **c)** $y = \dfrac{x - \sin x}{x^3}$

Solution The graphs in Fig. 7.24 allow us to predict that the three limits are 2, 0.5, and 0.17, respectively. Analytically, we first note that substituting $x = 0$ gives an indeterminate form in each case. Thus, we apply l'Hôpital's Rule:

a) $\displaystyle\lim_{x \to 0} \frac{3x - \sin x}{x} = \frac{3 - \cos x}{1}\bigg|_{x=0} = 2$

b) $\displaystyle\lim_{x \to 0} \frac{\sqrt{1+x} - 1}{x} = \frac{1/(2\sqrt{1+x})}{1}\bigg|_{x=0} = 0.5$

c) $\displaystyle\lim_{x \to 0} \frac{x - \sin x}{x^3} = \frac{1 - \cos x}{3x^2}\bigg|_{x=0} = \ ?$

For part (c), the first form of l'Hôpital's Rule does not tell us what the limit is because the derivative of $g(x) = x^3$ is zero at $x = 0$. However, a stronger form of l'Hôpital's Rule from advanced calculus says that whenever the rule gives 0/0 we can apply it again, repeating the process until we get a different result. We use this stronger rule:

$$\lim_{x \to 0} \frac{x - \sin x}{x^3} = \lim_{x \to 0} \frac{1 - \cos x}{3x^2} \qquad \text{Still } \frac{0}{0}; \text{ apply the rule again.}$$

$$= \lim_{x \to 0} \frac{\sin x}{6x} \qquad \text{Still } \frac{0}{0}; \text{ apply the rule again.}$$

$$= \lim_{x \to 0} \frac{\cos x}{6} = \frac{1}{6}. \qquad \text{A different result. Stop.}$$

Notice that 1/6 is 0.17 accurate to hundredths. ≣

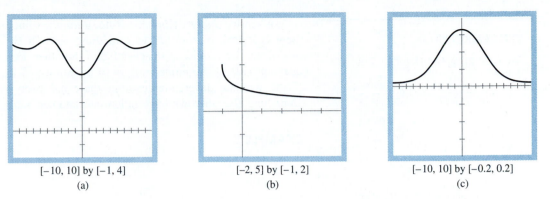

$$[-10, 10] \text{ by } [-1, 4] \qquad\qquad [-2, 5] \text{ by } [-1, 2] \qquad\qquad [-10, 10] \text{ by } [-0.2, 0.2]$$
$$\text{(a)} \qquad\qquad\qquad \text{(b)} \qquad\qquad\qquad \text{(c)}$$

7.24 Graphs that suggest the limit as $x \to 0$ (a) of $y = (3x - \sin x)/x$ is 2, (b) of $y = \left(\sqrt{1 + x} - 1\right)/x$ is 0.5, and (c) $y = (x - \sin x)/x^3$ is 0.17.

THEOREM 3 **L'Hôpital's Rule (Stronger Form)**

Suppose that $f(x_0) = g(x_0) = 0$ and that the functions f and g are both differentiable on an open interval (a, b) that contains the point x_0. Suppose also that $g' \neq 0$ at every point in (a, b) except possibly x_0. Then

$$\lim_{x \to x_0} \frac{f(x)}{g(x)} = \lim_{x \to x_0} \frac{f'(x)}{g'(x)},$$

provided that the limit on the right exists.

EXAMPLE 2

$$\lim_{x \to 0} \frac{\sqrt{1 + x} - 1 - x/2}{x^2} \qquad\qquad \text{Indeterminate form } \frac{0}{0}$$

$$= \lim_{x \to 0} \frac{(1/2)(1 + x)^{-1/2} - (1/2)}{2x} \qquad\qquad \text{Still } \frac{0}{0}$$

$$= \lim_{x \to 0} \frac{-(1/4)(1 + x)^{-3/2}}{2} = -\frac{1}{8} \qquad\qquad\qquad \blacksquare$$

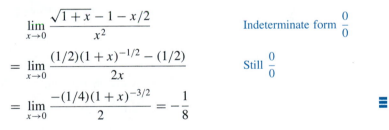

EXPLORATION 2

Where a Grapher Lacks Enough Precision

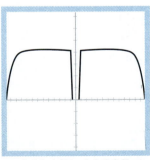

$$[-1.3, 1.3] \text{ by } [-0.5, 0.8]$$

7.25 This graph of

$$f(x) = \left(1 - \cos x^6\right)/x^{12}$$

incorrectly suggests that there is a gap near $x = 0$.

To end Section 2.1, we tried to view the graph of $f(x) = (1 - \cos x^6)/x^{12}$. (See Fig. 7.25.) Our grapher lacked the precision needed to produce a correct graph near $x = 0$. We can now, however, confirm the claim that $\lim_{x \to 0} f(x) = 0.5$.

1. Confirm that $\lim_{x \to 0} f(x)$ gives an indeterminate form.
2. Confirm that l'Hôpital's Rule can be applied. Then apply it.
3. $g(x) = (1 - \cos x^6)/(x^{18} + x^{12})$ is another function that baffles our grapher near $x = 0$. Confirm that l'Hôpital's Rule can be applied, then apply it as many times as necessary to find $\lim_{x \to 0} g(x)$.

EXPLORATION BIT

We tried to apply l'Hôpital's Rule as follows:

$$\lim_{x \to 0} \frac{1 - \cos x}{x + x^2} = \lim_{x \to 0} \frac{\sin x}{1 + 2x}$$

$$= \lim_{x \to 0} \frac{\cos x}{2}$$

$$= \frac{1}{2},$$

but our grapher supports another limit. Can you find what's wrong? What can we conclude?

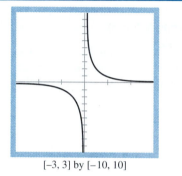

[−3, 3] by [−10, 10]

7.26 The graph of $f(x) = (\sin x)/x^2$ shows that $\lim_{x \to 0^+} f(x) = \infty$ and $\lim_{x \to 0^-} f(x) = -\infty$.

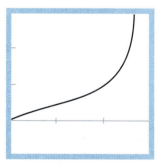

[0, 3π/4] by [−1, 3]

7.27 The graph of

$$y = \frac{\tan x}{1 + \tan x}$$

suggests that the discontinuity at $x = \pi/2$ is removable.

When we apply l'Hôpital's Rule, a limit's value is revealed only when there is a change from 0/0 to something else. If we reach a point where one of the derivatives approaches 0 and the other does not, then the limit in question is 0 (if the numerator → 0) or infinity (if the denominator → 0). If we continue differentiating after there has been a change from 0/0 to something else, then we have fallen into another trap.

EXAMPLE 3

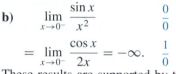

a) $\lim_{x \to 0^+} \dfrac{\sin x}{x^2}$ $\qquad \dfrac{0}{0}$

$= \lim_{x \to 0^+} \dfrac{\cos x}{2x} = \infty.$ $\qquad \dfrac{1}{0}$

b) $\lim_{x \to 0^-} \dfrac{\sin x}{x^2}$ $\qquad \dfrac{0}{0}$

$= \lim_{x \to 0^-} \dfrac{\cos x}{2x} = -\infty.$ $\qquad \dfrac{1}{0}$

These results are supported by the graph in Fig. 7.26.   ☰

The Indeterminate Forms ∞/∞, and ∞ · 0 and ∞ − ∞

In more advanced books, it is proved that l'Hôpital's Rule applies to the indeterminate form ∞/∞ as well as 0/0. If $f(x)$ and $g(x)$ both approach infinity as x approaches a, then

$$\lim_{x \to a} \frac{f(x)}{g(x)} = \lim_{x \to a} \frac{f'(x)}{g'(x)},$$

provided that the limit on the right exists. In the notation $x \to a$, a may be either finite or infinite.

EXPLORATION 3

The Indeterminate Form ∞/∞

1. Use Fig. 7.27 or your own viewing window to make a conjecture about

$$\lim_{x \to \pi/2} \frac{\tan x}{1 + \tan x}.$$

2. Use l'Hôpital's Rule to confirm your conjecture analytically. (Note that the limits from the left and right give two indeterminate forms, ∞/∞ and (−∞)/(−∞), so you will have to evaluate the limits from each side.)

The forms $\infty \cdot 0$ and $\infty - \infty$ can sometimes be handled by using algebra to get $0/0$ or ∞/∞ instead. Here again, we do not mean to suggest that there is a number $\infty \cdot 0$ any more than we mean to suggest that there is a number $0/0$ or ∞/∞. These forms are not numbers but descriptions of limits.

EXAMPLE 4

Show that $\lim\limits_{x \to \pm\infty} x \sin \dfrac{1}{x} = 1$.

Solution   The graph of $f(x) = x \sin(1/x)$ in Fig. 7.28 clearly suggests that $\lim\limits_{x \to \pm\infty} f(x) = 1$. Analytically,

$$\lim_{x \to \infty} x \sin \frac{1}{x} \qquad \infty \cdot 0$$

$$= \lim_{h \to 0^+} \frac{1}{h} \sin h \qquad \text{Let } h = 1/x.$$

$$= 1$$

In the exercises you will be asked to show that

$$\lim_{x \to -\infty} x \sin \frac{1}{x} = 1.$$

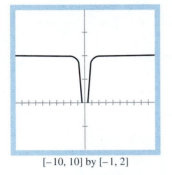

[−10, 10] by [−1, 2]

7.28   The graph suggests that

$$\lim_{x \to \pm\infty} x \sin \frac{1}{x} = 1.$$

This is confirmed analytically for $x \to \infty$ in Example 4. You are asked to confirm it analytically for $x \to -\infty$ as an exercise.

EXAMPLE 5

Evaluate $\lim\limits_{x \to 1} \left(\dfrac{1}{\ln x} - \dfrac{1}{x - 1} \right)$.

Solution   The graph in Fig. 7.29 suggests the limit to be $\dfrac{1}{2}$. Analytically,

$$\lim_{x \to 1} \left(\frac{1}{\ln x} - \frac{1}{x - 1} \right) \qquad \infty - \infty$$

$$= \lim_{x \to 1} \frac{x - 1 - \ln x}{(x - 1) \ln x} \qquad \frac{0}{0}$$

$$= \lim_{x \to 1} \frac{x - 1}{x \ln x + x - 1} \qquad \text{Using l'Hôpital's Rule and algebra; still } 0/0$$

$$= \lim_{x \to 1} \frac{1}{2 + \ln x} = \frac{1}{2}. \qquad \text{Using l'Hôpital's Rule again}$$

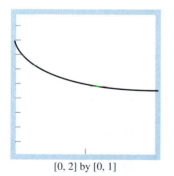

[0, 2] by [0, 1]

7.29 $\lim\limits_{x \to 1} \left(\dfrac{1}{\ln x} - \dfrac{1}{x - 1} \right)$ appears to be 0.5. This is confirmed analytically in Example 5.

The Indeterminate Forms 1^{∞}, 0^{0}, and ∞^{0}

The indeterminate forms 1^{∞}, 0^0, and ∞^0 can sometimes be handled by taking logarithms first. The idea is to calculate the limit of the logarithm and exponentiate that limit.

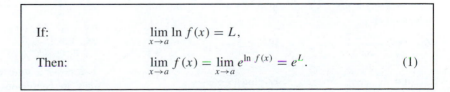

If: $\lim\limits_{x \to a} \ln f(x) = L,$

Then: $\lim\limits_{x \to a} f(x) = \lim\limits_{x \to a} e^{\ln f(x)} = e^{L}.$ (1)

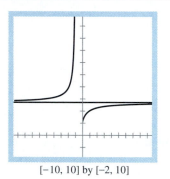

$[-10, 10]$ by $[-2, 10]$

7.30 Graphs of

$$y_1 = f(x) = (1 + \frac{1}{x})^x \quad \text{and} \quad y_2 = e.$$

In the exercises, we ask you to show that the graph of f is complete.

EXAMPLE 6

Let $f(x) = (1 + 1/x)^x$. Show that $\lim\limits_{x \to \pm\infty} f(x) = e$, that is, $y = e$ is a horizontal asymptote of $f(x)$. Draw a complete graph of f.

Solution The function f is defined outside of $[-1, 0]$. As $x \to \pm\infty$, $(1 + 1/x) \to 1$. Therefore, $f(x)$ approaches the indeterminate form 1^∞, so we take logarithms.

$$\ln f(x) = \ln \left(1 + \frac{1}{x}\right)^x = x \ln \left(1 + \frac{1}{x}\right) = \frac{\ln \left(1 + \frac{1}{x}\right)}{\frac{1}{x}}.$$

The latter expression gives the indeterminate form $0/0$, so we apply l'Hôpital's Rule:

$$\lim_{x \to \pm\infty} \ln f(x) = \lim_{x \to \pm\infty} \frac{\ln \left(1 + \frac{1}{x}\right)}{\frac{1}{x}} \qquad \frac{0}{0}$$

$$= \lim_{x \to \pm\infty} \left(\frac{\frac{1}{1 + \frac{1}{x}} \left(-\frac{1}{x^2}\right)}{-\frac{1}{x^2}} \right)$$

$$= \lim_{x \to \pm\infty} \frac{1}{1 + \frac{1}{x}} = 1.$$

Therefore,

$$\lim_{x \to \pm\infty} f(x) = e^1 = e. \qquad \text{Eq. (1) with } L = 1$$

A complete graph is shown in Fig. 7.30

Exercises 7.5

Find the limits in Exercises 1–8. Use l'Hôpital's Rule when the form is indeterminate. Support your answer graphically.

1. $\lim\limits_{x \to 2} \dfrac{x - 2}{x^2 - 4}$

2. $\lim\limits_{x \to 2} \dfrac{x^2 - 4}{x - 2}$

3. $\lim\limits_{x \to 1} \dfrac{x^3 - 1}{4x^3 - x - 3}$

4. $\lim\limits_{x \to 0} \dfrac{1 - \cos x}{x^2}$

5. $\lim\limits_{t \to 0} \dfrac{\sin t^2}{t}$

6. $\lim\limits_{x \to 0} \dfrac{\sin 5x}{x}$

7. $\lim\limits_{x \to \infty} \dfrac{3x^2 - 1}{2x^2 - x + 1}$

8. $\lim\limits_{t \to \infty} \dfrac{6t + 5}{3t - 8}$

Use graphs to find the limits in Exercises 9–20. Confirm your answers analytically. Use l'Hôpital's Rule when the form is indeterminate.

9. $\lim\limits_{x \to \pi/2} \dfrac{2x - \pi}{\cos x}$

10. $\lim\limits_{x \to 0} \dfrac{(1/2)^x - 1}{x}$

11. $\lim\limits_{x \to \infty} \dfrac{5x^2 - 3x}{7x^2 + 1}$

12. $\lim\limits_{t \to 0} \dfrac{\cos t - 1}{t^2}$

13. $\lim\limits_{x \to \pi/2} \dfrac{1 - \sin x}{1 + \cos 2x}$

14. $\lim\limits_{x \to \pi/2} \left(\dfrac{\pi}{2} - x\right) \tan x$

15. $\lim\limits_{x \to 0^+} \dfrac{2x}{x + 7\sqrt{x}}$

16. $\lim\limits_{x \to \infty} \dfrac{x - 2x^2}{3x^2 + 5x}$

17. $\lim\limits_{t \to 0} \dfrac{10(\sin t - t)}{t^3}$

18. $\lim\limits_{x \to 0} \dfrac{x(1 - \cos x)}{x - \sin x}$

19. $\lim\limits_{x \to 0} \left(\dfrac{1}{\sin x} - \dfrac{1}{x}\right)$

20. $\lim\limits_{x \to 0^+} \left(\dfrac{1}{x} - \dfrac{1}{\sqrt{x}}\right)$

Use the technique of Example 6 to find the limits in Exercises 21–26 analytically, and support your answer graphically.

21. $\lim\limits_{x \to 0^+} x^{(1/\ln x)}$

22. $\lim\limits_{x \to 0^+} x^{1/x}$

23. $\lim\limits_{x \to 0^+} (e^x + x)^{1/x}$

24. $\lim\limits_{x \to 1} x^{1/(x-1)}$

25. $\lim\limits_{x \to 0} \left(\dfrac{1}{x^2}\right)^x$

26. $\lim\limits_{x \to 0} \left(\ln \left|\dfrac{1}{x}\right|\right)^x$

Is it possible to extend the functions in Exercises 27–30 so that they are continuous at $x = 0$? If so, explain each extension.

27. $y = x^{\sqrt{2}}$

28. $y = x^{-\sqrt{3}}$

29. $y = x^{\ln x}$

30. $y = x^{x+1}$

31. Refer to Section 7.3, Example 9. Show that

  a) $f(x) = x^x$ has a removable discontinuity at $x = 0$.

  b) $\lim\limits_{x \to 0^+} f'(x) = -\infty$.

32. a) Show that $f(x) = |x|^x$ has a removable discontinuity at $x = 0$ and that

$$F(x) = \begin{cases} |x|^x, & x \neq 0 \\ 1, & x = 0 \end{cases}$$

  is the continuous extension at $x = 0$.

  b) Show that $F'(0)$ does not exist.

  c) ZOOM-IN at the point $(0, 1)$ on the graph of F until δx is no larger than 10^{-5}. Graph the final result in a square viewing window. Estimate the slope of the tangent line to F at $(0, 1)$ in this window. Why is the tangent line not vertical as suggested in part (b)? Is this a contradiction?

33. Which is correct, (a) or (b)? Explain.

  a) $\lim\limits_{x \to 3} \dfrac{x - 3}{x^2 - 3} = \lim\limits_{x \to 3} \dfrac{1}{2x} = \dfrac{1}{6}$

  b) $\lim\limits_{x \to 3} \dfrac{x - 3}{x^2 - 3} = \dfrac{0}{6} = 0$

34. Show that $\lim\limits_{x \to -\infty} x \sin \dfrac{1}{x} = 1$.

35. Refer to Example 6. Let $f(x) = \left(1 + \dfrac{1}{x}\right)^x$. Show that

  a) the domain of f is $(-\infty, -1) \cup (0, \infty)$.

  b) $\lim\limits_{x \to (-1)^-} f(x) = \infty$.

  c) $\lim\limits_{x \to 0^+} f(x) = 1$.

  d) the graph of f in Fig. 7.30 is complete using analytic methods.

36. Use graphs to determine

$$\lim\limits_{x \to (\pi/2)^-} \frac{\sec x}{\tan x} \quad \text{and} \quad \lim\limits_{x \to (\pi/2)^+} \frac{\sec x}{\tan x}.$$

  Try to confirm using l'Hôpital's Rule. Explain.

Draw complete graphs of the functions in Exercises 37–40 and confirm analytically.

37. $f(x) = \left(1 + \dfrac{2}{x}\right)^x$

38. $f(x) = \left(1 + \dfrac{3}{x}\right)^x$

39. $f(x) = x^{(1/\ln x)}$

40. $f(x) = (1 + x)^{1/x}$

Find the limits in Exercises 41 and 42 and draw a complete graph of the function in the specified interval. Is the function continuous in the interval? Explain.

41. $\lim\limits_{x \to 0} \dfrac{3^{\sin x} - 1}{x}, [-2\pi, 2\pi]$

42. $\lim\limits_{x \to 0} \dfrac{2^{\cos x} - 2}{x}, [-4\pi, 4\pi]$

43. Let $A(t)$ be the area of the region in the first quadrant enclosed by the coordinate axes, the curve $y = e^{-x}$, and the line $x = t > 0$. Let $V(t)$ be the volume of the solid generated by revolving the region about the x-axis.

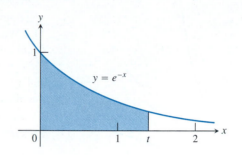

Find the following limits.

  a) $\lim\limits_{t \to \infty} A(t)$

  b) $\lim\limits_{t \to \infty} V(t)/A(t)$

  c) $\lim\limits_{t \to 0^+} V(t)/A(t)$

44. Let

$$f(x) = \begin{cases} x + 2, & x \neq 0 \\ 0, & x = 0 \end{cases}$$

$$g(x) = \begin{cases} x + 1, & x \neq 0 \\ 0, & x = 0 \end{cases}$$

Show that

$$\lim\limits_{x \to 0} \frac{f'(x)}{g'(x)} = 1 \quad \text{but} \quad \lim\limits_{x \to 0} \frac{f(x)}{g(x)} = 2.$$

Doesn't this contradict l'Hôpital's Rule?

45. *Continuously compounded interest continued.* The introduction to Section 2.1 says that the limit to how much you can earn in a year by investing \$100 at 6% compound interest is

$$\lim\limits_{k \to \infty} 100 \left(1 + \frac{0.06}{k}\right)^k \leq 106.20.$$

  a) Use the technique of Example 6 to confirm the limit that led to the continuous compound interest formula (Section 7.2, Eq. 9), namely, that

$$\lim\limits_{k \to \infty} A_0 \left(1 + \frac{r}{k}\right)^{kt} = A_0 e^{rt}.$$

  b) Find $100 e^{0.06}$ to three decimal places.

  c) Round $e^{0.06}$ to two decimal places, and then compute $1{,}000{,}000 e^{0.06}$.

  d) Round $1{,}000{,}000 e^{0.06}$ to two decimal places.

  e) Write a paragraph explaining the significance of the computations in parts (c) and (d).

7.6 The Rates at Which Functions Grow

This section shows how to extend our understanding about the end behavior of functions by comparing the rates at which functions of x grow as $|x|$ becomes large. We also introduce the so-called "little-oh" and "big-oh" notation that is sometimes used to describe the results of these comparisons.

We can restrict our attention to functions whose values eventually become and remain positive as $x \to \infty$. This can be done for two reasons. First, we can compare functions f and g as $x \to -\infty$ by comparing $f(-x)$ and $g(-x)$ as $x \to \infty$. Second, we can compare functions f and g with possible negative values by comparing $|f|$ and $|g|$.

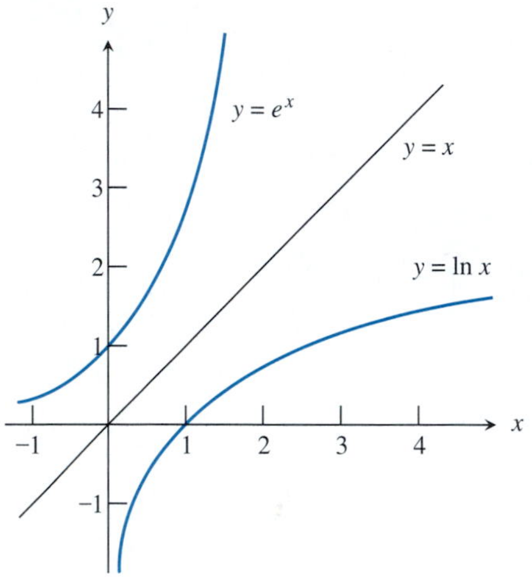

7.31 Graphs of $y = x$ and the inverse functions e^x and $\ln x$. Notice the symmetry: e^x and $\ln x$ are reflections of each other across the line $y = x$.

Relative Rates of Growth

For $a > 1$ and $x \to \infty$, the exponential function a^x grows so rapidly and the logarithmic function $\log_a x$ grows so slowly that they set the standards by which we can judge the growth of other functions. The familiar graphs (Fig. 7.31) of e^x, $\ln x$, and the identity function $f(x) = x$ (which show the inverse behavior of e^x and $\ln x$) suggest how rapidly and slowly e^x and $\ln x$, respectively, grow in comparison to $y = x$.

In fact, all the functions $a^x, a > 1$, grow faster (eventually) than any power x^n of x and hence faster (eventually) than any polynomial function.

EXPLORATION 1

Comparing Rates of Growth

If we define the *rate of growth* of a function f as f', compare graphically the rate of growth of e^x with the rate of growth of x^n for $n = 1, 2, 3, \ldots$. Over which intervals is one growing faster than the other?

Now, we want to consider a way to compare the rates of growth of functions for large values of x—which we call the *rates of growth* as $x \to \infty$—that does not employ the derivative.

To get a feeling for how rapidly the values of $y = e^x$ grow with increasing x, think of graphing the function on a large blackboard, with the axes scaled in centimeters. At $x = 1$ cm, the graph is $e^1 \approx 3$ cm above the x-axis. At $x = 6$ cm, the graph is $e^6 \approx 403$ cm ≈ 4 m high. (It is about to go through the ceiling if it hasn't done so already.) At $x = 10$ cm, the graph is $e^{10} \approx 22{,}026$ cm ≈ 220 m high, higher than most buildings. At $x = 24$ cm, the graph is more than halfway to the moon, and at $x = 43$ cm from the origin, approximately the distance across this book when it lies open on your desk, the graph is high enough to reach well past the nearest neighboring star, Proxima Centauri, which is about 4.3 light years away.

$$\begin{aligned} e^{43} &\approx 4.7 \times 10^{18} \text{ cm} \\ &= 4.7 \times 10^{13} \text{ km} \\ &\approx 1.58 \times 10^{8} \text{ light-seconds} \\ &\approx 5.0 \text{ light-years} \end{aligned}$$

Light travels at 300,000 km/sec in a vacuum.

In contrast, logarithmic functions like $y = \log_2 x$ and $y = \ln x$ grow more slowly as $x \to \infty$ than any positive power of x (Exercise 15). With axes scaled in centimeters, you have to go nearly 5 light-years out on the x-axis to find a point where the graph of $y = \ln x$ is even $y = 43$ cm high.

These important comparisons of exponential, polynomial, and logarithmic functions can be made precise by defining what it means for a function $f(x)$ to grow *faster than* another function $g(x)$ as $x \to \infty$.

DEFINITIONS   Rates of Growth as $x \to \infty$

1. f **grows faster than** g (and g **grows more slowly than** f) as $x \to \infty$ if $\displaystyle\lim_{x \to \infty} \frac{f(x)}{g(x)} = \infty$ or, equivalently, $\displaystyle\lim_{x \to \infty} \frac{g(x)}{f(x)} = 0.$

2. f and g **grow at the same rate** as $x \to \infty$ if

$$\lim_{x \to \infty} \frac{f(x)}{g(x)} = L \neq 0. \qquad L \text{ finite and not zero}$$

According to these definitions, $y = 2x$ does not grow faster than $y = x$ as $x \to \infty$. The two functions grow at the same rate because

$$\lim_{x \to \infty} \frac{2x}{x} = \lim_{x \to \infty} 2 = 2,$$

which is a finite nonzero limit. The reason for this apparent disregard of common sense is that we want "f grows faster than g" to mean that for large x-values, g is negligible in comparison to f.

This meaning can be illustrated rather dramatically with a grapher. By definition, $y_4 = e^x$ grows faster than $y_3 = x^2$, which, in turn, grows faster than $y_2 = 2x$ and $y_1 = x$. Figure 7.32 shows $y_1, y_2, y_3,$ and y_4 in a $[0, 10]$ by $[0, 100]$ viewing window. Change the view dimensions to $[0, 100]$ by $[0, 10{,}000]$, and compare visually the growth of the four functions.

If $L = 1$ in Definition 2, then f and g are right-end behavior models for each other, as we saw in Section 2.4. If f grows faster than g, then

$$\lim_{x \to \infty} \frac{f + g}{f} = \lim_{x \to \infty} \left(1 + \frac{g}{f}\right) = 1,$$

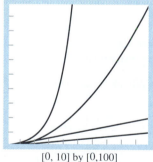

[0, 10] by [0,100]

7.32 Graphs of $y_1 = x, y_2 = 2x, y_3 = x^2,$ and $y_4 = e^x$.

so f is a right-end behavior model for $f + g$. This means that g can be ignored if we are looking for a simple right-end behavior model for $f + g$. This confirms why in Section 2.4 we could say that $y = a_n x^n$ is an end behavior model for $f(x) = a_n x^n + a_{n-1} x^{n-1} + \cdots + a_0$.

L'Hôpital's Rule can help us to compare rates of growth.

EXAMPLE 1

e^x grows faster than x^2 as $x \to \infty$ because

$$\lim_{x \to \infty} \frac{e^x}{x^2} = \infty,$$

as we can see by two applications of l'Hôpital's Rule:

$$\lim_{x \to \infty} \frac{e^x}{x^2} = \lim_{x \to \infty} \frac{e^x}{2x} = \lim_{x \to \infty} \frac{e^x}{2} = \infty.$$

EXPLORATION 2

Comparing Rates of Growth as $x \to \infty$

Sometimes specific examples suggest ways to prove more general statements.

1. Use the argument in Example 1 as a guide to

   a) show that any exponential function $a^x, a > 1$, grows faster than x^2.

   b) give a convincing argument that the exponential function e^x grows faster than any power function x^n.

2. Give a convincing argument that *any* exponential function $a^x, a > 1$, grows faster than any power function x^n.

3. Show that 3^x grows faster than 2^x.

4. Use your argument from part 3 to show that exponential functions never grow at the same rate. More precisely, show that for $a > b > 1, a^x$ grows faster than b^x.

5. Compare the rates of growth as $x \to \infty$ of the power functions, x^m, x^n, for positive m, n.

EXAMPLE 2

$\ln x$ grows more slowly than x as $x \to \infty$ because

$$\lim_{x\to\infty} \frac{\ln x}{x} = \lim_{x\to\infty} \frac{1/x}{1} \qquad \text{l'Hôpital's Rule}$$

$$= \lim_{x\to\infty} \frac{1}{x} = 0.$$

≡

EXPLORATION 3

The Slow-Growing Natural Log Function

Show that $\ln x$ grows more slowly than any power function $x^n, n > 0$. Use Example 2 as a guide.

EXAMPLE 3

In contrast to exponential functions, all logarithmic functions grow at the same rate as $x \to \infty$. For any bases a and b, (a and b greater than 1),

$$\lim_{x\to\infty} \frac{\log_a x}{\log_b x} = \lim_{x\to\infty} \frac{\ln x / \ln a}{\ln x / \ln b} = \frac{\ln b}{\ln a}.$$

The limiting ratio is always finite and never zero.

≡

"Growing at the Same Rate as" Is a Transitive Relation

If f grows at the same rate as g as $x \to \infty$ and g grows at the same rate as h as $x \to \infty$, then f grows at the same rate as h as $x \to \infty$. The reason is that

$$\lim_{x\to\infty} \frac{f}{g} = L_1 \qquad \text{and} \qquad \lim_{x\to\infty} \frac{g}{h} = L_2$$

together imply that

$$\lim_{x \to \infty} \frac{f}{h} = \lim_{x \to \infty} \frac{f}{g} \cdot \frac{g}{h} = L_1 L_2.$$

If L_1 and L_2 are finite and nonzero, then so is $L_1 L_2$.

EXAMPLE 4

Show that $y_1 = \sqrt{x^2 + 5}$ and $y_2 = (2\sqrt{x} - 1)^2$ grow at the same rate as $x \to \infty$.

Solution   We show that the functions grow at the same rate by showing that they both grow at the same rate as x:

$$\lim_{x \to \infty} \frac{\sqrt{x^2 + 5}}{x} = \lim_{x \to \infty} \sqrt{1 + \frac{5}{x^2}} = 1$$

$$\lim_{x \to \infty} \frac{(2\sqrt{x} - 1)^2}{x} = \lim_{x \to \infty} \left(\frac{2\sqrt{x} - 1}{\sqrt{x}} \right)^2 = \lim_{x \to \infty} \left(2 - \frac{1}{\sqrt{x}} \right)^2 = 4. \quad \blacksquare$$

EXPLORATION BITS

1. Show *directly* that y_1 and y_2 of Example 4 grow at the same rate.
2. The analytic solution in Example 4 suggests that $y = x$ is a right-end behavior model for y_1 and $y = 4x$ is a right-end behavior model for y_2. Explain. Find an end behavior model (right and left) for y_1 and for y_2. Viewing the graph of each might help.

Order of Magnitude as $x \to \infty$

Here we introduce the "little-oh" and "big-oh" notation, invented by number theorists a hundred years ago and now commonplace in mathematical analysis and computer science.

DEFINITION

A function f is **of smaller order than** g as $x \to \infty$ if $\lim_{x \to \infty} \frac{f(x)}{g(x)} = 0$. We indicate this by writing $f = o(g)$ ("f is little-oh of g").

Notice that saying that $f = o(g)$ as $x \to \infty$ is another way to say that f grows slower than g as $x \to \infty$.

EXAMPLE 5

$$\ln x = o(x) \text{ as } x \to \infty \qquad \text{because} \qquad \lim_{x \to \infty} \frac{\ln x}{x} = 0.$$

$$x^2 = o(x^3 + 1) \text{ as } x \to \infty \qquad \text{because} \qquad \lim_{x \to \infty} \frac{x^2}{x^3 + 1} = 0. \quad \blacksquare$$

If $f(x)/g(x)$ fails to approach zero as $x \to \infty$ but the ratio remains bounded, we say that f is at most the order of g as $x \to \infty$. What we mean, exactly, is that $f(x)/g(x)$ stays less than or equal to some integer M for x sufficiently large.

> **DEFINITION**
>
> A function f is **of at most the order of** g as $x \to \infty$ if there is a positive integer M for which
>
> $$\frac{f(x)}{g(x)} \le M$$
>
> for x sufficiently large. We indicate this by writing $\boldsymbol{f = O(g)}$ ("f is big-oh of g").

EXAMPLE 6

$x + \sin x = O(x)$ as $x \to \infty$ because $\dfrac{x + \sin x}{x} \le 2$ for x sufficiently large.

EXAMPLE 7

$$e^x + x^2 = O(e^x) \text{ as } x \to \infty \text{ because } \frac{e^x + x^2}{e^x} \to 1, \text{ as } x \to \infty.$$

EXAMPLE 8

$$x = O(e^x) \text{ as } x \to \infty \text{ because } \frac{x}{e^x} \to 0, \text{ as } x \to \infty.$$

Note that the "little oh" notation matches up with the notion of *growing more slowly* or *growing faster*. The "big oh" notation, however, does not exactly match with *growing at the same rate*. If we know, for example, that $f = O(g)$, there is nothing we can conclude about how the growth rates of f and g compare. Why?

If you look at the definitions again, you will see that $f = o(g)$ implies $f = O(g)$. Also, if f and g grow at the same rate, then $f = O(g)$ and $g = O(f)$, as you will be asked to confirm in Exercise 39.

Sequential versus Binary Search

Your graphing utility works according to algorithms programmed into it. Computer scientists sometimes measure the efficiency of an algorithm by counting the number of steps a computer has to take to use the algorithm to do something. There can be significant differences in how efficiently algorithms perform, even if they are designed to accomplish the same task. These differences are often described in big-oh notation. Here is an example.

Webster's *Third New International Dictionary* lists about 26,000 words that begin with the letter a. One way to look up a word or find out that it is not there is to read through the list one word at a time from the beginning until you either find your word or determine that it is not there. This method, called **sequential search**, makes no particular use of the words' alphabetical arrangement. You can be sure of getting an answer this way, but it might take 26,000 steps.

Another way to find the word or determine that it is not there is to go straight to the middle of the list (give or take a few words). If you do not find the word, then go to the middle of the half that contains it and forget about the half that does not. (You know which half contains it because you know that the list is ordered alphabetically.) This method eliminates roughly 13,000 words in a single step. If you do not find your word on the second try, then jump to the middle of the half that contains it. Continue this way until you have either found your word or divided the list in half so many times that there are no words left. How many times do you have to divide the list to find the word or determine that it is not there? At most 15, because

$$2^{14} < 26{,}000 < 2^{15}.$$

That certainly beats a possible 26,000.

For a list of length n, a sequential search algorithm takes on the order of n steps to find a word or determine that it is not in the list. A **binary search**, as the second algorithm is called, takes on the order of $\log_2 n$ steps. The reason is that if $2^{m-1} < n \leq 2^m$, then $m - 1 < \log_2 n \leq m$, and the number of bisections required to narrow the list to one word will be at most $m = \lceil \log_2 n \rceil$, the smallest integer greater than or equal to $\log_2 n$.

Big-oh notation provides a compact way to say all this. The number of steps in a sequential search of an ordered list is $O(n)$, while the number of steps in a binary search is $O(\log_2 n)$. In our example, there is a big difference between the two, and the difference can only increase with n because n grows faster than $\log_2 n$ as $n \to \infty$.

A sequential search takes $O(n)$ steps.

A binary search takes $O(\log_2 n)$ steps.

Exercises 7.6

1. Which of the following functions grow faster than e^x as $x \to \infty$? Which grow at the same rate as e^x? Which grow more slowly?

a) $x + 3$ **b)** $x^3 - 3x + 1$

c) $\sqrt{x}$ **d)** 4^x

e) $(5/2)^x$ **f)** $\ln x$

g) $\log_{10} x$ **h)** e^{-x}

i) e^{x+1} **j)** $(1/2)e^x$

2. Which of the following functions grow faster than e^x as $x \to \infty$? Which grow at the same rate as e^x? Which grow more slowly?

a) $10x^4 + 30x + 1$ **b)** $x \ln x - x$

c) $\sqrt{1 + x^4}$ **d)** x^{1000}

e) $(e^x + e^{-x})/2$ **f)** xe^x

g) $e^{\cos x}$ **h)** e^{x-1}

3. Which of the following functions grow faster than x^2 as $x \to \infty$? Which grow at the same rate as x^2? Which grow more slowly?

a) $x^2 + 4x$ **b)** $x^3 + 3$

c) x^5 **d)** $15x + 3$

e) $\sqrt{x^4 + 5x}$ **f)** $(x + 1)^2$

g) $\ln x$ **h)** $\ln(x^2)$

i) $\ln(10^x)$ **j)** 2^x

4. Which of the following functions grow faster than $\ln x$ as $x \to \infty$? Which grow at the same rate as $\ln x$? Which grow more slowly?

a) $\log_3 x$ **b)** $\log_2 x^2$

c) $\log_{10} \sqrt{x}$ **d)** $1/x$

e) $1/\sqrt{x}$ **f)** e^{-x}

g) x **h)** $5 \ln x$

i) 2 **j)** $\sin x$

5. Order the following functions from slowest-growing to fastest-growing as $x \to \infty$.

a) e^x **b)** x^x

c) $(\ln x)^x$ **d)** $e^{x/2}$

6. Order the following functions from slowest-growing to fastest-growing as $x \to \infty$.

a) 2^x **b)** x^2

c) $(\ln 2)^x$ **d)** e^x

7. Show that $\sqrt{10x + 1}$ and $\sqrt{x + 1}$ grow at the same rate as $x \to \infty$ by showing that they both grow at the same rate as $\sqrt{x}$ as $x \to \infty$.

8. Show that $\sqrt{x^4 + x}$ and $\sqrt{x^4 - x^3}$ grow at the same rate as $x \to \infty$ by showing that they both grow at the same rate as x^2 as $x \to \infty$.

9. Show that $\sqrt{x^4 + x}$ and $\sqrt[3]{x^6 + x}$ grow at the same rate as $x \to \infty$ by showing that they both grow at the same rate as x^2 as $x \to \infty$.

10. Show that $\sqrt[4]{x^6 + x}$ and $\sqrt{x^3 - 4x}$ grow at the same rate as $x \to \infty$ by showing that they both grow at the same rate as $x^{3/2}$ as $x \to \infty$.

11. True or false? As $x \to \infty$,

a) $x = o(x)$ **b)** $x = o(x + 5)$

c) $x = O(x + 5)$ **d)** $x = O(2x)$

e) $e^x = o(e^{2x})$ **f)** $x + \ln x = O(x)$

g) $\ln x = o(\ln 2x)$ **h)** $\sqrt{x^2 + 5} = O(x)$

12. True or false? As $x \to \infty$,

a) $\dfrac{1}{x + 3} = O\left(\dfrac{1}{x}\right)$ **b)** $\dfrac{1}{x} + \dfrac{1}{x^2} = o\left(\dfrac{1}{x}\right)$

c) $\dfrac{1}{x} - \dfrac{1}{x^2} = o\left(\dfrac{1}{x}\right)$ **d)** $2 + \cos x = O(2)$

e) $e^x + x = O(e^x)$ **f)** $x \ln x = o(x^2)$

g) $\ln(\ln x) = O(\ln x)$ **h)** $\ln(x) = o(\ln(x^2 + 1))$

13. Confirm that e^x grows faster as $x \to \infty$ than x^n for any positive integer n, even $x^{1,000,000}$. (*Hint*: What is the nth derivative of x^n?)

14. *The function e^x grows faster than any polynomial.* Show that e^x grows faster as $x \to \infty$ than any polynomial

$$a_n x^n + a_{n-1} x^{n-1} + \cdots + a_1 x + a_0.$$

15. a) Show that $\ln x$ grows more slowly as $x \to \infty$ than $x^{1/n}$ for any positive integer n, even $x^{1/1,000,000}$.

b) There are two values of x for which $\ln x$ and $x^{1/1,000,000}$ are equal. One is a little larger than 1, and the other is very large. Find the large value of x.

16. *The function $\ln x$ grows more slowly than any polynomial.* Show that $\ln x$ grows more slowly as $x \to \infty$ than any non-constant polynomial.

17. Show that $y = 4x$ is a right-end behavior model for $y = \left(2\sqrt{x} - 1\right)^2$.

18. Show that $y = |x|$ is an end behavior model for $y = \sqrt{x^2 + 5}$.

19. Show that e^x is a right-end behavior model and x^2 is a left-end behavior model for $e^x + x^2$.

20. Show that if f grows faster than g and $\lim_{x \to -\infty} \dfrac{g}{f} = 0$, then f is an end behavior model for $f + g$.

In Exercises 21–30, determine whether g is an end behavior model for f. If so, then say whether g is a right-end behavior model, a left-end behavior model, or both.

21. $g(x) = x$, $f(x) = x + \sin x$

22. $g(x) = x^2$, $f(x) = x^2 - \cos x$

23. $g(x) = 2x^3$, $f(x) = 2x^3 - 3x^2 + x - 1$

24. $g(x) = \dfrac{3}{2}x^2$, $f(x) = \dfrac{3x^4 - x^3 + x - 1}{2x^2 + x - 1}$

25. $g(x) = x$, $f(x) = 2^x + x$

26. $g(x) = 2^x$, $f(x) = 2^x + x$

27. $g(x) = x^2$, $f(x) = \sqrt{x^4 + 2x - 1}$

28. $g(x) = x^{3/2}$, $f(x) = \sqrt{x^3 + 1}$

29. $g(x) = x^{2/3}$, $f(x) = \sqrt[3]{x^2 - 2x - 1}$

30. $g(x) = x^{1/4}$, $f(x) = \sqrt[4]{x + 2}$

31. Suppose that f_1 and g_1 are right-end behavior models for f and g, respectively. Show that $f_1 g_1$, f_1/g_1, $|f_1|$, $\sqrt[n]{|g_1|}$ are right-end behavior models for fg, f/g, $|f|$, $\sqrt[n]{|g|}$, respectively.

32. Repeat Exercise 31 replacing "right" by "left."

33. Repeat Exercise 31 deleting the word "right."

34. Show g is an end behavior model for f if and only if g is both a left- and a right-end behavior model for f.

35. Suppose that the polynomial $p(x) = a_n x^n + \cdots + a_0$, $a_n \neq 0$, grows faster than x. What can you say about n?

36. Suppose that the polynomials $p(x)$ and $q(x)$ grow at the same rate. What can you say about $\lim_{|x| \to \infty} (p(x)/q(x))$?

37. Suppose you have four different algorithms for solving the same problem and each algorithm takes a number of steps equal to one of the functions listed here:

$$n, \qquad n \log_2 n, \qquad n^2, \qquad n(\log_2 n)^2.$$

Which of the algorithms, if any, is the most efficient in the long run?

38. Suppose you are looking for an item in an ordered list that is one million items long. How many steps might it take to find that item with a sequential search? A binary search?

39. Show that if functions f and g grow at the same rate as $x \to \infty$ then $f = O(g)$ and $g = O(f)$.

40. *Simpson's rule and the trapezoidal rule.* The definitions in the present section can be made more general by lifting the restriction that $x \to \infty$ and considering limits as $x \to a$ for any real number a. Show that the error E_s in the Simpson's rule approximation of a definite integral is $O(h^4)$ as $h \to 0$, while the error E_T in the trapezoidal rule approximation is $O(h^2)$. This gives another way to explain the relative accuracies of the two approximation methods.

7.7 — The Inverse Trigonometric Functions

The inverse trigonometric functions arise in problems that require finding angles from side measurements in triangles. They also provide antiderivatives for a wide variety of functions and hence appear in solutions to a number of differential equations that arise in mathematics, engineering, and physics. In this section, we complete the definitions started in Section 1.7. In the next section, we look at their derivatives and integrals.

The trigonometric functions are not one-to-one, so their inverses are not functions. However, we can restrict their domains so that the resulting inverses are functions.

The Arc Sine, Arc Cosine, Arc Tangent

We repeat the definitions given in Section 1.7.

DEFINITIONS

The **inverse sine function**, denoted by $y = \sin^{-1} x$, or $y = \arcsin x$, is the function with domain $[-1, 1]$ and range $[-\pi/2, \pi/2]$ such that $x = \sin y$ (Fig. 7.33a).

The **inverse cosine function**, denoted by $y = \cos^{-1} x$, or $y = \arccos x$, is the function with domain $[-1, 1]$ and range $[0, \pi]$ such that $x = \cos y$ (Fig. 7.33b).

The **inverse tangent function**, denoted by $y = \tan^{-1} x$, or $y = \arctan x$, is the function with domain $(-\infty, \infty)$ and range $(-\pi/2, \pi/2)$ such that $x = \tan y$ (Fig. 7.33c).

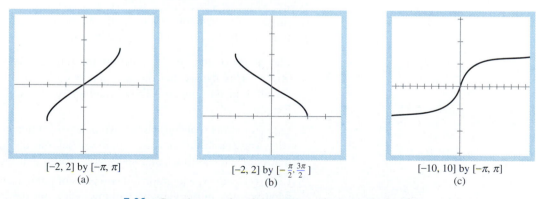

$[-2, 2]$ by $[-\pi, \pi]$
(a)

$[-2, 2]$ by $[-\frac{\pi}{2}, \frac{3\pi}{2}]$
(b)

$[-10, 10]$ by $[-\pi, \pi]$
(c)

7.33 Complete graphs of the inverse trigonometric functions:
a) $y = \sin^{-1} x$, $-1 \le x \le 1$, $-\pi/2 \le y \le \pi/2$,
b) $y = \cos^{-1} x$, $-1 \le x \le 1$, $0 \le y \le \pi$, and
c) $y = \tan^{-1} x$, $-\infty < x < \infty$, $-\pi/2 < y < \pi/2$.

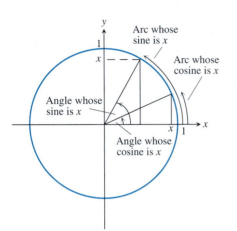

7.34 Arcs on the unit circle whose lengths represent $\sin^{-1} x$ and $\cos^{-1} x$.

EXPLORATION 1

The Inverse Trigonometric Functions

Most graphers can plot the inverse trigonometric functions in function mode and in parametric mode. It is instructive to try them both ways.

1. GRAPH $\sin^{-1} x$, $\cos^{-1} x$, and $\tan^{-1} x$ if they are library functions in your grapher. Use the definitions as a guide for setting [xMin, xMax] and [yMin, yMax].

2. With your grapher in parametric mode, set

$$x_1(t) = t, \qquad y_1(t) = \sin t,$$
$$x_2(t) = y_1(t), \quad y_2(t) = x_1(t)$$

with [tMin, tMax] = $[-\pi/2, \pi/2]$, and GRAPH. Explain what you see in larger [tMin, tMax] intervals. GRAPH $\cos^{-1} x$ and $\tan^{-1} x$ in a similar manner.

In case you are wondering what the "arc" is doing in the definitions, look at Fig. 7.34, which gives the geometric interpretation of $y = \sin^{-1} x$ and $y = \cos^{-1} x$ when y is positive. If $x = \sin y$, then y is the *arc* on the unit circle whose sine is x. For every value of x in the interval $[-1, 1]$, $y = \sin^{-1} x$ is the number in the interval $[-\pi/2, \pi/2]$ whose sine is x.

EXPLORATION 2

$f(x)$ and $f(-x)$

With your grapher in parametric mode, set

$$x_1(t) = t, \qquad y_1(t) = \sin t,$$
$$x_2(t) = y_1(t), \quad y_2(t) = x_1(t)$$

with [tMin, tMax] = $[-2\pi, 2\pi]$, t-step = 0.1, [xMin, xMax] = $[-2, 2]$, and [yMin, yMax]=$[-2\pi, 2\pi]$. GRAPH, and you will see part of the graph of the inverse sine relation. (Why is the graph not that of a function? What portion of it is the graph of the $\sin^{-1}$ function?)

1. The graph of the $\sin^{-1} t$ function (or relation) is symmetric about the origin. This means that the function is odd and $\sin^{-1}(-x) = -\sin^{-1} x$ for all x in its domain. Show graphically that $\sin^{-1}(-x) = -\sin^{-1} x$. Explain.

2. Let $y_1(t) = \tan t$, and graph the two pairs of parametric equations. Explain what you see in the viewing window. Pick out the graph of the function that is of interest to us, and show graphically that the function is odd.

3. Let $y_1(t) = \cos t$, and graph the two pairs of parametric equations. Explain what you see in the viewing window. Pick out the graph of the function that is of interest to us. Explain why $\cos^{-1} t$ is neither odd nor even. But then how are $\cos^{-1} t$ and $\cos^{-1}(-t)$ related? Use your grapher to help you make a conjecture. (*Hint:* Graph $-\cos^{-1} t$ and $\cos^{-1}(-t)$, and find a relationship between them.)

EXPLORATION BIT

Show how to graphically support the identity

$$\sin^{-1} x + \cos^{-1} x = \pi/2.$$

The relationship

$$\sin^{-1}(-x) = -\sin^{-1} x$$

from Exploration 2, part 1, and the relationship

$$\cos^{-1}(-x) = -\cos^{-1} x + \pi,$$

which you may have discovered in part 3, as well as the relationship

$$\sin^{-1} x + \cos^{-1} x = \pi/2$$

can be supported geometrically by using the unit circle as in Fig. 7.34. You will be asked to do this in the exercises.

WARNING ABOUT ARCSECANT

There is no general agreement about how to define $\sec^{-1} x$ for negative values of x. We would have chosen $\sec^{-1} x$ to lie between 0 and $\pi/2$ when x is positive and between $-\pi$ and $-\pi/2$ when x is negative (hence as a negative angle in the third quadrant). This has the advantage of simplifying the formula for the derivative of $\sec^{-1} x$ but the disadvantage of failing to satisfy the equation $\sec^{-1} x = \cos^{-1}(1/x)$ when x is negative.

The Inverses of sec x, csc x, and cot x

The other three basic trigonometric functions, $y = \sec x = \dfrac{1}{\cos x}$, $y = \csc x = \dfrac{1}{\sin x}$, and $y = \cot x = \dfrac{1}{\tan x}$, also have inverses when suitably restricted. Reviewing their graphs on a graphing utility or in Section 1.7 will help you to understand the following restrictions on the domains:

$$y = \sec x, \qquad\qquad 0 \le x \le \pi, \ x \ne \pi/2,$$

$$y = \csc x, \qquad\qquad -\pi/2 \le x \le \pi/2, \ x \ne 0,$$

$$y = \cot x, \qquad\qquad 0 < x < \pi.$$

The inverses of these functions are graphed in Fig. 7.35.

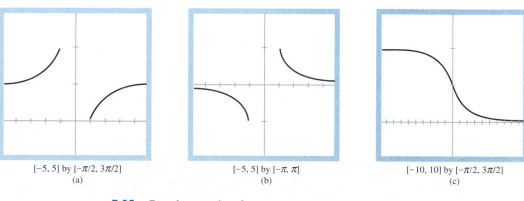

$[-5, 5]$ by $[-\pi/2, 3\pi/2]$	$[-5, 5]$ by $[-\pi, \pi]$	$[-10, 10]$ by $[-\pi/2, 3\pi/2]$
(a)	(b)	(c)

7.35 Complete graphs of

(a) $y = \sec^{-1} x = \cos^{-1}(1/x)$, $|x| \ge 1$, y in $[0, \pi/2) \cup (\pi/2, \pi]$,

(b) $y = \csc^{-1} x = \sin^{-1}(1/x)$, $|x| \ge 1$, y in $[-\pi/2, 0) \cup (0, \pi/2]$,

(c) $y = \cot^{-1} x = \pi/2 - \tan^{-1} x$, $-\infty < x < \infty$, $0 < y < \pi$.

Note how we use the $\cos^{-1}$, $\sin^{-1}$, and $\tan^{-1}$ functions to graph the $\sec^{-1}$, $\csc^{-1}$, and $\cot^{-1}$ functions, respectively.

EXPLORATION BIT

In Fig. 7.35, we note that

$$\sec^{-1} x = \cos^{-1} (1/x),$$

and

$$\csc^{-1} x = \sin^{-1} (1/x).$$

Explain why

$$\cot^{-1} x \neq \tan^{-1} (1/x).$$

Right-triangle Interpretations

The right-triangle interpretations of the inverse trigonometric functions in Fig. 7.36 can be useful in integration problems that require substitutions. We will use some of them in Chapter 8.

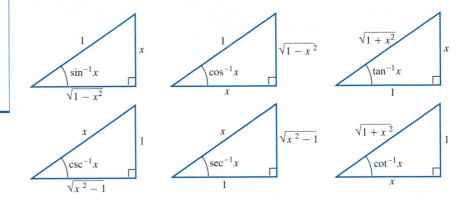

7.36   Right-triangle interpretations of the inverse trigonometric function angles (first-quadrant values).

EXAMPLE 1   Common Values of $\sin^{-1} x$.

$\sin^{-1} x$ is the angle whose sine is x, which can be written as the ratio $x/1$. We draw a right triangle, choose an acute angle, and mark the side opposite as x and the hypotenuse as 1. (Recall that for a right triangle, sine = opposite/hypotenuse.) The Pythagorean Theorem tells us that the third side is $\sqrt{1 - x^2}$.

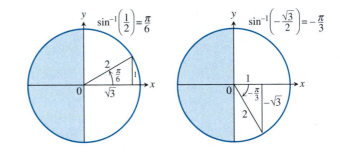

The angles come from the first and fourth quadrant because the range of $\sin^{-1} x$ is $[-\pi/2, \pi/2]$.   ≡

EXAMPLE 2   Common Values of $\sec^{-1} x$.

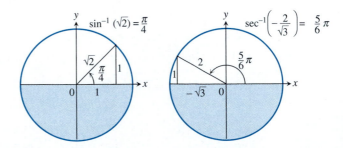

The angles come from the first and second quadrants because the range of $\sec^{-1} x$ is $[0, \pi/2) \cup (\pi/2, \pi]$.   ≡

EXPLORATION 3

Finding Angles

Use reference triangles to find the angles.

1. $\cos^{-1}\left(\dfrac{\sqrt{3}}{2}\right)$, $\cos^{-1}\left(\dfrac{1}{\sqrt{2}}\right)$, $\cos^{-1}\left(-\dfrac{1}{2}\right)$, $\cos^{-1}\left(-\dfrac{\sqrt{3}}{2}\right)$

2. $\tan^{-1}\left(\sqrt{3}\right)$, $\tan^{-1}(1)$, $\tan^{-1}\left(-\dfrac{1}{\sqrt{3}}\right)$, $\tan^{-1}\left(-\sqrt{3}\right)$

3. $\csc^{-1}\left(\sqrt{2}\right)$, $\csc^{-1}\left(\dfrac{2}{\sqrt{3}}\right)$, $\csc^{-1}(-2)$, $\csc^{-1}\left(-\dfrac{2}{\sqrt{3}}\right)$

4. $\cot^{-1}(1)$, $\cot^{-1}\left(\dfrac{1}{\sqrt{3}}\right)$, $\cot^{-1}\left(-\dfrac{1}{\sqrt{3}}\right)$, $\cot^{-1}\left(-\sqrt{3}\right)$

EXAMPLE 3

Find $\csc\alpha$, $\cos\alpha$, $\sec\alpha$, $\tan\alpha$, and $\cot\alpha$ if

$$\alpha = \sin^{-1}\frac{\sqrt{3}}{2}. \tag{1}$$

Solution   Equation (1) tells us that

$$\sin\alpha = \frac{\sqrt{3}}{2}.$$

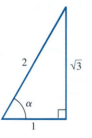

7.37 If $\alpha = \sin^{-1}(\sqrt{3}/2)$, then the values of the trigonometric functions of α can be read from this triangle.

We draw a reference triangle with vertical leg $\sqrt{3}$ and hypotenuse 2 (Fig. 7.37). The length of the remaining side is

$$\sqrt{(2)^2 - (\sqrt{3})^2} = \sqrt{4 - 3} = 1.$$

The values we want can be read as ratios of side lengths from the completed triangle:

$$\csc\alpha = \frac{2}{\sqrt{3}} = \frac{2\sqrt{3}}{3}, \qquad \cos\alpha = \frac{1}{2}, \qquad \sec\alpha = 2,$$

$$\tan\alpha = \frac{\sqrt{3}}{1} = \sqrt{3}, \qquad \cot\alpha = \frac{1}{\sqrt{3}} = \frac{\sqrt{3}}{3}.$$

EXAMPLE 4

Find $\cot\left[\sec^{-1}\left(-\dfrac{2}{\sqrt{3}}\right) + \csc^{-1}(-2)\right]$.

Solution   We work from inside out.

STEP 1: $\sec^{-1}\left(-\dfrac{2}{\sqrt{3}}\right) = \dfrac{5}{6}\pi$ Example 2

$\csc^{-1}(-2) = -\dfrac{\pi}{6}$ Exploration 3

STEP 2:
$$\cot\left[\sec^{-1}\left(-\frac{2}{\sqrt{3}}\right) + \csc^{-1}(-2)\right]$$

$$= \cot\left(\frac{5\pi}{6} - \frac{\pi}{6}\right)$$

$$= \cot\left(\frac{2\pi}{3}\right)$$

$$= -\frac{1}{\sqrt{3}}$$

EXAMPLE 5

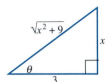

Find $\sec\left(\tan^{-1}\frac{x}{3}\right)$.

Solution Let $\theta = \tan^{-1}(x/3)$ and picture θ in a right triangle with opposite/adjacent = $x/3$, as shown at left. The length of the triangle's hypotenuse is $\sqrt{x^2+9}$. Thus, $\sec\left(\tan^{-1} x/3\right) = \sec\theta = \sqrt{x^2+9}/3$.

EXAMPLE 6 Drift Correction

During an airplane flight from Chicago to St. Louis, the navigator determines that the plane is 12 miles off course, as shown in Fig. 7.38. Find the angle a for a course parallel to the original, correct course, the angle b, and the correction angle $c = a + b$.

Solution

$$a = \sin^{-1}\frac{12}{180} = 0.067 \text{ radians } = 3.8°,$$

$$b = \sin^{-1}\frac{12}{62} = 0.195 \text{ radians } = 11.2°,$$

$$c = a + b = 15°.$$

7.38 Diagram for drift correction (Example 6), with distances rounded to the nearest mile (drawing not to scale).

Exercises 7.7

Use reference triangles like those in Figs. 7.36 and 7.37 to find the angles in Exercises 1–12.

1. a) $\tan^{-1} 1$ **b)** $\tan^{-1}\sqrt{3}$ **c)** $\tan^{-1}\left(\frac{1}{\sqrt{3}}\right)$

2. a) $\tan^{-1}(-1)$ **b)** $\tan^{-1}(-\sqrt{3})$ **c)** $\tan^{-1}\left(\frac{-1}{\sqrt{3}}\right)$

3. a) $\sin^{-1}\left(\frac{-1}{2}\right)$ **b)** $\sin^{-1}\left(\frac{-1}{\sqrt{2}}\right)$ **c)** $\sin^{-1}\left(\frac{-\sqrt{3}}{2}\right)$

4. a) $\sin^{-1}\left(\frac{1}{2}\right)$ **b)** $\sin^{-1}\left(\frac{1}{\sqrt{2}}\right)$ **c)** $\sin^{-1}\left(\frac{\sqrt{3}}{2}\right)$

5. a) $\cos^{-1}\left(\frac{1}{2}\right)$ **b)** $\cos^{-1}\left(\frac{1}{\sqrt{2}}\right)$ **c)** $\cos^{-1}\left(\frac{\sqrt{3}}{2}\right)$

6. a) $\cos^{-1}\left(\frac{-1}{2}\right)$ **b)** $\cos^{-1}\left(\frac{-1}{\sqrt{2}}\right)$ **c)** $\cos^{-1}\left(\frac{-\sqrt{3}}{2}\right)$

7. a) $\sec^{-1}(-\sqrt{2})$ **b)** $\sec^{-1}\left(\frac{-2}{\sqrt{3}}\right)$ **c)** $\sec^{-1}(-2)$

8. a) $\sec^{-1}\sqrt{2}$ **b)** $\sec^{-1}\left(\frac{2}{\sqrt{3}}\right)$ **c)** $\sec^{-1} 2$

9. a) $\csc^{-1}\sqrt{2}$ **b)** $\csc^{-1}\left(\frac{2}{\sqrt{3}}\right)$ **c)** $\csc^{-1} 2$

10. a) $\csc^{-1}(-\sqrt{2})$ **b)** $\csc^{-1}\left(\dfrac{-2}{\sqrt{3}}\right)$ **c)** $\csc^{-1}(-2)$

11. a) $\cot^{-1}(-1)$ **b)** $\cot^{-1}(-\sqrt{3})$ **c)** $\cot^{-1}\left(\dfrac{-1}{\sqrt{3}}\right)$

12. a) $\cot^{-1} 1$ **b)** $\cot^{-1} \sqrt{3}$ **c)** $\cot^{-1}\left(\dfrac{1}{\sqrt{3}}\right)$

13. Given that $\alpha = \sin^{-1}(1/2)$, find $\cos\alpha, \tan\alpha, \sec\alpha, \csc\alpha$.

14. Given that $\alpha = \cos^{-1}(-1/2)$, find $\sin\alpha, \tan\alpha, \sec\alpha, \csc\alpha$.

15. Given that $\alpha = \tan^{-1}(4/3)$, find $\sin\alpha, \cos\alpha, \sec\alpha, \csc\alpha$, and $\cot\alpha$.

16. Given that $\alpha = \sec^{-1}(-\sqrt{5})$, find $\sin\alpha, \cos\alpha, \tan\alpha, \csc\alpha$, and $\cot\alpha$.

Evaluate the expressions in Exercises 17–28.

17. $\sin\left[\cos^{-1}\dfrac{\sqrt{2}}{2}\right]$ **18.** $\sec\left[\cos^{-1}\dfrac{1}{2}\right]$

19. $\tan\left[\sin^{-1}\left(-\dfrac{1}{2}\right)\right]$ **20.** $\cot\left[\sin^{-1}\left(-\dfrac{\sqrt{3}}{2}\right)\right]$

21. $\csc\left(\sec^{-1} 2\right) + \cos\left(\tan^{-1}(-\sqrt{3})\right)$

22. $\tan\left(\sec^{-1} 1\right) + \sin\left(\csc^{-1}(-2)\right)$

23. $\sin\left(\sin^{-1}\left(-\dfrac{1}{2}\right) + \cos^{-1}\left(-\dfrac{1}{2}\right)\right)$

24. $\cot\left(\sin^{-1}\left(-\dfrac{1}{2}\right) - \sec^{-1} 2\right)$

25. $\sec\left(\tan^{-1} 1 + \csc^{-1} 1\right)$

26. $\sec\left(\cot^{-1}\sqrt{3} + \csc^{-1}(-1)\right)$

27. $\sec^{-1}\left(\sec\left(-\dfrac{\pi}{6}\right)\right)$

28. $\cot^{-1}\left(\cot\left(-\dfrac{\pi}{4}\right)\right)$

Evaluate the expressions in Exercises 29–40.

29. $\sec\left[\tan^{-1}\dfrac{x}{2}\right]$ **30.** $\sec\left(\tan^{-1} 2x\right)$

31. $\tan\left(\sec^{-1} 3y\right)$ **32.** $\tan\left[\sec^{-1}\dfrac{y}{5}\right]$

33. $\cos\left(\sin^{-1} x\right)$ **34.** $\tan\left(\cos^{-1} x\right)$

35. $\sin\left[\tan^{-1}\sqrt{x^2 - 2x}\right]$ **36.** $\sin\left[\tan^{-1}\dfrac{x}{\sqrt{x^2+1}}\right]$

37. $\cos\left[\sin^{-1}\dfrac{2y}{3}\right]$ **38.** $\cos\left[\sin^{-1}\dfrac{y}{5}\right]$

39. $\sin\left[\sec^{-1}\dfrac{x}{4}\right]$ **40.** $\sin\left[\sec^{-1}\dfrac{\sqrt{x^2+4}}{x}\right]$

Find the limits in Exercises 41–48. (If in doubt, look at the function's graph.)

41. $\displaystyle\lim_{x\to 1^-} \sin^{-1} x$ **42.** $\displaystyle\lim_{x\to -1^+} \cos^{-1} x$

43. $\displaystyle\lim_{x\to\infty} \tan^{-1} x$ **44.** $\displaystyle\lim_{x\to -\infty} \tan^{-1} x$

45. $\displaystyle\lim_{x\to\infty} \sec^{-1} x$ **46.** $\displaystyle\lim_{x\to -\infty} \sec^{-1} x$

47. $\displaystyle\lim_{x\to\infty} \csc^{-1} x$ **48.** $\displaystyle\lim_{x\to -\infty} \csc^{-1} x$

49. You are sitting in a classroom next to the wall looking at the blackboard at the front of the room. The blackboard is 12 ft long and starts 3 ft from the wall you are sitting next to. Show that your viewing angle is

$$\alpha = \cot^{-1}\dfrac{x}{15} - \cot^{-1}\dfrac{x}{3}$$

if you are x ft from the front wall.

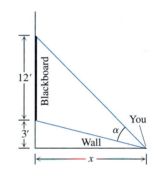

50. The region between the curve $y = \sec^{-1} x$ and the x-axis between $x = 1$ and $x = 2$ is revolved about the y-axis to generate a solid. Find the volume of the solid. (*Hint:* Use washers.)

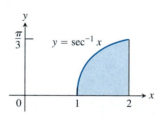

51. The cone has slant height 3 m. How large should the indicated angle be to maximize the cone's volume?

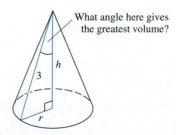

What angle here gives the greatest volume?

52. Find the value of α.

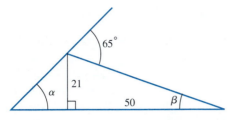

53. Find the values of $\cot^{-1} 2$, $\sec^{-1} (1.5)$, and $\csc^{-1} (1.5)$.

54. Here is an informal proof that

$$\tan^{-1} 1 + \tan^{-1} 2 + \tan^{-1} 3 = \pi.$$

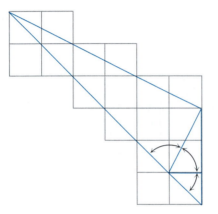

Explain this "picture proof." (*Source*: Edward M. Harris, *Behold! Sums of Arctan*, The College Mathematics Journal, Vol. 18, No. 2, Page 141.)

Confirm the identities in Exercises 55–58.

55. $\pi - \cos^{-1} x = \cos^{-1} (-x)$

56. $\sin^{-1} x + \cos^{-1} x = \pi/2$

57. $\sin^{-1} (-x) = -\sin^{-1} x$

58. $\tan^{-1} (-x) = -\tan^{-1} x$

59. Show that $y = \pi/2$ and $y = -\pi/2$ are right- and left-end behavior models, respectively, of $y = \tan^{-1} x$.

60. Determine the domain and range and draw a complete graph of $y = \sin\left(\sin^{-1} (x)\right)$.

Draw a complete graph of the functions in Exercises 61–66.

61. $y = \csc^{-1} (2x)$ **62.** $y = 3 \tan^{-1} x$

63. $y = 2 \sec^{-1} (3x)$ **64.** $y = \cot^{-1} (x + 2)$

65. $y = 3 + \cos^{-1} (x - 2)$ **66.** $y = \sin^{-1} (x - 3) - 2$

7.8

Derivatives of Inverse Trigonometric Functions; Related Integrals

In this section, we show how to confirm the derivatives of the inverse trigonometric functions, list the standard formulas for the derivatives, and discuss their companion integral formulas. As we will see, the restrictions on the domains of the inverse trigonometric functions show up in natural ways as restrictions on the domains of the derivatives.

EXPLORATION 1

Using Mental Checks for Support

As you proceed in mathematics, you acquire a great number of mental checks that you can use to support your work. We are about to show that

$$\frac{d}{dx} \sin^{-1} x = \frac{1}{\sqrt{1 - x^2}}.$$

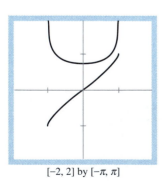

$[-2, 2]$ by $[-\pi, \pi]$

7.39   The graphs of $y_1 = \sin^{-1} x$ and $y_2 = $ NDER y_1.

The graphs of $y_1 = \sin^{-1} x$ and $y_2 = $ NDER y_1 are shown in Fig. 7.39. Let's check out mentally some things that we should already know.

1. What kind of symmetry does $\sin^{-1} x$ have? How do you know?

2. Knowing the type of symmetry of y_1, what type of symmetry should y_2 have? Does the graph of y_2 agree?

3. Thinking of y_1' as the slope of the graph of y_1, is y_1' ever positive? Negative? 0? Does the graph of y_2 agree?

4. When is the slope a minimum? Estimate its minimum value. Does the graph of y_2 agree?

5. What happens to the slope of y_1 near the endpoints of the domain? Does the graph of y_2 agree?

6. Which of parts 2–5 can you confirm mentally for the actual derivative given above?   ☘

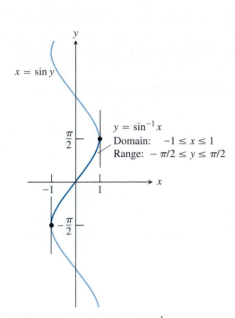

$x = \sin y$

$y = \sin^{-1} x$
Domain: $-1 \le x \le 1$
Range: $-\pi/2 \le y \le \pi/2$

7.40   The graph of $y = \sin^{-1} x$.

The Derivative of $y = \sin^{-1} u$

We know that the function $x = \sin y$ is differentiable in the open interval $-\pi/2 < y < \pi/2$ and that its derivative, the cosine, is positive there. The derivative rule for inverses in Section 7.2 therefore assures us that the inverse function $y = \sin^{-1} x$ is differentiable throughout the interval $-1 < x < 1$. We cannot expect it to be differentiable at $x = 1$ or $x = -1$, however, because the tangents to the graph are vertical at these points. (See Fig. 7.40).

To calculate the derivative of $y = \sin^{-1} x$, we differentiate both sides of the equation $\sin y = x$ with respect to x:

$$\sin y = x,$$

$$\frac{d}{dx} \sin y = 1,$$

$$\cos y \frac{dy}{dx} = 1.$$

We then divide through by $\cos y$ (> 0 for $-\pi/2 < y < \pi/2$) to get

$$\frac{dy}{dx} = \frac{1}{\cos y} = \frac{1}{\sqrt{1 - \sin^2 y}} = \frac{1}{\sqrt{1 - x^2}}.$$

The derivative of $y = \sin^{-1} x$ with respect to x is

$$\frac{d}{dx} \sin^{-1} x = \frac{1}{\sqrt{1 - x^2}}.$$

If u is a differentiable function of x with $|u| < 1$, we apply the Chain Rule in the form

$$\frac{dy}{dx} = \frac{dy}{du} \frac{du}{dx}$$

to $y = \sin^{-1} u$ to obtain

$$\frac{d}{dx} \sin^{-1} u = \frac{1}{\sqrt{1 - u^2}} \frac{du}{dx}.$$

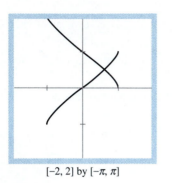

$[-2, 2]$ by $[-\pi, \pi]$

7.41 Graphs of the functions $y_1 = \sin^{-1} x$ and $y_2 = \cos^{-1} x$.

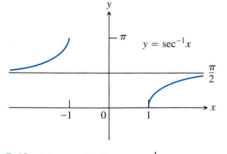

7.42 The graph of $y = \sec^{-1} x$.

EXPLORATION 2

Seeing Transformations

The graphs of the functions $y_1 = \sin^{-1} x$ and $y_2 = \cos^{-1} x$ appear to be congruent (Fig. 7.41).

1. What transformations can you apply to the graph of y_1 to obtain the graph of y_2?

2. What effect do these transformations have on the graph of the derivative of y_1?

3. Knowing that $y_1' = 1/\sqrt{1 - x^2}$, what is a reasonable prediction for y_2'?

The Derivative of $y = \sec^{-1} u$

We begin by differentiating both sides of the equation $\sec y = x$ with respect to x:

$$\sec y = x,$$

$$\frac{d}{dx} \sec y = 1,$$

$$\sec y \tan y \frac{dy}{dx} = 1,$$

$$\frac{dy}{dx} = \frac{1}{\sec y \tan y}.$$

To express the result in terms of x, we use the relations

$$\sec y = x \qquad \text{and} \qquad \tan y = \pm\sqrt{\sec^2 y - 1} = \pm\sqrt{x^2 - 1}.$$

Hence,

$$\frac{dy}{dx} = \pm\frac{1}{x\sqrt{x^2 - 1}}.$$

What do we do about the sign? A glance at Fig. 7.42 shows that the slope of the graph of $y = \sec^{-1} x$ is always positive. Therefore,

$$\frac{d}{dx} \sec^{-1} x = \begin{cases} \dfrac{1}{x\sqrt{x^2 - 1}} & \text{if } x > 1, \\ -\dfrac{1}{x\sqrt{x^2 - 1}} & \text{if } x < -1. \end{cases} \qquad (1)$$

With absolute values, we can write Eq. (1) as a single formula:

$$\frac{d}{dx} \sec^{-1} x = \frac{1}{|x|\sqrt{x^2 - 1}}, \qquad |x| > 1.$$

We can then apply the Chain Rule to obtain

$$\frac{d}{dx} \sec^{-1} u = \frac{1}{|u|\sqrt{u^2 - 1}} \frac{du}{dx}, \qquad |u| > 1,$$

where u is a differentiable function of x.

Derivative and Integral Formulas

Following is a list of the standard formulas for derivatives of inverse trigonometric functions and corresponding integrals that lead to inverse trigonometric functions. We have proved derivative formulas 1 and 5. You will be asked to verify the others in the exercises. The derivative formulas lead immediately to six new integration formulas, but only the three that matter are shown.

EXPLORATION BITS

1. Make a statement of how the derivative of the inverse of a function is related to the derivative of the inverse of its cofunction.

2. Explain why the three integration formulas given are "the only ones that matter."

Derivatives of Inverse Trigonometric Functions

1. $\dfrac{d(\sin^{-1} u)}{dx} = \dfrac{du/dx}{\sqrt{1 - u^2}}, \quad -1 < u < 1$

2. $\dfrac{d(\cos^{-1} u)}{dx} = -\dfrac{du/dx}{\sqrt{1 - u^2}}, \quad -1 < u < 1$

3. $\dfrac{d(\tan^{-1} u)}{dx} = \dfrac{du/dx}{1 + u^2}$

4. $\dfrac{d(\cot^{-1} u)}{dx} = -\dfrac{du/dx}{1 + u^2}$

5. $\dfrac{d(\sec^{-1} u)}{dx} = \dfrac{du/dx}{|u|\sqrt{u^2 - 1}}, \quad |u| > 1$

6. $\dfrac{d(\csc^{-1} u)}{dx} = \dfrac{-du/dx}{|u|\sqrt{u^2 - 1}}, \quad |u| > 1$

Integrals Leading to Inverse Trigonometric Functions

1. $\displaystyle\int \dfrac{du}{\sqrt{1 - u^2}} = \sin^{-1} u + C \qquad$ Valid for $u^2 < 1$

2. $\displaystyle\int \dfrac{du}{1 + u^2} = \tan^{-1} u + C \qquad$ Valid for all u

3. $\displaystyle\int \dfrac{du}{u\sqrt{u^2 - 1}} = \int \dfrac{d(-u)}{(-u)\sqrt{u^2 - 1}} = \sec^{-1} |u| + C = \cos^{-1} \left|\dfrac{1}{u}\right| + C$

Valid for $u^2 > 1$

SUPPORT

Support each calculation in Example 1. For example, for part (a), let $y_1 = \sin^{-1} x^2$, $y_2 = $ NDER y_1, and $y_3 = 2x/\sqrt{1 - x^4}$. Then GRAPH y_2 and y_3, and explain how your viewing window supports the calculated result.

EXAMPLE 1

a) $\dfrac{d}{dx} \sin^{-1} x^2 = \dfrac{1}{\sqrt{1 - (x^2)^2}} \cdot \dfrac{d}{dx}(x^2) = \dfrac{2x}{\sqrt{1 - x^4}}$

b) $\dfrac{d}{dx} \tan^{-1} \sqrt{x + 1} = \dfrac{1}{1 + (\sqrt{x + 1})^2} \cdot \dfrac{d}{dx}(\sqrt{x + 1})$

$= \dfrac{1}{x + 2} \cdot \dfrac{1}{2\sqrt{x + 1}} = \dfrac{1}{2\sqrt{x + 1}(x + 2)}$

c) $\dfrac{d}{dx} \sec^{-1} (3x) = \dfrac{1}{|3x|\sqrt{(3x)^2 - 1}} \cdot \dfrac{d}{dx}(3x)$

$= \dfrac{3}{|3x|\sqrt{9x^2 - 1}} = \dfrac{1}{|x|\sqrt{9x^2 - 1}}.$

EXPLORATION 3

Drawing a Complete Graph

1. Let $y_1 = f(x) = \sin^{-1} x^2$, $y_2 = \text{NDER } y_1$, and $y_3 = \text{NDER2 } y_1$. GRAPH y_1, y_2, and y_3 in appropriate viewing windows (Fig. 7.43).

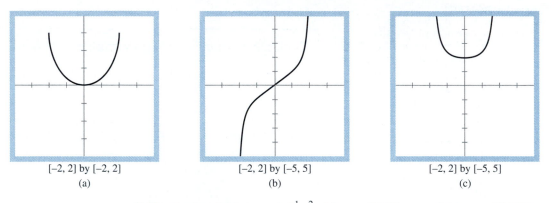

[−2, 2] by [−2, 2]   [−2, 2] by [−5, 5]   [−2, 2] by [−5, 5]

(a)   (b)   (c)

7.43 Graphs of (a) $y_1 = \sin^{-1} x^2$, (b) $y_2 = \text{NDER } y_1$, and (c) $y_3 = \text{NDER2 } y_1$. Information from viewing windows (b) and (c) help support that the graph of y_1 is complete.

2. Confirm that the graph of y_1 is complete. Use $y_2 = f'(x) = 2x/\sqrt{1 - x^4}$ (from Example 1), and confirm the intervals on which f is rising and falling. Determine the extreme values of f on $[-1, 1]$. Use y_3 to determine the intervals on which f is concave up and concave down. Confirm the intercepts of f.   ✢

EXAMPLE 2

Evaluate analytically. Support with an NINT computation.

a) $\displaystyle\int_0^1 \frac{dx}{1 + x^2}$ **b)** $\displaystyle\int_{2/\sqrt{3}}^{\sqrt{2}} \frac{dx}{x\sqrt{x^2 - 1}}$

Solution

a) $\displaystyle\int_0^1 \frac{dx}{1 + x^2} = \tan^{-1} x \Big]_0^1 = \tan^{-1} 1 - \tan^{-1} 0 = \frac{\pi}{4} - 0 = \frac{\pi}{4} = 0.785.$

Support: NINT $(1/(1 + x^2), 0, 1) = 0.785$.

b) $\displaystyle\int_{2/\sqrt{3}}^{\sqrt{2}} \frac{dx}{x\sqrt{x^2 - 1}} = \sec^{-1} x \Big]_{2/\sqrt{3}}^{\sqrt{2}} = \frac{\pi}{4} - \frac{\pi}{6} = \frac{\pi}{12} = 0.262.$ (See

Fig. 7.44). Support: NINT $(1/(x\sqrt{x^2 - 1}), 2/\sqrt{3}, \sqrt{2}) = 0.262$.   ☰

EXAMPLE 3

Evaluate

$$\int \frac{x^2 \, dx}{\sqrt{1 - x^6}}.$$

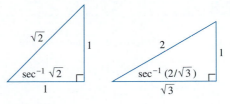

An isosceles right triangle:
$$\sec^{-1}\sqrt{2} = \frac{\pi}{4}$$

A 30-60-90 triangle:
$$\sec^{-1}\left(\frac{2}{\sqrt{3}}\right) = \frac{\pi}{6}$$

7.44 If you do not see right away how to evaluate $\sec^{-1}x$ at the limits of integration, use triangles like these or a calculator to help you.

Solution The resemblance between the given integral and the standard form

$$\int \frac{du}{\sqrt{1-u^2}} = \sin^{-1}u + C$$

suggests the substitution

$$u = x^3, du = 3x^2\,dx.$$

Indeed,

$$\int \frac{x^2\,dx}{\sqrt{1-x^6}} = \frac{1}{3}\int \frac{3x^2\,dx}{\sqrt{1-(x^3)^2}} = \frac{1}{3}\int \frac{du}{\sqrt{1-u^2}}$$

$$= \frac{1}{3}\sin^{-1}u + C = \frac{1}{3}\sin^{-1}(x^3) + C.$$

EXPLORATION 4

Do Analytically, Support Graphically

Evaluate $\int (1/\sqrt{9-x^2})\,dx$. Support graphically. *Hint:* To evaluate, you should recognize that this is the integral of arcsin x with a 9 in place of the 1. So a first step is to factor out a 9:

$$\sqrt{9-x^2} = \sqrt{9\left(1-\frac{x^2}{9}\right)} = 3\sqrt{1-\left(\frac{x}{3}\right)^2}.$$

Then compute

$$\int \frac{dx}{3\sqrt{1-(x/3)^2}}$$

by substituting $u = x/3$.

To support graphically, let $y_1 = 1/\sqrt{9-x^2}$, y_2 = NINT y_1, and y_3 = your computed result. GRAPH, compare y_2 and y_3, and explain.

Provide additional support by applying the toolbox program SLOPEFLD to y_1, for $-3 \le x \le 3$, and compare with the graph of y_3.

Exercises 7.8

Find dy/dx in Exercises 1–18 analytically, and support graphically.

1. $y = \cos^{-1} x^2$

2. $y = \cos^{-1} (1/x)$

3. $y = 5 \tan^{-1} 3x$

4. $y = \cot^{-1} \sqrt{x}$

5. $y = \sin^{-1} (x/2)$

6. $y = \sin^{-1} (1 - x)$

7. $y = \sec^{-1} 5x$

8. $y = (1/3) \tan^{-1} (x/3)$

9. $y = \csc^{-1} (x^2 + 1)$

10. $y = \cos^{-1} 2x$

11. $y = \csc^{-1} \sqrt{x} + \sec^{-1} \sqrt{x}$

12. $y = \csc^{-1} \dfrac{1}{x}, x > 0$

13. $y = \cot^{-1} \sqrt{x - 1}$

14. $y = x\sqrt{1 - x^2} - \cos^{-1} x$

15. $y = \sqrt{x^2 - 1} - \sec^{-1} x$

16. $y = \cot^{-1} \dfrac{1}{x} - \tan^{-1} x$

17. $y = 2x \tan^{-1} x - \ln (x^2 + 1)$

18. $y = \ln (x^2 + 1) - 2x + 2 \tan^{-1} x$

Evaluate the integrals in Exercises 19–30 analytically. Support with a NINT computation.

19. $\displaystyle\int_0^{1/2} \frac{dx}{\sqrt{1 - x^2}}$

20. $\displaystyle\int_{-1}^{1} \frac{dx}{1 + x^2}$

21. $\displaystyle\int_{\sqrt{2}}^{2} \frac{dx}{x\sqrt{x^2 - 1}}$

22. $\displaystyle\int_{-2}^{-\sqrt{2}} \frac{dx}{x\sqrt{x^2 - 1}}$

23. $\displaystyle\int_{-1}^{0} \frac{4\,dx}{1 + x^2}$

24. $\displaystyle\int_{\sqrt{3}/3}^{\sqrt{3}} \frac{6x}{1 + x^2}$

25. $\displaystyle\int_0^{\sqrt{2}/2} \frac{x\,dx}{\sqrt{1 - x^4}}$

26. $\displaystyle\int_0^{1/4} \frac{dx}{\sqrt{1 - 4x^2}}$

27. $\displaystyle\int_{1/\sqrt{3}}^{1} \frac{dx}{x\sqrt{4x^2 - 1}}$

28. $\displaystyle\int_0^{1} \frac{x}{1 + x^4}\,dx$

29. $\displaystyle\int_0^{1} \frac{4x\,dx}{\sqrt{4 - x^4}}$

30. $\displaystyle\int_0^{1} \frac{dx}{\sqrt{4 - x^2}}$

Evaluate the integrals in Exercises 31–38 analytically.

31. $\displaystyle\int \frac{dx}{\sqrt{9 - x^2}}$

32. $\displaystyle\int \frac{dx}{\sqrt{1 - 4x^2}}$

33. $\displaystyle\int \frac{dx}{17 + x^2}$

34. $\displaystyle\int \frac{dx}{9 + 3x^2}$

35. $\displaystyle\int \frac{dx}{x\sqrt{25x^2 - 2}}$

36. $\displaystyle\int \frac{dx}{x\sqrt{5x^2 - 4}}$

37. $\displaystyle\int \frac{y\,dy}{\sqrt{1 - y^4}}$

38. $\displaystyle\int \frac{\sec^2 y\,dy}{\sqrt{1 - \tan^2 y}}$

Use the indicated substitutions to evaluate the integrals in Exercises 39–44. Support your answer with an NINT computation.

39. $\displaystyle\int_{\sqrt[4]{2}}^{\sqrt{2}} \frac{2x\,dx}{x^2\sqrt{x^4 - 1}}, \qquad u = x^2$

40. $\displaystyle\int_0^{2} \frac{dx}{1 + (x - 1)^2}, \qquad u = x - 1$

41. $\displaystyle\int_1^{\sqrt{3}} \frac{2\,dx}{(1 + x^2) \tan^{-1} x}, \quad u = \tan^{-1} x$

42. $\displaystyle\int_0^{\ln\sqrt{3}} \frac{e^x\,dx}{1 + e^{2x}}, \qquad u = e^x$

43. $\displaystyle\int_2^{4} \frac{dx}{2x\sqrt{x - 1}}, \qquad u^2 = x$

44. $\displaystyle\int_{-\pi/2}^{\pi/2} \frac{2\cos x\,dx}{1 + \sin^2 x}, \qquad u = \sin x$

Use l'Hôpital's Rule to find the limits in Exercises 45–48.

45. $\displaystyle\lim_{x \to 0} \frac{\sin^{-1} x}{x}$

46. $\displaystyle\lim_{x \to 0} \frac{\sin^{-1} x}{x^3}$

47. $\displaystyle\lim_{x \to 0} \frac{\tan^{-1} x}{x}$

48. $\displaystyle\lim_{x \to 0} \frac{\tan^{-1} x}{x^3}$

49. Find the volume of this solid of revolution.

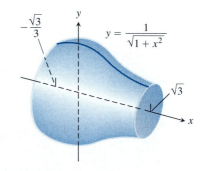

$$y = \frac{1}{\sqrt{1 + x^2}}$$

50. Find the length of the curve $y = \sqrt{1 - x^2}, -1/2 \le x \le 1/2$.

51. *Continuation of Exercise 49, Section 7.7.* You want to move your chair to a position along the wall that will maximize your viewing angle α. About how far from the front wall should you sit?

52. Find the center of mass of a thin plate of constant density $\delta = 1$ bounded by the curves $y = 1/(1 + x^2)$ and $y = -1/(1 + x^2)$ and by the lines $x = 0$ and $x = 1$.

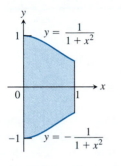

53. Find the value of x that maximizes θ in the figure below. How large is θ at that point? Begin by showing that

$$\theta = \pi - \cot^{-1} x - \cot^{-1} (2 - x).$$

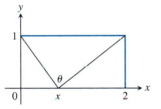

54. Find the linearizations of $\sin^{-1} x, \cos^{-1} x, \tan^{-1} x,$ and $\cot^{-1} x$ at $x = 0$.

Solve the initial-value problems in Exercises 55–58.

55. Differential equation: $\dfrac{dy}{dx} = \dfrac{1}{x\sqrt{x^2 - 1}}, x > 0$

   Initial condition: $y = \pi$ when $x = 2$

56. Differential equation: $\dfrac{dy}{dx} = \dfrac{1}{\sqrt{1 - x^2}}$

   Initial condition: $y = 1$ when $x = 0$

57. Differential equation: $\dfrac{dy}{dx} = -\dfrac{1}{\sqrt{1 - x^2}}$

   Initial condition: $y = \pi/2$ when $x = -\sqrt{2}/2$

58. Differential equation: $\dfrac{dy}{dx} = -\dfrac{1}{1 + x^2}$

   Initial condition: $y = \pi/2$ when $x = 0$

59. Use NINT to compute the value of

$$\sin^{-1} 0.6 = \int_0^{0.6} \frac{dx}{\sqrt{1 - x^2}}.$$

For reference, $\sin^{-1} 0.6 = 0.64350$ to five places.

60. Use NINT to compute the value of

$$\pi = 4 \int_0^1 \frac{1}{1 + x^2}\, dx.$$

61. Can the integrations in parts (a) and (b) both be correct? Explain.

   a) $\displaystyle \int \frac{dx}{\sqrt{1 - x^2}} = \sin^{-1} x + C$

   b) $\displaystyle \int \frac{dx}{\sqrt{1 - x^2}} = -\int -\frac{dx}{\sqrt{1 - x^2}} = -\cos^{-1} x + C$

62. a) Show that the functions

$$f(x) = \sin^{-1} \frac{x - 1}{x + 1} \quad \text{and} \quad g(x) = 2\tan^{-1} \sqrt{x},$$

   both defined for $x \geq 0$, have the same derivative and therefore that

$$f(x) = g(x) + C. \tag{2}$$

   b) Find C. (*Hint:* Evaluate both sides of Eq. (2) for a particular value of x.)

Verify the derivative formulas in Exercises 63–66.

63. $\dfrac{d(\cos^{-1} u)}{dx} = -\dfrac{du/dx}{\sqrt{1 - u^2}}, \quad -1 < u < 1$

64. $\dfrac{d(\tan^{-1} u)}{dx} = \dfrac{du/dx}{1 + u^2}$

65. $\dfrac{d(\cot^{-1} u)}{dx} = -\dfrac{du/dx}{1 + u^2}$

66. $\dfrac{d(\csc^{-1} u)}{dx} = \dfrac{-du/dx}{|u|\sqrt{u^2 - 1}}, \quad |u| > 1$

In Exercises 67–69, use NDER to provide graphical support for the derivative formulas for

67. $\sin^{-1} x$ and $\cos^{-1} x$.

68. $\tan^{-1} x$ and $\cot^{-1} x$.

69. $\sec^{-1} x$ and $\csc^{-1} x$.

Draw complete graphs of the functions in Exercises 70–73.

70. $y = \tan^{-1} \sqrt{x + 1}$ 71. $y = \sec^{-1}(3x)$

72. $y = \csc^{-1} \sqrt{x} + \sec^{-1} \sqrt{x}$

73. $y = \cot^{-1} \sqrt{x^2 - 1}$

7.9 _____   Hyperbolic Functions

It can be shown that every function defined on an interval centered at the origin can be written in a unique way as the sum of one even function and one odd function. For an arbitrary function f, the decomposition is

$$f(x) = \underbrace{\frac{f(x) + f(-x)}{2}}_{\text{Even}} + \underbrace{\frac{f(x) - f(-x)}{2}}_{\text{Odd}}.$$

If we write e^x this way, we get

$$e^x = \underbrace{\frac{e^x + e^{-x}}{2}}_{\text{Even}} + \underbrace{\frac{e^x - x^{-x}}{2}}_{\text{Odd}}.$$

The even and odd parts of e^x, called the hyperbolic cosine and hyperbolic sine of x, respectively, turn out to be important functions in their own right. They describe the motions of waves in elastic solids, the shapes of hanging electric power lines, and the temperature distributions in metal cooling fins. They even come up in the general theory of relativity. The designers of the Gateway Arch to the West in St. Louis used a hyperbolic cosine function to predict the arch's internal forces and then shaped the arch like an upside-down hyperbolic cosine curve.

READING THE NOTATION

The notation for the hyperbolic functions is read in different ways. For example, *cosh* may be read as "kosh" in *kosher* or to rhyme with *gosh*, and *sinh* may be pronounced as if spelled "cinch" or "shine."

You had an opportunity to investigate the hyperbolic functions in the exercises of Section 4.5 using NDER for the derivative. Now we are able to confirm information about these functions analytically.

Definitions and Identities

Function	Definition	Domain	Range
Hyperbolic cosine	$\cosh x = \dfrac{e^x + e^{-x}}{2}$	all real	$y \geq 1$
Hyperbolic sine	$\sinh x = \dfrac{e^x - e^{-x}}{2}$	all real	all real
Hyperbolic tangent	$\tanh x = \dfrac{\sinh x}{\cosh x} = \dfrac{e^x - e^{-x}}{e^x + e^{-x}}$	all real	$-1 < y < 1$
Hyperbolic secant	$\operatorname{sech} x = \dfrac{1}{\cosh x} = \dfrac{2}{e^x + e^{-x}}$	all real	$0 < y \leq 1$
Hyperbolic cosecant	$\operatorname{csch} x = \dfrac{1}{\sinh x} = \dfrac{2}{e^x - e^{-x}}$	$x \neq 0$	$y \neq 0$
Hyperbolic cotangent	$\coth x = \dfrac{\cosh x}{\sinh x} = \dfrac{e^x + e^{-x}}{e^x - e^{-x}}$	$x \neq 0$	$-\infty < y < -1,$ $1 < y < \infty$

Most graphers have cosh x, sinh x, and tanh x built in. The other three hyperbolic functions are reciprocals of these three. Complete graphs are given in Fig. 7.45. Notice that $y = \pm 1$ are horizontal asymptotes for $y = \tanh x$ and $y = \coth x$, and $y = 0$ is a horizontal asymptote for $y = \operatorname{sech} x$ and $y = \operatorname{csch} x$.

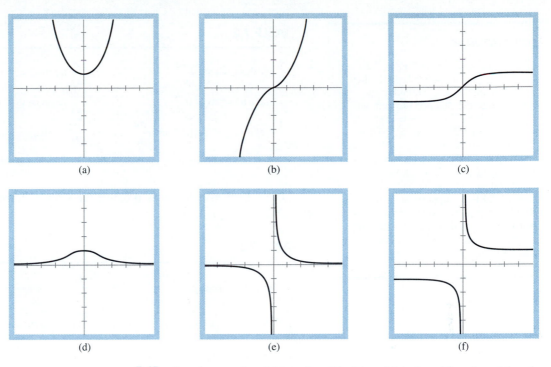

7.45 Complete graphs of (a) $\cosh x$, (b) $\sinh x$, (c) $\tanh x$, (d) $\operatorname{sech} x$, (e) $\operatorname{csch} x$, and (f) $\coth x$ in $[-5, 5]$ by $[-5, 5]$ viewing windows.

Hyperbolic functions satisfy the identities shown in Table 7.1. Except for differences in sign, these are the identities that we already know for trigonometric functions. We will say more about the analogy between hyperbolic and trigonometric functions in the exercises.

TABLE 7.1

$$\cosh^2 x - \sinh^2 x = 1$$

$$\cosh 2x = \cosh^2 x + \sinh^2 x \qquad \sinh 2x = 2 \sinh x \cosh x$$

$$\coth^2 x = 1 + \operatorname{csch}^2 x \qquad \tanh^2 x = 1 - \operatorname{sech}^2 x$$

$$\cosh^2 x = \frac{\cosh 2x + 1}{2} \qquad \sinh^2 x = \frac{\cosh 2x - 1}{2}$$

Derivatives and Integrals

The six hyperbolic functions, being rational combinations of the differentiable functions e^x and e^{-x}, have derivatives at every point at which they are defined. Finding formulas for the derivatives is a straightforward exercise in the sum, difference, and quotient rules of differentiation. Again, there are similarities with trigonometric functions. These derivatives are shown in Table. 7.2 on the following page.

EXPLORATION BIT

Each formula shown in Tables 7.2 and 7.3 can be supported using either NDER or NINT. For example, you could define the functions as follows:

$y_1 = \sinh x,$ $y_2 = \cosh x,$
$y_3 = $ NDER $y_1,$ $y_4 = $ NDER $y_2,$
$y_5 = $ NINT $y_1,$ $y_6 = $ NINT $y_2,$

and then compare graphs of the pairs
1. y_1 and y_4 **2.** y_2 and y_3
3. y_1 and y_6 **4.** y_2 and y_5.
Using NDER to support rather than NINT has at least two advantages. Do you know what they are? (Comparing y_2 and y_3 then y_2 and y_5 might remind you.)

TABLE 7.2

$$\frac{d}{dx}(\cosh u) = \sinh u \frac{du}{dx} \qquad \frac{d}{dx}(\sinh u) = \cosh u \frac{du}{dx}$$

$$\frac{d}{dx}(\tanh u) = \text{sech}^2 u \frac{du}{dx} \qquad \frac{d}{dx}(\coth u) = -\text{csch}^2 u \frac{du}{dx}$$

$$\frac{d}{dx}(\text{sech } u) = -\text{sech } u \tanh u \frac{du}{dx} \qquad \frac{d}{dx}(\text{csch } u) = -\text{csch } u \coth u \frac{du}{dx}$$

These derivative formulas produce the integral formulas shown in Table 7.3.

TABLE 7.3

$$\int \sinh u \, du = \cosh u + C \qquad \int \cosh u \, du = \sinh u + C$$

$$\int \text{sech}^2 u \, du = \tanh u + C \qquad \int \text{csch}^2 u \, du = -\coth u + C$$

$$\int \text{sech } u \tanh u \, du = -\text{sech } u + C \qquad \int \text{csch } u \coth u \, du = -\text{csch } u + C$$

EXAMPLE 1

EXPLORATION BITS

1. In what two ways can you support the result in Example 1?

2. To support the calculation in Example 2, compute NINT $((\sinh x)^2, 0, 1)$ and compare.

$$\int \coth 5x \, dx = \int \frac{\cosh 5x}{\sinh 5x} \, dx = \frac{1}{5} \int \frac{du}{u} \qquad u = \sinh 5x.$$

$$= \frac{1}{5} \ln |u| + C = \frac{1}{5} \ln |\sinh 5x| + C.$$

EXAMPLE 2

$$\int_0^1 \sinh^2 x \, dx = \int_0^1 \frac{\cosh 2x - 1}{2} \, dx \qquad \text{Table 7.1}$$

$$= \frac{1}{2} \int_0^1 (\cosh 2x - 1) \, dx = \frac{1}{2} \left[\frac{\sinh 2x}{2} - x \right]_0^1$$

$$= \frac{\sinh 2}{4} - \frac{1}{2} = 0.407$$

EXAMPLE 3

Evaluate

$$\int_0^1 \sinh x^2 \, dx.$$

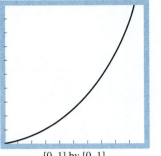

[0, 1] by [0, 1]

7.46 A graph of $y = \sinh(x^2)$ for $0 \le x \le 1$ suggests that

$$\int_0^1 \sinh x^2 \, dx < 0.5.$$

Solution We note that

$$\sinh x^2 = \frac{e^{x^2} - e^{-x^2}}{2}.$$

In Section 5.4 we stated that there were no explicit formulas for the antiderivatives of the function e^{-x^2}. None exists for e^{x^2} either. However,

$$\text{NINT}(\sinh x^2, 0, 1) = 0.358.$$

At present, we can't confirm this analytically. We can, however, provide additional grapher support by observing the graph of $y = \sinh x^2$ shown in a [0, 1] by [0, 1] viewing window (Fig. 7.46). The graph suggests that the value of the integral is a little less than 1/2 (why?) to support the value we found for NINT.

The Inverses of the Hyperbolic Functions

Notice in Fig. 7.45 that $y = \sinh x$, $y = \tanh x$, $y = \operatorname{csch} x$, and $y = \coth x$ are one-to-one, so their inverses are functions. The inverses are denoted by $y = \sinh^{-1} x$, $y = \tanh^{-1} x$, $y = \operatorname{csch}^{-1} x$, and $y = \coth^{-1} x$. The functions $y = \cosh x$ and $y = \operatorname{sech} x$ are not one-to-one. Restricting their domains to $x \ge 0$ produces one-to-one functions whose inverses are functions. These two inverses are also denoted by $y = \cosh^{-1} x$ and $y = \operatorname{sech}^{-1} x$ with the following understanding:

$$y = \cosh^{-1} x \text{ is the inverse of } y = \cosh x, x \ge 0. \tag{1}$$

$$y = \operatorname{sech}^{-1} x \text{ is the inverse of } y = \operatorname{sech} x, x \ge 0. \tag{2}$$

The domains and ranges of the inverse hyperbolic functions follow from the definitions of the hyperbolic functions and Eqs. (1) and (2) and are given in Table 7.4. Complete graphs of the inverse hyperbolic functions are given in Fig. 7.47.

TABLE 7.4

Inverse Function	Domain	Range
$y = \cosh^{-1} x$	$x \ge 1$	$y \ge 0$
$y = \sinh^{-1} x$	all real	all real
$y = \tanh^{-1} x$	$-1 \le x \le 1$	all real
$y = \operatorname{sech}^{-1} x$	$0 < x \le 1$	$y \ge 0$
$y = \operatorname{csch}^{-1} x$	$x \ne 0$	$y \ne 0$
$y = \coth^{-1} x$	$-\infty < x < -1, 1 < x < \infty$	$y \ne 0$

Most graphers have $\sinh^{-1} x$, $\cosh^{-1} x$, and $\tanh^{-1} x$ built in. The inverse hyperbolic secant, cosecant, and cotangent satisfy the three identities shown in Table 7.5. These identities come in handy when we have to calculate the values of, or draw complete graphs of, $\operatorname{sech}^{-1} x$, $\operatorname{csch}^{-1} x$, or $\coth^{-1} x$.

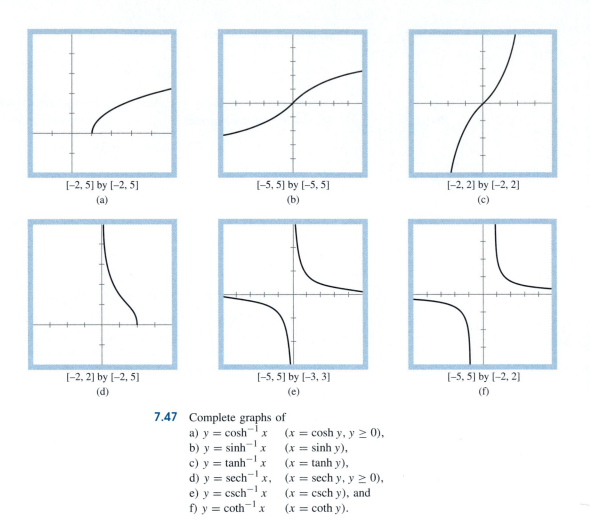

[–2, 5] by [–2, 5]
(a)

[–5, 5] by [–5, 5]
(b)

[–2, 2] by [–2, 2]
(c)

[–2, 2] by [–2, 5]
(d)

[–5, 5] by [–3, 3]
(e)

[–5, 5] by [–2, 2]
(f)

7.47 Complete graphs of
a) $y = \cosh^{-1} x$ $(x = \cosh y,\ y \geq 0)$,
b) $y = \sinh^{-1} x$ $(x = \sinh y)$,
c) $y = \tanh^{-1} x$ $(x = \tanh y)$,
d) $y = \operatorname{sech}^{-1} x$, $(x = \operatorname{sech} y,\ y \geq 0)$,
e) $y = \operatorname{csch}^{-1} x$ $(x = \operatorname{csch} y)$, and
f) $y = \coth^{-1} x$ $(x = \coth y)$.

TABLE 7.5

$$\operatorname{sech}^{-1} x = \cosh^{-1} \frac{1}{x} \qquad \operatorname{csch}^{-1} x = \sinh^{-1} \frac{1}{x} \qquad \coth^{-1} x = \tanh^{-1} \frac{1}{x}.$$

EXPLORATION 1

Viewing Inverses of Hyperbolic Functions

Let
$$x_1(t) = t, \qquad y_1(t) = t,$$
$$x_2(t) = t, \qquad y_2(t) = 1/\cosh t,$$
$$x_3(t) = y_2(t) \qquad y_3(t) = x_2(t).$$

1. Graph the three parametric equations simultaneously in a square viewing window that contains $0 \le x \le 6$, $0 \le y \le 4$. Set $t\text{Min} = 0$, $t\text{Max} = 6$, and t-step $= 0.05$. Explain what you see in the viewing window. Explain the domain of each function.

2. Let $x_4(t) = t$, $y_4(t) = \cosh^{-1}(1/t)$. Graph and compare (x_3, y_3) and (y_4, x_4). Predict what you should see, and explain what you do see. &

EXPLORATION BIT

If your grapher does not have cosh built in, you can still do Exploration 1. How?

Derivatives and Integrals

The chief merit of the inverse hyperbolic functions lies in their historical usefulness in integration. You will see why after we have derived the formulas for their derivatives in Table 7.6.

The restrictions $|u| < 1$ and $|u| > 1$ on the derivative formulas for $\tanh^{-1} u$ and $\coth^{-1} u$ come from the natural restrictions on the values of these functions. (See Table 7.4.) The distinction between $|u| < 1$ and $|u| > 1$ becomes important when we convert the derivative functions into integral formulas. If $|u| < 1$, the integral of $1/(1 - u^2)$ is $\tanh^{-1} u + C$. If $|u| > 1$, the integral is $\coth^{-1} u + C$. (See Table 7.6 on the following page.)

EXAMPLE 4

Show that if u is a differentiable function of x whose values are greater than 1, then

$$\frac{d}{dx} \cosh^{-1} u = \frac{1}{\sqrt{u^2 - 1}} \frac{du}{dx}.$$

Solution   First we find the derivative of $y = \cosh^{-1} x$ for $x > 1$:

$y = \cosh^{-1} x$

$x = \cosh y$ Invert.

$1 = \sinh y \dfrac{dy}{dx}$ Differentiate with respect to x.

$\dfrac{dy}{dx} = \dfrac{1}{\sinh y} = \dfrac{1}{\sqrt{\cosh^2 y - 1}}$ Since $x > 1$, $y > 0$ and $\sinh y > 0$.

$= \dfrac{1}{\sqrt{x^2 - 1}}$ $\cosh y = x$.

In short,

$$\frac{d}{dx} \cosh^{-1} x = \frac{1}{\sqrt{x^2 - 1}}.$$

The Chain Rule gives the final result:

$$\frac{d}{dx} \cosh^{-1} u = \frac{1}{\sqrt{u^2 - 1}} \frac{du}{dx}. \qquad \blacksquare$$

TABLE 7.6   **Derivatives of Inverse Hyperbolic Functions**

$$\frac{d(\cosh^{-1} u)}{dx} = \frac{1}{\sqrt{u^2 - 1}} \frac{du}{dx}, \qquad u > 1$$

$$\frac{d(\sinh^{-1} u)}{dx} = \frac{1}{\sqrt{1 + u^2}} \frac{du}{dx},$$

$$\frac{d(\tanh^{-1} u)}{dx} = \frac{1}{1 - u^2} \frac{du}{dx}, \qquad |u| < 1,$$

$$\frac{d(\operatorname{sech}^{-1} u)}{dx} = \frac{-du/dx}{u\sqrt{1 - u^2}}, \qquad 0 < u < 1,$$

$$\frac{d(\operatorname{csch}^{-1} u)}{dx} = \frac{-du/dx}{|u|\sqrt{1 + u^2}}, \qquad u \neq 0,$$

$$\frac{d(\coth^{-1} u)}{dx} = \frac{1}{1 - u^2} \frac{du}{dx}, \qquad |u| > 1.$$

Integrals Leading to Inverse Hyperbolic Functions

$$\int \frac{du}{\sqrt{u^2 - 1}} = \cosh^{-1} u + C, u > 1$$

$$\int \frac{du}{\sqrt{1 + u^2}} = \sinh^{-1} u + C$$

$$\int \frac{du}{1 - u^2} = \begin{cases} \tanh^{-1} u + C & \text{if } |u| < 1 \\ \coth^{-1} u + C & \text{if } |u| > 1 \end{cases}$$

$$\int \frac{du}{u\sqrt{1 - u^2}} = -\operatorname{sech}^{-1} |u| + C = -\cosh^{-1}\left(\frac{1}{|u|}\right) + C, \ 0 < |u| < 1$$

$$\int \frac{du}{u\sqrt{1 + u^2}} = -\operatorname{csch}^{-1} |u| + C = -\sinh^{-1}\left(\frac{1}{|u|}\right) + C, \ u \neq 0$$

EXAMPLE 5

EXPLORATION BIT

To how many decimal places do NINT $\left(2/\sqrt{1 + 4x^2}, 0, 1\right)$ and the exact result agree in Example 5?

$$\int_0^1 \frac{2\,dx}{\sqrt{1 + 4x^2}} = \int_0^2 \frac{du}{\sqrt{1 + u^2}} = \sinh^{-1} u]_0^2 \qquad u = 2x, \, du = 2dx.$$

$$= \sinh^{-1} 2 - \sinh^{-1} 0$$

$$= \sinh^{-1} 2 - 0 = \sinh^{-1} 2$$

Exercises 7.9 _____

Solve the equations in Exercises 1–6.

1. $\sinh x = -\dfrac{3}{4}$

2. $\operatorname{csch} x = \dfrac{4}{3}$

3. $\cosh x = 2$

4. $\tanh x = 0.5$

5. $\operatorname{sech} x = 0.7$

6. $\coth x = -2$

Rewrite the expressions in Exercises 7–12 in terms of exponentials, and simplify the results as much as you can. Support your answers graphically.

7. $2 \cosh (\ln x)$ **8.** $\sinh (2 \ln x)$

9. $\cosh 5x + \sinh 5x$ **10.** $(\sinh x + \cosh x)^4$

11. $\cosh 3x - \sinh 3x$

12. $\ln (\cosh x + \sinh x) + \ln (\cosh x - \sinh x)$

13. Starting with the identities

$$\sinh (x + y) = \sinh x \cosh y + \cosh x \sinh y,$$

$$\cosh (x + y) = \cosh x \cosh y + \sinh x \sinh y,$$

which you may assume, show that
a) $\sinh 2x = 2 \sinh x \cosh x.$

b) $\cosh 2x = \cosh^2 x + \sinh^2 x.$

14. Starting with the definitions of $\cosh x$ and $\sinh x$, show that

$$\cosh^2 x - \sinh^2 x = 1.$$

Draw complete graphs of the functions in Exercises 15–26, and confirm analytically.

15. $y = \sinh 3x$ **16.** $y = \dfrac{1}{2} \sinh (2x + 1)$

17. $y = 2 \tanh \dfrac{x}{2}$ **18.** $y = x - \tanh x$

19. $y = \ln (\text{sech } x)$ **20.** $y = \ln (\text{csch } x)$

21. $y = \sinh^{-1} 2x$ **22.** $y = 2 \cosh^{-1} \sqrt{x}$

23. $y = (1 - x) \tanh^{-1} x$ **24.** $y = (1 - x^2) \coth^{-1} x$

25. $y = x \, \text{sech}^{-1} \, x$ **26.** $y = x^2 \, \text{csch}^{-1} \, x^2$

Find dy/dx in Exercises 27–38.

27. $y = \ln (\text{csch } x + \coth x)$ **28.** $y = x \cosh x - \sinh x$

29. $y = \dfrac{1}{2} \ln |\tanh x|$ **30.** $y = \tan^{-1} (\sinh x)$

31. a) $y = \cosh^2 x$ **b)** $y = \sinh^2 x$ **c)** $y = \dfrac{1}{2} \cosh 2x$

32. a) $y = (x^2 + 1) \, \text{sech} (\ln x)$

     b) Evaluate dy/dx for $x = 1, 2, 5.$

     c) Express y in terms of exponentials and simplify. Then find dy/dx.

33. $y = \sinh^{-1} (\tan x)$

34. $y = \cosh^{-1} (\sec x), 0 < x < \pi/2$

35. $y = \tanh^{-1} (\sin x), -\pi/2 < x < \pi/2$

36. $y = \coth^{-1} (\sec x), -\pi/2 < x < \pi/2$

37. $y = \text{sech}^{-1} (\sin x), 0 < x < \pi/2$

38. $y = \text{csch}^{-1} (\tan x), 0 < x < \pi/2$

Evaluate the integrals in Exercises 39–50 analytically, and support with a NINT computation.

39. $\displaystyle\int_{-1}^{1} \cosh 5x \, dx$ **40.** $\displaystyle\int_{-1}^{0} \cosh (2x + 1) \, dx$

41. $\displaystyle\int_{-3}^{3} \sinh x \, dx$ **42.** $\displaystyle\int_{-\pi}^{\pi} \tanh 2x \, dx$

43. $\displaystyle\int_{0}^{1/2} 4e^x \cosh x \, dx$ **44.** $\displaystyle\int_{0}^{1/2} 4e^{-x} \sinh x \, dx$

45. $\displaystyle\int_{1}^{2} \dfrac{\cosh (\ln x)}{x} \, dx$ **46.** $\displaystyle\int_{0}^{\ln 2} \dfrac{\sinh x}{\cosh x} \, dx$

47. $\displaystyle\int_{0}^{\ln 3} \text{sech}^2 x \, dx$ **48.** $\displaystyle\int_{0}^{\ln 2} \tanh^2 x \, dx$

49. $\displaystyle\int_{0}^{4} \dfrac{\cosh \sqrt{x}}{\sqrt{x}} \, dx$ **50.** $\displaystyle\int_{\ln 2}^{\ln 3} \text{csch}^2 x \, dx$

Evaluate the integrals in Exercises 51–58 analytically.

51. $\displaystyle\int \sinh 2x \, dx$ **52.** $\displaystyle\int 4 \cosh (3x - \ln 2) \, dx$

53. $\displaystyle\int 2e^{2t} \cosh t \, dt$ **54.** $\displaystyle\int 8e^{-t} \sinh t \, dt$

55. $\displaystyle\int \tanh \dfrac{x}{7} \, dx$ **56.** $\displaystyle\int \coth \sqrt{2} x \, dx$

57. $\displaystyle\int \dfrac{\text{sech} \sqrt{t} \tanh \sqrt{t} \, dt}{\sqrt{t}}$ **58.** $\displaystyle\int \dfrac{\text{csch}(\ln t) \coth (\ln t) \, dt}{t}$

Evaluating Inverse Hyperbolic Functions and Related Integrals

When hyperbolic function keys are not available on a calculator, it is still possible to evaluate the inverse hyperbolic functions by expressing them first in terms of logarithms. The conversion formulas are listed in Table 7.7.

TABLE 7.7   Logarithm Formulas for Evaluating Inverse Hyperbolic Functions

$\sinh^{-1} x = \ln \left(x + \sqrt{x^2 + 1} \right),$	$-\infty < x < \infty$
$\cosh^{-1} x = \ln \left(x + \sqrt{x^2 - 1} \right),$	$x \geq 1$
$\tanh^{-1} x = \dfrac{1}{2} \ln \dfrac{1 + x}{1 - x},$	$|x| < 1$
$\text{sech}^{-1} x = \ln \left(\dfrac{1 + \sqrt{1 - x^2}}{x} \right),$	$0 < x \leq 1$
$\text{csch}^{-1} x = \ln \left(\dfrac{1}{x} + \dfrac{\sqrt{1 + x^2}}{|x|} \right),$	$x \neq 0$
$\coth^{-1} x = \dfrac{1}{2} \ln \dfrac{x + 1}{x - 1},$	$|x| > 1$

59. Use graphs to support the identities in Table 7.7 for $\sinh^{-1} x$, $\tanh^{-1} x$, and $\text{sech}^{-1} x$.

60. Use graphs to support the identities in Table 7.7 for $\cosh^{-1} x$, $\operatorname{csch}^{-1} x$, and $\coth^{-1} x$.

Evaluate the integrals in Exercises 61–68 in terms of
a) inverse hyperbolic functions,

b) natural logarithms using the formulas in Table 7.7. Support each answer with a NINT computation.

61. $\displaystyle\int_0^1 \frac{dx}{\sqrt{1+x^2}}$

62. $\displaystyle\int_{3/5}^{4/5} \frac{dx}{x\sqrt{1-x^2}}$

63. $\displaystyle\int_{5/4}^{5/3} \frac{dx}{\sqrt{x^2-1}}$

64. $\displaystyle\int_0^{1/2} \frac{dx}{1-x^2}$

65. $\displaystyle\int_{5/4}^{2} \frac{dx}{1-x^2}$

66. $\displaystyle\int_0^{2\sqrt{3}} \frac{dx}{\sqrt{4+x^2}}$ (Set $x = 2u$.)

67. $\displaystyle\int_1^{2} \frac{dx}{x\sqrt{4+x^2}}$ (Set $x = 2u$.)

68. $\displaystyle\int_0^{\pi} \frac{\cos x\, dx}{\sqrt{1+\sin^2 x}}$ (Set $u = \sin x$.)

Evaluate the integrals in Exercises 69–72.

69. $\displaystyle\int_1^3 \frac{\sinh x}{x}\, dx$

70. $\displaystyle\int_{-1}^3 \cosh(x^2)\, dx$

71. $\displaystyle\int_1^4 \frac{\cosh x - 1}{x}\, dx$

72. $\displaystyle\int_{-2}^{2} \frac{\sinh^{-1} x}{x}\, dx$

In Exercises 73–76, graph one of the antiderivatives of the function.

73. $y = \dfrac{\sinh x}{x}$

74. $y = \cosh(x^2)$

75. $y = \dfrac{\cosh x - 1}{x}$

76. $y = \dfrac{\sinh^{-1} x}{x}$

In Exercises 77 and 78, find the volumes of the solids generated by revolving the shaded regions about the x-axis.

77.

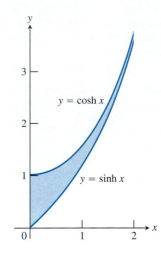

78.

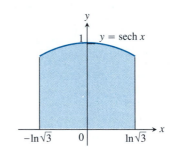

79. Find the centroid of the shaded region in Exercise 78.

80. Find the volume of the solid generated by revolving the shaded region about the line $y = 1$.

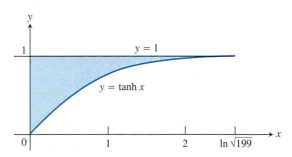

81. Show that if a function f is defined on an interval centered at the origin (so that $f(-x)$ is defined whenever $f(x)$ is defined), then

$$f(x) = \frac{f(x) + f(-x)}{2} + \frac{f(x) - f(-x)}{2}.$$

Then show that the functions defined by the formulas

$$(f(x) + f(-x))/2 \quad \text{and} \quad (f(x) - f(-x))/2$$

are even and odd, respectively.

82. What does the equation

$$f(x) = \frac{f(x) + f(-x)}{2} + \frac{f(x) - f(-x)}{2}$$

look like when f itself is (a) an even function, (b) an odd function?

83. *Skydiving.* If a body of mass m falling from rest under the action of gravity encounters an air resistance proportional to the square of the velocity, then the body's velocity t seconds into the fall satisfies the differential equation

$$m\frac{dv}{dt} = mg - kv^2,$$

where k is a constant that depends on the body's aerodynamic properties and the density of the air. (We assume that the fall is short enough that the variation in the air's density will not affect the outcome.)

a) Show that

$$v = \sqrt{\frac{mg}{k}} \tanh\left(\sqrt{\frac{gk}{m}}\,t\right)$$

satisfies the differential equation and the initial condition that $v = 0$ when $t = 0$.

b) Find the body's *limiting velocity*, $\lim\limits_{t\to\infty} v$.

c) For a 160-lb skydiver ($mg = 160$), and with time in seconds and distance in feet, a typical value for k is 0.005. What is the diver's limiting velocity?

84. *Accelerations whose magnitudes are proportional to displacement.* Suppose that the position of a body moving along a coordinate line at time t is
a) $s(t) = a\cos kt + b\sin kt$.

b) $s(t) = a\cosh kt + b\sinh kt$.

Show that in both cases, the acceleration d^2s/dt^2 is proportional to s but that in the first case, it is always directed toward the origin, while in the second case, it is directed away from the origin.

Hanging Cables and Hyperbolic Cosines

It can be shown that clotheslines, chains, telephone lines, and electric power cables that are strung from one support to another always hang in the shape of a hyperbolic cosine curve, an example of which is shown here.

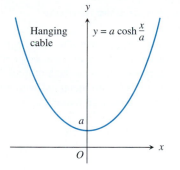

In contrast, the cables of suspension bridges, which do not hang freely but support a uniform load per horizontal foot, hang in parabolas. Hyperbolic cosine curves are sometimes called **chain curves** or **catenaries**, the latter term deriving from the Latin *catena*, for "chain."

85. Find the length of the curve

$$y = 10\cosh(x/10), \quad -10\ln 10 \le x \le 10\ln 10$$

86. Find the area between the curve in Exercise 85 and the x-axis.

87. *Minimal surfaces.* Find the area of the *minimal surface* swept out by revolving the curve $y = 2\cosh(x/2)$, $0 \le x \le \ln 8$, about the x-axis.

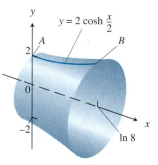

It can be shown that, of all continuously differentiable curves joining the points A and B in the figure above, the hyperbolic cosine curve $y = 2\cosh(x/2)$ generates the surface of least area. If you made a rigid wire frame of the circles through A and B and dipped it in a soap solution, the surface spanning the circles would be the one generated by the hyperbolic cosine.

88. Show that the function $y = a\cosh(x/a)$ solves the initial value problem

Differential equation: $\quad y'' = (1/a)\sqrt{1 + (y')^2}$,

Initial conditions: $\quad y'(0) = 0 \quad$ and $\quad y(0) = a$.

(By analyzing the forces in hanging cables, we can show that the curves they hang in always satisfy the differential equation and initial conditions given here. That is how we know that cables hang in hyperbolic cosines.)

The Hyperbolic in Hyperbolic Functions

In case you are wondering where the name *hyperbolic* comes from, here is the answer: Just as $x = \cos u$ and $y = \sin u$ are identified with points (x, y) on the unit circle, the functions $x = \cosh u$ and $y = \sinh u$ are identified with points (x, y) on the right-hand branch of the unit hyperbola, $x^2 - y^2 = 1$.

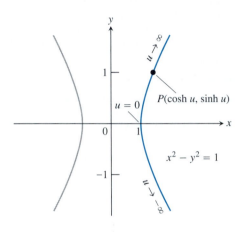

89. Another analogy between hyperbolic and circular functions is that the variable u in the coordinates $(\cosh u, \sinh u)$ for the points of the right-hand branch of the hyperbola $x^2 - y^2 = 1$ is twice the area of the sector AOP, just as u in $(\cos u, \sin u)$ is twice the area of sector AOP of the circle pictured. To see why this is so, carry out the steps that follow.

a) Show that the area $A(u)$ of sector AOP in the top diagram is given by the formula

$$A(u) = \frac{1}{2} \cosh u \sinh u - \int_1^{\cosh u} \sqrt{x^2 - 1}\, dx.$$

b) Differentiate both sides of the equation in part (a) with respect to u to show that

$$A'(u) = \frac{1}{2}.$$

c) Solve this last equation for $A(u)$. What is the value of $A(0)$?

Review Questions

1. What is the inverse of a one-to-one function? How do you tell when functions are inverses of one another? Give examples of functions that are inverses of one another; that are not inverses of one another.

2. What continuous functions have inverses?

3. How are the graphs, domains, and ranges of functions and their inverses related? Give an example.

4. How are the derivatives of functions and their inverses related? Give an example.

5. How is the function $y = \ln x$ defined? What are the rules for doing arithmetic with natural logarithms?

6. What is the derivative of $y = \ln x$? What integrals involve logarithms? Give examples.

7. What is the function $y = \exp(x)$? What is number e? How is the function $y = e^x$ defined?

8. Graph the functions $\ln x$ and e^x together. What are the domains and ranges of these functions? Can e^x ever equal zero?

9. What rules of exponents do exponential functions obey?

10. What derivatives and integrals are associated with the function $y = e^x$?

11. How are the functions a^x and $\log_a x$ defined? What are their derivatives? What integrals are associated with these functions? Give examples.

12. How is $\log_a x$ evaluated in terms of $\ln x$? How is the graph of $\log_a x$ related to the graph of $\ln x$?

13. What does the graph of the function a^x look like if $a > 1$? If $0 < a < 1$?

14. Describe some of the applications of base-10 logarithms.

15. Where does the equation $y = y_0 e^{kt}$ come from? Describe some of the applications of this equation.

16. How are limits found with l'Hôpital's Rule? Give examples. How do you know when to stop using the rule?

17. How do you compare the growth rates of functions of x as $x \to \infty$? Give examples.

18. Define right-end behavior model and left-end behavior model. How are right-end behavior model, left-end behavior model, and end behavior model related?

19. Explain "little-oh" and "big-oh" notation. Give examples.

20. Graph the six basic inverse trigonometric functions. Indicate their domains and ranges.

21. What are the derivatives of the inverse trigonometric functions? What integrals are associated with these functions? Give examples.

22. How are the six basic hyperbolic functions defined? Graph them, and indicate their domains and ranges.

23. What identities are associated with hyperbolic functions?

24. What are the derivatives of the six basic hyperbolic functions? What are the corresponding integrals? Give examples.

25. What is special about the hyperbolic cosine curves?

26. Graph the six inverse hyperbolic functions, and give their domains and ranges.

27. What are the derivatives of the inverse hyperbolic functions? Give examples of integrals that lead naturally to inverse hyperbolic functions.

Practice Exercises

In Exercises 1–22, find dy/dx. Support graphically.

1. $y = \ln \sqrt{x}$

2. $y = \ln \left(\dfrac{e^x}{2} \right)$

3. $y = \ln (3x^2 + 6)$

4. $y = \ln (1 + e^x)$

5. $y = \dfrac{1}{e^x}$

6. $y = xe^{-x}$

7. $y = e^{(1+\ln x)}$

8. $y = \displaystyle\int_1^x \ln t \, dt$

9. $y = \ln (\cos x)$

10. $y = \ln (\sin x)$

11. $y = \ln (\cos^{-1} x)$

12. $y = \ln (\sin^{-1} x)$

13. $y = \log_2(x^2)$

14. $y = \log_5(x - 7)$

15. $y = 8^{-x}$

16. $y = 9^x$

17. $y = \sin^{-1}(\sqrt{1 - x})$

18. $y = \tan^{-1}(\tan 2x)$

19. $y = \cos^{-1}(1/x) - \csc^{-1} x, \; x > 0$

20. $y = (1 + x^2) \cot^{-1} 2x$

21. $y = 2\sqrt{x - 1} \, \sec^{-1} \sqrt{x}$

22. $y = \csc^{-1}(\sec x), \; 0 < x < \pi/2$

23. Simplify:

   a) $\ln e^{2x}$ **b)** $\ln 2e$ **c)** $\ln \dfrac{1}{e}$

24. Simplify:

   a) $e^{2\ln 2}$ **b)** $e^{-\ln 4}$ **c)** $e^{\ln (\ln x)}$

Solve for y in Exercises 25–28.

25. $\ln (y^2 + y) - \ln y = x, \; y > 0$

26. $\ln (y - 4) = -4t$

27. $e^{2y} = 4x^2, \; x > 0$

28. $e^{-0.1y} = \dfrac{1}{2}$

Use logarithmic differentiation to find dy/dx in Exercises 29–34.

29. $y = \dfrac{2(x^2 + 1)}{\sqrt{\cos 2x}}$

30. $y = \sqrt[10]{\dfrac{3x + 4}{2x - 4}}$

31. $y = \left(\dfrac{(x + 5)(x - 1)}{(x - 2)(x + 3)} \right)^5$

32. $y = x^{\ln x}, \; x > 1$

33. $y = (1 + x^2)e^{\tan^{-1} x}$

34. $y = \dfrac{2x 2^x}{\sqrt{x^2 + 1}}$

Find the derivatives with respect to x of the functions whose formulas appear in Exercises 35–40.

35. $x - \coth x$

36. $x \sinh x - \cosh x$

37. $\ln (\operatorname{csch} x) + x \coth x$

38. $\ln (\operatorname{sech} x) + x \tanh x$

39. $\sin^{-1}(\tanh x)$

40. $\tan^{-1}(\sinh x)$

Find the derivatives with respect to x of the functions whose formulas appear in Exercises 41–46.

41. $\sqrt{1 + x^2} \sinh^{-1} x$

42. $\sqrt{x^2 - 1} \cosh^{-1} x$

43. $1 - \tanh^{-1}(1/x), |x| > 1$

44. $\coth^{-1}(\csc x), 0 < x < \pi/2$

45. $\operatorname{sech}^{-1}(\cos 2x), 0 < x < \pi/4$

46. $\operatorname{csch}^{-1}(\cot x), 0 < x < \pi/2$

Draw a complete graph of the functions in Exercises 47–50, and confirm analytically.

47. $y = e^{\tan^{-1} x}$

48. $y = e^{\cot^{-1} x}$

49. $y = x\tan^{-1} x - \dfrac{1}{2}\ln x$

50. $y = x\cos^{-1} x - \sqrt{1 - x^2}$

Evaluate the integrals in Exercises 51–72 analytically.

51. $\displaystyle\int_{-1}^{1} \frac{dx}{3x - 4}$

52. $\displaystyle\int_{1}^{e} \frac{\sqrt{\ln x}}{x}\,dx$

53. $\displaystyle\int_{\ln 3}^{\ln 4} e^x\,dx$

54. $\displaystyle\int_{0}^{\ln 3} e^{2x}\,dx$

55. $\displaystyle\int_{0}^{\pi/4} e^{\tan x}\sec^2 x\,dx$

56. $\displaystyle\int_{0}^{\pi/3} e^{\sec x}\sec x\tan x\,dx$

57. $\displaystyle\int_{0}^{\pi} \tan\frac{x}{3}\,dx$

58. $\displaystyle\int_{1/6}^{1/4} 2\cot \pi x\,dx$

59. $\displaystyle\int_{0}^{4} \frac{2x\,dx}{x^2 - 25}$

60. $\displaystyle\int_{-\pi/3}^{\pi/3} \frac{\sec x + \tan x}{\sec x}\,dx$

61. $\displaystyle\int_{0}^{\pi/4} \frac{\sec x\tan x + \sec^2 x}{\sec x + \tan x}\,dx \quad (u = \sec x + \tan x)$

62. $\displaystyle\int_{-\pi/2}^{\pi/2} \frac{\cos x}{2 - \sin x}\,dx$

63. $\displaystyle\int_{1}^{8} \frac{\log_4 x}{x}\,dx$

64. $\displaystyle\int_{1}^{e} \frac{8\ln 3\log_3 x}{x}\,dx$

65. $\displaystyle\int_{0}^{1} x3^{(x^2)}\,dx$

66. $\displaystyle\int_{0}^{\pi/4} 2^{\tan x}\sec^2 x\,dx$

67. $\displaystyle\int_{-1/2}^{1/2} \frac{3\,dx}{\sqrt{1 - x^2}}$

68. $\displaystyle\int_{1}^{1+(\sqrt{2}/2)} \frac{dx}{\sqrt{1 - (x - 1)^2}} \quad (u = x - 1)$

69. $\displaystyle\int_{-1}^{1} \frac{1}{1 + x^2}\,dx$

70. $\displaystyle\int_{1}^{3} \frac{2\,dx}{\sqrt{x}(1 + x)} \quad (u = \sqrt{x})$

71. $\displaystyle\int_{1/2}^{3/4} \frac{dx}{\sqrt{x}\sqrt{1 - x}} \quad (u = \sqrt{x})$

72. $\displaystyle\int_{\sqrt{2}/3}^{2/3} \frac{dx}{x\sqrt{9x^2 - 1}} \quad (u = 3x)$

Evaluate the integrals in Exercises 73–78 analytically. Support with a NINT computation.

73. $\displaystyle\int_{0}^{\ln 2} 4e^x\cosh x\,dx$

74. $\displaystyle\int_{0}^{\ln 2} \frac{\sinh x}{1 + \cosh x}\,dx$

75. $\displaystyle\int_{-\ln 3}^{\ln 3} 3\sqrt{\cosh 2x + 1}\,dx$

76. $\displaystyle\int_{1}^{2} \frac{5\operatorname{sech}^2(\ln x)}{x}\,dx$

77. $\displaystyle\int_{2}^{4} 10\operatorname{csch}^2 x\coth x\,dx$

78. $\displaystyle\int_{0}^{\ln\sqrt{2}} 4\operatorname{sech}^4 x\tanh x\,dx$

Evaluate the integrals in Exercises 79–86 analytically.

79. $\displaystyle\int \sec^2(x)e^{\tan x}\,dx$

80. $\displaystyle\int \csc^2(x)e^{\cot x}\,dx$

81. $\displaystyle\int \frac{\tan(\ln v)}{v}\,dv$

82. $\displaystyle\int \frac{dv}{v\ln v}$

83. $\displaystyle\int (x)3^{x^2}\,dx$

84. $\displaystyle\int 2^{\tan x}\sec^2 x\,dx$

85. $\displaystyle\int \frac{dy}{y\sqrt{4y^2 - 1}}$

86. $\displaystyle\int \frac{24\,dy}{y\sqrt{y^2 - 16}}$

Evaluate the integrals in Exercises 87–92 in terms of
a) inverse hyperbolic functions.
b) natural logarithms using the formulas in Table 7.7 in Exercises 7.9.

87. $\displaystyle\int_{0}^{\pi/2} \frac{\sin x\,dx}{\sqrt{1 + \cos^2 x}}$

88. $\displaystyle\int_{\sqrt{2}}^{\sqrt{17}} \frac{2x\,dx}{\sqrt{x^4 - 1}}$

89. $\displaystyle\int_{1/5}^{1/2} \frac{4\tanh^{-1} x}{1 - x^2}\,dx$

90. $\displaystyle\int_{\pi/6}^{\pi/4} \frac{\cos x\,dx}{\sin x\sqrt{1 + \sin^2 x}}$

91. $\displaystyle\int_{3/5}^{4/5} \frac{2\operatorname{sech}^{-1} x}{x\sqrt{1 - x^2}}\,dx$

92. $\displaystyle\int_{\sqrt{8}}^{\sqrt{3}} \frac{e^{\coth^{-1} x}}{1 - x^2}\,dx$

Evaluate the integrals in Exercises 93–96.

93. $\displaystyle\int_{1}^{6} \log_5 x\,dx$

94. $\displaystyle\int_{0}^{1} 3^{x^2}\,dx$

95. $\displaystyle\int_{1}^{5} \frac{e^{\sinh^{-1} x}}{1 + x^2}\,dx$

96. $\displaystyle\int_{2}^{5} \frac{e^{\cosh^{-1} x}}{1 + \sqrt{x}}\,dx$

Graph one of the antiderivatives of the functions in Exercises 97 and 98.

97. $y = \dfrac{\tanh x}{x}$

98. $y = x\coth x$

99. The function $f(x) = e^x + x$, being differentiable and one-to-one, has a differentiable inverse, $f^{-1}(x)$. Find the value of df^{-1}/dx at the point $f(\ln 2)$.

100. Find the area between the curve $y = 2(\ln x)/x$ and the x-axis from $x = 1$ to $x = e$.

101. The functions $\ln 5x$ and $\ln 3x$ differ by a constant. What constant?

102. Show that the area between the curve $y = 1/x$ and the x-axis from $x = 1$ to $x = 2$ is the same as the area under the curve from $x = 10$ to $x = 20$.

103. Find the area between each of the curves $y = (2\log_2 x)/x$ and $y = (2\log_4 x)/x$ and the interval $1 \le x \le e$ of the x-axis. What is the ratio of these areas?

104. Find the following numbers as accurately as your calculator will allow using only $\boxed{\text{LN}}$ and $\boxed{\div}$ (and the number keys, of course).

a) $\log_{10} 5$ **b)** $\log_2 3$ **c)** $\log_7 2$

105. What is the age of a sample of charcoal in which 90% of the carbon-14 that was originally present has decayed?

106. *Californium-252.* What costs $27 million per gram and can be used to treat brain cancer, analyze coal for its sulfur content, and detect explosives in luggage? The answer is Californium-252, a radioactive isotope so rare that only 8 g of it have been made in the Western world since its discovery by Glenn Seaborg in 1950. The half-life of the isotope is 2.645 years—long enough for a useful service life and short enough to have a high radioactivity per unit mass. One microgram of the isotope releases 170 million neutrons per second.

a) What is the value of k in the decay equation for this isotope?

b) What is the isotope's mean life? (See Section 7.4, Exercise 14.)

c) How long will it take 95% of a sample's radioactive nuclei to disintegrate?

107. *Appreciation.* A violin made in 1785 by John Betts, one of England's finest violin makers, cost $250 in 1924 and sold for $7500 in 1988. Assuming a constant rate of appreciation, what was that rate?

108. *Working under water.* The intensity $L(x)$ of light x feet beneath the surface of the ocean satisfies the differential equation

$$\frac{dL}{dx} = -kL.$$

As a diver, you know from experience that diving to 18 ft in the Caribbean Sea cuts the intensity in half. You cannot work without artificial light when the intensity falls below a tenth of the surface value. About how deep can you expect to work without artificial light?

109. *The purchasing power of the dollar.* How long will it take the purchasing power of the dollar to decrease to three-fourths of its present value if the annual inflation rate will be a constant 4%? (*Hint*: How long will it take the Consumer Price Index to increase to four thirds its present value at this inflation rate?)

110. *Inflation in West Germany.* The Consumer Price Index (1967 = 100) in West Germany was 175.8 in 1980 and 211.9 in 1986.

a) Assuming a constant rate of inflation, what was it?

b) What would you expect the index to be in 1996 if this rate were to continue?

111. *Inflation in Italy.* In contrast to West Germany (Exercise 110), the CPI in Italy was 295.5 in 1980 and 480.1 in 1986.

a) Assuming a constant rate of inflation, what was it?

b) What would you expect the index to be in 1996 if this rate were to continue?

112. The spread of flu in a certain school is given by

$$P(t) = \frac{150}{1 + e^{4.3 - t}},$$

where t is the number of days after students are exposed to infected students.

a) Estimate the initial number of students infected with the flu.

b) What is the maximum number of students who will get the flu?

c) Estimate the number of days it will take for 125 students to become infected.

d) When will the rate at which the flu spreads be a maximum? What is this rate?

113. *The Rule of 70 for depletion.* The Rule of 70 applies to deflation and depletion, too, to estimate how many years it will take something that decreases at a constant rate to reach one half its present value. To see why, solve the equation

$$A_0 e^{-kt} = \frac{1}{2} A_0$$

to show that $t = (\ln 2)/k$. Thus, if $k = i/100$ is given as $i\%$,

$$t = \frac{\ln 2}{(i/100)} = \frac{100 \ln 2}{i} \approx \frac{69.3}{i} \approx \frac{70}{i}.$$

As a rule, it will take about $70/i$ years for the amount in question to decline from A_0 to $A_0/2$.

114. *Continuation of Exercise 113.* Use the Rule of 70 to tell about how long it will take

a) for prices to decline to half their present level at a constant deflation rate of 5%;

b) for oil reserves to deplete to half their present volume at a constant depletion rate of 7%.

115. *Cooling a pie.* A deep-dish apple pie, whose internal temperature was 220°F when removed from the oven, was set out on a 40° breezy porch to cool. Fifteen minutes later, the pie's internal temperature was 180°F. How long did it take the pie to cool from there to 70°F?

116. *Transport through a cell membrane.** Under some conditions, the result of the movement of a dissolved substance across a cell's membrane is described by the equation

$$\frac{dy}{dt} = k\frac{A}{V}(c - y).$$

In this equation, y is the concentration of the substance inside the cell, and dy/dt is the rate with which y changes over time. The letters $k, A, V,$ and c stand for constants, k being the *permeability coefficient* (a property of the membrane), A the surface area of the membrane, V the cell's volume, and c the concentration of the substance outside the cell. The equation says that the rate at which the concentration changes within the cell is proportional to the difference between it and the outside concentration.

a) Solve the equation for $y(t)$, using y_0 to denote $y(0)$.

b) Find the steady-state concentration, $\lim\limits_{t\to\infty} y(t)$.

Find the limits in Exercises 117–122.

117. $\lim\limits_{t\to 0} \dfrac{t - \ln(1 + 2t)}{t^2}$

118. $\lim\limits_{x\to 0} \dfrac{\sin x}{e^x - x - 1}$

119. $\lim\limits_{x\to 0} \dfrac{x\sin x}{1 - \cos x}$

120. $\lim\limits_{x\to 1} \dfrac{\log_4 x}{\log_2 x}$

121. Show that $f(x) = \dfrac{2^{\sin x} - 1}{e^x - 1}$ has a removable discontinuity at $x = 0$, and draw a complete graph of f in $[-6, 3]$.

122. Show that $f(x) = x\ln x$ has a removable discontinuity at $x = 0$, and draw a complete graph of f in $(0,2]$.

Use logarithms and l'Hôpital's Rule to find the limits in Exercises 123–126.

123. $\lim\limits_{x\to\infty} x^{1/x}$

124. $\lim\limits_{x\to\infty} x^{1/x^2}$

125. $\lim\limits_{x\to\infty} \left(1 + \dfrac{3}{x}\right)^x$

126. $\lim\limits_{x\to 0} \left(1 + \dfrac{3}{x}\right)^x$

127. Compare the rates of growth as $x \to \infty$ of

a) x and $5x$;

b) $x + \dfrac{1}{x}$ and x;

c) $x^2 + x$ and $x^2 - x$.

128. Compare the rates of growth as $x \to \infty$ of

a) $\ln 2x, \quad \ln x^2,$ and $\ln(x + 2)$;

b) $x^{\ln x}$ and $x^{\log_2 x}$;

c) $\left(\dfrac{1}{2}\right)^x$ and $\left(\dfrac{1}{3}\right)^x$.

* Based on *Some Mathematical Models in Biology*, Revised Edition, R. M. Thrall, J. A. Mortimer, K. R. Rebman, R. F. Baum, eds., December 1967, PB–202 364, pp. 101–103; distributed by N.T.I.S., U.S. Department of Commerce.

129. True or false?

a) $\dfrac{1}{x^2} + \dfrac{1}{x^4} = O\left(\dfrac{1}{x^2}\right)$

b) $\dfrac{1}{x^2} + \dfrac{1}{x^4} = O\left(\dfrac{1}{x^4}\right)$

c) $\sqrt{x^2 + 1} = O(x)$

130. True or false?

a) $\ln x = o(x)$

b) $\ln\ln x = o(\ln x)$

c) $x = o(x + \ln x)$

In Exercises 131–136, determine whether g is a left-end behavior model, a right end behavior model, or an end behavior model for f.

131. $g(x) = x^{3/4}, f(x) = \sqrt[4]{x^3 - x}$

132. $g(x) = x^{2/3}, f(x) = \sqrt[3]{x^2 - 2x + 1}$

133. $g(x) = 0.5x, f(x) = \dfrac{2x^3 - x + 1}{4x^2 - x + 1}$

134. $g(x) = x^3, f(x) = \dfrac{x^5 - 1}{x^2 + x - 1}$

135. $g(x) = 2^x, f(x) = 2^x + 2^{-x}$

136. $g(x) = 2^{-x}, f(x) = 2^x + 2^{-x}$

137. Show that

$$\tan^{-1} x + \tan^{-1}\frac{1}{x} = \text{constant.}$$

Find the constant.

138. Use the diagram to show that

$$\int_0^{\pi/2} \sin x\, dx = \frac{\pi}{2} - \int_0^1 \sin^{-1} x\, dx.$$

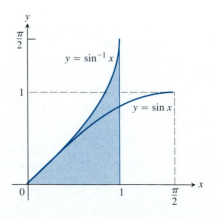

139. *Locating a solar station.* You are under contract to build a solar station at ground level on the east-west line between the two buildings shown.

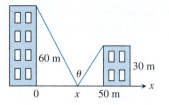

How far from the taller building should you place the station to maximize the number of hours it will be in the sun on a day when the sun passes directly overhead? Begin by observing that

$$\theta = \pi - \cot^{-1}\frac{x}{60} - \cot^{-1}\frac{50-x}{30}.$$

Then find the value of x that maximizes θ.

140. *The best branching angles for blood vessels and pipes.* When a smaller pipe branches off from a larger one in a flow system, we may want it to run off at an angle that is best from some energy-saving point of view. We might require, for instance, that energy loss due to friction be minimized along the section AOB shown in Fig. 7.48.

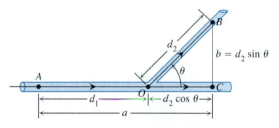

7.48 The smaller pipe OB branches away from the larger OAC at an angle θ that minimizes the friction loss along AO and OB. The optimum angle is found to be $\theta_c = \cos^{-1}(r^4/R^4)$, where r is the radius of the smaller pipe and R is the radius of the larger (Exercise 140).

In this diagram, B is a given point to be reached by the smaller pipe, A is a point in the larger pipe upstream from B, and O is the point where the branching occurs. A law due to Poiseuille states that the loss of energy due to friction in nonturbulent flow is proportional to the length of the path and inversely proportional to the fourth power of the radius. Thus, the loss along AO is $(kd_1)/R^4$, and that along OB is $(kd_2)/r^4$, where k is a constant, d_1 is the length of AO, d_2 is the length of OB, R is the radius of the larger pipe, and r is the radius of the smaller pipe. The angle θ is to be chosen to minimize the sum of these two losses:

$$L = k\frac{d_1}{R^4} + k\frac{d_2}{r^4}.$$

In our model, we assume that $AC = a$ and $BC = b$ are fixed. Thus, we have the relations

$$d_1 + d_2\cos\theta = a, \qquad d_2\sin\theta = b,$$

so that

$$d_2 = b\csc\theta$$

and

$$d_1 = a - d_2\cos\theta = a - b\cot\theta.$$

We can express the total loss L as a function of θ:

$$L = k\left(\frac{a-b\cot\theta}{R^4} + \frac{b\csc\theta}{r^4}\right).$$

a) Show that the critical value of θ for which $dL/d\theta$ equals zero is

$$\theta_c = \cos^{-1}\frac{r^4}{R^4}.$$

b) If the ratio of the pipe radii is $r/R = 5/6$, estimate to the nearest degree the optimal branching angle given in part (a).

The mathematical analysis described here is also used to explain the angles at which arteries branch in an animal's body. See Edward Batschelet's *Introduction to Mathematics for Life Scientists*, 2nd ed. (New York: Springer-Verlag, 1976).

141. *Tractor trailers and the tractrix.* When a tractor trailer turns into a cross street, its rear wheels follow a curve like the one in Fig. 7.49.

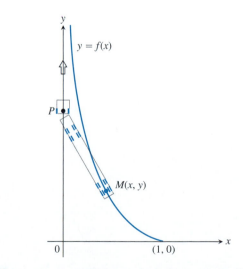

7.49 As P moves up the y-axis, it drags M after it at the end of a rod 1 unit long. The curve $y = f(x)$ traced by M is a *tractrix*. The goal of Exercise 141 is to find an equation for the curve.

(This is why the rear wheels sometimes ride up on the sidewalk.) We can find an equation for the curve if we picture the rear wheels as a mass M at the point $(1, 0)$ on the x-axis attached by a rod of unit length to a point P representing the cab at the origin. As the point P moves up the y-axis, it drags M along behind it. The curve traced by M, called a *tractix* from the Latin word *tractum* for drag, can be shown to be the graph of the function $y = f(x)$ that solves the initial value problem

Differential equation: $\quad \dfrac{dy}{dx} = -\dfrac{1}{x\sqrt{1 - x^2}} + \dfrac{x}{\sqrt{1 - x^2}},$

Initial condition: $\quad y = 0$ when $x = 1.$

Solve the initial value problem to find an equation for the curve. (You need an inverse hyperbolic function.)

8

Techniques of Integration

OVERVIEW   We have seen how integrals arise in modeling real phenomena and in measuring objects in the world around us, and we know in theory how integrals are evaluated with antiderivatives. The more sophisticated our models become, the more involved our integrals are likely to become. We can handle simple integral forms easily with pencil and paper. For more complicated integrals, we can use computer algebra systems (CAS) like *Derive*®, *Maple*®, or *Mathematica*® to find antiderivatives. And for the vast number of integral forms for which no explicit form of antiderivative exists, we can use numerical integration, NINT, to evaluate the definite integral and to see what the graph of the indefinite integral looks like.

In this chapter, we will work with integrals whose antiderivatives are not readily recognized, but that can be put into a form that will allow for pencil- and paper-solution, the second stage in Fig. 8.1. We will also touch on the CAS results (the third stage) and continue our work with the powerful NINT procedure on our graphing utility.

The Search for ∫ *f*(*x*) *dx*

| Antiderivatives easily recognized | Antiderivatives found by manipulation | Antiderivatives found by CAS | Antiderivatives do not exist in explicit form but can be viewed and evaluated by using NINT |

8.1   In this chapter, we travel across the methods for finding indefinite integrals.

It is a fair question to ask why we should study pencil-and-paper techniques to find explicit forms for integrals if CAS can do it so efficiently. Aside from a natural desire to learn, an answer is suggested by the analogy of arithmetic skills and the hand calculator. It is agreed that basic arithmetic skills will always be essential, as will be some understanding of arithmetic processes and their related algorithms. And we saw the extensive use of the electronic calculator only when it became small enough and inexpensive enough for universal use.

Similarly, basic integration skills will always be essential, and it is likely that CAS will soon fit inexpensively into the palms of our hands. Consequently, only a selection of the "most important" integration processes need be studied. Right now, we wrestle in the gray area of deciding how much time and energy should be devoted to the techniques of integration. We include in this book those that we feel are important at present. These will likely change in future editions, but we wait with you to find out how.

8.1　Formulas for Elementary Integrals

As we saw in Section 5.5, indefinite integration is the reverse of differentiation. So, to evaluate $\int f(x)\,dx$, we need to find an antiderivative function $F(x)$ whose derivative is $f(x)$. The list of integrals in Table 8.1 includes the integrals that we have worked with so far. You will find a more extensive list at the end of this book.

TABLE 8.1　　**A Summary of Standard Elementary Integral forms**

Definition	$\int du = u + C$
Constant multiple	$\int a\,du = au + C$
Sum	$\int (du + dv) = \int du + \int dv$

Powers
$$\int u^n\,du = \frac{u^{n+1}}{n+1} + C, n \neq -1 \qquad \int \frac{du}{u} = \ln |u| + C$$

Trigonometric
$$\int \cos u\,du = \sin u + C \qquad \int \sin u\,du = -\cos u + C$$
$$\int \tan u\,du = \ln |\sec u| + C \qquad \int \cot u\,du = \ln |\sin u| + C$$
$$\int \sec u\,du = \ln |\sec u + \tan u| + C \qquad \int \csc u\,du = -\ln |\csc u + \cot u| + C$$

Exponential
$$\int e^u\,du = e^u + C \qquad \int a^u\,du = \frac{1}{\ln a}a^u + C; a > 0, a \neq 1$$

Special algebraic forms
$$\int \frac{du}{1+u^2} = \tan^{-1} u + C \qquad \int \frac{du}{\sqrt{1-u^2}} = \sin^{-1} u + C$$
$$\int \frac{du}{u\sqrt{u^2-1}} = \sec^{-1} |u| + C$$

Historically, tables of integrals were used to find more complicated integrals like

$$\int \tan^5 x\,dx.$$

EXPLORATION BIT

If you can find a slide rule, find out *how* it works. Then figure out *why* it works. Prepare a demonstration and explanation.

Today, however, the fate of lengthy integral tables is tied to the growth of CAS. Just as we no longer use tables for approximate values of numbers like $\sin 20°$ or $\sqrt{\ln 55}$ because of the widespread availability of scientific calculators, it is likely that soon we will no longer need integral tables because of the accessibility of modern CAS. Indeed, the electronic "tables" provided by electronic calculators and CAS are better than *any* tables found in print.

SUPPORT REMINDER

Remember, for support that $F = \int f$, you can view either

$$F \text{ and NINT } f$$

or

$$\text{NDER } F \text{ and } f$$

on your grapher. What are the two possibilities for Example 1?

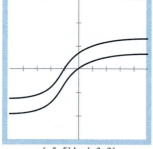

[–5, 5] by [–3, 3]

8.2 Graphs of

$$y_1 = \tan^{-1}(x + 1)$$

and

$$y_2 = \text{ NINT } (1/(t^2 + 2t + 2), 0, x).$$

The graphs are different. How do they support the result in Example 2?

Table 8.1 contains most of the basic integral forms, many of which we commit to memory because we use them so often. We note that two forms in the table, $\int \sec u \, du$ and $\int \csc u \, du$, are new to us. We include them to complete the list of integrals of the six basic trigonometric functions. (See Example 5.)

Algebraic Procedures and Trigonometric Identities

The most frequently used integration technique is *substitution*. We look at the function to be integrated and try to match it with a standard form.

EXAMPLE 1   Substitution

Evaluate

$$\int \frac{dx}{1 + 4x^2}.$$

Solution   The nearest standard form is

$$\int \frac{du}{1 + u^2} = \tan^{-1} u + C.$$

Since $4x^2 = u^2$ if $u = 2x$, we substitute

$$u = 2x, \qquad du = 2 \, dx, \qquad \text{and} \qquad dx = \frac{1}{2} du.$$

Then,

$$\int \frac{dx}{1 + 4x^2} = \int \frac{(1/2) \, du}{1 + u^2} = \frac{1}{2} \int \frac{du}{1 + u^2}$$

$$= \frac{1}{2} \tan^{-1} u + C = \frac{1}{2} \tan^{-1} 2x + C.$$

Examples 2–4 show how we can use some simple algebra steps to find integrands that match some standard forms.

EXAMPLE 2   Completing the Square

Evaluate

$$\int \frac{dx}{x^2 + 2x + 2},$$

and support graphically.

Solution   We complete the square of the quadratic in the denominator:

$$x^2 + 2x + 2 = \left(x^2 + 2x + 1\right) + (2 - 1) = (x + 1)^2 + 1.$$

The integrand then matches a standard form if we substitute $u = x + 1$:

$$\int \frac{dx}{x^2 + 2x + 2} = \int \frac{dx}{(x + 1)^2 + 1} = \int \frac{du}{u^2 + 1}$$

$$= \tan^{-1} u + C = \tan^{-1} (x + 1) + C.$$

To support graphically, we GRAPH $\tan^{-1}(x + 1)$ and NINT $(1/(t^2 + 2t + 2), 0, x)$. (See Fig. 8.2.)

EXAMPLE 3   Expanding a Power and Using a Trigonometric Identity

Evaluate

$$\int (\sec x + \tan x)^2 \, dx.$$

Solution   We expand the integrand and get

$$(\sec x + \tan x)^2 = \sec^2 x + 2 \sec x \tan x + \tan^2 x.$$

The first two terms on the right-hand side of this equation are old friends; we can integrate them at once. How about $\tan^2 x$? There is an identity that connects it with $\sec^2 x$:

$$\tan^2 x + 1 = \sec^2 x, \qquad \tan^2 x = \sec^2 x - 1.$$

We replace $\tan^2 x$ by $\sec^2 x - 1$ and get

$$\int (\sec x + \tan x)^2 \, dx = \int (\sec^2 x + 2 \sec x \tan x + \sec^2 x - 1) \, dx$$

$$= 2 \int \sec^2 x \, dx + 2 \int \sec x \tan x \, dx - \int 1 \, dx$$

$$= 2 \tan x + 2 \sec x - x + C. \qquad \equiv$$

We'll do more with trigonometric techniques in Sections 8.3 and 8.4.

EXAMPLE 4   Separating a Fraction

Evaluate

$$\int_0^{1/2} \frac{3x + 2}{\sqrt{1 - x^2}}$$

using analytic methods of calculus. Support the answer with an NINT computation.

Solution   First we treat this problem as an indefinite integration problem and find an antiderivative. A little trial and error suggests that the integrand does not seem to match any of the standard forms (Table 8.1). Next, we separate the integrand to get

$$\int \frac{3x + 2}{\sqrt{1 - x^2}} \, dx = 3 \int \frac{x \, dx}{\sqrt{1 - x^2}} + 2 \int \frac{dx}{\sqrt{1 - x^2}}.$$

In the first of these new integrals, we substitute

$$u = 1 - x^2, \qquad du = -2x \, dx, \qquad \text{and} \qquad x \, dx = -\frac{1}{2} du.$$

$$3 \int \frac{x \, dx}{\sqrt{1 - x^2}} = 3 \int \frac{(-1/2) \, du}{\sqrt{u}} = -\frac{3}{2} \int u^{-1/2} \, du$$

$$= -\frac{3}{2} \frac{u^{1/2}}{1/2} + C_1 = -3\sqrt{1 - x^2} + C_1.$$

The second of the new integrals is a standard elementary integral form,

$$2 \int \frac{dx}{\sqrt{1-x^2}} = 2 \sin^{-1} x + C_2.$$

Combining these results gives

$$\int \frac{3x+2}{\sqrt{1-x^2}}\, dx = -3\sqrt{1-x^2} + 2\sin^{-1} x + C.$$

Thus,

$$\int_0^{1/2} \frac{3x+2}{\sqrt{1-x^2}}\, dx = \left[-3\sqrt{1-x^2} + 2\sin^{-1} x \right]\Big|_0^{1/2}$$

$$= -3\sqrt{\frac{3}{4}} + 2\sin^{-1}(1/2) + 3$$

$$= 1.449.$$

Support:

$$\text{NINT}\left((3x+2)/\sqrt{1-x^2}, 0, 1/2 \right) = 1.449 \qquad \blacksquare$$

Example 5 evaluates $\int \sec x\, dx$ and suggests how to evaluate $\int \csc x\, dx$, the last two of the integrals of the six basic trigonometric functions. Basically, no "trick" is used in finding the antiderivative in Example 5, just some intelligent guessing and testing back and forth between derivatives and antiderivatives. Once we know the result, however, we are able to present the path to it in the fairly efficient manner shown here.

EXAMPLE 5   Multiplying by a Form of 1

Show that

$$\int \sec x\, dx = \ln |\sec x + \tan x| + C.$$

WRITING MATHEMATICS

The solution in Example 5 illustrates how a mathematical result that required lots of scratch paper to find can be written in a clear, concise manner.

Solution

$$\int \sec x\, dx = \int (\sec x)(1)\, dx = \int \sec x \frac{\sec x + \tan x}{\sec x + \tan x}\, dx$$

$$= \int \frac{\sec^2 x + \sec x \tan x}{\sec x + \tan x}\, dx$$

$$= \int \frac{du}{u} \qquad\qquad \begin{aligned} u &= \sec x + \tan x, \\ du &= (\sec^2 x + \sec x \tan x)dx. \end{aligned}$$

$$= \ln |u| + C = \ln |\sec x + \tan x| + C. \qquad \blacksquare$$

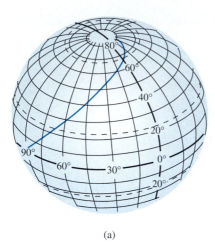

(a)

(b)

8.3 A course with a constant bearing of 45° E of N from the Galápagos Islands in the Pacific to Franz Josef Land in the Arctic Ocean as it appears (a) on the globe and (b) on a Mercator map.

EXPLORATION BITS

1. Suppose you want to mount the map of Example 6 on your wall. How wide would it be? What is the maximum latitude that the map could show? For what latitude would the map be nearly square?

2. Is the Mercator map of Fig. 8.3 accurately scaled?

With cosecants and cotangents in place of secants and tangents, the method of Example 5 leads to the companion formula

$$\int \csc x \, dx = -\ln |\csc x + \cot x| + C.$$

Mercator's World Map

The integral of the secant plays an important role in making maps for compass navigation. The easiest course for a sailor or pilot to steer is a course whose compass heading is constant. This might be a course of 45° (northeast), for example, or a course of 225° (southwest), or whatever. Such a course will be along a spiral that winds around the globe toward one of the poles (Fig. 8.3a) unless the course runs due north, south, east, or west.

In 1569, Gerhard Kremer, a Flemish surveyor and geographer known to us by his Latinized last name, Mercator, made a world map on which all spirals of constant compass heading appeared as straight lines (Fig. 8.3b). A sailor could then read the compass heading for a voyage between any two points from the direction of a straight line connecting them on Mercator's map.

If you look closely at Fig. 8.3(b), you will see that the vertical lines of longitude that meet at the poles on a globe have been spread apart to lie parallel on the map. The horizontal lines of latitude that are shown every 10° are also parallel, as they are on the globe, but they are not evenly spaced. The spacing between them increases toward the poles.

It can be shown that the scaling factor by which horizontal distances are increased at a fixed latitude $\theta°$ is precisely $\sec \theta$ and that this scaling factor increases with the latitude θ. If R is the radius of the globe being modeled (Fig. 8.4), then the distance on the map between the lines representing the equator and the latitude $\alpha°$ is R times the integral of the secant from zero to α:

$$D = R \int_0^\alpha \sec x \, dx.$$

Therefore, the map distance between the two latitude lines on the same side of the equator, say, at $\alpha°$ and $\beta°(\alpha < \beta)$, is

$$R \int_0^\beta \sec x \, dx - R \int_0^\alpha \sec x \, dx = R \int_\alpha^\beta \sec x \, dx = R \Big[\ln |\sec x + \tan x| \Big]_\alpha^\beta.$$

EXAMPLE 6

Suppose that the equatorial length of a Mercator map just matches the equator of a globe of radius 25 cm. Then the equation above gives the spacing on the map between the equator and latitude 20° north as

$$25 \int_0^{20°} \sec x \, dx = 25 \Big[\ln |\sec x + \tan x| \Big]_0^{20°} = 8.909 \text{ cm.}$$

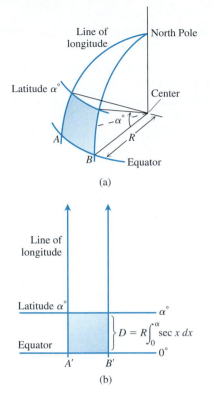

(a)

(b)

8.4 Lines of latitude and longitude (a) on a globe and (b) on a Mercator map.

The spacing between $60°$ and $80°$ north is given as

$$25 \int_{60°}^{80°} \sec x \, dx = 25 \Big[\ln |\sec x + \tan x| \Big]_{60°}^{80°} = 27.982 \text{ cm.}$$

As you can see, the map distance between latitude lines that are $20°$ apart is considerably greater near the pole than it is near the equator. The navigational properties of a Mercator map are achieved at the expense of a considerable distortion of distance. ≡

EXPLORATION 1

A Word of Caution about Using NINT

The view dimension [xMin, xMax] is like a closed interval in that it contains its endpoints—xMin and xMax identify the left and right columns of pixels in the viewing window. When we use NINT ($f(t)$, a, x), we have to be sure that both [a, x] and the "open" view dimension (xMin, xMax) contain no point of infinite discontinuity of f.

1. Consider $\int \sec x \, dx$ of Example 5. Let $y_1 = $ NINT ($\sec t$, 0, x) and $y_2 = \ln |\sec x + \tan x|$. GRAPH y_1 and y_2 in a $[-\pi/2, \pi/2]$ by $[-10, 10]$ viewing window. Everything should be okay. Is it?

2. Now GRAPH y_1 only in a $[0, \pi]$ by $[-10, 10]$ viewing window. Everything shouldn't be okay. Where does trouble occur?

3. GRAPH $y_1 = $ NINT ($\sec t$, -2, x) only in a $[-1.5, 1.5]$ by $[-10, 10]$ viewing window. Where does trouble occur? Why? ✿

Exercises 8.1

Evaluate the integrals in Exercises 1–12 by using substitution to match each integral to a standard elementary form. Support with a NINT computation.

1. $\displaystyle\int \frac{16x \, dx}{\sqrt{8x^2 + 1}}$

2. $\displaystyle\int \frac{3 \cos x \, dx}{\sqrt{1 + 3 \sin x}}$

3. $\displaystyle\int_0^1 \frac{16x \, dx}{8x^2 + 2}$

4. $\displaystyle\int_4^9 \frac{dx}{x - \sqrt{x}}$

5. $\displaystyle\int 4x \tan x^2 \, dx$

6. $\displaystyle\int \cot (3 - 7x) \, dx$

7. $\displaystyle\int_{-\pi}^{\pi} \sec \frac{x}{3} \, dx$

8. $\displaystyle\int x \sec (x^2 - 5) \, dx$

9. $\displaystyle\int_{3\pi/2}^{7\pi/4} \csc (x - \pi) \, dx$

10. $\displaystyle\int \frac{1}{x^2} \csc \frac{1}{x} \, dx$

11. $\displaystyle\int e^x \csc (e^x + 1) \, dx$

12. $\displaystyle\int \frac{\cot (3 + \ln x)}{x} \, dx$

Evaluate the integrals in Exercises 13–22. Support each with an NINT computation.

13. $\displaystyle\int_0^{\sqrt{\ln 2}} 2x e^{x^2} \, dx$

14. $\displaystyle\int_{\pi/2}^{\pi} \sin(x) e^{\cos x} \, dx$

15. $\displaystyle\int_{-1}^0 3^{x+1} dx$

16. $\displaystyle\int_1^2 \frac{2^{\ln x}}{x} \, dx$

17. $\displaystyle\int_1^3 \frac{6 \, dy}{\sqrt{y}(1 + y)}$

18. $\displaystyle\int_{-1}^0 \frac{4 \, dx}{1 + (2x + 1)^2}$

19. $\displaystyle\int_0^{1/6} \frac{dx}{\sqrt{1 - 9x^2}}$

20. $\displaystyle\int_0^1 \frac{dx}{\sqrt{4 - x^2}}$

21. $\displaystyle\int_{2/(5\sqrt{3})}^{2/5} \frac{6\,dx}{x\sqrt{25x^2-1}}$ **22.** $\displaystyle\int_{-6}^{-3\sqrt{2}} \frac{dx}{x\sqrt{x^2-9}}$

Evaluate the integrals in Exercises 23–28 by completing the square and using substitution to match each to a standard form.

23. $\displaystyle\int \frac{dx}{\sqrt{-x^2+4x-3}}$ **24.** $\displaystyle\int \frac{dx}{\sqrt{2x-x^2}}$

25. $\displaystyle\int_{1}^{2} \frac{8\,dx}{x^2-2x+2}$ **26.** $\displaystyle\int_{2}^{4} \frac{2\,dx}{x^2-6x+10}$

27. $\displaystyle\int \frac{dx}{(x+1)\sqrt{x^2+2x}}$ **28.** $\displaystyle\int \frac{dx}{(x-2)\sqrt{x^2-4x+3}}$

Evaluate the integrals in Exercises 29–32 using analytic methods. Support each with a NINT computation.

29. $\displaystyle\int_{\pi/4}^{3\pi/4} (\csc x - \cot x)^2\,dx$

30. $\displaystyle\int_{0}^{\pi/4} (\sec x + 4\cos x)^2\,dx$

31. $\displaystyle\int_{\pi/6}^{\pi/3} (\csc x - \sec x)(\sin x + \cos x)\,dx$

32. $\displaystyle\int_{0}^{\pi/2} (\sin 3x \cos 2x - \cos 3x \sin 2x)\,dx$

Evaluate the integrals in Exercises 33–36 by separating the fractions. Support each with a NINT computation.

33. $\displaystyle\int_{0}^{\sqrt{3}/2} \frac{1-x}{\sqrt{1-x^2}}\,dx$ **34.** $\displaystyle\int_{2}^{5} \frac{x+2\sqrt{x-1}}{2x\sqrt{x-1}}\,dx$

35. $\displaystyle\int_{0}^{\pi/4} \frac{1+\sin x}{\cos^2 x}\,dx$ **36.** $\displaystyle\int_{0}^{1/2} \frac{2-8x}{1+4x^2}\,dx$

37. Find the length of the curve $y = \ln(\cos x)$, $0 \le x \le \pi/3$.

38. Find the length of the curve $y = \ln(\sec x)$, $0 \le x \le \pi/4$.

39. Find the centroid of the region bounded by the x-axis, the curve $y = \sec x$, and the lines $x = -\pi/4$, $x = \pi/4$.

40. Find the area of the region bounded above by $y = 2\cos x$ and below by $y = \sec x$, $-\pi/4 \le x \le \pi/4$.

How far apart should the lines of latitude in Exercises 41 and 42 be on the Mercator map in Example 6?

41. Latitudes $30°$ and $45°$ north (New Orleans, Louisiana, and Minneapolis, Minnesota).

42. Latitudes $45°$ and $60°$ north (Salem, Oregon, and Seward, Alaska).

Evaluate the integrals in Exercises 43–58. Support each with a NINT computation.

43. $\displaystyle\int_{0}^{\pi/2} 3\sqrt{\sin x}\cos x\,dx$ **44.** $\displaystyle\int_{\pi/6}^{\pi/2} \cot^3 x \csc^2 x\,dx$

45. $\displaystyle\int_{-\pi}^{0} \frac{\sin x\,dx}{2+\cos x}$ **46.** $\displaystyle\int_{0}^{2} \frac{x\,dx}{4x^2+1}$

47. $\displaystyle\int_{0}^{1/4} \sec \pi x\,dx$ **48.** $\displaystyle\int_{1/4}^{3/4} \csc \pi x\,dx$

49. $\displaystyle\int_{0}^{\pi/3} e^{\tan x} \sec^2 x\,dx$ **50.** $\displaystyle\int_{\ln^2 2}^{\ln^2 3} \frac{e^{\sqrt{x}}\,dx}{\sqrt{x}}$

51. $\displaystyle\int_{1}^{4} \frac{2^{\sqrt{x}}\,dx}{2\sqrt{x}}$ **52.** $\displaystyle\int_{0}^{1} 10^{2x}\,dx$

53. $\displaystyle\int_{0}^{\sqrt{3}/3} \frac{9\,dx}{1+9x^2}$ **54.** $\displaystyle\int_{0}^{\ln\sqrt{3}} \frac{e^x\,dx}{1+e^{2x}}$

55. $\displaystyle\int_{0}^{1/4} \frac{2\,dx}{\sqrt{1-4x^2}}$ **56.** $\displaystyle\int_{0}^{1/\sqrt{2}} \frac{2x\,dx}{\sqrt{1-x^4}}$

57. $\displaystyle\int_{1/\sqrt{2}}^{1} \frac{dx}{x\sqrt{4x^2-1}}$ **58.** $\displaystyle\int_{\ln(2/\sqrt{3})}^{\ln 2} \frac{e^{-x}\,dx}{\sqrt{e^{2x}-1}}$

59. Find the value of C so that $\tan^{-1}(x+1) + C = $ NINT $\left(1/(t^2+2t+2), 0, x\right)$. (See Fig. 8.2 and Example 2.)

60. Find the value of C so that $(1/2)\tan^{-1} 2x + C = $ NINT $\left(1/(1+4t^2), 0, x\right)$. (See Example 1.)

8.2

Integration by Parts

Integration by parts is a technique used to simplify integrals of the form

$$\int f(x)g(x)\,dx.$$

A common type of integral of this form is one in which some higher-order derivative of $f(x)$ is 0 and repeated integrations beginning with $\int g(x)\,dx$ are done without difficulty. The integral

$$\int xe^x\,dx$$

is such an integral because, if we let $f(x) = x$, then $f''(x) = 0$, and repeated integrations beginning with $\int e^x \, dx$ are quite easy.

Integration by parts also applies to integrals like

$$\int e^x \cos x \, dx,$$

in which each part of the integrand reappears after repeated differentiation or integration.

In this section, we describe integration by parts and show how to apply it.

The Formula

The formula for integration by parts comes from the Product Rule,

$$\frac{d}{dx}(uv) = u\frac{dv}{dx} + v\frac{du}{dx}.$$

In its differential form, the rule becomes

$$d(uv) = u \, dv + v \, du,$$

which is then written as

$$u \, dv = d(uv) - v \, du$$

and integrated to give the following formula.

The Integration-by-Parts Formula

$$\int u \, dv = uv - \int v \, du.$$

The integration-by-parts formula expresses one integral, $\int u \, dv$, in terms of a second integral, $\int v \, du$. With a proper choice of u and v, the second integral may be easier to evaluate than the first. This is the reason for the importance of the formula. When faced with an integral that we cannot handle analytically, we can replace it by one with which we might have more success.

EXAMPLE 1

Evaluate

$$\int x \cos x \, dx.$$

Solution   We use the formula

$$\int u \, dv = uv - \int v \, du$$

with

$$u = x, \qquad dv = \cos x \, dx.$$

Then
$$du = dx, \qquad v = \sin x,$$

and
$$\int x \cos x \, dx = x \sin x - \int \sin x \, dx = x \sin x + \cos x + C.$$ ▤

EXPLORATION 1

Consider the Possibilities

The possible choices for u and dv in
$$\int u \, dv = \int x \cos x \, dx$$

of Example 1 are listed systematically below. Investigate each possibility in the integration-by-parts formula, and comment on where it leads you.

$u =$	$dv =$	$uv - \int v \, du =$	Comment
1	$x \cos x \, dx$	?	?
x	$\cos x \, dx$	?	?
$\cos x$	$x \, dx$	?	?
$x \cos x$	dx	?	?

Integration-by-parts summary   Keep in mind that the object is to go from the given integral ($\int u \, dv$) to a new integral ($\int v \, du$) that is simpler. (Integration by parts does not always work, so we can't always achieve that goal.)

EXAMPLE 2

Evaluate $\int_{-1}^{3} xe^{-x} dx$ analytically. Support with a NINT computation.

Solution   We use the formula $\int u \, dv = uv - \int v \, du$ with
$$u = x, \qquad dv = e^{-x} \, dx.$$

Then
$$du = dx, \qquad v = -e^{-x},$$

and
$$\int xe^{-x} \, dx = -xe^{-x} + \int e^{-x} \, dx = -xe^{-x} - e^{-x} + C.$$

Thus,
$$\int_{-1}^{3} xe^{-x} \, dx = \left(-xe^{-x} - e^{-x} \right) \Big|_{-1}^{3}$$
$$= -3e^{-3} - e^{-3} - (e - e) = -4e^{-3}.$$

Support: NINT $(xe^{-x}, -1, 3) = -0.199$, which agrees with $-4e^{-3}$ to three decimal places. ▤

EXAMPLE 3

Find the moment about the y-axis (Section 6.8) of a thin plate of constant density δ covering the region in the first quadrant bounded by the curve $y = e^x$ and the line $x = 1$ (Fig. 8.5).

Solution A typical vertical strip has

center of mass (c.m.): $\qquad (\tilde{x}, \tilde{y}) = \left(x, \dfrac{e^x}{2} \right),$

length: $\qquad\qquad\qquad e^x,$
width: $\qquad\qquad\qquad dx,$
area: $\qquad\qquad\qquad dA = e^x\, dx,$
mass: $\qquad\qquad\qquad dm = \delta\, dA = \delta e^x\, dx.$

The moment of the strip about the y-axis is therefore

$$\tilde{x}\, dm = x \cdot \delta e^x\, dx = \delta x e^x\, dx.$$

The moment of the plate about the y-axis is $M_y = \int \tilde{x}\, dm = \delta \int_0^1 x e^x\, dx.$
To evaluate this integral, we use the formula $\int u\, dv = uv - \int v\, du$ with

$$u = x, \qquad dv = e^x dx,$$
$$du = dx, \qquad v = e^x.$$

Then

$$\int x e^x\, dx = x e^x - \int e^x\, dx,$$

so

$$\int_0^1 x e^x\, dx = x e^x \Big]_0^1 - \int_0^1 e^x\, dx = e - \big[e^x\big]_0^1 = e - [e - 1] = 1.$$

The moment of the plate about the y-axis is

$$M_y = \delta \int_0^1 x e^x\, dx = \delta \cdot 1 = \delta. \qquad\qquad \blacksquare$$

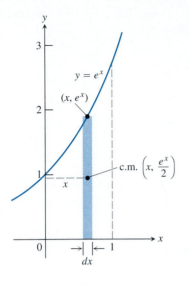

8.5 The moment of the strip about the y-axis is $x\delta\, dA = \delta x e^x\, dx.$

 EXPLORATION 2

($\int u\, dv = uv - \int v\, du$)

To integrate "by parts" requires practice in recognizing the parts that give you a simpler integral.

1. To evaluate $\int \ln x\, dx$ by parts, the possible choices for u and dv are listed. Comment on the two choices, and then complete the integration.

$u =$	$dv =$
1	$\ln x\, dx$
$\ln x$	dx

2. Support your results in part 1 graphically. Compare your graphs with our graphs in Fig. 8.6. How far apart are our graphs? Your graphs?

3. Evaluate $\int x^2 e^x dx$ by parts. Remember, you are looking for u and dv so that du and v make $\int v\, du$ simpler than $\int u\, dv$. Use a chart to systematically list the possible choices if necessary. (Completing this integration will require one additional small, insightful step on your part.)

4. Support your results in part 3 graphically. Comment on what you see in the viewing window. &

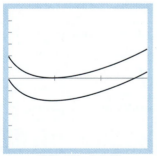

[0.01, 3] by [–3, 3]

8.6 Graphs of

$$y_1 = \text{the explicit form of } \int \ln x\, dx$$

and

$$y_2 = \text{NINT } (\ln t, 1, x)$$

for part 2 of Exploration 2.

Solving for the Unknown Integral

Integrals like the one in the next example occur in electrical engineering problems. Integration by parts must be applied two (or more) times to yield an equation in which the integral appears twice. We solve that equation for the value of the integral in an explicit form.

EXAMPLE 4

Evaluate

$$\int e^x \cos x \, dx.$$

Solution   We first use the formula

$$\int u \, dv = uv - \int v \, du$$

with

$$u = e^x, \qquad dv = \cos x \, dx,$$
$$du = e^x \, dx, \qquad v = \sin x.$$

Then

$$\int e^x \cos x \, dx = e^x \sin x - \int e^x \sin x \, dx. \tag{1}$$

The second integral is like the first except that it has $\sin x$ in place of $\cos x$. To evaluate it, we use integration by parts with

$$u = e^x \quad \text{(again)}, \qquad dv = \sin x \, dx,$$
$$du = e^x \, dx, \qquad\qquad v = -\cos x.$$

Then

$$\int e^x \cos x \, dx = e^x \sin x - \left(-e^x \cos x - \int (-\cos x)(e^x \, dx) \right)$$

$$= e^x \sin x + e^x \cos x - \int e^x \cos x \, dx.$$

The unknown integral now appears on both sides of the equation. Adding $\int e^x \cos x \, dx$ to both sides gives

$$2 \int e^x \cos x \, dx = e^x \sin x + e^x \cos x.$$

Dividing by 2 and adding a constant of integration give

$$\int e^x \cos x \, dx = \frac{e^x \sin x + e^x \cos x}{2} + C. \qquad\blacksquare$$

In Example 4, our choice of $u = e^x$ and $dv = \sin x \, dx$ in the second integration may have seemed arbitrary, but it wasn't. In theory, we could

---

EXPLORATION BIT

Now try to evaluate $\int e^x \sin x \, dx$. First, however, predict how many times you will have to use the integration-by-parts formula:

$$\int u \, dv = uv - \int v \, du.$$

have chosen $u = \sin x$ and $dv = e^x\, dx$. Doing so, however, would have turned Eq. (1) into

$$\int e^x \cos x\, dx = e^x \sin x - \left(e^x \sin x - \int e^x \cos x\, dx \right)$$

$$= \int e^x \cos x\, dx.$$

The resulting identity is correct but useless. *Moral:* Once you have decided on what to differentiate and integrate in circumstances like these, stick with them for the second integration.

General formulas for the integrals of $e^{ax} \cos bx$ and the closely related $e^{ax} \sin bx$ can be found in the integral table at the end of this book.

Tabular Integration

We have seen that integrals of the form $\int f(x)g(x)\, dx$, in which some higher-order derivative of $f(x)$ is zero and repeated integrations beginning with $g(x)\, dx$ are not difficult, are natural candidates for integration by parts. However, if many repetitions are required, the calculations can be cumbersome. In situations like this, there is a way to organize the calculations that saves a great deal of work. It is called *tabular integration* and is illustrated in the following examples.

EXAMPLE 5

Evaluate

$$\int x^2 e^x\, dx$$

by tabular integration.

Solution   With $f(x) = x^2$ and $g(x) = e^x$, we list

$f(x)$ and higher-order derivatives		$g(x)$ and subsequent integrals
x^2	$(+)$	e^x
$2x$	$(-)$	e^x
2	$(+)$	e^x
0		e^x

HOW TO DO TABULAR INTEGRATION

1. List the derivatives and integrals.
2. Link the terms with arrows.
3. Mark the operations $(+)$ and $(-)$ to alternate.
4. Combine the products of the linked terms according to the marked operations.

We combine the products of the functions connected by the arrows according to the operation sign along each arrow to obtain

$$\int x^2 e^x\, dx = x^2 e^x - 2xe^x + 2e^x + C.$$ ▤

EXAMPLE 6

Find the area under the curve $y = x^3 \sin x$ for $0 \le x \le 3$ numerically. Confirm analytically.

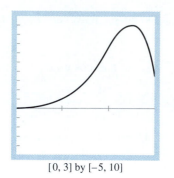

[0, 3] by [−5, 10]

8.7   The graph of $y = x^3 \sin x$ for $0 \le x \le 3$.

Solution   The desired area is given by $\int_0^3 x^3 \sin x \, dx$. (See Fig. 8.7.) A NINT computation gives $\int_0^3 x^3 \sin x \, dx = 11.873$. This numerical approximation can be confirmed by using analytic methods of calculus and the tabular integration technique. With $f(x) = x^3$ and $g(x) = \sin x$, we list

$f(x)$ and higher-order derivatives		$g(x)$ and subsequent integrals
x^3	$(+)$	$\sin x$
$3x^2$	$(-)$	$-\cos x$
$6x$	$(+)$	$-\sin x$
6	$(-)$	$\cos x$
0		$\sin x$

Combining the products of the functions connected by the arrows according to the operation signs, we obtain

$$\int x^3 \sin x \, dx = -x^3 \cos x + 3x^2 \sin x + 6x \cos x - 6 \sin x + C.$$

Thus, the area is

$$\int_0^3 x^3 \sin x = -x^3 \cos x + 3x^2 \sin x + 6x \cos x - 6 \sin x \big|_0^3$$
$$= -27 \cos 3 + 27 \sin 3 + 18 \cos 3 - 6 \sin 3$$
$$= 11.873 \quad \text{(square units).}$$

Exercises 8.2

Evaluate the integrals in Exercises 1–16. Support each with a NINT or NDER computation.

1. $\displaystyle\int x \sin x \, dx$

2. $\displaystyle\int x \cos 2x \, dx$

3. $\displaystyle\int x^2 \sin x \, dx$

4. $\displaystyle\int x^2 \cos x \, dx$

5. $\displaystyle\int x \ln x \, dx$

6. $\displaystyle\int x^3 \ln x \, dx$

7. $\displaystyle\int \tan^{-1} x \, dx$

8. $\displaystyle\int \sin^{-1} x \, dx$

9. $\displaystyle\int x \sec^2 x \, dx$

10. $\displaystyle\int 4x \sec^2 2x \, dx$

11. $\displaystyle\int x^3 e^x \, dx$

12. $\displaystyle\int x^4 e^{-x} \, dx$

13. $\displaystyle\int (x^2 - 5x) e^x \, dx$

14. $\displaystyle\int (x^2 + x + 1) e^x \, dx$

15. $\displaystyle\int x^5 e^x \, dx$

16. $\displaystyle\int x^2 e^{4x} \, dx$

Evaluate the integrals in Exercises 17–24. Support each with a NINT computation.

17. $\displaystyle\int_0^{\pi/2} x^2 \sin 2x \, dx$

18. $\displaystyle\int_0^{\pi/2} x^3 \cos 2x \, dx$

19. $\displaystyle\int_1^2 x \sec^{-1} x \, dx$

20. $\displaystyle\int_1^4 \sec^{-1} \sqrt{x} \, dx$

21. $\displaystyle\int_{-1}^3 e^x \sin x \, dx$

22. $\displaystyle\int_{-3}^2 e^{-x} \cos x \, dx$

23. $\displaystyle\int_{-2}^3 e^{2x} \cos 3x \, dx$

24. $\displaystyle\int_{-3}^2 e^{-2x} \sin 2x \, dx$

25. Find the area of the region enclosed by the x-axis and the curve $y = x \sin x$ for (a) $0 \le x \le \pi$, (b) $\pi \le x \le 2\pi$.

26. Use cylindrical shells to find the volume swept out by revolving the region bounded by $x = 0$, $y = 0$, and $y = \cos x$, $0 \le x \le \pi/2$, about the y-axis.

27. Find the volume swept out by revolving about the y-axis the region in the first quadrant bounded by the coordinate axes, the curve $y = e^{-x}$, and the line $x = 1$.

28. Find the volume swept out when the region shown in the first quadrant, bounded by the x-axis and the curve $y = x \sin x$, $0 \le x \le \pi$ is revolved (a) about the x-axis, (b) about the line $x = \pi$.

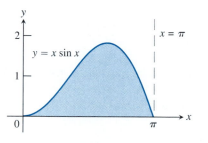

29. Find the moment about the y-axis of a thin plate of density $\delta = (1 + x)$ covering the region bounded by the x-axis and the curve $y = \sin x$, $0 \le x \le \pi$.

30. Although in integration by parts we usually drop the constant of integration when finding v from $\int dv$, choosing the constant to be different from zero can occasionally be helpful. As a case in point, evaluate

$$\int x \tan^{-1} x \, dx$$

with $u = \tan^{-1} x$ and $v = x^2/2 + 1/2$.

Reduction Formulas

Historically, *reduction formulas* were important practical integration techniques. Today, because of widespread use of CAS, they are more interesting as mathematical generalizations rather than as computational tools. Exercises 31 and 32 give examples of reduction formulas.

31. **a)** Show that for any positive integer n,

$$\int (\ln x)^n \, dx = x(\ln x)^n - n \int (\ln x)^{n-1} \, dx.$$

Hint: Use integration by parts.

b) Apply the reduction formula in part (a) to analytically determine the exact value of

$$\int_1^3 (\ln x)^4 \, dx.$$

Use a NINT computation to support your answer.

32. **a)** Show that for any positive integer n,

$$\int \tan^n x \, dx = \frac{1}{n-1} \tan^{n-1} x - \int \tan^{n-2} x \, dx.$$

Hint: Use $1 + \tan^2 x = \sec^2 x$.

b) Apply the reduction formula in part (a) to analytically determine the exact value of

$$\int_{\pi/6}^{\pi/3} \tan^4 x \, dx.$$

Use a NINT computation to support your answer.

33. **a)** Find the area bounded by $y = e^x$, $x = \ln 2$, $x = \ln 3$, $y = 0$.

b) Find the area bounded by $y = e^x$, $y = 2$, $y = 3$, $x = 0$.

c) How do the areas in parts (a) and (b) compare with the area of the rectangle determined by $x = 0$, $x = \ln 3$, $y = 0$, $y = 3$?

d) Show that for an appropriate choice of $u(x)$ and $v(x)$, your two integrals are $\int u \, dv$ and $\int v \, du$.

34. Let f be a one-to-one continuous function, $u = f(v)$, $v = f^{-1}(u)$, $u_1 = f(v_1)$, and $u_2 = f(v_2)$. Use an argument that involves area in the uv-plane to show that

$$\int_{v_1}^{v_2} u \, dv = u_2 v_2 - u_1 v_1 - \int_{u_1}^{u_2} v \, du.$$

(Assume that $0 < v_1 < v_2$.)

8.3 — Integrals Involving Trigonometric Functions

Trigonometric integrals involve algebraic combinations of the six basic trigonometric functions. In principle, we can always express such integrals in terms of sines and cosines, but it is often simpler to work with other functions, as in the integral

$$\int \sec^2 x \, dx = \tan x + C.$$

The general idea is to use identities to transform the integrals we are asked to evaluate into integrals that are easier to work with.

Products of Sines and Cosines, Part 1

We begin with integrals of the form

$$\int \sin^m x \cos^n x \, dx,$$

where m and n are nonnegative integers (positive or zero). We can divide the work into three cases.

    Case 1: m is odd.
    Case 2: m is even, n is odd.
    Case 3: m is even, n is even.

In each case, we can use a trigonometric identity to transform the integral into a more convenient form.

CASE 1: If m is odd, we write m as $2k + 1$ and use the identity $\sin^2 x = 1 - \cos^2 x$ to obtain

$$\sin^m x = \sin^{2k+1} x = (\sin^2 x)^k \sin x = (1 - \cos^2 x)^k \sin x.$$

Then we substitute $u = \cos x, du = - \sin x \, dx$ in the integrand.

EXAMPLE 1

Evaluate

$$\int \sin^3 x \cos^2 x \, dx.$$

Support graphically.

Solution

$$\int \sin^3 x \cos^2 x \, dx = \int \sin^2 x \cos^2 x \sin x \, dx$$

$$= \int (1 - \cos^2 x) \cos^2 x \sin x \, dx$$

$$= \int (1 - u^2)(u^2)(-du) \qquad \begin{array}{l} u = \cos x, du = - \sin x \, dx, \\ \text{so } \sin x \, dx = -du. \end{array}$$

$$= \int (u^4 - u^2) \, du$$

$$= \frac{u^5}{5} - \frac{u^3}{3} + C$$

$$= \frac{\cos^5 x}{5} - \frac{\cos^3 x}{3} + C.$$

SUPPORT: The graphs of

$$y_1 = \sin^3 x \cos^2 x \qquad \text{and} \qquad y_2 = \text{NDER}\left(\frac{\cos^5 x}{5} - \frac{\cos^3 x}{3}, x\right)$$

appear to be the same (Fig. 8.8), providing support for our analytic work.

$[-2\pi, 2\pi]$ by $[-0.3, 0.3]$

8.8 The graphs of

$$y_1 = \sin^3 x \cos^2 x$$

and

$$y_2 = \text{NDER}\left(\frac{\cos^5 x}{5} - \frac{\cos^3 x}{3}, x\right).$$

CASE 2: If m is even and n is odd in $\int \sin^m x \cos^n x \, dx$, we write $n = 2k + 1$ and use the identity $\cos^2 x = 1 - \sin^2 x$ to obtain

$$\cos^n x = \cos^{2k+1} x = (\cos^2 x)^k \cos x = (1 - \sin^2 x)^k \cos x.$$

Then we substitute $u = \sin x$, $du = \cos x \, dx$ in the integrand.

EXAMPLE 2

Evaluate

$$\int \cos^5 x \, dx.$$

Solution

$$\int \cos^5 x \, dx = \int \cos^4 x \, \cos x \, dx = \int (1 - \sin^2 x)^2 \cos x \, dx$$

$$= \int (1 - u^2)^2 \, du \qquad\qquad u = \sin x, \, du = \cos x \, dx.$$

$$= \int (1 - 2u^2 + u^4) \, du$$

$$= u - \frac{2}{3}u^3 + \frac{1}{5}u^5 + C = \sin x - \frac{2}{3}\sin^3 x + \frac{1}{5}\sin^5 x + C. \quad \blacksquare$$

CASE 3: If both m and n are even in $\int \sin^m x \cos^n x \, dx$, we substitute

$$\sin^2 x = \frac{1 - \cos 2x}{2}, \qquad \cos^2 x = \frac{1 + \cos 2x}{2}$$

to reduce the integrand to one in lower powers of $\cos 2x$.

EXAMPLE 3

Evaluate

$$\int \sin^2 x \cos^4 x \, dx.$$

Solution

$$\int \sin^2 x \cos^4 x \, dx = \int \left(\frac{1 - \cos 2x}{2} \right) \left(\frac{1 + \cos 2x}{2} \right)^2 dx$$

$$= \frac{1}{8} \int (1 - \cos 2x)(1 + 2\cos 2x + \cos^2 2x) \, dx$$

$$= \frac{1}{8} \int (1 + \cos 2x - \cos^2 2x - \cos^3 2x) \, dx. \qquad \begin{array}{l}\text{After some}\\\text{algebra}\end{array}$$

For the term involving $\cos^2 2x$, we use

$$\int \cos^2 2x \, dx = \frac{1}{2} \int (1 + \cos 4x) \, dx \qquad \text{Case 3}$$

$$= \frac{1}{2} \left(x + \frac{1}{4} \sin 4x \right). \qquad \begin{array}{l}\text{Omitting the constant of}\\\text{integration until the final result}\end{array}$$

For the $\cos^3 2x$ term, we have

$$\int \cos^3 2x \, dx = \int (1 - \sin^2 2x) \cos 2x \, dx \qquad \text{Case 2; then let } u = \sin 2x, \\ du = 2 \cos 2x \, dx.$$

$$= \frac{1}{2} \int (1 - u^2) \, du = \frac{1}{2} \left(\sin 2x - \frac{1}{3} \sin^3 2x \right). \qquad \text{Again omitting } C$$

Combining everything and simplifying, we get

$$\int \sin^2 x \cos^4 x \, dx = \frac{1}{16} \left(x - \frac{1}{4} \sin 4x + \frac{1}{3} \sin^3 2x \right) + C. \qquad \blacksquare$$

FLOWCHART 8.1 $\int \sin^m x \cos^n x \, dx$ (*m, n* nonnegative integers)

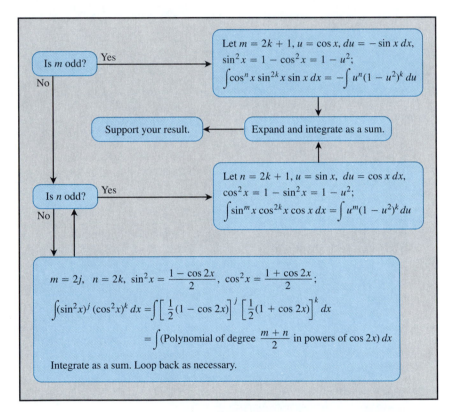

EXPLORATION 1

Eliminating Square Roots

The identity $2\cos^2 x = 1 + \cos 2x$ provides a way to eliminate a square root from an integrand. Evaluate

$$\int_0^{\pi/4} \sqrt{1 + \cos 4x} \; dx$$

numerically. Do you recognize the result? To confirm the result analytically, use the identity above to write $1 + \cos 4x$ as $2 \cos^2 2x$, simplify the integrand, and evaluate.   ⚛

Integrals of Powers of tan x and sec x

We know how to integrate the tangent and secant and their squares. To integrate higher powers, we use the identities $\tan^2 x = \sec^2 x - 1$ and $\sec^2 x = \tan^2 x + 1$ and integrate by parts when necessary to reduce the higher powers to lower powers.

EXAMPLE 4

Evaluate

$$\int \tan^4 x \, dx.$$

Support graphically.

Solution

$$\int \tan^4 x \, dx = \int \tan^2 x \cdot \tan^2 x \, dx = \int \tan^2 x \cdot (\sec^2 x - 1) \, dx$$

$$= \int \tan^2 x \sec^2 x \, dx - \int \tan^2 x \, dx$$

$$= \int \tan^2 x \sec^2 x \, dx - \int (\sec^2 x - 1) \, dx$$

$$= \int \tan^2 x \sec^2 x \, dx - \int \sec^2 x \, dx + \int dx.$$

In the first integral, we let

$$u = \tan x, \qquad du = \sec^2 x \, dx$$

and have

$$\int u^2 \, du = \frac{1}{3} u^3 + C'.$$

The remaining integrals are standard forms, so

$$\int \tan^4 x \, dx = \frac{1}{3} \tan^3 x - \tan x + x + C.$$

Figure 8.9 provides graphical support.   ≡

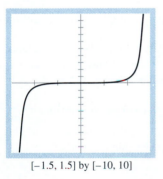

[−1.5, 1.5] by [−10, 10]

8.9 The graphs of

$$y_1 = \text{NINT} \, (\tan^4 t, 0, x)$$

and

$$y_2 = \frac{1}{3} \tan^3 x - \tan x + x$$

appear to be identical on $[-1.5, 1.5]$ to support the result in Example 4. What would happen if we used a wider [xMin, xMax] dimension for our viewing window?

EXAMPLE 5

Evaluate

$$\int \sec^3 x \, dx.$$

Solution   We integrate by parts, using

$$u = \sec x, \qquad\qquad dv = \sec^2 x \, dx,$$

$$du = \sec x \tan x \, dx, \qquad\qquad v = \tan x.$$

Then

$$\int \sec^3 x \, dx = \sec x \tan x - \int (\tan x)(\sec x \tan x \, dx)$$

$$= \sec x \tan x - \int (\sec^2 x - 1) \sec x \, dx \qquad \tan^2 x = \sec^2 x - 1.$$

$$= \sec x \tan x + \int \sec x \, dx - \int \sec^3 x \, dx.$$

Combining the two secant-cubed integrals gives

$$2 \int \sec^3 x \, dx = \sec x \tan x + \int \sec x \, dx,$$

so

$$\int \sec^3 x \, dx = \frac{1}{2} \sec x \tan x + \frac{1}{2} \ln |\sec x + \tan x| + C. \qquad \blacksquare$$

Products of Sines and Cosines, Part 2

The integrals

$$\int \sin mx \sin nx \, dx, \quad \int \sin mx \cos nx \, dx, \quad \text{and} \quad \int \cos mx \cos nx \, dx$$

arise in many applications of trigonometric functions. We can evaluate these integrals through integration by parts, but two such integrations are required in each case. It is simpler to use the identities

$$\sin mx \sin nx = \frac{1}{2}[\cos (m - n)x - \cos (m + n)x], \qquad \textbf{(1)}$$

$$\sin mx \cos nx = \frac{1}{2}[\sin (m - n)x + \sin (m + n)x], \qquad \textbf{(2)}$$

$$\cos mx \cos nx = \frac{1}{2}[\cos (m - n)x + \cos (m + n)x]. \qquad \textbf{(3)}$$

These come from combining the identities

$$\cos (A + B) = \cos A \cos B - \sin A \sin B, \qquad \textbf{(4)}$$

$$\cos (A - B) = \cos A \cos B + \sin A \sin B, \qquad \textbf{(5)}$$

$$\sin (A + B) = \sin A \cos B + \cos A \sin B, \qquad \textbf{(6)}$$

$$\sin (A - B) = \sin A \cos B - \cos A \sin B. \qquad \textbf{(7)}$$

For example, if we take $A = mx$ and $B = nx$ in Eqs. (4) and (5), add, and divide by 2, we get Eq. (3). We get Eq. (1) by subtracting (4) from (5) and dividing by 2. To get Eq. (2), we add Eqs. (6) and (7) and divide by 2.

EXPLORATION BIT

The integral

$$\int \sin mx \cos nx \, dx$$

resembles somewhat a form that we saw at the start of this section. Evaluate this integral or either of

$$\int \sin mx \sin nx \, dx$$

or

$$\int \cos mx \cos nx \, dx$$

by parts.

⬡ **EXPLORATION 2**

A Product of Sines and Cosines

Show graphically that

$$\int \sin 3x \cos 5x \, dx = -\frac{\cos 8x}{16} + \frac{\cos 2x}{4} + C.$$

Then match the integrand with one of the forms in Eqs. (1), (2), and (3) above, and confirm algebraically.

Definite Integrals of Even and Odd Functions

The definite integral of an even function $f(x)$ over $[-a, a]$ is twice the value of the definite integral of f over $[0, a]$ when this integral exists (Fig. 8.10). This is because the integral of f from $-a$ to 0 has the same value as the integral of f from 0 to a:

$$\int_{-a}^{0} f(x)\,dx = \int_{a}^{0} f(-u)\,(-du) \qquad x = -u, \;\; dx = -du.$$

$$= -\int_{a}^{0} f(u)\,du \qquad f \text{ even} \Leftrightarrow f(-u) = f(u).$$

$$= \int_{0}^{a} f(u)\,du.$$

This observation saves time when the antiderivative of f is more easily evaluated at 0 than at $-a$.

The definite integral of an odd function $f(x)$ over $[-a, a]$ is zero. This is because the integral of f from $-a$ to 0 has a value opposite that of the integral of f from 0 to a (Fig. 8.11).

Symmetric functions over symmetric intervals occur in a surprising number of applications.

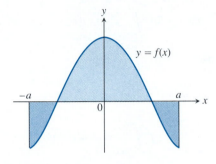

8.10 The graph of an even function is symmetric about the y-axis, so

$$\int_{-a}^{0} f(x)\,dx = \int_{0}^{a} f(x)\,dx$$

and

$$\int_{-a}^{a} f(x)\,dx = 2\int_{0}^{a} f(x)\,dx.$$

EXAMPLE 6

a) $\displaystyle \int_{-\pi/4}^{\pi/4} \cos x\,dx = 2\int_{0}^{\pi/4} \cos x\,dx = 2\,[\sin x]_{0}^{\pi/4} = 2\left[\frac{\sqrt{2}}{2} - 0\right] = \sqrt{2}.$

b) $\displaystyle \int_{-\pi}^{\pi} \sin x\,dx = -\cos x]_{-\pi}^{\pi} = -\cos \pi + \cos(-\pi) = -(-1) + (-1) = 0.$

$\equiv$

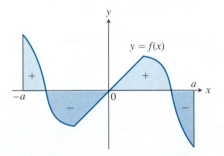

8.11 The graph of an odd function is symmetric about the origin, so

$$\int_{-a}^{a} f = \int_{-a}^{0} f + \int_{0}^{a} f = -\int_{0}^{a} f + \int_{0}^{a} f = 0.$$

The "+" and "−" signs in the picture suggest a geometric argument using area for why this is so.

EXPLORATION 3

Integrals of Even, Odd Functions

1. Tell whether f is even or odd, then evaluate both $\int_{-a}^{a} f$ and $\int_{0}^{a} f$ for the given value of a.

a) $f(x) = x^2 \cos x, \, a = \pi/2$ **b)** $f(x) = \sin^3 x, \, a = \pi/2$

c) $f(x) = \dfrac{\sin^2 x}{\cos x}, \, a = \pi/4$ **d)** $f(x) = \sqrt{\sin^2 x + 1}, \, a = \pi/2$

2. Suppose f is continuous on $[-a, a]$.

a) If $\int_{-a}^{a} f(x)\,dx = 2\int_{0}^{a} f(x)\,dx$, is f even? Give an example.

b) If $\int_{-a}^{a} f(x)\,dx = 0$, is f odd? Give an example.

Exercises 8.3

Evaluate the integrals in Exercises 1–14. Support each with a NINT or NDER computation.

1. $\displaystyle\int \sin^5 x\, dx$

2. $\displaystyle\int \sin^5 \frac{x}{2}\, dx$

3. $\displaystyle\int \cos^3 x\, dx$

4. $\displaystyle\int 3 \cos^5 3x\, dx$

5. $\displaystyle\int \sin^7 y\, dy$

6. $\displaystyle\int 7 \cos^7 t\, dt$

7. $\displaystyle\int 8 \sin^4 x\, dx$

8. $\displaystyle\int 8 \cos^4 2\pi x\, dx$

9. $\displaystyle\int 16 \sin^2 x \cos^2 x\, dx$

10. $\displaystyle\int 8 \sin^4 y \cos^2 y\, dy$

11. $\displaystyle\int 35 \sin^4 x \cos^3 x\, dx$

12. $\displaystyle\int \sin 2x \cos^2 2x\, dx$

13. $\displaystyle\int 8 \cos^3 2\theta \sin 2\theta\, d\theta$

14. $\displaystyle\int \sin^2 2\theta \cos^3 2\theta\, d\theta$

Evaluate the integrals in Exercises 15–38. Support each with a NINT computation.

15. $\displaystyle\int_0^{2\pi} \sqrt{\frac{1 - \cos x}{2}}\, dx$

16. $\displaystyle\int_0^{\pi} \sqrt{1 - \cos 2x}\, dx$

17. $\displaystyle\int_0^{\pi} \sqrt{1 - \sin^2 t}\, dt$

18. $\displaystyle\int_0^{\pi} \sqrt{1 - \cos^2 \theta}\, d\theta$

19. $\displaystyle\int_{-\pi/4}^{\pi/4} \sqrt{1 + \tan^2 x}\, dx$

20. $\displaystyle\int_{-\pi/4}^{\pi/4} \sqrt{\sec^2 x - 1}\, dx$

21. $\displaystyle\int_0^{\pi/2} \theta \sqrt{1 - \cos 2\theta}\, d\theta$

22. $\displaystyle\int_{-\pi}^{\pi} (1 - \cos^2 t)^{3/2}\, dt$

23. $\displaystyle\int_{-\pi/3}^{0} 2 \sec^3 x\, dx$

24. $\displaystyle\int_0^{\ln \pi/4} e^x \sec^3 e^x\, dx$

25. $\displaystyle\int_0^{\pi/4} \sec^4 \theta\, d\theta$

26. $\displaystyle\int_0^{\pi/12} 3 \sec^4 3x\, dx$

27. $\displaystyle\int_{\pi/4}^{\pi/2} \csc^4 \theta\, d\theta$

28. $\displaystyle\int_{\pi/2}^{\pi} 3 \csc^4 \frac{\theta}{2}\, d\theta$

29. $\displaystyle\int_0^{\pi/4} 4 \tan^3 x\, dx$

30. $\displaystyle\int_{-\pi/4}^{\pi/4} 6 \tan^4 x\, dx$

31. $\displaystyle\int_{\pi/6}^{\pi/3} \cot^3 x\, dx$

32. $\displaystyle\int_{\pi/4}^{\pi/2} 8 \cot^4 t\, dt$

33. $\displaystyle\int_{-\pi}^{0} \sin 3x \cos 2x\, dx$

34. $\displaystyle\int_0^{\pi/2} \sin 2x \cos 3x\, dx$

35. $\displaystyle\int_{-\pi}^{\pi} \sin 3x \sin 3x\, dx$

36. $\displaystyle\int_0^{\pi/2} \sin x \cos x\, dx$

37. $\displaystyle\int_0^{\pi} \cos 3x \cos 4x\, dx$

38. $\displaystyle\int_{-\pi/2}^{\pi/2} \cos x \cos 7x\, dx$

39. Which integrals are zero, and which are not? (You can do most of these without writing anything. Explain why.)

a) $\displaystyle\int_{-\pi}^{\pi} \sin x \cos^2 x\, dx$

b) $\displaystyle\int_{-L}^{L} \sqrt[3]{\sin x}\, dx$

c) $\displaystyle\int_{-\pi/4}^{\pi/4} x \sec x\, dx$

d) $\displaystyle\int_{-\pi/2}^{\pi/2} x \sin x\, dx$

e) $\displaystyle\int_{-a}^{a} \sin mx \cos mx\, dx$

f) $\displaystyle\int_{-\pi/2}^{\pi/2} \cos^3 x\, dx$

g) $\displaystyle\int_{-\ln 2}^{\ln 2} x(e^x + e^{-x})\, dx$

h) $\displaystyle\int_{-\pi/2}^{\pi/2} \sin x \sin 2x\, dx$

i) $\displaystyle\int_{-a}^{a} (e^x \sin x + e^{-x} \sin x)\, dx$

40. Which integrals are zero, and which are not? (You can do most of these without writing anything. Explain why.)

a) $\displaystyle\int_{-1}^{1} \sin 3x \cos 5x\, dx$

b) $\displaystyle\int_{-a}^{a} x \sqrt{a^2 - x^2}\, dx$

c) $\displaystyle\int_{-\pi/4}^{\pi/4} \tan^3 x\, dx$

d) $\displaystyle\int_{-\pi/2}^{\pi/2} x \cos x\, dx$

e) $\displaystyle\int_{-\pi}^{\pi} \sin^5 x\, dx$

f) $\displaystyle\int_{-\pi}^{\pi} \cos^5 x\, dx$

g) $\displaystyle\int_{-\pi/2}^{\pi/2} \sin^2 x \cos x\, dx$

h) $\displaystyle\int_{-\pi/4}^{\pi/4} \sec x \tan x\, dx$

i) $\displaystyle\int_{-1}^{1} \frac{\sin x\, dx}{e^x + e^{-x}}$

41. Which integrals in Exercise 39 have even integrands? Evaluate these integrals.

42. Which integrals in Exercise 40 have even integrands? Evaluate these integrals.

43. Show that

$$\int \csc x\, dx = -\ln |\csc x + \cot x| + C.$$

(*Hint:* Repeat the derivation in Section 8.1, Example 5, with cofunctions.)

44. Use the result in Exercise 43 to show that

$$\int \csc^3 x\, dx = -\frac{1}{2} \csc x \cot x - \frac{1}{2} \ln |\csc x + \cot x| + C.$$

45. Give geometric arguments using area that

$$\int_{-a}^{a} f(x)\, dx = 2 \int_0^{a} f(x)\, dx$$

when f is even and

$$\int_{-a}^{a} f(x)\,dx = 0$$

when f is odd.

46. Show analytically that

$$\int_{-a}^{0} f(x)\,dx = -\int_{0}^{a} f(x)\,dx$$

when f is an odd function. (*Hint:* Study the proof that

$$\int_{-a}^{0} f(x)\,dx = \int_{0}^{a} f(x)\,dx$$

for even functions.)

In Exercises 47–52, tell whether f is even or odd, then evaluate both $\int_{-a}^{a} f(x)\,dx$ and $\int_{0}^{a} f(x)\,dx$ for the given value of a.

47. $f(x) = x \sin x, a = \pi/2$

48. $f(x) = x^2 \sin x, a = \pi/2$

49. $f(x) = \cos x \sin 3x, a = \pi/2$

50. $f(x) = \dfrac{\sin x}{\cos^2 x}, a = \pi/4$

51. $f(x) = \cos(\sin x), a = \pi/2$

52. $f(x) = \sqrt{1 + \tan^2 x}, a = \pi/4$

8.4

Trigonometric Substitutions

We now embark on a three-step program that will enable us (in theory, at least) to integrate analytically all rational functions of x. The first step is to study substitutions that change binomials like $a^2 + x^2$, $a^2 - x^2$, and $x^2 - a^2$ into single squared terms. The second step will be to simplify integrals involving $ax^2 + bx + c$ by completing the square and then replacing the resulting sums and differences of squares by single squared terms. The third and last step, taken in Section 8.5, will be to express rational functions of x as sums of polynomials (which we already know how to integrate), fractions with linear-factored denominators (which become logarithms or fractions when integrated), and fractions with quadratic denominators (which we will be able to integrate by the techniques of the present section).

Trigonometric Substitutions for Combining Squares

Trigonometric substitutions enable us to replace the binomials

$$a^2 + x^2, \qquad a^2 - x^2, \qquad \text{and} \qquad x^2 - a^2$$

by single squared terms and thereby transform a number of important integrals into integrals with a more recognizable form. The most commonly used substitutions, $x = a \tan \theta$, $x = a \sin \theta$, and $x = a \sec \theta$, come from the reference triangles in Fig. 8.12.

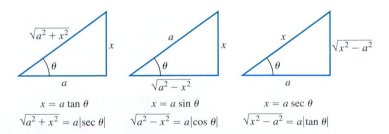

8.12 Reference triangles for trigonometric substitutions that change binomials into single squared terms.

With $x = a \tan\theta$, x can be any real number, and then

$$a^2 + x^2 = a^2 + a^2 \tan^2\theta = a^2 \left(1 + \tan^2\theta\right) = a^2 \sec^2\theta.$$

With $x = a \sin\theta$, we must have $-a \le x \le a$, and then

$$a^2 - x^2 = a^2 - a^2 \sin^2\theta = a^2 \left(1 - \sin^2\theta\right) = a^2 \cos^2\theta.$$

With $x = a \sec\theta$, we must have $x \le -a$ or $x \ge a$, and then

$$x^2 - a^2 = a^2 \sec^2\theta - a^2 = a^2 \left(\sec^2\theta - 1\right) = a^2 \tan^2\theta.$$

Trigonometric Substitutions

$x = a \tan\theta$ replaces $a^2 + x^2$ by $a^2 \sec^2\theta$.

$x = a \sin\theta$ replaces $a^2 - x^2$ (for $-a \le x \le a$)   by $a^2 \cos^2\theta$.

$x = a \sec\theta$ replaces $x^2 - a^2$ (for $x \le -a$ or $x \ge a$) by $a^2 \tan^2\theta$.

When we make substitutions, we always want them to be reversible so that we can change back to the original variables when we're done. For example, if $x = a \tan\theta$, we want to be able to set

$$\theta = \tan^{-1}\frac{x}{a}$$

after the integration takes place. If $x = a \sin\theta$, we want to be able to set

$$\theta = \sin^{-1}\frac{x}{a}$$

when we're done, and similarly for $x = a \sec\theta$.

As we know from Section 7.7, the functions in these substitutions have inverses only for selected values of θ (Fig. 8.13). For reversibility,

$$x = a \tan\theta \text{ requires } \theta = \tan^{-1}\frac{x}{a} \quad \text{with} \quad -\frac{\pi}{2} < \theta < \frac{\pi}{2},$$

$$x = a \sin\theta \text{ requires } \theta = \sin^{-1}\frac{x}{a} \quad \text{with} \quad -\frac{\pi}{2} \le \theta \le \frac{\pi}{2} \quad \text{if } -1 \le \frac{x}{a} \le 1,$$

$$x = a \sec\theta \text{ requires } \theta = \sec^{-1}\frac{x}{a} \quad \text{with} \quad \begin{cases} 0 \le \theta < \dfrac{\pi}{2} & \text{if } \dfrac{x}{a} \ge 1, \\[2mm] \dfrac{\pi}{2} < \theta \le \pi & \text{if } \dfrac{x}{a} \le -1. \end{cases}$$

The following examples of trigonometric substitutions show how reference triangles can be used at different places in a solution. Each time we use a reference triangle, we must remember to consider all the values that are possible for θ.

EXAMPLE 1

Evaluate

$$\int \frac{dx}{\sqrt{4 + x^2}}.$$

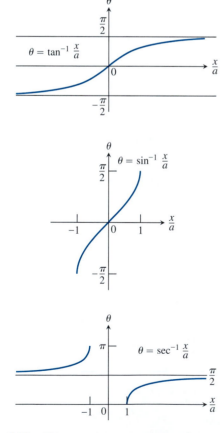

8.13 The arc tangent, arc sine, and arc secant of x/a, graphed as functions of x/a.

Solution We set

$$x = 2\tan\theta, \qquad dx = 2\sec^2\theta\,d\theta, \qquad -\frac{\pi}{2} < \theta < \frac{\pi}{2}.$$

From the triangle for $x = 2\tan\theta$ in Fig. 8.14, we have

$$\frac{\sqrt{4+x^2}}{2} = |\sec\theta|, \qquad \text{or} \qquad \sqrt{4+x^2} = 2|\sec\theta|.$$

Then,

$$\int \frac{dx}{\sqrt{4+x^2}} = \int \frac{2\sec^2\theta\,d\theta}{2|\sec\theta|}$$

$$= \int \sec\theta\,d\theta \qquad \sec\theta > 0 \text{ for } -\frac{\pi}{2} < \theta < \frac{\pi}{2}$$

$$= \ln|\sec\theta + \tan\theta| + C$$

$$= \ln\left|\frac{\sqrt{4+x^2}}{2} + \frac{x}{2}\right| + C \qquad \text{From Fig. 8.14}$$

$$= \ln\left|\sqrt{4+x^2} + x\right| + C'. \qquad \text{Taking } C' = C - \ln 2 \qquad \blacksquare$$

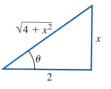

8.14 A reference triangle for $x = 2\tan\theta$ (Example 1). This allows us to express $\ln|\sec\theta + \tan\theta|$ in terms of x by reading the ratios

$$\sec\theta = \frac{\sqrt{4+x^2}}{2} \quad \text{and} \quad \tan\theta = \frac{x}{2}$$

directly from the triangle.

EXAMPLE 2

Evaluate

$$\int \frac{x^2\,dx}{\sqrt{9-x^2}}.$$

Support graphically.

Solution To replace $9 - x^2$ by a single squared term, we set

$$x = 3\sin\theta, \qquad dx = 3\cos\theta\,d\theta, \qquad -\frac{\pi}{2} < \theta < \frac{\pi}{2}.$$

Then

$$9 - x^2 = 9(1 - \sin^2\theta) = 9\cos^2\theta$$

and,

$$\int \frac{x^2\,dx}{\sqrt{9-x^2}} = \int \frac{9\sin^2\theta \cdot 3\cos\theta\,d\theta}{|3\cos\theta|}$$

$$= 9\int \sin^2\theta\,d\theta \qquad \cos\theta > 0 \text{ for } -\frac{\pi}{2} < \theta < \frac{\pi}{2}.$$

$$= 9\int \frac{1 - \cos 2\theta}{2}\,d\theta = \frac{9}{2}\left(\theta - \frac{\sin 2\theta}{2}\right) + C$$

$$= \frac{9}{2}(\theta - \sin\theta\cos\theta) + C$$

$$= \frac{9}{2}\left(\sin^{-1}\frac{x}{3} - \frac{x}{3} \cdot \frac{\sqrt{9-x^2}}{3}\right) + C \qquad \text{Fig. 8.15}$$

$$= \frac{9}{2}\sin^{-1}\frac{x}{3} - \frac{x}{2}\sqrt{9-x^2} + C.$$

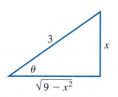

8.15 A reference triangle for $x = 3\sin\theta$ (Example 2).

Figure 8.16 provides graphical support.

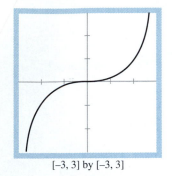

[−3, 3] by [−3, 3]

8.16   The graphs of

$$y_1 = \frac{9}{2} \sin^{-1} \frac{x}{3} - \frac{x}{2} \sqrt{9 - x^2}$$

and

$$y_2 = \text{NINT} \left(\frac{t^2}{\sqrt{9 - t^2}}, 0, x \right)$$

appear to be identical to support the results of Example 2.

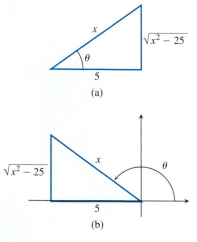

(a)

(b)

8.17   Reference triangles for $x = 5 \sec \theta$ for (a) $0 < \theta < \pi/2$ and (b) $\pi/2 < \theta < \pi$.

EXAMPLE 3

Evaluate

$$\int \frac{dx}{\sqrt{x^2 - 25}}.$$

Solution   To replace $x^2 - 25$ by a single squared term, we set

$$x = 5 \sec \theta, \qquad dx = 5 \sec \theta \tan \theta \, d\theta, \qquad \theta = \sec^{-1} \frac{x}{5}.$$

Then

$$x^2 - 25 = 25 \sec^2 \theta - 25 = 25(\sec^2 \theta - 1) = 25 \tan^2 \theta$$

and

$$\int \frac{dx}{\sqrt{x^2 - 25}} = \int \frac{5 \sec \theta \tan \theta \, d\theta}{\sqrt{25 \tan^2 \theta}}$$

$$= \int \frac{\sec \theta \tan \theta \, d\theta}{|\tan \theta|} \qquad \sqrt{\tan^2 \theta} = |\tan \theta|.$$

$$= \pm \int \sec \theta \, d\theta = \pm \ln |\sec \theta + \tan \theta| + C$$

$$= \pm \ln \left| \frac{x}{5} \pm \frac{\sqrt{x^2 - 25}}{5} \right| + C. \qquad \text{Fig. 8.17}$$

$$= \pm \ln \left| x \pm \sqrt{x^2 - 25} \right| + C'. \qquad C' = C \pm \ln 5.$$

What do we do about the signs? When $0 < \theta < \pi/2$, the tangent and secant are both positive and both signs are $(+)$. When $\pi/2 < \theta < \pi$, the tangent and secant are both negative, and both signs are $(-)$. Therefore,

$$\int \frac{dx}{\sqrt{x^2 - 25}} = \begin{cases} \ln \left| x + \sqrt{x^2 - 25} \right| + C' \\ \text{or} \\ -\ln \left| x - \sqrt{x^2 - 25} \right| + C'. \end{cases} \qquad (1)$$

Fortunately, we do not have to live with this two-line formula because the two logarithmic expressions on the right-hand side differ only by a constant (see Exploration 1). Therefore,

$$\int \frac{dx}{\sqrt{x^2 - 25}} = \ln \left| x + \sqrt{x^2 - 25} \right| + C.$$

EXPLORATION 1

Do Graphically, Confirm Analytically

1. Demonstrate graphically that

$$y_1 = \ln |x + \sqrt{x^2 - 25}| \qquad \text{and} \qquad y_2 = -\ln |x - \sqrt{x^2 - 25}|$$

(see Eq. (1)) differ by a constant. Explain your choice of [xMin, xMax], and account for any peculiar behavior over [xMin, xMax] in your viewing window.

2. Find a value for the difference $y_1 - y_2$. Call it k. Do you recognize k? (Probably not, but here is a hint: Since k is a difference of logs, it is reasonable that k may be the log of a more familiar number. To check, you could let $k = \ln c$, then ask what c is.)

3. Confirm analytically that $y_1 - y_2 = k$.

Two Useful Formulas

Integrals of the form $\int du/(u^2 + a^2)$ and $\int du/\sqrt{a^2 - u^2}$ arise so often in applications of integration that many people find that it saves time to memorize formulas for evaluating them.

$$\int \frac{du}{u^2 + a^2} = \frac{1}{a} \tan^{-1} \frac{u}{a} + C \qquad (2)$$

$$\int \frac{du}{\sqrt{a^2 - u^2}} = \sin^{-1} \frac{u}{a} + C \qquad (3)$$

We can derive Eq. (2) by substituting $u = a \tan\theta$ and Eq. (3) by substituting $u = a \sin\theta$.

SUPPORT

Explain how NDER can be used to support Eqs. (2) and (3). Then demonstrate using Example 4.

EXAMPLE 4

a)
$$\int \frac{dx}{(x+1)^2 + 4} = \frac{1}{2} \tan^{-1} \frac{x+1}{2} + C \qquad \text{Eq. (2) with } u = x + 1, \\ du = dx, a = 2$$

b)
$$\int \frac{dx}{\sqrt{3 - 4x^2}} = \frac{1}{2} \int \frac{du}{\sqrt{a^2 - u^2}} \qquad u = 2x, \frac{du}{2} = dx, a = \sqrt{3}.$$

$$= \frac{1}{2} \sin^{-1} \frac{u}{a} + C \qquad \text{Eq. (3)}$$

$$= \frac{1}{2} \sin^{-1} \frac{2x}{\sqrt{3}} + C$$

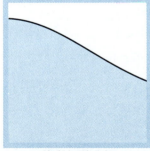

[0, 2] by [0, 1.2]

8.18 The region bounded by the curve

$$y_1 = \frac{4}{x^2 + 4},$$

the x-axis (at base of window), and the lines $x = 0$ (left side of window) and $x = 2$ (right side of window).

EXPLORATION 2

Confirm and Support

1. Find the area of the region bounded by the curve $y_1 = 4/(x^2 + 4)$, the x-axis, and the lines $x = 0$ and $x = 2$ (Fig. 8.18) using NINT. Do you recognize the result? Use Eq. (2), and confirm analytically.

2. Work through the following steps to find the volume of the solid generated by revolving the region of part 1 about the *x*-axis. Support numerically.

a) By the disk method, volume $= \int_0^2 \pi (y_1)^2 \, dx$. Use the disk method to write the integral that calculates the volume of the given solid, moving the constants to the outside of the integral.

b) To evaluate the integral, substitute $x = 2\tan\theta$. Why? What is dx? How does the integrand simplify? What are the restrictions on θ? How do the limits of integration change? Comment on how the restrictions on θ relate to the limits of integration. Don't forget, a reference triangle can be helpful.

c) Complete the integration to show that the volume $= (\pi/4)(\pi + 2)$.

d) Evaluate $(\pi/4)(\pi + 2)$ and NINT $(\pi y_1^2, 0, 2)$, and compare. &

Integrals Involving $ax^2 + bx + c, a \neq 0$

We handle these by first completing the square:

$$ax^2 + bx + c = a\left(x^2 + \frac{b}{a}x + \left(\frac{b}{2a} \right)^2 \right) + \left(c - \frac{b^2}{4a} \right)$$

$$= a\left(x + \frac{b}{2a} \right)^2 + \left(c - \frac{b^2}{4a} \right).$$

We then substitute

$$u = x + \frac{b}{2a}, \qquad x = u - \frac{b}{2a}, \qquad dx = du.$$

Example 5 illustrates the method.

EXAMPLE 5

Evaluate

$$\int \frac{dx}{\sqrt{2x - x^2}}.$$

Solution First we do the necessary algebra:

$$2x - x^2 = -(x^2 - 2x + 1) + 1 = 1 - (x - 1)^2.$$

Then we substitute $u = x - 1$ and $du = dx$ to get

$$\int \frac{dx}{\sqrt{2x - x^2}} = \int \frac{du}{\sqrt{1 - u^2}} = \sin^{-1} u + C = \sin^{-1}(x - 1) + C. \quad \blacksquare$$

EXPLORATION 3

Completing the Square

Show graphically, then confirm algebraically, that

$$\int \frac{dx}{4x^2 + 4x + 2} = \frac{1}{2} \tan^{-1}(2x + 1) + C.$$

Your analysis should involve a substitution after you complete a square. It may also involve a second substitution. &

Exercises 8.4

Evaluate the integrals in Exercises 1–16 using analytic methods. Support each with a NINT or NDER computation.

1. $\int_{-2}^{2} \dfrac{dx}{4 + x^2}$

2. $\int_{0}^{2} \dfrac{dx}{8 + 2x^2}$

3. $\int_{0}^{3/2} \dfrac{dx}{\sqrt{9 - x^2}}$

4. $\int_{0}^{1/(2\sqrt{2})} \dfrac{2\,dx}{\sqrt{1 - 4x^2}}$

5. $\int \dfrac{dx}{\sqrt{x^2 - 4}}$

6. $\int \dfrac{3\,dx}{\sqrt{9x^2 - 1}}$

7. $\int \sqrt{25 - x^2}\,dx$

8. $\int \sqrt{1 - 9x^2}\,dx$

9. $\int \dfrac{4x^2\,dx}{(1 - x^2)^{3/2}}$

10. $\int \dfrac{dx}{(4 - x^2)^{3/2}}$

11. $\int_{0}^{\sqrt{3}/2} \dfrac{2\,dy}{1 + 4y^2}$

12. $\int_{0}^{1/3} \dfrac{3\,dy}{\sqrt{1 + 9y^2}}$

13. $\int \dfrac{(x - 1)\,dx}{\sqrt{2x - x^2}}$

14. $\int \dfrac{(x - 2)\,dx}{\sqrt{5 + 4x - x^2}}$

15. $\int \dfrac{dx}{\sqrt{x^2 - 2x}}$

16. $\int \dfrac{dx}{\sqrt{x^2 + 2x}}$

Evaluate the integrals in Exercises 17–28 with a NINT computation. Confirm analytically.

17. $\int_{0}^{3\sqrt{2}/4} \dfrac{dx}{\sqrt{9 - 4x^2}}$

18. $\int_{0}^{5} \sqrt{25 - x^2}\,dx$

19. $\int_{1/\sqrt{3}}^{1} \dfrac{2\,dz}{z\sqrt{4z^2 - 1}}$

20. $\int_{8/\sqrt{3}}^{8} \dfrac{24\,dx}{x\sqrt{x^2 - 16}}$

21. $\int_{0}^{2} \dfrac{dx}{\sqrt{4 + x^2}}$

22. $\int_{0}^{1} \dfrac{x^3\,dx}{\sqrt{x^2 + 1}}$

23. $\int_{1}^{2} \dfrac{6\,dx}{\sqrt{4 - (x - 1)^2}}$

24. $\int_{1/2}^{1} \dfrac{\sqrt{1 - x^2}}{x^2}\,dx$

25. $\int_{1}^{3} \dfrac{dy}{y^2 - 2y + 5}$

26. $\int_{1}^{4} \dfrac{dy}{y^2 - 2y + 10}$

27. $\int_{-2}^{2} \dfrac{(x + 2)\,dx}{\sqrt{x^2 + 4x + 13}}$

28. $\int_{0}^{1} \dfrac{(1 - x)\,dx}{\sqrt{8 + 2x - x^2}}$

29. Find the area of the region in the first quadrant bounded by the coordinate axes, the line $x = 1$, and the curve $y = 2/(x^2 - 4x + 5)$ as shown.

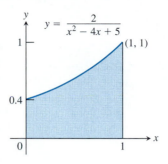

30. Find the volume of the solid generated by revolving the region bounded by the x-axis, the curve

$$y = 20/\sqrt{x^2 - 2x + 17},$$

and the lines $x = -2$ and $x = 11$, about the x-axis.

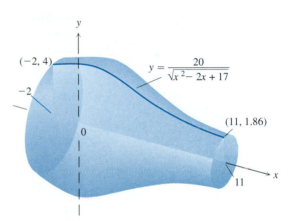

Evaluate the integrals in Exercises 31 and 32.

31. $\int \dfrac{4x^2\,dx}{(1 - x^2)^{3/2}}$

32. $\int_{0}^{1} \dfrac{4\,dx}{(4 - x^2)^{3/2}}$

33. Find the area of the region cut from the first quadrant by the elliptical curve $3y = \sqrt{9 - x^2}$.

34. Find the average value of the function $f(x) = 4/(x^2 - 4x + 8)$ over the interval $[2, 4]$.

35. Why can't $x = a \sin \theta$ be used to evaluate

$$\int_{4}^{5} \dfrac{dx}{9 - x^2}$$

analytically? How would you evaluate this definite integral analytically? Support your answer with a NINT computation.

36. Show that, except for the endpoints, the domain of the integrand $1/\sqrt{2x - x^2}$ of Example 5 and its antiderivative $\sin^{-1}(x - 1)$ are the same.

8.5 _____ Rational Functions and Partial Fractions

Now we are ready to take the third step that we described at the beginning of Section 8.4. We will express a rational function $f(x)$ as possibly the sum of a polynomial, of fractions with linear-factored denominators, and of fractions with denominators containing quadratic expressions.

If $f(x) = p(x)/h(x)$ is a rational function formed by the polynomials $p(x)$ and $h(x)$, then dividing $p(x)$ by $h(x)$ gives us our old friends $q(x)$ and $r(x)$, the quotient and remainder polynomials, respectively, where

$$f(x) = q(x) + \frac{r(x)}{h(x)} \qquad \text{with} \quad \deg r < \deg h.$$

EXAMPLE 1

Evaluate

$$\int \frac{(x-2)^3}{x^2-4}\, dx.$$

Solution The numerator and denominator have a common factor $x - 2$. The integrand is not defined for $x = 2$, so we can divide both numerator and denominator by $x - 2$ because it is not zero in the domain of the quotient:

$$\frac{(x-2)^3}{x^2-4} = \frac{(x-2)^3}{(x-2)(x+2)} = \frac{(x-2)^2}{x+2} = \frac{x^2-4x+4}{x+2}.$$

The result is a rational function $p(x)/h(x)$ in which $\deg p \geq \deg h$. We divide to get

$$\frac{x^2-4x+4}{x+2} = x - 6 + \frac{16}{x+2}.$$

Therefore,

$$\int \frac{(x-2)^3}{x^2-4}\, dx = \int \left(x - 6 + \frac{16}{x+2} \right) dx = \frac{x^2}{2} - 6x + 16 \ln |x+2| + C.$$

$$\equiv$$

$$
\begin{array}{r}
x \quad - \ 6 \\
x+2\overline{\smash{)}\,x^2 - 4x \quad + 4} \\
\underline{x^2 + 2x} \\
-6x \quad + 4 \\
\underline{-6x \quad -12} \\
+16
\end{array}
$$

Partial Fractions

A theorem from advanced algebra (mentioned later in more detail) says that every rational function, no matter how complicated, comes from adding simpler fractions. This suggests another algebraic technique that can help us to evaluate integrals "by hand." Namely, when we have a rational function integrand, we reverse the addition to get back to the simpler fractions—fractions that we can integrate with techniques we already know. This technique is the **method of partial fractions**. We practice it to help us review elementary integral forms.

EXAMPLE 2 An Example of the Kind of Addition We Want to Reverse

$$\frac{2}{x+1} + \frac{3}{x-3} = \frac{2(x-3) + 3(x+1)}{(x+1)(x-3)} = \frac{5x-3}{x^2-2x-3}$$

Adding the fractions on the left produces the fraction on the right. The reverse process consists of finding constants A and B such that

$$\frac{5x - 3}{x^2 - 2x - 3} = \frac{A}{x + 1} + \frac{B}{x - 3}. \tag{1}$$

(Pretend for a moment that we don't know that $A = 2$ and $B = 3$ will work.) We call the fractions $A/(x + 1)$ and $B/(x - 3)$ **partial fractions** because their denominators are only part of the original denominator $x^2 - 2x - 3$. We call A and B **undetermined coefficients** until proper values for them have been found.

To find A and B, we first clear Eq. (1) of fractions, obtaining

$$5x - 3 = A(x - 3) + B(x + 1) = (A + B)x - 3A + B. \tag{2}$$

This will be an identity in x if and only if the coefficients of like powers of x on the two sides are equal:

$$A + B = 5, \qquad -3A + B = -3.$$

Solving these equations simultaneously gives $A = 2$ and $B = 3$. ▤

EXAMPLE 3   Two Distinct Linear Factors in the Denominator

Evaluate

$$\int \frac{5x - 3}{(x + 1)(x - 3)} \, dx,$$

and support graphically

Solution   From Example 2,

$$\int \frac{5x - 3}{(x + 1)(x - 3)} \, dx = \int \frac{2}{x + 1} \, dx + \int \frac{3}{x - 3} \, dx$$

$$= 2 \ln |x + 1| + 3 \ln |x - 3| + C.$$

Figure 8.19 provides graphical support for our result. ▤

EXAMPLE 4   A Repeated Linear Factor in the Denominator

Express

$$\frac{6x + 7}{(x + 2)^2}$$

as a sum of partial fractions.

Solution   Since the denominator has a repeated linear factor, $(x + 2)^2 = (x + 2)(x + 2)$, we must express the fraction in the form

$$\frac{6x + 7}{(x + 2)^2} = \frac{A}{x + 2} + \frac{B}{(x + 2)^2}.$$

Clearing the equation of fractions gives

$$6x + 7 = A(x + 2) + B = Ax + (2A + B).$$

Matching coefficients of like terms gives $A = 6$ and

$$7 = 2A + B = 12 + B, \qquad \text{or} \qquad B = -5.$$

TWO METHODS FOR FINDING UNDETERMINED COEFFICIENTS

Method 1

1. Clear the equation of fractions.
2. Equate the coefficients of like powers of x.
3. Solve the system of equations that results to find the coefficients. (Remember, most graphers can help you solve a system of equations.)

Method 2

Use the fact that the equation cleared of fractions and containing the undetermined coefficients (Eq. (2) in Example 2) is an identity. This equation is true for all values of x including some that are particularly well chosen.

For example, in Eq. (2) let $x = 3$ to find $B = 3$. Then let $x = -1$ to find $A = 2$.

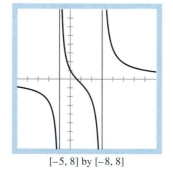

$$[-5, 8] \text{ by } [-8, 8]$$

8.19   The graphs of

$$y_1 = \frac{5x - 3}{(x + 1)(x - 3)}$$

and

$$y_2 = \text{NDER}\,(2 \ln |x + 1| + 3 \ln |x - 3|, x)$$

appear to be identical to support the result of Example 3 graphically.

Hence,
$$\frac{6x + 7}{(x + 2)^2} = \frac{6}{x + 2} - \frac{5}{(x + 2)^2}.$$

≡

EXPLORATION 1

Partial Fractions

See Example 4. Let

$$y_1 = \frac{6}{x + 2}, \qquad y_2 = \frac{-5}{(x + 2)^2}, \qquad y_3 = y_1 + y_2, \qquad y_4 = \frac{6x + 7}{(x + 2)^2}.$$

1. Show graphically that $y_4 = y_3$.

2. GRAPH the functions y_i. Using your viewing window as a guide, formulate and solve a problem of your choice that requires integration.

3. Let $y_5 = (2x^3 - 4x^2 - x - 3)/(x^2 - 2x - 3)$. Evaluate $\int y_5 \, dx$ analytically. Use NINT to support your result numerically and graphically.   ⌘

EXAMPLE 5   An Irreducible Quadratic Factor in the Denominator

Evaluate

$$\int \frac{3x^2 - 7x + 12}{(x - 2)(x^2 - 2x + 5)} \, dx.$$

Solution   The denominator has an irreducible quadratic factor as well as a linear factor, so we write

$$\frac{3x^2 - 7x + 12}{(x - 2)(x^2 - 2x + 5)} = \frac{A}{x - 2} + \frac{Bx + C}{x^2 - 2x + 5}.$$

Notice the numerator over $x^2 - 2x + 5$. For quadratic factors, we use first-degree numerators. Clearing the equation of fractions and equating coefficients of like terms gives

Coefficients of $x^2 : 3 = A + B$

Coefficients of $x^1 : -7 = -2A - 2B + C$

Coefficients of $x^0 : 12 = 5A - 2C$

Solving these equations simultaneously, we find that $A = 2, B = 1$, and $C = -1$. Thus,

$$\int \frac{3x^2 - 7x + 12}{(x - 2)(x^2 - 2x + 5)} \, dx = \int \frac{2}{x - 2} dx + \int \frac{(x - 1) \, dx}{x^2 - 2x + 5}$$

$$= 2 \ln |x - 2| + \frac{1}{2} \ln |x^2 - 2x + 5| + C. \ \equiv$$

General Description of the Method

Success in writing a rational function $f(x)/g(x)$ as a sum of partial fractions depends on two things:

1. *The degree of f(x) must be less than the degree of g(x).* (If it isn't, divide and work with the remainder term.)

2. *We must know the factors of* $g(x)$. (In theory, any polynomial with real coefficients can be written as a product of real linear factors and real quadratic factors. In practice, the factors may be impossible to find without CAS assistance.)

If these two conditions are met we can take the following steps.

STEP 1: Let $x - r$ be a linear factor of $g(x)$. Suppose $(x - r)^m$ is the highest power of $x - r$ that divides $g(x)$. Then assign the sum of m partial fractions to this factor, as follows:

$$\frac{A_1}{x - r} + \frac{A_2}{(x - r)^2} + \cdots + \frac{A_m}{(x - r)^m}.$$

Do this for each distinct linear factor of $g(x)$.

STEP 2: Let $x^2 + px + q$ be a quadratic factor of $g(x)$. Suppose

$$(x^2 + px + q)^n$$

is the highest power of this factor that divides $g(x)$. Then, to this factor, assign the sum of the n partial fractions:

$$\frac{B_1 x + C_1}{x^2 + px + q} + \frac{B_2 x + C_2}{(x^2 + px + q)^2} + \cdots + \frac{B_n x + C_n}{(x^2 + px + q)^n}.$$

Do this for each distinct quadratic factor of $g(x)$ that cannot be factored into linear factors with real coefficients.

STEP 3: Set the original fraction $f(x)/g(x)$ equal to the sum of all these partial fractions. Clear the resulting equation of fractions and arrange the terms in decreasing powers of x.

STEP 4: Equate the coefficients of corresponding powers of x and solve the resulting equations for the undetermined coefficients.

Proofs that $f(x)/g(x)$ can be written as a sum of partial fractions as described here are given in advanced algebra texts.

The Substitution $z = \tan(x/2)$

The substitution

$$z = \tan \frac{x}{2} \tag{3}$$

reduces the problem of integrating any rational function of $\sin x$ and $\cos x$ to a problem involving a rational function of z. This in turn can be integrated by partial fractions. Thus, the substitution (3) is theoretically a very powerful tool. For pencil-and-paper evaluation of integrals, however, it is very cumbersome and should be used only when simpler methods do not apply.

Figure 8.20 shows how $\tan(x/2)$ expresses a rational function of $\sin x$ and $\cos x$. To see the effect of the substitution, we calculate

$$\cos x = 2\cos^2 \frac{x}{2} - 1 = \frac{2}{\sec^2(x/2)} - 1 = \frac{2}{1 + \tan^2(x/2)} - 1 = \frac{2}{1 + z^2} - 1,$$

or

$$\cos x = \frac{1 - z^2}{1 + z^2}, \tag{4}$$

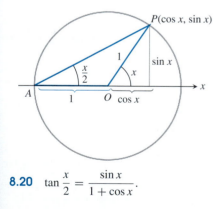

8.20 $\tan \dfrac{x}{2} = \dfrac{\sin x}{1 + \cos x}$.

and

$$\sin x = 2 \sin \frac{x}{2} \cos \frac{x}{2} = 2 \frac{\sin (x/2)}{\cos (x/2)} \cdot \cos^2 \frac{x}{2}$$

$$= 2 \tan \frac{x}{2} \cdot \frac{1}{\sec^2 (x/2)} = \frac{2 \tan (x/2)}{1 + \tan^2 (x/2)},$$

or

$$\sin x = \frac{2z}{1 + z^2}. \tag{5}$$

Finally, $x = 2 \tan^{-1} z$, so

$$dx = \frac{2 \, dz}{1 + z^2}. \tag{6}$$

EXAMPLE 6

$$\int_0^{\pi/2} \frac{1}{1 + \cos x} \, dx = \int_0^1 \frac{1 + z^2}{2} \cdot \frac{2 \, dz}{1 + z^2} \qquad \text{Eqs. (4) and (6) plus some algebra}$$

$$= \int_0^1 dz = z \Big]_0^1 = 1 \qquad\qquad \blacksquare$$

For a definite integral, the substitution $z = \tan(x/2)$ requires a change in the limits of integration. As in Example 6, when $x = 0$, $z = \tan 0 = 0$. When $x = (\pi/2)$, $z = \tan(x/2) = \tan(\pi/4) = 1$.

EXAMPLE 7

$$\int \frac{1}{2 + \sin x} \, dx = \int \frac{1 + z^2}{2 + 2z + 2z^2} \cdot \frac{2 \, dz}{1 + z^2} \qquad \text{Eqs. (5) and (6) plus some algebra}$$

$$= \int \frac{dz}{z^2 + z + 1} = \int \frac{dz}{(z + 1/2)^2 + 3/4} \qquad \text{Complete the square.}$$

$$= \int \frac{du}{u^2 + a^2} \qquad\qquad u = z + 1/2, \; a = \sqrt{3}/2.$$

$$= \frac{1}{a} \tan^{-1} \left(\frac{u}{a} \right) + C \qquad\qquad \text{Section 8.4, Eq. (2)}$$

$$= \frac{2}{\sqrt{3}} \tan^{-1} \left(\frac{2z + 1}{\sqrt{3}} \right) + C \qquad\qquad u = z + 1/2, \; a = \sqrt{3}/2.$$

$$= \frac{2}{\sqrt{3}} \tan^{-1} \left(\frac{1 + 2 \tan (x/2)}{\sqrt{3}} \right) + C \qquad z = \tan \frac{x}{2}.$$

Exercises 8.5

Expand the quotients in Exercises 1–8 by partial fractions.

1. $\dfrac{5x - 13}{(x - 3)(x - 2)}$

2. $\dfrac{5x - 7}{x^2 - 3x + 2}$

3. $\dfrac{x + 4}{(x + 1)^2}$

4. $\dfrac{2x + 2}{x^2 - 2x + 1}$

5. $\dfrac{x + 1}{x^2(x - 1)}$

6. $\dfrac{x}{x^3 - x^2 - 6x}$

7. $\dfrac{x^2 + 8}{x^2 - 5x + 6}$

8. $\dfrac{x^3 + 1}{x^2 + 4}$

Evaluate the integrals in Exercises 9–20 by the method of partial fractions. Support each with a NINT computation.

9. $\displaystyle\int_0^{1/2} \dfrac{dx}{1 - x^2}$

10. $\displaystyle\int_1^2 \dfrac{dx}{x^2 + 2x}$

11. $\displaystyle\int_4^8 \dfrac{y\,dy}{y^2 - 2y - 3}$

12. $\displaystyle\int_1^2 \dfrac{y + 4}{y^2 + y}\,dy$

13. $\displaystyle\int \dfrac{dt}{t^3 + t^2 - 2t}$

14. $\displaystyle\int \dfrac{t + 3}{2t^3 - 8t}\,dt$

15. $\displaystyle\int_0^{2\sqrt{2}} \dfrac{x^3\,dx}{x^2 + 1}$

16. $\displaystyle\int_0^1 \dfrac{x^4 + 2x}{x^2 + 1}\,dx$

17. $\displaystyle\int_0^{\sqrt{3}} \dfrac{5x^2\,dx}{x^2 + 1}$

18. $\displaystyle\int_1^5 \dfrac{y^3 + 4y^2}{y^3 + y}\,dy$

19. $\displaystyle\int \dfrac{dx}{\left(x^2 - 1\right)^2}$

20. $\displaystyle\int \dfrac{x^2\,dx}{(x - 1)\left(x^2 + 2x + 1\right)}$

Evaluate the integrals in Exercises 21–28 by the method of partial fractions. Support each with an NDER or NINT computation.

21. $\displaystyle\int \dfrac{x + 4}{x^2 + 5x - 6}\,dx$

22. $\displaystyle\int \dfrac{2x + 1}{x^2 - 7x + 12}\,dx$

23. $\displaystyle\int \dfrac{2\,dx}{x^2 - 2x + 2}$

24. $\displaystyle\int \dfrac{3\,dx}{x^2 - 4x + 5}$

25. $\displaystyle\int \dfrac{x^2 - 2x - 2}{x^3 - 1}\,dx$

26. $\displaystyle\int \dfrac{x^2 - 4x + 4}{x^3 + 1}\,dx$

27. $\displaystyle\int \dfrac{2x^4 + x^3 + 16x^2 + 4x + 32}{\left(x^2 + 4\right)^2}\,dx$

28. $\displaystyle\int \dfrac{x^4 - 3x^3 + 2x^2 - 3x + 1}{\left(x^2 + 1\right)^2}\,dx$

Evaluate the integrals in Exercises 29–32 using analytic methods. Support each with a NINT computation.

29. $\displaystyle\int_0^1 \dfrac{x}{x + 1}\,dx$

30. $\displaystyle\int_0^1 \dfrac{x^2}{x^2 + 1}\,dx$

31. $\displaystyle\int_{\sqrt{2}}^3 \dfrac{2x^3}{x^2 - 1}\,dx$

32. $\displaystyle\int_{-1}^3 \dfrac{4x^2 - 7}{2x + 3}\,dx$

33. Find the volume of the solid generated by revolving about the x-axis the region bounded by the x-axis, the lines $x = 1/2$ and $x = 5/2$, and the curve $y = 3/\sqrt{3x - x^2}$.

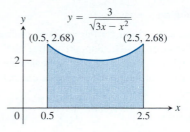

34. Find the length of the curve $y = \ln(1 - x^2), 0 \le x \le 1/2$.

35. *Social diffusion.* Sociologists sometimes use the phrase "social diffusion" to describe the way information spreads through a population. The information might be a rumor, a cultural fad, or news about a technical innovation. In a sufficiently large population, the number of people x who have the information is treated as a differentiable function of time t, and the rate of diffusion, dx/dt, is assumed to be proportional to the number of people who have the information times the number of people who do not. This leads to the equation

$$\dfrac{dx}{dt} = kx(N - x),$$

where N is the number of people in the population.

Suppose t is measured in days, $k = 1/250$, and two people start a rumor at time $t = 0$ in a population of $N = 1000$ people.

a) Find x as a function of t, and GRAPH $x(t)$.

b) When will half the population have heard the rumor? (This is when the rumor will be spreading the fastest.)

36. *Second-order chemical reactions.* Many chemical reactions are the result of the interaction of two molecules that undergo a change to produce a new product. The rate of the reaction typically depends on the concentrations of the two kinds of molecules. If a is the amount of substance A, b is the amount of substance B at time $t = 0$, and x is the amount of product at time t, then the rate of formation of x may be given by the differential equation

$$\dfrac{dx}{dt} = k(a - x)(b - x),$$

or

$$\dfrac{1}{(a - x)(b - x)} \dfrac{dx}{dt} = k$$

(k is a constant for the reaction).

a) Integrate both sides of this equation to obtain a relation between x and t (i) if $a = b$ and (ii) if $a \ne b$. Assume in each case that $x = 0$ when $t = 0$.

b) For $k = 2.35, a = 10$, and $b = 15$, find x as a function of t and GRAPH $x(t)$.

The Substitution $z = \tan(x/2)$

Use the substitution $z = \tan(x/2)$ to evaluate the integrals in Exercises 37–44. Support with a NINT computation. Integrals like these arise when we calculate the average angular velocity of the output shaft of a universal coupling.

37. $\displaystyle\int_0^{\pi/2} \frac{dx}{1 + \sin x}$

38. $\displaystyle\int_{\pi/3}^{\pi/2} \frac{\pi/2 \, dx}{1 - \cos x}$

39. $\displaystyle\int \frac{dx}{1 - \sin x}$

40. $\displaystyle\int \frac{dx}{2 + \cos x}$

41. $\displaystyle\int \frac{\cos x \, dx}{1 - \cos x}$

42. $\displaystyle\int \frac{dx}{1 + \sin x + \cos x}$

43. $\displaystyle\int \frac{dx}{\sin x - \cos x}$

44. $\displaystyle\int_{\pi/2}^{2\pi/3} \frac{dx}{\sin x + \tan x}$

8.6 Improper Integrals

The definite integral is defined for a function that is continuous on a closed interval. What about a function f that is continuous on an interval that is not closed? We can use our idea of definite integral to give meaning, for example, to $\int_a^c f$, where f is continuous on $[a, c)$ with c either a real number or ∞. The key to understanding this new type of integral is the idea of limit. (Where have we heard this before!)

Essentially, we first integrate the function over a closed interval *inside* the nonclosed interval; *then we take the limit of this integral as the closed interval expands to fill the nonclosed interval.* Figure 8.21 shows how to do this for a half-open interval $[a, c)$. It is conventional to call the resulting limit, whether it exists or not, the *improper integral* of the function over the nonclosed interval.

Here are four examples of this new type of integral:

$$\int_0^3 \frac{x+3}{x-3}, \qquad \int_0^1 \frac{dx}{x}, \qquad \int_{-\pi/2}^{\pi/2} \sec x \, dx, \qquad \int_0^\infty e^{-x^2} dx$$

(Can you see why each is improper?) Thus, for example,

$$\int_0^1 \frac{1}{x} \, dx = \lim_{b \to 0^+} \int_b^1 \frac{1}{x} \, dx.$$

Recall that a key question about a limit, and thus about an indefinite integral, is "Does it exist?" This is the basic question that we address in the rest of this section. Once the integral is known to exist, its value, if not immediately apparent, can be found by numerical or graphical methods.

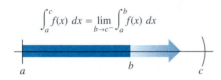

$$\int_a^c f(x) \, dx = \lim_{b \to c^-} \int_a^b f(x) \, dx$$

8.21 To integrate over a half-open interval $[a, c)$, we (1) integrate over a closed interval $[a, b]$ inside $[a, c)$ and (2) take the limit as $[a, b]$ expands to fill $[a, c)$.

EXPLORATION BIT

Evaluate

$$\int_{-2}^1 \frac{1}{x^2} \, dx$$

by the procedure of the Fundamental Theorem of Calculus, part 2. Does your result seem reasonable? Explain.

Convergence and Divergence

In lifting-line theory in aerodynamics, the function

$$f(x) = \sqrt{\frac{1+x}{1-x}}$$

needs to be integrated over the interval from $x = 0$ to $x = 1$. The function is not defined at $x = 1$, although it is defined and continuous everywhere else in $[0, 1]$. To integrate f from zero to one, we integrate f from zero to a positive number b less than 1 and take the limit of the resulting definite integral as b approaches 1. If the limit exists, we define the integral of f

INTEGRABLE FUNCTIONS

The definite integral is defined for a function that is continuous on a closed interval. In Section 5.2 we indicated that some discontinuous functions are integrable. If the discontinuity occurs at a single point and is a jump discontinuity or a removable discontinuity, then it can be proved that the limit of the Riemann sums exists and the function is *Riemann integrable.* (See Fig. 8.22.) It can also be proved that any increasing function or decreasing function or any bounded function with a finite number of discontinuities on a closed interval [a, b] is Riemann integrable.

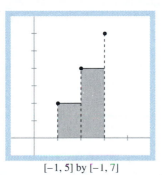

[−1, 5] by [−1, 7]

8.22 The function $y = 2[x]$ is not continuous on [1, 3], but the limit of Riemann sums exists and is 6. Thus,

$$\int_1^3 2[x]\,dx = 6.$$

EXPLORATION BIT

Do you see a more efficient way to do Exploration 1 using functions in parametric form?

from zero to 1 to be this value and write

$$\int_0^1 \sqrt{\frac{1+x}{1-x}}\,dx = \lim_{b \to 1^-} \int_0^b \sqrt{\frac{1+x}{1-x}}\,dx.$$

In this case, we also say that the integral

$$\int_0^1 \sqrt{\frac{1+x}{1-x}}\,dx$$

converges and say that the area under the curve $y = \sqrt{(1+x)/(1-x)}$ from 0 to 1 is the value of the integral. If the limit above does not exist, we say that the integral *diverges.*

EXAMPLE 1

Determine whether

$$\int_0^1 \sqrt{\frac{1+x}{1-x}}\,dx$$

converges or diverges. If it converges, compute its value.

Solution 1   See Fig. 8.23(a). Multiplying numerator and denominator by $\sqrt{1+x}$ gives

$$\int \sqrt{\frac{1+x}{1-x}}\,dx = \int \frac{1+x}{\sqrt{1-x^2}}\,dx = \int \frac{1}{\sqrt{1-x^2}}\,dx + \int \frac{x}{\sqrt{1-x^2}}\,dx$$

$$= \sin^{-1} x - \sqrt{1-x^2} + C.$$

Therefore,

$$\lim_{b \to 1^-} \int_0^b \sqrt{\frac{1+x}{1-x}}\,dx = \lim_{b \to 1^-} \left[\sin^{-1} x - \sqrt{1-x^2} \right]_0^b$$

$$= \lim_{b \to 1^-} [\sin^{-1} b - \sqrt{1-b^2} + 1]$$

$$= \sin^{-1} 1 - 0 + 1 = \pi/2 + 1.$$

The integral converges to $\pi/2 + 1$.

Solution 2   See Fig. 8.23(b).   If $y = \sqrt{(1+x)/(1-x)}$, then $x = (y^2 - 1)/(y^2 + 1)$, and we can integrate with respect to y. The portion of the area above the 1-by-1 square is

$$\int_1^\infty 1 - \frac{y^2 - 1}{y^2 + 1}\,dy = \int_1^\infty \frac{2}{y^2 + 1}\,dy$$

$$= \lim_{c \to \infty} \int_1^c \frac{2}{y^2 + 1}\,dy$$

$$= \lim_{c \to \infty} \left[2 \tan^{-1} y \right]_1^c$$

$$= \lim_{c \to \infty} 2 \tan^{-1} c - 2 \cdot \frac{\pi}{4} = 2 \cdot \frac{\pi}{2} - \frac{\pi}{2} = \frac{\pi}{2}.$$

Including the 1-by-1 square, the area is $\pi/2 + 1$, in agreement with our first calculation. Thus,

$$\int_0^1 \sqrt{\frac{1+x}{1-x}}\,dx = \frac{\pi}{2} + 1.$$

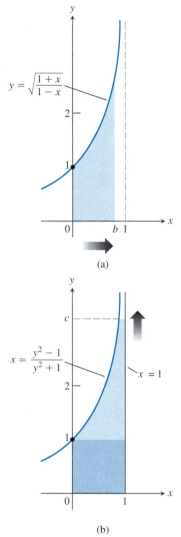

$$y = \sqrt{\frac{1+x}{1-x}}$$

$$x = \frac{y^2 - 1}{y^2 + 1}$$

$$x = 1$$

(a)

(b)

8.23 (a) To evaluate

$$\int_0^1 \sqrt{(1+x)/(1-x)} \, dx,$$

we evaluate

$$\int_0^b \sqrt{(1+x)/(1-x)} \, dx$$

and find the limit as b approaches 1 from below. (b) Equivalently, we can find $x = (y^2 - 1)/(y^2 + 1)$, integrate $1 - (y^2 - 1)/(y^2 + 1)$ with respect to y from 1 to c, find the limit as c approaches ∞, and then add 1 to account for the 1-by-1 square at the bottom.

EXPLORATION 1

Support for Improper Integrals

Example 1 illustrates the surprising fact that some *unbounded* regions can have *finite* area. We can support this using NINT. However, as we have already learned, when we use NINT $(f(t), a, x)$, we have to be sure that both $[a, x]$ and the open view dimension (xMin, xMax) contain no point of infinite discontinuity of f. (See Section 8.1, Exploration 1.)

Show four ways to support the result in Example 1 as suggested below.

Support Based on Solution 1   Let $y_1 = \sqrt{(1+x)/(1-x)}$. Explain how to use the *values* NINT $(y_1, 0, a)$ to provide numerical support.

Explain how to use the *function* NINT $(y_1, 0, x)$, and TRACE to provide graphical support.

Support Based on Solution 2   Let $y_2 = 1 - (x^2 - 1)/(x^2 + 1)$. Explain how to provide numerical support and graphical support. Which type of support do you prefer? Why?   ⚘

The notation

$$\int_a^b f(x) \, dx$$

for improper integrals is the same as the notation for definite integrals. In any given case, it is usually a simple matter to tell whether a particular integral is to be calculated as an ordinary definite integral or as a limit. If a and b are finite and f is continuous at every point of $[a, b]$, the integral is an ordinary definite integral. If f becomes infinite at one or more points in the interval of integration, or one or both of the limits of integration is infinite, then the designated integral is improper and is to be calculated as a limit.

DEFINITIONS

An integral with an infinite limit of integration or an integral of a function that becomes infinite at a point within the interval of integration is an **improper integral**. We evaluate such an integral as follows:

1. If f is continuous on $[a, \infty)$, then
$$\int_a^\infty f(x) \, dx = \lim_{b \to \infty} \int_a^b f(x) \, dx.$$

2. If f is continuous on $(-\infty, b]$, then
$$\int_{-\infty}^b f(x) \, dx = \lim_{a \to -\infty} \int_a^b f(x) \, dx.$$

3. If f is continuous on $(a, b]$, then
$$\int_a^b f(x) \, dx = \lim_{c \to a^+} \int_c^b f(x) \, dx.$$

4. If f is continuous on $[a, b)$, then
$$\int_a^b f(x) \, dx = \lim_{c \to b^-} \int_a^c f(x) \, dx.$$

If the limit involved exists, we say that the improper integral **converges** and that the limit is the value of the improper integral. If the limit fails to exist, we say that the improper integral **diverges**.

Two very important improper integrals are analyzed in Examples 2 and 3.

EXAMPLE 2

The integral $\int_1^\infty dx/x$ is improper because one of the limits of integration is ∞. We replace ∞ with b and recognize the natural log function from Chapter 7:

$$\int_1^b \frac{dx}{x} = \ln b.$$

To investigate the behavior of the integral as $b \to \infty$, we use known properties of definite integrals and of the natural logarithm. First we note that since $1/x \geq 1/2$ on $[1, 2]$ (see Fig. 8.24),

$$\ln 2 = \int_1^2 \frac{dx}{x} \geq \int_1^2 \frac{1}{2}\,dx = \frac{1}{2}. \qquad \text{Section 5.3, Rule 6}$$

Then, for any positive integer N,

$$2N \ln 2 \geq 2N \left(\frac{1}{2}\right) = N.$$

Thus,

$$\int_1^{2^{2N}} \frac{dx}{x} = \ln 2^{2N} \geq N,$$

which means that the increasing $\ln x$ function can take on values as great as desired (just choose $x = 2^{2N}$ with N large enough). In other words,

$$\lim_{b \to \infty} \ln b = \infty,$$

and the integral $\int_1^\infty dx/x$ diverges. ≡

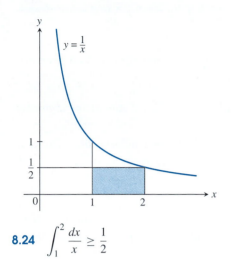

8.24 $\int_1^2 \dfrac{dx}{x} \geq \dfrac{1}{2}$

EXAMPLE 3

In the integral $\displaystyle\int_0^1 \frac{dx}{x}$, the function

$$f(x) = \frac{1}{x}$$

becomes infinite as $x \to 0^+$. We cut off the point $x = 0$ and start our integration at some positive number b between 0 and 1 (Fig. 8.25):

$$\int_b^1 \frac{dx}{x} = \ln x \bigg]_b^1 = \ln 1 - \ln b = \ln \frac{1}{b}.$$

We then investigate the behavior of the integral as b approaches zero from the right:

$$\lim_{b \to 0^+} \int_b^1 \frac{dx}{x} = \lim_{b \to 0^+} \left(\ln \frac{1}{b}\right) = \infty.$$

Thus, like $\int_1^\infty dx/x$, the integral $\int_0^1 dx/x$ also diverges. ≡

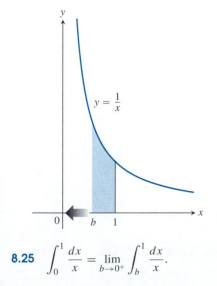

8.25 $\displaystyle\int_0^1 \frac{dx}{x} = \lim_{b \to 0^+} \int_b^1 \frac{dx}{x}.$

EXPLORATION 2

The Diverging $\int_0^\infty \dfrac{dx}{x}$

As Examples 2 and 3 show, the integral $\int dx/x$ diverges "at both ends" of the interval $(0, \infty)$.

1. Example 3 shows that $\lim_{x \to 0^+} \ln(1/x) = \infty$. Give a convincing argument that $\lim_{x \to 0^+} \ln x = -\infty$.

2. Investigate either limit of part 1 graphically and numerically. Do your investigations support the fact that $\int_0^1 dx/x$ diverges? Explain.   ☙

EXPLORATION BIT

Let $f(x) = 1/x$. Use the symmetry of the graph of f and an area argument to tell how

$$\int_1^\infty f(x)\,dx \quad \text{and} \quad \int_0^1 f(x)\,dx$$

are alike.

If a function f becomes infinite at an interior point c of an interval of integration $[a, b]$, then the improper integral $\int_a^b f$ is expressed as the sum of two improper integrals:

$$\int_a^b f = \int_a^c f + \int_c^b f.$$

If both $\int_a^c f$ and $\int_c^b f$ converge, then we say that $\int_a^b f$ converges and its value is the sum. Otherwise, we say that $\int_a^b f$ diverges.

EXAMPLE 4

The integral

$$\int_0^3 \frac{dx}{(x - 1)^{2/3}}$$

is improper because the integrand

$$f(x) = \frac{1}{(x - 1)^{2/3}}$$

becomes infinite at $x = 1$ in the interval of integration $[0, 3]$. Thus, we have two integrals to investigate,

$$\int_0^b \frac{dx}{(x - 1)^{2/3}} \quad \text{and} \quad \int_c^3 \frac{dx}{(x - 1)^{2/3}},$$

one on each side of $x = 1$. (See Fig. 8.26.) Taking limits, we find

$$\lim_{b \to 1^-} \int_0^b (x - 1)^{-2/3}\,dx = \lim_{b \to 1^-} [3(b - 1)^{1/3} - 3(0 - 1)^{1/3}] = 3$$

and

$$\lim_{c \to 1^+} \int_c^3 (x - 1)^{-2/3}\,dx = \lim_{c \to 1^+} [3(3 - 1)^{1/3} - 3(c - 1)^{1/3}] = 3\sqrt[3]{2}.$$

Since both limits exist and are finite, the integral of f converges, and its value is $3 + 3\sqrt[3]{2}$.   ≡

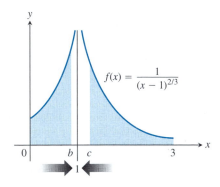

$f(x) = \dfrac{1}{(x - 1)^{2/3}}$

8.26

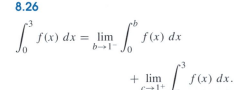

$$\int_0^3 f(x)\,dx = \lim_{b \to 1^-} \int_0^b f(x)\,dx$$

$$+ \lim_{c \to 1^+} \int_c^3 f(x)\,dx.$$

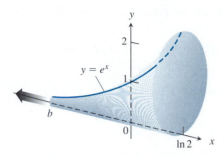

8.27 The calculations in Example 5 show that this infinite solid horn has a finite volume.

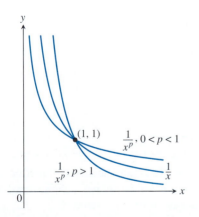

8.28 The function $1/x$ separates the functions $1/x^p$ on (0, 1] and [1, ∞). Those with graphs above the graph of $1/x$ on either interval have improper integrals that diverge. Those with graphs below the graph of $1/x$ have improper integrals that converge.

EXAMPLE 5

Each cross-section of the solid infinite "horn" in Fig. 8.27, cut by a plane perpendicular to the x-axis for $-\infty < x \le \ln 2$, is a circular disc with one diameter reaching from the x-axis to the curve $y = e^x$. Find the "volume" of the horn.

Solution We use the method of slicing from Section 6.9. The area of a typical cross-section is

$$A(x) = \pi (\text{radius})^2 = \pi \left(\frac{1}{2} y \right)^2 = \frac{\pi}{4} e^{2x}.$$

We define the *volume of the horn* to be the limit as $b \to -\infty$ of the volume of the portion from b to $\ln 2$. The volume of the portion from b to $\ln 2$ is

$$V = \int_b^{\ln 2} A(x) \, dx = \int_b^{\ln 2} \frac{\pi}{4} e^{2x} \, dx = \frac{\pi}{8} e^{2x} \Big]_b^{\ln 2} = \frac{\pi}{8} \left(e^{\ln 4} - e^{2b} \right)$$

$$= \frac{\pi}{8} \left(4 - e^{2b} \right).$$

As $b \to -\infty$, $e^{2b} \to 0$ and $V \to (\pi/8)(4 - 0) = \pi/2$. The volume of the horn is $\pi/2$. ≡

EXPLORATION 3

A View of Limits as $x \to \infty$

Because

$$\lim_{x \to \infty} f(x) = \lim_{x \to 0^+} f(1/x),$$

you can study what happens to $f(x)$ as $x \to \infty$ by studying what happens to $f(1/x)$ as $x \to 0^+$, an x-value that is easy to locate in the viewing window. (You may wish to review Section 1.3, Exploration 7 on composing functions.)

1. It is a fact that $\int_1^\infty dx/x^2$ converges. Find graphically what its value is by letting $f(x) = \int_1^x dt/t^2$ and viewing $f(1/x)$. Explain.

2. Now extend the above idea to give graphical support to the results of Example 5 by viewing a graph near $x = 0$. Describe the steps you take. ⚓

The Integrals $\int_0^1 dx/x^p$ and $\int_1^\infty dx/x^p$

We have seen that $\int dx/x$ diverges on both (0, 1] and [1, ∞). We can view $f(x) = 1/x$ as the function that separates converging and diverging functions of the form $g(x) = 1/x^p$ (Fig. 8.28).
If $p > 1$:

The graph of g lies below the graph of f on [1, ∞), and $\int_1^\infty dx/x^p$ converges.

The graph of g lies above the graph of f on (0, 1], and $\int_0^1 dx/x^p$ diverges.

If $0 < p < 1$:

> The graph of g lies above the graph of f on $[1, \infty)$, and $\int_1^\infty dx/x^p$ diverges.

> The graph of g lies below the graph of f on $(0, 1]$, and $\int_0^1 dx/x^p$ converges.

We ask you to confirm the above in Exercises 49 and 50.

EXPLORATION 4

Confirm Analytically

1. Confirm analytically that $\int_1^\infty dx/x^2$ converges. Does its value agree with your result from Exploration 3, part 1?

2. Show that the inequality $0 < p < 1$ above for integrals of $g(x) = 1/x^p$ on $(0, 1]$ and $[1, \infty)$ can be replaced by $p < 1$.

The Domination Test for Convergence and Divergence

Sometimes, we can determine whether an improper integral converges without having to evaluate it. Instead, we compare it to an integral whose convergence or divergence we already know. This is the case with the next example, an integral important in probability theory.

EXAMPLE 6

Determine whether the improper integral

$$\int_1^\infty e^{-x^2}\, dx$$

converges or diverges.

Solution Even though we cannot find any simpler expression for

$$\int_1^b e^{-x^2}\, dx,$$

we can show that it has a finite limit as $b \to \infty$.

The function $\int_1^b e^{-x^2}\, dx$ represents the area between the x-axis and the curve $y = e^{-x^2}$ from $x = 1$ to $x = b$. It is an increasing function of b. Therefore, it either becomes infinite as $b \to \infty$ or has a finite limit as $b \to \infty$. It does not become infinite because for every value of $b \ge 1$,

$$e^{-x^2} \le e^{-x} \qquad \text{(Fig. 8.29),}$$

so that

$$\int_1^b e^{-x^2}\, dx \le \int_1^b e^{-x}\, dx = e^{-1} - e^{-b} < e^{-1}. \qquad \text{Section 5.3, Rule 6}$$

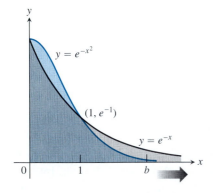

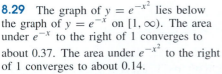

8.29 The graph of $y = e^{-x^2}$ lies below the graph of $y = e^{-x}$ on $[1, \infty)$. The area under e^{-x} to the right of 1 converges to about 0.37. The area under e^{-x^2} to the right of 1 converges to about 0.14.

Hence,

$$\int_1^\infty e^{-x^2}\, dx = \lim_{b\to\infty} \int_1^b e^{-x^2}\, dx$$

converges to a definite value.

≣

A positive function f **dominates** a positive function g as $x \to \infty$ if $g(x) \le f(x)$ for all values of x beyond some point a. For instance, $f(x) = e^{-x}$ dominates $g(x) = e^{-x^2}$ as $x \to \infty$ because $e^{-x^2} \le e^{-x}$ for all $x > a = 1$.

If f dominates g as $x \to \infty$, then

$$\int_a^b g(x)\, dx \le \int_a^b f(x)\, dx \qquad (b > a),$$

and from this it can be argued as in Example 6 that

$$\int_a^\infty g(x)\, dx \text{ converges if } \int_a^\infty f(x) \text{ converges.}$$

Turning this around says that

$$\int_a^\infty f(x)\, dx \text{ diverges if } \int_a^\infty g(x) \text{ diverges.}$$

We state these results as a theorem and then give examples.

THEOREM 1 **Domination Test for Convergence and Divergence of Improper Integrals**

Given $f(x) \ge g(x) \ge 0$ for all $x > a$:

1. If $\int_a^\infty f(x)\, dx$ converges, then $\int_a^\infty g(x)\, dx$ converges also.
2. If $\int_a^\infty g(x)\, dx$ diverges, then $\int_a^\infty f(x)\, dx$ diverges also.

Theorem 1 assumes that f and g are integrable over every finite interval $[a, b]$, which will be true, for instance, if the functions are continuous.

EXAMPLE 7

a) $\displaystyle\int_1^\infty \frac{1}{e^{2x}}\, dx$ converges because $\dfrac{1}{e^{2x}} < \dfrac{1}{e^x}$ and $\displaystyle\int_1^\infty \frac{1}{e^x}\, dx$ converges.

b) $\displaystyle\int_1^\infty \frac{1}{\sqrt{x}}\, dx$ diverges because $\dfrac{1}{\sqrt{x}} \ge \dfrac{1}{x}$ for $x \ge 1$ and $\displaystyle\int_1^\infty \frac{1}{x}\, dx$ diverges.

c) $\displaystyle\int_1^\infty \left(\frac{1}{x} + \frac{1}{x^2}\right) dx$ diverges because $\dfrac{1}{x} + \dfrac{1}{x^2} > \dfrac{1}{x}$ and $\displaystyle\int_1^\infty \frac{1}{x}\, dx$ diverges.

≣

The Limit Comparison Test

Another useful result, which we shall not prove, is the Limit Comparison

---

EXPLORATION BITS

1. For the two integrals in Example 7(a), find their values graphically, numerically, and analytically.

2. If you were to try the method of Exploration 3 on the integrals in Examples 7(b) and 7(c), what should you expect to see in the viewing windows?

Test for the convergence and divergence of improper integrals. It goes like this:

THEOREM 2   Limit Comparison Test for Convergence and Divergence of Improper Integrals

Suppose that $f(x)$ and $g(x)$ are positive functions and that

$$\lim_{x \to \infty} \frac{f(x)}{g(x)} = L \qquad (0 < L < \infty).$$

Then, $\int_a^\infty f(x)\,dx$ and $\int_a^\infty g(x)\,dx$ both converge or both diverge.

Like Theorem 1, Theorem 2 assumes f and g to be integrable over every finite interval $[a, b]$.

EXPLORATION 5

Limit Comparison Test

In the language of Section 7.6, Theorem 2 says that *if two functions grow at the same rate as $x \to \infty$, then their integrals from a to ∞ behave alike: they both converge or both diverge.*

1. The integral $\int_1^\infty dx/x^2$ converges. Why? Show that $\int_1^\infty dx/(1 + x^2)$ also converges by applying the Limit Comparison Test graphically, then analytically.

2. If the Limit Comparison Test shows that two integrals both converge, this does not mean that they converge to the same value. Verify this by evaluating the two integrals of part 1.

3. Determine the convergence or divergence of the two integrals

$$\int_1^\infty \frac{3}{e^x + 5}\,dx \qquad \text{and} \qquad \int_1^\infty \frac{1}{e^x}\,dx.$$

Include the Limit Comparison Test as part of your analysis.

EXPLORATION BIT

We define $\int_{-\infty}^\infty f$ as in Eq. (1). It also would seem reasonable to say that

$$\int_{-\infty}^\infty f = \lim_{b \to \infty} \int_{-b}^b f.$$

However, it is possible to find a function f for which this limit exists but the integrals in Eq. (1) diverge, which says that the two interpretations of $\int_{-\infty}^\infty f$ are *not* the same.

1. Can you find such a function f? (*Hint:* For what kind of function is $\int_{-b}^b f = 0$?) Confirm.

2. If all the integrals in Eq. (1) converge, can we correctly say that $\int_{-\infty}^\infty f = 2\int_0^\infty f$? Give a convincing argument.

Two Infinite Limits of Integration

In the mathematics underlying studies of light, electricity, and sound, we often encounter improper integrals, $\int_{-\infty}^\infty f$, that have two infinite limits of integration. We say that

$$\int_{-\infty}^\infty f = \int_{-\infty}^0 f + \int_0^\infty f \qquad (1)$$

if both integrals on the right-hand side converge. Otherwise, we say that $\int_{-\infty}^\infty f$ diverges.

EXAMPLE 8

$$\int_{-\infty}^{\infty} \frac{dx}{1+x^2} = \int_{-\infty}^{0} \frac{dx}{1+x^2} + \int_{0}^{\infty} \frac{dx}{1+x^2} \qquad \text{Eq. 1}$$

$$= \lim_{b \to -\infty} \left[\tan^{-1} x \right]_{b}^{0} + \lim_{c \to \infty} \left[\tan^{-1} x \right]_{0}^{c}$$

$$= \lim_{b \to -\infty} \left[\tan^{-1} 0 - \tan^{-1} b \right] + \lim_{c \to \infty} \left[\tan^{-1} c - \tan^{-1} 0 \right]$$

$$= 0 - \left(-\frac{\pi}{2} \right) + \frac{\pi}{2} - 0 = \pi.$$

We can interpret the value of this integral as the area of the unbounded region between the curve $y = 1/(1 + x^2)$ and the x-axis (Fig. 8.30).   ≡

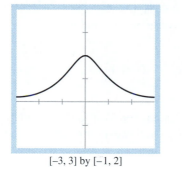

[−3, 3] by [−1, 2]

8.30 A complete graph of
$$y = \frac{1}{1 + x^2}.$$

EXPLORATION 6

Two Infinite Limits of Integration

Let $y_1 = 1/(1 + x^2)$, the integrand in Example 8. GRAPH $y_2 = $ NINT $(y_1, x, 0)$ for $x < 0$, $y_3 = $ NINT $(y_1, 0, x)$ for $x > 0$, and $y_4 = \pi/2$ in a $[-10, 10]$ by $[-2, 2]$ viewing window. Explain how your viewing window supports the result of Example 8.   ⌘

Probability

The domain of the *normal probability distribution function*

$$f(x) = \left(1/\sqrt{2\pi} \right) e^{-x^2/2}$$

is $(-\infty, \infty)$. The area under the curve (Fig. 8.31) is an unbounded region. However, the area must be 1 because, according to the theory of probability, the sum of all possible probabilities of an event must be 1. The integral involved is an improper integral.

EXAMPLE 9

Show that

$$\int_{-\infty}^{\infty} \frac{1}{\sqrt{2\pi}} e^{-x^2/2} \, dx = 1.$$

Solution   First,

$$\int_{-\infty}^{\infty} \frac{1}{\sqrt{2\pi}} e^{-x^2/2} \, dx = \int_{-\infty}^{0} \frac{1}{\sqrt{2\pi}} e^{-x^2/2} \, dx + \int_{0}^{\infty} \frac{1}{\sqrt{2\pi}} e^{-x^2/2} \, dx$$

$$= \lim_{a \to -\infty} \int_{a}^{0} \frac{1}{\sqrt{2\pi}} e^{-x^2/2} \, dx + \lim_{b \to \infty} \int_{0}^{b} \frac{1}{\sqrt{2\pi}} e^{-x^2/2} \, dx.$$

Now we must look to a graphical method because $\int_{a}^{0} (1/\sqrt{2\pi}) e^{-x^2/2} \, dx$ has no explicit antiderivative. The graphs in Figure 8.32 strongly sug-

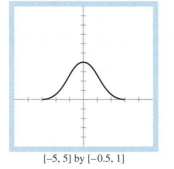

[−5, 5] by [−0.5, 1]

8.31 The graph of $f(x) = \dfrac{1}{\sqrt{2\pi}} e^{-x^2/2}$.

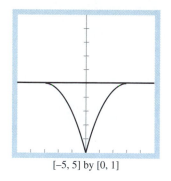

[-5, 5] by [0, 1]

8.32 The graphs of

$$y = 1/2,$$

$$f(x) = \text{NINT}\,(p(t), x, 0)$$

for $x < 0$, and

$$g(x) = \text{NINT}\,(p(t), 0, x)$$

for $x > 0$, where

$$p(t) = \frac{1}{\sqrt{2\pi}} e^{-t^2/2}.$$

EXPLORATION BIT

Interpret the integral

$$\int_0^\infty \frac{1}{x^2}\,dx,$$

and analyze it according to your meaning.

---

gest that the line $y = 1/2$ is an end behavior model for both functions $f(a) = \int_a^0 \left(1/\sqrt{2\pi}\right) e^{-x^2/2}\,dx$ and $g(b) = \int_0^b \left(1/\sqrt{2\pi}\right) e^{-x^2/2}\,dx$. Thus, it appears that both improper integrals converge to 1/2, and hence

$$\int_{-\infty}^{\infty} \left(1/\sqrt{2\pi}\right) e^{-x^2/2}\,dx = 1.$$

For Example 9, we could use the Domination Test to show analytically that $\int_{-\infty}^{\infty} (1/\sqrt{2\pi}) e^{-x^2/2}\,dx$ converges. Again, however, this test does not tell us what the limit is.

Concluding Remarks

We know that

$$\int_1^\infty \frac{1}{x^2}\,dx$$

converges, but what about integrals like

$$\int_2^\infty \frac{1}{x^2}\,dx \qquad \text{and} \qquad \int_{100}^\infty \frac{1}{x^2}\,dx?$$

The answer is that they converge, too. The existence of the limits

$$\lim_{b \to \infty} \int_2^b \frac{1}{x^2}\,dx \qquad \text{and} \qquad \lim_{b \to \infty} \int_{100}^b \frac{1}{x^2}\,dx$$

does not depend on the starting points $a = 2$ or $a = 100$, but only on the values of $1/x^2$ as x approaches infinity. Indeed, we find for any positive a that

$$\lim_{b \to \infty} \int_a^b \frac{1}{x^2}\,dx = \lim_{b \to \infty} \left(-\frac{1}{b} + \frac{1}{a}\right) = \frac{1}{a}.$$

The *value* of the limit depends on the value of a, but the *existence* of the limit does not. The limit exists for any positive number a. The convergence and divergence of integrals that are improper only because the upper limit is infinite never depend on the lower limit of integration.

Exercises 8.6

Evaluate the integrals in Exercises 1–10. Support each, if possible, with a graphical or numerical argument.

1. $\displaystyle\int_0^\infty \frac{dx}{x^2 + 4}$

2. $\displaystyle\int_0^1 \frac{dx}{\sqrt{x}}$

3. $\displaystyle\int_{-1}^1 \frac{dx}{x^{2/3}}$

4. $\displaystyle\int_1^\infty \frac{dx}{x^{1.001}}$

5. $\displaystyle\int_0^4 \frac{dx}{\sqrt{4 - x}}$

6. $\displaystyle\int_0^1 \frac{dx}{\sqrt{1 - x^2}}$

7. $\displaystyle\int_0^1 \frac{dx}{x^{0.999}}$

8. $\displaystyle\int_{-\infty}^2 \frac{dx}{4 - x}$

9. $\displaystyle\int_2^\infty \frac{2}{x^2 - x}\,dx$

10. $\displaystyle\int_0^\infty \frac{dx}{(1 + x)\sqrt{x}}$

In Exercises 11–44, determine whether the integrals converge or diverge. (In some cases, you may not need to evaluate the integrals to decide. Name any tests you use.) Support each, if possible, with a graphical or numerical argument.

11. $\displaystyle\int_1^\infty \frac{dx}{\sqrt[3]{x}}$

12. $\displaystyle\int_1^\infty \frac{dx}{x^3}$

13. $\displaystyle\int_1^\infty \frac{dx}{x^3 + 1}$

14. $\displaystyle\int_0^\infty \frac{dx}{x^3}$

15. $\displaystyle\int_0^\infty \frac{dx}{x^{3/2} + 1}$

16. $\displaystyle\int_0^\infty \frac{dx}{1 + e^x}$

17. $\displaystyle\int_0^{\pi/2} \tan x \, dx$

18. $\displaystyle\int_{-1}^1 \frac{dx}{x^2}$

19. $\displaystyle\int_{-1}^1 \frac{dx}{x^{2/5}}$

20. $\displaystyle\int_0^\infty \frac{dx}{\sqrt{x}}$

21. $\displaystyle\int_2^\infty \frac{dx}{\sqrt{x-1}}$

22. $\displaystyle\int_1^\infty \frac{5}{x} \, dx$

23. $\displaystyle\int_0^2 \frac{dx}{1 - x^2}$

24. $\displaystyle\int_2^\infty \frac{dx}{(x+1)^2}$

25. $\displaystyle\int_0^\infty \frac{dx}{\sqrt{x^6 + 1}}$

26. $\displaystyle\int_{-1}^1 \frac{dx}{\sqrt[3]{x}}$

27. $\displaystyle\int_0^\infty x^2 e^{-x} \, dx$

28. $\displaystyle\int_1^\infty \frac{\sqrt{x+1}}{x^2} \, dx$

29. $\displaystyle\int_\pi^\infty \frac{2 + \cos x}{x} \, dx$

30. $\displaystyle\int_1^\infty \frac{\ln x}{x} \, dx$

31. $\displaystyle\int_6^\infty \frac{1}{\sqrt{x+5}} \, dx$

32. $\displaystyle\int_1^\infty \frac{dx}{\sqrt{2x + 10}}$

33. $\displaystyle\int_2^\infty \frac{2}{x^2 - 1} \, dx$

34. $\displaystyle\int_1^\infty \frac{1}{e^{\ln x}} \, dx$

35. $\displaystyle\int_2^\infty \frac{1}{\ln x} \, dx$

36. $\displaystyle\int_1^\infty \frac{1}{\sqrt{e^x - x}} \, dx$

37. $\displaystyle\int_1^\infty \frac{1}{e^x - 2^x} \, dx$

38. $\displaystyle\int_2^\infty \frac{1}{x^3 - 5} \, dx$

39. $\displaystyle\int_0^\infty \frac{dx}{\sqrt{x + x^4}}$

40. $\displaystyle\int_0^\infty e^{-x} \cos x \, dx$

(*Hint for Exercise 39:* Compare the integral with $\int dx/\sqrt{x}$ for x near zero and with $\int dx/x^2$ for large x.)

41. $\displaystyle\int_0^\infty e^{-2x} \, dx$

42. $\displaystyle\int_{-5}^\infty e^{-3x} \, dx$

43. $\displaystyle\int_{-3}^\infty x^2 e^{-2x} \, dx$

44. $\displaystyle\int_{-2}^\infty x e^{-x} \, dx$

45. a) Draw a complete graph of

$$f(x) = \frac{x^2 - 1}{x - 1} \quad \text{for} \quad 0 \le x \le 3.$$

b) Evaluate $\displaystyle\int_0^3 \frac{x^2 - 1}{x - 1} \, dx$.

46. a) Draw a complete graph of

$$g(x) = \frac{x^3 + x^2 - 2x - 1}{x^2 - 2} \quad \text{for } 0 \le x \le 3.$$

b) Evaluate

$$\int_0^3 g(x) \, dx.$$

47. *Estimating the value of a convergent improper integral whose domain is infinite.* Show that

$$\int_3^\infty e^{-3x} \, dx = \frac{1}{3} e^{-9} < 0.000042,$$

and hence that

$$\int_3^\infty e^{-x^2} \, dx < 0.000042.$$

Therefore, $\displaystyle\int_0^\infty e^{-x^2} \, dx$ can be replaced by $\displaystyle\int_0^3 e^{-x^2} \, dx$ without introducing more than this much error. Evaluate this last integral by Simpson's Rule with $n = 6$. (This illustrates one method by which a convergent improper integral may be approximated numerically.)

48. *The infinite paint can, or Gabriel's horn.* The region bounded by the function $y = 1/x$, the x-axis, and the lines $x = 1$, $x = b$ sweeps out a solid of revolution about the x-axis as shown. Extending this solid to the right indefinitely produces an "infinite solid" with a finite volume and an infinite surface area.

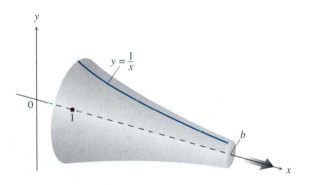

As Example 2 shows, the integral $\int_1^\infty dx/x$ diverges. This means that the integral

$$\int_1^\infty 2\pi \frac{1}{x} \sqrt{1 + \frac{1}{x^4}} \, dx,$$

which measures the surface area of the solid, diverges also because

$$\int_1^b 2\pi \frac{1}{x} \sqrt{1 + \frac{1}{x^4}} \, dx > \int_1^b \frac{1}{x} \, dx$$

for every finite value $b > 1$.

However, the integral

$$\int_1^\infty \pi \left(\frac{1}{x}\right)^2 dx$$

for the volume of the solid converges. Calculate it. This solid of revolution is sometimes described as a can that does not hold enough paint to cover its outside surface.

49. Show that

$$\int_1^\infty \frac{dx}{x^p} = \frac{1}{p-1} \quad \text{when} \quad p > 1$$

but that the integral is infinite when $p < 1$. Example 2 shows what happens when $p = 1$.

50. Show that

$$\int_0^1 \frac{dx}{x^p} = \frac{1}{1-p} \quad \text{when} \quad p < 1$$

but that the integral is infinite when $p > 1$.

In Exercises 51 and 52, find the values of p for which each integral converges.

51. $\int_1^2 \frac{dx}{x(\ln x)^p}$ **52.** $\int_2^\infty \frac{dx}{x(\ln x)^p}$

Exercises 53–56 are about the region in the first quadrant between the curve $y = e^{-x}$ and the x-axis.

53. Find the area of the region.

54. Find the centroid of the region.

55. Find the volume swept out by revolving the region about the y-axis.

56. Find the volume swept out by revolving the region about the x-axis.

57. Find the area of the region that lies between the curves $y = \sec x$ and $y = \tan x$ for $0 \le x \le \pi/2$.

58. Show that the area of the region between the curve $y = 1/(1 + x^2)$ and the entire x-axis is the same as the area of the unit disk, $x^2 + y^2 \le 1$.

Determine analytically which of the integrals in Exercises 59–62 converge and which diverge? Then support graphically.

59. $\int_{-\infty}^\infty \frac{dx}{\sqrt{x^2 + 1}}$ **60.** $\int_{-\infty}^\infty \frac{dx}{\sqrt{x^6 + 1}}$

61. $\int_{-\infty}^\infty \frac{dx}{e^x + e^{-x}}$ **62.** $\int_{-\infty}^\infty \frac{e^{-x} dx}{x^2 + 1}$

63. $\int_{-\infty}^\infty f(x)\, dx$ may not equal $\lim_{b\to\infty} \int_{-b}^b f(x)\, dx$. Show that $\int_0^\infty \frac{2x\, dx}{x^2+1}$ diverges and hence that $\int_{-\infty}^\infty \frac{2x\, dx}{x^2+1}$ diverges. Then show that

$$\lim_{b\to\infty} \int_{-b}^b \frac{2x\, dx}{x^2 + 1} = 0.$$

64. Use graphs to estimate the value of $\int_{-\infty}^\infty e^{-x^2} dx$.

8.7 Differential Equations

Differential equations can arise whenever we model the effects of change, motion, and growth. They can arise when we study moving particles, changing business conditions, oscillating voltages in neural networks, changing concentrations in chemical reactions, and flowing resources in a market economy. Our space program more or less runs on differential equations, and mathematical models of our environment involve huge systems of differential equations. The basic tool for solving differential equations is calculus. This section previews differential equations and some methods that we can use to solve them.

Separable First-Order Equations

We study differential equations for their mathematical importance and for what they tell us about reality. When we study unchecked bacterial growth, for example, and assume that the rate of increase in the population at any given time t is proportional to the number of bacteria then present, we are assuming that the population size y obeys the differential equation $dy/dt = ky$. If we also know that the size of the population is y_0 when $t = 0$, we can find a formula for the size of the population at any time $t > 0$ by solving the initial value problem

Differential equation: $\quad \dfrac{dy}{dt} = ky$

Initial condition: $\quad y = y_0 \quad$ when $t = 0$.

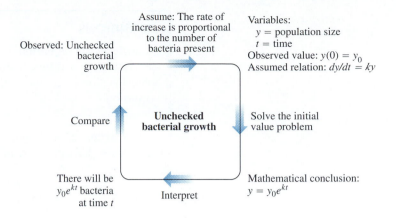

As we saw in Section 7.2, the general solution of the differential equation (the formula that gives all possible solutions) is

$$y = Ae^{kt} \qquad (A \text{ is an arbitrary constant}),$$

and the particular solution that satisfies the initial condition $y(0) = y_0$ is

$$y = y_0 e^{kt}.$$

We interpret this as saying that unchecked bacterial growth is *exponential*. There should be $y_0 e^{kt}$ bacteria present at time t. We know from experience that there are limits to all growth, and we do not expect this prediction to be accurate for large values of t. But for the early stages of growth, the formula is a good predictor and gives valuable information about the population's size.

Differential Equations and Integration

Solving differential equations is intimately connected with integration. For example, we solve $dy/dx = x^2$ by integrating both sides with respect to x:

$$\frac{dy}{dx} = x^2$$

$$\int \frac{dy}{dx}\, dx = \int x^2\, dx$$

$$\int dy = \int x^2\, dx$$

$$y + C_1 = \frac{x^3}{3} + C_2, \qquad \text{or simply } y = \frac{x^3}{3} + C.$$

Thus, solving the differential equation $dy/dx = x^2$ is equivalent to evaluating the indefinite integral $y = \int x^2\, dx$. Any solution y is a function whose derivative is x^2. We know that $y = x^3/3$ is *one* of infinitely many solutions of the form $x^3/3 + C$. $y = x^3/3 + C$ is called the general solution of the differential equation $dy/dx = x^2$.

The equations in this chapter are *ordinary differential equations.*

> **DEFINITIONS**
>
> A **differential equation** is an equation that contains one or more derivatives of a differentiable function. An equation with derivatives of a function of a single variable is called an **ordinary differential equation**.

The **order** of a differential equation is the order of the equation's highest-order derivative. A differential equation is **linear** if it can be put in the form

$$a_n(x)\frac{d^n y}{dx^n} + a_{n-1}(x)\frac{d^{n-1} y}{dx^{n-1}} + \cdots + a_1(x)\frac{dy}{dx} + a_0(x)y = F(x),$$

where y and the a's are functions of x.

EXAMPLE 1

First-order, linear: $\qquad \dfrac{dy}{dx} = 5y;\ 3\dfrac{dy}{dx} - \sin x = 0.$

Third-order, nonlinear: $\left(\dfrac{d^3 y}{dx^3}\right)^2 + \left(\dfrac{d^2 y}{dx^2}\right)^5 - \dfrac{dy}{dx} = e^x.$ $\qquad$ ☰

Almost all the differential equations that we have solved so far in this book have been linear equations of first or second order. This will continue to be true, but the equations will have more variety and will model a broader range of applications.

We call a function $y = f(x)$ a **solution** of a differential equation if y and its derivatives satisfy the equation. To test whether a given function solves a particular equation, we substitute the function and its derivatives into the equation. If the equation then reduces to an identity, the function solves the equation; otherwise, it does not.

EXAMPLE 2

Show that for any values of the arbitrary constants C_1 and C_2 the function $y = C_1 \cos x + C_2 \sin x$ is a solution of the differential equation

$$\frac{d^2 y}{dx^2} + y = 0.$$

Solution We differentiate twice to find $d^2 y/dx^2$:

$$y = C_1 \cos x + C_2 \sin x$$

$$\frac{dy}{dx} = -C_1 \sin x + C_2 \cos x$$

$$\frac{d^2 y}{dx^2} = -C_1 \cos x - C_2 \sin x.$$

We then substitute the expressions for y and $d^2 y/dx^2$ into the differential equation to see whether the left-hand side reduces to zero, which is the right-hand side. It does, because

$$\frac{d^2 y}{dx^2} + y = (-C_1 \cos x - C_2 \sin x) + (C_1 \cos x + C_2 \sin x) = 0.$$

So the function is a solution of the differential equation. $\qquad$ ☰

It can be shown that the formula $y = C_1 \cos x + C_2 \sin x$ gives all possible solutions of the equation $d^2y/dx^2 + y = 0$. A formula that gives all the solutions of a differential equation is called the **general solution** of the equation. In this sense, $y = C_1 \cos x + C_2 \sin x$ is the general solution of $d^2y/dx^2 + y = 0$. To *solve* a differential equation means to find its general solution.

Notice that the equation $d^2y/dx^2 + y = 0$ has order two and that its general solution has two arbitrary constants. The general solution of an nth-order ordinary differential equation can be expected to contain n arbitrary constants.

Separable Differential Equations

A method that sometimes works for solving a first-order differential equation involves treating the derivative dy/dx as a quotient of differentials and rearranging the equation to group the y terms alone with dy and the x terms alone with dx. We cannot always accomplish this, but when we can, the solution may then be found by separate integrations with respect to x and y.

DEFINITION

A first-order differential equation is **separable** if it can be put in the form

$$M(x) + N(y)\frac{dy}{dx} = 0 \qquad (1)$$

or in the equivalent differential form

$$M(x)\,dx + N(y)\,dy = 0.$$

When we write the equation this way, we say that we have **separated the variables.**

We can solve Eq. (1) by integrating both sides with respect to x to get

$$\int M(x)\,dx + \int N(y)\frac{dy}{dx}\,dx = C.$$

However, the second integral in this equation is equivalent to

$$\int N(y)\,dy,$$

and it usually saves time to integrate in the equivalent equation

$$\int M(x)\,dx + \int N(y)\,dy = C$$

instead. The result of the integration will express y either explicitly or implicitly as a function of x that solves Eq. (1).

Steps for Solving a Separable First-Order Differential Equation

1. Write the equation in the form $M(x)\,dx + N(y)\,dy = 0$.
2. Integrate M with respect to x and N with respect to y to obtain an equation that relates y and x.

EXAMPLE 3

Solve the differential equation

$$\frac{dy}{dx} = (1 + y^2)e^x.$$

Solution We use algebra to separate the variables and write the equation in the form $M(x)\,dx + N(y)\,dy = 0$, obtaining

$$e^x\,dx - \frac{1}{1+y^2}\,dy = 0.$$

We then integrate to get

$$\int e^x\,dx - \int \frac{1}{1+y^2}\,dy = C$$

$$e^x - \tan^{-1} y = C \qquad \text{Constants of integration combined}$$

$$\tan^{-1} y = e^x - C.$$

In this case, we can solve the resulting equation explicitly for y by taking the tangent of each side:

$$y = \tan(e^x - C). \qquad\qquad \blacksquare$$

> The general solution found in Example 3 can be easily confirmed by differentiating $y = \tan(e^x - C)$:
>
> $$\frac{dy}{dx} = e^x \sec^2(e^x - C) = (1 + y^2)e^x,$$
>
> since
>
> $$\sec^2 \theta = 1 + \tan^2 \theta.$$

Logistic Growth Model

In Chapter 7, we studied the fundamental growth model $dy/dx = ky$. That is, we made the assumption that the rate of increase of the population growth at any time t is directly proportional to the population. This model assumes unlimited (exponential) growth, which is usually a very unrealistic assumption. A more realistic model for growth assumes that growth is bounded by environmental factors (e.g., food) or economic factors (limited resources). In this model, the rate of increase of the population at any time t is directly proportional to *both* y and $M - y$, where M is the *maximum* possible size of the population. This leads to the differential equation

$$\frac{dy}{dt} = ky(M - y),$$

which is a more realistic growth model. This growth model is called the **logistic growth model**.

> **EXPLORATION BIT**
>
> Points of inflection are important in both biology and business. For example, the point of inflection of the bear growth graph is associated with the maximum bear population growth rate. Can you explain why?

EXAMPLE 4   Logistic Growth

A certain national park is known to be able to support at most 100 grizzly bears. Ten bears are known to be in the park at present. Assume that the bear population growth is logistic in nature. That is, we assume that the rate of growth is directly proportional to both the population y and the limiting factor $100 - y$ at any time t (years) with proportionality constant $k = 0.001$.

a) Determine the population of bears $y(t)$ as an explicit function of time t.

b) Draw a complete graph of $y(t)$.

c) What values of y and t make sense in the problem situation? Draw a graph of the problem situation.

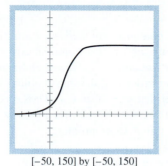

[−50, 150] by [−50, 150]

8.33 A complete graph of $y(t)$.

Be sure you understand why one symbol C can be used for the constant throughout the solution.

d) When should the bear population be 25? 50?

e) When should the bear population become constant?

Solution

a) We are given that

$$\frac{dy}{dt} = 0.001y(100 - y)$$

and that $y = 10$ when $t = 0$. Separating the variables leads to

$$1000 \int \frac{dy}{(100 - y)y} = \int dt$$

$$10 \left[\int \frac{dy}{y} + \int \frac{dy}{100 - y} \right] = \int dt \qquad \text{Partial fractions}$$

$$10 \left[\ln |y| - \ln |100 - y| \right] = t + C$$

$$\ln y - \ln (100 - y) = t/10 + C \qquad 0 < y < 100$$

$$\ln \frac{y}{100 - y} = t/10 + C$$

$$\frac{y}{100 - y} = Ce^{t/10} \qquad \text{Exponentiate.}$$

$$y = \frac{100Ce^{t/10}}{1 + Ce^{t/10}} \qquad \text{Solve for } y. \qquad (2)$$

Using Eq. (2) and the initial condition $y(0) = 10$, we find

$$10 = \frac{100C}{1 + C}$$

$$C = \frac{1}{9}.$$

So the bear population growth is given by the logistic function

$$y = \frac{(100/9)e^{t/10}}{1 + (1/9)e^{t/10}}.$$

b) See Figure 8.33 for a complete graph of $y(t)$.

c) Only $t \geq 0$ and $0 < y < 100$ make sense in the bear problem situation. See Figure 8.34 for a graph of the problem situation.

d) A TRACE of the graph in Fig. 8.34 shows that the bear population should be 25 in about 11 years and 50 in about 22 years.

e) TRACE farther along the graph in Fig. 8.34, and we see that the bear population should reach the park's capacity of 100 in about 75 years.

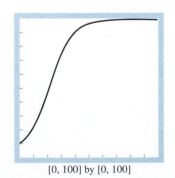

[0, 100] by [0, 100]

8.34 A graph of the bear problem situation.

Initial Value Problems

Example 4 is an illustration of an *initial value problem with one initial condition*. In general, initial value problems are nth-order differential equations with n initial conditions. For example, the equation $y'' + y = 0$ of Example 2 with $y(0) = 0$ and $y'(0) = 1$ is a second-order initial value problem with two initial conditions. Example 2 showed that $y = A \sin x + B \cos x$ is the general solution to $y'' + y = 0$. As an exercise, we ask you to show that $y = \sin x$ is the particular solution to this initial value problem.

Many practical problem situations involve initial value problems with differential equations whose solutions have no explicit algebraic form. Numerical methods and graphing provide the only method to solve such problems.

Numerical Methods for Solving Initial Value Problems

We complete our preview of differential equations by describing three numerical methods for solving the initial value problem $y' = f(x, y)$, $y(a) = y_0$ over an interval $a \le x \le b$. These methods do not produce a general solution of $y' = f(x, y)$; they produce tables of values of y for preselected values of x. Instead of being a drawback, however, this is a definite advantage, especially if we want to solve a differential equation like $y' = x^2 + y^2$ with no explicit algebraic solution. Also, in solving equations of motion in real situations like launching a rocket or intercepting one in orbit, we find that numerical answers are much more useful than algebraic expressions would be. We keep our examples simple, but the ideas can be extended to far more complicated equations and systems of equations. Naturally we will not attempt to solve such complicated systems with paper and pencil; instead, we will turn to technology.

We consider three numerical methods: first a method that dates back to Euler, then an improved Euler method, and finally an even more accurate method called the Runge–Kutta method. We use graphing calculators or computers to apply these methods. The programs that we use are EULERT, EULERG, IMPEULT, IMPEULG, RUNKUTT, and RUNKUTG. (Refer to the *Resource Manual*.) Note that each of these program names ends in T or G. The T programs produce a table of values while the G programs produce a graph of the solution.

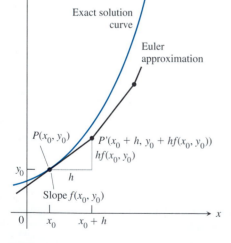

8.35 The Euler approximation to the solution of the initial value problem $y = f(x)$, $y = y_0$ when $x = x_0$. The errors involved may accumulate as we take more steps.

The Euler Method

The initial value problem

$$y' = f(x, y), \qquad y(x_0) = y_0,$$

provides us with a starting point, $P(x_0, y_0)$, and a slope $f(x_0, y_0)$. We know that the graph of the solution must be a curve through P with that slope (Fig. 8.35). If we use the tangent through P to approximate the actual solution curve, the approximation may be fairly good from $x_0 - h$ to $x_0 + h$ for small values of h. Thus, we might choose $h = 0.1$, say, and move along the tangent line from P to $P'(x_1, y_1)$, where $x_1 = x_0 + h$ and $y_1 = y_0 + hf(x_0, y_0)$.

If we think of P' as a new starting point, we can move from P' to $P''(x_2, y_2)$, where $x_2 = x_1 + h$ and $y_2 = y_1 + hf(x_1, y_1)$. If we replace h by $-h$, we move to the left from P instead of to the right. The process can be continued, but the errors are likely to accumulate as we take more steps.

EXAMPLE 5

Take $h = 0.1$, and investigate the accuracy of the Euler approximation method for the initial value problem

$$y' = 1 + y, \qquad y(0) = 1, \tag{3}$$

over the interval $0 \le x \le 1$ by letting

$$x_{n+1} = x_n + h, \qquad y_{n+1} = y_n + h(1 + y_n). \tag{4}$$

Solution The particular solution of Eqs. (3) is $y = 2e^x - 1$. Table 8.2 shows the results using Eqs. (4) and the exact results rounded to four decimals for comparison. By the time we get to $x = 1$, the error is about 5.6%. ≡

TABLE 8.2 **The Approximate Solution of $y' = 1 + y$, $y(0) = 1$, Using Euler's Method (Step Size 0.1)**

x	y (approx.)	y (exact)	Error = y (exact) $-$ y (approx.)
0	1	1	0
0.1	1.2	1.2103	0.0103
0.2	1.42	1.4428	0.0228
0.3	1.662	1.6997	0.0377
0.4	1.9282	1.9836	0.0554
0.5	2.2210	2.2974	0.0764
0.6	2.5431	2.6442	0.1011
0.7	2.8974	3.0275	0.1301
0.8	3.2872	3.4511	0.1639
0.9	3.7159	3.9192	0.2033
1.0	4.1875	4.4366	0.2491

The Improved Euler Method

This method first gets an estimate of y_{n+1}, as in the original Euler method, but calls the result z_{n+1}. We then take the average of $f(x_n, y_n)$ and $f(x_{n+1}, z_{n+1})$ in place of $f(x_n, y_n)$ in the next step. Thus,

$$z_{n+1} = y_n + hf(x_n, y_n),$$

$$y_{n+1} = y_n + \frac{h}{2}[f(x_n, y_n) + f(x_{n+1}, z_{n+1})].$$

If we apply this improved method to Example 5, again with $h = 0.1$, we get the following results at $x = 1$:

$$y \text{ (approx.)} = 4.428161693,$$

$$y \text{ (exact)} = 4.436563656,$$

$$\text{Error} = y \text{ (exact)} - y \text{ (approx.)} = 0.008401963,$$

and the error is less than 2/10 of 1%.

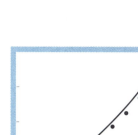

[0, 1] by [1, 5]

8.36 Plot of Table 8.2 for $0 \leq x \leq 1$ and $1 \leq y \leq 5$ with the graph of the analytic solution $y = 2e^x - 1$. Notice the error in the Euler method increases as x increases.

German scientist Carl Runge (1856–1927) was a mathematical physicist of Max Planck's caliber who developed numerical methods for solving the differential equations that arose in his studies of atomic spectra. He used so much mathematics in his research that physicists thought he was a mathematician, and he did so much physics that mathematicians thought he was a physicist. Neither group claimed him as their own, and it was years before anyone could find him a professorship. In 1904, Felix Klein finally persuaded his Göttingen colleagues to create for Runge Germany's only full professorship in applied mathematics. Runge was the professorship's first and only occupant, there being no one of his accomplishments to assume the post when he died.

EXPLORATION 1

Viewing Euler Approximations

1. Apply the EULERT graphing calculator program to the initial value problem of Example 5 to produce a table of values like Table 8.2. Compare your table with Table 8.2.

2. Apply the EULERG graphing calculator program to plot a graph of the Euler-approximate solution to the initial value problem. GRAPH the exact solution in the same viewing window. Compare your viewing window with Fig. 8.36.

3. Using the same view dimensions, apply the IMPEULG graphing calculator program to plot a graph of the solution of the initial value problem. Use $h = 0.1$. Compare your graph with the graph of the exact solution.

A Runge–Kutta Method

The Runge–Kutta method that we use (one of many) requires four intermediate calculations, as given in the following equations:

$$k_1 = hf(x_n, y_n), \qquad k_2 = hf\left(x_n + \frac{h}{2}, y_n + \frac{k_1}{2}\right),$$

$$k_3 = hf\left(x_n + \frac{h}{2}, y_n + \frac{k_2}{2}\right), \qquad k_4 = hf(x_n + h, y_n + k_3).$$

We then calculate y_{n+1} from the formula

$$y_{n+1} = y_n + \frac{1}{6}(k_1 + 2k_2 + 2k_3 + k_4).$$

When we apply this method to the problem of estimating $y(1)$ for the problem $y' = 1 + y$, $y(0) = 1$, still using $h = 0.1$, we get

$$y(1) = 4.436559488$$

$$(\text{recall, } y \text{ (exact)} = 4.436563657)$$

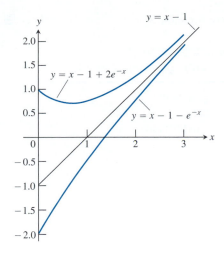

8.37 Graphs of the two solutions of $y' = x - y$ of Example 6. The upper graph,

$$y = x - 1 + 2e^{-x} (y(0) = 1),$$

is concave up and has a minimum when $x = y = \ln 2$. The lower graph,

$$y = x - 1 - e^{-x} (y(0) = -2),$$

is concave down, always rises as x increases, and crosses the x-axis at a value of x near 1.3. Both curves approach the line $y = x - 1$ as $x \to \infty$.

with an error 0.000004169, which is less than 1/10,000 of 1%. Runge–Kutta, in general, is clearly the most accurate of the three methods.

EXAMPLE 6

Consider the two initial value problems

a) $y' = x - y, y(0) = 1$ **b)** $y' = x - y, y(0) = -2.$

Table 8.3 shows the comparison of $y(x)$ as estimated by the Runge–Kutta method (RUNKUTT with $h = 0.1$) and the true value. Figure 8.37 shows the graphs of the Runge-Kutta approximations. (More points than are shown in the table were actually computed and plotted for the graphs.) ▤

Example 6 shows that the error in the Runge–Kutta approximation did not continue to increase as the process was continued. In fact, with $h = 0.1$, the difference between the exact solutions and the approximations remained less than 10^{-6} for the two initial value problems:

a) $y' = x - y, \quad y(0) = 1,$

b) $y' = x - y, \quad y(0) = -2.$
    The fact that the differential equation is linear in y is significant in discussing the accuracy of the Runge–Kutta approximation. Such accuracy will not be attained for the nonlinear differential equation of Example 7.

TABLE 8.3   **A Comparison of Runge–Kutta Approximations with Actual Values (Step Size 0.1)**

	x	y (Runge–Kutta)	y (true value)	Difference
a) $y' = x - y, y(0) = 1$	0	1	1	0
	0.5	0.713061869	0.713061319	5.50×10^{-7}
	1.0	0.735759549	0.735758882	6.67×10^{-7}
	1.5	0.946260927	0.946260320	6.07×10^{-7}
	2.0	1.270671057	1.270670566	4.91×10^{-7}
	2.5	1.664170369	1.664169997	3.72×10^{-7}
	3.0	2.099574407	2.099574137	2.70×10^{-7}
b) $y' = x - y, y(0) = -2$	0	-2	-2	0
	0.5	-1.106530934	-1.106530660	-2.74×10^{-7}
	1.0	-0.367879774	-0.367879441	-3.33×10^{-7}
	1.5	0.276869537	0.276869840	-3.03×10^{-7}
	2.0	0.864664472	0.864664717	-2.45×10^{-7}
	2.5	1.417914815	1.417915001	-1.86×10^{-7}
	3.0	1.950212796	1.950212932	-1.36×10^{-7}

EXAMPLE 7

Table 8.4 lists Runge–Kutta approximations ($h = 0.1$) for the initial value problem

$$y' = x^2 + y^2, \qquad y(0) = 0.$$

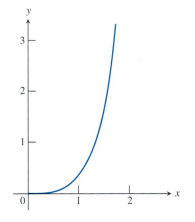

8.38 The graph of the Runge–Kutta solution of the initial value problem $y' = x^2 + y^2$, $y(0) = 0$, for $x > 0$.

Figure 8.38 shows the graph of y as a function of x.

x	y **(Runge–Kutta)**	y **(actual)**
	TABLE 8.4 **The Approximate Solution of** $y' = x^2 + y^2$, $y(0) = 0$ **(Step Size 0.1.)**	
0	0	0
0.5	0.041791288	0.041791146
1.0	0.350233742	0.350231844
1.5	1.517473414	1.517447544
2.0	71.57899545	317.224400
2.1	1.47001E+11	
2.2	1.66667E+**	(Meaning: "You broke the bank!")

For Example 7, the Runge–Kutta approximation to $y(2.1)$, using $h = 0.1$, is 1.47×10^{11}. For this problem, the value of y increases without bound as x increases just beyond 2. The solution curve has a vertical asymptote at just beyond 2. No matter how small we take h, we cannot assert any accuracy for our approximations as the curve approaches this asymptote.

EXPLORATION 2

Confirming a Solution

1. Confirm that $y = x - 1 + 2e^{-x}$ is the solution to the initial value problem $y' = x - y$, $y(0) = 1$, and that $y = x - 1 - e^{-x}$ is the solution to the initial value problem $y' = x - y$, $y(0) = -2$.

2. Draw the graphs of the line $y = x - 1$ and of $y = x - 1 + Ce^{-x}$ for $C = \pm 6, \pm 5, \ldots \pm 1$ in the same $[0, 5]$ by $[-5, 5]$ viewing window.

3. Could $y = x - 1 + Ce^{-x}$ be the general solution to $dy/dx = x - y$? Why? Must it be? Explain.   ❧

Exercises 8.7

In Exercises 1–4, show that each function $y = f(x)$ is a solution of the accompanying differential equation.

1. $xy'' - y' = 0$

   a) $y = x^2$ **b)** $y = 1$ **c)** $y = C_1 x^2 + C_2$

2. $y' + \dfrac{1}{x}y = 1$

   a) $y = \dfrac{x}{2}$ **b)** $y = \dfrac{1}{x} + \dfrac{x}{2}$ **c)** $y = \dfrac{C}{x} + \dfrac{x}{2}$

3. $2y' + 3y = e^{-x}$

   a) $y = e^{-x}$ **b)** $y = e^{-x} + e^{-(3/2)x}$

   c) $y = e^{-x} + Ce^{-(3/2)x}$

4. $yy'' = 2(y')^2 - 2y'$

   a) $y = 1$

   b) $y = \tan x$

In Exercises 5–8, show that each function $y = f(x)$ is a solution of the accompanying initial value problem.

Differential Equation	Initial Condition(s)	Solution Candidate
5. $y'' = -32$	$y(5) = 400,$ $y'(5) = 0$	$y = 160x - 16x^2$
6. $2\dfrac{dy}{dx} + 3y = 6$	$y(0) = 0$	$y = 2\left(1 - e^{-3x/2}\right)$
7. $y'' + 4y = 0$	$y(0) = 3,$ $y'(0) = -2$	$y = 3\cos 2x - \sin 2x$
8. $y'' - (y')^2 = 1$	$y(1) = 2,$ $y'(1) = 0$	$y = 2 - \ln\cos(x - 1)$

9. *Continuous compounding.* You have $1000 with which to open an account and plan to add $1000 per year. All funds in the account will earn 10% interest per year compounded continuously. If the added deposits are also credited to your account continuously, the number of dollars x in your account at time t (years) will satisfy the initial value problem

Differential equation: $\quad \dfrac{dx}{dt} = 1000 + 0.10x$

Initial condition: $\quad x(0) = 1000.$

a) Solve the initial value problem for x as a function of t.

b) About how many years will it take for the amount in your account to reach $100,000?

10. *Newton's Law of Cooling.* Newton's Law of Cooling assumes that the temperature T of a small hot object placed in a surrounding cooling medium of constant temperature T_s decreases at a rate proportional to $(T - T_s)$. An object cooled from $100°C$ to $40°C$ in 20 minutes when the surrounding temperature was $20°C$. How long did it take the temperature to reach $60°C$ on the way down?

In Exercises 11–20, determine an analytic solution to each initial value problem. Then draw a complete graph of the solution. Support graphically with IMPEULG.

11. $\dfrac{dy}{dx} = 0.017(y - 95), y(0) = 30$

12. $\dfrac{dx}{dt} = 0.013(x - 20), x(0) = 6$

13. $\dfrac{dy}{dx} = 0.00125y(350 - y), y(0) = 30$

14. $\dfrac{dx}{dt} = 0.005x(500 - x), x(0) = 20$

15. $\dfrac{dy}{dx} = e^{-2y}, y(0) = 0$

16. $\dfrac{dy}{dx} = e^{2x} - x, y(0) = 1$

17. $\dfrac{dy}{dx} = \dfrac{1}{x^2 + 1}, y(0) = 1$

18. $\dfrac{dy}{dx} = \sqrt{1 - y^2}, y(0) = 1$

19. $\dfrac{dy}{dx} = xe^x, y(0) = 0$

20. $\dfrac{dy}{dx} = \dfrac{1}{y\ln y}, y(0) = 1$

Solve the initial value problems in Exercises 21 and 22. (Refer to Example 2.)

21. $\dfrac{d^2y}{dx^2} + y = 0, y(0) = 0, y'(0) = 1$

22. $\dfrac{d^2y}{dx^2} + y = 0, y(0) = 1, y'(0) = 0$

23. Show that for any values of the arbitrary constants C_1 and C_2, the function $y = C_1 e^x \cos x + C_2 e^x \sin x$ is a solution of the differential equation

$$\dfrac{d^2y}{dx^2} - 2\dfrac{dy}{dx} + 2y = 0.$$

24. Solve the following initial value problem for the differential equation in Exercise 23.
a) $y(0) = 0, y'(0) = 1$ **b)** $y(0) = 1, y'(0) = 0$

25. Use the Euler method with $h = 1/5$ to estimate $y(1)$ if $y' = y$ and $y(0) = 1$. What is the exact value of $y(1)$?

26. Show that the Euler method leads to the estimate $\left(1 + 1/n\right)^n$ for $y(1)$ if $h = 1/n, y' = y$, and $y(0) = 1$. What is the limit as $n \to \infty$?

27. Use the improved Euler method with $h = 1/5$ to estimate $y(1)$ if $y' = y$ and $y(0) = 1$.

28. Use the Runge–Kutta method with $h = 1/5$ to estimate $y(1)$ if $y' = y$ and $y(0) = 1$.

29. Show that the solution of the initial value problem $y' = x^2 + y^2, y(0) = 1$ increases faster on the interval $0 \le x < 1$ than does the solution of the initial value problem $y' = y^2, y(0) = 1$. Solve the latter problem by separation of variables, and thus show that the solution of the original problem becomes infinite at a value of x not greater than 1. (In case you are interested, the value is about 0.969810654.)

30. Solve the initial value problem $y' = 1 + y^2, y(0) = 0$, by (a) separation of variables and (b) using the substitution $y = -u'/u$ and solving the equivalent problem for u. (Notice the similarity with the initial value problem $y' = x^2 + y^2, y(0) = 0$.)

Find numerical solutions (tables) and graphical solutions (graphs) of the initial value problems in Exercises 31–34. Each exercise gives a differential equation in the form $y' = f(x, y)$, a solution interval $a \le x \le b$, the initial value $y(a)$, the step size, and the number of steps. Use the EULER, IMPEUL, and RUNKUT programs.

31. $y' = x/7, a = 0, b = 4, y(0) = 1, h = 0.1, n = 20$

32. $y' = -y^2/x, a = 2, b = 4, y(2) = 2, h = 0.1, n = 20$

33. $y' = (x^2 + y^2)/2y, a = 0, b = 2, y(0) = 0.1, h = 0.2, n = 10$

34. Repeat Exercise 33, but with $h = 0.1$ and $n = 20$.

35. *Evaluating a nonelementary integral.* The value of the nonelementary integral

$$\int_0^1 \sin\left(t^2\right) dt$$

is the value of the function

$$y(x) = \int_0^x \sin\left(t^2\right) dt$$

at $x = 1$. The function, in turn, is the solution of the initial value problem

$$y' = \sin\left(x^2\right), \quad y(0) = 0.$$

Thus, by solving the initial value problem numerically on the interval $0 \le x \le 1$, we can find the value of the integral as the value of y that corresponds to $x = 1$. Use RUNKUT to estimate the integral's value by solving the initial value problem with (a) $n = 20$ steps and (b) $n = 40$ steps.

36. Use the method of Exercise 35 to estimate the value of

$$\int_0^1 x^2 e^{-x} \, dx$$

by solving an appropriate initial value problem on the interval $[0, 1]$ with (a) $n = 20$ steps, and (b) $n = 40$ steps. Compare your solution with the exact solution obtained by using integration by parts.

37. Show that $y = \int_a^x -2t \sec^2\left(a^2 - t^2\right) dt$, for a real, is a general solution to the differential equation $dy/dx = -2x\left(y^2 + 1\right)$. Use analytic methods.

38. Use IMPEULG to graph a solution to the initial value problem $dy/dx = -2x\left(y^2 + 1\right)$, where $y(1) = 1$.

39. Determine the value a so that $y = \int_a^x -2t \sec^2\left(a^2 - t^2\right) dt$ is the solution to the initial value problem in Exercise 38.

40. Show that $y = \int_x^\infty 2t/\left(1 + t^2\right)^2 dt$ is a solution to the differential equation $dy/dx = -2xy^2$. Use analytic methods.

41. Reproduce the values in Table 8.3 with step size 0.5. Compare with the step size 0.1 values shown in the table. Why are they different?

42. A 2000-gallon aquarium can support no more than 150 guppies. Six guppies are introduced into the aquarium. Assume that the rate of guppy population growth is directly proportional to the population y and limiting factor $150 - y$ at any time t (weeks) with proportionality constant $k = 0.0015$.
a) Determine the guppy population $y(t)$ as an explicit function of time t.

b) Draw a complete graph of $y(t)$.

c) What values of y and t make sense in the problem situation?

d) Draw a graph of the problem situation.

e) When is the guppy population 100? 125?

f) When should the guppy population reach the aquarium's capacity?

43. *Logistic growth.* A certain wild animal preserve is known to be able to support no more than 250 lowland gorillas. Twenty-eight gorillas were known to be in the preserve in 1970. Assume that the gorilla population growth from 1970 is logistic in nature. That is, we assume that the rate of growth is directly proportional to both the population y and the limiting factor $250 - y$ at any time t (years) with proportionality constant $k = 0.0004$.

a) Determine the population of gorillas $y(t)$ as an explicit function of time t.

b) Draw a complete graph of $y(t)$.

c) What values of y and t make sense in the problem situation?

d) Draw a graph of the problem situation.

e) When is the gorilla population 100? 200?

f) When should the gorilla population reach the preserve's capacity?

44. Prove that the unbounded area for $x > 0$ under the curve given by the witch of Agnesi,

$$y = \frac{a^3}{a^2 + x^2} \qquad (a > 0),$$

is finite. What is the area?

45. For Exercise 44, find the value(s) of a for which the unbounded area is exactly 1.

46. *The path of least time.* A steel ball is rolled down the decreasing curve paths given by the parametric equations $(x_i(t), y_i(t)), 0 \le t \le 1$, listed below. Each path starts at $(0, 0)$ and ends at $(\pi, -2)$. The time T it takes the steel ball to roll down the path is given by

$$T = \frac{1}{\sqrt{g}} \int_0^1 \sqrt{\frac{(x'(t))^2 + (y'(t))^2}{-2y(t)}} \, dt$$

(problem posed by Jerry Johnson, Oklahoma State University).

$$(x_1(t), y_1(t)) = (\pi t, -2t)$$

$$(x_2(t), y_2(t)) = (\pi t, -2\sqrt{t})$$

$$(x_3(t), y_3(t)) = (\pi t - \sin \pi t, -1 + \cos \pi t)$$

$$(x_4(t), y_4(t)) = (\pi t, 2(t - 1)^2 - 2)$$

$$(x_5(t), y_5(t)) = (\pi t, -2\sqrt{2t - t^2})$$

a) Verify that each path is decreasing for $0 \le t \le 1$. Also verify that $(x_i(0), y_i(0)) = (0, 0)$ and $(x_i(1), y_i(1)) = (\pi, -2)$ for each i.

b) Draw each path using a grapher.

c) Write the explicit time integral T for each path. Identify any integrals that are improper.

d) Evaluate T for path $(x_1(t), y_1(t))$ analytically.

e) Evaluate each time integral T using a NINT computation. Justify the use of NINT with the improper integrals. (*Hint:* Choose a very small lower integration limit.)

f) In advanced mathematics, it can be proven that path $(x_3(t), y_3(t))$, which is called a *brachistochrone*, is the "path of least time." Confirm *analytically* that the exact integral value for the least time is $\pi/\sqrt{g}$. In part (e), you should have discovered that other paths seem to produce

times less than $\pi/\sqrt{g}$. Give a geometric argument why some of the NINT computations give bad information in this case. (This is another example of why numerical methods will never replace the need for analytic methods!)

47. Devise a model and a related business application scenario that involve a logistic growth function and its derivative. Explain the derivative of the logistic growth function and the point of inflection of the logistic growth curve in terms of your business problem scenario.

8.8 _____ Computer Algebra Systems (CAS)

Computer algebra systems have powerful symbolic algorithms for evaluating complicated integrals when they have explicit antiderivatives.

Examples of CAS Output

Examples 1 and 2 display actual output from *Derive*® and *Mathematica*®. In the past, the integrals in these examples were typically evaluated by using integral tables.

EXAMPLE 1

Evaluate $\int x \sin^{-1} x \, dx$.

Solution   *Derive*®

1:   x ASIN (x)

2:   $\int$ x ASIN (x) dx

3:   $\left[x^2/2 - 1/4\right]$ ASIN (x) $+ \dfrac{x\sqrt{(1-x^2)}}{4}$

Mathematica®

In[11]:=

  Integrate[x ArcSin[x], x]

Out[11]=

  $\dfrac{\text{x Sqrt}[1 - x^2]}{4} - \dfrac{\text{ArcSin }[x]}{4} + \dfrac{x^2 \text{ ArcSin }[x]}{2}$

Notice that the CAS does not give the constant of integration.   ≡

In Example 1, the value of $\int x \sin^{-1} x \, dx$ is given in line 3 and OUT [11].

EXAMPLE 2

Evaluate $\int x \sin^4 x \, dx$.

In Example 2, the value of $\int x \sin^4 x \, dx$ is given in line 3 and OUT[12].

Solution *Derive*®

1: $x \; \text{SIN} \; (x)^4$

2: $\int x \; \text{SIN} \; (x)^4 dx$

3: $\dfrac{\text{COS} \; (x)^4}{16} - \dfrac{5 \; \text{COS} \; (x)^2}{16}$

$\qquad - \text{COS} \; (x) \left[\dfrac{x \; \text{SIN} \; (x)^3}{4} + \dfrac{3x \; \text{SIN} \; (x)}{8} \right] + \dfrac{3x^2}{16}$

Mathematica™

    In[12]:=

      Integrate[x(Sin[x])^4,x]

    Out[12]=

$\dfrac{3x^2}{16} - \dfrac{\text{Cos}[2 \; x]}{8} + \dfrac{\text{Cos}[4 \; x]}{128} - \dfrac{x \; \text{Sin}[2x]}{4} + \dfrac{x \; \text{Sin}[4x]}{32}$ ≡

EXPLORATION 1

Working with *Derive*® and *Mathematica*® Solutions

Notice the dramatically different symbolic form of the *Derive*® and Mathematica® solutions to Example 2.

1. Is $\dfrac{3x^2}{16} - \dfrac{\cos 2x}{8} + \dfrac{\cos 4x}{128} - \dfrac{x \sin 2x}{4} + \dfrac{x \sin 4x}{32} =$

$\dfrac{\cos^4 x}{16} - \dfrac{5 \cos^2 x}{16} - \cos x \left(\dfrac{x \sin^3 x}{4} + \dfrac{3x \sin x}{8} \right) + \dfrac{3x^2}{16}$? Explain.

2. Let $y_1 = x \sin^4 x$. Compute $\int_{-2}^{3} y_1 \, dx$ using an analytic result of Example 2. Compute NINT $(y_1, -2, 3)$. Compare the two values.

3. Let $f(x) = \int_0^x t \sin^4 t \, dt$, and draw its graph. Discuss the number of solutions to the equation $f(x) = 5$. Estimate each solution with error of at most 0.1.

4. Solve for x: $\int_1^x t \sin^4 t \, dt = 0$.

Nonelementary Integrals

The development of computer programs that evaluate indefinite integrals by symbolic manipulation has led to a renewed interest in determining which integrals can be expressed as finite algebraic combinations of elementary functions (the functions we have been studying) and which require infinite series (Chapter 9) or numerical methods (NINT) for their evaluation. Examples of the latter include the *error function*

$$\text{erf}\,(x) = \frac{2}{\sqrt{\pi}} \int_0^x e^{-t^2}\, dt$$

and integrals like

$$\int \sin x^2 \, dx \qquad \text{and} \qquad \int \sqrt{1 + x^4}\, dx$$

that arise in engineering and physics. These, and a number of others, like

$$\int \frac{e^x}{x}\, dx, \quad \int e^{(e^x)}\, dx, \quad \int \frac{1}{\ln x}\, dx, \quad \int \ln\,(\ln x)\, dx, \quad \int \frac{\sin x}{x}\, dx,$$

$$\int \sqrt{1 - k^2 \sin^2 x}\, dx, \quad (0 < k < 1),$$

look so easy that they tempt us to try them just to see how they turn out. *It can be proved, however, that there is no way to express these integrals as finite combinations of elementary functions. The same applies to integrals that can be changed into these by substitutions.* Thus, we do not know an explicit analytic form of any of the antiderivatives. However, we can GRAPH the antiderivative $y = \int_0^x f(t)\, dt + C$. For example, even though we cannot write down an antiderivative of $\sin x^2$, we can GRAPH an antiderivative accurately, and thus, in a real sense, we *know* the antiderivative.

EXAMPLE 3

a) GRAPH

$$y = \int_0^x \sin t^2 \, dt.$$

b) Solve the equation

$$\int_0^x \sin t^2 \, dt = 1$$

for $|x| \le 5$.

Solution Complete graphs of $y_1 = \text{NINT}\,(\sin t^2, 0, x)$ and $y_2 = 1$ are shown in Fig. 8.39 for $|x| \le 5$. Clearly, there is no solution to the equation for $|x| \le 5$ because the graphs of y_1 and y_2 never intersect.   ☰

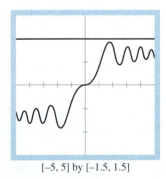

[−5, 5] by [−1.5, 1.5]

8.39 Complete graphs of $y_1 = \int_0^x \sin t^2 \, dt$ and $y_2 = 1$ for $|x| \le 5$.

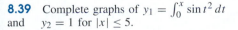

EXPLORATION 2

Sketching the Unknown

1. GRAPH $y = \sin x^2$ in $[0, 3]$.

2. Pretend you haven't seen the graph in Fig. 8.39. Based only on your knowledge of the graph of $y = \sin x^2$ in $[0, 3]$, use a sketch to predict the behavior of $f(x) = \int_0^x \sin t^2 \, dt$ for $0 \le x \le 3$.

3. Compare your sketch to Fig. 8.39.   &

Exercises 8.8

Use a computer algebra system or the integral formulas in the back of this book to evaluate analytically the integrals in Exercises 1–16. Support either method with NINT computations.

1. $\displaystyle\int_0^\infty e^{-x^2} \, dx$

2. $\displaystyle\int x \cos^{-1} x \, dx$

3. $\displaystyle\int_6^9 \frac{dx}{x\sqrt{x-3}}$

4. $\displaystyle\int_0^{1/2} x \tan^{-1} 2x \, dx$

5. $\displaystyle\int \frac{dx}{(9-x^2)^2}$

6. $\displaystyle\int_4^{10} \frac{\sqrt{4x+9}}{x^2} \, dx$

7. $\displaystyle\int_3^{11} \frac{dx}{x^2\sqrt{7+x^2}}$

8. $\displaystyle\int \frac{dx}{x^2\sqrt{7-x^2}}$

9. $\displaystyle\int_{-2}^{-\sqrt{2}} \frac{\sqrt{x^2-2}}{x} \, dx$

10. $\displaystyle\int_{-\pi/12}^{\pi/4} \frac{dx}{5+4\sin 2x}$

11. $\displaystyle\int \frac{dx}{4+5\sin 2x}$

12. $\displaystyle\int_3^6 \frac{x}{\sqrt{x-2}} \, dx$

(*Hint for 12:* The n in Formula 7 in the back of this book need not be an integer)

13. $\displaystyle\int x\sqrt{2x-3} \, dx$

14. $\displaystyle\int \frac{\sqrt{3x-4}}{x} \, dx$

15. $\displaystyle\int_0^\infty x^{10} e^{-x} \, dx$

16. $\displaystyle\int_0^1 x^2 \tan^{-1} x \, dx$

The following integrals in Exercises 17–24 have no explicit analytic antiderivatives. Determine a complete graph of each function in the specified interval. If you can determine that the function is even or odd, take advantage of the fact to shorten the calculations.

17. $y = \displaystyle\int_{0.01}^x \frac{2}{\sqrt{t}} e^{-t^2} \, dt, 0.01 \le x \le 5$

18. $y = \displaystyle\int_0^x \cos t^2 \, dt, |x| \le 5$

19. $y = \displaystyle\int_{0.01}^x \frac{e^t}{t} \, dt, 2 \le x \le 5$

20. $y = \displaystyle\int_{0.01}^x \frac{dt}{\ln t}, 0.01 \le x \le 0.99$

21. $y = \displaystyle\int_0^x \frac{3\sin t}{t} \, dt, |x| \le 5$

22. $y = \displaystyle\int_0^x \sqrt{1 - \frac{1}{2}\sin^2 t} \, dt, |x| \le 5$

23. $y = \displaystyle\int_0^x \sqrt{1+t^4} \, dt, |x| \le 5$

24. $y = \displaystyle\int_0^x t^2 \tan^5 t \, dt, |x| \le 1.5$

In Exercises 25–28, solve the equation $y = 1/2$ for $|x| \le 3$.

25. $y = \displaystyle\int_0^x \sin^{-1} \sqrt{t} \, dt$

26. $y = \displaystyle\int_{\pi/4}^x \sqrt{1+\sin^2 t} \, dt$

27. $y = \displaystyle\int_0^x \sqrt{1+t^4} \, dt$

28. $y = \displaystyle\int_0^x t^2 \tan^5 t \, dt$

29. Predict the behavior of $\int_0^x t e^{-t} \, dt$ for $x > 0$ based only on the graph of $y = xe^{-x}$. Support using NINT.

30. Predict the behavior of $\int_{0.1}^x \left(2 + \frac{1}{t} - \ln\left(25 - t^2\right)\right) dt$ for $x > 0$ based only on the graph of $y = 2 + \frac{1}{x} - \ln\left(25 - x^2\right)$. Support using NINT.

31. Find the centroid of the region cut from the first quadrant by the curve $y = 1/\sqrt{x+1}$ and the line $x = 3$.

32. A thin plate of constant density $\delta = 1$ occupies the region enclosed by the curve $y = 36/(2x+3)$ and the line $x = 3$ in the first quadrant. Find the moment of the plate about the y-axis.

Exercises 33–36 refer to the formulas at the back of this book by number.

33. Verify Formula 55 for $x > 0$ by differentiating the right side.

34. Verify Formula 76 by differentiating the right side.

35. Verify Formula 9 by integrating

$$\int \frac{x\,dx}{(ax+b)^2}$$

with the substitution $u = ax + b$.

36. Verify Formula 46 by integrating

$$\int \frac{dx}{x^2\sqrt{x^2 - a^2}}$$

with the substitution $x = a \sec u$.

Review Questions

1. Name some of the general methods for finding indefinite integrals?

2. What substitution(s) would you consider if the integrand contained the following terms?

   a) $\sqrt{x^2 + 9}$ **b)** $\sqrt{x^2 - 9}$

   c) $\sqrt{9 - x^2}$ **d)** $\sin^3 x \cos^2 x$

   e) $\sin^2 x \cos^2 x$

3. What analytic method(s) would you try if the integrand contained the following terms?

   a) $\sin^{-1} x$ **b)** $\ln x$

   c) $\sqrt{1 + 2x - x^2}$ **d)** $x \sin x$

   e) $\dfrac{2x + 3}{x^2 - 5x + 6}$ **f)** $\sin 5x \cos 3x$

   g) $\dfrac{1 - \sqrt{x}}{1 + \sqrt[4]{x}}$ **h)** $x\sqrt{2x + 3}$

4. Discuss several types of improper integrals. Define convergence and divergence of each type. How do you test for convergence and divergence of improper integrals? Give examples.

5. What is a differential equation?

6. Describe the method of separation of variables for solving first-order differential equations.

7. What is a solution of a differential equation?

8. Describe the Euler, improved Euler, and Runge–Kutta methods for solving the first-order initial value problem $y' = f(x, y)$, $y(a) = y_0$ over an interval $a \le x \le b$.

9. Explain how solutions to differential equations are related to integrals.

10. Explain how solutions to differential equations are related to initial value problems.

Practice Exercises

Evaluate the integrals in Exercises 1–42. Support with a NDER or NINT computation as appropriate.

1. $\displaystyle\int_0^{\pi/2} \frac{\cos x\,dx}{\sqrt{1 + \sin x}}$ **2.** $\displaystyle\int_{\pi^2/16}^{\pi^2/9} \frac{\tan\sqrt{x}}{2\sqrt{x}}\,dx$

3. $\displaystyle\int_{-1}^{1} \frac{2y}{y^4 + 1}\,dy$ **4.** $\displaystyle\int \sec x \tan x\, e^{\sec x}\,dx$

5. $\displaystyle\int_0^{\sqrt{2}/2} \frac{\sin^{-1} x}{\sqrt{1 - x^2}}\,dx$ **6.** $\displaystyle\int_1^e \frac{dx}{x(2 + \ln x)}$

7. $\displaystyle\int_{\pi/4}^{\pi/3} \frac{dx}{2 \sin x \cos x}$ **8.** $\displaystyle\int_0^{\pi/6} \frac{2\,dt}{\cos^2 t - \sin^2 t}$

9. $\displaystyle\int \frac{x + 4}{x^2 + 1}\,dx$ **10.** $\displaystyle\int \frac{x + 2}{\sqrt{1 - x^2}}\,dx$

11. $\displaystyle\int x^2 \ln x\,dx$ **12.** $\displaystyle\int_0^1 \ln(x + 1)\,dx$

13. $\displaystyle\int x^5 \sin x\,dx$ **14.** $\displaystyle\int \frac{2\tan^{-1} x}{x^2}\,dx$

15. $\displaystyle\int e^x \cos 2x\,dx$ **16.** $\displaystyle\int e^{-x} \sin x\,dx$

17. $\displaystyle\int \sin^3 y\,dy$ **18.** $\displaystyle\int \sin^3 y \cos^2 y\,dy$

19. $\displaystyle\int \sin^4 x \cos^2 x\,dx$ **20.** $\displaystyle\int \sin^3 x \cos^3 x\,dx$

21. $\displaystyle\int_0^{\pi} \sqrt{\frac{1 + \cos 2x}{2}}\,dx$ **22.** $\displaystyle\int_{\pi/4}^{3\pi/4} \sqrt{\cot^2 t + 1}\,dt$

23. $\displaystyle\int_0^{\pi/3} \tan^3 t\,dt$ **24.** $\displaystyle\int_{-\pi/4}^{\pi/4} 6 \sec^4 t\,dt$

25. $\displaystyle\int_0^3 \frac{dz}{(16 + z^2)^{3/2}}$ **26.** $\displaystyle\int_0^{\sqrt{3}} \frac{z^3 + z}{\sqrt{1 + z^2}}\,dz$

27. $\displaystyle\int \frac{dx}{x^2\sqrt{1 - x^2}}$ **28.** $\displaystyle\int \frac{x^2\,dx}{\sqrt{1 - x^2}}$

29. $\displaystyle\int_{5/4}^{5/3} \frac{12\,dx}{(x^2 - 1)^{3/2}}$ **30.** $\displaystyle\int_5^6 \frac{dx}{\sqrt{x^2 - 9}}$

31. $\displaystyle\int_{1/3}^{1} \frac{3\,dx}{9x^2 - 6x + 5}$

32. $\displaystyle\int_{-1}^{-1/2} \frac{dx}{\sqrt{-2x - x^2}}$

33. $\displaystyle\int_{0}^{1} \frac{dx}{(x+1)\sqrt{x^2 + 2x}}$

34. $\displaystyle\int_{-2}^{-1} \frac{2\,dx}{x^2 + 4x + 5}$

35. $\displaystyle\int_{2}^{6} \frac{x^3 + x^2}{x^2 + x - 2}\,dx$

36. $\displaystyle\int_{2}^{3} \frac{x^3 + 1}{x^3 - x}\,dx$

37. $\displaystyle\int \frac{x\,dx}{(x-1)^2}$

38. $\displaystyle\int \frac{8\,dx}{x^3(x+2)}$

39. $\displaystyle\int \frac{4\,dx}{x^3 + 4x}$

40. $\displaystyle\int_{0}^{1} \frac{x\,dx}{x^4 - 16}$

41. $\displaystyle\int_{3}^{\infty} \frac{2\,dx}{x^2 - 2x}$

42. $\displaystyle\int_{1}^{\infty} \frac{\ln x\,dx}{x^2}$

43. Find the moment of inertia about the x-axis of a thin plate of constant density $\delta = 2$ covering the region enclosed by the lines $x = 1$ and $y = 1$ and the curve $y = \ln x$ in the first quadrant.

44. Find the area of the region in the first quadrant bounded by the x-axis, the line $x = 2$, and the curve $y = \ln x$.

45. Find the volume of the solid generated by revolving about the y-axis the region bounded by the x-axis and the curve $y = 3x\sqrt{1 - x}$, $0 \le x \le 1$.

46. Find the length of the segment of the parabola $y = x^2$ that extends from $x = 0$ to $x = \sqrt{3}/2$.

47. Find the length of the curve $y = \sqrt{x^3 + 5x}$ from $x = 1$ to $x = 8$.

48. Find the centroid of the region bounded by the curves $y = \pm(1 - x^2)^{-1/2}$ and the lines $x = 0$ and $x = 1$.

49. The infinite region bounded by the coordinate axes and the curve $y = -\ln x$ in the first quadrant is revolved about the x-axis to generate a solid. Find the volume of the solid.

50. The region between the curve $y = x \ln x$ and the x-axis from $x = 0$ to $x = 2$ is revolved about the x-axis to generate a solid as shown. Find the volume of the solid.

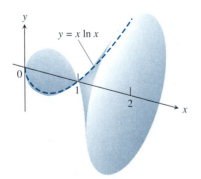

$y = x \ln x$

Which of the integrals in Exercises 51–54 converge, and which diverge?

51. $\displaystyle\int_{-\infty}^{\infty} \frac{4x\,dx}{x^2 + 1}$

52. $\displaystyle\int_{-\infty}^{\infty} \frac{8\,dx}{2x^2 + 3}$

53. $\displaystyle\int_{-\infty}^{\infty} \frac{e^{-x}\,dx}{e^{-x} + e^{x}}$

54. $\displaystyle\int_{-\infty}^{\infty} \frac{x^3\,dx}{1 + e^{x}}$

Use IMPEULG to graph the solution to the initial value problems in Exercises 55–62. Estimate the function value at the specified value of x.

55. $e^{y-2}\,dx - e^{x+2y}\,dy = 0$, $y(0) = -2$, $x = 2$

56. $y \ln y\,dx + (1 + x^2)\,dy = 0$, $y(0) = e$, $x = 2$

57. $\dfrac{dy}{dx} = \dfrac{x^2 + y^2}{2xy}$, $y(1) = -2$, $x = 6$

58. $\dfrac{dy}{dx} = \dfrac{y(1 + \ln y - \ln x)}{x(\ln y - \ln x)}$, $y(3) = 0.1$, $x = 5$

59. $(x^2 + y)\,dx + (e^{y} + x)\,dy = 0$, $y(0) = 0$, $x = 3$

60. $(e^{x} + \ln y)\,dx + \left(\dfrac{x + y}{y}\right)\,dy = 0$, $y(\ln 2) = 1$, $x = 1$

61. $(x + 1)\dfrac{dy}{dx} + 2y = x$, $y(0) = 1$, $x = 3$

62. $x\dfrac{dy}{dx} + 2y = x^2 + 1$, $y(1) = 1$, $x = 6$

Determine analytically the solution to the initial value problem in Exercises 63–66, and draw a complete graph of the solution. Support with a graphical solution using IMPEULG.

63. $y' = -0.15(y - 22)$, $y(0) = 50$

64. $\dfrac{dx}{dt} = 0.03(x - 55)$, $x(0) = 50$

65. $\dfrac{dP}{dt} = 0.002P(500 - P)$, $P(0) = 20$

66. $\dfrac{dP}{dt} = 0.0055P(200 - P)$, $P(0) = 50$

67. *Autocatalytic reactions.* The equation that describes the autocatalytic reaction in Section 4.3, Exercise 56, can be written as $dx/dt = kx(a - x)$. Read Exercise 56 for background (there is no need to do the exercise). Then solve the differential equation above to find x as a function of t. Assume that $x = x_0$ when $t = 0$.

Infinite Series

INFINITE POLYNOMIALS

A *finite polynomial* is an algebraic expression of the form

$$c_0 + c_1x + c_2x^2 + \cdots + c_nx^n, c_n \neq 0$$

(Section 1.5).
An *infinite polynomial* looks like this:

$$c_0 + c_1x + c_2x^2 + \ldots + c_nx^n + \ldots$$

We build to its meaning in the first several sections of this chapter.

OVERVIEW In this chapter, we study infinite polynomials called power series and, as a product of our study, develop Taylor's formula, one of the most remarkable formulas in all of mathematics. The formula does two things for us. It shows how to calculate the value of some infinitely differentiable functions like e^x at any point, just from its value and the values of its derivatives at the origin. As if that were not enough, the formula gives us *polynomial approximations of differentiable functions* of any order we want, along with their error formulas, all in a single equation. Power series have many additional uses. They provide an efficient way to evaluate integrals without explicit antiderivatives, and they solve differential equations that give insight into heat flow, vibration, chemical diffusion, and signal transmission. What you will learn here sets the stage for the roles played by series of functions of all kinds in science and mathematics.

9.1 Limits of Sequences of Numbers

This section describes what it means for an infinite sequence of numbers to have a limit and shows how to find the limits of many of the sequences that arise in mathematics and applied fields.

Informally, a sequence is an ordered collection of things, but in this chapter the things will usually be numbers. We have seen sequences before, such as sequences $x_0, x_1, \cdots, x_n, \cdots$ of numerical approximations generated by Newton's method and the sequence $A_3, A_4, \cdots, A_n, \cdots$ of areas of n-sided regular polygons used to define the area of a circle. These sequences have limits, but many equally important sequences do not. The sequence $1, 2, 3, \cdots, n, \cdots$ of positive integers has no limit, nor does the sequence $2, 3, 5, 7, 11, 13, \cdots$ of prime numbers. Frequently, we need to know when sequences do and do not have limits and how to find the limits when they exist. As with functions, we can use a grapher to suggest what a limiting value may be, and then we can confirm the limit analytically with theorems based on a formal definition. Formally, we will see a close parallel between what we do here and what we did with limits of functions in Chapter 2. Graphically, we will "see" limits of infinite sequences in a couple ways.

Definitions and Notation

Our starting point is a formal definition of sequence.

DEFINITION

An **infinite sequence** (or **sequence**) of numbers is a function whose domain is the set of integers greater than or equal to some integer n_1.

Usually, n_1 is 1, and the domain is the set of all positive integers. But sometimes we want to start our sequences elsewhere. We might take $n_1 = 0$, for instance, when we begin Newton's method or take $n_1 = 3$ when we work with n-sided polygons.

Sequences are defined the way other functions are, some typical functions being

$$a(n) = n - 1, \qquad a(n) = 1 - \frac{1}{n}, \qquad a(n) = \frac{\ln n}{n^2}$$

for the sequences

$$0, 1, 2, 3, \ldots, \qquad \frac{1}{2}, \frac{2}{3}, \frac{3}{4}, \ldots, \qquad 0, \frac{\ln 2}{4}, \frac{\ln 3}{9}, \frac{\ln 4}{16}, \ldots,$$

respectively.

To indicate that the domains are sets of integers, we use a letter like n from the middle of the alphabet for the independent variable, instead of the x, y, z, and t used so widely in other contexts. The formulas in the defining rules, however, like the ones above, are often valid for domains much larger than the set of positive integers. This can be an advantage, as we will see.

When $n_1 = 1$, the number $a(n)$ is the **nth term** of the sequence, or the **term with index n.** For example, if $a(n) = (n - 1)/n$, then the terms are

First term	Second term	Third term	nth term
$a(1) = 0,$	$a(2) = \dfrac{1}{2},$	$a(3) = \dfrac{2}{3}, \cdots,$	$a(n) = \dfrac{n-1}{n}.$

When we use the subscript notation a_n for $a(n)$, this sequence becomes

$$a_1 = 0, \qquad a_2 = \frac{1}{2}, \qquad a_3 = \frac{2}{3}, \cdots, \qquad a_n = \frac{n-1}{n}.$$

To describe sequences, we often write the first few terms as well as a formula for the nth term.

EXAMPLE 1

We write	For the sequence whose defining rule is
$0, 1, 2, \ldots, n-1, \ldots$	$a_n = n - 1$
$1, \dfrac{1}{2}, \dfrac{1}{3}, \cdots, \dfrac{1}{n}, \cdots$	$a_n = \dfrac{1}{n}$
$1, -\dfrac{1}{2}, \dfrac{1}{3}, -\dfrac{1}{4}, \cdots, (-1)^{n+1}\dfrac{1}{n}, \cdots$	$a_n = (-1)^{n+1}\dfrac{1}{n}$
$0, \dfrac{1}{2}, \dfrac{2}{3}, \dfrac{3}{4}, \cdots, \dfrac{n-1}{n}, \cdots$	$a_n = \dfrac{n-1}{n}$
$-\dfrac{1}{2}, \dfrac{2}{3}, -\dfrac{3}{4}, \dfrac{4}{5}, \cdots, (-1)^n \dfrac{n}{n+1}, \cdots$	$a_n = (-1)^n \dfrac{n}{n+1}$
$3, 3, 3, \ldots, 3, \ldots$	$a_n = 3$

ALTERNATING SIGNS

Notice in two of the sequences how the use of $(-1)^{n+1}$ and $(-1)^n$ in the defining rules give the alternating $+$ and $-$ signs of the terms.

Notation We refer to the sequence whose nth term is a_n with the notation $\{a_n\}$ ("the sequence a sub n"). The second sequence in Example 1 is $\{1/n\}$ ("the sequence 1 over n"); the last sequence is $\{3\}$ ("the sequence 3").

EXPLORATION 1

Graphing Sequences

We can investigate the behavior of sequences with graphs, just as we do for other functions. We use the sequence

$$\{a_n\} = \left\{\frac{n-1}{n}\right\} = 0, \frac{1}{2}, \frac{2}{3}, \frac{3}{4}, \cdots$$

to show three ways to visualize sequences. An important fourth way will be shown in Exploration 4.

In parametric mode, dot format

1. **a)** Let

$$x(t) = \frac{t-1}{t}, \qquad y(t) = c,$$

   and GRAPH with tMin = 1, t-step = 1. This plots the terms a_n on the line $y = c$ ($y = 2$ in Fig. 9.1a). Try it, and watch how the pixels appear in the viewing window.

   b) For a variation of this method, make the sequence plot along the line $x = c$.

2. Let

$$x(t) = t, \qquad y(t) = \frac{t-1}{t},$$

   and GRAPH with tMin = 1, t-step = 1 (Fig. 9.1b). Explain how this compares to the methods in parts 1(a) and 1(b).

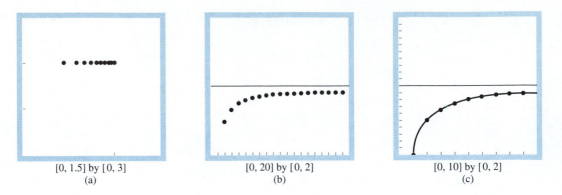

[0, 1.5] by [0, 3]   [0, 20] by [0, 2]   [0, 10] by [0, 2]
(a)   (b)   (c)

9.1   Three visualizations of the sequence $\{(n-1)/n\}$ as (a) points on the horizontal line $y = 2$, (b) points (n, a_n), and (c) points embedded in the graph of $y = (x-1)/x$ for $x = 1, 2, 3, \cdots$. We included the graph of $y = 1$ in (b) and (c), and we sketched in the dots in (c).

In function mode, connected format

3. GRAPH $y = (x-1)/x$ with x-scale $= 1$ (Fig. 9.1c). Explain how this compares to the method in part 2. In particular, explain where the graph of the sequence is in your viewing window.

TEST YOURSELF. Try to graph the sequence

$$\left\{ (-1)^{n+1} \frac{n-1}{n} \right\} = 0, \ -\frac{1}{2}, \ \frac{2}{3}, \ -\frac{3}{4}, \ \cdots$$

using each method above. Explain your results. Explain what happens when you TRACE on your graphs.   ♧

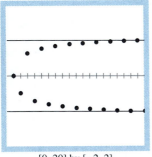

[0, 20] by [−2, 2]

9.2   A graph of the sequence

$$\left\{ (-1)^{n+1} \frac{n-1}{n} \right\}$$

and the lines $y = \pm 1$. Reproduce this on your grapher. TRACE the graph, and watch the terms alternate signs. Note that as n increases, the positive terms approach 1 and the negative terms approach -1. We say that the sequence $\{a_n\}$ diverges.

Convergence and Divergence

Your viewing windows for Exploration 1 (or ours in Fig. 9.1) suggest that the terms of the sequence $\{(n-1)/n\}$ approach a limiting value of 1 as n increases.

On the other hand, the terms of the sequence

$$\left\{ (-1)^{n+1} \frac{n-1}{n} \right\}$$

seem to accumulate near two different values -1 and 1. See Fig. 9.2. The dots in the viewing window seem to cluster around the horizontal lines $y = 1$ and $y = -1$ as $n \to \infty$.

To distinguish sequences that approach a unique limiting value L as n increases from those that do not, we say that those former sequences *converge*, according to the following definition.

What does it mean visually for a sequence to *converge* when we depict it as a set of (n, a_n) values in the coordinate plane? Look at Fig. 9.1(b). The most obvious visual message is that the dots are evenly spaced, representing the term-by-term increment of n. Less obvious, *but much more important*, is that the dots cluster about a horizontal line $y = c$ as $n \to \infty$. This is the visual suggestion that $a_n \to c$.

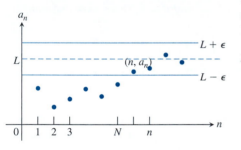

9.3 When $a_n \to L$, all the a_n's after a_N lie within ϵ of L.

DEFINITIONS

The sequence $\{a_n\}$ **converges** to the number L if to every positive number ϵ there corresponds an integer N such that for all n,

$$n > N \Rightarrow |a_n - L| < \epsilon.$$

If no such limit exists, we say that $\{a_n\}$ **diverges**.

If $\{a_n\}$ converges to L, we write $\lim_{n \to \infty} a_n = L$, or simply $a_n \to L$, and call L the **limit** of the sequence. (See Fig. 9.3.)

EXPLORATION 2

Limits of Sequences

For each of the four sequences whose nth term is given, construct a visualization suggested by Exploration 1. What do your viewing windows suggest about convergence for each sequence as n increases?

1. $a_n = 3$ **2.** $a_n = n - 1$

3. $a_n = \dfrac{1}{n}$ **4.** $a_n = \dfrac{(-1)^{n+1}}{n}$

Part 3 of Exploration 2 should suggest that $\{1/n\}$ converges to 0, or

$$\lim_{n \to \infty} \frac{1}{n} = 0.$$

To *support* this numerically, we must find, for a given number $\epsilon > 0$, an integer value for N so that all values of n greater than N will have their reciprocal values, $1/n$, within ϵ units of 0 (Exploration 3). To *confirm* the limit, we must find, for any value of $\epsilon > 0$, a representation for N so that $n > N$ will yield $1/n < \epsilon$ (Example 2).

To show that

$$\lim_{x \to c} f(x) = L,$$

we must find, for any ϵ-interval about L, a δ-interval about c such that x in the δ-interval implies that $f(x)$ is in the ϵ-interval. There is an analogy to limits at ∞. To show that

$$\lim_{x \to \infty} a_n = L,$$

we must find, for any ϵ-interval about L, a "δ-interval about ∞" such that x in the δ-interval implies that a_n is in the ϵ-interval. Can you tell exactly what is meant by a "δ-interval about ∞"?

EXPLORATION 3

Given ϵ, Find N

1. Represent the sequence $\{1/n\}$ parametrically with $x_1(t) = t$, $y_1(t) = 1/t$. GRAPH $\{1/n\}$ in dot format in a $[0, 20]$ by $[-0.1, 0.2]$ viewing window with tMin $= 1$ and t-step $= 1$.

2. Let $\epsilon = 0.1$. GRAPH $y = 0.1$ ($x_2(t) = t$, $y_2(t) = 0.1$) in the same viewing window as part 1. TRACE on the graph of $\{1/n\}$ to find the first value of n for which $1/n$ appears to be within ϵ-unit of 0.

3. Repeat part 2 with $\epsilon = 0.001$ and appropriate view dimensions.

EXAMPLE 2

Show analytically that $\lim_{n \to \infty} 1/n = 0$.

Solution We set $a_n = 1/n$ and $L = 0$ in the definition of convergence. To show that $1/n \to 0$, we must show that for any $\epsilon > 0$, there exists an

N such that

$$n > N \quad \Rightarrow \quad \left| \frac{1}{n} - 0 \right| < \epsilon.$$

This implication will hold for all n for which

$$\frac{1}{n} < \epsilon \qquad \text{or, equivalently,} \qquad n > \frac{1}{\epsilon}.$$

Thus, we choose N to be some integer greater than $1/\epsilon$ (the integer $[1/\epsilon]+1$, for example). Then any n greater than N will also be greater than $1/\epsilon$, and implication (1) will hold.   ≡

EXAMPLE 3

Show that if k is any number, then the sequence $\{k\}$ of constant values converges to k.

Solution   We set $a_n = k$ and $L = k$ in the definition of convergence. To show that $a_n \to k$, we must show that for any $\epsilon > 0$, there exists an integer N such that

$$n > N \quad \Rightarrow \quad |k - k| < \epsilon.$$

This implication holds for any integer N because $|k - k| = 0$ is less than every positive ϵ for all n.   ≡

The last activity of Exploration 1 suggests that the terms of the sequence $\{(-1)^{n+1}((n-1)/n)\}$ do not seem to have a particular limiting value. Example 4 confirms that this sequence diverges.

EXAMPLE 4

Confirm that $\{(-1)^{n+1}((n-1)/n)\}$ diverges.

Solution   Take a positive ϵ smaller than 1 so that the bands shown in Fig. 9.4 about the lines $y = 1$ and $y = -1$ do not overlap. Any $\epsilon < 1$ will do.

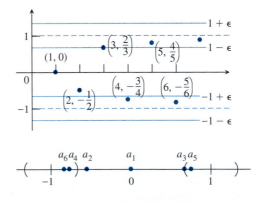

9.4   Neither the ϵ-interval about 1 nor the ϵ-interval about -1 contains a complete tail of the sequence

$$\left\{ (-1)^{n+1} \frac{n-1}{n} \right\}.$$

The sequence diverges.

Convergence to 1 would require every point of the graph beyond a certain index N to lie inside the upper band, but this will never happen. As soon as a point (n, a_n) lies within the upper band, every alternate point starting with $(n + 1, a_{n+1})$ will lie within the lower band. Hence, the sequence cannot converge to 1. Likewise, it cannot converge to -1. On the other hand, because the terms of the sequence get alternately closer to 1 and -1, they never accumulate near any other value. Therefore, the sequence diverges.

A **tail** of a sequence $\{a_n\}$ is the collection of all terms whose indices are greater than some integer N. In other words, a tail is one of the sets $\{a_n \mid n > N\}$. Another way to say that $a_n \to L$ is to say that every ϵ-interval about L contains a tail of the sequence. The convergence or divergence of a sequence has nothing to do with how a sequence starts out. It depends only on how the tails behave.

The behavior of $\{(-1)^{n+1}(n-1)/n\}$ is qualitatively different from that of $\{n - 1\}$, which diverges because it outgrows every real number L. To describe the behavior of $\{n - 1\}$, we write

$$\lim_{n \to \infty} (n - 1) = \infty.$$

In speaking of infinity as a limit of a sequence $\{a_n\}$, we do not mean that the difference between a_n and infinity becomes small as n increases. We mean that a_n becomes numerically large as n increases.

EXPLORATION BIT

Give an argument explaining why the sequence

$$\left\{ (-1)^{n+1} \frac{n - 1}{n} \right\}$$

does not converge to two different limits, namely, -1 and 1. Your explanation can parallel the argument for the general case as shown at the right.

Limits Are Unique

A sequence cannot converge to two different limits.

If a sequence $\{a_n\}$ converges, then its limit is unique.

The argument is as follows: If $\{a_n\}$ converged to two different limits L_1 and L_2, we could take ϵ to be the positive number $|L_1 - L_2|/2$. Because $a_n \to L_1$, there would exist an integer N_1 such that

$$n > N_1 \quad \Rightarrow \quad |a_n - L_1| < \epsilon.$$

There would also exist an integer N_2 such that

$$n > N_2 \quad \Rightarrow \quad |a_n - L_2| < \epsilon.$$

For n greater than both N_1 and N_2, we would then have

$$|L_1 - L_2| = |L_1 - a_n + a_n - L_2|$$

$$\leq |L_1 - a_n| + |a_n - L_2| < \epsilon + \epsilon \qquad \text{Triangle inequality}$$

$$= 2(|L_1 - L_2|/2) = |L_1 - L_2|.$$

But this is absurd; no number is less than itself. Hence, if a sequence converges, its limit is unique.

Subsequences

If the terms of one sequence occur in their given order among the terms of a second sequence, we call the first sequence a **subsequence** of the second.

EXAMPLE 5   Subsequences of the Sequence of Positive Integers

a) The sequence $2, 4, 6, \ldots, 2n, \ldots$ of even integers.

b) The sequence $1, 3, 5, \ldots, 2n - 1, \ldots$ of odd integers.

c) The sequence $2, 3, 5, 7, 11, \ldots$ of primes.   ≡

Subsequences are important for two reasons. The second reason is based on the contrapositive of the first.

---

1. If a sequence converges to a limit L, then all subsequences also converge to L.
2. If any subsequence of a sequence diverges, or if two subsequences have different limits, then the original sequence diverges.

Part 1 suggests that if we know that a sequence converges, it may be more efficient for us to approximate or confirm its limit by using a judiciously chosen subsequence. Part 2 says that a sequence like $\{(-1)^{n-1}\}$ diverges because the subsequence $1, 1, 1, \ldots$ of odd-numbered terms converges to 1 while the subsequence $-1, -1, -1, \ldots$ of even-numbered terms converges to a different limit.

Useful Theorems

The study of limits would be a cumbersome business if we had to answer every question about convergence by applying the definition directly. Fortunately, three theorems and our graphing utility will move us beyond having to rely on the definition. The first two theorems parallel familiar theorems from Chapter 2.

THEOREM 1   **Properties of Limits**

If A and B are real numbers, and $\lim_{n\to\infty} a_n = A$ and $\lim_{n\to\infty} b_n = B$, then as $n \to \infty$,

1. *Sum Rule:* $\lim \{a_n + b_n\} = A + B$
2. *Difference Rule:* $\lim \{a_n - b_n\} = A - B$
3. *Product Rule:* $\lim \{a_n \cdot b_n\} = A \cdot B$
4. *Constant Multiple Rule:* $\lim \{k \cdot b_n\} = k \cdot B$ (any number k)
5. *Quotient Rule:* $\lim \dfrac{a_n}{b_n} = \dfrac{A}{B}$ (provided that $B \neq 0$)

By combining Theorem 1 with Examples 2 and 3, we can proceed immediately to

$$\lim_{n\to\infty}\left(-\frac{1}{n}\right) = -1\cdot\lim_{n\to\infty}\frac{1}{n} = -1\cdot 0 = 0,$$

$$\lim_{n\to\infty}\left(\frac{n-1}{n}\right) = \lim_{n\to\infty}\left(1-\frac{1}{n}\right) = \lim_{n\to\infty}1 - \lim_{n\to\infty}\frac{1}{n} = 1-0 = 1,$$

$$\lim_{n\to\infty}\frac{5}{n^2} = 5\cdot\lim_{n\to\infty}\frac{1}{n}\cdot\lim_{n\to\infty}\frac{1}{n} = 5\cdot 0\cdot 0 = 0,$$

$$\lim_{n\to\infty}\frac{4-7n^6}{n^6+3} = \lim_{n\to\infty}\frac{(4/n^6)-7}{1+(3/n^6)} = \frac{0-7}{1+0} = -7.$$

A general consequence of Theorem 1 is that every nonzero multiple of a divergent sequence $\{a_n\}$ diverges. Suppose $\{ca_n\}$ were to converge for some number $c \neq 0$. Then $(1/c)\{ca_n\} = \{a_n\}$ would converge by the Constant Multiple Rule—but it doesn't.

The next theorem is the sequence version of the Sandwich Theorem of Section 2.3.

THEOREM 2   The Sandwich Theorem for Sequences

If $a_n \leq b_n \leq c_n$ for all n beyond some index N, and if $\lim a_n = \lim c_n = L$, then $\lim b_n = L$ also.

An immediate consequence of Theorem 2 is that, if $|b_n| \leq c_n$ and $c_n \to 0$, then $b_n \to 0$ because $-c_n \leq b_n \leq c_n$. We use this fact in the next example.

EXAMPLE 6

SUPPORT

Support the analytic results of Example 6 graphically.

Since $1/n \to 0$, we know that

a) $\dfrac{\cos n}{n} \to 0$ because $0 \leq \left|\dfrac{\cos n}{n}\right| = \dfrac{|\cos n|}{n} \leq \dfrac{1}{n};$

b) $\dfrac{1}{2^n} \to 0$ because $0 \leq \dfrac{1}{2^n} \leq \dfrac{1}{n};$

c) $(-1)^n\dfrac{1}{n} \to 0$ because $0 \leq \left|(-1)^n\dfrac{1}{n}\right| \leq \dfrac{1}{n}.$ ≡

The application of Theorems 1 and 2 is broadened by a theorem stating that applying a continuous function to a convergent sequence produces a convergent sequence. We state the theorem without proof.

THEOREM 3

If $a_n \to L$ and if f is a function that is continuous at L and defined at all the a_n's, then $f(a_n) \to f(L)$.

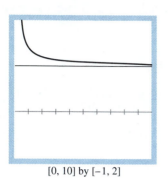

[0, 10] by [−1, 2]

9.5 The graphs of $y_3 = 1$ and $y_2 = \sqrt{y_1}$, where $y_1 = (x + 1)/x$. The graph of the sequence $\{\sqrt{(n + 1)/n}\}$ is embedded in the graph of y_2. The viewing window suggests that y_3 is a right-end behavior model for y_2, to support the fact that $\sqrt{(n + 1)/n} \to 1$.

LIMITS AT ∞

For a view of

$$\lim_{x \to \infty} f(x)$$

that corresponds to the view of

$$\lim_{n \to \infty} a_n$$

in Exploration 4, see Section 8.6, Exploration 3.

EXPLORATION BIT

Exploration 4 suggests how to learn about a limit as $n \to \infty$ by viewing a limit as $1/n \to 0^+$. This approach can be modified to have $-(1/n) \to 0^-$. Try it.

EXAMPLE 7

Show analytically that $\sqrt{(n + 1)/n} \to 1$. Support graphically.

Solution We know that $(n + 1)/n \to 1$. Taking $f(x) = \sqrt{x}$ and $L = 1$ in Theorem 3 therefore gives $\sqrt{(n + 1)/n} \to \sqrt{1} = 1$.

To support graphically, we use function mode, connected format. We let $y_1 = (x + 1)/x$, $y_2 = \sqrt{y_1}$, and $y_3 = 1$. We GRAPH y_2 and y_3 and see that y_3 appears to be a right-end behavior model for y_2 (Fig. 9.5). ≡

EXPLORATION 4

Another View of Limits as $n \to \infty$

If we regard the sequence $\{a(n)\}$ as a function $a(x)$ of a real variable x, Theorem 3 says that if $a(x) \to L$, then $f(a(x)) \to f(L)$ when f is continuous at L. Because

$$\lim_{x \to \infty} f(x) = \lim_{x \to 0^+} f(\frac{1}{x})$$

you can study what happens to $f(a(x))$ as $x \to \infty$ by studying what happens to $f(a(1/x))$ as $x \to 0^+$, an x-value that is easy to locate in the viewing window.

1. Support the result of Example 7 graphically by viewing $f(a_n)$ as $n \to \infty$ in parametric mode, dot format. Let

$$x_1(t) = t, \qquad y_1(t) = \frac{t + 1}{t},$$

$$x_2(t) = t, \qquad y_2(t) = \sqrt{y_1},$$

$$x_3(t) = t, \qquad y_3(t) = 1,$$

and graph (x_2, y_2) and (x_3, y_3). Explain what you see in the viewing window. (Figure 9.5, showing function mode, connected format, suggests the pattern you should see in your viewing window.)

2. Support the result of Example 7 graphically by viewing $f(a(1/x))$ as $x \to 0^+$ in function mode, connected format. Let

$$y_1 = \frac{1}{x}, \qquad y_2 = a(y_1) = \frac{y_1 + 1}{y_1}, \qquad y_3 = \sqrt{y_2},$$

and graph y_3 in a [0, 10] by [−1, 4] viewing window. Explain what you see in the viewing window. Explain how to use ZOOM-IN to give numerical support. ꝏ

EXAMPLE 8

Show that the successive square roots of 2, namely,

$$\sqrt{2}, \sqrt{\sqrt{2}}, \sqrt{\sqrt{\sqrt{2}}}, \dots,$$

converge to 1.

Solution We know that $1/n \to 0$. Taking $L = 0$ and $f(x) = 2^x$ in Theorem 3 gives $2^{1/n} \to 2^0 = 1$. The successive square roots of 2, namely, $2^{1/2}, 2^{1/4}, 2^{1/8}, \dots$, form a subsequence of the sequence $\{2^{1/n}\}$ and therefore converge to 1 also. ≣

The next theorem justifies a graphical check that we have used (function mode, connected format). It also enables us to use l'Hôpital's Rule to find the limits of some sequences. We state and prove the theorem first, then show how to apply it.

THEOREM 4

Suppose that $f(x)$ is a function defined for all $x \geq n_1$ and $\{a_n\}$ is a sequence such that $a_n = f(n)$ when $n \geq n_1$. If

$$\lim_{x \to \infty} f(x) = L, \qquad \text{then} \qquad \lim_{n \to \infty} a_n = L.$$

Proof Suppose that $\lim_{x \to \infty} f(x) = L$. Then for each positive number ϵ, there is a number M such that

$$x > M \qquad \Rightarrow \qquad |f(x) - L| < \epsilon.$$

Let N be an integer greater than M and greater than or equal to n_1. Then

$$n > N \Rightarrow a_n = f(n) \qquad \text{and} \qquad |a_n - L| = |f(n) - L| < \epsilon. \qquad ≣$$

EXAMPLE 9

Show that $\lim_{n \to \infty} (\ln n)/n = 0$.

Solution The function $(\ln x)/x$ is defined for all $x \geq 1$ and agrees with the given sequence at positive integers. Therefore, $\lim_{n \to \infty} (\ln n)/n$ will equal $\lim_{x \to \infty} (\ln x)/x$ if the latter exists. A single application of l'Hôpital's Rule shows that

$$\lim_{x \to \infty} \frac{\ln x}{x} = \lim_{x \to \infty} \frac{1/x}{1} = \frac{0}{1} = 0.$$

We conclude that $\lim_{n \to \infty} (\ln n)/n = 0$. ≣

When we use l'Hôpital's Rule to find the limit of a sequence, we often treat n as a continuous real variable and differentiate directly with respect to n. This saves us from having to rewrite the formula for a_n as we did in Example 9.

EXPLORATION BIT

Show graphically that

$$\lim_{n \to \infty} \frac{\ln n}{n} = 0$$

using function mode, connected format. Explain how Theorem 4 applies.

EXAMPLE 10

Find $\lim\limits_{n \to \infty} (2^n / 5n)$.

Solution Treating n as a continuous real variable and applying l'Hôpital's Rule, we have

$$\lim_{n \to \infty} \frac{2^n}{5n} = \lim_{n \to \infty} \frac{2^n \cdot \ln 2}{5} = \infty.$$

Recursive Definitions

So far, we have calculated each a_n directly from the value of n. But sequences are often defined **recursively** by giving

1. The value of the first term (or first few terms), and

2. A rule, called a **recursion formula**, for calculating any later term from terms that precede it.

Equation (1) used in Newton's method (Section 4.3) is a recursion formula. The terms of the *factorial sequence* 1, 2, 6, 24, ..., $n!$, ... are usually calculated recursively with the formula $(n+1)! = (n+1)n!$ and $1!$ defined to be 1.

Limits That Arise Frequently

The limits $n \to \infty$ and $1/n \to 0$ are basic. Other useful limits are shown in Table 9.1. The first limit is from Example 9. The others are derived in Appendix 5.

EXAMPLE 11   Limits Using the Equations in Table 9.1

1. $\dfrac{\ln (n^2)}{n} = \dfrac{2 \ln n}{n} \to 2 \cdot 0 = 0$ Eq. (1)

2. $\sqrt[n]{n^2} = n^{2/n} = (n^{1/n})^2 \to (1)^2 = 1$ Eq. (2)

3. $\sqrt[n]{3n} = 3^{1/n}(n^{1/n}) \to 1 \cdot 1 = 1$ Eq. (3) with $x = 3$, and Eq. (2)

4. $\left(-\dfrac{1}{2}\right)^n \to 0$ Eq. (4) with $x = -\dfrac{1}{2}$

5. $\left(\dfrac{n-2}{n}\right)^n = \left(1 - \dfrac{2}{n}\right)^n \to e^{-2}$ Eq. (5) with $x = -2$

6. $\dfrac{100^n}{n!} \to 0$ Eq. (6) with $x = 100$

TABLE 9.1   **Limits of Some Common Sequences**

1. $\lim\limits_{n \to \infty} \dfrac{\ln n}{n} = 0$ **2.** $\lim\limits_{n \to \infty} \sqrt[n]{n} = 1$

3. $\lim\limits_{n \to \infty} x^{1/n} = 1$ $(x > 0)$ **4.** $\lim\limits_{n \to \infty} x^n = 0$ $(|x| < 1)$

5. $\lim\limits_{n \to \infty} \left(1 + \dfrac{x}{n}\right)^n = e^x$ (any x)   **6.** $\lim\limits_{n \to \infty} \dfrac{x^n}{n!} = 0$ (any x)

In Eqs. (3)–(6), x remains fixed while $n \to \infty$.

Exercises 9.1

Each of Exercises 1–4 gives a formula for the *n*th term a_n of a sequence $\{a_n\}$. Find the values of $a_1, a_2, a_3,$ and a_4.

1. $a_n = \dfrac{1-n}{n^2}$

2. $a_n = \dfrac{1}{n!}$

3. $a_n = \dfrac{(-1)^{n+1}}{2n-1}$

4. $a_n = 2 + (-1)^n$

Each of Exercises 5–10 gives the first term or two of a sequence and a recursion formula for the remaining terms. Write out the first ten terms of each sequence.

5. $x_1 = 1, \quad x_{n+1} = x_n + (1/2^n)$

6. $x_1 = 1, \quad x_{n+1} = x_n/(n+1)$

7. $x_1 = 2, \quad x_{n+1} = x_n/2$

8. $x_1 = -2, \quad x_{n+1} = nx_n/(n+1)$

9. $x_1 = x_2 = 1, \quad x_{n+2} = x_n + x_{n+1}$

10. $x_1 = 1, \quad x_n = x_1 + x_2 + \cdots + x_{n-1}$

Which of the sequences $\{a_n\}$ in Exercises 11–26 converge, and which diverge? Find the limit of each convergent sequence analytically, and support graphically.

11. $a_n = 2 + (0.1)^n$

12. $a_n = 1 + (-1)^n$

13. $a_n = 5$

14. $a_n = \dfrac{10^{n+1}}{10^n}$

15. $a_n = \dfrac{n}{10}$

16. $a_n = 5^n$

17. $a_n = \dfrac{1-2n}{1+2n}$

18. $a_n = \dfrac{2n+1}{1-3n}$

19. $a_n = \dfrac{n^2 - 2n + 1}{n-1}$

20. $a_n = \dfrac{n+3}{n^2 + 5n + 6}$

21. $a_n = \dfrac{1-5n^4}{n^4 + 8n^3}$

22. $a_n = \dfrac{1-n^3}{70 - 4n^2}$

23. $a_n = \dfrac{n + (-1)^n}{n}$

24. $a_n = (-1)^n \left(1 - \dfrac{1}{n}\right)$

25. $a_n = \left(\dfrac{n+1}{2n}\right)\left(1 - \dfrac{1}{n}\right)$

26. $a_n = \left(2 - \dfrac{1}{2^n}\right)\left(3 + \dfrac{1}{2^n}\right)$

Find the limits of the sequences $\{a_n\}$ whose *n*th terms appear in Exercises 27–34.

27. $a_n = \dfrac{(-1)^{n+1}}{2n-1}$

28. $a_n = \left(-\dfrac{1}{2}\right)^n$

29. $a_n = \dfrac{\sin n}{n}$

30. $a_n = \dfrac{\sin^2 n}{2^n}$

31. $a_n = \sqrt{\dfrac{2n}{n+1}}$

32. $a_n = \sin\left(\dfrac{\pi}{2} + \dfrac{1}{n}\right)$

33. $a_n = \tan^{-1} n$

34. $a_n = \ln n - \ln (n+1)$

Find graphically the limits of the *converging* sequences whose *n*th terms appear in Exercises 35–58. Confirm analytically.

35. $a_n = \dfrac{n}{2^n}$

36. $a_n = \dfrac{3^n}{n^3}$

37. $a_n = \dfrac{\ln (n+1)}{n}$

38. $a_n = \dfrac{\ln n}{\ln 2n}$

39. $a_n = 8^{1/n}$

40. $a_n = (0.03)^{1/n}$

41. $a_n = \left(1 + \dfrac{7}{n}\right)^n$

42. $a_n = \left(1 - \dfrac{1}{n}\right)^n$

43. $a_n = \dfrac{1}{(0.9)^n}$

44. $a_n = \dfrac{(n+1)!}{n!}$

45. $a_n = \sqrt[n]{10n}$

46. $a_n = \sqrt[n]{n^2}$

47. $a_n = \left(\dfrac{3}{n}\right)^{1/n}$

48. $a_n = (n+4)^{1/(n+4)}$

49. $a_n = \dfrac{\ln n}{n^{1/n}}$

50. $a_n = \sqrt[n]{4^n n}$

51. $a_n = \left(\dfrac{1}{3}\right)^n$

52. $a_n = \dfrac{1}{\sqrt{2^n}}$

53. $a_n = \dfrac{1}{n!}$

54. $a_n = \dfrac{(-4)^n}{n!}$

55. $a_n = \left(\dfrac{1}{n}\right)^{1/\ln n}$

56. $a_n = \dfrac{n!}{2^n \cdot 3^n}$

57. $a_n = \dfrac{n!}{10^{6n}}$

58. $a_n = \dfrac{3^n \cdot 6^n}{2^{-n} \cdot n!}$

In Exercises 59–62, experiment with a graphing utility to identify a smallest value of N that will make the inequality hold for $n > N$.

59. $|\sqrt[n]{0.5} - 1| < 10^{-3}$

60. $|\sqrt[n]{n} - 1| < 10^{-3}$

61. $(0.9)^n < 10^{-3}$

62. $2^n/n! < 10^{-7}$

63. Support the analytic result of Example 8 graphically.

64. Support the analytic result of Example 9 graphically.

Use graphs to support the limits in Exercises 65 and 66.

65. Eqs. (1), (3), and (5) of Table 9.1. For those equations involving x, give support for two different values of x.

66. Eqs. (2), (4), and (6) of Table 9.1. For those equations involving x, give support for two different values of x.

67. Write an expression for a_{n+1} in terms of a_n to describe the recursively-defined sequence that you get when you solve $\sin x = x^2$ using Newton's method. (See Section 4.3.) Use Newton's method to solve $\sin x = x^2$ using the initial values $a_1 = 1, 2,$ and -1.

68. Write an expression for a_{n+1} in terms of a_n to describe the recursively-defined sequence that you get when you solve $\cos x = x^2$ using Newton's method. (See Section 4.3.) Use Newton's method to solve $\cos x = x^2$ using the initial values $a_1 = 1, -1$, and -2.

69. *A recursive definition of $\pi/2$.* In the diagram, the length $\pi/2$ of the circular arc is approximated by $x_n + \cos x_n$. If you define $\{x_n\}$ so that $x_1 = 1$ and $x_{n+1} = x_n + \cos x_n$, you generate a sequence that converges rapidly to $\pi/2$. Try it.

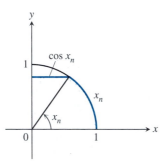

70. *A recursive solution of an equation* $\cos x = x$. If you define $\{x_n\}$ so that $x_1 = 1$ and $x_{n+1} = \cos x_n$, your sequence will converge to a solution of the equation $\cos x = x$, as the diagram suggests. Try it.

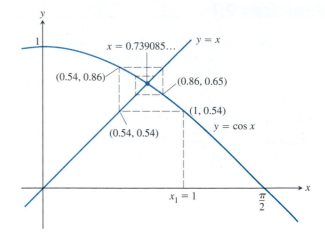

Find recursive solutions to the equations in Exercises 71 and 72.

71. The nonzero solutions of $\sin x = 0.5x$. *Hint:* To find the negative one, start with $x_1 = -1$.

72. The nonzero solutions to $\tan x = 2x$ in $(-\pi/2, \pi/2)$. *Hint:* See Exercise 71.

Sonya Kovalevsky

Although Sonya Kovalevsky (1850–1891) was the daughter of Russian nobility, her great achievements in science and mathematics were earned through hard work, determination, and sacrifice. As a nineteenth-century woman, she could not travel freely, could not attend public lectures, and had difficulty finding a job. She taught herself trigonometry and calculus as a young girl. At age 18 she left Russia to study in Germany, where she was tutored by several renowned mathematicians. She received her doctorate in 1874.

Later in her career, Kovalevsky worked with the famous mathematician Karl Weierstrass and received a life professorship at the University of Stockholm in Sweden. She is considered one of the greatest mathematical talents of the nineteenth century for her remarkable contributions in the areas of infinite series in mathematics, Abelian integrals, and partial differential equations, as well as in astronomy.

9.2 — Infinite Series

In mathematics and science, we often use infinite polynomials like

$$1 + x + x^2 + x^3 + \cdots + x^n + \cdots$$

to represent functions, evaluate nonelementary integrals, and solve differential equations. For any particular value of x, such a polynomial is an infinite sum of constants, a sum we call an *infinite series*. The goal of this and the next three sections is to learn to work with infinite series. Then, in the three sections after that, we will build on what we have learned to study infinite polynomials and the mathematical entities that they represent.

The Sequence of Partial Sums

Let us begin by asking how to assign meaning to an expression like

$$1 + \frac{1}{2} + \frac{1}{4} + \frac{1}{8} + \frac{1}{16} + \cdots.$$

The way to do so is not to try to add all the terms at once (we can't) but rather to add the terms one at a time from the beginning and look for a pattern in how these *partial sums* grow. When we do this, we find:

Partial sum		Value
First:	$s_1 = 1$	1
Second:	$s_2 = 1 + \dfrac{1}{2}$	$2 - \dfrac{1}{2}$
Third:	$s_3 = 1 + \dfrac{1}{2} + \dfrac{1}{4}$	$2 - \dfrac{1}{4}$
$\vdots$		
nth:	$s_n = 1 + \dfrac{1}{2} + \dfrac{1}{4} + \cdots + \dfrac{1}{2^{n-1}}$	$2 - \dfrac{1}{2^{n-1}}$ (after some algebra)

Indeed there *is* a pattern. The partial sums form a sequence whose nth term is

$$s_n = 2 - \frac{1}{2^{n-1}},$$

and this sequence converges to 2 (Fig. 9.6). We therefore say

"the sum of the infinite series $1 + \dfrac{1}{2} + \dfrac{1}{4} + \cdots + \dfrac{1}{2^{n-1}} + \cdots$ is 2."

Our knowledge of sequences and limits enables us to break away from the confines of finite sums.

In certain important ways, the idea of "sum of an infinite series" parallels that of the limit of a function at a point.

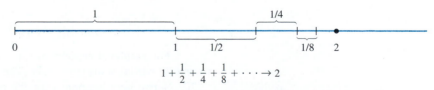

$$1 + \frac{1}{2} + \frac{1}{4} + \frac{1}{8} + \cdots \to 2$$

9.6 As the lengths $1, 1/2, 1/4, 1/8, \ldots$, are added one by one, the sums approach 2.

TERMINOLOGY

Stability on a grapher means there is partial sum agreement in decimals from left to right. We say that the partial sums s_n *stabilize* on the calculator if they are equal in all decimal places for all n greater than a particular n_1.

$\lim_{x \to c} f(x)$	Sum of an Infinite Series
• $f(x)$ is defined for values of x that are as close to c as we wish.	• The partial sums are defined for as many terms as we wish.
• We do not care whether f is defined at c.	• We do not know how to add infinitely many numbers.
• The values of f may approach a particular number as x approaches c.	• The partial sums may approach a particular number as more terms are added.
• The number approached by the values of f is given a name, *the limit of f at c.*	• The number approached by the partial sums is given a name, *the sum of the infinite series.*
• If the values of f do not approach a particular number, we say that the limit does not exist.	• If the partial sums do not approach a particular number, we say that the series diverges.

Finding a pattern for the partial sums of a series is difficult, often impossible. In this textbook, we assume that your grapher has the capability of adding the partial sums of a series either as a built-in feature or with a program. The *Resource Manual* contains two programs, PARTSUMT, which displays the partial sums in table form, and PARTSUMG, which displays the partial sums graphically.

EXAMPLE 1

Give numerical and graphical support for 2 as the sum of the infinite series

$$1 + \frac{1}{2} + \frac{1}{4} + \frac{1}{8} + \frac{1}{16} + \cdots.$$

Solution   For numerical support, we use the program PARTSUMT on our grapher to compute the values of several partial sums s_n. Some of the values are shown in the table below and give strong numerical support for 2 as the sum of the series. (In fact, on our grapher, s_n stabilizes at $s_{32} = 2$.)

n	s_n
5	1.9375
10	1.998046875
20	1.999998093
30	1.999999998

For graphical support, we use PARTSUMG to produce a graph of the first 20 partial sums (Fig. 9.7). The points plotted are (t, s_t) for $t = 1$ to $t = 20$. Notice how the points for the partial sums s_t appear to be approaching the line $y = 2$.

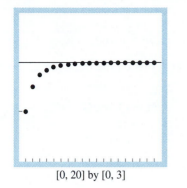

[0, 20] by [0, 3]

9.7   The graph of the first 20 partial sums s_t of the infinite series

$$1 + \frac{1}{2} + \frac{1}{4} + \frac{1}{8} + \frac{1}{16} + \cdots.$$

Notice how the s_t values appear to approach the line $y = 2$.

DEFINITIONS

Given a sequence of numbers $\{a_n\}$, an expression of the form

$$a_1 + a_2 + a_3 + \cdots + a_n + \cdots$$

is called an **infinite series.** The number a_n is called the **nth term** of the series. The sequence $\{s_n\}$ defined by

$$s_1 = a_1$$
$$s_2 = a_1 + a_2$$
$$\vdots$$
$$s_n = a_1 + a_2 + \cdots + a_n = \sum_{k=1}^{n} a_k$$

is the **sequence of partial sums** of the series, the number s_n being the **nth partial sum.** If the sequence of partial sums converges to a limit L, we say that the series **converges** and that its **sum** is L. In this case, we also write

$$a_1 + a_2 + \cdots + a_n + \cdots = \sum_{n=1}^{\infty} a_n = L.$$

If the sequence of partial sums of the series does not converge, we say that the series **diverges.**

NOTATION

When we study a series

$$a_1 + a_2 + \ldots + a_n + \ldots$$

for the first time, it is convenient to use sigma notation to write the series as

$$\sum_{n=1}^{\infty} a_n, \quad \sum_{k=1}^{\infty} a_k, \quad \text{or} \quad \sum a_n$$

without knowing whether it converges. These forms are read as "summation from n equals 1 to infinity of a_n," "summation from k equals 1 to infinity of a_k," and "summation a_n," respectively.

Geometric Series

Geometric series are series of the form

$$a + ar + ar^2 + \cdots + ar^{n-1} + \cdots = \sum_{n=1}^{\infty} ar^{n-1} \tag{1}$$

in which a and r are fixed real numbers and $a \neq 0$. The **ratio** r can be positive, as in

$$1 + \frac{1}{2} + \frac{1}{4} + \cdots + \frac{1}{2^{n-1}} + \cdots, \tag{2}$$

or negative, as in

$$1 - \frac{1}{3} + \frac{1}{9} - \cdots + (-1)^{n-1}\frac{1}{3^{n-1}} + \cdots. \tag{3}$$

If $r = 1$, the nth partial sum of the geometric series in (1) is

$$s_n = a + a(1) + a(1)^2 + \cdots + a(1)^{n-1} = na,$$

and the series diverges because $\lim_{n \to \infty} s_n = \pm\infty$. If $r \neq 1$, we can determine the convergence or nonconvergence of the series in the following way. We multiply the nth partial sum

$$s_n = a + ar + ar^2 + \cdots + ar^{n-1}$$

TERMINOLOGY

We call a series of the form (1) a geometric series.

The number r is called the *ratio* because it is the ratio of one term of the series to its preceding term. Check this out for the series (2) and (3).

by r, obtaining

$$rs_n = ar + ar^2 + \cdots + ar^{n-1} + ar^n.$$

We then subtract rs_n from s_n. Most of the terms on the right cancel, leaving only

$$s_n - rs_n = a - ar^n \qquad \text{or} \qquad s_n(1-r) = a(1-r^n).$$

We solve for s_n, obtaining

$$s_n = \frac{a(1-r^n)}{(1-r)} \qquad (r \neq 1).$$

If $|r| < 1$, then $r^n \to 0$ as $n \to \infty$ (Table 9.1, Eq. (4), in Section 9.1), and $s_n \to a/(1-r)$. In other words, the series converges to $a/(1-r)$. If $|r| > 1$, then $|r^n| \to \infty$, and the series diverges.

EXPLORATION BIT

You can support the results of Example 2 by using a second approach. Compute

$$\frac{1}{9}\left(1 + \frac{1}{3} + (\frac{1}{3})^2 + \ldots\right)$$

and

$$4\left(1 - \frac{1}{2} + \frac{1}{4} - \frac{1}{8} + \frac{1}{16} - \ldots\right),$$

then compare.

If $|r| < 1$, the geometric series converges, and

$$\sum_{n=1}^{\infty} ar^{n-1} = \frac{a}{1-r}. \qquad (4)$$

If $|r| \geq 1$, the series diverges.

EXAMPLE 2

Use Eq. (4) to find the sum of the following geometric series.

a) $\dfrac{1}{9} + \dfrac{1}{27} + \dfrac{1}{81} + \cdots$ **b)** $4 - 2 + 1 - \dfrac{1}{2} + \dfrac{1}{4} - \cdots$

Solution

a) For this series, $a = 1/9$, $r = 1/3$, and the sum

$$\frac{1}{9} + \frac{1}{27} + \frac{1}{81} + \cdots = \frac{1/9}{1 - 1/3} = \frac{1}{6}.$$

b) For this series, $a = 4$, $r = -1/2$, and the sum

$$4 - 2 + 1 - \frac{1}{2} + \frac{1}{4} - \cdots = \frac{4}{1 + 1/2} = \frac{8}{3}.$$

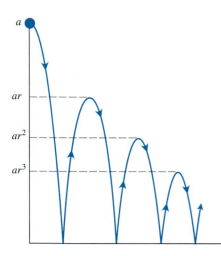

9.8 The height of each rebound is reduced to r times the previous height.

EXAMPLE 3

You drop a ball from a meters above a flat surface. Each time the ball hits after falling a distance h, it rebounds a distance rh, where r is positive but less than 1. Find the total distance the ball travels up and down (Fig. 9.8).

Solution   The total distance is

$$s = a + \underbrace{2ar + 2ar^2 + 2ar^3 + \cdots}_{\text{This sum is } 2ar/(1-r).} = a + \frac{2ar}{1-r} = a\frac{1+r}{1-r}.$$

If $a = 6$ m and $r = 2/3$, for instance, the distance is

$$s = 6\frac{1 + 2/3}{1 - 2/3} = 6\frac{5/3}{1/3} = 30 \text{ m.}$$

EXPLORATION 1

Repeating Decimals

We can use geometric series to find the fractional form of a rational number represented by an infinite repeating decimal.

1. Show how the infinite repeating decimal 5.23 23 23 ... can be represented by an infinite series. (The grouping of the decimal digits is a hint.)

2. Part of the infinite series is geometric with $|r| < 1$. Identify the geometric series, and find a, r, and $a/(1 - r)$.

3. Simplify your results to the fractional form for 5.23 23 23 ....

4. Try this process for an infinite decimal of your choice.

Sums of Other Convergent Series

For geometric series, we get an explicit form for the nth partial sum and, from that, the formula $s = a/(1 - r)$. Unfortunately, such formulas are rare. What we usually have to do is somehow test the series for convergence and then estimate the sums of the series that pass the test. Sometimes, a convergent series represents a known function like $\sin x$ or $\ln x$, but most of the time, we need a computer or calculator to estimate the sum.

The next example, however, is one of those pleasing, rare cases in which we can find the series' sum from a formula for s_n.

EXAMPLE 4

Determine whether $\sum_{n=1}^{\infty} \frac{1}{n(n + 1)}$ converges. If it does, find the sum.

Solution We look for a pattern in the sequence of partial sums that might lead us to an explicit expression for s_k. The key to success here, as in the integration

$$\int \frac{dx}{x(x + 1)} = \int \frac{dx}{x} - \int \frac{dx}{x + 1},$$

is partial fractions. The observation that

$$\frac{1}{k(k + 1)} = \frac{1}{k} - \frac{1}{k + 1}$$

permits us to write the partial sum

$$\sum_{n=1}^{k} \frac{1}{n(n + 1)} = \frac{1}{1 \cdot 2} + \frac{1}{2 \cdot 3} + \cdots + \frac{1}{k \cdot (k + 1)}$$

as

$$s_k = \left(\frac{1}{1} - \frac{1}{2}\right) + \left(\frac{1}{2} - \frac{1}{3}\right) + \cdots + \left(\frac{1}{k} - \frac{1}{k + 1}\right). \tag{5}$$

EXPLORATION BIT

There are at least two other ways to change an infinite decimal form to its equivalent fractional form. One is an algebraic method that you may have learned when you were studying algebra. Do you recall it? Another is through the use of technology. Some calculators will convert a decimal to its fractional form. If you have such a calculator, enter several digits of the infinite repeating decimal 5.23 23 23 ..., and see whether the calculator returns 518/99.

EXPLORATION BIT

Use your grapher to investigate the partial sums of

$$\sum_{n=1}^{\infty} \frac{1}{n(n + 1)}.$$

(One way is to apply the program PARTSUMG.) Do you think the series converges? If so, to what sum?

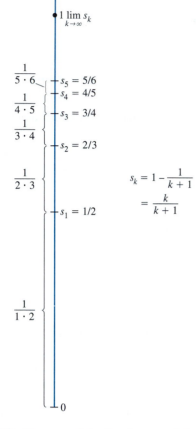

9.9 The sum of the first k terms of the series

$$\sum_{n=1}^{\infty} \frac{1}{n(n+1)}$$

is $k/(k+1)$, and the sum of the series is

$$\lim_{k \to \infty} \frac{k}{k+1} = 1.$$

TELESCOPING SERIES

If the partial sums of a series "collapse in the middle," as happens in Example 4 from Eq. (5) to Eq. (6), we informally refer to the series as a *telescoping series*.

Removing parentheses and canceling the terms of opposite sign collapses the sum to

$$s_k = 1 - \frac{1}{k+1}. \tag{6}$$

We then see that $s_k \to 1$ as $k \to \infty$. The series converges, and its sum is 1 (Fig. 9.9):

$$\sum_{n=1}^{\infty} \frac{1}{n(n+1)} = 1. \qquad\qquad \equiv$$

Other Divergent Series and a Test for Divergence

There are other series besides geometric series with $|r| \geq 1$ that diverge.

EXAMPLE 5

Both of these series diverge:

a) $\displaystyle\sum_{n=1}^{\infty} n^2 = 1 + 4 + 9 + \cdots + n^2 + \cdots$

b) $\displaystyle\sum_{n=1}^{\infty} \frac{n+1}{n} = \frac{2}{1} + \frac{3}{2} + \frac{4}{3} + \cdots + \frac{n+1}{n} + \cdots$

because the partial sums eventually exceed every preassigned number. Each term (beyond the first in series (a)) is greater than 1, and so the sum of n terms is greater than n. $\qquad \equiv$

There is a test for detecting the kinds of divergence that occur in Example 5.

The *n*th-Term Test for Divergence

If $\displaystyle\lim_{n \to \infty} a_n \neq 0$, or if $\displaystyle\lim_{n \to \infty} a_n$ fails to exist, then $\displaystyle\sum_{n=1}^{\infty} a_n$ diverges.

When we apply the *n*th-Term Test to the series in Example 5, we find that

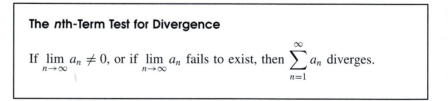

$\displaystyle\sum_{n=1}^{\infty} n^2 \qquad$ diverges because $n^2 \to \infty$,

$\displaystyle\sum_{n=1}^{\infty} \frac{n+1}{n} \qquad$ diverges because $\dfrac{n+1}{n} \to 1$,

$\displaystyle\sum_{n=1}^{\infty} (-1)^{n+1} \qquad$ diverges because $\displaystyle\lim_{n \to \infty} (-1)^{n+1}$ does not exist.

The nth-Term Test works because $\lim_{n\to\infty} a_n$ must equal zero if $\sum a_n$ converges. To see why, let

$$s_n = a_1 + a_2 + \cdots + a_n,$$

and suppose that $\sum a_n$ converges to S; that is, $s_n \to S$. When n is large, so is $n-1$, and both s_n and s_{n-1} are close to S. Their difference, a_n, must then be close to zero. More formally, $a_n = s_n - s_{n-1} \to S - S = 0$.

EXAMPLE 6

Apply the nth-Term Test for Divergence to these series.

a) $\displaystyle\sum_{n=1}^{\infty} \frac{n}{2n+5}$ **b)** $\displaystyle\sum_{n=1}^{\infty} \frac{5(-1)^n}{4^n}$

Solution

a) $\displaystyle\lim_{n\to\infty} \frac{n}{2n+5} = \frac{1}{2} \neq 0$
The series diverges by the nth-Term Test.

b) $\displaystyle\lim_{n\to\infty} \frac{5(-1)^n}{4^n} = 5 \lim_{n\to\infty} \left(\frac{-1}{4}\right)^n = 0$ Table 9.1, Eq. (4), Section 9.1

The nth-Term Test is inconclusive. The series may or may not converge.

We often express the nth-Term Test for Divergence using the equivalent contrapositive form:

If $\displaystyle\sum_{n=1}^{\infty} a_n$ converges, then $a_n \to 0$.

Caution: This does *not* mean that $\sum a_n$ converges if $a_n \to 0$. A series $\sum a_n$ may diverge even though $a_n \to 0$. Thus, $\lim a_n = 0$ is a *necessary* but *not a sufficient* condition for the series $\sum a_n$ to converge.

EXPLORATION 2

A Necessary Condition for Convergence

1. Explain why the statement above is the contrapositive of the statement of the nth-Term Test for Divergence.

2. Give a convincing argument that the series

$$\sum_{n=1}^{\infty} \sum_{i=1}^{n} \frac{1}{n}$$

diverges. (Be careful. This is not the series $1 + 1/2 + 1/3 + \cdots$.) Then write the series as an infinite sum having the form

$$a_1 + a_2 + a_3 + \cdots$$

where the $a_n \to 0$. Explain how this series is an example of the kind we cautioned you about above. &

A Useful Theorem

Whenever we have two convergent series, we can add them, subtract them, and multiply them by constants to make other convergent series. Theorem 5 gives the details.

THEOREM 5

If $\sum a_n = A$ and $\sum b_n = B$, then

1. *Sum Rule:* $\qquad \sum (a_n + b_n) = A + B$

2. *Difference Rule:* $\quad \sum (a_n - b_n) = A - B$

3. *Constant Multiple Rule:* $\quad \sum k a_n = k \sum a_n = kA \qquad$ (any number k)

As corollaries of Theorem 5, the following hold:

1. Every nonzero constant multiple of a divergent series diverges.

2. If $\sum a_n$ converges and $\sum b_n$ diverges, then $\sum (a_n + b_n)$ and $\sum (a_n - b_n)$ both diverge.

EXAMPLE 7

a) $\displaystyle\sum_{n=1}^{\infty} \frac{4}{2^{n-1}} = 4 \sum_{n=1}^{\infty} \frac{1}{2^{n-1}} = 4 \frac{1}{1 - \frac{1}{2}} = 4(2) = 8$

b) $\displaystyle\sum_{n=1}^{\infty} \frac{3^{n-1} - 1}{6^{n-1}} = \sum_{n=1}^{\infty} \frac{1}{2^{n-1}} - \sum_{n=1}^{\infty} \frac{1}{6^{n-1}} = 2 - \frac{1}{1 - \frac{1}{6}} = 2 - \frac{6}{5} = \frac{4}{5}$ ≣

We can always add a finite number of terms to a series or delete a finite number of terms from a series without altering its convergence or divergence because the series would have the same tails. If $\sum_{n=1}^{\infty} a_n$ converges, then $\sum_{n=k}^{\infty} a_n$ converges for any $k > 1$, and

$$\sum_{n=1}^{\infty} a_n = a_1 + a_2 + \cdots + a_{k-1} + \sum_{n=k}^{\infty} a_n.$$

Conversely, if $\sum_{n=k}^{\infty} a_n$ converges for any $k > 1$, then $\sum_{n=1}^{\infty} a_n$ converges. Thus,

$$\sum_{n=1}^{\infty} \frac{1}{5^n} = \frac{1}{5} + \frac{1}{25} + \frac{1}{125} + \sum_{n=4}^{\infty} \frac{1}{5^n},$$

and

$$\sum_{n=4}^{\infty} \frac{1}{5^n} = \sum_{n=1}^{\infty} \frac{1}{5^n} - \frac{1}{5} - \frac{1}{25} - \frac{1}{125} = \frac{1}{500}.$$

We can always shift the indexing of a series without altering its convergence. For example, we can write the geometric series $1 + 1/2 + 1/4 + \cdots = \sum_{n=1}^{\infty} 1/2^{n-1}$ as

$$\sum_{n=0}^{\infty} \frac{1}{2^n}, \quad \sum_{n=5}^{\infty} \frac{1}{2^{n-5}}, \quad \text{or even} \quad \sum_{n=-4}^{\infty} \frac{1}{2^{n+4}}.$$

> **EXPLORATION BIT**
>
> Confirm that
>
> $$\sum_{n=4}^{\infty} \frac{1}{5^n} = \frac{1}{500}.$$

Perpetuities

Suppose we want to give our favorite school or charity $1000 a year forever. This is an example of an *infinite income stream* called a **perpetuity**. We will assume that we can earn 8% annually on our money for the purposes of this example.

If a_n represents the amount that we need today to provide the nth $1000 payment, then

$$
\begin{aligned}
a_0 &= 1000 & &\text{The annual payout} \\
a_1 &= 1.08^{-1}(1000) & &\text{In 1 year, } a_1 \text{ grows to 1000.} \\
a_2 &= 1.08^{-1}a_1 & &\text{In 2 years, } a_2 \text{ grows to 1000.} \\
a_3 &= 1.08^{-1}a_2 & &\text{In 3 years, } a_3 \text{ grows to 1000.} \\
&\;\;\vdots & & \\
a_n &= 1.08^{-1}a_{n-1} & &\text{In } n \text{ years, } a_n \text{ grows to 1000.} \\
&\;\;\vdots & &
\end{aligned}
$$

The amount that we would have to set aside today to start the payments one year from today is called the *present value of the perpetuity* and is given by

$$\sum_{n=1}^{\infty} a_n.$$

In Example 8, we use the toolbox program PSUMRECT (see the *Resource Manual*) to compute the partial sums of an infinite series with the nth term given *recursively*. If you are using a CAS, there is probably a built-in operation for building such a table.

EXAMPLE 8

What is the present value of a $1000-per-year perpetuity earning interest at 8%. Make a conjecture. Then confirm the present value analytically.

Give a simple reason why $12,500 invested at 8% annually will provide $1,000 a year *forever*.

Solution We use PSUMRECT to compute the following table, where $a_n = 1.08^{-1}a_{n-1}$ and $a_0 = 1000$.

N	$\sum\limits_{n=1}^{N} a_n$
1	925.93
50	12,233.48
100	12,494.32
1000	12,500.00

The last line of the table shows that the present value that will provide $1000 for 1000 years is $12,500, not much more than it took to provide $1000 for 100 years. It is reasonable to conjecture that $12,500 will provide $1000 per year in perpetuity.

It can be shown that

$$a_n = (1.08^{-n})(1000). \quad \text{See the Exploration Bit.}$$

So the series $\sum\limits_{n=1}^{\infty} a_n$ is a geometric series with first term $a = (1000)(1.08^{-1})$ and ratio $r = 1.08^{-1}$. The sum of the infinite series is

$$\frac{a}{1 - r} = \frac{(1000)\left(1.08^{-1}\right)}{1 - 1.08^{-1}} = 12,500,$$

to confirm that $12,500 is the present value of the perpetuity. ≡

EXPLORATION BIT

Show that for the present value

$$\sum_{n=1}^{\infty} a_n$$

of the $1000 per year perpetuity earning interest at 8%,

$$a_n = (1.08^{-n})(1000).$$

Exercises 9.2

In Exercises 1–6, write the first five partial sums of each series, find a formula for the nth partial sum, and use it to find the series sum if the series converges. Support your result numerically and graphically.

1. $2 + \dfrac{2}{3} + \dfrac{2}{9} + \dfrac{2}{27} + \cdots + \dfrac{2}{3^{n-1}} + \cdots$

2. $\dfrac{9}{100} + \dfrac{9}{100^2} + \dfrac{9}{100^3} + \cdots + \dfrac{9}{100^n} + \cdots$

3. $1 - \dfrac{1}{2} + \dfrac{1}{4} - \dfrac{1}{8} + \cdots + (-1)^{n-1}\dfrac{1}{2^{n-1}} + \cdots$

4. $1 - 2 + 4 - 8 + \cdots + (-1)^{n-1}2^{n-1} + \cdots$

5. $\dfrac{1}{2 \cdot 3} + \dfrac{1}{3 \cdot 4} + \dfrac{1}{4 \cdot 5} + \cdots + \dfrac{1}{(n+1)(n+2)} + \cdots$

6. $\dfrac{5}{1 \cdot 2} + \dfrac{5}{2 \cdot 3} + \dfrac{5}{3 \cdot 4} + \cdots + \dfrac{5}{n(n+1)} + \cdots$

In Exercises 7–14, write the first five partial sums of each series. Then find the sum of the series. Support your result numerically and graphically.

7. $\sum\limits_{n=0}^{\infty} \dfrac{1}{4^n}$

8. $\sum\limits_{n=2}^{\infty} \dfrac{1}{4^n}$

9. $\sum\limits_{n=1}^{\infty} \dfrac{7}{4^n}$

10. $\sum\limits_{n=0}^{\infty} (-1)^n \dfrac{5}{4^n}$

11. $\sum\limits_{n=0}^{\infty} \left(\dfrac{5}{2^n} + \dfrac{1}{3^n} \right)$

12. $\sum\limits_{n=0}^{\infty} \left(\dfrac{5}{2^n} - \dfrac{1}{3^n} \right)$

13. $\sum\limits_{n=0}^{\infty} \left(\dfrac{1}{2^n} + \dfrac{(-1)^n}{5^n} \right)$

14. $\sum\limits_{n=0}^{\infty} \left(\dfrac{2^{n+1}}{5^n} \right)$

Use partial fractions to find the sums of the series in Exercises 15–18. Support with PARTSUMT computations.

15. $\sum\limits_{n=1}^{\infty} \dfrac{4}{(4n-3)(4n+1)}$

16. $\sum\limits_{n=1}^{\infty} \dfrac{1}{(4n-3)(4n+1)}$

17. $\sum\limits_{n=3}^{\infty} \dfrac{4}{(4n-3)(4n+1)}$

18. $\sum\limits_{n=1}^{\infty} \dfrac{2n+1}{n^2(n+1)^2}$

Which series in Exercises 19–32 converge, and which diverge?

If the series converges, find its sum.

19. $\displaystyle\sum_{n=0}^{\infty}\left(\frac{1}{\sqrt{2}}\right)^n$

20. $\displaystyle\sum_{n=1}^{\infty}\ln\frac{1}{n}$

21. $\displaystyle\sum_{n=1}^{\infty}(-1)^{n+1}\frac{3}{2^n}$

22. $\displaystyle\sum_{n=1}^{\infty}(\sqrt{2})^n$

23. $\displaystyle\sum_{n=0}^{\infty}\cos n\pi$

24. $\displaystyle\sum_{n=0}^{\infty}\frac{\cos n\pi}{5^n}$

25. $\displaystyle\sum_{n=0}^{\infty}e^{-2n}$

26. $\displaystyle\sum_{n=1}^{\infty}\frac{n^2+1}{n}$

27. $\displaystyle\sum_{n=1}^{\infty}(-1)^{n+1}n$

28. $\displaystyle\sum_{n=1}^{\infty}\frac{2}{10^n}$

29. $\displaystyle\sum_{n=0}^{\infty}\frac{2^n-1}{3^n}$

30. $\displaystyle\sum_{n=1}^{\infty}\left(1-\frac{1}{n}\right)^n$

31. $\displaystyle\sum_{n=0}^{\infty}\frac{n!}{1000^n}$

32. $\displaystyle\sum_{n=0}^{\infty}\frac{1}{x^n},\quad |x|>1$

The series in Exercises 33 and 34 are geometric series. Find a and r in each case to confirm the formulas.

33. $\displaystyle\frac{1}{1+x}=\sum_{n=0}^{\infty}(-1)^n x^n\quad(|x|<1)$

34. $\displaystyle\frac{1}{1+x^2}=\sum_{n=0}^{\infty}(-1)^n x^{2n}\quad(|x|<1)$

35. Find the smallest value for n for which s_n agrees with the sum of the geometric series in Examples 2(a) and 2(b) to nine digits.

36. For the infinite series of Example 4, find the smallest value of n for which $s_n > 0.99$.

37. A ball is dropped from a height of 4 m. Each time it strikes the pavement after falling from a height of h it rebounds to a height of $0.75h$. Find the total distance the ball travels up and down.

38. *Continuation of Exercise 37.* Find the total number of seconds the ball travels in Exercise 37. (*Hint:* The formula $s = 4.9t^2$ gives $t = \sqrt{s/4.9}$.)

39. Write the infinite repeating decimal $0.234\,234\,234\ldots$ as a geometric series, and express in fractional form the rational number it represents.

40. Express the rational number represented by the number $1.24\,123\,123\,123\ldots$, in which the 123 repeats, as the ratio of two integers.

41. The series in Exercise 5 can be written as

$$\sum_{n=1}^{\infty}\frac{1}{(n+1)(n+2)}\quad\text{and}\quad\sum_{n=-1}^{\infty}\frac{1}{(n+3)(n+4)}.$$

Write it as a sum beginning with (a) $n=-2$, (b) $n=0$, (c) $n=5$.

42. The series in Exercise 6 can be written as

$$\sum_{n=1}^{\infty}\frac{5}{n(n+1)}\quad\text{and}\quad\sum_{n=0}^{\infty}\frac{5}{(n+1)(n+2)}.$$

Write it as a sum beginning with (a) $n=-1$, (b) $n=3$, (c) $n=20$.

43. Write proofs of the two corollaries listed below Theorem 5.

44. Give convincing arguments for the three parts of Theorem 5.

In Exercises 45–48, use numerical and graphical methods to help you make a conjecture about the convergence or divergence of the infinite series $\sum_{n=0}^{\infty}a_n$ with the given nth term. If the series converges, make a conjecture about its sum.

45. $a_n = \dfrac{1}{2}a_{n-1},\ a_0 = 1$

46. $a_n = (1.05)^{-1}a_{n-1},\ a_0 = 2500$

47. $a_n = a_{n-1}+(2n-1),\ a_0 = 0$

48. $a_n = \dfrac{1}{n}a_{n-1},\ a_0 = 1$

49. Find the present value of a \$5000-per-year perpetuity at 7%.

50. Find the present value of a \$1000-per-year perpetuity at 5%.

Find the amount of money to set aside today for your favorite charity to start the payments indicated in Exercises 51 and 52 one year from today. Assume your money earns 7%.

51. \$500 a year for (a) 50 years, (b) 100 years.

52. \$2000 a year for (a) 20 years, (b) 100 years.

53. The diagram shows the first five of an infinite sequence of squares. The outermost square has an area of 4 m^2. Each of the other squares is obtained by joining the midpoints of the sides of the square before it. Find the sum of the areas of all the squares.

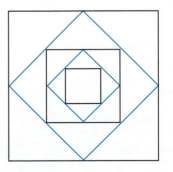

54. The figure on the following page shows the first three rows (and the beginning of the fourth row) of an infinite sequence of rows of semicircles. There are 2^n semicircles in the nth row, each of radius $1/2^n$. Find the sum of the areas of all the semicircles.

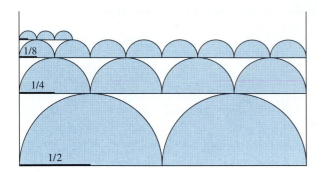

55. Make up an example of two divergent series whose term-by-term sum converges.

56. Show by example that $\sum(a_n/b_n)$ may diverge even though $\sum a_n$ and $\sum b_n$ converge and no b_n equals 0.

57. Show by example that $\sum(a_n/b_n)$ may converge to something other than A/B even when $A = \sum a_n$, $B = \sum b_n \neq 0$, and no b_n equals 0.

58. Show that if $\sum a_n$ converges and $a_n \neq 0$ for all n, then $\sum(1/a_n)$ diverges.

9.3   Series without Negative Terms: Comparison and Integral Tests

Given a series $\sum a_n$, we have two questions:

1. Does the series converge?

2. If it converges, what is its sum?

Much of the rest of this chapter is devoted to answering the first question. It would seem practical to answer the second question at the same time by computing the sum directly from nice formulas for the partial sums as we did for the geometric series. But in most cases, such formulas are not available. In their absence, we can turn to a grapher to approximate the sum when we need to know what it is.

In this section and the next, we study series that do not have negative terms. The reason for this restriction is that the partial sums of these series form nondecreasing sequences, and nondecreasing sequences *that are bounded from above* always converge. To show that a series of nonnegative terms converges, we need only show that there is some number beyond which the partial sums never go.

Nondecreasing Sequences

Suppose that $\sum a_n$ is an infinite series and that $a_n \geq 0$ for every n. Then, when we calculate the partial sums s_1, s_2, s_3, and so on, we see that each one is greater than or equal to its predecessor because $s_{n+1} = s_n + a_{n+1}$:

$$s_1 \leq s_2 \leq s_3 \leq \cdots \leq s_n \leq s_{n+1} \leq \cdots.$$

A sequence $\{s_n\}$ with the property that $s_n \leq s_{n+1}$ for every n is called a **nondecreasing sequence.**

There are two kinds of nondecreasing sequences—those that increase beyond any finite bound and those that don't. The former diverge to infinity, so we turn our attention to the other kind: those that do not grow beyond all bounds. Such a sequence is said to be **bounded from above,** and any number M such that $s_n \leq M$ for all n is called an **upper bound** of the sequence.

EXPLORATION BIT

Is the sequence

$$\{s_n\} = \left\{\frac{n}{n+1}\right\}$$

a nondecreasing sequence? Give a convincing argument.

EXAMPLE 1

If $s_n = n/(n + 1)$, then 1 is an upper bound, and so is any number greater than 1. No number smaller than 1 is an upper bound, so for this sequence, 1 is the **least upper bound**.   ≡

A nondecreasing sequence bounded from above always has a least upper bound, but we will not prove this fact. We prove that if L is the least upper bound, then the sequence converges to L. The following argument shows why L is the limit.

Suppose we plot the points $(1, s_1), (2, s_2), \ldots, (n, s_n)$ in the xy-plane (Fig. 9.10). If M is an upper bound of the sequence, all these plotted points will lie on or below the line $y = M$. The line $y = L$ is the lowest such line. None of the points (n, s_n) lies above $y = L$, but some do lie above any lower line $y = L - \epsilon$ if ϵ is a positive number. The sequence converges to L because

a) $s_n \leq L$ for *all* values of n and

b) given any $\epsilon > 0$, there exists an integer N for which $s_N > L - \epsilon$.

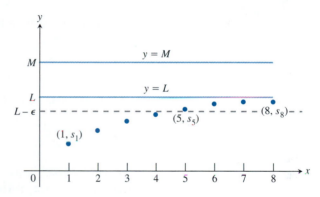

9.10 If the terms of a nondecreasing sequence have an upper bound M, they have a limit $L \leq M$.

The fact that $\{s_n\}$ is a nondecreasing sequence tells us further that

$$s_n \geq s_N > L - \epsilon \qquad \text{for all } n \geq N.$$

This means that *all* the numbers s_n beyond the Nth one lie within ϵ of L. This is precisely the condition for L to be the limit of the sequence s_n.

The facts for nondecreasing sequences are summarized in the following theorem.

THEOREM 6   The Nondecreasing Sequence Theorem

A nondecreasing sequence converges if and only if its terms are bounded from above. If all the terms are less than or equal to M, then the limit of the sequence is less than or equal to M as well.

Theorem 6 tells us that we can show that a series $\sum a_n$ of nonnegative terms converges if we can show that its partial sums are bounded from above. The question, of course, is how to find out in any particular instance whether the s_n's have an upper bound.

Sometimes we can show that the s_n's are bounded above by showing that each one is less than or equal to the corresponding partial sum of a series that is already known to converge. The next example shows how this can happen.

EXAMPLE 2

The series

$$\sum_{n=0}^{\infty} \frac{1}{n!} = 1 + \frac{1}{1!} + \frac{1}{2!} + \frac{1}{3!} + \cdots$$

converges because its terms are all positive and less than or equal to the corresponding terms of

$$1 + \sum_{n=0}^{\infty} \frac{1}{2^n} = 1 + 1 + \frac{1}{2} + \frac{1}{2^2} + \cdots.$$

To see how this relationship leads to an upper bound for the partial sums of $\sum_{n=0}^{\infty}(1/n!)$, let

$$s_n = 1 + \frac{1}{1!} + \frac{1}{2!} + \cdots + \frac{1}{n!},$$

and observe that, for each n,

$$s_n \leq 1 + 1 + \frac{1}{2} + \frac{1}{2^2} + \cdots + \frac{1}{2^n} < 1 + \sum_{n=0}^{\infty} \frac{1}{2^n} = 1 + \frac{1}{1 - (1/2)} = 3.$$

Thus, the partial sums of $\sum_{n=0}^{\infty}(1/n!)$ are all less than 3. Therefore, $\sum_{n=0}^{\infty}(1/n!)$ converges (but not necessarily to 3). ≡

EXPLORATION BIT

From the second term on in Example 2, we compare

$$\frac{1}{n!} \quad \text{with} \quad \frac{1}{2^{n-1}}.$$

Give a convincing argument that

$$\frac{1}{n!} \leq \frac{1}{2^{n-1}}$$

for all n.

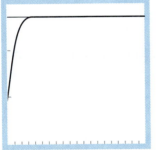

[0, 20] by [0, 3]

9.11 This viewing window shows the partial sums of the series

$$\sum_{n=0}^{\infty} \frac{1}{n!}$$

graphed in connected format (by PARTSUMG). The graph appears to approach the line $y = s_{20}$ (computed by PARTSUMT) quite rapidly as n increases. We will confirm the exact value of the sum in Section 9.7.

EXPLORATION 1

The Sum of a Series, Part 1

Example 2 tells us that the series $\sum_{n=0}^{\infty}(1/n!)$ converges. But what exactly is its sum?

1. Use PARTSUMT and your graphing utility to gather numerical evidence about the sum. Do you recognize what the sum is from s_{20}? (It turns out that the sum and s_{20} computed by PARTSUMT agree to 12 digits.)

2. Use PARTSUMG and your graphing utility to gather graphical evidence about the sum.
   Our results are shown in Fig. 9.11.

3. Compare the series $\sum_{n=0}^{\infty}(1/n!)$ and $\sum_{n=1}^{\infty}(1/n)$ term by term. On the basis of the convergence of the first series, can you make a conjecture about the convergence of the second? If necessary, use your grapher to gather numerical and graphical data about the second series. ☮

EXAMPLE 3 The Harmonic Series

The series

$$\sum_{n=1}^{\infty} \frac{1}{n} = 1 + \frac{1}{2} + \frac{1}{3} + \cdots + \frac{1}{n} + \cdots$$

is called the **harmonic series**. The series diverges because there is no upper bound for its sequence of partial sums. To see why, imagine grouping the terms of the series in the following way:

$$1 + \frac{1}{2} + \underbrace{\left(\frac{1}{3} + \frac{1}{4}\right)}_{> \frac{2}{4} = \frac{1}{2}} + \underbrace{\left(\frac{1}{5} + \frac{1}{6} + \frac{1}{7} + \frac{1}{8}\right)}_{> \frac{4}{8} = \frac{1}{2}} + \underbrace{\left(\frac{1}{9} + \frac{1}{10} + \cdots + \frac{1}{16}\right)}_{> \frac{8}{16} = \frac{1}{2}} + \cdots.$$

The sum of the first two terms is 1.5. The sum of the next two terms is $1/3 + 1/4$, which is greater than $1/4 + 1/4 = 1/2$ (because $1/3$ is greater than $1/4$). The sum of the next four terms is $1/5 + 1/6 + 1/7 + 1/8$, which is greater than $1/8 + 1/8 + 1/8 + 1/8 = 1/2$. The sum of the next eight terms is $1/9 + 1/10 + 1/11 + 1/12 + 1/13 + 1/14 + 1/15 + 1/16$, which is greater than $8/16 = 1/2$. The sum of the next 16 terms is greater than $16/32 = 1/2$, and so on. In general, the sum of the 2^n terms ending with $1/2^{n+1}$ is greater than $2^n/2^{n+1} = 1/2$. The sequence of partial sums is not bounded: If $n = 2^k$, the partial sum s_n is greater than $k/2$. The harmonic series diverges. (We will see later that the nth partial sum is slightly greater than $\ln(n+1)$.) ▤

Notice that the nth-Term Test for Divergence does not detect the divergence of the harmonic series. The nth term, $1/n$, goes to zero, but the series still diverges.

Comparison Test for Convergence

We established the convergence of the series in Example 2 by comparing it with a series that was already known to converge. This kind of comparison is typical of a procedure called the Comparison Test for convergence of series of nonnegative terms.

Comparison Test for Series of Nonnegative Terms

Let $\sum a_n$, $\sum c_n$ and $\sum d_n$ be series with no negative terms.

a) **Test for Convergence.** The series $\sum a_n$ converges if $\sum c_n$ converges and $a_n \le c_n$ for all $n > n_0$ for some positive integer n_0.

b) **Test for Divergence.** The series $\sum a_n$ diverges if $\sum d_n$ diverges and $a_n \ge d_n$ for all $n > n_0$ for some positive integer n_0.

In part (a), the partial sums of the series $\sum a_n$ are bounded above by

$$M = a_1 + a_2 + \cdots + a_{n_0} + \sum_{n=n_0+1}^{\infty} c_n.$$

WHY THE *HARMONIC* SERIES?

The terms in a harmonic series correspond to the stops on a vibrating string that produce multiples of the fundamental frequency. For example, 1/2 produces the first harmonic at a frequency that is twice the fundamental frequency, 1/3 produces a frequency that is three times the fundamental, and so on. This combination of frequencies produces musical harmony; hence the name of the series.

THE COMPARISON TEST

To apply the Comparison Test, we need to have on hand a list of series that we already know about. You may want to copy and add to this list as we go.

Convergent Series:

Geometric series with $|r| < 1$

Telescoping series like $\sum \dfrac{1}{n(n+1)}$

The series $\sum_{n=1}^{\infty} \dfrac{1}{n!}$

Divergent Series:

Geometric series with $|r| \ge 1$

Any series $\sum a_n$ with $\lim a_n \ne 0$

The harmonic series $\sum_{n=1}^{\infty} \dfrac{1}{n}$

EXPLORATION BIT

How "fast" does the harmonic series diverge?

1. Find the smallest value of n for which the partial sum $s_n \geq 5$.

2. Find the smallest value of n for which the partial sum $s_n \geq 8$.

The last paragraph of Example 3 gives a clue to the answers for parts 1 and 2. Does the clue agree with your results?

They therefore form a nondecreasing sequence with a limit L that is less than or equal to M.

In part (b), the partial sums for $\sum a_n$ are not bounded from above. If they were, the partial sums for $\sum d_n$ would be bounded by

$$M' = d_1 + d_2 + \cdots + d_{n_0} + \sum_{n=n_0+1}^{\infty} a_n,$$

and $\sum d_n$ would have to converge instead of diverge.

To apply the Comparison Test to a series, we do not have to include the early terms of the series. We can start the test with any index N, provided that we include all the terms of the series being tested from there on.

EXAMPLE 4

We can establish the convergence of the series

$$5 + \frac{2}{3} + 1 + \frac{1}{7} + \frac{1}{2} + \frac{1}{3!} + \frac{1}{4!} + \cdots + \frac{1}{k!} + \cdots$$

by ignoring the first four terms and comparing the remainder with the convergent geometric series

$$\sum_{n=1}^{\infty} \frac{1}{2^n} = \frac{1}{2} + \frac{1}{4} + \frac{1}{8} + \cdots.$$

The Integral Test

The Integral Test connects integral and series convergence. It will allow us to add to our list of convergent and divergent series. We introduce this test with a specific example, the series whose nth term is $1/n^2$. We demonstrate its convergence by relating it to the convergent improper integral, $\int_1^{\infty} 1/x^2 \, dx$.

EXAMPLE 5

Does the series

$$\sum_{n=1}^{\infty} \frac{1}{n^2} = 1 + \frac{1}{4} + \frac{1}{9} + \frac{1}{16} + \cdots + \frac{1}{n^2} + \cdots$$

converge or diverge?

Solution We will compare the partial sums of the series to the area under the graph of $f(x) = 1/x^2$. In Chapter 8, we learned that $\int_1^{\infty} 1/x^2 \, dx$ converges. Figure 9.12 suggests how we can use area under the graph of f to find an upper bound for

$$s_n = \frac{1}{1^2} + \frac{1}{2^2} + \frac{1}{3^2} + \cdots + \frac{1}{n^2} = f(1) + f(2) + f(3) + \cdots + f(n).$$

The first term of the series is $f(1) = 1$, which we can interpret as the area of a rectangle of height $1/1^2$ and base the length of the interval $[0, 1]$ on the x-axis. The next rectangle, over the interval $[1, 2]$, has area $f(2) = 1/2^2$. That rectangle lies below the curve $y = 1/x^2$, so

$$f(2) = \frac{1}{2^2} < \int_1^2 \frac{1}{x^2} \, dx.$$

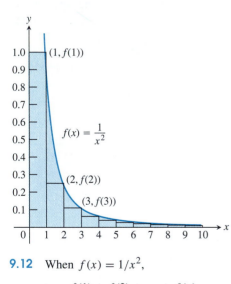

9.12 When $f(x) = 1/x^2$,

$$s_n = f(1) + f(2) + \cdots + f(n)$$

$$< f(1) + \int_1^n f(x) \, dx$$

$$= 1 + \left(\frac{1}{1} - \frac{1}{n} \right) < 2.$$

In the same way,

$$f(3) = \frac{1}{3^2} < \int_2^3 \frac{1}{x^2}\, dx, \qquad \cdots, \qquad f(n) = \frac{1}{n^2} < \int_{n-1}^n \frac{1}{x^2}\, dx.$$

Adding these inequalities, we get

$$f(2) + f(3) + \cdots + f(n) < \int_1^n (1/x^2)\, dx = -\frac{1}{x}\Big]_1^n = 1 - \frac{1}{n}.$$

To get s_n, we must add $f(1) = 1$. When we do so, we have

$$s_n = f(1) + f(2) + f(3) + \cdots + f(n) < 2 - \frac{1}{n} < 2.$$

The sequence of partial sums $\{s_n\}$ is an increasing sequence that is bounded above, so it has a limit, and the given series converges. ≣

EXPLORATION 2

The Sum of a Series, Part 2

Example 5 tells us that the series $\sum_{n=1}^{\infty} (1/n^2)$ converges. But what exactly is its sum?

1. Use your grapher, perhaps using a program such as PARTSUMT, to gather numerical evidence about the sum, S. When you believe that you have a good approximation, S_a, for S, compute $\sqrt{6\, S_a}$. Do you recognize the result? Make a conjecture about the exact value of S.

2. Using the exact value of S from part 1, determine the smallest value of n for which the partial sum s_n agrees with S to two digits; to three digits. Comment on whether s_n seems to converge to S rapidly or slowly.   ⌘

We now state and prove the Integral Test in more general terms.

Integral Test

Let $a_n = f(n)$, where $f(x)$ is a continuous, positive, decreasing function of x for all $x \geq 1$. Then the series $\sum a_n$ and the integral $\int_1^{\infty} f(x)\, dx$ both converge or both diverge.

Proof of the Integral Test   We start with the assumption that f is a decreasing function with $f(n) = a_n$ for every n. This leads us to observe that the rectangles in Fig. 9.13(a),
which have areas $a_1, a_2, \ldots, a_n$, collectively enclose more area than that under the curve $y = f(x)$ from $x = 1$ to $x = n + 1$. That is,

$$\int_1^{n+1} f(x)\, dx \leq a_1 + a_2 + \cdots + a_n.$$

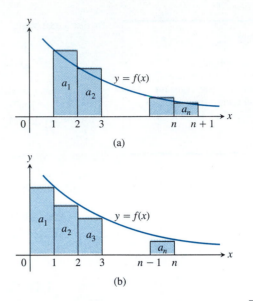

(a)

(b)

9.13 Subject to the conditions of the Integral Test, the series $\sum_{n=1}^{\infty} a_n$ and the integral $\int_1^{\infty} f(x)\, dx$ both converge or both diverge.

In Fig. 9.13(b), the rectangles have been faced to the left instead of to the right. If we momentarily disregard the first rectangle of area a_1, we see that

$$a_2 + a_3 + \cdots + a_n \leq \int_1^n f(x)\, dx.$$

If we include a_1, we have

$$a_1 + a_2 + \cdots + a_n \leq a_1 + \int_1^n f(x)\, dx.$$

Combining these results gives

$$\int_1^{n+1} f(x)\, dx \leq a_1 + a_2 + \cdots + a_n \leq a_1 + \int_1^n f(x)\, dx. \tag{1}$$

If the integral $\int_1^{\infty} f(x)\, dx$ is finite, the right-hand inequality shows that $\sum a_n$ is also finite. But if $\int_1^{\infty} f(x)\, dx$ is infinite, then the left-hand inequality shows that the series is also infinite.

Hence, the series and the integral are both finite or both infinite. ≡

EXPLORATION 3

The p-Series

The information about $\int_1^{\infty}(1/x^p)\, dx$ in Section 8.6 (proved in Exercises 8.6) and the Integral Test allow us to analyze the p-series for any real value of p. If p is a real constant, the series

$$\sum_{n=1}^{\infty} \frac{1}{n^p} = \frac{1}{1^p} + \frac{1}{2^p} + \frac{1}{3^p} + \cdots + \frac{1}{n^p} + \cdots$$

converges if $p > 1$ and diverges if $p \leq 1$.

1. Show that the p-series converges for $p > 1$ by showing that $\int_1^\infty (1/x^p)\, dx$ converges and then using the Integral Test.

2. Explain why the p-series diverges for $p = 1$.

3. Show why the p-series diverges for $p < 1$ by comparing its terms with the terms of the divergent harmonic series and then using the Comparison Test.   &

The Limit Comparison Test

We now present a more powerful form of the Comparison Test, known as the Limit Comparison Test. It is particularly handy when we deal with series in which a_n is a rational function of n. The next example will show you what we mean.

EXAMPLE 6

Determine analytically whether the following series converge or diverge.

a) $\displaystyle\sum_{n=2}^{\infty} \frac{2n}{n^2 - n + 1}$ **b)** $\displaystyle\sum_{n=2}^{\infty} \frac{2n^3 + 100n^2 + 1000}{(1/8)n^6 - n + 2}$

Solution In determining convergence or divergence, only the tails count. When n is very large, the highest powers of n in numerator and denominator are what count the most. So in part (a), we reason this way:

$$a_n = \frac{2n}{n^2 - n + 1}$$

behaves about like the end behavior model $2n/n^2 = 2/n$, and by comparison with $\sum 1/n$, we guess that $\sum a_n$ diverges.

To be more precise, we take

$$a_n = \frac{2n}{n^2 - n + 1} \qquad \text{and} \qquad d_n = \frac{1}{n}$$

and look at the ratio

$$\frac{a_n}{d_n} = \frac{2n^2}{n^2 - n + 1} = \frac{2}{1 - \left(\dfrac{1}{n}\right) + \left(\dfrac{1}{n^2}\right)}.$$

Clearly, as $n \to \infty$ the limit is 2: $\lim a_n/d_n = 2$.

This means that, in particular, if we take $\epsilon = 1$ in the definition of limit, we know that there is an integer N such that a_n/d_n is within 1 unit of this limit for all $n \geq N$:

$$2 - 1 \leq a_n/d_n \leq 2 + 1 \qquad (n \geq N).$$

Thus, $a_n \geq d_n$ for $n \geq N$. Therefore, by the Comparison Test, $\sum a_n$ diverges because $\sum d_n$ diverges.

In part (b), we reason that a_n will behave about like the end behavior model $2n^3/(1/8)n^6 = 16/n^3$, and by comparison with $\sum 1/n^3$, a p-series with $p = 3$, we guess that the series converges.

EXPLORATION BIT

Use your grapher, perhaps using PARTSUMG, to investigate the series in Example 6(a). Explain the difficulty with using this graphical information to predict convergence. The analytic method in Example 6 shows the power of algebra.

EXPLORATION BIT

The Comparison Tests require that the terms a_n be positive, or at least nonnegative, for $n \geq$ some n_0. For the terms a_n as given in Example 6(b), give a convincing argument that $a_n \geq 0$.

If we let $c_n = 1/n^3$, we can show that $\lim a_n/c_n = 16$. Taking $\epsilon = 1$ in the definition of limit, we can conclude that there is an index N such that

$$15 \leq \frac{a_n}{c_n} \leq 17 \qquad (n \geq N),$$

or $a_n \leq 17c_n$. Because $\sum c_n$ converges, so also does $\sum 17c_n$, and thus, so does $\sum a_n$. ≡

Our rather rough guesswork paved the way for successful choices of comparison series. We make all of this more precise in the following Limit Comparison Test.

Limit Comparison Test

a) **Test for Convergence.** If $a_n \geq 0$ for $n \geq n_0$ and there is a convergent series $\sum c_n$ such that $c_n > 0$ and

$$\lim \frac{a_n}{c_n} < \infty,$$

then $\sum a_n$ converges.

b) **Test for Divergence.** If $a_n \geq 0$ for $n \geq n_0$ and there is a divergent series $\sum d_n$ such that $d_n > 0$ and

$$\lim \frac{a_n}{d_n} > 0,$$

then $\sum a_n$ diverges.

A simpler version of the Limit Comparison Test combines parts (a) and (b) as follows.

Simplified Limit Comparison Test

If the terms of the two series $\sum a_n$ and $\sum b_n$ are positive for $n \geq n_0$, and the limit of a_n/b_n is finite and positive, then both series converge or both diverge.

The Simplified Limit Comparison Test is the one we use most often.

EXPLORATION 4

The Limit Comparison Test

Apply the Limit Comparison Test to determine whether each of the following series converges or diverges. For each series, follow the indicated steps.

a) $\dfrac{3}{4} + \dfrac{5}{9} + \dfrac{7}{16} + \dfrac{9}{25} + \cdots = \displaystyle\sum_{n=1}^{\infty} \dfrac{2n+1}{(n+1)^2}$

b) $\dfrac{101}{3} + \dfrac{102}{10} + \dfrac{103}{29} + \cdots = \displaystyle\sum_{n=1}^{\infty} \dfrac{100+n}{n^3+2}$

c) $\dfrac{1}{1} + \dfrac{1}{3} + \dfrac{1}{7} + \cdots = \displaystyle\sum_{n=1}^{\infty} \dfrac{1}{2^n-1}$

1. For the given nth term, a_n, determine an end behavior model b_n. Confirm that $\{b_n\}$ converges or diverges.

2. GRAPH $y = a_n/b_n$. Check that the limit of a_n/b_n is finite and positive. State whether $\{a_n\}$ is convergent or divergent.

3. Confirm the limit in part 2.

In Exploration 4, the series in parts (b) and (c) converge to a finite sum. But what exactly is the sum for each?

For the series $\sum_{n=1}^{\infty} 1/(2^n - 1)$ of part (c), we used PARTSUMT on our grapher and got the partial sum $s_t = 1.606695151$ for $t \geq 31$, strongly suggesting this to be the sum of the series to ten-digit accuracy.

For the series $\sum_{n=1}^{\infty}(100+n)/(n^3+2)$ in part (b), stability of values on the home screen is slower as $s_{400} = 51.63081059$ and $s_{800} = 51.63229194$. These values and the graph of the partial sums (Fig. 9.14) are compelling evidence that 51.6 is a reasonable estimate for the sum of the series.

We should be aware that there is some risk in drawing conclusions about sums of series from a graphing utility, especially if they converge very slowly. We will be able to have more confidence in our estimates when we can determine the possible error in our work.

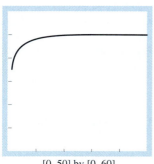

[0, 50] by [0, 60]

9.14 Shown in connected mode, the graph of the partial sums s_n of the series $\sum_{n=1}^{\infty}(100+n)/(n^3+2)$ suggests that the partial sums have stabilized.

Exercises 9.3

Which series in Exercises 1–24 converge, and which diverge? Give reasons for your answers. If a series converges, estimate its sum, and support your estimate graphically. Test the stability of the partial sums both numerically and graphically.

1. $\displaystyle\sum_{n=1}^{\infty} \dfrac{1}{10^n}$

2. $\displaystyle\sum_{n=1}^{\infty} -\dfrac{1}{8^n}$

3. $\displaystyle\sum_{n=1}^{\infty} \dfrac{n}{n+2}$

4. $\displaystyle\sum_{n=1}^{\infty} \dfrac{5}{n}$

5. $\displaystyle\sum_{n=1}^{\infty} \dfrac{\sin^2 n}{2^n}$

6. $\displaystyle\sum_{n=1}^{\infty} \dfrac{1+\cos n}{n^2}$

7. $\displaystyle\sum_{n=2}^{\infty} \dfrac{\ln n}{n}$

8. $\displaystyle\sum_{n=2}^{\infty} \dfrac{\sqrt{n}}{\ln n}$

9. $\displaystyle\sum_{n=1}^{\infty} \dfrac{1}{n\sqrt{n}}$

10. $\displaystyle\sum_{n=1}^{\infty} \dfrac{2^n}{3^n}$

11. $\displaystyle\sum_{n=0}^{\infty} \dfrac{-2}{n+1}$

12. $\displaystyle\sum_{n=1}^{\infty} \dfrac{1}{1+\ln n}$

13. $\displaystyle\sum_{n=1}^{\infty} \dfrac{1}{2n-1}$

14. $\displaystyle\sum_{n=1}^{\infty} \dfrac{2^n}{n+1}$

15. $\displaystyle\sum_{n=1}^{\infty} \left(1+\dfrac{1}{n}\right)^n$

16. $\displaystyle\sum_{n=1}^{\infty} \left(\dfrac{n}{3n+1}\right)^n$

17. $\displaystyle\sum_{n=1}^{\infty} \dfrac{n}{n^2+1}$

18. $\displaystyle\sum_{n=1}^{\infty} \dfrac{\sqrt{n}}{n^2+1}$

19. $\displaystyle\sum_{n=1}^{\infty} \dfrac{1}{\sqrt{n^3+2}}$

20. $\displaystyle\sum_{n=1}^{\infty} \dfrac{1}{n\sqrt[n]{n}}$

21. $\displaystyle\sum_{n=1}^{\infty} \dfrac{1+n}{n \cdot 2^n}$

22. $\displaystyle\sum_{n=1}^{\infty} \dfrac{1}{(\ln 2)^n}$

23. $\displaystyle\sum_{n=1}^{\infty} \dfrac{1}{3^{n-1}+1}$

24. $\displaystyle\sum_{n=1}^{\infty} \dfrac{10n+1}{n(n+1)(n+2)}$

25. Refer to Exploration 1. Find the smallest value of n such that s_n agrees with e to the full calculator display.

26. Later you will confirm that $\sum_{n=0}^{\infty} (1/n!) = e$. Use this fact to show that the sum of the series in Example 4 is $e + 101/21$. Support this answer using PARTSUMT.

27. There is absolutely no empirical evidence for the divergence of the harmonic series even though we know that it diverges. The partial sums, which satisfy the inequality

$$\ln(n+1) = \int_1^{n+1} \frac{1}{x}\, dx \leq 1 + \frac{1}{2} + \cdots + \frac{1}{n}$$

$$\leq 1 + \int_1^n \frac{1}{x}\, dx = 1 + \ln n$$

(Eq. 1), just grow too slowly. To see what we mean, suppose you had started with $s_1 = 1$ the day the universe was formed, 13 billion years ago, and added a new term every *second*. About how large would s_n be today?

28. There are no values of x for which $\sum_{n=1}^{\infty} (1/nx)$ converges. Why?

29. Show that if $\sum_{n=1}^{\infty} a_n$ is a convergent series of nonnegative numbers, then $\sum_{n=1}^{\infty} (a_n/n)$ converges.

30. Show that if $\sum a_n$ and $\sum b_n$ are convergent series with $a_n \geq 0$ and $b_n \geq 0$, then $\sum a_n b_n$ converges. (*Hint:* From some integer on, $0 \leq a_n$ and $b_n < 1$ so $a_n b_n \leq a_n + b_n$.)

31. *Nonincreasing sequences.* A sequence of numbers $\{s_n\}$ in which $s_n \geq s_{n+1}$ for every n is called a **nonincreasing sequence**. A sequence $\{s_n\}$ is bounded from below if there is a finite constant M with $M \leq s_n$ for every n. Such a number M is called a lower bound for the sequence. Deduce from Theorem 6 that a nonincreasing sequence that is bounded from below converges and that a nonincreasing sequence that is not bounded from below diverges.

32. *The Cauchy Condensation Test.* The Cauchy Condensation Test says: Let $\{a_n\}$ be a nonincreasing sequence ($a_n \geq a_{n+1}$ for all n) of positive terms that converges to 0. Then $\sum a_n$ converges if and only if $\sum 2^n a_{2^n}$ converges. For example, $\sum (1/n)$ diverges because $\sum 2^n \cdot (1/2^n) = \sum 1$ diverges. Use the Cauchy Condensation Test to show that

a) $\displaystyle\sum_{n=2}^{\infty} \frac{1}{n \ln n}$ diverges.

b) $\displaystyle\sum_{n=1}^{\infty} \frac{1}{n^p}$ converges if $p > 1$ and diverges if $p \leq 1$.

9.4

Series with Nonnegative Terms: Ratio and Root Tests

Convergence tests that depend on comparing one series with another series or with an integral are called *extrinsic* tests. They are very useful, but there are reasons to look for tests that do not require comparison. As a practical matter, we may not be able to find the series or function that we need to make a comparison work. And, in principle, all the information about a given series should be contained in its own terms. We therefore turn our attention to *intrinsic* tests—those that depend only on the series at hand.

The Ratio Test

Our first intrinsic test, the Ratio Test, measures the rate of growth (or decline) of a series by examining the ratio a_{n+1}/a_n. For a geometric series, this rate of growth is the constant ratio of the series, and the series converges if and only if the ratio is less than 1 in absolute value. Even if the ratio is not constant, we may be able to find a geometric series for comparison, as in the next example.

EXAMPLE 1

Let $a_1 = 1$, and define a_{n+1} to be $a_{n+1} = (n/(2n+1))a_n$. Does the series $\sum a_n$ converge or diverge? Estimate the sum if the series converges.

Solution   We begin by writing out a few terms of the series:

$$a_1 = 1, \quad a_2 = \frac{1}{3}a_1 = \frac{1}{3}, \quad a_3 = \frac{2}{5}a_2 = \frac{1 \cdot 2}{3 \cdot 5}, \quad a_4 = \frac{3}{7}a_3 = \frac{1 \cdot 2 \cdot 3}{3 \cdot 5 \cdot 7}.$$

Each term is somewhat less than $1/2$ the term before it because $n/(2n+1)$ is less than $1/2$. Therefore, the terms of the given series are less than or equal to the terms of the geometric series

$$1 + \left(\frac{1}{2}\right) + \left(\frac{1}{4}\right) + \cdots + \left(\frac{1}{2}\right)^{n-1} + \cdots$$

that converges to 2. So our series also converges, and its sum is less than 2. For $n \geq 30$, our grapher gives $s_n = 1.570796327$, a good estimate for the sum. ▤

EXPLORATION BIT

For $n \geq 30$, $s_n = 1.570796327$. Can you relate this number to one that is more familiar? If so, make a conjecture about the exact sum of the series.

In proving the Ratio Test, we will make a comparison with appropriate geometric series as in the example above, but when we *apply* it, we do not actually make a direct comparison.

EXPLORATION BITS

1. It part (b) of the Ratio Test, could $\rho = \infty$? Study the proof and decide.

2. In the Ratio Test, might

$$\lim_{n \to \infty} \frac{a_{n+1}}{a_n}$$

not exist for some reason other than the possibility mentioned in part 1 above? Give an example if you can. If this happens, what can you conclude from the Ratio Test?

3. Find two series, one convergent and one divergent, for which $\rho = 1$ in the Ratio Test.

The Ratio Test

Let $\sum a_n$ be a series with positive terms, and suppose that

$$\lim_{n \to \infty} \frac{a_{n+1}}{a_n} = \rho. \qquad \text{Greek letter rho}$$

Then

a) the series *converges* if $\rho < 1$,

b) the series *diverges* if $\rho > 1$,

c) the series *may converge or it may diverge* if $\rho = 1$. (The test provides no information.)

Proof of the Ratio Test

a) $\rho < 1$. Let r be a number between ρ and 1. Then the number $\epsilon = r - \rho$ is positive. Since

$$\frac{a_{n+1}}{a_n} \to \rho,$$

a_{n+1}/a_n must lie within ϵ of ρ when n is large enough, say, for all $n \geq N$. In particular,

$$\frac{a_{n+1}}{a_n} < \rho + \epsilon = r \qquad (n \geq N).$$

That is,

$$a_{N+1} < ra_N,$$

$$a_{N+2} < ra_{N+1} < r^2 a_N,$$

$$a_{N+3} < ra_{N+2} < r^3 a_N,$$

$$\vdots$$

$$a_{N+m} < ra_{N+m-1} < r^m a_N.$$

These inequalities show that the terms of our series, after the Nth term, approach zero more rapidly than the terms in a geometric series with ratio $r < 1$. More precisely, consider the series $\sum c_n$, where $c_n = a_n$ for $n = 1, 2, \ldots, N$ and $c_{N+1} = ra_N, c_{N+2} = r^2 a_N, \ldots, c_{N+m} = r^m a_N, \ldots$. Now $a_n \leq c_n$ for all n, and

$$\sum_{n=1}^{\infty} c_n = a_1 + a_2 + \cdots + a_{N-1} + a_N + ra_N + r^2 a_N + \cdots$$

$$= a_1 + a_2 + \cdots + a_{N-1} + a_N(1 + r + r^2 + \cdots).$$

The geometric series $1 + r + r^2 + \cdots$ converges because $|r| < 1$, so $\sum c_n$ converges. Since $a_n \leq c_n$, $\sum a_n$ also converges.

b) $\rho > 1$. From some index M on,

$$\frac{a_{n+1}}{a_n} > 1 \qquad \text{and} \qquad a_M < a_{M+1} < a_{M+2} < \cdots.$$

The terms of the series do not approach zero as n becomes infinite, and the series diverges by the nth-Term Test.

c) $\rho = 1$. The two series

$$\sum_{n=1}^{\infty} \frac{1}{n} \qquad \text{and} \qquad \sum_{n=1}^{\infty} \frac{1}{n^2}$$

show that some other test for convergence must be used when $\rho = 1$.

$$\text{For } \sum_{n=1}^{\infty} \frac{1}{n}: \qquad \frac{a_{n+1}}{a_n} = \frac{1/(n+1)}{1/n} = \frac{n}{n+1} \to 1.$$

$$\text{For } \sum_{n=1}^{\infty} \frac{1}{n^2}: \qquad \frac{a_{n+1}}{a_n} = \frac{1/(n+1)^2}{1/n^2} = \left(\frac{n}{n+1}\right)^2 \to 1^2 = 1.$$

In both cases, $\rho = 1$, yet the first series diverges while the second converges. $\equiv$

> **EXPLORATION BIT**
>
> Apply the Ratio Test to series whose convergence or divergence is already known to you.

The Ratio Test is often effective when the terms of the series contain factorials of expressions involving n or expressions raised to the nth power or combinations, as in the next example.

EXAMPLE 2

Use the Ratio Test to investigate the convergence or divergence of the following series. Estimate or confirm the sum if the series converges.

a) $\displaystyle\sum_{n=1}^{\infty} \frac{n!n!}{(2n)!}$ **b)** $\displaystyle\sum_{n=1}^{\infty} \frac{4^n n!n!}{(2n)!}$ **c)** $\displaystyle\sum_{n=0}^{\infty} \frac{2^n + 5}{3^n}$

Solution

a) If $a_n = n!n!/(2n)!$, then $a_{n+1} = (n+1)!(n+1)!/(2n+2)!$, and

$$\frac{a_{n+1}}{a_n} = \frac{(n+1)!(n+1)!(2n)!}{n!n!(2n+2)(2n+1)(2n)!}$$

$$= \frac{(n+1)(n+1)}{(2n+2)(2n+1)} = \frac{n+1}{4n+2} \rightarrow \frac{1}{4}.$$

The series converges because $\rho = 1/4$ is less than 1. To estimate the sum, we find (using PARTSUMT) that the partial sums stabilize at 0.7363998587.

b) If $a_n = 4^n n!n!/(2n)!$, then

$$\frac{a_{n+1}}{a_n} = \frac{4^{n+1}(n+1)!(n+1)!}{(2n+2)(2n+1)(2n)!} \times \frac{(2n)!}{4^n n!n!}$$

$$= \frac{4(n+1)(n+1)}{(2n+2)(2n+1)} = \frac{2(n+1)}{2n+1} \rightarrow 1.$$

Because the limit is $\rho = 1$, we cannot decide on the basis of the Ratio Test alone whether the series converges or diverges. However, when we note that $a_{n+1}/a_n = (2n+2)/(2n+1)$, we conclude that a_{n+1} is always greater than a_n because $(2n+2)/(2n+1)$ is always greater than 1. Therefore, all terms are greater than or equal to $a_1 = 2$, and the nth term does not go to zero as n tends to infinity. Hence, by the nth-Term Test, the series diverges.

c) For the series $\displaystyle\sum_{n=0}^{\infty}(2^n + 5)/3^n$,

$$\frac{a_{n+1}}{a_n} = \frac{(2^{n+1} + 5)/3^{n+1}}{(2^n + 5)/3^n} = \frac{1}{3} \cdot \frac{2^{n+1} + 5}{2^n + 5}$$

$$= \frac{1}{3} \cdot \left(\frac{2 + 5/2^n}{1 + 5/2^n} \right) \rightarrow \frac{1}{3} \cdot \frac{2}{1} = \frac{2}{3}.$$

The series converges because $\rho = 2/3$ is less than 1. We can confirm the sum to be 21/2:

$$\sum_{n=0}^{\infty} \frac{2^n + 5}{3^n} = \sum_{n=0}^{\infty} \left(\frac{2}{3}\right)^n + \sum_{n=0}^{\infty} \frac{5}{3^n} = \frac{1}{1 - 2/3} + \frac{5}{1 - 1/3} = \frac{21}{2}. \quad \equiv$$

The *n*th-Root Test

We return to the question "Does $\sum a_n$ converge?" When there is a simple formula for a_n, we can try one of the tests we already have. But consider the following example.

EXPLORATION BITS

1. Confirm that

$$\frac{\sqrt[n]{n}}{2} \to \frac{1}{2}.$$

2. Explain why a program such as PARTSUMT cannot be applied to the series of Example 3.

EXAMPLE 3

Let $a_n = f(n)/2^n$, where

$$f(n) = \begin{cases} n & \text{if } n \text{ is a prime number,} \\ 1 & \text{otherwise.} \end{cases}$$

Does $\sum a_n$ converge?

Solution   We write out several terms of the series:

$$\sum a_n = \frac{1}{2} + \frac{2}{4} + \frac{3}{8} + \frac{1}{16} + \frac{5}{32} + \frac{1}{64} + \frac{7}{128} + \cdots + \frac{f(n)}{2^n} + \cdots.$$

Clearly, this is not a geometric series. The nth term approaches zero as $n \to \infty$, so we don't know that the series diverges. The Integral Test doesn't look promising. The Ratio Test produces

$$\frac{a_{n+1}}{a_n} = \frac{1}{2}\frac{f(n+1)}{f(n)} = \begin{cases} \dfrac{1}{2} & \text{if neither } n \text{ nor } n+1 \text{ is a prime,} \\[2mm] \dfrac{1}{2n} & \text{if } n \text{ is a prime} \geq 3, \\[2mm] \dfrac{n+1}{2} & \text{if } n+1 \text{ is a prime} \geq 5. \end{cases}$$

The ratio is sometimes close to zero, sometimes very large, and sometimes $1/2$. It has no limit because there are infinitely many primes. A test that will answer the question (affirmatively—yes, the series does converge) is the nth-Root Test. To apply it, we consider the following:

$$\sqrt[n]{a_n} = \frac{\sqrt[n]{f(n)}}{2} = \begin{cases} \dfrac{\sqrt[n]{n}}{2} & \text{if } n \text{ is a prime,} \\[2mm] \dfrac{1}{2} & \text{otherwise.} \end{cases}$$

Therefore,

$$\frac{1}{2} \leq \sqrt[n]{a_n} \leq \frac{\sqrt[n]{n}}{2},$$

and $\lim \sqrt[n]{a_n} = 1/2$ by the Sandwich Theorem. Because this limit is less than 1, the nth-Root Test tells us that the given series converges, as we will now see.   ▤

EXPLORATION BITS

1. In part (b) of the nth-Root Test, could $\rho = \infty$? Study the proof and decide.

2. In the nth-Root Test, might

$$\lim_{n \to \infty} \sqrt[n]{a_n}$$

not exist for some reason other than the possibility mentioned in part 1? Give an example if you can. If this happens, what conclusion can you make from the nth-Root Test?

3. Find two series, one convergent and one divergent, for which $\rho = 1$ in the nth-Root Test.

The nth-Root Test

Let $\sum a_n$ be a series with $a_n \geq 0$ for $n \geq n_0$, and suppose that $\sqrt[n]{a_n} \to \rho$. Then

a) the series *converges* if $\rho < 1$,

b) the series *diverges* if $\rho > 1$, and

c) the test is *not conclusive* if $\rho = 1$.

Proof of the nth-Root Test

a) $\rho < 1$. Choose an $\epsilon > 0$ so small that $\rho + \epsilon < 1$ also. Since $\sqrt[n]{a_n} \to \rho$, the terms $\sqrt[n]{a_n}$ eventually get closer than ϵ to ρ. In other words, there exists an index $N \geq n_0$ such that

$$\sqrt[n]{a_n} < \rho + \epsilon \qquad (n \geq N).$$

Then it is also true that

$$a_n < (\rho + \epsilon)^n \qquad (n \geq N).$$

Now, $\sum_{n=N}^{\infty} (\rho + \epsilon)^n$, a geometric series with ratio $(\rho + \epsilon) < 1$, converges. By the Comparison Test, $\sum_{n=N}^{\infty} a_n$ converges, from which it follows that

$$\sum_{n=1}^{\infty} a_n = a_1 + \cdots + a_{N-1} + \sum_{n=N}^{\infty} a_n$$

converges.

b) $\rho > 1$. For all indices beyond some integer M, we have $\sqrt[n]{a_n} > 1$, so that $a_n > 1$ for $n > M$. The terms of the series do not converge to zero. The series diverges by the nth-Term Test.

c) $\rho = 1$. The series $\sum_{n=1}^{\infty}(1/n)$ and $\sum_{n=1}^{\infty}(1/n^2)$ shows that the test is not conclusive when $\rho = 1$. The first series diverges, and the second converges, but in both cases, $\sqrt[n]{a_n} \to 1$. ■

EXAMPLE 4

One of the following series converges, and the other diverges. Which does which?

a) $\displaystyle\sum_{n=1}^{\infty} \frac{n^2}{2^n}$ b) $\displaystyle\sum_{n=1}^{\infty} \frac{2^n}{n^2}$

EXPLORATION BIT

Apply the Ratio Test to the series in Example 4.

Solution   Series (a) converges because $\sqrt[n]{a_n} = \dfrac{\sqrt[n]{n^2}}{2} \to \dfrac{1}{2} < 1$. But series (b) diverges because $\sqrt[n]{b_n} = \dfrac{2}{\sqrt[n]{n^2}} \to 2 > 1$. ■

Exercises 9.4

Which series in Exercises 1–26 converge, and which diverge? Give reasons for your answers. Estimate the sum if the series converges.

1. $\displaystyle\sum_{n=1}^{\infty} 2\frac{n^2}{2^n}$

2. $\displaystyle\sum_{n=1}^{\infty} \frac{n!}{10^n}$

3. $\displaystyle\sum_{n=1}^{\infty} \frac{n^{10}}{10^n}$

4. $\displaystyle\sum_{n=1}^{\infty} n^2 e^{-n}$

5. $\displaystyle\sum_{n=1}^{\infty} n! \, e^{-n}$

6. $\displaystyle\sum_{n=1}^{\infty} \frac{(-2)^n}{3^n}$

7. $\displaystyle\sum_{n=1}^{\infty} \left(\frac{n-2}{n}\right)^n$

8. $\displaystyle\sum_{n=1}^{\infty} \frac{2 + (-1)^n}{1.25^n}$

9. $\displaystyle\sum_{n=1}^{\infty} \left(1 - \frac{3}{n}\right)^n$

10. $\displaystyle\sum_{n=1}^{\infty} \left(1 - \frac{1}{n^2}\right)^n$

11. $\displaystyle\sum_{n=1}^{\infty}\left(\frac{1}{n}-\frac{1}{n^2}\right)$

12. $\displaystyle\sum_{n=1}^{\infty}\frac{\ln n}{n^3}$

13. $\displaystyle\sum_{n=1}^{\infty}\frac{\ln n}{n}$

14. $\displaystyle\sum_{n=1}^{\infty}\frac{n\ln n}{2^n}$

15. $\displaystyle\sum_{n=1}^{\infty}\frac{(n+1)(n+2)}{n!}$

16. $\displaystyle\sum_{n=1}^{\infty}e^{-n}(n^3)$

17. $\displaystyle\sum_{n=1}^{\infty}\frac{(n+3)!}{3!n!3^n}$

18. $\displaystyle\sum_{n=1}^{\infty}\frac{1}{(2n+1)!}$

19. $\displaystyle\sum_{n=1}^{\infty}-\frac{n^2}{2^n}$

20. $\displaystyle\sum_{n=1}^{\infty}\frac{n!}{n^n}$

21. $\displaystyle\sum_{n=2}^{\infty}\frac{n}{(\ln n)^n}$

22. $\displaystyle\sum_{n=2}^{\infty}\frac{1}{(\ln n)^2}$

23. $\displaystyle\sum_{n=1}^{\infty}\frac{n!}{(n+2)!}$

24. $\displaystyle\sum_{n=1}^{\infty}\frac{n!}{(2n+1)!}$

25. $\displaystyle\sum_{n=1}^{\infty}\frac{3^n}{n^3 2^n}$

26. $\displaystyle\sum_{n=1}^{\infty}\frac{(n!)^n}{n^{(n^2)}}$

Do the series $\sum_{n=1}^{\infty}a_n$ defined by the formulas in Exercises 27–35 converge or diverge? Give reasons for your answers. Esti-

mate the sum using a grapher (and a program like PSUMRECT) if the series converges.

27. $a_1=2,\quad a_{n+1}=\dfrac{1+\sin n}{n}a_n$

28. $a_1=\dfrac{1}{3},\quad a_{n+1}=\dfrac{3n-1}{2n+5}a_n$

29. $a_1=3,\quad a_{n+1}=\dfrac{n}{n+1}a_n$

30. $a_1=2,\quad a_{n+1}=\dfrac{2}{n}a_n$

31. $a_1=-1,\quad a_{n+1}=\dfrac{1+\ln n}{n}a_n$

32. $a_1=\dfrac{1}{2},\quad a_{n+1}=\dfrac{n+\ln n}{n+10}a_n$

33. $a_n=\dfrac{2^n n!n!}{(2n)!}$

34. $a_n=\dfrac{(3n)!}{n!(n+1)!(n+2)!}$

35. $a_1=1,\quad a_{n+1}=\dfrac{n(n+1)}{(n+2)(n+3)}a_n$

(Hint: Write out several terms, see what factors cancel, and then generalize.)

36. Neither the Ratio Test nor the nth-Root Test helps with p-series. Try both tests on $\sum_{n=1}^{\infty}(1/n^p)$, and explain what happens.

9.5 Alternating Series and Absolute Convergence

A series in which the terms are alternately positive and negative is called an **alternating series.** Here are three examples:

$$1\frac{1}{2}+\frac{1}{3}-\frac{1}{4}+\frac{1}{5}-\cdots+\frac{(-1)^{n+1}}{n}+\cdots \tag{1}$$

$$-2+1-\frac{1}{2}+\frac{1}{4}-\frac{1}{8}+\cdots+\frac{(-1)^n 4}{2^n}+\cdots \tag{2}$$

$$1-2+3-4+5-6+\cdots+(-1)^{n-1}n+\cdots \tag{3}$$

Series (1), called the **alternating harmonic series,** converges, as we will see in a moment. Series (2), a geometric series with ratio $r=-1/2$, converges to $-2/(1+1/2)=-4/3$. Series (3) diverges because the nth term does not approach zero.

The Alternating Series Theorem

We prove the convergence of the alternating harmonic series by applying a general result known as the Alternating Series Theorem.

> **THEOREM 7** **The Alternating Series Theorem**
>
> The series
>
> $$\sum_{n=1}^{\infty}(-1)^{n+1}a_n = a_1 - a_2 + a_3 - a_4 + \cdots$$
>
> converges if all three of the following conditions are satisfied:
>
> **1.** The a_n's are all positive.
>
> **2.** $a_n > a_{n+1}$ for all n.
>
> **3.** $a_n \to 0$.

Proof If n is an even integer, say, $n = 2m$, then the sum of the first n terms is

$$S_{2m} = (a_1 - a_2) + (a_3 - a_4) + \cdots + (a_{2m-1} - a_{2m})$$

$$= a_1 - (a_2 - a_3) - (a_4 - a_5) - \cdots - (a_{2m-2} - a_{2m-1}) - a_{2m}.$$

The first equality shows that S_{2m} is the sum of m positive terms, since each term in parentheses is positive. Hence, $S_{2m+2} > S_{2m}$, and the sequence $\{S_{2m}\}$ is nondecreasing. The second equality shows that $S_{2m} < a_1$. Since $\{S_{2m}\}$ is nondecreasing and bounded from above, it has a limit, say,

$$\lim_{m \to \infty} S_{2m} = L. \tag{4}$$

If n is an odd integer, say, $n = 2m + 1$, then the sum of the first n terms is

$$S_{2m+1} = S_{2m} + a_{2m+1}.$$

Since $a_n \to 0$,

$$\lim_{m \to \infty} a_{2m+1} = 0,$$

and as $m \to \infty$,

$$S_{2m+1} = S_{2m} + a_{2m+1} \to L + 0 = L. \tag{5}$$

When we combine the results of (4) and (5), we get

$$\lim_{n \to \infty} S_n = L. \qquad \blacksquare$$

EXPLORATION BIT

Theorem 7 is also known as **Leibniz's Theorem**. The alternating series

$$1 - \frac{1}{3} + \frac{1}{5} - \frac{1}{7} + \frac{1}{9} - \cdots + \frac{(-1)^n}{2n+1} + \cdots$$

is **Leibniz's sum**. Can you find a value for Leibniz's sum?

WHY POSITIVE a_n's?

The first condition in Theorem 7 could say, "The a_n's are all nonnegative." But that condition and condition 2 would mean that if a particular $a_N = 0$, then all a_n "beyond" a_N would also be zero, and the series could be considered finite with a finite sum.

EXAMPLE 1

The *alternating harmonic series*

$$\sum_{n=1}^{\infty}(-1)^{n+1}\frac{1}{n} = 1 - \frac{1}{2} + \frac{1}{3} - \frac{1}{4} + \cdots$$

satisfies the three requirements of the theorem; therefore, it converges. $\blacksquare$

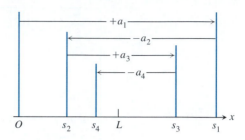

9.15 If an alternating series satisfies the hypotheses of Theorem 7, then its partial sums straddle their limit, denoted here as L.

We can use a visual model (Fig. 9.15) of the partial sums to gain insight into the way in which an alternating series converges to its limit L when the three conditions of Theorem 7 are satisfied.

Starting from the origin on a number line, we lay off the positive distance $s_1 = a_1$. To find the point corresponding to $s_2 = a_1 - a_2$, we must back up a distance equal to a_2. Since $a_2 < a_1$, we do not back up as far as O. Next we go forward a distance a_3 and mark the point corresponding to $s_3 = a_1 - a_2 + a_3$. Since $a_3 < a_2$, we go forward by an amount that is less than the previous backward step; that is, $s_3 < s_1$.

We continue in this seesaw fashion, backing up or going forward as the signs in the series demand. But each forward or backward step is shorter than the preceding step because $a_{n+1} < a_n$. And since the nth term approaches zero as n increases, the size of the steps we take forward or backward gets smaller and smaller.

We thus oscillate across the limit L, but the amplitude of oscillation decreases and approaches zero as its limit. The even-numbered partial sums $s_2, s_4, s_6, \ldots, s_{2m}$ increase toward L, while the odd-numbered sums $s_1, s_3, s_5, \ldots, s_{2m+1}$ decrease toward L. The limit L is between any two successive sums s_n and s_{n+1} and hence differs from s_n by an amount less than a_{n+1}.

EXPLORATION 1

The Alternating Harmonic Series, Part 1

Because the alternating harmonic series

$$\sum_{n=1}^{\infty} (-1)^{n+1} \frac{1}{n}$$

converges to a number L (Example 1), it must behave in the manner described above.

1. Use PARTSUMT to compute the partial sums s_n for $n = 10, 20, \ldots, 90$. Are these partial sums overestimates or underestimates of L? Why?

2. Use PARTSUMT to compute the partial sums s_n for $n = 9, 19, \ldots, 89$. Are these partial sums overestimates or underestimates of L? Why?

3. Compute s_{200} and s_{500}. Discuss the stability of the partial sums of the series. Does the series converge rapidly or slowly to its sum? Explain. (See Fig. 9.16.)

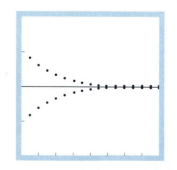

9.16 A graph of the partial sums of the alternating harmonic series and their limit.

4. Make a conjecture about the exact value of L. (*Hint:* Let $L = (s_{90} + s_{89})/2$, and compute e^L.)

Because

$$|L - s_n| < a_{n+1} \qquad \text{for every } n,$$

we can make useful estimates of the sums of convergent alternating series.

THEOREM 8   **The Alternating Series Estimation Theorem**

If the alternating series

$$\sum_{n=1}^{\infty} (-1)^{n+1} a_n$$

satisfies the three conditions of Theorem 7, then

$$s_n = a_1 - a_2 + \cdots + (-1)^{n+1} a_n$$

approximates the sum L of the series with an error whose absolute value is less than a_{n+1}, the numerical value of the first unused term. Furthermore, the remainder, $L - s_n$, has the same sign as the first unused term.

We leave the verification of the sign of the remainder for Exercise 45.

EXAMPLE 2

We first try Theorem 8 on an alternating series whose sum we already know, namely, the geometric series

$$\sum_{n=0}^{\infty} (-1)^n \frac{1}{2^n} = 1 - \frac{1}{2} + \frac{1}{4} - \frac{1}{8} + \frac{1}{16} - \frac{1}{32} + \frac{1}{64} - \frac{1}{128} + \frac{1}{256} - \cdots.$$

The theorem says that if we truncate the series after the eighth term, we throw away a total that is positive and less than 1/256. The sum of the first eight terms is 0.6640625. The sum of the series is

$$\frac{1}{1 - (-1/2)} = \frac{1}{3/2} = \frac{2}{3}.$$

DECIDING OVER OR UNDER

An easy way to decide whether s_n is an overestimate or underestimate of L is to think about the next term of the series. If it "adds" to the partial sum, then s_n is under L. If the next term "subtracts" from s_n, then s_n is over L.

The difference

$$\frac{2}{3} - 0.6640625 = 0.0026041666\ldots,$$

is positive and less than

$$\frac{1}{256} = 0.00390625.$$

EXPLORATION 2

The Alternating Harmonic Series, Part 2

Consider the alternating harmonic series

$$\sum_{n=1}^{\infty}(-1)^{n+1}\frac{1}{n}.$$

1. Show that s_{100} underestimates the sum of the series with an error no greater than 0.01.

2. Find an overestimate of the sum of the series with an error of at most 0.01.

3. How many terms of the series must be summed for the error to be at most 0.001?

4. Find an underestimate and an overestimate of the sum of the series, each with an error no greater than 0.001. Explain.

5. Repeat parts 3 and 4 with an error no greater than 0.0001.

Absolute Convergence

> ### DEFINITIONS
>
> A series $\sum a_k$ **converges absolutely** (is **absolutely convergent**) if the corresponding series of absolute values, $\sum |a_k|$, is convergent.
> A series that converges but does not converge absolutely **converges conditionally.**

EXAMPLE 3

a) The geometric series

$$1 - \frac{1}{2} + \frac{1}{4} - \frac{1}{8} + \cdots$$

converges absolutely because the corresponding series of absolute values

$$1 + \frac{1}{2} + \frac{1}{4} + \frac{1}{8} + \cdots$$

converges.

b) The alternating harmonic series, although it converges, does not converge absolutely because the corresponding series of absolute values is the (divergent) harmonic series. The alternating harmonic series converges conditionally.

The concept of absolute convergence is important because, first, we have many good tests for convergence of series of positive terms. Second, if a series converges absolutely, then it converges. That is the thrust of the next theorem.

EXPLORATION BIT

Being absolutely convergent is a *sufficient* condition for a series to be convergent, but it is not a *necessary* condition. A series can converge without necessarily being absolutely convergent. Example 3(b) shows one example. Find another.

THEOREM 9   The Absolute Convergence Theorem

If $\displaystyle\sum_{n=1}^{\infty} |a_n|$ converges, then $\displaystyle\sum_{n=1}^{\infty} a_n$ converges.

Proof of Theorem 9   For each n,

$$- |a_n| \leq a_n \leq |a_n|,$$

so

$$0 \leq a_n + |a_n| \leq 2|a_n|.$$

If $\sum_{n=1}^{\infty} |a_n|$ converges, then $\sum_{n=1}^{\infty} 2|a_n|$ converges, and by the Comparison Test, the nonnegative series $\sum_{n=1}^{\infty} (a_n + |a_n|)$ converges. The equality $a_n = (a_n + |a_n|) - |a_n|$ now lets us express $\sum_{n=1}^{\infty} a_n$ as the difference of two convergent series:

$$\sum_{n=1}^{\infty} a_n = \sum_{n=1}^{\infty} (a_n + |a_n| - |a_n|) = \sum_{n=1}^{\infty} (a_n + |a_n|) - \sum_{n=1}^{\infty} |a_n|.$$

Therefore, $\sum_{n=1}^{\infty} a_n$ converges.   ≡

We can rephrase Theorem 9 to say that *every absolutely convergent series converges*. However, the converse statement is false. Many convergent series do not converge absolutely. A reason is that the convergence of many series depends on the series' having infinitely many positive and negative terms arranged in a particular order.

EXAMPLE 4

For

$$\sum_{n=1}^{\infty} (-1)^{n+1} \frac{1}{n^2} = 1 - \frac{1}{4} + \frac{1}{9} - \frac{1}{16} + \cdots, \tag{6}$$

the corresponding series of absolute values is

$$\sum_{n=1}^{\infty} \frac{1}{n^2} = 1 + \frac{1}{4} + \frac{1}{9} + \frac{1}{16} + \cdots. \tag{7}$$

The series (6) converges absolutely and hence itself converges. By Theorem 8, $s_{49} = 0.82268$ is an overestimate, and $s_{50} = 0.82227$ is an underestimate (rounding safely) with an error of at most $0.82267 - 0.82227 = 0.0004$, or, more precisely, with an error at most $1/51^2 = 0.00039$ (rounding safely).

≡

EXAMPLE 5

For $\displaystyle\sum_{n=1}^{\infty} \frac{\sin n}{n^2} = \frac{\sin 1}{1} + \frac{\sin 2}{4} + \frac{\sin 3}{9} + \cdots$, the corresponding series of absolute values is

$$\sum_{n=1}^{\infty} \left| \frac{\sin n}{n^2} \right| = \frac{|\sin 1|}{1} + \frac{|\sin 2|}{4} + \cdots,$$

which converges by comparison with $\sum_{n=1}^{\infty} (1/n^2)$ because $|\sin n| \le 1$ for every n. The original series converges absolutely; therefore, it converges.

≡

Alternating p-series

When p is a positive constant, the sequence $\{1/n^p\}$ is a decreasing sequence with limit zero. Therefore, the alternating p-series

$$\sum_{n=1}^{\infty} \frac{(-1)^{n-1}}{n^p} = 1 - \frac{1}{2^p} + \frac{1}{3^p} - \frac{1}{4^p} + \cdots \qquad (p > 0)$$

converges.

For $p > 1$, the series converges absolutely. For $0 < p \le 1$, the series converges conditionally. For example,

$$1 - \frac{1}{\sqrt{2}} + \frac{1}{\sqrt{3}} - \frac{1}{\sqrt{4}} + \cdots$$

converges conditionally, while

$$1 - \frac{1}{2^{3/2}} + \frac{1}{3^{3/2}} - \frac{1}{4^{3/2}} + \cdots$$

converges absolutely.

Preview of Power Series

In the next section we will take up the subject of **power series**, series in which the nth term is a constant times x^n. We investigate the convergence of such series with the Ratio Test or nth-Root Test, as in the next example.

EXAMPLE 6

Find all values of x for which the series

$$x - \frac{x^2}{2} + \frac{x^3}{3} - \frac{x^4}{4} + \cdots + (-1)^{n-1}\frac{x^n}{n} + \cdots$$

converges.

Solution   The series converges absolutely for all values of x for which

$$\lim_{n\to\infty} \sqrt[n]{\left|(-1)^{n-1}\frac{x^n}{n}\right|} = \lim_{n\to\infty} \frac{|x|}{\sqrt[n]{n}} = |x|$$

is less than 1.

At $x = 1$, the series is the alternating harmonic series, which converges. At $x = -1$, the series is the negative of the harmonic series and diverges. For $|x| > 1$, the nth term of the series does not approach zero, and the series diverges. Hence, the series converges only for $-1 < x \le 1$. ≡

> **EXPLORATION BIT**
>
> Give a convincing argument that for $|x| > 1$, the nth term of the series in Example 6 does not approach zero.

EXAMPLE 7

The series

$$\sum_{n=1}^{\infty} (-1)^{n-1}\frac{x^n}{n}$$

converges for $x = 0.6$. Find an estimate for the sum with an error of at most 0.001.

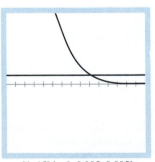

[1, 15] by [−0.005, 0.005]

9.17   Graphs of $y_1 = 0.6^x/x$ and $y_2 = 0.001$. Notice that the graph for y_1 drops below the graph for y_2 at about $x = 9.2$. In other words, $0.6^x/x \le 0.001$ for $x \ge 9.2$. We note that $a_9 = 0.0011$, so s_8 would *not* be a satisfactory estimate for the sum of the series. We note also that the sign accompanying a_{10} is negative, so s_9 is an overestimate of the sum.

Solution   By Theorem 8, the partial sum s_n is an estimate of the sum of this series with an error of at most $a_{n+1} = 0.6^{n+1}/(n+1)$. We need to determine n so that $0.6^{n+1}/(n+1) \le 0.001$.

We let $y_1 = f(x) = 0.6^x/x$, $y_2 = 0.001$, and we GRAPH y_1 and y_2 (Fig. 9.17). Their graphs suggest that $0.6^x/x \le 0.001$ for $x \ge 9.2$. Thus, to have $n + 1 \ge 9.2$, we can let $n = 9$. We find $s_9 = 0.470395$ as an estimate for the sum of the series with an error that should be at most 0.001. We find that $a_{10} = 0.6^{10}/10 = 0.0006$, which confirms our result. ≡

FLOWCHART 9.1 **Procedure for Determining Convergence**

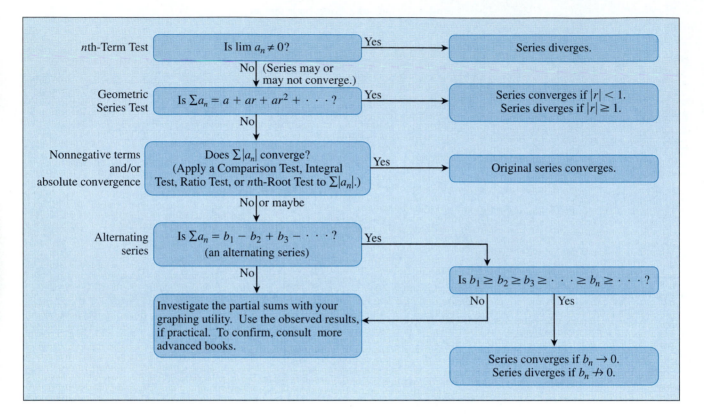

Exercises 9.5

Which of the alternating series in Exercises 1–10 converge, and which diverge? If the series converges, then given an underestimate and an overestimate of the sum L, each being within 0.001 of L. For each estimate, give the error value from Theorem 8.

1. $\sum_{n=1}^{\infty} (-1)^{n+1} \dfrac{1}{n^2}$

2. $\sum_{n=2}^{\infty} (-1)^{n+1} \dfrac{1}{\ln n}$

3. $\sum_{n=1}^{\infty} (-1)^{n+1}$

4. $\sum_{n=1}^{\infty} (-1)^{n+1} \dfrac{10^n}{n^{10}}$

5. $\sum_{n=1}^{\infty} (-1)^{n+1} \dfrac{\sqrt{n}+1}{n+1}$

6. $\sum_{n=1}^{\infty} (-1)^{n+1} \dfrac{1}{n^{3/2}}$

7. $\sum_{n=1}^{\infty} (-1)^{n+1} \dfrac{\ln n}{n}$

8. $\sum_{n=1}^{\infty} (-1)^{n+1} \dfrac{\ln n}{\ln n^2}$

9. $\sum_{n=1}^{\infty} (-1)^{n+1} \dfrac{3\sqrt{n}+1}{\sqrt{n}+1}$

10. $\sum_{n=1}^{\infty} (-1)^{n} \ln \left(1 - \dfrac{1}{n}\right)$

Which of the series in Exercises 11–36 converge absolutely, which converge conditionally, and which diverge?

11. $\sum_{n=1}^{\infty} (-1)^{n+1} (0.1)^n$

12. $\sum_{n=1}^{\infty} (-1)^{n+1} \dfrac{1}{\sqrt{n}}$

13. $\sum_{n=1}^{\infty} (-1)^{n+1} \dfrac{n}{n^3+1}$

14. $\sum_{n=1}^{\infty} \dfrac{n!}{2^n}$

15. $\sum_{n=1}^{\infty} (-1)^{n} \dfrac{1}{n+3}$

16. $\sum_{n=1}^{\infty} (-1)^{n+1} \dfrac{3+n}{5+n}$

17. $\displaystyle\sum_{n=1}^{\infty}(-1)^n\frac{\sin n}{n^2}$

18. $\displaystyle\sum_{n=2}^{\infty}(-1)^n\frac{1}{\ln n^3}$

19. $\displaystyle\sum_{n=1}^{\infty}(-1)^{n+1}\frac{1+n}{n^2}$

20. $\displaystyle\sum_{n=1}^{\infty}\frac{(-2)^{n+1}}{n+5^n}$

21. $\displaystyle\sum_{n=1}^{\infty}n^2(2/3)^n$

22. $\displaystyle\sum_{n=1}^{\infty}(-1)^{n+1}(\sqrt[n]{10})$

23. $\displaystyle\sum_{n=1}^{\infty}(-1)^n\frac{\tan^{-1}n}{n^2+1}$

24. $\displaystyle\sum_{n=2}^{\infty}(-1)^{n+1}\frac{1}{n\ln n}$

25. $\displaystyle\sum_{n=1}^{\infty}\left(\frac{1}{n}-\frac{1}{2n}\right)$

26. $\displaystyle\sum_{n=1}^{\infty}(-1)^{n+1}\frac{(0.1)^n}{n}$

27. $\displaystyle\sum_{n=1}^{\infty}(-1)^n\frac{n}{n+1}$

28. $\displaystyle\sum_{n=1}^{\infty}\frac{(-1)^n}{1+\sqrt{n}}$

29. $\displaystyle\sum_{n=1}^{\infty}\frac{-1}{n^2+2n+1}$

30. $\displaystyle\sum_{n=1}^{\infty}\frac{(-100)^n}{n!}$

31. $\displaystyle\sum_{n=1}^{\infty}(5)^{-n}$

32. $\displaystyle\sum_{n=2}^{\infty}(-1)^n\left(\frac{\ln n}{\ln n^2}\right)^n$

33. $\displaystyle\sum_{n=1}^{\infty}\frac{\cos n\pi}{n\sqrt{n}}$

34. $\displaystyle\sum_{n=1}^{\infty}\frac{\cos n\pi}{n}$

35. $\displaystyle\sum_{n=1}^{\infty}\frac{(-1)^n}{\sqrt{n}+\sqrt{n+1}}$

36. $\displaystyle\sum_{n=1}^{\infty}\frac{(-1)^{n+1}(n!)^2}{(2n)!}$

In Exercises 37–40, give the maximum error possible in using the sum of the first four terms to approximate the sum of the series.

37. $\displaystyle\sum_{n=1}^{\infty}(-1)^{n+1}\frac{1}{n}$

38. $\displaystyle\sum_{n=1}^{\infty}(-1)^{n+1}\frac{1}{10^n}$

39. $\displaystyle\ln(1.01)=\sum_{n=1}^{\infty}(-1)^{n+1}\frac{(0.01)^n}{n}$

40. $\displaystyle\frac{1}{1+t}=\sum_{n=0}^{\infty}(-1)^n t^n \quad (0<t<1)$

In Exercises 41 and 42, approximate the sums with an error of less than 5×10^{-6}.

41. $\displaystyle\sum_{n=0}^{\infty}(-1)^n\frac{1}{(2n)!}$ The sum is cos 1, the cosine of one radian.

42. $\displaystyle\sum_{n=0}^{\infty}(-1)^n\frac{1}{n!}$ The sum is $\dfrac{1}{e}$.

43. a) The series

$$\frac{1}{3}-\frac{1}{2}+\frac{1}{9}-\frac{1}{4}+\frac{1}{27}-\frac{1}{8}+\cdots+\frac{1}{3^n}-\frac{1}{2^n}+\cdots$$

does not meet one of the conditions of Theorem 7. Which one?

b) Find the sum of the series in part (a).

44. The limit L of an alternating series that satisfies the conditions of Theorem 7 lies between the values of any two consecutive partial sums. This suggests using the average

$$\frac{s_n+s_{n+1}}{2}=s_n+\frac{1}{2}a_{n+1}$$

to estimate L. Compute

$$s_{20}+\frac{1}{2}\cdot\frac{1}{21}$$

as an approximation to the sum of the alternating harmonic series. For a clue to the exact sum, see Exploration 1, part 4.

45. Show that whenever an alternating series is approximated by one of its partial sums, and the three conditions of Theorem 7 are satisfied, then the *remainder* (sum of the unused terms) has the same sign as the first unused term. (*Hint:* Group the terms of the remainder in consecutive pairs.)

46. Show that the sum of the first $2n$ terms of the series

$$1-\frac{1}{2}+\frac{1}{2}-\frac{1}{3}+\frac{1}{3}-\frac{1}{4}+\frac{1}{4}-\frac{1}{5}+\frac{1}{5}-\frac{1}{6}+\cdots$$

is the same as the sum of the first n terms of the series

$$\frac{1}{1\cdot2}+\frac{1}{2\cdot3}+\frac{1}{3\cdot4}+\frac{1}{4\cdot5}+\frac{1}{5\cdot6}+\cdots.$$

Do these series converge? What is the sum of the first $2n+1$ terms of the first series? If the series converge, what is the sum?

47. Consider the alternating series

$$\sum_{n=1}^{\infty}(-1)^{n-1}\frac{x^n}{n}$$

of Example 6. Give an overestimate and an underestimate of the sum of the series, each with error at most 0.001, for the value of x specified.

a) 0.2

b) 0.5

c) 0.8

d) 0.9

48. Consider the series

$$\sum_{n=1}^{\infty}(\sin n)/n^2$$

of Example 5.

a) Does Theorem 8 apply to this series? Explain.

b) Compute s_n for $n=100,500,1000,2000$.

c) Estimate the sum of the series.

d) How confident are you of your estimate in part (c)? Explain.

9.6

Power Series

Now that we know how to test infinite series for convergence, we can study the infinite polynomials that we mentioned at the beginning of Section 9.2. Once we know the values of x for which an infinite polynomial converges (its domain), we can use a grapher to approximate the values of the polynomial. We call these infinite polynomials *power series* because they are defined as infinite series of powers of some variable, in our case x. Like polynomials, power series can be added, subtracted, multiplied, differentiated, and integrated to give new power series. They can be divided, too, but we will not do that at this time.

Almost any function with infinitely many derivatives can be represented by a power series, as long as the derivatives don't become "too large." In fact, such power series representations provide methods by which some calculators are programmed to evaluate nonpolynomial functions. As the chapter continues, we will see what the series are for functions like e^x, $\sin x$, $\cos x$, $\ln(1+x)$, and $\tan^{-1} x$, and how these series enable us to approximate function values as accurately as we choose.

In this textbook, we assume that a grapher has the capability of graphing partial sums of a power series *over its domain* either as a built-in feature or with a program called GRAPHSUM. (See the *Resource Manual*.)

Power Series and Convergence

We begin with the formal definition. It shows two forms of a power series. Note that (1) is the special case of (2) with $a = 0$.

DEFINITIONS

A **power series** is a series of the form

$$\sum_{n=0}^{\infty} c_n x^n = c_0 + c_1 x + c_2 x^2 + \cdots + c_n x^n + \cdots \tag{1}$$

or

$$\sum_{n=0}^{\infty} c_n (x-a)^n = c_0 + c_1(x-a) + c_2(x-a)^2 + \cdots + c_n(x-a)^n + \cdots \tag{2}$$

in which the **center** a and the **coefficients** $c_0, c_1, c_2, \ldots c_n, \ldots$ are real constants.

EXAMPLE 1

Taking all the coefficients to be 1 in Eq. (1) gives the geometric power series with first term 1 and ratio x

$$\sum_{n=0}^{\infty} x^n = 1 + x + x^2 + \cdots + x^n + \cdots.$$

The series converges to $1/(1-x)$ when $|x| < 1$. We express this fact by writing

$$\frac{1}{1-x} = 1 + x + x^2 + \cdots + x^n + \cdots \qquad \text{for } -1 < x < 1. \qquad (3)$$

Figure 9.18 shows that the graph of $y = \sum\limits_{n=0}^{10} x^n$ approximates the graph of $y = 1/(1-x)$ quite nicely for most of $-1 < x < 1$. ≡

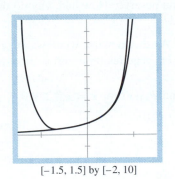

[−1.5, 1.5] by [−2, 10]

9.18 Graphs of

$$y_1 = \frac{1}{1-x}$$

and

$$y_2 = P_{10}(x) = s_{11}(x) = \sum_{n=0}^{10} x^n.$$

Note how alike y_1 and y_2 appear to be for $-1 < x < 1$. This viewing window (using just the 11th partial sum) provides strong graphical support that the function $1/(1-x)$ and the power series $\sum\limits_{n=0}^{\infty} x^n$ are identical on $(-1, 1)$, the domain of convergence of the power series.

We are now at another point in our study at which we can "see" with technology an idea seldom seen before in the study of calculus, simply because of the enormous amount of calculation involved. Using Eq. (3) as an example, we can think of the partial sums of the series on the right as nth-degree polynomials $P_n(x)$:

$$s_1(x) = P_0(x) = 1,$$
$$s_2(x) = P_1(x) = 1 + x,$$
$$s_3(x) = P_2(x) = 1 + x + x^2,$$
$$\vdots$$
$$s_n(x) = P_{n-1}(x) = 1 + x + x^2 + \cdots + x^{n-1}.$$

We can actually view how the P_n converge to their sum, the function $1/(1-x)$. Moreover, we can see that the convergence takes place on the interval $(-1, 1)$, exactly as confirmed by our analytical work.

EXPLORATION 1

Seeing a Power Series Converge

Consider the power series representation

$$y_1 = f(x) = \frac{1}{1-x} = \sum_{n=0}^{\infty} x^n \qquad \text{on } (-1, 1).$$

Let $P_n(x) = \sum\limits_{i=0}^{n} x^i.$

1. GRAPH the function $y_1 = 1/(1-x)$. Then, without actually graphing them, tell how the graphs of $P_0(x)$ and $P_1(x)$ compare to the graph of y_1. (Remember, a nonsquare viewing window distorts slope.)

2. GRAPH P_2, P_6, and P_{10}. Tell what you observe. (You may wish to use GRAPHSUM.)

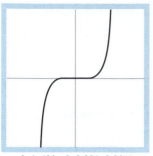

[−1, 1] by [−0.001, 0.001]

9.19 Graph of $y_3 = y_1 - y_2$, where

$$y_1 = \frac{1}{1 - x}$$

and

$$y_2 = P_{10}(x) = \sum_{n=0}^{10} x^n$$

(the two functions graphed in Fig. 9.18). TRACE shows that $|y_1 - y_2|$ is at most 0.0001 for $-0.43 < x < 0.39$.

3. GRAPH P_3, P_7, and P_{11}. Tell what you observe. How and why do these graphs differ from the graphs in part 2?

The interval of convergence is $(-1, 1)$. As the viewing windows suggest, the convergence is rapid for x close to 0 but requires more terms for x near -1 or 1.

4. Compare the values of $f(x)$, $P_3(x)$, and $P_{10}(x)$ for $x = \pm 0.2, \pm 0.5$.

5. How large must n be so that $|f(0.9) - P_n(0.9)| < 0.1$?

EXAMPLE 2

Estimate the range of values of x for which approximating $y = 1/(1 - x)$ by $P_{10}(x) = \sum_{n=0}^{10} x^n$ has an error of at most 0.0001.

Solution The graph of

$$f(x) = \frac{1}{1 - x} - P_{10}(x)$$

in Fig. 9.19 and TRACE can be used to estimate that $|f(x)| \leq 0.0001$ for $-0.43 < x < 0.39$. ▤

EXAMPLE 3

The power series

$$1 - \frac{1}{2}(x - 2) + \frac{1}{4}(x - 2)^2 + \cdots + \left(-\frac{1}{2}\right)^n (x - 2)^n + \cdots \qquad (4)$$

matches Eq. (2) with $a = 2, c_0 = 1, c_1 = -1/2, c_2 = 1/4, \ldots, c_n = (-1/2)^n$. This is a geometric series with first term 1 and ratio $r = -\frac{x - 2}{2}$. The series converges for $\left|\frac{x - 2}{2}\right| < 1$ or $0 < x < 4$ and not at $x = 0$ or $x = 4$. (Why?) The sum is

$$\frac{1}{1 - r} = \frac{1}{1 + \frac{x - 2}{2}} = \frac{2}{x},$$

so we write

$$\frac{2}{x} = 1 - \frac{(x - 2)}{2} + \frac{(x - 2)^2}{4} - \cdots + \left(-\frac{1}{2}\right)^n (x - 2)^n + \cdots \qquad (0 < x < 4).$$

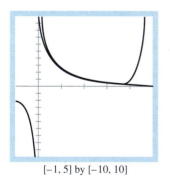

[−1, 5] by [−10, 10]

9.20 Graphs of

$$y_1 = \frac{2}{x}$$

and

$$y_2 = P_{10}(x) = \sum_{n=0}^{10} \left(\frac{-1}{2}\right)^n (x - 2)^n.$$

The graphs support $2/x$ as being an explicit form for the sum of the infinite series.

We use series (4) to generate polynomial approximations of $2/x$ (or $1/x$ if we divide by 2) for values of x near 2. We would keep the powers of $(x - 2)$ and write the approximating polynomials as

$$P_0(x) = 1, \quad P_1(x) = 1 - \frac{1}{2}(x - 2), \quad P_2(x) = 1 - \frac{1}{2}(x - 2) + \frac{1}{4}(x - 2)^2$$

and so on. The higher powers of $(x - 2)$ decrease rapidly as n increases when $|x - 2|$ is small, which means that the P_n approach $2/x$ rapidly near $x = 2$. Figure 9.20 suggests how well $P_{10}(x)$ approximates $2/x$ on $(0, 4)$. ▤

EXAMPLE 4

For what values of x does each of the following series converge?

a) $\displaystyle\sum_{n=1}^{\infty}(-1)^{n-1}\frac{x^n}{n} = x - \frac{x^2}{2} + \frac{x^3}{3} - \cdots$

b) $\displaystyle\sum_{n=1}^{\infty}(-1)^{n-1}\frac{x^{2n-1}}{2n-1} = x - \frac{x^3}{3} + \frac{x^5}{5} - \cdots$

c) $\displaystyle\sum_{n=0}^{\infty}\frac{x^n}{n!} = 1 + x + \frac{x^2}{2!} + \frac{x^3}{3!} + \cdots$

d) $\displaystyle\sum_{n=0}^{\infty}n!x^n = 1 + x + 2!x^2 + 3!x^3 + \cdots$

Solution   Apply the ratio test to the series $\sum |u_n|$, where u_n is the nth term of the series in question.

a) $\left|\dfrac{u_{n+1}}{u_n}\right| = \dfrac{n}{n+1}|x| \to |x|$. The series converges absolutely for $|x| < 1$, in agreement with the results of the nth-Root Test in Example 6 of Section 9.5. The rest of the analysis is the same as in that previous example; thus, series (a) converges for $-1 < x \leq 1$ and diverges elsewhere.

b) $\left|\dfrac{u_{n+1}}{u_n}\right| = \dfrac{2n-1}{2n+1}x^2 \to x^2$. The series converges absolutely for $x^2 < 1$. It diverges for $x^2 > 1$ because the nth term does not approach zero as $n \to \infty$. At $x = 1$, the series becomes $1 - 1/3 + 1/5 - 1/7 + \cdots$, which converges by the Alternating Series Theorem. It also converges at $x = -1$ because it again is an alternating series that satisfies the conditions for convergence. The value at $x = -1$ is the negative of the value at $x = 1$.

c) $\left|\dfrac{u_{n+1}}{u_n}\right| = \left|\dfrac{x^{n+1}}{(n+1)!}\cdot\dfrac{n!}{x^n}\right| = \dfrac{|x|}{n+1} \to 0$ for every x. The series converges absolutely for all x.

d) $\left|\dfrac{u_{n+1}}{u_n}\right| = \left|\dfrac{(n+1)!x^{n+1}}{n!x^n}\right| = (n+1)|x| \to \infty$ unless $x = 0$. The series diverges for all values of x except $x = 0$. The nth term does not approach zero unless $x = 0$. ≡

A PREVIEW

Although the series in parts (a), (b), and (c) of Example 4 converge for certain values of x, we do not yet have an explicit form for any of the sums in terms of elementary functions. As we proceed in this section, we learn explicit forms for the sums in parts (a) and (b). Later in the chapter, we will see that the series in part (c) converges to e^x.

AGREEMENT

Suppose that a power series

$$\sum_{n=1}^{\infty}a_n x^n$$

converges for values of x in a set D. We will use "GRAPH the sum of the series" to mean GRAPH a partial sum that approximates the exact sum reasonably well over a specified portion of D.

If we know the values of x for which a power series converges, and if we are able to estimate the error in approximating the sum by a partial sum polynomial, then we can identify those values of x over which the graph of a partial sum is a reasonably good graph of the exact sum, as the series from Example 4(a) illustrates in Example 5.

EXAMPLE 5

Graph the sum of the series

$$\sum_{n=1}^{\infty} (-1)^{n-1} \frac{x^n}{n} = x - \frac{x^2}{2} + \frac{x^3}{3} - \cdots .$$

Indicate a range of values for x for which the graph represents the sum reasonably accurately.

REASONABLY ACCURATE

A "reasonably accurate approximation" is as difficult to define as "good estimate." Suffice it to say that the meaning depends upon the situation and the needs of the moment. Thus, in the abstract, the best support you can give that an answer is reasonably accurate is to cite the maximum error possible.

   Visually, we can say that an approximation is reasonably accurate wherever its graph appears identical to the exact graph. Thus, for some [yMin, yMax], an approximation will be reasonably accurate, but for a smaller [yMin, yMax], the two graphs will separate, and the approximation can be considered not accurate enough.

Solution   From Example 4(a), we know that the series converges for $-1 < x \le 1$. For each x in $0 < x \le 1$, $\sum_{n=1}^{\infty} (-1)^{n-1} x^n / n$ is an alternating series. Notice that for $x < 0$, the series is *not* alternating. The Alternating Series Estimation Theorem assures us that the approximating polynomial $P_{10}(x) = \sum_{n=1}^{10} (-1)^{n-1} x^n / n$ is an underestimate of the sum of the series at x with an error of at most $x^{11}/11$. Notice that

$$\frac{0.5^{11}}{11} = 0.00004,$$

$$\frac{0.9^{11}}{11} = 0.02853,$$

$$\frac{0.999^{11}}{11} = 0.08991,$$

$$\frac{1^{11}}{11} = 0.09091.$$

These computations suggest that the error near $x = 1$ can be several times greater than the error near 0, but even at $x = 0.999$, we know that the approximation is within 0.09 of the exact value.

   Now, what about values of x for $-1 < x < 0$? We do not yet have a way to estimate the error in approximating the sum of the series for such x. We will find in Example 7, however, that the sum of this series is $\ln(1+x)$ on $(-1, 1]$. Figure 9.21(a) shows the graphs of $y_1 = P_{10}(x)$ and $y_2 = \ln(1+x)$. For this view, it appears that $P_{10}(x)$ is a reasonably accurate approximation of the exact sum for $-0.9 < x < 0.9$. Figure 9.21(b) suggests that we would have to shorten the interval for x if we make finer restrictions on $P_{10}(x)$ being a reasonably accurate approximation of $\ln(1 + x)$.   ▤

Example 4 illustrates how we usually test a power series for convergence and the kinds of results we get.

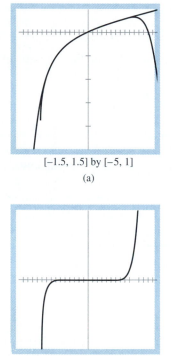

$$[-1.5, 1.5] \text{ by } [-5, 1]$$
(a)

$$[-1.5, 1.5] \text{ by } [-0.1, 0.1\,]$$
(b)

9.21 (a) The graph of

$$y_1 = P_{10}(x) = \sum_{n=1}^{10} (-1)^{n-1} \frac{x^n}{n}$$

appears to be a reasonable approximation of the graph of

$$y_2 = \ln{(1+x)} = \sum_{n=1}^{\infty} (-1)^{n-1} \frac{x^n}{n}$$

for $-0.9 < x < 0.9$. (Note that the x-scale $= 0.1$.) If, however, we are concerned about "finer" range values, the portion of the graph in (b) for $y_3 = y_2 - y_1$ and a smaller [yMin, yMax] suggests that y_1 is a reasonable approximation for y_2 over a smaller x-interval.

How to Test a Power Series for Convergence

STEP 1: Use the Ratio Test (or nth-Root Test) to find the x-values for which the series converges absolutely. Ordinarily, this is an open interval

$$|x - a| < h \qquad \text{or} \qquad a - h < x < a + h.$$

In some instances, as in Example 4(c), the series converges for all values of x. These instances are not uncommon. In rare cases, the series may converge at only a single point, as in Example 4(d).

STEP 2: If the interval of absolute convergence is finite, test for convergence or divergence at each of the two endpoints as in Example 3 and Examples 4(a) and 4(b). Neither the Ratio Test nor the nth-Root Test is useful at these points. Use a Comparison Test, the Integral Test, or the Alternating Series Theorem.

STEP 3: If the interval of absolute convergence is $a - h < x < a + h$, conclude that the series diverges (it does not even converge conditionally) for $|x - a| > h$, because for those values of x, the nth term does not approach zero.

In the next section, we will see how series in powers of $(x - a)$ are generated by the values of a function f and its derivatives at $x = a$. Since $1/x$, $\sqrt{x}$, and $\ln x$ exist and have derivatives of all orders at $x = 1$, the power series that we use to represent them can be in powers of $x - 1$. But they do not have power series representations in powers of x because they have no derivatives at $x = 0$.

To simplify the notation, the next theorem deals with the convergence of series of the form $\sum a_n x^n$. For series of the form $\sum a_n (x - a)^n$, we can apply a horizontal shift and replace $x - a$ by x' and apply the results to the series $\sum a_n (x')^n$.

THEOREM 10   **The Convergence Theorem for Power Series**

If

$$\sum_{n=0}^{\infty} a_n x^n = a_0 + a_1 x + a_2 x^2 + \cdots \qquad (5)$$

converges for $x = c$ ($c \neq 0$), then it converges absolutely for all $|x| < |c|$. If the series diverges for $x = d$, then it diverges for all $|x| > |d|$.

Proof   Suppose the series

$$\sum_{n=0}^{\infty} a_n c^n \qquad (6)$$

converges. Then

$$\lim_{n \to \infty} a_n c^n = 0.$$

Hence, there is an integer N such that $|a_n c^n| < 1$ for all $n \geq N$. That is,

$$|a_n| < \frac{1}{|c|^n} \qquad \text{for } n \geq N. \tag{7}$$

Now take any x such that $|x| < |c|$, and consider

$$|a_0| + |a_1 x| + \cdots + |a_{N-1} x^{N-1}| + |a_N x^N| + |a_{N+1} x^{N+1}| + \cdots. \tag{8}$$

There is only a finite number of terms before $|a_N x^N|$, and their sum is finite. Starting with $|a_N x^N|$ and beyond, the terms are less than

$$\left|\frac{x}{c}\right|^N + \left|\frac{x}{c}\right|^{N+1} + \left|\frac{x}{c}\right|^{N+2} + \cdots \tag{9}$$

because of (7). But the series (9) is a geometric series with ratio $r = |x/c|$, which is less than 1, since $|x| < |c|$. Hence, the series (9) converges, so the series (8) converges, and the original series (6) converges absolutely. This proves the first half of the theorem.

The second half of the theorem follows from the first. If the series diverges at $x = d$ and converges at a value x_0 with $|x_0| > |d|$, we may take $c = x_0$ in the first half of the theorem and conclude that the series converges absolutely at d. But the series cannot converge absolutely and diverge at one and the same time. Hence, if it diverges at d, it diverges for all $|x| > |d|$.
≡

The Radius and Interval of Convergence

The examples that we have looked at and the theorem that we just proved lead to the conclusion that a power series always behaves in exactly one of the following three ways.

Possible Behavior of $\sum c_n (x - a)^n$

1. The series converges at $x = a$ and diverges elsewhere.

2. There is a positive number h such that the series diverges for $|x - a| > h$ but converges absolutely for $|x - a| < h$. The series may or may not converge at either of the endpoints $x = a - h$ and $x = a + h$.

3. The series converges absolutely for every x.

In case 2, the set of points at which the series converges is a finite interval, called the **interval of convergence**. We know from past examples that the interval may be open, half-open, or closed, depending on the particular series. But no matter which kind of interval it is, h is called the **radius of convergence** of the series, and $a + h$ is the least upper bound of the set of points at which the series converges. The convergence is absolute at every point in the interior of the interval. If $a = 0$, the interval is centered at the origin. If a power series converges absolutely for all values of x (case 3), we say that its radius of convergence is infinite. If it converges only at $x = a$ (case 1), we say that the radius of convergence is zero.

Term-by-Term Differentiation of Power Series

A theorem from advanced calculus tells us that a power series can be differentiated term by term at each point in the interior of its interval of convergence.

THEOREM 11   The Term-by-Term Differentiation Theorem

If $\sum c_n(x - a)^n$ converges for $a - h < x < a + h$ for some $h > 0$, it defines a function f:

$$f(x) = \sum_{n=0}^{\infty} c_n(x - a)^n \qquad (a - h < x < a + h).$$

Such a function f has derivatives of all orders inside the interval of convergence. We can obtain the derivatives by differentiating the original series term by term:

$$f'(x) = \sum_{n=0}^{\infty} nc_n(x - a)^{n-1},$$

$$f''(x) = \sum_{n=0}^{\infty} n(n - 1)c_n(x - a)^{n-2},$$

and so on. Each of these derived series converges at every interior point of the interval of convergence of the original series.

Here is an example of how to apply term-by-term differentiation.

EXAMPLE 6

Is there an explicit form for the function f defined by the power series

$$f(x) = \sum_{n=1}^{\infty} (-1)^{n-1}\frac{x^{2n-1}}{2n - 1} = x - \frac{x^3}{3} + \frac{x^5}{5} - \cdots \qquad (-1 \le x \le 1)?$$

Solution   We differentiate the original series term by term and get

$$f'(x) = \sum_{n=1}^{\infty} (-1)^{n-1}\frac{(2n - 1)x^{2n-2}}{2n - 1}$$

$$= \sum_{n=1}^{\infty} (-1)^{n-1}x^{2n-2} = 1 - x^2 + x^4 - x^6 + \cdots.$$

This is a geometric series with first term 1 and ratio $-x^2$, so

$$f'(x) = \frac{1}{1 - (-x^2)} = \frac{1}{1 + x^2}.$$

EXPLORATION BITS

1. GRAPH

$$y_1 = P_{19}(x)$$

$$= \sum_{n=1}^{10} (-1)^{n-1} \frac{x^{2n-1}}{2n-1}$$

and

$$y_2 = \tan^{-1} x$$

in a $[-2, 2]$ by $[-\pi/2, \pi/2]$ viewing window. Explain what you see. (Also, do you understand why we used the symbol $P_{19}(x)$ and not $P_{10}(x)$?)

2. What can you say about the series

$$1 - x^2 + x^4 - x^6 + \ldots$$

when $x = 1$?

3. What can you say about the series

$$x - \frac{x^3}{3} + \frac{x^5}{5} - \frac{x^7}{7} + \cdots$$

when $x = 1$?

EXPLORATION BIT

Theorem 11 guarantees that the derived series converges *inside* the interval of convergence of the original series. Note that we get nothing more than this in Example 6 even though the interval of convergence of the original series includes its endpoints. Does this fact require that you fill in a couple of details in Example 6?

We can now integrate $f'(x) = 1/(1 + x^2)$ to get

$$f(x) = \int f'(x)\, dx = \int \frac{dx}{1 + x^2} = \tan^{-1} x + C.$$

When $x = 0$, $\tan^{-1} x = 0$ and the series for $f(x)$ is 0, so $C = 0$ also. Hence,

$$f(x) = x - \frac{x^3}{3} + \frac{x^5}{5} - \frac{x^7}{7} + \cdots = \tan^{-1} x \qquad (-1 \le x \le 1).$$

The original function f is the restriction of $\tan^{-1} x$ to the interval $[-1, 1]$.

≡

We were lucky in Example 6 both to find a formula for $f'(x)$ and then in turn to find a formula for the antiderivative of $f'(x)$. We may not always be able to quickly find a formula for f or f'. But sometimes our graphing utility can give a clue, as shown in Exploration 2.

EXPLORATION 2

Evaluating a Power Series

Consider the series

$$f(x) = \sum_{n=0}^{\infty} (-1)^n \frac{x^{2n+1}}{(2n+1)!}.$$

1. Apply the Ratio Test to show that the series converges absolutely for all x.

2. GRAPH the approximating polynomial

$$y_1 = P_{21}(x) = \sum_{n=0}^{10} (-1)^n \frac{x^{2n+1}}{(2n+1)!} \qquad \text{An 11-term polynomial of degree 21}$$

in a $[-2\pi, 2\pi]$ by $[-2, 2]$ viewing window. On the basis of your graph, make a conjecture about an explicit form for $f(x)$.

3. Support your conjecture by finding the series $f'(x)$ and then drawing a graph of $y_2 = y_1'$, the polynomial approximation of $f'(x)$. What do you expect the graph to look like? (What happens if you try to view $f'(x)$ by graphing NDER y_1?)

4. Support your conjecture further by finding the series $f''(x)$. How should it compare to the series $f(x)$?

5. Finally, GRAPH y_1 and $y_3 = \sin x$ in a $[-12, 12]$ by $[-2, 2]$ viewing window. Does this destroy all that you've conjectured? Explain.   ✂

Arctangents

In Example 6, we found a series for $\tan^{-1} x$ by differentiating to get

$$1 - x^2 + x^4 - x^6 + \cdots = \frac{1}{1 + x^2}$$

and integrating to get

$$x - \frac{x^3}{3} + \frac{x^5}{5} - \frac{x^7}{7} + \cdots = \tan^{-1} x.$$

$$\tan^{-1} x = x - \frac{x^3}{3} + \frac{x^5}{5} - \frac{x^7}{7} + \cdots \qquad (|x| \le 1). \qquad (10)$$

Figure 9.22 supports Eq. (10). When we put $x = 1$ and $\tan^{-1} 1 = \pi/4$ in Eq. (10), we get **Leibniz's formula:**

$$\frac{\pi}{4} = 1 - \frac{1}{3} + \frac{1}{5} - \frac{1}{7} + \frac{1}{9} - \cdots + \frac{(-1)^n}{2n + 1} + \cdots.$$

This series converges too slowly to be a good source of decimal approximations of π. It is better to use a formula like

$$\pi = 48 \tan^{-1} \frac{1}{18} + 32 \tan^{-1} \frac{1}{57} - 20 \tan^{-1} \frac{1}{239},$$

which uses values of x closer to zero. (See Exercise 5 4.)

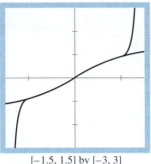

[−1.5, 1.5] by [−3, 3]

9.22 The polynomial

$$y_2 = P_{21}(x) = \sum_{n=0}^{10} (-1)^n \frac{x^{2n+1}}{2n + 1}$$

is a good approximation to $y_1 = \tan^{-1} x$ in $-1 \le x \le 1$.

Term-by-Term Integration of Power Series

Another theorem from advanced calculus states that a power series can also be integrated term by term throughout its interval of convergence.

THEOREM 12   **The Term-by-Term Integration Theorem**

If

$$f(x) = \sum_{n=0}^{\infty} c_n(x - a)^n,$$

where $\sum c_n(x-a)^n$ converges for $a - h < x < a + h$ for some $h > 0$, then the series

$$\sum c_n \frac{(x - a)^{n+1}}{n + 1}$$

converges for $a - h < x < a + h$ and

$$\int f(x)\, dx = \sum_{n=0}^{\infty} c_n \frac{(x - a)^{n+1}}{n + 1} + C$$

for $a - h < x < a + h$.

EXPLORATION BITS

Theorem 12 guarantees that the antiderivative series converges *inside* the interval of convergence of the original series. Note that in Example 7, we get nothing about convergence at the endpoints of the interval even though the interval of convergence of the antiderivative series includes an endpoint.

1. What can you say about the series

$$1 - t + t^2 - t^3 + \cdots$$

when $t = 1$?

2. What can you say about the series

$$x - x^2/2 + x^3/3 - x^4/4 + \cdots$$

when $x = 1$?

EXAMPLE 7

The series

$$\frac{1}{1 + t} = 1 - t + t^2 - t^3 \cdots$$

converges on the open interval $-1 < t < 1$. Therefore,

$$\ln(1+x) = \int_0^x \frac{1}{1+t}\, dt = t - \frac{t^2}{2} + \frac{t^3}{3} - \frac{t^4}{4} \cdots \Big]_0^x$$

$$= x - \frac{x^2}{2} + \frac{x^3}{3} - \frac{x^4}{4} + \cdots \qquad (-1 < x < 1).$$

As you know, the latter series also converges at $x = 1$ (Example 4a), but that was not guaranteed by the theorem. ≡

Multiplication of Power Series

Still another theorem from advanced calculus states that power series can be multiplied term by term.

THEOREM 13   **The Series Multiplication Theorem for Power Series**

If both $\sum a_n x^n$ and $\sum b_n x^n$ converge absolutely for $|x| < h$, and

$$c_n = a_0 b_n + a_1 b_{n-1} + a_2 b_{n-2} + \cdots + a_{n-1} b_1 + a_n b_0$$

$$= \sum_{k=0}^n a_k b_{n-k}, \qquad (11)$$

then the series $\sum c_n x^n$ also converges absolutely for $|x| < h$, and

$$\left(\sum a_n x^n\right) \cdot \left(\sum b_n x^n\right) = \sum c_n x^n. \qquad (12)$$

EXAMPLE 8

Multiply the geometric series

$$\sum_{n=0}^\infty x^n = 1 + x + x^2 + \cdots + x^n + \cdots = \frac{1}{1-x} \qquad (|x| < 1),$$

by itself to get the power series for $1/(1-x)^2$ for $|x| < 1$.

Solution   Let

$$A(x) = \sum_{n=0}^\infty a_n x^n = 1 + x + x^2 + \cdots + x^n + \cdots = 1/(1-x)$$

$$B(x) = \sum_{n=0}^\infty b_n x^n = 1 + x + x^2 + \cdots + x^n + \cdots = 1/(1-x)$$

and

$$c_n = \underbrace{a_0 b_n + a_1 b_{n-1} + \cdots + a_k b_{n-k} + \cdots a_n b_0}_{n+1 \text{ terms}} = \underbrace{1 + 1 + \cdots + 1}_{n+1 \text{ ones}} = n + 1.$$

EXPLORATION BIT

The product series of Example 8 seems tantalizingly close to the series

$$x + 2x^2 + 3x^3 + 4x^4 \cdots.$$

Can you find an explicit form for this series? Support graphically.

Then, by the Series Multiplication Theorem,

$$A(x) \cdot B(x) = \sum_{n=0}^{\infty} c_n x^n = \sum_{n=0}^{\infty} (n+1)x^n$$

$$= 1 + 2x + 3x^2 + 4x^3 + \cdots + (n+1)x^n + \cdots$$

is the series for $1/(1-x)^2$. The series all converge absolutely for $|x| < 1$.

≡

Exercises 9.6

Each of Exercises 1–14 gives a formula for the nth term of a series. For what values of x does the series (a) converge and (b) converge absolutely?

1. $(-1)^n (x+1)^n$

2. $\dfrac{nx^n}{(n+2)}$

3. $\dfrac{x^n}{n\sqrt{n}}$

4. $\dfrac{(x-1)^n}{\sqrt{n}}$

5. $\dfrac{x^{2n+1}}{n!}$

6. $\dfrac{(x-3)^{2n+1}}{n!}$

7. $\dfrac{x^n}{\sqrt{n^2+3}}$

8. $\dfrac{(-1)^n x^n}{\sqrt{n^2+3}}$

9. $\dfrac{nx^n}{n^2+1}$

10. $n(x-3)^n$

11. $\dfrac{\sqrt{n}x^n}{3^n}$

12. $\sqrt[n]{n}(x-1)^n$

13. $\left(1+\dfrac{1}{n}\right)^n x^n$

14. $(\ln n)x^n$

Each of Exercises 15–18 gives a formula for the nth term of a series. For what values of x does the series (a) converge and (b) converge absolutely? (c) Find the sum of the series. (d) Indicate a range of values for x for which P_{20} represents the sum with an error of at most 0.01.

15. x^n

16. $(x+5)^n$

17. $\dfrac{(x-2)^n}{10^n}$

18. $(2x)^n$

19. Consider the series $\sum_{n=1}^{\infty} (-1)^{n-1} x^n / n$ of Example 5.
   a) Compute $P_n(-1)$ for $n = 20, 30, 50, 100$.

   b) Estimate the maximum error in approximating the sum $\ln(1+x)$ of the series by $P_{10}(x)$ in $-0.9 < x < 0$.

   c) GRAPH $y = \ln(1+x)$ and $y = P_{35}(x)$ in a $[-1.5, 1.5]$ by $[-5, 1]$ viewing window. Compare this graph with Fig. 9.21(a).

20. The series $\sum_{n=1}^{\infty} (-1)^{n-1} x^{2n-1}/(2n-1)$ converges to $\tan^{-1} x$ in $-1 \le x \le 1$ (see Example 6). For each x in $-1 \le x \le 1$, the series is alternating.

a) Use PARTSUMT to support convergence at $x = 1$ and $x = -1$.

b) Use the Alternating Series Estimation Theorem to choose n so that $P_n(x)$ approximates $\tan^{-1} x$ with error at most 0.01 in $-1 \le x \le 1$.

c) Explain why the approximations at the endpoints of the interval of convergence are better than we were able to obtain in Example 5.

Each of Exercises 21–24 gives a formula for the nth term of a series. For what values of x does the series (a) converge and (b) converge absolutely? (c) Show that the series is alternating on half of the interval of convergence. (d) Find an approximating polynomial $P_n(x)$ that approximates the sum of the series on 90% of the interval in part (c) with an error of at most 0.01. (e) Compare the graph of the polynomial in part (d) with the actual sum over the interval of convergence. Discuss the error near the endpoints of the interval of convergence.

21. $\dfrac{(x-2)^n}{n}$ (See Exercise 33b.)

22. $\dfrac{(-1)^n(x+2)^n}{n}$ (See Exercise 33c.)

23. $n(x-3)^n$ (See Exercise 34c.)

24. $n(x-2)^n$ (The sum is $\dfrac{x-2}{(3-x)^2}$.)

25. Consider the series $\sum_{n=0}^{\infty} (-1)^n x^{2n+1}/(2n+1)!$ of Exploration 2 whose sum is $\sin x$.

   a) Compare $P_5(x)$, $P_{11}(x)$, and $P_{17}(x)$ for $\pm\pi/4, \pm\pi/2$, and $\pm 2\pi$.

   b) Estimate the range of values of x for which $P_{11}(x)$, $P_{21}(x)$, and $P_{31}(x)$ approximates the sum of the series with an error of at most 0.01.

26. Consider the series $\sum_{n=0}^{\infty} (-1)^n x^{2n}/(2n)!$.

   a) Use Theorem 11 and Exercise 25 to show that the sum of the series is $\cos x$.

   b) Confirm the values of x for which the series converges.

   c) Provide graphical support for your answer in part (b). Explain.

In Exercises 27–30, find (a) the values of x for which the series converges and (b) the range of values of x for which P_{20} approximates the sum with an error of at most 0.01.

27. $\displaystyle\sum_{n=0}^{\infty} \frac{x^n}{n!}$ (The sum is e^x.)

28. $\displaystyle\sum_{n=0}^{\infty} \frac{(-1)^n x^n}{n!}$ (Use Exercise 27 to find the sum.)

29. $\displaystyle\sum_{n=0}^{\infty} \frac{3^n x^n}{n}!$ (Use Exercise 27 to find the sum.)

30. $\displaystyle\sum_{n=0}^{\infty} (n+1)x^n$ (The sum is $\dfrac{1}{(1-x)^2}$.)

31. Consider the series $\sum_{n=1}^{\infty} x^n$.

a) Find the sum of this geometric series.

b) Use $\sum_{n=0}^{\infty} x^n = 1/(1-x)$ to derive the sum in part (a).

c) For what values of x does this series converge?

d) Estimate the range of values of x for which P_{20} is an approximation with an error of at most 0.01 of the sum.

32. Consider the series $\sum_{n=1}^{\infty} (-1)^{n+1} x^{2n+1}/(2n+1)!$.

a) For what values of x does the series converge?

b) Use $\sum_{n=0}^{\infty} (-1)^n x^{2n+1}/(2n+1)! = \sin x$ to find the sum of the series.

c) Estimate the range of values of x for which $P_{21}(x)$ is an approximation with an error of at most 0.01 of the sum.

33. Use $\sum_{n=1}^{\infty} (-1)^{n-1} x^n/n = \ln(1+x)$ to derive the following sums.

a) $\displaystyle\sum_{n=1}^{\infty} \frac{x^n}{n} = -\ln(1-x)$

b) $\displaystyle\sum_{n=1}^{\infty} \frac{(x-2)^n}{n} = -\ln(3-x)$

c) $\displaystyle\sum_{n=1}^{\infty} \frac{(-1)^n (x+2)^n}{n} = -\ln(3+x)$

34. a) Use $\sum_{n=1}^{\infty} (-1)^{n-1} x^n/n = \ln(1+x)$ and Theorem 11 to show that

$$x\frac{d^2}{dx^2}\ln(1+x) = -\frac{x}{(1+x)^2} = \sum_{n=1}^{\infty}(-1)^n n x^n.$$

b) Use part (a) to show that $x/(1-x)^2 = \sum_{n=1}^{\infty} nx^n$.

c) Use part (b) to show that $(x-3)/(4-x)^2 = \sum_{n=1}^{\infty} n(x-3)^n$.

d) For what range of values of x are the formulas in parts (a), (b), and (c) valid? Support your answers graphically.

35. Consider the series $\sum_{n=0}^{\infty} n! x^n$ of Example 4(d).

a) GRAPH $P_2(x)$ and $P_3(x)$ for $-2 \le x \le 2$. For what values of x is $|P_2(x) - P_3(x)| < 0.01$?

b) Repeat part (a) with $P_3(x)$ and $P_4(x)$.

c) Repeat part (a) with $P_9(x)$ and $P_{10}(x)$.

d) For what values of x does this series converge?

36. Describe how you would give a graphical argument that the radius of convergence of a power series is 0. (*Hint:* See Exercise 35.)

Each of Exercises 37 and 38 gives a formula for the nth term of a series. For what values of x does the series (a) converge and (b) converge absolutely? Support your answers graphically.

37. $n^n x^n$ **38.** $n!(x-4)^n$

39. For what values of x does the geometric series

$$1 - \frac{1}{2}(x-3) + \frac{1}{4}(x-3)^2 + \cdots + \left(-\frac{1}{2}\right)^n (x-3)^n + \cdots$$

converge? What is its sum? What series do you get if you differentiate the given series term by term? Where does that series converge? What is its sum? Support your answers graphically.

40. If you integrate the series of Exercise 39 term by term, what new series do you get? Where does the new series converge, and what is another name for its sum? Support your answers graphically.

41. The series for $\tan x$,

$$\tan x = x + \frac{x^3}{3} + \frac{2x^5}{15} + \frac{17x^7}{315} + \cdots,$$

converges for $-\pi/2 < x < \pi/2$.

a) Estimate the range of values of x for which $P_7(x)$ approximates the sum with an error of at most 0.01.

b) Find the first four terms of the series for $\ln|\sec x|$. For what values of x should the series converge? Support graphically.

c) Find the first three terms of the series for $\sec^2 x$. For what values of x should this series converge? Support graphically.

d) Check your result in part (c) by squaring the series given for $\sec x$ in Exercise 42.

42. The series for $\sec x$,

$$\sec x = 1 + \frac{x^2}{2} + \frac{5}{24}x^4 + \frac{61}{720}x^6 + \cdots,$$

converges for $-\pi/2 < x < \pi/2$.

a) Estimate the range of values of x for which $P_6(x)$ approximates the sum with an error of at most 0.01.

b) Find the first four terms of a series for the function $\ln|\sec x + \tan x|$. For what values of x should the series converge? Support graphically.

c) Find the first three terms of a series for $\sec x \tan x$. For what values of x should the series converge? Support graphically.

d) Check your result in part (c) by multiplying the series for $\sec x$ by the series given for $\tan x$ in Exercise 41.

43. The series $(4/\pi) \sum_{n=1}^{\infty} (-1)^{n+1} \cos((2n-1)x)/(2n-1)$ is known to converge to the function

$$f(x) = \begin{cases} -1, & -\pi < x < -\pi/2 \\ 1, & -\pi/2 < x < \pi/2 \\ -1, & \pi/2 < x < \pi \end{cases}$$

in the interval $-\pi < x < \pi$.

a) GRAPH the partial sums $s_{10}(x)$, $s_{20}(x)$, and $s_{30}(x)$. Describe their behavior near $x = \pm \pi/2$.

b) Estimate the range of values for x for which $s_{30}(x)$ approximates $f(x)$ with error of at most 0.01.

44. The series $(4/\pi) \sum_{n=1}^{\infty} \sin((2n-1)x))/(2n-1)$ is known to converge to the function

$$f(x) = \begin{cases} -1, & -\pi < x < 0 \\ 1, & 0 < x < \pi \end{cases}$$

in the interval $-\pi < x < \pi$.

a) GRAPH the partial sums $s_{10}(x)$, $s_{20}(x)$, and $s_{30}(x)$. Describe their behavior near $x = 0$.

b) Estimate the range of values for x for which $s_{30}(x)$ approximates $f(x)$ with an error of at most 0.01.

Exercises 45–51 (Group Project) refer to the series $\sum_{n=1}^{\infty} (\sin(n!x))/n^2$.

45. Use a comparison test to show that the series converges absolutely for all x.

46. GRAPH the partial sums $s_1(x), s_2(x), s_3(x), s_4(x), s_8(x)$, and $s_{12}(x)$ in $[-12.6, 12.6]$ by $[-2, 2]$. (*Note:* Graphing

the partial sums of this series for $h \geq 15$ may lead to erroneous information about the domain of the sum of the series on some graphers.)

47. Compare the values of $s_{12}(x)$ and $s_{13}(x)$ for several values of x in $-12.6 \leq x \leq 12.6$.

48. Show that $s_{13}(x)$ approximates the sum $s(x)$ of the series with an error of at most 0.075. (*Hint:* $|s(x) - s_{13}(x)| =$

$$\left| \sum_{n=14}^{\infty} \frac{\sin(n!x)}{n^2} \right| \leq \sum_{n=14}^{\infty} \left| \frac{\sin(n!x)}{n^2} \right| \leq \sum_{n=14}^{\infty} \frac{1}{n^2} < 0.075$$

using $\sum_{n=1}^{\infty} \frac{1}{n^2} = \frac{\pi^2}{6}$.)

49. ZOOM-IN on several points of $s_{13}(x)$. Describe what you see.

50. GRAPH NDER $s_{13}(x)$ on $[0, 12.6]$ by $[-10, 10]$. Describe what you see.

51. Make a conjecture about the derivative of $s(x)$.

52. According to the Alternating Series Estimation Theorem, how many terms of the power series for $\tan^{-1} 1$ would you have to add to be sure of finding $\pi/4$ with an error of at most 10^{-3}? Use PARTSUMT to support your answer.

53. Show that the power series for $\tan^{-1} x$ diverges for $|x| > 1$. How would you support this fact graphically?

54. About how many terms of the power series for $\tan^{-1} x$ would you have to use to evaluate each term on the right-hand side of the equation

$$\pi = 48 \tan^{-1} \frac{1}{18} + 32 \tan^{-1} \frac{1}{57} - 20 \tan^{-1} \frac{1}{239}$$

with an error of at most 10^{-6}? Then use PARTSUMT to calculate the corresponding three values and approximate π. Compare with your calculator value of π.

9.7 _____ Taylor Series and Maclaurin Series

Thus far, we have studied infinite series and have found some functions that match series sums on certain x-intervals. Now we address the other direction: Given a function, can we find an infinite series that agrees with that function on some x-interval? *Taylor series* provide an answer for functions that have infinitely many derivatives. Once we find the series for such a function, we will see how to analytically control the errors involved in using the partial sums of the series as polynomial approximations of the function. As a special case, we will at last see how to control the errors in linearizations.

As we look at series from this viewpoint, we point out that we have a situation here similar to what we had with integrals. It is a fact that the number of functions with explicit antiderivatives is small in comparison with the number of integrable functions. Likewise, the number of series in x that agree with explicit functions on x-intervals is small in comparison with the

An important feature of Taylor series is that they enable us to extend the domains of functions to include complex numbers, something we will look at briefly at the end of this section.

number of such series that converge on some x-interval. On graphing calculators, programs like GRAPHSUM can help us to study such series in much the same way that numerical integration, NINT, helps us to study indefinite integrals. On a CAS, the ability to study these series is often built in.

Taylor Polynomials

The linearization of a function f at a point a where f is differentiable is the polynomial

$$P_1(x) = f(a) + f'(a)(x - a).$$

If f has derivatives of higher order at a, then it has higher-order polynomial approximations as well, one for each available derivative. These polynomials are called the Taylor polynomials of f.

TERMINOLOGY

Here we speak of the polynomial $P_n(x)$ as the Taylor polynomial of *order n* rather than *degree n* so that $P_n(x)$ identifies the highest-order derivative (and the last term) in its definition. The degree of $P_n(x)$ may now be less than n because $f^n(a)$ may be zero. For example, the first two Taylor polynomials of $\cos x$ at $x = 0$ are $P_0(x) = 1$ and $P_1(x) = 1$. The first-order polynomial has degree zero, not one.

DEFINITION

Let f be a function with derivatives of order k for $k = 1, 2, \ldots, N$ in some interval containing a as an interior point. Then, for any integer n from 0 through N, the **Taylor polynomial** of order n generated by f at a is the polynomial

$$P_n(x) = f(a) + f'(a)(x - a) + \frac{f''(a)}{2!}(x - a)^2 + \cdots$$

$$+ \frac{f^{(k)}(a)}{k!}(x - a)^k + \cdots + \frac{f^{(n)}(a)}{n!}(x - a)^n.$$

EXPLORATION 1

What's Special about Taylor Polynomials?

Let f be a function with derivatives of order k for $k = 1, 2, \ldots, N$. Let

$$P(x) = \sum_{k=0}^{N} C_k(x - a)^k$$

be a polynomial of degree N satisfying

$$P(a) = f(a), \qquad P^{(k)}(a) = f^{(k)}(a) \qquad (k = 1, 2, \ldots, N).$$

1. Show that $C_0 = f(a)$.
2. Use $P'(x)$ to show that $C_1 = f'(a)$.
3. Show that $C_k = f^{(k)}(a)k!$ for $k = 2, 3, \ldots, N$.

From Exploration 1, you should have found the following.

The Taylor polynomial of order n and its first n derivatives have the same values as f and its corresponding derivatives at $x = a$.

A function that has derivatives of all orders at $x = a$ generates a Taylor polynomial for every $n \geq 0$.

EXAMPLE 1

Find the Taylor polynomials generated by $f(x) = e^x$ at $a = 0$.

Solution   Expressed in terms of x, the given function and its derivatives are

$$f(x) = e^x, \qquad f'(x) = e^x, \qquad \ldots, \qquad f^{(n)}(x) = e^x,$$

so

$$f(0) = e^0 = 1, \qquad f'(0) = 1, \qquad \ldots, \qquad f^{(n)}(0) = 1,$$

and

$$P_n(x) = 1 + x + \frac{x^2}{2!} + \frac{x^3}{3!} + \cdots + \frac{x^n}{n!}.$$

See Fig. 9.23.   ≡

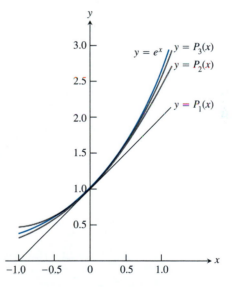

9.23   Graphs of $f(x) = e^x$ and its Taylor polynomials $P_1(x) = 1 + x$, $P_2(x) = 1 + x + x^2/2!$, and $P_3(x) = 1 + x + x^2/2! + x^3/3!$. Notice the close agreement near the center $x = 0$.

EXAMPLE 2

Find the Taylor polynomials generated by $f(x) = \cos x$ at $a = 0$.

EXPLORATION BIT

Use the method shown in Example 2 to find the Taylor polynomials for $\sin x$.

Solution   The cosine and its derivatives are

$$f(x) = \cos x, \qquad f'(x) = -\sin x,$$

$$f''(x) = -\cos x, \qquad f^{(3)}(x) = \sin x,$$

$$\vdots \qquad\qquad\qquad \vdots$$

$$f^{(2n)}(x) = (-1)^n \cos x, \qquad f^{(2n+1)}(x) = (-1)^{n+1} \sin x.$$

When $x = 0$, the cosines are 1 and the sines are 0, so

$$f^{(2n)}(0) = (-1)^n, \qquad f^{(2n+1)}(0) = 0.$$

Notice that the Taylor polynomials of order $2n$ and of order $2n + 1$ are identical:

$$P_{2n}(x) = P_{2n+1}(x) = 1 - \frac{x^2}{2!} + \frac{x^4}{4!} - \cdots + (-1)^n \frac{x^{2n}}{(2n)!}. \qquad \equiv$$

Figure 9.24 shows how well the Taylor polynomials of Example 2 approximate $y = \cos x$ near $x = 0$. Only the right-hand portions of the graphs are shown because the graphs are symmetric about the y-axis.

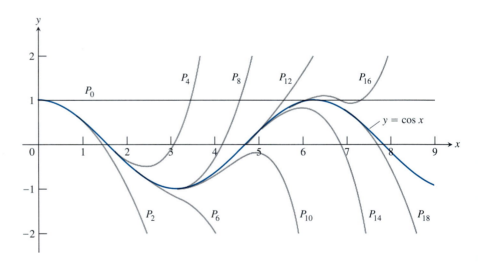

9.24   The Taylor polynomials

$$P_{2n}(x) = \sum_{k=0}^{n} (-1)^k x^{2k}/(2k)!$$

appear to converge to $\cos x$ as $n \to \infty$.

Taylor Series and Maclaurin Series

The natural extension of a Taylor polynomial to an infinite sum gives us a Taylor series. A special case is called the Maclaurin series.

Brook Taylor (1685–1731) did not invent Taylor series, and Maclaurin series were not developed by Colin Maclaurin (1698–1746). James Gregory was working with Taylor series when Taylor was only a few years old, and he published the Maclaurin series for $\tan x$, $\sec x$, $\tan^{-1} x$, and $\sec^{-1} x$ ten years before Maclaurin was born. At about the same time, Nicolaus Mercator discovered the Maclaurin series for $\ln(1 + x)$.

Taylor was not aware of Gregory's work when he published his book *Methodus incrementorum directa et inversa* in 1715, which contained what we now call Taylor series. Maclaurin quoted Taylor's work in a calculus book that he wrote in 1742. The book popularized series representations of functions, and although Maclaurin never claimed to have discovered them, Taylor series centered at $a = 0$ became known as Maclaurin series. History evened things up in the end. Maclaurin, a brilliant mathematician, was the original discoverer of the rule for solving systems of equations using determinants that we now call Cramer's rule.

DEFINITIONS

Let f be a function with derivatives of all orders throughout some interval containing a as an interior point. Then the **Taylor series generated by f at a** is

$$\sum_{k=0}^{\infty} \frac{f^{(k)}(a)}{k!}(x - a)^k = f(a) + f'(a)(x - a) + \frac{f''(a)}{2!}(x - a)^2$$

$$+ \cdots + \frac{f^{(n)}(a)}{n!}(x - a)^n + \cdots,$$

and the **Maclaurin series generated by f** is

$$\sum_{k=0}^{\infty} \frac{f^{(k)}(0)}{k!}x^k = f(0) + f'(0)x + \frac{f''(0)}{2!}x^2 + \cdots + \frac{f^{(n)}(0)}{n!}x^n + \cdots,$$

the Taylor series generated by f at $a = 0$.

Once we have found the Taylor series generated by the function f at a particular a, we can apply our usual tests to find where the series converges—usually in some interval $(a - h, a + h)$ or for all x.

EXAMPLE 3

Find the Taylor series generated by $f(x) = 1/x$ at $a = 2$. Where, if anywhere, does the series converge to $1/x$? Estimate the range of values of x for which the approximating partial sum polynomial of degree 10 agrees with $1/x$ with an error of at most 0.01.

Solution We need to compute $f(2)$, $f'(2)$, $f''(2)$, and so on. Taking derivatives, we get

$$f(x) = x^{-1}, \qquad\qquad f(2) \quad = 2^{-1} = \frac{1}{2},$$

$$f'(x) = -x^{-2}, \qquad\qquad f'(2) \quad = -\frac{1}{2^2},$$

$$f''(x) = 2!x^{-3}, \qquad\qquad \frac{f''(2)}{2!} \quad = 2^{-3} = \frac{1}{2^3},$$

$$f'''(x) = -3!x^{-4}, \qquad\qquad \frac{f'''(2)}{3!} \quad = -\frac{1}{2^4},$$

$$\vdots \qquad\qquad\qquad\qquad \vdots$$

$$f^{(n)}(x) = (-1)^n n! x^{-(n+1)}, \qquad \frac{f^{(n)}(2)}{n!} = \frac{(-1)^n}{2^{n+1}}$$

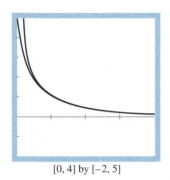

[0, 4] by [−2, 5]

9.25   The graphs of

$$y_1 = \frac{1}{x}$$

and

$$y_2 = P_{10}(x) = \sum_{n=0}^{10} (-1)^n \frac{(x-2)^n}{2^{n+1}}$$

closely approximate each other on $(0, 4)$. For $x > 2$, the series

$$\sum_{n=0}^{\infty} (-1)^n \frac{(x-2)^n}{2^{n+1}}$$

is alternating.

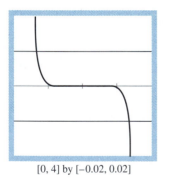

[0, 4] by [−0.02, 0.02]

9.26   The graphs of $y_3 = y_1 - y_2$ (of Fig. 9.25), $y_4 = 0.01$, and $y_5 = -0.01$ show that y_1 and y_2 are within 0.01 of each other for $0.73 < x < 3.47$.

The Taylor series is

$$f(2) + f'(2)(x-2) + \frac{f''(2)}{2!}(x-2)^2 + \frac{f'''(2)}{3!}(x-2)^3 +$$

$$\cdots + \frac{f^{(n)}(2)}{n!}(x-2)^n + \cdots$$

$$= \frac{1}{2} - \frac{(x-2)}{2^2} + \frac{(x-2)^2}{2^3} - \cdots + (-1)^n \frac{(x-2)^n}{2^{n+1}} + \cdots.$$

This is a geometric series with first term $1/2$ and ratio $r = -(x-2)/2$. It converges absolutely for $|x - 2| < 2$, and its sum is

$$\frac{1/2}{1 + (x-2)/2} = \frac{1}{2 + (x-2)} = \frac{1}{x}.$$

In this example, the Taylor series generated by $f(x) = 1/x$ at $a = 2$ converges to $1/x$ for $|x - 2| < 2$ or $0 < x < 4$.

Figure 9.25 shows graphically how $y_1 = 1/x$ and $y_2 = P_{10}(x) = \sum_{n=0}^{10} (-1)^n (x-2)^n / 2^{n+1}$ approximate each other for $0 < x < 4$. The error in approximating $1/x$ by $P_{10}(x)$ is suggested by $y_3 = y_1 - y_2$ (Fig. 9.26). We can use SOLVE to show that $P_{10}(x)$ approximates $1/x$ with an error of at most 0.01 for $0.73 < x < 3.47$.   ≡

EXPLORATION 2

Bringing Ideas Together—Group Project

Work together in groups of two or three outside the classroom. Bring together the ideas of Taylor polynomials, Taylor series, and what you learned in Section 9.6 about infinite series and their convergence. Develop your own Exploration activity based on Examples 1–3. You may get some ideas by reviewing Exploration 2 of Section 9.6. You may get some other ideas by thinking about the relationship between a *complete graph* of a Taylor *polynomial* of order (and degree) $2n$ for the cosine function and the "bumpiness" of the cosine function and why the graph of the Taylor polynomials of any order always, eventually, clearly separates from the graph of the cosine function.   ☙

Taylor's Theorem with Remainder

When the Taylor polynomials are defined for unbounded values of n, the Taylor polynomials are the partial sums of the Taylor series generated by f at a. The next theorem helps us to find out whether these sums converge to $f(x)$ in some interval $(a - h, a + h)$ or, perhaps, for all x.

When you study this statement of Taylor's Theorem, pay particular attention to c. It can get lost in the terminology of Taylor's Theorem, but its role is critical.

THEOREM 14a   Taylor's Theorem

If f and its first n derivatives f', f'', ..., $f^{(n)}$ are continuous on $[a, b]$ or on $[b, a]$ and $f^{(n)}$ is differentiable on (a, b) or on (b, a), then there exists a number c between a and b such that

$$f(b) = f(a) + f'(a)(b - a) + \frac{f''(a)}{2!}(b - a)^2 + \cdots$$

$$+ \frac{f^{(n)}(a)}{n!}(b - a)^n + \frac{f^{(n+1)}(c)}{(n + 1)!}(b - a)^{n+1}.$$

EXPLORATION BITS

1. The Alternating Series Estimation Theorem (Theorem 8) tells us whether the partial sum $P_{10}(x)$ of Example 3 is an overestimate or an underestimate of the exact sum. Which is it, and how can you tell from the graphs in Fig. 9.25?

2. In Fig. 9.26, why does the difference y_3 appear to become so great to the right of 3 when the graphs of y_1 and y_2 to the right of 3 (Fig. 9.25) are visually indistinguishable?

We prove the theorem assuming that $a < b$. The proof for the case $a > b$ is nearly the same.

Proof (for $a < b$)

The Taylor polynomial

$$P_n(x) = f(a) + f'(a)(x - a) + \frac{f''(a)}{2!}(x - a)^2 + \cdots + \frac{f^{(n)}(a)}{n!}(x - a)^n$$

and its first n derivatives match the function f and its first n derivatives at $x = a$. We do not disturb that matching if we add another term of the form $K(x - a)^{n+1}$, where K is any constant, because such a term and its first n derivatives are all equal to zero at $x = a$. The new function

$$\phi_n(x) = P_n(x) + K(x - a)^{n+1}$$

and its first n derivatives still agree with f and its first n derivatives at $x = a$.

We now choose the particular value of K that makes the curve $y = \phi_n(x)$ agree with the original curve $y = f(x)$ at $x = b$. This can be done; we need only satisfy

$$f(b) = P_n(b) + K(b - a)^{n+1} \qquad \text{or} \qquad K = \frac{f(b) - P_n(b)}{(b - a)^{n+1}}. \qquad (1)$$

With K defined by Eq. (1), the function

$$F(x) = f(x) - \phi_n(x)$$

measures the difference between the original function f and the approximating function ϕ_n for each x in $[a, b]$.

We now use Rolle's Theorem. First, because $F(a) = F(b) = 0$ and both F and F' are continuous on $[a, b]$, we know that

$$F'(c_1) = 0 \qquad \text{for some } c_1 \text{ in } (a, b).$$

Next, because $F'(a) = F'(c_1) = 0$ and both F' and F'' are continuous on $[a, c_1]$, we know that

$$F''(c_2) = 0 \qquad \text{for some } c_2 \text{ in } (a, c_1).$$

Rolle's Theorem, applied successively to F'', F''', ..., $F^{(n-1)}$ implies the existence of

$$c_3 \text{ in } (a, c_2) \qquad \text{such that } F'''(c_3) = 0,$$
$$c_4 \text{ in } (a, c_3) \qquad \text{such that } F^{(4)}(c_4) = 0,$$
$$\vdots \qquad\qquad\qquad \vdots$$
$$c_n \text{ in } (a, c_{n-1}) \qquad \text{such that } F^{(n)}(c_n) = 0.$$

Finally, because $F^{(n)}$ is continuous on $[a, c_n]$ and differentiable on (a, c_n) and $F^{(n)}(a) = F^{(n)}(c_n) = 0$, Rolle's Theorem implies that there is a number c_{n+1} in (a, c_n) such that

$$F^{(n+1)}(c_{n+1}) = 0. \tag{2}$$

If we differentiate

$$F(x) = f(x) - P_n(x) - K(x - a)^{n+1}$$

$n + 1$ times, we get

$$F^{(n+1)}(x) = f^{(n+1)}(x) - 0 - (n + 1)!K. \tag{3}$$

Equations (2) and (3) together give

$$K = \frac{f^{(n+1)}(c)}{(n + 1)!} \qquad \text{for some number } c = c_{n+1} \text{ in } (a, b). \tag{4}$$

Equations (1) and (4) give

$$\frac{f(b) - P_n(b)}{(b - a)^{n+1}} = \frac{f^{(n+1)}(c)}{(n + 1)!} \qquad \text{or} \qquad f(b) = P_n(b) + \frac{f^{(n+1)}(c)}{(n + 1)!}(b - a)^{n+1}.$$

This concludes the proof. ≡

When we apply Taylor's Theorem, we usually want to hold a fixed and treat b as an independent variable. *Taylor's formula* is easier to use in circumstances like these if we change b to x. Here is how the theorem reads with this change.

THEOREM 14b   Taylor's Theorem with Taylor's Formula

If f has derivatives of all orders in an open interval I containing a, then for each positive integer n and for each x in I,

$$f(x) = f(a) + f'(a)(x - a) + \frac{f''(a)}{2!}(x - a)^2 + \cdots$$
$$+ \frac{f^{(n)}(a)}{n!}(x - a)^n + R_n(x), \tag{5}$$

where

$$R_n(x) = \frac{f^{(n+1)}(c)}{(n + 1)!}(x - a)^{n+1} \qquad \text{for some } c \text{ between } a \text{ and } x. \tag{6}$$

When we state Taylor's Theorem this way, it says that for each x in I,

$$f(x) = P_n(x) + R_n(x).$$

Pause for a moment to think about how remarkable this equation is. For any value of n that we want, the equation gives both a polynomial approximation of f of that order and a formula for the error involved in using that approximation over the interval I.

Equation (5) is called **Taylor's formula**. The function $R_n(x)$ is called the **remainder of order** n or the **error term** for the approximation of f by $P_n(x)$ over I. It is sometimes called the *Lagrange form* of the remainder. When $R_n(x) \to 0$ as $n \to \infty$, for all x in I, we say that the Taylor series generated by f at $x = a$ **converges** to f on I, and we write

$$f(x) = \sum_{k=0}^{\infty} \frac{f^{(k)}(a)}{k!}(x - a)^k.$$

EXAMPLE 4   The Maclaurin series for e^x

Show that the Taylor series generated by $f(x) = e^x$ at $x = 0$ converges to $f(x)$ for every real value of x.

Solution   The function has derivatives of all orders throughout the interval $-\infty < x < \infty$. Equations (5) and (6) with $f(x) = e^x$ and $a = 0$ give

$$e^x = 1 + x + \frac{x^2}{2!} + \cdots + \frac{x^n}{n!} + R_n(x),$$ Polynomial from Example 1

and

$$R_n(x) = \frac{e^c}{(n+1)!}x^{n+1}   \quad \text{for some } c \text{ between } 0 \text{ and } x.$$

Since e^x is an increasing function of x, e^c lies between $e^0 = 1$ and e^x. When x is negative, so is c, and $e^c < 1$. When x is zero, $e^x = 1$, and $R_n(x) = 0$. When x is positive, so is c, and $e^c < e^x$. Thus,

$$|R_n(x)| \le \frac{|x|^{n+1}}{(n+1)!}   \quad \text{when } x \le 0,$$

and

$$|R_n(x)| < e^x \frac{x^{n+1}}{(n+1)!}   \quad \text{when } x > 0.$$

Finally, because

$$\lim_{n \to \infty} \frac{x^{n+1}}{(n+1)!} = 0   \quad \text{for every } x$$ Section 9.1, Table 9.1,

it is also true that $\lim_{n \to \infty} R_n(x) = 0$, and the series converges to e^x for every x.

$$e^x = \sum_{k=0}^{\infty} \frac{x^k}{k!} = 1 + x + \frac{x^2}{2!} + \cdots + \frac{x^k}{k!} + \cdots. \qquad (7)$$

Estimating the Remainder

It is often possible to estimate $R_n(x)$ as we did in Example 4. This method of estimation is so convenient that we state it as a theorem for future reference.

THEOREM 15   **The Remainder Estimation Theorem**

If there are positive constants M and r such that $|f^{(n+1)}(t)| \leq Mr^{n+1}$ for all t between a and x inclusive, then the remainder term $R_n(x)$ in Taylor's Theorem satisfies the inequality

$$|R_n(x)| \leq M\frac{r^{n+1}|x-a|^{n+1}}{(n+1)!}.$$

If these conditions hold for every n and all the other conditions of Taylor's Theorem are satisfied by $f(x)$, then the series converges to $f(x)$.

In the simplest examples, we can take $r = 1$ provided that f and all its derivatives are bounded by some constant M. But if $f(x) = 2 \cos(3x)$, each time we differentiate we get a factor of 3, so we could take $r = 3$ and $M = 2$.

There are examples in which finding an upper bound for $|R_n(x)|$ is very difficult, either because the n and r of the Remainder Estimation Theorem cannot be found or because finding the nth derivative of a function is tedious. Estimating the nth derivative graphically can be helpful, but this is best done with a computer algebra system because most graphing calculators at this time allow us to graph to only the fourth derivative without computing derivatives by hand. If you can use a computer algebra system, you should explore some of the examples and exercises of this section and the next section by finding upper bounds for the nth derivative graphically.

We are now ready to look at some examples of how the Remainder Estimation Theorem and Taylor's Theorem can be used together to settle questions of convergence. We will see how they can also be used to determine the accuracy with which a function is approximated by one of its Taylor polynomials.

EXAMPLE 5   The Maclaurin series for sin x

Show that the Maclaurin series for $\sin x$ converges to $\sin x$ for all x.

Solution   The function and its derivatives are

$$f(x) = \sin x, \qquad\qquad f'(x) = \cos x,$$

$$f''(x) = -\sin x, \qquad\qquad f'''(x) = -\cos x,$$

$$\vdots \qquad\qquad\qquad\qquad \vdots$$

$$f^{(2k)}(x) = (-1)^k \sin x, \qquad f^{(2k+1)}(x) = (-1)^k \cos x,$$

so

$$f^{(2k)}(0) = 0 \quad \text{and} \quad f^{(2k+1)}(0) = (-1)^k.$$

The series has only odd-powered terms, and for $n = 2k + 1$, Taylor's Theorem gives

$$\sin x = x - \frac{x^3}{3!} + \frac{x^5}{5!} - \cdots + \frac{(-1)^k x^{2k+1}}{(2k+1)!} + R_{2k+1}(x).$$

All the derivatives of $\sin x$ have absolute values less than or equal to 1, so we can apply the Remainder Estimation Theorem with $M = 1$ and $r = 1$ to obtain

$$|R_{2k+1}(x)| \leq 1 \cdot \frac{|x|^{2k+2}}{(2k+2)!}.$$

Since $\left(|x|^{2k+2} / (2k+2)! \right) \to 0$ as $k \to \infty$, whatever the value of x, $R_{2k+1}(x) \to 0$, and the Maclaurin series for $\sin x$ converges to $\sin x$ for every x. ≡

> Compare the Taylor polynomials for $\sin x$ with those for $\cos x$ in Example 2.

$$\sin x = \sum_{k=0}^{\infty} \frac{(-1)^k x^{2k+1}}{(2k+1)!} = x - \frac{x^3}{3!} + \frac{x^5}{5!} - \frac{x^7}{7!} + \cdots. \qquad (8)$$

EXPLORATION 3

The Maclaurin Series for cos x

Add the remainder term to the Taylor polynomial for $\cos x$ in Example 2 to obtain Taylor's formula for $\cos x$. Find values for M and r of the Remainder Estimation Theorem so that you can apply the Remainder Estimation Theorem. Give an argument that the remainder term $\to 0$, and conclude that the Maclaurin series for $\cos x$ converges to $\cos x$ for every x. Thus, we have

$$\cos x = \sum_{k=0}^{\infty} \frac{(-1)^k x^{2k}}{(2k)!} = 1 - \frac{x^2}{2!} + \frac{x^4}{4!} - \frac{x^6}{6!} + \cdots. \qquad (9)$$

> Historically, Taylor series gave us a way to compute the values of e^x, $\sin x$, and other functions by simply evaluating polynomials. Today, calculators have even more powerful methods built in.

EXAMPLE 6

Find the Maclaurin series for $\cos 2x$, and show that it converges to $\cos 2x$ for every value of x.

Solution We replace the x in Eq. (9) by $2x$ to obtain

$$\cos 2x = \sum_{k=0}^{\infty} \frac{(-1)^k (2x)^{2k}}{(2k)!} = 1 - \frac{(2x)^2}{2!} + \frac{(2x)^4}{4!} - \frac{(2x)^6}{6!} + \cdots.$$

The Maclaurin series for $\cos x$ converges to $\cos x$ for every value of x and therefore converges for every value of $2x$. ≡

Confirming the Truncation Error

The Remainder Estimation Theorem gives us a means to confirm analytically the *truncation error* that we can estimate graphically or numerically.

EXAMPLE 7

Estimate e with an error of at most 10^{-6}. Confirm the estimate analytically.

Solution We can use the result of Example 4 with $x = 1$ to write

$$e = \underbrace{1 + 1 + \frac{1}{2!} + \cdots + \frac{1}{n!}}_{P_n(1)} + \underbrace{e^c \frac{1}{(n+1)!}}_{R_n(1)}$$

$$= P_n(1) + R_n(1) \qquad \text{for some } c \text{ between } 0 \text{ and } 1.$$

Apply a grapher program such as GRAPHSUM to graph the partial sums of the series, incrementing from 0 to 20 (Fig. 9.27). Graphical analysis shows that the partial sums stabilize in six decimal places at 2.718281 from $n = 9$ on.

To confirm that $e = 2.718281$ to within $0.000001 = 10^{-6}$ for partial sums $P_9(1), P_{10}(1), \ldots$, we want to show that the remainder $|R_n(1)| < 10^{-6}$ when $n \geq 9$.

First, because $0 < c < 1$, we assume (with confidence) that $1 < e^c < 3$. Then, for $n \geq 9$,

$$0 < e^c \frac{1}{(n+1)!} = R_n(1) < 3\frac{1}{(9+1)!} = 3\frac{1}{10!} < 0.83 \times 10^{-6} < 10^{-6}.$$

≡

For an alternating series, we now have two analytical methods for confirming the possible error when we approximate the series sum by a partial sum, the Alternating Series Estimation Theorem (Section 9.5) and the Remainder Estimation Theorem. There are also a variety of ways—many yet undiscovered—of estimating the error with a grapher.

EXAMPLE 8

The Maclaurin series (8) for $\sin x$ is an alternating series for every value of x. For what values of x can we replace $\sin x$ by $x - (x^3/3!)$ with an error no greater than 3×10^{-4}? Confirm by two methods, and support graphically.

Solution According to the Alternating Series Estimation Theorem, the error in truncating

$$\sin x = x - \frac{x^3}{3!} + \frac{x^5}{5!} - \cdots$$

after $(x^3/3!)$ is no greater than

$$\left|\frac{x^5}{5!}\right| = \frac{|x|^5}{120}.$$

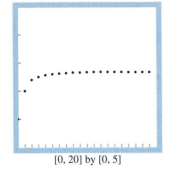

[0, 20] by [0, 5]

9.27 A GRAPHSUM plot of $(n, P_n(1))$.

Example 7 shows how we can use Theorem 15 to confirm error that we have been estimating graphically and numerically.

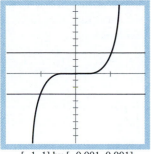

[−1, 1] by [−0.001, 0.001]

9.28 Graphs of

$$y_1 = \sin x - \left(x - \frac{x^3}{3!}\right),$$

$$y_2 = 3 \times 10^{-4},$$

and

$$y_3 = -3 \times 10^{-4}$$

show (using SOLVE) that $x - x^3/3!$ approximates $\sin x$ with an error of at most 3×10^{-4} in $-0.515 < x < 0.515$.

If we write

$$\sin x = x - \frac{x^3}{3!} + R_3,$$

then the Remainder Estimation Theorem with $M = r = 1$ gives

$$|R_3| \le 1 \cdot \frac{|x|^4}{4!} = \frac{|x|^4}{24},$$

which is not very good. But when we recognize that $x - x^3/3! = 0 + x + 0x^2 - x^3/3! + 0x^4$ is the Taylor polynomial of order 4 as well as of order 3, then we have

$$\sin x = x - \frac{x^3}{3!} + 0 + R_4,$$

and the Remainder Estimation Theorem also gives

$$|R_4| \le 1 \cdot \frac{|x|^5}{5!} = \frac{|x|^5}{120}.$$

Therefore, by either analytical method, the error will be less than or equal to 3×10^{-4} if

$$\frac{|x|^5}{120} < 3 \times 10^{-4} \qquad \text{or} \qquad |x| < \sqrt[5]{360 \times 10^{-4}} \approx 0.514. \qquad \text{Rounded down to be safe}$$

Figure 9.28 supports this result graphically. ≡

EXPLORATION 4

Sine Polynomials

1. Use the Maclaurin series for $\sin x$ to GRAPH polynomial approximations $P_{2k+1}(x)$ of the sine function for $k = 0, 1, 2, \cdots$.

2. Find the smallest value of k for which $P_{2k+1}(x)$ appears to be identical to $\sin x$ for one complete period to the right of the origin. Let $[y\text{Min}, y\text{Max}] = [-2, 2]$.

3. Devise a graphical method to determine the maximum error in using $P_{2k+1}(x)$ in part (2) to approximate $\sin x$ in $[0, 2\pi]$.

EXPLORATION BIT

Is the Maclaurin series for

$$\cos x + \sin x$$

an alternating series? What else can you say about this series?

Combining Taylor Series

On common intervals of convergence, Taylor series can be added, subtracted, and multiplied by constants just as other series can, and the results are once again Taylor series. The Taylor series for $f(x) + g(x)$ is the sum of the Taylor series for $f(x)$ and $g(x)$ because the nth derivative of $f + g$ is $f^{(n)} + g^{(n)}$, and so on. Thus, we obtain the Maclaurin series for $(1 + \cos 2x)/2$ by adding 1 to the Maclaurin series for $\cos 2x$ and dividing the combined results by 2, and the Maclaurin series for $\cos x + \sin x$ is the term-by-term sum of the Maclaurin series for $\cos x$ and $\sin x$.

Defining $e^{i\theta}$

As you may recall, a complex number is a number of the form $a + bi$, where a and b are real numbers and $i = \sqrt{-1}$. If we substitute $x = i\theta$ (θ real) in the Maclaurin series (7) for e^x and use the relations

$$i^2 = -1, \qquad i^3 = i^2 i = -i, \qquad i^4 = i^2 i^2 = 1, \qquad i^5 = i^4 i = i,$$

and so on, to simplify the result, we obtain

$$e^{i\theta} = 1 + \frac{i\theta}{1!} + \frac{i^2\theta^2}{2!} + \frac{i^3\theta^3}{3!} + \frac{i^4\theta^4}{4!} + \frac{i^5\theta^5}{5!} + \frac{i^6\theta^6}{6!} + \cdots$$

$$= \left(1 - \frac{\theta^2}{2!} + \frac{\theta^4}{4!} - \frac{\theta^6}{6!} + \cdots\right) + i\left(\theta - \frac{\theta^3}{3!} + \frac{\theta^5}{5!} - \cdots\right)$$

$$= \cos\theta + i\sin\theta.$$

This does not *prove* that $e^{i\theta} = \cos\theta + i\sin\theta$ because we haven't yet defined what it means to raise e to an imaginary power. But it does say how we ought to define $e^{i\theta}$ to be consistent with other things we know.

DEFINITION

For any real number θ,

$$e^{i\theta} = \cos\theta + i\sin\theta. \qquad (10)$$

EXPLORATION BIT

The number 2 is an important constant also, being the only even prime number and also the base for binary numbers, the key to how we talk with computers. Combine the six constants

$$0, 1, \pi, i, e, \text{ and } 2$$

into one equation so that each is used just once and no other number appears.

Equation (10), called **Euler's formula**, enables us to define e^{a+bi} to be $e^a \cdot e^{bi}$ for any complex number $a + bi$.

One of the amazing consequences of Euler's formula is the equation

$$e^{i\pi} = -1.$$

When written in the form $e^{i\pi} + 1 = 0$, this equation combines the five most important constants in mathematics: $0, 1, \pi, i,$ and e.

EXPLORATION 5

Extending Domains to Complex Numbers

Let $z = x + iy$ be any complex number.

1. Use Eq. (10) to define e^z and extend the domain of e^x to the complex numbers. Show that
2. $e^{-i\theta} = \cos\theta - i\sin\theta$
3. $\cos z = (e^{iz} + e^{-iz})/2$
4. $\sin z = \dfrac{e^{iz} - e^{-iz}}{2i}.$

Exercises 9.7

In Exercises 1–8, find the Taylor polynomials of orders 0, 1, 2, and 3 generated by f at a. Compare the graphs of the Taylor polynomials with the graph of f. Estimate graphically the range of values of x for which $P_3(x)$ approximates f with an error of at most 0.01.

1. $f(x) = \ln x, \quad a = 1$
2. $f(x) = \ln(1 + x), \quad a = 0$

3. $f(x) = 1/x, \quad a = 2$
4. $f(x) = 1/(x + 2), \quad a = 0$

5. $f(x) = \sin x, \quad a = \pi/4$
6. $f(x) = \cos x, \quad a = \pi/4$

7. $f(x) = \sqrt{x}, \quad a = 4$
8. $f(x) = \sqrt{x + 4}, \quad a = 0$

Find the Maclaurin series for the functions in Exercises 9–18. Estimate graphically the range of values of x for which $P_{10}(x)$ approximates f with an error of at most 0.01. Compare with the error bound obtained by using $R_{10}(x)$.

9. e^{-x}
10. $e^{x/2}$
11. $\sin 3x$

12. $5 \cos \pi x$
13. $\cos(-x)$
14. $x \sin x$

15. $\cosh x = (e^x + e^{-x})/2$

16. $\sinh x = (e^x - e^{-x})/2$

17. $x^2/2 - 1 + \cos x$

18. $\cos^2 x$ \quad (*Hint:* $\cos^2 x = (1 + \cos 2x)/2$.)

Quadratic Approximations

Write out Taylor's formula (Eq. 5) with $n = 2$ and $a = 0$ for the functions in Exercises 19–24. This will give you the quadratic approximations of these functions at $x = 0$ and the associated error terms.

19. $f(x) = \dfrac{1}{1 + x}$
20. $f(x) = \sqrt{1 + x}$

21. $f(x) = \ln(1 + x)$

22. $f(x) = (1 + x)^k$ (any number k)

23. $f(x) = \sin x$
24. $f(x) = \cos x$

25. Use the Taylor series generated by e^x at $x = a$ to show that

$$e^x = e^a \left(1 + (x - a) + \frac{(x - a)^2}{2!} + \cdots \right).$$

26. Find the Taylor series generated by e^x at $a = 1$. Compare your answer with the formula in Exercise 25.

Use analytic methods in Exercises 27–34, and support your answers graphically.

27. For approximately what values of x can you replace $\sin x$ by $x - x^3/6$ with an error of magnitude no greater than 5×10^{-4}?

28. If $\cos x$ is replaced by $1 - x^2/2$ and $|x| < 0.5$, what estimate can be made of the error? Does $1 - x^2/2$ tend to be too large or too small?

29. How far off is the approximation $\sin x = x$ when $|x| < 10^{-3}$? For which of these values of x is $x < \sin x$?

30. The estimate $\sqrt{1 + x} = 1 + x/2$ is used when x is small. Estimate the error when $|x| < 0.01$.

31. The approximation $e^x = 1 + x + x^2/2$ is used when x is small. Use the Remainder Estimation Theorem to estimate the error when $|x| < 0.1$.

32. When $x < 0$, the series for e^x is an alternating series. Use the Alternating Series Estimation Theorem to estimate the error that results from replacing e^x by $1 + x + x^2/2$ when $-0.1 < x < 0$. Compare with Exercise 31.

33. Estimate the error in the approximation $\sin x = x - x^3/3!$ when $|x| < 0.5$. (*Hint:* Use R_4, not R_3.)

34. When $0 \le h \le 0.01$, show that e^h may be replaced by $1 + h$ with an error no greater than 0.6% of h. Use $e^{0.01} = 1.01$.

Each of the series in Exercises 35 and 36 is the value of the Maclaurin series of a function $f(x)$ at some point. What function and what point? What is the sum of the series? Support your answer using a program such as PARTSUM.

35. $(0.1) - \dfrac{(0.1)^3}{3!} + \dfrac{(0.1)^5}{5!} - \cdots + \dfrac{(-1)^k (0.1)^{2k+1}}{(2k + 1)!} + \cdots$

36. $1 - \dfrac{\pi^2}{4^2 \cdot 2!} + \dfrac{\pi^4}{4^4 \cdot 4!} - \cdots + \dfrac{(-1)^k (\pi)^{2k}}{4^{2k} \cdot (2k!)} - \cdots$

Support your answers in Exercises 37-42 graphically.

37. Differentiate the Maclaurin series for $\sin x, \cos x$, and e^x term by term, and compare your results with the Maclaurin series for $\cos x, \sin x$, and e^x.

38. Integrate the Maclaurin series for $\sin x, \cos x$, and e^x term by term, and compare your results with the Maclaurin series for $\cos x, \sin x$, and e^x.

39. Multiply the Maclaurin series for e^x and $\sin x$ together to find the first five nonzero terms of the Maclaurin series for $e^x \sin x$.

40. Multiply the Maclaurin series for e^x and $\cos x$ together to find the first five nonzero terms of the Maclaurin series for $e^x \cos x$.

41. Use the Maclaurin series for $\sin x$ and the Alternating Series Estimation Theorem to show that

$$1 - \frac{x^2}{6} < \frac{\sin x}{x} < 1 \qquad \text{for} \qquad (0 < |x| < 1).$$

(This is the inequality in Section 2.3, Exercise 19.)

42. Use the Maclaurin series for $\cos x$ and the Alternating Series Estimation Theorem to show that

$$\frac{1}{2} - \frac{x^2}{24} < \frac{1 - \cos x}{x^2} < \frac{1}{2} \qquad (0 < |x| < 1).$$

(This is the inequality in Section 2.3, Exercise 21.)

43. Use Eq. (10) to write the following powers of e in the form $a + bi$.

a) $e^{-i\pi}$ b) $e^{i\pi/4}$ c) $e^{-i\pi/2}$

44. *Euler's identities.* Use Eq. (10) to show that

$$\cos\theta = \frac{e^{i\theta} + e^{-i\theta}}{2} \quad \text{and} \quad \sin\theta = \frac{e^{i\theta} - e^{-i\theta}}{2i}$$

(that is, $\cos\theta = \cosh(i\theta)$ and $\sin\theta = -i\sinh(i\theta)$).

45. Establish the equations in Exercise 44 by combining the formal Maclaurin series for $e^{i\theta}$ and $e^{-i\theta}$.

46. When a and b are real, we define $e^{(a+ib)x}$ to be $e^{ax}(\cos bx + i\sin bx)$. From this definition, show that

$$\frac{d}{dx}e^{(a+ib)x} = (a+ib)e^{(a+ib)x}.$$

47. Two complex numbers, $a+ib$ and $c+id$, are equal if and only if $a = c$ and $b = d$. Use this fact to evaluate

$$\int e^{ax}\cos bx\, dx \quad \text{and} \quad \int e^{ax}\sin bx\, dx$$

from

$$\int e^{(a+ib)x}\, dx = \frac{a-ib}{a^2+b^2}e^{(a+ib)x} + C,$$

where $C = C_1 + iC_2$ is a complex constant of integration.

9.8 Further Calculations with Taylor Series

This section introduces the binomial series, discusses how to choose the best center for a Taylor series, and shows how series are sometimes used to evaluate nonelementary integrals. At the end of the section, there is a table of important Maclaurin series.

The Binomial Series

The Maclaurin series generated by $f(x) = (1+x)^m$, when m is constant, is

$$1 + mx + \frac{m(m-1)}{2!}x^2 + \frac{m(m-1)(m-2)}{3!}x^3 + \cdots$$

$$+ \frac{m(m-1)(m-2)\cdots(m-k+1)}{k!}x^k + \cdots. \tag{1}$$

This series, called the **binomial series**, converges absolutely for $|x| < 1$. To derive the series, we first list the function and its derivatives:

$$f(x) = (1+x)^m,$$

$$f'(x) = m(1+x)^{m-1},$$

$$f''(x) = m(m-1)(1+x)^{m-2},$$

$$f'''(x) = m(m-1)(m-2)(1+x)^{m-3},$$

$$\vdots$$

$$f^{(k)}(x) = m(m-1)(m-2)\cdots(m-k+1)(1+x)^{m-k}.$$

We then evaluate these at $x = 0$ and substitute in the Maclaurin series formula to obtain

$$1 + mx + \frac{m(m-1)}{2!}x^2 + \cdots$$

$$+ \frac{m(m-1)(m-2)\cdots(m-k+1)}{k!}x^k + \cdots. \tag{2}$$

If m is an integer greater than or equal to zero, the series stops after $(m + 1)$ terms because the coefficients from $k = m + 1$ on are zero.

If m is not a positive integer or zero, the series is infinite and converges for $|x| < 1$. To see why, let u_k be the term involving x^k. Then apply the Ratio Test for Absolute Convergence to see that

$$\left| \frac{u_{k+1}}{u_k} \right| = \left| \frac{m - k}{k + 1} x \right| \to |x| \qquad \text{as } k \to \infty.$$

Our derivation of the binomial series shows only that it is generated by $(1 + x)^m$ and converges for $|x| < 1$. The derivation does not show that the series actually converges to $(1 + x)^m$. It does, but we will assume that part without proof. We will explore this series further in the exercises.

For $-1 < x < 1$,

$$(1 + x)^m = 1 + \sum_{k=1}^{\infty} \binom{m}{k} x^k, \tag{3}$$

where

$$\binom{m}{1} = m, \qquad \binom{m}{2} = \frac{m(m - 1)}{2!},$$

and

$$\binom{m}{k} = \frac{m(m - 1)(m - 2) \cdots (m - k + 1)}{k!} \qquad (k \geq 3).$$

EXAMPLE 1

Show that when $m = -1$, Eq. (3) gives the geometric series

$$\frac{1}{1 + x} = 1 - x + x^2 - x^3 + \cdots + (-1)^k x^k + \cdots. \tag{4}$$

Solution When $m = -1$,

$$\binom{-1}{1} = -1, \qquad \binom{-1}{2} = \frac{-1(-2)}{2!} = 1,$$

and

$$\binom{-1}{k} = \frac{-1(-2)(-3) \cdots (-1 - k + 1)}{k!} = (-1)^k \left(\frac{k!}{k!} \right) = (-1)^k.$$

With these coefficient values, Eq. (3) becomes

$$(1 + x)^{-1} = 1 + \sum_{k=1}^{\infty} (-1)^k x^k = 1 - x + x^2 - x^3 + \cdots + (-1)^k x^k + \cdots,$$

which is Eq. (4).

Choosing Centers for Taylor Series

Taylor's formula,

$$f(x) = f(a) + f'(a)(x-a) + \frac{f''(a)}{2!}(x-a)^2 + \cdots$$

$$+ \frac{f^{(n)}(a)}{n!}(x-a)^n + \frac{f^{(n+1)}(c)}{(n+1)!}(x-a)^{n+1}, \tag{5}$$

expresses the value of f at x in terms of f and its derivatives at a. In numerical computations, we therefore need a to be a point where we know the values of f and its derivatives. We also need a to be close enough to the values of x in which we are interested to make $(x-a)^{n+1}$ so small that we can neglect the remainder.

EXAMPLE 2

What value of a might we choose in Taylor's formula (Eq. 5) to compute $\sin 35°$ efficiently?

Solution   The radian measure for $35°$ is $35\pi/180$. We could choose $a = 0$ and use the series

$$\sin x = x - \frac{x^3}{3!} + \frac{x^5}{5!} - \cdots + (-1)^n \frac{x^{2n+1}}{(2n+1)!} + 0 \cdot x^{2n+2} + R_{2n+2}(x). \tag{6}$$

The remainder in Eq. (6) satisfies the inequality

$$|R_{2n+2}(x)| \le \frac{|x|^{2n+3}}{(2n+3)!},$$

which tends to zero as $n \to \infty$, no matter how large $|x|$ may be. We could therefore calculate $\sin 35°$ by placing

$$x = \frac{35\pi}{180} \approx 0.6108652$$

in the approximation

$$\sin x \approx x - \frac{x^3}{6} + \frac{x^5}{120} - \frac{x^7}{5040}.$$

This gives a truncation error of magnitude no greater than 3.3×10^{-8}, since

$$\left| R_8 \left(\frac{35\pi}{180} \right) \right| < \frac{(0.611)^9}{9!} < 3.3 \times 10^{-8}.$$

Alternatively, we could choose $a = \pi/6$ (which corresponds to $30°$) and use the series

$$\sin x = \sin \frac{\pi}{6} + \cos \frac{\pi}{6} \left(x - \frac{\pi}{6} \right) - \sin \frac{\pi}{6} \frac{(x - \pi/6)^2}{2!} - \cos \frac{\pi}{6} \frac{(x - \pi/6)^3}{3!}$$

$$+ \cdots + \sin \left(\frac{\pi}{6} + n \frac{\pi}{2} \right) \frac{(x - \pi/6)^n}{n!} + R_n(x).$$

By using this series, we could obtain equal accuracy with a smaller exponent n, but at the expense of introducing $\cos \pi/6$ as one of the coefficients. In this series, with $a = \pi/6$, we would take $x = 35\pi/180$, and the quantity $(x - a)$ would be

$$x - \frac{\pi}{6} = \frac{35\pi}{180} - \frac{30\pi}{180} = \frac{5\pi}{180} = 0.0872665,$$

which decreases rapidly as it is raised to higher powers. ≡

Evaluating Nonelementary Integrals

Maclaurin series are often used to express nonelementary integrals in terms of series.

EXAMPLE 3

Express $\int \sin x^2 \, dx$ as a power series.

Solution   From the series for $\sin x$, we obtain

$$\sin x^2 = x^2 - \frac{(x^2)^3}{3!} + \frac{(x^2)^5}{5!} - \frac{(x^2)^7}{7!} + \frac{(x^2)^9}{9!} - \cdots$$

$$= x^2 - \frac{x^6}{3!} + \frac{x^{10}}{5!} - \frac{x^{14}}{7!} + \frac{x^{18}}{9!} - \cdots.$$

Therefore,

$$\int \sin x^2 \, dx = C + \frac{x^3}{3} - \frac{x^7}{7 \cdot 3!} + \frac{x^{11}}{11 \cdot 5!} - \frac{x^{15}}{15 \cdot 7!} + \frac{x^{19}}{19 \cdot 9!} - \cdots.$$ ≡

EXPLORATION BIT

Use a series for $\int \sin x^2 \, dx$ shown in Example 3 to estimate $\int_0^1 \sin x^2 \, dx$ with an error of less than 10^{-5}.

EXAMPLE 4

Estimate $\int_0^1 \sin x^2 \, dx$ with an error of less than 1.1×10^{-9}.

Solution   From the indefinite integral in Example 3,

$$\int_0^1 \sin x^2 \, dx = \frac{1}{3} - \frac{1}{7 \cdot 3!} + \frac{1}{11 \cdot 5!} - \cdots$$

$$= \sum_{n=0}^{\infty} \frac{(-1)^n}{(4n+3)(2n+1)!}.$$

The sum of the first five terms is 0.3102683028. The series alternates, so the error in this approximation is at most

$$\frac{1}{(4 \cdot 5 + 3)(2 \cdot 5 + 1)!} < 1.1 \times 10^{-9}.$$

Compare this result with NINT $(\sin x^2, 0, 1)$. To guarantee this accuracy with the error formula for the Trapezoidal Rule would require using about 13,000 subintervals! ≡

EXAMPLE 5

Find a power series expansion for $\int_0^x \sin t/t \, dt$. Compare the graphs of $\int\limits_0^x \sin t/t \, dt$ and $P_{21}(x)$.

Solution From the series for $\sin x$, we obtain

$$\frac{\sin x}{x} = 1 - \frac{x^2}{3!} + \frac{x^4}{5!} - \cdots = \sum_{n=0}^{\infty} \frac{(-1)^n x^{2n}}{(2n+1)!}.$$

The series representation clearly shows that the discontinuity of $\sin x/x$ at $x = 0$ is removable. Integrating, we obtain

$$\int_0^x \frac{\sin t}{t} \, dt = \int_0^x \sum_{n=0}^{\infty} (-1)^n \frac{t^{2n}}{(2n+1)!} \, dt$$

$$= \sum_{n=0}^{\infty} (-1)^n \frac{t^{2n+1}}{(2n+1)(2n+1)!} \Big|_0^x$$

$$= \sum_{n=0}^{\infty} (-1)^n \frac{x^{2n+1}}{(2n+1)(2n+1)!}$$

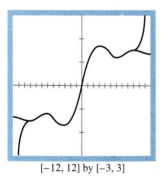

[−12, 12] by [−3, 3]

9.29 Graphs of

$$y_1 = \text{NINT}\left(\frac{\sin t}{t}, 0, x\right)$$

and its approximating Taylor polynomial

$$y_2 = P_{21}(x) = \sum_{n=0}^{10} (-1)^n \frac{x^{2n+1}}{(2n+1)(2n+1)!}.$$

Figure 9.29 shows the graph of $y = \int_0^x (\sin t)/t \, dt$ and the Taylor polynomial approximation $P_{21}(x) = \sum_{n=0}^{10} (-1)^n x^{2n+1}/((2n+1)(2n+1)!)$. ≡

Important Maclaurin Series

$$\frac{1}{1-x} = 1 + x + x^2 + \cdots + x^n + \cdots = \sum_{n=0}^{\infty} x^n \qquad (|x| < 1)$$

$$\frac{1}{1+x} = 1 - x + x^2 - \cdots + (-x)^n + \cdots = \sum_{n=0}^{\infty} (-1)^n x^n \qquad (|x| < 1)$$

$$e^x = 1 + x + \frac{x^2}{2!} + \cdots + \frac{x^n}{n!} + \cdots = \sum_{n=0}^{\infty} \frac{x^n}{n!} \qquad (|x| < \infty)$$

$$\sin x = x - \frac{x^3}{3!} + \frac{x^5}{5!} - \cdots + (-1)^n \frac{x^{2n+1}}{(2n+1)!} + \cdots = \sum_{n=0}^{\infty} \frac{(-1)^n x^{2n+1}}{(2n+1)!} \qquad (|x| < \infty)$$

$$\cos x = 1 - \frac{x^2}{2!} + \frac{x^4}{4!} - \cdots + (-1)^n \frac{x^{2n}}{(2n)!} + \cdots = \sum_{n=0}^{\infty} \frac{(-1)^n x^{2n}}{(2n)!} \qquad (|x| < \infty)$$

$$\ln(1+x) = x - \frac{x^2}{2} + \frac{x^3}{3} - \cdots + (-1)^{n-1} \frac{x^n}{n} + \cdots = \sum_{n=1}^{\infty} \frac{(-1)^{n-1} x^n}{n} \qquad (-1 < x \leq 1)$$

$$\ln \frac{1+x}{1-x} = 2 \tanh^{-1} x = 2\left(x + \frac{x^3}{3} + \frac{x^5}{5} + \cdots + \frac{x^{2n+1}}{2n+1} + \cdots\right) = 2\sum_{n=0}^{\infty} \frac{x^{2n+1}}{2n+1} \qquad (|x| < 1)$$

$$\tan^{-1} x = x - \frac{x^3}{3} + \frac{x^5}{5} - \cdots + (-1)^n \frac{x^{2n+1}}{2n+1} + \cdots = \sum_{n=0}^{\infty} \frac{(-1)^n x^{2n+1}}{2n+1} \qquad (|x| \leq 1)$$

Binomial Series

$$(1+x)^m = 1 + mx + \frac{m(m-1)x^2}{2!} + \frac{m(m-1)(m-2)x^3}{3!} + \cdots$$

$$+ \frac{m(m-1)(m-2)\cdots(m-k+1)x^k}{k!} + \cdots$$

$$= 1 + \sum_{k=1}^{\infty} \binom{m}{k} x^k \qquad (|x| < 1)$$

where

$$\binom{m}{1} = m, \qquad \binom{m}{2} = \frac{m(m-1)}{2!}, \qquad \binom{m}{k} = \frac{m(m-1)\cdots(m-k+1)}{k!} \qquad (k \geq 3).$$

Note: It is customary to define $\binom{m}{0}$ to be 1 and to take $x^0 = 1$ (even in the usually excluded case in which $x = 0$) in order to write the binomial series compactly as

$$(1+x)^m = \sum_{k=0}^{\infty} \binom{m}{k} x^k \qquad (|x| < 1).$$

If m is a *positive integer*, the series terminates at x^m, and the result converges for all x.

Exercises 9.8

What Taylor series would you use to represent the functions in Exercises 1–6 near the given values of x? (There may be more than one good answer.) Write out the first four nonzero terms of the series you choose. Estimate the error in using these terms to evaluate the function at the given value of x.

1. $\cos x$ near $x = 1$ **2.** $\sin x$ near $x = 6.3$

3. e^x near $x = 0.4$ **4.** $\ln x$ near $x = 1.3$

5. $\cos x$ near $x = 69$ **6.** $\tan^{-1} x$ near $x = 2$

7. Show that the Maclaurin series generated by $f(x) = (1+x)^3$ is the ordinary binomial expansion $1 + 3x + 3x^2 + x^3$ of $(1 + x)^3$.

8. If n is a positive integer, show that the Maclaurin series generated by $f(x) = (1 + x)^m$ is the ordinary binomial expansion of $(1 + x)^m$.

In Exercises 9–14, estimate the range of values of x for which $P_5(x)$ approximates $f(x) = (1+x)^m$ for the specified value of m with error at most 0.01. Then specify view dimensions for a window that shows where the graphs of f and $P_5(x)$ separate.

9. $m = 3/2$ **10.** $m = \sqrt{5}$

11. $m = -\sqrt{2}$ **12.** $m = -3/2$

13. $m = -3$ **14.** $m = -2$

15. Refer to Example 5. Estimate the error in approximating $\int_0^x (\sin t)/t \, dt$ by $P_{21}(x)$ for (a) $x = 5$, (b) $x = 8$, and (c) $x = 10$. (*Hint:* Notice that $P_{21}(x) = P_{22}(x)$.)

16. Estimate the range of values of x for which $P_{21}(x) = \sum_{n=0}^{10} (-1)^n x^{2n+1}/(2n + 1)$ approximates $\tan^{-1} x$ with an error of at most 0.001.

17. Express $\int_0^x (1 - \cos t)/t^2 \, dt$ as a power series. Compare the graphs of $\int_0^x (1 - \cos t)/t^2 \, dt$ and $P_{21}(x)$.

18. a) Express $\int_0^x (1 - e^{-t})/t \, dt$ as a power series.

b) Compare the graphs of $\int_0^x (1 - e^{-t})/t \, dt$ and $P_{10}(x)$.

c) Find n so that $P_n(1)$ approximates $\int_0^1 (1 - e^{-t})/t \, dt$ with an error of at most 10^{-6}.

Express each integral in Exercise 19–22 as an infinite series. Approximate the value of the integral using $P_{10}(x)$, and give an upper bound for the error. Support your answer with a NINT computation.

19. $\displaystyle\int_0^{0.1} \frac{\sin x}{x} \, dx$ **20.** $\displaystyle\int_0^{0.1} e^{-x^2} \, dx$

21. $\displaystyle\int_0^1 \frac{1 - \cos x}{x^2} \, dx$

22. $\displaystyle\int_0^{0.1} \sqrt{1 + x^4} \, dx$

(Use the binomial series with x^4 in place of x.)

23. Replace x by $-x$ in the Maclaurin series for $\ln(1 + x)$ to obtain a series for $\ln(1 - x)$. Then subtract this series from the Maclaurin series for $\ln(1 + x)$ to show that for $|x| < 1$,

$$\ln \frac{1+x}{1-x} = 2\left(x + \frac{x^3}{3} + \frac{x^5}{5} + \cdots\right).$$

Support this representation graphically.

24. How many terms of the Maclaurin series for $\ln(1+x)$ should you add to be sure of calculating $\ln(1.1)$ with an error of at most 10^{-8}? Use your graphing utility to support your answer.

Review Questions

1. Define infinite sequence (sequence), infinite series (series), and sequence of partial sums of a series.

2. Define convergence for (a) sequences, (b) series.

3. Which of the following statements are true, and which are false?

a) If a sequence does not converge, then it diverges.

b) If a sequence $\{a_n\}$ does not converge, then a_n tends to infinity as n does.

c) If a sequence $\{a_n\}$ converges, then so does the series
$$\sum_{n=1}^{\infty} a_n.$$

d) If a series does not converge, then its nth term does not

approach zero as n tends to infinity.

e) If the nth term of a series does not approach zero as n tends to infinity, then the series diverges.

f) If a sequence $\{a_n\}$ converges, then there is a number L such that a_n lies within 1 unit of L (i) for all values of n, (ii) for all but a finite number of values of n.

g) If all partial sums of a series are less than some constant L, then the series converges.

h) If a series converges, then its partial sums s_n are bounded (that is, $m \le s_n \le M$ for some constants m and M).

4. What tests do you know for the convergence and divergence of infinite series?

5. Under what circumstances do you know for sure that a bounded sequence converges?

6. Define absolute convergence and conditional convergence for infinite series. Give examples of series that (a) converge absolutely, (b) converge conditionally, (c) diverge.

7. What are geometric series? Under what circumstances do they converge? Diverge?

8. If a series converges, how confident can you be about partial sums computed with a grapher program such as PART-SUMT? What other information might increase your level of confidence?

9. What test would you apply to decide whether an alternating series converges? Give examples of convergent and divergent alternating series. How can you estimate the error involved in using a partial sum to estimate the sum of a convergent alternating series?

10. What is a power series? How do you test a power series for convergence? What kinds of results can you get? Give examples.

11. What are the basic facts about term-by-term differentiation and integration of power series? Give examples.

12. What is the Taylor series generated by a function $f(x)$ at a point $x = a$? What do you need to know about f to construct the series? Give an example.

13. What is the Maclaurin series generated by a function $f(x)$? Give an example.

14. What are Taylor polynomials, and what good are they? Give examples.

15. What is Taylor's formula, and what does Taylor's theorem say about it? When does the Taylor series generated by f at $x = a$ converge to f? What does it mean to support the sum graphically?

16. What does Taylor's formula tell us about the error in a linearization?

17. What are the Maclaurin series for e^x, $\sin x$, $\cos x$, $1/(1 + x)$, $1/(1 - x)$, $1/(1 + x^2)$, $\tan^{-1} x$, and $\ln(1 + x)$? How do you estimate the errors involved in replacing these series by their partial sums? Give examples.

18. What do you take into account when choosing a center for a Taylor series? Illustrate with examples.

19. How can you sometimes use series to evaluate nonelementary integrals? Give an example.

Practice Exercises

Determine which of the sequences $\{a_n\}$ in Exercises 1–12 converge and which diverge. Find the limit of each convergent sequence analytically, and support graphically.

1. $a_n = 1 + \dfrac{(-1)^n}{n}$

2. $a_n = \dfrac{1 - 2^n}{2^n}$

3. $a_n = \cos \dfrac{n\pi}{2}$

4. $a_n = \left(\dfrac{4}{n}\right)^{n/2}$

5. $a_n = \dfrac{\ln(n^2)}{n}$

6. $a_n = \left(\dfrac{n+5}{n}\right)^n$

7. $a_n = \sqrt[n]{\dfrac{3^n}{n}}$

8. $a_n = \dfrac{1}{3^{2n-1}}$

9. $a_n = \dfrac{(-4)^n}{n!}$

10. $a_n = \dfrac{\ln(2n+1)}{n}$

11. $a_n = \dfrac{(n+1)!}{n!}$

12. $a_n = \dfrac{n^2 - n}{2n^2 + n}$

Find the sums of the series in Exercises 13–18. Support with a grapher program such as PARTSUM.

13. $\displaystyle\sum_{n=1}^{\infty} \ln\left(\dfrac{n}{n+1}\right)$

14. $\displaystyle\sum_{n=2}^{\infty} \dfrac{-2}{n(n+1)}$

15. $\displaystyle\sum_{n=0}^{\infty} e^{-n}$

16. $\displaystyle\sum_{n=1}^{\infty} (-1)^n \dfrac{3}{4^n}$

17. $\displaystyle\sum_{n=0}^{\infty} a_n, a_0 = 125, a_n = 1.05^{-1} a_{n-1} \quad (n \geq 1)$

18. $\displaystyle\sum_{n=0}^{\infty} a_n, a_0 = 1, a_n = a_{n-1}/(1+n) \quad (n \geq 1)$

Which of the series in Exercises 19–30 converge absolutely, which converge conditionally, and which diverge? Give reasons for your answers. If the series converges, estimate the sum, and give an upper bound for the error if possible.

19. $\displaystyle\sum_{n=1}^{\infty} \dfrac{1}{\sqrt{n}}$

20. $\displaystyle\sum_{n=1}^{\infty} \dfrac{-5}{n}$

21. $\displaystyle\sum_{n=1}^{\infty} \dfrac{(-1)^n}{\sqrt{n}}$

22. $\displaystyle\sum_{n=1}^{\infty} \dfrac{1}{2n^3}$

23. $\displaystyle\sum_{n=1}^{\infty} \dfrac{(-1)^n}{\ln(n+1)}$

24. $\displaystyle\sum_{n=2}^{\infty} \dfrac{1}{n(\ln n)^2}$

25. $\displaystyle\sum_{n=1}^{\infty} \dfrac{(-1)^n}{n\sqrt{n^2+1}}$

26. $\displaystyle\sum_{n=1}^{\infty} \dfrac{(-1)^n 3n^2}{n^3+1}$

27. $\displaystyle\sum_{n=1}^{\infty} \dfrac{n+1}{n!}$

28. $\displaystyle\sum_{n=1}^{\infty} \dfrac{(-1)^n (n^2+1)}{2n^2+n-1}$

29. $\displaystyle\sum_{n=1}^{\infty} \frac{(-3)^n}{n!}$

30. $\displaystyle\sum_{n=1}^{\infty} \frac{2^n 3^n}{n^n}$

Each of Exercises 31–36 gives a formula for the nth term of a series. For what values of x does each series (a) converge and (b) converge absolutely? Graph the partial sum polynomial $s_{20}(x)$, and indicate a range of values of x for which the approximation has an error of at most 0.01, if possible.

31. $\dfrac{(x+2)^n}{3^n \cdot n}$

32. $\dfrac{x^n}{\sqrt{n}}$

33. $\dfrac{x^n}{n^n}$

34. $\dfrac{n+1}{2n+1}\dfrac{(x-1)^n}{2^n}$

35. $\dfrac{(-1)^{n-1}(x-1)^n}{n^2}$

36. $\dfrac{(x-1)^{2n-2}}{(2n-1)!}$

Each of the series in Exercises 37–42 is the value of the Maclaurin series of a function $f(x)$ at a particular point. What function and what point? What is the sum of the series? Support your answer with a PARTSUMT computation.

37. $1 - \dfrac{1}{4} + \dfrac{1}{16} - \cdots + (-1)^n \dfrac{1}{4^n} + \cdots$

38. $\dfrac{2}{3} - \dfrac{4}{18} + \dfrac{8}{81} - \cdots + (-1)^{n-1} \dfrac{2^n}{n\,3^n} + \cdots$

39. $\pi - \dfrac{\pi^3}{3!} + \dfrac{\pi^5}{5!} - \cdots + (-1)^n \dfrac{\pi^{2n+1}}{(2n+1)!} + \cdots$

40. $1 - \dfrac{\pi^2}{9 \cdot 2!} + \dfrac{\pi^4}{81 \cdot 4!} - \cdots + (-1)^n \dfrac{\pi^{2n}}{3^{2n}(2n!)} + \cdots$

41. $1 + \ln 2 + \dfrac{(\ln 2)^2}{2!} + \cdots + \dfrac{(\ln 2)^n}{n!} + \cdots$

42. $\dfrac{1}{\sqrt{3}} - \dfrac{1}{9\sqrt{3}} + \dfrac{1}{45\sqrt{3}} - \cdots$

$\quad + (-1)^{n-1} \dfrac{1}{(2n-1)(\sqrt{3})^{2n-1}} + \cdots$

In Exercises 43 and 44, find the first four nonzero terms of the Taylor series for the function $f(x)$ at the center $x = a$. Compare the graph of the Taylor polynomial with the graph of f. Estimate graphically the range of values of x for which the Taylor polynomial approximates f with an error of at most 0.01.

43. $f(x) = \sqrt{3+x^2}$ at $x = -1$

44. $f(x) = 1/(1-x)$ at $x = 2$

Use analytic methods in Exercises 45 and 46, and support your answers graphically.

45. Multiply the Maclaurin series for $\sin x$ and $2\cos x$ together to get the first four nonzero terms of the Maclaurin series for $\sin 2x$.

46. Use the formula $\sin^2 x = (1-\cos 2x)/2$ to get the Maclaurin series for $\sin^2 x$ from the Maclaurin series for $\cos 2x$.

Use series to approximate the values of the integrals in Exercises 47 and 48 with an error of at most 10^{-8}.

47. $\displaystyle\int_0^{1/2} e^{-x^3}\, dx$

48. $\displaystyle\int_0^{1/2} \dfrac{\tan^{-1} x}{x}\, dx$

49. *A convergent Taylor series that converges to its generating function only at its center.* It can be shown (though not simply) that the function f graphed below and defined by the rule

$$f(x) = \begin{cases} 0 & (x = 0) \\ e^{-1/x^2} & (x \neq 0) \end{cases}$$

has derivatives of all orders at $x = 0$ and that $f^{(n)}(0) = 0$ for all n.

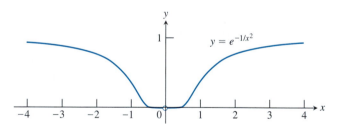

a) Find the Maclaurin series generated by f. At what values of x does the series converge? At what values of x does the series converge to $f(x)$?

b) Write out Taylor's formula (Section 9.7, Eq. 5) for f, taking $a = 0$ and assuming that $x \neq 0$. What does the formula tell you about the value of $R_n(x)$?

50. Use recursion to solve the equation $\sin x = x^3$.

Plane Curves, Parametrizations, and Polar Coordinates

OVERVIEW   The study of motion has been important since ancient times, and calculus gives us the mathematics that we need to describe it. In this chapter, we extend our ability to analyze motion by showing how to keep track of the position of a moving body as a function of time. We do this with parametric equations. We study them in the coordinate plane here and then extend our work to three dimensions in Chapters 11 and 12. We begin our study by developing equations for conic sections, since these are the paths traveled by planets, satellites, and other bodies whose motions are driven by inverse square forces. Planetary motion is best described in polar coordinates (another of Newton's inventions, although James Bernoulli usually gets the credit because he published first), so we will spend our remaining time finding out what curves, derivatives, and integrals look like in this new coordinate system.

10.1 _____ Conic Sections and Quadratic Equations

This section shows how the conic sections that originated in Greek geometry are described today as the graphs of quadratic equations in the coordinate plane. The Greeks of Plato's time described these curves as the curves formed by cutting a double cone with a plane (Fig. 10.1); hence the name "conic section." We begin by reviewing briefly the equation for circles, then continue on to parabolas, ellipses, and hyperbolas.

   The mathematics of conic sections is just what we need to describe the paths of planets, comets, moons, asteroids, satellites, or anything else that is moved through space by gravitational forces. Once we know that

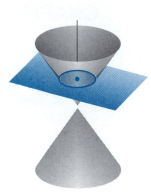

Circle: plane perpendicular
to cone axis

Ellipse

Parabola: plane parallel
to side of cone

Hyperbola: plane
parallel to cone axis

10.1   The standard conic sections—circles, ellipses, parabolas, and hyperbolas—are
the curves in which a plane cuts a double cone. Hyperbolas come in two parts,
called *branches*. Not shown here are three *degenerate* conic sections—a point,
a single line, and a pair of intersecting lines—each formed when a plane cuts
a double cone in a special way. Can you describe how these three special
cases could occur?

the path of a moving body is a conic section, we immediately have information about its velocity and the force that drives it, as we will see in
Chapter 12.

Equations from the Distance Formula—Circles

As we saw in Section 1.2, the distance between two points (x_1, y_1) and
(x_2, y_2) in the coordinate plane is

$$d = \sqrt{(x_2 - x_1)^2 + (y_2 - y_1)^2}.$$

We used this distance formula in Section 1.6 to derive the standard equations for circles centered at the origin and circles centered at any point
(h, k).

TO SIMPLIFY OUR WORK

In this chapter, most figures in the
coordinate plane will be centered
at the origin to simplify their
mathematical expressions. In practical
applications, any figure away from the
origin is simply a horizontal and/or
vertical shift of a congruent figure
that is centered at the origin.

The Standard Equations for a Circle

$$(x - h)^2 + (y - k)^2 = a^2 \qquad \text{Radius } a, \text{ centered at } (h, k)$$

$$x^2 + y^2 = a^2 \qquad \text{Radius } a, \text{ centered at } (0, 0)$$

Parabolas

Now we use the distance formula to derive standard equations for parabolas.

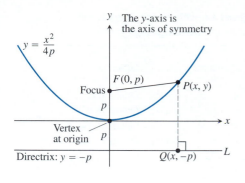

(a)

(b)

10.2   (a) The parabola $y = x^2/(4p)$.
(b) The parabola $y = -x^2/(4p)$. Notice that each parabola passes halfway between the directrix and the focus.

DEFINITIONS

A **parabola** is the set of points in a plane that are equidistant from a given fixed point and fixed line. The fixed point is the **focus** of the parabola. The fixed line is the **directrix** of the parabola.

Parabolas have simple equations when their foci (pronounced "FOE-sigh") and directrices ("di-*REC*-tri-seas") straddle the coordinate axes. For example, suppose the focus is the point $F(0,p)$ on the positive y-axis and the directrix is the line $y = -p$ (Fig. 10.2a). In the notation of the figure, a point $P(x, y)$ lies on the parabola if and only if

$$PF = PQ.$$

From the distance formula,

$$PF = \sqrt{(x - 0)^2 + (y - p)^2} = \sqrt{x^2 + (y - p)^2},$$

and

$$PQ = \sqrt{(x - x)^2 + (y - (-p))^2} = \sqrt{(y + p)^2}.$$

When we equate these expressions, square, and simplify, we get

$$y = \frac{x^2}{4p}. \qquad \text{Standard equation} \qquad (1)$$

If we did not already know from the geometry that the parabola opened upward, we could tell from the equation. The curve rises as x moves away from the origin. As an even function, the equation also reveals the parabola's symmetry about the y-axis. We call the y-axis the **axis** of the parabola (short for "axis of symmetry").

The point where a parabola crosses its axis, midway between the focus and the directrix, is called the **vertex** of the parabola. The vertex of the parabola $y = x^2/(4p)$ lies at the origin.

EXAMPLE 1

The standard equation of the parabola with vertex at the origin and focus at the point $(0, p) = (0, 1/4)$ is

$$y = \frac{x^2}{4(1/4)} \qquad \text{or} \qquad y = x^2. \qquad \blacksquare$$

If the parabola opens downward (Fig. 10.2b), with its focus at $F(0, -p)$ and its directrix the line $y = p$, then Eq. (1) becomes

$$y = -\frac{x^2}{4p}. \qquad \text{A parabola that opens downward}$$

Similar steps give us equations for parabolas that open to the left or right.

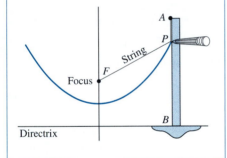

TABLE 10.1 The Standard Equations for Parabolas with Vertices at the Origin ($p > 0$)

Equation	Focus	Directrix	Axis	Direction
$y = x^2/(4p)$	$(0, p)$	$y = -p$	y-axis	Opens up
$y = -x^2/(4p)$	$(0, -p)$	$y = p$	y-axis	Opens down
$x = y^2/(4p)$	$(p, 0)$	$x = -p$	x-axis	Opens to right
$x = -y^2/(4p)$	$(-p, 0)$	$x = p$	x-axis	Opens to left

EXAMPLE 2

Find the focus and directrix of the parabola

$$y = \frac{x^2}{8}.$$

Solution   Find the value of p in the standard equation:

$$y = \frac{x^2}{8} \quad \text{is} \quad y = \frac{x^2}{4p} \quad \text{with} \quad 4p = 8, \quad \text{or} \quad p = 2.$$

Then find the focus and directrix for this value of p:

Focus: $(0, p)$ or $(0, 2)$,

Directrix: $y = -p$ or $y = -2$.

Ellipses

The equations that we use for ellipses come from the distance formula, too.

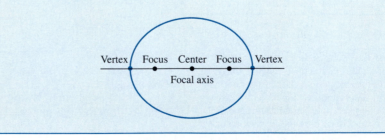

DEFINITIONS

An **ellipse** is the set of points in a plane whose distances from two fixed points in the plane have a constant sum.

   The two fixed points are the **foci** of the ellipse. The line through the foci is the **focal axis**. The point on this line halfway between the foci is the ellipse's **center**. The points where the focal axis and the ellipse cross are the ellipse's **vertices**.

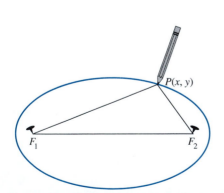

10.3   How to sketch an ellipse.

The quickest way to sketch an ellipse reasonably well uses the definition. Put a loop of string around two tacks F_1 and F_2, pull the string taut with a pencil point P, and move the pencil around to trace a curve (Fig. 10.3). The

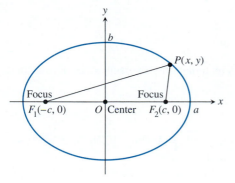

10.4   The ellipse defined by the equation

$$PF_1 + PF_2 = 2a$$

is the graph of the equation

$$\frac{x^2}{a^2} + \frac{y^2}{b^2} = 1.$$

EXPLORATION BITS

1. If the two fixed points (the foci) in the definition of an ellipse are the same ($c = 0$) and the constant sum ($2a$) is zero, the "ellipse" is one of the degenerate conic sections mentioned in the caption to Fig. 10.1. Identify it. (Henceforth in the text, we assume that $a > 0$.)

2. If the foci of an ellipse are the same, the "ellipse" is a special conic section. Identify it and explain. (Henceforth in the text, we assume that $c > 0$.)

3. If the constant sum ($2a$) in the definition of an ellipse is the same as the distance ($2c$) between the foci, the "ellipse" is a degenerate one that is *not* mentioned in the caption to Fig. 10.1. Describe it. (Henceforth in the text, we assume that $a > c > 0$.)

curve is an ellipse because the sum $PF_1 + PF_2$, being equal to the length of the loop minus the distance between the tacks, has a constant value. The foci of the ellipse lie at F_1 and F_2.

If the foci are $F_1(-c, 0)$ and $F_2(c, 0)$ (Fig. 10.4) and the sum of the distances $PF_1 + PF_2$ is denoted by $2a$, then the coordinates of a point P on the ellipse satisfy the equation

$$\sqrt{(x + c)^2 + y^2} + \sqrt{(x - c)^2 + y^2} = 2a.$$

To simplify this equation, we move the second radical to the right-hand side, square, isolate the remaining radical, and square again, obtaining

$$\frac{x^2}{a^2} + \frac{y^2}{a^2 - c^2} = 1. \tag{2}$$

Since the sum $PF_1 + PF_2$ is greater than the length F_1F_2 (triangle inequality for triangle PF_1F_2), the number $2a$ is greater than $2c$. Accordingly, a is greater than c, and the number $a^2 - c^2$ in Eq. (2) is positive.

The algebraic steps taken to arrive at Eq. (2) can be reversed to show that every point P whose coordinates satisfy an equation of this form with $0 < c < a$ also satisfies the equation $PF_1 + PF_2 = 2a$. Thus, a point lies on the ellipse if and only if its coordinates satisfy Eq. (2).

If we let b denote the positive square root of $(a^2 - c^2)$,

$$b = \sqrt{a^2 - c^2}, \tag{3}$$

then $a^2 - c^2 = b^2$, and Eq. (2) takes the more compact form

$$\frac{x^2}{a^2} + \frac{y^2}{b^2} = 1. \tag{4}$$

Equation (4) reveals that this ellipse is symmetric with respect to the origin and both coordinate axes. It lies inside the rectangle bounded by the lines $x = \pm a$ and $y = \pm b$. It crosses the axes at the points $(\pm a, 0)$ and $(0, \pm b)$. The tangents at these points are perpendicular to the axes because the slope

$$\frac{dy}{dx} = -\frac{b^2 x}{a^2 y}$$

is zero when $x = 0$ and infinite when $y = 0$.

Major and Minor Axes   The **major axis** of the ellipse described by Eq. (4) is the line segment of length $2a$ joining the intercepts $(\pm a, 0)$. The **minor axis** is the line segment of length $2b$ joining the intercepts $(0, \pm b)$. The number a itself is called the **semimajor axis**, and the number b the **semiminor axis**. The number c, which can be found from Eq. (3) to be

$$c = \sqrt{a^2 - b^2},$$

is sometimes called the **center-to-focus distance** of the ellipse (because that's what it is). When we interchange x and y in Eq. (4), we obtain the inverse

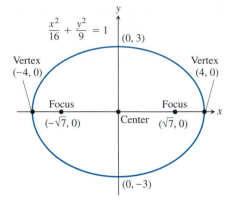

$$\frac{x^2}{16} + \frac{y^2}{9} = 1$$

(a)

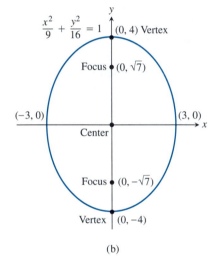

$$\frac{x^2}{9} + \frac{y^2}{16} = 1$$

(b)

10.5

Note that you can associate the larger denominator with the focal axis, the major axis, and the semimajor axis.

relationship

$$\frac{y^2}{a^2} + \frac{x^2}{b^2} = 1 \qquad \text{or} \qquad \frac{x^2}{b^2} + \frac{y^2}{a^2} = 1, \tag{5}$$

and the graphs of Eqs. (4) and (5) are reflections of each other across the line $y = x$.

EXAMPLE 3

To compare the two ellipses $x^2/16 + y^2/9 = 1$ (Fig. 10.5a) and $x^2/9 + y^2/16 = 1$ (Fig. 10.5b), we make a chart:

	$\dfrac{x^2}{16} + \dfrac{y^2}{9} = 1$	$\dfrac{x^2}{9} + \dfrac{y^2}{16} = 1$
Focal axis:	The x-axis	The y-axis
Major axis:	Horizontal	Vertical
Minor axis:	Vertical	Horizontal
Semimajor axis:	$a = \sqrt{16} = 4$	$a = \sqrt{16} = 4$
Semiminor axis:	$b = \sqrt{9} = 3$	$b = \sqrt{9} = 3$
Center-to-focus distance:	$c = \sqrt{16-9} = \sqrt{7}$	$c = \sqrt{16-9} = \sqrt{7}$
Foci:	$(\pm c, 0) = (\pm\sqrt{7}, 0)$	$(0, \pm c) = (0, \pm\sqrt{7})$
Vertices:	$(\pm a, 0) = (\pm 4, 0)$	$(0, \pm a) = (0, \pm 4)$
Center:	$(0, 0)$	$(0, 0)$

There is never any cause for confusion in analyzing equations like Eqs. (4) and (5). We simply find the intercepts on the coordinate axes; then we know which way the major axis runs because it is the longer of the two axes. The center always lies at the origin, and the foci always lie on the major axis.

Standard Equations for Ellipses Centered at the Origin

Foci on the x-axis: $\dfrac{x^2}{a^2} + \dfrac{y^2}{b^2} = 1 \qquad (a > b)$

Center-to-focus distance: $c = \sqrt{a^2 - b^2}$

Foci: $(\pm c, 0)$

Vertices: $(\pm a, 0)$

Foci on the y-axis: $\dfrac{x^2}{b^2} + \dfrac{y^2}{a^2} = 1 \qquad (a > b)$

Center-to-focus distance: $c = \sqrt{a^2 - b^2}$

Foci: $(0, \pm c)$

Vertices: $(0, \pm a)$

In each case, a is the semimajor axis, and b is the semiminor axis.

EXPLORATION 1

Graphing Ellipses, Part 1

When you graph conics, it is helpful to view them in a square viewing window.

1. Confirm that $\dfrac{x^2}{9} + \dfrac{y^2}{16} = 1$ is equivalent to $y = \pm\sqrt{16 - \dfrac{16}{9}x^2}$.

2. GRAPH $y_1 = \sqrt{16 - \dfrac{16}{9}x^2}$ and $y_2 = -y_1$ simultaneously in a $[-3, 3]$ by $[-4, 4]$ viewing window. Is the result what you expected?

3. Square the viewing window. Compare the new view with the previous view. Explain why square viewing windows are important when viewing graphs, particularly conics.   &

EXPLORATION BITS

1. Give a convincing argument why the vertices of an ellipse are $2a$ apart.

2. Give a convincing argument why, for an ellipse, the number $b = \sqrt{a^2 - c^2}$ happens also to be the distance from the origin to the y-intercepts.

Hyperbolas

Finally, we will see how the equations for hyperbolas also come from the distance formula.

DEFINITIONS

A **hyperbola** is the set of points in a plane whose distances from two fixed points in the plane have a constant difference.

The two fixed points are the **foci** of the hyperbola. The line through the foci is the **focal axis**. The point on this line halfway between the foci is the hyperbola's **center**. The points where the hyperbola and focal axis cross are the hyperbola's **vertices.**

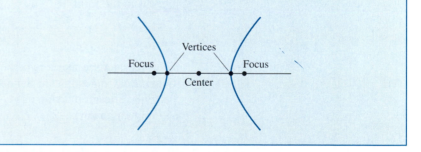

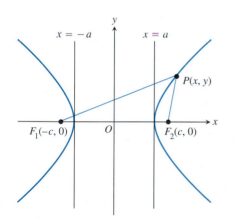

10.6 Hyperbolas have two branches. For points on the right-hand branch of the hyperbola shown here, $PF_1 - PF_2 = 2a$. For points on the left-hand branch, $PF_2 - PF_1 = 2a$.

If the foci are $F_1(-c, 0)$ and $F_2(c, 0)$ (Fig. 10.6) and the constant difference is $2a$, then a point (x, y) lies on the hyperbola if and only if

$$\sqrt{(x + c)^2 + y^2} - \sqrt{(x - c)^2 + y^2} = \pm 2a. \qquad (6)$$

To simplify this equation, we move the second radical to the right-hand side, square, isolate the remaining radical, and square again, obtaining

$$\frac{x^2}{a^2} + \frac{y^2}{a^2 - c^2} = 1. \qquad (7)$$

EXPLORATION BIT

Discuss the degenerate hyperbolas. (Henceforth in the text, we assume that the two foci are different ($c > 0$) and the constant difference, $2a$, is such that $0 < a < c$.)

So far, this is just like the equation for an ellipse. But now $a^2 - c^2$ is negative because $2a$, being the difference of two sides of triangle PF_1F_2, is less than $2c$, the third side.

The algebraic steps taken to arrive at Eq. (7) can be reversed to show that every point P whose coordinates satisfy an equation of this form with $0 < a < c$ also satisfies Eq. (6). Thus, a point lies on the hyperbola if and only if its coordinates satisfy Eq. (7).

If we let b denote the positive square root of $c^2 - a^2$,

$$b = \sqrt{c^2 - a^2}, \tag{8}$$

then $a^2 - c^2 = -b^2$, and Eq. (7) takes the more compact form

$$\frac{x^2}{a^2} - \frac{y^2}{b^2} = 1. \tag{9}$$

The only difference between Eq. (9) and the equation for an ellipse is the minus sign in the equation and the new relation

$$c^2 = a^2 + b^2$$

given by Eq. (8).

Like the ellipse of Eq. (4), the hyperbola of Eq. (9) is symmetric with respect to the origin and both coordinate axes. It crosses the x-axis at the points $(\pm a, 0)$. The tangents at these points are vertical because the derivative

$$\frac{dy}{dx} = \frac{b^2 x}{a^2 y}$$

is infinite when $y = 0$. The hyperbola has no y-intercepts; in fact, no part of the curve lies between the lines $x = -a$ and $x = a$. (Why?)

End Behavior and Asymptotes of Hyperbolas

Equation (9) for a hyperbola does not define a function. We can rewrite Eq. (9) in the form

$$y = \pm b \sqrt{\frac{x^2}{a^2} - 1},$$

however, and see that a hyperbola can be viewed as the graph of two functions. Both functions are defined for all $|x| \geq a$, so we can study the end behavior of each function as $x \to \infty$.

By comparing

$$f(x) = b \sqrt{\frac{x^2}{a^2} - 1} \qquad \text{to} \qquad g(x) = \frac{b}{a}|x|$$

and letting $|x|$ increase without bound, we find that $\lim_{x \to \pm \infty} (f/g) = 1$. Thus, g is an end behavior model for f. Similarly, we can show that $-(b/a)|x|$ is an end behavior model for $-b\sqrt{(x^2/a^2) - 1}$. Combined, we can call the lines

$$y = \pm \frac{b}{a} x$$

an *end behavior model for the hyperbola.* In fact, the hyperbola eventually appears to be indistinguishable from the lines $y = \pm (b/a)x$ as $x \to \infty$. So, as we did in Section 2.4, we call the lines $y = \pm(b/a)x$ the **asymptotes** of the hyperbola.

As with ellipses, when we interchange x and y in the equation of a hyperbola, we obtain the inverse relationship whose graphs are reflections of each other across the line $y = x$.

EXAMPLE 4

Compare the two hyperbolas $x^2/4 - y^2/5 = 1$ (Fig. 10.7a) and $y^2/4 - x^2/5 = 1$ (Fig. 10.7b).

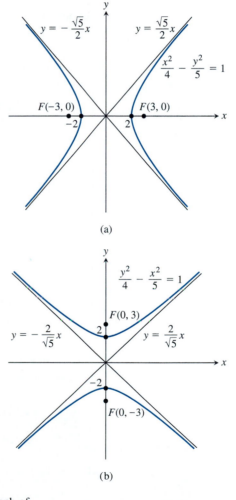

(a)

(b)

10.7 (a) The graph of

$$\frac{x^2}{4} - \frac{y^2}{5} = 1$$

and (b) the graph of

$$\frac{y^2}{4} - \frac{x^2}{5} = 1.$$

Note that you can associate the first term (the minuend) with the focal axis.

Solution Make a chart

	$\dfrac{x^2}{4} - \dfrac{y^2}{5} = 1$	$\dfrac{y^2}{4} - \dfrac{x^2}{5} = 1$
Focal axis:	The x-axis	The y-axis
Center-to-focus distance: $c = \sqrt{a^2 + b^2}$	$c = \sqrt{4 + 5} = 3$	$c = \sqrt{4 + 5} = 3$
Foci:	$(\pm c, 0) = (\pm 3, 0)$	$(0, \pm c) = (0, \pm 3)$
Vertices:	$(\pm a, 0) = (\pm 2, 0)$	$(0, \pm a) = (0, \pm 2)$
Center:	$(0, 0)$	$(0, 0)$
Asymptotes:	$y = \pm \dfrac{b}{a} x = \pm \dfrac{\sqrt{5}}{2} x$	$y = \pm \dfrac{a}{b} x = \pm \dfrac{2}{\sqrt{5}} x$

Standard Equations for Hyperbolas Centered at the Origin

Foci on the x-axis: $\dfrac{x^2}{a^2} - \dfrac{y^2}{b^2} = 1$

  Center-to-focus distance: $c = \sqrt{a^2 + b^2}$

  Foci: $(\pm c, 0)$

  Vertices: $(\pm a, 0)$

  Asymptotes: $y = \pm \dfrac{b}{a} x$

Foci on the y-axis: $\dfrac{y^2}{a^2} - \dfrac{x^2}{b^2} = 1$

  Center-to-focus distance: $c = \sqrt{a^2 + b^2}$

  Foci: $(0, \pm c)$

  Vertices: $(0, \pm a)$

  Asymptotes: $y = \pm \dfrac{a}{b} x$

EXPLORATION 2

Graphing Hyperbolas, Part 1

1. For the hyperbola $x^2/4 - y^2/9 = 1$, GRAPH

$$y_1 = 3\sqrt{\dfrac{x^2}{4} - 1} \quad \text{and} \quad y_2 = -3\sqrt{\dfrac{x^2}{4} - 1}.$$

Does y_1 give you one branch of the hyperbola and y_2 the other? Explain.

2. GRAPH y_1 and its end behavior model $y_3 = (3/2)|x|$.

3. Let $y_4 = -(3/2)|x|$, an end behavior model for y_2. GRAPH the hyperbola and its asymptotes. What other pair of equations besides those for y_3 and y_4 are possible for the asymptotes? Which pair do you prefer? Why?

Note the differences in the equations of the asymptotes. The asymptotes of

$$\frac{x^2}{a^2} - \frac{y^2}{b^2} = 1 \quad \text{are} \quad y = \pm\frac{b}{a}x.$$

The asymptotes of the inverse relation

$$\frac{y^2}{a^2} - \frac{x^2}{b^2} = 1$$

are found also by interchanging the x's and y's. Thus, the asymptotes of

$$\frac{y^2}{a^2} - \frac{x^2}{b^2} = 1$$

with foci on the y-axis are

$$x = \pm\frac{b}{a}y \quad \text{or} \quad y = \pm\frac{a}{b}x.$$

4. Describe how to use a, b, and the asymptotes to help you to *sketch* a graph of the hyperbola $x^2/a^2 - y^2/b^2 = 1$.   ⚘

Classifying Conic Sections by Eccentricity—The Focus–Directrix Equation

Although the center-to-focus distance c does not appear in the standard equation

$$\frac{x^2}{a^2} + \frac{y^2}{b^2} = 1 \quad (a > b)$$

for an ellipse, we may still determine the value of c from the equation

$$c = \sqrt{a^2 - b^2}.$$

If we keep a fixed and vary c over the interval $0 < c < a$, the resulting ellipses will vary in shape. They are close to being a circle when c is close to 0 (and a is close to b). They flatten as c increases. When c is close to a, the foci are close to the vertices, and the ellipse is close to a line segment.

We use the ratio of c to a to describe the various shapes possible for an ellipse. We call this ratio the ellipse's *eccentricity*. If the eccentricity is close to zero, the ellipse is close to being a circle. If the eccentricity is close to 1, the ellipse is long and thin.

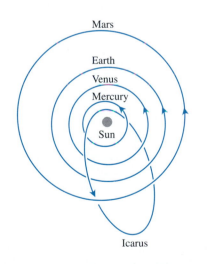

10.8 The orbit of the asteroid Icarus is highly eccentric. Earth's orbit is so nearly circular that both its foci lie inside the sun.

> **DEFINITION**
>
> The **eccentricity** of the ellipse $x^2/a^2 + y^2/b^2 = 1 (a > b)$ is the number
>
> $$e = \frac{c}{a} = \frac{\sqrt{a^2 - b^2}}{a}.$$

The planets in the solar system revolve around the sun in elliptical orbits with the sun at one focus. Most of the planets, including Earth, have orbits that are nearly circular, as can be seen from the eccentricities in Table 10.2. Pluto, however, has a fairly eccentric orbit, with $e = 0.25$, as does Mercury, with $e = 0.21$. Other members of the solar system have orbits that are even more eccentric. Icarus, an asteroid about 1 mile wide that revolves around the sun every 409 Earth days, has an orbital eccentricity of 0.83 (Fig. 10.8).

TABLE 10.2   **Eccentricities of Planetary Orbits**

Mercury	0.21	Mars	0.09	Uranus	0.05
Venus	0.01	Jupiter	0.05	Neptune	0.01
Earth	0.02	Saturn	0.06	Pluto	0.25

EXAMPLE 5

The orbit of Halley's comet is an ellipse 36.18 astronomical units long by 9.12 astronomical units wide. (One *astronomical unit* (AU) is the semimajor axis of Earth's orbit, about 92,600,000 miles.) Its eccentricity is

$$e = \frac{\sqrt{a^2 - b^2}}{a} = \frac{\sqrt{(36.18/2)^2 - (9.12/2)^2}}{(36.18/2)}$$

$$= \frac{\sqrt{(18.09)^2 - (4.56)^2}}{18.09} = 0.97.$$ ≡

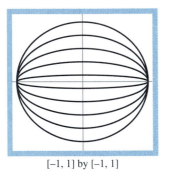
Whereas a parabola has one focus and one directrix, each ellipse has two foci and two directrices. These are the lines perpendicular to the major axis at distances $\pm a/e$ from the center. The parabola has the property that

$$PF = 1 \cdot PD \tag{10}$$

for any point P on it, where F is the focus and D is the point nearest P on the directrix. For an ellipse, it can be shown that the equations that replace Eq. (10) are

$$PF_1 = e \cdot PD_1, \qquad PF_2 = e \cdot PD_2. \tag{11}$$

Here, e is the eccentricity, P is any point on the ellipse, F_1 and F_2 are the foci, and D_1 and D_2 are the points on the directrices nearest P (Fig. 10.9).

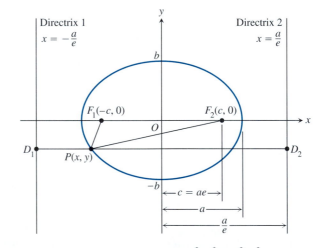

10.9   The foci and directrices of the ellipse $x^2/a^2 + y^2/b^2 = 1$. Directrix 1 corresponds to focus F_1, and Directrix 2 to focus F_2.

In Eqs. (11), the directrix and focus must correspond; that is, if we use the distance from P to F_1, we must also use the distance from P to the directrix at the same end of the ellipse. Thus, the directrix $x = -a/e$ corresponds to $F_1(-c, 0)$, and the directrix $x = a/e$ corresponds to $F_2(c, 0)$.

NOTATION

We will always be able to distinguish between the eccentricity e of an ellipse and the transcendental number $e = 2.71828\ldots$ by the context in which either is used. (Does this make you wonder what a hyperbola looks like for which $e = e$?)

EXPLORATION BIT

For an ellipse, we know that e near zero means "close to circular," and e near 1 means "long and flat." What does it mean for a hyperbola to have e near 1? $e \to \infty$?

Consider the equation

$$y = \pm b \sqrt{(x^2/a^2) - 1}.$$

Let $a = 1$,

$$y_1 = b\sqrt{x^2 - 1}, \quad \text{and} \quad y_2 = -y_1.$$

Then GRAPH y_1 and y_2 for $b = 0.25, 0.5, 1, 2, 4, 8, 16$ in a square viewing window. What do you predict should happen to the graphs in the viewing window as b increases from near zero to large values; that is, e increases from near 1 with c close to a, to large values with c "moving away" from a. (If your grapher has "list" capability, use it.)

We define the eccentricity of a hyperbola with the same formula that we use for the ellipse, $e = c/a$, but in this case, c equals $\sqrt{a^2 + b^2}$ instead of $\sqrt{a^2 - b^2}$. In contrast to the eccentricity of an ellipse, the eccentricity of a hyperbola is always greater than 1.

DEFINITION

The **eccentricity** of the hyperbola $x^2/a^2 - y^2/b^2 = 1$ is the number

$$e = \frac{c}{a} = \frac{\sqrt{a^2 + b^2}}{a}.$$

In both ellipse and hyperbola, the eccentricity is the ratio of the distance between the foci to the distance between the vertices (because $c/a = 2c/2a$).

$$\text{Eccentricity} = \frac{\text{Distance between foci}}{\text{Distance between vertices}}.$$

In an ellipse, the foci are closer together than are the vertices, and the ratio is less than 1. In a hyperbola, the foci are farther apart than are the vertices, and the ratio is greater than 1.

EXAMPLE 6

Find the eccentricity of the hyperbola $9x^2 - 16y^2 = 144$.

Solution   We divide both sides of the hyperbola's equation by 144 to put it in standard form, obtaining

$$\frac{9x^2}{144} - \frac{16y^2}{144} = 1 \quad \text{and} \quad \frac{x^2}{16} - \frac{y^2}{9} = 1.$$

With $a^2 = 16$ and $b^2 = 9$, we find that

$$c = \sqrt{a^2 + b^2} = \sqrt{16 + 9} = 5,$$

so

$$e = \frac{c}{a} = \frac{5}{4}. \qquad \qquad \equiv$$

As with the ellipse, it can be shown that the lines $x = \pm a/e$ act as directrices for the hyperbola and that

$$PF_1 = e \cdot PD_1 \quad \text{and} \quad PF_2 = e \cdot PD_2. \qquad (12)$$

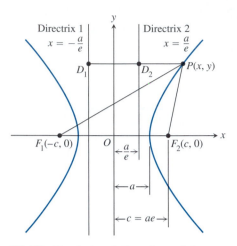

10.10   The foci and directrices of the hyperbola $x^2/a^2 - y^2/b^2 = 1$. No matter where P lies on the hyperbola, $PF_1 = e \cdot PD_1$ and $PF_2 = e \cdot PD_2$.

Here, P is any point on the hyperbola, F_1 and F_2 are the foci, and D_1 and D_2 are the points nearest P on the directrices (Fig. 10.10).

To complete the picture, we now define the eccentricity of a parabola to be $e = 1$. Equations (10), (11), and (12) then have the common form $PF = e \cdot PD$.

DEFINITION

The **eccentricity** of a parabola is $e = 1$.

The "focus–directrix" equation $PF = e \cdot PD$ unites the parabola, ellipse, and hyperbola in the following way. Suppose that the distance PF of a point P from a fixed point F (the focus) is a constant multiple of its distance from a fixed line (the directrix). That is, suppose

$$PF = e \cdot PD, \tag{13}$$

where e is the constant of proportionality. Then the path traced by P is

a) a *parabola* if $e = 1$,

b) an *ellipse* of eccentricity e if $e < 1$, and

c) a *hyperbola* of eccentricity e if $e > 1$.

Equation (13) may not look like much to get excited about. There are no coordinates in it, and when we try to translate it into coordinate form, it translates in different ways, depending on the numerical size of e. At least, that is what happens in the Cartesian plane. However, in the polar coordinate plane, as we will see in Section 10.7, the equation $PF = e \cdot PD$ translates into a single equation regardless of the size of e, an equation so simple that it has been the equation of choice for astronomers and space scientists for nearly 300 years.

Reflective Properties

The chief application of parabolas involves their use as reflectors of light and radio waves. Rays originating at a parabola's focus are reflected out of the parabola parallel to the parabola's axis. Similarly, rays coming in parallel to the axis are reflected toward the focus. We will see why this is so when we get into the calculus of conics.

This property is used in parabolic mirrors and telescopes, in automobile headlamps, in spotlights of all kinds, in parabolic radar and microwave antennas, and in television dish receivers. Parabolas are also used in bridge construction, wind tunnel photography, and submarine tracking.

Like the parabola, ellipses and hyperbolas have reflective properties that are important in medicine, science and engineering. If an ellipse is revolved about its major axis to generate a surface (the surface is called an *ellipsoid*)

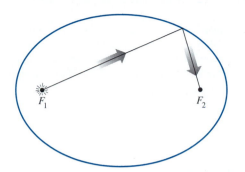

and the interior is silvered to produce a mirror, light from one focus will be reflected to the other focus (Fig. 10.11). Ellipsoids reflect sound the same way, and this property is used to construct *whispering galleries*, rooms in which a person standing at one focus can hear a whisper from the other focus. Ellipsoids also appear in instruments used to study aircraft noise in wind tunnels (sound at one focus can be received at the other focus with relatively little interference from other sources).

Light directed toward one focus of a hyperbolic mirror is reflected in a line through the other focus (Fig. 10.12). This property of hyperbolas is combined with the reflective properties of parabolas and ellipses in designing modern telescopes (Fig. 10.13).

10.11 An elliptical mirror (shown here in profile) reflects light from one focus to the other.

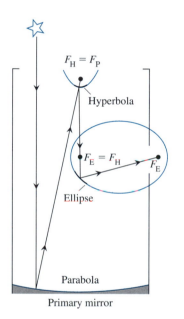

10.13 In this schematic drawing of a reflecting telescope, starlight reflects off a primary parabolic mirror toward the mirror's focus F_P. It is then reflected by a small hyperbolic mirror, whose focus is $F_H = F_P$, toward the second focus of the hyperbola, $F_E = F_H$. Since this focus is shared by an ellipse, the light is reflected by the elliptical mirror to the ellipse's second focus to be seen by an observer.

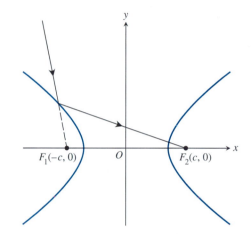

10.12 In this profile of a hyperbolic mirror, light coming toward focus F_1 is reflected toward focus F_2.

Other Applications

Ellipses appear in airplane wings (British Spitfire) and sometimes in gears designed for racing bicycles. Stereo systems often have elliptical styli, and water pipes are sometimes designed with elliptical cross-sections to allow for expansion when the water freezes. The triggering mechanisms in some lasers are elliptical, and stones on a beach become more and more elliptical as they are ground down by waves. There are also applications of ellipses to fossil formation. The ellipsolith, once thought to be a separate species, is now known to be an elliptically deformed nautilus.

Hyperbolic paths arise in Einstein's theory of relativity and form the basis for the (unrelated) LORAN radio navigation system. (LORAN is short for Long Range Navigation.) Hyperbolas also form the basis for a new system that the railroads are developing for using synchronized electronic signals from satellites to track freight trains. In a recent test in Minnesota, computers aboard locomotives were able to track trains to within one mile per hour of their speed and to within 150 feet of their actual location.

Exercises 10.1

In Exercises 1–6, write an equation for the parabola with the given focus and directrix.

1. Focus: $(0, 4)$ Directrix: $y = -4$

2. Focus: $\left(0, \dfrac{1}{4}\right)$ Directrix: $y = -\dfrac{1}{4}$

3. Focus: $(0, -3)$ Directrix: $y = 3$

4. Focus: $\left(0, -\dfrac{1}{2}\right)$ Directrix: $y = \dfrac{1}{2}$

5. Focus: $(-3, 0)$ Directrix: $x = 3$

6. Focus: $(2, 0)$ Directrix: $x = -2$

Find the foci and directrices of the parabolas in Exercises 7–10.

7. $y = 4x^2$

8. $y = x^2/3$

9. $y = -3x^2$

10. $y = -x^2/4$

Sketch a graph of each of the parabolas in Exercises 11–14.

11. $y = x^2/2$ **12.** $y = -x^2/6$

13. $x = y^2/8$ **14.** $x = -y^2/4$

Sketch a graph of each of the conic sections in Exercises 15–18. Support with a graphing utility.

15. $\dfrac{x^2}{4} + \dfrac{y^2}{9} = 1$ **16.** $\dfrac{x^2}{2} + y^2 = 1$

17. $\dfrac{y^2}{4} - x^2 = 1$ **18.** $\dfrac{x^2}{4} - \dfrac{y^2}{9} = 1$

Exercises 19–34 give equations for ellipses or hyperbolas. Put each equation in standard form, and find the eccentricity. Then sketch the conic. Include the foci in your sketch. Include the asymptotes when appropriate. Support your sketch with a graphing utility.

19. $64x^2 - 36y^2 = 2304$ **20.** $16x^2 + 25y^2 = 400$

21. $8y^2 - 2x^2 = 16$ **22.** $7x^2 + 16y^2 = 112$

23. $169x^2 + 25y^2 = 4225$ **24.** $y^2 - 3x^2 = 3$

25. $8x^2 - 2y^2 = 16$ **26.** $6x^2 + 9y^2 = 54$

27. $9x^2 + 10y^2 = 90$ **28.** $y^2 - x^2 = 8$

29. $x^2 - y^2 = 1$ **30.** $2x^2 + y^2 = 4$

31. $y^2 - x^2 = 4$ **32.** $2x^2 + y^2 = 2$

33. $3x^2 + 2y^2 = 6$ **34.** $9x^2 - 16y^2 = 144$

35. Make a chart similar to those in Examples 3 and 4, comparing the graphs of the two conics in Exercises 10 and 14.

36. Make a chart similar to those in Examples 3 and 4, comparing the graphs of the two conics in Exercises 21 and 25.

37. *Archimedes' formula for the volume of a parabolic solid.* The region enclosed by the parabola $y = (4h/b^2)x^2$ and the line $y = h$ is revolved about the y-axis to generate a solid as shown. Show that the volume of the solid is three-halves the volume of the corresponding cone.

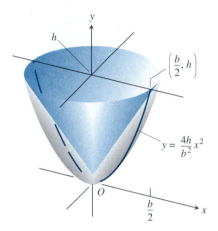

38. *Suspension bridge cables hang in parabolas.* The diagram below shows a cable of a suspension bridge supporting a uniform load of w pounds per horizontal foot. It can be shown that if H is the horizontal tension in the cable at the origin, then the curve of the cable satisfies the differential equation

$$\frac{dy}{dx} = \frac{w}{H}x.$$

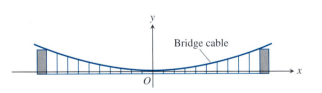

Show that the cable hangs in a parabola by solving this equation with the condition that $y = 0$ when $x = 0$.

39. Draw an ellipse of eccentricity 4/5.

40. Draw the orbit of Pluto to scale.

41. Find the dimensions of the rectangle of largest area that can be inscribed in the ellipse $x^2 + 4y^2 = 4$ with its sides parallel to the coordinate axes. What is the area of the rectangle?

42. Find the center of mass of a thin homogeneous plate that is bounded below by the x-axis and above by the ellipse $x^2/9 + y^2/16 = 1$.

43. The region bounded on the left by the y-axis, on the right by the hyperbola $x^2 - y^2 = 1$, and above and below by the lines $y = \pm 3$, is revolved about the x-axis to generate a solid. Find the volume of the solid.

44. The curve $y = \sqrt{x^2 + 1}\ \left(0 \le x \le \sqrt{2}\right)$, which is part of the upper branch of the hyperbola $y^2 - x^2 = 1$, is revolved about the x-axis to generate a surface. Find the area of the surface.

45. *The reflective property of parabolas.* The diagram below shows a typical point $P(x_0, y_0)$ on the parabola $y^2 = 4px$.

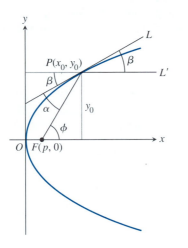

The line L is tangent to the parabola at P. The parabola's focus lies at $F(p, 0)$. The ray L' extending from P to the right is parallel to the x-axis. We show that light from F to P will be reflected out along L' by showing that β equals α. Establish this equality by taking the following steps.

a) Show that $\tan \beta = 2p/y_0$.

b) Show that $\tan \phi = y_0/(x_0 - p)$.

c) Use the identity

$$\tan \alpha = \frac{\tan \phi - \tan \beta}{1 + \tan \phi \tan \beta}$$

to show that $\tan \alpha = 2p/y_0$. Since the angles involved are both acute, $\tan \beta = \tan \alpha$ implies that $\beta = \alpha$.

46. In physics, circular waves are created in a ripple tank by touching the surface of the water, first at a point A and then

at a point B (see diagram below). As the waves expand, their point of intersection seems to trace a hyperbola. Does it really do that?

To find out, we can model the waves with circles centered at A and B. At time t, the point P is $r_A(t)$ units from A and $r_B(t)$ units from B. Since the radii of the circles increase at a constant rate, the rate at which the waves are traveling is

$$\frac{dr_A}{dt} = \frac{dr_B}{dt}.$$

Conclude from this equation that $r_A - r_B$ has a constant value, so P must lie on a hyperbola with foci at A and B.

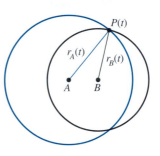

47. Let $x^2/a^2 - y^2/b^2 = 1$. Show that $y = \pm(b/a)\sqrt{x^2 - a^2}$.

48. Show that $g(x) = (b/a)\,|x|$ is an end behavior model for $f(x) = (b/a)\sqrt{x^2 - a^2}$.

49. Show that $g(x) = -(b/a)\,|x|$ is an end behavior model for $f(x) = -(b/a)\sqrt{x^2 - a^2}$.

50. a) Derive Eq. (1) for the parabola in Fig. 10.2(a).

    b) Show that the algebraic steps in part (a) can be reversed to prove that every point that satisfies Eq. (1) lies on the parabola.

51. a) Derive Eq. (4) for the ellipse in Fig. 10.4.

    b) Show that the algebraic steps in part (a) can be reversed to prove that every point that satisfies Eq. (4) lies on the ellipse.

52. a) Derive Eq. (9) for the hyperbola in Fig. 10.6.

    b) Show that the algebraic steps in part (a) can be reversed to prove that every point that satisfies Eq. (9) lies on the hyperbola.

10.2

The Graphs of Quadratic Equations in x and y

In this section, we establish one of the most amazing results in analytic geometry, which is that the Cartesian graph of any equation of the form

$$Ax^2 + Bxy + Cy^2 + Dx + Ey + F = 0 \qquad (1)$$

Equation (1) is called the **general quadratic equation.**

SUMMARY

A graph of the equation

$$Ax^2 + Cy^2 + Dx + Ey + F = 0$$

is

a) a *circle* if $A = C \neq 0$ (degenerate cases: the graph is a point, or there is no graph at all);

b) a *parabola* if the equation is quadratic in one variable and linear in the other ($A \neq 0, C = 0, E \neq 0$, *or* $A = 0, C \neq 0, D \neq 0$);

c) an *ellipse* if $A \neq C$ and A and C are both positive or both negative (degenerate cases: a single point or no graph at all);

d) a *hyperbola* if A and C have opposite signs (degenerate case: a pair of intersecting lines).

in which A, B, and C are not all zero is nearly always a conic section. The only exceptions are the case in which the graph consists of two parallel lines and the case in which there is no graph at all. It is conventional to call all graphs of Eq. (1), curved or not, **quadratic curves.**

You should already be familiar with Eq. (1) in the form

$$Ax^2 + Cy^2 + Dx + Ey + F = 0$$

where $B = 0$, so there is no xy term. The graph is a familiar conic section (possibly shifted, stretched, and/or shrunk) for most values of A, B, C, D, E, and F. Our final analysis of Eq. (1) involves the effect of the term Bxy when $B \neq 0$. It connects the conic sections to the one geometric transformation that we haven't studied yet, namely, rotations. To support this, check out what we can do with a graphing utility in the following Exploration. Pay particular attention to part 3.

EXPLORATION 1

Graphing Quadratic Curves

Assume that $C \neq 0$. Then Eq. (1) can be thought of as a quadratic equation in y:

$$Cy^2 + (Bx + E)y + Ax^2 + Dx + F = 0. \qquad (2)$$

Equation (2) can be solved for y by using the Quadratic Formula:

$$y = \frac{-(Bx + E) \pm \sqrt{(Bx + E)^2 - 4C(Ax^2 + Dx + F)}}{2C}.$$

This can be thought of as two functions y_1 and y_2 whose graphs, in general, we probably wouldn't want to *sketch*. We can, however, GRAPH y_1 and y_2 with ease.

1. Use y_1 and y_2 as described above. GRAPH

$$64x^2 + 144y^2 - 320x + 504y - 455 = 0 \qquad (3)$$

in a square viewing window. Note that Eq. (3) contains no Bxy term, so you should expect to see a conic section (do you know which?) with an axis that is horizontal or vertical. Also, to simplify your keystroking, you can let

$$y_3 = \sqrt{E^2 - 4C(Ax^2 + Dx + F)}.$$

Then,

$$y_1 = \frac{-E + y_3}{2C} \qquad \text{and} \qquad y_2 = \frac{-E - y_3}{2C}.$$

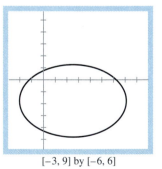

[-3, 9] by [-6, 6]

10.14   The graph of

$64x^2 + 144y^2 - 320x + 504y - 455 = 0.$

2. The graph from part 1 should be an ellipse (Fig. 10.14). TRACE to find its center and values for the semimajor and semiminor axes. (We selected the coefficients so that you should get familiar numbers.)

3. GRAPH $4x^2 - 2xy + y^2 - 3x + 6y - 16 = 0$ in a square viewing window. Note that now there is a Bxy term. Describe what you see.   ⚮

It is a worthwhile activity to organize information about the quadratic curves. One way is to consider the types of curves possible for various combinations of values of the coefficients in Eq. (1).

EXPLORATION 2

Organizing Information—Group Project

Work in groups of two or three. Organize examples of quadratic curves

$$Ax^2 + Bxy + Cy^2 + Dx + Ey + F = 0$$

into a chart. Be systematic, and show the effects of various combinations of values for A, B, C, D, E, and F. (For example, $A = C = 1$, $B = D = E = 0$, and $F = -r^2 (r > 0)$ is a circle with center $(0, 0)$ and radius r.)

You may wish to devote part of the chart to the degenerate conics.

> **EXPLORATION BIT**
>
> Here are some examples of equations whose graphs are degenerate quadratic curves. Give a geometric description of each graph.
>
> **1.** $x^2 = 0$
>
> **2.** $x^2 = -1$
>
> **3.** $y^2 = 1$
>
> **4.** $x^2 + y^2 = 0$
>
> **5.** $x^2 - 3x + 2 = 0$
>
> **6.** $xy + x - y - 1 = 0$
>
> (Is the graph of $x = 0$ a quadratic curve? Be careful!)

Translations of Ellipses and Hyperbolas

Your ellipse from part 1 of Exploration 1 (Fig. 10.14) has major and minor axes parallel to the coordinate axes. This ellipse is a shift, or translation, of an ellipse centered at the origin.

EXAMPLE 1

Find an ellipse centered at the origin that is congruent to the ellipse described by Eq. (3).

Solution   Use the method of completing the square to write Eq. (3),

$$64x^2 + 144y^2 - 320x + 504y - 455 = 0,$$

in the form

$$\frac{(x - x_1)^2}{a^2} + \frac{(y - y_1)^2}{b^2} = 1.$$

$$64x^2 + 144y^2 - 320x + 504y - 455 = 0$$

$$64(x^2 - 5x) + 144\left(y^2 + \frac{7}{2}y\right) = 455$$

$$64\left(x^2 - 5x + \frac{25}{4}\right) + 144\left(y^2 + \frac{7}{2}y + \frac{49}{16}\right) = 455 + 400 + 441 \qquad \text{Completing the square}$$

$$64\left(x - \frac{5}{2}\right)^2 + 144\left(y + \frac{7}{4}\right)^2 = 1296$$

$$\frac{\left(x - \frac{5}{2}\right)^2}{1296/64} + \frac{\left(y + \frac{7}{4}\right)^2}{1296/144} = 1$$

$$\frac{\left(x - \frac{5}{2}\right)^2}{(9/2)^2} + \frac{\left(y + \frac{7}{4}\right)^2}{3^2} = 1$$

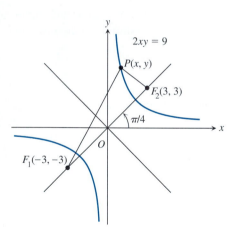

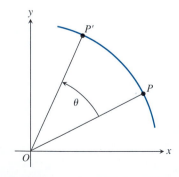

10.15 The focal axis of the hyperbola $2xy = 9$ makes an angle of $\pi/4$ radians with the positive x-axis.

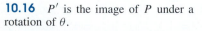

10.16 P' is the image of P under a rotation of θ.

This means that the ellipse

$$\frac{x^2}{(9/2)^2} + \frac{y^2}{3^2} = 1$$

centered at the origin will match the ellipse of Eq. (3) when shifted 5/2 units to the right and 7/4 units down.   ≡

The *Bxy* Term—Conics with Oblique Axes

The effect of the Bxy term, $B \neq 0$, in Eq. (1) is to rotate the conic section graph so that an axis is neither horizontal nor vertical. To see what happens when a conic section has such an *oblique* axis, let us write an equation for the hyperbola with $a = 3$ and foci at $F_1(-3, -3)$ and $F_2(3, 3)$ (Fig. 10.15). The equation $|PF_1 - PF_2| = 2a$ then becomes $|PF_1 - PF_2| = 2(3) = 6$, and

$$\sqrt{(x + 3)^2 + (y + 3)^2} - \sqrt{(x - 3)^2 + (y - 3)^2} = \pm 6.$$

When we transpose one radical, square, solve for the radical that still appears, and square again, this reduces to

$$2xy = 9, \tag{4}$$

which is a special case of Eq. (1) in which $B \neq 0$. The asymptotes of the hyperbola in Eq. (4) are the x- and y-axes, and the focal axis makes an angle of $\pi/4$ radians with the positive x-axis.

Because a nonzero Bxy term in general produces a "rotated" conic section, our task now becomes one of finding the conic section in *standard orientation* (that is, with horizontal and vertical axes) and a rotation that "turns" it to match the rotated conic.

Rotations

First, some terminology is needed. Points P and P' are related by a rotation θ (about the origin), provided that P and P' both lie on the same circle centered at the origin and angle POP' is θ. If θ is positive, the rotation is viewed as a counterclockwise rotation of the line segment OP through an angle of θ to the line segment OP' (Fig. 10.16). P' is called the image of P under a rotation of θ (about the origin).

Given $P = (x, y)$, what are the coordinates of $P'(x', y')$ in terms of x, y, and θ? Figure 10.17 will help to answer this question. Notice that

$$\frac{x'}{OP'} = \cos(\alpha + \theta) \qquad \text{and} \qquad \frac{y'}{OP'} = \sin(\alpha + \theta).$$

So

$$x' = OP' \cos\alpha \cos\theta - OP' \sin\alpha \sin\theta,$$

and

$$y' = OP' \sin\alpha \cos\theta + OP' \sin\theta \cos\alpha.$$

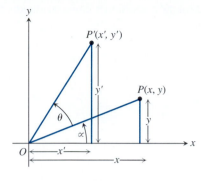

10.17 Two right triangles involved in rotating P to P'.

But

$$\frac{x}{OP} = \frac{x}{OP'} = \cos\alpha, \qquad \frac{y}{OP} = \frac{y}{OP'} = \sin\alpha, \qquad OP = OP'$$

so

$$x = OP'\cos\alpha, \qquad y = OP'\sin\alpha.$$

Therefore, by substitution,

$$x' = x\cos\theta - y\sin\theta,$$
$$y' = x\sin\theta + y\cos\theta. \tag{5}$$

A MATRIX CONNECTION

Equations (5) can be summarized in the matrix equation

$$\begin{bmatrix} x' \\ y' \end{bmatrix} = \begin{bmatrix} \cos\theta & -\sin\theta \\ \sin\theta & \cos\theta \end{bmatrix} \begin{bmatrix} x \\ y \end{bmatrix}.$$

This shows that a rotation in the xy-plane can be regarded as a function (of two variables) defined by a matrix and the operation of matrix multiplication. This is covered in some advanced mathematics courses.

How to Rotate a Graph

Equations (5) tell us how to rotate a graph. Moreover, they suggest how we can illustrate rotations on a graphing utility using parametric mode, the mode that we used to illustrate shifts, stretches, and shrinks in Chapter 1.

EXPLORATION 3

Seeing a Rotation

1. GRAPH the equation $y = x^3 - x$ using parametric mode with

$$x_1(t) = t, \qquad y_1(t) = t^3 - t.$$

Use a $[-6, 6]$ by $[-4, 4]$ viewing window with $[t\text{Min}, t\text{Max}] = [-6, 6]$ and t-step $= 0.1$. Is the graph a quadratic curve? Explain.

2. Now apply Eqs. (5) with $\theta = \pi/6$. Let

$$x_2(t) = x_1\cos\frac{\pi}{6} - y_1\sin\frac{\pi}{6},$$
$$y_2(t) = x_1\sin\frac{\pi}{6} + y_1\cos\frac{\pi}{6}.$$

GRAPH (x_1, y_1) and (x_2, y_2) in *sequential* format. Explain what you see.

3. In part 2, replace $\pi/6$ everywhere by θ. Exit to the home screen, and store $\pi/4$ in θ, then GRAPH. Repeat, using the following values for θ: $\pi/3, \pi/2, 2\pi/3, 3\pi/4, \pi, 3\pi/2$. (When $\theta = \pi$, you may wish to TRACE on your results.)

4. If your grapher has "list" capability, store the values in part 3 as a list in θ. Then GRAPH. Use TRACE to study your results.   ¢

Now we will show how to perform rotations algebraically. First we will show how to rotate a conic $Ax^2 + Cy^2 + F = 0$ in *standard position* (standard orientation with center at the origin). Then we will take a given conic $A'x^2 + B'xy + C'y^2 + F' = 0 \, (B' \neq 0)$ centered at the origin but displaced from standard orientation by a rotation and find a conic in standard position that rotates onto it (Fig. 10.18).

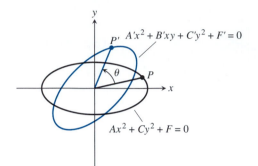

10.18 The conic $Ax^2 + Cy^2 + F = 0$ in standard position rotates through an angle θ about the origin onto the conic $A'x^2 + B'xy + C'y^2 + F' = 0$.

EXAMPLE 2

Find an equation whose graph is the graph of $3x^2 + y^2 - 12 = 0$ rotated $\pi/4$ radians (counterclockwise) about the origin. Support graphically.

Solution Let (x, y) be a point on the graph of $3x^2 + y^2 - 12 = 0$ and (x', y') its image under a rotation of $\pi/4$ radians about the origin. Eqs. (5) give

$$x' = x \cos\left(\frac{\pi}{4}\right) - y \sin\left(\frac{\pi}{4}\right) = \frac{x - y}{\sqrt{2}},$$

$$y' = x \sin\left(\frac{\pi}{4}\right) + y \cos\left(\frac{\pi}{4}\right) = \frac{x + y}{\sqrt{2}}.$$

> If the graph of a given equation in x and y is rotated counterclockwise through an angle θ about the origin, then an equation for the rotated graph can be found by solving Eqs. (5) for x and y and substituting the solutions into the given equation.

Solving the equations for x and y in terms of x' and y' gives

$$x = \frac{x' + y'}{\sqrt{2}}, \qquad y = \frac{-x' + y'}{\sqrt{2}}.$$

Substituting for x and y in $3x^2 + y^2 - 12 = 0$ gives

$$3\left(\frac{x' + y'}{\sqrt{2}}\right)^2 + \left(\frac{-x' + y'}{\sqrt{2}}\right)^2 - 12 = 0$$

$$\frac{3}{2}(x')^2 + 3x'y' + \frac{3}{2}(y')^2 + \frac{1}{2}(x')^2 - x'y' + \frac{1}{2}(y')^2 - 12 = 0$$

$$2(x')^2 + 2x'y' + 2(y')^2 - 12 = 0$$

$$(x')^2 + x'y' + (y')^2 - 6 = 0.$$

So the points (x', y') of the rotated conic satisfy the equation

$$x^2 + xy + y^2 - 6 = 0.$$

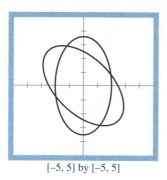

[-5, 5] by [-5, 5]

10.19   The viewing window supports the results of Example 2, namely, the graph of

$$x^2 + xy + y^2 - 6 = 0$$

appears to match the graph of

$$3x^2 + y^2 - 12 = 0$$

rotated through an angle of $\pi/4$.

CAUTION

If $B' \neq 0$ and $A' = C'$, use

$$\theta = \frac{1}{2} \cot^{-1} \frac{A' - C'}{B'},$$

or simply remember that $A' = C'$ means $\theta = \dfrac{\pi}{4}$.

EXPLORATION BITS

1. Show algebraically the result of rotating the conic

$$x^2 + xy + y^2 - 6 = 0$$

counterclockwise through an angle of $7\pi/4$.

2. Determine the conic that will match

$$x^2 + xy + y^2 - 6 = 0$$

when rotated through an angle of $-\pi/4$.

Figure 10.19 shows that the graph of $x^2 + xy + y^2 - 6 = 0$ appears to match the graph of $3x^2 + y^2 - 12 = 0$ rotated counterclockwise through an angle of $\pi/4$. The graphs were drawn by using the two-function method. (For example,

$$y_1 = \frac{-x + \sqrt{x^2 - 4(x^2 - 6)}}{2}, \qquad y_2 = \frac{-x - \sqrt{x^2 - 4(x^2 - 6)}}{2}$$

gives the graph of $x^2 + xy + y^2 - 6 = 0$.)

If θ is the angle through which the conic $Ax^2 + Cy^2 + F = 0$ can be rotated counterclockwise onto the conic $A'x^2 + B'xy + C'y^2 + F' = 0$, then it can be proved that

$$\theta = \frac{1}{2} \cot^{-1} \frac{A' - C'}{B'} = \frac{1}{2} \tan^{-1} \frac{B'}{A' - C'}. \tag{6}$$

EXAMPLE 3

Find the equation of a conic in standard position that, when rotated through an angle θ, will match the graph of $x^2 + xy + y^2 - 6 = 0$. Also, find θ.

Solution   The angle θ of rotation is given by Eq. (6) with $A' = B' = C' = 1$. Thus,

$$\begin{aligned} \theta &= \frac{1}{2} \cot^{-1} \left(\frac{1-1}{1} \right) = \frac{1}{2} \cot^{-1} \left(\frac{0}{1} \right) \\ &= \frac{1}{2} \cdot \frac{\pi}{2} \\ &= \frac{\pi}{4}. \end{aligned}$$

So the conic that we seek needs to be rotated $\pi/4$ radians to match the original conic $x^2 + xy + y^2 - 6 = 0$. This means that if the point $P = (x, y)$ lies on the conic we seek, then the rotated point $P' = (x', y')$ satisfies Eqs. (5),

$$\begin{aligned} x' &= x \cos\left(\frac{\pi}{4}\right) - y \sin\left(\frac{\pi}{4}\right), \\ y' &= x \sin\left(\frac{\pi}{4}\right) + y \cos\left(\frac{\pi}{4}\right), \end{aligned} \tag{7}$$

and lies on the conic

$$(x')^2 + x'y' + (y')^2 - 6 = 0. \tag{8}$$

Substituting Eqs. (7) into Eq. (8) and doing a little algebra, we find

$$3x^2 + y^2 - 12 = 0.$$

Thus, the graph of $3x^2 + y^2 - 12 = 0$ rotated through $\pi/4$ radians will match the graph of $x^2 + xy + y^2 - 6 = 0$, to agree with Example 2.

We can generalize: For *any* conic,

$$Ax^2 + Bxy + Cy^2 + Dx + Ey + F = 0, \tag{9}$$

the procedure of Example 3 can be used to find an angle θ and a congruent conic

$$A'x^2 + C'y^2 + D'x + E'y + F' = 0 \tag{10}$$

that will match the conic of Eq. (9) when rotated through the angle θ. The procedure of completing the square can then be used to find a congruent conic centered at the origin

$$A'x^2 + C'y^2 + F'' = 0 \tag{11}$$

that can be *shifted*, or *translated*, to match the conic of Eq. (9).

We rewrite Eq. (11) in the form

$$Ax^2 + Cy^2 + F = 0,$$

which we call the **standard form** of a conic, and we summarize our results as follows.

The graph of any (nondegenerate) conic

$$Ax^2 + Bxy + Cy^2 + Dx + Ey + F = 0$$

in which B is not zero is the translation of a graph of a conic in standard form followed by a rotation through an angle θ, where

$$\theta = \frac{1}{2}\cot^{-1}\frac{A-C}{B} = \frac{1}{2}\tan^{-1}\frac{B}{A-C}.$$

EXPLORATION 4

Rotating a Conic, Part 1

1. GRAPH

$$x_1(t) = t, \qquad\qquad y_1(t) = \sqrt{12 - 3t^2};$$
$$x_2(t) = x_1\cos\theta - y_1\sin\theta, \qquad y_2(t) = x_1\sin\theta + y_1\cos\theta,$$

with $\theta = \pi/4$. Explain what you see. How does this relate to Example 2?

2. Can you illustrate Example 2 completely in one viewing window? Explain.

3. The conic in part 3 of Exploration 1 results from the translation, then rotation of a conic in standard form. Describe the rotation. Describe the translation. (The algebra needed to produce the standard form is messy. Methods introduced in Section 10.7 will provide analytic tools that are easier to use.)   ꙮ

REMINDER

When $A = C$, we use

$$\frac{1}{2}\cot^{-1}\frac{A-C}{B}$$

for θ, and not

$$\frac{1}{2}\tan^{-1}\frac{B}{A-C}.$$

EXPLORATION BITS

1. Show that the discriminant is the coefficient of the x^2 term in the quadratic under the radical sign in Eq. (12). There is a connection between this fact and the information that the discriminant provides. Can you find the connection?

2. Show that the quantity $B^2 - 4AC$ is not changed by a rotation. That is, for any angle of rotation, $B^2 - 4AC = B'^2 - 4A'C'$.

3. The one case that we have not yet considered when we GRAPH the general quadratic is the case in which $C = 0$. In this case, Eqs. (12) do not apply. Analyze the general quadratic for $C = 0$, and tell how to GRAPH it. (Note that the discriminant gives you some immediate information.)

The Discriminant

By now and in general, we should be able to tell by studying the coefficients whether the graph of the general quadratic

$$Ax^2 + Bxy + Cy^2 + Dx + Ey + F = 0$$

is a parabola, an ellipse, or a hyperbola *when B is zero*. If B is not zero, there are two efficient ways to tell the shape of the graph. One is by evaluating the **discriminant**, $B^2 - 4AC$. The graph is a (real or degenerate)

a) *parabola* if $B^2 - 4AC = 0$,

b) *ellipse* if $B^2 - 4AC < 0$,

c) *hyperbola* if $B^2 - 4AC > 0$.

The other way is to actually GRAPH the two functions

$$y = \frac{-(Bx + E) \pm \sqrt{(Bx + E)^2 - 4C(Ax^2 + Dx + F)}}{2C} \qquad (12)$$

on a graphing utility.

EXAMPLE 4

a) $3x^2 - 6xy + 3y^2 + 2x - 7 = 0$ represents a parabola because

$$B^2 - 4AC = (-6)^2 - 4 \cdot 3 \cdot 3 = 36 - 36 = 0.$$

b) $x^2 + xy + y^2 - 1 = 0$ represents an ellipse because

$$B^2 - 4AC = (1)^2 - 4 \cdot 1 \cdot 1 = -3 < 0.$$

c) $xy - y^2 - 5y + 1 = 0$ represents a hyperbola because

$$B^2 - 4AC = (1)^2 - 4(0)(-1) = 1 > 0.$$

> **EXPLORATION BIT**
>
> Support the results of Example 4 graphically using Eqs. (12). As you proceed, look for a way to use (12) efficiently. Does the structure of (12) suggest a way to remember it, or at least how to derive it?

Exercises 10.2

Use the discriminant $B^2 - 4AC$ to decide whether the equations in Exercises 1–20 represent parabolas, ellipses, or hyperbolas. GRAPH each of them using the two-function method. For each conic that is not in standard form, list the angle of rotation that will match a conic with axes parallel to the coordinate axes to it.

1. $x^2 - y^2 - 1 = 0$

2. $25x^2 + 9y^2 - 225 = 0$

3. $y^2 - 4x - 4 = 0$

4. $x^2 + y^2 - 10 = 0$

5. $x^2 + 4y^2 - 4x - 8y + 4 = 0$

6. $2x^2 - y^2 + 4xy - 2x + 3y = 6$

7. $x^2 + 4xy + 4y^2 - 3x = 6$

8. $x^2 + y^2 + 3x - 2y = 10$

9. $xy + y^2 - 3x = 5$

10. $3x^2 + 6xy + 3y^2 - 4x + 5y = 12$

11. $x^2 - y^2 = 1$

12. $2x^2 + 3y^2 - 4x = 7$

13. $xy = 1$

14. $xy = -3$

15. $2x^2 + xy + x - y + 1 = 0$

16. $x^2 + 2xy - 2x + 3y - 3 = 0$

17. $x^2 - 3xy + 3y^2 + 6y = 7$

18. $25x^2 - 4y^2 - 350x = 0$

19. $6x^2 + 3xy + 2y^2 + 17y + 2 = 0$

20. $3x^2 + 12xy + 12y^2 + 435x - 9y + 72 = 0$

In Exercises 21–28, identify and GRAPH each conic using the two-function method. Find an equation of a conic in standard position that, when rotated through an angle θ, will match the graph of the given conic. Also find θ.

21. $xy = 2$

22. $x^2 + xy + y^2 = 1$

23. $x^2 - \sqrt{3}xy + 2y^2 = 1$

24. $x^2 - 2xy + y^2 = 2$

25. $x^2 - 3xy + y^2 = 5$

26. $xy - y - x + 1 = 0$

27. $3x^2 + 2xy + 3y^2 = 19$

28. $3x^2 + 4\sqrt{3}xy - y^2 = 7$

29. Solve Eqs. (5) for x and y in terms of x' and y'.

30. Show that an equation for the graph obtained by rotating the graph of $Ax^2 + Bxy + Cy^2 + Dx + Ey + F = 0$ by $\pi/4$ radians about the origin contains no xy term when $A = C$.

In Exercises 31–34, find an equation for the graph obtained by rotating the given conic counterclockwise about the origin through the angle θ.

31. $2x^2 - y^2 - 8 = 0$, $\theta = \pi/4$

32. $x^2 + 2y^2 - 10 = 0$, $\theta = \pi/6$

33. $4x^2 - y^2 - 10 = 0$, $\theta = \pi/3$

34. $2x^2 + 8y^2 - 7 = 0$, $\theta = \pi/4$

35. a) Derive Eq. (4) of the hyperbola in Fig. 10.15.
   b) Show that the algebraic steps in part (a) can be reversed to prove that every point that satisfies Eq. (4) lies on the hyperbola.

36. Give a reasonable explanation for

$$\frac{\pi}{4} = \frac{1}{2}\tan^{-1}\frac{B'}{A' - C'}$$

in Eq. (6) when $B' \neq 0$ and $A' = C'$.

10.3

Parametric Equations for Plane Curves

The parametric mode of a graphing utility allows us to draw graphs of relationships that are not functions. The grapher language that we have used is based in the mathematics of parametric equations.

PARAMETRIC GRAPHING

For parametric equations, the parameter interval need not be closed and it is often unbounded in one direction, such as $[5, \infty)$, or in both directions $(-\infty, \infty)$. However, for the practical purpose of using parametric mode on a grapher, the parameter interval is closed and bounded.

Also, the parameter interval is usually an interval of real numbers. When we GRAPH in parametric mode, however, we plot the graph using only a sample of the numbers in the interval, namely, the numbers

$$t\text{Min} + k \cdot t\text{-step}$$

for

$$k = 0, 1, 2, \dots, \left[\frac{t\text{Max} - t\text{Min}}{t\text{-step}}\right].$$

In parametric mode, the grapher settings control the initial point, the terminal point, how much of the curve is drawn, and the motion from the initial point to the terminal point.

DEFINITIONS

If x and y are given as functions

$$x = x_1(t), \qquad y = y_1(t)$$

over an interval of t-values, then the set of points $(x_1(t), y_1(t))$ defined by these equations is a **curve** (or **parametric curve**) in the coordinate plane. The equations are **parametric equations** for the curve, and we say that the curve has been *parametrized*. The variable t is the **parameter** of the curve, and its domain $I = [a, b] = [t\text{Min}, t\text{Max}]$ is the **parameter interval**. The point $(x_1(a), y_1(a))$ is the **initial point** of the curve, and $(x_1(b), y_1(b))$ is the **terminal point** of the curve.

When the path of a particle moving in the xy-plane is not the graph of a function, we try to express each of the particle's coordinates as a function of time t and describe the path with a pair of parametric equations, $x = x_1(t)$ and $y = y_1(t)$. Because these equations immediately tell us the particle's position at any time t, they are equations for the motion of the particle as well as equations for the path along which the motion takes place. We have seen how this enables us to calculate the particle's velocity and acceleration at any time t, something that we will look at again in Chapter 12. Now, instead of studying the rates of change in the motion of the particle, we are going to study the motion itself and the geometry of the curves that are produced.

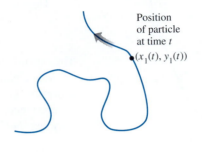

Position
of particle
at time t
$(x_1(t), y_1(t))$

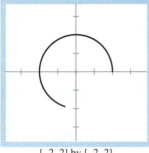

[–2, 2] by [–2, 2]

10.20 The parametric equations

$$x_1(t) = \cos t, \qquad y_1(t) = \sin t$$

plot the unit circle from $(1, 0)$ around to $(1, 0)$ as t increases from 0 to 2π. In this picture, a pixel has just lit for a value of t between π and $3\pi/2$.

After a parametric curve is drawn in a viewing window, we can TRACE to move the cursor along the path of the particle.

EXPLORATION BITS

1. In a parametrization of the circle

$$x^2 + y^2 = a^2,$$

can a be negative? Explain.

2. Investigate the parametrization

$$x = \cosh t, \qquad y = \sinh t.$$

EXPLORATION 1

Graphing Circles, Part 3

In applications of parametric equations, t can denote time or other quantities. For the parametric equations of a circle (see Section 1.6, Exploration 1),

$$x_1(t) = a \cos t, \qquad y_1(t) = a \sin t, \tag{1}$$

t can be thought of as representing time, angle measure, or distance of a particle from an initial point $(a \cos t_0, a \sin t_0)$ on the circle. (See Fig. 10.20.)

1. Remembering that the sine and cosine functions are periodic and have the same period, what size should be selected for a parameter interval so that the curve for Eqs. (1) is a circle? A semicircle?

2. What is the initial point of the curve if $t_0 = t\text{Min} = 0$? $\pi/2$? π? $3\pi/2$?

3. For each initial point in part 2, what is the terminal point of the curve when the parameter interval is of length $2\pi/3$? $4\pi/3$?

4. Use Eqs. (1) to GRAPH a semicircle with initial point $(1, 0)$ and terminal point $(-1, 0)$; with initial point $(0, 3)$ and terminal point $(0, -3)$.

5. Use Eqs. (1) to GRAPH a circle with initial point = terminal point. Experiment with different values of t-step. Describe the "motions" you view and how they are affected by t-step.

6. GRAPH the circle of part 5 so that it plots in the reverse order of how it plotted in part 5. In how many different ways can you reverse the order of the plot of a circle on your grapher? Explain each.

Now we move on to slightly less familiar parametric equations for the other conic sections.

EXAMPLE 1   Half a Parabola

The position $P(x, y)$ of a particle moving in the xy-plane is given by the equations

$$x = \sqrt{t}, \qquad y = t \qquad (t \geq 0). \qquad \text{The parameter interval is } [0, \infty).$$

Confirm the path traced by the parabola, and describe the motion.

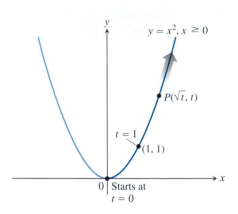

10.21 The parametric equations $x = \sqrt{t}$, $y = t$ of Example 1 with parameter interval $[0, \infty)$ describe the motion of a particle that traces the right-hand half of the parabola $y = x^2$.

EXPLORATION BIT

Give graphic support to Examples 1 and 2. Explain why you cannot demonstrate either example completely.

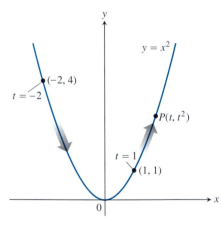

10.22 The parametric equations $x = t$, $y = t^2$ of Example 2 with parameter interval $(-\infty, \infty)$ define the complete parabola $y = x^2$. This example illustrates the straightforward parametrization,

$$x(t) = t, \qquad y(t) = f(t),$$

of a function $y = f(x)$.

Solution In general, we try to identify and confirm a parametric curve by eliminating t between the two parametric equations to produce an explicit relationship between x and y.

Here we find that

$$y = t = \left(\sqrt{t}\right)^2 = x^2.$$

This means that the particle's position coordinates satisfy the equation $y = x^2$, so the particle moves along the parabola $y = x^2$. The parameter interval, however, has an endpoint ($t = 0$), and for $t > 0$, the particle's x-coordinate is never negative. Thus, the particle starts at $(0, 0)$ and traces completely only that *half* of the parabola that is in the first quadrant (Fig. 10.21). ≡

EXAMPLE 2   A Complete Parabola

The position $P(x, y)$ of a particle moving in the xy-plane is given by the equations

$$x = t, \quad y = t^2 \quad (-\infty < t < \infty). \qquad \text{The parameter interval is } (-\infty, \infty).$$

Confirm the particle's path, and describe the motion.

Solution We identify the path by eliminating t between the equations $x = t$ and $y = t^2$, obtaining

$$y = (t)^2 = x^2.$$

The particle's position coordinates satisfy the equation $y = x^2$, so the particle moves along this curve.

In contrast to Example 1, however, the parameter interval is unbounded in both directions, and the particle traverses the complete parabola from left to right as t increases from $-\infty$ to ∞ (Fig. 10.22). ≡

Parametric equations for an ellipse and a hyperbola are based on some familiar trigonometric identities, as suggested in Examples 3 and 4.

EXAMPLE 3   Parametric Equations for the Ellipse $x^2/a^2 + y^2/b^2 = 1$

Confirm the path and describe the motion of a particle whose position $P(x, y)$ at time t is given by the equations

$$x = a \cos t, \qquad y = b \sin t \qquad (0 \le t \le 2\pi).$$

Solution We find a Cartesian equation for the particle's coordinates by eliminating t between the equations

$$\cos t = \frac{x}{a}, \qquad \sin t = \frac{y}{b}.$$

We accomplish this with the identity $\cos^2 t + \sin^2 t = 1$, which yields

$$\left(\frac{x}{a}\right)^2 + \left(\frac{y}{b}\right)^2 = 1 \qquad \text{or} \qquad \frac{x^2}{a^2} + \frac{y^2}{b^2} = 1.$$

The particle's coordinates (x, y) satisfy the equation $x^2/a^2 + y^2/b^2 = 1$, so the particle moves along this ellipse. When $t = 0$, the particle's coordinates are

$$x = a \cos (0) = a, \qquad y = b \sin (0) = 0,$$

so the motion starts at $(a, 0)$. As t increases, the particle rises and moves toward the left, moving counterclockwise. It traverses the ellipse once, returning to its starting position $(a, 0)$ at time $t = 2\pi$ (Fig. 10.23). ▤

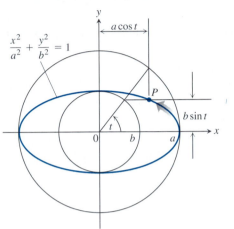

10.23 The coordinates of P are $x = a \cos t$, $y = b \sin t$.

EXPLORATION 2

Graphing Ellipses, Part 2

1. Graphically support Example 3. Explain your support.

2. Recall what you know about ellipses. Tell what must be true for $a > b$, $a = b$, $a < b$.

3. Recall what you know about parametric graphing. Tell how your choice of tMin, tMax, and t-step affect what you see in the viewing window when you graph an ellipse.

4. Figure 10.23 suggests a logo that is seen frequently on television. Can you reproduce it on your grapher? If successful, do you prefer graphing the logo in sequential format or simultaneous format? Why? ✄

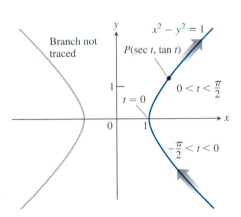

10.24 The equations $x = \sec t$, $y = \tan t$ are parametric equations for the hyperbola $x^2 - y^2 = 1$. When the parameter interval is $-\pi/2 < t < \pi/2$, the equations represent the right-hand branch only. Can you describe a parameter interval for the complete hyperbola?

EXAMPLE 4 Parametric Equations for the Right-hand Branch of the Hyperbola $x^2 - y^2 = 1$

Confirm the path and describe the motion of the particle whose position $P(x, y)$ at time t is given by the equations

$$x = \sec t, \qquad y = \tan t \qquad \left(-\frac{\pi}{2} < t < \frac{\pi}{2} \right).$$

Solution We find a Cartesian equation for the coordinates of P by eliminating t between the equations

$$\sec t = x, \qquad \tan t = y.$$

We accomplish this with the identity $\sec^2 t - \tan^2 t = 1$, which yields

$$x^2 - y^2 = 1.$$

Since the particle's coordinates (x, y) satisfy the equation $x^2 - y^2 = 1$, the motion takes place somewhere on this hyperbola. As t runs between $-\pi/2$ and $\pi/2$, $x = \sec t$ remains positive, and $y = \tan t$ runs between $-\infty$ and ∞, so P traverses the hyperbola's right-hand branch. It comes in along the branch's lower half as $t \to 0^-$, reaches $(1, 0)$ at $t = 0$, and moves out into the first quadrant as t increases toward $\pi/2$ (Fig. 10.24). ▤

EXPLORATION 3

Graphing Hyperbolas, Part 2

1. Graphically support Example 4. Explain. Note particularly that the parameter interval $(-\pi/2, \pi/2)$ has to be open. (Why?) Explain how you deal with this for [tMin, tMax].

2. Can you use parametric graphing to graph both branches of the hyperbola? Explain.

3. Explain how to parametrically GRAPH the hyperbolas $x^2/a^2 - y^2/b^2 = 1$ for various values of a and b.

4. Watch one branch of the graph of a hyperbola being drawn in parametric mode. Why does the "particle" tracing out the graph slow down near a vertex? (You may have to adjust your view dimensions to see this well.)

5. For some hyperbolas, a grapher may show a line approximating an asymptote. Explain why. How could it be removed? (*Hint:* Recall the meanings of *dot* and *connected* formats.) ☮

In addition to the conic sections, there are numerous curves that can be modeled geometrically and defined parametrically. One of the more interesting is the *cycloid*.

EXAMPLE 5   Cycloids

A wheel of radius a rolls along a horizontal line without slipping. Find parametric equations for the path (called a **cycloid**) traced by a point P on the wheel's edge.

Solution   We take the line to be the x-axis and suppose that the wheel rolls to the right, P being at the origin when the turn angle $t = 0$. Figure 10.25 shows the wheel after it has turned t radians. The base of the wheel is at units from the origin. The wheel's center is at (at, a), and the coordinates of P are

$$x = at + a\cos\theta, \qquad y = a + a\sin\theta.$$

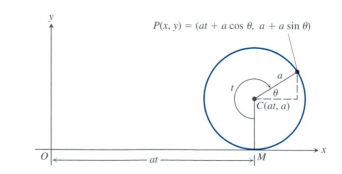

10.25   The position of $P(x, y)$ on the edge of the wheel when the wheel has turned t radians.

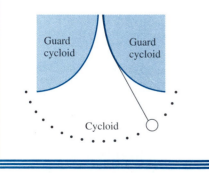

PENDULUM CLOCKS

One problem with a pendulum clock whose bob swings in a circular arc is that the frequency with which the pendulum swings changes with the amplitude of the swing. The wider the swing, the longer it takes the bob to return to the center.

This does not happen if the bob can be made to swing in a cycloid, and in 1673, Christiaan Huygens (1629–1695), the Dutch mathematician, physicist, and astronomer who discovered the rings of Saturn, designed a pendulum clock whose bob would swing in a cycloid. He hung the bob from a fine wire constrained by "guards" that caused it to draw up as it swung. How were the guards shaped? They were cycloids, too.

To express θ in terms of t, we observe that $t + \theta = 3\pi/2 + 2k\pi$ for some integer k, so

$$\theta = \frac{3\pi}{2} - t + 2k\pi.$$

This makes

$$\cos\theta = \cos\left(\frac{3\pi}{2} - t + 2k\pi\right) = -\sin t,$$

$$\sin\theta = \sin\left(\frac{3\pi}{2} - t + 2k\pi\right) = -\cos t.$$

The equations that we seek are

$$x = at - a\sin t, \qquad y = a - a\cos t.$$

These are usually written with the a factored out:

$$x = a(t - \sin t), \qquad y = a(1 - \cos t).$$

Figure 10.26 shows the first arch of the cycloid and part of the next.

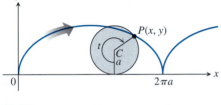

10.26 The cycloid $x = a(t - \sin t)$, $y = a(1 - \cos t)$, shown for $t \geq 0$.

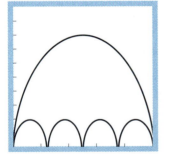

10.27 Five arches!

EXPLORATION 4

Graphing Cycloids

1. With your grapher in parametric mode, GRAPH

$$x(t) = a(t - \sin t), \qquad y(t) = a(1 - \cos t)$$

for $A = 0.5$, 1, and 1.5.

2. Relate part 1 to the wheel model of Example 5. How big are the "wheels" you are using? What should your parameter intervals be for each wheel to make three complete turns? What effect does t-step have on the model? What effect does a square viewing window have on the model?

3. Find the parametric equations and make the other adjustments on your grapher to show the graphs of Fig. 10.27. Interpret your viewing window in terms of the wheel model.

4. Model the movement of two points,

   a) opposite each other on the edge of one wheel.

   b) in similar positions on the edges of the two wheels on the same side of an automobile.

   c) on the same automobile wheel, one on the edge of the tire and one on the edge of the hubcap.

5. Predict what will happen if you use $a < 0$ in the parametric equations of part 1, then test your prediction.

THE WITCH OF AGNESI

The first text to include differential and integral calculus along with analytic geometry, infinite series, and differential equations was written in the 1740s by the Italian mathematician Maria Gaetana Agnesi (1718–1799). Agnesi, a gifted scholar and linguist whose Latin essay defending higher education for women was published when she was only nine years old, was a well-published scientist by age 20, and an honorary faculty member of the University of Bologna by age 30.

Today, Agnesi is remembered chiefly for a bell-shaped curve called *the witch of Agnesi*. This name, found only in English texts, is the result of a mistranslation. Agnesi's own name for the curve was *versiera* or "turning curve." John Colson, a noted Cambridge mathematician, probably confused versiera with *avversiera*, which means "wife of the devil" and translated it into "witch." You can find out more about the witch by doing Exercise 36.

Summary: Standard Parametric Equations

Circle $x^2 + y^2 = a^2$:
$$x = a\cos t, \qquad y = a\sin t \qquad (0 \le t \le 2\pi)$$

Parabola $y = ax^2$ or $x = ay^2$:
$$x = t, y = at^2 \qquad \text{or} \qquad x = at^2, y = t \qquad (-\infty < t < \infty)$$

Ellipse $\dfrac{x^2}{a^2} + \dfrac{y^2}{b^2} = 1$:
$$x = a\cos t, \qquad y = b\sin t \qquad (0 \le t \le 2\pi)$$

Hyperbola $\dfrac{x^2}{a^2} - \dfrac{y^2}{b^2} = 1$ or $\dfrac{y^2}{a^2} - \dfrac{x^2}{b^2} = 1$:
$$x = a\sec t, \ y = b\tan t \ \text{ or } \ x = b\tan t, \ y = a\sec t \ \ (0 \le t \le 2\pi)$$

Cycloid generated by a circle of radius a:
$$x = a(t - \sin t), \qquad y = a(1 - \cos t) \qquad (-\infty < t < \infty)$$

Exercises 10.3

Exercises 1–24 give parametric equations for the motion of a particle in the xy-plane. Graph the particle motion. Determine the equivalent Cartesian equation of the particle motion of each by analytic means. Show on a sketch the portion of the Cartesian graph that represents the particle motion. Also show the direction of the motion.

1. $x = \cos t, \quad y = \sin t \ (0 \le t \le \pi)$

2. $x = \cos 2t, \quad y = \sin 2t \ (0 \le t \le \pi)$

3. $x = \sin 2\pi t, \quad y = \cos 2\pi t \ (0 \le t \le 1)$

4. $x = \cos(\pi - t), \quad y = \sin(\pi - t) \ (0 \le t \le \pi)$

5. $x = 4\cos t, \quad y = 2\sin t \ (0 \le t \le 2\pi)$

6. $x = 4\sin t, \quad y = 2\cos t \ (0 \le t \le \pi)$

7. $x = 4\cos t, \quad y = 5\sin t \ (0 \le t \le \pi)$

8. $x = 4\sin t, \quad y = 5\cos t \ (0 \le t \le 2\pi)$

9. $x = 3t, \quad y = 9t^2 \ (-\infty < t < \infty)$

10. $x = -\sqrt{t}, \quad y = t \ (t \ge 0)$

11. $x = t, \quad y = \sqrt{t} \ (t \ge 0)$

12. $x = \sec^2 t - 1, \quad y = \tan t \ (-\pi/2 < t < \pi/2)$

13. $x = -\sec t, \quad y = \tan t \ (-\pi/2 < t < \pi/2)$

14. $x = \csc t, \quad y = \cot t \ (0 < t < \pi)$

15. $x = 2t - 5, \quad y = 4t - 7 \ (-\infty < t < \infty)$

16. $x = 1 - t, \quad y = 1 + t \ (-\infty < t < \infty)$

17. $x = t, \quad y = 1 - t \ (0 \le t \le 1)$

18. $x = 3t, \quad y = 2 - 2t \ (0 \le t \le 1)$

19. $x = t, \quad y = \sqrt{1 - t^2} \ (-1 \le t \le 1)$

20. $x = t, \quad y = \sqrt{4 - t^2} \ (0 \le t \le 2)$

21. $x = t^2, \quad y = \sqrt{t^4 + 1} \ (t \ge 0)$

22. $x = \sqrt{t + 1}, \quad y = \sqrt{t} \ (t \ge 0)$

23. $x = \cosh t, \quad y = \sinh t \quad (-\infty < t < \infty)$

24. $x = 2\sinh t, \quad y = 2\cosh t \quad (-\infty < t < \infty)$

25. Find parametric equations for the motion of a particle that starts at $(a, 0)$ and traces the circle $x^2 + y^2 = a^2$
   a) once clockwise.
   b) once counterclockwise.
   c) twice clockwise.
   d) twice counterclockwise.
   (There are lots of ways to do these, so your answers may not be the same as the ones in the back of the book.)

Closed Curves

A curve defined parametrically is **closed** if its initial point and terminal point are the same. For example, the circle $x = \cos t, y = \sin t, 0 \le t \le 2\pi$ is closed because $(\cos 0, \sin 0) = (1, 0) = (\cos 2\pi, \sin 2\pi)$. The semicircle $x = \cos t, y = \sin t, 0 \le t \le \pi$ is not closed.

In Exercises 26–33, find a value of a for $0 \le t \le a$ that makes each curve closed.

26. $x = 5\sin t, \ y = 5\sin 2t$

27. $x = 5\sin 2t, \ y = 5\sin 3t$

28. $x = (5\sin 3t)\cos t, \ y = (5\sin 3t)\sin t$

29. $x = (5\sin 2t)\cos t, \ y = (5\sin 2t)\sin t$

30. $x = (5\sin 2.5t)\cos t, \ y = (5\sin 2.5t)\sin t$

31. $x = (5\sin 1.5t)\cos t, \ y = (5\sin 1.5t)\sin t$

32. $x = t\cos t, \ y = t\sin t$

33. $x = t\sin t, \ y = t\cos t$

34. Find parametric equations for the motion of a particle that starts at $(a, 0)$ and traces the ellipse $x^2/a^2 + y^2/b^2 = 1$
   a) once clockwise.
   b) once counterclockwise.
   c) twice clockwise.
   d) twice counterclockwise.
   (As in Exercise 25, there are many correct answers.)

35. *The involute of a circle.* If a string wound around a fixed circle is unwound while being held taut in the plane of the circle, its end P traces an *involute* of the circle as suggested by the diagram below. In the diagram, the circle is the unit circle in the xy-plane, and the initial position of the tracing point is the point $(1, 0)$ on the x-axis. The unwound portion of the string is tangent to the circle at Q, and t is the radian measure of the angle from the positive x-axis to segment OQ. Derive parametric equations for the involute by expressing the coordinates x and y of P in terms of t. GRAPH the unit circle and its involute in a viewing window.

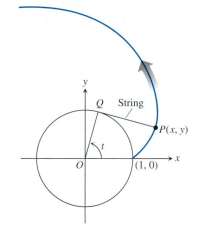

36. *The witch of Maria Agnesi.* The bell-shaped witch of Maria Agnesi can be constructed in the following way. Start with a circle of radius 1, centered at the point $(0, 1)$ on the y-axis as shown in the diagram.

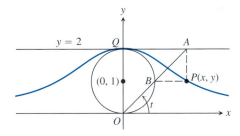

Choose a point A on the line $y = 2$, and connect it to the origin with a line segment. Call the point where the segment crosses the circle B. Let P be the point where the vertical line through A crosses the horizontal line through B. The witch is the curve traced by P as A moves along the line $y = 2$. Find (and GRAPH) parametric equations for the witch by expressing the coordinates of P in terms of t, the radian measure of the angle that segment OA makes with the positive x-axis. The following equalities (which you may assume) will help:
   (1) $x = AQ$
   (2) $y = 2 - AB\sin t$
   (3) $AB \cdot OA = (AQ)^2$

37. Find the point on the parabola $x = t, y = t^2$ closest to the point $(2, 1/2)$. (*Hint:* Minimize the square of the distance as a function of t.)

38. Find the point on the ellipse $x = 2\cos t, y = \sin t, 0 \le t \le 2\pi$ closest to the point $(3/4, 0)$. (*Hint:* Minimize the square of the distance as a function of t.)

39. *Parametric equations for lines in the plane.*
   a) Show that the equations
   $$x = x_0 + (x_1 - x_0)t, \quad y = y_0 + (y_1 - y_0)t, \quad (-\infty < t < \infty)$$
   are parametric equations for the line through the points

(x_0, y_0) and (x_1, y_1). In the diagram, the arrow shows the direction of increasing t.

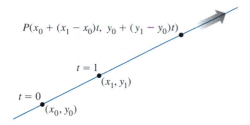

$P(x_0 + (x_1 - x_0)t, \ y_0 + (y_1 - y_0)t)$

$t = 1$

(x_1, y_1)

$t = 0$

(x_0, y_0)

b) Write parametric equations for the line through a point (x_1, y_1) and the origin.

c) Write parametric equations for the line through $(-1, 0)$ and $(0, 1)$. GRAPH in parametric mode to support your result.

GRAPH the equations in Exercises 40–44.

40. *Ellipse.* $x = 4 \cos t, \ y = 2 \sin t$, over
  a) $0 \le t \le 2\pi$
  b) $0 \le t \le \pi$
  c) $-\pi/2 \le t \le \pi/2$

41. *Hyperbola branch.* $x = \sec t, \ y = \tan t$ over
  a) $-1.5 \le t \le 1.5$
  b) $-0.5 \le t \le 0.5$
  c) $1.6 \le t \le 4.6$

42. *Parabola.* $x = 2t + 3, \ y = t^2 - 1 \ (-2 \le t \le 2)$

43. *Cycloid.* $x = t - \sin t, \ y = 1 - \cos t$, over
  a) $0 \le t \le 2\pi$
  b) $0 \le t \le 4\pi$
  c) $\pi \le t \le 3\pi$

44. *Hypocycloid.* For a *cycloid*, the generating circle rolls along a line. For a *hypocycloid*, the generating circle rolls around a larger circle while inside it. The parametric equations to graph are

$$x = \cos^3 t, \quad y = \sin^3 t, \quad \text{over}$$

  a) $0 \le t \le 2\pi$
  b) $-\pi/2 \le t \le \pi/2$
  c) On the basis of the graph in part (a), sketch the generating circle and the circle that contains it. Justify your sketch.

45. *A nice curve (a deltoid).*   GRAPH

$$x = 2 \cos t + \cos 2t,$$

$$y = 2 \sin t - \sin 2t,$$

$$0 \le t \le 2\pi.$$

What happens if you replace 2 with -2 in the equations for x and y? GRAPH the new equations, and find out.

46. *An even nicer curve.*   GRAPH

$$x = 3 \cos t + \cos 3t,$$

$$y = 3 \sin t - \sin 3t,$$

$$0 \le t \le 2\pi.$$

What happens if you replace 3 with -3 in the equations for x and y? GRAPH the new equations, and find out.

47. *Projectile motion.*   GRAPH

$$x = (64 \cos \alpha)t,$$

$$y = -16t^2 + (64 \sin \alpha)t,$$

$$0 \le t \le 4 \sin \alpha,$$

for the following launch angles of a projectile:
  a) $\alpha = \pi/4$ **b)** $\alpha = \pi/6$ **c)** $\alpha = \pi/3$ **d)** $\alpha = \pi/2$

48. *Five beautiful curves.*   GRAPH each of the following:
  a) *Epicycloid:*

$$x = 9 \cos t - \cos 9t$$

$$y = 9 \sin t - \sin 9t$$

$$0 \le t \le 2\pi$$

  b) *Hypocycloid:*

$$x = 8 \cos t + 2 \cos 4t$$

$$y = 8 \sin t + 2 \sin 4t$$

$$0 \le t \le 2\pi$$

  c) *Hypotrochoid:*

$$x = \cos t + 5 \cos 3t$$

$$y = 6 \cos t - 5 \sin 3t$$

$$0 \le t \le 2\pi$$

  d) *A rose:*

$$x = (8 \sin 2t) \cos t$$

$$y = (8 \sin 2t) \sin t$$

$$0 \le t \le 2\pi$$

  e) *A trillium:*

$$x = -(\sin 3t) \cos t$$

$$y = -(\sin 3t) \sin t$$

$$0 \le t \le 2\pi$$

Use parametric mode to GRAPH each equation in Exercises 49–56. Then rotate the graph counterclockwise with respect to the positive x-axis through the given angle. Finally, find an equation for the rotated curve.

49. $y = x^2 + 1, \theta = 30°$

50. $y = x^2 + 1, \theta = 90°$

51. $y^2 = x, \theta = \dfrac{\pi}{2}$

52. $y^2 = x, \theta = \dfrac{\pi}{4}$

53. $\dfrac{x^2}{4} + \dfrac{y^2}{9} = 1, \theta = \dfrac{\pi}{3}$

54. $\dfrac{x^2}{4} + \dfrac{y^2}{9} = 1, \theta = \dfrac{\pi}{4}$

55. $\dfrac{x^2}{16} - \dfrac{y^2}{25} = 1, \theta = 60°$

(*Hint:* GRAPH $x = 4 \sec t, y = 5 \tan t$, over $0 \le t \le 2\pi$.)

56. $\dfrac{x^2}{16} - \dfrac{y^2}{25} = 1, \theta = 30°$

10.4 ——————————— The Calculus of Parametric Equations

This section shows how to calculate slopes, lengths, and surface areas associated with *differentiable* parametric curves.

> **DEFINITIONS**
>
> A parametrized curve $x = f(t)$, $y = g(t)$ is said to be **differentiable at** $t = t_0$ if f and g are differentiable at $t = t_0$. The curve is **differentiable** if it is differentiable at every parameter value.

Slopes of Parametrized Curves

The restriction that y be a differentiable function of x is not, in general, an obstacle to our work. Most curves that are defined parametrically can be viewed as combinations of nicely behaved functions. Complications usually occur only at their points of nondifferentiability.

At a point on a differentiable parametrized curve where y is also a differentiable function of x, the derivatives dx/dt, dy/dt, and dy/dx are related by the Chain Rule:

$$\frac{dy}{dt} = \frac{dy}{dx} \cdot \frac{dx}{dt}$$

(Fig. 10.28). If $dx/dt \ne 0$, we may divide both sides of this equation by dx/dt to solve for dy/dx.

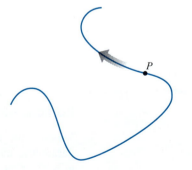

10.28 When the three first derivatives dx/dt, dy/dt, and dy/dx exist at a point P on a parametrized curve, they are related by the Chain Rule:

$$\frac{dy}{dt} = \frac{dy}{dx} \cdot \frac{dx}{dt}.$$

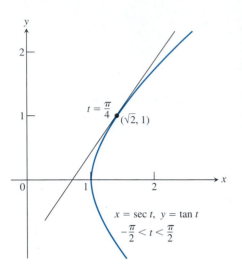

10.29   The hyperbola branch in Example 1. Equation (1) applies for every point on the graph except (1,0). Can you state why Eq. (1) fails at (1,0)?

> **Formula for Finding dy/dx as a function of t from dy/dt and dx/dt ($dx/dt \neq 0$)**
>
> $$\frac{dy}{dx} = \frac{dy/dt}{dx/dt} \tag{1}$$

If you want to think in terms of differentials, think of the two dt's canceling out to produce dy/dx. Equation (1) is the equation that we use to calculate the slopes of parametrized curves.

EXAMPLE 1

Find the tangent to the right-hand hyperbola branch

$$x = \sec t, \qquad y = \tan t \qquad \left(-\frac{\pi}{2} < t < \frac{\pi}{2} \right)$$

at the point $\left(\sqrt{2}, 1 \right)$, where $t = \pi/4$ (Fig. 10.29).

Solution   The slope of the curve at t is

$$\frac{dy}{dx} = \frac{dy/dt}{dx/dt} \qquad\qquad \text{Eq. (1)}$$

$$= \frac{\sec^2 t}{\sec t \tan t} = \frac{\sec t}{\tan t} = \csc t.$$

Setting t equal to $\pi/4$ gives

$$\left. \frac{dy}{dx} \right|_{t=\pi/4} = \csc \frac{\pi}{4} = \sqrt{2}.$$

The point-slope equation of the tangent is

$$y - y_0 = m(x - x_0)$$
$$y - 1 = \sqrt{2}(x - \sqrt{2})$$
$$y = \sqrt{2}x - 2 + 1$$
$$y = \sqrt{2}x - 1. \qquad\qquad \blacksquare$$

> **EXPLORATION BIT**
>
> Use NDER to support this result.

Particle Motion and the Tangent Line

Suppose the motion of a particle in the xy-plane is represented by certain parametric equations. The tangent line models the *direction* of the particle at a given point. Locally (that is, for values of t that give points near the point of tangency), the tangent line is a very good approximation for the path of the particle (Fig. 10.30). This is the same property of local straightness that we encountered when we studied nonparametrized differentiable functions.

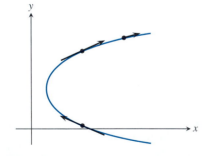

10.30   The tangent line models the direction of the particle at a given point.

The Parametric Formula for d^2y/dx^2

If the parametric equations for a curve define y as a twice-differentiable function of x, we may calculate d^2y/dx^2 as a function of t in the following way:

$$\frac{d^2y}{dx^2} = \frac{d}{dx}(y')$$

$$= \frac{dy'/dt}{dx/dt} \qquad \text{Eq. (1) with } y \text{ replaced by } y'$$

Formula for Finding d^2y/dx^2 as a function of t from dx/dt and $y' = dy/dx$ $(dx/dt \neq 0)$

$$\frac{d^2y}{dx^2} = \frac{dy'/dt}{dx/dt}$$

The following example illustrates the steps that we take to evaluate d^2y/dx^2.

To find d^2y/dx^2 in terms of t,

1. express $y' = dy/dx$ in terms of t,
2. find dy'/dt, and
3. divide dy'/dt by dx/dt.

EXAMPLE 2

Find d^2y/dx^2 if $x = t - t^2$ and $y = t - t^3$.

Solution

STEP 1: Express y' in terms of t:

$$y' = \frac{dy}{dx} = \frac{dy/dt}{dx/dt} = \frac{1 - 3t^2}{1 - 2t} \qquad \text{Eq. (1) with } x = t - t^2, y = t - t^3$$

STEP 2: Differentiate y' with respect to t:

$$\frac{dy'}{dt} = \frac{d}{dt}\left(\frac{1 - 3t^2}{1 - 2t}\right) = \frac{2 - 6t + 6t^2}{(1 - 2t)^2} \qquad \text{Quotient Rule}$$

STEP 3: Divide dy'/dt by dx/dt:

$$\frac{dx}{dt} = \frac{d}{dt}(t - t^2) = 1 - 2t$$

$$\frac{dy'/dt}{dx/dt} = \frac{2 - 6t + 6t^2}{(1 - 2t)^2} \cdot \frac{1}{1 - 2t}$$

$$= \frac{2 - 6t + 6t^2}{(1 - 2t)^3}.$$

EXPLORATION BIT

Use NDER to support this result.

The Length of a Parametric Curve

We can find an integral for the length of a differentiable parametric curve whose derivative is continuous

$$x = f(t), \qquad y = g(t) \qquad (a \leq t \leq b)$$

by rewriting the integral $L = \int ds$ from Section 6.4 in the following way:

$$\text{Length} = \int_{t=a}^{t=b} ds$$

$$= \int_a^b \sqrt{dx^2 + dy^2} \qquad \text{Because } ds = \sqrt{dx^2 + dy^2}$$

$$= \int_a^b \sqrt{\left(\frac{dx^2}{dt^2} + \frac{dy^2}{dt^2}\right) dt^2}$$

$$= \int_a^b \sqrt{\left(\frac{dx}{dt}\right)^2 + \left(\frac{dy}{dt}\right)^2} \, dt.$$

The only requirement besides the continuity of the integrand is that the point $P(x, y)$ not trace out any portion of the curve more than once as t moves from a to b.

Parametric Formula for Arc Length

If the functions $x = f(t)$ and $y = g(t)$ have continuous first derivatives with respect to t for $a \le t \le b$, and if the point $P(x, y)$ traces the curve defined by these equations exactly once as t moves from $t = a$ to $t = b$, then the length of the curve is given by the formula

$$\text{Length} = \int_a^b \sqrt{\left(\frac{dx}{dt}\right)^2 + \left(\frac{dy}{dt}\right)^2} \, dt. \tag{2}$$

What if we have two different parametrizations for a curve whose length we want to find—does it matter which one we use? The answer, from advanced calculus, is no. As long as the parametrization that we choose meets the conditions preceding Eq. (2), the formula gives the correct length.

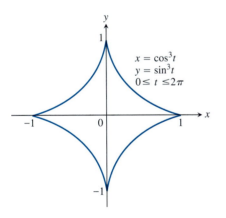

10.31 The astroid in Example 3.

EXAMPLE 3

Find the length of the astroid (Fig. 10.31)

$$x = \cos^3 t, \qquad y = \sin^3 t \qquad (0 \le t \le 2\pi).$$

Solution We find the length of the first-quadrant portion, $0 \le t \le \pi/2$, and multiply by 4. We have

$$x = \cos^3 t, \qquad y = \sin^3 t$$

$$\left(\frac{dx}{dt}\right)^2 = [3 \cos^2 t (-\sin t)]^2 = 9 \cos^4 t \sin^2 t$$

$$\left(\frac{dy}{dt}\right)^2 = [3 \sin^2 t (\cos t)]^2 = 9 \sin^4 t \cos^2 t$$

$$\sqrt{\left(\frac{dx}{dt}\right)^2 + \left(\frac{dy}{dt}\right)^2} = \sqrt{9\cos^2 t \sin^2 t \underbrace{(\cos^2 t + \sin^2 t)}_{1}}$$

$$= \sqrt{9\cos^2 t \sin^2 t}$$

$$= 3|\cos t \sin t|$$

$$= 3\cos t \sin t. \quad \cos t \sin t \geq 0 \text{ for } 0 \leq t \leq \pi/2.$$

Therefore,

Length of first-quadrant portion

$$= \int_0^{\pi/2} 3\cos t \sin t \, dt$$

$$= 3\int_{u=0}^{u=1} u \, du \qquad \begin{aligned} u &= \sin t, du = \cos t \, dt, \\ u(0) &= 0, u(\pi/2) = 1. \end{aligned}$$

$$= 3\left[\frac{u^2}{2}\right]_0^1$$

$$= \frac{3}{2}.$$

The length of the complete curve is four times this:

$$\text{Length of complete curve } = 4\left(\frac{3}{2}\right) = 6.$$

EXAMPLE 4

Find the length of a Spiral of Archimedes, $x = t\cos t$, $y = t\sin t$, for $0 \leq t \leq 20$.

Solution Let

$$x_1(t) = t\cos t, \qquad y_1(t) = t\sin t \qquad (0 \leq t \leq 20);$$

$$x_2(t) = t, \qquad y_2(t) = \sqrt{(\text{NDER }(x_1(t), t))^2 + (\text{NDER }(y_1(t), t))^2};$$

$$x_3(t) = t, \qquad y_3(t) = \text{NINT }(y_2(t), 0, t).$$

GRAPH (x_1, y_1) to see this Spiral of Archimedes (Fig. 10.32). Then GRAPH (x_3, y_3) to see the lengths for the values of t. At $t = 20$, the arc length is 202.095.

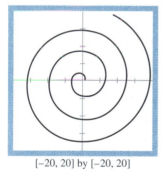

[−20, 20] by [−20, 20]

10.32 The Spiral of Archimedes,

$$x = t\cos t, \qquad y = t\sin t,$$

of Example 4 for $0 \leq t \leq 20$. Other such spirals may be found by graphing the equations

$$x = at\cos t, \qquad y = at\sin t$$

for different values of a.

The Area of a Surface of Revolution

The formula $S = \int 2\pi\rho \, ds$ developed in Section 6.5 for the area of the surface swept out by revolving a differentiable curve, with continuous derivative, about an axis translates into $S = \int 2\pi y \, ds$ if the axis is the x-axis and into $S = \int 2\pi x \, ds$ if the axis is the y-axis. With $ds = \sqrt{(dx/dt)^2 + (dy/dt)^2} \, dt$, these lead to the following formulas.

Parametric Formulas for the Area of a Surface of Revolution

If the functions $x = f(t)$ and $y = g(t)$ have continuous first derivatives with respect to t for $a \le t \le b$, and if the point $P(x, y)$ traces the curve defined by the equations exactly once as t moves from $t = a$ to $t = b$, then the areas of the surfaces generated by revolving the curve about the coordinate axes are the following:

1. Revolution about the x-axis ($y \ge 0$):

$$\text{Area} = \int_a^b 2\pi y \sqrt{\left(\frac{dx}{dt}\right)^2 + \left(\frac{dy}{dt}\right)^2} \, dt \tag{3}$$

2. Revolution about the y-axis ($x \ge 0$):

$$\text{Area} = \int_a^b 2\pi x \sqrt{\left(\frac{dx}{dt}\right)^2 + \left(\frac{dy}{dt}\right)^2} \, dt \tag{4}$$

As with length, we can calculate surface area from any convenient parametrization that meets the criteria stated above.

EXAMPLE 5

The standard parametric equations for the circle of radius 1 centered at the point $(0, 1)$ in the xy-plane are

$$x = \cos t, \qquad y = 1 + \sin t \qquad (0 \le t \le 2\pi).$$

Use these equations to find the area of the surface swept out by revolving this circle about the x-axis (Fig. 10.33).

Solution   We use Eq. (3) with

$$y = 1 + \sin t, \qquad \frac{dx}{dt} = -\sin t, \qquad \frac{dy}{dt} = \cos t$$

to obtain

$$\text{Area} = \int_a^b 2\pi y \sqrt{\left(\frac{dx}{dt}\right)^2 + \left(\frac{dy}{dt}\right)^2} \, dt \qquad \text{Eq. (3)}$$

$$= \int_0^{2\pi} 2\pi (1 + \sin t) \underbrace{\sqrt{(-\sin t)^2 + (\cos t)^2}}_{1} \, dt$$

$$= 2\pi \int_0^{2\pi} (1 + \sin t) \, dt$$

$$= 2\pi [t - \cos t]_0^{2\pi}$$

$$= 2\pi [(2\pi - 1) - (0 - 1)]$$

$$= 4\pi^2.$$

Remember, once we see a method of solution for a problem like any in this section, we have two basic ways to calculate our way there. One

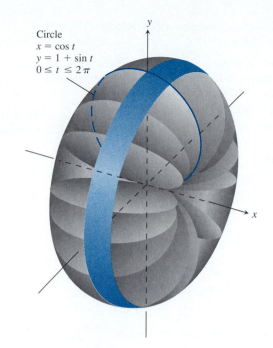

Circle
$x = \cos t$
$y = 1 + \sin t$
$0 \le t \le 2\pi$

10.33 The surface in Example 5. Using a grapher, we find

$$2\pi \cdot \text{NINT } (1 + \sin t, 0, 2\pi) = 39.478,$$

which agrees with the analytic result of $4\pi^2$.

is analytical, the other numerical using a grapher. A third way involves a combination of approaches. With much practice, we acquire the wisdom to choose the best way.

Exercises 10.4

In Exercises 1–12, find an equation for the line tangent to the curve given at the point defined by the given value of t. Also, find the value of d^2y/dx^2 at this point. GRAPH each curve. Overlay the graph of the tangent line at the specified point to support your analytic work.

1. $x = 2\cos t, \quad y = 2\sin t \quad (t = \pi/4)$

2. $x = \sin 2\pi t, \quad y = \cos 2\pi t \quad (t = -1/6)$

3. $x = 4\sin t, \quad y = 2\cos t \quad (t = \pi/4)$

4. $x = \cos t, \quad y = \sqrt{3}\cos t \quad (t = 2\pi/3)$

5. $x = \sec^2 t - 1, \quad y = \tan t \quad (t = -\pi/4)$

6. $x = \sec t, \quad y = \tan t \quad (t = \pi/6)$

7. $x = t, \quad y = \sqrt{t} \quad (t = 1/4)$

8. $x = -\sqrt{t+1}, \quad y = \sqrt{3t} \quad (t = 3)$

9. $x = 2t^2 + 3, \quad y = t^4 \quad (t = -1)$

10. $x = 1/t, \quad y = -2 + \ln t \quad (t = 1)$

11. $x = t - \sin t, \quad y = 1 - \cos t \quad (t = \pi/3)$

12. $x = \cos t, \quad y = 1 + \sin t \quad (t = \pi/2)$

GRAPH each curve in Exercises 13–18. Find the length of each curve by analytic means. Support with a NINT computation.

13. $x = \cos t, \quad y = t + \sin t \quad (0 \le t \le \pi)$

14. $x = t^3, \quad y = 3t^2/2 \quad (0 \le t \le \sqrt{3})$

15. $x = t^2/2, \quad y = (2t + 1)^{3/2}/3 \quad (0 \le t \le 4)$

16. $x = (2t + 3)^{3/2}/3, \quad y = t + t^2/2 \quad (0 \le t \le 3)$

17. $x = 8\cos t + 8t\sin t, \quad y = 8\sin t - 8t\cos t \quad (0 \le t \le \pi/2)$

18. $x = \ln(\sec t + \tan t) - \sin t, \quad y = \cos t, \quad (0 \le t \le \pi/3)$

(For a bigger picture of what the relationship looks like, GRAPH on the interval $-\pi/2 < t < \pi/2$.)

Use analytic means to find the areas of the surfaces generated by revolving the curves in Exercises 19–22 about the indicated axes. Support with a NINT computation.

19. $x = \cos t, \quad y = 2 + \sin t \quad (0 \le t \le 2\pi)$
about the x-axis

20. $x = (2/3)t^{3/2}, \quad y = 2\sqrt{t} \quad (0 \le t \le \sqrt{3})$
about the y-axis

21. $x = t + \sqrt{2}, \quad y = t^2/2 + \sqrt{2}t \quad (-\sqrt{2} \le t \le \sqrt{2})$
about the y-axis

22. $x = \ln(\sec t + \tan t) - \sin t, \quad y = \cos t \quad (0 \le t \le \pi/3)$
about the x-axis

23. *A cone frustum.* The line segment joining the points $(0, 1)$ and $(2, 2)$ is revolved about the x-axis to generate a frustum of a cone. Find the surface area of the frustum using the parametric equations for the segment:

$$x = 2t, \quad y = t + 1 \quad (0 \le t \le 1).$$

Check your result with the geometry formula

$$\text{Area} = \pi(r_1 + r_2)(\text{slant height}).$$

24. *A cone.* The line segment joining the origin to the point (h, r) is revolved about the x-axis to generate a cone of height h and base radius r. Find the cone's surface area using the parametric equations for the segment:

$$x = ht, \quad y = rt \quad (0 \le t \le 1).$$

Check your result with the geometry formula

$$\text{Area} = \pi(r)(\text{slant height}).$$

25. *Length is independent of parametrization.* To illustrate the fact that the numbers that we get for length do not depend on how we parametrize our curves (except for the mild restrictions mentioned earlier), calculate the length of the semicircle $y = \sqrt{1 - x^2}$ with these two different parametrizations:
a) $x = \cos 2t, \quad y = \sin 2t \quad (0 \le t \le \pi/2)$
b) $x = \sin \pi t, \quad y = \cos \pi t \quad (-1/2 \le t \le 1/2)$

26. *Elliptic integrals.* The length of the ellipse

$$x = a \cos t, \quad y = b \sin t \quad (0 \le t \le 2\pi)$$

turns out to be

$$\text{Length} = 4a \int_0^{\pi/2} \sqrt{1 - e^2 \cos^2 t}\, dt,$$

where e is the ellipse's eccentricity. The integral in this formula, called an *elliptic integral*, is nonelementary except when $e = 0$ or 1.
a) Use the Trapezoidal Rule with $n = 10$ to estimate the length of the ellipse when $a = 1$ and $e = 1/2$. Support with a NINT computation.

b) Use the fact that the absolute value of the second derivative of $f(t) = \sqrt{1 - e^2 \cos^2 t}$ is less than 1 to find an upper bound for the error in the estimate that you obtained in part (a).

27. Find the length of the curve given by $x = (8 \sin 2t) \cos t$, $y = (8 \sin 2t) \sin t$, for $0 \le t \le \pi/2$ (one "rose petal").

28. Find the length of the curve given by $x = (6 \sin 2.5t) \cos t$, $y = (6 \sin 2.5t) \sin t$, for $0 \le t \le a$, where a determines the first "rose petal."

29. Find the length of the Spiral of Archimedes, $x = 2t \cos t$, $y = 2t \sin t$, for $0 \le t \le 50$.

30. A particle moves on a curve given by $x = 8t \sin t$, $y = 8 \cos t$, for $0 \le t \le 3$. What is the farthest distance the particle travels from its starting position?

31. Find the area of the solid of revolution determined by revolving the curve $x = 3 \sin t$, $y = 5 + 3 \sin 2t$, for $0 \le t \le 2\pi$, about the y-axis. Sketch an appropriate figure.

32. Find the area of the solid of revolution determined by revolving the curve $x = 5 + 3 \sin t$, $y = 3 \sin 2t$, for $0 \le t \le 2\pi$ about the y-axis. Sketch an appropriate figure.

In Exercises 33 and 34, draw the *Bowditch curves* or *Lissajous figures*. In each case, find the point in the first quadrant where the tangent to the curve is horizontal, and find the equations of the two tangents at the origin. Draw the tangent lines. Support with a graphing utility.

33. $x = \sin t, \quad y = \sin 2t$, for $0 \le t \le 2\pi$.

34. $x = \sin 2t, \quad y = \sin 3t$, for $0 \le t \le 2\pi$.

Find closed curves for the parametric equations in Exercises 35–41. State the range of t values that produce a closed curve. The curves are all Bowditch curves (Lissajous figures), their general formula being

$$x = a \sin(mt + d), \quad y = b \sin nt$$

with m and n integers.

35. $x = \sin 2t, \quad y = \sin t$

36. $x = \sin 3t, \quad y = \sin 4t$

37. $x = \sin t, \quad y = \sin 4t$

38. $x = \sin t, \quad y = \sin 5t$

39. $x = \sin 3t, \quad y = \sin 5t$

40. $x = \sin(3t + \pi/2), \quad y = \sin 5t$

41. $x = \sin(3t + \pi/4), \quad y = \sin 5t$

"Spirograph" Patterns

Let

$$x = a \cos t + b \cos(ct),$$

$$y = a \sin t + b \sin(ct).$$

42. For $a = 5$ and $b = 10$, draw a closed curve for $c = 2, 4, 6$, and 8.

43. For $a = 12$ and $b = 6$, draw a closed curve for $c = 2, 4, 6$, and 8.

44. For $c = 8$ in Exercise 42, find the length of the curve for $0 \le t \le 2\pi$.

45. For $c = 8$ in Exercise 43, find the length of the curve for $0 \le t \le 2\pi$.

10.5 _____ Polar Coordinates

In this section, we define polar coordinates and study their relation to Cartesian coordinates. One of the distinctions between polar and Cartesian coordinates is that while a point in the plane has just one pair of Cartesian coordinates, it has infinitely many pairs of polar coordinates. This has interesting consequences for graphing, as we will see in the next section. Polar coordinates enable us to describe all conic sections with a single equation, as we will see in Section 10.7. The calculus that we have done in rectangular coordinates carries over to this new system as well.

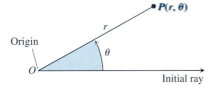

10.34 To define polar coordinates for the plane, we start with an origin and an initial ray.

Definition of Polar Coordinates

To define polar coordinates, we first fix an **origin** O and an **initial ray** from O (Fig. 10.34). Then each point P can be located by assigning to it a **polar coordinate pair** (r, θ), in which the first number, r, gives the directed distance from O to P and the second number, θ, gives the directed angle from the initial ray to the segment OP:

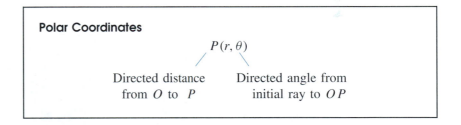

Polar Coordinates

$$P(r, \theta)$$

Directed distance from O to P Directed angle from initial ray to OP

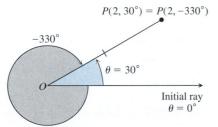

10.35 The ray $\theta = 30°$ is the same as the ray $\theta = -330°$.

Multiple Representations of Points

As in trigonometry, the angle θ is positive when measured counterclockwise and negative when measured clockwise. But the angle associated with a given point is not unique. For instance, the point that is 2 units from the origin along the ray $\theta = 30°$ has polar coordinates $r = 2, \theta = 30°$. It also has coordinates $r = 2, \theta = -330°$ and $r = 2, \theta = 390°$ (Fig. 10.35).

There are occasions when we wish to allow r to be negative. That is why the term "directed distance" is used. The ray $\theta = 30°$ and the ray $\theta = 210°$ together make a complete line through O (Fig. 10.36). The point

$P(2, 210°)$ that is 2 units from O on the ray $\theta = 210°$ has polar coordinates $r = 2, \theta = 210°$. It can be reached by turning $210°$ counterclockwise from the initial ray and going forward 2 units. It can also be reached by turning $30°$ counterclockwise from the initial ray and going *backward* two units. So we say that the point also has polar coordinates $r = -2$, $\theta = 30°$.

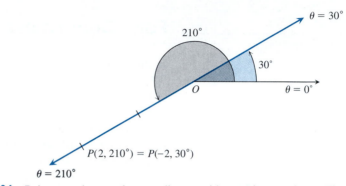

10.36 Points can have polar coordinates with negative r-values. Thus, P could have polar coordinates $(2, 210°)$ or $(-2, 30°)$. It could also have coordinates $(2, -150°)$ or $(-2, -330°)$. Other pairs are possible for P. Name some of them.

Whenever the angle formed by two rays is $180°$, the rays make a line. We then say that each ray is the **opposite** of the other. Points on the ray θ have polar coordinates (r, θ) with $r \geq 0$. Points on the opposite ray, the ray $\theta + 180°$, have coordinates (r, θ) with $r \leq 0$. (See Fig. 10.37.)

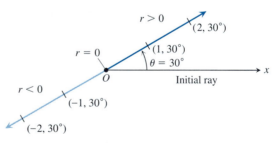

10.37 The ray $\theta = 30°$ and its opposite, the ray $\theta = 210°$, form a line.

RADIAN MEASURE

Although nothing in the definition of polar coordinates requires the use of radian measure, we must remember to have all angles in radians when we differentiate and integrate trigonometric functions of θ. (Do you recall why?)

EXAMPLE 1

Find all the polar coordinates of the point $P(2, 30°)$. Express the angles in radians as well as in degrees.

Solution We sketch the initial ray of the coordinate system, draw the ray through the origin that makes a $30°$ angle with the initial ray, and mark the point $P(2, 30°)$ (Fig. 10.38).

Names for P in Example 1 can be generated in the following ways:

1. For $r = 2$,

Degrees	Radians
$(2, 30°)$	$(2, \pi/6)$
$(2, 30° \pm 360°)$	$(2, \pi/6 \pm 2\pi)$
$(2, 30° \pm 720°)$	$(2, \pi/6 \pm 4\pi)$
$\vdots$	$\vdots$
$(2, 30° \\ \pm n \cdot 360°)$	$(2, \pi/6 \pm n \cdot 2\pi)$

2. For $r = -2$,

Degrees	Radians
$(-2, -150°)$	$(-2, -5\pi/6)$
$(-2, -150° \\ \pm 360°)$	$(-2, -5\pi/6 \\ \pm 2\pi)$
$(-2, -150° \\ \pm 720°)$	$(-2, -5\pi/6 \\ \pm 4\pi)$
$\vdots$	$\vdots$
$(-2, -150° \\ \pm n \cdot 360°)$	$(-2, -5\pi/6+ \\ \pm 2\pi) \\ n.2\pi$

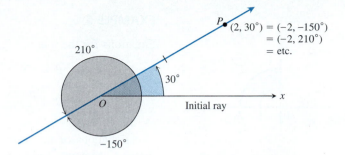

10.38 The point $P(2, 30°)$ has infinitely many different polar coordinates.

We then find formulas for the coordinate pairs in which $r = 2$ and $r = -2$. These are shown in the margin. We can combine the last lines in parts 1 and 2 to obtain the following representations for the polar coordinates of $P(2, 30°)$ in terms of the first-quadrant 30° angle.

Degrees $\qquad\qquad$ Radians

$$((-1)^n \cdot 2, 30° \pm n \cdot 180°) \qquad ((-1)^n \cdot 2, \pi/6 \pm n \cdot \pi) \qquad (n = 0, 1, 2, \ldots)$$

Elementary Polar Equations and Inequalities

If r is a constant nonzero value a, then any point $P(r, \theta) = P(a, \theta)$ is $|a|$ units from the origin. As θ varies over any interval of length 2π radians, the points P form a circle of radius $|a|$ centered at the origin (Fig. 10.39).

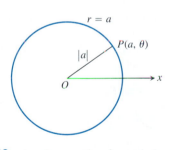

10.39 A polar equation for a circle of radius $|a|$ centered at the origin is $r = a$.

Polar Equation for Circles Centered at the Origin

$$r = a$$

EXAMPLE 2

The equation $r = 1$ is an equation for the circle of radius 1 centered at the origin. So is the equation $r = -1$.

If θ is a constant value θ_0 and r takes on all values between $-\infty$ and ∞, the points $P(r, \theta)$ form a line through the origin that makes an angle θ_0 with the initial ray.

Polar Equation for Lines through the Origin

$$\theta = \theta_0$$

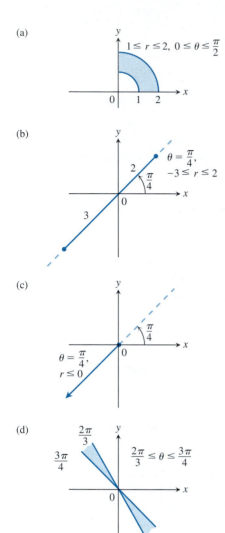

(a)

$1 \le r \le 2, \ 0 \le \theta \le \dfrac{\pi}{2}$

(b)

$\theta = \dfrac{\pi}{4},$ $-3 \le r \le 2$

(c)

$\theta = \dfrac{\pi}{4},$ $r \le 0$

(d)

$\dfrac{2\pi}{3} \le \theta \le \dfrac{3\pi}{4}$

10.40 The graphs of typical inequalities in r and θ.

EXAMPLE 3

Equations for the line in Fig. 10.38 include

$$\theta = 30°, \qquad \theta = 210°, \qquad \theta = -150°,$$

$$\theta = \frac{\pi}{6}, \qquad \theta = \frac{7\pi}{6}, \qquad \theta = \frac{-5\pi}{6}.$$

Equations of the form $r = a$ and $\theta = \theta_0$ can be combined to define regions, segments, and rays.

EXAMPLE 4

Graph the sets of points whose polar coordinates satisfy the following conditions.

a) $1 \le r \le 2$ and $0 \le \theta \le \dfrac{\pi}{2}$

b) $-3 \le r \le 2$ and $\theta = \dfrac{\pi}{4}$

c) $r \le 0$ and $\theta = \dfrac{\pi}{4}$

d) $\dfrac{2\pi}{3} \le \theta \le \dfrac{3\pi}{4}$ (no restriction on r)

Solution   The graphs are shown in Fig. 10.40.

Relating Cartesian and Polar Coordinates

When we use both polar and Cartesian coordinates in a plane, we place the two origins together and take the initial polar ray to be the positive x-axis. The ray $\theta = \pi/2, r > 0$ becomes the positive y-axis (Fig. 10.41). The two sets of coordinates are then related by the following equations:

$$x = r \cos \theta, \qquad y = r \sin \theta, \tag{1}$$

$$x^2 + y^2 = r^2, \qquad \frac{y}{x} = \tan \theta \tag{2}$$

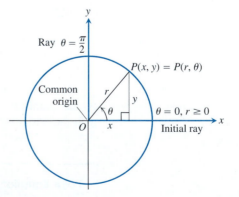

10.41 A diagram like this suggests the relationships between polar and Cartesian coordinates given in Eqs. (1) and (2).

EXPLORATION BIT

Show how Eqs. (2) follow from Eqs. (1).

EXPLORATION 1

The Polar-Cartesian Connection, Part 1

Equations (1) are particularly important because they allow us to display polar graphs in our Cartesian viewing windows just as they would appear in a polar viewing window.

1. In parametric mode, GRAPH

$$x(t) = 6\cos t, \qquad y(t) = 6\sin t$$

for $0 \le t \le 2\pi$ in a *square* viewing window containing $-6 \le x \le 6$, $-6 \le y \le 6$. Describe what you see.

2. We know that $r = a$ is the polar equation for a circle of radius $|a|$ centered at the origin. Give the polar equation for the graph in part 1. ⚭

Equations (1) and Exploration 1 suggest a general procedure for graphing polar curves parametrically in a Cartesian viewing window. If $r = f(\theta)$ is a polar equation, then its graph in the xy-plane is given by the parametric equations

$$x(t) = f(t)\cos t, \qquad y(t) = f(t)\sin t. \qquad \text{Here } \theta = t.$$

EXAMPLE 5

Graph $r = 1 + 2r\cos\theta$ by graphing an equivalent parametric representation.

Solution We know that the polar graph is also the parametric graph

$$x(t) = f(t)\cos t, \qquad y(t) = f(t)\sin t,$$

where $r = f(t)$. Here, $r = 1 + 2r\cos t$ (remember that $t = \theta$). So

$$r(1 - 2\cos t) = 1,$$

or

$$r = \frac{1}{1 - 2\cos t}.$$

So, the parametric equations

$$x(t) = \frac{\cos t}{1 - 2\cos t}, \qquad y(t) = \frac{\sin t}{1 - 2\cos t}$$

with $0 \le t \le 2\pi$ will produce the graph of the polar equation $r = 1 + 2r\cos\theta$ (Fig. 10.42). ≣

EXPLORATION 2

The Polar-Cartesian Connection, Part 2

1. Use the equations

$$x = r\cos\theta, \qquad y = r\sin\theta$$

to transform one equation to the other.

Sidebar (left column)

EXPLORATION BIT

GRAPH

$$x(t) = 6\sin t, \qquad y(t) = 6\cos t$$

for the same values of t, x, and y as in part 1 of Exploration 1. How does this graph differ from the graph in Exploration 1?

POLAR GRAPHING, POLAR GRIDS

Some graphers have built-in polar graphing capability that will graph a polar equation on a *polar grid*. A sheet of *polar graph paper* will show you what a polar grid looks like. It is also instructive to sketch some graphs of polar equations on polar graph paper. To find out whether your grapher can graph in *polar mode*, consult the *Resource Manual* or your *Owner's Guide*. If polar graphing is available, feel free to use it as needed.

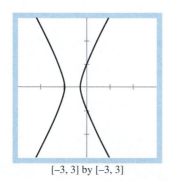

[−3, 3] by [−3, 3]

10.42 A complete graph of

$$r = 1 + 2r\cos\theta.$$

Try showing this graph using polar mode if this option is available on your grapher.

	Polar Equation	Cartesian Equation
a)	$r \cos \theta = 2$	$x = 2$
b)	$r^2 \cos \theta \sin \theta = 4$	$xy = 4$
c)	$r^2 \cos^2 \theta - r^2 \sin^2 \theta = 1$	$x^2 - y^2 = 1$
d)	$r = 1 + 2r \cos \theta$	$y^2 - 3x^2 - 4x - 1 = 0$
e)	$r = 1 - \cos \theta$	$x^4 + y^4 + 2x^2 y^2 + 2x^3 + 2xy^2 - y^2 = 0$

2. If possible, support each of the above graphically. For example, for 1(d), let

$$x_1(t) = \frac{\cos t}{1 - 2\cos t}, \qquad y_1(t) = \frac{\sin t}{1 - 2\cos t}, \qquad \text{Why?}$$

$$x_2(t) = t, \qquad y_2(t) = \sqrt{3t^2 + 4t + 1}, \qquad \text{Why?}$$

$$x_3(t) = t, \qquad y_3(t) = -y_2(t), \qquad \text{Why?}$$

and GRAPH for $-3 \le t \le 3$, t-step = 0.1, in a *square* viewing window containing $-3 \le x \le 3$, $-2 \le y \le 2$. Explain the support.

3. Can you support 1(d) in polar mode? Explain.

4. With some curves, we are better off with polar coordinates; with others, we are better off with Cartesian coordinates. Explain. &

To *find* a Cartesian equation corresponding to a given polar equation requires familiarity with trigonometric identities—and some luck. The polar graph in the xy-plane can give a clue as to the nature of the Cartesian relationship.

EXAMPLE 6

Find a Cartesian equation for the curve

$$r \cos \left(\theta - \frac{\pi}{3} \right) = 3.$$

Solution We graph

$$x(t) = \frac{3}{\cos (t - \pi/3)} \cos t, \qquad y(t) = \frac{3}{\cos (t - \pi/3)} \sin t$$

(Fig. 10.43) and see that the relationship appears to be linear. We use the identity

$$\cos (A - B) = \cos A \cos B + \sin A \sin B$$

with $A = \theta$ and $B = \pi/3$:

$$r \cos \left(\theta - \frac{\pi}{3} \right) = 3,$$

$$r \left(\cos \theta \cos \frac{\pi}{3} + \sin \theta \sin \frac{\pi}{3} \right) = 3,$$

$$r \cos \theta \cdot \frac{1}{2} + r \sin \theta \cdot \frac{\sqrt{3}}{2} = 3, \qquad \text{Luck!}$$

$$\frac{1}{2}x + \frac{\sqrt{3}}{2}y = 3,$$

$$x + \sqrt{3}y = 6.$$

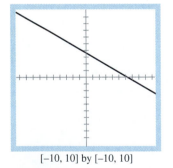

[-10, 10] by [-10, 10]

10.43 The graph of the polar equation of Example 6 (in either parametric mode or polar mode—try each) appears to be a line with a small negative slope and y-intercept between 3 and 4. The results of Example 6 confirm that it is.

EXPLORATION 3

The Polar-Cartesian Connection, Part 3

GRAPH the polar equation, and make a conjecture about the corresponding Cartesian equation. Then confirm.

1. $r \cos \theta = -4$

2. $r^2 = 4r \cos \theta$

3. $r = 4/(2 \cos \theta - \sin \theta)$ (The Cartesian equation will have some removable discontinuities. Can you identify them?)

Exercises 10.5

Note: All angles are in radians.

1. Pick out the polar coordinate pairs that label the same point.
- **a)** $(3, 0)$
- **b)** $(-3, 0)$
- **c)** $(-3, \pi)$
- **d)** $(-3, 2\pi)$
- **e)** $(2, 2\pi/3)$
- **f)** $(2, -\pi/3)$
- **g)** $(2, 7\pi/3)$
- **h)** $(-2, \pi/3)$
- **i)** $(2, -\pi/3)$
- **j)** $(2, \pi/3)$
- **k)** $(-2, -\pi/3)$
- **l)** $(-2, 2\pi/3)$
- **m)** (r, θ)
- **n)** $(r, \theta + \pi)$
- **o)** $(-r, \theta + \pi)$
- **p)** $(-r, \theta)$

2. Find the Cartesian coordinates of the points whose polar coordinates are given in parts (a)–(l) of Exercise 1.

3. Plot the following points (given in polar coordinates). Then find all the polar coordinates of each point.
- **a)** $(2, \pi/2)$
- **b)** $(2, 0)$
- **c)** $(-2, \pi/2)$
- **d)** $(-2, 0)$

4. Plot the following points (given in polar coordinates). Then find all the polar coordinates of each point.
- **a)** $(3, \pi/4)$
- **b)** $(-3, \pi/4)$
- **c)** $(3, -\pi/4)$
- **d)** $(-3, -\pi/4)$

5. Find the Cartesian coordinates of the following points (given in polar coordinates).
- **a)** $(\sqrt{2}, \pi/4)$
- **b)** $(1, 0)$
- **c)** $(0, \pi/2)$
- **d)** $(-\sqrt{2}, \pi/4)$
- **e)** $(-3, 5\pi/6)$
- **f)** $(5, \tan^{-1}(4/3))$
- **g)** $(-1, 7\pi)$
- **h)** $(2\sqrt{3}, 2\pi/3)$

6. Find all polar coordinates of the origin.

Sketch a graph of each set of points whose polar coordinates satisfy each equation or inequality in Exercises 7–22.

7. $r = 2$

8. $0 \leq r \leq 2$

9. $r \geq 1$

10. $1 \leq r \leq 2$

11. $0 \leq \theta \leq \pi/6, \quad r \geq 0$

12. $\theta = 2\pi/3, \quad r \leq -2$

13. $\theta = \pi/3, \quad -1 \leq r \leq 3$

14. $\theta = 11\pi/4, \quad r \geq -1$

15. $\theta = \pi/2, \quad r \geq 0$

16. $\theta = \pi/2, \quad r \leq 0$

17. $0 \leq \theta \leq \pi, \quad r = 1$

18. $0 \leq \theta \leq \pi, \quad r = -1$

19. $\pi/4 \leq \theta \leq 3\pi/4, \quad 0 \leq r \leq 1$

20. $-\pi/4 \leq \theta \leq \pi/4, \quad -1 \leq r \leq 1$

21. $-\pi/2 \leq \theta \leq \pi/2, \quad 1 \leq r \leq 2$

22. $0 \leq \theta \leq \pi/2, \quad 1 \leq |r| \leq 2$

Replace the polar equations in Exercises 23–34 by equivalent Cartesian equations by analytic means. Then identify the graph of each. Support with a grapher.

23. $r \cos \theta = 2$

24. $r \sin \theta = -1$

25. $r \sin \theta = 4$

26. $r \cos \theta = 0$

27. $r \sin \theta = 0$

28. $r \cos \theta = -3$

29. $r \cos \theta + r \sin \theta = 1$

30. $r \sin \theta = r \cos \theta$

31. $r^2 = 1$

32. $r^2 = 4r \sin \theta$

33. $r = \dfrac{5}{\sin \theta - 2 \cos \theta}$

34. $r = 4 \tan \theta \sec \theta$

Replace the Cartesian equations in Exercises 35–44 by equivalent polar equations by analytic means. Identify the graph of each. Support with a grapher.

35. $x = 7$

36. $y = 1$

37. $x = y$

38. $x - y = 3$

39. $x^2 + y^2 = 4$

40. $x^2 - y^2 = 1$

41. $\dfrac{x^2}{9} + \dfrac{y^2}{4} = 1$

42. $xy = 2$

43. $y^2 = 4x$

44. $x^2 - y^2 = 25\sqrt{x^2 + y^2}$

45. Show that the Cartesian graph of the polar equation $r = f(\theta)$ is given by the graph of the parametric equations

$$x = f(\theta) \cos \theta, \quad y = f(\theta) \sin \theta.$$

10.6

Graphing in Polar Coordinates

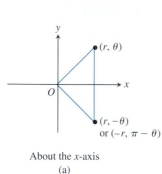

About the *x*-axis

(a)

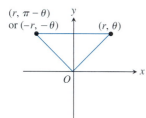

About the *y*-axis

(b)

The graph of the equation

$$F(r, \theta) = 0$$

consists of the points whose polar coordinates in some form satisfy the equation. We say "in some form" because it is a sad fact that some coordinate pairs of a point on the graph may not satisfy the equation even when others do.

To speed our work, we look for symmetries, for values of θ at which the curve passes through the origin, and for points at which r takes on extreme values. When the curve passes through the origin, we also try to calculate the curve's slope there. Coordinates of points that appear in this section are polar coordinates unless specified otherwise.

Symmetry and Slope

The three parts of Fig. 10.44 illustrate the standard polar coordinate tests for symmetry.

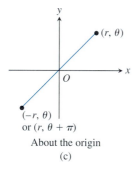

About the origin

(c)

10.44 Three tests for symmetry.

Symmetry Tests for Graphs

1. *Symmetry about the x-axis:* If the point (r, θ) lies on the graph, the point $(r, -\theta)$ or $(-r, \pi - \theta)$ lies on the graph (Fig. 10.44a).

2. *Symmetry about the y-axis:* If the point (r, θ) lies on the graph, the point $(r, \pi - \theta)$ or $(-r, -\theta)$ lies on the graph (Fig. 10.44b).

3. *Symmetry about the origin:* If the point (r, θ) lies on the graph, the point $(-r, \theta)$ or $(r, \theta + \pi)$ lies on the graph (Fig. 10.44c).

If a curve has any two of the symmetries listed here, it also has the third (as you will be invited to show in Exercise 48). Thus, if two of the tests are positive, there is no need to apply the third test.

The slope of a polar curve $r = f(\theta)$ is given not by the derivative $r' = df/d\theta$, but by a different formula. To see why, and what the formula is, think of the graph of f as the graph of the parametric equations

$$x = r \cos \theta = f(\theta) \cos \theta,$$

$$y = r \sin \theta = f(\theta) \sin \theta.$$

If f is a differentiable function of θ, then so are x and y, and when

$dx/d\theta \neq 0$, we may calculate dy/dx from the parametric formula

$$\frac{dy}{dx} = \frac{dy/d\theta}{dx/d\theta} \qquad \text{Section 10.4, Eq. (1), with } t = \theta$$

$$= \frac{\dfrac{d}{d\theta}(f(\theta) \cdot \sin\theta)}{\dfrac{d}{d\theta}(f(\theta) \cdot \cos\theta)}$$

$$= \frac{\dfrac{df}{d\theta}\sin\theta + f(\theta)\cos\theta}{\dfrac{df}{d\theta}\cos\theta - f(\theta)\sin\theta} \qquad \text{Product Rule for Derivatives}$$

$$= \frac{r'\sin\theta + r\cos\theta}{r'\cos\theta - r\sin\theta}. \qquad r = f, r' = df/d\theta.$$

Slope of a Polar Curve

If $r = f(\theta)$ is differentiable and $dx/d\theta \neq 0$, then the slope dy/dx at the point (r, θ) on the graph of f is given by the formula

$$\text{Slope at } (r, \theta) = \frac{r'\sin\theta + r\cos\theta}{r'\cos\theta - r\sin\theta}. \tag{1}$$

If $r = 0$ when $\theta = \theta_0$, then Eq. (1) reduces to

$$\text{Slope at } (0, \theta_0) = \frac{r'\sin\theta_0}{r'\cos\theta_0} = \frac{\sin\theta_0}{\cos\theta_0} = \tan\theta_0.$$

Slope at the Origin

If the graph of $r = f(\theta)$ passes through the origin at the value $\theta = \theta_0$, then the slope of the curve there is

$$\text{Slope at } (0, \theta_0) = \tan\theta_0.$$

EXPLORATION BITS

1. GRAPH $r = 8\sin 3(\theta + c)$ in a square viewing window for different values of c. Study the graphs for $c = 0, \pm\pi/6, \pm\pi/3$, and so on, then try to find a value for c so that the graph matches the GNA Exploration logo used in this book.

2. Recall how horizontal shifts in the graph of $y = f(x)$ occur in Cartesian coordinates. Describe an analogy in polar coordinates. Illustrate with the equation from part 1. (It will have to be put into a slightly different form.)

We say "slope at $(0, \theta_0)$" and not just "slope at the origin" because a polar curve may pass through the origin more than once, with different slopes at different θ-values. This will not be the case in our first example, however.

EXAMPLE 1   A Cardioid

Sketch a complete graph of the curve

$$r = 1 - \cos\theta.$$

Support with a grapher.

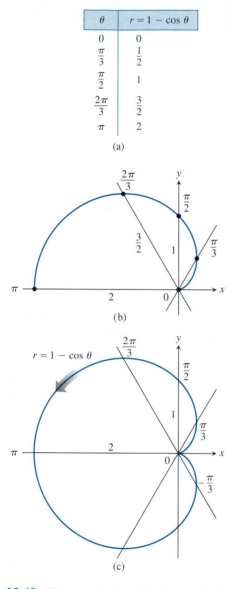

θ	$r = 1 - \cos\theta$
0	0
$\dfrac{\pi}{3}$	$\dfrac{1}{2}$
$\dfrac{\pi}{2}$	1
$\dfrac{2\pi}{3}$	$\dfrac{3}{2}$
π	2

(a)

(b)

$r = 1 - \cos\theta$

(c)

10.45 The steps in sketching the graph of the cardioid $r = 1 - \cos\theta$. The arrow shows the direction of increasing θ.

Solution   The curve is symmetric about the x-axis because (r, θ) on the graph

$$\Rightarrow \quad r = 1 - \cos\theta$$

$$\Rightarrow \quad r = 1 - \cos(-\theta) \qquad \cos\theta = \cos(-\theta).$$

$$\Rightarrow \quad (r, -\theta) \text{ on the graph.}$$

As θ increases from 0 to π, $\cos\theta$ decreases from 1 to -1, and $r = 1 - \cos\theta$ increases from a minimum value of 0 to a maximum value of 2. As θ continues on from π to 2π, $\cos\theta$ increases from -1 back to 1 and r decreases from 2 back to 0. The curve starts to repeat when $\theta = 2\pi$ because the cosine has period 2π. The curve leaves the origin with slope $\tan(0) = 0$ and returns to the origin with slope $\tan(2\pi) = 0$.

We make a table of values from $\theta = 0$ to $\theta = \pi$, plot the points, sketch a smooth curve through them with a horizontal tangent at the origin, and reflect the curve across the x-axis to complete the graph (Fig. 10.45). The curve is called a *cardioid* because of its heart shape.

The graph of $r = 1 - \cos\theta$ in polar mode for $0 \le \theta \le 2\pi$, or of the equivalent parametric equations

$$x(t) = (1 - \cos t)\cos t, \qquad y(t) = (1 - \cos t)\sin t,$$

for $0 \le t \le 2\pi$ in a *square* viewing window supports the sketch. Try either one.   ≣

EXPLORATION 1

Complete Polar Graphs

A polar graph on a grapher should be *complete* in the same sense that a Cartesian graph is complete. The viewing window should suggest all the important features of the graph. There should be no hidden behavior.

Many polar graphs that we study that are periodic are nicely behaved and can be shown completely by graphing them over an interval whose length is the *greater* of

a) 2π, so we have covered all possible θ values at least once, or

b) the period of the function.

1. Draw the polar graph of $r = \sin 2\theta$ for $0 \le \theta \le \pi$. Note that π is the period of $\sin 2\theta$. Then draw the graph for $0 \le \theta \le 2\pi$. Explain what you observe.

2. Draw the polar graph of the *lemniscate*, $r = \sqrt{\sin 2\theta}$ for $0 \le \theta \le \pi/2$; for $0 \le \theta \le \pi$; for $0 \le \theta \le 3\pi/2$; for $0 \le \theta \le 2\pi$. Explain what you observe.

3. Find a complete graph of $r = 8\sin 2.5\theta$. Explain how parts (*a*) and (*b*) above fail in this case.

4. Find other polar functions for which parts (*a*) and (*b*) above fail.

If a graph (either polar or Cartesian) suggests hidden or confusing behavior, we now have the other mode in which to view it to perhaps understand it a bit better.

5. Draw the polar graph of $r = \cos\theta^2$ for $0 \le \theta \le 100$, θ-step $= 0.1$. (Note that $\cos\theta^2$ is not periodic.) Is your graph complete? Is there hidden behavior? Compare with the Cartesian graph of $y = \cos x^2$.   ✼

EXAMPLE 2

Draw a complete graph of the curve $r^2 = 4\cos\theta$. Confirm the apparent symmetries in the graph.

Solution   The square viewing window shows the graph of

$$x_1(t) = \left(2\sqrt{\cos t}\right)\cos t,$$

$$y_1(t) = \left(2\sqrt{\cos t}\right)\sin t$$

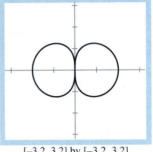

$$[-3.2, 3.2] \text{ by } [-3.2, 3.2]$$

and

$$x_2(t) = -x_1(t),$$

$$y_2(t) = -y_1(t)$$

for $-\pi/2 \le t \le \pi/2$. We chose the parameter interval so that the radicand remains positive, and we could have chosen *any* such interval because of the periodicity of the sine and cosine functions.

We note that the curve is closed; its initial point ($t = \theta = -\pi/2$) and terminal point ($t = \theta = \pi/2$) are the same—the origin. To confirm the symmetries, we see that the curve is symmetric about the x-axis because

$$(r, \theta) \text{ on the graph} \quad \Rightarrow r^2 = 4\cos\theta$$

$$\Rightarrow r^2 = 4\cos(-\theta) \qquad \begin{array}{l}\cos\theta = \\ \cos(-\theta).\end{array}$$

$$\Rightarrow \quad (r, -\theta) \quad \text{on the graph.}$$

The curve is also symmetric about the origin because

$$(r, \theta) \text{ on the graph} \quad \Rightarrow \quad r^2 = 4\cos\theta$$

$$\Rightarrow \quad (-r)^2 = 4\cos\theta$$

$$\Rightarrow \quad (-r, \theta) \quad \text{on the graph.}$$

Together, these two symmetries imply symmetry about the y-axis.   ∎

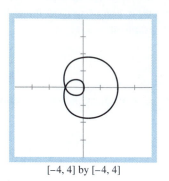

[−4, 4] by [−4, 4]

10.46 A complete graph of

$$r = 1 + \cos\frac{\theta}{2}.$$

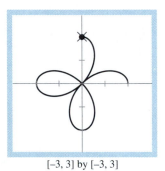

[−3, 3] by [−3, 3]

10.47 The graph of

$$r = 2\cos 2\theta, \quad 0 \le \theta \le 2\pi,$$

is shown here illuminating the pixel for $(-2, 3\pi/2)$. Another polar name for $(-2, 3\pi/2)$ is $(2, \pi/2)$, but when $\theta = \pi/2$, the pixel for $(-2, \pi/2)$ at the tip of the bottom "leaf" illuminates. The pair $(2, \pi/2)$ does not satisfy the equation $r = 2\cos 2\theta$.

EXPLORATION BIT

If your grapher does not have simultaneous graphing capability, you can identify the points of intersection that correspond to simultaneous solutions by using TRACE. Explain how.

EXAMPLE 3

Draw a complete graph of the curve

$$r = 1 + \cos\frac{\theta}{2}.$$

Solution   Because the period of $\cos\theta$ is 2π, the period of $\cos(\theta/2)$ is 4π. We let θ run from 0 to 4π to produce the graph in Fig. 10.46.   ▤

Finding the Points Where Curves Intersect

The fact that we can represent a point in different ways in polar coordinates makes extra care necessary in deciding which ordered pair causes a pixel to illuminate on a graph. It is possible, for example, to have (r, θ) satisfy an equation and $(-r, \theta + \pi)$ not satisfy the same equation. The pixel on the graph, while representing both (r, θ) and $(-r, \theta + \pi)$, would be lit only because of (r, θ).

EXAMPLE 4

$(2, \pi/2)$ and $(-2, 3\pi/2)$ are polar coordinates for the same point (Fig. 10.47), but $(-2, 3\pi/2)$ satisfies $r = 2\cos 2\theta$:

$$-2 = 2\cos 2(3\pi/2) = 2\cos 3\pi = 2(-1),$$

and $(2, \pi/2)$ does not:

$$2 \ne 2\cos 2(\pi/2) = 2\cos\pi = 2(-1) = -2.$$ ▤

If the curves for two polar equations intersect, the (r, θ) pair that illuminates the point of intersection for one equation may be different from the (r, θ) pair that illuminates the point for the other equation. In other words, a point of intersection of two graphs does not necessarily represent a simultaneous solution to the equations.

Thus, for any two polar equations and their graphs,

a) the simultaneous solutions of the equations identify points of intersection of the graphs, but there could be others;

b) the points of intersection of the graphs include all simultaneous solutions of the equations, but some points of intersection may have coordinates that are not simultaneous solutions.

 ### EXPLORATION 2

Simultaneous Solutions

The *simultaneous format* of a graphing utility gives new meaning to the *simultaneous solution* of a pair of equations. A simultaneous solution occurs only where the two graphs "collide" while they are drawing and not where one graph draws across the other at a point that had been illuminated earlier. GRAPH the polar equations.

$$r = \cos 2\theta \quad \text{and} \quad r = \sin 2\theta$$

in simultaneous format with $0 \le \theta < 2\pi$, θ-step = 0.1, and view dimensions $[-1, 1]$ by $[-1, 1]$.

1. *While the graphs are drawing on the screen,* count the number of times the two graphs illuminate a single pixel simultaneously. Explain why these points of intersection of the two graphs correspond to simultaneous solutions of the equations. (You may find it helpful to slow down the graphing by making θ-step smaller, θ-step = 0.05, for example.)

2. How many points of intersection are there in all for the two graphs?

3. How can you use the Cartesian graphs of $y = \cos 2x$ and $y = \sin 2x$ to support the number of simultaneous solutions?

EXAMPLE 5

When a motorcycle drill team performs, the paths of the motorcycles cross but the motorcycles do not crash because of the time differential.

Find the points of intersection of the curves

$$r^2 = 4\cos\theta \qquad \text{and} \qquad r = 1 - \cos\theta.$$

Solution Drawing the graphs simultaneously suggests there are two simultaneous solutions and two other points of intersection, $(0, 0)$ and $(2, \pi)$. (See Fig. 10.48.) Solving simultaneously, we substitute $\cos\theta = 1 - r$ into the equation $r^2 = 4\cos\theta$ and get

$$r^2 = 4(1 - r)$$

$$r^2 = 4 - 4r$$

$$r^2 + 4r - 4 = 0$$

$$r = -2 \pm 2\sqrt{2}. \quad \text{Quadratic formula}$$

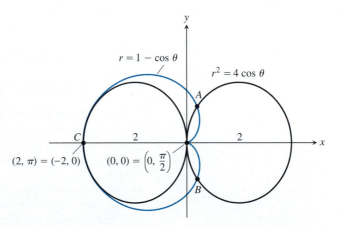

10.48 Drawing $r_1 = 1 - \cos\theta$ and $r_2 = 2\sqrt{\cos\theta}$ simultaneously suggests that simultaneous solutions correspond to points A and B and that the point $(0, 0) = (0, \pi/2)$ is a point of intersection of the graphs but not a simultaneous solution. Drawing $r_1 = 1 - \cos\theta$ and $r_3 = -2\sqrt{\cos\theta} = -r_2$ simultaneously suggests no other simultaneous solutions but one other point of intersection of the two graphs, $(2, \pi) = (-2, 0)$.

The value $r = -2 - 2\sqrt{2}$ has too large an absolute value to give a point on either curve. The values of θ corresponding to $r = -2 + 2\sqrt{2}$ are

$$\theta = \cos^{-1}(1 - r) \qquad \text{From } r = 1 - \cos\theta$$

$$= \cos^{-1}(1 - (2\sqrt{2} - 2)) \qquad \text{Set } r = 2\sqrt{2} - 2.$$

$$= \cos^{-1}(3 - 2\sqrt{2})$$

$$= \pm 1.398 \text{ radians} = \pm 80.12°$$

We have thus identified two points of intersection:

$$(r, \theta) = \left(2\sqrt{2} - 2, \pm 1.398\right) = \left(2\sqrt{2} - 2, \pm 80.12°\right)$$

The points $(0, 0)$ and $(2, \pi)$ are on both graphs but are not simultaneous solutions of the equations because they are not reached at the same value of θ. On the curve $r = 1 - \cos\theta$, the point $(2, \pi)$ is reached when $\theta = \pi$. On the curve $r^2 = 4\cos\theta$, it is reached when $\theta = 0$, where it is identified not by the coordinates $(2, \pi)$, which do not satisfy the equation, but by the coordinates $(-2, 0)$, which do. Similarly, the cardioid reaches the origin when $\theta = 0$, but the curve $r^2 = 4\cos\theta$ reaches the origin when $\theta = \pi/2$.   ≡

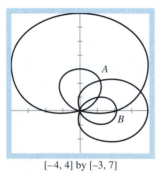

[−4, 4] by [−3, 7]

10.49   Complete graphs of

$r_1 = 1 + 3\cos\theta$ and $r_2 = -2 + 5\sin\theta$.

Graphing r_1 and r_2 simultaneously suggests simultaneous solutions of the equations occur at points with polar coordinates $A(2.41, 1.08)$ and $B(-2, \pi)$.

EXAMPLE 6

Let $r_1 = 1 + 3\cos\theta$ and $r_2 = -2 + 5\sin\theta$. Find all points of intersection of the two curves.

Solution   Complete graphs (Fig. 10.49) can be obtained for $0 \le \theta \le 2\pi$. (Why?) A careful examination of the graphs shows that there are five points of intersection. Two of them are simultaneous solutions with polar coordinates $A(2.41, 1.08)$ and $B(-2, \pi)$. The rectangular coordinates of the other three points of intersection are $(0, 0)$, $(0.96, 0.83)$, and $(3.54, 1.37)$.   ≡

EXPLORATION 3

Colliding Particles

Suppose the two polar equations in Example 6 represent two particles moving as a function of time. (Think of θ as t.) The simultaneous solutions of the equations correspond to the points in time at which the two particles are in the same position—in other words, when they collide.

1. Find when and where the two particles collide for the first time. Assuming that they survive the first collision, find where they collide a second time. If they survive the second collision and continue moving with time, will they collide a third time? Explain.

2. Using parametric mode and graphing in the xy-plane, let

$$x_1(t) = (1 + 3\cos t)\cos t, \qquad y_1(t) = (1 + 3\cos t)\sin t,$$

$$x_2(t) = (-2 + 5\sin t)\cos t, \qquad y_2(t) = (-2 + 5\sin t)\sin t.$$

The parametric equations that show the distance between points $(x_1(t), y_1(t))$ and $(x_2(t), y_2(t))$ at time t are

$$x_3(t) = t, \qquad y_3(t) = \sqrt{(x_1(t) - x_2(t))^2 + (y_1(t) - y_2(t))^2}.$$

GRAPH $(x_3(t), y_3(t))$, and explain what you see. What points in the graph indicate particle collisions? Do these agree with part 1? When are the particles 4 units apart? Explain.

3. What could you do to the time t to improve the accuracy in the above activities?

4. Confirm at least one simultaneous solution.

Exercises 10.6 _____

Sketch the curves in Exercises 1–10. Support with a grapher.

1. $r = 1 + \cos\theta$

2. $r = 2 - 2\cos\theta$

3. $r = 1 - \sin\theta$

4. $r = 1 + \sin\theta$

5. $r = 2 + \sin\theta$

6. $r = 1 + 2\sin\theta$

7. $r^2 = 4\cos 2\theta$

8. $r^2 = 4\sin\theta$

9. $r = \theta$

10. $r = \sin(\theta/2)$

11. *Four-leaved rose.* Determine a complete graph of $r = 8\cos 2\theta$. What range of values of θ is necessary to close the curve? In $r = 8\cos 2\theta$, what is the effect of the number 8? The number 2?

12. *Another rose.* Determine a complete graph of $r = 8\cos 3\theta$. How many rose petals are there? What range of values of θ is necessary to close the curve? In $r = 8\cos 3\theta$, what is the effect of the number 8? The number 3?

Use Eq. (1) to find the slopes of the curves in Exercises 13–16 at the given points. Sketch the curves along with their tangents at these points. Support with a grapher.

13. *Cardioid.* $r = -1 + \cos\theta, \theta = \pm\pi/2$

14. *Cardioid.* $r = -1 + \sin\theta, \theta = 0, \pi/2, \pi$

15. *Four-leaved rose.* $r = \sin 2\theta, \theta = \pm\pi/4, \pm 3\pi/4$, and the values of θ at which the curve passes through the origin.

16. *Four-leaved rose.* $r = \cos 2\theta, \theta = 0, \pm\pi/2, \pi$, and the values of θ at which the curve passes through the origin.

Draw complete graphs of the lemniscates in Exercises 17 and 18. What range of θ values is necessary to close each curve?

17. $r^2 = 4\cos 2\theta$

18. $r^2 = 4\sin 2\theta$

Find complete graphs of the limaçons in Exercises 19–22. Limaçon ("LEE-ma-sahn") is Old French for "snail." You will see why the name is appropriate when you graph the limaçons in Exer-

cise 19. Equations for limaçons have the form $r = a \pm b\cos\theta$ or $r = a \pm b\sin\theta$. There are four basic shapes.

19. *Limaçons with an inner loop.*
  a) $r = \dfrac{1}{2} + \cos\theta$
  b) $r = \dfrac{1}{2} + \sin\theta$

20. *Cardioids.*
  a) $r = 1 - \cos\theta$
  b) $r = -1 + \sin\theta$

21. *Dimpled limaçons.*
  a) $r = \dfrac{3}{2} + \cos\theta$
  b) $r = \dfrac{3}{2} - \sin\theta$

22. *Convex limaçons.*
  a) $r = 2 + \cos\theta$
  b) $r = -2 + \sin\theta$

23. Sketch the region defined by the inequality $0 \le r \le 2 - 2\cos\theta$.

24. Sketch the region defined by the inequality $0 \le r^2 \le \cos\theta$.

25. Show that the point $(2, 3\pi/4)$ lies on the curve $r = 2\sin 2\theta$.

26. Show that the point $(-1, 3\pi/2)$ lies on the curve $r = -\sin(\theta/3)$.

Find the points of intersection of the pairs of curves in Exercises 27–34.

27. $r = 1 + \cos\theta, \quad r = 1 - \cos\theta$

28. $r = 1 + \sin\theta, \quad r = 1 - \sin\theta$

29. $r = 1 - \sin\theta, \quad r^2 = 4\sin\theta$

30. $r^2 = \sqrt{2}\sin\theta, \quad r^2 = \sqrt{2}\cos\theta$

31. $r^2 = \sin 2\theta, \quad r^2 = \cos 2\theta$

32. $r = 1 + \cos\dfrac{\theta}{2}, \quad r = 1 - \sin\dfrac{\theta}{2}$

33. $r = 1,$ $r = 2 \sin 2\theta$

34. $r = 1,$ $r^2 = \sin 2\theta$

Period of a Polar Graph

We define the *period* of a polar graph to be the length of the smallest interval of values of θ that is necessary to close a complete graph. For example, the rose $r = 8 \sin 2\theta$ has period 2π, while the rose $r = 8 \sin 3\theta$ has period π (based on Exercises 11 and 12). GRAPH and determine the period of the graphs in Exercises 35–44.

35. $r = 5 \sin \theta.$ **36.** $r = 5 \cos \theta.$

37. $r = 5 \sin 2\theta.$ **38.** $r = 5 \cos 2\theta.$

39. $r = 5 \sin 5\theta.$ **40.** $r = 5 \cos 5\theta.$

41. $r = 5 \sin (5\theta/2).$ **42.** $r = 5 \cos (5\theta/2).$

43. $r = 5 \sin (3\theta/2).$ **44.** $r = 5 \cos (3\theta/2).$

45. *A rose within a rose.* Find a complete graph of the equation $r = 1 - 2 \sin 3\theta$. What is its period?

46. Determine a complete graph of *the nephroid of Freeth:*

$$r = 1 + 2 \sin \frac{\theta}{2}.$$

What is its period? On the basis of the graph (in a square viewing window), guess what *nephroid* might mean.

47. *Spirals.* Polar coordinates are just the thing for defining spirals. Graph the following spirals:
  a) *A logarithmic spiral:* $r = e^{\theta/10}$
  b) *A hyperbolic spiral:* $r = 8/\theta$
  c) *Equilateral hyperbola:* $r = \pm 10/\sqrt{\theta}$

48. Show that a curve with any two of the symmetries listed at the beginning of the section automatically has the third.

49. How would you describe the period of the graph of $r = \theta$?

50. Predict the shape of a complete graph of

$$r = a \sin \left(\frac{m}{n} \theta \right),$$

where m and n are *relatively prime* positive integers (that is, m and n have no common factors). (*Hint:* Study the patterns in Exercises 41 and 43, and then do more grapher explorations.)

Videocassette Tape Length

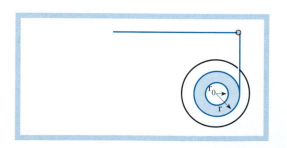

The length of a tape wound onto a take-up reel (shown above) of a videocassette is given by

$$S = \int_0^\alpha \sqrt{r^2 + \left(\frac{b}{2\pi} \right)^2} \, d\theta,$$

where b is the tape thickness and

$$r = r_0 + \left(\frac{\alpha}{2\pi} \right) b$$

is the radius of the tape on the take-up reel. The initial radius of the tape on the take-up reel is r_0, and α is the angle in radians through which the wheel has turned.

51. Simulate the tape accumulating on the take-up reel using polar graphing with $r_0 = 1.75$ cm and $b = 0.06$ cm.

52. Confirm analytically the formula for S, the length of the tape.

53. Determine the length of tape on the take-up reel if the reel has turned through an angle of 80π and $r_0 = 1.75$ cm, $b = 0.06$ cm.

54. Assume that b is very small in comparison to r at any time.
  a) Show analytically that

$$S_a = \int_0^\alpha r \, d\theta$$

  is an excellent approximation to the exact value of S.
  b) For the values given in Exercise 53, compare S_a with S.

55. Let n be the number of complete turns the take-up reel has made.
  a) Find a formula for n in terms of S, the tape length.
  b) When a VCR operates, the tape moves past the heads at a constant speed. Describe the speed of the take-up reel as time progresses.

56. Suppose in Exercise 55 that the VCR tape counter is the number, n, of complete turns of the take-up reel. Describe the counter values as a function of time t.

57. Support the formula for S by the following experiment. Take a roll of wide ribbon and a paper tube. Determine the radius of the paper tube, r_0. (How can you do this without measuring from the center?) Wrap the ribbon around the tube carefully, and count the number of times that it wraps (to find α). Determine the thickness of the ribbon, b, by dividing the radial length between r_0 and the outer ribbon by the layers of ribbon. Calculate S by the formula, measure the length of the ribbon, and compare values.

58. The first Exploration Bit of this section suggests a method to *rotate* polar graphs.

a) Determine a formula that you believe will rotate the graph of $r = f(\theta)$ through an angle of α radians about the origin.

b) Support your conjecture in part (a) with $r = 1 - \cos\theta$ and $\alpha = \pi/6, \pi/2$, and $2\pi/3$.

10.7 —————————— Polar Equations of Conic Sections

Polar coordinates are important in astronomy and astronautical engineering because the ellipses, parabolas, and hyperbolas along which satellites, moons, planets, and comets move can all be described with one general polar equation. In Cartesian coordinates, the equations of conics have different forms, but not so here. This section develops the general equation, along with special equations for lines and circles.

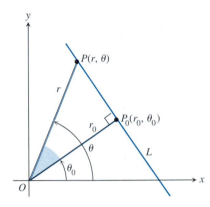

10.50 We can obtain a polar equation for line L by reading the relation $r_0/r = \cos(\theta - \theta_0)$ from triangle OP_0P.

Lines

Suppose the perpendicular from the origin to line L meets L at the point $P_0(r_0, \theta_0)$, with $r_0 \geq 0$ (Fig. 10.50). Then, if $P(r, \theta)$ is any other point on L, the points P, P_0, and O are the vertices of a right triangle, from which we can read the relation

$$\frac{r_0}{r} = \cos(\theta - \theta_0)$$

or

$$r\cos(\theta - \theta_0) = r_0.$$

The coordinates of $P_0(r_0, \theta_0)$ satisfy this equation as well.

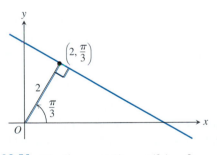

10.51 The line $r\cos(\theta - \pi/3) = 2$.

The Standard Polar Equation for Lines

If the point $P_0(r_0, \theta_0)$ is the foot of the perpendicular from the origin to the line L, and $r_0 \geq 0$, then an equation for L is

$$r\cos(\theta - \theta_0) = r_0. \tag{1}$$

EXAMPLE 1

The standard polar equation for the line shown in Fig. 10.51 is

$$r\cos\left(\theta - \frac{\pi}{3}\right) = 2.$$

≡

EXAMPLE 2

Use the identity $\cos(A - B) = \cos A \cos B + \sin A \sin B$ to find a Cartesian equation for the line in Example 1.

Solution

$$r \cos \left(\theta - \frac{\pi}{3}\right) = 2$$

$$r \left(\cos\theta \cos\frac{\pi}{3} + \sin\theta \sin\frac{\pi}{3}\right) = 2$$

$$\frac{1}{2} r \cos\theta + \frac{\sqrt{3}}{2} r \sin\theta = 2$$

$$\frac{1}{2}x + \frac{\sqrt{3}}{2}y = 2$$

$$x + \sqrt{3}y = 4$$

EXPLORATION 1

A Line in Polar Form

1. GRAPH the line

$$L_1 : \qquad 2y - 6x = 18$$

in a square viewing window. Find analytically the line L_2 through the origin that is perpendicular to L_1.

2. Find analytically θ_0 and r_0 in Eq. (1). Note that θ_0 is between $\pi/2$ and π.

3. In polar or parametric mode, GRAPH the line

$$L_3 : \qquad r \cos(\theta - \theta_0) = r_0$$

for the values of θ_0 and r_0 from part 2. Compare the graphs of L_1 and L_3.

4. Explain how you could have done parts 1–3 graphically. (*Hint:* Use the DRAW features of your grapher to construct L_2 and then estimate the coordinates of its point of intersection with L_1.)

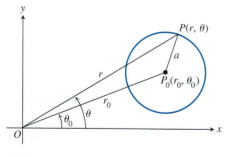

10.52 We can get an equation for this circle by applying the Law of Cosines to triangle $O P_0 P$.

Circles

To find a polar equation for the circle of radius a centered at $P_0(r_0, \theta_0)$, we let $P(r, \theta)$ be a point on the circle and apply the Law of Cosines to triangle $O P_0 P$ (Fig. 10.52). This gives

$$a^2 = r_0{}^2 + r^2 - 2r_0 r \cos(\theta - \theta_0). \qquad (2)$$

If the circle passes through the origin, then $r_0 = a$, and Eq. (2) simplifies somewhat:

$$a^2 = a^2 + r^2 - 2ar \cos(\theta - \theta_0) \qquad \text{Eq. (2) with } r_0 = a$$

$$r^2 = 2ar \cos(\theta - \theta_0)$$

$$r = 2a \cos(\theta - \theta_0). \qquad (3)$$

If the circle's center lies on the positive x-axis, so $\theta_0 = 0$, Eq. (3) becomes

$$r = 2a \cos\theta. \qquad (4)$$

If, instead, the circle's center lies on the positive y-axis, $\theta_0 = \pi/2$, $\cos(\theta - \pi/2) = \sin\theta$, and Eq. (3) becomes

$$r = 2a\sin\theta. \tag{5}$$

Equations for circles through the origin centered on the negative x- and y-axes can be obtained from Eqs. (4) and (5) by replacing r with $-r$.

Polar Equations for Circles through the Origin Centered on the x- and y-axes, Radius $a > 0$

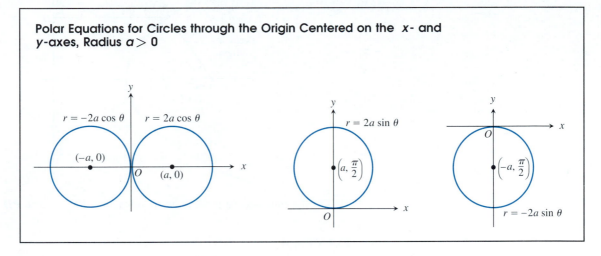

EXAMPLE 3 Circles through the Origin

SUPPORT

Support Example 3 graphically. Remember that each center is given in polar coordinates.

Radius	Center (polar coordinates)	Equation
3	$(3, 0)$	$r = 6\cos\theta$
2	$(2, \pi/2)$	$r = 4\sin\theta$
1/2	$(-1/2, 0)$	$r = -\cos\theta$
1	$(-1, \pi/2)$	$r = -2\sin\theta$

Ellipses, Parabolas, and Hyperbolas

To find polar equations for ellipses, parabolas, and hyperbolas, we first assume that the conic has one focus at the origin (for the parabola, its only focus) and that the corresponding directrix is the vertical line $x = k$ lying to the right of the origin (Fig. 10.53). This makes

$$PF = r$$

and

$$PD = k - FB = k - r\cos\theta.$$

The conic's focus–directrix equation $PF = e \cdot PD$ then becomes

$$r = e(k - r\cos\theta),$$

which can be solved for r to obtain

$$r = \frac{ke}{1 + e\cos\theta}. \tag{6}$$

This equation represents an ellipse if $0 < e < 1$, a parabola if $e = 1$, and a hyperbola if $e > 1$. And there we have it—ellipses, parabolas, and hyperbolas all with the same basic equation!

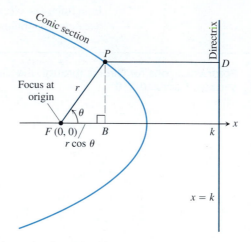

10.53 If a conic section is put in this position, then $PF = r$, and $PD = k - r \cos \theta$

EXAMPLE 4

Typical conics from Eq. (6) are shown in Fig. 10.54.

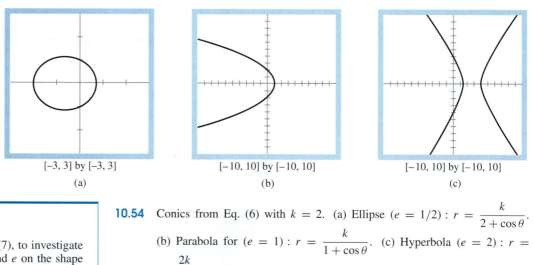

[–3, 3] by [–3, 3]

(a)

[–10, 10] by [–10, 10]

(b)

[–10, 10] by [–10, 10]

(c)

EXPLORATION BITS

1. Use Eqs. (6) and (7), to investigate the effects of k and e on the shape of each type of conic section. Summarize your conclusions.

2. When you graph Eq. (6) or (7), explain the sequence in which the graph appears for each of $e < 1, e = 1, e > 1$.

10.54 Conics from Eq. (6) with $k = 2$. (a) Ellipse $(e = 1/2)$: $r = \dfrac{k}{2 + \cos \theta}$. (b) Parabola for $(e = 1)$: $r = \dfrac{k}{1 + \cos \theta}$. (c) Hyperbola $(e = 2)$: $r = \dfrac{2k}{1 + 2 \cos \theta}$.

≡

You may see some variations of Eq. (6) from time to time, depending on the location of the directrix. If the directrix is the line $x = -k$ to the left of the origin (the origin is still the focus), the equation that we get in place of Eq. (6) is

$$r = \frac{ke}{1 - e \cos \theta}. \tag{7}$$

EXPLORATION BIT

Exploration 2 of Section 10.2 suggested organizing the quadratic curves on the basis of the equation

$$Ax^2 + Bxy + Cy^2 + Dx + Ey + F = 0.$$

You may wish to extend your presentation to the polar form, including degenerate cases.

The denominator now has a $-$ instead of a $+$. If the directrix is either of the lines $y = k$ or $y = -k$, the equations that we get have sines in them instead of cosines, as shown in Table 10.3.

EXAMPLE 5

Find an equation for the hyperbola with eccentricity 3/2 and directrix $x = 2$.

Solution We use Eq. (6) with $k = 2$ and $e = 3/2$ to get

$$r = \frac{2(3/2)}{1 + (3/2)\cos\theta} \quad \text{or} \quad r = \frac{6}{2 + 3\cos\theta} \qquad \equiv$$

EXAMPLE 6

Predict the shape of the graph of $r = 25/(10 + 10\cos\theta)$. Identify the directrix and eccentricity of the conic section. Then draw the graph.

Solution We divide the numerator and denominator by 10 to put the equation into a form shown in Table 10.3:

$$r = \frac{5/2}{1 + \cos\theta}.$$

TABLE 10.3   Equations for conic sections

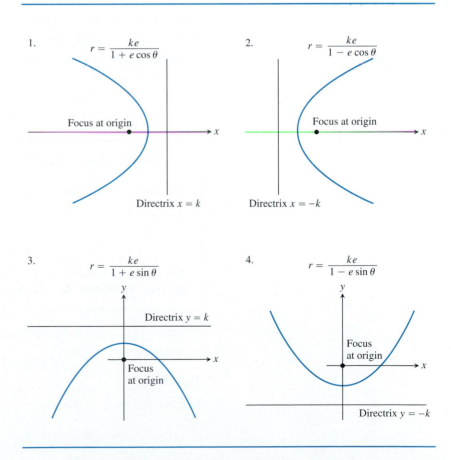

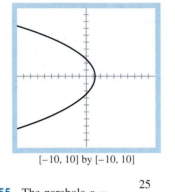

[−10, 10] by [−10, 10]

10.55 The parabola $r = \dfrac{25}{10 + 10\cos\theta}$.

This is the equation

$$r = \frac{ke}{1 + e\cos\theta} \qquad \text{Table 10.3, part 1}$$

with $k = 5/2$ and $e = 1$. The graph is a parabola ($e = 1$), and the directrix ($x = k$) is $x = 5/2$. See Fig. 10.55. ≡

Rotations

A conic is in *standard polar form* if its equation has a form shown in Table 10.3. One of the foci is the origin, and an axis of the conic is coincident with a coordinate axis.

One of the great advantages of conics in polar form is that the equation for a rotated conic requires nothing more than "tweaking" the standard form.

EXPLORATION 2

Rotating a Conic, Part 2

On the home screen, store 5 in K, 0.8 in E, and 0 in A. Using polar (or parametric) mode, enter

$$r = \frac{KE}{1 - E\cos(\theta - A)}.$$

1. Use Table 10.3 to predict what the graph will look like, then GRAPH.

2. Store 0.2 in A, then GRAPH. Repeat for $A = 0.4, 0.6, 1, 1.5, 2,$ and 2.5.

3. Predict what you will see in the viewing window for $A = \pi/4, \pi/3, \pi/2, 2\pi/3, 3\pi/4, \pi, 3\pi/2,$ and 2π. Test your predictions. ☙

Exploration 2 suggests the formula for rotating a conic.

The Rotation of a Conic Through an Angle α

The conic in polar form $r = \dfrac{ke}{1 \pm e\cos\theta}$ rotated through angle α has equation

$$r = \frac{ke}{1 \pm e\cos(\theta - \alpha)}.$$

The center of rotation is one of the foci. A similar result holds with the sine function in place of the cosine function.

We can use this method to graph a conic *anywhere* in the xy-plane, as the next example suggests.

EXAMPLE 7

Describe the graph of the conic

$$r = \frac{4(0.6)}{1 - 0.6\sin\theta}. \tag{8}$$

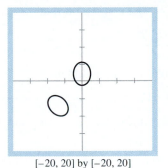

[−20, 20] by [−20, 20]

10.56 The two conics of Example 7. The one with focus at the origin is rotated through an angle $\pi/3$, then shifted left 5 units and down 10 units to match the other.

Find an analytic representation of the conic formed by rotating this conic $\pi/3$ about its focus $(0, 0)$ and then shifting the conic so that the corresponding focus is $(-5, -10)$.

Solution Equation (8) matches the form in part 4 of Table 10.3. Its graph is an ellipse (because $e = 0.6 < 1$) with focus $(0, 0)$, directrix $y = -4$ and major axis coincident with the y-axis. The parametric equations

$$x_1(t) = \frac{2.4}{1 - 0.6 \sin t} \cos t, \qquad y_1(t) = \frac{2.4}{1 - 0.6 \sin t} \sin t$$

generate its graph. The parametric equations

$$x_2(t) = \frac{2.4}{1 - 0.6 \sin(t - \pi/3)} \cos t, \qquad y_2(t) = \frac{2.4}{1 - 0.6 \sin(t - \pi/3)} \sin t$$

rotate the graph of (x_1, y_1) counterclockwise by $\pi/3$ radians. The parametric equations

$$x_3(t) = x_2(t) - 5, \qquad y_3(t) = y_2(t) - 10$$

shift the graph of (x_2, y_2) left 5 units and down 10 units to one with focus $(-5, -10)$. The graphs of (x_1, y_1) and (x_3, y_3) are shown in Fig. 10.56. ≣

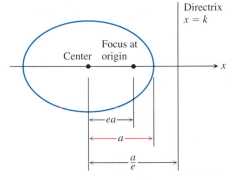

10.57 In an ellipse with semimajor axis a, the focus–directrix distance is $k = a/e - ea$, so $ke = a(1 - e^2)$.

From the ellipse diagram in Fig. 10.57, we see that k is related to the eccentricity e and the semimajor axis a by the equation

$$k = \frac{a}{e} - ea.$$

From this, we find that $ke = a(1 - e^2)$. Replacing ke by $a(1 - e^2)$ in Eq. (6) gives the standard polar equation for an ellipse with eccentricity e and semimajor axis a.

Ellipse with Eccentricity e and Semimajor Axis a (one focus at the origin)

$$r = \frac{a(1 - e^2)}{1 + e \cos \theta} \qquad (0 < e < 1) \qquad (9)$$

EXPLORATION BIT

Notice that if $e = 0$, Eq. (9) becomes $r = a$, the polar equation for a circle. Equation (9) was derived from Eq. (6). Analyze Eq. (6) for $e = 0$. Explain your findings. What happens if $e = 1$ in Eq. (9)? Explain.

Equation (9) is the starting point for calculating planetary orbits in astronomy.

EXAMPLE 8

Find a polar equation for an ellipse with semimajor axis 39.44 AU (astronomical units) and eccentricity 0.25. This is the approximate size of Pluto's orbit around the sun. Draw its graph with 1 scale unit = 1 AU.

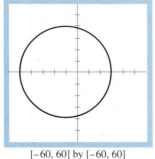

[−60, 60] by [−60, 60]

10.58 The graph of Pluto's orbit around the sun. One focus (origin) is the sun. Each scale mark represents 10 AUs. For comparison purposes, Earth is 1 AU from the sun.

Solution   We use Eq. (9) with $a = 39.44$ and $e = 0.25$ to find

$$r = \frac{39.44 \left(1 - (0.25)^2\right)}{1 + 0.25 \cos \theta}.$$

At its point of closest approach (perihelion), $\theta = 0$, and Pluto is

$$r = 29.58 \text{ AU}$$

from the sun. At its most distant point (aphelion), $\theta = \pi$, and Pluto is

$$r = 49.3 \text{ AU}$$

from the sun. A graph of Pluto's orbit is shown in Fig. 10.58.   ▤

Exercises 10.7

Find polar and Cartesian equations for the lines in Exercises 1 and 2.

1.

2.

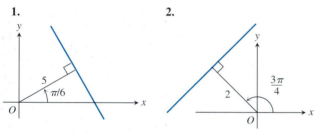

Sketch the lines in Exercises 3 and 4, and find Cartesian equations for them.

3. $r \cos \left(\theta - \dfrac{\pi}{4}\right) = \sqrt{2}$ **4.** $r \cos \left(\theta - \dfrac{2\pi}{3}\right) = 3$

Find polar equations for the circles in Exercises 5 and 6.

5.   **6.**

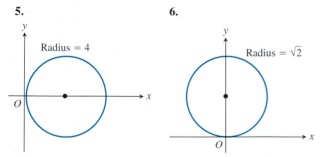

Radius = 4   Radius = $\sqrt{2}$

Sketch the circles in Exercises 7 and 8. Give polar coordinates for their centers, and identify their radii.

7. $r = 4 \cos \theta$ **8.** $r = 6 \sin \theta$

Exercises 9–16 give the eccentricities of conic sections with one focus at the origin, along with the directrix corresponding to that focus. Find a polar equation for each conic section.

9. $e = 1, x = 2$ **10.** $e = 1, y = 2$

11. $e = 2, x = 4$ **12.** $e = 5, y = -6$

13. $e = 1/2, x = 1$ **14.** $e = 1/4, x = -2$

15. $e = 1/5, y = -10$ **16.** $e = 1/3, y = 6$

Sketch the parabolas and ellipses in Exercises 17–24. Include the directrix that corresponds to the focus at the origin. Label the vertices with appropriate polar coordinates. Label the centers of the ellipses as well. Support with a grapher.

17. $r = \dfrac{1}{1 + \cos \theta}$ **18.** $r = \dfrac{6}{2 + \cos \theta}$

19. $r = \dfrac{25}{10 - 5 \cos \theta}$ **20.** $r = \dfrac{4}{2 - 2 \cos \theta}$

21. $r = \dfrac{400}{16 + 8 \sin \theta}$ **22.** $r = \dfrac{12}{3 + 3 \sin \theta}$

23. $r = \dfrac{8}{2 - 2 \sin \theta}$ **24.** $r = \dfrac{4}{2 - \sin \theta}$

25. In the same viewing window, GRAPH the conics

$$r = \frac{e}{1 + e \cos \theta}$$

for $e = 0.1, 0.2, 0.3, 0.4, 0.5, 0.6, 0.7, 0.8,$ and 0.9. Explain the effect of the parameter e.

26. In the same viewing window, GRAPH the conics

$$r = \frac{e}{1 + e \cos \theta}$$

for $e = 1.1, 1.2, 1.3, 1.4, 1.5, 2, 2.5, 3, 10$. Explain the effect of the parameter e.

27. In the same viewing window, GRAPH the conics

$$r = \frac{k}{1 + \cos\theta}$$

for $k = -8, -4, -3, -2, -1, 0, 1, 2, 3, 4, 8$. Explain the effect of the parameter k.

28. In the same viewing window, GRAPH the conics

$$r = \frac{k}{1 - \sin\theta}$$

for $k = -8, -4, -3, -2, -1, 0, 1, 2, 3, 4, 8$. Explain the effect of the parameter k.

GRAPH the lines and conic sections in Exercises 29–38.

29. $r = 3\sec(\theta - \pi/3)$ **30.** $r = 4\sin\theta$

31. $r = 8/(4 + \cos\theta)$ **32.** $r = 1/(1 - \sin\theta)$

33. $r = 1/(1 + 2\sin\theta)$ **34.** $r = 4\sec(\theta + \pi/6)$

35. $r = -2\cos\theta$ **36.** $r = 8/(4 + \sin\theta)$

37. $r = 1/(1 + \cos\theta)$ **38.** $r = 1/(1 + 2\cos\theta)$

Sketch the regions defined by the inequalities in Exercises 39 and 40.

39. $0 \le r \le 2\cos\theta$ **40.** $-3\cos\theta \le r \le 0$

41. *Perihelion and aphelion.* A planet travels about its sun in an ellipse whose semimajor axis has length a.

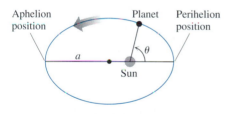

Aphelion position · Planet · Perihelion position · a · θ · Sun

a) Show that $r = a(1 - e)$ when the planet is closest to the sun (perihelion) and that $r = a(1 + e)$ when the planet is farthest from the sun (aphelion).

b) Use the data in Table 10.4 to find how close each planet in our solar system comes to the sun and how far away each planet gets from the sun.

42. *Planetary orbits.* In Example 8, we drew the orbit of Pluto.

a) Use the data in Table 10.4 to find polar equations for the orbits of the other planets. Draw the orbits in the same viewing window for Mercury through Mars. Then draw the orbits in the same viewing window for Jupiter through Pluto. Explain why it was necessary to use two viewing windows.

TABLE 10.4   Semimajor Axes and Eccentricities of the Planets in Our Solar System

Planet	Semimajor axis (astronomical units)	Eccentricity
Mercury	0.3871	0.2056
Venus	0.7233	0.0068
Earth	1.000	0.0167
Mars	1.524	0.0934
Jupiter	5.203	0.0484
Saturn	9.539	0.0543
Uranus	19.18	0.0460
Neptune	30.06	0.0082
Pluto	39.44	0.2481

b) Study the orbits of Neptune and Pluto together. The right side of the viewing window suggests an astronomical event that occurred on January 21, 1979, will occur next on March 14, 1999, and then won't happen again until September 2226. The behavior that better identifies the event is hidden. Make a conjecture about what the event is. Use ZOOM-IN to help you decide that the event is indeed possible.

43. a) Find Cartesian equations for the curves $r = 2\sin\theta$ and $r = \csc\theta$.

b) Sketch the curves together, and label their points of intersection in both Cartesian and polar coordinates.

44. Repeat Exercise 43 for $r = 2\cos\theta$ and $r = \sec\theta$.

45. Find a polar equation for the parabola whose focus lies at the origin and whose directrix is the line $r\cos\theta = 4$.

46. Find a polar equation for the parabola whose focus lies at the origin and whose directrix is the line $r\cos(\theta - \pi/2) = 2$.

47. Generalize the procedure in Exploration 1 of this section to find a polar equation for the line $ax + by = c$.

In Exercises 48–51, find a polar equation of the line using the procedure in Exploration 1 and Exercise 47. Support with a grapher.

48. $2x - y = -5$ **49.** $3x + 2y = 6$

50. $2x - y = 4$ **51.** $4x + 3y = -12$

52. Draw a diagram similar to Fig. 10.57 for a hyperbola with focus at the origin and semimajor axis a. Then show that Eq. (9) also gives the graph of a hyperbola with eccentricity e, semimajor axis a, and one focus at the origin.

53. Derive the four equations in Table 10.3.

10.8 _____ Integration in Polar Coordinates

This section shows how to calculate areas of plane regions, lengths of curves, and areas of surfaces of revolution in polar coordinates. The general methods for setting up the integrals are the same as for Cartesian coordinates, although the resulting formulas are somewhat different. For anyone who is interested in finding centroids, we have included the formulas for doing so at the end of the exercise set.

Area in the Plane

The region AOB in Fig. 10.59 is bounded by the rays $\theta = \alpha, \theta = \beta$ and the curve $r = f(\theta)$. We divide angle AOB into n parts and approximate a typical sector POQ by a *circular* sector of radius r and central angle $\Delta\theta$ (Fig. 10.60). Then

$$\text{Area of } POQ \approx \frac{1}{2}r^2 \,\Delta\theta,$$

and

$$A = \text{ Area } AOB \approx \sum_{\theta=\alpha}^{\beta} \frac{1}{2}r^2 \,\Delta\theta.$$

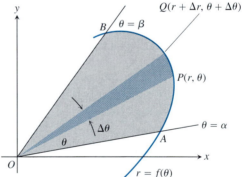
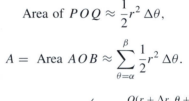
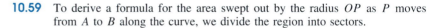

10.59   To derive a formula for the area swept out by the radius OP as P moves from A to B along the curve, we divide the region into sectors.

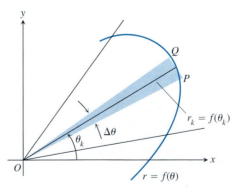
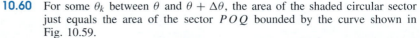

10.60   For some θ_k between θ and $\theta + \Delta\theta$, the area of the shaded circular sector just equals the area of the sector POQ bounded by the curve shown in Fig. 10.59.

If the function $r = f(\theta)$ is continuous for $\alpha \leq \theta \leq \beta$, then there is a θ_k between θ and $\theta + \Delta\theta$ such that the circular sector of radius

$$r_k = f(\theta_k)$$

and central angle $\Delta\theta$ gives the *exact* area of POQ (Fig. 10.60). Then the entire area is given exactly by

$$A = \sum \frac{1}{2} r_k{}^2 \, \Delta\theta = \sum \frac{1}{2} (f(\theta_k))^2 \Delta\theta.$$

If we let $\Delta\theta \to 0$, we see that

$$A = \lim_{\Delta\theta \to 0} \sum \frac{1}{2} (f(\theta_k))^2 \, \Delta\theta = \int_\alpha^\beta \frac{1}{2} f^2(\theta) \, d\theta = \int_\alpha^\beta \frac{1}{2} r^2 \, d\theta.$$

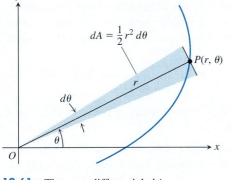

10.61 The area differential dA.

Area Between the Origin and $r = f(\theta)$ $(\alpha \leq \theta \leq \beta)$

$$A = \int_\alpha^\beta \frac{1}{2} r^2 \, d\theta \qquad (1)$$

This is the integral of the **area differential** (Fig. 10.61)

$$dA = \frac{1}{2} r^2 \, d\theta.$$

EXAMPLE 1

Find the area of the region enclosed by the cardioid $r = 2(1 + \cos\theta)$. Support with a NINT computation.

Solution We graph the cardioid (Fig. 10.62) and determine that the radius OP sweeps out the region exactly once as θ runs from 0 to 2π. The area is therefore

$$\int_{\theta=0}^{\theta=2\pi} \frac{1}{2} r^2 \, d\theta = \int_0^{2\pi} \frac{1}{2} \cdot 4(1 + \cos\theta)^2 \, d\theta$$

$$= \int_0^{2\pi} 2(1 + 2\cos\theta + \cos^2\theta) \, d\theta$$

$$= \int_0^{2\pi} \left(2 + 4\cos\theta + 2\frac{1 + \cos 2\theta}{2} \right) \, d\theta$$

$$= \int_0^{2\pi} (3 + 4\cos\theta + \cos 2\theta) \, d\theta$$

$$= 3\theta + 4\sin\theta + \frac{\sin 2\theta}{2} \Big|_0^{2\pi}$$

$$= 6\pi - 0 = 6\pi.$$

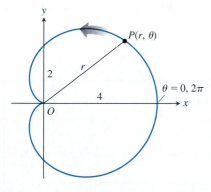

10.62 The cardioid $r = 2(1 + \cos\theta)$.

Support: NINT $(2(1 + \cos x)^2, 0, 2\pi) = 18.850$, which is 6π to three decimal places.

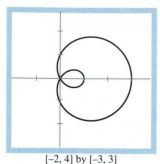

[−2, 4] by [−3, 3]

10.63 The limaçon

$$r = 2\cos\theta + 1 \quad (0 \le \theta \le 2\pi).$$

Once we know the θ-interval for the inner loop (Exploration 1), we can show that the area inside it is $\pi - (3\sqrt{3}/2)$ analytically and 0.544 using NINT. How do these two values compare?

EXPLORATION BIT

Find the area inside one loop of the GNA Exploration logo used in this book. Use the equation $r = 3\cos 3\theta$. First, find and justify the limits of integration. Compute the area using NINT. Then confirm the area analytically. (Also, explain the orientation of your graph compared to that of the logo.)

EXPLORATION 1

Finding Area

Figure 10.63 shows a complete graph of the limaçon

$$r = 2\cos\theta + 1$$

that appears for $0 \le \theta \le 2\pi$.

1. Watching the graph appear, it seems that the "top" curve, the entire inner loop, and the "bottom" curve each take about the same time to complete. If that is true, what would be a reasonable guess for a θ-interval that would draw the inner loop. Support your guess graphically, and confirm analytically.

2. Use Eq. (1) and NINT to find the area inside the inner loop. What are your limits of integration?

3. Confirm part 2 analytically. Use $\cos^2\theta = (1 + \cos 2\theta)/2$ to eliminate the $\cos^2\theta$ term from your integrand.

4. The graph suggests symmetry about the x-axis. Confirm this. What does symmetry tell you about the regions above and below the x-axis inside the inner loop? How does this suggest that you could change the integration used in finding the area? Show that this change works using NINT. Confirm analytically that this new integration gives the same area for the inner loop.   ⚭

To find the area of a region like the one in Fig. 10.64, which lies between two polar curves from $\theta = \alpha$ to $\theta = \beta$, we subtract the integral of $(1/2)r_1^2\,d\theta$ from the integral of $(1/2)r_2^2\,d\theta$. This leads to the following formula.

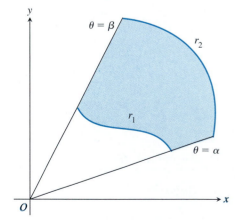

10.64 The area of the shaded region is calculated by subtracting the area of the region between r_1 and the origin from the area of the region between r_2 and the origin.

Area of the Region $r_1(\theta) \le r \le r_2(\theta)$ $(\alpha \le \theta \le \beta)$

$$A = \int_\alpha^\beta \frac{1}{2}r_2^2\,d\theta - \int_\alpha^\beta \frac{1}{2}r_1^2\,d\theta = \int_\alpha^\beta \frac{1}{2}(r_2^2 - r_1^2)\,d\theta \qquad (2)$$

EXAMPLE 2

Find the area of the region that lies inside the circle $r = 1$ and outside the cardioid $r = 1 - \cos\theta$.

Solution   We graph the region to determine its boundaries and find the limits of integration (Fig. 10.65). The outer curve is $r_2 = 1$, the inner curve is $r_1 = 1 - \cos\theta$, and θ runs from $-\pi/2$ to $\pi/2$. The area, from

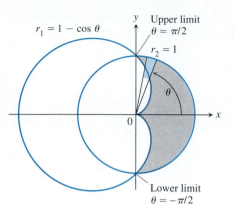

$r_1 = 1 - \cos\theta$

Upper limit $\theta = \pi/2$

$r_2 = 1$

θ

Lower limit $\theta = -\pi/2$

10.65 The region and limits of integration in Example 2.

Eq. (2), is

$$A = \int_{-\pi/2}^{\pi/2} \frac{1}{2}(r_2^2 - r_1^2)\, d\theta$$

$$= 2\int_0^{\pi/2} \frac{1}{2}(r_2^2 - r_1^2)\, d\theta \qquad \text{Symmetry}$$

$$= \int_0^{\pi/2} (1 - (1 - 2\cos\theta + \cos^2\theta))\, d\theta$$

$$= \int_0^{\pi/2} (2\cos\theta - \cos^2\theta)\, d\theta = \int_0^{\pi/2}\left(2\cos\theta - \frac{\cos 2\theta + 1}{2}\right) d\theta$$

$$= \left[2\sin\theta - \frac{\sin 2\theta}{4} - \frac{\theta}{2}\right]_0^{\pi/2} = 2 - \frac{\pi}{4}.$$

EXPLORATION BIT

For Example 2, you can show just the region on your grapher by graphing r_1 and r_2 from $-\pi/2$ to $\pi/2$. Is the area of the region the same as the area inside the graph of $r_2 - r_1$ from $-\pi/2$ to $\pi/2$? Your grapher can suggest the answer. You can confirm it.

POLAR COORDINATE INTEGRALS

For arc length, we can convert an integral in Cartesian coordinates to one in polar coordinates by substitution of variables. This is not the case for the area integral in Eq. (1) because we do not have a corresponding area integral in Cartesian coordinates.

The Length of a Curve

We calculate the length of a curve $r = f(\theta), \alpha \le \theta \le \beta$ by expressing the differential $ds = \sqrt{dx^2 + dy^2}$ in terms of θ and integrating from α to β. To express ds in terms of θ, we first write dx and dy as

$$dx = d(r\cos\theta) = -r\sin\theta\, d\theta + \cos\theta dr,$$

$$dy = d(r\sin\theta) = r\cos\theta\, d\theta + \sin\theta dr.$$

We then square and add (algebra omitted) to obtain

$$ds = \sqrt{dx^2 + dy^2} = \sqrt{r^2 d\theta^2 + dr^2}. \qquad (3)$$

Think of ds as the hypotenuse of a right triangle whose sides are $r\, d\theta$ and dr (Fig. 10.66). For the purpose of evaluation, Eq. (3) is usually written with $d\theta$ factored out:

$$ds = \sqrt{r^2 + \left(\frac{dr}{d\theta}\right)^2}\, d\theta.$$

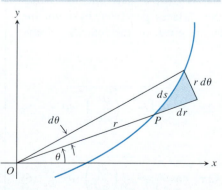

10.66 For arc length, $ds^2 = r^2 d\theta^2 + dr^2$.

Length of a Curve

If $r = f(\theta)$ has a continuous first derivative for $\alpha \le \theta \le \beta$ and if the point $P(r, \theta)$ traces the curve $r = f(\theta)$ exactly once as θ runs from α to β, then the length of the curve is given by the formula

$$\text{Length} = \int_\alpha^\beta \sqrt{r^2 + \left(\frac{dr}{d\theta}\right)^2}\, d\theta. \qquad (4)$$

EXAMPLE 3

Find the length of the curve $r = 2\theta + \sqrt{\theta} - 2\sin\theta + 1$ for $0.1 \le \theta \le 5$.

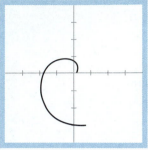

[−20, 20] by [−20, 20]

10.67   The curve

$$r = 2\theta + \sqrt{\theta} - 2\sin\theta + 1$$

for $0.1 \le \theta \le 5$. The graph suggests that there are no problems with differentiability (away from zero) and the curve is traced exactly once, so Eq. (4) can be applied. Is the NINT value that we obtain in Example 3 reasonable? Give a convincing argument.

Solution   We GRAPH the curve and see that the necessary conditions to apply Eq. (4) appear to be met (Fig. 10.67). Therefore,

$$\text{Length} = \int_{0.1}^{5} \sqrt{r^2 + \left(\frac{dr}{d\theta}\right)^2}\, d\theta$$

$$= \int_{0.1}^{5} \sqrt{(2\theta + \sqrt{\theta} - 2\sin\theta + 1)^2 + \left(2 + \frac{1}{2\sqrt{\theta}} - 2\cos\theta\right)^2}\, d\theta.$$

A NINT computation yields 39.009.   ≡

EXPLORATION 2

Finding Curve Length

Follow these steps to find the length of the cardioid $r = 1 - \cos\theta$.

1. GRAPH the cardioid to find the limits of integration.
2. Show that the integrand is $\sqrt{2 - 2\cos\theta}$.
3. Compute the length using NINT.
4. Confirm the length analytically. You may find the identity
   $1 - \cos\theta = 2\sin^2(\theta/2)$ helpful.

The Area of a Surface of Revolution

The formula for the area of a surface of revolution is $S = \int 2\pi\rho\, ds$, just as in rectangular coordinates, but now we express the radius function ρ and the arc length differential ds in terms of r and θ.

Area of a Surface of Revolution

If $r = f(\theta)$ has a continuous first derivative for $\alpha \le \theta \le \beta$ and if the point $P(r, \theta)$ traces the curve $r = f(\theta)$ exactly once as θ runs from α to β, then the areas of the surfaces generated by revolving the curve about the x- and y-axes are given by the following formulas:

1. Revolution about the x-axis:

$$\text{Area} = \int_{\alpha}^{\beta} 2\pi y\, ds = \int_{\alpha}^{\beta} 2\pi r \sin\theta \sqrt{r^2 + \left(\frac{dr}{d\theta}\right)^2}\, d\theta \qquad (5)$$

2. Revolution about the y-axis:

$$\text{Area} = \int_{\alpha}^{\beta} 2\pi x\, ds = \int_{\alpha}^{\beta} 2\pi r \cos\theta \sqrt{r^2 + \left(\frac{dr}{d\theta}\right)^2}\, d\theta \qquad (6)$$

EXPLORATION 3

Finding Surface Area

Follow these steps to find the area of the surface (Fig. 10.68b) generated by revolving the right-hand loop (Fig. 10.68a) of the lemniscate $r^2 = \cos 2\theta$ about the y-axis.

1. GRAPH the lemniscate to find the limits of integration. Give a convincing argument that your limits are correct.

2. Show that the integrand in Eq. (6) is $2\pi \cos \theta$ by showing that

$$r\sqrt{r^2 + \left(\frac{dr}{d\theta}\right)^2} = 1.$$

3. Compute the surface area using NINT.

4. Confirm the surface area analytically.

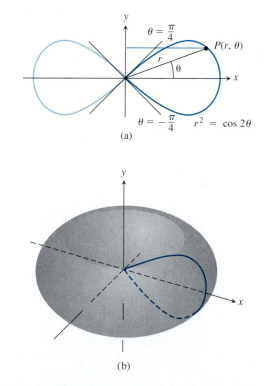

10.68   (a) The right-hand half of a lemniscate is revolved about the y-axis to generate (b) a surface whose area can be found in Exploration 3.

Exercises 10.8

Draw graphs for Exercises 1–18, then find the areas of the regions. Support each with a NINT computation.

1. Inside the circle $r = \cos \theta$ between the rays $\theta = 0$ and $\theta = \pi/4$

2. Enclosed by the rays $\theta = 0$ and $\theta = \ln 25$ and the curve $r = e^\theta, 0 \le \theta \le \ln 25$

3. Inside the convex limaçon $r = 4 + 2\cos\theta$

4. Inside the cardioid $r = a(1 + \cos\theta)$

5. Inside one leaf of the four-leaved rose $r = \cos 2\theta$

6. Inside the circle $r = 2a\sin\theta$

7. Inside the lemniscate $r^2 = 2a^2\cos 2\theta$

8. Inside one loop of the lemniscate $r^2 = 4\sin 2\theta$

9. Inside the six-leaved rose $r^2 = 2\sin 3\theta$

10. Shared by the circles $r = 2\cos\theta$ and $r = 2\sin\theta$

11. Shared by the circles $r = 1$ and $r = 2\sin\theta$

12. Shared by the circle $r = 2$ and the cardioid $r = 2(1 - \cos\theta)$

13. Shared by the cardioids $r = 2(1 + \cos\theta)$ and $r = 2(1 - \cos\theta)$

14. Inside the lemniscate $r^2 = 6\cos 2\theta$ and outside the circle $r = \sqrt{3}$

15. Inside the circle $r = 3a\cos\theta$ and outside the cardioid $r = a(1 + \cos\theta)$

16. Inside the circle $r = -2\cos\theta$ and outside the circle $r = 1$

17. a) Inside the outer loop (and including the inner loop) of the limaçon $r = 2\cos\theta + 1$ in Exploration 1
    b) Inside the outer loop and outside the inner loop

18. Inside the circle $r = 6$ above the line $r\sin\theta = 3$

Draw the curves in Exercises 19–23, and find the lengths of the curves. Support with a NINT computation.

19. The spiral $r = \theta^2$, $0 \le \theta \le \sqrt{5}$

20. The curve $r = e^\theta/\sqrt{2}$, $0 \le \theta \le \pi$

21. The curve $r = \sec\theta$, $0 \le \theta \le \pi/4$

22. The curve $r = \csc\theta$, $\pi/6 \le \theta \le \pi/2$

23. The cardioid $r = 1 + \cos\theta$

24. As usual, when faced with a new formula, it is a good idea to try it out on familiar objects to be sure that it gives results consistent with past experience. Use the length formula in Eq. (4) to calculate the circumferences of the following circles. Compare with the exact values.
    a) $r = a$
    b) $r = a\cos\theta$
    c) $r = a\sin\theta$

Find the areas of the surfaces generated by revolving the curves in Exercises 25–28 about the indicated axes. Support each with a NINT computation.

25. $r = \sqrt{\cos 2\theta}$, $0 \le \theta \le \pi/4$, y-axis

26. $r = \sqrt{2}e^{\theta/2}$, $0 \le \theta \le \pi/2$, x-axis

27. $r^2 = \cos 2\theta$, x-axis

28. $r = 2a\cos\theta$, y-axis

In Exercises 29 and 30, let

$$y_1(x) = \sqrt{\left(2x + \sqrt{x} - 2\sin x + 1\right)^2 + \left(2 + \frac{1}{2\sqrt{x}} - 2\cos x\right)^2},$$

the integrand of the arc length integral of Example 3.

29. a) Draw the graph of y_1 in $[0, 10]$ by $[0, 20]$ and $[0, 0.001]$ by $[0, 200]$ viewing windows.
    b) Describe the behavior of y_1 as $x \to 0^+$.
    c) Is $\int_0^5 y_1(x)\,dx$ an improper integral? Explain.

30. a) Draw the graph of $f(x) = \text{NINT}\,(y_1(t), x, 5)$ for $0 \le x \le 5$. Investigate the behavior of f as $x \to 0^+$.
    b) Describe the behavior of f in the interval $0 \le x \le 5$.
    c) On the basis of the graphical evidence in parts (a) and (b), make a conjecture about $\lim\limits_{x \to 0^+} \int_x^5 y_1(t)\,dt$. Give a value for the limit if you believe that it exists.

Centroids

Since the centroid of a triangle is located on each median, two-thirds of the way from the vertex to the opposite base, the lever arm for the moment about the x-axis of the thin triangular region in Fig. 10.69 is about $(2/3)r\sin\theta$. Similarly, the lever arm for

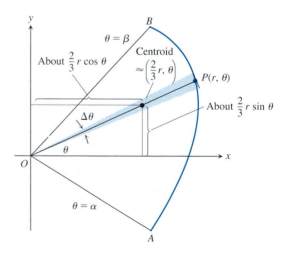

10.69 The moment of the thin triangular sector about the x-axis is approximately

$$\frac{2}{3}r\sin\theta\,dA = \frac{2}{3}r\sin\theta \cdot \frac{1}{2}r^2\,d\theta = \frac{1}{3}r^3\sin\theta\,d\theta.$$

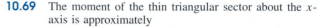

the moment of the triangular region about the y-axis is about $(2/3)r\cos\theta$. These approximations improve as $\Delta\theta \to 0$ and lead to the following formulas for the coordinates of the centroid of region AOB:

$$\bar{x} = \frac{\int \frac{2}{3}r\cos\theta \cdot \frac{1}{2}r^2\,d\theta}{\int \frac{1}{2}r^2\,d\theta} = \frac{\frac{2}{3}\int r^3\cos\theta\,d\theta}{\int r^2\,d\theta},$$

$$\bar{y} = \frac{\int \frac{2}{3}r\sin\theta \cdot \frac{1}{2}r^2\,d\theta}{\int \frac{1}{2}r^2\,d\theta} = \frac{\frac{2}{3}\int r^3\sin\theta\,d\theta}{\int r^2\,d\theta},$$

with limits $\theta = \alpha$ to $\theta = \beta$ on all integrals.

31. Find the centroid of the region enclosed by the cardioid $r = a(1 + \cos\theta)$.

32. Find the centroid of the semicircular region $0 \le r \le a$, $0 \le \theta \le \pi$.

Review Questions

1. Name the conic sections. Where does the term *conic* come from?

2. What reflective properties do parabolas, ellipses, and hyperbolas have? What applications use these properties?

3. How are circles defined in terms of distance? Give typical equations for circles. Graph one of the equations, and include the circle's center.

4. How are parabolas defined in terms of distance? Give typical equations of parabolas. Graph one of the equations, and include the parabola's vertex, focus, and directrix.

5. How are ellipses defined in terms of distance? Give typical equations for ellipses. Graph one of the equations, and include the ellipse's vertices and foci. How is the eccentricity of an ellipse defined? Sketch an ellipse whose eccentricity is close to 1 and another whose eccentricity is close to 0.

6. How are hyperbolas defined in terms of distance? Give typical equations for hyperbolas. Graph one of the equations, and include the hyperbola's vertices, foci, axes, and asymptotes. What values can the eccentricities of hyperbolas have?

7. Explain the equation $PF = e \cdot PD$.

8. What can be said about the graph of the equation

$$Ax^2 + Bxy + Cy^2 + Dx + Ey + F = 0$$

if A, B, and C are not all zero? What can you tell from the number $B^2 - 4AC$?

9. Explain the two-function method for graphing a conic $Ax^2 + Bxy + Cy^2 + Dx + Ey + F = 0$. Give examples. Does the method always work?

10. Give parametric equations for a circle, a parabola, and an ellipse. In each case, give a parameter domain that covers the conic exactly once.

11. Give parametric equations for a hyperbola. What is the appropriate domain for the parameter if the curve is to be traced out exactly once by your equations?

12. What is a cycloid? What are typical parametric equations for a cycloid?

13. How do you find dy/dx and d^2y/dx^2 when a curve in the xy-plane is given by parametric equations? Give examples.

14. How do you find the length of a parametric curve? Give an example.

15. How do you find the area of a surface of revolution generated by a parametric curve? Give an example.

16. Make a diagram to show the standard relations between Cartesian coordinates (x, y) and polar coordinates (r, θ). Express each set of coordinates in terms of the other kind.

17. If a point has polar coordinates (r_0, θ_0), what other polar coordinates does the point have?

18. How do you test the graph of the equation $F(r, \theta) = 0$ for symmetry about the origin? About the x-axis? About the y-axis? Give examples.

19. Describe how to rotate a curve given in parametric form through an angle θ with respect to the positive x-axis. Give examples.

20. What are the standard polar equations for lines, circles, ellipses, parabolas, and hyperbolas? Give examples.

21. How do you find the areas of plane regions in polar coordinates? Give examples.

22. How do you find the length of a curve in polar coordinates? Give an example.

23. How do you find the area of a surface of revolution in polar coordinates? Give an example.

Practice Exercises

Sketch the parabolas in Exercises 1 and 2. Include the parabola's focus and directrix in each sketch. Support with a grapher.

1. $x = y^2/8$ **2.** $y = -x^2/4$

Sketch the ellipses and hyperbolas in Exercises 3–6. Include the vertices, foci, and directrices in each sketch. Support with a grapher.

3. $16x^2 + 7y^2 = 112$ **4.** $x^2 + 2y^2 = 4$

5. $3x^2 - y^2 = 3$ **6.** $2y^2 - 8x^2 = 16$

Use the discriminant to decide whether the equations in Exercises 7–10 represent parabolas, ellipses, or hyperbolas. Draw a complete graph of each conic.

7. $x^2 + xy + y^2 + x + y + 1 = 0$

8. $x^2 + 3xy + 2y^2 + x + y + 1 = 0$

9. $x^2 + 4xy + 4y^2 + x + y + 1 = 0$

10. $x^2 + 2xy - 2y^2 + x + y + 1 = 0$

Identify the conic sections in Exercises 11 and 12. Determine an angle of rotation that will produce the given conic from a conic in standard position. Graph the equivalent conic in standard position and the given conic.

11. $2x^2 + xy + 2y^2 - 15 = 0$

12. $x^2 + 2\sqrt{3}xy - y^2 - 4 = 0$

13. Find the eccentricity of the hyperbola $xy = 2$. (*Hint:* What is an equivalent conic in standard form?)

14. Find an equation for the hyperbola with eccentricity 2 and vertices $(2, 0)$ and $(-2, 0)$.

15. Find the volume of the solid generated by revolving the region enclosed by the ellipse $9x^2 + 4y^2 = 36$ about the
  a) x-axis   **b)** y-axis

16. The "triangular" region in the first quadrant bounded by the x-axis, the line $x = 4$, and the hyperbola $9x^2 - 4y^2 = 36$ is revolved about the x-axis to generate a solid. Find the volume of the solid.

17. Find the points on the parabola $x = 2t, y = t^2$ closest to the point $(0, 3)$.

18. *Archimedes's area formula for parabolic arches.* Show that the area of the shaded region in the diagram is $A = (2/3)bh$. Thus, the area under a parabolic arch is two-thirds the base times the height.

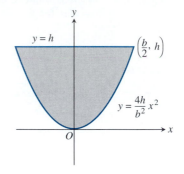

Exercises 19–24 give parametric equations for the motion of a particle in the xy-plane. Identify the particle's path by finding a Cartesian equation for it. Graph the Cartesian equation, and indicate the direction of motion and the portion traced by the particle. Support with a grapher.

19. $x = t/2, y = t + 1 \ (-\infty < t < \infty)$

20. $x = \sqrt{t}, y = 1 - \sqrt{t} \ (t \geq 0)$

21. $x = (1/2)\tan t, y = (1/2)\sec t \ (-\pi/2 < t < \pi/2)$

22. $x = -2\cos t, y = 2\sin t \ (0 \leq t \leq \pi)$

23. $x = -\cos t, y = \cos^2 t \ (0 \leq t \leq \pi)$

24. $x = 4\cos t, y = 9\sin t \ (0 \leq t \leq 2\pi)$

25. Find parametric equations for the motion of a particle in the xy-plane that traces the ellipse $16x^2 + 9y^2 = 144$ once counterclockwise. (There are many ways to do this, so your answer might not be the same as the one in the back of the book.)

26. Find parametric equations for the motion of a particle that starts at the point $(-2, 0)$ in the xy-plane and traces the circle $x^2 + y^2 = 4$ three times clockwise. (There are many ways to do this.)

In Exercises 27 and 28, find an equation for the line in the xy-plane that is tangent to the curve at the point corresponding to the given value of t. Also, find the value of d^2y/dx^2 at this point.

27. $x = (1/2)\tan t, y = (1/2)\sec t \ (t = \pi/3)$

28. $x = 1 + 1/t^2, y = 1 - 3/t \ (t = 2)$

Find the lengths of the curves in Exercises 29 and 30. Graph the curves, and find their lengths using analytic methods. Support each with a NINT computation.

29. $x = e^{2t} - \dfrac{t}{8}, y = e^t \ (0 \leq t \leq \ln 2)$

30. The closed loop of the graph of $x = t^2, y = t^3/3 - t$.

Find the areas of the surfaces generated by revolving the curves in Exercises 31 and 32 about the indicated axes.

31. $x = t^2/2,\ y = 2t,\ 0 \le t \le \sqrt{5}$; x-axis

32. $x = t^2 + 1/(2t),\ y = 4\sqrt{t},\ 1/\sqrt{2} \le t \le 1$; y-axis
(*Hint:* $(dx/dt)^2 + (dy/dt)^2$ is a perfect square.)

Determine a complete graph of the polar curves in Exercises 33–44. Specify the period of each curve. Name each curve (four-leaved rose, spiral, limaçon, lemniscate, circle, cardioid, parabola, or hyperbola).

33. $r = \cos 2\theta$ **34.** $r \cos \theta = 1$

35. $r = \dfrac{6}{1 - 2\cos\theta}$ **36.** $r = \sin 2\theta$

37. $r = \theta$ **38.** $r^2 = \cos 2\theta$

39. $r = 1 + \cos\theta$ **40.** $r = 1 - \sin\theta$

41. $r = \dfrac{2}{1 - \cos\theta}$ **42.** $r^2 = \sin 2\theta$

43. $r = -\sin\theta$ **44.** $r = 2\cos\theta + 1$

Sketch the lines given by the polar equations in Exercises 45 and 46, and find Cartesian equations for them. Support with a grapher.

45. $r \cos\left(\theta - \dfrac{\pi}{3}\right) = 2\sqrt{3}$

46. $r \cos\left(\theta - \dfrac{3\pi}{4}\right) = \sqrt{2}/2$

Find the centers and radii of the circles given by the polar equations in Exercises 47 and 48.

47. $r = 2\sin\theta$ **48.** $r = -4\cos\theta$

Sketch the regions defined by the polar inequalities in Exercises 49 and 50.

49. $0 \le r \le 6\cos\theta$

50. $-4\sin\theta \le r \le 0$

Find the points of intersection of the curves given by the polar equations in Exercises 51–54. Use analytic methods, and support with a grapher.

51. $r = \sin\theta,\ r = 1 + \sin\theta$

52. $r = \cos\theta,\ r = 1 - \cos\theta$

53. $r = 1 + \sin\theta,\ r = -1 + \sin\theta$

54. $r = 1 + \cos\theta,\ r = -1 - \cos\theta$

Sketch the conic sections whose polar equations are given in Exercises 55–58. Give polar coordinates for the vertices and, in the case of ellipses, for the centers as well. Support each with a grapher.

55. $r = \dfrac{2}{1 + \cos\theta}$ **56.** $r = \dfrac{8}{2 + \cos\theta}$

57. $r = \dfrac{6}{1 - 2\cos\theta}$ **58.** $r = \dfrac{12}{3 + \sin\theta}$

Exercises 59–62 give the eccentricities of conic sections with one focus at the origin of the polar coordinate plane, along with

the directrix for that focus. Find a polar equation for each conic section.

59. $e = 2,\ r\cos\theta = 2$ **60.** $e = 1,\ r\cos\theta = -4$

61. $e = 1/2,\ r\sin\theta = 2$ **62.** $e = 1/3,\ r\sin\theta = -6$

Find equations for the lines that are tangent to the polar coordinate curves in Exercises 63 and 64 at the origin.

63. The lemniscate $r^2 = \cos 2\theta$

64. The limaçon $r = 2\cos\theta + 1$

65. Find polar equations for the lines that are tangent to the tips of the petals of the four-leaved rose $r = \sin 2\theta$. Support with a grapher.

66. Find polar equations for the lines that are tangent to the cardioid $r = 1 + \sin\theta$ at the points where it crosses the x-axis. Support with a grapher.

Graph the regions in the polar coordinate plane described in Exercises 67–70, and find the area of each.

67. Enclosed by the limaçon $r = 2 - \cos\theta$

68. Enclosed by one leaf of the three-leaved rose $r = \sin 3\theta$

69. Inside the two-leaved rose $r = 1 + \cos\theta$ and outside the circle $r = 1$

70. Inside the cardioid $r = 2(1 + \sin\theta)$ and outside the circle $r = 2\sin\theta$

Find the areas of the surfaces generated by revolving the polar curves in Exercises 71 and 72 about the indicated axes. Support each with a NINT computation.

71. $r = \sqrt{\cos 2\theta},\ 0 \le \theta \le \pi/4$, about the x-axis

72. $r^2 = \sin 2\theta$, about the y-axis

GRAPH the curves given by the polar equations in Exercises 73–76, and find the length of each.

73. $r = -1 + \cos\theta$

74. $r = \sin\theta,\ 0 \le \theta \le \pi$

75. $r = \cos^3(\theta/3),\ 0 \le \theta \le \pi/4$

76. $r = \sqrt{1 + \sin 2\theta}$

Use parametric mode to GRAPH the equations in Exercises 77 and 78, then rotate each graph counterclockwise with respect to the positive x-axis through the given angle. Finally, find an equation for the rotated curve.

77. $y = x^2 - 1,\ \theta = 30°$

78. $\dfrac{x^2}{9} + \dfrac{y^2}{4} = 1,\ \theta = \dfrac{\pi}{3}$

79. *Average value.* The average value of the polar coordinate r over the curve $r = f(\theta)\ (\alpha \le \theta \le \beta)$ with respect to θ is given by the formula

$$r_{av} = \frac{1}{\beta - \alpha} \int_\alpha^\beta f(\theta)\,d\theta.$$

Use this formula to find the average value of r with respect to θ over the following curves.

a) The cardioid $r = a(1 - \cos\theta)$

b) The circle $r = a$

c) The circle $r = a\cos\theta, -\pi/2 \le \theta \le \pi/2$

80. *Archimedes's spiral.* The graph of an equation of the form $r = a\theta$, where a is a nonzero constant, is called an *Archimedes spiral*. Show that such a spiral cuts any ray from the origin into congruent segments. In other words, show that the width between successive turns of the spiral remains the same.

81. *The space engineer's formula for eccentricity.* The space engineer's formula for the eccentricity of an elliptical orbit is

$$e = \frac{r_{max} - r_{min}}{r_{max} + r_{min}},$$

where r is the distance from the space vehicle to the focus of its elliptical path. Why does the formula work? (*Hint:* You do not need to use calculus. Just think about the definition of eccentricity.)

82. *A satellite orbit.* A satellite is in an orbit that passes over the North and South Poles of Earth. When it is over the South Pole, it is at the highest point of its orbit, 1000 miles above the planet's surface. Above the North Pole, it is at the lowest point of its orbit, 300 miles above the surface.

a) Assuming that the orbit (with reference to Earth) is an ellipse with one focus at the center of the planet, find its eccentricity. (Take the diameter of Earth to be 8000 miles.)

b) Using the north-south axis of the earth as the x-axis and the center of the earth as origin, find a polar equation for the orbit.

83. Show that the length of the polar curve $r = 2f(\theta), \alpha \le \theta \le \beta$, is twice the length of the curve $r = f(\theta), \alpha \le \theta \le \beta$.

84. Show that the area of the surface generated by revolving the polar curve $r = 2f(\theta), \alpha \le \theta \le \beta$, about the x-axis is four times the area of the surface generated by revolving the curve $r = f(\theta), \alpha \le \theta \le \beta$, about the x-axis.

Vectors and Analytic Geometry in Space

OVERVIEW   This chapter introduces vectors and three-dimensional coordinate systems for space. Just as the coordinate plane is the natural place to study functions of a single variable, coordinate space is the place to study functions of two variables (or even more). We establish coordinates in space by adding an axis that measures distance above and below the xy-plane. This builds on what we already know without forcing us to start over again.

Equations in three variables define surfaces in space as equations in two variables define curves in the plane. We use these surfaces to graph functions (not in this chapter, but later), define regions, bound solids, describe walls of containers, and so on. In short, we use them to do all the things we do in the plane, but stepped up one dimension.

Once in space, we can model motion in three dimensions and track the positions of moving bodies with vectors. We can also calculate the directions and magnitudes of their velocities and accelerations and predict the effects of the forces that we see working on them. As we will see in the next chapter, coordinates and vectors are a powerful combination. Coordinates tell us where a moving body is, and vectors tell us what is happening to it.

11.1   Vectors in the Plane

Some of the things that we measure are completely determined by their magnitudes. To record mass, length, or time, for example, we need only write down a number and indicate an appropriate unit of measure. But we need more than that to describe a force, displacement, or velocity, because these quantities have direction as well as magnitude. To describe a force, we need to record the direction in which it acts as well as how large it is. To describe a body's displacement, we have to say in what direction it moved as well as how far it moved. To describe a body's velocity at any given time, we have to know where the body is headed as well as how fast it is going.

Quantities that have direction as well as magnitude are usually represented by arrows that point in the direction of the action and whose lengths represent the magnitude of the action in terms of a suitably chosen unit. When we describe these arrows abstractly, as directed line segments in the plane or in space, we call them *vectors*.

DEFINITIONS

A **vector** in the plane is a directed line segment. Two vectors are **equal** or **the same** if they have the same length and direction.

In print, vectors are usually represented by single boldface roman (nonitalic) letters, as in **v** ("vector v"). A vector may be defined by the directed line segment from point A to point B, however, and written as $\overrightarrow{AB}$ ("vector AB").

The arrows that we use when we draw vectors are understood to represent the same vector if they have the same length and point the same direction (Fig. 11.1).

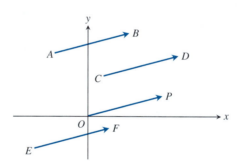

11.1 The four arrows shown here have the same length and direction. They therefore represent the same vector. We write

$$\overrightarrow{AB} = \overrightarrow{CD} = \overrightarrow{OP} = \overrightarrow{EF}.$$

Scalars and Scalar Multiples

We scale vectors by multiplying them by real numbers. To double a vector's length, we multiply the vector by 2. To increase a vector's length 50%, we multiply by 1.5. To reverse a vector's direction and double its length at the same time, we multiply by -2.

If c is a number and **v** is a vector, the direction of $c\mathbf{v}$ agrees with **v** if c is positive and is opposite to that of **v** if c is negative (Fig. 11.2). Since real numbers work like scaling factors in this context, we tend to call them **scalars** and to call multiples like $c\mathbf{v}$ **scalar multiples**.

Geometric Addition—The Parallelogram Law

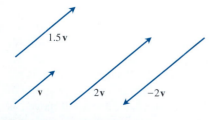

11.2 Scalar multiples of **v**.

Two vectors $\mathbf{v}_1$ and $\mathbf{v}_2$ may be added geometrically by drawing a representative of $\mathbf{v}_1$, say, from A to B in Fig. 11.3, and then a representative of $\mathbf{v}_2$

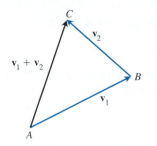

11.3 The sum of $\mathbf{v}_1$ and $\mathbf{v}_2$.

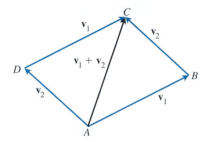

11.4 The parallelogram law of addition. Quadrilateral $ABCD$ is a parallelogram because its opposite sides have equal lengths. Aristotle used the law to describe the combined action of two forces.

starting from the terminal point B of $\mathbf{v}_1$. In Fig. 11.3, $\mathbf{v}_2 = \overrightarrow{BC}$. The sum $\mathbf{v}_1 + \mathbf{v}_2$ is then the vector from the initial point A of $\mathbf{v}_1$ to the terminal point C of $\mathbf{v}_2$. That is, if

$$\mathbf{v}_1 = \overrightarrow{AB} \qquad \text{and} \qquad \mathbf{v}_2 = \overrightarrow{BC},$$

then

$$\mathbf{v}_1 + \mathbf{v}_2 = \overrightarrow{AB} + \overrightarrow{BC} = \overrightarrow{AC}.$$

This description of addition is sometimes called the **Parallelogram Law** of addition because $\mathbf{v}_1 + \mathbf{v}_2$ is given by the diagonal of the parallelogram determined by $\mathbf{v}_1$ and $\mathbf{v}_2$ (Fig. 11.4).

EXPLORATION 1

Properties of Vector Operations—Group Project

We can add any two vectors; therefore, we can say that vector addition is a *binary operation* on the set of vectors. Whenever you meet a new binary operation in mathematics, it is good practice to ask about certain properties, specifically those suggested by the words *commutative*, *associative*, *identity*, and *inverse*. In this chapter, you will meet several binary operations that may be new to you. We would like you to work together in groups of two or three to answer questions about their properties. The Exploration Bits will alert you to the operations and an occasional new question that you can explore.

To begin, interpret each of the following questions, and then answer the question with a convincing geometric argument.

1. Is vector addition a commutative operation?

2. Is vector addition an associative operation?

3. Is there an identity element for vector addition?

4. Do vectors have vector-addition inverses?

Components

Whenever a vector $\mathbf{v}$ can be written as a sum

$$\mathbf{v} = \mathbf{v}_1 + \mathbf{v}_2,$$

the vectors $\mathbf{v}_1$ and $\mathbf{v}_2$ are said to be **components** of $\mathbf{v}$. We also say that we have *represented* or *resolved* $\mathbf{v}$ in terms of $\mathbf{v}_1$ and $\mathbf{v}_2$.

The most common algebra of vectors is based on representing each vector in terms of components parallel to the Cartesian coordinate axes and writing each component as an appropriate multiple of a **basic unit** vector. The basic unit vector in the positive x direction is the vector $\mathbf{i}$ that runs from $(0, 0)$ to $(1, 0)$. The basic unit vector in the positive y direction is the vector $\mathbf{j}$ from $(0, 0)$ to $(0, 1)$. Then $a\mathbf{i}$, a being a scalar, represents a vector of length $|a|$ parallel to the x-axis, pointing to the right if a is positive and to the left if a is negative. Similarly, $b\mathbf{j}$ is a vector of length $|b|$ parallel to the y-axis, pointing up if b is positive and down if b is negative. Figure 11.5 shows a

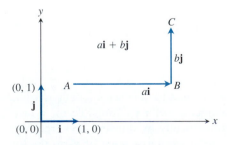

11.5 The basic vectors $\mathbf{i}$ and $\mathbf{j}$. Any vector $\overrightarrow{AC}$ can be expressed as a multiple of $\mathbf{i}$ plus a multiple of $\mathbf{j}$.

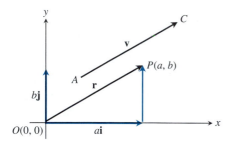

11.6 Any vector $v = \overrightarrow{AC}$ is equal to an appropriate position vector $r = \overrightarrow{OP}$.

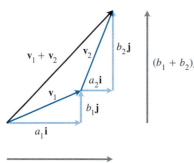

11.7 If $\mathbf{v}_1 = a_1\mathbf{i} + b_1\mathbf{j}$ and $\mathbf{v}_2 = a_2\mathbf{i} + b_2\mathbf{j}$, then $\mathbf{v}_1 + \mathbf{v}_2 = (a_1 + a_2)\mathbf{i} + (b_1 + b_2)\mathbf{j}$.

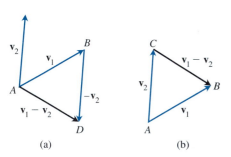

11.8 Two ways to draw $\mathbf{v}_1 - \mathbf{v}_2$: (a) as $\mathbf{v}_1 + (-\mathbf{v}_2)$ and (b) as the vector from the tip of $\mathbf{v}_2$ to the tip of $\mathbf{v}_1$.

vector $\mathbf{v} = \overrightarrow{AC}$ resolved into its $\mathbf{i}$- and $\mathbf{j}$-components as the sum

$$\mathbf{v} = a\mathbf{i} + b\mathbf{j}.$$

Figure 11.6 shows vector $\mathbf{v}$ and also the **position vector $\mathbf{r}$** $= \overrightarrow{OP}$ from the origin O to the point $P(a, b)$. Notice that $\mathbf{v} = a\mathbf{i} + b\mathbf{j} = \mathbf{r}$.

> **DEFINITION**
>
> If $\mathbf{v} = a\mathbf{i} + b\mathbf{j}$, the vectors $a\mathbf{i}$ and $b\mathbf{j}$ are the **basic vector components of v**. The numbers a and b are the **(basic) scalar components of v**, respectively.

Basic components give us a way to define equality of vectors algebraically.

> **DEFINITION** **Equality of Vectors (Algebraic)**
>
> $$a\mathbf{i} + b\mathbf{j} = a'\mathbf{i} + b'\mathbf{j} \qquad \Leftrightarrow \qquad a = a' \quad \text{and} \quad b = b'$$

That is, two vectors are equal if and only if their scalar components are equal.

Algebraic Addition

Two vectors may be added algebraically by adding their corresponding scalar components, as shown in Fig. 11.7.

> If $\mathbf{v}_1 = a_1\mathbf{i} + b_1\mathbf{j}$ and $\mathbf{v}_2 = a_2\mathbf{i} + b_2\mathbf{j}$, then
> $$\mathbf{v}_1 + \mathbf{v}_2 = (a_1 + a_2)\mathbf{i} + (b_1 + b_2)\mathbf{j}.$$

EXAMPLE 1

$$(2\mathbf{i} - 4\mathbf{j}) + (5\mathbf{i} + 3\mathbf{j}) = (2 + 5)\mathbf{i} + (-4 + 3)\mathbf{j} = 7\mathbf{i} - \mathbf{j}. \qquad \blacksquare$$

Subtraction

The negative of a vector $\mathbf{v}$ is the vector $-\mathbf{v}$ that has the same length as $\mathbf{v}$ but points in the opposite direction. To subtract a vector $\mathbf{v}_2$ from a vector $\mathbf{v}_1$, we add $-\mathbf{v}_2$ to $\mathbf{v}_1$. This may be done geometrically by drawing $-\mathbf{v}_2$ from the tip of $\mathbf{v}_1$ and then drawing the vector from the initial point of $\mathbf{v}_1$ to the tip of $-\mathbf{v}_2$, as shown in Fig. 11.8(a), where

$$\overrightarrow{AD} = \overrightarrow{AB} + \overrightarrow{BD} = \mathbf{v}_1 + (-\mathbf{v}_2) = \mathbf{v}_1 - \mathbf{v}_2.$$

Another way to draw $\mathbf{v}_1 - \mathbf{v}_2$ is to draw $\mathbf{v}_1$ and $\mathbf{v}_2$ with a common initial point and then draw $\mathbf{v}_1 - \mathbf{v}_2$ as the vector from the tip of $\mathbf{v}_2$ to the tip of $\mathbf{v}_1$. This is illustrated in Fig. 11.8(b), where

$$\overrightarrow{CB} = \overrightarrow{CA} + \overrightarrow{AB} = -\mathbf{v}_2 + \mathbf{v}_1 = \mathbf{v}_1 - \mathbf{v}_2.$$

Thus, $\overrightarrow{CB}$ is the vector that when added to $\mathbf{v}_2$ gives $\mathbf{v}_1$:

$$\overrightarrow{CB} + \mathbf{v}_2 = (\mathbf{v}_1 - \mathbf{v}_2) + \mathbf{v}_2 = \mathbf{v}_1.$$

We subtract vectors by subtracting the corresponding scalar components.

EXPLORATION BITS

1. Now that you have an algebraic definition of vector addition, you should be able to give convincing algebraic arguments for parts 1–4 of the Group Project in Exploration 1. Try it.

2. Give a convincing argument that

$$(-1)\mathbf{v} = -\mathbf{v}.$$

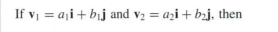

If $\mathbf{v}_1 = a_1\mathbf{i} + b_1\mathbf{j}$ and $\mathbf{v}_2 = a_2\mathbf{i} + b_2\mathbf{j}$, then

$$\mathbf{v}_1 - \mathbf{v}_2 = (a_1 - a_2)\mathbf{i} + (b_1 - b_2)\mathbf{j}$$

EXAMPLE 2

$$(6\mathbf{i} + 2\mathbf{j}) - (3\mathbf{i} - 5\mathbf{j}) = (6 - 3)\mathbf{i} + (2 - (-5))\mathbf{j} = 3\mathbf{i} + 7\mathbf{j}.$$

We find the components of the vector from a point $P_1(x_1, y_1)$ to a point $P_2(x_2, y_2)$ by subtracting the components of $\overrightarrow{OP_1} = x_1\mathbf{i} + y_1\mathbf{j}$ from the components of $\overrightarrow{OP_2} = x_2\mathbf{i} + y_2\mathbf{j}$.

The vector from $P_1(x_1, y_1)$ to $P_2(x_2, y_2)$ is

$$\overrightarrow{P_1P_2} = (x_2 - x_1)\mathbf{i} + (y_2 - y_1)\mathbf{j}.$$

EXAMPLE 3

The vector from $P_1(3, 4)$ to $P_2(5, 1)$ is

$$\overrightarrow{P_1P_2} = (5 - 3)\mathbf{i} + (1 - 4)\mathbf{j} = 2\mathbf{i} - 3\mathbf{j}.$$

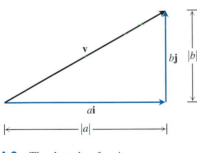

11.9 The length of $\mathbf{v}$ is $\sqrt{|a|^2 + |b|^2} = \sqrt{a^2 + b^2}$.

The Length of a Vector

We calculate the length of $\mathbf{v} = a\mathbf{i} + b\mathbf{j}$ by representing $\mathbf{v}$ as the hypotenuse of a right triangle with sides $|a|$ and $|b|$ (Fig. 11.9) and applying the Pythagorean Theorem to get

$$\text{Length of } \mathbf{v} = \sqrt{|a|^2 + |b|^2} = \sqrt{a^2 + b^2}.$$

The usual symbol for the length is $|\mathbf{v}|$, which is read "the length of $\mathbf{v}$" or "the magnitude of $\mathbf{v}$," the latter being more common in applied fields. The bars are the same as the ones we use for absolute values.

The **length** or **magnitude** of $\mathbf{v} = a\mathbf{i} + b\mathbf{j}$ is $|\mathbf{v}| = \sqrt{a^2 + b^2}$. (1)

Scalar Multiplication

In terms of components, scalar multiplication takes the following form.

If c is a scalar and $\mathbf{v} = a\mathbf{i} + b\mathbf{j}$ is a vector, then
$$c\mathbf{v} = c(a\mathbf{i} + b\mathbf{j}) = (ca)\mathbf{i} + (cb)\mathbf{j}. \qquad (2)$$

To check that the length of $c\mathbf{v}$ is $|c|$ times the length of $\mathbf{v}$ when we do scalar multiplication this way, we can calculate the length with Eq. (1):

$$|c\mathbf{v}| = |(ca)\mathbf{i} + (cb)\mathbf{j}| \qquad \text{Eq. (2)}$$
$$= \sqrt{(ca)^2 + (cb)^2} \qquad \text{Eq. (1) with } ca \text{ and } cb \text{ in place of } a \text{ and } b$$
$$= \sqrt{c^2(a^2 + b^2)}$$
$$= \sqrt{c^2}\sqrt{a^2 + b^2}$$
$$= |c||\mathbf{v}|.$$

If c is a scalar and $\mathbf{v}$ is a vector, then
$$|c\mathbf{v}| = |c||\mathbf{v}|. \qquad (3)$$

EXAMPLE 4

If $c = -2$ and $\mathbf{v} = -3\mathbf{i} + 4\mathbf{j}$, then

$$|\mathbf{v}| = |-3\mathbf{i} + 4\mathbf{j}| = \sqrt{(-3)^2 + (4)^2} = \sqrt{9 + 16} = \sqrt{25} = 5$$

$$|-2\mathbf{v}| = |(-2)(-3\mathbf{i} + 4\mathbf{j})| = |6\mathbf{i} - 8\mathbf{j}| = \sqrt{(6)^2 + (-8)^2} = \sqrt{36 + 64}$$
$$= \sqrt{100} = 10 = |-2|\,5 = |c||\mathbf{v}|. \qquad \equiv$$

The Zero Vector

We define

$$\mathbf{0} = 0\mathbf{i} + 0\mathbf{j}$$

as the **zero vector**. It is a special vector because it has no direction (or infinitely many directions) and its length is zero:

$$|a\mathbf{i} + b\mathbf{j}| = \sqrt{a^2 + b^2} = 0 \qquad \Leftrightarrow \qquad a = b = 0.$$

Unit Vectors

Any vector $\mathbf{u}$ whose length is 1 is called a **unit vector**. The vectors $\mathbf{i}$ and $\mathbf{j}$ are unit vectors:

$$|\mathbf{i}| = |1\mathbf{i} + 0\mathbf{j}| = \sqrt{1^2 + 0^2} = 1, \qquad |\mathbf{j}| = |0\mathbf{i} + 1\mathbf{j}| = \sqrt{0^2 + 1^2} = 1.$$

If $\mathbf{u}$ is the unit vector obtained by rotating $\mathbf{i}$ through an angle θ in the positive direction (Fig. 11.10), then **u** has a horizontal component $\cos\theta$ and a vertical component $\sin\theta$ so that

$$\mathbf{u} = (\cos\theta)\mathbf{i} + (\sin\theta)\mathbf{j}. \tag{4}$$

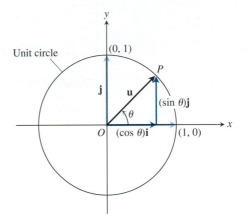

11.10 The unit vector that makes an angle of measure θ with the positive x-axis. Every unit vector has the form

$$\mathbf{u} = (\cos\theta)\mathbf{i} + (\sin\theta)\mathbf{j}$$

for some θ.

If we allow the angle θ in Eq. (4) to vary from 0 to 2π, the point P in Fig. 11.10 traces the unit circle $x^2 + y^2 = 1$. Since this takes in all possible directions, every unit vector in the plane is given by Eq. (4) for some value of $\theta\,(0 \le \theta \le 2\pi)$.

Direction as a Vector

It is common in subjects like classical electricity and magnetism, which use vectors a great deal, to define the **direction** of a nonzero vector **v** to be the unit vector obtained by dividing **v** by its own length.

HANDWRITTEN NOTATION

In handwritten work, it is common to denote unit vectors with small "hats," as in $\hat{u}$ (pronounced "u hat"). In hat notation, **i** and **j** become $\hat{i}$ and $\hat{j}$.

> **DEFINITION**
>
> If $\mathbf{v} \ne \mathbf{0}$, the **direction** of **v** is the unit vector $\dfrac{\mathbf{v}}{|\mathbf{v}|}$.

Notice that instead of just saying that $\mathbf{v}/|\mathbf{v}|$ *represents* the direction of **v**, we say that it *is* the direction of **v**.

To see that $\mathbf{v}/|\mathbf{v}|$ really is a unit vector, we can calculate its length directly:

$$\text{Length of } \frac{\mathbf{v}}{|\mathbf{v}|} = \left| \frac{\mathbf{v}}{|\mathbf{v}|} \right|$$

$$= \frac{1}{|\mathbf{v}|}|\mathbf{v}| \qquad \text{Eq. (3) with } c = \frac{1}{|\mathbf{v}|}$$

$$= 1.$$

Any nonzero vector can be expressed in terms of its length and direction by using the equation

$$\mathbf{v} = |\mathbf{v}| \cdot \frac{\mathbf{v}}{|\mathbf{v}|} = (\text{length of } \mathbf{v}) \cdot (\text{direction of } \mathbf{v}). \qquad 5)$$

EXAMPLE 5

Express $\mathbf{v} = 3\mathbf{i} - 4\mathbf{j}$ in terms of its length and direction.

Solution

Length of $\mathbf{v}$: $\qquad |\mathbf{v}| = \sqrt{(3)^2 + (-4)^2} = \sqrt{9 + 16} = 5$

Direction of $\mathbf{v}$: $\qquad \dfrac{\mathbf{v}}{|\mathbf{v}|} = \dfrac{3\mathbf{i} - 4\mathbf{j}}{5} = \dfrac{3}{5}\mathbf{i} - \dfrac{4}{5}\mathbf{j}$

$$\mathbf{v} = 3\mathbf{i} - 4\mathbf{j} = 5 \left(\frac{3}{5}\mathbf{i} - \frac{4}{5}\mathbf{j} \right)$$

$$\underset{\text{Length of } \mathbf{v}}{\diagup} \qquad \underset{\text{Direction of } \mathbf{v}}{\diagdown}$$

≡

> **EXPLORATION BIT**
>
> Demonstrate that the direction of (the direction of $\mathbf{v}$) is the same as the direction of $\mathbf{v}$ (as we should hope that it would be).

It follows from the definition of direction of a vector that nonzero vectors $\mathbf{A}$ and $\mathbf{B}$ have the *same* direction if and only if

$$\frac{\mathbf{A}}{|\mathbf{A}|} = \frac{\mathbf{B}}{|\mathbf{B}|} \qquad \text{or} \qquad \mathbf{A} = \frac{|\mathbf{A}|}{|\mathbf{B}|}\mathbf{B}.$$

Thus, if $\mathbf{A}$ and $\mathbf{B}$ have the same direction, then $\mathbf{A}$ is a positive scalar multiple of $\mathbf{B}$. Conversely, if $\mathbf{A} = k\mathbf{B}$, $(k > 0)$, then

$$\frac{\mathbf{A}}{|\mathbf{A}|} = \frac{k\mathbf{B}}{|k\mathbf{B}|} = \frac{k}{|k|}\frac{\mathbf{B}}{|\mathbf{B}|} = \frac{k}{k}\frac{\mathbf{B}}{|\mathbf{B}|} = \frac{\mathbf{B}}{|\mathbf{B}|}.$$

Therefore, nonzero vectors $\mathbf{A}$ and $\mathbf{B}$ have the same direction if and only if $\mathbf{A}$ is a positive scalar multiple of $\mathbf{B}$.

We say that two nonzero vectors $\mathbf{A}$ and $\mathbf{B}$ have *opposite* directions if

$$\frac{\mathbf{A}}{|\mathbf{A}|} = -\frac{\mathbf{B}}{|\mathbf{B}|}.$$

From this it follows that $\mathbf{A}$ and $\mathbf{B}$ have opposite directions if and only if $\mathbf{A}$ is a negative scalar multiple of $\mathbf{B}$.

EXAMPLE 6

a) Same direction: $\mathbf{A} = 3\mathbf{i} - 4\mathbf{j}$ and $\mathbf{B} = \dfrac{3}{2}\mathbf{i} - 2\mathbf{j} = \dfrac{1}{2}\mathbf{A}$

$\qquad$ ($\mathbf{B}$ is a positive scalar multiple of $\mathbf{A}$.)

b) Opposite directions: $\mathbf{A} = 3\mathbf{i} - 4\mathbf{j}$ and $\mathbf{B} = -9\mathbf{i} + 12\mathbf{j} = -3\mathbf{A}$

$\qquad$ ($\mathbf{B}$ is a negative scalar multiple of $\mathbf{A}$.)

≡

Slopes, Tangents, and Normals

Two vectors are said to be **parallel** if they are either positive or negative scalar multiples of one another or, equivalently, if the line segments representing them are collinear or parallel. Similarly, a vector is parallel to a line if some segment that represents the vector is parallel to the line.

Two vectors are **normal** (*perpendicular* and *orthogonal* are sometimes used) if the line segments representing them belong to two perpendicular lines. A vector is normal to a line if it has a representative line segment that is perpendicular to the line.

The **slope** of a vector that is not parallel to the y-axis is the slope shared by the lines parallel to the vector. Thus, when $a \neq 0$, the vector $\mathbf{v} = a\mathbf{i} + b\mathbf{j}$ has a well-defined slope that can be calculated from the components of $\mathbf{v}$ as the number b/a (Fig. 11.11).

When we say that a vector is **tangent** or **normal** to a curve at a point, we mean that the vector is parallel or normal, respectively, to the line that is tangent to the curve at the point. The next example shows how to find such vectors.

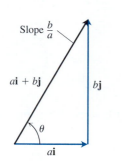

11.11 If $a \neq 0$, the vector $a\mathbf{i} + b\mathbf{j}$ has slope $b/a = \tan\theta$.

EXAMPLE 7

Find unit vectors tangent and normal to the curve

$$y = \frac{x^3}{2} + \frac{1}{2}$$

at the point $(1, 1)$.

Solution   We find the two unit vectors that are parallel and normal to the curve's tangent line at the point $(1, 1)$, shown in Fig. 11.12.

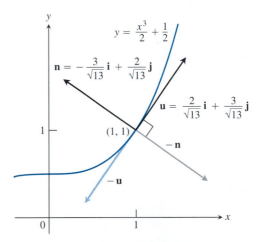

11.12 The unit tangent and unit normal vectors at the point $(1, 1)$ on the curve
$$y = \frac{x^3}{2} + \frac{1}{2}.$$

The slope of the line tangent to the curve at $(1, 1)$ is

$$y' = \frac{3x^2}{2}\bigg|_{x=1} = \frac{3}{2}.$$

We find a unit vector with this slope. The vector $\mathbf{v} = 2\mathbf{i} + 3\mathbf{j}$ has slope 3/2, as does every nonzero multiple of $\mathbf{v}$. To find a multiple of $\mathbf{v}$ that is a unit vector, we divide $\mathbf{v}$ by its length,

$$|\mathbf{v}| = \sqrt{2^2 + 3^2} = \sqrt{13}.$$

This produces the unit vector

$$\mathbf{u} = \frac{\mathbf{v}}{|\mathbf{v}|} = \frac{2}{\sqrt{13}}\mathbf{i} + \frac{3}{\sqrt{13}}\mathbf{j}.$$

The vector $\mathbf{u}$ is tangent to the curve at $(1, 1)$ because it has the same slope as $\mathbf{v}$. For the same reason, the vector

$$-\mathbf{u} = -\frac{2}{\sqrt{13}}\mathbf{i} - \frac{3}{\sqrt{13}}\mathbf{j},$$

which points in the opposite direction, is also tangent to the curve at $(1, 1)$. Without some additional requirement, there is no reason to prefer one of these vectors to the other.

To find unit vectors normal to the curve at $(1, 1)$, we look for unit vectors whose slopes are the negative reciprocal of the slope of $\mathbf{u}$. This is quickly done by interchanging the scalar components of $\mathbf{u}$ and changing the sign of one of them. We obtain

$$\mathbf{n} = -\frac{3}{\sqrt{13}}\mathbf{i} + \frac{2}{\sqrt{13}}\mathbf{j} \qquad \text{and} \qquad -\mathbf{n} = \frac{3}{\sqrt{13}}\mathbf{i} - \frac{2}{\sqrt{13}}\mathbf{j}.$$

Again, either one will do. The vectors have opposite directions, but both are normal to the curve at the point $(1, 1)$. ≡

> If $\mathbf{v} = a\mathbf{i} + b\mathbf{j}$, then $\mathbf{p} = -b\mathbf{i} + a\mathbf{j}$ and $\mathbf{q} = b\mathbf{i} - a\mathbf{j}$ are perpendicular to $\mathbf{v}$ because their slopes are both $-a/b$, the negative reciprocal of the slope of $\mathbf{v}$.

Exercises 11.1

1. The vectors $\mathbf{A}$, $\mathbf{B}$, and $\mathbf{C}$ shown below lie in a plane. Copy them on a sheet of paper.

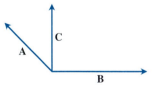

Then, by arranging vectors head to tail, as in Figs. 11.4, 11.7, and 11.8, sketch

a) $\mathbf{A} + \mathbf{B}$ b) $\mathbf{A} + \mathbf{B} + \mathbf{C}$

c) $\mathbf{A} - \mathbf{B}$ d) $\mathbf{A} - \mathbf{C}$

2. The vectors $\mathbf{A}$, $\mathbf{B}$, and $\mathbf{C}$ shown lie in a plane. Copy them on a sheet of paper.

Then, by arranging vectors head to tail, as in Figs. 11.4, 11.7, and 11.8, sketch

a) $\mathbf{A} - \mathbf{B}$ b) $\mathbf{A} - \mathbf{B} + \mathbf{C}$

c) $2\mathbf{A} - \mathbf{B}$ d) $\mathbf{A} + \mathbf{B} + \mathbf{C}$

Let $\mathbf{u} = 3\mathbf{i} - 2\mathbf{j}$ and $\mathbf{v} = -2\mathbf{i} + 5\mathbf{j}$. Find the (basic) scalar components of the vectors in Exercises 3–8.

3. $3\mathbf{u}$ **4.** $-2\mathbf{v}$

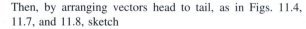

5. $\mathbf{u} + \mathbf{v}$

6. $\mathbf{u} - \mathbf{v}$

7. $2\mathbf{u} - 3\mathbf{v}$

8. $-2\mathbf{u} + 5\mathbf{v}$

In Exercises 9–16, express the vectors in the form $a\mathbf{i} + b\mathbf{j}$, and sketch them as arrows in the coordinate plane.

9. $\overrightarrow{P_1 P_2}$ if P_1 is the point $(1, 3)$ and P_2 is the point $(2, -1)$

10. $\overrightarrow{OP_3}$ if O is the origin and P_3 is the midpoint of the vector $\overrightarrow{P_1 P_2}$ joining $P_1(2, -1)$ and $P_2(-4, 3)$

11. The vector from the point $A(2, 3)$ to the origin

12. The sum of the vectors $\overrightarrow{AB}$ and $\overrightarrow{CD}$, given the four points $A(1, -1)$, $B(2, 0)$, $C(-1, 3)$, and $D(-2, 2)$

13. The unit vectors $\mathbf{u} = (\cos\theta)\mathbf{i} + (\sin\theta)\mathbf{j}$ for $\theta = \pi/6$ and $\theta = 2\pi/3$. Include the circle $x^2 + y^2 = 1$ in your sketch.

14. The unit vectors $\mathbf{u} = (\cos\theta)\mathbf{i} + (\sin\theta)\mathbf{j}$ for $\theta = -\pi/4$ and $\theta = -3\pi/4$. Include the circle $x^2 + y^2 = 1$ in your sketch.

15. The unit vector obtained by rotating $\mathbf{j}$ $120°$ clockwise about the origin

16. The unit vector obtained by rotating $\mathbf{i}$ $135°$ counterclockwise about the origin

Find the magnitude of each vector in Exercises 17–22.

17. $2\mathbf{i} - 3\mathbf{j}$

18. $3\mathbf{i} + 4\mathbf{j}$

19. $\dfrac{3}{5}\mathbf{i} + \dfrac{4}{5}\mathbf{j}$

20. $\dfrac{4}{5}\mathbf{i} - \dfrac{3}{5}\mathbf{j}$

21. $-\dfrac{5}{13}\mathbf{i} + \dfrac{12}{13}\mathbf{j}$

22. $\dfrac{8}{17}\mathbf{i} + \dfrac{15}{17}\mathbf{j}$

Find a unit vector in the same direction as each vector in Exer-

cises 23–28.

23. $3\mathbf{i} + 4\mathbf{j}$

24. $4\mathbf{i} - 3\mathbf{j}$

25. $12\mathbf{i} - 5\mathbf{j}$

26. $-15\mathbf{i} + 8\mathbf{j}$

27. $2\mathbf{i} + 3\mathbf{j}$

28. $5\mathbf{i} - 2\mathbf{j}$

In Exercises 29–34, use Eq. (5) to express the vectors in terms of their lengths and directions.

29. $\mathbf{i} + \mathbf{j}$

30. $2\mathbf{i} - 3\mathbf{j}$

31. $\sqrt{3}\mathbf{i} + \mathbf{j}$

32. $-2\mathbf{i} + 3\mathbf{j}$

33. $5\mathbf{i} + 12\mathbf{j}$

34. $-5\mathbf{i} - 12\mathbf{j}$

35. Show that $\mathbf{A} = 3\mathbf{i} + 6\mathbf{j}$ and $\mathbf{B} = -\mathbf{i} - 2\mathbf{j}$ have opposite directions. Sketch $\mathbf{A}$ and $\mathbf{B}$ together.

36. Show that $\mathbf{A} = 3\mathbf{i} + 6\mathbf{j}$ and $\mathbf{B} = (1/2)\mathbf{i} + \mathbf{j}$ have the same direction.

In Exercises 37–42, find the unit vectors (four vectors in all) that are tangent and normal to the curve at the given point. Show the curve and vectors together in a sketch.

37. $y = x^2$, $(2, 4)$

38. $y = e^x$, $(\ln 2, 2)$

39. $x^2 + 2y^2 = 6$, $(2, 1)$

40. $4x^2 + y^2 = 16$, $(\sqrt{2}, -2\sqrt{2})$

41. $y = \tan^{-1} x$, $(1, \pi/4)$

42. $y = \sin^{-1} x$, $(\sqrt{2}/2, \pi/4)$

43. Let $\mathbf{v}$ be a vector in the plane not parallel to the y-axis. Does

$$\text{slope of } -\mathbf{v} = \text{slope of } \mathbf{v},$$

or does

$$\text{slope of } -\mathbf{v} = -(\text{slope of } \mathbf{v})?$$

Justify your answer.

44. Is the zero vector the only vector whose length is zero? Give reasons for your answer.

11.2 Cartesian (Rectangular) Coordinates and Vectors in Space

Our goal now is to describe the three-dimensional Cartesian coordinate system and learn our way around in space. This means defining distance, practicing with the arithmetic of vectors (same rules as in the plane but with an extra term), and making connections between sets of points in space and equations and inequalities. One reason that inexpensive graphing utilities are not quite ready yet for handling graphs in three-space is because of the difficulty of representing a three-dimensional picture on a two-dimensional surface. The difficulty extends to sketching as well, so we have included some sketching tips in this chapter.

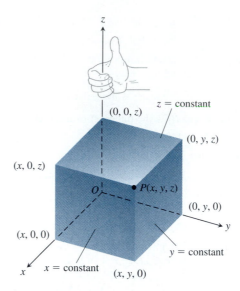

11.13 The Cartesian coordinate system is right-handed.

Cartesian Coordinates

The *Cartesian coordinates* for space are often called *rectangular coordinates* because the axes that define them meet at right angles. To locate points in space, we use these three mutually perpendicular coordinate axes arranged as in Fig. 11.13.

The axes Ox, Oy, and Oz shown there make what is known as a *right-handed* coordinate frame. When you hold your right hand so that the fingers curl from the positive x-axis toward the positive y-axis, your thumb points along the positive z-axis.

The Cartesian coordinates (x, y, z) of a point P in space are the numbers where the planes through P perpendicular to the three axes cut the axes.

Points that lie on the x-axis have their y- and z-coordinates equal to zero. That is, they have coordinates of the form $(x, 0, 0)$. Similarly, points on the y-axis have coordinates of the form $(0, y, 0)$, and points on the z-axis have coordinates of the form $(0, 0, z)$.

The points in a plane perpendicular to the x-axis all have the same x-coordinate, the number at which that plane cuts the x-axis. Similarly, the points in a plane perpendicular to the y-axis have a common y-coordinate, and the points in a plane perpendicular to the z-axis have a common z-coordinate. It is therefore easy to write equations for these planes—we just name the common coordinate's value. The equation $x = 2$ is an equation for the plane perpendicular to the x-axis at $x = 2$. The equation $y = 3$ is an equation for the plane perpendicular to the y-axis at $y = 3$. The equation $z = 5$ is an equation for the plane perpendicular to the z-axis at $z = 5$. Figure 11.14 shows the planes $x = 2$, $y = 3$, and $z = 5$, together with their intersection point $(2, 3, 5)$.

The planes $x = 2$ and $y = 3$ in Fig. 11.14 intersect in a line that...

The planes $x = 2$ and $y = 3$ in Fig. 11.14 intersect in a line that runs parallel to the z-axis. This line is described by the *pair* of equations $x = 2, y = 3$. A point (x, y, z) lies on the line if and only if x equals 2 and y equals 3. Similarly, the line of intersection of the planes $y = 3$ and $z = 5$ is described by the equation pair $y = 3, z = 5$. This line runs parallel to the x-axis. The line of intersection of the planes $x = 2$ and $z = 5$, parallel to the y-axis, is described by the equation pair $x = 2, z = 5$.

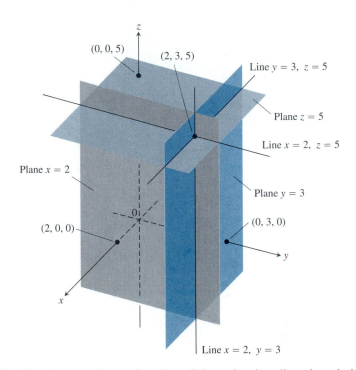

11.14 The planes $x = 2$, $y = 3$, and $z = 5$ determine three lines through the point $(2, 3, 5)$.

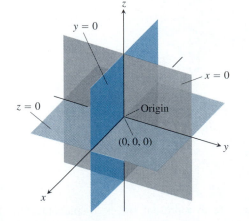

11.15 The planes $x = 0$, $y = 0$, and $z = 0$ are the planes determined by the coordinate axes. They divide space into eight cells called octants.

The planes determined by the three coordinate axes are the **xy-plane**, whose standard equation is $z = 0$; the **yz-plane**, whose standard equation is $x = 0$; and the **xz-plane**, whose standard equation is $y = 0$. They meet in the point $(0, 0, 0)$, which is called the **origin** of the coordinate system (Fig. 11.15).

The three **coordinate planes** $x = 0$, $y = 0$, and $z = 0$ divide space into eight cells called **octants**. The octant in which the point coordinates are all positive is called the **first octant**, but there is no conventional numbering for the remaining seven octants.

In the following examples, we match a number of coordinate equations and inequalities with the sets of points they define in space.

EXAMPLE 1

Defining Equations and Inequalities	Verbal Description
$z \geq 0$	The half-space consisting of the points on and above the xy-plane.
$x = -3$	The plane perpendicular to the x-axis at $x = -3$. This plane lies parallel to the yz-plane and 3 units behind it.
$z = 0, x \leq 0, y \geq 0$	The second quadrant of the xy-plane, including the positive axes.
$x \geq 0, y \geq 0, z \geq 0$	The first octant, including the positive axes.
$-1 \leq y \leq 1$	The slab-like region between the planes $y = -1$ and $y = 1$ (planes included).
$y = -2, z = 2$	The line in which the planes $y = -2$ and $z = 2$ intersect. Alternatively, the line through the point $(0, -2, 2)$ parallel to the x-axis.

EXAMPLE 2

Identify the set of points $P(x, y, z)$ whose coordinates satisfy the two equations

$$x^2 + y^2 = 4 \qquad \text{and} \qquad z = 3.$$

Solution The points lie in the horizontal plane $z = 3$ and, in this plane, make up the circle $x^2 + y^2 = 4$. We call this set of points "the circle $x^2 + y^2 = 4$ in the plane $z = 3$" or, more simply, "the circle $x^2 + y^2 = 4, z = 3$" (Fig. 11.16).

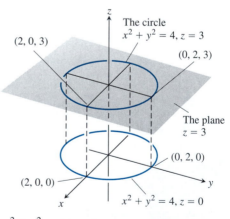

11.16 The circle $x^2 + y^2 = 4, z = 3$.

EXPLORATION BIT

Describe the interior of a cylinder of radius 2 and height h with inequalities.

Vectors in Space

The directed line segments that we use to represent forces, displacements, and velocities in space are called vectors, just as they are in the plane. The same rules of addition, subtraction, and scalar multiplication apply.

The vectors from the origin to the points $(1,0,0)$, $(0,1,0)$, and $(0,0,1)$ are the **basic unit vectors**. We denote them by **i**, **j**, and **k**. The **position vector**

r from the origin O to the typical point $P(x, y, z)$ is

$$\mathbf{r} = \overrightarrow{OP} = x\mathbf{i} + y\mathbf{j} + z\mathbf{k}.$$

The values x, y, z are the scalar components of the position vector.

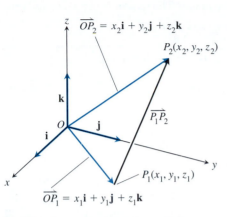

$\overrightarrow{OP_2} = x_2\mathbf{i} + y_2\mathbf{j} + z_2\mathbf{k}$

$P_2(x_2, y_2, z_2)$

$\overrightarrow{P_1P_2}$

$P_1(x_1, y_1, z_1)$

$\overrightarrow{OP_1} = x_1\mathbf{i} + y_1\mathbf{j} + z_1\mathbf{k}$

11.17   The vector from P_1 to P_2 is
$\overrightarrow{P_1P_2} = (x_2 - x_1)\mathbf{i} + (y_2 - y_1)\mathbf{j} + (z_2 - z_1)\mathbf{k}.$

Addition, Subtraction, and Scalar Multiplication for Vectors in Space

For any vectors $\mathbf{A} = a_1\mathbf{i} + a_2\mathbf{j} + a_3\mathbf{k}$ and $\mathbf{B} = b_1\mathbf{i} + b_2\mathbf{j} + b_3\mathbf{k}$ and for any scalar c,

$$\mathbf{A} + \mathbf{B} = (a_1 + b_1)\mathbf{i} + (a_2 + b_2)\mathbf{j} + (a_3 + b_3)\mathbf{k},$$
$$\mathbf{A} - \mathbf{B} = (a_1 - b_1)\mathbf{i} + (a_2 - b_2)\mathbf{j} + (a_3 - b_3)\mathbf{k},$$
$$c\mathbf{A} = (ca_1)\mathbf{i} + (ca_2)\mathbf{j} + (ca_3)\mathbf{k}.$$

The Vector between Two Points

Often, we want to express the vector $\overrightarrow{P_1P_2}$ from the point $P_1(x_1, y_1, z_1)$ to the point $P_2(x_2, y_2, z_2)$ in terms of the coordinates of P_1 and P_2. To do so, we first observe that $\overrightarrow{P_1P_2} = \overrightarrow{P_1O} + \overrightarrow{OP_2}$, as in Fig. 11.17. We then write $\overrightarrow{P_1O}$ as $-\overrightarrow{OP_1}$ and express the results in terms of $\mathbf{i}$, $\mathbf{j}$, and $\mathbf{k}$:

$$\overrightarrow{P_1P_2} = \overrightarrow{P_1O} + \overrightarrow{OP_2}$$
$$= \overrightarrow{OP_2} - \overrightarrow{OP_1}$$
$$= (x_2\mathbf{i} + y_2\mathbf{j} + z_2\mathbf{k}) - (x_1\mathbf{i} + y_1\mathbf{j} + z_1\mathbf{k})$$
$$= (x_2 - x_1)\mathbf{i} + (y_2 - y_1)\mathbf{j} + (z_2 - z_1)\mathbf{k}.$$

The vector from $P_1(x_1, y_1, z_1)$ to $P_2(x_2, y_2, z_2)$ is

$$\overrightarrow{P_1P_2} = (x_2 - x_1)\mathbf{i} + (y_2 - y_1)\mathbf{j} + (z_2 - z_1)\mathbf{k}.$$

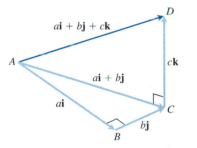

D

$a\mathbf{i} + b\mathbf{j} + c\mathbf{k}$

A

$c\mathbf{k}$

$a\mathbf{i} + b\mathbf{j}$

$a\mathbf{i}$

C

$b\mathbf{j}$

B

11.18   The length of the vector $\overrightarrow{AD}$ can be determined by applying the Pythagorean Theorem to the right triangles ABC and ACD.

Length and Direction

As in the plane, the important features of a vector are its length and direction. The length of a vector $a\mathbf{i} + b\mathbf{j} + c\mathbf{k}$ is calculated by applying the Pythagorean Theorem twice. In the notation of Fig. 11.18, we have

$$|\overrightarrow{AC}| = |a\mathbf{i} + b\mathbf{j}| = \sqrt{a^2 + b^2},$$

from triangle ABC, and then

$$|a\mathbf{i} + b\mathbf{j} + c\mathbf{k}| = |\overrightarrow{AD}| = \sqrt{|\overrightarrow{AC}|^2 + |\overrightarrow{CD}|^2} = \sqrt{a^2 + b^2 + c^2},$$

from triangle ACD.

The **length** of the vector $\mathbf{A} = a\mathbf{i} + b\mathbf{j} + c\mathbf{k}$ is

$$|\mathbf{A}| = |a\mathbf{i} + b\mathbf{j} + c\mathbf{k}| = \sqrt{a^2 + b^2 + c^2}. \tag{1}$$

Read Eq. (2) as "The length of $c\mathbf{A}$ is the absolute value of c times the length of $\mathbf{A}$."

EXAMPLE 3

The length of $\mathbf{A} = \mathbf{i} - 2\mathbf{j} + 3\mathbf{k}$ is

$$|\mathbf{A}| = \sqrt{(1)^2 + (-2)^2 + (3)^2} = \sqrt{1 + 4 + 9} = \sqrt{14}.$$

If we multiply a vector $\mathbf{A} = a_1\mathbf{i} + a_2\mathbf{j} + a_3\mathbf{k}$ by a scalar c, the length of $c\mathbf{A}$ is $|c|$ times the length of $\mathbf{A}$, just as in the plane. The reason is the same, as well:

$$c\mathbf{A} = ca_1\mathbf{i} + ca_2\mathbf{j} + ca_3\mathbf{k},$$

so

$$|c\mathbf{A}| = \sqrt{(ca_1)^2 + (ca_2)^2 + (ca_3)^2} = \sqrt{c^2a_1^2 + c^2a_2^2 + c^2a_3^2}$$

$$= |c|\sqrt{a_1^2 + a_2^2 + a_3^2} = |c||\mathbf{A}|. \tag{2}$$

EXAMPLE 4

If $\mathbf{A}$ is the vector of Example 3, then the length of

$$2\mathbf{A} = 2(\mathbf{i} - 2\mathbf{j} + 3\mathbf{k}) = 2\mathbf{i} - 4\mathbf{j} + 6\mathbf{k}$$

is

$$\sqrt{(2)^2 + (-4)^2 + (6)^2} = \sqrt{4 + 16 + 36} = \sqrt{56}$$

$$= \sqrt{4 \cdot 14} = 2\sqrt{14} = 2|\mathbf{A}|.$$

Again as with vectors in the plane, vectors of unit length are called **unit vectors**. The vectors $\mathbf{i}$, $\mathbf{j}$, and $\mathbf{k}$ are unit vectors because

$$|\mathbf{i}| = |1\mathbf{i} + 0\mathbf{j} + 0\mathbf{k}| = \sqrt{1^2 + 0^2 + 0^2} = 1,$$

$$|\mathbf{j}| = |0\mathbf{i} + 1\mathbf{j} + 0\mathbf{k}| = \sqrt{0^2 + 1^2 + 0^2} = 1,$$

$$|\mathbf{k}| = |0\mathbf{i} + 0\mathbf{j} + 1\mathbf{k}| = \sqrt{0^2 + 0^2 + 1^2} = 1.$$

The *direction* of a nonzero vector $\mathbf{A}$ is the unit vector obtained by dividing $\mathbf{A}$ by its length $|\mathbf{A}|$.

If $A \neq 0$, the **direction** of $\mathbf{A}$ is the unit vector $\dfrac{\mathbf{A}}{|\mathbf{A}|}$.

As in the plane, we can use the equation

$$\mathbf{A} = |\mathbf{A}| \cdot \frac{\mathbf{A}}{|\mathbf{A}|} \tag{3}$$

to express any nonzero vector as a product of its length and direction.

EXAMPLE 5

Express $\mathbf{A} = \mathbf{i} - 2\mathbf{j} + 3\mathbf{k}$ as a product of its length and direction.

Solution

$$\mathbf{A} = |\mathbf{A}| \cdot \frac{\mathbf{A}}{|\mathbf{A}|} \qquad \text{Eq. 3}$$

$$= \sqrt{14} \cdot \frac{\mathbf{i} - 2\mathbf{j} + 3\mathbf{k}}{\sqrt{14}} \qquad \text{Length from Example 3}$$

$$= \sqrt{14}\left(\frac{1}{\sqrt{14}}\mathbf{i} - \frac{2}{\sqrt{14}}\mathbf{j} + \frac{3}{\sqrt{14}}\mathbf{k}\right) = \text{(length of } \mathbf{A}) \cdot \text{(direction of } \mathbf{A})$$

The vector $a\mathbf{v}$ is in the direction of $\mathbf{v}$ if $a > 0$, and is in the opposite direction of $\mathbf{v}$ if $a < 0$.

EXAMPLE 6

Find a unit vector $\mathbf{u}$ in the direction of the vector from $P_1(1, 0, 1)$ to $P_2(3, 2, 0)$.

Solution   The vector that we want is the direction of $\overrightarrow{P_1P_2}$. To find it, we divide $\overrightarrow{P_1P_2}$ by its own length:

$$\overrightarrow{P_1P_2} = (3 - 1)\mathbf{i} + (2 - 0)\mathbf{j} + (0 - 1)\mathbf{k} = 2\mathbf{i} + 2\mathbf{j} - \mathbf{k},$$

$$|\overrightarrow{P_1P_2}| = \sqrt{(2)^2 + (2)^2 + (-1)^2} = \sqrt{4 + 4 + 1} = \sqrt{9} = 3,$$

$$\mathbf{u} = \frac{\overrightarrow{P_1P_2}}{|\overrightarrow{P_1P_2}|} = \frac{2\mathbf{i} + 2\mathbf{j} - \mathbf{k}}{3} = \frac{2}{3}\mathbf{i} + \frac{2}{3}\mathbf{j} - \frac{1}{3}\mathbf{k}.$$

EXAMPLE 7

Find a vector 6 units long in the direction of $\mathbf{A} = 2\mathbf{i} + 2\mathbf{j} - \mathbf{k}$.

Solution   The vector that we want is

$$6\frac{\mathbf{A}}{|\mathbf{A}|} = 6\frac{2\mathbf{i} + 2\mathbf{j} - \mathbf{k}}{\sqrt{2^2 + 2^2 + (-1)^2}} = 6\frac{2\mathbf{i} + 2\mathbf{j} - \mathbf{k}}{3} = 4\mathbf{i} + 4\mathbf{j} - 2\mathbf{k}.$$

Distance in Space

To find the distance between two points P_1 and P_2 in space, we find the length of $\overrightarrow{P_1P_2}$ (see Fig. 11.17). Equation (4) is the resulting formula:

The Distance between $P_1(x_1, y_1, z_1)$ and $P_2(x_2, y_2, z_2)$

$$|\overrightarrow{P_1P_2}| = \sqrt{(x_2 - x_1)^2 + (y_2 - y_1)^2 + (z_2 - z_1)^2} \qquad (4)$$

EXAMPLE 8

The distance between

$$P_1(2, 1, 5) \qquad \text{and} \qquad P_2(-2, 3, 0)$$

is

$$|\overrightarrow{P_1P_2}| = \sqrt{(-2 - 2)^2 + (3 - 1)^2 + (0 - 5)^2} = \sqrt{16 + 4 + 25}$$
$$= \sqrt{45} = 6.708.$$

Tips for Sketching

Some tips for making pictures of three-dimensional objects look three-dimensional are presented on the following pages. Sketching with a pencil is safer than sketching with a pen because you can erase.

≡ *D R A W I N G   L E S S O N*

How to Sketch Three-dimensional Objects to Look Three-dimensional

1. **Break lines. When one line passes behind another, break it to show that it doesn't touch and that part of it is hidden.**

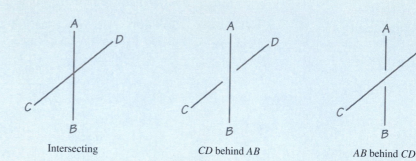

Intersecting   *CD* behind *AB*   *AB* behind *CD*

2. **Make the angle between the positive *x*-axis and the positive *y*-axis large enough.**

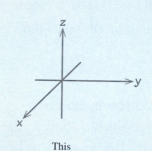
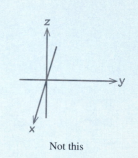

This   Not this

3. **Sketch planes parallel to the coordinate planes as if they were rectangles (sketched as parallelograms) with sides parallel to the coordinate axes.**

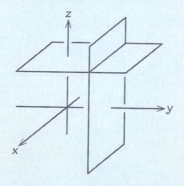

4. **Dot or omit hidden portions of lines. Don't let the line touch the boundary of the parallelogram that represents the plane unless the line lies in the plane.**

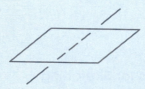

Line below plane   Line above plane   Line *in* plane

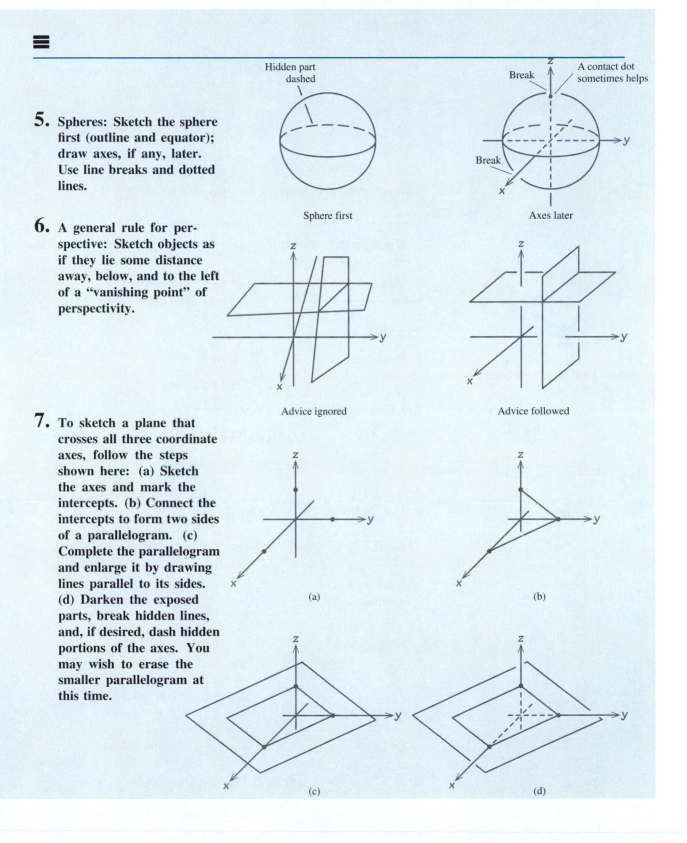

5. Spheres: Sketch the sphere first (outline and equator); draw axes, if any, later. Use line breaks and dotted lines.

Hidden part dashed

Sphere first

A contact dot sometimes helps

Break

Break

Axes later

6. A general rule for perspective: Sketch objects as if they lie some distance away, below, and to the left of a "vanishing point" of perspectivity.

Advice ignored

Advice followed

7. To sketch a plane that crosses all three coordinate axes, follow the steps shown here: (a) Sketch the axes and mark the intercepts. (b) Connect the intercepts to form two sides of a parallelogram. (c) Complete the parallelogram and enlarge it by drawing lines parallel to its sides. (d) Darken the exposed parts, break hidden lines, and, if desired, dash hidden portions of the axes. You may wish to erase the smaller parallelogram at this time.

(a)

(b)

(c)

(d)

Spheres

We can use Eq. (4) to write equations for spheres. Since a point $P(x, y, z)$ lies on the sphere of radius a centered at $P_0(x_0, y_0, z_0)$ if and only if it lies a units from P_0, it lies on the sphere if and only if

$$|\overrightarrow{P_0P}| = a$$

or

$$(x - x_0)^2 + (y - y_0)^2 + (z - z_0)^2 = a^2.$$

This equation is the *standard equation* for a sphere (Fig. 11.19).

The Standard Equation for the Sphere of Radius a with Center (x_0, y_0, z_0)

$$(x - x_0)^2 + (y - y_0)^2 + (z - z_0)^2 = a^2 \tag{5}$$

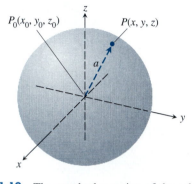

11.19 The standard equation of the sphere of radius a centered at the point (x_0, y_0, z_0) is

$$(x - x_0)^2 + (y - y_0)^2 + (z - z_0)^2 = a^2.$$

EXAMPLE 9

Find the center and radius of the sphere

$$x^2 + y^2 + z^2 + 2x - 4y = 0.$$

Solution Complete the squares in the given equation to obtain

$$x^2 + 2x + 1 + y^2 - 4y + 4 + z^2 = 0 + 1 + 4$$

$$(x + 1)^2 + (y - 2)^2 + z^2 = 5.$$

This is Eq. (5) with $x_0 = -1$, $y_0 = 2$, $z_0 = 0$, and $a = \sqrt{5}$. The center is $(-1, 2, 0)$, and the radius is $\sqrt{5}$. ≡

EXAMPLE 10   Sets Bounded by Spheres or Portions of Spheres

Defining Equations and Inequalities	Description
a) $x^2 + y^2 + z^2 < 4$	The interior of the sphere $x^2 + y^2 + z^2 = 4$.
b) $x^2 + y^2 + z^2 \leq 4$	The solid ball bounded by the sphere $x^2 + y^2 + z^2 = 4$. Alternatively, the sphere $x^2 + y^2 + z^2 = 4$ together with its interior.
c) $x^2 + y^2 + z^2 > 4$	The exterior of the sphere $x^2 + y^2 + z^2 = 4$.
d) $x^2 + y^2 + z^2 = 4, z \leq 0$	The lower hemisphere cut from the sphere $x^2 + y^2 + z^2 = 4$ by the xy-plane (the plane $z = 0$). ≡

EXPLORATION 1

Cross-sections of Spheres

1. Find the cross-section of the sphere $x^2 + y^2 + z^2 = 16$ formed by its intersection with the planes $z = 0, z = -2,$ and $z = 3$.

2. GRAPH the three curves in part 1 in a two-dimensional coordinate system.

3. Show how to set up your grapher to graph the cross section of the sphere for any value of z. If your grapher has list capability, GRAPH the cross-sections for several values of z in one viewing window. (*Suggestion:* Try the list $\{-3.5, -2.5, -1.5, 0, 2, 3, 3.3, 3.7, 3.8, 3.9, 3.95, 3.99\}$, and predict what the viewing window will show.)

4. For each value of z, what is the area, A, of the cross-section? Draw a graph of A plotted against z. ☙

Midpoints of Line Segments

The coordinates of the midpoint M of the line segment joining two points $P_1(x_1, y_1, z_1)$ and $P_2(x_2, y_2, z_2)$ are found by averaging the coordinates of P_1 and P_2. That is,

$$M = \left(\frac{x_1 + x_2}{2}, \frac{y_1 + y_2}{2}, \frac{z_1 + z_2}{2} \right). \tag{6}$$

To see why, we observe that the vector between P_1 and M is $(1/2)\,\overrightarrow{P_1P_2}$ (see Fig. 11.20), so the position vector

$$\overrightarrow{OM} = \overrightarrow{OP_1} + \frac{1}{2}(\overrightarrow{P_1P_2}) = \overrightarrow{OP_1} + \frac{1}{2}(\overrightarrow{OP_2} - \overrightarrow{OP_1})$$

$$= \frac{1}{2}(\overrightarrow{OP_1} + \overrightarrow{OP_2})$$

$$= \frac{x_1 + x_2}{2}\mathbf{i} + \frac{y_1 + y_2}{2}\mathbf{j} + \frac{z_1 + z_2}{2}\mathbf{k}.$$

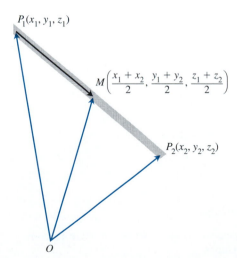

11.20 The coordinates of the point halfway between P_1 and P_2 are found by averaging the coordinates of P_1 and P_2.

Because the scalar components of the position vector $\overrightarrow{OM}$ are identical to the coordinates of M, we have confirmed Eq. (6).

EXAMPLE 11

The midpoint of the segment joining $P_1(3, -2, 0)$ and $P_2(7, 4, 4)$ is

$$\left(\frac{3+7}{2}, \frac{-2+4}{2}, \frac{0+4}{2} \right) = (5, 1, 2).$$

Exercises 11.2

In Exercises 1–12, give a geometric description of each set of points in space whose coordinates satisfy the given pair of equations. Sketch a graph of each set.

1. $x = 2, y = 3$
2. $x = -1, z = 0$

3. $y = 0, z = 0$
4. $x = 1, y = 0$

5. $x^2 + y^2 = 4, z = 0$
6. $x^2 + y^2 = 4, z = -2$

7. $x^2 + z^2 = 4, y = 0$
8. $y^2 + z^2 = 1, x = 0$

9. $x^2 + y^2 + z^2 = 1, x = 0$

10. $x^2 + y^2 + z^2 = 25, y = -4$

11. $x^2 + y^2 + (z+3)^2 = 25, z = 0$

12. $x^2 + (y-1)^2 + z^2 = 4, y = 0$

In Exercises 13–18, describe the sets of points in space whose coordinates satisfy the given inequalities or combinations of equations and inequalities.

13. a) $x \geq 0, y \geq 0, z = 0$

    b) $x \geq 0, y \leq 0, z = 0$

14. a) $0 \leq x \leq 1$

    b) $0 \leq x \leq 1, 0 \leq y \leq 1$

    c) $0 \leq x \leq 1, 0 \leq y \leq 1, 0 \leq z \leq 1$

15. a) $x^2 + y^2 + z^2 \leq 1$

    b) $x^2 + y^2 + z^2 > 1$

16. a) $x^2 + y^2 \leq 1, z = 0$

    b) $x^2 + y^2 \leq 1, z = 3$

    c) $x^2 + y^2 \leq 1$, no restriction on z

17. a) $x^2 + y^2 + z^2 = 1, z \geq 0$

    b) $x^2 + y^2 + z^2 \leq 1, z \geq 0$

18. a) $x = y, z = 0$

    b) $x = y$, no restriction on z

In Exercises 19–28, describe the given set with a single equation or with a pair of equations.

19. The plane perpendicular to the
    a) x-axis at $(3, 0, 0)$,

    b) y-axis at $(0, -1, 0)$,

    c) z-axis at $(0, 0, -2)$

20. The plane through the point $(3, -1, 2)$ perpendicular to the
    a) x-axis,   **b)** y-axis,   **c)** z-axis

21. The plane through the point $(3, -1, 1)$ parallel to the
    a) xy-plane, **b)** yz-plane, **c)** xz-plane

22. The circle of radius 2 centered at $(0, 0, 0)$ and lying in the
    a) xy-plane, **b)** yz-plane, **c)** xz-plane

23. The circle of radius 2 centered at $(0, 2, 0)$ and lying in the
    a) xy-plane, **b)** yz-plane, **c)** plane $y = 2$

24. The circle of radius 1 centered at $(-3, 4, 1)$ and lying in a plane parallel to the
    a) xy-plane, **b)** yz-plane, **c)** xz-plane

25. The line through the point $(1, 3, -1)$ parallel to the
    a) x-axis,   **b)** y-axis,   **c)** z-axis

26. The set of points in space equidistant from the origin and the point $(0, 2, 0)$

27. The circle in which the plane through the point $(1, 1, 3)$ perpendicular to the z-axis meets the sphere of radius 5 centered at the origin

28. The set of points in space that lie 2 units from both point $(0, 0, 1)$ and, point $(0, 0, -1)$

Write inequalities to describe the sets in Exercises 29–34. Sketch a graph of each set.

29. The slab bounded by the planes $z = 0$ and $z = 1$ (planes included)

30. The solid cube in the first octant bounded by the planes $x = 2, y = 2$, and $z = 2$

31. The half-space consisting of the points on and below the xy-plane

32. The upper hemisphere of the sphere of radius 1 centered at the origin

33. The (a) interior and (b) exterior of the sphere of radius 1 centered at the point $(1, 1, 1)$

34. The closed region bounded by the spheres of radius 1 and radius 2 centered at the origin. (*Closed* means that the spheres are to be included. Had we wanted the spheres left out, we would have asked for the *open* region bounded by the spheres. This is analogous to the way we use "closed" and "open" to describe intervals: "Closed" means endpoints included, "open" means endpoints left out. Closed sets include boundaries; open sets leave them out.)

Find the lengths and directions of the vectors in Exercises 35–46.

35. $2\mathbf{i} + \mathbf{j} - 2\mathbf{k}$

36. $3\mathbf{i} - 6\mathbf{j} + 2\mathbf{k}$

37. $\mathbf{i} + 4\mathbf{j} - 8\mathbf{k}$

38. $9\mathbf{i} - 2\mathbf{j} + 6\mathbf{k}$

39. $5\mathbf{k}$

40. $6\mathbf{i}$

41. $-4\mathbf{j}$

42. $\dfrac{3}{5}\mathbf{i} + \dfrac{4}{5}\mathbf{k}$

43. $-\dfrac{1}{3}\mathbf{j} + \dfrac{1}{4}\mathbf{k}$

44. $\dfrac{1}{\sqrt{2}}\mathbf{i} - \dfrac{1}{\sqrt{2}}\mathbf{k}$

45. $\dfrac{1}{\sqrt{6}}\mathbf{i} - \dfrac{1}{\sqrt{6}}\mathbf{j} - \dfrac{1}{\sqrt{6}}\mathbf{k}$

46. $\dfrac{\mathbf{i}}{\sqrt{3}} + \dfrac{\mathbf{j}}{\sqrt{3}} + \dfrac{\mathbf{k}}{\sqrt{3}}$

In Exercises 47–52, find

a) the distance between points P_1 and P_2,

b) the direction of $\overrightarrow{P_1 P_2}$,

c) the midpoint of line segment $P_1 P_2$.

47. $P_1(1, 1, 1)$,　$P_2(3, 3, 0)$

48. $P_1(-1, 1, 5)$,　$P_2(2, 5, 0)$

49. $P_1(1, 4, 5)$,　$P_2(4, -2, 7)$

50. $P_1(3, 4, 5)$,　$P_2(2, 3, 4)$

51. $P_1(0, 0, 0)$,　$P_2(2, -2, -2)$

52. $P_1(5, 3, -2)$,　$P_2(0, 0, 0)$

In Exercises 53 and 54, find the vectors whose lengths and directions are given. Do the calculations mentally. Sketch the vector.

53.

	Length	Direction
a)	2	$\mathbf{i}$
b)	$\sqrt{3}$	$-\mathbf{k}$
c)	$\dfrac{1}{2}$	$\dfrac{3}{5}\mathbf{j} + \dfrac{4}{5}\mathbf{k}$
d)	7	$\dfrac{6}{7}\mathbf{i} - \dfrac{2}{7}\mathbf{j} + \dfrac{3}{7}\mathbf{k}$

54.

	Length	Direction
a)	7	$-\mathbf{j}$
b)	$\sqrt{2}$	$-\dfrac{3}{5}\mathbf{i} - \dfrac{4}{5}\mathbf{k}$
c)	$\dfrac{13}{12}$	$\dfrac{3}{13}\mathbf{i} - \dfrac{4}{13}\mathbf{j} - \dfrac{12}{13}\mathbf{k}$
d)	$a > 0$	$\dfrac{1}{\sqrt{2}}\mathbf{i} + \dfrac{1}{\sqrt{3}}\mathbf{j} - \dfrac{1}{\sqrt{6}}\mathbf{k}$

55. Find a vector of magnitude 7 in the direction of $\mathbf{A} = 12\mathbf{i} - 5\mathbf{k}$.

56. Find a vector $\sqrt{5}$ units long in the direction of $\mathbf{A} = \mathbf{i} + \mathbf{j} + \mathbf{k}$.

57. Find a vector 5 units long in the direction opposite to the direction of $\mathbf{A} = 2\mathbf{i} - 3\mathbf{j} + 6\mathbf{k}$.

58. Find a vector of magnitude 3 in the direction opposite to the direction of $\mathbf{A} = (1/2)\mathbf{i} - (1/2)\mathbf{j} - (1/2)\mathbf{k}$.

Spheres and Distance

59. Find the centers and radii of the following spheres.

a) $(x + 2)^2 + y^2 + (z - 2)^2 = 8$

b) $\left(x + \dfrac{1}{2}\right)^2 + \left(y + \dfrac{1}{2}\right)^2 + \left(z + \dfrac{1}{2}\right)^2 = \dfrac{21}{4}$

c) $(x - \sqrt{2})^2 + (y - \sqrt{2})^2 + (z + \sqrt{2})^2 = 2$

d) $x^2 + \left(y + \dfrac{1}{3}\right)^2 + \left(z - \dfrac{1}{3}\right)^2 = \dfrac{29}{9}$

60. Find equations for the spheres whose centers and radii are given here.

	Center	Radius
a)	$(1, 2, 3)$	$\sqrt{14}$
b)	$(0, -1, 5)$	2
c)	$(-2, 0, 0)$	$\sqrt{3}$
d)	$(0, -7, 0)$	7

Find the centers and radii of the spheres in Exercises 61–64.

61. $x^2 + y^2 + z^2 + 4x - 4z = 0$

62. $2x^2 + 2y^2 + 2z^2 + x + y + z = 9$

63. $x^2 + y^2 + z^2 - 2z = 0$

64. $3x^2 + 3y^2 + 3z^2 + 2y - 2z = 9$

65. a) Find the intersection of the sphere $x^2 + y^2 + z^2 = 9$ with the planes $y = -2$, $y = 0$, and $y = 1$.

b) GRAPH the three curves in part (a) in a two-dimensional coordinate system, and compare the areas bounded by the curves.

66. a) Find the intersection of the sphere $x^2 + y^2 + z^2 = 9$ with the planes $x = -1$, $x = 0$, and $x = 2$.

b) GRAPH the three curves in part (a) in a two-dimensional coordinate system, and compare the areas bounded by the curves.

67. a) Find the intersection of the sphere in Exercise 61 with the planes $z = 2$ and $z = 4$.

b) GRAPH the two curves in part (a) in a two-dimensional coordinate system.

68. a) Find the intersection of the sphere in Exercise 61 with the planes $x = -2$ and $x = 0$.

b) GRAPH the two curves in part (a) in a two-dimensional coordinate system.

69. Find a formula for the distance from the point $P(x, y, z)$ to the

a) x-axis,   **b)** y-axis,   **c)** z-axis.

70. Find a formula for the distance from the point $P(x, y, z)$ to the

a) xy-plane,   **b)** yz-plane,   **c)** xz-plane.

11.3 _____   Dot Products

We now introduce the *dot product* of two vectors, the first of the two methods that we will learn for multiplying vectors together. Our motivation is the need to calculate the work done by a constant force in displacing a mass. When the force and displacement are represented as vectors, the dot product of the two vectors gives the work done by the force during the displacement.

Dot products, also called *scalar products* because the resulting products are numbers and not vectors, have applications in mathematics as well as in engineering and physics. In this section, we present the algebraic and geometric properties on which many of these applications depend. The second kind of vector product, the *cross product*, will be described in the next section.

DEFINITION

The **dot product A · B** ("**A** dot **B**") or **scalar product** of two vectors **A** and **B** is the number

$$\mathbf{A} \cdot \mathbf{B} = |\mathbf{A}||\mathbf{B}| \cos \theta, \tag{1}$$

where θ measures the angle ($0 \leq \theta \leq \pi$) made by representative **A** and **B** vectors with the same endpoint (as in Fig. 11.21).

In words, the dot product of **A** and **B** is the length of **A** times the length of **B** times the cosine of the angle between **A** and **B**. The product is a scalar, not a vector. It is called the dot product because of the dot in the notation **A** · **B**.

From Eq. (1), we see that the dot product of two vectors is positive when the angle between them is acute, zero when the angle is right, and negative when the angle is obtuse.

Since the angle that a vector **A** makes with itself is zero and the cosine of the angle is 1,

$$\mathbf{A} \cdot \mathbf{A} = |\mathbf{A}||\mathbf{A}| \cos 0 = |\mathbf{A}||\mathbf{A}|(1) = |\mathbf{A}|^2 \qquad \text{or} \qquad |\mathbf{A}| = \sqrt{\mathbf{A} \cdot \mathbf{A}}. \tag{2}$$

This gives us a handy way to calculate a vector's length, as we will see.

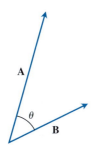

11.21 **A** · **B** is the number $|\mathbf{A}||\mathbf{B}| \cos \theta$.

Calculation

To calculate **A** · **B** from the components of **A** and **B**, we let

$$\mathbf{A} = a_1\mathbf{i} + a_2\mathbf{j} + a_3\mathbf{k},$$
$$\mathbf{B} = b_1\mathbf{i} + b_2\mathbf{j} + b_3\mathbf{k},$$

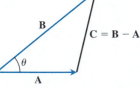

11.22 Equation (3) is obtained by applying the law of cosines to a triangle whose sides represent **A**, **B**, and **C** = **B** − **A**.

and

$$\mathbf{C} = \mathbf{B} - \mathbf{A} = (b_1 - a_1)\mathbf{i} + (b_2 - a_2)\mathbf{j} + (b_3 - a_3)\mathbf{k}.$$

Then we apply the law of cosines to a triangle (Fig. 11.22) whose sides represent the vectors **A**, **B**, and **C** and obtain

$$|\mathbf{C}|^2 = |\mathbf{A}|^2 + |\mathbf{B}|^2 - 2|\mathbf{A}||\mathbf{B}|\cos\theta,$$

$$|\mathbf{A}||\mathbf{B}|\cos\theta = \frac{|\mathbf{A}|^2 + |\mathbf{B}|^2 - |\mathbf{C}|^2}{2}.$$

The left side of this equation is **A** · **B**, and we may evaluate the right side by applying Eq. (1) of Section 11.2 to find the lengths of **A**, **B**, and **C**. The result of this algebra (see Exercise 43) is the formula

$$\mathbf{A} \cdot \mathbf{B} = a_1b_1 + a_2b_2 + a_3b_3. \tag{3}$$

Thus, to find the dot product of two given vectors, we multiply their corresponding **i**, **j**, and **k** components and add the results. In particular, from Eq. (2), we have

$$|\mathbf{A}| = \sqrt{\mathbf{A} \cdot \mathbf{A}} = \sqrt{a_1^2 + a_2^2 + a_3^2}. \tag{4}$$

When we solve Eq. (1) for θ, we get a formula for finding the angle between two nonzero vectors.

The Angle between Two Vectors

The angle between two nonzero vectors **A** and **B** is

$$\theta = \cos^{-1}\left(\frac{\mathbf{A} \cdot \mathbf{B}}{|\mathbf{A}||\mathbf{B}|}\right). \tag{5}$$

EXAMPLE 1

Find the angle between $\mathbf{A} = \mathbf{i} - 2\mathbf{j} - 2\mathbf{k}$ and $\mathbf{B} = 6\mathbf{i} + 3\mathbf{j} + 2\mathbf{k}$.

Solution We use Eq. (5):

$$\mathbf{A} \cdot \mathbf{B} = (1)(6) + (-2)(3) + (-2)(2) = 6 - 6 - 4 = -4$$

$$|\mathbf{A}| = \sqrt{\mathbf{A} \cdot \mathbf{A}} = \sqrt{(1)^2 + (-2)^2 + (-2)^2} = \sqrt{9} = 3$$

$$|\mathbf{B}| = \sqrt{\mathbf{B} \cdot \mathbf{B}} = \sqrt{(6)^2 + (3)^2 + (2)^2} = \sqrt{49} = 7$$

$$\theta = \cos^{-1}\left(\frac{\mathbf{A} \cdot \mathbf{B}}{|\mathbf{A}||\mathbf{B}|}\right) \qquad \text{Eq. 5}$$

$$= \cos^{-1}\left(\frac{-4}{(3)(7)}\right)$$

$$= \cos^{-1}\left(-\frac{4}{21}\right)$$

$$= 100.981° = 1.762 \text{ radians}$$

Laws of Multiplication

From the equation $\mathbf{A} \cdot \mathbf{B} = a_1b_1 + a_2b_2 + a_3b_3$, we can see right away that

$$\mathbf{A} \cdot \mathbf{B} = \mathbf{B} \cdot \mathbf{A}. \qquad (6)$$

In other words, dot multiplication is commutative. We can also see from Eq. (3) that if c is any number, then

$$(c\mathbf{A}) \cdot \mathbf{B} = \mathbf{A} \cdot (c\mathbf{B}) = c(\mathbf{A} \cdot \mathbf{B}).$$

If $\mathbf{C} = c_1\mathbf{i} + c_2\mathbf{j} + c_3\mathbf{k}$ is any third vector, then

$$\mathbf{A} \cdot (\mathbf{B} + \mathbf{C}) = a_1(b_1 + c_1) + a_2(b_2 + c_2) + a_3(b_3 + c_3)$$

$$= (a_1b_1 + a_2b_2 + a_3b_3) + (a_1c_1 + a_2c_2 + a_3c_3)$$

$$= \mathbf{A} \cdot \mathbf{B} + \mathbf{A} \cdot \mathbf{C}.$$

Hence, dot products obey a distributive law:

$$\mathbf{A} \cdot (\mathbf{B} + \mathbf{C}) = \mathbf{A} \cdot \mathbf{B} + \mathbf{A} \cdot \mathbf{C}. \qquad (7)$$

EXPLORATION BIT

Equation (7) is an answer to part (3) of the Exploration Bit at the start of this section. There is, however, one special fact about the operations in Eq. (7) that makes this distributive law different from other familiar distributive laws. Do you know what it is? Comment on how this fact relates also to Eqs. (8) and (9).

If we combine this with the commutative law, Eq. (6), it is also evident that

$$(\mathbf{A} + \mathbf{B}) \cdot \mathbf{C} = \mathbf{A} \cdot \mathbf{C} + \mathbf{B} \cdot \mathbf{C}. \qquad (8)$$

Equations (7) and (8) together permit us to multiply sums of vectors by the familiar laws of algebra. For example,

$$(\mathbf{A} + \mathbf{B}) \cdot (\mathbf{C} + \mathbf{D}) = \mathbf{A} \cdot \mathbf{C} + \mathbf{A} \cdot \mathbf{D} + \mathbf{B} \cdot \mathbf{C} + \mathbf{B} \cdot \mathbf{D}. \qquad (9)$$

Orthogonal Vectors

For the vectors that we study, "orthogonal" means the same as "perpendicular." In some scientific contexts in which the word "orthogonal" is used, there is no such geometric interpretation.

Two vectors whose dot product is zero are said to be *orthogonal*.

DEFINITION

Vectors $\mathbf{A}$ and $\mathbf{B}$ are **orthogonal** if $\mathbf{A} \cdot \mathbf{B} = 0$.

The zero vector $\mathbf{0} = 0\mathbf{i} + 0\mathbf{j} + 0\mathbf{k}$ is orthogonal to every vector because its dot product with every vector is zero. When neither $|\mathbf{A}|$ nor $|\mathbf{B}|$ is zero, the equation $\mathbf{A} \cdot \mathbf{B} = |\mathbf{A}||\mathbf{B}| \cos\theta$ tells us that $\mathbf{A} \cdot \mathbf{B}$ is zero if and only if $\cos\theta$ is zero, that is, when θ equals $\pi/2$.

EXAMPLE 2

The vectors $\mathbf{A} = 3\mathbf{i} - 2\mathbf{j} + \mathbf{k}$ and $\mathbf{B} = 2\mathbf{j} + 4\mathbf{k}$ are orthogonal because

$$\mathbf{A} \cdot \mathbf{B} = (3)(0) + (-2)(2) + (1)(4) = 0.$$ ≡

Vector Projections and Scalar Components

The vector we get by *projecting* a vector $\mathbf{B}$ onto the line through a vector $\mathbf{A}$ is called the **vector projection of B onto A**, sometimes denoted

$$\text{proj}_{\mathbf{A}}\mathbf{B} \qquad (\text{"the vector projection of } \mathbf{B} \text{ onto } \mathbf{A}\text{"}).$$

If $\mathbf{B}$ represents a force, then the vector projection of $\mathbf{B}$ onto $\mathbf{A}$ represents the effective force in the direction of $\mathbf{A}$ (Fig. 11.23).

If the angle between $\mathbf{B}$ and $\mathbf{A}$ is acute, the length of the vector projection of $\mathbf{B}$ onto $\mathbf{A}$ is $|\mathbf{B}| \cos\theta$. If the angle is obtuse, its cosine is negative, and the length of the vector projection of $\mathbf{B}$ onto $\mathbf{A}$ is $-|\mathbf{B}| \cos\theta$. In either case, the number $|\mathbf{B}| \cos\theta$ is called the **scalar component of B in the direction of A** (Fig. 11.24).

The scalar component of $\mathbf{B}$ in the direction of $\mathbf{A}$ can be found by dividing both sides of the equation $\mathbf{A} \cdot \mathbf{B} = |\mathbf{A}||\mathbf{B}| \cos\theta$ by $|\mathbf{A}|$. This gives

$$|\mathbf{B}| \cos\theta = \frac{\mathbf{A} \cdot \mathbf{B}}{|\mathbf{A}|} = \mathbf{B} \cdot \frac{\mathbf{A}}{|\mathbf{A}|}. \tag{10}$$

Equation (10) says that the scalar component of $\mathbf{B}$ in the direction of $\mathbf{A}$ can be obtained by "dotting" $\mathbf{B}$ with the direction of $\mathbf{A}$.

The vector projection of $\mathbf{B}$ onto $\mathbf{A}$ is the scalar component of $\mathbf{B}$ in the direction of $\mathbf{A}$ times the direction of $\mathbf{A}$. If the angle between $\mathbf{A}$ and $\mathbf{B}$ is acute, the vector projection has length $|\mathbf{B}| \cos\theta$ and direction $\mathbf{A}/|\mathbf{A}|$. If the angle is obtuse, the vector projection has length $-|\mathbf{B}| \cos\theta$, and direction $-\mathbf{A}/|\mathbf{A}|$. In either case, we have the following.

$$\text{proj}_{\mathbf{A}}\mathbf{B} = (|\mathbf{B}| \cos\theta)\frac{\mathbf{A}}{|\mathbf{A}|} \tag{11}$$

Equations (10) and (11) together give a useful way to find $\text{proj}_{\mathbf{A}}\mathbf{B}$:

$$\text{proj}_{\mathbf{A}}\mathbf{B} = (|\mathbf{B}| \cos\theta)\frac{\mathbf{A}}{|\mathbf{A}|} \qquad \text{Eq. (11)}$$

$$= \left(\frac{\mathbf{A} \cdot \mathbf{B}}{|\mathbf{A}|}\right)\frac{\mathbf{A}}{|\mathbf{A}|} \qquad \text{Eq. (10)}$$

$$= \frac{\mathbf{A} \cdot \mathbf{B}}{\mathbf{A} \cdot \mathbf{A}}\mathbf{A}. \qquad |\mathbf{A}|^2 = \mathbf{A} \cdot \mathbf{A} \tag{12}$$

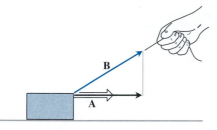

11.23 Pulling on the box has the effect of moving the box in the direction of $\mathbf{A}$. The effective force in this direction is represented by the vector projection of $\mathbf{B}$ onto $\mathbf{A}$.

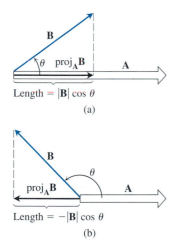

11.24 (a) When θ is acute, the length of the vector projection of $\mathbf{B}$ onto $\mathbf{A}$ is $|\mathbf{B}| \cos\theta$, the scalar component of $\mathbf{B}$ in the direction of $\mathbf{A}$. (b) When θ is obtuse, the scalar component of $\mathbf{B}$ in the direction of $\mathbf{A}$ is negative, and the length of the vector projection of $\mathbf{B}$ onto $\mathbf{A}$ is $-|\mathbf{B}| \cos\theta$.

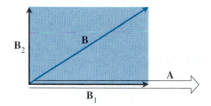

11.25 The vector **B** as the sum of vectors parallel and orthogonal to **A**. Vectors **A** and **B** can be thought of as determining one plane then the vectors **B**$_1$ parallel to **A** and **B**$_2$ orthogonal to **A** are a resolution of **B** in that plane.

EXAMPLE 3

Find the vector projection of $\mathbf{B} = 6\mathbf{i} + 3\mathbf{j} + 2\mathbf{k}$ onto $\mathbf{A} = \mathbf{i} - 2\mathbf{j} - 2\mathbf{k}$ and the scalar component of **B** in the direction of **A**.

Solution   We find $\text{proj}_{\mathbf{A}}\mathbf{B}$ from Eq. (12):

$$\text{proj}_{\mathbf{A}}\mathbf{B} = \frac{\mathbf{A} \cdot \mathbf{B}}{\mathbf{A} \cdot \mathbf{A}}\mathbf{A} = \frac{6 - 6 - 4}{1 + 4 + 4}(\mathbf{i} - 2\mathbf{j} - 2\mathbf{k})$$

$$= -\frac{4}{9}(\mathbf{i} - 2\mathbf{j} - 2\mathbf{k}) = -\frac{4}{9}\mathbf{i} + \frac{8}{9}\mathbf{j} + \frac{8}{9}\mathbf{k}.$$

We find the scalar component of **B** in the direction of **A** from Eq. (10):

$$|\mathbf{B}| \cos\theta = \mathbf{B} \cdot \frac{\mathbf{A}}{|\mathbf{A}|} = (6\mathbf{i} + 3\mathbf{j} + 2\mathbf{k}) \cdot \left(\frac{1}{3}\mathbf{i} - \frac{2}{3}\mathbf{j} - \frac{2}{3}\mathbf{k}\right)$$

$$= 2 - 2 - \frac{4}{3} = -\frac{4}{3}.$$

≡

Writing a Vector as a Sum of Orthogonal Vectors

In mechanics, we often want to express a vector **B** as a sum of a vector **B**$_1$ parallel to a vector **A** and a vector **B**$_2$ orthogonal to **A** (Fig. 11.25). We can do this by writing **B** as a sum of its vector projection onto **A** plus whatever is left over, because the leftover part will automatically be orthogonal to **A**.

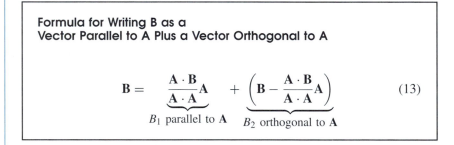

**Formula for Writing B as a
Vector Parallel to A Plus a Vector Orthogonal to A**

$$\mathbf{B} = \underbrace{\frac{\mathbf{A} \cdot \mathbf{B}}{\mathbf{A} \cdot \mathbf{A}}\mathbf{A}}_{B_1 \text{ parallel to } \mathbf{A}} + \underbrace{\left(\mathbf{B} - \frac{\mathbf{A} \cdot \mathbf{B}}{\mathbf{A} \cdot \mathbf{A}}\mathbf{A}\right)}_{B_2 \text{ orthogonal to } \mathbf{A}} \qquad (13)$$

The vector **B**$_1$, being the vector projection of **B** onto **A**, is parallel to **A**, while **B**$_2$ can be seen to be orthogonal to **A** because $\mathbf{A} \cdot \mathbf{B}_2$ is zero:

$$\mathbf{A} \cdot \mathbf{B}_2 = \mathbf{A} \cdot \left(\mathbf{B} - \frac{\mathbf{A} \cdot \mathbf{B}}{\mathbf{A} \cdot \mathbf{A}}\mathbf{A}\right) = \mathbf{A} \cdot \mathbf{B} - \frac{\mathbf{A} \cdot \mathbf{B}}{\mathbf{A} \cdot \mathbf{A}}\mathbf{A} \cdot \mathbf{A} = \mathbf{A} \cdot \mathbf{B} - \mathbf{A} \cdot \mathbf{B} = 0.$$

EXAMPLE 4

Express $\mathbf{B} = 2\mathbf{i} + \mathbf{j} - 3\mathbf{k}$ as the sum of a vector parallel to $\mathbf{A} = 3\mathbf{i} - \mathbf{j}$ and a vector orthogonal to **A**.

Solution   We use Eq. (13). With

$$\mathbf{A} \cdot \mathbf{B} = 6 - 1 = 5 \qquad \text{and} \qquad \mathbf{A} \cdot \mathbf{A} = 9 + 1 = 10,$$

Eq. (13) gives

$$\mathbf{B} = \frac{\mathbf{A} \cdot \mathbf{B}}{\mathbf{A} \cdot \mathbf{A}}\mathbf{A} + \left(\mathbf{B} - \frac{\mathbf{A} \cdot \mathbf{B}}{\mathbf{A} \cdot \mathbf{A}}\mathbf{A}\right) = \frac{5}{10}(3\mathbf{i} - \mathbf{j}) + \left(2\mathbf{i} + \mathbf{j} - 3\mathbf{k} - \frac{5}{10}(3\mathbf{i} - \mathbf{j})\right)$$

$$= \left(\frac{3}{2}\mathbf{i} - \frac{1}{2}\mathbf{j}\right) + \left(\frac{1}{2}\mathbf{i} + \frac{3}{2}\mathbf{j} - 3\mathbf{k}\right).$$

parallel to A   orthogonal to A

Check: The first vector in the sum is parallel to **A** because it is $(1/2)\mathbf{A}$. The second vector in the sum is orthogonal to **A** because

$$\left(\frac{1}{2}\mathbf{i} + \frac{3}{2}\mathbf{j} - 3\mathbf{k}\right) \cdot (3\mathbf{i} - \mathbf{j}) = \frac{3}{2} - \frac{3}{2} = 0. \qquad \equiv$$

Lines in the Plane and Distances from Points to Lines

Dot products give a new understanding of the equations that we write for lines in the plane and a quick way to calculate distances from points to lines.

Suppose L is a line through the point $P_0(x_0, y_0)$ perpendicular to a vector $\mathbf{N} = A\mathbf{i} + B\mathbf{j}$ (Fig. 11.26). Then a point $P(x, y)$ lies on L if and only if

$$\mathbf{N} \cdot \overrightarrow{P_0P} = 0 \qquad \text{or} \qquad A(x - x_0) + B(y - y_0) = 0.$$

When rearranged, the second equation becomes

$$Ax + By = Ax_0 + By_0.$$

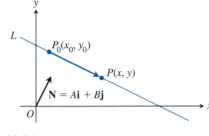

11.26 A point $P(x, y)$ lies on the line through P_0 perpendicular to **N** if and only if $\mathbf{N} \cdot \overrightarrow{P_0P} = 0$.

The Line through $P(x_0, y_0)$ Perpendicular to $\mathbf{N} = A\mathbf{i} + B\mathbf{j}$

$$Ax + By = C, \qquad C = Ax_0 + By_0 \qquad (14)$$

Notice how the components of **N** become coefficients in the equation $Ax + By = C$.

EXAMPLE 5

Find an equation for the line through $P_0(3, 5)$ perpendicular to $\mathbf{N} = \mathbf{i} + 2\mathbf{j}$.

Solution We use Eq. (14) with $A = 1$, $B = 2$, and $C = (1)(3) + (2)(5) = 13$ to get

$$x + 2y = 13. \qquad \equiv$$

EXAMPLE 6

Find the distance from the point $S(4, 4)$ to the line L: $x + 3y = 6$.

Solution We find a point P on the line and calculate the distance as the length of the vector projection of $\overrightarrow{PS}$ onto a vector **N** perpendicular to the line (Fig. 11.27). Any point on the line will do for P, and we can find **N** from the coefficients of $x + 3y = 6$ as $\mathbf{N} = \mathbf{i} + 3\mathbf{j}$. With $P = (0, 2)$, say, we

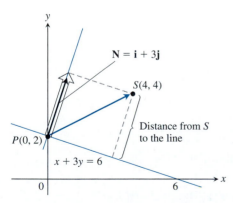

11.27 The distance from point S to the line is the length of the vector projection of $\overrightarrow{PS}$ onto **N**.

To find the distance from a point S to the line L: $Ax + By = C$, find

1. a point P on L,

2. $\overrightarrow{PS}$, and

3. the direction of $\mathbf{N} = A\mathbf{i} + B\mathbf{j}$.

Then calculate the distance as

$$\left| \overrightarrow{PS} \cdot \frac{\mathbf{N}}{|\mathbf{N}|} \right|.$$

then have

$$\overrightarrow{PS} = (4 - 0)\mathbf{i} + (4 - 2)\mathbf{j} = 4\mathbf{i} + 2\mathbf{j}.$$

$$\text{Distance from } S \text{ to } L = |\text{proj}_{\mathbf{N}} \overrightarrow{PS}| = \left| \overrightarrow{PS} \cdot \frac{\mathbf{N}}{|\mathbf{N}|} \right|$$

$$= \left| (4\mathbf{i} + 2\mathbf{j}) \cdot \frac{\mathbf{i} + 3\mathbf{j}}{\sqrt{(1)^2 + (3)^2}} \right| = \frac{4 + 6}{\sqrt{10}} = \sqrt{10}. \quad \blacksquare$$

Work

To return to the situation mentioned at the beginning of this section, in mechanics, the work done by a constant force $\mathbf{F}$ when the point of application undergoes a displacement $\overrightarrow{PQ}$ (Fig. 11.28) is defined to be the dot product of $\mathbf{F}$ with $\overrightarrow{PQ}$.

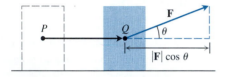

11.28 The work done by a constant force $\mathbf{F}$ during a displacement $\overrightarrow{PQ}$ is $(|\mathbf{F}| \cos \theta)|\overrightarrow{PQ}| = F \cdot \overrightarrow{PQ}$.

DEFINITION

The **work** done by a constant force $\mathbf{F}$ acting through a displacement $\overrightarrow{PQ}$ is

$$\text{Work} = \mathbf{F} \cdot \overrightarrow{PQ} = |\mathbf{F}||\overrightarrow{PQ}| \cos \theta. \tag{15}$$

EXAMPLE 7

If $|\mathbf{F}| = 40$ newtons (about 9 pounds), $|\overrightarrow{PQ}| = 3$ m, and $\theta = 50°$, the work done by $\mathbf{F}$ in acting from P to Q is

$$\begin{aligned}
\text{Work} &= |\mathbf{F}||\overrightarrow{PQ}| \cos \theta & \text{Eq. (15)} \\
&= (40)(3) \cos 50° & \text{Given values} \\
&= 77.135 \text{ newton-meters.} & \blacksquare
\end{aligned}$$

Exercises 11.3

In Exercises 1–12, find $\mathbf{A} \cdot \mathbf{B}$, $|\mathbf{A}|$, $|\mathbf{B}|$, the cosine of the angle between $\mathbf{A}$ and $\mathbf{B}$, the scalar component of $\mathbf{B}$ in the direction of $\mathbf{A}$, and the vector projection of $\mathbf{B}$ onto $\mathbf{A}$.

1. $\mathbf{A} = 3\mathbf{i} + 2\mathbf{j}$, $\mathbf{B} = 5\mathbf{j} + \mathbf{k}$

2. $\mathbf{A} = \mathbf{i}$, $\mathbf{B} = 5\mathbf{j} - 3\mathbf{k}$

3. $\mathbf{A} = 3\mathbf{i} - 2\mathbf{j} - \mathbf{k}$, $\mathbf{B} = -2\mathbf{j}$

4. $\mathbf{A} = -2\mathbf{i} + 7\mathbf{j}$, $\mathbf{B} = \mathbf{k}$

5. $\mathbf{A} = 5\mathbf{j} - 3\mathbf{k}$, $\mathbf{B} = \mathbf{i} + \mathbf{j} + \mathbf{k}$

6. $\mathbf{A} = \dfrac{1}{\sqrt{2}}\mathbf{i} + \dfrac{1}{\sqrt{3}}\mathbf{j} + \dfrac{1}{\sqrt{6}}\mathbf{k}$, $\mathbf{B} = \dfrac{1}{\sqrt{2}}\mathbf{j} - \mathbf{k}$

7. $\mathbf{A} = -\mathbf{i} + \mathbf{j}$, $\mathbf{B} = \sqrt{2}\mathbf{i} + \sqrt{3}\mathbf{j} + 2\mathbf{k}$

8. $\mathbf{A} = \mathbf{i} + \mathbf{k}$, $\mathbf{B} = \mathbf{i} + \mathbf{j} + \mathbf{k}$

9. $\mathbf{A} = 2\mathbf{i} - 4\mathbf{j} + \sqrt{5}\mathbf{k}$, $\mathbf{B} = -2\mathbf{i} + 4\mathbf{j} - \sqrt{5}\mathbf{k}$

10. $\mathbf{A} = -5\mathbf{i} + \mathbf{j}$, $\mathbf{B} = 2\mathbf{i} + \sqrt{17}\mathbf{j} + 10\mathbf{k}$

11. $\mathbf{A} = 10\mathbf{i} + 11\mathbf{j} - 2\mathbf{k}$, $\mathbf{B} = 3\mathbf{j} + 4\mathbf{k}$

12. $\mathbf{A} = 2\mathbf{i} + 10\mathbf{j} - 11\mathbf{k}$, $\mathbf{B} = 2\mathbf{i} + 2\mathbf{j} + \mathbf{k}$

13. Write $\mathbf{B} = 3\mathbf{j} + 4\mathbf{k}$ as the sum of a vector parallel to $\mathbf{A} = \mathbf{i} + \mathbf{j}$ and a vector orthogonal to $\mathbf{A}$.

14. Write $\mathbf{B} = \mathbf{j} + \mathbf{k}$ as the sum of a vector parallel to $\mathbf{A} = \mathbf{i} + \mathbf{j}$ and a vector orthogonal to $\mathbf{A}$.

15. Write $\mathbf{B} = 8\mathbf{i} + 4\mathbf{j} - 12\mathbf{k}$ as the sum of a vector parallel to $\mathbf{A} = \mathbf{i} + 2\mathbf{j} - \mathbf{k}$ and a vector orthogonal to $\mathbf{A}$.

16. $\mathbf{B} = \mathbf{i} + (\mathbf{j} + \mathbf{k})$ is already the sum of a vector parallel to $\mathbf{i}$ and a vector orthogonal to $\mathbf{i}$. If you use Eq. (13) with $\mathbf{A} = \mathbf{i}$, do you get $\mathbf{B}_1 = \mathbf{i}$ and $\mathbf{B}_2 = \mathbf{j} + \mathbf{k}$? Try it and find out.

In Exercises 17–20, find an equation for the line in the xy-plane that passes through the given point perpendicular to the given vector. Then sketch the line. Include the vector in your sketch as a vector starting at the origin (as in Fig. 11.26).

17. $P(2, 1), \quad \mathbf{i} + 2\mathbf{j}$

18. $P(-2, 1), \quad \mathbf{i} - \mathbf{j}$

19. $P(-1, 2), \quad -2\mathbf{i} - \mathbf{j}$

20. $P(-1, 2), \quad 2\mathbf{i} - 3\mathbf{j}$

In Exercises 21–24, find the distance in the xy-plane from the point to the line.

21. $S(2, 8), \quad x + 3y = 6$ **22.** $S(0, 0), \quad x + 3y = 6$

23. $S(2, 1), \quad x + y = 1$ **24.** $S(1, 3), \quad y = -2x$

25. Show that the vectors

$$\mathbf{A} = \frac{1}{\sqrt{3}}(\mathbf{i} - \mathbf{j} + \mathbf{k}), \quad \mathbf{B} = \frac{1}{\sqrt{2}}(\mathbf{j} + \mathbf{k}),$$

$$\mathbf{C} = \frac{1}{\sqrt{6}}(-2\mathbf{i} - \mathbf{j} + \mathbf{k})$$

are orthogonal to one another.

26. Find the vector projections of $\mathbf{D} = \mathbf{i} + \mathbf{j} + \mathbf{k}$ onto the vectors $\mathbf{A}$, $\mathbf{B}$, and $\mathbf{C}$ of Exercise 25. Then show that $\mathbf{D}$ is the sum of these vector projections.

27. *Cancellation in dot products is risky.* In real-number multiplication, if $ab_1 = ab_2$ and a is not zero, we can safely cancel the a and conclude that $b_1 = b_2$. Not so for vector multiplication—if $\mathbf{A} \cdot \mathbf{B}_1 = \mathbf{A} \cdot \mathbf{B}_2$ and $\mathbf{A} \neq \mathbf{0}$, it is not safe to conclude that $\mathbf{B}_1 = \mathbf{B}_2$. See whether you can come up with an example. Keep it simple: Experiment with $\mathbf{i}$, $\mathbf{j}$, and $\mathbf{k}$.

28. *Sums and differences.* In the diagram below, it looks as if $\mathbf{v}_1 + \mathbf{v}_2$ and $\mathbf{v}_1 - \mathbf{v}_2$ are orthogonal. Is this mere coincidence, or are there circumstances under which we may expect the sum of two vectors to be orthogonal to their difference? Find out by expanding the left-hand side of the equation

$$(\mathbf{v}_1 + \mathbf{v}_2) \cdot (\mathbf{v}_1 - \mathbf{v}_2) = 0.$$

Give an example of $\mathbf{v}_1$ and $\mathbf{v}_2$ to support your conclusion.

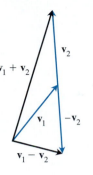

29. Find the angles of the triangle ABC whose vertices are $A(-1, 0, 2)$, $B(2, 1, -1)$, and $C(1, -2, 2)$.

30. Find the angle between $\mathbf{A} = 2\mathbf{i} + 2\mathbf{j} + \mathbf{k}$ and $\mathbf{B} = 2\mathbf{i} + 10\mathbf{j} - 11\mathbf{k}$.

31. Find the angle between the diagonal of a cube and the diagonal of one of its faces. (*Hint:* Use a cube whose edges represent $\mathbf{i}$, $\mathbf{j}$, and $\mathbf{k}$.)

32. Find the angle between the diagonal of a cube and one of the edges it meets at a vertex.

Work

33. Find the work done by a force $\mathbf{F} = -5\mathbf{k}$ (magnitude 5 newtons) in moving an object along the line from the point $(1, 1, 1)$ to the origin (distance in meters).

34. A locomotive exerts a constant force of 60,000 newtons (N) on a freight train while pulling it 1 km along a straight track. How much work does the locomotive do?

35. How much work does it take to slide a crate 20 m along a loading dock by pulling on it with a 200-N force at an angle of 41° from the horizontal?

36. The wind passing over a boat's sail exerts a 4000-N force $\mathbf{F}$ as shown below. If the force vector makes a 60° angle with the line of the boat's forward motion, how much work does the wind perform in moving the boat forward 1 km?

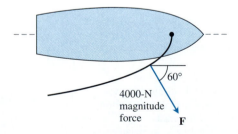

4000-N
magnitude
force $\mathbf{F}$

Angles between Curves

The angles between two differentiable curves at a point of intersection are the angles between the curves' tangents at that point. Find the angles between the curves in Exercises 37–42 at their points of intersection.

37. $3x + y = 5$, $2x - y = 4$

38. $y = 2x - 1$, $y = 3 - 5x$

39. $y = x^2 - 2$, $y = \sqrt{1-x}$

40. $y = x^3$, $y = \sqrt{x}$ (two points of intersection)

41. $y = x^2 + 2x + 1, y = \cos x$

42. $y = x^2 - 2x + 1, y = \sin x$

43. Derive Eq. (3). (*Hint:* $|\mathbf{A}|^2 = \mathbf{A} \cdot \mathbf{A} = a_1{}^2 + a_2{}^2 + a_3{}^2$ for $\mathbf{A} = a_1\mathbf{i} + a_2\mathbf{j} + a_3\mathbf{k}$.)

11.4 _____   Cross Products

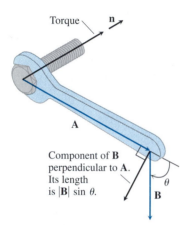

11.29   The torque vector describes the tendency of the force **B** to drive the bolt forward.

When we turn a bolt by applying a force to a wrench (Fig. 11.29), the *torque* that we produce acts along the axis of the bolt to drive the bolt forward. The magnitude of the torque depends on how far along the wrench the force is applied and on how much of the force is actually perpendicular to the wrench at that point. The number that we use to measure the torque's magnitude is a product made up of the length of the vector arm **A** and the scalar component of the force **B** perpendicular to **A**. In the notation of Fig. 11.29,

$$\text{Magnitude of the torque vector} = |\mathbf{A}||\mathbf{B}| \sin\theta.$$

If we let **n** be a unit vector along the axis of the bolt in the direction of the torque, then the complete description of the torque vector is

$$\text{Torque vector} = \mathbf{n}|\mathbf{A}||\mathbf{B}| \sin\theta.$$

We call the torque the **vector product** of **A** and **B**.

Vector products are widely used to describe the effects of forces in studies of electricity, magnetism, fluid mechanics, and planetary motion. The goal of this section is to acquaint you with the mathematical properties of vector products that account for their use in these fields. We will see a number of these applications ourselves in Chapters 12 and 15, where we study motion in space and the integrals associated with fluid flow. In the next section, we will also see how vector products are combined with scalar products to produce equations for lines and planes in space.

EXPLORATION BITS

1. Suppose vectors **A** and **B** belong to *skew* lines in space. Explain what is meant by "the plane of **A** and **B**."

2. As part of Exploration 1 of Section 11.1, discuss the vector product as a binary operation. Which of the ideas

   commutative   associative
   identity   inverse

make sense intuitively for the vector product? Give supporting geometric arguments.

The Vector Product of Two Vectors in Space

When we define vector products in mathematics, we start with two nonzero vectors **A** and **B** in space without requiring them to have any particular physical interpretation. If **A** and **B** are not parallel, they determine a plane. We select a unit vector **n** perpendicular to the plane by the *right-hand rule*. This means that we choose **n** to be the unit normal vector that points the way your thumb points when the fingers of your right hand curl through the angle θ from **A** to **B** (Fig. 11.30).

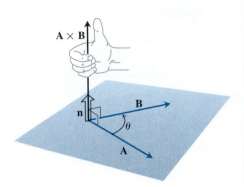

11.30 The construction of $\mathbf{A} \times \mathbf{B}$. Note that $\mathbf{A} \times \mathbf{B}$ is a vector, in contrast to the dot product $\mathbf{A} \cdot \mathbf{B}$ which is a scalar.

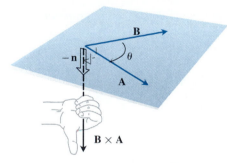

11.31 The construction of $\mathbf{B} \times \mathbf{A}$.

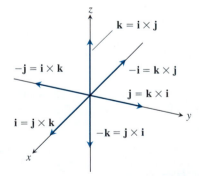

11.32 The pairwise cross products of $\mathbf{i}$, $\mathbf{j}$, and $\mathbf{k}$ follow from the definition and can be read from this diagram. For example, by definition, we have

$$\mathbf{i} \times \mathbf{j} = \mathbf{k}|\mathbf{i}||\mathbf{j}|\sin\frac{\pi}{2} = \mathbf{k}(1)(1)(1) = \mathbf{k}.$$

DEFINITION

The **vector product** or **cross product**, $\mathbf{A} \times \mathbf{B}$ (pronounced "A cross B"), of two vectors $\mathbf{A}$ and $\mathbf{B}$ is the vector

$$\mathbf{A} \times \mathbf{B} = \mathbf{n}|\mathbf{A}||\mathbf{B}|\sin\theta, \tag{1}$$

where θ measures the angle $(0 \leq \theta \leq \pi)$ made by representative $\mathbf{A}$ and $\mathbf{B}$ vectors with the same endpoint (as in Fig. 11.30) and $\mathbf{n}$ is a unit vector chosen perpendicular to the plane of $\mathbf{A}$ and $\mathbf{B}$ by the *right-hand rule*.

Since $\mathbf{A} \times \mathbf{B}$ is a scalar multiple of $\mathbf{n}$, and $\mathbf{n}$ is perpendicular to both $\mathbf{A}$ and $\mathbf{B}$, $\mathbf{A} \times \mathbf{B}$ is perpendicular to both $\mathbf{A}$ and $\mathbf{B}$.

If θ approaches $0°$ or $180°$ in Eq. (1), the length of $\mathbf{A} \times \mathbf{B}$ approaches zero. We therefore define $\mathbf{A} \times \mathbf{B}$ to be $\mathbf{0}$ if $\mathbf{A}$ and $\mathbf{B}$ are parallel (and fail to determine a plane). This is consistent with our torque interpretation as well. If the force $\mathbf{B}$ in Fig. 11.29 is parallel to the wrench, meaning that we are trying to turn the bolt by pushing or pulling straight along the handle of the wrench, the torque produced is $\mathbf{0}$.

If one or both of $\mathbf{A}$ and $\mathbf{B}$ is zero, we define $\mathbf{A} \times \mathbf{B}$ to be zero as well. Thus, the cross product of two vectors $\mathbf{A}$ and $\mathbf{B}$ is zero if and only if $\mathbf{A}$ and $\mathbf{B}$ are parallel or one or both of them is the zero vector.

$\mathbf{A} \times \mathbf{B}$ versus $\mathbf{B} \times \mathbf{A}$

Reversing the order of the factors in a nonzero vector product reverses the direction of the resulting vector. When the fingers of our right hand curl through the angle θ from $\mathbf{B}$ to $\mathbf{A}$, our thumb points the opposite way, and the unit vector that we choose in forming $\mathbf{B} \times \mathbf{A}$ is the negative of the one that we choose in forming $\mathbf{A} \times \mathbf{B}$ (Fig. 11.31). Thus,

$$\mathbf{B} \times \mathbf{A} = -(\mathbf{A} \times \mathbf{B}) \tag{2}$$

for all vectors $\mathbf{A}$ and $\mathbf{B}$. Unlike the dot product, the cross product is not a commutative operation.

When we apply the definition of the cross product to the unit vectors $\mathbf{i}$, $\mathbf{j}$, and $\mathbf{k}$, we find (Fig. 11.32)

$$\mathbf{i} \times \mathbf{j} = \mathbf{k}, \qquad \mathbf{j} \times \mathbf{i} = -\mathbf{k},$$
$$\mathbf{j} \times \mathbf{k} = \mathbf{i}, \quad \text{and} \quad \mathbf{k} \times \mathbf{j} = -\mathbf{i},$$
$$\mathbf{k} \times \mathbf{i} = \mathbf{j}, \qquad \mathbf{i} \times \mathbf{k} = -\mathbf{j},$$

and

$$\mathbf{i} \times \mathbf{i} = \mathbf{j} \times \mathbf{j} = \mathbf{k} \times \mathbf{k} = 0$$

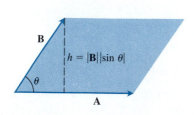

11.33 The parallelogram determined by **A** and **B** has area

$$|\mathbf{A} \times \mathbf{B}| = |\mathbf{A}||\mathbf{B}||\sin\theta|.$$

What value of θ will give the greatest value of $|\mathbf{A} \times \mathbf{B}|$? Why?

$|\mathbf{A} \times \mathbf{B}|$ Is the Area of a Parallelogram

Because **n** is a unit vector, the magnitude of $\mathbf{A} \times \mathbf{B}$ is

$$|\mathbf{A} \times \mathbf{B}| = |\mathbf{n}||\mathbf{A}||\mathbf{B}||\sin\theta| = |\mathbf{A}||\mathbf{B}||\sin\theta|. \qquad (3)$$

This is the area of the parallelogram determined by **A** and **B** (Fig. 11.33), $|\mathbf{A}|$ being the base of the parallelogram and $|\mathbf{B}||\sin\theta|$ the height.

The Magnitude of a Torque

Equation (3) is the equation that we use to calculate magnitudes of torques.

EXAMPLE 1

Force **F** is applied at Q as shown. The magnitude of the torque produced at the pivot point P is

$$|\overrightarrow{PQ} \times \mathbf{F}| = |\overrightarrow{PQ}||\mathbf{F}|\sin 70° \qquad \text{Eq. (3)}$$
$$= (3)(20)\sin 70°$$
$$= 56.382 \text{ ft-lb.}$$

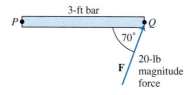

EXPLORATION BIT

Provide convincing geometric arguments that

$$(\mathbf{A} \times \mathbf{B}) \times \mathbf{C}$$

lies in the plane of **A** and **B** and

$$\mathbf{A} \times (\mathbf{B} \times \mathbf{C})$$

lies in the plane of **B** and **C**. Use a sketch if it would be helpful.

The Associative and Distributive Laws

As a rule, cross-product multiplication is not associative because $(\mathbf{A} \times \mathbf{B}) \times \mathbf{C}$ lies in the plane of **A** and **B**, whereas $\mathbf{A} \times (\mathbf{B} \times \mathbf{C})$ lies in the plane of **B** and **C**. However, the **scalar distributive law**

$$(r\mathbf{A}) \times (s\mathbf{B}) = (rs)\mathbf{A} \times \mathbf{B} \qquad (4)$$

does hold, as do the **vector distributive laws**

$$\mathbf{A} \times (\mathbf{B} + \mathbf{C}) = \mathbf{A} \times \mathbf{B} + \mathbf{A} \times \mathbf{C} \qquad (5)$$

and

$$(\mathbf{B} + \mathbf{C}) \times \mathbf{A} = \mathbf{B} \times \mathbf{A} + \mathbf{C} \times \mathbf{A}. \qquad (6)$$

As a special case of (4), we have

$$(-\mathbf{A}) \times \mathbf{B} = \mathbf{A} \times (-\mathbf{B}) = -(\mathbf{A} \times \mathbf{B}).$$

This allows us to write

$$-(\mathbf{A} \times \mathbf{B}) = -\mathbf{A} \times \mathbf{B}.$$

The scalar distributive law can be verified by applying Eq. (1) to the products on both sides of Eq. (4) and comparing the results. The vector distributive law in Eq. (5) is not easy to prove, however, and we will assume it here. Equation (6) follows from Eq. (5); just multiply both sides of Eq. (5) by -1 and reverse the orders of the products.

The Determinant Formula for A × B

Our next objective is to show how to calculate the basic vector components of $\mathbf{A} \times \mathbf{B}$ from the basic vector components of $\mathbf{A}$ and $\mathbf{B}$.

Suppose

$$\mathbf{A} = a_1\mathbf{i} + a_2\mathbf{j} + a_3\mathbf{k}, \qquad \mathbf{B} = b_1\mathbf{i} + b_2\mathbf{j} + b_3\mathbf{k}.$$

Then the distributive laws and the rules for multiplying $\mathbf{i}$, $\mathbf{j}$, and $\mathbf{k}$ tell us that

$$
\begin{aligned}
\mathbf{A} \times \mathbf{B} &= (a_1\mathbf{i} + a_2\mathbf{j} + a_3\mathbf{k}) \times (b_1\mathbf{i} + b_2\mathbf{j} + b_3\mathbf{k}) \\
&= a_1b_1\mathbf{i} \times \mathbf{i} + a_1b_2\mathbf{i} \times \mathbf{j} + a_1b_3\mathbf{i} \times \mathbf{k} \\
&\quad + a_2b_1\mathbf{j} \times \mathbf{i} + a_2b_2\mathbf{j} \times \mathbf{j} + a_2b_3\mathbf{j} \times \mathbf{k} \\
&\quad + a_3b_1\mathbf{k} \times \mathbf{i} + a_3b_2\mathbf{k} \times \mathbf{j} + a_3b_3\mathbf{k} \times \mathbf{k} \\
&= (a_2b_3 - a_3b_2)\mathbf{i} + (a_3b_1 - a_1b_3)\mathbf{j} + (a_1b_2 - a_2b_1)\mathbf{k}. \qquad (7)
\end{aligned}
$$

The terms at the end of Eq. (7) are the same as the terms in the expansion of the symbolic *determinant*

$$
\begin{vmatrix}
\mathbf{i} & \mathbf{j} & \mathbf{k} \\
a_1 & a_2 & a_3 \\
b_1 & b_2 & b_3
\end{vmatrix}.
$$

We may therefore use the following rule to calculate $\mathbf{A} \times \mathbf{B}$.

If $\mathbf{A} = a_1\mathbf{i} + a_2\mathbf{j} + a_3\mathbf{k}$ and $\mathbf{B} = b_1\mathbf{i} + b_2\mathbf{j} + b_3\mathbf{k}$, then

$$
\mathbf{A} \times \mathbf{B} =
\begin{vmatrix}
\mathbf{i} & \mathbf{j} & \mathbf{k} \\
a_1 & a_2 & a_3 \\
b_1 & b_2 & b_3
\end{vmatrix}. \qquad (8)
$$

EXAMPLE 2

Find $\mathbf{A} \times \mathbf{B}$ and $\mathbf{B} \times \mathbf{A}$ if

$$\mathbf{A} = 2\mathbf{i} + \mathbf{j} + \mathbf{k}, \qquad \mathbf{B} = -4\mathbf{i} + 3\mathbf{j} + \mathbf{k}.$$

Solution We use Eq. (8) to find $\mathbf{A} \times \mathbf{B}$:

$$
\mathbf{A} \times \mathbf{B} =
\begin{vmatrix}
\mathbf{i} & \mathbf{j} & \mathbf{k} \\
2 & 1 & 1 \\
-4 & 3 & 1
\end{vmatrix}
=
\begin{vmatrix} 1 & 1 \\ 3 & 1 \end{vmatrix}\mathbf{i} -
\begin{vmatrix} 2 & 1 \\ -4 & 1 \end{vmatrix}\mathbf{j} +
\begin{vmatrix} 2 & 1 \\ -4 & 3 \end{vmatrix}\mathbf{k}
$$

$$= -2\mathbf{i} - 6\mathbf{j} + 10\mathbf{k}.$$

Equation (2) then gives $\mathbf{B} \times \mathbf{A}$:

$$\mathbf{B} \times \mathbf{A} = -(\mathbf{A} \times \mathbf{B}) = 2\mathbf{i} + 6\mathbf{j} - 10\mathbf{k}. \qquad \blacksquare$$

DETERMINANT FORMULAS

(For more information, see Appendix 7.)

$$
\begin{vmatrix} a & b \\ c & d \end{vmatrix} = ad - bc
$$

EXAMPLE

$$
\begin{vmatrix} 2 & 1 \\ -4 & 3 \end{vmatrix} = (2)(3) - (-4)(1)
$$

$$= 6 + 4 = 10 \qquad \blacksquare$$

$$
\begin{vmatrix}
a_{11} & a_{12} & a_{13} \\
a_{21} & a_{22} & a_{23} \\
a_{31} & a_{32} & a_{33}
\end{vmatrix}
$$

$$
= a_{11}\begin{vmatrix} a_{22} & a_{23} \\ a_{32} & a_{33} \end{vmatrix} - a_{12}\begin{vmatrix} a_{21} & a_{23} \\ a_{31} & a_{33} \end{vmatrix}
$$

$$
+ a_{13}\begin{vmatrix} a_{21} & a_{22} \\ a_{31} & a_{32} \end{vmatrix}
$$

EXAMPLE

$$
\begin{vmatrix}
-5 & 3 & 1 \\
2 & 1 & 1 \\
-4 & 3 & 1
\end{vmatrix}
$$

$$
= (-5)\begin{vmatrix} 1 & 1 \\ 3 & 1 \end{vmatrix} - (3)\begin{vmatrix} 2 & 1 \\ -4 & 1 \end{vmatrix}
$$

$$
+ (1)\begin{vmatrix} 2 & 1 \\ -4 & 3 \end{vmatrix}
$$

$$= -5(1 - 3) - 3(2 + 4) + 1(6 + 4)$$

$$= 10 - 18 + 10 = 2 \qquad \blacksquare$$

Note: The Exploration Bit at the end of this section suggests how to evaluate a determinant using a graphing utility.

EXAMPLE 3

Find a vector perpendicular to the plane of $P(1, -1, 0)$, $Q(2, 1, -1)$, and $R(-1, 1, 2)$.

Solution The vector $\overrightarrow{PQ} \times \overrightarrow{PR}$ is perpendicular to the plane because it is perpendicular to both vectors. In terms of components,

$$\overrightarrow{PQ} = (2 - 1)\mathbf{i} + (1 + 1)\mathbf{j} + (-1 - 0)\mathbf{k} = \mathbf{i} + 2\mathbf{j} - \mathbf{k},$$

$$\overrightarrow{PR} = (-1 - 1)\mathbf{i} + (1 + 1)\mathbf{j} + (2 - 0)\mathbf{k} = -2\mathbf{i} + 2\mathbf{j} + 2\mathbf{k},$$

$$\overrightarrow{PQ} \times \overrightarrow{PR} = \begin{vmatrix} \mathbf{i} & \mathbf{j} & \mathbf{k} \\ 1 & 2 & -1 \\ -2 & 2 & 2 \end{vmatrix}$$

$$= \begin{vmatrix} 2 & -1 \\ 2 & 2 \end{vmatrix} \mathbf{i} - \begin{vmatrix} 1 & -1 \\ -2 & 2 \end{vmatrix} \mathbf{j} + \begin{vmatrix} 1 & 2 \\ -2 & 2 \end{vmatrix} \mathbf{k}$$

$$= 6\mathbf{i} + 6\mathbf{k}. \qquad \blacksquare$$

EXAMPLE 4

Find the area of the triangle with vertices $P(1, -1, 0)$, $Q(2, 1, -1)$, and $R(-1, 1, 2)$ (Fig. 11.34).

Solution The area of the parallelogram determined by $\overrightarrow{PQ}$ and $\overrightarrow{PR}$ is

$$|\overrightarrow{PQ} \times \overrightarrow{PR}| = |6\mathbf{i} + 6\mathbf{k}| \qquad \text{Values from Example 3}$$

$$= \sqrt{(6)^2 + (6)^2} = \sqrt{2 \cdot 36} = 8.485.$$

The triangle's area is half of this, or 4.243. $\qquad \blacksquare$

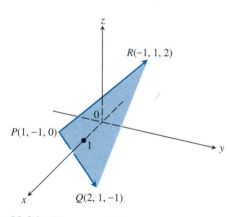

11.34 The area of triangle PQR is half of $\left|\overrightarrow{PQ} \times \overrightarrow{PR}\right|$.

EXAMPLE 5

Find a unit vector perpendicular to the plane of $P(1, -1, 0)$, $Q(2, 1, -1)$, and $R(-1, 1, 2)$.

Solution Since $\overrightarrow{PQ} \times \overrightarrow{PR}$ is perpendicular to the plane, its direction is a unit vector, n, perpendicular to the plane. In terms of components,

$$\mathbf{n} = \frac{\overrightarrow{PQ} \times \overrightarrow{PR}}{|\overrightarrow{PQ} \times \overrightarrow{PR}|} \qquad \text{The direction of } \overrightarrow{PQ} \times \overrightarrow{PR}$$

$$= \frac{6\mathbf{i} + 6\mathbf{k}}{6\sqrt{2}} \qquad \text{Values from Examples 3 and 4}$$

$$= \frac{1}{\sqrt{2}}\mathbf{i} + \frac{1}{\sqrt{2}}\mathbf{k}. \qquad \blacksquare$$

The Test for Parallelism

Since the sines of $0°$ and $180°$ are both zero, the cross product $\mathbf{A} \times \mathbf{B} = \mathbf{n}|\mathbf{A}||\mathbf{B}| \sin \theta$ of two nonzero vectors will be the zero vector if and only if vectors A and B are parallel.

> Nonzero vectors **A** and **B** are parallel if and only if
>
> $$\mathbf{A} \times \mathbf{B} = \mathbf{0}.$$

EXAMPLE 6

Parallel: $\mathbf{A} = \mathbf{i} + 2\mathbf{j} - \mathbf{k}$ and $\mathbf{B} = -2\mathbf{i} - 4\mathbf{j} + 2\mathbf{k}$.

The test: $\mathbf{A} \times \mathbf{B} = \begin{vmatrix} \mathbf{i} & \mathbf{j} & \mathbf{k} \\ 1 & 2 & -1 \\ -2 & -4 & 2 \end{vmatrix}$

$$= \begin{vmatrix} 2 & -1 \\ -4 & 2 \end{vmatrix} \mathbf{i} - \begin{vmatrix} 1 & -1 \\ -2 & 2 \end{vmatrix} \mathbf{j} + \begin{vmatrix} 1 & 2 \\ -2 & -4 \end{vmatrix} \mathbf{k}$$

$$= (4 - 4)\mathbf{i} - (2 - 2)\mathbf{j} + (-4 + 4)\mathbf{k}$$

$$= 0\mathbf{i} - 0\mathbf{j} + 0\mathbf{k} = \mathbf{0}$$

Not parallel: $\mathbf{A} = \mathbf{i} + 2\mathbf{j} - \mathbf{k}$ and $\mathbf{C} = -2\mathbf{i} + 2\mathbf{j} + 2\mathbf{k}$

The test: $\mathbf{A} \times \mathbf{C} = 6\mathbf{i} + 6\mathbf{k} \neq \mathbf{0}$ Example 3   ≡

The Triple Scalar or Box Product

The product $(\mathbf{A} \times \mathbf{B}) \cdot \mathbf{C}$ is called the **triple scalar product** of **A**, **B**, and **C** (in that order). As you can see from the formula

$$(\mathbf{A} \times \mathbf{B}) \cdot \mathbf{C} = |\mathbf{A} \times \mathbf{B}||\mathbf{C}| \cos\theta,$$

the product is the volume of the *parallelepiped* (parallelogram-sided box) determined by **A**, **B**, and **C** (Fig. 11.35). The number $|\mathbf{A} \times \mathbf{B}|$ is the area of the base parallelogram, and $|\mathbf{C}| \cos\theta$ is the height of the tip of **C** above the plane of **A** and **B**. If θ is greater than $90°$, then $\cos\theta$ is negative, and we must take the absolute value of $(\mathbf{A} \times \mathbf{B}) \cdot \mathbf{C}$ to get the volume. Because of the geometry of this situation, $(\mathbf{A} \times \mathbf{B}) \cdot \mathbf{C}$ is often called the **box product** of **A**, **B**, and **C**.

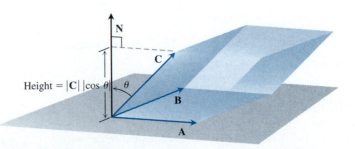

11.35 Except perhaps for sign, the number $(\mathbf{A} \times \mathbf{B}) \cdot \mathbf{C}$ is the volume of the parallelepiped (parallelogram-sided box) shown here.

By treating the planes of **B** and **C** and of **C** and **A** as the base planes of the parallelepiped determined by **A**, **B**, and **C**, we can see that

$$(\mathbf{A} \times \mathbf{B}) \cdot \mathbf{C} = (\mathbf{B} \times \mathbf{C}) \cdot \mathbf{A} = (\mathbf{C} \times \mathbf{A}) \cdot \mathbf{B}. \qquad (9)$$

Since the dot product is commutative, Eq. (9) also gives

$$(\mathbf{A} \times \mathbf{B}) \cdot \mathbf{C} = \mathbf{A} \cdot (\mathbf{B} \times \mathbf{C}). \qquad (10)$$

In other words, the dot and cross products may be interchanged in a triple scalar product without altering its value.

The triple scalar product can be evaluated as a determinant in the following way:

$$\mathbf{A} \cdot (\mathbf{B} \times \mathbf{C}) = \mathbf{A} \cdot \left[\begin{vmatrix} b_2 & b_3 \\ c_2 & c_3 \end{vmatrix} \mathbf{i} - \begin{vmatrix} b_1 & b_3 \\ c_1 & c_3 \end{vmatrix} \mathbf{j} + \begin{vmatrix} b_1 & b_2 \\ c_1 & c_2 \end{vmatrix} \mathbf{k} \right]$$

$$= a_1 \begin{vmatrix} b_2 & b_3 \\ c_2 & c_3 \end{vmatrix} - a_2 \begin{vmatrix} b_1 & b_3 \\ c_1 & c_3 \end{vmatrix} + a_3 \begin{vmatrix} b_1 & b_2 \\ c_1 & c_2 \end{vmatrix}$$

$$= \begin{vmatrix} a_1 & a_2 & a_3 \\ b_1 & b_2 & b_3 \\ c_1 & c_2 & c_3 \end{vmatrix}$$

EXPLORATION BIT

Use your grapher's MATRIX menu to compute the determinant of a matrix. In particular, support the result of Example 7.

Show also that *any* arrangement of the three rows of the matrix in Example 7 results in the same volume for the parallelepiped. Relate this fact to the discussion about Eqs. (9) and (10).

The number $\mathbf{A} \cdot (\mathbf{B} \times \mathbf{C})$ can be calculated with the formula

$$\mathbf{A} \cdot (\mathbf{B} \times \mathbf{C}) = \begin{vmatrix} a_1 & a_2 & a_3 \\ b_1 & b_2 & b_3 \\ c_1 & c_2 & c_3 \end{vmatrix}.$$

EXAMPLE 7

Find the volume of the parallelepiped determined by $\mathbf{A} = \mathbf{i} + 2\mathbf{j} - \mathbf{k}$, $\mathbf{B} = -2\mathbf{i} + 3\mathbf{k}$, and $\mathbf{C} = 7\mathbf{j} - 4\mathbf{k}$.

Solution   The volume is the absolute value of

$$\mathbf{A} \cdot (\mathbf{B} \times \mathbf{C}) = \begin{vmatrix} 1 & 2 & -1 \\ -2 & 0 & 3 \\ 0 & 7 & -4 \end{vmatrix}$$

$$= \begin{vmatrix} 0 & 3 \\ 7 & -4 \end{vmatrix} - 2 \begin{vmatrix} -2 & 3 \\ 0 & -4 \end{vmatrix} - \begin{vmatrix} -2 & 0 \\ 0 & 7 \end{vmatrix}$$

$$= -21 - 16 + 14 = -23.$$

So the volume is 23.   ≡

Exercises 11.4

In Exercises 1–8, find the length and direction (when defined) of $\mathbf{A} \times \mathbf{B}$ and $\mathbf{B} \times \mathbf{A}$.

1. $\mathbf{A} = 2\mathbf{i} - 2\mathbf{j} - \mathbf{k}, \quad \mathbf{B} = \mathbf{i} - \mathbf{k}$

2. $\mathbf{A} = 2\mathbf{i} + 3\mathbf{j}, \quad \mathbf{B} = -\mathbf{i} + \mathbf{j}$

3. $\mathbf{A} = 2\mathbf{i} - 2\mathbf{j} + 4\mathbf{k}, \quad \mathbf{B} = -\mathbf{i} + \mathbf{j} - 2\mathbf{k}$

4. $\mathbf{A} = \mathbf{i} + \mathbf{j} - \mathbf{k}, \quad \mathbf{B} = \mathbf{0}$

5. $\mathbf{A} = 2\mathbf{i}, \quad \mathbf{B} = -3\mathbf{j}$

6. $\mathbf{A} = \mathbf{i} \times \mathbf{j}, \quad \mathbf{B} = \mathbf{j} \times \mathbf{k}$

7. $\mathbf{A} = -8\mathbf{i} - 2\mathbf{j} - 4\mathbf{k}, \quad \mathbf{B} = 2\mathbf{i} + 2\mathbf{j} + \mathbf{k}$

8. $\mathbf{A} = \dfrac{3}{2}\mathbf{i} - \dfrac{1}{2}\mathbf{j} + \mathbf{k}, \quad \mathbf{B} = \mathbf{i} + \mathbf{j} + 2\mathbf{k}$

In Exercises 9–14, sketch the coordinate axes and then include the vectors **A**, **B**, and **A** × **B** as vectors coming out from the origin.

9. $\mathbf{A} = \mathbf{i}, \quad \mathbf{B} = \mathbf{j}$

10. $\mathbf{A} = \mathbf{i} + \mathbf{k}, \quad \mathbf{B} = \mathbf{j}$

11. $\mathbf{A} = \mathbf{i} - \mathbf{k}, \quad \mathbf{B} = \mathbf{j} + \mathbf{k}$

12. $\mathbf{A} = 2\mathbf{i} - \mathbf{j}, \quad \mathbf{B} = \mathbf{i} + 2\mathbf{j}$

13. $\mathbf{A} = \mathbf{i} + 3\mathbf{j} + 2\mathbf{k}, \quad \mathbf{B} = \mathbf{k}$

14. $\mathbf{A} = \mathbf{i} + 2\mathbf{j}, \quad \mathbf{B} = 2\mathbf{j} + \mathbf{k}$

In Exercises 15–18:

a) Find a vector **N** perpendicular to the plane of the points P, Q, and R.

b) Find the area of triangle PQR.

c) Find a unit vector perpendicular to plane PQR.

15. $P(1, -1, 2), \quad Q(2, 0, -1), \quad R(0, 2, 1)$

16. $P(1, 1, 1), \quad Q(2, 1, 3), \quad R(3, -2, 1)$

17. $P(2, -2, 1), \quad Q(3, -1, 2), \quad R(3, -1, 1)$

18. $P(-2, 2, 0), \quad Q(0, 1, -1), \quad R(-1, 2, -2)$

19. Let $\mathbf{A} = 5\mathbf{i} - \mathbf{j} + \mathbf{k}, \mathbf{B} = \mathbf{j} - 5\mathbf{k}, \mathbf{C} = -15\mathbf{i} + 3\mathbf{j} - 3\mathbf{k}$. Which vectors, if any, are (a) perpendicular, (b) parallel?

20. Let $\mathbf{A} = \mathbf{i} + 2\mathbf{j} - \mathbf{k}, \mathbf{B} = -\mathbf{i} + \mathbf{j} + \mathbf{k}, \mathbf{C} = \mathbf{i} + \mathbf{k}, \mathbf{D} = \mathbf{i} + 2\mathbf{j} - \mathbf{k}$. Which vectors, if any, are (a) perpendicular, (b) parallel?

21. If $\mathbf{A} = 2\mathbf{i} - \mathbf{j}$ and $\mathbf{B} = \mathbf{i} + 3\mathbf{j} - 2\mathbf{k}$, find $\mathbf{A} \times \mathbf{B}$. Then calculate $(\mathbf{A} \times \mathbf{B}) \cdot \mathbf{A}$ and $(\mathbf{A} \times \mathbf{B}) \cdot \mathbf{B}$.

22. Is $(\mathbf{A} \times \mathbf{B}) \cdot \mathbf{A}$ always zero? Explain. What about $(\mathbf{A} \times \mathbf{B}) \cdot \mathbf{B}$?

23. Given vectors **A**, **B**, and **C**, use dot-product and cross-product notation to describe the following vectors:

a) The vector projection of **A** onto **B**

b) A vector orthogonal to **A** and **B**

c) A vector with the length of **A** and the direction of **B**

d) A vector orthogonal to **A** × **B** and **C**

e) A vector in the plane of **B** and **C** perpendicular to **A**

24. *Cancellation in cross products is risky, too.* Find an example to show that $\mathbf{A} \times \mathbf{B} = \mathbf{A} \times \mathbf{C}, \mathbf{A} \neq \mathbf{0}$, need not imply that **B** equals **C**.

In Exercises 25 and 26, find the magnitude of the torque exerted by **F** on the bolt at P if $|\overrightarrow{PQ}| = 8$ in. and $|\mathbf{F}| = 30$ lb. Answer in foot-pounds.

25.   **26.**

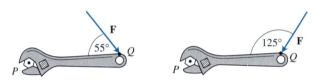

Triple Scalar Products

In Exercises 27–30, verify that $(\mathbf{A} \times \mathbf{B}) \cdot \mathbf{C} = (\mathbf{B} \times \mathbf{C}) \cdot \mathbf{A} = (\mathbf{C} \times \mathbf{A}) \cdot \mathbf{B}$ and find the volume of the parallelepiped determined by **A**, **B**, and **C**.

	A	B	C
27.	$2\mathbf{i}$	$2\mathbf{j}$	$2\mathbf{k}$
28.	$\mathbf{i} - \mathbf{j} + \mathbf{k}$	$2\mathbf{i} + \mathbf{j} - 2\mathbf{k}$	$-\mathbf{i} + 2\mathbf{j} - \mathbf{k}$
29.	$2\mathbf{i} + \mathbf{j}$	$2\mathbf{i} - \mathbf{j} + \mathbf{k}$	$\mathbf{i} + 2\mathbf{k}$
30.	$\mathbf{i} + \mathbf{j} - 2\mathbf{k}$	$-\mathbf{i} - \mathbf{k}$	$2\mathbf{i} + 4\mathbf{j} - 2\mathbf{k}$

11.5 _____   Lines and Planes in Space

This section shows how to use scalar and vector products to write equations for lines, line segments, and planes in space.

Equations for Lines and Line Segments

Suppose L is a line in space that passes through a point $P_0(x_0, y_0, z_0)$ and lies parallel to a vector $\mathbf{v} = A\mathbf{i} + B\mathbf{j} + C\mathbf{k}$. Then L is the set of all points $P(x, y, z)$ for which the vector $\overrightarrow{P_0P}$ is parallel to $\mathbf{v}$ (Fig. 11.36). That is, P lies on L if and only if

$$\overrightarrow{P_0P} = t\mathbf{v} \tag{1}$$

for some number t.

When we write Eq. (1) in terms of components,

$$(x - x_0)\mathbf{i} + (y - y_0)\mathbf{j} + (z - z_0)\mathbf{k} = t(A\mathbf{i} + B\mathbf{j} + C\mathbf{k}),$$

and equate the corresponding components on the two sides, we get three equations in the parameter t:

$$x - x_0 = tA, \qquad y - y_0 = tB, \qquad z - z_0 = tC.$$

When rearranged, these become the standard parametric equations of line L:

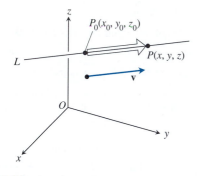

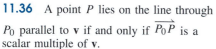

11.36 A point P lies on the line through P_0 parallel to $\mathbf{v}$ if and only if $\overrightarrow{P_0P}$ is a scalar multiple of $\mathbf{v}$.

Parametric Equations for the Line through $P_0(x_0, y_0, z_0)$ Parallel to $\mathbf{v} = A\mathbf{i} + B\mathbf{j} + C\mathbf{k}$

$$x = x_0 + tA, \qquad y = y_0 + tB, \qquad z = z_0 + tC \tag{2}$$

EXPLORATION BIT

How are Eqs. (2) and the equations in Exercise 39 of Section 10.3 related?

EXAMPLE 1

Find parametric equations for the line through the point $(-2, 0, 4)$ parallel to the vector $\mathbf{v} = 2\mathbf{i} + 4\mathbf{j} - 2\mathbf{k}$.

Solution   With $P_0(x_0, y_0, z_0) = (-2, 0, 4)$ and

$$A\mathbf{i} + B\mathbf{j} + C\mathbf{k} = 2\mathbf{i} + 4\mathbf{j} - 2\mathbf{k},$$

Eqs. (2) become

$$x = -2 + 2t, \qquad y = 4t, \qquad z = 4 - 2t.$$

The line for these equations is shown in this diagram along with vector $\mathbf{v}$. The arrows show the direction of increasing t.

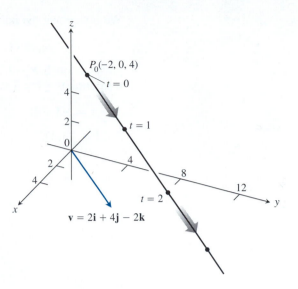

EXAMPLE 2

Find parametric equations for the line through the points $P(-3, 2, -3)$ and $Q(1, -1, 4)$.

Solution The vector

$$\overrightarrow{PQ} = 4\mathbf{i} - 3\mathbf{j} + 7\mathbf{k}$$

is parallel to the line, and Eqs. (2) with $(x_0, y_0, z_0) = (-3, 2, -3)$ give

$$x = -3 + 4t, \qquad y = 2 - 3t, \qquad z = -3 + 7t. \tag{3}$$

We could equally well have chosen $Q(1, -1, 4)$ as the "base point" and written

$$x = 1 + 4t, \qquad y = -1 - 3t, \qquad z = 4 + 7t \tag{4}$$

as equations for the line. The parametrizations in Eqs. (3) and (4) place you at different points for a given value of t, but each set of equations covers the line completely as t runs from $-\infty$ to ∞.

To find equations for the line segment joining two points, we first find equations for the line determined by the points. We then find the t-values for which the line passes through the points and restrict t to lie within the closed interval bounded by these values. The line equations, together with this added restriction, are called equations for the line segment.

EXAMPLE 3

Find parametric equations for the line segment joining the points $P(-3, 2, -3)$ and $Q(1, -1, 4)$.

Solution We begin with equations for the line through P and Q, taking them, in this case, from Example 2:

$$x = -3 + 4t, \qquad y = 2 - 3t, \qquad z = -3 + 7t. \tag{5}$$

We notice that the point

$$(x, y, z) = (-3 + 4t, 2 - 3t, -3 + 7t)$$

passes through $P(-3, 2, -3)$ at $t = 0$ and $Q(1, -1, 4)$ at $t = 1$. We add the restriction $0 \le t \le 1$ to Eqs. (5) to get equations for the segment:

$$x = -3 + 4t, \qquad y = 2 - 3t, \qquad z = -3 + 7t, \qquad (0 \le t \le 1).$$

See Fig. 11.37.

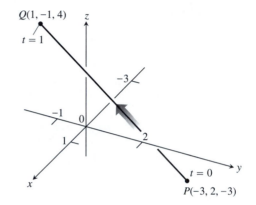

11.37   The arrow shows the direction of increasing t in line segment PQ for the parametrization found in Example 3.

The Distance from a Point to a Line

Finding the distance from a point to a line involves two steps.

To find the distance from a point P to a line L,

STEP 1: Find the point Q on L closest to P.

STEP 2: Calculate the distance from P to Q.

EXAMPLE 4

Find the distance from the point $P(1, 1, 5)$ to the line

$$x = 1 + t, \qquad y = 3 - t, \qquad z = 2t.$$

Solution

STEP 1: We find the point on the line closest to P. The coordinates of a typical point Q on the line are $Q(1 + t, 3 - t, 2t)$.

We want the value of t that minimizes the distance from P to Q. To avoid working with square roots, we find this value of t by minimizing the *square* of the distance instead. The formula for the square of the distance from P to Q is

$$f(t) = (1 + t - 1)^2 + (3 - t - 1)^2 + (2t - 5)^2$$
$$= (t)^2 + (2 - t)^2 + (2t - 5)^2$$
$$= t^2 + 4 - 4t + t^2 + 4t^2 - 20t + 25$$
$$= 6t^2 - 24t + 29.$$

The derivative,

$$\frac{df}{dt} = 12t - 24,$$

equals 0 when $t = 2$. This value of t gives a minimum value for f because $d^2 f/dt^2 = 12$ is positive. The point Q closest to P is $Q(1 + t, 3 - t, 2t)_{t=2} = Q(3, 1, 4)$.

STEP 2: We find the distance from $P(1, 1, 5)$ to $Q(3, 1, 4)$:

$$d(P, Q) = \sqrt{(1 - 3)^2 + (1 - 1)^2 + (5 - 4)^2}$$
$$= \sqrt{(-2)^2 + (0)^2 + (1)^2} = \sqrt{4 + 1} = \sqrt{5}.$$

The distance from P to the line is $\sqrt{5}$.   ≡

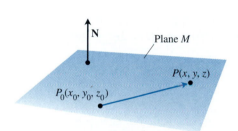

11.38 The standard equation for a plane in space is defined in terms of a vector normal to the plane: A point P lies in the plane through P_0 normal to $\mathbf{N}$ if and only if $\mathbf{N} \cdot \overrightarrow{P_0 P} = 0$.

Equations for Planes

Suppose M is a plane in space that passes through a point $P_0(x_0, y_0, z_0)$ and is perpendicular (normal) to the nonzero vector $\mathbf{N} = A\mathbf{i} + B\mathbf{j} + C\mathbf{k}$. Then M consists of all points $P(x, y, z)$ for which the vector $\overrightarrow{P_0 P}$ is orthogonal to $\mathbf{N}$ (Fig. 11.38). That is, P lies on M if and only if

$$\mathbf{N} \cdot \overrightarrow{P_0 P} = 0$$

or

$$A(x - x_0) + B(y - y_0) + C(z - z_0) = 0.$$

When rearranged, this becomes

$$Ax + By + Cz = Ax_0 + By_0 + Cz_0.$$

Equation for the Plane through $P_0 (x_0, y_0, z_0)$ Perpendicular to $\mathbf{N} = A\mathbf{i} + B\mathbf{j} + C\mathbf{k}$

$$Ax + By + Cz = D, \qquad \text{where} \qquad D = Ax_0 + By_0 + Cz_0 \qquad (6)$$

Notice how the components of **N** become coefficients in the equation

$$Ax + By + Cz = D.$$

EXAMPLE 5

Compare Eq. (6) and Eq. (14) of Section 11.3.

Find an equation for the plane through $P_0(-3, 0, 7)$ perpendicular to $\mathbf{N} = 5\mathbf{i} + 2\mathbf{j} - \mathbf{k}$.

Solution We use Eq. (6) to find

$$D = 5(-3) + 2(0) - 1(7) = -15 - 7 = -22$$

and

$$5x + 2y - z = -22. \qquad\qquad \blacksquare$$

EXAMPLE 6

Find an equation for the plane through the points $A(0, 0, 1)$, $B(2, 0, 0)$, and $C(0, 3, 0)$.

Solution We find a vector normal to the plane and use it with one of the points to write an equation for the plane.

The cross product

$$\overrightarrow{AB} \times \overrightarrow{AC} = \begin{vmatrix} \mathbf{i} & \mathbf{j} & \mathbf{k} \\ 2 & 0 & -1 \\ 0 & 3 & -1 \end{vmatrix} = 3\mathbf{i} + 2\mathbf{j} + 6\mathbf{k}$$

is normal to the plane. We substitute the components of this vector and the coordinates of the point $(0, 0, 1)$ into Eq. (6) to get $D = 3(0)+2(0)+6(1) = 6$ and

$$3x + 2y + 6z = 6$$

as an equation for the plane. $\qquad\qquad \blacksquare$

EXAMPLE 7

Find the point in which the line

$$x = \frac{8}{3} + 2t, \qquad y = -2t, \qquad z = 1 + t$$

meets the plane $3x + 2y + 6z = 6$.

Solution The point

$$\left(\frac{8}{3} + 2t, -2t, 1 + t \right)$$

will lie in the plane if its coordinates satisfy the equation of the plane, that is, if

$$3 \left(\frac{8}{3} + 2t \right) + 2(-2t) + 6(1 + t) = 6$$

$$8 + 6t - 4t + 6 + 6t = 6$$

$$8t = -8$$

$$t = -1.$$

The point of intersection is

$$(x, y, z)_{t=-1} = \left(\frac{8}{3} - 2, 2, 1 - 1\right) = \left(\frac{2}{3}, 2, 0\right).$$

≡

EXAMPLE 8

Find the distance from the point $S(1, 1, 3)$ to the plane $3x + 2y + 6z = 6$.

Solution   We use the same approach that we used in Section 11.3 to find the distance from a point to a line: We find a point P in the plane and calculate the length of the vector projection of $\overrightarrow{PS}$ onto a vector **N** normal to the plane (Fig. 11.39).

To find the distance from a point S to a plane $Ax + By + Cz = D$, find
1. a point P on the plane,

2. $\overrightarrow{PS}$, and

3. the direction of $\mathbf{N} = A\mathbf{i} + B\mathbf{j} + C\mathbf{k}$.
  Then calculate the distance as

$$\left| \overrightarrow{PS} \cdot \frac{\mathbf{N}}{|\mathbf{N}|} \right|.$$

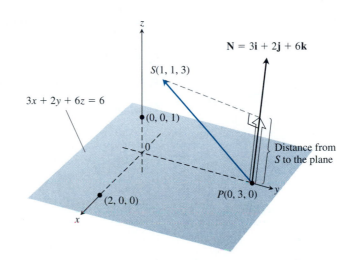

11.39   The distance from S to the plane is the length of the vector projection of $\overrightarrow{PS}$ onto **N**.

The coefficients in the equation $3x + 2y + 6z = 6$ give

$$\mathbf{N} = 3\mathbf{i} + 2\mathbf{j} + 6\mathbf{k}.$$

The points on the plane that are easiest to find from the plane's equation are the intercepts. If we take P to be the y-intercept $(0, 3, 0)$, then

$$\overrightarrow{PS} = (1 - 0)\mathbf{i} + (1 - 3)\mathbf{j} + (3 - 0)\mathbf{k}$$

$$= \mathbf{i} - 2\mathbf{j} + 3\mathbf{k}.$$

$$|\mathbf{N}| = \sqrt{(3)^2 + (2)^2 + (6)^2}$$

$$= \sqrt{49}$$

$$= 7.$$

$$\text{Distance from } S \text{ to the plane} \;=\; \left| \overrightarrow{PS} \cdot \frac{\mathbf{N}}{|\mathbf{N}|} \right|$$

$$= \left| (\mathbf{i} - 2\mathbf{j} + 3\mathbf{k}) \cdot \left(\frac{3}{7}\mathbf{i} + \frac{2}{7}\mathbf{j} + \frac{6}{7}\mathbf{k} \right) \right|$$

$$= \left| \frac{3}{7} - \frac{4}{7} + \frac{18}{7} \right|$$

$$= \frac{17}{7}.$$

Angles between Planes; Lines of Intersection

The angle between two intersecting planes is defined to be the acute (or right) angle formed by their normal vectors (Fig. 11.40).

EXAMPLE 9

Find the angle between the planes $3x - 6y - 2z = 15$ and $2x + y - 2z = 5$.

Solution   The vectors

$$\mathbf{N}_1 = 3\mathbf{i} - 6\mathbf{j} - 2\mathbf{k}, \qquad \mathbf{N}_2 = 2\mathbf{i} + \mathbf{j} - 2\mathbf{k}$$

are normals to the planes. The angle between them is

$$\theta = \cos^{-1}\left(\frac{\mathbf{N}_1 \cdot \mathbf{N}_2}{|\mathbf{N}_1||\mathbf{N}_2|} \right) \qquad \text{Eq. (5), Section 11.3}$$

$$= \cos^{-1} \frac{4}{21}$$

$$= 79.019° = 1.379 \text{ radians}$$

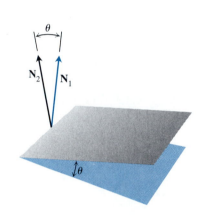

11.40   The angle between two planes is obtained from an angle formed by their normal vectors.

EXPLORATION 1

The Line of Intersection of Two Planes

The planes $3x - 6y - 2z = 15$ and $2x + y - 2z = 5$ intersect. Find parametric equations for the line of intersection by the following steps.

STEP 1: Let $\mathbf{N}_1$ and $\mathbf{N}_2$ be normals to the two planes. Explain why $\mathbf{N}_1 \times \mathbf{N}_2$ is parallel to the line of intersection of the two planes.

STEP 2: Find a point common to the two planes.

STEP 3: Use $\mathbf{N}_1 \times \mathbf{N}_2$ and the point in Step 2 to write the desired equations.

EXAMPLE 10

Find a vector parallel to the line of intersection of the planes $3x - 6y - 2z = 15$ and $2x + y - 2z = 5$.

Solution   Any vector that is parallel to the line of intersection will be parallel to both planes and therefore perpendicular to their normals. Conversely, any vector that is perpendicular to the planes' normals will be parallel to both planes and hence parallel to their line of intersection. These requirements are met by the vector

$$\mathbf{v} = \mathbf{N}_1 \times \mathbf{N}_2$$

$$= \begin{vmatrix} \mathbf{i} & \mathbf{j} & \mathbf{k} \\ 3 & -6 & -2 \\ 2 & 1 & -2 \end{vmatrix}$$

$$= 14\mathbf{i} + 2\mathbf{j} + 15\mathbf{k}.$$

Any nonzero scalar multiple of $\mathbf{v}$ will do as well.   ▤

EXAMPLE 11

Find parametric equations for the line in which the planes $3x - 6y - 2z = 15$ and $2x + y - 2z = 5$ intersect.

Solution   We find a vector parallel to the line and a point on the line and use Eqs. (2).

Example 10 gives a vector parallel to the line, namely, $\mathbf{v} = 14\mathbf{i} + 2\mathbf{j} + 15\mathbf{k}$. To find a point on the line, we find a point common to the two planes. Substituting $z = 0$ in the plane equations and solving for x and y simultaneously gives the point $(3, -1, 0)$. The line is

$$x = 3 + 14t, \qquad y = -1 + 2t, \qquad z = 15t.$$ ▤

EXPLORATION BIT

Compare the result of Example 11 with your result from Exploration 1. If the two results are different, explain why. If they are the same, explain why they could be different.

Exercises 11.5

Find parametric equations for the lines in Exercises 1–14.

1. The line through $P(3, -4, -1)$ parallel to the vector $\mathbf{i} + \mathbf{j} + \mathbf{k}$

2. The line through $P(2, 3, -1)$ parallel to the vector $4\mathbf{i} - 2\mathbf{j} + 3\mathbf{k}$

3. The line through $P(1, 2, -1)$ and $Q(-1, 0, 1)$

4. The line through $P(-2, 0, 3)$ and $Q(3, 5, -2)$

5. The line through $P(1, 2, 0)$ and $Q(1, 1, -1)$

6. The line through $P(3, \pi, -2)$ and $Q(e, \sqrt{2}, 7)$

7. The line through the origin parallel to the vector $2\mathbf{j} + \mathbf{k}$

8. The line through the point $(3, -2, 1)$ parallel to the line $x = 1 + 2t, y = 2 - t, z = 3t$

9. The line through $(1, 1, 1)$ parallel to the z-axis

10. The line through $(2, 4, 5)$ perpendicular to the plane $3x + 7y - 5z = 21$

11. The line through $(0, -7, 0)$ perpendicular to the plane $x + 2y + 2z = 13$

12. The line through $(2, 3, 0)$ perpendicular to the vectors $\mathbf{A} = \mathbf{i} + 2\mathbf{j} + 3\mathbf{k}$ and $\mathbf{B} = 3\mathbf{i} + 4\mathbf{j} + 5\mathbf{k}$

13. The x-axis

14. The z-axis

Find parametric equations for the line segment joining the points in each of Exercises 15–22. Draw coordinate axes and sketch the segment, indicating the direction of increasing t for your parametrization.

15. $(0, 0, 0), \quad (1, 1, 1)$

16. $(0, 0, 0), \quad (1, 0, 0)$

17. $(1, 0, 0), \quad (1, 1, 0)$

18. $(1, 1, 0), \quad (1, 1, 1)$

19. $(0, -1, 1), \quad (0, 1, 1)$

20. $(3, 0, 0), \quad (0, 2, 0)$

21. $(2, 2, 0), \quad (1, 2, -2)$

22. $(1, -1, -2), \quad (0, 2, 1)$

Find equations for the planes in Exercises 23–28.

23. The plane through $P_0(0, 2, -1)$ perpendicular to
$\mathbf{N} = 3\mathbf{i} - 2\mathbf{j} - \mathbf{k}$

24. The plane through $(1, -1, 3)$ parallel to the plane
$3x + y + z = 7$

25. The plane through $(1, 1, -1)$, $(2, 0, 2)$, and $(0, -2, 1)$

26. The plane through $(2, 4, 5)$, $(1, 5, 7)$, and $(-1, 6, 8)$

27. The plane through $P_0(2, 4, 5)$ perpendicular to the line
$$x = 5 + t, \quad y = 1 + 3t, \quad z = 4t$$

28. The plane through $A(1, -2, 1)$ perpendicular to the vector
from the origin to A

In Exercises 29–32, find the distance from the point to the line.

29. $x = 4t, \quad y = -2t, \quad z = 2t; \quad P(0, 0, 12)$

30. $x = 5 + 3t, \quad y = 5 + 4t, \quad z = -3 - 5t; \quad P(0, 0, 0)$

31. $x = 2 + 2t, \quad y = 1 + 6t, \quad z = 3; \quad P(2, 1, 3)$

32. $x = 2t, \quad y = 1 + 2t, \quad z = 2t; \quad P(2, 1, -1)$

In Exercises 33–38, find the distance from the point to the plane.

33. $(2, -3, 4), \quad x + 2y + 2z = 13$

34. $(0, 0, 0), \quad 3x + 2y + 6z = 6$

35. $(0, 1, 1), \quad 4y + 3z = -12$

36. $(2, 2, 3), \quad 2x + y + 2z = 4$

37. $(0, -1, 0), \quad 2x + y + 2z = 4$

38. $(1, 0, -1), \quad -4x + y + z = 4$

In Exercises 39–42, find the point in which the line meets the plane.

39. $x = 1 - t, \quad y = 3t, \quad z = 1 + t; \quad 2x - y + 3z = 6$

40. $x = 2, \quad y = 3 + 2t, \quad z = -2 - 2t; \quad 6x + 3y - 4z = -12$

41. $x = 1 + 2t, \quad y = 1 + 5t, \quad z = 3t; \quad x + y + z = 2$

42. $x = -1 + 3t, \quad y = -2, \quad z = 5t; \quad 2x - 3z = 7$

Find the angle between each pair of planes in Exercises 43–48.

43. $x + y = 1, \quad 2x + y - 2z = 2$

44. $5x + y - z = 10, \quad x - 2y + 3z = -1$

45. $2x + 2y + 2z = 3, \quad 2x - 2y - z = 5$

46. $x + y + z = 1, \quad z = 0$ (the xy-plane)

47. $2x + 2y - z = 3, \quad x + 2y + z = 2$

48. $4y + 3z = -12, \quad 3x + 2y + 6z = 6$

Find parametric equations for the line in which each pair of
planes in Exercises 49–52 intersect.

49. $x + y + z = 1, \quad x + y = 2$

50. $3x - 6y - 2z = 3, \quad 2x + y - 2z = 2$

51. $x - 2y + 4z = 2, \quad x + y - 2z = 5$

52. $5x - 2y = 11, \quad 4y - 5z = -17$

53. Find an equation for the plane that contains both the point
$(1, 2, 3)$ and the line $x = 3t - 1, y = 6, z = t + 2$.

54. Find an equation for the plane that contains intersecting lines
$$L_1 : x = 3t - 1, y = t + 2, z = t - 1, -\infty < t < \infty$$
$$L_2 : x = 5s - 1, y = 2s + 2, z = -s - 1, -\infty < s < \infty$$

55. Find the points in which the line $x = 1 + 2t, y = -1 - t, z = 3t$ meets the coordinate planes. Describe the reasoning behind your answer.

56. Find equations for the line in the plane $z = 3$ that makes
a $30°$ angle with $\mathbf{i}$ and a $60°$ angle with $\mathbf{j}$. Describe the
reasoning behind your answer.

57. Is the line $x = 1 - 2t, y = 2 + 5t, z = -3t$ parallel to the
plane $2x + y - z = 8$? Give reasons for your answer.

58. How can you tell when two planes
$$A_1 x + B_1 y + C_1 z = D_1 \quad \text{and} \quad A_2 x + B_2 y + C_2 z = D_2$$
are parallel? Perpendicular? Give reasons for your answers.

11.6 _____ Surfaces in Space

Just as we call the graph of an equation $F(x, y) = 0$ in the plane a curve, we call
the graph of the equation $F(x, y, z) = 0$ in space a **surface**. We use surfaces
to describe boundaries of solids, to model the membranes across which fluids
flow, to describe plates over which electrical charges are distributed, and to
define the walls of containers that are subjected to pressures of various kinds.
We will see some of these applications in the chapters to come.

Our goal in the present section is to become acquainted with the surfaces
that are most commonly used in the theory and application of the calculus
of functions of more than one variable. This includes finding out what the
surfaces look like, what their equations are, and how to sketch them.

Cylinders

Besides planes, probably the simplest surfaces to sketch and write equations for are cylinders.

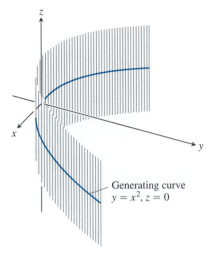

11.41 The cylinder formed by lines parallel to the z-axis and passing through the parabola $y = x^2$ in the xy-plane.

> ### DEFINITION
>
> A **cylinder** is a surface composed of all the lines parallel to a given line that pass through a given plane curve. The curve is the **generating curve** of the cylinder.

In solid geometry, *cylinder* often means *right circular cylinder*. We will assume *right* for our cylinders with the generating curve always in one of the coordinate planes and the lines that form the cylinder parallel to the third axis. This means, for example, that $(x_0, y_0, 0)$ is a point on a generating curve in the xy-plane if and only if (x_0, y_0, z) is a point on the cylinder for any value of z. On the other hand, we will not assume *circular*; we will not even assume that the generating curve is closed. We allow our cylinders to have cross-sections of *any* kind, such as in Example 1, in which each cross-section is a parabola.

EXAMPLE 1   A Parabolic Cylinder

Find an equation for the cylinder made by the lines parallel to the z-axis that pass through the parabola $y = x^2, z = 0$ (Fig. 11.41).

Solution   If a point Q on the cylinder has coordinates (x_0, y_0, z), then $(x_0, y_0, 0)$ lies on the generating curve. Hence, $y_0 = x_0^2$. Conversely, any point Q with coordinates (x_0, x_0^2, z) lies on a line parallel to the z-axis (or on the z-axis itself) that passes through $P_0(x_0, x_0^2, 0)$, a point on the generating curve (Fig. 11.42). Hence, Q is on the cylinder, and we have proven that the points of the cylinder are precisely those for which $y = x^2$. This makes $y = x^2$ an equation for the cylinder. ≡

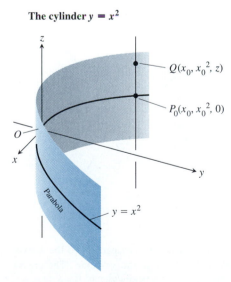

11.42 Every point in the cylinder of Example 1 has coordinates of the form (x_0, x_0^2, z).

As Example 1 suggests, any curve $f(x, y) = C$ in the xy-plane defines a cylinder parallel to the z-axis whose equation is also $f(x, y) = C$. The equation $x^2 + y^2 = 1$ defines the *circular cylinder* made by the lines parallel to the z-axis and passing through the circle $x^2 + y^2 = 1$ in the xy-plane. The equation $x^2 + 4y^2 = 9$ defines the *elliptic cylinder* made by the lines parallel to the z-axis and passing through the ellipse $x^2 + 4y^2 = 9$ in the xy-plane, and so on.

In a similar manner, any curve $g(x, z) = C$ in the xz-plane defines a (right) cylinder parallel to the y-axis whose equation is also $g(x, z) = C$, and any curve $h(y, z) = C$ defines a (right) cylinder parallel to the x-axis whose equation is also $h(y, z) = C$.

In short, an equation in any two of the three Cartesian coordinates defines a cylinder made of lines parallel to the axis of the third coordinate. See Figs. 11.43 and 11.44.

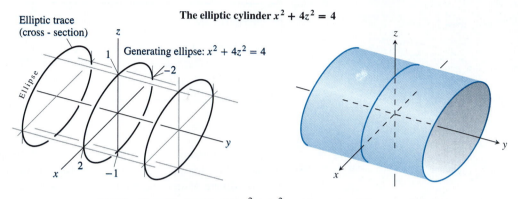

11.43   The elliptic cylinder $x^2 + 4z^2 = 4$ is made of lines parallel to the y-axis and passing through the ellipse $x^2 + 4z^2 = 4$ in the xz-plane. The cross-sections or "traces" of the cylinder in planes perpendicular to the y-axis are ellipses congruent to the generating ellipse. The cylinder extends along the entire y-axis, but we can draw only a finite portion of it.

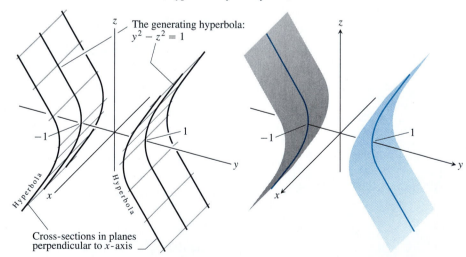

11.44   The hyperbolic cylinder $y^2 - z^2 = 1$ is made of lines parallel to the x-axis and passing through the hyperbola $y^2 - z^2 = 1$ in the yz-plane. The cross-sections or "traces" of the cylinder in planes perpendicular to the x-axis are hyperbolas congruent to the generating hyperbola.

☰ *SKETCHING LESSON*

HOW TO SKETCH CYLINDERS PARALLEL TO THE COORDINATE AXES

1. Sketch all three coordinate axes very *lightly*.

2. Sketch the trace of the cylinder in the coordinate plane of the two variables that appear in the cylinder's equation. Sketch *very lightly*.

3. Sketch traces in parallel planes on either side (again, lightly).

4. Add parallel outer edges to give the shape definition.

5. If more definition is required, darken the parts of the lines that are exposed to view. Leave the hidden parts light. Use line breaks when you can.

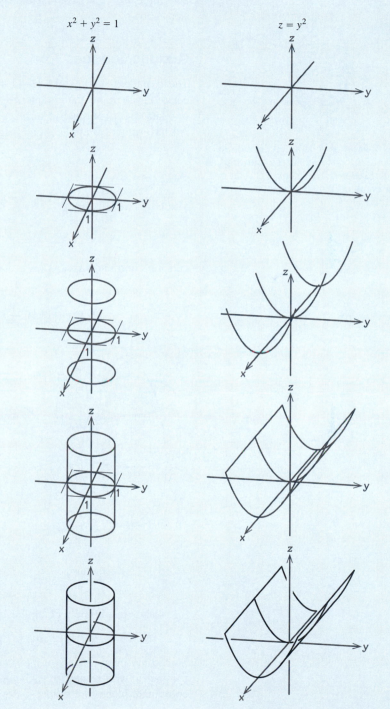

$x^2 + y^2 = 1$

$z = y^2$

Sketching Cylinders

On the preceding page is some advice about sketching cylinders. As always, pencil is safer than pen because you can erase more easily. For each sketch, determine first the axis that the cylinder is parallel to. Then carry out the steps shown.

Quadric Surfaces

The cylinders in Figs. 11.42–11.44 are **quadric surfaces**, surfaces whose equations combine quadratic terms with linear terms and constants. A number of other important quadric surfaces are described in the examples that follow. All have practical applications (Fig. 11.45).

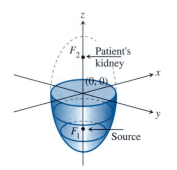

11.45 A *lithotripter* is an application in medicine of the quadric surface known as an ellipsoid. It emits UHF shock waves at one focus to break up kidney stones at the other focus.

EXAMPLE 2

The **ellipsoid**

$$\frac{x^2}{a^2} + \frac{y^2}{b^2} + \frac{z^2}{c^2} = 1$$

cuts the coordinate axes at $(\pm a, 0, 0)$, $(0, \pm b, 0)$, and $(0, 0, \pm c)$ (Fig. 11.46).

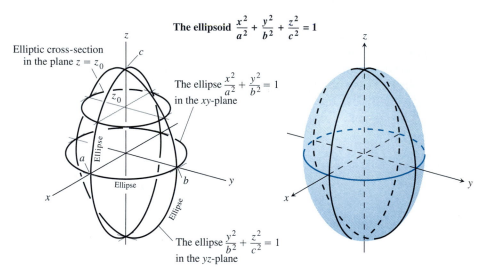

The ellipsoid $\dfrac{x^2}{a^2} + \dfrac{y^2}{b^2} + \dfrac{z^2}{c^2} = 1$

Elliptic cross-section in the plane $z = z_0$

The ellipse $\dfrac{x^2}{a^2} + \dfrac{y^2}{b^2} = 1$ in the xy-plane

The ellipse $\dfrac{y^2}{b^2} + \dfrac{z^2}{c^2} = 1$ in the yz-plane

11.46 A typical ellipsoid.

It lies inside the rectangular box (we can assume a, b, c are positive)

$$|x| \le a, \qquad |y| \le b, \qquad |z| \le c.$$

Since only even powers of x, y, and z occur in the equation, this surface is symmetric with respect to each coordinate plane. The sections cut out by the coordinate planes are ellipses. For example,

$$\frac{x^2}{a^2} + \frac{y^2}{b^2} = 1 \qquad \text{when} \quad z = 0.$$

Each section cut out by a plane

$$z = z_0, \qquad |z_0| < c,$$

is an ellipse

$$\frac{x^2}{a^2[1 - z_0^2/c^2]} + \frac{y^2}{b^2[1 - z_0^2/c^2]} = 1.$$

When two of the three semiaxes a, b, and c are equal, the surface is an ellipsoid of revolution; when all three are equal, it is a sphere.   ≡

EXAMPLE 3

The **elliptic paraboloid**

$$\frac{x^2}{a^2} + \frac{y^2}{b^2} = \frac{z}{c} \qquad (c > 0) \tag{1}$$

is symmetric with respect to the planes $x = 0$ and $y = 0$ (Fig. 11.47). The only intercept on the axes is at the origin. Except for this point, the surface lies above the xy-plane because z is positive whenever either x or y is different from zero. The sections cut by the coordinate planes are

$$x = 0: \qquad \text{the parabola } z = \frac{c}{b^2}y^2,$$

$$y = 0: \qquad \text{the parabola } z = \frac{c}{a^2}x^2,$$

$$z = 0: \qquad \text{the point } (0, 0, 0).$$

Each plane $z = z_0$ above the xy-plane cuts the surface in the ellipse

$$\frac{x^2}{a^2} + \frac{y^2}{b^2} = \frac{z_0}{c}.$$ ≡

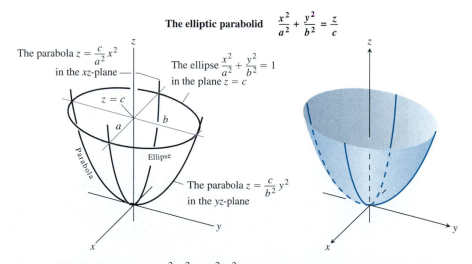

The elliptic parabolid $\dfrac{x^2}{a^2} + \dfrac{y^2}{b^2} = \dfrac{z}{c}$

The parabola $z = \dfrac{c}{a^2}x^2$ in the xz-plane

The ellipse $\dfrac{x^2}{a^2} + \dfrac{y^2}{b^2} = 1$ in the plane $z = c$

$z = c$

Ellipse

Parabola

The parabola $z = \dfrac{c}{b^2}y^2$ in the yz-plane

11.47 The surface $x^2/a^2 + y^2/b^2 = z/c$ is an elliptic paraboloid. The cross-sections perpendicular to the z-axis above the xy-plane are ellipses. The cross-sections in planes that contain the z-axis are parabolas.

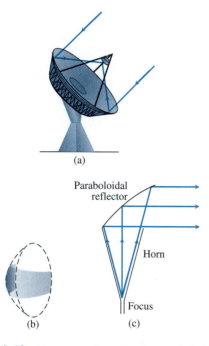

(a)

Paraboloidal
reflector

Horn

Focus

(b)   (c)

11.48 Shapes cut from circular paraboloids are used in many antennas such as satellite trackers and cellular phone links. (a) Radio telescopes use the same principles as do optical telescopes. (b) A "rectangular-cut" radar reflector. (c) The profile of a horn antenna in a microwave radio link.

EXAMPLE 4

The **circular paraboloid** or **paraboloid of revolution**

$$\frac{x^2}{a^2} + \frac{y^2}{a^2} = \frac{z}{c} \qquad (c > 0)$$

is obtained by taking $b = a$ in Eq. (1) for the elliptic paraboloid. The cross-sections of the surface by planes perpendicular to the z-axis are circles centered on the z-axis. The cross-sections by planes containing the z-axis are congruent parabolas with a common focus at the point $(0, 0, a^2/4c)$. Many antennas are shaped like pieces of paraboloids of revolution. (See Fig. 11.48.)

EXAMPLE 5

The **elliptic cone**

$$\frac{x^2}{a^2} + \frac{y^2}{b^2} = \frac{z^2}{c^2} \qquad (2)$$

is symmetric with respect to the three coordinate planes (Fig. 11.49).

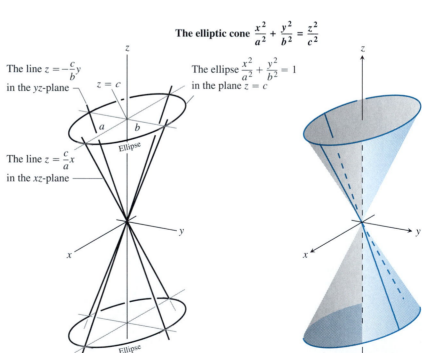

11.49 The surface $x^2/a^2 + y^2/b^2 = z^2/c^2$ is an elliptic cone. Planes perpendicular to the z-axis cut the cone in ellipses above and below the xy-plane. Vertical planes that contain the z-axis cut it in pairs of intersecting lines.

The sections cut by the coordinate planes are

$$x = 0: \qquad \text{the lines } z = \pm\frac{c}{b}y, \qquad (3)$$

$$y = 0: \qquad \text{the lines } z = \pm\frac{c}{a}x, \qquad (4)$$

$$z = 0: \qquad \text{the point } (0, 0, 0).$$

The sections cut by planes $z = z_0$ above and below the xy-plane are ellipses whose centers lie on the z-axis and whose vertices lie on the lines in Eqs. (3) and (4).

If $a = b$, the cone is a circular cone. ≡

EXAMPLE 6

The **hyperboloid of one sheet**

$$\frac{x^2}{a^2} + \frac{y^2}{b^2} - \frac{z^2}{c^2} = 1 \qquad (5)$$

is symmetric with respect to each of the three coordinate planes (Fig. 11.50).

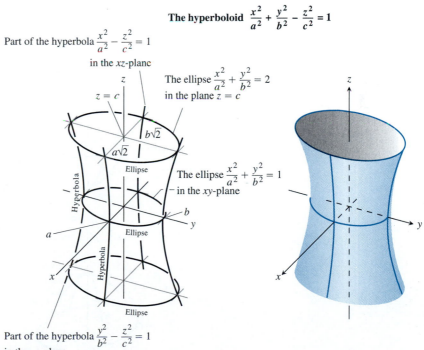

The hyperboloid $\dfrac{x^2}{a^2} + \dfrac{y^2}{b^2} - \dfrac{z^2}{c^2} = 1$

Part of the hyperbola $\dfrac{x^2}{a^2} - \dfrac{z^2}{c^2} = 1$ in the xz-plane

The ellipse $\dfrac{x^2}{a^2} + \dfrac{y^2}{b^2} = 2$ in the plane $z = c$

The ellipse $\dfrac{x^2}{a^2} + \dfrac{y^2}{b^2} = 1$ in the xy-plane

Part of the hyperbola $\dfrac{y^2}{b^2} - \dfrac{z^2}{c^2} = 1$ in the yz-plane

11.50 The surface $x^2/a^2 + y^2/b^2 - z^2/c^2 = 1$ is an (elliptic) hyperboloid of one sheet. Planes perpendicular to the z-axis cut it in ellipses. Vertical planes containing the z-axis cut it in hyperbolas.

The sections cut out by the coordinate planes are

$$x = 0: \qquad \text{the hyperbola } \frac{y^2}{b^2} - \frac{z^2}{c^2} = 1,$$

$$y = 0: \qquad \text{the hyperbola } \frac{x^2}{a^2} - \frac{z^2}{c^2} = 1, \qquad (6)$$

$$z = 0: \qquad \text{the ellipse } \frac{x^2}{a^2} + \frac{y^2}{b^2} = 1.$$

> This shape (and the one in Example 7) is sometimes called an *elliptic hyperboloid*.

The plane $z = z_0$ cuts the surface in an ellipse with center on the z-axis and vertices on one of the hyperbolas in Eqs. (6).

The surface is connected, meaning that it is possible to travel from any point on it to any other point on it without leaving the surface. For this reason, it is said to have *one* sheet, in contrast to the hyperboloid in the next example, which has two sheets.

If $a = b$, the hyperboloid is a surface of revolution.   ≡

EXAMPLE 7

The **hyperboloid of two sheets**

$$\frac{z^2}{c^2} - \frac{x^2}{a^2} - \frac{y^2}{b^2} = 1 \qquad (7)$$

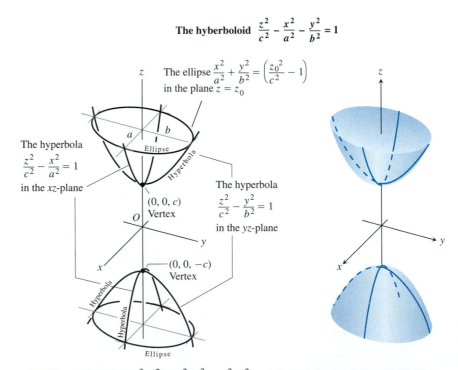

The hyberboloid $\dfrac{z^2}{c^2} - \dfrac{x^2}{a^2} - \dfrac{y^2}{b^2} = 1$

The ellipse $\dfrac{x^2}{a^2} + \dfrac{y^2}{b^2} = \left(\dfrac{z_0^2}{c^2} - 1\right)$ in the plane $z = z_0$

The hyperbola $\dfrac{z^2}{c^2} - \dfrac{x^2}{a^2} = 1$ in the xz-plane

The hyperbola $\dfrac{z^2}{c^2} - \dfrac{y^2}{b^2} = 1$ in the yz-plane

$(0, 0, c)$ Vertex

$(0, 0, -c)$ Vertex

11.51 The surface $z^2/c^2 - x^2/a^2 - y^2/b^2 = 1$ is an (elliptic) hyperboloid of two sheets. Planes perpendicular to the z-axis above and below the vertices cut it in ellipses. Vertical planes containing the z-axis cut it in hyperbolas.

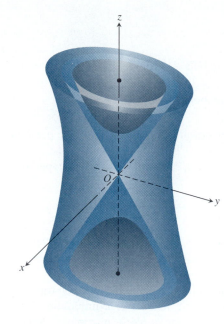

11.52 The cone is asymptotic to both hyperboloids.

is symmetric with respect to the three coordinate planes (Fig. 11.51). The plane $z = 0$ does not intersect the surface; in fact, for a horizontal plane to intersect the surface, we must have $|z| \geq c$. The hyperbolic sections

$$x = 0: \qquad \frac{z^2}{c^2} - \frac{y^2}{b^2} = 1,$$

$$y = 0: \qquad \frac{z^2}{c^2} - \frac{x^2}{a^2} = 1$$

have their vertices and foci on the z-axis. The surface is separated into two portions, one above the plane $z = c$ and the other below the plane $z = -c$. This accounts for its name.

Equations (5) and (7) have different numbers of negative terms. The number in each case is the same as the number of sheets of the hyperboloid. If we compare with Eq. (2), we see that replacing the 1 on the right side of either Eq. (5) or (7) by zero gives the equation of a cone. This cone (Fig. 11.52) is *asymptotic* to both of the hyperboloids (5) and (7) in the same way that the lines

$$\frac{x^2}{a^2} - \frac{y^2}{b^2} = 0$$

are asymptotic to the two hyperbolas

$$\frac{x^2}{a^2} - \frac{y^2}{b^2} = \pm 1$$

in the xy-plane.   ☰

EXAMPLE 8

The **hyperbolic paraboloid**

$$\frac{y^2}{b^2} - \frac{x^2}{a^2} = \frac{z}{c} \qquad (c > 0)$$

has symmetry with respect to the planes $x = 0$ and $y = 0$ (Fig. 11.53).

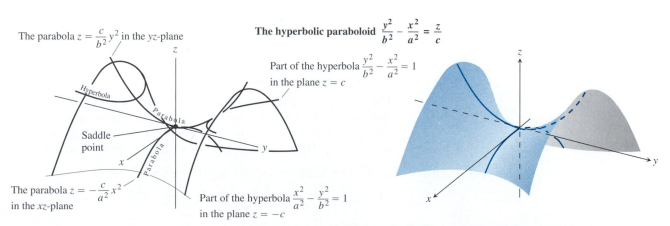

11.53 The surface $y^2/b^2 - x^2/a^2 = z/c$ is a hyperbolic paraboloid. The cross-sections in planes perpendicular to the x- or y-axis are parabolas. The cross-sections in planes perpendicular to the z-axis above and below the xy-plane are hyperbolas. What is the cross-section in the xy-plane itself?

The sections in these planes are

$$x = 0: \qquad \text{the parabola } z = \frac{c}{b^2}y^2, \qquad (8)$$

$$y = 0: \qquad \text{the parabola } z = -\frac{c}{a^2}x^2. \qquad (9)$$

In the plane $x = 0$, the parabola opens upward from the origin. The parabola in the plane $y = 0$ opens downward.

If we cut the surface by a plane $z = z_0 > 0$, the section is a hyperbola,

$$\frac{y^2}{b^2} - \frac{x^2}{a^2} = \frac{z_0}{c},$$

with its focal axis parallel to the y-axis and its vertices on the parabola in (8). If z_0 is negative, the focal axis is parallel to the x-axis, and the vertices lie on the parabola in (9).

Near the origin, the surface is shaped like a saddle. To a person traveling along the surface in the yz-plane, the origin looks like a minimum. To a person traveling in the xz-plane, the origin looks like a maximum. Such a point is called a **minimax** or **saddle point** of a surface. We shall discuss maximum and minimum points on surfaces in Chapter 13. ▤

EXPLORATION 1

Cross-sections of Quadric Surfaces

1. Find the cross-section of the elliptic cone $x^2/9 + y^2/16 = z^2/4$ formed by its intersection with the given planes. Then GRAPH each cross-section in a two-dimensional coordinate system. Explain the cross-section graphs in terms of the graph of the elliptic cone.

   a) The planes $y = 0$, $y = \pm 4$, and $y = \pm 10$.

   b) The planes $x = 0$, $x = \pm 3$, and $x = \pm 10$.

   c) The planes $z = 0$, $z = \pm 2$, and $z = \pm 8$.

2. Choose appropriate planes parallel or identical to the coordinate planes, and repeat part 1 for the following surfaces. If your grapher has list capability, graph several cross-sections in the same viewing window.

   a) $x^2/9 + y^2/16 + z^2/25 = 1$ **b)** $x^2/25 + y^2/9 - z^2/16 = 1$

   c) $x^2/16 + y^2/25 = z/2$ **d)** $x^2/9 + y^2/9 = z/4$

   e) $z^2/16 - x^2/25 - y^2/9 = 1$ **f)** $y^2/4 - x^2/25 = z/3$

▤ *S K E T C H I N G   L E S S O N*

How to Sketch Quadric Surfaces

$$x^2 + \frac{y^2}{4} + z^2 = 1 \qquad\qquad z = 4 - x^2 - y^2$$

1. Lightly sketch the three coordinate axes.

2. Decide on a scale and mark the intercepts on the axes.

3. Sketch cross-sections in the coordinate planes and in a few parallel planes, but don't clutter the picture. Use tangent lines as guides.

4. If more is required, darken the parts exposed to view. Leave the rest light. Use line breaks when you can.

Exercises 11.6

Sketch the surfaces in Exercises 1–50.

Cylinders

1. $x^2 + y^2 = 4$
2. $x^2 + z^2 = 4$
3. $y^2 + z^2 = 1$
4. $z = y^2/4$
5. $z = y^2 - 1$
6. $x = y^2$
7. $z = 4 - x^2$
8. $x = 4 - y^2$
9. $y = x^2$
10. $y = x^2 - 2$
11. $y^2 + 4z^2 = 16$
12. $4x^2 + y^2 = 36$
13. $z^2 + 4y^2 = 9$
14. $y^2 - z^2 = 4$
15. $z^2 - y^2 = 1$
16. $yz = 1$

Ellipsoids

17. $9x^2 + y^2 + z^2 = 9$
18. $4x^2 + 4y^2 + z^2 = 16$
19. $x^2 + y^2 + z^2 = 4$
20. $9x^2 + 4y^2 + z^2 = 36$
21. $4x^2 + 9y^2 + 4z^2 = 36$
22. $9x^2 + 4y^2 + 36z^2 = 36$

Paraboloids

23. $x^2 + y^2 = z$
24. $x^2 + z^2 = y$
25. $x^2 + 4y^2 = z$
26. $z = x^2 + 9y^2$
27. $z = 8 - x^2 - y^2$
28. $z = 18 - x^2 - 9y^2$
29. $x = 4 - 4y^2 - z^2$
30. $y = 1 - x^2 - z^2$
31. $z = x^2 + y^2 + 1$
32. $z = 4x^2 + y^2 - 4$

Cones

33. $x^2 + y^2 = z^2$
34. $y^2 + z^2 = x^2$
35. $x^2 + z^2 = y^2$
36. $4x^2 + 9y^2 = z^2$
37. $9x^2 + 4y^2 = 36z^2$
38. $4x^2 + 9z^2 = 9y^2$

Hyperboloids

39. $x^2 + y^2 - z^2 = 1$
40. $y^2 + z^2 - x^2 = 1$
41. $y^2/4 + z^2/9 - x^2/4 = 1$
42. $x^2/4 + y^2/4 - z^2/9 = 1$
43. $x^2/4 + y^2 - z^2 = 1$

44. $z^2 - x^2 - y^2 = 1$
45. $z^2 - x^2/4 - y^2 = 1$
46. $y^2/4 - x^2/4 - z^2 = 1$
47. $x^2 - y^2 - z^2/4 = 1$
48. $x^2/4 - z^2/4 - y^2 = 1$

Hyperbolic Paraboloids

49. $y^2 - x^2 = z$
50. $x^2 - y^2 = z$

Identify and sketch the surfaces in Exercises 51-64. Support with graphs of cross-sections in a two-dimensional coordinate system.

51. $9x^2 + 36y^2 + 4z^2 = 36$
52. $3y^2 + 3z^2 = 2x$
53. $9x^2 + 36z^2 = 4y^2$
54. $16y^2 - 9x^2 - 144z^2 = 144$
55. $9x^2 + 16z^2 = 72y$
56. $-4x^2 + 9y^2 + 36z^2 = 36$
57. $z^2 - 9y^2 = 3x$
58. $36x^2 + 4y^2 + 9z^2 = 36$
59. $2x^2 + 2z^2 = 3y$
60. $12y^2 + 3z^2 = 16x$
61. $9x^2 - 36y^2 - 4z^2 = 36$
62. $x^2 - 4z^2 = 2y$
63. $25x^2 - 100y^2 + 4z^2 = 100$
64. $16y^2 + z^2 = 4x^2$

Computer 3D Grapher

If you have access to a computer 3D grapher, try graphing the surfaces in Exercises 65–71.

65. $z = y^2$
66. $z = 1 - y^2$
67. $z = x^2 + y^2$
68. $z = x^2 + 2y^2$
69. $z = \sqrt{1 - x^2}$ (upper half of a circular cylinder)
70. $z = \sqrt{1 - y^2/4}$ (upper half of an elliptical cylinder)
71. $z = \sqrt{x^2 + 2y^2 + 4}$ (one sheet of an elliptic hyperboloid)
72. The shape of a lithotripter (Fig. 11.45) is generated by rotating the portion of an ellipse below its (horizontal) minor axis about its major axis. The major diameter (length of the major axis) is 26 inches, and the minor diameter is 10 inches. Where should the shock-wave source and patient be placed for maximum effect? Give the appropriate measurements.

11.7 Cylindrical and Spherical Coordinates

In this section, we introduce two new systems of coordinates for space: the cylindrical coordinate system and the spherical coordinate system. In the cylindrical coordinate system, cylinders whose axes lie along the z-axis and planes that contain the z-axis have especially simple equations. In the spherical coordinate system, spheres centered at the origin and single cones

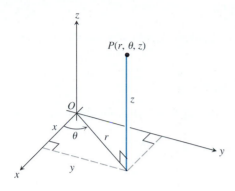

11.54 The cylindrical coordinates of a point in space are $r, \theta,$ and z.

at the origin whose axes lie along the z-axis have especially simple equations. When your work involves these shapes and you need a coordinate system, these may be the best systems to use, as we will see in later chapters.

Cylindrical Coordinates

We obtain cylindrical coordinates for space by combining polar coordinates in the xy-plane with the usual z-axis. This assigns to every point in space one or more coordinate triples of the form (r, θ, z), as shown in Fig. 11.54.

The values of $x, y, r,$ and θ in cylindrical coordinates are related by the usual equations:

$$x = r\cos\theta, \quad y = r\sin\theta, \quad r^2 = x^2 + y^2, \quad \tan\theta = y/x. \qquad (1)$$

We will use cylindrical coordinates to study planetary motion in Section 12.5.

In cylindrical coordinates, the equation $r = a$ describes not just a circle in the xy-plane but an entire circular cylinder about the z-axis (Fig. 11.55a). The z-axis itself is given by the equation $r = 0$. The equation $\theta = \theta_0$ describes the plane that contains the z-axis and makes an angle of θ_0 radians with the positive x-axis.

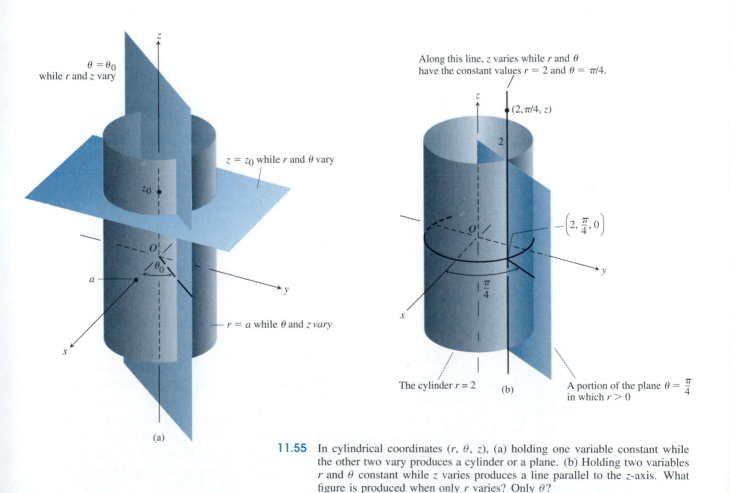

11.55 In cylindrical coordinates (r, θ, z), (a) holding one variable constant while the other two vary produces a cylinder or a plane. (b) Holding two variables r and θ constant while z varies produces a line parallel to the z-axis. What figure is produced when only r varies? Only θ?

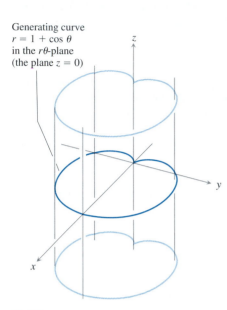

Generating curve
$r = 1 + \cos\theta$
in the $r\theta$-plane
(the plane $z = 0$)

11.56 The cylindrical coordinate equation $r = 1 + \cos\theta$ defines a cylinder in space whose cross-sections perpendicular to the z-axis are cardioids.

EXAMPLE 1

Describe the points in space whose cylindrical coordinates satisfy the equations

$$r = 2, \qquad \theta = \frac{\pi}{4}.$$

Solution These points make up the line in which the cylinder $r = 2$ cuts the portion of the plane $\theta = \pi/4$ in which r is positive (Fig. 11.55b). This is the line through the point $(2, \pi/4, 0)$ parallel to the z-axis. ▤

EXAMPLE 2

Sketch the surface $r = 1 + \cos\theta$.

Solution The equation involves only r and θ; the coordinate variable z is missing. Therefore, the surface is a cylinder of lines that pass through the cardioid $r = 1 + \cos\theta$ in the $r\theta$-plane and are parallel to the z-axis. The rules for sketching the cylinder are the same as always: Sketch the x-, y-, and z-axes, draw a few perpendicular cross-sections, connect the cross-sections with parallel lines, and darken the exposed parts (Fig. 11.56). ▤

EXAMPLE 3

Find a Cartesian equation for the surface $z = r^2$, and identify the surface.

Solution From Eqs. (1), we have $z = r^2 = x^2 + y^2$. The surface is the circular paraboloid $x^2 + y^2 = z$. ▤

EXAMPLE 4

Find an equation for the circular cylinder $4x^2 + 4y^2 = 9$ in cylindrical coordinates.

Solution The cylinder consists of the points whose distance from the z-axis is $\sqrt{x^2 + y^2} = 3/2$. The corresponding equation in cylindrical coordinates is $r = 3/2$. ▤

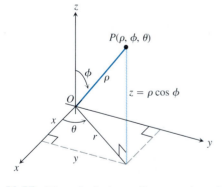

$P(\rho, \phi, \theta)$

$z = \rho\cos\phi$

11.57 The spherical coordinates ρ, ϕ, and θ and their relation to x, y, z, and r.

Spherical Coordinates

Spherical coordinates locate points in space with two angles and a distance, as shown in Fig. 11.57.

The first coordinate, $\rho = |\overrightarrow{OP}|$, is the point's distance from the origin. Unlike r, the variable ρ is never negative. The second coordinate, ϕ, is the angle that the vector $\overrightarrow{OP}$ makes with the positive z-axis. It is required to lie in the interval $[0, \pi]$. The third coordinate is the angle θ as it was found for cylindrical coordinates.

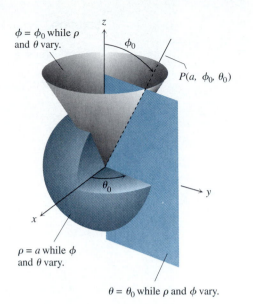

$\phi = \phi_0$ while ρ and θ vary.

ϕ_0

$P(a, \phi_0, \theta_0)$

$\rho = a$ while ϕ and θ vary.

$\theta = \theta_0$ while ρ and ϕ vary.

11.58 Spheres whose centers are at the origin, single cones at the origin whose axes lie along the z-axis, and half-planes "hinged" along the z-axis have constant-coordinate equations in spherical coordinates.

Selected Equations Relating Cartesian (Rectangular), Cylindrical, and Spherical Coordinates

$$r = \rho \sin\phi, \qquad x = r\cos\theta = \rho\sin\phi\cos\theta,$$

$$z = \rho\cos\phi, \qquad y = r\sin\theta = \rho\sin\phi\sin\theta, \qquad (2)$$

$$\rho = \sqrt{x^2 + y^2 + z^2} = \sqrt{r^2 + z^2}$$

The equation $\rho = a$ describes the sphere of radius a centered at the origin (Fig. 11.58).

The equation $\phi = \phi_0$ describes a single cone whose vertex lies at the origin and whose axis lies along the z-axis. If $0 < \phi_0 < \pi/2$, the cone $\phi = \phi_0$ opens upward. If $\pi/2 < \phi_0 < \pi$, it opens downward. The equations $\phi = 0$ and $\phi = \pi$ describe the positive z-axis and negative z-axis, respectively (degenerate cones). The equation $\phi = \pi/2$ describes the xy-plane—the "cone" flattened out.

A few books give spherical coordinates in the order (ρ, θ, ϕ), with the θ and ϕ reversed. Watch out for this when you read elsewhere.

EXAMPLE 5

Find a spherical coordinate equation for the sphere

$$x^2 + y^2 + (z - 1)^2 = 1$$

with center $(0, 0, 1)$ and radius 1.

Solution   From Eqs. (2), we find that the *left side* of the equation is

$$\rho^2 \sin^2 \phi (\cos^2 \theta + \sin^2 \theta) + \rho^2 \cos^2 \phi - 2\rho \cos \phi + 1 = \rho^2 - 2\rho \cos \phi + 1.$$

Hence, the original equation transforms into

$$\rho^2 - 2\rho \cos \phi + 1 = 1,$$
$$\rho^2 = 2\rho \cos \phi,$$
$$\rho = 2 \cos \phi.$$

Exercises 11.7

The following table gives the coordinates of specific points in space in one of three coordinate systems. In Exercises 1–10, find coordinates for each point in the other two systems. There may be more than one right answer because points in cylindrical and spherical coordinates may have more than one coordinate triple.

Rectangular (x, y, z)	Cylindrical (r, θ, z)	Spherical (ρ, ϕ, θ)
1. $(0, 0, 0)$		
2. $(1, 0, 0)$		
3. $(0, 1, 0)$		
4. $(0, 0, 1)$		
5.	$(1, 0, 0)$	
6.	$(\sqrt{2}, 0, 1)$	
7.	$(1, \pi/2, 1)$	
8.		$(\sqrt{3}, \pi/3, -\pi/2)$
9.		$(2\sqrt{2}, \pi/2, 3\pi/2)$
10.		$(\sqrt{2}, \pi, 3\pi/2)$

In Exercises 11–30, translate the equations from the given coordinate system (rectangular, cylindrical, spherical) into equations in the other two systems. Also, describe the set of points defined by the equation.

11. $r = 0$

12. $x^2 + y^2 = 5$

13. $z = 0$

14. $z = -2$

15. $\rho \cos \phi = 3$

16. $\sqrt{x^2 + y^2} = z$

17. $\rho \sin \phi \cos \theta = 0$

18. $\tan^2 \phi = 1$

19. $x^2 + y^2 + z^2 = 4$

20. $x^2 + y^2 + \left(z - \dfrac{1}{2}\right)^2 = \dfrac{1}{4}$

21. $\rho = 2 \sin \theta$

22. $\rho = 6 \cos \phi$

23. $r = \csc \theta$

24. $r = -3 \sec \theta$

25. $x^2 + y^2 + (z - 1)^2 = 1, z \leq 1$

26. $x^2 + y^2 + z^2 = 3, 0 \leq z \leq \sqrt{3}/2$

27. $\rho = 2, 0 \leq \phi \leq \pi/3$

28. $r^2 + z^2 = 4, z \leq -\sqrt{2}$

29. $\phi = 4\pi/3, 0 \leq \rho \leq 2$

30. $\phi = \pi/2, 0 \leq \rho \leq \sqrt{7}$

In Exercises 31–38, describe the sets of points in space whose cylindrical coordinates satisfy the given equations or pairs of equations. Sketch.

31. $r = 4$

32. $r^2 + z^2 = 1$

33. $r = 1 - \cos \theta$

34. $r = 2 \cos \theta$

35. $r = 2, \quad z = 3$

36. $\theta = \pi/6, \quad z = r$

37. $r = 3, \quad z = \theta/2$

38. $r = 1, z = 2\pi - \theta$

In Exercises 39–46, describe the sets of points in space whose spherical coordinates satisfy the given equations or pairs of equations. Sketch.

39. $\rho = 2, \phi = \pi/2$

40. $\rho = 6, \quad \phi = \pi/4$

41. $\theta = \pi/4, \quad \phi = \pi/4$

42. $\theta = \pi/2, \phi = 3\pi/4$

43. $\theta = \pi/2, \quad \rho = 4 \sin \phi$

44. $\rho = \sin \phi$

45. $\rho = \cos \phi$

46. $\rho = 1 - \cos \phi$ (*Hint:* The absence of θ indicates symmetry with respect to the z-axis. What is the trace of the surface in the yz-plane?)

Review Questions

1. When are two vectors equal?

2. How are vectors added and subtracted?

3. How are the length and direction of a vector calculated?

4. If a vector is multiplied by a scalar, how is the result related to the original vector? What if the scalar is zero? Negative?

5. Define the *dot product* or *scalar product* of two vectors. Which algebraic laws (commutative, associative, distributive) are satisfied by dot products, and which, if any, are not? Give examples. When is the dot product of two vectors equal to zero?

6. What is the vector projection of a vector **B** onto a vector **A**? How do you write **B** as the sum of a vector parallel to **A** and a vector orthogonal to **A**?

7. Define the *cross product* or *vector product* of two vectors. Which algebraic laws (commutative, associative, distributive) are satisfied by cross products, and which are not? Give examples. When is the cross product of two vectors equal to zero?

8. What is the determinant formula for evaluating the cross

product of two vectors? Use it in an example.

9. How are cross and dot products used to find equations for lines, line segments, and planes? Give examples.

10. How can vectors be used to calculate the distance between a point and a line? A point and a plane? Give examples.

11. What is the geometric interpretation of $(\mathbf{A} \times \mathbf{B}) \cdot \mathbf{C}$ as a volume? How may the product be calculated from the components of **A**, **B**, and **C**?

12. What is a cylinder? Give examples of equations that define cylinders in Cartesian coordinates; in cylindrical coordinates. What advice can you give about drawing cylinders?

13. Give examples of ellipsoids, paraboloids, cones, and hyperboloids (equations and sketches). What advice can you give about sketching these surfaces?

14. How are cylindrical and spherical coordinates defined? Draw diagrams that show how cylindrical and spherical coordinates are related to rectangular coordinates. Describe graphs that have constant-coordinate equations (like $x = 1, r = 1$, or $\phi = \pi/3$) in the three coordinate systems.

Practice Exercises

1. Draw the unit vectors $\mathbf{u} = (\cos\theta)\mathbf{i} + (\sin\theta)\mathbf{j}$ for $\theta = 0$, $\pi/2, 2\pi/3, 5\pi/4$, and $5\pi/3$, together with the coordinate axes and unit circle.

2. Find the unit vector obtained by rotating

   a) **i** clockwise $45°$ b) **j** counterclockwise $120°$.

In Exercises 3 and 4, find the unit vectors that are tangent and normal to the curve at point P.

3. $y = \tan x$, $P(\pi/4, 1)$ 4. $x^2 + y^2 = 25$, $P(3, 4)$

Express the vectors in Exercises 5–8 in terms of their lengths and directions.

5. $\sqrt{2}\mathbf{i} + \sqrt{2}\mathbf{j}$

6. $-\mathbf{i} - \mathbf{j}$

7. $2\mathbf{i} - 3\mathbf{j} + 6\mathbf{k}$

8. $\mathbf{i} + 2\mathbf{j} - \mathbf{k}$

9. Find a vector 2 units long in the direction of $\mathbf{A} = 4\mathbf{i} - \mathbf{j} + 4\mathbf{k}$.

10. Find a vector 5 units long in the direction opposite to the direction of $\mathbf{A} = (3/5)\mathbf{i} + (4/5)\mathbf{k}$.

In Exercises 11 and 12, find $|\mathbf{A}|, |\mathbf{B}|, \mathbf{A} \cdot \mathbf{B}, \mathbf{B} \cdot \mathbf{A}, \mathbf{A} \times \mathbf{B}$, $\mathbf{B} \times \mathbf{A}, |\mathbf{A} \times \mathbf{B}|$, the acute angle between **A** and **B**, the scalar component of **B** in the direction of **A**, and the vector projection of **B** onto **A**.

11. $\mathbf{A} = \mathbf{i} + \mathbf{j}$,
    $\mathbf{B} = 2\mathbf{i} + \mathbf{j} - 2\mathbf{k}$

12. $\mathbf{A} = 5\mathbf{i} + \mathbf{j} + \mathbf{k}$,
    $\mathbf{B} = \mathbf{i} - 2\mathbf{j} + 3\mathbf{k}$

In Exercises 13 and 14, write **B** as the sum of a vector parallel to **A** and a vector orthogonal to **A**.

13. $\mathbf{A} = 2\mathbf{i} + \mathbf{j} - \mathbf{k}$,   14. $\mathbf{A} = \mathbf{i} - 2\mathbf{j}$,

    $\mathbf{B} = \mathbf{i} + \mathbf{j} - 5\mathbf{k}$ $\mathbf{B} = \mathbf{i} + \mathbf{j} + \mathbf{k}$

In Exercises 15 and 16, draw coordinate axes, and then sketch **A**, **B**, and $\mathbf{A} \times \mathbf{B}$ as vectors at the origin.

15. $\mathbf{A} = \mathbf{i}$, $\mathbf{B} = \mathbf{i} + \mathbf{j}$ 16. $\mathbf{A} = \mathbf{i} - \mathbf{j}$, $\mathbf{B} = \mathbf{i} + \mathbf{j}$

In Exercises 17 and 18, find the distance from the point to the line in the xy-plane.

17. $(3, 2)$, $3x + 4y = 2$ 18. $(-1, 1), 5x - 12y = 9$

In Exercises 19 and 20, find the distance from the point to the plane.

19. $(6, 0, -6)$, $x - y = 4$

20. $(3, 0, 10)$, $2x + 3y + z = 2$

21. Find an equation for the plane that passes through the point $(3, -2, 1)$ normal to the vector $\mathbf{N} = 2\mathbf{i} + \mathbf{j} - \mathbf{k}$.

22. Find an equation for the plane that passes through the point $(-1, 6, 0)$ perpendicular to the line $x = -1 + t$, $y = 6 - 2t, z = 3t$.

In Exercises 23 and 24, find an equation for the plane through points P, Q, and R.

23. $P(1, -1, 2)$, $Q(2, 1, 3)$, $R(-1, 2, -1)$

24. $P(1, 0, 0)$, $Q(0, 1, 0)$, $R(0, 0, 1)$

25. Find parametric equations for the line that passes through the point $(1, 2, 3)$ parallel to the vector $\mathbf{v} = -3\mathbf{i} + 7\mathbf{k}$.

26. Find the points in which the line $x = 1 + 2t$, $y = -1 - t$, $z = 3t$ meets the three coordinate planes.

27. Find the point in which the line through the origin perpendicular to the plane $2x - y - z = 4$ meets the plane $3x - 5y + 2z = 6$.

28. Find parametric equations for the line segment joining the points $P(1, 2, 0)$ and $Q(1, 3, -1)$.

29. Find parametric equations for the line in which the planes $x + 2y + z = 1$ and $x - y + 2z = -8$ intersect.

30. Find the angle between the planes $x + y = 1$ and $y + z = 1$.

In Exercises 31 and 32, find the distance from the point to the line.

31. $(2, -1, -10)$; $x = 4$, $y = 4 + 2t$, $z = 4t$

32. $(-1, 4, 3)$; $x = 10 + 4t$, $y = -3$, $z = 4t$

33. *Work.*   Find the work done in pushing a car 800 ft with a force of magnitude 40 lbs directed $28°$ downward from the horizontal against the back of the car.

34. *Torque.*   The operators manual for the *Toro* 21-in. lawn-mower says, "Tighten the spark plug to 15 ft-lb (20.4 N · m)." If you are installing the plug with a 10.5-in. socket wrench that places the center of your hand 9 in. from the axis of the spark plug, about how hard should you pull? Answer in pounds.

In Exercises 35 and 36, find (a) the area of the parallelogram determined by vectors $\mathbf{A}$ and $\mathbf{B}$, (b) the volume of the parallelepiped determined by the vectors $\mathbf{A}$, $\mathbf{B}$, and $\mathbf{C}$.

35. $\mathbf{A} = \mathbf{i} + \mathbf{j} - \mathbf{k}$, $\mathbf{B} = 2\mathbf{i} + \mathbf{j} + \mathbf{k}$, $\mathbf{C} = -\mathbf{i} - 2\mathbf{j} + 3\mathbf{k}$

36. $\mathbf{A} = \mathbf{i} + \mathbf{j}$, $\mathbf{B} = \mathbf{j}$, $\mathbf{C} = \mathbf{i} + \mathbf{j} + \mathbf{k}$

37. Which of the following are *always true*, and which are *not always true*?

a) $|\mathbf{A}| = \sqrt{\mathbf{A} \cdot \mathbf{A}}$ b) $\mathbf{A} \cdot \mathbf{A} = |\mathbf{A}|$

c) $\mathbf{A} \times \mathbf{0} = \mathbf{0} \times \mathbf{A} = \mathbf{0}$ d) $\mathbf{A} \times (-\mathbf{A}) = \mathbf{0}$

e) $\mathbf{A} \times \mathbf{B} = \mathbf{B} \times \mathbf{A}$

f) $\mathbf{A} \times (\mathbf{B} + \mathbf{C}) = \mathbf{A} \times \mathbf{B} + \mathbf{A} \times \mathbf{C}$

g) $(\mathbf{A} \times \mathbf{B}) \cdot \mathbf{B} = 0$ h) $(\mathbf{A} \times \mathbf{B}) \cdot \mathbf{C} = \mathbf{A} \cdot (\mathbf{B} \times \mathbf{C})$

38. Which of the following are *always true*, and which are *not always true*?

a) $\mathbf{A} \cdot \mathbf{B} = \mathbf{B} \cdot \mathbf{A}$ b) $\mathbf{A} \times \mathbf{B} = -(\mathbf{B} \times \mathbf{A})$

c) $(-\mathbf{A}) \times \mathbf{B} = -(\mathbf{A} \times \mathbf{B})$

d) $(c\mathbf{A}) \cdot \mathbf{B} = \mathbf{A} \cdot (c\mathbf{B}) = c(\mathbf{A} \cdot \mathbf{B})$ (any number c)

e) $c(\mathbf{A} \times \mathbf{B}) = (c\mathbf{A}) \times \mathbf{B} = \mathbf{A} \times (c\mathbf{B})$ (any number c)

f) $\mathbf{A} \cdot \mathbf{A} = |\mathbf{A}|^2$ g) $(\mathbf{A} \times \mathbf{A}) \cdot \mathbf{A} = 0$

h) $(\mathbf{A} \times \mathbf{B}) \cdot \mathbf{A} = \mathbf{B} \cdot (\mathbf{A} \times \mathbf{B})$

The equations in Exercises 39–48 define sets both in the plane and in three-dimensional space. Identify both sets for each equation.

Rectangular Coordinates

39. $x = 0$ **40.** $x + y = 1$

41. $x^2 + y^2 = 4$ **42.** $x^2 + 4y^2 = 16$

43. $x = y^2$ **44.** $y^2 - x^2 = 1$

Cylindrical Coordinates

45. $r = 1 - \cos\theta$ **46.** $r = \sin\theta$

47. $r^2 = 2\cos 2\theta$ **48.** $r = \cos 2\theta$

Describe the sets defined by the spherical-coordinate equations and inequalities in Exercises 49–54.

49. $\rho = 2$ **50.** $\theta = \pi/4$

51. $\phi = \pi/6$ **52.** $\rho = 1$, $\phi = \pi/2$

53. $\rho = 1$, $0 \leq \phi \leq \pi/2$ **54.** $1 \leq \rho \leq 2$

The following table gives the coordinates of points in space in one of the three standard coordinate systems. In Exercises 55–60, find coordinates for each point in the other two systems. There may be more than one right answer because cylindrical and spherical coordinates are not unique.

Rectangular (x, y, z)	Cylindrical (r, θ, z)	Spherical (ρ, ϕ, θ)
55.		$(1, 0, 0)$
56.	$\left(1, \dfrac{\pi}{2}, 0\right)$	
57.		$\left(\sqrt{2}, \dfrac{\pi}{4}, \dfrac{\pi}{2}\right)$
58.		$\left(2, \dfrac{5\pi}{6}, 0\right)$
59. $(1, 1, 1)$		
60. $(0, -1, 1)$		

In Exercises 61–66, translate the equations from the given coordinate system (rectangular, cylindrical, spherical) into the other two systems. Identify the set of points defined by the equation.

Rectangular

61. $z = 2$

62. $z = \sqrt{x^2 + y^2}$

Cylindrical

63. $z = r^2$

64. $r = \cos \theta$

Spherical

65. $\rho = 4$

66. $\rho \cos \phi = 1$

Identify and sketch the quadric surfaces in Exercises 67–78. Support with graphs of cross-sections in a two-dimensional coordinate system.

67. $x^2 + y^2 + z^2 = 4$

68. $x^2 + (y - 1)^2 + z^2 = 1$

69. $4x^2 + 4y^2 + z^2 = 4$

70. $36x^2 + 9y^2 + 4z^2 = 36$

71. $z = -(x^2 + y^2)$

72. $y = -(x^2 + z^2)$

73. $x^2 + y^2 = z^2$

74. $x^2 + z^2 = y^2$

75. $x^2 + y^2 - z^2 = 4$

76. $4y^2 + z^2 - 4x^2 = 4$

77. $y^2 - x^2 - z^2 = 1$

78. $z^2 - x^2 - y^2 = 1$

12

Vector-valued Functions, Parametrizations, and Motion in Space

OVERVIEW When a body travels through space, the parametric equations $x = f(t)$, $y = g(t)$, and $z = h(t)$ may be used to model the body's motion and path. Using vector notation, we can condense these three equations into a single equation

$$\mathbf{r} = f(t)\mathbf{i} + g(t)\mathbf{j} + h(t)\mathbf{k}$$

that gives the body's position as a *vector function* of t.

Thus, we have two easily interchangeable ways to represent position and motion of a body traveling in space. In this chapter, we use calculus to differentiate and integrate vector functions and to study the paths, velocities, and accelerations of bodies in the plane and in space. We apply graphing utility techniques in parametric mode to reinforce our findings about the vector functions.

12.1   Vector-valued Functions and Curves in Space; Derivatives and Integrals

When a particle moves along a path in space, we can keep track of it by watching the vector $\mathbf{r}$ that runs from the origin out to the particle (Fig. 12.1). If the particle's position coordinates are twice-differentiable functions of time, then so is $\mathbf{r}$, and we can find the particle's velocity and acceleration vectors as functions of time by differentiating $\mathbf{r}$. Conversely, if we know the particle's velocity or acceleration vector as a function of time, then under the right conditions, we can find $\mathbf{r}$ by integration. This section gives the details.

913

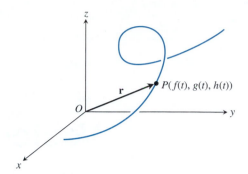

12.1   The position vector $\mathbf{r} = \overrightarrow{OP}$ of a particle moving through space is a function of time.

For efficiency in our written work, we can represent a path in space with one of three forms:

Parametrically with the points

$$(x(t),\ y(t),\ z(t))$$

or with vector equation

$$\mathbf{r} = x(t)\mathbf{i} + y(t)\mathbf{j} + z(t)\mathbf{k}$$

or

$$\mathbf{r} = f\mathbf{i} + g\mathbf{j} + h\mathbf{k}.$$

Given any form, we should be able to shift easily to another as needed.

Definitions

When a particle moves through space as a function of time, each of its rectangular coordinates is also a function of time. We may describe the particle's path with a triple of equations

$$x = f(t), \qquad y = g(t), \qquad z = h(t) \qquad (t \in I),$$

where I is the time interval in question. The set of points (x, y, z) defined by these equations is called a **curve in space**. The equations are **parametric equations** for the curve, t being the **parameter** and I the **parameter interval**. The point $P(f(t), g(t), h(t))$ is called the particle's **position** at time t.

The vector

$$\mathbf{r} = \overrightarrow{OP} = f(t)\,\mathbf{i} + g(t)\,\mathbf{j} + h(t)\,\mathbf{k} \qquad (1)$$

is called the particle's **position vector**, and we think of the particle's path as the curve determined by $\mathbf{r}$. We use Eq. (1) to name the curve as "the curve $\mathbf{r} = f(t)\,\mathbf{i} + g(t)\,\mathbf{j} + h(t)\,\mathbf{k}$" or "the curve $\mathbf{r} = f\mathbf{i} + g\mathbf{j} + h\mathbf{k}$."

In general, a **vector function** or **vector-valued function** is any set of ordered pairs whose second components are vectors and that satisfies the requirement for being a function. The first components of the ordered pairs form the domain of the function. In this chapter, each vector function that we study can usually be defined by a vector equation such as Eq. (1), and its domain will be an interval of real numbers. In Chapter 15, we will study vector functions whose domains are regions of a plane, or space, and call such functions *vector fields*.

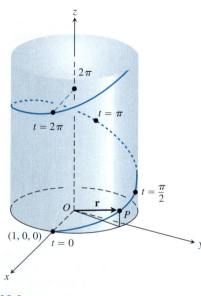

12.2   The helix $\mathbf{r} = (\cos t)\,\mathbf{i} + (\sin t)\,\mathbf{j} + t\,\mathbf{k}$ spirals up from the xy-plane as t increases from zero.

EXAMPLE 1   A Helix

The curve

$$\mathbf{r} = (\cos t)\,\mathbf{i} + (\sin t)\,\mathbf{j} + t\,\mathbf{k},$$

called a *helix* ("HEE-lix") from an old Greek word for spiral, winds around the cylinder $x^2 + y^2 = 1$. (See Fig. 12.2 for this helix and Fig. 12.3 for other helices.) We know that the curve lies on the cylinder because the x-

and y-coordinates of the point $P(\cos t, \sin t, t)$, satisfy the cylinder's equation:

$$x^2 + y^2 = (\cos t)^2 + (\sin t)^2 = 1.$$

Because $z = t$, the curve rises steadily as t increases and completes one turn around the cylinder each time t increases by 2π. The parametric equations for the helix are

$$x = \cos t, \qquad y = \sin t, \qquad z = t.$$

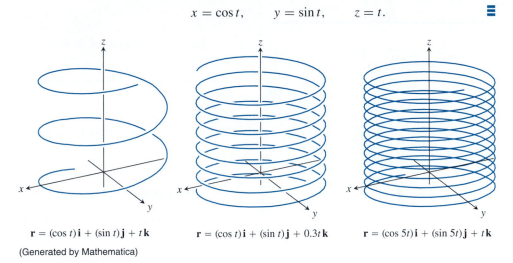

$$\mathbf{r} = (\cos t)\,\mathbf{i} + (\sin t)\,\mathbf{j} + t\,\mathbf{k} \qquad\qquad \mathbf{r} = (\cos t)\,\mathbf{i} + (\sin t)\,\mathbf{j} + 0.3t\,\mathbf{k} \qquad\qquad \mathbf{r} = (\cos 5t)\,\mathbf{i} + (\sin 5t)\,\mathbf{j} + t\,\mathbf{k}$$

(Generated by Mathematica)

12.3 Helices ("HEE-lih-sees") drawn by computer.

Limits and Continuity

We define limits of vector functions in terms of their scalar components.

> **DEFINITION**
>
> If $\mathbf{L} = L_1\,\mathbf{i} + L_2\,\mathbf{j} + L_3\,\mathbf{k}$ is a vector and $\mathbf{r}$ is a vector-valued function of t defined by the rule $\mathbf{r}(t) = f(t)\,\mathbf{i} + g(t)\,\mathbf{j} + h(t)\,\mathbf{k}$, then $\mathbf{r}$ **has limit L as t approaches t_0,**
>
> $$\lim_{t \to t_0} \mathbf{r}(t) = \mathbf{L},$$
>
> if
>
> $$\lim_{t \to t_0} f(t) = L_1, \qquad \lim_{t \to t_0} g(t) = L_2, \qquad \text{and} \qquad \lim_{t \to t_0} h(t) = L_3.$$

EXAMPLE 2

If $\mathbf{r}(t) = (\cos t)\,\mathbf{i} + (\sin t)\,\mathbf{j} + t\,\mathbf{k}$, then

$$\lim_{t \to \pi/4} \mathbf{r}(t) = \left(\lim_{t \to \pi/4} \cos t \right)\mathbf{i} + \left(\lim_{t \to \pi/4} \sin t \right)\mathbf{j} + \left(\lim_{t \to \pi/4} t \right)\mathbf{k}$$

$$= \frac{\sqrt{2}}{2}\,\mathbf{i} + \frac{\sqrt{2}}{2}\,\mathbf{j} + \frac{\pi}{4}\,\mathbf{k}.$$

EXPLORATION BIT

Which properties in the Properties of Limits Theorem (Theorem 1(c) in Section 2.1) extend to vector-valued functions? Give convincing arguments. Are any of the properties used in Example 2?

Continuity is defined the same way for vector functions as it is for scalar functions.

DEFINITION

The vector function **r** is **continuous at a point** $t = t_0$ if

$$\lim_{t \to t_0} \mathbf{r}(t) = \mathbf{r}(t_0).$$

It is a **continuous function** if it is continuous at every point in its domain.

The fact that limits of vector functions are defined in terms of components leads to the following test for continuity.

Component Test for Continuity at a Point

A vector function **r** defined by the rule $\mathbf{r}(t) = f(t)\,\mathbf{i} + g(t)\,\mathbf{j} + h(t)\,\mathbf{k}$ is continuous at $t = t_0$ if and only if f, g, and h are continuous at t_0.

EXPLORATION BITS

1. Sketch or describe a graph of the function in Example 4. Refer to the types of discontinuities listed in Section 2.2, and suggest a name for the type of discontinuity illustrated by Example 4.

2. Describe some other types of discontinuities that could occur in space graphs of vector-valued functions. (Can you think of any type that is different from those suggested by the list in Section 2.2?) If possible, define vector functions to illustrate.

EXAMPLE 3

The function

$$\mathbf{r} = (\cos t)\,\mathbf{i} + (\sin t)\,\mathbf{j} + t\,\mathbf{k}$$

is continuous because the component functions $\cos t$, $\sin t$, and t are continuous. ≡

EXAMPLE 4

The function

$$\mathbf{r} = (\cos t)\,\mathbf{i} + (\sin t)\,\mathbf{j} + \frac{1}{t}\,\mathbf{k}$$

is not continuous at $t = 0$ because the third component function, $h(t) = 1/t$, is not continuous at $t = 0$. ≡

Derivatives and Motion

Suppose that

$$\mathbf{r} = f(t)\,\mathbf{i} + g(t)\,\mathbf{j} + h(t)\,\mathbf{k}$$

is the position vector of a particle moving along a curve in space and that f, g, and h are differentiable functions. Then the difference between the particle's positions at time t and a nearby time $t + \Delta t$ can be expressed as the vector difference

$$\Delta \mathbf{r} = \mathbf{r}(t + \Delta t) - \mathbf{r}(t).$$

(See Fig. 12.4.) In terms of components,

$$\Delta \mathbf{r} = \mathbf{r}(t + \Delta t) - \mathbf{r}(t)$$

$$= [f(t + \Delta t)\,\mathbf{i} + g(t + \Delta t)\,\mathbf{j} + h(t + \Delta t)\,\mathbf{k}] - [f(t)\,\mathbf{i} + g(t)\,\mathbf{j} + h(t)\,\mathbf{k}]$$

$$= [f(t + \Delta t) - f(t)]\,\mathbf{i} + [g(t + \Delta t) - g(t)]\,\mathbf{j} + [h(t + \Delta t) - h(t)]\,\mathbf{k}.$$

As Δt approaches zero, three things happen simultaneously. First, Q approaches P along the curve. Second, the secant line PQ approaches a limiting position tangent to the curve at P. Third, the quotient $\Delta \mathbf{r}/\Delta t$ approaches the limit

$$\lim_{\Delta t \to 0} \frac{\Delta \mathbf{r}}{\Delta t} = \left[\lim_{\Delta t \to 0} \frac{f(t + \Delta t) - f(t)}{\Delta t} \right] \mathbf{i} + \left[\lim_{\Delta t \to 0} \frac{g(t + \Delta t) - g(t)}{\Delta t} \right] \mathbf{j}$$

$$+ \left[\lim_{\Delta t \to 0} \frac{h(t + \Delta t) - h(t)}{\Delta t} \right] \mathbf{k}$$

$$= \left[\frac{df}{dt} \right] \mathbf{i} + \left[\frac{dg}{dt} \right] \mathbf{j} + \left[\frac{dh}{dt} \right] \mathbf{k}.$$

We are therefore led by past experience to the following definition.

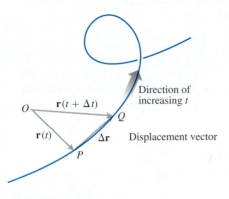

12.4 Between time t and time $t + \Delta t$, the particle moving along the path shown here undergoes the displacement $\overrightarrow{PQ} = \Delta \mathbf{r}$. The vector sum $\mathbf{r}(t) + \Delta \mathbf{r}$ gives the new position, $\mathbf{r}(t + \Delta t)$.

DEFINITION   **Derivative of a Vector Function of a Real Variable**

If f, g, and h are differentiable scalar functions of t and $\mathbf{r} = f\mathbf{i} + g\mathbf{j} + h\mathbf{k}$, then the vector

$$\mathbf{r}' = \frac{d\mathbf{r}}{dt} = \frac{df}{dt}\mathbf{i} + \frac{dg}{dt}\mathbf{j} + \frac{dh}{dt}\mathbf{k}$$

is the **derivative** of $\mathbf{r}$ with respect to t. The vector function $\mathbf{r} = f\mathbf{i} + g\mathbf{j} + h\mathbf{k}$ is said to be **differentiable at a point** $t = t_0$ if f, g, and h are differentiable at t_0 and is said to be **differentiable** if it is differentiable at every point in its domain.

Look once again at Fig. 12.4. We drew the figure for Δt positive so that $\Delta \mathbf{r}$ points in the direction of increasing t. The vector $\Delta \mathbf{r}/\Delta t$, not shown, also points the same direction as $\Delta \mathbf{r}$. If Δt had been negative, $\Delta \mathbf{r}$ would have pointed backward, against the direction of motion. The quotient $\Delta \mathbf{r}/\Delta t$, however, having a sign opposite to the sign of $\Delta \mathbf{r}$, would still have pointed forward. In either case, then, $\Delta \mathbf{r}/\Delta t$ points forward, and we expect the vector $d\mathbf{r}/dt = \lim_{\Delta t \to 0} \Delta \mathbf{r}/\Delta t$, when different from $\mathbf{0}$, to do the same. This means that $d\mathbf{r}/dt$ has all the makings of a velocity vector—it gives the rate of change of position with respect to time, and it points forward as we would expect.

SUPPORT

We can support the vector-analytic results on our graphers by

1. applying NDER to the component scalar-valued functions to help find **v**.

2. computing the scalar value $|\mathbf{v}(t)|$ for different values of t.

3. applying NDER2 to the component scalar-valued functions to help find **a**.

4. computing the values of t that give orthogonal vectors, such as finding when $\mathbf{v}(t) \cdot \mathbf{a}(t) = 0$.

DEFINITIONS

If the position vector **r** of a particle moving in space is a differentiable function of time t, then the vector

$$\mathbf{v} = \frac{d\mathbf{r}}{dt}$$

is the particle's **velocity vector**. At any time t, the direction of **v** is the **direction of motion**, the magnitude of **v** is the particle's **speed**, and the derivative $\mathbf{a} = d\mathbf{v}/dt$, when it exists, is the particle's **acceleration vector**. In short,

1. Velocity is the derivative of position: $\mathbf{v} = \dfrac{d\mathbf{r}}{dt}$.

2. Speed is the magnitude of velocity: Speed $= |\mathbf{v}|$.

3. Acceleration is the derivative of velocity: $\mathbf{a} = \dfrac{d\mathbf{v}}{dt} = \dfrac{d^2\mathbf{r}}{dt^2}$.

4. The vector $\mathbf{v}/|\mathbf{v}|$ is the direction of motion at time t.

We use the formula $\mathbf{A} = |\mathbf{A}| \cdot (\mathbf{A}/|\mathbf{A}|)$ from Section 11.2 to express the velocity of a moving particle as the product of its speed and direction.

$$\text{Velocity} = |\mathbf{v}| \cdot \frac{\mathbf{v}}{|\mathbf{v}|} = (\text{speed}) \cdot (\text{direction})$$

EXAMPLE 5

The vector function

$$\mathbf{r} = (3 \cos t)\,\mathbf{i} + (3 \sin t)\,\mathbf{j} + t^2\,\mathbf{k}$$

gives the position of a moving body at time t. Find the body's speed and direction when $t = 2$. At what times, if any, are the body's velocity and acceleration orthogonal?

Solution

$$\mathbf{r} = (3 \cos t)\,\mathbf{i} + (3 \sin t)\,\mathbf{j} + t^2\,\mathbf{k}$$

$$\mathbf{v} = \frac{d\mathbf{r}}{dt} = -(3 \sin t)\,\mathbf{i} + (3 \cos t)\,\mathbf{j} + 2t\,\mathbf{k}$$

$$\mathbf{a} = \frac{d^2\mathbf{r}}{dt^2} = -(3 \cos t)\,\mathbf{i} - (3 \sin t)\,\mathbf{j} + 2\,\mathbf{k}$$

At $t = 2$, the body's speed and direction are

Speed: $\qquad |\mathbf{v}(2)| = \sqrt{(-3 \sin 2)^2 + (3 \cos 2)^2 + (4)^2} = 5$

Direction: $\qquad \dfrac{\mathbf{v}(2)}{|\mathbf{v}(2)|} = -\left(\dfrac{3}{5} \sin 2\right)\mathbf{i} + \left(\dfrac{3}{5} \cos 2\right)\mathbf{j} + \dfrac{4}{5}\,\mathbf{k}.$

To find t when $\mathbf{v}$ and $\mathbf{a}$ are orthogonal, we find t for which

$$\mathbf{v} \cdot \mathbf{a} = 9 \sin t \cos t - 9 \cos t \sin t + 4t = 4t = 0.$$

The only value is $t = 0$.

Differentiation Rules

The rules for differentiating vector functions resemble the rules for differentiating scalar functions. This is due in large part to the derivative of a vector function being defined in terms of components.

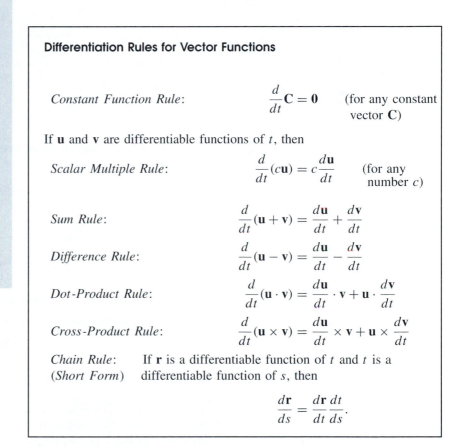

Differentiation Rules for Vector Functions

Constant Function Rule:
$$\frac{d}{dt}\mathbf{C} = \mathbf{0} \qquad \text{(for any constant vector } \mathbf{C})$$

If $\mathbf{u}$ and $\mathbf{v}$ are differentiable functions of t, then

Scalar Multiple Rule:
$$\frac{d}{dt}(c\mathbf{u}) = c\frac{d\mathbf{u}}{dt} \qquad \text{(for any number } c)$$

Sum Rule:
$$\frac{d}{dt}(\mathbf{u} + \mathbf{v}) = \frac{d\mathbf{u}}{dt} + \frac{d\mathbf{v}}{dt}$$

Difference Rule:
$$\frac{d}{dt}(\mathbf{u} - \mathbf{v}) = \frac{d\mathbf{u}}{dt} - \frac{d\mathbf{v}}{dt}$$

Dot-Product Rule:
$$\frac{d}{dt}(\mathbf{u} \cdot \mathbf{v}) = \frac{d\mathbf{u}}{dt} \cdot \mathbf{v} + \mathbf{u} \cdot \frac{d\mathbf{v}}{dt}$$

Cross-Product Rule:
$$\frac{d}{dt}(\mathbf{u} \times \mathbf{v}) = \frac{d\mathbf{u}}{dt} \times \mathbf{v} + \mathbf{u} \times \frac{d\mathbf{v}}{dt}$$

Chain Rule: If $\mathbf{r}$ is a differentiable function of t and t is a
(Short Form) differentiable function of s, then

$$\frac{d\mathbf{r}}{ds} = \frac{d\mathbf{r}}{dt}\frac{dt}{ds}.$$

We will prove the Dot-Product Rule and the Chain Rule but leave the rules for constants, scalar multiples, sums, differences, and the Cross-Product as exercises.

Proof of the Dot-Product Rule

Suppose that $\qquad \mathbf{u} = u_1(t)\,\mathbf{i} + u_2(t)\,\mathbf{j} + u_3(t)\,\mathbf{k}$

and $\qquad\qquad \mathbf{v} = v_1(t)\,\mathbf{i} + v_2(t)\,\mathbf{j} + v_3(t)\,\mathbf{k}.$

Then

$$\frac{d}{dt}(\mathbf{u} \cdot \mathbf{v}) = \frac{d}{dt}(u_1 v_1 + u_2 v_2 + u_3 v_3)$$

$$= \underbrace{u_1' v_1 + u_2' v_2 + u_3' v_3}_{\mathbf{u}' \cdot \mathbf{v}} + \underbrace{u_1 v_1' + u_2 v_2' + u_3 v_3'}_{\mathbf{u} \cdot \mathbf{v}'}.$$ ▤

Proof of the Chain Rule   Suppose that $\mathbf{r} = f(t)\mathbf{i} + g(t)\mathbf{j} + h(t)\mathbf{k}$ is a differentiable vector function of t and that t is a differentiable scalar function of some other variable s. Then f, g, and h are differentiable functions of s, and the Chain Rule for differentiable real-valued functions gives

$$\frac{d\mathbf{r}}{ds} = \frac{df}{ds}\mathbf{i} + \frac{dg}{ds}\mathbf{j} + \frac{dh}{ds}\mathbf{k}$$

$$= \frac{df}{dt}\frac{dt}{ds}\mathbf{i} + \frac{dg}{dt}\frac{dt}{ds}\mathbf{j} + \frac{dh}{dt}\frac{dt}{ds}\mathbf{k}$$

$$= \left(\frac{df}{dt}\mathbf{i} + \frac{dg}{dt}\mathbf{j} + \frac{dh}{dt}\mathbf{k}\right)\frac{dt}{ds}$$

$$= \frac{d\mathbf{r}}{dt}\frac{dt}{ds}.$$ ▤

Derivatives of Vector Functions of Constant Length

We might at first think that a vector whose length remains constant as time passes would have to have a zero derivative, but this is not the case. Think of a clock hand. Its length remains constant as time passes, but its direction still changes. What we *can* say about the derivative of a vector of constant length is that it is always orthogonal to the vector. Direction changes, as it were, take place at right angles.

If $\mathbf{u}$ is a differentiable vector function of t of constant length, then

$$\mathbf{u} \cdot \frac{d\mathbf{u}}{dt} = 0. \qquad (2)$$

To see why Eq. (2) holds, suppose that $\mathbf{u}$ is a differentiable function of t and that $|\mathbf{u}|$ is constant. Then $\mathbf{u} \cdot \mathbf{u} = |\mathbf{u}|^2$ is constant, and we may differentiate both sides of this equation to get

$$\frac{d}{dt}(\mathbf{u} \cdot \mathbf{u}) = \frac{d}{dt}|\mathbf{u}|^2 = 0$$

$$\frac{d\mathbf{u}}{dt} \cdot \mathbf{u} + \mathbf{u} \cdot \frac{d\mathbf{u}}{dt} = 0 \qquad \text{\color{blue}Dot-Product Rule with } \mathbf{v} = \mathbf{u}$$

$$2\mathbf{u} \cdot \frac{d\mathbf{u}}{dt} = 0 \qquad \text{\color{blue}Dot multiplication is commutative.}$$

$$\mathbf{u} \cdot \frac{d\mathbf{u}}{dt} = 0.$$

EXPLORATION BIT

In general, a vector function could be one of the following types:

a) constant in length and direction,

b) constant in length and varying in direction,

c) varying in length and constant in direction,

d) varying in length and direction.

We discuss types a, b, and d in this chapter and leave the analysis of type c for you to explore here. (*Hint:* Relate this type of vector function to another type of function that you have seen before.)

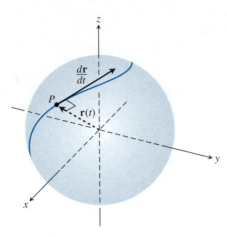

12.5 The velocity vector of a particle *P* that moves on the surface of a sphere is tangent to the sphere.

EXAMPLE 6

Suppose a particle moves on a sphere centered at the origin in such a way that its position vector $\mathbf{r}(t)$ is differentiable (Fig. 12.5). Show that the particle's velocity is always orthogonal to $\mathbf{r}$.

Solution   The length of $\mathbf{r}$ is constant (it equals the radius of the sphere). Hence, $\mathbf{r} \cdot (d\mathbf{r}/dt) = 0$.   ▬

EXPLORATION 1

The Tethered Particle

Suppose the position of a particle in space is a function of time *t* as follows:

$$\mathbf{u}(t) = (\sin t)\,\mathbf{i} + (\cos t)\,\mathbf{j} + \sqrt{3}\,\mathbf{k}.$$

1. Describe a geometric model of the path of the particle. Why does this model suggest that the particle remains a fixed distance from the origin?
2. *Confirm* that the particle remains a fixed distance from the origin.
3. Verify Eq. (2) by computing $d\mathbf{u}/dt$ and then $\mathbf{u} \cdot d\mathbf{u}/dt$.
4. Interpret the meaning of Eq. (2) in terms of your model.

Integrals of Vector Functions

If the scalar components of a vector function $\mathbf{r}$ are continuous functions of *t* throughout an interval $a \leq t \leq b$, we define the definite integral of $\mathbf{r}$ from *a* to *b* by the following formula.

DEFINITION

The **definite integral** of the function $\mathbf{r}$ defined by the rule $\mathbf{r}(t) = f(t)\,\mathbf{i} + g(t)\,\mathbf{j} + h(t)\,\mathbf{k}$ over the interval $a \leq t \leq b$ is

$$\int_a^b \mathbf{r}(t)\,dt = \left(\int_a^b f(t)\,dt\right)\mathbf{i} + \left(\int_a^b g(t)\,dt\right)\mathbf{j} + \left(\int_a^b h(t)\,dt\right)\mathbf{k}.$$

We call a differentiable vector function $\mathbf{F}$ an **antiderivative** of a vector function $\mathbf{f}$ if $d\mathbf{F}/dt = \mathbf{f}$. If $\mathbf{F}$ is an antiderivative of $\mathbf{f}$, it can be shown, working one component at a time, that every antiderivative of $\mathbf{f}$ has the form $\mathbf{F}+\mathbf{C}$ for some constant vector $\mathbf{C}$. We call the set of all antiderivatives of $\mathbf{f}$ the **indefinite integral** of $\mathbf{f}$ and denote it with an integral sign in the usual way.

DEFINITION

The **indefinite integral** of $\mathbf{f}$ with respect to *t* is the set of all antiderivatives of $\mathbf{f}$, denoted by $\displaystyle\int \mathbf{f}(t)\,dt$. If $\mathbf{F}$ is any antiderivative of $\mathbf{f}$, then

$$\int \mathbf{f}(t)\,dt = \mathbf{F}(t) + \mathbf{C}.$$

We can use the Integral Existence Theorem for continuous scalar functions to show that every continuous vector function has an antiderivative. All the usual arithmetic rules for definite and indefinite integrals apply, and grapher support can be provided for analytic results by applying NINT to the three component scalar functions.

EXAMPLE 7

$$\int_0^\pi ((\cos t)\,\mathbf{i} + (\sin t)\,\mathbf{j} + t\,\mathbf{k})\,dt = \left(\int_0^\pi \cos t\ dt\right)\mathbf{i} + \left(\int_0^\pi \sin t\ dt\right)\mathbf{j}$$

$$+ \left(\int_0^\pi t\ dt\right)\mathbf{k}$$

$$= \left[\sin t\right]_0^\pi \mathbf{i} + \left[-\cos t\right]_0^\pi \mathbf{j} + \left[\frac{t^2}{2}\right]_0^\pi \mathbf{k}$$

$$= [0-0]\mathbf{i} + [-(-1)+(1)]\mathbf{j} + \left[\frac{\pi^2}{2}-0\right]\mathbf{k}$$

$$= 2\,\mathbf{j} + \frac{\pi^2}{2}\,\mathbf{k}.$$
≡

EXAMPLE 8

The velocity of a particle moving in space is

$$\frac{d\mathbf{r}}{dt} = (\cos t)\,\mathbf{i} - (\sin t)\,\mathbf{j} + \mathbf{k}.$$

Find the particle's position as a function of t if $\mathbf{r} = 2\,\mathbf{i} + \mathbf{k}$ when $t = 0$.

Solution   Our goal is to solve the initial value problem that consists of

The differential equation: $\dfrac{d\mathbf{r}}{dt} = (\cos t)\,\mathbf{i} - (\sin t)\,\mathbf{j} + \mathbf{k},$

The initial condition: $\mathbf{r} = 2\,\mathbf{i} + \mathbf{k}$ when $t = 0$.

To solve it, we first use what we know about derivatives to find the general solution of the differential equation. Integrating both sides with respect to t gives

$$\mathbf{r} = (\sin t + C_1)\,\mathbf{i} + (\cos t + C_2)\,\mathbf{j} + (t + C_3)\,\mathbf{k}$$

$$= (\sin t)\,\mathbf{i} + (\cos t)\,\mathbf{j} + t\,\mathbf{k} + \underbrace{C_1\,\mathbf{i} + C_2\,\mathbf{j} + C_3\,\mathbf{k}}_{\mathbf{C}}$$

$$= (\sin t)\,\mathbf{i} + (\cos t)\,\mathbf{j} + t\,\mathbf{k} + \mathbf{C}.$$

We then use the initial condition to find the value of $\mathbf{C}$:

$$(\sin 0)\,\mathbf{i} + (\cos 0)\,\mathbf{j} + (0)\mathbf{k} + \mathbf{C} = 2\,\mathbf{i} + \mathbf{k} \qquad \text{The initial condition is } \mathbf{r}(0) = 2\,\mathbf{i} + \mathbf{k}.$$

$$\mathbf{j} + \mathbf{C} = 2\,\mathbf{i} + \mathbf{k}$$

$$\mathbf{C} = 2\,\mathbf{i} - \mathbf{j} + \mathbf{k}$$

The particle's position as a function of t is

$$\mathbf{r}(t) = (\sin t + 2)\,\mathbf{i} + (\cos t - 1)\,\mathbf{j} + (t + 1)\,\mathbf{k}.$$

As a check, we can see from this formula that

$$\frac{d\mathbf{r}}{dt} = (\cos t + 0)\,\mathbf{i} + (-\sin t - 0)\,\mathbf{j} + (1 + 0)\,\mathbf{k} = (\cos t)\,\mathbf{i} - (\sin t)\,\mathbf{j} + \mathbf{k}$$

and that

$$\mathbf{r}(0) = (\sin 0 + 2)\,\mathbf{i} + (\cos 0 - 1)\,\mathbf{j} + (0 + 1)\,\mathbf{k} = 2\,\mathbf{i} + \mathbf{k}. \qquad \equiv$$

EXPLORATION 2

Motion in the Plane

When a particle moves in the xy-plane, its z-coordinate is 0, and its position vector function reduces to

$$\mathbf{r} = \mathbf{f}(t)\,\mathbf{i} + \mathbf{g}(t)\,\mathbf{j}.$$

1. Recall the parametric equations for the function $y = x^2$:

$$x(t) = t, \qquad y(t) = t^2.$$

Use the parametric equations to write the vector function for a particle moving along the curve $y = x^2$ in the xy-plane.

2. Find the velocity and acceleration vector functions for the particle.

3. Give an analytic and geometric interpretation of the three vectors shown in Fig. 12.6 based on your results above.

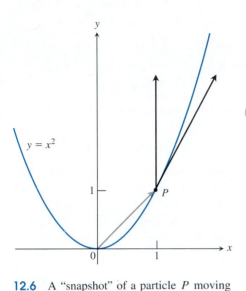

12.6 A "snapshot" of a particle P moving along the curve $y = x^2$ in the xy-plane. Use your findings from Exploration 2 to interpret the three vectors shown.

Exercises 12.1

In Exercises 1–8, **r** is the position of a particle in space at time t. Find the particle's **(a)** velocity vector and **(b)** acceleration vector. Find the particle's **(c)** speed and **(d)** direction of motion at the given value of t, and **(e)** write the velocity at that time as the product of the particle's speed and direction.

1. $\mathbf{r} = (2\cos t)\,\mathbf{i} + (3\sin t)\,\mathbf{j} + 4t\,\mathbf{k}, \quad t = \pi/2$

2. $\mathbf{r} = (t + 1)\,\mathbf{i} + (t^2 - 1)\,\mathbf{j} + 2t\,\mathbf{k}, \quad t = 1$

3. $\mathbf{r} = (\cos 2t)\,\mathbf{j} + (2\sin t)\,\mathbf{k}, \quad t = 0$

4. $\mathbf{r} = e^t\,\mathbf{i} + \dfrac{2}{9}e^{2t}\,\mathbf{j}, \quad t = \ln 3$

5. $\mathbf{r} = (\sec t)\,\mathbf{i} + (\tan t)\,\mathbf{j} + \dfrac{4}{3}t\,\mathbf{k}, \quad t = \pi/6$

6. $\mathbf{r} = (2\ln(t + 1))\,\mathbf{i} + t^2\,\mathbf{j} + \dfrac{t^2}{2}\,\mathbf{k}, \quad t = 1$

7. $\mathbf{r} = (e^{-t})\,\mathbf{i} + (2\cos 3t)\,\mathbf{j} + (2\sin 3t)\,\mathbf{k}, \quad t = 0$

8. $\mathbf{r} = (1 + t)\,\mathbf{i} + \dfrac{t^2}{\sqrt{2}}\,\mathbf{j} + \dfrac{t^3}{3}\,\mathbf{k}, \quad t = 1$

In Exercises 9–12, **r** is the position of a particle in space at time t. Find the angle between the velocity and acceleration vectors at time $t = 0$.

9. $\mathbf{r} = (3t + 1)\,\mathbf{i} + \sqrt{3}t\,\mathbf{j} + t^2\,\mathbf{k}$

10. $\mathbf{r} = \left(\dfrac{\sqrt{2}}{2}t\right)\mathbf{i} + \left(\dfrac{\sqrt{2}}{2}t - 16t^2\right)\mathbf{j}$

11. $\mathbf{r} = (\ln(t^2 + 1))\,\mathbf{i} + (\tan^{-1}t)\,\mathbf{j} + \sqrt{t^2 + 1}\,\mathbf{k}$

12. $\mathbf{r} = \dfrac{4}{9}(1 + t)^{3/2}\,\mathbf{i} + \dfrac{4}{9}(1 - t)^{3/2}\,\mathbf{j} + \dfrac{1}{3}t\,\mathbf{k}$

In Exercises 13 and 14, **r** is the position of a particle in space at time t. Find the time or times in the given time interval when the velocity and acceleration vectors are orthogonal.

13. $\mathbf{r} = (t - \sin t)\,\mathbf{i} + (1 - \cos t)\,\mathbf{j} \quad (0 \le t \le 2\pi)$

14. $\mathbf{r} = (\sin t)\,\mathbf{i} + t\,\mathbf{j} + (\cos t)\,\mathbf{k} \quad (t \ge 0)$

Evaluate the integrals in Exercises 15–20 analytically. Support with NINT computations.

15. $\displaystyle\int_0^1 [t^3\,\mathbf{i} + 7\,\mathbf{j} + (t + 1)\,\mathbf{k}]\,dt$

16. $\displaystyle\int_1^2 \left[(6 - 6t)\,\mathbf{i} + 3\sqrt{t}\,\mathbf{j} + \left(\dfrac{4}{t^2}\right)\mathbf{k}\right]\,dt$

17. $\int_{-\pi/4}^{\pi/4} [(\sin t)\,\mathbf{i} + (1 + \cos t)\,\mathbf{j} + (\sec^2 t)\,\mathbf{k}]\,dt$

18. $\int_{0}^{\pi/3} [(\sec t \tan t)\,\mathbf{i} + (\tan t)\,\mathbf{j} + (2 \sin t \cos t)\,\mathbf{k}]\,dt$

19. $\int_{1}^{4} \left[\frac{1}{t}\,\mathbf{i} + \frac{1}{5 - t}\,\mathbf{j} + \frac{1}{2t}\,\mathbf{k}\right] dt$

20. $\int_{0}^{1} \left[\frac{2}{\sqrt{1 - t^2}}\,\mathbf{i} + \frac{\sqrt{3}}{1 + t^2}\,\mathbf{k}\right] dt$

Exercises 21–24 give the position vectors of particles moving along various curves in the xy-plane. In each case, find the particle's velocity and acceleration vectors at the stated times, and sketch them as vectors on the curve.

21. *Motion on the circle $x^2 + y^2 = 1$.*

$$\mathbf{r} = (\sin t)\,\mathbf{i} + (\cos t)\,\mathbf{j} \quad (t = \pi/4 \text{ and } \pi/2)$$

22. *Motion on the circle $x^2 + y^2 = 16$.*

$$\mathbf{r} = \left(4 \cos \frac{t}{2}\right)\mathbf{i} + \left(4 \sin \frac{t}{2}\right)\mathbf{j} \quad (t = \pi \text{ and } 3\pi/2)$$

23. *Motion on the cycloid $x = t - \sin t, \ y = 1 - \cos t$.*

$$\mathbf{r} = (t - \sin t)\,\mathbf{i} + (1 - \cos t)\,\mathbf{j} \quad (t = \pi \text{ and } 3\pi/2)$$

24. *Motion on the parabola $y = x^2 + 1$.*

$$\mathbf{r} = t\,\mathbf{i} + (t^2 + 1)\,\mathbf{j} \quad (t = -1, 0, \text{ and } 1)$$

Solve analytically each initial value problem in Exercises 25–30 for $\mathbf{r}$ as a vector function of t. Support each indefinite integral computation by using NINT on the three component scalar functions.

25. Differential equation: $\dfrac{d\mathbf{r}}{dt} = -t\,\mathbf{i} - t\,\mathbf{j} - t\,\mathbf{k}$

   Initial condition: $\mathbf{r} = \mathbf{i} + 2\,\mathbf{j} + 3\,\mathbf{k}$ when $t = 0$

26. Differential equation: $\dfrac{d\mathbf{r}}{dt} = (180t)\,\mathbf{i} + (180t - 16t^2)\,\mathbf{j}$

   Initial condition: $\mathbf{r} = 100\,\mathbf{j}$ when $t = 0$

27. Differential equation: $\dfrac{d\mathbf{r}}{dt} = \dfrac{3}{2}(t + 1)^{1/2}\,\mathbf{i} + e^{-t}\,\mathbf{j} + \dfrac{1}{t + 1}\,\mathbf{k}$

   Initial condition: $\mathbf{r} = \mathbf{k}$ when $t = 0$

28. Differential equation: $\dfrac{d\mathbf{r}}{dt} = (t^3 + 4t)\,\mathbf{i} + t\,\mathbf{j} + 2t^2\,\mathbf{k}$

   Initial condition: $\mathbf{r} = \mathbf{i} + \mathbf{j}$ when $t = 0$

29. Differential equation: $\dfrac{d^2\mathbf{r}}{dt^2} = -32\,\mathbf{k}$

   Initial condition: $\mathbf{r} = 100\mathbf{k}$ and $\dfrac{d\mathbf{r}}{dt} = 8\mathbf{i} + 8\mathbf{j}$ when $t = 0$

30. Differential equation: $\dfrac{d^2\mathbf{r}}{dt^2} = -(\mathbf{i} + \mathbf{j} + \mathbf{k})$

   Initial conditions: $\mathbf{r} = 10\,\mathbf{i} + 10\,\mathbf{j} + 10\,\mathbf{k}$ and $\dfrac{d\mathbf{r}}{dt} = \mathbf{0}$ when $t = 0$

31. A particle moves on a cycloid in the xy-plane in such a way that its position at time t is

$$\mathbf{r} = (t - \sin t)\,\mathbf{i} + (1 - \cos t)\,\mathbf{j}.$$

Find the maximum and minimum values of $|\mathbf{v}|$ and $|\mathbf{a}|$ analytically. (*Hint:* Find the extreme values of $|\mathbf{v}|^2$ and $|\mathbf{a}|^2$ first, and take square roots later.) Support graphically.

32. A particle moves around the ellipse $(y/3)^2 + (z/2)^2 = 1$ in the yz-plane in such a way that its position at time t is

$$\mathbf{r} = (3 \cos t)\,\mathbf{j} + (2 \sin t)\,\mathbf{k}.$$

Find the maximum and minimum values of $|\mathbf{v}|$ and $|\mathbf{a}|$ analytically. (*Hint:* Find the extreme values of $|\mathbf{v}|^2$ and $|\mathbf{a}|^2$ first, and take square roots later.) Support graphically.

33. *The Constant Function Rule.* Prove that if $\mathbf{f}$ is the vector function with the constant value $\mathbf{C}$, then $d\mathbf{f}/dt = \mathbf{0}$.

34. *The Scalar Multiple Rule.* Prove that if $\mathbf{f}$ is a differentiable function of t and c is any real number, then

$$\frac{d(c\mathbf{f})}{dt} = c\frac{d\mathbf{f}}{dt}.$$

35. *The Sum and Difference Rules.* Prove that if $\mathbf{u}$ and $\mathbf{v}$ are differentiable functions of t, then

$$\frac{d}{dt}(\mathbf{u} + \mathbf{v}) = \frac{d\mathbf{u}}{dt} + \frac{d\mathbf{v}}{dt}$$

and

$$\frac{d}{dt}(\mathbf{u} - \mathbf{v}) = \frac{d\mathbf{u}}{dt} - \frac{d\mathbf{v}}{dt}.$$

36. *The Cross-Product Rule.* Prove that if $\mathbf{u}$ and $\mathbf{v}$ are differentiable functions of t, then

$$\frac{d}{dt}(\mathbf{u} \times \mathbf{v}) = \frac{d\mathbf{u}}{dt} \times \mathbf{v} + \mathbf{u} \times \frac{d\mathbf{v}}{dt}.$$

12.2 Modeling Projectile Motion

When we launch a projectile (e.g., a baseball or a golf ball) into the air, we often want to know beforehand how far it will go (will it reach the home run fence or the green?), how high it will rise (will it clear the fence?),

and when it will land (when do we get results?). We can get this information from parametric equations that represent the motion of the particle and simulations using graphers. The equations are based on the direction and magnitude of the projectile's initial velocity vector, described
in terms of the angle and speed at which the projectile is launched. The
equations come from combining calculus and Newton's Second Law of
Motion in vector form. In this section, we derive these equations and
show how to use them to get the information that we want about projectile
motion.

The Vector and Parametric Equations for Ideal Projectile Motion

To derive equations for *ideal* projectile motion, we assume that the projectile
behaves like a particle moving in a vertical coordinate plane and that the only
force acting on the projectile during its flight is the constant force of gravity,
which always points straight down. In practice, neither of these assumptions
really holds. There are wind factors, the ground moves beneath the projectile
as Earth turns, the air creates a frictional force that varies with the projectile's
speed and altitude, and the force of gravity changes as the projectile moves
along. All this must be taken into account by applying "corrections" to
the ideal equations that we are about to derive. We will illustrate how
this is done by applying air resistance (*drag*) and wind corrections in this
section.

We assume that our projectile is launched from the origin at time $t = 0$
into the first quadrant with an initial velocity $\mathbf{v}_0$ (Fig. 12.7a). If α is the
angle that $\mathbf{v}_0$ makes with the horizontal, then

$$\mathbf{v}_0 = (|\mathbf{v}_0| \cos \alpha)\,\mathbf{i} + (|\mathbf{v}_0| \sin \alpha)\,\mathbf{j}.$$

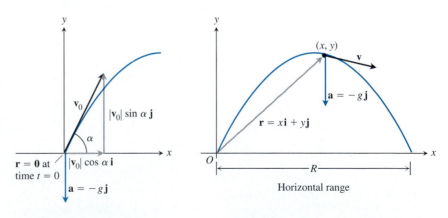

(a) Position, velocity, acceleration,
and launch angle at $t = 0$.

(b) Position, velocity, and acceleration
at a later time t.

12.7 Ideal projectile motion for a projectile with an initial velocity $\mathbf{v}_0$ and launch
angle α.

If we use the simpler notation v_0 for the initial speed $|\mathbf{v}_0|$, then

$$\mathbf{v}_0 = (v_0 \cos \alpha)\,\mathbf{i} + (v_0 \sin \alpha)\,\mathbf{j}. \tag{1}$$

We let

$$\mathbf{r} = x(t)\,\mathbf{i} + y(t)\,\mathbf{j} = x\,\mathbf{i} + y\,\mathbf{j}$$

denote the projectile's position vector at time t (Fig. 12.7b), so the projectile's initial position at $t = 0$ is

$$\mathbf{r}_0 = 0\,\mathbf{i} + 0\,\mathbf{j} = \mathbf{0}. \tag{2}$$

If the only force acting on the projectile during its flight is a constant downward acceleration of gravity of magnitude g, then

$$\frac{d^2\mathbf{r}}{dt^2} = -g\,\mathbf{j}. \tag{3}$$

We can find the projectile's position as a function of time t by solving the following initial value problem:

Differential equation: $\quad \dfrac{d^2\mathbf{r}}{dt^2} = -g\,\mathbf{j}$

Initial conditions: $\quad \mathbf{r} = \mathbf{0} \quad$ and $\quad \dfrac{d\mathbf{r}}{dt} = \mathbf{v}_0 \quad$ when $t = 0$.

The first integration gives

$$\frac{d\mathbf{r}}{dt} = -(gt)\,\mathbf{j} + \mathbf{v}_0.$$

A second integration gives

$$\mathbf{r} = -\frac{1}{2}gt^2\,\mathbf{j} + \mathbf{v}_0 t + \mathbf{r}_0. \tag{4}$$

Substituting the values of $\mathbf{v}_0$ and $\mathbf{r}_0$ from Eqs. (1) and (2) gives

$$\mathbf{r} = -\frac{1}{2}gt^2\,\mathbf{j} + \underbrace{(v_0 \cos \alpha)t\,\mathbf{i} + (v_0 \sin \alpha)t\,\mathbf{j}}_{\mathbf{v}_0 t} + \mathbf{0}$$

$$= (v_0 \cos \alpha)t\,\mathbf{i} + \left((v_0 \sin \alpha)t - \frac{1}{2}gt^2 \right)\mathbf{j}.$$

> Note that when $\alpha = 90°$, the second of Eqs. (6) below becomes the familiar
> $$y = v_0 t - (1/2)gt^2.$$

Thus, we have the following equations for ideal (no corrections) projectile motion. If we measure distance in meters and time in seconds, then g is 9.8 m/sec^2, and Eqs. (6) give x and y in meters. If we measure distance in feet and time in seconds, then g is 32 ft/sec^2, and Eqs. (6) give x and y in feet.

Equations for the Ideal Motion of a Projectile Launched from the Origin at $t = 0$

Vector Form: $\quad \mathbf{r} = (v_0 \cos \alpha) t\, \mathbf{i} + \left((v_0 \sin \alpha) t - \frac{1}{2} g t^2 \right) \mathbf{j} \quad$ (5)

Parametric Form: $\quad x(t) = (v_0 \cos \alpha) t, \quad y(t) = (v_0 \sin \alpha) t - \frac{1}{2} g t^2 \quad$ (6)

where α is the **launch angle** and v_0 is the **initial speed**.

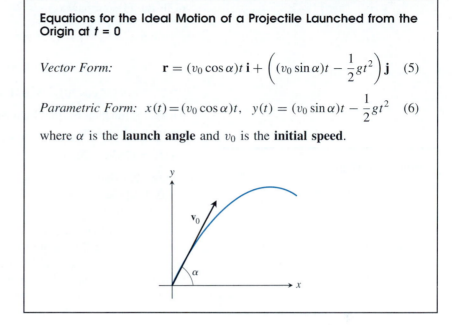

EXAMPLE 1

A baseball is hit when it is 3 feet above the ground. It leaves the bat with initial velocity of 152 feet per second, making an angle of 20° with the ground. A 6 mi/h (8.8 ft/sec) wind is blowing in the horizontal direction, directly opposite the direction the ball is taking toward the outfield.

a) Determine algebraic representations for the flight of the baseball in both vector function and parametric forms.

b) Simulate the motion of the baseball with a graphing utility.

c) Use the simulation to answer the following questions.

- How high does the baseball go, and when does it reach maximum height?

- When is the baseball 35 feet high?

- Assuming that it is not caught, how far (ground distance) does the baseball travel from home plate in the air, and when does it hit the ground?

Solution

a) The launch velocity of the baseball is given by Eq. (1) less the 8.8 ft/sec horizontal effect of the wind. Thus,

$$\mathbf{v}_0 = (v_0 \cos \alpha)\, i + (v_0 \sin \alpha)\, \mathbf{j} - (\text{wind speed})\, \mathbf{i}$$

$$= (152 \cos 20°) \mathbf{i} + (152 \sin 20°) \mathbf{j} - 8.8\, \mathbf{i}$$

$$= (152 \cos 20° - 8.8) \mathbf{i} + (152 \sin 20°) \mathbf{j}.$$

The initial position is

$$\mathbf{r}_0 = 0\, \mathbf{i} + 3\, \mathbf{j}.$$

With the wind taken into account, the only other force acting on the baseball in flight is gravity (Eq. 3), which yields Eq. (4). Using $\mathbf{v}_0$ and

$\mathbf{r}_0$ in Eq. (4), we have

$$\mathbf{r} = -\frac{1}{2}gt^2\,\mathbf{j} + \mathbf{v}_0 t + \mathbf{r}_0$$

$$= -16t^2\,\mathbf{j} + (152\cos 20° - 8.8)t\,\mathbf{i} + (152\sin 20°)t\,\mathbf{j} + 3\,\mathbf{j}$$

$$= (152\cos 20° - 8.8)t\,\mathbf{i} + (3 + (152\sin 20°)t - 16t^2)\,\mathbf{j},$$

the vector function for the flight of the baseball. The corresponding parametric equations are

$$x(t) = (152\cos 20° - 8.8)t, \qquad y(t) = 3 + (152\sin 20°)t - 16t^2. \quad (7)$$

b) Figure 12.8 shows a graph of the parametric equations (7) in a [0, 500] by [0, 100] viewing window. You should also view the graph in a square viewing window.

c) A TRACE on the graph (Fig. 12.8a) suggests that the baseball's maximum height is a little more than 45 ft, occurring about 1.7 seconds after being hit. Figure 12.8(b) suggests that the baseball is about 35 ft. high in 2.4 seconds. If allowed to land, the baseball would hit the ground about 442 ft from home plate in about 3.3 seconds (Fig. 12.8c).

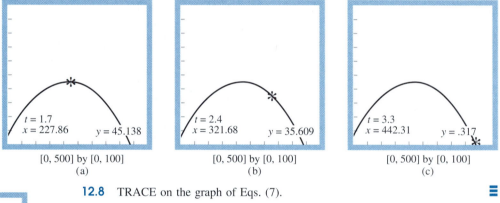

[0, 500] by [0, 100]　　　[0, 500] by [0, 100]　　　[0, 500] by [0, 100]
(a)　　　　　　　　　　(b)　　　　　　　　　　(c)

12.8 TRACE on the graph of Eqs. (7).

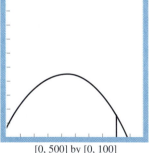

12.9 The home-run fence. Has the batter hit a home run?

EXPLORATION 1

A Home Run—Or Is It?

The baseball in Example 1 was hit toward the "400 ft" sign on the right-centerfield fence, which shows the distance from the batter to the fence. If it clears the fence, it will be a home run. The fence is 15 ft high.

1. Use the parametric equations from Example 1 to simulate the flight of the baseball. Use another pair of parametric equations, or the LINE command, to simulate the 15-ft outfield fence, 400 ft from home plate (Fig. 12.9).

2. Has the batter hit a home run? Give a convincing argument.

3. Simulate the flight of the baseball with *no wind*. GRAPH the no-wind flight and the wind-resisted flight in the same viewing window. Compare the distances, heights, and times of both flights.

4. FOR BASEBALL FANS: Make up a baseball problem of your own that can be solved with vector functions and parametric equations. Figure 12.15 in the exercises may suggest some ideas.

⚘

Height, Flight Time, and Range

Equations (6) enable us to answer most questions analytically about the ideal motion of a projectile launched from the origin.

The projectile reaches its highest point when its vertical velocity component is zero, that is, when

$$\frac{dy}{dt} = v_0 \sin\alpha - gt = 0 \qquad \text{or} \qquad t = \frac{v_0 \sin\alpha}{g}.$$

For this value of t, the value of y is

$$y_{max} = (v_0 \sin\alpha)\left(\frac{v_0 \sin\alpha}{g}\right) - \frac{1}{2}g\left(\frac{v_0 \sin\alpha}{g}\right)^2 = \frac{(v_0 \sin\alpha)^2}{2g}.$$

To find when the projectile lands, when it is launched over a horizontal surface, we set y equal to zero in Eqs. (6) and solve for t:

$$(v_0 \sin\alpha)t - \frac{1}{2}gt^2 = 0$$

$$t\left(v_0 \sin\alpha - \frac{1}{2}gt\right) = 0$$

$$t = 0, \qquad t = \frac{2v_0 \sin\alpha}{g}.$$

Since 0 is the time the projectile is launched, $(2v_0 \sin\alpha)/g$ must be the time when the projectile strikes the ground.

To find the projectile's range R, the distance from the origin to the landing point on the horizontal surface, we find the value of x when $t = (2v_0 \sin\alpha)/g$:

$$x = (v_0 \cos\alpha)t$$

$$R = (v_0 \cos\alpha)\left(\frac{2v_0 \sin\alpha}{g}\right) = \frac{v_0^2}{g}(2\sin\alpha\cos\alpha) = \frac{v_0^2}{g}\sin 2\alpha.$$

Notice that the range is largest when $\sin 2\alpha = 1$ or $\alpha = 45°$.

EXPLORATION BITS

FOR BASKETBALL FANS: Give a grapher analysis of "hang time" for the leap of a basketball player.

FOR FOOTBALL FANS: Give a grapher analysis of "hang time" of a punt.

For either of the above, include a discussion on how to maximize hang time.

Height, Flight Time, and Range

For ideal projectile motion when an object is launched from the origin over a horizontal surface:

Maximum height: $\qquad y_{max} = \dfrac{(v_0 \sin\alpha)^2}{2g}$ (8)

Flight time (launch to landing): $\qquad t = \dfrac{2v_0 \sin\alpha}{g}$ (9)

Range (horizontal distance, launch to landing): $R = \dfrac{v_0^2}{g}\sin 2\alpha$ (10)

EXAMPLE 2

Find the maximum height, flight time, and range of a projectile launched from the origin over a horizontal surface at an initial speed of 500 m/sec and a launch angle of 60°. Support with a grapher simulation.

Solution

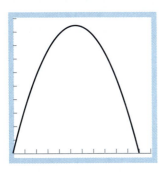

Maximum height (Eq. 8): $y_{max} = \dfrac{(v_0 \sin \alpha)^2}{2g} = \dfrac{(500 \sin 60°)^2}{2(9.8)}$

$= 9566.327$ m

Flight time (Eq. 9): $t = \dfrac{2v_0 \sin \alpha}{g} = \dfrac{2(500) \sin 60°}{9.8}$

$= 88.370$ sec

Range (Eq. 10): $R = \dfrac{v_0{}^2}{g} \sin 2\alpha = \dfrac{(500)^2 \sin 120°}{9.8}$

$= 22{,}092.485$ m

GRAPH as in Fig. 12.10, and TRACE to support these results.

12.10   A graph for Example 2. What viewing dimensions do you think we used?

Launching from (x_0, y_0)

If we fire our ideal projectile from the point (x_0, y_0) instead of from the origin (Fig. 12.11), the equations that replace Eqs. (6) are

$$x = x_0 + (v_0 \cos \alpha)t, \qquad y = y_0 + (v_0 \sin \alpha)t - \frac{1}{2}gt^2, \qquad (11)$$

as you will be invited to show in Exercise 25.

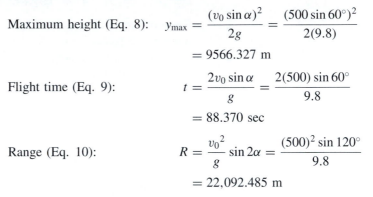

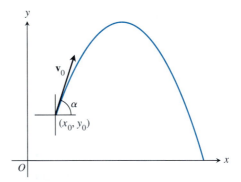

12.11   The path of a projectile launched from (x_0, y_0) with an initial velocity v_0 at an angle of α degrees with the horizontal.

EXPLORATION 2

Golf Shots

Figure 12.12 illustrates a recent golf shot that we made. We hit the golf ball horizontally off a level spot on a 20-ft hill to a green 20 ft below on level ground and 120 ft away (horizontally).

1. Assuming ideal projectile motion, give the parametric equations that model the flight of the golf ball. Use Eqs. (11) and coordinate axes as suggested by Fig. 12.12. (*Note:* $g = 32$ ft/sec^2.)

2. Confirm how long the ball was in flight.

3. Confirm the initial velocity that we imparted to the golf ball.

4. GRAPH the path of the ball to support all the information above. Compare your graph with our sketch.

5. FOR GOLFERS: What club might we have used? What club should we have used? If you are avid about the game, you may find it interesting to research the slopes of club faces and show the relative effect of each using your graphing utility.

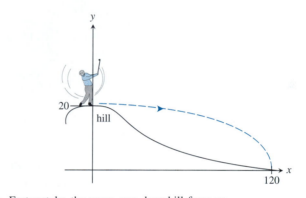

12.12 Fortunately, the green was downhill from us.

CONSIDER AN EXTREME CASE

To "see" that the water from a hose does not trace a parabola, you have to know what to look for. Consider an "extreme case" in which the water can fall a long way. Air resistance will eventually slow the forward progress of the water to a standstill. Gravity continues to move the water vertically, so the path eventually becomes one that is straight down, clearly different from the path of a parabola. Thus, to see that the path is not parabolic, look at the water that is headed toward the ground. To see even better, point the hose higher.

Ideal Trajectories Are Parabolic

It is often claimed that water from a hose traces a parabola in the air, but this isn't so. Air resistance slows down the water and distorts its path to one that is different from a parabola.

What is really being claimed is that *ideal* projectile motion is along parabolas, and this we can see from Eqs. (6). If we substitute $t = x/(v_0 \cos \alpha)$ from the first of Eqs. (6) into the second, we obtain the Cartesian coordinate equation

$$y = -\left(\frac{g}{2v_0{}^2 \cos^2 \alpha}\right) x^2 + (\tan \alpha)x.$$

This equation is quadratic in x and linear in y, so its graph is a parabola.

Projectile Motion with Air Resistance

For many nonspinning projectiles, the main factor affecting the path of the projectile, other than gravity, is a slowing down force due to air resistance. This factor is called the drag force and it acts in a direction *opposite* to the velocity of the projectile (Fig. 12.13).

It is known that for some high speed projectiles, the drag force is proportional to different powers of the speed over different velocity ranges. Such

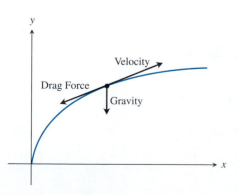

12.13 Projectile motion with air resistance.

a drag force is *nonlinear*. For projectiles moving through the air at relatively low speeds, however, the drag force is (very nearly) proportional to the speed (to the first power) and so is called *linear*. This is summarized in the following theorem.

THEOREM 1

Using the *linear drag model*, the position at time t of a projectile is

Vector Form:

$$\mathbf{r} = \frac{v_0}{k}(1 - e^{-kt})(\cos\alpha)\,\mathbf{i} + \left(\frac{v_0}{k}(1 - e^{-kt})(\sin\alpha)\right. $$

$$\left. + \frac{32}{k^2}(1 - kt - e^{-kt})\right)\mathbf{j}, \tag{12}$$

Parametric Form:

$$x(t) = \frac{v_0}{k}(1 - e^{-kt})(\cos\alpha),$$

$$y(t) = \frac{v_0}{k}(1 - e^{-kt})(\sin\alpha) + \frac{32}{k^2}(1 - kt - e^{-kt}), \tag{13}$$

where k is a positive constant representing the air density and v_0 and α are the initial velocity and launch angle, respectively, of the projectile.

Proof Because drag is linear, directly proportional to the particle speed, and acting in the opposite direction of the velocity vector, we have the following initial value problem for $\mathbf{r} = x(t)\mathbf{i} + y(t)\mathbf{j}$:

Differential equation:
$$\frac{d^2\mathbf{r}}{dt^2} = -g\mathbf{j} - k\mathbf{v}$$

$$= -g\mathbf{j} - k\frac{d\mathbf{r}}{dt} \qquad \text{Compare with Eq. (3).} \tag{14}$$

Initial conditions: $\mathbf{r} = \mathbf{0}$ and $\mathbf{v}_0 = (v_0\cos\alpha)\mathbf{i} + (v_0\sin\alpha)\mathbf{j}$ when $t = 0$.

From Eq. (14), we have

$$\frac{d^2\mathbf{r}}{dt^2} = -g\mathbf{j} - k\left(\frac{dx}{dt}\mathbf{i} + \frac{dy}{dt}\mathbf{j}\right)$$

$$= -k\frac{dx}{dt}\mathbf{i} + \left(-g - k\frac{dy}{dt}\right)\mathbf{j}.$$

Comparing components of the two sides, we have

$$\frac{d^2x}{dt^2} = -k\frac{dx}{dt} \tag{15}$$

and

$$\frac{d^2y}{dt^2} = -g - k\frac{dy}{dt}. \tag{16}$$

We integrate Eq. (15) and use $x = 0$ and $\dfrac{dx}{dt} = v_0\cos\alpha$ when $t = 0$ to obtain

$$\frac{dx}{dt} = -kx + v_0\cos\alpha. \tag{17}$$

To solve Eq. (17), we separate the variables (Section 8.7) and obtain

$$\frac{dx}{v_0 \cos \alpha - kx} = dt.$$

Integrating both sides, we obtain

$$-\frac{1}{k} \ln \ |v_0 \cos \alpha - kx| = t + C$$

$$\ln \ |v_0 \cos \alpha - kx| = -kt + C$$

$$|v_0 \cos \alpha - kx| = Ce^{-kt}$$

$$v_0 \cos \alpha - kx = Ce^{-kt} \tag{18}$$

$$v_0 \cos \alpha - kx = (v_0 \cos \alpha)e^{-kt} \qquad C = v_0 \cos \alpha \text{ when } t = 0 \text{ and } x = 0.$$

$$kx = v_0 \cos \alpha - (v_0 \cos \alpha)e^{-kt}$$

$$x = \frac{v_0}{k}(1 - e^{-kt})(\cos \alpha).$$

To solve Eq. (16), we set $w = \dfrac{dy}{dt}$ and separate the variables:

$$\frac{dw}{dt} = -g - kw$$

$$\frac{dw}{-g - kw} = dt$$

$$\frac{-1}{k} \ln \ |-g - kw| = t + C$$

$$\ln \ |-g - kw| = -kt + C$$

$$-g - kw = Ce^{-kt} \tag{19}$$

Now when $t = 0$, $w = \dfrac{dy}{dt} = v_0 \sin \alpha$, so $C = -g - kv_0 \sin \alpha$. Equation (19) becomes

$$-g - kw = (-g - k \ v_0 \sin \alpha)e^{-kt}$$

$$-g - k\frac{dy}{dt} = (-g - k \ v_0 \sin \alpha)e^{-kt}. \tag{20}$$

Integrating both sides of Eq. (20), solving for the constant of integration, and then solving for y, we obtain

$$y = \frac{v_0}{k}(1 - e^{-kt})(\sin \alpha) + \frac{32}{k^2}(1 - kt - e^{-kt})$$

to complete the proof.   ≡

We can now model projectile-motion problems more realistically, as the next example illustrates.

EXAMPLE 3

A baseball is hit as it was in Example 1, this time with linear drag to slow it down but no wind (a very calm day). For the temperature and humidity on game day, it has been established that the constant of proportionality k

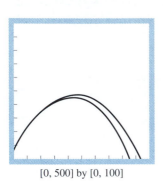

[0, 500] by [0, 100]

12.14 The baseball problem with and without drag (no wind).

(called the *resistance* or *drag coefficient*) is 0.05. Solve parts (a)–(c) given in Example 1 for these new conditions.

Solution

a) Using Eqs. (13), the parametric representation of the position of the baseball as a function of time t, assuming drag coefficient 0.05, is

$$x(t) = \frac{152 \cos 20°}{0.05} \left(1 - e^{-0.05t}\right),$$

$$y(t) = 3 + \frac{152 \sin 20°}{0.05} \left(1 - e^{-0.05t}\right) + \left(\frac{32}{0.05^2}\right) \left(1 - 0.05t - e^{-0.05t}\right).$$

We leave the vector representation for you to write.

b) Figure 12.14 shows simulations of the flight of the baseball (no wind) with and without the drag factor. The inner graph is the flight of the baseball with air resistance (drag).

c) We leave it to you to answer the questions about height, range, and flight times. ▤

Exercises 12.2

Projectile flights in the following exercises are to be treated as ideal unless stated otherwise. All launch angles are assumed to be measured from the horizontal. All projectiles are assumed to be launched from the origin over a horizontal surface unless stated otherwise. Solve all problems analytically or with a graphing utility, and then support or confirm, respectively, if possible.

1. A projectile is launched at a speed of 840 m/sec at an angle of 60°. How long will it take to get 21 km downrange?

2. Find the muzzle speed of a gun whose maximum range is 24.5 km.

3. A projectile is launched over level ground with an initial speed of 500 m/sec at a launch angle of 45°.

   a) When and how far away will the projectile land?

   b) How high overhead will the projectile be when it is 5 km downrange?

   c) What is the highest the projectile will go?

4. A baseball is thrown from the stands 32 ft above the field at a launch angle of 30°. When and how far away will the ball strike the ground if its initial speed is 32 ft/sec?

5. Show that a projectile launched at an angle of α degrees $(0 < \alpha < 90)$, has the same range as a projectile launched at the same speed and an angle of $(90 - \alpha)$ degrees.

6. What two launch angles will enable a projectile to reach a target 16 km downrange if the projectile's initial speed is 400 m/sec?

7. A spring gun at ground level launches a golf ball at an angle of 45°. The ball lands 10 m away. What was the

ball's initial speed? For the same initial speed, find the two launch angles that make the range 6 m.

8. An electron in a TV tube is beamed horizontally at a speed of 5×10^6 m/sec toward the face of the tube 40 cm away. About how far will the electron drop before it hits?

9. Laboratory tests designed to find how far golf balls of different hardness go when hit with a driver showed that a 100-compression, Surlyn-covered, two-piece ball hit with a club-head speed of 100 mi/h at a launch angle of 9° carried 248.8 yd. What was the launch speed of the ball? It was not 100 mi/h. (The data are from "Does Compression Really Matter?" by Lew Fishman, *Golf Digest*, August 1986, pp. 35–37.)

10. Show that doubling the initial speed of a projectile multiplies its maximum height and range by 4. By about what percentage should you increase the initial speed to double the height and range?

11. Show that a projectile attains three-quarters of its maximum height in half the time it takes to reach the maximum height.

12. A human cannonball is to be fired with an initial speed of $v_0 = 80\sqrt{10}/3$ ft/sec. The circus performer (of the right caliber, naturally) hopes to land on a special cushion located 200 ft downrange. The circus is being held in a large room with a flat ceiling 75 ft high. Can the performer be fired to the cushion without striking the ceiling? If so, what should the cannon's angle of elevation be?

13. A golf ball is hit from the tee to an elevated green with an initial speed of 116 ft/sec and a launch angle of 45°, as

shown in the diagram. Will the ball reach the pin?

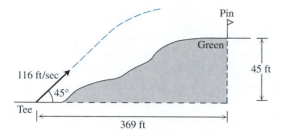

14. A golf ball leaves the ground at a 30° angle at a speed of 90 ft/sec. Will it clear the top of a 30-ft tree 135 ft away?

15. A baseball hit by a Boston Red Sox player at a 20° angle from 3 ft above the ground just cleared the left end of the "Green Monster," the left-field wall in Fenway Park (Fig. 12.15). The wall there is 37 ft high and 315 ft from home plate. About how fast was the ball going? How long did it take the ball to reach the wall?

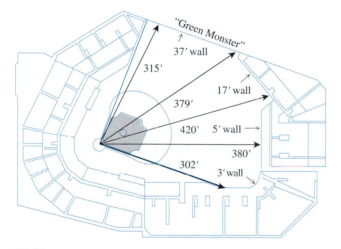

12.15 The Green Monster, the left-field wall at Fenway Park in Boston.

16. A baseball is hit when it is 2.5 feet above the ground and leaves the bat with initial velocity of 145 ft/sec at a launch angle of 23°. A 10-mi/h wind is blowing in the horizontal direction against the ball. A 15-foot-high fence is 300 feet from home plate in the direction of the flight.

a) Determine algebraic representations for the flight of the baseball in both vector function and parametric forms.

b) Simulate the motion of the baseball with a graphing utility. Show the fence in the viewing window.

c) Use the simulation to answer the following questions.

i) How high does the baseball go, and when does it reach maximum height?

ii) How far (ground distance) could the baseball travel from home plate in the air, and when would it hit the ground?

iii) When is the baseball 20 feet high? How far (ground distance) is the baseball from home plate at that height? Give a complete solution.

iv) Has the batter hit a home run?

17. A volleyball is hit when it is 4 feet above the ground and 12 feet from a 6 foot net. It leaves the point of impact with an initial velocity of 35 ft/sec at an angle of 27°.

a) Determine algebraic representations for the flight of the volleyball in both vector function and parametric forms.

b) Simulate the motion of the volleyball with a graphing utility. Show the net and both ends of a 50-foot court in the viewing window.

c) Use the simulation to answer the following questions.

i) How high does the volleyball go, and when does it reach maximum height?

ii) How far (ground distance) could the volleyball travel in the air from its launch, and when would it hit the ground?

iii) When is the volleyball 7 feet above the ground? How far (ground distance) is it from where it would land? Give a complete solution.

iv) Suppose the net is raised to 8 feet? Does this change things? Explain.

18. Complete Example 3, part (c). Also, discuss how far away the outfield fence would have to be to prevent a home run. The fence's height is at least 12 feet and at most 15 feet.

19. Consider the baseball problem of Example 3 again. Assume a drag coefficient of 0.12.

a) How high does the baseball go, and when does it reach maximum height?

b) How far (ground distance) could the baseball travel from home plate in the air, and when would it hit the ground?

c) When is the baseball 30 feet high? How far (ground distance) is it from home plate at that height? Give a complete solution.

d) A 10-foot outfield fence is 340 feet from home plate in the direction of the flight of the baseball. The outfielder can jump and catch any ball up to 11 feet off the ground to stop it from going over the fence. Has the batter hit a home run?

20. Consider the baseball problem of Example 3 again. This time, assume a drag coefficient of 0.08 *and* a wind blowing at 12 mi/h in the horizontal direction, directly opposite the direction the ball is taking toward the outfield.

a) Determine algebraic representations for the flight of the baseball in both vector function and parametric forms.

b) Simulate the motion of the baseball with and without wind.

c) Use the simulation to answer the following questions.

i) How high does the baseball go, and when does it reach maximum height?

ii) How far (ground distance) could the baseball travel from home plate in the air, and when would it hit the ground?

iii) When is the baseball 35 feet high? How far (ground distance) is the baseball from home plate at that height? Give a complete solution.

iv) A 20-foot outfield fence is 380 feet from home plate in the direction of the flight of the baseball. Has the batter hit a home run? If "yes," what wind speed would have kept the ball in the park? If "no," what wind speed would have allowed it to be a home run?

21. Consider the baseball problem of Example 3 again. In the same viewing window, simulate the motion of the baseball for drag coefficients of $k = 0.01, 0.02, 0.05, 0.1, 0.15, 0.20,$ and 0.25. Use the simulations to answer these questions for each value of k.

a) How high does the baseball go, and when does it reach maximum height?

b) How far (ground distance) could the baseball travel from home plate in the air, and when would it hit the ground?

c) As $k \to 0$, show that the right-hand members of Eqs. (13) approach the right-hand members of Eqs. (6). Explain why this makes sense.

22. *Modeling Projectile Motion Launched on a Side of a Hill.* We assume that the hill can be modeled by an inclined plane that makes an angle β with the horizontal. From point A on the inclined plane, a projectile is launched with an initial speed v_0 at an angle α from the horizontal as pictured. Its flight direction is directly toward the top of the hill, and the only force acting on it in flight is gravity; there is no wind, spin, or drag.

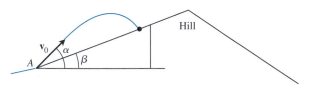

a) If $v_0 = 95$ ft/sec, $\alpha = 42°$, and $\beta = 25°$, what is the time of flight (t_f) before the projectile hits the side of the hill? Assume that the projectile hits the same side of the hill from which it is launched. Simulate the hillside and the flight with a grapher.

b) What is the range of the flight (r_f) on the hillside; that is, what is the distance from launch point to landing point on the hillside?

c) Show algebraically that

$$t_f = \frac{2v_0 \sin (\alpha - \beta)}{g \cos \beta},$$

where g is 32 ft/sec². Compare with the graphical solution to part (a) for the given values of $v_0, \alpha,$ and β.

d) Determine a formula for r_f in terms of $v_0, \alpha,$ and β. Compare with the graphical solution to part (b) for the given values of $v_0, \alpha,$ and β.

23. The diagram below shows an experiment with two marbles. Marble A is launched toward marble B at an initial speed v_0. The launch angle is α. At the same instant, marble B, suspended directly over a spot R units downrange from A, is released from a height of $R \tan \alpha$. The marbles appear to collide regardless of the value of v_0. Explain why this happens. Must it always happen? Why?

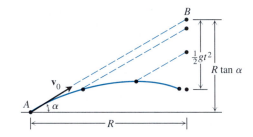

24. A projectile is launched down an inclined plane as shown below in profile. Show that the ideal maximum downhill range is achieved when the initial velocity vector bisects the angle AOR. If the projectile were launched uphill instead of down, what launch angle would maximize its range?

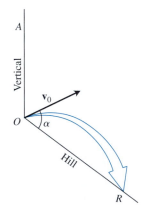

25. Derive Eqs. (11) in the text:

$$x = x_0 + (v_0 \cos \alpha)t,$$

$$y = y_0 + (v_0 \sin \alpha)t - \frac{1}{2}gt^2$$

by solving the following initial value problem for a vector **r** in the plane.

Differential equation: $\dfrac{d^2\mathbf{r}}{dt^2} = -g\mathbf{j}$

Initial conditions: $\mathbf{r} = x_0\mathbf{i} + y_0\mathbf{j}$ and

$$\frac{d\mathbf{r}}{dt} = (v_0\cos\alpha)\mathbf{i} + (v_0\sin\alpha)\mathbf{j}$$

when $t = 0$.

26. Follow the suggestions provided near the end of the proof of Theorem 1 to solve Eq. (20) and complete the proof of the theorem.

12.3 _____ Directed Distance and the Unit Tangent Vector **T**

As you can imagine, differentiable curves, especially those with continuous first and second derivatives, have been subjects of intense study, for their mathematical interest as well as for their applications to motion in space. In this section and the next, we describe some of the features that account for the importance of these curves.

Distance Along a Curve

One of the special features of space curves whose coordinate functions have continuous first derivatives is that, like plane curves, they are smooth enough to have a measurable length. This enables us to locate points along these curves by giving their directed distance s along the curve from some **base point**, the way we locate points on coordinate axes by giving their directed distance from the origin (Fig. 12.16).

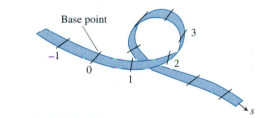

12.16 Differentiable curves with continuous first derivatives can be scaled like coordinate axes or like tape measures that include negative numbers as well as positive numbers. The coordinate given to each point is its directed distance from the base point.

Although time is the natural parameter for describing a moving body's velocity and acceleration, the **directed distance coordinate** s is the natural parameter for studying a curve's geometry. The relationships between these parameters play an important role in calculations of space flight.

In the plane, we define the length of a parametrized curve $x = f(t)$, $y = g(t)$ $(a \le t \le b)$ by the formula

$$\text{Length} = \int_a^b \sqrt{\left(\frac{dx}{dt}\right)^2 + \left(\frac{dy}{dt}\right)^2}\, dt.$$

In space calculations, the analogous formula has an additional term for the third coordinate.

DEFINITION

If $x = f(t)$, $y = g(t)$, and $z = h(t)$ have continuous first derivatives with respect to t for $a \leq t \leq b$ and the position vector $\mathbf{r} = f(t)\mathbf{i} + g(t)\mathbf{j} + h(t)\mathbf{k}$ defined by these equations traces its curve exactly once as t moves from $t = a$ to $t = b$, then the length of the curve is given by the formula

$$\text{Length} = \int_a^b \sqrt{\left(\frac{dx}{dt}\right)^2 + \left(\frac{dy}{dt}\right)^2 + \left(\frac{dz}{dt}\right)^2}\, dt. \qquad (1)$$

Just as for plane curves, we can calculate the length of a curve in space from any convenient parametrization that meets the stated conditions. Again, we will omit the proof.

Notice that the square root in Eq. (1) is $|\mathbf{v}|$, the length of the velocity vector $d\mathbf{r}/dt$. This lets us write the formula for length a shorter way.

Length Formula (Short Form)

$$\text{Length} = \int_a^b |\mathbf{v}|\, dt \qquad (2)$$

EXAMPLE 1

Find the length of one turn of the helix

$$\mathbf{r} = (\cos t)\,\mathbf{i} + (\sin t)\,\mathbf{j} + t\,\mathbf{k}.$$

Solution Since the sine and cosine both have periods of 2π, the helix makes one full turn as t runs from 0 to 2π (Fig. 12.17). The length of this portion of the curve is

$$\begin{aligned}
\text{Length} &= \int_a^b |\mathbf{v}|\, dt \quad \text{Eq. (2)} \\
&= \int_0^{2\pi} \sqrt{(-\sin t)^2 + (\cos t)^2 + (1)^2}\, dt \\
&= \int_0^{2\pi} \sqrt{2}\, dt \\
&= 2\pi\sqrt{2}. \qquad \blacksquare
\end{aligned}$$

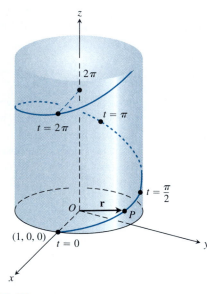

12.17 The helix

$$\mathbf{r} = (\cos t)\,\mathbf{i} + (\sin t)\,\mathbf{j} + t\,\mathbf{k}$$

makes a complete turn during the time interval from $t = 0$ to $t = 2\pi$. We calculate the distance traveled by P by integrating $|\mathbf{v}|$ from $t = 0$ to $t = 2\pi$ (Example 1).

EXPLORATION BIT

The length, $2\pi\sqrt{2}$, of one turn of the helix is $\sqrt{2}$ times the length of the circle in the xy-plane over which the helix stands. Does this suggest a geometric argument in support of the length being $2\pi\sqrt{2}$? (In what more familiar geometric figure is one length $\sqrt{2}$ times another length?)

If $x = f(t)$, $y = g(t)$, and $z = h(t)$ have continuous first derivatives with respect to t and we choose a base point $P(t_0)$ on the curve $\mathbf{r} = f(t)\mathbf{i} + g(t)\mathbf{j} + h(t)\mathbf{k}$ (Fig. 12.18), the integral of $|\mathbf{v}|$ from t_0 to t gives the directed distance along the curve from $P(t_0)$ to $P(t)$. The distance is a function of t, and we denote it by $s(t)$.

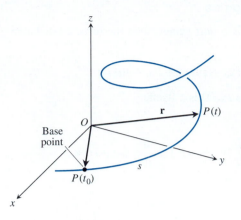

12.18 The directed distance along the curve from $P(t_0)$ to any point $P(t)$ is

$$s = \int_{t_0}^{t} |\mathbf{v}(\tau)| \, d\tau.$$

We use the Greek letter τ (tau) as the variable of integration to distinguish it from t.

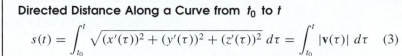

> **Directed Distance Along a Curve from t_0 to t**
>
> $$s(t) = \int_{t_0}^{t} \sqrt{(x'(\tau))^2 + (y'(\tau))^2 + (z'(\tau))^2} \, d\tau = \int_{t_0}^{t} |\mathbf{v}(\tau)| \, d\tau \quad (3)$$

The value of s is positive if t is greater than t_0 and negative if t is less than t_0.

EXAMPLE 2

If $t_0 = 0$, the directed distance along the helix

$$\mathbf{r} = (\cos t)\,\mathbf{i} + (\sin t)\,\mathbf{j} + t\,\mathbf{k}$$

from t_0 to t is

$$
\begin{aligned}
s(t) &= \int_{t_0}^{t} |\mathbf{v}(\tau)| \, d\tau \qquad \text{Eq. (3)} \\
&= \int_{0}^{t} \sqrt{2} \, d\tau \qquad \text{Value from Example 1} \\
&= \sqrt{2}\,t.
\end{aligned}
$$

Thus, $s(2\pi) = 2\pi\sqrt{2}$, $s(-2\pi) = -2\pi\sqrt{2}$, and so on (Fig. 12.19).

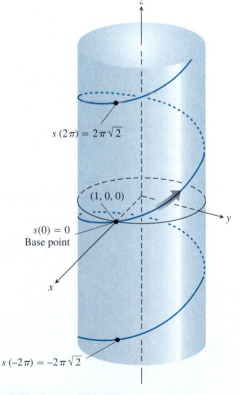

12.19 Directed distances along the helix

$$\mathbf{r} = (\cos t)\,\mathbf{i} + (\sin t)\,\mathbf{j} + t\,\mathbf{k}$$

from the base point $(1, 0, 0)$.

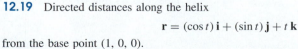

EXAMPLE 3   Distance along a Line

Show that if $\mathbf{u} = u_1\mathbf{i} + u_2\mathbf{j} + u_3\mathbf{k}$ is a *unit* vector, then the directed distance along the line

$$\mathbf{r} = (x_0 + tu_1)\,\mathbf{i} + (y_0 + tu_2)\,\mathbf{j} + (z_0 + tu_3)\,\mathbf{k}$$

from the point $P_0(x_0, y_0, z_0)$ where $t = 0$ is t itself.

Solution

$$\mathbf{v} = \frac{d}{dt}(x_0 + tu_1)\,\mathbf{i} + \frac{d}{dt}(y_0 + tu_2)\,\mathbf{j} + \frac{d}{dt}(z_0 + tu_3)\,\mathbf{k}$$

$$= u_1\,\mathbf{i} + u_2\,\mathbf{j} + u_3\,\mathbf{k} = \mathbf{u},$$

so

$$s(t) = \int_0^t |\mathbf{v}|\,d\tau = \int_0^t |\mathbf{u}|\,d\tau = \int_0^t 1\,d\tau = t. \qquad \blacksquare$$

If the derivatives beneath the radical in Eq. (3) are continuous, part 1 of the Fundamental Theorem of Calculus (Section 5.4) tells us that s is a differentiable function of t whose derivative is

$$\frac{ds}{dt} = |\mathbf{v}(t)|. \qquad (4)$$

As we expect, the speed with which a particle moves along a path is the magnitude of $\mathbf{v}$.

Notice that although the base point $P(t_0)$ plays a role in defining s in Eq. (3), it plays no role in Eq. (4). The rate at which a moving particle covers the distance along its path does not depend on how far away the base point is.

Notice also that as long as $|\mathbf{v}|$ is different from zero, *as we will assume it to be in all examples from now on*, ds/dt is positive, and s is an increasing function of t.

The Unit Tangent Vector **T**

Since ds/dt is positive for the curves we are considering from now on, s is one-to-one and has an inverse that gives t as a differentiable function of s (Section 7.2). The derivative of the inverse function is

$$\frac{dt}{ds} = \frac{1}{ds/dt} = \frac{1}{|\mathbf{v}|}.$$

This makes $\mathbf{r}$ a differentiable function of s whose derivative can be calculated with the Chain Rule for vector-valued functions to be

$$\frac{d\mathbf{r}}{ds} = \frac{d\mathbf{r}}{dt}\frac{dt}{ds} = \mathbf{v}\cdot\frac{1}{|\mathbf{v}|} = \frac{\mathbf{v}}{|\mathbf{v}|}. \qquad (5)$$

This tells us that $d\mathbf{r}/ds$ has the constant length

$$\left|\frac{d\mathbf{r}}{ds}\right| = \frac{1}{|\mathbf{v}|}|\mathbf{v}| = 1. \qquad (6)$$

Together, Eqs. (5) and (6) say that $d\mathbf{r}/ds$ is a unit vector that points in the direction of **v**. We call $d\mathbf{r}/ds$ the *unit tangent vector* of the curve traced by **r**, and we denote it by **T** (Fig. 12.20).

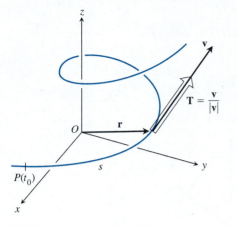

12.20 We get the unit tangent vector **T** by dividing **v** by $|\mathbf{v}|$.

DEFINITION

The **unit tangent vector** of a differentiable curve $\mathbf{r} = \mathbf{f}(t)$ is

$$\mathbf{T} = \frac{d\mathbf{r}}{ds} = \frac{d\mathbf{r}/dt}{ds/dt} = \frac{\mathbf{v}}{|\mathbf{v}|}.$$

The unit tangent vector **T** is a differentiable function of t whenever **v** is a differentiable function of t. As we will see in the next section, **T** is one of three unit vectors in a traveling reference frame that is used to describe the motion of space vehicles and other bodies moving in three dimensions.

EXAMPLE 4

Find the unit tangent vector of the helix

$$\mathbf{r} = (\cos t)\,\mathbf{i} + (\sin t)\,\mathbf{j} + t\,\mathbf{k}.$$

Solution

$$\mathbf{v} = (-\sin t)\,\mathbf{i} + (\cos t)\,\mathbf{j} + \mathbf{k}$$

$$|\mathbf{v}| = \sqrt{(-\sin t)^2 + (\cos t)^2 + (1)^2} = \sqrt{2}$$

$$\mathbf{T} = \frac{\mathbf{v}}{|\mathbf{v}|} = -\frac{\sin t}{\sqrt{2}}\,\mathbf{i} + \frac{\cos t}{\sqrt{2}}\,\mathbf{j} + \frac{1}{\sqrt{2}}\,\mathbf{k}$$ ≡

EXAMPLE 5   The Involute of a Circle

Find the unit tangent vector of the curve

$$\mathbf{r} = (\cos t + t\sin t)\,\mathbf{i} + (\sin t - t\cos t)\,\mathbf{j} \qquad (t > 0).$$

(See Fig. 12.21.)

Solution

$$\mathbf{v} = \frac{d\mathbf{r}}{dt} = (-\sin t + \sin t + t\cos t)\,\mathbf{i} + (\cos t - \cos t + t\sin t)\,\mathbf{j}$$

$$= (t\cos t)\,\mathbf{i} + (t\sin t)\,\mathbf{j}$$

$$|\mathbf{v}| = \sqrt{t^2\,\cos^2 t + t^2\,\sin^2 t} = \sqrt{t^2} = |t| = t \qquad \begin{array}{l}|t| = t \\ \text{because } t > 0.\end{array}$$

$$\mathbf{T} = \frac{\mathbf{v}}{|\mathbf{v}|} = \frac{\mathbf{v}}{t} = (\cos t)\,\mathbf{i} + (\sin t)\,\mathbf{j}$$ ≡

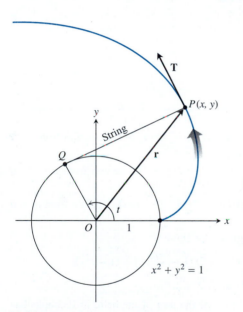

12.21 The *involute* of a circle is the path traced by the endpoint P of a string unwinding from a circle, here the unit circle in the xy-plane. The position vector of P can be shown to be

$$\mathbf{r} = (\cos t + t\sin t)\,\mathbf{i} + (\sin t - t\cos t)\,\mathbf{j},$$

where t is the angle from the positive x-axis to Q. Example 5 derives a formula for the curve's unit tangent vector **T**. How is **T** related to the vector $\overrightarrow{OQ}$?

EXAMPLE 6

For the counterclockwise motion

$$\mathbf{r} = (\cos t)\,\mathbf{i} + (\sin t)\,\mathbf{j}$$

around the unit circle,

$$\mathbf{v} = (-\sin t)\,\mathbf{i} + (\cos t)\,\mathbf{j}$$

is already a unit vector, so

$$\mathbf{T} = \mathbf{v}.$$

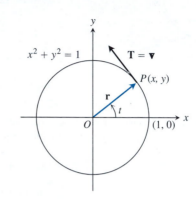

Exercises 12.3

In Exercises 1–8, find the curve's unit tangent vector. Also, find the length of the indicated section of the curve.

1. $\mathbf{r} = (2\cos t)\,\mathbf{i} + (2\sin t)\,\mathbf{j} + \sqrt{5}t\mathbf{k}$, section from $t = 0$ to $t = \pi$

2. $\mathbf{r} = (6\sin 2t)\,\mathbf{i} + (6\cos 2t)\,\mathbf{j} + 5t\mathbf{k}$, section from $t = 0$ to $t = \pi$

3. $\mathbf{r} = t\mathbf{i} + (2/3)t^{3/2}\mathbf{k}$, section from $t = 0$ to $t = 8$

4. $\mathbf{r} = (\cos^3 t)\,\mathbf{j} + (\sin^3 t)\,\mathbf{k}$, section from $t = 0$ to $t = \pi/2$

5. $\mathbf{r} = (2 + t)\,\mathbf{i} - (t + 1)\,\mathbf{j} + t\,\mathbf{k}$, section from $t = 0$ to $t = 3$

6. $\mathbf{r} = 6t^3\mathbf{i} - 2t^3\mathbf{j} - 3t^3\mathbf{k}$, section from $t = -1$ to $t = 1$

7. $\mathbf{r} = (t\cos t)\,\mathbf{i} + (t\sin t)\,\mathbf{j} + (2\sqrt{2}/3)t^{3/2}\,\mathbf{k}$, section from $t = 0$ to $t = \pi$

8. $\mathbf{r} = (t\sin t + \cos t)\,\mathbf{i} + (t\cos t - \sin t)\,\mathbf{j}$, section from $t = 0$ to $t = \sqrt{2}$

In Exercises 9–12, use analytic methods and then support with NINT computations. Find the directed distance along the curve from point t_0 to point t by evaluating the integral

$$s(t) = \int_{t_0}^{t} |\mathbf{v}(\tau)|\, d\tau.$$

Then find the length of the indicated section of the curve.

9. $\mathbf{r} = (4\cos t)\,\mathbf{i} + (4\sin t)\,\mathbf{j} + 3t\mathbf{k}$, section from $t_0 = 0$ to $t = \pi/2$

10. $\mathbf{r} = (\cos t + t\sin t)\,\mathbf{i} + (\sin t - t\cos t)\,\mathbf{j}$, section from $t_0 = \pi/2$ to $t = \pi$

11. $\mathbf{r} = (e^t\cos t)\,\mathbf{i} + (e^t\sin t)\,\mathbf{j} + e^t\mathbf{k}$, section from $t_0 = 0$ to $t = \ln 4$

12. $\mathbf{r} = (1 + 2t)\,\mathbf{i} + (1 + 3t)\,\mathbf{j} + (6 - 6t)\,\mathbf{k}$, section from $t_0 = 0$ to $t = 1$

13. Find the length of the curve

$$\mathbf{r} = (\sqrt{2}t)\,\mathbf{i} + (\sqrt{2}t)\,\mathbf{j} + (1 - t^2)\,\mathbf{k}$$

from $(0, 0, 1)$ to $(\sqrt{2}, \sqrt{2}, 0)$.

14. The length $2\pi\sqrt{2}$ of the turn of the helix in Example 1 is also the length of the diagonal of a square 2π units on a side. Show how to obtain this square by cutting away and flattening a portion of the cylinder around which the helix winds.

15. *Length is independent of parametrization.* To illustrate the fact that the length of a smooth curve in space does not depend on the parametrization that we use to compute it, as long as the parametrization meets the conditions given with Eq. (1), calculate the length of one turn of the helix in Example 1 with the following parametrizations.
  a) $\mathbf{r} = (\cos 4t)\,\mathbf{i} + (\sin 4t)\,\mathbf{j} + 4t\mathbf{k}$ $(0 \le t \le \pi/2)$
  b) $\mathbf{r} = (\cos(t/2))\,\mathbf{i} + (\sin(t/2))\,\mathbf{j} + (t/2)\,\mathbf{k}$ $(0 \le t \le 4\pi)$
  c) $\mathbf{r} = (\cos t)\,\mathbf{i} - (\sin t)\,\mathbf{j} - t\mathbf{k}$ $(-2\pi \le t \le 0)$

12.4

Curvature, Torsion, and the **TNB** Frame

In this section, we define a *frame* of mutually orthogonal unit vectors that always travels with a vehicle moving along a curve in space (Fig. 12.22). The frame has three vectors. The first is **T**, the unit tangent vector. The second is **N**, the unit vector that gives the direction of $d\mathbf{T}/ds$. The third is $\mathbf{B} = \mathbf{T} \times \mathbf{N}$. These vectors and their derivatives, when available, give useful information about a vehicle's orientation in space and how the vehicle's path turns and twists as the vehicle moves along.

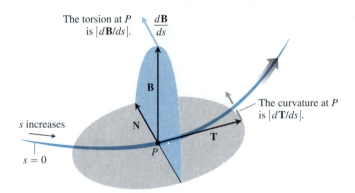

The torsion at P is $|d\mathbf{B}/ds|$. $\dfrac{d\mathbf{B}}{ds}$

The curvature at P is $|d\mathbf{T}/ds|$.

s increases

$s = 0$

12.22 Every moving vehicle travels with a **TNB** frame that describes how the vehicle moves.

For example, the magnitude of the derivative $d\mathbf{T}/ds$ tells how much a vehicle's path turns to the left or right as it moves along; it is called the *curvature* of the vehicle's path. The number $|d\mathbf{B}/ds|$ tells how much a vehicle's path rotates or twists as the vehicle moves along; it is called the *torsion* of the vehicle's path. Look at Fig. 12.22 again. If P is a train climbing up a curved track, the rate at which the headlight beam turns per unit distance is the same as the curvature of the track. The rate at which the engine rotates about its longitudinal axis (the line of **T**) is the torsion.

We begin with curves in the plane.

The Curvature of a Plane Curve

As we move along a differentiable curve in the plane, the unit tangent vector **T** turns as the curve bends. We measure the rate at which **T** turns by measuring the change in the angle ϕ that **T** makes with **i** (Fig. 12.23). At each point P, the absolute value of $d\phi/ds$, stated in radians per unit of length along the curve, is called the **curvature** at P. If $|d\phi/ds|$ is large, **T** is turning sharply as we pass through P, and the curvature at P is large. If $|d\phi/ds|$ is close to zero, **T** is turning slowly, and the curvature at P is small. On circles and lines, the curvature is constant, as we will see in Examples 1 and 2. On other curves, the curvature can vary from place to place along the curves. If $x(t)$ and $y(t)$ are twice-differentiable

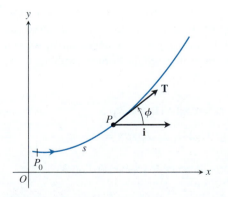

12.23 The value of $|d\phi/ds|$ at P is called the *curvature* of the curve at P.

functions of t, we can derive a formula for the curvature of the curve $r = x(t)\,\mathbf{i} + y(t)\,\mathbf{j}$ in the following way. In Newton's dot notation, in which $\dot{y}$ ("y dot") means dy/dt, $\ddot{y}$ ("y double dot") means d^2y/dt^2, and so on, we have

$$\tan\phi = \frac{dy}{dx} = \frac{dy/dt}{dx/dt} = \frac{\dot{y}}{\dot{x}} \quad \text{and} \quad \phi = \tan^{-1}\left(\frac{\dot{y}}{\dot{x}}\right).$$

Hence,

$$\frac{d\phi}{ds} = \frac{d\phi}{dt}\frac{dt}{ds} = \frac{1}{1+(\dot{y}/\dot{x})^2}\frac{d}{dt}\left(\frac{\dot{y}}{\dot{x}}\right)\frac{1}{(\dot{x}^2+\dot{y}^2)^{1/2}} \qquad \frac{dt}{ds} = \frac{1}{|\mathbf{v}|}$$

$$= \frac{\dot{x}^2}{(\dot{x}^2+\dot{y}^2)^{3/2}}\frac{\dot{x}\ddot{y}-\dot{y}\ddot{x}}{\dot{x}^2}$$

$$= \frac{\dot{x}\ddot{y}-\dot{y}\ddot{x}}{|\mathbf{v}|^3}.$$

The curvature is therefore

$$\left|\frac{d\phi}{ds}\right| = \frac{|\dot{x}\ddot{y}-\dot{y}\ddot{x}|}{|\mathbf{v}|^3}. \tag{1}$$

The observation that $|\dot{x}\ddot{y}-\dot{y}\ddot{x}|$ is the magnitude of the vector

$$\mathbf{v}\times\mathbf{a} = \begin{vmatrix} \mathbf{i} & \mathbf{j} & \mathbf{k} \\ \dot{x} & \dot{y} & 0 \\ \ddot{x} & \ddot{y} & 0 \end{vmatrix}$$

enables us to write Eq. (1) in the following compact vector form.

The traditional symbol for the curvature function is the Greek letter κ (kappa).

Curvature

$$\kappa = \frac{|\mathbf{v}\times\mathbf{a}|}{|\mathbf{v}|^3} \tag{2}$$

Equation (2) calculates the curvature, a geometric property of the curve, from the velocity and acceleration of any vector representation of the curve in which $|\mathbf{v}|$ is different from zero. Take a moment to think about how remarkable this really is: From any formula for motion along a curve, no matter how variable the motion may be (as long as $\mathbf{v}$ is never zero), we can calculate a physical property of the curve that seems to have nothing to do with the way the curve is traversed.

EXAMPLE 1   The Curvature of a Straight Line Is Zero

On a straight line, ϕ is constant (Fig. 12.24), so $d\phi/ds = 0$. ≡

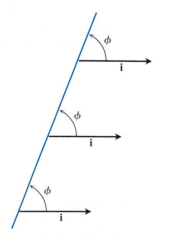

12.24   Along a line, the angle ϕ stays the same from point to point, and the curvature, $d\phi/ds$, is zero.

EXAMPLE 2   The Curvature of a Circle of Radius a Is $1/a$

To see why, parametrize the circle as

$$\mathbf{r} = (a\cos t)\,\mathbf{i} + (a\sin t)\,\mathbf{j}.$$

Then

$$\mathbf{v} = -(a \sin t)\mathbf{i} + (a \cos t)\mathbf{j},$$

$$\mathbf{a} = -(a \cos t)\mathbf{i} - (a \sin t)\mathbf{j}.$$

Then

$$\mathbf{v} \times \mathbf{a} = \begin{vmatrix} \mathbf{i} & \mathbf{j} & \mathbf{k} \\ -a \sin t & a \cos t & 0 \\ -a \cos t & -a \sin t & 0 \end{vmatrix} = (a^2 \sin^2 t + a^2 \cos^2 t)\mathbf{k} = a^2\mathbf{k},$$

$$|\mathbf{v}|^3 = \left[\sqrt{(-a \sin t)^2 + (a \cos t)^2}\right]^3 = a^3,$$

and

$$\kappa = \frac{|\mathbf{v} \times \mathbf{a}|}{|\mathbf{v}|^3} = \frac{|a^2\,\mathbf{k}|}{a^3} = \frac{a^2|\mathbf{k}|}{a^3} = \frac{1}{a}.$$

Circle of Curvature and Radius of Curvature

The **circle of curvature** or **osculating circle** at a point P on a plane curve where $\kappa \neq 0$ is the circle in the plane of the curve that

1. is tangent to the curve at P (has the same tangent that the curve has);
2. has the same curvature the curve has at P; and
3. lies toward the concave or inner side of the curve (as in Fig. 12.25).

The **radius of curvature** of the curve at P is the radius of the circle of curvature, which, according to Example 2, is

$$\text{Radius of curvature} = \rho = \frac{1}{\kappa}.$$

To calculate ρ, we calculate κ and take its reciprocal.

The **center of curvature** of the curve at P is the center of the circle of curvature.

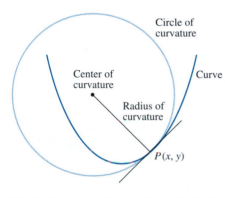

12.25 The osculating circle or circle of curvature at $P(x, y)$ lies toward the inner side of the curve. Note that for greater curvature κ, the circle (radius $\rho = 1/\kappa$) is smaller. For smaller values of κ, the circle is larger.

The Principal Unit Normal Vector for Curves in the Plane

The vectors $d\mathbf{T}/ds$ and $d\mathbf{T}/d\phi$ are related by the Chain Rule equation

$$\frac{d\mathbf{T}}{ds} = \frac{d\mathbf{T}}{d\phi}\frac{d\phi}{ds}.$$

Furthermore, since $\mathbf{T}$ is a vector of constant length, $d\mathbf{T}/ds$ and $d\mathbf{T}/d\phi$ are both orthogonal to $\mathbf{T}$. They have the same direction if $d\phi/ds$ is positive and opposite directions if $d\phi/ds$ is negative. Now

$$\mathbf{T} = (\cos\phi)\mathbf{i} + (\sin\phi)\mathbf{j},$$

and

$$\frac{d\mathbf{T}}{d\phi} = -(\sin\phi)\mathbf{i} + (\cos\phi)\mathbf{j} = \cos\left(\phi + \frac{\pi}{2}\right)\mathbf{i} + \sin\left(\phi + \frac{\pi}{2}\right)\mathbf{j}$$

is the unit vector obtained by rotating $\mathbf{T}$ counterclockwise through $\pi/2$ radians. Therefore, if we stand at a point on the curve facing in the direction of **T** (see Fig. 12.26), the vector $d\mathbf{T}/ds = (d\mathbf{T}/d\phi)(d\phi/ds)$ will point toward the left if $d\phi/ds$ is positive and toward the right if $d\phi/ds$ is negative. In other words, $d\mathbf{T}/ds$ will point toward the concave side of the curve.

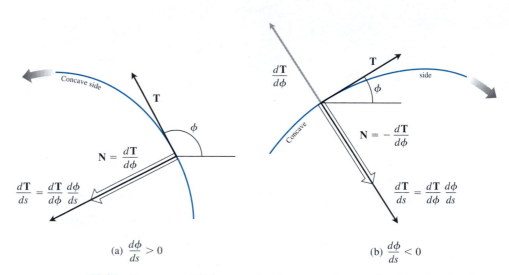

12.26 The vector $d\mathbf{T}/ds$, normal to the curve, always points inward. The principal unit normal vector **N** is the direction of $d\mathbf{T}/ds$.

Since $d\mathbf{T}/d\phi$ is a unit vector, the magnitude of $d\mathbf{T}/ds$ at any point on the curve is the curvature at that point, as we can see from the equation

$$\left| \frac{d\mathbf{T}}{ds} \right| = \left| \frac{d\mathbf{T}}{d\phi} \right| \left| \frac{d\phi}{ds} \right| = (1)(\kappa) = \kappa.$$

When $d\mathbf{T}/ds \neq \mathbf{0}$, its direction is given by the unit vector

$$\mathbf{N} = \frac{d\mathbf{T}/ds}{|d\mathbf{T}/ds|} = \frac{1}{\kappa} \frac{d\mathbf{T}}{ds}.$$

Since **N** points the same way $d\mathbf{T}/ds$ does, **N** is always orthogonal to **T** and directed toward the concave side of the curve. The vector **N** is called the **principal unit normal vector** of the curve.

Because the directed distance on a curve $\mathbf{r} = x(t)\mathbf{i} + y(t)\mathbf{j}$ is defined with ds/dt positive, the Chain Rule gives

$$\mathbf{N} = \frac{d\mathbf{T}/ds}{|d\mathbf{T}/ds|} = \frac{(d\mathbf{T}/dt)(dt/ds)}{|d\mathbf{T}/dt||dt/ds|}$$

$$= \frac{d\mathbf{T}/dt}{|d\mathbf{T}/dt|}. \tag{3}$$

This formula enables us to find **N** without having to find ϕ, κ, or s first.

EXAMPLE 3

Find **T** and **N** for the circular motion

$$\mathbf{r} = (\cos 2t)\,\mathbf{i} + (\sin 2t)\,\mathbf{j}.$$

Solution   We first find $\mathbf{T}$

$$\mathbf{v} = -(2\sin 2t)\,\mathbf{i} + (2\cos 2t)\,\mathbf{j},$$

$$|\mathbf{v}| = \sqrt{4\sin^2 2t + 4\cos^2 2t} = 2,$$

$$\mathbf{T} = \frac{\mathbf{v}}{|\mathbf{v}|} = -(\sin 2t)\,\mathbf{i} + (\cos 2t)\,\mathbf{j}.$$

From this we find

$$\frac{d\mathbf{T}}{dt} = -(2\cos 2t)\,\mathbf{i} - (2\sin 2t)\,\mathbf{j},$$

$$\left|\frac{d\mathbf{T}}{dt}\right| = \sqrt{4\cos^2 2t + 4\sin^2 2t} = 2,$$

and

$$\mathbf{N} = \frac{d\mathbf{T}/dt}{|d\mathbf{T}/dt|} = -(\cos 2t)\,\mathbf{i} - (\sin 2t)\,\mathbf{j}. \qquad \text{Eq. (3)}$$

Curvature for Curves in Space

In space, there is no natural way to find an angle like ϕ with which to measure the change in $\mathbf{T}$ along a differentiable curve. But we still have s, the directed distance along the curve, and can define the curvature to be

$$\kappa = \left|\frac{d\mathbf{T}}{ds}\right|,$$

as it worked out to be for curves in the plane. The formula

$$\kappa = \frac{|\mathbf{v} \times \mathbf{a}|}{|\mathbf{v}|^3} \qquad (4)$$

still holds, as you will see if you do the calculation in Exercise 23.

EXAMPLE 4

How do the values of a and b control the curvature of the helix (Fig. 12.27)

$$\mathbf{r} = (a\cos t)\,\mathbf{i} + (a\sin t)\,\mathbf{j} + bt\mathbf{k}? \qquad (a, b \geq 0, \text{ not both } 0)$$

Solution   We calculate the curvature with Eq. (4):

$$\mathbf{v} = -(a\sin t)\,\mathbf{i} + (a\cos t)\,\mathbf{j} + b\mathbf{k},$$

$$\mathbf{a} = -(a\cos t)\,\mathbf{i} - (a\sin t)\,\mathbf{j},$$

$$\mathbf{v} \times \mathbf{a} = \begin{vmatrix} \mathbf{i} & \mathbf{j} & \mathbf{k} \\ -a\sin t & a\cos t & b \\ -a\cos t & -a\sin t & 0 \end{vmatrix} = (ab\sin t)\,\mathbf{i} - (ab\cos t)\,\mathbf{j} + a^2\mathbf{k},$$

$$\kappa = \frac{|\mathbf{v} \times \mathbf{a}|}{|\mathbf{v}|^3} = \frac{\sqrt{a^2 b^2 + a^4}}{(a^2 + b^2)^{3/2}} = \frac{a\sqrt{a^2 + b^2}}{(a^2 + b^2)^{3/2}} = \frac{a}{a^2 + b^2}. \qquad (5)$$

From Eq. (5), we see that increasing b for a fixed a decreases the curvature. Decreasing a for a fixed b eventually decreases the curvature as well. In other words, stretching a spring tends to straighten it.

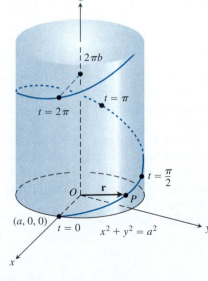

12.27   The helix

$$\mathbf{r} = (a\cos t)\,\mathbf{i} + (a\sin t)\,\mathbf{j} + bt\,\mathbf{k},$$

drawn with a and b positive (Example 4).

If $b = 0$, the helix reduces to a circle of radius a, and its curvature reduces to $1/a$, as it should. If $a = 0$, the helix becomes the z-axis, and its curvature reduces to 0, again as it should.

N for Curves in Space

To define the principal unit normal vector of a curve in space, we use the same definition that we use for a curve in the plane:

$$\mathbf{N} = \frac{d\mathbf{T}/ds}{|d\mathbf{T}/ds|} = \frac{1}{\kappa}\frac{d\mathbf{T}}{ds} = \frac{d\mathbf{T}/dt}{|d\mathbf{T}/dt|}.$$

As before, $d\mathbf{T}/ds$ and its unit-vector direction $\mathbf{N}$ are orthogonal to $\mathbf{T}$ because $\mathbf{T}$ has constant length, in this case 1.

EXPLORATION BIT

Give a geometric interpretation of the $\mathbf{N}$ found in Example 5.

EXAMPLE 5

Find $\mathbf{N}$ for the helix in Example 4.

Solution Using values from Example 4, we have

$$\mathbf{T} = \frac{\mathbf{v}}{|\mathbf{v}|} = \frac{-(a\sin t)\,\mathbf{i} + (a\cos t)\,\mathbf{j} + b\mathbf{k}}{\sqrt{a^2 + b^2}}$$

$$\frac{d\mathbf{T}}{dt} = -\frac{a}{\sqrt{a^2 + b^2}}((\cos t)\,\mathbf{i} + (\sin t)\,\mathbf{j})$$

$$\left|\frac{d\mathbf{T}}{dt}\right| = \frac{a}{\sqrt{a^2 + b^2}}\sqrt{\cos^2 t + \sin^2 t} = \frac{a}{\sqrt{a^2 + b^2}}$$

$$\mathbf{N} = \frac{d\mathbf{T}/dt}{|d\mathbf{T}/dt|} = -(\cos t)\,\mathbf{i} - (\sin t)\,\mathbf{j}.$$

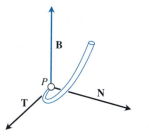

12.28 The vectors $\mathbf{T}$, $\mathbf{N}$, and $\mathbf{B}$ (in that order) make a right-handed frame of mutually orthogonal unit vectors in space. It is called the *Frenet* ("fren A") *frame* (after Jean-Frédéric Frenet, 1816–1900) or the *TNB frame*.

Torsion and the Binormal Vector

The **binormal vector** of a curve in space is the vector $\mathbf{B} = \mathbf{T} \times \mathbf{N}$, a unit vector orthogonal to both $\mathbf{T}$ and $\mathbf{N}$ (Fig. 12.28). Together, $\mathbf{T}$, $\mathbf{N}$, and $\mathbf{B}$ define a moving *right-handed* vector frame that plays a significant role in calculating the flight paths of space vehicles.

The three planes determined by $\mathbf{T}$, $\mathbf{N}$, and $\mathbf{B}$ are shown in Fig. 12.29. The curvature $\kappa = |d\mathbf{T}/ds|$ can be thought of as the rate at which the normal plane *turns* as the point P moves along the curve. Similarly, the **torsion** $\tau = |d\mathbf{B}/ds|$ is the rate at which the osculating plane *lifts* as P moves along the curve. It is a measure of how much the curve twists.

The most widely used formula for torsion, derived in more advanced texts, is

$$\tau = \pm\frac{\begin{vmatrix} \dot{x} & \dot{y} & \dot{z} \\ \ddot{x} & \ddot{y} & \ddot{z} \\ \dddot{x} & \dddot{y} & \dddot{z} \end{vmatrix}}{|\mathbf{v} \times \mathbf{a}|^2} \qquad \text{(if } \mathbf{v} \times \mathbf{a} \neq \mathbf{0}\text{).} \tag{6}$$

This formula calculates the torsion directly from the derivatives of the component functions $x = x(t), y = y(t), z = z(t)$ that make up $\mathbf{r}$. The determinant's first row comes from $\mathbf{v}$, the second row comes from $\mathbf{a}$, and the

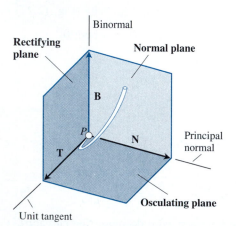

12.29 The three planes determined by $\mathbf{T}$, $\mathbf{N}$, and $\mathbf{B}$.

third row comes from $\dot{\mathbf{a}}$. The sign ($\pm$) is chosen to keep τ from being negative. Exercise 24 gives a formula that you can use to calculate τ if one or more of the third derivatives fails to exist (but that will not be a problem in calculations in this book).

EXAMPLE 6

Find the torsion of the helix

$$\mathbf{r} = (\cos t)\,\mathbf{i} + (\sin t)\,\mathbf{j} + t\mathbf{k}.$$

Solution We evaluate Eq. (6). We find the entries in the determinant by differentiating $\mathbf{r}$:

$$\mathbf{v} = -(\sin t)\,\mathbf{i} + (\cos t)\,\mathbf{j} + \mathbf{k}$$

$$\mathbf{a} = -(\cos t)\,\mathbf{i} - (\sin t)\,\mathbf{j}$$

$$\dot{\mathbf{a}} = (\sin t)\,\mathbf{i} - (\cos t)\,\mathbf{j}$$

Then

$$\tau = \frac{\begin{vmatrix} \dot{x} & \dot{y} & \dot{z} \\ \ddot{x} & \ddot{y} & \ddot{z} \\ \dddot{x} & \dddot{y} & \dddot{z} \end{vmatrix}}{|\mathbf{v} \times \mathbf{a}|^2} = \frac{\begin{vmatrix} -\sin t & \cos t & 1 \\ -\cos t & -\sin t & 0 \\ \sin t & -\cos t & 0 \end{vmatrix}}{\left\| \begin{vmatrix} \mathbf{i} & \mathbf{j} & \mathbf{k} \\ -\sin t & \cos t & 1 \\ -\cos t & -\sin t & 0 \end{vmatrix} \right\|^2}$$

$$= \frac{\cos^2 t + \sin^2 t}{|(\sin t)\,\mathbf{i} - (\cos t)\,\mathbf{j} + \mathbf{k}|^2} = \frac{1}{2}.$$

The Tangential and Normal Components of Acceleration

When a moving body is accelerated by gravity, brakes, a combination of rocket motors, or some other force, we usually want to know how much of the acceleration acts to move the body straight ahead in the direction of motion, that is, in the tangential direction $\mathbf{T}$. We can find out if we use the Chain Rule to rewrite $\mathbf{v}$ as

$$\mathbf{v} = \frac{d\mathbf{r}}{dt} = \frac{d\mathbf{r}}{ds}\frac{ds}{dt} = \mathbf{T}\frac{ds}{dt}$$

and differentiate both ends of this string of equalities to get

$$\mathbf{a} = \frac{d\mathbf{v}}{dt} = \frac{d}{dt}\left(\mathbf{T}\frac{ds}{dt} \right) = \frac{d^2s}{dt^2}\mathbf{T} + \frac{ds}{dt}\frac{d\mathbf{T}}{dt}$$

$$= \frac{d^2s}{dt^2}\mathbf{T} + \frac{ds}{dt}\left(\frac{d\mathbf{T}}{ds}\frac{ds}{dt} \right) = \frac{d^2s}{dt^2}\mathbf{T} + \frac{ds}{dt}\left(\kappa\mathbf{N}\frac{ds}{dt} \right)$$

$$= \frac{d^2s}{dt^2}\mathbf{T} + \kappa\left(\frac{ds}{dt} \right)^2 \mathbf{N}.$$

This is a remarkable equation. There is no **B** in it. No matter how the path of the moving body that we are watching may appear to twist and turn, the acceleration **a** always lies in the plane of **T** and **N** orthogonal to **B**. The equation also tells us exactly how much of the acceleration takes place tangent to the motion (d^2s/dt^2) and how much takes place normal to the motion $(\kappa(ds/dt)^2)$.

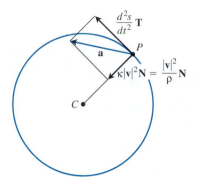

12.30 The tangential and normal components of the acceleration of a body speeding up as it moves counterclockwise around a circle of radius ρ.

The **tangential** and **normal** scalar components of acceleration are

$$a_\mathrm{T} = \frac{d^2s}{dt^2} = \frac{d}{dt}|\mathbf{v}| \quad \text{and} \quad a_\mathrm{N} = \kappa\left(\frac{ds}{dt}\right)^2 = \kappa|\mathbf{v}|^2. \tag{7}$$

That is,

$$\mathbf{a} = \frac{d^2s}{dt^2}\mathbf{T} + \kappa\left(\frac{ds}{dt}\right)^2\mathbf{N}. \tag{8}$$

If a body moves in a circle at a constant speed, d^2s/dt^2 is zero, and the acceleration points along **N** toward the circle's center. Notice that the normal scalar component of the acceleration is the curvature times the square of the speed. This explains why you have to have firm control of the steering wheel when your car enters a sharp curve (large κ), its speed is excessive (large $|\mathbf{v}|$), or both.

It the body is speeding up or slowing down while moving in a circle, **a** has a nonzero tangential component (Fig. 12.30).

To calculate a_N, we usually use the formula

$$a_\mathrm{N} = \sqrt{|\mathbf{a}|^2 - a_\mathrm{T}^2}, \tag{9}$$

which comes from solving the equation $|\mathbf{a}|^2 = \mathbf{a} \cdot \mathbf{a} = a_\mathrm{T}^2 + a_\mathrm{N}^2$ for a_N. With this formula, we can find a_N without having to calculate κ first.

EXAMPLE 7

Without finding **T** and **N**, write the acceleration of the motion

$$\mathbf{r} = (\cos t + t \sin t)\,\mathbf{i} + (\sin t - t \cos t)\,\mathbf{j}, \qquad (t > 0)$$

in the form $\mathbf{a} = a_\mathrm{T}\mathbf{T} + a_\mathrm{N}\mathbf{N}$. (The path of the motion is the involute of the circle in Fig. 12.31.)

Solution We use the first of Eqs. (7) to find a_T:

$$\mathbf{v} = (t \cos t)\,\mathbf{i} + (t \sin t)\,\mathbf{j} \qquad \text{Value from Section 12.3, Example 5}$$

$$|\mathbf{v}| = \sqrt{t^2 \cos^2 t + t^2 \sin^2 t} = \sqrt{t^2} = |t| = t \qquad (t > 0)$$

$$a_\mathrm{T} = \frac{d}{dt}|\mathbf{v}| = \frac{d}{dt}(t) = 1 \qquad \text{Eq. (7)}$$

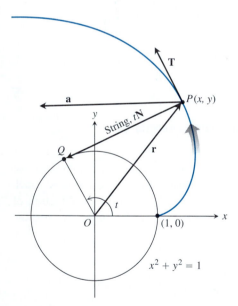

12.31 The tangential and normal components of the motion

$$\mathbf{r} = (\cos t + t \sin t)\,\mathbf{i} + (\sin t - t \cos t)\,\mathbf{j}$$

shown here are **T** and t **N** (Example 7). As you redraw the graphs in Exploration 1, you will see that, while a point travels around the circle at a constant speed, the corresponding point on the involute of the circle is visibly accelerating.

Knowing a_T, we use Eq. (9) to find a_N:

$$\mathbf{a} = (\cos t - t \sin t)\,\mathbf{i} + (\sin t + t \cos t)\,\mathbf{j}$$

$$|\mathbf{a}|^2 = t^2 + 1 \qquad \text{(after some algebra)}$$

$$a_\mathrm{N} = \sqrt{|\mathbf{a}|^2 - a_\mathrm{T}^2} = \sqrt{(t^2 + 1) - (1)} = \sqrt{t^2} = t.$$

We then use Eq. (8) to find **a**:

$$\mathbf{a} = a_{\mathrm{T}}\mathbf{T} + a_{\mathrm{N}}\mathbf{N} = (1)\,\mathbf{T} + (t)\,\mathbf{N} = \mathbf{T} + t\mathbf{N}.$$ ≣

EXPLORATION 1

Unwinding a Circle

Let

$$x_1(t) = \cos t, \qquad y_1(t) \qquad\quad = \sin t,$$

$$x_2(t) = \cos t + t\sin t, \qquad y_2(t) = \sin t - t\cos t,$$

and $\mathbf{r} = (x_2(t),\, y_2(t))$.

1. Compute $\mathbf{a} = \dfrac{d^2\mathbf{r}}{dt^2}$.

2. GRAPH the unit circle $(x_1(t),\, y_1(t))$, **r**, and **a** simultaneously in a square viewing window containing $[-4,\ 4]$ by $[-4,\ 4]$ for $0 \le t \le 5$.

3. Sketch the vectors in Fig. 12.31.   &

Motion Formulas ($|v| \ne 0$)

Unit tangent vector:	$\mathbf{T} = \dfrac{\mathbf{v}}{|\mathbf{v}|}$
Principal unit normal vector:	$\mathbf{N} = \dfrac{d\mathbf{T}/dt}{|d\mathbf{T}/dt|}$
Binormal vector:	$\mathbf{B} = \mathbf{T} \times \mathbf{N}$
Curvature:	$\kappa = \left|\dfrac{d\mathbf{T}}{ds}\right| = \dfrac{|\mathbf{v} \times \mathbf{a}|}{|\mathbf{v}|^3}$
Torsion:	$\tau = \left|\dfrac{d\mathbf{B}}{ds}\right| = \pm\dfrac{\begin{vmatrix} \dot{x} & \dot{y} & \dot{z} \\ \ddot{x} & \ddot{y} & \ddot{z} \\ \dddot{x} & \dddot{y} & \dddot{z} \end{vmatrix}}{|\mathbf{v} \times \mathbf{a}|^2}$
Tangential and normal scalar components of acceleration:	$\mathbf{a} = a_{\mathrm{T}}\mathbf{T} + a_{\mathrm{N}}\mathbf{N}$ $a_{\mathrm{T}} = \dfrac{d}{dt}|\mathbf{v}|, \qquad a_{\mathrm{N}} = \sqrt{|\mathbf{a}|^2 - a_{\mathrm{T}}^2}$

Exercises 12.4

Find **T**, **N**, and κ for the plane curves in Exercises 1–4.

1. $\mathbf{r} = t\mathbf{i} + (\ln\cos t)\,\mathbf{j}$ $(-\pi/2 < t < \pi/2)$

2. $\mathbf{r} = (\ln\sec t)\,\mathbf{i} + t\mathbf{j}$ $(-\pi/2 < t < \pi/2)$

3. $\mathbf{r} = (2t + 3)\,\mathbf{i} + (5 - t^2)\,\mathbf{j}$

4. $\mathbf{r} = (\cos t + t\sin t)\,\mathbf{i} + (\sin t - t\cos t)\,\mathbf{j}$ $(t > 0)$

Find **T**, **N**, **B**, κ, and τ for the space curves in Exercises 5–8.

5. $\mathbf{r} = (3\sin t)\,\mathbf{i} + (3\cos t)\,\mathbf{j} + 4t\mathbf{k}$

6. $\mathbf{r} = (\cos t + t\sin t)\,\mathbf{i} + (\sin t - t\cos t)\,\mathbf{j} + 3\mathbf{k}$

7. $\mathbf{r} = (e^t \cos t)\,\mathbf{i} + (e^t \sin t)\,\mathbf{j} + 2\mathbf{k}$

8. $\mathbf{r} = (6 \sin 2t)\,\mathbf{i} + (6 \cos 2t)\,\mathbf{j} + 5t\mathbf{k}$

In Exercises 9–12, write $\mathbf{a}$ in the form $\mathbf{a} = a_T\mathbf{T} + a_N\mathbf{N}$ without finding $\mathbf{T}$ and $\mathbf{N}$.

9. $\mathbf{r} = (2t + 3)\,\mathbf{i} + (t^2 - 1)\,\mathbf{j}$

10. $\mathbf{r} = \ln(t^2 + 1)\,\mathbf{i} + (t - 2\tan^{-1} t)\,\mathbf{j}$

11. $\mathbf{r} = (a \cos t)\,\mathbf{i} + (a \sin t)\,\mathbf{j} + bt\mathbf{k}$

12. $\mathbf{r} = (1 + 3t)\,\mathbf{i} + (t - 2)\,\mathbf{j} - 3t\mathbf{k}$

In Exercises 13–16, write $\mathbf{a}$ in the form $\mathbf{a} = a_T\,\mathbf{T} + a_N\,\mathbf{N}$ at the given value of t without finding $\mathbf{T}$ and $\mathbf{N}$. You can save yourself some work by evaluating the vectors $\mathbf{v}$ and $\mathbf{a}$ at the given value of t *before* finding their lengths.

13. $\mathbf{r} = (t + 1)\,\mathbf{i} + 2t\mathbf{j} + t^2\mathbf{k}$ $(t = 1)$

14. $\mathbf{r} = (t \cos t)\,\mathbf{i} + (t \sin t)\,\mathbf{j} + t^2\mathbf{k}$ $(t = 0)$

15. $\mathbf{r} = t^2\mathbf{i} + (t + (1/3)t^3)\,\mathbf{j} + (t - (1/3)t^3)\,\mathbf{k}$ $(t = 0)$

16. $\mathbf{r} = (e^t \cos t)\,\mathbf{i} + (e^t \sin t)\,\mathbf{j} + \sqrt{2}e^t\mathbf{k}$ $(t = 0)$

In Exercises 17 and 18, find $\mathbf{r}$, $\mathbf{T}$, $\mathbf{N}$, and $\mathbf{B}$ at the given value of t. Then find equations for the osculating, normal, and rectifying planes at that value of t.

17. $\mathbf{r} = (\cos t)\,\mathbf{i} + (\sin t)\,\mathbf{j} - \mathbf{k}$ $(t = \pi/4)$

18. $\mathbf{r} = (\cos t)\,\mathbf{i} + (\sin t)\,\mathbf{j} + t\mathbf{k}$ $(t = 0)$

19. Show that if a particle's speed is constant, its acceleration is always normal to its path.

20. *Formula for the curvature of the graph of a function in the xy-plane.*

a) The graph of a function $y = f(x)$ in the xy-plane automatically has the parametrization $x = t$, $y = f(t)$ and the vector formula $\mathbf{r} = x\mathbf{i} + y\mathbf{j}$. Use this formula to show that if f is a twice-differentiable function of x, then

$$\kappa = \frac{|y''|}{(1 + (y')^2)^{3/2}}.$$

b) Use the curvature formula in part (a) to find the curvature of the curve $y = \ln(\cos x)$, $-\pi/2 < x < \pi/2$.

c) GRAPH κ parametrically, and discuss its behavior in terms of the curvature of f.

21. Find an equation for the circle of curvature of the curve $\mathbf{r} = t\mathbf{i} + (\sin t)\,\mathbf{j}$ at the point $(\pi/2, 1)$. GRAPH the curve and the circle of curvature.

22. Find an equation for the circle of curvature of the curve $\mathbf{r} = (2 \ln t)\,\mathbf{i} - (t + (1/t))\,\mathbf{j}$ in the xy-plane at the point $(0, -2)$ (the point where $t = 1$). GRAPH the curve and the circle of curvature.

23. *How to derive the formula $\kappa = |\mathbf{v} \times \mathbf{a}|/|\mathbf{v}|^3$ for curves in space.* To derive the formula, carry out these two steps:

STEP 1: Use the equations $\mathbf{v} = \mathbf{T}(ds/dt)$ and $\mathbf{a} = (d^2s/dt^2)\mathbf{T} + \kappa(ds/dt)^2\mathbf{N}$ to find a formula for $|\mathbf{v} \times \mathbf{a}|$.

STEP 2: Solve the resulting equation for κ, assuming that $|\mathbf{v} \times \mathbf{a}| \neq 0$.

24. *Formula that calculates torsion directly from $\mathbf{B}$ and $\mathbf{v}$.* If we start with the definition $\tau = |d\mathbf{B}/ds|$ and apply the Chain Rule to rewrite $d\mathbf{B}/ds$ as

$$\frac{d\mathbf{B}}{ds} = \frac{d\mathbf{B}}{dt} \cdot \frac{dt}{ds} = \frac{d\mathbf{B}}{dt} \cdot \frac{1}{|\mathbf{v}|},$$

we arrive at the formula

$$\tau = \frac{|d\mathbf{B}/dt|}{|\mathbf{v}|}.$$

The advantage of this formula over the one in Eq. (6) is that it is easier to derive and state. The disadvantage is that it can take a lot of work to find $|d\mathbf{B}/dt|$. Use the new formula to find the torsion of the helix in Example 6.

25. *Maximizing the curvature of a helix.* In Example 4, the curvature of the helix $\mathbf{r} = (a \cos t)\,\mathbf{i} + (a \sin t)\,\mathbf{j} + bt\mathbf{k}$ $(a, b \geq 0,$ not both 0) is $\kappa = a/(a^2 + b^2)$. What is the largest value that κ can have for a given value of b?

12.5 Planetary Motion and Satellites

This section introduces the standard vector equations for motion in polar and cylindrical coordinates and discusses Kepler's three laws of planetary motion.

Vector Equations for Motion in Polar and Cylindrical Coordinates

When a particle moves along a curve in the polar coordinate plane, we express its position, velocity, and acceleration in terms of the *moving*

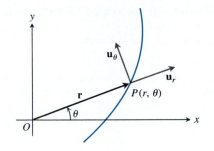

12.32 The length of the position vector **r** is the positive polar coordinate r of the point P. Thus, $\mathbf{u}_r$, which is $\mathbf{r}/|\mathbf{r}|$, is also $\mathbf{r}/r$. Equations (1) express $\mathbf{u}_r$ and $\mathbf{u}_\theta$ in terms of **i** and **j**.

unit vectors

$$\mathbf{u}_r = (\cos\theta)\,\mathbf{i} + (\sin\theta)\,\mathbf{j}, \qquad \mathbf{u}_\theta = -(\sin\theta)\,\mathbf{i} + (\cos\theta)\,\mathbf{j} \qquad (1)$$

shown in Fig. 12.32. The vector $\mathbf{u}_r$ points along the position vector $\overrightarrow{OP}$, so $\mathbf{r} = r\mathbf{u}_r$. The vector $\mathbf{u}_\theta$, orthogonal to $\mathbf{u}_r$, points in the direction of increasing θ.

We find from Eqs. (1) that

$$\frac{d\mathbf{u}_r}{d\theta} = -(\sin\theta)\,\mathbf{i} + (\cos\theta)\,\mathbf{j} = \mathbf{u}_\theta,$$

$$\frac{d\mathbf{u}_\theta}{d\theta} = -(\cos\theta)\,\mathbf{i} - (\sin\theta)\,\mathbf{j} = -\mathbf{u}_r.$$

When we differentiate $\mathbf{u}_r$ and $\mathbf{u}_\theta$ with respect to t to find how they change with time, the Chain Rule gives

$$\frac{d\mathbf{u}_r}{dt} = \frac{d\mathbf{u}_r}{d\theta}\frac{d\theta}{dt} = \frac{d\theta}{dt}\mathbf{u}_\theta, \qquad \frac{d\mathbf{u}_\theta}{dt} = \frac{d\mathbf{u}_\theta}{d\theta}\frac{d\theta}{dt} = -\frac{d\theta}{dt}\mathbf{u}_r. \qquad (2)$$

Hence,

$$\mathbf{v} = \frac{d\mathbf{r}}{dt} = \frac{d}{dt}(r\mathbf{u}_r) = \frac{dr}{dt}\mathbf{u}_r + r\frac{d\mathbf{u}_r}{dt} = \frac{dr}{dt}\mathbf{u}_r + r\frac{d\theta}{dt}\mathbf{u}_\theta. \qquad (3)$$

See Fig. 12.33.

The acceleration is

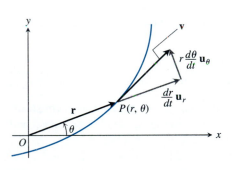

12.33 In polar coordinates, the velocity vector is

$$\mathbf{v} = \frac{dr}{dt}\,\mathbf{u}_r + r\frac{d\theta}{dt}\,\mathbf{u}_\theta.$$

$$\mathbf{a} = \frac{d\mathbf{v}}{dt} = \left(\frac{d^2r}{dt^2}\mathbf{u}_r + \frac{dr}{dt}\frac{d\mathbf{u}_r}{dt}\right) + \left(r\frac{d^2\theta}{dt^2}\mathbf{u}_\theta + \frac{dr}{dt}\frac{d\theta}{dt}\mathbf{u}_\theta + r\frac{d\theta}{dt}\frac{d\mathbf{u}_\theta}{dt}\right).$$

When Eqs. (2) are used to evaluate $d\mathbf{u}_r/dt$ and $d\mathbf{u}_\theta/dt$ and the components are separated, the equation for acceleration becomes

$$\mathbf{a} = \left(\frac{d^2r}{dt^2} - r\left(\frac{d\theta}{dt}\right)^2\right)\mathbf{u}_r + \left(r\frac{d^2\theta}{dt^2} + 2\frac{dr}{dt}\frac{d\theta}{dt}\right)\mathbf{u}_\theta.$$

To extend the equations of motion to space, we add $z\mathbf{k}$ to the right-hand side of the equation $\mathbf{r} = r\mathbf{u}_r$. Then, in cylindrical coordinates,

$$\mathbf{r} = r\mathbf{u}_r + z\mathbf{k},$$

$$\mathbf{v} = \frac{dr}{dt}\mathbf{u}_r + r\frac{d\theta}{dt}\mathbf{u}_\theta + \frac{dz}{dt}\mathbf{k},$$

$$\mathbf{a} = \left(\frac{d^2r}{dt^2} - r\left(\frac{d\theta}{dt}\right)^2\right)\mathbf{u}_r + \left(r\frac{d^2\theta}{dt^2} + 2\frac{dr}{dt}\frac{d\theta}{dt}\right)\mathbf{u}_\theta + \frac{d^2z}{dt^2}\mathbf{k}.$$

The vectors $\mathbf{u}_r$, $\mathbf{u}_\theta$, and $\mathbf{k}$ make a right-handed frame (Fig. 12.34) in which

$$\mathbf{u}_r \times \mathbf{u}_\theta = \mathbf{k}, \qquad \mathbf{k} \times \mathbf{u}_r = \mathbf{u}_\theta, \qquad \mathbf{u}_\theta \times \mathbf{k} = \mathbf{u}_r.$$

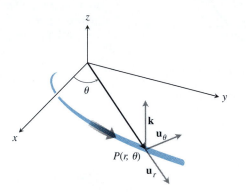

12.34   Motion in space with cylindrical coordinates.

Coordinates for Planetary Motion; Initial Conditions

Newton's Law of Gravitation says that if $\mathbf{r}$ is the radius vector from the center of a sun of mass M to the center of a planet of mass m, then the force $\mathbf{F}$ of the gravitational attraction between the planet and sun is given by the equation

$$\mathbf{F} = -\frac{GmM}{|\mathbf{r}|^2}\frac{\mathbf{r}}{|\mathbf{r}|} \qquad (4)$$

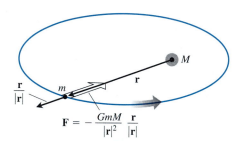

$$\mathbf{F} = -\frac{GmM}{|\mathbf{r}|^2}\frac{\mathbf{r}}{|\mathbf{r}|}$$

12.35   The force of gravity is directed along the line joining the centers of mass.

(Fig. 12.35). The constant G is called the (universal) **gravitational constant.** If we measure the masses m and M in kilograms, time in seconds, and distance in meters, then G is about 6.6720×10^{-11} Nm^{2}kg^{-2}.

Combining Eq. (4) with Newton's Second Law of Motion, $\mathbf{F} = m\,d^2\mathbf{r}/dt^2$, gives

$$m\frac{d^2\mathbf{r}}{dt^2} = -\frac{GmM}{|\mathbf{r}|^2}\frac{\mathbf{r}}{|\mathbf{r}|},$$

$$\frac{d^2\mathbf{r}}{dt^2} = -\frac{GM}{|\mathbf{r}|^3}\mathbf{r}. \qquad (5)$$

This is typical of a **central force**, one that points toward a fixed center at all times—in this case, the center of the sun.

Equation (5) says that $d^2\mathbf{r}/dt^2$ is a scalar multiple of $\mathbf{r}$ and hence tells us that

$$\mathbf{r} \times \frac{d^2\mathbf{r}}{dt^2} = \mathbf{0}. \qquad (6)$$

THE GRAVITATIONAL CONSTANT

The gravitational constant,

$$G \approx 6.6720 \times 10^{-11}\frac{Nm^2}{kg^2}.$$

When you use this value of G in a calculation, remember to express mass in kilograms, distance in meters, and time in seconds.

The left-hand side of this equation is the derivative of $\mathbf{r} \times d\mathbf{r}/dt$:

$$\frac{d}{dt}\left(\mathbf{r} \times \frac{d\mathbf{r}}{dt}\right) = \frac{d\mathbf{r}}{dt} \times \frac{d\mathbf{r}}{dt} + \mathbf{r} \times \frac{d^2\mathbf{r}}{dt^2} \qquad \text{\color{blue}Product Rule}$$

$$= \mathbf{0} + \mathbf{r} \times \frac{d^2\mathbf{r}}{dt^2} \qquad \text{\color{blue}$\mathbf{A} \times \mathbf{A} = \mathbf{0}$ for any vector.}$$

$$= \mathbf{r} \times \frac{d^2\mathbf{r}}{dt^2}.$$

Thus, Eq. (6) is equivalent to

$$\frac{d}{dt}\left(\mathbf{r} \times \frac{d\mathbf{r}}{dt}\right) = \mathbf{0},$$

which, in turn, says that

$$\mathbf{r} \times \frac{d\mathbf{r}}{dt} = \mathbf{C} \qquad\qquad (7)$$

for some constant vector $\mathbf{C}$.

Equation (7) tells us that $\mathbf{r}$ and $d\mathbf{r}/dt$ always lie in a plane perpendicular to $\mathbf{C}$. Hence, the planet moves in a fixed plane through the center of its sun.

We now adjust the coordinate system for space, if necessary, to place the origin at the sun's center of mass and make the plane of the planet's motion the polar coordinate plane. This makes $\mathbf{r}$ the planet's polar coordinate position vector. We also position the z-axis in a way that makes $\mathbf{k}$ the direction of $\mathbf{C}$ and makes the planet's motion counterclockwise when viewed from the positive z-axis. This makes θ increase with t, so $d\theta/dt$ is positive. Finally, we rotate the polar coordinate plane about the z-axis, if necessary, to make the initial ray coincide with the direction that $\mathbf{r}$ has when the planet is closest to the sun. This runs the initial ray through the **perihelion** position of the planet (Fig. 12.36).

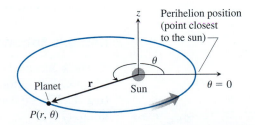

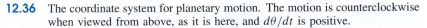

12.36 The coordinate system for planetary motion. The motion is counterclockwise when viewed from above, as it is here, and $d\theta/dt$ is positive.

If we now measure time so that $t = 0$ when the planet is at perihelion, we have the following initial conditions for the planet's motion:

1. $r = r_0$, the minimum radius, when $t = 0$.

2. $\dfrac{dr}{dt} = 0$ when $t = 0$ (because r has a minimum value then).

3. $\theta = 0$ when $t = 0$.

4. $|\mathbf{v}| = v_0$ when $t = 0$.

Since

$$v_0 = |\mathbf{v}|_{t=0} \qquad \text{Standard notation}$$

$$= \left| \frac{dr}{dt}\mathbf{u}_r + r\frac{d\theta}{dt}\mathbf{u}_\theta \right|_{t=0} \qquad \text{Eq. (3)}$$

$$= \left| r\frac{d\theta}{dt}\mathbf{u}_\theta \right|_{t=0} \qquad \frac{dr}{dt} = 0 \text{ when } t = 0.$$

$$= \left| r\frac{d\theta}{dt} \right| |\mathbf{u}_\theta|_{t=0}$$

$$= \left| r\frac{d\theta}{dt} \right|_{t=0} \qquad |\mathbf{u}_\theta| = 1.$$

$$= \left(r\frac{d\theta}{dt} \right)_{t=0}, \qquad r \text{ and } d\theta/dt \text{ are both positive.}$$

we also know that

5. $r\dfrac{d\theta}{dt} = v_0$ when $t = 0$.

The German astronomer, mathematician, and physicist Johannes Kepler (1571–1630) was the first, and until Descartes the only, scientist to demand physical (as opposed to theological) explanations of celestial phenomena. His three laws of motion, the results of a lifetime of work, changed the course of astronomy forever and played a crucial role in the development of Newton's physics.

Statement of Kepler's First Law (The Conic-section Law)

Kepler's First Law says that a planet's path is a conic section with the sun at one focus. The eccentricity of the conic is

$$e = \frac{r_0 v_0{}^2}{GM} - 1, \tag{8}$$

and the polar equation is

$$r = \frac{(1+e)r_0}{1 + e\cos\theta}. \tag{9}$$

The derivation is involved, and we will not give it here.

Kepler's Second Law (The Equal-area Law)

Kepler's Second Law says that the radius vector from a sun to a planet (the position vector **r** in our model) sweeps out equal areas in equal times (Fig. 12.37). To derive this law, we use Eq. (3) to evaluate the cross product in Eq. (7):

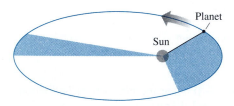

12.37 The line joining a planet to its sun sweeps over equal areas in equal times.

$$\mathbf{C} = \mathbf{r} \times \frac{d\mathbf{r}}{dt} \qquad \text{Eq. (7)}$$

$$= \mathbf{r} \times \mathbf{v} \qquad \mathbf{v} = d\mathbf{r}/dt.$$

$$= r\mathbf{u}_r \times \left(\frac{dr}{dt}\mathbf{u}_r + r\frac{d\theta}{dt}\mathbf{u}_\theta \right) \qquad \text{Eq. (3)}$$

$$= r\frac{dr}{dt}\underbrace{\mathbf{u}_r \times \mathbf{u}_r}_{\mathbf{0}} + r\left(r\frac{d\theta}{dt} \right)\underbrace{\mathbf{u}_r \times \mathbf{u}_\theta}_{\mathbf{k}}$$

$$= r\left(r\frac{d\theta}{dt} \right)\mathbf{k}. \tag{10}$$

Setting t equal to zero then shows that

$$\mathbf{C} = \left[r \left(r \frac{d\theta}{dt} \right) \right]_{t=0} \mathbf{k} = r_0 v_0 \, \mathbf{k}.$$

Substituting this value for $\mathbf{C}$ in Eq. (10) gives

$$r_0 v_0 \mathbf{k} = r^2 \frac{d\theta}{dt} \mathbf{k} \quad \text{or} \quad r^2 \frac{d\theta}{dt} = r_0 v_0.$$

Here is where the area comes in. From Section 10.8, the area differential in polar coordinates is

$$dA = \frac{1}{2} r^2 \, d\theta.$$

Accordingly, dA/dt has the constant value

$$\frac{dA}{dt} = \frac{1}{2} r^2 \frac{d\theta}{dt} = \frac{1}{2} r_0 v_0,$$

which is Kepler's Second Law.

For Earth, r_0 is about 150,000,000 km, v_0 is about 30 km/sec, and dA/dt is about 2,250,000,000 km^2/sec. Every time your heart beats, Earth advances 30 km along its orbit, and the line joining Earth to the sun sweeps out 2,250,000,000 square kilometers.

Statement of Kepler's Third Law (The Time–Distance Law)

The time T that it takes a planet to go around its sun once is the planet's **orbital period.** The semimajor axis a of the planet's orbit is the planet's **mean distance** from the sun. Kepler's Third Law says that T and a are related by the equation

$$\frac{T^2}{a^3} = \frac{4\pi^2}{GM}. \tag{11}$$

Since the right-hand side of this equation is constant within a given solar system, the ratio of T^2 to a^3 is the same for every planet in the system. The derivation of Eq. (11) can be found, along with the derivation of Kepler's First Law, in more advanced texts.

Kepler's Third Law is the starting point for working out the size of our solar system. It allows the semimajor axis of each planetary orbit to be expressed in astronomical units, Earth's semimajor axis being one unit. The distance between any two planets at any time can then be predicted in astronomical units, and all that remains is to find one of these distances in kilometers. This can be done by bouncing radar waves off Venus, for example. The astronomical unit is now known, after a series of such measurements, to be 149,597,870 km.

Astronomical Data

Although Kepler discovered his laws empirically and stated them only for the six planets that were known at the time, the modern derivations of Kepler's laws show that they apply to any body that is driven by a force that obeys an inverse square law. They apply to Halley's comet and the asteroid Icarus. They apply to the moon's orbit about Earth, and they applied to the orbits of the Apollo spacecraft about the moon. Charged particles fired at the nuclei of atoms scatter by an inverse square law force along hyperbolic paths.

Tables 12.1–12.3 give additional data for planetary orbits and for the orbits of seven of Earth's artificial satellites (Fig. 12.38). Vanguard 1 was nicknamed "Grapefruit." The data that Vanguard sent back revealed differences between the levels of Earth's oceans and provided the first determination of the precise locations of some of the more isolated Pacific islands. The data also verified that the gravitation of the sun and moon would affect the orbits of Earth's satellites and that solar light could exert enough pressure to deform an orbit.

Syncom 3 is one of a series of U.S. Department of Defense telecommunications satellites. Tiros 11 (for "television infrared observation satellite") is one of a series of weather satellites. GOES 4 (for "geostationary operational environmental satellite") is one of a series of satellites designed to gather information about Earth's atmosphere. Its orbital period, 1436.2 minutes, is nearly the same as Earth's rotational period of 1436.1 minutes, and its orbit is nearly circular ($e = 0.0003$). Intelsat 5 is a heavy-capacity commercial telecommunications satellite.

TABLE 12.1 Data on Earth and the Sun

The Sun's mass: 1.99×10^{30} kg
Earth's mass: 5.975×10^{24} kg
Equatorial radius of Earth: 6378.533 km
Polar radius of Earth: 6356.912 km
Earth's rotational period: 1436.1 min
Earth's orbital period: 1 year = 365.256 days

TABLE 12.2 Values of a, e, and T for the Major Planets

Planet	Semimajor Axis a^*	Eccentricity, e	Period, T
Mercury	57.95	0.2056	87.967 days
Venus	108.11	0.0068	224.701 days
Earth	149.57	0.0167	365.256 days
Mars	227.84	0.0934	1.8808 years
Jupiter	778.14	0.0484	11.8613 years
Saturn	1427.0	0.0543	29.4568 years
Uranus	2870.3	0.0460	84.0081 years
Neptune	4499.9	0.0082	164.784 years
Pluto	5909	0.2481	248.35 years

*Millions of kilometers

TABLE 12.3   Data on Earth's Satellites

Name	Launch Date	Time or Expected Time Aloft	Weight at Launch (kg)	Period (min)	Perigee Height (km)	Apogee Height (km)	Semimajor Axis a (km)	Eccentricity
Sputnik 1	Oct. 1957	57.6 days	83.6	96.2	215	939	6,955	0.052
Vanguard 1	Mar. 1958	300 years	1.47	138.5	649	4,340	8,872	0.208
Syncom 3	Aug. 1964	$> 10^6$ years	39	1436.2	35,718	35,903	42,189	0.002
Skylab 4	Nov. 1973	84.06 days	13,980	93.11	422	437	6,808	0.001
Tiros 11	Oct. 1978	500 years	734	102.12	850	866	7,236	0.001
GOES 4	Sept. 1980	$> 10^6$ years	627	1436.2	35,776	35,800	42,166	0.0003
Intelsat 5	Dec. 1980	$> 10^6$ years	1,928	1417.67	35,143	35,707	41,803	0.007

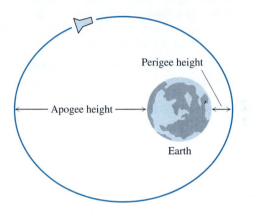

12.38   The orbit of an Earth satellite: $2a$ = diameter of Earth + perigee height + apogee height.

Circular Orbits

For circular orbits, e is zero, $r = r_0$ is a constant, and Eq. (8) gives

$$r = r_0 = \frac{GM}{{v_0}^2},$$

which reduces to

$$r = \frac{GM}{v^2}$$

because v is constant as well (Exercise 12). Kepler's Third Law becomes

$$\frac{T^2}{r^3} = \frac{4\pi^2}{GM}$$

because $a = r$.

Exercises 12.5

Reminder: When a calculation involves the gravitational constant G, express distance in meters, mass in kilograms, and time in seconds.

1. Since the orbit of Skylab 4 had a semimajor axis of $a = 6808$ km, Kepler's Third Law with M equal to Earth's mass should give the period. Calculate it. Compare your result with the value in Table 12.3.

2. Earth's distance from the Sun at perihelion is approximately 149,557,000 km, and the eccentricity of Earth's orbit about the sun is 0.0167. Compute the velocity v_0 of Earth in its orbit at perihelion. (Use Eq. 8).

3. In July 1965, the Soviet Union launched Proton I, weighing 12,200 kg (at launch), with a perigee height of 183 km, an apogee height of 589 km, and a period of 92.25 minutes. Using the relevant data for the mass of Earth and the gravitational constant G, compute the semimajor axis a of the orbit from Eq. (11). Compare your answer with the number that you get by adding the perigee and apogee heights to the diameter of Earth.

4. a) The Viking 1 orbiter, which surveyed Mars from August 1975 to June 1976, had a period of 1639 min. Use this and the fact that the mass of Mars is 6.418×10^{23} kg to find the semimajor axis of the Viking 1 orbit.

   b) The Viking 1 orbiter was 1499 km from the surface of Mars at its closest point and 35,800 km from the surface at its farthest point. Use this information together with the value that you obtained in part (a) to estimate the average diameter of Mars.

5. The Viking 2 orbiter, which surveyed Mars from September 1975 to August 1976, moved along an ellipse whose semimajor axis was 22,030 km. What was the orbital period? (Express your answer in minutes.)

6. If a satellite is to hold a geostationary orbit, what must the semimajor axis of its orbit be? Compare your result with the semimajor axes of the satellites in Table 12.3.

7. The mass of Mars is 6.418×10^{23} kg. If a satellite revolving about Mars is to hold a stationary orbit (have the same period as the period of Mars's rotation, which is 1477.4 min), what must the semimajor axis of its orbit be?

8. The period of the moon's rotation about Earth is 2.36055×10^6 sec. About how far away is the moon?

9. A satellite moves around Earth in a circular orbit. Express the satellite's speed as a function of the orbit's radius.

10. If T is measured in seconds and a in meters, what is the value of T^2/a^3 for planets in our solar system? For satellites orbiting Earth? For satellites orbiting the moon? (The moon's mass is 7.354×10^{22} kg.)

11. For what values of v_0 in Eq. (8) is the orbit in Eq. (9) a circle? An ellipse? A hyperbola?

12. Show that a planet in a circular orbit moves with a constant speed. (*Hint:* This is a consequence of one of Kepler's laws.)

13. *Practice with* $\mathbf{u}_r$ *and* $\mathbf{u}_\theta$. A ball is placed in a frictionless tube that is pivoted at one end and rotates with constant angular velocity $d\theta/dt = 2$. In polar coordinates, the position of the ball at time t is $\theta = 2t, r = \cosh\theta$. Show that the $\mathbf{u}_r$ component of acceleration is always zero.

Review Questions

1. What are the rules for differentiating and integrating vector functions? Give examples.

2. How do you find the velocity, speed, direction of motion, and acceleration of a body moving along a smooth curve in space? Give examples.

3. What is special about the derivatives of vectors of constant length? Give an example.

4. What are the vector and parametric equations for ideal projectile motion? How do you find a projectile's maximum height, flight time, and range? Give examples.

5. How do you define and calculate the length of a segment of a curve in space? Give an example. What mathematical assumptions are involved in the definition?

6. How do you measure distance along a curve in space from a preselected base point? Give an example of a directed-distance function.

7. What is a curve's unit tangent vector? Give an example.

8. How do you define the curvature of a curve in the plane? Give an example.

9. What is a plane curve's principal normal vector? Which way does it point? Give an example.

10. How do you define $\mathbf{N}$ and κ for curves in space? How are these quantities related? Give examples.

11. What is a curve's binormal vector? Give an example. What is the relation of this vector to the curve's torsion? Give an example.

12. What formulas are available for writing a moving body's acceleration as a sum of its tangential and normal components? Give an example. Why might one want to write the acceleration this way? What if the body moves at a constant speed? At a constant speed around a circle?

13. State Kepler's laws. To what do they apply?

Practice Exercises

GRAPH the curves in Exercises 1 and 2, and sketch their velocity and acceleration vectors at the given values of t.

1. $\mathbf{r} = (4\cos t)\mathbf{i} + (\sqrt{2}\sin t)\mathbf{j}$, $t = 0$ and $\pi/4$

2. $\mathbf{r} = (\sqrt{3}\sec t)\mathbf{i} + (\sqrt{3}\tan t)\mathbf{j}$, $t = 0$ and $\pi/6$

Evaluate the integrals in Exercises 3 and 4. Support with NINT computations.

3. $\displaystyle\int_0^1 [(3 + 6t)\mathbf{i} + (4 + 8t)\mathbf{j} + (6\pi\cos\pi t)\mathbf{k}]\,dt$

4. $\displaystyle\int_e^{e^2} \left[\frac{2\ln t}{t}\mathbf{i} + \frac{1}{t\ln t}\mathbf{j} + \frac{1}{t}\mathbf{k}\right]\,dt$

Solve the initial value problems in Exercises 5–8 analytically. Support each indefinite integral computation by using NINT on the three component scalar functions.

5. $\dfrac{d\mathbf{r}}{dt} = -(\sin t)\mathbf{i} + (\cos t)\mathbf{j} + \mathbf{k}$, $\mathbf{r} = \mathbf{j}$ when $t = 0$

6. $\dfrac{d\mathbf{r}}{dt} = \dfrac{1}{t^2 + 1}\mathbf{i} - \dfrac{1}{\sqrt{1 - t^2}}\mathbf{j} + \dfrac{t}{\sqrt{t^2 + 1}}\mathbf{k}$, $\mathbf{r} = \mathbf{j} + \mathbf{k}$
when $t = 0$

7. $\dfrac{d^2\mathbf{r}}{dt^2} = 2\mathbf{j}$, $\dfrac{d\mathbf{r}}{dt} = \mathbf{k}$ and $\mathbf{r} = \mathbf{i}$ when $t = 0$

8. $\dfrac{d^2\mathbf{r}}{dt^2} = -2\mathbf{i} - 4\mathbf{j}$, $\dfrac{d\mathbf{r}}{dt} = 4\mathbf{i}$ and $\mathbf{r} = 3\mathbf{i} + 3\mathbf{j}$ when $t = 1$

Find the lengths of the curves in Exercises 9 and 10.

9. $\mathbf{r} = (2\cos t)\mathbf{i} + (2\sin t)\mathbf{j} + t^2\mathbf{k}(0 \le t \le \pi/4)$

10. $\mathbf{r} = (3\cos t)\mathbf{i} + (3\sin t)\mathbf{j} + 2t^{3/2}\mathbf{k}$ $(0 \le t \le 3)$

11. Find $\mathbf{T}$, $\mathbf{N}$, $\mathbf{B}$, κ, and τ at $t = 0$ if
$\mathbf{r} = \dfrac{4}{9}(1 + t)^{3/2}\,\mathbf{i} +$
$\dfrac{4}{9}(1 - t)^{3/2}\,\mathbf{j} + \dfrac{1}{3}t\,\mathbf{k}.$

12. Find $\mathbf{T}$, $\mathbf{N}$, $\mathbf{B}$, κ, and τ as functions of t if
$\mathbf{r} = (\sin t)\,\mathbf{i} + (\sqrt{2}\cos t)\,\mathbf{j} + (\sin t)\,\mathbf{k}.$

In Exercises 13 and 14, write $\mathbf{a}$ in the form $\mathbf{a} = a_\mathrm{T}\mathbf{T} + a_\mathrm{N}\mathbf{N}$ at $t = 0$ without finding $\mathbf{T}$ and $\mathbf{N}$.

13. $\mathbf{r} = (2 + 3t + 3t^2)\mathbf{i} + (4t + 4t^2)\mathbf{j} - (6\cos t)\mathbf{k}$

14. $\mathbf{r} = (2 + t)\mathbf{i} + (t + 2t^2)\mathbf{j} + (1 + t^2)\mathbf{k}$

15. The position of a particle in the plane at time t is

$$\mathbf{r} = \frac{1}{\sqrt{1 + t^2}}\mathbf{i} + \frac{t}{\sqrt{1 + t^2}}\mathbf{j}.$$

Find the particle's greatest speed.

16. Suppose $\mathbf{r} = (e^t\cos t)\mathbf{i} + (e^t\sin t)\mathbf{j}$. Show that the angle between $\mathbf{r}$ and $\mathbf{a}$ never changes. What *is* the angle?

17. At what times in the interval $0 \le t \le \pi$ are the velocity and acceleration vectors orthogonal for the motion
$\mathbf{r} = \mathbf{i} + (5\cos t)\mathbf{j} + (3\sin t)\mathbf{k}$?

18. The position of a particle moving in space at time $t \ge 0$ is

$$\mathbf{r} = 2\mathbf{i} + \left(4\sin\frac{t}{2}\right)\mathbf{j} + \left(3 - \frac{t}{\pi}\right)\mathbf{k}.$$

Find the first time that $\mathbf{r}$ is perpendicular to the vector $\mathbf{i} - \mathbf{j}$.

19. *Shot-put.* A shot leaves the thrower's hand 8 ft above the ground at a $45°$ angle at 44 ft/sec. Where is it 3 sec later?

20. *Javelin.* A javelin leaves the thrower's hand 7 ft above the ground at a $45°$ angle at 80 ft/sec. How high does it go?

21. *The Dictator.* The Civil War mortar *Dictator* weighed so much (17,120 lb) that it had to be mounted on a railroad car. It had a 13-in. bore and fired a 200-lb shell with a 20-lb powder charge. It was made by Mr. Charles Knapp in his ironworks in Pittsburgh, Pennsylvania, and was used by the Union Army in 1864 in the siege of Petersburg, Virginia. How far did it actually shoot? Here we have a difference of opinion. The ordnance manual claimed 4325 yards, while field officers claimed 4752 yards. Assuming a $45°$ firing angle, what muzzle speeds are involved here?

22. *The world's record for popping a champagne cork.* The world's record for popping a champagne cork is 109 feet, 6 inches, held by Captain Michael Hill of the British Royal Artillery (of course). Assuming that Capt. Hill held the bottle neck at ground level at a $45°$ angle, how fast was the cork going as it left the bottle, assuming ideal projectile flight?

23. A baseball is hit when it is 4 feet above the ground. It leaves the bat with an initial velocity of 155 feet per second, making an angle of $18°$ with the ground. An 8-mi/h wind is blowing in the horizontal direction against the ball. A 10-foot-high fence is 380 feet from home plate in the direction of the flight.

a) Determine algebraic representations for the flight of the baseball in both vector function and parametric forms.

b) Simulate the motion of the baseball with a graphing utility. Show the fence in the viewing window.

c) Use the simulation to answer the following questions.

  i) How high does the baseball go, and when does it reach maximum height?

  ii) How far (ground distance) could the baseball travel from home plate in the air, and when would it hit the ground?

  iii) When is the baseball 25 feet high? How far (ground distance) is the baseball from home plate at that height? Give a complete solution.

  iv) Has the batter hit a home run?

24. Consider the baseball problem of Exercise 23 again. This time, assume a linear drag model with a drag coefficient of 0.09. Use a grapher simulation to answer the following questions.

a) How high does the baseball go, and when does it reach maximum height?

b) When is the baseball 30 feet high? How far (ground distance) is the baseball from home plate at that height? Give a complete solution.

c) How far (ground distance) could the baseball travel from home plate in the air, and when would it hit the ground?

d) Has the batter hit a home run? If "yes," find a drag coefficient that would have prevented a home run. If "no," find a drag coefficient that would have allowed it to be a home run. Give a complete solution.

25. Find parametric equations for the line that is tangent to the curve $\mathbf{r} = e^t\mathbf{i} + (\sin t)\mathbf{j} + \ln(1-t)\mathbf{k}$ at $t = 0$.

26. Find parametric equations for the line tangent to the helix $\mathbf{r} = (\sqrt{2}\cos t)\mathbf{i} + (\sqrt{2}\sin t)\mathbf{j} + t\mathbf{k}$ at the point where $t = \pi/4$.

27. At point P, the velocity and acceleration of a particle moving in the plane are $\mathbf{v} = 3\mathbf{i} + 4\mathbf{j}$ and $\mathbf{a} = 5\mathbf{i} + 15\mathbf{j}$. Find the curvature of the particle's path at P.

28. Find the normal plane of the curve $\mathbf{r} = t\mathbf{i} + t\mathbf{j} + (2/3)t^{3/2}\mathbf{k}$ at the point $(1, 1, 2/3)$.

29. *The torsion of a helix.* Find the torsion of the helix $\mathbf{r} = (a\cos t)\mathbf{i} + (a\sin t)\mathbf{j} + bt\mathbf{k}$ of Example 4, Section 12.4. For a given value of a, what is the largest value the torsion can have?

30. *A useful formula for the curvature of a plane curve.* Equation (1) in Section 12.4 says that the curvature of a plane curve $\mathbf{r} = f(t)\mathbf{i} + g(t)\mathbf{j}$ in which the coordinate functions $x = f(t)$, $y = g(t)$ are twice differentiable is

$$\kappa = \frac{|\dot{x}\ddot{y} - \dot{y}\ddot{x}|}{(\dot{x}^2 + \dot{y}^2)^{3/2}}.$$

Use this formula to find the curvature of curve $\mathbf{r} = \ln(\sec t)\mathbf{i} + t\mathbf{j}$ $(-\pi/2 < t < \pi/2)$.

31. *The view from Skylab 4.* What percentage of Earth's surface area could the astronauts see when Skylab 4 was at its apogee height, 437 km above the surface? To find out, first model the visible surface as the surface generated by revolving about the y-axis the circular arc GT. See the diagram.

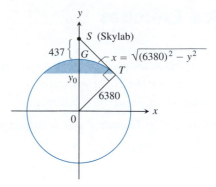

Then carry out these steps:

a) Use similar triangles to show that

$$y_0/6380 = 6380/(6380 + 437).$$

Solve for y_0.

b) Calculate the visible area as

$$VA = \int_{y_0}^{6380} 2\pi x \; ds.$$

c) Express the result as a percentage of the earth's surface area.

32. Use analytic calculus to deduce from the orbit equation

$$r = \frac{(1+e)r_0}{1 + e\cos\theta}$$

that a planet is closest to its sun when $\theta = 0$, and show that $r = r_0$ at that time.

33. *A Kepler equation.* The problem of locating a planet in its orbit at a given time and date eventually leads to solving "Kepler" equations of the form

$$f(x) = x - 1 - \frac{1}{2}\sin x = 0.$$

a) Show that this particular equation has a solution between $x = 0$ and $x = 2$.

b) Use Newton's method to find the solution to as many places as you can. Support graphically.

A.1 ——————— Formulas from Precalculus Mathematics

Algebra

1. Laws of Exponents

$$a^m a^n = a^{m+n}, \quad (ab)^m = a^m b^m, \quad (a^m)^n = a^{mn}, \quad a^{m/n} = \sqrt[n]{a^m}$$

If $a \neq 0$, $\qquad \dfrac{a^m}{a^n} = a^{m-n}, \quad a^0 = 1, \quad a^{-m} = \dfrac{1}{a^m}.$

2. Zero   Division by zero is not defined.

If $a \neq 0$: $\qquad \dfrac{0}{a} = 0, \quad a^0 = 1, \quad 0^a = 0$

For any number a: $\qquad a \cdot 0 = 0 \cdot a = 0$

3. Fractions

$$\frac{a}{b} + \frac{c}{d} = \frac{ad + bc}{bd}, \quad \frac{a}{b} \cdot \frac{c}{d} = \frac{ac}{bd}, \quad \frac{a/b}{c/d} = \frac{a}{b} \cdot \frac{d}{c}, \quad \frac{-a}{b} = -\frac{a}{b} = \frac{a}{-b},$$

$$\frac{(a/b) + (c/d)}{(e/f) + (g/h)} = \frac{(a/b) + (c/d)}{(e/f) + (g/h)} \cdot \frac{bdfh}{bdfh} = \frac{(ad + bc)fh}{(eh + fg)bd}$$

4. The Binomial Theorem
For any positive integer n,

$$(a + b)^n = a^n + na^{n-1}b + \frac{n(n-1)}{1 \cdot 2}a^{n-2}b^2$$

$$+ \frac{n(n-1)(n-2)}{1 \cdot 2 \cdot 3}a^{n-3}b^3 + \cdots + nab^{n-1} + b^n.$$

For instance, $(a + b)^1 = a + b$,

$$(a + b)^2 = a^2 + 2ab + b^2,$$

$$(a + b)^3 = a^3 + 3a^2b + 3ab^2 + b^3,$$

$$(a + b)^4 = a^4 + 4a^3b + 6^2b^2 + 4ab^3 + b^4.$$

5. Difference of Like Integers Powers, $n > 1$

$$a^n - b^n = (a - b)(a^{n-1} + a^{n-2}b + a^{n-3}b^2 + \cdots + ab^{n-2} + b^{n-1})$$

For instance, $a^2 - b^2 = (a - b)(a + b),$

$$a^3 - b^3 = (a - b)(a^2 + ab + b^2),$$

$$a^4 - b^4 = (a - b)(a^3 + a^2b + ab^2 + b^3).$$

6. Completing the Square
If $a \neq 0$, we can rewrite the quadratic $ax^2 + bx + c$ in the form $au^2 + C$ by a process called completing the square:

$$ax^2 + bx + c = a\left(x^2 + \frac{b}{a}x\right) + c \qquad \text{Factor } a \text{ from the first two terms.}$$

$$= a\left(x^2 + \frac{b}{a}x + \frac{b^2}{4a^2} - \frac{b^2}{4a^2}\right) + c \qquad \text{Add and subtract the square of half the coefficient of } x.$$

$$= a\left(x^2 + \frac{b}{a}x + \frac{b^2}{4a^2}\right) + a\left(-\frac{b^2}{4a^2}\right) + c \qquad \text{Bring out the } -b^2/4a^2.$$

$$= a\underbrace{\left(x^2 + \frac{b}{a}x + \frac{b^2}{4a^2}\right)}_{\text{This is } \left(x + \frac{b}{2a}\right)^2.} + \underbrace{c - \frac{b^2}{4a}}_{\text{Call this part } C.}$$

$$= au^2 + C \qquad u = x + b/2a$$

7. The Quadratic Formula
By completing the square on the first two terms of the equation

$$ax^2 + bx + c = 0$$

and solving the resulting equation for x (details omitted), we obtain the formula

$$x = \frac{-b \pm \sqrt{b^2 - 4ac}}{2a}.$$

This equation is called the **quadratic formula**.
The solutions of the equation $2x^2 + 3x - 1 = 0$ are

$$x = \frac{-3 \pm \sqrt{(3)^2 - 4(2)(-1)}}{2(2)} = \frac{-3 \pm \sqrt{9 + 8}}{4}$$

or $x = \dfrac{-3 + \sqrt{17}}{4}$ and $x = \dfrac{-3 - \sqrt{17}}{4}.$

The solutions of the equation $x^2 + 4x + 6 = 0$ are

$$x = \frac{-4 \pm \sqrt{(4)^2 - 4 \cdot 1 \cdot 6}}{2} = \frac{-4 \pm \sqrt{16 - 24}}{2}$$

$$= \frac{-4 \pm \sqrt{-8}}{2} = \frac{-4 \pm 2\sqrt{2}\sqrt{-1}}{2} = -2 \pm \sqrt{2}i.$$

The solutions are the complex numbers $-2 + \sqrt{2}i$ and $-2 - \sqrt{2}i$. Appendix A.6 has more on complex numbers.

Geometry

(A = area, B = area of base, C = circumference, S = lateral area or surface area, V = volume)

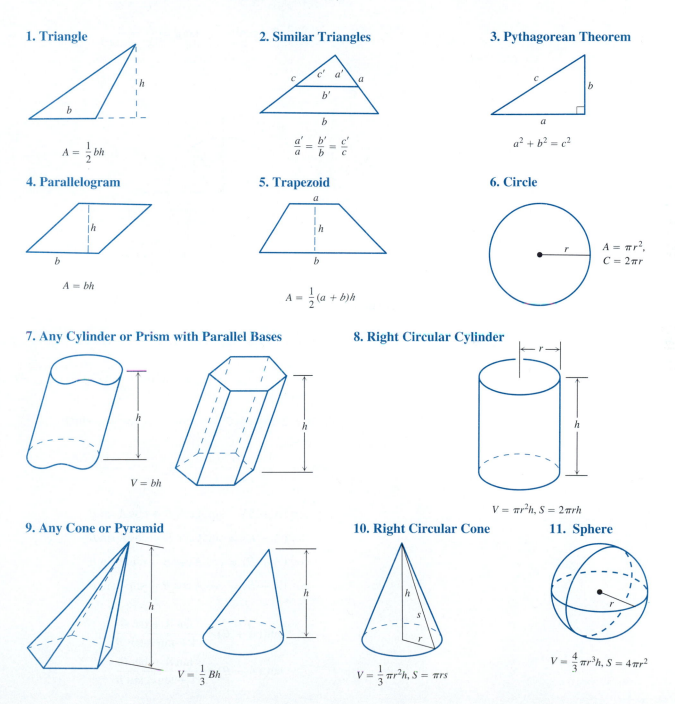

1. Triangle

$A = \dfrac{1}{2}bh$

2. Similar Triangles

$\dfrac{a'}{a} = \dfrac{b'}{b} = \dfrac{c'}{c}$

3. Pythagorean Theorem

$a^2 + b^2 = c^2$

4. Parallelogram

$A = bh$

5. Trapezoid

$A = \dfrac{1}{2}(a + b)h$

6. Circle

$A = \pi r^2,$
$C = 2\pi r$

7. Any Cylinder or Prism with Parallel Bases

$V = bh$

8. Right Circular Cylinder

$V = \pi r^2 h, S = 2\pi rh$

9. Any Cone or Pyramid

$V = \dfrac{1}{3}Bh$

10. Right Circular Cone

$V = \dfrac{1}{3}\pi r^2 h, S = \pi rs$

11. Sphere

$V = \dfrac{4}{3}\pi r^3 h, S = 4\pi r^2$

Trigonometry

1. Definitions and Fundamental Identities

$$\text{Sine:} \qquad \sin\theta = \frac{y}{r} = \frac{1}{\csc\theta}$$

$$\text{Cosine:} \qquad \cos\theta = \frac{x}{r} = \frac{1}{\sec\theta}$$

$$\text{Tangent:} \qquad \tan\theta = \frac{y}{x} = \frac{1}{\cot\theta}$$

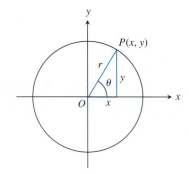

2. Identities

$$\sin(-\theta) = -\sin\theta, \qquad \cos(-\theta) = \cos\theta$$

$$\sin^2\theta + \cos^2\theta = 1, \qquad \sec^2\theta = 1 + \tan^2\theta, \qquad \csc^2\theta = 1 + \cot^2\theta$$

$$\sin 2\theta = 2\sin\theta\cos\theta, \qquad \cos 2\theta = \cos^2\theta - \sin^2\theta$$

$$\cos^2\theta = \frac{1 + \cos 2\theta}{2}, \qquad \sin^2\theta = \frac{1 - \cos 2\theta}{2}$$

$$\sin(A + B) = \sin A \cos B + \cos A \sin B$$
$$\sin(A - B) = \sin A \cos B - \cos A \sin B$$
$$\cos(A + B) = \cos A \cos B - \sin A \sin B$$
$$\cos(A - B) = \cos A \cos B + \sin A \sin B$$

$$\tan(A + B) = \frac{\tan A + \tan B}{1 - \tan A \tan B}$$

$$\tan(A - B) = \frac{\tan A - \tan B}{1 + \tan A \tan B}$$

$$\sin\left(A - \frac{\pi}{2}\right) = -\cos A, \qquad \cos\left(A - \frac{\pi}{2}\right) = \sin A$$

$$\sin\left(A + \frac{\pi}{2}\right) = \cos A, \qquad \cos\left(A + \frac{\pi}{2}\right) = -\sin A$$

$$\sin A \sin B = \frac{1}{2}\cos(A - B) - \frac{1}{2}\cos(A + B)$$

$$\cos A \cos B = \frac{1}{2}\cos(A - B) + \frac{1}{2}\cos(A + B)$$

$$\sin A \cos B = \frac{1}{2}\sin(A - B) + \frac{1}{2}\sin(A + B)$$

$$\sin A + \sin B = 2\sin\frac{1}{2}(A + B)\cos\frac{1}{2}(A - B)$$

$$\sin A - \sin B = 2\cos\frac{1}{2}(A + B)\sin\frac{1}{2}(A - B)$$

$$\cos A + \cos B = 2\cos\frac{1}{2}(A + B)\cos\frac{1}{2}(A - B)$$

$$\cos A - \cos B = -2\sin\frac{1}{2}(A + B)\sin\frac{1}{2}(A - B)$$

3. Common Reference Triangles

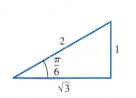

4. Angles and Sides of a Triangle

Law of cosines: $c^2 = a^2 + b^2 - 2ab\cos C$

Law of sines: $\dfrac{\sin A}{a} = \dfrac{\sin B}{b} = \dfrac{\sin C}{c}$

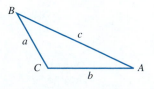

$$\text{Area} = \frac{1}{2}bc\sin A = \frac{1}{2}ac\sin B = \frac{1}{2}ab\sin C$$

A.2

Proofs of the Limit Theorems in Chapter 2

This appendix furnishes the ϵ–δ proofs of the limit theorems in Chapter 2.

THEOREM 1(C)

If L_1 and L_2 are real numbers and $\lim_{x \to x_0} f(x) = L_1$ and $\lim_{x \to x_0} g(x) = L_2$, then

1. **Sum Rule:** $\qquad\qquad\quad \lim (f((x) + g(x)) = L_1 + L_2$
2. **Difference Rule:** $\qquad\;\; \lim (f(x) - g(x)) = L_1 - L_2$
3. **Product Rule:** $\qquad\quad\;\; \lim f(x) \cdot g(x) = L_1 \cdot L_2$
4. **Constant Multiple Rule:** $\lim k \cdot f(x) = k \cdot L_1 \quad$ (any number k)
5. **Quotient Rule:** $\qquad\quad\;\; \lim \dfrac{f(x)}{g(x)} = \dfrac{L_1}{L_2}$ (provided that $L_2 \neq 0$.)

The limits are all to be taken as $x \to x_0$ in the appropriate domain (D).

We proved the Sum Rule in Section 2.6, and we obtain the Difference Rule by replacing $g(x)$ by $(-g(x))$ and L_2 by $(-L_2)$ in the Sum Rule. The Constant Multiple Rule is the special case $f(x) = k$ of the Product Rule. This leaves only the Product and Quotient Rules to prove.

Proof of the Limit Product Rule   We need to show that for any $\epsilon > 0$, there exists a $\delta > 0$ such that for all x in D,

$$0 < |x - x_0| < \delta \quad \Rightarrow \quad |f(x)g(x) - L_1L_2| < \epsilon. \tag{1}$$

Suppose then that ϵ is a positive number, and write $f(x)$ and $g(x)$ as

$$f(x) = L_1 + (f(x) - L_1) \qquad \text{and} \qquad g(x) = L_2 + (g(x) - L_2).$$

Multiply these expressions, and subtract L_1L_2:

$$\begin{aligned}
f(x) \cdot g(x) - L_1L_2 &= (L_1 + (f(x) - L_1))(L_2 + (g(x) - L_2)) - L_1L_2 \\
&= L_1L_2 + L_1(g(x) - L_2) + L_2(f(x) - L_1) \\
&\quad + (f(x) - L_1)(g(x) - L_2) - L_1L_2 \\
&= L_1(g(x) - L_2) + L_2(f(x) - L_1) + (f(x) - L_1) \\
&\qquad (g(x) - L_2). \tag{2}
\end{aligned}$$

Since f and g have limits L_1 and L_2 as $x \to x_0$, there exist positive numbers $\delta_1, \delta_2, \delta_3,$ and δ_4 such that for all x,

$$\begin{aligned}
0 < |x - x_0| < \delta_1 &\quad \Rightarrow \quad |f(x) - L_1| < \sqrt{\epsilon/3}, \\
0 < |x - x_0| < \delta_2 &\quad \Rightarrow \quad |g(x) - L_2| < \sqrt{\epsilon/3}, \\
0 < |x - x_0| < \delta_3 &\quad \Rightarrow \quad |f(x) - L_1| < \epsilon/(3(1 + |L_2|)), \\
0 < |x - x_0| < \delta_4 &\quad \Rightarrow \quad |g(x) - L_2| < \epsilon/(3(1 + |L_1|)). \tag{3}
\end{aligned}$$

All four of the inequalities on the right-hand side of (3) will hold for $0 < |x - x_0| < \delta$ if we take δ to be the smallest of the numbers δ_1 through δ_4. Therefore, for all x, $0 < |x - x_0| < \delta$ implies that

$$|f(x) \cdot g(x) - L_1 L_2|$$

$$\leq |L_1||g(x) - L_2| + |L_2||f(x) - L_1| + |f(x) - L_1||g(x) - L_2|$$

Triangle inequality applied to Eq.(2)

$$\leq (1 + |L_1|)|g(x) - L_2| + (1 + |L_2|)|f(x) - L_1| + |f(x) - L_1||g(x) - L_2|$$

$$\leq \frac{\epsilon}{3} + \frac{\epsilon}{3} + \sqrt{\frac{\epsilon}{3}}\sqrt{\frac{\epsilon}{3}} = \epsilon. \qquad \text{Values from (3)}$$

This completes the proof of the Limit Product Rule. ▮

Proof of the Limit Quotient Rule   We show that

$$\lim_{x \to x_0} \frac{1}{g(x)} = \frac{1}{L_2}.$$

Then we can apply the Limit Product Rule to show that

$$\lim_{x \to x_0} \frac{f(x)}{g(x)} = \lim_{x \to x_0} f(x) \cdot \frac{1}{g(x)} = \lim_{x \to x_0} f(x) \cdot \lim_{x \to x_0} \frac{1}{g(x)} = L_1 \cdot \frac{1}{L_2} = \frac{L_1}{L_2}.$$

To show that $\lim_{x \to x_0}(1/g(x)) = 1/L_2$, we need to show that for any $\epsilon > 0$, there exists a $\delta > 0$ such that for all x,

$$0 < |x - x_0| < \delta \quad \Rightarrow \quad \left|\frac{1}{g(x)} - \frac{1}{L_2}\right| < \epsilon.$$

Since $|L_2| > 0$, there exists a positive number δ_1 such that for all x,

$$0 < |x - x_0| < \delta_1 \quad \Rightarrow \quad |g(x) - L_2| < \frac{|L_2|}{2}. \tag{4}$$

For any numbers A and B, it can be shown that $|A| - |B| \leq |A - B|$ and $|B| - |A| \leq |A - B|$, from which it follows that

$$||A| - |B|| \leq |A - B|. \tag{5}$$

With $A = g(x)$ and $B = L_2$, this gives

$$||g(x)| - |L_2|| \leq |g(x) - L_2|,$$

which we can combine with the right-hand inequality in (4) to get, in turn,

$$||g(x)| - |L_2|| < \frac{|L_2|}{2},$$

$$-\frac{|L_2|}{2} < |g(x)| - |L_2| < \frac{|L_2|}{2},$$

$$\frac{|L_2|}{2} < |g(x)| < \frac{3|L_2|}{2},$$

$$|L_2| < 2|g(x)| < 3|L_2|,$$

$$\frac{1}{|g(x)|} < \frac{2}{|L_2|} < \frac{3}{|g(x)|}. \tag{6}$$

Therefore, $0 < |x - x_0| < \delta_1$ implies that

$$\left| \frac{1}{g(x)} - \frac{1}{L_2} \right| = \left| \frac{L_2 - g(x)}{L_2 g(x)} \right| \leq \frac{1}{|L_2|} \cdot \frac{1}{|g(x)|} \cdot |L_2 - g(x)|$$

$$< \frac{1}{|L_2|} \cdot \frac{2}{|L_2|} \cdot |L_2 - g(x)|. \qquad \text{Eq. 6}$$

Suppose now that ϵ is an arbitrary positive number. Then $\frac{1}{2}|L_2|^2 \epsilon > 0$, so there exists a number $\delta_2 > 0$ such that for all x

$$0 < |x - x_0| < \delta_2 \quad \Rightarrow \quad |L_2 - g(x)| < \frac{\epsilon}{2}|L_2|^2. \qquad (7)$$

The conclusions in (6) and (7) both hold for all x such that $0 < |x - x_0| < \delta$ if we take δ to be the smaller of the positive values δ_1 and δ_2. Combining (6) and (7) then gives

$$0 < |x - x_0| < \delta \quad \Rightarrow \quad \left| \frac{1}{g(x)} - \frac{1}{L_2} \right| < \epsilon. \qquad (8)$$

This concludes the proof of the Limit Quotient Rule.   ≡

THEOREM 2   The Sandwich Theorem

Suppose that $g(x) \leq f(x) \leq h(x)$ for all $x \neq x_0$ in some interval about x_0 and that $\lim_{x \to x_0} g(x) = \lim_{x \to x_0} h(x) = L$. Then $\lim_{x \to x_0} f(x) = L$.

Proof for Right-hand limits   Suppose $\lim_{x \to x_0^+} g(x) = \lim_{x \to x_0^+} h(x) = L$. Then for any $\epsilon > 0$, there exists a $\delta > 0$ such that for all x, the inequality $x_0 < x < x_0 + \delta$ implies that

$$L - \epsilon < g(x) < L + \epsilon \quad \text{and} \quad L - \epsilon < h(x) < L + \epsilon \qquad (9)$$

These inequalities combine with the inequality $g(x) \leq f(x) \leq h(x)$ to give

$$L - \epsilon < g(x) \leq f(x) \leq h(x) < L + \epsilon, \qquad (10)$$

$$L - \epsilon < f(x) < L + \epsilon,$$

$$-\epsilon < f(x) - L < \epsilon.$$

Therefore, for all x, the inequality $x_0 < x < x_0 + \delta$ implies that $|f(x) - L| < \epsilon$.   ≡

Proof for Left-hand Limits   Suppose that $\lim_{x \to x_0^-} g(x) = \lim_{x \to x_0^-} h(x) = L$. Then for any $\epsilon > 0$, there exists a $\delta > 0$ such that for all x, the inequality $x_0 - \delta < x < x_0$ implies that

$$L - \epsilon < g(x) < L + \epsilon \quad \text{and} \quad L - \epsilon < h(x) < L + \epsilon.$$

We conclude as before that for all x the inequality $x_0 - \delta < x < x_0$ implies $|f(x) - L| < \epsilon$.   ≡

Proof for Two-sided Limits   If $\lim_{x \to x_0} g(x) = \lim_{x \to x_0} h(x) = L$, then $g(x)$ and $h(x)$ both approach L as $x \to x_0^+$ and as $x \to x_0^-$; so $\lim_{x \to x_0^+} f(x) = L$, and $\lim_{x \to x_0^-} f(x) = L$. Hence, $\lim_{x \to x_0} f(x)$ exists and equals L.   ≡

Exercises A.2

1. Suppose that functions $f_1(x)$, $f_2(x)$, and $f_3(x)$ have limits L_1, L_2, and L_3, respectively, as $x \to x_0$. Show that their sum has limit $L_1 + L_2 + L_3$. Use mathematical induction (Appendix A.4) to generalize this result to the sum of any finite number of functions.

2. Use mathematical induction and the Limit Product Rule in Theorem 1 to show that if functions $f_1(x)$, $f_2(x)$, ..., $f_n(x)$ have limits L_1, L_2, ..., L_n as $x \to x_0$, then $\lim_{x \to x_0} f_1(x) f_2(x) \cdot \ldots \cdot f_n(x) = L_1 \cdot L_2 \cdot \ldots \cdot L_n$.

3. Use the fact that $\lim_{x \to x_0} x = x_0$ and the result of Exercise 2 to show that $\lim_{x \to x_0} x^n = x_0^n$ for any integer $n > 1$.

4. Use the fact that $\lim_{x \to x_0}(k) = k$ for any number k together with the results of Exercises 1 and 3 to show that $\lim_{x \to x_0} f(x) = f(x_0)$ for any polynomial function
$$f(x) = a_0 x^n + a_1 x^{n-1} + \cdots + a_{n-1} x + a_n.$$

5. Use Theorem 1 and the result of Exercise 4 to show that if $f(x)$ and $g(x)$ are polynomial functions and $g(x_0) \neq 0$, then
$$\lim_{x \to x_0} \frac{f(x)}{g(x)} = \frac{f(x_0)}{g(x_0)}.$$

6. Figure A.1 gives the diagram for a proof that the composite of two continuous functions is continuous. Reconstruct the proof from the diagram. The statement to be proved is this: If g is continuous at $x = x_0$ and f is continuous at $g(x_0)$, then $f \circ g$ is continuous at x_0.

Assume that x_0 is an interior point of the domain of g and that $g(x_0)$ is an interior point of the domain of f. This will make the limits involved two-sided. (The arguments for the cases that involve one-sided limits are similar.)

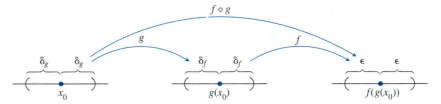

A.1   The diagram for a proof that the composite of two continuous functions is continuous. The continuity of composites holds for any finite number of functions. The only requirement is that each function be continuous where it is applied. In the figure, g is to be continuous at x_0, and f is to be continuous at $g(x_0)$.

A.3 _____

A Proof of the Chain Rule for Functions of a Single Variable

Our goal is to show that if $f(u)$ is a differentiable function of u and $u = g(x)$ is a differentiable function of x, then the composite $y = f(g(x))$ is a differentiable function of x. More precisely, if g is differentiable at x_0 and f is differentiable at $g(x_0)$, then the composite is differentiable at x_0, and

$$\left.\frac{dy}{dx}\right|_{x=x_0} = f'(g(x_0)) \cdot g'(x_0). \tag{1}$$

Suppose that Δx is an increment in x and that Δu and Δy are the corresponding increments in u and y. The following figure is drawn with the increments all positive, but Δx could be negative, and Δu and Δy could be negative or even zero.

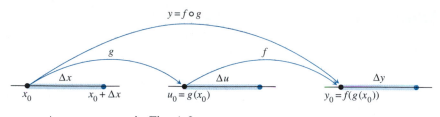

As you can see in Fig. A.2,

$$\frac{dy}{dx}\bigg|_{x=x_0} = \lim_{\Delta x \to 0} \frac{\Delta y}{\Delta x},$$

so our goal is to show that this limit is $f'(g(x_0)) \cdot g'(x_0)$.

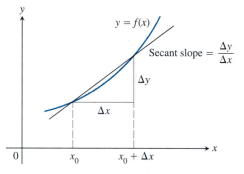

A.2 The graph of y as a function of x. The derivative of y with respect to x at $x = x_0$ is $\lim_{\Delta x \to 0} \Delta y / \Delta x$.

By virtue of Eq. (4) in Section 3.8,

$$\Delta u = g'(x_0)\Delta x + \epsilon_1 \Delta x = [g'(x_0) + \epsilon_1]\Delta x, \qquad (2)$$

where $\epsilon_1 \to 0$ as $\Delta x \to 0$. Similarly,

$$\Delta y = f'(u_0)\Delta u + \epsilon_2 \Delta u = [f'(u_0) + \epsilon_2]\Delta u, \qquad (3)$$

where $\epsilon_2 \to 0$ as $\Delta u \to 0$. Combining the equation for Δu with the one for Δy gives

$$\Delta y = [f'(u_0) + \epsilon_2][g'(x_0) + \epsilon_1]\Delta x, \qquad (4)$$

so

$$\frac{\Delta y}{\Delta x} = f'(u_0)g'(x_0) + \epsilon_2 g'(x_0) + f'(u_0)\epsilon_1 + \epsilon_2 \epsilon_1.$$

Since ϵ_1 and ϵ_2 go to zero as Δx goes to zero, the three terms on the right vanish in the limit, and

$$\lim_{\Delta x \to 0} \frac{\Delta y}{\Delta x} = f'(u_0)g'(x_0) = f'(g(x_0)) \cdot g'(x_0).$$

This concludes the proof.

A.4

Mathematical Induction

Many formulas like

$$1 + 2 + \cdots + n = \frac{n(n+1)}{2},$$

can be shown to hold for every positive integer n by applying an axiom called the *mathematical induction principle.* A proof that uses this axiom is called a *proof by mathematical induction* or a *proof by induction.*

The steps in proving a formula by induction are as follows.

STEP 1: Check that it holds for $n = 1$.

STEP 2: Prove that if it holds for any positive integer $n = k$, then it also holds for $n = k + 1$.

Once these steps are completed (the axiom says), we know that the formula holds for all positive integers n. By Step 1, it holds for $n = 1$. By Step 2 it holds for $n = 2$, and therefore by Step 2, it holds also for $n = 3$, for $n = 4$, and so on. If the first domino falls, and the kth domino always knocks over the $(k + 1)$st when it falls, all the dominos fall.

From another point of view, suppose we have a sequence of statements

$$S_1, S_2, \ldots, S_n, \ldots,$$

one for each positive integer. Suppose we can show that assuming any one of the statements to be true implies that the next statement in line is true. Suppose that we can also show that S_1 is true. Then we may conclude that the statements are true from S_1 on.

EXAMPLE 1

Show that for every positive integer n,

$$1 + 2 + \cdots + n = \frac{n(n + 1)}{2}.$$

Solution   We accomplish the proof by carrying out the two steps of mathematical induction.

STEP 1: The formula holds for $n = 1$ because

$$1 = \frac{1(1 + 1)}{2}.$$

STEP 2: If the formula holds for $n = k$, does it also hold for $n = k + 1$? The answer is yes, and here's why: If

$$1 + 2 + \cdots + k = \frac{k(k + 1)}{2},$$

then

$$1 + 2 + \cdots + k + (k + 1) = \frac{k(k + 1)}{2} + (k + 1) = \frac{k^2 + k + 2k + 2}{2}$$

$$= \frac{(k + 1)(k + 2)}{2} = \frac{(k + 1)((k + 1) + 1)}{2}$$

The last expression in this string of equalities is the expression $n(n+1)/2$ for $n = (k + 1)$.

The mathematical induction principle now guarantees the original formula for all positive integers n.

Notice that all *we* have to do is carry out Steps 1 and 2. The mathematical induction principle does the rest. ≡

EXAMPLE 2

Show that for all positive integers n,

$$\frac{1}{2^1} + \frac{1}{2^2} + \cdots + \frac{1}{2^n} = 1 - \frac{1}{2^n}.$$

Solution   We accomplish the proof by carrying out the two steps of mathematical induction.

STEP 1: The formula holds for $n = 1$ because

$$\frac{1}{2^1} = 1 - \frac{1}{2^1}.$$

STEP 2: If

$$\frac{1}{2^1} + \frac{1}{2^2} + \cdots + \frac{1}{2^k} = 1 - \frac{1}{2^k},$$

then

$$\frac{1}{2^1} + \frac{1}{2^2} + \cdots + \frac{1}{2^k} + \frac{1}{2^{k+1}} = 1 - \frac{1}{2^k} + \frac{1}{2^{k+1}} = 1 - \frac{1 \cdot 2}{2^k \cdot 2} + \frac{1}{2^{k+1}}$$

$$= 1 - \frac{2}{2^{k+1}} + \frac{1}{2^{k+1}} = 1 - \frac{1}{2^{k+1}}.$$

Thus, the original formula holds for $n = k + 1$ whenever it holds for $n = k$.

With these two steps verified, the mathematical induction principle now guarantees the formula for every positive integer n.   ≡

Other Starting Integers

Instead of starting at $n = 1$, some induction arguments start at another integer. The steps for such an argument are

STEP 1: Check that the formula holds for $n = n_1$ (whatever the appropriate first integer is).

STEP 2: Prove that if the formula holds for any integer $n = k \geq n_1$, then it also holds for $n = k + 1$.

Once these steps are completed, the mathematical induction principle will guarantee the formula for all $n \geq n_1$.

EXAMPLE 3

Show that $n! > 3^n$ if n is large enough.

Solution   How large is large enough? We experiment:

n	1	2	3	4	5	6	7
$n!$	1	2	6	24	120	720	5040
3^n	3	9	27	81	243	729	2187

It looks as if $n! > 3^n$ for $n \geq 7$. To be sure, we apply mathematical induction. We take $n_1 = 7$ in Step 1 and try for Step 2.

Suppose $k! > 3^k$ for some $k \geq 7$. Then

$$(k + 1)! = (k + 1)(k!) > (k + 1)3^k > 7 \cdot 3^k > 3^{k+1}.$$

Thus, for $k \geq 7$.

$$k! > 3^k \quad \Rightarrow \quad (k+1)! > 3^{k+1}.$$

The mathematical induction principle now guarantees that $n! \geq 3^n$ for all $n \geq 7$. ▬

Exercises A.4

1. Assuming that the triangle inequality $|a + b| \leq |a| + |b|$ holds for any two numbers a and b, show that

$$|x_1 + x_2 + \cdots + x_n| \leq |x_1| + |x_2| + \cdots + |x_n|$$

for any n numbers.

2. Show that if $r \neq 1$, then

$$1 + r + r^2 + \cdots + r^n = \frac{1 - r^{n+1}}{1 - r}$$

for all positive integers n.

3. Use the Product Rule

$$\frac{d}{dx}(uv) = u\frac{dv}{dx} + v\frac{du}{dx}$$

and the fact that

$$\frac{d}{dx}(x) = 1$$

to show that

$$\frac{d}{dx}(x^n) = nx^{n-1}$$

for all positive integers n.

4. Suppose that a function $f(x)$ has the property that $f(x_1 x_2) = f(x_1) + f(x_2)$ for any two positive numbers x_1 and x_2. Show that

$$f(x_1 x_2 \cdots x_n) = f(x_1) + f(x_2) + \cdots + f(x_n)$$

for the product of any n positive numbers $x_1, x_2 \ldots, x_n$.

5. Show that

$$\frac{2}{3^1} + \frac{2}{3^2} + \cdots + \frac{2}{3^n} = 1 - \frac{1}{3^n}$$

for all positive integers n.

6. Show that $n! > n^3$ if n is large enough.

7. Show that $2^n > n^2$ if n is large enough.

8. Show that $2^n \geq 1/8$ for $n \geq -3$.

9. Show that the sum of the squares of the first n positive integers is $n(n+1)(2n+1)/6$.

10. Show that the sum of the cubes of the first n positive integers is $(n(n+1)/2)^2$.

11. Show that the following finite-sum rules hold for every positive integer n.

   a) $\displaystyle\sum_{k=1}^{n}(a_k + b_k) = \sum_{k=1}^{n} a_k + \sum_{k=1}^{n} b_k$

   b) $\displaystyle\sum_{k=1}^{n}(a_k - b_k) = \sum_{k=1}^{n} a_k - \sum_{k=1}^{n} b_k$

   c) $\displaystyle\sum_{k=1}^{n} ca_k = c \cdot \sum_{k-1}^{n} a_k \qquad$ (any number c)

   d) $\displaystyle\sum_{k=1}^{n} a_k = n \cdot c \quad$ if a_k has the constant value c.

A.5 _____ Limits That Arise Frequently

This appendix verifies the limits in Table 9.1 of Section 9.1.

1. $\displaystyle\lim_{n \to \infty} \frac{\ln n}{n} = 0$

2. $\displaystyle\lim_{n \to \infty} \sqrt[n]{n} = 1$

3. $\displaystyle\lim_{n \to \infty} x^{1/n} = 1 \quad (x > 0)$

4. $\displaystyle\lim_{n \to \infty} x^n = 0 \quad (|x| < 1)$

5. $\displaystyle\lim_{n \to \infty} \left(1 + \frac{x}{n}\right)^n = e^x \quad$ (any x)

6. $\displaystyle\lim_{n \to \infty} \frac{x^n}{n!} = 0 \quad$ (any x)

In Eqs. (3)–(6) above, x remains fixed while $n \to \infty$.

1. $\displaystyle\lim_{n\to\infty} \frac{\ln n}{n} = 0$ We proved this in Section 9.1, Example 9.

2. $\displaystyle\lim_{n\to\infty} \sqrt[n]{n} = 1$ Let $a_n = n^{1/n}$. Then

$$\ln a_n = \ln n^{1/n} = \frac{1}{n} \ln n \to 0. \tag{1}$$

Applying Theorem 3, Section 9.1, with $f(x) = e^x$ gives

$$n^{1/n} = a_n = e^{\ln a_n} = f(\ln a_n) \to f(0) = e^0 = 1. \tag{2}$$

3. **If $x > 0$, $\displaystyle\lim_{n\to\infty} x^{1/n} = 1$**   Let $a_n = x^{1/n}$. Then

$$\ln a_n = \ln x^{1/n} = \frac{1}{n} \ln x \to 0 \tag{3}$$

because x remains fixed as $n \to \infty$. Applying Theorem 3, Section 9.1, with $f(x) = e^x$ gives

$$x^{1/n} = a_n = e^{\ln a_n} \to e^0 = 1. \tag{4}$$

4. **If $|x| < 1$, $\displaystyle\lim_{n\to\infty} x^n = 0$**   We need to show that to each $\epsilon > 0$, there corresponds an integer N so large that $|x^n| < \epsilon$ for all n greater than N. Since $\epsilon^{1/n} \to 1$ and $|x| < 1$, there exists an integer N for which

$$\epsilon^{1/N} > |x|. \tag{5}$$

In other words,

$$|x^N| = |x|^N < \epsilon. \tag{6}$$

This is the integer that we seek because, if $|x| < 1$, then

$$|x^n| < |x^N| \qquad \text{for all } n > N. \quad \text{By induction} \tag{7}$$

Combining (6) and (7) produces

$$|x^n| < \epsilon \qquad \text{for all } n > N, \tag{8}$$

and we're done.

5. **For any number x, $\displaystyle\lim_{n\to\infty} \left(1 \mathcal{C} \frac{x}{n}\right)^n = e^x$**   Let

$$a_n = \left(1 + \frac{x}{n}\right)^n.$$

Then $\qquad \ln a_n = \ln \left(1 + \frac{x}{n}\right)^n = n \ln \left(1 + \frac{x}{n}\right) \to x,$

as we can see by the following application of l'Hôpital's Rule, in which we differentiate with respect to n:

$$\lim_{n\to\infty} n \ln \left(1 + \frac{x}{n}\right) = \lim_{n\to\infty} \frac{\ln(1 + x/n)}{1/n}$$

$$= \lim_{n\to\infty} \frac{\left(\frac{1}{1+x/n}\right) \cdot \left(-\frac{x}{n^2}\right)}{-1/n^2} = \lim_{n\to\infty} \frac{x}{1 + x/n} = x.$$

Apply Theorem 3, Section 9.1, with $f(x) = e^x$ to conclude that

$$\left(1 + \frac{x}{n}\right)^n = a_n = e^{\ln a_n} \to e^x.$$

6. For any number x, $\lim\limits_{n\to\infty} \dfrac{x^n}{n!} = 0$ Since

$$-\frac{|x|^n}{n!} \le \frac{x^n}{n!} \le \frac{|x|^n}{n!},$$

all we need to show is that $|x|^n/n! \to 0$. We can then apply the Sandwich Theorem for Sequences (Section 9.1, Theorem 2) to conclude that $x^n/n! \to 0$.

The first step in showing that $|x|^x/n! \to 0$ is to choose an integer $M > |x|$ so that

$$\frac{|x|}{M} < 1 \qquad \text{and} \qquad \left(\frac{|x|}{M}\right)^n \to 0. \qquad \text{By part 4}$$

We then restrict our attention to values of $n > M$. For these values of n, we can write

$$\frac{|x|^n}{n!} = \frac{|x|^n}{1 \cdot 2 \cdots \cdot M \cdot \underbrace{(M+1)(M+2)\cdots \cdot n}_{(n-M) \text{ factors}}}$$

$$\le \frac{|x|^n}{M!\,M^{n-M}} = \frac{|x|^n M^M}{M!\,M^n} = \frac{M^M}{M!}\left(\frac{|x|}{M}\right)^n.$$

Thus, $$0 \le \frac{|x|^n}{n!} \le \frac{M^M}{M!}\left(\frac{|x|}{M}\right)^n.$$

The constant $M^M/M!$ does not change as n increases. Thus, the Sandwich Theorem tells us that

$$\frac{|x|^n}{n!} \to 0 \qquad \text{because} \qquad \left(\frac{|x|}{M}\right)^n \to 0.$$

A.6   Complex Numbers

Complex numbers are numbers of the form $a + bi$, where a and b are real numbers and $i = \sqrt{-1}$. The number a is the **real part** of $a + bi$, and b is the **imaginary part.**

Two complex numbers $a + bi$ and $c + di$ are equal if and only if $a = c$ and $b = d$. In particular, $a + bi = 0$ if and only if $a = 0$ and $b = 0$, or, equivalently, if and only if $a^2 + b^2 = 0$.

We add and subtract complex numbers by adding and subtracting their real and imaginary parts:

$$(a + bi) + (c + di) = (a + c) + (b + d)i,$$

$$(a + bi) - (c + di) = (a - c) + (b - d)i.$$

We multiply complex numbers the way we multiply other binomials, using the fact that $i^2 = -1$ to simplify the final result:

$$(a + bi)(c + di) = ac + adi + bci + bdi^2$$

$$= ac + adi + bci - bd \qquad i^2 = -1.$$

$$= (ac - bd) + (ad + bc)i. \qquad \text{Real and imaginary parts combined}$$

To divide a complex number $c + di$ by a nonzero complex number $a + bi$, multiply the numerator and denominator of the quotient by $a - bi$ (the number $a - bi$ is the **complex conjugate** of $a + bi$):

$$\frac{c + di}{a + bi} = \frac{c + di}{a + bi} \cdot \frac{a - bi}{a - bi}$$

Multiply the numerator and denominator by the complex conjugate of $a + bi$.

$$= \frac{ac - bci + adi - bdi^2}{a^2 - abi + abi - b^2 i^2}$$

$$= \frac{(ac + bd) + (ad - bc)i}{a^2 + b^2}$$

$$= \frac{ac + bd}{a^2 + b^2} + \frac{ad - bc}{a^2 + b^2} i.$$

EXAMPLE 1

a) $(2 + 3i) + (6 - 2i) = (2 + 6) + (3 - 2)i = 8 + i$

b) $(2 + 3i) - (6 - 2i) = (2 - 6) + (3 - (-2))i = -4 + 5i$

c) $(2 + 3i)(6 - 2i) = (2)(6) + (2)(-2i) + (3i)(6) + (3i)(-2i)$
$$= 12 - 4i + 18i - 6i^2 = 12 + 14i + 6 = 18 + 14i$$

d) $\dfrac{2 + 3i}{6 - 2i} = \dfrac{2 + 3i}{6 - 2i} \cdot \dfrac{6 + 2i}{6 + 2i}$

$$= \frac{12 + 4i + 18i + 6i^2}{36 + 12i - 12i - 4i^2}$$

$$= \frac{6 + 22i}{40} = \frac{3}{20} + \frac{11}{20}i$$ ≡

A.7 _____   Determinants and Cramer's Rule

A rectangular array of numbers like

$$A = \begin{bmatrix} 2 & 1 & 3 \\ 1 & 0 & -2 \end{bmatrix}$$

is called a **matrix.** We call A a 2 by 3 matrix because it has two rows and three columns. An m by n matrix has m rows and n columns, and the **entry** or **element** (number) in the ith row and jth column is often denoted by a_{ij}:

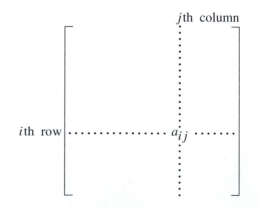

The matrix
$$A = \begin{bmatrix} 2 & 1 & 3 \\ 1 & 0 & -2 \end{bmatrix}$$

has
$$a_{11} = 2, \qquad a_{12} = 1, \qquad a_{13} = 3,$$
$$a_{21} = 1, \qquad a_{22} = 0, \qquad a_{23} = -2.$$

A matrix with the same number of rows as columns is a **square matrix.** It is a **matrix of order** n if the number of rows and columns is n.

With each square matrix A, we associate a number det A or $|a_{ij}|$, called the **determinant** of A, calculated from the entries of A in the following way. (The vertical bars in the notation $|a_{ij}|$ do not mean absolute value.) For $n = 1$ and $n = 2$, we define

$$\det [a] = a, \tag{1}$$

$$\det \begin{bmatrix} a_{11} & a_{12} \\ a_{21} & a_{22} \end{bmatrix} = a_{11}a_{22} - a_{21}a_{12}. \tag{2}$$

For a matrix of order 3, we define

$$\det A = \det \begin{bmatrix} a_{11} & a_{12} & a_{13} \\ a_{21} & a_{22} & a_{23} \\ a_{31} & a_{32} & a_{33} \end{bmatrix} = \begin{array}{l} \text{Sum of all signed products} \\ \text{of the form } \pm a_{1i} \cdot a_{2j} \cdot a_{3k}, \end{array} \tag{3}$$

where i, j, k is a permutation of 1, 2, 3 in some order. There are $3! = 6$ such permutations, so there are six terms in the sum. Half of these have plus signs and the other half have minus signs, according to the index of the permutation, where the index is the number that we define next. The sign is positive when the index is even and negative when the index is odd.

DEFINITION Index of a Permutation

Given any permutation of the numbers $1, 2, 3, \ldots, n$, denote the permutation by $i_1, i_2, i_3, \ldots, i_n$. In this arrangement, some of the numbers following i_1 may be less than i_1, and however many of these there are is called the **number of inversions** in the arrangement pertaining to i_1. Likewise, there are a number of inversions pertaining to each of the other i's; it is the number of indices that come after that particular one in the arrangement and are less than it. The **index** of the permutation is the sum of all of the numbers of inversions pertaining to the separate indices.

EXAMPLE 1

For $n = 5$, the permutation
$$5 \quad 3 \quad 1 \quad 2 \quad 4$$
has

four inversions pertaining to the first element, 5,

two inversions pertaining to the second element, 3,

and no further inversions, so the index is $4 + 2 = 6$.

The following table shows the permutations of 1,2,3, the index of each permutation, and the signed product in the determinant of Eq. (3).

Permutation	Index	Signed Product
1 2 3	0	$+a_{11}a_{22}a_{33}$
1 3 2	1	$-a_{11}a_{23}a_{32}$
2 1 3	1	$-a_{12}a_{21}a_{33}$
2 3 1	2	$+a_{12}a_{23}a_{31}$
3 1 2	2	$+a_{13}a_{21}a_{32}$
3 2 1	3	$-a_{13}a_{22}a_{31}$

(4)

The sum of the six signed products is

$$a_{11}(a_{22}a_{33} - a_{23}a_{32}) - a_{12}(a_{21}a_{33} - a_{23}a_{31}) + a_{13}(a_{21}a_{32} - a_{22}a_{31})$$

$$= a_{11}\begin{vmatrix} a_{22} & a_{23} \\ a_{32} & a_{33} \end{vmatrix} - a_{12}\begin{vmatrix} a_{21} & a_{23} \\ a_{31} & a_{33} \end{vmatrix} + a_{13}\begin{vmatrix} a_{21} & a_{22} \\ a_{31} & a_{32} \end{vmatrix} = \begin{vmatrix} a_{11} & a_{12} & a_{13} \\ a_{21} & a_{22} & a_{23} \\ a_{31} & a_{32} & a_{33} \end{vmatrix}. \quad (5)$$

The formula

$$\begin{vmatrix} a_{11} & a_{12} & a_{13} \\ a_{21} & a_{22} & a_{23} \\ a_{31} & a_{32} & a_{33} \end{vmatrix} = a_{11}\begin{vmatrix} a_{22} & a_{23} \\ a_{32} & a_{33} \end{vmatrix} - a_{12}\begin{vmatrix} a_{21} & a_{23} \\ a_{31} & a_{33} \end{vmatrix} + a_{13}\begin{vmatrix} a_{21} & a_{22} \\ a_{31} & a_{32} \end{vmatrix} \quad (6)$$

reduces the calculation of a 3 by 3 determinant to the calculation of three 2 by 2 determinants.

Many people prefer to remember the following scheme for calculating the six signed products in the determinant of a 3 by 3 matrix:

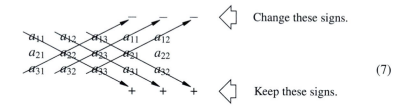

(7)

Minors and Cofactors

The second order determinants on the right-hand side of Eq. (6) are called the **minors** (short for minor determinant) of the entries that they multiply. Thus,

$$\begin{vmatrix} a_{22} & a_{23} \\ a_{32} & a_{33} \end{vmatrix} \quad \text{is the minor of } a_{11},$$

$$\begin{vmatrix} a_{21} & a_{23} \\ a_{31} & a_{33} \end{vmatrix} \quad \text{is the minor of } a_{12},$$

and so on. The minor of the element a_{ij} in a matrix A is the determinant of the matrix that remains after we delete the row and column containing a_{ij}:

$$\begin{vmatrix} a_{11} & a_{12} & a_{13} \\ a_{21} & a_{22} & a_{23} \\ a_{31} & a_{32} & a_{33} \end{vmatrix} \qquad \text{The minor of } a_{22} \text{ is} \qquad \begin{vmatrix} a_{11} & a_{13} \\ a_{31} & a_{33} \end{vmatrix}.$$

$$\begin{vmatrix} a_{11} & a_{12} & a_{13} \\ a_{21} & a_{22} & a_{23} \\ a_{31} & a_{32} & a_{33} \end{vmatrix} \qquad \text{The minor of } a_{23} \text{ is} \qquad \begin{vmatrix} a_{11} & a_{12} \\ a_{31} & a_{32} \end{vmatrix}.$$

The **cofactor** A_{ij} of a_{ij} is $(-1)^{i+j}$ times the minor of a_{ij}. Thus,

$$A_{22} = (-1)^{2+2} \begin{vmatrix} a_{11} & a_{13} \\ a_{31} & a_{33} \end{vmatrix} = \begin{vmatrix} a_{11} & a_{13} \\ a_{31} & a_{33} \end{vmatrix},$$

$$A_{23} = (-1)^{2+3} \begin{vmatrix} a_{11} & a_{12} \\ a_{31} & a_{32} \end{vmatrix} = - \begin{vmatrix} a_{11} & a_{12} \\ a_{31} & a_{32} \end{vmatrix}.$$

The effect of the factor $(-1)^{i+j}$ is to change the sign of the minor when the sum $i + j$ is odd. There is a checkerboard pattern for remembering these sign changes.

$$\begin{matrix} + & - & + \\ - & + & - \\ + & - & + \end{matrix}$$

In the upper left corner, $i = 1$, $j = 1$, and $(-1)^{1+1} = +1$. In going from any cell to an adjacent cell in the same row or column, we change i by 1 or j by 1, but not both, so we change the exponent from even to odd or from odd to even, which changes the sign from $+$ to $-$ or from $-$ to $+$.

When we rewrite Eq. (6) in terms of cofactors, we get

$$\det A = a_{11}A_{11} + a_{12}A_{12} + a_{13}A_{13}. \tag{8}$$

EXAMPLE 2

Find the determinant of the matrix

$$A = \begin{bmatrix} 2 & 1 & 3 \\ 3 & -1 & -2 \\ 2 & 3 & 1 \end{bmatrix}.$$

Solution 1   The cofactors are

$$A_{11} = (-1)^{1+1} \begin{vmatrix} -1 & -2 \\ 3 & 1 \end{vmatrix}, \qquad A_{12} = (-1)^{1+2} \begin{vmatrix} 3 & -2 \\ 2 & 1 \end{vmatrix},$$

$$A_{13} = (-1)^{1+3} \begin{vmatrix} 3 & -1 \\ 2 & 3 \end{vmatrix}.$$

To find $\det A$, we multiply each element of the first row of A by its cofactor and add:

$$\det A = 2 \begin{vmatrix} -1 & -2 \\ 3 & 1 \end{vmatrix} + (-1) \begin{vmatrix} 3 & -2 \\ 2 & 1 \end{vmatrix} + 3 \begin{vmatrix} 3 & -1 \\ 2 & 3 \end{vmatrix}$$

$$= 2(-1 + 6) - 1(3 + 4) + 3(9 + 2) = 10 - 7 + 33 = 36.$$

Solution 2   From (7), we find

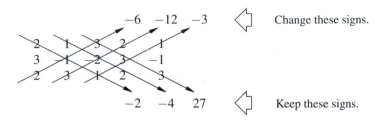

Change these signs.

Keep these signs.

$$\det A = -(-6) - (-12) - 3 + (-2) + (-4) + 27 = 36.$$ ≡

Expanding by Columns or by Other Rows

The determinant of a square matrix can be calculated from the cofactors of any row or any column.

If we were to expand the determinant in Example 2 by cofactors according to elements of its third column, say, we would get

$$+3 \begin{vmatrix} 3 & -1 \\ 2 & 3 \end{vmatrix} - (-2) \begin{vmatrix} 2 & 1 \\ 2 & 3 \end{vmatrix} + 1 \begin{vmatrix} 2 & 1 \\ 3 & -1 \end{vmatrix}$$

$$= 3(9 + 2) + 2(6 - 2) + 1(-2 - 3) = 33 + 8 - 5 = 36.$$

Useful Facts About Determinants

FACT 1: If two rows (or columns) of a matrix are identical, the determinant is zero.

FACT 2: Interchanging two rows (or columns) of a matrix changes the sign of its determinant.

FACT 3: The determinant of a matrix is the sum of the products of the elements of the ith row (or column) by their cofactors for any i.

FACT 4: The determinant of the **transpose** of a matrix is the same as the determinant of the original matrix. ("Transpose" means to write the rows as columns.)

FACT 5: Multiplying each element of some row (or column) of a matrix by a constant c multiplies the determinant by c.

FACT 6: If all elements of a matrix above the main diagonal (or all below it) are zero, the determinant of the matrix is the product of the elements on the main diagonal. (The **main diagonal** is the diagonal from upper left to lower right.)

EXAMPLE 3

$$\begin{vmatrix} 3 & 4 & 7 \\ 0 & -2 & 5 \\ 0 & 0 & 5 \end{vmatrix} = (3)(-2)(5) = -30.$$ ≡

FACT 7: If the elements of any row of a matrix are multiplied by the cofactors of the corresponding elements of a different row and these products are summed, the sum is zero.

EXAMPLE 4

If A_{11}, A_{12}, A_{13} are the cofactors of the elements of the first row of $A = (a_{ij})$, then the sums

$$a_{21}A_{11} + a_{22}A_{12} + a_{23}A_{13}$$

(elements of second row times cofactors of elements of first row) and

$$a_{31}A_{11} + a_{32}A_{12} + a_{33}A_{13}$$

are both zero.

FACT 8: If the elements of any column of a matrix are multiplied by the cofactors of the corresponding elements of a different column and these products are summed, the sum is zero.

FACT 9: If each element of a row of a matrix is multiplied by a constant c and the results added to a different row, the determinant is not changed. A similar result holds for columns.

EXAMPLE 5

If we start with

$$A = \begin{bmatrix} 2 & 1 & 3 \\ 3 & -1 & -2 \\ 2 & 3 & 1 \end{bmatrix}$$

and add -2 times row 1 to row 2 (subtract 2 times row 1 from row 2), we get

$$B = \begin{bmatrix} 2 & 1 & 3 \\ -1 & -3 & -8 \\ 2 & 3 & 1 \end{bmatrix}.$$

Since det $A = 36$ (Example 2), we should find that det $B = 36$ as well. Indeed we do, as the following calculation shows:

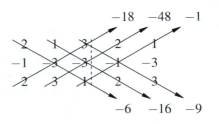

$$\det B = -(-18) - (-48) - (-1) + (-6) + (-16) + (-9)$$
$$= 18 + 48 + 1 - 6 - 16 - 9 = 67 - 31 = 36.$$

EXAMPLE 6

Evaluate the fourth order determinant

$$D = \begin{vmatrix} 1 & -2 & 3 & 1 \\ 2 & 1 & 0 & 2 \\ -1 & 2 & 1 & -2 \\ 0 & 1 & 2 & 1 \end{vmatrix}.$$

Solution   We subtract 2 times row 1 from row 2 and add row 1 to row 3 to get

$$D = \begin{vmatrix} 1 & -2 & 3 & 1 \\ 0 & 5 & -6 & 0 \\ 0 & 0 & 4 & -1 \\ 0 & 1 & 2 & 1 \end{vmatrix}.$$

We then mulitply the elements of the first column by their cofactors to get

$$D = \begin{vmatrix} 5 & -6 & 0 \\ 0 & 4 & -1 \\ 1 & 2 & 1 \end{vmatrix} = 5(4+2) - (-6)(0+1) + 0 = 36. \qquad \blacksquare$$

Cramer's Rule

If the determinant

$$D = \det A = \begin{vmatrix} a_{11} & a_{12} \\ a_{21} & a_{22} \end{vmatrix}$$

of the coefficient matrix of the system

$$
\begin{aligned}
a_{11}x + a_{12}y &= b_1, \\
a_{21}x + a_{22}y &= b_2
\end{aligned}
\tag{9}
$$

of linear equations is 0, the system has either infinitely many solutions or no solution at all. The system

$$
\begin{aligned}
x + y &= 0, \\
2x + 2y &= 0
\end{aligned}
$$

whose determinant is

$$D = \begin{vmatrix} 1 & 1 \\ 2 & 2 \end{vmatrix} = 2 - 2 = 0$$

has infinitely many solutions. We can find an x to match any given y. The system

$$
\begin{aligned}
x + y &= 0, \\
2x + 2y &= 2
\end{aligned}
$$

has no solution. If $x + y = 0$, then $2x + 2y = 2(x + y)$ cannot be 2.

If $D \neq 0$, the system (9) has a unique solution, and Cramer's Rule states that it may be found from the formulas

$$
x = \frac{\begin{vmatrix} b_1 & a_{12} \\ b_2 & a_{22} \end{vmatrix}}{D}, \qquad y = \frac{\begin{vmatrix} a_{11} & b_1 \\ a_{21} & b_2 \end{vmatrix}}{D}.
\tag{10}
$$

The numerator in the formula for x comes from replacing the first column in A (the x-column) by the column of constants b_1 and b_2 (the b-column). Replacing the y-column by the b-column gives the numerator of the y-solution.

EXAMPLE 7

Solve the system

$$
\begin{aligned}
3x - y &= 9, \\
x + 2y &= -4.
\end{aligned}
$$

Solution   We use Eqs. (10). The determinant of the coefficient matrix is

$$D = \begin{vmatrix} 3 & -1 \\ 1 & 2 \end{vmatrix} = 6 + 1 = 7.$$

Hence,

$$x = \frac{\begin{vmatrix} 9 & -1 \\ -4 & 2 \end{vmatrix}}{D} = \frac{18 - 4}{7} = \frac{14}{7} = 2,$$

$$y = \frac{\begin{vmatrix} 3 & 9 \\ 1 & -4 \end{vmatrix}}{D} = \frac{-12 - 9}{7} = \frac{-21}{7} = -3.$$

Systems of three equations in three unknowns work the same way. If the determinant

$$D = \det A = \begin{vmatrix} a_{11} & a_{12} & a_{13} \\ a_{21} & a_{22} & a_{23} \\ a_{31} & a_{32} & a_{33} \end{vmatrix}$$

of the system

$$a_{11}x + a_{12}y + a_{13}z = b_1,$$
$$a_{21}x + a_{22}y + a_{23}z = b_2, \tag{11}$$
$$a_{31}x + a_{32}y + a_{33}z = b_3$$

is zero, the system has either infinitely many solutions or no solution at all. If $D \neq 0$, the system has a unique solution, given by Cramer's Rule:

$$x = \frac{1}{D}\begin{vmatrix} b_1 & a_{12} & a_{13} \\ b_2 & a_{22} & a_{23} \\ b_3 & a_{32} & a_{33} \end{vmatrix}, \quad y = \frac{1}{D}\begin{vmatrix} a_{11} & b_1 & a_{13} \\ a_{21} & b_2 & a_{23} \\ a_{31} & b_3 & a_{33} \end{vmatrix}, \quad z = \frac{1}{D}\begin{vmatrix} a_{11} & a_{12} & b_1 \\ a_{21} & a_{22} & b_2 \\ a_{31} & a_{32} & b_3 \end{vmatrix}.$$

The pattern continues in higher dimensions.

Exercises A.7

Evaluate each determinant in Exercises 1–4.

1. $\begin{vmatrix} 2 & 3 & 1 \\ 4 & 5 & 2 \\ 1 & 2 & 3 \end{vmatrix}$

2. $\begin{vmatrix} 2 & -1 & -2 \\ -1 & 2 & 1 \\ 3 & 0 & -3 \end{vmatrix}$

3. $\begin{vmatrix} 1 & 2 & 3 & 4 \\ 0 & 1 & 2 & 3 \\ 0 & 0 & 2 & 1 \\ 0 & 0 & 3 & 2 \end{vmatrix}$

4. $\begin{vmatrix} 1 & -1 & 2 & 3 \\ 2 & 1 & 2 & 6 \\ 1 & 0 & 2 & 3 \\ -2 & 2 & 0 & -5 \end{vmatrix}$

Evaluate each determinant in Exercises 5–8 by expanding according to the cofactors of (a) the third row and (b) the second column.

5. $\begin{vmatrix} 2 & -1 & 2 \\ 1 & 0 & 3 \\ 0 & 2 & 1 \end{vmatrix}$

6. $\begin{vmatrix} 1 & 0 & -1 \\ 0 & 2 & -2 \\ 2 & 0 & 1 \end{vmatrix}$

7. $\begin{vmatrix} 1 & 1 & 0 & 0 \\ 0 & 0 & -2 & 1 \\ 0 & -1 & 0 & 7 \\ 3 & 0 & 2 & 1 \end{vmatrix}$

8. $\begin{vmatrix} 0 & 1 & 0 & 0 \\ 0 & 1 & 1 & 0 \\ 1 & 1 & 1 & 1 \\ 1 & 1 & 0 & 0 \end{vmatrix}$

Solve each system of equations in Exercises 9–16 by Cramer's Rule.

9. $x + 8y = 4$
   $3x - y = -13$

10. $2x + 3y = 5$
   $3x - y = 2$

11. $4x - 3y = 6$
   $3x - 2y = 5$

12. $x + y + z = 2$
   $2x - y + z = 0$
   $x + 2y - z = 4$

13. $2x + y - z = 2$
   $x - y + z = 7$
   $2x + 2y + z = 4$

14. $2x - 4y = 6$
   $x + y + z = 1$
   $5y + 7z = 10$

15. $x - z = 3$
   $2y - 2z = 2$
   $2x + z = 3$

16. $x_1 + x_2 - x_3 + x_4 = 2$
   $x_1 - x_2 + x_3 + x_4 = -1$
   $x_1 + x_2 + x_3 - x_4 = 2$
   $x_1 + x_3 + x_4 = -1$

17. Find values of h and k for which the system

$$2x + hy = 8,$$
$$x + 3y = k$$

has (a) infinitely many solutions, (b) no solution at all.

18. For what value of x will

$$\begin{vmatrix} x & x & 1 \\ 2 & 0 & 5 \\ 6 & 7 & 1 \end{vmatrix} = 0?$$

19. Suppose u, v, and w are twice-differentiable functions of x that satisfy the relation $au + bv + cw = 0$, where a, b, and c are constants, not all zero. Show that

$$\begin{vmatrix} u & v & w \\ u' & v' & w' \\ u'' & v'' & w'' \end{vmatrix} = 0.$$

20. *Partial fractions.* Expanding the quotient

$$\frac{ax+b}{(x-r_1)(x-r_2)}$$

by partial fractions calls for finding the values of C and D that make the equation

$$\frac{ax+b}{(x-r_1)(x-r_2)} = \frac{C}{x-r_1} + \frac{D}{x-r_2}$$

hold for all x.

a) Find a system of linear equations that determines C and D.

b) Under what circumstances does the system of equations in part (a) have a unique solution? That is, when is the determinant of the coefficient matrix of the system different from zero?

A.8

Lagrange Multipliers with Two Constraints

In many applied problems, we need to find the extreme values of a function $f(x, y, z)$ whose variables are subject to two constraints. This appendix continues the discussion in Section 13.9 to show how to find the extreme values with Lagrange multipliers if the functions involved have continuous first partial derivatives.

If the constraints on x, y, and z are

$$g_1(x, y, z) = 0 \qquad \text{and} \qquad g_2(x, y, z) = 0,$$

we find the constrained local maxima and minima of f by introducing two Lagrange multipliers λ and μ (mu, pronounced "mew"). That is, we locate the points $P(x, y, z)$ where f takes on its constrained extreme values by finding the values of x, y, z, λ, and μ that simultaneously satisfy the equations

$$\nabla f = \lambda \nabla g_1 + \mu \nabla g_2, \qquad g_1(x, y, z) = 0, \qquad g_2(x, y, z) = 0. \quad (1)$$

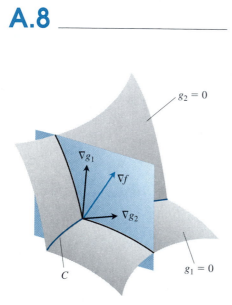

A.3 The vectors ∇g_1 and ∇g_2 lie in a plane perpendicular to the curve C because ∇g_1 is normal to the surface $g_1 = 0$ and ∇g_2 is normal to the surface $g_2 = 0a$.

The equations in (1) have a nice geometric interpretation. The surfaces $g_1 = 0$ and $g_2 = 0$ (usually) intersect in a differentiable curve, say C (Fig. A.3), and along this curve we seek the points where f has local maximum and minimum values relative to its other values on the curve. These are the points where ∇f is normal to C, as we saw in Section 13.9, Theorem 7. But ∇g_1 and ∇g_2 are also normal to C at these points because C lies in the surfaces $g_1 = 0$ and $g_2 = 0$. Therefore, ∇f lies in the plane determined by ∇g_1 and ∇g_2, which means that

$$\nabla f = \lambda \nabla g_1 + \mu \nabla g_2$$

for some λ and μ. Since the points that we seek also lie in both surfaces, their coordinates must satisfy the equations $g_1(x, y, z) = 0$ and $g_2(x, y, z) = 0$, which are the remaining requirements in Eqs. (1).

EXAMPLE 1

The plane $x + y + z = 1$ cuts the cylinder $x^2 + y^2 = 1$ in an ellipse (Fig. A.4). Find the points on the ellipse that lie closest to and farthest from the origin.

Solution   We model this as a Lagrange multiplier problem in which we find the extreme values of

$$f(x, y, z) = x^2 + y^2 + z^2$$

(the square of the distance from (x, y, z) to the origin) subject to the two constraints

$$g_1(x, y, z) = x^2 + y^2 - 1 = 0, \qquad (2)$$

$$g_2(x, y, z) = x + y + z - 1 = 0. \qquad (3)$$

The gradient equation in (1) then gives

$$\nabla f = \lambda \nabla g_1 + \mu \nabla g_2 \qquad \text{Eq. (1)}$$

$$2x\mathbf{i} + 2y\mathbf{j} + 2z\mathbf{k} = \lambda(2x\mathbf{i} + 2y\mathbf{j}) + \mu(\mathbf{i} + \mathbf{j} + \mathbf{k}) \qquad (4)$$

$$2x\mathbf{i} + 2y\mathbf{j} + 2z\mathbf{k} = (2\lambda x + \mu)\mathbf{i} + (2\lambda y + \mu)\mathbf{j} + \mu\mathbf{k} \qquad (5)$$

or

$$2x = 2\lambda x + \mu, \qquad 2y = 2\lambda y + \mu, \qquad 2z = \mu. \qquad (6)$$

The scalar equations in (6) yield

$$2x = 2\lambda x + 2z \quad \Rightarrow \quad (1 - \lambda)x = z,$$
$$2y = 2\lambda y + 2z \quad \Rightarrow \quad (1 - \lambda)y = z. \qquad (7)$$

Equations (7) are satisfied simultaneously either if $\lambda = 1$ and $z = 0$ or if $\lambda \neq 1$ and $x = y = z/(1 - \lambda)$.

If $z = 0$, then solving Eqs. (2) and (3) simultaneously to find the corresponding points on the ellipse gives (algebra omitted) the two points $(1, 0, 0)$ and $(0, 1, 0)$. This makes sense when you look at Fig. A.4.

If $x = y$, then Eqs. (2) and (3) give

$$\begin{array}{ll} x^2 + x^2 - 1 = 0 & x + x + z - 1 = 0, \\ 2x^2 = 1 & z = 1 - 2x, \\ x = \pm\dfrac{\sqrt{2}}{2} & z = 1 \mp \sqrt{2}. \end{array} \qquad (8)$$

The corresponding points on the ellipse are

$$\left(\frac{\sqrt{2}}{2}, \frac{\sqrt{2}}{2}, 1 - \sqrt{2}\right) \qquad \text{and} \qquad \left(-\frac{\sqrt{2}}{2}, -\frac{\sqrt{2}}{2}, 1 + \sqrt{2}\right). \qquad (9)$$

Again this makes sense when you look at Fig. A.4.

The points on the ellipse closest to the origin are $(1, 0, 0)$ and $(0, 1, 0)$. The points on the ellipse farthest from the origin are the two points displayed in (9). ≡

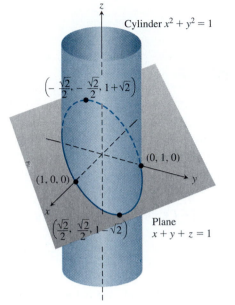

A.4   On the ellipse where the plane and cylinder meet, what are the points closest to and farthest from the origin (Example 1)?

Exercises A.8

1. Find the point closest to the origin on the line of intersection of the planes $y + 2z = 12$ and $x + y = 6$.

2. Find the maximum value that $f(x, y, z) = x^2 + 2y - z^2$ can have on the line of intersection of the planes $2x - y = 0$ and $y + z = 0$.

3. Find extreme values of $f(x, y, z) = x^2yz + 1$ on the intersection of the plane $z = 1$ with the sphere $x^2 + y^2 + z^2 = 10$.

4. a) Find the maximum value of $w = xyz$ on the line of intersection of the two planes $x + y + z = 40$ and $x + y - z = 0$.

b) Give a geometric argument to support your claim that you have found a maximum, and not a minimum, value of w.

5. Find the extreme values of the function $f(x, y, z) = xy + z^2$ on the circle in which the plane $y - x = 0$ intersects the sphere $x^2 + y^2 + z^2 = 4$.

A.9

Path Independence of $\int \mathbf{F} \cdot d\mathbf{r}$ Implies That $\mathbf{F} = \nabla f$

This appendix completes the proof of Theorem 1 in Section 15.7 by showing that, under the mathematical hypothesis discussed there, path independence of the integral $\int \mathbf{F} \cdot d\mathbf{r}$ throughout a domain D implies that $\mathbf{F}$ is the gradient of a scalar function f defined on D.

We need to show that if $\mathbf{F} = M\mathbf{i} + N\mathbf{j} + P\mathbf{k}$ there exists a scalar function f defined on D for which

$$\frac{\partial f}{\partial x} = M, \qquad \frac{\partial f}{\partial y} = N, \qquad \text{and} \qquad \frac{\partial f}{\partial z} = P. \tag{1}$$

To define f, we choose an arbitrary basepoint A in D. We define the value of f at A to be zero and the value of f at every other point B of D to be the value of the integral of $\mathbf{F} \cdot d\mathbf{r}$ along some smooth path in D from A to B. Such paths exist because D is a connected open region, and all such paths assign the same value to $f(B)$ because the integral of $\mathbf{F} \cdot d\mathbf{r}$ is path independent in D. Since the integral of $\mathbf{F} \cdot d\mathbf{r}$ from A to A is zero, the values of f are all defined by the single equation

$$f(B) = \int_A^B \mathbf{F} \cdot d\mathbf{r}. \tag{2}$$

One consequence of this definition is that for any points B and B_0 in D,

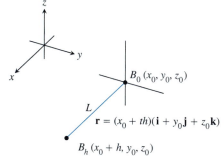

A.5 The line segment L from B_0 to B_h lies parallel to the x-axis. The partial derivative with respect to x of the function

$$f(x, y, z) = \int_A^{(x,y,z)} \mathbf{F} \cdot d\mathbf{r}$$

at B_0 is the limit of $(1/h) \int_L \mathbf{F} \cdot d\mathbf{r}$ as $h \to 0$.

$$f(B) - f(B_0) = \int_A^B \mathbf{F} \cdot d\mathbf{r} - \int_A^{B_0} \mathbf{F} \cdot d\mathbf{r} = \int_{B_0}^A \mathbf{F} \cdot d\mathbf{r} + \int_A^B \mathbf{F} \cdot d\mathbf{r} = \int_{B_0}^B \mathbf{F} \cdot d\mathbf{r}. \tag{3}$$

Let us now show that f has partial derivatives that give the components of $\mathbf{F}$ at every point $B_0(x_0, y_0, z_0)$ in D.

Since D is open, B_0 is the center of a solid sphere that lies entirely in D. We can therefore find a positive number h (it might be small) for which the point $B_h(x_0 + h, y_0, z_0)$ and the line segment L joining B_0 to B_h both lie in D (Fig. A.5). When we calculate the value of the integral of $\mathbf{F} \cdot d\mathbf{r}$ along L with the parametrization

$$\mathbf{r} = (x_0 + th)\mathbf{i} + y_0\mathbf{j} + z_0\mathbf{k} \quad (0 \le t \le 1), \tag{4}$$

we find that

$$\int_{B_0}^{B_h} \mathbf{F} \cdot d\mathbf{r} = \int_0^1 (M\mathbf{i} + N\mathbf{j} + P\mathbf{k}) \cdot (h \, dt\mathbf{i}) = h \int_0^1 M(x_0 + th, y_0, z_0) \, dt. \tag{5}$$

The difference quotient for defining $\partial f/\partial x$ at B_0 is therefore

$$\frac{f(B_h) - f(B_0)}{h} = \frac{f(x_0 + h, y_0, z_0) - f(x_0, y_0, z_0)}{h}$$

$$= \frac{1}{h} \int_{B_0}^{B_h} \mathbf{F} \cdot d\mathbf{r} \qquad \text{Eq. (3)}$$

$$= \int_0^1 M(x_0 + th, y_0, z_0) \, dt. \qquad \text{Eq. (5)}$$

Since M is continuous, given any $\epsilon > 0$ there exists a $\delta > 0$ such that

$$|th| < \delta \quad \Rightarrow \quad |M(x_0 + th, y_0, z_0) - M(x_0, y_0, z_0)| < \epsilon.$$

This implies that whenever $|h| < \delta$,

$$\left| \int_0^1 M(x_0 + th, y_0, z_0) \, dt - M(x_0, y_0, z_0) \right|$$

$$= \left| \int_0^1 (M(x_0 + th, y_0, z_0) - M(x_0, y_0, z_0)) \, dt \right|$$

$$\le \int_0^1 \left| M(x_0 + th, y_0, z_0) - M(x_0, y_0, z_0) \right| \, dt < \int_0^1 \epsilon \, dt = \epsilon.$$

We can find such a positive δ for every positive ϵ, so

$$\left. \frac{\partial f}{\partial x} \right|_{(x_0, y_0, z_0)} = \lim_{h \to 0} \int_0^1 M(x_0 + th, y_0, z_0) \, dt = M(x_0, y_0, z_0). \qquad (6)$$

In other words, the partial derivative of f with respect to x exists at B_0 and equals M at B_0. Since B_0 was chosen arbitrarily, we can conclude that $\partial f/\partial x$ exists and equals M at every point of D.

The equations $\partial f/\partial y = N$ and $\partial f/\partial z = P$ are derived in a similar way. This concludes the proof.

Glossary

Absolute Value: The absolute value of a number x, denoted by $|x|$, is defined by the formula

$$|x| = \begin{cases} x \text{ if } x \geq 0 \\ -x \text{ if } x < 0 \end{cases}$$

Acceleration: The derivative of velocity.

Alternating harmonic series:

$$\sum_{n=1}^{\infty} (-1)^{n+1} \frac{1}{n} = 1 - \frac{1}{2} + \frac{1}{3} + \frac{1}{4} + \ldots$$

Alternating series: A series in which the terms are alternately positive and negative.

Angle of inclination: Of a line crossing the x-axis, the smallest angle obtained measuring counterclockwise from the x-axis around the point of intersection.

Antiderivative: An antiderivative of function f is the function whose derivative is f.

Approximating sum: The sum $S = \sum_a^b f(c_k)\Delta x_k$ that approximates the integral $\int_a^b f(x)dx$.

Arc cosine of x: $y = \cos^{-1} x$.

Argument: See Independent variable.

Asymptote: See horizontal, vertical, and slant asymptotes.

Average daily inventory: The average number of items on hand each day.

Average rate of change: The amount of change in a quantity over a period of time divided by the length of the time period.

Average velocity: The distance a body moves in a straight line divided by the time it takes to move that distance.

Axis of a parabola: The axis of symmetry of the parabola.

Basic vectors: Vectors of unit length along the coordinate axes in terms of which other vectors in the plane may be expressed.

Big-oh notation: f is big-oh of g (written $f = 0(g)$) as $x \to \infty$ if there exists a positive integer M such that $\frac{f(x)}{g(x)} \leq M$ for x sufficiently large.

Binomial series: The series for the function $f(x) = (1 + x)^m$.

Cardioid: The graph of the curve $r = a(1 - \cos \theta)$.

Cartesian coordinate: The numerical coordinate assigned to a point in a plane that measures the distance of the point from either of two intersecting straight-line axes.

Center of a circle: The fixed point in the circle's definition.

Center of an ellipse: The point of intersection of the axes of the ellipse.

Center of gravity: Center of mass.

Center of a hyperbola: The point of intersection of the hyperbola's axes of symmetry.

Center of mass: In many cases, physical objects behave as if their mass were concentrated at a single point, called the center of mass. Article 5.8 shows how this point is located.

Centroid: Center of mass.

Circle: The set of points in a plane whose distance from a given fixed point in the plane is a constant.

Closed: Refers to intervals that contain both endpoints.

Common logarithms: Base 10 logarithms.

Complex number: A number consisting of real and imaginary components, of the form $a + bi$, where a and b are real numbers and i is defined as $\sqrt{-1}$.

Composite: The composite $g \circ f$ of two functions g and f is the function whose inputs are the inputs of f and whose outputs are the values of $g(f(x))$.

Concave down: Refers to the graph of a differentiable function $y = f(x)$ on an interval where y' decreases.

Concave up: Refers to the graph of a differentiable function $y = f(x)$ on an interval where y' increases.

Conic sections: Parabolas, ellipses, and hyperbolas.

Connected: If the graph of f over interval I consists of a single unbroken curve, it is said to be connected.

Constant of integration: The constant C in the formula $F(x) + C$ that gives all possible antiderivatives of the function $f = dF / dx$.

Continuous compound interest formula: A formula that gives the amount of interest continuously compounded in an account over time as a function of the original amount, the interest rate, and the time elapsed.

Continuous extension: A function that extends the domain of a function f to include more points where it is continuous.

Continuous function: A function $f(x)$ is continuous at each point c in its domain if $\lim_{x \to c} f(x) = f(c)$. A function is continuous if it is continuous at each point of its domain.

Converges: The sequence $\{a_n\}$ converges to the number L if to every positive ε there corresponds an integer N such that for all n, $n > N \Rightarrow |a_n - L| < \varepsilon$.

Coordinate Pair: The x- and y-coordinates (x,y) of a point in the Cartesian plane.

Critical Point: A point in the domain of a function for which the function's first derivative is zero or does not exist.

Definite integral: The Riemann integral.

Dependent variable: The output variable of a function. The variable y in the function $y = f(x)$.

Derivative: The function f' derived from f whose value at x is defined by the equation

$$f'(x) = \lim_{\Delta x \to 0} \frac{f(x + \Delta x) - f(x)}{\Delta x} \text{ whenever the limit exists.}$$

Differentiable at x: A function is differentiable at a point x if it has a derivative at x.

Differentiable curve: The curve traced by a body whose position vector $\mathbf{R}(t)$ is a differentiable function of t. Also, the graph of the differentiable function $y = f(x)$.

Differentiable function: A function that is differentiable at every point of its domain.

Differential: The differential df of a function $f(x)$ is the change in the linearization of f that results from a change dx in x.

Differential calculus: The branch of mathematics that deals with derivatives.

Differential equation: An equation that contains one or more derivatives.

Directrix of a parabola: The fixed line in a parabola.

Discontinuity: A point where a function fails to be continuous. For example, a point where a step function jumps from one value to another without taking on any of the intermediate values.

Diverges: The sequence $\{a_n\}$ diverges if it does not converge.

Domain: The set of first entries of a set of ordered pairs or a relation.

Dominant terms: Terms whose values give most of a function's output for selected portions of the function's domain. For the function $y = x + 1 / x$, the term $1 / x$ is dominant for values of x close to zero.

Eccentricity: The number e in the equation $PF = e \cdot PD$ that defines a conic section. In an ellipse e indicates the degree of departure from circularity.

Ellipse: The set of points in a plane whose distances from two fixed points in the plane have a constant sum.

End behavior asymptote: Let $f(x) = p(x) / h(x)$ be a rational function, and $q(x)$ and $r(x)$ be the quotient and remainder when $p(x)$ is divided by $h(x)$. That is

$$f(x) = q(x) + \frac{r(x)}{h(x)} \text{ with deg } r < \text{deg } h$$

Then the graph of q is called the end behavior of asymptote of f.

End behavior model: The function g is an end behavior model of f if:

1. $\lim\limits_{x \to \pm\infty} f / g = 1$ when $g(x) \neq 0$ for $|x|$ large, or

2. $\lim\limits_{x \to \pm\infty} f(x) = 0$ when $g(x) = 0$.

Euler's formula: For any real number θ, $e^{i\theta} = \cos \theta + i \sin \theta$.

Equation for a line: An equation that is satisfied only by the coordinates of the points that lie on the line.

Error function: The function $\text{erf}(x) = \dfrac{2}{\sqrt{\pi}} \displaystyle\int_o^x e^{-t^2} \, dt$.

Even function: A function in which $f(-x) = f(x)$ for every x in the function's domain.

Exponential function with base e and exponent x: $y = e^x$.

Extrinsic text for convergence: Test that compares one series with another series or with an integral.

Focal axis: The axis of symmetry that passes through the focus of a parabola or the foci of an ellipse or hyperbola.

Focus (pl: foci): The fixed point (or points) in the definition of a parabola, ellipse, or hyperbola.

Formal power series: The result of connecting the terms of a general sequence of constants times powers of x with plus signs to yield

$$\sum_{n=0}^{\infty} a_n x^n = a_o + a_1 x + a_2 x^2 + a_3 x^3 + \cdots + a_n x^n + \cdots$$

Function: Any rule in mathematics that assigns to each element in one set a unique element from another set. The sets may be sets of numbers, sets of number pairs, sets of points, or sets of objects of any kind.

General linear equation: The equation $Ax + By = C$, in which A, B, and C are constants and A and B are not both zero. Its graph is always a line.

General solution: A formula that gives all the solutions of a differential equation.

Geometric series: A series of the form

$$a + ar + ar^2 + ar^3 + \cdots = ar^{n+1} + \cdots$$

Grade: The slope of a roadbed.

Graph: The points in a plane whose coordinate pairs are the input-output pairs of a function.

Half-life: In a radioactive element, the time required for half of the radioactive nuclei present in a sample to decay.

Half-open: Refers to intervals that contain one endpoint but not both.

Harmonic series:

$$\sum_{n=1}^{\infty} \frac{1}{n} = 1 + \frac{1}{2} + \frac{1}{3} + \cdots + \frac{1}{n} + \cdots$$

Heaviside "cover-up method": A shortcut used to integrate rational functions of x whose denominators have been factored.

Homogeneous equation: Any differential equation of the form

$$\frac{dy}{dx} = F\left(\frac{y}{x}\right)$$

Homogeneous material: Material of constant density.

Hooke's Law: A law stating that the force F required to stretch or compress a spring from its natural length can be approximated by the equation $F = kx$, where x is the amount the spring has displaced from its unstressed length, and k is a constant characteristic of the spring.

Horizontal asymptote: A line $y = b$ is a horizontal asymptote of the graph of a function $y = f(x)$ if either the $\lim_{x \to \infty} f(x) = b$ or $\lim_{x \to -\infty} f(x) = b$.

Horizontal line: A line parallel to the x-axis.

Hydrostatic force: The total force exerted by a standing liquid against a containing wall.

Hyperbola: The set of points in a plane whose distances from two fixed points in the plane have a constant difference.

Identity function: The function that assigns each number to itself.

Implicit differentiation: The differentiation of implicitly defined functions.

Implicitly: When an equation $F(x,y) = 0$ defines f on (a,b) even though it does not define f by giving y explicitly in terms of x.

Improper integral: The integral of a function over a non-closed interval.

Increment: The net change in a variable.

Indefinite integral: The set of all antiderivatives of a function.

Independent variable: The variable x in a function $y = f(x)$. The input variable of a function.

Indeterminate forms: Expressions of the form $0/0$, $\infty \cdot 0$, ∞/∞, and $\infty - \infty$, etc. that are used to describe limits even though the expressions themselves are not numbers.

Infinite intervals: Intervals of the form (a,∞), $[a,\infty)$, $(-\infty,b)$, $(-\infty,b]$, and $(-\infty, \infty)$.

Infinite series: Given a sequence of numbers $\{a_n\}$, an expression of the form $a_1 + a_2 + a_3 + \cdots + a_n + \cdots$

Infinity: A symbol, ∞, used to describe the limiting behavior of functions whose values grow beyond all numerical bounds.

Initial condition: The condition that $y = y_0$ when $x = x_0$.

Initial value problem: The problem of finding a function y of x given its derivative $\frac{dy}{dx} = f(x)$ and its value y_o at a particular x_o.

Instantaneous velocity: The derivative of position or distance.

Integral sign: The symbol $\int$.

Integrand: In an integral, the function being integrated.

Integrating factor: The factor $\rho = e^{\int P\,dx}$ used to integrate a first order differential equation.

Intrinsic convergence test: Test that depends only on the series under consideration.

Inverse of f: The function defined by reversing a one-to-one function.

Inverse relation: The relation R^{-1} whose ordered pairs (b,a) are such that the ordered pairs (a,b) belong to the relation R.

Isocline: An isocline of the differential equation $y' = f(x,y)$ is a curve with equation $f(x,y) = $ constant.

Law of exponential change: The equation $y = y_0 e^{kt}$. Used when a quantity y grows or declines at a rate that at any time t is proportional to the amount that is present.

Law of refraction: See Snell's law.

Least upper bound: The smallest number that is an upper bound for an increasing sequence.

Left-hand derivative: The derivative defined by a left-hand limit.

Left-hand limit: The limit of function $f(t)$ as t approaches a number c from the left, denoted by $\lim_{t \to c^-} f(t)$.

Leibniz's theorem: The series

$$\sum_{n=1}^{\infty} (-1)^{n+1} a_n = a_1 - a_2 + a_3 - a_4 + \cdots$$

converges if the a's are all positive, $a \geq a_{n+1}$ for all n, and $a_n \to 0$

L'Hôpital's rule: A rule for calculating limits of fractions whose numerators and denominators both approach zero.

Limit: The number L as t approaches c of function $f(t)$.

Limits of integration: The numbers a and b of an integral $\int_a^b f(x)dx$.

Linearization: The linearization L of f at a is $L(x) = f(a) + f'(a)(x - a)$, if $y = f(x)$ is differentiable at $x = a$.

Little-oh notation: f is little-oh of g (written $f = o(g(x))$ as $x \to \infty$ if $\lim_{x \to \infty} \dfrac{f(x)}{g(x)} = 0$.)

Local maximum value of f: A function $f(x)$ defined at $x = c$ has a local maximum value at c if $f(x) \geq f(c)$ for all x in some interval about c.

Local minimum value of f: A function $f(x)$ defined at $x = c$ has a local maximum value at c if $f(x) \leq f(c)$ for all x in some interval about c.

Logarithmic differentiation: Taking the logarithm of both sides of an equation before differentiating.

Logarithm of x to the base a: The inverse of the function $y = a^x$. It exists when $a > 0$ and $a \neq 1$.

Maclaurin series generated by f: A Taylor series generated by f at $a = 0$.

Magnitude: Absolute value.

Magnitude of a vector: Its length.

Major axis: The longer of the two axes of an ellipse.

Method of undetermined coefficients: The method of first guessing the form of the solution up to certain predetermined constants and then determining the values of these constants by using the differential equation.

Minor axis: The shorter of the two axes of an ellipse.

Moment about the origin: For a system of masses, m_i, located at the points x_i on the x-axis, the number $M_0 = \Sigma\, m_i x_i$. For a slender rod of density $\delta(x)$ along the interval $[a,b]$ on the x-axis, the number $M_0 = \displaystyle\int_a^b x\delta\,dx$.

Moment about the x-axis: Of a thin plate in the xy-plane, the number

$$M_x = \int \bar{y}\,dm.$$

Moment about the y-axis: Of a thin plate in the xy-plane, the number

$$M_x = \int \bar{y}\,dm.$$

Monomial in x: A single term of the form cx^n, where c is a constant and n is zero or a positive integer.

Natural logarithm: Of a positive number x the natural logarithm, denoted by $\ln x$, is defined to be the value of the integral of the function $1/t$ from $t = 1$ to $t = x$.

Negative x-axis: The x-axis to the left of the origin.

Negative y-axis: The y-axis below the origin.

Newton's law of cooling: The rate at which the temperature of an object changes is proportional to the difference between the temperature of the object and the temperature of the surrounding medium.

Nondecreasing sequence: A sequence of numbers $\{S_n\}$ in which $S_n \leq S_{n+1}$ for every n.

Normal line: A line perpendicular to the tangent to a curve at a point of tangency.

Norm of a partition: The length of a partion's largest subinterval, denoted $\|P\|$ (for partion P).

nth-degree polynomial: A polynomial of the form $a_n x^n + a_{n-1}x^{n-1} + \cdots + a_1 x + a_0$

nth term: The number of index n in an infinite sequence or series, often denoted by a_n.

Odd function: A function in which $f(-x) = -f(x)$ for every x in the function's domain.

One-sided limit: A limit $\lim_{x \to c} F(t)$ in which x lies to one particular side of c.

One-to-one function: A function that always gives different outputs for different inputs.

Open: Refers to intervals that do not contain either endpoint.

Ordinary differential equation: An equation with derivatives of a function of a single variable.

Origin: The intersection of coordinate axes in a plane Cartesian coordinate system.

Parabola: A set of points in a plane that are equidistant from a given fixed point and fixed line in the plane.

Parallelogram law: Describes the geometric addition of two vectors.

Parameter: The variable t in the parametric equations $x = f(t), y = g(t)$.

Parametric equations: The equations for x and y are given as functions of a third variable t.

Partial fraction: The partial fractions of a rational function $f(x) / g(x)$, where the degree of $f(x)$ is less than the degree of $g(x)$, are the rational functions whose denominators are factors of $g(x)$, and which, when added together, equal $f(x) / g(x)$.

Periodic function: A function $f(x)$ such that $f(x + p) = f(x)$ for every value of x, where p is a positive number called the period.

Point of inflection: A point on a curve where the concavity changes.

Point-slope equation: The equation $y - y_1 = m(x - x_1)$ for the line through the point (x_1, x_2) with slope m.

Polar coordinate pair: A coordinate pair (r, θ) for a point P in which r gives the directed distance from the origin O to

P and θ gives the directed angle from the initial ray from O to the ray OP.

Polynomial in x: Any sum of a finite number of monomials in x.

Position vector: The radius vector that vies the position at time t of a moving particle.

Positive x-axis: The x-axis to the right of the origin.

Positive y-axis: The y-axis above the origin.

Power series: A series of the form

$$\sum_{n=0}^{\infty} c_n x^n = c_0 + c_1 x + \cdots + c_n x^n + \cdots$$

or

$$\sum_{n=0}^{\infty} c_n (x - a)^n$$
$$= c_0 + c_1 (x - a) + \cdots + c_n (x - a)^n + \cdots$$

in which a and c_n are real constants.

Quadrant: One of the four regions into which the coordinate axes divide the Cartesian plane.

Quadratic curves: Graphs of equations of the form
$$Ax^2 + Bxy + Cy^2 + Dx + Ey + F = 0.$$

Radian measure: The radian measure of an angle at the center of the unit circle equals the length of the arc that the angle cuts from the unit circle.

Radioactive: The property of being an element whose atoms decay radioactively.

Radioactive decay: The process whereby a radioactive atom emits some of its mass as radiation and reforms to make an atom of some new substance.

Radius of a circle: The constant distance from a circle to its center.

Range: The set of second entries of a set of ordered pairs or a relation.

Rational function of x: The ratio or quotient of two polynomials in x.

Rational number: A real number that can be expressed as the quotient $\dfrac{a}{b}$ where a and b are integers, and b is not zero.

Reaction rate: The rate which a product forms in a chemical reaction.

Real-valued functions of a real number: Functions that have domains and ranges that are sets of real numbers.

Recursion formula: A rule by which any later term in a sequence can be calculated from the terms preceding it.

Relation: Any mathematical rule that assigns the members of one set to the members of a second set.

Relative maximum: See Local maximum.

Relative minimum: See Local minimum.

Remainder: See Truncation error.

Riemann integral: The integral of a function over a closed interval.

Riemann sum: Any expression S_P of the form

$$S_P = \sum_{k=1}^{n} f(c_k)\Delta x_k$$

where f is a function defined on the closed interval $[a,b]$, P is a partition of $[a,b]$, and c_k is in the kth subinterval of P.

Right-hand derivative: A derivative defined by a right-hand limit.

Right-hand limit: The limit of function $f(t)$ as t approaches a number c from the right, denoted by $\lim_{t \to c^+} f(t)$.

Rise: The vertical distance covered from one point to another of a nonvertical line in a plane.

Run: The horizontal distance covered from one point to another of a nonvertical line in a plane.

Second derivative: The derivative of the first derivative.

Semimajor axis: Half the length of the major axis of an ellipse.

Semiminor axis: Half the length of the minor axis of an ellipse.

Separating the variables: In a differential equation, combining all the y terms with dy/dx on one side of the equation and putting all the x terms on the other side.

Sequence: A function whose domain is the set of positive integers.

Sequence of partial sums: Of an infinite series $\Sigma\, a_n$, the sequence $\{s_n\}$ defined by

$$s_n = a_1 + a_2 \cdots + a_n + \cdots = \sum_{k=1}^{n} a_k.$$

Sinusoid: A function that can be written in the form $f(x) = a\sin(bx + c) + d$ where a, b, c, and d are real numbers.

Slant asymptote: A linear asymptote that is not parallel to either coordinate axis.

Slope of a nonvertical line: The ratio of the distance a line rises or falls to the distance it runs horizontally.

Slope field: Of a differential equation, $y' = f(x,y)$, assigns to each point $P(x,y)$ in the domain of f a short line segment whose slope is $f(x,y)$.

Slope-intercept equation: The equation $y = mx + b$ of a line with slope m and y-intercept b.

Slope of a curve: At a point, $(x_0, f(x_0))$, on a differentiable curve $y = f(x)$, the number $f'(x_0)$.

Slope of a vector: The slope shared by the lines parallel to the vector.

Snell's law: $\dfrac{\sin \theta_1}{c_1} = \dfrac{\sin \theta_2}{c_2}$

Solid of revolution: A solid whose shape can be generated by revolving regions in a plane about axes.

Solution of a differential equation: A function $F(x)$ is a solution of the differential equation $dy/dx = f(x)$ over the interval I if F is differentiable at every point of I and if

$$\frac{d}{dx}F(x) = f(x)$$

at every point of I.

Solution curve of a differential equation: The graph of the solution of a differential equation.

Speed: The absolute value of velocity.

Spring constant: The constant k in Hooke's Law.

Step function: A function that jumps from one value to another without taking on any of the intermediate values.

Subsequence: A sequence whose terms occur in a given order among the terms of a second sequence.

Surface of revolution: A surface whose shape can be generated by revolving a curve in a plane about axes.

Tail of a sequence: The collection of all the terms whose indices are greater than some index N.

Tangent to a curve: At a point $P(x_0, f(x_0))$ where the curve $y = f(x)$ is differentiable, the line through P with slope $f'(x_0)$.

Taylor series generated by f at a:

$$\sum_{k=0}^{\infty} \frac{f(k)(a)}{k!}(x - a)^k =$$

$$f(a) + f'(a)(x - a) + \frac{f''(a)}{2!}(x - a)^2 +$$

$$\cdots + \frac{f^{(n)}(a)}{n!}(x - a)^n + \cdots$$

where f is a function with derivatives of all orders throughout some interval containing a as an interior point.

Telescoping series: A series whose middle terms cancel each other.

Translation of axes: The coordinate transformation $x' = x - h$, $y' = y - k$ that shifts the coordinate axes to the positions formerly occupied by the lines $x = h$, $y = k$.

Truncation error: The difference between the sum of a series and its nth partial sum.

Two-sided limit: A function $f(t)$ has a limit at point c if and only if the right-hand and left-hand limits at c exist and are equal. When the side from which $x \to c$ is unrestricted, the limit $\lim_{x \to c} f(x)$ is sometimes called a two-sided limit.

Unit circle: The graph of $x^2 + y^2 = 1$.

Unit hyperbola: The graph of $x^2 - y^2 = 1$.

Unit vector: Any vector whose length is equal to the unit of length used along the coordinate axes.

Upper bound: For a sequence, any number M such that $s_n < M$ for all n.

Variable of integration: The variable x in the integral $\int f(x)dx$.

Variation of parameters: A method for finding the solution of a nonhomogeneous differential equation once the general solution of the corresponding homogeneous equation is known.

Vector: A directed line segment. Used to describe a quantity that has direction as well as magnitude.

Vertex (*pl*: vertices): The point of a parabola or the points of an ellipse or hyperbola that cross the focal axis.

Vertical asymptote: A line $x = a$ of a graph of a function $y = f(x)$ if either $\lim_{x \to a^-} f(x) = \pm\infty$ or $\lim_{x \to a^+} f(x) = \pm\infty$.

Vertical line: A line parallel to the y-axis.

Work: Moving a mass over a distance by exerting a force.

Zero of f: A point at which $f(x) = 0$.

Zero vector: The vector $\mathbf{0} = 0\mathbf{i} + 0\mathbf{j}$, whose length is zero.

Index of Explorations

Chapter 1

Section 1.1

1. $(0,6)$, $(5,-5)$ **3.** One choice: $[-17,21]$ by $[-12,76]$
5. One choice:
x Scl $= 5$, y Scl $= 5$ gives many scale marks at convenient points which are still distinguishable. **7.** One choice:
x Scl $= 0.05$ y Scl $= 10$ **9.** Intercepts: $x = -1$, $y = 1$.

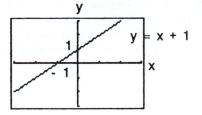

11. $y = -x^2$; $(0,0)$ gives the only intercepts. **13.** $x = -y^2$; $(0,0)$ gives the only intercepts.

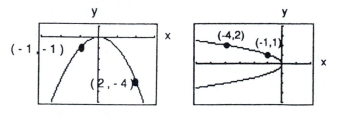

15. (e) **17.** (e) **19.** (e) **21.** A complete graph of $y = 3x - 5$ may be viewed in the rectangle $[-2,4]$ by $[-10,5]$. x-intercept: $x = 5/3$. y-intercept: $y = -5$.
23. A graph of $y = 10 + x - 2x^2$ may be viewed in the rectangle $[-6,6]$ by $[-30,15]$. Intercepts: $x = -2, 5/2$; $y = 10$. **25.** A graph of $y = 2x^2 - 8x + 3$ may be viewed in the rectangle $[-7,10]$ by $[-20,70]$. x-intercepts: $x \approx 0.42, 3.58$. y-intercept: $y = 3$. **27.** Using TRACE on the graph of $y = x^2 + 4x + 5 = (x+2)^2 + 1$ in the rectangle $[-8,4]$ by $[-2,20]$, we see there are no x-intercepts, $y = 5$ is the y-intercept and $(-2,1)$ is the low point. **29.** We graph $y = 12x - 3x^3$ in the rectangle $[-4,4]$ by $[-15,15]$. $x = \pm 2, 0$ are the x-intercepts, $y = 0$ is the y-intercept. Using TRACE, $(-1.14, -9.24)$ and $(1.14, 9.24)$ are the approximate local low and high points, respectively. **31.** A complete graph of $y = -x^3 + 9x - 1$ can be obtained in the viewing rectangle $[-5,5]$ by $[-35,35]$. The y-intercept is -1. Use of TRACE yields the following approximations. x-intercepts: $-3.02, 0.08, 2.93$; low point: $(-1.75, -11.39)$; high point: $(1.75, 9.39)$. **33.** An idea of a complete graph of $y = x^3 + 2x^2 + x + 5$ can be obtained by using the viewing rectangles $[-3,2]$ by $[-2,10]$ and $[-2,1]$ by $[4,6]$. The y-intercept is 5 and using TRACE in the first rectangle, we obtain -2.44 as the approximate x-

intercept. In the second rectangle, we obtain $(-1,5)$ and $(-0.33, 4.85)$ as the approximate local high and low points. **37.** With $N = 127$, $M = 63$, $a = 116$, $b = 52$. **39.** With $(N, M) = (127, 63)$, $a = 53$, $b = 114$. **41.** Yes

Section 1.2

1. $\Delta x = -2$, $\Delta y = -3$ **3.** $\Delta x = -5$, $\Delta y = 0$ **5.** Slope is 3; slope of lines perpendicular is $-1/3$. **7.** $m = 0$. The perpendicular lines are vertical and have no slope. **9.** $\sqrt{2}$ **11.** 6 **13.** $\sqrt{a^2 + b^2}$ **15.** 3 **17.** 5 **19.** 6 **21.** a) $x = 2$ b) $y = 3$ **23.** a) $x = 0$ b) $y = -\sqrt{2}$ **25.** $y = x$ **27.** $y = x + 2$ **29.** $y = 2x + b$
31. $y = \frac{3}{2}x$ **33.** $x = 1$ **35.** $3x + 4y + 2 = 0$ **37.** $y = 3x - 2$ **39.** $y = x + \sqrt{2}$ **41.** $y = -5x + 2.5$ **43.** x-intercept is 4, y-intercept is 3 **45.** x-intercept: $x = 3$, y-intercept: $y = -4$ **47.** x-intercept: $x = -2$, y-intercept: $y = 4$ **49.** $x = 3$, $y = 4$ **51.** $x = a$, $y = b$ **53.** Perpendicular: $y = x$. $y = -x + 2$ and $y = x$ may be viewed in the rectangle $[-6,6]$ by $[-4,4]$. Distance $= \sqrt{2}$. **55.** Perpendicular: $y = 2x$. $y = -\frac{1}{2}x + \frac{3}{2}$ and $y = 2x$ may be viewed in the rectangle $[-6,6]$ by $[-4,4]$. Distance $= 2\sqrt{5}/5$. **57.** Perpendicular: $y = x + 3$. $y = -x + 3$ and $y = x + 3$ may be viewed in the rectangle $[-7,8]$ by $[-3,7]$. Distance $= 3\sqrt{2}$. **59.** $y = x - 1$. $y = x + 2$ and $y = x - 1$ may graphed in $[-10,10]$ by $[-10,10]$. **61.** $y = -2x + 2$ **63.** a) $|x - 3|$ b) $|x + 2|$ **65.** The distance between 5 and x is 1. **67.** a) $a = -2$ b) $a \geq 0$ **69.** a) $-2.5°$/in b) $-16.1°$/in c) $-8\frac{1}{3}°$/in
71. 5.97 atmospheres **73.** $-40°C$ is equivalent to $-40°F$ **75.** a) $d(t) = 45t + d_0$ where d_0 is the distance of the car from the point P at time $t = 0$. c) $m = 45$ d) If $t = 0$ corresonds to a specific time, say 1:00 p.m., then negative values of t would correspond to times before 1:00 p.m. e) The initial distance (from point P) $d_0 = 30$ miles. **77.** The possibilities for the third vertex are $(-1,4)$, $(-1,-2)$, $(5,2)$.

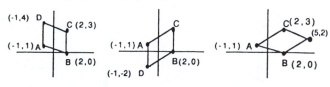

79. $x = 1$

Section 1.3

1. Not a function. There are x-values that correspond to two y-values. **3.** This is a function. No value of x corresponds to two y-values. **5.** a) $(3,-1)$ b) $(-3,1)$ c) $(-3,-1)$ **7.** a) $(-2,-1)$ b) $(2,1)$ c) $(2,-1)$ **9.** a) $(1,\sqrt{2})$ b) $(-1,-\sqrt{2})$ c) $(-1,\sqrt{2})$

11. a) $(0, -\pi)$ b) $(0, \pi)$ c) $(0, -\pi)$ **13.** Domain: $[1, \infty)$, range: $[2, \infty)$ **15.** Domain: $(-\infty, 0]$, range: $(-\infty, 0]$ **17.** Domain: $(-\infty, 3]$, range: $[0, \infty)$ **19.** Domain: $(-\infty, 2) \cup (2, \infty)$, range: $(-\infty, 0) \cup (0, \infty)$ **21.** Domain: $(-\infty, \infty)$, range: $[-9, \infty)$. Symmetric about the y-axis.   **23.** Domain = range = $(-\infty, \infty)$. No symmetry.   **25.** Domain = range = $(-\infty, \infty)$. No symmetry.   **27.** Domain = range = $(-\infty, 0) \cup (0, \infty)$. Symmetric about the origin.   **29.** Domain: $(-\infty, 0) \cup (0, \infty)$, range: $(-\infty, 1) \cup (1, \infty)$. No symmetry.   **31.** a) No   b) No   c) $(0, \infty)$ **33.** Odd   **35.** Neither   **37.** Even   **39.** Even   **41.** Odd   **43.** Symmetric about the y-axis. Graph $y = -x^2$ in the viewing rectangle $[-10, 10]$ by $[-10, 10]$.   **45.** Symmetric about the y-axis. Graph $y = 1/x^2$ in the viewing rectangle $[-5, 5]$ by $[0, 3]$.   **47.** Symmetric about the origin. Graph $y = 1/x$ in the viewing rectangle $[-4, 4]$ by $[-4, 4]$.   **49.** $x^2y^2 = 1$ has graph symmetric about both axes and the origin. Graph $y = 1/|x|$ and $y = -1/|x|$ in the viewing rectangle $[-4, 4]$ by $[-4, 4]$.   **51.** Graph $y = |x + 3|$ in the viewing rectangle $[-7, 1]$ by $[0, 4]$.   **53.** Graph $y = \dfrac{|x|}{x}$ in the viewing rectangle $[-2, 2]$ by $[-2, 2]$. There is no point on the graph when $x = 0$.   **55.** $y = x$ when $x \leq 0$ and $y = 0$ when $x \geq 0$.   **57.** a) Graph $y = 3 - x + 0\sqrt{1 - x}$ and $y = 2x + 0\sqrt{x - 1}$ in the viewing rectangle $[-10, 10]$ by $[0, 20]$.   b) $f(0) = 3$, $f(1) = 2$, $f(2.5) = 5$.   **59.** a) Graph $y = 1 + 0\sqrt{5 - x}$ in the viewing rectangle $[-3, 10]$ by $[0, 2]$. It is understood that the x-axis for $x \geq 5$ is part of the graph and the point $(5, 1)$ is not.   b) $f(0) = 1$, $f(5) = 0$, $f(6) = 0$.   **61.** a) Graph $y = 4 - x^2 + 0\sqrt{1 - x}$, $y = \dfrac{3}{2}x + \dfrac{3}{2} + 0\sqrt{x - 1} + 0\sqrt{3 - x}$ and $y = x + 3 + 0\sqrt{x - 3}$ in the viewing rectangle $[-3, 7]$ by $[-5, 10]$.   b) $f(0.5) = 3.75$, $f(1) = 3$, $f(3) = 6$, $f(4) = 7$.

63. a) $1 - |x - 1|$, $0 \leq x \leq 2$,   b) $f(x) = \begin{cases} 2, & 0 \leq x < 1 \\ 0, & 1 \leq x < 2 \\ 2, & 2 \leq x < 3 \\ 0, & 3 \leq x \leq 4 \end{cases}$

65. a) $0 \leq x < 1$ b) $-1 < x \leq 0$ **67.** a) Graph of $y = x - \lfloor x \rfloor$, $-3 \leq x \leq 3$ b) Graph of $y = \lfloor x \rfloor - \lceil x \rceil$, $-3 \leq x \leq 3$

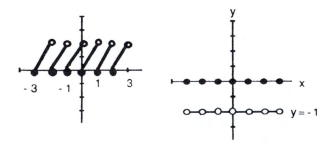

69. Graph $y = abs(x + 1) + 2abs(x - 3)$ in the viewing rectangle $[-2, 5]$ by $[0, 12]$.

$$f(x) = \begin{cases} -3x + 5, & x \leq -1 \\ -x + 7, & -1 < x \leq 3 \\ 3x - 5, & x > 3 \end{cases}$$

71. Graph $y = |x| + |x - 1| + |x - 3|$ in the viewing rectangle $[-1, 4]$ by $[0, 7]$. $y = -3x + 4$, $x \leq 0$, $y = -x + 4$, $0 < x \leq 1$, $y = x + 2$, $1 < x \leq 3$, $y = 3x - 4$, $x > 3$.   **73.** $f(x) = x$, domain: $(-\infty, \infty)$. $g(x) = \sqrt{x - 1}$, domain: $[1, \infty)$. $f(x) + g(x) = x + \sqrt{x - 1}$, domain $(f + g)$: $[1, \infty)$, complete graph in $[0, 5]$ by $[0, 10]$. $f(x) - g(x) = x - \sqrt{x - 1}$, domain $(f - g)$: $[1, \infty)$, complete graph in $[0, 5]$ by $[0, 4]$. $f \circ g(x) = f(g(x)) = f(\sqrt{x - 1}) = \sqrt{x - 1}$, domain $f \circ g$: $[1, \infty)$, complete graph in $[0, 5)$ by $[0, 2]$. $f(x)/g(x) = x/\sqrt{x - 1}$, domain: $(1, \infty)$, complete graph in $[0, 10]$ by $[0, 5]$. $g(x)/f(x) = \sqrt{x - 1}/x$, domain: $[1, \infty)$, complete graph in $[0, 10]$ by $[0, 0.5]$, graph starts at $(1, 0)$.   **75.** a) 2   b) 22   c) $x^2 + 2$ d) $x^2 + 10x + 22$ e) 5   f) -2 g) $x + 10$ h) $x^4 - 6x^2 + 6$ **77.** a) $\sqrt{x - 7}$ b) $3(x + 2)$ c) x^2 d) x e) $\dfrac{1}{x - 1}$ f) $\dfrac{1}{x}$ **79.** a) $C(10) = 72$ b) $C(30) - C(20)$ is the increase of cost if the production level is raised from 20 to 30 items daily.   **81.** The two functions are identical.   **83.** $g(x) = \sqrt{x}$ is one possibility.   **85.** Graph $y = abs(x + 3) + abs(x - 2) + abs(x - 4)$ in the viewing rectangle $[-4, 5]$ by $[0, 15]$. We see that $d(x)$ is minimized when $x = 2$ so you would put the table next to Machine 2.   **87.** $d(x)$ now has minimum value 17 when $x = 2$. The table should be placed next to Machine 3.

Section 1.4

1. a) $y = (x + 4)^2$ b) $y = (x - 7)^2$ **3.** a) Position 4   b) Position 1   c) Position 2   d) Position 3   **5.** Shift the graph of $|x|$ to the left 4 units. Then shift the resulting graph down 3 units.   **7.** Reflect the graph of $y = \sqrt{x}$ over the y-axis and then stretch the resulting graph vertically by a factor of 3.   **9.** Stretch the graph of $y = \dfrac{1}{x}$ vertically by a factor of 2 and shift the resulting graph down 3 units.   **11.** Shift the graph of $y = x^3$ right 3 units. Shrink the resulting graph by a factor of 0.5. Reflect the last graph over the x-axis and shift the last graph up 1 unit.   **13.** Shift the graph of $y = \dfrac{1}{x}$ to the right 2 units (obtaining the graph of $\dfrac{1}{x - 2}$). Shift the resulting graph up 3 units.   **15.** Graph the function in the viewing rectangle: $[-8, 12]$ by $[-15, 5]$. Domain = range = $(-\infty, \infty)$.   **17.** Check your result by graphing y in $[-5, 5]$ by $[-10, 10]$. Domain = range = $(-\infty, \infty)$.

19. Check your result by graphing $y = -(x+3)^{(-2)} + 2$ in the viewing rectangle $[-10, 5]$ by $[-3, 3]$. Domain $= (-\infty, -3) \cup (-3, \infty)$, range $= (-\infty, 2)$. **21.** Check your result by graphing $y = -2((x-1)^{(1/3)})^2 + 1$ in the viewing rectangle $[-1, 3]$ by $[-2, 1]$. Domain $= (-\infty, \infty)$, range $= (-\infty, 1]$. **23.** Check your result by graphing $y = 2[1-x] = 2\,\text{Int}(1-x)$ in the viewing window $[-3, 4]$ by $[-6, 8]$. (Graph this in Dot Mode if possible.) Domain $= (-\infty, \infty)$, range $= \{2n : n = 0, \pm1, \pm2, \ldots\}$. **25.** $y = 3x^2 + 4$ **27.** $y = 0.2(\frac{1}{x} - 2)$ **29.** $y = 3|x+2| + 5$ **31.** $y = -0.8(x-1)^3 - 2$

33. $y = 5\sqrt{-(x+6)} + 5$ **35.** $y = \sqrt{\frac{3}{2}x + 1}$ **37.** The resulting curve is the same in both cases. Let $c > 1$ be given. Applied to $y = x^n$, a vertical stretch by c has the same end result as a horizontal shrink by $\frac{1}{\sqrt[n]{c}}$. **39.** In #25 and #26 we obtain, respectively, $y = 3x^2 + 4$ and $y = 3(x^2 + 4) = 3x^2 + 12$. We obtain different geometric results by reversing the order of the transformations. The second graph is 8 units above the first graph. **41.** a) Reflect across the y-axis. b) Reflect across the x-axis. c) They are the same: $y = \sqrt[3]{-x} = -\sqrt[3]{x}$. **43.** $y = mx + b$ **45.** $(2, 3), (3, 2), (4, 3)$. **47.** $(-1, -2), (0, 0), (1, -2)$. **49.**

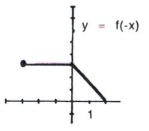

$y = f(-x)$

51.

$y = f(x-2)$

53.

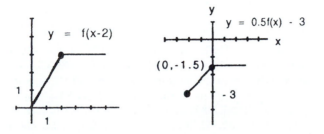

$y = 0.5f(x) - 3$

$(0, -1.5)$

55.

$y = -2f(x+1) + 3$

$(-1, -3)$

57.

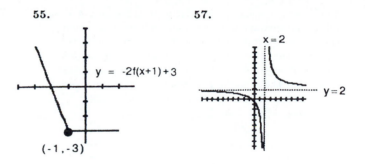

$x = 2$

$y = 2$

59. $y = -2(x-3)^2 + 7$, vertex $(3, 7)$, axis of symmetry: $x = 3$. Check your result by graphing $y = -2x^2 + 12x - 11$ in the viewing rectangle $[-1, 7]$ by $[-16, 7]$. **61.** Vertex: $(-\frac{5}{2}, -6)$, axis of symmetry: $x = -\frac{5}{2}$. Check your result by graphing the function in $[-5, -0]$ by $[-6, 10]$. **63.** $x = (y-3)^2 + 2$. Shift up 3, shift right 2. Check your sketch by graphing $y = 3 + \sqrt{x-2}$ and $y = 3 - \sqrt{x-2}$ in the viewing rectangle $[2, 20]$ by $[-5, 11]$. In parametric mode use $x = t^2 - 6t + 11$, $y = t$, $-5 \leq t \leq 11$. **65.** $x = 2(y+1)^2 - 1$. Shift down 1, horizontal stretch by 2, shift left 1. Check your sketch by graphing $y = -1 + \sqrt{\frac{x+1}{2}}$ and $y = -1 - \sqrt{\frac{x+1}{2}}$ in $[-1, 10]$ by $[-4, 2]$. In parametric mode $x = 2t^2 + 4t + 1$, $y = t$, $-4 \leq t \leq 2$. **67.** $x = -2(y-3)^2 + 5$. Shift up 3, horizontal stretch by 2, reflection across y-axis, shift right 5. Check your sketch by graphing $y = 3 + \sqrt{\frac{5-x}{2}}$ and $y = 3 - \sqrt{\frac{5-x}{2}}$ in $[-9, 5]$ by $[0, 7]$. In parametric mode $x = -2t^2 + 12t - 13$, $y = t$, $0 \leq t \leq 7$. **69.** Two.

Section 1.5

1. $\{-3/2, 2/3\}$ **3.** $\{1 - \sqrt{5}/2, 2, 1 + \sqrt{5}/2\}$ **5.** $\{-4, \frac{1}{2}\}$

7. $\{-0.5, -0.41, 1.08\}$ or $\{-\frac{1}{2}, \frac{1 \pm \sqrt{5}}{3}\}$ **9.** We give a sequence for the smallest solution only. $[-2, -1]$ by $[-1, 1]$, $x\,Scl = 0.1$; $[-2, -1.9]$ by $[-0.1, 0.1]$, $x\,Scl = 0.01$; $[-1.94, -1.93]$ by $[-0.01, 0.01]$, $x\,Scl = 0.001$; $[-1.931, -1.930]$ by $[-0.001, 0.001]$, $x\,Scl = 0.0001$. **11.** No real solution **13.** $\{3, \frac{1 \pm \sqrt{7}}{2}\}$ **15.** $\{0.74, 7.56, 12.70\}$ **17.** We assume that the number

x satisfies $2 < x < 6$. The statement a) $0 < x < 4$ about x does not contain precise enough information about x to be able to determine whether the statement is true or false. The same goes for the statement g) which is equivalent to $-2 < x < 6$. Use of rules of inequalities shows that the remaining statements are all equivalent to $2 < x < 6$ whence they are true. **19.** $\{-2, 2\}$ **21.** $\{-9/2, -1/2\}$ **23.** $\{-1/3, 17/3\}$ **25.** $-1 \le y \le 3$ **27.** $5/3 < y < 3$ **29.** $0.9 < y < 1.1$ **31.** $|x - 6| <$ 3 **33.** $|x + 1| < 4$ **35.** $3 < x < 7$ **37.** $(-\infty, -1] \cup [3, \infty)$ **39.** $-8 \le x \le 2$ **41.** $(-\infty, -5) \cup (-1, \infty)$ **43.** $(-2/3, 2)$ **45.** $[-5, 2]$ **47.** $(-\infty, 2 - \sqrt{7}] \cup [2, 2 + \sqrt{7}]$ **49.** $(0, 1.9) \cup (2.1, \infty)$ **51.** $(15, 225)$ **53.** b) If the y-range is too large, the graphing utility cannot distinguish between very close values of y. **55.** We may multiply both sides of the equation by the least common multiple of the denominators of the coefficients. This produces an equivalent equation with integer coefficients. For the given example the l.c.m. is 6 and we obtain the equation $12x^3 + 3x^2 - 4x + 6 = 0$. The possible rational roots of this equation are $\pm\dfrac{p}{q}$ where $p = 1, 2, 3$ or 6 and $q = 1, 2, 3, 4, 6$ or 12. **57.** a) $A(x) = x(50 - x)$ b) Check your sketch by graphing $y = x(50 - x)$ in $[-10, 60]$ by $[-100, 700]$ c) domain $= (-\infty, \infty)$, range $= (-\infty, 625]$ d) $0 < x < 50$ e) Using TRACE in b) leads to the approximation 13.8ft by 36.2ft. The exact dimensions are $25 - 5\sqrt{5}$ft by $25 + 5\sqrt{5}$ft f) $\{0 < x < 25 - 5\sqrt{5}\} \cup \{25 + 5\sqrt{5} < x < 50\}$ **59.** a) $A(x) = (8.5 - 2x)(11 - 2x)$ b) Graph $y = A(x)$ in $[0, 10]$ by $[-2, 50]$ c) Domain $= (-\infty, \infty)$, range $= [-1.5625, \infty)$ d) $0 < x < 4.25$ only makes sense. This part of the graph is the graph of the problem situation. e) 0.95in. **61.** a) $V(x) = 0.1x + 0.25(50 - x)$ b) Graph $y = V(x)$ in $[-20, 100]$ by $[-2, 20]$ c) Domain $=$ range $= (-\infty, \infty)$ d) $0 \le x \le 50$, x an integer e) 22 dimes, 28 quarters f) There is no integral solution. **63.** a) $V = x(20 - 2x)(25 - 2x)$ b) Graph $y = V$ in $[-2, 17]$ by $[-100, 900]$. c) Domain $=$ range $= (-\infty, \infty)$ d) $0 < x < 10$. Graphing $y = V$ in $[0, 10]$ by $[0, 900]$ gives a graph of the problem situation. e) $x = 3.68$in., maximum volume is 820.53in^3. **65.** a) $L = 5.5 - x$. b) $V(x) = x(5.5 - x)(8.5 - 2x)$ c) Graph V in $[-1, 7]$ by $[-10, 50]$. The domain and range are both the set of all real numbers. d) $0 < x < 4.25$. Graph V in $[0, 4.25]$ by $[-10, 50]$. e) $x = 0.753$in or 2.592in. f) $x = 1.585$in. $V_{\max} = V(1.585) = 33.074$in^3.

Section 1.6

1. Not one-to-one **3.** One-to-one **5.** Your graph should be the line through $(0, -3)$ and $(2, 0)$.

7.

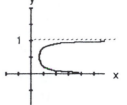

9. $x^2 + (y - 2)^2 = 4$. Graph $y = 2 + \sqrt{4 - x^2}$ and $y = 2 - \sqrt{4 - x^2}$ in $[-3.4, 3.4]$ by $[0, 4]$. **11.** $(x - 3)^2 + (y + 4)^2 = 25$. Graph $y = -4 + \sqrt{25 - (x - 3)^2}$ and $y = -4 - \sqrt{25 - (x - 3)^2}$ in $[-5.5, 11.5]$ by $[-9, 1]$. **13.** $x^2 + y^2 = 4$ **15.** $(x - 3)^2 + (y - 3)^2 = 9$ **17.** $C = (3, -4)$, $r = 3$. Graph $x = 3 + 3\cos t$, $y = -4 + 3\sin t$, $0 \le t \le 2\pi$, in $[-2.1, 8.1]$ by $[-7, -1]$. Domain $= [0, 6]$, range $= [-7, -1]$. **19.** $C = (-2, -3)$, $r = \sqrt{5}$. Domain $= [-2 - \sqrt{5}, -2 + \sqrt{5}]$, range $= [-3 - \sqrt{5}, -3 + \sqrt{5}]$. Graph $x = -2 + \sqrt{5}\cos t$, $y = -3 + \sqrt{5}\sin t$, $0 \le t \le 2\pi$, in $[-8.15, 4.15]$ by $[-5.7, 1.5]$. **21.** Graph $y = \dfrac{4}{3}\sqrt{x^2 - 9}$ and $y = -\dfrac{4}{3}\sqrt{x^2 - 9}$ in $[-10, 10]$ by $[-10, 10]$. The calculator screen shows incorrect gaps in the curve when the slopes of tangent lines become numerically very large. **23.** a) All points outside the boundary of the unit circle. b) All points in the interior of the circular disk with center $(0, 0)$ and radius 2. c) The ring between the two circles. **25.** $(x + 2)^2 + (y + 1)^2 < 6$ **27.** The graph of the inverse relation can be viewed by graphing $x_1 = 3/(t - 2) - 1$, $y_1 = t$, $-10 \le t \le 10$, Tstep $= .1$ in $[-10, 10]$ by $[-10, 10]$. It is a function. **29.** Graph $x = t^3 - 4t + 6$, $y = t$, $-10 \le t \le 10$, tstep $= 0.1$, in $[-10, 20]$ by $[-15, 15]$. This is not a function. **31.** For the inverse graph $y = e^{x/2}$ and $y = -e^{x/2}$ in $[-2, 5]$ by $[-10, 10]$. This is not a function. **33.** $f^{-1}(x) = (x - 3)/2$. **35.** $f^{-1}(x) = \sqrt[3]{x + 1}$ **37.** $f^{-1}(x) = \sqrt{x - 1}$ **39.** $f^{-1}(x) = 2 - \sqrt{-x}$, $x \le 0$ **41.** $f^{-1}(x) = \dfrac{1}{\sqrt{x}}$ **43.** $f^{-1}(x) = \dfrac{1 - 3x}{x - 2}$, $x \ne 2$ **45.** Check your sketch by graphing $y = 2[\ell n(x - 4)/\ell n(3)] - 1$ in $[3, 14]$ by $[-10, 4]$ noting that $x = 4$ is a vertical asymptote. **47.** Check your sketch by graphing $y = -3[\ell n(x + 2)/\ell n(0.5)] + 2$ in $[-2, 4]$ by $[-18, 10]$. **49.** Graph $y = 5(e^{3x}) + 2$ in $[-2, 1]$ by $[0, 20]$. **51.** Graph $y = -2(3^x) + 1$ in $[-4, 2]$ by $[-10, 2]$. **53.** The graph of $\log x = \log_{10} x$ has the same shape as that in Figure 1.91. We start with the graph of $y = \log x$, reflect it through the x-axis and stretch vertically by a factor of 3. (We now have the graph of $y = -3\log x$.) We then shift left 2 and then up 1. The domain comes from $x + 2 > 0$ so it is $(-2, \infty)$

and the range is $(-\infty, \infty)$.   **55.** Start with the graph of $y = 3^x$, shift left 1 $(y = 3^{x+1})$, reflect across the y-axis $(y = 3^{-x+1})$, stretch vertically by a factor of 2 and shift up 1.5. Domain $= (-\infty, \infty)$, range $= (1.5, \infty)$.   **57.** Shift the graph of $x^2 + y^2 = 9$ three units left and 5 units up. Domain $= [-6, 0]$, range $= [2, 8]$.   **59.** Graph $y_1 = 2^x$, $y_2 = \dfrac{\ell n\, x}{\ell n\, 2}$, $y_3 = x$ in $[-7.5, 14.6]$ by $[-5, 8]$.

61. Graph $y_1 = \log_3 x = \dfrac{\ell n\, x}{\ell n\, 3}$, $y_2 = 3^x$, $y_3 = x$ in

$[-7.5, 14.6]$ by $[-5, 8]$.   **63.** $\left\{ \ell n(\dfrac{3 - \sqrt{5}}{2}), \ell n(\dfrac{3 + \sqrt{5}}{2}) \right\}$

65. $\{2 - \sqrt{3}, 2 + \sqrt{3}\}$

Section 1.7

1. 8.901 rad   **3.** -0.73 **5.** $355.234°$ **7.** $-114.592°$

9. $\dfrac{5\pi}{4}$ **11.** 18   **13.** $\dfrac{1}{2}$

15. a) $\sin \dfrac{\pi}{3} = \dfrac{\sqrt{3}}{2}$,   b)$\sin(-\dfrac{\pi}{3}) = -\dfrac{\sqrt{3}}{2}$,

$\cos \dfrac{\pi}{3} = \dfrac{1}{2}$, $\cos(-\dfrac{\pi}{3}) = \dfrac{1}{2}$,

$\tan \dfrac{\pi}{3} = \sqrt{3}$, $\tan(-\dfrac{\pi}{3}) = -\sqrt{3}$,

$\cot \dfrac{\pi}{3} = \dfrac{\sqrt{3}}{3}$, $\cot(-\dfrac{\pi}{3}) = -\dfrac{\sqrt{3}}{3}$,

$\sec \dfrac{\pi}{3} = 2$, $\sec(-\dfrac{\pi}{3}) = 2$,

$\csc \dfrac{\pi}{3} = \dfrac{2\sqrt{3}}{3}$ $\csc(-\dfrac{\pi}{3}) = -\dfrac{2\sqrt{3}}{3}$

17. a) $\sin(6.5) = 0.2151$ b) $\sin(-6.5) = -0.2151$

$\cos(6.5) = 0.9766$ $\cos(-6.5) = 0.9766$

$\tan(6.5) = 0.2203$ $\tan(-6.5) = -0.2203$

$\cot(6.5) = 4.5397$ $\cot(-6.5) = -4.5397$

$\sec(6.5) = 1.0240$ $\sec(-6.5) = 1.0240$

$\csc(6.5) = 4.6486$ $\csc(-6.5) = -4.6486$

19. a) $\sin \dfrac{\pi}{2} = 1$ b) $\sin \dfrac{3\pi}{2} = -1$

$\cos \dfrac{\pi}{2} = 0$ $\cos \dfrac{3\pi}{2} = 0$

$\tan \dfrac{\pi}{2}$ is undefined $\tan \dfrac{3\pi}{2}$ is undefined

$\cot(\dfrac{\pi}{2}) = 0$ $\cot \dfrac{3\pi}{2} = 0$

$\sec \dfrac{\pi}{2}$ is undefined $\sec \dfrac{3\pi}{2}$ is undefined

$\csc \dfrac{\pi}{2} = 1$ $\csc \dfrac{3\pi}{2} = -1$

21. $\dfrac{\pi}{6}$, $30°$ **23.** -1.3734, $-78.6901°$ **25.** In parametric mode graph $x_1(t) = \cos t$, $y_1(t) = \sin t$, t Min $= 0$, t Max $= 2\pi$, t Step $= 0.1$. Then use TRACE: For a given t value, the displayed $x = \cos t$ and $y = \sin t$.   **27.** $[-\pi, 2\pi]$ by $[-1, 1]$, $[-\pi, 2\pi]$ by $[-1, 1]$, $[-1.5\pi, 1.5\pi]$ by $[-2, 2]$, respectively.   **29.** $[-270°, 450°]$ by $[-3, 3]$, $[-360°, 360°]$ by $[-3, 3]$, $[-180°, 180°]$ by $[-3, 3]$,

respectively.   **31.** Check by graphing the functions in $[-\pi, \pi]$ by $[-3, 3]$.   **33.** Graph the functions in $[0, 2\pi]$ by $[-1, 1]$.   **35.** Amplitude $= 2$, period $= 6\pi$, horizontal stretch by a factor of 3. To see one period of the function graph it in $[0, 6\pi]$ by $[-2, 2]$.   **37.** $y = \cot\left(2x + \dfrac{\pi}{2}\right) = \cot\left[2(x + \dfrac{\pi}{4})\right]$. A horizontal shrinking by a factor of $\dfrac{1}{2}$ is applied to the graph of $y = \cot x$ followed by a horizontal shift left $\dfrac{\pi}{4}$ units. The period is $\dfrac{\pi}{2}$. Graphing the function in $[-\dfrac{\pi}{4}, \dfrac{3\pi}{4}]$ by $[-2, 2]$ shows two periods of the function.   **39.** Period $= 2\pi/3$, domain is all real numbers except $n\pi/3$, n an integer, range $= (-\infty, -5) \cup (1, \infty)$. One period of the graph may be viewed in $[-\dfrac{\pi}{3}, \dfrac{\pi}{3}]$ by $[-11, 7]$. Start with the graph of $y = \csc x$, shrink horizontally by a factor of $1/3$, shift horizontally left $\pi/3$ units, stretch vertically by a factor of 3, shift vertically downward 2 units.   **41.** Period $= \pi/3$, domain is all real numbers except odd multiples of $\pi/6$, range is all real numbers. One period of the graph may be viewed in $[-\dfrac{\pi}{6}, \dfrac{\pi}{6}]$ by $[-11, 15]$. Start with the graph of $y = \tan x$, shrink horizontally by a factor of $1/3$, shift horizontally left $\pi/3$ units, stretch vertically by a factor of 3, reflect across the x-axis, shift vertically upward 2 units.   **43.** $\{\pm \cos^{-1}(-0.7) + 2n\pi\}$.   **45.** $\{(\tan^{-1} 4) + n\pi\}$ **47.** $\{-2.596, 0, 2.596\}$.   **49.** 2.219.   **51.** $(-\infty, -2.596) \cup (0, 2.596)$.   **53.** Graph $x_1 = \sqrt{5} \cos t$, $y_1 = \sqrt{5} \sin t$, $0 \le t \le 2\pi$ in $[-3.8, 3.8]$ by $[-\sqrt{5}, \sqrt{5}]$.   **55.** Graph $x_1 = 2 + 3 \cos t$, $y_1 = -3 + 3 \sin t$, $0 \le t \le 2\pi$, in $[-3.1, 7.1]$ by $[-6, 0]$.   **57.** $y = \sqrt{13} \sin(x + \alpha)$, where $\alpha = \sin^{-1}(3/\sqrt{13}) = 0.9828$.

59. $y = \sqrt{2} \sin\left(2x + \dfrac{\pi}{4}\right)$.   **61.** b) and d);   c) and e)

65. Graphs of cosine, sine and tangent are symmetric about the y-axis, the origin and the origin, respectively.

67. a) yes,   b) $-1 \le \cos 2x \le 1$,   c) $0 \le \dfrac{1 + \cos 2x}{2} \le 1$,

d) The domain is the set of all reals; the range is the interval $[0, 1]$.   **69.** a) 37   b) 365   c) 101 units to the right   d) 25 units upward   **71.** We obtain $\cos(A - A) = \cos A \cos A + \sin A \sin A$ or $1 = \cos^2 A + \sin^2 A$ for the first equation. For the second equation we obtain $\sin(A - A) = \sin A \cos A - \cos A \sin A$ or $0 = 0$.

73. $\cos(A + \dfrac{\pi}{2}) = \cos A \cos \dfrac{\pi}{2} - \sin A \sin \dfrac{\pi}{2} = -\sin A$.

If we start the cosine curve, reflect it across the x-axis and shift horizontally $\dfrac{\pi}{2}$ units to the left, we obtain the sine curve. Similarly, $\sin(A + \dfrac{\pi}{2}) = \cos A$.   **75.** $\dfrac{\sqrt{6} + \sqrt{2}}{4}$

77. $\dfrac{\sqrt{2} + \sqrt{6}}{4}$ **79.** $\dfrac{2 + \sqrt{2}}{4}$ **81.** $\dfrac{2 - \sqrt{3}}{4}$ **87.** a) Let

$f(x) = \cot x.$ $f(-x) = \dfrac{\cos(-x)}{\sin(-x)} = \dfrac{\cos x}{-\sin x} = -f(x),$

proving $f(x)$ is odd. b) Let $g(x) = \dfrac{h(x)}{k(x)}$ where $h(x)$ is

even and $k(x)$ is odd. $g(-x) = \dfrac{h(-x)}{k(-x)} = \dfrac{h(x)}{-k(x)} = -g(x)$

proving $g(x)$ is odd where defined. c) The graph of $y = \cot(-x) = -\cot x$ can be obtained by reflecting the graph of $y = \cot x$ across the x-axis. **89.** Use the method indicated in the solution of Exercise 87. **91.** Fundamental period $= \dfrac{\pi}{30}$. Graph $y = \sin(60x)$ in the window $[0, \dfrac{\pi}{30}]$ by $[-1, 1]$. **95.** $41.186°$ **97.** $18.435°$

Chapter 1 Practice Exercises

1. a) $(1, -4)$ b) $(-1, 4)$ c) $(-1, -4)$ **3.** a) $(-4, -2)$ b) $(4, 2)$ c) $(4, -2)$ **5.** a) origin only b) y-axis only **7.** a) both axes and the origin b) none of the mentioned symmetries **9.** $x = 1, y = 3$ **11.** $x = 0, y = -3$ **13.** $y = 2x - 1$. Intercepts: $x = \dfrac{1}{2}, y = -1$ **15.** $y = -x + 1$. Intercepts: $x = 1, y = 1$ **17.** $y + 6 = 3(x - 1)$ or $y = 3x - 9$. Intercepts: $x = 3, y = -9$. **19.** $y - 2 = -\dfrac{1}{2}(x + 1)$ or $y = -\dfrac{1}{2}x + \dfrac{3}{2}$. Intercepts: $x = 3, y = \dfrac{3}{2}$. **21.** $3y = 5x + 4$ **23.** $y = \dfrac{5}{2}x - 6$ **25.** $y = \dfrac{1}{2}x + 2$ **27.** $y = -2x - 1$ **29.** a) $2x - y = 12$ b) $x + 2y = 6$ c) $14\sqrt{5}/5$ **31.** a) $4x + 3y = -20$ b) $y + 12 = \dfrac{3}{4}(x - 4)$ c) $\dfrac{32}{5}$ **33.** Domain $=$ range $= (-\infty, \infty)$ **35.** Check your sketch by graphing $y = 2\,\mathrm{abs}(x - 1) - 1$ in $[-2, 4]$ by $[-1, 5]$. Domain $= (-\infty, \infty)$, range $= [-1, \infty)$ **37.** Domain $= (-\infty, \infty)$, range $= [-1, 1]$. **39.** Domain $= (-\infty, 0]$, range $= (-\infty, \infty)$. Check your sketch by graphing $y = -\sqrt{-x}$ and $y = \sqrt{-x}$ in $[-9, 0]$ by $[-3, 3]$. **41.** Domain $=$ range $= (-\infty, \infty)$. Graph $f(x)$ in $[-9, 4]$ by $[-100, 100]$. **43.** Domain $= (1, \infty)$, range $= (-\infty, \infty)$. Graph $y = \ell n(x - 1)/\ell n(7) + 1$ in $[1, 3]$ by $[-3, 3]$, recalling that $x = 1$ is a vertical asymptote. **45.** Domain $= (-\infty, \infty)$, range $= [5, \infty)$. Graph $y = \mathrm{abs}(x - 2) + \mathrm{abs}(x + 3)$ in $[-5, 4]$ by $[4, 9]$. **47.** Stretch vertically by a factor of 2, reflect across the x-axis, shift horizontally right one unit, shift vertically 5 units upward. **49.** Stretch vertically by a factor of 3, shrink horizontally by a factor of $1/3$, shift horizontally $\pi/3$ units left. **51.** $y = -2(x - 2)^2 + 3$ **53.** $y = \dfrac{3}{x + 2} + 5$

55.

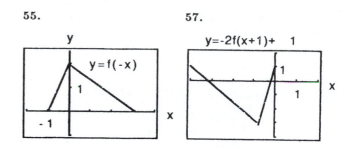
y = f(-x)

57.

y = -2f(x+1) + 1

59. Vertex is $(2, 3)$. $x = 2$ is the line of symmetry. Check your sketch by graphing y in $[-2, 5]$ by $[-6, 3]$. **61.** all are even **63.** a) even b) odd c) odd **65.** a) even b) odd c) odd **67.** Graph y in $[-2, 2]$ by $[0, 2]$. The function is periodic of period 1. **69.** Graph $y_1 = \sqrt{-x}$ and $y_2 = \sqrt{x}$ at the same time in $[-4, 4]$ by $[0, 2]$. **71.** The graph consists of one period of the sine function on $[0, 2\pi]$ together with all points on the x-axis larger than 2π. **73.** $y = \begin{cases} 1 - x, & 0 \le x < 1 \\ 2 - x, & 1 \le x \le 2 \end{cases}$ **75.** For $f(x)$, domain $=$ range $=$ all real numbers except 0. For the remaining functions, domain $=$ range $=$ all positive real numbers. **77.** $(x - 1)^2 + (y - 1)^2 = 1$ **79.** $(x - 2)^2 + (y + 3)^2 = \dfrac{1}{4}$ **81.** $(3, -5), 4$ **83.** $(-1, 7), \sqrt{11}$ **85.** a) $x^2 + y^2 < 1$ b) $x^2 + y^2 \le 1$ **87.** $\left\{\dfrac{1}{2}, \dfrac{3}{2}\right\}$ **89.** $\{-20, 15\}$ **91.** $-\dfrac{5}{2} \le x \le -\dfrac{3}{2}$ **93.** $-\dfrac{1}{5} < y < 1$ **95.** $\{0.19, 2.47, 4.34\}$ **97.** $\left\{\dfrac{15 - \sqrt{5}}{6}, \dfrac{15 + \sqrt{5}}{6}\right\}$ or $\{2.127, 2.873\}$ **99.** $-1 < x < 2$ **101.** $(-\infty, -1) \cup (5, \infty)$ **103.** $(-\infty, 0.19) \cup (2.47, 4.34)$ **105.** a) $\dfrac{\pi}{6}$ b) 0.122π c) -0.722π d) $-\dfrac{5\pi}{6}$ **107.** a) $0.891, 0.454, 1.965, 0.509, 2.205, 1.122$ b) $-0.891, 0.454, -1.965, -0.509, 2.205, -1.122$ c) $\sqrt{3}/2, -1/2, -\sqrt{3}, -\sqrt{3}/3, -2, 2\sqrt{3}/3$ d) $-\sqrt{3}/2, -1/2, \sqrt{3}, \sqrt{3}/3, -2, -2\sqrt{3}/3$ **109.** Graph the functions in $[0, 2\pi]$ by $[-1, 2]$. **111.** $\dfrac{3}{4}$ **113.** $f^{-1}(x) = \dfrac{2 - x}{3}$ **115.** The inverse relation is not a function. Its graph may be obtained using $x_1 = t^3 - t, y_1 = t, -2 \le t \le 2$ in $[-6, 6]$ by $[-2, 2]$. **117.** $0.775, 44.427°$ **119.** Graph $y = |\cos x|$ in $[-\dfrac{\pi}{2}, \dfrac{\pi}{2}]$ by $[0, 1]$. The graph is complete because the function has period π. **121.** For x in the interval $\left[-\dfrac{\pi}{2}, \dfrac{3\pi}{2}\right]$,

$y = \begin{cases} 0, & -\dfrac{\pi}{2} \le x < \dfrac{\pi}{2} \\ -\cos x, & \dfrac{\pi}{2} \le x \le \dfrac{3\pi}{2} \end{cases}.$

Graphing this part gives a complete graph because

the function has period 2π.   **123.** a) $A(x) = (\frac{x}{4})^2 +$ $(\frac{100 - x}{4})^2$ b) Graph this function in $[-50, 150]$ by $[300, 1000]$ c) Domain $= (-\infty, \infty)$, range $= [312.5, \infty)$ d) $0 < x < 100$ e) $50 - 10\sqrt{7}$in and $50 + 10\sqrt{7}$in f) The maximum $(\frac{100}{4})^2 = 625$in^2 cannot be attained. The minimum of 312.5in^2 is attained if both pieces are 50in.   **125.** The graph is the square with vertices $(1, 0), (0, 1), (-1, 0), (0, -1)$.

Chapter 2

Section 2.1

1. 4   **3.** 2   **5.** 9   **7.** 1   **9.** -15 **11.** -2 **13.** 0
15. Limit does not exist.   **17.** Function is not defined at $x = 0$ so substitution cannot be used. The limit does not exist because the left-hand limit -1 does not equal the right-hand limit 1.   **19.** $\frac{1}{2}$ **21.** $-\frac{1}{2}$ **23.** $\frac{1}{2}$
25. $-\frac{1}{4}$ **27.** $\lim_{x \to 0} x \sin \frac{1}{x} = 0$ **29.** The limit does not exist.   **31.** $\lim_{x \to 0} \frac{2^x - 1}{x} = 0.693\ldots$ We will later be able to show that the limit is $\ell n\,2$.   **33.** Limiting value of investment $= \$106.18$. There is not much advantage here in compounding more frequently than about 8 times.   **35.** There appear to be no points of the graph very near to $(0, 4)$. The actual graph is the graph of $y = x + 4$ with the point $(0, 4)$ missing.   **37.** $\lim_{x \to 2^+} f(x) = \infty$, $\lim_{x \to 2^-} f(x) = -\infty$ **39.** a), b), d), e), f)
41. a)

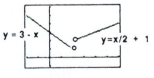

y = 3 - x   y=x/2 + 1

b) $\lim_{x \to 2^+} f(x) = 2$, $\lim_{x \to 2^-} f(x) = 1$
c) Does not exist because right-hand and left-hand limits are not equal.

43. a) A complete graph of $f(x)$ can be obtained on a graphing calculator by graphing both $y = (x - 1)^{-1} + 0\sqrt{(1 - x)}$ and $y = x^3 - 2x + 5 + 0\sqrt{(x - 1)}$ in the viewing rectangle $[-3, 5]$ by $[-25, 25]$.   b) $\lim_{x \to 1^+} f(x) = 4$ and $\lim_{x \to 1^-} f(x)$ does not exist.   c) No. For this limit to exist, the two limits in b) must be equal to the same finite number.   **45.** $a = 15$

47. a)

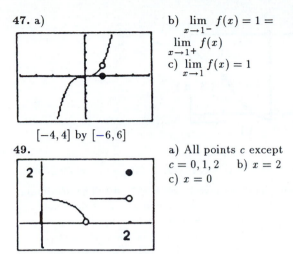

$[-4, 4]$ by $[-6, 6]$

b) $\lim_{x \to 1^-} f(x) = 1 = \lim_{x \to 1^+} f(x)$
c) $\lim_{x \to 1} f(x) = 1$

49.

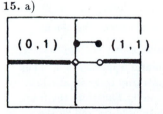

a) All points c except $c = 0, 1, 2$ b) $x = 2$
c) $x = 0$

51. 0   **53.** 0   **55.** 1   **57.** 1   **59.** a) 10   b) 20   **61.** a) 4
b) -21 c) -12 d) $-7/3$ **63.** 0   **65.** 0   **67.** 0
69. $\lim_{x \to 0} (1 + x)^{4/x} = 54.598$ with error less than
0.01.   **71.** $\lim_{x \to 0} \frac{2^x - 1}{x} = 0.6931$ with error less than
0.0001.   **75.** Graph the function in $[1.49\pi, 1.51\pi]$ by $[-0.00001, 0.00001]$ and use TRACE. This suggests the limit is about 9.536×10^{-7}. This is close to the actual limit which can be shown to be $\frac{1}{2^{20}}$.   **77.** $b \to \frac{1}{2}$

Section 2.2

1. a) Yes, $f(-1) = 0$.   b) Yes, $\lim_{x \to -1^+} f(x) = 0$.   c) Yes
d) Yes   **3.** a) No   b) No   **5.** a) $\lim_{x \to 2} f(x) = 0$ b) Define $g(x) = f(x)$, $x \neq 2$, $g(2) = 0$. Then g is an extension of f which is continuous at $x = 2$.   **7.** $f(x)$ is continuous at all points of $[-1, 2]$ except $x = 0$ and $x = 1$.   **9.** $f(x)$ is continuous at all points except $x = 2$.   **11.** $f(x)$ is continuous at all points except $x = 1$.   **13.** $f(x)$ is continuous at all points except $x = 1$.
15. a)

(0, 1)   **(1, 1)**

b) All points except $x = 0$ and $x = 1$

17. $x = 2$ **19.** $x = 1$ and 3   **21.** $x = \pm 1$ **23.** There are no points of discontinuity.   **25.** $x = 0$ **27.** All x with $x < -3/2$ **29.** No points of discontinuity
31. $\lim_{x \to 1} \frac{x^2 - 1}{x - 1} = \lim_{x \to 1} (x + 1) = 2$. Hence $f(1) = \lim_{x \to 1} f(x)$ and f is continuous at $x = 1$.   **33.** $h(2) = 7$
35. $g(4) = 8/5$

37. $a = 4/3$

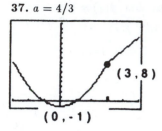

(3,8)

(0,-1)

39. $\lim_{x \to 0} \sec x = 1$ **41.** 1 **43.** 1 **45.** $-\sqrt[3]{2} \approx -1.2599$
47. Both methods give 1.324717957 **49.** The maximum value of f occurs when $x = 2$ and $x = 3$. f does not take on its minimum value 0 but only approaches it arbitrarily closely. Theorem 7 is not contradicted because of the discontinuities. **51.** The maximum value 1 is not attained but is only approached as x approaches ± 1. The minimum value 0 is attained at $x = 0$. Theorem 7 is not contradicted because the interval $(-1, 1)$ is not closed. **53.** We are given $f(0) < 0 < f(1)$. By Theorem 8 there exists some c in $[0, 1]$ such that $f(c) = 0$. A possible graph is

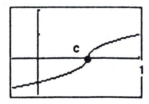

c

1

Section 2.3

1. 1 **3.** $\dfrac{1}{2}$ **5.** 1 **7.** 2 **9.** 1 **11.** -1 **13.** 0 **15.** 4

17. a) Approximately 0.6 b) Very close c) $\dfrac{3}{5}$ **19.** 1

21. a) We graph $y_1 = \dfrac{1}{2} - \dfrac{x^2}{24}$, $y_2 = \dfrac{1 - \cos x}{x^2}$ and
$y_3 = \dfrac{1}{2}$ (y_1 and y_2 are extremely close near $x = 0$) in the window $[-6, 6]$ by $[-0.1, 0.6]$ supporting the inequality. b) $\dfrac{1}{2}$ **23.** The numerator $\cos x$ approaches 1 as $x \to 0$ while the denominator approaches 0. Thus the fraction can be made arbitrarily large in absolute value if x is sufficiently close to 0. Therefore the fraction cannot approach any finite number and so the limit does not exist. **25.** $\dfrac{\pi}{180}$ **27.** $f(x) = \dfrac{\tan 3x}{x}$.

x	± 0.1	± 0.01	± 0.001
$f(x)$	3.0934	3.0009	3.0000

Conjecture: $\lim_{x \to 0} f(x) = 3$. **29.** $f(x) = \dfrac{x - \sin x}{x^2}$.

x	-0.01	0.01	-0.001	0.001
$f(x)$	-0.00167	0.00167	-0.00017	0.00017

Conjecture: $\lim_{x \to 0} f(x) = 0$.

Section 2.4

1. a) $\dfrac{2}{5}$ b) $\dfrac{2}{5}$ **3.** a) 0 b) 0 **5.** a) ∞ b) $-\infty$ **7.** a) 0
b) 0 **9.** a) $-\dfrac{2}{3}$ b) $-\dfrac{2}{3}$ **11.** a) -1 b) -1 **13.** ∞
15. ∞ **17.** ∞ **19.** $-\infty$ **21.** $y = 0$ is the end behavior asymptote. Vertical asymptotes at $x = -\dfrac{5}{2}$ and $x = 1$.
23. EBA: $y = 3$; VA: $x = \pm 2$ **25.** EBA: $y = x - 4$;
VA: $x = -2$ **27.** EBA: $y = x^2 + 2x + 2$; VA: $x = 2$
29. EBA: $y = 1$; VA: $x = \dfrac{1 \pm \sqrt{5}}{2}$ **31.** a) ∞ b) $-\infty$
c) $-\infty$ d) ∞ **33.** a) ∞ b) $-\infty$ **35.** 0 **37.** 1 **39.** ∞
41. $-\infty$ **43.** $0, -\infty, -1, -1$, respectively **45.** 2 **47.** 2
49. 0 **51.** Both limits are equal to 2. **53.** Each graph satisfies $y \to \infty$ as $x \to \infty$ and $y \to -\infty$ as $x \to -\infty$. As the power of x increases, the vertical steepness of the graph increases for $|x| > 1$. **55.** Carrying out the hint proves that $y = -\dfrac{1}{7}$ is an end behavior model for $f(x)$ by definition of end behavior model. **57.** One such
function is $f(x) = \begin{cases} x + 1, & x \le 2 \\ \dfrac{1}{5 - x}, & 2 < x < 5 \\ -1, & x \ge 5 \end{cases}$. Graph $y_1 =$
$x + 1 + 0\sqrt{2 - x}$, $y_2 = \dfrac{1}{5 - x} + 0\sqrt{x - 2} + 0\sqrt{5 - x}$ and
$y_3 = -1 + 0\sqrt{x - 5}$ in $[-5, 10]$ by $[-10, 10]$. **59.** Does not exist, 0, 1, respectively. **61.** Does not exist, 0, 0, respectively. **63.** Let $f(x) = x^3$, $g(x) = \dfrac{1}{x^2}$. Then
$\lim_{x \to 0} (fg) = \lim_{x \to 0} x = 0$. Let $f(x) = 5x^2$, $g(x) = \dfrac{1}{x^2}$.
Then $\lim_{x \to 0} (fg) = \lim_{x \to 0} 5 = 5$. Let $f(x) = x^2$, $g(x) = \dfrac{1}{x^4}$.
Then $\lim_{x \to 0} (fg) = \lim_{x \to 0} \dfrac{1}{x^2} = \infty$. **65.** $\lim_{x \to \pm\infty} \dfrac{f(x)}{a_n x^n} =$
$\lim_{x \to \pm\infty} \left(1 + \dfrac{a_{n-1}}{a_n}\dfrac{1}{x} + \cdots + \dfrac{a_1}{a_n}\dfrac{1}{x^{n-1}} + \dfrac{a_0}{a_n}\dfrac{1}{x^n}\right) = 1 + 0 +$
$\cdots + 0 = 1$ **67.** Using graphs, support $\left(1 + \dfrac{1}{x}\right)^x \to e$
as $x \to \pm\infty$. **69.** Using graphs, we can support $(1 + \dfrac{0.07}{x})^x \to e^{0.07}$ as $x \to \pm\infty$. **71.** $xe^{-x} = \dfrac{x}{e^x} \to 0$ as
$x \to \infty$. $xe^{-x} \to -\infty$ as $x \to -\infty$. **73.** $y = xe^x \to \infty$ as $x \to \infty$. $y \to 0$ as $x \to -\infty$.

Section 2.5

1. All are equivalent except a), e) and g). **3.** g)
5. e) **7.** h) **9.** i) **11.** $-3 \le y \le 7$ **13.** $2 < y < 3$
15. $0 \le y \le 4$ **17.** $\dfrac{9}{5} < y < \dfrac{11}{5}$ **19.** $\left|x - \dfrac{9}{2}\right| < \dfrac{7}{2}$

21. $\left|x + \dfrac{3}{2}\right| < \dfrac{5}{2}$ **23.** $-1.22 < x < -0.71$ **25.** $0.93 < x < 1.36$ **27.** $-1.36 < x < -0.93$ **29.** $9.995 < x < 10.004$ rounding to thousandths appropriately
31. $22.21 < x < 23.81$ **33.** $20 < x < 30$ **35.** $-\dfrac{1}{9} < x < \dfrac{1}{11}$ **37.** $3.94 < x < 4.06$ **39.** $-2.68 < x < -2.66$ rounding to hundredths appropriately **41.** $\ell n\,0.4 < x < \ell n\,0.6$ or rounding to hundredths appropriately $-0.91 < x < -0.52$ **43.** $|x - 3| < 0.5$ **45.** $|x - 1| < 0.04$ **47.** $\lim\limits_{x \to 1} x^2 = 1$, $\lim\limits_{x \to 1} x^2 = 1$, $\lim\limits_{x \to \pi/6} \sin x = 0.5$, $\lim\limits_{x \to 3} \dfrac{x+1}{x-2} = 4$, respectively. **49.** $\lim\limits_{x \to 10} x^2 = 100$, $\lim\limits_{x \to -10} x^2 = 100$, $\lim\limits_{x \to 23} \sqrt{x-7} = 4$, $\lim\limits_{x \to 10} \sqrt{19-x} = 3$, $\lim\limits_{x \to 24} \left(\dfrac{120}{x}\right) = 5$, $\lim\limits_{x \to 1/4} \left(\dfrac{1}{4x}\right) = 1$, $\lim\limits_{x \to 0} \dfrac{3-2x}{x-1} = -3$, $\lim\limits_{x \to -3} \dfrac{3x+8}{x+2} = 1$, respectively. **51.** $3.384 < x < 3.387$ or, in symmetric form, $|x - x_0| < 0.001$. **53.** $x > 702$ **55.** $x > 50 + \sqrt{1504} \approx 88.781$ **57.** Let $x_1 = 0.65241449628$, $x_2 = 0.93923517764$. x must satisfy $x_1 + 2n\pi < x < x_2 + 2n\pi$ or $(2n+1)\pi - x_2 < x < (2n+1)\pi - x_1$. **59.** $x < -175.5$ **61.** $x < -\dfrac{1097}{9}$

Section 2.6

1. $\delta = 2$ **3.** $\delta = \dfrac{1}{2}$ **5.** $\delta = 0.1$ **7.** $\delta = 0.23$ **9.** $\delta = \dfrac{7}{16}$ **11.** $L = 5$. $\delta = 0.005$. **13.** $L = 4$. $\delta = 0.05$.
15. $L = 2$. $\delta = 0.0399$ or any other smaller positive number. **17.** $L = 2$. $\delta = 1/3$. **19.** $\delta = \epsilon$ in each case
21. $\lim\limits_{x \to \infty} f(x) = 1$ **23.** $\lim\limits_{x \to -1+} \dfrac{x+2}{x+1} = \infty$ **25.** $L = \sin 1 \approx 0.84$. $\delta = \epsilon/0.54 = 1.85\epsilon$, using the method of Example 6 and rounding down δ to hundredths to be safe. **27.** 1.17ϵ **29.** 0.30ϵ **31.** 1.78ϵ **33.** $I = (5, 5 + \epsilon^2)$. $\lim\limits_{x \to 5+} \sqrt{x - 5} = 0$.

35.

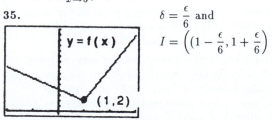

$\delta = \dfrac{\epsilon}{6}$ and $I = \left(\left(1 - \dfrac{\epsilon}{6}, 1 + \dfrac{\epsilon}{6}\right)\right)$

$[-2, 3]$ by $[-1, 16]$

37. $\lim\limits_{x \to 2} f(x) = 5$ means corresponding to any radius $\epsilon > 0$ about 5, there exists a radius $\delta > 0$ about 2 such that $0 < |x - 2| < \delta$ implies $|f(x) - 5| < \epsilon$. **39.** $\delta = \sqrt{4 + \epsilon} - 2$. $\lim\limits_{x \to 2} x^2 = 4$ or $\lim\limits_{x \to 2} (x^2 - 4) = 0$. $\delta \to 0$ as

$\epsilon \to 0$. The graph of δ as a function of ϵ can be viewed by graphing $y = \sqrt{4 + x} - 2$ in the rectangle $[0, 4]$ by $[0, 1]$ and excluding the endpoints.
45. a)

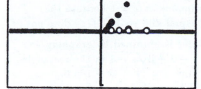

b) $\lim\limits_{x \to 0} f(x) = 0$

$[-2, 2]$ by $[-1, 1]$

Chapter 2 Practice Exercises

1. Exists **3.** Exists **5.** Exists **7.** Continuous at $x = a$ **9.** Not continuous at $x = c$ **11.** -4 **13.** 0 **15.** -1 **17.** Does not exist **19.** 2 **21.** $\dfrac{1}{5}$ **23.** 3 **25.** $\dfrac{2}{5}$ **27.** 0 **29.** $-\infty$ **31.** ∞ **33.** ∞ **35.** $\dfrac{1}{2}$ **37.** 8 **39.** a) 0.78 b) all close to 0.78 c) f appears to have a minimal value at $x = 0$. d) $7/9$ **41.** a) ∞ b) $-\infty$
43. a)

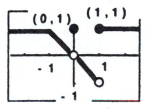

b) $\lim\limits_{x \to -1+} f(x) = 1$, $\lim\limits_{x \to -1-} f(x) = 1$, $\lim\limits_{x \to 0+} f(x) = 0$, $\lim\limits_{x \to 0-} f(x) = 0$, $\lim\limits_{x \to 1+} f(x) = 1$, $\lim\limits_{x \to 1-} f(x) = -1$ c) $\lim\limits_{x \to -1} f(x) = 1$, $\lim\limits_{x \to 0} f(x) = 0$ but $\lim\limits_{x \to 1} f(x)$ does not exist because the right-hand and left-hand limits of f at 1 are not equal. d) Only at $x = -1$ **45.** a) A graph of f may be obtained by graphing the functions $y = abs(x^3 - 4x) + 0\sqrt{1 - x}$ and $y = x^2 - 2x - 2 + 0\sqrt{x - 1}$ in the viewing rectangle $[-5, 7]$ by $[-4, 10]$.

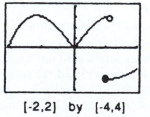

$[-2, 2]$ by $[-4, 4]$

b) $\lim\limits_{x \to 1+} f(x) = \lim\limits_{x \to 1+} (x^2 - 2x - 2) = -3$. $\lim\limits_{x \to 1-} f(x) = \lim\limits_{x \to 1-} |x^3 - 4x| = 3$ c) f does not have a limit at $x = 1$

because the right-hand and left-hand limits at $x = 1$ are not equal.   d) $x^3 - 4x$ is continuous by 2.2 Example 5 and $|x|$ is continuous by 2.2 Example 8. Thus $|x^3 - 4x|$ is continuous by Theorem 5 and so f is continuous for $x < 1$. For $x > 1$, $f(x) = x^2 - 2x - 2$, a polynomial, is continuous. Thus $f(x)$ is continuous at all points except $x = 1$.   e) f is not continuous at $x = 1$ because the two limits in b) are not equal and so $\lim_{x \to 1} f(x)$ does not exist.   **47. a)** A graph of f is obtained by graphing $y = -x + 0\sqrt{1 - x}$ and $y = x - 1 + 0\sqrt{x - 1}$ in the rectangle $[-2, 4]$ by $[-2, 4]$.   b) $\lim_{x \to 1+} f = \lim_{x \to 1+} x - 1 = 0$. $\lim_{x \to 1-} f = \lim_{x \to 1-} -x = -1$.   c) No value assigned to $f(1)$ makes f continuous at $x = 1$.   **49.** $x = \pm 2$ **51.** $y = 0$ **53.** $x^2 - x$ **55. a)** -21 b) 49 c) 0 d) 1 e) 1 f) 7 **57.** 0 **59.** 1 **61.** 0 **63.** Set $k = 8$ **65. a)** $\lim_{x \to 0-} f(x) = 0$ b) $\lim_{x \to 0+} f(x) = \infty$ c) The limit does not exist. To exist both one-sided limits must exist and be equal.   **67.** This is not a contradiction because $0 < x < 1$ is not a *closed* interval.   **69.** True because $0 = f(1) < 2.5 < f(2) = 3$ and so by Theorem 7, $2.5 = f(c)$ for some c in $[1, 2]$.   **73.** $\lim_{x \to 0} \dfrac{\sin x}{x} = 1$ means given any radius $\epsilon > 0$ about 1 there exists a radius $\delta > 0$ about 0 such that for all x $0 < |x - 0| < \delta$ implies $|f(x) - 1| < \epsilon$.   **75.** This "definition" puts a requirement on $f(x)$ but not necessarily for x near x_0. Thus for any given $\epsilon > 0$ there is an x such that $|x^2 - 0| < \epsilon$ but this does not imply $\lim_{x \to x_0} x^2 = 0$ in general.   **77.** $-\dfrac{1}{2} < x < \dfrac{7}{2}$, $\left| x - \dfrac{3}{2} \right| < 2$ **79.** $2.31 < x < 2.35$ or taking $x_0 = 7/3$, $|x - 7/3| < 0.02$.   **81.** $-0.280 < x < -0.228$. $|x + 0.254| < 0.026$ **83.** $0 < \delta \le \epsilon$ **85.** $\lim_{x \to \frac{1}{3}+} \dfrac{1 - 2x}{3x - 1} = \infty$ **87.** $L = 0$, $\delta = 0.01$ **89.** $L = 1$, $\delta = 3/8$ **91.** 3.12ϵ **93.** $65° < t < 75°$

Chapter 3

Section 3.1

1. a) 0 b) -4 **3. a)** 1 b) -0.75 **5. a)** April 15; $\dfrac{2}{3}$ degree per day.   b) Yes. Near January 1 and July 1.   c) Positive from mid-January until about July 1, negative from about July 2 until mid-December.   **7.** $f'(x) = 4x$. When $x = 3$, $m = 12$ and the tangent is $y = 12x - 23$.   **9.** $f'(x) = 4x - 13$. $f'(3) = -1$. Tangent line: $y = -x - 13$.   **11.** $f'(x) = -2/x^2$. At $x = 3$ the slope is $-2/9$ and the tangent has equation $2x + 9y = 12$.   **13.** $f'(x) = \dfrac{1}{(x + 1)^2}$. $f'(3) = \dfrac{1}{16}$. Tangent line: $16y - x = 9$.   **15.** $f'(x) = 1 - \dfrac{9}{x^2}$. At $x = 3$ the slope is 0

and the tangent has equation $y = 6$.   **17.** $f'(x) = \dfrac{1}{2\sqrt{x}}$. $f'(3) = \dfrac{1}{2\sqrt{3}}$. Tangent line: $y - (1 + \sqrt{3}) = \dfrac{1}{2\sqrt{3}}(x - 3)$.   **19.** $f'(x) = \dfrac{1}{\sqrt{2x}}$. At $x = 3$ the slope is $\dfrac{1}{\sqrt{6}}$ and the tangent has equation $\sqrt{6}y = x + 3$.   **21.** Tangent line: $y = 2x + 5$. Graph $y_1 = 4 - x^2$ and $y_2 = 2x + 5$ in $[-5, 5]$ by $[-10, 10]$.   **23.** Tangent line: $y = \dfrac{1}{2}(x + 1)$. Graph $y_1 = \sqrt{x}$ and $y_2 = 0.5x + 0.5$ in $[-3, 5]$ by $[-1, 3]$.   **25.** $f'\left(\dfrac{1}{2}\right) = 0$ **27.** $f'(-1) = -1$ **29.** $f'(4) = -\dfrac{1}{16}$ **31.** The right-hand and left-hand derivatives at $x = 0$ are, respectively, 1 and 0. Since these are unequal, the function is not differentiable at $x = 0$.   **33.** Left- and right-hand derivatives at $x = 1$ are, respectively, $\dfrac{1}{2}$ and 2. Since they are not equal, f is not differentiable at $x = 1$.   **35. a)** $f'(x) = -2x$ b) Graph $y_1 = -x^2$ and $y_2 = -2x$ in $[-3, 3]$ by $[-10, 5]$.   c) $y' > 0$ for $x < 0$ and $y' < 0$ for $x > 0$. $f'(0) = 0$.   d) y increases on $(-\infty, 0)$ and decreases on $(0, \infty)$. The interval on which $y' > 0$ $(y' < 0)$ is the interval on which y increases (decreases).   **37.** $f'(x) = x^2$ b) Graph $y_1 = \dfrac{x^3}{3}$ and $y_2 = x^2$ in $[-3, 3]$ by $[-5, 5]$.   c) $f' > 0$ on $(-\infty, 0)$ and $0, \infty)$. $f'(0) = 0$.   d) f is increasing on the same intervals.

Section 3.2

1. $y = 4x - 3$. The graphs of $y = x^2 + 1$ and $y = 4x - 3$ can be viewed in the rectangle $[-10, 10]$ by $[-10, 20]$.   **3.** $y = \sqrt{3} + 0.58(x + 1)$. The graphs can be viewed in the rectangle $[-6, 6]$ by $[-4, 4]$.   **5.** $y = 0.8(x - 2)$. The graphs can be viewed in the rectangle $[-8, 8]$ by $[-8, 8]$.   **7. a)** Only. We can draw the graph without lifting our pencil, so the function is continuous. But the function is not differentiable at each of the points which are peaks or low points. At these points the left-hand and right-hand derivatives are not equal (there is not a *unique* tangent line).   **9.** c) $x = 0$ is not a point of the domain. At every other point there is a unique tangent line so the funciton is both continuous and differentiable.

11 through 18. In these exercises one may evaluate $D(h)$ and $S(h)$ directly using a calculator, or one may first algebraically simplify $D(h)$ and $S(h)$. If the calculator is used, meaningful results may not be obtained when $h = \pm 10^{-15}$ due to the limits of machine accuracy. This answers part c) of these exercises.

11. b) Conjectures: $f'(2) = 10$, $f'(0) = -2$. $S(h)$ is closer in both cases (in fact exact).   **13. b)** $f'(2) = -0.25$, $S(h)$ is closer. $f'(0)$ and $D(h)$ for $a = 0$ are not defined but $S(h) = 0$ for all h.   **15. a)** If $a = 2$, $D(h) = S(h) = 1$ for all the h's considered. If $a = 0$, $D(h) = $

$\begin{cases} -1, & \text{if } h < 0 \\ 1, & \text{if } h > 0 \end{cases}$ and $S(h) = 0$ for all h.   b) $f'(2) = 1$;

both $S(h)$ and $D(h)$ are exact. $f'(0)$ does not exist.
17. a) If $a = 2$, $D(h) \to 0$ as $h \to 0$ while $S(h) = 0$
for those h's considered. If $a = 0$, $D(h)$ is undefined
for $h < 0$ while $D(h) \to \infty$ as $h \to 0^+$.   b) $f'(2) = 0$,
$S(h)$ is closer. $f'(0)$ does not exist.   **19.** Even though
the derivative may not exist, the values of $S(h)$ may
be defined giving meaningless approximations of the
derivative.   **21.** See the comment for Exercise 19.
25. a) We agree that $f'(a) = \text{NDER}(f(x), a)$ (with
$h = 0.01$) rounded to two decimal places and obtain
$f'(-1) = 0.14$, $f'(0) = -14.94$, $f'(1.5) = -9.69$ and
$f'(3.5) - 0.13$.   b) $\text{NDER}(f(x), a)$ may give results
even though $f'(a)$ does not exist. This was the case for
$f'(0)$.   **27.** b) y_1' exists for all x.   e) y_1 is increasing
over the interval $(-\infty, 0)$ where y_1' is positive. y_1 is
decreasing over the interval $(0, \infty)$ where y_1' is negative.
29. a) Graph $y_1 = \sqrt[3]{x - 2}$ and $y_2 = \text{NDER } y_1$ in $[0, 3]$
by $[-2.2, 3.8]$.   b) y_1' does not exist at $x = 2$ because
the tangent line is vertical there. This is suggested by
the graph because $y_2 \to \infty$ as $x \to 2$. y_1' is positive,
tangent lines have positive slope and y_1 is increasing
for all x except $x = 2$.   **31.** a) Graph $y_1 = \sqrt{1 - x}$ and
$y_2 = \text{NDER } y_1$ in $[-2, 1]$ by $[-2.2, 2.5]$.   b) y_1' does
not exist for $x > 1$ because $(1, \infty)$ is not part of the
domain of y_1. y_1' does not exist at $x = 1$ because the
graph has a vertical tangent there. This is suggested by
the graph because $y_1 \to -\infty$ as $x \to 1^-$. y_1' is negative,
slopes of tangent lines are negative and y_1 is decreasing
on the interval $(-\infty, 1)$.   **33.** $[0.25895206, 0.25895208]$
by $[0.135, 0.145]$ is one possibility.   **35.** a) Let $y_1 =$
$-x^2(x < 0) + (4 - x^2)(x \geq 0)$. Graph $\text{NDER}(y_1, x)$ in
the given window.   b) We see the graph of $y = -2x$,
$x \neq 0$.   c) With $h = 0.01$, $\text{NDER}(f(x), 0) = 200$.

$f'(0)$ does not exist.   **37.** a) Let $y_1 = -\dfrac{x^2}{2} + 0\sqrt{-x}$,

$y_2 = \dfrac{x^2}{2} + 0\sqrt{x}$, $y_3 = \text{NDER}(y_1, x)$, $y_4 = \text{NDER}(y_2, x)$.

Graph y_3, y_4 in the given viewing window.   b) This is

the graph of $y = \begin{cases} -x, & x < 0 \\ x, & x \geq 0 \end{cases}$.   c) $\text{NDER}(f(x), 0) =$

0.005 if $h = 0.01$ but $f'(0)$ does not exist.   **39.** Graph
$y_1 = (x^3 + 6x^2 + 12x)(x < 0) + (-x^2)$ $(x > 0)$ and
$y_2 = \text{NDER}(y_1, x)$ in $[-5, 5]$ by $[-10, 10]$. Since the
two one-sided derivatives at $x = 0$ are unequal, $f'(0)$

does not exist.   **41.** Graph $y_1 = -5(\text{abs } x)^{1/3}$ and
$y_2 = \text{NDER}(y_1, x)$ in $[-5, 5]$ by $[-10, 10]$. $f'(0)$ does
not exist.   **43.** b) $D_x(\sin x) = \cos x$ **45.** By Example
6 of 3.1, $f'(0)$ does not exist. But $\text{NDER}(|x|, 0) = 0$.
$\text{NDER}(f(x), a)$ may exist even if $f(a)$ does not exist.

Section 3.3

1. $1, 0$ **3.** $-2x, -2$ **5.** $2, 0$ **7.** $x^2 + x + 1, 2x + 1$
9. $4x^3 - 21x^2 + 4x, 12x^2 - 42x + 4$ **11.** $8x - 8, 8$
13. $y' = 2x - 1$, $y'' = 2$, $y^{(n)} = 0$ for $n \geq 3$ **15.** $y' =$
$2x^3 - 3x - 1$, $y'' = 6x^2 - 3$, $y''' = 12x$, $y^{(4)} = 12$, $y^{(n)} =$
0 for $n \geq 5$ **17.** $3x^2 + 2x + 1$ **19.** $3x^2$ **21.** $12x + 13$

23. $5x^4 - 2x$ **25.** $\dfrac{8}{(x + 7)^2}$ **27.** $\dfrac{2x^3 - 7}{x^2}$ **29.** $\dfrac{3}{x^4}$

31. $\dfrac{x^2 - 2x - 1}{(1 + x^2)^2}$ **33.** $\dfrac{x^4 + 2x}{(1 - x^3)^2}$ **35.** $\dfrac{-5}{\sqrt{x}(\sqrt{x} - 4)^2}$

37. $\dfrac{1}{\sqrt{x}(\sqrt{x} + 1)^2}$ **39.** $-\dfrac{4x^3 + 3x^2 - 1}{(x^2 - 1)^2(x^2 + x + 1)^2}$

41. $y' = -\dfrac{6}{x^3}$, $y'' = \dfrac{18}{x^4}$ **43.** $y' = -\dfrac{20}{x^5}$, $y'' = \dfrac{100}{x^6}$

45. $y' = 1 - \dfrac{1}{x^2}$, $y'' = \dfrac{2}{x^3}$ **47.** $y - 3^{-0.2} = 0.63(x - 1)$.

The result is confirmed by viewing $y = x3^{-0.2x}$ and
$y = 3^{-0.2} + 0.63(x - 1)$ in the rectangle $[-10, 10]$
by $[-10, 10]$.   **49.** $y = (3/5) + 0.44x$. The result is
confirmed by graphing $y = f(x)$ and $y = (3/5) + 0.44x$
in the rectangle $[-1, 1]$ by $[0.3, 0.9]$.   **51.** The graphs can
be viewed in the two rectangles $[-10, 10]$ by $[-10, 10]$
and $[-50, 50]$ by $[-50, 50]$.   **53.** $y = f'(x)$ or $y =$
$\text{NDER}(f(x))$ can be viewed in the rectangle $[-2, 8]$
by $[-4, 4]$. $y = f''(x)$ or $y = \text{NDER2}(f(x))$ can be
viewed in $[-2, 10]$ by $[-4, 10]$.   **55.** $y = \text{NDER}(f(x))$
can be viewed in the rectangle $[-8, 8]$ by $[-2, 0]$. $y =$
$\text{NDER2}(f(x))$ can be viewed in $[-6, 8]$ by $[-1, 1]$.
57. The graph of $y = \text{NDER}(f(x))$ oscillates, appears to
cross the x-axis infinitely often and to be symmetric with
respect to the origin. The three solutions of $f'(x) = 0$
of the smallest absolute value are $-2.029, 0, 2.029$. The
graph of $y - \text{NDER2}(f(x))$ appears to cross the x-axis
infinitely often. The solution set of $f''(x) > 0$ consists
of an infinite sequence of intervals. The three closest
to the origin are $(-6.578, -3.644)$, $(-1.077, 1.077)$ and
$(3.644, 6.578)$ rounding appropriately.   **59.** We use zoom-
in and graphs of Exercise 53. $f'(x) = 0$ for $x = -0.313$
and $x = 3.198$. $f''(x) > 0$ for x in the set $(-\infty, -1) \cup$
$(1, \infty)$.   **61.** $f'(x) = 0$ has no solution. $f''(x) > 0$
has solution set $(-3, -0.333) \cup (5, \infty)$.   **63.** $f''(x) =$
$6(6x^3 - 45x^2 - 24x + 20)/(3x^2 + 4)^3$. We cannot
solve $f''(x) > 0$ exactly by a convenient method. The
approximate solution set is $(-0.911, 0.460) \cup (7.950, \infty)$.

67. a) 13 b) -7 c) $\dfrac{7}{25}$ d) 20 **69.** c)   **71.** $x + 9y =$

29 **73.** $(-1, 27)$, $(2, 0)$ **75.** $y = 4x$, $y = 2$ **77.** $\dfrac{dP}{dV} =$

$-\dfrac{nRT}{(V - nb)^2} + \dfrac{2an^2}{V^3}$ **79.** $\dfrac{dR}{dM} = CM - M^2$ **83.** a) $x =$

$1.442695\ldots$ b) $x = (1.4427, 0.5307)$ c) very close

85. a) 0.0424 b) $y_2 = \dfrac{y_1(x + 0.01) - y_1(x - 0.01)}{0.02}$

c) About 0.01. NOTE: The answers for a) and c) depend on the h used in the NDER algorithm. We used $\delta = 0.01$ here.

Section 3.4

1. a) The particle first moves left then right. b) $(2,3)$ at $t = 0$, $(0,3)$ at $t = 1$, $(0,3)$ at $t = 2$, $(2,3)$ at $t = 3$. c) The particle changes direction at $(-0.25, 3)$ when $t = 1.5$. When $t = 1.5$, $v = 0$ and $a = 2$. d) 14.5 meters f) The particle is at rest when $t = 1.5$ sec. **3.** We use Tstep 0.05 in the viewing rectangle $[-10, 10]$ by $[-15, 25]$. a) The particle first moves to the right, then to the left and then to the right again. b) $(-3, 3)$, $(-1, 3)$, $(-5, 3)$, $(-9, 3)$ c) $(-0.70, 3)$ when $t = 0.7$, $v = 0.07$, $a = -7.8$; $(-9.30, 3)$ when $t = 3.3$, $v = 0.07$, $a = 7.8$ (TRACE approximations) d) 27.2 meters f) Approximately when $t = 0.7$ and $t = 3.3$ sec. **5.** We use $x_1 = t\sin t$, $y_1 = 3$, $x_2 = t$, $y_2 = \text{NDER}(x_1, t, t)$, $x_3 = t$, $y_3 = \text{NDER}(y_2, t, t)$ with $0 \le t \le 15$, Tstep 0.05 in $[-15, 15]$ by $[-15, 15]$ and TRACE approximations. a) The particle moves right, left, right, left, right, and finally left slightly. b) $(0, 3)$, $(0.84, 3)$, $(1.82, 3)$, $(0.42, 3)$ c) $(1.82, 3)$ when $t = 2.05$, $v = -0.06$, $a = -2.7$; $(-4.814, 3)$ when $t = 4.90$, $v = -0.07$, $a = 5.19$; $(7.91, 3)$ when $t = 8$, $v = -0.17$, $a = -8.2$; $(-11.04, 3)$ when $t = 11.1$, $v = 0.16$, $a = 11.2$; $(14.17, 3)$ at $t = 14.2$, $v = 0.106$, $a = -14.3$ d) 69.75 meters f) $t = 0$, 2.05, 4.9, 8, 11.1 and 14.2. **7.** 4.46 sec on Mars, 0.73 sec on Jupiter. **9.** One possibility: Graph $x_1(t) = t$, $y_1(t) = 24t - 0.8t^2$, $0 \le t \le 30$, in $[0, 30]$ by $[0, 180]$. Then use TRACE and zoom-in if more accuracy is desired. **11.** 320 sec, 52 sec **13.** a) 10^4 per hour b) 0 c) -10^4 per hour **15.** a) The average cost of one washing machine when producing the first 100 washing machines is $\dfrac{c(100)}{100} = \$110$. During production of the first 100 machines the average increase in producing one more machine is: average increase $= \dfrac{c(100) - c(0)}{100 - 0} = \$90$; the fixed cost $c(0) = \$2000$ is omitted with this method. b) \$80 c) \$79.90 **17.** $a = -6\text{m/sec}^2$ when $t = 1$ sec and $a = 6\text{m/sec}^2$ when $t = 3$ sec. **19.** a) We use $-12 \le t \le 12$, Tstep 0.05 in the viewing rectangle $[-35, 35]$ by $[-3, 10]$. b) For this graph we can use $0 \le t \le 6.29 \approx 2\pi$, Tstep 0.05 in $[-3, 8]$ by $[-3, 3]$. c) This line segment can be viewed using $0 \le t \le 6.29$, Tstep 0.05 in $[-6, 10]$ by $[-6, 2]$. **21.** a) All have derivative $3x^2$. c) The result of a) suggests that the family consists of all functions of the form $x^3 + C$ where C can be any constant. d) Yes, $f(x) = x^3$. e) Yes, $g(x) = x^3 + 3$. **23.** a) 190 ft/s b) 2 c) At $8s$ when $v = 0$ d) At $10.8s$ when it was falling at 90 ft/s e) From $t = 8s$ to $t = 10.8s$, i.e., $2.8s$ f) Just before burnout, i.e., just

before $t = 2s$. The acceleration was constant from $t = 2s$ to $t = 10.8s$ during free fall. **25.** a) 0, 0 b) 1700, 1400 c) Rabbits per day and foxes per day **27.** (b) **29.** (d) **31.** a)
b) $x = 0, 2, 4, 5$

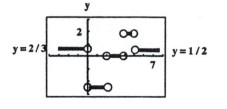

33. $t = 2.832$
35.

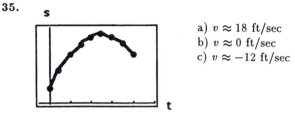

a) $v \approx 18$ ft/sec
b) $v \approx 0$ ft/sec
c) $v \approx -12$ ft/sec

Section 3.5

1. $1 + \sin x$ **3.** $-\dfrac{1}{x^2} + 5\cos x$ **5.** $-\csc x \cot x - 5$ **7.** $\sec x(x \tan x + 1)$ **9.** $x(2\cot x - x\csc^2 x)$ **11.** $3 + x\sec^2 x + \tan x$ **13.** $\sec^2 x$ **15.** 0 **17.** $4\sec x \tan x$ **19.** $-\dfrac{x\sin x + \cos x}{x^2}$ **21.** $\dfrac{1 + \cos x + x\sin x}{(1 + \cos x)^2}$ **23.** $-\dfrac{\csc^2 x}{(1 + \cot x)^2}$ **25.** $\csc x(\csc^2 x + \cot^2 x)$ **27.** Tangent: $y = x$. Normal: $y = -x$. We may graph the three functions $y = \sin x$, $y = x$ and $y = -x$ in the viewing rectangle $[-3, 3]$ by $[-2, 2]$. **29.** Tangent: $y = 2\sin^2 2 + 4\sin 2\cos 2(x - 2)$. Normal: $y = 2\sin^2 2 - (4\sin 2\cos 2)^{-1}(x - 2)$. We may view $y = 2(\sin x)^2$, the tangent and the normal in $[0, 4]$ by $[0, 2.7]$. **33.** $(\tan x)' = \sec^2 x = \dfrac{1}{\cos^2 x}$ and $(\cot x)' = -\csc^2 x = -\dfrac{1}{\sin^2 x}$ cannot be 0 for any value of x. **35.** $y' = 0$ is equivalent to $\cos x = -2$. Since the latter equation has no solution, the graph has no horizontal tangent. **37.** There are horizontal tangents at the points $\left(\dfrac{\pi}{6}, \dfrac{\pi}{6} + \sqrt{3}\right)$, $\left(\dfrac{5\pi}{6}, \dfrac{5\pi}{6} - \sqrt{3}\right)$. **39.** $y = -x + 1 + \dfrac{\pi}{4}$, $y = x + 1 - \dfrac{\pi}{4}$ **41.** $y = -1$ is the tangent line at $\left(\dfrac{\pi}{4}, -1\right)$, the only point at which the tangent is horizontal. **43.** The graph of $\tan x$ and its derivative $\sec^2 x$, $-\dfrac{\pi}{2} < x < \dfrac{\pi}{2}$ may be viewed in the rectangle $[-1.57, 1.57]$ by $[-5, 5]$. **45.** $1/2$

Section 3.6

1. $3\cos(3x+1)$ **3.** $-\frac{1}{3}\sin\left(\frac{x}{3}\right)$ **5.** $(2-3x^2)\sec^2(2x-x^3)$ **7.** $1+2x\sec(x^2+\sqrt{2})\tan(x^2+\sqrt{2})$ **9.** $(2x+7)\csc(x^2+7x)\cot(x^2+7x)$ **11.** $\frac{10}{x^2}\csc^2\left(\frac{2}{x}\right)$

13. $[-\sin(\sin x)]\cos x$ **15.** $y=(2x+1)^5$. $y'=5(2x+1)^4(2)=10(2x+1)^4$. We support this result by graphing $10(2x+1)^4$ and $\text{NDER}((2x+1^5),x)$ in $[-2,1]$ by $[0,810]$ and seeing that the two graphs coincide.

17. $-6x(x^2+1)^{-4}$ **19.** $\left(1-\frac{x}{7}\right)^{-8}$ **21.** $y'=4\left(\frac{x^2}{8}+x-\frac{1}{x}\right)^3\left(\frac{x}{4}+1+\frac{1}{x^2}\right)$ **23.** $\csc x(\csc x+\cot x)^{-1}$ **25.** $2\sin x(2\sin^2 x\cos x+\cos^{-3}x)$

27. $x^2(2x-5)^3(14x-15)$ **29.** $(4x+3)^3(x+1)^{-4}[-3(4x+3)+16(x+1)]=(4x+3)^3(x+1)^{-4}(4x+7)$ **31.** $\frac{2\sin x}{(1+\cos x)^2}$ **33.** $3\left(\frac{x-1}{x}\right)^2\frac{1}{x^2}$

35. $\sin^2 x(4\sin x\sec^2 4x+3\cos x\tan 4x)$ **37.** $\frac{\cos x}{2\sqrt{\sin x}}$

39. $2\sec x\sqrt{\sec x+\tan x}$ **41.** $-\frac{3}{(2x+1)^{3/2}}$

43. $\frac{3x+7}{\sqrt{x+5}}$ **45.** $3\sin\left(\frac{\pi}{2}-3t\right)=3\cos 3t$

47. $\frac{4}{\pi}(\cos 3t-\sin 5t)$ **49.** $-\sec^2(2-\theta)$

51. $\frac{\theta\cos\theta+\sin\theta}{2\sqrt{\theta\sin\theta}}$ **53.** $6\sin(3x-2)\cos(3x-2)$ **55.** $-4(\sin 2x)(1+\cos 2x)$ **57.** $-2[\sin(2x-5)]\cos(\cos(2x-5))$ **59.** $-\frac{\csc^2\sqrt{2x}}{\sqrt{2x}}$ **61.** $2(\sec^2 x)\tan x$

63. $18\csc^2(3x-1)\cot(3x-1)$ **65.** $\frac{5}{2}$ **67.** $-\frac{\pi}{4}$ **69.** 0

71. In both cases we get $-6\sin(6x+2)$. **73.** $\frac{dy}{dx}=1$ in both cases. **75.** 5 **77.** $\frac{1}{2}$ **79.** Tangent: $y-2=\pi(x-1)$. Normal: $y-2=-\frac{1}{\pi}(x-1)$. **81.** a) $\frac{2}{3}$ b) $2\pi+5$ c) $15-8\pi$ d) $\frac{37}{6}$ e) -1 f) $\frac{\sqrt{2}}{24}$ g) $\frac{5}{32}$ h) $\frac{-5\sqrt{17}}{51}$

83. $s=A\cos(2\pi bt)$, $V=s'=-2\pi bA\sin(2\pi bt)$ and $a=v'=-4\pi^2 b^2 A\cos(2\pi bt)$. Now let $s_1=A\cos[2\pi(2b)t]=A\cos(4\pi bt)$. Then the new velocity and acceleration are given by $v_1=4\pi bA\sin(4\pi bt)=2(-2\pi bA)\sin(4\pi bt)$ and $a_1=-16\pi^2 b^2 A\cos(4\pi bt)=4(-4\pi^2 b^2 A)\cos(4\pi bt)$. Thus the amplitude of v is doubled and the amplitude of a is quadrupled. **85.** a) Graph y in $[0,365]$ by $[-12,62]$. b) When $t=101$ or April 12 in a non-leap year. c) About $0.637°\mathrm{F}$ per day.

Section 3.7

1. $\frac{9}{4}x^{5/4}$ **3.** $\frac{1}{3}x^{-2/3}=\frac{1}{3\sqrt[3]{x^2}}$ **5.** $-(2x+5)^{-3/2}$

7. $\frac{2x^2+1}{\sqrt{x^2+1}}$ **9.** $-\frac{(2xy+y^2)}{x^2+2xy}$ **11.** $\frac{1-2y}{2x+2y-1}$

13. $\frac{x(1-y^2)}{y(x^2-1)}$ **15.** $\frac{1}{y(x+1)^2}$ **17.** $\frac{-1}{4\sqrt{x}\sqrt{1-\sqrt{x}}}$

19. $-\left(\frac{9}{2}\right)\csc^{3/2}x\cot x$ **21.** $\cos^2 y$ **23.** $-\frac{\cos^2(xy)+y}{x}$

25. $y'=y/\left[\frac{1}{y}\cos\left(\frac{1}{y}\right)-\sin\left(\frac{1}{y}\right)-x\right]$ **27.** b), c), d).

29. $-\frac{x}{y}$, $-\frac{y^2+x^2}{y^3}=-\frac{1}{y^3}$ **31.** $\frac{x+1}{y}$, $\frac{y^2-(x+1)^2}{y^3}$

33. $\frac{\sqrt{y}}{\sqrt{y}+1}$, $\frac{1}{2(\sqrt{y}+1)^3}$ **35.** a) $7x-4y=2$ b) $4x+7y=29$ **37.** a) $y-3x=6$ b) $x+3y=8$ **39.** a) $y=\frac{6}{7}(x+1)$ b) $y=-\frac{7}{6}(x+1)$ **41.** a) $y=-\frac{\pi}{2}x+\pi$ b) $y=\frac{2}{\pi}x+\frac{\pi^2-4}{2\pi}$ **43.** a) $y=2\pi(x-1)$ b) $y=-\frac{1}{2\pi}(x-1)$

45. $-\frac{\pi}{2}$ **47.** a) At $\left(\frac{\sqrt{3}}{4},\frac{\sqrt{3}}{2}\right)$ the slope is -1 at $\left(\frac{\sqrt{3}}{4},\frac{1}{2}\right)$ it is $\sqrt{3}$. b) Graph $x_1=\sqrt{t^2-t^4}$, $y_1=t$ and $x_2=-x_1$, $y_2=t$, $-1\le t\le 1$ in $[-0.5,0.5]$ by $[-1,1]$. **49.** b) $\frac{3}{2}$ **51.** x-intercepts are $\pm\sqrt{7}$. The tangents at $(\pm\sqrt{7},0)$ both have slope -2. **53.** Graph $y_1=\sqrt{\frac{5-2x^2}{3}}$, $y_2=-y_1$, $y_3=x^{1.5}$, $y_4=-y_3$, $y_5=-\left(\frac{2}{3}\right)x+\frac{5}{3}$, $y_6=\left(\frac{3}{2}\right)x-\frac{1}{2}$, $y_7=\left(\frac{2}{3}\right)x-\frac{5}{3}$, $y_8=-\left(\frac{3}{2}\right)x+\frac{1}{2}$ in $[-6,6]$ by $[-3.5,3.5]$. **55.** $a=\frac{k^2}{2}$

57. b) $xy=\sin^{-1}(-y^5)+2k\pi$ or $xy=(2k+1)\pi+\sin^{-1}(y^5)$. c) Domain: $\left[-\frac{\pi}{2},0\right)$, range of graphed relation: $[-1,0)\cup(0,1]$. d) Domain: $\left(-\infty,\frac{-5\pi}{2}\right]\cup\left[\frac{3\pi}{2},\infty\right)$, range: $[-1,0)\cup(0,1]$. **59.** $\left(-2\sqrt{\frac{7}{3}},-\sqrt{\frac{7}{3}}\right)$, $\left(2\sqrt{\frac{7}{3}},\sqrt{\frac{7}{3}}\right)$

Section 3.8

1. $4x-3$ **3.** $2(x-1)$ **5.** $\frac{x}{4}+1$ **7.** $2x$ **9.** -5

11. $2+\left(\frac{1}{12}\right)(x-8)$ **13.** $L(x)=x$ **15.** $L(x)=\pi-x$

17. $L(x)=1+2\left(x-\frac{\pi}{4}\right)$ **19.** a) $1+2x$ b) $1-5x$

c) $2(1+x)$ d) $1-6x$ e) $3+x$ f) $1-\left(\frac{1}{2}\right)x$

21. $L(x) = 1 + \left(\frac{3}{2}\right)x$ is the sum of the linearizations of $\sqrt{x+1}$ and $\sin x$. **23.** c), d) the sequence of numbers always approaches 1. **25.** a) 0.21 b) 0.2 c) 0.01 **27.** a) 0.231 b) 0.2 c) 0.031 **29.** a) $-\frac{1}{3}$ b) $-\frac{2}{5}$ c) $\frac{1}{15}$ **31.** $4\pi r_0^2\, dr$ **33.** $3x_0^2\, dx$ **35.** $2\pi r_0 h\, dr$ **37.** a) $0.08\pi \approx 0.2513 m^2$ b) 2.000% **39.** 3% **41.** 3% **43.** 1/3 of 1% **45.** The variation of the radius should not exceed 1/2000 of its ideal value, that is, 0.05% of the ideal value. **49.** b) $x = 28\sqrt{3} - 45$ c) $g(28\sqrt{3} - 45) \approx -0.009$ d) $x = 3.156$ with error at most 0.01 e) $\frac{225}{64}$

51. $3(x^2 - 1)dx$ **53.** $\frac{2(1-x^2)dx}{(1+x^2)^2}$ **55.** $y = \frac{x}{1+x}$, $dy = \frac{dx}{(1+x)^2}$ **57.** $5\cos(5x)dx$ **59.** $2\sec^2\left(\frac{x}{2}\right)dx$

61. $\csc\left(1 - \frac{x}{3}\right)\cot\left(1 - \frac{x}{3}\right)dx$ **63.** b) The slope of the straight line should be $f'(0) = 0.5$ or very close to it.

Chapter 3 Practice Exercises

1. $5x^4 - \frac{1}{4}x + \frac{1}{4}$ **3.** $2(x+1)(2x^2 + 4x + 1)$

5. $2\cos 2x$ **7.** $\frac{1}{(x+1)^2}$ **9.** $-4x^2(x^3 + 1)^{-7/3}$

11. $2\sin(1 - 2x)$ **13.** $3(x^2 + x + 1)^2(2x + 1)$

15. $\frac{4(x+2)}{\sqrt{4x^2 + 16x + 15}}$ **17.** $\frac{-y}{x + 2y}$ **19.** $\frac{5 - 2x - y}{x + 2y}$

21. $\frac{-1}{3(xy)^{1/5}}$ **23.** $\frac{1}{2y(x+1)^2}$

25. $\frac{(5x+1)(5x^2 + 2x)^{1/2}}{2y}$ **27.** $\frac{x-1}{2x\sqrt{x}}$ **29.** $3\sec(1 + 3x)\tan(1 + 3x)$ **31.** $-2x(\csc x^2)^2$

33. $\frac{x^2 - 2x - 1}{(1-x)^{1/2}(1 + x^2)^{3/2}}$

35. a)

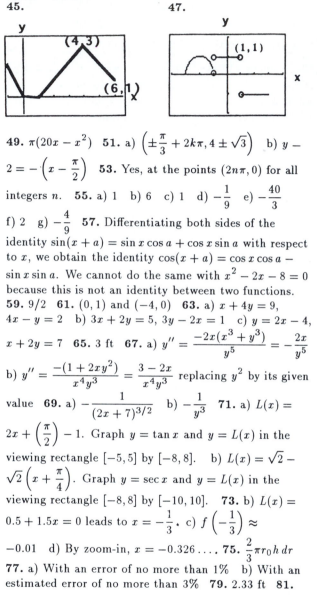

b) Yes c) f is not differentiable at $x = 1$ because its left-hand derivative (1) is not equal to its right-hand derivative (−1) at $x = 1$.

37. $(-1, 27)$, $(2, 0)$ **39.** b) $5\sqrt{2}$ c) $-10, 10$ d) at -10, $v = 0$, $a = 10$; at 10, $v = 0$, $a = -10$. e) The particle first reaches the origin at $t = \pi/4$. At that time velocity $= -10$, speed $= 10$, acceleration $= 0$. **41.** a) $\frac{4}{7}$ sec; 280

cm/sec b) 560 cm/sec; 980 cm/sec^2 **43.** a)(iii) b) (i) c) (ii)

45.   **47.**

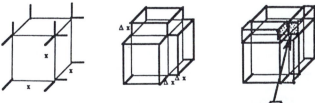

49. $\pi(20x - x^2)$ **51.** a) $\left(\pm\frac{\pi}{3} + 2k\pi, 4 \pm \sqrt{3}\right)$ b) $y - 2 = -\left(x - \frac{\pi}{2}\right)$ **53.** Yes, at the points $(2n\pi, 0)$ for all integers n. **55.** a) 1 b) 6 c) 1 d) $-\frac{1}{9}$ e) $-\frac{40}{3}$ f) 2 g) $-\frac{4}{9}$ **57.** Differentiating both sides of the identity $\sin(x + a) = \sin x \cos a + \cos x \sin a$ with respect to x, we obtain the identity $\cos(x + a) = \cos x \cos a - \sin x \sin a$. We cannot do the same with $x^2 - 2x - 8 = 0$ because this is not an identity between two functions. **59.** 9/2 **61.** $(0, 1)$ and $(-4, 0)$ **63.** a) $x + 4y = 9$, $4x - y = 2$ b) $3x + 2y = 5$, $3y - 2x = 1$ c) $y = 2x - 4$, $x + 2y = 7$ **65.** 3 ft **67.** a) $y'' = \frac{-2x(x^3 + y^3)}{y^5} = -\frac{2x}{y^5}$ b) $y'' = \frac{-(1 + 2xy^2)}{x^4y^3} = \frac{3 - 2x}{x^4y^3}$ replacing y^2 by its given value **69.** a) $-\frac{1}{(2x + 7)^{3/2}}$ b) $-\frac{1}{y^3}$ **71.** a) $L(x) = 2x + \left(\frac{\pi}{2}\right) - 1$. Graph $y = \tan x$ and $y = L(x)$ in the viewing rectangle $[-5, 5]$ by $[-8, 8]$. b) $L(x) = \sqrt{2} - \sqrt{2}\left(x + \frac{\pi}{4}\right)$. Graph $y = \sec x$ and $y = L(x)$ in the viewing rectangle $[-8, 8]$ by $[-10, 10]$. **73.** b) $L(x) = 0.5 + 1.5x = 0$ leads to $x = -\frac{1}{3}$. c) $f\left(-\frac{1}{3}\right) \approx -0.01$ d) By zoom-in, $x = -0.326\ldots$ **75.** $\frac{2}{3}\pi r_0 h\, dr$ **77.** a) With an error of no more than 1% b) With an estimated error of no more than 3% **79.** 2.33 ft **81.**

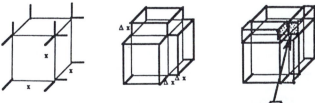

Chapter 4

Section 4.1

1. $4/3, -2$ **3.** 0 **5.** $-1, 1, 3$ **7.** 0 **9.** a) Between
two zeros of a function lies a zero of its derivative.
11. Between zeros of $f(x) = \sin x$ there is a zero of
$f'(x) = \cos x$. **13.** a) Use the window $[-5, 5]$ by $[-1, 15]$.
b) no c) no d) Local maxima at $(\pm 1.73, 10.39)$;
local minima at $(\pm 3, 0)$, $(0, 0)$. f) Increasing on
$[-3, -1.73] \cup [0, 1.73] \cup [3, \infty)$; decreasing on $(-\infty, -3] \cup$
$[-1.73, 0] \cup [1.73, 3]$ **15.** Local maximum at $(5/2, 6.25)$;
increasing on $[0, 5/2]$, decreasing on $[5/2, 6]$; absolute
maximum at $(5/2, 6.25)$, absolute minimum at $(6, -6)$.
Local minimum at $(0, 0)$. **17.** Absolute maximum at
$(8, 2)$; absolute minimum at $(4, 0)$; increasing on $[4, 8]$.
19. Local and absolute minima at $(\pm\sqrt{5}, -16)$; absolute
maximum at $(0, 9)$, local maxima at $(\pm 3, 0)$. Increasing
on $[-\sqrt{5}, 0] \cup [\sqrt{5}, 3]$, decreasing on $[-3, -\sqrt{5}] \cup$
$[0, \sqrt{5}]$. **21.** Absolute minimum at $(-5, -115.567)$,
absolute maximum at $(5, 114.433)$; local minimum at
$(0.559, -2.639)$, local maximum at $(-1.126, -0.362)$.
Increasing on $[-5, -1.126] \cup [0.559, 5]$; decreasing on
$[-1.126, 0.559]$. **23.** $1/2$ **25.** 1 **27.** $s'(c) = v(c) = 79.5$
29. $v(c) = 7.66$ knot
31.

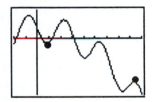

33. $0, \pm 2.029$ **35.** 1 **37.** $y' = -1/x^2 < 0$. By Cor 1, y
is decreasing. **43.** $f(-2) = 11$, $f(-1) = -1$. $f'(x) \neq 0$
on $(-2, -1)$. **45.** $f(1) = -1$; $f(3) = 7/3$. $f'(x) =$
$1 + \dfrac{2}{x^2}$, which is never zero. **47.** Select $b \neq 3$. By
the M.V.T. $f(b) - 3 = 0 \cdot (b - 0) = 0$. **49.** a) $(2, 5)$
b) $(-2, 2)$ c) $x = 2$ **51.** $f(x) = \sqrt{(x - a)(b - x)}$, for
example
53.

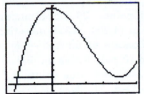

55. a) 1.000 b)

a	$f(a)$	b	area
0.5	0.303	1.762	0.382
0.8	0.359	1.238	0.157
1.0	0.368	1.0	0
1.2	0.361	0.817	0.138
1.5	0.335	0.627	0.292

c) near 0.3, 0.459

Section 4.2

1. $f' > 0$ on $(-\infty, -1) \cup (1, \infty)$; $f' < 0$ on $(-1, 1)$; $f' = 0$
for $x = \pm 1$. **3.** Rising on $(-\infty, -2]$ and $[0, 2]$; falling
on $[-2, 0]$ and $[2, \infty)$. Local maxima at $x = -2$, $x = 2$;
local minimum at $x = 0$. **5.** Falling on $(-\infty, 1/2]$,
rising $[1/2, \infty)$: concave up everywhere, local minimum
at $(1/2, -5/4)$. **7.** Local maximum at $(1, 5)$; local
minimum at $(3, 1)$, inflection point at $(2, 3)$. Rising
$(-\infty, 1] \cup [3, \infty)$, falling on $[1, 3]$, concave down for
$x < 2$, concave up for $x > 2$. **9.** Local minima $(\pm 1, -1)$,
local maximum $(0, 1)$, inflection points $(\pm 1/\sqrt{3}, -1/9)$.
Rising on $[-1, 0] \cup [1, \infty)$; falling on $(-\infty, -1] \cup [0, 1]$;
concave down on $(-1/\sqrt{3}, 1/\sqrt{3})$; concave up otherwise.
11. Decreasing between $(0.667, 11.037)$ and $(1, 11)$.
Inflection point at $(5/6, 11.019)$. Falling on $[2/3, 1]$, rising
elsewhere. **13.** Local minima at $\pm 1/\sqrt{6}, -121/12$,
local maximum at $(0, -10)$. Inflection points occur
when $x = \pm 1\sqrt{18})$. Concave up for $x < -1/\sqrt{18}$
and $x > 1/\sqrt{18}$, concave down on $(-1/\sqrt{18}, 1/\sqrt{18})$.
Falling on $(-\infty, -1/\sqrt{6}] \cup [0, 1/\sqrt{6}]$; rising elsewhere.
15. Inflection point (π, π); always rising, minimum
at $(0, 0)$, maximum at $(2\pi, 2\pi)$; concave down for $0 <$
$x < \pi$, concave up $\pi < x < 2\pi$. **17.** Local minima at
$(-2.115, -22.236)$ and $(1.861, -6.268)$; local maximum
at $(0.254, 2.504)$; inflection points at $(-2/\sqrt{3}, -11.508)$
and $(2/\sqrt{3}, -2.270)$. Concave down on $(-2/\sqrt{3}, 2/\sqrt{3})$,
concave up elsewhere. Falling on $(-\infty, -2.135] \cup$
$[0.254, 1.861]$, rising elsewhere. **19.** Local maximum
at $(-1.263, 2.435)$; inflection points at $(-\sqrt{1/3}, 0.288)$

and $(\sqrt{1/3}, -3.176)$. Rising on $(-\infty, -1.263]$, falling thereafter; concave up on $(-1/\sqrt{3}, 1/\sqrt{3})$, concave down elsewhere.   **21.** Graph is always rising; inflection point at $(0, 3)$; concave up for $x < 0$, concave down for $x > 0$. **23.** Defined for $x \geq 0$; always rising and concave down; local minimum at $(0, -1)$.   **25.** The graph is always rising; concave up for $x < 2$; inflection point at $(2, 2.5)$. **27.** Local minimum at $(0, 1)$. Always rising, concave down for $x < 0$, concave up for $x > 0$, but $x = 0$ is not an inflection point.   **29.** Local minimum at $(0, 0)$; local maximum at $(1.200, 2.033)$, inflection point at $(-0.6, 2.561)$. Concave up for $x < -0.6$, concave down on $(-0.6, 0)$ and $(0, \infty)$. Rising on $[0, 1.2]$.   **31.** Local minimum at $(1, -3)$; inflection points at $(-2, 7.560)$ and $(0, 0)$; falling on $(-\infty, 1]$, rising thereafter.
33.

time	$a(t) = 2$	$v(t) = 2t - 4$	speed	direction
$t < 2$	pos	neg, inc	slowing	to the left
$t = 2$	pos	0		stopped
$2 < t$	pos	pos, inc	gaining	to the right

35.

time	$a(t)$	$v(t)$	speed	direction
$t < -1$	neg	pos, dec	dec	to the right
$t = -1$	neg	0	0	stopped
$-1 < t < 0$	neg	neg, dec	inc	to the left
$0 < t < 1$	pos	neg, inc	dec	to the left
$1 < t$	pos	pos	inc	to the right

37. Local minimum at $x = 2$; inflection points at $x = 1$, $5/3$.   **39.** Local maximum at $(-1, -2)$; local minimum at $(1, 2)$.   **41.** No, f might have an inflection point.
43. True   **45.** a) $t = 2, 6, 9.5$ b) $t = 4, 8, 11.5$
47.   **49.**

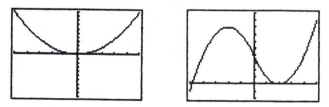

51. Viewing rectangle: $[-1, 1]$ by $[-1, 1]$.   **53.** Viewing rectangle: $[-0.5, 0.5]$ by $[-5, 5]$.   **55.** $y_2 = (-2a^3) + (3a^2 - 9)x$; locally y_2 is below y_1 for $x > 0$.   **57.** Graph in $[0, 300]$ by $[-20, 150]$ using $0 \leq t \leq 10$.   **59.** $75.964°$

Section 4.3

1. $0.618033988, -1.618033989$ **3.** 1.16403514, -1.452626879 **5.** $0.6301153962, 2.5732719864$
7. $3.216451347, -1.564587289$ **9.** 1.89207115 **11.** If $f'(x_0) \neq 0$, all $x_n = x_0$.

13. $x_0 = h > 0 \Rightarrow x_1 = x_0 - \dfrac{\sqrt{x_0}}{1/(2\sqrt{x_0})} = -h$

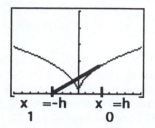

15. Always rising; concave down $x < 1$, concave up $x > 1$. Inflection point $(1, -1)$. One real root.
17. Rising to local maximum $\left(\dfrac{2 - \sqrt{13}}{3}, 8.879\right)$,

then falling to local minimum at $\left(\dfrac{2 + \sqrt{13}}{3}, 1.9354\right)$,

then rising; inflection point at $\left(\dfrac{2}{3}, 5.407\right)$; concave down for $x < 2/3$, concave up for $x > 2/3$. One real root.   **19.** Local minimum $(-1/\sqrt{12}, 11.808)$; local maximum at $(1/\sqrt{12}, 12.192)$, inflection point at $(0, 12)$; one real root. Rising on $[-1/\sqrt{12}, 1/\sqrt{12}]$, falling elsewhere. Concave up for $x < 0$, concave down for $x > 0$.   **21.** Use $[-10, 10]$ by $[-60, 60]$. Graph is always falling; inflection point $(0, 20)$. Concave up for $x < 0$, concave down for $x > 0$. One real root.   **23.** Falling to local minimum at $(-0.383, 7.785)$, then rising; concave up everywhere; no real roots.   **25.** Use $[-1, 6]$ by $[-30, 10]$. Local maximum $(0.259, 0.001)$; inflection point $(1.417, -12.412)$; local minimum $(2.547, -24.825)$; one zero. Rising on $(-\infty, 0.259] \cup [2.574, \infty)$, falling on $[0.259, 2.574]$; concave down for $x < 1.417$, concave up for $x > 1.417$.   **27.** Use $[-5, 5]$ by $[-17, 17]$. Graph rises to a local maximum at $(2.000, 15.333)$, then falls. Concave down for $x < 1.333$, then concave up. Inflection points $(0, 10)$ and $(1.333, 13.160)$; two real roots.   **29.** Use $[-3, 3]$ by $[0, 34]$. Falling to a local minimum at $(-1.107, 17.944)$; rising to a local maximum at $(0.270, 20.130)$; falling to a local minimum at $(0.837, 19.927)$; rising thereafter. Concave down between inflection points $(-0.577, 18.867)$ and $(0.577, 20.022)$; concave up elsewhere. No real roots.   **31.** Use $[-8, 4]$ by $[-20, 35]$. Rising to a local maximum at $(-4.023, 32.012)$; falling to a local minimum at $(-1.514, -0.189)$, rising to a local maximum at $(0.287, 13.095)$. Concave up between inflection points $(-3, 18)$ and $(-0.5, 7.063)$; concave down elsewhere. Four real roots.   **33.** Falling to local minimum at $(-2.601, -7.580)$; rising to local maximum at $(-1.097, 21.196)$; falling to local minimum at $(0.534, -0.495)$; rising to $(2.364, 52.006)$, falling thereafter. Concavity: up until $(-2.016, 4.530)$, down until $(-0.266, 10.206)$, up until $(1.681, 31.029)$, down thereafter. Five real roots.   **35.** Use $[-2, 2]$ by $[-50, 50]$. Always rising; inflection point at $(-0.288, -2.922)$ where concavity changes from concave down to concave up. One real root. **37.** 0 and 20 **39.** $5/3 \times 14/3 \times 35/3$ inches **41.** a) $y = 1 - x$ b) $A(x) = 2x(1-x)$ c) $1/2$ **43.** a) $v(0) = 100$ ft/sec b) 356.25 feet c) -150.997 feet/sec **45.** $18 \times 18 \times 36$ inches **47.** a) $x = 12$, $y = 6$ b) $x = 12$, $y = 6$ **49.** 0.653 **51.** a) $a = -3$, $b = -9$ b) $a = -3$, $b = -24$ **53.** $32\pi/3$ **55.** 6 inches deep by $6\sqrt{3}$ inches wide **57.** $p = K/2$ **59.** $M = c/2$ **61.** Answer depends on technology used; 1.003 is common. **63.** c) $1.879, -0.347, -1.532$ **65.** Use $x - (\tan x) \times (\cos x)^2 \to x$ **67.** b) $v = 2x(24 - 2x)(18 - 2x)$ d) $0 < x < 9$ e) $x = 3.394$ in, $v = 1309.955$ in^3 f) $x = 2$ or $x = 5$ **69.** Square cut from 15-in end: $x = 1.962$ in, $v_{\max} = 66.019$ in^3. **71.** f and g will have extrema at the same values of x although their nature may be reversed. **73.** 0 **75.** Solve $3aL - 2b = 0$ and $-aL^3 + bL^2 = H$ for a and b.

Section 4.4

1. No extrema or inflection points; always rising; concave up if $x < 0$, concave down if $x > 0$. $[-5, 5]$ by $[5, 5]$ shows a complete graph. **3.** Local minima at $[\pm 1, 2)$. $[-5, 5]$ by $[0, 5]$ shows a complete graph. Always concave up. Rising on $[-1, 0) \cup [1, \infty)$, falling on $(-\infty, -1] \cup (0, 1]$. **5.** Concavity: down $(-\infty, -2)$, up $(-2, 0)$, down $(0, 2)$, up $(2, \infty)$. Falling on $(-\infty, -2) \cup (-2, 2) \cup (2, \infty)$. **7.** $[-5, 5]$ by $[-5, 5]$ shows a complete graph. $(0, -1)$ is a local maximum. No inflection points. Concave up for $|x| > 1$, concave down on $(-1, 1)$. Rising on $(-\infty, -1) \cup (-1, 0]$. Falling on $[0, 1) \cup (1, \infty)$. **9.** Local maximum at $(0, -2)$. The graph is that of the negative of #7, increased by 1. $[-5, 5]$ by $[-5, 5]$ shows a complete graph. **11.** $[-10, 10] \times [-10, 10]$ shows a complete graph. Concave up for $x < 1$, rising on $(-\infty, 1) \cup (1, \infty)$. **13.** $[-5, 4]$ by $[-10, 25]$ shows a complete graph. Local minimum at $(0.575, 0.144)$. Inflection point at $(-3, 11)$. Concave up on $(-\infty, -3) \cup (-2, \infty)$, concave down on $(-3, -2)$; falling on $(-\infty, -2) \cup (-2, 0.575]$, rising on $[0.575, \infty)$. **15.** Use $[-4, 4]$ by $[-0.5, 1.5]$. $(-2.414, -0.207)$ is a local minimum, $(-0.268, 0.683)$ is an inflection point, $(0.414, 1.207)$ is a local maximum, $(1, 1)$ is an inflection point. Rising on $(-2.414, 0.414]$, falling on $(-\infty, 2.414]$ and $[0.414, \infty)$; concave down on $(-0.268, 1)$, concave up elsewhere. **17.** $[-4, 4]$ by $[-10, 10]$ shows a complete graph. $(-0.475, -3.331)$ local maximum, $(0.490, 0.800)$ local minimum. Rising on $(-\infty, -2) \cup [-2, -0.475] \cup [0.490, \infty)$; falling on $[-0.475, 0) \cup (0, 0.490]$. Concave up on $(-\infty, -2) \cup (0, \infty)$; concave down on $(-2, 0)$. **19.** Graph needs both $[-2, 6]$ by $[-10, 30]$ and $[-2, 2]$ by $[-2, 2]$ to be seen completely. Local minima at $(0.243, -1.589)$ and $(2.543, 18.459)$; local maximum at $(1.214, -0.869)$; inflection point at

$(0.855, -1.158)$. Rising on $[0.243, 1.214] \cup [2.543, \infty)$; falling on $[-\infty, 0.243] \cup [1.214, 2) \cup (2, 2.543]$. Concave up on $(-\infty, 0.855) \cup (2, \infty)$, concave down on $(0.855, 2)$. **21.** $[-5, 5]$ by $[-15, 15]$ shows a complete graph. Inflection points at $(0, -0.5)$ and $(-3.005, 10.792)$; local minimum at $(1.666, 2.884)$. Falling on $(-\infty, -2) \cup (-2, 1) \cup (1, 1.666]$; rising on $[1.666, \infty)$; concave down on $(-3.005, -2)$ and $(0, 1)$, concave up on $(-\infty, -3.005)$, $(-2, 0)$ and $(1, \infty)$. **23.** Three views are necessary: $[-5, 5]$ by $[-100, 100]$, $[-3.2, -2.8]$ by $[-100, 100]$ and $[2.8, 3.2]$ by $[-100, 100]$. Local maximum at $(2.919, 45.572)$, and a local minimum at $(3.077, 62.540)$. Inflection points at $(-3.257, -67.885)$, $(0.004, -0.222)$ and $(2.727, 39.274)$. Rising on $(-\infty, -3) \cup (-3, 2.919] \cup [3.077, \infty)$, falling on $[2.919, 3) \cup (3, 3.077)$; concave up for $-3.257x < -3$, $0.440 < x < 2.727$, and $x > 3$; concave down on $(-\infty, -3.257) \cup (-3, 0.004) \cup (2.919, 3)$. **25.** $[-4, 4]$ by $[-1, 2]$ shows a complete graph. Inflection points at $(\pm 1.155, 1.5)$, maximum at $(0, 2)$. Rising on $(-\infty, 0]$, falling on $[0, \infty)$; concave down for $|x| < 1.155$, concave up for $|x| > 1.55$. **27.** 16 **29.** 12×18 m, 72 m **31.** $x = 15$ ft, $y = 5$ ft **33.** $h = r = \dfrac{10}{\sqrt[3]{\pi}}$ cm **35.** Minimum value occurs when $x = 51/8$; minimum $L = 11.04$ inches. **37.** a) 16 b) -1 **39.** $y \geq \dfrac{147}{9} > 0$ **41.** $50 + \dfrac{c}{2}$ **43.** $(2km/h)^{1/2}$ **45.** Maximum of p is $p(3) = 0$ **47.** The denominators of f, f', f'' have the same zeros. **49.** Diameter and height are the same.

Section 4.5

Exercises 1–48. For all problems, local extrema and inflection points are given. If intervals for rising, falling, and/or concavity are not specified and there are no discontinuities, the graph rises as y goes from a local minimum to a local maximum, etc. Concavity changes at an inflection point. Normally, if y is differentiable at a local maximum (minimum) $x = c$, the graph is concave down (up) on a neighborhood of c. **1.** Always rising; concave up for $x < 0$, concave down for $x > 0$; inflection point at $(0, 0)$. **3.** Local minimum at $(0, 0)$, increasing and concave up for $x \geq 0$. **5.** For $x > -3/2$, y is rising and concave down; local minimum at $(-1.5, 0)$. **7.** The interval $-0.144 \leq x \leq 1.999$ (between $3x + 5 = 3\pi/2$ and $3x + 5 = 7\pi/2$) is one period; $y = 1$, a minimum at the endpoints. $y = 5$, a maximum, at $x = 0.951$. Inflection points at $(0.428, 3)$, $(1.475, 3)$. **9.** Local maximum at $(\pi/12, \sqrt{2})$, inflection point $(3\pi/12, 0)$, local minimum $(5\pi/12, -\sqrt{2})$, inflection point $(7\pi/12, 0)$, local maximum $(9\pi/12, \sqrt{2})$. **11.** Always increasing, concave down for $x < 0$, inflection point at $x = 0$, concave up for $x > 0$. **13.** Concave down and increasing from a minimum at $(3/2, 0)$. **15.** Use

$[0, 50]$ by $[-3, 4]$. Increasing from a minimum of $(4, -2)$, concave down. **17.** Local minimum at $(3\pi/4, -2)$; local maximum at $(\pi/4, -8)$. Not defined at $x = \pi/2$; concave down on $(0, \pi/2)$, concave up on $(\pi/2, \pi)$. **19.** Inflection points at $(0, 0)$ and $(0.268, 0.716)$. Local minimum at $(-0.341, -0.582)$. **21.** y always increasing; $x \le -2 \Rightarrow y$ increasing; concavity changes from up to down at $(-2, 0)$. **23.** Graph on $[-\pi, \pi]$ by $[-2, 2]$. Minimum at $(0, 1)$, maximum at $(\pi/2, -1)$; concave up on $(-\pi/4, \pi/4)$, concave down on $(\pi/4, 3\pi/4)$. **25.** Use $[-5, 5]$ by $[0, 100]$; always rising and concave up. **27.** Graph $y = 3(ln(x + 1))/ln2$ in $[-5, 5]$ by $[-10, 10]$. Rising for $x > -1$, concave down. **29.** Use $[-5, 5]$ by $[0, 4]$; concave down where defined, local minima at $(0, 0)$ and $(2, 0)$. **31.** Use $[-\pi/2, 3\pi/2]$ by $[-8, 8]$; $(0, 0)$ and $(\pi, 0)$ are inflection points; rising on $(-\pi/2, \pi/2)$, falling on $(\pi/2, 3\pi/2)$, concave up on $(0, \pi/2) \cup (\pi/2, \pi)$. **33.** Use $[-\pi/2, 3\pi/2]$ by $[-3, 3]$, $x\,\mathrm{Scl} = \pi/16$;

Local minima at	Local maxima at	Inflection points at
$(-0.968, -1.906)$	$(0.216, 1.216)$	$(-0.413, -0.408)$
$(1.228, -0.223)$	$(1.914, 0.223)$	$(0.673, 0.542)$
$(2.925, -1.216)$	$(4.109, 1.906)$	$(1.571, 0)$
		$(2.469, -0.542)$
		$(3.554, 0.408)$
		$(4.712, 0)$

35. Use $[-\pi, \pi]$ by $[-6, 6]$; $(-1.298, -4.132)$ is a local minimum, $(-0.858, -3.890)$ inflection point, $(-0.578, -3.718)$ local maximum, $(0.578, 3.718)$ local minimum, $(0.86, 3.90)$ inflection point, $(1.298, 4.132)$ local maximum. **37.** Minimum at $(0.368, -0.368)$, analytically at $(1/e, -1/e)$. **39.** Local minima at $(\pm 0.60, -0.18) = (\pm\sqrt{(1/e)}, -0.18)$. **41.** Minimum at $(0, 1/2)$. **43.** Minimum/maximum at $(\mp 0.850, \mp 0.515)$; inflection points at $(\mp 1.471, \mp 0.328)$ and $(0, 0)$. **45.** Maxima at $(\pm 4.493, -4.603)$, $(\pm 10.904, -10.950)$; minima at $(\pm 7.725, 7.790)$. Concavity changes $x = K\pi$, $K = \pm 1, \pm 2 \dots$. Not defined at $x = 0$. **47.** Inflection points at $(-2.626, -0.500)$, $(1.090, 0.089)$ and $(5.333, 0.182)$. Not defined at $x = 0$. Concave down $-2.626 < x < -1$ and $1.090 < x < 5.333$; concave up $x < -2.626$, $-1 < x < 1.090$, $5.333 < x$. Concave down $x < -1$ and $x > 1.090$. **49.** Min $y = (-5)^{3/5} = -2.627$; max $y = 2.627$. **51.** Maximum $= 7.460$, minimum $= -172.64$. **53.** Use $[-3, 3]$ by $[-3, 3]$; inflection point at $(0, 0)$. **55.** Use $[-3, 3]$ by $[-3, 3]$; concavity changes at $x = 0$. **57.** Use $[-3, 3]$ by $[-3, 3]$. **59.** 6.25 cm^2 **61.** 5 **63.** $\pi/6$ **65.** 2 **67.** No; minimum value is 0. **69.** 0.873 miles from point opposite the boat. **71.** $A = 8 \sin\theta \cos\theta$, 4 **73.** 29.925

Section 4.6

1. $\dfrac{dA}{dt} = 2\pi r \dfrac{dr}{dt}$ **3.** $\dfrac{dV}{dt} = \left(\dfrac{2}{3}\right)\pi r h \dfrac{dr}{dt}$

5. a) $\dfrac{dP}{dt} = I^2 \dfrac{dR}{dt} + 2IR\dfrac{dI}{dt}$ b) $\dfrac{dP}{dt} = -2IR\dfrac{dI}{dt}$

7. $\dfrac{ds}{dt} = \left(x\dfrac{dx}{dt} + y\dfrac{dy}{dt} + z\dfrac{dz}{dt}\right)/s$ **9.** π cm^2/min

11. a) 14 cm^2/sec, increasing   b) 0 cm/sec, constant c) $-14/13$ cm/sec, decreasing **13.** -680 miles/hr; $\dfrac{-520(x + y)}{\sqrt{x^2 + y^2}}$ mph **15.** decreasing 0.06366 mm/min **17.** a) $1125/32\pi$ cm/min   b) $1500/32\pi$ cm/min **19.** a) -1.326 cm/min   b) $r = \sqrt{26y - y^2}$ c) -0.553 cm/min **21.** 1 ft/min, 40π ft^2/min **23.** 11 ft/sec **25.** 80% per minute **27.** 1 rad/sec, 0 rad/sec

29. a) $\dfrac{dc}{dt} = 0.3$, $\dfrac{dr}{dt} = 0.9$, $\dfrac{dp}{dt} = 0.6$ b) $\dfrac{dc}{dt} = -1.567$, $\dfrac{dr}{dt} = 3.5$, $\dfrac{dp}{dt} = 5.063$ **31.** -1500 ft/sec **33.** $\dfrac{-5}{72\pi}$ in/min, $\dfrac{-10}{3}$ in^2/min **35.** 7.1 in/min **37.** 29.5 knots

Section 4.7

1. a) $x^2 + C$ b) $\dfrac{x^3}{3} + C$ c) $\dfrac{x^3}{3} = x^2 + x + C$

3. a) $x^{-3} + C$ b) $-\dfrac{x^{-3}}{3} + C$ c) $-\dfrac{x^{-3}}{3} + x^2 + 3x + C$ **5.** a) $-\dfrac{1}{x} + C$ b) $-\dfrac{5}{x} + C$ c) $2x + \dfrac{5}{x} + C$ **7.** a) $x^{3/2} + C$ b) $(8/3)x^{3/2} + C$ c) $(1/3)x^3 - (8/3)x^{3/2} + C$ **9.** a) $x^{2/3} + C$ b) $x^{1/3} + C$ c) $x^{-1/3} + C$ **11.** a) $(1/3)\cos(3x) + C$ b) $-3\cos x + C$ c) $-3\cos x + (1/3)\cos(3x) + C$ **13.** a) $\tan x + C$ b) $\tan(5x) + C$ c) $(1/5)\tan(5x) + C$ **15.** a) $\sec x + C$ b) $\sec 2x + C$ c) $2\sec 2x + C$ **17.** $x + (1/2)\cos 2x + C$ **19.** a) $-\sqrt{x} + C$ b) $x + C$ c) $\sqrt{x} + C$ d) $-x + C$ e) $x - \sqrt{x} + C$ f) $-3\sqrt{x} - 2x + C$ g) $x^2/2 - \sqrt{x} + C$ h) $-3x + C$ **21.** b **23.** $y = x^2 - 7x + 10$ **25.** $y = x^3/3 + x + 1$ **27.** $y = 5/x + 2$ **29.** $y = x^3 + x^2 + x - 3$ **31.** $y = x + \sin x + 4$ **33.** $y = -x^3 + x^2 + 4x + 1$ **35.** $s = 4.9t^2 + 10$ **37.** $s = 16t^2 + 20t$ **39.** $y = 2x^{3/2} - 50$ **41.** $x = x^3 - 3x^2 + 12x$ **43.** 48 m/sec downwards **45.** -14 m/sec **47.** $y = (C - kt)^2/4$ **49.** a) 24 seconds   c) 45.821, 4.109 **51.** d **53.** $y = x^2 + C$ **55.** $y = x - x^3 + C$ **57.** Increasing, inflection point at $(0, 1)$; from concave down to up. **59.** Minimum at $(0, 0)$; $y \to$ constant as $x \to \pm\infty$.

61.

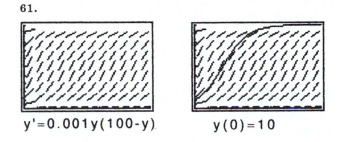

$y'=0.001y(100-y)$ $y(0)=10$

Chapter 4 Practice Exercises

1. $y' = 1/(x+1)^2 > 0$ **3.** $y' > 0 \Rightarrow y$ always increasing
5. 12   **7.** Local minimum at $x = -1$, inflection points at
$x = 0$ and $x = 2$.   **9.** a) T b) P
Exercises **11–30.** See comment at beginning of answers
to Exercises, Section 4.5.
11. Use $[-6, 6]$ by $[-4, 4]$; inflection point at $(-1, 0)$.
13. Use $[-6, 6]$ by $[-4, 4]$; inflection point at $(2, 0)$.
15. Use $[-6, 6]$ by $[-4, 4]$; maximum at $(0.385, 1.215)$,
concave down.   **17.** Use $[-6, 6]$ by $[-50, 50]$. Since
$y' \leq 0$ everywhere, there are no hidden exrema. Always
falling. Concavity changes from up to down at inflection
point $(-1/2, -12.333)$.   **19.** Use $[-6, 6]$ by $[-8, 3]$
and $[-0.5, 0.5]$ by $[0.99, 1.01]$ to show the behavior
of the graph. Local minima at $(-0.118, 0.9976)$ and
$(2.118, -6.456)$, local maximum at $(0, 1)$, inflection
points at $(-0.598, 0.9988)$ and $(1.393, -3.414)$.   **21.** Use
$[-5, 5]$ by $[-15, 25]$. Local minimum at $(0.578, 0.972)$,
inflection point at $(1.079, 13.601)$, local maximum
at $(1.692, 20.517)$.   **23.** Use $[-5, 5]$ by $[-15, 20]$;
inflection point at $(3.710, -3.420)$, local maximum
at $(0.215, -2.417)$.   **25.** Graph $y = (\ell n|x|)/\ell n\, 3$ in
$[-1, 4]$ by $[-5, 5]$, not defined at $x = 0$, always concave
down.   **27.** Use $[1.9, 4]$ by $[-5, 5]$; defined for $x > 2$,
always rising, concave down on $(2, 4)$, concave up on
$(4, \infty)$.   **29.** Use $[-1, 2]$ by $[-1, 2]$. $(0.500, 0.707)$ is
a local maximum; local minima at $(0, 0)$ and $(1, 0)$.
31. 2.1958...   **33.** 0.828...   **37.** Mimimum is $f(-6) = -74$; maximum is $f(-4.550) = 16.25$ **39.** $r = 25$,
$s = 50$ **41.** Height = 3 ft, side of base = 6 ft   **43.** $h = 2$, $r = \sqrt{2}$ **45.** $x = 48/\sqrt{7}$ miles, $y = 36/\sqrt{7}$ miles
47. 276 Grade A tires, 553 Grade B   **49.** $12 \times 6 \times 2$
inches, 144 in^3 **51.** Particle starts at $(5, 0)$, moves left
until it reaches $(0.94, 0)$ then moves right.   **53.** -40
m^2/sec   **55.** 1 cm/min   **57.** b) $-125/144\pi$ ft/min
59. -2 rad/sec   **61.** All such functions are represented
by the formula.   **63.** a) C b) $x + C$ c) $\dfrac{x^2}{2} + C$
d) $\dfrac{x^3}{3} + C$ e) $\dfrac{x^{11}}{11} + C$ f) $-x^{-1} + C$ g) $-\dfrac{x^{-4}}{4} + C$
h) $\dfrac{2}{7}x^{7/2} + C$ i) $\dfrac{3}{7}x^{7/3} + C$ j) $\dfrac{4}{7}x^{7/4} + C$ k) $\dfrac{2}{3}x^{3/2} + C$ l) $2x^{1/2} + C$ m) $\dfrac{7}{4}x^{4/7} + C$ n) $\dfrac{-3}{4}x^{-4/3} + C$

65. $x^3 + (5/2)x^2 - 7x + C$ **67.** $(2/3)x^{3/2} + 2x^{1/2} + C$
69. $(3/5)\sin 5x + C$ **71.** $\tan 3x + C$ **73.** $(1/2)x - \sin x + C$ **75.** $3\sec(x/3) + 5x + C$ **77.** $\tan x - x + C$
79. $x - (1/2)\sin 2x + C$ **81.** $y = 1 + x + x^2/2 + x^3/6$ **83.** $y = x - 1/x - 1$ **85.** $y = \sin x$ **87.** $y = x$
89. Duck!
91.

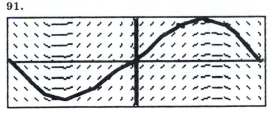

Chapter 5

Section 5.1

1. a)

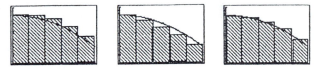

b) $\text{LRAM}_5(6 - x^2) = (6 - 0^2)\Delta x + (6 - (0.4)^2)\Delta x + (6 - (0.8)^2)\Delta x + (6 - (1.2)^2)\Delta x + (6 - (1.6)^2)\Delta x = 10.08$ (where $\Delta x = 0.4$).
$\text{RRAM}_5(6 - x^2) = (6 - 0.4^2)\Delta x + (6 - 0.8^2)\Delta x + (6 - 1.2^2)\Delta x + (6 - 1.6^2)\Delta x + (6 - 2^2)\Delta x = 8.48$.
$\text{MRAM}_5(6 - x^2) = (6 - 0.2^2)\Delta x + (6 - 0.6^2)\Delta x + (6 - 1^2)\Delta x + (6 - 1.4^2)\Delta x + (6 - 1.8^2)\Delta x = 9.36$.
3. a)

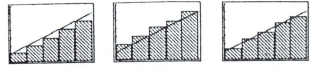

b) $\text{LRAM}_5(x+1) = [(0+1)+(1+1)+(2+1)+(3+1) +(4+1)]\Delta x = 15$ (where $\Delta x = 1$).
$\text{RRAM}_5(x+1) = [(1+1)+(2+1)+(3+1)+(4+1)+ (5+1)]\Delta x = 20$.
$\text{MRAM}_5(x+1) = [(0.5+1)+(1.5+1)+(2.5+1)+ (3.5+1)+(4.5+1)]\Delta x = 17.5$.

5. a)

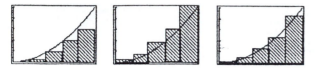

b) $\text{LRAM}_5(2x^2) = [2(0^2)+2(1^2)+2(2^2)+2(3^2)+ 2(4^2)]\Delta x = 60$ (where $\Delta x = 1$).
$\text{RRAM}_5(2x^2) = [2(1^2)+2(2^2)+2(3^2)+2(4^2)+ 2(5^2)]\Delta x = 110$.
$\text{MRAM}_5(2x^2) = [2(.5^2)+2(1.5^2)+2(2.5^2)+2(3.5^2)+ 2(4.5^2)]\Delta x = 82.5$.

7.

n	$\text{LRAM}_n f$	$\text{RRAM}_n f$	$\text{MRAM}_n f$
10	12.645	14.445	13.4775
100	13.41045	13.59045	13.499775
1000	13.4910045	13.5090045	13.49999775

9.

n	$\text{LRAM}_n f$	$\text{RRAM}_n f$	$\text{MRAM}_n f$
10	268.125	393.125	325.9375
100	321.28125	333.78125	327.48438..
1000	326.87531..	328.12531..	327.49984..

11.

n	$\text{LRAM}_n f$	$\text{RRAM}_n f$	$\text{MRAM}_n f$
10	1.98352..	1.98352..	2.00825..
100	1.99984..	1.99984..	2.00008..
1000	1.99999..	1.99999..	2.00000..

13.

n	$\text{LRAM}_n f$	$\text{RRAM}_n f$	$\text{MRAM}_n f$
10	1.77264..	1.77264..	1.77227..
100	1.77245..	1.77245..	1.77245..
1000	1.77245..	1.77245..	1.77245..

15. 17.5, 83, 13.5, 327.5, 2, respectively. **19.** $\displaystyle\sum_{k=1}^{4} \frac{1}{k} =$

$\dfrac{1}{1} + \dfrac{1}{2} + \dfrac{1}{3} + \dfrac{1}{4} = \dfrac{25}{12}$ **21.** $\displaystyle\sum_{k=1}^{3}(k+2) = (1+2)+(2+$

$2)+(3+2) = 12$ **23.** $\displaystyle\sum_{k=0}^{4} \frac{k}{4} = \frac{0}{4}+\frac{1}{4}+\frac{2}{4}+\frac{3}{4}+\frac{4}{4} = \frac{5}{2}$

25. $\displaystyle\sum_{k=1}^{4} \cos k\pi = \cos(1\cdot\pi)+\cos 2\pi+\cos 3\pi+\cos 4\pi = 0$

27. $\displaystyle\sum_{k=1}^{4}(-1)^k = (-1)^1+(-1)^2+(-1)^3+(-1)^4 = 0$

29. All **31.** $\displaystyle\sum_{k=1}^{6} k$ **33.** $\displaystyle\sum_{k=1}^{4} \frac{1}{2^k}$ **35.** $\displaystyle\sum_{k=1}^{5}(-1)^{k+1}\frac{k}{5}$

37. 55 **39.** -91 **41.** -20 **43.** 1,000,000 **45.** a) -15
b) 1 c) 1 d) -11 e) 16 **47.** $\dfrac{6\cdot 1}{1+1} + \dfrac{6\cdot 2}{2+1} + \dfrac{6\cdot 3}{3+1} +$

$\dfrac{6\cdot 4}{4+1} + \dfrac{6\cdot 5}{5+1}$. $\displaystyle\sum_{k=1}^{100} \frac{6k}{k+1} = 574.816..$

49. $1(1-1)(1-2) + 2(2-1)(2-2) + 3(3-1)(3-2) +4(4-1)(4-2) + 5(5-1)(5-2)$.
$\displaystyle\sum_{k=1}^{500} k(k-1)(k-2) = 15,562,437,750$ **51.** 78

57. $\text{RRAM}_n f < A_a^b f < \text{LRAM}_f^n$ **63.** $\text{RRAM}_n(2x^2) = \dfrac{125}{3}\dfrac{(n+1)(2n+1)}{n^2} \to \dfrac{250}{3}$ as $n \to \infty$. **65.** $\text{RRAM}_n f = \dfrac{9}{2}\left(\dfrac{n+1}{n}\right)\left(\dfrac{2n+1}{n}\right) - \dfrac{9}{2}\left(\dfrac{n+1}{n}\right) + 9 \to \dfrac{27}{2}$

as $n \to \infty$. **67.** $\text{RRAM}_n f = 4 + 24\left(\dfrac{n+1}{n}\right) - \dfrac{64}{3}\left(\dfrac{n+1}{n}\right)\left(\dfrac{2n+1}{n}\right) + 64\left(\dfrac{n+1}{n}\right)^2 \to \dfrac{148}{3}$ as $n \to \infty$.

73. $A_0^x(t^3) = \dfrac{x^4}{4}$ **75.** The explanation is below the picture.

Section 5.2

1. a) 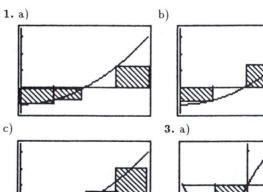   b)

c)   **3.** a)

b)

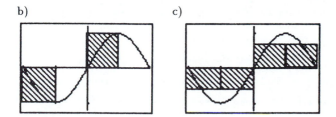

c)

5. 0.8   **7.** $\int_0^2 x^2 dx$ **9.** $\int_{-7}^5 (x^2 - 3x)dx$

11. $\int_2^3 \frac{1}{1-x} dx$ **13.** $\int_0^4 \cos x\, dx$ **15.** $A = -\int_0^2 (x^2 -$

$4)dx = 5.333$ **17.** $A = \int_0^5 \sqrt{25 - x^2}\, dx = 19.634$

19. $A = \int_0^{\pi/4} \tan x\, dx = 0.346$

21.

	$\text{LRAM}_n f$	$\text{RRAM}_n f$	$\text{MRAM}_n f$
$n = 100$	0.44290145	0.44293477	0.44291878
$n = 1000$	0.44291688	0.44292022	0.44291856
$\text{NINT}(f)$		0.442918559	

23.

	$\text{LRAM}_n f$	$\text{RRAM}_n f$	$\text{MRAM}_n f$
$n = 100$	0.75272914	0.67210056	0.71218935
$n = 1000$	0.71629745	0.70823459	0.71226377
$\text{NINT}(f)$		0.71226452	

25.

	$\text{LRAM}_n f$	$\text{RRAM}_n f$	$\text{MRAM}_n f$
$n = 100$	2.0282123	2.0575260	2.0427050
$n = 1000$	2.0412951	2.0442265	2.0427592
$\text{NINT}(f)$		2.0427597	

27. $\text{NINT}(2 - x - 5x^2, x, -1.3) = -42.666$.

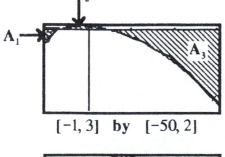

$[-1, 3]$ **by** $[-50, 2]$

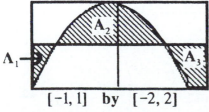

$[-1, 1]$ **by** $[-2, 2]$

The definite integral has the value $A_2 - A_1 - A_3$. Since A_1 and A_3 are below the x-axis, they each contribute a negative value to the integral.
29. $\text{NINT}(\sin(x^2), x, 0, 2\pi) = 0.642$. Graph $y = \sin(x^2)$ in $[0, 2\pi]$ by $[-1, 1]$. The integral is the sum of the signed areas (positive if above the x-axis, negative if below the x-axis) between the x-axis and the curve.

31. $\int_{-1}^1 \sqrt{1 - x^2}\, dx = \frac{\pi}{2}$. $\text{NINT}(\sqrt{1 - x^2}, x, -1.1) = 1.571$.   **33.** $\int_{-1}^1 (1 - |x|)dx = 1 = \text{NINT}(1 - |x|, x, -1.1)$.

35. For $[0, 5]$, $\text{MRAM}_n f = \frac{625}{8}\left(2 - \frac{1}{n^2}\right) \to \frac{625}{4}$ as $n \to \infty$. For $[0, a]$, $\text{MRAM}_n f = \frac{a^4}{8}\left(2 - \frac{1}{n^2}\right) \to \frac{a^4}{4}$ as $n \to \infty$.   **37.** Graph $y_1 = 1 + 0\sqrt{x}$ and $y_2 = -1 + 0\sqrt{-x}$ in $[-2, 3]$ by $[-2, 2]$. $x = 0$ is the only discontinuity. Integral $= 1$.   **39.** Graph $y = \frac{x^2 - 1}{x + 1}$ in $[-3, 4]$ by $[-5, 4]$. $x = -1$ is the only discontinuity. Integral $= -3.5$.   **41.** a) $g(x) = \frac{\sin x}{x}$, $x \neq 0$, $g(0) = 1$
b) 1.848   c) 3.697   **45.** Conjecture: $\text{NINT}(y_i(x), 0, 2) = \text{NINT}(y_i(x), 0, 1) + \text{NINT}(y_i(x), 1, 2)$.   **47.** $a = c = -1$, $b = 1$ **49.** $S_n = \text{RRAM}_n(x^2)$ on $[0, 1]$ and $\lim_{n\to\infty} S_n = 0.333$.   **51.** $S_n = \text{RRAM}(1 + x)$ on $[0, 1]$; $\lim_{n\to\infty} S_n = 1.5$.   **53.** $S_n = \text{RRAM}_n((2 + x)^2)$ on $[0, 1]$; $\lim_{n\to\infty} S_n = 6.333$.

Section 5.3

1. $\frac{8}{3}$ **3.** $\frac{16}{3}$ **5.** $\frac{8}{3}$ **7.** $\frac{27}{4}$ **9.** 1   **11.** 1   **13.** $\frac{2}{\pi}$ **15.** $1 + \sqrt{3}$ **17.** $\frac{b^{n+1}}{n + 1}$ **19.** a) 0   b) -8 c) -12 d) 10   e) -2 f) 16   **21.** a) 5   b) -5 c) -3 **23.** $\frac{-13}{3}$ **25.** $\frac{2}{\sqrt{3}}$

27. $-0.475\ldots$ **29.** $\frac{1}{2} \leq \int_0^1 \frac{1}{1 + x^2} dx \leq 1$ **31.** $c = \frac{1}{\sqrt[3]{4}}$.
Graph $y_1 = x^3 + 1$ and $y_2 = f(c) = \frac{5}{4}$ in $[0, 1]$ by $[-1, 2]$.

33. $c = \sqrt{\frac{4}{\pi} - 1}$. Graph $y_1 = \frac{1}{x^2 + 1}$, $y_2 = f(c) = \frac{\pi}{4}$ in $[0, 1]$ by $[-0.5, 1]$.   **35.** $\frac{15}{4}$ **37.** This is an immediate consequence of the Mean Value Theorem for Definite Integrals.

Section 5.4

1. 3   **3.** 1   **5.** $\frac{5}{2}$ **7.** 2   **9.** $2\sqrt{3}$ **11.** 0   **13.** $\frac{8}{3}$ **15.** $\frac{5}{2}$ **17.** $\frac{1}{2}$ **19.** $-\frac{5}{2}$ **21.** The integral does not exist.

23. 2.551.. **25.** $\dfrac{5}{6}$ **27.** π **29.** $F(x) = \dfrac{x^2}{2} - 2x$. The
two graphs appear to be the same in $[-10, 10]$ by
$[-10, 10]$. $F(0.5) = -0.875$ and we get the same value
after zooming in. $F(1) = -1.5$ compared to -1.516
as one approximation. $F(1.5) = -1.875$ compared to
-1.879 as one approximation. $F(2) = -2$ compared to
-2. $F(5) = 2.5$ compared to 2.53 as one approximation.

31. $F(x) = \dfrac{x^3}{3} - \dfrac{3}{2}x^2 + 6x$. The two graphs are
indistinguishable in the viewing rectangle $[-15, 15]$ by
$[-1000, 1000]$. The values of $F(x)$ and $\text{NINT}(f(t), t, 0, x)$
agree when accurately calculated. **33.** Graph $y =$
$\text{NINT}(t^2 \sin t, t, 0, x)$ in the viewing window $[-3, 3]$ by
$[0, 9]$. **35.** Graph $y = \text{NINT}(5e^{-0.3t^2}, t, 0, x)$ in the
viewing window $[0, 5]$ by $[0, 10]$. **37.** The two graphs are
identical. **39.** $K = -\dfrac{3}{2}$ **41.** $x = 0.699$ is the solution
to $\displaystyle\int_0^x e^{-t^2} dt = 0.6$. **43.** $\sqrt{1 + x^2}$ **45.** $\dfrac{\sin x}{2\sqrt{x}}$ **47.** d)
49. b) **51.** $x = a$ **53.** $f(4) = 1$ **57.** a) \$9 b) \$10
59. $I_{av} = 300$. Average daily holding cost = \$6 per day.
61. a) Compare your drawing with the result of graphing
$y = (\cos x)/x$ in $[-15, 15]$ by $[-1, 1]$. The x- and y-axes
are asymptotes. b) Graph $y = \text{NINT}((\cos t)/t, t, 1, x)$
in $[0, 15]$ by $[-1, 1]$. c) Because $f(0)$ is undefined.
d) For $x > 0$, $g(x)$ and $h(x)$ have the same derivative
$f(x)$ and so they differ by an additive constant. This
is confirmed if one graph can be obtained from the
other by a vertical shift. Along with the function
in b), graph $y = \text{NINT}((\cos t)/t, t, 0.5, x)$ in $[0.01, 3]$
by $[-3, 3]$ to see that this is the case. Alternatively,
$$\int_{0.5}^x f(t)dt = \int_{0.5}^1 f(t)dt + \int_1^x f(t)dt \approx 0.5 + \int_1^x f(t)dt.$$
63. a) $[-1, 1]$ b) $F'(x) = 2x\sqrt{1 - x^4}$ c) Zeros of
$F'(x): x = 0, \pm 1$. F' increasing on $\left[-\dfrac{1}{\sqrt[4]{3}}, \dfrac{1}{\sqrt[4]{3}}\right]$,
decreasing on $\left[-1, -\dfrac{1}{\sqrt[4]{3}}\right]$ and $\left[\dfrac{1}{\sqrt[4]{3}}, 1\right]$. Local extrema
of F' at $\pm 1, \pm\dfrac{1}{\sqrt[4]{3}}$. d) The zeros of $F'(x)$ tell us where
the graph has a horizontal tangent and possible local
extremes. The graph of F is concave up where F' is
increasing and concave down where F' is decreasing. The
local extrema of F' at $x = \pm\dfrac{1}{\sqrt[4]{3}}$ correspond to inflection
points of F. Graph $\text{NINT}(\sqrt{1 - t^2}, t, 1, x^2)$ in $[-1, 1]$ by
$[-2, 1]$. **65.** a) $\left(-\dfrac{1}{2}, \dfrac{1}{2}\right)$. Later work will show that F
has a continuous extension to a function with domain
$\left[-\dfrac{1}{2}, \dfrac{1}{2}\right]$. b) $\dfrac{2}{\sqrt{1 - 4x^2}}$ c) F' is decreasing on $\left(-\dfrac{1}{2}, 0\right)$,

increasing on $\left[0, \dfrac{1}{2}\right)$, and has a local minimum at $x = 0$,
the only local extreme on $\left(-\dfrac{1}{2}, \dfrac{1}{2}\right)$.
d) c) tells us that the graph of F is concave down
on $\left(-\dfrac{1}{2}, 0\right)$, concave up on $\left(0, \dfrac{1}{2}\right)$, and that F has
an inflection point at $x = 0$. In parametric mode (to
save time) graph $x = t$, $y = \text{NINT}\left(\dfrac{1}{\sqrt{1 - s^2}}, s, 1, 2t\right)$,
$-0.499 \le t \le 0.499$, t step $= 0.05$ in $[-0.499, 0.499]$ by
$[-\pi, 0]$. **67.** $F'(x) = 3x^2 \cos(2x^3) - 2x\cos(2x^2)$. This
is supported by graphing, in parametric mode, $x_1 = t$,
$y_1 = \text{NDER}(\text{NINT}(\cos(2s), s, t^2, t^3), t, t)$ and $x_2 = t$,
$y_2 = 3t^2 \cos(2t^3) - 2t\cos(2t^2)$, $-1.5 \le t \le 1.5$.
t Step $= 0.05$, in $[-1.5, 1.5]$ by $[-12, 12]$.

Section 5.5

1. $\displaystyle\int x^3 dx = \dfrac{x^4}{4} + C$. $\left(\dfrac{x^4}{4} + C\right)' = x^3$ **3.** $\dfrac{x^2}{2} + x + C$
5. $2x^{3/2} + C$ **7.** $\dfrac{3}{2}x^{2/3} + C$ **9.** $\dfrac{5}{3}x^3 + x^2 + C$
11. $\dfrac{1}{2}x^4 - \dfrac{5}{2}x^2 + 7x + C$. We graph the integrand
and $\text{NDER}\left(\dfrac{1}{2}x^4 - \dfrac{5}{2}x^2 + 7x, x\right)$ in $[-3, 3]$ by $[-50, 50]$
and see that two graphs are identical. **13.** $2\sin x + C$
15. $-3\cos\dfrac{x}{3} + C$ **17.** $-3\cot x + C$ **19.** $-\dfrac{1}{2}\csc x + C$
21. $4\sec x - 2\tan x + C$ **23.** $-\dfrac{1}{2}\cos 2x + \cot x + C$
25. $2y - \sin 2y + C$ **27.** $-\dfrac{1}{4}\cos 2x + C$ **29.** $\tan\theta + C$
31. $g(x) = \dfrac{1 - \cos x}{x^2}$, $-10 \le x \le 10$. $x \ne 0$, $g(0) = \dfrac{1}{2}$.
Graph $y = \displaystyle\int_0^x f(t)dt = \text{NINT}(f(t), t, 0, x)$ in $[-10, 10]$ by
$[-2, 2]$. For quicker results, use parametric mode, $x = t$,
$y = \text{NINT}((1 - \cos s)/s^2, s, 0, t)$, t step $= 0.3$, $-10 \le t \le$
10, in the above window. **37.** Only c) is right. **39.** $y =$
$2x^{3/2} - 50$ **41.** $y = \displaystyle\int_0^x 2^t \, dt + 2 = \dfrac{1}{\ln 2} \cdot 2^x - \dfrac{1}{\ln 2} + 2$.
Graph $y = 2 + \text{NINT}(2^t, t, 0, x)$ in $[-7.5, 9.5]$ by $[0, 10]$.
43. $y = 2x$ **45.** $y = \dfrac{x^3}{16}$ **47.** 16 ft/sec^2 **49.** 1.24 sec
55. A) $\dfrac{\sqrt{\pi^2 + 4}}{\sqrt{g}}$ C) $\dfrac{\pi}{\sqrt{g}}$

Section 5.6

1. $-\dfrac{1}{3}\cos 3x + C$ **3.** $\dfrac{1}{2}\sec 2x + C$
5. $(7x - 2)^4 + C$ **7.** $-6\sqrt{1 - r^3} + C$
9. a) $-\dfrac{\cot^2 2\theta}{4} + C$ b) $-\dfrac{\csc^2 2\theta}{4}$ **11.** $\dfrac{3}{16}$ **13.** $\dfrac{1}{2}$ **15.** 0

17. $\dfrac{2}{3}$ **19.** $\dfrac{1}{1-x} + C$ **21.** $\tan(x+2) + C$

23. $3(r^2-1)^{4/3} + C$ **25.** $\sec\left(\theta + \dfrac{\pi}{2}\right) + C$

27. $2(1+x^4)^{3/4} + C$ **29.** a) $\dfrac{14}{3}$ b) $\dfrac{2}{3}$ **31.** a) $\dfrac{1}{2}$

b) $-\dfrac{1}{2}$ **33.** a) $\dfrac{1}{2}(\sqrt{10}-3)$ b) $\dfrac{1}{2}(3-\sqrt{10})$ **35.** a) $\dfrac{45}{8}$

b) $-\dfrac{45}{8}$ **37.** a) $\dfrac{1}{6}$ b) $\dfrac{1}{2}$ **39.** a) 0 b) 0 **41.** $2\sqrt{3}$

43. 0 **45.** 8 **47.** $\dfrac{38}{3}$ **49.** $\dfrac{16}{3}$

51. $s = (3t^2-1)^4 - 1$ **53.** $s = -6\cos(t+\pi) - 6$ **55.** 1
57. All integrations are correct. The graph of any one
of the antiderivatives can be obtained from the graph of
any other antiderivative by a vertical shift verifying that
they differ by an additive constant.

Section 5.7

1. a) 2 b) 2 c) 2 **3.** a) 4.25 b) 4 c) 4

5. a) $5.146\ldots$ b) $5.252\ldots$ c) $\dfrac{16}{3} = 5.333\ldots$

7. $\displaystyle\int_{-1}^{3} e^{-x^2}\,dx$

n	TRAP	SIMP	LRAM
10	1.62316	1.63322	1.69671
100	1.63293	1.63303150	1.64029
1000	1.6330305	1.63303148	1.63377

n	RRAM	MRAM
10	1.54961	1.63799
100	1.62558	1.63308
1000	1.63229	1.63303

NINT yields 1.63303148105.

9. $\displaystyle\int_{-5}^{5} x\sin x\,dx$

n	TRAP	SIMP	LRAM
10	−4.682	−4.73	−4.68
100	−4.7537	−4.754469	−4.7547
1000	−4.75446	−4.7544704038	−4.75446

n	RRAM	MRAM
10	−4.68	−4.79
100	−4.7537	−4.7549
1000	−4.75446	−4.754474

NINT yields −4.75447040396.

11. $\displaystyle\int_{3\pi/4}^{4.5} \dfrac{\tan x}{x}\,dx$

n	TRAP	SIMP	LRAM
10	0.257	0.246	0.101
100	0.244	0.243771	0.228
1000	0.2437718	0.2437703542	0.242

n	RRAM	MRAM
10	0.413	0.238
100	0.259	0.244
1000	0.245	0.2437696

NINT yields 0.243770354155.

13. $|E_T| \leq \dfrac{1}{600} = 0.0016666\ldots$ **15.** a) $n = 1$
b) $n = 2$ **17.** a) $n = 283$ b) $n = 2$ **19.** a) $n = 76$
b) $n = 12$ **21.** 3.1379, 3.14029 **23.** 1.3669, 1.3688
25. a) $0.057\ldots$ and $0.0472\ldots$ b) Let $y_1 = \sin x / x$,

$y_2 = \text{der}2(y_1, 2, 2)$, $y_3 = \left(\dfrac{1.5\pi}{12}\right)\left(\dfrac{1.5\pi}{10}\right)^2 absy_2$, $y_4 =$

$\text{NDER}(\text{NDER}(y_2, x, x), x, x)$, $y_5 = \dfrac{1.5\pi}{180}\left(\dfrac{1.5\pi}{10}\right)^4 absy_4$.

We graph y_3 in $\left[\dfrac{\pi}{2}, 2\pi\right]$ by $[0, 0.03]$ and y_5 in $\left[\dfrac{\pi}{2}, 2\pi\right]$

by $[0, 3 \times 10^{-4}]$. Max $y_3 = 2.168\ldots \times 10^{-2}$, max $y_5 =$
$2.26\ldots \times 10^{-4}$ c) Max $E_T(x) = 5.42\ldots \times 10^{-3}$, max
$E_S(x) = 1.41\ldots \times 10^{-5}$. d) Max $E_T(x) = 8.67 \times 10^{-4}$,
max $E_S(x) = 3.62\ldots \times 10^{-7}$. e) We cannot find
the exact value, but we can approximate the integral
as closely as we like by increasing n. With $n = 50$,
Simpson's Rule gives the value $0.0473894\ldots$. By d)
the error is at most 3.62×10^{-7}. **27.** Refer to the
method of Exercise 25. a) $3.6664\ldots$ and $4.65348218\ldots$
b) Max $E_T = 0.2466\ldots$, max $E_S = 2.55\ldots \times 10^{-4}$.
c) Max $E_T = 6.16\ldots \times 10^{-2}$, max $E_S = 1.59\ldots \times 10^{-5}$.
d) Max $E_T = 9.86\ldots \times 10^{-3}$, max $E_S = 4.08\ldots \times 10^{-7}$.
e) Simpson's Rule with $n = 50$ yields $3.6534844\ldots$ with
error at most $4.08\ldots \times 10^{-7}$. **29.** $\text{LRAM}_{50}f = 3.2591$,
$\text{RRAM}_{50}f = 3.3205$ and $T_{50}f = 3.2898$. Yes, $2T_{50}f =$
$\text{LRAM}_{50}f + \text{RRAM}_{50}f$. **31.** 1013 **35.** $466.66\ldots$ in^2
37. Using the odd-numbered hours, we get 56.86 kwh
per customer. **39.** In parametric mode graph $x(t), y(t)$,
$0 \leq t \leq 1$, t step $= 0.05$ in $[0, \pi]$ by $[0, 2]$. Domain $=$
$[0, \pi]$, range $= [0, 2]$. **41.** $A \approx 1.596507\ldots$ **43.** $g(x) =$
$\cos^{-1}(x-1) - \sqrt{2x - x^2}$. $\text{NINT}(g(x), x, 0, 2) =$
1.57079603391.

Chapter 5 Practice Exercises

1. a)

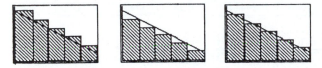

b) 20, 15, 17.5

3.

	$n = 10$	$n = 100$	$n = 1000$
$LRAM_n$	22.695	23.86545	23.9865045
$RRAM_n$	25.395	24.13545	24.0135045
$MRAM_5$	23.9775	23.999775	23.99999775

5.

	$n = 10$	$n = 100$	$n = 1000$
$LRAM_n$	3.9670	3.99967	3.9999967
$RRAM_n$	3.9670	3.99967	3.9999967
$MRAM_n$	4.0165	4.00016	4.0000016

7. a) 75 **b)** -10 **9. a)** 0 **b)** 60 **11. a)** $\sum_{k=0}^{3} 2^k$

b) $\sum_{k=0}^{4} \frac{1}{3^k}$ **c)** $\sum_{k=1}^{5}(-1)^{k+1}k$ **d)** $\sum_{k=1}^{3} \frac{5}{2^k}$ **13.** $\int_{0}^{1} e^x \, dx$

15. $RRAM_n f = \frac{625}{2}\left(1 + \frac{1}{n}\right) + \frac{75}{2}\left(1 + \frac{1}{n}\right)$.

$\lim_{n \to \infty} RRAM_n f = 350$ **17. a)** π **b)** $-\pi$ **c)** -3π

19. $\pi - 2$ **21.** 10 **23.** 16 **25.** 2 **27.** 1 **29.** $\frac{2}{5}$

31. $\sqrt{3}$ **33.** $6\sqrt{3} - 2\pi$ **35.** 8 **37.** -1 **39.** 2

41. $0.6931\ldots$ **43.** $0.2938\ldots$ **45.** $\frac{17}{3}$, $NINT((x - 2/x)(x + 2/x), x, 2, 3) = 5.6666\ldots$ **47.** 8 **49.** -2

55. Graph $NINT((\sin(3t))/t, t, 1, x)$ in $[0.01, 8]$ by $[-2, 0.004]$. There is no explicit elementary formula for this integral. **57.** $-3.091, 1.631$ **59.** No solution

61. a) and c) **63.** $f(x) = \cos x$, $0 \le x \le 1$ **65.** $F(1) - F(0)$ **67.** 10 m **69.** 21.511 ft/sec^2 **71. a)** $(V^2)_{av} = \frac{(V_{max})^2}{2}$ **b)** 339.411 volts **75.** $n = 16$, $h = \frac{1}{8}$ **77.** T and S both agree with π up to the limits of calculator accuracy. **79.** By Simpson's Rule the approximate area of the lot is 6059 ft^2. The job cannot be done for $11,000.

Chapter 6

Section 6.1

1. $\frac{125}{6}$ **3.** $\frac{125}{6}$ **5.** $\frac{\pi}{2}$ **7.** $\frac{32}{3}$ **9.** $\frac{32}{3}$ **11.** $\frac{1}{6}$

13. 26.15341 **15.** $\frac{9}{2}$ **17.** 0.22016 **19.** $\frac{1}{6}$ **21.** 15.68376

23. $\frac{1}{12}$ **25.** 1 **27.** $\sqrt{2} - 1$ **29.** $c = 2^{4/3}$ **31. a)** $A(t) = \frac{2}{3}t^3\left(1 - \frac{1}{\sqrt{2}}\right)$ **c)** $0.195\ldots$ **33.** 4

Section 6.2

1. $\frac{8\pi}{3}$ **3.** 36π **5.** $\frac{128\pi}{7}$ **7.** π **9.** $\frac{32\pi}{3}$ **11.** 2π **13.** 4π

15. $4\pi \ln 4$ **17.** $\frac{2\pi}{3}$ **19.** $\frac{128\pi}{5}$ **21.** $\frac{117\pi}{5}$ **23.** $\pi^2 - 2\pi$

25. $\frac{4\pi}{3}$ **27.** 8π **29.** $\frac{500\pi}{3}$ **31.** $\pi^2 - 2\pi$ **33.** $\frac{\pi}{3}$

35. a) 8π **b)** $\frac{32\pi}{5}$ **c)** $\frac{8\pi}{3}$ **d)** $\frac{224\pi}{15}$ **37. a)** $\frac{16\pi}{15}$

b) $\frac{56\pi}{15}$ **c)** $\frac{64\pi}{15}$ **39.** $\frac{\pi r^2 h}{3}$ **41. a)** $x = -2$, $x = 0.59375\ldots$ **b)** 76.8153067 **43.** $1053\pi \approx 3.3L$

Section 6.3

1. 8π **3.** $\frac{3\pi}{2}$ **5.** 3π **7.** $\frac{4\pi}{3}$ **9.** $\frac{16\pi}{3}$ **11.** $\frac{8\pi}{3}$ **13.** $\frac{14\pi}{3}$

15. $\frac{6\pi}{5}$ **17. a)** $\frac{5\pi}{3}$ **b)** $\frac{4\pi}{3}$ **19. a)** $\frac{11\pi}{15}$ **b)** $\frac{97\pi}{105}$

c) $\frac{121\pi}{210}$ **d)** $\frac{23\pi}{30}$ **21. a)** $\frac{512\pi}{21}$ **b)** $\frac{832\pi}{21}$ **23. a)** $\frac{\pi}{6}$

b) $\frac{\pi}{6}$ **25. a)** 20.637π **b)** $2\pi(11.610)$ **27.** 94.782

29. b) $\frac{\pi}{4}$ **c)** 5.033

Section 6.4

1. 12 **3.** $\frac{14}{3}$ **5.** $\frac{53}{6}$ **7.** $\frac{123}{32}$ **9.** $\frac{3}{2}$ **11.** 3.1385

13. $e^3 - e^{-3}$ **15.** 6 **17.** $y = x^{1/2}$, $0 \le x \le 4$ **19.** $y = \ln x + 3$ **21.** $\int_{0}^{1/2} \frac{1 + x^2}{1 - x^2} dx = 0.5986\ldots$ **23.** 21.068

inches **25.** $9.033\ldots$ million miles **27.** $1089.89\ldots$

Section 6.5

1. $4\pi\sqrt{5}$ **3.** $3\pi\sqrt{5}$ **5.** $\frac{98\pi}{81}$ **7.** $\frac{28\pi}{3}$ **9.** $\frac{49\pi}{3}$

11. $\frac{2\pi(2\sqrt{2} - 1)}{3}$ **13.** $\frac{253\pi}{20}$ **15.** $\frac{x}{27}\left[\left(\frac{5}{4}\right)^3 - 1\right]$

17. 4591π **19.** 452.390 L of each color **21.** 11,900 tiles

Section 6.6

1. 400 ft·lb **3.** 925 $N \cdot m$ **5.** 64,800 ft·lb **7.** 1.2 $N \cdot m$

9. a) $104\frac{1}{6}$ ft·lb b) 312.5 ft·lb **11.** 245,436.926 ft·lb
13. 7,238,229.473 ft·lb **15.** a) 1,500,000 ft·lb b) 100
minutes **17.** 21,446,605.85 $N \cdot m$ **19.** Through the
valve (84823 ft·lb vs 98960 ft·lb) **21.** 967,610.537 ft·lb;
cost is \$4838.05, yes. **23.** $5.1441 \times 10^{10} N \cdot m$

Section 6.7

1. 2812.5 lb **3.** 375 lb. No. **5.** 1166.67 lb **7.** 41.67 lb
9. $F = 1309$ lb/in^3 **11.** 1161 lb **13.** 1034.16 ft^3 **15.** A
plate $h \times w$ with top d below the surface experiences a
force of $\omega w \left[\dfrac{h^2 + 2dh}{2}\right] = \dfrac{\omega d + \omega(d + h)}{2} hw$

Section 6.8

1. 4 ft **3.** $M_0 = 8, \bar{x} = 1$ **5.** $M_0 = \dfrac{68}{3}, \bar{x} = \dfrac{17}{7}$
7. $\left(0, \dfrac{2}{3}\right)$ **9.** $\left(\dfrac{16}{105}, \dfrac{8}{15}\right)$ **11.** $\left(\dfrac{3}{5}, 1\right)$ **13.** $\left(0, \dfrac{\pi}{8}\right)$
15. $\left(1, -\dfrac{2}{5}\right)$ **17.** $\left(\dfrac{2}{4 - \pi}, \dfrac{2}{4 - \pi}\right)$ **19.** $(0, 1)$
21. $\left(\dfrac{a}{3}, \dfrac{a}{3}\right)$ **23.** $\left(\dfrac{14}{9}, \dfrac{\ln 4}{3}\right)$ **25.** $V = 32\pi; S = 32\sqrt{2}\pi$
27. $4\pi^2$ **29.** $\left(0, \dfrac{2a}{\pi}\right)$ **31.** $\left(0, \dfrac{4a}{3\pi}\right)$ **33.** $\dfrac{\pi a^3}{6}(4 + 3\pi)\sqrt{2}$

Section 6.9

1. 16 **3.** $\dfrac{16}{3}$ **5.** $\pi \ln 2$ **7.** $\dfrac{8}{3}$ **9.** $\dfrac{4\pi r^3}{3}$ **11.** $s^2 h, s^2 h$
13. b) 20 m c) 0 m **15.** b) 6 m c) 2 m **17.** a) 245 m
b) 0 m **19.** b) 6 m c) 4 m **21.** a) $d = 2, s = 2$
b) $d = 4, s = 0$ c) $d = 4, s = 4$ d) $d = 2, s = 2$
23. approximately 1.73π instead of 2π

Chapter 6 Practice Exercises

1. 1 **3.** $\dfrac{9}{2}$ **5.** 18 **7.** $\dfrac{9}{8}$ **9.** $\dfrac{\pi^2}{32} + \dfrac{\sqrt{2}}{2} - 1$ **11.** 4
13. $\dfrac{8\sqrt{2} - 7}{6}$ **15.** 3 **17.** a) 2π b) π **19.** a) 8π
b) $\dfrac{1088\pi}{15}$ c) $\dfrac{512\pi}{15}$ **21.** $\pi\left(\sqrt{3} - \dfrac{\pi}{3}\right)$ **23.** $\pi \ln 16$
25. $\dfrac{28\pi}{3}$ **27.** $2\sqrt{3}$ **29.** $\dfrac{92}{9}$ **31.** $\dfrac{2\pi}{3}[26^{3/2} - 2^{3/2}] \approx$
86.5π **33.** 3π **35.** $\dfrac{3\pi}{2}$ **37.** $4560 N \cdot m$ **39.** 10 ft·lb,
30 ft·lb **41.** $6400\pi\omega/3$ ft·lb, where $\omega = 62.5$ **43.** 333.3
lb **45.** 2200 lb **47.** $\left(0, \dfrac{8}{5}\right)$ **49.** $\left(\dfrac{3}{2}, \dfrac{12}{5}\right)$ **51.** $\left(0, \dfrac{8}{15}\right)$
53. $\left(7, \dfrac{1}{3}\ln 2\right)$ **55.** $\dfrac{9\pi}{280}$ **57.** π^2 **59.** 18 **61.** b) 5 ft
c) 3 ft **63.** b) 15 ft c) -5 ft

Chapter 7

Section 7.1

1. $2(\ln 2 - \ln 3)$ **3.** $-\ln 2$ **5.** $2\ln 3 - \ln 2$ **7.** $\ln 3 +$
$\dfrac{1}{2}\ln 2$ **9.** $\dfrac{2}{x}$. The result is supported by graphing
$\dfrac{2}{x}$ and $\text{NDER}(\ln(x^2), x, x)$ in $[-5, 5]$ by $[-5, 5]$.
11. $-\dfrac{1}{x}$ **13.** $\dfrac{1}{x + 2}$ **15.** $\dfrac{\sin x}{2 - \cos x}$ **17.** $\dfrac{1}{x \ln x}$
19. $\dfrac{2x + 1}{2\sqrt{x(x + 1)}}$ **21.** $\dfrac{\sin x + 2(x + 3)\cos x}{2\sqrt{x + 3}}$
23. $3x^2 + 6x + 2$ **25.** $\dfrac{x + 5}{x \cos x}\left[\tan x - \dfrac{5}{x(x + 5)}\right]$
27. $\dfrac{x\sqrt{x^2 + 1}}{(x + 1)^{2/3}}\left[\dfrac{1}{x} + \dfrac{x}{x^2 + 1} - \dfrac{2}{3(x + 1)}\right]$
29. $\dfrac{2(x^2 + x - 1)}{3x(x - 2)(x^2 + 1)}\sqrt[3]{\dfrac{x(x - 2)}{x^2 + 1}}$ **31.** $\ln 2 - \ln 3$
33. $\ln 2 - \ln 5$ **35.** $-\ln 2$ **37.** $2\ln(0.8)$ **39.** $\ln 4$
41. $\ln 3$ **43.** $(\ln 2)^2$ **45.** $\ln 2$ **47.** $3\ln 3$ **49.** $\ln x^2 + C$
51. $\dfrac{1}{2}\ln(x^2 + 4) + C$ **53.** $-3\ln\left|\cos\dfrac{x}{3}\right| + C$ **55.** $-\infty$
57. $-\infty$ **59.** $\ln 2$ **63.** a) Graph $y_1 = \sqrt{x + 3}\sin x$ in
$[-4, 20]$ by $[-5, 5]$. b) In the same viewing rectangle
graph $y_2 = \text{NDER}(y_1, x)$ and $y_3 = \dfrac{dy}{dx}$ (found in Exercise
21). The fact that y_2 and y_3 coincide supports that
y_3 is valid where $y < 0$. **65.** Graph $y_1 = \dfrac{x + 5}{x \cos x}$ in
$[-20, 20]$ by $[-20, 20]$. Proceed now as in Exercise 23.
67. Graph $y_1 = (x(x - 2)/(x^2 + 1))^{1/3}$ in $[-20, 20]$ by
$[-1.1, 1.2]$. Proceed now as in exercise 63. **69.** $\dfrac{dy}{dx} =$
$\dfrac{(df/dx)g - f(dg/dx)}{g^2}$ **71.** a) Graph $y = f(x) =$
$\text{NINT}(\cot t, t, \dfrac{\pi}{2}, x)$ in $[0, \pi]$ by $[-5, 5]$. (Use $tol = 1$.)
Conjecture: $\lim\limits_{x \to 0^+} f(x) = \lim\limits_{x \to \pi^-} f(x) = -\infty$. b) Near
$x = 4$ there is rather wild oscillating behavior due to
the discontinuity of the integrand at $x = \pi$. c) Except
for the vertical asymptote at $x = \pi$ the graph of the
antiderivative $\ln|\sin x|$ is smooth. The two graphs are
identical on $(0, \pi)$.

Section 7.2

1. 7 **3.** 2 **5.** $3e^2$ **7.** $\ln 2$ **9.** $\dfrac{\ln 2}{10}$ **11.** $\dfrac{\ln 2}{\ln(3/2)}$ **13.** 0
15. $-10\ln 3$ **17.** $t = 0$ is the only solution **19.** $y =$
$e^{2t + 4}$ **21.** $y = 40 + e^{5t}$ **23.** $y = e^{(2x^2 + 1 - 5)}$ **25.** $2e^x$
27. $-e^{-x}$ **29.** $\dfrac{2}{3}e^{2x/3}$ **31.** $(xe^2 - e^x)' = e^2 - e^x$

33. $\dfrac{e^{\sqrt{x}}}{2\sqrt{x}}$ **35.** 2 **37.** 1 **39.** 8 **41.** $\dfrac{1}{2}(e^2 + 2e - 3)$

43. $\ln(3/2)$ **45.** 0 **47.** $e^2 - e$ **49.** $2\sin(e^x) +$
C **51.** $\ln(1 + e^x) + C$ **53.** $-2\ln|\cos(\sqrt{x})| + C$

55. a) $f^{-1}(x) = \dfrac{x-3}{2}$ b) We may graph $y = 2x + 3$
and $y = \dfrac{x-3}{2}$ together in $[-10,10]$ by $[-10,10]$.

57. a) $f^{-1}(x) = f(x) = \dfrac{1-2x}{x+2}$ b) The graph of
$f(x)$ is symmetric about the line $y = x$ confirming
that f is self-inverse. **59.** $\dfrac{1}{4}$ **63.** ∞ **65.** $\ln 2$

67. $\{0.679\ldots, 1.3086\ldots\}$ **69.** $1 + 5x$ **71.** e and e^{-1}
73. $y = 2 + x + \ln|x|$ **79.** 1250 **81.** a) The amount
in the account after t years is $A(t) = A_0 e^t$. b) 1.0986
years (rounded) c) e times the original amount
83. $4.875\ldots\%$ **85.** a) 14 years, 10 years b) 14%,
3.5% **87.** August, 2015 **89.** About 3.9% **91.** $\dfrac{100\ln 2}{r}$
years **93.** a) $p(x) = 20.09e^{1-0.01x}$ b) (rounded) \$49.41,
\$22.20 d) Graph $y = 20.09xe^{1-0.01x}$ in $[0,200]$ by
$[0,2100]$ **95.** a) $0.04\ln 10$ b) 109.65 (rounded) c) We
will always get the same result in b) and c). **97.** a) 8
b) 1000 c) 8.03 months, impossible d) 7.082 months,
173.919 rabbits/month **99.** a) $D_x F^{-1}(x) \approx 6$.

Section 7.3

1. $\pi x^{\pi-1}$ **3.** $-\sqrt{2}x^{-\sqrt{2}-1}$ **5.** $8^x \ln 8$
7. $-\csc x \cot x (3^{\csc x})\ln 3$ **9.** $2\left(\dfrac{\ln x}{x}\right)x^{\ln x}$ **11.** $(x +$
$1)^x\left[\dfrac{x}{x+1} + \ln(x+1)\right]$ **13.** $x^{\sin x}\left[\dfrac{\sin x}{x} + (\cos x)\ln x\right]$
15. $\dfrac{1}{x\ln 2}$ **17.** $\dfrac{3}{(\ln 3)(3x+1)}$ **19.** $-\dfrac{1}{x\ln 2}, x >$
0 **21.** $\dfrac{1}{x}$ **23.** $\dfrac{1}{\ln 10}$ **25.** $\dfrac{3}{\sqrt{3}+1} = \dfrac{3}{2}(\sqrt{3}-1)$
27. $\dfrac{4}{\ln 5}$ **29.** $\dfrac{1}{\ln 4}$ **31.** $\dfrac{3}{2}$ **33.** $\dfrac{1}{\ln 2}$ **35.** $\dfrac{\ln 10}{2}$
37. $\dfrac{3}{2}\ln 2$ **39.** $\ln 10$ **41.** $\lim_{x\to 0} f(x) = \ln 3$. $\dfrac{2^{\sin x}}{\ln 2} + C$
43. $\dfrac{(\ln(x-2))^2}{2\ln 3} + C$ **45.** No local extrema or inflection
points. The gaph is falling and concave up on $(0,\infty)$.
Check your graph by graphing y in $[0,10]$ by $[0,3]$.
47. Graph $y = x^{\sqrt{x}}$ in $[-1,3]$ by $[-2,8]$. y is decreasing
on $(0, e^{-2}]$, has a relative minimum at $x = e^{-2}$, and is
increasing on $[e^{-2},\infty)$. The graph of y is concave up
for all $x > 0$. **49.** Period 2π. Graph y in $\left[-\dfrac{\pi}{2}, \dfrac{3\pi}{2}\right]$
by $[0,4]$. Rel. min. at $(0,2)$, rel. max. at $\left(\pi, \dfrac{1}{2}\right)$. Let
$v = 1.90392136$. Inflection points at $(v, 0.12)$ and

$(2\pi - v, 0.12)$. Rising on $\left[0, \dfrac{\pi}{2}\right)$ and $\left(\dfrac{\pi}{2}, \pi\right)$, falling on
$\left(-\dfrac{\pi}{2}, 0\right]$ and $\left[\pi, \dfrac{3\pi}{2}\right)$. Concave up on $\left(-\dfrac{\pi}{2}, \dfrac{\pi}{2}\right), \left(\dfrac{\pi}{2}, v\right)$
and $\left(2\pi - v, \dfrac{3\pi}{2}\right)$. It is concave down on $(v, 2\pi - v)$.
51. Period 2π. Graph y in $[0, 2\pi]$ by $[-3, 0]$. Rel. max.
at $\left(\dfrac{\pi}{2}, 0\right)$. No inflection point. Rising on $\left(0, \dfrac{\pi}{2}\right]$,
falling on $\left[\dfrac{\pi}{2}, \pi\right)$. Concave down on $\left(0, \dfrac{\pi}{2}\right) \cup \left(\dfrac{\pi}{2}, \pi\right)$.
53. 6.052 **55.** 3.591 **57.** 12 **59.** The solution set
is $\{(-0.77, 0.58), (2, 4), (4, 16)\}$. **61.** a) ∞ b) $-\infty$
63. a) ∞ b) 0 **65.** a) The ratio and its limit are
both $\ln 10$. b) The ratio and its limit are both $\dfrac{\ln 3}{\ln 2}$.
67. Check your result by graphing $y = x^{\sin x}$ in $[0, 40]$
by $[0, 40]$. **69.** In each case we have $x^{\beta} < x^{\sqrt{3}} < x^{\alpha}$
for $0 < x < 1$ and $x^{\alpha} < x^{\sqrt{3}} < x^{\beta}$ for $x > 1$ where
$0 < \alpha < \sqrt{3} < \beta$. The closer α and β are, the more we
must zoom in to distinguish the curves. For $x > 0$, $(1, 1)$
is the only point of intersection. **73.** 3.63×10^{-8} and
4.27×10^{-8} **75.** 10

Section 7.4

1. a) $k = \dfrac{\ln 0.99}{1000}$ b) 10,483 years c) about 82%
3. 54.88 grams **5.** 0.59 day **7.** We use $m = 66 +$
$7 = 73$. a) 168.46 m b) 41.1 sec **9.** a) 17.53 min.
longer b) About 13.26 min. **11.** $-3°$ C **13.** 92.10 sec
15. About 6658.30 years **17.** 41.22 years **19.** b) Never
21. 13.768 years from now

Section 7.5

1. $\dfrac{1}{4}$ **3.** $\dfrac{3}{11}$ **5.** 0 **7.** $\dfrac{3}{2}$ **9.** -2 **11.** $\dfrac{5}{7}$ **13.** $\dfrac{1}{4}$ **15.** 0
17. $-\dfrac{5}{3}$ **19.** 0 **21.** e **23.** e^2 **25.** 1 **27.** If we define
$y = 0$ when $x = 0$, then y is continuous on $[0, \infty)$.
29. No. $y \to \infty$ as $x \to 0^+$ **31.** a) The function
$F(x)$ defined by $F(x) = f(x)$, $x > 0$ and $F(0) = 1$ is
continuous on $[0, \infty)$. **33.** b) is correct. a) is incorrect
because L'Hôpital's rule does not apply to the limit
form $\dfrac{0}{6}$; it is not an indeterminate form. **37.** Graph f
in $[-10, 10]$ by $[0, 20]$. The graph resembles the one in
Fig. 7.30. **39.** $f(x) = e$ for $x > 0$, $x \neq 1$ **41.** Graph
in $[-2\pi, 2\pi]$ by $[-1, 2]$. The function is continuous
except at $x = 0$ where it has a removable discontinuity.
43. a) 1 b) $\dfrac{\pi}{2}$ c) π **45.** b) 106.184 c) 1,060,000
d) 1061836.55

Section 7.6

1. Function grows a) slower than b) slower than c) slower than d) faster than e) slower than f) slower than g) slower than h) slower than i) at the same rate as j) at the same rate as e^x as $x \to \infty$.   **3.** The function grows a) at the same rate as b) raster than c) faster than d) slower than e) at the same rate as f) at the same rate as g) slower than h) slower than i) slower than j) faster than x^2 as $x \to \infty$.   **5.** $e^{x/2}$, e^x, $(\ln x)^x$, x^x **11.** Only c), d), e), f), h) are true.   **15.** b) $x = e^{e^u}$ where $u = 16.6265089014$ **21.** g is an end behavior model for f.   **23.** g is an end behavior model for f.   **25.** g is only a left end behavior model for f.   **27.** g is an end behavior model for f.   **29.** g is an end behavior model for f.   **35.** $n > 1$ **37.** The first algorithm.

Section 7.7

1. a) $\dfrac{\pi}{4}$ b) $\dfrac{\pi}{3}$ c) $\dfrac{\pi}{6}$ **3.** a) $-\dfrac{\pi}{6}$ b) $-\dfrac{\pi}{4}$ c) $-\dfrac{\pi}{3}$
5. a) $\dfrac{\pi}{3}$ b) $\dfrac{\pi}{4}$ c) $\dfrac{\pi}{6}$ **7.** a) $\dfrac{3\pi}{4}$ b) $\dfrac{5\pi}{6}$ c) $\dfrac{2\pi}{3}$
9. a) $\dfrac{\pi}{4}$ b) $\dfrac{\pi}{3}$ c) $\dfrac{\pi}{6}$ **11.** a) $\dfrac{3\pi}{4}$ b) $\dfrac{5\pi}{6}$ c) $\dfrac{2\pi}{3}$
13. $\cos\alpha = \dfrac{\sqrt{3}}{2}$, $\tan\alpha = \dfrac{1}{\sqrt{3}}$, $\sec\alpha = \dfrac{2}{\sqrt{3}}$, $\csc\alpha = 2$
15. $\sin\alpha = \dfrac{4}{5}$, $\cos\alpha = \dfrac{3}{5}$, $\sec\alpha = \dfrac{5}{3}$, $\csc\alpha = \dfrac{5}{4}$ and $\cot\alpha = \dfrac{3}{4}$ **17.** $\dfrac{\sqrt{2}}{2}$ **19.** $-\dfrac{1}{\sqrt{3}}$ **21.** $\dfrac{2}{\sqrt{3}} + \dfrac{1}{2}$ **23.** 1
25. $-\sqrt{2}$ **27.** $\dfrac{\pi}{6}$ **29.** $\dfrac{\sqrt{x^2+4}}{2}$ **31.** $\tan(\sec^{-1} 3y) =$
$\begin{cases} -\sqrt{9y^2 - 1} & \text{if } 3y \le -1 \\ \sqrt{9y^2 - 1} & \text{if } 3y \ge 1 \end{cases}$ **33.** $\sqrt{1 - x^2}$
35. $\dfrac{\sqrt{x^2 - 2x}}{|x - 1|}$ **37.** $\dfrac{\sqrt{9 - 4y^2}}{3}$ **39.** $\dfrac{\sqrt{x^2 - 16}}{|x|}$
41. $\dfrac{\pi}{2}$ **43.** $\dfrac{\pi}{2}$ **45.** $\dfrac{\pi}{2}$ **47.** 0   **51.** 0.955 radian or
$54.736°$ **53.** 0.464, 0.841, and 0.730 in radian measure
61. Graph $y = \sin^{-1}(1/2x) = \sin^{-1}((2x)^{-1})$ in
$[-3.5, 3.5]$ by $\left[-\dfrac{\pi}{2}, \dfrac{\pi}{2}\right]$ **63.** Graph $y = 2\cos^{-1}(1/3x) = 2\cos^{-1}((3x)^{-1})$ in $[-3, 3]$ by $[0, 2\pi]$.   **65.** Graph $y = 3 + \cos^{-1}(x - 2)$ in $[1, 3]$ by $[3, 3 + \pi]$

Section 7.8

1. $-\dfrac{2x}{\sqrt{1 - x^4}}$. To confirm graphically we graph this last result and $\text{NDER}(y, x)$ in $[-1, 1]$ by $[-5, 5]$ and see that the two graphs match.   **3.** $\dfrac{15}{1 + 9x^2}$
5. $\dfrac{1}{\sqrt{4 - x^2}}$ **7.** $\dfrac{1}{|x|\sqrt{25x^2 - 1}}$ **9.** $\dfrac{-2x}{(x^2 + 1)\sqrt{x^4 + 2x^2}}$

11. $y = \dfrac{\pi}{2}$, $y' = 0$ **13.** $-\dfrac{1}{2x\sqrt{x - 1}}$ **15.** $\dfrac{x|x| - 1}{|x|\sqrt{x^2 - 1}}$
17. $2\tan^{-1}x$ **19.** $\dfrac{\pi}{6}$ **21.** $\dfrac{\pi}{12}$ **23.** π **25.** $\dfrac{\pi}{12}$ **27.** $\dfrac{\pi}{6}$
29. $\dfrac{\pi}{3}$ **31.** $\sin^{-1}\dfrac{x}{3} + C$ **33.** $\dfrac{1}{\sqrt{17}}\tan^{-1}\dfrac{x}{\sqrt{17}} + C$
35. $\sec^{-1}\left(\dfrac{5}{\sqrt{2}}|x|\right) + C$ **37.** $\dfrac{1}{2}\sin^{-1}(y^2) + C$ **39.** $\dfrac{\pi}{12}$
41. $2\ln(4/3)$ **43.** $\dfrac{\pi}{12}$ **45.** 1   **47.** 1   **49.** $\dfrac{\pi^2}{2}$ **51.** $3\sqrt{5}$
ft ≈ 6.71 ft   **53.** $x = 1$, $\theta = \dfrac{\pi}{2}$ **55.** $y = \sec^{-1}x + \dfrac{2\pi}{3}$
57. $y = \cos^{-1}x - \dfrac{x}{4}$ **61.** Both answers can be correct because they differ by a constant: $\sin^{-1}x = \dfrac{x}{2} - \cos^{-1}x$.

71. Graph $y = \sec^{-1}(3x) = \cos^{-1}\left(\dfrac{1}{3x}\right)$ in $[-5, 5]$ by $[0, \pi]$. Other windows can show that $y = 0$ when $x = \dfrac{1}{3}$ and $y = \pi$ when $x = -\dfrac{1}{3}$.   **73.** Graph $y = \cot^{-1}\sqrt{x^2 - 1} = \dfrac{\pi}{2} - \tan^{-1}\sqrt{x^2 - 1}$ in $[-10, 10]$ by $[-1, 2]$.

Section 7.9

1. -0.693 or $-\ln 2$ **3.** ± 1.317 **5.** ± 0.896 **7.** $x + \dfrac{1}{x}$, $x > 0$ **9.** e^{5x} **11.** e^{-3x} **15.** Graph $y = \sinh 3x$ in $[-3, 3]$ by $[-3, 3]$. This graph may be obtained from the graph of $y = \sinh x$ by horizontally shrinking it by a factor of $\dfrac{1}{3}$.   **17.** Graph $y = 2\tanh\dfrac{x}{2}$ in $[-4, 4]$ by $[-2, 2]$. The graph can be obtained from the graph of $y = \tanh x$ by stretching vertically and horizontally by a factor of 2.   **19.** Graph in $[-10, 10]$ by $[-10, 0]$. Graph rises on $(-\infty, 0]$, falls on $[0, \infty)$ and is concave down for all x.   **21.** Graph $y = \sinh^{-1}(2x)$ in $[-5, 5]$ by $[-5, 5]$. The graph may be obtained from the graph of $y = \sinh^{-1}x$ by shrinking horizontally by a factor of $\dfrac{1}{2}$.   **23.** Graph in $[-1, 1]$ by $[-4, 0.5]$. The domain is $(-1, 1)$. $\lim\limits_{x\to x+} = -\infty$ and $\lim\limits_{x\to 1} y = 0$. y is rising on $(-1, 0.564]$ and falling on $[0.564, 1)$. The curve is concave down on $(-1, 1)$.   **25.** Graph y in $[0, 1]$ by $[0, 1]$. y has domain $(0, 1]$. It is rising for $0 < x \le x_0$ and falling for $x_0 \le x \le 1$ where $x_0 = 0.552$. The curve is concave down on $(0, 1)$.   **27.** $-\text{csch } x$ **29.** $\text{csch}(2x)$ **31.** All three have derivative $\sinh(2x)$.   **33.** $|\sec x|$
35. $\sec x$ **37.** $-\csc x$ **39.** $\dfrac{2\sin 5}{5}$ **41.** 0   **43.** e **45.** $\dfrac{3}{4}$
47. $\dfrac{4}{5}$ **49.** $2\sinh 2$ **51.** $\dfrac{\cosh 2x}{2} + C$ **53.** $\dfrac{e^{3t}}{3} +$
$e^t + C$ **55.** $7\ln\left(\cosh\dfrac{x}{7}\right) + C$ **57.** $-2\,\text{sech}\sqrt{t} + C$

61. a) $\sinh^{-1}(1)$ b) $\ln(1+\sqrt{2})$ **63.** a) $\cosh^{-1}\left(\frac{5}{3}\right) - \cosh^{-1}\left(\frac{5}{4}\right)$ b) $\ln\frac{3}{2}$ **65.** a) $\coth^{-1}2 - \coth^{-1}\left(\frac{5}{4}\right)$

b) $-\frac{\ln 3}{2}$ **67.** a) $\frac{1}{2}\left[\sinh^{-1}2 - \sinh^{-1}(1)\right]$

b) $\frac{1}{2}\ln\frac{2+\sqrt{5}}{1+\sqrt{2}}$ **69.** 3.916 **71.** 7.589 **73.** Graph $y = $ NINT $\left(\frac{\sinh t}{t}, t, 1, x\right)$ in $[-4, 4]$ by $[-10, 10]$. **75.** Graph NINT $\left(\frac{\cosh t - 1}{t}, t, 1, x\right)$ in $[-5, 5]$ by $[-1, 10]$. **77.** 2π

79. $(0, 0.477)$ **83.** b) $\sqrt{\frac{mg}{k}}$ c) 178.885 ft/sec **85.** 99

87. $\pi\left(\frac{63}{8} + 2\ln 8\right)$ **89.** c) $A(0) = 0$ so $A(u) = \frac{1}{2}u + C = \frac{1}{2}u$ and $u = 2A(u)$.

Chapter 7 Practice Exercises

1. $\frac{dy}{dx} = \frac{1}{2x}$. Support by graphing NDER$(\ln\sqrt{x}, x)$ in $[0, 2]$ by $[0, 10]$. **3.** $\frac{dy}{dx} = \frac{2x}{x^2+2}$ **5.** $\frac{dy}{dx} = -\frac{1}{e^x}$

7. $\frac{dy}{dx} = e$ **9.** $-\tan x$ **11.** $-\frac{1}{\sqrt{1-x^2}\cos^{-1}x}$

13. $\frac{2}{(\ln 2)x}$ **15.** $-(\ln 8)8^{-x}$ **17.** $\frac{-1}{2\sqrt{x(1-x)}}$

19. $\frac{2}{x\sqrt{x^2-1}}, x > 0$ **21.** $\frac{1}{x} + \frac{\sec^{-1}\sqrt{x}}{\sqrt{x-1}}$ **23.** a) $2x$

b) $1 + \ln 2$ c) -1 **25.** $y = e^x - 1$ **27.** $y = \ln 2x$

29. $\frac{2(x^2+1)}{\sqrt{\cos 2x}}\left(\frac{2x}{x^2+1} + \tan 2x\right)$

31. $5\left(\frac{(x+5)(x-1)}{(x-2)(x+3)}\right)^5 \cdot$
$\left[\frac{1}{x+5} + \frac{1}{x-1} - \frac{1}{x-2} - \frac{1}{x+3}\right]$ **33.** $e^{\tan^{-1}x} * 2x + $ 1) **35.** $\coth^2 x$ **37.** $-x\,\mathrm{csch}^2 x$ **39.** $\mathrm{sech}\,x$ **41.** $1 + \frac{x\sinh^{-1}x}{\sqrt{1+x^2}}$ **43.** $\frac{1}{x^2-1}$ **45.** $2\sec 2x$ **47.** Graph y in $[-30, 30]$ by $[0, 5]$. **49.** Graph in $[0, 4]$ by $[0, 5]$. Minimum at $(0.544, 0.578)$. **51.** $-\frac{\ln 7}{3}$ **53.** 1

55. $e - 1$ **57.** $\ln 8$ **59.** $\ln(9/25)$ **61.** $\ln(1+\sqrt{2})$

63. $\frac{(\ln 8)^2}{2\ln 4} = \frac{9}{8}\ln 4 = \frac{9}{4}\ln 2$ **65.** $\frac{1}{\ln 3}$ **67.** π **69.** $\frac{\pi}{2}$

71. $\frac{\pi}{6}$ **73.** $3 + \ln 4$ **75.** $8\sqrt{2}$ **77.** $5(\mathrm{csch}^2 - \mathrm{csch}^2 4)$

79. $e^{\tan x} + C$ **81.** $-\ln|\cos(\ln v)| + C$ **83.** $\frac{3^{x^2}}{2\ln 3} + C$ **85.** $\sec^{-1}|2y| + C$ **87.** a) $\sinh^{-1}1$ b) $\ln(1 + \sqrt{2})$ **89.** a) $2\left[\left(\tanh^{-1}\left(\frac{1}{2}\right)\right)^2 - \left(\tanh^{-1}\left(\frac{1}{5}\right)\right)^2\right]$

b) $\frac{1}{2}(\ln 2)\ln\frac{9}{2}$ **91.** a) $\left(\mathrm{sech}^{-1}\frac{3}{5}\right)^2 - \left(\mathrm{sech}^{-1}\frac{4}{5}\right)^2$

b) $\left(\ln\frac{3}{2}\right)\ln 6$ **93.** $\frac{6\ln 6 - 5}{\ln 5}$ **95.** 2.714 **97.** Graph $\int_1^x \frac{\tanh t}{t}dt = $ NINT $\left(\frac{\tanh t}{t}, t, 1, x\right)$ in $[-10, 10]$ by $[-5, 3]$. **99.** $\frac{1}{3}$ **101.** $\ln\frac{5}{3}$ **103.** Areas: $\frac{1}{\ln 2}$ and $\frac{1}{\ln 4}$. Ratio: 2 **105.** 18935 years **107.** About 5.3% **109.** About 7.19 years **111.** a) About 8% b) About 771.8 using the 8% of a) **115.** 91.943 minutes **117.** The limit is ∞ as $t \to 0^-$ and $-\infty$ as $t \to 0^+$. **119.** 2 **121.** Define $f(0) = \ln 2$ **123.** 1 **125.** e^3 **127.** In all three cases the functions grow at the same rate as $x \to \infty$ because $\lim_{x\to\infty}\frac{f(x)}{g(x)} = L$, L finite and not 0. **129.** a) True b) False c) True **131.** g is a right end behavior model for f. **133.** g is an end behavior model for f. **135.** g is a right end behavior model for f. **137.** $f(x) = \frac{\pi}{2}$ for $x > 0$ and $f(x) = -\frac{\pi}{2}$ for $x < 0$. **139.** $100 - 20\sqrt{17} \approx 17.538$ m **141.** $y = \mathrm{sech}^{-1}|x| - \sqrt{1-x^2} = \cosh^{-1}\left(\frac{1}{|x|}\right) - \sqrt{1-x^2}$

Chapter 8

Section 8.1

1. $2(8x^2 + 1)^{1/2} + C$ **3.** $\int_2^{10}\frac{1}{u}du \approx 1.69044$

5. $\ln(\cos^2 x^2) + C$ **7.** $3\int_{-\pi/3}^{\pi/3}\sec u\,du \approx 7.902$

9. 0.881 **11.** $-\ln|\csc(e^x + 1) + \cot(e^x + 1)| + C$ **13.** $\int_0^{\ln 2}e^u\,du = 1$ **15.** $\int_0^1 3^u\,du \approx 1.82048$

17. $12\int_1^{\sqrt{3}}\frac{du}{1-u^2} \approx 3.14159(\pi)$ **19.** $\frac{\pi}{18}$

21. $6\int_{2/\sqrt{3}}^{\sqrt{3}}du/(u\sqrt{u^2-1}) \approx 3.14159(\pi)$ **23.** $u = x - 2$; $\sin^{-1}(x-2) + C$ **25.** $u = x - 1$; 6.28319 (2π) **27.** $u = x + 1$; $\sec^{-1}|x+1| + C$ **29.** $4 - \pi/2$ **31.** 0 **33.** $\frac{\pi}{3} - \frac{1}{2}$

35. $\sqrt{2}$ **37.** $\int_0^{\pi/3}\sec x\,dx \approx 1.31696$ **39.** $\bar{x} = 0$, $\bar{y} = \frac{1}{2\ln(1+\sqrt{2})}$ **41.** 8.30169 cm **43.** $u = \sin x$; 2 **45.** $\ln(1/3)$ **47.** $u = \pi x$; $\frac{1}{\pi}\ln(1+\sqrt{2})$ **49.** $u = \tan x$; $e^{\sqrt{3}} - 1$ **51.** $u = \sqrt{x}$; $\frac{2}{\ln 2}$ **53.** $u = 3x$; π

55. $u = 2x$; $\dfrac{\pi}{6}$ **57.** $u = 2x$; $\dfrac{\pi}{12}$

59. $C = -\dfrac{\pi}{4}$

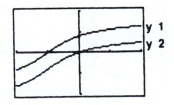

Section 8.2

1. $-x \cos x + \sin x + C$ **3.** $-x^2 \cos x + 2x \sin x +$

$2 \cos x + C$ **5.** $\dfrac{x^2}{2} \ln x - \dfrac{1}{4}x^2 + C$ **7.** $x \tan^{-1} x -$

$\dfrac{1}{2} \ln(1 + x^2) + C$ **9.** $x \tan x + \ln|\cos x| + C$ **11.** $x^3 e^x -$

$3x^2 e^x + 6x e^x - 6e^x + C$ **13.** $(x^2 - 7x + 7)e^x + C$

15. $(x^5 - 5x^4 + 20x^3 - 60x^2 + 120x - 120)e^x + C$

17. $\dfrac{x^2 - 4}{8}$ **19.** $\dfrac{2\pi}{3} - \dfrac{\sqrt{3}}{2}$ **21.** 11.614 **23.** 18.186

25. a) π b) 3π (area is positive)   **27.** $2\pi\left(1 - \dfrac{2}{e}\right)$

29. $\pi^2 + \pi - 4$ **31.** Exact answer is $3[(\ln 3)^4 -$

$4(\ln 3)^3 + 12(\ln 3)^2 - 24 \ln 3 + 16]$ **33.** a) 1 b) $\ln \dfrac{27}{4} -$

1, $(a) + (b) + 2\ln(2) = 3 \ln 3$ d) let $u = y = e^x$, $v = x$

Section 8.3

1. $-\cos x + 2\dfrac{\cos^3 x}{3} - \dfrac{\cos^5 X}{5} + C$ **3.** $-\dfrac{\sin^3 x}{3} +$

$\sin x + C$ **5.** $-\cos y + \dfrac{3\cos^3 y}{3} - \dfrac{3\cos^5 y}{5} + \dfrac{\cos^7 y}{7} + C$

7. $3x - 2\sin 2x + \dfrac{1}{4}\sin 4x + C$ **9.** $2\left[x - \dfrac{\sin 4x}{4}\right] + C$

11. $35\left[\dfrac{\sin^5 x}{5} - \dfrac{\sin^7 x}{7}\right] + C$ **13.** $-\cos^4 20 + C$ **15.** 4

17. 2 **19.** $\ln(3 + 2\sqrt{2})$ **21.** $\sqrt{2}$ **23.** $2\sqrt{3} + \ln(2 - \sqrt{3})$

25. $\dfrac{4}{3}$ **27.** $\dfrac{4}{3}$ **29.** $2 - \ln 4$ **31.** $\dfrac{4}{3} - \ln\sqrt{3}$ **33.** $-\dfrac{6}{5}$

35. π **37.** 0 **39.** (a), (b), (c), (e), (g), and (i) are zero.

(d), (f), and (h) are not zero.   **41.** (d) 2, (f) $\dfrac{4}{3}$, (h) $\dfrac{4}{3}$

45. a)

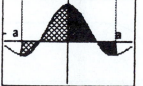

b)

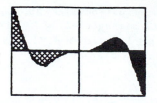

47. even, 2, 1   **49.** odd, 0, 0.5   **51.** even, 2.404, 1.202

Section 8.4

1. $\dfrac{\pi}{4}$ **3.** $\dfrac{\pi}{6}$ **5.** $\ln|x + \sqrt{x^2 - 4}| + C$

7. $\left[\dfrac{25}{2}\sin^{-1}\left(\dfrac{x}{5}\right) + \dfrac{x}{2}\sqrt{25 - x^2}\right] + C$

9. $4\left[\dfrac{x}{\sqrt{1 - x^2}} - \sin^{-1} x\right] + C$ **11.** $\dfrac{\pi}{3}$

13. $-\sqrt{2x - x^2} + C$ **15.** $\ln|x - 1 + \sqrt{x^2 - 2x}| + C$

17. $\dfrac{\pi}{8}$ **19.** $\dfrac{\pi}{3}$ **21.** $\ln(\sqrt{2} + 1)$ **23.** π **25.** $\dfrac{\pi}{8}$ **27.** 2

29. $0.6435011\ldots$ **31.** $4\left[\dfrac{x}{\sqrt{1 - x^2}} - \sin^{-1} x\right] + C$

33. $\dfrac{3\pi}{4}$ **35.** $x = 3\sin\theta$ would mean $\sin\theta = \dfrac{4}{3}$ at the left-hand endpoint. Use $x = 3\sec\theta$.

Section 8.5

1. $\dfrac{2}{x - 3} + \dfrac{3}{x - 2}$ **3.** $\dfrac{1}{x + 1} + \dfrac{3}{(x + 1)^2}$ **5.** $-\dfrac{2}{x} -$

$\dfrac{1}{x^2} + \dfrac{2}{x - 1}$ **7.** $1 + \dfrac{17}{x - 3} + \dfrac{-12}{x - 2}$ **9.** $\dfrac{1}{2}\ln 3$

11. $\dfrac{1}{2}\ln 15$ **13.** $\dfrac{1}{6}\ln\left|\dfrac{(t + 2)(t - 1)^2}{t^3}\right| + C$ **15.** $4 - \ln 3$

17. $5\left[\sqrt{3} - \dfrac{\pi}{3}\right]$ **19.** $-\dfrac{1}{4}\ln|x - 1| - \dfrac{1}{4}\dfrac{1}{(X - 1)} +$

$\dfrac{1}{4}\ln|x + 1| - \dfrac{1}{4}\dfrac{1}{(x + 1)} + C$ **21.** $\dfrac{1}{7}\ln|(x + 6)^2(x -$

$1)^5| + C$ **23.** $2\tan^{-1}(x - 1) + C$ **25.** $\ln\left|\dfrac{x^2 + x + 1}{x - 1}\right| +$

C **27.** $2x + \dfrac{1}{2}\ln(x^2 + 4) + C$ **29.** $1 - \ln 2$ **31.** $7 +$

$\ln 8$ **33.** $3\pi\ln 25$ **35.** a) $x(T) = \dfrac{1000e^{4t}}{499 + e^{4t}}$ **37.** 1

39. $\dfrac{2}{1 - \tan(x/2)} + C$ **41.** $-\cot\left(\dfrac{x}{2}\right) - x - C$

43. $\dfrac{1}{\sqrt{2}}\ln\left|\dfrac{\tan(x/2) + 1 - \sqrt{2}}{\tan(x/2) + 1 + \sqrt{2}}\right| + C$

Section 8.6

1. $\dfrac{\pi}{4}$ **3.** 6 **5.** 4 **7.** 1000 **9.** $\ln 4$ **11.** diverges

13. converges   **15.** converges   **17.** diverges   **19.** converges   **21.** diverges   **23.** diverges   **25.** converges

27. converges   **29.** diverges   **31.** diverges   **33.** con-

verges **35.** diverges **37.** converges **39.** converges
41. converges **43.** converges **45.** b) 7.5 **47.** 0.8862
51. converges if $p < 1$, diverges if $p \geq 1$. **53.** 1 **55.** 2π
57. $\ln 2$ **59.** diverges **61.** converges

Section 8.7

9. a) $x = 1000(11e^{0.10t} - 10)$ b) 23.026 years

11. $y = 95 - 65e^{0.017x}$ **13.** $y = \dfrac{3.350e^{04375t}}{32 + 3e^{0.4375t}}$ **15.** $y =$

$\dfrac{1}{2}\ln(2x + 1)$ **17.** $y = \arctan x + 1$ **19.** $y = xe^x - e^x + 1$

21. $y = \sin x$ **25.** $y(1) \approx y_5 = 2.0736$; $y(1) = e^1 =$

$2.71828\ldots$ **27.** $2.7027\ldots$ **29.** $y = \dfrac{1}{1 - x}$ is the solution

to $y' = y^2$, $y(0) = 1$
31. All three methods find $y(2) = 1.28571428571\ldots$

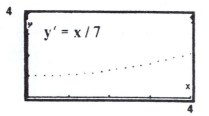

33. The computed values of $y(2)$ are:
Euler, 2.07334340632;
Improved Euler, 2.21157126441;
Runge-Kutta, 2.20283462521.

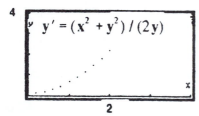

35. a) 0.310268270416   b) 0.310268299767
37. Differentiate $y = \tan(a^2 - x^2)$.
39. $\sqrt{\pi/4 + 1} \approx 1.336$
41.

	x	y(R–K)	y(true)	Difference
$y' - x - y$				
$y(0) = 1$	3	2.09981094687	...	2.37×10^{-4}
$y' = x - y$				
$y(0) = -2$	3	1.9009452656	...	-1.18×10^{-4}

The errors are much greater because the step size is five
times as large.

43. a) $y = \dfrac{3500e^{0.1t}}{111 + 14e^{0.1t}}$
e) 16.65 years, 34.57 years
f) 92 years

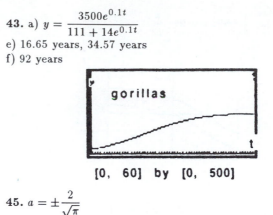

[0, 60] by [0, 500]

45. $a = \pm\dfrac{2}{\sqrt{\pi}}$

Section 8.8

1. (Using *Mathematica*)
In[2]: = Integrate[Exp[−x^2],{x,0,Infinity}]
Out[2] = $\dfrac{\text{Sqrt[Pi]}}{2}$

3. In[5]: = Integrate[1/(x Sqrt[x−3]),{x,6,9}]
Out[5] = $\dfrac{-\text{Pi}}{2\,\text{Sqrt[3]}} + \dfrac{2\,\text{ArcTan[Sqrt[2]]}}{\text{Sqrt[3]}}$

5. In[7]: = Integrate[1/(9−x^2k)^2,x]
Out[7] = $\dfrac{-\text{x}}{18(-9+\text{x})} - \dfrac{\text{Log}[-3 + \text{x}]}{108} + \dfrac{\text{Log}[3 + \text{x}]}{108}$

7. In[9]: = Integrate[1/(x^2 Sqrt[7+x^2]),{x,3,11}]
Out[9] = $\dfrac{4}{21} - \dfrac{8\,\text{Sqrt[2]}}{77}$

9. In[11]: = Integrate[(Sqrt[x^2−2])/x,{x,−2,−Sqrt[2]}]
Out[11] = $\dfrac{\text{Pi}}{\text{Sqrt[2]}} - \dfrac{4 + \text{Pi}}{2\,\text{Sqrt[2]}}$

11. In[13]: = Integrate[1/(4+5Sin[2x]),x]
Out[13] = $\dfrac{-\text{Log}[2\,\text{Cos[x]} + \text{Sin[x]}]}{6}$
$+ \dfrac{\text{Log}[\text{Cos[x]} + 2\,\text{Sin[x]}]}{6}$

13. In[15]: = Integrate[x Sqrt[2x−3],x]
Out[15] = $\text{Sqrt}[-3 + 2\text{x}]\left(-\dfrac{3}{5} - \dfrac{\text{x}}{5} + \dfrac{2\text{x}^2}{5}\right)$

15. In[17]: = Integrate[x^10 Exp[−x],{x,0,Infinity}]
Out[17] = 3628800

17.

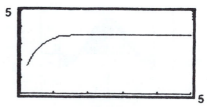

19.

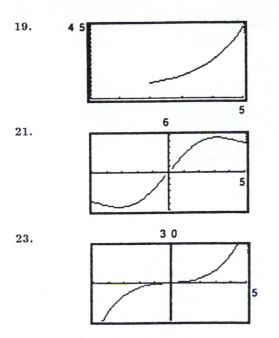

21.

23.

25. $x = 0.7705$ **27.** $x = 0.4987$ **29.** f always increasing, concave up on $(0, 1)$; inflection point at 1, $f \to$ constant as $x \to \infty$. **31.** $\bar{x} = 1.33333333143$, $\bar{y} = 1.38629436004$

35. $\dfrac{1}{a^2}\left[\ln|ax+b| + \dfrac{b}{ax+b}\right] + C$

Chapter 8 Practice Exercises

1. $2[\sqrt{2} - 1]$ **3.** 0 **5.** $\dfrac{\pi^2}{32}$ **7.** $\dfrac{\ln 3}{4}$ **9.** $\dfrac{1}{2}\ln(x^2+1) +$

$4\tan^{-1} x + C$ **11.** $\dfrac{x^3}{3}\ln x - \dfrac{x^3}{9} + C$ **13.** $-x^5\cos x +$

$5x^4\sin x + 20x^3\cos x - 60x^2\sin x - 120x\cos x +$

$120\sin x + C$ **15.** $e^x[\cos 2x + 2\sin 2x]/5 + C$

17. $-\cos y + \dfrac{\cos^3 y}{3} + C$ **19.** $\dfrac{\sin^5 x \cos x}{6} -$

$\dfrac{\sin^3 x \cos x}{24} + \dfrac{x}{16} - \dfrac{\sin 2x}{32} + C$ **21.** 2

23. $\dfrac{3}{2} - \ln 2$ **25.** $\dfrac{3}{80}$ **27.** $-\dfrac{\sqrt{1-x^2}}{x} + C$ **29.** 5 **31.** $\dfrac{\pi}{8}$

33. $\dfrac{\pi}{3}$ **35.** $16 + \dfrac{\ln 400}{3}$ **37.** $\ln|x-1| - \dfrac{1}{x-1} + C$

39. $\ln\left[\dfrac{|x|}{\sqrt{x^2+4}}\right] + C$ **41.** $\ln 3$ **43.** 1 **45.** $\dfrac{32\pi}{35}$

47. $s = 22.254$ **49.** 2π **51.** Diverges **53.** Diverges

55. $y(2) = -1.1377$ **57.** $y(6) = -7.349$

59. $y(3) = -2.691$ **61.** $y(3) = 0.907$

63. $y = 28e^{-0.15x} + 22$ **65.** $P = 500\dfrac{e^t}{24 + e^t}$

67. $x = \dfrac{ax_0}{x_0 + (a - x_0)e^{-akt}}$

Chapter 9

Section 9.1

1. $a_1 = 0$, $a_2 = -\dfrac{1}{4}$, $a_3 = -\dfrac{2}{9}$, $a_4 = -\dfrac{3}{16}$ **3.** $a_1 = 1$,

$a_2 = -\dfrac{1}{3}$, $a_3 = \dfrac{1}{5}$, $a_4 = -\dfrac{1}{7}$ **5.** 1, $\dfrac{3}{2}$, $\dfrac{7}{4}$, $\dfrac{15}{8}$, $\dfrac{31}{16}$, $\dfrac{63}{32}$,

$\dfrac{127}{64}$, $\dfrac{255}{128}$, $\dfrac{511}{256}$, $\dfrac{1023}{512}$ **7.** $2, 1, \dfrac{1}{2}, \dfrac{1}{4}, \ldots, x_{10} = \dfrac{1}{2^8}$

9. $x_1 = 1$, $x_2 = 1$, $x_3 = 2$, $x_4 = 3$, $x_5 = 5$, $x_6 = 8$,

$x_7 = 13$, $x_8 = 21$, $x_9 = 34$, $x_{10} = 55$ **11.** Converges to 2

13. Converges to 5 **15.** Diverges **17.** Converges to -1

19. Diverges **21.** Converges to -5 **23.** Converges

to 1 **25.** Converges to $1/2$ **27.** 0 **29.** 0 **31.** $\sqrt{2}$

33. $\dfrac{\pi}{2}$ **35.** Converges to 0 **37.** Converges to 0

39. Converges to 1 **41.** e^7 **43.** Diverges **45.** Converges

to 1 **47.** Converges to 1 **49.** Diverges **51.** Converges

to 0 **53.** Converges to 0 **55.** Converges to $1/e$

57. Converges to 0 **59.** $n = 693$ **61.** $n = 66$

63.

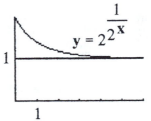

$$y = 2^{2^{\frac{1}{x}}}$$

67. $a_1 = 1 \to 0.877$, $a_1 = 2 \to 0.877$, $a_1 = 0 \to 0$

71. $\pm 1.895494\ldots$

Section 9.2

1. $s_n = \dfrac{2(1 - 1/3)^n)}{1 - 1/3}$, 3 **3.** $s_n = \dfrac{1 - (1/2)^n}{1 - (-1/2)}$, $\dfrac{2}{3}$

5. $s_n = \dfrac{1}{2} - \dfrac{1}{n+2}$, $\dfrac{1}{2}$ **7.** $s_5 = 1.3320\ldots$, $s_n \to \dfrac{4}{3}$

9. $s_5 = 2.33105\ldots$, $s_n \to \dfrac{7}{3}$ **11.** $s_n \to 11.5$

13. $s_n \to \dfrac{17}{6}$ **15.** $a_n = \dfrac{1}{4n-3} - \dfrac{1}{4n+1}$, $s_n \to 1$

17. $s_n \to \dfrac{1}{9}$ **19.** Converges to $2 + \sqrt{2}$ **21.** Converges

to 1 **23.** Diverges, $a_n \nrightarrow 0$ **25.** Converges to $\dfrac{e^2}{e^2 - 1}$

27. Diverges, $a_n \nrightarrow 0$ **29.** Converges to $\dfrac{3}{2}$ **31.** Diverges,

$a_n \nrightarrow 0$ **33.** $a - 1$, $r = -x$ **35.** a) 18 terms needed

b) 31 terms needed **37.** 28 m **39.** $\dfrac{234}{999} = \dfrac{26}{111}$

41. $a_n =$ a) $\dfrac{1}{(n+4)(n+5)}$, b) $\dfrac{1}{(n+2)(n+3)}$,

c) $\dfrac{1}{(n-3)(n-2)}$ **45.** Appears to converge to 2

47. Diverges **49.** \$71,428.58 **51.** a) \$6,900.38

b) $7,134.63 **53.** 8 m^2 **55.** $a_n = n$, $b_n = -n$
57. $\sum 1/3^n = 3/2$, $\sum 1/2^n = 2$, $\sum \left(\frac{2}{3}\right)^n = 3 \neq 3/4$

Section 9.3

1. Converges to 10/9; geometric series **3.** Diverges,
$a_n \not\to 0$ **5.** Converges by the comparison with $\frac{1}{2^n}$;
sum ≈ 0.637 **7.** Diverges, by comparison test
9. Converges by the integral test **11.** Diverges by limit
comparison test **13.** Diverges by limit comparison test
15. Diverges, $a_n \not\to 0$ **17.** Diverges by comparison with
$\sum \frac{1}{n}$ **19.** Converges by comparison with $\sum \frac{1}{n^{3/2}}$
21. Converges **23.** Converges, geometric series **25.** 14
27. Between 40.5 and 41.6 **29.** Compare $\sum \left(\frac{a_n}{n}\right)$ with
$\sum a_n$ **31.** $\{-s_n\}$ is nondecreasing and either bounded
above or not

Section 9.4

1. Converges, by the ratio test, sum ≈ 6. **3.** Converges,
by the ratio test, to approximately 376.179.
5. Diverges, by the ratio test. **7.** Diverges; $a_n \not\to 0$.
9. Diverges, $a_n \not\to 0$. **11.** Diverges **13.** Diverges, by
comparison test. **15.** Converges, by the ratio test;
17.0279727... **17.** Converges, by the ratio test: 4.0625.
19. Converges, by the ratio test; -6. **21.** Converges,
by the ratio test; 8.25271035... **23.** Converges by
comparison with $\sum \frac{1}{n^2}$; 0.5. **25.** Diverges by the ratio
test. **27.** Converges, by the ratio test; 2.680118...
29. Diverges, $a_n = \frac{3}{n}$. **31.** Converges, by the ratio
test; $-2.11952700...$ **33.** Converges, by the ratio test;
2.5707963... **35.** Converges by comparison with $\sum \frac{12}{n^4}$;
1.26079119...

Section 9.5

1. Converges; $s_{32} = 0.82199... < L < 0.82297... = s_{31}$
3. Diverges **5.** Converges; $s_{1,000,000}$ is within 0.001.
7. Converges. $|L - s_{10,000}| < 0.001$. **9.** Diverges
11. Converges absolutely **13.** Converges absolutely
15. Converges conditionally **17.** Converges absolutely
19. Converges conditionally **21.** Converges absolutely
23. Converges absolutely **25.** Diverges **27.** Diverges
29. Converges absolutely **31.** Converges absolutely
33. Converges absolutely **35.** Converges conditionally
37. |error| < 0.2 **39.** |error| $< 2 \times 10^{-11}$ **41.** 0.540302
43. a) $|a_n|$ not strictly decreasing b) $= -\frac{1}{2}$
45. $(a_{n+1} - a_{n+2}) + (a_{n+3} - a_{n+4}) + \dots$ has same sign
as a_{n+1}. **47.** a) $0.18226 < $ sum $ < 0.18233$
b) $0.4053 < $ sum $ < 0.4055$

c) $0.5875 < $ sum $ < 0.5881$
d) $0.64147 < $ sum $ < 0.64218$

Section 9.6

1. a) $-2 < x < 0$ b) $-2 < x < 0$
3. a) $-1 \leq x \leq 1$ b) $-1 \leq x \leq 1$
5. a) All x b) All x
7. a) $-1 \leq x < 1$ b) $-1 < x < 1$
9. a) $-1 \leq x < 1$ b) $-1 < x < 1$
11. a) $-3 < x < 3$ b) $-3 < x < 3$
13. a) $-1 < x < 1$ b) $-1 < x < 1$
15. a) $(-1, 1)$ b) $(-1, 1)$ c) $\frac{1}{1-x}$ d) $-0.8 < x < 0.7$
17. a) $(-8, 12)$ b) $(-8, 12)$ c) $\frac{10}{12-x}$ d) $-6 \leq x \leq 9$
19. a) $P_{20} = -3.597...$, $P_{100} = -5.187...$, b) 0.19
21. a) Convergent, $[1, 3)$ b) Absolutely convergent,
$(1, 3)$ c) Alternating on $(1, 2)$ d) P_{30} on $[1.1, 2]$,
error < 0.01 **23.** a) Convergent, $(2, 4)$ b) Absolutely
convergent, $(2, 4)$ c) Alternating on $(2, 3)$ d) P_{110} on
$[2.1, 3]$, error < 0.01 **25.** b) $[-8.4, 8.4]$, $[-15.9, 15.9]$,
$[-23.3, 23.3]$ **27.** All x; $[-7, 7]$ **29.** All x; $[-\frac{7}{3}, \frac{7}{3}]$
31. d) $-0.80 < x < 0.74$ **33.** b) Use $-\ln(3 - x) =$
$-\ln(1 - (2 - 2))$ and part a) **35.** a) $(-0.38, 0.38)$
b) $(-0.698, 0.698)$ c) $(-2.85, 2.85)$ d) All x **37.** $x = 0$
39. Converges for $1 < x < 5$ to $\frac{2}{x-1}$
41. a) $-0.873 < x < 0.873$ b) $\frac{x^2}{2} + \frac{x^4}{12} + \frac{x^6}{45} +$
$\frac{17x^8}{(8) \cdot (315)} + \dots$, $-\frac{\pi}{2} < x < \frac{\pi}{2}$
c) $1 + x^2 + \frac{2x^4}{3} + \dots$ **43.** b) $|s_{30} - f(x)| < 0.01$ when
$|x| < 1.172$ or $1.97 < |x| < \pi$ **45.** Compare $|a_n|$ with
$\frac{1}{n^2}$ **49.** A smooth peak becomes an intricate "mountain
range." **51.** Conjecture: $s(x)$ is not differentiable
53. Diverges, by the ratio test

Section 9.7

1. $P_3 = (x - 1) - \frac{(x-1)^2}{2} + \frac{(x-1)^3}{3}$; $0.60 < x < 1.47$
3. $P_3 = \frac{1}{2} - \frac{1}{4}(x - 2) + \frac{1}{8}(x - 2)^2 - \frac{1}{16}(x - 2)^3$; $1.34 < x < 2.78$
5. $P_3 = \frac{1}{\sqrt{2}}\left[1 + \left(x - \frac{\pi}{4}\right) - \frac{(x - \pi/4)^2}{2} - \frac{(x - \pi/4)^3}{6}\right]$;
$-0.008 < x < 1.515$ **7.** $P_3 = 2 + \frac{x-4}{4} - \frac{(x-4)^2}{64} +$
$\frac{(x-4)^3}{512}$; $1.91 < x < 6.60$ **9.** $\sum_{n=0}^{\infty} (-x)^n / n!$, $(-5, 2.5)$

11. $\sin 3x = \sum_{0}^{\infty} \frac{(-1)^n (3x)^{2n+1}}{(2n+1)!}$; graphically, when

$|x| < 1.083$; using R_{10} when $|x| < 0.95$

13. $\sum_{n=0}^{\infty} \frac{(-1)^n x^{2n}}{(2n)!}$; graphically, when $|x| < 3.624$;

analytically, when $|x| < 3.22$ **15.** $\sum_{n=0}^{\infty} \frac{x^{2n}}{(2n)!}$; graphically,

when $|x| < 3.581$; analytically, when $|x| < 2.734$

17. $\sum_{n=2}^{\infty} \frac{(-1)^n x^{2n}}{(2n)!}$; graphically, when $|x| < 3.624$;

analytically, when $|x| < 3.603$ **19.** $[1 - x + x^2] -$

$\frac{x^3}{(1+c)^4}$, c between 0 and x **21.** $\left[x - \frac{x^2}{2}\right] +$

$\frac{x^3}{3(1+c)^3}$, c between 0 and x **23.** $(x) - \frac{x^3 \cos c}{3!}$

25. $\sum_{n=0}^{\infty} \frac{(x-a)^n e^a}{n!} = e^a \left[1 + (x-a) + \frac{(x-a)^2}{2!} + \infty \right]$

27. $|x| \le 0.56$ **29.** $|\text{error}| < 1.67 \times 10^{-10}$, $x < 0$
31. $|R_2| < 1.84 \times 10^{-4}$ **33.** 2.6×10^{-4} **35.** $\sin 0.1 =$
$0.09978334 \dots$ **39.** $x + x^2 + \left(\frac{1}{3}\right) x^3 = \left(\frac{1}{30}\right) x^5 -$
$\left(\frac{1}{90}\right) x^6 + \dots$ **43.** a) -1 b) $\frac{1}{\sqrt{2}}(1 + i)$ c) $-i$

Section 9.8

1. $\cos 1 - (x-1)\sin 1 - \frac{(x-1)^2}{2}\cos 1 + \frac{(x-1)^3}{6}\sin 1$;
$\frac{|x-1|^4}{4!}$

3. $e^{0.4}\left[1 + (x-0.4) + \frac{(x-0.4)^2}{2} + \frac{(x-0.4)^3}{6}\right]$;
$\frac{|x-0.4|e^{0.4}}{4!}$

5. $\cos 69 - (x-69)\sin 69 - \frac{(x-69)^2 \cos 69}{2} +$
$\frac{(x-69)^3 \sin 69}{2}$; $\frac{|x-69|^4}{4!}$

	x-Range	Viewing Window
9.	$(-0.88, 1.14)$	$(-2, 6)$ by $(-2, 15)$
11.	$(-0.35, 0.4)$	$(-1, 6)$ by $(0, 3)$
13.	$(-0.27, 0.26)$	$(-1, 4)$ by $(0, 1.5)$

15. a) $2.005 . 10^{-8}$ b) $9.93 . 10^{-4}$ c) 0.169

17. $\sum_{k=1}^{\infty} \frac{(-1)^{k+1} x^{2k-1}}{(2k-1)(2k)!}$

19. $\sum_{k=1}^{\infty} \frac{(-1)^{k+1}(0.1)^{2k-1}}{(2k-1)(2k-1)!}$; $\int_0^{0.1} P_{10}(x)dx = 0.999444612$

21. $\sum_{n=1}^{\infty} \frac{(-1)^{n+1} x^{2n-1}}{(2n-1)(2n)!}$; $\int_0^{0.1} P_{10}(x)dx = 0.49986114 \dots$

Chapter 9 Practice Exercises

1. Converges to 1   **3.** Diverges   **5.** Converges to 0
7. Converges to 3   **9.** Converges to 0   **11.** Diverges
13. Diverges   **15.** $\frac{e}{e-1}$ **17.** 2625   **19.** Diverges
21. Conditionally convergent, approx value of S is 0.6
23. Conditionally convergent; -0.92 is within 0.5 of S
25. Absolutely convergent; -0.55...   **27.** Converges,
$2e - 1$ **29.** Convergent, $e^{-3} - 1$ **31.** a) $-5 \le x < 1$
b) $-5 < x < 1$, -4.78 **33.** all x, all x

35. a) $0 \le x \le 2$ b) $0 \le x \le 2$ **37.** $f(x) = \frac{1}{1+x}$; at
$x = \frac{1}{4}$, sum is 0.8   **39.** $f(x) = \sin x$; at $x = \pi$, sum is 0
41. $f(x) = e^x$; at $x = \ln 2$, sum is 2   **43.** $2 - \frac{1}{2}(x+1) +$
$\frac{3}{16}(x+1)^2 + \frac{9}{192}(x+1)^3 + \dots$; $-2.143 < x < 0.238$
45. $(2x) - \frac{(2x)^3}{3!} + \frac{(2x)^5}{5!} - \frac{(2x)^7}{7!} + \dots$ **47.** 0.48491714

49. a) $\sum_{n=0}^{\infty} 0\frac{x^n}{n!}$, converges for all x, converges to f only

at $x = 0$ b) for $x \ne 0$, $R_n(x) = e^{-1/x^2}$

Chapter 10

Section 10.1

1. $y = \frac{x^2}{16}$ **3.** $y = -\frac{x^2}{12}$ **5.** $x = -\frac{y^2}{12}$ **7.** $F\left(0, \frac{1}{16}\right)$,
$y = -\frac{1}{16}$ **9.** Focus: $\left(0, -\frac{1}{12}\right)$. Directrix: $y = \frac{1}{12}$

11. Graph $y = \frac{x^2}{2}$ in $[-10.6, 10.6]$ by $[0, 12.5]$ to check
your result. (We have used the "screen-squaring"
feature of our calculator to help determine the viewing
rectangle.)   **13.** Graph $y = \sqrt{8x}$ and $y = -\sqrt{8x}$
together in $[-9.4, 14.4]$ by $[-7, 7]$ to check your result.
15. Graph $y = \frac{3}{2}\sqrt{4 - x^2}$ and $y = -\frac{3}{2}\sqrt{4 - x^2}$ in
$[-5.1, 5.1]$ by $[-3, 3]$.   **17.** Graph $y_1 = 2\sqrt{1 + x^2}$ and
$y_2 = -y_1$ in $[-13.6, 13.6]$ by $[-8, 8]$.   **19.** $\frac{x^2}{6^2} - \frac{y^2}{8^2} = 1$,
$e = \frac{5}{3}$. Foci: $(\pm 10, 0)$, asymptotes: $y = \pm \frac{4}{3}x$. Graph
$y_1 = (4/3)\sqrt{x^2 - 36}$, $y_2 = -y_1$, $y_3 = (4/3)x$ and $y_4 = -y_3$ in $[-34, 34]$ by $[-20, 20]$.   **21.** $\frac{y^2}{2} - \frac{x^2}{8} = 1$, $e = \sqrt{5}$. Foci: $(0, \pm\sqrt{10})$, asymptotes: $y = \pm\frac{x}{2}$. Graph
$y = \frac{\sqrt{x^2 + 8}}{2}$, $y = -\frac{\sqrt{x^2 + 8}}{2}$, $y = \frac{x}{2}$ and $y = -\frac{x}{2}$ in

$[-8.5, 8.5]$ by $[-5, 5]$.   **23.** $\dfrac{x^2}{25} + \dfrac{y^2}{169} = 1$, $e = \dfrac{12}{13}$.

Foci: $(0, \pm 12)$. Graph $y_1 = \dfrac{13}{5}\sqrt{25 - x^2}$ and $y_2 = -y_1$

in $[-22, 22]$ by $[-13, 13]$.   **25.** $\dfrac{x^2}{2} - \dfrac{y^2}{8} = 1$, $e = \sqrt{5}$.

Foci: $(\pm\sqrt{10}, 0)$, asymptotes: $y = \pm 2x$. Graph $y = 2\sqrt{x^2 - 2}$, $y = -2\sqrt{x^2 - 2}$, $y = 2x$ and $y = -2x$ in $[-17, 17]$ by $[-10, 10]$.   **27.** $\dfrac{x^2}{10} + \dfrac{y^2}{9} = 1$, $e = \dfrac{1}{\sqrt{10}}$.

Foci: $(\pm 1, 0)$. Graph $y_1 = \dfrac{3\sqrt{10 - x^2}}{\sqrt{10}}$, and $y_2 = -y_1$ in

$[-5.1, 5.1]$ by $[-3, 3]$.   **29.** $x^2 - y^2 = 1$, $e = \sqrt{2}$. Foci: $(\pm\sqrt{2}, 0)$, asymptotes: $y = \pm x$. Graph $y = \sqrt{x^2 - 1}$, $y = -\sqrt{x^2 - 1}$, $y = x$ and $y = -x$ in $[-8.2, 8.2]$ by $[-4.9, 4.9]$.   **31.** $\dfrac{y^2}{4} - \dfrac{x^2}{4} = 1$, $e = \sqrt{2}$. Foci: $(0, \pm 2\sqrt{2})$. Graph $y = \sqrt{x^2 + 4}$, $y = -\sqrt{x^2 + 4}$, $y = x$ and $y = -x$ in $[-8.2, 8.2]$ by $[-4.9, 4.9]$.   **33.** $\dfrac{x^2}{2} + \dfrac{y^2}{3} = 1$, $e = \dfrac{1}{\sqrt{3}}$. Foci: $(0, \pm 1)$. Graph $y = \sqrt{1.5}\sqrt{2 - x^2}$ and $y = -\sqrt{1.5}\sqrt{2 - x^2}$ in $[-2.9, 2.9]$ by $[-\sqrt{3}, \sqrt{3}]$.

35.

	$y = -x^2/4$	$x = -y^2/4$
Focal axis:	The y-axis	The x-axis
Focus:	$(0, -1)$	$(-1, 0)$
Vertex:	$(0, 0)$	$(0, 0)$
Directrix:	$y = 1$	$x = 1$

39. One example is the graph of $\dfrac{x^2}{25} + \dfrac{y^2}{9} = 1$.

41. Dimensions: $2\sqrt{2}$ (horizontal) by $\sqrt{2}$. Area $= 4$.

43. 24π

Section 10.2

1. Hyperbola. Graph $y_1 = \sqrt{x^2 - 1}$ and $y_2 = -y_1$ in $[-17, 17]$ by $[-10, 10]$.

3. Parabola. Graph $y_1 = 2\sqrt{x + 1}$ and $y_2 = -y_1$ in $[-5.8, 7.8]$ by $[-4, 4]$. Shift the graph of $x = \dfrac{y^2}{4}$ horizontally left one unit.

5. Ellipse. Let $y_1 = \sqrt{64 - 16(x^2 - 4x + 4)}$. Graph $y_2 = \dfrac{8 + y_1}{8}$ and $y_3 = \dfrac{8 - y_1}{8}$ in $[0, 4]$ by $[-3.8, 2.3]$.

7. Parabola. Let $y_1 = \sqrt{(4x)^2 - 16(x^2 - 3x - 6)}$. Graph $y_2 = \dfrac{-4x + y_1}{8}$ and $y_3 = \dfrac{-4x - y_1}{8}$ in $[-8.4, 13.6]$ by $[-9.1, 3.9]$. $\theta = -0.464$.   **9.** Hyperbola. Let $y_1 = \sqrt{x^2 + 12x + 20}$. Graph $y_2 = \dfrac{-x + y_1}{2}$ and $y_3 = \dfrac{-x - y_1}{2}$ in $[-49, 43]$ by $[-27.4, 26.8]$. $\theta = -\dfrac{\pi}{8}$.

11. Hyperbola. Graph $y = \sqrt{x^2 - 1}$ and $y = -\sqrt{x^2 - 1}$ in $[-8.5, 8.5]$ by $[-5, -5]$.   **13.** Hyperbola. Graph $y = \dfrac{1}{x}$ in $[-5.1, 5.1]$ by $[-3, 3]$. $\theta = \dfrac{\pi}{4}$.

15. Hyperbola. Graph $y = -\dfrac{2x^2 + x + 1}{x - 1}$ in dot format in $[-8, 8.7]$ by $[-19, 9.1]$. $\theta = 0.232$.

17. Ellipse. Let $y_1 = \sqrt{(6 - 3x)^2 - 4(3)(x^2 - 7)}$. Graph $y_2 = \dfrac{-(6 - 3x) + y_1}{6}$ and $y_2 = \dfrac{-(6 - 3x) - y_1}{6}$ in $[-20.3, 8.2]$ by $[-12.9, 3.9]$. $\theta = 0.491$.

19. Ellipse. Let $y_1 = \sqrt{(3x + 17)^2 - 4(2)(6x^2 + 2)}$. Graph $y_2 = \dfrac{-(3x + 17) + y_1}{4}$ and $y_3 = \dfrac{-(3x + 17) - y_1}{4}$ in $[-11.9, 14.4]$ by $[-12.6, 2.9]$. $\theta = 0.322$.   **21.** Graph the hyperbola $y = \dfrac{2}{x}$ in $[-8.5, 8.5]$ by $[-5, 5]$. $\theta = \dfrac{\pi}{4}$. $x^2 - y^2 = 4$.   **23.** Let $y_1 = \sqrt{3x^2 - 4(2)(x^2 - 1)}$. Graph $y_2 = \dfrac{\sqrt{3}x + y_1}{4}$ and $y_3 = \dfrac{\sqrt{3}x - y_1}{4}$ in $[-2.6, 2.1]$ by $[-1.5, 1.3]$ obtaining an ellipse. $\theta = \dfrac{\pi}{6}$. $x^2 + 5y^2 = 2$.

25. Let $y_1 = \sqrt{9x^2 - 4(x^2 - 5)}$. Graph $y_2 = \dfrac{3x + y_1}{2}$ and $y_3 = \dfrac{3x - y_1}{2}$ in $[-17, 17]$ by $[-10, 10]$ obtaining a hyperbola. $\theta = \dfrac{\pi}{4}$. $5y^2 - x^2 = 10$.   **27.** Ellipse. Let $y_1 = \sqrt{4x^2 - 12(3x^2 - 19)}$. Graph $y_2 = \dfrac{-2x + y_1}{6}$ and $y_3 = \dfrac{-2x - y_1}{6}$ in $[-5.1, 5.1]$ by $[-3, 3]$. $\theta = \dfrac{\pi}{4}$. $4x^2 + 2y^2 = 19$.   **29.** $x = x'\cos\theta + y'\sin\theta$, $y = -x'\sin\theta + y'\cos\theta$.   **31.** $x^2 + 6xy + y^2 - 16 = 0$.   **33.** $x^2 + 10\sqrt{3}xy + 11y^2 - 40 = 0$.

Section 10.3

1. Graph $x_1 = \cos t$, $y_1 = \sin t$, $0 \le t \le \pi$ in $[-1.7, 1.7]$ by $[-1, 1]$. The upper half of the unit circle, $x^2 + y^2 = 1$, is traced out in the counterclockwise direction from $(1, 0)$ to $(-1, 0)$.   **3.** Graph $x_1 = \sin 2\pi t$, $y_1 = \cos 2\pi t$, $0 \le t \le 1$ in $[-1.7, 1.7]$ by $[-1, 1]$. The unit circle is traced out once in the clockwise direction starting at $(0, 1)$.   **5.** $x = 4\cos t$, $y = 2\sin t$, $0 \le t \le 2\pi$. Graph this with t-step $= 0.1$ in $[-4, 4]$ by $[-2.4, 2.4]$. The ellipse $\dfrac{x^2}{16} + \dfrac{y^2}{4} = 1$ is traced out once in the counterclockwise direction.   **7.** Graph in $[-8.5, 8.5]$ by $[-5, 5]$. The upper half of the ellipse $\dfrac{x^2}{16} + \dfrac{y^2}{25} = 1$ is traced out in the counterclockwise direction.   **9.** Graph for $-1 \le t \le 1$ in $[-7.6, 7.6]$ by $[0, 9]$. The parabola $y = x^2$ is traced out from left to right.   **11.** Graph for $0 \le t \le 10$ in $[0, 10]$ by $[-1.4, 4.5]$. The upper half of the parabola $y^2 = x$ is traced out from left to right. **13.** Graph in $[-17, 17]$ by $[-10, 10]$. The left branch of the hyperbola $x^2 - y^2 = 1$ is traced out from bottom to top. **15.** Graph for $-5 \leq t \leq 5$ in $[-16.6, 17.4]$ by $[-7, 13]$. The graph of the line $y = 2x + 3$ is traced out from left to right. **17.** Graph in $[-2, 3]$ by $[-1, 2]$. The line segment from $(0, 1)$ to $(1, 0)$ is traced out from left to right. It is part of $y = 1 - x$. **19.** Graph in $[-1, 1]$ by $[-0.1, 1.1]$. The upper half of the unit circle, $y = \sqrt{1 - x^2}$, is traced out from left to right. **21.** Graph for $0 \leq t \leq 4$ in $[-6.9, 21.9]$ by $[0, 17]$. The top half of the hyperbola $y^2 - x^2 = 1$ for $x \geq 0$ is traced out from left to right. **23.** Graph for $-3 \leq t \leq 3$ in $[-17, 17]$ by $[-10, 10]$. The right branch of the hyperbola $x^2 - y^2 = 1$ is traced out from bottom to top. **25.** a) $x = a \cos t$, $y = -a \sin t$, $0 \leq t \leq 2\pi$ b) $x = a \cos t$, $y = a \sin t$, $0 \leq t \leq 2\pi$ c) $x = a \cos(2t)$, $y = -a \sin(2t)$, $0 \leq t \leq 2\pi$ d) $x = a \cos t$, $y = a \sin t$, $0 \leq t \leq 4\pi$ **27.** $a = \pi$. $a = 2\pi$ is required for the complete graph. **29.** $a = \dfrac{\pi}{2}$. For a complete graph use $a = 2\pi$.

31. $a = \dfrac{\pi}{1.5}$. A complete graph is obtained if $a = 4\pi$. **33.** No such a exists. **35.** $x = \cos t + t \sin t$, $y = \sin t - t \cos t$. Graph this and $x_2 = \cos t$, $y_2 = \sin t$, $0 \leq t \leq 2\pi$ in $[-10.2, 10.2]$ by $[-6, 6]$. **37.** $(1, 1)$ **39.** b) $x = x_1 t$, $y = y_1 t$ c) $x = -1 + t$, $y = t$. Graph in $[-2, 2]$ by $[-1, 3]$, $-1 \leq t \leq 3$. **41.** a) Use $[-18, 32]$ by $[-15, 15]$. b) Use $[-0.13, 2]$ by $[-0.55, 0.55]$. c) Use $[-50, 1]$ by $[-55, 15]$. **43.** The three graphs can be compared in the viewing window $[0, 4\pi]$ by $[-2.7, 4.7]$. **45.** Graph in $[-5.1, 5.9]$ by $[-3.2, 3.2]$. The new equations amount to $x = -2 \cos t + \cos(2t)$, $y = -2 \sin t + \sin(2t)$. Graph in the same window. The original curve had three cusps. The new curve appears to be cardioid. **47.** Graph a), b), c) in $[0, 128]$ by $[-21, 54]$. In d) the curve is part of the y-axis traced from $(0, 0)$ to $(0, 64)$ and back down to $(0, 0)$. **49.** Graph $x_1 = t$, $y_1 = t^2 + 1$,

$$x_2 = \frac{\sqrt{3}t - (t^2 + 1)}{2}, \; y_2 = \frac{t + \sqrt{3}(t^2 + 1)}{2}, \; -4 \leq t \leq 4$$

in $[-16.1, 16.1]$ by $[-2, 17]$. $3x^2 + 2\sqrt{3}xy + y^2 + 2x - 2\sqrt{3}y + 4 = 0$. **51.** The rotated curve has equation $y = x^2$. Graph $x_1 = t^2$, $y_1 = t$, $x_2 = t$, $y_2 = t^2$, $-3 \leq t \leq 3$ in $[-15.3, 15.3]$ by $[-9, 9]$. **53.** Graph $x_1 = 2 \cos t$, $y_1 = 3 \sin t$, $x_2 = \cos t - \left(\dfrac{3\sqrt{3}}{2}\right) \sin t$, $y_2 = \sqrt{3} \cos t + \dfrac{3}{2} \sin t$,

$0 \leq t \leq 2\pi$ in $[-5.1, 5.1]$ by $[-3, 3]$. The rotated curve has equation $21x^2 + 10\sqrt{3}xy + 31y^2 = 144$. **55.** Graph $x_1 = 4 \sec t$, $y_1 = 5 \tan t$, $x_2 = \dfrac{4 \sec t - 5\sqrt{3} \tan t}{2}$, $y_2 =$

$\dfrac{4\sqrt{3} \sec t + 5 \tan t}{2}$, $0 \leq t \leq 2\pi$ in $[-34, 34]$ by $[-20, 20]$ (dot format may help). The rotated curve has equation $-23x^2 + 82\sqrt{3}xy + 59y^2 = 1600$.

Section 10.4

1. Tangent line: $y = -x + 2\sqrt{2}$. $\dfrac{d^2 y}{dx^2}\Big|_{t=\pi/4} = -\sqrt{2}$. Graph $x_1 = 2 \cos t$, $y_1 = 2 \sin t$, $x_2 = t$, $y_2 = -t + 2\sqrt{2}$, $-3.5 \leq t \leq 9$, T step $= 0.05$ in $[-10.5, 10.5]$ by $[-6.1, 6.1]$. **3.** $y = -\dfrac{1}{2}x + 2\sqrt{2}$. $\dfrac{d^2 y}{dx^2}\Big|_{t=\pi/4} = -\dfrac{\sqrt{2}}{4}$. Graph $x_1 = 4 \sin t$, $y_1 = 2 \cos t$, $x_2 = t$, $y_2 = -\dfrac{1}{2}x + 2\sqrt{2}$, $-10.5 \leq t \leq 10.5$ in $[-10.5, 10.5]$ by $[-6.1, 6.1]$. **5.** $y = -\dfrac{1}{2}x - \dfrac{1}{2}$. $\dfrac{d^2 x}{dx^2}\Big|_{t=-\pi/4} = \dfrac{1}{4}$. Graph $x_1 = \sec^2 t - 1$, $y_1 = \tan t$, $x_2 = t$, $y_2 = -\dfrac{1}{2}t - \dfrac{1}{2}$, $-7 \leq t \leq 15$ in $[-6.2, 14.2]$ by $[-8, 4]$. **7.** $y = x + \dfrac{1}{4}$. $\dfrac{d^2 x}{dx^2}\Big|_{t=1/4} = -2$. Graph $x_1 = t$, $y_1 = \sqrt{t}$, $x_2 = t$, $y_2 = t + 0.25$, $-10.5 \leq t \leq 10.5$ in $[-10.5, 10.5]$ by $[-6.1, 6.1]$. **9.** $y = x - 4$. $\dfrac{d^2 y}{dx^2} = \dfrac{1}{2}$ for all $t \neq 0$. Graph $x_1 = 2t^2 + 3$, $y_1 = t^4$, $x_2 = t$, $y_2 = t - 4$, $-2 \leq t \leq 20$ in $[-13.2, 24.2]$ by $[-6, 16]$. **11.** $y = \sqrt{3}x + 2 - \dfrac{\sqrt{3}}{3}\pi$. $\dfrac{d^2 y}{dx^2}\Big|_{t=\pi/3} = -4$. Graph $x_1 = t - \sin t$, $y_1 = 1 - \cos t$, $x_2 = t$, $y_2 = \sqrt{3}t + 2 - \dfrac{\sqrt{3}}{3}\pi$, $-17 \leq t \leq 17$ in $[-17, 17]$ by $[-10, 10]$. **13.** Graph in $[-1, 1]$ by $[0, \pi]$. $L = 4$ **15.** Graph in $[0, 8]$ by $[1/3, 9]$. $L = 12$ **17.** Graph in $[8, 4\pi]$ by $[0, 8]$. $L = \pi^2$ **19.** $8\pi^2$ **21.** $\dfrac{52\pi}{3}$ **23.** $3\sqrt{5}\pi$ **25.** a) π b) π **27.** 19.377 **29.** 2505.105 **31.** 159.485 **33.** Graph $x_1 = \sin t$, $y_1 = \sin 2t$, $0 \leq t \leq 2\pi$ in $[-1.86, 1.86]$ by $[-1.1, 1.1]$. Horizontal tangent at $\left(\dfrac{\sqrt{2}}{2}, 1\right) \left(t = \dfrac{\pi}{4}\right)$. Tangents at origin: $y = \pm 2x$. To confirm, we graph x_1, y_1, $x_2 = t$, $y_2 = 2t$, $x_3 = t$, $y_3 = -2t$, $-2\pi \leq t \leq 2\pi$ in the viewing rectangle given above.

35 through 41. For each of these we may use $0 \leq t \leq 2\pi$ in $[-1.86, 1.86]$ by $[-1.1, 1.1]$. For 38 and 39, $\dfrac{\pi}{2} \leq t \leq \dfrac{3\pi}{2}$ suffices. **43.** Graph $x = 12 \cos t + 6 \cos(ct)$, $y = 12 \sin t + 6 \sin(ct)$, $0 \leq t \leq 2\pi$ in $[-34, 34]$ by $[-20, 20]$ for each $c = 2, 4, 6, 8$. **45.** 306.324

Section 10.5

1. $\{a, c\}$, $\{b, d\}$, $\{e, k\}$, $\{f, h\}$, $\{g, j\}$, $\{i, l\}$, $\{m, 0\}$, $\{n, p\}$

3.

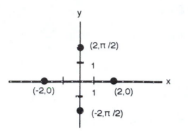

a) $\left(2, \dfrac{\pi}{2} + 2n\pi\right), \left(-2, -\dfrac{\pi}{2} + 2n\pi\right)$ b) $(2, 2n\pi)$,

$(-2, (2n+1)\pi)$ c) $\left(-2, \dfrac{\pi}{2} + 2n\pi\right), \left(2, -\dfrac{\pi}{2} + 2n\pi\right)$

d) $(-2, 2n\pi), (2, (2n+1)\pi)$. $n = 0, \pm1, \pm2, \dots$

5. a) $(1, 1)$ b) $(1, 0)$ c) $(0, 0)$ d) $(-1, -1)$

d) $\left(\dfrac{3\sqrt{3}}{2}, -\dfrac{3}{2}\right)$ f) $(3, 4)$ g) $(1, 0)$ h) $(-\sqrt{3}, 3)$

7. Graph $r = 2, 0 \le \theta \le 2\pi$ is a square window containing $[-2, 2]$ by $[-2, 2]$ in polar mode.

9.

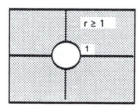

11.

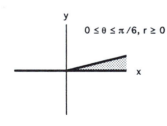

13.

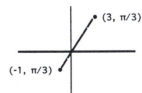

15. The graph consists of the origin and the positive y-axis.

17. The graph consists of the upper half of the unit circle including $(-1, 0)$ and $(1, 0)$.

19.

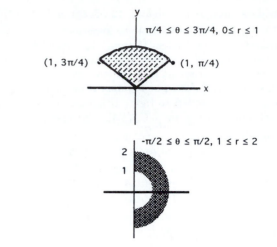

21.

$-\pi/2 \le \theta \le \pi/2, 1 \le r \le 2$

23. $x = 2$. Vertical line consisting of all points with x-coordinate 2. **25.** $y = 4$, the horizontal line through $(1, 4)$. **27.** $y = 0$, the x-axis. **29.** $x + 1 = 1$, the line through $(1, 0)$ and $(0, 1)$. **31.** $x^2 + y^2 = 1$, the unit circle. **33.** $y = 2x + 5$, line with slope 2 through $(0, 5)$. **35.** $r = 7 \sec \theta, -\dfrac{\pi}{2} < \theta < \dfrac{\pi}{2}$, the vertical line through $(7, 0)$. **37.** $\theta = \dfrac{\pi}{4}$ **39.** $r = 2$, circle **41.** $r = \dfrac{\pm 6}{\sqrt{4\cos^2 \theta + 9\sin^2 \theta}}$, ellipse **43.** $r = \dfrac{4\cos\theta}{\sin^2 \theta}$, parabola

Section 10.6

1. In Exercises 1 through 12 the student should use the method of Example 1 including a table of values, use of symmetries and Equation (4) as a guide as to how the curve goes into and out of the origin. The student's results can then be checked on a grapher. One way of carrying out this check is given in the answers. It is assumed that a graphing utility with a polar graphing mode and with a screen "squaring" function is being used.

Graph $r = 1 + \cos\theta, 0 \le \theta \le 2\pi, \theta$ Step $= 0.1$ in a square rectangle containing $[-0.25, 2]$ by $[-1.3, 1.3]$, for example $[-1.33, 3.08]$ by $[-1.3, 1.3]$. **3.** Graph $r = 1 - \sin\theta$, $0 \le \theta \le 2\pi$ in $[-2.4, 2.4]$ by $[-2.2, 0.7]$. **5.** Graph $r = 2 + \sin\theta, 0 \le \theta \le 2\pi$ in $[-3.5, 3.6]$ by $[-1.2, 3]$.

7. Graph $r = 2\sqrt{\cos 2\theta}$ and $r = -2\sqrt{\cos 2\theta}$ in $[-2, 2]$ by $[-1.2, 1.2]$. Use θ Step $= 0.01, -\dfrac{\pi}{4} \le \theta \le \dfrac{\pi}{4}$. **9.** Graph $r = \theta$ in $[-33, 33]$ by $[-18.4, 20]$ first using $0 \le \theta \le 20$ then $-20 \le \theta \le 0$ and then $-20 \le \theta \le 20$. **11.** Graph $r = 8\cos 2\theta, 0 \le \theta \le 2\pi$ in $[-13.6, 13.6]$ by $[-8, 8]$. A complete graph requires a minimum of 2π for the range of θ. The factor 8 stretches the graph of $r = \cos 2\theta$ away from the origin by a factor of 8. The range, $\dfrac{\pi}{4} \le \theta \le \dfrac{3\pi}{4}$, for example, produces a closed curve. Replacing θ by 2θ

produces 3 more leaves.   **13.** $m = \pm 1$. Graph $r = -1 + \cos\theta$, $r = -(\sin\theta + \cos\theta)^{-1}$ and $r = (\sin\theta - \cos\theta)^{-1}$, $-\pi \le \theta \le \pi$ in $[-2.6, 4.3]$ by $[-2, 2]$.   **15.** $\theta = \dfrac{\pi}{4}$, $m = -1$; $\theta = -\dfrac{\pi}{4}$, $m = 1$; $\theta = \dfrac{3\pi}{4}$, $m = 1$; $\theta = -\dfrac{3\pi}{4}$, $m = -1$. $\theta = 0$, $m = 0$; $\theta = \dfrac{\pi}{2}$, $m = \infty$; $\theta = \pi$, $m = 0$; $\theta = \dfrac{3\pi}{2}$, $m = -\infty$. Graph $r = \sin 2\theta$, $r = \pm\dfrac{\sqrt{2}}{\sin\theta + \cos\theta}$, $r = \pm\dfrac{\sqrt{2}}{\sin\theta - \cos\theta}$, $0 \le \theta \le 2\pi$ in $[-1.9, 1.9]$ by $[-1.2, 1.2]$ and regard the x- and y-axes as tangent lines also. **17.** Graph $r = \pm 2\sqrt{\cos 2\theta}$, $-\dfrac{\pi}{4} \le \theta \le \dfrac{\pi}{4}$, θ Step $= 0.01$ in $[-2, 2]$ by $[-1.2, 1.2]$. $\dfrac{\pi}{2}$ is a minimum range of θ, but it must be over an interval in which $\cos 2\theta$ is non-negative.   **19.** a) Graph $r = \dfrac{1}{2} + \cos\theta$, $0 \le \theta \le 2\pi$, θ Step $= 0.1$ in $[-0.9, 2.3]$ by $[-0.94, 0.94]$.   b) Graph $r = 0.5 + \sin\theta$, $0 \le \theta \le 2\pi$ in $[-1.5, 1.5]$ by $[-0.2, 1.6]$. **21.** a) Graph $r = 1.5 + \cos\theta$, $0 \le \theta \le 2\pi$ in $[-2, 4]$ by $[-1.8, 1.9]$.   b) Graph $r = 1.5 - \sin\theta$, $0 \le \theta \le 2\pi$ in $[-3.1, 3.1]$ by $[-2.7, 1]$.   **23.** Graph $r = 2 - 2\cos\theta$, $0 \le \theta \le 2\pi$ in $[-6.5, 3.2]$ by $[-2.8, 2.8]$. The region consists of this closed curve (a cardioid) and every point inside it.   **27.** The origin, $\left(1, \dfrac{\pi}{2}\right)$ and $\left(1, \dfrac{3\pi}{2}\right)$ **29.** $(2(\sqrt{2} - 1), \sin^{-1}(3 - 2\sqrt{2}))$, $(2(\sqrt{2} - 1), \pi - \sin^{-1}(3 - 2\sqrt{2}))$, the origin and $\left(2, \dfrac{3\pi}{2}\right) = \left(-2, \dfrac{\pi}{2}\right)$. **31.** $\left(\pm 2^{-1/4}, \dfrac{\pi}{8}\right)$ and the origin   **33.** $\left(1, \dfrac{\pi}{12}\right)$, $\left(1, \dfrac{5\pi}{12}\right)$, $\left(-1, \dfrac{19\pi}{12}\right)$, $\left(-1, \dfrac{23\pi}{12}\right)$, $\left(1, \dfrac{13\pi}{12}\right)$, $\left(1, \dfrac{17\pi}{12}\right)$, $\left(-1, \dfrac{7\pi}{12}\right)$, $\left(-1, \dfrac{11\pi}{12}\right)$ **35.** Graph $y = 5\sin\theta$, $0 \le \theta \le \pi$ in $[-5, 5]$ by $[-0.44, 5.44]$. Period π.   **37.** Graph $r = 5\sin 2\theta$, $0 \le \theta \le 2\pi$ in $[-8.5, 8.5]$ by $[-5, 5]$. Period 2π. **39.** Graph $r = 5\sin 5\theta$, $0 \le \theta \le \pi$ in $[-8.5, 8.5]$ by $[-5, 5]$. Period π.   **41.** Graph $r = 5\sin(2.5\theta)$, $0 \le \theta \le 4\pi$ in $[-8.5, 8.5]$ by $[-5, 5]$. Period 4π.   **43.** Graph $r = 5\sin 1.5\theta$, $0 \le \theta \le 4\pi$ in $[-8.5, 8.5]$ by $[-5, 5]$. Period 4π.   **45.** Graph $r = 1 - 2\sin 3\theta$, $0 \le \theta \le 2\pi$ in $[-4.7, 4.7]$ by $[-2.1, 3.4]$. Period 2π.   **47.** a) Graph $r = e^{\theta/10}$, $-20 \le \theta \le 10$ in $[-4.3, 3.6]$ by $[-2, 2.7]$.   b) Graph $r = \dfrac{8}{\theta}$, $-20 \le \theta \le 20$ in the same window.   c) Graph $r = \dfrac{10}{\sqrt{\theta}}$ and $r = -\dfrac{10}{\sqrt{\theta}}$, $0 \le \theta \le 200$, θ Step $= 0.5$ in the same window.   **49.** Infinite period   **51.** Graph $r = 1.75 + \left(\dfrac{0.06}{2\pi}\right)\theta$, $0 \le \theta \le 10\pi$ in $[-3, 3]$ by $[-3, 3]$.

53. 741.420 cm   **55.** a) $n = \dfrac{-2\pi + \sqrt{4\pi^2 + 2bS_a}}{b}$.
b) The speed of the take-up reel steadily decreases.

Section 10.7

1. $r\cos\left(\theta - \dfrac{\pi}{6}\right) = 5$, $\sqrt{3}x + y = 10$

3. Graph $r = \dfrac{\sqrt{2}}{\cos[\theta - (\pi/4)]}$, $0 \le \theta \le \pi$ in $[-5.8, 7.8]$ by $[-3, 5]$. $x + y = 2$.   **5.** $r = 8\cos\theta$ **7.** Check your sketch by graphing $r = 4\cos\theta$, $0 \le \theta \le \pi$ in $[-1.4, 5.4]$ by $[-2, 2]$. Center: $(2, 0)$, radius $= 2$.   **9.** $r = \dfrac{2}{1 + \cos\theta}$ **11.** $r = \dfrac{8}{1 + 2\cos\theta}$ **13.** $r = \dfrac{1}{2 + \cos\theta}$ **15.** $r = \dfrac{10}{5 - \sin\theta}$ **17.** Directrix: $x = 1$, vertex: $\left(\dfrac{1}{2}, 0\right)$. Graph for $-\pi \le \theta \le \pi$ in $[-7.4, 2.8]$ by $[-3, 3]$. Include the directrix $r = 1/\cos\theta$.   **19.** Graph $r_1 = \dfrac{5}{2 - \cos\theta}$ and $r_2 = \dfrac{-5}{\cos\theta}$, $-\pi \le \theta \le \pi$ in $[-8.5, 8.5]$ by $[-5, 5]$. Directrix concerned is $x = -5$. Vertices: $\left(\dfrac{5}{3}, \pi\right)$, $(5, 0)$, center: $\left(\dfrac{5}{3}, 0\right)$. **21.** Graph $r = \dfrac{25}{1 + 0.5\sin\theta}$, $0 \le \theta \le 2\pi$ in $[-93, 9]$ by $[-50, 60]$. Also include $r = \dfrac{50}{\sin\theta}$. Vertices: $\left(\dfrac{50}{3}, \dfrac{\pi}{2}\right)$, $\left(-50, \dfrac{\pi}{2}\right)$, center: $\left(-\dfrac{50}{3}, \dfrac{\pi}{2}\right)$.   **23.** Graph $r = \dfrac{4}{1 - \sin\theta}$ and $r = \dfrac{-4}{\sin\theta}$, $0 \le \theta \le 2\pi$ in $[-13, 13]$ by $[-5, 12]$. Vertex: $\left(2, \dfrac{3\pi}{2}\right)$.   **25.** Graph the ellipses sequentially in $[-2, 1.4]$ by $[-1, 1]$, $0 \le \theta \le 2\pi$, θ Step $= 0.1$. The last two require a larger rectangle. As e increases, the center moves to the left, the ellipse increases in size. The ellipse also flattens out horizontally as can be seen by graphing in $[-11.3, 2.3]$ by $[-4, 4]$.   **27.** Graph these sequentially in $[-27, 27]$ by $[-16, 16]$. As k becomes more negative, the parabola opens up wider and wider to the right. As k becomes more and more positive, the parabola opens to the left wider and wider.   **29.** Graph $r = 3\sec\left(\theta - \dfrac{\pi}{3}\right) = \dfrac{3}{\cos(\theta - \pi/3)}$, $0 \le \theta \le 2\pi$ in $[-11, 19]$ by $[-6, 11.7]$. $x + \sqrt{3}y = 6$ in rectangular coordinates. **31.** Graph $r = \dfrac{8}{4 + \cos\theta}$, $0 \le \theta \le 2\pi$ in $[-4.9, 3.9]$ by $[-2.6, 2.6]$.   **33.** Graph $r = \dfrac{1}{1 + 2\sin\theta}$, $0 \le \theta \le 2\pi$ in $[-2.7, 2.7]$ by $[-0.93, 2.27]$ in dot format, θ Step $= 0.01$.   **35.** Graph $r = -2\cos\theta$, $0 \le \theta \le \pi$ in $[-2.7, 0.7]$ by $[-1, 1]$.   **37.** Graph $r = \dfrac{1}{1 + \cos\theta}$, $0 \le \theta \le 2\pi$, θ Step $= 0.1$ in $[-7.7, 5.9]$ by $[-4, 4]$. **39.** Graph $r = 2\cos\theta$, $0 \le \theta \le \pi$ in $[-0.7, 2.7]$ by $[-1, 1]$. The region consists of the circle and all points within it.

41. b)

Planet	$a(1-e)AU$	$a(1+e)AU$
Mercury	0.3075	0.4667
Venus	0.7184	0.7282
Earth	0.9833	1.017
Mars	1.382	1.666
Jupiter	4.951	5.455
Saturn	9.021	10.057
Uranus	18.30	20.06
Neptune	29.81	30.31
Pluto	29.65	49.23

43. a) $x^2 + (y-1)^2 = 1$, $y = 1$ b) Graph $r = 2\sin\theta$, $r = \dfrac{1}{\sin\theta}$, $0 \le \theta \le \pi$ in $[-1.7, 1.7]$ by $[0, 2]$. Label the points of intersection $(1,1)$, $\left(\sqrt{2}, \dfrac{\pi}{4}\right)$ and $(-1, 1)$, $\left(\sqrt{2}, \dfrac{3\pi}{4}\right)$.   **45.** $r = \dfrac{4}{1 + \cos\theta}$ **47.** Assume a and b are non-zero. Then $P_0 = \left(\dfrac{ca}{a^2+b^2}, \dfrac{cb}{a^2+b^2}\right)$ and $r = \dfrac{|c|}{\sqrt{a^2+b^2}\cos(\theta - \theta_0)}$ where $\theta_0 = \tan^{-1}\dfrac{b}{a}$ if P_0 is in the 1st or 4th quadrant and $\theta_0 = \pi + \tan^{-1}\dfrac{b}{a}$ if P_0 is in the 2nd or 3rd quadrant.   **49.** $r = \dfrac{6}{\sqrt{13}\cos(\theta - \tan^{-1}(2/3))}$

51. $r = \dfrac{12}{5\cos(\theta - (\pi + \tan^{-1}(3/4)))}$

Section 10.8

1. Graph $r = \cos\theta$, $0 \le \theta \le \dfrac{\pi}{4}$ in $[0, 1]$ by $[0, 0.5]$. Then draw a line segment connecting $(0, 0)$ and the rectangular point $\left(\dfrac{1}{2}, \dfrac{1}{2}\right)$. $A = \dfrac{\pi + 2}{16}$.   **3.** Graph $r = 4 + 2\cos\theta$, $0 \le \theta \le 2\pi$ in $[-8.2, 12.2]$ by $[-6, 6]$. $A = 18\pi$.   **5.** Graph $r = \cos 2\theta$, $-\dfrac{\pi}{4} \le \theta \le \dfrac{\pi}{4}$ in $[-0.09, 1.12]$ by $[-0.35, 0.35]$ for one leaf. $A = \dfrac{\pi}{8}$.   **7.** For the purpose of graphing let $a = 2$. For the entire graph use $r = \sqrt{8\cos 2\theta}$, $0 \le \theta \le 2\pi$ in $[-3, 3]$ by $[-1.7, 1.7]$. $A = 2a^2$.   **9.** Graph $r_1 = \sqrt{2\sin 3\theta}$ and $r_2 = -r_1$, $0 \le \theta \le 2\pi$ in $[-2.5, 2.5]$ by $[-1.5, 1.5]$. $A = 4$.   **11.** Graph $r = 1$ and $r = 2\sin\theta$, $0 \le \theta \le 2\pi$ in $[-2.5, 2.5]$ by $[-1, 2]$. $A = \dfrac{2\pi}{3} - \dfrac{\sqrt{3}}{2}$.   **13.** Graph $r = 2(1 + \cos\theta)$ and $r = 2(1 - \cos\theta)$, $0 \le \theta \le 2\pi$ in $[-6.8, 6.8]$ by $[-4, 4]$. $A = 6\pi - 16$.   **15.** Graph $r = 3a\cos\theta$ and $r = a(1 + \cos\theta)$, with $a = 2$, $0 \le \theta \le 2\pi$ in $[-3, 8.3]$ by $[-3.3, 3.3]$. $A = \pi a^2$.   **17.** a) $2\pi + \dfrac{3\sqrt{3}}{2}$ b) $\pi + 3\sqrt{3}$ **19.** Graph $r = \theta^2$, $0 \le \theta \le \sqrt{5}$ in $[-4.9, 2.5]$ by $[0, 4.4]$. $L = \dfrac{19}{3}$.   **21.** The

graph is the line segment from $(1, 0)$ to $(1, 1)$ which has length 1.   **23.** Graph $r = 1 + \cos\theta$, $0 \le \theta \le 2\pi$ in $[-1.7, 3.4]$ by $[-1.5, 1.5]$. $L = 8$.   **25.** $\sqrt{2}\pi$ **27.** $2(2 - \sqrt{2})\pi$ **29.** b) $y_1 \to \infty$ as $x \to 0^+$ c) The integral is improper because y_1 has a vertical asymptote at the endpoint $x = 0$.   **31.** $((0.8333\ldots)a, 0) = \left(\dfrac{5a}{6}, 0\right)$

Chapter 10 Practice Exercises

1. Graph $y = \pm 2\sqrt{2x}$ in $[-11.8, 16.8]$ by $[-8.4, 8.4]$. Use the line-drawing feature to include the directrix $x = -2$. Focus: $(2, 0)$.   **3.** Vertices $(0, \pm 4)$, foci: $(0, \pm 3)$. Graph $y = \pm 4\sqrt{1 - \dfrac{x^2}{7}}$, $y = \pm\dfrac{16}{3}$ (directrices) in $[-10.2, 10.2]$ by $[-6, 6]$.   **5.** Vertices: $(\pm 1, 0)$, foci: $(\pm 2, 0)$, directrices: $x = \pm\dfrac{1}{2}$. Graph $y = \pm\sqrt{3x^2 - 3}$ in $[-15, 15]$ by $[-10, 10]$.   **7.** $B^2 - 4AC = -3 < 0$ indicates ellipse but there is no solution and no graph.   **9.** Parabola. Graph $y = \dfrac{-(4x+1) \pm \sqrt{-8x - 15}}{8}$ in $[-13.5, 3.5]$ by $[0, 10]$.   **11.** Ellipse, $\theta = \dfrac{\pi}{4}$. Equivalent conic: $\dfrac{x^2}{6} + \dfrac{y^2}{10} = 1$. Graph $y = \pm 3\sqrt{10\left(1 - \dfrac{x^2}{6}\right)}$ and $y = \dfrac{-x \pm \sqrt{x^2 - 8(2x^2 - 15)}}{4}$ in $[-7.5, 7.5]$ by $[-4, 5]$.   **13.** $e = \sqrt{2}$ **15.** a) 24π b) 16π **17.** $(\pm 2, 1)$ **19.** $y = 2x + 1$. The entire line is traced out from left to right.   **21.** $\dfrac{y^2}{(0.5)^2} - \dfrac{x^2}{(0.5)^2} = 1$, hyperbola. The graph is the upper branch traced out from left to right. This may be confirmed by graphing in parametric mode in $[-25.5, 25.5]$ by $[-10, 20]$.   **23.** The portion of the parabola $y = x^2$ determined by $-1 \le x \le 1$ is traced out from left to right.   **25.** $x = 3\cos t$, $y = 4\sin t$, $0 \le t \le 2\pi$ **27.** $y - 1 = \left(\dfrac{\sqrt{3}}{2}\right)\left(x - \dfrac{\sqrt{3}}{2}\right)$. At $t = \dfrac{\pi}{3}$, $\dfrac{d^2y}{dx^2} = \dfrac{1}{4}$.   **29.** Graph in $[-0.16, 4.4]$ by $[-0.26, 2.5]$. Length $= 3 + \ln 2/8$.   **31.** $\dfrac{76\pi}{3}$ **33.** Graph $r = \cos 2\theta$, $0 \le \theta \le 2\pi$ in $[-1.7, 1.7]$ by $[-1, 1]$. Period $= 2\pi$. Four-leaved rose.   **35.** Graph $r = \dfrac{6}{1 - 2\cos\theta}$, $0 \le \theta \le 2\pi$ in $[-30, 22]$ by $[-13, 18]$ in dot format. Period $= 2\pi$. Hyperbola.   **37.** Graph $r = \theta$, $-4\pi \le \theta \le 4\pi$ in $[-18.6, 18.6]$ by $[-12.3, 9.7]$. Infinite period. Spiral.   **39.** Graph $r = 1 + \cos\theta$, $0 \le \theta \le 2\pi$ in $[-1.6, 3.2]$ by $[-1.4, 1.4]$. Period 2π. Cardioid.   **41.** Graph $r = \dfrac{2}{1 - \cos\theta}$, $0 \le \theta \le 2\pi$ in $[-10, 17]$ by $[-8, 8]$. Period 2π. Parabola.   **43.** Graph $r = -\sin\theta$, $0 \le \theta \le 2\pi$ in $[-1, 1]$ by $[-1.1, 0.1]$. Period

π. Circle. **45.** Graph $r = \dfrac{2\sqrt{3}}{\cos(\theta - \pi/3)}$ $0 \le \theta \le 2\pi$

in $[-17, 17]$ by $[-5, 15]$. $y = -\dfrac{1}{\sqrt{3}}x + 4$. **47.** Center

$(0, 1)$, radius $= 1$ **49.** The region consists of all points on and within the circle with center $(3, 0)$ and radius 3. Graph the circle $r = 6\cos\theta$, $0 \le \theta \le \pi$ in $[-2.1, 8.1]$ by $[-3, 3]$. **51.** The origin is the only point of intersection. **53.** All points on the curve since the two curves are

identical. **55.** Graph the parabola $r = \dfrac{2}{1 + \cos\theta}$, $0 \le$

$\theta \le 2\pi$ in $[-26.5, 7.5]$ by $[-10, 10]$. $(1, 0)$ is the vertex. **57.** Graph r, $0 \le \theta \le 2\pi$ in $[-13, 4]$ by $[-5, 5]$ in dot

format. Vertices: $(-6, 0)$, $(2, \pi)$. **59.** $r = \dfrac{4}{1 + 2\cos\theta}$

61. $r = \dfrac{2}{2 + \sin\theta}$ **63.** $\theta = \dfrac{\pi}{4}$ and $\theta = \dfrac{3\pi}{4}$ or $y = \pm x$

65. $r\cos(\theta - \theta_0) = 1$ where $\theta_0 = \dfrac{\pi}{4}, \dfrac{3\pi}{4}, \dfrac{5\pi}{4}$ and $\dfrac{7\pi}{4}$

67. Graph $r = 2 - \cos\theta$, $0 \le \theta \le 2\pi$ in $[-4.7, 2.7]$

by $[-2.2, 2.2]$. $A = \dfrac{9\pi}{2}$ **69.** Graph $r = 1 + \cos 2\theta$

and $r = 1$, $0 \le \theta \le 2\pi$ in $[-2, 2]$ by $[-1.2, 1.2]$. $A =$

$2 + \dfrac{\pi}{4}$ **71.** $(2 - \sqrt{2})\pi$ **73.** 8 **75.** $\dfrac{\pi + 3}{8}$ **77.** Graph

$x_1 = t$, $y_1 = t^2 - 1$, $x_2 = x_1 \cos\left(\dfrac{\pi}{6}\right) - y_1 \sin\left(\dfrac{\pi}{6}\right)$,

$y_2 = x_1 \sin\left(\dfrac{\pi}{6}\right) + y_1 \cos\left(\dfrac{\pi}{6}\right)$, $-3 \le t \le 3$ in $[-6.6, 3]$

by $[-1, 8.4]$. $3x^2 + 2\sqrt{3}xy + y^2 + 2x - 2\sqrt{3}y - 4 =$

0 **79.** a) a b) a c) $\dfrac{2a}{\pi}$ **81.** We use Fig. 10.13.

$\dfrac{r_{max} - r_{min}}{r_{max} + r_{min}} = \dfrac{(c + a) - (a - c)}{(c + a) + (a - c)} = \dfrac{2c}{2a} = \dfrac{c}{a} = e.$

Chapter 11

Section 11.1

1. a)

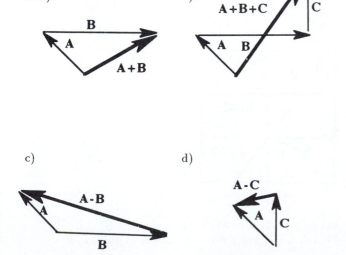

b)

c)

d)

3. 9 and 6 **5.** 1 and 3 **7.** 12 and -19
9. $i - 4j$

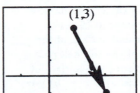

11. $-2i - 3j$

13. $\theta = \dfrac{\pi}{6} : \dfrac{\sqrt{3}}{2}i + \dfrac{1}{2}j;$

$\theta = \dfrac{2\pi}{3} : -\dfrac{1}{2}i + \dfrac{\sqrt{3}}{2}j$

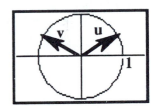

15. $\dfrac{\sqrt{3}}{2}i - \dfrac{1}{2}j$ **17.** $\sqrt{13}$ **19.** 1 **21.** 1 **23.** $\dfrac{3}{5}i + \dfrac{4}{5}j$

25. $\dfrac{12}{13}i - \dfrac{5}{13}j$ **27.** $\dfrac{2}{\sqrt{13}}i + \dfrac{3}{\sqrt{13}}j$

29. $\sqrt{2}\left(\dfrac{1}{\sqrt{2}}i + \dfrac{1}{\sqrt{2}}j\right)$ **31.** $2\left(\dfrac{\sqrt{3}}{2}i + \dfrac{1}{2}j\right)$

33. $13\left[\dfrac{5}{13}i + \dfrac{12}{13}j\right]$ **35.** $\dfrac{A}{|A|} = \dfrac{1}{\sqrt{5}}i + \dfrac{2}{\sqrt{5}}j; \dfrac{B}{|B|} =$

$-\dfrac{1}{\sqrt{5}}i - \dfrac{2}{\sqrt{5}}j$ **37.** $\pm\dfrac{1}{\sqrt{5}}(i + 4j)$ are unit tangent

vectors; $\pm\dfrac{1}{\sqrt{5}}(4i - j)$ are unit normal vectors. **39.** Unit

tangent vectors are $\pm\dfrac{1}{\sqrt{2}}(i - j)$; unit normal vectors are

$\pm\dfrac{1}{\sqrt{2}}(i + j)$. **41.** Unit tangent vectors are $\pm\dfrac{1}{\sqrt{5}}(2i + j)$;

unit normal vectors are $\pm\dfrac{1}{\sqrt{5}}(i - 2j)$. **43.** The slopes

are the same.

Section 11.2

1. Line through $(2, 3, 0)$ parallel to the z-axis. **3.** The x-axis **5.** Circle in xy-plane, center at $(0, 0, 0,)$, radius 2. **7.** Circle in the xz-plane, center at $(0, 0, 0)$, radius 2. **9.** Circle in the yz-plane, center at $(0, 0, 0)$, radius

1. **11.** Circle in the xy-plane, center at $(0,0,0)$, radius 4. **13.** a) The first quadrant in the xy-plane b) The fourth quadrant in the xy-plane **15.** a) The interior and surface of the unit sphere (center at $(0,0,0)$) b) All of 3-space <u>but</u> the interior and surface of the unit sphere **17.** a) The surface of the top half ($z \geq 0$) of the unit sphere b) The interior and surface of the top half of the unit sphere **19.** a) $x = 3$ b) $y = -1$ c) $z = -2$ **21.** a) $z = 1$ b) $x = 3$ c) $y = -1$ **23.** a) $x^2 + (y - 2)^2 = 4, z = 0$ b) $(y - 2)^2 + z^2 = 4, x = 0$ c) $x^2 + z^2 = 4, y = 2$ **25.** a) $y = 3, z = -1$ b) $x = 1, z = -1$ c) $x = 1, y = 3$ **27.** $x^2 + y^2 = 16, z = 3$ **29.** $0 \leq z \leq 1$ **31.** $z \leq 0$ **33.** a) $(x - 1)^2 + (y - 1)^2 + (z - 1)^2 < 1$ b) $(x - 1)^2 + (y - 1)^2 + (z - 1)^2 > 1$ **35.** Length 3, direction $\frac{2}{3}\mathbf{i} + \frac{1}{3}\mathbf{j} - \frac{2}{3}\mathbf{k}$ **37.** $9, \frac{1}{9}\mathbf{i} + \frac{4}{9}\mathbf{j} - \frac{8}{9}\mathbf{k}$ **39.** $5, \mathbf{k}$ **41.** $4, -\mathbf{j}$ **43.** $\frac{5}{12}, -\left(\frac{4}{5}\right)\mathbf{i} + \left(\frac{3}{5}\right)\mathbf{j}$ **45.** $\frac{1}{\sqrt{2}}, \left(\frac{1}{\sqrt{3}}\right)\mathbf{i} - \left(\frac{1}{\sqrt{3}}\right)\mathbf{j} - \left(\frac{1}{\sqrt{3}}\right)\mathbf{k}$ **47.** a) 3 b) $\frac{2}{3}\mathbf{i} + \frac{2}{3}\mathbf{j} - \frac{1}{3}\mathbf{k}$ c) $\left(2, 2, \frac{1}{2}\right)$

49. a) 7 b) $\frac{3}{7}\mathbf{i} - \frac{6}{7}\mathbf{j} + \frac{2}{7}\mathbf{k}$ c) $\left(\frac{5}{2}, 1, 6\right)$ **51.** a) $2\sqrt{3}$ b) $\frac{1}{\sqrt{3}}\mathbf{i} - \frac{1}{\sqrt{3}}\mathbf{j} - \frac{1}{\sqrt{3}}\mathbf{k}$ c) $(1, -1, -1)$ **53.** a) $2\mathbf{i}$ b) $-\sqrt{3}\mathbf{k}$ c) $\frac{3}{10}\mathbf{j} + \frac{4}{10}\mathbf{k}$ d) $6\mathbf{i} - 2\mathbf{j} + 3\mathbf{k}$ **55.** $\frac{84}{13}\mathbf{i} - \frac{35}{13}\mathbf{j}$ **57.** $-\frac{10}{7}\mathbf{i} + \frac{15}{7}\mathbf{j} - \frac{30}{7}\mathbf{k}$ **59.** a) $C(-2, 0, 2)$, radius $= \sqrt{8} = 2\sqrt{2}$ b) $C\left(-\frac{1}{2}, -\frac{1}{2}, -\frac{1}{2}\right)$, radius $= \frac{\sqrt{21}}{2}$ c) $C(\sqrt{2}, \sqrt{2}, -\sqrt{2})$, radius $= \sqrt{2}$ d) $C\left(0, -\frac{1}{3}, \frac{1}{3}\right)$, radius $= \frac{\sqrt{29}}{3}$ **61.** Center $(-2, 0, 2)$, radius $\sqrt{8}$

63. Center $(0, 0, 1)$, radius 1 **65.** a) $x^2 + z^2 = 5$ in the plane $y = -2$, $x^2 + z^2 = 9$ in the plane $y = 0$, $x^2 + z^2 = 8$ in the plane $y = 1$ b) the areas are $5\pi, 9\pi, 8\pi$

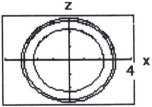

67. a) In $z = 2$, $(x + 2)^2 + y^2 = 8$; in $z = 4$, $(x + 2)^2 + y^2 = 4$ b)

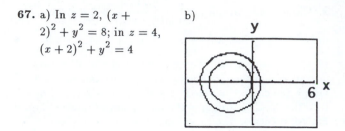

69. a) $\sqrt{y^2 + z^2}$ b) $\sqrt{x^2 + z^2}$ c) $\sqrt{x^2 + y^2}$

Section 11.3

	$\mathbf{A} \cdot \mathbf{B}$	$|\mathbf{A}|$	$|\mathbf{B}|$	$\cos \theta$	$|\mathbf{B}| \cos \theta$
1.	10	$\sqrt{13}$	$\sqrt{26}$	$\frac{10}{13\sqrt{2}}$	$\frac{10}{\sqrt{13}}$
3.	4	$\sqrt{14}$	2	$\frac{2}{\sqrt{14}}$	$\frac{4}{\sqrt{14}}$
5.	2	$\sqrt{34}$	$\sqrt{3}$	$\frac{2}{\sqrt{3}\sqrt{34}}$	$\frac{2}{\sqrt{34}}$
7.	$\sqrt{3} - \sqrt{2}$	$\sqrt{2}$	3	$\frac{\sqrt{3} - \sqrt{2}}{2}$	$\frac{\sqrt{3} - \sqrt{2}}{2}$
9.	-25	5	5	-1	-5
11.	25	15	5	$\frac{1}{3}$	$\frac{5}{3}$

	$\text{Proj}_{\mathbf{A}} \mathbf{B} = \dfrac{\mathbf{A} \cdot \mathbf{B}}{\mathbf{A} \cdot \mathbf{A}} \mathbf{A}$
1.	$\frac{10}{13}[3\mathbf{i} + 2\mathbf{j}]$
3.	$\frac{2}{7}[3\mathbf{i} - 2\mathbf{j} - \mathbf{k}]$
5.	$\frac{1}{17}[5\mathbf{j} - 3\mathbf{k}]$
7.	$\frac{\sqrt{3} - \sqrt{2}}{2}[-\mathbf{i} + \mathbf{j}]$
9.	$-2\mathbf{i} + 4\mathbf{j} - \sqrt{5}\mathbf{k}$
11.	$\frac{1}{9}[10\mathbf{i} + 11\mathbf{j} - 2\mathbf{k}]$

13. $\left[\frac{3}{2}\mathbf{i} + \frac{3}{2}\mathbf{j}\right] + \left[-\frac{3}{2}\mathbf{i} + \frac{3}{2}\mathbf{j} + 4\mathbf{k}\right]$ **15.** $\frac{14}{3}[\mathbf{i} + 2\mathbf{j} - \mathbf{k}] + \frac{2}{3}[5\mathbf{i} - 8\mathbf{j} - 11\mathbf{k}]$ **17.** $x + 2y = 4$ **19.** $2x + y = 0$

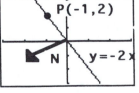

21. $2\sqrt{10}$ **23.** $\sqrt{2}$ **25.** $\mathbf{A} \cdot \mathbf{B} = \mathbf{A} \cdot \mathbf{C} = \mathbf{B} \cdot \mathbf{C} = 0$ **27.** $\mathbf{i} \cdot \mathbf{j} = \mathbf{i} \cdot \mathbf{k}$ but $\mathbf{j} \neq \mathbf{k}$ **29.** $< A = 71.068°, < B =$

$37.864°$, $< C = 71.068°$ **31.** $35.264°$ **33.** $-5 \; N \cdot m$
35. $3018.838 \; N \cdot m$ **37.** $45°$ or $135°$ **39.** $59.156°$
41. $63.4°$ at $x = 0$ and $83.7°$ at $x = 1$ **43.** $|A|^2 + |B|^2 -$
$|C|^2 = a_1^2 + a_2^2 + a_3^2 + b_1^2 + b_2^2 + b_3^2 - [(b_1 - a_1)^2 + (b_2 -$
$a_2)^2 + (b_3 - a_3)^2] = 2(a_1 b_1 + a_2 b_2 + a_3 b_3)$

Section 11.4

1. $A \times B$: length $= 3$, direction $= \left[\frac{2}{3}i + \frac{1}{3}j + \frac{2}{3}k\right]$;

$B \times A$: length $= 3$, direction $= -\left[\frac{2}{3}i + \frac{1}{3}j + \frac{2}{3}k\right]$

3. $A \times B = B \times A = 0$, both have length 0 and no
direction   **5.** $A \times B$: length $= 6$, direction $= -k$; $B \times A$
has same length, opposite direction   **7.** $A \times B$ has length
$6\sqrt{5}$, direction $\frac{1}{\sqrt{5}}(i - 2k)$; $B \times A$ has same length,

opposite direction

9. $A \times B = k$ **11.** $A \times B = i - j + k$

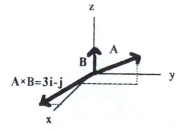

13. $A \times B = 3i - j$

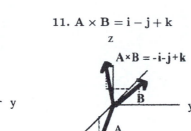

15. a) $\pm(8i + 4j + 4k)$ b) $2\sqrt{6}$ c) $\pm\frac{1}{\sqrt{6}}(2i + j + k)$

17. a) $\pm(-i + j)$ b) $\frac{1}{\sqrt{2}}$ c) $\pm\frac{1}{\sqrt{2}}(-i + j)$ **19.** a) none

b) A and C **21.** $2i + 4j + 7k$, $0, 0$ **23.** a) $\frac{A \cdot B}{B \cdot B}B$

b) $A \times B$ c) $\frac{\sqrt{A \cdot A}}{\sqrt{B \cdot B}}B$ d) $(A \times B) \times C$ e) $(B \times$

$C) \times A$ **25.** 16.383 foot-pounds

	27.	29.
$A \times B$	$4k$	$i - 2j - 4k$
$(A \times B) \cdot C$	8	-7
$B \times C$	$4i$	$-2i - 3j + k$
$(B \times C) \cdot A$	8	-7
$C \times A$	$4j$	$-2i + 4j + k$
$(C \times A) \cdot B$, Vol.	$8, 8$	$-7, 7$

Section 11.5

1. $x = 3 + 1 \cdot t$, $y = -4 + 1 \cdot t$, $z = -3 + 1 \cdot t$ **3.** $x =$
$1 - 2t$, $y = 2 - 2t$, $z = 1 + 2t$ **5.** $x = 1$, $y = 2 - t$, $z =$
$-t$ **7.** $x = 0$, $y = 2t$, $z = t$ **9.** $x = 1$, $y = 1$, $z = 1 + t$
11. $x = t$, $y = -7 + 2t$, $z = 2t$ **13.** $x = t$, $y = 0$, $z = 0$
15. $x = 5$, $y = t$, $z = t$,   **17.** $x = 1$, $y = t$, $z = 0$,
   $0 \le t \le 1$, $t = 0$ $0 \le t \le 1$
   corresponds to $(0, 0, 0)$

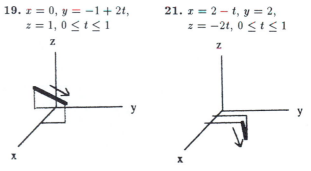

19. $x = 0$, $y = -1 + 2t$,   **21.** $x = 2 - t$, $y = 2$,
   $z = 1$, $0 \le t \le 1$ $z = -2t$, $0 \le t \le 1$

23. $3x - 2y - z = -3$ **25.** $7x - 5y - 4z = 6$ **27.** $x +$
$3y + 4z = 34$ **29.** $\sqrt{120}$ **31.** 0 **33.** 3 **35.** $\frac{19}{5}$ **37.** $\frac{5}{3}$
39. $\left(\frac{3}{2}, -\frac{3}{2}, \frac{1}{2}\right)$ **41.** $(1, 1, 0)$ **43.** $45°$ **45.** $101.096°$
47. $47.124°$ **49.** $x = -t$, $y = 2 + t$, $z = -1$ **51.** $x =$
4, $y = 1 + 6t$, $z = 3t$ **53.** $4x - y - 12z = -34$
55. $(1, -1, 0)$, $\left(0, -\frac{1}{2}, -\frac{3}{2}\right)$, $(-1, 0, 3)$ **57.** No,
$N \cdot (-2i + 5j - 3k) \ne 0$

Section 11.6

1.

$x^2 + y^2 = 4$

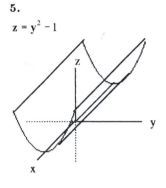

3. $y^2 + z^2 = 1$

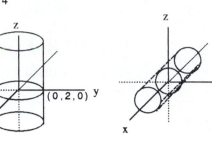

5.

$z = y^2 - 1$

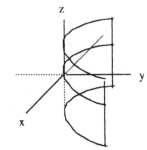

7. $z = 4 - x^2$

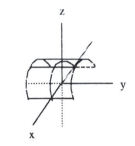

9. $y = x^2$

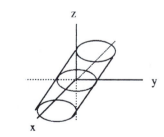

11. $y^2 + 4z^2 = 16$

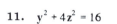
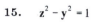
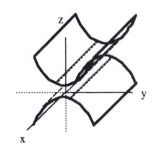

13. $z^2 + 4y^2 = 9$

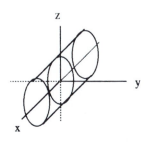

15. $z^2 - y^2 = 1$

17. $9x^2 + y^2 + z^2 = 9$

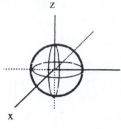

19. $x^2 + y^2 + z^2 = 4$

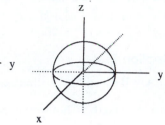

21.
$4x^2 + 9y^2 + 4z^2 = 36$

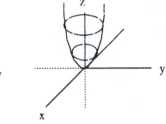

23. $x^2 + y^2 = z$

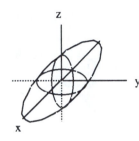

25. $x^2 + 4y^2 = z$

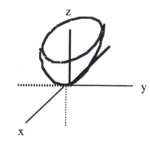

27. $z = 8 - x^2 - y^2$

29. $x = 4 - 4y^2 - z^2$

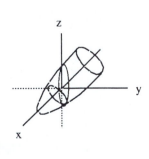

31. $z = x^2 + y^2 + 1$

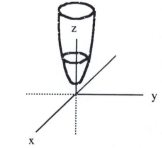

33.
$$x^2 + y^2 = z^2$$

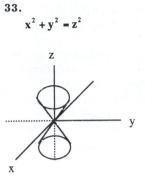

35.
$$x^2 + z^2 = y^2$$

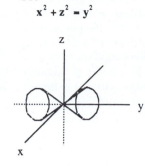

49.
$$y^2 - x^2 = z$$

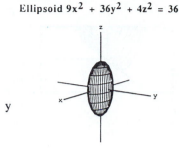

51.
Ellipsoid $9x^2 + 36y^2 + 4z^2 = 36$

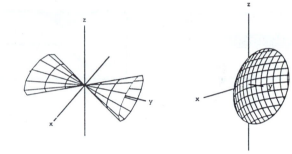

37.
$$9x^2 + 4y^2 = 36z^2$$

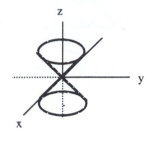

39.
$$x^2 + y^2 - z^2 = 1$$

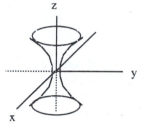

53.
Cone $9x^2 + 36z^2 = 4y^2$

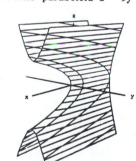

55.
Paraboloid $9x^2 + 16z^2 = 72y$

41.
$$(y^2/4) + (z^2/9) - (x^2/4) = 1$$

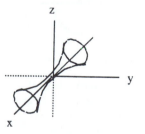

43.
$$(x^2/4) + y^2 - z^2 = 1$$

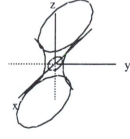

57.
Hyperbolic paraboloid $z^2 - 9y^2 = 3x$

59.
Paraboloid $2x^2 + 2z^2 = 3y$

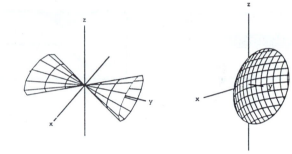

45.
$$z^2 - (x^2/4) - y^2 = 1$$

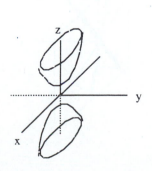

47.
$$x^2 - y^2 - (z^2/4) = 1$$

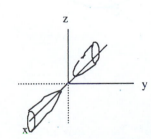

61.
Hyperboloid of two sheets
$9x^2 - 36y^2 - 4z^2 = 36$

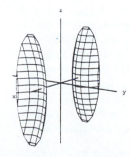

63.
Hyperboloid of one sheet
$25x^2 - 100y^2 + 4z^2 = 100$

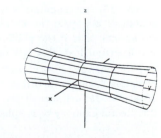

65.

$z = y^2$

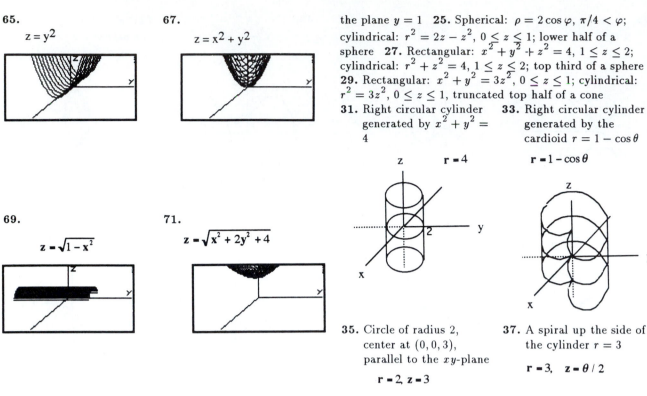

67.

$z = x^2 + y^2$

69.

$z = \sqrt{1 - x^2}$

71.

$z = \sqrt{x^2 + 2y^2 + 4}$

the plane $y = 1$ **25.** Spherical: $\rho = 2\cos\varphi$, $\pi/4 < \varphi$; cylindrical: $r^2 = 2z - z^2$, $0 \le z \le 1$; lower half of a sphere **27.** Rectangular: $x^2 + y^2 + z^2 = 4$, $1 \le z \le 2$; cylindrical: $r^2 + z^2 = 4$, $1 \le z \le 2$; top third of a sphere **29.** Rectangular: $x^2 + y^2 = 3z^2$, $0 \le z \le 1$; cylindrical: $r^2 = 3z^2$, $0 \le z \le 1$, truncated top half of a cone **31.** Right circular cylinder generated by $x^2 + y^2 = 4$ **33.** Right circular cylinder generated by the cardioid $r = 1 - \cos\theta$

$r = 4$

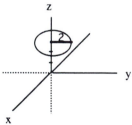

$r = 1 - \cos\theta$

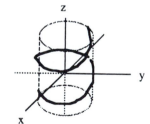

35. Circle of radius 2, center at $(0, 0, 3)$, parallel to the xy-plane   **37.** A spiral up the side of the cylinder $r = 3$

$r = 2, \ z = 3$

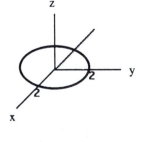

$r = 3, \quad z = \theta/2$

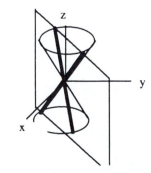

39. A circle in the xy-plane   **41.** The intersection of the cone $\varphi = \pi/4$ and the plane $\theta = \pi/4$; intersecting lines

Section 11.7

	Rectangular	Cylindrical	Spherical
	(x, y, z)	(r, θ, z)	(ρ, ϕ, θ)
1.	$(0, 0, 0)$	$(0, 0^*, 0)$	$(0, 0^*, 0^*)$
3.	$(0, 1, 0)$	$\left(1, \frac{\pi}{2}, 0\right)$	$\left(1, \frac{\pi}{2}, \frac{\pi}{2}\right)$
5.	$(1, 0, 0)$	$(1, 0, 0)$	$\left(1, \frac{\pi}{2}, 0\right)$
7.	$(0, 1, 1)$	$\left(1, \frac{\pi}{2}, 1\right)$	$\left(\sqrt{2}, \cos^{-1}\left(\frac{1}{\sqrt{2}}\right), \frac{\pi}{2}\right)$
9.	$(0, -2\sqrt{2}, 0)$	$\left(2\sqrt{2}, \frac{3\pi}{2}, 0\right)$	$\left(2\sqrt{2}, \frac{\pi}{2}, \frac{3\pi}{2}\right)$

0^* can be any angle

11. Rectangular: $x = 0$, $y = 0$; spherical: $\phi = 0$ or $\phi = \pi$; z-axis   **13.** Rectangular: $z = 0$; cylindrical: $z = 0$; spherical: $\phi = \pi/2$; the xy-plane   **15.** Rectangular, cylindrical: $z = 3$; the plane $z = 3$ **17.** Rectangular: $x = 0$; cylindrical: $\theta = \pi/2$ or $3\pi/3$; the yz-plane **19.** Cylindrical: $r^2 + z^2 = 4$; spherical: $\rho = 2$; sphere of radius 2, center at origin   **21.** Rectangular: $x^2 + y^2 + z^2 = 2y$; cylindrical: $r^2 + z^2 = 2r\sin\theta$; a sphere **23.** Rectangular: $y = 1$; spherical: $\rho\sin\varphi\sin\theta = 1$;

43. Circle in yz-plane:
$(y-2)^2 + z^2 = 4$

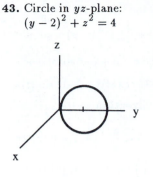

45. There will be symmetry about the z-axis. In the yz-plane, $\rho = \cos\varphi \Rightarrow \rho^2 = \rho\cos\varphi \Rightarrow y^2 + z^2 = z$ or $y^2 + \left(z - \frac{1}{2}\right)^2 = \frac{1}{4}$. When revolved about the z-axis, this becomes a sphere.

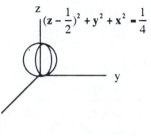

$$(z - \tfrac{1}{2})^2 + y^2 + x^2 = \tfrac{1}{4}$$

Chapter 11 Practice Exercises

1.

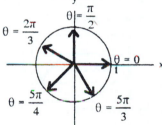

3. tangents: $\pm\frac{1}{\sqrt{5}}(\mathbf{i} + 2\mathbf{j})$; normals: $\pm\frac{1}{\sqrt{5}}(2\mathbf{i} - \mathbf{j})$ **5.** 2, $\frac{1}{\sqrt{2}}(\mathbf{i} + \mathbf{j})$ **7.** 7, $\frac{1}{7}(2\mathbf{i} - 3\mathbf{j} + 6\mathbf{k})$ **9.** $\frac{2}{\sqrt{33}}(4\mathbf{i} - \mathbf{j} + 4\mathbf{k})$ **11.** $|\mathbf{A}| = \sqrt{2}$, $|\mathbf{B}| = 3\mathbf{A} \cdot \mathbf{B} = 3 = \mathbf{B} \cdot \mathbf{A}$, $\mathbf{A} \times \mathbf{B} = -2\mathbf{i} + 2\mathbf{j} - \mathbf{k}$, $\mathbf{B} \times \mathbf{A} = 2\mathbf{i} - 2\mathbf{j} + \mathbf{k}$, $|\mathbf{A} \times \mathbf{B}| = 3$, $\theta = \frac{\pi}{4}$, $|\mathbf{B}| \cos\theta = \frac{3}{\sqrt{2}}$, $\text{proj}_\mathbf{A}\mathbf{B} = 3(\mathbf{i} + \mathbf{j})$ **13.** $\frac{4}{3}(2\mathbf{i} + \mathbf{j} - \mathbf{k}) - \frac{1}{3}(5\mathbf{i} + \mathbf{j} + 11\mathbf{k})$

15.

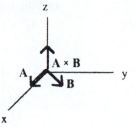

17. 3   **19.** $\sqrt{2}$ **21.** $2x + y - z = 3$ **23.** $-9x + y + 7z = 4$ **25.** $x = 1 - 3t$, $y = 2$, $z = 3 + 7t$ **27.** $\left(\frac{4}{3}, -\frac{2}{3}, -\frac{2}{3}\right)$ **29.** $x = 10 - 15t$, $y = 3t$, $z = -9 + 9t$ **31.** 2   **33.** 28254.323 ft·lb   **35.** a) $\sqrt{14}$ b) 1   **37.** (b) and (e) are not always true; others are always true.   **39.** y-axis in plane, yz-plane in 3-space.   **41.** In plane: circle; in 3-space: cylinder generated by the circle with axis parallel to z-axis.   **43.** Parabola opening to the right, cylinder generated by the parabola.   **45.** Cardioid with dimple on right; cylinder generated by the cardioid.   **47.** Horizontal lemniscate; cylinder generated by the lemniscate.   **49.** Surface of sphere centered at origin with radius 2.   **51.** The upper nappe of a cone whose surface makes an angle of $\pi/6$ with the z-axis.   **53.** The upper hemisphere of the unit sphere.

	Rectangular	Cylindrical	Spherical
55.	$(1,0,0)$	$(1,0,0)$	$\left(1, \frac{\pi}{2}, 0\right)$
57.	$(0,1,1)$	$\left(1, \frac{\pi}{2}, 1\right)$	$\left(\sqrt{2}, \frac{\pi}{4}, \frac{\pi}{2}\right)$
59.	$(1,1,1)$	$\left(\sqrt{2}, \frac{\pi}{4}, 1\right)$	$\left(\sqrt{3}, \cos^{-1}\left(\frac{1}{\sqrt{3}}\right), \frac{\pi}{4}\right)$

61. Cylindrical: $z = 2$, spherical: $\rho\cos\phi = 2$; plane parallel to xy-plane.   **63.** Rectangular: $z = x^2 + y^2$; spherical: $\rho = \cos\phi / \sin^2\phi$; paraboloid opening up.   **65.** Rectangular: $x^2 + y^2 + z^2 = 16$, cylindrical $r^2 + z^2 = 16$; sphere of radius 4.

67. sphere

$$x^2 + y^2 + z^2 = 4$$

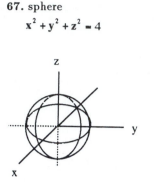

69. ellipsoid

$$4x^2 + 4y^2 + z^2 = 4$$

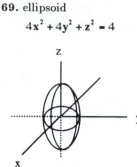

71. circular paraboloid

$$z = -(x^2 + y^2)$$

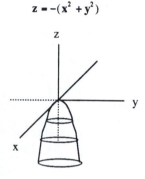

73. cone about z-axis

$$x^2 + y^2 = z^2$$

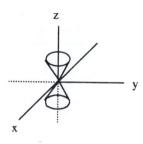

75. hyperboloid of one sheet

$$x^2 + y^2 - z^2 = 4$$

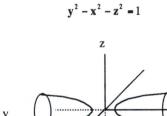

77. hyperboloid of two sheets

$$y^2 - x^2 - z^2 = 1$$

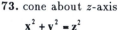

Chapter 12

Section 12.1

1. a) $\mathbf{v} = (-\sin t)\mathbf{i} + (3\cos t)\mathbf{j} + 4\mathbf{k}$ b) $\mathbf{a} = (-2\cos t)\mathbf{i} + (-3\sin t)\mathbf{j}$ c) At $t = \dfrac{\pi}{2}$: speed $= 2\sqrt{5}$ d) direction $= \dfrac{-2\mathbf{i} + 4\mathbf{k}}{2\sqrt{5}}$ e) $\mathbf{v} = 2\sqrt{5}\left(-\dfrac{1}{\sqrt{5}}\mathbf{i} + \dfrac{2}{\sqrt{5}}\mathbf{k}\right)$ **3.** a) $\mathbf{v} = (-2\sin 2t)\mathbf{j} + (2\cos t)\mathbf{k}$ b) $\mathbf{a} = (-4\cos 2t)\mathbf{j} + (-2\sin t)\mathbf{k}$ c) At $t = 0$: speed $= 2$ d) direction $= \mathbf{k}$ e) $\mathbf{v} = 2\mathbf{k}$ **5.** a) $\mathbf{v} = (\sec t \tan t)\mathbf{i} + (\sec^2 t)\mathbf{j} + \dfrac{4}{3}\mathbf{k}$ b) $\mathbf{a} =$

$(\sec t \tan^2 t + \sec^3 t)\mathbf{i} + (2\sec^2 t \tan t)\mathbf{j}$ c) At $t = \dfrac{\pi}{6}$, speed $= 2$ d) direction $= \dfrac{1}{3}\mathbf{i} + \dfrac{2}{3}\mathbf{j} + \dfrac{2}{3}\mathbf{k}$ e) $\mathbf{v} = 2\left[\dfrac{1}{3}\mathbf{i} + \dfrac{2}{3}\mathbf{j} + \dfrac{2}{3}\mathbf{k}\right]$ **7.** a) $\mathbf{v} = (-e^{-t})\mathbf{i} + (-6\sin 3t)\mathbf{j} + (6\cos 3t)\mathbf{k}$ b) $\mathbf{a} = (e^{-t})\mathbf{i} + (-18\cos 3t)\mathbf{j} + (-18\sin 3t)\mathbf{k}$ c) At $t = 0$: speed $= \sqrt{37}$ d) direction $= \dfrac{-\mathbf{i} + 6\mathbf{k}}{\sqrt{37}}$ e) $\mathbf{v} = \sqrt{37}\left(\dfrac{-\mathbf{i} + 6\mathbf{k}}{\sqrt{37}}\right)$ **9.** $\dfrac{\pi}{2}$ **11.** $\dfrac{\pi}{2}$ **13.** $t = 0,\ \pi,\ 2\pi$

15. $\dfrac{1}{4}\mathbf{i} + 7\mathbf{j} + \dfrac{3}{2}\mathbf{k}$ **17.** $\left(\dfrac{\pi}{2} + \sqrt{2}\right)\mathbf{j} + 2\mathbf{k}$ **19.** $(\ln 4)\mathbf{i} + (\ln 4)\mathbf{j} + (\ln 2)\mathbf{k}$

21. $\mathbf{v}\left(\dfrac{\pi}{4}\right) = \dfrac{\sqrt{2}}{2}(\mathbf{i} - \mathbf{j})$, $\mathbf{a}\left(\dfrac{\pi}{4}\right) = -\dfrac{\sqrt{2}}{2}(\mathbf{i} + \mathbf{j})$, $\mathbf{v}\left(\dfrac{\pi}{2}\right) = -\mathbf{j}$, $\mathbf{a}\left(\dfrac{\pi}{2}\right) = -\mathbf{i}$

23. $\mathbf{v}(\pi) = 2\mathbf{i}$, $\mathbf{a}(\pi) = -\mathbf{j}$, $\mathbf{v}\left(\dfrac{3\pi}{2}\right) = \mathbf{i} - \mathbf{j}$, $\mathbf{a}\left(\dfrac{3\pi}{2}\right) = -\mathbf{i}$.

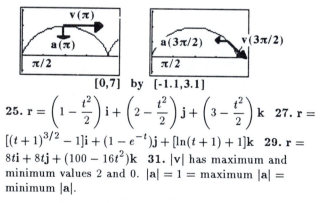

$[0,7]$ by $[-1.1, 3.1]$

25. $\mathbf{r} = \left(1 - \dfrac{t^2}{2}\right)\mathbf{i} + \left(2 - \dfrac{t^2}{2}\right)\mathbf{j} + \left(3 - \dfrac{t^2}{2}\right)\mathbf{k}$ **27.** $\mathbf{r} = [(t+1)^{3/2} - 1]\mathbf{i} + (1 - e^{-t})\mathbf{j} + [\ln(t+1) + 1]\mathbf{k}$ **29.** $\mathbf{r} = 8t\mathbf{i} + 8t\mathbf{j} + (100 - 16t^2)\mathbf{k}$ **31.** $|\mathbf{v}|$ has maximum and minimum values 2 and 0. $|\mathbf{a}| = 1 = $ maximum $|\mathbf{a}| = $ minimum $|\mathbf{a}|$.

Section 12.2

1. 50 sec **3.** a) 72.15 sec, 25.51 km b) 4.02 km c) 6377.55 m **7.** 9.9 m/sec; 18.43° and 71.57° **9.** 189.56 mph **13.** In flight, the ball passes just above the pin. **15.** It takes $t_1 \approx 2.245$ sec to reach the wall. The initial speed is $v_0 \approx 149.31$ ft/sec. The speed at t_1 is about 141.83 ft/sec neglecting all forces except gravity. **17.** a) $\mathbf{r} = [13 + (35\cos 27°)t]\mathbf{i} + [4 + (35\sin 27°)t - 16t^2]\mathbf{j}$. $x = 13 + (35\cos 27°)t$, $y = 4 + (35\sin 27°)t - 16t^2$ b) Graph x, y above, $0 \le t \le 1.2$ in $[0, 51]$ by $[-9.7, 19.7]$. Also draw in the net: Line $(25, 0, 25, 6)$. c) I) $y_{max} \approx 7.9$ ft occurs at $t \approx 0.5$ sec. II) It travels 37.4 ft and hits the ground at $t = 1.2$ sec approximately. III) $t \approx 0.25$ sec and 0.75 sec at 7.8 ft and 23.4 ft from the point of impact. IV) The ball hits the net. **19.** a) $y_{max} = 40.435$ ft at $t = 1.48$ sec b) It travels about 373 ft, hitting ground at $t = 3.13$ sec c) $y = 30$ ft at about $t = 0.69$ sec and $t = 2.3$ sec, 94.59 ft and 287.08 ft from home plate, respectively. d) Yes

21.

k	y_{max}	Time for y_{max}	Flight Distance	Flight Time
0.01	44.77 ft	1.6 sec	463.66 ft	3.3 sec
0.02	44.34	1.6	456.13	3.3
0.05	43.05	1.6	422.38	3.2
0.1	41.15	1.5	391.15	3.2
0.15	39.39	1.5	354.10	3.1
0.20	37.85	1.4	322.22	3.0
0.25	36.41	1.4	294.62	2.9

c) As the air density diminishes to 0, the air resistence to the motion of the projectile diminishes to 0, as was assumed in Eq. (6).

Section 12.3

1. $T = -\frac{2}{3}\sin t\,i + \frac{2}{3}\cos t\,j + \frac{\sqrt{5}}{3}k$. $L = 3\pi$. **3.** $T = \frac{1}{\sqrt{1+t}}i + \sqrt{\frac{t}{1+t}}k$. $L = \frac{52}{3}$. **5.** $T = \frac{1}{\sqrt{3}}(i - j + k)$. $L = 3\sqrt{3}$. **7.** $T = \left(\frac{\cos t - t\sin t}{t+1}\right)i + \left(\frac{\sin t + t\cos t}{t+1}\right)j + \left(\frac{\sqrt{2}t}{t+1}\right)k$. $L = \frac{\pi^2}{2} + \pi$. **9.** $s(t) = 5(t - t_0)$. $L = \frac{5\pi}{2}$. **11.** $s(t) = \sqrt{3}(e^t - e^{t_0})$. $L = 3\sqrt{3}$. **13.** $\sqrt{2} + \ln(1 + \sqrt{2})$.

Section 12.4

1. $T = \cos i - \sin t\,j$, $N = -\sin t\,i - \cos t\,j$, $\kappa = \cos t$ **3.** $T = \frac{i - tj}{\sqrt{1+t^2}}$, $N = -\frac{1}{\sqrt{t+t^2}}i - \frac{1}{\sqrt{1+t^2}}j$, $\kappa = \frac{1}{2(1+t^2)^{3/2}}$ **5.** $T = \frac{3}{5}\cos t\,i - \frac{3}{5}\sin t\,j + \frac{4}{5}k$, $N = -\sin t\,i - \cos t\,j$, $B = \frac{4}{5}\cos t\,i - \frac{4}{5}\sin t\,j - \frac{3}{5}k$, $\kappa = \frac{3}{25}$, $\tau = \frac{4}{25}$ **7.** $T = \left(\frac{\cos t - \sin t}{\sqrt{2}}\right)i + \left(\frac{\cos t + \sin t}{\sqrt{2}}\right)j$, $N = -\left(\frac{\sin t + \cos t}{\sqrt{2}}\right)i + \left(\frac{\cos t - \sin t}{\sqrt{2}}\right)j$, $B = k$, $\kappa = \frac{1}{\sqrt{2}e^t}$, $\tau = 0$ **9.** $a = \frac{2t}{\sqrt{1+t^2}}T + \frac{2}{\sqrt{1+t^2}}N$ **11.** aN, assuming $a > 0$ **13.** $a = \frac{4}{3}T + \frac{2\sqrt{5}}{3}N$ **15.** $a = 2N$ **17.** At $t = \frac{\pi}{4}$, $r = \frac{\sqrt{2}}{2}i + \frac{\sqrt{2}}{2}j - k$, $T = -\frac{\sqrt{2}}{2}i + \frac{\sqrt{2}}{2}j$,

$N = -\frac{\sqrt{2}}{2}i - \frac{\sqrt{2}}{2}j$, $B = k$. Osculating plane: $z = -1$. Normal plane: $y = x$. Rectifying plane: $y = -x$. **21.** $\left(x - \frac{\pi}{2}\right)^2 + y^2 = 1$. Graph $x_1(t) = t$, $y_1(t) = \sin t$ and $x_2(t) = \frac{\pi}{2} + \cos t$, $y_2(t) = \sin t$, $-2\pi \le t \le 2\pi$, t step $= 0.1$ in $[-2\pi, 2\pi]$ by $[-4.2, 3.2]$. **25.** $\frac{1}{2b}$

Section 12.5

1. 93.17 min. compared to 93.11 min. in the table **3.** 6763 km compared with 6765 km **5.** 1655 min **7.** 20420 km **9.** $\frac{1.9967 \times 10^7}{\sqrt{r}}$ m/sec **11.** Circle: $v_0 = \sqrt{\frac{GM}{r_0}}$. Ellipse: $\sqrt{\frac{GM}{r_0}} < v_0 < \sqrt{\frac{2GM}{r_0}}$. Hyperbola: $v_0 > \sqrt{\frac{2GM}{r_0}}$.

Chapter 12 Practice Exercises

1.

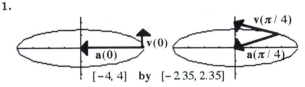

$[-4, 4]$ by $[-2.35, 2.35]$

3. $6i + 8j$ **5.** $r = (\cos t - 1)i + (\sin t + 1)j + tk$ **7.** $r = i + t^2j + tk$ **9.** $\frac{\pi\sqrt{16 + \pi^2}}{16} + \ln\left(\frac{\pi + \sqrt{16 + \pi^2}}{4}\right) = 1.7199\ldots$ **11.** At $t = 0$: $T = \frac{2}{3}i - \frac{2}{3}j + \frac{1}{3}k$, $N = \frac{i + j}{\sqrt{2}}$, $B = -\frac{1}{3\sqrt{2}}i + \frac{1}{3\sqrt{2}}j + \frac{4}{3\sqrt{2}}k$, $\kappa = \frac{\sqrt{2}}{3}$, $\tau = \frac{1}{6}$ **13.** $a(0) = 10T(0) + 6N(0)$ **15.** 1 **17.** $t = 0, \frac{\pi}{2}, \pi$ **19.** $x = 66\sqrt{2}$ ft, $y = -42.66$ ft so it must be on the level ground. **21.** 644.36 ft/sec and 675.42 ft/sec **23.** a) $x = (155\cos 18°)t = \left(\frac{176}{15}\right)t$, $y = 4 + (155\sin 18°)t - 16t^2$ and with the same x, y, $r = xi + yj$. b) Graph x, y in a), $0 \le t \le 4$, t step $= 0.05$ in $[0, 500]$ by $[0, 100]$. Also include LINE$(380, 0, 380, 10)$. c) With the set-up in b) we use the TRACE function. (c-i) $y_{max} = 39.85$ ft at $t = 1.5$ sec; (c-ii) $y = 0$ at about $t - 3.1$ sec and $x = 420.61$ ft; (c-iii) $y = 25$ ft at about $t = 0.55$ sec and $t = 2.45$ sec when $x = 74.62$ ft and 332.42 ft, respectively; (c-iv) Yes. **25.** $x = 1 + t$, $y = t$, $z = -t$ **27.** 1/5 **29.** $\tau = b/(a^2 + b^2)$ has maximum value $1/2a$ for a given value of $a > 0$. **31.** a) 5971 km b) $VA = (5.2188 \times 10^6)\pi$ km^2 c) 3.21% **33.** b) 1.49870113352. Graph f in the window $[-2, 3]$ by $[-2, 2]$.

Appendixes

Appendix A.4

7. Statement true for $n \geq 5$

Appendix A.7

1. -5 **3.** 1 **5.** a) -7 b) -7 **7.** a) 38 b) 38 **9.** $x = -4$, $y = 1$ **11.** $x = 3$, $y = 2$ **13.** $x = 3$, $y = -2$, $z = 2$
15. $x = 2$, $y = 0$, $z = -1$ **17.** a) $h = 6$, $k = 4$ b) $h = 6$, $k \neq 4$

Appendix A.8

1. $(2, 4, 4)$ **3.** $f(\sqrt{6}, \sqrt{3}, 1) = f(-\sqrt{6}, \sqrt{3}, 1) = 1 + \sqrt{6}$, max; $f(\sqrt{6}, -\sqrt{3}, 1) = f(-\sqrt{6}, -\sqrt{3}, 1) = 1 - \sqrt{6}$, min **5.** $f(0, 0, 2) = f(0, 0, -2) = 4$, max; $f(2, 2, 0) = f(-2, -2, 0) = 2$, min

Index

A Brief Table of Integrals

1. $\displaystyle\int u\,dv = uv - \int v\,du$

2. $\displaystyle\int a^u\,du = \frac{a^u}{\ln a} + C, \qquad a \neq 1, \qquad a > 0$

3. $\displaystyle\int \cos u\,du = \sin u + C$

4. $\displaystyle\int \sin u\,du = -\cos u + C$

5. $\displaystyle\int (ax + b)^n\,dx = \frac{(ax + b)^{n+1}}{a(n + 1)} + C, \qquad n \neq -1$

6. $\displaystyle\int (ax + b)^{-1}\,dx = \frac{1}{a}\ln|ax + b| + C$

7. $\displaystyle\int x(ax + b)^n\,dx = \frac{(ax + b)^{n+1}}{a^2}\left[\frac{ax + b}{n + 2} - \frac{b}{n + 1}\right] + C, \qquad n \neq -1, -2$

8. $\displaystyle\int x(ax + b)^{-1}\,dx = \frac{x}{a} - \frac{b}{a^2}\ln|ax + b| + C$

9. $\displaystyle\int x(ax + b)^{-2}\,dx = \frac{1}{a^2}\left[\ln|ax + b| + \frac{b}{ax + b}\right] + C$

10. $\displaystyle\int \frac{dx}{x(ax + b)} = \frac{1}{b}\ln\left|\frac{x}{ax + b}\right| + C$

11. $\displaystyle\int \left(\sqrt{ax + b}\right)^n\,dx = \frac{2}{a}\frac{\left(\sqrt{ax + b}\right)^{n+2}}{n + 2} + C, \qquad n \neq -2$

12. $\displaystyle\int \frac{\sqrt{ax + b}}{x}\,dx = 2\sqrt{ax + b} + b\int \frac{dx}{x\sqrt{ax + b}}$

13. (a) $\displaystyle\int \frac{dx}{x\sqrt{ax + b}} = \frac{2}{\sqrt{-b}}\tan^{-1}\sqrt{\frac{ax + b}{-b}} + C, \qquad$ if $\;b < 0$

 (b) $\displaystyle\int \frac{dx}{x\sqrt{ax + b}} = \frac{1}{\sqrt{b}}\ln\left|\frac{\sqrt{ax + b} - \sqrt{b}}{\sqrt{ax + b} + \sqrt{b}}\right| + C, \qquad$ if $\;b > 0$

14. $\displaystyle\int \frac{\sqrt{ax + b}}{x^2}\,dx = -\frac{\sqrt{ax + b}}{x} + \frac{a}{2}\int \frac{dx}{x\sqrt{ax + b}} + C$

15. $\displaystyle\int \frac{dx}{x^2\sqrt{ax + b}} = -\frac{\sqrt{ax + b}}{bx} - \frac{a}{2b}\int \frac{dx}{x\sqrt{ax + b}} + C$

16. $\displaystyle\int \frac{dx}{a^2 + x^2} = \frac{1}{a}\tan^{-1}\frac{x}{a} + C$

17. $\displaystyle\int \frac{dx}{(a^2 + x^2)^2} = \frac{x}{2a^2(a^2 + x^2)} + \frac{1}{2a^3}\tan^{-1}\frac{x}{a} + C$

18. $\displaystyle\int \frac{dx}{a^2 - x^2} = \frac{1}{2a}\ln\left|\frac{x + a}{x - a}\right| + C$

19. $\displaystyle\int \frac{dx}{(a^2 - x^2)^2} = \frac{x}{2a^2(a^2 - x^2)} + \frac{1}{2a^2}\int \frac{dx}{a^2 - x^2}$

20. $\displaystyle\int \frac{dx}{\sqrt{a^2 + x^2}} = \sinh^{-1}\frac{x}{a} + C = \ln\left(x + \sqrt{a^2 + x^2}\right) + C$

21. $\displaystyle\int \sqrt{a^2 + x^2}\,dx = \frac{x}{2}\sqrt{a^2 + x^2} + \frac{a^2}{2}\ln\left(x + \sqrt{a^2 + x^2}\right) + C$

22. $\displaystyle\int x^2\sqrt{a^2 + x^2}\,dx = \frac{x}{8}(a^2 + 2x^2)\sqrt{a^2 + x^2} - \frac{a^4}{8}\ln\left(x + \sqrt{a^2 + x^2}\right) + C$

23. $\displaystyle \int \frac{\sqrt{a^2 + x^2}}{x}\,dx = \sqrt{a^2 + x^2} - a \ln \left| \frac{a + \sqrt{a^2 + x^2}}{x} \right| + C$

24. $\displaystyle \int \frac{\sqrt{a^2 + x^2}}{x^2}\,dx = \ln \left(x + \sqrt{a^2 + x^2} \right) - \frac{\sqrt{a^2 + x^2}}{x} + C$

25. $\displaystyle \int \frac{x^2}{\sqrt{a^2 + x^2}}\,dx = -\frac{a^2}{2} \ln \left(x + \sqrt{a^2 + x^2} \right) + \frac{x\sqrt{a^2 + x^2}}{2} + C$

26. $\displaystyle \int \frac{dx}{x\sqrt{a^2 + x^2}} = -\frac{1}{a} \ln \left| \frac{a + \sqrt{a^2 + x^2}}{x} \right| + C$

27. $\displaystyle \int \frac{dx}{x^2\sqrt{a^2 + x^2}} = -\frac{\sqrt{a^2 + x^2}}{a^2 x} + C$

28. $\displaystyle \int \frac{dx}{\sqrt{a^2 - x^2}} = \sin^{-1} \frac{x}{a} + C$

29. $\displaystyle \int \sqrt{a^2 - x^2}\,dx = \frac{x}{2}\sqrt{a^2 - x^2} + \frac{a^2}{2} \sin^{-1} \frac{x}{a} + C$

30. $\displaystyle \int x^2\sqrt{a^2 - x^2}\,dx = \frac{a^4}{8} \sin^{-1} \frac{x}{a} - \frac{1}{8} x\sqrt{a^2 - x^2}\,(a^2 - 2x^2) + C$

31. $\displaystyle \int \frac{\sqrt{a^2 - x^2}}{x}\,dx = \sqrt{a^2 - x^2} - a \ln \left| \frac{a + \sqrt{a^2 - x^2}}{x} \right| + C$

32. $\displaystyle \int \frac{\sqrt{a^2 - x^2}}{x^2}\,dx = -\sin^{-1} \frac{x}{a} - \frac{\sqrt{a^2 - x^2}}{x} + C$

33. $\displaystyle \int \frac{x^2}{\sqrt{a^2 - x^2}}\,dx = \frac{a^2}{2} \sin^{-1} \frac{x}{a} - \frac{1}{2} x\sqrt{a^2 - x^2} + C$

34. $\displaystyle \int \frac{dx}{x\sqrt{a^2 - x^2}} = -\frac{1}{a} \ln \left| \frac{a + \sqrt{a^2 - x^2}}{x} \right| + C$

35. $\displaystyle \int \frac{dx}{x^2\sqrt{a^2 - x^2}} = -\frac{\sqrt{a^2 - x^2}}{a^2 x} + C$

36. $\displaystyle \int \frac{dx}{\sqrt{x^2 - a^2}} = \cosh^{-1} \frac{x}{a} + C = \ln \left| x + \sqrt{x^2 - a^2} \right| + C$

37. $\displaystyle \int \sqrt{x^2 - a^2}\,dx = \frac{x}{2}\sqrt{x^2 - a^2} - \frac{a^2}{2} \ln \left| x + \sqrt{x^2 - a^2} \right| + C$

38. $\displaystyle \int \left(\sqrt{x^2 - a^2} \right)^n dx = \frac{x\left(\sqrt{x^2 - a^2} \right)^n}{n + 1} - \frac{na^2}{n + 1} \int \left(\sqrt{x^2 - a^2} \right)^{n-2} dx, \qquad n \neq -1$

39. $\displaystyle \int \frac{dx}{\left(\sqrt{x^2 - a^2} \right)^n} = \frac{x\left(\sqrt{x^2 - a^2} \right)^{2-n}}{(2 - n)a^2} - \frac{n - 3}{(n - 2)a^2} \int \frac{dx}{\left(\sqrt{x^2 - a^2} \right)^{n-2}}, \qquad n \neq 2$

40. $\displaystyle \int x\left(\sqrt{x^2 - a^2} \right)^n dx = \frac{\left(\sqrt{x^2 - a^2} \right)^{n+2}}{n + 2} + C, \qquad n \neq -2$

41. $\displaystyle \int x^2\sqrt{x^2 - a^2}\,dx = \frac{x}{8}(2x^2 - a^2)\sqrt{x^2 - a^2} - \frac{a^4}{8} \ln \left| x + \sqrt{x^2 - a^2} \right| + C$

42. $\displaystyle \int \frac{\sqrt{x^2 - a^2}}{x}\,dx = \sqrt{x^2 - a^2} - a \sec^{-1} \left| \frac{x}{a} \right| + C$

43. $\displaystyle \int \frac{\sqrt{x^2 - a^2}}{x^2}\,dx = \ln \left| x + \sqrt{x^2 - a^2} \right| - \frac{\sqrt{x^2 - a^2}}{x} + C$

44. $\displaystyle \int \frac{x^2}{\sqrt{x^2 - a^2}}\,dx = \frac{a^2}{2} \ln \left| x + \sqrt{x^2 - a^2} \right| + \frac{x}{2}\sqrt{x^2 - a^2} + C$

45. $\displaystyle \int \frac{dx}{x\sqrt{x^2 - a^2}} = \frac{1}{a} \sec^{-1} \left| \frac{x}{a} \right| + C = \frac{1}{a} \cos^{-1} \left| \frac{a}{x} \right| + C$

46. $\displaystyle \int \frac{dx}{x^2\sqrt{x^2 - a^2}} = \frac{\sqrt{x^2 - a^2}}{a^2 x} + C$

47. $\displaystyle \int \frac{dx}{\sqrt{2ax - x^2}} = \sin^{-1} \left(\frac{x - a}{a} \right) + C$

48. $\displaystyle \int \sqrt{2ax - x^2}\,dx = \frac{x - a}{2}\sqrt{2ax - x^2} + \frac{a^2}{2} \sin^{-1} \left(\frac{x - a}{a} \right) + C$

49. $\displaystyle \int \left(\sqrt{2ax - x^2} \right)^n dx = \frac{(x - a)\left(\sqrt{2ax - x^2} \right)^n}{n + 1} + \frac{na^2}{n + 1} \int \left(\sqrt{2ax - x^2} \right)^{n-2} dx$

50. $\displaystyle \int \frac{dx}{\left(\sqrt{2ax - x^2} \right)^n} = \frac{(x - a)\left(\sqrt{2ax - x^2} \right)^{2-n}}{(n - 2)a^2} + \frac{(n - 3)}{(n - 2)a^2} \int \frac{dx}{\left(\sqrt{2ax - x^2} \right)^{n-2}}$

51. $\displaystyle \int x\sqrt{2ax - x^2}\,dx = \frac{(x + a)(2x - 3a)\sqrt{2ax - x^2}}{6} + \frac{a^3}{2} \sin^{-1} \left(\frac{x - a}{a} \right) + C$

52. $\displaystyle \int \frac{\sqrt{2ax - x^2}}{x}\,dx = \sqrt{2ax - x^2} + a \sin^{-1} \left(\frac{x - a}{a} \right) + C$

53. $\displaystyle \int \frac{\sqrt{2ax - x^2}}{x^2}\,dx = -2\sqrt{\frac{2a - x}{x}} - \sin^{-1} \left(\frac{x - a}{a} \right) + C$

54. $\displaystyle \int \frac{x\,dx}{\sqrt{2ax - x^2}} = a \sin^{-1} \left(\frac{x - a}{a} \right) - \sqrt{2ax - x^2} + C$

55. $\displaystyle\int \frac{dx}{x\sqrt{2ax - x^2}} = -\frac{1}{a}\sqrt{\frac{2a - x}{x}} + C$

56. $\displaystyle\int \sin ax\, dx = -\frac{1}{a}\cos ax + C$

57. $\displaystyle\int \cos ax\, dx = \frac{1}{a}\sin ax + C$

58. $\displaystyle\int \sin^2 ax\, dx = \frac{x}{2} - \frac{\sin 2ax}{4a} + C$

59. $\displaystyle\int \cos^2 ax\, dx = \frac{x}{2} + \frac{\sin 2ax}{4a} + C$

60. $\displaystyle\int \sin^n ax\, dx = -\frac{\sin^{n-1} ax \cos ax}{na} + \frac{n-1}{n}\int \sin^{n-2} ax\, dx$

61. $\displaystyle\int \cos^n ax\, dx = \frac{\cos^{n-1} ax \sin ax}{na} + \frac{n-1}{n}\int \cos^{n-2} ax\, dx$

62. (a) $\displaystyle\int \sin ax \cos bx\, dx = -\frac{\cos (a+b)x}{2(a+b)} - \frac{\cos (a-b)x}{2(a-b)} + C, \qquad a^2 \neq b^2$

(b) $\displaystyle\int \sin ax \sin bx\, dx = \frac{\sin (a-b)x}{2(a-b)} - \frac{\sin (a+b)x}{2(a+b)} + C, \qquad a^2 \neq b^2$

(c) $\displaystyle\int \cos ax \cos bx\, dx = \frac{\sin (a-b)x}{2(a-b)} + \frac{\sin (a+b)x}{2(a+b)} + C, \qquad a^2 \neq b^2$

63. $\displaystyle\int \sin ax \cos ax\, dx = -\frac{\cos 2ax}{4a} + C$

64. $\displaystyle\int \sin^n ax \cos ax\, dx = \frac{\sin^{n+1} ax}{(n+1)a} + C, \quad n \neq -1$

65. $\displaystyle\int \frac{\cos ax}{\sin ax}\, dx = \frac{1}{a}\ln |\sin ax| + C$

66. $\displaystyle\int \cos^n ax \sin ax\, dx = -\frac{\cos^{n+1} ax}{(n+1)a} + C, \quad n \neq -1$

67. $\displaystyle\int \frac{\sin ax}{\cos ax}\, dx = -\frac{1}{a}\ln |\cos ax| + C$

68. $\displaystyle\int \sin^n ax \cos^m ax\, dx = -\frac{\sin^{n-1} ax \cos^{m+1} ax}{a(m+n)} + \frac{n-1}{m+n}\int \sin^{n-2} ax \cos^m ax\, dx, n \neq -m$

(If $n = -m$, use No. 86.)

69. $\displaystyle\int \sin^n ax \cos^m ax\, dx = \frac{\sin^{n+1} ax \cos^{m-1} ax}{a(m+n)} + \frac{m-1}{m+n}\int \sin^n ax \cos^{m-2} ax\, dx, m \neq -n$

(If $m = -n$, use No. 87.)

70. $\displaystyle\int \frac{dx}{b + c \sin ax} = \frac{-2}{a\sqrt{b^2 - c^2}}\tan^{-1}\left[\sqrt{\frac{b-c}{b+c}}\tan\left(\frac{\pi}{4} - \frac{ax}{2}\right)\right] + C, \qquad b^2 > c^2$

71. $\displaystyle\int \frac{dx}{b + c \sin ax} = \frac{-1}{a\sqrt{c^2 - b^2}}\ln\left|\frac{c + b \sin ax + \sqrt{c^2 - b^2}\cos ax}{b + c \sin ax}\right| + C, \qquad b^2 < c^2$

72. $\displaystyle\int \frac{dx}{1 + \sin ax} = -\frac{1}{a}\tan\left(\frac{\pi}{4} - \frac{ax}{2}\right) + C$

73. $\displaystyle\int \frac{dx}{1 - \sin ax} = \frac{1}{a}\tan\left(\frac{\pi}{4} + \frac{ax}{2}\right) + C$

74. $\displaystyle\int \frac{dx}{b + c \cos ax} = \frac{2}{a\sqrt{b^2 - c^2}}\tan^{-1}\left[\sqrt{\frac{b-c}{b+c}}\tan\frac{ax}{2}\right] + C, \qquad b^2 > c^2$

75. $\displaystyle\int \frac{dx}{b + c \cos ax} = \frac{1}{a\sqrt{c^2 - b^2}}\ln\left|\frac{c + b \cos ax + \sqrt{c^2 - b^2}\sin ax}{b + c \cos ax}\right| + C, \qquad b^2 < c^2$

76. $\displaystyle\int \frac{dx}{1 + \cos ax} = \frac{1}{a}\tan\frac{ax}{2} + C$

77. $\displaystyle\int \frac{dx}{1 - \cos ax} = -\frac{1}{a}\cot\frac{ax}{2} + C$

78. $\displaystyle\int x \sin ax\, dx = \frac{1}{a^2}\sin ax - \frac{x}{a}\cos ax + C$

79. $\displaystyle\int x \cos ax\, dx = \frac{1}{a^2}\cos ax + \frac{x}{a}\sin ax + C$

80. $\displaystyle\int x^n \sin ax\, dx = -\frac{x^n}{a}\cos ax + \frac{n}{a}\int x^{n-1}\cos ax\, dx$

81. $\displaystyle\int x^n \cos ax\, dx = \frac{x^n}{a}\sin ax - \frac{n}{a}\int x^{n-1}\sin ax\, dx$

82. $\displaystyle\int \tan ax\, dx = \frac{1}{a}\ln |\sec ax| + C$

83. $\displaystyle\int \cot ax\, dx = \frac{1}{a}\ln |\sin ax| + C$

84. $\displaystyle\int \tan^2 ax\, dx = \frac{1}{a}\tan ax - x + C$

85. $\displaystyle\int \cot^2 ax\, dx = -\frac{1}{a}\cot ax - x + C$

86. $\displaystyle\int \tan^n ax\, dx = \frac{\tan^{n-1} ax}{a(n-1)} - \int \tan^{n-2} ax\, dx, \quad n \neq 1$

87. $\displaystyle\int \cot^n ax\, dx = -\frac{\cot^{n-1} ax}{a(n-1)} - \int \cot^{n-2} ax\, dx, \quad n \neq 1$

88. $\int \sec ax \, dx = \frac{1}{a} \ln |\sec ax + \tan ax| + C$

89. $\int \csc ax \, dx = -\frac{1}{a} \ln |\csc ax + \cot ax| + C$

90. $\int \sec^2 ax \, dx = \frac{1}{a} \tan ax + C$

91. $\int \csc^2 ax \, dx = -\frac{1}{a} \cot ax + C$

92. $\int \sec^n ax \, dx = \frac{\sec^{n-2} ax \tan ax}{a(n-1)} + \frac{n-2}{n-1} \int \sec^{n-2} ax \, dx, \qquad n \neq 1$

93. $\int \csc^n ax \, dx = -\frac{\csc^{n-2} ax \cot ax}{a(n-1)} + \frac{n-2}{n-1} \int \csc^{n-2} ax \, dx, \qquad n \neq 1$

94. $\int \sec^n ax \tan ax \, dx = \frac{\sec^n ax}{na} + C, \qquad n \neq 0$

95. $\int \csc^n ax \cot ax \, dx = -\frac{\csc^n ax}{na} + C, \qquad n \neq 0$

96. $\int \sin^{-1} ax \, dx = x \sin^{-1} ax + \frac{1}{a} \sqrt{1 - a^2 x^2} + C$

97. $\int \cos^{-1} ax \, dx = x \cos^{-1} ax - \frac{1}{a} \sqrt{1 - a^2 x^2} + C$

98. $\int \tan^{-1} ax \, dx = x \tan^{-1} ax - \frac{1}{2a} \ln (1 + a^2 x^2) + C$

99. $\int x^n \sin^{-1} ax \, dx = \frac{x^{n+1}}{n+1} \sin^{-1} ax - \frac{a}{n+1} \int \frac{x^{n+1} \, dx}{\sqrt{1 - a^2 x^2}}, \qquad n \neq -1$

100. $\int x^n \cos^{-1} ax \, dx = \frac{x^{n+1}}{n+1} \cos^{-1} ax + \frac{a}{n+1} \int \frac{x^{n+1} \, dx}{\sqrt{1 - a^2 x^2}}, \qquad n \neq -1$

101. $\int x^n \tan^{-1} ax \, dx = \frac{x^{n+1}}{n+1} \tan^{-1} ax - \frac{a}{n+1} \int \frac{x^{n+1} \, dx}{1 + a^2 x^2}, \qquad n \neq -1$

102. $\int e^{ax} \, dx = \frac{1}{a} e^{ax} + C$

103. $\int b^{ax} \, dx = \frac{1}{a} \frac{b^{ax}}{\ln b} + C, \qquad b > 0, \quad b \neq 1$

104. $\int xe^{ax} \, dx = \frac{e^{ax}}{a^2} (ax - 1) + C$

105. $\int x^n e^{ax} \, dx = \frac{1}{a} x^n e^{ax} - \frac{n}{a} \int x^{n-1} e^{ax} \, dx$

106. $\int x^n b^{ax} \, dx = \frac{x^n b^{ax}}{a \ln b} - \frac{n}{a \ln b} \int x^{n-1} b^{ax} \, dx, \qquad b > 0, \quad b \neq 1$

107. $\int e^{ax} \sin bx \, dx = \frac{e^{ax}}{a^2 + b^2} (a \sin bx - b \cos bx) + C$

108. $\int e^{ax} \cos bx \, dx = \frac{e^{ax}}{a^2 + b^2} (a \cos bx + b \sin bx) + C$

109. $\int \ln ax \, dx = x \ln ax - x + C$

110. $\int x^n (\ln ax)^m \, dx = \frac{x^{n+1} (\ln ax)^m}{n+1} - \frac{m}{n+1} \int x^n (\ln ax)^{m-1} \, dx, \qquad n \neq -1$

111. $\int x^{-1} (\ln ax)^m = \frac{(\ln ax)^{m+1}}{m+1} + C, \qquad m \neq -1$

112. $\int \frac{dx}{x \ln ax} = \ln |\ln ax| + C$

113. $\int \sinh ax \, dx = \frac{1}{a} \cosh ax + C$

114. $\int \cosh ax \, dx = \frac{1}{a} \sinh ax + C$

115. $\int \sinh^2 ax \, dx = \frac{\sinh 2ax}{4a} - \frac{x}{2} + C$

116. $\int \cosh^2 ax \, dx = \frac{\sinh 2ax}{4a} + \frac{x}{2} + C$

117. $\int \sinh^n ax \, dx = \frac{\sinh^{n-1} ax \cosh ax}{na} - \frac{n-1}{n} \int \sinh^{n-2} ax \, dx, \qquad n \neq 0$

118. $\int \cosh^n ax \, dx = \frac{\cosh^{n-1} ax \sinh ax}{na} + \frac{n-1}{n} \int \cosh^{n-2} ax \, dx, \qquad n \neq 0$

119. $\displaystyle\int x \sinh ax \, dx = \frac{x}{a} \cosh ax - \frac{1}{a^2} \sinh ax + C$

120. $\displaystyle\int x \cosh ax \, dx = \frac{x}{a} \sinh ax - \frac{1}{a^2} \cosh ax + C$

121. $\displaystyle\int x^n \sinh ax \, dx = \frac{x^n}{a} \cosh ax - \frac{n}{a} \int x^{n-1} \cosh ax \, dx$

122. $\displaystyle\int x^n \cosh ax \, dx = \frac{x^n}{a} \sinh ax - \frac{n}{a} \int x^{n-1} \sinh ax \, dx$

123. $\displaystyle\int \tanh ax \, dx = \frac{1}{a} \ln (\cosh ax) + C$ 　　　　　**124.** $\displaystyle\int \coth ax \, dx = \frac{1}{a} \ln |\sinh ax| + C$

125. $\displaystyle\int \tanh^2 ax \, dx = x - \frac{1}{a} \tanh ax + C$ 　　　　**126.** $\displaystyle\int \coth^2 ax \, dx = x - \frac{1}{a} \coth ax + C$

127. $\displaystyle\int \tanh^n ax \, dx = -\frac{\tanh^{n-1} ax}{(n-1)a} + \int \tanh^{n-2} ax \, dx, \qquad n \neq 1$

128. $\displaystyle\int \coth^n ax \, dx = -\frac{\coth^{n-1} ax}{(n-1)a} + \int \coth^{n-2} ax \, dx, \qquad n \neq 1$

129. $\displaystyle\int \operatorname{sech} ax \, dx = \frac{1}{a} \sin^{-1}(\tanh ax) + C$ 　　　**130.** $\displaystyle\int \operatorname{csch} ax \, dx = \frac{1}{a} \ln \left| \tanh \frac{ax}{2} \right| + C$

131. $\displaystyle\int \operatorname{sech}^2 ax \, dx = \frac{1}{a} \tanh ax + C$ 　　　　**132.** $\displaystyle\int \operatorname{csch}^2 ax \, dx = -\frac{1}{a} \coth ax + C$

133. $\displaystyle\int \operatorname{sech}^n ax \, dx = \frac{\operatorname{sech}^{n-2} ax \tanh ax}{(n-1)a} + \frac{n-2}{n-1} \int \operatorname{sech}^{n-2} ax \, dx, \qquad n \neq 1$

134. $\displaystyle\int \operatorname{csch}^n ax \, dx = -\frac{\operatorname{csch}^{n-2} ax \coth ax}{(n-1)a} - \frac{n-2}{n-1} \int \operatorname{csch}^{n-2} ax \, dx, \qquad n \neq 1$

135. $\displaystyle\int \operatorname{sech}^n ax \tanh ax \, dx = -\frac{\operatorname{sech}^n ax}{na} + C, \qquad n \neq 0$

136. $\displaystyle\int \operatorname{csch}^n ax \coth ax \, dx = -\frac{\operatorname{csch}^n ax}{na} + C, \qquad n \neq 0$

137. $\displaystyle\int e^{ax} \sinh bx \, dx = \frac{e^{ax}}{2} \left[\frac{e^{bx}}{a+b} - \frac{e^{-bx}}{a-b} \right] + C, \qquad a^2 \neq b^2$

138. $\displaystyle\int e^{ax} \cosh bx \, dx = \frac{e^{ax}}{2} \left[\frac{e^{bx}}{a+b} + \frac{e^{-bx}}{a-b} \right] + C, \qquad a^2 \neq b^2$

139. $\displaystyle\int_0^\infty x^{n-1} e^{-x} \, dx = \Gamma(n) = (n-1)!, \qquad n > 0$ 　　　**140.** $\displaystyle\int_0^\infty e^{-ax^2} \, dx = \frac{1}{2} \sqrt{\frac{\pi}{a}}, \qquad a > 0$

141. $\displaystyle\int_0^{\pi/2} \sin^n x \, dx = \int_0^{\pi/2} \cos^n x \, dx = \begin{cases} \dfrac{1 \cdot 3 \cdot 5 \cdots (n-1)}{2 \cdot 4 \cdot 6 \cdots n} \cdot \dfrac{\pi}{2}, & \text{if } n \text{ is an even integer} \geq 2 \\[2ex] \dfrac{2 \cdot 4 \cdot 6 \cdots (n-1)}{3 \cdot 5 \cdot 7 \cdots n}, & \text{if } n \text{ is an odd integer} \geq 3 \end{cases}$